Summer Biomechanics, Bioengineering and Biotransport Conference 2017

Tucson, Arizona, USA
21 - 24 June 2017

Volume 1 of 2

ISBN: 978-1-5108-5993-7

Printed from e-media with permission by:

Curran Associates, Inc.
57 Morehouse Lane
Red Hook, NY 12571

Some format issues inherent in the e-media version may also appear in this print version.

Printed by Curran Associates, Inc. (2018)

Summer Biomechanics, Bioengineering and Biotransport Organizing Committee
201 Waterfront St
National Harbor, MD 20745
USA

info@sb3c.org

Additional copies of this publication are available from:

Curran Associates, Inc.
57 Morehouse Lane
Red Hook, NY 12571 USA
Phone: 845-758-0400
Fax: 845-758-2633
Email: curran@proceedings.com
Web: www.proceedings.com

TABLE OF CONTENTS

VOLUME 1

HEART VALVE MECHANICS (SOLIDS)

SPINE MECHANICS (SOLIDS)

CARDIOVASCULAR GROWTH, REMODELING, & REPAIR (SOLIDS)

BONE BIOMECHANICS (SOLIDS)

THROMBOSIS (FLUIDS)

BIOTRANSPORT AND MICROFLUIDICS (BTR)

INSTRINSIC AND EXTRINSIC REGULATION OF CELLULAR MECHANOTRANSDUCTION (CTE)

SOFT TISSUE CHARACTERIZATION AND MODELING (SOLIDS)

CELEBRATION FOR LARY TABER: GROWTH AND REMODELING IN DEVELOPMENT AND DISEASE (SOLIDS/CTE)

ATHEROSCLEROSIS (FLUIDS)

BIOFLUIDS (FLUIDS)

STRATEGIES TO IMPROVE REHABILITATION TREATMENTS (DDR/IAB)

MICRO-ENGINEERED PHYSIOLOGIC SYSTEMS (CTE)

MICROENVIRONMENTAL CONTROL OF TISSUE FORMATION AND CELL FUNCTION (CTE)

MULTISCALE ANALYSIS OF CARTILAGE AND INTERVERTEBRAL DISC (SOLIDS)

SOFT TISSUE MECHANOBIOLOGY (SOLIDS)

HEART VALVE FLOW AND FUNCTION (FLUIDS/SOLIDS)

CARDIOVASCULAR DEVICE (FLUIDS/SOLIDS)

ACTIVE LEARNING IN BIOMECHANICAL ENGINEERING EDUCATION (EDU)

HYPERTHERMIA, CRYOTHERAPY, AND CRYOPRESERVATION (BTR)

DISEASE MODELS AND ENGINEERED THERAPIES (CTE)

MECHANICS AND MODELING OF MUSCULOSKELETAL SOFT TISSUES (SOLIDS)

MECHANICAL REGULATION OF REMODELING AND REPAIR (CTE)

THERAPEAUTIC MATERIALS FOR REPAIR AND REGENERATION (CTE)

PHD PAPER COMPETITION: IMAGING, BIOFLUID MECHANICS, AND BIOTRANSPORT

PHD PAPER COMPETITION: TISSUE MECHANICS AND CHARACTERIZATION

PHD PAPER COMPETITION: CELLULAR MECHANICS AND MECHANOBIOLOGY

PHD PAPER COMPETITION: DISEASES, INJURY, AND REMODELING

PHD PAPER COMPETITION: EXTRACELLULAR MATRIX BIOMECHANICS

PHD PAPER COMPETITION: BIOMATERIALS AND MATERIAL-CELLULAR INTERACTION

UNDERGRADUATE DESIGN COMPETITION

UPPER AND LOWER EXTREMITY JOINT MECHANICS (SOLIDS)

HEAD INJURY & INJURY BIOMECHANICS 1 (SOLIDS)

VASCULAR MECHANICS (SOLIDS)

IMAGING AND DIAGNOSTICS (FLUIDS)

NANO AND MIRCROTHERAPEUTICS (BTR)

MECHANICAL REGULATION OF MORPHOGENESIS (CTE)

HEAD INJURY & INJURY BIOMECHANICS 2 (SOLIDS)

REPRODUCTIVE BIOMECHANICS (SOLIDS)

TENDON TISSUE ENGINEERING AND REGENERATION (CTE/SOLIDS)

ANEURYSM (FLUIDS)

EXPERIMENTAL MODELING FOR CLINICAL SURGICAL APPLICATIONS (DDR/IAB)

VASCULAR, LYMPHATIC, AND OCULAR TRANSPORT (BTR)

MEASURING AND MODELING CELL MECHANCIS AND THE MICROENVIRONMENT (CTE)

HEAD INJURY & INJURY BIOMECHANICS 3 (SOLIDS)

REPRODUCTIVE, OCULAR, AND GASTROINTESTINAL BIOMECHANICS (SOLIDS)

ANEURYSM MECHANICS (SOLIDS)

VOLUME 2

TENDON MECHANICS AND STRUCTURE (SOLIDS/CTE)

PEDIATRIC FLOW (FLUIDS)

SURGICAL DEVICE DESIGN APPLICATIONS (DDR/IAB)

MULTI-SCALE MEASURES AND MODELS OF ENGINEERED MATERIALS AND TISSUES (CTE)

POSTER SESSIONS

BACHELORS LEVEL STUDENT PAPER COMPETITION I – DYNAMICS & INJURY, DEVICES, AND IMAGING

BACHELORS LEVEL STUDENT PAPER COMPETITION II – FLUIDS & MICROFLUIDICS, CELLULAR & TISSUE MECHANICS, PHYSIOLOGY & DISEASES

MASTERS LEVEL STUDENT PAPER COMPETITION I – PHYSIOLOGY & DISEASES, CELLULAR & TISSUE MECHANICS, DEVICES

MASTERS LEVEL STUDENT PAPER COMPETITION II – DYNAMICS & INJURY, FLUIDS & MICROFLUIDICS, BIOTRANSPORT & HEAT TRANSFER

BIOTRANSPORT POSTERS

EDUCATION POSTERS

DESIGN, DYNAMICS, AND REHABILITATION POSTERS

CELL AND TISSUE ENGINEERING POSTERS – MECHANOBIOLOGY AND THE MICROENVIRONMENT

CELL AND TISSUE ENGINEERING POSTERS – TISSUE ENGINEERING AND DISEASE MODELS

CELL AND TISSUE ENGINEERING POSTERS – MEASUREMENTS AND MODELING IN CELL AND TISSUE ENGINEERING

SOLIDS POSTERS – BONE AND CARTILAGE

SOLIDS POSTERS - MUSCULOSKELETAL

SOLIDS POSTERS – TISSUE MECHANICS

SOLIDS POSTERS - CARDIOVASCULAR

FLUIDS POSTERS – CARDIOVASCULAR DEVICES, VALVES, AND FLOWS

FLUIDS POSTERS – CARDIOVASCULAR DIAGNOSTICS AND FLOW ANEURYSMS

SB^{3}C2017
Summer Biomechanics, Bioengineering and Biotransport Conference
June 21 – 24, Tucson, AZ, USA

ALTERATIONS IN MECHANICAL PROPERTIES AND *IN VIVO* GEOMETRY OF THE MITRAL VALVE FOLLOWING MYOCARDIAL INFARCTION

Bruno V. Rego (1), Salma Ayoub (1), Amir H. Khalighi (1), Andrew Drach (1), Joseph H. Gorman III (2), Robert C. Gorman (2), Michael S. Sacks (1)

(1) *Center for Cardiovascular Simulation*
Institute for Computational Engineering and Sciences
Department of Biomedical Engineering
The University of Texas at Austin
Austin, TX, USA

(2) *Gorman Cardiovascular Research Group*
Perelman School of Medicine
University of Pennsylvania
Philadelphia, PA, USA

INTRODUCTION

Each year, more than 40,000 people in the United States undergo surgery for mitral valve (MV) repair, frequently as a treatment for mitral regurgitation brought on by myocardial infarction (MI) [1]. Currently, the preferred method for repairing the MV is through the insertion of an annuloplasty ring, which constricts the mitral annulus sufficiently for the leaflets to coapt and thus for valve function to be restored. Long-term efficacy of this repair procedure remains a major challenge, however. A full 30% of patients experience recurrence of ischemic regurgitation within 6 months of surgery, and more than 60% have regurgitation within 5 years [2]. Repair failure is primarily attributed to changes in the geometry of the MV annulus, whose normal saddle shape is severely flattened post-MI and during ring implant [3]. The boundary conditions on the MV are further perturbed by permanent displacement of the papillary muscles (PMs), which coincides with post-MI left ventricular distention. These dramatic changes cause stress overload throughout the valve apparatus, leading to annuloplasty ring suture rupture, irreversible leaflet tissue damage, and ultimately the return of ischemic regurgitation [4,5].

In addition to altered loading, growth and remodeling in the MV following MI and after surgery undoubtedly also play a major role in determining the success or failure of repair. Recent work has uncovered evidence of significant cell activation and matrix turnover in the MV leaflets following MI [6]. Despite its importance, however, the mechanism through which the MV remodels in response to MI—as well as what mediates this process—remains poorly understood. In the present study, we examined how the mechanical properties and *in vivo* geometry of the MV change after MI, in an effort to gain a deeper understanding of the driving factors behind post-MI valvular remodeling. This foundation is an essential prerequisite for future efforts to design, optimize, and simulate novel MV repair devices and surgical strategies.

METHODS

Posterolateral MI was induced in three adult male Dorsett sheep, following established procedures [7]. Echocardiography was performed to obtain 3D ultrasound scans of the left ventricle and the MV prior to infarction, immediately following infarction, and at 8 weeks post-MI. At 8 weeks, animals were sacrificed and MV tissues were explanted for mechanical, microscopic, and histological analysis. The belly region of the anterior leaflet from each MV specimen was isolated for multiple protocols of biaxial mechanical testing up to a peak stress of 200 kPa. The same belly region specimen from each animal was then partitioned and fixed into subsamples for histological, microscopic, and biochemical analyses. To evaluate changes in the architecture of the extracellular matrix owing to MI, samples were imaged using transmission electron microscopy (TEM).

Ultrasound scans from diastole and systole (open and closed valvular states, respectively) were manually traced to separately label the MV annulus, anterior leaflet, posterior leaflet, coapted region, and commissure points (Fig. 1a, panels 1 and 3). The ultrasound tracings were then computationally analyzed to identify significant changes in the *in vivo* shape and function of the MV following MI. In addition to examining the geometry of the valve directly, we have developed a framework to estimate *in vivo* strains throughout the leaflet by integrating the ultrasound tracings into a finite element (FE) modeling approach, in which we exploit the measured mechanical properties of each specimen (Fig. 1a). This methodology will be validated using a database of *in vitro* MV closure images, acquired using high-resolution computed tomography (CT) (Fig. 1b) [8].

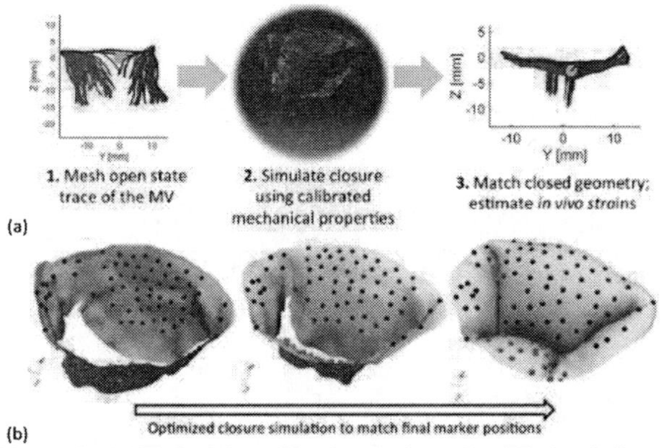

Figure 1: (a) Estimation of *in vivo* strains using FE simulations. (b) Validation of FE approach using *in vitro* micro-CT data.

RESULTS

Biaxial testing results showed that at 8 weeks post-MI, the mechanical properties of the MV anterior leaflet are severely disturbed. While normal MV leaflets are much more extensible in the radial direction than in the circumferential direction (Fig. 2a), the anisotropy of post-MI specimens was the opposite (Fig. 2b). This reversal is consistent with a large radial permanent set resulting from ventricular distention and subsequent PM displacement, and indicates that the leaflet's fiber structure undergoes dramatic changes post-MI. TEM images of MV leaflet tissue supported this conclusion, showing visible damage to the collagen network following infarction (Fig. 3). Unlike healthy tissue, the post-MI extracellular matrix exhibited a collection of shorter-than-normal fibrils that were largely unbound, possibly as a result of overload-induced collagen rupture. Ultrasound scans showed that, *in vivo*, the MV annulus is substantially dilated post-MI (Fig. 4), consistent with previous findings [9]. Moreover, the annulus loses its normal saddle shape in diastole (Fig. 5a,b) and the leaflets no longer exhibit tenting in closure (Fig. 5c,d).

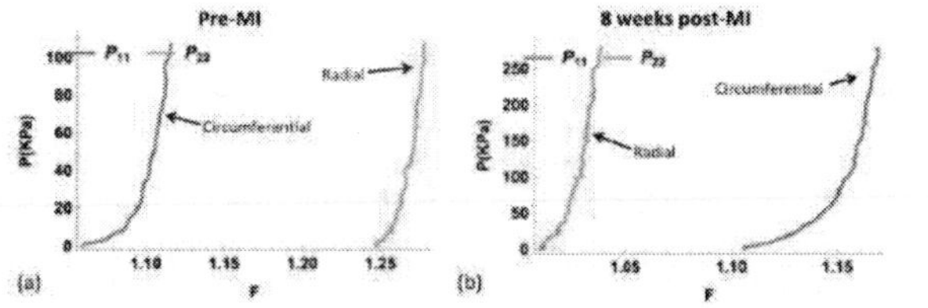

Figure 2: Equibiaxial response of (a) a healthy MV and (b) an MV after MI. Ventricular distention causes a reversal of anisotropy.

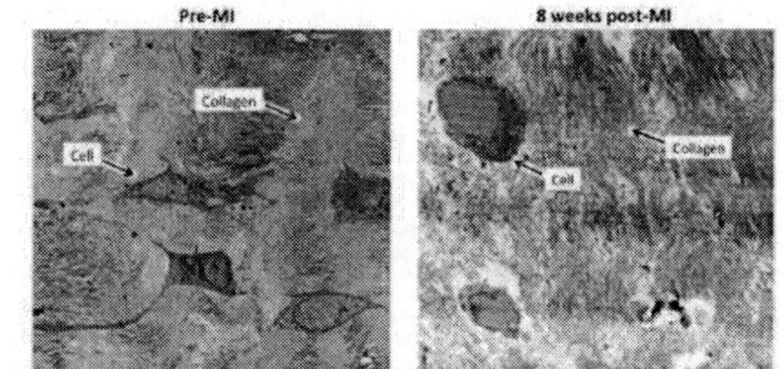

Figure 3: TEM images of (a) healthy and (b) post-MI MV leaflet tissue. While healthy collagen is tightly packed, this architecture is disturbed post-MI, resulting in a loose network of shorter fibrils.

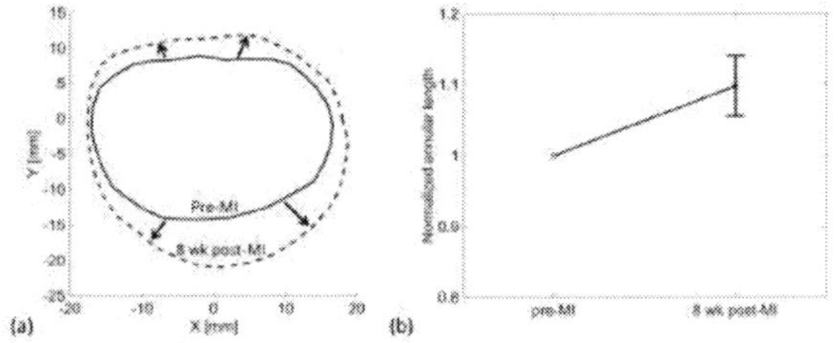

Figure 4: (a) Top view of the annulus pre- and post-MI, showing notable dilation. (b) Relative increase in annular length post-MI.

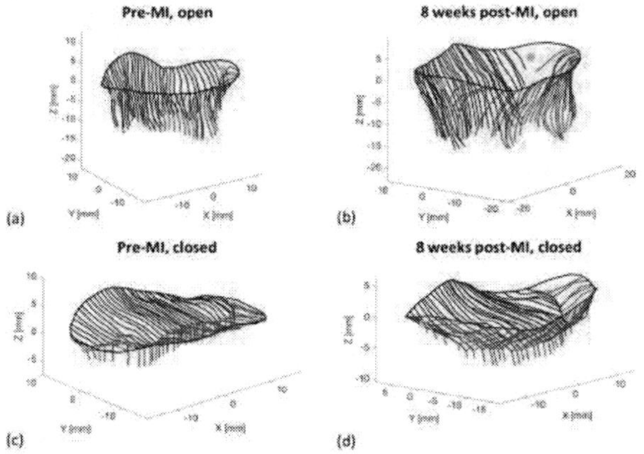

Figure 5: Changes in the *in vivo* geometry of the MV following MI. In diastole (a,b), the annulus loses its normal saddle shape. In systole (c,d), the leaflets lose their normal tenting behavior.

DISCUSSION

Our results shed significant light on the post-MI mechanical and geometric properties of the MV. These findings are highly relevant to the design of MV repair devices and the optimization of surgical strategies, as they directly elucidate the state of the valve at each time of imaging as well as at the time of explant. While current MV repair endeavors largely seek to return the valve to its pre-MI state, our results suggest that this approach may not be favorable, especially given the substantial changes in mechanical properties. Instead, an effort to place the MV in an alternative homeostatic state may lead to decreased repair failure. This overarching goal is a main driver in our ongoing work to (1) estimate how strains experienced by the MV leaflets change over time after MI, and (2) draw connections between tissue-level deformations and cellular behavior *in vivo*.

ACKNOWLEDGEMENTS

This work is supported by the National Institutes of Health grant no. R01-HL119297 to M.S.S. and J.H.G., and by the National Science Foundation grant no. DGE-1610403 to B.V.R.

REFERENCES

[1] Glower, DD et al., *J Am Coll Cardiol*, 64:172-181, 2014.
[2] McGee, EC et al., *J Thorac Cardiovasc Surg*, 128:916-924, 2004.
[3] Gorman, JH et al., *Ann Thorac Surg*, 77:544-548, 2004.
[4] Gillinov, AM et al., *Ann Thorac Surg*, 72:20-27, 2001.
[5] Go, AS et al., *Circulation*, 129:28-292, 2014.
[6] Dal-Bianco, JP et al., *J Am Coll Cardiol*, 67:275-287, 2016.
[7] Robb, JD et al., *Eur J Cardiothorac Surg*, 40:1501-1507, 2011.
[8] Khalighi, AH et al., *Func Imag Model Heart*, 164-171, 2015.
[9] Daimon, M et al., *Echocardiography*, 27:1069-1077, 2010.

SB^{3}C2017
Summer Biomechanics, Bioengineering and Biotransport Conference
June 21 – 24, Tucson, AZ, USA

BLOCKING CADHERIN-11 PREVENTS CALCIFIC AORTIC VALVE DISEASE IN NOTCH1$^{+/-}$ MICE

Cynthia R. Clark, Meghan A. Bowler, J. Caleb Snider, and W. David Merryman

Biomedical Engineering
Vanderbilt University
Nashville, TN, USA

INTRODUCTION

Calcific aortic valve disease (CAVD) affects approximately up to one third of the elderly population accounting for ~15 000 patient deaths per year in the United States. Currently, therapeutic intervention does not exist for patients until severe stenosis requires surgical replacement of the valve. Studies to identify the molecular mechanism of CAVD are essential to the discovery of a nonsurgical strategy.

Mutations in *Notch1* cause heritable CAVD in humans and mice [1]. We have previously demonstrated an enhanced myofibroblast phenotype in the aortic valve interstitial cells (AVICs) of *Notch1*$^{+/-}$ mice [2]. *Notch1*$^{+/-}$ AVICs have increased cadherin-11 (CDH11) and are hypersensitive to mechanical strain as demonstrated by the formation of calcific nodules [3]. More recently, Hadji et al demonstrated that increased expression of long noncoding (lnc) RNA *H19* decreased *Notch1* expression and leads to idiopathic CAVD [4]. Additionally, knockdown of *H19* in human VICs decreased *CDH11* by ≈30%. These observations have led us to speculate that CDH11 may be the common mediator of Notch1 receptor dysfunction that results in both heritable and idiopathic CAVD [5]. Here we sought to determine if a CDH11 blocking antibody, SYN0012, would slow or prevent aortic valve pathophysiology in *Notch1*$^{+/-}$ mice.

METHODS

Notch1$^{+/-}$ mice were given a high fat/ high cholesterol (HFD) chow at 10 weeks. At 4 months, the mice began a regimen of 10 mg/kg SYN0012 or IgG2a by intraperitoneal (IP) injection once a week for 8 weeks. *Notch1*$^{+/-}$ and *Notch1*$^{+/-}$;*Cdh11*$^{+/-}$ mice were introduced the HFD at 6 months and aged to 12 months. Aortic valve velocity (V_{max}) and ejection fraction (EF) was measured from aortic PW Dopplar Mode and parasternal short axis M-Mode echocardiographic images, respectively. Valves were dissected and immediately embedded in OCT or flash frozen on dry ice and placed at -80°C. 7μm sections were used to measure topography and stiffness by atomic force microscopy (AFM) and for alizarin red staining. RNA was isolated for cDNA synthesis and qPCR.

RESULTS

Notch1$^{+/-}$ mice administered IgG2a showed a significant increase in their aortic valve peak velocity from 4 months to 6 months; whereas, mice treated with SYN0012 did not change (Fig. 1A). Moreover, the EF/V_{max} ratio was decreased in IgG2a treated mice compared to no changed in the SYN0012 group over the treatment cycle (Fig. 1B); note that a decrease in EF/V_{max} indicates a valve dysfunction. Taking a transgenic approach, *Notch1*$^{+/-}$;*Cdh11*$^{+/-}$ mice had decreased peak velocity (Fig. 1C) and increased EF/V_{max} (Fig. 1D) relative to aged matched *Notch1*$^{+/-}$ mice. These data suggests that decreased CDH11 function prevented aortic valve stenosis.

Histology revealed hyperplastic leaflets in mice treated with IgG2a. Conversely, SYN0012 treated mice had thin leaflets, indicative of healthier valve morphology (Fig. 2). AFM of unfixed tissue sections demonstrated that SYN0012 leaflets were significantly less stiff than the IgG2a group (Fig. 3). Additionally, expression of genes associated with inflammation and fibrosis, *Ykl40* and *Il6*, were decreased in the valves of the SYN0012 group (Fig. 4). These analyses demonstrate that administering SYN0012 stopped the aortic valve pathological phenotype normally observed in *Notch1*$^{+/-}$ mice. Blocking CDH11 appears to have prevented inflammation and fibrosis, thus reducing the thickening and stiffening of the valve leaflets.

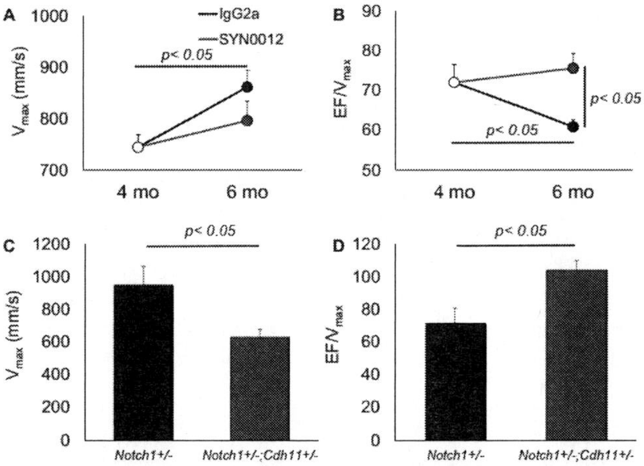

Figure 1: CDH11 blocking antibody (SYN0012) slowed the progression of V_{max} from 4-6 months (A). Note the IgG2a treatment is significantly increased at 6 mo versus 4 mo, but the SYN0012 is not different and SYN0012 also prevented the decrease in EF/V_{max} which is an indicator of CAVD (B). Similarly, mice with both *Notch1* and *Cdh11* mutations have decreased aortic velocity (C) and increased EF/V_{max} (D) versus mice with *Notch1* mutations alone.

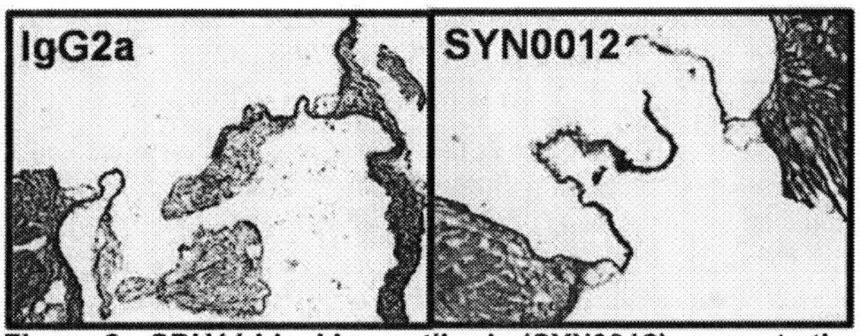

Figure 2: CDH11 blocking antibody (SYN0012) prevents the characteristic thickening observed in Notch1+/- mice treated with IgG2a.

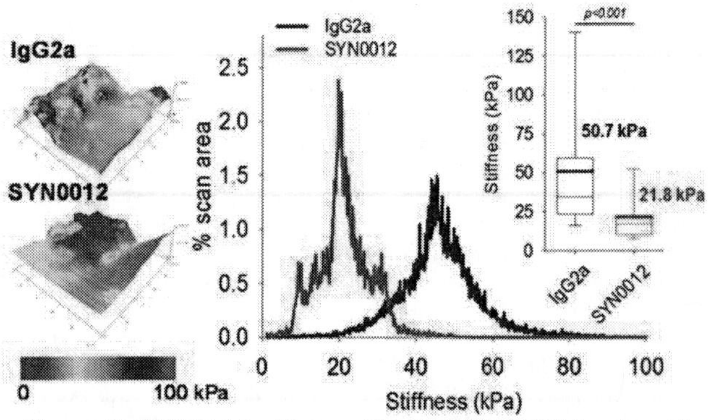

Figure 3: CDH11 blocking antibody (SYN0012) treatment leads to decreased tissue stiffness as measured by AFM.

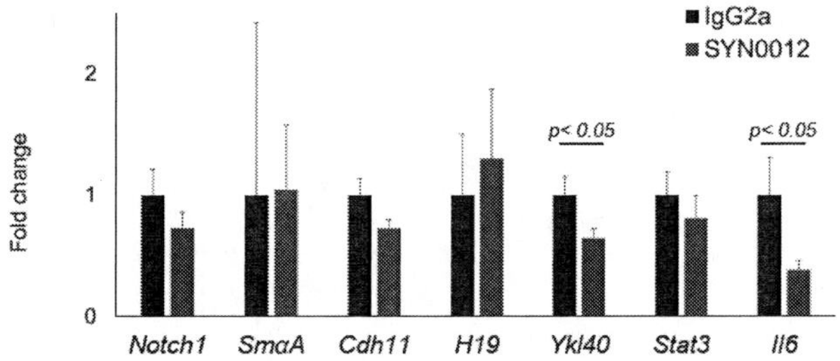

Figure 4: qPCR results of leaflets following SYN0012 treatment for two months. Inflammatory markers, Ylk40 and Il6 were decreased due to SYN0012 treatment. Other hallmarks of CAVD were not changed.

DISCUSSION

Here we report that an anti-CDH11 blocking antibody prevented or slowed the progression of CAVD in $Notch1^{+/-}$ mice and suggests that CDH11 may be a novel target for non-surgical treatment in human CAVD patients. We used a transgenic model of decreased *Cdh11* expression to confirm our results were specific to the inhibition of CDH11. $Notch1^{+/-}$;$Cdh11^{+/-}$ mice fed HFD and aged to one year showed notable differences in their aortic valve hemodynamic functions compared to $Notch1^{+/-}$ mice. Reduced *Cdh11* appears preserve the health of the tissue while no signs of stenosis were observed. Although the molecular mechanism driving myofibroblast differentiation and tissue calcification, hallmarks of CAVD, in aortic valve leaflets is unclear, several sources of data have implicated CDH11 as a key molecule in the pathology of the disease. *CDH11* is upregulated in the valves of human CAVD patients [3] and overexpression causes aortic valve calcification in mice [6], emphasizing the importance of understanding this mechanism. Ongoing studies are pursuing how CDH11 localization and signaling changes the mechanobiology of AVICs to a pathological phenotype. Taken together, these data reveal an exciting new pharmacological strategy that should be further investigated in hopes of preventing CAVD in humans.

ACKNOWLEDGEMENTS

This research was supported by the NIH (R01 HL115130 and R35 HL135790). We acknowledge the Vanderbilt Cardiovascular Physiology Core for performing echoes, and Dominik Hartl and Uwe Junker from Roche for supplying the blocking antibody, SYN0012.

REFERENCES

[1] Garg, V, et al., *Nature*, 437:270-4, 2005.
[2] Chen, J, et al, *ATVB*, 35:1597-605, 2015.
[3] Hutcheson, JD, et al, *ATVB*, 33:114-20, 2013.
[4] Hadji, F, et al, *Circulation*, 134:1848-62, 2016.
[5] Clark, CR and Merryman, WD, *Circulation*, 134:1863-5, 2016.
[6] Sung, DC, et al, *ATVB*, 36:1627-37, 2016.

SB³C2017
Summer Biomechanics, Bioengineering and Biotransport Conference
June 21 – 24, Tucson, AZ, USA

COAPTATION ZONE IN TRICUSPID ANNULUS CINCHING: A PILOT STUDY USING AN EX-VIVO PORCINE MODEL

Ashley Thomas (1*), Paola Diaz-Portela (1*), Edward Y. Sako (2), Shamik Bhattacharya (1)

(1) School of Science, Engineering and Technology
St. Mary's University
San Antonio, Texas, United States

(2) Cardiothoracic Surgery
University of Texas Health Science Center at San Antonio
San Antonio, Texas, United States

(*) The first two authors have made equal contributions to this study

INTRODUCTION

The tricuspid valve (TV) is a one-way valve made of three leaflets that forms the boundary between the right ventricle and the right atrium of the heart. A fully functioning TV maintains the direction of blood flow from the atrium into the ventricle. However, an enlarged TV could result in tricuspid regurgitation where the blood flows back into the atrium [1], often associated with heart failure. To correct this pathology, Tricuspid annuloplasty (TAP) is currently the most preferred surgical technique. In TAP, the dilated annulus is septa-laterally cinched to reduce regurgitation. Reports of recurring regurgitation, septal dilation and ring dehiscence after TAP indicates leaflet malcoaptation [2]. We hypothesize that the coaptation (Figure 1) of the tricuspid valve is changed as the annulus is cinched, thereby affecting the valve performance. The length of the coaptation zone (CZ) is a key indicator for leaflet coaptation and may be more important than the degree of valve area reduction. Previous studies support that an increased cinching force results in annulus area reduction [3]. However, the change in CZ due to cinching has not yet been investigated.

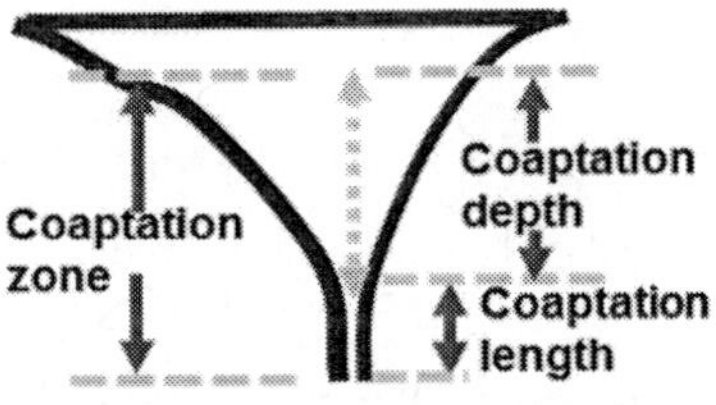

Figure 1 – Coaptation Zone

METHODS

This experiment is based on a previous model of tricuspid annulus cinching [3]. Porcine hearts with intact aortas are collected from a local abattoir and are cryopreserved until the day of the experiment upon which the hearts were thawed in a warm water bath at 37.4°C. The heart is dissected to expose the right ventricle and the TV. Sonomicrometry transducers consisting of small piezo-electric crystals that both receive and emit ultrasonic pulses is used to quantify distances within the TV. These distances can be continuously traced using SonoLAB™ and analyzed using SonoXYZ™ to quantify or develop an image of the data. Before the heart is placed in the chamber, eight Sonometric crystals are sutured on the septal and anterior leaflets (Figure 4). A suture is inserted into tissue adjacent to the tricuspid annulus. One end of the suture is tied and anchored at the anteroseptal commissure and the other end is exteriorized through the posteroseptal commissure (Figure 2).

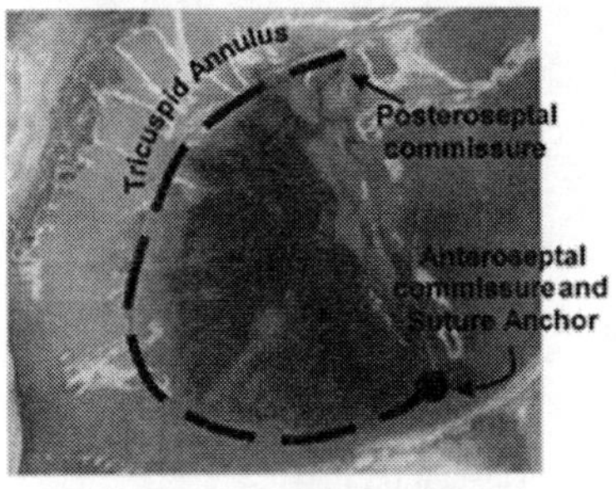

Figure 2- Suture Positioning

The free-end (Figure 2) of the suture is kept parallel to the septa-lateral direction without touching the tissue and is attached to a lever and force transducer mounted on a slider system (Figure 3A). The slider moves the force transducer in a given direction at regular intervals as shown (Figure 3A). The tricuspid valve is closed by pressurizing the right ventricle through the pulmonary valve, using a custom built static pressure head system (Figure 2 & 3A). The system is immersed in a Plexiglas chamber filled with physiological saline (Figure 3A). The suture will be pulled at 5 mm increments to cinch after the sonomicrometry crystals are implanted in the leaflets (Figure 3B & Figure 4).

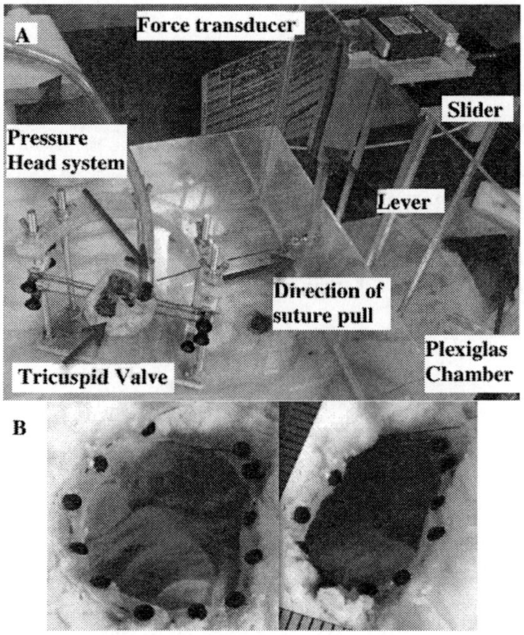

Figure 3- A) Experimental Setup B) Image Of Tricuspid Annulus Showing Reduction In Area Due To Cinching After Pressurization Of The Ventricle [3].

A plane on each of the leaflets is created with the crystals (Figure 4) to trace the length of the coaptation zone (CZ) using the Sonometrics software and analyze to form an image in SonoXYZ (Figure 4). SonoXYZ gives, X, Y, and Z coordinates in reference to the crystals set as the origin. Three are placed on the septal leaflet, with two at the top and one at the bottom, and four are placed on the anterior leaflet (Figure 4). One crystal is placed on the septum between the right and left ventricle where the heart will not move so that it may serve as the origin in the 3-D coordinate system for the analysis. When the leaflets coaptate, the planes will also meet and CZ is calculated (Figure 5).

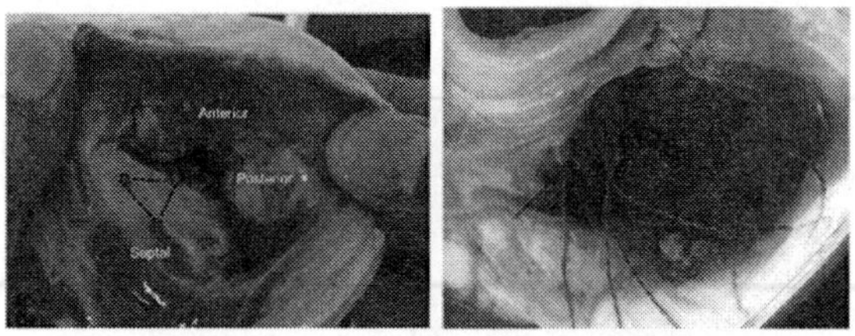

Figure 4 – Sonometric Crystal Positioning

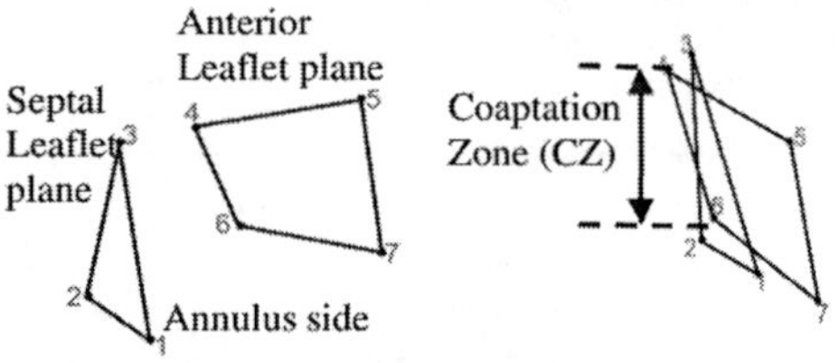

Figure 5– Plane Formation and Coaptation Zone (CZ) Measurement

RESULTS

The preliminary results of the experiment found the CZ of a normal porcine heart (267 g, 30 mm ring sizer) to be 8.78 mm based on the image generated by SonoXYZ (Figure 6). The CZ is the length formed by overlapping of two planes (septal and posterior) We used the marker method (Figure 7) which is commonly used by surgeons as a **'thumb rule'** to verify this. The length $\overline{AB}$ = 9.58 mm and length $\overline{CD}$ = 7.31 mm. Here the length is measured using ImageJ software. This is a preliminary study to get the length of coaptaion zone (CZ) using sonocrystals. We are not able to create two perfect parallel planes which overlapped each other but it seems we are pretty close. We will continue to improve the technique by doing more experiments.

In addition to this result, we are able to cinch the annulus, and we observed a change in CZ as we cinch the annulus using the (Figure 3B) experimental set up. At present we are trying different configurations for optimal positioning of the Sonometric crystals and the different intervals of cinching along with the cinching force.

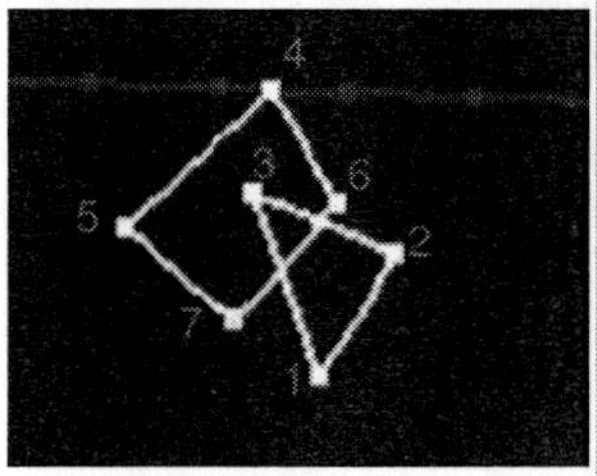

Figure 6 – Overlapping of the Planes of Normal TV during Coaptation from SonoXYZ

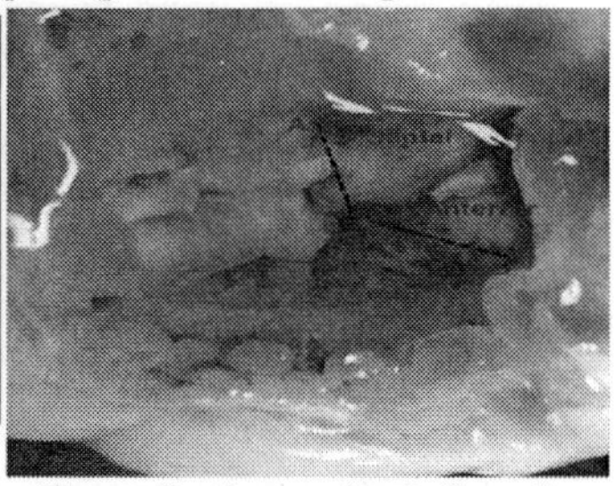

Figure 7 – Approximate Location of the Coaptation Zone using the Marker Technique

DISCUSSION

While feasible, there is not enough data to show that 8.78 mm is the best measurement of the CZ. It is expected that the length of coaptation zone (CZ) will vary with change in annulus area (Figure 3B) and cinching force. Therefore, an optimized CZ relevant to physiological conditions can be derived by correlating the CZ with other parameters like leakage, area reduction and cinching force. The optimal CZ, can be an important decision making tool for surgeons performing TAP and contribute in the development of percutaneous technique along with the cinching force.

ACKNOWLEDGEMENTS

This work is supported by the San Antonio Area Foundation Biomedical Research grant. We sincerely thank Dr. Gus Kalogeros from the Sonometrics Corporation for all his help and are grateful for all the support given by Dr. Tim Raabe, the Associate Dean of the School of Science Engineering and Technology at St. Mary's University. Many thanks to Dr. Morten Olgaard Jensen (Biomedical Engineering, University of Arkansas, Fayetteville) for his important suggestions, which improved this abstract.

REFERENCES

[1] Matsunaga, Akira, and Carlos MG Duran. "Progression of tricuspid regurgitation after repaired functional ischemic mitral regurgitation." *Circulation* (112.9 suppl):I-452, 2005.

[2] Calafiore, Antonio Maria, et al. "A single-size band, 50 mm long, for tricuspid annuloplasty." *European Journal of Cardio-Thoracic Surgery* 34.3:677-679, 2008.

[3] Adkins, Amy, et al. "Force Required to Cinch the Tricuspid Annulus: An Ex-Vivo Study." *J Hearth Valve Dis* 24.5:644-652, 2015.

EFFECT OF RESTRICTING MITRAL VALVE ANNULAR CONTRACTION ON ANTERIOR LEAFLET STRAIN: AN IN VITRO STUDY

Thomas F. Easley (1,2), Vinay Bhal (3), Charles H. Bloodworth (1,2), Ajit P. Yoganathan (1,3)

(1) Parker H. Petit Institute for Bioengineering and Bioscience
Georgia Institute of Technology
Atlanta, GA, USA

(2) George W. Woodruff School of Mechanical Engineering
Georgia Institute of Technology
Atlanta, GA, USA

(3) Wallace H. Coulter Department of Biomedical Engineering
Georgia Institute of Technology
Atlanta, GA, USA

INTRODUCTION

In mitral valve (MV) disease, poor valvular geometry (e.g. annular or ventricular dilation) or tissue degeneration, results in improper coaptation of the mitral valve leaflets leading to mitral regurgitation (MR). This systolic backflow of blood from the ventricle to the atrium is commonly repaired with a restrictive annuloplasty surgical procedure. The annuloplasty ring is implanted into the MV to reduce the annular area by bringing the MV leaflets closer together to provide better leaflet coaptation during systole and reduce MR [1]. Implantation of an annuloplasty ring restricts the motion of the mitral valve throughout the cardiac cycle, with stiffer rings imparting more severe restriction. Although this surgical repair is an effective treatment, studies have shown MR recurrence as high as 30% in patients with ischemic MR within the first 6 months after operation [2]. Assessment of healthy annular dynamics and its effect on the MV leaflets could help in further understanding long-term MV repair outcomes of the annuloplasty procedure.

A recently developed *in vitro* left heart simulator provided a dynamically contracting MV annulus capable of replicating the MV annular contraction over the cardiac cycle [3]. Using this simulator, we were able to study the effects of a contracting MV annulus on anterior leaflet strain compared to a static MV annulus, i.e. resulting from a rigid annuloplasty repair.

METHODS

Experimental Setup

Healthy mitral valves (n=8) were excised from fresh ovine hearts (Superior Farms, Denver, CO). During excision, the annuli and subvalvular structures (i.e. chordae tendineae and papillary muscles) were preserved. The MVs were then mounted into the left heart chamber by suturing the MV annulus to the annulus cuff and attaching the papillary muscles (PM) to the PM positioning rods (Figure 1).

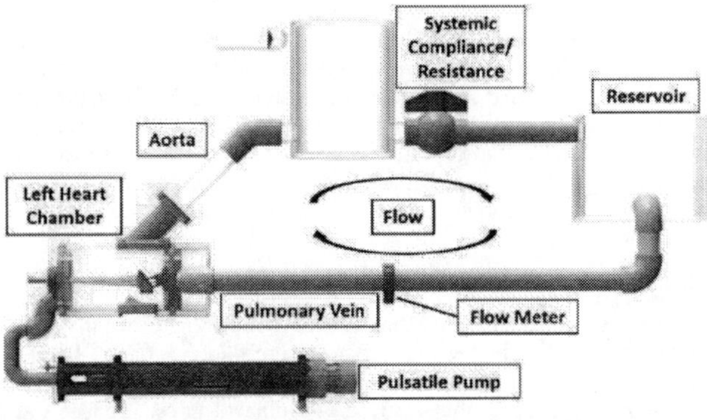

Figure 1. Schematic of left heart simulator complete flow loop.

Before mounting the chamber into the flow loop, a grid of markers was manually placed on the MV leaflets using tissue dye (Fisher Scientific, Hampton, NH) (Figure 2).

Experimental Testing

A novel contracting annulus was employed, where two linear actuators (HAD2-2, RobotZone, Winfield, KS) were used in conjunction to generate a cyclic annular area change, as previously reported [4]. This area change was matched to that of healthy human subjects, as measured by 3D echocardiography [4]. Three states of annular function were tested: 1) healthy, 2) undersized, and 3) true-sized. To simulate a true-sized restrictive annuloplasty, the static, peak-diastolic state was chosen as it would mimic the annular size of a flaccid heart (annular area = 5.5 cm²). In addition, to simulate an under-sized restrictive annuloplasty, the static, peak-systolic state was chosen as it would mimic the reduction in annular size (annular area = 4.5 cm²). Both states were the maximum and minimum annular area of the healthy dynamic contraction used as the healthy state, respectively.

The left heart simulator was tuned to normal hemodynamics for all testing conditions (5 L/min cardiac output, 70 beats/min, 35/65 systole/diastole ratio, 120 mmHg peak systolic pressure). In order to control the linear actuators and pulsatile pump (Vivitro Labs, Victoria, BC, CA), and record simulator hemodynamics, a custom code was written in LabVIEW (2015, National Instruments, Austin, TX).

Stereophotogrammetry was used to track a 3x3 marker grid on the central A2 scallop (Figure 2) of the MV using two high-speed A504k cameras (Basler Inc., Exton, PA) and XCAP video acquisition software (EPIX, Inc.; Buffalo Grove, IL) at 250 Hz.

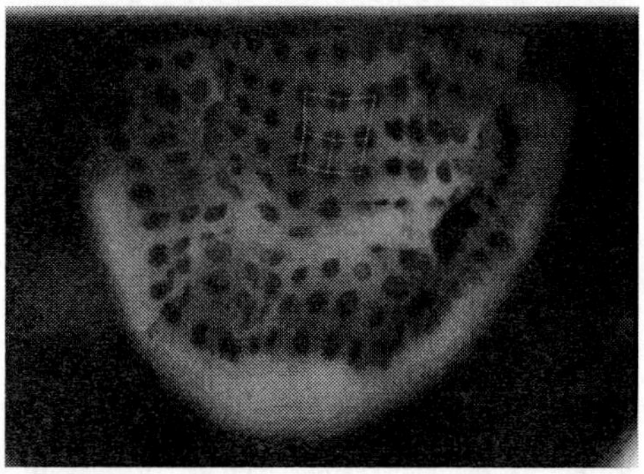

Figure 2. Mitral valve leaflet marked with tissue dye for stereophotogrammetry.

A custom code was written in MATLAB (R2016a, MathWorks, Natick, MA) for computing Green areal strain between peak systole and peak diastole for each state of annular function. For MV visualization, 4D echocardiography via ie33 xMatrix ultrasound system and x7-2 probe (Philips Healthcare, Andover, MA) was performed. From 4D echocardiography leaflet coaptation for each state was measured, in addition to annular area change being measured to ensure proper annular contraction.

RESULTS

The mean ± SEM anterior leaflet strain of the healthy, rigid undersized, and rigid true-sized states were 0.321 ± 0.061, 0.368 ± 0.061, and 0.412 ± 0.050, respectively (Figure 3). The healthy state significantly reduced leaflet strain versus both undersized and true-sized (each $p<0.05$, using Wilcoxon signed-ranked test). There was no significant difference in leaflet midline coaptation length between the three states.

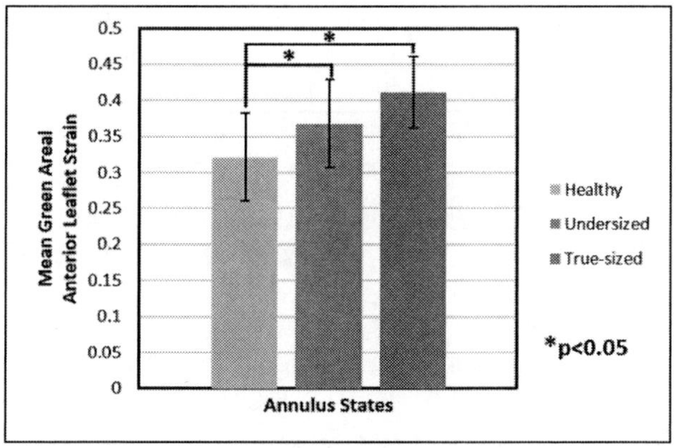

Figure 3. Resultant mean ± SEM of Green areal strain between annular states: healthy, rigid undersized, and rigid true-sized.

DISCUSSION

It was shown *in vitro* that simulated rigid true-sized and rigid undersized annuloplasties increased mitral anterior leaflet strain compared to a healthy contraction. Previous *in vivo* animal studies have shown similar results where rigid true-sized annuloplasty rings increase anterior mitral leaflet strains [5]. Our work also shows that rigid undersized rings can lead to increased strain in the anterior leaflet. However, we did not find a significant difference between rigid true-sized and rigid undersized anterior leaflet strain. Clinical studies are needed to investigate if anterior leaflet strain is increased in actual patients with restrictive annuloplasty rings, and, if so, how it impacts repair failure.

Additionally, there was no significant difference seen in midline coaptation length between the three states. This study used healthy mitral valves and geometry for all experiments. Thus it was not expected that there would be any differences in coaptation between the true-sized and undersized annular areas, in contrast to having an ischemic MV with improper coaptation that is being repaired.

Future *in vitro* work in this area will focus on the effects of diseased contraction pattern, e.g. ischemic, on the biomechanics of the mitral valve compared to that of healthy/normal contraction.

ACKNOWLEDGEMENTS

This work was supported by the National Heart Lung and Blood Institute under Grant R01HL119297. The authors would also like to thank Eric Pierce for his assistance in reviewing the submission.

REFERENCES

[1] Braun, J et al., *Ann Thorac Surg*, 85.2:430-437, 2008.
[2] McGee, E et al., *JTCVS*, 128.6:916-925, 2004.
[3] Easley, T et al, *SB³C2016*, C702:592-593, 2016.
[4] Levack, M et al., *Circ*, 126.11 suppl 1:S183-S188, 2012.
[5] Bothe, W et al., *Circ*, 124.11 suppl 1:S81-S96, 2011.

SB[3]C2017
Summer Biomechanics, Bioengineering and Biotransport Conference
June 21 – 24, Tucson, AZ, USA

IN VITRO LEFT HEART SYSTEM WITH 7T MRI PROVIDES HIGH RESOLUTION MITRAL VALVE 3D IMAGING DATASETS FOR COMPUTATIONAL MODELING

Sam E. Stephens, BS (1); Mariana R. Maisonnette (2); Serguei Liachenko, MD, PhD (3)
Jonathan F. Wenk, PhD (4); Morten O. Jensen, PhD, DrMed (2)

(1) Department of Mechanical Engineering
University of Arkansas
Fayetteville, AR, USA

(2) Department of Biomedical Engineering
University of Arkansas
Fayetteville, AR, USA

(3) Bioimaging Laboratory
Division of Neurotoxicology
FDA National Center for Toxicology Research
Jefferson, AR, USA

(4) Department of Mechanical Engineering
University of Kentucky
Lexington, KY, USA

INTRODUCTION

Computational models of the heart have the potential ability to simulate and predict the performance of medical devices and intervention techniques. This allows early feedback in the design process, ultimately reducing development time and improving patient outcome through personalized procedures. The overall accuracy and the reliability of such computational models however require biomechanical validation. Models of the mitral valve (MV) are based on in vivo/vitro imaging and bench-top measured or assumed material properties without converging with actual force measurements. Including force measurements in a precise replica of individual valves will provide optimal tissue properties. The highest forces on the valve occur at peak left ventricular pressure. Focusing on that time point simplifies imaging and force measurements, allowing for higher imaging quality and more comprehensive force measurements.

The work presented here focuses on MV imaging. While in vitro imaging of MV has advanced tremendously over the last decade [1,2], a number of deficiencies still need to be addressed. The positioning and orientation of each papillary muscle (PM) and annular geometry are vital to proper function. It has been shown that leaflet strain and the chordal force balance are significantly impacted by the physiologically appropriate shape of the annulus [3,4,5] and PM position [6,7,8]. In addition to the overall position of the PMs, the orientation of the muscle tip with respect to the annulus can also have a similar effect since the chordal tendineae (CT) do not all insert into the PMs at a common point, rather having a spatial distribution whose position with respect to the annulus varies with tip angle. As there is substantial variation between individual MVs, controlling PM position/orientation and annular shape to the specific native conditions for the individual valve will maintain native leaflet strain and proper coaptation. The importance of CT distribution [9] is included in current computational models [10], but the location and orientation of the PMs as well as the 3-dimensional annular shape in relation to PM position do not necessarily reflect *in vivo* conditions. Pre-mortem measurements are required to correlate the *in vitro* conditions to that of the native valve in order to validate computational models against realistic geometries. Additionally, depending upon the medium, the valve may also be subject to gravitational forces which distort the shape, most significantly in open valve and low-pressure configurations. The use of a fluid environment keeps the tissue in a more natural and neutrally buoyant configuration.

The following is a description of a left heart apparatus and process by which a high-resolution 3-dimensional image of the mitral valve is obtained with a fluid environment utilizing precise papillary muscle positioning and orientation as well as a saddle-shaped annulus holder. Furthermore, the process is designed such that the apparatus reflects *in vivo* conditions within a particular animal by employing pre-mortem ultrasound imaging.

METHODS

The apparatus for magnetic resonance imaging (MRI) consists of three main components; a clamp for holding the valve annulus, a holder for supporting and positioning the papillary muscles, and a case to mount the clamp and muscle holder and apply fluid pressure (Fig 1). The annulus clamp is customizable for an individual MV using annular dimensions obtained via pre-mortem ultrasonic imaging (Fig 2 a+b). Automated adjustments are made through customized computer aided design software. The clamp utilizes sutures to guide the tissue during implantation and mechanical pressure applied by bolts to hold the valve in place (Fig. 3 a+b). The PM holder configuration is adjusted in a similar manner as the clamp, making use of an automated algorithm for determining several key dimensions of the holder (Fig

2c). The MV dimensions required for the hardware customizations are shown in (Fig 2d). The annular shape of the clamp is governed by the commissure-commissure width, septal-lateral distance, and annular height. The positioning of the PMs is determined through measurements from the posterior commissure to each muscle tip, anterior commissure to each muscle tip, and the mid-anterior annulus to each muscle tip. The PM holder orientation is controlled such that the muscle tips are directed towards the quarter-commissure points, mimicking natural PM orientation.

Both the clamp and muscle holder are 3D printed using a stereolithographic printer (Fig 1). The apparatus is intended to operate a low trans-MV pressures (<10mmHg). The fluid pressure is controlled throughout the imaging procedure to within ±1mmHg to ensure a steady, closed valve.

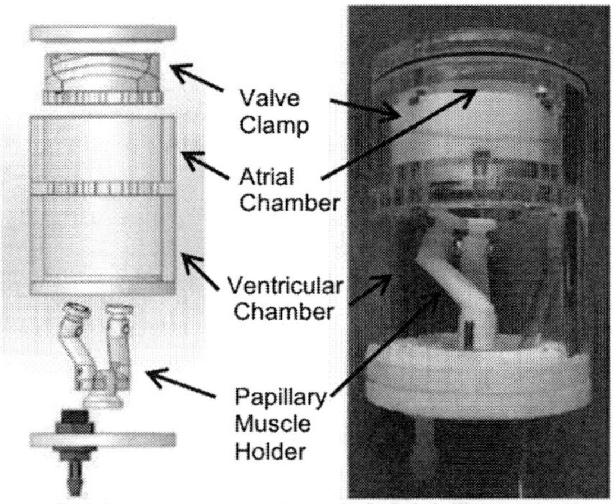

Figure 1: Left heart apparatus in exploded (left illustration) and assembled (right image) views

The apparatus is designed such that once the valve has been imaged, it can be taken from the left heart chamber without removing it from the clamp and PM holder, to be inserted into different apparatus for force measurements [11]. A 7T MRI machine with a 30µm voxel resolution will be used (Bruker, Billerica, MA).

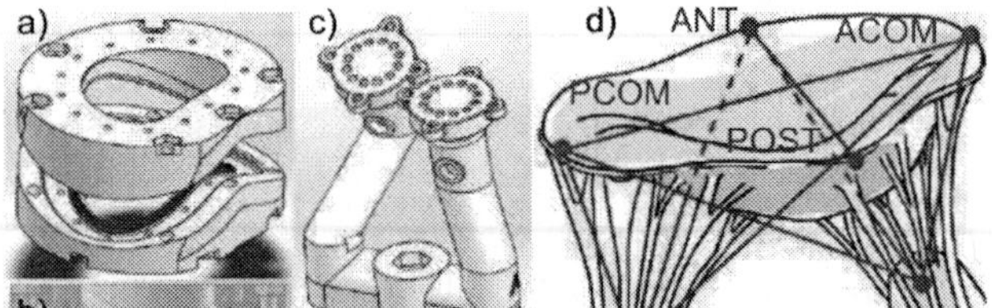

Figure 2: Mitral valve clamp: (a) 3D view, (b) cross section and (c) papillary muscle holder, (d) in vivo ultrasound measurements. APM: Anterior Papillary Muscle. PPM: Posterior Papillary Muscle. PCOM: Posterior Commissure. ACOM: Anterior Commissure. ANT: Anterior Annulus. POST: Posterior Annulus

RESULTS

The materials used were tested in the 7T MRI system and did not distort images. The clamp was tested for accessibility, ease of mounting, and overall strength and rigidity. It was found that the clamp holds the mitral annulus with sufficient pressure to prevent any slipping of the valve with negligible deflection of the clamp. Images of the mounted MV are shown in Fig 3a+b. A pilot 7T MRI scan to evaluate the imaging time and quality was performed, shown in Fig 3c, with attachment points for the mounting system identified. We expect to present a full dataset at SB3C. The valve was completely immersed in a bath of Fomblin that can match the density of blood (Solvay, Brussels, Belgium).

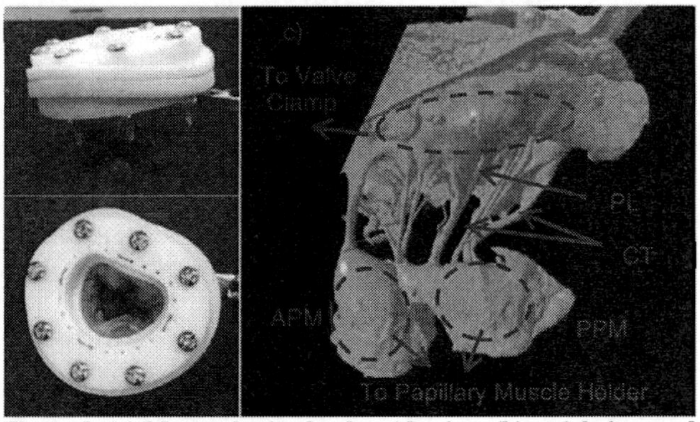

Figure 3: (a) Mounted mitral valve side-view, (b) atrial view, and (c) Preliminary 7T MRI scan of MV segment in fluid with mounting points indicated by dashed regions. APM: Anterior Papillary Muscle. PPM: Posterior Papillary Muscle. PL: Posterior Leaflet. CT: Chordae Tendineae

DISCUSSION

The high resolution image, realistic positioning and conditions, and subsequent detailed static force measurements stand to serve as valuable tools in validating and informing computational models of the MV. This system accurately positions and orients the annulus and PMs to specific *in vivo* positions of the animal from which the valve was removed. In addition, the spatial locations of the PM tips are combined with the correct orientation, as the PM holder maintains appropriate tip angle. This system is designed to use fluid to generate hydrostatic pressure across the valve, preventing tissue drying, folding, and buoyancy issues. Hence, the tissue can be imaged in its natural state without altering tissue mechanical properties and geometry prior to imaging. Finally, the system is designed to be modular, allowing the valve mounting components to be removed and used in another apparatus without first dismounting the valve.

ACKNOWLEDGEMENTS

Funding was provided by the Arkansas Research Alliance, NCTR protocol #C17009, and NSF Grant # CMMI-1538754.

REFERENCES

[1] Rabbah J, Yoganathan A, et al, *ABME,* 2013; 41:305-315
[2] Bloodworth IV C, Yoganathan A, et al, *ABME,* 2016;1-12
[3] Jimenez J, Yoganathan A, et al, *JTCVS,* 2007;134:1592-1598
[4] Jensen MO, Nielsen S, et al, *JTCVS,* 2011;142,3:697-703
[5] Salgo I, Edmunds H, et al, Circulation. 2002; 106:711-717
[6] Jimenez J, Yoganathan A, et al, *JHVD,* 2005;14:295-302
[7] He S, Yoganathan A, et al, *JHVD,* 2000;9:495-501
[8] Cochran R, Kunzelman K, et al, *ANS,* 1998; 66:S155-161
[9] Khalighi AH, Sacks MS, et al, *ABME,* 2016; Dec 19 *[EPUB]*
[10] Toma M, Kunzelman K, et al, *ABME,* 20116; 44:942-953
[11] Rahmani A, Jensen M, *JHVD,* 2013; 22:28-35

SB³C2017
Summer Biomechanics, Bioengineering and Biotransport Conference
June 21–24, 2017, Tucson, Arizona, USA

LONG-TERM GROWTH OF CALCIFIC AORTIC VALVE DISEASE: A MECHANOBIOLOGY MODEL

Amirhossein Arzani (1), Kristyn S. Masters (2), Mohammad R. K. Mofrad (1)

(1) Department of Bioengineering
University of California, Berkeley
Berkeley, CA, USA

(2) Department of Biomedical Engineering
University of Wisconsin
Madison, WI, USA

INTRODUCTION

Calcific aortic valve disease (CAVD) is a common form of aortic stenosis where calcific nodules stiffen the valve leaflets, impairing valve functionality. It is well known that CAVD primarily occurs on the aortic side of the valve where disturbed hemodynamic conditions exist. The progression of CAVD is a multiscale process. The biochemical events take place on the biomolecular and cellular scales, whereas the hemodynamics cover a larger scale. Temporally, notable calcification progression happens during months/years, whereas the hemodynamics time scale is a cardiac cycle. Furthermore, CAVD initiation and progression is a multi-stage disease [1]. The initiation phase shares some similarities with atherosclerosis where low-density lipoprotein (LDL) and monocyte infiltration and the subsequent inflammation play a key role. Two mechanisms have been proposed for the calcification phase of the disease. In the osteogenic pathway, osteogenic differentiation of valvular interstitial cells (VICs) leads to bone-like calcification. In the myofibroblastic pathway, the VICs differentiate to activated myofibroblasts, which lead to the creation of apoptotic bodies providing nucleation sites for calcification. The myofibroblastic pathway is the focus of the current study. A systems biology model is developed to quantify the long-term progression of calcification. Wall shear stress (WSS) and mechanical strain are accounted in the model and are updated based on calcification progression.

METHODS

An overview of the proposed model is shown in Figure 1. In the atherosclerosis-mimicking phase, injury to the endothelial cells (ECs) enhances LDL penetration into the subendothelium. Conversion of LDL into oxidized LDL (ox-LDL) promotes monocyte arrest by the ECs. Infiltrated monocytes differentiate to activated macrophages and foam cells. The subsequent inflammatory environment leads to secretion of cytokines such as transforming growth factor-β (TGF-β) that cause differentiation of VICs to myofibroblasts. Finally, nodules are deposited creating a safe environment for calcification to occur [2]. The cascade of biochemical events is influenced by the hemodynamics. WSS influences infiltration of LDL and monocytes, and circumferential strain enhances dystrophic calcification.

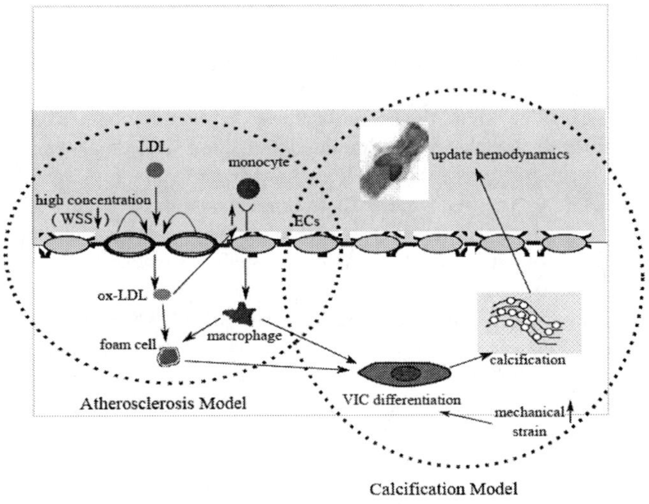

FIGURE 1: AN OVERVIEW OF THE MECHANO-SENSITIVE BIOCHEMICAL EVENTS LEADING TO CALCIFICATION

LDL: $$\dot{C}_{LDL} = \underbrace{f_L(\tau)}_{\text{LDL influx}} - \underbrace{d_{LDL}C_{LDL}}_{\text{LDL conversion to ox-LDL}} - \underbrace{d_L C_{LDL}}_{\text{LDL diffusing out}}$$

ox-LDL: $$\dot{C}_{oxLDL} = \underbrace{d_{LDL}C_{LDL}}_{\text{LDL conversion to ox-LDL}} - \underbrace{k_L C_{oxLDL} C_M}_{\text{ox-LDL reacting with macrophages}}$$

Monocyte: $$\dot{C}_m = \underbrace{f_m(\tau, C_{oxLDL})}_{\text{monocyte capturing by ECs}} - \underbrace{d_m C_m}_{\text{differentiation to macrophages}} - \underbrace{m_d C_m}_{\text{monocyte apoptosis}}$$

Macrophage: $$\dot{C}_M = \underbrace{d_m C_m}_{\text{differentiation to macrophages}} - \underbrace{\alpha k_L C_{oxLDL} C_M}_{\text{ox-LDL macrophage reaction}}$$

Foam cell: $$\dot{C}_f = \underbrace{\alpha k_L C_{oxLDL} C_M}_{\text{ox-LDL macrophage reaction}} - \underbrace{k_f C_f}_{\text{foam cell efferocytosis and apoptosis}}$$

Latent TGF-β $$\dot{\mathrm{TGF}} = \underbrace{f_t(C_F)}_{\text{TGF-}\beta\text{ production by macrophages}} - \underbrace{\lambda \mathrm{TGF}}_{\text{TGF-}\beta\text{ apoptosis}} - \underbrace{a_F C_F \mathrm{TGF}}_{\text{TGF-}\beta\text{ activation}}$$

$$C_F = C_f + C_M$$

Active TGF-β $$\dot{\mathrm{TGF}}_{act} = \underbrace{a_F C_F \mathrm{TGF}}_{\text{TGF-}\beta\text{ activation}} - \underbrace{\lambda_a \mathrm{TGF}_{act}}_{\text{TGF-}\beta\text{ apoptosis}}$$

Calcification: $$\dot{Ca} = \underbrace{\gamma k_c \mathrm{TGF}\,(1 + h(\epsilon))}_{\text{TGF-}\beta\text{ induced calcification promoted by strain}}$$

WSS: $$\tau = \frac{36}{Ca + 181}\ [Pa]$$

Strain: $$\epsilon = \frac{0.2\,Ca + 30}{Ca + 333.33}$$

The aforementioned cascade of biochemical events is modeled with a system of ordinary differential equations (ODEs) shown in the boxed equations. The model quantifies the temporal evolution of the biochemical concentrations involved. Calcification is quantified in Agatston units. A pathological aging model is developed to account for the age-dependent increase in LDL influx during CAVD [3]:

$$C_{LDL_{in}} = \beta \bar{C}_{LDL_{in}} \begin{cases} \frac{6.959\,t - 130.8}{t + 34.8} & \text{if t} \geq 18.8 \\ 0 & \text{if t} < 18.8 \end{cases},$$

where t is the age in years and β is a parameter used to simulate different levels of LDL influx. The atherosclerosis phase of the model is built similar to prior atherosclerosis models [4], and the calcification phase is developed based on in-vitro data reported in the literature [5]. Calcification progression increasingly stiffens the valve leaflets, therefore, affecting the hemodynamics. To account for this effect, the hemodynamics (WSS and circumferential strain) are modeled with simplified functions such that when no calcification is present (Ca = 0) the anticipated value for a healthy tricuspid valve is recovered. These functions are further tuned such that the anticipated range and trend of hemodynamics are modeled with calcification progression (manifested in increased stiffness, compliance mismatch, and incomplete opening/closure). In future work, we will be replacing these functions with finite element simulations coupled to the biological reactions to build a multiscale model of CAVD. This will increase the accuracy of our model and enable quantification of spatial calcification patterns.

RESULTS

The system of ODEs is solved and the long-term evolution of the biochemicals in the model are quantified. To validate the predictive capabilities of our model, calcification progression is compared with longitudinal clinical data reported in the literature (Fig. 2). In Fig. 2a, the thick green line and the scattered points are the data reported by Owens *et al.* during an average 2.4 year followup [6]. In Fig. 2b, the red dashed lines are calcification progression during 3.8 years followup, reported by Messika-Zeitoun *et al.* [7]. Good agreement is seen between the model and longitudinal clinical data.

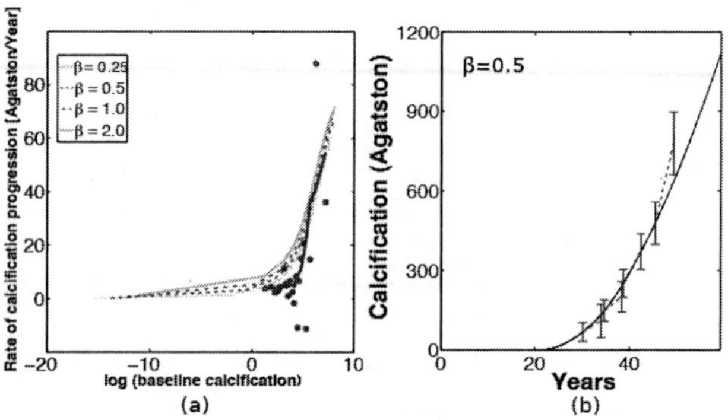

FIGURE 2: CALCIFICATION PROGRESSION VALIDATED WITH CLINICAL DATA IN (A) OWENS *et al.* [6] (B) MESSIKA-ZEITOUN *et al.* [7]

Recent clinical trials have shown a failure in the effectiveness of statin therapy to prevent CAVD progression. To evaluate this, an idealized statin therapy simulation is performed where LDL influx is set to zero after a certain age. The corresponding results are shown in Figure 3. An age-dependent response is observed where the effectiveness of statin therapy depends on the age that the therapy has started (or more accurately on the baseline calcification level). To evaluate the uncertainty in the parameters estimated in our model, the temporal evolution of control coefficients (a measure of normalized sensitivity) is computed. The peak control coefficients were in an acceptable range.

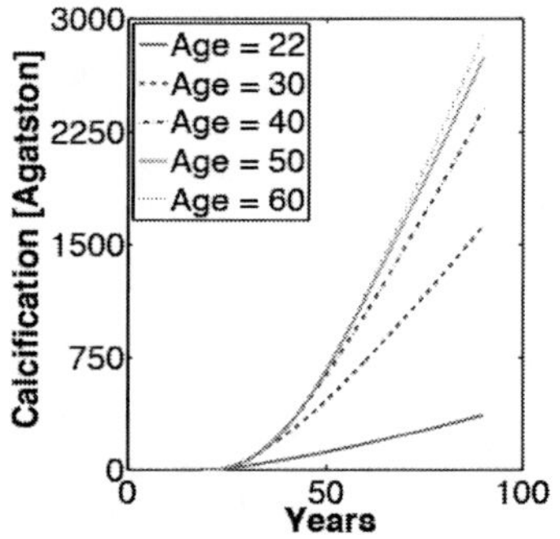

FIGURE 3: CALCIFICATION PROGRESSION WITH STATIN THERAPY STARTING AT A CERTAIN AGE.

DISCUSSION

We have developed a predictive model of long-term CAVD progression. Our model accounts for the prominent biochemical events involved in the myofibroblastic pathway of CAVD. The biochemical events are mediated by mechanical forces. Low WSS increases the EC permeability to LDL and increases monocyte adhesion rate. An increase in mechanical stress and the subsequent increase in circumferential strain promotes VIC differentiation and calcification. Our model is built based on the hypothesis that the nodules formed during the myofibroblastic pathway create a safe environment for calcification [2]. A complete model of aortic valve calcification needs to also account for the osteogenic calcification of VICs, however, suitable data does not exist in the literature. In future work, we will be performing in-vitro experiments to build a mechano-sensitive osteogenic calcification model. The simplified hemodynamics estimates will be replaced by finite element simulations within a multiscale framework to increase the accuracy of the model.

In summary, our model is a first step towards a comprehensive predictive model of long-term CAVD progression, a clinically valuable tool that can guide surgeons in valve replacement decision makings.

ACKNOWLEDGEMENTS

We acknowledge a grant from the American Heart Association.

REFERENCES

[1] Pawade, T. A . *et al. Journal of the American College of Cardiology*, 2015.

[2] Chen, J. H. and Simmons, C. A. *Circulation Research*, 2011.

[3] Sell, S. and Scully, R. E. *The American Journal of Pathology*, 1965.

[4] Cilla, M. *et al. Journal of The Royal Society Interface*, 2014.

[5] Fisher, C. I. *et al. Biomechanics and Modeling in Mechanobiology*, 2013.

[6] Owens, D. S. *et al. The American Journal of Cardiologyy*, 2010.

[7] Messika-Zeitoun, D. *et al. Arteriosclerosis, Thrombosis, and Vascular Biology*, 2007.

BIOMECHANICAL DIFFERENCES BETWEEN MALE AND FEMALE SACROILIAC JOINTS IMPLANTED WITH THREE DIFFERENT SACROILIAC IMPLANT SYSTEMS: STRESS ANALYSES

Amin Joukar (1), Anoli Shah (1), Ali Kiapour (1), Ardalan Seyed Vosoughi (1), Anand K. Agarwal (1), Hossein Elgafy (1), Nabil Ebraheim (1), Vijay K. Goel (1)

(1) Engineering Center for Orthopaedic
Research Excellence (E-CORE), University of
Toledo, Toledo, Ohio

INTRODUCTION

About 90% of adults are suffering from low back pain (LBP) at some stage of their lives. Recent studies have considered that 15-30% of LBP is originating from sacroiliac joint (SIJ), while the main reason for the pain is believed to arise from the lumbar region [1]. Understanding the SIJ mechanics is of a great value much like the other spine regions. The anatomical differences between male and female SIJs are well established in the literature [1].

Increased perception of SIJ as a frequent source of LBP, has upsurge the clinical suspicion about the treatment planning, specifically when surgical arthrodesis is considered as the only solution. When non-surgical treatments to alleviate the pain and discomfort of patients with suspected SIJ dysfunction fail, surgical interventions would be considered as an option. There are various implant devices to fuse SIJ using minimally invasive surgery (MIS).

A few *in silico* studies have been performed to calculate the stress values of different SIJ implants. Bruna-Rosso et al. [2] used finite element method to analyze SIJ biomechanics under RI-ALTO fusion implant under compression load. The proximally insertion of implant which was farther from SIJ center of rotation was taking more stresses than distally insertion of implant.

Ivanov et al. [3] have reported that prior lumbar fusion can directly increase in angular motion and stress across the patient's SIJ, and the magnitude of both of these parameters is strongly correlated to the specific lumbar levels fused as well as the number of segments fused

There is no biomechanical data to show the amount of stress which was taken by different SIJ implants. The objective of this study is to assess the biomechanics of the three types of SIJ fusion implants, triangular implants, half threaded screw, and fully threaded screw, in terms of stress values and their effect on each gender using finite element simulation.

METHODS

CT scans of a male and a female patients were used to develop the ligamentous L1- pelvis finite element. Cortical and cancellous bone and nucleus, annulus fibrosis, and ligaments were simulated using Isotropic elastic, hyperelastic, and hypoelastic material properties as detailed for the male model elsewhere [3]. In each model, SIJ was instrumented using three implants. Titanium alloy material properties were used for all three implants used in the constructs. While fully thread screw was implanted posteriorly, triangular implant and half threaded screw were inserted laterally into the SIJ. Implants and bone were tied together. The model was fixed at the hip joint. To simulate various motions of flexion-extension, lateral bending and axial rotation, 10 Nm moment plus 400 N follower load were applied to the models. The von Mises stress at the sacroiliac joint were calculated for both models instrumented with all three fusion devices and compared to each other, Figure 1. Using previous literature data, female and male intact spinopelvic models were validated. All implants were placed into the left sacroiliac joint of both male and female models fusing the left SIJ and leaving right SIJ unfused.

RESULTS

The von Mises stress of the female and male SIJs under instrumented conditions were calculated. Figures 2, 3, 4 and 5 show the maximum von Mises stress values at the sacrum and ilium for male and female models under different loading modes for intact and three instrumented conditions.

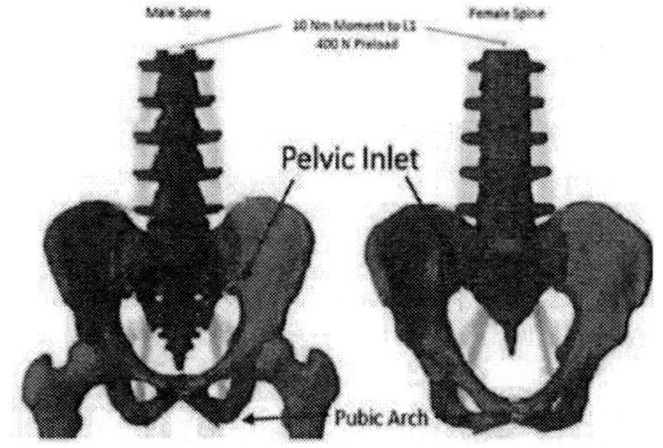

Figure 1: Finite element models of male and female sacroiliac joints.

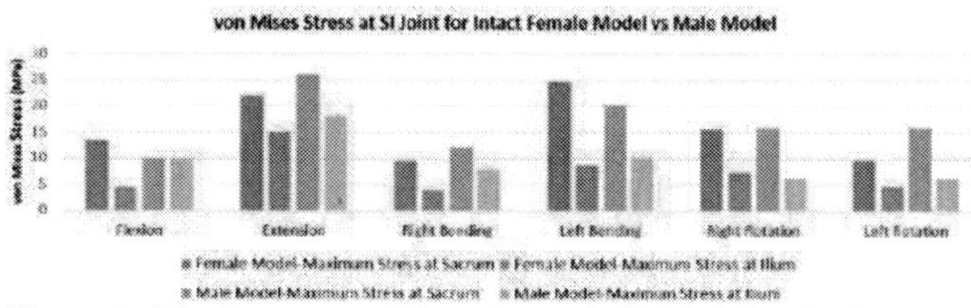

Figure 2: Maximum von Mises stress at SI joint for the intact model: female vs male model

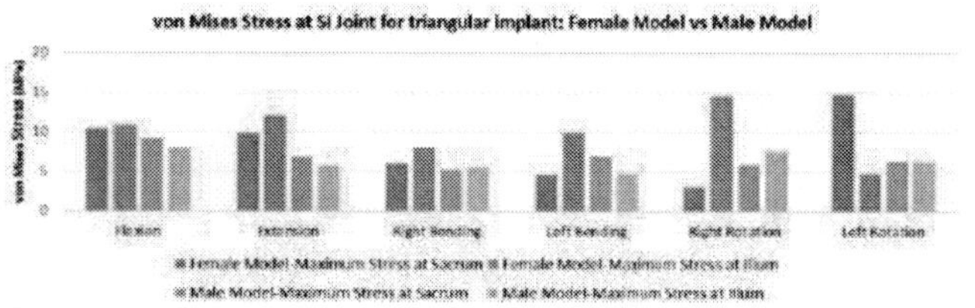

Figure 3: Maximum von Mises stress at SI joint for triangular implant: female vs male model

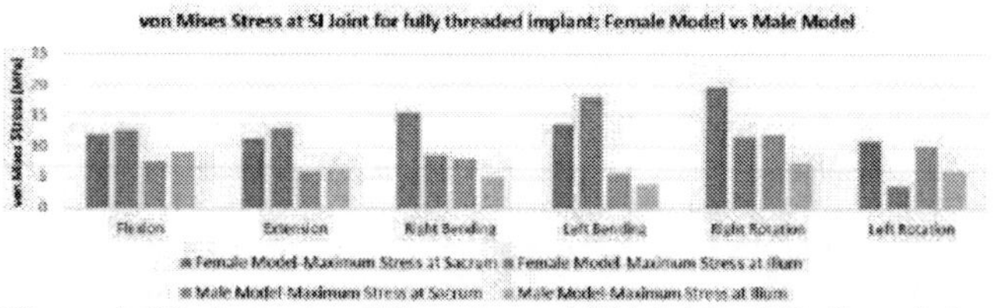

Figure 4: Maximum von Mises stress at SI joint for fully threaded screw: female vs male model

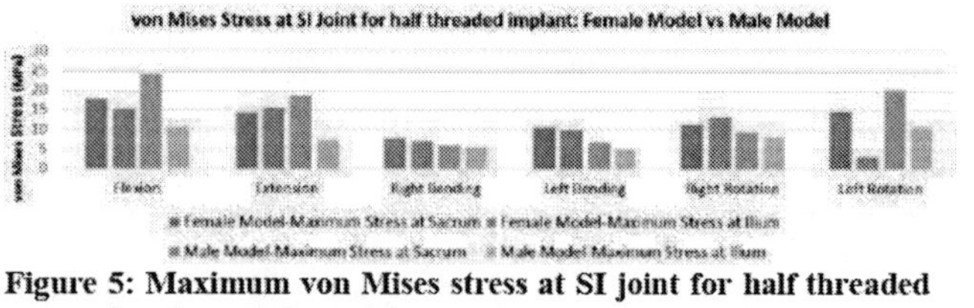

Figure 5: Maximum von Mises stress at SI joint for half threaded screw: female vs male model

It was shown that the maximum von Mises values at the SIJ in the female model were almost higher than male model up to 5 times for all three instrumented devices. In some motions, sacrum was experiencing higher stress values than ilium up to 3 times, and in others, ilium was taking higher stresses up to 4 times.

DISCUSSION

The stress values for the female model were mostly higher than the male model in all instrumented conditions. The range of motion (ROM) at the SIJ for the female model was higher compared to the male model and this could be one of the reason for higher stresses at the SIJ in the female model. Triangular implants produced low stress at bone for both genders. Half threaded and fully threaded screws caused higher stresses at bone. This might be due to the triangular design of this implant and not having threads. Some gender-related differences in the SIJ can lead to a higher rate of SIJ misalignment in young women [1]. Having greater pubic angle facilitates the parturition in females, followed by ligaments laxity which leads to increased mobility. This unique aspect of the SIJ provides females with the ability to give birth and it may also predispose females to a greater risk of experiencing pelvic pain. One factor that plays an important role in determining the severity of this predisposition involves the laxity of the female SI joints during pregnancy [1]. Due to the anatomical differences and effect of pregnancy, women are more susceptible to develop pelvic girdle pain, and are therefore at greater risk of experiencing low back pain. One limitation of the present study was using similar material properties for female and male models, which was mainly due to lack of experimental data.

ACKNOWLEDGEMENTS

The work was supported by NSF Industry/University Cooperative Research Center at The University of California at San Francisco, CA and The University of Toledo, Toledo, OH (www.nsfcdmi.org).

REFERENCES

[1] Vleeming, A et al, J. Anat. Journal of Anatomy, 221(6):537-567, 2012.
[2] Bruna-Rosso, C et al, Int J Spine Surg, 10:16, 2016.
[3] Ivanov, A et al, Spine, 34.5, 2009.

THE OCCIPITOATLANTAL CAPSULAR LIGAMENTS ARE THE PRIMARY STABILIZER OF THE ADULT CRANIOCERVICAL JUNCTION

Rinchen Phuntsok (1, 2),
Douglas L. Brockmeyer (3), Andrew T. Dailey (3), Michael R. Herron (1, 2),
Kenneth L. Smith (1, 2), Benjamin J. Ellis (1, 2)

(1) Department of Bioengineering
University of Utah
Salt La□e City, Utah, USA

(2) Scientific Computing and Imaging Institute
University of Utah
Salt La□e City, Utah, USA

(3) Department of Neurosurgery
University of Utah
Salt La□e City, Utah, USA

INTRODUCTION

The adult craniocervical junction (CCJ), which includes the bony portion of occiput to C2, is stabilized by several important ligamentous structures. These structures include the occipitoatlantal capsular ligaments (OACL), tectorial membrane (TM) and transverse ligament (TL) [1]. Damage to one or more of these ligaments can result in CCJ instability. The diagnoses of CCJ instability by clinicians requires an in-depth understanding of the role played by each of these ligaments. However, previous studies investigating the relative role of these structures have resulted in contradictory conclusions. For example, Werne et al. concluded that TM restricts extension while Oda et al. reported that it limits flexion [2, 3]. Tubbs et al. concluded the TM prevents impingement of the cervical canal by the odontoid [4]. Radcliff et al. on the other hand concluded that the TL is the main restraint to instability across the CCJ [5]. A parametric finite element (FE) study by Brolin et al. concluded that capsular ligaments play a major role in joint motion [6]. Contrary to this, Child et al. ascribed an "all-or-none" phenomenon where the threshold to instability is a result of the confluence of the fibers between TM, capsular ligaments and alar ligaments [7]. However, each of these studies performed sequential sectioning of ligamentous structures, inadequately simulating isolated injury scenarios. To date, no FE study has examined the relative role of the CCJ ligaments in subject specific FE models.

In order to address these issues, we have created 3 subject specific FE models of the adult CCJ. The goal of this study is to identify the key stabilizers of the adult CCJ and assess their relative roles. Isolated and combined injury scenarios of the OACL, TL and TM were simulated.

METHODS

Three adult FE models of the CCJ were used for normal and injury simulations. The models were: an open source model (Adult 1), a model created from the NIH visible human data set (Adult 2) and a model by Puttlitz et al. (Adult 3) [8, 9, 10]. The meshes from these models were imported into FEBio's preprocessor, PreView (http://febio.org/). Linear adult material properties were used for ligamentous structures (e.g. OACL, TL and TM) as described in our previous work [11]. The normal models were reevaluated by comparing the flexion-extension, lateral bending and axial rotation ROM data obtained under physiological loads to Panjabi et al. (Table 1).

Sensitivity Studies (Ligamentous injury models)

The intact normal FE models were modified to evaluate the sensitivity of the models when damage to OACL, TL and TM were simulated. Seven injury scenarios were simulated with flexion-extension, axial rotation and lateral bending applied moments. These include 1) OACL injury (i.e. reducing the stiffness of the OACL by 3 orders of magnitude from normal) 2) removal of TL, 3) removal of the TM, 4) removal of both TM and TL, 5) OACL injury with TM removal, 6) OACL injury with TL removal, 7) and OACL injury with both TM and TL removed. Boundary conditions used for the models are the same as reported previously [6, 11]. A moment of 1.5 Nm was applied to the adult FE models for validation, but only 1 Nm was applied to the injury models because the injury models behaved non-physiologically beyond this level of torque. FEBio and its post processor Postview were used for all analysis and post processing, respectively. C0-C2, C0-C1 and C1-C2 ROM were predicted for flexion-extension, axial rotation and lateral bending. Averaged percent increases and standard deviation (SD) from normal are reported for the injury models. Axial rotation and lateral bending data are averaged for left and right.

RESULTS

Flexion ROMs for all three normal models were within one SD of previously reported data (Table 1) [12]. Extension ROM was also within one SD of previously reported data except for Adult 2 [12]. Axial rotation and lateral bending ROMs were below 1 SD of previously reported data [12, 13].

Isolated OACL injury caused the largest percent increase from normal across all ROMs. Flexion increased by 21.9±14.8% (Fig.1 Top) while extension increased by 26.3±18.9% (Fig.1 Bottom) at C0-C1 for isolated OACL injury. Axial rotation increased by 107± 23.8% (Fig. 2 Top) and lateral bending increased by 96±35.4% (Fig. 2 Bottom) at C0-C1 for the isolated OACL injury scenario. OACL injury combined with TM-TL injury further increased all ROMs by <6% from the isolated OACL injury case, except lateral bending where there was >50% increase (Fig 1 and 2) at C0-C1. Isolated TM and TL, and combined TM-TL removal scenarios had <5% increase for axial rotation and lateral bending for all levels (Fig. 2). Combined TM-TL injury increased flexion by < 18% and extension by < 26% for all levels. Isolated TM injury had < 8% increase for both flexion and extension at all levels.

Table 1. C0-C2 ROM data in degrees for normal adult FE models. Axial and lateral rotation data are averaged for left and right side.

FE Models	Adult 1 (26,F)	Adult 2 (59, F)	Adult 3 (64, F)	Panjabi [12,13]
Flex-Ext	32.8 - 20	29 - 17	33 - 28	27.1(5.8) – 24.9(6.8)
Axial	29.2	21	20.9	40.35 (10.1)
Lateral	7.3	7.8	10.5	16.15 (5.9)

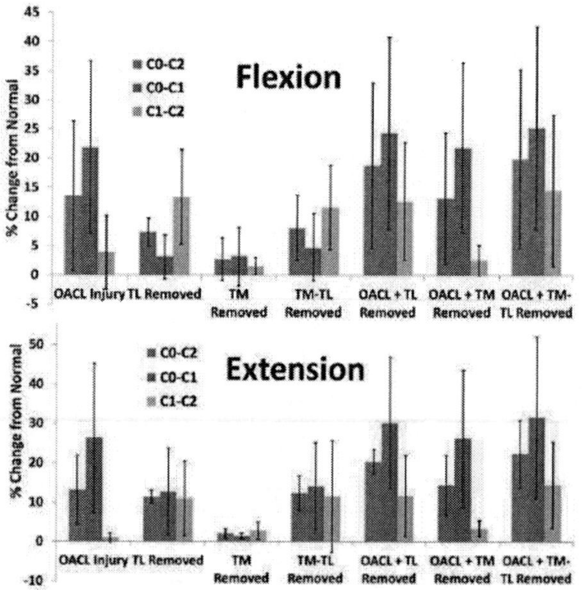

Figure 1: Top: Percent increase in Flexion from normal. Bottom: Percent increase in Extension from normal.

DISCUSSION

The results of this study show that the OACLs are the primary stabilizers of the adult CCJ. The results from the isolated OACL injury showed the largest increase in ROM in all loading scenarios. Contrary to a previous study, isolated TM injury showed the least increase in ROM in all loading scenarios and did not restrict flexion or extension ROMs when the OACL was injured [2, 3]. Similarly, an isolated TL injury did not cause a large increase in axial rotation and lateral bending. Isolated TL injury resulted in increased flexion and extension ROM; however that was considerably smaller than the increase due to an isolated OACL injury. In contrast to Child et al's study, the combined injury of multiple ligaments, in our case OACL, TM and TL also did not lead to a drastic increase from the isolated OACL injury case. We did not see an 'all-or-none' phenomenon as Child et al [7].The results of this sensitivity study challenge the traditional belief that the TL is the primary stabilizer of the CCJ and that the OACL play a secondary role [5]. This study also overcomes the biases in the results caused due to sequential sectioning of ligamentous structures in previous studies [5]. Linear material properties were used in this study to save computational time and obtain converged results. A similar approach has been used previously [14]. The stiffer results in normal (Table 1) axial rotation and lateral bending could be attributed to the use of linear material properties.

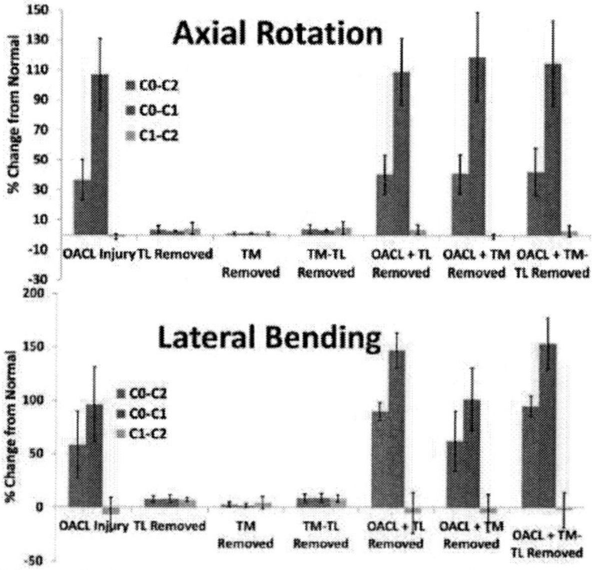

Figure 2: Top: Percent increase in Axial Rotation from normal. Bottom: Percent increase in Lateral bending from normal.

ACKNOWLEDGEMENTS

K2M, Inc. and the Marion L. Walker, M.D. Endowment are gratefully acknowledged for funding this research. We also thank the NIH, OpenHBM, and Christian Puttlitz, PhD, Colorado State University for providing the initial data for our models.

REFERENCES

[1] Tubbs, R. S. et al.,*J Neurosurg Spine,*14: 697-709, 2011.[2] Werne, S.,*Acta Orthop Scand,*23:1-150,1957. [3] Oda, T. et al.,*Clin Biomech,*7: 201-207,1992. [4] Tubbs, R. S. et al.,*Clin Anat,*20:382-386,2007. [5] Radcliff, K. E. et al.,*Spine J,*15:1618-1628,2015.[6] Brolin, K. et al.,*Spine(Phila Pa 1976),*29: 376-385, 2004. [7] Child, Z. et al.,*Spine J,*16:1116-1123, 2016. [8] Osth, J. et al.,*J Biomech Eng,*138: 061005, 2016. [9] Ackerman, M. J.,*Yearb Med Inform*:115-117,2002. [10] Leahy, PD. PhD Thesis,Colorado State University,2012. [11] Phuntsok, R. et al.,*J Neurosurg Pediatr*: 1-7, 2015. [12] Panjabi, M. et al.,*J.Spine.Disord,*4:157-167,1991.[13] Panjabi, M. et al., *J.Ortho.Res,* 9:584-593,1991. [14] Kuo, C. S. et al.,*BMC Musculoskelet Disord,*11: 151, 2010.

TOWARDS ENHANCING THE CONSISTENCY OF VERTEBRAL KINEMATICS IN A RAT DISLOCATION SPINAL CORD INJURY MODEL

Stephen Mattucci (1,2), Jie Liu (2), Paul Fijal (1,2), Wolfram Tetzlaff (2), Thomas Oxland (1,2)

(1) Departments of Orthopaedics and Mechanical Engineering
University of British Columbia
Vancouver, BC, Canada

(2) International Collaboration on Repair Discoveries (ICORD)
University of British Columbia
Vancouver, BC, Canada

INTRODUCTION

Dislocation is the most common clinically observed spinal cord injury (SCI) mechanism, occurring in 45% of SCI cases [1], however, there are few pre-clinical models [2, 3]. Despite demonstrating different patterns of tissue damage and behavioural function compared to more common contusion injury models [4], the dislocation models that do exist have issues with repeatability, making it challenging to investigate treatment options, as differences may be masked by the inherent injury variability. Since the dislocation model involves injury clamps which grip and displace the spine, a potential cause of the variability may exist at this interface, where there may be relative motion between the injury clamps and the vertebrae.

The objectives of this study were to: i) design new injury clamps that pivot and self-align to the vertebrae, ii) measure intervertebral kinematics during a high-speed dislocation injury in an in vivo rat model using the existing and redesigned clamps; and iii) quantify slippage (i.e. relative motion) at the vertebrae-clamp interface to determine which clamps provide the most rigid connection.

METHODS

The existing clamps perform a dislocation injury by attaching to the spine, where the rostral clamp grips the two vertebrae rostral to the level of the dislocation and holds both together and stationary, while the caudal clamp grips the two vertebrae caudal to the dislocation level, to be displaced dorsally [2]. The existing clamps act as two parallel ridges to grab two adjacent vertebrae; however, different widths of adjacent vertebrae would mean the clamp could only be tightened to the wider of the two, allowing the narrower vertebra to move independently.

The design criterion for the new dislocation clamps was to enable clamps to grip both vertebrae independently for both the rostral and caudal pair. New clamps were designed to pivot, and self-align (PSA) as it is tightened via a custom saddle washer resting against a rounded outer surface. This ensures that each clamp holds both intended vertebrae.

The dislocation injury was performed using both the existing and redesigned clamps, where 400 μm radiopaque tantalum beads were fixed to the vertebrae and the clamps to act as fiducial markers (Figure 1) to track relative motion.

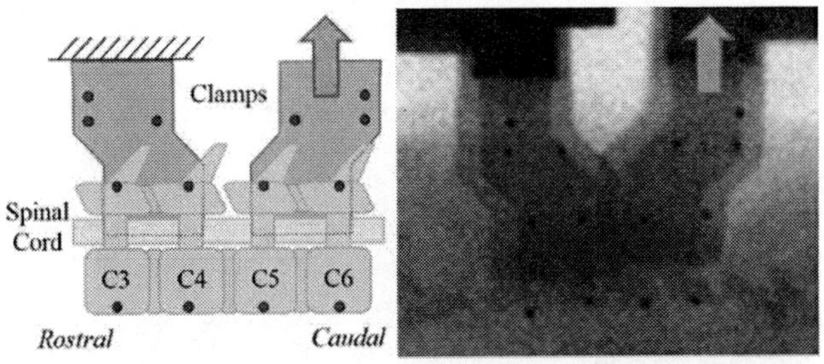

Figure 1: Schematic of the location of the fiducial markers (red dots) on the vertebrae and injury clamps, with a corresponding high-speed x-ray image. The clamps are represented as outlines on the schematic to visualize the marker locations on the vertebrae. The rostral clamp is held rigid while the caudal clamp is dislocated dorsally.

Sprague-Dawley rats (n = 17 per group, mean weight 315 g), were deeply anesthetized before the spine was surgically exposed and the clamps were attached to C3/C4 and C5/C6 (Figure 1). The caudal clamp dislocated C5/C6 dorsally by 2.3 mm at a velocity of 700 mm/s.

The injury was performed within a high-speed x-ray system, where the camera recorded at 8000 frames per second and 512 x 512 image resolution, with a spatial resolution of 0.12 mm/pixel. The motion of the clamps and vertebrae were tracked using TEMA motion analysis.

RESULTS

The cases of relative motion analyzed were: C5 relative to the caudal clamp in translation (Figure 2) and rotation, C4 relative to the rostral clamp in translation and rotation, and C5 relative to C4 in translation.

The relative motion of C5 with respect to the caudal clamp was significantly less for both translation and rotation for the PSA clamps ($p < 0.05$) (Table 1). The motion of C4 with respect to the rostral clamp was not statistically different between the different clamps.

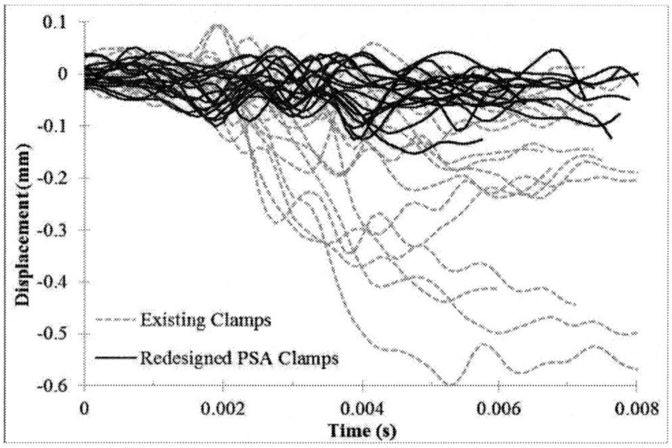

Figure 2. Relative motion of C5 with respect to the caudal clamp compared between the existing clamps and redesigned PSA clamps. Importantly slipping only occurred in some instances for the existing clamps, compared to the PSA clamps where relative motion never exceeded 0.15 mm.

DISCUSSION

This was the first time the motion of the vertebrae with respect to the clamps has been measured in a dislocation model. These results help to identify and solve a potential cause of the variability present in the dislocation model: the mechanical interface where the clamps gripped the vertebrae. Relative motion has been significantly reduced, and further studies can be conducted with confidence that the rat vertebrae are moving rigidly with respect to the PSA dislocation injury clamps.

It is still unknown if reducing the kinematic variability will correlate to reducing the histological and behavioural variability in the injury outcomes. The injury was performed at a severe level to exaggerate any relative motion, however this lead to several animals perishing immediately after injury, allowing the injury to progress to different stages in each animal post-injury. Logistical limitations at our centre prevented any survival studies to be conducted in conjunction with the high-speed x-ray apparatus due to radiation exposure.

This study has clearly demonstrated that relative motion (or slippage) was occasionally present between the existing dislocation injury clamps and the vertebrae, and the degree of variability in spine kinematics during a dislocation injury in a rat model. By redesigning the injury clamps, relative motion between the vertebrae and injury clamps has been reduced, and thus produced more repeatable spine kinematics during injury. These improvements will serve to address a potential source of variability in the injury model and progress toward a more repeatable rat dislocation model, better suited for investigating SCI interventions, and further shed light on the importance of injury mechanism.

ACKNOWLEDGEMENTS

The authors wish to acknowledge the Canadian Institutes for Health Research (CIHR) and the Blusson Integrated Cures Partnership (BICP) for financial support.

REFERENCES

[1] Sekhon, L et al., *Spine,* 26, 24 Suppl: S2-12, 2001.
[2] Choo, A et al., *J Neurosurgery: Spine,* 6, 3:255-66, 2007.
[3] Fiford, R et al., *J Neurotrauma*, 21, 4:451-8, 2004.
[4] Chen, K et al., *J Neurotrauma,* 33:1-18, 2016.

Table 1. Average of maximum, absolute relative motion between clamps and vertebrae

		Existing clamps Mean (SD)	PSA clamps Mean (SD)	P value
C4 w.r.t. rostral clamp	Y translation (mm)	0.11 (0.06)	0.11 (0.06)	0.95
	Rotation (degrees)	2.88 (1.50)	2.99 (1.82)	0.84
C5 w.r.t. caudal clamp	Y translation (mm)	0.25 (0.17)	0.09 (0.03)	0.00*
	Rotation (degrees)	4.85 (3.23)	1.56 (0.54)	0.00*

*Relative motion between vertebra and clamps was significantly different ($p < 0.05$)

ELECTRO-MECHANICAL ACTUATOR FOR HIGH FREQUENCY MAGNETIC RESONANCE ELASTOGRAPHY IN-VIVO OF THE SPINE

Sean M. Rothenberger (1), Thomas Neuberger (2), Corina S. Drapaca (3), Daniel H. Cortes (1, 2)

(1) Department of Mechanical and Nuclear Engineering
Pennsylvania State University
University Park, PA, USA

(2) Department of Biomedical Engineering
Pennsylvania State University
University Park, PA, USA

(3) Department of Engineering Science and Mechanics
Pennsylvania State University
University Park, PA, USA

INTRODUCTION

Intervertebral discs degeneration is closely correlated to lower back pain, which affects nearly 85% of people at some point in their lives [1]. The long-term effects of this degeneration are easily recognizable with current diagnosing methods using disc morphology, proposed by Pfirrmann et al. [2]. However, to more closely study the underlying mechanisms behind disc degeneration and the effects of possible treatments, it is necessary to first have the ability to quantify the degree of degeneration at earlier stages.

Our previous study showed intervertebral disc degeneration strongly correlates to a decrease in the shear modulus of the nucleus pulposus (NP) [3]. Magnetic Resonance Elastography (MRE) was adapted within this previous study to measure the NP shear modulus in intact human cadaveric disc segments. MRE is a technique where small-amplitude mechanical waves are excited and mechanical properties are calculated from measurements of wave propagation. Previous tests on ex-vivo spine segments indicate an optimum frequency range of 500 to 1200 Hz [3]. It is necessary to determine if this optimal frequency range still holds true in-vivo. Unfortunately, commercial MRE actuators can only operate up to 200 Hz. As a result, the objective of this study is to design an actuator for the spine which operates in-vivo at high frequencies.

METHODS

Proposed Electro-Mechanical Actuator

The purpose of the actuator is to induce shear waves within the intervertebral discs which will later allow for the calculation of mechanical properties present within these tissues of interest. The approach taken in this study was to design an electro-mechanical actuator similar to the one used for spine samples, but adapted for living human subjects. Shear waves are produced through the rocking motion of the electro-mechanical actuator against the patient's lumbar spine. Subjects are positioned on top of the device within the core of the MRI machine (Figure 1). An AC current source is supplied to electromagnetic coils (located within the actuator) by an external function generator. This current induces magnetic fields which interact with the main magnetic field of the MRI machine producing the oscillating motion of the actuator. The actuator built for this study is shown in Fig. 2.

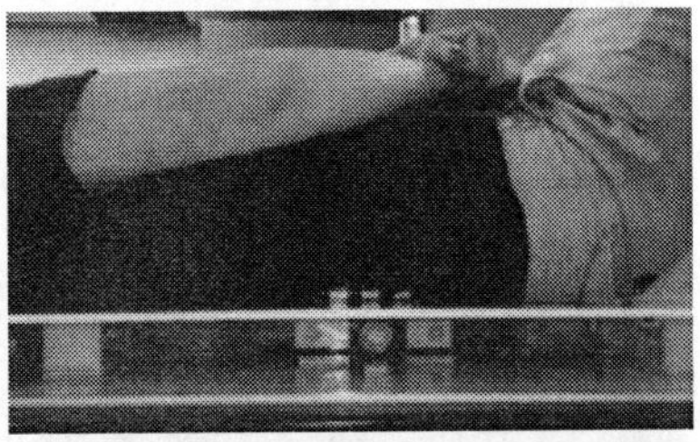

Figure 1: Electro-mechanical Actuator Placement

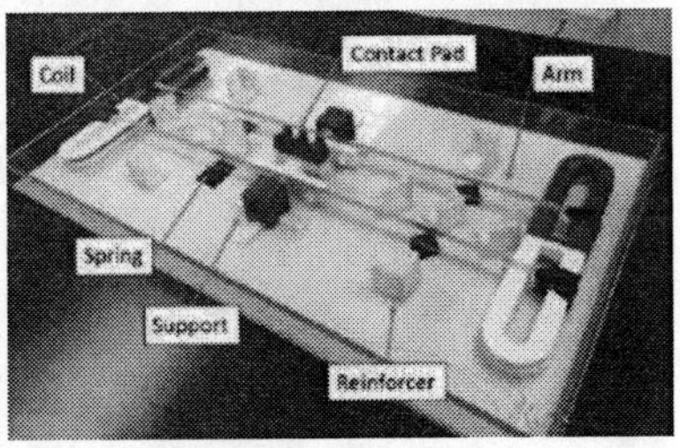

Figure 2: Constructed Actuator Interior

Electrical coil design and optimization

Two electrical coils located on each side of the rocking arms were used to generate motion. The electrical diagram of the actuator is presented in Fig. 3. To insure that both coils produce the same magnitude of magnetic field, each coil was designed to have the same amount of current running through it, along with the same number of loops. The current within the coils was maximized by minimizing the AC system's respective impedance (Z). Once the reactive component of the circuit could be eliminated, the impedance was at its lowest value and the current was subsequently at its highest value. To do this, the capacitance needed in the circuit at a given frequency was:

$$C = {}^{1}/_{(2\pi f)^2 L} \quad (1)$$

Where L is the inductance of the coils and f is the operating frequency. The capacitor is inserted into the circuit in between the actuator and the amplifier, as seen in Fig. 3.

Measurements and Simulations

The temperature increase of the coils was measured for safety purposes. Given that the actuator produces its own magnetic field while within the MRI machine, there is potential that this could cause interference with the main magnetic field in the region being imaged. The magnetic field produced by the actuator was simulated using Finite Element Method Magnetics (FEMM).

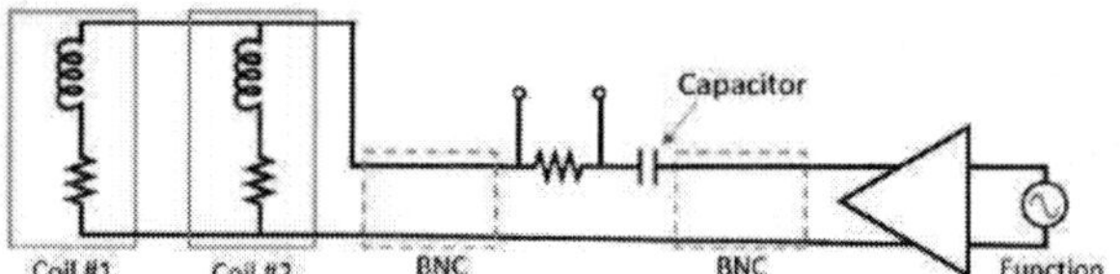

Figure 3: Circuit Diagram

The actuator was tested in one healthy subject (male, age 21 years) at 100, 200, 250, 500 and 1000 Hz. Displacement data was measured using a 2D MRI sequence similar to that used in our previous study [3]. Measurements were taken in the middle plane of an L4-L5 disc. Displacement data was then analyzed using MREWave software.

RESULTS

The coil's temperature increases by less than four degrees Celsius within a 16-minute period. Therefore, it was determined that the coil will not generate enough heat to produce any dangerous or damaging outcomes. The magnitude of the magnetic field produced by the actuator within the imaging region was approximately 12.5 μT (Figure 4). Comparison of images with the actuator on and off produced no noticeable artifacts in the images. The actuator was successfully tested at frequency values including 100, 200, 250 and 500 Hz. However, no measurable displacements were generated at 1000 Hz. An example of a wave image resulting from MRE is shown in Fig. 5a. This information was then used to generate an elastogram (Figure 5b).

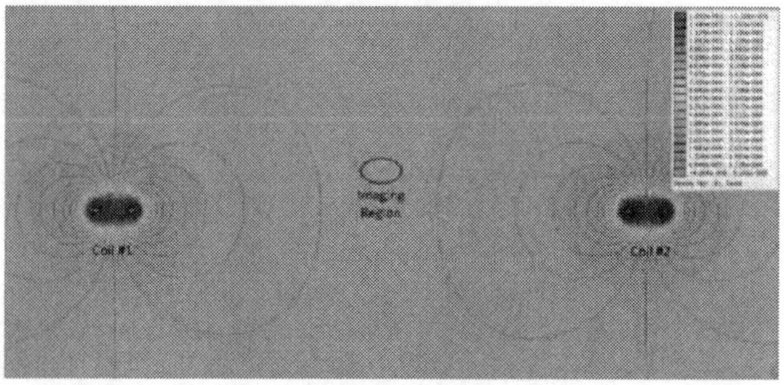

Figure 4: Magnetic Field Density

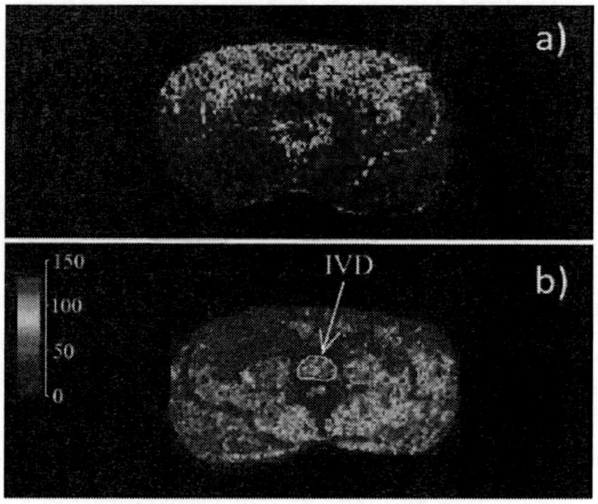

Figure 5: Elastography of the torso at 500 Hz a) wave field b) shear modulus (kPa)

For 100, 200, 250 and 500 Hz the average shear moduli for the IVD were 8.0, 9.3, 18.3 and 87.9 kPa, respectively.

DISCUSSION

This study showed the design and testing of an electro-mechanical actuator for high frequency elastography measurements of the intervertebral disc. Displacements were induced and measurable up to 500 Hz. To the best of our knowledge, this is the highest frequency ever used for MRE in human in-vivo.

Intervertebral disc degeneration produces significant changes in mechanical properties of disc tissues. Therefore, mechanical properties may serve as sensitive biomarker for disc degeneration. Previous studies using disc samples highlighted the need of high frequencies (~1000 Hz) for MRE in the intervertebral disc [1,3]. A previous study applied MRE to the intervertebral disc at frequencies up to 80 Hz [4]. Our proposed set-up extended the range of testable frequency to 500 Hz.

This actuator will serve as a valuable tool to further study the use of MRE as a clinical method to quantify the degree of degeneration of the intervertebral disc. Future investigations will utilize this actuator to evaluate the degree of degeneration of intervertebral discs in-vivo along with the effect of frequency on the measured values of stiffness.

ACKNOWLEDGEMENTS

This study was financed using start-up funds provided by the Pennsylvania State University.

REFERENCES

[1] Ben-Abraham, E et al., Magnetic Resonance Imaging 2015.
[2] Pfirrmann, CW et al., *Spine* 26:1873–1878 2001.
[3] Cortes, D et al., *Mag. Resonance in Medicine*, 00:4–8 2013.
[4] Streitberger KJ et al. *Mag. Resonance in Medicine*, 74(5):1380-7 2015.

MRI QUANTIFICATION OF IN VIVO HUMAN DISC DIURNAL COMPRESSION AND INDUCED FLEXION

Kyle D. Meadows (1), John M. Peloquin (1), Edward J. Vresilovic (2), Dawn M. Elliott (1)

(1) Biomedical Engineering
University of Delaware
Newark, DE, USA

(2) Department of Orthopaedics and Rehabilitation
Pennsylvania State University
Hershey, PA, USA

INTRODUCTION

In vivo measures of intervertebral disc mechanical function are important for identification of abnormal discs and for translating results from cadaveric studies to the in vivo context. Intervertebral disc mechanics are an important factor in spine function, degenerative disc disease, and low back pain. Disc degeneration is currently classified using qualitative integer scales based on appearance in T2 weighted MRI. This approach is acknowledged as insufficient to identify symptomatic discs for treatment. In vivo measurements of mechanical function may improve identification of problem discs. This study developed methods to quantify in vivo function by using MRI to measure activity-dependent changes in disc geometry and T2 relaxation time. A key focus was quantification of intra-subject reliability, which was unknown. Subject-specific classification and modeling requires data that is consistent from day to day. Young, healthy volunteers were used to characterize baseline healthy disc function, providing a foundation for future degeneration studies.

METHODS

Volunteers aged 23–28 years with healthy discs were recruited in accordance with human subjects guidelines. Each subject was assigned to one of two protocols: diurnal loading (n=2) or induced flexion (n=1). The diurnal protocol consisted of two supine MRI sessions, the first < 1 hour after waking (AM) and the second in the late afternoon (PM). Between scans, subjects wore an ~10 kg backpack to magnify diurnal changes and went about their normal routines. Each subject repeated the AM–PM diurnal protocol on two different days. The induced flexion protocol consisted of an AM MRI session in supine position immediately followed by a session in a flexion position supported by pillows [1]. Each MRI session lasted approximately 1 hour and consisted of 4 lumbar spine scans: sagittal and coronal T2-weighted (T2-w) turbo spin echo (TSE) (3D, 0.5x0.5x5.0 mm, TR/TE=4540/124), mid-sagittal T2 map (2D, 0.6x0.6x5.0 mm, TR=3000, TE=13.6, 27.2,...81.6), and sagittal T1-weighted (T1-w) FLASH (3D, 0.5 x 0.5 x 3.0 mm TR/TE= 9.6/3.65).

Disc geometry (volume, height, wedge angle) and spine posture (sagittal Cobb angle) were measured using the T1-w FLASH images. A representative L4–L5 pair of AM and PM images is shown in Figure 1. Discs were segmented and 3D disc volume was calculated. Disc height was measured by dividing disc area by anterior–posterior width in the mid-sagittal image [2]. Wedge angle was measured as the angle between superior and inferior endplates, and Cobb angle was measured as the angle between the inferior L1 endplate and the inferior L5 endplate in the mid-sagittal image. T2 maps were calculated by fitting an exponential decay to the series of TE times [3]. Average nucleus pulposus (NP) T2 was calculated from a circular region centered on the NP, with diameter = 3/4 the central disc height. The T2-w image was used to confirm that the subjects had non-degenerate discs. A representative T2-w image of entire lumbar spine in supine and flexed position is shown in Figure 2. All levels were healthy (grade 1 or 2). Measurements were qualitatively evaluated for the AM starting point and between conditions (diurnal: AM vs. PM; flexion: supine vs. flexed) due to the sample size.

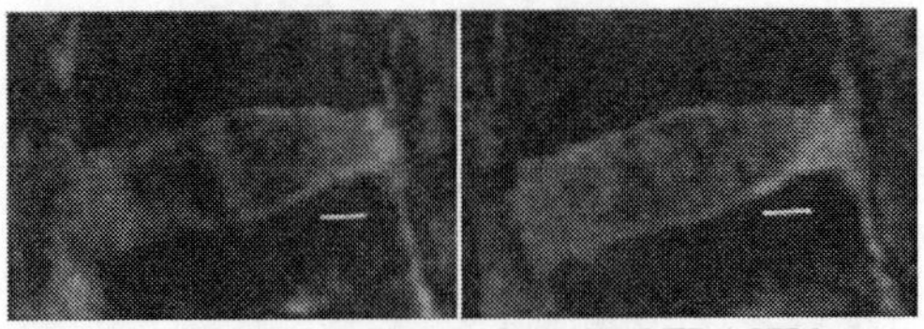

Figure 1: Representative mid-sagittal FLASH image of L4-L5 during AM (left) and PM (right) scans. Scale = 5 mm.

RESULTS

AM (Fig 3A): Disc geometry and NP T2 times taken in the AM were quite consistent within subjects. Height and NP T2 were consistent across the three subjects for each level (Fig 3A). AM disc height and volume increased from L1 to L5 for all subjects, as expected, and in similar fashion between subjects. Within each subject AM disc height and volume were very consistent across different days. Together these are promising findings for using these parameters in a statistical model for normal disc variation with level. Between subjects, NP T2 values were also consistent with level from L1 through L5. Interestingly T2 times were unchanged from L1–L2 through L3–L4, but higher at L4-L5 for all three subjects. Wedge angle had more spread across subjects, although seemed to be more closely grouped at L4-L5.

Diurnal (Fig 3B): Diurnal changes from AM to PM, unlike AM parameters, were much more variable both within subjects and across subjects (Fig 3B). In almost all cases the disc height decreased AM to PM. Similarly, in most cases disc volume decreased AM to PM, but one subject's volume increased across levels for one trial. On a matched basis the changes in height and volume were not correlated, suggesting in some cases height loss was not all due to fluid flow out of the disc, but in some cases fluid redistributed and contributed to disc bulge. The T2 times, which are related to hydration (although other components such as proteoglycan also contribute) support this notion, as the change in T2 time from AM to PM was variable between the subjects and even within a subject on different days. The change in wedge angle was only 2–4 degrees from AM to PM and likely not an important parameter for diurnal effects.

Flexion (Fig 3C) Almost all of the forced flexion occurred in the lower lumbar spine as measured by sagittal Cobb angle (33° supine vs 20° flexed for the lower lumbar spine; an unchanged 14° for the upper lumbar spine). The disc wedge angle was unchanged for L1–L2 through L3–L4, but increased by 5° at L3–L4 and 10° at L4–L5. Disc height, volume, and T2 time were unchanged with flexion.

DISCUSSION

The diurnal and flexion protocols both produced meaningful changes in disc geometry and T2. However, the same-subject variation in disc height, volume, wedge angle, and NP T2 for diurnal AM–PM changes was large, comparable to the differences between subjects (Fig 3B). This kind of variation makes subject-specific modeling difficult, and fitting a population average more efficient. For classification of disc health, diurnal changes will be useful if the magnitude of degeneration-induced changes exceeds day-to-day same-subject variability. The same-subject consistency in the AM disc geometry and T2 (Fig 3A) makes it preferable to measure those properties specifically in the morning, especially if the disc properties are being used for diagnosis or as a reference state for modeling. Although AM disc state is consistent, it is static, and thus has limited potential to provide information about a disc's mechanical properties.

The flexion protocol showed promise as a more controlled protocol to perturb the disc in vivo and measure its mechanical function. It induced flexion specifically in the L3–L4 and L4–L5 discs (Fig 3C). The negligible change in disc height, volume, and T2 time is consistent with the limited time for water exudation. Because the flexion is artificially imposed it can in principle be adjusted during imaging to achieve a specific value. Both protocols require additional study with more subjects with both healthy and degenerate discs. Quantification of in vivo disc mechanics has great potential to improve our ability to identify, characterize, and treat degenerative disc disease.

ACKNOWLEDGEMENTS

NIH R01AR05005; UD Center for Biomedical and Brain Imaging

REFERENCES

[1] Fazey, P et al., *Clin Biomech*, 21(5):538-542, 2001.
[2] O'Connell, G et at., *Spine*, 32(3):328-333, 2007.
[3] Wang, Y et al., *Eur Radiol*, 23(1):228-234, 2013.

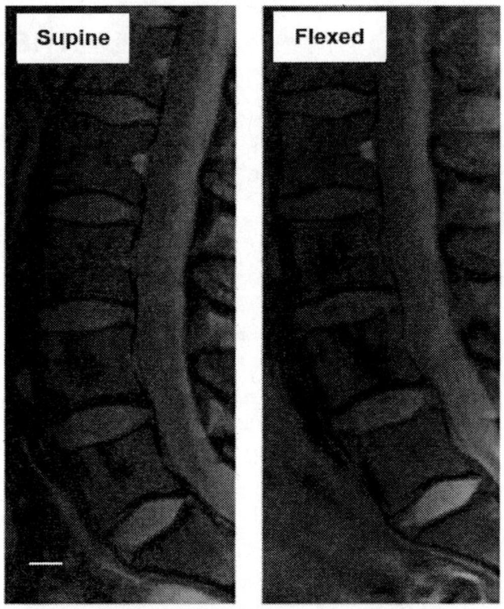

Figure 2: Supine and flexed mid-sagittal T2-w images of the lumbar spine. Scale = 10 mm.

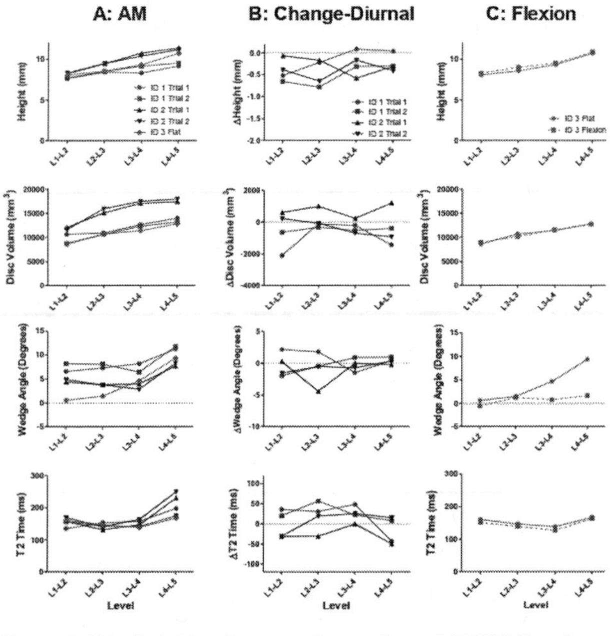

Figure 3: Disc height, volume, wedge angle, and NP T2 time by lumbar level for (A) AM starting point, (B) diurnal changes from AM to PM, and (C) supine compared with forced flexion.

EFFECT OF SPINAL FUSION ON BIOMECHANICS OF ADJACENT SEGMENT DISCS – AN IN-VIVO PATIENT STUDY

Kamran Z. Khan (1), Thomas D. Cha (1), Louis G. Jenis (1), James D. Kang (2), Kirkham B. Wood (3), Guoan Li (1)*

(1) Department of Orthopaedic Surgery
Harvard Medical School & Massachusetts General Hospital
Boston, MA, United States.

(2) Department of Orthopaedic Surgery
Harvard Medical School & Brigham and Women's Hospital
Boston, MA, United States.

(3) Department of Orthopaedic Surgery,
Stanford University Medical Center.
Redwood City, CA, United States.

INTRODUCTION

Low back pain (LBP) is the number one cause of disability in individuals, with an estimated annual total cost exceeding $100 billion in the United States [1]. Degenerative Disc Disease (DDD) causes structural and mechanical failure of the intervertebral discs (IVD) and is a major cause of LBP [2, 3]. Furthermore, spinal fusion surgery is a commonly performed surgical procedure for the treatment of DDD with severe pain. However, longitudinal radiographic studies have demonstrated that spinal fusion surgery accelerates disc degeneration in adjacent vertebral segments. Up to 80% of patients develop adjacent segment degeneration (ASD) as early as 1 year postoperatively, with an estimated 17-36% of patients requiring reoperation within 5-10 years.

ASD is a multifactorial process, including biomechanical and biological factors [4, 5], but clear scientific evidence demonstrating the mechanisms that cause accelerated ASD after a fusion surgery is lacking [6, 7]. Specifically, it is assumed that mechanical factors initiate disc degeneration by exposing the adjacent segment disc tissue to abnormal and excessive loads, which alters the vertebral kinematics and accelerates ASD. However, no clear in-vivo data has been reported that shows effects of spinal fusion on adjacent segment discs pre and post-fusion surgery [8, 9]. Thus, it is critically important to identify the biomechanical etiology of accelerated lumbar spine ASD in order to prevent its occurrence and potentially open up new modalities of diagnosis and treatment.. This study investigated adjacent segment disc deformation under weight bearing conditions in patients with low back pain and DDD before and after lumbar fusion surgeries and compared the data with healthy subjects.

METHODS

Ten patients with disk degenerative disease at L4-S1 were MRI scanned and 3D anatomical models were constructed for each (Fig. 1a). Subjects were then imaged using a dual fluoroscopic imaging system at weight bearing standing, flexion and extension positions before undergoing a fusion surgery at the L4-S1 segments (Fig. 1b). Five patients returned 3 years after undergoing fusion surgery for a post-op analysis and were imaged again using the same methodology. . Disc heights and translations of the adjacent segment of L3-4 were calculated using the changes in geometry between the endplates of L3 and L4 vertebra (Fig. 2), using the disc dimensions measured during MRI as the reference. Patient data was compared with 8 healthy control subjects whom were investigated the same way as the patients. We analyzed 9 points of interest on the endplate (Fig. 2e).

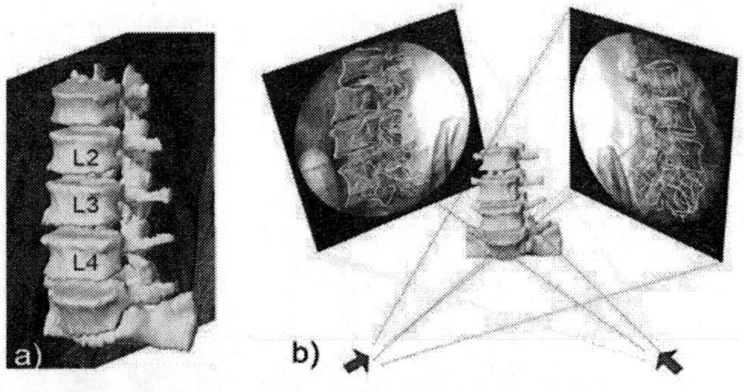

Figure 1. a) Construction of 3D vertebrae from parallel MR images. b) Reproduction of in-vivo vertebrae positions by matching 3D model projections to 2D osseous contours.

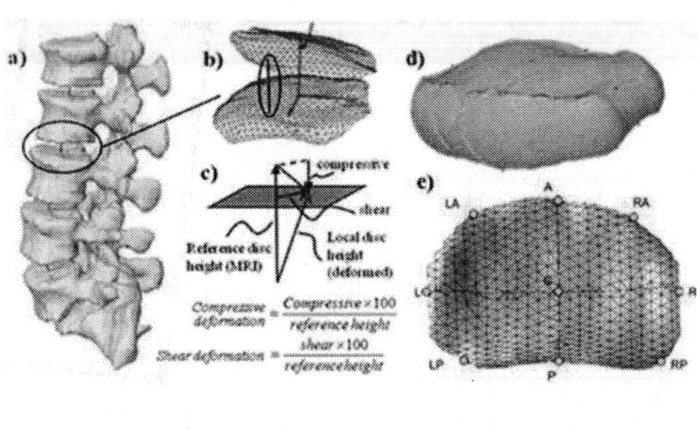

Fig. 2. a) Coordinate systems at endplates of lumbar vertebrae for determination of intervertebral kinematics. b) Determination of disc heights for calculating disc deformation. c) Shear deformation and other kinematic calculation method. d) Typical 3D rendering of intervertebral disc. e) 9 points of interest were studied on the endplate.

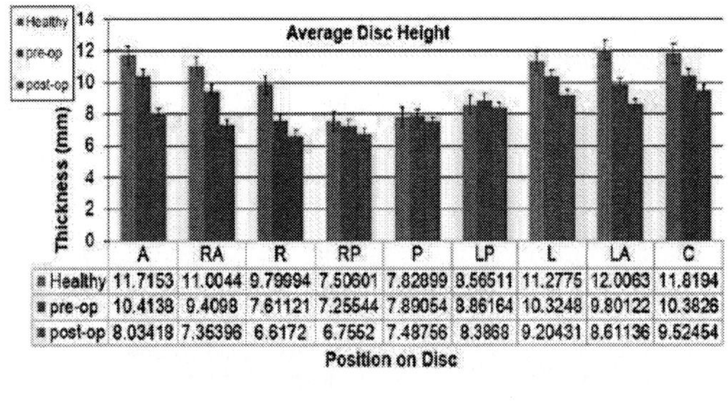

	A	RA	R	RP	P	LP	L	LA	C
Healthy	11.7153	11.0044	9.79994	7.50601	7.82899	8.56511	11.2775	12.0063	11.8194
pre-op	10.4138	9.4098	7.61121	7.25544	7.89054	8.86164	10.3248	9.80122	10.3826
post-op	8.03418	7.35396	6.6172	6.7552	7.48756	8.3868	9.20431	8.61136	9.52454

Fig. 3. Average disc height decreased post-operatively in each of the 9 positions studied when compared to pre-op disc height.

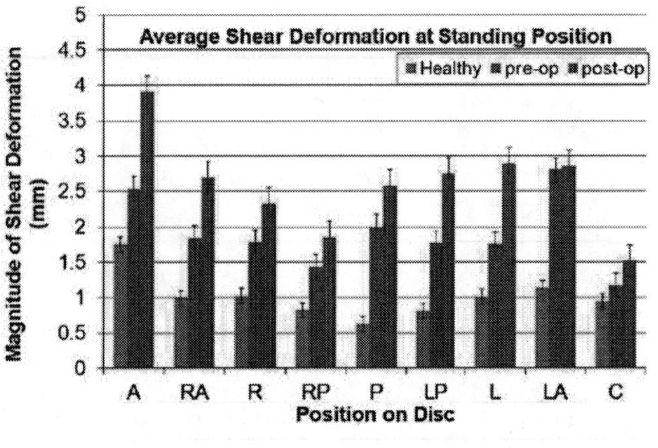

Figure 4: Average disc translations changed before and after the fusion surgery during the weightbearing standing position.

RESULTS

The adjacent disc experienced a significant decrease in disc height (Fig. 3) and an increase in translations (Fig.4), both before and after the fusion surgery when compared to the normal, healthy subjects at the weight bearing standing position. At the anterior position of the disc, on average, the disc height was reduced by 1.3 mm before fusion surgery, compared to the healthy control subjects. After the fusion surgery, the disc height was further reduced by 2.3 mm, compared to pre-op disc height. Nearly all the 9 studied points on the disc experienced loss of disc height after fusion surgery. Furthermore, the disc translations were increased by 0.78 mm, on average, at the anterior portion of the disc and 1.38 mm at the posterior portion of the disc before undergoing fusion surgery. Post-fusion surgery, the disc translations were further increased by 1.36 mm at the anterior portion of the disc and 0.57 mm at the posterior portion of the disc. The aforementioned data was collected during weight bearing standing position.

DISCUSSION

This study indicates that for patients with DDD, the adjacent segment disc experiences higher changes in disc dimensions (disc heights and translations) compared to healthy subjects. However, fusion surgery further increased the changes in disc dimensions at the adjacent level. The anterior portion of the disc experienced greater changes than other parts of the disc.

These data provide interesting clinical implications. Most current fusion techniques aim to restore disc heights to restore intersegmental stability. However, this surgery may cause interference to adjacent levels based on this experimental data. The observed increases in disc dimensions at the adjacent levels can further cause abnormal motion and instability at these levels. As adjacent segment disease has been widely reported after fusion surgery, the increased deformation post-surgery could therefore be a possible biomechanical mechanism that correlates to post-operative adjacent segment disease. Therefore, fusion surgery should aim to restore dynamic biomechanical stability of the index levels to minimize the interferences to adjacent segments and to improve long term clinical outcomes.

In summary, this study found that fusion surgery dramatically increased disc dimensions at the adjacent segment discs. A future study is warranted to investigate early adjacent segment disc degeneration after a fusion surgery using advanced MR imaging technique, such as T1ρ and T2 mapping. This will enable us to quantitatively evaluate disc degeneration and its association with altered disc deformation.

ACKNOWLEDGEMENTS

We would like to acknowledge and thank the NIH and MGH for supporting this research.

REFERENCES

[1] Katz, JN, J Bone Joint Surg Am, 2006. 88 Suppl 2: p. 21-4.
[2] Adams, MA et al., Spine, 2000. 25(13): p. 1625-36.
[3] Adams, MA et al., Spine, 2006. 31(18): p. 2151-61.
[4] Kumar, MN et al., Eur Spine J, 2001. 10(4): p. 309-13.
[5] Schulte, TL et al., Eur Spine J, 2007. 16(12): p. 2152-8.
[6] Lee, CK, Spine, 1988. 13(3): p. 375-7.
[7] Ekman, P et al., Eur Spine J, 2009. 18(8): p. 1175-86.
[8] Untch, C et al., Spine, 2004. 29(21): p. 2376-81.
[9] Cunningham, BW et al., Spine, 1997. 22(22): p. 2655-63.

COMPUTATIONAL MODELING OF REMODELING FOLLOWING RELIEF OF HEMODYNAMIC OVERLOAD IN A BIVENTRICULAR CANINE HEART MODEL

A. Nikou (1), K. Yoshida (2), C. M. Witzenburg (2), A. D. McCulloch (1,3), J. H. Omens (1,3), J. W. Holmes (2,4)

(1) Department of Bioengineering
University of California, San Diego
La Jolla, CA, USA

(2) Department of Biomedical Engineering
University of Virginia
Charlottesville, VA, USA

(3) Department of Medicine
University of California, San Diego
La Jolla, CA, USA

(4) Department of Medicine
University of Virginia
Charlottesville, VA, USA

INTRODUCTION

Computational models with the ability to predict growth and remodeling of the heart could have useful clinical applications. As an example, surgeons planning repair of congenital heart defects must balance two conflicting goals: delaying surgery often allows the patient to better tolerate the procedure and/or the repair to last longer without revision. Continued heart remodeling, however, risks passing a "point-of-no-return" after which full cardiac function cannot be recovered. Models with the ability to predict the time-course of remodeling for individual patients could thus be of clinical utility.

Current cardiac growth models can predict responses to different hemodynamic perturbations used to study ventricular remodeling in experimental animals, including pressure overload (PO, induced by aortic banding) and volume overload (VO, induced by mitral regurgitation or an arteriovenous fistula). More commonly though, the major clinical questions center on how – or whether – the heart will *reverse* remodel following medical therapy or surgical repair. To date, cardiac growth models have not been comprehensively evaluated for their ability to predict reverse remodeling, typically a decrease in myocyte and heart size and restoration of a more physiological ventricular shape. Accordingly, we employed a growth law, previously used by our group, to simulate 18 days of experimental aortic banding in dogs followed by removal of the band, and compared the model results with experiments reported by Sasayama et al. [1].

METHODS

The coupled cardiovascular and heart model was previously published by Kerckhoffs et al. for normal and overload conditions [2]. Briefly, a finite element model of the beating canine left and right ventricles (LV and RV) with realistic myofiber structure was coupled to a nonlinear lumped-parameter model of systemic and pulmonary circulation including atria and valves [3]. Active and passive mechanics were modeled by Hill-type and Fung-type models [4,5]. The anisotropic strain-based growth law was also previously developed [6]. Briefly, this law uses changes in maximum fiber strain (which occurs at or near end diastole) to predict growth in the fiber direction and changes in minimum first principal strain in the plane perpendicular to the fibers (which occurs near end systole) to predict growth in the radial and crossfiber directions. The normal heart model (baseline) was coupled to the circulatory model and run to reach a hemodynamic steady state. In order to simulate acute aortic stenosis (PO), the area of the open aortic valve was decreased in the circulatory model and re-run to reach a new steady state. For each growth step, growth stretches were calculated and applied at end diastole, then the new grown configuration was coupled to the circulatory model (with altered aortic valve size) and run to steady state. In order to simulate reverse remodeling, after 12 forward growth steps (equivalent to 18 days) the size of the aortic valve was returned to that of the normal heart and 3 reverse growth steps were simulated. Compared with our published study of forward growth during PO [6], here we readjusted the severity of the aortic stenosis to match maximum LV pressure after banding in the published experiments [1]. The only other difference was a lower value for the systemic peripheral resistance in this study to better match the experimental hemodynamics.

RESULTS

The computational model matched the experimentally reported increase in maximum LV pressure with banding and decrease following unloading. The model also captured the reported increase in LV end-diastolic pressure (EDP) following banding, but simulated EDP was lower after unloading than in the experiments (Figure 1). Model-predicted changes in LV dimensions matched the experimental results closely. In response to banding, the model predicted approximately 10% radial growth on average across elements in the LV freewall, and the resulting changes in posterior wall thickness fell within 1 SD of Sasayama's reported mean values at all time points (Figure 2B). By contrast, the model predicted less than 4% growth on average in the fiber direction, and little change in anterior-posterior cavity diameter, in agreement with Sasayama's data (Figure 2A). Following simulated relief of PO, the model predicted very little reverse remodeling despite a return to near-normal hemodynamics.

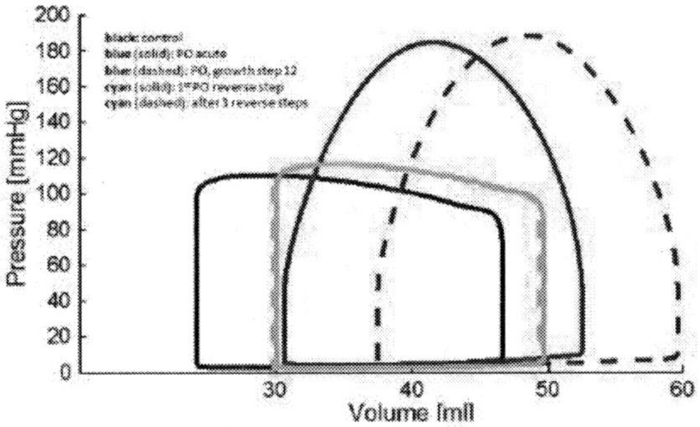

Figure 1: Left ventricular pressure-volume loops

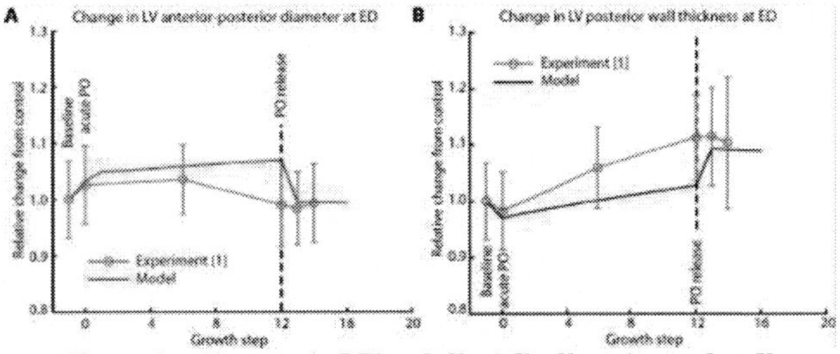

Figure 2: Changes in LV end diastolic diameter and wall thickness during forward and reverse remodeling

DISCUSSION

We employed a computational cardiovascular model previously shown to correctly predict hypertrophy ("forward remodeling") during experimental pressure and volume overload [2] to investigate growth responses following relief of pressure overload ("reverse remodeling"). Following simulated relief of PO after 18 days of forward remodeling, the model predicted very little reverse remodeling. Instead, maps of the stimuli that drive growth in the model (computed as differences between various strain components and their baseline normal values) showed that relief of PO returned strains to their normal baseline values across most of the heart, eliminating the stimulus for further remodeling.

While this result matched the experimental study used for comparison, it was initially surprising. The approach used here decomposes total deformation into an elastic deformation and a growth deformation: $\mathbf{F}_{eg} = \mathbf{F}_e \cdot \mathbf{F}_g$. Thus, if we prescribe a total axial stretch λ_{eg} in the fiber direction and simulate the growth response, λ_g will be >1 and the growth will gradually reduce the elastic stretch $\lambda_e = \lambda_{eg} / \lambda_g$ (physically, this simulates series addition of sarcomeres gradually returning sarcomere lengths to a homeostatic value). If the muscle is then restored to its original length, elastic stretches will drop below their baseline values, stimulating reversal of the original growth. Our simulations reveal the importance of the coupling between the ventricles and the circulation. In our study, relieving PO restored pressures to baseline values, but not global stretches; as one indication of this distinction, the cavity volumes in Figure 1 remained elevated even after relief of PO.

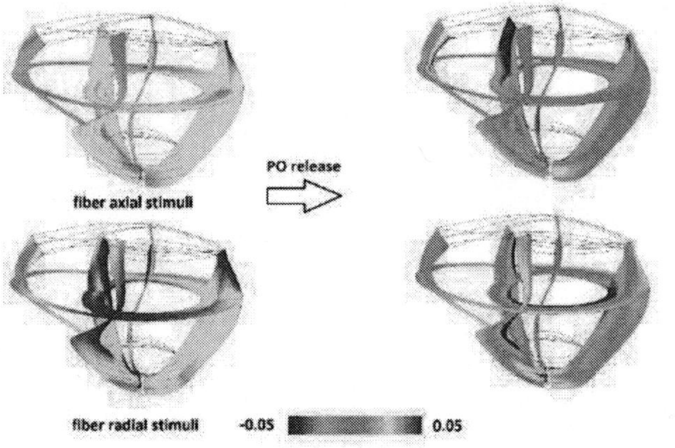

Figure 3: Fiber axial and fiber radial stimuli color map 18 days after inducing PO and two days after releasing PO

The experimental study used for comparison here followed animals for a very short time after relief of PO. Clinical studies in patients with aortic stenosis followed by valve replacement suggest that over longer time periods, some reverse remodeling does occur, but it is remarkably slow. For example, Monrad et al. found that eight years following surgical correction of aortic stenosis, LV mass normalized but LV wall-thickness-to-radius ratio remained elevated [7]. The response to hemodynamic unloading also likely depends on the nature and duration of the forward remodeling. In some situations, such as pregnancy or athletic training, cardiac growth and remodeling induced by altered hemodynamic load appear to reverse completely once normal hemodynamics are restored. In others, some features of remodeling appear irreversible. For example, we reported that unloading end-stage failing hearts with left ventricular assist devices (LVADs) did produce a dramatic reduction in myocyte size, but did not restore normal radius-to-wall-thickness ratios in the LV, suggesting that myocyte shape changes did not reverse completely [8].

In conclusion, the growth model employed here provided a good match to experiments following experimental pressure overload by aortic banding and relief of that overload. Model results suggest that realistic hemodynamic loading of the heart model is essential to capturing observed growth responses. Future work includes examining responses over much longer durations of overloading and unloading for comparison to available clinical data.

ACKNOWLEDGEMENTS

This study was supported by the National Institute of Health grant U01 HL-120862 (to J. W. Holmes).

REFERENCES

[1] Sasayama S. et al., Circ Res 38: 172-178, 1976.
[2] Kerckhoffs, R.C.P. et al., Circ Heart Fail 3:528-536, 2010.
[3] Arts T et al., Am J Physiol 288: 1943–1954, 2005.
[4] Lumens J. et al., Ann Biomed Eng 37:2234–2255, 2009.
[5] Guccione J.M. et al., J Biomech Eng 113:42-55, 1991.
[6] Kerckhoffs R.C.P. et al., Mech Res Commun 42:40-50, 2012.
[7] Monrad E. S. et al., Circ 77:1345-1355, 1988.
[8] Barbone A. et al., Circ 104 (suppl): I229-I232, 2001.

TARGETING SEROTONIN 2B RECEPTOR TO IMPROVE CARDIAC FUNCTION FOLLOWING MYOCARDIAL INFARCTION

J. Caleb Snider (1), Qinkun Zhang (2), Hind Lal (2), and W. David Merryman (1)

(1) Biomedical Engineering
Vanderbilt University
Nashville, TN, USA

(2) Department of Cardiovascular Medicine
Vanderbilt University
Nashville, TN, USA

INTRODUCTION

Nearly 1.5 million Americans experience a myocardial infarction (MI) annually, accounting for one in every four deaths. MI leads to cardiomyocyte death. The inflammatory response removes cellular debris, and granulation tissue creates a scaffold for mature scar formation. The complex cellular response creates a heterogeneous mechanical environment altering cellular activity and tissue morphology. Scar formation is vital for mechanical stability to prevent rupture of the ventricular wall. However, local stiffening increases mechanical strain on surrounding tissue, causing scar expansion and ultimately, compromised output and cardiac failure. Cardiac fibroblasts (CFs) are the primary cells responsible for scar formation. Following injury, factors such as mechanical strain and transforming growth factor β1 (TGF-β1) signaling result in myofibroblast activation of CFs. Myofibroblasts remodel the microenvironment through hypersecretion of collagen and compaction of the extracellular matrix (ECM) [1]. The regulation of myofibroblast activity must be understood to modulate scar formation and limit fibrosis.

Serotonin signaling is highly upregulated during cardiac injury. The 5-HT_{2B} receptor has been implicated in cardiac hypertrophy and right ventricle fibrosis [2-3]. In noncardiomyocytes, it is required for increased expression of inflammatory and profibrotic markers like interleukin-6 and TGF-β [1]. It is also known to have mitogenic effects, control collagen synthesis of fibroblasts, and influence myofibroblast activation [4]. 5-HT_{2B} involvement in ECM regulation makes it a potential therapeutic target for regulating scar formation after MI. The objective of this work is to clarify the role of 5-HT_{2B} in myofibroblast activity and scar formation. We hypothesize antagonism of the 5-HT_{2B} receptor will limit post-MI remodeling by downregulating myofibroblast activity and modulating scar formation.

METHODS

MI was induced in wild type mice by ligating the left anterior descending branch of the left coronary artery. Echocardiograms were taken prior to ligation, as well as 7 and 21 days following the procedure. Measurements were made on m-mode images of the short axis of hearts captured on the Vevo2100 system to compare mice treated with the 5-HT_{2B} antagonist RS-127445 hydrochloride (RS) or DMSO. 21 days after MI, mice were euthanized. Hearts were excised and either flash frozen or frozen in OCT. Hearts were cryosectioned at 10 μm thickness and subjected to Masson's trichrome staining or atomic force microscopy (AFM) analysis. BioScope Catalyst AFM (Bruker AXS, Santa Barbara CA) was operated in Peak Force – Quantitative Nanomechanical Mapping mode in fluid. 10 x 10 μm areas were scanned using an MLCT-Bio tip with approximate spring constant of 0.03 N/m and a tip radius of 70 nm. mRNA was extracted from flash frozen hearts using TRIzol. qPCR was performed to detect transcriptional differences in tissue exposed to RS vs. DMSO.

RESULTS

Transcription of 5-HT_{2B}, but not 5-HT_{2A}, drastically increased and peaked 5 days following MI (**1A**). This increase coincided with an increase in α-smooth muscle actin (αSMA), TGF-β1, and collagen-1 (**1B-C**). Therefore, the effects of antagonizing 5-HT_{2B} on cardiac function, gene transcription, and scar formation were investigated following MI.

Ejection fraction in both RS and vehicle treated groups was decreased by at least 20% at day 7 and was not significantly different

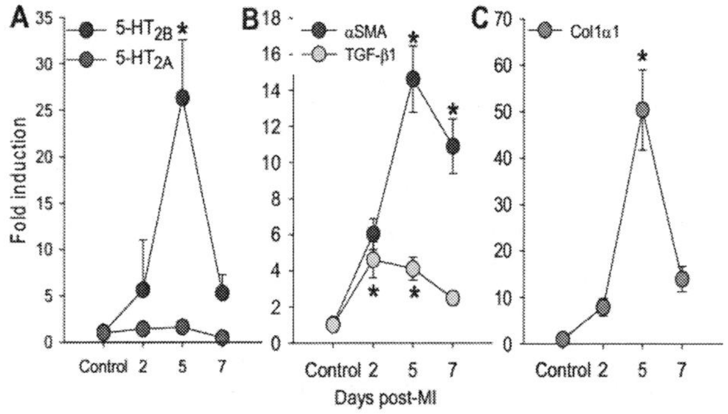

Fig. 1: 5-HT$_{2B}$ and myofibroblast response to MI. Fold induction after MI determined by qPCR. * indicates significant difference from control animals ($p<0.05$)

between the groups, indicating similar initial cardiomyocyte damage. 21 days following infarction, 5-HT$_{2B}$ antagonism preserved the ejection fraction and fractional shortening compared to vehicle treated mice (**2A-B**). Increasing systolic dilation of the left ventricle inner diameter and volume was observed in vehicle treated mice but was diminished by RS treatment (**2C-D**).

Hearts were excised and subjected to Masson's trichrome staining and AFM analysis to determine physical properties of the scar tissue. No difference was observed in scar size. Vehicle treated mice had stiffer scars than RS treated mice (**3A**). qPCR analysis revealed that RS treatment significantly reduced transcript levels of 5-HT$_{2B}$, TGF-β type 2 receptor (TβRII), and smooth muscle protein 22α. Collagen-1 transcription was also reduced (**3B**).

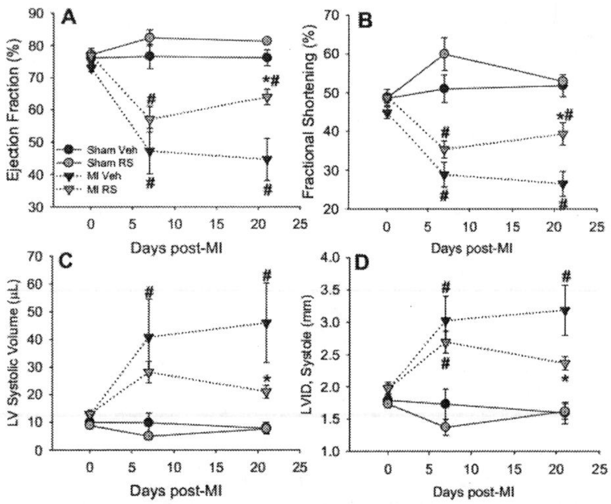

Fig. 2: Functional effect of 5-HT$_{2B}$ antagonism in response to MI. Combined echocardiogram data in WT mice treated with vehicle or RS (n=6). * indicates significant difference between Veh and RS treated animals. # indicates significant difference compared to sham ($p<0.05$)

DISCUSSION

Our initial findings implicate the serotonin receptor 5-HT$_{2B}$ in the process of scar formation following MI. The transcriptional increase of this receptor, and not the 5-HT$_{2A}$ receptor, points to a specific therapeutic target. Since it correlates with the increase in αSMA, TGF-β1, and collagen-1, we believe modulating scar formation through 5-HT$_{2B}$ is a viable approach to limiting damage after MI.

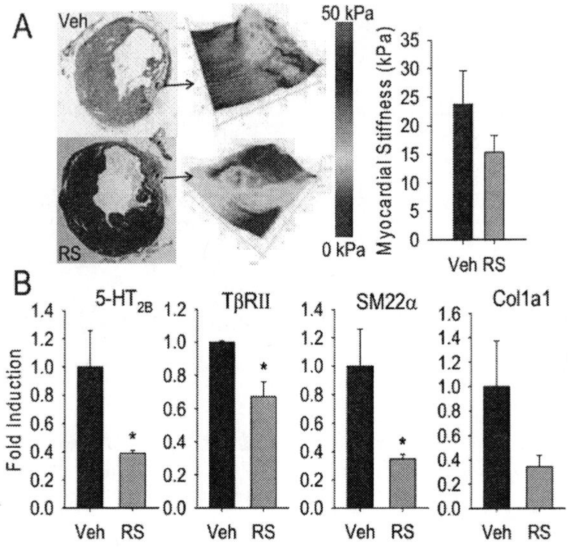

Fig. 3: A) Biophysical (n=2) and B) transcriptional ($n \geq 3$) effects of 5-HT$_{2B}$ antagonism in vivo. * indicates $p<0.05$

We have shown 5-HT$_{2B}$ antagonism following MI preserves cardiac output by restricting expansion of the left ventricle. The cardiac function of vehicle treated mice continues to decrease over 21 days while being maintained in RS treated mice.

While no difference in scar size was observed, 5-HT$_{2B}$ antagonism resulted in a softer scar. The decrease in scar stiffness following MI suggests that surrounding cardiomyocytes experience less strain. This would abrogate reactive fibrosis, limiting tissue damage beyond the initial insult. The scar was still mechanically sound enough to prevent rupture of the ventricle wall. We expect the vehicle treated mice to continue to worsen into cardiac failure, while the RS treated mice will maintain their cardiac function.

Finally, we have identified decreased gene transcription in genes known to be involved in fibrosis. 5-HT$_{2B}$ antagonism decreases transcription of TβRII, SM22α, and collagen-1. TGF-β signaling is known to resolve inflammation and instigate a fibrotic response in injured tissue. By limiting signaling of this receptor, we are potentially curbing the fibrotic response by limiting activation of myofibroblasts, (seen by the decrease in the smooth muscle protein SM22α) which would decrease the amount of collagen secretion and compaction.

Further exploration into the mechanical effects and signaling pathways involved in 5-HT$_{2B}$ antagonism following MI will provide insight into the utility of targeting 5-HT$_{2B}$ signaling as a therapeutic approach for improving outcomes following MI.

ACKNOWLEDGEMENTS

This work was supported by the NIH (R01 HL115103 and R35 HL135790). We thank the Vanderbilt Cardiovascular Physiology Core for performing echoes for this study.

REFERENCES

[1] Frangogiannis, NG. *Nat Rev Cardiol.* 11(5):255-265, 2014.
[2] Jaffré, F. et al., *Circ Res.,* 104(1):113-123, 2009.
[3] Janssen, W. Et al., *Biomed Res Int.* 2015:1-8, 2015.
[4] Hutcheson, J. Et al., *J. Mol Cell Cardiol.* 53(5):707-714,2012.

SB[3]C2017
Summer Biomechanics, Bioengineering and Biotransport Conference
June 21 – 24, Tucson, AZ, USA

THE INTERPLAY OF GROWTH AND REMODELING IN HUMAN HEART VALVES DURING SOMATIC GROWTH

Pim J.A. Oomen (1,2,3), Carlijn V.C. Bouten (1,2), Ellen Kuhl (3), Sandra Loerakker (1,2,3)

(1) Department of Biomedical Engineering
Eindhoven University of Technology
Eindhoven, The Netherlands

(2) Institute for Complex Molecular Systems
Eindhoven University of Technology
Eindhoven, The Netherlands

(3) Department of Mechanical Engineering
Stanford University
Stanford, CA, United States

INTRODUCTION

Cardiovascular tissues are widely known to adapt to changes in hemodynamic demands, in order to maintain or restore mechanical homeostasis [1]. Growth and remodeling play eminent roles in preserving homeostasis, although it is often unclear which mechanical parameter(s) drive(s) these processes and define(s) the homeostatic state. In recent studies, we established an extensive data set of human heart valves of different ages [2,3], and computational models derived from these data suggested that human heart valves appear to maintain a stretch homeostasis [3].

Yet, a mechanistic insight into *how* this stretch homeostasis is preserved on a local level is unclear. The goals of the present study are to use this extensive experimental data set to: (1) establish a computational framework that can locally predict growth (change in mass or volume) and remodeling (change in tissue structure or material properties) of human heart valves, and (2) elucidate the relative roles of growth and remodeling in preserving mechanical homeostasis with age.

METHODS

Experimental data

The geometrical, structural, and material properties of human aortic and pulmonary heart valves of different ages (infant to adult origin, n=32) were determined via numerical-experimental techniques [2,3]. To provide input for the computational framework, all data were divided into three categories (infant, adolescent, adult) to assess the average changes in annulus size (boundary conditions), leaflet geometry (growth), and material properties (remodeling) with age.

Computational model

Finite element models were developed for each age group, to predict the growth and remodeling processes between the different time points. The valve geometry for each age was based on measurements of the valve radius and leaflet thickness, changes in annulus size served as displacement boundary conditions, and hemodynamic loading conditions were chosen in accordance with each age group. A hyperelastic fiber-reinforced material model was used to represent the material behavior of the valve leaflets, where the material parameters at each age were obtained from the experimental data (indentation testing). Leaflet growth was incorporated via the theory of finite growth, where the total deformation gradient tensor $\boldsymbol{F}$ is multiplicatively decomposed into an elastic part $\boldsymbol{F}_e$ and a growth part $\boldsymbol{F}_g$:

$$\boldsymbol{F} = \boldsymbol{F}_e \cdot \boldsymbol{F}_g \,, \tag{1}$$

where $\boldsymbol{F}_g$ was derived from the experimental data on valve geometry. Tissue remodeling was implemented via a linear transition (with $0 \leq W \leq 1$) from the material behavior of one age group to the other:

$$\sigma_{i,j} = (1 - W_j)\sigma_{i,\text{age1}} + W_j\sigma_{i,\text{age2}}, \qquad i = circ, rad. \tag{2}$$

RESULTS

Experimental results

Both aortic and pulmonary leaflets demonstrated transversely isotropic growth (Fig. 1A), where the total leaflet area increased monotonically with age (Fig. 1B). In terms of leaflet thickness (Fig. 1C), the average thickness increased in the aortic valve between the infant and adolescent stage, whereas a subsequent decrease was present during maturation towards the adult stage. The average thickness of the pulmonary valve appeared to be rather stable with age.

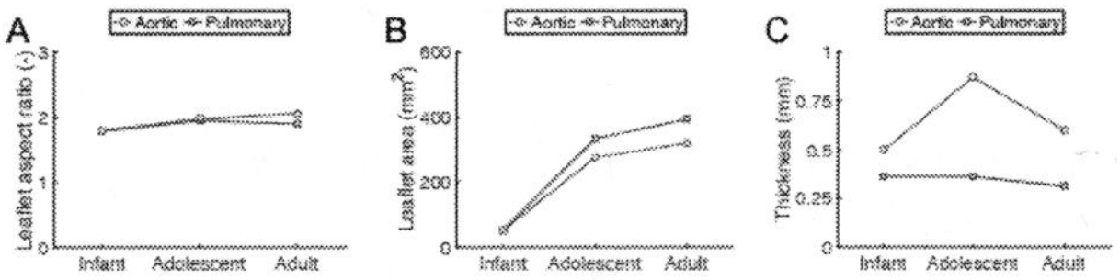

Figure 1: Changes in leaflet aspect ratio (A), area (B) and thickness (C) as a function of age.

With regard to tissue remodeling, the stiffness in the circumferential direction appeared to decrease with age during early development (Fig. 2A), while a subsequent increase in stiffness was evident during the later stages of development (Fig. 2B).

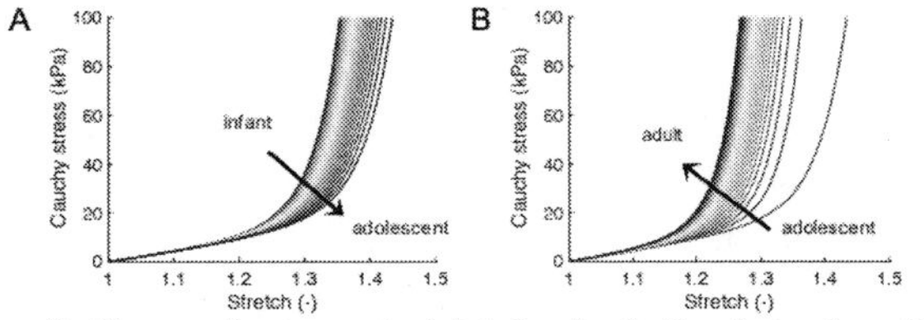

Figure 2: Changes in stress-stretch behavior in the circumferential direction during development from the infant to the adolescent stage (A), and from the adolescent to the adult stage (B).

Computational results

When the valve geometry, material properties, and hemodynamic loading conditions corresponding with each age group were applied, the area stretches were predicted to be similar in valves representing the average adolescent and adult aortic case (Fig. 3), supporting the hypothesis that growth and remodeling in human heart valves aim to restore a stretch homeostasis [3].

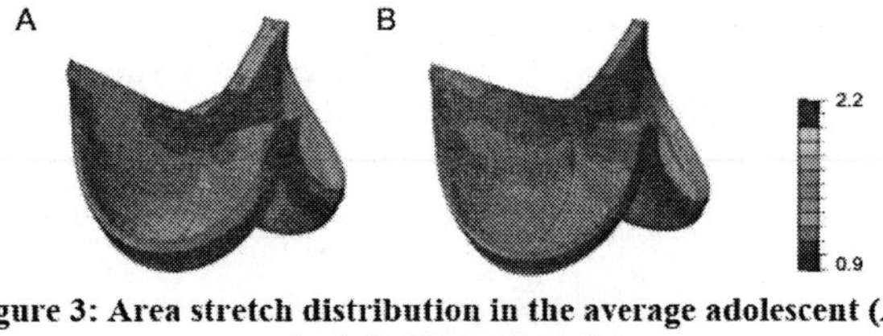

Figure 3: Area stretch distribution in the average adolescent (A) and adult (B) aortic valve.

To elucidate the relative roles of growth and remodeling in the preservation of stretch during the evolution of the aortic valve from the adolescent to the adult stage, systematic analyses were performed where the effects of growth and remodeling were studied in isolation or combined. Starting from the adolescent valve, increasing the annulus size and hemodynamic load towards their corresponding adult values (representing the maturation of the surrounding tissues and hemodynamic conditions) resulted in a considerable increase in stretch in the valve (Fig. 4A), indicating that valve adaptation is necessary to restore stretches to their homeostatic values. Interestingly though, applying the growth law that was derived from the experimental data in combination with the changes in boundary conditions, appeared to only induce further increases in stretch (Fig. 4B). Hence, volumetric growth during this time span is not likely to contribute to restoring mechanical homeostasis. Conversely, when remodeling was applied instead of growth, area stretches clearly decreased towards the homeostatic values (Fig. 4C). An even closer resemblance was obtained when growth and remodeling were applied in combination (Fig. 4D), although the difference compared to remodeling only is rather limited.

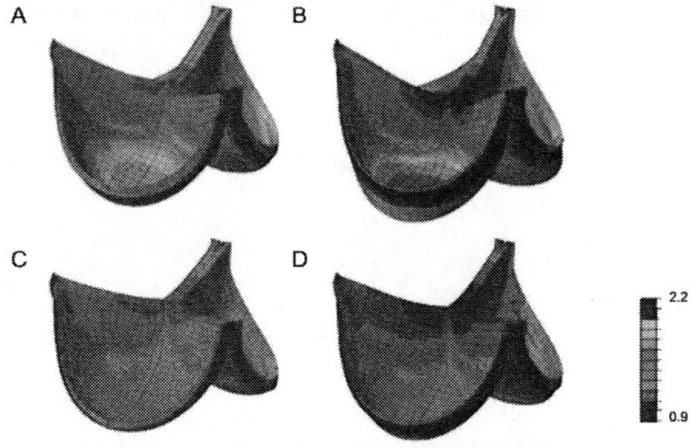

Figure 4: Area stretch distribution in the adolescent aortic valve subjected to: (A) adult boundary conditions (increase in annulus size and hemodynamic loads), (B) adult boundary conditions and growth, (C) adult boundary conditions and remodeling, (D) adult boundary conditions and both growth and remodeling.

DISCUSSION

In the present study, we have established a computational framework to predict growth and remodeling of human heart valves during somatic growth, based on an extensive experimental data set. Considering the evolution of the aortic valve from the adolescent to the adult stage, our results indicated that remodeling is the dominant mechanism for restoring mechanical homeostasis. For the pulmonary valve, remodeling was found to be of similar importance for restoring homeostasis. For the transition from the infant to the adolescent stage, however, we expect that growth will appear as the major determinant in preserving homeostasis. In future studies, hypotheses for growth and remodeling will be implemented to further increase our mechanistic understanding of valve growth and remodeling. These insights are indispensable for understanding healthy valve development, and valve pathologies, as well as for creating tissue-engineered valves that can provide long-term growth and remodeling corresponding with the patient's demands.

ACKNOWLEDGEMENTS

This project has received funding from the European Union's Horizon 2020 research and innovation programme under the Marie Sklodowska-Curie grant agreement No 654513. We further acknowledge the support from the European Union's Seventh Framework Program (grant agreement No 604514), and the Netherlands CardioVascular Research Initiative (CVON2012-01): the Dutch Heart Foundation, Dutch Federation of University Medical Centers, the Netherlands Organization for Health Research and Development, and the Royal Netherlands Academy of Sciences.

REFERENCES

[1] Humphrey JD et al., *Nat Rev Mol Cell Biol*, 15:802-812, 2014.
[2] Van Geemen D et al., *PLoS One*, 11:e0149020, 2016.
[3] Oomen, PJA et al., *Acta Biomater*, 29:161-169, 2016.

REGIONAL VARIATIONS IN CELL-MATRIX MECHANO-ADAPTATION DRIVE HYPERTENSIVE VASCULAR REMODELING

Matthew R. Bersi (1,2), Ramak Khosravi (2), Anna Wujciak (2), Alexander W. Caulk (2), David G. Harrison (3), Jay D. Humphrey (2,4)

(1) Department of Biomedical Engineering
Vanderbilt University
Nashville, Tennessee, USA

(2) Department of Biomedical Engineering
□ale University
New Haven, Connecticut, USA

(3) Departments of Medicine and Pharmacology
Vanderbilt University
Nashville, Tennessee, USA

(4) Vascular Biology & Therapeutics Program
□ale University
New Haven, Connecticut, USA

INTRODUCTION

The aorta is the primary conduit for blood flow in mammals and its unique structure and function – to elastically store and release energy [1] - play critical roles in health; aortic disease is yet responsible for significant morbidity and mortality. Regional variations exist along the aorta, including embryonic cell lineage [2], extracellular matrix organization, local geometry, and hemodynamic loading [3]. Nevertheless, it remains unclear why certain diseases manifest differentially along the aorta. Toward this end, angiotensin II (AngII), the primary component of the renin-angiotensin system, is a critical mediator of arterial homeostasis and contributes to arterial stiffening, atherosclerosis, and the progression of aortic aneurysms and dissections in central arteries [4].

Using the AngII infusion model of induced hypertension [5], we characterized the evolving cell-matrix content and biaxial wall mechanics in four regions of the aorta, from proximal thoracic to distal abdominal. Our analyses reveal time-varying changes in cellular and matrix composition that lead to altered biomechanical behaviors, implicating (i) an inability of intramural cells to regulate homeostatic material stiffness in the aneurysmal propensity of the ascending aorta (ii) a mechano-inflammatory driven thickening of the wall in excessive fibrosis of the descending aorta, and (iii) a loss of medial stress shielding by degrading adventitial collagen in dissection propensity of the suprarenal aorta. In the absence of such maladaptations, smooth muscle contractility helps to maintain homeostatic mechanical properties in the infrarenal aorta despite elevated blood pressures by lowering wall stress. Importantly, mechanical stimuli precede, and likely drive, the inflammatory infiltration that ultimately leads to differential maladaptive regional remodeling. This indicates a need for more systematic assessments of aortic disease progression, not simply a singular focus on a particular condition or region.

METHODS

Animal Model - Hypertension was induced in 19.4±0.02 week old, male, apolipoprotein-E null (*ApoE*$^{-/-}$) mice by Ang II infusion (1000 ng/kg/min) for up to 28 days. Conscious blood pressure was measured at 0, 4, 7, 14, 21, and 28 days post-implantation using a tail-cuff system. At each experimental time point, four regions of the aorta – ascending thoracic (ATA), descending thoracic (DTA), suprarenal abdominal (SAA) and infrarenal abdominal (IAA) – were excised and prepared for testing and analysis.

Mechanical Testing and Contractile Assessment - Samples were cleaned of perivascular tissue and cannulated onto a custom computer-controlled mechanical testing device [6]. Specimens were placed in a warmed (37°C) Hanks buffered physiologic solution and subjected to cyclic preconditioning near the *in vivo* axial stretch. Following estimation of the unloaded geometry and *in vivo* axial stretch, vessels were subjected to 3 cyclic pressure-diameter protocols (from 10 to 140 mmHg at 0.95, 1.0, and 1.05 times the *in vivo* axial stretch) and 4 cyclic axial force-length protocols (from 0 to f_{max} at 10, 60, 100, and 140 mmHg, where f_{max} was up to 65 mN, 40mN, 40 mN, or 30 mN for the ATA, DTA, SAA, and IAA, respectively) [7]. The continuous acquisition of loads (pressure and axial force) and geometry (outer diameter and axial length) during mechanical testing allowed for material parameter identification using the constitutive equation W:

$$W(\mathbf{C},\mathbf{M}^i) = \frac{c}{2}(I_\mathbf{C} - 3) + \sum_{i=1}^{4} \frac{c_1^i}{4c_2^i}\left\{\exp\left[c_2^i(IV_\mathbf{C}^i - 1)^2\right] - 1\right\}, \quad (1)$$

W describes the elastin-dominated behavior and four families of locally parallel collagen fibers to the transmurally averaged material properties of the aortic wall, whereby stress and material stiffness can be computed from first and second derivatives of W, respectively, with respect to an appropriate measure of strain.

Baseline contractility was assessed by measuring changes in outer diameter and circumferential stress in response to either 100 mM KCl, 1 μM phenylephrine (PE), or 1 μM AngII. All vessels were activated while held at the *in vivo* axial stretch and pressure of 90 mmHg.

Histology – Tested aortas were fixed overnight in a 10% neutral buffered formalin solution prior to staining for extracellular matrix proteins (e.g., collagen and elastin), matrix metalloproteases (MMPs), or inflammatory cell markers (e.g., CD45). Image analysis was performed in order to extract wall composition, constituent area fractions, and (positively staining) cell counts.

Statistical Analysis – Mechanical metrics and layer-specific expression were compared using either one- or two-way ANOVA, respectively, with post-hoc Tukey HSD test for multiple comparisons; potential correlations between measured variables were assessed using the Pearson product-moment correlation coefficient.

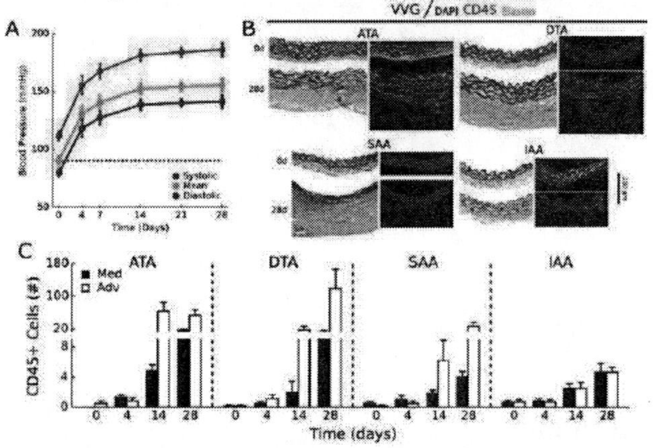

Figure 1: (A) Blood pressure and (B) histological images showing wall remodeling and (C) inflammatory ($CD45^+$) cell expression

RESULTS

AngII infusion increased systolic blood pressure from 112±4 mmHg at baseline to 186±6 mmHg at 28 days ($p < 0.001$). The associated pressure-driven wall thickness increase led to a reduction in distensibility in all regions ($p < 0.001$), albeit more pronounced proximally (ATA: 55% vs. IAA: 37% reduction). While the change in thickness, primarily a result of adventitial collagen deposition, in the ATA and IAA mimicked the blood pressure change, the DTA and SAA showed a delayed exuberant thickness increase, consistent with inflammatory cell recruitment to the adventitia.

With increasing pressure, all regions increased circumferential stiffness, albeit to varying degrees (ATA: 232% vs. IAA: 44% increase). Whereas the DTA, SAA, and IAA restored circumferential stiffness back to near-baseline values by 28 days, the ATA alone did not return to baseline (remaining 1.8-fold higher), indicating aneurysmal propensity [8].

All regions showed a reduction in energy storage that was apparent after either 4, 7, or 14 days in the ATA, DTA, or SAA, respectively. The sustained increase in energy storage, particularly in the SAA, was a result of increases in blood pressure and collagen degradation as revealed by increased MMP-13 staining (day 4 and 14, $p < 0.01$). This biomechanical feature, in addition to transient reductions in adventitial area at 4 days that coincided with increased MMP levels, potentially renders the SAA vulnerable to dissection.

The time course of remodeling gave rise to an early peak in wall stress (up to 68% in the SAA), resulting in modest increases in wall thickness that preceded, and were independent of, inflammatory cell infiltration. While the IAA showed the least remodeling, it was also determined to be the most contractile region, suggesting further that the ability to attenuate the early mechanical response (by reducing wall stress [9]) may also attenuate the maladaptive inflammatory response. Indeed, $CD45^+$ cell expression revealed a strong positive correlation with wall thickness ($R > 0.8$, $p < 0.001$), a strong negative correlation with stored energy ($R < -0.65$, $p < 0.01$), and a weak correlation with circumferential stiffness ($R < 0.25$).

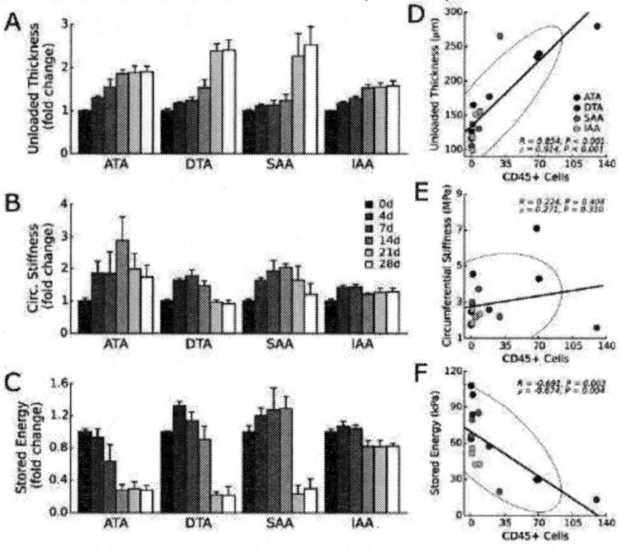

Figure 2: (A-C) Fold changes and (D-F) correlation with $CD45^+$ cells for wall thickness, circumferential stiffness, and stored energy from all regions over 28 days of infusion.

DISCUSSION

Pressure-induced increases in mechanical stress stimulate matrix deposition [10] and matrix degradation [11] as well as expression of chemokines that influence inflammatory cell recruitment and activity [12]. Ultimately the local balance between matrix deposition and degradation dictates whether the response is homeostatic or pathologic. Here, we show that aortic remodeling in response to the same hypertensive stimulus differs by region and includes aneurysmal propensity (ATA), excessive fibrosis and structural stiffening (DTA), a propensity to dissection (SAA), and a homeostatic response (IAA). We show, too, that adventitial inflammation leads to maladaptive structural stiffening and it follows, rather than precedes, blood pressure-driven responses to increased wall stresses. Inflammation must be targeted to mitigate the positive feedback between hypertension and arterial stiffening.

ACKNOWLEDGEMENTS

This work was supported, in part, by NIH grants R01 HL086418, R01 HL105297, and U01 HL116323.

REFERENCES

[1] Bersi, MR et al., *Hypertension*, 67(5):890-896, 2016. **[2]** Majesky, M., *Arterioscler Thromb Vasc Biol*, 27:1248-1258, 2007. **[3]** Cuomo, F et al., *Ann Biomed Eng*, 43(7): 1555-1570, 2015. **[4]** Van Thiel, B et al., *Eur J Pharmacol*, 763: 3-14, 2015. **[5]** Daugherty, A et al., *J Clin Invest*, 105(11):1605-1612, 2000. **[6]** Gleason, RL et al., *J Biomech Eng*, 126(6):787, 2004. **[7]** Ferruzzi, J et al., *Ann Biomed Eng*, 41(7): 1311-1330, 2013. **[8]** Bellini, C et al., J Biomech, 49(12): 2383-2389, 2016. **[9]** Humphrey, JD, Wilson, E, *J Biomech*, 36(11):1595-1601, 2003. **[10]** Bardy et al., *Circ Res*, 79(1):70-78, 1996. **[11]** Lehoux et al., *Circulation*, 109(8):1041-1047, 2004. **[12]** Capers et al., *Hypertension*, 30(6):1397-1402, 1997.

STRAIN MEDIATED ENZYME DEGRADATION OF ARTERIAL TISSUE; IMPLICATIONS IN DISEASE AND MEDICAL DEVICE DESIGN

Robert Gaul (1,2), Caitríona Lally (1,2)

(1) Trinity Centre for Bioengineering
Trinity College Dublin
Dublin, Ireland.

(2) Mechanical and Manufacturing Engineering,
Trinity College Dublin,
Dublin, Ireland.

INTRODUCTION

Collagen fibre architecture is of critical importance in healthy arterial function. It is believed that maladaptive remodelling of this architecture may play a role in the development and progression of arterial disease [1]. To date, literature has focused on load induced fibre reorganisation with a clear lack of information available on collagen production and degradation in response to load in arterial tissue.

Previous studies on other soft tissues, such as tendon and bovine pericardium (BP), have shown conflicting results on whether load inhibits [2] or enhances [3] enzymatic activity. In addition, dynamic loading has been shown to exhibit an accelerated enzymatic response in BP [3]. It is hypothesised that mechanical loading may also influence the enzymatic activity of collagenase in arteries, although further investigation is needed to decipher these mechanisms.

Small angle light scattering (SALS) is a technique which lends itself well to investigating load induced degradation as it allows structural information to be acquired non-destructively for fibrous test specimens. In fact, SALS has previously been used to identify the preferential degradation of unloaded collagen fibres in collagenase treated corneal tissue in real time [4].

The aim of the present study is to use a SALS system to measure collagen fibre changes in intact arterial tissue and to explore the influence of load on collagen degradation in arterial tissue.

METHODS

Porcine common carotid arteries were dissected into 5 mm ring samples before being cut longitudinally to obtain planar samples. The adventitial layer was carefully removed prior to testing to allow adequate light transmission for future SALS analysis. Samples were placed in a custom stretching device at 0%, 5% and 25% circumferential strain levels, representative of stiff, healthy and weakened tissue respectively. Strained samples were incubated in bacterial collagenase (Clostridium histolyticum, Sigma Aldrich) at 37°C.

Samples were analysed at 1-hour intervals for a total of 4 hours. Measurements were taken at the same 64 interrogation regions at each time point. Testing was carried out using an in-house SALS system consisting of a 5mW HeNe laser (λ = 632.8 nm) and single focusing lens (f_l = 75 mm). Each sample was interrogated sequentially using a 150 µm beam diameter and a 250 µm scanning resolution, controlled using LabVIEW.

Fibre alignment was determined based on the eccentricity of the scattered light distribution using Eq. 1. Histological analysis was also carried out on the same samples post collagenase treatment to support SALS data.

$$E = 2\frac{\sqrt{\left(\frac{Major\ Axis}{2}\right)^2 - \left(\frac{Minor\ Axis}{2}\right)^2}}{Major\ Axis} \quad (1)$$

Technical Presentation #17

RESULTS

SALS images of a sample exposed to 5% strain and collagenase for 4 hours showed increased fibre alignment based on scattered light eccentricity, see Figure 1a) and 1b). Figure 1c) shows the relative change in SALS eccentricity under different loading conditions in the presence of collagenase for 4 hours (Two outliers have been removed based on the ROUT method [5]). The relative change in fibre alignment is also shown at each time point for a single sample at 5% and 25% strain in Figure 1d). Whilst variations exist in the initial fibre angle and the degree of fibre eccentricity, the change in fibre eccentricity is measured relative to the initial fibre eccentricity of each sample prior to collagenase treatment.

Histological images of the same samples also show contrasting fibre organisation based on the loading conditions applied to the vessel wall, see Figure 2.

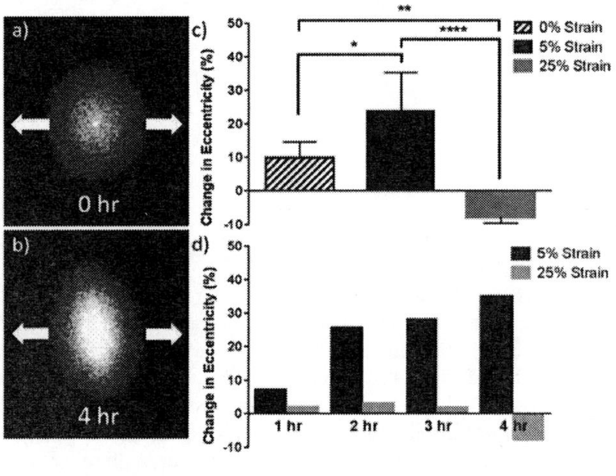

Figure 1: a) and b) SALS light distribution after 4 hours at 5% strain subjected to collagenase. Relative change in SALS eccentricity in the presence of collagenase c) after 4 hours and d) at each time point. N=7, * $P \leq 0.05$, ** $P \leq 0.01$, **** $P \leq 0.0001$

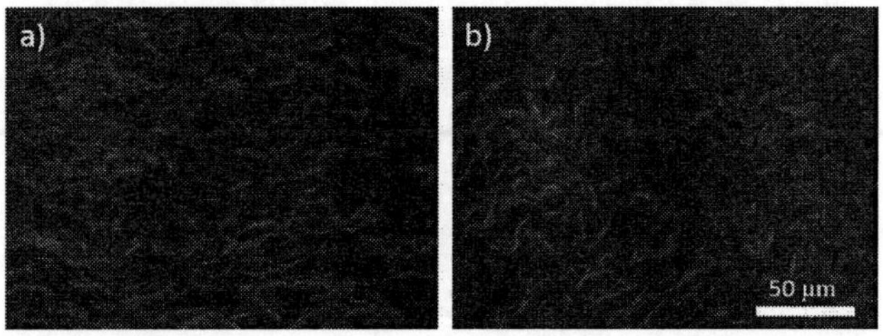

Figure 2: Polarised light images of picrosirius red stained arterial tissue showing collagen fibre organisation in arterial tissue subjected to a) 5% strain and b) 25% strain.

DISCUSSION

Results shown in Figure 1 demonstrate that strain mediated degradation mechanisms exist in arterial tissue which are dependent on the level of strain experienced by the tissue.

The large relative increase in SALS eccentricity at 5% strain indicates increased fibre alignment over time and therefore preferential degradation of unloaded fibres. In contrast, a 25% strain resulted in a reduction in eccentricity suggesting less fibre alignment over time and consequently, the preferential degradation of loaded fibres. The relatively small increase in alignment in the 0% strain condition was attributed to a greater signal to noise ratio as sample density decreases over time with a more isotropic collagen degradation. These findings are supported by histological analysis where a clear lack of collagen organisation is evident in tissue subjected to 25% strain (Figure 2).

In earlier published work by Huang and Yannes [6], a similar degradation response was identified in reconstituted bovine collagen, in the form of a tape. A protective mechanism was identified at low strain levels (< 4%) whilst accelerated collagen degradation was observed at higher strain levels (> 6%) [5]. This study demonstrates that collagen under load within arterial tissue responds to collagen degrading enzymes in a similar fashion. This load–mediated degradation process may have implications in disease where weakening of the vessel wall could result in higher strains which induce further collagen degradation through a positive feedback mechanism.

Future work aims to explore a range of different loading regimes, including physiological and diseased conditions, to more fully investigate load mediated collagen degradation and production in response to mechanical stimuli. These results not only provide important information on the development and progression of arterial disease, but also its treatment, where the mechanical environment may be artificially altered, e.g. stenting.

ACKNOWLEDGEMENTS

This project has received funding from the Irish Research Council (GOIPG/2014/515).

REFERENCES

[1] Creane, A *et al.*, Biomech Model Mechanobiol., 10:831-843, 2011.
[2] Wyatt, K *et al.*, J Biomech. Eng., 131:1-11, 2009.
[3] Ellsmere, J *et al.*, Biomaterials, 20:1143-1150, 1999.
[4] Robitaille, M *et al.*, Interface focus, 1:767-776, 2011.
[5] Motulsky and Brown, BMC Bioinformatics, 7:123, 2006
[6] Huang, C and Yannes, I, J Biomed Mater Res., 11:137-154. 1977.

MODELLING THE DISSECTION OF ARTERIAL TISSUE

Brian FitzGibbon (1,2), Niamh Hynes (2), Sherif Sultan (2), Peter E. McHugh (1), Patrick McGarry (1)

(1) Discipline of Biomedical Engineering, College of Engineering & Informatics National University of Ireland, Galway

(2) The Western Vascular Institute, Vascular & Endovascular Surgery, Galway University Hospitals / The Galway Clinic Galway, Ireland

INTRODUCTION

Aortic Dissection (AD) occurs due to intimal injury which causes an initial separation of the arterial layers. The separation is thought to occur most commonly between the intima and media[1]. This intima-medial tear is the pre-cursor for crack propagation, which is thought to occur between the lamellar elastic layers of the media [2,3]. The propagation of this crack/buckle commonly results in the development of pathological blood flow between the medial layers. This blood flow exerts an internal intra- media pressure which not only causes further separation of the layers, but also expansion of this pathological conduit known as the "False lumen" [4]. This is often problematic as it results in a reduction in blood flow to the major arteries which stem from the aorta; the celiac, the superior mesenteric (SMA), the renals, the inferior mesenteric (IMA), the intercostal, and the spinal arteries. If these arteries are occluded/malperfused the outcome for the patient can be catastrophic (paraplegia, renal failure, stroke, etc.) [5].

The potential for separation of the arterial layers is highly influenced by the strength of the inter-lamellar bonds. If there is any portion of the media in which the morphology of elastin and collagen is pathologically altered, as seen in hypertension-induced obstruction of the vasa vasorum [6-8], the strength of the bridge fibers between those lamellar layers is also affected.

Cohesive zone models can be used to describe the delamination between two surfaces [9-10]. AD is fundamentally a biomechanics problem involving an initial material failure and subsequent unstable buckling between the layers of elastic laminae throughout the aorta extending from proximal to distal sections.

To date no *in silico* models have been developed to simulate fracture propagation leading to AD. In this study a cohesive zone modelling approach is developed to simulate AD. A fundamental examination of the role of intraluminal septum thickness and residual stress in AD development is presented.

MODEL DEVELOPMENT

An aortic ring geometry was created consisting of two anatomical layers; a Neo-Hookean adventitial-medial layer with an elevated stiffness (to simulate mild hypoxia), and a Neo-Hookean intimo-medial layer. The Neo-Hookean parameters were calibrated from uniaxial tensile tests of ovine aortic specimens. The interface between the layers was then divided into 2 sections; a portion of weakened aortic wall, and a portion of healthy aortic wall (AD is a local phenomenon, therefore, initiation is assumed to depend on some underlying local aortic weakness). The weakened wall was implemented through decreasing the strength of the interface between the layers. A cohesive zone model (McGarry et al. [10]) was then implemented at the boundary between the arterial layers. The normal and tangential interface tractions are defined in Eq.'s 1&2 below:

$$T_n(\Delta_n, \Delta_t) = \sigma_{max} * \exp(1)\left(\frac{\Delta_n}{\partial_n}\right) exp\left(-\sqrt{\frac{\Delta_n^2}{\delta_n^2}+\frac{\Delta_t^2}{\delta_t^2}}\right) \quad (1)$$

$$T_t(\Delta_n, \Delta_t) = \tau_{max} * \exp(1)\left(\frac{\Delta_t}{\partial_t}\right) exp\left(-\sqrt{\frac{\Delta_n^2}{\delta_n^2}+\frac{\Delta_t^2}{\delta_t^2}}\right) \quad (2)$$

Debonding occurs at the interface if the calculated traction (T_n *or* T_t) at a given increment exceeds the prescribed maximum allowable traction before separation. σ_{max} is the maximum traction in the normal direction without normal separation and τ_{max} is the maximum traction in the tangential direction. Traction in a given direction is a function of Δ_n and Δ_t and the strength in the given traction direction. The model predicts identical behavior in mode I and mode II fracture if $\sigma_{max} = \tau_{max}$.

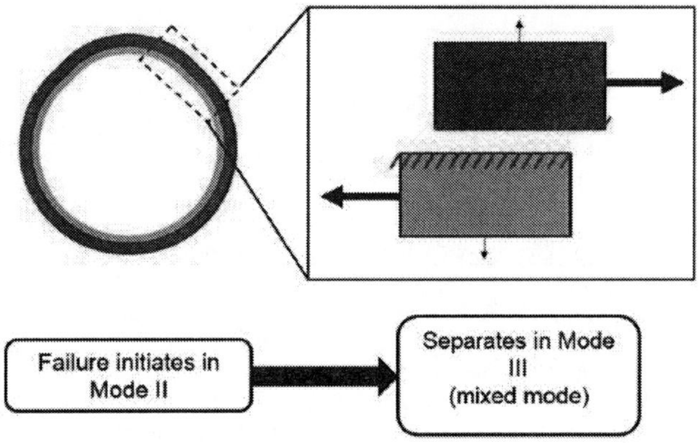

Figure 1: Schematic illustrating the mode of delamination observed in arterial dissection.

A slight geometric asymmetry was introduced in the ring geometry to represent a bleb in the artery wall. A minor stress concentration occurs due to such a geometric imperfection. A residual stress was applied to the intima-medial layer of the artery.

RESULTS

As shown in Figure 2, simulations predict that the greater the initial residual stress in the intimo-medial layer, the greater the subsequent cross-sectional area reduction of the true lumen. The thickness of the intraluminal septum as a percentage of the total aortic thickness also was also seen to influence the amount of true lumen loss.

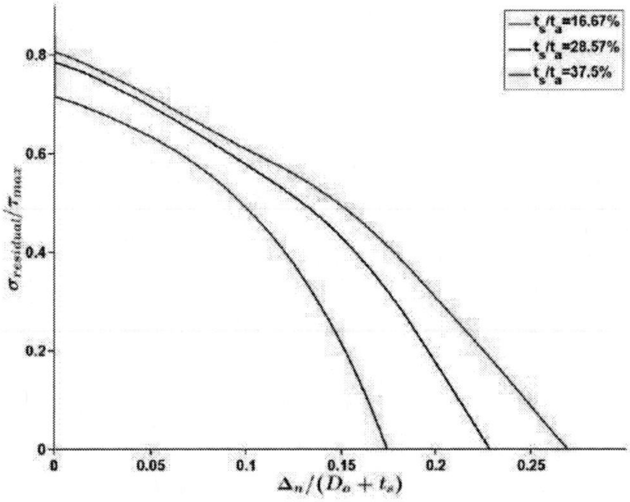

Figure 2: Relationship between normalized residual stress in artery wall ($\sigma_{residual}/\tau_{max}$) and the length of the false lumen ($\Delta_n/(D_o + t_s)$) for a range of values of septum thickness (t_s). (t_a) is the thickness of the intact aortic wall.

Figure 3 shows the predicted size of the true lumen following crack propagation in the artery wall. An in-vivo image of an AD is shown for comparison. Significantly higher levels of dissection are observed in-vivo than computed by our model, suggesting that residual stress is not the only driving force in the propagation of arterial dissection. However, the contribution of residual stress to AD development is significant, reducing the cross sectional area of the lumen by 17.4%-26.9%.

DISCUSSION

The clinical implications of the role the intraluminal septum thickness plays in the propagation of an intra-lamellar tear may be profound; should a patient present with an acute AD and has a thick intraluminal septum, this may indicate that the false lumen is likely to reduce the relative size of the true lumen due to the residual stress present.

New imaging modalities such as time-resolved flow MRI will allow for the quantification of arterial wall motion. The detection of localized regions of deformation adjacent to localized regions of elevated tissue stiffness may indicate a high risk of the initiation and propagation of AD.

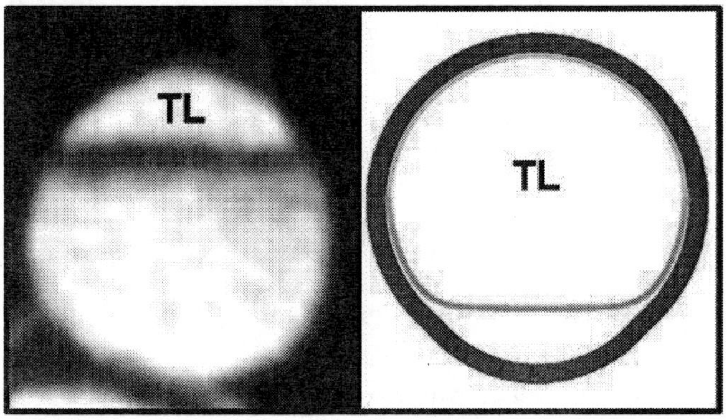

Figure 3: In-vivo image from a computed tomography scan of a dissected aorta (left). The true lumen is depicted by a **"TL" in the image.** Aortic dissection computed using a cohesive zone model is shown for comparison (right).

ACKNOWLEDGEMENTS

B.F acknowledges funding from the Irish Research Council Enterprise Partnership Scheme (RCS1563) and the enterprise partner, The Western Vascular Institute.

REFERENCES

[1] Criado, F. J. Tex. Heart Inst. J. 38, 694–700, 2011
[2] Mikich, B. & Mikich, M. Heart 89, 6–8, 2003
[3] Sommer,G.et al. J. Biomech. (2016) **49,** 2374–2382, 2016
[4] Carson, M. W. & Roach, M. R. J. Biomech. 23, 579–588, 1990
[5] Fattori, R. et al. J. Am. Coll. Cardiol. 61, 1661–78, 2013
[6] Angouras, D. Eur. J. Cardio-Thoracic Surg. 17, 468–473, 2000
[7] Stefanadis, C. et al. Circulation 91, 1995
[8] Heistad, D. D. et al. J. Clin. Invest. 62, 133–140, 1978
[9] Xu, X.-P. et al. Model. Simul. Mater. Sci. Eng. 1, 111–132, 1993
[10] McGarry, J. P. et al. J. Mech. Phys. Solids 63, 336–362, 2014

EFFECTS OF DAILY AND CYCLIC PARATHYROID HORMONE (PTH) TREATMENT REGIMENS ON BONE IN OVARIECTOMIZED RATS

Hongbo Zhao, Wei-Ju Tseng, Wonsae Lee, Yang Liu, Yihan Li, Chantal de Bakker, X Sherry Liu*

McKay Orthopaedic Research Laboratory,
Department of Orthopaedic Surgery
University of Pennsylvania,
Philadelphia, PA, USA

INTRODUCTION

Intermittent parathyroid hormone (PTH) is currently the only FDA-approved anabolic agent for postmenopausal osteoporosis. Unlike the anti-resorptive agents which suppress bone resorption through inhibition of osteoclast activities, PTH can initiate bone formation by activation of osteoblast activities. However, in clinical practice the recommended PTH treatment duration is limited to 18-24 months. Furthermore, bone mineral density rapidly decreases upon withdrawal from PTH treatment [1]. However, postmenopausal osteoporosis is a life-long chronic condition, thus, a more sustained, long-term treatment effect is needed. A cyclic treatment regimen with repeated cycles of on and off daily injection of PTH has been proposed and proven to be more efficient for improving bone mass than the traditional daily injection regime [2-4]. Additionally, our previous study discovered a continuous anabolic window upon early withdrawal from PTH treatment, which offers a new mechanism in support of the cyclic PTH treatment regime (Fig 1) [5]. After a 3-week PTH treatment in ovariectomized (OVX) rats, there was a trend of continued improvement in bone volume fraction (BV/TV, Fig 1), trabecular thickness (Tb.Th), and structure model index (SMI) after one week of withdrawal from PTH treatment. However, such trends disappeared after two weeks of withdrawal from PTH treatment and the treatment benefits diminished after three weeks of withdrawal (Fig 1). Therefore, the objective of this study was to investigate the effects of cyclic PTH treatment regimes on bone microarchitecture in estrogen-deficient rats.

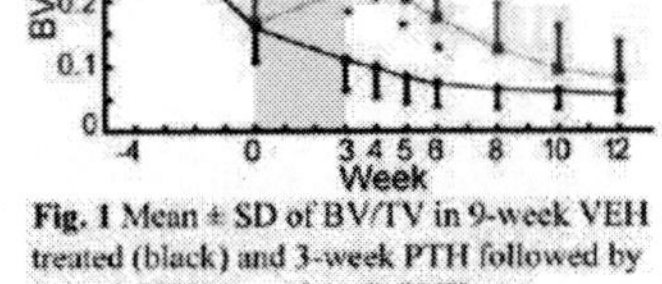

Fig. 1 Mean ± SD of BV/TV in 9-week VEH treated (black) and 3-week PTH followed by 6-week VEH treated (red) OVX rats.
* Significant difference between groups ($p<0.05$)

METHODS

PTH treatment study: 21 female Sprague Dawley rats received bilateral OVX surgery at age of 16-17 weeks and developed osteopenia for 4 weeks. Rats were then assigned to 4 groups: P9V9 (n=6, PTH injected 5 days/week for 9 weeks at 40μg/kg followed by saline injected 5 days/week for 9 weeks), VEH (n=3, saline injected 5 days/week for 18 weeks), C3P3V3 (n=7, 3 repeated cycles of 3 weeks of PTH injected 5 days/week at 40μg/kg, followed by 3 weeks of saline injected 5 days/week, for a total of 18 weeks over 3 cycles), and C3P3V2 (n=5, 3 repeated cycles of 3 weeks of PTH injected 5 days/week at 40μg/kg followed by 2 weeks of saline injected 5 days/week, for a total of 15 weeks over 3 cycles) (Fig 2).

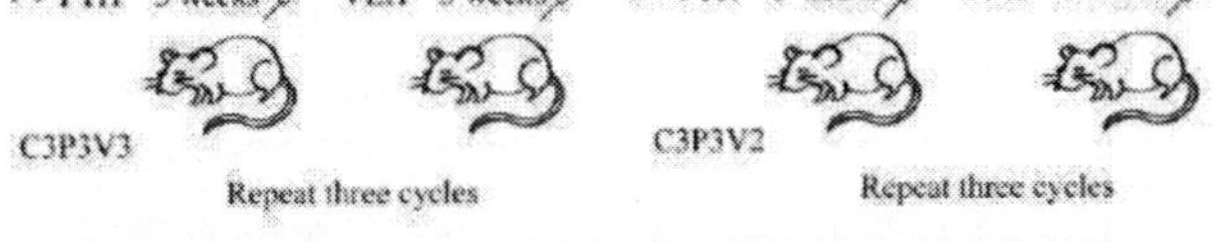

Fig. 2 Experiment design of cyclic PTH treatment: (A) C3P3V3 and (B) C3P3V2.

μCT Imaging: A 4-mm region distal to the proximal tibial growth plate was scanned by an *in vivo* μCT system (VivaCT 40, Scanco Medical AG, Brüttisellen, Switzerland) using a customized holder at 10.5 mm voxel size [6]. Rats in the P9V9, VEH and C3P3V3 groups were scanned at treatment weeks -4 (OVX surgery), 0, 3, 6, 9, 12, 15 and 18, whereas rats in the C3P3V2 group were scanned at treatment weeks -4, 0, 3, 5, 8, 10, 13 and 15. Three dimensional (3D) image registration was applied to identify the same volume of interest (VOI) in all scans [6]. Standard trabecular bone structural parameters including bone volume fraction (BV/TV), trabecular thickness (Tb.Th), trabecular spacing (Tb.Sp), trabecular number (Tb.N), structure model index (SMI), and connectivity density (Conn.D) were measured.

Statistics: Longitudinal comparisons were made using 2-way, repeated-measures ANOVA, adjusted for baseline values. Bonferroni corrections were applied to all *post hoc* tests. Differences were considered significant when $p<0.05$.

RESULTS

4-week osteopenia development caused 52%, 11%, and 36% decreases in BV/TV, Tb.Th, and Tb.N, respectively, and a 123% increase in SMI (Fig 3A-D, $p<0.05$). Bone loss continued in VEH rats for 18 weeks (Fig 3). In contrast, BV/TV and Tb.Th increased and SMI decreased in P9V9 rats for 9-week PTH treatment; however, these improvements were reversed upon withdrawal from PTH. At week 18, there was no difference between P9V9 and VEH groups in BV/TV, Tb.Th or Tb.N (Fig 3A-C), and BV/TV and Tb.N in both P9V9 and VEH groups were significantly lower than at week 0. On the other hand, cyclic PTH treatment (C3P3V3 group) effectively alleviated the withdrawal effect when compared to P9V9 group. During the first cycle of treatment in the C3P3V3 group, 3-week PTH treatment caused 97% and 32% increases in BV/TV and Tb.Th (Fig 3AB), and a 32% decrease in the SMI (Fig 3D) at week 3 compared to baseline (week 0). 3 weeks of withdrawal (week 6) did not cause significant decline in any of trabecular bone parameters compared to week 3 (Fig 3 A-D). During the second cycle of PTH treatment (week 6-9), Tb.Th tended to increase ($p=0.053$) over the treatment period while no change was found in the other parameters. During the withdrawal period of the second cycle (week 9-12), no changes were observed in any of trabecular bone parameters (Fig 3A-D). Trabecular bone responses during the third cycle (week 12-18) were similar to those in the second cycle. At the end of cyclic PTH treatment (week 18), BV/TV and Tb.Th in C3P3V3 rats were significantly greater than both P9V9 and VEH rats (Fig 3AB). Additionally, for rats in the C3P3V3 group, BV/TV at week 18 was not different and Tb.Th had increased 66% when compared to week 0. The C3P3V2 group also showed attenuated PTH treatment withdrawal effects. Since PTH treatment in the C3P3V2 group was restarted one week earlier than the C3P3V3 group, the overall reduction in BV/TV was lower and overall increase in Tb.Th was greater in C3P3V2 group than in C3P3V3 group (Fig 4).

DISCUSSION

Our results showed that significant bone loss and bone microarchitecture deterioration occurred in OVX animals after discontinuation of PTH treatment, similar to previous clinical findings [1]. 9 weeks after withdrawal of PTH, the treatment benefits were completely reversed, with no difference in trabecular bone parameters between P9V9 and VEH groups. The cyclic administration regime in both C3P3V3 and C3P3V2 groups alleviated bone deterioration and extended the total duration of treatment. However, despite continued increases in Tb.Th, these treatment strategies did not continue to improve bone mass (reflected by BV/TV) or trabecular architecture (reflected by SMI). Because of the 1-week anabolic window after early withdrawal from PTH in rats [5], C3P3V2 group showed a slower rate of bone mass decline compared with C3P3V3 group over the three cycles. Therefore, a cyclic PTH treatment regime with 2-weeks off and 3-weeks on may have better efficacy over longer time periods. In fact, extrapolation of the bone loss/gain trends reported here beyond the 3 treatment cycles leads us to predict that the BV/TV of the C3P3V2 group would decline to baseline (week 0) values after 5 additional treatment cycles (~43 weeks of treatment) while the BV/TV of the C3P3V3 group would decline to baseline values after only 1 additional treatment cycles (~24 weeks of treatment). In addition, during the PTH treatment period of the second and third cycles, there was no improvement in BV/TV despite increases in Tb.Th, suggesting that the PTH treatment effects reached a plateau at this period due to high cellular activities not only in osteoblasts, but also in osteoclasts. In future studies, anti-resorptive agents may be given during the off-PTH treatment period, in order to maintain the treatment benefits. Therefore, a cyclic and sequential treatment regime of PTH and anti-resorptive agents may provide a better strategy to treat chronic osteoporosis.

ACKNOWLEDGEMENTS

This study is supported by NIH/NIAMS P30-AR050950, and NIH/NIAMS K01-AR066743.

REFERENCES

[1] Black, D.M., et al., *N Engl J Med*, 353(6):555-565, 2005.
[2] Cosman, F., et al., *N Engl J Med*, 353(6):566-575, 2005.
[3] Cosman, F., et al., *J. Clin. Endocrinol. Metab.*, 100(7):2769-2776, 2015.
[4] Iida-Klein, A., et al., *Bone*, 40(2): 391-398, 2007.
[5] Tseng, W.-J., et al., *ASBMR Annual Meeting, Atlanta, GA*, 2016.
[6] Lan, S., et al., *Bone*, 56(1):83-90, 2013.

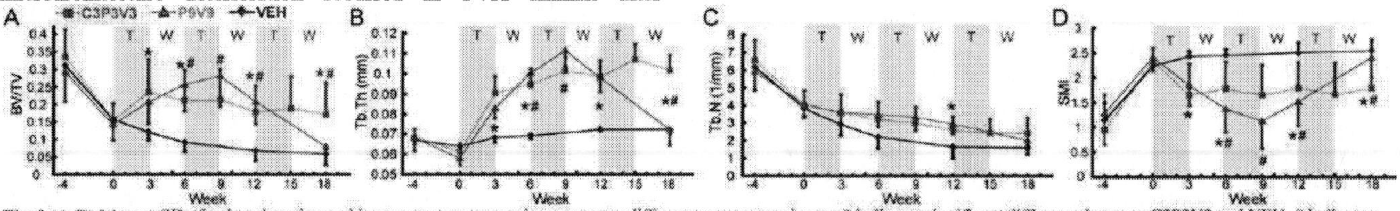

Fig. 3 (A-D) Mean ± SD of trabecular microarchitecture measurements in response to different treatment regimens. * indicates significant difference between C3P3V3 and VEH, # indicates significant difference between C3P3V3 and P9V9 ($p<0.05$).

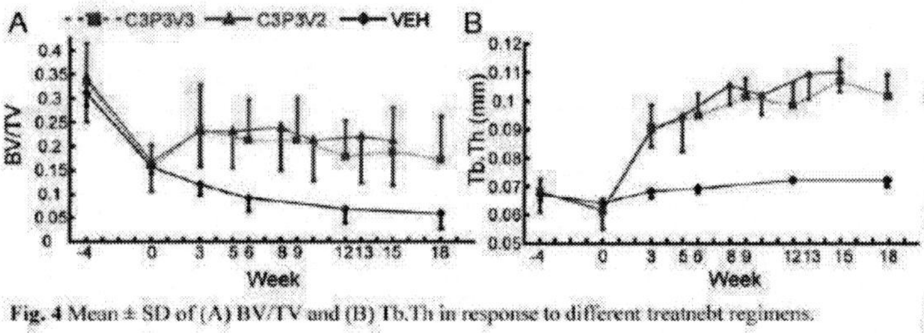

Fig. 4 Mean ± SD of (A) BV/TV and (B) Tb.Th in response to different treatnebt regimens.

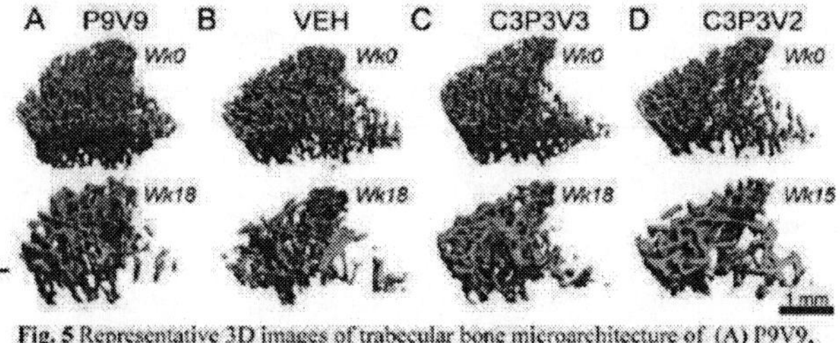

Fig. 5 Representative 3D images of trabecular bone microarchitecture of (A) P9V9, (B) VEH, (C) C3P3V3, and (D) C3P3V2 at the baseline and end of treatment.

TART CHERRY POTENTIALLY PREVENTS BONE LOSS THROUGH INHIBITION OF RANKL IN TNF-OVEREXPRESSING MICE

Linda A. Effiong[1], Nicolas Moon[2], Saqib Nizami[1], Thomas R. Gardner[1], Do Y. Soung[1]

(1) Department of Orthopedic Surgery
Columbia University
New □ork, N□, USA

(2) Department of Medicine
Case Western Reserve University
Cleveland, OH, USA

INTRODUCTION

Osteoporosis is a state of low bone mineral density and breakdown of bone homeostasis through the loss of mechanical and structural properties. This bone loss leads to increased skeletal fragility with the concomitant fractures and poor healing. Hormonal changes and inflammation are major factors in sustaining osteoporotic states. Standard treatments for osteoporotic conditions seek to mitigate the effects rather than address those causes. Anti-resorptive agents, such as bisphosphonates or anabolic agents such as parathyroid hormone, have shown great efficacy in halting bone breakdown and improving fracture healing. An approach that would address the inflammation that drives osteoporosis is needed.

Polyphenolic compounds, such as quercetin, proanthocyanidins, and anthocyanidins have been shown to be effective in attenuating a chronic inflammatory state like rheumatoid arthritis (RA), which creates an osteoporotic bone environment. The Tart Cherry (TC) is a fruit rich in polyphenols with effective antioxidant and anti-inflammatory properties under consideration as a food to combat the inflammatory osteoporotic disease environment. TC consumption has been shown to lower serum levels of TNFα, a major driver of inflammation, in overweight human subjects and obesity-prone rats fed a high-fat diet. The study investigates the efficacy of Tart Cherry to prevent or reverse the activity of an inflammatory bone loss disease state.

METHODS

Animals and Diet: Twenty-six mice, 8 female wild type (WT) and 20 female TNF transgenic (TG) were used in this study. Mice aged 3.5-5weeks-old, were grouped as: Group 1 (control group, n=8) 5-week-old WT (wild type), fed a 0% TC diet for 5 weeks. Groups 2, 3, and 4 was age-matched TG randomly assigned to feed on a diet of 0% TC (n=5), 5% TC (n=4), and 15% TC (n=5). Group 5 (n=4) TG were fed a 0% TC diet and treated with an intraperitoneal dose of 10 mg/kg infliximab twice per week as a positive control for standard rheumatoid arthritis (RA) treatment.

Micro Computed Tomography: To assess if the TC impacted the cortical and trabecular structure of the bone, micro-CT analysis was performed. Femora of the mice were harvested and stored at -80°C and thawed prior to testing. The femora were scanned in PBS using a Skyscan 1272 Micro CT system (Bruker). The bones were scanned with a 10.4 μm resolution at 60kVP, with a rotation step of 0.4mm and an aluminum filter of 0.25mm. The scans were calibrated using 0.25g/cm^3 and 0.75g/cm^3 phantoms. Reconstructed files were analyzed using 200 slices. Properties of the cortical bone were taken at the mid-diaphysis and the trabecular bone properties were taken from the metaphysis. Cortical thicknesses and cortical bone fractions were calculated. Bone volume fraction (BV/TV), trabecular thickness (TB.Th), trabecular number (Tb.N), trabecular separation (Tb.Sp) measuring the thickness of the spaces, connectivity (Conn.D) were computed. The bone mineral density and tissue mineral density of the cortical and trabecular bone, respectively, were computed. The inner and outer diameters along with the cortical mean thicknesses of the diaphyseal region of the bones were also calculated (1).

Mechanical Testing: Following micro-CT imaging, bone quality and bone integrity were assessed by three-point bending (2) the right femora of the mice (Instron Microtester 5848, Norwood, MA). A three point bending test was performed in the anterior-posterior direction using a 100N load cell (accuracy, ±0.025N). A constant ramp displacement (0.05 mm/s) was used to apply a load at the midpoint of the femur until failure. Stiffness, maximum failure, failure deflection and failure energy were measured. The elastic moduli were determined using Classical Beam Theory. Dimensional values of the mouse femora were obtained from morphometric Micro-CT data.

Statistical Analysis. One-way ANOVA (GLM) were performed for micro CT and mechanical testing data using SAS 9.4 (Cary, NC). Statistical significance was defined as P<.05. A Tukey range test was performed to compare between levels. Data are presented as a mean ± SD.

RESULTS

Micro-CT: Overall, WT mice were found to have a significantly larger cortical thickness when compared to TG+0%TC (p<.05).

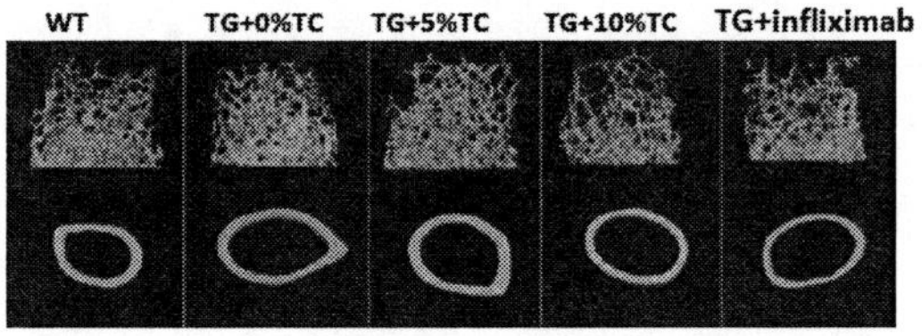

Figure 1: Micro-CT scans of WT, TG +0%TC, TG+5%TC, TG+10%TC and TG+Infliximab, First row represents a 1mm slice of trabecular bone taken below the growth plate. Second row represents a 1mm slice of cortical bone taken at the midshaft.

The cortical area fraction (Ct.Ar/Tt.Ar) is the ratio of cross sectional area to the total cross area section. It was found that WT mice had significantly larger cortical area fraction in relation to TG+5%TC (p<.05). The bone volume fraction (BV/TV) is the volume of the region segmented as trabecular bone. There was significantly a larger volume of trabecular bone in the TG+ infliximab mice when compared to the TG+0% (p<.05). Connectivity of the tissue was significantly higher in the infliximab treated femurs compared to the TG+0%.

Mechanical Testing. WT mice had a higher stiffness (p<.0.0066) and a greater maximum load to failure when compared to TG mice (p<.0355). No differences in mechanical properties were found between the 5 experimental groups

DISCUSSION

Micro-CT findings showed differences in the cortical thicknesses and the cortical fraction attributed to the TG mice having established RA derived osteoporosis. This is further supported by a higher cortical fraction in the WT mice when compared to Groups 2,3,5. The TG+10%TC had a higher cortical fraction than the TG+0% indicating that the TC had an effect on amount of cortical bone growth. The infliximab treatment on the TG mouse femurs exhibited higher trabecular bone volume than the TG mice with TC treatment. Infliximab is a TNF inhibitor that stops the increase of osteoclasts, subsequently decreasing the bone volume. This increase in trabecular bone volume reflected an increase in the connectivity of the tissue. No differences were found in inner diameter, outer diameter and mean thicknesses of the bone between experimental groups which is consistent with a lack of differences in the mechanical properties. However, when WT versus TG were compared, WT had higher failure loads and a higher stiffness, which is expected for non-osteoporotic bones as compared to the TG mice.

Due to the small sample size for the TC groups and short 4 week treatment time, it is unclear if TC has an effect the bone mechanical properties indicating a need for a larger sample size. Future studies will include a robust sample size, and the inclusion of a WT control with TC treatments of different dosages and prolonged treatment times.

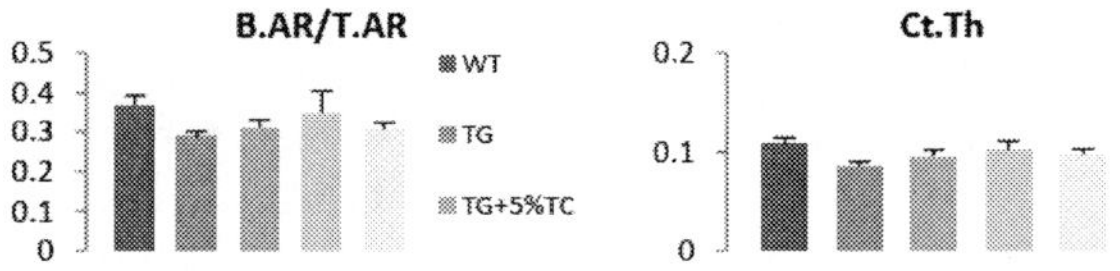

Figure 2: Micro-CT parameters calculated for cortical bone.

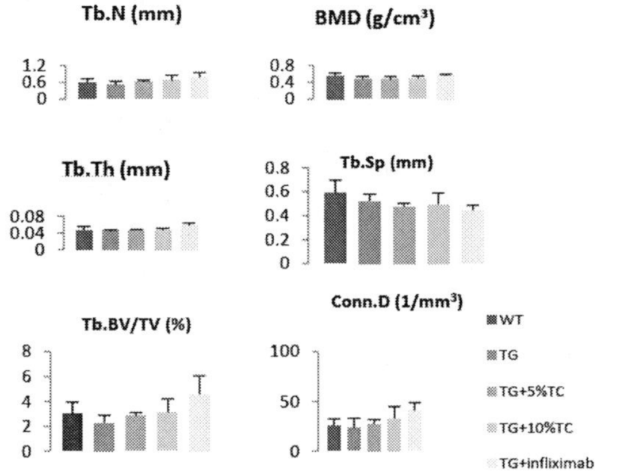

Figure 3: Micro-CT parameters for trabecular bone.

	WT	TG+ 0% TC	TG+ 5%TC	TG+ 10%TC	TG+ Infliximab
Failure Load (N)	14.7±2.7	11.8±1.7	12.0±2.7	13.1±2.2	13.0±1.3
Max Deflection (mm)	0.4±0.1	0.4±0.1	0.5±0.1	0.5±0.0	0.6±0.3
Stiffness (N/mm)	93.0±20.0	56.5±18.3	77.2±16.7	70.8±20.7	63.0±19.2
Failure Energy (mJ)	3.6±0.8	3.2±0.7	3.9±0.8	4.0±0.6	3.2±0.7
Modulus (Gpa)	0.7±0.1	0.5±0.2	0.7±0.1	0.6±0.2	0.5±0.1

Table 1: Means and STD of femora mechanical properties.

ACKNOWLEDGEMENTS

Special thanks for Dr. Lee Song and Dr. Adam Abraham at the Department of Orthopedics at Columbia University for their assistance with this study.

REFERENCES

1. Bouxsein, M. L., et al. (2010). "Guidelines for assessment of bone microstructure in rodents using micro–computed tomography." Journal of Bone and Mineral Research **25**(7): 1468-1486.

2. Tascau, L., et al. (2015). "Activation of Protein Kinase A in Mature Osteoblasts Promotes a Major Bone Anabolic Response." Endocrinology **157**(1): 112-126.

IN VIVO BONE STRAIN AND CORTICAL BONE RESPONSE TO MECHANICAL LOAD IN THE MOUSE TIBIA

Kari A. Verner (1), Haisheng Yang (2), Russell P. Main (1,2)

(1) Weldon school of Biomedical Engineering
Purdue University
W. Lafayette, IN, USA

(2) Basic Medical Sciences
Purdue University
W. Lafayette, IN, USA

INTRODUCTION

The skeleton is a fascinating biological structure that has the ability to adapt its bone-specific architecture to optimize for daily mechanical function, and can also repair itself when mechanical failure does occur. However, several factors can reduce bone's ability to adapt and maintain structural integrity in response to increased mechanical strains. Failure of mechanobiological mechanisms can allow skeletal resorption to progress unabated and could eventually lead to structural failure. Skeletal diseases such as osteoporosis, occur due to compromised mechanobiological regulation of bone quality. These diseases affect millions of people and require billions of dollars in medical treatments every year[1]. Enhancing the understanding of how skeletal biology is regulated in response to mechanical perturbation is critical for developing innovative diagnostic tools and treatments.

Development of the non-invasive axial loading mouse model has provided extensive opportunities for exploring the cellular, genetic, and whole-organ response of the skeleton to mechanical stimuli in both cortical and cancellous bone[2-4]. In addition to assessing the response of the bone to an applied load stimuli, characterizing the load-induced strain environment is fundamental to interpreting the mechanobiological response.

The first goal of this work was to assess *in vivo* physiologic strains and develop an applied mechanical load:strain calibration for the mouse tibia such that the relative increase in bone strains induced via a controlled, applied load compared to the *in vivo* locomotor strain environment is well characterized. Second, long term applied loading studies were completed to determine the tissue-level response to a specific increase in strain on the medial midshaft surface. Finally, locations of active mineralization during the loading study will be related to cross-sectional strain distributions.

METHODS

Tibia from male, 16 week old C57Bl/6 mice were used for all experiments. To assess physiologic strains *in vivo* and develop an applied mechanical load:strain calibration, one single element strain gauge was surgically attached to the medial midshaft surface of the left tibia (N=7). One hour after surgery, mice were run on a small treadmill at maximum sustainable speeds for brief bouts while strain data was collected simultaneously. Immediately following, mice were anesthetized and the gauged limb was loaded non-invasively in axial compression incrementally from -4 to -14.5N. Physiologic and load calibration strain data were analyzed to determine peak longitudinal strains during running and at each applied load level, respectively.

Specimen-specific finite element models were generated for the gauged limbs (N=6). Dissected tibiae were scanned by microcomputed tomography (microCT), and the images were reconstructed and meshed using a series of custom MATLAB codes. Meshed 3D models were then imported into ABAQUS for boundary condition and load definitions. Each model was validated by matching *in vivo* measured strains from the gauge location to modeled strains at the same location. Validated models were used to assess the applied load:strain relationship at the gauge location as well as maximum strains at non-instrumented locations, such as the midshaft cross-section.

Long term loading studies were conducted to assess the tissue-level response of the tibia to a load inducing 2.5x physiological strain at the medial midshaft. A load level of -6.4N was determined from the load:strain calibration as well as the finite elements models. A loading protocol of 216 cycles at a peak load of -6.4N was applied at 4Hz to the left tibia (which allowed the right tibia to serve as a contralateral control) was conducted 5 days/week for two (N=5) or four (N=10) weeks (Figure 1). At specific time points throughout both groups, animals received IP injections of bone flourochromes. Three days after the final load application, mice were euthanized and both tibiae extracted. Loaded and control limbs were scanned by microCT. Cortical bone morphometry was analyzed for a bone volume corresponding to 2.5% of bone length centered at the midshaft, and significant differences between loaded and control limbs were identified by paired, one-tail T-tests. After scanning, bones were processed for mineralized tissue histology to assess cortical mineralization dynamics.

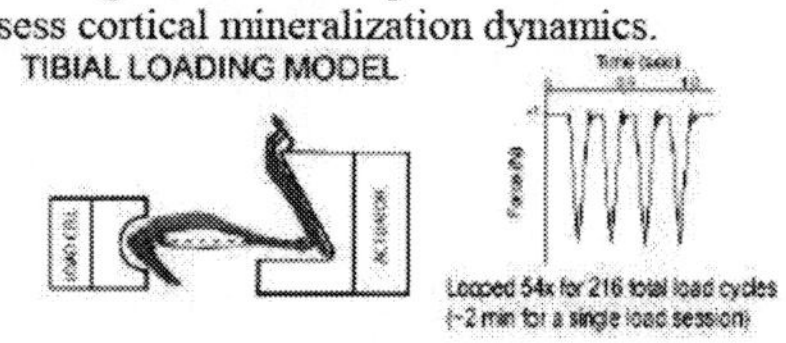

Figure 1: Schematic illustration of the mouse hind limb in the mechanical loading device, and the triangular waveform and timing of the applied load.

RESULTS

The maximum speed mice could maintain on the treadmill was 0.5 m/s. At this speed, the average peak longitudinal strains on the medial midshaft surface were 236±61με (Figure 2A). An applied load inducing 590με was, therefore, necessary to induce 2.5x peak physiologic strain on the medial midshaft. Based on the stiffness relationship developed from the applied mechanical load:strain calibration, the appropriate load level is required to induce 2.5x the physiological strains was -6.5N (Figure 2B).

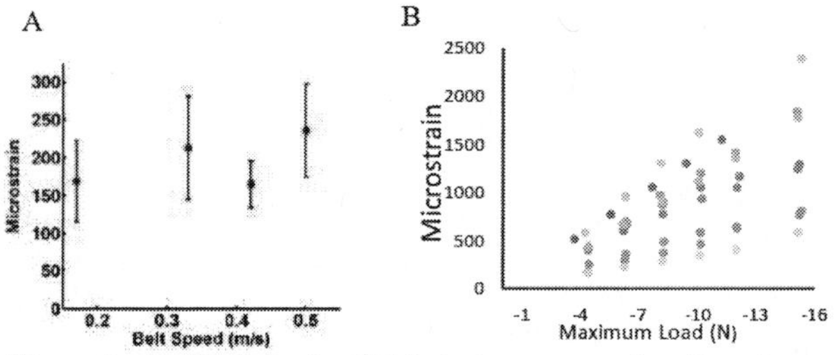

Figure 2: (A) Peak *in vivo* tibial strains measured as the mice ran on a motorized treadmill at increasing speeds. Mean ± 1 standard deviation. (B) Applied mechanical load:strain plot for (N=8) mice.

Validated finite element models predicted that a load of -6.4N would induce approximately 590με at the gauge location on the medial midshaft surface of the tibia. Models were also used to predict maximum tensile and compressive strains, which occurred on the posterior and anterior surfaces at the midshaft (Table 1, Figure 3), respectively.

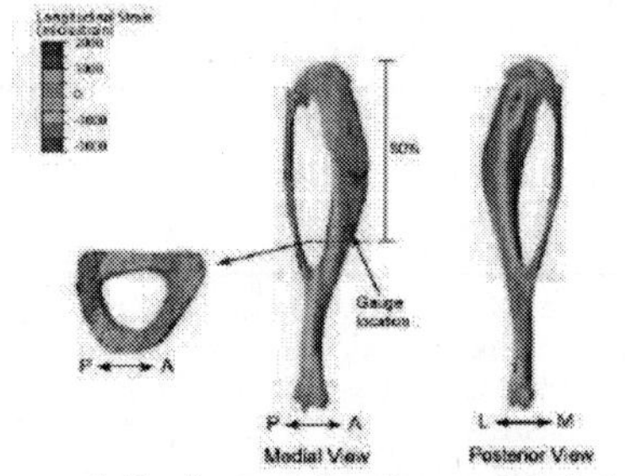

Figure 3: Representative images showing modeled strains induced throughout the tibia during applied mechanical loading.

Table 2: Predicted peak principal strains at the tibial midshaft resulting from a -6.4N applied load.

		Strain (με)
Tensile	95th percentile	1024
	Mean	452
Compressive	95th percentile	-1601
	Mean	-663

Cortical morphometry analyses of loaded versus contralateral control tibiae showed increases in maximum and minimum moments of inertia, cortical bone area, and bone volume after two weeks, but not after four weeks of applied loading (Figure 4).

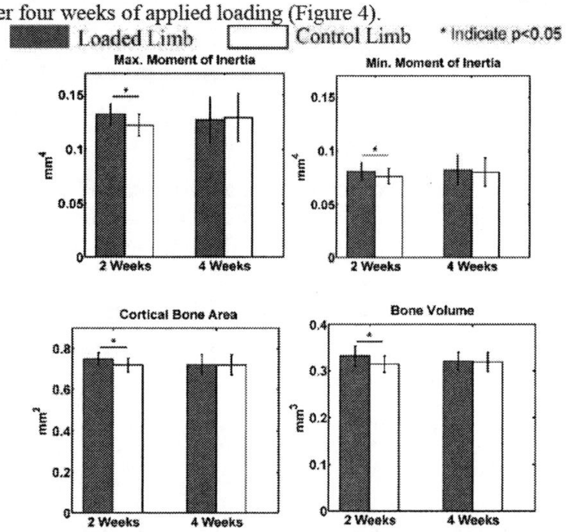

Figure 4: Morphometry results for 2 vs 4 week applied loading studies. Bars represent ±1 standard deviation.

DISCUSSION

Based on *in vivo* measurements and finite element models, an applied axial compressive load of -6.4N would induce 2.5x peak physiological strain on the medial midshaft surface of the mouse tibia. The models predicted that during applied loading, peak principal strains occur on the posterior and anterior surfaces, and are 2-3x greater than strains on the medial surface. When -6.4N was applied 5 days/week for two or four weeks, an adaptive response was measured in the loaded limb after two weeks but not after four. Future work will assess mineral apposition and growth rates with the completion of the mineralized tissue histology. Those results will help elucidate the difference between the 2 and 4 week loading groups, and will also be related to local periosteal and endosteal strain measures from FEA-modeled cross-sections.

ACKNOWLEDGEMENTS

This work was supported by an NSF GRFP to KV (DGE-1333468) and an NSF grant to RPM (CMMI 1463523). We would like to thank the Purdue Bone and Body Composition Core of the Indiana CTSI for providing microCT scanning.

REFERENCES

[1] Burge, R et al., *J Bone Miner Res*, 22(3):465-75, 2007.
[2] Fritton, J et al., *Bone*, 36(6):1030-8, 2005.
[3] De Souza, R et al., *Bone*, 37(6):810-8, 2005.
[4] Lynch, M. & Main, R., *Bone*, 49(3):439-46, 2011.

The Relationship Between Pore Morphology and Cortical Bone Mechanics

Lydia P. Bakalova (1), Jesper S. Thomsen (2), Christina M. Andreasen (3), Annemarie Brüel (2), Ellen M. Hauge (2), Birgitte Jul Kill (2), Thomas Levin Andersen (3), Mariana E. Kersh (1)

(1) Department of Mechanical Science and Engineering
University of Illinois at Urbana-Champaign
Urbana, Illinois, USA

(2) Department of Biomedicine
Aarhus University
Aarhus, Denmark

(3) Dept. of Clinical Cell Biology (KCB), Vejle Hospital – Lillebælt Hospital
Institute of Regional Health Research, University of Southern Denmark
Vejle, Denmark

INTRODUCTION

Osteoporosis is a bone disease that affects over 200 million people worldwide, and is characterized by an accelerated loss of bone mass and microarchitecture deterioration of both trabecular and cortical bone. This bone loss is the result of a dysfunctional bone remodeling process, in which bone resorption and formation becomes uncoupled and imbalanced, and is suggested to alter its strength and lead to an increased risk of fractures [1,2]. Currently, bone strength is clinically assessed based on dual-energy X-ray absorptiometry (DEXA) measurements of bone mineral density (BMD) [3]. Although BMD is linked to mechanical strength and fracture risk, it is not a sufficient factor to assess bone strength [4]. Although bone porosity and microarchitecture have been previously shown to be essential to the mechanical competence of bone [7], they are not included in DXA measurements.

High-resolution peripheral-Quantiative Computed Tomography (hr-pQCT) data may be a viable option for including porosity measurements, but which measurements should be taken and how they relate to bone strength remains unclear. Non-invasive experimental measurements of human bone strength in-vivo are not possible, leaving a gap in the understanding of the relationship between micro-architecture and bone mechanics. Alternatively, the use of patient-specific models generated from micro computed tomography (μCT) data and analyzed with the finite element method (FE) can be used to better understand the relationship between bone microstructure and strength [4-6].

The aim of this study was to investigate the relationship between the strain energy within the cortical bone surrounding pores and pore morphology. Identifying how pores change the mechanical response of bone to load will assist in identifying relevant measures for improving the assessment of bone strength.

METHODS

Bone Samples and Imaging: Human cortical bone specimens (n=23) harvested from the fibular diaphysis were acquired and scanned using μCT data, (spatial resolution of 6μm x 6μm x 6μm), to create high resolution micro-architectural finite element (FE) models. The μCT images were binarized and connected regions less than 20 voxels were removed using custom software (Matlab R2015-Academic, Mathworks, Natick, MA, USA) (Figure 1). Bone fragments that were not connected to the main cortical structure were also removed and final images were exported as PNG files.

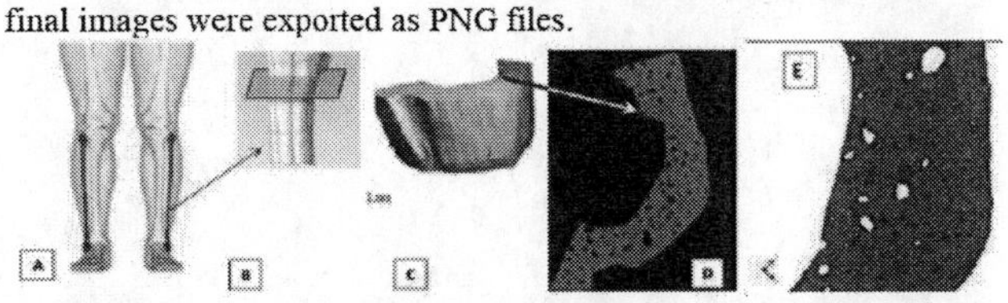

Figure 1: (A) Region of leg used for analysis [4]; (B) Region of fibula bone analyzed for porosity; (C) micro-CT scanning reconstruction of bone section; (D) cross-sectional images of human cortical bone from Fibula bone; (E) Finite element mesh

Finite Element Models: The processed image stacks was then reconstructed into a finite element mesh (ScanIP, Simpleware, Exeter, UK). Each mesh was generated using quadratic tetrahedral volume elements and was assigned homogeneous, isotropic, elastic material properties (Young's Modulus, E = 17.5 GPa, Poisson's ratio, ν = 0.3) [8]. For consistency, each model was limited to 100 μCT images. Three nodes on the bottom face of the mesh were fully constrained, and the remaining nodes of this same face were constrained in the vertical direction. A unit compressive force (1 N) was uniformly applied to the top face of the mesh. The strain energy within the bone was calculated using a linear stress analysis (Abaqus v6.14; Dassault Systems Simulia Corp., Pawtucket, RI, USA).

Pore Morphology and Strain Energy Analysis: The binarized μCT image corresponding to the center of the finite element mesh was used to calculate pore size and circularity. Pore area was defined as the total number of voxels within each pore. Pore circularity was defined as the ratio of the area of the largest circle that can be inscribed in the pore to the actual area of the pore [9].

Elements within 6 microns of the central portion of the finite element model were isolated, and the nearest border elements surrounding each pore were identified. The total strain energy around each pore normalized to the total strain energy of the respective bone segment was calculated. Differences in strain energy between various sized and shaped pores were assessed using a Wilcoxon Rank-Sum Test ($p < 0.05$).

RESULTS

Among the 21 bone samples, 3085 pores were identified and analyzed. Across all samples, most pores were small and circular (Figure 2). There were more than four times the number of small pores (Figure 3, right axis) than larger pores per unit area and the circularity of most pores was 60-80%.

When analyzed individually, the contribution of each pore to the total strain energy increased with increasing pore size ($p < 0.001$) (Figure 3, 4, A and C). Irregularly shaped pores contributed to more of the total strain energy compared to circular pores. However, within each group of size and shape, there was a noticeable distribution of strain energy.

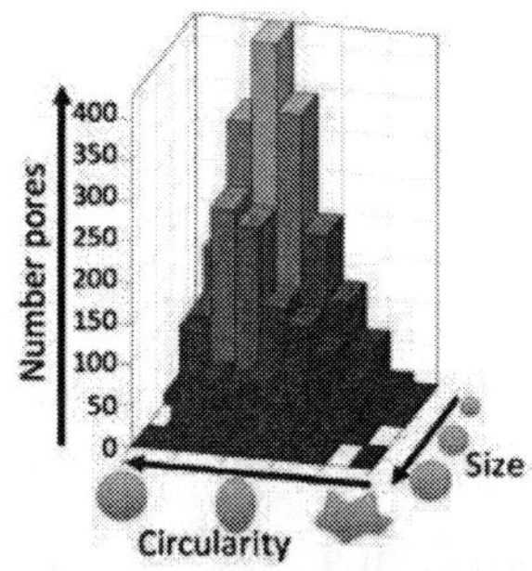

Figure 2: Number of pores across all samples and their size and shape distributions.

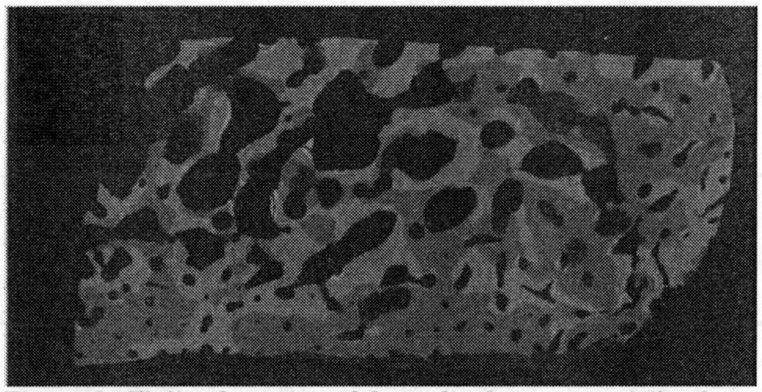

Figure 3: Finite element model results for representative bone sample. Larger pores resulted in higher strain energy compared to smaller pores.

The irregular shaped pores (circularity 0-20%) contributed the most to the strain energy across all samples ($p < 0.001$), and circularity values above 40% resulted in negligible strain energy (Figure 4, B and D).

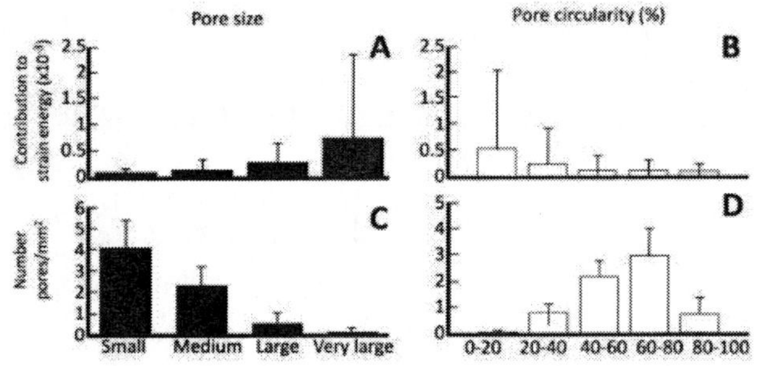

Figure 4: (A) Contribution of pores of different size to strain energy and (C). the number of pores per unit area across all specimens. (B) Contribution of pores of different shape and (D) the number of pores by shape per unit area within each.

DISCUSSION

The purpose of this study was to investigate the influence of cortical porosity to the strength of bone. Our results demonstrate that strength was sensitive to pore size and pore shape. There was an inverse relationship between the pore size and contribution to strain energy. It should be noted that we did not include all of the smallest pores, and the Nyquist theorem did not allow us to identify pores small than 12 μm. However, for those pores we could identify, while the individual contribution of small pores was small, their net sum contribution should be considered when assessing bone strength since there are many more of these small pores. While HR-pQCT data provides more information than DEXA or clinical-level computed tomography data, our results indicate that the spatial resolution of HR-pQCT is currently insufficient to detect these small pores.

Future work includes experimental validation of our finite element model predictions using in vitro based full field strain measurements of these samples. Our model results suggest that the strain distribution within cortical bone is complex - we observed a large deviation of strain energy values within a given pore size and shape. This may indicate that cortical strength is sensitive to other factors including pore clustering and spatial location relative to the endosteal and periosteal surfaces.

In conclusion, we have shown that various porosity metrics affect bone strength. Large pores and less-circular pores have significant contribution to the total strain energy of cortical bone.

ACKNOWLEDGEMENTS

We would like to thank, Gete Ester Toft Eschen, from the Aarhus University Hospital in Aarhus, Denmark for obtaining and CT images of the bone samples used in this research. Thanks to Jette Barlach, Kaja Laursen and Birgit MacDonald, whom assisted with the handling of the tissues.

REFERENCES

[1] Andersen, TL., et al. *Am J Pathology* 183.1 (2013) 235-246.
[2] Hirano, T et al., *Bone*, 27 (2000)13–20
[3] Reginster, J et al., *Bone*, 38 (2006) S4-S9;
[4] Mueller, T et al., *Bone*, 48 (2011) 1232-1238
[5] MacNeil, J et al, *Bone* 42 (2008) 1203–1213
[6] Crawford, R et al., *Bone*, 33 (2003);33:744–50
[7] Nirody, J et al., *Bone*, 75 (2015) 88–95
[8] Varghese, B et al, *Journal of biomechanics*, *44*(7), 1374-1379.
[9] Birdal, Tolga (2011). MATLAB Central File Exchange. *Retrieved 11 Mar 2016*

EXPERIMENTAL MEASUREMENT OF MULTIAXIAL STRAIN STATES AND MULTIAXIAL YIELDING OF TRABECULAR BONE DURING VERTEBRAL FAILURE

Johnfredy Loaiza (1), Amira I Hussein (2), Elise F. Morgan (1,2,3)

(1) Mechanical Engineering
Boston University
Boston, MA, USA

(2) Orthopaedic Surgery
Boston University School of Medicine
Boston, MA, USA

(3) Biomedical Engineering
Boston University
Boston, MA, USA

INTRODUCTION

Vertebral fractures occur in 12-20 % of patients over the age of 50 [1]. Across all ages, approximately one third of all vertebral fractures occur from falling and one sixth from lifting or traffic accidents, whereas in one half of the cases, the cause cannot be tied to a single traumatic event [2]. Therefore, to the extent that the loading conditions associated with vertebral fracture are known, these conditions are complex. Moreover, results of finite element (FE) simulations have indicated that the trabecular bone within the vertebral body is subjected to multiaxial stress and strain states during fracture [3]. These results suggest that the multiaxial failure behavior of trabecular bone is of high clinical relevance and is essential to understanding how vertebrae fail and how the incidence of vertebral fractures can be reduced. However, these results have not been verified fully due to a lack of available experimental data on the stress or strain states experienced by trabecular bone *in situ* within the vertebra.

The overall goal of this study was to determine the strains and the ensuing failure that occur *in situ* in the trabecular compartment during vertebral fracture. Time-lapse series of micro-computed tomography (μCT) images captured during mechanical testing of vertebrae were analyzed by digital volume correlation (DVC) [4] to compute the strains that developed in the trabecular compartment. Then, these strains were evaluated using a multiaxial yield criterion developed for trabecular bone [5], to estimate which regions experienced yield. Our specific objectives in this study were to determine: (1) the nature of the multiaxial loading occurring in vertebral trabecular bone; and (2) where local yielding commences and propagates in the vertebral body.

METHODS

Fresh-frozen human thoracolumbar spine segments (n=21; 41-91 years; 9 male, 12 female; NDRI, Philadelphia, PA) were obtained by a dissection process that involved making transverse cuts above the T12 inferior endplate and below the L2 superior endplate. These segments underwent mechanical testing and μCT imaging as part of a prior study [4]. Each segment was potted at each end in bone cement and then placed in a custom-built, radiolucent loading device. After reducing viscoelastic behavior of the sample through a preconditioning process of ten cycles to 400 N, the vertebrae were imaged with μCT (μCT 80, Scanco Medical, Brüttisellen, Switzerland; 37 μm/voxel). Following the initial scan, the sample was axially compressed (1 mm/step at 0.25 mm/sec) and then scanned again at the same settings. This stepwise loading and imaging continued until a drop in force was observed in the force-displacement curve. Global yield for the sample was defined as the increment at which the initial linear slope declined markedly. The strains in the L1 vertebra at each increment were measured via DVC [4]. Briefly, this technique involves creating a mesh that subdivides the vertebra of interest into hexahedral regions with ~5mm side length. The displacement of each node in this mesh is estimated using a Gauss-Newton optimization method to register the image intensities before and after deformation. The (2nd-order) Green strain tensor $\mathbf{E}$ is then calculated for each hexahedral region. An error threshold corresponding to the mean ± two standard deviations in strain error [4] (0.00115 ± 0.00172) was applied to filter out values of strain that could not be robustly distinguished from zero.

The yield criterion employed in this study, the fabric-based Tsai-Wu (FBTW) criterion [5], defines the yield surface g in strain space as

$$g(\mathbf{E}) := \mathbf{G}:\mathbf{E} + \mathbf{E}:\mathbb{G}\mathbf{E} = 1 \qquad (1)$$

where $\mathbf{G}$ and $\mathbb{G}$ are 2nd- and 4th-order coefficient tensors, respectively.

Both **G** and $\mathbb{G}$ are functions of the (known) uniaxial tensile and compressive yield strains [5] and the fabric tensor, a quantitative measure of trabecular architecture [6]. For each region at each loading increment, the **E** computed by DVC and the fabric tensor measured for that region [7] were used to compute g, and regions with $g \geq 1$ were defined as yielded ("locally yielded") for that increment. The peripheral-most regions of the vertebral body were not analyzed for the occurrence of yielding, because the combination of cortical shell and trabecular bone in these regions prohibited calculation of a single fabric tensor in any of these regions.

RESULTS

Multiaxial strain states occurred in the trabecular centrum. Averaged across all samples, and across all loading increments following the global yield point, triaxial strain states were found in 4.54% (standard deviation among samples = 14.53%) of the regions of the centrum (Table 1). This percentage corresponded to roughly one quarter of the regions in which any strain was detected. The nature of the biaxial strain states that were found to occur differed among samples but often involved shear strain (Figure 1). Moreover, the magnitudes of the biaxial strains were sufficiently large to cause local yielding, as defined by the FBTW criterion, in nine of the L1 vertebrae.

Table 1: Percentage of regions that experienced no detectable strain, uniaxial or biaxial strain, and triaxial strain.

Strain State	Mean (STD) %
No Strain	81.91 (23.14)
Uniaxial	5.74 (5.47)
Biaxial	7.81 (13.26)
Triaxial	4.54 (14.53)

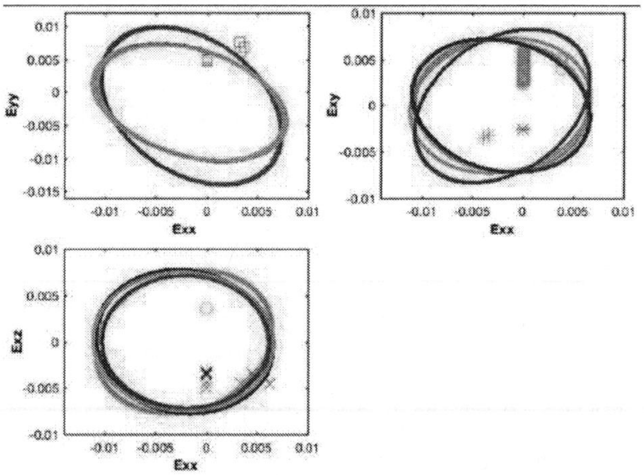

Figure 1: Strains in regions of trabecular bone in the L1 vertebra of four spine segments plotted in one normal biaxial plane (E_{yy}-E_{xx}; see coordinate frame in Figure 2 for reference) and two shear-normal biaxial planes (E_{xy}-E_{xx} and E_{xz}-E_{xx}), along with the FBTW yield envelopes in these planes. The data correspond to the loading increment at which local yielding was first detected. Each pair of color and symbol represents one L1 vertebra.

Qualitative examination revealed that local yielding commenced superiorly and posteriorly in seven of the nine samples for which local yielding was detected. The yielded regions were located at or just below the endplate. Local yielding appeared to propagate inferiorly from the superior endplate and outward in the transverse plane (Figure 2).

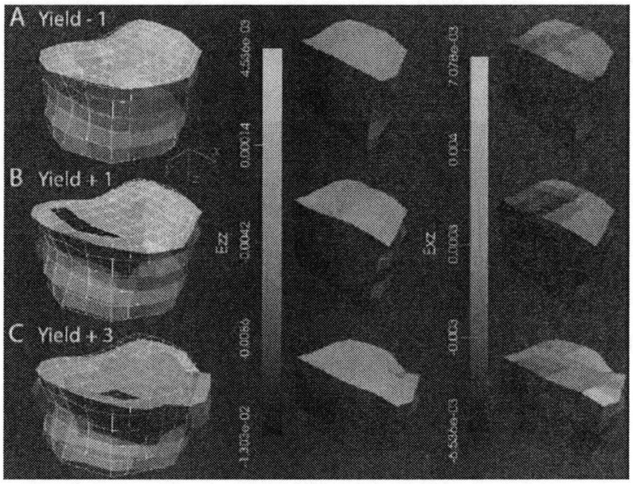

Figure 2: Display of the location of yielded regions (red shading in left column) and, on sagittal half-sections, the distribution of E_{zz} (middle column) and E_{xz} (right column) in one sample for three loading increments (A) one prior, (B) one following, and (C) three following the occurrence of the global yield point. In the left column, the wireframe represents the L1 vertebra's original shape, the translucent gray rendering represents a 25x magnification of the deformed vertebra.

DISCUSSION

The results of this study indicate that trabecular bone in the vertebral centrum is subjected to some degree of multiaxial loading even when the global loading conditions applied to the spine segment are simple axial compression. To our knowledge, this study is the first to use an experimental approach to measure the multiaxial nature of the strains and the ensuing yielding that develops *in situ* in trabecular bone.

Surprisingly, local yielding was detected in only nine out of the 21 spine segments, despite the fact that a global yield point was observed in all 21. This result may be due to the stringent error threshold that we applied to the measured strains and to potentially limited accuracy of the FBTW criterion in predicting yield. The error threshold may be responsible for the result that no strain was detected in 82% of the regions (Table 1). Nevertheless, given the nature of the multiaxial strain states that were detected, the findings of this study suggest that shear strains may play a substantial role in vertebral failure under axial compression.

ACKNOWLEDGEMENTS

NIH AR054620; Dr. Paul Barbone;
Boston University Micro-Computed Tomography Core Facility

REFERENCES

[1] Melton, L et al., *Osteoporosis Int*, 3(3):113-119, 1993.
[2] Silva, M, *Injury*, 4(3):143-154, 2007.
[3] Clouthier, AL, *Med Eng Phys*, 37(6):599-604, 2015.
[4] Hussein, A et al., *Procedia IUTAM*, 4:116-125, 2012.
[5] Wolfram, U et al., *J Mech Behav Biomed Mater*, 15: 218-228, 2012.
[6] Harrigan, T, Mann R, *J Mat Sci*, 19(3): 761-767, 1984.
[7] Doube, M et al., *Bone*, 47(6): 1076-1079, 2010.

DEVELOPMENT AND VALIDATION OF SUBJECT-SPECIFIC PROXIMAL FEMUR AND LUMBAR SPINE FINITE ELEMENT MODELS OF OBESE, OLDER ADULTS TO EVALUATE THE EFFECTS OF WEIGHT LOSS ON BONE STRENGTH

Ashley A. Weaver (1), Samantha L. Schoell (1), Dan P. Beavers (2), Leon Lenchik (3), W. Jack Rejeski (4), Joel D. Stitzel (1), Kristen M. Beavers (4)

(1) Wake Forest School of Medicine,
Department of Biomedical Engineering,
Winston-Salem, NC, USA

(2) Wake Forest School of Medicine,
Department of Biostatistical Sciences,
Winston-Salem, NC, USA

(3) Wake Forest School of Medicine,
Department of Radiology,
Winston-Salem, NC, USA

(4) Wake Forest University,
Department of Exercise Science,
Winston-Salem, NC, USA

INTRODUCTION

Osteoporosis is a significant public health concern with fracture at the hip and spine presenting as the primary clinical outcomes. Hip fractures are serious, costly, and disproportionately affect older adults [1]. Compression fracture of the spine is also common, especially in older adults, and can significantly contribute to back pain, depression, and reduced quality of life [2]. The increase in fracture risk is generally attributed to a loss in areal bone mineral density (aBMD) with age, although up to half of all fractures occur in individuals with normal aBMD, suggesting that other factors contribute to fracture risk [3]. Emerging data implicate weight and weight change as well as bone quality in fracture risk prediction. Previous studies have found significant decreases in total hip aBMD associated with diet-induced weight loss which can increase fracture risk [4]. To improve the assessment of fracture risk, bone strength as measured by finite element (FE) models derived from computed tomography (CT) data has emerged as a robust noninvasive tool. FE model development can include other bone quality measures including volumetric BMD (vBMD) and cortical thickness to better predict bone strength and fracture risk. To date, no study has assessed the changes in bone strength in older adults undergoing intentional weight loss.

The objective of this study is to develop and validate subject-specific proximal femur and lumbar spine FE models using morphing techniques to examine the effect of intentional weight loss over 18 months in obese, older adults on bone quality measures including vBMD, cortical thickness, and bone strength derived from clinical CT scan data of the total hip, femoral neck, and lumbar spine.

METHODS

Clinical CT scans of the proximal femur and lumbar spine of 25 (19 F + 6 M) and 30 (21 F + 9 M) obese, older adults, respectively, undergoing an 18-month intentional weight loss intervention were obtained at baseline and post-intervention. Image segmentation of the femur and lumbar spine CT scans was performed using Mimics (Materialise, Plymouth, MI) using bone thresholding techniques and manual editing. The segmentation masks were converted to 3D triangulated surface models. Quantitative measures of vBMD of the total hip, femoral neck, and lumbar spine were obtained using N-vivo software and elasticity-density relationships from the literature were used to derive subject-specific material properties (Figure 1) (Image Analysis, Columbia, KY).

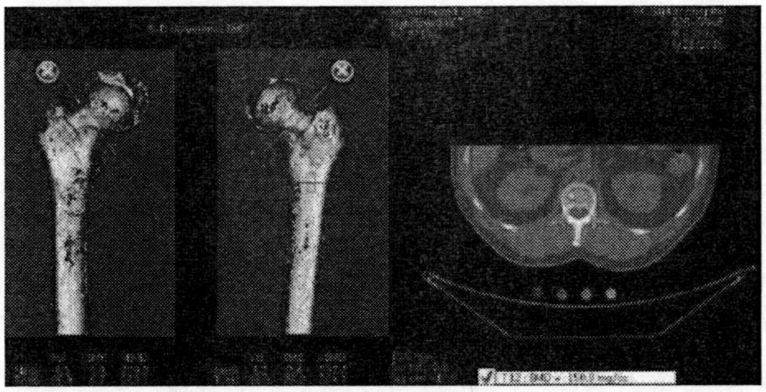

Figure 1: vBMD collection using a calibrated bone mineral phantom (N-Vivo v10.0.45862, Image Analysis, Columbia, KY).

High resolution cortical bone thickness measurements across the entire surface of the proximal femur and lumbar spine were obtained by applying a validated density-based algorithm [5]. This algorithm assumes a constant cortex density to estimate the cortical bone thickness from the clinical CT scans allowing for accurate estimation of cortex thickness down to approximately 0.3 mm. A cortical thickness mapping algorithm was developed to apply the cortical thickness measurements directly to the subject-specific FE models (Figure 2).

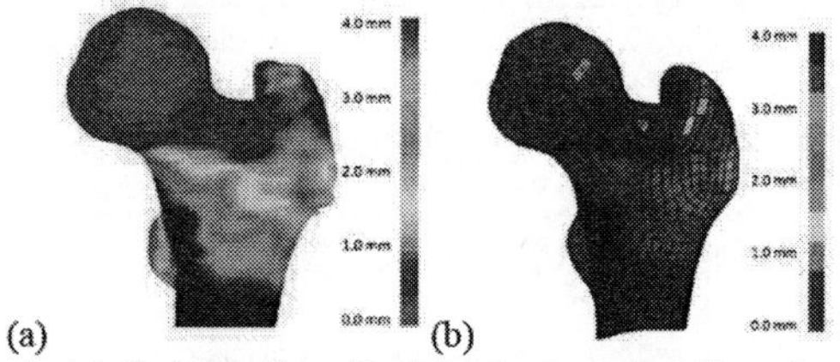

Figure 2: (a) Output of cortical thickness estimation algorithm and (b) resulting subject-specific FE model with mapped cortical thickness measurements.

Model development of the subject-specific FE models involved homologous landmark collection and model morphing techniques to accelerate the development of the models. Homologous landmark collection involves image segmentation, atlas development, and image registration techniques to derive atlas and subject point clouds. The atlas models used include the Global Human Body Models Consortium (GHBMC) M50-O v4.4 femur and the Total HUman Model for Safety (THUMS) AM50 v4.01 lumbar spine. Model morphing was performed using thin-plate spline interpolation to accurately capture the subject-specific geometry. Bone strength was estimated through simulation of a single-limb stance and sideways fall configuration for the femur and quasi-static uniaxial compression test for the spine based on experimental tests [6,7] (Figure 3). The peak fracture force or bone strength was defined as the peak force recorded between the impactor and femoral head or vertebral body. Statistical analyses were performed to obtain baseline values and correlations between weight change and change in bone quality parameters.

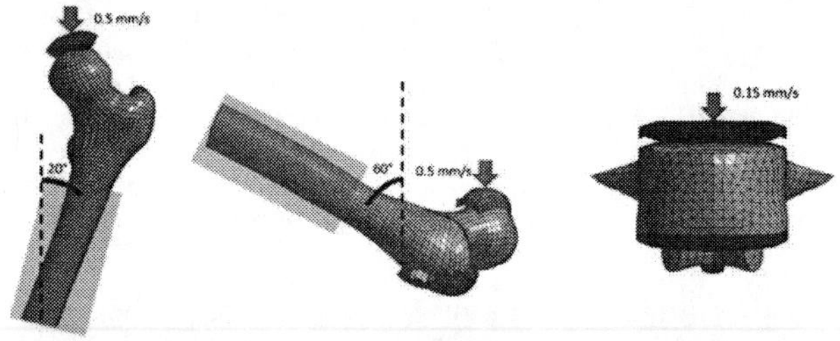

Figure 3: FE simulation setup for femur and lumbar spine.

RESULTS

Participants with proximal femur scans (n=25, 65.6 average age, 76% female, 20% African American) lost an average of 9.8% of their body weight over 18 months. Participants with lumbar spine scans (n=30, 65.9 average age, 70% female, 20% African American) lost an average of 10.4% of their body weight over 18 months. In analyses adjusted for baseline bone quality parameters and sex, correlations were observed between weight change and change in bone quality measures (r range: -0.30-0.60). Analyses show a significant correlation between weight change and change in total hip and femoral neck vBMD and cortical thickness. For every 1 kilogram of weight loss, vBMD in the total hip decreased by 0.003 g/cm^3 and in the femoral neck decreased by 0.004 g/cm^3. In addition, for every 1 kilogram of weight loss, cortical thickness in the total hip and femoral neck decreased by 0.003 mm, each. No significant correlations were present for the strength data for the femur or any bone quality measures for the lumbar spine.

Table 1: Correlations and parameter estimates between weight change (kg) and change in bone quality measures, adjusted for baseline quality parameters and gender. *p-value <0.05.

Bone Quality Measure	Adjusted r	B (SE)
Δ vBMD (g/cm^3)		
Total Hip	0.46*	0.003 (0.001)
Femoral Neck	0.45*	0.004 (0.002)
Lumbar Spine	0.22	0.548 (0.481)
Δ Cortical Thickness (mm)		
Total Hip	0.60*	0.003 (0.001)
Femoral Neck	0.53*	0.003 (0.001)
Lumbar Spine	-0.20	-0.002 (0.002)
Δ Estimated Strength (kN)		
Femoral Stance	-0.16	-0.004 (0.005)
Femoral Fall	-0.30	-0.003 (0.002)
Lumbar Spine	-0.13	-0.007 (0.010)

DISCUSSION

In this study, subject-specific FE models of the proximal femur and lumbar vertebrae of obese, older adults were developed to study the effect of intentional weight loss over 18 months on bone quality parameters. Clinical CT scan data was used to derive subject-specific bone quality measures at baseline and 18 month post weight loss intervention of vBMD, cortical thickness, and bone strength predicted by subject-specific FE models. Pilot data suggests modest associations between weight change and vBMD and cortical thickness changes, particularly in the proximal femur.

The methods presented offer the ability to accurately account for variations in geometry, cortical thickness, and material properties which can result in more reliable FE models to predict bone strength and fracture risk. The techniques used can be applied to a larger sample population to improve the understanding of the complex relationship between morphologic, compositional, and material changes of the elderly and obese undergoing intentional weight loss. Integration of advanced measures of bone quality derived from clinical CT with conventional aBMD can help improve the assessment and prediction of osteoporosis and fracture risk. These tools can be used to further understand the effects of intentional weight loss on bone health and the help aid clinicians in designing optimal weight loss strategies.

ACKNOWLEDGEMENTS

We thank Divya Jain and Elizabeth Lopez for their assistance with data collection and analysis. Funding sources: National Institutes of Health (K01 AG047921, R18 HL076441, and P30 AG21332), Wake Forest School of Medicine Translational Science Institute, Wake Forest University Translational Science Center and the National Science Foundation Research Experiences for Undergraduates (REU) under Award No. 1559700. Views expressed are those of the authors and do not represent the views of the sponsors.

REFERENCES

[1] Burge, R et al., *J Bone Miner Res*, 22: 465-475, 2007.
[2] Al-Sari, U et al., *Osteoporis Int*, 27:2891-2900, 2016.
[3] Nguyen, ND et al., *J Clin Endocrinol Metab*, 92:955-962, 2007.
[4] Zibellini, J et al., *J Bone Miner Res*, 30(12):2168-2178, 2015.
[5] Treece, GM et al., *Med Image Anal*, 16(5): 952-965, 2012.
[6] Keyak, JH et al., *J Biomech*, 31:125-133, 1998.
[7] Crawford, RP et al., *Bone*, 33:744-750, 2003.

Limitations of the Scalar Stress for Predicting Hemolysis in Complex Flows

Mohammad M. Faghih, M. Keith Sharp

Department of Mechanical Engineering,
University of Louisville, Louisville, KY, USA

INTRODUCTION

Mechanical circulatory support devices represent promising solutions for the 5.7 million patients in the US with congestive heart failure, but clinical complications, including blood damage issues (*e.g.*, mechanical blood damage, platelet activation and thrombosis) remain to be minimized. Since the introduction of blood-contacting devices, there has been a long history of research on hemolysis (red blood cell damage), including developing hemolysis prediction models. Such models, if universal across a wide range of flows, could be used during the design and development stage for new cardiovascular devices to reduce development expense and improve device performance. To the extent that platelet activation and thrombosis are also induced by fluid stress on cells, similar mechanisms may apply.

Earlier studies mainly focused on fluid shear stress and exposure time as the main factors of mechanical damage, as exemplified by the well-known power-law hemolysis model [Giersiepen, *et al.*, 1990], which was developed for pure laminar shear flow. Many improvements and modifications have been made to the power-law model to address issues of reliable predictions [Grigioni, *et al.*, 2006], however, the basis for the models remains a purely empirical connection between fluid stress and the cell membrane mechanics that lead to release of hemoglobin. As an alternative to these fluid-stress-based models, membrane-strain-based models base hemolysis predictions on a yield criterion for the membrane [*e.g.*, Chen & Sharp 2011, Chen, *et al.* 2013], which introduces a mechanistic relationship between fluid stresses and membrane tensions. Chen & Sharp [2011] calibrated their model to their own new data (also for shear flow only), while most other models of both types are based on the shear flow measurements underlying the power-law model.

It is clear, then, that both types of models are limited by the lack of validation data for flows other than laminar shear. Nonetheless, these models have been applied in more complex laminar flows, and even in turbulent flows, by substituting a von-Mises-like scalar stress [Bludszuweit 1995] in place of shear stress in their algorithms.

The purpose of this project is to evaluate whether the von-Mises-like stress, borrowed from solid mechanics, adequately scales the cell membrane tension that leads to membrane rupture and hemoglobin release as a predictor of hemolysis in flows of different hydrodynamic conditions. In particular, membrane tension will be compared for three different fluid stress types, *i.e.*, laminar shear, laminar extension and turbulent Reynolds stresses, each having the same scalar stress.

METHODS

The two-factor regression power-law equation is

$$\Delta PfHb / Hb = C\tau^{a}t^{b} \qquad (1)$$

where $\Delta PfHb$ is the increase in plasma-free hemoglobin, Hb is total hemoglobin (both intracellular and extracellular), τ is fluid shear stress, t is exposure time, and a, b and C are experimentally determined constants. Typical values for a, b and C are 2.416, 0.785 and 3.62×10^{-7}, respectively [Giersiepen, *et al.*, 1990].

The state of fluid stress at a point in a laminar flow is defined by a symmetric tensor having nine components. The scalar stress for such a flow is [Faghih & Sharp, 2016]

$$\sigma = \left[\frac{1}{2}\sigma'_{ij} : \sigma'_{ij}\right]^{1/2} \qquad (2)$$

where $\sigma'_{ij} = \sigma_{ij} - \frac{1}{3}\delta_{ij}\sigma_{kk}$ is the deviatoric stress, σ_{ij} is the full stress, δ_{ij} is the Kronecker delta and i, j and k are tensor indices. For turbulent flow, two similar tensors describe the mean viscous and Reynolds stresses, respectively.

For laminar shear (Fig. 1a), the scalar stress equals the fluid shear stress. Membrane tension in an ellipsoidal cell tank-treading in this flow can be calculated with techniques by Tran-Son-Tay, *et al.* [1987].

For laminar extensional flow, Chen & Sharp [2011] derived an equation for the maximum membrane tension around the middle of an ellipsoidal cell (the dotted line in Fig. 1b).

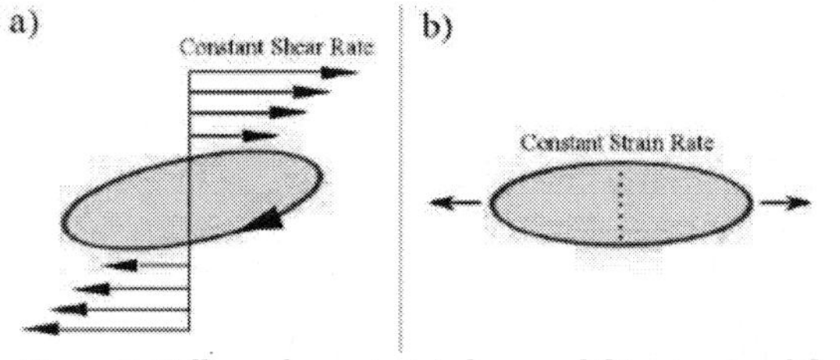

Figure 1. Cell membrane in (a) shear and (b) extensional flows.

For turbulent flow, Reynolds stress was measured in flow downstream of a prosthetic heart valve [Liu, *et al.*, 2000]. While a number of models have been proposed to scale the fluid stress exerted on cells in turbulent flows, the model of Antiga & Steinman [2009] is used here. This model is based on assumptions of turbulent eddies of length scale 100 μm and characteristic frequency of ~ 17 kHz, between which the cell is sheared at low Reynolds number (based on cell size). To estimate cell membrane tension, the cell is assumed to tank-tread and the Tran-Son-Tay, *et al.* model is again used. Since the cells would undergo chaotic motion in a turbulent flow, it is challenging to characterize the motion of cells within eddies of different time and length scales. Therefore, for simplicity, it is assumed that the cell is immediately trapped between another pair of eddies of the same characteristics after the previous eddies break down. Thus, the cell experiences continuous tank-treading motion for the purposes of this comparison. Laminar shear and extensional flow conditions were selected to match the Liu, *et al.* Reynolds stress so that the scalar stresses are the same for all three cases.

RESULTS

According to data of Liu, *et al.*, Reynolds stress 7.8 mm downstream on the centerline in the wake of a St. Jude valve is ~ 530 dyn/cm^2. This value is used as the common scalar stress for the three cases here, consistent with its use as the scalar stress by many authors in the power-law model. However, based on the analysis of Antiga & Steinman, for the same location in the wake, a red cell between co-rotating eddies of length scale 100 μm would experience a viscous shear stress of ~ 160 dyn/cm^2. Laminar shear and extensional flow cases were both evaluated with scalar stress equal to 530 dyn/cm^2, the same as for the turbulent case.

The resulting cell membrane tension for laminar extensional stress is greater than the mean for laminar shear by about an order of magnitude (Fig. 2). Mean tension for the turbulent case is about three times lower than the mean for laminar shear. In addition, for both laminar and turbulent shear flows, membrane tension is periodic, due to the tank-treading motion of the cell. The period of the tension fluctuations is about three times longer for turbulent shear. For a characteristic vortex lifetime of ~ 0.06 ms, the cell membrane completes only a fraction of a circulation between a particular pair of vortices. For extensional flow, the membrane does not circulate, so local tension in the membrane is constant.

DISCUSSION

Since it was reported in 1995, the scalar stress has become widely-used for extending the power-law and other hemolysis prediction methods to complex flows. The scalar stress hypothesis has, however, never been validated across a range of controlled flows. For the three flow cases evaluated here, the power law (equation 1) and any other method utilizing the scalar stress concept all give the same hemolysis prediction. The estimates shown in Fig. 2, however, strongly suggest that different types of flow cause distinctly different magnitudes and waveforms of membrane tension and, thus, can be expected to lead to different levels of hemolysis. The potential influence of different mean tensions in causing membrane rupture is obvious, however, the significance of membrane tension periodicity (in spite of constant fluid stress for laminar flow) remains to be explored. Periodicity may, for instance, produce fatigue failure more readily than constant tension of the same level.

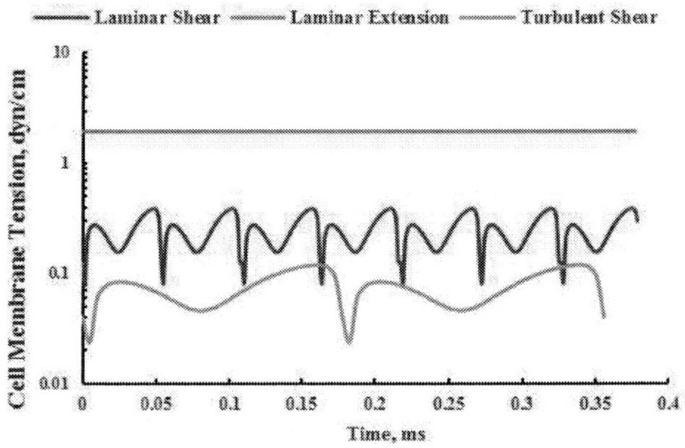

Figure 2. Tension of the cell membrane in turbulent, laminar extensional and laminar shear flows.

It appears, then, that the scalar stress hypothesis should be reevaluated and, in particular, needs to be properly validated for the types of flows that occur in blood-contacting devices, which includes laminar shear and extensional flows, as well as turbulent flows with ranges of combinations of viscous and Reynolds stresses.

Considerable uncertainty exists in the appropriate scaling of turbulent stresses for hemolysis prediction. The red cell motion proposed by Antiga & Steinman and used as the basis for the turbulent flow case in Fig. 2 has not been verified. Indeed, it has also been proposed that hemolysis scales directly with Reynolds stresses, as well as with turbulent viscous stresses alone.

Microscopic observations of red cell motion and deformation have thus far been performed mostly in laminar flow with elevated plasma viscosity, at cell concentrations much lower than normal hematocrit and at fluid stress levels well below the threshold for hemolysis. Fundamental discoveries of critical importance to the development of more mechanistic hemolysis models may await such direct imaging of cells in flow conditions relevant to cardiovascular devices.

REFERENCES

Antiga L, Steinman DA. *Biorheology* 46:77-81, 2009.
Bludszuweit C. *Artif Organs* 19:7:590-6, 1995.
Chen Y, Sharp MK. *Artif Organs* 35:145-56, 2011.
Chen Y, Kent TL, Sharp MK. *Artif Organs* 37:3:256-66, 2013.
Faghih MM, Keith Sharp M. *J Biomech Eng* 138: 124504-1-4, 2016.
Giersiepen M, *et al. Int J Artif Organs* 13:300-6, 1990.
Grigioni M, *et al. Artif Organs* 28:467-75, 2004.
Liu JS, Lu PC, Chu SH. *J Biomech Eng* 122:118-24, 2000.
Tran-Son-Tay R, *et al. Biophys J.* 51:915-24, 1987.

SB^{3}C2017
Summer Biomechanics, Bioengineering and Biotransport Conference
June 21 – 24, Tucson, AZ, USA

3D RECONSTRUCTION OF THE HEMOSTATIC PLUG TRANSPORT MICROENVIRONMENT

Mehran Mirramezani (1), Maurizio Tomaiuolo (2), Timothy J. Stalker (2) Shawn C. Shadden (1)

(1) Department of Mechanical Engineering
University of California, Berkeley
Berkeley, California, USA

(2) Department of Medicine
University of Pennsylvania
Philadelphia, Pennsylvania, USA

INTRODUCTION

Numerous studies have shown that mass transport affects the hemostatic process. Biochemical and biological processes that define platelet activation and coagulation reactions are deeply intertwined with the physical characteristics of the environment where such processes occur. The flow velocity is one of the major factors that determine the rate of delivery and removal of soluble agonists and cellular components. It has remained unclear however how clot microstructure affects such transport processes.

Computational studies have explicitly considered the plasma velocity in the gaps between platelets in 2D clot models [1], or in a single 3D structure informed by fluorescence microscopy [2]. While both studies found that plasma velocity is slow, interpretation of the results is difficult. The 2D model cannot consider the influence of tortuous paths accessible to flow in a 3D geometry. In the 3D model study, only one structure was examined, thus there was no repeatability or generalizability of the results. The goal of this study is to examine the nature of a platelet aggregate microenvironment and integrate its architectural features in a data-driven single-platelet-resolution 3D hemostatic plug reconstruction procedure that is repeatable, reproducible, and can be procedurally applied to any hemostatic mass on which there is sufficient data.

METHODS

Vascular injury in the mouse cremaster muscle microcirculation: A thrombus can be formed following penetrating laser injury. A laser was used to make a defect in the mouse cremaster artery large enough to allow red cells as well as plasma to escape. Using the confocal z-stack of fluorescent channels, the 3D volume of the hemostatic platelet plug is visualized (Figure 1A) in vivo during clot formation. Highly activated platelets, which form a less porous core, are captured by antibody tagging and differentiated from sparsely activated platelets, which form a more porous shell region [3].

3D platelet aggregates microenvironment: Once the morphology of the thrombotic mass is parameterized by interrogation of the z-stack, a jammed disordered packing of non-intersecting ellipsoids is used to fill the volume. This method utilizes a hard-particle molecular dynamics algorithm to producing dense jammed disordered packing given the ratio between the three semi-axis (1, 1, and 0.4 μm was used to represent the *nominal* dimensions of human platelets) and the number of particles in a unit cube with periodic boundary conditions. The ellipsoids are then assigned random initial position and velocities, their motion is followed as they collide and uniformly expand until a jammed state is reached and the density reaches a maximal value [4]. Then this packing configuration is used repeatedly to fill the in-vivo plug volume (Figure 1B). Based on experimental data, different scaling factors for the nominal platelet semi-axis dimensions mentioned above were employed for the core and shell regions to produce desired porosities in these respective regions.

Computational fluid dynamics: A computational model was used to study the flow characteristics inside the 3D reconstruction of the hemostatic plug. Briefly, the reconstructed platelet aggregate was combined with a cylindrical vascular domain having the same dimensions of the vessel injured during the experiment. The entire intraluminal volume was meshed using Abaqus. The mesh is highly resolved to capture the intraclot microarchitecture so that a no-slip velocity on each platelet surface is well resolved. A finite element flow solver was developed using the open-source FEniCS package to solve the incompressible Stokes equation in the fluid domain. A parabolic flow profile was imposed at the inlet, with a centerline velocity of 2 mm/s (Figure 1C).

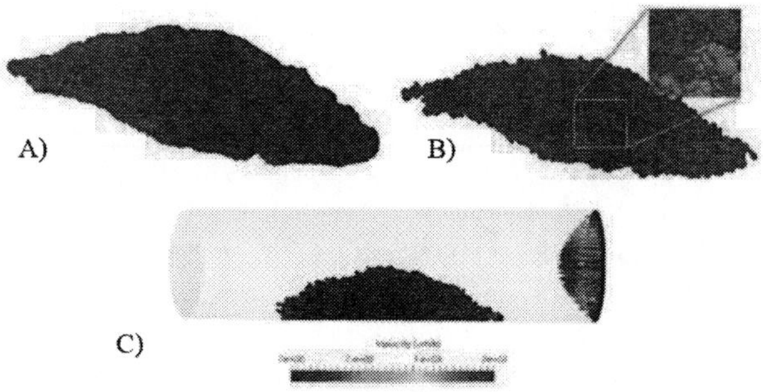

Figure 1 A) The volume of a hemostatic plug from the cremaster muscle injury model; blue (shell), red (core) B) The hemostatic plug volume is filled with a jammed ellipsoid packing and meshed C) The hemostatic plug is positioned in the modeled vessel

RESULTS

The microstructure of the clot can be parameterized by the distribution of pore gap size. The gap size distribution was calculated for the various jammed disordered packings resulting from the different semi-axis scaling factors, as well as for the actual experimental data obtained from ex-vivo SEM images of the thrombus. All images were converted into binary versions for analysis via a thresholding method with a custom Matlab code to obtain gap size distributions. A comparison of the gap size distributions from each scaling factor packing configurations with that of experimental data shows that the microenvironment of a platelet aggregate can be reproduced when values of the scaling factor is between 1.1 and 1.3 (Figure 2B).

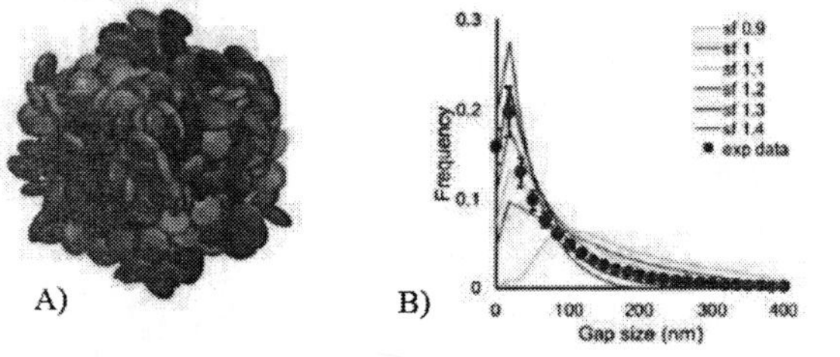

Figure 2 A) An example of a randomly jammed ellipsoid packing B) Gap size distributions computed from ellipsoid packings with different scaling factors compared with experimental data

Interplatelet plasma velocity was studied adopting a parametric approach. The first parameter considered is mean gap size. The volume of an in-vivo hemostatic platelet plug was reconstructed and the volume of each ellipsoid in it was modified by multiplying each semi-axis by a scaling factor from 0.7 to 1.3. The results indicate that large variations in gap size correspond to small variations in the mean interplatelet velocity (Figure 3A). The results show that the plasma velocity between the platelets is in the microns-per-second range, approximately three orders of magnitude slower than the blood flow in the lumen vessel.

Molecular movement between the platelet gaps occurs because of convection, diffusion, or a combination of the two. The Peclet number specifies the relative contribution of each for a particular substance in a given porous media. The Peclet number was computed using the reported diffusion coefficient of thrombin [5], the mean gap size, and the mean computed velocity. Even with a variation spanning an order of magnitude in mean gap size the value of the Peclet number was always well below 1, indicating the dominant nature of diffusion in the hemostatic plug microenvironment (Figure 3B).

The effects of maximum centerline velocity in the vessel were also considered. When the inlet centerline velocity was increased from 2 to 10 mm/s, the interplatelet plasma velocity values remained negligibly small (Figure 3C). The interplatelet velocity across the hemostatic plug in the wall-normal direction was also investigated by sampling points selected at different distances from the base of the

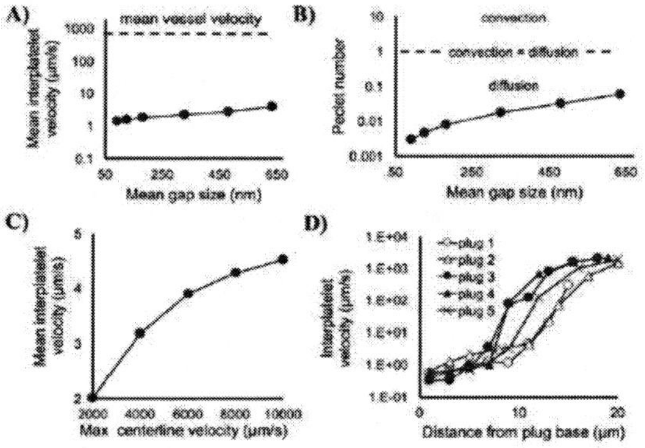

Figure 3 A) Effect of mean gap size. B) Peclet number versus mean gap size. C) Effect of centerline velocity. D) Fluid velocity measured in wall-normal direction across and just outside the clot (clot height ~10 micron)

plug (vessel wall). Starting from the outside of a plug the velocity is about 1 mm/s, however there is a sudden drop of three orders of magnitude just inside the clot (Figure 3D).

DISCUSSION

The microenvironmental features of a blood clot, such as porosity and permeability, can affect the delivery and removal of both cells and coagulation proteins necessary for hemostasis. Here, a combined computational-experimental approach was used to reconstruct the geometric and fluid dynamic microenvironment inside a realistic clot.

Our results suggest that interplatelet velocity is low and invariant to the mean gap size, the outside blood velocity, or the specific location within a hemostatic plug. In all cases diffusion becomes responsible for molecular movement within the hemostatic plug. Furthermore, a sheltered environment promotes the residence time and the bioactivity of relevant molecules and proteins, but also isolates the processes. From this perspective diffusion-driven processes become a necessary but not sufficient condition for hemostasis during clot growth. Because interplatelet velocity is insensitive to increasing gap size, it appears that platelets are positioned to satisfy both selection constraints of slow plasma velocity and narrow gap sizes.

ACKNOWLEDGEMENTS

This work was supported by the American Heart Association (11SDG5720011) and NIH (R01-HL-103419).

REFERENCES

[1] Tomaiuolo M et al., *Blood,* 124(11):1816-1823, 2014.
[2] Voronov R.S et al., *Ann Biomed Eng,* 41(6):1297-1307, 2013.
[3] Stalker T.J et al., *Blood,* 121(10):1875-1885, 2013.
[4] Donev AI et al, *Science,* 303(5660):990-993, 2004.
[5] Hubbell JA et al, *Biophys J,* 50:937-945, 1986.

AN INVESTIGATION OF THE RELATIONSHIP BETWEEN PLATELET ADHESION AND SURFACE TOPOGRAPHY IN THE PENN STATE PEDIATRIC VAD

A. Mueser (1), C. A. Siedlecki (1,2), W. J. Weiss (1,2), K. B. Manning (1,2)

(1) Department of Biomedical Engineering
Pennsylvania State University
University Park, PA, USA

(2) Department of Surgery
Penn State Hershey Medical Center
Hershey, PA, USA

INTRODUCTION

Ventricular assist devices are primarily used to bridge the gap of heart complications and waiting for a heart transplant. With cardiovascular disease being the leading cause of death in the United States, creating a sustainable, effective device is necessary [1]. Children are in grave need of a device as shown by the 321 children listed on the heart transplant list as of January 2015. In 2015, only 385 children received a heart transplant, leaving the current 321 children on the list to continue waiting. [2] Thrombus formation remains a challenge with current pediatric ventricular assist devices (PVADs) and could lead to complications such as stroke and heart attack. To alleviate these issues, determining the underlying mechanisms of clot formation is necessary.

One mechanism that is thought to play a role in clot formation in PVADs is surface topography. Previous studies have been conducted to study mechanisms of clot formation in adult left ventricular assist devices [3-5]; however, studies on PVADs specifically have yet to be conducted. The Penn State PVAD uses a segmented poly-(ether polyurethane urea) (SPEUU) blood sac within the device. A thorough surface analysis was performed to understand how surface imperfections of the blood sac could be correlated to specific depositions.

METHODS

To determine if a correlation between surface topography and clot formation exists, various forms of microscopy were completed following device explantation at the Penn State Hershey Medical Center for five blood sacs. Throughout the surface analysis, the only animal study identification provided was the sac number. No other information about the study, including anticoagulants, animal sex, etc., was provided. Outlined below are the steps that were followed for the analysis.

- At Penn State Hershey Medical Center, the PVAD was implanted into sheep for 30-60 days. (IACUC#46372)
- Following explantation, eight samples from specific regions of each sac were fluorescently labeled for platelets and fibrin.
- To assess platelet and fibrin depositions, the samples were imaged using confocal microscopy and an environmental scanning electron microscope (ESEM).
- The sample was cleared of all biologic material using a combination of hydrochloric acid and pepsin in a 3:1 ratio. Optical profilometry was used to collect data regarding the surface topography.

Once these steps were completed, the optical profilometry data were analyzed using MetroPro® 8.3.5 software (Zygo Corporation, Middlefield, CT) to obtain various surface roughness parameters. These parameters were the roughness average (Ra) and the root mean square (RMS). The roughness average was chosen due to its statistical stability and the ease of replication for surfaces using a controlled process for fabrication (Eq. 1).

$$R_a = \frac{1}{L}\int_0^L |z(x)|\, dx \tag{1}$$

The root mean square was chosen as a parameter because it is sensitive to peaks and valleys than the average (Eq. 2).

$$RMS = \sqrt{\frac{1}{L}\int_0^L z^2(x)dx} \tag{2}$$

Following data analysis, comparisons were performed between locations that showed macroscopic clot depositions, those that are visible to the human eye (millimeters in size), and microscopic clot depositions, those that are only visible through microscopy (microns in size). Between the *in vivo* thrombus and the *in vitro* microscopy, an initial analysis was completed to determine a relationship between surface topography and platelet deposition to have a better understanding of the impact surfaces have on pulsatile VADs.

RESULTS

Through confocal microscopy and ESEM, regions where biologic depositions occur were identified. Platelets were labeled red and fibrin was labeled green. In a macroscopic clot, or larger microscopic clots, the fluorescence was seen as yellow meaning there was a mixture of platelets and fibrin in that area. Sample images of macroscopic and microscopic clots taken from confocal microscopy are shown in Figure 1. The left image shows a large amount of fluorescence, implying a large amount of deposit in that region, hence a larger clot. The image on the right shows minimal fluorescence, implying only a few platelets and minimal fibrin deposits in that region.

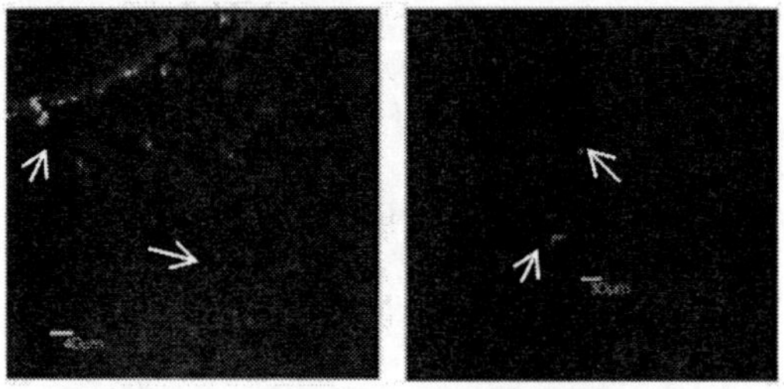

Figure 1: Sample images of macroscopic clot (left) and microscopic clot (right) from confocal microscopy.

From the animal studies, two of the sacs resulted in macroscopic clots and all five of the sacs had at least one microscopic deposition. All of the clots are in the top half of the sac with a majority of them developing near the inlet and outlet ports on the diaphragm (moving) side. The locations of each deposit are shown in Figure 2 where macroscopic clots are orange and microscopic clots are blue.

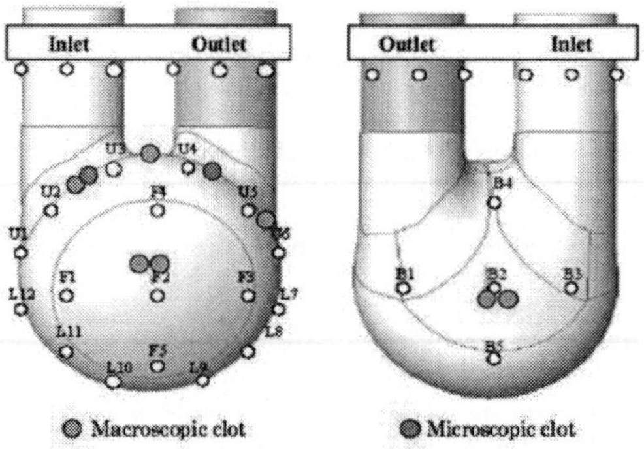

Figure 2: Map of macroscopic clot locations (orange) and microscopic clot locations (blue).

After determining the locations of platelet and fibrin deposits and degrading them from the surface, the profilometry data were analyzed to compare the surface roughness between the locations of macroscopic and microscopic clots. As shown in Figure 3, the macroscopic clot locations demonstrate higher surface topography parameters than the microscopic clots. The roughness average (Ra) is nearly double for macroscopic clots when compared to microscopic clots. The root mean square (RMS) is also approximately double for the macroscopic clot.

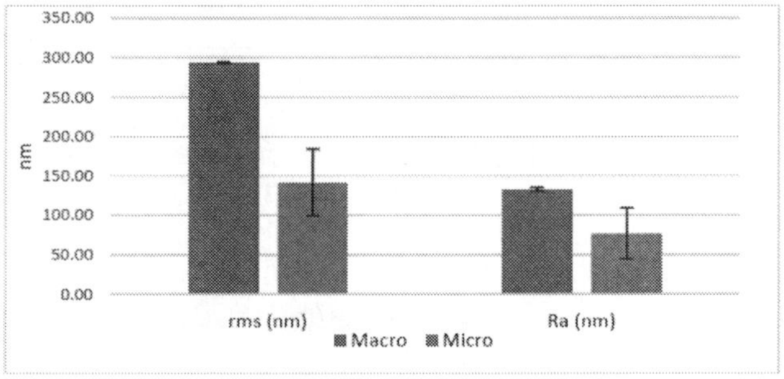

Figure 3: Surface roughness evaluation comparing surface topography between macroscopic and microscopic clot locations.

DISCUSSION

The surface topography was rougher under macroscopic clots than under the microscopic clots, leading to the conclusion that surface topography does play a role in clot formation in the Penn State PVAD. The roughness of a surface can impact clot formation due to the ability for a platelet to attach to a divot or scratch on a surface. When a platelet adheres to a surface, it triggers the clotting cascade. Previous studies have supported this conclusion through creation of submicron topographic structures on various polymers [6, 7]. Fluid mechanics may also play a role in the clot formation near the inlet and outlet ports. One possible explanation for the deposits in these areas is the way the valve is placed within the device. There would be little to no washing in the area of the clots due to the strut in the valve, giving potential for a clot to form.

CONCLUSIONS & FUTURE WORK

Further analysis needs to be performed on six remaining blood sacs to gain further significance to our findings. Other factors, including fluid flow and anticoagulants, have yet to be considered. Following the completion of the blood sac analysis, particle image velocimetry data previously collected on the Penn State PVAD will be examined against the deposit locations to determine if there is a correlation between clot formation and shear rate. The animal studies will also be examined to ensure there were no complications with the animals that could have led to the platelet and fibrin deposits. Once all of these analyses are complete, a stronger statement will be able to be made regarding the mechanism for clot formation in the Penn State PVAD.

ACKNOWLEDGEMENTS

This work was supported through National Institutes of Health, HL108123. We also thank Kaitlin Saloky for her aid with lab work and Nick Ceneviva for the creation of the protocol.

REFERENCES

[1] Heron, M., *Natl Vital Stat Rep,* 60(6):1-94, 2012.
[2] OPTN: Organ Procurement and Transplantation Network, 2015.
[3] Topper, S., *et al., Cardiovasc Eng Technol,* 5(1): 54–69, 2014.
[4] Yamanaka, H., *et al., ASAIO J,* 52(3): 257-265, 2006.
[5] Navitsky, M., *et al., ABME,* 41(1): 4-16, 2013.
[6] Milner, K., *et al., ASAIO J,* 52(2): 12A, 2006.
[7] Koh, L., *et al., Biomaterials,* 31(7): 1533-1545, 2010.

A PREDICTIVE MULTISCALE MODE FOR SIMULATING FLOW-INDUCED PLATELET ACTIVATION AND AGGREGATION: CORRELATING WITH IN VITRO RESULTS

Peng Zhang (1), Jawaad Sheriff (1), Prachi Gupta (2), Marvin J. Slepian (3), Yuefan Deng (2), Danny Bluestein (1)

(1) Biomedical Engineering Department
Stony Brook University
Stony Brook, N□, United States

(2) Applied Mathematics Department
Stony Brook University
Stony Brook, N□, United States

(3) Departments of Medicine and Biomedical Engineering
University of Ari□ona
Tucson, AZ, United States

INTRODUCTION

The coagulation cascade of blood may be initiated by flow-induced platelet activation, which prompts clot formation in prosthetic cardiovascular devices and vascular disease processes. Upon activation, platelets undergo complex morphological changes. Activated platelets polymerize fibrinogen into a fibrin network that enmeshes red blood cells. Continuum methods fail to capture the molecular mechanisms such as filopodia formation during platelet activation, while utilizing molecular dynamics is computationally prohibitive. A multiscale approach offers a means to bridge the gap between macroscopic flow and the cellular scales.

We present a multiscale model for simulating platelets activation and aggregation in viscous blood flows. This model incorporates a Dissipative Particle Dynamics (DPD) model of viscous flow that interfaces with Coarse Grained Molecular Dynamics (CGMD) model of mechanobiology-based platelets, to simulate their activation via mechanotransduction pathways. This model bridges the gap between macroscopic transport flow-induced platelet activation and aggregation scales and the ensuing molecular events, and is further validated through the use of several shear-based techniques that allow observation of platelet shape change and motion over a wide range of shear stresses and exposure durations.

METHODS

Our multiple spatiotemporal model employs a modified DPD formula to describe viscous blood flow in the cases of stenosis and microchannel [1], and an electrical-free CGMD formula to model the intra-platelet constituents and study the mechanotransduction process of hemodynamic in platelets [2]. Spatially, the DPD-CGMD interface was established by imposing a hybrid force field [3]. In the spatial interface, the Lennard–Jones term helps the cytoskeleton-confined shape and the incompressibility of platelets against the applied shear stress of circumfluent flow; and the dissipative and random terms maintain the flow local thermodynamic and mechanical properties and exchange momentum to express interactions between the platelet and the flow particles. Temporally, a 4-level multiple time-stepping (MTS) scheme was used [4] as an efficient and ultra-scalable numeric solver on top supercomputers [5].

In the mesoscale, we enhanced the conventional DPD formula by adding a Morse-based repulsive term for favorably producing Poiseuille flow of an incompressible fluid through a stenosis where the compressibility becomes a problem for DPD [1, 6]. We conducted a comparative study to investigate the fluid flow properties of DPD-Morse and DPD fluids through a 67% stenotic microchannel. Our DPD-Morse can extend the conventional DPD to sustain large compression caused by a stenosis.

In the nanoscale, we built a mechanobiology-based platelet model by describing key constituents and biophysical properties [2, 7]. We modeled a bilayer membrane, ellipsoid based discoid shape, rigid filamentous core, gel-like cytoplasm using the Morse potential [6], and filamentous actin. An α-helix structure was used to mimic a protrusible actin filament.

The platelet's membrane in our model has 67,004 particles. αIIbβ3 is the major platelet integrin (receptor) with 50,000 to 80,000 copies per platelet [8]. Thus, we use a membrane particle to model one αIIbβ3 receptor in activated state, where two particles represent the two heads of the receptor (Fig. 1). The force field between the member particles of different platelets is modeled as follows:

$$\boldsymbol{F}_{ij} = D_0\left(e^{-2\alpha(r_{ij}-r_0)} - 2e^{-\alpha(r_{ij}-r_0)}\right)\boldsymbol{e}_{ij} + f^A(t_{ij})\left(1-\frac{r_{ij}}{d_c}\right)\boldsymbol{e}_{ij}$$

where D_0 is the well depth of repulsive force, α is the scaling factor, r_0 is the zero-force length. $r_{ij} = |\boldsymbol{r}_{ij}|$ and $\boldsymbol{e}_{ij} = \boldsymbol{r}_{ij}/r_{ij}$. f^A is the adhesive force strength coefficient and depends on time t_{ij} that denotes time duration of adhesion, and d_c is the force relaxation distance that is equal to two times the physical radius of the platelets.

We validated our numerical method by correlating numerical simulations and model predictions with in vitro platelet activation and aggregation results. Purified platelets [7] were prepared from whole blood obtained from consenting healthy donors of both genders. For evaluation of shear-mediated activation response, platelets at a count of 20,000/µl were exposed in a Hemodynamic Shearing Device to different shear stress-exposure time combinations up to 50 dyne/cm^2 and 4 min, respectively. Platelet samples were fixed with 2% glutaraldehyde, mounted on poly-L-lysine coated glass slides, and dehydrated through an ethanol series prior to imaging with scanning electron microscopy. In particular, we examined the average number of visible filopods per platelet, circularity of activated platelets, and the position, length and the aspect ratio of the filopods.

For aggregation experiments, purified platelets were prepared as previously described, diluted to 150,000/µl, treated with 0.5 mg/ml fibrinogen (Enzyme Research Laboratories, South Bend, IN), and perfused through 100 µm× 1 mm off-the-shelf (µ-Slide III0.1, ibidi USA, Inc., Madison, WI) and 50 µm × 200 µm PDMS microchannels using a syringe pump (Fig. 2a). The maximum centerline velocities of platelets through the channels was defined as 200 cm/min, corresponding to maximum wall shear stresses of 4 and 40 dyne/cm^2, respectively. Aggregation events were observed at 100× magnification under a DIC microscope (Ti-E, Nikon, Melville, NY) at up to 420 fps (Neo 5.5 sCMOS, Andor Technology Ltd., Belfast, UK). We measured the length of the bond and contact time between 2 aggregating platelets (Fig. 2b).

RESULTS

Our multiscale model describes the biophysical properties for platelets down to the nm-length and ps-time scales [2]. Membrane Young's modulus is 31.2 µN/m and shear elastic modulus is 33.0±9.0 µN/m. Cytoplasm viscosity is 4.1 mPa·s. Actin filament stiffness is 56.3±1.0 pN/nm.

Using this model, we simulated the flow-induced platelet activation. When platelets were flipping in the viscous blood flow, dynamic stresses were mapped on the membrane and transmitted to the cytoskeleton, with simultaneous accumulation of the mapped stresses over the period. This emulated the effect of mechanotransduction process of dynamics stresses in platelets. Filopodial formation is applied to actin filaments that are exposed to the highest stresses. Regardless of the shear exposure, most platelets had 2 or 3 filopods, with mean length of 1.29 ± 0.17 µm (n=945 filopods), circularity of 0.99 ± 0.01 (n=331 platelets), and median filopod number of 3 (n=339 platelets). These experimental observations correlate well with numerical predictions.

In addition, we mimicked the aggregation of two platelets under a shear stress rate of 8000 s^{-1} in a 35.6 µm × 16 µm × 35.6 µm channel. The contact time was between 0.2 and 1.0 ms. The contact area was 0.0~0.2 µm^2, which decreases almost linearly with an increase in the microchannel width. The time integral of contact area was 0.0~0.08 µm^2·ms [9].

DISCUSSION

Our approach is the first computationally affordable numerical method for simulating the platelet activation and aggregation by highly resolved mapping of mechanical stresses on the cytoskeleton in dynamic flow. Biophysical properties of a platelet are accurately described down to nm-length and ps-time scales. The viscous flows are described at µm-length and ns-time scales. Phenomena of filopodia formation and platelet aggregation are mimicked and correlate well with ongoing experimental results. This model can be further employed to describe the initiation of thrombosis in blood flow and study the effects of modulating platelet properties to enhance their shear resistance via mechanotransduction pathways.

ACKNOWLEDGEMENTS

This publication was made possible by grants from the National Institute of Health: NHLBI R21HL096930-01A2, NIBIB Quantum Award Implementation Phase II-U01 EB012487-0 and NIH U01 HL13105201A1 (PI: Danny Bluestein). The simulations in this work used the XSEDE computing resource award DMS150011 at the SDSC Comet supercomputer (PI: Peng Zhang).

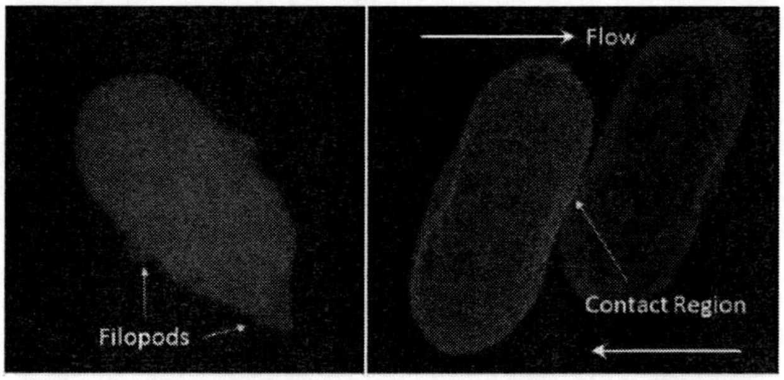

Figure 1: Aggregation of adjacent activated/quiescent platelets

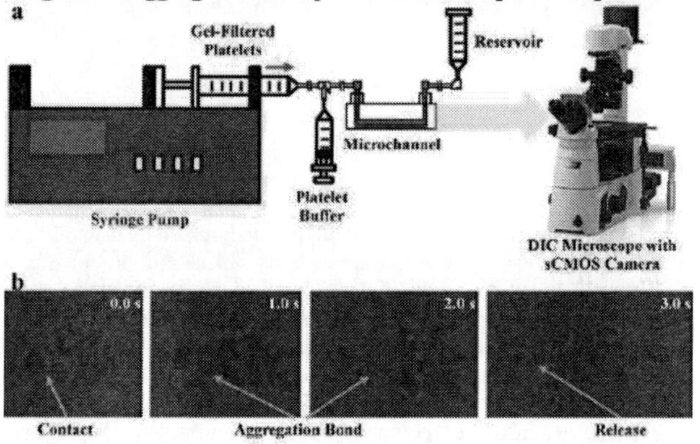

Figure 2: (a) Set-up for aggregation experiments. b) Sequential imaging of platelet aggregation/disaggregation under flow.

REFERENCES

[1] Gao, C et al., Reducing the Effects of Compressibility in DPD-based Blood Flow Simulations through Severe Stenotic Microchannel, *J Comput Phys*, 2017. In press
[2] Zhang, P et al., *J Biomech*, 50:26-33, 2017.
[3] Zhang, P et al., *Cell Mol Bioeng*, 7:552-574, 2014.
[4] Zhang, P et al., *J Comput Phys*, 284:668-686, 2015.
[5] Zhang, P et al., *Comput Phys Commun*, 204:132-140, 2015.
[6] Zhang, N et al., *J Comput Phys*, 257(A):726-736, 2014.
[7] Pothapragada, S et al., *Int J Numer Method Biomed Eng*, 31:1-16, 2015.
[8] Michelson, AD, ed. *Platelets*, 3rd ed. Academic Press, 2013.
[9] Mody, N.A. et al, *Biophys J*, 95:2539-2555, 2008.

CLOT FORMATION IN A MODEL INTRACRANIAL ANEURYSM IS MODULATED BY ENDOVASCULAR COIL SHAPE AND ARRANGEMENT

Brittany J. Earnest (1), Avery J. Evans, M.D. (2), Brian P. Helmke, Ph.D. (1)

(1) Department of Biomedical Engineering
University of Virginia
Charlottesville, VA, USA

(2) Department of Radiology
University of Virginia
Charlottesville, VA, USA

INTRODUCTION

Intracranial aneurysms (IA) are a serious and prevalent condition. IA rupture accounts for 3-5% of new strokes, and 1 in 50 people in the United States has an unruptured IA[1]. Therefore, it is necessary to treat IAs before rupture and in a way that prevents recurrence. The most common treatment is endovascular coiling; however approximately 30% of cases require additional intervention. In order to design and screen improved coils, the mechanism by which coiled IAs heal needs to be well understood and simulated in test models.

Ex vivo and animal studies demonstrate that the basic mechanism for IA healing begins when intra-aneurysmal hemodynamics, facilitated by coils, drive thrombosis formation within the clot. The thrombus provides an initial scaffold for cellularization and matrix reorganization within the IA sac, or dome, and opening, or "neck". Still, it is not clear why bioactive coils and other treatments fail in patients, which indicates that our understanding of this mechanism is incomplete. *In vitro* models provide a supplement to *in vivo* and *ex vivo* studies and give new insight into human mechanism by using human-derived materials, providing analysis of the system at early time points and allowing control over variables in the system that cannot be controlled *in vivo*.

The goal of this study is to design an *in vitro* model of coiled bifurcation IA thrombosis in order to better understand the progression and structure of IA blood clots and their dependencies on coil arrangement, coil design and intra-aneurysmal hemodynamics. With this new information, novel coils that improve rate and extent of IA thrombus formation and stabilization can be developed and screened efficiently.

METHODS

Polycarbonate parallel plate flow chambers were designed and fabricated to simulate a small bifurcation IA. One chamber was designed to simulate hemodynamics experienced by a coiled IA neck, where a triangular section in the middle of the chamber supports a single 1 cm section of coil (see Figure 1). A second chamber was designed to simulate flow profile throughout a bifurcation IA dome, where a circular opening 5 mm in diameter was machined into the triangular section, creating a "sac" with a 3 mm neck opening. Coils were arranged in this "sac". Chambers, tubing and reservoirs were incubated with 2.5% bovine serum albumin and 30 U/mL heparin before coiling to prevent clotting in the absence of coils.

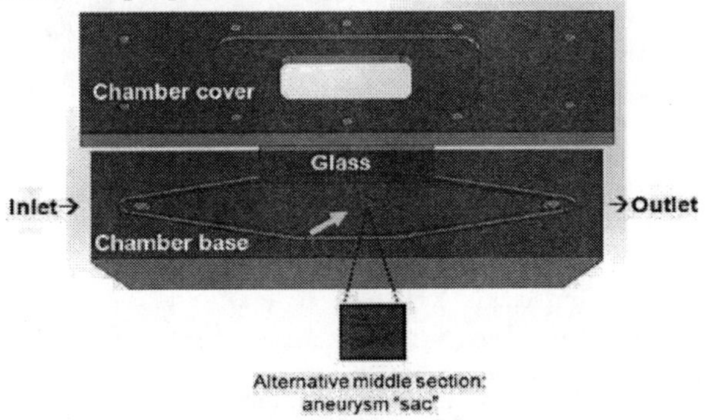

Figure 1: Diagram of single coil flow chamber. Arrow indicates coil placement. The middle section was modified in the aneurysm "sac" model.

Whole blood from healthy volunteers was drawn into 30-cc syringes, according to approved IRB protocol, diluted to 37% in saline and immediately centrifuged at 100xg for 20 min. Platelet-rich plasma (PRP) was collected. This dilution allowed for extended clot times and clot progression analysis without anticoagulant. Oregon green-tagged fibrinogen was added to PRP at 1:100 dilution, and DiI-labeled red blood cells (RBCs) were added to PRP as flow tracers. PRP was immediately transferred to the flow loop and recirculated (or kept stagnant) for 17 to 20 minutes at 200 s^{-1} shear rate. Fibrin formation and tracer flow were recorded with fluorescence confocal microscopy.

Chambers were immediately perfused with saline, then 5 mL 4% paraformaldehyde, allowed to fix for 20 minutes, and rinsed in saline. Chambers were imaged using confocal microscopy and then disassembled. Coils were removed and dehydrated using ethanol solutions of increasing concentration and hexamethyldisilazane (HMDS). Coils were imaged using scanning electron microscopy (SEM). Measurements of RBC flow profile, clot coverage of aneurysm and coils, mixing time, platelet spreading and fibrin structure were performed in ImageJ[2]. Two-way ANOVA and student's t-tests were used to evaluate variability within and between groups, respectively.

RESULTS

In single coil chambers (Figure 1), clots formed on coils under flow with 200 s^{-1} shear rate had morphology similar to those observed *ex vivo*, whereas clots formed on coils without flow were larger, more homogenous, and contained fibrin fibers with thinner bundle diameters. Also in these single coil chambers, the structure of clots formed under flow was modified by coil design. Coils constructed by braiding flat wires that have rectangular cross-sections (braided coils) were coated with dense regions of fibrin fibers and few spread, adhered platelets at 20 minutes. Coils constructed with coiled smooth wire (coiled coils) presented surfaces with many spread platelets and fewer fibrin fibers in the same time frame (Figure 2).

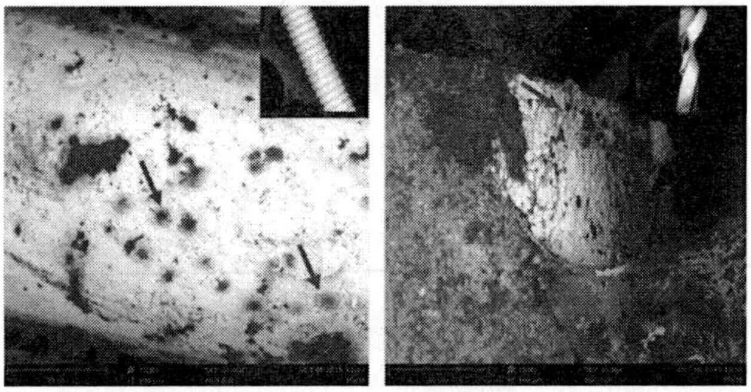

Figure 2: SEM images of clotted coils after 20 minutes of flow in single coil chambers. Platelets spread and adhered (arrows) but few fibrin fibers formed on coiled coils (left). In contrast, dense fibrin fiber networks (arrows) containing few platelets formed on braided coils (right). Inset shows coil geometry, scale bar = 30 μm.

In the aneurysm sac chamber, coil design and arrangement determined flow pattern within the aneurysm dome in a way that affected the clot formation rate and structure. Coiled coils placed in a "circumferential" arrangement generated flow tracer streamlines that are curved and divergent (Figure 3). Flow tracers moved more slowly and more closely followed the edges of the coils where coiled coils were placed side-by-side in a "linear" arrangement. Live imaging of clot formation suggests that these hemodynamics dictate the location of clot progression, and that the rate of clot formation is significantly slower in a sac filled with braided coils than coiled coils. Live imaging of the back of the aneurysm dome showed that a sac filled with coiled coils requires a significantly longer time to become well-mixed than a sac filled with braided coils (Figure 4). This indicates that the effective permeability of the aneurysm packed with coiled coils is less than that of an aneurysm packed with braided coils.

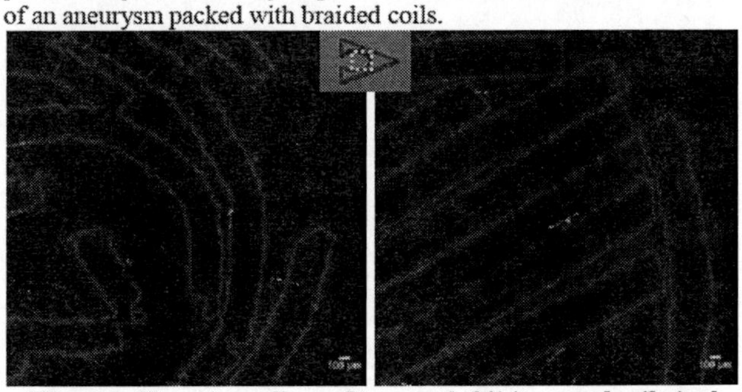

Figure 3: Flow streamlines of tracers (white) around coils (red outline) constructed from 2-D projections of 1 min time stacks. Inset shows field of view location in the sac, scale bar = 100 μm.

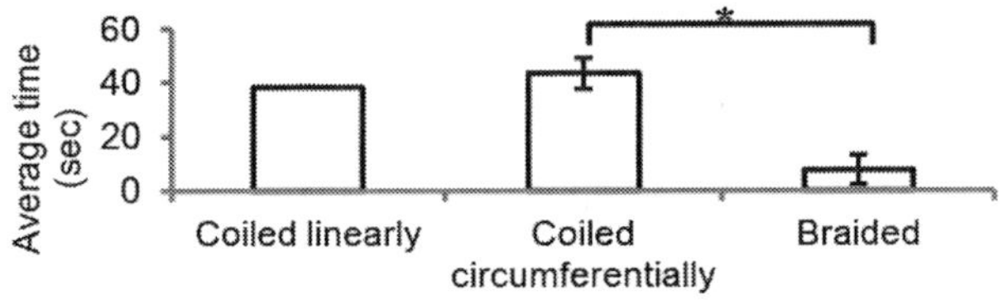

Figure 4: Average time until a well-mixed field of view was observed at back of the dome. (* indicates p<0.05)

DISCUSSION

The difference in clot morphology between those formed on coils without flow and with flow indicate that arterial shear is necessary to replicate IA thrombosis. Clot formation in both single coil and aneurysm sac chambers suggests that braided coils promote more fibrous clotting at the IA neck, but do not create a low enough effective IA sac permeability to promote space-filling clot within the IA dome. Flow streamline and clot progression data indicate that the hemodynamics dictated by coil arrangement subsequently facilitate the pattern of clot progression.

These data show how IA coil design and placement dictate flow patterns and, in turn, clot rate and structure in an *in vitro* model. Such characterization cannot be directly measured *in vivo*, demonstrating the importance of *in vitro* models to a full understanding of IA healing. Initial clot formation provides the scaffold for cellularization and tissue formation in IAs in patients; therefore, if clot formation is improved via coil design and arrangement, then tissue formation and subsequent healing may also be improved.

ACKNOWLEDGEMENTS

The authors would like to thank the Cardiovascular Research Center Training Grant (NHLBI: T32 HL007284), AHA Predoctoral Grant (14PRE19580020), and a UVa-Coulter Foundation ROPe grant.

REFERENCES

[1] Brain Aneurysm Statistics and Facts – Brain Aneurysm Foundation. http://www.bafound.org/.

[2] Rasband, W.S., ImageJ, US National Institute of Health, Bethesda, MD, USA, http://rsb.info.nih.gov/ij/, 1997-2016.

EVALUATION OF A NEAR-WALL RESIDENCE TIME MODEL FOR THROMBOGENIC POTENTIAL

Kirk B. Hansen and Shawn C. Shadden

Department of Mechanical Engineering
University of California, Berkeley
Berkeley, CA, US

INTRODUCTION

Thrombus formation is a complex process involving the transport and reaction of dozens of biochemical species, the activation, adhesion, and aggregation of platelets, interactions between the flow domain and the vessel wall, and fibrin network formation [1]. Previous researchers have incorporated many of these factors into comprehensive computational models of the thrombosis process [2]. Due to the complexity of these models, however, their application has typically been limited to either simplified or reduced geometries. They also typically require the knowledge of a large number of parameters, many of which are virtually impossible to determine experimentally. As such, while these models have provided valuable insight into the underlying mechanisms of thrombus formation, they have been of little use in the determination of localized thrombogenic potential on a patient-specific basis. Simplified models that enable the estimation of this thrombogenic potential could have valuable clinical relevance.

In many diseased conditions, blood stasis is thought to play a key role in thrombus formation. As such, residence time measures have been used as surrogates for thrombogenic potential in coronary artery aneurysms [3], ventricular assist devices [4], and the left ventricle [5]. However, the thrombotic process typically initiates and progresses at a thrombogenic surface, and thus *near-wall* residence time may play a more important role than these previously utilized *global* residence time measures. Near-wall measures may also be conducive to a previously proposed reduced-dimensional model, significantly reducing the computational cost of resolving near-wall transport [6]. Previous researchers have proposed surrogates for near-wall residence time such as relative residence time [7] and wall shear stress exposure time [8], but these measures do not provide actual time scales that can be compared to those of e.g. the coagulation cascade.

In this work, we propose a new model for stasis-driven thrombogenic potential. This model (hereafter called the near-wall residence time model) approximates residence time by tracking the concentration of a single species released from the vessel wall. The time scales of this process are evaluated in an idealized abdominal aortic aneurysm and compared to those obtained with a more standard global residence time calculation.

METHODS

Near-wall Residence Time Model: The near-wall residence time model is based on the conclusion that the initial stages of coagulation can be approximated as the concentration of a single species released from a diseased vessel wall at a constant flux [9]. Herein, the advection-diffusion equation was solved for the concentration c of this species:

$$\frac{\partial c(\mathbf{x},t)}{\partial t} + \mathbf{u}(\mathbf{x},t)\cdot\nabla c(\mathbf{x},t) = D\Delta c(\mathbf{x},t), \tag{1}$$

where $\mathbf{u}$ is a precomputed velocity field and D is diffusivity. Boundary conditions were set to $c = 0$ on the inlet and outlet surfaces, $D\nabla c\cdot\mathbf{n} = q$ at the reactive surface (where q is the arbitrary flux into the domain), and $\nabla c\cdot\mathbf{n} = 0$ on all other surfaces, and the initial condition was set to $c(\mathbf{x},0) = 0$.

The concentration field solution of Eq. (1) is arbitrary, and cannot be directly linked to any thrombotic processes. In order to use this model to determine thrombogenic potential, it is useful to establish a near-wall residence time that approximates how far along a wall-bound reaction can progress before the products are flushed downstream. This measure can then be compared to the time scales of thrombotic processes such as the initiation phase of the coagulation cascade. To do so, we consider the one-dimensional problem corresponding to the advection-free case of diffusion in the wall-normal direction, which is a reasonable approximation in the thin near-wall concentration boundary layers where diffusion is the dominant transport mechanism:

$$\frac{\partial c(x,t)}{\partial t} = D\frac{\partial^2 c(x,t)}{\partial x^2}, \tag{2a}$$

$$D\frac{\partial c}{\partial x}(0,t) = -q, \tag{2b}$$

$$\frac{\partial c}{\partial x}(\infty,t) = 0, \tag{2c}$$

$$c(x,0) = 0. \tag{2d}$$

This problem has an analytical solution for $c(x,t)$ that can be used to compute the concentration at the wall ($x = 0$) as a function of time. By inverting this relationship, we obtain a near-wall residence time measure

$$\tau_{\mathrm{NW}} = \frac{\pi c_{\mathrm{wall}}^2 D}{4q^2} \tag{3}$$

Technical Presentation #30

Thus, by substituting into Eq. (3) the solution at the wall from Eq. (1), the near-wall residence time can be computed on the reactive surface and compared to the time scales of thrombotic processes and the global residence time model.

Global Residence Time Model: Global residence time τ_G was computed using established methods [3, 4, 5]:

$$\frac{\partial \tau_G(\mathbf{x},t)}{\partial t} + \mathbf{u}(\mathbf{x},t) \cdot \nabla \tau_G(\mathbf{x},t) = D \Delta \tau_G(\mathbf{x},t) + 1. \quad (4)$$

Boundary conditions were set to $\tau_G = 0$ on the inlet and outlet surfaces and $\nabla \tau_G \cdot \mathbf{n} = 0$ on all other surfaces, and the initial condition was set to $\tau_G(\mathbf{x}, 0) = 0$.

RESULTS AND DISCUSSION

The idealized abdominal aortic aneurysm from [6, 10] was used as a test case (Fig. 1). The unsteady incompressible Navier-Stokes equations were solved using a stabilized finite element method to obtain an unsteady velocity field based on the inlet flow rate given in [11]. The near-wall and global residence time models were then solved until they reached periodic convergence, also using a stabilized finite element method. For the near-wall model, the reactive surface was chosen to be the surface of the aneurysm ($z = 7\,\text{cm}$ to $z = 15\,\text{cm}$). In both models, diffusivity was set to $D = 5 \cdot 10^{-6}\,\text{cm}^2/\text{s}$ to represent a typical value for the shear-enhanced diffusivity of large proteins [12].

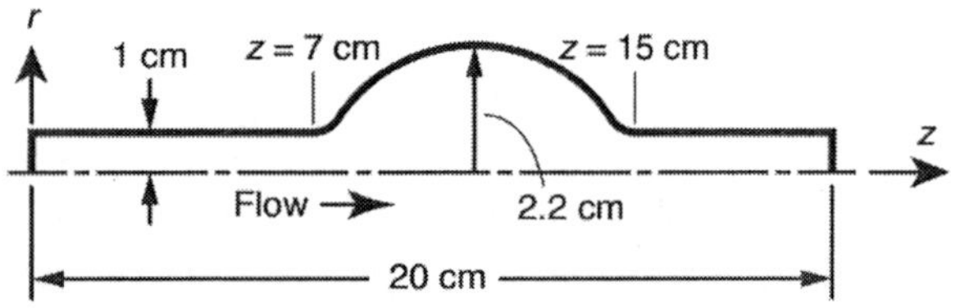

Figure 1: Idealized abdominal aortic aneurysm geometry. Figure taken from [6].

Results from the global and near-wall residence time models show similar trends, but some quantitative differences (Fig. 2). Both models show peaks at approximately $z = 8.5\,\text{cm}$ and $z = 13.5\,\text{cm}$, which is consistent with previous works investigating near-wall transport in this geometry [6]. The near-wall residence time results show sharper peaks and valleys, however. This can be explained by examining the difference between the concentration field from the near-wall results and the residence time field from the global results (Fig. 3). The concentration field shows thin near-wall boundary layer behavior as is typical of wall-bound reactions in high-Péclet-number cardiovascular flows [6, 8]. The global residence time field shows high values throughout the domain and is therefore likely subject to greater levels of mixing.

Further investigation is necessary to conclusively determine which residence time model is a more appropriate surrogate for stasis-driven thrombogenic potential. Specifically, these authors plan to perform numerical experiments to compare the results from the global and near-wall residence time models to those obtained by solving for the complete coagulation cascade, in both the idealized geometry presented herein and in more complex, patient-specific geometries. Because the near-wall model more closely represents the underlying physics of a wall-bound reaction, however, we expect it to better approximate thrombogenic potential in these types

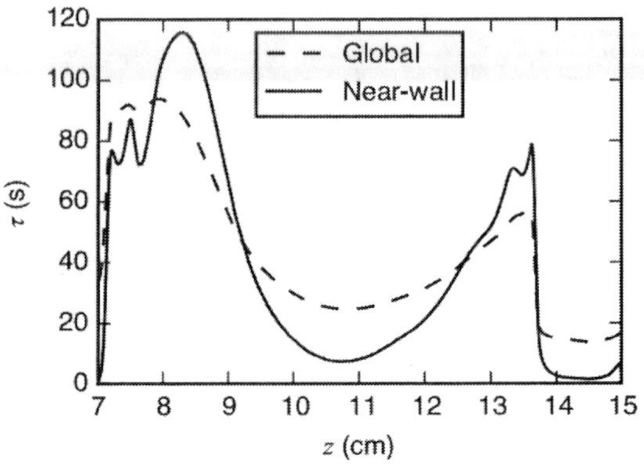

Figure 2: Results of global and near-wall residence time computations on the aneurysm surface.

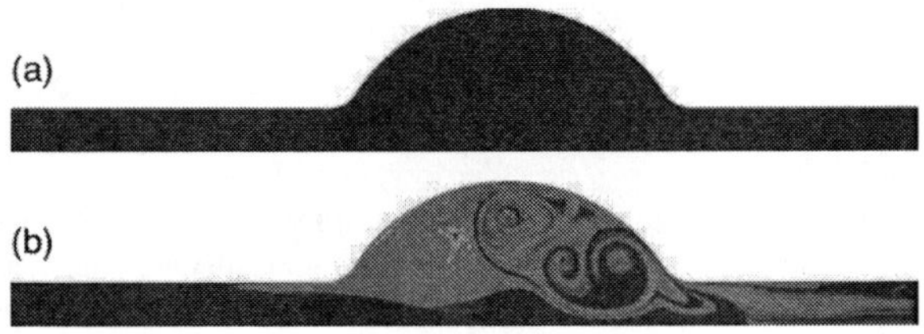

Figure 3: Values of (a) concentration from near-wall computation and (b) residence time from global computation, normalized to their respective peak values. Concentrations from the near-wall computation are confined to very thin near-wall boundary layers that are barely visible.

of problems.

ACKNOWLEDGMENTS

This work was supported by the NSF GRFP and NSF Career Award #1150947.

REFERENCES

[1] Tomaiuolo, M et al., *Intervent Cardiol Clin*, 6(1):1–12, 2016.
[2] Cito, S et al., *Thromb Res*, 131(2):116–124, 2013.
[3] Sengupta, D et al., *Biomech Model Mechan*, 13(6):1261–1276, 2014.
[4] Long, CC et al., *Comput Mech*, 54(4):911–919, 2014.
[5] Rossini, L et al., *J Biomech*, 41(11):2152–2161, 2016.
[6] Hansen, KB and SC Shadden, *Biomech Model Mechan*, 15(3):713–722, 2016.
[7] Lee, SW et al., *J Biomech Eng*, 131(6):061013-1–061013-7, 2009.
[8] Arzani, A et al., *Biomech Model Mechan*, doi:10.1007/s10237-016-0853-7, 2016.
[9] Papadopoulos, K et al., *Med Eng Phys*, 36(2):196–204, 2014.
[10] Biasetti et al., *Front Phys*, 3:2661-266-16, 2012.
[11] Biasetti et al., *Ann Biomed Eng*, 38(2):380–390, 2009.
[12] Sorensen et al., *Ann Biomed Eng*, 27(4):436–448, 1999.

EFFICIENT CAPTURE OF CIRCULATING TUMOR CELLS IN A MICROFLUIDIC DEVICE

Yaling Liu[1,2,*], Shunqiang Wang,[1] Wentao Shi[2]

(1) Department of *Mechanical Engineering and Mechanics, Lehigh University, Bethlehem, PA 18015*

(2) *Bioengineering Program, Lehigh University, Bethlehem, PA 18015, USA.*

INTRODUCTION

Isolation and quantification of circulating tumor cells (CTCs) in patient blood is crucial for both cancer early diagnosis and prognosis. Various techniques have been developed in the past decade while microfluidic chips attract significant attention due to its better controlled microenvironment. This unique characteristic in combination with advanced nanotechnologies provides a possibility to further improve the CTC isolation with better performance. In this article, by extending the wavy-herringbone concept developed earlier in our group, we developed a hierarchical micro/nanostructured microfluidic chip by introducing a uniform coating of nanoparticles coated with anti-EpCAM. This hierarchical structured platform not only keeps the high capture efficiency and purity brought by the wavy-herringbone structure, but further improves the capture efficiency due to nanoparticles with increased surface area. Our results demonstrated an almost 100% capture efficiency at low shear rate (i.e., 60/s) and an up to 70% purity. Especially at higher shear rate, i.e., 400/s, compared to pure wavy-herringbone structures, the hierarchical micro/nanostructures demonstrated an enhancement of up to ~3-fold and ~1.5-fold for capture efficiency and purity, respectively. With these promising results, this hierarchical structured platform can further advance CTC isolation and cancer care.

METHODS

The hierarchical micro/nanostructured microfluidic chip was fabricated in two steps. First, a PMDS slab with micron wavy-HB structures was made by using the previously developed reflow process [1]. Briefly, an SU8 master with traditional grooved-HB was first fabricated, followed by a replication of PDMS molds. A second SU8 master was then replicated by pressing the PDMS mold on an uncured photoresist layer. After removing the PDMS mold, the sharp grooved-HB structures were melted into wavy-HB structures by increasing the temperature. The fabricated SU8 master was then silane treated and used for PDMS replications. Another PDMS cover layer was also made through the traditional photolithography method, followed by hole punching for the inlet and outlet.

To deposit NPs on the wavy-HB structures, a layer-by-layer deposition approach was developed which utilized the electrostatic interaction between oppositely charged polymers [2]. First, negative charges were introduced to the surface of the above-mentioned PDMS slab with wavy-HB structures by a plasma treatment. The PDMS slab was then immersed in a 0.01 mol/L Polyarylamine hydrochloride (PAH) solution for 5 mins to form a layer of positively charged polymers followed by a rinse with DI water. Negatively charged silica NPs with a diameter of 100 nm were deposited by immersing the PDMS slab in 1% NP solution in ethanol for 5 s. To control the deposition consistency for each deposition, an automated manipulator was used to fix the pulling velocity at 5 mm/min [3]. The deposited PDMS slab was dried overnight under an enclosed environment of ethanol to reduce the evaporation induced convection effect, which could lead to a non-uniform particle coating [4,5]. The micro/nanostructures were then characterized by SEM scanning. Lastly, the hierarchical micro/nanostructured PDMS slab was assembled with the PDMS cover layer to make a microfluidic chip.

RESULTS

Fig. 1 illustrated the fabrication process and the NP deposited on the wavy surface.

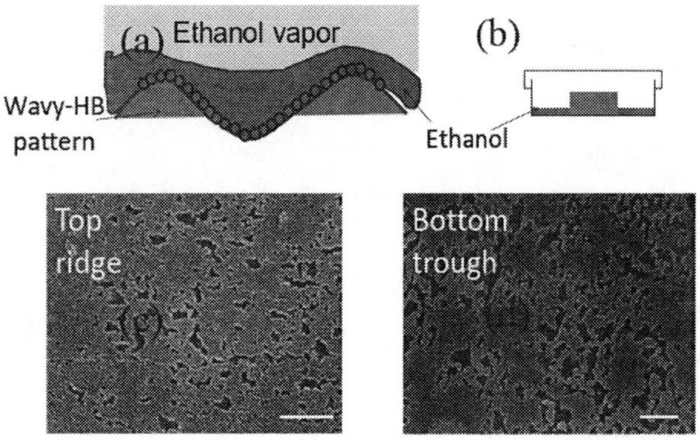

Fig. 1 (a) Illustrative image of the NP deposition under the (b) improved evaporation setting. SEM images of deposited NPs on (c) top ridge and (d) bottom trough after the deposition process.

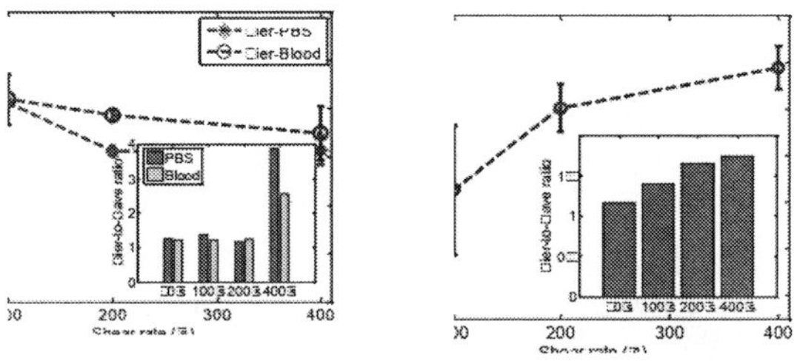

Fig. 2 CTC performance of hierarchical wavy-HB chip and its comparison with the pure wavy-HB chip. (a) Capture efficiency vs. shear rate in the hierarchical wavy-HB chip in both PBS buffer solution and blood. (b) Purity vs. shear rate in the hierarchical wavy-HB chip

Capture efficiency of cancer cells in the hierarchical wavy-HB chip was compared to that in the pure wavy-HB chip, as shown in Fig. 2(a). The overall trend of capture efficiency over shear rate in the hierarchical wavy-HB chip agrees well with that for the pure wavy-HB chip for both PBS and whole blood cases, which is mainly due to the increased shear detachment force under the increased shear rate. However, it was noticed that the hierarchical wavy-HB chip outweighs the pure wavy-HB chip in capture efficiency (see the inset). Especially at higher shear rate, i.e. 400/s, the enhancement reaches up to around 2 times in the whole blood case (69.8% vs. ~27.3%). This enhancement was speculated from the increased total surface area brought by the integration of NPs, which agrees well with the findings in other works in literature [6,7]. Fig. 2(b) shows that the purity is enhanced with increasing shear rate. Interestingly, the inset in Fig.2(b) shows that the hierarchical wavy-HB chip owns a better purity performance compared to the pure wavy chip. The exact mechanism behind this phenomenon is not clear yet, but it is speculated that there might be a geometry mismatch between NPs and WBCs which unexpectedly reduce the WBC non-specific binding. Based on results shown in Fig. 2(a-b), the hierarchical wavy-HB chip can lead to better capture efficiency and purity than the pure wavy-HB chip. This is extremely important in lots of CTC post-analysis applications like DNA analysis of isolated CTCs where both high capture efficiency and high purity are required. In this case, the hierarchical wavy-HB chip can serve as a suitable candidate to be tested at high shear rates like 200/s and 400/s.

DISCUSSION

A microfluidic CTC isolation chip with hierarchical micro/nanostructures were developed. By extending the concept of previously developed wavy-HB chip, the integration of nanoparticles enhanced CTC capture due to the increased overall surface area. A previously developed particle deposition approach was used to fabricate the hierarchical patterns. In comparison to the pure wavy-HB chip and the hallmark grooved-HB chip, the wavy-HB chip with hierarchical micro/nanostructures demonstrated a capture efficiency up to almost 100% at low shear rate (i.e. 60/s) and a high purity performance (up to ~70% and only ~680 WBCs per 1 mL blood). SEM images were also used to further characterize the interaction between captured cancer cells and micro/nanostructures.

ACKNOWLEDGEMENTS

This work was supported in part by National Science Foundation (NSF) grant CBET-1264808, DMS-1516236 and National Institute of Health (NIH) grant EB015105 (to Yaling Liu).

REFERENCES

1. Wang, S.T., Liu, K., Liu, J.A., Yu, Z.T.F., Xu, X.W., Zhao, L.B., Lee, T., Lee, E.K., Reiss, J., Lee, Y.K., Chung, L.W.K., Huang, J.T., Rettig, M., Seligson, D., Duraiswamy, K.N., Shen, C.K.F. & Tseng, H.R. Highly Efficient Capture of Circulating Tumor Cells by Using Nanostructured Silicon Substrates with Integrated Chaotic Micromixers. *Angew Chem Int Edit* **50**, 3084-3088 (2011).
2. Decher, G. Fuzzy nanoassemblies: Toward layered polymeric multicomposites. *Science* **277**, 1232-1237 (1997).
3. Lin, P.C. & Yang, S. Mechanically switchable wetting on wrinkled elastomers with dual-scale roughness. *Soft Matter* **5**, 1011-1018 (2009).
4. Kumnorkaew, P. & Gilchrist, J.F. Effect of Nanoparticle Concentration on the Convective Deposition of Binary Suspensions. *Langmuir* **25**, 6070-6075 (2009).
5. Fischer, B.J. Particle convection in an evaporating colloidal droplet. *Langmuir* **18**, 60-67 (2002).
6. Chen, L., Liu, X.L., Su, B., Li, J., Jiang, L., Han, D. & Wang, S.T. Aptamer-Mediated Efficient Capture and Release of T Lymphocytes on Nanostructured Surfaces. *Adv Mater* **23**, 4376-+ (2011).
7. Chen, W.Q., Weng, S.N., Zhang, F., Allen, S., Li, X., Bao, L.W., Lam, R.H.W., Macoska, J.A., Merajver, S.D. & Fu, J.P. Nanoroughened Surfaces for Efficient Capture of Circulating Tumor Cells without Using Capture Antibodies. *Acs Nano* **7**, 566-575 (2013).

MICROFLUIDIC SORTING OF CELL VIABILITY BASED ON STIFFNESS FOR APPLICATIONS IN REGENERATIVE MEDICINE

Muhymin Islam1, Hannah Brink2, Syndey Blanche2, Caleb DiPrete2, Tom Bongiorno1, Nicholas Stone1, Anna Liu2, Anisha Philip3, Gonghao Wang1, Wilbur Lam2,3, Alexander Alexeev1, Edmund K. Waller3, and Todd Sulchek1,2*

(1) George W. Woodruff School of Mechanical Engineering, Georgia Institute of Technology, 801 Ferst Drive, Atlanta, GA, 30332-0405, USA

(2) Wallace H. Coulter Department of Biomedical Engineering, Georgia Institute of Technology, 313 Ferst Drive, Atlanta, GA, 30332-0535, USA

(3) Winship Cancer Institute, Emory School of Medicine, 1365 Clifton NE Rd Atlanta, GA 30322

INTRODUCTION

The enrichment of viable cells is an essential step to obtain effective products for cell therapy. The presence of nonviable cells in cell suspension is a ubiquitous problem therapies and assays. In some sensitive embryonic stem cell cultures, the presence of necrotic cells release factors that negatively impact the health of the culture. Also the presence of nonviable cells can impact cell therapies, for example delaying of engraftment time of hematopoietic stem cell transplantation.

While procedures exist to characterize the viability of cells, most methods to exclude nonviable cells require the use of density gradient centrifugation, other methods that modify the cells. For example flow cytometry with promidium iodide staining and monoclonal antibody-magnetic bead pull down assay can be used but require time consuming processing. Consequently, label-free techniques that are high throughput and accurate are needed as an alternative.

We report a label-free microfluidic technique to separate live and dead cells that exploits differences in cellular stiffness. The device uses a channel with repeated, diagonal ridges. Stiff nonviable cells directed through the channel are translated orthogonally to the channel length, while soft live cells follow hydrodynamic flow and squeeze under the ridges. Jurkat cells are enriched to high purity of viable cells by a factor of 185-fold. To highlight the utility for hematopoietic stem cell transplantation, frozen samples of cord blood were thawed and the purity of viable nucleated cells was increased from 65% to over 94% with a recovery of 73% of the viable cells.

METHODS

The microfluidic device was fabricated using replica molding of polydimethylsiloxane (PDMS) on a SU-8 patterned silicon wafer1,2. The ridge gap was chosen to be small enough to compress the cells without clogging the device. Three and five outlet devices were designed to understand how the ability to fractionate the output can improve cell purity. The mold pattern was transferred to PDMS. An optical micrograph of the device is shown in Fig. 1.

Jurkat cells were cultured and maintained in medium and FBS. The nonviable cell samples were prepared by heating to 60 C in a water bath for 2 hours. Discarded samples of cryopreserved cord blood were collected from Emory University Hospital under an IRB approved study. The cord blood samples had been previously frozen in media containing 10% v/v DMSO and stored in liquid nitrogen. The blood was thawed using a warm water bath for 10 minutes.

RESULTS

The balance of these forces results in stiff cells being directed to upper outlets and softer cells to lower outlets as illustrated in Fig. 1(a).

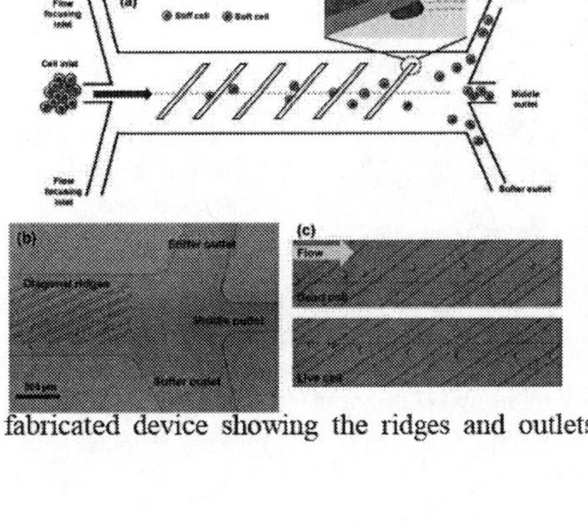

Fig. 1: Ridge based microfluidic device for viable cell sorting. (a) Schematic of the device showing cells will squeeze through ridges and flow towards different outlets based on their stiffness; (b) optical micrograph of the fabricated device showing the ridges and outlets; (c) representative

micrograph of overlaid video of dead and live cells moving towards opposite directions in the device.

Live and dead cells were separately flowed through the device at various flow rates to observe their trajectories. Video microscopy showed that the live cells moved perpendicular to the ridges consistent with softness and resulted in a net negative transverse displacement with respect to the direction of fluid flow of -6.01 ± 2.76 μm/ridge (Fig. 2(b)). Dead cells moved in the direction parallel to the ridges consistent with the higher stiffness, resulting in a net positive transverse displacement with respect to the direction of fluid flow of 8.84 ± 3.14 μm/ridge (Fig.2(b)).

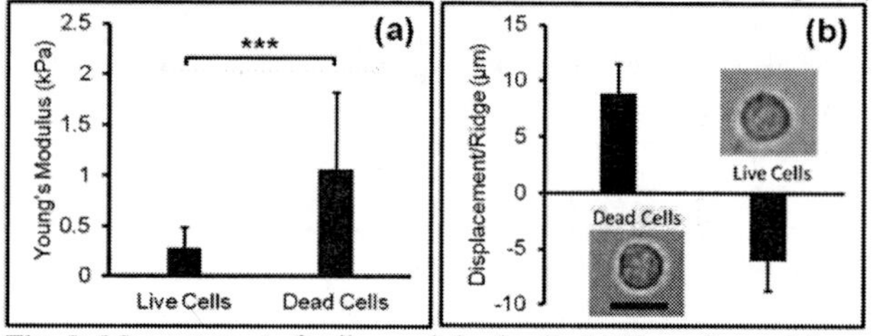

Fig. 2: Measurements for live and dead cells (a) Young's modulus of live and dead cells obtained from AFM measurement; (b) average displacement of dead and live cells per ridge in the microfluidic device (N = 50); optical micrographs of live and dead cells are shown.

To study the sorting of live and dead cells, we generated mixtures of live and dead cells and evaluated the outlets for live and dead cells using flow cytometry. The optical micrograph of the device is shown in Fig. 3(a) and the simulated streamline in the device is shown in Fig. 3(b). This device was designed to have two soft (soft 1 and soft 2), one middle, and two stiff (stiff 1 and stiff 2) outlets. The experiment was performed with a starting ratio of 1:1 for live and dead. The flow cytometry results obtained from different outlets are shown in Fig. 3(c). The enrichment for live cells in soft 1 outlet was 185-fold leading to purity of over 99%. However, this enrichment was associated with recovery of only 28.8% of live cells in a single outlet. Combining the cells in both soft outlets resulted in recovery of 84.0% live cells with 94.9% purity.

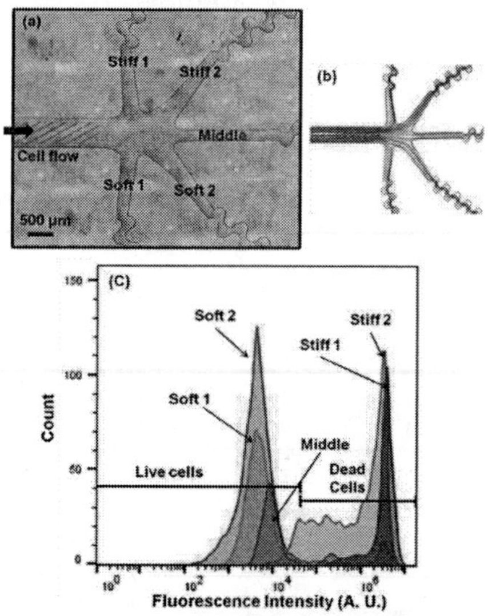

Fig.3: (a) Five outlets device; (b) streamline of fluid flow; (c) flow cytometry results obtained from different outlets of the device.

To validate the utility of viability sorting in a therapeutic context, microfluidic sorting was used to improve the purity of viable cells from samples of thawed cord blood that were rich in hematopoietic stem cells. The Young's modulus of nucleated cells from thawed cord blood samples and heat-treated, nonviable samples were measured separately using AFM. Fig. 4(a) shows the stiffness histogram of the samples. We note that the heat-treated cells sample is uniformly stiffer than the thawed sample, and that there is a region of overlap, consistent with the existence of a nonviable subpopulation within the thawed sample. The microfluidic device was then implemented to sort the thawed cord blood sample for viability. The thawed cells sent to the inlet and sorted cells collected from all five outlets were stained with EthD-1 and flow cytometry was performed to calculate the percentage of viable nucleated cells and used to identify nucleated cells and red blood cells (Fig. 4(b)). The percentage of viable nucleated cells in the blood was determined to be 65.2% post-thaw. The combination of both soft outlets recovered 73.3% live cells with 94.8% purity and 9.77-fold enrichment. The recovery of the live cells can be increased to 97.5% with a purity of 76.1% by including cells from middle outlet.

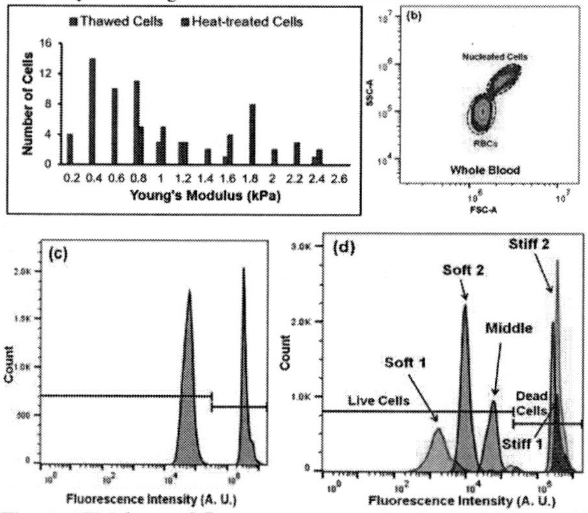

Fig. 4: AFM data and flow cytometry results obtained from cord blood (a) distribution of Young's modulus for nucleated cord blood cells; (b) scatter plot of cord blood to distinguish nucleated cells from red blood cells; (c) staining of live and dead nucleated cells in the cord blood; (d) viability of nucleated cells collected from different outlets after sorting.

DISCUSSION

The unique repeated skew ridge microchannel design is effective to sort live and dead cells based upon differences in cell stiffness with high purity and high collection efficiency. We observed enrichment over 185-fold for live cells with purity of over 99% from the mixture of live and dead cells. Sorting viable cells with high purity and recovery was achieved with no labeling of the cells, which offers significant practical advantages over the existing label-based sorting methods. This simple device can be implemented within the laboratory much like a filter or directly integrated in a microfluidic lab-on-a-chip to purify cell samples to higher viability to improve the quality of downstream analyses. With the emerging importance of cell therapies, the microfluidics approach may be effective to improve the surety of the product and obtain consistent potency and safe cell therapies.

ACKNOWLEDGEMENTS

This research was supported by National Institute of Health (1R21EB020977-01).

REFERENCES

1. Wang, G.; Crawford, K.; Turbyfield, C.; Lam, W.; Alexeev, A.; Sulchek, T. Lab on a Chip 2015, 15, (2), 532-540.
2. Wang, G.; Mao, W.; Byler, R.; Patel, K.; Henegar, C.; Alexeev, A.; Sulchek, T. PLoS One 2013, 8, (10), e75901.

PTEN DELETION IN PANCREATIC CANCER ASSOCIATED FIBROBLASTS DECREASES HYDRAULIC PERMEABILITY IN A 3D MICROFLUIDIC MODEL OF THE TUMOR STROMA

Alex Avendano (1), Jonathan Chang (1), Christina Ennis (2), Amanda Stratton (3), Jason R. Pitarresi (4,5), Michael C. Ostrowski (5), Jonathan W. Song (1,5)

(1) Mechanical and Aerospace Engineering
The Ohio State University
Columbus, Ohio, US

(2) Neuroscience
Kenyon College
Gambier, Ohio, US

(3) Bioengineering
Lehigh University
Bethlehem, Pennsylvania, US

(4) Division of Gastroenterology
University of Pennsylvania
Philadelphia, Pennsylvania, US

(5) The Comprehensive Cancer Center
The Ohio State University
Columbus, Ohio, US

INTRODUCTION

Pancreatic ductal adenocarcinoma (PDAC) is one of the deadliest malignant solid tumors, with an overall 5 year survival of 5% and no established effective therapies. PDAC is unique when compared to other tumors in that the majority of cells composing the tumor are not cancer cells but other cell types such as fibroblasts, immune cells, and endothelial cells that exist in areas surrounding the tumor (known as the tumor stroma). The tumor stroma has emerged as an important mediator of pancreatic cancer growth and metastasis to distant sites of the body **[1]**. In particular, loss of the tumor suppressor gene phosphatase and tensin homolog (PTEN) in pancreatic cancer associated fibroblasts (CAFs) within the tumor stroma has been shown to accelerate tumor progression and therapeutic resistance, resulting in a more aggressive form of PDAC **[2]**. However, knowledge of how loss of PTEN in CAFs function to produce this effect is limited, due in part to difficulties of current in vitro approaches to model the tumor stroma and in vivo approaches to isolate and quantify the individual effects of CAFs.

Using a microfluidic 3D model of the tumor stroma, this study set out to characterize how loss of PTEN in CAFs enable these cells to alter two biophysical properties of the tumor microenvironment: i) collagen fiber alignment and ii) hydraulic permeability or the ability of fluid to flow through a porous medium. Aligned collagen fibers have been shown to provide contact guidance for tumor cells to metastasize while hydraulic permeability is an indicator of how well flow and drugs can penetrate into the tumor and thus provide important parameters to profile the effects of CAFs in PDAC. We hypothesized that loss of PTEN would result in more aligned collagen fibers and decreased hydraulic permeability, characteristic of aggressive PDAC.

METHODS

To evaluate our hypothesis, a microfluidic model of the tumor stroma was designed, fabricated, and optimized to measure collagen fiber alignment and hydraulic permeability as a function of PTEN deletion in CAFs (Fig 1a). The microfluidic model consisted of Polydimethylsiloxane (PDMS) devices fabricated using soft lithography**[3]**. The devices consisted of a single straight channel (5 mm long, 500 um wide, and 1mm tall) with 4mm inlet and outlet ports. Pancreatic CAFs, isolated from human pancreatic tumors, were then suspended in a rat tail type I collagen hydrogel, injected into the channel, and cultured under static (no flow) conditions between 2 days for hydraulic permeability measurements and 4 days for collagen fiber

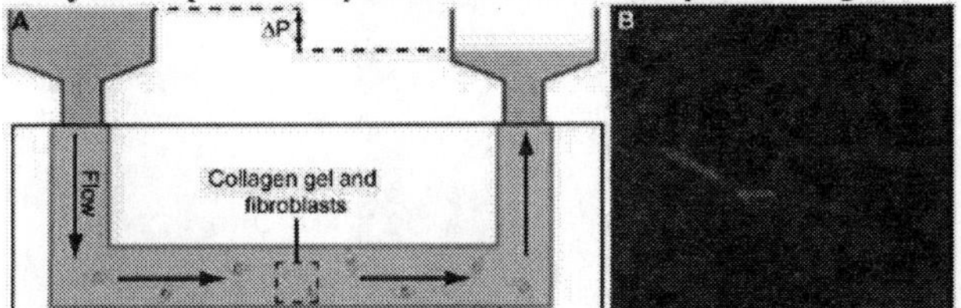

Fig.1 A) Channel design schematic. B) Confocal reflectance image of a CAF (red) suspended in a porous collagen gel (green). C) Bulk BSA-TRITC dye flow tracking inside channel. D) Estimation of velocity from timelapse images, bulk dye x positions are plotted vs time to determine the velocity.

alignment measurements. The conditions tested were control pancreatic CAFs (shnc), pancreatic CAFs with PTEN silenced with shRNA (shPTEN), and acellular collagen Type I hydrogel. Additional studies were also conducted to enzymatically degrade hyaluronan with hyaluronidase (HAdase, 0.5mg/mL) and inhibit and phosphorylated Akt (pAkt) with the pharmacological inhibitor MK2206 (4.5μM) in cell culture medium, respectively. Fibroblasts in the microdevice were maintained for 48 hours prior to measurements by placing ~400 μL of medium at the ports with the medium being refreshed every 24 hours.

Collagen fiber alignment was assessed with confocal reflectance microscopy images of the collagen fibers within the device **(Fig. 1b)**. These images were post processing with a custom Fast Fourier Transform (FFT) Matlab algorithm following previous guidelines **[4]**. The algorithm calculated an alignment index, a ratio between the frequency of the most aligned region of the image and the ideally random case. With this criteria, alignment indices closer to 1 indicate random alignment while higher values indicate more aligned collagen fibers.

The hydraulic permeability was determined by flowing a bovine serum albumin-tetramethylrhodamine (BSA-TRITC) dye through the channel and measuring the average fluid velocity through the channel. The velocity was measured by tracking the bulk of the dye over time inside the channel using time lapse microscopy **(Fig 1c-d)**. Image were acquired every 5 seconds for 20 minutes or until dye flowed completely through the field of view. Darcy's Law for flow through porous medium was then used to calculate the hydraulic permeability **[5]**.

RESULTS

We detected no significant differences in collagen fiber alignment when comparing shnc, shPTEN, and acellular collagen conditions (data not shown). In contrast, the hydraulic permeability decreased significantly for the shPTEN condition when compared to both shnc and acellular collagen conditions **(Fig. 2)**. Furthermore, application of HAdase and MK2206 rescued shPTEN hydraulic permeability to acell/shnc control levels **(Fig. 3)**. Application of HAdase and MK2206 to acell and shnc did not alter the hydraulic permeability (results not shown).

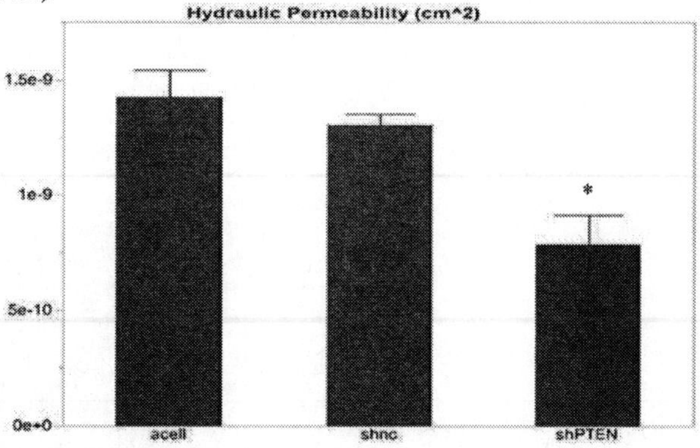

Fig. 2 Hydraulic permeability decreases significantly due to PTEN deletion in CAFs decreases hydraulic permeability. Statistical testing was done using ANOVA with Tukey's HSD post testing. (n = 4-8 across all conditions, α set to 0.05 for statistical significance). Error bars indicate the standard error of the mean. (* P<0.05)

DISCUSSION

The decrease in the hydraulic permeability can explain why loss of PTEN in CAFs result in more aggressive pancreatic tumors. Decreased stromal hydraulic permeability in tumors has been shown using both

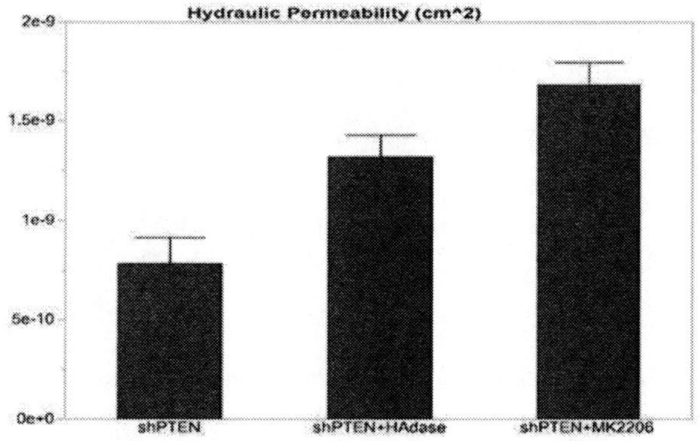

Fig. 3 HA and pAKT inhibition rescues shPTEN hydraulic permeability. Statistical testing was as described in Fig. 3. Error bars indicate the standard error of the mean

computational and experimental approaches to enhance tumor growth and impede delivery of therapeutics **[6,7]**. Moreover, since fiber alignment was not observed to be impacted by expression of PTEN in CAFs, we suspect that this decrease in hydraulic permeability is mediated by the deposition of extra-fibrillar matrix molecules that increase the hydraulic resistance through the semi-porous medium **[8]** (e.g non-fibrillar collagen, hyaluronan, or other glycosaminoglycan (GAG) molecules).

Furthermore, the application of HAdase and the subsequent rescue of hydraulic permeability suggest that increases in hydraulic permeability by CAFs with PTEN deleted is mediated by deposition of hyaluronan. However, future studies should be completed to assess the presence of other molecules that can modulate hydraulic permeability such as other GAG molecules.

pAKT is known to be upregulated in PTEN deleted CAFs **[3]**. Our results demonstrate that inhibition of pAKT with MK2206 also rescued the hydraulic permeability to control levels. Taken together, these results support a hypothesis that a CAF mediated mechanism of enhanced tumor growth and therapeutic resistance occurs by decreasing stromal hydraulic permeability by depositing excess hyaluronan and signaling through pAkt. Future studies will apply our established tumor stroma model as well as *in vivo* models to validate this new hypothesis. Moreover, a more complete understanding of the mechanisms that regulate hydraulic permeability can lead to new therapeutic strategies to contain PDAC and improve the poor patient survival outcome of this disease.

ACKNOWLEDGEMENTS

Funding to JWS from: American Cancer Society (IRG6700350), American Heart Association (15SDG25480000), and Ohio State University Institute for Materials Research. CE/JWS acknowledge support from Pelotonia.

REFERENCES

[1] Hidalgo, M and Von Hoff, D, *Clin Cancer Res*, 18:4249-4256, 2012.
[2] Pitarresi et al., *submitted*, 2017.
[3] Qin, D, and Xia, Y, *Nat Protocol*, 5:491-502, 2010.
[4] Ng, C.P and Swartz, M, *Ann Biomed Eng*, 34(3):446-454, 2006.
[5] Wiig, H, and Swartz, M, *Physiol Rev*, 92:1005-1060, 2012.
[6] Wu, M et al., *J. Theor. Biol.*, 335:194-207, 2014.
[7] Netti, P.A et al., *Cancer Res*, 34:2497-2503, 2000.
[8] Levick, J.R., *Q. J. Exp. Physiol*, 34:859-871, 2001.

DEVELOPMENT OF VASCULARIZED 3D MICROFLUIDIC TUMOR PLATFORMS TO STUDY PARTICLE TRANSPORT

Manasa Gadde (1), Rhys Michna (2), Marissa N Rylander (3)

(1) Biomedical Engineering
University of Texas at Austin
Austin, TX, USA

(2) Mechanical Engineering
University of Texas at Austin
Austin, TX, USA

(3) Mechanical Engineering
University of Texas at Austin
Austin, TX, USA

INTRODUCTION

Physiologically representative tumor platforms have not been applied to elucidating transport kinetics of nanoparticles in response to variations in their properties and perturbations in the microenvironment. Although 2D cell monolayers provide the basis for initial testing of nanoparticle efficacy, these systems cannot recapitulate critical features of the tumor microenvironment including hemodynamics, matrix mechanics, and elevated interstitial pressure thereby inhibiting prediction of accurate cellular response to therapeutics. While *in vivo* models provide physiologic fidelity, they can be highly variable - confounding determination of the biologic response to nanoparticles and the large number of animals needed to characterize toxicity and biodistribution can be cost prohibitive. Furthermore, isolating the influence of specific nanoparticle properties and features of the tumor microenvironment on nanoparticle transport are impossible due to the complexity of animal models thereby limiting optimization of the nanotherapeutic to overcome transport barriers.

3D *in vitro* platforms with physiological fidelity capable of characterizing the transport and therapeutic efficacy of nanoparticles would provide critical insight into strategies for therapy optimization and reduce the expense associated with nanoparticle clinical translation. There have been significant advancements in creation of microfluidic tumor models and several studies have been published related to uncovering the role of flow on the tumor microenvironment. However, these systems have not been exploited for studying nanoparticle transport in a physiological tumor microenvironment.

We present platform designs and methods used to develop *in vitro* platforms that replicate the vasculature of the tumor microenvironment. Features of these platforms include a continuous, aligned endothelium that allows for cell-cell interactions between vasculature and tumor cells. A novel platform for fabrication of a single endothelialized microchannel encased within a collagen platform hosting breast cancer cells was developed and utilized to study the influence of cellular interaction on transport phenomenon through vasculature in a hyperpermeable tumor microenvironment. This platform relied on subtractive tissue engineering fabrication techniques and we have confirmed the platform's ability to recapitulate physiological vasculature in a tumor microenvironment as well as demonstrated the influence of tumor endothelial interactions on transport of particles of varying size. Additionally, a second platform capable of combining lithographic techniques with additive tissue engineering methods was used to create more intricate endothelialized microfluidic networks that capture the more complex geometries of tumor microvasculature representative of patient tumors. By modeling microvascular networks after *in vivo* tumors we are able to create patient specific *in vitro* platforms that can be used to develop personalized patient treatments

METHODS

Collagen was extracted using a protocol developed by Buchanan et. al. [1]. Briefly, collagen was extracted from rat tail tendons and dissolved in 0.01 M HCl overnight, centrifuged, and lyophilized. To create the perfusable tumor platforms, 14 mg/ml collagen stock solution was mixed with 10x DMEM, 1N NaOH, and DMEM containing human breast adenocarcinoma cells, MDA-MB-231 (ATCC) to bring the final solution to 7 mg/ml collagen. The mixture was then injected into PDMS mold consisting of single or multichannels with a 22G needle inserted to create a vessel lumen in the center. Telomerase immortalized human microvascular endothelial cells (TIME) cells were then added to the inner vessel lumen and a 72 hr graded flow rate protocol previously developed [1] was utilized in order to maintain endothelial integrity and establish a confluent monolayer (Figure 1C). To create the vascularized network platforms, a lithographically produced PDMS stamp mimicking vasculature data from patient tumors was used (Figure 3). Live dead staining was performed using calcein-AM and propidium iodide to determine the viability of the cells under long term culture and F-actin staining was used to observe cellular morphology. Vessel permeability was quantified by flowing the channel with 10μg/ml fluorescently labeled 70 kDa dextran at a rate of 26 μl/min to investigate the barrier function of the endothelium. The vascularized microfluidic platforms were perfused with 70 kDa dextran for 2 hours with images being taken every five minutes for evaluation of transport and diffusion. The last five consecutive data points from the two hours of flow were used to calculate P_d using the following formula:

$$P_d = \frac{1}{I_1 - I_b}\left(\frac{I_2 - I_1}{\Delta t}\right)\frac{d}{4} \qquad (1)$$

where I_b is the background intensity, I_1 is the average initial intensity, I_2 is the average intensity after recovery time interval Δt, and d is the diameter of the microchannel. To determine the ability of the 3D vascularized multichannel microfluidic system to be used to measure nanoparticle transport, 100nm blue polystyrene nanoparticles were introduced into the system at a rate of 26 μl/min.

RESULTS

Calcein-AM and propidium iodide tests revealed that the PDMS housing does not affect the viability of the MDA-MB-231 cells as seen in Figure 2a. The dominant green fluorescence reveals that the majority of the cells are alive with a few dead cells (red signal). F-actin staining revealed a confluent endothelium with TIME cells elongated and oriented in the direction of flow (Figure 1 b-d). Also, the F-actin staining revealed that the endothelium is much more confluent and tighter when the tumor cells are absent (Figure 1b). In the presence of the tumor cells, the endothelium had holes (Figure 1d).

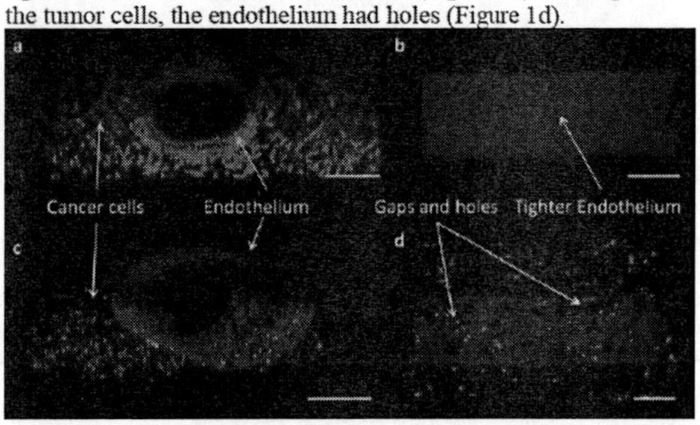

Figure 1: Viability and F-actin staining of the vascularized microfluidic tumor platform. Scale bars are 500 μm. (a) Front view of viability staining. (b) Top view of f-actin stained vascularized microfluidic tumor platform containing only TIME cells. (c) Isometric 3D view of endothelium surrounded by cancer cells. (d) Top view of f-actin stained co-culture vascularized microfluidic tumor platform with MDA-MB-231 (green) and TIME (red) cells.

Vessel permeability data reinforced the conclusions from the F-actin staining that co-culture of tumor and endothelial cells results in a leaky vasculature. Effective permeability data for vascularized microfluidic tumor platforms under three conditions: acellular, TIME endothelium only and co-culture of TIME and MDA-MB-231 cells is shown in Figure 2. The acellular platforms without the presence of an endothelium exhibited the highest diffusion coefficient average, 26±3 nm/sec, followed by the co-culture platform with a diffusion coefficient 25±1.7 nm/sec, and finally the mono-culture of TIME endothelium with a diffusion coefficient 16±1.9 nm/sec.

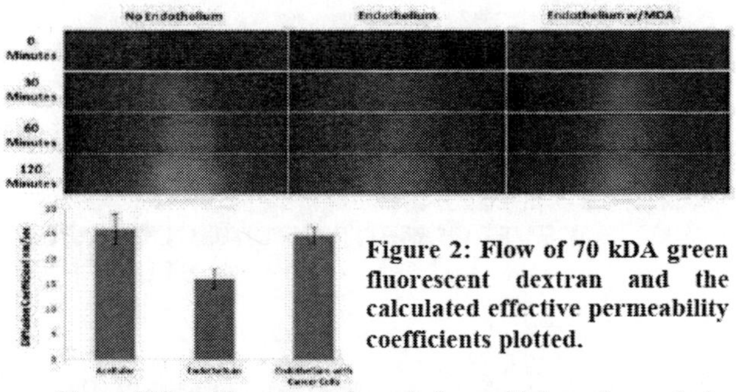

Figure 2: Flow of 70 kDA green fluorescent dextran and the calculated effective permeability coefficients plotted.

Flow of blue microsphere through the multichannel vascularized co-culture platform revealed the capability of the platform to support dynamic monitoring of particle flow as shown in Figure 3a, b. Additionally, we were able to replicate vasculature from patient tumor data (Figure 3b) using lithographic methods (Figure 3c).

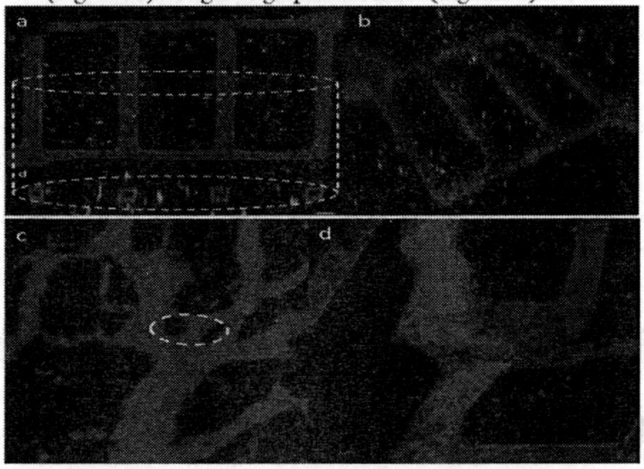

Figure 3: In vitro vascularized tumor platforms. (a,b) Multichannel vascularized co-culture platform showing particle flow (blue) (c) Two-photon image showing human colon carcinoma vasculature in mice at day 0. (d) Microfluidic tumor microenvironment capturing the geometry of an in vivo tumor.

DISCUSSION

In this study, vascularized tumor platforms were created with embodiments of scalable single channel, dual-channel, and networks. The platforms enable sustained transport of macromolecules and the recreation of hemodynamic forces that are integral to tumor growth. We have shown that a confluent endothelium can be formed and that co-culture of MDA MB 231 cancer cells in the platform creates a leakier endothelium as is typical for tumors *in vivo*. The leakier endothelium can affect the transport of particles such as nanoparticles or chemotherapeutics and play a role in their effectiveness. Finally, the geometry of complex *in vivo* tumor vasculature was reproduced *in vitro*. By incorporating patient data into the design of our microvascular network we can create tumor platforms that are physiologically representative of patients. These platforms form a foundation for the future study of transport, chemical gradients, and cell behavior in a physiologically relevant tumor microenvironment. They can be expanded upon to incorporate immune cells, stromal cells, and lymphatic vessels to create a complete tumor microenvironment.

REFERENCES

[1] Buchanan, C et al., *Tissue Eng C*, 20:64-75, 2013.

RELATION BETWEEN ACCURACY AND PERSISTENCE OF CANCER CELL MIGRATION UNDER CHEMICAL GRADIENT

Hye-ran Moon (1), Julien Varennes (2), Andrew J Mugler (2), Bumsoo Han (1, 3)

(1) School of Mechanical Engineering
Purdue University
West Lafayette, IN, USA

(2) Department of Physics and Astronomy
Purdue University
West Lafayette, IN, USA

(3) Purdue Center for Cancer Research
Purdue University
West Lafayette, IN, USA

INTRODUCTION

Cancer cell migration is one of the key steps of cancer metastasis, which is the major cause of cancer-related death. During this step, multiple environmental cues are imposed on cells migrating and often guide their movements. Concentration gradient of growth factors is one of the cues whose significance has been recognized [1]. However, tumor microenvironment is very complex where multiple cues with distinct characteristics co-exist including biophysical cues by extracellular matrix (ECM) and neighboring cells. It is unclear how cancer cells sense and respond to these complex cues. Specifically, it is poorly understood how the "accuracy" and "persistence" of the cellular migratory behaviors are affected by the environmental cues. The "accuracy" and "persistence" has been defined as chemotactic index (CI) and chemotactic ration (CR) [2] as illustrated in **Figure 1.**

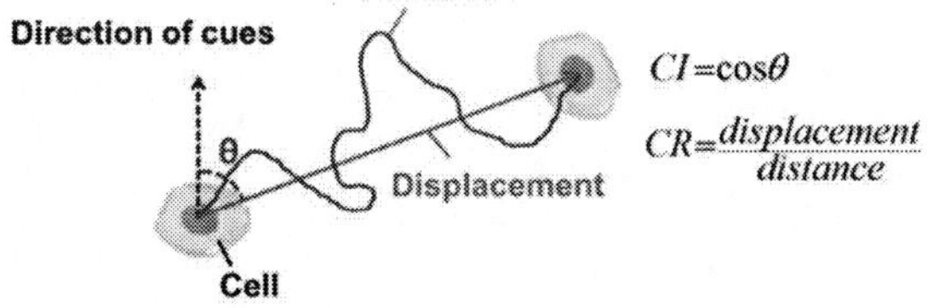

Figure1: Definition of CI and CR

Among the environmental cues, chemotaxis has been considered a major driving force of cell migration. For example, *Wang S-J et al* reported that the breast cancer cell, MDA-MB-231, responded to epidermal growth factor (EGF) gradient and migrated directionally [3]. Although cell migration has been explained as a single-cell motile behavior, invasion of cancer cells occur as collective migration, which is multicellular movement with intercellular communications [4][5]. The previous studies have shown the migration behavior to address the effect of environmental cues, but were mostly performed in 2D which doesn't reflect 3D microenvironment. In spite of these studies, due to lack of understanding of cellular response to complex cues, prediction of cancer cell migratory behavior is significantly limited.

In the present study, we performed an integrated theoretical and experimental study to postulate a hypothesis of inter-relation between "accuracy" and "persistence" of the cancer cell migration. With human breast cancer cells (MDA-MB-231) within 3D collagen matrix under a concentration gradient of EGF, we quantify their migratory behavior and propose a theoretical framework to explain the cellular migratory behavior.

METHODS

Cells and reagents

Human breast cancer cell line (MDA-MB-231) was cultured in Advanced DMEM/F12 basal medium (Invitrogen) supplemented with 5% Fatal bovine serum (FBS), 1% L-glutamine (L-glu), and 1% Pen/Strep. MDA-MB-231 cells were harvested using 0.025% trypsin when confluence reached 80-85%. A starvation medium with 1% FBS was used prior to EGF treatment.

In vitro chemotaxis platform

A microfluidic device was fabricated by soft lithography technique. There were three channels in the device, which were side channels (Up-channel, UC and Down-channel, DC) and a center channel (CC) as shown in **Figure 2A**. Ends of the channels were connected to reservoirs through ports drilled by a biopsy punch.

Harvested MDA-MB-231 and type I collagen solution (2mg/ml) were mixed and the mixture was slowly loaded to the CC. After a 48-hour culture, UC was filled by EGF containing medium (800nM) and the starvation medium was infused to DC. Consequently, EGF gradient was generated through the collagen matrix in CC because of the concentration difference between side channels. The prepared device was placed on an incubation stage. Finally, a time-lapse image was taken by a CCD camera (Q-imaging) with an auto-inverted microscopy (Olympus, IX71).

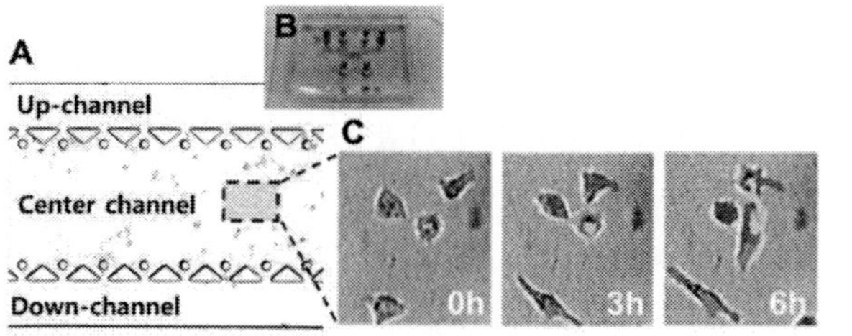

Figure 2: Experimental Setup. (A) Top-view microscopy image of the chip. (B) Photograph of the fabricated chip. (C) Microscopy analysis of cell migration.

Migration analysis

Single cell migration was observed with and without EGF gradient. Cell movement was taken by the microscopy every 15 minutes for more than an 8hour duration. The time-lapse images were further analyzed to quantify cell migration. The results were displayed as cell trajectories originating from the initial point to (0, 0). 41 trajectories were analyzed for both control and EGF gradient group. Also, the average cell speed, CI and CR were statistically analyzed. All p-values were determined by student's t test.

RESULTS AND DISCUSSION

Cell trajectories on **Figure 3** show a response to the EGF gradient in cell speed and directions of cell migration. The lengths of trajectories under the gradient are observed as larger than without the gradient. Accordingly, the cell speed under the gradient can be considered stimulated. In addition to the cell speed, directionality also changed depending on the presence of the gradient. Roughly, the majority of cells under the EGF gradient migrate in a gradient direction unlike the control group.

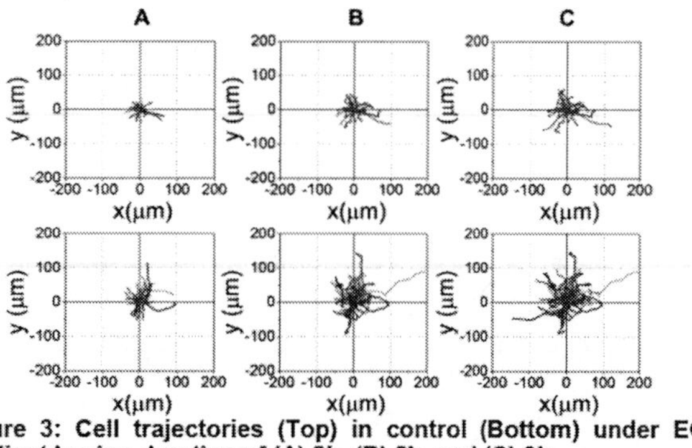

Figure 3: Cell trajectories (Top) in control (Bottom) under EGF gradient having duration of (A) 3h, (B) 6h, and (C) 8h.

Figure 4 shows quantitative analysis of cell speed, CI, and CR. As expected, EGF gradient significantly stimulated average cell speed. To compare this directional response, we analyzed CI which implicates the direction of final destination with respect to the gradient direction. Consequently, the parameter addresses how accurately the biased motility of the cells is following the gradient. **Figure 4C** depicts significant increase of CI on exposure to the EGF gradient. In the control group without EGF gradient, CI was 0.006±0.09 while CI of cell trajectories under EGF gradient was 0. 277±0.11. Although the results from CI clearly shows the biased response to EGF gradient, no remarkable difference was observed on the persistence represented by CR. These results depict that the gradient can affect the accuracy of biased migration under the EGF gradient but hardly evoke a consistent migration tendency to the direction in the case of single cells.

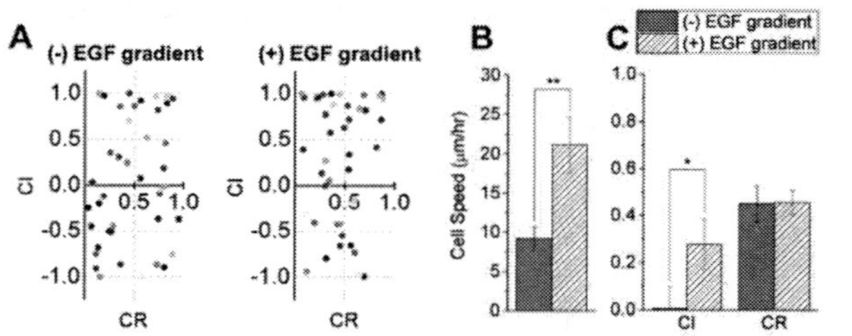

Figure 4: (A) CI-CR distribution of each cells (B) Cell speed (C) CI and CR (*: p<0.05, **: p<0.001)

Based on the present observation, we postulate a hypothesis that the cellular migratory behavior responding to chemical gradients with the presence of other environmental cues can be illustrated by the quadrants of the map as shown below. Although cancer migration is a key phenomenon during cancer metastasis, mechanisms of migration are still unclear largely due to the complexity of tumor microenvironment. Further research is warranted, but our hypothesis may quantitatively explain the role of the various microenvironment cues in determining the way that cells execute the fundamental behavior of gradient tracking and migration.

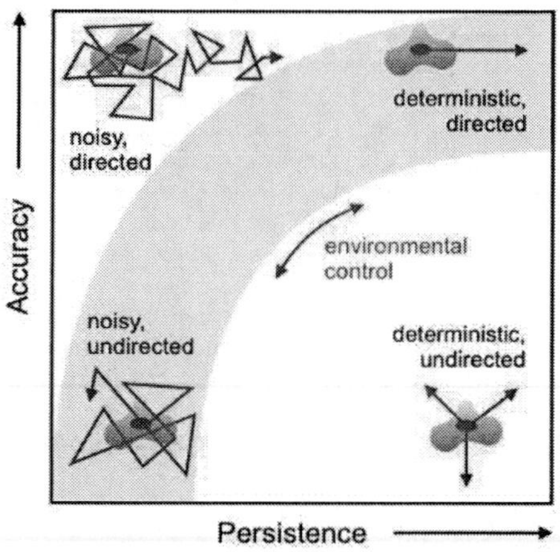

Figure 5: Proposed cellular migratory behavior under complex environmental cues.

ACKNOWLEDGEMENTS

The present work was partially supported by a grant from Showalter Research Trust.

REFERENCES

1. Roussos ET, *et al*: *Nat Rev Cancer* 2011, **11**(8):573-587.
2. McCutcheon M: *Annals of NY Acd. Sci.* 1955, **59**(5):941-944.
3. Wang S-J, et al: *Exp. Cell Res.* 2004, **300**(1):180-189.
4. Sahai E: *Curr. Opinion Gen. & Devel.* 2005, **15**(1):87-96.
5. Ellison D, *et al*: *PNAS* 2016, **113**(6):E679-E688.

THE IMPACT OF CXCL-12 ISOFORMS ON BREAST CANCER INVASION

Sarah M. Bushman (1), Talia Arcieri (1), Ayush Garg (2) Jonathan W. Song (2,3)

(1) Biomedical Engineering
Ohio State University
Columbus, Ohio, USA

(2) Mechanical and Aerospace Engineering
Ohio State University
Columbus, Ohio, USA

(3) The Comprehensive Cancer Center
The Ohio State University
Columbus, OH, USA

INTRODUCTION

CXCL-12, also known as SDF1, is an inflammatory cytokine which is known to be overexpressed in breast cancer patients. Overexpression of CXCL-12 has been shown to promote tumor growth, metastasis, angiogenesis, and tumor malignancy [1,3]. Areas in the body which express a high level of CXCL-12, including bone and brain, are common secondary sites for metastasis [2]. As a result, CXCL-12 has emerged as a prospective target for therapy. CXCL-12 works through binding to its cognate receptor CXCR4 [3], while CXCR7 is believed to be another cell surface receptor for CXCL-12 that scavenges or sequesters CXCL-12, thereby generating gradients of CXCL-12 that can lead to increased signaling by CXCR4. [4]

While CXCL-12 has been widely studied, the majority of experiments have focused on a combination of CXCL-12 isoforms. As a result, differences between CXCL-12 isoforms remain relatively unstudied. In humans, CXCL-12 is known to exist in six isoforms (α, β, γ, ε, φ, δ) [5]. Clinical research has shown that levels of CXCL-12α, β, and γ are often elevated during malignant stage breast tumors [1]. CXCL-12γ specifically has been shown to be present during advanced stages of breast cancer [3]. In addition, different isoforms demonstrate different binding affinities with CXCR4 and the extracellular matrix (ECM). CXCL-12α has been shown to have the strongest receptor binding affinity followed by β and γ. The isoform matrix binding affinities follow the opposite trend. Previous research has shown that CXCL-12α relied highly on CXCR7 interactions while CXCL-12γ and CXCL-12β were able to be scavenged from the ECM. This result is significant as the α isoform has been shown to have substantial effects on bulk cellular migration assays such as boyden chambers while CXCL-12γ has typically not been studied [6].

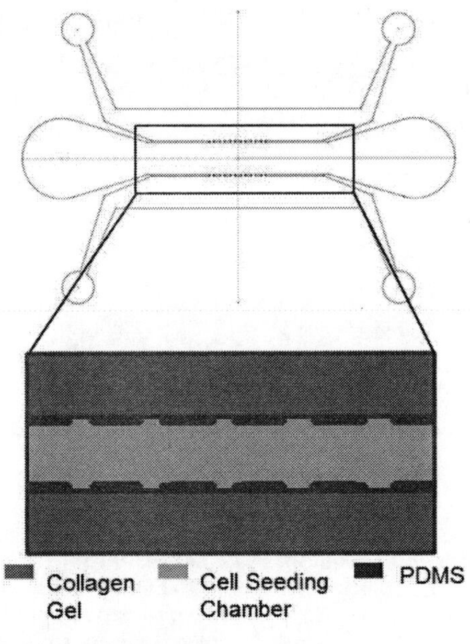

Figure 1: Device Schematic

The above design has a height of 50um. The Apertures are a width of 100um. Distance between apertures is 500um. The cell seeding chamber has a width of 2.1mm and the collagen seeding chamber has a width of 1mm.

In addition to being an indicator of tumor progression CXCL-12 expression has been shown to be clinically treatable with AMD-3100, a competitive inhibitor of CXCR4. However, it remains unclear if AMD-3100 has variable effectiveness with different CXCL-12 isoforms. This study uses microfluidics to study how the matrix binding properties affects breast cancer invasion and inhibitor binding dynamics [7].

METHODS

In order to study the effect of CXCL-12 two different cell lines were used, MCF10As and MCF10IVs. MCF10As are a non-transformed cell line while MCF104s are a Ras-transformed variant of MCF10As. In order to visualize cellular invasion, a microfluidic device was used. This device features a central cell seeding chamber and two side chambers for containing localized collagen as an ECM space for cancer cell invasion. The device was formed using standard photo and soft lithography techniques. 24 hours after device formation 10μl of 3mg/ml collagen was injected into the collagen chamber of the device. Following collagen polymerization, a fibronectin solution (100μg/ml) was injected into the cell port and allowed to incubate for 45min. The device was washed twice with experimental media to ensure excess fibronectin was removed. Cells were seeded at a density of 50,000 cells/μl and allowed to attach. Images were acquired every 24 hours for 72 hours. To induce cell invasion, each CXCL-12 isoform was added to the experimental media at a concentration of 100 ng/ml. The media was refreshed every 24 hours during the course of the experiment. AMD-3100 was also applied at a concentration of 100ng/ml. The invasion area was measured using ImageJ software.

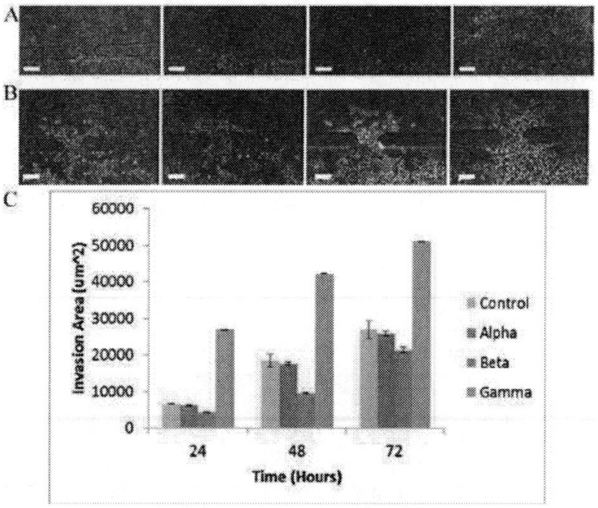

Figure 2: MCF10IV Invasion. A. Representative images of MCF10A invasion at 72 hours. From Left to right: Control, 100ng/ml Alpha, 100ng/ml Beta, 100ng/ml Gamma. B. Representative images of MCF10IV invasion at 72 hours. From Left to right: Control, 100ng/ml Alpha, 100ng/ml Beta, 100ng/ml gamma. Scale bars are 100um. C. Average MCF10IV invasion.

RESULTS

The results of both MCF10A and MCF104 invasion show that CXCL-12γ is the most effective CXCL-12 isoform which potentiates invasion. During control experiments MCF10A cells did not invade the collagen matrix. Similarly, CXCL-12α and CXCL-12β conditions showed no invasion. In contrast, CXCL-12γ stimulation resulted in an average invasion of 5,899, 16,478, and 59,194μm^2 at 24, 48, and 72 hours respectively. Standard deviation values were 71, 753, and 2,315 respectfully. MCF10IV cells showed a similar invasion trend. Control, CXCL-12α, and CXCL-12β stimulated invasion of MCF10IV cells showed insignificant differences in invasion area. CXCL-12γ stimulation of MCF10IVs showed a 2-4 fold increase in invasion. The MCF10IV invasion trends can be seen in Figure 2. Preliminary results with AMD-3100 show a decrease in invasion back down to control levels for all conditions and no change with AMD-3100 alone. The number of replicates for each experimental condition was n>3.

DISCUSSION

The findings from this study exhibit the differential effects of CXCL-12 isoforms on the invasion of malignant and non-malignant epithelia cells. MCF10As are a well-studied non-invasive epithelial cell. Without stimulation by CXCL-12 isoforms, MCF10A cells form a stable monolayer that does not invade collagen. The Ras-transformation in the MCF10IV cells result in the loss of a cellular monolayer and spontaneous cellular invasion without CXCL-12 stimulation. Upon stimulation with CXCL-12α or CXCL-12β our results show that these isoforms do not significantly affect invasion levels for both MCF10A and MCF10IV cells compared to unstimulated controls. In contrast, CXCL-12γ significantly increases invasion of both MCF10A and MCF10IV cell lines. These results suggest that CXCL-12γ selectively potentiates the invasion of both MCF10IV and MCF10A cells, and that this effect is independent the Ras mutation present in the MCF10IV cells and absent in the MCF10A cells.

Differences in the potency of CXCL-12 isoforms on invasion suggest that that the affinity of CXCL-12 binding to the matrix is more significant than the affinity of receptor binding for cancer invasion. One possible explanation for this outcome could be due to an increased availability of CXCL-12γ at the invasive cell front due to the matrix binding properties. This notion will be further explored in future studies that combine stimulation by different CXCL-12 isoforms with mechanical stimulation due to interstitial flow.

ACKNOWLEDGEMENTS

We acknowledge funding from the OSU Pelotonia Foundation and the Institute for Materials Research. In addition, we thank the Ganju Lab for the generous gift of MCF10IV cells and the Ostrowski lab for the generous gift of MCF10A cells.

REFERENCES

[1] B.A. Teicher, S.P. Fricker, *Clinical Cancer Research. 2010*
[2] R.R. Langley, I. J. Fidler. *Int. J. Cancer* 2012
[3] I. Kryczek et. al. *American Journal of Physiology* 2006
[4] S.P. Cavner et. al. *Integrated Biology*. 2014
[5] S. Zhao et.al. *Translational Oncology* 2014
[6] H. Zheng et. al. *Cardiovascular Pharmacology* 2007
[7] P. Ray et. al. *Neoplasia* 2011

NUCLEAR ENVELOPE WRINKLING AND CONNECTIVITY REGULATES MSC MECHANO-ADAPTATION AND YAP/TAZ TRANSLOCATION

Brian D. Cosgrove (1,2,3), Tristan P. Driscoll (1,2), Eric N. Dai (1,2), Su-Jin Heo (2), Jason A. Burdick (1,3), Robert L. Mauck (1,2,3)

(1) Department of Bioengineering
University of Pennsylvania
Philadelphia, PA, 19104

(2) McKay Orthopaedic Research Labs
University of Pennsylvania
Philadelphia, PA, 19104

(3) Translational Musculoskeletal Research Center
Philadelphia VA Medical Center
Philadelphia, PA, 19104

INTRODUCTION. Mechanical forces present in the cellular microenvironment are rapidly transmitted through the actin cytoskeleton to the nucleus, where they can deform chromatin and alter gene expression [1]. The nuclear envelope (NE) functions as a critical element in the translation of these mechanical cues [1], and can also act as a regulator of the translocation of cytosolic molecules into the nucleus. One such protein complex, YAP/TAZ, moves to the nucleus with increasing force, where it acts as part of a transcription factor complex that initiates transcriptional activities associated with substrate stiffness [2]. Two important elements for cytoskeleton connectivity to the NE are the nesprin giant isoforms, which connect F-actin to the nuclear envelope, and the lamin-based intermediate filament network inside of the nuclear membrane, which provides mechanical rigidity to the nucleus and plays an important role in chromatin organization [3]. Additionally, lamin-A/C is a dynamic mechanosensory element, with its expression tightly regulated throughout stem cell development and differentiation [4,5]. Despite the important role that the nucleus plays in cellular mechanotransduction, there remain open questions relating to how the structure of the NE in stem cells responds to time-varying properties of the microenvironment. In this work, we sought to determine how the nuclear envelope architecture of mesenchymal stem cells changes during mechano-adaptation to variation in internal and external biophysical inputs.

METHODS. Hyaluronic Acid Hydrogels: Methacrylate-modified hyaluronic acid was synthesized, thin hydrogels were polymerized, and fibronectin adhesive domain peptide sequences (GCGYG**RGD**SSPG-) were conjugated to methacrylate groups at 1mM as described previously [6]. Cell Culture, Immunofluorescence Staining, and Confocal Microscopy: Juvenile bovine mesenchymal stem cells (MSCs) were isolated from bone marrow and cultured for one passage before being plated on gels for 18 hours prior to fixation and staining. Antibodies used included α-YAP/TAZ (Santa Cruz) and α-LaminAC, α-Laminβ1, and α-Nucleoporin (Abcam). YAP/TAZ nuclear-to-cytoplasmic ratios were determined by measuring the average fluorescent intensity in each domain. Pharmacologic regulators of the contractile state were applied as follows: LPA (50μM for 30m) and ML7 (25 μM for 60m). Confocal imaging was performed on a Nikon A1R+ laser scanning scope with a 100x 1.4NA objective (82 nm/px XY, 200 nm/px Z). Wrinkle Index: The FeatureJ plugin was used to calculate the contrast boundaries of wrinkles in the NE using a Sobel edge detection method. Edges were binarized and the total boundary pixels were normalized to the nuclear area. Nesprin Knockdown: To knock down nesprin-1 giant, an inhibitory microRNA ('TGCCGAGGACCTTCATCTTCT') was delivered using lentivirus. Virus (media supernatant) was added to MSCs in culture at a ratio of 1:8 for 24hr, with infection confirmed by analysis of GFP+ cells. MSCs were seeded onto gels 48hrs post-infection. Statistics: Analysis was performed using GraphPad. All graphs are shown as mean ± SE. *=$p<0.05$, **=$p<0.01$, ****=$p<0.0001$ using either student's t-test or one-way ANOVA with Sidak's post-hoc testing.

RESULTS. Juvenile bovine MSCs were seeded onto thin 2D methacrylated hyaluronic acid hydrogels (MeHA) that were functionalized with RGD and UV polymerized to a substrate stiffness of 10kPa (as verified by AFM) unless indicated otherwise. Nuclear envelope (NE) wrinkles were noted across the nucleus, as evidenced by confocal imaging of Lamin-A/C (**Fig. 1A**), Lamin-β1 (data not shown), and nuclear pore complexes (data not shown). XZ confocal slices highlighted the multi-planar nature of these wrinkles, with bright foci likely representing the NE folding onto itself (**Fig 1A**). To

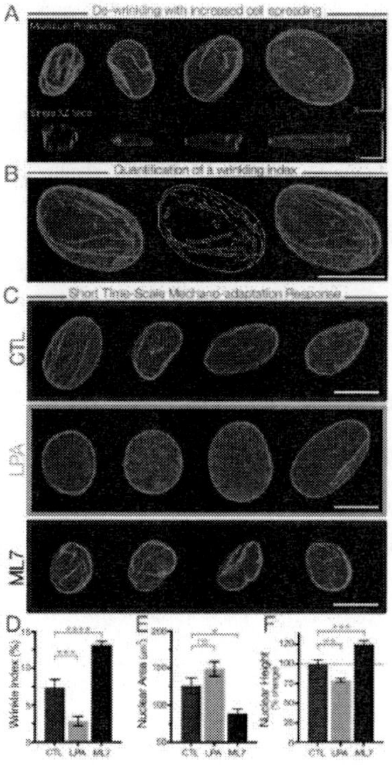

Figure 1- (A) Maximum projection z-stacks and single XZ slices of Lamin-A/C staining show de-wrinkling with increased cell spreading. SB = 5 μm. (B) Iterative wrinkle index procedure (C) Representative Lamin-A/C max projection images for MSCs on 10 kPa MeHA gels on CTL cells or following treatment with LPA for 30m and ML7 for 60m. Quantifications of (D) wrinkle index, (E) spread area, and (F) percent change in height of nuclei in these perturbations. n>19 nuclei/group. SB = 10 μm.

quantify wrinkling, we used a contrast based edge-detection method to find wrinkle boundaries from maximum intensity projections of Lamin-A/C z-stacks (**Fig. 1B**). From this, we computed a "Wrinkle Index", which represents the fraction of the nuclear envelope surface occupied by a wrinkle. Across all datasets this metric ranged from 0-20%. We first queried how NE wrinkling changed over short timescales following treatment with pharmacologic agents that alter cell contractility (**Fig. 1C**). Contractility was increased using lysophosphatidic acid (LPA) for 30m, or decreased with ML7 treatment (a myosin light chain kinase inhibitor) for 60m. These two pharmacologic agents do not decrease nuclear-actin connectivity [6]. The short timescales probed allowed for determination of the influence of contractility, independent of the NE remodeling that may occur over longer timescales [7]. Treatment with LPA significantly reduced nuclear wrinkling (**Fig. 1D**), with a corresponding increase in nuclear cross-sectional area (CSA) and decrease in nuclear height (**Fig. 1E/F**). Treatment with ML7 resulted in the opposite responses (**Fig. 1D-F**). These data suggest that NE state wrinkling depends on cytoskeletal contractility, independent of other regulators of NE structure [7].

Next, we analyzed wrinkling indices from z-stacks of >100 MSCs cultured across five different physiologic substrate stiffness conditions. This analysis showed a correlation in which wrinkling index decreased as cell spread area increased (**Fig. 2**). Cell spreading is thought to be driven by increased contractility, and in accordance with this, analysis of YAP/TAZ ratios from the same dataset similarly showed a robust mechanosensitive nuclear translocation only under conditions where there was low nuclear envelope wrinkling (**Fig. 2**). Bilinear curve-fits identified two distinct regions of the wrinkling response: an *unwrinkling regime* where there are large changes in NE wrinkling with only small changes in cellular spread area/cytoskeletal tension, and a *nuclear prestress regime,* where there is low nuclear wrinkling with increases in cellular spreading and cytoskeletal tension. The transition point between these regimes occurred at a wrinkling index of ~1.8%. Together, this suggests that the generation of cytoskeletal stress and subsequent YAP/TAZ nuclear translocation, occurs only after NE wrinkling is removed via increased actomyosin tension. In a further set of studies, we examined how reduction of cytoskeletal to nuclear connectivity via Nesprin-1G knockdown (N1G) altered nuclear YAP/TAZ translocation and NE wrinkling. Traction force generation by N1G knockdown cells increased by ~300% compared to miRNA negative controls (**Fig. 3A**). Despite this increase in contractility, N1G connectivity was required for normal YAP/TAZ translocation with increased substrate stiffness (**Fig. 3B**). Knockdown of N1G increased baseline levels of NE wrinkling compared to miRNA negative controls, and reduced changes in nuclear wrinkling and shape that occur with addition of LPA (**Fig. 3C/D**).

DISCUSSION. This work highlights the importance of pre-stress and laxity of the nuclear envelope in dictating the mechanoresponsivity of stem cells to biophysical inputs. Nuclear wrinkling appears to act as a strain sink [9], where NE wrinkles must first be smoothed out before pre-stress can be built up in both the NE and the cell. NE wrinkling thus creates a "toe region" in the cellular response to exogenous force, providing a buffer to pre-stress generation while also potentially enabling rapid changes in chromatin architecture in response to physiologic mechanical loading. In addition, we found that Nesprin-1G is essential for proper mechanosensing and YAP/TAZ translocation, and that this nuclear connectivity regulates the NE wrinkling state. This provides further insight into why N1G connectivity is an essential regulator of YAP/TAZ translocation. It is additionally possible that NE wrinkling directly influences nuclear import mechanisms and/or chromatin dynamics, making YAP/TAZ-to-TEAD binding more amenable. Future experiments will further elucidate how the wrinkled NE architecture regulates stem cell mechanobiology.

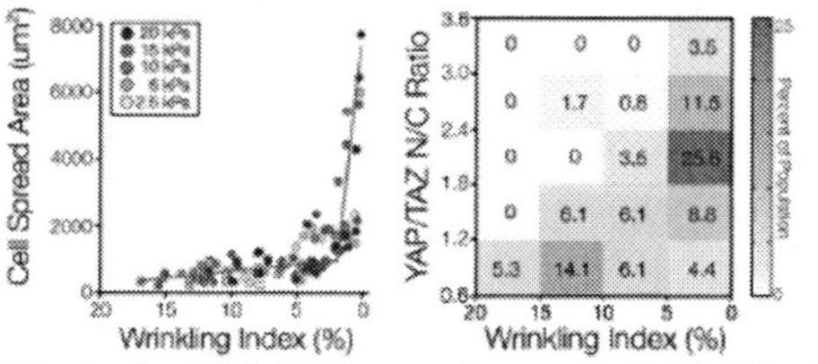

Figure 2- (*left*) Cell spread area and (*right*) YAP/TAZ ratios compared to wrinkle indices (plotted as decreasing, from 20 to 0%) of MSCs cultured across five different stiffness MeHA hydrogels. Red line indicates bilinear fit of unwrinkling and prestress regimes. n=101 nuclei over five stiffness conditions.

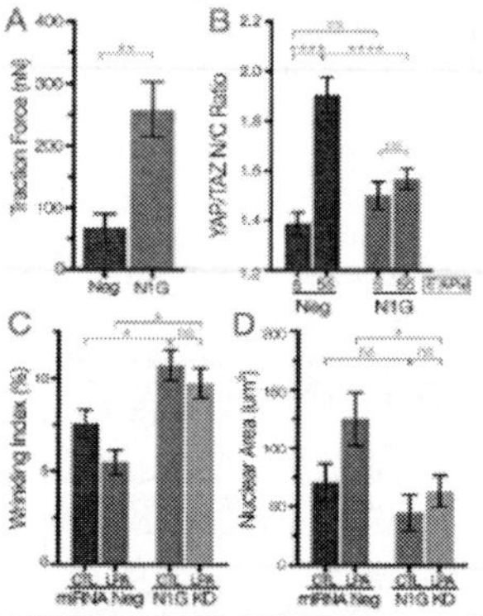

Figure 3- (A) Traction force generation in Neg control and N1G knockdown MSCs on 5kPa PA substrates n=11-23 cells/group. (B) YAP/TAZ nuclear ratios of MSCs on 5 and 55kPa PA substrates, n>45 cells /group. Nuclear wrinkling indices (C) and spread area (D) of Neg control and N1G knockdown MSCs cultured on 10kPa MeHA hydrogels before and after 30m LPA treatment, n>34cells/group.

ACKNOWLEDGEMENTS. This work was supported by the NIH (R01 EB008722) and the Penn Center for Musculoskeletal Disorders.
REFERENCES. [1] Tajik+, *Nat Mat*, 2016. [2] Dupont+, *Nature*, 2011. [3] Aureille+, *Cur Opin Cell Bio*, 2016. [4] Swift+, *Science*, 2013. [5] Burdick+, *BioMacroMol*, 2005. [6] Driscoll+, *Biophys J*, 2015. [7] Buxboim+, *Cur Bio*, 2014. [8] Heo+, *eLife*, 2016.

MODULATING SUBSTRATE STIFFNESS, CELL MORPHOLOGY AND OXYGEN AVAILABILITY IN 3D HYDROGELS DIRECT THE CHONDROGENIC AND MYOGENIC DIFFERENTIATION OF MESENCHYMAL STEM CELLS

Paola Aprile (1,2), Binulal N. Sathy (1,2), Daniel J. Kelly (1,2,3)

(1) Trinity Center for Bioengineering,
Trinity College Dublin
Dublin, Ireland

(2) Department of Mechanical and
Manufacturing Engineering
Trinity College Dublin
Dublin, Ireland

(3) Advanced Materials and Bioengineering
Research Centre (AMBER)
Royal College of Surgeons in Ireland and
Trinity College Dublin
Dublin, Ireland

INTRODUCTION

Local environmental cues such as substrate stiffness [1], cell morphology [2] and oxygen availability [3] can direct different biological processes, from cell migration and differentiation to tissue homeostasis. In particular, mesenchymal stem cells (MSC) can differentiate into distinct cell lineages, including chondrocytes, osteocytes, adipocytes and myocites [4] when exposed to specific physical cues present in the local tissue environment. *In vivo*, cells are surrounded by their extracellular matrix (ECM), which acts as a 3D substrate that resists cell-traction forces and thereby regulates tissue development by altering physical force distributions and hence biological functions. Although many studies confirm a role for such biophysical cues in regulating MSC differentiation *in vitro*, such biological processes are poorly understood in physiologically relevant 3D environments. The overall objective of this study was to investigate the interplay between cell shape, oxygen availability and substrate stiffness in regulating the differentiation of MSCs along the chondrogenic or myogenic pathway, and to explore specific molecular mechanisms of mechanotransduction in such systems. To this end, we developed an interpenetrating network (IPN) hydrogel system consisting of type I collagen and alginate, which allowed us to tailor both the stiffness of the hydrogel as well as the shape a MSC would adopt in this environment.

METHODS

Three-dimensional IPN gels were produced using final concentrations of 1.5% low molecular weight alginate (Pronova) and 1.3 mg/ml of rat-tail collagen type-I (Corning). This blend was mixed with $5x10^6$ cells/ml passage 2 porcine MSCs and transferred to a custom-made mould to obtain gels of cylindrical shape (5mmx2mm thickness). To obtain a spread cell morphology, cells embedded in the IPN blend were initially incubated for 4 hours at 37°C to allow MSC to adhere and bind to the collagen network, then 20mM $CaCl_2$ was added for 2.5 hours to crosslink the alginate network and hence obtain a spread cell in a stiff gel. 500 mM of sodium citrate was added to obtain soft gels with a spread cell morphology. In contrast, using the same method, but avoiding the initial incubation time allowed the formations of both soft and stiff gels with a round cell morphology.

Samples were maintained in a chemically defined media supplemented with 10ng/ml TGF-β3 at 20%O_2 or 5%O_2. Cell viability was assessed at day 0 (Live/Dead kit, Cambridge Bioscience) as well as mechanical properties using a Zwick materials testing machine with a 5N load-cell. Gene expression of day 3 and 7 samples was analysed with qRT-PCR for Sox9, Col2a1, ACAN, CNN1, ACTA2, MYH1 normalized to B2M (SYBR-green®) using the ΔΔCt method. Immunolocalization of actin, MyoD, Yap/TAZ and Smad2/3 (Santa Cruz) was performed on day 7 samples, while biochemical quantification of DNA (PicoGreen®) and glycosaminoglycans (Blyscan Biocolor) were carried on day 21 samples. Statistical analyses were performed using GraphPad Prism (5.00 GraphPad software). Results, both numerical and graphical, are presented as mean ± SEM. A level of $p<0.05$ was considered significant.

RESULTS

The alginate-collagen IPN allowed the production of four experimental groups, namely: soft round, stiff round, soft spread and stiff spread. Cell viability, mechanical properties and cell morphology of day 0 samples were analyzed to validate the experimental setting

(Fig. 1A, B and C respectively). The stiffness of the IPN hydrogels spanned from approximately 15 kPa to 4 kPa by decreasing $CaCl_2$ mediated cross-linking of the alginate network (Fig 1B). Moreover, all 3D-IPN maintained >90% cell viability (Fig.1A) and supported the development of distinct cell morphologies, as demonstrated by the actin circularity (Fig.1C).

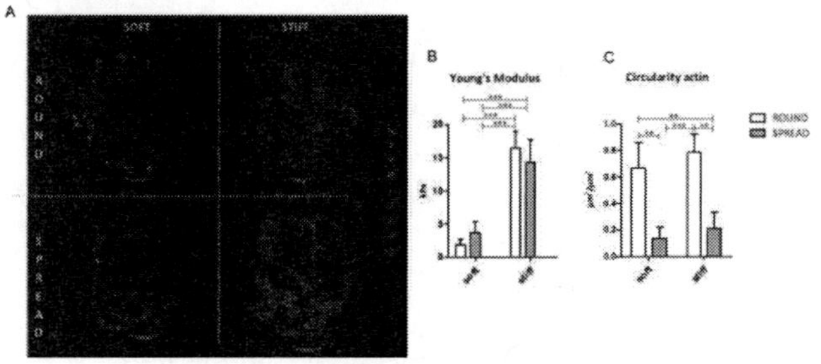

Figure 1: Validation of IPN system. A) Live/Dead analysis. B) Young's Modulus of IPN. C) Confocal analysis of actin circularity.

The expression of all chondrogenic markers was higher at physioxic (5% O_2) compared to hyperoxic (21% O_2) in all 4 IPNs (data not shown). Chondrogenesis of MSCs was enhanced by soft substrates and a rounded cell morphology (Fig.2).

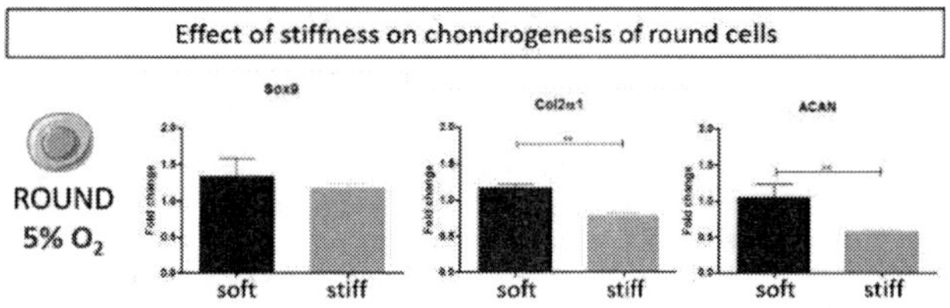

Figure 2: The effect of substrate stiffness on chondrogenic genes expression in MSCs after 7 days in 3D-IPN.

Having demonstrated that soft substrates are more chondro-inductive, we next sought to examine how altering MSCs shape in these IPNs would alter their fate. Sox9, Col2α1 and ACAN expression all decreased when MSCs were allowed to spread in these IPN in physioxic conditions (Fig.3).

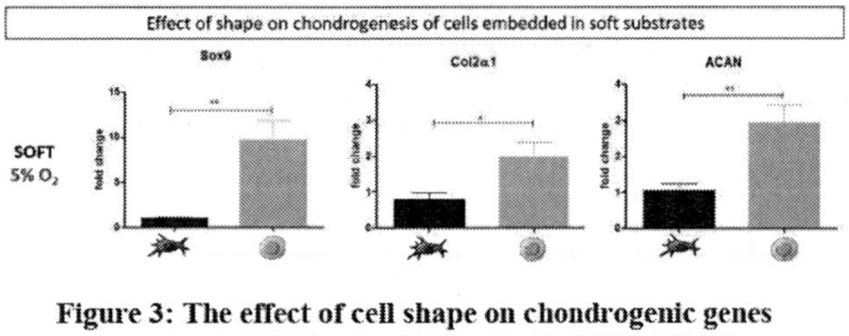

Figure 3: The effect of cell shape on chondrogenic genes expression in MSCs after 7 days in 3D-IPN at 5% O_2.

Localization of transcriptional regulators in MSCs were next investigated in these 3D IPNs. While YAP remained cytosolic in MSCs irrespective of substrate stiffness and cell morphology, MyoD nuclear translocation was observed as the stiffness of the IPNs were increased (Fig. 4A). This correlated with an increase in the expression of myogenic markers as the stiffness of the gel increased (Fig.4B). Unexpectedly, allowing the MSCs to adopt a more spread morphology was found to suppress the expression of myogenic markers (Fig. 4C).

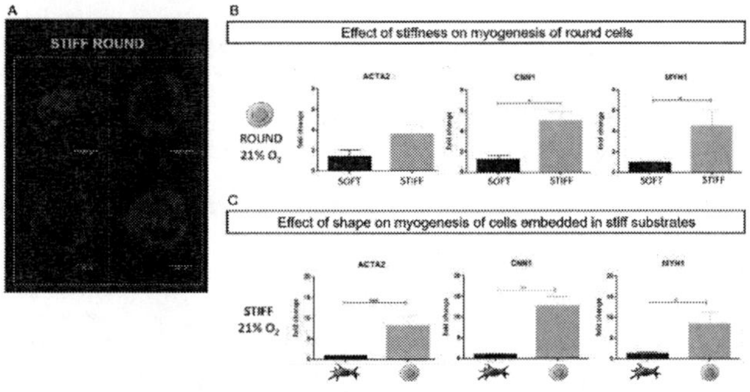

Figure 4: Myogenic profile of MSC in 3D-IPN. A) Immunofluorescence for MyoD (green), actin (red), nuclei (blue). B) Effect of substrate stiffness on gene expression in MSCs at 21%O_2. C) Effect of cell shape on gene expression in MSCs at 21% O_2.

DISCUSSION

Mimicking the complexity of ECM adhesion elements is required to reproduce the biological microenvironment in 3D hydrogels. 3D-IPN can satisfy this requirement and furthermore can be used to modulate the morphology and contractility of MSCs. To the best of our knowledge, this is the first attempt to systematically investigate the effect of oxygen tension, cell shape and substrate stiffness on the differentiation of MSCs in a 3D environment. Our findings confirms that a rounded cell morphology and a soft matrix is beneficial to MSC chondrogenesis in a 3D environment, confirming the findings of 2D studies. Stiffer environments and higher levels of oxygen availability were found to support a more myogenic phenotype, which could be explained as a dimensionality effect [5], however further analysis are required to confirm this hypothesis. To conclude, this IPN system enables the exploration of MSC mechanobiology in a physiological relevant 3D environment, which has important implications for the development of novel tissue engineering strategies.

ACKNOWLEDGEMENTS

This study was supported by Science Foundation Ireland (12/IA/1554; 12/US/I2489). We wish to confirm that there are no known conflicts of interest.

REFERENCES

[1] Engler, AJ et al., *Cell*, 126(4): 677-689, 2006
[2] McBeath, R et al., *Developmental Cell*, 6(4):483-495, 2004
[3] Holwarth, C et al., *BMC Cell Biol*, 11:11, 2010
[4] Pittenger, MF et al., *Science*, 284:143-147, 1999
[5] Caliari, S et al., *Biomaterials*, 103:314-323, 2016

THE ROLE OF CADHERIN-11 IN MEDIATING MECHANICAL CUES IN FIBROBLASTS

Meghan A. Bowler (1), Matthew R. Bersi (1), Rachel J. Jerrell (2), Aron Parekh (2), and W. David Merryman (1)

(1) Biomedical Engineering
Vanderbilt University
Nashville, TN, USA

(2) Otolaryngology
Vanderbilt University
Nashville, TN, USA

INTRODUCTION

Cadherin-11 (CDH11) is a mechanosensitive transmembrane protein involved in cell-cell adhesion, which is thought to be responsible for mediating much of the intercellular tension integral to the process of dystrophic calcification [1]. While CDH11 is known to be upregulated in calcific aortic valve disease, and a blocking strategy against CDH11 is currently in clinical trials for treatment of rheumatoid arthritis, its mechanism of action is poorly understood – as is cadherin mediated signaling and mechanotransduction in general.

CDH11 forms homotypic bonds between neighboring cells that are twice as strong as other mesenchymal cadherins, such as cadherin-2 (CDH2) [2]. This indicates a particularly important function of CDH11 in signaling initiated by tension between neighboring cells. CDH11 is also known to complex with catenins, thereby interacting with the Wnt pathway, and to regulate inflammatory cytokines, specifically IL-6 in synovial fibroblasts [3]. How these pathways are interrelated in the context of aortic valve interstitial cells (AVICs) will prove useful in treatment of calcific aortic valve disease, among other fibroblast-like-mediated diseases.

METHODS

Murine aortic valves were excised from immortalized CDH11 wild-type (WT), heterozygous (CDH11$^{+/-}$), and null (CDH11$^{-/-}$) mice and digested in 500 μL 600 U/mL collagenase for 30 minutes at room temperature. Following digestion, valves were seeded on 0.1% gelatin and expanded in DMEM supplemented with 10% FBS and 1% PenStrep to yield AVICs.

WT AVICs were infected with pLL-CDH11-mCherry and processed via fluorescence-assisted cell sorting to create a pure population of AVICs overexpressing CDH11 tagged with mCherry (CDH11-OX).

In order to quantify the traction forces generated by cells from each population, CDH11-OX, WT, CDH11$^{+/-}$, and CDH11$^{-/-}$ AVICs were seeded on 1023 Pa, fibronectin-coated polyacrylamide gels overnight and imaged before and after disruption from the gel. The associated displacement and traction fields were reconstructed using the FTTC method. Modifications to the Tractions for All code [4] were made to find a mean traction force per cell,

$$|\mathrm{F}| = \int\int \sqrt{(T_x^2(x,y) + T_y^2(x,y))}dxdy \qquad (1)$$

and the work done by each cell,

$$\mathrm{W} = \frac{1}{2}\int \vec{T}\,(\vec{r}) \cdot \vec{u}\,(\vec{r})\, d^2r \qquad (2)$$

Where T_x and T_y are the in-plane components of the traction vector $\vec{T}$, and $\vec{u}$ represents the displacement field and each position $\vec{r}$. Each metric was normalized to cell area.

CDH11-OX, WT, CDH11$^{+/-}$, and CDH11$^{-/-}$ AVICs were seeded on collagen gels at 40,000/gel and imaged at least every 24 hours for 4 days to assay contractility. CDH11-OX, WT, CDH11$^{+/-}$, and CDH11$^{-/-}$ AVICs were plated at 10,000/cm^2 on tissue culture plastic and grown for 24h before media was harvested. An ELISA for IL-6 (R&D Systems DY406-05) was then performed with that supernatant. Strain experiments were performed using the Flexcell FX-4000 Tension System. Cells were plated onto the flexible bottom of Flexcell plates coated with pronectin and exposed to 24 hours of 10% or 15% equibiaxial strain to recapitulate the mechanical environment of the valves. Samples were analyzed for CDH2, CDH11, Chi3l1, IL-6, STAT3, and GAPDH mRNA levels via qPCR. Significance is defined as different from WT unless otherwise noted, and * = $p<0.05$.

RESULTS

Loss of CDH11 results in a decrease in the amount of mean force (**1A**) and work (**1B**) AVICs are able to generate. The displacement of gels by single cells (**1C**) was measured via traction force microscopy (TFM). At the population level, as measured via collagen gel contraction assay, we see a similar decrease in contractility with loss of CDH11. However, blocking CDH11 homotypic binding with a targeted antibody, SYN0012, did not modify contraction (**1D**).

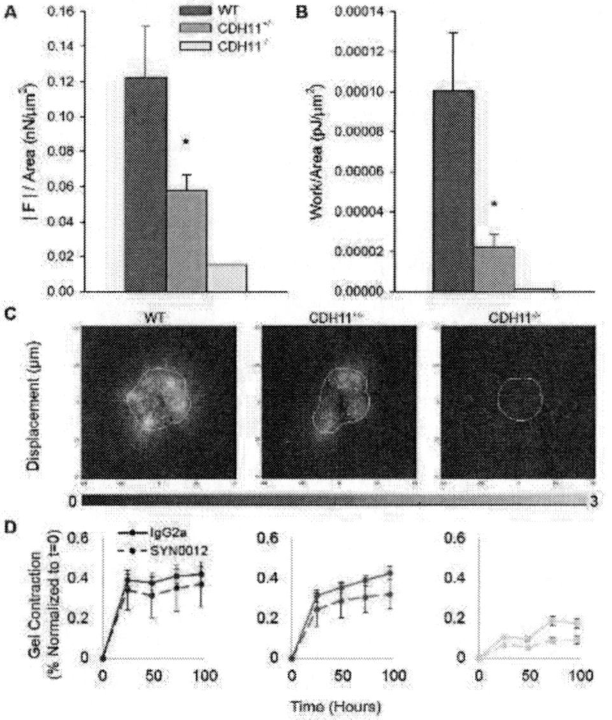

Figure 1: AVICs lacking CDH11 are not able to produce as much force (A) or work (B), as measured by TFM (C). AVICs with less CDH11 are able to contract less as a population, and blocking CDH11 interactions does not affect their contraction (D).

Since SYN0012 treatment of fibroblasts is known to affect the production of IL-6, an inflammatory cytokine, we investigated how the expression of CDH11 affects IL-6 secretion (**2B**) and transcription (**2C**). While IL-6 transcription is correlated with CDH11 transcription (**2A**), the CDH11-OX AVICs secreted less IL-6 than WT AVICs (**2B**). Interestingly, collagen gel contractility of the CDH11-OX AVICs is lower than that of WT AVICs, correlating well with secreted IL-6.

Further investigation of the IL-6 signaling cascade led us to assay the transcription of STAT3, and one of its transcriptional targets, Chi3l1. qPCR demonstrates a decrease in STAT3 transcription with loss of CDH11 (**2D**) and a decrease in Chi3l1 transcription that is very highly correlated with CDH11 transcription (**2E**). Chi3l1, or YKL-40 in humans, is known to modulate CDH2 expression and bond stability, and we discovered an inverse relationship between Chi3l1 and CDH2 transcription (**2F**).

Subjecting the AVICs to 10% or 15% equibiaxial strain demonstrates a dependence of IL-6 (**3A**) and Chi3l1 (**3B**) transcription on CDH11 expression. As CDH11 is lost, IL-6 and Chi3l1 transcription are less dramatically modified when cells are strained.

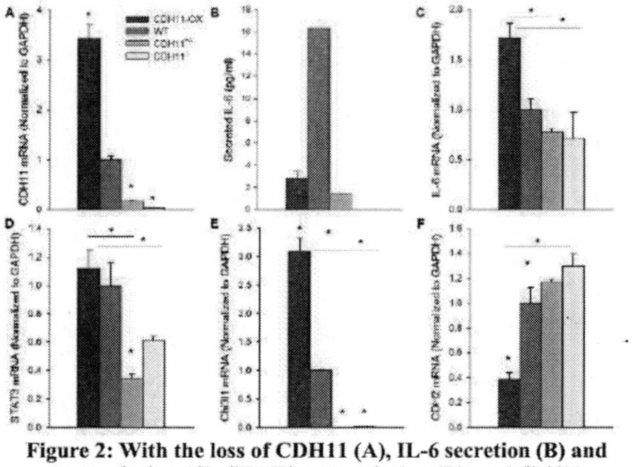

Figure 2: With the loss of CDH11 (A), IL-6 secretion (B) and transcription (C), STAT3 transcription (D), and Chi3l1 transcription (E) decrease, but CDH2 transcription (F) increases.

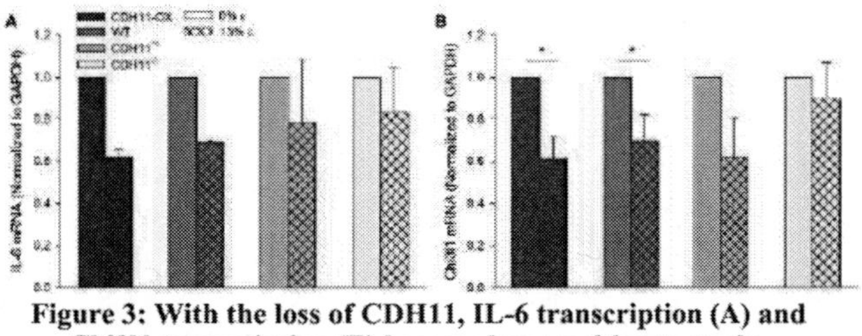

Figure 3: With the loss of CDH11, IL-6 transcription (A) and Chi3l1 transcription (B) become less sensitive to strain.

DISCUSSION

While CDH11 expression regulates AVIC contraction, it is clear that this is a result of more than just homotypic bond interactions. Noting that CDH11 and IL-6 are well-correlated in AVICs and that CDH11 regulates IL-6 in other fibroblasts, the IL-6 pathway was a good candidate for further research. Indeed, we found that STAT3 and Chi3l1 are correlated with CDH11 expression and that CDH2 is inversely correlated. This could be a pathway connecting CDH11 with CDH2 and providing an explanation for switching of CDH11 to CDH2 under strain. Of note, in human aortic valves, Chi3l1, or YKL-40 in humans, is 35-fold higher in stenotic samples than healthy samples [5]. Therefore, this pathway provides a connection from CDH11 upregulation in calcific aortic valve disease to a serum biomarker, YKL-40, potentially allowing for minimally invasive disease monitoring.

ACKNOWLEDGEMENTS

This work was supported by the NIH (R01-HL115103 and R35 HL135790) and the NSF (Graduate Research Fellowship: 2013170175).

REFERENCES

[1] Hutcheson, JD et al., *ATVB*, 33:114-20, 2013.
[2] Pittet, P et al, *J Cell Sci*, 121:877-86, 2008.
[3] Chang, SK et al, *PNAS*, 108:8402-7, 2011.
[4] Marinković, A et al, *AJPLCMP*, 303:L169-80, 2012.
[5] Bossé, Y et al, *Circ Cardiovasc Genet*, 2:489-98, 2009.

SB³C2017
Summer Biomechanics, Bioengineering and Biotransport Conference
June 21 – 24, Tucson, AZ, USA

MODELLING THE INFLUENCE OF CELL SHAPE AND OTHER MECHANICAL CUES ON DIFFERENTIATION

Hamsini Suresh (1), Siamak S. Shishvan (1), Vikram S. Deshpande (1)

(1) Department of Engineering
University of Cambridge
Cambridge, United Kingdom

INTRODUCTION

The extent of spreading of a cell confined to a square island depends on the dimensions of the island[1]. Cells confined to smaller islands spread to a limited extent and are more likely to form adipocytes, while cells confined to larger islands spread out more and usually tend to form osteoblasts[2]. In this study, we propose a model that essentially utilizes the differences in the cytoskeletal arrangements in cells confined to adhesive islands of different sizes, to predict the corresponding differentiation trend observed in the experiments, and subsequently extend the model to investigate experimental observations involving the impact of other mechanical 'cues' on cell spreading and differentiation[4,5].

MODEL OVERVIEW: A STATISTICAL MECHANICS APPROACH

Consider a thermodynamic system that consists of the cell and the adhesive square island, interacting with the external, thermo-nutrient reservoir. A given cell shape is characterized by a 'spread micro-state', which is defined as the connection of material points on the cell membrane to those on the island.

At every spread micro-state, the cell is at mechanical and thermal equilibrium with its surroundings. Defining the free-energy of the cell G^i_{tot} in micro-state i as

$$G^i_{\text{tot}} = G^i_{\text{cyto}} + G^i_{\text{elas}} + G^i_{\text{wink}} \quad (1)$$

where G^i_{cyto} is the free-energy associated with the cytoskeletal configuration in the cell in micro-state i (as defined by Vigliotti et al[3]), G^i_{elas} is the contribution from the passive elasticity of the cell (modified hyperelastic Ogden model) and G^i_{wink} is the free-energy associated with the winkler foundation used to approximate the adhesion between the cell and the island. We assume that the islands are printed on rigid substrates, so that the free-energy associated with the deformation of the substrate is zero, and hence does not feature in eqn. (1).

The external reservoir imposes a homeostatic constraint on the cell such that the cell fluctuates between a variety of different micro-states with their associated shapes and free-energies, while ensuring

$$\sum_i G^i_{\text{tot}} = G_0 \quad (2)$$

where the summation denotes the ensemble average of the total free-energy of the cell, while G_0 denotes the free-energy associated with the homeostasis constrained imposed by the external reservoir.

With the equilibrium free-energy of the cell in a micro-state given by eqn. (1), the probability P^i that the cell is observed in micro-state i (with total free-energy G^i_{tot}) can be calculated by minimizing the free-energy of the micro-states with respect to the constraint in eqn. (2) to yield

$$P^i \propto e^{-\beta G^i_{\text{tot}}} \quad (3)$$

where β is the parameter chosen to enforce the homeostatic condition imposed on the system by the thermo-nutrient reservoir.

Since the cell can take a wide variety of shapes on each island, the cell is allowed to explore this enormous phase-space through a few million Markov Chain Monte Carlo (MCMC) steps on each island. A small random displacement is given to the cell in each MCMC walk, for which the total-free energy of the cell is calculated using eqn. (1) and the resulting configuration (micro-state $i+1$) is accepted with the probability $e^{-\beta(G^{i+1}_{\text{tot}} - G^i_{\text{tot}})}$.

RESULTS

Simulations were performed with cells confined to square islands of a range of sizes and the extent of spreading on each island was observed. Fig. 1 depicts a few shapes (or spread micro-states) of cells observed on islands of two different sizes. Notice that the island area (A^p) is normalized with A^e – the area of cell under zero elastic strain.

Figure 1: Few of the spread micro-states observed in cells on (a) a small island (A^p/A^e=1), and (b) a large island (A^p/A^e=8)

Our hypothesis is that studying the fluctuations of cells between different cytoskeletal configurations when confined to an island of a given size plays a key role in predicting the trends in cell spreading and differentiation. Fig. 2(a) depicts the G_{cyto} distribution as a function of patch size. In order to correlate the information contained in the cytoskeletal free-energy distribution to cell differentiation, we develop a differentiation model where the central parameter is the mean cytoskeletal free-energy $\bar{G}_{\text{cyto}}$ calculated over a long period of time period for a cell confined to a given island. For a cell fluctuating on an island over N_c spread micro-states,

$$\bar{G}_{\text{cyto}} = \sum_{i=1}^{N_c} P^i G^i_{\text{cyto}} \quad (4)$$

For large N_c, central limit theorem can be applied to eqn. (4) to express $\bar{G}_{\text{cyto}}$ as a normal distribution $N(\mu, \bar{\sigma})$, where $\bar{\sigma} = \frac{\sigma}{\sqrt{N_c}}$, μ and σ are the mean and variance of the cytoskeletal free-energy distribution of the cell on the given island. Fig. 2(b) depicts the distribution of $\bar{G}_{\text{cyto}}$ corresponding to the G_{cyto} distributions illustrated in Fig. 2(a).

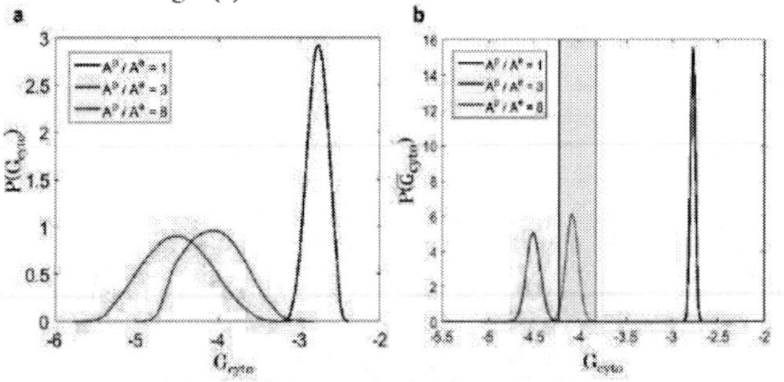

Figure 2: (a) G_{cyto} and (b) $\bar{G}_{\text{cyto}}$ distributions for three different island sizes (x-axis normalized with G_0)

Let us assume that the cells undergo differentiation when their mean cytoskeletal free-energy $\bar{G}_{\text{cyto}}$ calculated over a long period of time period attains a value over a specific range. For example, stem cells differentiate into osteoblasts when $\bar{G}_{\text{cyto}}$ lies within the interval $[G_{\text{osteo}} - \Delta_{\text{osteo}}, G_{\text{osteo}} + \Delta_{\text{osteo}}]$ (this interval is illustrated as the yellow box in Fig. 2(b)). Therefore, the probability that a cell differentiates into an osteoblast can be written as

$$P_{\text{osteo}} = \int_{G_{\text{osteo}}-\Delta_{\text{osteo}}}^{G_{\text{osteo}}+\Delta_{\text{osteo}}} p(\bar{G}_{\text{cyto}})\, d\bar{G}_{\text{cyto}} \quad (5)$$

An expression similar to eqn. (5) can be written for other cell types, with cell-specific differentiation parameters. In this study, we focus on two types of cells formed – osteoblasts and adipocytes. Therefore, G_{osteo}, Δ_{osteo}, G_{adipo}, Δ_{adipo} and N_c are the different parameters that determine the number of cells that commit to the two lineages on each of the adhesive islands. For a particular choice of the five parameters, we plot the differentiation as a function of island size as predicted by the simulation (2D) in Fig. 3(a), and compare it with the experimental observation (3D) in Fig. 3(b).

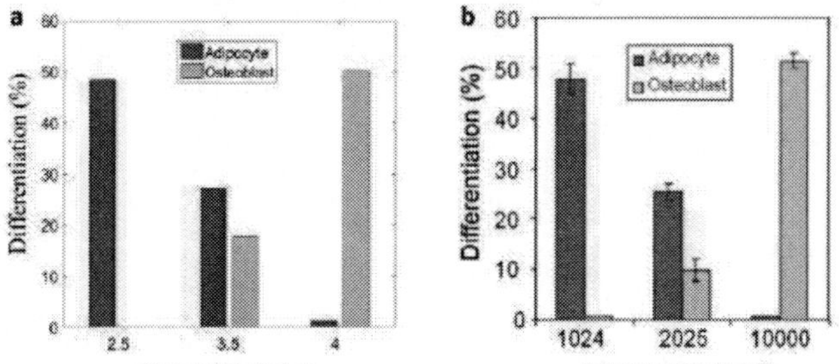

Figure 3: Differentiation as a function of island size, from (a) simulation and, (b) experiment.

DISCUSSION

We observe that the statistical mechanics model, coupled with the differentiation model, is able to capture the essential trend observed in the differentiation of stem cells as a function of the size of the adhesive island. The range of areas of islands over which the differentiation trend is predicted by the simulation is lesser than the range observed in the experiments. This can be attributed to the fact that the spreading of cells is significantly higher in 3D, but the model used in the current study only captures the 2D effects, since the model's primary aim is to demonstrate its versatility while being computationally efficient.

The versatility of the model stems from the fact that it can be used to study different types of mechanical 'cues', such as island size, island shape and substrate stiffness. With the same underlying statistical mechanics model and the different parameters that dictate the differentiation behavior, the current study is extended to predict the differentiation of cells as a function of substrate rigidity (as reported in Engler et al[4]) and the aspect ratio of the adhesive island (as reported in Kilian et al[5]). The contractility of the actomyosin bundles can also be varied to study the impact of actin or myosin inhibitors such as cyto-D or blebbistatin on the extent of differentiation[2,4,6].

ACKNOWLEDGEMENTS

HS is funded by the Commonwealth Scholarship Commission UK.

REFERENCES

[1] Chen, CS et et al., *Biochem Biophys Res Commun*, 307:355–361, 2003.
[2] McBeath, R et al., *Dev Cell*, 6:483–495, 2004.
[3] Vigliotti, A et al., *Biomech Model Mechanobiol*, 15:761-89, 2015.
[4] Engler, AJ et al., *Cell*, 126:677-689, 2006.
[5] Kilian, KA et al., *Proc Natl Acad Sci*, 107(11):4872–4877, 2010.
[6] Huang, S et al., *Mol Biol Cell*, 9:3179–3193, 1998.

ANISOTROPIC YAP MECHANOTRANSDUCTION

Wen-Cih Wen and Pen-Hsiu Grace Chao

Institute of Biomedical Engineering
National Taiwan University
Taipei, Taiwan

INTRODUCTION

YAP is the main transcription factor for the Hippo signaling pathway, which controls growth and development. Recent studies have found YAP to relay mechanical signals into the nucleus, such as stiffness, deformation, and shape [1-3]. Actin structure, cell tension, and JNK signaling have all been demonstrated as regulators of YAP mechanotransduction [1, 2, 4]. As orientation of uniaxial stretch have been shown to regulate cytoskeleton organization and gene expression [5, 6], we hypothesize that anisotropic mechanical loading will result in a direction-dependent YAP response.

METHODS

Microfabrication PDMS membranes were treated with 2 mg/ml dopamine hydrochloride (pH=8.5) and microcontact printed with 30 µg/ml fibronectin. Uncoated regions were blocked with 1% BSA. Linear and elliptic patterns were 17 µm in width and the elliptic pattern had an area of ~1250 µm² with aspect ratio of 6.

Cell Culture Human bone marrow-derived mesenchymal stem cells (hMSCs) were seeded at 4000 cells/cm² on the patterned PDMS membranes in serum-free medium for 5 hours.

Mechanical Stimulation A custom bioreactor applied 10% uniaxial strain at 0.5 Hz for 30 minutes either parallel or perpendicular to the long axis of the cells (strain rate = 0.1/sec).

Imaging and Analysis Cytoskeleton organization was examined by fluorescently labeling f-actin with Alexa Fluor 488-conjugated phalloidin (Invitrogen) and the nuclei were labeled with by DAPI. YAP localization was assessed by immunostaining with YAP primary antibody (Santa Cruz). YAP location was quantified by a custom MATLAB program.

Statistical Analysis SPSS was used to perform ANOVA analysis with post-hoc tests (LSD, α=0.05). Chi-square tests with Bonferroni correction was used to compare proportions. Error bars represent standard deviations or standard errors of a proportion.

RESULTS

MSCs conformed to the pattern shapes as illustrated in Figure 1, where unpatterned cells (random) had a significantly larger spread area compared to cells confined in linear patterns (53% of the random group, Figure 2A). Elliptic patterns further constrained cell area compared with the linear patterns (79% of the linear group) where cells could freely elongate, resulting in a higher aspect ratio (Figure 2AB). Dynamic stretch for 30 minutes reduced cell area in the unpatterned cells but did not influence cell area in the patterned cells. However, a small reduction in cell aspect ratio was observed in the patterned and loaded cells. YAP translocated to the nucleus in the randomly oriented cells upon dynamic stretch (Figure 3). In the linearly patterned cells, only parallel loading increased YAP nuclear localization, whereas no YAP translocation was observed in the loaded elliptic cells.

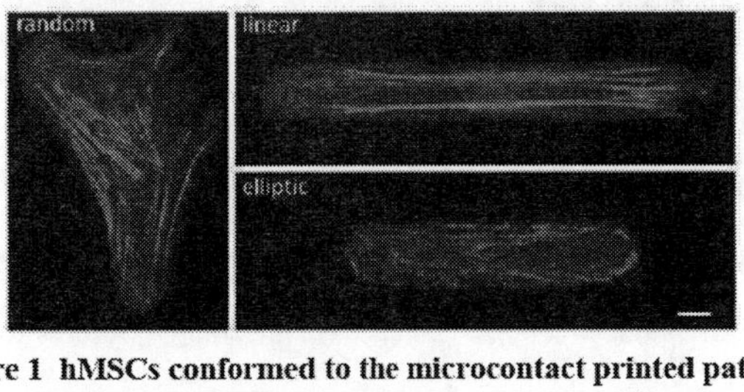

Figure 1 hMSCs conformed to the microcontact printed patterns (Green: phalloidin, blue: DAPI, scale bar=10 µm)

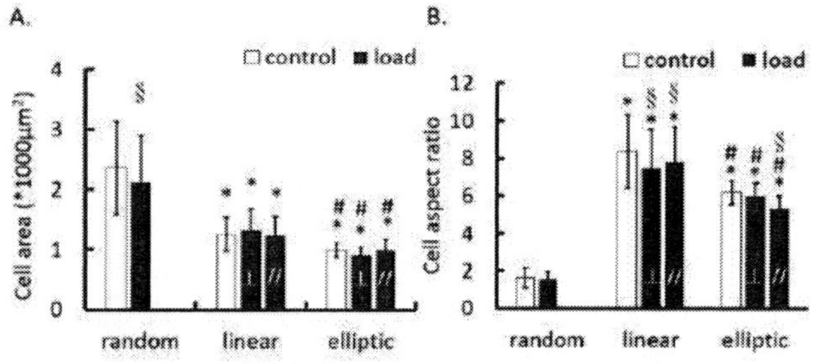

Figure 2 Effects of cell shape and anisotropic stretching on cell area (A) and aspect ratio (B) (n=65-107, *p<0.05 vs random, § p<0.05 vs control, #p<0.05 vs linear, ⊥ represents perpendicular and // represents parallel loading)

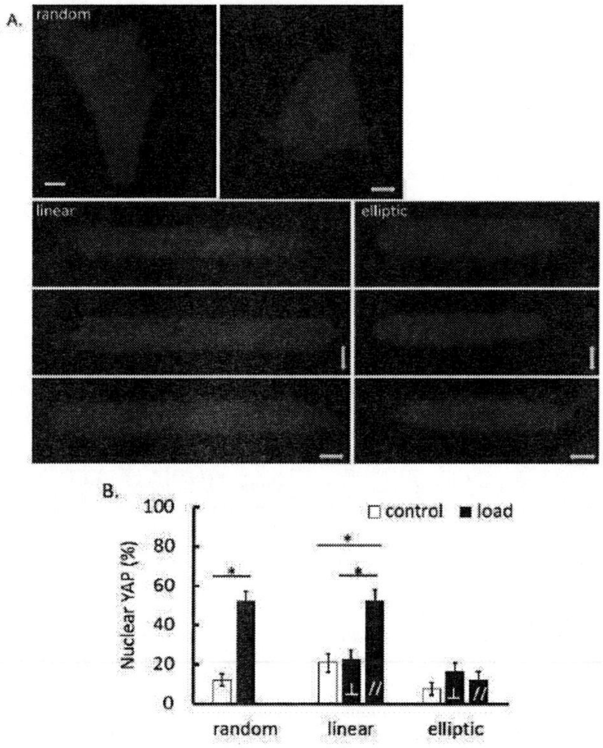

Figure 3 Direction-dependent YAP localization in response to uniaxial dynamic stretch (A) Representative images of YAP localization (Scale bar=10 μm, yellow arrows indicate strain direction) (B) Quantitative analysis of nuclear YAP localization. (n=65-107, *p<0.005 vs indicated groups, ⊥ represents perpendicular and // represents parallel loading)

DISCUSSION

Cells in situ reside in a variety of matrices and exhibit a wide variety of phenotypic morphologies, which is reflected in cytoskeletal structures. Perturbations to the cytoskeleton via deformational load can influence global cellular functions, including proliferation and phenotype [7]. Parallel and perpendicular uniaxial loading (relative to the cell long axis) have been shown to result in differential phenotype [6-8]. Few studies, however, have examined the mechanistic interactions between cell geometry and mechanical regulation [9]. In this study, we investigate the anisotropic YAP nuclear localization response to dynamic uniaxial loading. In randomly oriented cells, dynamic cyclic loading induces YAP nuclear translocation, as demonstrated previously [1]. In cells cultured in linear patterns, where cells can elongate freely, YAP translocates to the nucleus only when dynamic stretch is applied parallel to cell orientation. When cell elongation is limited by the elliptic patterns, no YAP translocation is observed with mechanical loading. Our results demonstrate the anisotropic mechanotransduction of YAP signaling, which is dependent on cell area and/or aspect ratio. Anisotropic loading have been shown to result in direction-dependent cytoskeleton and cell tension changes [6, 10]. At the same time, cell tension and cytoskeleton integrity have been shown to regulate YAP signaling [4]. Future studies will delineate the effects of cell spread area and aspect ratio, as well as the underlying mechanisms for the anisotropic YAP signaling.

ACKNOWLEDGEMENTS

The work was supported by the Taiwanese MOST (105-2221-E-002-006) and NHRI (EX106-10411EI)

REFERENCES

[1] Codelia VA, Sun G, Irvine Kenneth D. Regulation of YAP by Mechanical Strain through Jnk and Hippo Signaling. Current Biology. 2014;24:2012-7.

[2] Aragona M, Panciera T, Manfrin A, Giulitti S, Michielin F, Elvassore N, et al. A Mechanical Checkpoint Controls Multicellular Growth through YAP/TAZ Regulation by Actin-Processing Factors. Cell. 2013;154:1047-59.

[3] Dupont S, Morsut L, Aragona M, Enzo E, Giulitti S, Cordenonsi M, et al. Role of YAP/TAZ in mechanotransduction. Nature. 2011;474:179-83.

[4] Das A, Fischer RS, Pan D, Waterman CM. YAP Nuclear Localization in the Absence of Cell-Cell Contact Is Mediated by a Filamentous Actin-dependent, Myosin II- and Phospho-YAP-independent Pathway during Extracellular Matrix Mechanosensing. Journal of Biological Chemistry. 2016;291:6096-110.

[5] Kurpinski K, Chu J, Hashi C, Li S. Anisotropic mechanosensing by mesenchymal stem cells. Proceedings of the National Academy of Sciences. 2006;103:16095-100.

[6] Shao Y, Mann JM, Chen W, Fu J. Global architecture of the F-actin cytoskeleton regulates cell shape-dependent endothelial mechanotransduction. Integrative Biology. 2014;6:300-11.

[7] Lee C-F, Haase C, Deguchi S, Kaunas R. Cyclic stretch-induced stress fiber dynamics – Dependence on strain rate, Rho-kinase and MLCK. Biochemical and Biophysical Research Communications. 2010;401:344-9.

[8] Lee CH, Shin HJ, Cho IH, Kang Y-M, Kim IA, Park K-D, et al. Nanofiber alignment and direction of mechanical strain affect the ECM production of human ACL fibroblast. Biomaterials. 2005;26:1261-70.

[9] Chen Y, Pasapera AM, Koretsky AP, Waterman CM. Orientation-specific responses to sustained uniaxial stretching in focal adhesion growth and turnover. Proceedings of the National Academy of Sciences. 2013;110:E2352-E61.

[10] Wen S-M, Chao PG. Role of Cell Tension on Anisotropic Mechanosensing. SB3C. Snowbird, UT2015.

CELL CYCLE SYNCHRONIZATION MODULATES CHONDROGENESIS AND MECHANOTRANSDUCTION OF MESENCHYMAL STEM CELLS

Andrea R. Tan (1), Eben G. Estell (1), Alfonso Martin-Peña (2), J. Chloe Bulinski (3), Clark T. Hung (1)

(1) Department of Biomedical Engineering
Columbia University
New York, NY, USA

(2) Department of Orthopaedics and Rehabilitation
University of Florida
Gainesville, FL, USA

(3) Department of Biological Sciences
Columbia University
New York, NY, USA

INTRODUCTION

Mesenchymal stem cells (MSCs) are an attractive cell source due to their increased availability and intrinsic ability to differentiate down desired lineages. Their success in tissue engineering, however, has been limited by variation in differentiation potential and proliferation capacity, and there is poor understanding how to select the most potent cells for clinical application. In particular, while MSCs isolated from adipose tissue (ADSCs) are among the easiest to extract, previous reports have found inferior potential for proliferation and chondrogenesis [1]. To increase utility of these cells for cartilage tissue engineering, here we applied a method of synchronizing cells to a particular cell cycle phase in 2D that has previously shown to be successful at enhancing subsequent 3D chondrogenic differentiation of bone marrow derived MSCs [2]. As such, two sequential studies were performed. In study 1, we explored the ability to use our synchronization technology to enhance chondrogenesis of ADSCs, hypothesizing that modulating the %S phase cells directly impacts the glycosaminoglycan (GAG) production of these cells.

In study 2, we investigated the potential role of primary cilia in mounting this chondrogenic response. The primary cilium has been suggested to play a regulatory role in cell cycle progression [3, 4] and is considered a critical effector of cell mechanotransduction and chondrogenesis [5,6]. Briefly, mechanosensitive cells sense perturbations that are transduced to biochemical signals that regulate proliferation, differentiation, and gene expression in stem cells [7]. Furthermore, maintenance of the phenotype of chemically induced differentiated MSCs was found to rely on the presence of primary cilia [8]. As such, here we synchronized cells and characterized primary cilia (incidence rate and cilia length) and the subsequent calcium signaling response of these cell populations to fluid shear. We hypothesized that synchronized cells would exhibit an increased cilia incidence rate and decreased cilia length due to their advanced position in the cell cycle. We also hypothesized that the increase in cilia presence would facilitate a more robust response to external stimuli.

METHODS

Cell isolation and transduction Following euthanization for other studies (IACUC-University of Florida), adipose tissue was harvested from healthy, skeletally mature horses (2-5 years). ADSCs were isolated and plated for two passages before being transduced with dual gene lentiviral vector construct encoding a secreted *metridia* luciferase reporter, as previously described [9]. Cell Synchronization Following transduction, cells were returned to monolayer culture for 2 additional passages and used immediately (asynchronous control) or synchronized via methylcellulose suspension and 2D re-plating, as described in [2]. Cell cycle phase was confirmed via flow cytometry (Figure 1a). *Study 1*: After 24 hours (post-methylcellulose (PMC) 24) and 27 hours (PMC 27), cells were trypsinized from monolayer culture to produce populations composed of varying %S phase cells. Cells from each group were then used to create cell pellets (0.5×10^6 cells) and cultured for 35 days in chondrogenic medium supplemented with 10 ng/mL TGF-β3 and 10 ng/mL BMP-6. At days 3, 14, and 35, pellet samples were harvested and evaluated for GAG and DNA content, as described [10]. *Study 2*: Cells were re-plated for 27 hours (PMC 27) for cilia staining and Ca^{2+} fluid shear experiments. Cilia staining and imaging Cells were plated on coverslip-glass bottom dishes and fixed with 4% paraformaldehyde after 27 hours. Immunohistochemistry was performed to assess the presence and length of primary cilium (labeled with Alexa-Fluor-488 conjugated acetylated alpha-tubulin and counterstained with TRITC-conjugated phalloidin and diamidino-2-

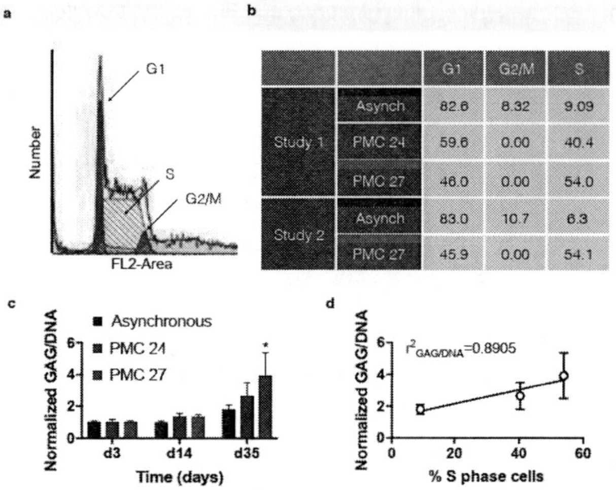

		G1	G2/M	S
Study 1	Asynch	82.6	8.32	9.09
	PMC 24	59.6	0.00	40.4
	PMC 27	46.0	0.00	54.0
Study 2	Asynch	83.0	10.7	6.3
	PMC 27	45.9	0.00	54.1

Figure 1. (a) Representative flow cytometry plot of % cell cycle phase, summarized in (b) table for each synchronization group in Studies 1 and 2. (c) Normalized GAG/DNA over culture time (n=4-6/group, *p<0.05 vs. asynchronous group). (d) Linear regression of GAG output vs. %S phase cells.

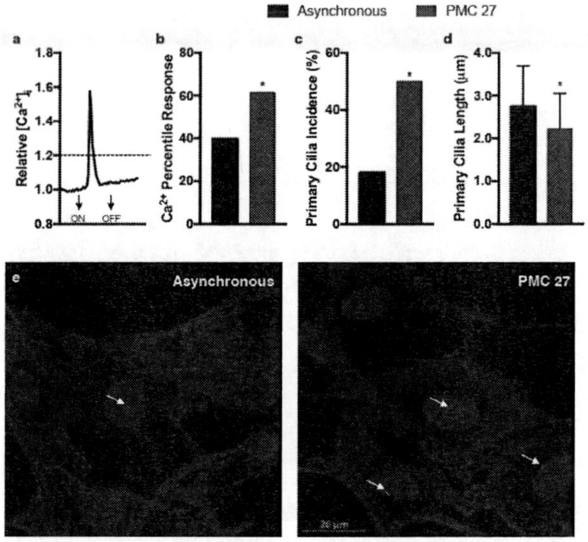

Figure 2. (a) Representative Ca^{2+} signal in response to 0.5 dyne/cm² fluid shear stress over time. Dotted line = response threshold. Arrows indicate the onset (120 sec) and offset of flow (240 sec). (b) Percentile response of cells mounting a calcium response for asynchronous and synchronized cells, n=100 cells, *p<0.05 vs. asynchronous group. (c) Primary cilia incidence (% of total cells) and (d) primary cilia length for control and PMC 27 cells, n=250 cells, *p<0.05 vs. asynchronous group. (e) Representative images demonstrating increased incidence and shorter primary cilia lengths for PMC 27 cells. Primary cilium (green, Alexa-Fluor 488 conjugated acetyl α-tubulin, identified by arrow) co-labeled with actin cytoskeleton (red, rhodamine phalloidin) and nucleus (blue, DAPI).

phenylindole (DAPI) for primary cilia, cytoskeleton and nuclear visualization, respectively). Calcium imaging and analysis Cells were plated in silicone wells on glass slides (50 x 10^3 cells/well) and imaged after 27 hours for fluid shear response characterization, as described in [11]. Briefly, changes in intracellular calcium ($[Ca^{2+}]_i$) were tracked using the calcium-sensitive fluorescent indicator Fura Red-AM, and a custom Matlab code was employed to calculate percent responding cells (n=100 cells/slide, pooled across 3 slides/group). Statistics Two-way ANOVA (α=0.05) with Tukey's HSD post-hoc tests was used to compare groups for biochemical output. Linear regression was performed to assess the relationship between GAG output and %S phase cells. Fisher's Exact Test ($p<0.05$) was used to assess changes in Ca^{2+} response and cilia measurements.

RESULTS

Cell synchronization of equine ADSCs via methylcellulose suspension successfully arrested cells in the G1 phase, and re-plating these synchronized cells triggered cells to re-enter the cell cycle, yielding distinct cell populations with varying composition of %S phase cells (Figure 1b). *Study 1*: The biochemical composition of cell pellets was strongly correlated with the %S phase cells. By day 35, PMC 24 cells were slightly more biosynthetically active than asynchronous pellets (GAG/DNA normalized to d3 values, 1.47-fold increase, $p=0.32$) while PMC 27 cells were significantly more biosynthetically active (2.2-fold increase, $p<0.0001$, Figure 1c), yielding a strong positive linear relationship between GAG output and %S phase cells (r^2 = 0.8905, Figure 1d). *Study 2*: The calcium signaling response to fluid shear, as characterized by the increase in intracellular Ca^{2+} (Figure 2a), was enhanced for synchronized cells; percentile response increased significantly from 40% to 62% ($p<0.05$, Figure 2b). Cilia incidence was significantly increased in PMC 27 cells (50% vs. 18%, $p<0.05$, Figure 2c, 2e) with commensurate decrease in cilia length (2.22 ± 0.83 µm vs. 2.75 ± 0.94 µm, $p<0.05$, Figure 2d, 2e).

DISCUSSION

Synchronization of equine ADSCs produced populations of cells with enhanced cilia incidence that mounted a significantly more robust calcium signaling response to fluid shear stress. This may be due to the increased incidence rate as well as the shortened cilia length. There is evidence that an optimal length of cilium may exist for mechanotransduction, where lengthening of the structure may inhibit or decrease associated Hedgehog signaling that regulates differentiation [12]. Toward this end, our findings support an increasing role for cilia during chondrogenesis in 3D culture, as suggested by others [6,8]. Together, we accept our hypotheses that cell synchronization in the manner described directly modulates chondrogenesis by increasing GAG biosynthetic output and mechanotransduction by modulating the response to fluid shear. While numerous efforts have focused on the manipulation of the cilia through biological and chemical means, here we have shown effectively that similar results can be attained through synchronization of the cell cycle to control cells in the S phase. Furthermore, we present here a powerful technique for optimization of strategies for functional tissue engineering of articular cartilage. Future work will continue to explore the potential to control and select the most relevant stem cells for clinical application as well as their response to exogenous factors. Finally, careful control of cell cycle phase and cilia may provide additional insight into cartilage physiology and pathology.

ACKNOWLEDGEMENTS

Funded in part by the Coulter Foundation, Columbia Technology Ventures (WHCF CU12-0369), and NIH 5P41EB002520.

REFERENCES

[1] Yoshimura+, *Cell Tissue Res*, 327:449-462, 2007, [2] Tan+, *Trans SB³C 2016*, [3] Tucker+, *Cell*, 17: 527-535, 1979, [4] Pan+, *Cell*, 7:1255-1257, 2007, [5] Muhammad+ *Cell Mol Life Sci*, 69:2101-2107, 2012, [6] Deren+, *Int J Mol Sci*, 2:188-198, 2016, [7] Wang+ *Sports Med Arthrosc Rehabil Ther Technol* 2:16, 2010, [8] Tummala+, *Cell Mol Bioeng*, 3:207-212, 2010, [9] Martin-Peña+ *Trans ORS 2016*, [10] Kelly+, *J Biomech*, 39:1489-1497, 2006, [11] Estell+, *Trans ORS 2017*, [12] Breslin+, *Cell Cycle*, 13:2773-2779, 2014.

SB³C2017
Summer Biomechanics, Bioengineering and Biotransport Conference
June 21 – 24, Tucson, AZ, USA

EXPERIMENTAL CHARACTERIZATION OF AIRWAY TISSUE EXHIBITS PRONOUNCED DIRECTIONAL AND REGIONAL MECHANICAL PROPERTY VARIATIONS

Mona Eskandari, Alberto L. Arvayo, Ellen Kuhl, Marc E. Levenston

Department of Mechanical Engineering
Stanford University
Stanford, CA, USA

INTRODUCTION

Lung disease is ranked as the third leading cause of death in the United States and has a higher fatal incidence elsewhere. Yet, pulmonary mechanics research—particularly airway obstruction mechanics—is still in its infancy. Diseases causing airway collapse, such as asthma and bronchitis, require an understanding of how the tissue deforms in order to inform diagnoses, design interventions, and formulate preventative measures. To date, only a handful of studies have documented the stiffness response of the airway walls, and most have been restricted to the trachea, the largest, most accessible passageway [1, 2, 3]. Computational models have been limited by the availability of airway-specific constitutive relations and the inference of measured material properties of the trachea to predict the mechanics of diseases affecting the small bronchi and bronchioles [4]. Additionally, despite viscoelasticity being documented for nearly all other biological structures, it has been overlooked in the airways [5]. Consequently, the objective of this study was to establish the stiffness and viscous mechanical response in the proximal and distal airway tree to investigate anisotropic and region-dependent behavior.

METHODS

Fresh lungs were obtained from five pigs (>6months) and used to prepare six sets of test specimens (~30 samples/set) from three regions (trachea, large bronchi, and small bronchi) and two directions (axial and circumferential) as shown in Figure 1. An equal number of samples were excised from the right and left lung lobe of each pig.

Landmark airways (the trachea splitting into the right and left main bronchus) were excised from parenchymal lobes and preserved in the specimen's blood. The cylindrical airways, composed of stiff outermost cartilage with innermost soft tissue, were cut axially and circumferentially while avoiding branching regions. To evade artifacts due to tissue damage, we did not separate soft tissue and cartilage layers [1, 2], but rather tested the intact, full thickness samples to determine the effective mechanical properties of the airway wall. The width and thickness of each sample was measured in the unloaded state, and used to calculate the Piola stress from the measured load during testing.

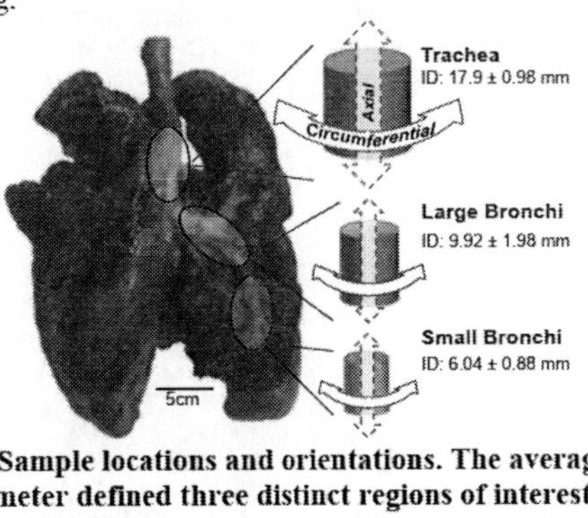

Figure 1: Sample locations and orientations. The average inner diameter defined three distinct regions of interest.

An Instron 5848 Microtester and 10N load cell was used for uniaxial tensile tests. The sample was preconditioned with a triangular waveform five times to 35% elongation at 1%/s [2, 3]. Preliminary tests detected no difference between 0.2%/s, 1%/s or 5%/s. This was immediately followed by a ramp to 35% elongation at 1%/s that was held for 300s before unloading at the same rate. The grip-to-grip distance was treated as the sample's reference length and care was taken to avoid applying stress during mounting. Biological samples typically apply a tare load preceding data collection to exclude the convex portion of the stress-strain response [3]. However, samples drastically differ in their regional and directional behavior, and imposing an identical tare load was unsuitable. After data collection a tare load was added to ensure consistent sample treatment.

For each sample, a stiffness modulus was derived from the linear portion of the stress-strain curve up to 15% and a two-term exponential function was fit to the 300s stress relaxation. The percent relaxation was defined by reduction in stress over the 300s hold divided by the peak stress. All data were subject to optimal Box-Cox transformation and analyzed via multifactor general linear models ($p<0.05$) with donor included as a random variable and Bonferroni's test for pairwise comparisons. Presented values are mean ± standard error of the mean.

RESULTS

Uniaxial tensile tests revealed substantial regional and directional variations in airway stiffness and viscous response as illustrated in representative stress-time curves (Figure 2).

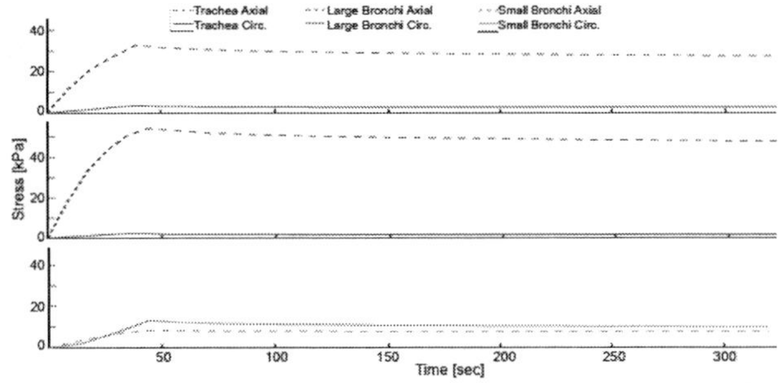

Figure 2: Representative stress-time curves for ramp-and-hold strain profiles for each region and orientation.

The stiffness modulus (Figure 3A) was significantly greater in axial versus circumferential specimens ($p<0.001$), averaging nearly an order of magnitude greater in the axial orientation. The modulus of axial samples did not significantly vary among regions. The modulus of circumferential samples was significantly greater for the small bronchi than for the trachea ($p=0.007$) or large bronchi ($p=0.021$).

In Figure 3B circumferential samples relaxed significantly more than axial samples ($p<0.001$), and the tracheal samples relaxed more than the large bronchi ($p=0.005$) and small bronchi ($p<0.001$).

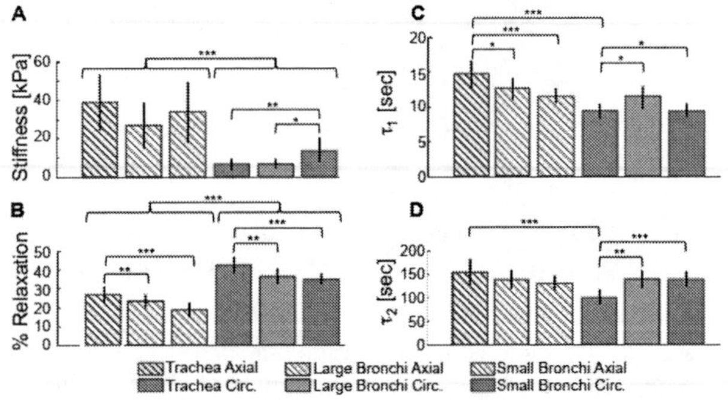

Figure 3: Stiffness modulus (A), and relaxed stress percentage (B), and viscous time constants τ_1 (C) and τ_2 (D).

The first (shorter) relaxation time constant (τ_1, Figure 3C) for trachea samples was greater for axial than for circumferential samples ($p<0.001$), with no significant anisotropy for small or large bronchi. Among axial samples, τ_1 was greater for the trachea than for the large bronchi ($p=0.040$) and small bronchi ($p<0.001$). Among circumferential samples, τ_1 was smaller for the trachea than for the large bronchi ($p=0.037$) and small bronchi ($p=0.030$). The second (longer) time constant (τ_2, Figure 3D) was lower in trachea circumferential samples than in trachea axial samples ($p<0.001$) or circumferential samples from large bronchi ($p=0.002$) and small bronchi ($p<0.001$).

DISCUSSION

While the use of porcine tissue limits direct extrapolation to human airways, the similarity between human and animal lung structure supports drawing qualitative inferences. In this fundamental study we have introduced an understanding of mechanical stiffness and viscous behavior informed by airway orientation and location. While prior works have established the anisotropic response in human and animal tracheal tissues, none have explored the stiffness or viscous behaviors of the smaller airways embedded within the parenchyma, highlighting an important void in lung mechanics research.

Airway tissue exhibits mechanical behavior similar to many well-established biological structures [5]. This study agrees with previous works claiming strong material anisotropy in the trachea [1], and indicates that similar patterns of anisotropy hold for distal airways as well. Moreover, viscous anisotropy is also observed, both proximally and distally.

The airways exhibit a region-dependent mechanical response. Preliminary studies comparing the circumferential stiffness response of the trachea to the immediate adjacent large bronchi found no difference in stress-strain response, in agreement with our findings [2]; however, dissimilar material properties are revealed when comparing the trachea and large bronchi to small bronchi: small airways are circumferentially stiffer than other larger regions, which was previously unaccounted for. Correspondingly, the axial response of the tissue does not significantly vary based on tissue location.

Stress relaxation and time constants exhibited significant differences between airway sample subsets, with circumferential tissue relaxing more, and the trachea relaxing the most. However, initial correlations allowed us to conclude the observed trends are not simply due to differences in the geometry of the trachea, large bronchi, and small bronchi, but rather the inherent composition of the airways that vary in accordance with orientation and location.

In conclusion, the observed directional and regional stiffness variations emphasize the non-uniform and anisotropic nature of the bronchial tree. The results of this study determine airway-specific material behavior, informing future computational simulations, which currently utilize generic, isotropic, biomechanical models and parameters. The novel stiffness and viscous documentation of anisotropic and region-dependent airway mechanics directly impacts clinical questions regarding the damage seen throughout the airway tree, moving beyond the trachea, to explain life-threatening airway complications such as ventilated-induced lung injury.

ACKNOWLEDGEMENTS

This study was supported by the DARE Doctoral Fellowship to Mona Eskandari and by the Ford Fellowship to Alberto L. Arvayo.

REFERENCES

[1] Codd, SL et al., *J Appl Physiol*, 76.6: 2627-2635, 1994.
[2] Noble, P et al., *J Appl Physiol*, 99.6: 2061-2066, 2005.
[3] Teng, Z et al., *J Biomech*, 45.9: 1717-1723, 2012.
[4] Eskandari, M et al., *J Theor Biol*, 403: 209-218, 2016.
[5] Fung, Y, *Biomechanics: Mechanical Properties of Living Tissues*, 2013.

FATIGUE FAILURE OF SIMULATED NETWORKS: EFFECT OF NETWORK ARCHITECTURE ON ITS FATIGUE BEHAVIOR

Rohit Y. Dhume (1), Victor H. Barocas (2)

(1) Department of Mechanical Engineering
University of Minnesota,
Minneapolis, MN, USA

(2) Department of Biomedical Engineering
University of Minnesota,
Minneapolis, MN, USA

INTRODUCTION

Structures composed of fiber networks are extremely common in nature. Several consumer products, such as paper, are composed of networks of polymeric elements with diameters on the order of micrometers. Networks of interconnected filaments or fibers are also prevalent in the biological world, forming hierarchical structures of varying sizes in many connective tissues (e.g., sclera, adventitia, facet capsular ligament).

Being widespread in both natural and manufactured load-bearing structures, networks have received much research attention with a tremendous amount of work focused on constitutive equations governing the macroscale mechanics of materials [1]. Recent developments in multiscale modeling have further motivated research to relate macroscale material behavior to parameters of the underlying microscale network, e.g., [2]. In spite of the advances, fatigue failure of networks, especially networks in biological tissues, is still not well understood. Tissues undergo many cycles of varying loads over a person's lifespan, e.g., tendons in the body experience fluctuating tensile stress during daily activities. Even though tissues remodel over time, fatigue can still become increasingly relevant in certain tissues with age as the metabolic rate of turnover of macromolecular fibers, e.g., collagen, slows down [3]. This tendency to fatigue can lead to injuries and a reduction in the quality of life, e.g., [4].

Obviously, the fatigue behavior of a fiber network must depend on the individual constituent fiber properties. Additionally, geometric effects arising due to the networks architecture can potentially affect its fatigue behavior. Thus, the objective of this work was to determine how the fatigue behavior of a network depends on its geometry and how it differs from that of its individual fibers. This question was examined in the context of simulated networks subjected to cyclic uniaxial loading until failure.

METHODS

Network Generation and Structure: Three-dimensional fiber networks were represented as trusses composed of non-linear elastic cylindrical members (fibers) connected by means of pin-joints (nodes) allowing free rotation. Geometrically different networks were generated from random seed points using three different algorithms: Delaunay (D), Voronoi (V), and Erdös-Rényi (ER). Fiber networks thus generated were isotropic and were trimmed to fit within a cube of unit dimension in the computational domain. The physical size of the network was related to the dimensions of the fibers and their desired volume fraction within the network using the relation [5]

$$X = \sqrt{\frac{\pi r^2 \sum l}{\varphi}} \tag{1}$$

where X is the conversion factor relating the physical and computational domains, r is the fiber radius, Σl is the total fiber length in the network, and φ is the fiber volume fraction.

Network Mechanics and Failure: Network fibers were modeled as one-dimensional non-linear springs with a constitutive equation [6],

$$f = \frac{EA}{B}\left(e^{\left(\frac{\lambda^2-1}{2}\right)} - 1\right) \tag{2}$$

where f is the force generated within the fiber, A is its cross section area, E controls the fiber stiffness, B is a non-linearity parameter, and λ is the fiber stretch. Failure was introduced into the model by removing individual fibers if the fiber stretch exceeded a critical value or if the cumulative fatigue damage fraction calculated from Miner's rule [7] exceeded 1. Since the primary focus of this work was a comparison between fiber and network behavior, any reasonable choice of fiber fatigue properties would suffice. Thus, fibers were assigned an exponentially decreasing stress-life (SN) curve with parameters selected such that fiber fatigue failure occurred within a computationally acceptable number of cycles.

At the network scale, uniaxial deformation due to an external load was simulated as shown in the flowchart (Figure 1).

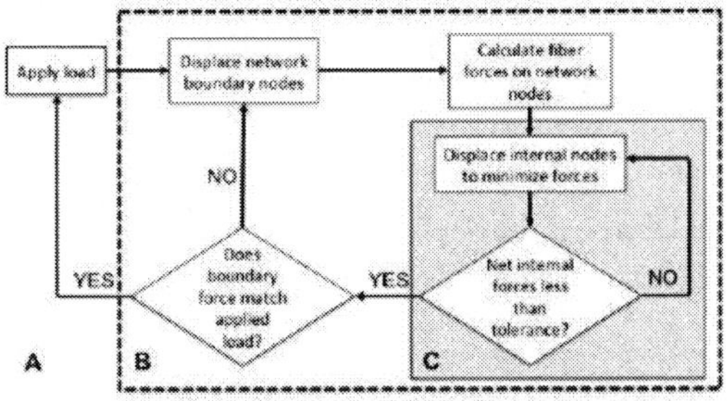

Figure 1: Flowchart of the network model. The outer loop (A) specifies the external load. The intermediate loop (B) updates the boundary displacements to match the external load. The inner loop (C) calculates the equilibrium positions of the internal nodes for a specified boundary displacement.

Failure of the network was defined as a drop in the network stress for any further increase in network stretch mathematically represented as

$$\frac{d\sigma^{net}}{d\lambda^{net}} < 0 \quad (3)$$

where σ^{net} and λ^{net} are the stress and stretch of the overall network respectively, measured in the direction of deformation. Ten networks each of the different network architectures were subjected to a cyclic load-to-failure test to simulate fatigue and to obtain an SN curve.

RESULTS

The physical dimension of the networks was on the order of 10μm. The networks were confirmed to be sufficiently resolved, i.e., the network stress-strain response did not change with a further increase in the number of nodes and fibers. Table 1 shows the average network parameters for each of the three network types.

Table 1: Average network parameters

Type	Total fiber length	Mean fiber length	Mean nodal degree	Number of nodes	Number of fibers	Physical network size, μm
D	127.90	0.23	15	397	560	8.90
V	68.40	0.11	4	430	612	6.50
ER	366.38	0.60	17	590	660	15.10

Figure 2 shows mean SN data for simulations of each network type. The amplitude of the oscillating external load applied on the networks was varied to produce network failure from 1 to 30 cycles. The semi-log SN curves for all three network types fit well to a line in the high stress-low cycle region with some deviation in the low stress-high cycle limit.

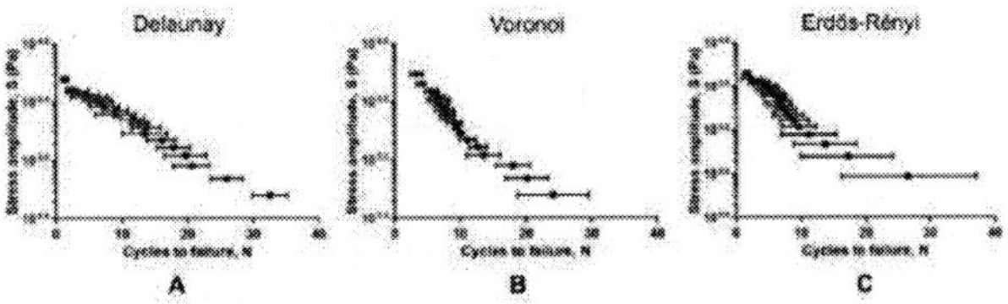

Figure 2: Mean SN data for the three network types.

A comparison between the SN curves of the different network types and that of the fiber is shown in Figure 3. The fiber SN curve was scaled by a factor of $\frac{\varphi}{3}$ to account for the difference in the magnitude of stresses experienced by the fiber and the network due to differences in their cross-section areas. The scaling factor was calculated by analyzing a lattice network under uniaxial loading in which all fibers share the applied load equally.

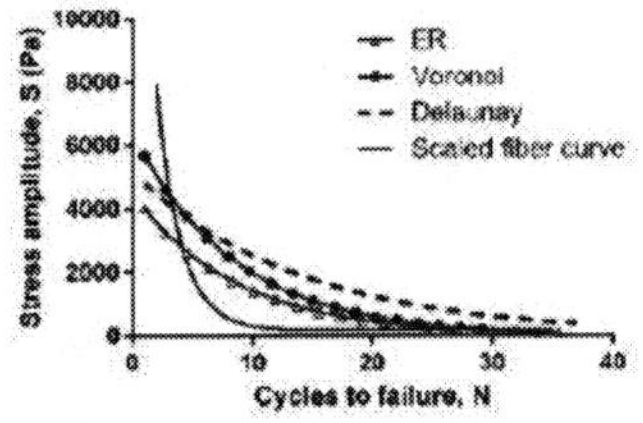

Figure 3: Comparison between fiber and network SN curves.

Figure 4 shows the cumulative distribution of fiber stresses in the networks at two different external loads. The Voronoi networks had more loaded fibers, but most of them were at lower stresses than in the other two network types.

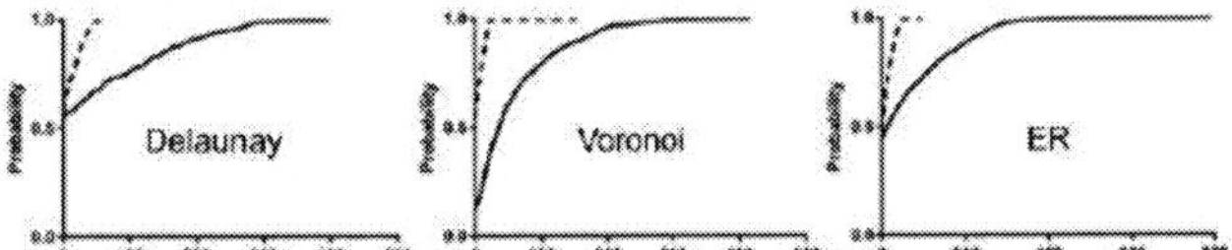

Figure 4: Cumulative distribution plots of fiber stresses for two applied loads, 300 N (dashed), and 1500 N (solid).

DISCUSSION

Fatigue loading until failure was simulated on three distinct fiber network geometries. SN curves obtained for the different network types were compared with that of the individual fiber. The SN curves demonstrated that the network structure was mechanically more endurant compared to the fiber. At high loads, in addition to fatigue, fiber failure in the network also occurred due to fibers overstretching which resulted in the networks' N values falling behind those of the fiber (Figure 3).

Among the different network geometries, the ER networks consistently performed poorly on the fatigue tests. At a given network load, the longer average length of fibers in the ER networks resulted in the fibers experiencing larger stretches, and consequently larger stresses, compared to the other two network types (Figure 4C), reducing the overall life of the ER network.

The Delaunay networks fared better at low to moderate loads while the Voronoi networks lasted longer at higher loads. Failing to satisfy the Maxwell stability criterion, which requires a stable truss-like network to have an average nodal degree greater than 6, the Voronoi networks, at low loads, had only a few fibers bearing the entire stress (Figure 4B). At higher loads, however, the stresses within the Voronoi networks were distributed over a greater number of fibers, which resulted in the Voronoi networks outlasting the Delaunay.

ACKNOWLEDGEMENTS

The work was supported by the National Institutes of Health (R01EB005813) and by the Minnesota Supercomputing Institute.

REFERENCES

[1] Flory, P.J. and Rehner, J. Jr., *J. Chem. Phys.*, 11:512, 1943. [2] Suki, B. et al., *Pulm Pharmacol Ther*, 25(4):268-275, 2012. [3] Freeman, M.A.R., *J. Orthop. Res.*, 17:3-8, 1999. [4] Andarawis-Puri, N. et al., *J. Musculoskelet. Neuronal Interact.*, 11:106-114, 2011. [5] Chandran, P.L. et al., *J. Biomech. Eng.*, 128, 2006. [6] Billiar, K.L. and Sachs, M.S., *J. Biomech Eng.*, 122:327-335, 2000. [7] Miner, M.A., *J. Appl. Mech.*, 67:159-164, 1945.

SB³C2017
Summer Biomechanics, Bioengineering and Biotransport Conference
June 21 – 24, Tucson, AZ, USA

BIAXIAL MECHANICAL PROPERTIES OF VENOUS VALVE LEAFLET TISSUES

Jiaqi Lu (1), Adam A. Benson (2), Hsiao-Ying Shadow Huang (3)

(1) Mechanical and Aerospace Engineering
North Carolina State University
Raleigh, NC, USA

(2) Mechanical and Aerospace Engineering
North Carolina State University
Raleigh, NC, USA

(3) Mechanical and Aerospace Engineering
North Carolina State University
Raleigh, NC, USA

INTRODUCTION

Chronic venous insufficiency (CVI) is widely accepted as the seventh most common chronic debilitating disease and has a huge impact on human health all over the world. In the United States, 10% to 35% of adults are affected by CVI [1]. Venous valve incompetence is a major cause of this disease. When the venous valves become incompetent, blood reflux can occur, which results in blood pooling in the lower extremities and therefore causes CVI.

Tissue engineered prosthetic venous valve is a potential effective treatment of the incompetent valves and the mechanical properties and biochemical components of native venous valve tissues are critical inputs to the proper design of functional prosthetic valves. To date, only uniaxial tensile studies have been reported on venous valve leaflet tissue mechanical properties [2]. However, it was inadequate to describe the complex mechanical behaviors of venous valve leaflets, whose native mechanical loading state is intrinsically multi-axial. In our study, we conducted equibiaxial tensile tests and collagen assay on bovine jugular and saphenous vein valve leaflet tissues to better understand their biaxial nonlinear and anisotropic mechanical properties and quantitatively analyze their collagen concentration.

METHODS

Bovine jugular and saphenous venous valves were used in lieu of human tissues due to their relatively large size, ease to access, and the previous usage in prosthetic studies [1]. The jugular venous valves (JV) had the dimension of 25~30mm in the circumferential direction, with 8~15mm in the radial direction. The saphenous venous valves (SV) had the dimension of 10~15mm in the circumferential direction and 3~5mm in the radial direction.

The leaflets were collected from the veins and put into biaxial mechanical testing (Fig. 1) and collagen assay. The central belly region of each leaflet specimen was isolated and mounted onto the BioTester. The specimens were stretched equibiaxially in both circumferential and radial directions to 60% strain at the rate of 1%/s under displacement control. Corresponding displacement and force data were recorded through the load cells.

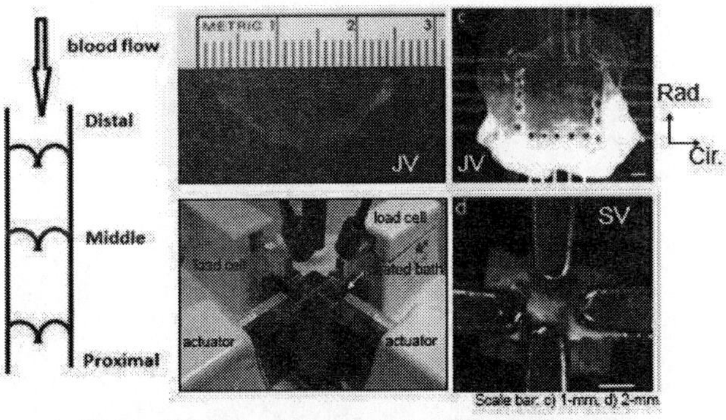

Figure 1: Mechanical testing for JV and SV tissues

After mechanical testing, the specimens were then frozen in preparation for collagen assay. The specimens were weighed and put into collagen extraction solution for 120h. Then the collagen concentration of each sample was obtained by adding dye reagent and analyzing the 550nm light absorbance of each sample.

RESULTS

Bovine jugular valves had non-linear anisotropic mechanical properties (Fig. 2). Tangent moduli of elasticity in the circumferential direction were larger than those in the radial direction for all the samples ($p<0.05$), which indicated the venous valves were stiffer in the circumferential direction than in the radial direction (Table 1).

For JV tissues, specimens from different locations of the vein were separated into three groups, as the Proximal, which was closer to the heart, the Middle, and the Distal. For SV tissues, the specimens were separated into the Proximal and the Distal groups. By comparing the leaflet properties from different locations, we observed that all JV and SV Distal tissues showed anisotropic mechanical properties, while SV Proximal tissues showed isotropic properties. Moreover, the JV tissues showed an increasing trend in collagen concentration from the Proximal to the Distal groups, while the SV tissues showed an decreasing trend (Fig. 3). The difference among the leaflets from different locations indicated that inter-valvular variabilities existed among the venous valve leaflet tissues. As the blood flows back to the heart from the distal end to the proximal end, leaflets from different locations are exposed to different pressure levels, which might have caused such inter-valvular variability.

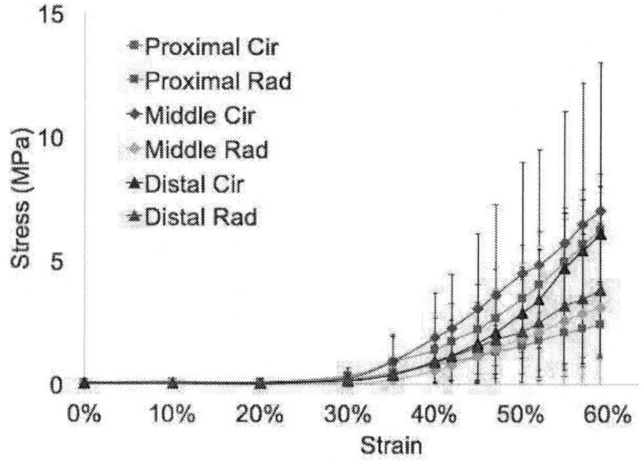

Figure 2 JV stress-strain curves

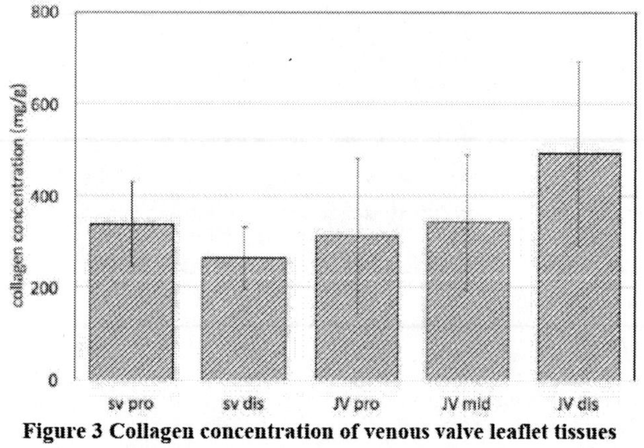

Figure 3 Collagen concentration of venous valve leaflet tissues

During dissection process, the JV valves were observed as either bicuspid (i.e. had two leaflets), or tricuspid (i.e. had three leaflets), and the SV valves were all bicuspid. By comparing the properties of the leaflets from the same valve location, we observed that they did not possess the same properties. One of the leaflet appeared to have stronger mechanical properties than the other one, which indicated that intra-valvular variability existed among the leaflet tissues.

CONCLUSION

Bovine jugular venous valves could be either bicuspid or tricuspid, while all bovine saphenous venous valves were bicuspid. All of the jugular valve leaflet and distal saphenous valve tissues possessed anisotropic non-linear mechanical properties. The leaflets appeared to be stiffer in the circumferential direction than in the radial direction. Proximal valves from the saphenous veins possessed isotropic mechanical properties. The collagen concentration of the tissues showed an increasing trend from the proximal end to the distal end in jugular veins, while the opposite in saphenous veins. By comparing the mechanical properties, valve leaflets from the saphenous veins appeared to be much stiffer than those from the jugular veins (Table 1). The difference in blood pressure levels might have caused such variabilities. To the best of the author's knowledge, this is the first study reporting the biaxial mechanical properties of venous valve leaflet tissues and thus contributes toward refining our collective understanding of valvular tissue biomechanics.

	JV Proximal		JV Middle		JV Distal	
	Cir	Rad	Cir	Rad	Cir	Rad
Tangent modulus of elasticity (MPa)	**26.98 ±9.36**	**8.89 ±5.57**	**23.49 ±20.21**	**10.35 ±16.26**	**29.72 ±6.18**	**15.42 ±9.46**

	SV Proximal		SV Distal	
	Cir	Rad	Cir	Rad
Tangent modulus of elasticity (MPa)	**62.66 ±34.88**	**41.61 ±16.01**	**92.08 ±27.86**	**31.76 ±13.87**

Table 1: Tangent moduli of elasticity for venous valve leaflet tissues

ACKNOWLEDGEMENTS

This research was supported by NSF CAREER GARDE 1553430.

REFERENCES

[1] C. Zervides. *J. Endovasc. Ther.*, vol. 19, no. 2, pp. 281–90, Apr. 2012.

[2] J. S. Ackroyd, et al. *Br. J. Surg.*, vol. 72, no. 2, pp. 117–119, 1985.

IN VIVO COMPARISION OF MYELIN AND STIFFNESS MAPS IN THE HUMAN BRAIN

E. Ozkaya (1), M. Wintermark (2), M. Kurt (1)

(1) Department of Mechanical Engineering
Stevens Institute of Technology
Hoboken, NJ, USA

(2) Department of Radiology
Stanford University
Stanford, CA, USA

INTRODUCTION

Myelin, whose dry mass is made up of 70-85% lipids and 15-30% protein, consists of 40% water [1]. It covers the axonal nerves to enhance the signal propagation. In humans, it is seen that myelination process extends at least through the first 20 years of life but its peak occurs during the first year of life [2,3,4]. In addition to enhancing signal propagation, myelin changes the overall stiffness of the brain tissue by dramatically changing the tissue composition. Previously, it has been demonstrated through histological and mechanical indentation tests on bovine brain tissue that, in different regions of cerebral white matter, white matter stiffness increases with myelin content [1]. However, no direct evidence of the relationship between stiffness and myelin content has been acquired *in vivo* up-to-date.

Recently, advanced imaging tools have been developed that can help establish this relationship: **1)** Non-invasive imaging tools based on MRI have been proposed for *in vivo* myelin mapping, such as conventional T1-weighted (T1-w) and T2-weighted (T2-w) imaging, magnetization transfer imaging, diffusion tensor imaging (DTI), fluid-attenuated inversion recovery (FLAIR), multi-component T2-relaxation imaging (MCRI), and multi-component Driven Equilibrium Single Pulse Observation of T1 and T2 (mcDESPOT) [5]. Among these methods, T1-w gives a level of brightness depending on the myelin-bond cholesterol and T2-w gives less intense hydrophobic lepidic bilayer. **2)** Magnetic resonance elastography (MRE) has been gaining attention as a non-invasive means for *in vivo* mechanical measurement of biological tissues, especially the brain tissue.

In this study, we aim to clarify the ambiguity about the relation between myelin content and brain stiffness in human brain through *in vivo* measurements. The correlation between myelin content and brain stiffness might result in an improved clinical tool for early diagnosis of neurodegenerative diseases arising from demyelination. In other words, accurate assessment of myelin-stiffness relation in vivo is extremely important for a comprehensive understanding of human neurodevelopment.

METHODS

In this study, 3T MR Scanner (GE Healthcare) and multi-frequency magnetic resonance elastography (Mayo Clinic, Rochester, MN) was used to investigate the correlation between myelin content and stiffness. We varied the vibration frequencies of the MRE actuator in an acoustic range from 40 to 80 Hz to create a database of brain viscoelasticity with respect to varying frequencies of shear vibrations in the brain tissue. The analysis conducted in this abstract was done with a 7-year-old male volunteer.

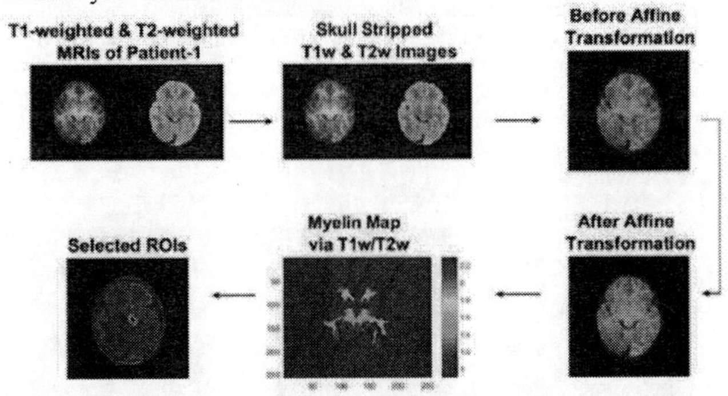

Figure 1: Analysis workflow. ROI-1: Telencephalon left within blue boundary. ROI-2: Frontal lobe right within orange boundary. ROI-3: Thalamus right within yellow boundary. ROI-4: Leniform nucleus right within purple boundary.

T1-weighted gradient images (BRAVO) were acquired for 160 slices and T2-weighted scans were acquired for 48 slices.

Analysis workflow, as shown in Figure 1, starts with skull stripping to avoid any correlation increase that can be caused by overlapping skull portions. This is followed up by affine transformation to overlay the T1-w and T2-w images. Then the whole myelin mapping is obtained by using T1-w/T2-w technique [5,6].

Regions of interest (ROIs) are selected manually on a T2-w image to correspond to different regions in the white matter. Anatomical names for the chosen four ROIs in Figure 1 are; (1) Telencephalon left within blue boundary, (2) Frontal lobe right within orange boundary, (3) Thalamus right within yellow boundary and (4) Leniform nucleus right within orange boundary.

For these selected ROIs, the correlations between T1-w/T2-w image and MRE scans (*i.e,* storage moduli, loss moduli and stiffness maps) with at frequencies 40 Hz, 60 Hz and 80 Hz are investigated as depicted in Figure 2. We used the MATLAB function corr2 to calculate the corresponding correlation coefficients for each ROI.

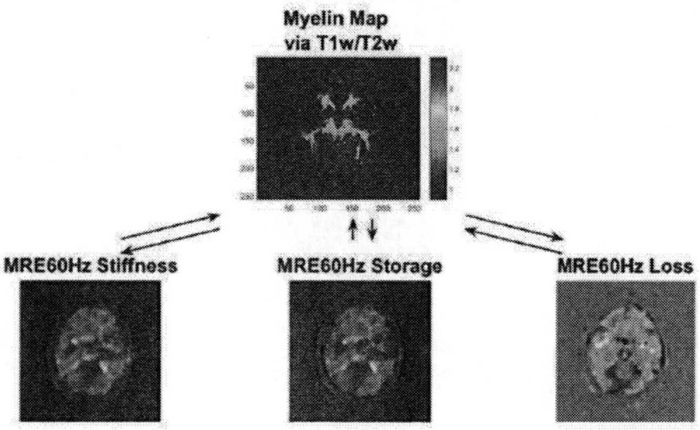

Figure 2: Correlation analysis is carried out among T1w/T2w and MRE60Hz images, stiffness, storage and loss moduli maps respectively.

RESULTS

The corresponding correlation results between T1-w/T2-w and MRE stiffness maps are reported in Table 1. Surprisingly, the correlation values between these two metrics are very high, especially for the elastic component, storage modulus (Figure 3). Most of the correlation values for storage and stiffness maps are above 0.80. Although we observe some low correlation values between the T1w/T2w and Loss Modulus as given in Table 1, some of these might have to do with the relative sensitivity of the loss moduli measurements to the transmitted and attenuated shear waves at low MRE frequencies.

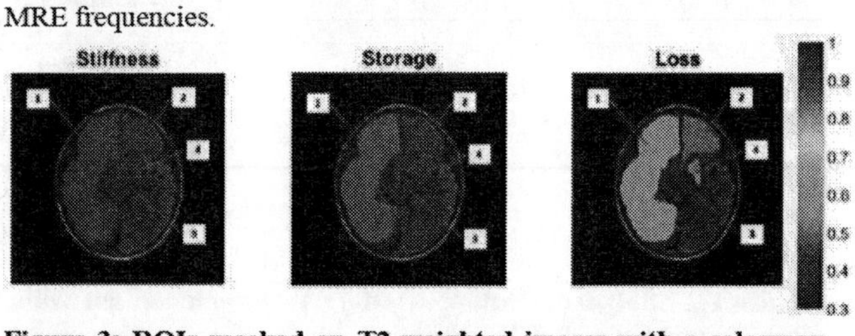

Figure 3: ROIs masked on T2-weighted images with a colormap depending on obtained correlation values among T1-w/T2-w and MRE60Hz images.

Table1: Correlation with the T1-w/T2-w values of each ROIs with respect to different MRE frequencies and modalities.

Experiment		ROI-1	ROI-2	ROI-3	ROI-4
MRE40Hz	**Stiffness**	0.8685	0.9265	0.9529	0.8611
	Storage	0.8639	0.9274	0.9528	0.8387
	Loss	0.4278	0.6747	NA	0.2156
MRE60Hz	**Stiffness**	0.8836	0.9205	0.9622	0.9220
	Storage	0.8618	0.9226	0.9666	0.9096
	Loss	0.5985	0.8337	NA	0.8082
MRE80Hz	**Stiffness**	0.8797	0.9110	0.9526	0.9409
	Storage	0.8624	0.9099	0.9525	0.9411
	Loss	0.7038	0.8663	0.3665	0.8194

DISCUSSION

Performed correlation analysis showed that myelin content could contribute to high stiffness as in the ROIs selected. Correlation values were relatively insensitive to the MRE actuation frequency for storage modulus and stiffness (Table 1). Correlations between the qualitative myelin measure and storage and stiffness values were noticeably higher than loss modulus. These lower correlation values for the loss moduli merits further review since the effect of myelin on the dissipative properties of the brain tissue has not been investigated in the prior art.

One main limitation of our study was the qualitative myelin content measure obtained by using the T1-w/T2-w method, which might not be ideal for a detailed investigation of the effect of myelin on tissue stiffness. To overcome the limitation caused by T1w/T2w method, alternative myelin measuring methods can be proposed. In [2], serial block face imaging by focused ion beam milling coupled to scanning electron microscopy (FIB-SEM), together with cyro-preperation methods, such as high-pressurized freezing, was suggested to be used to follow up myelin formation during its development. Alternatively, more advanced myelin imaging techniques such as quantitative magnetization transfer imaging (q-MT) can be utilized for future studies for more robust correlation analysis [10].

Causes of high and low correlations among regions of interest (ROI) can be investigated to find out the underlying reasons of correlation changes. One reason of these changes might be due to the different fiber directions inside the chosen ROIs, which effect the transmission of the shear waves through MRE and effectively change the stiffness measurements.

ACKNOWLEDGEMENTS

This work has been funded by Lucile Packard Children's Hospital through a Postdoctoral Research Grant. Dr. Kurt is the recipient of the grant. The authors would also like to acknowledge GE Healthcare for the generous support in scan time.

REFERENCES

[1] Weickenmeier, J et al., *Acta Biomateriala*, 42:265-272, 2016.
[2] Snaidero, N et al., *Journal of Cell Science*, 127:2999-3004, 2014.
[3] Fields , R D, *Trends Neurosci.*, 31:21-43, 2008.
[4] Paus, T et al., *Science*, 283:1908-1911, 1999.
[5] Ganzetti, M et al., *Front Hum. Neurosci.*, 8/671:1-14, 2014.
[6] Ganzetti, M et al.,G et al., *Neuroradiology*, 57:917-928, 2015.
[7] Haast, R A M et al., Frontiers in Neuroanatomy, 10/12:1-17.
[8] Kruse, S A et al., *NeuroImage*, 39:231-237, 2008.
[9] Murphy, M C et al., *PLoSONE*, 8/12:1-14, 2013.
[10] Samsanov A., *NeuroImage*, 62:1390-1398, 2012.

INFLUENCE OF SIZE AND SHAPE ON THE BIOMECHANICAL ENVIRONMENT OF THE HUMAN LAMINA CRIBROSA: A STUDY ON RACIOETHNIC DISPARITY

Hirut Kollech (1), Reza Behkam (2) Jonathan P. Vande Geest (1,2,3,4)

(1) Computational Modeling and Simulation, University of Pittsburgh, 326 Eberly Hall, Pittsburgh PA 15260

(2) Department of Bioengineering, University of Pittsburgh, 300 Technology Drive, Pittsburgh PA 15219

(3) Louis J. Fox Center for Vision Restoration, University of Pittsburgh, Pittsburgh, PA 15219

(4) McGowan Institute for Regenerative Medicine, University of Pittsburgh, Pittsburgh, PA 15219

INTRODUCTION

Across multiple prospective clinical studies of patients with glaucoma and ocular hypertension, age and ethnicity appear as significant risk factors for the development and progression of glaucomatous optic neuropathy [1-2]. While the reasons for these disparities are multifactorial [3, 4], there is sufficient evidence to suggest that differences in the biomechanical properties of the load bearing connective tissues of the optic nerve and posterior sclera between these racial and ethnic groups contribute to this increased risk [5-6]. The biomechanical model of glaucoma considers IOP-related stress and resultant strain on load bearing connective tissues of the optic nerve as one major causative influence that effects cellular, vascular and axonal components of the optic nerve [6]. Within the context of this model, differences in the ONH microstructure and mechanical properties of at-risk populations may provide a mechanism for increased susceptibility to optic nerve injury at all levels of IOP. Our research is focused on identifying if differences in the deformationally-dependent ONH microstructure and biomechanical behavior are different between at-risk minority groups using a combined nonlinear optical microscopy and finite element modeling approach. These differences may explain the observed greater predilection to IOP- related injury in these groups.

Our research team has identified differences in the scleral microstructure of those of African Descent (AD) as compared to those of European Descent (ED) [Yan et al., 2011]. This microstructural data suggested that the posterior sclera of AD populations may be at risk for larger increases in scleral canal opening as compared age matched donors of ED. We have also demonstrated that the lamina cribrosa (LC) itself behaves differently in AD and Hispanic Ethnicity (HE, also at higher risk for glaucoma) as compared to those of ED (Figure 1). These data all suggest that the microstructural and biomechanical environment of the human LC may play a role in the higher prevalence of glaucoma clinically seen in those high risk groups.

The purpose of this abstract is to use finite element modeling to better understand if differences in LC size and shape within the optic nerve head of at risk groups, as measured using nonlinear optical microscopy, lead to increased mechanical strains. Such information will connect anatomically measured LC geometry in HE, AD, and ED with mechanical strains in the optic nerve head.

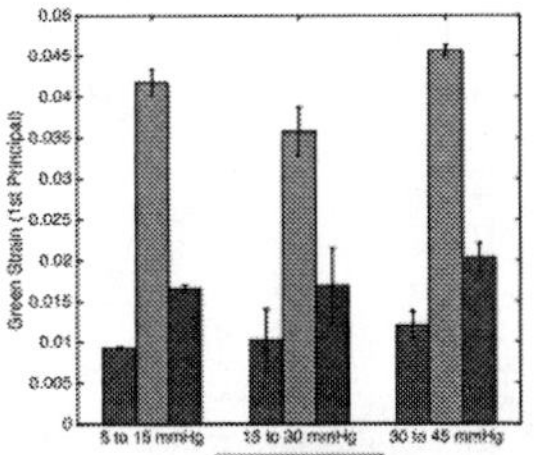

Figure 1. Principal strain differences in the LC of AD, ED, and HIS donors.

METHODS

All samples obtained followed an approved procurement method as outlined by the University of Arizona Institutional Review Board. Twelve human donor eye samples (post-two photon imaging) were used; four (4) each from African Decent(AD), European Decent (ED) and Hispanic Ethnicity (HE). The process began by isolating the LC; followed by a short mechanical and chemical digestion. Finally, the LC was mounted into a pressure chamber where four (4) different pressures (5, 15, 30, and 45 mmHg) were applied and imaged anteriorly with a two (2) photon microscope. Two photon imaging was performed using a Ti:Saph (Mira 900, Coherent, Inc.) tunable 120 fs pulsed light source. The excitation wavelength of the laser was λ= 780 nm and collection was done in the epidirection using a dichroic mirror (405 nm), a 377/50 nm bandpass filter (SHG), and a 460/80 nm bandpass filter. Digital images were acquired using a 4X objective (NA = 0.3) resulting in a pixel size of 2.5μm in the x- and y-direction and 5μm in the z-direction (Figure 2).

The maximum intensity projection for each image set was used to draw a spline around the rim of the sclera and fit an ellipse using MATLAB regionprops (Figure 2). The geometry of this ellipse was later used for the lamina cribrosa dimension for simulation.

For the finite element simulation, shell elements with anisotropic hyperelastic Holzapfel model were used with the following parameters. Lamina C10=50,000 Pa, D=10^{-6} Pa^{-1}, k1=50,000 Pa, k2=225 and kappa=0.3 and Sclera C10=300000 Pa, D=10^{-6} Pa^{-1}, k1=50,000 Pa, k2=50 and kappa=0.05. The sclera and the LC were assumed to be nearly incompressible.

An average thickness of 300 microns was used for the lamina while the mean thickness values calculated from MicroCT measurements were used for each subgroups as follows (AD=0.000743 m, CD=0.001022 m, HE=0.000821 m).

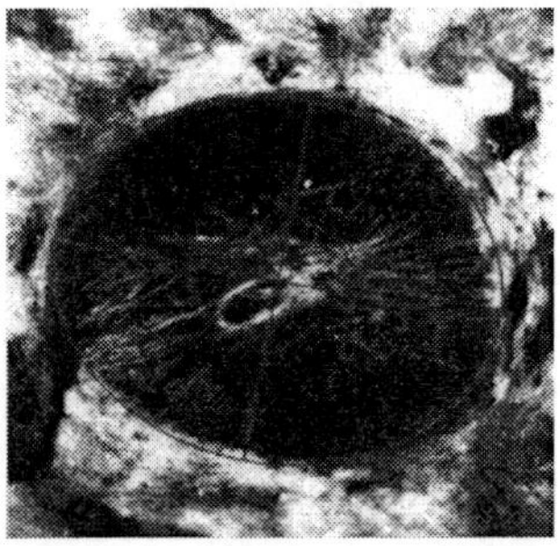

Figure 2. Maximum intensity projection of the two photon second harmonic image with a typical ellipse fit

RESULTS

The mean LC logarithmic strain values for HE, AD, and ED samples were 0.017875+/-0.00125, .021375+/-.0017017 , and .0125 +/-.00297, in first axis (LE11) The mean LC logarithmic strain values for HE, AD, and ED samples were .000372+/-.0018, 0.01375 +\-0.0033, and 0.00089+/-.00327 in the second axis (LE22).

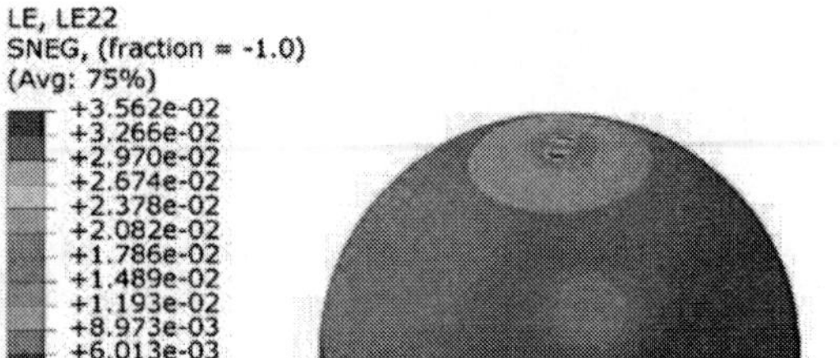

Figure 3. Finite element model used to investigate racioethnic differences in LC strain.

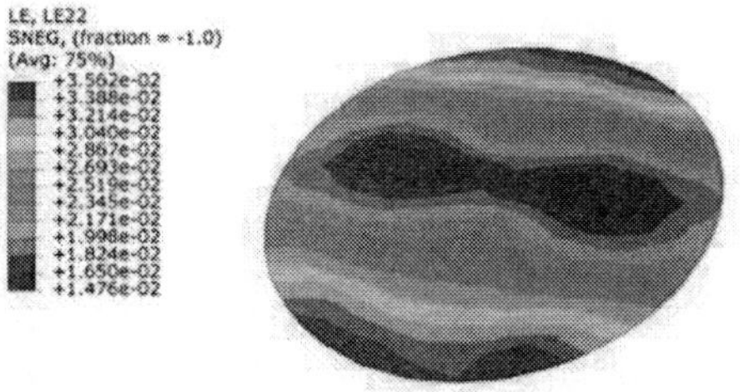

Figure 4. Representative contour plot of logarithmic strain on a human LC

DISCUSSION

One way ANOVA performed to evaluate statistical differences in strain across the subgroups show significant differences in LE22 (p value=.001202), LE11 (p value=0.00071) and the maximum principal strain (p value =0.043526). Further analysis using t-test within the subgroups groups shows the statistical difference between AD and CD (p value=0.003453) and CD and HE (p value =0.019274) in logarithmic strain direction 11. Similarly in the second direction, the logarithmic strain among AD and CD (p value =0.003517), AD and HE (p value=0.016099) and AD and CD (p value=0.028995. Finally the Maximum Principal strain was significantly different between AD and C (p value =0.045661). It is noteworthy to see that LE11 has significantly different results within each subgroup. For this test only 12 samples were used. Future testing should include more samplings.

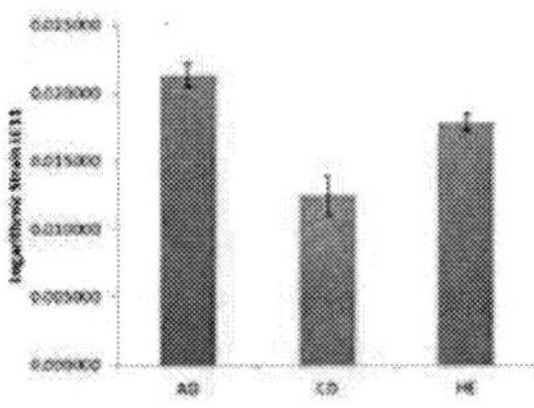

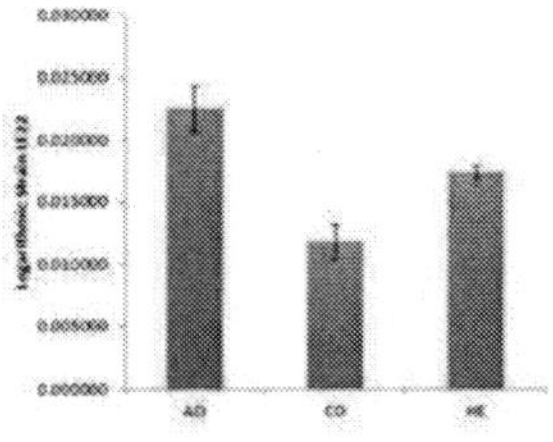

ACKNOWLEDGEMENTS

Research reported in this abstract was supported by the National Eye Institute of the National Institutes of Health under award number R01EY020890 to JPVG.

REFERENCES

1. Drance, S., D.R. Anderson, and M. Schulzer, Risk factors for progression of visual field abnormalities in normal-tension glaucoma. Am J Ophthalmol, 2001. 131(6): p. 699-708.
2. Tielsch, J.M., A. Sommer, J. Katz, R.M. Royall, H.A. Quigley, and J. Javitt, Racial variations in the prevalence of primary open-angle glaucoma. The Baltimore Eye Survey. Jama, 1991. 266(3): p. 369-74.
3. Girkin, C.A., Primary open-angle glaucoma in African Americans. Int Ophthalmol Clin, 2004. 44(2): p. 43-60
4. Racette, L., M.R. Wilson, L.M. Zangwill, R.N. Weinreb, and P.A. Sample, Primary open-angle glaucoma in blacks: a review. Surv Ophthalmol, 2003. 48(3): p. 295-313.
5. Urban, Z., O. Agapova, V. Hucthagowder, P. Yang, B.C. Starcher, and M.R. Hernandez, Population differences in elastin maturation in optic nerve head tissue and astrocytes. Invest Ophthalmol Vis Sci, 2007. 48(7): p. 3209-15. .
6. Seider, M.I., R.Y. Lee, D. Wang, M. Pekmezci, T.C. Porco, and S.C. Lin, Optic disk size variability between African, Asian, white, Hispanic, and Filipino Americans using Heidelberg retinal tomography. J Glaucoma, 2009. 18(8): p. 595-600

THE IMPACT OF CEREBROSPINAL FLUID PRESSURE ON OPTIC NERVE HEAD DEFORMATION

Andrew J. Feola (1), Baptiste Coudrillier (1,2), John Mulvihill (1,3), Diogo M. Geraldes (4), Nghia T. Vo (5), Julie Albon (6,7), Richard L. Abel (8), Brian Samuels (9), C. Ross Ethier (1)

(1) Department of Biomedical Engineering Georgia Institute of Technology/Emory University, Atlanta GA, USA

(2) E□ponent Inc□Menlo, California

(3) School of Engineering/Bernal Institute, University of Limerick, Limerick, Ireland

(4) Biomechanics Group, Department of Mechanical Engineering, Imperial College London, London, United Kingdom

(5) Diamond Light Source, Didcot, United Kingdom

(6) Optic Nerve Group, School of Optometry and Vision Sciences, Cardiff University, Cardiff, Wales, United Kingdom

(7) Cardiff Institute of Tissue Engineering and Repair, Cardiff University, Cardiff, Wales, United Kingdom

(8) Department of Surgery and Cancer, Imperial College, London, United Kingdom

(9) Department of Ophthalmology, U. Alabama at Birmingham, Birmingham, AL

INTRODUCTION

It is thought that biomechanics plays an important role in various ocular pathologies including glaucoma, visual impairment and intracranial pressure (VIIP) syndrome, and idiopathic intracranial hypertension (IIH). The optic nerve head (ONH) experiences a complex biomechanical environment due to exposure to several pressures, including: blood pressure, intraocular pressure (IOP), and cerebrospinal fluid pressure (CSFp) [1]. Previous work has focused on the impact of IOP on ONH biomechanics, with a focus on the lamina cribrosa (LC). This is because the LC is the primary site of retinal ganglion cell (RGC) loss in glaucoma; RGCs transmit visual information to the brain from the eye. Changes in CSFp were shown by optical imaging to also have a profound impact on ONH deformations [2]. However, due to light penetration issues, this study was unable to assess deformation within the LC or retrolaminar neural tissue (RLNT).

Compared to light-based methods, X-ray imaging methods such as micro-computed tomography (μCT) offer better tissue penetration. In traditional μCT, contrast arises from variations in X-ray attenuation by different tissue compositions. To overcome this limitation, phase-contrast techniques indirectly measure the phase-shift that X-rays undergo when passing through a tissue. In soft tissues, X-ray phase-shift signals are typically orders of magnitude larger than X-ray absorption signals. Therefore, PC μCT is an attractive way to noninvasively image the full thickness of the LC and RLNT at high spatial resolution. Here, we used phase-contrast micro-computed tomography (PC μCT) implemented at the beamline I12-JEEP, Diamond Light Source, UK [3], to measure CSFp-induced deformations of the LC and RLNT.

METHODS

The posterior eye and RLNT of porcine eyes (n=3) were imaged using PC μCT from synchrotron radiation. We collected eyes with an intact optic nerve and an optic nerve sheath longer than 2 cm. Samples were prepared following a previously established protocol [4]. In brief, after the posterior eye was isolated and the cornea removed, the posterior scleral shell was sealed onto a resin holder (Figure 1). The resin holders were secured onto pressure chambers filled with PBS in-line with a hydrostatic reservoir to control IOP. To simulate changes in CSFp, we inserted the tip of a 23G blunt cannula between the optic nerve and optic nerve sheath. The cannula was secured by a silk suture tied around the outside of the optic nerve sheath so the tissue was compressed against the cannula.

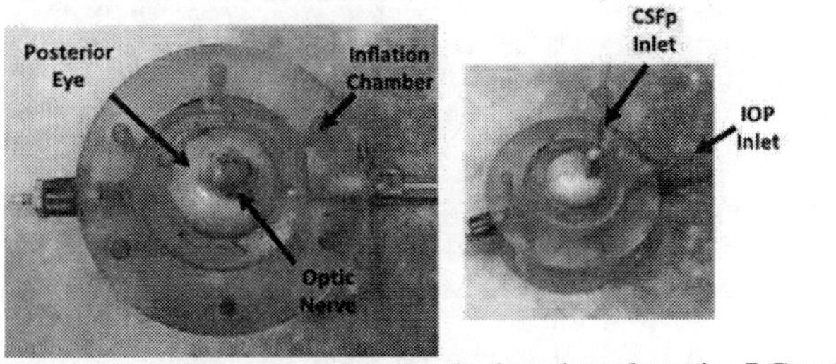

Figure 1: Experimental preparation for imaging of porcine LC and RLNT using PC μCT. IOP was applied within the inflation chamber while CSFp was applied within the optic nerve sheath

This approach allowed us to independently control IOP and CSFp. In this experiment, we held IOP constant at 15 mmHg (average physiological value) and increased CSFp from 4 mmHg to 10, 20 and 30 mmHg. 3D PC μCT of the lamina cribrosa and RLNT were acquired at each CSFp after 20 minutes of equilibration to minimize the influence of viscoelastic effects. We used these images to determine deformation within the LC and RLNT using digital image volume correlation (DVC; DaVis 8.1.3, LaVision, GmbG, Goettingen, Germany) on sub-volumes (20x20x20 voxels) with a 75% overlap. These settings have been previously established on a similar dataset of the porcine LC [4].

We focused on the 1st and 3rd principal strains determined relative to a CSFp of 4 mmHg as a 3D measure of tissue deformation. These represent the maximum magnitudes of strains experienced at the LC and RLNT, which we assume are related to the biomechanical insult experienced by RGCs in the ONH.

RESULTS

We were able to clearly identify the LC, as distinct from the surrounding neural and connective tissue, and the RLNT in the ONH (Figure 2). Our results showed that increasing CSFp caused the optic nerve sheath to expand by distending the dura mater (Figure 3).

We found that increasing CSFp increased the strain in both the LC and RLNT. Overall, the strains induced by higher magnitudes of CSFp were more pronounced in the RLNT; however, we found that CSFp still had an impact on LC deformation (Figure 4). Relative to a baseline CSFp of 4 mmHg, at a CSFp of 30 mmHg the LC experienced mean 1st and 3rd principal strains of 4.4 and -3.5%, respectively. Within the RLNT, elevating CSFp from 4 to 30 mmHg resulted in mean 1st and 3rd principal strains of 9.5% and -9.1%.

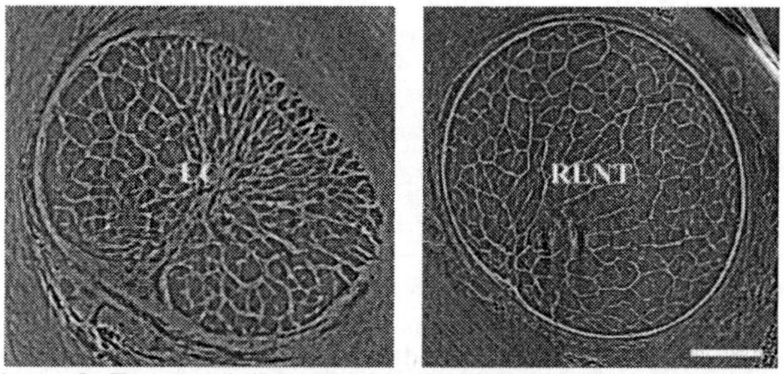

Figure 2: Representative cross-sectional images from our 3D PC μCT images of the LC (left) and RLNT (right). Scale bar = 1 mm

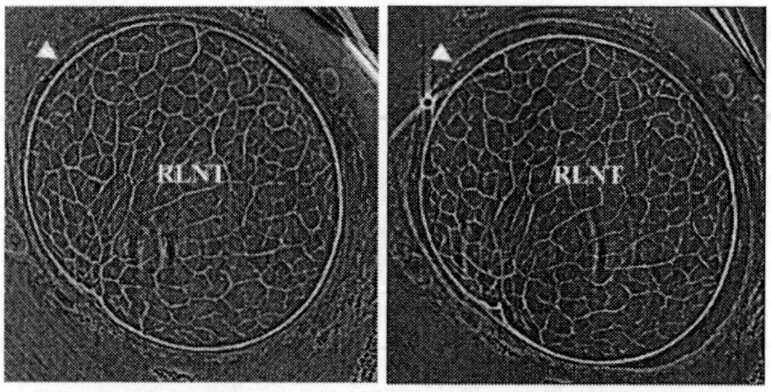

Figure 3: The impact of CSFp elevations on the optic nerve sheath. We can see the optic nerve sheath (arrowheads) expanding as CSFp increased from 4 mmHg (left) to 30 mmHg (right).

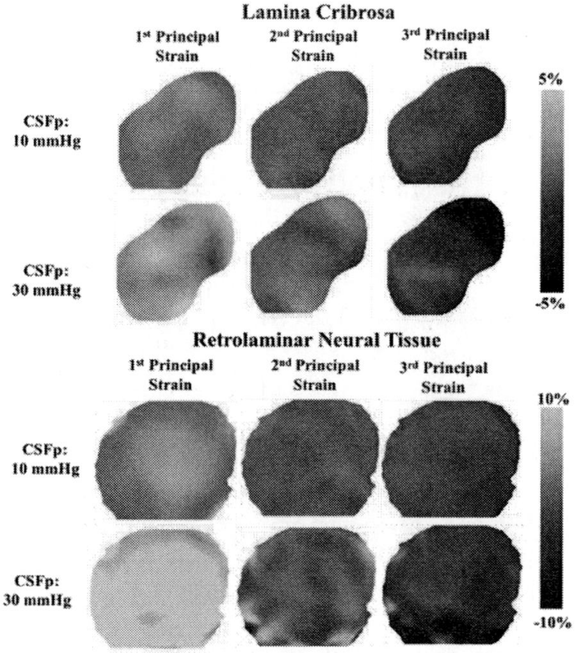

Figure 4: DVC-computed strain contour maps of the LC (top panel) and RLNT (bottom panel). Strains are relative to a CSFp of 4 mmHg. Note the different scale between the LC and RLNT.

DISCUSSION

PC μCT is a valuable tool to assess 3D CSFp-induced deformations of the LC and RLNT. We found that CSFp significantly impacted the strains within the LC and RLNT. Our approach allows us to directly assess the impact of CSFp on the LC and RLNT in 3D with high resolution and accuracy. These results are in qualitative agreement with Morgan et al., who found that elevating CSFp directly impacted optic nerve disc deformation [2].

Major limitations of our study are the limited sample size and no examination of the interactions between IOP and CSFp. However, due to the limited access to a synchrotron light source and length of experiments, we were forced to focus this initial study on investigating the effects of CSFp alone. Fortunately, we found small inter-specimen variation, increasing confidence in our results. Additional studies will examine the interplay between IOP and CSFp.

Overall, elevations in CSFp were positively correlated with increasing deformations in each region, and may play a role in ocular pathologies that are linked to changes in CSFp.

ACKNOWLEDGEMENTS

The authors would like to thank Robert Atwood, Michael Drakapoulos, Kaz Wanelik, and Christina Reinhard from Diamond for assisting on experiments using synchrotron PC μCT imaging and analysis. This work is supported by Science & Technology Facilities Council Grants EE8491, EE9825, and EE11407; the Georgia Research Alliance (CRE); and NASA grant NNX13AP91G.

REFERENCES

[1] Burgoyne, C., *Exp. Eye Res,* 93:120-132. 2011.
[2] Morgan, W.H., et al., *IOVS*, 36:1163-1172, 1995.
[3] Drakopoulos, M., et al., *J. of Synch. Rad.* 2015;22(3):828-838
[4] Coudrillier, B., et al., *IOVS* 57:2666-77, 2016.

SB³C2017
Summer Biomechanics, Bioengineering and Biotransport Conference
June 21 – 24, Tucson, AZ, USA

VASCULAR SMOOTH MUSCLE CELL MECHANO-ADAPTATION DEPENDS ON EXTRACELLULAR MECHANICAL PROPERTIES

Kerianne E. Steucke (1), Kamilah Y. Amen (1), Patrick W. Alford (1)

(1) Biomedical Engineering
University of Minnesota
Minneapolis, MN, USA

INTRODUCTION

Vascular diseases, such as hypertension and atherosclerosis, can perturb the mechanical environment of vascular smooth muscle cells (VSMCs) leading to maladaptive physiological responses. In response to these mechanical changes VSMCs undergo mechano-adaptation, or grow and remodel themselves and their surrounding environment, to maintain vessel integrity and function. Growth theory suggests that VSMCs grow and remodel to maintain a 'target stress', which can be characterized in one dimension by a growth law given by,

$$\dot{G} = A(\sigma - \sigma_o), \tag{1}$$

where $\dot{G}$ is the rate of growth, A is a constant, σ is the cell stress, and σ_o is the target stress [1,2].

Recent findings suggest that changes in the extracellular matrix (ECM) stiffness, as seen in vascular disease, could shift the target stress [3]. However, the impact of an evolving substrate modulus on VSMC mechano-adaptation is unknown. **Here, we experimentally measured VSMC temporal stress evolution after mechanically perturbing both the cell and its environment. These data were then used to generate a substrate-dependent mechano-adaptation law to capture the effects of ECM remodeling on VSMC function in vascular disease progression.**

METHODS

Cell Culture and Construct Fabrication. Human umbilical artery VSMCs were purchased from Lonza and Cell Applications. Micropatterned polyacrylamide gels with fluorescent microspheres were formed onto elastomer membranes fixed between steel clamps as previously published (Fig. 1A) [4].

Traction Force Microscopy Experiments. A uniaxial step change in strain (0%, 10%, 20%, or 30%) was applied to the elastomer membrane and VSMCs (Fig. 1A). Then, bright field and fluorescent images for each cell and the bead layer immediately below, respectively, were taken for each time point (0h, 4h, 8h, and 24h) and after cell lysis (Fig. 1B).

Cell Stress Generation Analysis. Cell induced substrate displacement was determined by comparing the fluorescent bead images when the cell is exerting force and after the cell is lysed using particle image velocimetry software (Fig. 1B) [6]. Traction stress was calculated from the substrate displacements and known substrate properties (E = 5kPa, 13.5kPa, or 100kPa; ν = 0.5) using Fourier transform traction cytometry algorithm software (Fig. 1B) [5, 6].

Substrate-Dependent Mechano-Adaptation Law Determination. Cells were assumed to be Neo-Hookean and undergo growth proportional to the perturbation of their stress away from the target stress (Eq. 1) (Fig. 2A) [4]. Experimental cell stress data were used to fit the substrate-dependent mechano-adaptation law using custom modeling in MATLAB (Fig. 2). The model was adapted to capture the temporal stress evolution of VSMCs exposed to varying chronic strain as previously published and incorporates the effect of a changing substrate modulus on the target stress [3, 4]. ECM production is assumed to be a function of the perturbation of the cell stress away from its target stress, such that the ECM stiffens due to incremented production under increased tension.

$$W_{ecm} = \frac{\alpha}{\beta}\left(e^{\beta(I_1-3)} - 1\right) \tag{2}$$

$$\dot{\alpha} = B(\sigma - \sigma_o) \tag{3}$$

Where α and ß are material parameters, I_1 is the first strain invariant, σ_o is the cell target stress, σ is the cell stress, and B is a constant. The target stress is assumed to be a linear function of ECM stiffness with a slope m_σ.

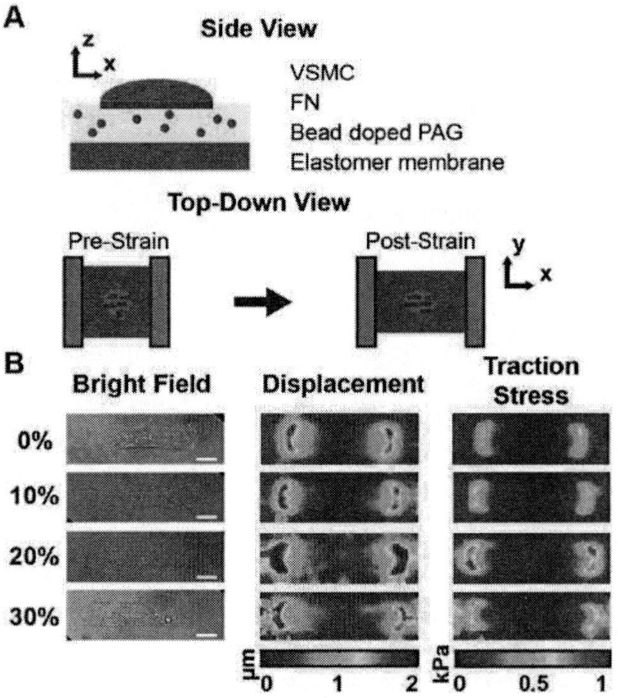

Figure 1. A. Construct Fabrication. Side View: A VSMC is constrained to the micropatterned FN island printed onto a florescent-bead doped polyacrylamide gel. The polyacrylamide gel is bonded to an elastic membrane. Top-Down View: The elastic membrane is secured between two steel clamps and uniaxial chronic strain is then applied. **B. VSMC Stress Determination.** Bright field images after 0%, 10%, 20%, or 30% strain is applied. Scale Bar: 25µm. Average substrate displacement and average traction stress heat maps for 0h post applied strain (n=19-92).

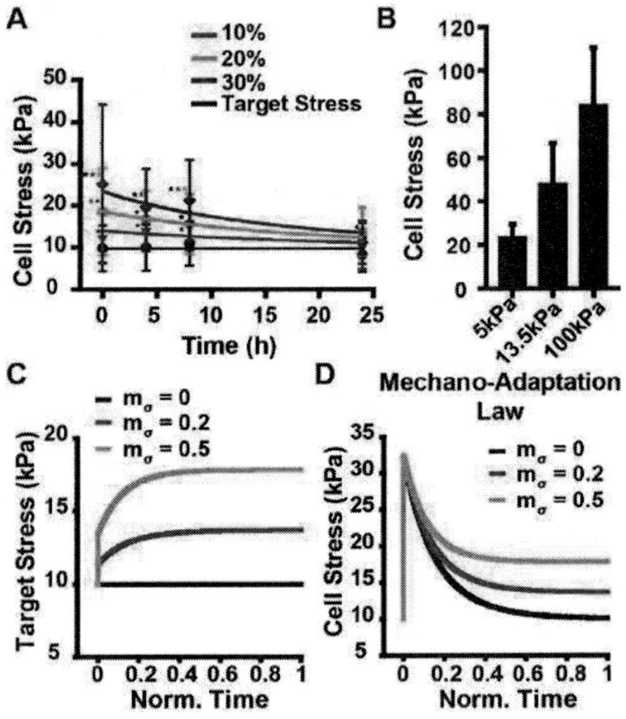

Figure 2. A. VSMC Temporal Stress. VSMC Cauchy stress for 0, 4, 8, and 24h post applied strain (10%, 20%, or 30%). Experimental data (points) are fit with a growth law (solid lines). Mean±STD. *Represents $p<0.05$. **B. Substrate Modulus Affects VSMC Contractility.** VSMC Cauchy stress for cells cultured on a 5, 13.5, or 100kPa substrate modulus. Mean±STD. **C. VSMC Target Stress.** Temporal target stress as a result of varying the strength of correlation, m_σ, between substrate modulus and target stress. **D. Substrate Dependent Mechano-Adaptation Law.** Temporal VSMC Cauchy stress response to changes in the mechanical environment for different correlation strengths (m_σ) of target stress and substrate modulus.

RESULTS

We found that the temporal stress evolution for VSMCs exposed to chronic applied strain can be fit using a simple growth law (Eq. 1), where the cell stress increases immediately following applied strain and, over time, approaches control values (Fig. 2A). The VSMC target stress increases with increasing substrate modulus (Fig. 2B).

The substrate-dependent VSMC mechano-adaptation law incorporates the changes in target stress due to evolving extracellular mechanical properties. The greater the correlation between substrate modulus and target stress the greater the target stress becomes during the mechano-adaptive process (Fig. 2C). When the substrate modulus dependent target stress was accounted for, we found that the stress increases immediately and, over time, approached a steady state. However, the steady state reached is dependent on the correlation coefficient (m_σ), which defines relationship strength between the target stress and the changing substrate modulus (Eq. 2 and 3) (Fig. 2D).

DISCUSSION

VSMCs are known to adapt their own function as well as their local environment in response to mechanical changes. Here, we measured the VSMC temporal stress following a mechanical perturbation of the VSMC and its environment and used that data to generate a substrate-dependent mechano-adaptation law. By improving our understanding of how mechanical changes affect VSMC function, we can better understand vascular pathology.

ACKNOWLEDGEMENTS

We are grateful for our funding support by the National Science Foundation (PWA) and the American Heart Association (KES).

REFERENCES

[1] Fung YC, Biomech: Mech Prop of Living Tissues, *Sptringer*, 2004.
[2] Taber LA, *J Biomech Eng*, 120:348-354, 1998.
[3] Steucke KE, et al., *J Biomech*, 48(12):3044-3051, 2015.
[4] Steucke KE, et al., *JBME*, Submitted, 2017.
[5] Kraning-Rush, CM, et al., *Methods Cell Bio*, 110:139-178, 2012.
[6] Tseng Q, et al., *PNAS*, 109:1506-1511, 2012.

"BEYOND THE FORCE" - MECHANICS OF EARLY DEVELOPMENT IN THE FROG ILLUSTRATE FUNDAMENTAL DESIGN PRINCIPLES OF GROWTH AND DEVELOPMENT.

Lance A. Davidson[1, 2, 3]

(1) Bioengineering, University of Pittsburgh, Pittsburgh PA, 15213

(2) Developmental Biology, University of Pittsburgh, Pittsburgh PA, 15260

(3) University of Pittsburgh, Pittsburgh PA, 15213

ABSTRACT: The form and function of the embryo is shaped by forces generated at the cellular scale directed against complex structures and materials of the early embryo. In this presentation I will describe our current understanding of the mechanics of amphibian development and how it relates to development in other animals. Like most vertebrates, amphibian embryos begin development as a remarkable, 'active' viscoelastic material with a bulk ultra-low residual elastic moduli less than 10 Pa. As zygotic gene expression ramps up, tissues begin the process of differentiation with each germ layer, endoderm, mesoderm, and ectoderm, displaying specific material properties and biophysical behaviors. Material properties and active force production at early stages are principally determined by actomyosin contractility with little contribution from the extracellular matrix. From this beginning, programs of early development integrate physical mechanics with 'biological-programs' that allow the embryo to adapt to variations in maternal contributions and the environment. To enable these robust programs of development mechano-biological feedback loops endow the embryo with the ability to adjust both force production, tissue mechanical properties, and structure. As we dissect these processes we reveal a new set of design principles and suggest new technologies that will enable future engineers to design tissues and organs.

FGF-MEDIATED TENSIONAL GRADIENTS DRIVE MORPHOGENESIS OF THE AVIAN HINDGUT

Nandan L. Nerurkar (1), L. Mahadevan (2), Cliff J. Tabin (1)

(1) Department of Genetics
Harvard Medical School
Boston, MA, USA

(2) Departments of Applied Math,
Organismal and Evolutionary Biology, and
Physics
Harvard University

ABSTRACT*

The dramatic changes in tissue shape that underlie embryogenesis necessitate a precise, stereotyped regulation of physical forces. To ensure that these forces arise at the right location and time with the right orientation and magnitude, they must ultimately be under genetic control. Here we combine in vivo imaging, biophysics, and mathematical modeling with molecular and embryological approaches to investigate the molecular control of forces underlying morphogenesis of the avian gut tube. Internalization of endoderm to form the gut tube is a fundamental, yet still poorly understood aspect of establishing the vertebrate body plan. We find that hindgut formation is driven by a previously unappreciated posterior epithelial movement resulting from a contractile gradient in the hindgut-forming endoderm. The force gradient is established through graded cell shape changes, in response to a morphogenic gradient of FGF8, providing mechanistic insight into how developmental signals can be translated into large-scale physical forces shaping the embryo.

Please note that this abstract has been submitted for inclusion in the special session honoring the retirement of Dr. Larry Taber. The session will feature talks from Dr. Taber's trainees and collaborators, highlighting his important contributions to the field.

SB[3]C2017
Summer Biomechanics, Bioengineering and Biotransport Conference
June 21 – 24, Tucson, AZ, USA

BUCKLING MORPHOGENESIS OF THE EMBRYONIC AIRWAY EPITHELIUM

Victor D. Varner

Department of Bioengineering
The University of Texas at Dallas
Richardson, TX, USA

INTRODUCTION

Tree-like networks form the basic architecture for many organs in the body, including the lung, kidney, and mammary gland [1]. In the embryo, these structures originate as simple epithelial tubes, which are elaborated by a sequence of branching events. In most cases, branching involves reciprocal signaling between the branching epithelium and surrounding mesenchyme [2].

In the developing mouse lung, airway branching is highly stereotyped [3] and regulated in part by fibroblast growth factor (FGF) signaling [4]. New epithelial branches emerge adjacent to focal regions of FGF10 expression in the mesenchyme, and it is thought that this pre-pattern of FGF10 specifies the locations of new branches [5]. But when the mesenchyme is removed, and reciprocal signaling is disrupted, isolated epithelial explants still branch in culture. In the absence of a mesenchymal template of FGF, it is unclear how the epithelial branching pattern is specified.

Here, using a combination of ex vivo culture experiments and mathematical modeling, we show that a physical instability, produced by constrained epithelial growth, drives the branching morphogenesis of airway epithelial explants in the absence of mesenchyme [6]. These results suggest that simple physical mechanisms can control the formation of patterned epithelial structures during embryonic development.

METHODS

Embryos were harvested from timed-pregnant CD1 mice (Charles River Laboratories) at embryonic day E12.5. Whole lung explants were dissected in phosphate-buffered saline (PBS) using fine forceps. To isolate the airway epithelium, explants were incubated in dispase (Invitrogen) for 15 min, and the mesenchyme was removed manually using fine tungsten needles. Isolated epithelial explants were then embedded in three-dimensional (3D) gels of reconstituted basement membrane protein (Figure 1). Explants were cultured ex vivo at 37°C using culture medium supplemented with recombinant FGF. Time-lapse movies were collected using a Nikon Ti-U inverted microscope customized with a spinning disk (BD Biosciences).

Proliferating cells were detected using the Click-iT EdU Imaging Kit (Invitrogen). Fixed explants were then incubated with primary antibody against E-cadherin and, subsequently, AlexaFluor-conjugated secondary antibodies to simultaneously label the airway epithelium. Explants were dissected from the gel, and confocal stacks were captured. 3D reconstructions of the airway epithelium and proliferating cells were generated using Imaris (Bitplane).

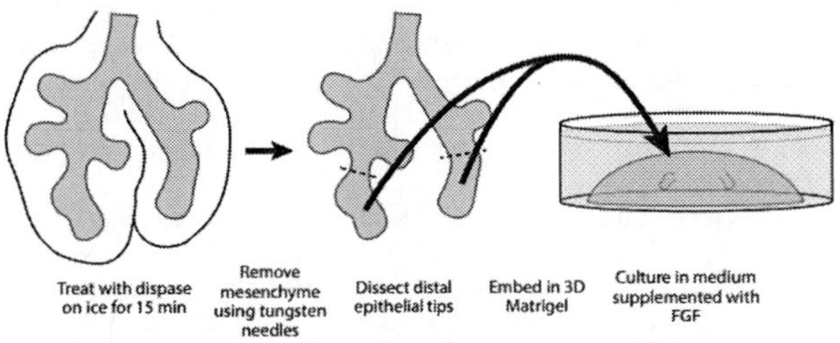

Figure 1: Schematic of mesenchyme-free culture assay.

A simple theoretical model consisting of a growing epithelium supported by a viscoelastic foundation (representing the surrounding

Technical Presentation #52

gel) was used to investigate the mechanics of epithelial branching. As a first approximation, we assumed small deformations and ignored the effects of initial epithelial curvature. Growth was included by decomposing the overall strain in the epithelium (ϵ) into a component due to growth (ϵ_g) and a component due to elastic deformation (ϵ^*) by $\epsilon = \epsilon^* + \epsilon_g$ [7]. The mechanical stresses depended only ϵ^* and were given by $\sigma = E(\epsilon - \epsilon_g)$, where E represents the elastic modulus of the epithelium. Assuming uniform epithelial growth and a sinusoidal deflection of the epithelium, the governing equation was given by

$$B(p)\frac{h^3}{12}\frac{d^4v}{dx^4} + N_g h\frac{d^2v}{dx^2} + \frac{\gamma B_f(p)}{2}v = 0,$$

where $v(x)$ is the deflection, γ represents the wave number of the assumed sinusoidal deflection, $N_g = bEh\epsilon_g$ is the in-plane force due to growth, b and h represent the height and width of the epithelium, respectively, and $B(p)$ and $B_f(p)$ are differential operators that define the viscoelastic properties of the epithelium and underlying foundation, respectively.

RESULTS

Time-lapse culture of mesenchyme-free explants revealed that new epithelial branches form simultaneously, with the airway epithelium cast into a folded geometry of characteristic wavelength λ (Figure 2). During this process, the airway epithelium was highly proliferative, but clear spatial patterns of proliferation did not become apparent until branches had already emerged.

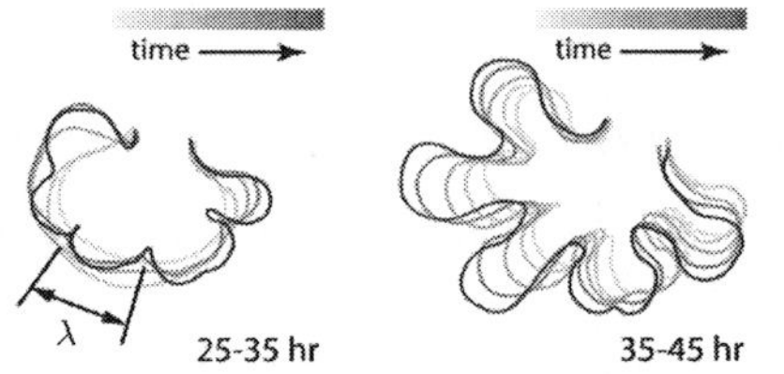

Figure 2: Traced contours of the airway epithelium during mesenchyme-free branching.

To investigate the mechanics of this process, we created a theoretical model for a growing epithelial layer on a viscoelastic foundation. The results of our model indicated that a growth-induced, mechanical instability underlies epithelial branching in the absence of mesenchyme. The dominant wavelength of this instability defines the overall branching pattern, and is controlled by the epithelial growth rate. To test this model prediction, we cultured mesenchyme-free explants in gels supplemented with different concentrations of FGF to modulate rates of epithelial proliferation. These experiments confirmed that branch wavelengths could be controlled by tuning the amount of epithelial growth.

We then extended our model to predict how folds might emerge from the initially unpatterned epithelium. The initial epithelial geometry was taken to consist of random, infinitesimally small deflections $v_0(x)$ [8]. This "noise" was then decomposed into Fourier components at different spatial frequencies (Figure 3). From these initial conditions, the deflection $v(x,t)$ increased exponentially in time t according to $v(x,t) = v_0 e^{kt}$, where the exponent k depended upon how closely the wavelength of each Fourier component corresponded to the dominant wavelength of the instability.

In the model, these initial irregularities were sufficient to initiate a folding instability in the growing epithelial layer, and regular folds of characteristic wavelength emerged spontaneously from the noisy initial conditions. Our simulations predicted shorter branch wavelengths at increased values of ϵ_g, consistent with our experimental results. The computed branch wavelengths were robust to variations in the initial irregularities in the epithelium.

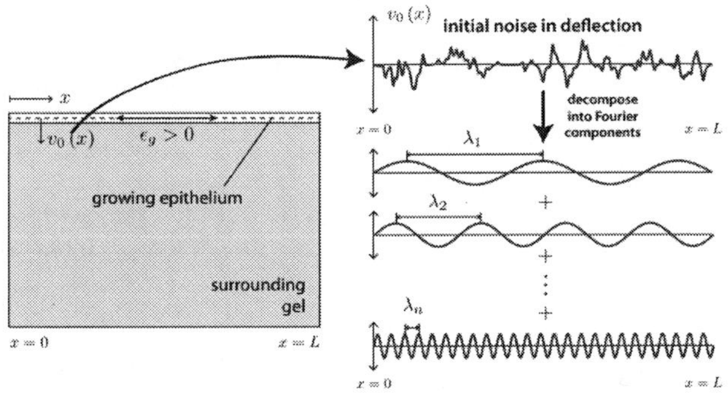

Figure 3: Schematic for the epithelial time-history of folding model. Noisy infinitesimal fluctuations in the initial geometry of the epithelial layer were decomposed into (spatial) Fourier components. The amplitudes of the components that corresponded to the dominant wavelength of the instability were preferentially amplified in time.

DISCUSSION

Epithelial morphogenesis is thought to be regulated primarily by biochemical signaling networks. Here, using a mesenchyme-free culture assay, we disrupted the biochemical pre-pattern thought to specify the locations of new branches in the developing mouse lung. We found that a physical instability, produced by constrained epithelial growth, is sufficient to drive buckling morphogenesis of the embryonic airway epithelium. These results suggest that mechanical forces, in addition to the well-established role of biochemical signals, can control the formation of patterned epithelial structures during embryonic development.

REFERENCES

[1] Varner VD & Nelson CM, *Development*, 141:2750-2759, 2014.
[2] Affolter M, et al., *Nat Rev Mol Cell Biol*, 10(12):831-842, 2009.
[3] Metzger RJ, et al., *Nature*, 453(7196):745-750, 2008.
[4] Shannon JM & Hyatt BA, *Annu Rev Physiol*, 66:625-645, 2004.
[5] Metzger RJ & Krasnow MA, *Science*, 284:1635-1639, 1999.
[6] Varner VD, et al., *PNAS*, 112(30):9230-9235, 2015.
[7] Taber LA, *Appl Mech Rev*, 48:1-59, 1995.
[8] Biot MA, *Geol Soc Am Bull*, 72(11):1595–1620, 1961.

SB^{3}C2017
Summer Biomechanics, Bioengineering and Biotransport Conference
June 21 – 24, Tucson, AZ, USA

MODELING MECHANICAL REGULATION OF GENE EXPRESSION IN MOUSE VENTRICULAR MYOCYTES

Andrew D. McCulloch (1,2), Kyle S. Buchholz (1), Philip M. Tan (3), Jeffrey H. Omens (2), Jeffrey J. Saucerman (3)

(1) Department of Bioengineering
University of California San Diego
La Jolla, CA, USA

(2) Department of Medicine
University of California San Diego
La Jolla, CA, USA

(3) Department of Biomedical Engineering
University of Virginia
Charlottesville, VA, USA

INTRODUCTION

Cardiac myocytes are sensitive to stress and strain, and they grow and remodel in response to altered load *in vivo*. Different hemodynamic conditions, such as pressure *vs.* volume overload give rise to distinct hypertrophic responses. Continuum models of growth and remodeling suggest that differential myocyte hypertrophy responses to transverse *vs.* longitudinal myofiber strain may be sufficient to explain these distinct remodeling patterns.

Cardiac myocytes are striated and contain structures that have a defined organization with respect to the longitudinal axis of the myofilament proteins. Many organelles and macromolecular complexes that have been implicated in myocyte mechanotransduction are localized at distinct sites both at the cell membrane and in the cytoskeleton: ion channels, costameres (focal adhesions), the dystrophin-dystroglycan complex, adherens junctions, and Z-disks of the sarcomere.

The micropatterned neonatal mouse ventricular myocyte culture preparation has been seen to replicate many of the important morphological features of ventricular myocytes *in vivo* including the cell aspect ratio, gap junctional coupling and striated myofibrils.

The goal of this study was to formulate a new network model of myocyte stretch signaling and gene expression and use it analyze the effects of transverse and longitudinal strain on gene expression to discriminate signaling pathways activated by each type of loading.

METHODS

We cultured micropatterned neonatal mouse ventricular myocytes on laminin-coated elastomeric substrates, and compared the effects of different non-equibiaxial strains applied to the myocytes. Here we used an elliptical stretch device to apply non-equibiaxial strains with maximum components aligned parallel or transverse to the myofibril axis. Transcriptomes measured by RNA-seq after 30 minutes to 4 hours of static strain exposure were then used to test a new network model of myocyte stretch signaling and gene expression. The network model was implemented as a system of logic-based ODEs representing 9 mechanosensors, 6 signaling pathways, 11 transcription factors and 645 target genes.

RESULTS

After 30 minutes of stretch, 53 and 168 genes were differentially expressed (DE) by transverse and longitudinal stretch, respectively. After 4 hours, the number of DE genes increased to 795 with longitudinal stretch while it decreased to 35 with transverse stretch. Gene ontology term (GO) analysis indicated enrichment of transcription factor and protein kinase activity by both types of stretch; whereas longitudinal but not transverse stretch caused expression of genes involved in sarcomere organization and cytoskeletal protein binding. Comparing these results with the model analysis suggested that longitudinal stretch activates protein kinase C which induces gene expression *via* the transcription factors serum response factor (SRF) and myocyte enhancer factor-2 (MEF2). In contrast, cAMP response element-binding protein (CREB) and its target genes were activated by both longitudinal and transverse stretch.

DISCUSSION

This approach to modeling myocyte mechanotransduction may provide a new link between the mechanosensitive cellular pathways that regulate myocyte hypertrophy and continuum models of ventricular growth and remodeling stimulated by anisotropic strain.

ACKNOWLEDGEMENTS

Supported by NIH grants P41 GM103426 and R01 HL105242

Technical Presentation #53

WHAT DRIVES CORTICAL FOLDING IN THE BRAIN?

Philip V. Bayly (1) and Gang Xu (2)

(1) Mechanical Engineering and Materials Science
Washington University
Saint Louis, Missouri, USA

(2) Engineering and Physics
University of Central Oklahoma
Edmond, Oklahoma, USA

INTRODUCTION

The folded shape of the human brain allows the cerebral cortex, the thin (2-4 mm) outer layer of neurons and their associated processes, to attain a very large surface area (~1600 cm²) relative to brain volume (~1400 cm³). Cortical folding, or gyrification, is critical to human brain development and function. Abnormal cortical folding has been associated with cognitive or emotional problems, including retardation, epilepsy, autism, and schizophrenia. In humans, cortical folding normally occurs in utero, during the third trimester (Fig. 1).

Large mammals (cows, pigs) and small carnivorous mammals (cats, ferrets; Fig. 1) also have folded brains; although rodents and birds (Fig. 2) do not. Disturbances of cortical folding in humans are clinically important and offer clues to the underlying mechanisms of normal development. Outward, or convex, folds are known as gyri and inward folds are called sulci. Despite decades of study (1), the mechanical basis of folding remains a topic of active research. In subsequent sections we summarize key experimental observations and leading theories of folding mechanics, noting the important contributions of Larry Taber.

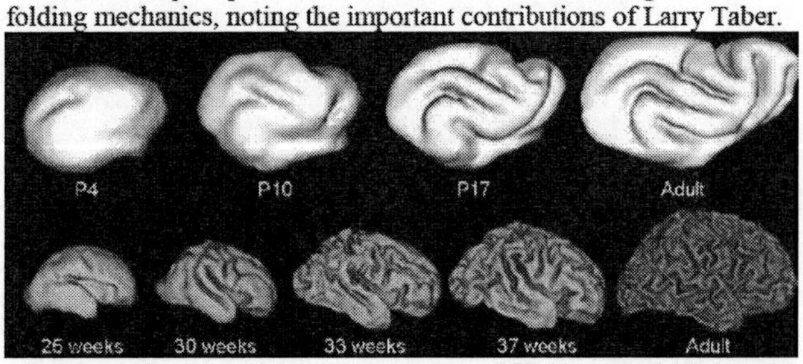

Figure 1. Surface representations from MR images of (*A*) ferret brains at postnatal day 4, 10, 17, and adult, and (*B*) human brains at 25, 30, 33, and 39 weeks gestation and in the adult. From (2).

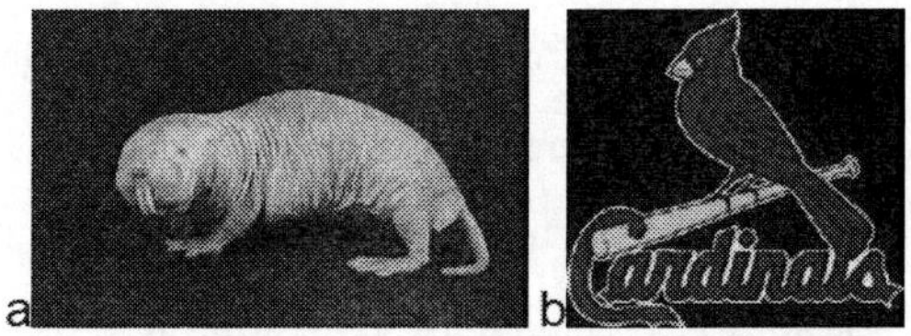

Figure 2. Rodents, such as the naked mole rat (a), and birds, such as the cardinal (b), have smooth (lissencephalic) brains. Ironically, both have been studied in detail by Larry Taber.

THE AXONAL TENSION HYPOTHESIS

Van Essen (3) has proposed that *axonal tension* produces folding by drawing sides of gyri together. This hypothesis suggests three questions that can be answered by direct experimentation. (i) Can axons in vitro sustain mechanical tension? (ii) Are axons in the intact brain under tension? (iii) Does axonal tension act to draw the walls of gyri together?

Several studies have addressed the first question. Dennerll et al. (4) showed that sensory neurites from the chick embryo grew in response to tension, and contracted to re-establish tension after the fiber was allowed to slacken. Neural fibers from the brain itself also grew in vitro in response to applied tension, but maintained much lower levels of tension and did not re-establish tension after slackening (5).

Dissection experiments (6, 7) convincingly showed that white matter in mammalian brains is under tension, including the white matter in the adult mouse brain (6) and in both mature and developing ferret brains (7) (**Fig. 4**). Incisions through cortical gray matter did not open, but cuts through white matter normal to the fiber direction opened, indicating stress along the fiber axis. Most axonal fibers in gyri are radial, so the stress within each gyrus is dominated by axon tension.

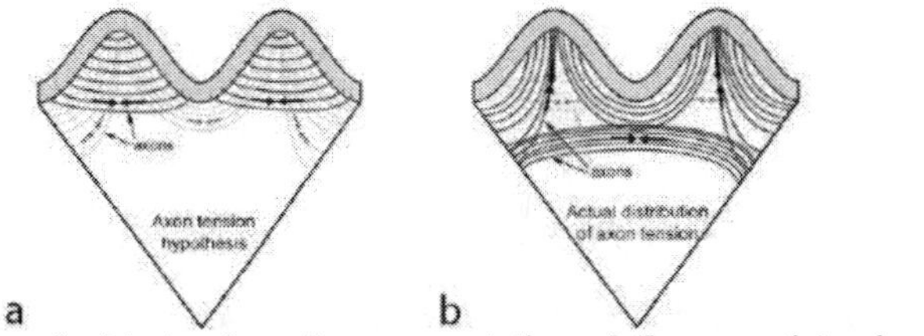

Figure 3. (a) A schematic representation of the *axonal tension hypothesis* of cortical folding (3). Tension (black arrows) in axons that connect two cortical regions is hypothesized to pull them, forming an outward fold. From Xu et al. (7) following Van Essen (3). (b) Actual distributions of axon tension based on dissection and histology data (7). Tension is circumferential in subcortical white matter and radial in the cores of gyri. Circumferential tension (grey arrows) was not detected between the walls of outward folds (7).

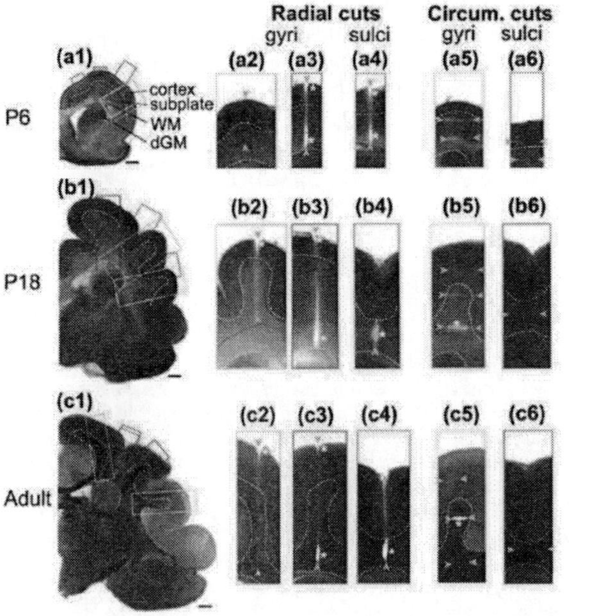

Figure 4. Microdissection study of the ferret brain. (a1-c1) Coronal brain sections at postnatal day (P) 6, 18 and adult. Dashed curves outline the boundaries between cortex, subplate, subcortical white matter (WM), and deep grey matter (dGM). Scale bars represent 1 mm. Color-coded rectangles indicate the regions shown in the close-ups to the right of each section. Cuts of various depths (indicated by pairs of arrowheads) are made either radially (a2-a4, b2-b4, and c2-c4) or circumferentially (a5-a6, b5-b6, and c5-c6) on developing gyri and sulci as noted. Asterisks indicate significant openings. From (7).

VISCOELASTIC INSTABILITY / STRESS-DRIVEN GROWTH

The theory of multiplicative growth due to Rodriguez et al. (8) has been widely used to model the development and growth of biological tissues including brain (6, 7, 9). The time scale of folding is long enough for the brain to grow significantly in response to stress. Such behavior of the deeper layers would resemble the response of a viscoelastic material, specifically a Maxwell fluid. Theoretical predictions for compressive folding instabilities in viscoelastic media were developed by Biot (10). Bayly et al. (11) extended Biot's analysis to the model the growth of the cortex. According to this analysis, Bayly et al. (11) predicted that the wavelength of folds, λ, (normalized by thickness h) will depend on the ratio of cortical (G_0) to sub-cortical (R_f) growth rates, and the ratio of the modulus of the outer cortex (μ) to that of deeper layers (μ_f).

The stress-driven growth model can explain both (i) variations in wavelength of folds and (ii) the stress fields observed in the developing brain. Simulations (COMSOL Multiphysics, Burlington, MA) confirmed that more rapid cortical expansion leads to shorter wavelengths (**Fig. 5**). Simulated wavelengths also scale directly with cortical thickness. Initial shape imperfections, as well as spatial variations in tangential expansion itself, may also affect the final shape.

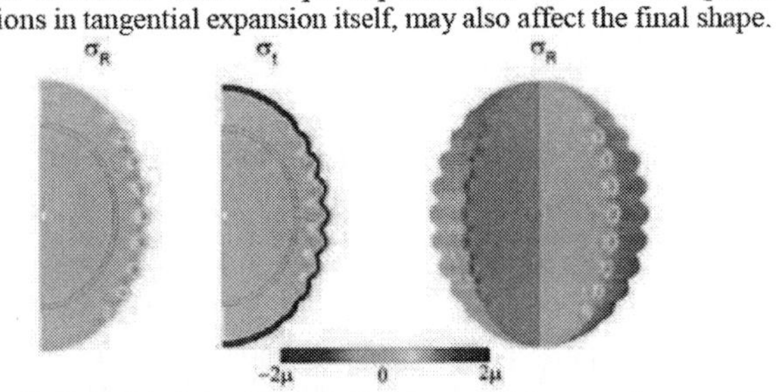

Figure 5. Radial (σ_r), and tangential stress (σ_t) in an axisymmetric model of cortical growth and stress-driven subcortical growth (11).

CONCLUSION

Measurements and modeling indicate that cortical folding is due to viscoelastic instability of stress-driven growth, not by axonal tension.

ACKNOWLEDGEMENTS

Funding from NIH grant R01 NS070918 (Taber), R21 EB005834 (Bayly), NSF grant DMS0540701 (Taber). We are deeply grateful for the mentorship, friendship, and collaboration of Larry A. Taber.

REFERENCES

1. Le Gros Clark WE (1945) Deformation patterns in the cerebral cortex. *Essays on Growth and Form*, eds Le Gros Clark WE & Medawar PB (Oxford University Press, London), pp 1-22.
2. Barnette AR, *et al.* (2009) Characterization of brain development in the ferret via MRI. *Pediatr Res* 66(1):80-84.
3. Van Essen DC (1997) A tension-based theory of morphogenesis and compact wiring in the central nervous system. *Nature* 385(6614):313-318.
4. Dennerll TJ, *et al.* (1989) The cytomechanics of axonal elongation and retraction. *J Cell Biol* 109(6 Pt 1):3073-3083.
5. Chada S, *et al.* (1997) Cytomechanics of neurite outgrowth from chick brain neurons. *J Cell Sci* 110 (Pt 10):1179-1186.
6. Xu G, Bayly PV, & Taber LA (2009) Residual stress in the adult mouse brain. *Biomech Model Mechanobiol* 8(4):253-262.
7. Xu G, *et al.* (2010) Axons pull on the brain, but tension does not drive cortical folding. *J Biomech Eng* 132(7):071013.
8. Rodriguez EK, Hoger A, & McCulloch AD (1994) Stress-dependent finite growth in soft elastic tissues. *J Biomech* 27(4):455-467.
9. Filas BA, Xu G, & Taber LA (2013) Mechanisms of brain morphogenesis. *Computer Models in Biomechanics*, eds Holzapfel GA & Kuhl E (Springer, New York), pp 337-349.
10. Biot MA (1957) Folding instability of a layered viscoelastic medium under compression. *Proceedings of the Royal Society of London - Series A* 242:444-454.
11. Bayly PV, Okamoto RJ, Xu G, Shi Y, & Taber LA (2013) A cortical folding model incorporating stress-dependent growth explains gyral wavelengths and stress patterns in the developing brain. *Phys Biol* 10(1):016005.

PREDICTION OF POST STENOTIC FLOW INSTABILITIES IN A PATIENT SPECIFIC COMMON CAROTID ARTERY MODEL

Mancini V. (1), Bergersen A. (2), Segers P. (1), Valen-Sendstad K. (2)

(1) IBiTech - bioMMeda
Ghent University
9000 Ghent, Belgium

(2) Scientific Computing
Simula Research Laboratory
1325 Lysaker, Norway

INTRODUCTION

In silico medicine shows great potential for revealing mechanisms of atherosclerotic plaque progression in the carotid bifurcation by modelling patient-specific local hemodynamic forces [1]. It is well known that the pulsatile nature of the flow in combination with narrowing of the lumen can cause post stenotic flow instabilities, which presents additional challenges from a flow modelling point of view. The three most common modelling approaches for unstable or turbulent flows are Reynolds-averaged Navier-Stokes (RANS), large eddy simulation (LES), and direct numerical simulation (DNS). RANS is the computationally cheapest one, but has shown to be directly misleading for predicting transitional flows [2]. DNS is in contrast accurate, but on the other hand extremely computationally expensive. LES can potentially offer an ideal balance between accuracy and computational cost [3]. The aim of this study was to explore the parameter space of the numerical simulation, and establish a DNS reference solution for stenotic carotid bifurcation, and that will be used in the future to assess whether LES can accurately predict post stenotic flow instabilities.

METHODS

Medical images of a carotid bifurcation with severe stenosis (82% by area) located in the internal carotid artery (ICA) of a 75 years old man were obtained from computed tomography angiography and segmented using *3D Slicer* [4]. The *Vascular Modelling Tool Kit* was used to create meshes with a local refinement in the stenotic- and downstream region, resulting in meshes ranging from 2 to 50 million (M) linear tetrahedral cells. Flow simulations were performed using the open-source CFD solver *Oasis* [5], where special care was taken to ensure a kinetic energy preserving and minimally dissipative numerical solution. The fluid was set to mimic water with kinematic viscosity of $\nu = 1 \cdot 10^{-6}$ m²/s, and a Reynolds number was 1400 at the inlet, to allow for direct comparison with *in vitro* experiments at a later point. The flow field was evaluated at peak systolic flow rate of 585.52 ml/min, thus, enabling rigorous assessment of the temporal and spatial resolution with respect to the smallest scales, however at the cost of an artificial flow condition. A parabolic velocity profile was prescribed as inlet condition, no-slip along the walls, and a resistance condition to meet a flow split between the ICA and the external carotid artery at 68.2:31.8, to obtain physiologically plausible flow split [6]. The initial condition was a fully developed flow from the 2M simulation that was projected onto each mesh.

We performed a spatial refinement study to obtain a reference solution, with a fixed time step (Δt) of $2 \cdot 10^{-5}$ s. Using the least computationally expensive mesh that gave adequate results, a temporal refinement study was performed, with Δt ranging from $1 \cdot 10^{-4}$ to $5 \cdot 10^{-6}$ s.

The instantaneous velocity (u) was decomposed into the mean ($\bar{u}$), and fluctuating components (u'), respectively ($u = \bar{u} + u'$), to compute the turbulent kinetic energy (TKE) $k = 0.5(u'^2 + v'^2 + w'^2)$. The analyses were based on the time interval from 0.1 to 2.0 s, to allow for initial transients to wash out and statistics to converge. Figure 1 (left) shows the region of interest and four cut planes (A to D) where we computed the time-averaged velocity. In addition, the power spectral density (PSD) of u' downstream of the stenosis (point P in Figure 1) was computed by using the Welch method with Hann windowing with data from 0.1 to 0.5 s.

RESULTS

Focusing first of qualitative results, Figure 1 (right) shows the volumetric TKE for the 22M simulation, and we can clearly observe flow instabilities already upstream of the stenosis, which increases in intensity at the center and downstream of the stenosis. However, the flow instabilities quickly dissipate further downstream.

The time-averaged velocity is shown in cut planes A to D from the spatial and temporal refinement study in Figure 2, left and right respectively. The 22M mesh was found to offer an optimal balance between accuracy and computational cost: the time-averaged velocity obtained from the finest 50M mesh are comparable with the 22M mesh, whereas it differs for the 2M and 6M simulation. Therefore, the 22M mesh was used for the temporal refinement study.

From the time-averaged velocity in Figure 2 (right) we find that the three largest time steps are phenotypically similar, although the smallest time steps show large differences, in particular at C1-C2.

The spectral distribution of energy in Figure 3 (left) confirms the ability of the 22M to sufficiently replicate the finest mesh results. Furthermore, from Figure 3 (right) we can observe adequate agreement between the different temporal resolutions.

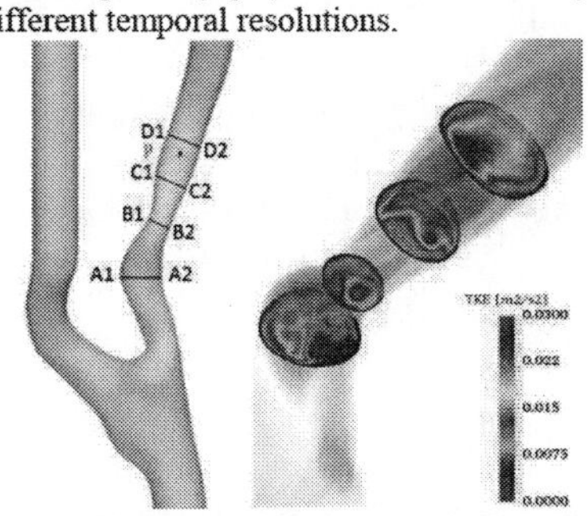

Figure 1: Region of interest with planes A to D and point P (left), and a volume rendering of the TKE and in planes A to D (right).

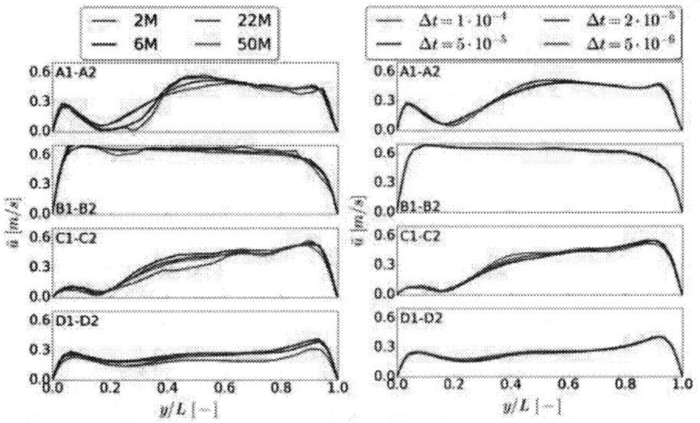

Figure 2: Time-averaged velocity along the four lines of interest for the spatial (left) and temporal (right) refinement study.

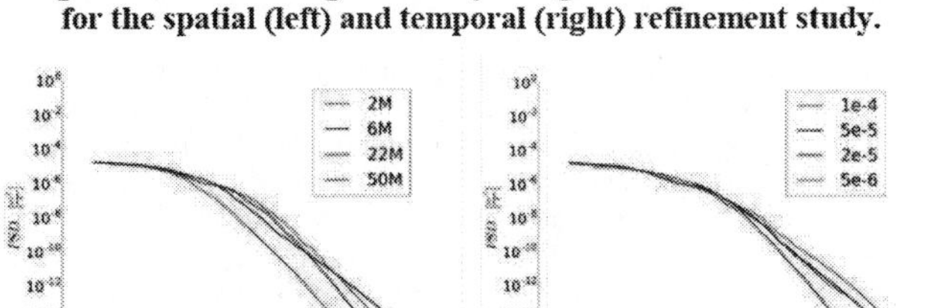

Figure 3: Power spectral density of the velocity field at point P, located 1.7 cm (2.2 diameters) downstream of the stenosis.

DISCUSSION

The aim of this study was to explore the parameter space relative to a DNS reference solution, a 22M mesh with a time step of $1 \cdot 10^{-4}$ s was found to be an optimal choice between computational cost and accuracy, and will be used in future studies on turbulence modeling for stenotic carotid bifurcations. That being said, there is admittedly a minor difference between 22M and 50M in Figure 2 (left). To assess if the 50M simulation was close enough to a proper DNS we compared the temporal (Δt) and spatial (Δx) scales in the numerical simulation to the Kolmogorov time scale (τ), and length scale (η), respectively.

The results of spatial assessment can be found in Table 1, displaying the mean Δx, the smallest Kolmogorov length scale, and the maximum ratio. There are two things of note, *first* the Kolmogorov length scale converges as the mesh is refined. *Second,* the ratio of the two finest meshes are below 10, typically sufficient to capture > 95 % of the dissipation. If the simulations, from a numerical point of view, were truly converged, the ratio needs to be unity. However, as observed in bottom right of Table 1, that would require a mesh with ~7^3 times 50M cells, if uniformly refined. We therefore consider the 50M sufficiently refined from a pragmatic point of view, and by extension the 22M mesh for biomedical applications. Furthermore, the difference in CPU hours is substantial, the 22M simulation with $\Delta t = 1 \cdot 10^{-4}$ used 1420 CPU hours on 96 cores, while the 50M simulation with $\Delta t = 2 \cdot 10^{-5}$ s spent 16 496 CPU hours on 128 cores to simulate 2 physical seconds, over an 11-fold difference.

Table 1: Comparison between Kolmogorov length scale and Δx

Num. of cells	2M	6M	22M	50M
Δx_{mean}[m]	3.19E-4	2.14E-4	1.38E-4	1.05E-4
η_{min} [m]	9.81E-6	9.30E-6	8.91E-6	8.58E-6
$(\Delta x/\eta)_{max}$ [-]	21.93	14.04	9.31	7.49

The temporal assessment of the flow simulation can be found in Table 2, showing that the temporal resolution was below the Kolmogorov scales. Furthermore, we can observe from the PSD in Figure 3 (left) that there was, for all practical purposes, no energy in the fluctuating component above 5000 Hz. We can therefore attribute the discrepancies observed in Figure 2 (lower) to an inadequate data for time-averaging (0 – 0.7 s). In fact, an even larger time step might have been an adequate temporal resolution, however $\Delta t = 1 \cdot 10^{-4}$ s is very close to the stability criteria of the solver.

Table 2: Comparison between olmogorov time scale and Δt

Δt [s]	1.00E-4	5.00E-5	2.00E-5	5.00E-6
τ_{min} [s]	7.96E-5	7.96E-5	7.93E-5	7.92E-5
$\Delta t/\tau_{min}$ [-]	1.2563	0.6281	0.2522	0.0631

The simulations were performed with rigid walls, and the effect of a compliant model remains to be assessed. Furthermore, the fluid used for these simulations was water instead of blood to allow for direct comparison against *in vitro* experiments, which will be the focus of future work. As part of a larger consortium, the final aim of this line of investigation is to build a non-contact device to diagnose severe stenosis in the carotid arteries through the analysis of neck's skin displacement. We will therefore compute the sensitivity of the flow split and of noise in the inflow pulse through *in silico* experiments, since both factors can be challenging to control in the *in vitro* experiments.

ACKNOWLEDGEMENTS

Funding: EU Horizon 2020 programme (CARDIS; ICT-26-2014).

REFERENCES

[1] Steinman DA. at al., *Annals of Biomed Engi*, 30:483-497 2002.
[2] Varghese S. at al., J. *Fluid Mech.*, 582:281-318, 2007.
[3] Aspden A. et al., *Commun Appl Math Comput Sci*, 3:103-26, 2009.
[4] Iannaccone F. et al., *Int J Art Org*, 37: 928-939, 2014.
[5] Mortensen M. et al., *Comp Phys Comm*, 188:177-188, 2015.
[6] Groen H. C. et al., *J Biomechs*, 43: 2332-8, 2010.

COMPREHENSIVE CHARACTERIZATION OF RABBIT AORTIC ARCH HEMODYNAMICS FROM 4D PC-MRI DERIVED CFD

David S. Molony (1), Lei Zhou (1), Jaekeun Park (1), Candace Fleischer (1), John N. Oshinski (1, 2), Xiaoping Hu (3), Habib Samady (1), Amir Rezvan (1), Don P. Giddens (2)

(1) Department of Medicine,
Emory University
Atlanta, GA, USA

(2) Wallace H Coulter Department of
Biomedical Engineering,
Georgia Institute of Technology
Atlanta, GA, USA

(3) Department of Bioengineering,
University of California, Riverside
Riverside, CA, USA

INTRODUCTION

Animal models have widely been used to gain a better understanding of the mechanisms involved in atherosclerotic plaque formation. These models offer an opportunity to investigate mechanistic pathways through gene or surgical manipulation. Small animals such as mice and rats have been most commonly used due to the ease of alterations and short time frame for disease formation [1]. However, their small size results in a hemodynamic environment far removed from that of humans. Larger animals such as pigs and rabbits are less common due to their significant expense; though on the other hand the vessel caliber is much closer to that of humans. In particular, the rabbit model has been extensively used to investigate the association between hemodynamics and atherosclerotic lesion formation in the descending aorta [2, 3].

Previous studies of rabbit hemodynamics have assumed generic inlet velocity profiles or ignored the full range of velocity components [2-4]. Morbiducci et al., (2013) stressed the importance of incorporating all 3D velocity components as neglecting these can lead to a misleading representation of human aortic hemodynamics [5]. Several metrics have been applied to examine the hemodynamics of arteries. These can be divided into near wall parameters and bulk flow parameters. Near wall parameters such as time-averaged wall shear stress (TAWSS) are associated with plaque formation while bulk flow parameters such as helicity are believed to play an atheroprotective role through reduced luminal surface LDL concentration [6]. Recently the role of multi-directional flow has been assessed and was found to have better correspondence with lesion formation in the descending rabbit aorta than TAWSS and oscillatory shear index (OSI) [2]. This parameter along with flow helicity may help better explain lesion patterns in the rabbit aorta. The aims of this research were to characterize the hemodynamics of the rabbit aortic arch through 4D PC-MRI derived CFD towards understanding the role of blood flow in lesion formation.

METHODS

Male New Zealand White Rabbits on a high fat diet underwent MRI imaging between 1 and 3 months after either left carotid partial ligation (n=8) or sham ligation (n=4). Anatomical data was acquired using a MRA sequence and flow data with a 4D PC-MRI sequence with the following respective parameters (acquisition matrix 400x400 pixels, 256x176 pixels; field of view - 16x16 cm, 7.5x5.15 cm; slice thickness - 2.5cm, 2cm). A venc of 120 cm/s was used for all velocity components. A 5x5 pixel median filter was applied to remove spurious vectors (figure 1).

Images were segmented using Segment (Medviso) by a level set approach. CFD models were meshed using ICEM-CFD and the solver Fluent was used. Rabbit specific branch outflows determined from PC-MRI measurements were assigned as outlet boundary conditions and the 3 measured velocity components were set as a velocity inlet. Post processing was performed using MATLAB and the near wall parameters TAWSS, transWSS and cross flow index (CFI) were assessed [7]. The helicity of the flow was assessed over the entire vessel by means of time-averaged helicity (h1), helicity intensity (h2), helical strength direction (h3), helical strength (h4) [5]. Pearson correlation was used to test for correlation between helical flow indices and mean vessel diameter, centerline length, tapering, curvature and Reynolds number. Continuous variables are reported as mean±standard deviation.

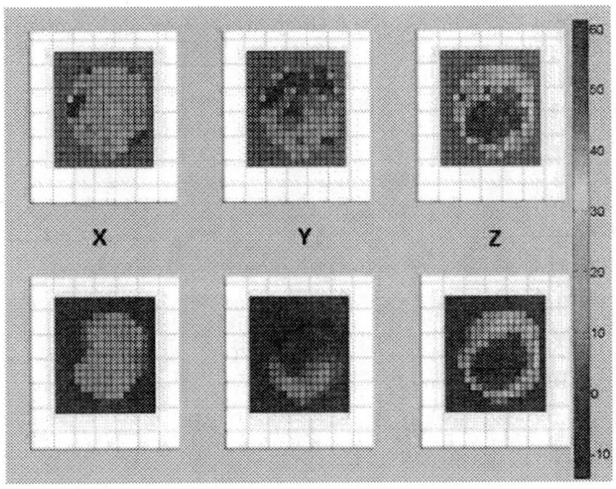

Figure 1: Example instantaneous 3D velocity components (cm/s) before and after median filtering.

RESULTS

The mean centerline length, median diameter, tapering and Reynolds number were 50.9±8.95mm, 2.51±0.35mm, 5.75±1.97°, 309±80 respectively. The Reynolds number was significantly correlated ($p<0.05$) with helicity intensity (h2). There were no other significant correlations between the geometric parameters with any of the helical indices. During systole helical flow was predominantly right handed and was observed to extend from the ascending aorta to the descending aorta (figure 2). In some rabbits this was accompanied with a left handed helical flow extending into the descending aorta. Throughout the cardiac cycle the sign of the helical strength direction (h3) was most frequently negative indicating a favoring towards right handed helical flow.

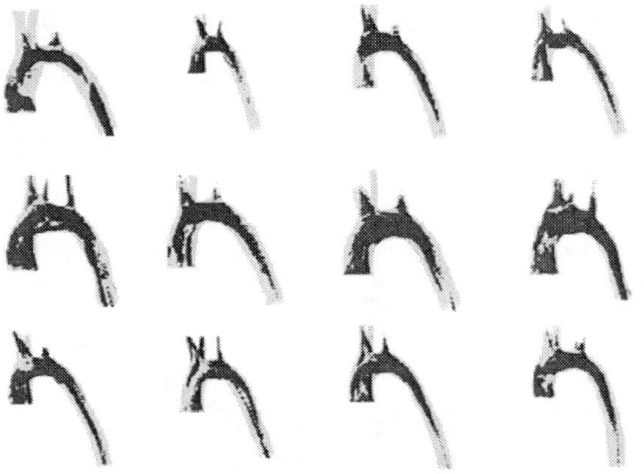

Figure 2: Isosurfaces of local normalized helicity (LNH) greater than 0.8 (red) and greater than -0.8 (blue) during peak systole. LNH>0.8 corresponds to left handed helical structures and LNH>-0.8 to right handed helical structures.

A specific pattern of near wall flow parameters was generally observed in all rabbits (figure 3). Low TAWSS was observed on the outer curvature of the ascending aorta and high TAWSS on the inner curvature of the ascending aorta. A region of low TAWSS was also generally seen on the inner curvature of the descending aorta. High values of CFI and transWSS were observed on the lateral walls of the descending aorta. CFI in particular showed a markedly consistent pattern.

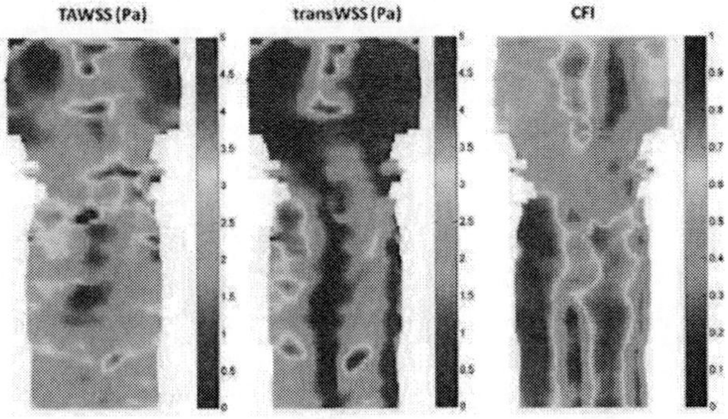

Figure 3: En face plots for TAWSS, transWSS and CFI for rabbit 3. The vessel has been unwrapped along the outer curvature.

DISCUSSION

4D PC-MRI provides a foundation for high spatial and temporal resolution simulations of the rabbit aortic arch allowing both bulk and near wall flow parameters to be quantified with high accuracy. Throughout the studied population consistent flow patterns were observed including distinct high CFI patterns in the descending aorta. This indicates a large degree of multi-directional flow on the lateral walls. Interestingly this appeared to occur in regions of low and high TAWSS, explaining the greater heterogeneity in transWSS compared to CFI. During systole a right handed helicity was dominant (blue) which dissipated further into the cardiac cycle. The poor correlation between rabbit geometry and helical flow metrics illustrate the complexity in these flows. More robust metrics such as statistical shape models may better explain these helical flow patterns. Future work will investigate rabbit plaque formation and it's correlation to hemodynamic parameters and statistical shape models of the aortic arch.

ACKNOWLEDGEMENTS

This work was funded by the Program of Excellence in Nanotechnology (PEN).

REFERENCES

[1] Getz, G et al., *Arterioscler Thromb Vasc Biol*, 32:1104-1115, 2012.
[2] Pfeiffer, V et al., *J Biomech*, 46:2651-2658, 2013.
[3] Vincent, P et al., *J R Soc Inferface*, 65: 2011:1708-1719, 2011.
[4] Menon, A et al., *J Pharm Tox Meth*, 65:18-28, 2012.
[5] Morbiducci, U et al., *J Biomech*, 46:102-109, 2013.
[6] Liu, X et al., *A J Physiol*, 297:H163-H170, 2009.
[7] Mohamied, Y et al., *J Biomech*, 50:102-109, 2017.

HEMODYNAMIC RISK IN CORONARY BIFURCATIONS: A COMPUTATIONAL EXPLORATION

D. Gallo (1), C. Chiastra (2,3), P. Tasso (1), F. Iannaccone (4), F. Migliavacca (2), J.J. Wentzel (3), U. Morbiducci (1)

(1) Department of Mechanical and Aerospace Engineering
Politecnico di Torino
Turin, Italy

(2) Department of Cardiology, Biomedical Engineering
Erasmus MC
Rotterdam, The Netherlands

(3) Laboratory of Biological Structure Mechanics, Department of Chemistry, Materials and Chemical Engineering Giulio Natta, Politecnico di Milano, Milan, Italy

(4) FEops NV
Ghent, Belgium

INTRODUCTION

The so-called "hemodynamic hypothesis" suggests that local hemodynamics is a main factor of the onset and progression of lesions at arterial bifurcations [1]. Local hemodynamics is mainly determined by the underlying anatomical features of an arterial bifurcation. For example in carotid arteries, computational fluid dynamics (CFD) studies showed a correlation between geometry and disturbed shear [2]. Furthermore, local geometry of carotid bifurcation has been reported to be significantly correlated with peculiar helical flow structures [3], which in turn have been proven to reduce the likelihood of flow disturbances at the bifurcation [4]. While the interplay between geometry, near-wall and intravascular flow features has been widely studied in carotid bifurcations, less is known about diseased coronary bifurcations. Indeed, several studies focused on the influence on near-wall descriptors attributable to anatomical features, such as the presence of stenosis at varying locations and with different severity/extension, bifurcation angle, curvature [5], and tortuosity [6]. The aim of the present study is to extend the investigation of the influence of peculiar coronary bifurcation anatomical features on both near-wall and intravascular flow features. In particular, the impact of stenosis, bifurcation angle, and curvature on local hemodynamics is analyzed by performing CFD simulations on a population-based, idealized model of coronary bifurcation. Such an idealized model-based approach will enable, by varying one specific geometrical feature at time while keeping the others constant, to clearly identify whether and to which extent specific anatomic features promote atherosensitive hemodynamic phenotypes.

METHODS

A parametric model of a coronary bifurcation (Fig.1) representing the left anterior descending (LAD) coronary artery with its diagonal branch was created, as detailed elsewhere [7]. Briefly, the diameter of the proximal main branch (P-MB) is 3.30 mm while the diameters of the distal main branch (D-MB) and the side branch (SB) are 2.77 mm and 2.10 mm, respectively [8].

The distal bifurcation angle was varied within the physiological range. The cardiac curvature was taken into account by bending the model on a sphere of radius 'R'. A physiological cardiac curvature radius was considered (R=56.3 mm), as well as two extreme values (R=∞, i.e. absence of curvature, and R=16.5 mm). Stenosed (diseased) and unstenosed (healthy) bifurcation models were analyzed. Stenosed models are characterized by 60% diameter stenosis in each branch (Medina classification 1,1,1). The lesion is 12 mm long and eccentric, with plaque located at the inner arc of the vessel (Fig. 1).

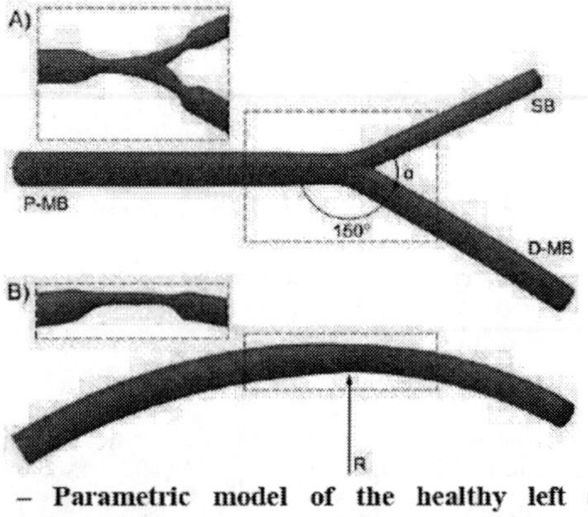

Figure 1 – Parametric model of the healthy left anterior descending / first diagonal coronary bifurcation: (A) top and (B) lateral views. Details of the diseased model are shown in the boxes.

In summary, 10 coronary bifurcation models were investigated by combining 3 distal angles (40°, 55°, and 70°) and 3 curvature radii, both for the healthy and the diseased coronary bifurcation models. The governing equations of unsteady fluid motion were solved by using the finite volume method. A typical human LAD flow waveform was imposed at the inlet as a plug velocity profile. A flow-split of 0.65:0.35 for the D-MB and SB, respectively, was applied at outlets. The flow-split was maintained constant in the diseased models in order to focus the analysis only on the impact that geometrical features have on local hemodynamics. Blood was modeled as a non-Newtonian fluid using the Carreau model. The flow was assumed as laminar (max Reynolds number ~ 330 at the P-MB stenosis of diseased models). Intravascular fluid structures were investigated in terms of helical flow topology and content. Helicity intensity h_2 and helical rotation balance h_4 were calculated, according to [4]:

$$h_2 = \frac{1}{TV}\int_T \int_V |\vec{v}\cdot\vec{\omega}|dVdt \quad (1) \qquad h_4 = \frac{\left|\int_T \int_V \vec{v}\cdot\vec{\omega}\, dVdt\right|}{\int_T \int_V |\vec{v}\cdot\vec{\omega}|dVdt} \quad (2)$$

where $\vec{v}$, $\vec{\omega}$ are the velocity and vorticity vectors, respectively, V is the volumetric fluid domain of interest, T is the cardiac cycle. Helicity intensity h_2 is an indicator of the total amount of helical flow, while h_4 measures the strength of relative rotations of helical flow structures. Near-wall hemodynamics was evaluated in terms of time-averaged WSS (TAWSS), oscillatory shear index (OSI), and relative residence time (RRT). In particular, the fraction of the luminal surface area exposed to TAWSS < 0.4 Pa, OSI > 0.2, and RRT > 4.17 Pa^{-1} (in consequence of values set for TAWSS and OSI) was calculated.

RESULTS AND DISCUSSION

The analysis was carried out on the entire bifurcation and single branches. Findings related to the hemodynamic descriptors over the entire bifurcation are summarized in Tables 1 and 2. Generation and transport of helical flow structures is influenced by the curvature radius. Smaller curvature radius is associated with higher helicity intensity in both unstenosed and stenosed cases. In healthy cases, helical flow topology is driven mainly by the curvature of the vessel. In diseased cases, the impact of curvature on helical flow structures piles up with the helicity generated because of the lumen reduction (Fig. 2A). Consequently, helicity intensity h_2 of diseased models is one order of magnitude higher than healthy cases. Globally, helical flow structures are symmetrical, as demonstrated by the nearly 0 values of the helical rotation balance h_4. The surface areas exposed to atherosensitive RRT values can be observed in diseased cases at the distal MB and at the SB, close to the reattachment point of the recirculation regions (Fig. 2B). The curvature radius moderately affects the near-wall hemodynamics of the diseased cases. In particular, smaller curvature radius leads to larger lumen area exposed to low TAWSS and smaller lumen area exposed to high OSI and RRT. The bifurcation angle has a minor effect on the calculated hemodynamic variables. Helicity intensity is not dependent from the bifurcation angle in both healthy and diseased cases. Furthermore, in the healthy cases, the fraction of the lumen surface area exposed to low TAWSS slightly decreases with increasing bifurcation angle, while the exposure to high OSI and RRT is negligible. In the diseased cases, the surface area exposed to low TAWSS increases to ~3.3%. Surface areas exposed to high OSI and RRT show a poor dependence on the bifurcation angle.

In conclusion, the approach proposed in this study provides a controlled benchmark to investigate the effect of various geometrical features on local hemodynamics, highlighting the complex interplay between anatomy and intricate fluid structures in coronary bifurcations.

Table 1 – Near-wall hemodynamic descriptors for cases with different curvature radius 'R'.

Stenosis	R [mm]	Relative surface area [%]			h_2 [m/s^2]	h_4
		TAWSS<0.4Pa	OSI>0.2	RRT>4.17Pa^{-1}		
No	∞	0.00	0.00	0.00	0.20	0.00
	56.3	0.07	0.00	0.00	0.43	0.02
	16.5	0.07	0.00	0.00	1.12	0.00
Yes	∞	2.74	2.08	2.12	4.56	0.01
	56.3	3.21	2.04	2.01	5.13	0.06
	16.5	3.81	1.11	1.56	6.60	0.08

Table 2 – Near-wall hemodynamic descriptors for cases with different bifurcation angles 'α'.

Stenosis	α	Relative surface area [%]			h_2 [m/s^2]	h_4
		TAWSS<0.4Pa	OSI>0.2	RRT>4.17Pa^{-1}		
No	40°	0.42	0.00	0.00	0.43	0.00
	55°	0.07	0.00	0.00	0.43	0.02
	70°	0.00	0.00	0.00	0.44	0.02
Yes	40°	3.35	1.58	2.04	4.58	0.07
	55°	3.31	2.04	2.01	5.13	0.06
	70°	3.24	2.07	2.06	4.85	0.01

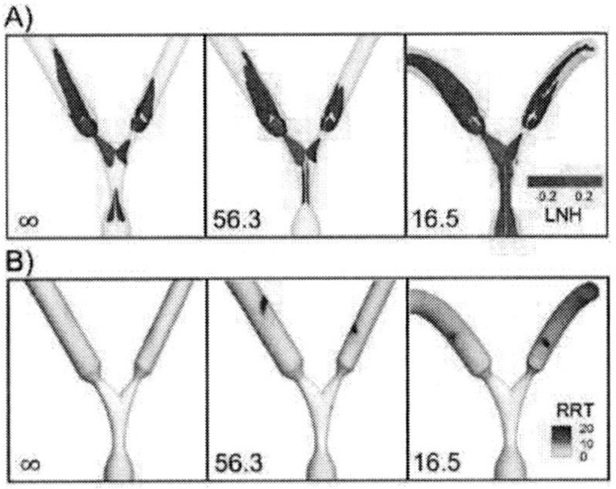

Figure 2 – Hemodynamics results of the diseased cases with different curvature radii: A) Isosurfaces of local normalized helicity (LNH); positive and negative values of LNH indicate counter-rotating flow structures. B) Contour maps of relative residence time (RRT, Pa^{-1}).

ACKNOWLEDGEMENTS

C. Chiastra and J.J. Wentzel are supported by the ERC starting grant (No. 310457, BioCCora).

REFERENCES

[1] Morbiducci, U et al., *Thromb Haemost*, 115(3):484-92, 2016.
[2] Lee, SW et al., *Stroke*, 39:2341–2347, 2008.
[3] Gallo, D et al., *Ann Biomed Eng*, 43:68–81, 2015.
[4] Gallo, D et al., *J Biomech*, 45:2398–2404, 2012.
[5] Peng, C et al., *PLoS One*, 11, e0157490, 2016.
[6] Vorobtsova, N et al., *Ann Biomed Eng*, 44(7):2228-39, 2016.
[7] Chiastra, C et al., *Biomed Eng Online*, 15(1):91, 2016.
[8] Finet, G et al., *EuroIntervention*, 3:490–8, 2008.

BLOCKING CADHERIN-11 DECREASES ATHEROSCLEROTIC PLAQUE DEVELOPMENT

Camryn L. Johnson (1), MacRae F. Linton (2), and W. David Merryman (1)

(1) Biomedical Engineering
Vanderbilt University
Nashville, TN, USA

(2) Department of Pharmacology
Vanderbilt University
Nashville, TN, USA

INTRODUCTION

Atherosclerosis is an inflammatory disease that involves the build-up of plaques in the coronary vessel walls. During disease progression, low-density lipoprotein (LDL) passes through the endothelial layer into the intima, where an immune response is initiated. Immature macrophages present in the blood, known as monocytes, are recruited to the vessel wall and invade the plaque via the endothelial monolayer. Monocytes differentiate into macrophages within the plaque and uptake oxidized-LDL, becoming foam cells, which make up the majority of the cellularity of the plaque [1]. Vascular smooth muscle cell proliferation and extracellular matrix deposition promote the construction of the fibrous cap [2]. The most serious clinical concern with atherosclerosis is thrombosis, which occurs if the fibrous cap is disrupted, exposing the inside of the plaque to the clotting factors in the blood and leading to occlusion of the vessel and, often, myocardial infarction.

Cadherin-11 (CDH11) is a calcium-dependent cell-cell adhesion protein, and it has been implicated in several fibrotic and inflammatory diseases. Previous work has shown that calcified aortic valves contain higher levels of CDH11, indicating that it may play an important role in the calcification process, possibly due to the increased cellular tension that it provides resulting in increased contractility and nodule formation [3]. Another study demonstrated that cardiac CDH11 levels spike after myocardial infarction (MI), and treatment with a blocking antibody against CDH11 improves post-MI tissue remodeling and recovery in mice. This same study posited that blocking CDH11 improved outcomes due to a disruption in the interaction between macrophages and cardiac fibroblasts, resulting in a decreased inflammatory response [4]. Therefore, CDH11 is a promising target for several cardiovascular diseases, but its role in the progression of the atherosclerosis or the inflammatory component of plaque formation has not yet been examined.

METHODS

ApoE-/- male mice were purchased (Jackson Laboratory) and at 4 weeks of age, mice either continued on a regular diet or were fed on a high fat, high cholesterol diet (HFD) for 10 weeks. Mice in the treatment groups were given intraperitoneal injections of either a CDH11 blocking antibody (SYN0012) or an isotype antibody control (IgG2a) every 10 days at a dose of 10 mg/kg for the 10 week period. Blood was collected either retro-orbitally or via post-mortem cardiac puncture. Monocyte content was assessed using fluorescence-activated cell sorting (FACS), using CD45.2, F480, and CD11b as markers. Mice were sacrificed at 14 weeks, and the aortas were taken for either mRNA harvesting or for analysis of the lesion area. Relative amounts of mRNA were quantified using qPCR. The percent lesion area was quantified using *en face* analysis. Aortas were split open, stained for lipids using Sudan IV, pinned down to a black dissecting pan, and imaged. Images were analyzed using MATLAB to calculate lesion area as a percentage of total aorta area.

RESULTS

In both the ApoE-/- mice fed the regular diet and those fed the HFD, CDH11 and MMP12 mRNA levels were higher in the aortic arch compared to the distal aorta (**Figure 1A-B**). The arch is known to be a region with a greater predilection for atherosclerotic plaque formation, and MMP12 levels are correlated with advanced plaque progression and macrophage invasion [2]. Additionally, CHI3L1 mRNA levels, also correlated with atherosclerosis [5], were higher in the arch of ApoE-/- mice fed the HFD, and CHI3L mRNA levels were correlated with CDH11 mRNA levels in both the arch and the distal aorta of the ApoE-/- mice fed the HFD. These results could indicate a possible role for CDH11 in lesion development.

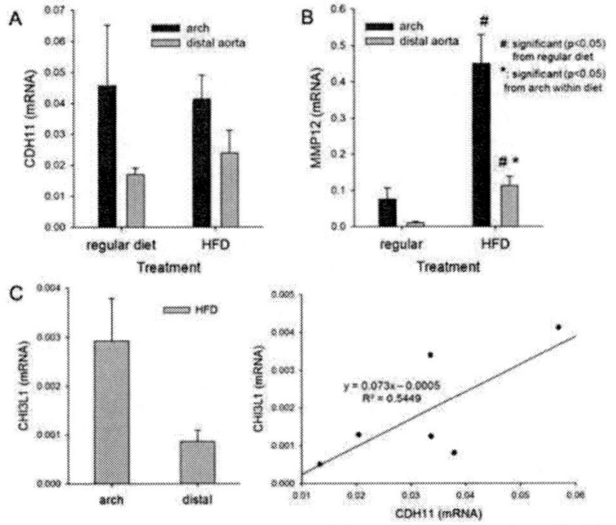

Figure 1: Relative mRNA levels of CDH11 (A) and MMP12 (B) are higher in the aortic arch when compared to the distal aorta in ApoE-/- mice fed on either a regular or a HFD for 10 weeks. C) Relative mRNA levels of CHI3L1 are higher in the aortic arch, and expression of CHI3L1 correlates with CDH11 in both the arch and the distal aorta.

ApoE-/- mice fed a HFD for 10 weeks receiving a CDH11 blocking antibody (SYN0012) treatment demonstrated decreased lesion area as a percent of total aorta area compared to the ApoE-/- mice receiving an isotype control antibody (IgG2a) (**Figure 2**). These results further indicate a role for CDH11 in atherosclerotic plaque formation and suggest that blocking CDH11 reduces disease progression.

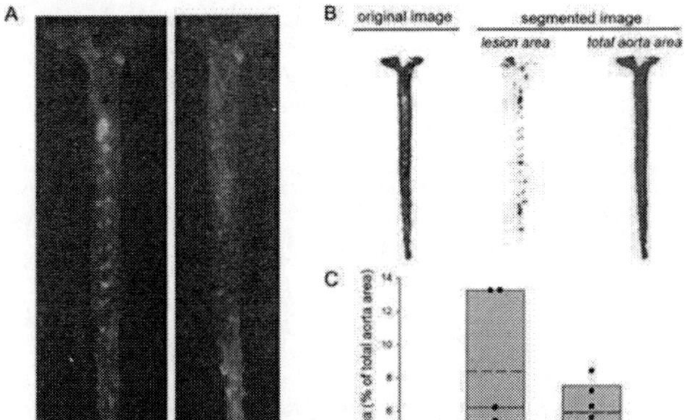

Figure 2: CDH11 blocking antibody (SYN0012) reduced lesion area in ApoE-/- mice fed a HFD for 10 weeks. A) *En face* images of the aorta with lipids stained red. B) Image segmenting strategy in MATLAB. C) Lesion area of both treatment groups represented as the percentage of total aorta area (solid line = median; dashed line = mean).

ApoE-/- mice fed a HFD for 10 weeks receiving a CDH11 blocking antibody (SYN0012) treatment contained higher monocyte content in peripheral blood compared to the ApoE-/- mice receiving an isotype control antibody (IgG2a) (**Figure 3**). These results indicate that the CDH11 blocking antibody treatment is preventing macrophages from migrating from the peripheral blood and could suggest a decrease in macrophage invasion of atherosclerotic plaques.

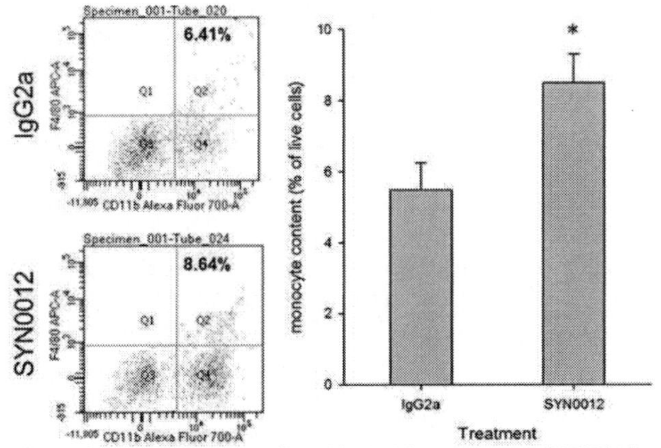

Figure 3: Treatment with CDH11 blocking antibody (SYN0012) for 10 weeks increased monocyte content in the peripheral blood of ApoE-/- mice fed a HFD for 10 weeks. * indicates significance of $p<0.05$.

DISCUSSION

These results combined indicate a role for CDH11 in the disease progression of atherosclerosis. While the mechanism remains unclear, preliminary evidence suggests that blocking CDH11 may decrease plaque burden by preventing macrophage migration into the plaque. This is supported by previous work which has suggested that blocking CDH11 decreases inflammation post-MI by disrupting an interaction between macrophages and cardiac fibroblasts [4]. Another possible explanation could be found in contractility. Other studies have demonstrated a link between cell contractility and endothelial permeability [6], and CDH11 has been shown to play an important role in cell contraction, particularly in cardiovascular diseases in which cell contractility is a primary mechanism [3]. Therefore, the CDH11 blocking antibody could be preventing macrophage infiltration by decreasing contraction and reducing endothelial permeability. More work is needed to be done in order to fully elucidate the mechanism, but there is a clear connection between CDH11 and atherosclerotic plaque development.

ACKNOWLEDGEMENTS

This work was supported by the NIH (R01 HL115013 and R35 HL135790) and the National Science Foundation (DGE-1445197). We acknowledge Dominik Hartl and Uwe Junker from Roche for supplying the blocking antibody, SYN0012.

REFERENCES

[1] Hansson, G. K., et al. *Nature Reviews Immunology* 6.7 (2006): 508-519.
[2] Schwanekamp, J. A., et al. *ATVB* 36.1 (2016): 60-68.
[3] Hutcheson, J. D., et al. " *ATVB* 33.1 (2013): 114-120.
[4] Schroer, Alison. Diss. Vanderbilt University. 2016.
[5] Rathcke, C.N., et al. *PLoS One* 7.8 (2012): e43498.
[6]Huynh, J., et al. *Science translational medicine* 3.112 (2011): 112ra122-112ra122.

TEMPORAL AND SPATIAL CORRELATION OF WALL SHEAR STRESS TO PLAQUE COMPOSITION IN ATHEROSCLEROTIC MICE DURING PLAQUE PROGRESSION

R Xing (1), A. Moerman (1), R.Y. Ridwan (2), K. Van der Heiden (1), F.J.H. Gijsen (1)

(1) Department of Biomedical Engineering and Cardiology, Erasmus Medical Center, Rotterdam, The Netherlands

(2) Department of Genetics, Erasmus Medical Center, Rotterdam, The Netherlands

INTRODUCTION

Wall shear stress (WSS) is the frictional force that blood flow exerts on the endothelium. It is known to initiate atherosclerotic plaque growth, which is the accumulation of lipids, fibrous tissue and inflammatory cells in the arterial wall.

As the disease advances, plaques can progress into two types: a stable or a vulnerable plaque. Vulnerable plaques are characterized by thin fibrous cap, large lipid pool and increased plaque inflammation. Rupture of a vulnerable plaque is the main cause of stroke and myocardial infarction. However, factors determining plaque progression and plaque vulnerability remain unclear.

We hypothesized that WSS plays a role in atherosclerotic plaque progression and plaque stability [1]. We want to study this hypothesis using a WSS-manipulated atherosclerotic mouse model, in which both a stable and a vulnerable plaque are present in the same vessel [2]. In this study, we demonstrate for the first time how WSS patterns change over time during plaque progression in this mouse model and how these WSS patterns are related to plaque composition.

METHODS

A tapering cast was placed around the right common carotid artery (RCCA) of female ApoE-/- mice (n=7). The cast has a tapering lumen (diameter from 400 µm to 200 µm over 1.5 mm) which changed the blood flow in the RCCA. The cast induces a low WSS region upstream of the cast, high WSS in the cast, and oscillatory WSS area downstream of the cast. It was previously shown that vulnerable plaques develop in the region upstream of the cast [2]. IN this study, we will focus on the upstream region. We followed plaque progression over time at 5, 7, 9 weeks after cast placement.

Blood velocity and vessel diameter in the RCCA was measured with Doppler Ultrasound (Vevo 2100). The two measurements were captured at the same location to determine the flow rate. Next, RCCA vessel geometry was obtained using the contrast-enhanced micro-CT imaging system (Quantum FX) with a resolution of 40 µm. RCCA vessel surface was then reconstructed using an in-house developed segmentation protocol [3]. Finally, computational fluid dynamics (Fluent 14.5) was applied to compute the velocity field and WSS distribution in the RCCA at three time points during plaque

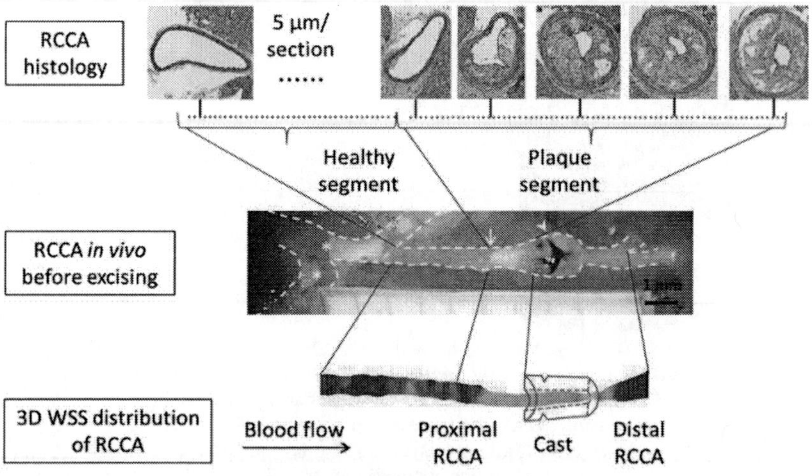

Figure 1: the histological images (top panel) are registered to the 3D shear stress map (bottom panel) by using the en-face imaging performed before excising the RCCA (central panel).

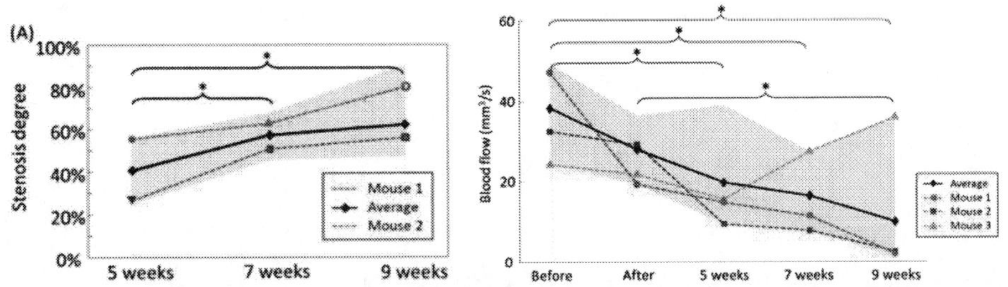

Figure 2: the area stenosis based on the CT (left) and the ultrasound flow measurements (left). The shaded area indicates the range of the measurements, and the asterix statistical significance.

progression. Plaque composition was evaluated at week 9 using standard histological techniques. Histology was matched to the WSS maps by quantifying the shrinkage of the RCCA during the processing steps (figure 1).

RESULTS

Geometry and flow: Plaques were present immediately upstream of the cast (see figure 1) at week 5 and continue to develop at week 7 and 9. The average degree of stenosis increases from 41 to 62% (figure 2, left panel). The flow reduced immediately after cast placement and a continuous decrease over time was observed. On average, a flow reduction from 38 mm^3/s to 10 mm^3/s was observed (figure 2, right panel). The individual data illustrate the variability between the animals, e.g. the increased flow for mouse 3 from week 5 on.

Shear stress data: Figure 3A illustrates the evolution of 3D WSS distributions of mouse 1 over time. At 5 weeks, the average WSS in the healthy segment was 2.2 ± 0.1 Pa and in the plaque segment it was 5.0 ± 2.2 Pa. The WSS decreased over time due to a gradual decrease in blood flow, finally resulting in average WSS values of 0.3 ± 0.0 Pa in the healthy segment and 1.7 ± 1.0 Pa in segment plaque segment. Figure 3B illustrates the differences observed between the animals. In mouse 2, flow only marginally decreased over time, and the increased lumen intrusion induced a peak in the WSS in the plaque region at week 7. Furthermore, the circumferentially averaged WSS distribution showed significant heterogeneity (figure 3, central panels).

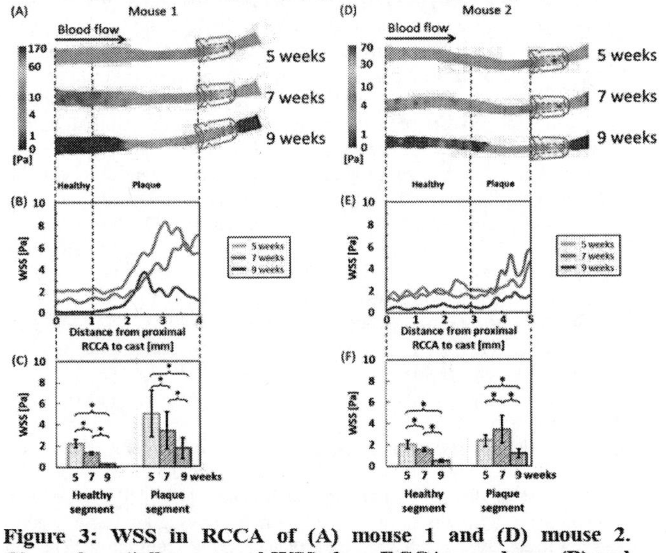

Figure 3: WSS in RCCA of (A) mouse 1 and (D) mouse 2. Circumferentially-averaged WSS along RCCA was shown (B) and (E). Averaged WSS in the healthy and plaque segments are shown in panels (C) and, and (F).

Histology data: An example of the histological results is given in the top row of figure 1. We always observed a single continuous plaque in RCCA upstream to the cast. The proximal part of the plaque was smaller and eccentric. There was rich accumulation of foam cells and macrophages which occupied 24.6% ± 18.9% of the plaque area. Cholesterol crystals were clearly seen as needle-like structures. Closer to the cast, the plaques were larger and concentric. Large necrotic cores were observed in proximity of the media layer while smaller cholesterol crystals lied superficially. Macrophages were less abundant with 13.9% ± 11.2%. However, macrophages accumulated superficially to the vessel lumen. Fibrous caps, covering the lipid pools, were slightly thinner in the concentric plaque region compared to the eccentric plaque region.

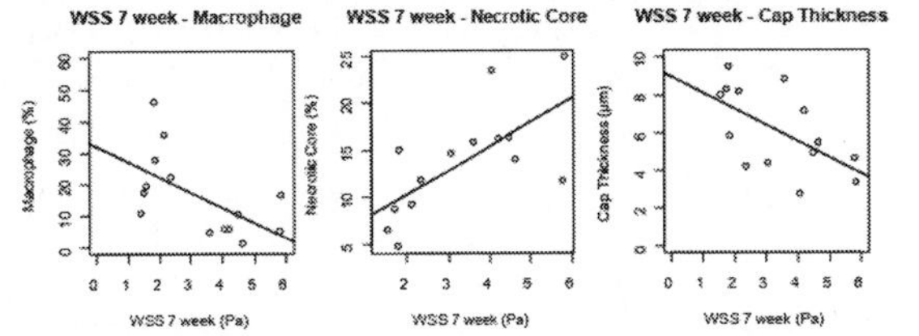

Figure 4: correlation scatter plots for WSS vs macrophages, necrotic core size and cap thickness for Mouse 1.

WSS vs composition: We found a negative correlation between WSS and the amount of macrophages: low WSS is correlated to increased inflammation (Figure 4). This correlation was observed in 6 out of 7 mice. We found a positive correlation between WSS and necrotic core size and a negative correlation between WSS and cap thickness, indicating that high WSS is associated with vulnerability. This correlation was observed in 4 out of 7 mice. The correlations between WSS and plaque composition were equally strong for the 7 and 9 week data.

DISCUSSION

Different plaque progression profiles were observed among animals. In general, blood flow decreased over time yet not in all mice. We combined geometrical and flow data specific to each animal and computed 3D WSS maps. Our study was the first to describe the evolution of WSS over a growing plaque in an *in vivo* atherosclerotic mouse model with a cast. We found that the healthy vessel segment was exposed to a homogeneous and decreasing level of WSS over time. However, the evolution of WSS in the plaque segment showed various profiles among animals, depending on the local geometrical and blood flow data. Also, the spatial distribution of WSS along the plaque segment was heterogeneous. Moreover, by matching histological sections to the RCCA geometry reconstructed from micro-CT, we were able to correlate WSS to plaque composition. In the majority of the mice, low WSS was associated with elevated levels of inflammation, which mainly localized proximally within small eccentric plaques. In the larger concentric plaques, high WSS was associated with increased relative necrotic core area and a thinner cap.

REFERENCES

[1] Wentzel,et al., Cardiovascular Research, 2012, 92(2)
[2] Cheng et al, Circulation, 2006 Jun 13;113(23)
[3] Xing et al., Biomedical Engineering Online, 2016. 15(15)

BASELINE RIGHT VENTRICULAR FUNCTION DOES NOT PREDICT SUDDEN DEATH IN SICKLE CELL MICE

David A. Schreier (1), Diana Tabima (1), Timothy A. Hacker (2) and Naomi C. Chesler (1,2)

(1) Department of Biomedical Engineering, University of Wisconsin-Madison, 2146 ECB, 1550 Engineering Dr., Madison, WI 53706

(2) Department of Medicine, University of Wisconsin-Madison, 1685 Highland Avenue, 5158 Medical Foundation Centennial Building, Madison, WI 53705

INTRODUCTION

Patients with sickle cell disease (SCD)-related pulmonary hypertension (PH) have a substantially increased risk of sudden death compared to those with SCD alone, as high as a 50% within 2 years [1, 2]. The prevalence of PH in adults with SCD is high--about 30%, while the percentage of SCD patients with evidence of PH via lung histology at the time of death is even higher--75% [3-5]. Upwards of 70% of SCD patients have evidence of pulmonary vascular disease by autopsy [5], but the effects of SCD on right ventricular (RV) function are not well understood. The major cause of death in PH is RV failure, and investigating RV function in patients with SCD and PH is merited.

SCD affects RV function through a number of mechanisms. A recent study demonstrated that both coronary microvascular disease and myocardial fibrosis, a result of chronic anemia, are associated with SCD [6], both of which impair RV function. Decreased RV ejection fraction, RV dilatation and RV hypertrophy are also common in SCD patients [7], but the connection to PH has not been made. The underlying condition that causes the symptoms of SCD are the SCD red blood cells (RBC), which lyse easily, are more rigid, dense and frequently trapped in the microcirculation, compared to healthy RBCs. As a consequence, SCD RBC interrupt blood flow and cause ischemia, thrombotic events, and reperfusion injury. SCD also increases oxidative stress [8], which has been shown to occur in patients with RV failure alone [9]. In addition, SCD can and often does impact hemodynamics including increasing systemic arterial wave reflections [10], and increasing pulmonary vascular resistance (PVR) [8], which has been shown to affect the interaction between the RV and pulmonary vasculature negatively [11].

In order to measure how SCD affects the RV, we propose to use a SCD mouse model that has been shown to successfully replicate the genetic, hematologic, and histopathologic features found in human SCD, including increased blood viscosity and chronic anemia [12]. In addition, these mice have shown to develop PH at 5 months of age [13].

The goal of this study is to quantify how SCD and its consequences affect pulmonary vascular hemodynamics, RV function, and the efficiency of interactions between the RV and the pulmonary vasculature. We chose to use pressure-volume (PV) loops to measure these changes as it provides a comprehensive view of right ventricular contractility, systolic and diastolic function, as well as the efficiency of the interaction between the pulmonary vasculature and RV, also referred to as ventricular vascular coupling (VVC). We hypothesized that SCD mice will have poor RV function and impaired ventricular-vascular coupling (VVC) compared to healthy control mice.

To test this hypothesis, we first measured PV loops in age-matched SCD and control mice, then exposed the animals to 5 minutes of 10% O_2 ventilation, which has been shown to lower blood oxygen levels in mice [14], after which we again measured pressure-volume loops. Our results demonstrate that while at baseline SCD mice have RV function and VVC comparable to control mice, acute hypoxia causes sudden death, likely a result of sickle cell crisis.

METHODS

6 male C57Bl6 mice (CTL) and 8 male Berkeley SCD mice (SCD), 20-24 weeks old, were obtained from Jackson Laboratory (Bar Harbor, ME). Using established techniques [14, 15], mice were anesthetized with an urethane solution (1000 mg/kg body weight, intraperitoneally), intubated, and placed on a ventilator. The left carotid artery was cannulated with a 1.2 F catheter tipped pressure transducer (Scisense, London, Ontario, Canada) that was advanced into the ascending aorta to measure systemic blood pressure. Subsequently, a 1.2 F admittance PV catheter (Scisense) was

introduced into the apex of the RV using a 20 gauge needle, leaving the pericardium intact. After instrumentation was established and initial RV pressure-volume measurements were obtained, the inferior vena cava was isolated and briefly occluded to obtain alterations in venous return for determination of end-systolic and end-diastolic pressure relations. This vena cava occlusion (VCO) was limited to a few seconds to avoid reflex responses. VCO was performed at least three times.

Measurements were obtained under normoxic ventilation conditions and, subsequently, under acutely hypoxic ventilation conditions. The pressure-volume signals were visually checked for quality and recorded for later analysis (Figure 1). At least ten consecutive cardiac cycles free of extra systolic beats were selected and used for the analysis. Standard hemodynamic variables including heart rate (HR), RV peak systolic pressure (RVSP), and RV function parameters such as stroke volume (SV), ejection fraction (EF), cardiac output (CO), end-diastolic volume (Ved), and effective arterial elastance (E_a=SV/PP) were calculated. RV contractile function was quantified as the slope of the ESPVR (E_{es}). Finally, VVC was calculated as E_{es}/E_a.

Statistical analysis was performed using a one-way ANOVA for condition (CTL vs. SCD) or generalized least squares with multiple comparisons for condition and ventilation (room air vs. acute hypoxia).

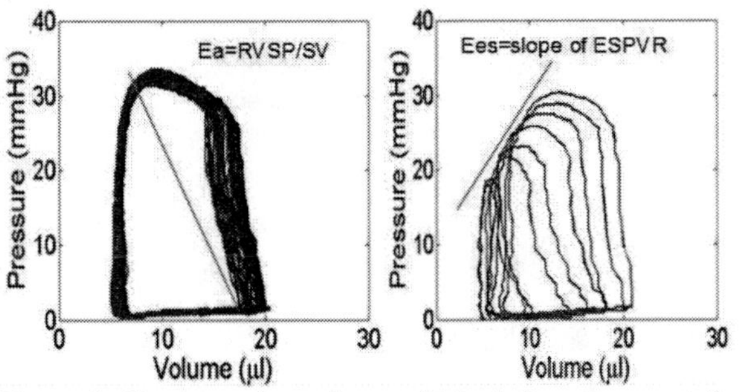

Figure 1. Pressure-volume loops obtained in a representative (CTL) mouse right ventricle showing both standard loops (left) and those collected during a vena cava occlusion (VCO; right). E_a and E_{es} are obtained as shown.

RESULTS

SCD did not change HR, RVSP, or RVEDP compared to CTL at baseline (Table 1). SCD mice did have trends of dilated RV, decreased SV, CO and PVR compared to CTL mice but no differences were statistically significant at baseline (Table 1). The major changes observed in this study were the results from exposure to 5 minutes of acute hypoxia (10% oxygen ventilation) after baseline measurements were obtained. 5 of 8 SCD mice died within 2 minutes of exposure, 2 of 8 died within the 5 minute period (we obtained limited data from these mice), and only 1 mouse survived the full 5 minute duration of acute hypoxia. In comparison, all CTL mice survived exposure to acute hypoxia and hemodynamic data were obtained (Table 1).

DISCUSSION

RV hemodynamics in SCD mice were not significantly different from the healthy control mice under baseline, normoxic conditions (21% oxygen ventilation). In contrast to the prior study showing PH in these SCD mice at 5 months of age [13], no PH was found. With no PH at baseline, no changes in RV function at baseline would be expected.

Table 1. Hemodynamic values calculated from pressure-volume loops in room air and acute hypoxia (AH; for CTL only) conditions.

	CTL	SCD	CTL AH
HR (beats/min)	550 ± 27	578 ± 32	525 ± 34
RVSP (mmHg)	24.6 ± 2.1	22.9 ± 3.6	26.9 ± 3.4
SV (µL)	21.4 ± 0.8	20.2 ± 3.9	19.7 ± 1.5
CO (ml/min)	11.9 ± 0.9	12.0 ± 2.7	10.1 ± 2.1
EF (%)	58.5 ± 6.2	42.3 ± 7.0	54.2 ± 9.0
Ved	37.9 ± 4.3	47.2 ± 6.1	34.5 ± 4.0
Ea (mmHg min/ml)	1.2 ± 0.2	1.7 ± 0.8	2.2 ± 1
Ees (mmHg min/ml)	1.4 ± 0.5	0.9 ± 0.3	1.3 ± 0.3
VVC	1.1 ± 0.3	0.8 ± 0.3	1.1 ± 0.7

Note: values are ±SE, n=6 for CTL, n=8 SCD. *P<0.05 vs. CTL.

However, upon exposure to acute hypoxia, SCD mice died suddenly. Erythrocyte sickling, which irreversibly sickles RBC in this mouse model [12], leads to hemolysis, thrombi formation and vaso-occlusion [16]. Our results suggest that acute changes in SCD RBC cause sudden death SCD, possibly via acute RV failure.

Future work includes treating SCD mice with a drug that increases fetal hemoglobin and increases resistance to RBC sickling to determine if these animals are more resilient to the stress of acute hypoxia.

ACKNOWLEDGEMENTS

This work was supported by NIH grant 1R21HL129047 (NCC), and NIH F31 HL128088-01 (DAS). The authors also would like to acknowledge Gaoussou Diarra for his surgical expertise.

REFERENCES

1. Castro, O., M. Hoque, and B.D. Brown, Blood, 2003. **101**(4): p. 1257-61.
2. Charache, S., J Lab Clin Med, 1994. **124**(4): p. 473-4.
3. Ataga, K.I., et al., Am J Med, 2004. **117**(9): p. 665-9.
4. De Castro, L.M., et al., Am J Hematol, 2008. **83**(1): p. 19-25.
5. Haque, A.K., et al., Hum Pathol, 2002. **33**(10): p. 1037-43.
6. Desai, A.A., et al., Circ Cardiovasc Imaging, 2014. **7**(3): p. 430-7.
7. Junqueira, F.P., et al., J Cardiovasc Magn Reson, 2013. **15**: p. 83.
8. Machado, R.F. and H.W. Farber, Clin Chest Med, 2013. **34**(4): p. 739-52.
9. Redout, E.M., et al., Cardiovasc Res, 2007. **75**(4): p. 770-81.
10. Aessopos, A., et al., Atherosclerosis, 2007. **191**(2): p. 427-32.
11. Wang, Z., et al., Physiol Rep, 2013. **1**(7): p. e00184.
12. Paszty, C., et al., Science, 1997. **278**(5339): p. 876-8.
13. Hsu, L.L., et al., Blood, 2007. **109**(7): p. 3088-98.
14. Schreier, D., et al., J Biomech Eng, 2013. **135**(2): p. 021018.
15. Tabima, D.M., T.A. Hacker, and N.C. Chesler, Am J Physiol Heart Circ Physiol, 2010. **299**(6): p. H2069-75.
16. Sun, K. and Y. Xia, Curr Opin Hematol, 2013. **20**(3): p. 215-21.

SB³C2017
Summer Biomechanics, Bioengineering and Biotransport Conference
June 21–24, 2017, Tucson, Arizona, USA

INTER-SUBJECT VARIABILITY TO INHALED AEROSOLS

Jessica M. Oakes[1]

(1) Department of Bioengineering, Northeastern University, Boston, MA, USA

INTRODUCTION

Despite wide variations in airway morphometry and respiration patterns, most computational models predict aerosol fate in either a single representative patient or in an idealized airway geometry. Therefore, until inter-patient variability is accounted for, it is impossible to translate these predictive models into clinical practice or public-health settings. Furthermore, correlation of deposition patterns to patient-specific airway morphometry will aid in not only understanding which geometric parameters are most important, but also provide a link to morphometric abnormalities and atypical distribution of particles in the lung. To answer to this need, the main goals of this study are to (1) assess variability in deposited particle concentrations in image-based airway geometries and (2) correlate airway morphometry to flow disturbances and deposition hotspots.

METHODS

Unsteady airflow and particle transport simulations are performed in three subject-specific pulmonary airway geometries to predict variability in ventilation, airway resistances, and deposited particle concentrations across three subjects of similar age. Airway geometries, airflow simulations, and particle tracking methods are performed with our previously-developed and validated frameworks [5, 6]. Conducting airway geometric models of the three subjects are constructed from clinically-obtained thoracic CT images. Spanning from the trachea to the most distal conducting airways distinguishable on the images, 3D conducting airway geometries of each subject are created manually with the open source software, SimVascular [9]. All three subjects are female and their ages, at the time of the CT scan, are 25.8, 35.7, and 37.2 for subjects 1, 2, and 3 respectively.

Airflow throughout the respiration cycle is calculated by solving the Navier-Stokes equations with physiologically realistic boundary conditions, assuming that the air is incompressible (fluid density: $\rho_f = 1.2E-6g-mm^{-3}$), Newtonian (viscosity: $\mu = 1.81E{-}5g{-}(mm{-}s)^{-1}$), and that the walls are rigid. Boundary condition descriptions at the distal faces ($\Gamma_{i,j}$) is not straightforward, as the spatial description of the time-dependent flow or pressure cannot be experimentally measured. Thus, it necessary to choose distal boundary conditions carefully, in order to best represent physiological conditions. To mimic the negative, relative to atmosphere, pressure that derives air in and out of the lungs, we choose to represent the lung periphery by a resistor and capacitor connected in series, driven by the time-dependent driving pressure (P_D, Figure 1), described by:

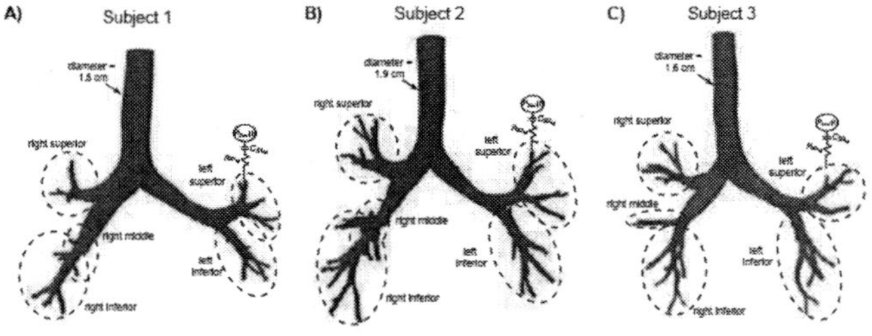

FIGURE 1: 3D image-based airway geometries of Subject 1 (panel A), Subject 2 (panel B), and Subject 3 (panel C). Terminal airways leading to each lobe are outlined. Note: not shown to scale.

$$P_D(t) = R_{global}Q(t) + \frac{V(t)}{C_{global}} - P_{atm} \quad (1)$$

where $Q(t) = \frac{dV(t)}{dt}$, and R_{global} and C_{global} are the respiratory resistance and compliance, respectively and are estimated based on data collected in anesthetized and paralyzed adults [?]. Assuming a sinusoidal respiration waveform, $V(t)$ is defined as:

$$V(t) = -\frac{1}{2}\left[TVcos\Big(\frac{2\pi}{T_B}t\Big) + TV\right], \quad (2)$$

where T_B is the total time for one breath, $\frac{1}{RR}$, and TV is the tidal volume; values were chosen to represent an average healthy adult. The coupling between the 3D and 0D models is completed by passing flow rate ($Q_{\Gamma_{i,j}}$) to the LPN [4]. For each subject, distal resistances and compliances ($R_{i,j}$ and $C_{i,j}$) are estimated, assuming proportionality to the subtending lobe volume and terminal airway's cross-sectional area. Here, we assume that the lobar fractions are 0.25, 0.20, 0.25, 0.09, and 0.21 for the left inferior, left superior, right inferior, right middle, and right superior lobes, respectively [2].

The incompressible Navier-Stokes equations are solved with a stabilized Galerkin finite element method with a custom linear solver that incorporates a combination of GMRES and conjugate gradient methods. The second order generalized α-method is employed for the time integration. To avoid flow divergence at $\Gamma_{trachea}$ and $\Gamma_{i,j}$, which may occur when the relationship between pressure and flow is applied as a boundary condition, and thus a velocity profile is not described, we apply a backflow stabilization framework with $\beta = 0.1$ [3].

Following the airflow simulations, spherical particles ($d_p = 3\mu m$) are tracked in the 3D models by solving a reduced form of the Maxey-Riley equation [6, 7]. Gravity was positioned parallel to the trachea and particles were assumed to be deposited once they penetrated the wall. Time-dependent concentration and air volume exiting each of the models was recorded and passed as a boundary condition for 1D advection-diffusion models where transport is solved distal to the 3D airways. Throughout expiration, particles are passed from the 1D models to the 3D geometry. Distal to the 3D model, defined for each lobe by Yeh et al. [10] were lumped together to solve the trumpet models [8], which included airway evolution throughout the respiration cycle based on mass conservation and a loss term to model deposition in the bronchioles and alveoli.

RESULTS AND DISCUSSION

Airway geometries, with the five lobes outlined, are shown in Figure 1. Tracheal diameter is smallest in subject 1 and largest in subject 2. The trachea and main bronchi are more elliptical in subjects 2 and 3 than in subject 1, with the greatest ellipticity seen in subject 3. Subject 2's left main bronchi's cross-sectional area reduces at a greater rate between the main carina and the bifurcation leading to the left inferior and superior lobes, compared to the other two models.

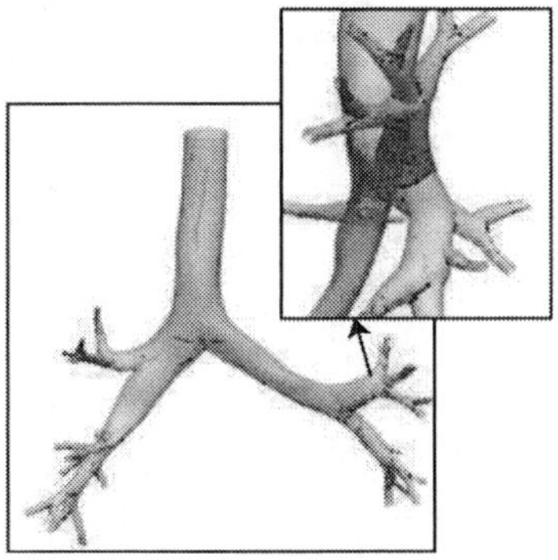

FIGURE 2: Deposited particle locations ($d_p = 3\ \mu m$) for Subject 1.

The global resistance and compliance are estimated as $R_{global} = 0.007\frac{cmH_2O-s}{ml}$ and $C_{global} = 59\frac{mL}{cmH_2O}$ and the tidal volume is set to 500 mL and the respiratory rate to 0.25 $\frac{1}{s}$. As the same lobar fractions are applied for each subject, the time dependent lobar flow rate and volume is the same between subjects.

For the 3 micron diameter particles considered here, deposition was mainly localized to the bifurcation regions for the larger airways and on the gravitationally-dependent side of the smaller airways. A representative particle concentration map is shown for Subject 1 in Figure 2. Subject 3's left and right main bronchi are more elliptical than Subject 1 and 2, resulting in more particles being deposited in these regions of Subject 3, compared to the other two subjects. Despite all three subjects having the same amount of air delivered to each lobe lobe, the distribution of particles between the subjects varied (Figure 3). Indeed, a value of 1 would indicate that the delivered fraction of particles is proportional to the lobe volume fraction. Subject 2 had more particle delivered to the left inferior lobe than the other two subjects.

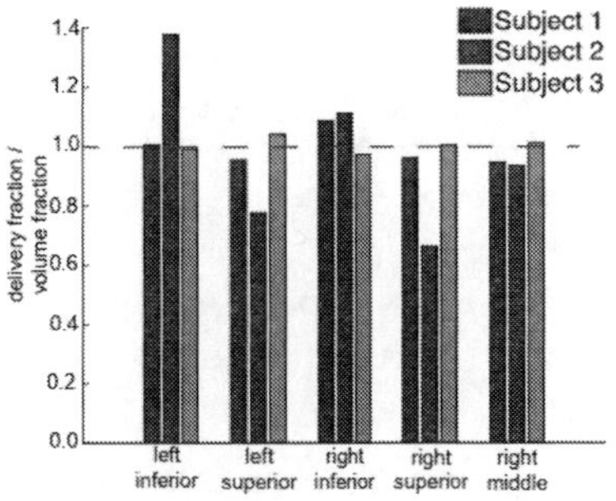

FIGURE 3: Fraction of particles delivered to each lobe normalized by the volume fraction.

By coupling distal respiratory mechanics to 3D image-based conducting airways of healthy adults, we are able to predict variations in particle deposition across three female subjects of similar age. Geometric differences, including the level of airway ellipticity and cross-sectional area, dictate differences between subjects on the location of deposited particles. Future studies shall focus on coupling experimentally derived ventilation to incorporate subject-specific distal ventilation.

ACKNOWLEDGMENTS

We would like to thank Dr. Jeff Feinstein for providing us with the thoracic CT images and the Information Technology Services, Research Computing at Northeastern University for providing high performance computing and storage.

REFERENCES

[1] D'Angelo, E. et al. *Journal of Applied Physiology*, 67: 2256-2564, 1989.
[2] Horsfield, K. et al. *J. Applied Physiology*, 31: 207-217, 1972.
[3] Esmaily Moghadam, M. et al. *Comp. Mech.*, 52: 1141-1152, 2013.
[4] Esmaily Moghadam, M. et al. *Comp. Physics.*, 244: 63-79, 2013.
[5] Oakes, J.M. et al. *Annals of Biomed. Eng.*, 42: 899-914, 2014.
[6] Oakes, J.M. et al. *International Journal For Numerical Methods in Biomedical Engineering*, DOI:10.1002/cnm.2847.
[7] Shadden. (2010) FlowVC [Computer software]. Retrieved from http://shaddenlab.berkeley.edu/software/.
[8] Taulbee et al. *J. of Applied Physiology*, 38: 77-85, 1975.
[9] Updegrove A. et al. *Annals of Biomed. Eng.*, DOI:10.1007/s10439-016-1762-8.
[10] Yeh, H. and Schum, G. *Bulletin of Math. Biology*, 42: 461-480, 1980.

MEASUREMENT OF THE DIFFUSION COEFFICIENT OF OXYGEN IN THE VITREOUS HUMOR

Anita N. Penkova (1,5), Komsan Rattanakijsuntorn (6), Anahid Khoobyaar (1), Karthik Murali (2,4), Mark S. Humayun (3,4), Satwindar S. Sadhal (1,5)

(1) Department of Aerospace & Mechanical Engineering
(2) D-Health Lab
(3) Department of Ophthalmology
(4) Department of Biomedical Engineering
University of Southern California
Los Angeles, California, USA

(5) SAIRC, Saban Research Center,
Children's Hospital Los Angeles,
Los Angeles, California, USA

(6) Department of Mechanical Engineering
Ubon Ratchathani University,
Ubonratchathani, Thailand

INTRODUCTION

Ischemic retinal diseases such as diabetic retinopathy (DR) and retinal vein occlusions (RVO) are leading causes of blindness [1]. Although they have different underlying pathophysiology, the common end result is inner retinal ischemia. Currently, the standard of care for ischemic retinal disease treatment involves using intravitreal medications, laser photocoagulation (focal or Pan Retinal Photocoagulation (PRP)) and/or Pars Plana Vitrectomy (PPV). Intravitreal medications have been shown to decrease the VEGF levels and vascular permeability but need to be administered monthly and may increase the risk of retinal ischemia [2]. Moreover, a recent publication has shown that only 22% of patients treated for diabetic macular edema with ranibizumab and 8% of patient treated with only laser gained 15 or more letters after a 12-month period of treatment [3] resulting in approximately 50% of patients without significant visual acuity improvement. Laser therapy is also destructive, leading to side effects such as constricted peripheral visual field, delayed dark adaptation, central scotomas, and scarring [4, 5]. PPV itself increases preretinal O_2 but those changes are insufficient to supply ischemic retina and can divert oxygen from the anterior structures, leading to iris and angle neovascularization and subsequent hyphema and elevated intraocular pressure [6]. Intravitreal oxygen therapy has been proposed as a treatment for retinal ischemia [7, 8]. For the development of these types of treatments, detailed understanding of intravitreal oxygen transport is necessary, and accurate measurements of the transport parameters are necessary. In particular, for mathematical modeling of the transport processes, the diffusion coefficient is a crucial physical property that is used in the determination of the O_2 distribution. It is of interest to ascertain how much O_2 reaches the retina and the lens as the diffusion and convection processes progress. The information about reaching the lens is particularly important because of adverse effects of oxygen in triggering cataracts.

METHODS

We have devised a technique to measure this diffusion coefficient, based on the classical theory of bubble dissolution, developed from the diffusion equation. We have conducted test experiments with bovine vitreous which is extracted from fresh bovine eyes (Manning Beef, Pico Rivera, CA) and placed in a clear plastic container. An oxygen bubble, 1-4 mm in diameter, is injected into the vitreous. The bubble gets trapped in the gelatinous mesh (collagen and hyaluronic acid) and does not rise with buoyancy. The free surface of the vitreous is covered with peanut oil to prevent contact with air. The system is slightly pressurized with 180 cm of hydrostatic head with the same oil. A preliminary low-cost setup for the experiment is shown in Figure 1. The O_2 bubble starts to dissolve in the vitreous, and the size history is recorded with a camera over 10-20 hours until the bubble completely disappears. The size history together with the gas diffusion theory will be applied to obtain the diffusion coefficient. Based on the work of Epstein & Plesset [9], the bubble radius $R(t)$ is described as a function of time by the nonlinear ordinary differential equation,

$$\frac{dR(t)}{dt} = \frac{D(c_\infty - c_s)}{\rho_\infty + 2M\sigma/(3R(t)\,\mathrm{RT})}\left\{\frac{1}{R(t)} + \frac{1}{(\pi D t)^{1/2}}\right\}, \qquad (1)$$

which results from the solution of the diffusion equation. The various symbols are given in Table 1. The surface concentration c_s is given by Henry's law, i.e., $c_s = Kp_b$ where $p_b = p_l + 2\sigma/R$, with

c_∞	initial O_2 concentration	R	universal gas constant
c_s	O_2 concentration at bubble surface	$R(t)$	bubble radius
D	O_2 diffusion coefficient	t	time
M	O_2 molecular mass	ρ_∞	O_2 density at STP
		σ	surface tension

Table 1: Nomenclature

Figure 1: Bovine vitreous humor in a clear container. The free surface of the vitreous is covered in peanut oil, and connected by tubing to an oil reservoir 180 cm above. An O_2 bubble was injected from the bottom, and stays trapped in the vitreous.

p_l being the liquid-side pressure at the bubble surface. Further adjustment is needed for O_2 depletion due to reaction with ascorbates. However, we wait sufficiently long before the start of the experiment (4-5 hours) by which time both the O_2 and the ascorbates have depleted. The experiment was conducted at 20°C, and the bubble collapse history was recorded with a camera set to photograph the bubble every 30 minutes (see Figure 2). Complete collapse was seen after 11.5 hours.

RESULTS

Equation (2) was solved numerically using the following data□ initial radius, $R_0 = 2$ mm; $c_\infty = 0$; $\rho_\infty = 1.331 kg/m^3$□ $\sigma = 0.065$ N/m; hydrostatic head, $h = 180$ cm□Henry's law constant for O_2, $K = 0.001384$ moles/l-atm. The numerical integration was conducted in a non-dimensional time domain $\tau = \pi D t / R_0^2$, with $\Delta\tau = 0.00157$ based on a time interval of 1s using $D = 2\times10^{-9} \mathrm{m}^2/\mathrm{s}$ for O_2 in water. The theory predicted complete collapse at $\tau_c = 59.57$. With at $t_c = 11.57$ h, we obtained $D = \tau_c R_0^2 / \pi \tau_c = 1.82\times10^{-9} \mathrm{m}^2/\mathrm{s}$. The experimental data was plotted against the theoretical prediction in Figure 3, and shows good agreement. The experimental collapse rate is a bit slow at early times, and it is possible that there is some residual O_2 in the vitreous.

DISCUSSION

The bubble-collapse method is an accurate approach for O_2 diffusion coefficient measurement in the vitreous humor. For the ex vivo eye, this is particularly convenient, especially since bubble suspension in the vitreous is not difficult. The data from the measurements are very useful for future mathematical modeling of O_2 transport through the vitreous humor, and developing O_2-based therapies for retinal diseases. Earlier measurements on feline vitreous have been conducted by Linsenmeier et al □10□who gave D values in the range (1.7 - 2.3)10^{-9} m^2/s at 38°C. While this agrees well with our result, further measurements at 37-38°C temperature range need to be conducted. Additional improvements in the experiment include developing a high-pressure system that would facilitate bubble collapse over a much shorter time period, as well as measurements at different temperatures. Investigations on effect of the presence of ascorbates are also needed, especially for in vivo application of the results.

REFERENCES

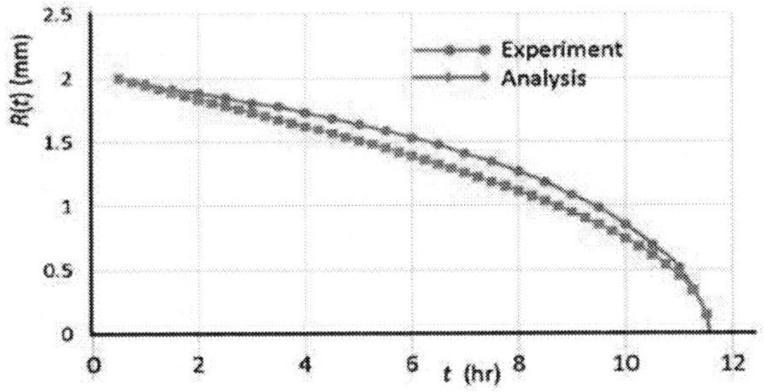

Figure 3: Plots of $R(t)$ from experimental measurements and numerical calculations.

[1] J. H. Kempen, et al *Arch. Ophthalmol.*, vol. 122, no. 4, p. 552–563, Apr. 2004.

[2] National Eye Institute, Bethesda, MD, *Diabetic Retinopathy, https://www.nei.nih.gov/eyedata/diabetic#4,*□ 2015.

[3] K. Manousaridis and J. Talks, *Br. J. Ophthalmol.*, vol. 96, no. 2, p. 179–184, Feb. 2012.

[4] M. Landers III, E. Stefansson and M. Wolbarsht, *Retina*, vol. 2, no. 3, p. 167–175, 1982.

[5] E. Stef□nsson, et al. *Am. J. Ophthalmol.*, vol. 113, no. 1, p. 36–38, Jan 1992.

[6] I. A. Barba□etto, et al. *Exp. Eye Res.*, vol. 78, no. 5, p. 917–924, May 2004.

[7] N. Scianmarello, et al. *IEEE 29th Int. Conf. on Micro Electro Mechanical Systems (MEMS)*, Shanghai, *24-28 Jan.*, 2016.

[8] K. Murali, et al. *PLOS One*, doi□10.1371/journal.pone.0149961.g003, 1-13, 2016.

[9] T. Epstein and M. Plesset, *J. Chem. Phys.*, vol. 18, no. 11, pp. 1501-1509, 1950.

[10] R. A. Linsenmeier, et al. *Exp. Eye Res.*, vol. 32, pp. 369-379, 1981.

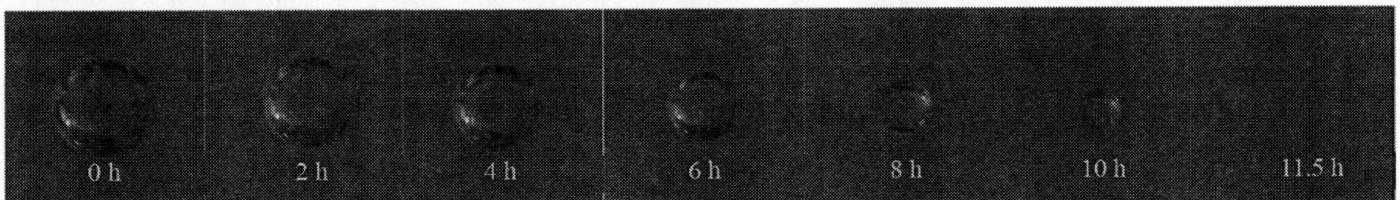

Figure 2: Images of bubble collapse history in an ex vivo bovine vitreous at 2-hour intervals. The initial bubble diameter is 4 mm (image at 0 h).

A SIMULATION FRAMEWORK OF MULTISCALE FLOW IN LYMPHATIC VESSEL NETWORKS

Lowell T. Edgar, Christopher J. Morris, James E. Moore Jr.

Department of Bioengineering
Imperial College London
South Kensington Campus, London,
United Kingdom

INTRODUCTION

The lymphatic system is an extensive vessel network that plays a vital role in maintaining fluid homeostasis by transporting approximately 4 L/day of viscous lymph from the interstitial space to the venous blood return. Unlike the blood circulation, there is no central pump; rather, lymph flow is generated against gravity and adverse pressure conditions by vessel contractions. These occur through extrinsic pumping mechanisms (such as motion of the surrounding tissue), and intrinsic pumping due to active contraction of individual vessel segments (known as lymphangions). Both mechanisms rely on bi-leaflet valves throughout the vessel network to prevent backflow.

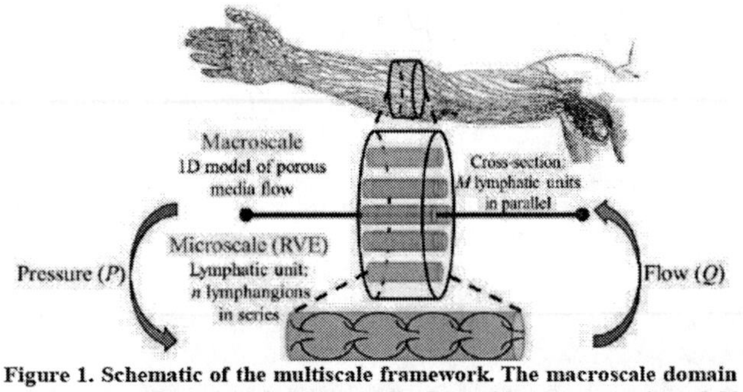

Figure 1. Schematic of the multiscale framework. The macroscale domain consists of a 1D model of porous media flow with a cross section of *M* lymphatic units in parallel at each integration point. Each lymphatic unit consists of *n* lymphangions in series (microscale RVE). Macro pressure (*P*) is passed down to the microscale to set up the boundary value problem, and the solution is passed back up to calculate macro flow (*Q*).

Disruption of lymphatic flow results in a debilitating condition known as lymphedema, which affects hundreds of millions of people worldwide. The direct causes of lymphedema are poorly understood, but thought to be initiated by disruption of lymph flow pathways and/or pumping dysfunction. The phenomenon of lymphatic pumping is multiscale in nature: lymphangions at the micron scale (~50-500 μm in diameter) move macroscopic volumes of fluid (L). However, the exact mechanisms as to how microscopic pumping integrates to produce macroscopic flow, and how dysfunction at the individual lymphangion level disrupts this flow, remain poorly understood. Towards resolving this issue, our goal is to develop a simulation framework that allows for a more in-depth and systematic investigation into multiscale lymphatic flow with the hope of providing insight into both healthy and diseased lymphatic function, as well as new therapeutic targets and treatment options for the millions of patients suffering from lymphedema.

METHODS

Lymphangions, defined as the segment of vessel between two valves, are the fundamental pumping unit of the lymphatic system. We have previously designed a computational ODE model to study lymphangion pumping using a lumped parameter approach. This model is useful for simulating tens of lymphangions in series and parallel, and produces results that correspond well with experiments. The model takes inlet and outlet pressure as input parameters and returns vessel diameter and flow as outputs. Further details on this model can be found in our previous publications [1, 2].

To extend the model to the whole tissue level, we have constructed a multiscale simulation framework consisting of a macroscale and microscale domain (Fig. 1). We represented flow through the lymphatic network at the macroscale using a continuous 1D model of porous media flow. The macroscale domain (X) was portioned into N

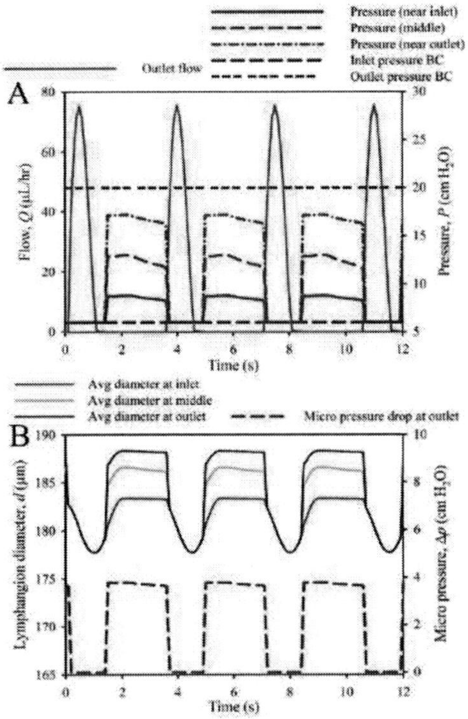

Figure 2. Simulation of a large lymphatic vessel. (A) Macro flow and pressure. On the left axis, macro flow *Q* at the outlet the vessel (red). On the right axis macro pressure *P* at the middle of the vessel, as well as the inlet and outlet pressure boundary conditions. (B) Lymphangion diameter and pressure drop. On the left axis, average diameter of the RVE at the inlet of the vessel (green), at the middle of the vessel (orange), and at the outlet (blue). On the right axis, the microscale pressure drop across the last RVE near the outlet.

finite elements, and each midpoint was assigned a microscale representative volume element (RVE), consisting of n lymphangions in series.

The computational homogenization was loosely based on the heterogeneous multiscale methods proposed by Chu et al. [3]. Macro pressure, P, was stored at each node and macro flow, Q, was evaluated at each mid-point. Pressure boundary conditions were passed down to each RVE to set up a boundary value problem based on a first-order homogenization scheme through which the pressure drop across the lymphangion chain is calculated using the macro pressure gradient,

$$\Delta p = \frac{\partial P}{\partial X}\delta, \quad (1)$$

where δ is the length of the RVE lymphangion chain. The ODE model was then solved using these pressure boundary conditions and the time-averaged flow passed back as macro flow,

$$Q = q_{avg}. \quad (2)$$

This formulation results in a nonlinear system of equations which we solved for the unknown macro nodal pressures $\{P\}$ using quasi-Newton iterations,

$$[\mathrm{K}]\{P\} = \{H\}, \quad (3)$$

where $[\mathrm{K}]$ is the tangent permeability matrix ($-\partial Q/\partial P$) and $\{H\}$ is the solution array used to enforce the conservation of mass by restricting the macro flow gradient to zero. Pressure and flow across the macro domain was calculated over time using quasi-steady analysis, in which the Newton iterations were solved until convergence criteria were met for each time step.

We have included a simulation of a large lymphatic vessel network to demonstrate the multiscale framework. The vessel was 0.3 m, consisting of 100 lymphangions, with $N = 10$ elements and an RVE chain of $n = 3$. The inlet pressure was held constant at 6 and the outlet pressure at 20 cm H_2O, and flow was simulated over 12 s at 0.1 s intervals.

RESULTS

Flow at the outlet of the macro domain vessel was periodic, reaching a peak of 75.5 μL/hr at 0.5 s with a period of 3.5 s (Fig. 2A, left axis in red). By integrating outlet flow over time we calculate that the vessel pumped 0.05 μL of lymph against the adverse pressure difference over the 12 s time interval. Pressure throughout the macro domain tended to remain low, near the inlet pressure value, but periodically swelled with a waveform that preceded flow (Fig. 2A, right axis in blue). During this swell, pressure built up linearly from the inlet and outlet and also shifted downward slightly during the peak, with the most shift occurring at the middle of the vessel.

Lymphangion contraction was synchronous with flow, and the macro pressure waveform preceded each contractile event (Fig. 2B, left axis). Although all lymphangions reached the same minimum diameter while contracting, the resting diameter increased nonlinearly from the inlet to the outlet. During the resting period, the pressure drop across the RVE was positive, but it dropped to slightly negative values during lymphangion contraction (Fig. 2B, right axis).

DISCUSSION

This framework successfully simulated a large lymphatic vessel pumping lymph against a significantly adverse pressure gradient. As the lymphangions begin to contract, the pressure drop over the RVE drops to slightly negative values, the valves are open and forward flow is produced. Once the contraction begins to release, the pressure drop over the RVE becomes adverse, the valves close and flow is negative although very small due to the high resistance of the valves. Note that the macro pressure converges to an adverse pressure gradient if flow is negative, and a favorable pressure difference if flow is positive. Also note that the macro pressure P is sustainably different magnitude than the RVE pressure difference Δp (Fig. 2).

Using this framework we modeled the behavior of 100 lymphangions by only calculating for 30 lymphangions and applying the homogenization scheme. These methods could easily be extended to larger numbers of lymphangions, both in series and parallel, but we have already begun to see interesting effects occurring across scales, such as the lag in the macro pressure waveform. Although this initial implementation was successful, there are numerous improvements to be made and features to be added. We would like to effectively scale our pressure-diameter structural relationship in the ODE model to account for lymphatic vessels of different diameters. We have begun to consider a second-order homogenization scheme, as the RVE lymphangion chains are long enough such that second-order pressure effects may not be negligible. We also consider a flow source and flow boundary conditions to simulate lymph entering the system via initial lymphatics, and externally applied pressure such as in massage therapy and compression bandage treatments currently used to treat lymphedema.

ACKNOWLEDGEMENTS

Funding from the US National Institute of Health U01 HL123420-01 is gratefully acknowledged.

REFERENCES

[1] Jamalian S et al., *Am J Physiol Heart Circ Physiol*, 2013, [2] Jamalian S et al., *PLoS One*, 2016, [3] Chu et al., *Multiscale Model Simul*, 2012.

CONTRACTION OF COLLECTING LYMPHATICS: ORGANIZATION OF PRESSURE-DEPENDENT RATE FOR MULTIPLE LYMPHANGIONS

Christopher D. Bertram (1), Charlie Macaskill (1), Michael J. Davis (2)

(1) School of Mathematics and Statistics
University of Sydney
New South Wales, Australia

(2) Dept. Medical Pharmacology & Physiology
University of Missouri School of Medicine
Columbia, Missouri, U.S.A.

INTRODUCTION

The lymphatic vascular system scavenges interstitial fluid that has leaked out of capillaries and returns it to the blood vascular system at the subclavian veins. In the main, its vessels comprise two functional types: non-muscular initial lymphatics which admit fluid from interstitial space, and muscular collecting lymphatics which pump it onward. The latter vessels are divided into short segments (typically some 3–10 diameters long) called lymphangions by non-return valves. We have previously developed models of multiple lymphangions [1] that prescribed fixed but adjustable values for both the contraction rate and the phase delay between adjacent lymphangions.

The rate of contraction of collecting lymphatics is in reality responsive to many stimuli. In particular, Scallan et al. [2] have shown how the contraction frequency is sensitive to small changes in transmural pressure, and may increase ten-fold over the pressure range 0–5 cmH_2O.

In the context of a single-lymphangion model, we have recently [3] introduced rate-dependence into our modelling. At the end of each systole, the algorithm measures the average transmural pressure over the preceding beat and calculates a relaxation period that will provide a contraction rate corresponding to data on rate vs. pressure measured in isobaric contractions *ex vivo*.

For application to a model of a vessel consisting of multiple lymphangions, it is necessary to consider not just the local pressure but also how adjacent lymphangions interact. We here describe a model that incorporates both the responsiveness of lymphangions to their local pressure and the propagation of contraction activation between lymphangions. We base our model on observations made in an isolated segment of rat mesenteric lymphatic vessel, which illuminate the principles that underlie coordination *in vivo*.

METHODS

Perfused segment

The mesenteric segment included three non-return valves. It was isolated and mounted for perfusion on micropipettes in a chamber viewed through an inverted microscope as described previously [4]. Each micropipette led to a reservoir of physiological saline solution (PSS) with albumin, in which the pressure was under the control of a servo system. The depth of PSS outside the vessel imposed a constant external pressure of 0.5 cmH_2O. After the cannulated vessel had resumed spontaneous contractions, it was subjected to step variations in each of the end pressures. Starting with both reservoirs at 3 cmH_2O, the downstream pressure was raised first to 7 cmH_2O, then to 11 cmH_2O, then to 15 cmH_2O, before being returned to its initial value. Each step was maintained for about 0.75 min. When downstream pressure exceeded that upstream, the valves prevented flow. After a 4 min pause for the segment to recuperate, the upstream pressure was raised in three similar but smaller steps, to 5, 7 and 9 cmH_2O, causing step increases in the flow-rate, before being returned to baseline.

The vessel was continuously imaged at 30 frames/s by video camera. Subsequently, vessel inside diameter was tracked from the video recordings [5] at three approximately equidistant axial sites, two between valves 1 and 2 and the third between valves 2 and 3. By similar means [4], the times when each valve was open or closed were tracked.

The diameter vs. time traces thus procured were analysed to detect the start of each contraction. The automated analysis steps included moving-average filtering, differentiation, and thresholding; each detected start was then inspected and if necessary adjusted manually.

Numerical model

The principles adduced from the results of the isolated-segment experiment (see below) were incorporated in an algorithm to control the time of the start of the contraction of each lymphangion in a multi-lymphangion model. As initially implemented the algorithm recognized three states (refractory, relaxed, and contracting) and two instants (end-diastole and end-systole). These were stored in a 5-level array keeping track of each lymphangion's state over time; see Figure 1. Later an alternative method of achieving the same ends was devised which dispensed with this large array.

This controlling process determined the start time of contractions. These are modeled as a prescribed smooth waveform $M(t)$ of fixed duration. $M(t)$ was earlier used to vary the active tension; we now [3] use it to vary the shape of a curve relating diameter and transmural pressure between limits representing the peak of isobaric twitch contractions and the inter-twitch relaxation.

RESULTS

In the isolated lymphatic segment, when upstream pressure was stepped up, contractions propagated in the mean direction of lymph flow as enforced by the valves. When downstream pressure was stepped up, the direction of this propagation was reversed. When upstream and downstream pressure were the same, contractions propagated in the antegrade direction.

We deduced the following principles from the *ex-vivo* experiment. (1) Each segment, including the cut part-lymphangions connected to micropipettes, is capable of initiating contractions at a rate depending on local pressure. (2) Signals corresponding to the start of contraction are propagated electrically within the vessel wall to adjacent lymphangions. (3) There is an absolute refractory period starting at end-diastole and ending at a time later than end-systole during which a further contraction cannot be provoked, and arriving activation signals are extinguished.

These principles were encompassed digitally as shown in Fig. 1.

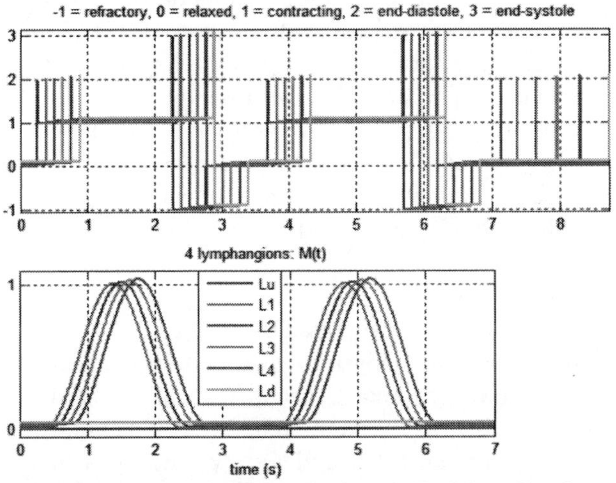

Figure 1: Four-lymphangion rate control. Each lymphangion would contract at a rate dependent on its local pressure if it were not entrained after a propagation delay by one of its neighbours. Here, transmural pressure decreases along the chain of lymphangions, so the non-contractile cut part-lymphangion upstream of the first valve controls rate for the whole chain. Only 7s is here simulated; the upper panel shows the look-ahead times after 7s at which the individual lymphangions would next contract in the absence of propagated signals.

The algorithm allows the coordinated rate of contraction of all the lymphangions to vary with the average transmural pressure over the preceding cycle experienced in the lymphangion with the highest transmural pressure. If another lymphangion subsequently experiences higher cycle-average transmural pressure, it has the opportunity to take over the pacing, subject only to whether the existing pattern of contractions and consequent refractory periods leaves it in a position to act on its local pressure; see Figure 2.

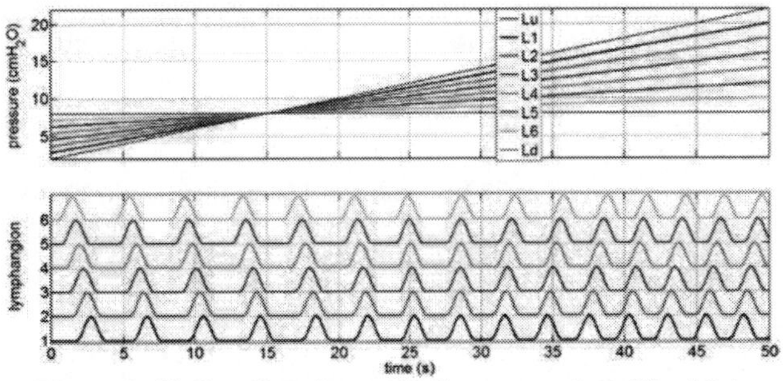

Figure 2: Six lymphangions, with linear ramping of the inlet pressure from an initial value below to a final value above the outlet pressure. Once inlet pressure rises above outlet pressure ($t > 15$s), the order in which the six lymphangions contract reverses. The phase changeover process is gradual, with all lymphangions in phase at 28s. Meanwhile, the contraction rate is constant until the maximum pressure starts to rise at 15s, then relaxation time decreases, and the rate smoothly increases.

We are in process of applying this scheme to the control of contractions in our multi-lymphangion model of lymphatic pumping.

DISCUSSION

Based on our observations in an isolated 2-lymphangion segment of rat mesenteric collecting lymphatic vessel, we believe that the algorithm we have devised accurately reflects how transmural pressure and wall-conducted activation signals interact. The method can potentially be extended to vary the magnitude of activation with transmural pressure. However, there is also a rate-of-change aspect to the lymphatic myogenic response to stretch or transmural pressure [6], which is not captured here. Flow has an inhibitory influence on lymphatic contractions [7,8], and this is also absent from the model put forward here.

ACKNOWLEDGEMENTS

All authors acknowledge support from U.S. National Institutes of Health (NIH) grant U01-HL-123420. MJD's laboratory was supported by NIH grant R01-HL-120867.

REFERENCES

[1] Bertram, CD et al., *Am J Physiol*, 310:H847-H860, 2016.
[2] Scallan, JP et al., *Am J Physiol*, 303:H809-H824, 2012.
[3] Bertram, CD et al., *Biomech Model Mechanobiol* (submitted).
[4] Davis, MJ et al., *Am J Physiol*, 301:H48-H60, 2011.
[5] Davis, MJ, *Microcirculation* 12:361-372, 2005.
[6] Davis, MJ et al., *J Physiol*, 587:165-182, 2009.
[7] Gashev, AA et al., *J Physiol*, 450:1023-1037, 2002.
[8] Kornuta, JA et al., *Am J Physiol*, 309:R1122-R1134, 2015.

PULSATILITY DICTATES LYMPH FLOW IN VIVO

Akshay Pujari (1), Daniel T. Sweet (2), Mark L Kahn (2), Juan M. Jiménez (1)

(1) Department of Mechanical & Industrial Engineering, University of Massachusetts, Amherst, MA 01003, USA

(2) Department of Medicine and Division of Cardiology, University of Pennsylvania, Philadelphia, PA 19104, USA

INTRODUCTION

The lymphatic system not only plays a role in immunity, but also in the delivery of interstitial fluid and adsorbed lipids to the blood circulatory system. Lymph forms as cell waste and toxin containing interstitial fluid is collected by lymphatic capillaries and transported via pre-collector and collecting lymphatic vessels. Due to the absence of a central pumping organ like the heart, lymph transport relies on the combined pressure gradients generated by fluid pressure head, local tissue motion, valves and lymphatic vessel contractility. The lack of synergy amongst these pumping mechanisms generates temporal variations in lymph flow rate and directionality[1]. Lymph flow irregularity makes difficult elucidating lymph fluid flow characteristics and their role in the transport of cells and macromolecules in the lymphatic system. The flow field is further complicated by the presence of lymphatic valves, which yield spatial and temporal flow gradients further complicating lymph transport. Here we employ 4-week-old wild type mice to study the transport of fluorescent tracers in efferent lymph vessels carrying lymph from subdermal inguinal lymph nodes. Using particle tracking velocimetry (PTV) we characterized the lymphatic flow field determining 1) wall shear stress, vorticity, and normal/shear strains, 2) lymph flow regulation via vessel contractility and coordination with neighboring lymphangions, and the 3) role of lymphatic valve efficiency in increasing momentum and decreasing static pressure across the valve to regulate lymph directionality.

METHODS

Mice were anesthetized using 2,2,2-Tribromoethanol (Sigma-Aldrich) dissolved in tert-amyl alcohol (Sigma-Aldrich) (0.5 mg/g). Body temperature was maintained at 37 °C via an animal temperature controller and warming pad (World Precision Instruments, ATC2000). Thoracic incision allowed peeling back of ventral skin and inguinal lymph node exposure. Five μL FluoSpheres [580/605] 1 μm latex particles were injected with a 32 gauge needle at a concentration of 1×10^8 beads/mL into the inguinal lymph. Particle displacement through efferent lymphatic vessels was imaged with an Olympus MVX10 dissecting microscope and recorded with a Nikon D5200 camera. Particle displacement was tracked using ImageJ and Imaris software. Tracks were checked manually to ensure accurate particle tracking. Algorithms were written in the Matlab software to determine velocity, vorticity, strain, wall shear stress and vessel wall motion. Lymphatic endothelial Prox1-GFP marker enabled lymphatic vessel edge detection and vessel tracking motion. Vessel diameter, D^{*}, was calculated by integrating the cross sectional imaged area of the vessel divided by the average of upper and lower lymphangion wall length. Results for two mice are shown here.

RESULTS

Figure 1 shows pathlines of fluorescent particles in the collecting vessel of mouse 1. The valves promote flow separation and recirculation zones. The delineation of the lymphatic vessel was outlined by the edge detection algorithm. Results from vessel motion tracking demonstrate asynchronous contractility of lymphatic vessels. Three different lymphangions for Mouse 1 show vessel diameter contractility variation between 0 and 16 percent (Figure 2) with a pumping period of 17.6 seconds. In contrast, mouse 2 demonstrates a maximum contraction of 10 percent with temporally varying lymphatic pumping periods from 2.5 to 3.9 seconds (Figure 3). Retrograde flow varies in duration and is observed randomly throughout the tracked periods. Phase difference between neighboring lymphangions varied between 5 and 73 degrees (Table 1). Although the net lymph flow direction is forward, lymph flow reverses frequently (Figure 3).

DISCUSSION

These results reveal some of the *in-vivo* complexities of lymphatic fluid flow. Sequential organized pumping of neighboring lymphangion contractions was identified as a necessary condition for significant increases in lymph transport, while lack of sequence served to reverse or arrest lymph flow. Given the absence of a centralized pumping mechanism, the transport of lymph is highly irregular and its implications to disease, drug transport and development are still being elucidated.

Table 1. Comparison of contractility phase difference between neighboring lymphangion average diameter for three neighboring lymphangions for mouse 2. Duration of flow reversal varied from cycle to cycle.

Cycle number		1	2	3	4	5	6	7	8
Phase difference (degrees)		41	39	12	25	18	10	7	19
Reverse flow (%)	Left lymphangion	38	17	0	39	55	66	0	0
	Right lymphangion	32	23	0	35	58	67	0	0

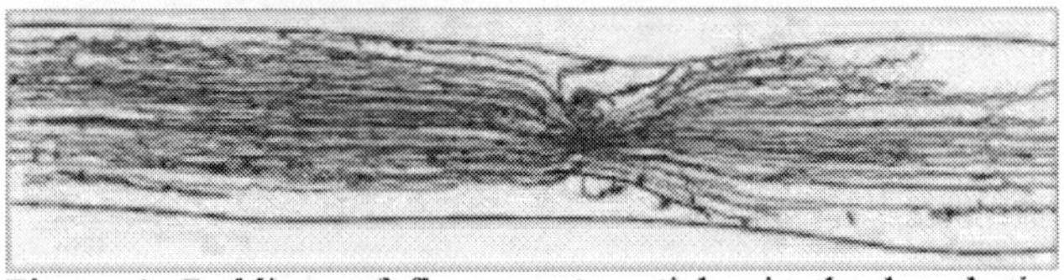

Figure 1. Pathlines of fluorescent particles in the lymphatic vessel of a live mouse.

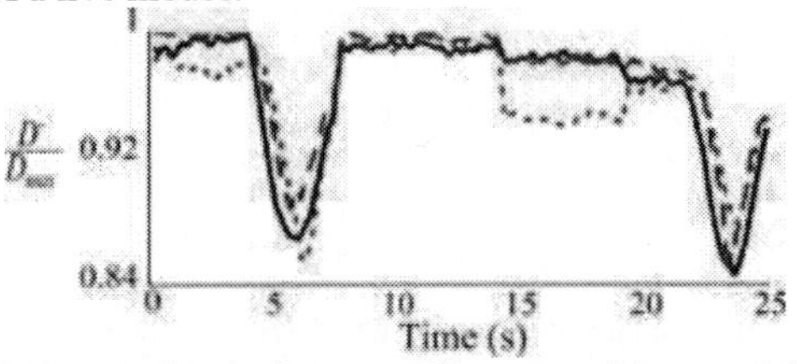

Figure 2. Comparison of average diameter of three neighboring lymphangions (---left, — — — middle, —— right) for mouse 1.

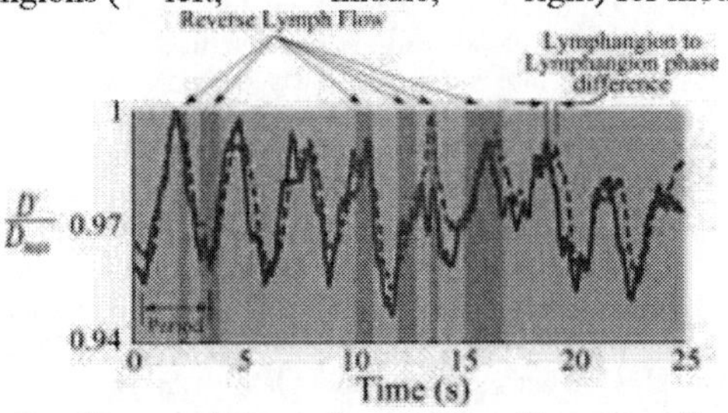

Figure 3. Comparison of average diameter for three neighboring lymphangions. Reverse flow is common along with varying period lengths.

ACKNOWLEDGEMENTS

Supported by NHLBI grants HL107617, HL007439 and HL103432

REFERENCES

[1] Rasmussen, J.C., et al., Translational oncology, 3(6): p. 362-72, 2010.

REPETITIVE SMALL-ANGLE FLEXION MAY INCREASE INJURY RISK: AN EX-VIVO STUDY

Nicole C. Corbiere-Gale (1), Stacey L. Zeigler (2), Christopher Towler (2), Kathleen A. Issen (1), Arthur J. Michalek (1), and Laurel Kuxhaus (1)

(1) Mechanical and Aeronautical Engineering Department
Clarkson University
Potsdam, N□, USA

(2) Physical Therapy Department
Clarkson University
Potsdam, N□, USA

INTRODUCTION

The likelihood of acute lower back injury is highly dependent on trunk stability. Stability is achieved *in-vivo* via complex interaction between the flexible spinal column and surrounding musculature. Each intervertebral disc in the spinal column exhibits nonlinear torque-rotation behavior in flexion-extension, centered around a low stiffness neutral zone [1]. Increased neutral zone length and decreased neutral zone thickness are indicators of increased spinal column flexibility, and potential risk factors of acute hyperflexion injury. Neutral zone behavior of the disc is affected by a number of factors, primarily water content and spatial distribution. While the neutral zone for individual discs has been quantified, the neutral zone length and slope for a multiple vertebra motion segment are unknown, as is the recovery period between repetitive flexion and load bearing activities [2].

We hypothesized that high cycle, low amplitude, cyclic flexion would increase lumbar motion segment laxity, indicating a decrease in column stability. This hypothesis was tested by computing the neutral zone of load-displacement data generated from the application of low-load, low-angle cyclic flexion to 5-vertebra motion segments. Specimens were obtained from white tailed deer (*O. virginianus*), aged 1.5-3 years, with a similar degree of skeletal maturity and range of motion [3] to adolescent humans. Human cadaver specimens (adults aged 59-72) were also used to verify that cyclic flexion would increase laxity under the same loading conditions.

METHODS

Load-displacement data from cyclically-loaded 5-vertebra motion segments were analyzed. To generate these data, seven cervine (2 male, 5 female; 2.4 ± 0.2 years) and five human cadaver (1 male, 4 female; 64.2 ± 4.8 years) spines were dissected into 5-vertebra motion segments (L1-L5) and potted using a previously described method [4]. BMD of the cervine specimens was 1.13 ± 0.20g/cm^2; the BMD of the cadaver specimens was 0.91 ± 0.24g/cm^2. Four of the five cadaver specimens had observable disc damage and osteophyte formation.

Motion segments were cyclically loaded from a neutral position to 15° of flexion [5] (as measured by the regional centroid angle, described below) for 20,000 cycles. Testing was performed at ambient temperature (approximately 22°C), specimens were kept moist by frequent application of physiologic saline. The eccentric cyclic loading with pinned end conditions at 0.5Hz cycled between displacements corresponding to 0° and 15° of flexion to simulate a low-flexion activity of daily living (ADL) such as bending slightly forward to reach an object. The specimens were preconditioned (four cycles); a fifth cycle was used for analysis as described below. Periodically (after cycles 0, 1,000, 3,000, 6,000, 10,000, 15,000 and 19,995), the loading frequency was reduced to 0.01Hz for five imaging cycles. Load-displacement data, sampled at 50Hz, from the fifth of these cycles were analyzed.

The initial displacement required to achieve 15° of flexion was determined for each specimen using the regional centroid angle as measured with a Qualisys motion capture system (Qualisys AB, Sweden). These bounds were used for all 20,000 cycles. Data were post-processed to apply a 4th order Butterworth filter with a cutoff frequency of 0.01Hz using a custom MATLAB code (The Mathworks Inc., Natick, MA). From the filtered data, hysteresis loops, hysteresis loop area, and neutral zone length and slope were computed with custom MATLAB code.

The filtered data points during the loading side (not unloading) of each analyzed hysteresis loop were extracted and used to calculate the neutral zone length and slope (Figure 1). The process began by choosing a point (displacement location) that was, by visual inspection, clearly in the neutral zone. A window of 1,000 neighboring data points was then fit with a line to the selected portion of the data. Once a line was fit to this portion of data, a 95% confidence interval was calculated. The next iteration re-defined the window to include all of the data points within the 95% confidence interval. Ten more iterations further refined the amount of load-displacement data in the neutral zone and also provided enough iterations for convergence (to the nearest 0.0001mm or N/mm). The neutral zone was the region of data points that were within the 95% confidence interval. The neutral zone slope was then extracted from the linear equation that defines the first order polynomial line fitting the neutral zone using "polyfit" in MATLAB. Neutral zone length and slope for each specimen were normalized by the specimen-specific neutral zone length or slope at cycle 5.

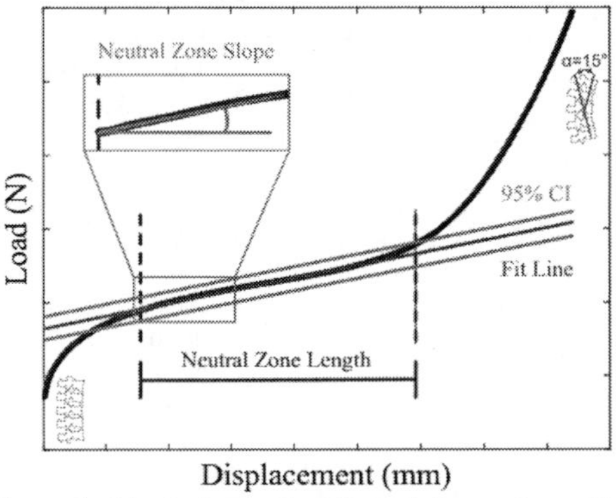

Figure 1: Neutral zone length and slope schematic.

RESULTS

After 20,000 of cyclic flexion, the general trend for neutral zone length (Figure 2) and neutral zone slope (Figure 3) show that both the cervine and cadaver specimens behave similarly during cyclic loading. The neutral zone length increases and stiffness (slope) decreases as the number of cycles increases indicating an increase in laxity. There appears to be more variability in the cadaver neutral zone lengths, perhaps due to the variability in specimen condition; the cervine specimens were a more homogeneous population. Four of the five cadaver specimens had osteophytes at every vertebral level which could alter the motion segment laxity.

From Figure 3, it is apparent that the neutral zone slope continues to decrease (although at a small rate) for a long period of time, perhaps

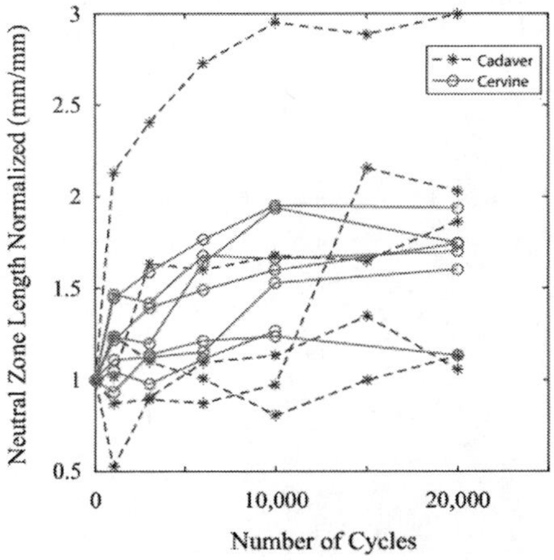

Figure 2: Neutral zone length normalized by the length at cycle 5.

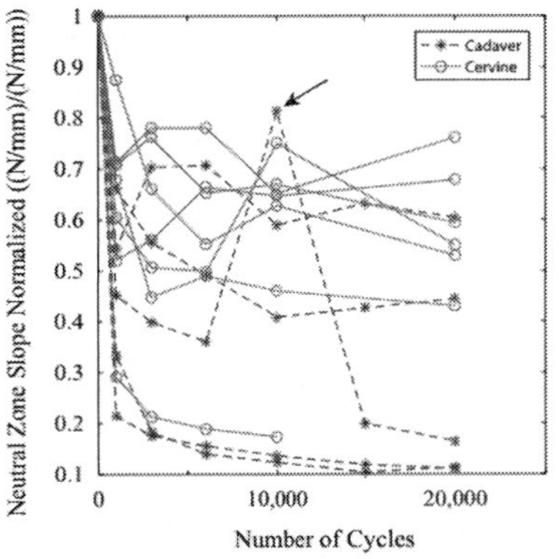

Figure 3: Neutral zone slope normalized by the slope at cycle 5. (The arrow indicates the 2nd day of testing for this specimen.)

longer than this test allows for. One cervine specimen test was halted after cycle 10,005 due to equipment failure. One cadaver specimen's cyclic loading was interrupted by unexpected failure of the potting material; this specimen resumed cyclically loaded on a later date and appears as an outlier in cycle 10,005 (arrow in Figure 3). Note that after cyclic loading resumed, the original trend continued.

DISCUSSION

Our results show that low amplitude cyclic flexion increases lumbar spinal laxity, as evident by increased neutral zone length and decreased neutral zone stiffness of isolated cadaveric motion segments. This increase in laxity may result in a clinically-significant increase in the risk of hyperflexion injury following prolonged cyclic flexion activities. We see that the cervine and cadaver specimens' neutral zone slopes behave similarly, despite differences in species, disc height, and neutral zone length. This offers further evidence of similar behavior of cervine and cadaveric specimens during cyclic loading as suggested by previously published [6] and ongoing work in our laboratory.

Limitations include a small sample size, neglecting surrounding muscles, and the use of quadruped cervine spines [6-7]. Future work includes comparisons of loading and unloading neutral zone behavior, modeling the effects of muscle forces on spinal column laxity, and investigation of the neutral zone during lateral bending.

Based on these results, we recommend that health care providers consider repetitive small movement ADLs in lower back injury prevention education. In particular, our work adds to growing evidence [2] that a sufficient recovery period should occur between repetitive flexion and heavy load bearing activities. This work will ultimately provide the appropriate medical evidence for advanced injury risk guidance and improve rehabilitation techniques.

ACKNOWLEDGEMENTS

The authors have no conflicts of interest. We thank Mark J. Hedgeland and Mario J. Ciani for technical assistance.

REFERENCES

[1] Wilke H, et al. *Eur Spine J*, 6:129-137, 1997.
[2] Callaghan J, et al. *Clinical Biomechanics*, 16;28-37, 2001.
[3] Lingle S. *Canada Journal of Zoology*, 71:708-724, 1993.
[4] Corbiere N, et al. *JBME*, 126:064504-1-4, 2014.
[5] Cobian D, et al. *J of ASTM Int*, 9:1-10, 2011.
[6] Corbiere N, et al. *J Biomech*, 49:1477-1481, 2016.
[7] Kumar N, et al. *Anatomical Record*, 260:189-203, 2000.

WALKING SPEED CHANGES IN RESPONSE TO USER-DRIVEN TREADMILL CONTROL

Nicole T. Ray[1], Brian A. Knarr[2], Jill S. Higginson[1]
[1]Department of Mechanical Engineering, University of Delaware, Newark, DE, USA
[2]Department of Biomechanics, University of Nebraska at Omaha, Omaha, NE, USA
Contact E-mail□nray□ udel.edu

INTRODUCTION

Stroke is the 3rd leading cause of death and the leading cause of long-term disability in the US, affecting approximately 795,000 people annually[1]. After a stroke, unilateral weakness, called hemiparesis, can change muscle activation and movement patterns[2], and this weakness in the ankle plantar and dorsiflexor muscles often leads to decreased walking speeds[3,4]. This decrease in walking speed corresponds to a diminished quality of life, and therefore, many poststroke motor rehabilitation programs aim to increase walking speed[5,6].

To increase walking speed, poststroke motor rehabilitation typically involves treadmill-based gait training. The patient walks on a treadmill while one physical therapist guides their weakened limb and another guides and lifts the torso to teach them the proper mechanics[7]. Since this method is subject to human error and often affected by therapist fatigue, researchers at the University of Delaware developed a new poststroke gait training program called FastFES. FastFES combines fast treadmill walking and functional electrical stimulation (FES) of the ankle plantar and dorsiflexor muscles of the weakened limb to increase walking speed and improve quality of life[8]. With FastFES, increased anterior ground reaction force (AGRF) and increased trailing limb angle (TLA) contribute to increased walking speed[9,10]. However, despite the success of the FastFES protocol, subject-specific programs with active training components may further improve functional outcomes and promote both motor and cognitive learning.

Active training programs, which require user input, improve both cognitive learning and user function[11]. Therefore, active training addresses both the cognitive and motor deficits caused by stroke. To utilize active training principles in poststroke gait training, we have developed an adaptive, user-driven treadmill controller that adjusts the treadmill's speed in real time to match the user's walking speed. However, before we implement this user-driven treadmill control in poststroke rehabilitation programs, we will isolate its effect on healthy adults to ensure that it promotes proper gait mechanics and actively engages the user.

To utilize the user-driven treadmill in poststroke rehabilitation, we must first characterize the mechanical response of users to this new treadmill control scheme. Therefore, we will determine the effect of the user-driven treadmill control on the gait mechanics of healthy adults to isolate the effect of the treadmill controller and determine its feasibility in a poststroke gait training rehabilitation scheme. Since previous work has shown the importance of increased peak push-off force, measured as the anterior ground reaction force (AGRF), and trailing limb angle (TLA) in the increased walking speed of stroke survivors[10], we focused on these variables for the following analysis (Figure 1). We hypothesized that users would choose similar self-selected (SS) walking speeds regardless of the treadmill controller and therefore, their gait mechanics were not expected to differ significantly between the 2 treadmill conditions.

METHODS

Data Collection

13 healthy adults (7M, 6F, 21.46 ± 2.88 years, 1.75 ± 0.11m, 71.41 ± 11.32kg) with no history of lower limb musculoskeletal injury were recruited for this study. The evaluation session was held in the Neuromuscular Biomechanics Laboratory at the University of Delaware. Kinetic and motion capture data were collected using an instrumented split-belt treadmill (Bertec Corp., Columbus, OH) at 1080Hz and an 8 camera motion capture system (Motion Analysis Corp, Santa Rosa, CA) at 60 Hz. The overground SS walking speed of each participant was measured using a 6m walking task.

During this evaluation, trials were performed in a random order and each participant walked at their SS speed on the treadmill in both its normal and user-driven modes while data was collected for 1 minute. For the normal mode, the treadmill was started 0.5m/s less than the overground walking speed from the 6m walking task. Then the treadmill speed was increased until the participant indicated that they were moving at their comfortable SS walking speed.

The user-driven treadmill controller adjusts the belt speeds in tandem based on the user's distance from the front of the treadmill, their push-off force with each step, and step cadence. The speed adjustments are averaged between belts and smoothed so the belt accelerations feel natural. Participants were given at least 5 minutes to acclimate to the user-driven treadmill control before data was collected. In this mode, the treadmill started at a fixed speed and ran in the normal mode for 10 seconds, then the user-driven control took over. For the measurement trials, the treadmill started at 0.5 m/s below the participant's overground walking speed and the user adjusted the speed independently until the subject indicated that they had reached their comfortable SS walking speed. Then data was collected for 1 minute.

Data Analysis

Motion capture data was processed using Cortex software (Motion Analysis Corp., Santa Rosa CA), and a low pass Butterworth filter was applied to motion capture data at 6Hz and force data at 30Hz. Kinetic and kinematic calculations were performed using Visual3D (C-Motion, Inc., Germantown, MD). Analysis began with walking speed, peak AGRF, and peak TLA (Figure 1). AGRF is the push-off force measured by the force plates during terminal stance phase and TLA is the angle between the vertical axis of the lab and the vector connecting the greater trochanter and 5th metatarsal. The peak values of AGRF and TLA from terminal stance phase were averaged over the 1 minute trial and

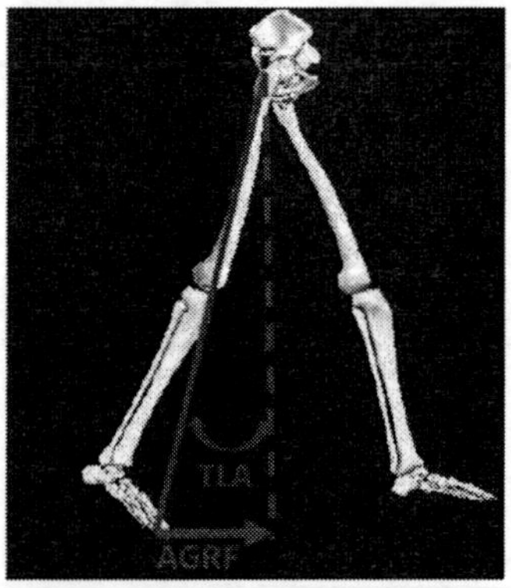

Figure 1: Definition of AGRF and TLA

then again between sides. These overall averages were used for the remainder of the analysis.

Paired T-tests were used to determine if the treadmill control strategy caused participants to choose significantly different walking speeds and subsequently different average peak AGRF and TLA (α=0.05). Linear models were used to evaluate changes in AGRF and TLA versus changes in walking speed. Since AGRF and TLA indicate changes in walking speed after stroke, subsequent linear models were adjusted for walking speed to determine the effect of the treadmill control strategy on AGRF and TLA.

RESULTS

For each subject, the set speed of the normal mode treadmill was compared to the average steady state walking speed from the user-driven mode treadmill (Figure 2). A paired T-test indicated that subjects chose significantly different SS walking speeds on the normal and user-driven treadmill (α=0.05, $p<0.0001$).

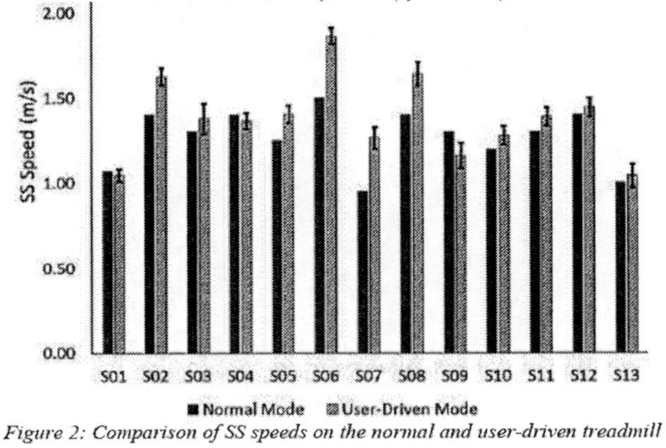

Figure 2: Comparison of SS speeds on the normal and user-driven treadmill

Each subject's AGRF and TLA increased or decreased in the same direction as their SS walking speed from the normal to user-driven modes (Figure 3).

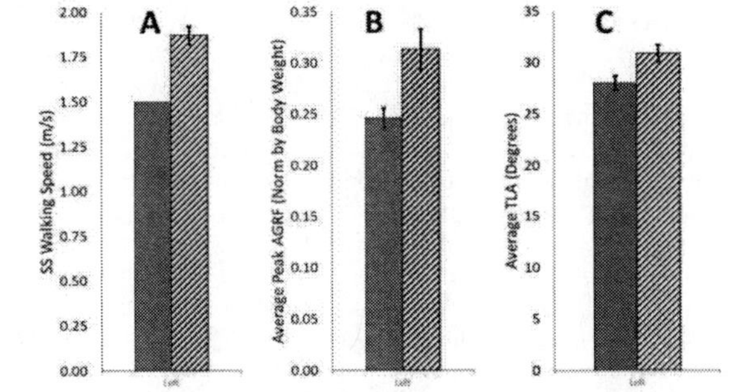

*Figure 3: A) Average Speed, B) Average AGRF normalized by body weight, and C) Average TLA at the SS walking speed for S06. *Note: solid bars = normal mode and striped bars = user-driven treadmill*

The unadjusted linear models showed strong relationships between SS speed and AGRF and TLA for both modes (Table 1). The adjusted linear models also highlighted the relationship between walking speed and AGRF and TLA regardless of the treadmill mode (Table 2).

Table 1: Parameters from linear model comparing AGRF and TLA versus SS walking speed

Independent Variable	Dependent Variable	Treadmill Mode	Model Parameters		
			Slope	Y-intercept	R^2
SS Speed (m/s)	AGRF	Normal	0.14	0.05	0.79
		User-Driven	0.12	0.08	0.84
	TLA	Normal	8.76	14.95	0.63
		User-Driven	10.47	12.86	0.70

Table 2: Parameters for adjusted linear models

Independent Variable (IV)	Dependent Variable	Covariate (CV)	Model Parameters			
			IV Coefficient	CV Coefficient	Y-intercept	R^2
TM Mode	AGRF	SS Speed (m/s)	0.006	0.130	0.007	0.85
	TLA		0.161	9.428	14.256	0.66

DISCUSSION

This study aims to determine the mechanical response of healthy adults to user-driven treadmill control. The primary variables of focus for this analysis were SS walking speed, AGRF, and TLA. Paired T-tests indicated that participants chose significantly different walking speeds on the normal and user-driven treadmill modes. In addition, they used significantly different peak AGRF and TLA for the two modes. However, adjusted linear models showed these differences are primarily due to the changes in walking speed rather than treadmill mode. We expect the treadmill modes have an indirect effect on these parameters by allowing the users to choose these different speeds.

When users walking on a normal and user-driven treadmill are provided visual flow in a virtual reality environment, there is no significant difference in spatiotemporal parameters, and kinetic and kinematic joint variables[12]. However, rehabilitation often seeks to change gait mechanics so they are more similar to an accepted standard. Therefore, the differences presented here may help stroke survivors further benefit from gait training. In conjunction with the analysis of other joint angles, kinetics, and kinematics, further analysis will quantify the effect of the user-driven treadmill and its potential for use as a clinical rehabilitation tool.

This study asked participants to indicate their SS walking speeds overground as well as on the normal and user-driven mode treadmill. Future work will compare the speeds of all 3 conditions to determine if the user-driven treadmill provides a training environment more similar to overground walking that may improve retention after gait training for rehabilitation.

Since active training improves cognitive learning and neuromuscular rehabilitation after stroke, the user-driven treadmill presents the unique opportunity to redefine the standard of care for gait training after stroke. This study shows that participants chose different SS walking speeds on the normal and user-driven mode treadmill. Therefore, the user-driven treadmill may allow patients in gait training rehabilitation to achieve improved functional outcomes by increasing their walking speed and improving cognitive learning throughout the training process.

ACKNOWLEDGEMENTS

The authors would like to thank NIH P30 103333 and the University of Delaware Helwig Mechanical Engineering Fellowship for their support of this work as well as Kelley Kempski for her assistance with data collection and processing.

REFERENCES

1. http://www.strokecenter.org/patients/about-stroke/stroke-statistics/.
2. http://www.stroke.org/understand-stroke/what-stroke.
3. Mayo NE, et al. Disabil Rehabil. 1999
4. Kesar TM, et al. Gait & Pos. 2011.
5. Abellan van Kan G, et al. J Nutr Health Aging. 2009
6. Bohannon RW, et al. Int J Rehabil Res. 1991
7. Sullivan KJ, et al. Arch Phys Med Rehabil. 2002
8. Kesar TM, et al. Stroke. 2009
9. Hsiao H, et al. J Neuroeng Rehabil. 2015
10. Hsiao H, et al. J Neuroeng Rehabil. 2016
11. Bulea TC, et al. IEEE Eng Med Biol Soc Annu Conf. 2014
12. Sloot LH, et al. Gait & Pos. 2014

IMPROVEMENTS IN GAIT AFTER COMBINATIONAL TREATMENT STRATEGY IN CONTUSED RATS

Alexander Herman (1), Rebecca Gomezrueda (2), Jennifer Kadlowec (1), Andrea J Vernengo (3), Anita Singh (1,2)

(1) Mechanical Engineering,
Rowan University,
Glassboro, NJ, USA

(2) Biomedical Engineering,
Widener University,
Chester, PA, USA

(3) Chemical Engineering,
Rowan University,
Glassboro, NJ, USA

INTRODUCTION

Spinal cord injury (SCI) is a devastating and debilitating condition that affects an estimated 227,080 to 300,938 persons in the United States with approximately 12,000 new cases occurring each year [1-3]. Current treatment strategies include activity based rehabilitation therapies, such as body weight supported treadmill training (BWSTT) that utilizes the uninjured descending pathways in incomplete SCI patients and spinal circuits below the level of injury in complete SCI patients. If the descending inputs are minimal or absent, recovery in walking over-ground is never observed in SCI patients. Since the amount of spared descending input strongly regulates the extent of recovery after BWSTT, combining transplantation treatment strategies that increase the number of descending inputs by inducing neuroprotective and regenerative environment around the injured spinal cord holds most promise to further enhance the functional recovery after BWSTT. Transplantation strategies using cells genetically engineered to deliver neurotrophins, which is a neuroprotective and neuroregenerative agent, limit spinal tissue loss, promote regeneration/sprouting of injured axons, bridge the site of injury and result in some functional recovery in animal SCI models [4-6]. Clinical translation of these techniques poses several problems, including rejection and issues with regulation of release, requiring more invasive approaches with additional problems associated with potential rejection and regulation of release of bioactive compounds. To overcome these limitations, we propose using bioengineered scaffold poly N-(isopropylacrylamide) – poly (ethylene glycol) (PNIPAAm-PEG), which has shown to be highly biocompatible and when functionalized to secrete a neurotropic factor can promote regeneration and functional recovery in a surgically induced SCI animal model. Furthermore by incorporating BWSTT in these animals, our collaborative group will for the first time report the combinatorial effect of scaffold+neurotrophins+BWSTT in a clinically relevant contusion SCI model.

METHODS

Female Sprague-Dawley rats were divided in five groups (Figure 1). All animals, except sham received a moderate contusion injury using an impactor device, in which a 10-g, 2.0-mm-diameter rod was dropped from a height of 25mm onto the exposed T9/10 level spinal cord and immediately removed. One week post-injury, animals in neurotrophin groups (Groups 4 and 5) received neurotrophin transplants using a positive displacement pipette (approximately 5 ml of PNIPAAm-g-PEG loaded with BDNF (0.5 x 10^6/5 µl) and NT-3 (0.5 x10^6/5 µl). One week post-injury or transplant, animals in BWSTT groups (Groups 3 and 5) began BWSTT for 1000 steps/day, 5 days/wk for 8 wks at 75% body weight support (BWS) and 7 cm/s speed using the custom-built Treadmill and Body Weight Support setup. All procedures were carried out in accordance with protocols approved by the Institutional Animal Care and Use Committee, following the NIH Guidelines for the Care and Use of Laboratory Animals.

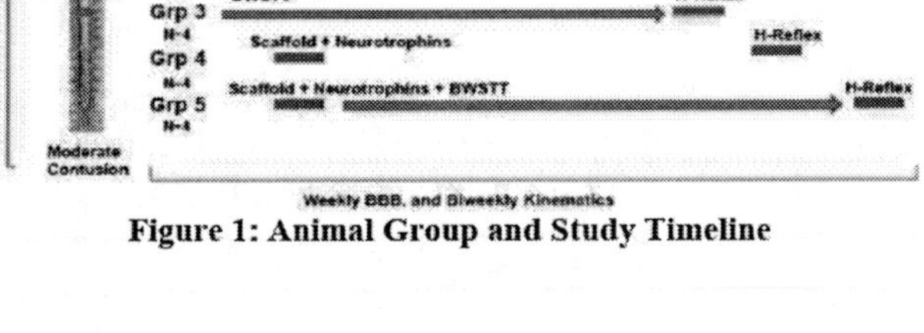

Figure 1: Animal Group and Study Timeline

Behavior Tests: *a) Treadmill Testing:* Biweekly, hind limbs of the animals were shaved and reflective markers were placed on the hip (greater trochanter and iliac crest), knee (femoral condyle) and ankle (lateral malleolus) and the fifth metatarsal of the hind limbs. Three high-speed OptiTrack (Natural Point, OR) cameras were used to record marker movements as the animals walked on the treadmill at 100 FPS. Stepping was first tested at 75% weight support and 7 cm/s speed. Post processing of data was performed using Motive (Natural Point, OR) and MATLAB (Mathworks Inc., MA). Data analysis included measurements of the kinematic characteristics, i.e., the length of the step, number of steps and hip, knee and ankle joint excursions.

b) Overground locomotor BBB Test: Hind limb motor function was assessed weekly using the open field [4] scoring system. Hind limb function was scored from 0 (no observable movements) to 21 (normal locomotion). Individual rats were placed in an elliptical enclosure, allowed to accommodate to the enclosure, and then taped for 4 minutes by a trained observer. At a later time point, another observer who was blinded to the group identity of the animals scored the tests.

RESULTS

Kinematic Analysis: The average number of steps performed in each group is presented in Figure 2 at 75% BWS and at 7 cm/s. Hip, knee and ankle excursion, as shown in Figure 3 were similar to normal uninjured rats when treated with BWSTT, or implant or combination treatment. As shown in Figure 4, all treated animals exhibited improvement in stride length when compared to injured animals.

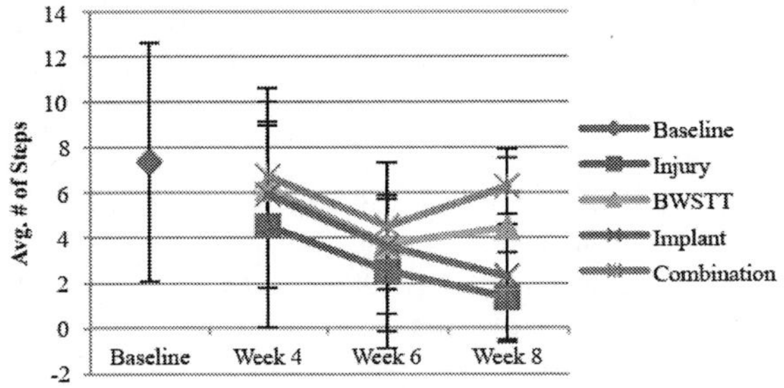

Figure 2: A line plot of the average number of steps per week per group for 75% BWS and at 7 cm/s.

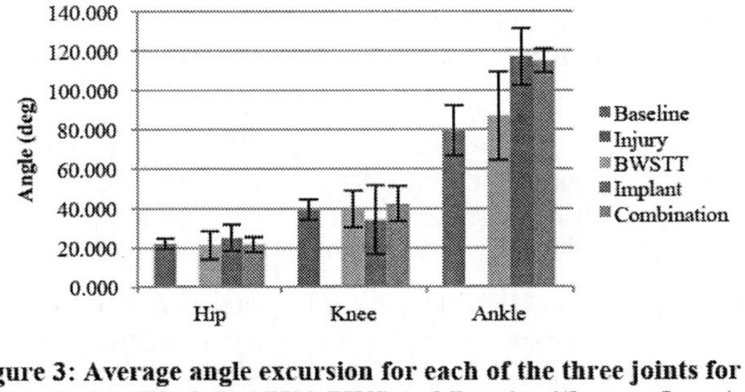

Figure 3: Average angle excursion for each of the three joints for rats walked in Week 8 at 75% BWS and 7 cm/s with error bars to 1 standard deviation. No rat in the injury group performed at these settings in this week.

BBB Test: A significant improvement in BBB scores is observed in BWSTT and combinational groups at Weeks 6-8 as showing in Figure 5.

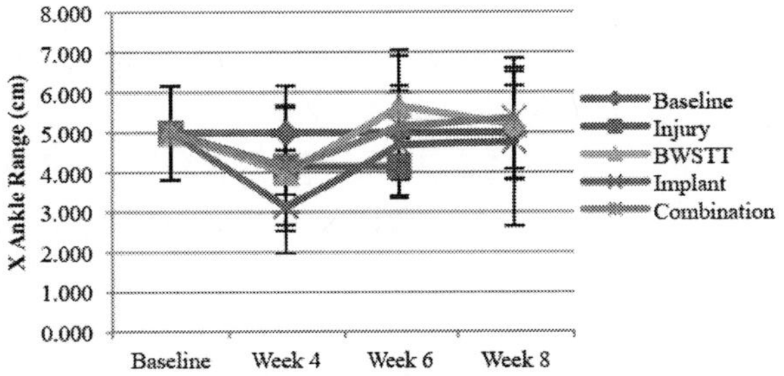

Figure 4: Line plot of the average stride length over the course of 8 weeks measured from the x component of the motion capture subject's ankle. Data is represented at 75% BWS and 7 cm/s. No rat in the injury group performed at these settings at Week 8.

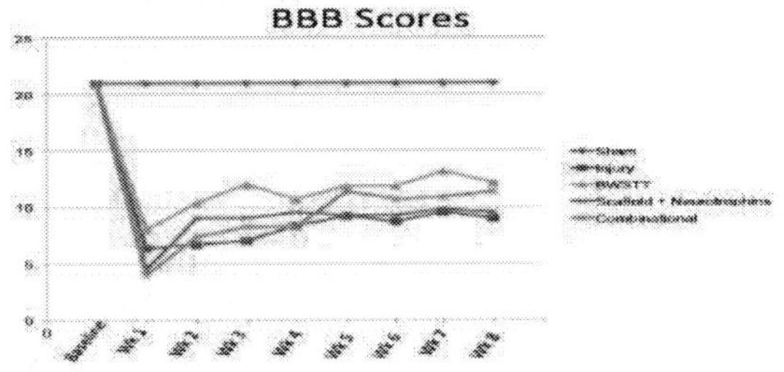

Figure 5: BBB Scores of the different groups over course of training.

DISCUSSION

BBB data and kinematic data analysis exhibit functional recovery in animals that received BWSTT, implants and combinational treatment. Statistical analysis is currently being performed to determine significant different between various groups. In summary, successful findings from this study will help develop a unique tissue-engineering approach with promising clinical application in incomplete SCI patients. By adding other growth promoting agents to this bioengineered scaffold, future studies can explore the beneficial effects of the proposed combinatorial treatment strategy in complete SCI animal model.

ACKNOWLEDGEMENTS

This study was supported by the New Jersey Commission on Spinal Cord Research – Exploratory Research Grant # CSCR14ERG001.

REFERENCES

[1] Hutchinson, K.J., Gomez-Pinilla, F., Crowe, M.J., Ying, Z., and Basso, D.M., Brain, 2004. **127**(Pt 6): p. 1403-14.

[2] Dobkin, B., Apple, D., Barbeau, H., Basso, M., Behrman, A., Deforge, D., Ditunno, J., Dudley, G., Elashoff, R., Fugate, L., Harkema, S., Saulino, M., and Scott, M., Neurology, 2006. **66**(4): p. 484-93.

[3] Hubli, M. and Dietz, V., J Neuroeng Rehabil, 2013. **10**: p. 5.

[4] Himes, B.T., Liu, Y., Solowska, J.M., Snyder, E.Y., Fischer, I., and Tessler, A., J Neurosci Res, 2001. **65**(6): p. 549-64.

[5] Jin, Y., Tessler, A., Fischer, I., and Houle, J.D., Neurorehabil Neural Repair, 2000. **14**(4): p. 311-7.

[6] Mendell, L.M., Munson, J.B., and Arvanian, V.L.,. J Physiol, 2001. **533**(Pt 1): p. 91-7.

SOFT ROBOTIC DEVICES FOR HAND REHABILITATION: A NARRATIVE REVIEW

Rita M. Patterson (1) and Chia-Ye Chu (2)

(1) Family Medicine
University of North Texas Health Science Center
Fort Worth, TX USA

(2) Texas College of Osteopathic Medicine
University of North Texas Health Science Center
Fort Worth, TX USA

INTRODUCTION

The debilitating effects on hand function from a number of a neurologic disorders has given rise to the development of rehabilitative robotic devices aimed at restoring hand function in these patients. To combat the shortcomings of previous traditional robotics, soft robotics are rapidly emerging as an alternative due to their inherent safety, less complex designs, and increased potential for portability and efficacy. While several groups have begun designing devices, there are few devices that have progressed enough to provide clinical evidence of their design's therapeutic abilities. Therefore, a global review of devices that have been previously attempted could facilitate the development of new and improved devices in the next step towards obtaining clinical proof of the rehabilitative effects of soft robotics in hand dysfunction.

METHODS

A literature search was performed in SportDiscus, Pubmed, Scopus, and Web of Science for articles related to the design of soft robotic devices for hand rehabilitation. A framework was also developed which listed the key design components reported in the literature. After obtaining primary search results from each database, the following inclusion and exclusion criteria were used to further narrow the literature search:

Inclusion criteria:

- The device described was a soft robot (no rigid components on the robot-human interface)
- The device intends to facilitates the movement of at least one finger joint in the hand
- The paper was a scientific article written in the English language and accessible to the authors

Exclusion criteria:

- The device contained rigid components on the robot-human interface
- The device focused on other joints without including the fingers

In order to guide the analysis of a relatively large number of devices, a general framework was developed (Figure 1) that conceptualizes the essential components of each device and the basic interactions between these components.

Our discussion of the schematic is broken into two parts: the first part deals with the design of the robotic device while the second part deals with the pro-active planning for interaction with the environment.

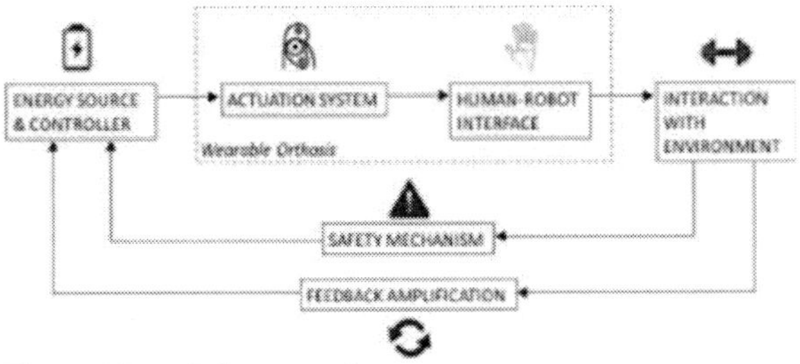

Figure 1Search framework

RESULTS

The final search entries for this review were completed on August 1st, 2016. A total of twenty-seven articles, representing twenty unique devices were identified for this overview. A summary of the results of the literature search are presented in Table 1 which include the database, search query, number of articles found and the total number of unique devices that were considered for this review. Six additional articles were identified as potentially fulfilling the search criteria, but were international articles, inaccessible to the authors and so were excluded in our results.

A brief summary of the main results is as follows:

- Energy source: no device fully portable or operational without a technician.
- Wearable orthosis: most were pneumatic and only one independently controlled each digit.
- Interaction with the environment: most were continuous passive motion machines.
- Feedback: less than half of the devices had any form of feedback.
- Safety: only 6 devices discussed specific safety mechanisms.

DISCUSSION

Our results have shown that there has been a rapid progression of the development of soft robotic devices for hand rehabilitation, with a majority of articles being published within the past three years. Consequently, most devices are still in preclinical prototype and research phase.

As previously noted, the number of independent actuators and their configuration have major implications on the types of movements supported during rehabilitation. Most devices focused simply on flexion and/or extension of the index through middle fingers, with some attention given to the thumb.

While there is a specific set of hand motions that should be achieved through rehabilitation, it follows that there should be a uniform way of evaluating this success across devices. In terms of evaluating performance, these preclinical devices have demonstrated a wide heterogeneity in methods of testing and reporting results. However, before clinical trials can be performed, there needs to be an established consensus on the method of evaluating the efficacy of therapy. This will facilitate comparisons between devices to determine which designs are more adept at rehabilitation.

CONCLUSIONS

The past three years has seen a rapid increase in the development of soft robotic devices for hand rehabilitative applications. These preclinical research prototypes display a wide range of technical solutions that will likely soon enter clinical trials. Consensuses should first be made about specific hand motions that devices should mimic as well as the evaluations methods used in testing them. It is our goal that this review will guide future developers through the various design considerations in order to develop better devices for patients with hand impairments.

Database	Search query	No. articles returned	No. of excluded articles	No. of relevant articles	Total No. of unique devices
SportDiscus	TX "soft robot*"	2	2	0	0
Pubmed	("soft robot*" [All Fields] OR ("Robotics" [MeSH] AND "soft" [All Fields])) AND (hand OR finger OR thumb OR glove [All fields])	30	28	2	2
Scopus	TITLE-ABS-KEY(Robot*) AND TITLE-ABS-KEY(Soft) AND TITLE-ABS-KEY("hand" OR "finger" OR "thumb" OR "glove")	773	750	23	14
Web of Science	TI = (soft robot*) AND TI = (hand OR finger OR thumb OR glove)	26	23	3	0
Referenced articles	-	-	-	-	4
Total		831	804	**27***	**20**

Table 1 Summary of literature search results
*This value represents the total number of unique relevant articles because some articles were found in multiple databases

LOADING PATTERNS ASSOCIATED WITH POSTURAL CHANGE

Justin Scott, Kelly Patterson, Lindsay Hoard, Michael Drost, and Tamara Reid Bush

Department of Mechanical Engineering
Michigan State University
East Lansing, MI, USA

INTRODUCTION

Many people sit in a chair for part of their day, getting up and adjusting themselves without a thought. It is so automatic that individuals may shift their weight several times in the short time it takes to read two pages of text. However, this is not the case for all individuals. In particular people with limited mobility, such as those in wheelchairs, may not be able to adjust their body, creating loads that occur over a long duration. This situation can lead to pressure ulcers (PUs). PUs are penetrating wounds that essentially leave a hole in the skin and deeper tissues, sometimes penetrating to the bone. They are caused by necrosis due to lack of nutrients to tissue and can take months to heal properly. Because of this, the average cost of a hospital visit to treat a PU is $48,000 [1]. This does not take into account lost work time or deterioration of quality of life for an individual with a PU. Wheelchair bound individuals are at an especially high risk of developing PUs because their buttocks and lower back are constantly loaded. High external pressure contributes to PUs by diminishing blood flow to tissues, resulting in malnourished tissue and eventually tissue death. The 282,000 people [2] who currently have a spinal cord injury (SCI) in the United States are highly prone to the formation of PUs. Further complicating the care of SCI patients is the fact that they may not be able to feel pressure points. They are not able to feel the discomfort after long periods of intense loading, nor can they adjust to remove this loading, making the formation of an ulcer more likely.

The goal of this work is to develop an articulating chair that moves the body so that loading is shifted from the ischial tuberosities, one of the regions where it is common for PUs to develop, to another body region. With such an apparatus, it will be possible to prescribe a regime of postural changes that will cyclically load and unload tissues throughout the day, reducing the chance of a PU in any specific region.

METHODS

Office chairs have a range of recline of 14 to 21 degrees from vertical, and wheelchairs can recline, but neither currently has the ability to articulate the spine [3]. To add this function, we designed a chair with independently rotating pelvic and thoracic supports, allowing us to simulate kyphotic (slouched), lordotic (arched), and neutral postures in addition to reclining. Tekscan pressure mats and Six-Axis AMTI load cells were used to measure the pressures and overall forces exerted on each face of the chair. The maximum magnitude and location of pressure was tracked statically for three individuals and dynamically for one person. While the participants were sitting with their body as far into the chair as possible, neutral, kyphotic, and lordotic postures were tested. Snapshots of the mat on the seat pan were taken for each posture while the recline angle increased from 0 degrees (upright) to 30 degrees (full recline), in increments of 5 degrees. For the static tests, the individual stood up to allow for posture change and recline angle increase, while the participant stayed in the chair for the posture changes during dynamic testing. Chair recline and articulation were used to shift the loading points under the ITs of the seated occupants.

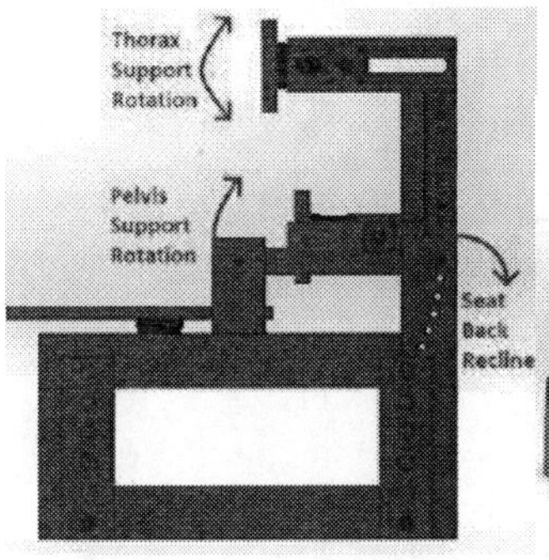

Figure 1. CAD design of the testing apparatus.

RESULTS

A posterior shift in the center of pressure (COP) with increasing recline angle was predicted and observed, as is seen in Table 1. However the 27.5 mm shift is not considered large enough to relieve the pressure on the IT. In terms of posture, the COP was more affected by the lordotic position. The lordotic posture produced a 12.7 mm anterior shift in the COP, while the COP position was almost the same for the neutral and kyphotic postures, shown in Table 2.

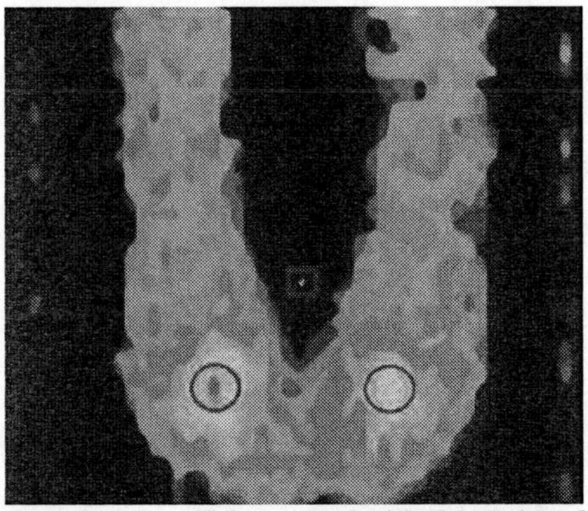

Figure 2. Snapshot from Tekscan mat with the center of pressure (COP) in the box and ischial tuberosities circled.

DISCUSSION

The location measurements for the COP are currently taken from the front of the seat pan. But it may be more appropriate to take such measurements from different landmarks in the future, such as the back of the seat pan or the point of contact closest to the back of the seat pan. The latter might account for variations in how far a participant can push back into the seat. Recline and spinal articulation are two motions that may shift the COP, and others are being considered for future prototypes. They include a pelvic cradle, which will move the seat pan and pelvic support as one, and a footrest that changes the position of the legs. Altering the thigh and shank angles has the potential to provide anterior and posterior shifts in the COP. Additionally raising or lowering the foot level relative to the seat pan can change the area of contact and the COP in turn.

Our initial testing has shown that recline angle and maximum pressure on the seat pan have an inverse relationship. This means that the load is shifted increasingly onto the pelvic and thoracic supports as the individual reclined, as expected. This may be an effective method to relieve pressure on the ITs, but it might not be possible to have a large angle of recline for long periods during the day. So the additional movements, mentioned before, need to be explored.

Pressure ulcers affect people across various populations, and anyone confined to a wheelchair has a higher risk of developing one. Ulcers, by definition, take a long time to heal; common health problem in geriatric or diabetic patients complicate wound healing. Procedures that ameliorate PUs before they form would save people time and suffering as well as money. These can be prescribed with an articulating chair that promotes blood perfusion. Our research to this point has been limited to one subject reclining and changing posture through the movement of the pelvic and thoracic supports. Future endeavors will a larger and more varied subject population.

Table 1. Average maximum pressure for the three dynamically tested torso articulations (slouched, neutral, lordotic) at each recline angle. Center of pressure measured from the front of the seat pan.

Recline Angle (°)	Average Max Pressure (kPa)	COP Position (mm)
0	16.00±.41	220.2±10.5
5	18.51±.19	225.7±7.0
10	19.68±.07	229.9±6.5
15	18.98±.13	233.5±6.3
20	16.96±.24	238.7±7.5
25	15.64±.24	239.4±8.2
30	15.11±.23	248.3±6.2

Table 2. Average maximum pressure of each position during the dynamics recline.

Posture	Average Max Pressure (kPa)	COP Position(mm)
Neutral	16.52±.26	237.8±10.0
Kyphotic	17.49±.24	238.0±8.4
Lordotic	17.80±.31	225.3±10.4

ACKNOWLEDGEMENTS

The work is supported by NSF CBET, grant number 1603646.

REFERENCES

[1] *National Spinal Cord Injury Statistical Center*, 2016.
[2] Brem, H., et al. *American Journal of Surgery*, 200(4): 473-477, 2010.
[3] Bush TR, et al. *Proceedings of the Human Factors and Ergonomics Society* **43**, 1999.

THE PSYCHOPHYSICAL EFFECTS OF HAPTIC FEEDBACK IN THE PERCEPTUAL AWARENESS OF A POWERED TRANSFEMORAL LIMB

J. Miles Canino (1), Kevin B. Fite (1)

(1) Mechanical and Aeronautical Engineering
Clarkson University
Potsdam, NY, USA

INTRODUCTION

Unfavorable rehabilitation outcomes in patients with lower limb amputations are increasing due to increased incidence of confounding dysvascular etiologies. The amount of time that it takes for proper rehabilitation using conventional techniques often leads to marked progression of underlying disease states [1, 2].

The use of vibrotactile haptic feedback had been shown to significantly shorten the rehabilitation time by training proper weight-bearing technique in persons with lower limb amputations [2].

Prior work conducted in our lab also successfully demonstrated that haptic feedback can successfully convey meaningful information in the completion of stepwise and sinusoidal motion tracking tasks with a powered transfemoral prosthesis [3,4]. The investigation of these complex haptic communications presented an unexpected observation wherein visual feedback seemed to be detrimental to motion tracking performance when compared to haptic feedback alone. It was then desired to understand if there was a meaningful relationship between visual feedback and motion tracking performance, and if so, what implications this could have for future use of haptic feedback in the rehabilitation process.

METHODS

With IRB approval (CU# 14-34), three informed and consenting able-bodied adults were tasked with following randomly generated sinusoidal one-degree of freedom non-weight bearing motion trajectories with an electromyography (EMG)-controlled powered-knee transfemoral prosthesis. These tasks were completed with and without visual feedback and with and without haptic feedback as detailed in [4].

Two haptic feedback devices were designed and fabricated, one providing quasi-static pressure feedback and one providing vibrotactile feedback [3]. These devices were affixed to the subject's outer thigh with a Velcro harness, extending distally from the greater trochanter region. The control system for these haptic feedback devices would convey two possible schemes - a corrective scheme and a pacemaking scheme. The corrective scheme sought to coax the user into flexion or extension as a means of getting them back on target while the pacemaking scheme would provide the user with a rhythmic pattern that they could use to extend and flex in a rhythmic manner required to follow the repetitive motion trajectory [4].

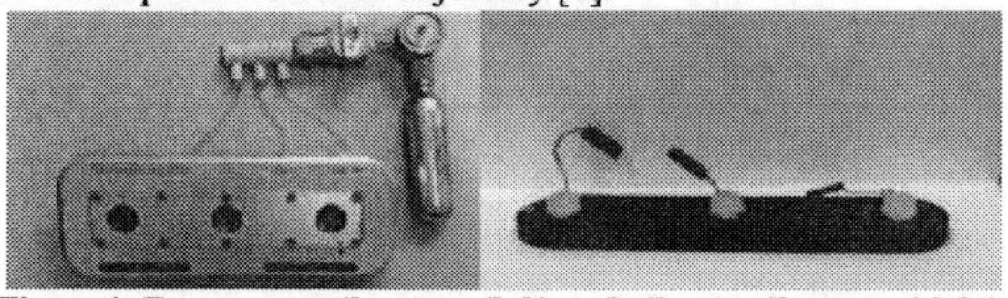

Figure 1: Pressure tactile array (left) and vibrotactile array (right).

The EMG-controlled transfemoral powered knee prosthesis used in this study is described by Hoover *et al.*, while the control architecture for the powered prosthesis is as described by Dawley *et al.* [5, 6]. In short, the control torque about the knee of the powered limb is volitionally controlled via EMG interface.

The periodic trajectories followed during the tracking tasks were randomly generated and band limited with the mean value of the sinusoid constrained between 30 and 50 degrees, the amplitude constrained between 10 and 40 degrees, and the frequency was constrained between .05 and 1.5 Hz. If the generated trajectory would go outside of the mechanical bounds of powered limb motion, it would be re-generated until it fell within these bounds [4].

Torque, knee angle, EMG, knee stiffness, and motor current data were recorded by Matlab at a 1 kHz sample rate during testing. The motion tracking performance was analyzed using Analysis of Variance (ANOVA) of the Magnitude-Squared Coherence (MSC) between the desired signal frequency and the user-generated signal frequency [4].

RESULTS

Haptic feedback of both quasi-static pressure and vibrotactile modalities offered a significant improvement in tracking performance ($p = 0.0001$ and $p = 0.0005$, respectively), yet there were unexpected observations during the testing of subjects [4]. While testing, it was observed that tests which incorporated visual feedback seemed to have greater performance instability than tests that contained haptic feedback alone. This can be best visualized in two example tests shown in Figure 2. The test on the left shows typical motion tracking performance while only provided visual feedback, while the test on the right shows typical performance when only provided haptic feedback. It is readily evident from these plots that haptic feedback alone demonstrates a marked improvement over the former in tracking. This phenomenon was briefly conjectured by Canino *et al.*, but furthered analysis was not conducted [4]. Looking at the plot of mean correlation data presented in Figure 3, one can observe that in all cases except one (circled), visual feedback caused a larger variance in frequency tracking. Furthered analysis of the frequency data, specifically cycle to cycle variation, corroborated this observation. It was found that visual feedback provided significantly greater variation for two out of three instances ($p = 0.0003$ for visual alone, $p = 0.006$ for visual with vibrotactile, and $p = 0.984$ for visual with pressure). In the latter data set, where greater variation was not observed, it is likely because subjects were observed using a test-taking strategy wherein they would close their eyes or avert their gaze from the motion trajectories, effectively eliminating the detrimental effects of visual feedback. This observation was noted by Canino et al. and, based on the analysis presented herein, it is suggested that this test-taking strategy had a meaningful skewing effect on the data analyzed [4].

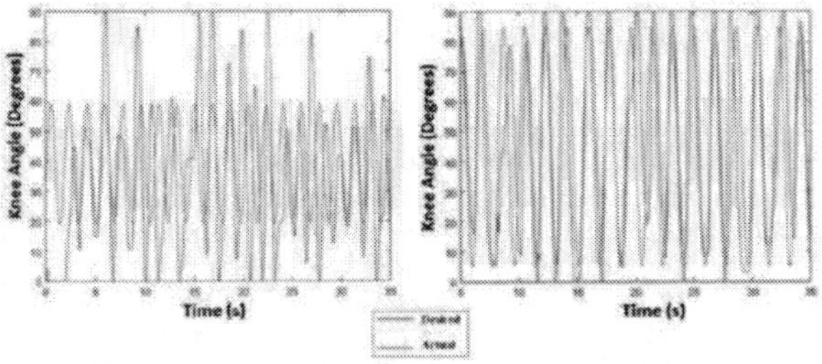

Figure 2: Tracking data for visual feedback alone (left) and pressure feedback alone (right).

DISCUSSION

This psychophysical phenomenon, what we describe as Detrimental Task Focus (DTF), is frequently observed and anecdotally explained. Examples often provided include athletes who have performed a sports-related task thousands of times before suddenly finding that they are unable to perform during the actual game, colloquially described as "choking." The rationale behind such behavior is often described as a conflicting schism between one's autonomous capabilities and their focused capabilities, usually brought on by stress, as described by Baumeister's self-focus theory [7].

The question then became, why was DTF observed in the accomplishment of these more complex motion tasks only when visual feedback was provided but not with haptic feedback alone? Other researchers have documented similar phenomenon [8, 9]. Lee et al. purported that this observation was due to the body's functional inability to effectively perform autonomous, monotonous, or otherwise patterned tasks while in a "focused state," wherein a focused state naturally comes from one's visual perception of a task. That is to say, actively trying to use visual feedback while performing a patterned, monotonous, or autonomous task inherently brings a detrimental focused state [8]. Ankarali et al. suggested that the observed detrimental effects of visual feedback stem from the user being more inclined to attempt to over-correct perceived instability in the performance of a rhythmic task. These researchers observed a notable reduction in cycle-to-cycle variability when only haptic feedback was provided, similar to the observations made herein [9].

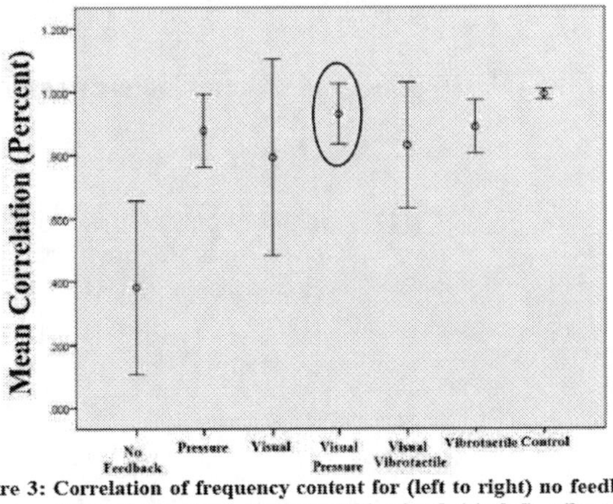

Figure 3: Correlation of frequency content for (left to right) no feedback, pressure alone, visual alone, pressure with visual (circled), vibrotactile with visual, vibrotactile alone, and control.

The observation of DTF and haptic feedback's seeming ability to successfully communicate with the autonomous functionality of the body has important implications in the rehabilitative potential of our haptic feedback apparatus. While conventional means of rehabilitation focus on the user slowly developing proper gait and weight-bearing techniques via repetition and the ultimate formation of newly ingrained motor control, our results suggest that haptic feedback could enhance and facilitate this process. Haptic feedback has demonstrated its ability to somehow convey rhythmic information to users in an intuitive way that dissuades DTF. This means that haptic feedback could lead to better rehabilitation outcomes by directly training the body's autonomous state, altogether bypassing the body's focused state. One could then develop a haptic feedback apparatus and program that allows for this more intuitive rehabilitation process.

Given that haptic feedback seems to provide direct access to the body's autonomous perception, the conveyance of real-time kinetic and kinesthetic limb information to a prosthetic user could also lead to a greatly enhanced prosthetic limb experience. This could even help to impart a sense of artificial proprioception, where neural plasticity eventually causes the prosthetic limb to become an autonomous extension of the user.

REFERENCES

[1] Dillingham, T, et al, Phys. Med. and Rehab, 89:1038-1045, 2008
[2] Fu, M, et al., Orthopedics, 37: E993-E998, 2014
[3] Canino, J.M, et al. IEEE Proceedings of EMBS: 1-4, 2016
[4] Canino, J.M, et al. ASME Proceedings of DSCC: 1-8, 2016
[5] Hoover, C., et al., J. of Med Dev, 6: 011005-1-011005-12, 2012
[6] Dawley, J., et al.," Proceedings of ICORR:1-6, 2013
[7] Baumeister, R. J. of Personality & Soc. Psych, 46(3) 610-620 1984
[8] Lee, T.G., et al. J. of Neuroscience, 33(32): 13233-13237, 2013
[9]Ankarali, M., et al. J. of Neurophysiology, 11.6; 1286-1299, 2014

SKIN-ON-A-CHIP: A MICROENGINEERED PLATFORM FOR STUDIES IN SKIN MECHANOBIOLOGY

Megan J. Farrell (1), Thomas F. Seykora (1), Susan W. Volk (2), George Cotsarelis (3), Dongeun Huh (1)

(1) Department of Bioengineering
University of Pennsylvania
Philadelphia, PA, USA

(2) Department of Clinical Studies
University of Pennsylvania
School of Veterinary Medicine
Philadelphia, PA, USA

(3) Department of Dermatology
University of Pennsylvania
Perelman School of Medicine
Philadelphia, PA, USA

INTRODUCTION

Skin disorders and pathologies are commonly associated with complex organ-level responses (i.e. inflammation or mechanical regulation of wound healing) or structure/function relationships (barrier function). However, *in vitro* models often lack critical components, such as vasculature, perfusion, and mechanical loading, needed to more completely model these processes. One promising approach to tackling these critical challenges is to leverage the recent rapid advancement in the development of *in vitro* tissue and organ models within the growing field of organs-on-chips. By synergistically combining technologies derived from tissue engineering, microfabrication, and microfluidics, organs-on-chips mimic the complex structural organization of multiple tissue types and their essential biological processes in functional units of human organs. Studies have shown that these microengineered models provide unprecedented capabilities for real-time monitoring and quantitative analysis of complex physiological and pathophysiological organ-level responses. Here we demonstrate the potential of this emerging technology for modeling the three-dimensional (3D) structure and dynamic physiological microenvironment of the human skin. Specifically, we developed a compartmentalized microfluidic system for co-culture of human keratinocytes, dermal fibroblasts, and vascular endothelial cells. This multilayered tissue achieves physiological spatial arrangement that resembles the epidermis and vascularized dermis. Furthermore, this model was integrated with a computer-controlled mechanical actuation system to mimic stretching of the skin and concomitant tissue deformation. Our preliminary data illustrate the capacity of our microdevice to model the structural and functional complexity of the skin in ways that have not been possible using traditional cell culture techniques. This microsystem may serve as a novel platform to mechanistically examine the mechanobiology of the skin during health and disease.

METHODS

Compartmentalized 3D microfluidic devices were fabricated via standard soft lithography methods from polydimethylsiloxane (PDMS). Skin-on-a-chip devices consisted of upper and lower channels separated by a thin porous (10 µm pores, 10 µm thickness) PDMS membrane (**Fig. 1 Left**). Human adult keratinocytes (Lonza) were seeded into the upper 'epidermis' chamber to form a confluent monolayer (maintained in CnT-07, CellnTec). To reconstitute the dermis in the lower channel, a dermal fibroblast (Fb, Lonza)-laden type 1 collagen hydrogel (4mg/mL) was cultured in the center of the channel. Two flanking channels ran along the gel channel for fluidic access and feeding of the dermis (FGM-2, Lozna). Devices were monitored for successful spatial patterning, cell morphology, and viability.

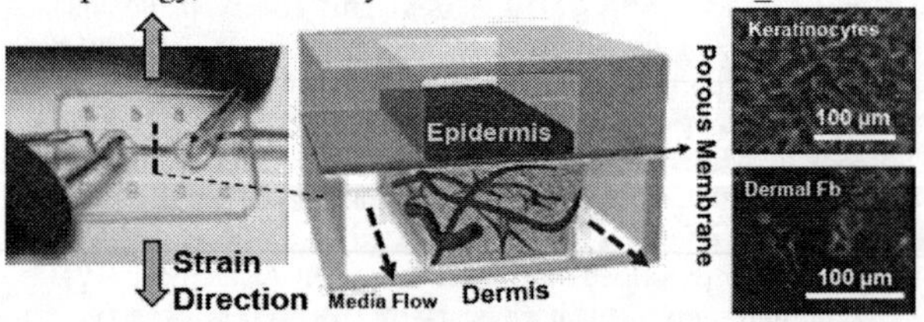

Figure 1: Left) Skin-on-a-Chip with cross-sectional illustration of multilayered design. Right) Epidermal keratinocyte monolayer (membrane) and dermal fibroblast (3D collagen gel) morphology.

For vascularized models, Fbs (Lonza) and human umbilical vein endothelial cells (HUVECs) were co-cultured in a 2 mg/mL type 1 collagen/ 10 mg/mL fibrin composite gel in a dermis only device. These devices were cultured over the course of 7 days under continuous flow conditions (syringe pump, 150 µl/hr) in endothelial growth medium supportive of vasculogenesis (EGM-2MV, Lonza). Robust network formation was confirmed with CD31 immunostaining, and the presence of open lumens verified in histological cross-sections.

To mechanically actuate the microtissues in these devices, we integrated our device with a custom computer-controlled mechanical stretcher/bioreactor capable of defined cyclic or static loading protocols. Keratinocytes were cultured as a monolayer on the membrane of multilayered devices. Strain transmission across the PDMS membrane to the epidermal cells was verified with strain mapping (digital image correlation, Vic2D) and analysis of nuclear aspect ratio and orientation. Average Lagrangian strain of the epidermal layer was determined and compared to expected strain levels computed with finite element analysis (FEA; COMSOL). To confirm the capacity to generate strain transfer to the cells cultured in the 3D hydrogel of the dermis, devices with fibroblast-laden collagen gels were stretched (20% grip-to-trip strain) and individual cells tracked (Calein-AM live cell staining). Cell shape was compared before and after device stretching.

RESULTS

Our device enabled the physiological 3D spatial patterning and co-culture of skin cells including keratinocytes and Fbs (**Fig. 1 Right**). With the inclusion of HUVECs in the dermal compartment, robust vascular networks (CD31 positive) were able to form via vasculogenesis over the course of 7 days (**Fig. 2**), similar to [1,2]. Open lumens of these networks were observable in histological cross-sections of the tissues cultured in the devices.

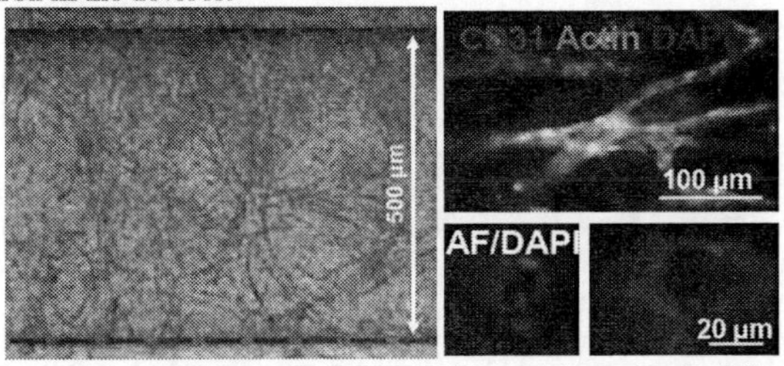

Figure 2: Left) Dermal Fb/HUVEC co-culture. Dotted lines denote channel edges. Top Right) Vascular networks stain positive for CD31. Bottom Right) Open lumens observable in tissue cross sections (AF=autofluorescence).

With mechanical stretch, cell elongation was observed in both the dermal and epidermal tissues. Digital image correlation and strain mapping mirrored FEA predictions with strain at the membrane approximately 2-fold higher than grip-to-grip applied strain due to cross-sectional geometry and channel bowing (**Fig. 3A**). Strain transmission to live cells in the epidermal layer was confirmed with a shift in both nuclear orientation along the direction of stretch and a positive change in nuclear aspect ratio of individually tracked cells (**Fig 3B**). Furthermore, to verify cells in the 3D hydrogel are in fact being stretched, confocal stacks of individually tracked Fbs were acquired at 0% and 20% grip-to-grip strain. Maximum intensity projections were generated along the imaging depth (**Fig. 3C**), and cell body shape compared (excluding extended processes). Positive increases in cellular aspect ratio in the direction of stretch confirms that we can generate tissue and cellular level strains by directly stretching the PDMS device.

DISCUSSION

While increasingly complex bioengineered models better replicate the physiological environment, models for skin [3,4] lack critical components of skin physiology, including defined vascular networks and mechanical stretch. To this end, we have developed a multi-layered, vascularized skin-on-a-chip model for applications in studies of skin mechanobiology. We have confirmed the capacity to culture epidermal and dermal cell types with physiological spatial patterning. More so, we generated vascular networks with open lumens in a dermal fibroblast/endothelial cell co-culture. In addition to microfluidic model development, we were able to actuate our devices to achieve strain transfer at the tissue level using a custom bioreactor system. This versatile bioreactor enables the generation of a number of loading protocols (static or cyclic) with continuous media perfusion (syringe pump) and extended time in culture. In future studies, this model will serve as a platform to study the role of mechanical stretch in the pathophysiology of keloid disease and hypertrophic scarring of the skin.

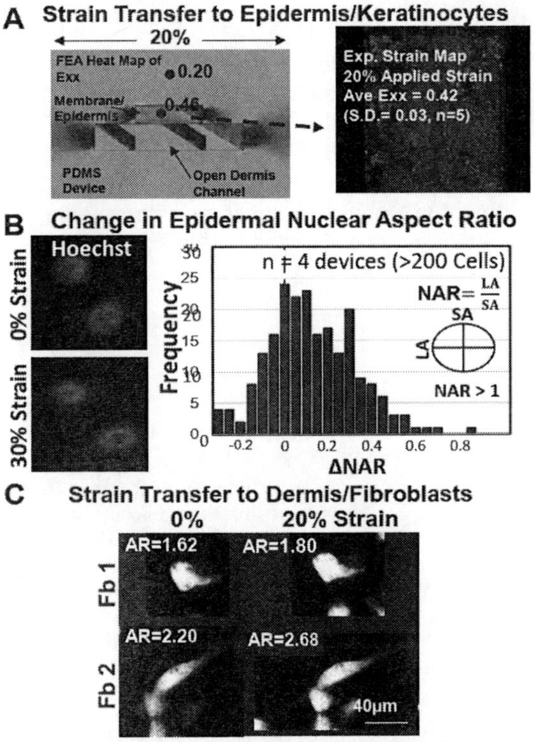

Figure 3: A) FEA predicts strain levels achieved experimentally at epidermal monolayer. B) Increased keratinocyte nuclear aspect ratio indicates successful strain transfer to cells on membrane. C) Elongation of Fbs in the direction of stretch confirms capacity to generate cellular strains through the hydrogel when applying stretch directly to the microfluidic device.

ACKNOWLEDGEMENTS

This research was funded through the UPenn Institute for Regenerative Medicine and the National Institutes of Health (F32 AR068838-02). Resources through the Penn Center for Musculoskeletal Disorders were used for histological processing and strain analyses.

REFERENCES

[1] Whisler, JA et al., *Tissue Eng Part C*, 20(7):543-552, 2014.
[2] Moya, ML et al., *Tissue Eng Part C Methods*, 19(9):730-737, 2013.
[3] Abaci, HE et al., *Lab Chip*, 15(3):882-888, 2015.
[4] Wufuer, M et al., *Scientific Reports*, 6: doi10.1038, 2016.

VALVE INTERSTITIAL CELL MECHANICS WITHIN A 3-D POLY(ETHYLENE GLYCOL) HYDROGEL ENVIRONMENT

Alex C. Khang (1), Andrea G. Rodriguez (2), Megan Schroeder (3), Kristi Anseth (2,3,4), and Michael S. Sacks (1)

(1) Center for Cardiovascular Simulation
Department of Biomedical Engineering
University of Texas at Austin
Austin, TX, USA

(2) Department of Chemical and Biological Engineering
University of Colorado at Boulder
Boulder, CO, USA

(3) Department of Materials Science and Engineering
University of Colorado at Boulder
Boulder, CO, USA

(4) Howard Hughes Medical Institute and the Biofrontiers Institute
University of Colorado at Boulder
Boulder, CO, USA

INTRODUCTION

The aortic valve is comprised of three distinct layers (fibrosa, ventricularis, spongiosa) that feature different amounts of elastin, collagen, and glycoaminoglycans. Consistent within all layers is the presence of valve interstitial cells (VICs) that maintain the extracellular matrix (ECM) [1]. VICs are dynamic with respect to phenotype and have fibroblast-like characteristics when healthy (quiescent) [2]. In diseased or injured valves, VICs switch to a myofibroblast-like phenotype (activated) and become contractile. Activated VICs display prominent αSMA stress fibers and an increase in ECM remodeling [2].

3-D hydrogel encapsulation is an increasingly popular technique for studying VICs [3-6]. Previously our group has studied VIC biomechanical states within the native tissue, but a need still exists for the investigation of VIC mechanics within 3-D hydrogels [7, 8]. In this study, we employed poly(ethylene glycol) (PEG) gels for VIC encapsulation to study their mechanical response to the surrounding hydrogel stiffness and to varying levels of adhesion. Cell contraction was elicited through chemical treatments and the resulting mechanical properties of the constructs were measured by end-loading flexural deformation testing.

METHODS

VIC-Hydrogel Fabrication and Modulation. The hydrogel chemistry used in this study was adopted from the Anseth Research Group and utilizes a thiol-ene, radical polymerization reaction to create a random network of the constituents (Fig. 1) [4-6]. In short, a solution containing norbornene-functionalized poly(ethylene glycol) (PEGnb), MMP-degradable peptide crosslinks (KCGPQGIWGQCK), RGDS adhesive peptides, and lithium phenyl-2,4,6-trimethylbenzoylphosphinate (LAP, photoinitiator) was mixed with VICs (isolated from porcine aortic valves) suspended in phosphate buffered saline (PBS). The solution was then pipetted into silicone molds and polymerized through UV light exposure (2.5 mW/cm^2, 365 nm light) resulting in rectangular VIC-hydrogels (~10 mm x 5 mm x 1 mm). VIC-hydrogels were allowed to incubate for 5 days in proliferation limiting media (1% FBS) before being subjected to end-loading flexural testing.

The thiol-ene ratio was adjusted to fabricate VIC-hydrogels with stiffnesses of 5 kPa, 10 kPa, and 15 kPa in order to study the response of VICs within environments with varying mechanical cues. In addition, separate 5 kPa VIC-hydrogels containing 0 mM, 0.5 mM, and 1.0 mM of RGDS peptides were fabricated to investigate VICs with respect to varying levels of adhesion availability.

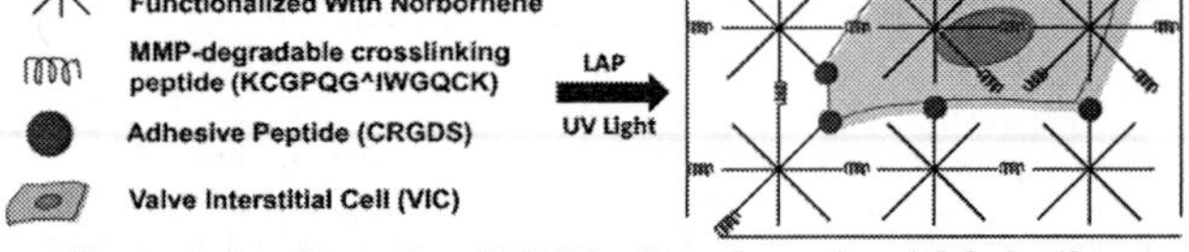

Figure 1: A schematic of VIC-hydrogel construct fabrication.

Visualization of VICs Within Hydrogels. Immunostaining was used to visualize the nucleus, αSMA stress fibers, and F-actin filaments of the VICs within the hydrogel environment.

End-Loading Flexural Testing Method. Small sections of Teflon tubing were attached to the shorter sides of the VIC-hydrogels and 5 evenly spaced red tracking beads were applied along the long side using cyanoacrylate glue (Fig. 2 A). The Teflon tubing was used to attach the VIC-hydrogels onto a moving post and a calibrated bending bar (Fig. 2 A). The moving post was fixed to a linear actuating stage that moved the VIC-hydrogel towards the bending bar, flexing the VIC-hydrogels as a result of end-loading (Fig. 2 B&C).

Protocols from Merryman et al. were adapted for end-loading flexural testing of the VIC-hydrogels [7]. Initially, the samples were allowed to incubate in a 5 mM potassium chloride solution (KCL, normal) for 10 minutes before end-loading testing. The testing media was then changed to a 90 mM KCL solution (contractile) and the VIC-hydrogels were allowed to incubate for 10 minutes before testing again. Finally, the testing media was exchanged for a 20 μM Cytochalasin D solution (CytoD, non-contractile) and the VIC-hydrogel was tested after 10 minutes of incubation.

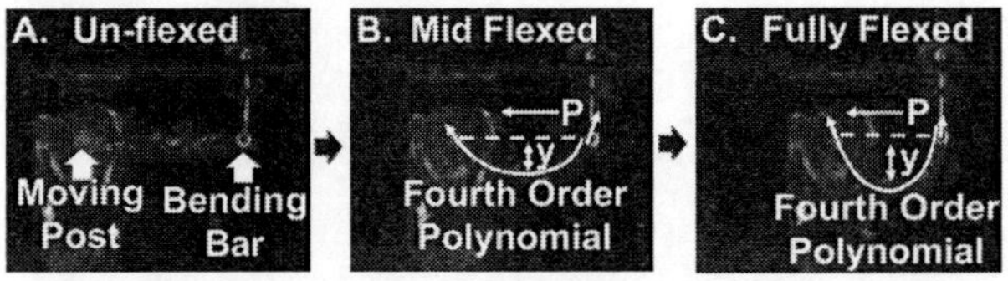

Figure 2: (A) VIC-hydrogel attached to a moving post and a bending bar. (B) Mid-flexed and (C) fully flexed VIC-hydrogel with axial force (P) and deflection of the center marker (y) labeled.

<u>Calculation of Effective VIC-Hydrogel Stiffness.</u> The displacement and known stiffness of the bending bar (Fig. 2 A) was used to calculate the axial force (P) exerted onto the VIC-hydrogels (Fig. 2 B&C). The 5 dots placed along the VIC-hydrogels were fit to a fourth order polynomial function (Fig. 2.1 B & C) and its curvature (κ) was calculated using:

$$\kappa = \frac{y''}{(1+(y')^2)^{\frac{3}{2}}} \tag{1}$$

The applied moment (M) was calculated using:

$$M = P * y \tag{2}$$

where y is the deflection of the center marker in the vertical direction (Fig. 2 B&C). The second moment of inertia (I) was calculated using:

$$I = \frac{1}{12t^3w} \tag{3}$$

where t and w are the thickness and width of the VIC-hydrogel respectively. M/I vs Δκ (κ-κo) plots were generated from the end-loading tests and the effective stiffness (E_{eff}) of the VIC-hydrogels were computed directly from the rate of change of those plots using the Euler-Bernoulli relationship [9]:

$$M = E_{eff} I \Delta k \tag{4}$$

RESULTS

<u>VIC-Hydrogel Matrix Interactions.</u> After 5 days of inoculation, the VICs began to spread within the hydrogel by adhering onto the RGDS peptides (Fig. 3). The VIC-hydrogels in the contractile group displayed higher amounts of αSMA stress fibers compared to the normal group. The non-contractile group displayed the least amount of αSMA stress fibers. No significant visual differences are observed for F-actin or DAPI among the three treatments conditions.

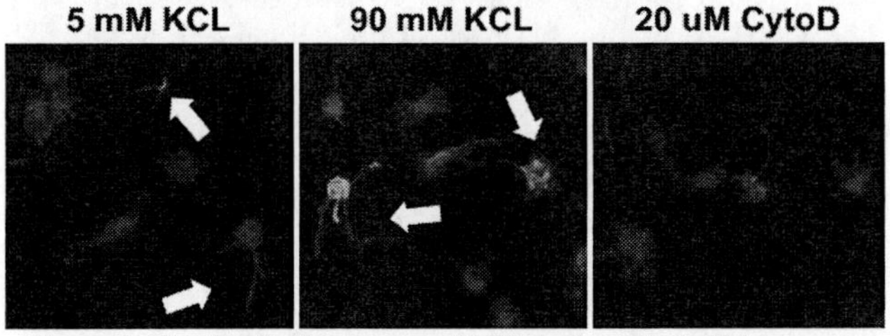

Figure 3: Confocal microscopy of VIC-hydrogels (αSMA-green, DAPI-blue, F-actin-red).

<u>Analyses of the Non-Linear Flexural Response.</u> The VIC-hydrogel M/I vs Δκ mechanical response was found to be non-linear for all treatment conditions (Fig. 4 A). This behavior was described by using a technique adapted from Mirnajafi et al. to fit the data to a second order polynomial [10]. The derivative of the polynomial was then computed, resulting in a linear relationship between E_{eff} and Δκ through utilization of equation 4 (Fig. 4 B).The contractile treatment produced higher M/I values with respect to Δκ compared to the normal treatment (Fig. 4 A). The non-contractile treatment resulted in the lowest M/I values among all the conditions. As the change in curvature (Δκ) increased, the rate of change of M/I (E_{eff}) decreased for all contractile levels (Fig. 4 B). In addition, a larger axial force was required to deform the VIC-hydrogels after 10 minutes of incubation within the 90 mM KCL solution (Fig. 4 C). The opposite was found to be the case after incubation within the 20 μM CytoD solution.

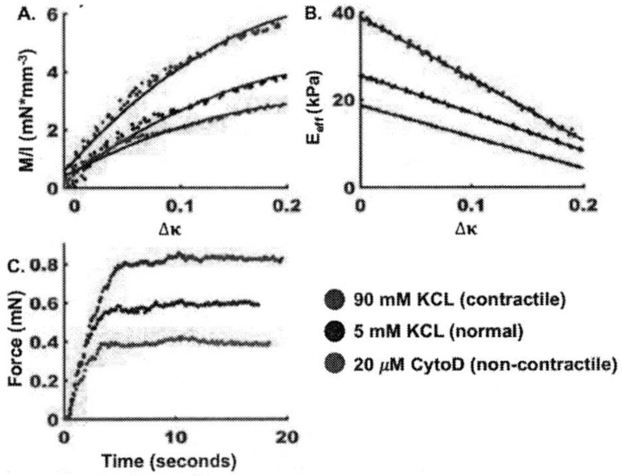

Figure 4: (A) Representative M/I vs Δκ plots for a VIC-hydrogel under normal, contractile, and non-contractile states. (B) The effective stiffness of the VIC-hydrogel plotted against change in curvature for all three conditions. (C) The axial force exerted on the VIC-hydrogel with respect to the duration of the end-loading tests for all three treatments.

DISCUSSION

Construct modulation was demonstrated to be an effective means of controlling VIC contraction [7, 8]. However, further studies are needed to accurately elucidate VIC biomechanical properties within a 3-D hydrogel environment. Future work will focus on adapting our established macro-micro finite element model of the aortic valve to estimate VIC connectivity, contractility, and stiffness within hydrogels [8].

ACKNOWLEDGEMENTS

This work was funded by the NHLBI grant R01 HL-119297.

REFERENCES

[1] Taylor, PM, et al., *The Interational Journal of Biochemistry and Cell Biology*, 35:113-118, 2003.[2] Jr., RHM, et al., *Journal of Surgical Research*, 57:1-21, 1994.[3] Duan, B, et al., *Acta Biomaterialia*, 9:7640-7650, 2013.[4] Benton, JA, et al., *Biomaterials*, 30:6593-6603, 2009.[5] Gould, ST, et al., *Journal of Tissue Engineering and Regenerative Medicine*, 10:2016.[6] Gould, ST, et al., *Acta Biomaterialia* 8:3201-3209, 2012.[7] Merryman, DW, et al., *Journal of Biomechanics* 39:88-96, 2006.[8] Buchanan, RM, http://hdl.handle.net/2152/39463 *(Doctoral Thesis), 2016.*[9] Frisch-Fay, R, *Flexible Bars*, 539, 1962.[10] Mirnajafi, A, et al., *Biomaterials*, 26:795-804, 2005.

INTERPLAY OF MULTI-TPYED HEPATIC CELLS UNDER SHEAR FLOW

Yu Du, Hao Yang, Ning Li, and Mian Long

Key Laboratory of Microgravity (National Microgravity Laboratory), Center of Biomechanics and Bioengineering, and Beijing Key Laboratory of Engineered Construction and Mechanobiology
Institute of Mechanics, Chinese Academy of Sciences
Beijing, China.

INTRODUCTION

Liver is the largest organ in human body with specific microcirculation due to its complicated structure of sinusoidal nodes, in which multiple types of hepatic and hemapoietic cells interact with each other under blood flow. The liver sinusoid is a specialized capillary network with narrow luminal diameter (7-15 μm), slow blood flow (0.1-1 dyn/cm²), and consists of four major types of cells, including fenestrated selectin-deficient liver sinusoid endothelial cells (LSECs), liver resident macrophage Kuppfer cells (KCs), hepatic stellate cells (HSCs), and hepatocytes (HCs)[1-2]. Polymorphonuclear leukocytes (neutrophils, PMN) recruitment in the liver sinusoid is widely observed in almost all liver diseases and contributes to pathogen clearance or tissue damage[3]. While selectin-mediated PMN tethering and rolling unlikely appears in liver sinusoids and Mac-1 or CD44 is assumed to play dominant roles during *in vivo* local fMLF- or systematic LPS-induced inflammatory[4], the regulating mechanisms of PMN adhesion and crawling dynamics are still obscure from those *in vivo* studies. To address this issue, *in vitro* two-dimensional (2D) and 3D models were developed to recapitulate key features of the liver microenvironment and to investigate the roles of distinct adhesive molecules and hepatic cellular interactions in the liver-specific PMN recruitment.

METHODS

To decipher the multistep process of PMN recruitment under physiological flow in liver microvasculature *in vitro*, we established a 2D live-cell flow chamber system (Fig. 1*A*) to perfuse PMNs under shear flow and record the interplay between PMNs and LSECs or LSECs co-cultured with KCs. PMN adhesion and crawling were quantified for fMLF-activated PMNs on TNF-α-stimulated LSECs. Related molecular regulations of LFA-1, Mac-1, and CD44 and the impact of KCs were also analyzed.

To further mimic liver microenvironment with well-organized cell composition and physical interactions, an *in vitro* 3D liver model (Fig. 2*A*) was developed. Two microfabricated polydimethylsiloxane (PDMS) layers were bonded together with a polyethylene (PE) porous membrane between them, forming two fluidic channels separated by a permeable membrane, which is highly similar to liver sinusoids with an endothelium separating microvenule and Disse Space. To replicate the physiological structure and cellular composition, the 3D microfluidic device was constructed by integrating LSEC monolayer with sparsely distributed KCs and HSCs on either side of extracellular matrix (ECM)-coated PE membrane into HC monolayer immobilized on lower PDMS substrate. Typical liver specific functions in protein secretion, drug metabolism and immune responses were analyzed.

RESULTS

We first tested fMLF-activated PMN adhesion and crawling on TNF-α-stimulated LSEC monolayer using 2D *in vitro* liver sinusoid model (Fig. 1*A*). PMN adhesion on LSECs was reduced significantly by LFA-1 blocking but remained unchanged by Mac-1 or CD44 blocking compared with non-blocking PMNs, indicating that LFA-1, but not Mac-1 and CD44, is positively correlated with PMN adhesion on LESCs (Fig. 1*B*). Median crawling speed was significantly increased by Mac-1 blocking but not by LFA-1 or CD44 blocking, as compared to that for non-blocking control (Figure 1*C*). The presence of KCs only down-regulated the directness of PMN crawling and this difference disappeared in LFA-1 blocking group but remained unchanged in other three groups (Figure 1*D*), implying that LFA-1s on PMN interacting with its ligands on KC might mediate highly randomized crawling to the flow direction [5].

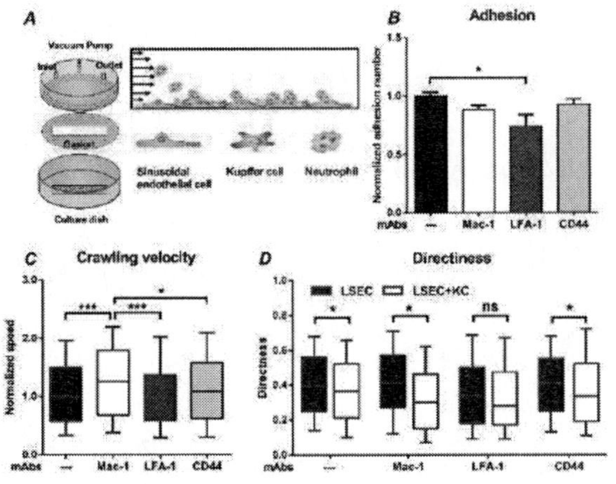

Figure 1: PMN adhesion and crawling dynamics were tested in a 2D *in vitro* liver sinusoid model. (*A*) Schematic of the experimental system. Normalized PMN adhesion number (*B*) and crawling speed (*C*) on LSECs were plotted for the intact and LFA-1-, Mac-1- or CD44-blocked PMNs. (*D*) Comparison of PMN crawling directness on LSECs alone and co-cultured with KCs. Data are presented as the mean ± SEM. $^{*}p < 0.05$, $^{**}p < 0.01$, $^{***}p < 0.001$.

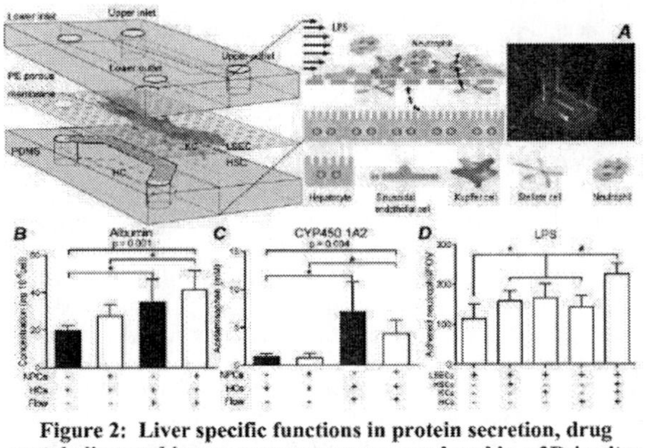

Figure 2: Liver specific functions in protein secretion, drug metabolism and immune responses were analyzed in a 3D *in vitro* liver sinusoid model. (*A*) Schematic of the experimental system. (*B*) ALB secretion was determined from the supernatant collected at 24-h. (*C*) Metabolic products of phenacetin (acetaminophen, for CYP1A2 activity) were determined by liquid chromatography-tandem mass spectrometry from the supernatant collected at 24-h. (*D*) Number of adhered neutrophils when LSECs were cultured alone, with respective HSCs, KCs, or HCs, or together with the other three cell types in the liver model. Data are presented as the mean ± SD. $^{*}p < 0.05$.

We next tested liver-specific functions in protein secretion, drug metabolism, and neutrophil accumulation using a 3D *in vitro* liver sinusoid model (Fig. 2*A*). Shear flow enhanced albumin secretion to 1.5- and 1.78-fold for HCs alone or HCs co-cultured with nonparenchymal cells (NPCs) compared with those under static condition (Fig. 2*B*). Combination of NPC co-culture with flow exposure yielded even higher ALB secretion (2.12-fold increased from HCs alone under static condition), implying that co-culture and shear flow could enhance albumin secretion cooperatively (Fig. 2*B*). Shear flow enhanced dramatically CYP1A2 activity to 6.3- and 3.42-fold when HCs were cultured alone or co-cultured with NPCs, while co-culture with NPCs had no effects on CYP activities under either static or flow condition (Figure 2*C*). Under LPS stimulations, co-culture of LSECs with HSCs, KCs, HCs, or all these three cell types increased PMN accumulation to 1.39-, 1.47-, 1.27- and 1.98-fold compared with LSECs alone (Fig. 2*D*), implying that each type of hepatic cells may contribute to PMN recruitment differently [6].

DISCUSSION

This work establishes *in vitro* 2D and 3D platforms to understand the biomechanics of PMNs interacting with hepatic cells in the sinusoids. The 2D model provides an effective way to understand the dynamics of shear-induced PMN adhesion and crawling on primary LSECs and elucidate the roles of adhesive molecules involved. The 3D model replicates the key architecture of liver sinusoids by integrating four major types of cells into two separated flow channels, which could serve as a functional platform to investigate the short-duration functional maintenance, hepatic cellular interactions, cytotoxic metabolism, and inflammatory cascade under physiologically-like microenvironment. From these models, we are able to analyze the molecular mechanism of PMN recruitment in the liver sinusoids and the distinct impact of each type of hepatic cells.

ACKNOWLEDGEMENTS

This work was supported by National Natural Science Foundation of China grants 31230027, 91642203, 31661143044 and 31110103918, and CAS Strategic Priority Research Program grant XDA01030604 and XDB22040101.

REFERENCES

[1] Dash, A et al., *Am J Physiol Cell Physiol*, 304:C1053-1063, 2013.
[2] Vollmar, B et al., *Physiol Rev*, 89:1269-1339, 2009.
[3] Heymann, F et al., *Nat Rev Gastroenterol Hepatol*, 13:88-110, 2016.
[4] Menezes, GB et al., *J Immunol*, 183:7757-7568, 2009.
[5] Yang, Y et al., *Exp Cell Res*, dio: 10.1016/j.yexcr.2017.01.002.
[6] Du, Y et al., *Lab on a Chip*, dio: 10.1039/C6LC01374K.

A CHEMOGENETIC TOOL TO CONTROL CHONDROCYTE ACTIVTY *IN VITRO*

Ryan C. McDonough (1), Janty Shoga (2), Christopher Price (1,2)

(1) Department of Biomedical Engineering
University of Delaware
Newark, DE, USA

(2) Department of Biomechanics and Movement Science
University of Delaware
Newark, DE, USA

INTRODUCTION

Calcium (Ca^{2+}) is a ubiquitous second messenger and critical component of mechanotransduction signaling cascades within chondrocytes. Several distinct calcium signaling mechanisms and pathways have been implicated with regards to chondrocyte mechanotransduction.[1] However, the majority of techniques employed to evoke/regulate intracellular calcium ($[Ca^{2+}]_i$) signaling within chondrocytes involve either mechanical perturbation or exogenous application of endogenous soluble factors.[2-4] However, these approaches often lack specificity in the mechanisms by which they initiate Ca^{2+} signaling; mechanical forces can target several mechanosensitive components simultaneously and soluble factors can have alternate or off target effects *in vitro* and *in vivo*. Furthermore, these activators can converge upon multiple signaling pathways that regulate Ca^{2+} signaling through a variety of mechanisms.[1] Thus, there is a need for well-controlled and precise Ca^{2+} signaling activation tools to aide in dissecting the role of $[Ca^{2+}]_i$ and cell activation in chondrocyte mechanobiology and homeostasis. Recent developments in molecular biology have resulted in a tool, generated via directed molecular evolution, that enables such desired control. The hM3Dq receptor, a 2nd generation DREADD (Designer Receptor Exclusively Activated by Designer Drugs), is a G-protein-coupled receptor (GPCR) that has been selected to be solely activated by clozapine n-oxide (CNO), a pharmacologically inert compound that only exhibits activity on this receptor.[5] This DREADD has been primarily examined in studies of neurological behavior and can regulate Ca^{2+} activation in neurons via a G_q-protein-coupled receptor-mediated PLC-IP3-DAG pathway (**Fig 1**).[6,7] The purpose of this study is to establish this chemogenetic tool as a targeted, specific, and well-controlled synthetic activator of Ca^{2+}signaling within chondrocytes, specifically as an instrument to dissect the role of Ca^{2+} related pathways in chondrocyte mechanotransduction and homeostasis.

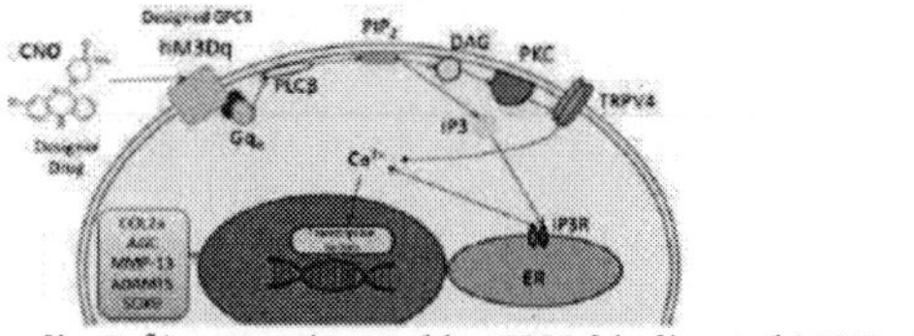

Figure 1. Cytosolic Ca^{2+} entry triggered by CNO binding to hM3Dq is facilitated through a GPCR signaling pathway ($G_{q\alpha}$) involving ER-mediated release of Ca^{2+} and possibly TRPV4-mediated Ca^{2+} entry.

METHODS

CELL CULTURE: ATDC5 cells were plated in glass-bottom dishes (~15,000 cells/cm^2) and cultured at 37°C, 5% CO_2 in complete growth media (1:1 Dulbecco's Modified Eagle Medium + Ham's F12 Nutrient Mixture, 5% fetal bovine serum, 1x Pen/Strep). After 24hr cells were transiently transfected with the hM3Dq-HA plasmid (Addgene; Plasmid #45547)[5] using jetPRIME (Polyplus-Transfection) at a reagent-to-DNA ratio of 3:1 following the manufacturer's protocol for 4hr prior to being replaced with complete growth media for 48hr. Transfection reagent and ratio selection was determined in a pilot study (data not shown). Cells were then stained with 10μM Fluo-8 AM, a fluorescent $[Ca^{2+}]_i$ indicating dye, in HBSS, washed, and then transferred to the microscope and allowed to equilibrate for 5min prior to $[Ca^{2+}]_i$ imaging.

IMAGE ACQUISITION: Imaging experiments were performed on a Zeiss AxioObserver.Z1 Apotome microscope with a 20X 0.8 NA objective. Experiments were performed at 25°C to limit the extent of spontaneous signaling due to TRPV4, which is known to be activated by temperatures above 25°C.[8] A characteristic group of cells was identified and a 15 min imaging protocol was initiated. The first 7.5 min were allotted for establishing the baseline level of spontaneous calcium activation within the cells. After the baseline period, cells were exposed to CNO diluted in HBSS to concentrations of 100-, 250-, 500-, 750nM, or 1μM. Additional samples were exposed to 1- or 100nM GSK101, or 50% hypotonic shock, both of which have been shown to cause up-regulation of $[Ca^{2+}]_i$ entry.[9] One frame was acquired every three sec to reduce the effects of photobleaching. Following imaging, cells were fixed in 4% paraformaldehyde and subject to immunocytochemistry to visualize the expression and localization of the HA-tagged hM3Dq protein.

IMAGE ANALYSIS: A custom MATLAB program was developed to identify and extract $[Ca^{2+}]_i$ signals from the imaging data. For each identified cell in the frame, its fluorescence was normalized to that of the first frame. Calcium peaks were defined as increases in fluorescence intensity greater than 5 standard deviations above the baseline signal. Population response was defined as the percentage of cells exhibiting at least one calcium signal following the baseline period. Multiple peaks were expressed as the percentage of cells with

>1 peak. The temporal kinetics of each signal (duration, rise, and fall times) were also determined. Statistical analysis was performed using one-way ANOVA with Tukey's post-hoc test; significance was set at $p < 0.05$.

RESULTS

The administration of CNO resulted in an immediate and coordinated transient calcium response within hM3Dq-transfected ATDC5 cells; only low-level 'spontaneous' $[Ca^{2+}]_i$ signaling was observed in wild-type CNO-treated controls (**Fig 2**). Increasing the concentration of CNO resulted in a dose-dependent increase in population response ($p<0.001$), with the highest CNO concentration (1µM) exhibiting a response similar to that of 100nM GSK101-activated samples (**Fig 3A**). Increasing CNO concentration did not alter the fraction of cells eliciting multiple $[Ca^{2+}]_i$ peaks, nor alter the health or viability of the cells. There were no significant differences between the temporal kinetics of CNO and hypotonic shock signaling; however, both were significantly faster than those elicited by 1- and 100nM GSK101 activation ($p<0.01$) (**Fig 3B**). No differences in baseline signaling were observed due to transfection with hM3Dq. Immunocytochemistry (ICC) staining revealed hM3Dq localization to the plasma membrane of transfected samples (**Fig 4**); no staining was observed in non-transfected or ICC control samples. hM3Dq transfection efficiency (determined by ICC) was ~40% (data not shown).

DISCUSSION

This study established the successful use of a novel chemogenetic tool for the targeted, well-controlled activation of calcium signaling within ATDC5 cells *in vitro*. To our knowledge, this is the first time that DREADDs have been utilized within a musculoskeletal cell line, and allowed us to precisely and synthetically activate signaling cascades that evoke transient $[Ca^{2+}]_i$ responses in chondrocytes.

Following the administration of CNO, hM3Dq-transfected ATDC5 cells exhibited immediate, rapid and coordinated transient $[Ca^{2+}]_i$ signals that were concentration dependent. Non-transfected cells were irresponsive to CNO. CNO also did not have a significant impact on multiple signaling events, unlike GSK101, suggesting the potential ability to downregulate hM3Dq-mediated cellular activation; such mechanisms may include inhibiting the re-association of the G_α-protein to its receptor or receptor internalization and warrant further study. The fact that the percentage of cells that responded to 1µM CNO (~70%) was greater than the number transfected (~40%) suggests that complete transfection is not necessary to induce large population responses, as autocrine/paracrine signaling mechanisms may be triggered to activate Ca^{2+}signaling in non-transfected cells.[10] The temporal kinetics also identified a similarity between CNO- and hypotonic shock-induced Ca^{2+} signaling, suggesting that the DREADD-based system may share similar Ca^{2+}-specific mechanobiological pathways to those activated in response to hypotonic shock; however, the role of TRPV4 in this DREADD response requires further study.

Limitations to this study include the fact that visualization of the DREADD protein currently requires ICC staining of the HA tag. This prevents direct co-localization of DREADD expression and Ca^{2+} signaling in real-time. Future studies will be performed using a fluorescently tagged DREADD receptor to enable such co-localization. Additionally, DREADs for G_s and G_i pathways exist and can be used in conjunction with mechanical perturbation and pathway inhibition studies to identify how these pathways integrate/synergize with mechanical activation to regulate chondrocyte mechanobiology. Thus, we believe that DREADDs represent a novel tool for exploring the role of controlled Ca^{2+} signaling in chondrocyte biology, cartilage pathology, and the improvement of cartilage tissue engineering.

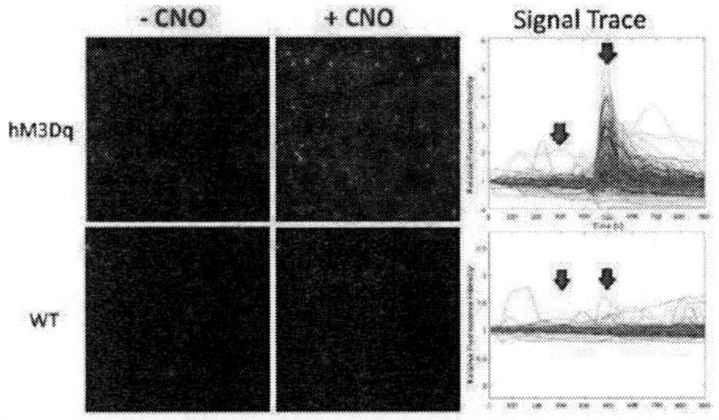

Figure 2. Ca^{2+} signaling following 1µM CNO administration in hM3Dq transfected ATDC5 cells and wild type (WT) controls. Signal traces showed the rapid and coordinated induction of $[Ca^{2+}]_i$ signals in hM3Dq cells, while no such events were detected in WT cells. Red arrow = 5 min (baseline). Blue arrow = 8 min (30 sec following addition).

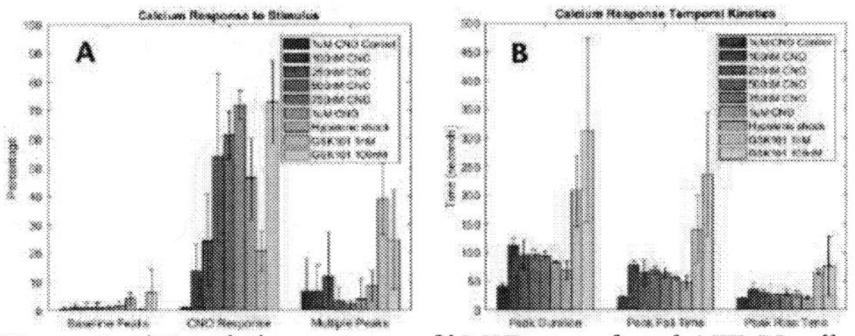

Figure 3. A) Population response of hM3Dq-transfected ATDC5 cells to stimuli. hM3Dq transfection had no effect on baseline spontaneous $[Ca^{2+}]_i$. signaling. Increasing CNO increased the percentage of cells exhibiting single, but not multiple, peaks; however, GSK101 administration significantly increased the number of cells exhibiting multiple peaks ($p<0.05$). B) Temporal kinetics of calcium signals showed a significant decrease in signaling duration between CNO and GSK101 ($p<0.05$). but not between CNO and hypotonic shock.

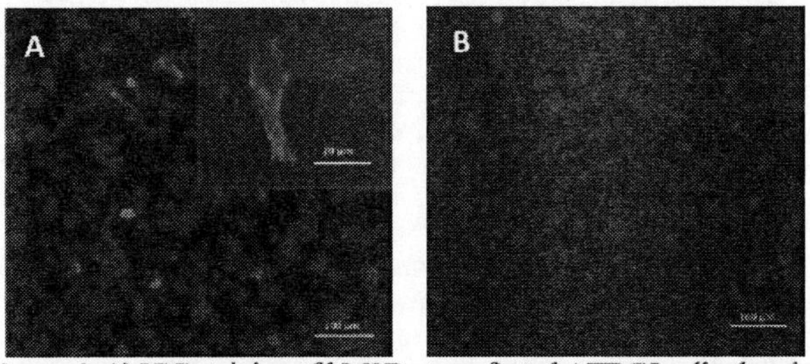

Figure 4. A) ICC staining of hM3Dq-transfected ATDC5 cells showing localization of the GPCR to the plasma membrane. B) ICC staining of non-transfected ATDC5 cells showing no staining. Main image 20X; inset image 63X. Green = hM3Dq. Blue = DAPI.

REFERENCES

1. Matta C, et al. *Front Biosci (Schol Ed)*. 2013;5(January):305-324. **2.** Weber JF, et al. *Biomech Model Mechanobiol*. 2014;13(6):1387-1397. **3.** Guilak F, et al. *J Orthop Res*. 17(3):421-429. **4.** Degala S, et al. *Arch Biochem Biophys*. 2011;505(1):112-117. **5.** Armbruster BN, et al. *PNAS*. 2007;104(12):5163-5168. **6.** Thiel G, et al. *Biol Chem*. 2013;394(12):1615-1622. **7.** Rogan SC, et al. *Pharmacol Rev*. 2011;63(2). **8.** Hurd L, et al. *Am J Med Genet*. 2015;167A:2286-2293. **9.** O'Conor CJ, et al. *PNAS*. 2014;111(4):1316-1321. **10.** D'andrea P, et al. *Biochem J*. 1998;329:681-687.

DEVELOP A HIGH THROUGHPUT PLATFORM BASED ON 3D NEOTISSUE ARRAYS AND FLUDIC SYSTEM FOR CONTROLLED STEM CELL GROWTH ACTIVITY

Ansha. Zhao(1,2), Yonghui. Ding(1), Michael Floren(1), Cameron Morley(1), Wei. Tan(1)*

(1) Department of Mechanical Engineering
University of colorado
Boulder, CO, United State

(2)Department of Material Science and Engineering, Southwest Jiaotong University
Chengdu, Sichuan, China

INTRODUCTION

Nowadays, pluripotency of stem cells have received much attention to be used as clinically therapeutic purpose for regenerative medicine and biomedical application because of their ability of self-renew or to differentiate into specialized progeny. Insights into the complicate stem cell niche have identified the micro-enviroments are critical cues regulating stem cell fate via cell–cell, cell- extracellular matrix, and cell-soluble factors (such as cytokine or growth factors) interactions. In order to create these complex microenvironments, researchers have developed a variety of bioactive materials to induce stem cell commitment. Traditional platforms to screen and optimize these complex factors are based on 2d cell culture system that requires a tremendous effort and the results are not reliable. Here, we represent a kind of high throughput platforms integrated by protein arrays and fluidic systems to indentify the biosynthetic vascular environments to allow efficient commitment in stem cells from numerous different substrates.

METHODS

Duke University Medical Center lead by Whitney O. Lane has published an article describing an established method of designing a parallel plate flow device. This method will be replicated and modified to obtain laminar flow over the sandwich structure. The modified sandwich structure can be seen in Figure 2. In Fig. 3. The co-culture flow device will meet the following requirements: stabilization of the membrane, laminar flow, and allow for separate flow access. There will be uniform laminar flow with a shear stress of 10-100 dyne/cm^2 to the exposed surface of the membrane. Separate flow access to both sides of the support will be necessary in order to account for the different flow profiles the membrane will endure. To simulate a physiological environment, the cell-seeded slide, the printed microarray, and the membrane will experience physiological flow. Electrospun PEGDM matrices were deposited onto standard glass slides that were pretreated with TMPMA to presentmethacrylate groups that can bond the matrices to the glass. Scanning electron microscopy was used to examine the microstructure of the electrospun PEGDM substrates in both dry and hydrated states. Protein arrays were fabricated by using stock solutions of collagen I at a concentration of 250 mg/ml. Samples were deposited on the fibrous PEGDM matrix using Aushon 2470 arrayer with 185 micron pins (AushonBioSystems, Billerica, MA), to achieve dots with a nominal diameter of 250 microns. Individual spots with 7 replicates were deposited with a 500 mm pitch distance onto the PEGDM matrices. Approximately twenty collagen I microarrays could be deposited simultaneously in this method within ~1hr. Prepared collagen I microarrays were stored at 4 °C in a humid environment for 24 h before use.

The collagen I microarrays were rinsed by water and sterilized by 70% ethanol for 1hr before cell seeding. Rat mesenchymal stem cells (MSCs) with passage from 3-8 were culture on the collagen I microarrays for 24h in a static condition before these samples were stored in flow device with the flow velocity of 10dynes/cm^2 for 16h. Following cell culture, immunofluorescent staining was explored to observe cell adhesion and spreading with immunofluorescent dye DAPI and Alexa488-phallodin.

RESULTS

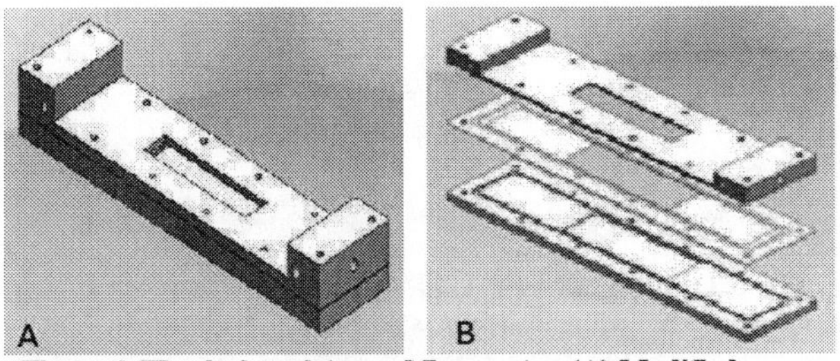

Figure 1:The design picture of flow system(A) Modified sandwich structure of the parallel-plate flow device.
(B) Exploded view of modified sandwich structure of the parallel-plate flow device

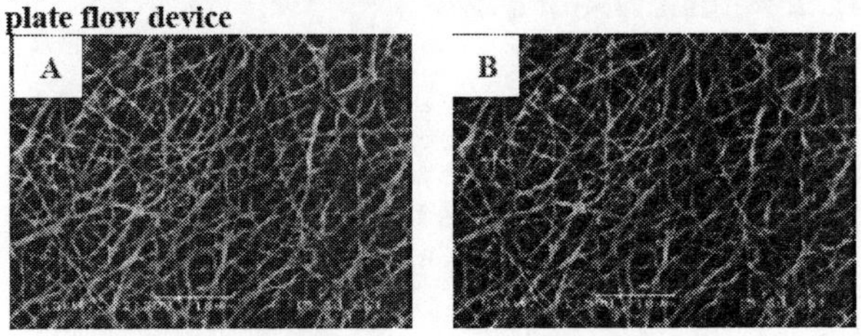

Figure2: The morphology of PEGDM fibrous architecture (A)dry condition (B)wet condition

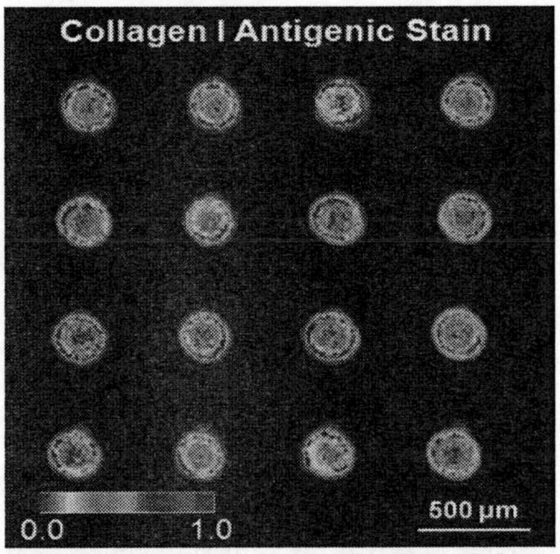

Figure3: Confocal microscopy image of collagen I deposition on PEGDM substrate illustrating 3-dimensional dot presentation.

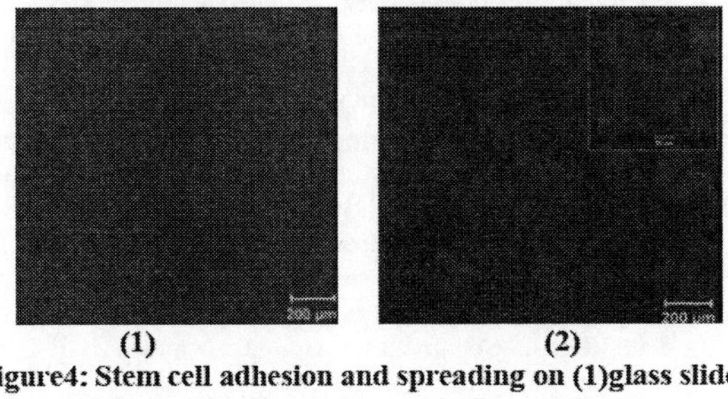

Figure4: Stem cell adhesion and spreading on (1)glass slides in static condition and collagen I array in flow system

DISCUSSION

The electrospinning process has been shown to house a better environment for cells to survive. But for vascular engineering application, the growth activity of stem cell on electrospun matrix under physiologically flow condition is still worth to study further. The flow system (Fig1.) developed in this paper will allow for the study of interaction between micoarrays and any variety of cells (endothelial cells, stem cells, etc.).

High-throughput tissue microarray for cell culture study to print onto a glass slide which enables an assortment of proteins with a variety of concentration gradients to be used in studying protein and cell interactions. Scanning electron microscopy results show the copolymerized electrospun PEGDM matrix have a highly porous structure which is a three-dimensional interconnected fibrous network with a fiber diameter of 0.5-1um. The fiber will swell and increase its diameter after the hydration occurs (Fig.2). Collagen I dots can decorate the matrix in a micro scale (Fig.3) where MSC prefer to attach and spread homogenously under flow condition (Fig.4). In conclusion, this system combined protein assays and flow system enables the production of engineered microenvironments to control stem cell growth behavior. And also it can be used as a high throughput platform to optimize the composition of extracellular along with the changeable compound of protein dots.

ACKNOWLEDGEMENTS
This work is finically supported through NIH funding sources K25HL097246 (W.T.) and R01 HL119371 (W.T.).

REFERENCES

[1] Whitney O. Lane, A. E.-H. (2012, January 17). Parrallel-pate Flow Chamber and Continuous Flow Circuit to Evaluate Endothelial Progenitor Cells under Laminar Flow Shear Stress. Retrieved October 2015, from Jove: http://www.jove.com/video/3349/parallel-plate-flow-chamber-continuous-flow-circuit-to-evaluate.
[2] Floren, Michael, and Wei Tan. Three-dimensional, Soft Neotissue Arrays as High Throughput Platforms for the Interrogation of Engineered Tissue Environments. Biomaterials 59 (2015): 39-52.
[3] Hwan D. Kim,etal. High throughput approaches for controlled stem cell differentiation. Acta Biomaterialia 34 (2016) 21–29.
[4] Asha K. Patel. High throughput screening for discovery of materials that control stem cell fate. Current Opinion in Solid State and Materials Science 20 (2016) 202–211.
[5] Naveen Nagiah. Highly Compliant Vascular Grafts with Gelatin-Sheathed Coaxially Structured Nanofibers. Langmuir. 2015 December 1; 31(47): 12993–13002

SB^3C2017
Summer Biomechanics, Bioengineering and Biotransport Conference
June 21 – 24, Tucson, AZ, USA

SPROUTING LYMPHANGIOGENESIS REGULATED BY COMBINED BIOCHEMICAL AND MECHANICAL STIMULATION IN A 3-D MICROFLUIDIC DEVICE

Chia-Wen Chang (1), Pawan Kumar (2), Jonathan W. Song (2,3)

(1) Chemical and Biomolecular Engineering
The Ohio State University
Columbus, OH, USA

(2) The Comprehensive Cancer Center
The Ohio State University
Columbus, OH, USA

(3) Mechanical and Aerospace Engineering
The Ohio State University
Columbus, OH, USA

INTRODUCTION

Lymphatic vessels comprise an essential physiological system that regulates many critical functions such as coordinating immunological responses and maintaining interstitial fluid pressure by acting as an interconnected drainage system to prevent accumulation of excess fluid within the tissue spaces [1, 2]. Moreover, several pathologies are believed to be related with impaired lymphatic vessel function. For instance, lymphedema is a consequence of the accumulation of tissue fluid. Another example is cancer, which can co-opt lymphatic vessels to enable metastases to lymph nodes.

While lymphangiogenesis, or the formation of new lymphatic vessels, is necessary to prevent lymphedema from arising, aberrant levels of lymphangiogenesis is associated with cancer and other pathologies such as inflammation. Several signaling molecules, including certain growth factors and cytokines, are believed to be associated with increased lymphangiogenesis in both *in vivo* and *in vitro* settings. Furthermore, while blood vessel angiogenesis is one of the established hallmarks of cancer, much less is known about what governs lymphangiogenesis. One reason for the relatively deficient understanding of lymphangiogenesis compared to angiogenesis is a lack of established techniques that integrate the combined mechanical and biochemical stimulation present within the lymphangiogenesis microenvironment. Here we assess lymphangiogenesis using a 3-D microfluidic device, which enables us to apply biochemical and mechanical stimulations on the vessels analogues comprised of lymphatic endothelial cells. We show that stimulation by interelukin-6 (IL-6) and vascular endothelial growth factor C (VEGF-C), whether applied alone or in combination, promotes lymphangiogenesis. These molecules were used for this study because they have been found to be elevated in cancer patient samples and during inflammation. In addition to biochemical stimulation with IL-6 and VEGF-C, we assessed the effect of interstitial fluid flow (IF) on lymphangiogenesis to determine whether IF or IF combined with IL-6 and VEGF-C promotes lymphangiogenesis. Finally, we used our microfluidic approach to assess for lymphatic vessel permeability to help establish a correlation between lymphangiogenesis and altered lymphatic permeability.

METHODS

The microfluidic device for studying lymphangiogenesis was comprised of a single PDMS layer sealed against a glass slide. The PDMS layer was fabricated using established soft lithography replica molding techniques (MSL). The master mold was fabricated by conventional photolithography method. The PDMS precursor was prepared by mixing the base and curing agents with 10:1 ratio and pouring onto the master mold. The device was cured in 65 °C oven overnight and inlets and outlets were punched using biopsy punch. For the PDMS layer, three parallel channels were patterned using replica molding. To assemble the device, plasma treatment was employed to irreversibly seal the PDMS layer onto the glass slide. Prior to cell seeding, Type I collagen gel added to the middle channel of microfluidic device, which was directly adjacent to the two lateral channels and subsequently lined with lymphatic endothelial cells.

Type I collagen gel isolated from rat tail was purchased from Corning company and injected into middle channel at a concentration of 3 mg/ml. Lymphatic endothelial cells were purchased from Lonza and cultured in EGM-2MV medium. Both IL-6 and VEGF-C were from Peperotech Inc. For chemical treatment, IL-6 and VEGF-C were mixed with culture medium individually or combinatory.

Fibronectin solution was coated onto cell channel surfaces for 20 minutes to improve biocompatibility prior to cell seeding. Lymphatic endothelial cells were seeding into both two lateral channels of microfluidic devices by pipetting and keep culturing 1 day in the humidified incubator at 37 °C for better cell attachment. Both chemical and mechanical stimulations were applied after 1 day cell culture in the devices. Phase contrast and epi-fluorescence microscopy was employed to capture the sprouting of lymphatic endothelial cells into collagen gel regions. NIH ImageJ software was used to quantify the sprouting area for different time periods. To measure the vascular permeability, fluorescently-conjugated tracer dye (Texas red-Dextran, 70 K M.W) was passively pumped into the lymphatic endothelial cell lined channels and time-lapse was employed to track the movement of dye-containing media from the cell lined channels into the collagen gel for 12 hours. The fluorescent intensity at various time points was analyzed by ImageJ.

We quantified lymphangiogenesis or sprouting of lymphatic endothelial cells in our microfluidic device in response to different conditions using a sprouting ratio of the area of the lymphatic sprouts to a reference area within the microfluidic device (Figure 1). Lymphatic sprouting was monitored in response to stimulation with IL-6, VEGF-C and their combinations (concentration of 50 ng/ml). In addition, we combined these biochemical conditions with interstitial flow and compared to static conditions (Figure 1).

RESULTS

To verify the biocompatibility of proposed microfluidic device, lymphatic endothelial cells were cultured in the device for 4 days. The results showed that cell viability was compatible with those cultured on tissue culture dishes and no significant cell death during culture period was observed. Furthermore, the collagen gel was stable without delaminating or contraction during experiments, indicating that this microfluidic device possesses the necessary biocompatibility for endothelial cell biology study. These results are in accordance with previous results with blood endothelial cells [3].

Under no biochemical stimulation and static conditions, we observed negligible amounts of lymphatic sprouting. In contrast, stimulation with IL-6 and VEGF-C under static conditions significantly increased lymphatic sprouting compared to control. These results suggest that stimulation with IL-6 and VEGF-C promotes lymphangiogenesis. Also, we applied IL-6 and VEGF-C together and compared the Sprouting Ratio to IL-6 alone and VEGF-C treatment conditions. We observed that while the Sprouting Ratio for IL-6+VEGF-C together is greater than for IL-6 and VEGF-C alone, the effect does not appear to be additive (Figure 1).

When we monitored sprouts in response to interstitial flow, we observed that interstitial flow alone in the absence of biochemical treatment increases the Sprouting Ratio 5-fold compared to static and unstimulated control (Figure 1). This result suggests that mechanical stimulation by interstitial flow alone significantly enhances lymphangiogenesis. We also monitored sprouts in response to interstitial flow combined with IL-6 and VEGF-C. The combination of interstitial flow with IL-6 increases the Sprouting Ratio by ~ 10-fold compared IL-6 under static conditions. . Similarly, interstitial flow combined with VEGF-C increases Sprouting Ratio by ~3-fold compared VEGF-C under static conditions (Figure 1). Finally interstitial flow combined with IL-6+VEGF-C increases Sprouting Ratio by ~ 2.5-fold compared to IL-6+VEGF-C under static condition (Figure 1). Collectively, these results suggest that either biochemical (IL-6 or VEGF-C) or mechanical (interstitial flow) stimulations on their own or applied together promote lymphatic sprouting.

In addition to lymphatic sprouting, we observed that both VEGF-C and IL-6 potently increase lymphatic vessel permeability (data not shown). Thus, the presence of theses biomolecules combined with interstitial flow stimulation potently upregulates lymphangiogenesis and lymphatic vessel leakiness, which in turn can have important implications for the pathogenesis of certain diseases such as cancer metastasis or inflammation

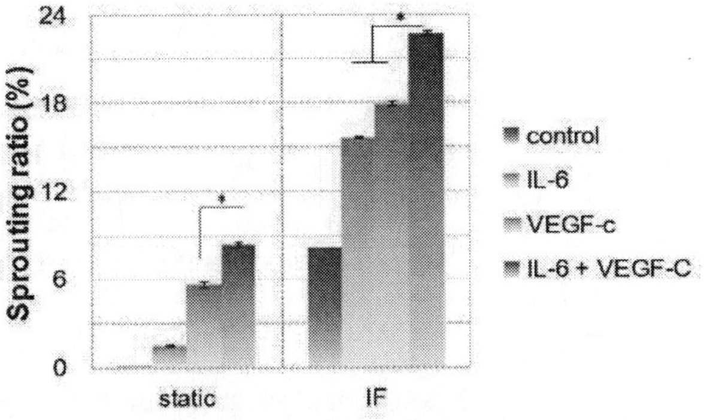

Figure 1: Lymphatic vessel sprouting ratio of various biochemical stimulations in static and interstitial flow (IF) conditions: control, IL-6, VEGF-C and combination of IL-6 and VEGF-C.
*** represents p-value < 0.05 using one-way ANOVA analysis by Origin lab software.**

DISCUSSION

Aberrant lymphangiogenesis is a critical determinant of certain pathologies. For instance, during cancer metastasis, invasive cancer cells arising from certain tumor types (e.g. breast cancer) are prone to disseminate to distant organs through interconnected lymphatic vessel system. Therefore, it is of high physiological importance to uncover the regulators of lymphangiogenesis. The microenvironment of tumors are known to overexpress certain signaling molecules (e.g. IL-6 and VEGF-C) while also residing at elevated interstitial flow levels. Our microfluidic system successfully reconstitutes these parameters *in vitro* by enabling different combinations of biochemical and mechanical stimulation. Our results demonstrated that lymphangiogenesis is highly associated with the IL-6 and VEGF-C stimulation as both molecules alone and in combination promote lymphatic sprouting. Moreover, interstitial flow on its own or in combination with IL-6 and VEGF-C potently enhances the lymphatic vessel sprouting. Our results suggest that the increased lymphatic sprouting arises due to increased lymphatic vessel permeability. Furthermore, this study demonstrates that our 3-D microfluidic device can serve as an *in vitro* platform for increasing our understanding of lymphangiogenesis, with applications for further study pertaining to cancer and inflammation responses.

ACKNOWLEDGEMENTS

Funding was provided by the American Heart Association (15SDG25480000) and the Pelotonia Junior Investigator Award.

REFERENCES

[1] Huethorst, Eline, et al., *TISSUE ENGINEERING: Part B* Volume 22, Number 1, 2016

[2] Hirakawa, Satoshi, et al., *Advanced Drug Delivery Reviews,* 74, 2014, 12–18

[3] Song, Jonathan, et al., *PNAS*, 2011, 108, 37, 15342-15347

INTERFACIAL MECHANICS DETERMINE TISSUE ARCHITECTURE OF NORMAL AND DISEASED BREAST

V. Srivastava (1), J.C. Garbe (1,2), M.A. LaBarge (3), Z.J. Gartner (1)

(1) Department of Pharmaceutical Chemistry
University of California San Francisco
San Francisco, California, USA

(2) Life Sciences Division
Lawrence Berkeley National Lab
Berkeley, California, USA

(3) Department of Population Sciences
City of Hope National Medical Center
Duarte, California, USA

INTRODUCTION

Breast cancer is the second leading cause of cancer death in women. It arises within the human mammary epithelium, a tissue with a bilayered structure, where a layer of contractile myoepithelial cells (MEPs) surrounds the milk-secreting luminal epithelial cells (LEPs). In addition to its importance in the function of secreting and pumping milk, the bilayered structure of the mammary gland also plays a critical role in suppressing breast cancer progression [1]. Progressive architectural breakdown can remove tumor suppressive MEPs from the tissue-ECM boundary, allowing transformed LEPs to contact the surrounding stroma, facilitating their invasion and metastasis [2] **(Fig. 1)**.

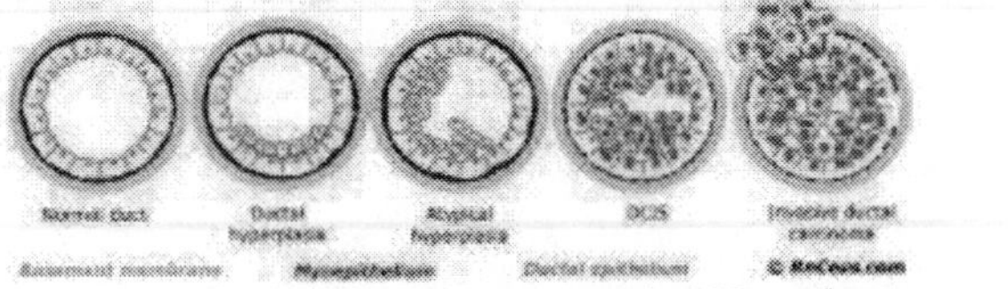

Figure 1: Stages for breast cancer progression. Normal mammary duct has a bilayered structure that is maintained through ductal carcinoma in situ. Transition to invasive ductal carcinoma is marked by a breakdown in the bilayered structure allowing transformed cells to contact stroma and metastasize.

The bilayered architecture of the mammary gland is maintained by the ability of mammary epithelial cells to couple their interfacial mechanical properties with each other and the extracellular matrix (ECM) [3,4]. Using in vitro reconstitution of primary human mammary epithelial cells (HMECs), we demonstrated that HMEC organoids can self-organize with in vivo-like cell positioning through a process that is governed by the interfacial mechanics of the mammary epithelium and its ECM [3] **(Fig. 2a)**. Other systems such as gastrulating embryos and prostate glands also undergo similar differential interfacial mechanics-driven cell sorting, suggesting this is a general property of tissues [3,5]. More specifically, the unique ability of MEPs to strongly interact with the ECM is central to robust glandular self-organization **(Fig. 2b)**.

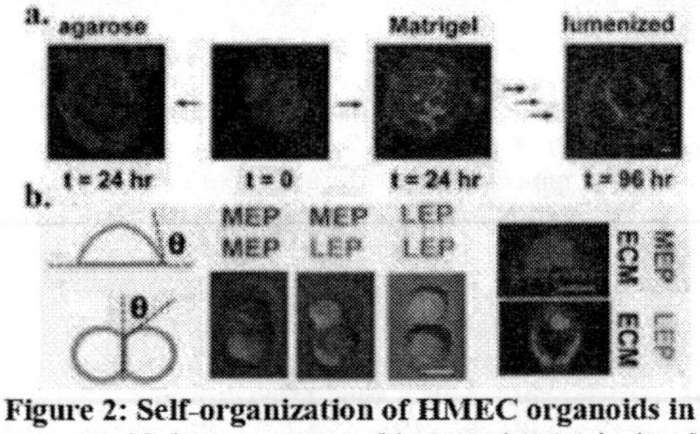

Figure 2: Self-organization of HMEC organoids in vitro. a. HMEC organoids have correct architecture in Matrigel and inverted architecture in agarose. **b.** MEPs have strongest cell-cell adhesion and uniquely interact with ECM based on contact angle measurements [3].

<u>We hypothesize that breast cancer drivers alter the physical properties of the mammary epithelium and ECM, which lead to the progressive loss of the mammary gland's ability to recover from architectural disruption, thus accelerating the progression to invasive cancers.</u> While changes in the physical properties of both tumor cells and surrounding matrix during cancer progression have been reported [6], a direct mechanistic link between oncogenes, altered interfacial mechanics, and a breakdown in tissue architecture has not been made. Therefore, we are using interfacial tension measurements of HMECs to develop a mechanical model for tissue self-organization. We have

gained important insight into the balance of these interfacial tensions through perturbation of different contractile and adhesive molecules. Next, we are leveraging our self-organization model to characterize the mechanisms by which single cell-level physical changes result in architectural changes associated with the malignant disease in humans.

METHODS

We used self-organization of organoids of primary HMECs to characterize the physical determinants of mammary tissue architecture. The organoids were made by mechanical aggregation of FACS-sorted LEPs and MEPs in agarose microwells, which were allowed to self-organize in Matrigel for 24-48 h [3]. Self-organization was quantified by visually scoring the organoids as correct, split, incorrect or scrambled. The cell-cell and cell-ECM interfacial tensions were measured using micropipette aspiration and contact angle analysis [3,7].

For studying the effect of breast cancer drivers, primary HMECs were transformed with lentiviral constructs prior to FACS. As most cancers involve transformations only in the LEP population, we assembled driver-modified LEPs with normal MEPs. The self-organization and biophysical measurements were compared to empty vector-transformed LEPs cultured identically. A chi-square test on the frequency distributions of normal and transformed tissues was used for determining statistical significance. For tension measurements, Student's T-test or ANOVA for Gaussian distributions and Mann-Whitney's non-parametric test for non-Gaussian data sets were used to calculate statistical significance

RESULTS

Normal HMECs self-organize by taking mechanical cues from cell-cell and cell-ECM interactions: In the presence of mechanical cues from the ECM, the organoids achieve the in vivo-like organization with a LEP core surrounded by MEPs. In contrast, in the inert agarose wells, self-organization is solely guided by cell-cell interactions, resulting in inverted structures (**Fig. 2**) [3]. Our measurements for cell-cell and cell-ECM interfacial tensions are consistent with this differential interfacial tension-based model for tissue self-organization [5] (**Fig. 3**).

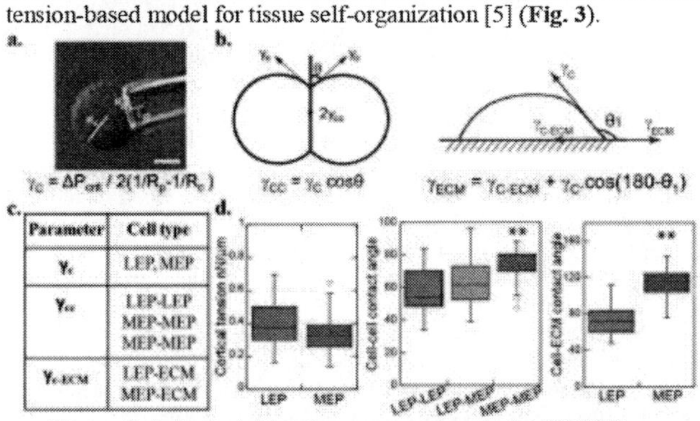

Figure 3: Interfacial tension measurements for HMECs. a. Micropipette aspiration for measuring cell-medium cortical tension. **b.** Cell contact angle analysis for cell-cell and cell-ECM interfacial tension measurement. **c.** List of interfacial parameters. **d.** Cortical tension, cell-cell and cell-ECM contact angles for normal HMECs.

Further, the depletion in MEPs of talin1, an adaptor protein linking integrins to the actin cytoskeleton, reduces MEP-ECM affinity leading to disorganized tissues (**Fig. 4**) [3]. In contrast, inhibiting cell-cell adhesion by depleting p120 catenin does not affect tissue architecture as it does not alter the dominant MEP-ECM interaction (**Fig. 4**) [3].

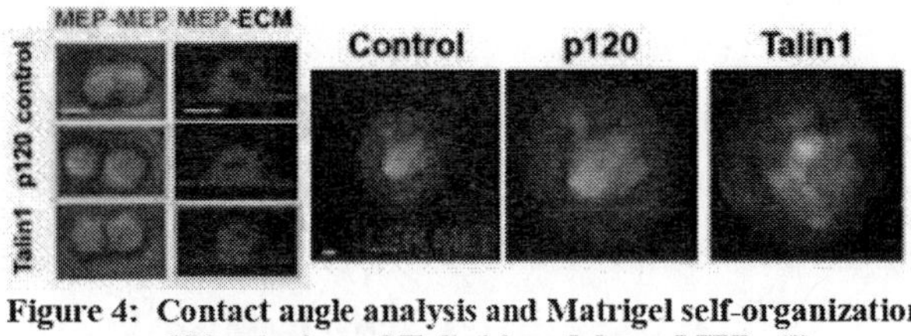

Figure 4: Contact angle analysis and Matrigel self-organization of p120 catenin and Talin1 knockdown MEPs (3).

Cyclin D1 expression does not affect self-organization of HMECs: we expressed activated cyclin D1 in LEPs, which is an important cell cycle regulator and is overexpressed in more than 50% of breast cancers [8]. We found that both acute (3 days) and chronic (10 days) cyclin D1 expression in LEPs did not significantly affect tissue architecture (**Fig. 5**). However, sustained expression of cyclin D1 in post-stasis cells (>15 days) [9] resulted in disorganized tissues, likely due to increased ECM interactions arising from extended cell culture.

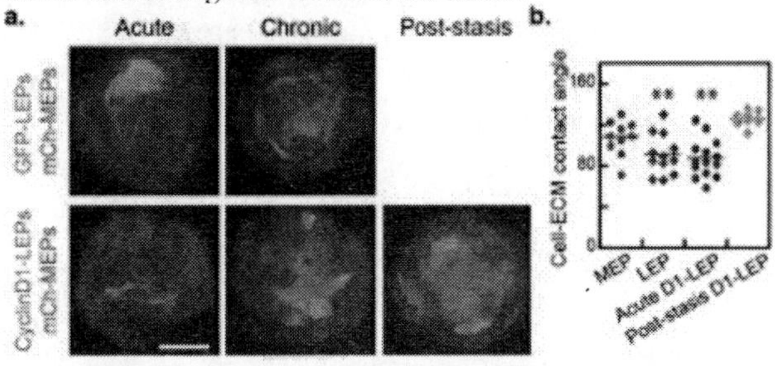

Figure 5: Impact of cyclin D1 on self-organization. a. Reconstituted organoids with GFP- or cyclin D1-expressing LEPs and mCherry MEPs at 24 hours. **b.** Cell-cell and cell-ECM contact angles for normal HMECs and cyclin D1-expressing LEPs. Scare bar: 50 µm.

DISCUSSION

We have established HMEC organoids as a powerful model to quantitatively characterize the impact of breast cancer drivers on cellular biophysical properties and tissue architecture. We are currently applying this system to identify which oncogenic changes alter tissue architecture, and the underlying mechanisms for progression to invasive breast cancers. Mechanistic insight into the biophysical regulation of human mammary architecture will help identify molecular targets to stabilize tissue architecture in a transformed mammary gland.

ACKNOWLEDGEMENTS

We thank members of the Gartner Lab for their valuable suggestions and support. We also thank Department of Defense Era of Hope Expansion Award (W81XWH-13-1-0221) for funding.

REFERENCES

1. Adriance, MC et al. *Breast Cancer Res*, 7(5):190-197, 2005
2. Nelson, CM et al., *Annu Rev Cell Dev Biol*, 22:287-309, 2006
3. Cerchiari, AE et al., *Proc Natl Acad Sci USA* 112(7):2287-92, 2015
4. Chanson, L et al., *Proc Natl Acad Sci USA*, 108(8):3264-9, 2011
5. Brodland GW. *J Biomech Engg*, 124(2):188-97, 2002
6. Kumar, S et al., *Cancer and Metastasis Rev*, 28(1-2):113-27, 2009
7. Schiffhauer, ES et al., *Curr Biol*, 26:1-7, 2016
8. Barnes, DM et al., *Breast Cancer Res Treat*, 52(1):1-15 1998
9. Lee, JK et al., *Front Cell Dev Biol*, 3:13, 2015

SB^{3}C2017
Summer Biomechanics, Bioengineering and Biotransport Conference
June 21 – 24, Tucson, AZ, USA

A COMPUTATIONAL MICROSTRUCTURAL NETWORK MODEL TO TEST DUNN'S HYPOTHESES OF CONTACT GUIDANCE

Victor K. Lai (1), Rohit Y. Dhume (2), Lauren M. Bersie (3), Victor H. Barocas (3), Robert T. Tranquillo (3,4)

(1) Department of Chemical Engineering
University of Minnesota – Duluth
Duluth, MN, USA

(2) Department of Mechanical Engineering
University of Minnesota – Twin Cities
Minneapolis, MN, USA

(3) Department of Biomedical Engineering
University of Minnesota – Twin Cities
Minneapolis, MN, USA

(4) Department of Chemical Engineering and Materials Science
University of Minnesota – Twin Cities
Minneapolis, MN, USA

INTRODUCTION

Contact guidance, or the propensity of cells to align and migrate along the direction of anisotropic environmental features, such as aligned fibrils or micropatterned substrates, is a well-established phenomenon [1, 2]. However, despite its significance in a variety of applications such as wound healing, tissue engineering, and cancer, the underlying mechanisms of contact guidance remain unidentified.

In his seminal work on the subject [3], Dunn presented the hypotheses that contact guidance can result from a cell sensing anisotropies of adhesion, porosity, and viscoelasticity present within the environment. Over three decades later, the dominant signal of the three has yet to be determined, largely due to the interconnected nature of the chemical, steric, and mechanical anisotropies within a fiber network. Experimental manipulation of just one parameter often confounds results, as all the anisotropies are concurrently affected. One such example is the method of varying collagen gel density in order to vary fiber stiffness, employed by many studies in the literature to observe contact guidance and cell protrusion dynamics [4, 5]. This method provides varying degrees of mechanical anisotropy as desired; however, it simultaneously changes the other two anisotropies, failing to isolate and thus elucidate the primary contact guidance signal in an aligned fiber network. Consequently, developing a physics-based modeling framework for separately examining Dunn's hypotheses may facilitate an understanding of contact guidance.

We present here the use of a previously validated microstructural network model [6, 7] as a computational means for investigating the three possible anisotropies within a network postulated by Dunn and their effect on contact guidance. Through the simulation of the interactions between a pseudopod and a fiber network, this work aims to elucidate the principal signal responsible for contact guidance.

METHODS

Network Generation: Representative Volume Elements (RVEs) comprised of Voronoi networks are used to represent the collagen extracellular matrix (ECM), as previously described [8]. To generate networks of varying degrees of alignment, a larger network (2x2x2 in dimension) is first generated and stretched in the x-direction to varying extents. Subsequently, the RVEs used for these simulations are obtained by extracting a 1x1x1 element from these stretched networks, while ensuring that the total fiber length in each RVE is approximately constant. In total, 21 networks varying in alignment in the x-direction from Ω_{xx} = 0.33 (isotropic) to 0.76 were generated. The actual dimension of each RVE side is approximately 10μm, based on an ECM concentration of 3mg/mL.

Mechanical (stiffness) Anisotropy: The force exerted by each fiber within the RVE upon deformation is represented by an exponential constitutive equation, and the average stress in the network calculated via volume-averaging, as previously described [6]. Each network is stretched in both the x- and y- directions separately to obtain stress-strain curves in both directions. The effect of alignment on mechanical anisotropy is quantified using the modulus ratio:

$$\frac{E_x}{E_y} = \frac{tangent\ modulus\ at\ \lambda{=}1.1\ of\ network\ stretched\ in\ x{-}direction}{tangent\ modulus\ at\ \lambda{=}1.1\ of\ network\ stretched\ in\ y{-}direction} \quad (1)$$

Adhesion (Chemical) Anisotropy: A cylinder spanning the entire RVE length is used to represent a pseudopod in the collagen network. The cylinder dimensions relative to the RVE were chosen based on experimental data of pseudopod measurements of human dermal fibroblasts embedded in an ECM at short time points (3h – 7h), which showed pseudopods of average diameter ~5μm and length ~10μm (unpublished). For each network, the cylinder is placed in both the x-

(aligned) and y-directions (Figure 1). In each case, the contact length of the pseudopod with the collagen fibers is approximated by the sum of intersection lengths of fibers with the cylinder. These fiber intersections represent lengths available for adhesions with the pseudopod after the pseudopod has pushed them to its surface. The effect of network alignment on adhesion anisotropy is quantified as follows:

$$\frac{L_x}{L_y} = \frac{\text{total intersection length with protrusion in the } x\text{-direction}}{\text{total intersection length with protrusion in the } y\text{-direction}} \qquad (2)$$

In addition, fibers that intersect within 20% of the cylinder axis are considered to be broken and hence disregarded. This threshold value is calculated based on a fiber failure stretch ratio of 1.4 from [9], assuming that the fiber gets pushed to the cylinder surface to form the arc length.

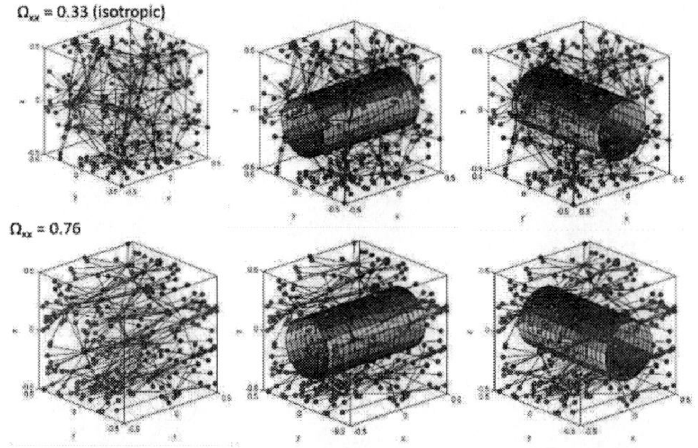

Figure 1. RVEs showing the network only (left), the pseudopod inserted in the x-direction (middle), and the pseudopod inserted in the y-direction. The top row shows an isotropic network (Ω_{xx} = 0.33), the bottom row shows the most aligned network (Ω_{xx} = 0.76).

RESULTS

Mechanical Anisotropy: A plot of E_x/E_y versus Ω_{xx} is shown in Figure 2 below. As expected, when the network is close to isotropic (Ω_{xx} ~ 0.33), the modulus ratio is around 1. This ratio increases rapidly as the network becomes more aligned in the x-direction, which confirms mechanical anisotropy in aligned networks exhibiting stiffer behavior in the direction of alignment.

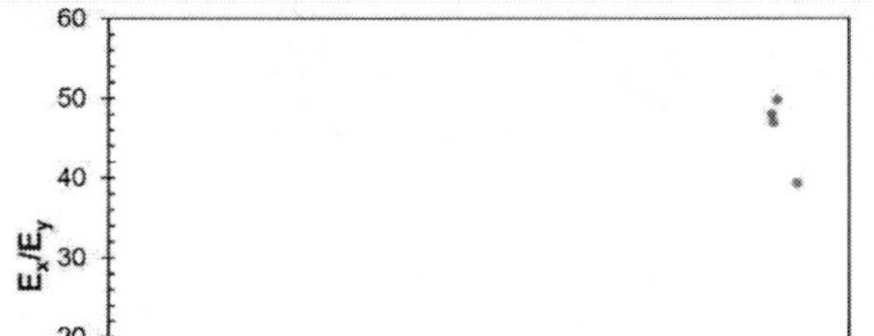

Figure 2. Modulus ratio versus network orientation. The stiffness ratio increases rapidly with orientation, with the networks exhibiting stiffer behavior in the aligned direction.

Adhesion Anisotropy: Figure 3 shows a plot of the intersection length ratio, L_x/L_y,versus Ω_{xx}. No definitive trend of intersection length ratio with degree of alignment is observed. While a pseudopod extended along direction of alignment showed longer intersection lengths per fiber as degree of alignment increased, this effect was offset by fewer number of fibers intersected (data not shown).

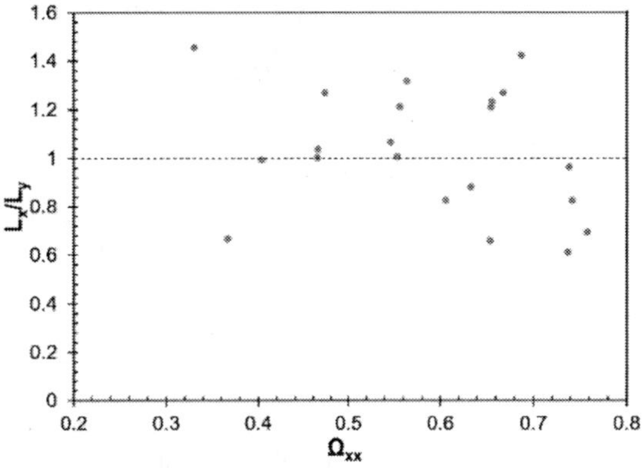

Figure 3. Intersection length ratio versus network orientation. Scattered data around an intersection length ratio of 1 suggests that degree of network alignment does not affect the total fiber contact length with a pseudopod available for focal adhesion.

DISCUSSION

This microstructural network model framework affords a zeroth-order assessment of Dunn's hypotheses that has never been conducted. Although the modulus ratio curve will change quantitatively with the stretch ratio chosen to define the moduli, the qualitative nature of the curve in Fig. 2 for $\lambda = 1.1$ suggests that stiffness anisotropy is a viable signal for how cells sense fiber alignment. This conclusion cannot be made for adhesion anisotropy for the limited sets of networks analyzed to date (Fig. 3). Topological constraints to reorganization of the network to accommodate the pseudopods (cylinders) may have directional dependence and yield a small anisotropy in the fiber contact length that equates to adhesion anisotropy in this simple view of pseudopod-fiber adhesion. The model can also be used to assess the porosity (steric) anisotropy hypothesis by computing the cumulative force needed to incrementally insert a pseudopod (cylinder) into the network and push the fibers to its surface, for pseudopods aligned with and against the fibers (Fig. 1).

ACKNOWLEDGEMENTS

This work is supported by the NIH grant [RO1 EB005813].

REFERENCES

[1] Weiss, P, *J Exp Zool,* 100:353-386, 1945.
[2] Dickinson, RB et al., *Ann Biomed Eng,* 22(4):342-356, 1994.
[3] Dunn, G.A., *Cell Behaviour,* 247-280, 1982.
[4] Riching, KM et al., *Biophys J,* 107:2546–2558, 2014.
[5] Zaman MH et al., *Proc Natl Acad Sci USA,* 89(2):1389-1397, 2005.
[6] Chandran, PL, Barocas, VH, *J Biomech Eng,* 128(2):259-270, 2006.
[7] Lai VK et al., *J Biomech Eng,* 134(1):011004, 2012
[8] Lai VK et al., *J Biomech Eng, 135(7):71007, 2013*
[9] Hadi MF et al., *J Biomech Eng, 134(9):091005, 2012*

A COUPLED CHEMO-MECHANICAL CELL-MATRIX MODEL TO PREDICT MECHANICAL FEEDBACK BETWEEN CELLS AND EXTRACELLULAR MATRICES

F. Alisafaei (1), M. Hall (2), M. Wu (2), V. Shenoy (1)

(1) Department of Materials Science and Engineering
University of Pennsylvania
Philadelphia, PA, USA

(2) Department of Biological and Environmental Engineering
Cornell University
Ithaca, N□, USA

INTRODUCTION

Reciprocal mechanical interactions between cells and their surrounding extracellular matrices (ECMs) regulate various physiological and pathological processes such as cell motility, and tumor growth. Our recent study [1] reveals a mechanical cross-talk (feedback loop) between breast tumor cells and fibrous ECMs. Contractile cells pulling on their surrounding ECMs reorganize, align, and subsequently stiffen the fibrous collagen matrices in their immediate vicinity. In return, the matrix stiffening leads to greater cell force generation and cell stiffening. Here, we propose a novel nonlinear chemo-mechanical cell model, coupled with a fiber network-inspired continuum matrix model, to link the mechanics of the cell (stiffness and architecture) to its surrounding extracellular matrix.

METHODS

As explained in our previous work [1], we engineer an array of type 1 collagen matrices with different microstructures modulated by altering; (i) collagen concentration, (ii) polymerization temperature, and (iii) cross-linking via preglycation with ribose. As shown in Figure 1 (B), at the same collagen concentration, an increase in pore diameter is observed in collagen matrices when they are (i) polymerized at low temperatures, and (ii) preglycated with ribose. Using a parallel-plate oscillatory shear rheometer, we also measure the initial shear modulus (shear modulus at shear strain $\gamma \cong 0$), and strain stiffening (shear modulus at higher shear strains $0 < \gamma \leq 0.4$) of all nine collagen matrices. The average pore diameter and the initial shear modulus respectively range from 1.8 to 5.1 μm and 6 to 900 Pa in the collagen matrices shown in Figure 1 (B). Note that normal and malignant breast tissues have shear moduli of about 60 and 1300 Pa, respectively. To investigate the interaction between cells and matrices, MDA-MB-231 breast cancer cells are embedded within the collagen matrices (see Figure 1 (A)) covalently bonded to fluorescent marker beads. Using a 3D particle tracking microscopy from the laboratory of M.W. and coworkers [1], the 3D displacements of ~10,000 beads are measured in the vicinity of each cell.

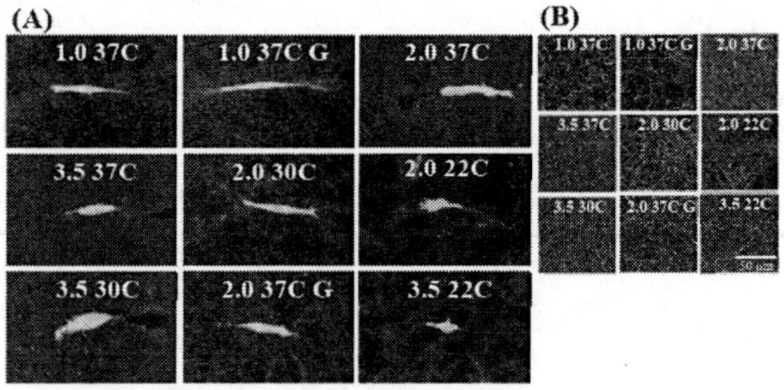

Figure 1: Modulating collagen architecture using gel concentration and cross linking strength. (A) Confocal images of cells (yellow) and collagen matrices (cyan). (B) Collagen matrices with various initial stiffness and strain stiffening. The first and second numbers at the top of each image represent collagen concentration (1.0, 2.0, and 3.5 mg/ml), and polymerization temperature (37, 30, and 22 °C), respectively, and G denotes preglycation.

RESULTS

Figure 2 depicts the experimentally measured displacement fields exerted by the cells within the collagen matrices. As reported in [1] and shown in Figure 2, collagen matrices with larger pore sizes exhibit greater fiber alignment accompanied by an increased range of displacement propagation. Furthermore, Figure 3 (A) shows that actin filaments are more aligned (along the cell's major axis) in stiffer matrices (2 mg/mL 30 °C and 3.5 mg/mL 22 °C) than in the softer matrix (1 mg/mL 37 °C) indicating that cell body stiffness increases with matrix stiffness (see also [1]); cells actively respond to their stiffened surrounding by alignment of f-actin filament along the long axis of the cell.

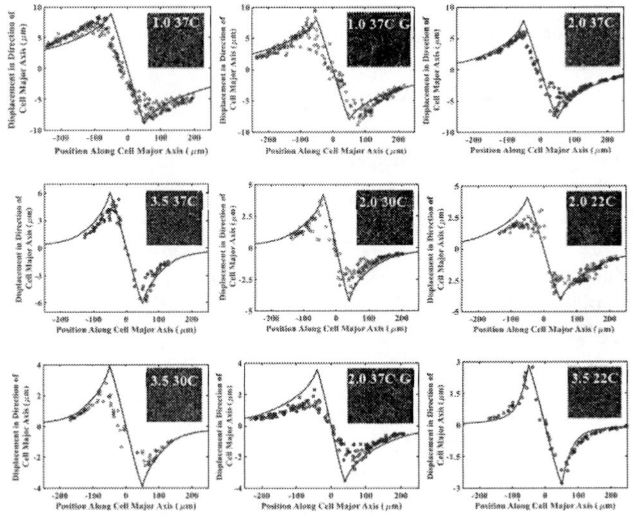

Figure 2: Cell induced collagen alignment promotes long range mechanical signal transmission. Blue circles and red solid lines represent the experimentally measured bead displacements and the predicted displacement fields from the cell-matrix model, respectively.

To capture the cell-matrix interaction, we propose a 3D chemo-mechanical cell model and couple it with the fiber network-inspired continuum matrix model of [1, 2]. The cell model is composed of a linear element connected in series to a nonlinear part. The linear part contains an active component (representing the cell actomyosin network) that acts in parallel with a linear spring [3]. As the nonlinear part of the cell model, we use a highly nonlinear spring to represent the actin filament network (shown in Figure 3(A)) and its stiffening with matrix stiffness. To train the cell model and uniquely determine the chemo-mechanical parameters, we use the displacement field of 1 mg/mL 37 °C collagen matrix (or any of the nine displacement fields of Figure 2) together with the stiffness data of [4]. In other words, the cell fitting parameters are determined such that (i) the resulting displacement field from the cell-matrix model captures the displacement field experimentally measured in 1 mg/mL 37 °C collagen matrix, and (ii) the average cell stiffness calculated from the model matches the cell stiffness data experimentally measured from atomic force indentations of the same cell type [4].

Figure 2 illustrates that the coupled cell-matrix model can successfully predict the displacement fields experimentally measured from the 3D particle tracking microscopy. Maximum principal stress σ_{max} is plotted in Figure 3 (B) for a polarized cell embedded within the stiffest (3.5 mg/mL 22 °C) and the softest (1 mg/mL 37 °C) collagen matrices. Figure 3 (B) depicts that σ_{max} (representing the actin fiber network of Figure 3 (A)) is aligned along the cell's polarization axis and significantly increases with increasing matrix stiffness. Furthermore, our coupled nonlinear cell-matrix model shows that σ_{max} is more tensile near the cell's tips as depicted in Figure 3 (B). We also compute cell-generated forces in all nine collagen matrices. We find that cell-generated force along the cell's major axis increases with initial collagen matrix stiffness as shown in Figure 3 (C). It has been reported that cells, including MDA-MB-231 breast cancer cells, promote more traction forces with increasing substrate stiffness [5].

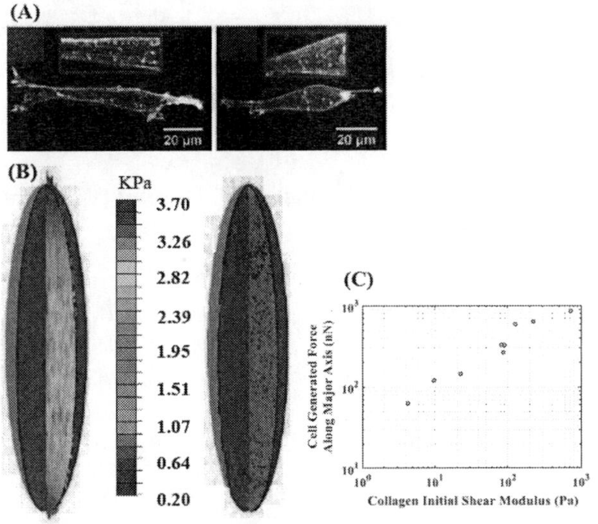

Figure 3: Cell body stiffness and cell force generation increase with matrix stiffness. (A) Actin fiber alignment along the cell's major axis in 2 mg/mL 30 °C collagen matrix (left), and a diffuse actin filament network without a preferential alignment direction in 1 mg/mL 37 °C collagen matrix (right). (B) Maximum principal stress values and their directions in 3.5 mg/mL 22 °C (left) and 1 mg/mL 37 °C (right) collagen matrices. (C) Cell-generated forces measured from the cell-matrix model; stiffer collagen matrices promote greater cell-generated forces.

DISCUSSION

We propose a coupled cell-matrix model to explain the physical mechanism that cells use to interact with their surrounding matrices. By implementing the constitutive model into a 3D finite element framework, we qualitatively predict (i) displacement fields induced by pulling cells within their surrounding matrices, (ii) cell-generated forces, actin fiber alignment, and cell stiffening as a function of matrix stiffness, and (iii) strain-induced fiber alignment of fibrous matrices and its associated long-range displacement propagation.

ACKNOWLEDGEMENTS

The theoretical part of the work was supported by National Cancer Institute Grants U01CA202177 and U54CA193417 (to V.B.S.), NIH Grant R01EB017753 (to V.B.S.), and National Science Foundation Grant 1548571 (to V.B.S.). The experimental part of the work was mainly supported by National Center for Research Resources, NIH Grant R21RR025801 (to M.W.), and National Institute of General Medical Sciences, NIH Grant R21GM103388 (to M.W.).

REFERENCES

[1] Hall, M et al., *Proc Natl Acad Sci USA*, 113:14043-14048, 2016.
[2] Wang, H et al., *Biophys J*, 107: 2592-2603, 2014.
[3] Shenoy, V et al., *Interface Focus*, 6: 20150067, 2016.
[4] Staunton, J et al., *Sci Rep*, 6:19686, 2016.
[5] Kraning-Rush, C et al., *PloS One*, 7:e32572, 2012.

IDENTIFICATION OF TRPV4 AS A PRESSURE MECHANOSENSOR IN THE DEVELOPING LUNG

Joshua T. Morgan (1), Wade G. Stewart (1), Jason P. Gleghorn (1)

(1) Biomedical Engineering
University of Delaware
Newark, DE, USA

INTRODUCTION

Branching morphogenesis of the lung is highly regulated and stereotyped, leading to an architecture that is conserved within a given species and optimized for gas exchange [1]. The lung develops within a dynamic mechanical environment; liquid is pushed throughout the airways via periodic peristaltic contractions of airway smooth muscle causing transient expansion of the distal tips. Clinical observations and animal models suggest a critical role for the (dynamic) regulation of pressure in the lung, with dysregulation resulting in an underbranched airway known as lung hypoplasia. Further research quantifying the role of physical forces such as pressure, as well as connecting these physical events to cellular mechanotransduction systems, is an essential step towards a more complete understanding of lung morophogenesis.

Whereas many mechanotransduction systems have been identified to date, and several are likely to play a role in development, a review of literature suggests a prime candidate is transient receptor potential cation channel subfamily V member 4 (TRPV4). TRPV4 is a calcium permeable cation channel that has been previously implicated in stretch mechanotransduction in multiple cell and tissue systems [2]. Importantly, peristaltic airway smooth muscle contractions dynamically move fluid within the airways that distend the distal airway branches and provide a stretch stimulus to tissue [1]. Indeed, TRPV4 is expressed in mature lung and has been implicated in mediating the response to ventilator-induced barotrauma [3]; however, TRPV4 has been minimally investigated in fetal lungs [4] and has not been investigated for its role as a pressure transducer in development.

To directly quantify the role of pressure and provide a culture model for studies of mechanotransduction in lung development, we developed a microfluidic "chest cavity" suitable for the culture of embryonic murine lungs *ex vivo* and used it to quantify the impact of pressure on development. Further, to elucidate the role of TRPV4 in lung development, we employed pharmacological agents to activate or inhibit TRPV4 signaling in *ex vivo* culture to delineate its role as a mechanosensor in the developing lung.

METHODS

Microfluidic Culture: We designed microfluidic "chest cavities" that enable us to apply dynamic control of pressure during the *ex vivo* culture of embryonic murine whole lung explants (Figure 1). Lung explants were intubated on a pulled glass microneedle connecting two fluidic chambers - one that surrounded the organ (pleural chamber) and one that was continuous with the airways via the intubated trachea (lumenal chamber). Pressure was regulated in both chambers to result in a transmural pressure difference across the lung, ΔP ($\Delta P = P_{lumen} - P_{pleural}$). Timed pregnant CD-1 mice were humanely euthanized using CO2 asphyxiation under an Animal Use Protocol approved by the University of Delaware IACUC. Lung explants (embryonic day 12; E12) were isolated and cultured for 48 h at controlled transmural pressures. Following culture, lungs were fixed and stained for E-cadherin to mark the airways.

Role of TRPV4: Western blotting and qPCR was performed to investigate expression levels of TRPV4 in the murine lung, while whole mount imaging was utilized to determine localization of TRPV4 in the lung. In order to determine the role of TRPV4 signaling in lung development we employed pharmacological agents to activate (GSK1016790A) or inhibit (GSK205) TRPV4. Lung explants were cultured in the presence of the pharmacological agents for 24 h. Following culture, lungs were stained for markers of the airway (E-cadherin), smooth muscle (αSMA), and vasculature (PECAM) to assess the impact of TRPV4 modulation of lung morphology.

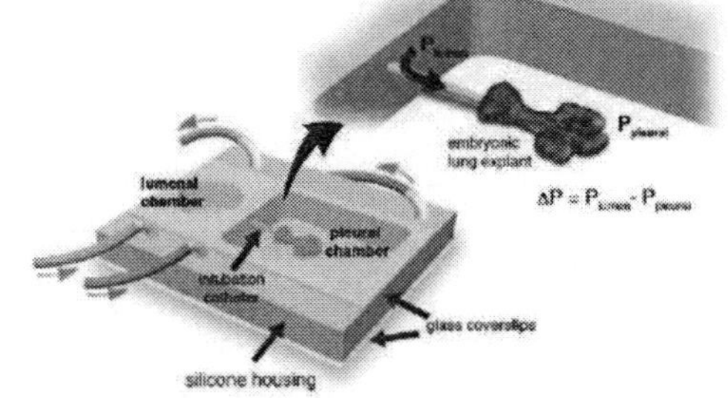

Figure 1: Microfluidic "chest cavity" culture model of lung development. Embryonic murine lungs can be cultured *ex vivo* under defined pressures.

RESULTS

Utilizing the microfluidic culture, lumenal and pleural pressures were regulated and airway architecture was imaged in real-time over 48 hours of culture. We have demonstrated that airway branching events throughout the lung are synchronized and the timing of branching events is accelerated by transmural pressure (Figure 2). Furthermore, the hierarchical branching pattern of individual explant cultures closely mapped the *in vivo* architecture, regardless of pressure, demonstrating that transmural pressure regulates the rate of airway morphogenesis and not the sites of branching.

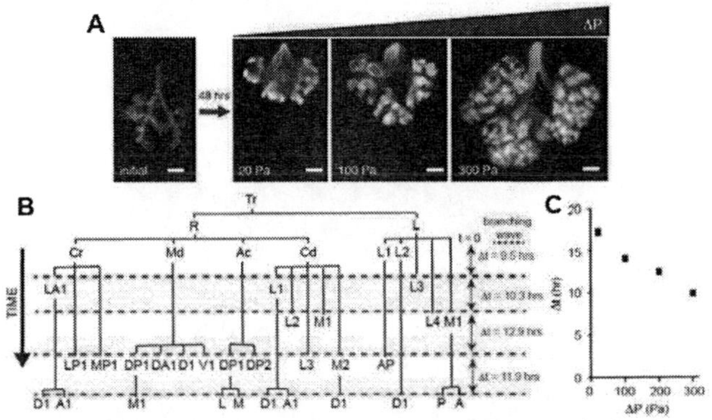

Figure 2: (A) Pressure increases the rate of lung growth. (B) Branching occurs at regularly spaced time intervals (Δt) in cultured murine lung explants. Letters refer to specific lung branches as defined by Metzger and colleagues [5]. (C) The length of this interval is regulated by the transmural pressure (ΔP). n = 4 explants/press. Error bars smaller than the markers.

TRPV4 expression was confirmed in E12 and E13 lungs at the protein and transcript levels and whole mount immunofluorescence revealed strong TRPV4 localization to the epithelium, subepithelial mesenchyme, and major vasculature. Pharmacological inhibition of TRPV4 resulted in decreased smooth muscle, near-complete ablation of the lung vasculature, and decreased airway branching (Figure 3A). Conversely, TRPV4 activation led to increases in the vascular density throughout the lung as well as prominent increases in smooth muscle (Figure 3B). Changes in airway morphology were quantified by counting the fold change in terminal buds over the culture period (Figure 3C). Vascular and smooth muscle morphology was quantified as percent stain coverage of the projected lung (Figure 3D-E).

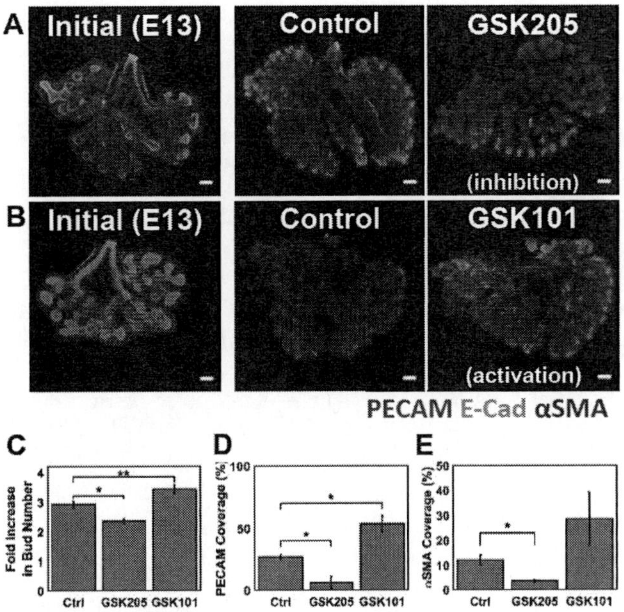

Figure 3: TRPV4 plays a critical role in lung development. (A) Inhibiting TRPV4 using 10 µM GSK205 results in dramatic decreases in vascular (red) and smooth muscle (blue) staining as well as decreased airway (green) branching compared to control cultures. (B) TRPV4 activation using 100 nM GSK1016790A results in increased vascular and smooth muscle staining. (C-E) Quantification of results. Scale bars are 50 µm.

DISCUSSION

In this work, we have demonstrated a novel culture model, allowing for quantitative studies of pressure during lung development, and utilized it to determine the role of pressure on the branch timing, but not organization. Further, we have identified a potential mechanosensor, TRPV4. Our studies demonstrate that TRPV4 activity has profound impact on the morphogenesis of the airway, vasculature and airway smooth muscle. Previous work has strongly implicated smooth muscle mediated stretch at the distal tips as a key mechanical cue in branching. Our work suggests TRPV4 may mediate the cellular response to these events, likely through Ca^{2+} signaling known to be important in lung development [1]. Continued studies will further define the pathways that underpin the linkages between molecular, genetic, and physical mechanisms of lung development.

ACKNOWLEDGEMENTS

Funding was provided by National Science Foundation (1537256), a Basil O'Connor Starter Scholar Award from the March of Dimes, and the National Institutes of Health (1P30GM110759-01).

REFERENCES

[1] D. Warburton *et al.*, *Curr. Top. Dev. Biol.* **90**, 73–158 (2010).
[2] W. Liedtke, *Handb. Exp. Pharmacol.*, 473–487 (2007).
[3] K. Hamanaka *et al.*, *Am. J. Physiol. Lung Cell. Mol. Physiol.* **299**, L353-362 (2010).
[4] P. S. Nayak *et al.*, *Respir. Res.* **16**, 60 (2015).
[5] R. J. Metzger, O. D. Klein, G. R. Martin, M. A. Krasnow, *Nature.* **453**, 745–750 (2008).

THE INFLUENCE OF MATRIX STIFFNESS ON DIRECTED CELL MIGRATION IN ALIGNED FIBROUS MICROENVIRONMENTS

William Y. Wang and Brendon M. Baker

Department of Biomedical Engineering
University of Michigan
Ann Arbor, MI, USA

INTRODUCTION

Studies performed on flat surfaces coated with extracellular matrix (ECM) proteins have provided a foundational understanding of the biology of cell migration, a process essential to growth and development, tissue homeostasis, and cancer metastasis [1]. While these studies have identified biochemical factors that trigger migration, elucidated migration-specific signaling pathways, and generated a paradigm of the cell migration cycle, it is now widely accepted that in many physiologic settings, the physical structure of the surrounding ECM critically defines diverse cell migration mechanisms and strategies [2]. Previous work correlated ECM fibril alignment and the direction of neural crest cell migration from the neural tube during early development, implicating contact guidance provided by the ECM's physical structure in developmental migration events [3]. More recently, the surrounding stroma of breast tumors has been shown to undergo marked remodeling and alignment, and tumor cell escape and intravasation is thought to be facilitated by collagen fibril alignment as observed in *in vitro* tumor explant models using collagen gels [4]. In native settings such as the tumor stroma, as well as in purified biopolymers such as collagen gels, mechanical properties such as a modulus or stiffness are integrally linked to fiber density and alignment. Indeed, tumors and the adjacent stroma are known to simultaneously stiffen and become increasingly organized with cancer progression. Towards isolating how individual features of the ECM influence cell migration, we implemented a recently established synthetic material system that affords independent control over several mechanical and architectural features of the ECM [5]. In particular, we investigated how cell migration is influenced by matrix stiffness in aligned fibrous matrices that model the metastatic breast tumor stroma, identifying a new role for cell force-mediated matrix deformation in promoting rapid migratory events.

METHODS

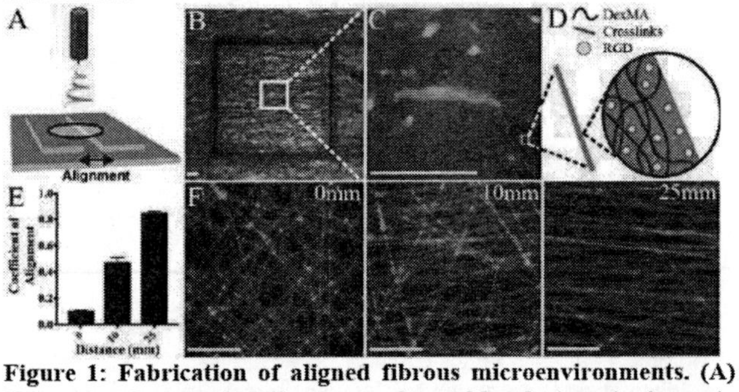

Figure 1: Fabrication of aligned fibrous microenvironments. (A) Networks of suspended fibers were formed by electrospinning onto microfabricated PDMS substrates seated on a platform containing parallel electrodes. (B) Brightfield image of DexMA fibers suspended over a microfabricated square well. (C) Fluorescent image of a 3T3 fibroblast on an aligned fiber network. (D) Arginylglycylaspartic acid (RGD) is coupled to the material to enable cell adhesion and migration. (E) Coefficient of alignment as a function of electrodes spacing. (F) Matrices formed with electrode distances of 0, 10, and 25mm. Scale bars: 100μm.

Fiber network fabrication: To generate suspended 3D fiber networks, methacrylated dextran (DexMA) was dissolved at 0.45 g/ml in a 1:1 mixture of water:dimethylformamide with 0.005% Irgacure 2959 photoinitiator and electrospun at 25^{0}C and 35% relative humidity in an environment-controlled glovebox (Fig. 1A-D). Samples were stabilized by a primary crosslinking under UV light (100mW/cm^2), hydrated in a solution of 1mg/ml Irgacure 2959 photoinitiator and exposed to varying durations of UV light to control the degree of crosslinking and resulting stiffness [5]. Microfabrication: Poly(dimethylsiloxane) (PDMS) substrates fabricated by soft lithography possessed arrays of square wells that suspend electrospun fibers, removing the mechanical influence of the underlying glass surface (Fig. 1). Cell culture: NIH-3T3 fibroblasts were cultured in high-glucose DMEM containing 10% bovine serum, 1% penicillin/streptomycin, and 1% L-glutamine. Cell nuclei were labeled with 3μg/ml Hoechst33342 for migration studies. Microscopy: Timelapse images were collected at 10 minute intervals over 8 hours at 10x magnification on a Nikon Eclipse Ti with custom incubation chamber. The positions of cell nuclei were tracked with custom Matlab software. Statistics: Significance was determined by ANOVA and data is presented as a mean ± standard deviation. *p-value < 0.05.

RESULTS

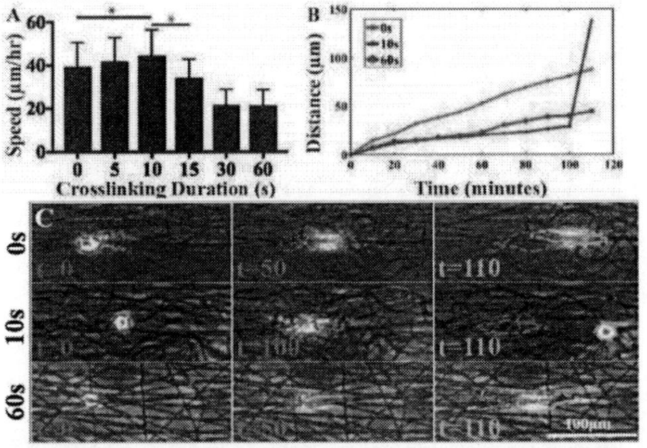

Figure 2: Matrix stiffness influences cell migration. (A) A biphasic relationship exists between cell migration speed and fibrous matrix stiffness. (B) Cell migration distance trajectories for cells migrating on aligned fibers of crosslinked for (C) 0s (top), 10s (middle), and 60s (bottom). Inset labels indicate time in minutes.

Control over fiber alignment of suspended networks was achieved by spacing parallel electrodes at the collecting surface (Fig. 1A, E). Subsequent studies were conducted with the most aligned matrices (0.8 coefficient of alignment), generated from a 25mm spacing (Fig. 1F). Holding other parameters constant (fiber diameter, density, and cell-adhesive ligand density), UV crosslinking duration was modulated to control the resulting stiffness of aligned fiber matrices. Quantifying 3T3 fibroblast migration over an 8h period via timelapse imaging, we observed that migration speed overall was slower at higher matrix stiffnesses (Fig. 2A). Interestingly, peak migration speeds were achieved at an intermediate degree (10s) of crosslinking. Cells in aligned fiber matrices demonstrated two distinct migration mechanisms. The more common mode, observed across all stiffnesses investigated, was steady migration where cells maintained a constant speed (Fig. 2B red and purple lines, Fig. 2C top and bottom). At intermediate stiffnesses, these periods of steady migration were interrupted by less continuous migration that involved the cells' ability to deform and reorganize its local microenvironment by recruiting nearby fibers towards the cell body (Fig. 2C middle). While limited net movement occurs during the matrix reorganization phase, upon release of the cell's trailing edge, the stretched matrix snapped back to its previous configuration and the cell was propelled forward in the direction of fiber alignment (Fig. 2B). This matrix-mediated migration mechanism did not occur in stiffer matrices where no deformations were observed.

The cell's contractile machinery is critical in migration and in generating traction forces that deform the local extracellular environment [1], so we next tested whether myosin and ROCK inhibition influenced migration in this setting. A 10μM dose of blebbistatin significantly decreased cell migration speed on softer matrices (10s crosslinking), whereas speed on stiff matrices (60s crosslinking) significantly decreased only at a higher dose of 30μM (Fig. 2A). Treatment with 10μM of Y-27632 significantly lowered migration speed on soft and stiff fibers (Fig. 2B).

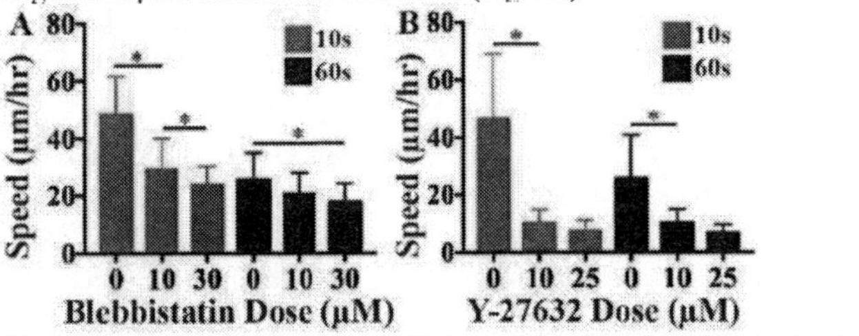

Figure 3: Actomyosin contractility is required for fast directed cell migration in aligned ECMs. (A) Cell migration speed with increasing concentrations of blebbistatin (A) and Y-27632 (B) on aligned fibrous matrices crosslinked for 10s and 60s.

DISCUSSION

We employed a synthetic material system to model fibrous ECM, while orthogonally defining mechanical properties and architectural features that may be important factors in cell migration. Within aligned fibrous matrices, we identified maximum cell migration speeds at an intermediate stiffness, supporting findings with smooth muscle cell and neutrophil migration on elastic ECM-coated polyacrylamide gels [6, 7]. These studies suggested that focal adhesion growth/disassembly are regulated by stiffness, and that intermediate stiffnesses yield optimal adhesion turnover for maximum migration. Differences in focal adhesion turnover remain to be characterized in this system, however our study suggests a novel role for cell-mediated matrix reorganization during directed cell migration. As in other ECMs, matrix stiffness defines the deformations resulting from cell forces. However, in aligned fibrous contexts, the elastic energy generated by this deformation can profoundly affect cell migration, allowing cells to achieve effective migration speeds of 348μm/hr (5.8μm/min) which is over five-fold greater than values previously reported [8]. Supporting a dependency for actomyosin-generated contractility in this phenomenon, inhibition of myosin and ROCK activity abrogated matrix deformations, diminishing the frequency of these rapid migratory events and reducing migration speeds to similar levels as stiff matrices where deformations were absent. These results suggest the cell's ability to physically reorganize its local ECM is critical in migration through fibrous ECMs. Given the ubiquity of fibrous tissues within the body, an understanding of how matrix structure and mechanics guide cell migration could improve approaches to promote cell migration to wound sites or inhibit metastatic cells from invading the surrounding tumor stroma.

ACKNOWLEDGEMENTS

This work was supported by a University of Michigan Rackham Merit Fellowship Program and the NIH (K99HL124322).

REFERENCES

[1] Ridley, AJ et al., Science, 302:1704-1709, 2003. [2] Charras, G and Sahai E, Nat Rev Mol Cell Biol, 15:813-824, 2014. [3] Lofberg, Jan et al., Dev Biol, 75:148-167, 1980. [4] Provenzano, PP et al., BMC Med, 4:38, 2006. [5] Baker, BM et al., Nat Mater 14:1262-1268, 2015. [6] Peyton, SR et al., J Cell Physiol, 204:198-209, 2005. [7] Stroka, KM et al., Cell Motil Cytoskeleton, 66:328-341, 2009. [8] Doyle, AD et al., J. Cell Biol, 184:481-490, 2009.

DEVELOPMENT OF THREE-DIMENSIONAL SOFT MATERIALS ELASTOGRAPHY BASED ON MAGNETIC RESONANCE IMAGING AND TOPOLOGY OPTIMIZATION

Luyao Cai (1), Claus B.W. Pedersen (2), Corey P. Neu (1,3)

(1) Weldon School of Biomedical Engineering
Purdue University
West Lafayette, IN, US

(2) Dassault Systèmes Deutschland
GmbH, Hamburg, Germany

(3) Department of Mechanical Engineering
University of Colorado Boulder
Boulder, CO, US

INTRODUCTION

Magnetic resonance imaging (MRI) enables the nondestructive imaging of internal tissue structure, in addition to other physical phenomena like flow and diffusion. One specialized MRI technique, termed displacements under applied loading by MRI (dualMRI), was developed to measure displacements and strain in musculoskeletal tissues, hydrogels and engineered constructs. However, deformation information does not directly describe spatial distributions of tissue properties (e.g. stiffness), which is critical to the understanding of disease progression [1,2]. In the pathogenesis of osteoarthritis, the stiffness of articular cartilage decreases [3], as measured using explanted tissues and invasive methods. While ideally MRI would assess cartilage stiffness noninvasively, traditional magnetic resonance elastography (MRE) is not possible due to the dissipation of high frequency shear waves in the human body, which are required to measure properties in thin and stiff cartilage, and deep within articular joints [4]. To avoid these limitations, we aimed to develop and validate a three-dimensional (3D) inverse modeling workflow that combined dualMRI, to directly measure intratissue deformation, with topology optimization in the application of heterogeneous (layered) materials representative of the complex gradient architecture of articular cartilage. The significance and potential of this approach was highlighted for the description of tissue degeneration, repair and complex material properties.

METHODS

Topology optimization, which is a non-parametric optimization and design method, is widely used to design stiff, durable and light-weight structures [5]. Here, we applied topology optimization to determine the stiffness distribution of layered agarose materials as a model for articular cartilage. Instead of maximizing the structure stiffness, as is used in traditional applications, we minimized the absolute difference between material deformation and the deformation of a base model [5]. Generally, the workflow was solved using finite element method software (Abaqus) [6] and optimization software Tosca (Dassault Systèmes) [7], to iteratively update the stiffness of each element representing the material.

To first validate our workflow in two dimensions, ideal displacements were derived from forward modeling in Abaqus and used for stiffness calculation (Figure 1). A bi-layer cylindrical model was created (top/bottom layers: E=1000Pa/500Pa), such that displacement patterns calculated were consistent with MRI data [8]. With the bottom edge of the model fully constrained, the top edge was indented to 15% of the thickness, and resulting displacements were calculated, with linear elasticity and plane stress assumptions.

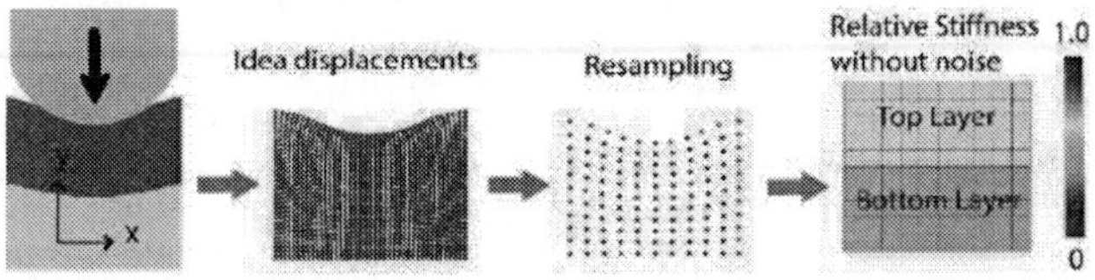

Figure 1. Validation of inverse modeling using known displacements calculated from forward simulation in Abaqus. The displacement data of a two-layer cylinder (with stiffness ratio top: bottom=2:1), with no noise added, restored the relative stiffness to known values.

The workflow bias and precision of stiffness measures was evaluated using Monte Carlo simulations (Figure 2). The pixel number, model dimensions, and data noise (stdev=0.1mm), were set according to displacement-encoded MRI data [8]. Simulations were repeated 100 times with random noise, and bias and precision were calculated as the root mean square error and pooled standard deviation of stiffness values, respectively, of each element [9].

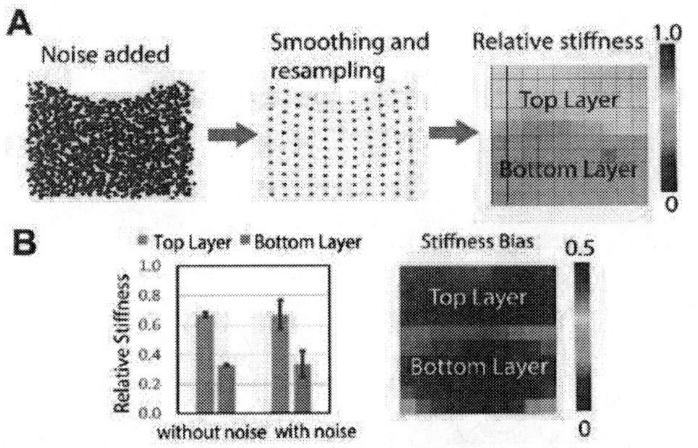

Figure 2. Noise typical of displacement-encoded MRI data introduced heterogeneity in stiffness distribution, but regional patterns are robust to noise. (A) Known data with added noise was smoothed and resampled, and relative stiffness (i.e. E1:E2=2:1) with noise was calculated. (B) Monte Carlo simulations allowed us to estimate error of the technique.

To further validate our workflow, dualMRI was used to measure internal displacements and strain within layered agarose gels to establish baseline deformation and noise levels commonly observed in small materials or explanted tissues [8] (Figure 3). Briefly, finite displacements were determined using dualMRI [2] under cyclic compressive loading (0.23 Hz, 0.67N) with a DENSE-FISP imaging sequence on both 7T and 14.1T MRI systems (Bruker). The relative stiffness was calculated in Figure 3.

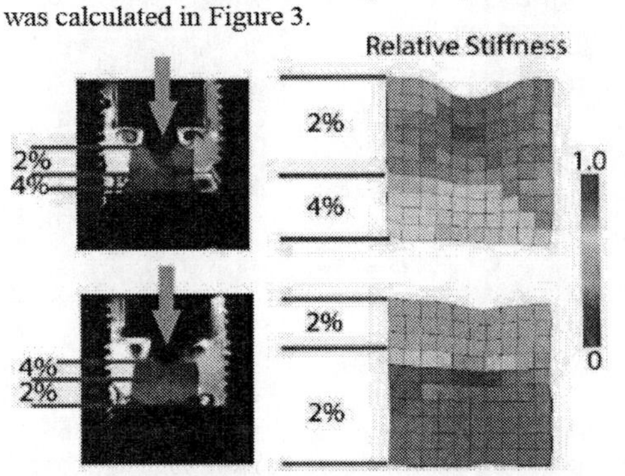

Figure 3. Relative stiffness calculated from experimental displacement data of two bilayer agarose gel models.

A sensitivity analysis (data not shown) identified the 2D assumption as the most important factor, which indicated the deficiency of the 2D assumption and the importance of the out-of-plane strains measurements. In order to correct the bias of this factor, a validation study was completed in three dimension using forward simulation displacements from Abaqus (Figure 4). Similar with the two dimensional model, a bi-layer cylinder was indented coaxial or non-coaxial to consider malalignments observed in experiments.

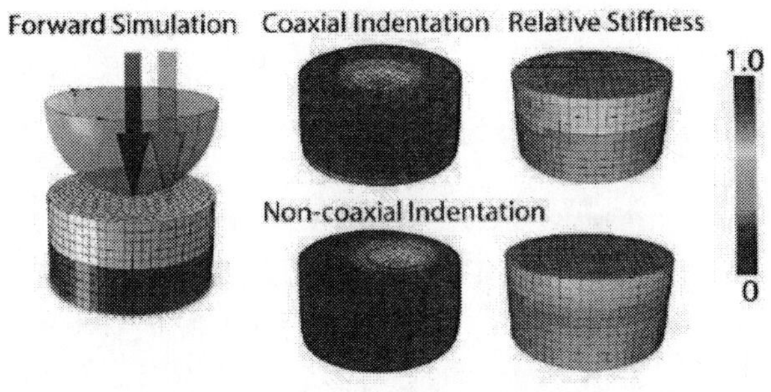

Figure 4. Validation of inverse modeling in three dimensions using "ideal" displacements from forward simulation in Abaqus.

RESULTS

Our inverse modeling approach restored original stiffness patterns from forward simulations (Figure 1). Monte Carlo simulations identified the stiffness ratio between top/bottoms material layers to be 2.019, which was very close to the known value of 2.0 (Figure 2). The bias and precision of stiffness values were 0.092 and 0.066, respectively. The inverse modeling was also sensitive to boundary geometries and smoothing process. The validation results of three dimensions showed the bi-layer configuration was restored both in coaxial and non-coaxial indentation.

DISCUSSION

The purpose of this study was to define a workflow to enable inverse modeling-based stiffness calculations that take advantage of displacement-encoded MRI data. Our inverse modeling approach produced a distinctive bilayer configuration of stiffness ratio 2:1 between two material layers, which was robust to noise. Some systematic bias persisted (Figure 2B), especially near material interfaces and edges. The interface bias could be due to the smoothing process and regularization constraints that blur out the gradient.

To improve our model under complex out-of-plane deformation, a 3D cylinder model was created and its stiffness configuration was successfully restored. The non-coaxial loading showed the ability of the algorithm to deal with non-ideal loading conditions. We noticed that the interface between two layers in the non-coaxial case was slightly blurred due to the filter applied. This effect can be relieved if more nodal displacements were input as design responses.

In summary, our inverse modeling workflow, which was built on displacement-encoded MRI, can provide a more intuitive result of the stiffness distribution inside the tissue during conditions of large deformation both in 2D and 3D. For articular cartilage, inverse modeling may be useful to analyze softening observed in osteoarthritis or provide a unique imaging biomarker for tissue repair.

REFERENCES

[1] Chan DD, et al., *Sci Rep*, 6:19220, 2016.
[2] Neu, CP, *Osteoarthritis and Cartilage* 22.10: 1349-1359, 2014.
[3] Waldstein W, et al., *J Orthop Res* 34(1):135-140, 2016.
[4] Hardy P, et al. *Magn Reson Med*, 53.5: 1065-1073, 2005.
[5] Bendsøe M and Sigmund O, Springer-Verlag, Berlin, 2003.
[6] Abaqus Manual, Dassault Systèmes, www.simulia.com
[7] TOSCA Structure, Dassault Systèmes, www.simulia.com.
[8] Griebel A, et al., *Comput Meth Biomech Biomed Eng*, 47.9:2149-2156, 2014.
[9] Chan DD, et al., *Comput Meth Biomech Biomed Eng*, 16.8:852-860, 2013.

ROLE OF PRICELLULAR MATRIX IN MODULATING CHONDROCYTE STRAINS IN HEALTHY AND OSTEOARTHERITIC CARTILAGE

Mehdi Khoshgoftar[1,2], Peter A. Torzilli[1], Suzanne A. Maher[1,2]

[1]Orthopaedic Soft Tissue Research Program
[2]Department of Biomechanics
Hospital for Special Surgery
New York, New York, United States

INTRODUCTION

Osteoarthritis (OA) is a degenerative joint disease characterized by pain, disability, and progressive changes in joint tissue. Altered mechanical properties of articular cartilage at the macroscale-tissue level and microscale-cellular levels have been quantified at all stages of OA [1]. In particular, it has been noted that the elastic modulus and permeability of the extracellular matrix (ECM) and the pericellular matrix (PCM) are inhomogeneous and change significantly with OA [1]. But the effect of changes in matrix properties on the micro-mechanical strains experienced by the cell and their resulting biological response are not understood.

The objective of this study was to understand how the relationship between the spatial variation in the material properties of ECM and PCM in healthy and OA cartilage affects cell deformation [1]. To satisfy this objective we built a multi-scale, biphasic, swelling, fiber-reinforced, nonlinear computational model that incorporates healthy and OA-like macroscale and microscale properties. Our hypothesis was that the role of the PCM in modulating cell deformation will be changed in OA vs. healthy cartilage.

METHODS

An axisymmetric multi-scale finite element model was generated in Abaqus v6.12 (Dassault Systemes, 2015), **Fig. 1**. Macroscale Level Model: A cartilage explant (radius 2.8 mm; height 1 mm) was modeled with healthy and OA-like ECM content and mechanical properties [2-3]. The base of the explant was restricted from displacements in all directions, to represent cartilage attachment to the subchondral bone. A nominal 10% compression was applied in 0.5 seconds to the explants, resulting in a reaction force of 286 N at the loading platen. Microscale Model: The boundary conditions of the microscale model were derived from the solution of the macroscale model. The microscale model included a chondrocyte located in the middle zone of the cartilage explant, surrounded by PCM and ECM.

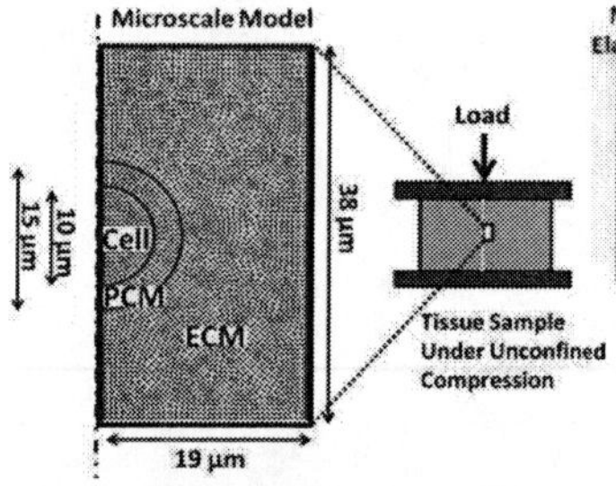

Figure 1: Multi-scale finite element model of cartilage that links tissue level unconfined compression loading to cell-level deformations.

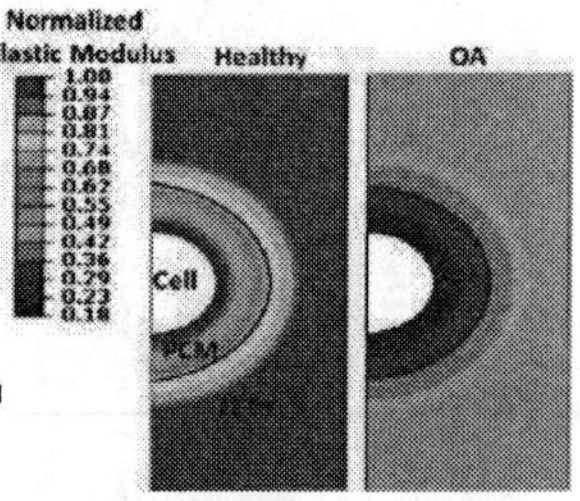

Figure 2: Variation in elastic modulus (normalized to that of healthy ECM) from PCM inner edge to ECM.

A validated, composition-based, fiber-reinforced, poroviscoelastic, swelling cartilage model that accommodates for effects of collagen reinforcement and swelling by proteoglycans in ECM and PCM was used. The governing stress equation is [2]:

$$\boldsymbol{\sigma}_{tot} = -\mu^{f}\mathbf{I} + \frac{n_{s,0}}{J}\left(\left(1 - \sum_{i=1}^{totf} n_{col}^{i}\right)\boldsymbol{\sigma}_{pgm} + \sum_{i=1}^{totf} n_{col}^{i}\,\boldsymbol{\sigma}_{col}^{i}\right) - \Delta\pi\mathbf{I}$$

where μ^f is the water chemical potential, $\boldsymbol{I}$ the unit tensor, $n_{s,0}$ the initial solid phase volume fraction, J the volumetric deformation, $\boldsymbol{\sigma}_{pgm}$ the stress in the non-fibrillar PG's solid matrix substrate, σ^i_{col} the fibril stress in the *ith* fibril direction, n^i_{col} the volume fraction of the collagen fibrils with respect to the total solid volume, *totf* the total number of the fibrils and $\Delta\pi$ the osmotic pressure gradient. For further details of the material model the reader is referred to [2].

Material properties and content of healthy and OA chondrocyte, ECM and PCM were derived from the literature [2-3]. ECM content and elastic modulus of the solid substrate of Grade 3 OA cartilage were modeled to be 40% and 55% less than those of healthy cartilage, respectively. Permeability of OA cartilage was modeled to be twice that of healthy cartilage (1.767×10^{-17} m^4/N s). At the cellular level, the chondrocyte was modeled as a homogeneous biphasic material [3] with elastic modulus of 0.7 kPa, Poisson's ratio of 0.4, solid fraction of 0.17, and permeability of 0.5×10^{-14} m^4/N s [3]. Based on recent AFM elastic mapping of the PCM and ECM around chondrocytes [1], the elastic modulus of the PCM and ECM was increased from the inner edge of the PCM towards the far field ECM (**Fig. 2**).

RESULTS

Higher chondrocyte and PCM strains occurred in OA tissue compared to healthy tissue: peak axial compressive strain increased from 24% to 32% peak lateral tensile strain increased from 12% to 17% (**Fig. 3**), and peak shear strain from 18% to 24% (**Fig. 4**). The distribution of strain inhomogeneities throughout the microscale model were consistent for healthy and OA tissue for axial and lateral strain. Specifically, (i) compressive and tensile strains were amplified within the chondrocyte, (ii) the intersecting boundary region between the chondrocyte and PCM experienced the highest strains, and (iii) the central region of the chondrocyte experienced less compressive and tensile strains compared to the periphery of the chondrocyte.

The distribution of shear strain inhomogeneities was different between healthy and OA tissue: Peak shear strains in healthy cartilage were concentrated within the PCM region. In OA cartilage, part of the internal lateral periphery of the chondrocyte also experienced high shear strains of 18%.

Of note, across all models, PCM axial tensile strains of up to 2% occurred, which was attributed to the internal stretch induced by the internal swelling pressure of the tissue.

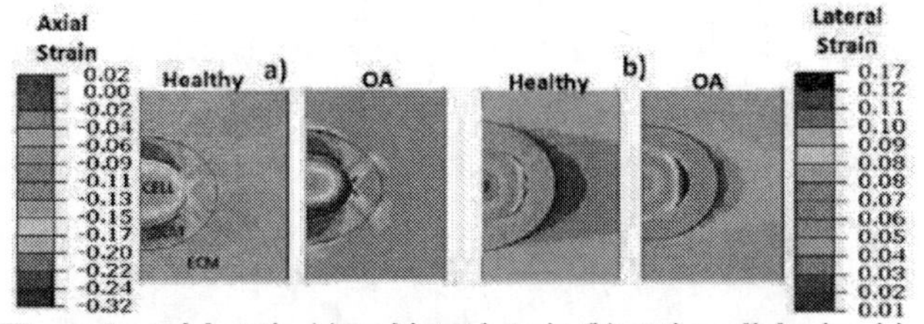

Figure 3: Axial strain (a) and lateral strain (b) at the cellular level in healthy and OA cartilage.

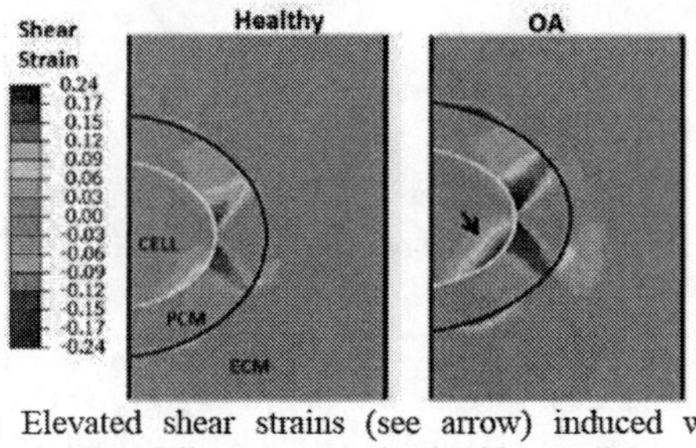

Figure 4: Elevated shear strains (see arrow) induced within the chondrocyte in OA cartilage compared to healthy cartilage.

DISCUSSION

By way of a multi-scale, biphasic, swelling, fiber-reinforced, nonlinear computational model, we found that chondrocytes in OA cartilage are exposed to increased mechanical deformations compared to that in healthy cartilage. The distribution of *axial* and *lateral* strain throughout the cell-PCM-ECM complex was not affected by simulating OA-like microscale properties. However *shear* strains distributions were different between health and OA cartilage. Specifically, peak shear strains were concentrated within the PCM region in healthy cartilage, while in OA cartilage the internal lateral periphery of the chondrocyte also experienced high shear strains.

There are limited reports in literature about the effects of distortional deformations on the biological response of chondrocytes. For example, it is reported that increased shear strains applied to isolated human chondrocytes result in increased glycosaminoglycan synthesis, longer chondroitin sulfate glycosaminoglycan chains and elevated expression of interleukin-6 (markers typical of osteoarthritic cartilage) [4-5]. To truly understand the mechano-biological response of chondrocytes, the complexity of the micro-mechanical environment of the cell must be defined. Our multiscale model incorporated spatially varying macrostructural and microstructural mechanical properties of the PCM and ECM to reflect both healthy and OA cartilage. Consistent with previous studies [3], we found that the PCM amplifies ECM compressive and tensile strains. However our study also indicated that shear strains are absorbed by the PCM in healthy cartilage, which helps to shield the chondrocytes from distortional deformations. Yet, in OA cartilage, the lower modulus and higher permeability of the PCM resulted in elevated shear strains within the chondrocytes. It is possible that these elevated strains lead to disrupted mechano-biological signals to the cell and lead to further tissue disruption.

In summary, our computational models allow for a more detailed analysis of the strain distribution within the cell-PCM-ECM complex, and with the help of parallel physical experiments, will help to unearth the mechano-biological mechanisms of cartilage degradation as a function of stage of OA. Models incorporating sub-cellular components, such as the nucleus and cytoskeletal fibers, may provide further insights into the role of extra and intracellular mechanotransduction mechanisms.

ACKNOWLEDGEMENTS

This study was funded by NIAMS, of NIH (AR066635, Mechanobiological Risk Factors for Initiation of Post Traumatic Osteoarthritis). We thank the Russell Warren Chair of Tissue Engineering and the Clark Foundation for their support.

REFERENCES

[1] Wilusz, RE, et al., *Osteoarth Cartilage*, 21(12):1895-1903, 2013.
[2] Quiroga, JP, et al., *Biomech Model Mechanobio*, 1-8, 2016.
[3] Alexopoulos, LG, et al., *Acta Biomaterialia,* 1(3): 317-325, 2005.
[4] Smith, RL, et al., *Biorheology*, 37(1): 95-107, 2000.
[5] Mohtai, M, et al., *J Orthop Res,* 14 (1): 67-73, 1996.

EFFECT OF BIPHASIC PARAMETERS AND FIBRIL ORIENATION ON TRANSIENT CARTILAGE MECHANICS IN THE HIP

Jocelyn N. Todd (1,2), Huashan Zou (1,2), Travis G. Maak (3), Jeffrey A. Weiss (1,2,3)

(1) Department of Bioengineering
University of Utah
Salt Lake City, UT, USA

(2) Scientific Computing and Imaging Institute
University of Utah
Salt Lake City, UT, USA

(3) Department of Orthopaedics
University of Utah
Salt Lake City, UT, USA

INTRODUCTION

Hip pathoanatomies such as dysplasia and femoroacetabular impingement result in overload and damage to the articular cartilage and labrum, leading to early osteoarthritis (OA). Finite element (FE) analysis of patient populations offers a structured approach to study the effects of alterations in anatomy and biophysical properties of the cartilage layers on load support by the fluid phase and stress in the solid matrix, yielding insight into injury mechanisms and the best treatment options. The magnitude of transient fluid depressurization and its sensitivity to various biphasic material parameters, as well as the importance of accurate representation of the anisotropy and heterogeneity of collagen fibril structure in the hip during simulations of physiological loading are unknown. Studies incorporating FE simulations of hip mechanics have often represented the articular cartilage as elastic, based on the assumption that there is no significant loss of fluid pressurization in cartilage of the hip during activities of daily living. Further, collagen fibrils within the articular cartilage exhibit a changing orientation through the thickness and a tension-compression non-linearity that has not been accurately represented in previous studies of cartilage mechanics in the hip. The objectives of this study were to examine the influences of anisotropy and inhomogeneity in collagen fibril orientation and assumptions related to permeability on the transient mechanics of the solid and fluid phases of articular cartilage during walking and single-leg stance.

METHODS

An idealized FE model of the articular layers of the hip was developed to provide a realistic standardized geometry of the articular layers for parametric analysis and decreased computational time (~75% decrease from patient-specific to idealized). The femoral and acetabular cartilage layers were constructed by fitting ellipsoidal surfaces to the articular layers of a normal hip (Fig. 1A). The resulting surfaces produced an excellent fit consistent with other parameterizations.[1] Labrum geometry was constructed by projecting a cross section of the labrum from the normal hip about the circumference of the idealized cartilage geometry. The resulting contact pattern and area of the model compared well to the native hip model with an applied load. (Fig. 1B) A mesh convergence study was performed to determine optimal discretization of the cartilage layers, resulting in a final mesh of 94,195 quadratic tetrahedral elements and an average element volume of 0.11 mm³.

Cartilage layers were represented as anisotropic biphasic materials. The solid phase consisted of a continuous ellipsoidal

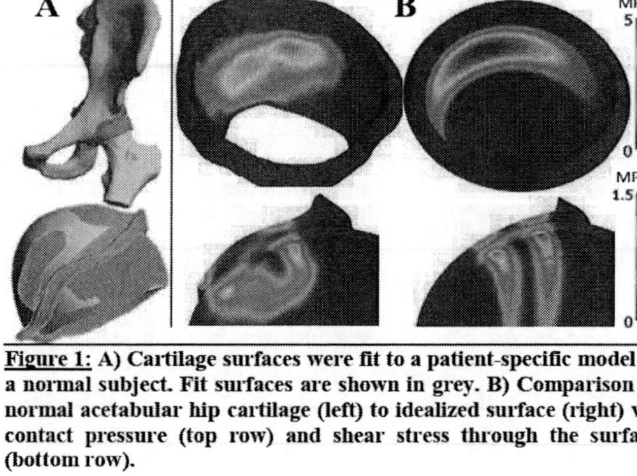

Figure 1: A) Cartilage surfaces were fit to a patient-specific model of a normal subject. Fit surfaces are shown in grey. B) Comparison of normal acetabular hip cartilage (left) to idealized surface (right) via contact pressure (top row) and shear stress through the surface (bottom row).

distribution of tension-only collagen fibrils embedded in an isotropic matrix to capture tension-compression nonlinearity, initial anisotropy and strain-induced anisotropy.[2, 3] The baseline model used a physiologically realistic distribution of fibril orientation through the thickness of the cartilage layers, where fibrils were aligned tangent to articular surface in superficial zone, spherical in middle zone, and perpendicular to the osteochondral interface in the deep zone. A continuous variation through the thickness was achieved with a custom FEBio plugin that allows scaling of the contribution from a local element axis as a function of surface depth. Permeability was represented as strain-dependent and anisotropic to align fluid flow to the direction of the fibrils[3]. The results of this baseline model were compared to three cases of simplification: Case 1) isotropic strain-dependent permeability, Case 2) constant isotropic permeability, and Case 3) homogeneous (spherical) fibril distribution.

The labrum was represented as a biphasic transversely-isotropic material with an isotropic ground matrix and fibers aligned circumferentially.[3] Osteochondral interfaces were tied to rigid bodies representing the femur and pelvis and were non-permeable. All other surfaces were free-draining. During simulations, the femoral cartilage was loaded into the acetabular cartilage through the femur while the pelvis was held in all degrees of displacement and rotation.

Single-leg stance (SLS) and gait were applied using three-dimensional loading and kinematics from Bergmann et al.[4] All models were analyzed with FEBio (www.febio.org). SLS was simulated by ramping to the maximum load and analyzing the response under creep for 400 s, as pilot simulations indicated that the variation of dependent variables was minimal after this time. A full cycle of gait was simulated to assess its effect on the same variables and variations in material parameters. Maximum tensile strain on the articular surface (E_1), maximum shear stress at the osteochondral interface (τ_{max}), fluid load support (FLS, calculated as the ratio of total fluid force to total applied force summed over the articular surface of the acetabular cartilage), and maximum fluid pressure (p_{max}) were examined within the acetabular cartilage as dependent variables.

We assessed sensitivity of the dependent variables to material coefficients. For Case 1, the strain-dependent coefficient M was varied[5]. For Case 2, the permeability coefficient k was varied (0.00075, 0.001 as baseline, 0.0025, 0.005, 0.0075 mm^4/Ns)[6,7].

RESULTS

After 400 s of SLS (long-time response), simplifications to the permeability model (Cases 1 and 2) under-predicted the fluid depressurization and decrease in FLS by ~5% (Fig. 2A). These simplifications also produced slightly reduced relaxation of τ_{max} over the long-time response. While variation of the strain-dependent constant M resulted in only slight changes to dependent variables during SLS, perturbing the permeability value resulted in large variation of p_{max} and FLS. Increasing the permeability constant from the baseline of k=0.001 to k=0.0075 resulted in a 43% increase of fluid depressurization at 400 s, resulting in a much larger decrease in FLS (Fig. 2B). However, this had little effect on τ_{max} and E_1.

At the beginning of SLS, changes in dependent variables were seen only when simplifying the physiological fibril orientation to homogeneous (continuous spherical). The response was only reflected in the τ_{max} result, which was 20% higher at t=0 s between physiological and homogeneous orientation (Fig. 2C). Analysis of the incremental fibril orientation models demonstrated that tangential alignment of the surface fibrils alone produced about the same degree of increase in τ_{max} at loading (17%) as the simplification to homogeneous fibrils, but there was a negligible difference compared

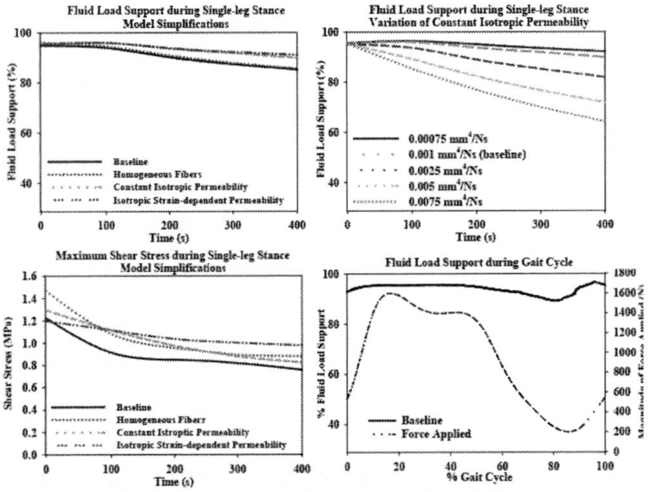

Figure 2: Effect of model simplification and variation of constant isotropic permeability on FLS and τ_{max} and FLS during SLS and gait

to the baseline physiological model when the fibrils at the osteochondral interface were aligned perpendicularly.

During gait, dependent variables did not vary widely based on material simplification. FLS varied during loading by 6%, but remained above 89% for the duration of the cycle (Fig. 2D).

DISCUSSION

The simplifications of Cases 1 and 3 produced only small variations in dependent variables (FLS, E_1, τ_{max}) in comparison to the most physiologically relevant baseline model during SLS and gait. However, there are several notable points. First, changing from anisotropic permeability to more simplified representations resulted in slower depressurization. Examination of the results revealed that anisotropy resulted in more tangentially directed fluid flow, toward unloaded areas, rather than normal to the surface, allowing the cartilage layers to depressurize more quickly. Regarding Case 2, the strong dependence on the constant isotropic permeability value reveals that this key parameter must be chosen carefully.

Analysis of the fibril orientation models demonstrated that the perpendicularly oriented fibrils near the osteochondral interface offered a greater contribution to stress and strain conditions within the joint compared to the contribution of the tangentially aligned fibrils on the articular surface. For simulations of chondrolabral mechanics that focus on the dependent variables used in our study, it may be sufficient to only represent anisotropy near the osteochondral surface.

The FLS during gait reported by this study is consistent with other literature in that FLS by hip cartilage remains above ~90% for various activities.[8] However, these studies do not include the variation of FLS during the activities studied. The present study indicates that FLS varied to a slight degree throughout the cycle and is restored to the original ratio by the end of the cycle.

REFERENCES

[1] Anderson, et al., *Journal of Biomechanics*, 43:1351-7, 2010.
[2] Ateshian, et al., *J Biomech Eng*, 131:061003, 2009.
[3] Henak, et al., *J Biomech Eng*, 136:021021, 2014.
[4] Bergmann, et al., *Journal of Biomechanics*, 34:859-71, 2001.
[5] Makela, et al., *Osteoarthritis Cartilage*, 20:1268-77, 2012.
[6] Athanasiou, et al., *Journal of orthopaedic research : official publication of the Orthopaedic Research Society*, 12:340-9, 1994.
[7] Haemer, et al., *Journal of Biomechanics*, 45:1450-6, 2012.
[8] Li, et al., *Proc Inst Mech Eng H*, 228:556-63, 2014.

Technical Presentation #87

FULLY AUTOMATED, HEXAHEDRAL MESHING OF PATIENT-SPECIFIC CARTILAGE STRUCTURES: DATA FROM THE OSTEOARTHRITIS INITIATIVE

Borja Rodriguez-Vila (1,2), David M. Pierce (3,4)

(1) Bioengineering and Telemedicine Centre
Universidad Polit□cnica de Madrid
Madrid, Spain

(2) Networking Research on Bioengineering
Biomaterials and Nanomedicine (CIBER-BBN)
Madrid, Spain

(3) Department of Mechanical Engineering
University of Connecticut
Storrs, CT, USA

(4) Department of Biomedical Engineering
University of Connecticut
Storrs, CT, USA

INTRODUCTION

Articular cartilage in diarthrodial joints must provide (i) a compliant, low-friction surface between the relatively rigid bones; (ii) a long-wearing and resilient surface; and (iii) a means to distribute the contact pressure to the underlying bone structure. Osteoarthritis (OA) is a disease of the synovial joint, with degeneration and loss of articular cartilage as one hallmark change. OA is a debilitating disease that afflicts nearly 20% of people in the US, costing more than $185.5 billion a year (in 2007 dollars), and its prevalence is projected to increase by about 40% in the next 25 years [1,2].

Fortunately, we can access large cohort databases on the progression of OA, e.g. the Osteoarthritis Initiative (OAI), with a cohort of 4,796 participants. The OAI is a multicenter, longitudinal, observational study of knee OA funded in part by the NIH [5]. This study selects men and women from the general population who likely either have preexisting OA or are at high risk as indicated by weight, knee symptoms, or history of knee injuries. The OAI public database supports investigations of knee OA onset and progression using traditional measures of disease and biomarkers developed from the study. Additionally, 3.0 T Siemens Trio Magnetic Resonance Images (MRIs) are available, including: localizer (3-plane), intermediate-weighted turbo spin echo, 3-D dual-echo in steady-state (DESS), intermediate-weighted turbo spin echo, T1-weighted 3-D flash and T2.

Despite the multifactorial nature of OA, mechanical stresses play a key role in the destructive evolution of the disease [3,4]. Both overloading (e.g. trauma) and reduced loading (e.g. immobilization) of cartilage induce molecular and microstructural changes that lead to mechanical softening, fibrillation, and erosion. Experiments can be used to quantify mechanical properties and biology of tissues, and imaging can be used to estimate tissue structure and even strains; however, only computational models can estimate intra-tissue stresses in human joints because the required *in vivo* experiments are impossible or unsafe. Finite element (FE) models are well-established at the macro (e.g. joint) scale as a means to estimate stress distributions.

Generating an appropriate computational mesh is prerequisite for applying many numerical techniques, including FE analyses, to patient-specific questions. Such meshes represent the geometry of interest using a set of polyhedral elements, commonly tetrahedra (a minimum of four connected nodes creating four triangular faces) or hexahedra (a minimum of eight connected nodes creating six quadrilateral faces). Many fast and robust methods exist for automatically generating tetrahedral meshes of arbitrary geometries, cf. [6,7].

Building computational meshes from hexahedra is far more restrictive, and a fully automatic algorithm for generating hexahedral meshes of arbitrary geometries has not yet been achieved. However, for a wide range of applications, hexahedral-based meshes are preferred. Among many reasons, FE meshes require far more tetrahedral elements (relative to hexahedral elements) to achieve the same solution accuracy for a given analyses, and this leads to higher computational costs (both time and memory) [8]. Additionally, when the aim is to apply FE analyses, tetrahedral meshes produce acceptable displacement results but are relatively inaccurate for predicting stresses [9].

In this work we propose a specific algorithm designed to work with patient-specific geometries of the knee obtained from MRI. Indeed, we establish a fully automated methodology for hexahedral meshing of knee joint structures, i.e. femoral cartilage, tibial plateau and menisci.

METHODS

MRI data sets from the OAI include 160 slices of 384x384 pixels^2, for a final voxel resolution of 0.365x0.365x0.7 mm^3. We label anatomical structures of the knee in each patient by manually segmenting sagittal slices of the fat-suppressed 3D dual-echo in steady state (DESS) from the OAI database (**Figure 1a**). DESS appears to be the best sequence to capture quantitative cartilage morphometry and to provide the best universal cartilage discrimination. From the manual segmentations we generate triangular surfaces of each structure using a marching cubes algorithm (**Figure 1b**).

We start from triangular surfaces of the structures of interest as inputs to our fully automatic, hexahedral meshing methodology for knee joint structures, i.e. femoral cartilage, medial/lateral tibial plateau and medial/lateral menisci. Our methodology uses the minimum bounding box of each structure to perform specific and customized sweeping algorithms, and while minimizing the use of collapsed elements.

In every case, the process pipeline is similar: (1) generate a low-element-density model with a custom sweeping algorithm which takes into account the structure's geometry, (2) apply Laplacian smoothing, (3) refine the model by subdividing hexahedrons, (4) expand the model to fit the original (triangular surface) segmentation, (5) smooth the model to eliminate sharp borders, and (6) optimize the quality of the elements. The mesh density is fully adjustable and we form meshes for femoral and tibial cartilages using three different layers (superficial/middle/deep) to allow use of different material parameters.

We demonstrate the quality of our patient-specific meshes of in the human knee using the element scaled Jacobian $\hat{J}$, a common quality metric for the analyses of solid structures [10]. Each hexahedral element has a center, $k = 0$, and nodes, $k \in [1, \ldots, 8]$, as well as edges (e_{k1}, e_{k2}, e_{k3}) connected to each node. We evaluate the scaled Jacobian of an element as the minimum of each nodal Jacobian divided by the length of the three corresponding edges using

$$\hat{J} = \min_{k \in [0,\ldots,8]} \left[\frac{e_{k1} \cdot (e_{k2} \times e_{k3})}{\|e_{k1}\| \|e_{k2}\| \|e_{k3}\|} \right], \quad (1)$$

where the scaled Jacobian takes the range $[-1, 1]$ for a hexahedral element, with -1 corresponding to the worst possible elements and $+1$ the best possible ones (NB: only elements with a nonzero, positive scaled Jacobian are suitable for FE analysis [10]).

RESULTS

Using baseline MRIs from six patients in the OAI database we manually segmented the cartilages and menisci, reconstructed and smoothed them in 3-D, and individually meshed them with hexahedral FEs. In all cases our methodology obtains the patient-specific meshes of interest from the triangular surfaces in less than 4 minutes running a MATLAB implementation on a common PC (**Tables 1,2**). Scaled Jacobian values of the meshes range from 0.5 to 1, where ~ 90% of the elements have values >0.8, and only 1.2% have values from 0.5 to 0.6.

Table 1: Mesh quality metrics (M±SD) by anatomical structure (Fem.=Femoral, Cart.=Cartilage, Tib.=Tibial, M.=Medial, L.=Lateral).

Structure	Elements (#)	Time (s)	$\hat{J} > 0.8$ (%)	$0.6 > \hat{J} > 0.5$ (%)
Fem. Cart.	24502±2758	80±7	94.67±0.93	0.96±0.14
M. Tib. C.	3816±0	11±1	93.27±3.74	0.24±0.30
L. Tib. C.	3906±0	11±1	94.62±2.04	0.14±0.09
M. Men.	1880±98	44±18	55.01±7.10	6.84±3.32
L. Men.	1700±34	57±21	42.89±2.55	11.33±1.6

Table 2: Mesh quality metrics organized by OAI patient ID [5].

OAI ID	Elements (#)	Time (s)	$\hat{J} > 0.8$ (%)	$0.6 > \hat{J} > 0.5$ (%)
9932809	37712	188	90.44	1.49
9948792	39562	221	89.29	1.51
9961728	37662	174	90.90	1.61
9977985	35366	226	90.40	1.25
9988421	31640	200	88.60	1.83
9988820	33482	226	90.38	1.82

The hexahedral meshes we obtained using our methodology (**Figure 1c**) correctly match the initial triangular surfaces (**Figure 1b**), with a RMS value (comparing the input triangular surfaces to the final hexahedral meshes) below 1 mm. We use collapsed elements in only the meshes of the femoral cartilages, resulting in a maximum of 12 wedges for an individual patient.

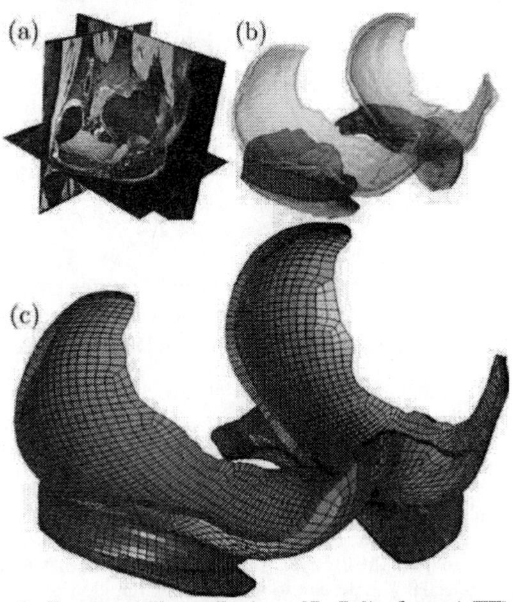

Figure 1: Representative patient-specific finite element (FE) mesh established from baseline magnetic resonance images (MRIs) from the OAI database. (a) Segmentation of cartilages and menisci. (b) 3-D reconstruction of femoral (yellow) and tibial (blue) cartilages, medial (green) and lateral (red) menisci. (c) Corresponding mesh of 20-node hexahedral FEs.

DISCUSSION

Our preliminary, fully hexahedral meshing methodology preserves tissue volume and produces high-quality elements. To the best of our knowledge, no other automatic hexahedral meshing for patient-specific knee structures has been proposed in the literature.

We hope to provide a fundamentally different means to test hypotheses on the mechanisms of disease progression at the patient level by integrating our patient-specific FE meshes and analysis framework, cf. [11], with data from individual patients or from natural history databases, e.g. the OAI. In the longer term, simulation-based, predictive medicine combined with medical imaging will likely improve the health, well-being, and quality of life for patients with OA.

REFERENCES

[1] Lawrence, RC et al., *Arthritis Rheum*, 58:26-35, 2008.
[2] Kotlarz, H et al., *Arthritis Rheum*, 60:3546-3553, 2009.
[3] Goldring MB, Marcu KB, *Arthritis Res Ther*, 11:224, 2009.
[4] Loeser, RF et al., *Arthritis Rheum*, 64:1697-1707, 2012.
[5] Nevitt, M et al., *The OAI: Protocol for the Cohort Study*, 2006.
[6] Löhner R, Parikh P, *Int J Num Method Fluids*, 8:1135-49, 1988.
[7] Yerry MA, Shephard MS, *Int J Num Method Eng*, 20:1965-90, 1984.
[8] Ramos A, Simões JA, *Med Eng Phys*, 28:916-924, 2006.
[9] Puso MA, Solberg J, *Int J Num Methods Eng*, 67:841-867, 2006.
[10] Tarjuelo-Gutierrez J et al., *Med Biol Eng Comp*, 52:159-168, 2014.
[11] Pierce D et al., *Biomech Model Mechanobiol*, 15:229-244, 2016.

DOES REGULAR PHYSICAL ACTIVITY HELP MITIGATE CARTILAGE STRAINS?

Axel C. Moore (1), Brian T. Graham (2), Christopher Price (1,2), David L. Burris (1,2)

(1) Biomedical Engineering
University of Delaware
Newark, Delaware, USA

(2) Mechanical Engineering
University of Delaware
Newark, Delaware, USA

INTRODUCTION

It is now well accepted that regular physical activity has a significant and beneficial effect on overall health [1]. As a result, the CDC suggests that 150 min of physical activity per week is a necessary component of a healthy lifestyle. Additionally, more recent studies have shown it is equally important to minimize the duration of continuous inactivity using even brief bouts of movement. For example, interrupting prolonged deskwork with hourly stepping activities (*e.g.* walking to the restroom) had significant and positive effects on cardiovascular health [2]. These studies have already translated to practice with companies like Apple using their fitness trackers to remind users to stand and move at least once per hour.

Archard's wear law, which states that wear increases with load and distance slid [3], suggests that regular physical activity promotes overall health at the expense of joint health. Fortunately, the many studies on the subject consistently show no additional risk of joint failure in active populations; in fact, the most recent and largest study to date showed that risk of joint failure decreased significantly with increased weekly physical activity [4].

The mechanisms underlying these unexpected benefits of activity on joint health remain unresolved. Furthermore, it is unclear if recommendations to regularly interrupt deskwork with a trip to the restroom or water cooler have implications for joint health. Recently, our team discovered a phenomenon, which we call tribological rehydration [5], that provides a unique opportunity for insights. Under static conditions, cartilage thins as it loses interstitial fluid, which disrupts lubrication and homeostasis. Tribological rehydration uses sliding to drive fluid back into the articular surface (Figure 1B). We propose that tribological rehydration promotes joint health and homeostasis by mitigating the loss of fluid (strain) and the corresponding loss of mechanical, tribological, and biological functions. This study isolates the tribological rehydration mechanism to test the hypothesis that activity frequency has a significant effect on cartilage's maximum strain and its associated tissue functions.

METHODS

Mature bovine stifle joints were acquired from a local abattoir. Large osteochondral cylinders (19 mm diameter, with a convex articular surface) were extracted from the femoral condyles (~6-12 cylinders per joint) using a coring bit. Tissue specimens were loaded on a uni-directional pin-on-disc materials tester (Figure 1A) with *in-situ* strain measurements to quantify the loss and recovery of interstitial water. Samples remained submerged in phosphate buffered saline (PBS) for the duration of testing. The applied load and rotational speed of the disc are controlled while the friction force and compressive strain responses of the sample are measured. The contact geometry is referred to as a convergent stationary contact area because it contains a convergence zone for hydrodynamic effects while keeping the contact area stationary relative to the cartilage.

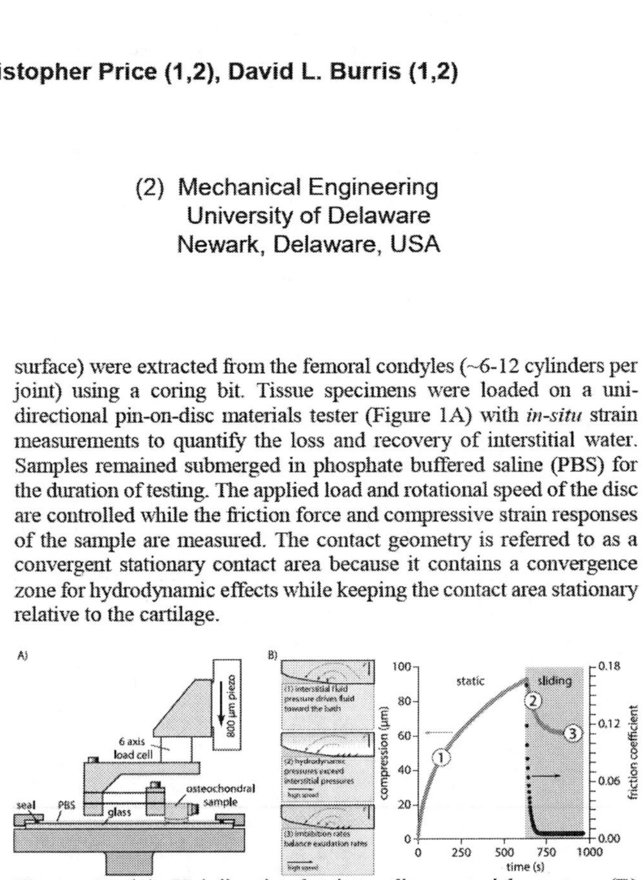

Figure 1: (A) Uni-directional pin-on-disc materials tester. (B) Tribological rehydration hypothesis schematic. It is proposed that hydrodynamic pressures developed during sliding drive fluid back into cartilage recovering strain and friction. Image is adapted from ref [5].

In-situ optical observation of representative samples compressed against a glass flat under a 5 N compressive load revealed a contact diameter of 5 ± 0.4 mm and resulting pressures of 0.25 ± 0.05 MPa [5]. This contact diameter and pressure are both about ¼ those assessed *in-vivo* [6]. Contact area and exudation time constant are known to be proportional; thus, to simulate one 'day' of *in-vivo* loading, a proportionally 'equivalent day' of 150 minutes was used. This equivalent day was broken into a 90 min 'awake' period and a 60 min 'sleep' period as illustrated in Figure 2A. A constant 5 N load was applied during the awake period, to represent load bearing, and reduced to a constant 0.1 N load during the sleep period to model the load applied during non-weight bearing activities. The awake period was broken into activity (sliding) and inactivity (static), during which the sliding speed was fixed at 100 mm/s and 0 mm/s, respectively. The total active (sliding at 100 mm/s) time was fixed at 30 min/day, which was administered in 1, 3, 6, or 15 equally distributed doses summing to 30 min total. An entirely inactive awake period (0 doses) was used as a control. A single simulated day was used to 'run-in' each sample.

RESULTS

The marked effects of activity dosing on peak strain are illustrated in Figure 2B. The inactive control (0) exceeded 18% strain at the conclusion of the 'awake' period and recovered to only 16% during the 'sleep' period. With one activity cycle, strain transitioned from 17% to 7.7%, falling to within the 95% confidence interval for active equilibrium strain. Following 'sleep', tissue strain was further reduced to 6.6% and dipped into the 95% confidence interval for resting equilibrium within 5 minutes. The number of activity doses in a day had no effect on the active or resting equilibria strains. However, increasing activity doses reduced the maximum strain as illustrated by Figure 3A.

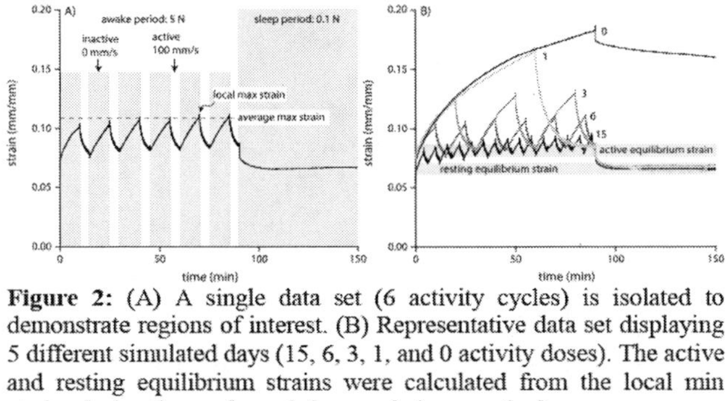

Figure 2: (A) A single data set (6 activity cycles) is isolated to demonstrate regions of interest. (B) Representative data set displaying 5 different simulated days (15, 6, 3, 1, and 0 activity doses). The active and resting equilibrium strains were calculated from the local min strains during the awake and sleep period, respectively.

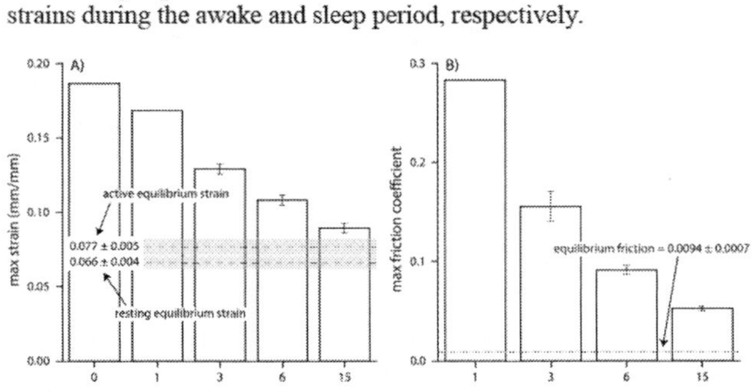

Figure 3: (A) Max strain and (B) max friction coefficient are shown as a function of activity cycles. The max strain and max friction were calculated from the mean of the local max strains (Figure 2A) and max friction coefficients. The equilibrium friction was calculated from the local min friction coefficient/s during the awake period.

Because strain and friction are fundamentally linked to interstitial hydration, the friction data produced similar trends. Like the maximum strain (Figure 3A), the maximum friction coefficients decreased with increased activity dosing (Figure 3b). Likewise, regardless of the dosing protocol, the equilibrium friction coefficient was 0.0094 ± 0.0007.

DISCUSSION

The CDC recommends 150 min of activity per week. Recent studies have shown compelling evidence that the regularity of these activities is as important to overall health as the weekly quantity. Although the mechanisms are unclear, most studies suggest that physical activity serves primarily to disrupt the detrimental effects of prolonged inactivity. This study demonstrated an analogous mechanism by which activity promotes joint function by limiting the amount of interstitial fluid lost during inactivity.

Interstitial fluid is important for biochemical, mechanical, and tribological functions of cartilage. During inactivity, the loss of interstitial fluid over time has significant detrimental effects on tissue function [7]. The results of this study clearly show that more frequent activity dosing helps mitigate the loss of interstitial fluid by periodically reversing the exudation process; moreover, the recovery process is far quicker than the exudation process (Fig 2A). Fifteen doses largely maintained hydration throughout the day relative to active equilibrium; interestingly, this dosing corresponds to, once scaled, ~1 interruption per hour *in-vivo* [2]. Thus, intermittent tribological rehydration helps maintain optimal joint biomechanics, tribology, function, and health by periodically reversing the exudation process.

Presently, we have no direct evidence that tribological rehydration participates in the maintenance of joint health. However, despite the vast differences between the model contacts in this study and those *in-vivo*, the similarities between the strain and frictional responses are compelling. During normal activities *in-vivo*, Eckstein *et al.* showed an 'active equilibrium' strain of approximately 2-6% above the 'resting equilibrium' strain [6]. Remarkably, the active equilibrium strain in this study was only 1% larger than the resting equilibrium strain despite the 50x larger applied normal force. Tribological rehydration enabled interstitial fluid pressure to carry nearly all of the load [5] and this would undoubtedly influence the biological functions and homeostasis of the tissue.

These data also suggest that cartilage contacts are unable to recover from rest alone (Fig. 2B). Comparing the recovery rate for 1 dose at 5 N to that of 0 doses at 0.1 N (Fig. 2B) suggests that passive recovery rates are exceedingly slow relative to tribological rehydration rates. Similar recovery rate trends have been observed *in-vivo* [6]. Our results suggest tribological rehydration is ~3X faster than exudation. This ratio should increase in larger joints because exudation rates scale with contact area but rehydration rates ought to scale with contact diameter; based on these results, we expect tribological rehydration rates to exceed exudation rates by ~12x in large joints like the knee. This may explain how relatively small strains are maintained in healthy populations [6] despite the fact that even 'active' people are sedentary across 90% of an average day [8].

ACKNOWLEDGEMENTS

This work was funded by the NSF BMMB 1635536, NIH S10 OD016361, and INBRE P20 GM103446.

REFERENCES

[1] AF Manly, *Surgeon General Report CDC*, 1996
[2] GN Healy et al., *Euro Heart J*, 2015
[3] JF Archard, *J Appl Phy*, 24:981-988, 1953
[4] P Williams, *Med and Sci in Sports*, 45:1292-1297, 2013
[5] AC Moore and DL Burris, *Osteo and Cart*, 25:99-107, 2016
[6] F Eckstein et al., *J Anat*, 208:491-512, 2006
[7] G Ateshain, *J Biomech*, 42:1163-1176, 2009
[8] T Catrine et al., *Appl Physiol Nur Metab*, 38: 100-114, 2013

HUMAN DISC NUCLEOTOMY: ANNULUS FIBROSUS INTERNAL DEFORMATIONS ARE ONLY ALTERED AT LOW LOADS

Amy A. Claeson (1), Brent L. Showalter (2), Edward J. Vresilovic (3)
Alexander C. Wright (4), James C. Gee (4), Neil R. Malhotra (4), Dawn M. Elliott (1)

(1) Department of Biomedical Engineering
University of Delaware
Newark, DE, USA

(2) Department of Bioengineering
University of Pennsylvania
Philadelphia, PA, USA

(3) Department of Orthopaedics and Rehabilitation
Pennsylvania State University
Hershey, PA, USA

(4) Department of Radiology
University of Pennsylvania
Philadelphia, PA, USA

INTRODUCTION

Nucleotomy is a surgical procedure following herniation and has been used ex vivo to study the effect of reduced nucleus pulposus (NP) pressure that occurs with disc degeneration [1-4]. Previous, studies of nucleotomized discs in axial compression show an elongated toe region (low load effect) that is likely due to reduced NP pressure, but in the linear region (high load effect), the disc compresses more and the stiffness is equivalent to the intact disc [1], likely because the annulus fibrosus (AF) and remaining NP become pressurized with the large compression loads and associated height and volume loss. In addition, following nucleotomy, the outward bulge of the outer AF increases in compression compared to intact [3,4]. Together, these effects have implications for AF internal strain, particularly radial deformations, as alterations in normal outward bulging may lead to fissures or delamination and advance disc degeneration. Therefore, it is important to evaluate disc mechanics at low loads.

We have shown that nucleotomy alters internal AF strains at high loads (in the linear region of compressive loading) [5], but no study quantifies how the reduced NP pressure at low loads affects internal AF mechanics. Here we use our previously developed non-invasive technique [6] that uses mechanical testing, MRI imaging, and image registration to quantify internal AF deformations. We hypothesize (1) a reduction in NP pressure after nucleotomy disrupts internal AF mechanics at low loads, but (2) NP pressure is not a significant factor in AF loading at high loads that are representative of physiological activities (e.g., 1-2X body weight).

METHODS

<u>Specimen Preparation</u>: Human cadaveric L3-L4 discs (grade II, n=8) were analyzed in their intact state and after a partial nucleotomy that removed 30-50% of the NP through a left posterolateral incision (Fig. 1, arrow). Specimens were hydrated unconfined in a 1X PBS bath overnight at 4°C prior to mechanical testing. After nucleotomy, remaining NP reconstituted during hydration filling the void (Fig. 1).

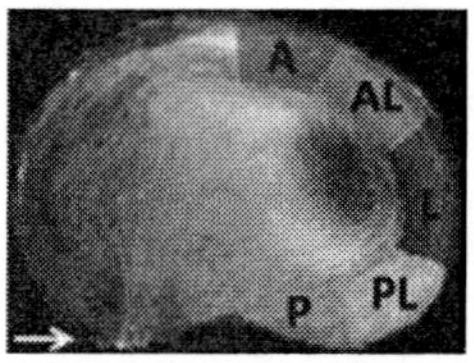

Figure 1. Midaxial MRI image of a partial nucleotomized disc after hydration. Left posterolateral AF incision site denoted by arrow. The right AF was unaffected by the incision and regions of interest defined in *A*: Anterior, *L*: Lateral, *P*: Posterior

<u>Mechanical Testing and Image Acquisition</u>: Each disc was subjected to axial compression stress-relaxation tests to a Low load (50N peak) and a High load (1180N peak) in the intact condition and after nucleotomy as follows (Fig. 2). After overnight hydration, intact discs were placed in an MRI compatible chamber and loaded in an Instron testing system to a Low load of 50N (Fig 2: A_1). The resultant disc height was held (DH_{A1}=DH_{A2}) while the load relaxed for 8 hours (A_2). The chamber was locked at A_2 and transferred to the MRI for Low load imaging (MRI_{A2}). Next, the disc and chamber were returned to the Instron, and the disc was compressed to a High Load (B_1) corresponding to 10% strain (average resultant load 1180N). The load relaxed for 3 hours in the Instron (B_2), then continued for another 8 hours (B_3, a second-order power function was used to determine this final load). The chamber was locked and the High load MRI image was acquired (MRI_{B3}). After nucleotomy, discs underwent the same hydration, stress-relaxation, and imaging protocols at the Low load state (MRI_{C2}) and High load state (MRI_{D3}). Discs were imaged in a 7T MRI scanner at 0.3mm isotropic resolution with a T2-weighted turbo spin echo sequence that emphasizes lamella contrast (Fig. 1).

<u>Image Processing</u>: Advanced Normalization Tools [7] was used with established methods [6] to create an average disc shape template at MRI_{A2} and to calculate 3D disc deformation tensors, using images

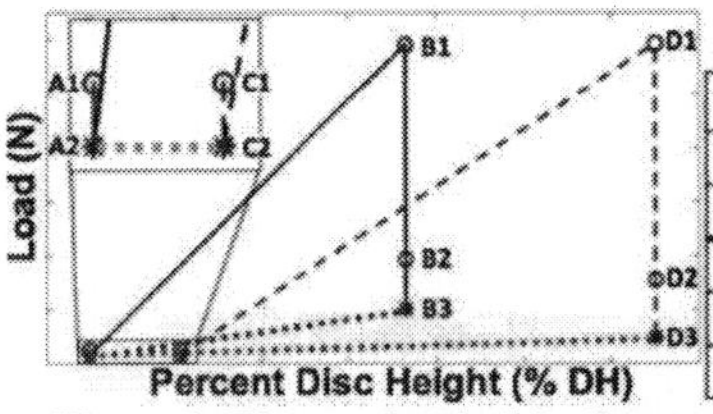

* MRI	Low Load		High Load	
INTACT	A1	A2	B1	B3
Load (N)	50	-	1180	203
% DH	100.		92.7	
NUCL	C1	C2	D1	D3
Load (N)	50	-	1180	105
% DH	98.0		87.0	

Figure 2. The mechanical testing profile included two stress-relaxation protocols in axial compression at Low and High load. The protocol was first applied to intact discs and each peak load was matched after nucleotomy.

of each disc, with respect to the intact reference condition (MRI_{A2}), for three states: Intact at High load (Intact-High, at MRI_{B3}, purple), and Nucleotomy at Low load (Nucl-Low, at MRI_{C2}, grey) and High load (Nucl-High, at MRI_{D3}, green) (Fig 2). Each registration output a 3D displacement tensor from which displacement fields and strains (axial (E_Z), circumferential (E_C), and radial (E_R)) were calculated and then morphed to the average disc shape template. Strain values and directional components of the displacement vector field were then averaged in regions of interest in the middle third of the disc height (to avoid bone boundary effects) and in five circumferential regions (Fig. 1, only *A*, *L*, *P* reported here) for each disc and each deformation state.

Data Analysis: To discern how nucleotomy affects internal disc strain and internal disc deformation, comparisons were made between deformation states (Intact-High, Nucl-Low, Nucl-High) and across regions (*A*, *L*, *P*). Data were compared with a repeated measure ANOVA with a Holm-Sidak post-hoc test at $p<0.05$ significance.

RESULTS

Intact-High: Peak load (B_1) averaged 1180±445N as a result of the applied displacement and relaxed to 203±85N (B_3) after 8 hours. With a 7.3±3.4% loss in disc height, axial strain averaged -18.5±1.2% and did not vary by region as indicated by the homogeneous strain field (Fig. 3A, 3D). Radial strains were more heterogeneous (Fig. 4A) but did not vary across region. Displacement vectors pointed outward (Fig. 4A), and show normal AF bulging due to NP pressure.

Nucl-Low: Nucleotomized disc height decreased by 2.0±2.9% to match the Low load ($C_1=A_1=50N$), with an average axial strain of -2.1±3.5%. Strain was inhomogeneous (Fig. 3B) and varied widely, especially in *A* and *L* (range [-18.5 32.7%]), yielding no regional statistical differences. Displacement vectors in region *A* oriented circumferentially, and *P* vectors oriented inward (Fig. 4B). Both regions were significantly different from Intact-High (Fig. 4D).

Nucl-High: Peak load ($D_1=B_1=1180N$) was achieved at a 13.0±3.6% loss of disc height and relaxed to 105±46N (D_3). Axial strain was homogeneous (Fig. 3C-D) and was increased in magnitude across all regions (average -22.5±1.5%), though not significantly different from Intact-High. Displacement vectors oriented outward in *A* and *P* (Fig. 4C), indicating a reengaged AF exhibiting normal bulging, because these vectors are not significantly different from those of Intact-High (Fig 4D).

DISCUSSION

We show that (1) a reduction in NP pressure after nucleotomy disrupts internal AF mechanics at Low load, but (2) AF strains were restored at High load, suggesting NP pressure is not a significant factor in AF loading at High load. Nucl-Low axial strain varied widely in *A* and *L*, but less in *P*. We speculate this variability appeared only in *A* and *L* regions because more NP was likely removed near right *A*, *AL* and *L* regions based on the trajectory from the left posterolateral incision site. At Nucl-Low, displacement vectors oriented circumferentially in *A* and *L*, while *P* vectors were variable and indicated either internal AF buckling or reduced outward pressure,

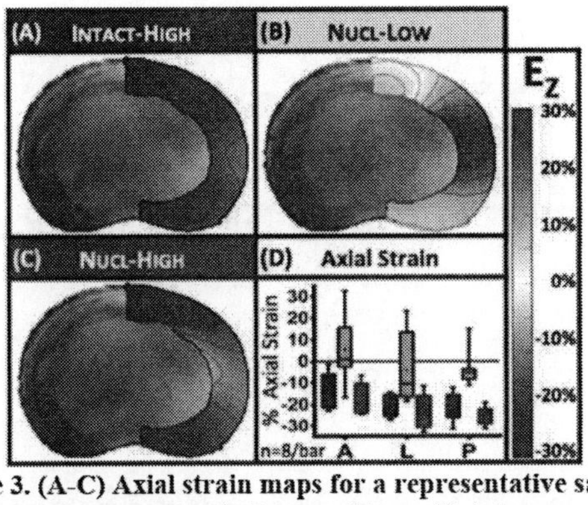

Figure 3. (A-C) Axial strain maps for a representative sample. Strain is (A) homogeneous in Intact-High, (B) inhomogeneous in Nucl-Low, and (C) again homogeneous in Nucl-High. (D) Higher magnitude strain occurred in Nucl-High across regions, though not significantly different from Intact-High.

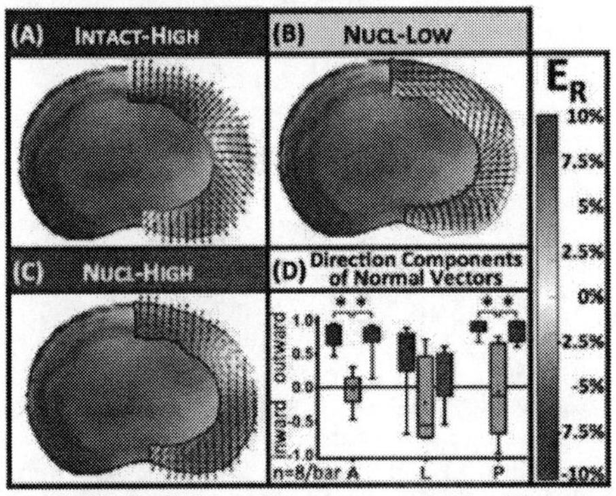

Figure 4. (A-C) Radial strain maps overlaid with normalized displacement vectors for a representative sample. (A) Internal pressurization yields outward facing displacement vectors, (B) that are disrupted at Low loads after nucleotomy and (C) restored upon reaching High loads (D) in *A* and *P* regions.

consistent with findings from previous studies that compromised disc structure by measuring AF strain using internal beads [4] or markers [3]. Diminished AF radial strain and increased compliance (indicated by buckling) may cause AF fissures or delamination that can lead to disc fatigue and further to degeneration. Nucl-High disc height decreased by an additional 5.7% to match the Intact-High peak load (B_1), and this change was mirrored in AF axial strain, though the change was not significant. We found this additional loss in disc height (elongated toe region) serves to reengage the compliant AF and overcome buckling at Low load. In conclusion, we show decreased NP pressure at Low load alters AF internal deformations, and based on our finding that Intact-High and Nucl-High strains are not significantly different, together with previous studies [1], we show that decreased NP pressure does not affect AF mechanics at High load.

ACKNOWLEDGEMENT

This study was funded by the NIH R01AR050052.

REFERENCES

[1] Showalter, BL et al., *J Biomech*, 47:2633-2640, 2014. [2] O'Connell, GD et al., *Spine*, 36(21):1765-1771, 2011. [3] Meakin, JR et al., *Clin Biomech*, 16(2)121-128, 2001. [4] Seroussi, RE et al., *J Orthop Res*, 7:122-131, 1989. [5] Claeson, AA et al., SB^3C, 2016. [6] Yoder, JH et al., *J Biomech Eng*, 136(11):1-9, 2014. [7] Lancaster, JL et al., *Hum Brain Mapp*, 28(11):1194-205, 2007.

THE ROLE OF MECHANICAL FORCES ON HEMISPHERE DIVISION IN THE EMBRYONIC FOREBRAIN

Kara E. Garcia (1), Larry A. Taber (1,2)

(1) Department of Biomedical Engineering
Washington University in St. Louis
St. Louis, MO, USA

(2) Department of Mechanical Engineering
Washington University in St. Louis
St. Louis, MO, USA

INTRODUCTION

During normal embryonic development, the anterior end of the brain grows into two distinct lobes, or telencephalic hemispheres, that become the right and left sides of the adult cerebrum. However, incomplete hemisphere division occurs in an estimated 1 out of every 250 human pregnancies, a condition called holoprosencephaly [1]. Of cases that survive to term, structural abnormalities range from mild to severe, and associated defects include decreased brain size (microcephaly), enlarged brain ventricles (hydrocephalus), facial abnormalities, and developmental delays [1].

Studies suggest that proper cleavage depends on morphogenetic signaling patterns that regulate differential growth [2]. As depicted in Figure 1, anterior growth factors stimulate proliferation while roof plate signals inhibit growth between hemispheres. Furthermore, floor plate signals encourage neural differentiation and contribute to growth factor production. Genetic and chemical perturbations of these signals alter growth to produce a range of holoprosencephalic phenotypes [2].

Despite an abundance of information regarding molecular signals, the mechanical forces involved in hemisphere division have received little attention. In this study, we combine computational modeling with experimental approaches to determine the forces needed to shape the hemispheres. Since both mechanical and molecular feedback have been shown to influence epithelial proliferation [3], we hypothesize that mechanical stress may also contribute to normal and abnormal telencephalon development. We specifically focus on the role of embryonic cerebrospinal fluid (eCSF), which is known to inflate the embryonic brain and stimulate proliferation [4].

METHODS

Experimental Methods: To determine the role of eCSF pressure during hemisphere morphogenesis, pressure was experimentally altered in chicken embryos during the period of initial hemisphere division.

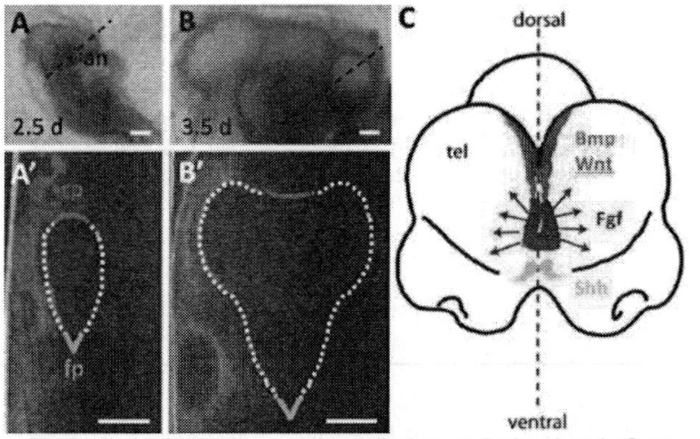

Figure 1. (A-B) Embryonic brain before and after hemisphere formation. (A'-B') OCT cross-sections at location of dotted lines in A-B. (C) Schematic of signals directing growth, adapted from [11]. Scale bars=300 μm, an=anterior, rp=roof plate, fp=floor plate.

Fertilized white Leghorn chicken embryos were incubated 2-2.5 days, extracted onto filter paper, and cultured at least 24 h using the EC method [5]. To reduce pressure, a glass tube was surgically inserted into the midbrain, equalizing pressure between the internal lumen and external fluid. For sham controls, a closed glass tube was used instead [6]. To increase pressure, embryos were pharmacologically treated with 4mM beta-D-xyloside (BDX, 25uL), which has been previously shown to increase lumen pressure (~50%) via increased eCSF osmolarity [4].

For each case, three-dimensional brain geometries were recorded over the course of development using optical coherence tomography (OCT) and analyzed using ImageJ software. To separate the effects of mechanical stretch and tissue growth, geometries were recorded with

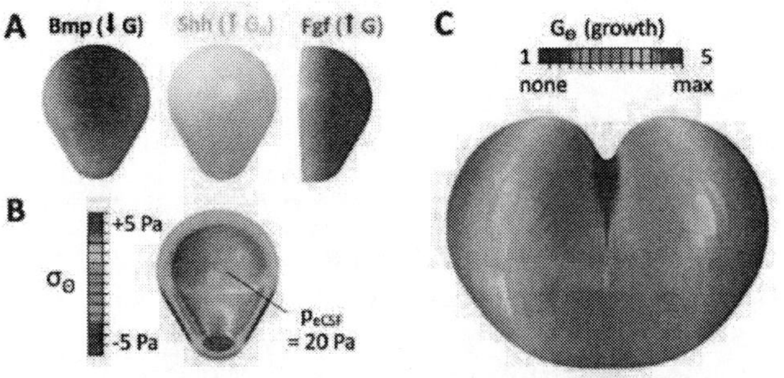

Figure 2. 3D model of hemisphere growth and division. (A) Molecular signals for growth, G. (B) Initial mechanical stress due to eCSF pressure. (C) Final geometry and growth.

and without eCSF pressure. To visualize cell proliferation, nuclei undergoing DNA synthesis were labeled using the Click-iT EdU assay (Invitrogen). Cell death was labeled with Nile Blue.

Computational Methods: A three-dimensional model of the telencephalon was created using ABAQUS/Standard (D. S. Simulia), incorporating growth via a user-defined material (UMAT) [7]. Initial geometry was matched to OCT measurements, while nonlinear material properties and eCSF pressures were taken from previous literature [8,9]. Total deformation was defined such that $\mathbf{F}=\mathbf{F}^{*}\bullet\mathbf{G}$, where $\mathbf{F}^{*}$ represents elastic deformation and $\mathbf{G}$ represents growth [10]. Tangential and radial growth components (G_{Θ}, G_{Φ}, G_R) were defined as a function of predefined signaling gradients (Figure 2A) and evolving Cauchy stress such that growth rate increases with tension (Figure 2B).

RESULTS

When eCSF pressure was pharmacologically increased, the telencephalon failed to cleave normally (Figure 3A, n=8-10 per group). Instead of invaginating, the roof plate bulged outward and became significantly wider than that of controls ($P<0.01$). Interestingly, this shape was similar to holoprosencephalic morphology observed due to upregulation of Shh or downregulation of Bmp4 (both leading to a net increase in proliferation) [2]. Growth also appeared higher throughout the brain, as estimated from OCT cross-sections by measuring change in total circumference without pressure ($P<0.1$).

The telencephalon also failed to cleave when eCSF pressure was surgically decreased (Figure 3B, n=10 per group). In this case, unloaded hemispheres were significantly smaller than those exposed to normal eCSF pressure ($P<0.01$). Similar effects have also observed with downregulation of Fgf8 or upregulation of Bmp4 (both leading to a net decrease in proliferation) [2].

To explain these results, a computational model was considered incorporating realistic initial geometry, approximate signaling patterns, and growth dependent on signal concentration and mechanical feedback. As shown in Figure 2C, the resulting differential growth was sufficient to produce hemisphere cleavage under normal pressure. EdU labeling was generally consistent with this pattern, revealing decreased proliferation at the midline for control embryos (n=6, not shown). As expected for a stress-dependent model, total growth increased under increased pressure and decreased under decreased pressure.

DISCUSSION

Here we find that, while molecular signals provide a template for differential growth, mechanical factors also contribute to the final growth and shape of the telencephalon. Through computational modeling, we see that differential growth is sufficient to cleave the telencephalic hemispheres under normal signaling and pressure. Through experimental perturbations, however, we observe that increased or decreased eCSF pressure can result in incomplete division of the prospective cerebral hemispheres.

Prior studies have suggested a link between mechanical feedback and growth [3,4,6], and genetic models have examined dysregulation of eCSF pressure at similar stages [12]. In the low-pressure case, we propose that reduced tension decreases growth to a level insufficient for cleavage. Alternately, the high-pressure case suggests a fundamental change in the growth distribution. Future work should test our computational model against altered pressure cases, comparing geometry and proliferation patterns for each loading condition. While our current model uses simple, approximate relationships to define growth, a refined growth law or signaling interactions may be necessary to explain this observed morphology.

Since this study is limited by confounding factors associated with existing interventions (eg. removal of eCSF growth factors under intubation), future studies should consider a microfluidics approach to precisely control pressure. Such studies, along with the results presented here, may elucidate the mechanics of normal brain development and highlight potential sources of abnormal brain structure.

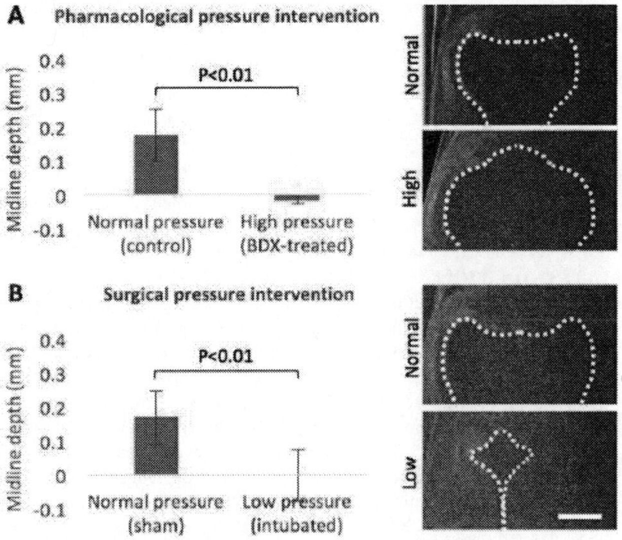

Figure 3. Effects of eCSF pressure on telencephalic hemispheres. All images to scale, scale bar=300 μm.

ACKNOWLEDGEMENTS

This work was funded by the National Institutes of Health Grants T32 EB018266 and R01 NS070918.

REFERENCES

[1] Orioli, I et al., Am J Med Gen C Sem Med Gen, 154C:13–21, 2010.
[2] Ohkubo Y et al., Neuroscience, 111(1):1-17, 2002.
[3] Wyatt TP et al., Proc Natl Acad Sci, 112(18):5726-5731, 2015.
[4] Desmond ME et al. Int J Dev Biol, 58:35-43, 2014.
[5] Chapman SC et al., Dev Dyn 220:284-289, 2001.
[6] Desmond ME and Jacobson, Dev Biol, 57(1), 188-198, 1977.
[7] Young T et al. J Biomech Eng, 132:104505, 2010.
[8] Xu G et al., J Biomech, 132(1):011005, 2010.
[9] Jelinek R and Pexieder, Physiol Bohemoslov, 17:297-305, 1968.
[10] Taber LA, Phil. Trans. R. Soc. A, 367:3555-3583, 2009.
[11] Stevens, H et al., Frontiers in Neuroscience, 4:1-59, 2010.
[12] Doganli C, J Biol Chem, 288(13):8862-8874, 2013.

PRESTRAIN REGULATES CELL SENSING OF TOPO-MECHANICAL CUES TO DIRECT ANNULUS FIBROSUS MECHANOBIOLOGY

Edward D. Bonnevie (1), Dawn M. Elliott (2), Robert L. Mauck (1,3)

(1) Department of Orthopaedic Surgery
University of Pennsylvania
Philadelphia, PA, USA

(2) Department of Biomedical Engineering
University of Delaware
Newark, DE, USA

(3) Department of Bioengineering
University of Pennsylvania
Philadelphia, PA, USA

INTRODUCTION

Mechanical function of the intervertebral disc arises from its composite structure. The inner, gel-like nucleus pulposus contains a high concentration of fixed negative charges, due to the abundance of proteoglycans, which in turn generates a substantial osmotic swelling pressure, on the order of 400 kPa [1]. This NP swelling is confined by the fibrous outer region of the disc, the annulus fibrosus (AF), which is typically in a prestrained state as it resists disc expansion [2].

One hallmark of early disc degeneration is proteoglycan loss from the NP and a corresponding decrease in osmotic pressure. This change in pressure not only decreases disc height, but can also influence the mechanical microenvironment of the AF. That is, given that the AF is a nonlinear strain stiffening material, loss of prestrain would lead to an effectively softer microenvironment. The aberrant remodeling that follows would compound these changes by promoting disorganization in the AF fiber network in pathological discs. These altered microenvironmental cues (both micromechanics and topography) likely affect cell phenotype and homeostasis [3,4], but little is known regarding how AF cells respond to these cues.

The goal of this study was to develop an experimental platform through which to query the impact of these microenvironmental cues on AF cell mechanobiology. To accomplish this, we utilized scaffold fabrication methods to generate aligned and nonaligned fibrous scaffolds that mimic the organized healthy AF and the disorganized degenerate AF. We analyzed the mechanical function and organization of these tissue analogs with applied prestrain. Additionally, we determined AF cell response to these scaffolds under free swelling conditions and with seeding onto prestrained scaffolds. Our findings indicate that cells sense both mechanical and topographical changes in these prestrained fiber networks, and demonstrate that this experimental platform can serve as a basis for the study of cell responses to tissue changes reflective of those that occur during disc degeneration and after injury (i.e., softened or aberrantly remodeled, disorganized tissue).

METHODS

Electrospun scaffolds were prepared as previously described [5]. Briefly, 14% w/v ε-polycaprolactone (PCL) was extruded at 2.5 mL/hr from a spinneret charged to 14 kV onto a grounded rotating mandrel. The mandrel was rotated with a surface speed of 14.5 m/s (aligned) or 2.7 m/s (nonaligned). Mechanical and structural analyses were conducted to determine how scaffold mechanics and topography change with strain. Mechanical analyses were conducted using an Instron 5848 in uniaxial tension to failure (in the presumptive fiber direction) at a strain rate of 0.01/min. To determine the role of tensile strain on topography, a custom tensioning device was developed to perform *in situ* straining within a scanning electron microscope (SEM, FEI Quanta 600). Aligned and nonaligned scaffolds were positioned within the tensioning device and sequentially imaged at 0%, 4.5%, 9%, and 13.5% strain. Fiber angle distributions were determined based on Fast Fourier Transforms in ImageJ and an anisotropy index (AI) was calculated based on a Gaussian distribution fit of fiber angles (AI = 1, random distribution; AI = 0, perfectly aligned).

Bovine annulus fibrosus cells were isolated from caudal discs via overnight digestion, expanded through passage two, and seeded at a density of 500 cells/mm² onto PCL scaffolds coated with fibronectin. Constructs were cultured for 24 hours at 37°C and 5% CO_2 in chemically defined medium to allow for cell attachment and spreading. Scaffolds were either free-swelling (no strain) or were subjected to 9% strain prior to cell seeding (prestrain, **Fig 3A**). At 24 hours, cells were stained with Hoechest to identify cell nuclei, and were imaged using an inverted fluorescent microscope (**Fig 3B**). At least 30 cells per scaffold were assessed for nuclear aspect ratio (NAR, the ratio of long and short axis of the nucleus based on elliptical fit) and nuclear angle, where 0° coincided with the dominant fiber and stress direction. Statistics were conducted based on a 2-way ANOVA.

RESULTS

Both aligned and nonaligned electrospun scaffolds exhibited nonlinear stress-strain behavior with modulus increasing as a function of strain applied in the fiber direction **(Fig 1A, B)**. Modulus increased within the first 10% strain, indicating the presence of a toe region, comparable to native fibrous tissues. For aligned scaffolds, modulus reached a maximum of 12 MPa, compared to 4 MPa for nonaligned scaffolds. SEM on scaffolds stretched *in situ* revealed conformational and topographic changes as a function of applied strain **(Fig 2)**. For aligned scaffolds, both fiber straightening and realignment in the direction of strain were evident, with the anisotropy index decreasing slightly as a function of applied strain **(Fig 2G)**. Fibers in nonaligned scaffolds reoriented in the direction of strain to a much larger magnitude; however, even at 13.5% strain, they were still more disorganized than aligned scaffolds **(Fig 2G)**.

For cells seeded onto scaffolds, fiber alignment and pre-strain dictated nuclear morphology. In terms of nuclear orientation **(Fig 3C-F)**, fiber organization promoted nuclear alignment along the fiber/strain direction ($p = 0.02$), but prestrain did not have a large effect in reorienting cell nuclei ($p = 0.5$). While prestrain did not significantly alter nuclear alignment, prestrain had a larger effect on nuclear aspect ratio **(Fig 3G-H)**. For nonaligned scaffolds, prestrain marginally increased NAR from 1.28 to 1.37 ($p = 0.1$), while for aligned scaffolds prestrain increased NAR from 1.43 to 1.71 ($p= 0.02$).

DISCUSSION

This study demonstrates that prestrain in fiber networks is a potent regulator of cellular perception of topo-mechanical cues. Cell phenotype is sensitive to both topographical and mechanical cues – both of which are altered by IVD injury and degeneration. Matrix elasticity, or more specifically, the microenvironmental stiffness of a substrate that a cell senses, regulates numerous activities, including proliferation, migration, and differentiation [3,6]. Our data indicate that prestrain of both aligned and nonaligned fibrous networks promotes stiffening of the network **(Fig 1)**, straightening of fibers, and realignment of the fibers in the direction of strain **(Fig 2)**. Thus, without altering the inherent material composition of the network, prestrain altered topo-mechanical cues as sensed by AF cells **(Fig 3)**. With no further mechanical perturbation after cell seeding, both nuclear orientation and aspect ratio of AF cells were sensitive to fiber alignment and prestress. Building upon these results, future studies will utilize this platform to determine AF signaling responses to these altered cues. Determining the roles that fiber network stiffness and organization play to initiate aberrant tissue remodeling is pivotal to understanding the initiation and progression of IVD degeneration.

ACKNOWLEDGEMENTS

This study was supported by NIH R01 EB02425 and P30 AR050950.

REFERENCES

[1] Guehring *et al. Eur Spine J,* 2006,
[2] Gardener-Morse *et al. J Orthop Res,* 2003,
[3] Baker *et al. Nature Mat,* 2015,
[4] Klingberg *et al. J Cell Bio,* 2014,
[5] Nerurkar *et al. J Orthop Res,* 2007,
[6] Engler *et al. Cell,* 2006,

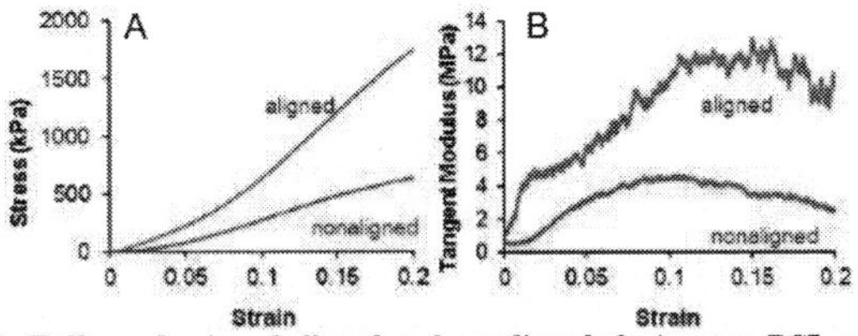

Figure 1: Bulk mechanics of aligned and nonaligned electrospun PCL scaffolds. (A) Aligned scaffolds are more than 100% stiffer than nonaligned scaffolds. (B) Significant Strain stiffening is evident for both types of scaffolds with a region reminiscent of the toe region or neutral zone of native tissue (n = 2).

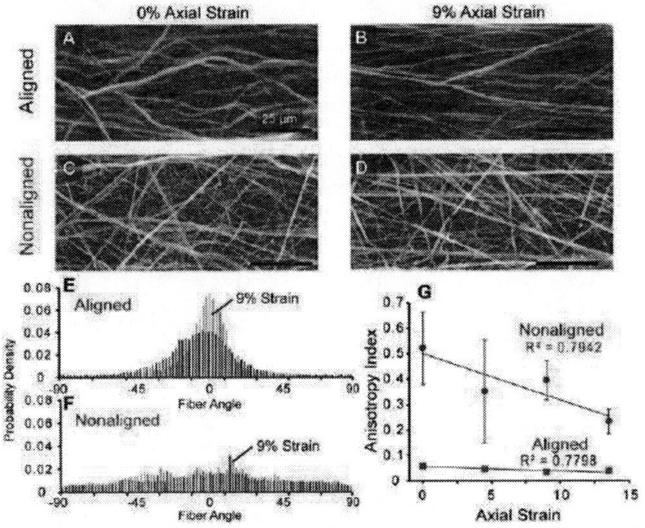

Figure 2: Topographical changes as a function of axial strain in aligned and nonaligned scaffolds. (A-D) SEM images of both nonstrained and prestrained (ε = 9%) PCL revealed fiber straightening and reorienting as a function of strain. (E-F) Fiber orientation distributions for nonstrained and prestrained PCL. (G) Anisotropy decreases for both nonaligned and aligned PCL scaffolds as a function of prestrain (n = 3).

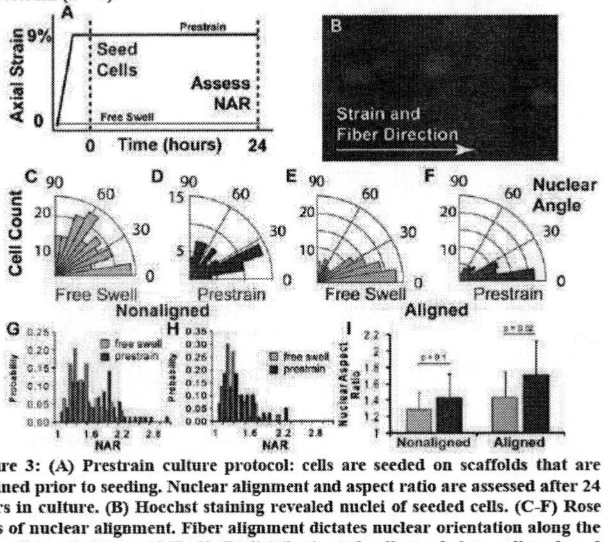

Figure 3: (A) Prestrain culture protocol: cells are seeded on scaffolds that are strained prior to seeding. Nuclear alignment and aspect ratio are assessed after 24 hours in culture. (B) Hoechst staining revealed nuclei of seeded cells. (C-F) Rose plots of nuclear alignment. Fiber alignment dictates nuclear orientation along the stress/fiber direction. (GH) NAR distribution of cells seeded on aligned and nonaligned scaffolds. (H) Both fiber alignment and prestrain dictated increases in NAR. (n = 2 scaffolds, >30 cells per scaffold.

LOCAL MECHANICAL PROPERTIES OF 3D COLLAGEN HYDROGELS ASSESSED VIA OPTICAL MAGNETIC TWISTING CYTOMETRY

Jacopo Ferruzzi (1), Haiyue Li (2), Atena Irani Shemirani (1), Yanhang Zhang (1,2), Muhammad H. Zaman (1,3)

(1) Department of Biomedical Engineering
Boston University
Boston, MA, USA

(2) Department of Mechanical Engineering
Boston University
Boston, MA, USA

(3) Howard Hughes Medical Institute
Boston University
Boston, MA, USA

INTRODUCTION

The extracellular matrix (ECM) constitutes a complex network of structural proteins that are dynamically remodeled by resident cells. Cellular behaviors such as adhesion and migration are in turn regulated by the structural and mechanical properties of the ECM [1]. In particular, the fibrillar organization of matrix proteins has been found to impact cellular morphology more than the biochemical nature of ECM adhesion molecules [2]. Type I collagen is the predominant fibrous constituent of the ECM and reconstituted collagen hydrogels are commonly used as substrates for 3D cell cultures. The mechanical properties of collagen gels are commonly assessed using global testing techniques such as shear rheometry and uniaxial tensile testing [3]. The global stiffness measured from these tests, in the order of Pa-kPa, is much lower than the stiffness of individual collagen fibrils, which is in the order of hundreds of MPa [4], and is more likely to be sensed by cultured cells. Therefore, a better understanding of cell-matrix interactions *in vitro* can only be achieved by monitoring the spatio-temporal evolution of collagen structure and mechanics locally.

In this study, we propose an approach that allows micromechanical testing of the 3D fibrous collagen matrix by employing optical magnetic twisting cytometry (OMTC), a technique that was originally developed – and used so far only in 2D settings – to test the mechanical properties of the cellular cytoskeleton [5]. This technique uses a high number of covalently bound ferromagnetic microbeads that are embedded within the gel during polymerization and, by means of two perpendicular magnetic fields, are subjected to coupled rotations and displacements. Here, we report the feasibility of such approach by using OMTC to measure the properties of 3D collagen gels. By measuring local values of apparent shear moduli (including both elastic and viscous components) at different planes along the gel thickness, we captured the local mechanical behavior of homogeneous acellular gels using OMTC.

METHODS

Cylindrical wells made of polydimethylsiloxane (PDMS), shown in Figure 1-a, were used to house the gels and avoid bulk motions caused by magnetic stimulation. Briefly, 3 to 3.5g of a liquid PDMS mixture (Dow Corning, Midland, MI), with a 10:1 weight ratio of silicon elastomer to curing agent, was deposited on 35mm glass-bottom dishes (MatTek Corporation, Ashland, MA) and cured at 50°C for 2 hours. Up to three cylindrical wells per dish were carved out of the PDMS layer using a punch with an outer diameter of 5mm. Wells were coated with poly-D-lysine and glutaraldehyde (Sigma-Aldrich, St. Loius, MO) to provide an anchoring layer for collagen and to avoid gel floating. Collagen gels were prepared using rat tail type I collagen (BD Biosciences, Franklin Lakes, NJ) at a concentration of 2mg/mL using established protocols [6]. Streptavidin-coated ferromagnetic beads with a diameter of ~4.2μm (Spherotec Inc., Lake Forest, IL) were added at a concentration of 2×10^7 beads/mL. Finally, 50μL of the collagen-bead mixture was deposited in each well. Collagen fibrillogenesis was allowed in an incubator at 37°C and 50% CO_2 for 2 hours, before adding 3mL of 1× phosphate buffered saline (PBS).

The microstructural organization of collagen was assessed using a scanning confocal microscope (Olympus FV1000) in reflection mode with a 60× water immersion lens. A 488nm laser at low intensity (8%) was used to excite collagen, while scattered light was collected in the range 485-495nm [6]. Image stacks (10μm thick with 0.5μm step size) were acquired at a distance of 100μm from the glass. OMTC measurements where performed at distances between 100 and 600μm from the glass. Magnetic twisting was induced in the beads by applying a strong magnetizing pulse horizontally (within the imaging plane) followed by a weak vertical field oscillating at 1Hz. The specific torque T causes a bead displacement d that was detected using a CCD camera mounted on an inverted microscope and a bead-tracking algorithm [5].

The specific torque is defined as a torque per unit bead volume, with dimensions of Pa, and is given by

$$T = cH\cos\theta \quad (1)$$

where H is the magnetic twisting field, θ is the angle between the direction of magnetization and the direction of the bead's magnetic moment, and c is the bead constant [7]. Finally, the complex shear modulus $G(f)$ is obtained via Fourier transformation as

$$G(f) = T(f)/d(f) = G' + iG'' \quad (2)$$

where G' and G'' are respectively the apparent storage and loss moduli, with dimensions of Pa/nm.

RESULTS

Bead calibration requires mixing 20µL of 1% w/v ferromagnetic beads in a silicone fluid of known viscosity (Brookfield Engineering, Middleborough, MA) and tracking the decay of the remnant magnetic field in the direction of magnetization after applying twisting fields of increasing intensities (H = 10, 30, 50, 70G). Linear fits to Equation 1 led to an average bead constant of $c = 0.68 \pm 0.05$ Pa/G for our beads. When beads were embedded within 3D collagen gels we observed consistent sinusoidal bead displacements upon the application of oscillating twisting fields. Equations 1 and 2 allowed us to compute the storage and loss moduli in different imaging planes (along the z axis). Overall, the mechanical behavior of collagen hydrogels was found to be predominantly elastic with the values of storage moduli one order of magnitude larger than the respective loss moduli (Figure 1-b). The apparent stiffness moduli, averaged within each imaging plane, were nearly homogeneous across the thickness of the gel, thus reflecting the homogeneous organization of the underlying collagen network. Such structural homogeneity was confirmed via confocal imaging, which also displayed qualitatively similar structural features between normal collagen gels and collagen-beads mixtures (Figure 2). Streptavidin-coated beads were found to localize around collagen fibrils hence suggesting that the measured displacements, and the resulting stiffness moduli, arise from discrete interactions between beads and fibrils.

DISCUSSION

Altered cell-matrix interactions are implicated in the initiation and progression of multiple pathologies. In particular, malignant tumors are characterized by increased collagen deposition, alignment, and stiffness, and such fibrotic features have been shown to play a causative role in cancer invasion [8]. Novel insights in the basic mechanisms of invasion and metastasis can be achieved if we gain a better understanding of ECM mechanics using realistic *in vitro* models.

To the best of our knowledge, our approach is the first to employ OMTC as a tool to probe the mechanical properties of 3D collagen gels. Prior studies of 3D collagen micromechanics adopted tools such as optical tweezers (i.e., displacement of one bead at a time) [9] and standard magnetic twisting (i.e., cumulative rotation of all embedded beads) [10]. In contrast, OMTC allowed us to track displacements of more than 100 beads per imaging plane, thus displaying the capability of monitoring local mechanical changes due to cellular adhesion and migration. Our findings demonstrate the feasibility of this approach but the physical interpretation of our results is limited by the measurement of an *apparent stiffness*. In OMTC, the apparent modulus can be transformed into a traditional shear modulus by introducing a geometric factor that corrects for the degree of bead embedding. However, in our experiment beads are fully embedded within the matrix and theoretical results from linear elasticity predict no displacements associated with a bead rotation within an infinite homogeneous elastic material. The observed sinusoidal displacements suggest therefore that beads interact with the collagen network only at discrete locations by pivoting around binding sites. A detailed interpretation of our experimental results requires computational modeling of the physiochemical interactions between the collagen network and rotating beads. Future studies will include embedding single cells and multicellular clusters to quantify the mechanical heterogeneities generated by cell-matrix interactions.

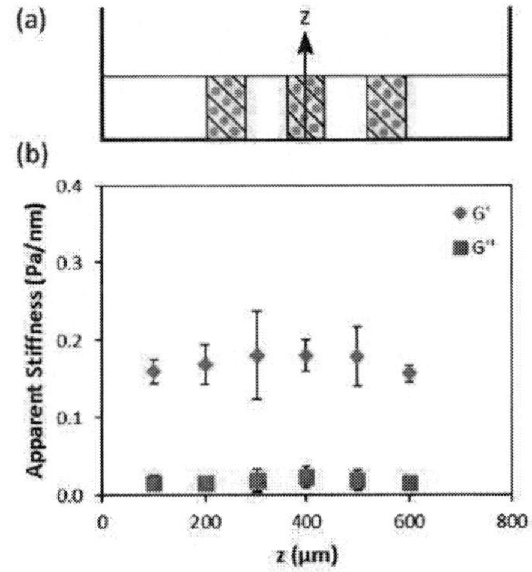

Figure 1: (a) PDMS wells in a 35mm dish housing the mixture of fibrous collagen (dashed lines) and magnetic beads (green circles). (b) Representative OMTC results show a predominantly elastic behavior of collagen and nearly homogeneous properties along z.

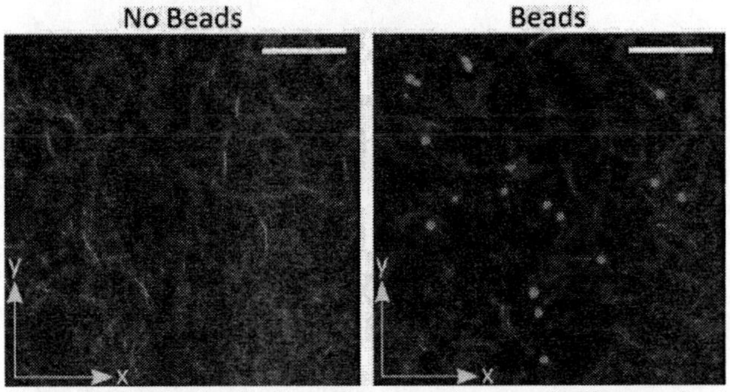

Figure 2: Confocal reflection microscopy images of collagen gels at z = 100µm without beads (left) and with beads embedded in 3D at a 2×10^7 beads/mL concentration (right). Scale bar represents 50µm.

ACKNOWLEDGEMENTS

This work was funded by NIH grant U01 CA202123 (MHZ) and NSF grant CMMI 0954825 (YZ). We acknowledge Professor Jeffrey J. Fredberg and Dr. Chan Y. Park (Harvard T.H. Chan School of Public Health) for insightful discussions and assistance with bead calibration.

REFERENCES

[1] Charras, G. and Sahai, E., *Nat Rev Mol Cell Biol*, 15: 813-24, 2014.
[2] Baker, B.M. et al., *Nat Mater*, 14: 1262-68, 2015.
[3] van Oosten, A.S., *Sci Rep*, 6: 19270, 2016.
[4] Liu, Y. et al., *Interface Focus*, 6: 20150080, 2016.
[5] Fabry, B. et al., *J Appl Physiol*, 91: 986-94, 2001.
[6] Harjanto, D. et al., *PLoS One*, 6: e24891, 2011.
[7] Fabry, B. et al., *Phys Rev E*, 68: 041914, 2003.
[8] Pickup, M.W. et al., *Cancer Res*, 73: 5336-46, 2013.
[9] Velegol, D. and Lanni, F., *Biophys J*, 81: 1786-92, 2001.
[10] Leung, L.Y. et al., *FASEB J*, 21: 2064-73, 2007.

SB³C2017
Summer Biomechanics, Bioengineering and Biotransport Conference
June 21 – 24, Tucson, AZ, USA

MECHANOBIOLOGY OF HEALING: MODELING THE COORDINATION BETWEEN COLLAGEN DEPOSITION AND WOUND CONTRACTION

A. Buganza Tepole (1)

(1) Mechanical Engineering
Purdue University
West Lafayette, IN, USA

INTRODUCTION

Perfect wound healing is challenging due to the difficulty in steering the interplay among different cell types, cell signaling networks and mechanical feedback loops evolving in space and time. We propose a computational framework to study healing dynamics starting from a continuum mechanics description of the tissue supplemented with a computational systems biology approach for cell behavior and signaling network dynamics [1]. At the tissue scale we consider tissue deformations that satisfy mechanical equilibrium, and cells and chemical signal concentrations which obey reaction-diffusion laws. At the cell scale we consider ordinary differential equations for cell behavior. This theoretical framework and its numerical implementation with finite elements allows quantitative reasoning of the healing dynamics, hopefully enabling paths to improve the design of wound management strategies.

We focus here on the aspects of cutaneous wound healing that pertain to restoration of mechanical function. Collagen is the main structural component, it appears in a wavy fiber network in the dermis and with a preferred orientation [2]. The main cell types in the dermis are the fibroblasts and myofibroblasts which are in charge of monitoring and tuning the microstructure to achieve the desired macroscopic mechanical properties [2].

Wound healing consists of four overlapping stages: hemostasis, inflammation, proliferation, and remodeling. We focus on the proliferation and remodeling phases in the dermis. At the beginning of the proliferative phase two main chemical signals are present in the wound and start diffusing outwards: transforming growth factor beta (TGF-β) and platelet derived growth factor (PDGF). These chemokines direct migration of fibroblasts into the wound and potentiate collagen deposition. Fibroblasts respond also to mechanical cues such as strain by producing TGF-β and increasing collagen deposition. TGF-β serves an additional role, it promotes differentiation from fibroblast to myofibroblast. Myofibroblasts are a contractile phenotype capable of compacting the collagen network. As healing progresses, the wound closes and remodels by two main mechanisms: collagen deposition by fibroblasts and wound contraction by myofibroblasts [3].

METHODS

We consider the tissue scale first. In a two-dimensional domain we define a cell density field ρ, a chemokine field c, and a displacement field $\boldsymbol{u}$. The cell density field accounts for fibroblasts and myofibroblasts and the chemokine field corresponds to TGF-β and PDGF. Locally, the tissue microstructure is characterized by the collagen mass fraction ϕ, permanent volume changes λ_a^p, λ_s^p, and collagen alignment $\boldsymbol{a}, \kappa$. Mechanical equilibrium follows from the balance of linear momentum

$$\nabla \cdot \boldsymbol{\sigma} = \boldsymbol{0} \tag{1}$$

While the non-equilibrium time-evolution of the biological fields obeys a usual transport law [4]

$$\dot{\rho} + \nabla \cdot \boldsymbol{q}_\rho = s_\rho, \tag{2}$$

$$\dot{c} + \nabla \cdot \boldsymbol{q}_c = s_c \,, \tag{3}$$

where $\boldsymbol{q}_\rho, \boldsymbol{q}_c$ are flux terms and s_ρ, s_c are source terms.

We include an active stress to model how myofibroblasts exert traction on the tissue, driving wound contraction [5],

$$\sigma^{act} = \phi\rho\left(\tau_f + \frac{\tau_{mc}c}{K_{\tau c}+c}\right) \tag{4}$$

with parameters $\tau_f, \tau_{mc}, K_{\tau c}$. Fibroblasts migrate following Keller-Segel chemotaxis [6]

$$\boldsymbol{q}_\rho = -D^{\rho\rho}\nabla\rho - \rho D^{\rho c}\nabla c \tag{5}$$

with parameters $D^{\rho\rho}, D^{\rho c}$. Cells proliferate following logistic growth and upregulation with Michaelis-Menten kinetics due to the inflammatory signal

$$s_\rho = \left(p_{\rho n} + \frac{p_{\rho c}c}{K_{\rho c}+c} + p_{\rho e}\hat{H}(\theta^e)\right)\left(1 - \frac{\rho}{K_{\rho\rho}}\right)\rho \tag{6}$$

with parameters $p_{\rho n}, p_{\rho c}, p_{\rho e}, K_{\rho\rho}, K_{\rho c}$, and where $\hat{H}$ is a monotonically increasing function of elastic strain which accounts for mechanotransduction.

Locally, the microstructure is remodeled. Collagen production by fibroblasts is dependent on TGF-β in a manner analogous to cell proliferation

$$\dot{\phi} = \left(p_{\phi,n} + \frac{p_{\phi,c}c}{K_{\phi,c}+c} + p_{\phi,e}\hat{H}(\theta^e - \vartheta^e)\right)\left(\frac{\rho}{K_{\phi,\rho}+\phi}\right) - d(c)\phi\rho \tag{7}$$

with the addition of a decay term dependent on the inflammation signal. Lastly, as the myofibroblasts pull on the tissue and the fibroblasts deposit collagen, the contracted configuration is made permanent. Therefore, we define the permanent contracture laws

$$\lambda^p_{a,s} = \frac{\dot{\phi}}{T_{a,s}}(\lambda^e_{a,s} - 1) \tag{8}$$

where $\lambda^e_{a,s}$ are the elastics stretches in the fiber and orthogonal directions respectively, and $T_{a,s}$ are time scale parameters. We solve the equations numerically with the finite element method.

RESULTS

We model a square domain of 100mm×100mm . We create a circular wound of 20mm in diameter at the center. The wound is defined by lack of collagen and fibroblasts; instead, the wound is filled with inflammatory chemokines. Parameters of the simulation are available in [1]. As time progresses, cells migrate into the wound (Fig. 1 top row) while the chemokine diffuses outwards (Fig. 1 second row). Fibroblasts deposit collagen restoring the tissue mechanical response (Fig. 1 third row). The active stress driving wound contraction depends on the myofibroblast density, collagen content, and the chemical signal; thus, the wound contracts primarily in a ring in the periphery of the disrupted tissue (Fig. 1 bottom row). We explore the effect of wound geometry. An elliptical wound oriented orthogonal to the fiber direction (Fig. 2 middle) shows larger contracture with respect to a wound oriented parallel the to the fibers (Fig. 2 bottom).

DISCUSSION

This model describes coupled chemo-bio-mechanical systems which are a unique tool to understand the dynamics of wounds and remodeling processes based on mechanistic considerations. For example, we show the preeminent role of the surrounding tissues in wound contraction compared to the wound bed at early phases of wound healing, a feature highlighted in recent experiments [7]. Further work is needed for calibration against experimental data and extension to more cell types and more sophisticated regulatory networks.

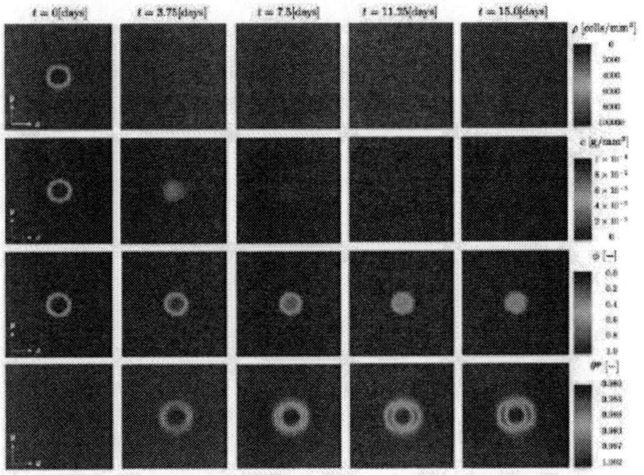

Figure 1: cells (top row) migrate into the wound, chemokine (second row) diffuses out, collagen (third row) is deposited, and contracture (bottom row) originates as a ring (from [1]).

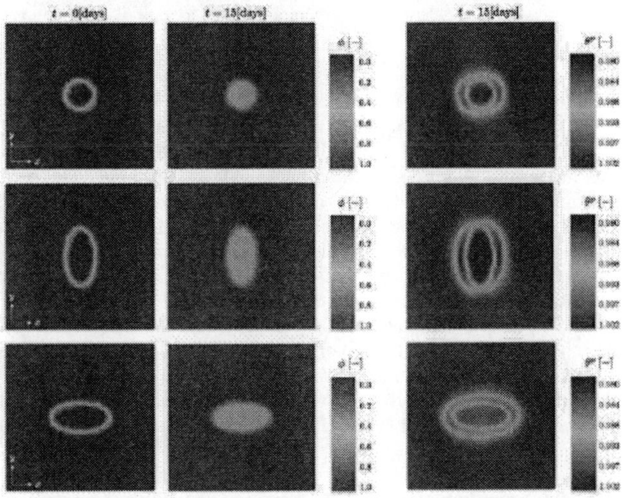

Figure 2: Three different wound shapes with respect to the fiber direction which is aligned with the x axis (from [1]).

REFERENCES

[1] Buganza, A et al., *Comput Method Appl Mech Eng*, 314:46-70, 2017.
[2] Li, B et al., *J Tissue Viability*, 20:108-120, 2011.
[3] Wong, V et al., *J Invest Dermatol*, 131:2186-96, 2011.
[4] Tranquillo, R et al., *Biomech Model Mechanobiol*,158:361-71, 1992.
[5] Loerakker, S et al., *Biomech Model Mechanobiol*, 13:985-01, 2013.
[6] Keller EF, et al, *J Theor Biol*, 30:225-34, 1971.
[7] Schmidt BA, Horsley V, *Development*, 140:1517-2, 2013.

EVALUATION OF STRAIN ENERGY FUNCTIONS FOR THE DEVELOPMENT OF A GROWTH AND REMODELING MODEL OF AGE-SPECIFIC MURINE PATELLAR TENDON HEALING

Akinjide R. Akintunde (1), Kristin S. Miller (1)

(1) Department of Biomedical Engineering
Tulane University
New Orleans, Louisiana, USA

INTRODUCTION

Tendon injuries are common, debilitating, and painful disorders characterized by altered composition, structure, and function. Existing treatment strategies often fail to restore tendon to pre-injury functional capacity [1, 2]. The underlying extracellular matrix (ECM) dynamics that drive post-injury mechanical response and influence the healing process, however, are not fully elucidated [1]. Towards this end, the capability of growth and remodeling (G&R) models to predict (patho)physiologic processes by delineating key mechanisms of matrix dynamics is well documented [cf. 3]. Such models afford time- and cost-efficient frameworks to test hypotheses regarding clinical interventions, as well as the critical timing of these interventions to improve clinical outcomes. However, such G&R models require strain energy functions (SEFs) with microstructural implications, capable of describing salient mechanisms of age-specific tendon healing (such as ECM production/repair, removal/damage, and organization) [4]. Microstructurally-motivated SEFs have the capability to capture the salient response of tendons under normal and (patho)physiologic conditions, thereby aiding our understanding of underlying mechanisms through which tendons attempt to restore homeostasis under altered loading conditions. Several SEFs have been adapted for tendons and ligaments, however, most of them are phenomenological [5]. Phenomenological model parameters are subject to the investigator's interpretation and cannot be measured experimentally, thereby impeding validation efforts and limiting clinical relevance. This study evaluates descriptive capability of two potential SEFs for murine patellar tendon healing, these are: the tendon-specific microstructural "Shearer (SHR)" model [5], and the microstructurally-motivated "Gasser-Ogden-Holzapfel (GOH)" model, which was adapted from cardiovascular tissue [6]. We hypothesize that, the material and structural parameters of each model will decline with age, and that structural parameters, such as fiber alignment, will fail to restore to control values following injury, thereby reproducing prior experimental findings [7-10].

METHODS

Uniaxial load-displacement and cross-sectional area data were obtained from published (IACUC-approved) studies [7-10] based on an established murine patellar tendon injury model [11]. While details can be found in the respective publications, briefly, pre-, and post-injury (at 3 and 6 weeks) ramp-to-failure data for mature (120d), aging (270d), and aged (540d) samples were obtained from excised tibia-patellar tendon-patella units. Load-displacement data were converted to Cauchy axial stress and strain using the cross-sectional area, and undeformed gauge length respectively, while assuming the tendon is incompressible. Guided by estimated normal physiologic strain range of 1-4% stretch in tendons, with acute stress leading to up to ~ 8% stretch [12], a 10% stretch cut-off was applied to the stress-stretch data. The Cauchy axial stress equation for the GOH model is:

$$\sigma_{zz}^{GOH} = C_{iso}\left(\lambda^2 - \frac{1}{\lambda}\right) + \chi^{GOH}\left[\kappa\left(\lambda^2 - \frac{1}{\lambda}\right) + (1-3\kappa)\left(\lambda^2\cos^2\xi - \frac{\sin^2\xi}{2\lambda}\right)\right] \quad (1)$$

where, λ is the axial stretch measured experimentally, $\chi^{GOH} = 4C_1^c v e^{C_2^c v^2}$, $v = [\kappa I_1 + (1-3\kappa)I_4 - 1]$, and the strain tensor 4th invariant, $I_4 = \frac{1}{\lambda}\sin^2\xi + \lambda^2\cos^2\xi$. C_{iso} is the ground-state shear modulus of the non-collagenous matrix when isolated. C_1^c and C_2^c are collagen-related stress-like and dimensionless parameters respectively. ξ, a structural parameter, is the angle from the tendon longitudinal axis to fiber mean direction. The model uniquely accounts for collagen fiber dispersion via structural parameter, κ, which equals 0 when fibers are perfectly aligned and 1/3 when distributed isotropically about the fiber mean direction. The SHR model Cauchy axial stress is:

$$\sigma_{zz}^{SHR} = (1-\phi)\mu\left(\lambda^2 - \frac{1}{\lambda}\right) + 2\chi^{SHR}\left(\lambda^2\cos^2\psi - \frac{\sin^2\psi}{2\lambda}\right) \quad (2),$$

$$\text{where, } \chi^{SHR} = \begin{cases} 0 & I_4 < 1 \\ \frac{\phi E \cos\alpha}{6\sqrt{I_4}\sin^2\theta_0}\left(2 - \frac{3\Gamma+1}{\Gamma\sqrt{\Gamma}}\right) & 1 \le I_4 \le \lambda^{*2} \\ \frac{\phi E \cos\alpha}{2\sqrt{I_4}}\left(\beta - \frac{1}{\sqrt{\Gamma}}\right) & I_4 > \lambda^{*2} \end{cases}, \quad \beta = \frac{2(1-\cos^3\theta_o)}{3\sin^2\theta_o},$$

ϕ is the collagen volume fraction, $(1-\phi)\mu$ is same as c_{iso}, previously defined. ϕE is fibril Young's modulus, θ_0 is the crimp angle of outermost fibrils, critical stretch, $\lambda^* = \frac{1}{\cos\alpha}\sqrt{\frac{1}{\cos^2\theta_o} - \sin^2\alpha}$ and $I_4 = \frac{1}{\lambda}\sin^2\psi + \lambda^2\cos^2\psi$. $\Gamma = \sin^2\alpha + I_4\cos^2\alpha$, α is the fibril helix angle and ψ is the fascicle helix angle. Existing and potential experimental techniques to measure the model parameters are described in Shearer's article [5].

To identify the best-fit material parameters for each model, the function *lsqnonlin* in MATLAB was employed to minimize the objective function, $f(\vec{x}) = \sigma_{zz}^{Theor}(\vec{x}) - \sigma_{zz}^{Exp}(\vec{x})$. σ_{zz}^{Theor} and $\sigma_{zz}^{Exp}(\vec{x})$ are the theoretical and experimental Cauchy axial stresses respectively. $\vec{x}$ is a vector of the model parameters, each component subjected to lower and upper bounds (guided by available experimental data). While c_1^c and $c_2^c \ge 0$, using published [13] shear and Young's moduli for 150d old mice patellar tendons, bounds for $(1-\phi)\mu$ and C_{iso} were set as mean ± 2(Std. Dev.) of the measured values, and $0 \le \phi E \le 1$ GPa. Due to lack of experimental/imaging data on the structural parameters, bounds were set theoretically as follows: $\kappa \in [0, 1/3]$, θ_0, α, and $\psi \in [0, \pi/2]$.

RESULTS

Figures 1 and 2 show the models' fits to group average experimental data during healing and aging.

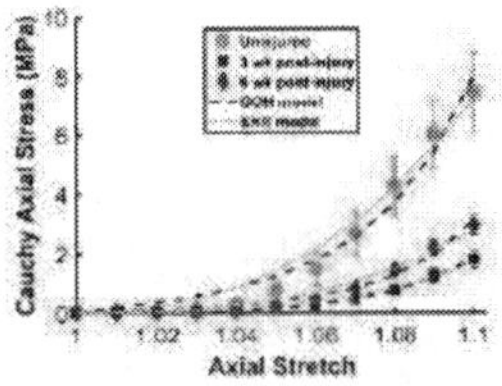

Figure 1: Healing: Data (closed symbols: avg ± SEM) fit to GOH and SHR models for mature (120d) group.

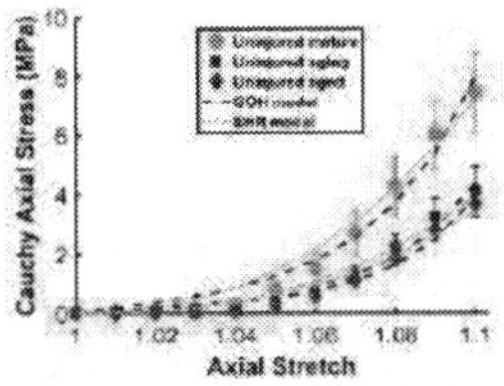

Figure 2: Aging: Data (avg ± SEM) fit to GOH and SHR models for mature, aging (270d) and aged (540d).

GOH and SHR model average parameters for each age-group and injury time point are listed in **Tables 1 and 2** respectively.

Table 1: Average model parameters (Mean ± Std. Error of the Mean) for the GOH model.

	C_{iso} (kPa)	C_1^c (MPa)	C_2^c	κ ±0.010	ξ (°)
120 - Ctl	8.0±1.7	3.7±0.9	40±7	0.020	8±2
- 3wk	3.8±0.2	0.5±0.1	48±4	0.013	5±3
- 6wk	4.4±0.6	1.4±0.5	54±6	0.021	4±2
270 - Ctl	4.1±0.5	2.2±0.7	59±5	0.032	7±2
- 3wk	5.3±0.5	1.4±0.3	60±10	0.024	10±4
- 6wk	4.5±0.5	1.5±0.4	76±12	0.027	14±4
540 - Ctl	5.0±0.6	1.5±0.4	58±9	0.015	8±3
- 3wk	4.4±0.7	0.2±0.1	56±5	0.014	4±2
- 6wk	3.9±0.3	1.4±0.5	54±7	0.010	10±5

Table 2: Average model parameters for the SHR model.

	$(1\text{-}\phi)\mu$ (kPa)	ϕE (MPa)	θ_o (°)	α (°)	ψ (°)
120 - Ctl	7.5±1.6	701±97	32±15	16±11	45±11
- 3wk	6.2±1.0	747±72	14±3	9±3	49±3
- 6wk	7.8±1.1	648±80	19±4	16±4	46±4
270 - Ctl	8.3±1.1	596±80	20±4	11±4	42±3
- 3wk	6.6±0.7	626±88	12±3	9±4	35±5
- 6wk	6.8±0.9	669±74	17±4	10±4	48±4
540 - Ctl	7.3±0.9	630±77	16±3	8±3	42±4
- 3wk	4.7±0.9	814±89	12±3	4±3	50±3
- 6wk	8.0±1.3	586±119	8±3	4±3	36±7

To evaluate our hypothesis, a 2-way ANOVA was performed, its results are shown in **Tables 3 and 4**.

Table 3: Results of 2-way ANOVA on GOH parameters

factor	C_{iso}	c_1^c	c_2^c	κ	ξ
Age	0.32	0.24	0.42	*0.04	0.17
Injury	#0.06	0.72	0.47	0.81	0.68
Age + Injury	*0.002	0.26	0.49	0.95	0.55

Table 4: Results of 2-way ANOVA on SHR parameters

factor	$(1\text{-}\phi)\mu$	ϕE	θ_o	α	ψ
Age	0.9	0.64	*0.04	#0.06	0.45
Injury	0.11	0.36	*0.02	0.45	0.91
Age + Injury	0.59	0.69	0.56	0.87	#0.09

*$p \le 0.05$ (statistical significance) and #$0.05 < p \le 0.1$ (statistical trend)

DISCUSSION

Both models exhibited acceptable fits to experimental data, although GOH model ($0.97 \le R^2 \le 0.99$) exhibited a relatively better fit than SHR ($0.94 \le R^2 \le 0.97$) with lower average root mean square error range of 0.05 - 0.35 MPa to SHR's 0.09 - 0.47 MPa. Unexpectedly, none of the material parameters for either model were statistically significant in aging or injury, except for C_{iso} which showed a trend during age-specific healing. C_{iso} showed significant interaction between aging and injury. Statistical significance of κ indicates the altered fiber alignment which has been observed post-injury in the patellar tendon [7-10]. For the SHR model, crimp angle, θ_o was significant in aging and injury. The fibril helix angle, α, exhibited trend with respect to age, and fascicle helix angle, ψ showed trend for interaction between age and injury.

The results of this study suggest that the structural parameters exhibited the highest significance, e.g., collagen organization at different architectural levels (fibril to fascicle), and that cells may preferentially reorganize the ECM in response to altered stimuli brought about by injury (and aging) rather than producing or removing collagen (i.e., compositional changes). Ongoing work (guided by histological, imaging data and physiology over statistical dictates) is focused on assessing model over-parameterization and then sensitivity analysis. This work also highlights the need for further experiments in order to provide physiologic bounds for microstructural model parameters to ensure that the relevant ECM and potential (patho)physiological mechanisms are captured.

ACKNOWLEDGEMENTS

We would like to thank Lou Soslowsky for the experimental data used in this study.

REFERENCES

[1] Andarawis-Puri, N. et al., J Orthop Res, 33:780-784, 2015.
[2] Docheva, D. et al., Adv Drug Deliv Rev, 84:222-239, 2015.
[3] Ateshian & Humphrey, Annu Rev Biomed Eng, 14:97-111, 2012.
[4] Humphrey & Rajagopal, Math Model Meth Appl Sci, 12:407-430, 2002.
[5] Shearer, T., J Biomech, 48:3017-3025, 2015.
[6] Gasser, T. et al., J Royal Society Interface, 3:15-35, 2006.
[7] Dunkman, A. et al., Matrix Biol, 32:3-13, 2013.
[8] Dunkman, A. et al., Matrix Biol, 35:232-238, 2014.
[9] Dunkman, A. et al., Ann Biomed Eng, 42:619-630, 2014.
[10] Mienaltowski et al., J Orthop Res, 34:1256-1263, 2016.
[11] Beason, D. et al., J Biomech, 45:1550-1553, 2012.
[12] Kannus, P., Scand J Med Sci Sports, 10:312-320, 2000.
[13] Dourte, L et al., J Orthop Res, 31:1430-1437, 2013.

BIOMECHANICAL COMPARISON OF ANATOMICAL OSTEOCHONDRAL ALLOGRAFT VS. "SNOWMAN" CONFIGURATION

Ferris M. Pfeiffer (1,2), Aaron M. Stoker (1), James P. Stannard (1), James L. Cook (1)

(1) Department of Orthopaedic Surgery
University of Missouri
Columbia, MO, USA

(2) Department of Bioengineering
University of Missouri
Columbia, MO, USA

INTRODUCTION

The majority of focal articular cartilage defects in the femoral condyles requiring surgery cannot be effectively treated using a single cylindrical osteochondral allograft (OCA). In these situations, multiple (2-4) cylindrical OCAs are used to fill the entire defect in a "snowman" configuration. While this technique can replace the damaged area with viable, functional tissue, it has numerous shortcomings. First, the "snowman" technique is technical demanding with increased risk for complications. The use of multiple OCAs also introduces additional interfaces (e.g. between each OCA) that can compromise healing, integration and functional performance of the entire construct. Additionally, this technique results in suboptimal restoration of the natural curvature of the condyle due to the use multiple independent curved surfaces. Finally, for repair of non-circular articular defects, using cylindrical OCAs can require removal of significant healthy tissue. Thus, an alternative that can overcome these shortcomings involves the use of a non-circular or anatomical OCA that can resurface large cartilage defects with a single anatomically curved section. Such an OCA will benefit from a tapered interface that will minimize damage at insertion, but provide proper fixation strength. Thus, we hypothesize that a single anatomically shaped osteochondral allograft will provide superior fixation strength and graft congruency as compared to multiple circular cross-sectioned grafts.

METHODS

Human tissue:
Fresh (n=8) and frozen (n=10) human femoral condyles were acquired from cadaveric tissue sources including Musculoskeletal Transplant Foundation (MTF, Edison, NJ) and Science Care (Phoenix, AZ). Anatomical (n=9) (figure 1) and snowman (n=9) (figure 2) osteochondral allografts were harvested from each condyle. Prior to graft transplantation and following graft harvest, site preparation, and graft implantation, the recipient site was scanned using a 3D laser scanning system (David Laser, Germany) to compare the pre-implant geometry with the post implanted geometry. Surface images were evaluated using Matlab (Mathworks, Natik, MA) to quantify congruency. Following implantation, condyles were mounted in a materials test machine (Textural Analyzer, Stable Micro Systems,

Figure 1. Example harvest procedure for human anatomical osteochondral allograft including a),b) guide placement, c) site preparation, d) anatomical graft, and e) implanted graft.

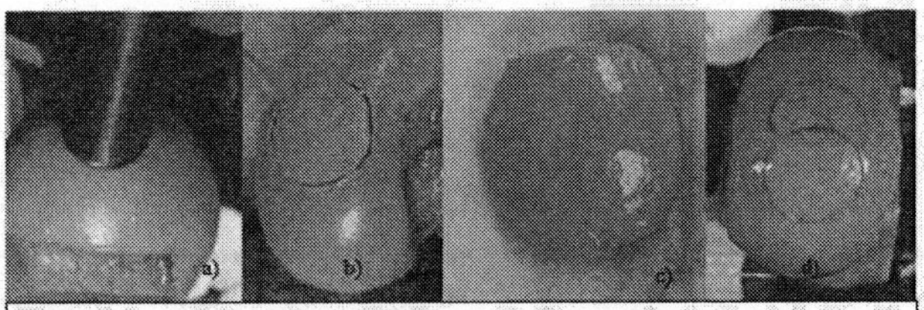

Figure 2. Example harvest procedure human for "snowman" osteochondral allograft including a) trephine harvest, b),c) single harvested graft, and d) implanted grafts.

London) and placed against an opposing tibial plateau. Contact pressure was measured using a pressure mapping system (Tekscan, Boston, MA) by placing a human condyle sensor between the implanted condyle and the opposing tibial plateau. Contact pressure was measured immediately following implantation and following 500 cycles at 1Hz. Finally, the extraction strength was measured by applying an extraction force to the underside of the OCA using a rigid rod placed through a pre-drilled hole under the OCA.

Canine tissue:
Fresh (n=10) canine femoral condyles and tibial plateaus were acquired from animals sacrificed for reasons unrelated to this study. Anatomical (n=5) (figure 3) and snowman (n=5) osteochondral allografts were harvested from each condyle.

Figure 3. Example harvest procedure for canine anatomical osteochondral allograft including a),b) guide placement, c) site preparation, d) anatomical graft, and e) implanted graft.

Tibial plateau with attached meniscus (n=10) (figure 4) were harvested from each Missouri Osteochondral Preservation System (MOPS) preserved donor tibial plateau and implanted into a canine cadaver limb from an animal sacrificed for reasons unrelated to this study.

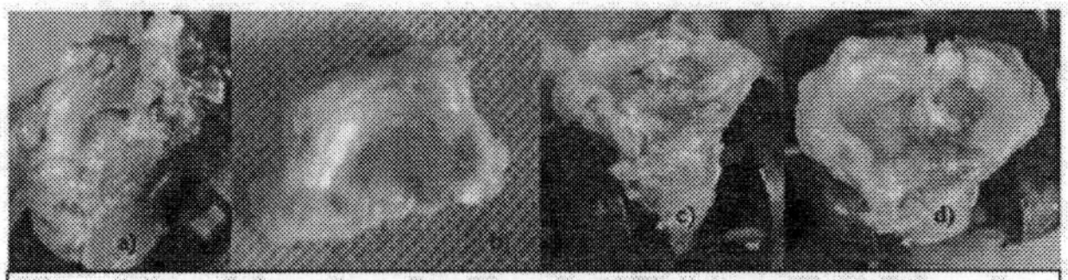

Figure 4. Example harvest procedure for canine tibial plateau with attached meniscus allograft including a) MOPS preserved tissue,b) prepared graft, c),d) implanted graft

Prior to graft transplantation and following graft harvest, site preparation, and graft implantation, the recipient site was scanned using a 3D laser scanning system (David Laser, Germany) to compare the pre-implant geometry with the post implanted geometry. Surface images

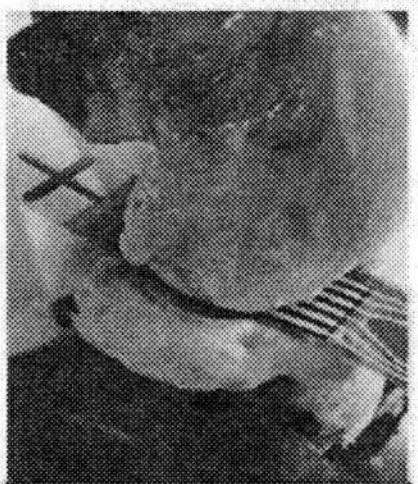

Figure 5. Experimental setup for contact pressure testing in canine biojoint samples.

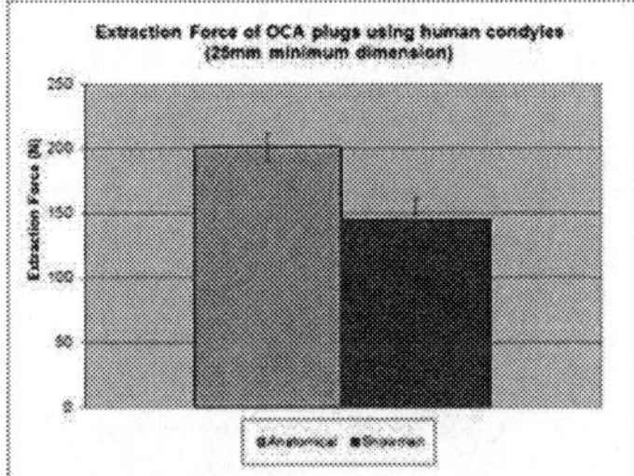

Figure 6. Graft extraction force for a single(25mm minimum dimension) human anatomical OCA vs 2 (25mm diameter) "snowman" grafts.

were evaluated using Matlab (Mathworks, Natik, MA) to quantify congruency. Following implantation, condyles were mounted in a materials test machine (Textural Analyzer, Stable Micro Systems, London) and placed against an opposing tibial plateau. Contact pressure was measured using a pressure mapping system (Tekscan, Boston, MA) by placing a canine condyle sensor between the implanted condyle and the opposing tibial plateau (figure 5).

Contact pressure was measured immediately following implantation and following 500 cycles at 1Hz. Finally, the extraction strength was measured by applying an extraction force to the underside of the OCA using a rigid rod placed through a pre-drilled hole under the OCA.

RESULTS

Human Tissue:
Anatomical graft extraction force was significantly higher for the single anatomical graft vs. the multiple "snowman" grafts (figure 6).

Graft congruency
Pre- and post-transplantation surface scanned images were compared using a custom Matlab program to quantify deviations in surface congruency.

Canine Tissue:
Anatomical graft extraction force was higher for the single anatomical graft vs. multiple "snowman" grafts (figure 7). More variability was seen in the extraction strength of the small size canine tissue grafts as compared to the larger grafts using human tissue. Such variability may be in part due to the technical difficulty of creating a reproducible interface at a smaller scale.

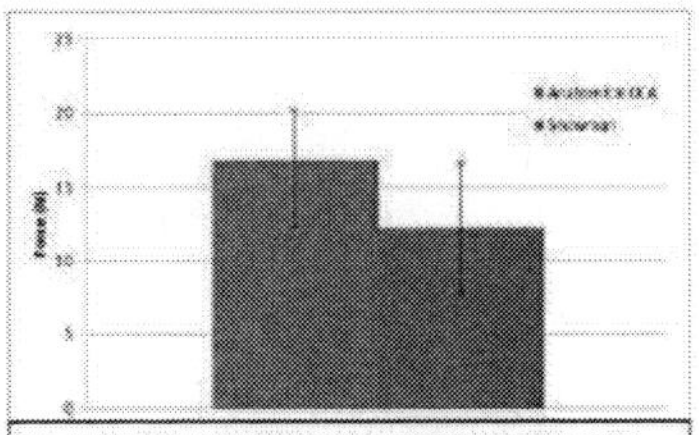

Figure 7. Graft extraction force for a single(6mm minimum dimension) canine anatomical OCA vs 2 (10mm diameter) "snowman" grafts.

Contact pressure mapping:
A marked difference was observed between contact pressures in the joints with a single anatomical graft (figure 8) vs. joints with multiple "snowman" grafts (figure 9). In general, the single anatomical grafts had a more evenly distributed pressure and demonstrated fewer "hot spots" as compared to the multiple grafts of the "snowman" procedure. Efforts are ongoing to quantify the magnitude of the difference.

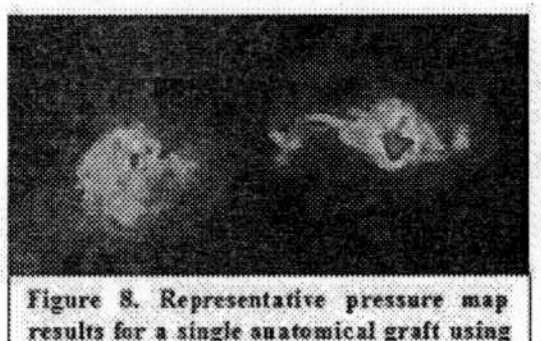

Figure 8. Representative pressure map results for a single anatomical graft using canine tissue.

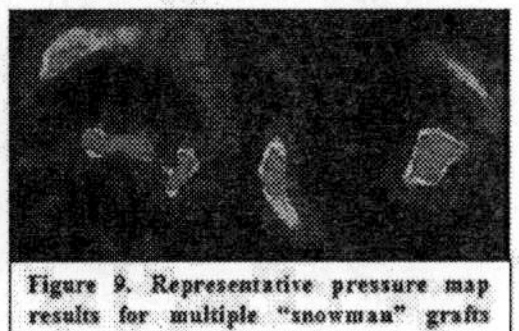

Figure 9. Representative pressure map results for multiple "snowman" grafts using canine tissue.

DISCUSSION

The results of this study demonstrate superior extraction strength properties and improved graft congruency for a single anatomical osteochondral allograft as compared to multiple circular cross section OCAs.

ACKNOWLEDGEMENTS
This project was supported by the DOD CDMRP Award W81XWH-15-1-0567

SB³C2017
Summer Biomechanics, Bioengineering and Biotransport Conference
June 21 – 24, Tucson, AZ, USA

PRESSURE-INDUCED CHANGES IN THE REGIONAL STRUCTURAL ARCHITECTURE OF THE PORCINE TRICUSPID VALVE LEAFLETS

Vineet S. Thomas (1), Anup D. Pant (1), Anthony Black (1), Taylor Verba (1), Rouzbeh Amini (1)

(1) Department of Biomedical Engineering
The University of Akron
Akron, Ohio, USA.

INTRODUCTION

The underlying arrangement of the extracellular matrix (ECM) components of the tricuspid valve (TV) leaflets influences their mechanical responses. In addition, ECM structural remodeling occurs in response to the modified mechanical environment induced by physiological changes (e.g. pregnancy), diseases (e.g. pulmonary hypertension), and/or TV repair/replacement surgeries. In this work, we have examined how increasing ventricular pressure alters the regional structural properties of the porcine TV leaflets.

METHODS

Fresh porcine hearts were obtained from a local slaughterhouse (3-D Meats, Dalton, Ohio) immediately after the animals were slaughtered. In previous studies, cardiac valves were isolated to remount pressurization setups [1]. However, the annular restriction of the atrioventricular valves could significantly change the mechanical responses of the leaflets [2]. As such, in our study, the *entire* ventricles were hydrostatically pressurized via an inlet through the pulmonary artery. Eight hearts were fixed using glutaraldehyde (0.5% v/v) under 25 mmHg ventricular pressure for 6-8 hours. Another group of eight samples were fixed without any pressure for the same period of time. The anterior, posterior, and septal leaflets of the valves were then excised and dehydrated using glycerol solutions as described previously by Joyce et al. [1]. The microstructural architecture of the leaflets was obtained using a custom-made small angle light scattering (SALS) equipment (Fig. 1a).

To minimize the effects of geometric variability among the samples, every scanned image of the leaflets was mapped to a nine-node biquadratic computational domain (Fig 1). Using nine equally spaced regions in the computational domain (Fig 1b), we divided each leaflet to nine regions in a curvilinear manner (Fig 1a).

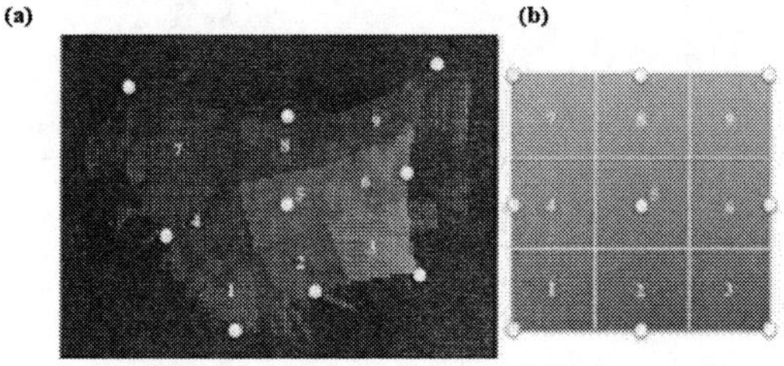

Figure 1: (a) Regions 1 to 9 used for statistical comparison on the physical domain of a typical TV posterior leaflet; (b) nine-node computational domain used for mapping.

From the SALS scans, the main fiber distribution μ and the normalized orientation index (NOI, a measure of ECM network splay [3]) were available for data points on each region with a resolution of 250 μm x 250 μm. The ECM fiber network for each data point was expressed using von Mises distribution [4]:

$$R(\theta;\kappa,\mu)=\frac{1}{\pi I_0(\kappa)}\exp(\kappa\cos(2(\theta-\mu)))\text{....}\quad(1)$$

with, $I_0(\kappa)$ being the modified Bessel function of the first kind of zero order and κ being a measure of network splay calculated from experimentally obtained NOI. To obtain a single value of μ and NOI for each region on each specimen, the individual von Mises distributions of each data point were augmented and normalized to combine a new distribution function. A non-linear least square fitting scheme was used to fit the combined distribution to a new von Mises distribution (Fig. 2). The main fiber distribution of the fitted function was used as the average μ' of the region and κ of the fitted distribution was employed to calculate the average NOI′ of the region.

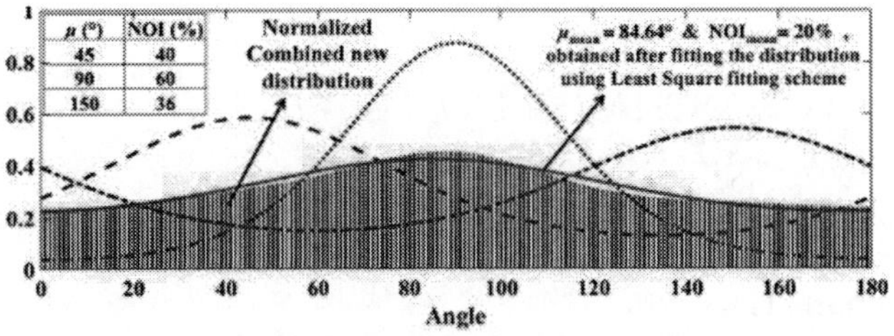

Figure 2: Computing NOI′ and μ' using augmentation and normalization of regional von-Mises distributions of three typical data points.

The two-tailed Student's t-test was then performed to compute the statistical significance ($p<0.05$) of average fiber structural parameters between the pressurized and non-pressurized samples in each leaflet. After the the averaged values *within* each region of *each* sample (i.e. μ' and NOI′) were obtained, the following equations were used to obtain the average value of each region *among all samples* [5, 6]:

$$a = \frac{\sum_{i=1}^{N} f(\mu_i')\cos(2\mu_i')}{\sum_{i=1}^{N} f(\mu_i')} \quad (2) \qquad b = \frac{\sum_{i=1}^{N} f(\mu_i')\sin(2\mu_i')}{\sum_{i=1}^{N} f(\mu_i')} \quad (3)$$

$$\mu_{mean} = \frac{1}{2}\tan^{-1}\left(\frac{b}{a}\right) \text{ if } a>0 \; ; \mu_{mean} = \frac{1}{2}\left[\pi + \tan^{-1}\left(\frac{b}{a}\right)\right] \text{ if } a<0 \quad (4)$$

where, $f(\mu_i')$ was the ratio of fibers oriented along μ_i'. The average value of NOI was calculated using arithmetic mean and standard deviation.

$$NOI_{mean} = \frac{\sum_{i=1}^{n} NOI_i'}{n} \quad (5) \qquad \sigma_{NOI} = \sqrt{\frac{\sum_{i=1}^{n}\left(NOI_i' - NOI_{mean}\right)^2}{n}} \quad (6)$$

NOI_{mean} and μ_{mean} were also computed using augmentation and normalization of the regional von Mises distributions.

RESULTS

Fig. 3 shows the relationship between the NOI_{mean}, obtained using the two different methods described above. Linear regression of the data led to $y = 1.05x+10.17$ with the R^2 value of 0.72, which indicated that the two methods produced relatively similar mean values.

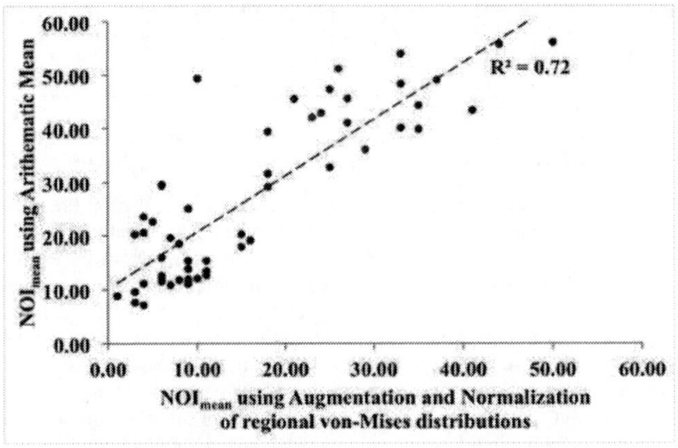

Figure 3: Comparison between the two averaging methods.

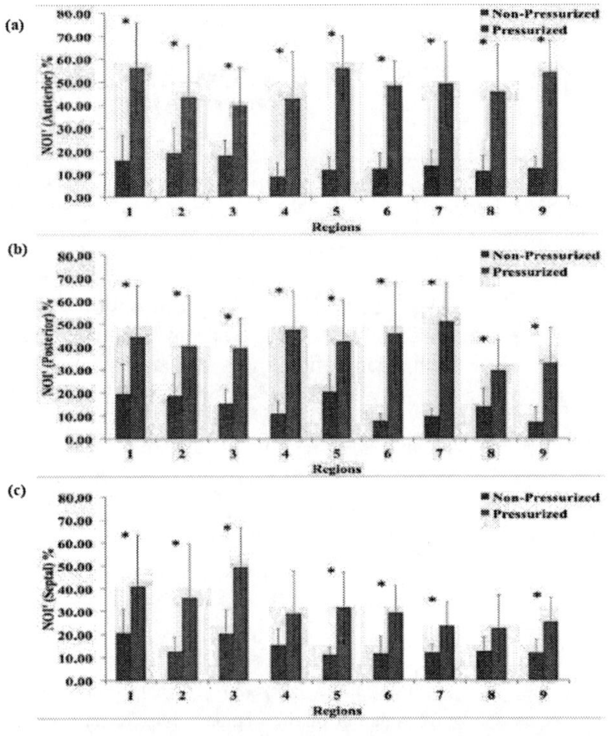

Figure 4: Statistical comparison of regions 1 to 9 for (a) anterior, (b) posterior and (c) septal leaflets in the non-pressurized state and the pressurized state (n = 8, bars are standard errors); asterisks show data that are significantly different.

Fig. 4a-4c demonstrates how ventricular pressurization increased ECM network organization for each of the nine regions of TV leaflets. The average network NOI was significantly higher in regions 1 to 9 of the anterior and posterior leaflets and in regions 1,2,3,5,6,7, and 9 of the septal leaflet.

DISCUSSION

Although the thickness, mechanical properties, and potentially regional loading environment are different among the three TV leaflets [7], it was observed that the ECM became more organized in all of them when the ventricular pressure increased. Such a pressure-induced alteration in ECM could provide insight into the normal and abnormal multi-scale biomechanics of the valve. For example, ECM remodeling could play an important role in the secondary TV regurgitation caused by pulmonary hypertension. The experimental results are also useful for validation of multi-scale computational models that provide predication of network deformation [8].

ACKNOWLEDGEMENTS

Funding for this project was provided in part by American Heart Association Grant No. 16SDG27770088.

REFERENCES

[1] Joyce, E. M. et al., *Ann of Thor Surgery*. 87.4:1240-1249, 2004.
[2] Amini, R. et al., *Ann of Biomed Eng*. 40.7:1455-1467, 2001.
[3] Sacks, M. S. et al., *Ann of Biomed Eng*. 25.4: 678-689, 1997.
[4] Gouget, L.M.G. et al., *J Biomech Mechanobio*. 11:475-482, 2012.
[5] Sander, E.A. et al., *J Biomed Mat Research*. 88A:322-331, 2009.
[6] Mardia K.V., Jupp P.E, In Directional Statistics. NY: Wiley; 2000.
[7] Aimini Khoiy, K. et al., *J of Biomech Eng*. 138.11:111006, 2016.
[8] Amini, R. et al., *J of Biomech Eng*. 136.3: 031003, 2014.

EFFECT OF GEOMETRIC REMODELING ON MITRAL VALVE LEAFLET MECHANICS: AN *EX VIVO* INVESTIGATION

Charles H. Bloodworth IV (1), Eric L. Pierce (1), Nancy J. Deaton (1), Michael S. Sacks (2), and Ajit P. Yoganathan (1)

(1) Wallace H. Coulter Department of Biomedical Engineering, Georgia Institute of Technology and Emory University, Atlanta, GA, USA

(2) Institute for Computational Engineering and Sciences, University of Texas at Austin, Austin, TX, USA

INTRODUCTION

In ischemic mitral regurgitation (IMR), MV insufficiency is *primarily* caused by remodeling of the papillary muscles and mitral annulus due to coronary artery blockage, which leads to improper coaptation of the mitral leaflets. To treat IMR, the MV can either be repaired or replaced altogether. Despite the widespread use of restrictive annuloplasty (RA) in repairing IMR [1], recurrence of MR following this repair is a persistent clinical problem. Gelsomino reported 33% recurrence of moderate to severe MR at three-year follow-up [2].

In this setting, numerous clinical investigations report a strong correlation of MR recurrence with left ventricle sphericity index, indicating that when displacement of papillary muscles (PMs) persists after RA, leaflet function remains impaired [3]. Recent studies also suggest that IMR is not solely a consequence of the structures *surrounding* the MV, but that leaflet remodeling at the micro-level occurs during disease progression, and after repair [4]. Alteration of the physiological environment of the MV, specifically, that of tissue loading is implicated as a possible cause for adverse leaflet remodeling, and, thereby, a contributor to MR recurrence [4].

Using an *ex vivo* left heart simulator, we sought to replicate MV geometric distortions characteristic of IMR and RA, and to describe the impact of these geometries on the leaflets' response to loading. Using an internally-controlled model, leaflet strain and radius of curvature (ROC) were computed at peak systole for each valve and in each geometry. Increased curvature of MV leaflets (ie smaller ROC) has been shown to reduce leaflet stress [5]. We hypothesized that IMR and RA would exacerbate leaflet strain and ROC, relative to the healthy state, because of the flattening of the annulus and displacement of the PMs.

METHODS

Pulse Duplicator. Novel *ex vivo* data collection and imaging methods for this study have been published by our group [6]. Briefly, ovine MVs (n=9) were excised and mounted within a cylindrical left heart simulator (CLHS). Left heart pressures and flows (5 liters/min, 100 mmHg peak MV gradient, 70 beats/min) were replicated within the simulator using a custom LabVIEW (National Instruments, Austin, TX) controlled pulse duplicator and flow loop. A novel device was used to control the mitral annular geometry in a static fashion, and papillary muscle (PM) positions were also adjustable.

On each MV, three geometric configurations were replicated. First, healthy MVs had a saddle shaped annular geometry, and PMs were positioned for healthy MV closure. Second, a geometry representative of IMR was replicated by dilating and flattening the annulus, and displacing the PMs. PM displacement methods are further described in Bloodworth *et al* [6]. Finally, the annulus was downsized and kept flat, with preserved PM displacement to replicate a repair with flat restrictive annuloplasty. The annular geometries corresponding to these configurations, referred to as 'healthy', 'IMR' and 'repaired', were derived from previously published data for healthy/IMR ovine subjects [7,8], as well as the dimensions of the Carpentier-Edwards Classic annuloplasty ring (**Table 1**).

	Healthy	IMR	Repaired
A/P dist. (mm)	24	30	24
C/C dist. (mm)	30	33	30
AHCWR (%)	15 (saddle)	0 (flat)	0 (flat)
PM Position	Normal	Displaced	Displaced

Table 1. Annular and papillary muscle geometries. A/P – Anterior/Posterior width, C/C – intercommissural width, AHCWR – Annular height to commissural width ratio.

Acquisition of 3D Echo and ROC Measurement. Within the pulse duplicator, each valve was imaged using 3D gated clinical echocardiography (3D echo) in each geometric configuration. Philips QLab software was used to analyze 3D echo images. Cut-planes were positioned to transect the MV at the A1/P1, A2/P2 or A3/P3 plane. At each of these cut-planes, landmark points were positioned at the annulus, belly, coaptation start, and edge of each leaflet. ROC was measured by defining a circle using three landmarks.

MicroCT Scans and Strain Measurement. Following acquisition of 3D echo for each of the geometric states, the CLHS chamber was drained of fluid and detached from the pulse duplicator in preparation for scanning by micro-computed tomography (microCT). A mesh of microCT-compatible fiducial markers were then attached to the mitral leaflet surfaces using previously published techniques [9]. This enables discrete one-to-one marker tracking between microCT scans taken under different geometric and loading conditions. Following the application of markers, the CLHS was pressurized using humidified air to a static MV gradient of 100 mmHg, and scanned once for each geometric configuration. These pressurized scans faithfully simulate the systolic closure of the valve in a static fashion for the duration of the 7 minute microCT scan. Finally, each MV was then submerged in static water and scanned for an unloaded reference state in its healthy configuration. Areal green strains were calculated using the high-contrast leaflet markers as previously reported [6,9].

Data Analysis. All data was analyzed using MatLab. ROC and strain measurements were classified by state (healthy, IMR, or repaired), leaflet (anterior or posterior), zone (coaptation, or belly), and segment (A1/P1, A2/P2, or A3/P3). These classifications were treated as ANOVA factors, and the significance of each individual factor was assessed using a multi-way ANOVA followed by a Tukey-Kramer test for significance. The analysis of interactions between factors, and ROC/strain correlations are not included in this brief report, but will be presented at SB3C. All data are presented below as mean ± 95% CI.

RESULTS

Leaflet ROC results are shown in **Figure 1**. ROC was found to be significantly larger in the belly region of leaflets than in the coaptation zone (1.08±.05 versus 0.54±.05 cm, $p<0.001$). Furthermore, anterior leaflet (AL) ROC was significantly larger than that of the posterior leaflet (PL) (1.11±.05 versus 0.51±.05 cm, $p<0.001$). However, overall radius of curvature did not change significantly with MV state.

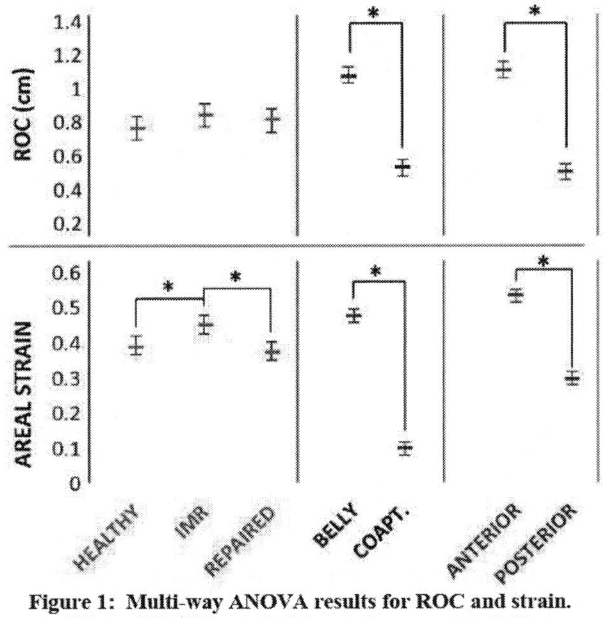

Figure 1: Multi-way ANOVA results for ROC and strain. Mean±95%CI is plotted for each group. * $p<0.05$

Leaflet areal strain results are shown in the lower panel of **Figure 1**. Strain magnitude on the anterior leaflet was in the range of previously published results [10]. Measured strain was significantly greater in the belly region of leaflets than in the coaptation zone (47±2% versus 10±2%, $p<0.001$), and AL strain was significantly greater than PL strain (53±2% versus 29±2%, $p<0.001$). Additionally, leaflet strain in the IMR state was significantly greater than that of both the healthy and repaired states (45±3% versus 39±3% and 38±3%, $p<0.001$). Strain patterns can be observed in **Figure 2**, which shows an averaged strain map of all nine MVs on a representative leaflet geometry.

DISCUSSION

In this study, we observed mitral leaflet curvature and strain in the setting of geometric MV remodeling on a cohort of nine excised ovine MVs. Across different leaflet segments and zones, ROC and strain were found to vary greatly. This is evidenced by the large differences found between the belly and coaptation zones, and between the anterior and posterior leaflets. ROC and strain differences were less pronounced across the different states.

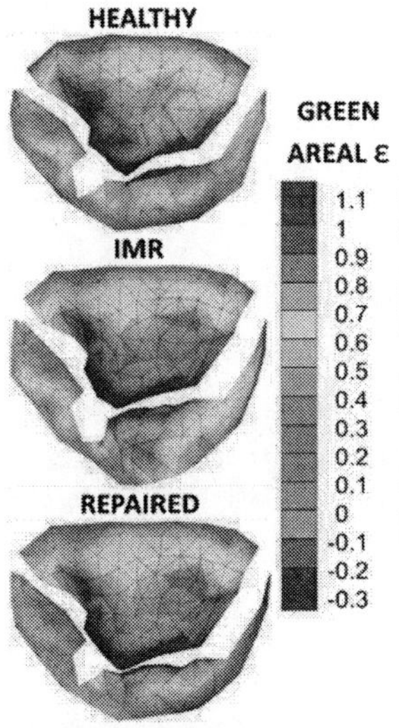

Figure 2: Local leaflet strain averaged across all MVs (n=9) and shown on a single representative MV for all states.

However, interesting trends can be qualitatively observed when examining the strain maps in **Figure 2**. Annular dilatation along the posterior aspect appears to cause increased strain in the posterior leaflet in the IMR state. Furthermore, an increase in strain on the right side of the anterior leaflet in both the IMR and repaired states is apparent. This could be due to the larger displacement of the posteromedial PM.

MV deformation was measured using 3D echo during dynamic MV function, and with novel microCT techniques under static loading. In combination, a detailed picture of MV loading was produced as a function of geometric distortion. Our in-depth analysis continues; examining the interaction of factors in our statistical model will allow us to more quantitatively evaluate patterns, and reveal more nuanced differences of the observed deformations. An understanding of the changes in MV loading that take place in IMR and after repair may lead to improved ability to restore a physiological mechanical environment, addressing the issue of long-term recurrence of MR after repair.

ACKNOWLEDGEMENTS

This work was partially supported by the National Science Foundation Graduate Research Fellowship (ELP) under Grant DGE-1148903, as well as by the NHLBI under Grant R01HL119297. The authors would also like to acknowledge Vinay Bhal, Camille Johnson, and Sriram Srirangam for their assistance.

REFERENCES

[1] Gillinov, A et al., *J Thorac Cardiovasc Surg,* 122:1125–1141, 2001.
[2] Gelsomino, S et al., *Eur Heart J,* 29:231-240, 2008.
[3] Kron, I et al., *J Thorac Cardiovasc Surg,* 149(3):752-761, 2015.
[4] Connell, P et al, *ABME,* 44(4):954-967, 2011.
[5] Salgo, I et al., *Circulation,* 106:7 11-717, 2002.
[6] Bloodworth, C et al, *ABME,* 2016 doi:10.1007/s10439-016-1734-z
[7] Daimon, M et al, *Circulation,* 111:2183-2189, 2005.
[8] Vergnat, M et al., *Ann Thorac Surg,* 91:157–164, 2011.
[9] Pierce, E et al., *ABME,* 44(7):2273-81, 2016.
[10] Sacks, M et al., *ABME,* 30:1281–1290, 2002.

HEMODYNAMIC PERFORMANCE OF VALVE-IN-VALVE IN CALCIFIED BIOPROSTHETIC VALVES IS SIGNIFICANTLY DIFFERENT THAN IN NON-CALCIFIED VALVES

Hoda Hatoum (1), Jennifer Dollery (2), Pablo Maureira (3), Juan A. Crestanello (2), Lakshmi P. Dasi (1)

(1) Department of Biomedical Engineering
The Ohio State University
Columbus, Ohio, USA

(2) Division of Cardiac Surgery
The Ohio State University
Columbus, Ohio, USA

(3) Department of Cardiovascular Surgery
Lorraine University Hospital of Nancy
Nancy, France

INTRODUCTION

Bioprosthetic surgical aortic valves (SAV) are increasingly preferred over mechanical valves because of the lower thrombosis risk associated with them. However, due to the limited durability of SAV [1], patients need to undergo a redo aortic valve replacement (AVR) surgery after the implanted SAV fails – often due to severe calcification of the fixed tissue leaflets. However, many of these patients needing a redo AVR are at high-risk for mortality from open heart surgery. A valve-in-Valve (ViV) procedure consists of implanting a transcatheter aortic valve (TAV) into the failed bioprosthesis and constitutes a less risky procedure than a redo AVR surgery. Unfortunately, adverse consequences such as elevated gradients and residual stenosis persist with the ViV procedure [2]. To better understand the outcomes of the ViV several in-vitro studies have been conducted with the hope of developing guidelines or best practices to ensure best hemodynamic performance [3,4]. However, all these studies modeled the failed bioprosthesis using a non-calcified SAV. Given that non-calcified SAVs and calcified SAVs can have dramatically different geometric characteristics, it is not clear if the knowledge gained from these prior studies provide meaningful recommendations for future ViV procedures. This study aims to determine the significance and validity of using a non-calcified SAV vs. using a calcified SAV to establish guidelines for ViV implantation.

METHODS

A 23 mm Medtronic Evolut TAV was implanted in a severely degenerated 23 mm Carpentier-Edwards Perimount Magna Ease SAV extracted from a patient who underwent a redo surgery. In the same manner, the 23 mm Medtronic Evolut TAV was implanted in a non-degenerated 23 mm Carpentier-Edwards Perimount 2800 SAV as a control. The TAV was deployed at 4 different axial positions, namely +6.0, 0, -6.2 and -9.8 mm relative to the bioprosthesis annulus, where the negative axial positions denote a sub-annular deployment and the positive a supra-annular deployment. The hemodynamic performance of the ViV was assessed in a flow loop that simulates a left heart under physiological pressure and flow. The flow loop includes a reservoir tank, mitral valve, pump, valve holder with flow development region, systemic compliance chamber and resistance valve. The working fluid used was approximately 60/40 water/glycerin mixture by volume, a density of 1090 Kg/m^3 and a kinematic viscosity of 3.88 cSt. The pressure and the flow rate across the aortic valve were regulated by a compliance chamber, a resistance valve, and a pump that is controlled by an in-house LabVIEW program. The flow rate was measured with an ultrasonic Transonic flow probe located immediately upstream of the aortic valve, and aortic and ventricular pressures were measured using Validyne pressure transducers. Flow and pressure waveforms were recorded in LabVIEW. The flow was seeded with fluorescent PMMA-Rhodamine B particles with diameters ranging from 1 to 20 μm. The dynamic models of the ViV were assessed using high resolution particle image velocimetry (PIV) which includes illuminating a region of interest in the flow domain using a laser sheet created by a Nd:YLF single cavity diode pumped solid state, high repetition rate laser coupled with external spherical and cylindrical lenses. The spatial and temporal resolutions were 0.07 mm/pixel and 1000 Hz respectively.

RESULTS

The average EOAs normalized by the inflow areas of each SAV are plotted versus the axial positions (Figure 1a). The EOAs for the non-degenerated SAV showed greater values than those with the calcified SAV as the axial deployment of the TAV increases. At -9.8 mm, the EOAs for both valves were not significantly different. The maximum difference in EOA as the axial positioning varies from -9.8 mm to +6.0

mm for the non-calcified SAV is larger than that for the calcified SAV as it is 0.161 versus 0.02, respectively. In addition, the maximum difference in EOA for the two cases occurred at the 0 axial position with the difference settling to 0.218.

Velocity vectors and vorticity contours of the main jet are displayed (Figure 1b) in peak systole at -9.8 mm and 0 mm axial positions. The distance between the parallel shear layers (red and blue contours) for the degenerated SAV case at 0 mm was shown to be 5 mm compared to 7.3 mm for the -9.8 mm case. This indicates that for both cases the jet is far narrower than the physical distance available for the flow at the exiting orifice defined as the most superior edge of the leaflets. En-face high speed imaging videos of the opening and closing of the valves showed significantly stronger and high frequency leaflet fluttering throughout the cardiac cycle (Figure 2) for the calcified SAV cases.

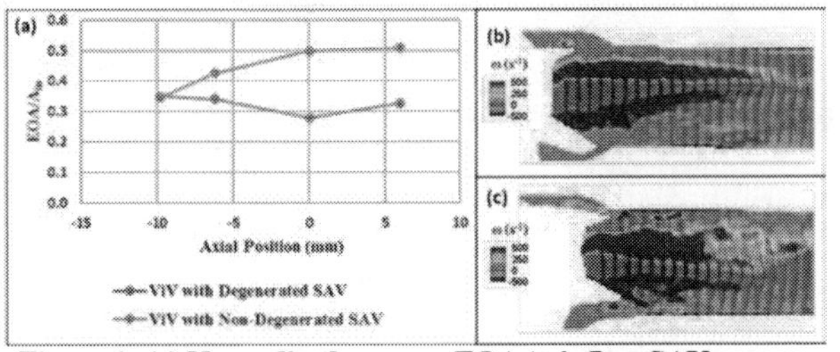

Figure 1: (a) Normalized average EOA to inflow SAV area vs axial position (b) Averaged velocity vectors and vorticity contours at peak systole at -9.8 mm (c) and at 0 mm axial positions for the ViV with degenerated SAV.

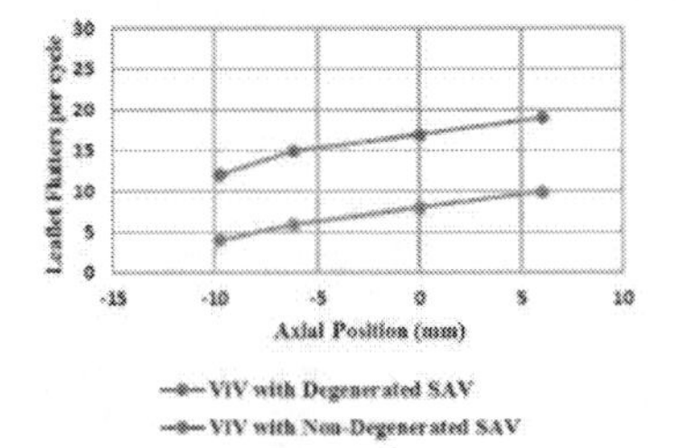

Figure 2: Leaflet Flutters per cardiac cycle

DISCUSSION

In-vitro assessment of ViV using both non-degenerated and degenerated SAVs has shown that calcified SAVs yield different hemodynamic performance and parameters compared with non-calcified valves. For non-calcified SAV ViV cases, supra-annular deployment of the TAV yields better EOA compared with lower axial positions in agreement with previous literature [3,4]. However, in the case of the ViV in a severely calcified SAV, the EOA does not seem to undergo a significant improvement as the axial deployment of the TAV varies from sub-annular positions to supra-annular positions. This result is clinically relevant given that most ViV procedures are performed in severely calcified bioprosthetic SAVs. Non-monotonic behavior of EOA vs. axial position is noted with the apparent drop in EOA for the 0 axial position case with a slight recovery of EOA with a supra-annular positioning. While this improvement in EOA does agree with in-vivo data presented [5], the overall behavior is contrasting. These observations can be explained by fluid dynamic effects imposed by the irregular geometry of the calcified SAV's inflow region due to both tissue ingrowth and subsequent calcific nodule formation. When the TAV is deployed inside a non-degenerated SAV (Figure 3), the smoothness of the SAV inflow region offers a smooth entry with minimal disturbances. The flow remains attached inside the available lumen of the ViV and eventually separates at the edge of the TAV leaflets ultimately issuing a central jet. However, for the calcified SAV case, the irregularities at the inflow region from the tissue ingrowth and calcium nodules force the flow to separate at the inflow orifice. The resulting turbulent jet is much narrower and flows through the length of the ViV configuration, also causing the leaflets to flutter in response to self-excited oscillations between the jet and the flexible TAV leaflets. Figure 3 illustrates this phenomenon which is confirmed both from the low jet width measurements (figure 1b) and the increased flutter noticed in calcified SAV cases. For these reasons a higher EOA is seen for the ViV with non-degenerated SAV compared with the degenerated SAV.

SUMMARY

In-vitro ViV hemodynamic studies using non-calcified SAVs do not faithfully represent the in-vivo hemodynamic assessments. We have highlighted new fluid dynamic mechanisms that occur for the case of ViV in calcified SAVs which can compromise EOA.

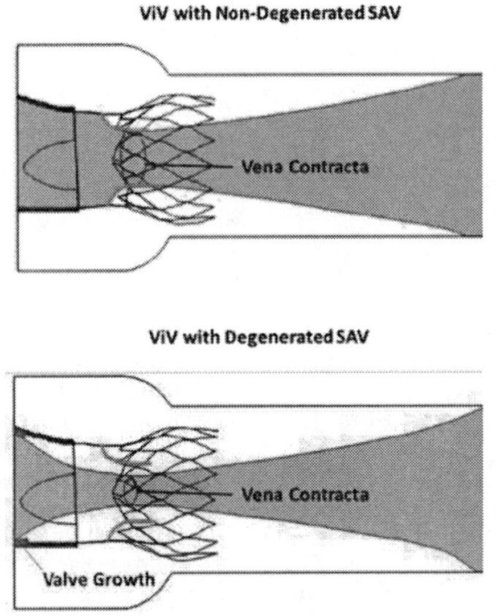

Figure 3: Flow fields upstream and downstream of ViV with the non-degenerated and the degenerated SAVs during the forward flow phase.

ACKNOWLEDGEMENTS

The research done was supported by National Institutes of Health (NIH) under Award Number R01HL119824.

REFERENCES

[1] Gao, G. et al., *JACC*, 44(2):384-388, 2004.
[2] Dasi, L.P., et al., *Annals of Biomedical Engineering, 1-22, 2016*
[3] Midha, P.A., et al., *JACC*, 9(15):1618-1628, 2016.
[4] Simonato, M., et al., *Circulation: Cardiovascular Interventions*, 9(6):e003651
[5] Eggebrecht, H., et al. *JACC-Cardiovascular Interventions*, 4(11):1218-1227

EXPERIMENTAL INVESTIGATION OF 3D LEFT VENTRICULAR FLOW USING A NOVEL MULTIPLANE SCANNING STEREO PIV SETUP

Hicham Saaid (1), Patrick Segers (1), Tom Claessens (2), Pascal Verdonck (1)

(1) Institute Biomedical Technology, Ghent University, Ghent, Belgium

(2) Department of Materials Science and Engineering, Ghent University, Ghent, Belgium

INTRODUCTION

The spatiotemporal characteristics of the intraventricular flow field are of (patho)physiological and clinical interest [1]. This is shown in several recent in vivo studies which describe and evaluate cardiac flow using modern medical imaging techniques (such as echocardiography, 4D-MRI and echocardiographic PIV [2, 3]). Intraventricular flow is also a topic of biofluid mechanic research, in silico as well as in vitro. This is exemplified by several particle image velocimetry (PIV) studies that have been performed in hydraulic bench models to investigate the flow field inside a left ventricle (LV) replica in a 2D plane [4], or to reconstruct the 3D flow structures numerically from 2D velocity data [5]. However, as the intraventricular flow has a complex 3D and unsteady structure, a dedicated PIV system allowing for dynamic 3D flow measurements in the LV is highly desired.

The aim of this project is to design and develop a novel system that allows to quantitatively study the flow in experimental in vitro PIV models, for instance vascular segments, heart valves and left ventricle replicas. In this study we present an application which permits 3D volume reconstruction of the flow field by means of phase-locked stereo-PIV in a transparent pulsatile LV membrane model.

METHODS

The setup (figure 1) was primarily designed and developed to facilitate consecutive stereoscopic measurements without repeating the complex and time consuming stereo calibration. The proposed system is partly based on a previous study of Yagi and co-workers [6], and is built around two hexagonal transparent tanks. The shape of the two tanks is chosen such that optical access can be established orthogonally through the walls. Both tanks were filled entirely with the same refractive-index matched (n = 1.49) liquid as a working fluid.

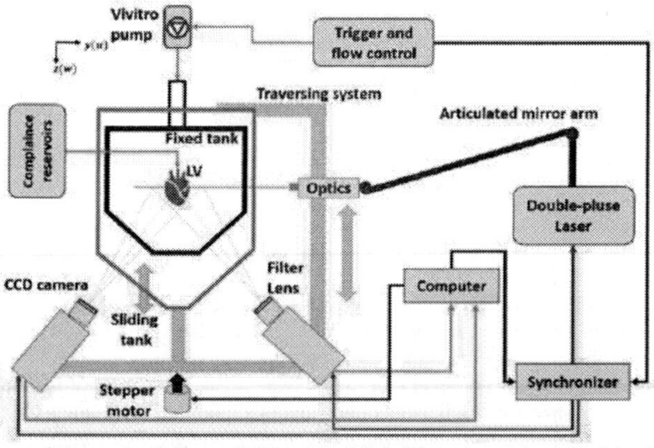

Figure 1: Schematic layout of the setup: two hexagonal tanks, CCD cameras, laser sheet, optical lenses, hydraulic components and the control system components.

To eliminate the need to perform the calibration in each measurement plane separately, the external tank, CCD cameras and the optical lenses have been rigidly connected to a traversing mechanism equipped with a stepper motor. Throughout the scanning sequence, the whole stereo-PIV apparatus translates simultaneously with the external tank, while the internal tank, which functions as the LV housing, remains fixed. This way, the relative optical positions between the cameras and the measurement plane are secured. The 45° stereo-PIV system consists of two CCD sensor cameras (PCO.2000, PCO, Germany) equipped with a long-pass filter at 550 nm, with a maximum resolution of 2048*2048 pixels and mounted on tilt adapters to achieve the Scheimpflug condition. A double-cavity pulsed NdYAG laser (Twins BSL 140, Quantel, USA delivering 140 mJ/pulse with a repetition rate of 15-30 Hz) and compound lenses were used to generate and shape a uniform laser light sheet (thickness 1.5-2 mm).

A commercially available cardiovascular simulator (Vivitro Systems, Inc., Victoria, Canada) was used to impose (and monitor) different physiological pressure and flow conditions. Compliance and resistance were carefully adjusted to achieve a physiological pressure wave form.

A pair of 25 mm mechanical bileaflet valves were mounted in the aortic and mitral positions. A three-component liquid consisting of sodium iodide, glycerol and distilled water (volume fraction 79:20:1) was used as a blood-mimicking liquid and seeded with fluorescent particles (ϕ10 μm) as tracers. Special attention has been paid to ensure the appropriate particle number and density in the interrogation window for reliable cross-correlation analysis of stereo PIV images. To avoid discoloration, a small amount (0.1% by weight) of sodium thiosulfate was added.

RESULTS AND DISCUSSION

Figure 2B shows a reconstruction of the 3D flow in the LV during early diastole, obtained after interpolating the velocity data from several parallel planes (distance 2 mm) measured in a phase-locked manner through 80 consecutive cycles. Stroke volume was set to 50 ml, for a heart rate of 70 beats/minute. The figure 2B displays an asymmetric transmitral jet flow passing through a bileaflet mechanical mitral valve (25mm), one jet toward the apical direction, whereas the second one drifted parallel to the LV wall. As a result of a relatively large LV model diameter (70mm) compared with an anatomical one, the velocity field map and the corresponding 3D flow reconstruction report low velocity values.

The CCD cameras and the double-cavity laser were triggered via the synchronizer (ILA GmbH). The onset of the measurement sequence was triggered by the cardiovascular simulator. For the stereo PIV calibration and image processing, a commercial PivView3c (PIVTEC GmbH) software was used. Tecplot was then used to visualize the resulting velocity data.

To the best of our knowledge, this is the first experimental stereoscopic PIV-setup that is capable of reconstructing the complex 3D left ventricle flow field at any given time instant in different controllable and repeatable physiologically relevant hydrodynamic conditions, in a time-efficient way. We believe that this experimental setup will serve as a reference and benchmark for future in vivo and numerical studies of left ventricle flow dynamics. In the near future, we plan to develop a more realistically shaped LV model, based on medical imaging.

ACKNOWLEDGEMENTS

This project was funded by the BOF (project 05V01713)

REFERENCES

[1] Bermejo J. et al., *Annu. Rev. Fluid Mech*, 47:315-342, 2015
[2] Muñoz D.R. et al., *Eur Heart J C Imaging*, 14, 1029-1038, 2013
[3] Hong G. R. et al., *J Cardiovasc Ultrasound*, 21 (4), 155-162, 2013
[4] Falahatpisheh A. et al., *Eur. J. Mech. B. Fluids*, 35, pp.2–8, 2012
[5] Fortini S. et al., *Exp Fluids*, 54:1609, 2013
[6] Yagi T. et al., *J R Soc Interface*, 10.82, 2013

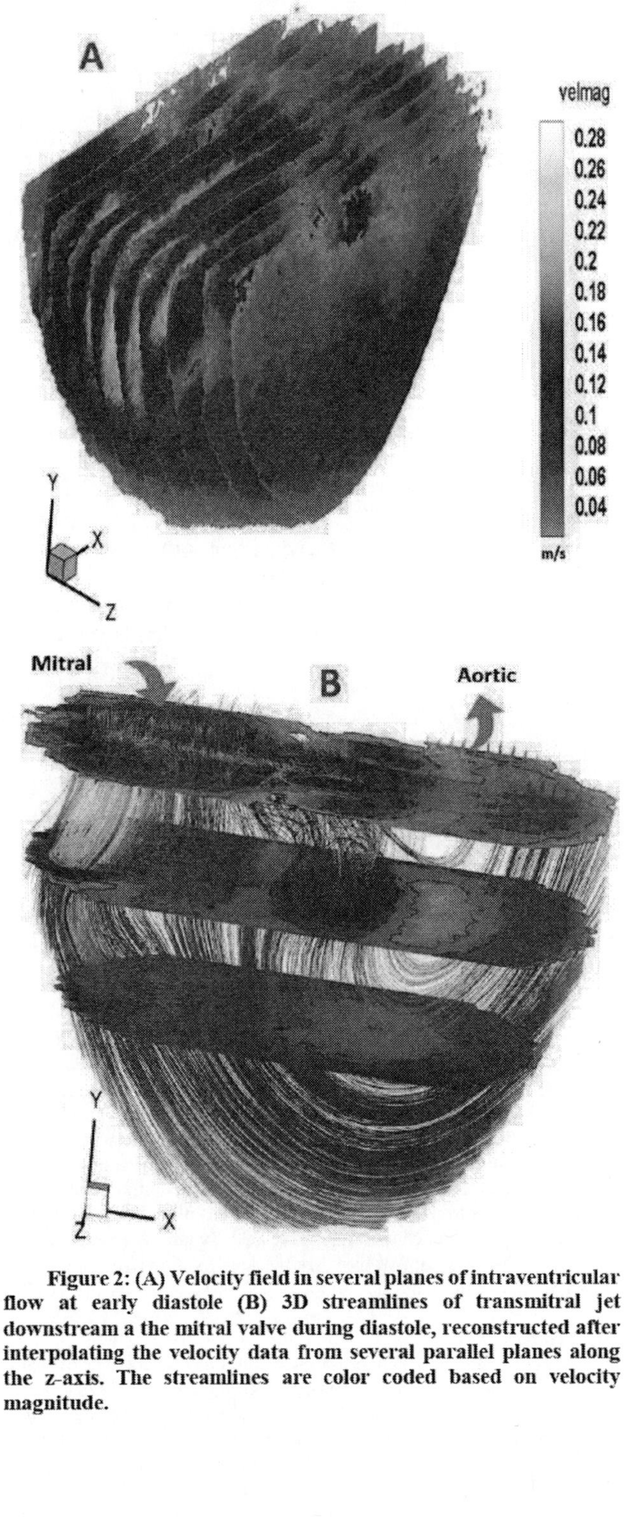

Figure 2: (A) Velocity field in several planes of intraventricular flow at early diastole (B) 3D streamlines of transmitral jet downstream a the mitral valve during diastole, reconstructed after interpolating the velocity data from several parallel planes along the z-axis. The streamlines are color coded based on velocity magnitude.

MRI-BASED FLUID STRUCTURE INTERACTION OF THE AORTIC VALVE: ALTERATION OF NONLINEAR VALVE PROPERTIES TO SIMULATE CALCIFICATION AND BICUSPID AORTIC VALVE

A. Gilmanov (1), A. J. Barker (2), H. Stolarski (3), F. Sotiropoulos (4)

(1) Saint Anthony Falls Laboratory
University of Minnesota Minneapolis
Minneapolis, MN, USA

(2) Dept. of Radiology
Northwestern University
Chicago, IL, USA

(3) Dept. of Civil, Environmental and Geo-engineering
University of Minnesota Minneapolis
Minneapolis, MN, USA

(4) Dept. of Civil Engineering
Stony Brook University
Stony Brook, N□, USA

INTRODUCTION

The bicuspid aortic valve (BAV) has attracted significant attention due to controversy surrounding whether genetics or hemodynamics are associated with concomitant aortopathy. Given that 4D flow MRI can simultaneously measure complex 3D anatomy and blood flow patterns, it is an ideal modality to investigate the hemodynamic hypothesis for aortopathy[1, 2]. Nonetheless, limitations of temporal and spatial resolution prevent MRI from visualizing boundary layers and fluid structure interaction (FSI) occurring immediately adjacent to the valve leaflets as well as at the aortic wall downstream from the valve.

Increasing computational power combined with image-based patient specific anatomy and boundary conditions (available from MRI) has enabled reliable computational fluid dynamic (CFD) approaches which incorporate moving boundary conditions via FSI[3, 4]. These numerical approaches have the potential to elucidate complex and dynamic blood and leaflet behavior previously not seen with imaging alone. In addition, the approach allows for simulation of parallel disease processes, such as aortic valve calcification. In this study, we present the initial efforts to perform subject specific simulations, incorporating a thin shell FSI approach to mimic valve cusp dynamics. In addition, the thickness of the valve cusps was parametrically varied to investigate the effect of valve calcification.

METHODS

4D flow MRI and standard bSSFP cine imaging was performed in a healthy volunteer with no history of cardiovascular disease and a normal tricuspid aortic valve (TAV). Additionally, two patients with a suspected right-left coronary cusp fusion (RL-BAV) and a right-noncoronary cusp fusion (RN-BAV) underwent CMR, including 4D flow MRI and bSSFP cine imaging. The morphology of the valves was confirmed using cine images. The aorta of the healthy subject was subsequently segmented in 3D to facilitate the FSI/CFD simulations. Additionally, 3 valve models were created to replicate the cusp morphology of the TAV, RN-BAV, and RL-BAV valves, as determined from the corresponding bSSFP images (inset, Figure 1A). A Curvilinear Immersed Boundary Finite Element Fluid Structure Interaction (CURVIB-FE-FSI) method[5] was employed. The leaflets of the aortic valve are modelled as a thin shell using a rotational free FE formulation[6]. A nonlinear anisotropic constitutive model of an aortic valve[7, 8] was used to describe the native tissue of the heart valve. All valve simulations were performed using the geometry of the healthy aorta. A flow waveform created from the MRI flow velocities was used as the inlet flow conditions for all models. In order to simulate the effect of calcification, the cusp thicknesses were varied (h_o= 0.6, 1, and 2mm).

RESULTS

Figure 1A shows MRI data from the healthy volunteer with normal TAV morphology. A valve model was constructed from the bSSFP MRI data and is shown in Figure 1B. The downstream flow pattern for the FSI model is shown in Figure 1C with different degrees of calcification. Figure 2 shows the evolution of the geometric orifice area and velocity at the valve over the cardiac cycle. As expected, the increased thickness (i.e. calcification) resulted in a shorter opening duration, reduced area, and increased velocity. Figure 3A shows geometric alterations to the healthy volunteer valve in order to mimic a RL-BAV valve morphology. The flow patterns in the aorta between the TAV and BAV morphotypes are characterized by markedly different transvalvular jets. The outflow jets are indicative of those seen to increase wall shear stress in the proximal outer curvature of the RL-BAV model. Similar patterns are seen for the patient MRI scans, even given the different left ventricular outflow tract and aortic geometry (Figure 3B).

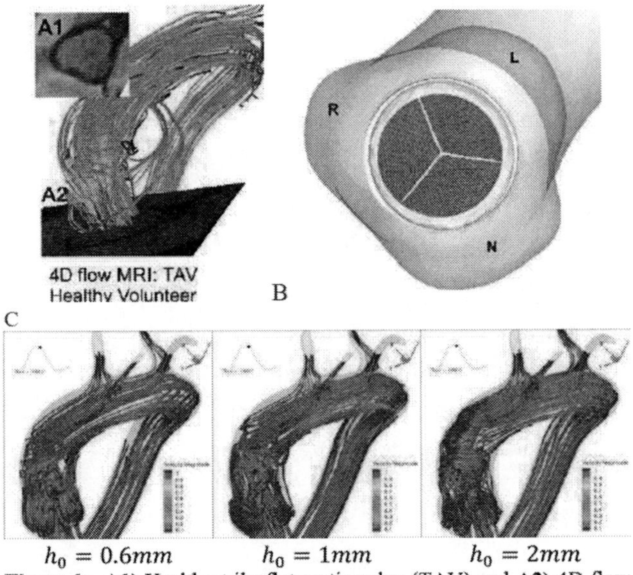

Figure 1: A1) Healthy trileaflet aortic valve (TAV) and **A2)** 4D flow MRI. **B)** Valve model for FSI/CFD simulation. **C)** Comparative analysis of blood flow distribution for the healthy patient with different values of "calcifications" (h_0 - thickness) at two time moments: $t_i = 800\,(0.1\,sec)$ and $1160\,(0.18\,sec)$. It is clearly seen that such "calcification" obstructs blood flow by causing flow acceleration.

A

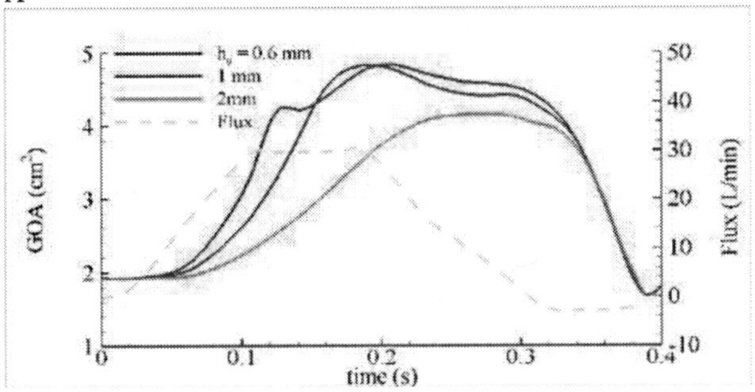

B

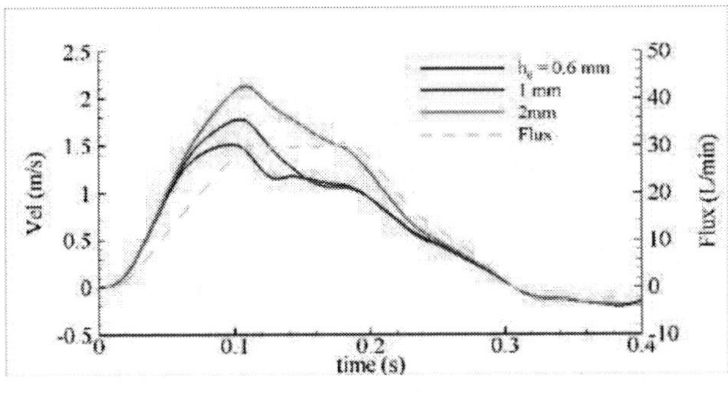

Figure 2. 2 Time history of (a) geometric orifice area (GOA) and (b) mean velocity across the section of GOA. Red solid line is $h_0 = 0.6\ mm$, blue is $1\ mm$, and green is $2\ mm$. Inflow volume rate (Flux) is shown as dashed gray line.

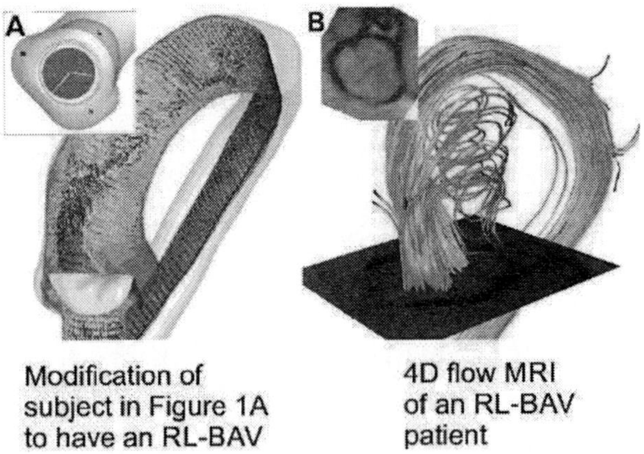

Figure 3. (A) An unobstructed RL-BAV patient measured via CMR. **(B)** The healthy subject from Figure 1A was parametrically altered to mimic RL-BAV leaflet morphology (systolic outflow is shown at systole). Note that even given the different left ventricular outflow tract and aortic geometry, the outflow jet remains along the outer curvature of the proximal aorta.

DISCUSSION

Blood flow patterns (such as the magnitude of flow velocities and the forces exerted on the aorta wall) are important factors leading endothelial mechanotransduction and maladaptive extracellular matrix remodeling. These mechanisms are strongly dependent on the action of the heart valve, and its intrinsic properties. Thus, we set out to develop realistic FSI simulations of aortic cusp motion and its downstream effects on blood flow in the aorta. The combination of two approaches allows for parametric variation and simulation of disease processes and intervention. An example of parametric variation is shown by altering the valve 'calcification' in the healthy patient to mimic the onset of calcific aortic stenosis. In addition, the valve cusp morphologies were altered to mimic the flow behaviors seen in two valve phenotypes observed in BAV patients. Future modifications to these models will be used to simulate the changes to the BAV during the onset of concomitant aortic stenosis.

ACKNOWLEDGEMENTS

Research reported in this publication is supported in part by the National Heart, Lung, And Blood Institute of the National Institutes of Health under Award Number K25HL119608.

REFERENCES

[1] Mahadevia, R, *et al.* Circulation, 2014. 129(6): p. 673-82.
[2] Guzzardi, DG, *et al.* J. Am. Coll. Cardiol., 2015. 66(8): p. 892-900.
[3] Chandran, KB. Cardiovasc. Eng. Technol., 2010. 1(1): p. 18-38.
[4] Sotiropoulos, F, *et al.* Progress in Aero. Sciences, 2014. 65: p. 1-21.
[5] Gilmanov, A, *et al.* J. Comput. Phys., 2015. 300(C): p. 814-843.
[6] Stolarski, H, *et al.* International Journal for Numerical Methods in Engineering, 2013. 95(9): p. 740-770.
[7] May-Newman, K, *et al.* J. Biomech. Eng., 1998. 120(1): p. 38-47.
[8] Gilmanov, A, *et al.* J. Biomech., 2017. 50: p. 56-62.

ASSESSMENT OF THROMBOSIS POTENTIAL OF A TRANSCATHETER HEART VALVE USING A NOVEL SINGLE-CAMERA VOLUMETRIC PIV TECHNIQUE

Christopher Clifford (1), Vrishank Raghav (1,2), Prem Midha (2), Ikechukwu Okafor (2,3), Camille Johnson (2), Brian Thurow (1), Ajit Yoganathan (2)

(1) Auburn University
Auburn, AL, USA

(2) Georgia Institute of Technology
Atlanta, GA, USA

(3) Exponent Inc.
Philadelphia, PA, USA

INTRODUCTION

Transcatheter aortic valve replacement (TAVR) has emerged as a safe and effective treatment for severe, symptomatic aortic stenosis in intermediate or greater surgical risk patients [1]. However, despite excellent short-term outcomes, improved imaging and awareness has led to the identification of leaflet thrombosis on the aortic side of the prosthesis in as many as 40% of patients [2]. When deployed into a native aortic valve, the open-celled stent frame of a transcatheter heart valve (THV) becomes enclosed in native aortic valve leaflet tissue. The displaced native leaflets divide the native sinus into two regions - a smaller anatomical sinus and a neo-sinus.

Understanding the flow patterns within the neo-sinus may provide insight into the occurrence and severity of THV thrombosis. However, due to extremely limited optical access through the THV stent, imaging the neo-sinus of a commercial valve is difficult, if not impossible. In addition, flow around cardiac valves has been demonstrated to be highly three dimensional, yet traditional imaging techniques used to capture these complex flows are often prohibitively costly and/or cumbersome to implement in a relevant experimental setup.

A novel solution to this problem is to use a plenoptic camera for rapid acquisition and reconstruction of 3D images. While a conventional camera maps all of the light from a single point on the focal plane to a single pixel on the image sensor, a plenoptic camera maps light from a point on the focal plane to distinct pixels on the image sensor based on the angle entering the main lens [3]. This can be used to render images from a variety of perspective views and focal planes during the post-processing period, as shown in Figure 1. On the left, two perspectives obscure/reveal the 8♥; on the right, focus is shifted from the K♣ to the 8♥. The Advanced Flow Diagnostics Laboratory (AFDL) at Auburn University has constructed several plenoptic cameras for a wide variety of applications, including plenoptic PIV.

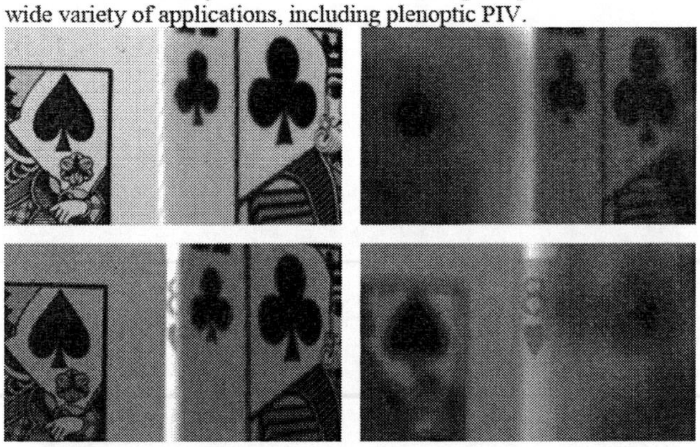

Figure 1: Demonstration of plenoptic imaging capabilities

The aims of this study are two-fold: a) introduce plenoptic PIV as an alternative novel solution to measure three-dimensional flow fields for biomedical applications and b) understand the flow patterns within the native sinus and neo-sinus using plenoptic PIV in two different THV models – the SAPIEN XT and a transparent THV model, the GT-TAVR.

METHODS

Left Heart Simulator and Flow Conditions: The Georgia Tech Left Heart Simulator (illustrated in Figure 2) capable of reproducing physiological flow rates and pressures [4] was used for this study. A balloon expandable THV (21 mm SAPIEN XT) was mounted in the test chamber for the validation experiments to study the flow field downstream and in the vicinity of the THV. For these experiments, data were acquired at two cardiac output conditions of 5 and 2 liters/min at a heart rate of 70 beats/min, a mean arterial pressure of 100 mmHg, and 7 time points across the cardiac cycle (5 in systole, 2 in diastole).

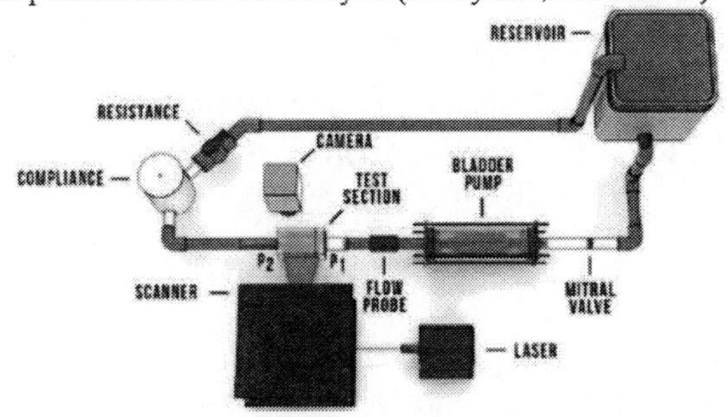

Figure 2: Illustration of the Georgia Tech Left Heart Simulator

To study the three-dimensional flow field in the neo-sinus region of the THV, an indigenously developed THV with transparent walls was used. In this case, data were acquired at a cardiac output of 5 liters/min, a heart rate of 70 beats/min, a mean arterial pressure of 100 mmHg, and 21 time points across the cardiac cycle (19 in systole, 2 in diastole). Our objective in the second set of experiments was to describe the temporal evolution of the flow field by acquiring velocity field data with better time resolution.

PIV Instrumentation

Plenoptic PIV: The plenoptic camera used in this study is a modified Imperx Bobcat ICL-B6620, which has a 29 megapixel (MP) CCD sensor with 14-bit interline readout. The hexagonally-packed microlens array has a focal length of 308 microns and a microlens pitch of 77 microns such that there are approximately 16×16 pixels per microlens. Spatial calibration was achieved by imaging a series of "dot card" calibration targets. Vector fields were obtained using an iterative multi-pass, multi-grid window deformation technique known as WIDIM [5]. The plenoptic PIV data used four passes with a final interrogation window of 24^3 voxels and 75% overlap. The resultant vector field had a spatial resolution of about 0.3 mm per vector.

Planar PIV: The planar (2D) PIV instrumentation used in this study including the details of the measurement uncertainties is well documented in our previous work [4].

RESULTS

SAPIEN XT: Figure 3 illustrates 5 slices sampled from the 3D velocity field across the diameter of the valve at peak systole. The figure also illustrates a comparison of the velocity profile at the sino-tubular junction in the central plane between plenoptic-PIV and 2D-PIV.

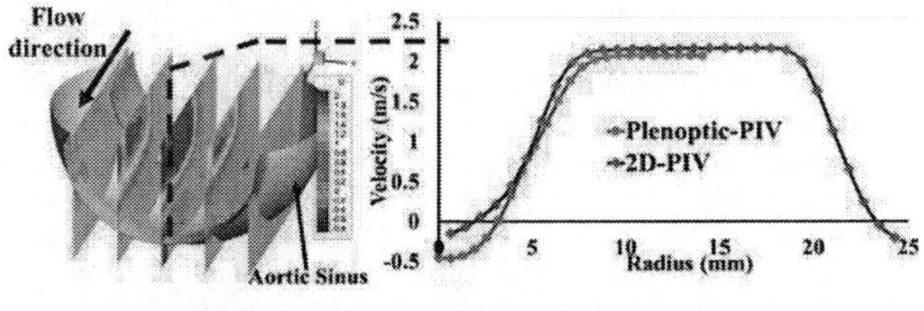

Figure 3: Illustration of 3D velocity field downstream of the SAPIEN XT valve and the velocity profile comparison between plenoptic and 2D PIV at peak systole.

Figure 4 illustrates the three-dimensional swirling strength field overlaid with streamlines downstream of the SAPIEN XT during systolic acceleration. The three-dimensional vortex shed from the leaflet tips is clearly visible as labeled in the figure.

GT-TAVR: The velocity field in the neo-sinus region was measured in the transparent GT-TAVR model as illustrated in Figure 5. We observed that the flow at the *base* of the neo-sinus region was stagnated (zero velocity) throughout systole and diastole (not shown).

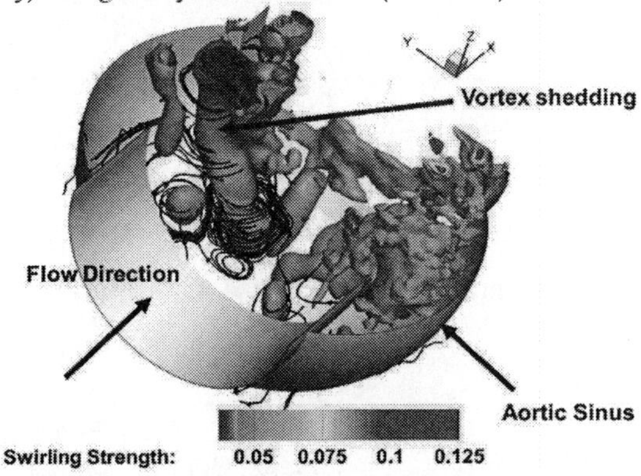

Figure 4: Three-dimensional swirling strength iso-contours overlaid on streamlines downstream of the SAPIEN XT illustrating the shed vortex during systolic acceleration.

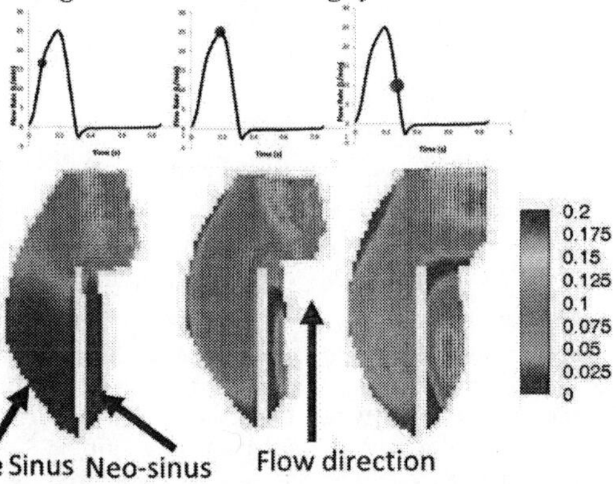

Figure 5: Illustration of the velocity field in the native and the neo-sinus regions of the transparent GT-TAVR valve model.

DISCUSSION

Plenoptic PIV is a feasible method to measure three-dimensional flow fields for biomedical applications. The quantitative validation against 2D PIV demonstrated that the difference in measured peak forward flow velocity was less than 3.5%. However, the difference in measured velocity in the reverse flow region was much larger, and these issues are currently being addressed. 3D velocity field downstream of the valve was spatially well resolved and capable of highlighting crucial flow events such as vortex shedding.

Stagnation was observed at the base of the neo-sinus region through the cardiac cycle which indicates that flow stagnation could be the possible mechanism for the recently discovered issue of thrombus in THVs.

ACKNOWLEDGEMENTS

The authors acknowledge the assistance of the members of the Cardiovascular Fluid Mechanics Laboratory at Georgia Tech and the Advanced Flow Diagnostics Laboratory at Auburn University.

REFERENCES

[1] Leon, MB et al., *N Engl J Med.*, 374:1609–1620, 2016.
[2] Piazza N., *JACC Cardiovasc Interv.*, 6:443–451, 2013.
[3] Fahringer, TW et al., Meas. Sci. and Tech., 26:201, 2015.
[4] Midha, P et al., *Ann. Biomed. Engg., 10:1-8, 2016.*
[5] Scarano, F., *Exp. Fluids, 29:51-60, 2000.*

INFLUENCE OF INLET BOUNDARY CONDITIONS ON THE EVALUATION OF AORTIC WALL SHEAR STRESS FOR PATIENTS WITH ABNORMAL AORTIC VALVES

Selene Pirola (1), Omar A. Jarral (2), Declan P. O'Regan (3), Thanos Athanasiou (2), Xiao Y. Xu (1)

(1) Department of Chemical Engineering
Imperial College London
London, UK

(2) Department of Surgery and Cancer
St. Mary's Hospital
Imperial College London
London, UK

(3) Institute of Clinical Science
Hammersmith Hospital
Imperial College London
London, UK

INTRODUCTION

Over the last three decades, evidence has been presented linking hemodynamic wall shear stress (WSS) to the initiation and development of cardiovascular diseases [1]. Image-based computational modelling has been widely adopted to evaluate patient-specific WSS, as it has been shown to be more reliable than deriving WSS from phase-contrast magnetic resonance imaging (PC-MRI) alone [2]. However, computational evaluation of WSS is highly dependent on the employed boundary conditions (BCs) [3]. PC-MRI allows in vivo velocity profiles to be obtained that can be imposed as inlet BC for computational models of the aorta, providing a more realistic evaluation of vascular hemodynamics. These profiles can be 1D (through-plane/axial velocity component) or 3D (through-plane and in-plane velocity components). However, as the acquisition of in-plane velocity requires a longer scanning time, in many cases inlet velocity profiles are not or only partially available. Morbiducci et al. [3] investigated this issue on a normal aorta, showing that using the axial velocity component alone may be sufficient for a correct hemodynamic evaluation. However, this may not be true for cases with abnormal aortic valves, as the presence of a pathological or prosthetic valve may increase the magnitude and influence of the in-plane velocity components.

The aim of this study is, therefore, to investigate if using the axial velocity component alone as inlet BC would be sufficient for an accurate evaluation of the aortic hemodynamics and WSS even in the presence of an abnormal valve.

METHODS

MR images were acquired at Hammersmith Hospital (London, UK) from a patient with a mechanical aortic valve, who gave his informed consent. A 1.5T Philips Achieva system was employed to acquire multi-slice sagittal anatomic images of the thoracic aorta and proximal vessels. Two PC-MRI planes were acquired normal to the aortic axis, respectively at the aortic annulus and in the proximal descending aorta. Three velocity components were acquired with encoding parameters set 10% above the expected peak velocity. Retrospective cardiac gating was employed to obtain 100 time points per average cardiac cycle. A pressure measurement was performed on the patient 30 min prior to the scan.

Patient-specific 3D geometry of the thoracic aorta and aortic branches was reconstructed by Mimics v18.0. An in-house MATLAB tool was used for image noise reduction and to manually segment the PC-MR images. A structured mesh of approximately 2.1M hexahedral elements was generated using ANSYS ICEM v15.0.

The model inlet was located in the aortic root, where the PC-MRI mapping plane was placed. This was used to obtain the 3D time-varying velocity profiles which were used as inlet BCs for the reference case (case 3D), and to extract 1D through-plane velocity profiles (case TP). These were obtained by the scalar product of the vector normal to the acquisition plane and the acquired velocity vectors. For both cases, the velocity profiles were mapped to the 3D global coordinates of the computational model. 3-element Windkessel model was applied as outlet BC for all model outlets, with its coefficients being calculated employing the acquired PC-MRI data and pressure waveform [4].

Numerical solutions were obtained with ANSYS CFX (v15.0). The simulations continued until a periodic solution was reached and only the last cycles were used for the analysis. Flow was assumed to be laminar and blood was considered as a Newtonian fluid (viscosity of $4 \cdot 10^{-3}$ Pa·s, density of 1060 kg·m^{3}). Time-averaged WSS (TAWSS) and oscillatory shear index (OSI) were obtained using CEI Ensight (v10.1), and maps of absolute and weighted-percentage differences were obtained for OSI and TAWSS respectively, with the 3D case assumed as reference case. In particular, the weighted percentage difference was obtained as follows:

$$100 \cdot \left|\frac{TAWSS_{3D}(x,y,z)-TAWSS_{TP}(x,y,z)}{TAWSS_{3D}(x,y,z)}\right| \cdot \frac{TAWSS_{3D}(x,y,z)}{TAWSS_{3D}^{MAX}} \quad (1)$$

which was defined to avoid misleadingly high percentage values in regions where the magnitude of TAWSS is close to 0.

RESULTS

Velocities at the model inlet were compared between the two cases (3D and TP), showing that case TP had a 25% lower mean value and 29.8% lower maximum velocity, compared to case 3D. On average the magnitude of in-plane velocity components was 32% of the total velocity and 48% of the through-plane velocity (Figure 1).

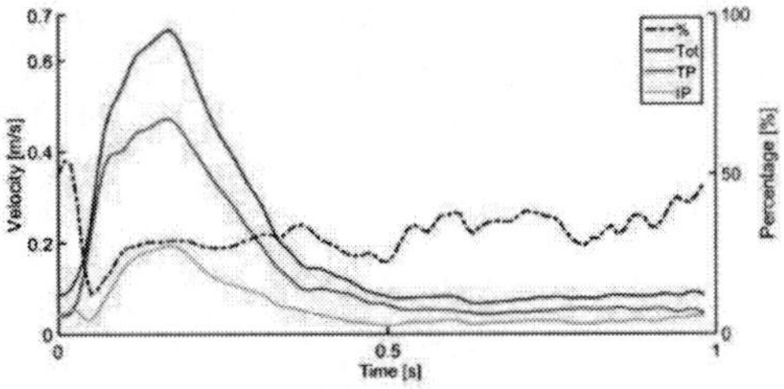

Figure 1: Total inlet velocity (Tot), through-plane (TP) and in-plane (IP) velocity components, along with the percentage of IP with respect to the total velocity.

TAWSS mean values and ranges were 1.24 (0.41-8.66) Pa and 1.65 (0.27-14.34) Pa, for TP and 3D respectively. The main differences in TAWSS and OSI distributions were observed in the ascending aorta, aortic arch and arch branches roots. Weighted percentage differences were up to 50% in the ascending aorta and on the inner brachiocephalic artery root (Figure 2). Absolute differences (Figure 2) close to the maximum OSI value were found in the ascending aorta and aortic arch, with TP tending to underestimate OSI in the ascending aorta, while overestimating it in the aortic arch. For both indices, the differences seem to be accentuated in the ascending aorta, where an angled velocity jet originating from the valve impinges the wall.

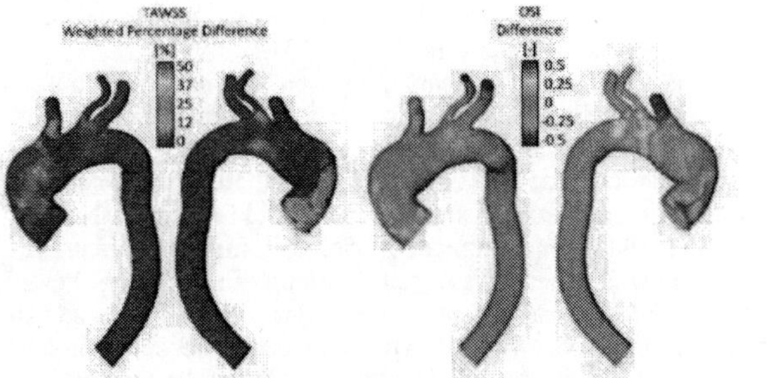

Figure 2: Maps of weighted-percentage differences (TAWSS, left) and absolute difference (OSI, right) between the two cases.

To further investigate the highlighted differences in TAWSS and OSI, their distributions were analyzed and compared between the two cases at a selected location along the ascending aorta (Figure 3). Case TP correctly captured the location of the peak TAWSS at about 0.4, while at distances 0.1 and 0.7 case TP predicted local peaks which were not observed in case 3D. Differences in OSI distributions were more pronounced: a peak at about 0.2 was correctly captured by case TP, while between distances 0.5 and 1 TP and 3D distributions were out of phase.

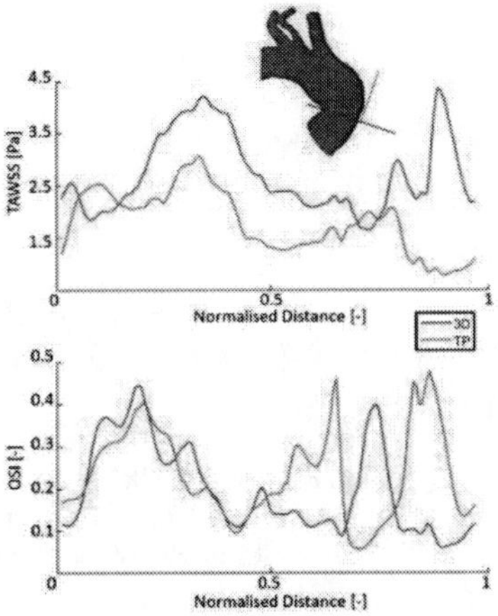

Figure 3: TAWSS (top) and OSI (bottom) distributions along the wall at a selected location (top) in the ascending aorta.

DISCUSSION

The importance of WSS in the process of vascular wall remodeling is well-known to the research community. However, the evaluation of this parameter still presents some uncertainties as it is dependent on the imposed boundary conditions [3], amongst other factors.

In this work, we modelled the aortic hemodynamics in a patient with a mechanical valve implanted and compared the results obtained by imposing the axial velocity component alone (TP) and full 3D velocity profiles (3D), which were extracted from PC-MR images, as inlet BCs. Preliminary results showed that the in-plane velocity components were non-negligible, with a higher mean ratio to the axial component than that observed in a healthy aorta (48% vs 32%) [3]. Comparisons of TAWSS and OSI in the two cases underlined the presence of a remarkable difference both in their magnitude and distributions. These results show evidence that the imposition of 1D axial velocity profiles may not provide sufficiently accurate estimation of WSS in the ascending aorta and aortic arch. However, the influence of the in-plane components seems to be negligible after the third aortic branch.

The comparison was so far conducted on a single case only, so no general conclusions can be drawn. However, the study is still ongoing and more cases of abnormal aortic valves will be analyzed soon.

Overall, our preliminary results suggest that the in-plane velocity components may play an important role in the evaluation of patient-specific WSS and, consequently, in the study of aortic flow-driven wall remodeling.

ACKNOWLEDGEMENTS

This work was supported by the European Commission within the Horizon 2020 Framework through the MSCA-ITN-ETN European Training Networks (project number 642458).

REFERENCES

[1] Mohamied, Y et al., *Ann Biomed Eng*, 43:16-25, 2015.
[2] Piatti, F, Pirola, S et al., *J Biomech*, 50:93-101,2017
[3] Morbiducci, U et al., *J Biomech*, 34:859-871, 2001.
[4] Xiao N. et al., *Int J Numer Method Biomed Eng*; 30:204-231, 2014

EVALUATION OF NOVEL POLYMERIC TRANSCATHETER AND SURGICAL AORTIC VALVES WITH FLUID-STRUCTURE INTERACTION MODELS AND EXPERIMENTAL ANALYSIS

Ram P. Ghosh (1), Gil Marom (1), Oren M. Rotman (1), Matteo Bianchi (1), Saurabh Prabhakar (2), Marc Horner (3), Marvin J. Slepian (1, 4), Danny Bluestein (1)

(1) Department of Biomedical Engineering,
Stony Brook University,
Stony Brook, N□, USA

(2) ANS□S Fluent India Pvt. Ltd,
Pune, India

(3) ANS□S, Inc.,
Evanston, IL, USA

(4) Sarver Heart Center,
University of Ari□ona,
Tucson, AZ, USA

INTRODUCTION

One out of four people above 65 years of age in the U.S. suffer from heart valve diseases, making prosthetic valves' structural and fluid mechanics a great interest among the researchers [1]. The prosthetic valves have been studied since 1950 and despite of significant supporting evidences that bioprosthetic valves are biocompatible and have similar hemodynamics to native valves, the material of the xenograft leaflets was adapted but not specifically designed or optimized for prosthetic use. Our research group has developed a novel polymeric prosthetic valve, which was optimized to reduce stresses on the leaflets, improve hemodynamics and has thrombogenic performance similar to that of a tissue valve [2].

There are currently two types of prosthetic aortic valve procedures for end stage patients with calcific aortic valve, surgical aortic valve replacement (SAVR) and transcatheter aortic valve replacement (TAVR). SAVR requires open heart surgery while TAVR is a minimally invasive procedure for patients who cannot undergo open heart surgery. The mechanics of these devices was previously studied with numerical tools. However, to the best of our knowledge, no such study compared the combined structural and fluid mechanics of TAVR and SAVR valves under identical conditions.

The aims of this study are to compare our polymeric TAVR and SAVR valves using a fluid-structure interaction (FSI) simulation with Arbitrary Lagrangian-Eulerian (ALE) method and to evaluate experimentally the TAVR valve's hemodynamics. Specifically, the model compares valves' hemodynamics parameters and mechanical stresses. The FSI simulation combines structural and fluid dynamics and can accurately capture the valve kinematics by transferring momentum between leaflets and blood flow [5, 6]. The experimental bench tests are used for thorough hemodynamic evaluation of the fabricated valves according to ISO 5840 and for validating the FSI simulation.

METHODS

The valves are made of cross-linked *Styrene-block-IsoButylene-block-Styrene* (xSIBS; Innovia LLC, Miami, FL) polymer and the geometries were obtained from Polynova cardiovascular Inc. (Stony Brook, NY) [2]. The initial leaflets design was utilized in the SAVR valve and later optimized and adapted for a TAVR valve by adding a sleeve sutured to a self-expandable nitinol stent, and by offsetting the leaflets nominal position to be semi-open for reduction of structural stress accumulation over the cardiac cycle. The stent was drawn using SolidWorks 2015 (Dassault Systemes, Concord, MA, USA). The fluid domain was extracted from the ViVitro Pulse Duplicator's aortic root geometry (Vivitro System Inc. Victoria, BC, CA) [6].

The fluid governing equations were solved using finite volume discretizing method while the structural equations were solved using an implicit displacement based finite element approach. The flow solution employed a coupled pressure-velocity method to solve continuity and momentum equations. An unsteady Reynold averaged Navier-Stokes (URANS) with k-ω turbulent model was implemented. A diffusion based dynamic mesh was used for remeshing highly skewed fluid domain cells. In order to keep the fluid domain continues and to prevent small-volume cells, virtual walls were added between the leaflets to maintaining a microscopic gap between the leaflets on the structural domain. Zero penetration was enforced by a Normal Lagrange contact formulation between the leaflets and the virtual wall surfaces.

The fluid and structural solvers were coupled using ANSYS system coupling where the time step size varied from 0.1 to 1.0 ms during the cardiac cycle. The simulations were ran for a cardiac cycle with a duration of 0.854 s. The blood was modeled as Newtonian fluid with a dynamic viscosity of 0.0035 Pa·s and a density of 1060 kg/m^3. The leaflets material property was obtained via a uniaxial tensile machine and fitted to Mooney-Rivlin isotropic hyperplastic model. The

ALE type FSI models were used because of their accurate boundary condition implementation, wall shear stress (WSS) calculation, and incorporation of valves' leaflet thickness, hence ensuring accurate calculation of flowrates, orifice areas and mechanical stresses. The 2-way iteratively implicit FSI simulation was solved separately for the fluid and solid equations in transient ANSYS Fluent and Structural 17.1 (ANSYS Inc., Canonsburg, PA), respectively.

RESULTS

The TAVR and SAVR valves show similar qualitative flow features and their flow rates, orifice areas, and WSS during systole and mechanical stress magnitudes during diastole can be quantitatively compared. The flow acceleration during systole initiates with a central jet flow and symmetrically forming vortex rings near the sinuses (Fig. 1 A, D). Then the sinuses' central flow aid in the velocity elevation with the central orifice jet. The vortex rings become weaker during peak systole and travels upwards towards the aorta (Fig. 1 B, E). During peak flow phase, the TAVR valve's central jet flow was expanding more extensively than in the SAVR valve. This is followed by flow deceleration phase and the beginning of diastole. During diastole, the leaflets closure was aided by the flow from the sinuses. The leaflets closure causes a flow depression and is responsible for a fluid suction. The flow from the upper region then travels to the sinuses and creates a counter rotating vortex to fill the sinuses.

The valves show the largest geometric orifice area (GOA; Fig. 2 A), effective orifice area (EOA), flow rates (Fig. 2 B), and maximum WSS on the leaflets during their systolic phase. The TAVR valve was found to have higher GOA, EOA and flowrate but lower WSS than the SAVR valve. The TAVR and SAVR valves calculated EOAs were 2.33 and 1.85 cm^2 respectively. The TAVR valves experienced lower WSS (72.25 Pa) compare to the SAVR valve (92.06 Pa) during peak systole. The valves experienced the highest mechanical stresses during peak diastole (Fig. 1 C, F). The peak equivalent stress magnitudes for the TAVR and SAVR valves are 8.84 and 4.44 MPa respectively. In both cases, the highest stress magnitudes were observed near the leaflets commissural region where the leaflets are attached.

DISCUSSION

This study compared numerically polymeric TAVR and SAVR

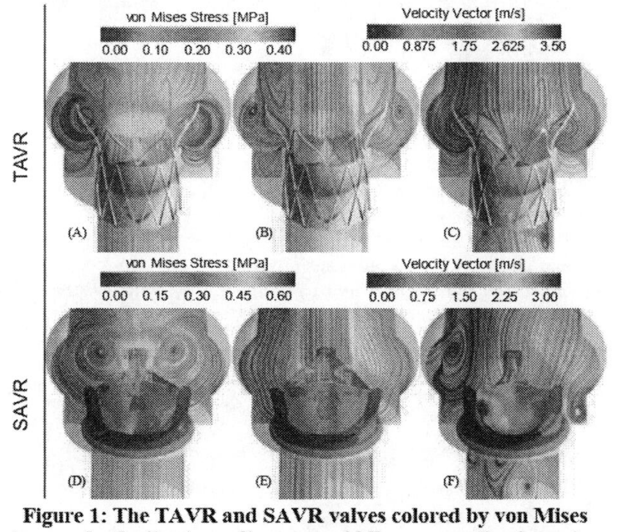

Figure 1: The TAVR and SAVR valves colored by von Mises stress and velocity streamline on a middle cross-section in various instances during the cardiac cycle: beginning of systole (left), peak systole (center), and peak diastole (right)

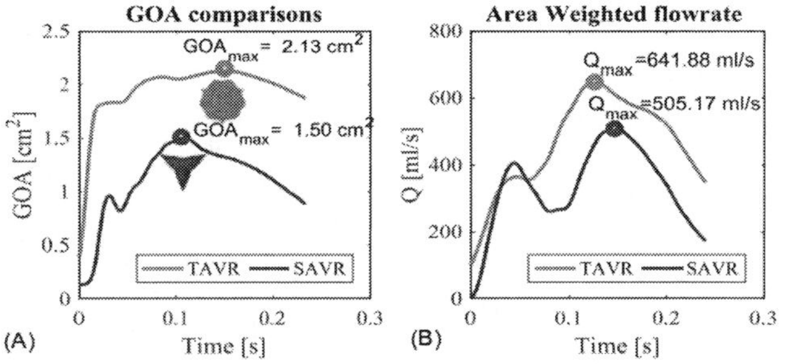

Figure 2: The valves' GOA (A) and area weighted flow rate (B) during systolic phase.

valves leaflets kinematics. TAVR valve had higher GOA, EOA, and flowrate and lower WSS than SAVR valve suggesting that TAVR valve is an improvement of the exiting SAVR valve. The TAVR valve, however, experienced higher mechanical stresses than the SAVR valve. The leaflets attachment in the commissural region can explain these higher stresses. The influence of these results on the durability of the SAVR and TAVR valves will be experimentally validated as described below.

The hemodynamics during diastole was not compared because the flow blockage model in the virtual gap has not been implemented yet, resulting in unphysical leakage. This study assumes an ideal circular configuration of the valves, however, deployed TAVR valve will be either under-expanded or in elliptical shape [7]. This interrupts the valve coaptation and valve functionality which in turns have an impact on leaflets stresses.

The polymeric TAVR valves are now being fabricated and bench tested in our facility. The fabrication is done by compression molding using an EDM-machined S7 tool steel mold (Fig. 3). Vacuum is applied to the mold throughout the molding process for efficient removal of air bubbles and for maximizing the molded valve quality. The valves are then sutured to a nitinol stent for completion of the process. The valves' durability is now being assessed using a Vivitro Hi-Cycle System (Vivitro Labs, Inc. Victoria, BC) according to ISO 5840. Hydrodynamic performance is being evaluated using the Vivitro left heart simulator (Vivitro Labs, Inc. Victoria, BC). This is also used for validation of the FSI simulation results in terms of the valves' EOA, GOA, and flow rates. Flow-induced platelet activation is measured in-vitro in comparison to a gold standard Carpentier-Edwards Perimount Magna Ease aortic valve.

Figure 3: The polymeric TAVR valve mold and the stented valve.

ACKNOWLEDGEMENTS

ANSYS Inc. is in an academic partnership with Dr. Bluestein.

REFERENCES

[1] Bavo, AM, et al., *PloS one*, 11:e0154517, 2016.
[2] Claiborne, TE, et al., *ASAIO J*, 59:275-83, 2013.
[3] Villablanca, PA, et al., *Int J Cardiol*, 225:234-243, 2016.
[4] Sedrakyan, A, et al., *JAMA Intern Med*, 2016.
[5] Marom, G, *Arch Computat Methods Eng*, 1-26, 2014.
[6] Piatti, F, et al., *J Biomech*, 2015.
[7] Martin, C, and Sun, W, *J Biomech*, 48:3026-3034, 2015.

PATIENT-SPECIFIC MITRAL VALVE ANNULOPLASTY REPAIR: THE OPTIMAL RING DESIGN FOR TREATING ISCHEMIC MITRAL REGURGITATION

Amir H. Khalighi, Andrew Drach, Michael S. Sacks

Center for Cardiovascular Simulation
Institute for Computational Engineering and Sciences
Department of Biomedical Engineering
The University of Te□as at Austin
Austin, T□, USA

INTRODUCTION

Mitral valve (MV) regurgitation is the most common heart valve disease, afflicting more than 2.5 million people in the U.S. alone [1] with a prevalence rate of 1% among the population [2]. Surgical procedures to treat MV regurgitation are either valve replacement or repair techniques which seek to restore a healthy valvular configuration. In the mitral annuloplasty (AP) technique to repair leaky valves, a prosthetic ring is placed on the MV annulus to shrink the dilated mitral orifice [3]. Although earlier outcome assessments predicted excellent long-term results for annuloplasty repair [4], more recent follow-up investigations have revealed a high rate of repair failure and regurgitation recurrence [5]. Furthermore, some studies have shown that patient-specific valvular geometry greatly impacts repair outcomes [6], though this structure-function relationship is not yet clearly understood. In this work, we developed multi-resolution models of five healthy MVs enriched with detailed structural information. Then, these high-fidelity models were used to simulate several repair scenarios. Our results provide insight into the efficacy of each AP ring design, and elucidates factors that should be considered in repair procedures.

METHODS

We used extant imaging data that we had previously collected [7]. In this work, we analyzed five imaging datasets to broaden our scope to a population level. First, the end-systolic state of the valve was simulated and imaged to capture chordae tendineae with high resolution (Figure 1). Next, the valve was imaged in the open state [8] to acquire leaflet geometry with high fidelity [9] (Figure 2). In both states, fiducial markers placed on the valve were used to register the open and closed scans through hyperelastic warping and to build highly detailed full MV models (Figure 3).

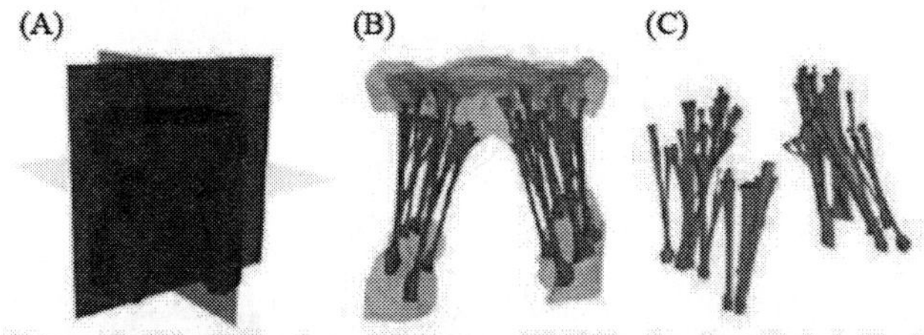

Figure 1: (A) a 3D micro-CT image of MV in the closed state that was sculpted (B) to extract chordae tendineae (C).

To build a structurally informed finite element (FE) model of the MV, fiber architecture data was collected [10] and mapped on the FE mesh using a spline-based method [11]. The mechanical behavior of leaflet tissue was simulated using a nearly incompressible transversely isotropic model [12]. To simulate physiologic loading, we applied a constant trans-valvular pressure of 100 mmHg on the MV leaflets. The contact interaction between the anterior and posterior leaflets was

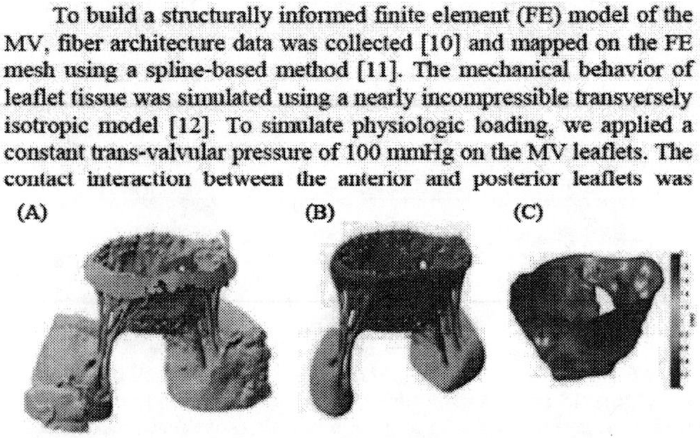

Figure 2: (A) a 3D micro-CT image of MV in the open state that was analyzed (B) to build a shell model for the MV leaflets (C).

modeled as a frictionless self-contact boundary condition (BC). Prior to simulating different repair scenarios, we imposed displacement BCs on the annulus and chordae origins to mimic mitral orifice dilation and papillary muscle relocation induced by ischemic MV regurgitation. This simulated "diseased" state was then used as the base to test and assess different AP ring designs.

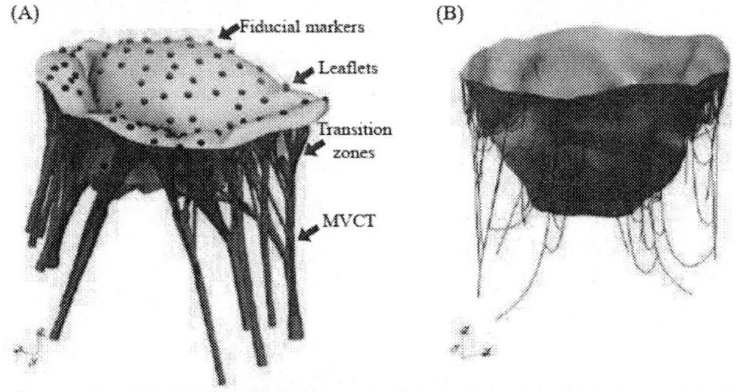

Figure 3: (A) MV leaflets morphed into the closed state, and full MV morphed back to the open state (B).

To quantify the effects of AP ring design on MV response, we applied annular BCs ranging from fully flat to nominal saddle shape. Moreover, the MV opening size was changed from 20% undersized to 20% oversized of the normal annular area (Figure 4). The complete array of all candidate repair designs was then tested through FE modeling in ABAQUS 6.14.1 (SIMULIA, Dassault Systèmes, Providence, RI, USA), configured to use a nonlinear (Newton-Raphson) quasi-static solver with direct time integration (Hilber-Hughes-Taylor implicit scheme) and automatic time stepping. The convergence criteria were set to 0.01 in relative displacements and 0.0001 in relative residuals.

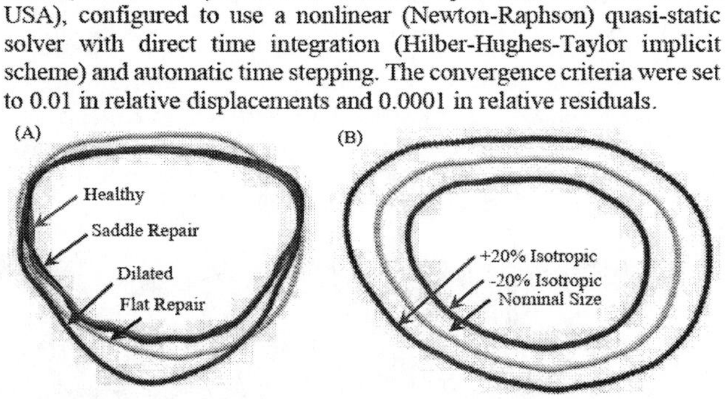

Figure 4: Studied annular configurations in terms of shape (A) and size (B) are shown.

RESULTS

We simulated the MV closing behavior under healthy, dilated, and multiple repair scenarios for five ovine MVs. For brevity, we have only presented the results from one valve throughout this abstract. To assess valvular response, we computed the pointwise field of total fiber recruitment (TFR) defined as

$$\mathbf{TFR} = \int_{-\pi/2}^{\pi/2} \Gamma(\theta) \left[\int_{0}^{E_{ens}(\theta)} D(\chi)\, d\chi \right] \mathbf{n}(\theta) \otimes \mathbf{n}(\theta)\, d\theta \qquad (1)$$

where $\Gamma(\theta)$ is the fiber ensemble orientation function, $D(\chi)$ is the fiber ensemble recruitment function, and E_{ens} is the fiber ensemble strain. The magnitude of TFR is a normalized index of local structural changes in the leaflet tissue, which varies between the slack tissue with full structural reserve (TFR=0) and fully recruited tissue with depleted structural reserve (TFR=1). We observed large differences in the TFR distribution between the considered states, especially in the central region of the anterior leaflet (Figure 5). Furthermore, our results have revealed that saddle-shaped rings with undersized orifice area better restore the normal valvular state.

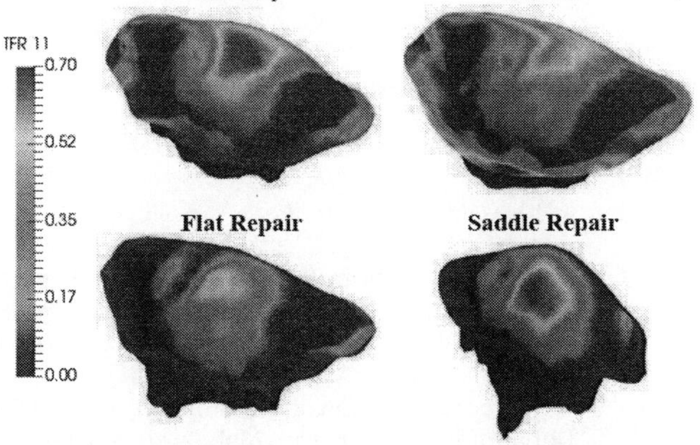

Figure 5: Fiber recruitment measure is analyzed under different conditions which indicated tissue sensitivity to annular states.

DISCUSSION

We developed attribute-rich computational models of the MV apparatus and demonstrated the utility of predicting biomechanical response of MV under different physiological states. The simulations results for the healthy and dilated cased were validated against *in vitro* measurements. The computed structural tensor index (TFR) revealed significant differences in the tissue response to dilation and various repair scenarios. This information was then used to direct the design of AP rings toward restoring the normal state of the leaflet tissue. In our ongoing work, we aim to simulate a more extensive collection of AP ring designs, with the objective of specifically finding an optimal set of geometries that succeeds in restoring the normal MV state.

This work is the first step towards our long-term goal of reducing the number of negative outcomes associated with MV regurgitation repairs. We believe that our study will lead to a precise patient-specific simulation tool for personalized AP ring designs with a minimized level of IMR recurrence. Ultimately, our framework should be extended to model the post-repair MV remodeling process, which would provide a platform to directly evaluate, compare, and optimize the long-term efficacy of a variety of repair approaches, including AP techniques.

ACKNOWLEDGEMENTS

Research reported in this publication was supported by National Heart, Lung, and Blood Institute of the National Institutes of Health under award number R01HL119297.

REFERENCES

[1] Enriquez-Sarano, M et al. *The Lancet*, *373*(9672), 1382-1394, 2009.
[2] Thom T. et al., *Circulation*, 113(6):e85–e151, 2006
[3] Rausch, M. et al., *Ann Biomed. Eng.*, 39(6), 1690-1702, 2013.
[4] David, T.E., et al., *Circulation*, 112.000699, 2013.
[5] Acker, M.A. et al., *N Engl J Med*, 370(1): p. 23-32, 2014.
[6] Salgo, I.S. et al., *Circulation*, 106(6): p. 711-717, 2002.
[7] Khalighi, A.H et al., *Ann Biomed. Eng.*, 1-16, 2016.
[8] Bloodworth, C.H et al., *Ann Biomed. Eng.*, 1-12, 2016.
[9] Khalighi, A.H. et al. *Lecture Notes in Comput. Sci.*, 164-171, 2015.
[10] Sacks, M.S et al., *Ann Biomed. Eng.*, 25 (4), 678-689, 1997.
[11] Aggrawal, A et al. *Lecture Notes in Comput. Sci.*, 141-149, 2013.
[12] Fan, R. et al. *J of Biomech.* 47 (9), 2043-2054, 2014.

POSITIONING OF A DEDICATED STENT FOR CORONARY BIFURCATIONS: AN IN SILICO STUDY

Claudio Chiastra (1), Maik J. Grundeken (2), Francesco Migliavacca (1), Gabriele Dubini (1), Patrick W. Serruys (3), Robbert J. de Winter (2), Joanna J. Wykrzykowska (2), Ender A. Finol (4), Wei Wu (4)

(1) Laboratory of Biological Structure Mechanics
Department of Chemistry, Materials and Chemical Engineering “Giulio Natta”, Politecnico di Milano
Milan, Italy

(2) The Heart Center, Academic Medical Center
University of Amsterdam
Amsterdam, The Netherlands

(3) International Centre for Circulatory Health
NHLI, Imperial College London
London, United Kingdom

(4) Department of Mechanical Engineering
University of Texas at San Antonio
San Antonio, Texas, USA

INTRODUCTION

Atherosclerotic lesions at coronary bifurcations represent 15-20% of total coronary lesions [1]. Their treatment is challenging for interventional cardiologists and results in lower procedural success rate and higher risk of long term cardiac events than non-bifurcated lesions [1]. Over the last decade, many stents specifically designed for coronary bifurcations have been developed. However, most of them remained prototypes and are not used in routine clinical practice. The Tryton stent (Tryton Medical, Inc., USA) is one of few dedicated devices that underwent large clinical trials [1, 2]. It is CE marked and is currently under evaluation by the Food and Drug Administration (FDA) for approval in the United States.

The Tryton stent is implanted in the bifurcation side branch (SB). Its design is characterized by fewer struts in the proximal portion and facilitates the implantation of an additional conventional stent in the main branch (MB). The stent is positioned according to four markers. Although the manufacturer recommends deploying the stent so that the carina is positioned 1/3 the distance from the distal middle marker (Fig. 1A), wrong positioning of the device is frequent.

In this context, the aim of this study is to investigate the effects of wrong positioning of the Tryton stent in terms of geometrical and biomechanical aspects by using a virtual bench testing approach. Indeed, modeling techniques based on the finite element method have emerged as a valuable tool for the assessment of geometrical and mechanical variables that are hardly detectable *in vitro* or *in vivo* [3].

METHODS

An idealized model of the left anterior descending / first diagonal coronary bifurcation was created with a distal bifurcation angle of 45°. Realistic diameters were set for the lumen [4]: 3.5 mm for the proximal MB, 2.8 mm for the distal MB, and 2.4 mm for the SB. The arterial wall thickness (proximal MB of 1.05 mm, distal MB of 0.83 mm and SB of 0.72 mm) was defined according to experimental measurements and an isotropic hyperelastic constitutive law was assigned to the three different layers that compose the arterial wall [5].

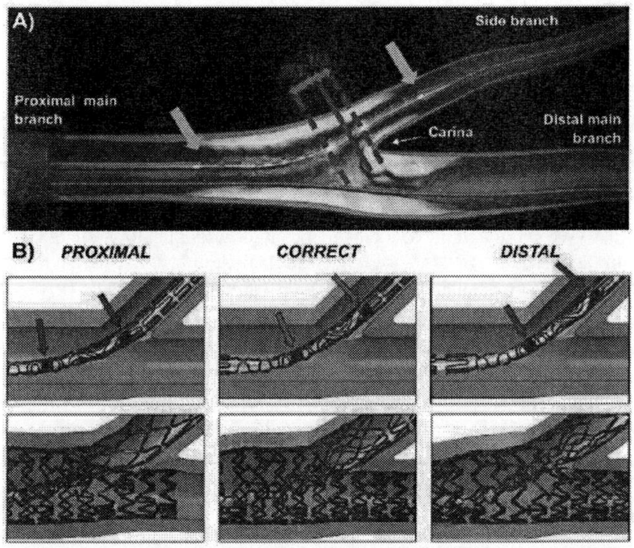

Figure 1: A) Manufacturer's recommendation for correct Tryton stent positioning [6]; B) Details of the three finite element models under investigation: (top) Tryton stent positioning; (bottom) final geometrical configurations after stents' deployment.

A previously validated model of the Tryton stent mounted on a stepped balloon was used [7]. Briefly, the cobalt-chromium alloy that characterizes the stent was described by a Von Mises-Hill plasticity model with isotropic hardening. The polymeric material of the balloon was described using an elastic linear isotropic model. Two different Young's moduli were assigned to the proximal and distal parts of the balloon to comply with the manufacturer's pressure-diameter relationship by using a calibration procedure.

All steps of the so-called 'Tryton-based culotte technique' [2] were simulated as a quasi-static process using Abaqus/Explicit (Dassault Systèmes Simulia Corp., USA): (1) expansion of the Tryton stent in the SB; (2) proximal optimization technique (i.e. inflation of a short balloon in the proximal MB prior to the SB); (3) opening of the MB access; (4) expansion of the conventional Xience V stent (Abbott Vascular, USA) in the MB; (5) kissing balloon inflation (i.e. simultaneous inflation of two balloons in both bifurcation branches); (6) proximal optimization technique.

The following three scenarios were investigated (Fig. 1B):

- 'Correct' Tryton stent positioning.
- 'Proximal' Tryton stent positioning: the distal end of the central stent zone is placed precisely at the level of the carina.
- 'Distal' Tryton stent positioning: the stent is placed so that the carina is at 2/3 from the distal middle marker, instead of 1/3.

RESULTS

Several geometrical variables were computed to compare the three scenarios after stent deployment. In particular, the ostial area stenosis was calculated to assess the SB opening after intervention (see Table 1). This variable is defined as the ratio of the difference between the SB ostium area and the area free of stent struts to the SB ostium area [8]. The worst scenario was represented by the 'proximal' case, which exhibited the highest ostial area stenosis. Conversely, the 'correct' case showed the lowest (and hence better) ostial area stenosis. The Tryton stent diameter at the carina was also calculated, as shown in Fig. 2A. The 'correct' Tryton stent positioning resulted in the largest diameter. In both the 'proximal' and 'distal' cases, the Tryton stent was slightly squeezed after deployment at the bifurcation region. The distal bifurcation angle decreased after stenting in all cases (see Table 1). The more distal the Tryton stent the larger distal bifurcation angle change obtained. Furthermore, stent malapposition was evaluated by calculating the percent area of malapposed struts (i.e. struts not in contact with the lumen surface) with respect to the total area of the abluminal stent surface [9]. It has been reported that malapposed struts can alter the local hemodynamics, potentially inducing stent thrombosis. The 'proximal' case had the highest percentage of malapposed struts, while the 'distal' case had the lower percentage, as shown in Table 1.

Table 1: Geometrical analysis after virtual stenting for the three cases investigated.

Case	Ostial area stenosis	Distal angle change	Stent malapposition
Proximal	74.3%	-4.8°	36.1%
Correct	44.8%	-6.9°	28.4%
Distal	51.5%	-8.4°	25.5%

The arterial wall stress was also computed for the three scenarios after intervention. High values of maximum principal stress were located in all cases at the proximal MB (see Fig. 2B). Larger areas with high stresses were found in the 'proximal' case. The peak stress occurred at the proximal MB in the vicinity of the SB ostium, opposite the carina (see Fig. 2B, black arrows). The 'proximal' Tryton stent positioning yielded the highest peak wall stress.

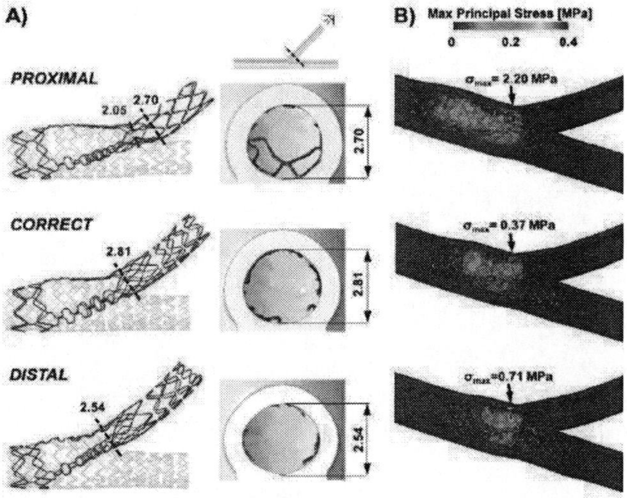

Figure 2: A) Tryton stent diameter at the carina for the three investigated cases. Dimensions are in mm. B) Maximum principal stress in the arterial wall after intervention for the three analyzed cases. Black arrows indicate the location with peak stress.

DISCUSSION

Overall, the 'correct' Tryton stent positioning is the best scenario resulting in the lowest ostial area stenosis, highest diameter at the carina, lower malapposition, and the lowest peak arterial wall stress. The 'distal' positioning ensures the lowest stent malapposition but induces the lowest Tryton stent diameter at the carina. The 'proximal' positioning should be avoided by interventional cardiologists as it represents the worst scenario with the highest ostial area stenosis, stent malapposition, and peak arterial wall stress.

The results of the present work explain and confirm the benefits of 'correct' Tryton stent positioning, as recommended by the manufacturer. The study also shows that virtual bench testing can be successfully used to compare different stenting scenarios in terms of geometrical and mechanical variables. The same procedure can be applied to compare different stenting techniques or stent designs. Furthermore, the results provide a foundation for future fluid dynamic analyses on Tryton stent configurations.

REFERENCES

[1] Lassen, JF et al., *EuroIntervention*, 12:38-46, 2016.
[2] Grundeken, MJ et al. *Expert Rev Med Devices*, 10:707-716, 2013
[3] Antoniadis, AP et al., JACC Cardiovasc Interv, 8:1282-1296, 2015.
[4] Finet, G et al., *EuroIntervention*, 3:490-498, 2007
[5] Holzapfel, G et al., *Am J Physiol Heart Circ Physiol*, 289:H2048–H2058, 2005.
[6] http://www.trytonmedical.com
[7] Chiastra, C et al. *EuroIntervention*, 10:e1-e3, 2015.
[8] Ormiston, J et al. *JACC Cardiovasc Interv*, 1:351-7, 2008.
[9] Chiastra, C et al. *J Biomech*, 48:2102-2111, 2016.

VERY SHORT PERIPHERAL CATHTER FOR REDUCTION OF CATHETER-RELATED THROMBOPHLEBITIS

D. Weiss (1), O. Rotman (2), U. Zaretsky (1), S. Einav (1, 2)

(1) Department of Biomedical Engineering, Tel Aviv University, Tel Aviv, Israel

(2) Department of Biomedical Engineering, Stony Brook University Stony Brook, NY, USA

INTRODUCTION

Short peripheral catheter thrombophlebitis (SPCT) is the major complication associated with the use of short peripheral catheters (SPCs) in intravenous procedures affecting up to 80% of all patients receiving IV theraphy. Recently, we demonstrated that SPCT is initiated by a prolonged mechanical interaction between the catheter cannula and the endothelial layer of the catheterized vein [1, 2, 3, 4]. Here we examine a novel design of a SPC- The very short peripheral catheter (VSPC). The novel VSPC has a significant shorter cannula to minimize the contact between the cannula and the vein wall and is also equipped with a small protrusion on the outer surface of the cannula as an anchoring element in order to avoid the short invasive part from accidently slipping out of the vein (Figure 1). The development of SPCT resulting from indwelling of VSPC vs. a commercial SPC was assessed and quantified in swine model using previously reported MRI tool [2] followed by histopathology analysis.

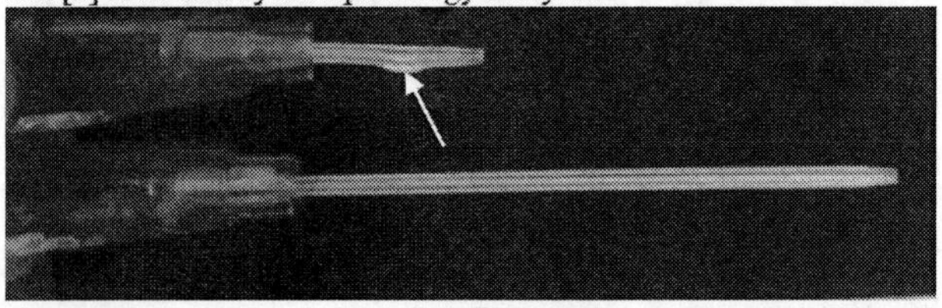

Figure 1: On top: our novel catheter prototype (VSPC) consisting of rounded protrusion as an anchoring element (white arrow) and a shorter cannula (10mm) compared to a matcing size commercial catheter shown at the bottom (Insyte, BD).

METHODS

The effect of catheters indwelling on the evolution of SPCT was studied through swine ear veins model. The severity of SPCT was quantitatively evaluated *in vivo* by means of Magnetic Resonance Imaging (MRI) and compared to blinded histopathology analysis. Five Female swine (landrace breed, Lahav, Israel), weighting 65 kg used in the study. Each swine ear was randomly assigned to either our 10mm length VSPC (Figure1) or the 20G commercially used SPC (30 mm length, Insyte, BD), both correspond to outer cannula diameter of ~1.1mm. All catheters were inserted by an authorized veterinarian into the lateral aspect of the marginal ear vein (under anesthesia), then flushed with heparinized saline (for vein patency test), glued to the surrounding tissue using designated tissue adhesive (hystoacryl) and finally secured to the surrounding tissue with transparent film dressing (Tegaderm™; 3M), full roll of gauze and leucoplast (All procedures were approved by the Helsinki committee of Assaf Harofeh Medical Center). For the following consecutive 12 days the swine was freed in the habitat (CRO facility with OECD-GLP; Assaf Harofeh Medical Center) with free access to diet and water. Each animal underwent two MRI scans through the study- 4 and 12 days following catheters insertion procedure. MRI was performed in a 3 tesla scanner (Magnetom Skyra, Siemens, Germany) using 64 channels head coil. The animal was placed in a prone position and the head of the animal was tilted to one side at a time in order to focus on the ear vein. MRI scanning included the following sequences: quantitative T2 mapping, and T1 pre and post gadolinium-DTPA (GD) injection, with and without fat saturation. The site of the catheter was marked using fiducial markers (Beekley Corporation). Image analysis was performed using Matalb (Mathworks, Natick, MA, USA). At the end of the second MRI scan the animal was sacrificed and its ears (along with the catheters) were harvested and fixed in 10% buffered formaldehyde for a comprehensive blinded histopathology analysis. Histopathological evaluation compared the relative local tissue changes in both ears. Biological response parameters evaluated by histopathology were scored using semi-quantitative grading of five grades (0-4), taking into consideration the severity and/or extension of the response (0=none, 1=slight, 2=mild, 3=moderate, 4=severe).

The samples were diagnosed for: Blood vessel inflammation, Thrombus formation, Exfoliation of endothelium, Intimal Edema, Intimal necrosis and Hemorrhage.

RESULTS

Analysis of the MRI results suggests that the T2 values of the SPC groups were higher than the VSPC ones (with average of 65ms Vs. 54ms) for all the tested animals (Figure 2). This finding imply on higher extent of edema identified in the region of the SPC site. It should be noticed that the difference is mainly evident in the second MRI scan which as mentioned, is carried out 12 days following procedure.

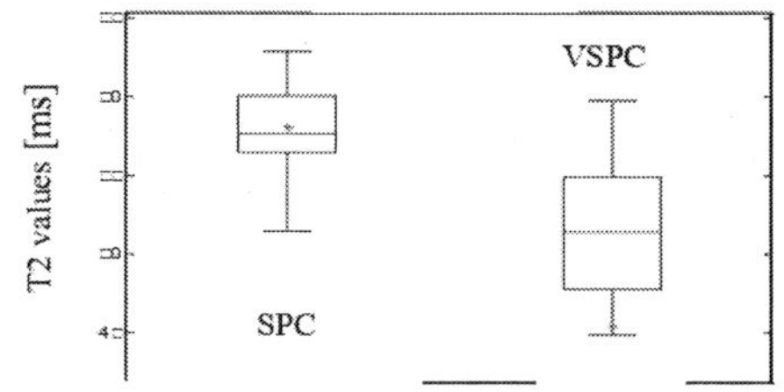

Figure 2: Statistics of the T2 value for each of the tested groups. Green asterisks denote the average and red lines indicate the median. The tops and bottoms of each box are the 25th and 75th percentiles of the samples, respectively.

The T1- weighted imaging revealed the most significant difference between the catheters groups as the averaged signal intensity in each time point was about 1.5 times higher in the SPC samples. As can be seen in Figure 3, "hot" intensity colors are mainly distributed surrounding the vein catheterized by the SPC. These MRI findings suggest on more severe extent of inflammation and edema in the SPC-related samples.

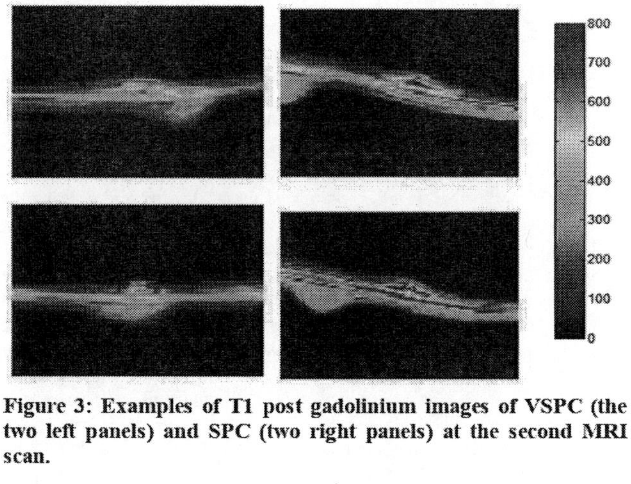

Figure 3: Examples of T1 post gadolinium images of VSPC (the two left panels) and SPC (two right panels) at the second MRI scan.

The histological findings showed marked stretching and focal damage along the vein which was catheterized with the SPC catheter.

In the specimens from swine#1-swine #4, all morphological parameters indicative of vascular integrity were lower in the VSPC-derived samples, when comparing to the SPC-derived samples. The singular parameter of "hemorrhage" was minimally higher in the "VSPC" group of swine#3, however, as the very minor increase in the "hemorrhage" was not accompanies with any other changes which may suggest loss of cells (i.e., necrosis, inflammation and thrombus formation), the increase grade of hemorrhage is considered of no biological significance. For swine #5, both the thrombus formation and intimal edema parameters found to be higher in the VSPC group comparing to the SPC one.

DISCUSSION

The MRI and the histopathological findings suggest that the VSPC catheter type is safer to use when comparing to the SPC catheter. Both the MRI parameters (T2 mapping and T1 post gadolinium) were found significantly higher in the SPC group indicating on higher extent of edema and inflammation compared to the VSPC group. The indwelling of the SPC caused mark stretching of the vein wall and severe inflammation in the surrounding tissue. The VSPC was able to dwell within the vein for all the period of the experiment with no evidence for catheter slipping out of the vein.

The thrombus formation in the VSPC group of swine #5 was identified in the insertion site and was assumed to be the result of an external trauma which most likely happened during the swine habitation. This kind of potential trauma usually accompanied by intimal edema thus explaining the higher scoring of these parameters comparing to the SPC samples in this animal.

The position of each catheter within the cannulated vein was further validated using ultrasound (data is not shown). The position of the VSPC answered our design requirements. It avoided contact with the opposite vein wall and was also allowed opened flow regions between its lower wall part and the vein wall, as opposed to the commercial SPC which was only allowed flow to its sides and constantly contacted the vein wall.

ACKNOWLEDGEMENTS

This study was supported by the Nadal Colton Fund (Grant no. 590046).

REFERENCES

[1] Weiss, D et al., Mechanical compression effects on the secretion of vWF and IL-8 by cultured human vein endothelium. *PLOS ONE*, 2016.

[2] Weiss, D et al., Quantitative T2 mapping for detection and quantification of thrombophlebitis in a rabbit model. *Journal of Biomechanics*, 2016.

[3] Weiss, D et al., Modeling catheter-vein biomechanical interactions during an intravenous procedure." *Computer methods in biomechanics and biomedical engineering*.,8,1-10,2015.

[4] Rotman, Oren Moshe, et al., Biomechanical aspects of catheter-related thrombophlebitis." *Journal of Biomedical Science and Engineering*, 2013.

FLUID-STRUCTURE INTERACTION MODELING OF THE PENN STATE PEDIATRIC VENTRICULAR ASSIST DEVICE: PRELIMINARY COMPUTATIONAL STUDIES

B. Good (1), P. Crompton (1), K. Manning (1,2)

(1) Department of Biomedical Engineering
Pennsylvania State University
University Park, PA, USA

(2) Department of Surgery
Penn State Hershey Medical Center
Hershey, PA, USA

INTRODUCTION

Children with congenital heart disease face the highest wait-list mortality rate in all of transplantation medicine with 23% dying within 6 months of being added [1]. Despite this, only recently has a pediatric ventricular assist device (PVAD), the Berlin Heart EXCOR, been approved for use in the United States [2]. As part of developing a reliable PVAD, it is necessary to accurately predict the hemodynamics within the device due to higher hemolysis and thrombosis rates in PVADs compared to adult VADs [3]. Very little work has been done on the modeling of blood pumps [4] due primarily to the moving components within the pump and the strong fluid-solid interactions that occur. For pulsatile devices, this includes the interactions of both air and blood with a moving membrane.

A current fluid-structure interaction (FSI) framework exists within the OpenFOAM Extend Project [5] that has been validated in simple geometries. The solver defines independent dynamic meshes for the fluid and solid regions and contains inner loops for strong coupling. However, this solver is only designed for one incompressible Newtonian fluid and one linear elastic solid coupled through a single FSI interface. Therefore, to model the complex interactions within a pulsatile pneumatic blood pump like the Penn State PVAD, a new solver must be developed to handle three independent meshes (two fluid meshes, air and blood, and one solid mesh, membrane) with two independent FSI interfaces (membrane-to-air and membrane-to-blood) and material specific constitutive models for air, blood and membrane.

METHODS

The proposed FSI solver consists of three partial solvers for air, blood and membrane. For air, a transient incompressible Newtonian solver with a standard k-epsilon model is used. For blood, our previously validated viscoelastic solver for pediatric blood [6] is used. For this study, a blood-analog solution was created consisting of 55% water, 45% glycerin and 0.01% xanthan gum by weight. The fluid's viscoelastic properties were measured with a Vilastic-3 viscometer and fit to the pediatric blood model (Fig. 1) (η=8.54 cP, η_1=10.6 cP, μ=1.621 cP/s and ρ= 1.13 g/ml).

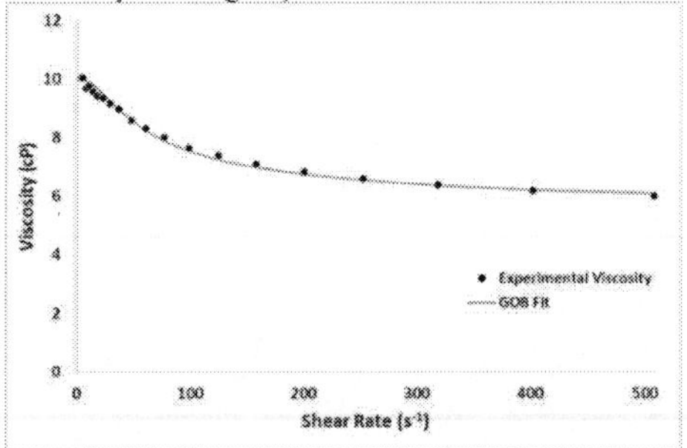

Figure 1: Viscosity vs. shear rate data for the FSI blood-analog and its resultant GOB viscoelastic model fit.

The Penn State PVAD membrane is made from a segmented poly(ether urethane) urea (SPEUU) material that exhibits hyperelasticity and hysteresis [7] that a simple linear elastic model cannot replicate. Therefore, a hyper-viscoelastic finite-volume solid mechanics solver is developed to model the SPEUU membrane. Material characterization was done with dog-bone samples cut to ASTM D638 standards (Fig. 2A) and both uniaxial tension and stress relaxation experiments were performed to determine the respective hyperelastic and viscoelastic models. For this material, it was shown that a Mooney-Rivlin hyperelastic model (Fig. 2B) along with a 3-term Prony series (Fig. 2B) accurately described the SPEUU.

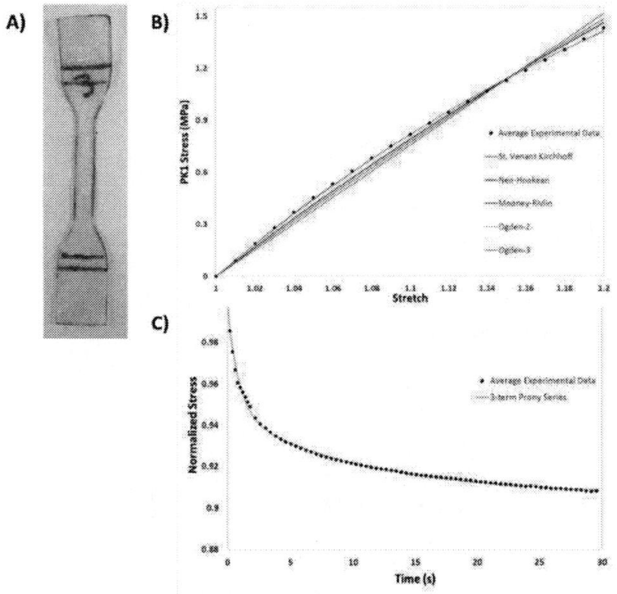

Figure 2: A) Dog-bone SPEUU sample for material characterization. B) Stress vs. strain loading data and hyperelastic model fit. C) Stress-relaxation data and Prony series fit.

The SPEUU membrane solver was validated using the uniaxial experimental data described previously (results not included) and also in a mock PVAD experimental set-up (Fig. 3A). The acrylic model is split into air (left) and blood-analog (right) chambers separated by a modified SPEUU membrane. Both steady and pulsatile pressures are applied with a pneumatic driver and markers on the membrane's surface are tracked using a two-camera system and a direct linear transformation (DLT) Matlab code (Fig. 3B). Both 2D and 3D models were created using Pointwise and steady and pulsatile experimental pressure waveforms were replicated computationally for validation.

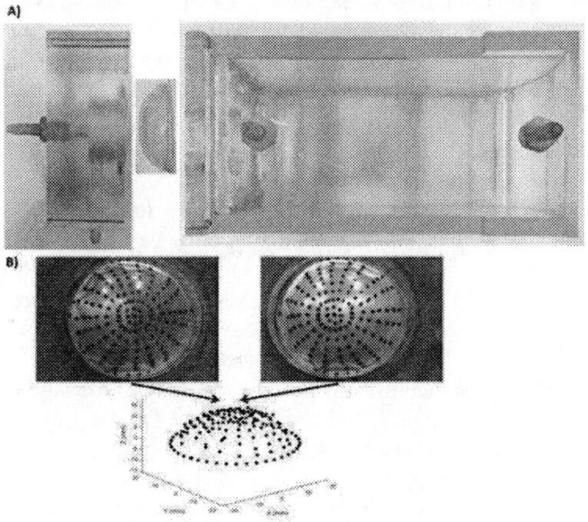

Figure 3: A) Experimental mock PVAD loop. B) Membrane tracking and 3D reconstruction with DLT.

RESULTS

Four steady pressures were applied experimentally (Fig. 4B) and computationally (Fig. 4C) in the 2D mock PVAD geometry (Fig. 4A). For the highest pressure (3730 Pa, Case D), the model under predicts peak displacement by approximately 12% while at a mid-pressure (934 Pa, Case C) the model is within 5% at every marker position.

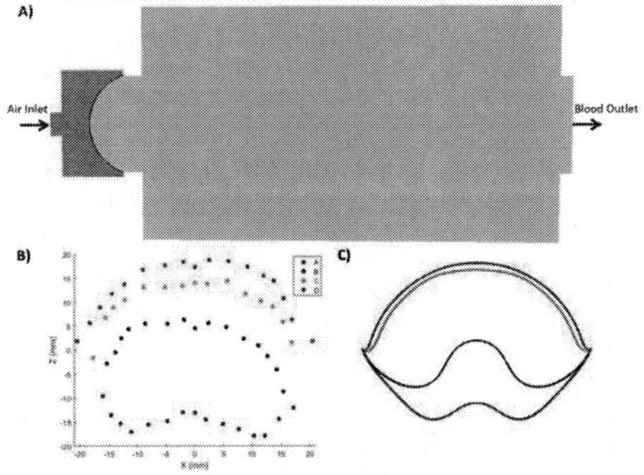

Figure 4: A) Computational FSI model. B) Steady pressure experimental results. C) Steady pressure computational results.

Two pulsatile pressure waveforms were applied (60 and 75 bpm (Fig. 5)) and markers tracked at 5 time points during the pulsatile cycle. This data along with standard deviations for each marker point will be compared to transient computational simulations.

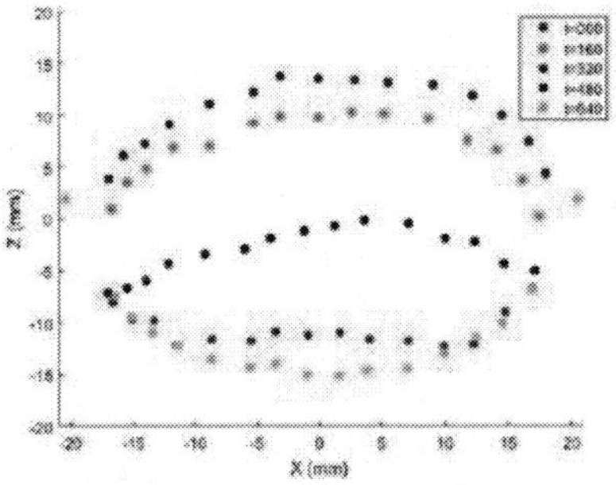

Figure 5: Pulsatile pressure experimental results.

DISCUSSION

To accurately predict hemodynamics inside the Penn State PVAD, a tightly-coupled FSI solver is developed that accurately models the interactions between air, a visco-hyperelastic SPEUU membrane and viscoelastic blood in a simplified model. The solver is validated with both steady and pulsatile 2D computational models and compared to experimental membrane tracking results. This FSI solver will be validated further in full 3D models and finally used to model the actual hemodynamics inside the Penn State PVAD.

ACKNOWLEDGEMENTS

This work was supported by NIH NHLBI HL108123.

REFERENCES

[1] Mah D, *et al. J Heart Lung Transplant.* 28(12):1292-1298. 2009.
[2] Almond C, *et al. Circulation.* 127(16):717-727. 2015.
[3] Bachmann C, *et al. Artif Organs.* 24(5):362-372. 2000.
[4] Long C, *et al. Comput Mech.* 52(5):971-981. 2013.
[5] Ponweiser T, *et al. PRACE.* 2015.
[6] Good B, *et al. Ann Biomed Eng.* 44(4):1019-1035. 2016.
[7] Qi H and Boyce M. *Mech Mater.* 37:817-839. 2005.

REAL-WORLD PROBLEM SOLVING AND VALUE CREATION IN THE BIOMECHANICS CLASSROOM

Laurel Kuxhaus (1), Karen L. Troy (2)

(1) Mechanical & Aeronautical Engineering
Clarkson University
Potsdam, N□, USA

(2) Biomedical Engineering
Worcester Polytechnic Institute
Worcester, MA, USA

INTRODUCTION

It is well-established that project- and problem-based learning engage students in the classroom. Biomechanics classes offer students the opportunity to build technical proficiency by applying mechanics principles to problems involving living organisms. However, technical proficiency is not the only skill necessary for a successful career; mindset also matters. [1]

"Entrepreneurial Mindset Learning" (EML), as championed by the Kern Entrepreneurial Engineering Network, offers a framework for faculty to consider ancillary outcomes (beyond technical proficiency). These include stimulating curiosity, making connections, and creating value. These outcomes are synergistic with bioengineering education, which by nature must teach students about biocompatibility, ethical decision making, and other issues unique to working with humans and animals. Bioengineering problems have the potential to include realistic and contextual constraints and to spark excitement in students who can immediately see the real-world application of their coursework. Using the Entrepreneurial Mindset as a framework when developing course projects offers faculty a structure to enrich content and benefit students.

Here we present a description of the instructors' thought processes behind using EML as a guide to enrich biomechanics class projects dealing with bone fracture. In each project, specific EML principles were identified with the goal of incorporation into a primarily technical project.

METHODS

The authors systematically enhanced biomechanics class projects with EML tasks. In each case, the technical nature of the project was preserved and enhanced. Specific highlights from EML are described below.

Project 1: Vertebral fracture project, "The Elevator Problem".
(KLT at Worcester Polytechnic Institute)
An earlier version of this project was implemented by KLT at the University of Illinois at Chicago. This enhanced version to emphasize EML principles is being implemented in Spring, 2017.
Overall Summary of the Project: Students were given the following scenario: "You are a biomechanics expert witness for an accident case. Mrs. Albert was riding in an elevator when it suddenly screeched to a halt between floors. A second later it free-fell 6 feet and abruptly came to a halt. Mrs. Albert felt sore and received an x-ray later that day, which showed no fracture. One year later she was diagnosed with multiple vertebral fractures and opted to sue the elevator manufacturer." The final goal of the assignment was to write a one-paragraph expert opinion, supported with data, summarizing whether it was more likely than not that Mrs. Albert's vertebral fractures were caused by the elevator accident. The enhanced version of the project includes the "twist" in which half the students were "hired" by the elevator company's attorney, while the other half were "hired" by Mrs. Albert's attorney. All groups were competing for the attorney's business, giving an incentive to produce coherent, thorough work.

Students work in assigned groups of four for the project. They are prompted in class to discuss mechanisms of injury, and to identify what questions they needed to ask and answer to determine whether Mrs. Albert had a case. Upon request they are provided with results from Mrs. Albert's most recent bone density scan, as well has her age, height, weight, occupation, and health history. Students are instructed to provide context, rationale, and citations (if available) for any values they assumed. They are advised to consider the problem in terms of upper and lower limits.

The EML-enhanced version of the project includes additional contextual questions designed to encourage exploration of related material, such as: How much should Mrs. Albert sue for? How could this injury be prevented in the future, given that elevators are necessary?

Final expert opinions will be presented to the class orally and sorted by potential employer, to determine whether results were skewed in favor of the employing attorney. This provides a context for discussing ethics and the need to produce unbiased results. The degree to which students have learned EML will be assessed using a structured self-assessment that prompts students to identify and reflect on specific aspects of the project designed to emphasize various principles.

EML synergies and enhancements: Throughout the development of the project, the Entrepreneurial Mindset was considered [2]. In addition to traditional engineering project goals of *performing analysis* and *validating* results, this particular project meets entrepreneurially-minded learning goals in the following ways:

- Students must *evaluate the technical plausibility, economic and societal impacts* of both the accident and the lawsuit.
- To determine whether Mrs. Albert had a viable case, they need to *assess policy and regulatory issues*.
- Students understand vertebral structure and function, and the engineering design of elevator braking systems. This allowed them to *analyze their solutions* in a real-world context.
- By writing an expert opinion, students learn to *communicate their engineering analysis in terms of economic and societal benefits* (or costs, in this project).

Project 2: Cervine cancellous bone microstructure analysis.
(LK at Clarkson University)

Overall Summary of the Project: Students (26 total) worked in assigned groups of 3 to analyze the microstructural parameters of cervine vertebral cancellous bone, including parameters such as trabecular thickness, trabecular number, bone volume fraction, and connectivity density. The overall research question posed to students was whether cervine vertebral cancellous bone was appropriate for use in biomechanical studies; it is important to note that this question remains unanswered in the literature.

Each group was assigned a particular focus of investigation and specific vertebrae to analyze (for example: comparing cranial vs. caudal parameters, or L2 to L4 vertebrae.) MicroCT scans of cervine lumbar vertebrae and rudimentary instructions for using ImageJ [3] and the BoneJ plugin were provided to students. Each group had to design their own analysis (How many samples of bone would they analyze? How many times would they analyze each vertebrae? Would they analyze any additional parameters beyond those required?) Each group delivered a report, structured similar to a journal paper, and reports were peer reviewed in class. Students then could revise their reports for additional credit.

To earn a grade of an "A", students had to exceed the minimum requirements by completing additional analysis of their choosing and compare to literature beyond that provided. Finally, all students' results were compiled into a spreadsheet, and each individual student was tasked with making an interesting observation about the data set and presenting it in a single-slide format. Students completed a peer-evaluation survey in CATME [4]. Group report scores were scaled by the individual factor from the CATME survey.

EML synergies and enhancements: This particular project meets entrepreneurially-minded learning goals in the following ways:

- Students were forced to consider the value of an ex-vivo model for laboratory research. This required them to *investigate the (research) market,* and *evaluate the societal benefits* and *sustainability* of ex-vivo animal models for biomechanical studies.
- Students developed curiosity about the use of cadaver specimens in research, and learned about relevant *policy and regulatory issues*.
- Students learned about intellectual property surrounding ImageJ (freely-available research-grade software).
- Students *communicated* the *societal benefits* of their conclusion in *economic terms* in their reports.

RESULTS

These projects have been implemented in undergraduate biomechanics classes.

Project #1 was originally implemented in a tissue biomechanics course over a four-week period, and received excellent feedback from students, who enjoyed the context, as well as "being forced to research" and think about their answers. The revised version presented here includes two classroom periods of worktime, plus one classroom period for discussion of the final product.

Project #2 was implemented in an introductory biomechanics class and included 5 classroom periods of work time, plus 2 classroom periods for peer review. Anecdotal observations indicate that students were highly engaged throughout the process; informal reporting suggested that many teams spent approximately 20 hours outside of class time working on the project. While students initially appeared frustrated by the unknown nature of the project and by the gaps in their own knowledge of regulatory issues and the research process, they displayed confidence in their results by the end, as evidenced by dialogue between groups about choice of methods and realism of results. Students' observations about the collection of data generated by all groups indicated a high level of curiosity about potential outliers and trends. Some indicated that the experience was meritous of discussion with prospective employers.

DISCUSSION

Taken together, these biomechanics class projects provide two examples of how EML can be incorporated into otherwise straightforward technical analysis projects. Others have shown that enhancing classroom projects to include skills within the Entrepreneurial Mindset may increase student engagement in the classroom. These experiences can inspire students to build *connections* between their classroom experience and the real world, look *creatively* at problems, and *create value* through their work.

The projects described here required relatively minor edits from previous instantiations to incorporate EML. It is important to note that EML is one framework that can be used to enhance student engagement in the classroom. Many elements of EML are synergistic with fundamental principles behind problem-based learning, flipped classrooms, and other contemporary pedagogical techniques. That is, incorporating elements of EML into existing classroom activities, such as projects, can add value and context to students with only a small effort from the instructor.

ACKNOWLEDGEMENTS

The authors thank the Kern Family Foundation awards to WPI and Clarkson University for supporting this work.

REFERENCES

[1] Osterwelder et al., (2014). *Value Proposition Design.* Wiley.
[2] http://engineeringunleashed.com
[3] Schneider et al., (2012) *Nature methods* **9(7)**: 671-675
[4] Loughry et al. (2007) Ed and Psy Meas, **67**, 505-24.

USING PROJECT-BASED PHYSICAL COMPUTING TO TEACH PROGRAMMING CONCEPTS TO BIOMEDICAL ENGINEERS

Trevor R. Ham (1), Rouzbeh Amini (1)

(1) Department of Biomedical Engineering
The University of Akron
Akron, Ohio, USA

INTRODUCTION

Biomedical engineering is a relatively new discipline of engineering with an extremely wide technical scope. Biomedical engineers are expected to have functional knowledge of concepts from disparate fields, such as mechanics, transport, circuits, and materials chemistry. As such, compared to many other engineering disciplines, undergraduate biomedical engineering students often have a relatively short time to acquire new concepts and incorporate them into their skillset as engineers. Problem [1] and project-based [2] learning has attracted much attention lately as professors recognize the need for more active forms of education. With such a crucial need for educational expediency within biomedical engineering, incorporating an engineering project, one that includes a design statement and where students are expected to meet specific deliverables, can be a useful method to encourage students to move to the highest level of Bloom's taxonomy (i.e., producing new or original work). [3]

Teaching computer programming to biomedical engineers is not an exception, either. As the concepts in programming can sometimes become too theoretical, we have chosen more hands-on projects using physical computing for this subject matter. In particular, since spring 2014, we have been using affordable Arduino microprocessors as the basis for projects in a 200-level programming course for biomedical engineers (Biomedical Computing). The focus of the course has traditionally been on learning MATLAB as a scripting language for simple engineering tasks. Over the past 3 years, the scope of the project has evolved to become a large part of the instructional material, which helps the students get hands-on experience working in MATLAB to solve a practical engineering problem that involves human-machine interactions. Herein, we describe the evolution of the project and how it has augmented student engagement and become an integral part of the course.

METHODS

Spring 2014 – Fall 2015: Arduino UNO microprocessors were introduced as the main hardware for the course project for the first time in the Spring 2014 semester. For two consecutive semesters, the students had the option of choosing the scope of their project as long as the instructor approved the feasibility of the project to be conducted within a single semester. The students had to either pick sensors, collect data from them, and further analyze the data, or upload previously saved data files, process them, and send them to actuators. During these two semesters, students successfully conducting a wide range of projects including quantifying angles using potentiometers, playing music synchronized with LED lights using the notes stored in a .xlsx file, and continuous graphical display of loads acquired by a pressure transducer.

Spring 2016: Beginning with the Spring 2016 semester, we substantially overhauled the way the project was administered. Rather than allowing the students to choose their project scope, we obtained 10 pulse sensors (a simple integrated circuit with a photoplethysmograph, (PPG), and an op-amp) and required all of the groups to build a heart rate monitor using the Arduino microprocessor and PPG (Fig 1.). We gave the students detailed structure, consisting of the following: (1) a detailed introductory lecture regarding how to use Arduinos, (2) a full-length walkthrough that outlined the code they should write and what features it should have, (3) and links to helpful code in online repositories. Students were required to validate the performance of their heart rate monitor and compare it to the beats per minute (BPM) observed when taking their radial pulse (see Table 1). We monitored their progress through the mandatory submission of weekly updates. To understand the extent to which the project helped them improve their comprehension of the material, we studied their perceptions of the project (via a survey) and the effect of the project on their final grade. We split the class into two halves, the first of which completed the

Figure 1: Arduino UNO (blue) with breadboard (white). This image also shows the pulse sensor (arrow) and green LED connected to the microprocessor. This is the setup that the students used in Spring and Fall 2016.

project in the MATLAB scripting language and the second in the Arduino Integrated Development Environment (AIDE), which closely resembles C syntax. We compared the results (surveys and final grades) between the two groups using Student's t-test (α=0.05).

Fall 2016: Based on the results from the Spring semester, we implemented changes in the format of the project. Namely, we gave the students extra time to complete it and added two in-class work sessions. We again measured student perceptions and final project grade. Only one programming language (MATLAB) was used.

RESULTS

Although for the first two semesters (i.e. Spring 2014 and Fall 2015) we did not collect any pre- and post- project surveys, many students acknowledged the educational advantages of hands-on projects in the written questionnaire of the course evaluation. They stated that conducting a project involving physical computing initiated/intensified their interests in computer programming. The students' feedback was used to modify the project scope and format for the following semesters.

Table 1: A comparison of beats per minute (BPM) as calculated by taking a radial pulse and by the Arduino-based monitor. Validation data are from a representative project report.

Group Member	*Radial BPM*	*PPG Monitor BPM*
Student 1	80	81
Student 2	60	62
Student 3	80	90

In general, the Arduino-based heart rate monitor was found to be accurate when compared to manual observation (Table 1). We found a number of interesting outcomes after revising the nature of the project for Spring 2016. First and foremost, we demonstrated that we were able to conduct the project in two different programming languages with no negative effects on final exam grade (p=0.622). In addition, we received valuable student feedback and used that to implement the second version of the heart rate monitor project in Fall 2016. In the Fall semester, we observed improvements in the number of students who reported that the project helped them prepare for the final exam and in the number of students who stated that they enjoyed the project (Table 2). We believe that this improvement was primarily due to the in-class work sessions, as the students were able to overcome hurdles with the help of the instructor that they would have otherwise been unable to. We observed that few, if any, groups made use of the extra time allowed (monitored via the weekly progress updates). Importantly, we noticed a "flattening" effect on the final report grade. While the average was slightly lower (see Table 2), the standard deviation was also substantially reduced. This flattening effect was likely a result of the in-class work sessions, as the students were able to ask questions on the final project report. In the Spring semester, where the groups were mostly left to their own devices, a few outstanding groups produced stellar work while some groups were unable to even finish the assignment. In the Fall semester, every group produced a working heart rate monitor and the quality of the reports was much more consistent.

Table 2: Survey results from Spring and Fall semesters, 2016.

	Spring 2016	*Fall 2016*
Percent reporting that the project helped prepare for the final	71%	78%
Percent reporting that they enjoyed the project	79%	83%
Final project grade (mean ± SD)	95.6 ± 8.77%	92.4 ± 2.08%

DISCUSSION

Through the use of both quantitative (t-test) and qualitative (surveys) measures, we were able to implement active, project-based learning with the goal of improving student comprehension. In addition to the results shown in Table 2, one of the comments that we commonly received from students was that they enjoyed how strongly the project was related to Biomedical Engineering. We are interested in performing future studies to test whether a project's relation to the field of study has an impact on how well students perform and retain the knowledge they learned.

Another outcome of this project was that students improved their confidence and "soft skills" regarding programming and computational problem-solving. Many of the students who took this class expressed a lack of confidence in their math or programming skills before the class began. One of our goals over each semester was to demonstrate to students that there is no secret to mastering programming or mathematics, but that they must do so by working hard and spending time with the material. After completing the project, many students expressed enthusiasm for the material and increased confidence in their ability to program and understand advanced mathematical concepts, despite the actual program for the heart rate monitor being relatively simple. In our opinion, self-efficacy and confidence in programming are extremely important for undergraduate students, especially when considering how fundamental coding skills are to all engineering disciplines.

ACKNOWLEDGEMENTS

The authors would like to acknowledge the Department of Biomedical Engineering and the Doctoral Teaching Program in Engineering at The University of Akron for providing a teaching fellowship for Trevor Ham.

REFERENCES

[1] Clyne AM, Billiar KL, *J Biomech Eng*, 138:070804, 2016.
[2] Savage RN, Chen KC, Vanasupa L, *J STEM Education*, 8:15-27, 2007.
[3] Krathwohl DR, *J Theor Into Pract*, 41: 212-218, 2002.

COURSE BASED UNDERGRADUATE RESEARCH EXPERIENCES IN BIOMECHANICAL ENGINEERING

Alisa Morss Clyne (1)

(1) Mechanical Engineering and Mechanics
Drexel University
Philadelphia, PA, USA

INTRODUCTION

Numerous studies have shown that undergraduate students gain positive outcomes from research experiences, including learning to think like a scientist, finding research exciting, and intending to pursue graduate school or science careers [1, 2]. These research experiences may have additional benefits for women and underrepresented minorities, since they offer critical support from professors and peers to students at risk for leaving STEM fields [3-5]. However, research experiences for undergraduates (REUs) and research internships are resource intensive, and few institutions have the resources to offer these experiences to a large number of undergraduate students. Since Drexel University is a co-operative education school, our students cannot participate in REUs or summer research internships, and therefore an even smaller number of engineering students engage in research experiences.

An alternative approach is a Course Based Undergraduate Research Experience (CURE), in which whole classes of students address a research question that is of interest to the scientific community. A CURE integrates five key elements: use of scientific practices, discovery of new knowledge, relevant work to the scientific community, collaboration, and iteration [6]. The major factors that distinguish a CURE from a typical inquiry based laboratory class is that students conduct research in which the outcome is both not known and is of interest in the field. Through CUREs, many students can be engaged in research, including those who would not seek out a research program since it serves all students who enroll in the course [7]. Students who participate in CUREs report many of the same gains as students who participate in research internships [8]. In addition, students who participated in a CURE as compared to a traditional or inquiry-based laboratory course improved their view of STEM as a creative process and continued to have increased understanding of how to do research and confidence in their abilities 2-3 years after the course [9].

METHODS

A Biofluid Mechanics course was adapted from its current problem-based learning format to focus on the multiscale biomechanics and mechanobiology of congenital heart defects. In the first part of the course, students learned cardiovascular fluid mechanics at the organism level, including cardiac and vascular anatomy, cardiovascular coupling, and mechanical assist devices. Students worked in teams to study how varying the flow parameters of an intravascular axial flow blood pump implanted in the inferior vena cava will affect pulmonary circulation. In the second part of the course, students learned cardiovascular fluid mechanics at the tissue and cell level, including vascular wall shear stress and endothelial cell mechanotransduction. Student teams learned how to use a cone and plate device to expose endothelial cells to varying levels of mean shear stress, pulsatile index, or oscillation frequency. The students then analyzed cell glucose metabolism under these varied mechanical conditions by mass spectroscopy.

The Biofluid Mechanics CURE was assessed using the Undergraduate Research Student Self-Assessment (URSSA) tool, which is a validated approach that gives reliable measures of gains perceived to be achieved by undergraduates participating in research [1, 10]. This tool is currently being used on Drexel's campus to assess the effectiveness of a traditional mentored research program, called the STAR Scholars Program [11]. The STAR Scholars data will serve as important baseline data for the assessment of the Biofluid Mechanics CURE, which will allow us to understand whether this more streamlined approach to research led to the same benefits seen from a traditional mentored research experience in the context of the same institution.

RESULTS

Undergraduate students in mechanical and biomedical engineering learned biofluid mechanics in the context of the multiscale biomechanics and mechanobiology of congenital heart defects. While course assessment is still ongoing, students did successfully modify flow parameters in a computational fluid dynamics model of an intravascular axial flow pump, test endothelial cell response to steady laminar and oscillating disturbed flow in a cone and plate device (Figure 1), and analyze glucose mass spectroscopy data (Figure 2).

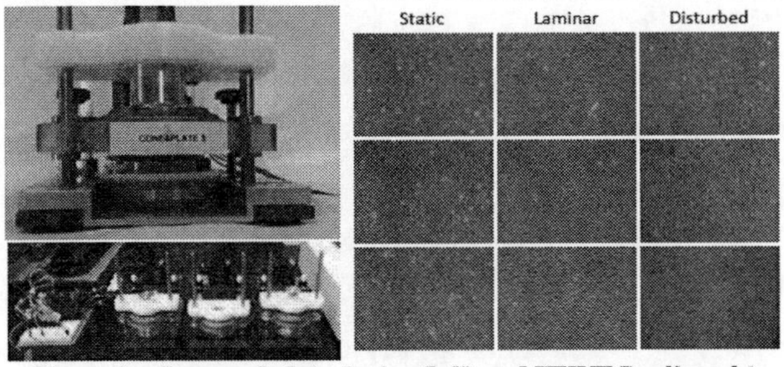

Figure 1: Cone and plate device (left) and HUVECs aligned to steady but not oscillating flow (phase contrast microscopy).

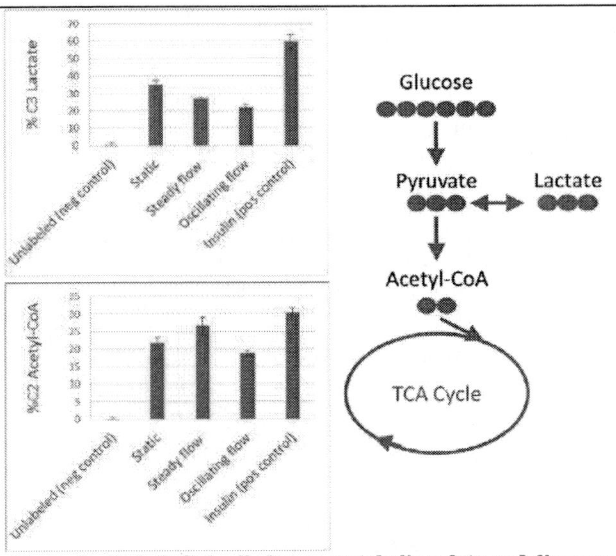

Figure 2: Endothelial cell glucose metabolism data and diagram for cells exposed to static conditions, steady laminar flow, and oscillating disturbed flow.

DISCUSSION

Drexel undergraduate students in biomedical and mechanical engineering were able to experience biofluid mechanics research through the CURE. This teaching method exposed more students to the research process than could otherwise be achieved with limited resources. By using existing data sets, the time commitment for both instructors and students was decreased. We hope that other faculty will implement the CURE teaching approach in biomechanics courses, in particular introductory courses so that interested students can later pursue additional research experiences.

ACKNOWLEDGEMENTS

We thank Dr. Jennifer Stanford for assistance in assessing the Biofluid Mechanics CURE.

REFERENCES

[1] Laursen S, Hunter A-B, Seymour E, Thiry E, Melton G. Undergraduate Research in the Sciences: Engaging Students in Real Science. San Francisco: Jossey-Bass; 2010.

[2] Lopatto D, Tobias S. Science in Solution: the impact of undergraduate research on student learning. In: Research CoU, editor. Washington, DC, 2010.

[3] Barlow AEL, Villarejo M. Making a difference for minorities: Evaluation of an educational enrichment program. J Res Sci Teach. 2004;41(9):861-81.

[4] Nagda BA, Gregerman SR, Jonides J, von Hippel W, Lerner JS. Undergraduate student-faculty research partnerships affect student retention. Rev High Educ. 1998;22(1):55.

[5] Eagan MK, Sharkness J, Hurtado S, Mosqueda CM, Chang MJ. Engaging Undergraduates in Science Research: Not Just About Faculty Willingness. Res High Educ. 2011;52(2):151-77..

[6] Auchincloss LC, Laursen SL, Branchaw JL, Eagan K, Graham M, Hanauer DI, et al. Assessment of Course-Based Undergraduate Research Experiences: A Meeting Report. Cbe-Life Sci Educ. 2014;13(1):29-40.

[7] Rowland SL, Lawrie GA, Behrendorff JBYH, Gillam EMJ. Is the undergraduate research experience (URE) always best?: The power of choice in a bifurcated practical stream for a large introductory biochemistry class. Biochem Mol Biol Edu. 2012;40(1):46-62.

[8] Russell CB, Weaver GC. A comparative study of traditional, inquiry-based, and research-based laboratory curricula: impacts on understanding of the nature of science. Chem Educ Res Pract. 2011;12(1):57-67.

[9] Szteinberg GA, Weaver GC. Participants' reflections two and three years after an introductory chemistry course-embedded research experience. Chem Educ Res Pract. 2013;14(1):23-35.

[10] Hunter A-B, Weston TJ, Laursen SL, Thiry H. URSSA: Evaluating student gains from undergraduate research in the sciences. CUR Quarterly. 2009;3:15-9.

[11] Stanford JS, Rocheleau SE, Smith KPW, Mohan J. Early undergraduate research experiences lead to similar learning gains for STEM and Non-STEM undergraduates. Studies in Higher Education. 2015:1-15.

A New Approach to Teaching Biomechanics by Bridging the Gap Between Classroom and Clinic

Anita Singh (1), Dawn Ferry (2)

(1) Biomedical Engineering
Widener University
Chester, PA, USA

(2) School of Nursing
Widener University
Chester, PA, USA

INTRODUCTION

Teaching biomechanics by itself is challenging because the field integrates two difficult bodies of knowledge: the complexity of human anatomy and physiology with the mechanics of the body and the external forces it encounters while being stationary or during motion. Furthermore, teaching the course to a team of multidisciplinary students warrants needs of teaching techniques that enhances student learning in the classroom. This paper reports the outcome of a project based learning approach by integrating clinic immersion experience in a classroom-oriented Biomechanics course.

METHODS

A senior level interdisciplinary Biomechanics course was offered to a group of mechanical and biomedical engineering students. Students were provided with power point lectures during regular class hours. Topics covered in the class included: Introduction to biomechanics, biomechanical models, kinematics, kinetics, bioinstrumentation, tissue biomechanics of bone, muscle and soft tissue and medical devices – hip, knee and spine, injury biomechanics and sports biomechanics. Wide range of topics covered in one semester course often poses several challenges, including the assessment of what was taught and what student learned in the class. To overcome this challenge, we utilized a project based learning approach (Fig 1). The process started with exposing the students to clinical settings where they had an opportunity to make observations and explored to identify an unmet need. Students were taken to the Center for Simulation and Computerized Testing located in the School of Nursing within the University. They met with the Director of the center and were exposed to the existing simulation mannequins including the SimMen, Noelle (a birthing simulator), two baby Hal simulators, pediatric simulator, and iStan simulator. After having a thorough understanding of the facility, the students were tasked with identifying a limitation in the mannequin and propose an engineering solution to overcome this limitation utilizing knowledge obtained in the course (Table 1). This approach not only helped enhance their critical thinking, but also exposed them to the components of creativity, innovation, ethical decision-making, global awareness, self-directed research, and life-long learning. These are some of the most important entrepreneurial skill sets needed in the workforce along with technical aptitude.

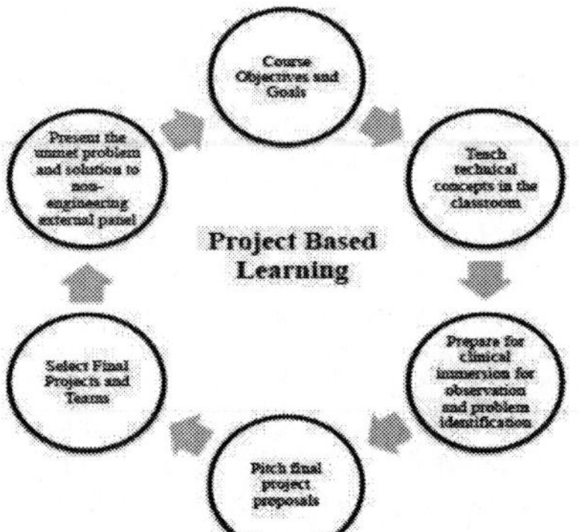

Figure 1: Project based learning approach to integrate clinical immersion and problem identification phase in the design process to meet the course objectives and outcomes.

Table 1: Details of Project Based Learning Projects for the Biomechanics Course

Pitched Proposals	Final Proposals	Proposal Design
Electromechanical lever to lift up the manikin upper body A new Obese SimMan	Electromechanical Lever to Lift Upper Body in Obese Patients	
SimMan: Fever and Hypothermic Conditions	SimMan: Fever and Hypothermic Conditions	
SimMan Improvement Proposal: The Need for Human Temperature Emulation	SimMan Improvement: The Need for Human Temperature Emulation	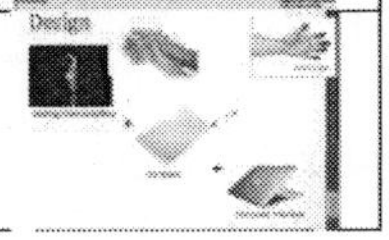

RESULTS

Student evaluations and informal interviews with the students have revealed successful implementation of this strategy. As an indirect measure, we used student evaluation of the course as shown in Table 2 and their comments that were as follows:

- "I have never gained more from a single course in my entire life. I feel pulled towards biomedical engineering as my future for graduate studies."
- "I really enjoyed the class! The instructor really used the resources on campus well (our final project proposal with the Nursing department, class in the PT department) and really tried to get us engaged in class and helped us understand how important biomechanics is and how it is applied in the real world."
- "This class was my favorite class this semester. The course material was fascinating and was presented in a clear, concise manner. The class was challenging in a fun way. I felt inclined to study more, and do more out of class research so that I was prepared and ready to go for the next class."

Table 2: Indirect Assessment of Student Evaluation of the Course

Questions:	High	Low	Mean
The instructor accomplished the goals of the course.	5	4	4.75
The course material was presented in a clear manner.	5	4	4.67
After taking this course, I understand why this subject is important to Engineering	5	4	4.92
After taking this course, I understand the subject better.	5	4	4.75

DISCUSSION

Several teaching techniques have been investigated while teaching Biomechanics course. Use of computer-assisted instruction (CAI) in introductory biomechanics had no significant effect, even though the student attitudes about CAI were positive [1]. Research using the biomechanics course inventory (BCI) has identified course and instructor characteristics to account for much smaller variance (2-5%) in learning [2] than student characteristics and behaviors (14-40%) do [3-4]. Another important factor that plays a key role in student learning of biomechanical concepts is student's perception of career relevance, and their interest in the subject. Implementing student-initiated projects in the course increases student motivation due to ownership and high level of interest that sets them for a desire to succeed. Furthermore, since the capstone course often limits the identification phase, having students learn about it in other courses in the curriculum helps expand their knowledge of the design process.

In conclusion, project based learning with problem identification phase enhances student interest in engineering, and motivate learning in upper division engineering science courses. Our future studies will aim to investigate if this approach also enhances student retention in engineering programs, and enhance performance in capstone design courses and experiences.

ACKNOWLEDGEMENTS

We wish to acknowledge the Simulation Lab staff at Widener University, Chester, PA

REFERENCES

1. McPherson, M. N., & Guthrie, B. M. (1991) In J. D. Wilkerson (Ed.) Proceedings of the Third National Symposium on Teaching Kinesiology and Biomechanics in Sports. 73-76.
2. Knudson, D., Bauer, J., & Bahamonde, R. (2009) Perceptual and Motor Skills, 108, .499-504.
3. Hsieh, C., & Knudson, D. (2008) Sports Biomechanics, 7(3), 398-402.
4. Hsieh, C., Smith, J.D., Bohne, M., & Knudson, D. (2010) Practice and Evidence of Scholarship of Teaching and Learning in Higher Education.

ENCOURAGING AN ENTREPRENEURIAL MINDSET IN BIOMECHANICS

Kristen L. Billiar

Department of Biomedical Engineering
Worcester Polytechnic Institute
Worcester, MA USA

INTRODUCTION

Through technical coursework, undergraduate engineering students gain the skills to solve well-defined problems in their area of study, yet the world is full of ill-defined problems. Thus many innovative educators have developed curricula ripe with open-ended problems which challenge students to apply their knowledge to come up with multiple solutions, and, in the best of cases, assess which is most reasonable based on literature and/or hands-on validation. Nevertheless, a gap in learning remains… there is often an opportunity for the students to come up with an innovative solution to an even broader problem if they are challenged to examine the problem, constraints, and opportunities more thoroughly. And sometimes the students, if in the proper mindset to examine the overall situation more deeply, will find that the problem being solved is not really the problem that needs to be solved.

An entrepreneurial mindset has been shown to aid students in using their technical skills to identify critical problems and develop innovative solutions "outside of the box." According to the KEEN (Kern Entrepreneurial Engineering Network) definition, the entrepreneurial mindset has three critical components: curiosity, connections, and creating value [1]. In short, the students should be empowered to be curious and investigate the highly dynamic world; they should be encouraged to make connections between different types of information and integrate knowledge; and they must be trained to create solutions which meet the needs of others and add meaningful value.

In this work, the implementation of a static analysis project in a sophomore-level biomechanics course is described in which an entrepreneurial mindset is encouraged. The students' performance on the team project was assessed using a detailed rubric, and their aptitude in creating value beyond solving the technical problem was qualitatively evaluated. The students' performance on straight-forward static analysis and conceptual understanding are also assessed and compared to previous offerings of the course. Further, the student feedback on the project was recorded and analyzed qualitatively.

METHODS

In place of two standard homework assignments covering forces, moments, and 2D static analysis, utilized in previous offerings of the course, the students were assigned a project.

To encourage their curiosity, orthopedic treatment of dogs in pain was chosen as the "hook." The students were told that veterinarians are reporting higher-than-acceptable cracking of the plastic portion of the acetabular cups in a total hip replacement products causing undue pain and suffering to many dogs, especially large breeds. The 75 students, mostly sophomores, were randomly assigned into three-person teams and challenged to analyze the mechanics of the canine hip to determine the force that the ball (replacing the femoral head) applies to the acetabular cup. To encourage the creation of greater value, the students were given extra credit if they could come up with a better design or alternative solution beyond the calculation of forces in the hip.

The project was assigned in two parts. In the first part, each team provided schematics of the problem, free body diagrams, the equations, and an initial back-of-the-envelope calculation. For the second part, each team was required to create a two-page report with multi-page appendices containing individual students' diagrams and calculations. A detail rubric was provided for the report after the first part was handed in to clarify what was being asked for and how the report would be graded.

The teaching assistant and instructor both graded the projects using the rubric, and the alternative design solutions, if provided, were recorded in a spreadsheet.

To assess technical knowledge gained from the lectures and project, a quiz was administered which had three parts: force and moment analysis, static equilibrium analysis, and conceptual questions. The quiz, designed to assess the students' ability to perform basic static analyses and describe static equilibrium concepts in writing, was equivalent in content and difficulty to the quizzes administered in the past three course offerings. To obtain feedback from students, an anonymous mid-course evaluation was administered electronically through the course site.

Results from the this offering and the three previous offerings were compared using a one-way ANOVA with $p<0.05$ considered significant. Differences between groups were determined using post hoc analysis.

RESULTS

The average score of the projects (99%) were higher than those of the homework assignments in the previous offering (<90%) (Figure 1). In both cases (project and homework assignments), the scores were high since the students could work together in teams and ask the instructor and TA for help. The students were also encouraged to utilize the rubric to assess their own work while completing the project which led to higher scores, and they could earn up to 10 points extra credit for offering additional impact demonstrating an entrepreneurial mindset, also leading to higher scores than the homework assignments.

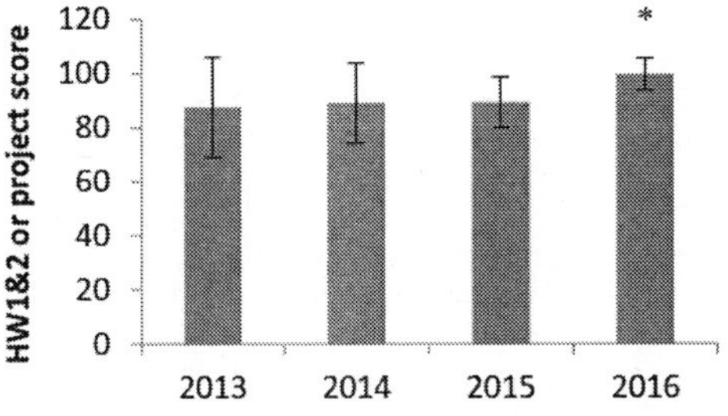

Figure 1: Comparison of combined scores for homework assignments 1 and 2 from 2013 to 2015 to overall project score (with extra credit) for 2016. *p<0.001.

Unexpectedly, the scores on question 1 (forces and moments) and question 2 (rotational static equilibrium) from the quiz were significantly lower for the students completing the project compared to those completing the homework assignments in the previous offerings (Figure 2). However, the students did significantly better in describing the concepts of static analysis following the project compared to doing homework assignments (Figure 3).

In terms of the entrepreneurial mindset, over 75% of the teams completed the extra credit. The majority of the teams suggested using more durable materials such as ceramics, metals, and higher molecular weight plastics (most of the students in the course are in the "Biomaterials and Tissue Engineering Track", not the "Biomechanics and Biofluids" Track in our program). A few teams suggested mechanical solutions such as changing the geometry, patient specific implants, and even use of an exoskeleton.

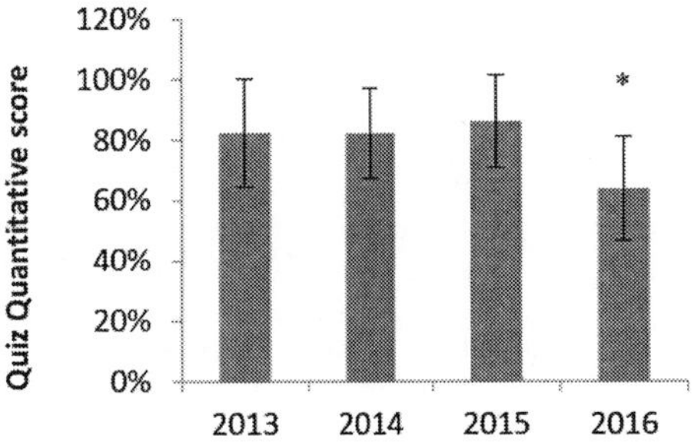

Figure 2: Combined scores on quantitative question 1 (forces and moments) and question 2 (static analysis). . *p<0.001.

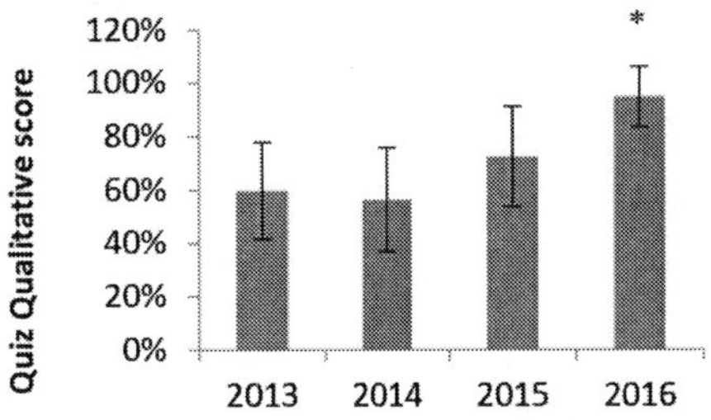

Figure 3: Combined scores for 5 or 6 qualitative questions in the third portion of the quiz. *p<0.001.

From mid-term feedback, the students generally liked the project, although many commented that it was too open-ended and assigned too early in the term (it was completed in parallel with learning static analysis) thus they had difficulty making connections with the course materials. In regards to encouraging curiosity, some students commended that analysis of humans would have been more relevant, and some found it difficult to find anatomical information about dogs.

DISCUSSION

In this work, a team project was developed for a sophomore-level biomechanics course in lieu of two standard homework assignments. In hindsight, it is not surprising that not requiring the students to do the same type of problems as assessed in the quiz resulted in lower scores on those types of problems. Alternatively, the high scores on the conceptual questions relative to previous offerings without the project was surprising. Our findings suggest that even limited project-based learning can have a positive impact on conceptual understanding (or at least the students' ability to describe the concepts in writing). Additional metrics of learning are needed to quantitatively assess the students' progress towards and entrepreneurial mindset.

ACKNOWLEDGEMENTS

The authors thank the Kern Family Foundation for the KEEN grant to WPI to support educational workshops and implementation into the engineering curriculum.

REFERENCES

[1] http://engineeringunleashed.com/keen/em101/ accessed 1/18/17.

UAB SOLUTION STUDIOS™ - A COLLABORATION BETWEEN NURSING, BIOMEDICAL ENGINEERING, AND HONORS

Joel L. Berry, Ph.D. (1), Nancy P. Wingo, Ph.D. (2), Kristen Noles, RN, MSN, CNL, (3) and Alan Eberhardt, Ph.D. (1)

(1) Department Biomedical Engineering
The University of Alabama at Birmingham
Birmingham, AL, USA

(2) School of Nursing
The University of Alabama at Birmingham
Birmingham, AL, USA

(3) UAB Hospital
The University of Alabama at Birmingham
Birmingham, AL, USA

INTRODUCTION

Innovative ideas originating in the clinical setting are often met with seemingly insurmountable obstacles, impeding their translation into improved patient care and outcomes. Clinicians may not know how to advance their ideas from concept to application, particularly when a solution requires expertise in other fields such as basic science or engineering. Additionally, clinicians may have little time to devote to nurturing a good idea. Concurrently, biomedical engineering (BME) students and students in other STEM disciplines are charged with solving clinical problems in classroom settings, yet they often do not receive enough exposure to the healthcare setting to learn to define problems and generate viable solutions. The challenge remains to bring together these stakeholders to accelerate the translation of innovation out of the clinical setting into healthcare practice.

UAB Solution Studios™ is an interprofessional education program that developed as a partnership between the UAB School of Nursing, the UAB Department of Biomedical Engineering, the UAB Honors College, and UAB Hospital to facilitate interaction between students and clinicians. The program promotes the Stanford d.school 5-step design thinking process [1] for students in clinical nursing disciplines, undergraduate Biomedical Engineering (BME) programs, and the Science and Technology Honors Program.

In this process, BME and Honors STEM students access the Solution Studios website in the Canvas learning management system to explore descriptions of clinical problems posted by graduate nursing students and UAB Hospital clinicians. After selecting problems that interest them, students meet with the clinical partners and form teams around unsolved clinical problems. Teams initially meet online, and then arrange for shadowing opportunities to learn more about the problem in the clinical setting so that they can pursue solutions.

The program provides a convenient and robust digital space for students, faculty, and clinicians to collaborate, share ideas and work through steps to find solutions to unsolved clinical problems.

METHODS

From January to August of 2016, graduate students in the Clinical Nurse Leader program at the UAB School of Nursing and clinicians from UAB Hospital were approached to provide unsolved clinical problems in cardiovascular, neurosurgical and surgical intensive care units, emergency department, orthopedics, general medicine/infectious diseases, gastroenterology and dermatology. We asked that the problems be significant enough that, if solved, a viable solution would be impactful to their practice. The participating clinicians were given a PowerPoint template with 6 slides and asked to provide the following information:

1. Short biography of clinician
2. Who are the patients?
3. What is the problem?
4. What is currently being used?
5. What is needed?
6. What would be the benefit of a solution?

The clinicians responded with the requested information and then submitted their PowerPoint slides to the website manager for posting to the UAB Solution Studios Canvas website.

In the Fall of 2016, UAB Solution Studios became accessible to two undergraduate classes: a sophomore level honors clinical innovation class (12 students) as well as a senior level biomedical engineering senior capstone class (48 students). At the start of the semester, all students accessed the site in the Canvas learning management system to explore problem descriptions and biographies posted by clinical partners (currently graduate nursing students and UAB Hospital nurse clinicians). After selecting problems that interested them, students met with clinical partners, shadowed them in their practices, and formed teams for problem solving. Teams initially met online and then arranged for shadowing opportunities for BME and Honors students to learn more about the problem and pursue solutions.

The honors clinical innovation class was given 2 weeks to shadow the clinicians to allow them ample time to understand the problem from the point of view of the clinician as well the patients (empathy phase). Five teams formed in this stage. The rest of the semester was devoted to background literature searches for relevant journal articles and patents describing the problem and potential solutions (definition). The students then entered the ideation phase, prototyping and testing over the remaining semester. Prototyping of devices was accomplished with crude materials for fabrication (cardboard, tape, wire, etc.) and prototyping of solutions to problems dealing with process issues was accomplished using storyboarding. The prototypes were then shown to the clinicians for feedback and refinement. All teams presented their final projects in oral and written form at the end of the semester.

Of a class of 48 senior BME capstone students, 9 participated in UAB Solution Studios and formed 2 teams. These students adhered to the Stanford Biodesign process [2]. In this approach, students were required to demonstrate engineering design aspects of their projects. Engineering topics included developing the problem statement and establishing design constraints in terms of engineering criteria. Brainstorming followed which required evaluation of potential solutions in terms of the design constraints. The desired outcome at this phase was concept refinement into a formal design proposal. As part of brainstorming, the students are asked to establish Evaluation Matrices (Pugh method), by which different designs are evaluated for their ability to meet constraints, as well as feasibility, cost, time to develop, etc. During this period, the students developed engineering CAD drawings & performed basic engineering calculations, completed circuit diagrams and explored prototype development. By the seventh week of the 15-week course, the students must submit a formal Design Proposal. The proposal must have signature approval from their client and a professional engineer.

RESULTS

Outcomes from the sophomore clinical innovation class produced 4 device solutions and 1 process solution. The device solutions were

1. Improved data acquisition from a pill camera used for endoscopic tracking of GI bleeds thus reducing human intervention.
2. A wire and tubing conduit to secure ECG pacing leads and catheter tubing away from the skin of ICU patients to reduce skin ulcerations.
3. Improvements to ostomy bags to reduce infection and facilitate management of bag emptying and cleaning
4. ICU hospital bed modification to facilitate movement of bed with peripheral devices (monitors and pumps) to CT and MRI scanners.
5. Process for increasing hospital staff awareness of patient drug abuse risk through intravenous catheter ports.

Outcomes from the first semester BME capstone students yielded prototype concepts for

1. Continuous passive motion device for patients recovering from total knee replacement surgery that improved comfort and patient compliance
2. A transport system for safely moving obese patients from hospital transport bed to surgical table

All students in all classes were required to submit UAB intellectual property disclosures for their proposed solutions.

DISCUSSION

The number of unsolved clinical problems in a large academic medical center like UAB is overwhelming. In many cases, these problems are being managed by clinical staff performing suboptimal workarounds that frequently compound the morbidity of patients. In parallel, BME and STEM students at UAB develop technical competence that could be channeled to address some of these problems Moreover, many of these students are pre-medical students and are highly motivated to pursue clinical settings as part of their undergraduate experience.

Our pilot semester of UAB Solution Studios provided some important information that will serve as the basis for improving and expanding the program. One of the most important findings of this work is that our students prefer to enter the clinical domain through the digital environment of UAB Solution Studios as opposed to the logistical challenge of arranging shadowing experiences by email, phone or face to face meetings. The clinicians favor the convenience of posting their problems online and then meeting with teams of students who display real interest in helping to solve their problem. Another important finding is that the undergraduate students feel empowered by entering the clinical setting and knowing they may be able to contribute to the healthcare enterprise. The clinicians feel empowered because a team is actively working to help them improve their delivery of patient care.

The limitations of this approach were apparent in 2 ways. First, there were many more problems posted to the website than there were students available to work on them. Some clinicians felt discouraged by providing problems only to have them rejected for consideration by the students. This discouragement was mitigated by first telling the clinicians that their problems may not be selected. Second, not all solution ideas represented viable candidates for further development. Students were always informed of this possibility at the beginning of the semester. Any discouragement felt by the students was observed to be offset by their enthusiasm for the experience of participating in a clinical environment.

Despite the limitations of this approach, UAB Solution Studios represents a path forward to empower students and clinicians in the delivery of patient care. The use of a digital platform offers increased convenience to all the participants and accelerates the timeline of clinical problem solving.

ACKNOWLEDGEMENTS

We would like to acknowledge the enthusiasm and participation of all students and clinicians involved in this process.

REFERENCES

[1] Retrieved from http://dschool.stanford.edu/use-our-methods/

[2] Zenios, S et al., Biodesign: the process of innovating medical technologies, 2010.

STUDY OF FREEZING INDUCED RADIOFREQUENCY ABLATION HEATING PATTERN CHANGE

Kangwei Zhang (1), Jincheng Zou (1), Aili Zhang (1), Lisa X. Xu (1,2)

(1) School of Biomedical Engineering
Shanghai Jiaotong University
Shanghai, China

(2) Med-□ Institute
Shanghai Jiaotong University
Shanghai, China

INTRODUCTION

Image-guided RFA and cryosurgery have been widely accepted for non-surgical treatment of hepatocellular carcinoma and metastatic liver tumors. However, there are still some concerns associated with these two therapies. For RFA, its therapeutic effect decreases with the tumor volume [1]. Moreover, incomplete damage exists due to tissue cooling by the neighboring blood vessels [2]. Meanwhile, cryotherapy is also restricted due to the peculiar systemic complications and the high rates of local failures [3]. In order to deal with these restrictions, some scientists have proposed to combine cryotherapy with RFA for a better therapeutic effect. It has been found that an alternative freezing and heating treatment of the tumor can not only directly kill the cancer cells but can also destroy the tumor immune-suppression environment and stimulate anti-tumor immunity through animal studies [4]. This outcome is closely related to the thermal history. During the combined treatment, the freezing process before heating may have greatly changed the properties of the tissue, make a profound influence on the followed heating process. Thus for better planning of the treatment, more accurately evaluating the outcome as well as better understanding the treatment mechanisms underlying the combined treatment, it is necessary to look into the thermal history change and find out the possible causes of the change. In this paper, both in vitro liver experiments and in vivo rabbit liver experiments are carried out to compare the heating pattern for tissues experiencing with or without freezing. A finite element method is also setup to analyze the contributions of potential changed properties such as blood perfusion rate, thermal conductivity and electrical conductivity.

METHODS

Three freshly excised porcine livers and two locations in each liver are selected for treatment, one for pure RF heating and the other one for alternately freezing and heating. For pure heating treatment, a commercial RF bipolar needle (Olympus-CELON) is used and the voltage is controlled at a constant data 30V by a laboratory–made RF generator. The frequency used is 460 kHz and the heating duration is 3 minutes. Four thermocouple probes are placed locally to record the temperature during RF heating as shown in Figure 1-a. For the cryo-thermal treatment, the liver is first frozen with a self-made cryo-system based on vaporization of liquid nitrogen [5]. The front of the freezing probe is attached to the tissue surface and an ice ball with diameter of 2.5 cm is produced. After rewarming, the RF needle is inserted into the center of the frozen area, and the heating process is the same as the pure heating group. For in vivo rabbit liver experiments, a total of ten ablations are performed on the livers of five rabbits. After an adequate state of anesthesia is achieved, the rabbit is fixed onto the bench with bottom up. A midline laparotomy incision was performed from the xiphisternum to the umbilicus to expose the liver. The liver is then treated by the same treatment protocol as in vitro groups. For each rabbit, both a heating lesion and an alternative freezing and heating lesion are created.

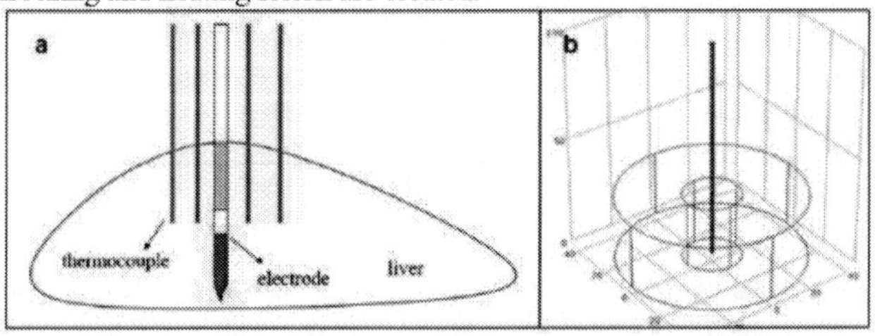

Figure 1: (a) Four thermocouples were set at the position away from the needle 3mm, 5mm, 8mm and 10mm separately. (b) The sketch of the numerical domain.

A finite element model is established for parametric analysis of all the affecting properties in the in vivo experiments. The rabbit and its liver is simplified to be a cylinder as shown in Figure 1-b. The inner cylinder with a diameter of 25 mm is the region that has been frozen. If there is only one single-heating, the whole cylinder has the properties of a normal liver tissue. The heat transfer in the tissue can be described by Pennes bioheat transfer equation [6].

The blood perfusion rate is set to be zero in the tissue that has been previously frozen due to damage of the blood vessels. Due to lack of direct measured value of the liver thermal conductivity and the electrical conductivity after frozen, Different values are used for parametric analysis and the data are shown in Table 1. COMSOL Multiphysics 5.2 was used for analysis.

Table 1: The properties used in the numerical model

	Unfrozen	Frozen
Blood perfusion rate (ml/s · ml)	0.0064 [7]	0
Electrical conductivity (S/m)	0.216 [8]	0.216/ 0.35 / 0.5
Thermal conductivity (W/°C · m)	0.494 [8]	0.4 / 0.494 / 0.8

RESULTS

Figure 2-a shows the temperature curves in vitro experiments. The results show that there is no significant difference in the heating rate between the heating of the frozen tissue and unfrozen tissue. However, in the in vivo experiments, the temperature rising rate of the alternately freezing and heating group is significantly higher than that of the single heating group as shown in Figure 2-b. It can be found that at a distance of 5 mm from the needle, the temperature of the cryo-thermal group is about 10 degrees higher than that of the direct heating group after heating for one minute. The difference between in vitro and in vivo experimental results may be attributed to the role of blood flow. For living tissues, due to the thermal stresses during the process, the blood vessels were broken resulting in excessive bleeding. The possible influence may include: (1) heat dissipation reduced due to the blood flow stopping after damage; (2) change of tissue thermophysical properties, such as electrical conductivity and thermal conductivity.

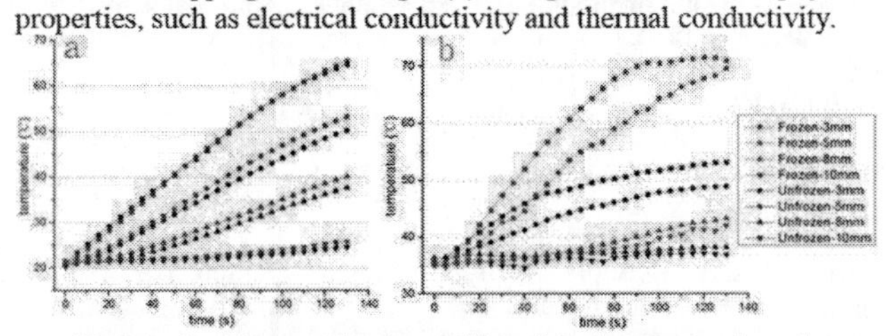

Figure 2: Temperature profiles for frozen and unfrozen tissues. (a) in vitro experiments; (b) in vivo experiments.

The simulated results and the experimental results for the pure RF heating process are shown in Figure 3. The results show that the simulated heating curve is in good agreement with the experiment at the beginning of the heating period. For the cryo-thermal model, the blood perfusion rate is set to zero. The simulated heating curves with different thermal conductivities are shown in Figure 4-a with comparison to the experimental results. The effect of electrical conductivity is shown in Figure 4-b. The simulated heating curves are also compared with the experimental results. The results show that the increase of conductivity leads to a significant increase in the RF heating capacity. And when the conductivity is set to 0.35 S/m, the simulated heating rate agrees well with the experimental results as shown in Figure 4-b. The comparison of the exepreimental results with the simulated suggests that only the increase in electrical conductivity after freezing can account for the significant increase of the heating rate inside the in vivo tissue that has been frozen before heating.

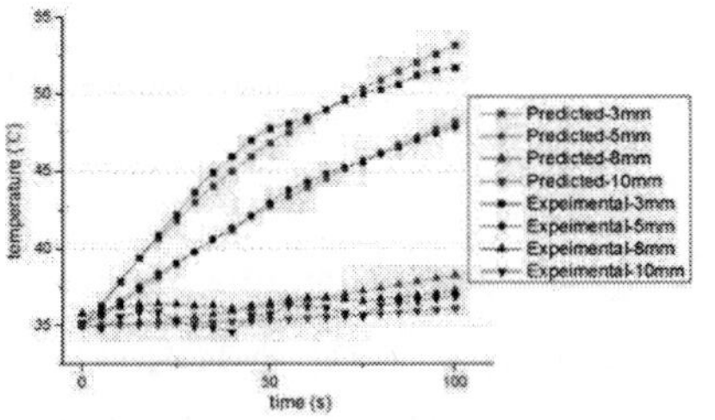

Figure 3: Comparison of experimental and simulated results for pure heating model.

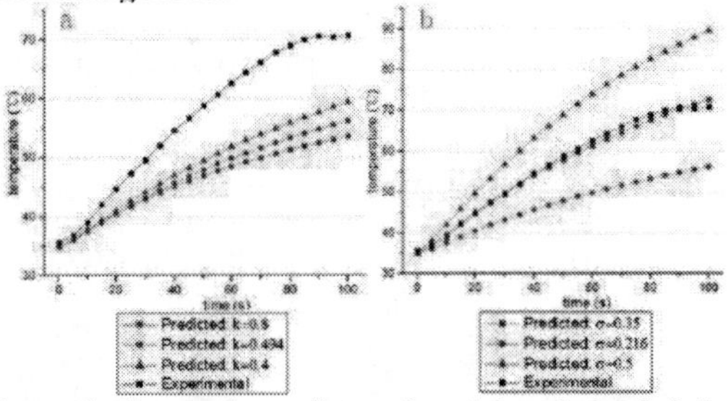

Figure 4: Comparison of experimental and simulated results for cryo-thermal model. The location selected is 3mm. (a) The thermal conductivity was set to 0.4, 0.494, and 0.8 (W/℃·m) in turn; (b) the electrical conductivity was set to 0.216, 0.35, and 0.5 (S/m) in turn.

DISCUSSION

This research has studied the treatment which combines cryosurgery and RF ablation from the view of heat transfer. It is found that the heating rate during the RF ablation is significantly increased for the living tissue if a previous freezing is imposed. The parametric study finds that the electrical conductivity plays a more important role than thermal conductivity. The simulated and experimental features will have many positive impacts on the RF treatment. Other than slowdown of the blood convection of the RF heating energy, the freezing induced significant increase of the electrical property of the tissue would shorten the ablation time for the same amount of the target tissue. Besides, there should also be a sharper thermal gradient at the edge of the tumor. Further study is necessary to quantify the increase of the therapeutic effect from either theoretical and experimental point of view.

ACKNOWLEDGEMENTS

This work was supported by National Natural Science Foundation of China (NSFC 51476101) and Ministry of Science and Technology of China (2016YFC0106200).

REFERENCES

[1] Chu K F, et al, *Nature Reviews Cancer*, 14(3):199-208, 2014.
[2] Adam R, Hagopian E J, Linhares M, et al, *Archives of Surgery*, 137(12):1332-1339, 2002.
[3] Bhardwaj, N., et al, *Cryobiology*, 61.1:1-9, 2010.
[4] Jun Zhu, Yan Zhang, Aili Zhang, Kun He, Ping Liu, and Lisa Xu, *Scientific Reports*, 6:27136 | DOI: 10.1038/srep27136, 2016.
[5] Sun J, et al, *International Journal of Heat & Mass Transfer*, 51(23–24):5478-5485, 2008.
[6] Berjano E J, *Biomedical Engineering Online*, 5(5):24, 2006.
[7] Chang I, *Biomedical Engineering Online*, 2(1):1-18, 2003.
[8] Duck F A, *Ta Egerev Sv Lyamshev Lm Naugol'nykh Ka Pashin Ae*, 1990.

BEHAVIOR OF INTERSTITIAL FLUID PRESSURE IN TUMORS WITH ENHANCED BLOOD PERFUSION

Timothy Munuhe, Myo Min Zaw, Liang Zhu, Ronghui Ma

Department of Mechanical Engineering
University of Maryland Baltimore County
Baltimore, MD, USA

INTRODUCTION

Elevated interstitial fluid pressure (IFP) is an important characteristics of solid tumors. Due to the high density and leaky nature of tumor vasculature and lack of functional lymphatic system, the interstitial fluid pressure is typically elevated throughout the tumor except the region near the tumor periphery. Such an elevated IFP hinders transvascular transport of therapeutic agents and constitutes a major obstacle for drug delivery to tumors. Regulation of IFP using various strategies has attracted substantial research interest in the past years [1-3]. Netti et al. [2] performed experimental and theoretical study to investigate the transient response of IFP in solid tumors to an increased blood perfusion. Both the experimental results and theoretical simulation show that the IFP increases substantially in response to a higher blood perfusion in the tumor within a time period of 100 seconds. For such a short period of time (100 second), the hydraulic conductivity of the tumor was considered a constant value despite of tumor deformation in the theoretical analysis. However, it is unclear whether the interstitial flow will behave differently if a higher blood perfusion rate is maintained for sufficiently long time so that the tissue pore size will be enlarged due to tissue deformation. It is hypothesized that the enlarged pore size will increase the tissue hydraulic conductivity, which in turn, decreases the IFP.

The objective of this study is to develop a steady state poroelastic model to study the mechanical response of a tumor and its subsequent influence on interstitial fluid flow during elevated blood perfusion. Specifically, the tumor porosity and hydraulic conductivity are considered to be dependent on tissue deformation for elevated blood perfusion, and their effect on interstitial fluid pressure in a solid tumor is predicted. Simulation results demonstrate that for tumors with a typical set of mechanical properties, reducing interstitial fluid pressure is possible if tumor blood perfusion is elevated and maintained for a long period of time.

METHODS

The poroelastic model consists of the equations for fluid flow through tumor and tumor tissue deformation [1-3]. We assume that the porous structure is homogenous before blood perfusion is elevated, and the solid phase of the medium is Hookian material, incompressible, isotropic, and fully saturated with fluid. The fluid flow in the tumor and tissue deformation have reached steady state after an increase of the blood perfusion rate. The steady state interstitial fluid transport in the extracellular matrix is governed by mass conservation and momentum balance in both solid and fluid phases [2], which are:

$$\nabla\cdot(\varepsilon\mathbf{V}_f)=\phi_b-\phi_l \qquad and \qquad \varepsilon\mathbf{V}_f=-K\nabla p \qquad (1)$$

where ε is the porosity, $\mathbf{V}_f$ is the interstitial velocity vector, ϕ_b and ϕ_l are, respectively, the rate of fluid extravasation from blood vessels per unit tissue volume and the rate of the lymphatic drainage per unit tissue volume, ∇p is the interstitial pressure gradient, and K is the hydraulic conductivity of tissue. The lymphatic drainage can be considered as zero in a solid tumor. The extravasation from blood vessels is governed by the Starling's law [2], which is

$$\phi_b=\frac{L_pS}{V}[P_B-p-\sigma_s(\pi_B-\pi_i)], \qquad (2)$$

where P_B and π are, respectively, the hydrostatic blood pressure and osmotic pressure, L_p is the hydraulic permeability of the microvessel wall in the tumor, and S/V is the vasculature surface area per unit tissue volume.

The constitutive equation for tissue deformation is

$$G\nabla^2\mathbf{u}+(\lambda+G)\nabla(\nabla\cdot\mathbf{u})=\nabla(p), \qquad (3)$$

where **u** is the displacement vector, and G and λ are Lamé constants. It should be noted that the porosity ε and permeability K in Eqs. (1) – (3) are spatially varying in the presence of tissue deformation; both are the functions of dilatation e, which is calculated by $e = \nabla \cdot \mathbf{u}$. When the volume of the solid phase of the porous medium is assumed not to change for a small deformation, the porosity ε is computed by the expression [4]

$$\varepsilon = \frac{\varepsilon_0 + e}{1+e}, \quad (4)$$

where ε_0 is the tissue porosity in the absence of deformation. Various empirical relationships have been proposed to quantify permeability K as a function of the dilatation e [2,4]. In this study, we employed a commonly used relationship which prescribes an exponential increase in tissue permeability with dilatation e [4]

$$K = K_0 \exp(Me), \quad (5)$$

where K_0 is the permeability in the absence of deformation and M is a material constant that governs the variation of the permeability with the deformation. Typically M is related to tissue properties, densities of the cells, and extracellular matrix [4]. Through curve-fitting of the experimental data, previous studies suggest that M varies in the range of 0 to 5 [4]. We used a constant $M = 5$ in this study. For spherical tumor of homogeneous properties, the tumor deformation and fluid flow can be considered as a one-dimensional problem. Eq. 3 can be expressed in terms of tumor dilatation e, which is

$$K(2G+\lambda)\nabla^2 e + \phi_b = 0 \quad (6)$$

The boundary conditions for a spherical tumor with free moving boundary are given as follows:

$$\frac{\partial p}{\partial r} = 0 \quad at \quad r = 0, \quad p = 0 \quad at \quad r = R \quad (7)$$

$$\frac{\partial e}{\partial r} = 0 \quad at \quad r = 0, \quad e = 0 \quad at \quad r = R \quad (8)$$

RESULTS

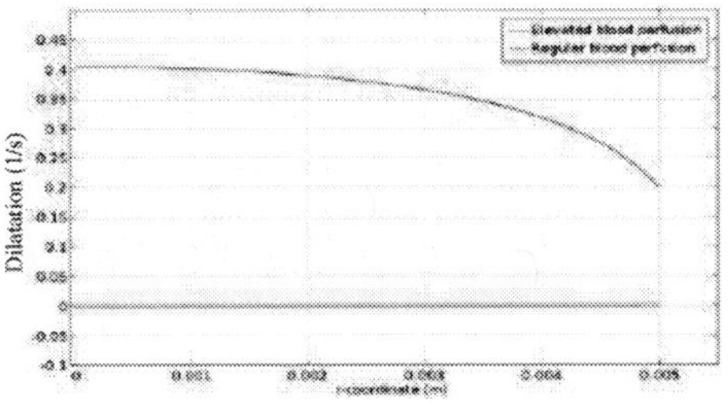

Figure 1. Dilatation (e) distributions for regular and elevated blood perfusion

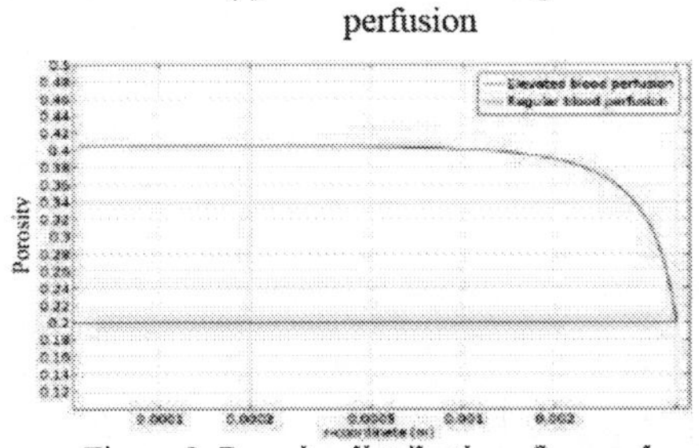

Figure 2. Porosity distributions for regular and elevated blood perfusion.

The poroelastic model was used to simulate the steady state IFP distributions with regular blood perfusion and elevated blood perfusion. For the latter case, it is assumed that the blood perfusion rate is increased by 10% for sufficient amount of time so that a steady state IFP is re-established. The simulation parameters used in the study are given in Table 1. The tumor tissue dilatation due to the elevated blood perfusion is shown in Fig. 1, and the tumor porosity after deformation is shown in Fig. 2. It can be seen that the central region experiences more deformation than the peripheral region. The enlarged porosity subsequently increases the hydraulic conductivity of the extracellular matrix and expedites interstitial fluid flow. As a result, the IFP decreases as shown in Fig. 3.

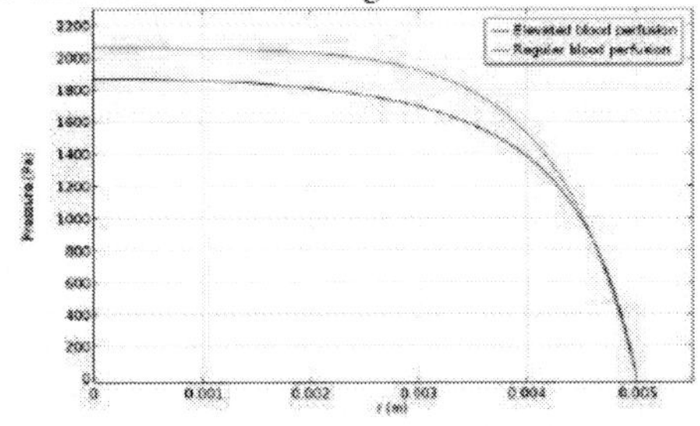

In summary, the simulation results demonstrated that it is possible to regulate IFP by elevating blood perfusion in tumors for extended period of time. The decreased IFP will enhance extravascular transport of drug. Mild regional or whole body hyperthermia has caused increase in blood perfusion in tumors for hours and decrease in IFPs has been reported in some studies [5]. One limitation of the current study is the assumption that the tumor is free to expand. While this is relevant to tumors planted under skin, it is not applicable to tumors embedded in stiff tissue. Also, it is uncertain how long it will take for tumors to reach steady state after the blood perfusion is increased. Further experimental and theoretical studies will be continued to investigate the effect of boundary condition on IFP and the transient behavior of tumor during protracted blood perfusion elevation.

Table 1. Simulation parameters [2,4]

Properties and Parameters	Values
L_p (cm/mmHg s)	2.8e-7
S/V (tumor) cm^{-1}	200
P_B ($mmHg^{-1} s^{-1}$)	15.6
σs (π_B - π_i) mmHg	4.1
Initial tissue porosity ε	0.4)
Initial tumor porosity ε	0.2
Initial tumor permeability K	5×10^{-16} m^2
Initial tissue permeability K	1×10^{-14} m^2
Tissue/tumor Poisson's ratio v	0.35
Tumor diameter	1 *cm*
Lame constant G	2026 pa
Lame constant λ	91190.9 pa

ACKNOWLEDGEMENTS

This study was supported in part by an NSF research grant (CBET-1335958) in partial fulfilment of the requirements for the PhD degree from UMBC by Myo Min Zaw.

REFERENCES

1. Netti, P.A., Baxter, L.T., Boucher, Y., Skalak, R., Jain, R.K., Cancer Research vol. 55, p. 5451 – 5458, 1995
2. Truskey, G.A., Fan, Y., katz, D.F., Transport Phenomena in Biological Systems, 2ed, Pearson Prentice Hall, New Jersey, 2009
3. Samtinoranont, M., Rooney, F., Ferrari, M., Anals of Biomedical Engineering, vol. 31, p. 327 – 335, 2003
4. Su, D. Ma, R., and Zhu, L., *Medical and Biological Engineering & Computing*, vol. 49(11), pp.1233-40, 2011
5. Sen, A., Capitano, M.L., Spernyak, J.A., Schueckler, J.T., Thomas, S., Singh, A.K., Evans, S.S., Hylander, B.L., Repasky, E.A. Cancer Res. Vol. 71(11): 3872–3880, 2011

TEMPERATURE FIELD MEASUREMENT OF OPTICAL THERMOCAVITATION FOR ENHANCED SKIN SURFACE COOLING

Vicente Robles[1], Darren Banks[1], Seyed Mahdi Akbarimoosavi[1], Luis Felipe Devia-Cruz[1], Santiago Camacho-López[2], Guillermo Aguilar[1].

(1) Department of Mechanical Engineering, University of California, Riverside, 92521, Riverside, CA, USA.
(2) Departamento de □ptica, Centro de Investigaci□n Cient□fica y de Educaci□n Superior de Ensenada, Carretera Ensenada-Tijuana No. 3918, Zona Playitas, C.P. 22860,

INTRODUCTION

Optical cavitation is a process in which a laser is focused into an absorptive liquid, causing it to superheat and leading to the growth and collapse of a vapor bubble. The collapse is succeeded by a shockwave emission powerful enough to damage titanium films [1]. The study of cavitation has evolved from learning how to prevent a damaging phenomenon to developing applications for biomedicine and micro-fluid controls [2,3].

New applications are using single-phase heat transfer using coolants to remove excess heat because traditional structures such as fans and heatsinks are limited by spatial conditions. Cryogen sprays are used in the treatment of port wine birthmarks to prevent burning of the epidermis as a laser treats blood vessels underneath the dermis. Tetrafluoroethane was traditionally used for spray cooling and is being replaced with less-effective, but far more environmentally friendly fluids [4]. This shift requires improved cooling techniques to offset the lower effectiveness. Another existing issue is that a film of coolant can buildup on the skin surface leading to discomfort such as frostbite and blistering. The buildup of coolant also impedes efficient heat transfer as formation of a boundary layer prevents full contact of fresh coolant to the skin surface. Our team utilizes this motivation to study the effects of continuous wave (CW) OTC to destroy the thermal boundary layer of a substrate and allowing for increased effectiveness of thermal management applications in the biomedical field.

This study shows that focusing a CW laser into an absorptive solution can periodically induce cavitation that can enhance local convection during the collapsing phase. Figure 1 is an image demonstrating the agitation of liquid resulting in enhanced convection.

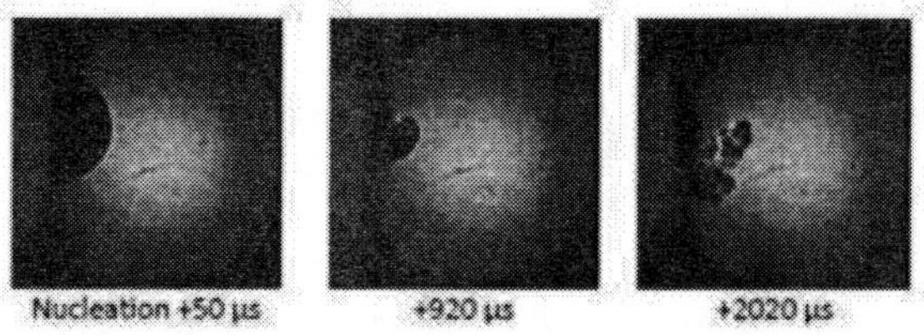

Figure 1: This figure is a phase contrast image sequence showing the ejection of local liquid and agitated mixing [5].

METHODS

OTC bubbles produced are approximately in the length scales of 1 mm and lifetime of about 100µs. These bubbles form at the substrate interaction due to the methods of inducing cavitation with a continuous laser as opposed to the traditional pulsed lasers which also result in potential thermal damage from plasma formation. We have successfully characterized bubbles (rate of formation and diameter) produced under various conditions such as focal position, irradiant power, and ambient pressure.

To advance the application for skin-cooling, the temperature field of the cavitation site must be accurately determined. The resulting shockwave makes sensors vulnerable to damage requiring a different method to accurately measure temperature throughout the bubble evolution. This study focuses on describing the development of planar-laser induced fluorescence (PLIF) as a method of non-intrusively measuring the temperature field of a single bubble.

As shown in Figure 2, a continuous wave 810nm laser induces cavitation in a copper nitrate solution while a 440nm planar laser concurrently excites rhodamine B dye to fluoresce in the solution. The intensity of fluorescence is recorded by a Miro 310 high speed camera and correlated to the liquid temperature.

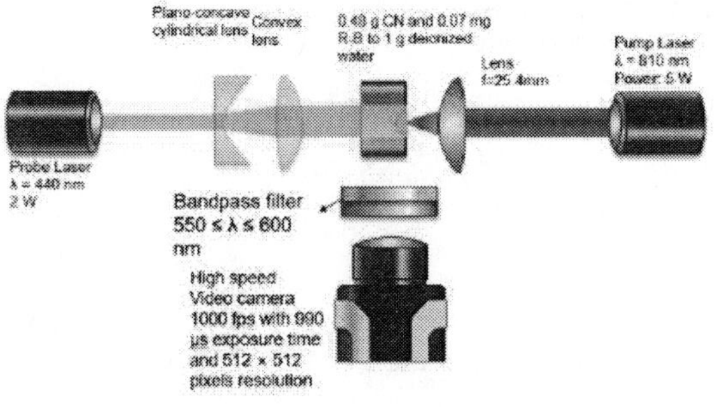

Figure 2: Optical setup for planar-laser induced fluorescence temperature measurement.

RESULTS

Figure 3, shows the temperature field throughout the bubble lifetime. While the bubble begins growth (~260ms), the formation of the boundary layer is seen. This boundary layer is destroyed after the collapse of the bubble. As the bubble grows to its maximum diameter (~1.5mm) in about 263ms, the temperature appears to surpass 100°C. While this may be true, it is important to note an absence of Rhodamine B solution inside the bubble may provide a false measurement.

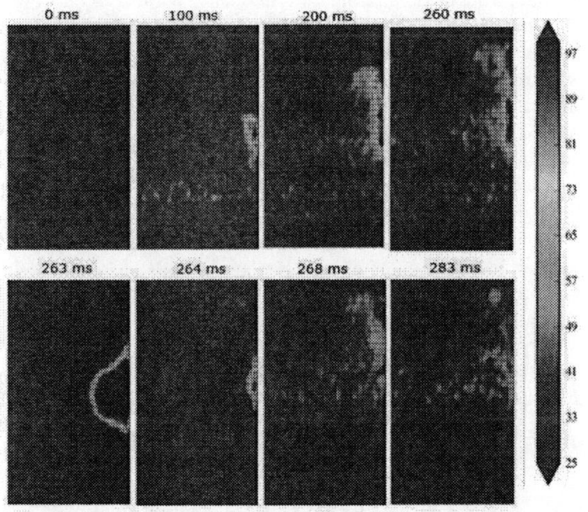

Figure 3: Temperature field during bubble lifetime. Maximum temperature exceeding 100°C is drastically decreased after the collapse of the bubble.

Ultimately, the maximum temperature during a bubble's lifetime at 100°C decreases drastically after the collapse to ~35°C, nearly decreasing back to ambient temperature. This is quantitative indication of strong agitation of parcel of fluid at the cavitation site. This is a large temperature jump especially when compared to previous measurements of cavitation induced by pulsed lasers which only differ by 12°C, effectively proving pulsed cavitation to be a thermally insulated process [7]; not effective for enhanced convective cooling.

DISCUSSION

The phenomenon of optical cavitation produces strong and rapid changes in mechanical, optical and thermal dynamics within the cavitation site. These resulting observations give potential for cavitation bubbles to be used for applications such as turbulent mixing in microfluidic devices or convective cooling in skin during tissue laser treatment.

The temperature field was measured throughout the lifetime of the bubble, showing a reduction of sensible heat upon the collapsing stage. Utilizing the information obtained, the resulting energy conversion was estimated to be ~0-50%, effectively making optical thermocavitation a viable method of cooling enhancement for surfaces in question.

Copper nitrate (CuNO4) is used to induce cavitation due to its high absorption however, it is highly corrosive and not biocompatible for applications such as skin poration for drug delivery [6]. Future work includes utilizing a biocompatible solution of gold nanorods dispersed in water. Furthermore, the PLIF technique will also be used characterize the 3-D temperature field surrounding a bubble.

ACKNOWLEDGEMENTS

The authors would like to thank Zhang, B., Martinez, I. for their laboratory assistance in this investigation. Discussions with Dr. Juan-Pablo Padilla Martinez, Dr. Monica Martinez, Dr. Marko Princevac, and Dr. Lorenzo Mangolini have been invaluable in advancing this work.

This work is supported by NSF-CBET grant 1403508 and UCMEXUS..

REFERENCES

[1] Ramirez-San-Juan, J., et al., Optical Materials Express, 2011. 1(4): 598-604.
[2] Padilla-Martinez, J., et al., Photonics and Lasers in Medicine, 2012. 1(3): 183-193.
[3] Aguilar, G., et al., Lasers in Surgery and Medicine, 2003. 32:152–159
[4] Banks, D. UC Riverside, 2015.
[5] J.P. Padilla-Martinez, et al., Photon Lasers Med, 2012. 1(3):183-193
[6] Padilla-Martinez, et al. "Towards the enhancement of transdermal drug delivery through thermocavitation" Photonics in Lasers and Medicine, 1:3, 2012.
[7]Quinto-Su, P.A., Suzuki, M. & Ohl, C.-D. Fast temperature measurement following single laser-induced cavitation inside a microfluidic gap. Sci. Rep. 4, 5445; DOI:10.1038/srep05445 (2014).

MEDIUM-TERM STABILITY OF CANCER BIOMARKERS IN HUMAN SERA STORED BY ISOTHERMAL VITRIFICATION

Morwena J. Solivio and Alptekin Aksan

Biostabilization and Bioencapsulation Laboratories, Department of Mechanical Engineering,
University of Minnesota,
Minneapolis, MN 55455, USA

INTRODUCTION

Millions of "archival" biospecimens are collected from patients and are stored in biorepositories and biobanks across the world **(1)**. These biospecimens are routinely stored for future research, including biomarker discovery and verification, where the availability of high quality biospecimens is an important pre-requisite.

Cryogenic storage is the current method of choice for preservation of biospecimens for future biomarker research **(2)**. This method however has several disadvantages. Exposure of biospecimens to mechanical and chemical stresses during freezing can cause damage by ice to many susceptible biomolecules. Furthermore, freezer farms and liquid nitrogen storage facilities are both expensive and vulnerable to failures, subjecting biospecimens to freeze-thaw, which is detrimental to many proteins. Additionally, biorepositories and biobanks receive large sample volumes of biospecimens, which necessitate thawing, aliquoting and re-freezing, all of which can incur damage to many vulnerable biomarkers **(3-4)**. Modifications to biospecimens through the above-mentioned processes could readily impede the discovery of robust biomarkers, consequently hindering development for biomarker-facilitated personalized medicine, risk assessment, disease diagnosis, management and treatment. This slow progress is evident by the approval rate of the Food and Drug Administration (FDA), where less than two-dozen biomarkers are approved for clinical use, despite thousands of molecular signatures being evaluated and over a hundred thousand studies already published in literature **(6)**.

This project aims to eliminate the requirement for the frozen state storage for archival serum and plasma samples by stabilizing and storing them at room temperature by isothermal vitrification. To accomplish this goal, we have developed a novel biopreservation sponge (an electrospun adsorbing/dissolving matrix), made from a lyoprotectant cocktail containing excipients to stabilize cancer-related proteinaceous biomarkers in the human serum during isothermal vitrification and room temperature storage.

We have previously shown that the current matrix and methodology was able to stabilize 5 biomarkers: **LDH**, **PSA**, **CRP**, **C3a**, and **MMP-7**, with > 90% recovery after four hours of desiccation **(6)**. The glass transition temperature ($\mathbf{T_g}$) of these samples was >4°C, allowing for storage in a standard refrigerator. This work aims to increase T_g such that storage at room temperature (RT) is possible.

METHODS

To produce the sugar-based electrospun matrix **(V1EX)**, a solution containing 1.2 M of trehalose (TRE) and 0.025 M dextran (DEX), 1.5% (v/v) glycerol, 0.1% (v/v) Tween 20, 1% (w/v) PEG, 0.3% (w/v) gluconic acid and 0.2% (w/v) glucamine was rigorously mixed for 24 hours. The solution was then electrospun at a flowrate of 0.003 mL/min through a 15 kV gradient across a 15 cm gap. The environment was kept at room temperature and 50% RH. Electrospun woven matrices were then formed into disks, placed in a petri dish, 6, 24 or 96-well plates **(Figure 1)**, and stored in a vacuum chamber before use.

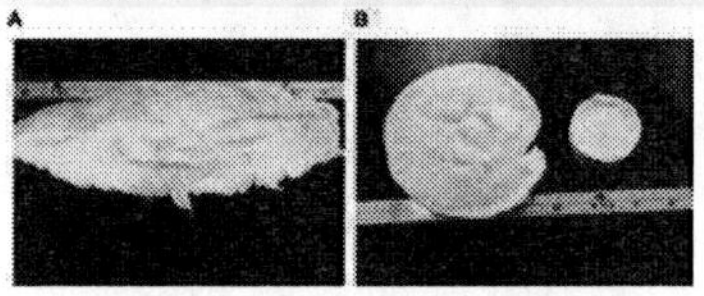

Figure 1. Electrospun woven matrix (A) are formed into disks (B) to be placed in a petri-dish or in 6-, 24- or 96-well plates before use.

A 150 µL volume of serum samples spiked with the biomarkers, **LDH** (8 mg/mL), **PSA** (4 ng/mL), **CRP** (1 mg/mL), **C3a** (9 µg/mL) or **MMP-7** (24 ng/mL) was added to 50 mg **V1EX**.

The samples were then dried in a vacuum chamber overnight prior to storage, re-suspension, and subsequent enzymatic activity analysis or sandwich ELISA. Enzyme activity assay was used to measure LDH stability. ELISA was used to measure CRP, PSA, C3a and MMP-7 recovery after overnight desiccation and storage at room temperature. Differential scanning calorimetry was used to measure T_g.

The results are reported as the mean± SD (n = 3). To determine variability in biomarker recovery in response to storage time, a one-way ANOVA was conducted. P-values ≤ 0.05 were considered statistically significant. Note that the P-values were obtained by comparing the experimental group with the untreated (fresh) control.

RESULTS

Human serum spiked with selected biomarkers (PSA, CRP, C3a and MMP-7) was added to the electrospun matrix (V1EX) and desiccated in a vacuum chamber overnight, reaching a $\mathbf{T_g}$ of 50±6°C. Complete recovery after resuspension was achieved with PSA at 98±3% (P = 0.844), CRP at 103±3% (P = 0.699) and C3a at 102±16% (P=0.860) **(Figure 2)**. However, only 64±5% (P=0.0009) of MMP-7 was recovered.

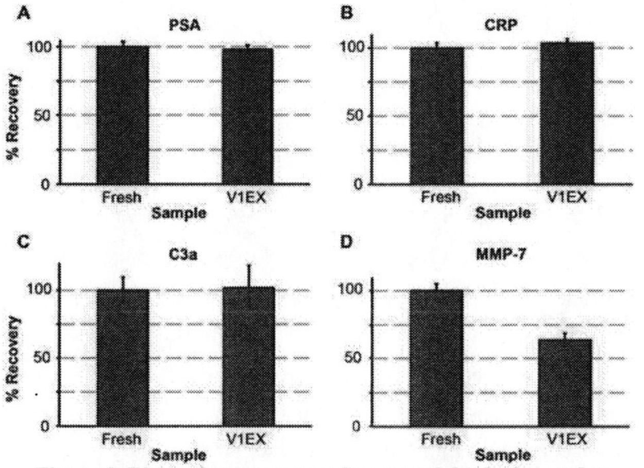

Figure 2. Biomarker recovery after overnight drying and re-suspension. Samples spiked with the biomarkers (A) PSA, (B) CRP, (D) C3a, and (C) MMP-7 were mixed with V1EX and desiccated overnight. Recovery for PSA after re-suspension is 98±3%, for CRP is 103±3%, C3a is 102±16%, and MMP-7 is 65±5%.

We have then conducted medium-term (1, 2 weeks) stability studies with serum samples spiked with LDH and PSA. The samples were added to the matrix and dried overnight in a vacuum chamber. Desiccated samples were stored at RT and re-suspended at predetermined time points. Full recovery of LDH (99±4%; P=0.800) **(Figure 3A)** and PSA (97±6%; P=0.599) **(Figure 3B)** was accomplished after 2 weeks of storage.

DISCUSSION

Cryogenic storage is the most ubiquitously used method for biospecimen preservation in biobanks and biorepositories for future biomarker research and verification. However, even when best practices are observed, cryogenic storage and freeze-thaw alone can impose harsh chemical and mechanical stresses that can compromise sample quality **(3-5)**. Additionally, cryogenic storage is costly, with

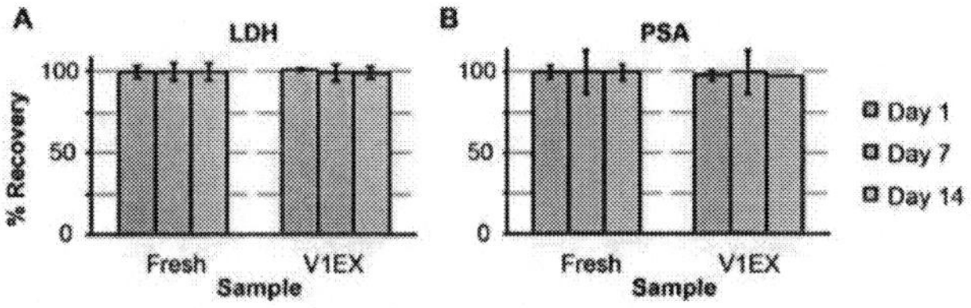

Figure 3. Medium-term stability studies. Storage for 14 days at RT resulted to 99±4% recovery for LDH (A) 97±6% for PSA (B)

the estimated cost of storing over 600 million biospecimens to be $100 million annually **(2)**.

This research aims to develop a room temperature stabilization technology for storing achievable serum and plasma samples. A sugar based electrospun matrix is designed to adsorb and isothermally vitrify liquid biospecimens. When vitrified, all biochemical reactions in the biospecimen are halted without exposure to extreme stresses induced by cryogenic storage and freeze-thaw **(7)**. This also makes the samples less vulnerable to power shortage, equipment breakdown and human error that can occur in cryogenic storage facilities. Removal of water during isothermal vitrification significantly reduces sample volume, further reducing storage and transportation costs.

In this work, we aimed to increase the T_g of the samples such that RT storage is achievable, and conducted medium-term stability studies to demonstrate the suitability of the method for stabilizing liquid biospecimens for extended periods. Increasing the desiccation time to 16 hours decreased the water content of the samples to 10%, and increased the T_g to 50±6°C. Complete post-re-suspension recovery was achieved for LDH, PSA, CRP and C3a following desiccation, while MMP-7 showed sensitivity to overnight drying. MMP-7 is a known freeze-thaw sensitive biomarker **(8)**. There is currently no available information in the literature on the sensitivity of this protein to desiccation. Medium-term stability studies on serum samples spiked with LDH and PSA, desiccated overnight and stored at RT resulted to complete recovery of the biomarkers after 14 days. Long-term stability studies for LDH, PSA, CRP and C3a are currently ongoing. The results shown in this work indicate the potential of this method for stabilizing an expanded number of protein biomarkers and as an excellent alternative for cryogenic storage.

The next step will be the determination of the stability of 300+ biomarkers in the Myriad RBM Human Discovery Platform, distinguishing the biomarkers that are unstable, and enhancing stability by identifying similarities among the destabilized proteins and modifying the matrix and method accordingly.

ACKNOWLEDGEMENTS

We would like to thank Miss Sampreeti Jena for her valuable assistance with DSC experiments and T_g determination. This research is funded by NIH IMAT grants CA157298 and CA204510.

REFERENCES

[1] Vaught, J et al., *J Natl Cancer Inst Monogr*, 42:24-31, 2011
[2] Shabihkhani, M et al., *Clin Biochem*, 47:258-266, 2014
[3] Twomey, A et al., *J of Phys Chem B*, 117:7889-7897, 2013.
[4] Twomey, A et al., *Int J Pharm*, 487: 91-100, 2015.
[5] Poste, G, *Nature*, 469: 156-157, 2011
[6] Solivio, MJ et al., *Scientific Reports*, 6: 24186, 2016
[7] Less, R et al., *Cryobiology*, 2: 176-185, 2013
[8] Kisand, K et al., *Clin Chem Lab*, 2: 229-235, 2011

Heat Shock Protein Expression during Short Pulse Laser Therapy

Neda Parchami (1), Amanda A Oliveira (2), Kenia P Nunes (2), Eric Guisbert (2), Kunal Mitra (1)

(1) Department of Biomedical Engineering
Florida Institute of Technology
Melbourne, FL, USA

(2) Department of Biological Sciences
Florida Institute of Technology
Melbourne, FL, USA

INTRODUCTION

The heat shock proteins (HSPs), originally identified as heat-inducible gene products, are a highly conserved family of proteins that respond to a wide variety of stressful conditions such as fever, alcohol, inflammation, necrosis, and cancer [1]. Present in all types of living cells, HSPs can be found in the nucleus as well in the cytoplasm. Induction of heat shock protein expression appears to improve healing of damaged tissues in animal models and in humans. Although the induction of HSPs have been studied extensively in cultured cells, their expression in *in vivo* organisms remains unclear [2]. Based on previous study, it is evident that in some tissues such as brain, lung, and skin, induction of HSP 70 is directly related to temperature of heat stress with the maximal expression in one hour; however, in some tissues such as liver temperature has no relation with expression of HSPs [3]. Lasers are extensively being used for therapeutic applications. The interaction of biological tissue with short-pulse lasers can provide important information regarding the material properties of the tissues, and it can be used as high source of energy to precisely cut and ablate the tissue. Previous studies analyzed the expressions of heat shock proteins 70 and 47 at different time-points post laser irradiation on a rat model having dermal tumors. It has been well documented that HSP 70 is active in the damage zone to destruct the target cell while HSP 47 is a part of recovery process of the thermally induced damage [4]. In addition, some previous studies have been reported analyzing if prior induction of HSP synthesis would be protective against laser induced injury. The effects of high temperature exposure for short durations such as for the case of short pulse lasers can lead to significantly different responses than traditional CW laser sources. The goal of this study is to first investigate the effects of heat stress in controlled thermal conditions using tissue samples from live anesthetized rat and euthanized rats. Finally the effects of short-pulse laser irradiation on the production of heat shock response through HSP70 and HSP47 expression using live anesthetized rat model.

MATERIALS AND METHODS

The difference in HSP expression is studied using male rats (3 weeks, 250g) for three different experimental conditions in order to elucidate the effects of each condition. The first group of rats was fully anesthetized for one hour using Isoflurane 3% and scarified after one hour. The second group of rats was anesthetized and then euthanized with dosage of 5% Isoflurane after exposure for one hour. Finally, the last group of rats was euthanized without anesthesia. For the first and second group of rats, tissue samples were collected immediately after euthanasia and stored in liquid nitrogen. For the last group of rats, tissue samples were preserved in Krebs's solution (130 mM NaCl, 4,7mM KCl, 1,18 mM KHPO4, 1,17 mM MgSO4, 1,6 mM CaCl2, 2H2O, 14,9 mM NaHCO3, 5,5 mM dextrose e 0,03 Mm $CaNa_2$ EDTA). For each group of rats, experiments were performed under two different thermal conditions - room temperature (control group) and at 40^{0}C. After euthanasia, hair was shaved using an electric hair trimmer and the skin cleaned with dry gauze, and approximately 1 cm^2 diameter region of the dorsal skin were collected and processed as aforementioned. Biopsy sections were flash frozen in liquid nitrogen and stored at -800°C. Tissue sections were cut into weight equivalent pieces (approximately100 mg) and homogenized in 400 μl ice-cold extraction buffer (T-PER: protease inhibitor, 1:10μL). Samples were electrophoresed by 10% SDS-PAGE, and transferred to a nitrocellulose membrane. The membrane was blocked with 5% non-fat dry milk in PBS-Tween 20 (PBST) buffer for 1 hour at room temperature. An anti-HSP70 (rabbit's polyclonal IgG anti-HSP70) and anti-HSP47 (rabbit polyclonal IgG anti-HSP47) monoclonal antibody (SantaCruz) diluted 1:500 in 5%non-fat dry milk was added and then incubated for 3 hours at room temperature. The blot was washed with PBST and incubated for 1 hour at room temperature with a conjugated goat anti-rabbit IgG secondary antibody diluted 1:1000 (SantaCruz). Blots were washed, developed with ECL-plus, and scanned. All experimental procedures on rats were approved by Florida Institute of Technology's Institutional Animal Care and Use Committee (IACUC).

RESULTS

Preliminary results have demonstrated expression of HSP70 and HSP47 in rat's skin after short-pulse laser irradiation at different times (Figure 1). Western blot analysis showed that the peak of both HSP47 and HSP70 expression occurred 4 hours post laser irradiation. In order to provide another measure of the extent of thermal effects following laser irradiation, the kinetic expression of HSP70and HSP47 as indicators of damage and healing procedure at various times following laser injury to the skin is analyzed as shown in Figure 2. In addition, after 2 hours post-irradiation the HSP 70 expressed immediately surrounding the damage zone (Figure 3).

DISCUSSIONS

In order to understand the roles of the heat shock proteins HSP47 and HSP70 during short-pulse laser irradiation, HSPs are used as marker of biochemical change in the tissue surrounding the laser irradiated zone. Effects of anesthesia and euthanasia in rats on HSP expression at different temperatures during laser irradiation will be studied. The response to temperature under different experimental conditions for anesthetized, euthanized, and tissue samples will be beneficial in understanding the role of short pulse laser parameters as a therapeutic tool.

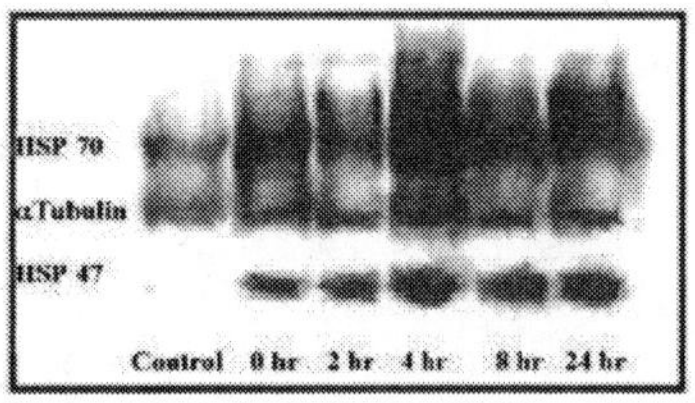

Figure 1. Western blot analysis of HSP70 and 47 expressions in rat skin following laser irradiation. Non-irradiated skin tissue biopsies from anesthetized rat are used as control samples. Western blot analysis of HSP70 and HSP47 at 0, 2, 4, 8, 24 h post-irradiation.

(A) (B) (C

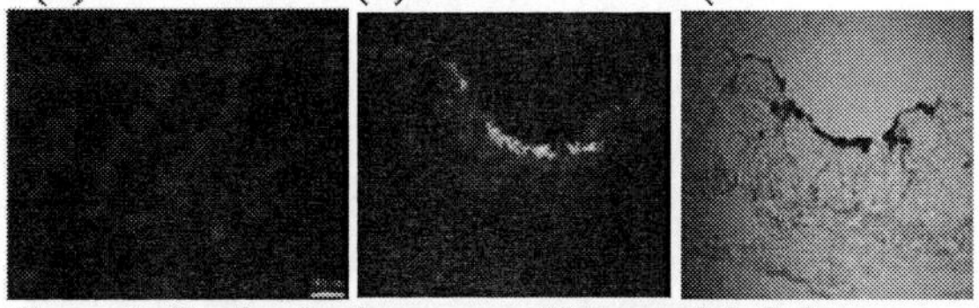

Figure 2. Confocal microscopy of immunolabeled cross-sections of rat skin section immediately after irradiation (time = 0). DNA is labeled with DAPI (blue); HSP70 is immunolabeled to appear red; HSP47 is immunolabeled to appear green. Non-irradiated skin tissue (A); laser-irradiated skin tissue (B); visible light image (C)

Figure 3: 3-D Image of skin (thickness = 20 μm) lesion induced by short-pulse laser visualized with confocal microscope. Optical scans are obtained at 1-μm intervals through a 20 μm thick tissue section. Green and red arrows show the regions where HSP47 and 70 are expressed in the tissue. HSP70 and 47 are visualized at 1 h.

REFERENCES

[1] Caitlin, E et al., *Opt,* 13(3), 030501). 10.1117/1.29046652, 2008.

[2] Atalay, M et al., *Curr Protein Pept Sci*, 10(1): 85–95, 2009.

[3] Belake, M et al., *J Biological Chemistry,* 265:15275-15279, 1990.

[4] Sajjadi, J et al., *Medical Engineering & Physics,* 87: 1406– 1414, 2013.

INITIAL STUDIES: A NOVEL APPROACH TO TREATING ACUTE PANCREATITIS USING THERAPEUTIC HYPOTHERMIA

Daniel P. Meckes (2), Matthew J. Skinner (3,4), Keith T. Wilkins (2), Gregory M. Donatelli (2), Christopher C. Thompson (3,4), Jennifer E. Mitchell (2), Thomas L. Merrill (1,2)

(1) Department of Mechanical Engineering
Rowan University
Glassboro, NJ, USA

(2) FocalCool, LLC
Mullica Hill, NJ, USA

(3) Harvard Medical School
Boston, MA, USA

(4) Department of Interventional Endoscopy
Brigham and Women's Hospital
Boston, MA, USA

INTRODUCTION

Acute pancreatitis (AP) is an inflammatory disorder of the pancreas. The disease can be deadly.[1,2] Treatment options are limited to none present. The disease carries significant burden to the patient, healthcare provider, and health economy in general.[3] If existing preventative and treatment options are not realized, the predictive cost of AP in the United States will exceed an underestimated $8.3 billion by the year 2025.

Four recent small-animal studies suggest that therapeutic hypothermia (TH) reduces cytokine (TNF-α, IL-1β, and IL-6) proliferation, blunts cellular injury, and confers a significant survival benefit.[4-8] Current cooling technology including whole-body and intravascular approaches are likely too invasive for treatment of AP in man. To address these problems we propose a minimally invasive alternative. Endoscopically-placed cooling catheter leverages *endoscopic retrograde cholangiopancreatography* (ERCP) procedures and devices. We hypothesize that intraductal pancreatic cooling provides a life saving treatment while limiting complications associated with previous generation cooling technology.

We tested the hypothesis that intraductal cooling via the pancreatic duct (PD) is technically feasible in both *ex vivo* and *in vivo* porcine models. We investigated intraductal conductive cooling depth and rate. First we proved our computational model numerically. Then we compared these findings using an *ex vivo* porcine model. Lastly, we tested our technology in a human-scaled *in vivo* porcine model. Cooling was least effective in the computational and numerical models. However, the *ex vivo* and *in vivo* models demonstrated promising results. Both models showed cooling effects moving away from the PD. Cooling depth was observed as deep as 20 and 18 millimeters in the *ex vivo* and *in vivo* models respectively. In the *in vivo* model, TH targets were achieved in less then 20 minutes.

METHODS

A geometrically uniform pancreas model was generated using SolidWorks. Human tissue thermal and physical properties were assigned according to those provided by IT'IS database of thermal and electromagnetic parameters of biological tissue. The pancreas model had a mass of m_{model} = 86.17g and an OD=2.67cm. The OD of the model was maintained at a BC of 37°C. These properties are comparable to those found in man.[9] A flow rate of 60ml/min was determined using Pennes bioheat equation with a desired treatment time of 30 minutes. 4°C coolant was passed through the flow channels of the catheter.

Before conducting our *in vivo* test, we created an *ex vivo* experiment that simulates similar heating conditions found in a living system. A porcine pancreas having mass = 79g was enclosed in a latex prophylactic and submerged in a warm water bath. The water bath was maintained at constant temperature of 38±1°C using a submersion water heater. Our cooling technology was placed inside the PD. Continuous tissue temperature was recorded using four type-K thermocouples at 2, 8, 15 and 20mm away from the device. Flow rate was initiated with 0°C (coolant) and a flow of 25ml/min. The experiment was reproduced using three different specimens.

A live Yorkshire pig weighing approximately 40kg was studied to evaluate the effects of tissue heat generation and circulatory heat transfer (Figure 1). Due to porcine anatomy, the procedure was done surgically instead of ERCP.[10,11] The PD was accessed using a standard 0.025" endoscopic guide wire. The PD was cannulated by the cooling catheter, which was passed over the guide wire. 0°C coolant (saline) was passed through the flow channels inside the catheter and at a flow rate of 120ml/min. Continuous tissue temperature was recorded using four type-T thermocouples at 3, 8, 12 and 18mm from the PD. A fifth thermocouple was placed in the liver to detect any systemic cooling effects.

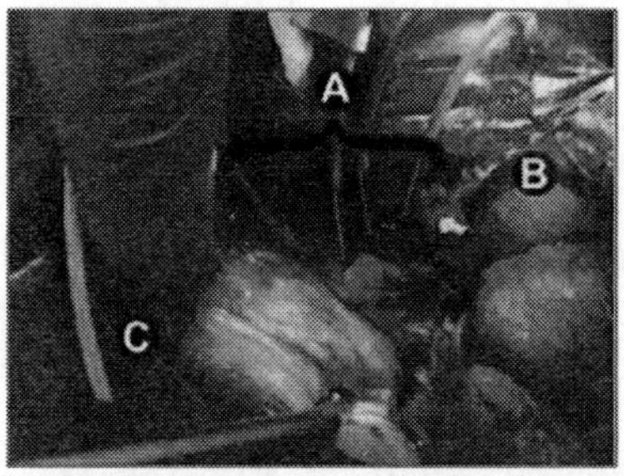

Figure 1. In vivo device placement: (A) four type-T thermocouples, (B) cooling catheter and guide wire, and (C) fifth thermocouple inserted into liver.

RESULTS

The computational model and numerical predictions showed cooling defects as deep as 12mm and a tissue temperature of 35°C. Cooling could not be observed beyond 14mm from the PD. Heat transfer steady-state analysis proved our computational model.

The *ex vivo* porcine pancreas was successfully cooled by 15°C with depreciating effects moving away from the PD (Figure 2). At 15mm a temperature of 35°C was reached in 5 minutes.

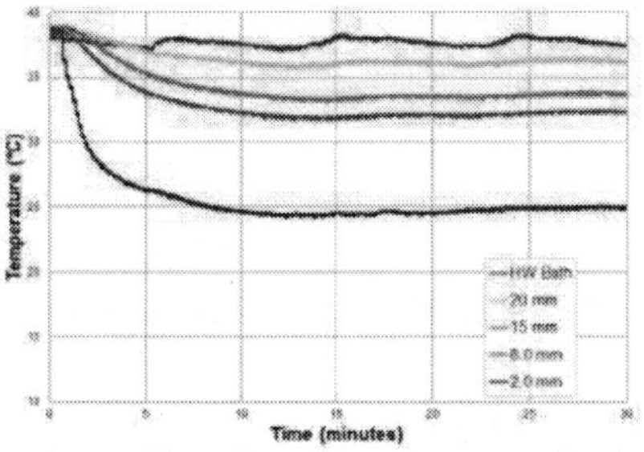

Figure 2: ex vivo cooling of a porcine pancreas maintained thermally at a human body temperature.

The *in vivo* pancreas experienced a maximal 10°C effect (Figure 3). In 4.65 minutes a dT of 2°C was observed at 12mm. These results compare favorably to those found in the *ex vivo* model. At 30 minutes, continuous cooling was observed at 18mm having a dT of 1.6°C. During this study the initial depth of the catheter insertion was 6cm, but ending depth was 4.1cm. The catheter's migration outwards occurred at roughly 7 minutes after initiation of cooling. Due to operating room conditions, the initial body temperature of the pig during the *in vivo* study was below normal.

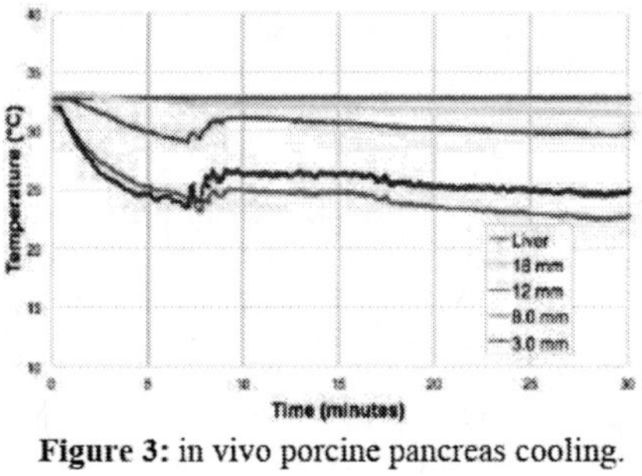

Figure 3: in vivo porcine pancreas cooling.

DISCUSSION

Focal cooling of the pancreas is technically feasible in both *ex vivo* and *in vivo* porcine models with a dT of 2°C observable at 12-15mm in 5 minutes. Mild therapeutic hypothermia (MTH) of 33-35°C demonstrates cardioprotective and neuroprotective behaviors; however, the optimal temperature range for treatment of AP remains unclear. For example, Rochas-Santos et al. observed no noticeable changes to pancreatic pathophysiology using profound temperatures of 4.6°C for treatment of ischemic-reperfusion AP injury in rats. To the best of our knowledge, no such study to date has evaluated the effects of the graded hypothermia (22-35°C) observed in this current report on tissue histopathology. Future studies should investigate the effects of graded hypothermia and histopathologic changes including vacuolization of acinar cells, pancreatic necrosis, and lymphocytic and neutrophilic infiltration. One lingering question is whether or not hypothermia impedes translocation of bacteria causing infection and increased severity.

There is evidence that AP severity is mediated by cytokines with serum levels of TNF-α and IL-6 increasing early on during the course of the disease.[4] Endoscopically-placed cooling catheter can rapidly induce localized hypothermia intraductally—roughly three times faster then conventional cooling approaches and technologies. Rapid localized cooling is just one advantage of the proposed device. Localized pancreatic cooling is not only more comfortable to the patient but it also significantly limits systemic cooling of supportive systems, which are attacked during systemic inflammatory response syndrome (SIRS). Our technology leverages a minimally invasive procedure. Even though our design makes more sense clinically, TH may not be a viable therapy option due to disease progression and the timing when a patient generally presents to the emergency room with AP. In this case, alternative strategies may be necessary to halt disease progression further along in this destructive cascade of events. Further survived animal studies monitoring cytokine profiles in the setting of induced pancreatitis as well as imaging changes to monitor evolution of pancreatitis are needed to determine the clinical effects of TH.

ACKNOWLEDGEMENTS

Our work was supported in part by NIH's National Institute of Neurological Diseases and Stroke (NINDS), grant #1R43NS095573-01A1. Authors: Meckes, Wilkins and Donatelli also thank the National Institutes of Biomedical Imaging and Bioengineering (NIBIB) and VentureWell for additional support. The content is solely the responsibility of the authors and does not necessarily represent the official view of the NIH.

REFERENCES

[1] Lowenfels, A et al., *Current gastroenterology reports* 11.2 (2009): 97-103.

[2] Nathens, A et al. *Critical care medicine* 32.12 (2004): 2524-2536.

[3] Fagenholz, P et al., *Pancreas* 35.4 (2007): 302-307.

[4] Rocha-Santos, V et al., *Pancreas* 43.6 (2014): 946-950.

[5] Matsuoka, et al., *Pancreas* 26.1 (2003): e12-e17.

[6] Wang, X et al., *Pancreas* 30.4 (2005): e80-e86.

[7] Fujimoto, K et al., *Pancreas* 37.2 (2008): 176-181.

[8] da Silva, I et al., *Pancreas* 42.3 (2013): 544-545.

[9] Caglar, V et al., *Forensic Medicine and Anatomy Research* (2014): 63-69

[10] Swindle, M et al. *Veterinary Pathology Online* 49.2 (2012): 344-356.

[11] Noar, M. *Endoscopy* 53 27(1995): 77-80

[12] Hirano, T et al., *British journal of surgery* 79.8 (1992): 803-806.

[13] Vanwoerkom, R et al., *Hospital Physician* (2009): 9-19

GLIAL CELL ANALYSIS IN THE BRAIN TUMOR MICROENVIRONMENT ELUCIDATES CONTRIBUTIONS TO GLIOBLASTOMA PATIENT PROGRESSION

R. Chase Cornelison (1), Jessica X. Yuan (1), Bethany J. Horton (2), Jennifer M. Munson (1)

(1) Department of Biomedical Engineering
University of Virginia
Charlottesville, VA, US

(2) Department of Public Health Sciences
University of Virginia
Charlottesville, VA, US

INTRODUCTION

Glioblastoma (GBM) is the most common and deadly brain tumor. Despite an aggressive standard of care, the average survival of GBM patients is only 15 months. Much of this poor response to therapy is due to the invasiveness of GBM cells, and better understanding factors that contribute to GBM invasion could have significant therapeutic implications. We have shown that interstitial fluid flow increases invasion of glioma cell lines both *in vitro* and *in vivo* (1). To date, this effect has only been examined in cultures of cancer cells alone. Recent evidence in multiple cancers suggests that the cellular tumor microenvironment can also contribute to cancer progression (2). Nonetheless, the GBM microenvironment is relatively under-characterized, and links between the microenvironment, interstitial flow, and GBM malignancy are currently unknown.

Tissue engineered models of cancer offer a means to dissect the contributions of individual components on cancer cell invasion in a controlled and reproducible manner. However, no current in vitro model incorporates the complex brain tumor microenvironment and interstitial fluid flow. By examining patient samples, we have developed a 3D culture model that incorporates both cellular and biophysical components in a physiologically-relevant context to examine individual and synergistic impacts on GBM cell invasion.

We histologically quantified the cellular elements of the brain microenvironment at the invasive edges of GBM patient sections (n=33, median age=62.5, 54% male) to obtain a physiologically-relevant ratio of cancer cells to glial cells. This cellular ratio was incorporated into a 3D in vitro model to examine invasion of patient-derived glioblastoma stem cells (GSC) from three different patients in the presence or absence of the brain tumor microenvironment as well as interstitial fluid flow. We found that GSCs from different patients responded differently to the addition of cellular components, in some cases increasing invasion while decreasing it in others. Each GSC line was found to be responsive to interstitial flow, increasing invasion, and the cellular effects were exacerbated by flow in one GSC line. Furthermore, trends in invasion correlated with microglial activation but not astrocytic activation.

METHODS

A quantitative immunohistochemical strategy was developed to assess important components of the brain microenvironment at the invasive edges of patient GBM sections (n=33, median age=62.5, 54% male). Quantitative analysis, using ImageJ, of the stained samples in regions at the invasive front of tumors provided composition concentration of non-cancer cell populations, including astrocytes (ALDH1L1), microglia (Iba1), and extracellular matrix components by pentachrome staining for general composition. A proportional hazards model was used to model overall survival (3). Additional tests were performed to ensure appropriateness of the proportional hazards assumption. Survival modeling and testing the proportional hazards assumption were performed using the "phreg" procedure in SAS (version 9.4). Survival curve comparisons were conducted using LogRank (Mantel-Cox) test and log-rank hazard ratios (HRs) are given as HR ¼ A/B (95% CI).

An average ratio of unstained cancer cells to stained astrocytes or microglia was also calculated from this histology for incorporation into a 3D *in vitro* model of the invasive tumor border Three patient-derived GSC lines were used for our experiments (G2, G34, and G528) derived and maintained from three individual patients, as previously described (4). Primary human astrocytes (Sciencell) and immortalized human microglia (Applied Biological Materials, Inc.) were used to model elements of the brain tumor microenvironment. 50,000 GSCs were seeded into thiolated hyaluronic acid-collagen gels atop 8μm-pore tissue culture inserts (1). When appropriate, glial cells labeled with

CellTracker dyes (Invitrogen) were added at a ratio of 6:1:1 cancer cells: astrocytes:microglia. A pressure head was applied to yield a superficial velocity of 1 μm/s through the cell/gel compartment for flow, or media held level for static conditions. After 18h, tissue inserts were imaged using an EVOS FL fluorescence microscope, and DAPI$^+$ nuclei were counted as transmigrated cells (reported as % migrating of total cells seeded).

To assess glial activation, some gels were fixed after 18 hours in culture using 4% paraformaldehyde and labeled for either glial fibrillary acidic protein (GFAP, Abcam ab7260) or CD68 (Abcam ab201845). Hydrogels containing only glial cells was used as an activation control. Images were acquired in five random locations in each gel (n=3) using a Zeiss 710 confocal microscope, and GFAP$^+$ labeled astroctyes and CD68$^+$ labeled microglia were manually quantified.

RESULTS

When quantitatively analyzing the cellular microenvironment, the percentage of area coverage by positive staining for markers of microglia (Iba-1) and astrocytes (ALDH1L1) were significantly different ($p<0.01$) in patient GBM samples within the invasive edges compared to the tumor bulk. Therefore, we developed a proportional hazards ratio model to determine the contribution of the microenvironment on patient survival. This model revealed that the balance of percent area coverage of astrocytes and microglia between the tumor bulk and invasive edges was particularly predictive of survival. Specifically, higher proportions of microglia in the invasive edges indicated better prognosis (HR = 0.45 [0.22, 0.91]; $p = 0.026$) while higher proportions of astrocytes in the invasive edges indicated poorer prognosis (HR = 1.81 [1.03, 3.19]; $p = 0.04$) compared to our whole patient cohort.

We then created a 3D *in vitro* model of the invasive tumor border (**Fig. 1A**) using a 6:1:1 ratio of cancer cells:astrocytes:microglia using this same histological data. Invasion of three different patient-derived GSC lines was evaluated in the presence or absence of the cellular and biophysical tumor microenvironment. G34 increased invasion with the addition of glial cells (**Fig. 1B**) while G528 decreased invasion. However, the latter was only a significant effect in the presence of

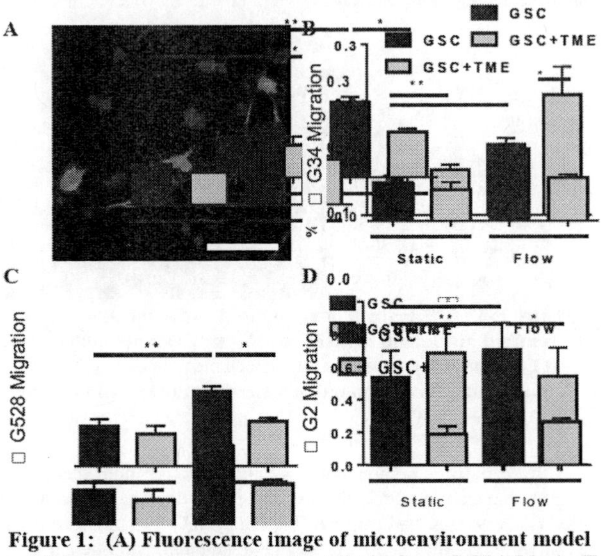

Figure 1: (A) Fluorescence image of microenvironment model containing astrocytes (green), microglia (red), and G34 (blue). (B-D) Invasion quantification +/- glia +/- flow for patient-derived GSC lines G34 (B), G528 (C), and G2 (D).

interstitial flow (**Fig. 1C**). The presence of glial cells had no apparent effect on invasion of G2 cells (**Fig. 1D**). Each cancer cell line was flow responsive, increasing invasion in the presence of interstitial flow.

Because glial cell area in the invasive border were predictive of patient survival, we also evaluated glial cell activation in our *in vitro* model (**Table 1**). Each GSC line was cultured in 3D with CellTracker-labeled astrocytes and microglia for 18 hours under static conditions. The tri-culture hydrogels were then fixed and stained for activation markers in astrocytes (GFAP) and microglia (CD68). Hydrogels containing only glial cells was used as an activation control.

Table 1: Quantification of glial cell activation markers before (Control) and after addition of G34, G528, and G2 cancer cells.

	Control	+G34	+G528	+G2
%GFAP$^+$ Astrocytes	10.06	27.51	20.01	9.11
%CD68$^+$ Microglia	43.77	77.98	100	98.18

Preliminary analysis revealed a trend in microglial activation that inversely corresponded with trends in the observed GSC invasion. For example, G34 caused the least activation of microglia, and this cell line significantly increased invasion in the presence of the TME (**Fig. 1B**). Conversely, G528 activated microglia the most and decreased invasion when combined with the TME (**Fig. 1C**).

DISCUSSION

Glioma cell invasion is a central reason for poor patient prognosis. Reactive astrocytes and microglia make up a large proportion of the GBM tumor bulk and adjacent regions of the brain, lending credence to the idea that they are involved in invasion of cancer cells into healthy parenchyma. While each cell type has individually been shown to increase cancer cell invasion *in vitro*, contributions from the combination of these two cells has not been examined. Furthermore, we and others have previously shown interstitial flow to stimulate cancer cell invasion, but this has not been studied in the context of the complex cellular tumor microenvironment.

Using patient samples, we found that the percent area of glial cells adjacent to the tumor bulk indeed correlated with patient survival in a proportional hazards ratio model. Higher microglia area indicated better survival while higher astrocytes indicated poorer survival. The trend for microglia could be replicated *in vitro* using a patient-defined model of the adjacent invasive region. The GSCs that induced the highest microglial activation decreased invasion upon adding glial cells, while the inverse was true for the line with least activation. These effects were occasionally dependent on the presence of interstitial flow, indicating that both cellular and biophysical cues in the tumor microenvironment can be important for evaluating cancer outcomes.

ACKNOWLEDGEMENTS

This work was supported by the University of Virginia Coulter Foundation (JMM), an American Cancer Society Institutional Research Grant (JMM), and an NIH Pre-doctoral Cancer Training Grant Award (NCI T32CA009109) (JXY). The authors would also like to acknowledge assistance from the Biorepository and Tissue Research Facility and the Cardiovascular Research Center Histology Core at the University of Virginia for assistance with obtaining patient samples and histological methods, respectively.

REFERENCES

[1] Munson, J et al., *Cancer Res,* 73:1536–4, 2013.
[2] Swartz, M et al., *Cancer Res*, 72:2473-2480, 2012.
[3] Yuan, J et al., *J Neuropathol Exp Neurol*, 75:1110-1123, 2016.
[4] Lee, J et al., *Cancer Cell,* 9:391–403, 2006.

FEASIBILITY OF A "SAME-DAY" AUTOLOGOUS TISSUE-ENGINEERED VASCULAR GRAFT REMODELING IN A SEEDED ELASTOMERIC SCAFFOLD

Darren G Haskett (1,3), Kamiel A Saleh (2), Jeffrey T Krawiec (2,3), Justin S Weinbaum (3,4), Antonio D'Amore (1,2,3), William R Wagner (1,2,3), Lauren E Kokai (3,4), Kacey G Marra (3,4), J Peter Rubin (3,4), David A Vorp (1,2,3,4,5)

(1) Department of Surgery
University of Pittsburgh
Pittsburgh, PA, USA

(2) Department of Bioengineering
University of Pittsburgh
Pittsburgh, PA, USA

(3) McGowan Institute of Regenerative Medicine
University of Pittsburgh
Pittsburgh, PA, USA

(4) Department Plastic Surgery
University Pittsburgh
Pittsburgh, PA, USA

(5) Department of Cardiothoracic Surgery
University of Pittsburgh
Pittsburgh, PA, USA

INTRODUCTION

Tissue engineered vascular grafts (TEVGs) containing adipose derived mesenchymal stem cells (ADMSCs) offer an alternative to small diameter vascular grafts currently used in cardiac and lower extremity revascularization procedures [1]. ADMSC infused TEVGs have been shown to promote remodeling and vascular homeostasis *in vivo* and offer a possible treatment solution for those suffering from cardiovascular disease [2-4]. Unfortunately, the time needed to cultivate ADMSCs remains a large hurdle for TEVGs as a treatment option.

An alternative TEVG treatment option bypasses the time needed for culture making use of cells taken directly from the stromal vascular fraction (SVF) (known to contain progenitor cells) and obtained directly after liposuction or panniculectomy procedures [5]. However, the ability of an uncultured, "same day" SVF seeded TEVG to remain patent and initiate remodeling *in vivo* and cell fate after implantation remain unknown.

The purpose of this study was to determine if SVF seeded TEVGs would remain patent *in vivo* and remodel after a "same day" implantation. Additionally, the duration of SVF presence within the implanted construct will be measured by using two different length implantations.

METHODS

SVF, obtained from adult adipose tissue, was seeded within 4 hours after acquisition from the patient onto poly(ester urethane)urea (PEUU) bilayered scaffolds using a customized rotational vacuum seeding device. These constructs were then placed for a short period of time in culture media allowing the cells to be incorporated into the scaffold during preparation for the surgical procedure. Constructs were then surgically implanted as abdominal aortic interposition grafts in Lewis rats (**Figure 1**) as described previously [2-4]. Implanted scaffolds were allowed to remodel for 5 days or 8 weeks. Patency was documented using angiography and gross inspection, while vascular components, along with the presence of any remaining implanted cells, were detected using immunofluorescent chemistry (IFC).

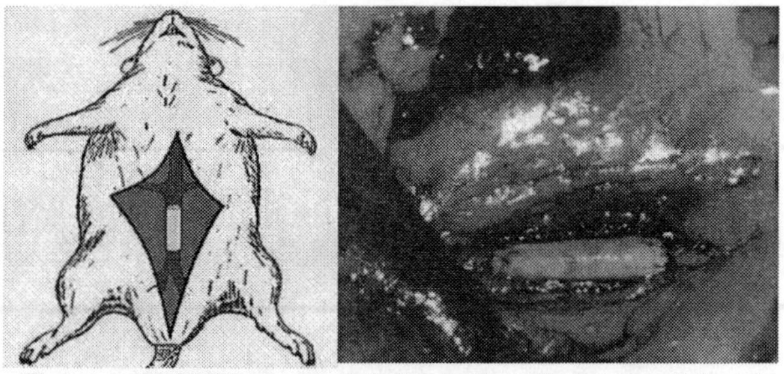

Figure 1: Left) Diagram of abdominal aortic interpostitional scaffold placement in a rat [4]. Right) Example image of TEVG implant immediately after suturing in place and having blood flow resumed in a Lewis rat (Forceps indicate

RESULTS

Initial findings demonstrated patency through both angiography and gross inspection after a short period (i.e. 5 days, n=1) and a longer period (i.e. 8 weeks, n=1) *in vivo* (**Figure 2**). IFC for human cells using the human progenitor cell marker, CD90, and a human nuclear antigen (HNA) was positive for cells remaining in the inner layer of the scaffold at 5 days but was negative at 8 weeks with greater HNA staining than CD90 (**Figure 3**). Further IFC analysis was positive for von Willebrand

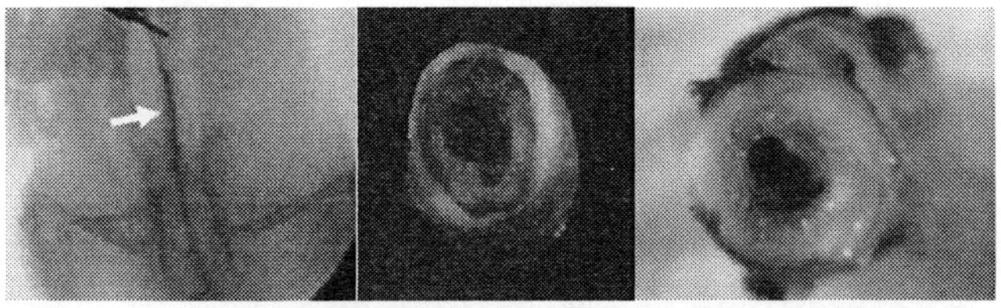

Figure 2: A) Angiogram displaying flow through an SVF seeded TEVG (arrow) after implantation. B) Gross inspection of explanted TEVG revealing patentcy from a 5-day implantation (Right) and 8-week implantation (Left).

Factor (vWF) indicating the presence of an endothelium, and smooth muscle alpha actin (αSMA suggesting the presence of smooth muscle cells in the neotissue. (**Figure 4**).

DISCUSSION

Initial findings show that a "same day" SVF-seeded TEVG remains patent after implantation *in vivo*, with remodeling and neo-tissue formation occurring by 8 weeks. Initial findings also demonstrate

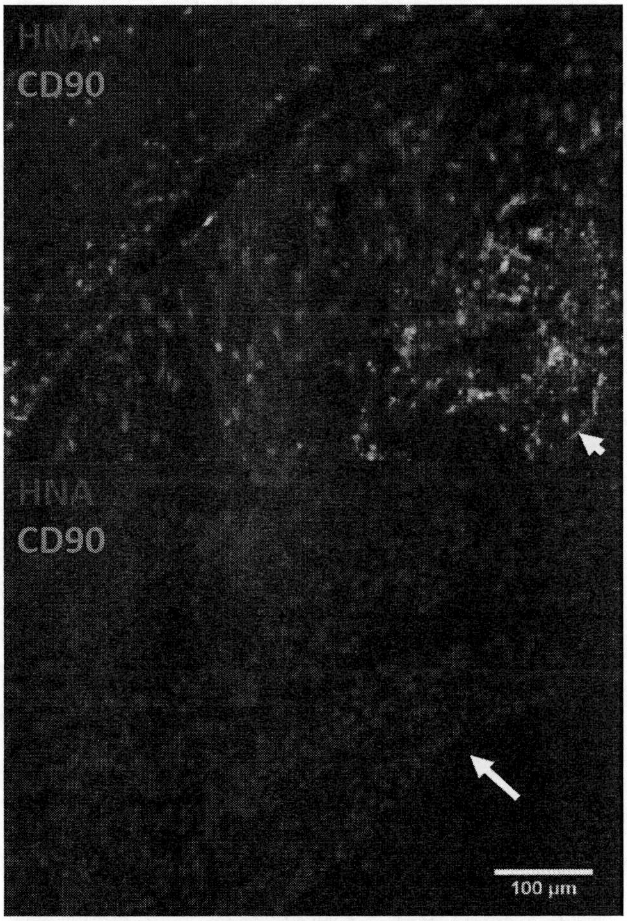

Figure 3: IFC on explanted TEVGs for HNA and human CD90 (red and green respectively) reveals presence of implanted cells after 5 days (Top) but not at 8 weeks (Bottom). Arrows indicate luminal edge. Counterstained with DAPI (blue)

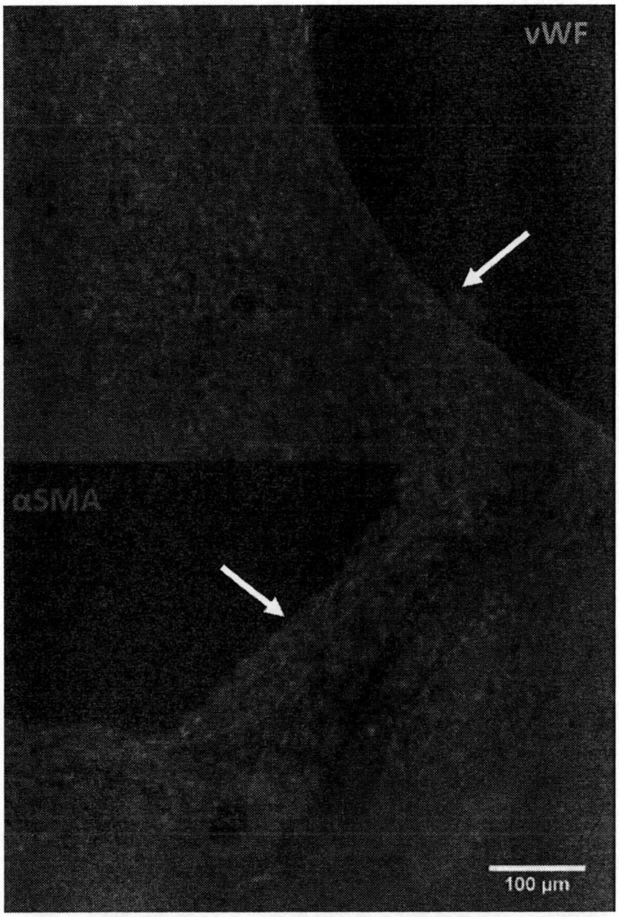

Figure 4: IFC staining for vWF (Top, green) and αSMA (Bottom, red) of a patent TEVG explanted after 8 weeks. Arrows indicate luminal edge. Counterstained with DAPI (blue)

the presence of human markers within the implanted TEVG at 5 days that are no longer detected at 8 weeks. This is comparable to previous work showing after a week, seeded cells are present in greatly reduced numbers compared to initial seeding, and merely the presence of stem cells promoted ingrowth and cellular migration [6], though as of yet the fate of the SVF cells during *in vivo* remodeling remains unknown.

Ultimately, understanding the fate of SVF cells and their ability to initiate *in* vivo remodeling will advance TEVG technology towards a "same-day" implementation and greatly enhance its appeal for clinical use.

ACKNOWLEDGEMENTS

The authors would like to acknowledge their sources of support: NIH Vascular Training Grant to DGH (T32 HL098036; Dr. Edith Tzeng, PI) and NIH R21 grant to DAV (R21 EB016138).

REFERENCES

[1] Weintraub, W, et al., *Am J of Cardiology*, **73**(2): 103-112, 1994.
[2] He, W, et al., *Car Eng and Tech*, **2**(3): p. 149-159, 2011.
[3] He, W, et al., *Biomaterials*, **31**(32): p. 8235-8244, 2010.
[4] Nieponice, A, et al., *Tissue Eng Part A*, **16**(4): p. 1215-23, 2010.
[5] Krawiec, J, et al., *J Vasc Surg*, 2016.
[6] Roh, J, et al., *Pro Nat Sciences US*, **107**(10): p. 4669-4674, 2010.

IN VIVO MATURATION AND INTEGRATION OF ENGINEERED ENPLATE-MODIFIED DISC-LIKE ANGLE PLY STRUCTURES (EDAPS)

SE. Gullbrand (1,2), DH. Kim (1,2), BG. Ashinksy (1,2), JT. Martin (1,2) LJ. Smith (1,2), DM. Elliott (3), HE. Smith (1,2), RL. Mauck (1,2)

(1) McKay Orthopaedic Research Laboratory
University of Pennsylvania
Philadelphia, PA, USA

(2) Translational Musculoskeletal Research Center
Philadelphia VA Medical Center
Philadelphia, PA, USA

(3) Department of Biomedical Engineering
University of Delaware
Newark, DE, USA

INTRODUCTION

Intervertebral disc degeneration, the most common cause of low back pain, is a progressive cascade of cellular, compositional, structural and mechanical changes [1]. The standard surgical treatment for disc degeneration involves fusion of the degenerated motion segment, which does not restore native disc structure or function, and may exacerbate degeneration of adjacent discs [2]. A promising alternative to fusion is total disc replacement with a cellular, engineered disc construct, which may restore normal structure and mechanics to the spine [3]. To this end, our group developed disc-like angle ply structures (DAPS) that mimic the structure and function of the native disc by combining cell-seeded hydrogels for the nucleus pulposus (NP) with an electrospun nanofibrous annulus fibrosus (AF) [4]. Our previous work showed that, following implantation in the rat caudal disc space for 5 weeks with external fixation, DAPS did not integrate well with the native tissue and exhibited progressive reduction in NP proteoglycan content and MRI T2 signal [5]. In this study, to improve construct integration and promote maintenance of construct composition *in vivo*, acellular porous polymer foam endplates were incorporated with the DAPS to form an endplate DAPS (eDAPS) implant. We hypothesized that the engineered endplates would promote DAPS integration, allowing for remobilization of the implanted motion segment. Here we report on the long-term *in vivo* performance of the eDAPS implant and the effects of prolonged external fixation versus remobilization.

METHODS

eDAPS Fabrication and Culture: DAPS sized for the rat caudal disc space were fabricated by concentrically wrapping aligned, angled strips of electrospun poly(ε-caprolactone) (PCL) nanofibers to form the AF region, and filling the center with a hyaluronic acid hydrogel to form the NP region [5]. Both regions were seeded with bovine disc cells ($2x10^6$ cells/AF and $6x10^5$ cells/NP) and cultured separately for two weeks in chemically defined media containing TGF-β3. After two weeks, the AF and NP regions were combined, and acellular porous PCL endplates (4 mm diameter, 1.5 mm high) fabricated via salt leaching, were apposed to each side of the DAPS to form the eDAPS (Figure 1A). eDAPS were cultured for an additional three weeks for a total of 5 weeks of preculture. *Implantation:* 18 Athymic male retired breeder rats were anesthetized, and Kirschner wires were passed through the C8 and C9 caudal vertebral bodies (VB) allowing the placement of a ring-type external fixator [5]. eDAPS were implanted following removal of the C8-C9 disc and a partial corpectomy of the adjacent vertebral bodies (Figure 1B). At five weeks post-implantation, seven rats were euthanized for analysis (5W Fixed (F) group). In the remaining animals, the external fixator was either removed to remobilize (R) the tail (5WF+5WR, n=6) or left in place to maintain fixation (10W F, n=5) for an additional five weeks.

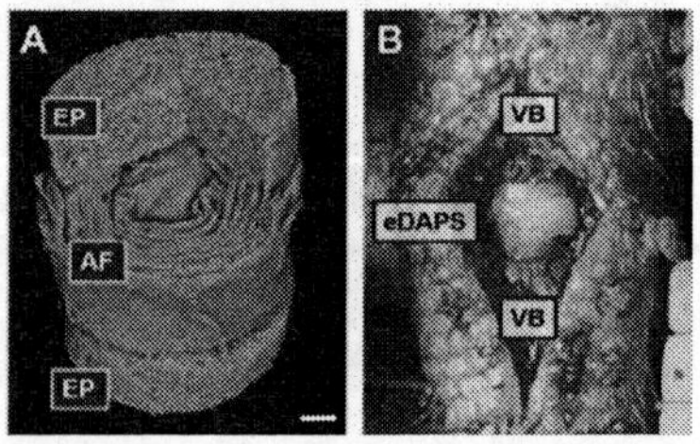

Figure 1: (A) eDAPS structure as visualized via μCT. (B) Intraoperative image of eDAPS in the rat caudal disc space.

Magnetic Resonance Imaging: T2 mapping of the implants was performed at 4.7T (16 echoes, TE/TR = 7.84 ms/2,000 ms, FOV = 15x15 mm^2). Average T2 maps were generated for each experimental group using a custom MATLAB code [6]. *µCT Imaging:* Following MRI, eDAPS implanted motion segments were subjected to µCT scanning (SCANCO µCT 50) at 3µm resolution to identify boney integration of the PCL foam endplates. *Mechanical Testing and Biochemistry:* Four vertebra-eDAPS-vertebra segments in each experimental group, and four native rat tail vertebra-disc-vertebra segments were subjected to ~3 MPa compression testing (20 cycles, 0 to 3N, 0.05 Hz). Displacement was tracked optically via a custom texture tracking MATLAB code. The 20th cycle was used to calculate the toe and linear region compressive moduli, and transition and maximum strains via a bilinear fit. After compression testing, a scalpel was used to transect adjacent tendons and muscle for complete soft tissue release around the implanted eDAPS, and a tension test to failure was performed. Following mechanical testing, the eDAPS were separated into NP, AF and EP regions, and digested with proteinase-K. Glycosaminoglycan (GAG) content was quantified via the DMMB assay and compared to the pre-implantation GAG content of eDAPS cultured for 5 weeks. Significant differences in quantitative outcomes were assessed via a one-way ANOVA with Tukey's post-hoc test.

RESULTS

NP T2 values for eDAPS were maintained from 5 to 10 weeks *in vivo*, and were not significantly different from the NP T2 of native rat tails discs at either time point. NP T2 values of the eDAPS at all time points were significantly higher than the T2 values of DAPS implanted without endplates for 5 weeks (50 ms, dashed line, Figure 1A, [5]). NP T2 values did not change with remobilization.

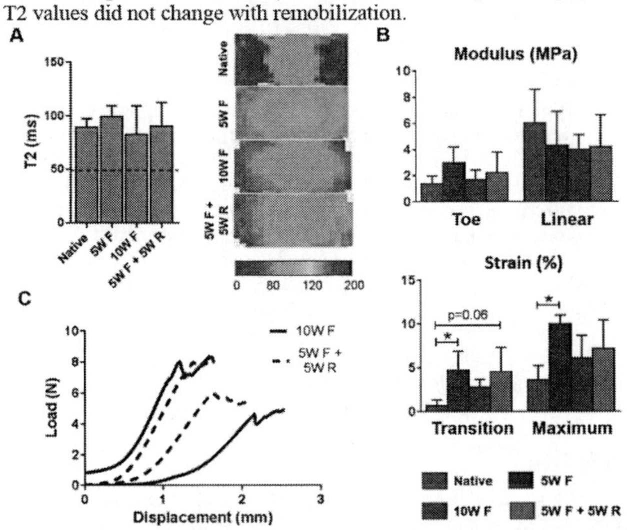

Figure 1: (A) NP T2 values and composite T2 maps for each experimental group. (B) Compressive modulus and strain compared to native rat tail discs (* = p<0.05). (C) Tension to failure curves of each sample tested in the 10 week groups.

The toe and linear region moduli of eDAPS implanted motion segments at 5 weeks and 10 weeks were not different from native rat tail motion segments (Figure 1B). Transition and maximum strain were significantly increased in eDAPS implanted for 5 weeks compared to native discs. However, by 10 weeks, there were no significant differences in transition or maximum strain between native discs and eDAPS implants with fixation. There was a trend (p=0.06) towards increased transition strain in the remobilized group compared to native.

Tension to failure tests were not possible at the 5 week time point due to insufficient construct integration; at 10 weeks, tensile testing was performed on 4 eDAPS implanted motion segments (2 10W F, 2 5WF + 5WR). Tensile load at failure ranged from 4 to 8 N (Figure 1C), with failure occurring at the DAPS- endplate interface (rather than the bone-endplate interface).

µCT scans showed new bone formation at the interface between the PCL foam endplates and the native vertebral body in the 10 week implanted groups (yellow arrow, Figure 2A). NP GAG content was not significantly different from pre-implantation values in any post-implantation group. NP GAG content in the 10 week remobilized group was significantly higher than the 5 week group. GAG content in the AF was maintained at all post-implantation time points. Significant increases in GAG content in the EP region (compared to pre-implantation) were noted at 10 weeks. EP GAG content in the remobilized group was higher than in the fixed group at 10 weeks.

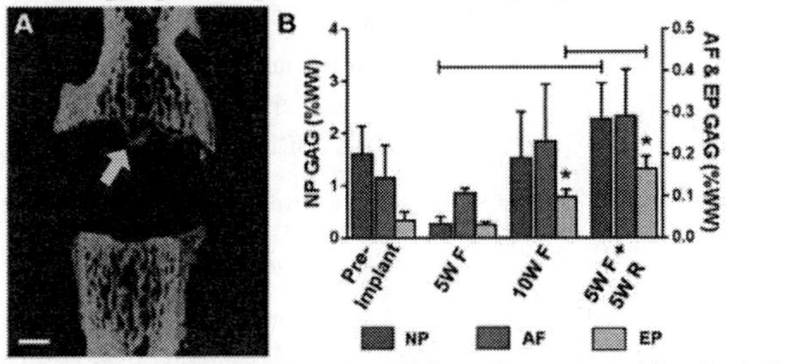

Figure 2: (A) µCT slice of a rat caudal motion segment implanted with an eDAPS for 10 weeks. (B) NP, AF and EP GAG content for each group compared to pre-implantation values (bars denote p<0.05, * = p<0.05 compared to pre-implantation and 5W F groups).

DISCUSSION

Taken together, these data suggest that long-term *in vivo* implantation of eDAPS implants results in functional integration and continued construct maturation. While eDAPS transition and maximal strains were super-physiologic after 5 weeks implantation, strains approached native values after 10 weeks. Furthermore, functional integration of the eDAPS with the native tissue occurred by 10 weeks, as evidenced by non-zero tensile strength and the evidence new bone formation within the PCL endplate region on µCT imaging. Biochemical assays illustrated the eDAPS either maintained or produced additional GAGs with longer *in vivo* implantation periods. Remobilization did not have a significant effect on NP T2 values or on mechanical properties of the implanted motion segments. However, remobilization appeared to have a positive effect on eDAPS GAG content in the NP and EP regions. Additional study of longer remobilization times is warranted. Overall, this data suggests that a tissue-engineered intervertebral disc with endplates is capable of integrating with native tissue and undergoing functional maturation in the *in vivo* space, particularly following remobilization of the implanted disc. Ongoing work is focused on furthering the clinical translation of the eDAPS by evaluating these constructs in a large animal model.

ACKNOWLEDGEMENTS

This work was supported by the Department of Veterans' Affairs and the Penn Center for Musculoskeletal Disorders.

REFERENCES

[1] Hadjipavlou A, et al, *JBJS*, 90:1261-70, 2008. [2] Ghiselli G, et al., *JBJS*, 86A:1497-503, 2004. [3] Bowles R, et al, *PNAS*, 108:13106-11, 2011. [4] Nerurkar N, et al, *Spine*, 35:867-73, 2010. [5] Martin J, et al, 2016 ORS Proceedings. [6] Martin J, et al, *JOR*, 33:140-148, 2015.

DENSIFICATION OF TYPE I COLLAGEN MATRICES AS A MODEL FOR CARDIAC FIBROSIS

Benjamin Seelbinder (1), Logan J. Worke (2), Jeanne E. Barthold (1), Tyler Novak (2), Russell P. Main (2,3), Corey P. Neu (1,2)

(1) Mechanical Engineering
University of Colorado Boulder
Boulder, CO, USA

(2) Biomedical Engineering
Purdue University
West Lafayette, IN, USA

(3) Department of Basic Medical Sciences
Purdue University
West Lafayette, IN, USA

INTRODUCTION

Cardiac fibrosis is a disease state characterized by excessive collagenous matrix accumulation within the myocardium and is a key element of many cardiac pathologies such as myocardial infarction and hypertrophic cardiomyopathy[1]. Even more systemic conditions like diabetes, obesity and aging are associated with significant cardiac fibrosis[2],[3]. The range of etiologies for cardiac fibrosis is indicative of its complexity and it is due to in part to this complexity that current treatment options are severely lacking. To close this gap, *in vitro* models that recapitulate the defining features of the fibrotic environment are in need. Type I collagen is a major cardiac ECM protein and the defining component of fibrotic depositions, therefore making it an attractive choice for a fibrosis model; however, it demonstrates poor mechanical strength due to solubility limits. Recently, plastic compression has been shown to overcome these issues and significantly increase the mechanical properties of collagen matrices[4]. Additionally, improvements in collagen extraction have yielded oligomeric collagen materials that more closely mimic the cross-linking behavior seen *in vivo*[5]. Here, we utilized confined compression of oligomeric, type I collagen matrices to resemble defining hallmarks seen in fibrotic tissue. We then populated these matrices with cardiomyocytes (CM) and performed functional and gene expression analysis to assess whether compressed collagen substrates are able to induce fibrotic symptoms such as decline in contractile properties and the alterations in gene expression.

METHODS

Type I oligomeric collagen from porcine dermis was obtained as previously described[5]. Collagen was solubilized at 4.25 mg/mL and neutralized with 1.7 M phosphate buffered saline (PBS). Glass-bottom 8-well plates (Ibidi) with 1 cm^2/well surface areas were filled with neutralized collagen to a final thickness of 1 mm (uncompressed) or 5 mm (compressed, Figure 1). Substrates were polymerized at 37°C for 1 hour. Thick (5mm) gels were densified in confined compression within the well using a custom-fit Teflon stamp. Compression was controlled using a Bose ElectroForce 5500 mechanical testing system at a 0.1%/s strain rate to 80% final strain (1 mm final height). Collagen density was analyzed via collagen autofluorescence, bulk compression via Bose ElectroForce 5500 (load cell: 50 N), microscale compression via AFM (Novascan) and nanoscale analysis via NanoSEM (FEI).

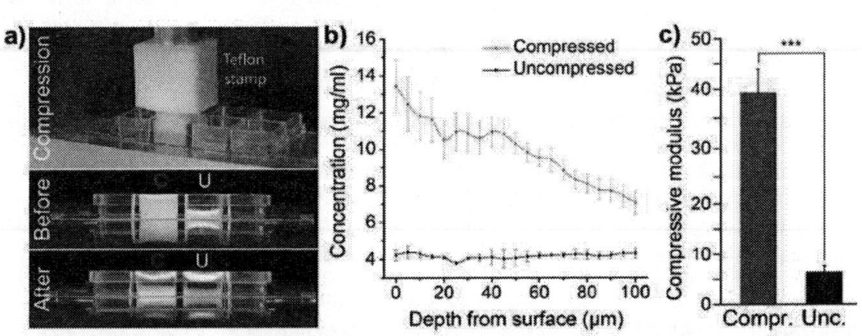

Figure 1. Compression and bulk properties of oligomeric, type I collagen matrices. a) Compression procedure of collagen after polymerization in 1 cm^2 wells. **b)** Collagen concentration in relation to surface depth. **c)** Bulk compressive properties of compressed (C) and uncompressed (U) matrices; *** $p<0.001$.

Cardiomyocytes (CM) were derived from mice embryos (E18.5). $0.5 \cdot 10^6$ CMs were cultured per matrix and cultured for 2d before fluo4 Ca^{2+} and gene expression analysis was performed.

RESULTS

Compressed matrices showed significantly higher collagen content, with concentrations being highest at the surface, and increased bulk compressive modulus compared to uncompressed (39.1 kPa vs. 6.3 kPa, Figure 1). Nanoscale analysis revealed that mean fiber thickness (1.14±0.10 vs. 0.83±0.09, p=0.045) and area fraction of collagen fibers (55.5±1.8% vs. 49.3±1.3%, p<0.005) was significantly higher in compressed versus uncompressed matrices. In contrast to bulk measurements, AFM analysis of the compressive modulus at microscale showed no significant difference between compressed and uncompressed samples (2.93 ± 0.25 kPa vs. 2.82 ± 0.14 kPa).

Beating performance of CMs cultured on compressed matrices severely declined: amplitudes were a magnitude lower, frequency was higher and broader and inter-beat variations were distinctly higher compared to uncompressed matrices (Figure 2). Average beat durations (FWHM) only marginally differed. Further analysis also revealed poor cell-to-cell synchronization (data not shown). Gene expression analysis of fibrosis-relevant markers showed expression change of three fibrosis marker in accord with current literature (stripped bars) while maturation was increased and inflammation decreased on compressed collagen matrices.

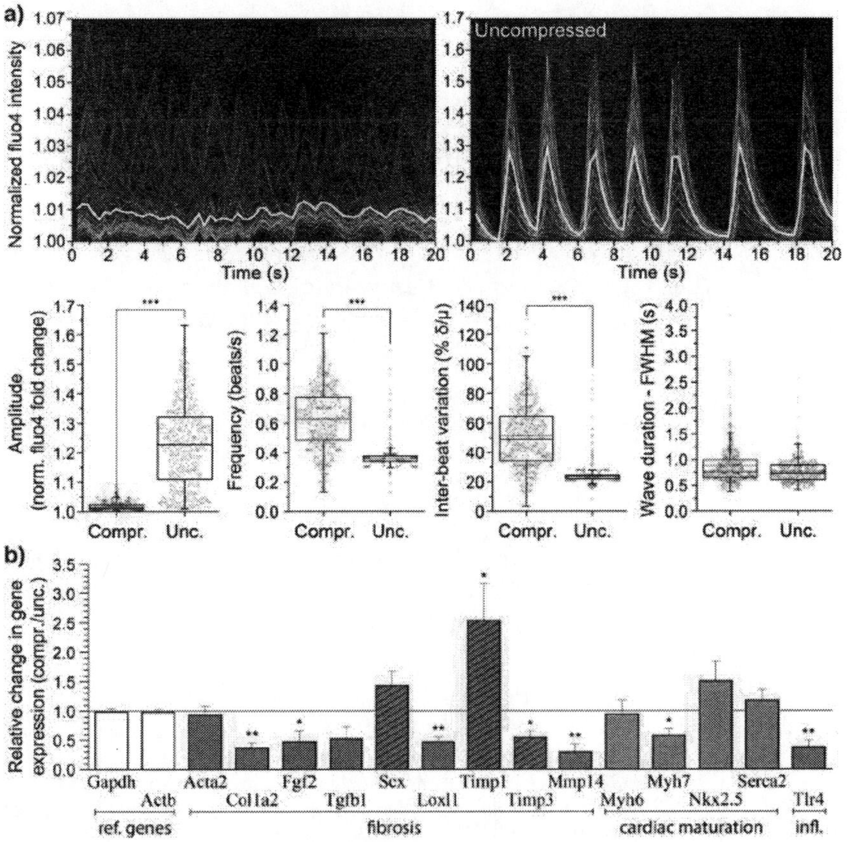

Figure 2. Ca^{2+} activity and gene expression analysis of CMs on collagen matrices. a) Representative fluo4-derived Ca^{2+} signals from one area (820x820 μm) imaged over 20s at 3.3 fps; colored lines=cells (compr. n=131; unc. n=147); thick white line=mean. **b)** CM beating properties determined from Ca^{2+} analysis; dots=cell from 4 areas (compr. n=547; unc. n=651). Statistics between area averages. **c)** Gene expression analysis; ref.=reference; infl.=inflammation; error bars=SEM; * p>0.01, ** p<0.01, *** p<0.001.

DISCUSSION

Confined compression of collagen matrices recapitulated defining features of ECM alterations in fibrosis: Type I collagen content was elevated, especially at the surface; bulk compressive moduli was comparable with healthy and disease states between uncompressed and compressed matrices (healthy: 6.32 kPa embryo or 10-15 kPa adult vs. fibrotic: > 35 kPa); and at the fibril level, compression resulted in increased mean fibril thickness and density which was in accord with findings in fibrotic tissues[6].

Embryonic CMs displayed a strong phenotypic response with beating patterns on compressed matrices showing all the hallmarks of CMs beating in fibrotic tissue: weak calcium amplitudes are indicative of decreased systolic output, low beat consistency is symptomatic for arrhythmias and poor cell-to-cell synchronization is reflective of a disturbed excitation coupling between cells[1]. Even elevated frequency is typically a symptom observed in patients with cardiac fibrosis.

After observing a strong phenotype, it was surprising to find that most fibrosis gene markers showed either no or the opposite response as expected from literature. A reason for this discrepancy was likely the use of embryonic CMs. This notion is supported through the observed increase in CM maturation markers potentially triggered by higher mechanical resistance imposed by compressed collagen similar to the postnatal increase in work load which is known to simulate cardiac maturation. The maturation cascade could suppress stress (Tgfb1) and inflammation cascades (Tlr4). Another reason for the unexpected gene expression results could be the lack of increased cross-linking or other missing ECM components of our collagen matrices compared to native fibrotic environments.

While more improvements in cell culture and ECM recapitulation need to be made to fully model all *in vivo* aspects, compressed collagen substrates prove effective to study distinct aspects of cardiac fibrosis, such as the observed abrogation of cardiac contractile properties.

ACKNOWLEDGEMENTS

The authors acknowledge funding from NSF CAREER 1349735.

REFERENCES

[1] Kong, P et al. *Cell. Mol. Life Sci.* 71:549–574, 2014
[2] Asbun, J et al. *J. Am. Coll. Cardiol.* 47:693–700, 2006
[3] Biernacka, A et al. *Aging Dis.* 2:158–173, 2011
[4] Novak, T et al. *Adv. Funct. Mater.* 26:2617–2628, 2016
[5] Bailey, JL et al. *Biopolymers* 95:77–93, 2011
[6] Jalil, JE et al. *Circ. Res.* 64:1041–1050, 1989

ABSENCE OF DECORIN ACCELERATES CARTILAGE FIBRILLATION AND AGGRECAN DEPLETION IN POST-TRAUMATIC OSTEOARTHRITIS

Q. Li (1), L. Ouyang (1), B. Doyran (1), L. Fan (1), W. Tong (2), W. J. Tseng (2), X. S. Liu (2), R. V. Iozzo (3), L. Qin (2), D. E. Birk (4), L. Han (1)

(1) School of Biomedical Engineering, Science and Health Systems
Drexel University
Philadelphia, PA, United States.

(2) McKay Orthopaedic Research Laboratory
Department of Orthopaedic Surgery
University of Pennsylvania
Philadelphia, PA, United States.

(3) Department of Pathology, Anatomy and Cell biology
Thomas Jefferson University
Philadelphia, PA, United States.

(4) Department of Molecular Pharmacology and Physiology, University of South Florida
Tampa, FL, United States.

INTRODUCTION

Post-traumatic osteoarthritis (PTOA) is the prevalent OA form in young adults [1]. Currently, the pathogenesis of chondrocytes and extracellular matrix (ECM) molecules in PTOA is not well understood. This study aims to define the roles of decorin, a critical cartilage matrix regulatory molecule, in PTOA. In normal cartilage, decorin can bind to collagens II [2] and VI [3], aggrecan core protein [4] and transforming growth factor-β (TGF-β) [5]. Our recent study showed that decorin-null ($Dcn^{-/-}$) murine cartilage has substantially reduced glycosaminoglycan (GAG) content and weakened mechanical properties [6]. In early OA, decorin is markedly up-regulated [7], and in late OA, it becomes increasingly fragmented [8]. These evidences suggest that decorin plays an important role not only in the proper functioning of normal cartilage, but also in OA pathogenesis. To this end, we examined the progression of PTOA in wild-type (WT) and $Dcn^{-/-}$ mice. The destabilization of the medial meniscus (DMM) surgery was performed to induce PTOA in mice, followed by an analysis of ECM morphology, nanostructure and the mechanical properties of cartilage during OA progression. Here, the more severe cartilage degradation in $Dcn^{-/-}$ compared to WT mice highlighted the critical roles of decorin in PTOA.

METHODS

Surgery: DMM surgery was performed on the right hind knees of skeletally mature, 12-week old, male C57BL/6 WT and $Dcn^{-/-}$ mice, with the Sham control on contralateral left knees, following established procedures [9]. All animal work was approved by the IACUC. Mice were euthanized at 2 and 8 weeks post-surgery for mechanical and structural tests of early and late stage of OA. *AFM-nanoindentation* was performed on the intact surface of medial condyle cartilage (n = 5) at 10 μm/s rate using a microspherical tip (R ≈ 5 μm, k ≈ 8.9 N/m, NanoAndMore) and a Dimension Icon AFM [10]. For each condyle cartilage, ≥ 15 locations were tested up to an indentation depth of ~1 μm to account for heterogeneity. Effective indentation modulus, E_{ind}, was calculated from the loading portion of each indentation force-depth curve via the Hertz model. *Histology and immunohistochemistry (IHC):* Whole joints of both knees were harvested from additional mice (n ≥ 3), fixed, decalcified, and embedded in paraffin. 6-μm-thick sagittal sections were cut across the medial side and two sections within every consecutively six sections of the medial side of the knee were stained with Safranin-O/Fast Green for GAGs and scored via the modified Mankin Score by two masked observers [11]. Serial sections adjacent to those with representative safranin O-stained images were stained with decorin antibody. To characterize the loss of articular cartilage, cartilage thickness (t_{total}) was quantified (Fig. 1c) [12]. *SEM:* After indentation, samples were treated with Karnovsky's fixative, dehydrated in hexamethyldisilazane and imaged under Zeiss Supra 50VP SEM. *μCT:* For joints used for histology, prior to decalcification, the fixed joints were scanned by ex vivo μCT (Scanco microCT 35) at 6 μm resolution to characterize subchondral bone morphology.

RESULTS

In WT mice, IHC demonstrated an up-regulation of decorin after DMM, indicating the active involvement of decorin in OA (Fig. 1a). At early stage of OA (2 weeks): there is upregulation of aggrecan concentration on DMM side comparing with Sham for $Dcn^{-/-}$ cartilage (Fig. 1b). Aggrecan depletion was observed on cartilage surface (Fig. 1c). While there is no observable cartilage damage for $Dcn^{-/-}$ mice (Fig. 2a), the significant increase of modified Mankin score shows detectable microscopic OA signs as early as 2 weeks (Fig. 2b). Surface collagen fibrils became more aligned along the mediolateral orientation in $Dcn^{-/-}$ Cartilage indicating collagen reorganization, which was absent in WT mice (Fig. 3b). Similar mechanical properties of cartilage were observed despite significant decreasing of indentation modulus for WT mice (Fig. 3a). At late stage of OA (8 weeks): more severe OA was observed in both genotypes, illustrated by the presence of chondrocyte hypertrophy and hypercellularity (Fig. 1c). $Dcn^{-/-}$ cartilage surface was depleted of GAGs and became completely dominated by highly aligned collagen fibrils (Fig. 3b), a clear sign of surface fibrillation. Significant reduction of total cartilage thickness (Fig. 2a) and escalation of modified Mankin score (Fig. 2b) demonstrate cartilage erosion. The modulus decrease was significant for both genotypes (Fig. 3a), illustrating cartilage dysfunction. Furthermore, $Dcn^{-/-}$ mice showed marked subchondral bone plate thickening (Fig. 4), which was absent in the WT.

DISCUSSION

The cartilage surface fibrillation is a typical OA-induced cartilage structural change in advanced OA [13]. In the DMM model, it is absent in WT, since DMM induces mild-to-moderate OA in the tested time frame of 8 weeks [12]. The salient fibrillation on $Dcn^{-/-}$ cartilage surface thus suggests that during PTOA, decorin can inhibit collagen fibril aggregation and alignment, preventing the formation of fibrocartilage that is mechanically inferior to healthy hyaline cartilage [7]. In $Dcn^{-/-}$ mice, as a result of more severe fibrillation and cartilage degradation, subchondral bone thickening occurs at 8 weeks, another sign of advanced OA [13]. These evidences illustrate the active involvement of decorin in PTOA and its important role in preventing cartilage fibrillation, aggrecan loss and OA progression. In addition, at 2 weeks post-DMM, although WT cartilage exhibits significant modulus reduction while $Dcn^{-/-}$ cartilage does not, such results do not suggest that $Dcn^{-/-}$ mice are more resistant to early OA (Fig. 3). As this could be (1) a temporary artifact due to the initiation of surface fibrillation (Fig. 3b) and cartilage thinning (Fig. 2a), (2) anabolic activity of chondrocytes for rescuing knee joint dysfunction (Fig. 1c). It should be noted, however, that in $Dcn^{-/-}$ mice, cartilage has a number

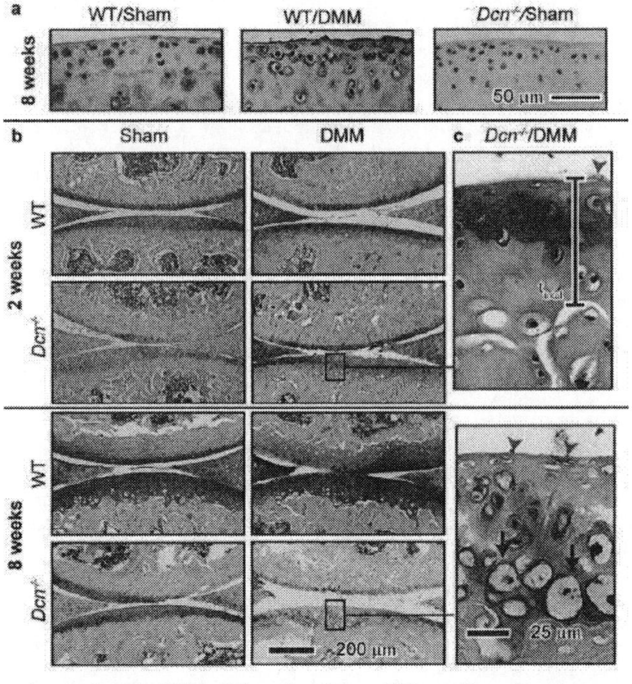

Figure 1: **(a)** IHC of decorin shows WT cartilage has increased decorin at 8 weeks after DMM. $Dcn^{-/-}$/*Sham* is shown as a negative control. **(b-c)** Safranin-O/Fast Green histology (red arrow head: GAG depletion, black arrow: Chondrocyte hypertrophy in $Dcn^{-/-}$ mice).

of developmental defects, such as reduced sGAG content and altered collagen fibril diameter, as well as weakened mechanical properties before DMM [6]. These defects can also contribute to the observed more severe OA in $Dcn^{-/-}$ mice. In order to isolate decorin activities in skeletal development versus in OA, our ongoing studies are using inducible decorin knockout mice to provide the temporal specificity of decorin gene ablation.

ACKNOWLEDGEMENTS

This work was supported by the National Institutes of Health (AR066824) and the Penn Center for Musculoskeletal Diseases (PCMD) (NIH P30 AR050950).

REFERENCES

[1] Helmick, CG et al., *Arthritis Rheum.*, 58:15-25, 2008.
[2] Douglas, T et al., *Biomacromolecules*, 7:2388-2393, 2006.
[3] Wiberg, C et al., *J. Biol. Chem.*, 276:18947-18952, 2001.
[4] Wiberg, C et al., *J. Biol. Chem.*, 278:37698-37704, 2003.
[5] Hildebrand, A et al., *Biochem. J.*, 302:527-534, 1994.
[6] Doyran, B et al., *Trans.Orthop. Res. Soc.*, 62:491, 2016.
[7] van der Kraan, PM et al., *J.Rheumatol.*, 24:543-9, 1997.
[8] Bock, HC et al., *Osteoarthritis Cartilage*, 9:654-663, 2001.
[9] Glasson, SS et al., *Osteoarthritis Cartilage*, 15:1061-1069, 2007.
[10] Batista, MA et al., *Matrix Biol.*, 38:84-90, 2014.
[11] McNulty, MA et al., *Osteoarthritis Cartilage*, 20:949-956, 2012.
[12] Doyran, B et al., *Osteoarthritis Cartilage*, 25:108-117, 2017.
[13] Buckwalter, JA et al., *Instr. Course Lect.*, 54:465-480, 2005.

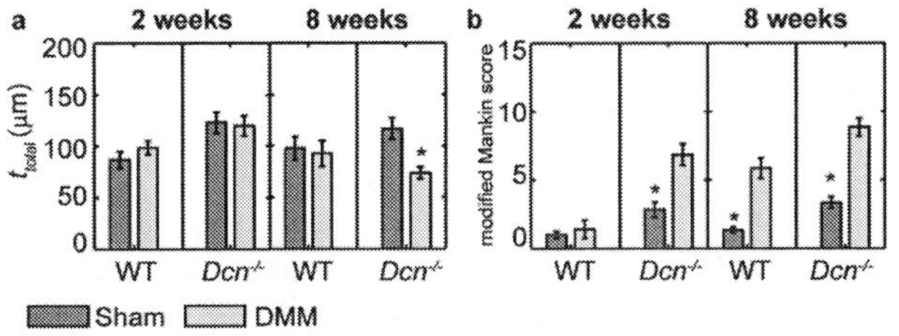

Figure 2: **(a)** cartilage total thickness of DMM vs Sham knees for 2 weeks and 8 weeks post DMM. **(b)** modified Mankin score of DMM vs. Sham knees for 2 weeks and 8 weeks post DMM (*: $p < 0.05$ via Wilcoxon signed rank test, mean ± 95% CI, $n \geq 3$ joints each).

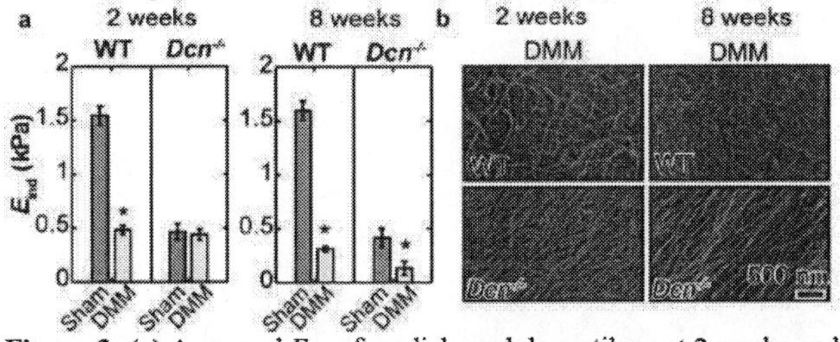

Figure 3: **(a)** Averaged E_{ind} of medial condyle cartilage at 2 weeks and 8 weeks post DMM and **(b)** cartilage surface fibril structure. (*: $p < 0.05$ via Wilcoxon signed rank test, mean ± 95% CI, $n \geq 5$ joints each. Red arrow: collagen alignment along mediolateral direction).

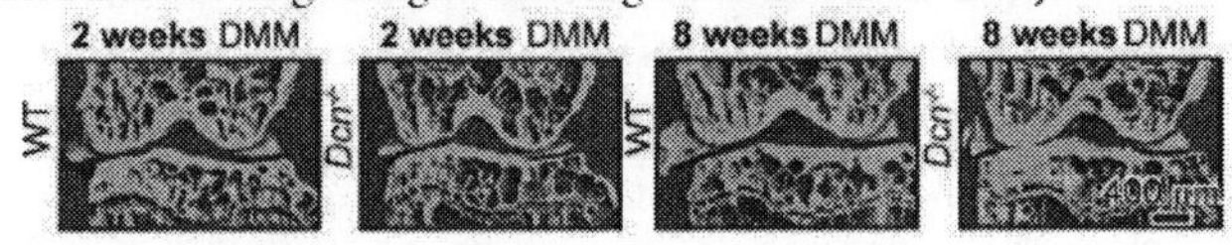

Figure 4: Subchondral trabecular bone frontal plane image via μCT (red arrowhead: subchondral bone thickening and osteophyte formation)

CONTROLLING AND MEASURING THE SPATIAL EXTENT OF CHONDROCYTE DEATH IN A NON-INVASIVE MURINE POST-TRAUMATIC OSTEOARTHRITIS MODEL

Alexander Kotelsky (1), Edward F. Ruppel (1), Mark R. Buckley (1)

(1) Department of Biomedical Engineering
University of Rochester
Rochester, NY, USA

INTRODUCTION

Osteoarthritis (OA), the leading cause of disability in the United States [1], is a whole joint disease involving degeneration of articular cartilage (AC). Approximately 12% of all cases of OA are caused by known joint trauma and are therefore classified as post-traumatic osteoarthritis (PTOA) [2]. Due to the low proliferative capacity of chondrocytes – the sole cells residing in AC – cartilage homeostasis may be compromised by chondrocyte death resulting from traumatic injuries. Moreover, immediate, necrotic chondrocyte death following mechanical trauma leads to the uncontrolled release of intracellular contents, possibly inducing aberrant behavior in surrounding regions of cartilage and surrounding tissues [3]. Thus, cell death has been implicated as a possible tipping point in the pathogenesis of PTOA. Nevertheless, the role of chondrocyte death in PTOA remains controversial; for example, a recent study induced expression of diphtheria toxin to kill superficial zone chondrocytes in mice, and chondrocyte death did not cause or exacerbate PTOA [4]. However, the effects of diphtheria-associated cell death and mechanical (necrotic) cell death on the joint may be distinct due to different spatial patterns and different effects on surrounding cells.

A key step towards understanding the role of *mechanically-mediated* cell death in PTOA is to assess how the extent of cell death impacts PTOA progression. Murine models provide a flexible, low cost research platform where PTOA is easily induced through surgical interventions [5, 6] or controlled external mechanical loading [7-9]. However, many of these powerful models do not allow for control over the extent of the initial injury (and the extent of cell death). Moreover, assessing the extent of cell death in these models generally requires time-consuming serial histological sectioning [6]. In one *in vivo* study, PTOA was induced non-invasively by compressing the femur against the tibial plateau with the knee at a controlled angle in a materials testing device [8]. We aimed to investigate whether a model inspired by this study can be used to investigate whether spatial extent of cell death can be controlled and easily measured without serial histological sectioning. Additionally, we sought to assess the impact of age on the susceptibility of chondrocytes to mechanical cell death following traumatic impact. We hypothesize that the area of chondrocyte death is governed by the energy of impact, and that younger mice will exhibit lower susceptibility to impact-induced chondrocyte death.

METHODS

Mechanical testing platform design. A custom modular testing platform was designed and assembled in Solidworks for manufacture by 3D printing on an Object 30 with FullCure 810, VeroClear UV sintered resin (Fig. 1 a-b). The device consists of three main parts: 1) the support base, 2) the body spacers, and 3) the impacter guide block. Parts 1 and 2 were 3D printed, whereas part 3 was machined out of white delrin to provide a low friction impact.

The support base includes mounting points for the body spacers and a dovetail linear guide for an adjustable compression boot. The boot consists of a foot cavity and a tibia alignment brace used to fix a mouse's knee joint in place. A set screw knee alignment block fixes the knee joint in place concentrically to the impacter (Fig. 1b).

The body spacers - one mobile and two stationary - raise the impacter guide block to provide space for the mouse body without exerting unintended loading (Fig. 1 a, c). The mobile spacer is alternated between left and right front mounting points depending on which leg is tested.

The impacter guide block includes a reamed bore through which a 1N cylindrical weight is allowed to free fall onto the knee from different heights, resulting in impacts of different controlled energies (Fig. 1 a, c).

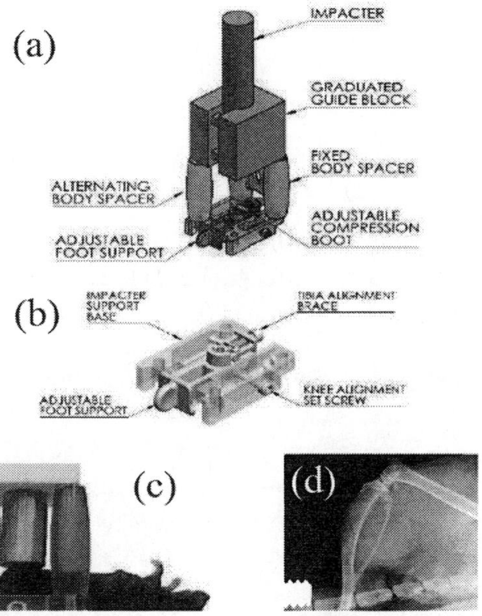

Figure 1: (a, b) Schematic representation of the mechanical testing platform used for murine knee impact experiments. (c) 3-D printed testing device used to secure the mouse tibia in place and load the knee joint with a cylindrical impacter. (d) X-ray image of murine knee joint secured in the testing device.

Impact experiment. Knee joints of C57BL6 mice in two age groups (16-19 and 38-41 week old male and female mice) were mounted in the device (Fig. 1c-d). To evaluate the knee flexion angle and ensure consistency of mounting, X-ray images were acquired and analyzed using ImageJ. The two age groups were divided into three loading subgroups: 1) sham (n=3 16-19 week old mice, n=8 38-41 week old mice), 2) impact with 406.7mJ (n=4 16-19 week old mice, n=5 38-41 week old mice) and 3) impact with 813.4mJ (n=4 16-19 week old mice, n=4 38-41 week old mice) of kinetic energy. In the sham group, the mouse was placed and secured into the testing device, but the knee was not impacted.

Analysis of cell death. Following the mechanical testing, distal femurs with intact articular cartilage were carefully dissected and stained with 10 μM calcein AM and 40 μg/mL propidium iodide (PI), indicators of live and dead cells (respectively). The posterior femoral articular surface (hydrated in HBSS, pH = 7.4, 302 mOsm) was imaged under fluorescent illumination with a 4X dry lens (NA = 0.13) on an Olympus IX-81 inverted microscope such that both femoral condyles were visible. Cells that lost calcein AM fluorescence and became PI-positive were considered to be dead. Cell death areas were quantified using ImageJ and compared across age and loading groups using a two-way analysis of variance (ANOVA) with Sidak post-hoc analysis. The significance level α was set to 0.05.

RESULTS

The knee flexion angle of mice secured in the impacting device was measured to be 78.4 ± 6.9 degrees (Fig. 1d). Cell death in impacted specimens was expansive, and generally observed away from the edges of the joint (Fig. 2b). Cell death area increased significantly with both age and energy (main effects). According to post hoc analysis, in 38-41 week old mice, cell death area increased from sham to 406.7 mJ and 813.34 mJ, and from 406.7 mJ to 813.34 mJ. In addition, cell death area was higher in 38-41 week old mice for the 831.4 mJ group (Fig. 2c).

DISCUSSION

In this study, we have established a testing platform inspired by a previously established non-invasive murine PTOA model [8] that can induce controlled, quantifiable chondrocyte death through impaction of the knee with different kinetic energies. Importantly, using this model, the spatial extent of chondrocyte death on the femoral articular surface can be measured quickly and easily without sectioning the cartilage.

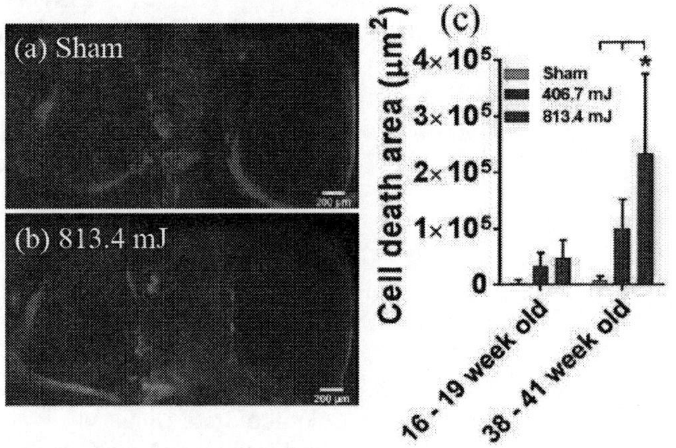

Figure 2: Representative fluorescent images from (a) 16 - 19 week old sham group and (b) 38 - 41 week old 813.4 mJ group. Green and red fluorescence represent live and dead cells, respectively. (c) Area of cell death on the articular surface of distal femoral condyles. Data are mean + standard deviation. Brackets denote $p < 0.05$ for comparison across impact intensities; * denotes $p < 0.05$ vs. 16 - 19 week old group.

The results of the study confirm our hypotheses and demonstrate that chondrocytes are less able to withstand impact loading with increased age and load intensity. Interestingly, the area of cell death was generally confined to the middle of the joint, away from the region of meniscal coverage (Fig. 2b). This observation is consistent with the known chondroprotective effect of menisci [6].

One limitation of the current study is that impact-induced damage to other soft tissues surrounding the articular cartilage (e.g., ligaments and menisci) was not assessed. Furthermore, chondrocyte death was analyzed only on the femoral articular surface and was not evaluated deeper in the tissue or on the tibial plateau.

Taken together, the results suggest that age- and energy-associated increase in traumatic chondrocyte death may lead to a faster onset and progression of post-traumatic osteoarthritis. Since the model used in this study is known to induce PTOA [8], future studies will 1) correlate the extent of chondrocyte death with PTOA progression across age groups and impact intensities and 2) assess cell and matrix changes that occur in and around the area of cell death.

ACKNOWLEDGEMENTS

We thank Arthur Salo for his assistance with machining of parts.

REFERENCES

[1] Neogi, *Osteoarthritis Cartilage,* 21:1145-53, 2013. [2] Brown et al., *J Orthop Trauma*, 20:739-744, 2006. [3] Levin et al., *Osteoarthritis Cartilage,* 9: 702-711, 2001. [4] Zhang et al., *J Clin Invest,* 126(8):2893–2902, 2016. [5] Kamekura et al., *Osteoarthritis Cartilage,* 13:632-641, 2005. [6] David et al., *J Orthop Res,* 1-11, 2016. [7] Furman et al., *J Orthop Res,* 578-592, 2007. [8] Poulet et al., *Arthritis& Rheumatism*, 63: 137-147, 2011. [9] Christiansen et al., *Osteoarthritis and Cartilage,* 20: 773-782, 2012.

DYNAMIC COMPRESSION OF HUMAN AND OVINE MENISCAL TISSUE COMPARED TO A BLOCK COPOLYMER MATERIAL FOR POTENTIAL MENISCAL REPLACEMENT

Kristine M. Fischenich (1), Katie Boncella (2), Travis S. Bailey (1,3), Tammy L. Haut Donahue (1,4)

(1) School of Biomedical Engineering
Colorado State University
Fort Collins, CO, USA

(2) Health and Exercise Science
Colorado State University
Fort Collins, CO, USA

(3) Chemical and Biological Engineering
Colorado State University
Fort Collins, CO, USA

(4) Mechanical Engineering
Colorado State University
Fort Collins, CO, USA

INTRODUCTION

The structure of menisci allows for stabilization of the knee and load distribution within the joint[1]. The presence and function of these fibrocartilaginous tissues is vital in maintaining the health of the underlying articular cartilage. Due to their avascular nature and limited healing capacity, damage has been shown to lead to early onset osteoarthritis. Currently, there are few technologies to replace damaged menisci, and standard treatment is to leave asymptomatic damage alone or perform partial meniscectomies[2]. A number of synthetic replacements have been investigated but limitations include poor durability, mechanics, and biocompatibility[3]. Likewise many degradable scaffolds have been shown to have limited mechanical integrity and lead to cartilage damage once tested *in vivo*[4,5]. With the rising number of elderly persons and the increase of meniscal injuries across young athletes, there is a greater need than ever for increased investigation into potential meniscal replacements. Many previous studies have focused on larger animal models including ovine models for their similarity to human meniscus both with respect to size[6], mechanics[7] and structural composition[8]. However, these studies have focused mechanical comparisons of instantaneous and equilibrium responses of the tissues rather than the repetitive cyclic loading. The objective of this study was to investigate the response of both human and ovine meniscal tissue to repeated compressive loading, and compare these findings to a potential hydrogel meniscal replacement.

METHODS

Menisci from seven human cadaveric specimens (ages 60±21) and 8 mature ovine cadaveric specimens were sectioned into medial and lateral, anterior and posterior segments. Plugs were created using 5 mm diameter biopsy punch taken from the proximal to distal end and stored at 1.6°C in 1x PBS for 24 hours. A sizing apparatus was used to cut the samples 3 mm thick from the mid-belly of the plug. Ten samples for all regions were tested for both human and sheep with the exception of the medial posterior region in human where only 9 samples were tested due to limited tissue. Polystyrene polyethylene oxide (PS-PEO) polymer was synthesized as previously described[9], packed between two Kapton sheets, melt pressed in a Carver Press (150°C, 500psi, 10 minutes), and allowed to swell in 1x PBS for 24. Swollen cylindrical plugs where 3 mm thick and 5mm in diameter. A total of 10 hydrogel samples were used for testing. All samples were fixed distally with cyanoacrylate between polished aluminum platens in a heated 1x PBS bath (96-99°F) and tested using a servo hydraulic testing system (Bionic Model 370.02 MTS) equipped with an axial 2lb load cell (Futek LSB200). The average thickness was measured, a 200mN preload applied and then samples were compressed to 12% strain for 5000 cycles at 1 Hz representing physiological strains, frequency, and the average daily steps taken by Americans[10,11]. Following the first cyclic test, samples were stored at 1.6°C in 1x PBS for 24 hours after which the testing protocol was repeated resulting in two cyclic tests for each plug. Modulus values were determined from a linear fit of the 2-10% stress-strain curve of each loading cycle. A second order power fit of the modulus vs cycle graph was used to assess relaxation. A one way analysis of variance (ANOVA) with a post hoc Tukey's test was used to assess differences between the two tests across all regions of interest as well as differences between regions with values of interest including cycle 1, cycle 10, cycle 25, cycle 50, and the final 5000th cycle and the three coefficients from the power fit (coefficients A, B, and C). Significance was set at $p<0.05$.

RESULTS

No differences were found between the two cyclic tests for each sample, so the two tests were averaged. After averaging the two tests, statistical assessments were performed to assess differences between regions of both human and sheep menisci and none were found with respect to modulus or the coefficients from the second order power fit. Since no regional differences were found in the native meniscal tissue, all regions were combined for comparison with the hydrogel plugs (Figure 1). Rapid relaxation occurred in both the human and sheep menisci with equilibrium typically reached prior to 1000 cycles. Once averaged an ANOVA was used to determine differences across cycles of interest within groups (human, sheep, hydrogel) as well as differences across groups within cycles for both modulus values and coefficients of the power fit (Figure 2).

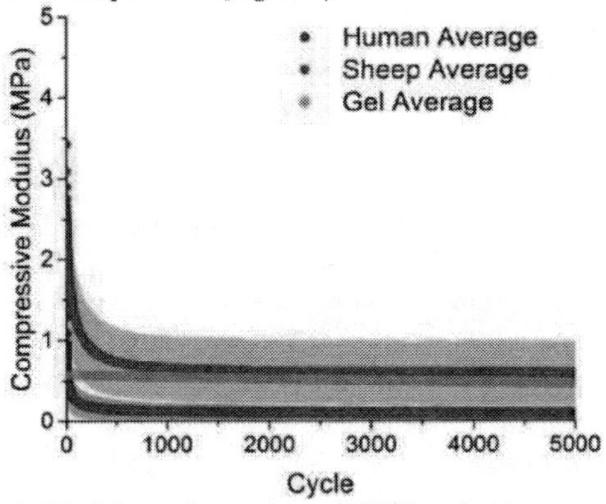

Figure 1: Modulus values across all 5000 cycles (average ± std)

DISCUSSION

This study is one of the first to investigate the mechanical response of multiple meniscal tissues as well as a potential replacement material under repeated cyclic compression similar to what is experienced *in vivo*. When allowed to rest for 24 hours, no significant differences were seen between runs in any of the samples with respect to the initial response, relaxation profile, and equilibrium response. Additionally, even though others have shown regional differences in single compression studies, no regional differences were found in either the human or the sheep menisci. Sheep menisci were found to have both a higher initial (cycle 1) and equilibrium (cycle 5000) response as well as a slightly longer relaxation time compared to the human samples. The overall decrease in modulus of the sheep menisci (approximately 82% from cycle 1 to cycle 5000) however was similar to that of the human samples (95%). The hydrogel material, although often found to be statistically different, had a modulus value between that of the human and sheep for the majority of the 5000 cycle test. The greatest difference between the hydrogel and the two native meniscal tissues was the limited relaxation observed in the hydrogel. Values reported in this study are similar to those previously reported for human and sheep menisci both instantaneously and at equilibrium[7,12,13]. It has been well documented meniscal tissue has an initial response vastly different from the equilibrium response likely due to the fluid flow within the tissue when compressed.

Although the sheep menisci were found to be statistically different from human they do relax similarly to human tissue. This in combination with previous literature supports the use of sheep as a large animal model for the human meniscus condition. The polystyrene-polyethylene oxide block copolymer hydrogel material tested was significantly different from both native tissues under most assessments, but did present with material properties within the range of the two and matched the values seen in the sheep model for the majority of the 5000 cycle test and should remain a viable materials option for meniscal replacement.

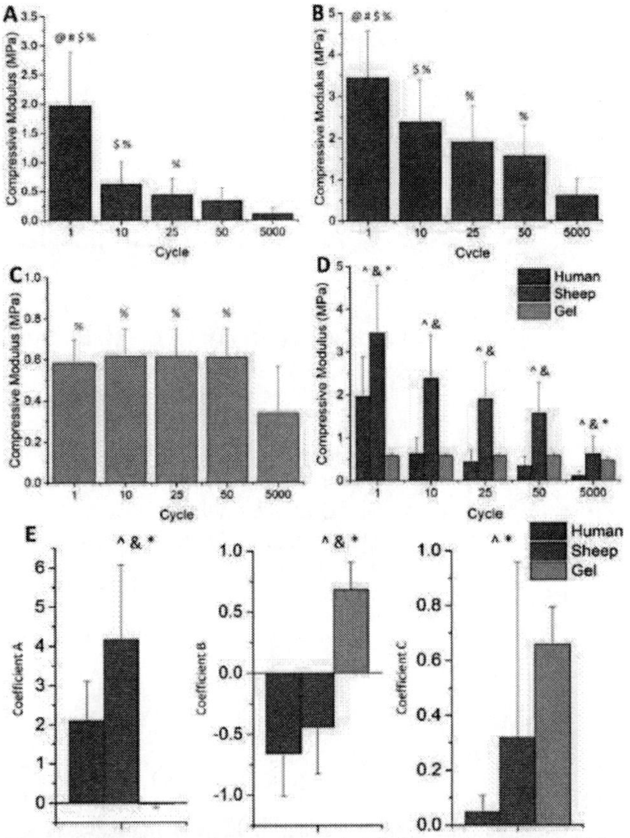

Figure 2: Modulus values (average ± std) for cycles of interest for human (A) sheep (B) and hydrogel (C) as well as a comparison of all three groups (D). Coefficients from power fit (average ± std) for all groups (E). Statistical significance denoted as the following @ sig diff from cycle 10, # sig diff from 25, $ sig diff from 50, % sig diff from 5000, ^ sig diff between human sheep, & sig diff between sheep and gel, * sig diff between human and gel.

ACKNOWLEDGEMENTS

Research reported in this publication was supported by the National Institute of Arthritis and Musculoskeletal and Skin Diseases of the National Institutes of Health under Award Number R21 AR069826. The content is solely the responsibility of the authors and does not necessarily represent the official views of the National Institutes of Health.

REFERENCES

[1] Allen, AA., et al., *Oper. Tech. Orthop.*, 5(1):2–9, 1995 [2] Katz, JN., et al., *Arthritis Rheum.* 60(3):633–5, 2009 [3] Vrancken, CT., et al., *Knee Surg. Sports Traumatol.*, 2015 [4] Tienen, TG., et al. *Biomat.*, 24:2541–2548., 2003 [5] Welsing, RTC., et al. *J. Sports Med.*, 36(10):1978–89., 2008 [6] Takroni, T., et al. *J. Exp. Orthop.* 3(1):35, 2016 [7] Joshi, MD., et al. *J. Biomed. Mater. Res.* 29(7):823–8., 1995 [8] Chevrier, A et al., J. Orthop. Res. 27(9):1197–203, 2009 [9] Guo, C. et al., *Soft Matter* 6(19):4807, 2010 [10] Tissakht, M. et al., *J. Biomech.* 28(4):411–22, 1995 [11] Choi, BCK et al. *Clin. Investg. Med.* 30(3):146–151, 2007 [12] Bursac, P et al., *Biorheology* 46(3):227–37, 2009 [13] Chia, HN et al., J. Orthop. Res. 26(7):951–6, 2008

ROLES OF TYPE V COLLAGEN IN THE STRUCTURE AND MECHANICAL PROPERTIES OF MANDIBULAR CONDYLE CARTILAGE

P. Chandrasekaran (1), Q. Li (1), M. Sun (2), L. J. Soslowsky (3), D. E. Birk (2), L. Han (1)

(1) School of Biomedical Engineering, Science and Health Systems
Drexel University
Philadelphia, PA, United States

(2) Department of Molecular Pharmacology and Physiology
University of South Florida
Tampa, FL, United States

(3) McKay Orthopaedic Research Laboratory
University of Pennsylvania
Philadelphia, PA, United States

INTRODUCTION

The mandibular condyle cartilage in the temporomandibular joint (TMJ) has a bi-layered layout. It is composed of a fibrocartilage superficial layer dominated by type I collagen fibers aligned along the anteroposterior orientation, covering a secondary hyaline cartilage layer of type II collagen and proteoglycans [1]. This unique hybrid structure endows the mandibular cartilage with its specialized anisotropic biomechanical properties [2]. However, to this day, there is little molecular-level understanding of how such hybrid structure is assembled during the skeletal development of TMJ.

This study aims to examine the roles of type V collagen in the development and functioning of the TMJ mandibular cartilage. Type V collagen is a fibril-forming collagen that plays critical roles in regulating the fibril nucleation during type I collagen fibrillogenesis (Fig 1) [3]. The importance of type V collagen in type I collagen-dominated connective tissues have been illustrated by the salient developmental defects in tendon [4-6] and cornea [7, 8]. Furthermore, the human genetic disease of classical Ehlers-Danlos Syndrome (EDS) is due to the deficiency of Col5A1 gene [3, 6, 9]. The resultant abnormal fibril nucleation and dysfunctional fibril growth leads to a number of defects in type I collagen-dominated tissues, including skin hyperextensibility, abnormal healing, joint hypermobility, and atrophic scarring [9]. Meanwhile, while the importance of type V collagen to the mandibular cartilage function is implied by frequent reports of TMJ subluxation in patients with classical form EDS [10], there is little knowledge of how type V collagen impacts the functions of the mandibular cartilage. To this end, we hypothesize that type V collagen plays critical roles in regulating the collagen fibrillar structure and mechanical properties of the TMJ mandibular cartilage. To test this hypothesis, we compared the nanostructure and mechanical properties of the mandibular cartilage superficial layer between the EDS mice ($Col5a1^{+/-}$) and wild-type ($Col5a1^{+/+}$) control.

METHODS

Sample preparation. Murine TMJs were harvested at 3 months (P90) and 6 months (P180) of age. The harvested tissues were stored in PBS with protease inhibitors for less than 48 hrs to minimize post-mortem degradation before AFM tests. Note that we did not evaluate null mice ($Col5a1^{-/-}$) since they are perinatal lethal [3].

AFM-nanoindentation. (Dimension Icon) was performed with a microspherical tip ($R \approx 5\mu$m, $k \approx 2$ N/m, μMasch) [11]. For each sample, at least 15-20 locations were tested in the central region of the mandibular cartilage surface. At each location, effective indentation modulus, E_{ind}, was calculated via the Hertz model. After nanoindentation, each sample was fixed with Karnovsky's fixative and dehydrated using gradient concentration of Hexamethyldisilazane to preserve the collagen architecture [12].

Structural analysis. Scanning electron micrographs (**SEM**, Zeiss Supra 50 VP) were acquired to visualize collagen fibril nanostructure. For histological analyses, the skulls were harvested and demineralized for 3-4 weeks before preparing the tissues for sectioning and Hemotoxylin and Eosin (H&E) staining was performed.

Statistical test. The Wilcoxon signed rank test was used to detect structural and mechanical differences between control and $Col5a1^{+/-}$ mice, with a significance level of $\alpha = 0.05$.

RESULTS

From SEM studies, there were significantly thicker collagen fibrils on the surfaces of $Col5a1^{+/-}$ (white arrows, Fig 2b) as compared to $Col5a1^{+/+}$. In addition, H&E histology (Fig 2d) showed altered gross-level shapes of the whole mandibular condyle surface in the *Col5a1*$^{+/-}$ mice.

In line with the salient structural phenotype, nanoindentation also showed significant reduction in the modulus of the mandibular cartilage surface in *Col5a1*$^{+/-}$ mice (Fig 3). Such reduction was present not only in mature mice at 3 months of age, but also in aging mice at 6 months of age.

DISCUSSION

Upon the reduction of Col5a1 expression in *Col5a1*$^{+/-}$ mice, the increase in both the average and heterogeneity (variance) (Fig 2c) of the surface fibril diameter is consistent with the activities that collagen V co-assemble with collagen I to form heterotypic fibrils, which limits the lateral growth of collagen I fibrils. As a result of the abnormal fibril growth, the mandibular condyle also develops abnormal gross-level morphology, which can lead to altered contact kinematics within the TMJ. Meanwhile, the significant reduction in E_{ind} for both mature and aged mice confirms that type V collagen is essential to the mechanical functioning of the mandibular cartilage. Similarly, in classical EDS, it is likely that the mutation causes haploinsufficiency of COL5A1 encoding pro-α1(V) chains[3], thereby reducing the capability of forming heterotypic fibrils.

This study directly illustrates the critical roles of type V collagen in the normal development and functioning of the TMJ mandibular cartilage. This work provides a foundation for future studies on the coordinated impacts of type V and XI collagens on the development of mandibular cartilage, as well as the development of the type I collagen-dominated TMJ disc. As the first study on the biomechanical properties of murine TMJ, this work provides a new experimental infrastructure for understanding the development and disease-associated degeneration of the TMJ in various transgenic and induced TMJ OA murine models [13, 14].

ACKNOWLEDGEMENTS

This work was supported by the National Institute of Health (AR066824), and the Penn Center for Musculoskeletal Diseases (PCMD) (NIH P30AR050950).

REFERENCES

[1] Singh, M et al., *J. Biomech. Eng.*, 130:011009, 2008.
[2] Allen, KD et al., *J. Biomech.*, 39:312-322, 2006.
[3] Wenstrup, RJ et al., *J. Biol. Chem.*, 279:53331-53337, 2004.
[4] Zhang, G et al., *J. Musculoskelet. Neuronal. Interact.*, 5(1):5-21, 2005.
[5] Connizzo, BK et al., *J. Ortho. Res.*, 34:2154-2161, 2016.
[6] Sun, M et al., *Am. J. Pathol.*, 185:1436-1447, 2015.
[7] Zhang, G et al., *J. Biol. Chem.*, 284:8888-8897, 2009.
[8] Sun, M et al., *J. Cell Sci.*, 124:4096, 2011.
[9] Malfait, F et al., *Genet Med.*, 12:597-605, 2010.
[10] Norton, LA et al., *Am. J. Orthod. Dentofacial. Orthop.*, 111:75-84, 1997.
[11] Sanchez-Adams, J et al., *J. Orthop. Res.*, 31:1218-1225, 2013.
[12] Bray, DF et al., *Microsc. Res. Tech.*, 26:489-495, 1993.
[13] Koyama, E et al., *J. Dent. Res.*, 93:663-670, 2014.
[14] Wadhwa, S et al., *Cells Tissues Organs.*, 181:136-143, 2005.

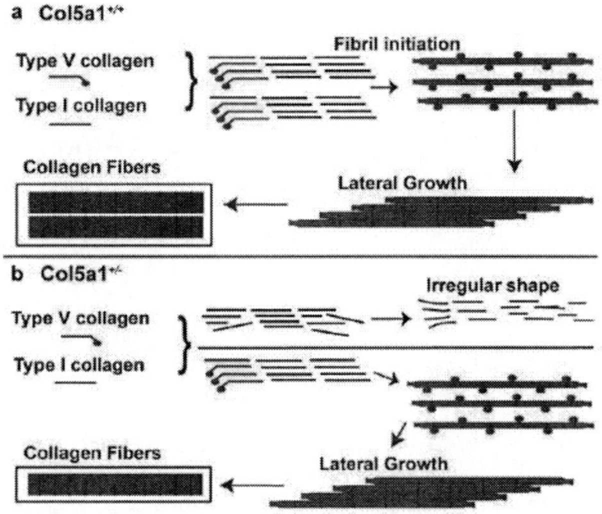

Fig 1. a) Representative figure of collagen formation in *Col5a1*$^{+/+}$ and **b)** in EDS affected mandibular condyle cartilage (*Col5a1*$^{+/-}$) [3].

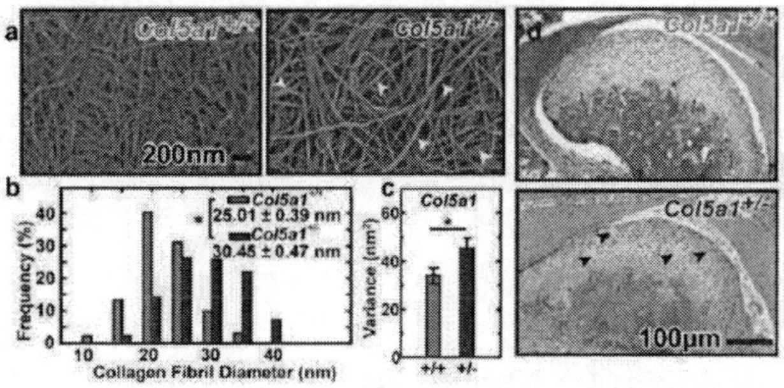

Fig 2. a) SEM of mandibular condyle surface of *Col5a1*$^{+/+}$ and *Col5a1*$^{+/-}$ respectively. **b)** Histogram frequency distribution of diameter. (*: $p < 0.05$ between *Col5a1*$^{+/+}$ and *Col5a1*$^{+/-}$, mean ± 95% CI, M=3) and **c)** diameter variance with 95% CI. **d)** H&E histological images of mandibular condyles of *Col5a1*$^{+/+}$ and *Col5a1*$^{+/-}$ mice.

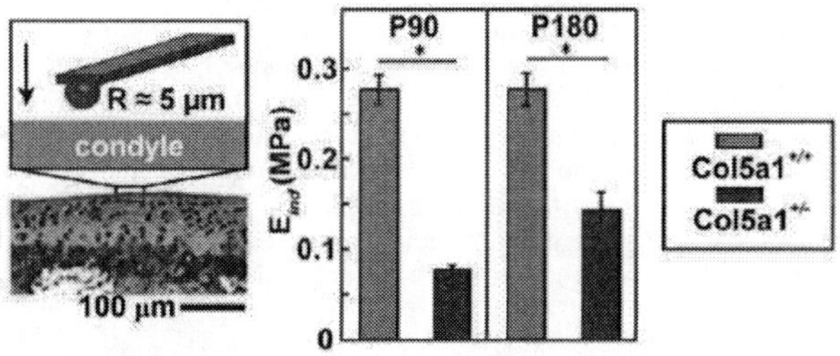

Fig 3. AFM based nanoindentation modulus on condyle surfaces (*: $p < 0.05$ between *Col5a1*$^{+/+}$ and *Col5a1*$^{+/-}$, mean ± SEM of n ≥ 15 from M=3 animals)

MICROMECHANICAL HETEROGENEITY OF THE TEMPOROMANDIBULAR JOINT DISC AND CONDYLE CARTILAGE SURFACES

L. Ouyang (1, 2), C. Wang (1), Q. Li (1), X. L. Lu (3), L. Han (1)

(1) School of Biomedical Engineering, Science and Health Systems
Drexel University
Philadelphia, PA, United States.

(2) Department of Orthopedics
Huazhong University of Science and Technology
Wuhan, Hubei, China.

(3) Department of Mechanical Engineering
University of Delaware
Newark, DE, United States

INTRODUCTION

In temporomandibular joint (TMJ), the disc and mandibular condyle work synergistically to sustain the high frequency tensile, compressive and shear loading specialized to the joint [1]. Due to the unique kinematics of the TMJ, the stress and strain experienced by both the disc and condyle are heterogeneous and nonlinear. In TMJ osteoarthritis (TMJ OA), the most prevalent degenerative TMJ disease [2], which has been reported to affect 10-30% of the population [3], and can be present without clinical symptoms [4], both the disc and condyle cartilage undergo irreversible degradation, leading to abnormal joint contact, craniomandibular pain and limited jaw motion [5]. To this day, there is very limited understanding of the structural and mechanical heterogeneity of the TMJ tissues. It is thus challenging to develop effective strategies that can document disease progression, repair and/or restore the native tissue functions [6]. This study aims to define the nanostructural and micromechanical heterogeneity of the disc and condyle at the scale of collagen fibril diameters, with a focus on their surfaces, where the two tissues are in direct contact and load transmission.

METHODS

Sample preparation. Disc and condyle were harvested from 3-month old porcine TMJs. 5 regions (anterior, posterior, central, medial and lateral) as thickness of 1mm, were punched out from disc and condyle, respectively (Fig. 1b, d).

Histology. Whole disc and condyle were harvested, fixed in 4% paraformaldehyde. Serial 6-μm-thick sagittal sections were cut, and stained for sulfated glycoaminoglycans (sGAGs) by Safranin-O/Fast Green.

AFM-nanoindentation was performed on the intact surfaces of both articular disc and condyle cartilage in PBS using a microspherical tip ($R \approx 5$ μm, $k \approx 0.03$ N/m, μMasch) and Dimension Icon AFM, following established procedures [7, 8].For the disc, the test was performed on both the superior and interior surfaces. On each surface, we tested five different anatomical regions: the anterior. For each disc and condyle, at least 10 locations were tested on the intact surface to account for spatial heterogeneity. The effective indentation modulus, E_{ind}, was calculated for each location via Hertz model .

Structural analysis. Samples were treated with hyaluronidase and trypsin for 24h to remove proteoglycans, fixed with Karnovsky's fixative for 3 hours in room temperature, then dehydrated in a series of graded ethanol and hexamethyldisilazane mixture to retain the 3D architecture [9]. The surface collagen fibrils were then visualized by scanning electron microscope (SEM, Zeiss Supra 50 VP).

Statistical test. One-way ANOVA followed by Tukey- Kramer post-hoc was used to examine the regional variations of E_{ind} and d_{col} on each of the disc and condyle surface. The significance level was set as $\alpha = 0.05$.

RESULTS

Safranin-O/Fast Green staining showed little proteoglycan in the disc (Fig. 1b) and the top layer of condyle cartilage, illustrating the fibrocartilage nature of these tissues. The secondary layer of the condyle cartilage is marked with appreciable sGAG staining, a characteristic of hyaline cartilage (Fig. 1d). Nanoindentation detected significant regional variations on all three tested surfaces. On the superior surface of the disc, posterior region showed significantly higher E_{ind} compared to all other four regions; on the inferior side, however, the highest modulus was detected at the lateral end. On condyle cartilage surface, the highest modulus, however, was detected at the anterior end (Fig. 2). Meanwhile, SEM also detected significant heterogeneity in the collagen fibril diameters on all three surfaces. On the superior side of the disc surface, no significant differences in the diameter were observed, while on the inferior side, the anterior end showed thinnest fibrils. On the condyle surface, the posterior end had thickest collagen fibril diameters compared to other regions (Fig. 3).

DISCUSSION

The mandibular condyle surfaces are much softer compared to the disc surfaces, which could be associated with the differences in their functions. The disc mainly sustains anteroposterior tensile stresses, while the condyle provides both compressive load bearing and shock absorption [10]. Here, the more compliant surface of the condyle cartilage can likely increase the disc-condyle congruency, and effectively reduce the contact stresses.

The salient micromechanical heterogeneity in E_{ind} highlights the complexity of both the articular disc and the mandibular cartilage. While the presence of heterogeneity is consistent with tissue-level studies, the overall trend appears to be different. On the superior side of the disc, the highest modulus at the posterior end is in agreement with tissue-level unconfined compression studies [11]. In contrast, on the inferior side, the lateral and medial ends have higher moduli, which are different from tissue-level results showing higher modulus at the anterior and posterior ends [11]. Similarly, on the mandibular cartilage surface, the fact that the anterior end has highest modulus is also different from observations at the tissue level, which had the highest modulus at the posterior end [12]. These variations can be attributed to the differences in tested length scales. In this study, AFM-nanoindentation measures the properties strictly reflecting the top ~ 10 μm surface of the superficial layer, while compression test includes the integrated properties of both the surface the interior.

On both disc and condyle cartilage surfaces, the region with highest modulus are not in line with the region with thickest collagen fibril diameters. It is possible that in addition to the fibril diameter, other structural factors, such as the fibril alignment and inter-fibrillar cross-linking are also important factors determining the modulus of the surface. Our ongoing studies are further examining the nanostructural features of these surfaces to reveal the structure-mechanics relationships.

In conclusion, this study is the first that defines the nanostructural and micromechanical heterogeneity of TMJ cartilage surfaces. Such knowledge will lay the ground for our ongoing studies on the elastic and viscoelastic micromechanics of the interior of both tissues, leading to a comprehensive understanding of the structure-mechanics relationship of the TMJ tissues at multiple scales.

ACKNOWLEDGEMENTS

This work was supported by the National Institutes of Health AR066824 (LH).

REFERENCES

[1] Buijs, GJ et al., *J. Oral Maxillofac. Surg.*, 67:782-7, 2009.
[2] Das, SK, *J. Oral Biol. Craniofac. Res.*, 3:109-10, 2013.
[3] Rando, C et al., *Am. J. Phys. Anthropol.*, 148:45-53, 2012.
[4] Bertram, S et al., *J. Am. Dent. Assoc.*, 132:753-61, 2001.
[5] LeResche, L, *Crit. Rev. Oral Biol. Med.*, 8:291-305, 1997.
[6] Amini, AR et al., *Crit. Rev. Biomed. Eng.*, 40:363-408, 2012.
[7] Batista, MA et al., *Matrix Biol*, 38:84-90, 2014.
[8] Li, Q et al., *J. Biomech.*, 48:1364-70, 2015.
[9] Bray, DF et al., *Microsc. Res. Tech.*, 26:489-495, 1993.
[10] MacBarb, RF et al., *J. Tissue Eng.*, 20:3290-302, 2014.
[11] Allen, KD et al., *Ann. Biomed. Eng.*, 33:951-62, 2005.
[12] Singh, M et al., *J. Biomech.*, 42:405-417, 2009.

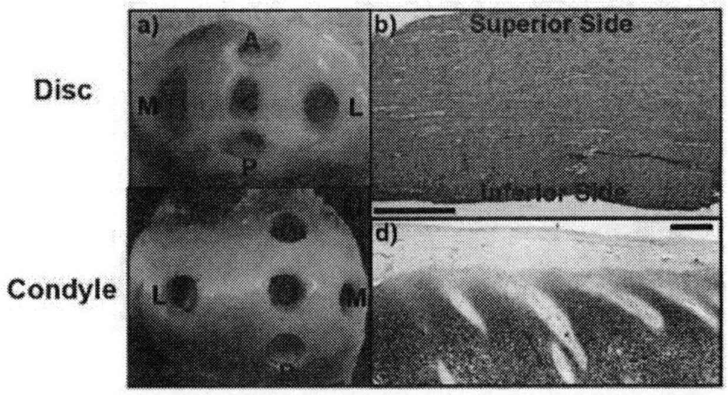

Figure 1: The representative transverse view of TMJ **a)** disc and **c)** condyle. Central sagittal section of TMJ **b)** disc and **d)** condyle stained with Safranin-O/Fast Green. Scale bar: 500 μm. A: anterior, P: posterior, C: central, M: medial, L: lateral,

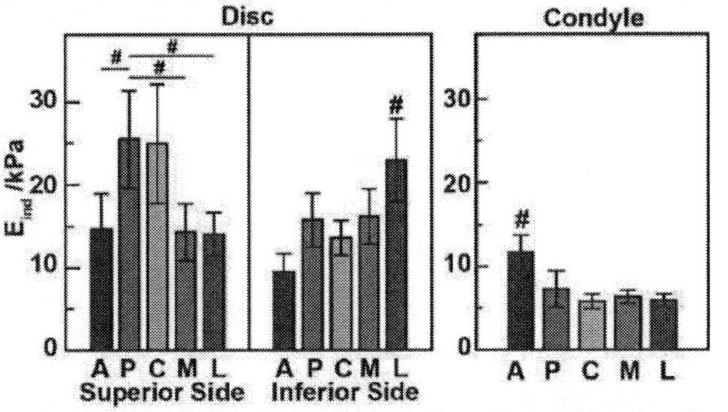

Figure 2: Nano-indentation modulus on disc and condyle surface (box plot, $n = 3$). A: anterior, P: posterior, C: central, M: medial, L: lateral. #: $p < 0.05$.

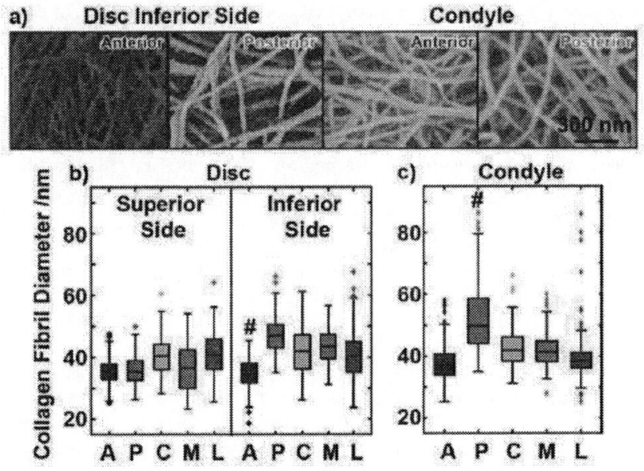

Figure 3: **a)** Nanostructure images, boxplot of TMJ **b)** disc and **c)** condyle surface collagen fibril diameter (≥ 158 fibrils from $n = 3$ animals). A: anterior, P: posterior, C: central, M: medial, L: lateral. #: $p < 0.05$.

MENISCAL ENTHESIS COLLAGEN FIBER ORIENTATION IS ALTERED WITH OSTEOARTHRITIS

Hannah M. Pauly (1), Tammy L. Haut Donahue (1,2)

(1) School of Biomedical Engineering
Colorado State University
Fort Collins, CO, USA

(2) Department of Mechanical Engineering
Colorado State University
Fort Collins, CO, USA

INTRODUCTION

The menisci are c-shaped fibrocartilaginous structures located in the knee joint and aid in stability and force distribution during mechanical loading [1]. Each meniscus is attached to the underlying tibial plateau via two ligamentous insertions: the anterior enthesis and the posterior enthesis. These entheses function to anchor the menisci firmly to the underlying bone and transmit tensile stresses applied to the menisci during loading. Integrity of the meniscal entheses is vital for proper joint function [2].

Meniscal entheses consist of four distinct tissue regions (Fig 1) [3]. The ligamentous (LI) region extends from the mainbody of the meniscus and transitions to a region of uncalcified fibrocartilage (UCFC). The UCFC consists of type I and II collagen fibers which are continuous across the tidemark where the region of calcified fibrocartilage (CFC) begins [4]. Finally the CFC interfaces with the underlying subchondral bone (SCB) at an interdigitated cement line. The orientation of collagen fibers throughout the meniscal enthesis is thought to be an adaptation to the applied loads [5].

Osteoarthritis (OA) is a debilitating joint disease that is commonly characterized by degradation of cartilage and other tissues in the knee. It is hypothesized that integrity of the meniscal entheses is important for preventing excessive transverse meniscal extrusion that could exacerbate cartilage damage [6]. Osteoarthritic meniscal entheses exhibit numerous morphological and mechanical changes including a breakdown of tidemark organization, micro-cracks, increased mineralization and increased mechanical compliance [7]. These changes could potentially contribute to meniscal extrusion and accelerate OA progression. Thus there is interest in determining whether structural changes to the collagen fibers also occur during osteoarthritis, as a potential indicator of further functional changes.

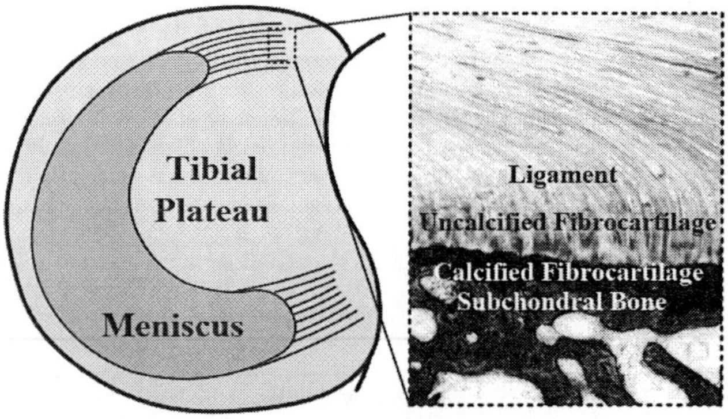

Figure 1: Meniscal enthesis structure.

METHODS

End-stage osteoarthritic tissue was obtained from three patients undergoing total knee arthroplasty. All meniscal entheses (medial anterior (MA), medial posterior (MP), lateral anterior (LA), and lateral posterior (LP)) were excised and sectioned along the main fiber axis. One half of each sectioned enthesis was fixed and embedded in methyl methacrylate. Histological sections (30 μm thick) were made using standard cutting and grinding techniques and sections were stained with a 0.1% Picrosirius red solution.

Polarized light analysis was used to determine the angles of collagen fibers present throughout the insertion. Images of stained sections were taken using a polarized microscope at 10° increments between 45° and 135°, with and without a compensator in place. After capture, all images were rotated such that the tidemark was horizontal.

A custom MATLAB graphical user interface was used to determine the collagen fiber extinction angle, and thus the orientation relative to the tidemark. The software was used to manually select the region of analysis, the LI, UCFC, CFC, or SCB region based on proximity to the tidemark and visual appearance. A grid of measurement points was overlaid throughout the selected region (Fig 2) and the mean collagen fiber angle and angular deviation, a measure of dispersion, was calculated. Student's t-tests were used to compare the mean angle and angular deviation of the anterior and posterior entheses of each menisci, within the designated regions. $P<0.10$ was considered to be statistically significant.

Figure 2: Example of grid of points overlaid on LI region (blue points) and CFC region (green points).

RESULTS

Structural analysis of the collagen fiber orientation relative to the tidemark demonstrated differences between the medial and lateral meniscal entheses. In the lateral meniscus the angle of collagen fiber insertion tended to increase when moving from the LI region of the enthesis to the SCB (Fig 3). In contrast in the medial meniscus collagen fiber angle remained relatively constant throughout all enthesis regions.

When the anterior and posterior entheses of medial and lateral menisci are compared the only significant differences were observed in the CFC of the lateral meniscus. The collagen fibers inserted at significantly smaller (shallower) angles in the LA enthesis compared to the LP enthesis.

The fiber deviation in the lateral meniscal entheses tended to be slightly higher compared to the medial meniscal entheses, suggesting more disperse collagen fiber insertion angles in the lateral meniscal entheses (Fig 4). The fiber angles in the CFC of the LA enthesis were significantly less dispersed than the fiber angles in the LP enthesis.

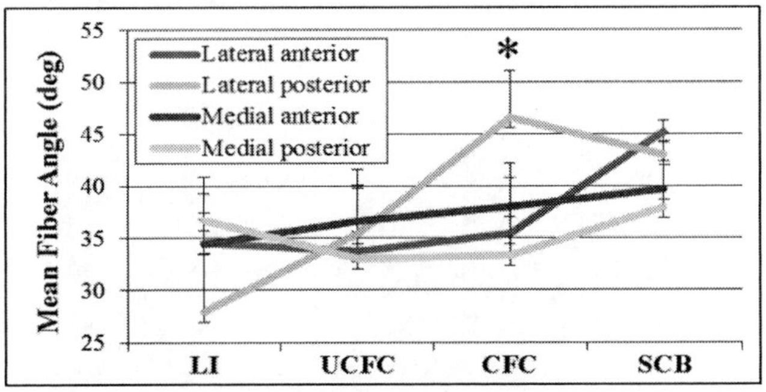

Figure 3: Meniscal enthesis mean fiber angles. * Indicates significant difference between lateral anterior and lateral posterior entheses.

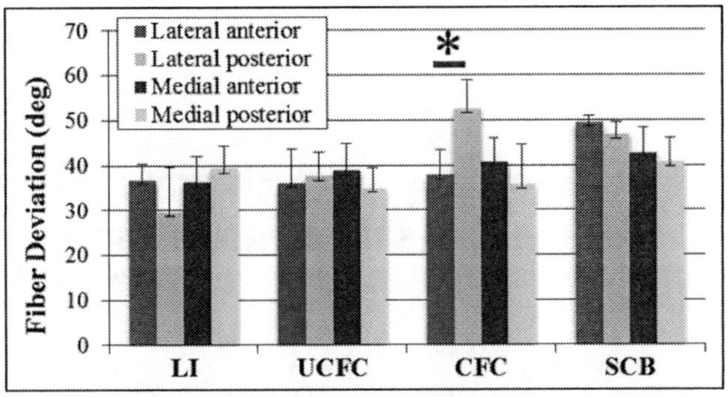

Figure 4: Meniscal enthesis fiber deviations. * Indicates significant difference between lateral anterior and lateral posterior entheses.

DISCUSSION

Assessment of collagen fiber angles throughout osteoarthritic meniscal entheses revealed that the CFC collagen fiber angle and fiber angle deviation were lower in the LA enthesis compared to the LP enthesis. This difference may be based on enthesis mobility, as the anterior horns tend to undergo more displacement [10].

The mean fiber angles and angular deviations obtained from osteoarthritic tissue in this study can be compared to healthy tissue, as assessed by Abraham et al [9]. The collagen fiber insertion angles were generally ~10-15° lower in osteoarthritic tissue, indicating overall shallower insertion angles compared to healthy tissue. These shallower insertion angles could permit more excessive translation of the menisci during knee loading which could contribute cartilage damage. Additionally in healthy tissue the collagen fibers of the MA enthesis inserted into the bone at shallower angles than in the MP enthesis. In osteoarthritic tissue this difference was absent, suggesting a "normalization" of collagen fiber angles in the medial meniscal entheses. Changes in collagen fiber structure between healthy and osteoarthritic tissue could indicate changes in functionality.

In osteoarthritic entheses the average fiber deviation was ~20° higher than in healthy entheses. The increased fiber dispersion in diseased tissue could suggest that a breakdown of collagen fibers is occurring, which could also explain the increased mechanical compliance observed in the osteoarthritic entheses [9].

One limitation to this study is that because the tissue was obtained from patients with end stage OA it is impossible to elucidate the timeline for the progression of structural changes observed. Overall the results of this study suggest that significant structural changes occur in the collagen fibers of osteoarthritic meniscal entheses.

ACKNOWLEDGEMENTS

The authors would like to thank Dr. Adam Abraham for his assistance with acquiring the images.

REFERENCES

[1] Messner, K et al., J Anat, 193:161-78, 1998.
[2] Haut Donahue, TL et al., J Biomech, 36:19-34, 2003.
[3] Benjamin, M et al., J Anat. 177:127-34, 1991.
[4] Gao, J et al., J Anat. 185:663-7, 1994.
[5] Abraham, A et al., Acta Biomat, 9:6322-9, 2012.
[6] Jones, A et al., Australas Radiol, 50:306-13, 2006.
[7] Abraham, A et al., Osteoarthr Cartilage, 22:275-283, 2014.
[10] Yao, J et al., J Orthop Res, 26:673-84, 2008.

RULE-BASED APPROACH FOR ASSIGNMENT OF MYOFIBER DISTRIBUTION TO HUMAN TONGUE MODELS

Arnold David Gomez (1), Nahla Elsaid (2), Jiachen Zhuo (2), Maureen L. Stone (3), Jerry L. Prince (1)

(1) Electrical and Computer Engineering Dept.
Johns Hopkins University
Baltimore, MD, USA

(2) Dept. of Diagnostic Radiology and Nuclear Medicine
University of Maryland School of Medicine
Baltimore, MD, USA

(3) Department of Neural and Pain Sciences
University of Maryland School of Dentistry
Baltimore, MD, USA

INTRODUCTION

The overall quality of human life is directly dependent on (and affected by) processes involving motion of the tongue, including production of sounds for speech, mastication of food, and breathing. For this reason, oral cancer has deep negative consequences on individuals despite not having a high mortality rate [1-2]. Because the incidence of this type of cancer is growing, there is an increasing need for effective therapeutic techniques, not only in terms of tumor removal, but also in preserving tongue function afterwards [2]. Mechanical modeling offers insights that can aid in the design of surgical techniques. For example: simulations enable relatively rapid assessment of functional changes due to changes in geometry or stiffness that can result from postsurgical tissue remodeling. As a rule, the input to computer simulations must faithfully represent the system under investigation in order to produce accurate and reliable results. This is especially true for the tongue's myofiber orientation, because it is intimately related to muscle contraction. Structural assessment is often achieved via diffusion-tensor MRI (DT-MRI) [3], but this is often insufficient to extract fiber directions (due to noise), or is unavailable altogether. While manual approximation is possible, it can introduce user variability and can be increasingly difficult in intricate geometries.

This study explores the creation of synthetic fiber directionality information using the rule-based algorithm, which has previously been used to construct cardiac models [4]. The main tenet of the approach is to apply a priori anatomical knowledge to the boundary of a subject-specific domain and resolve the interior via Laplace's equation. In the heart, the strategy consists of applying conditions to the endo- and epicardium, where some aspects of fiber directionality are generally known. While the myofiber distribution in the tongue is complex (including crossing fiber families), it follows a distinctive pattern [5]. Thus, it is possible to describe the overall pattern from subject-specific anatomy to generate an approximation of fiber distribution. In particular, we hypothesized that mechanical simulations with approximate fiber orientations (via the aforementioned approach) would be functionally similar to those obtained using experimental measurements from high-resolution DT-MRI.

METHODS

The experiments herein focus on the genioglossus (GG), which is the largest tongue muscle by volume [6]. Therefore, it has a large influence in the organ's motion, as well as a well-defined diffusion signal, which enables its structural characterization in vivo [6-3]. To test the hypothesis, finite-element (FE) biomechanical simulations (Fig. 1) of GG contraction with either approximated, or experimentally measured, fiber distributions were compared against each other.

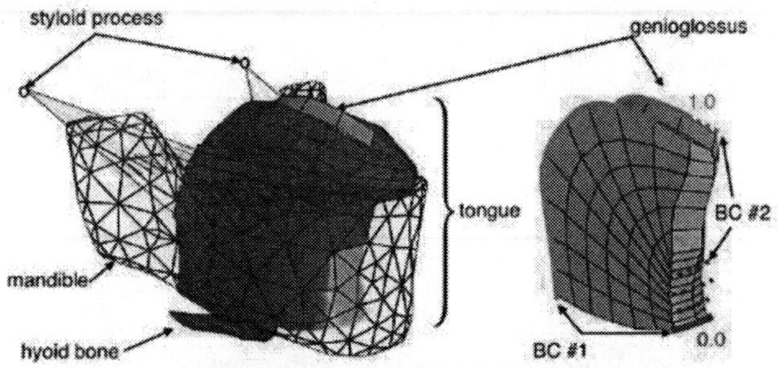

Figure 1: Tongue model. The FE mesh (left) was used to simulate GG contraction. Fiber distribution was obtained via DT-MRI, or via Laplace's Equation using the boundary conditions (BC, right).

The model consisted of the hyoid and mandibular bones, which were considered rigid and constrained to be stationary, and the tongue geometry, which was based on a published shape [7]. The material constitutive model was assumed to be an isotropic neo-Hookean solid with coefficients based on literature values [8].

The model was matched (via manual affine registration) to the imaging domain. DT-MRI of a healthy adult male was performed to extract fiber directions using 64 diffusion directions (2 averages), 80x80x80 matrix size, 3 mm isotropic voxel size, and a b-value 500 s/mm^2. Diffusion tensor reconstruction was obtained using DSI Studio [9]. Fiber information was extracted from the first eigenvalue of the diffusion tensor via nearest-neighbor interpolation.

While simulations with imaging-derived fibers used as the benchmark, two approximated fiber distributions were constructed by solving Laplace's equation. These differed in their boundary conditions (BC), which consisted of setting nodal values to 1.0 or 0.0. The first consisted of simply placing known values at the inferior-anterior corner of the GG (Fig. 1, right) and along the longitudinal surface. In the second, anterior values were extended to the entre insertion to the mandibular bone, and discarded towards the anterior tip of the tongue. These changes appear in Fig. 1 with dotted lines. Comparisons were quantified via displacement percent error.

Contractile stress in the GG was set to 25% of its maximum capacity of 0.35 Mpa [8]. An additional simulation of 25% activation in the styloglossus (SG) muscle was performed to evaluate error when GG motion was passively generated. Biomechanical simulations were carried out using FEBIO Software Suite [10].

RESULTS

The solutions to Laplace's equation (potential distributions) for the two BC sets appear in Figure 2. As described, approximate fiber distributions relate to these solutions via gradients.

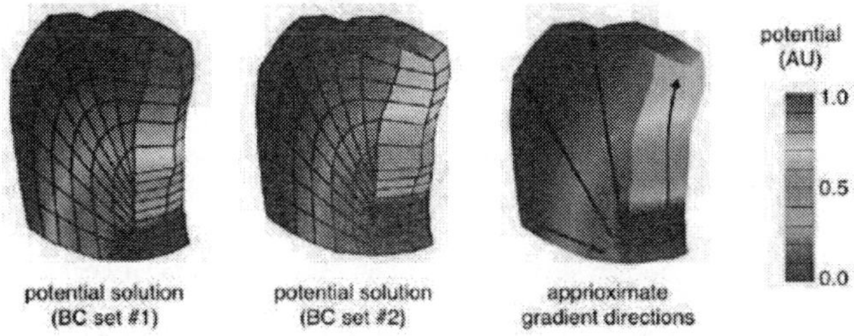

Figure 2: Differences between boundary conditions sets. The potential distributions (in arbitrary units, AU) are equivalent to the solution of Laplace's equation with the different BC sets.

Despite the similarities in their potential distributions, the resulting fiber orientations are noticeably different (Figure 3). Fibers from BC set #2 are visually similar to DT-MRI results across a greater area than BC set #1, but some discrepancies are noticeable on the superior (top) portion. Some noise is present on DT-MRI results.

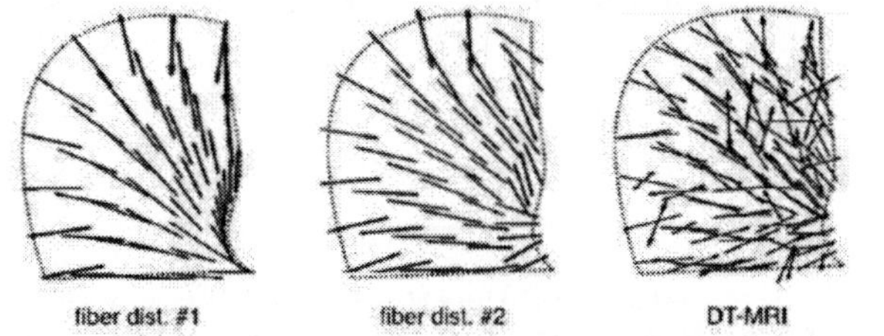

Figure 3: Fiber distributions. Approximate fiber distributions (left, center) are distinct and less noisy compared to DT-MRI.

Error maps between simulated GG activation with either of the approximations, against imaging-based results are shown in Figure 4. Using fiber distribution #1 (Fig. 3) results in relatively large displacement error particularly above the mandible insertion of the GG. The mean error across the tongue was 49%, with an SEM of 2%. In contrast, application of fiber distribution #2 results in a visibly reduced error, although error was still concentrated in the same area. The mean error across the tongue in this case was 28%, with an SEM of 3%. Displacement percent error during SG contraction was 8.8×10^{-3} using fiber distribution #1 and 1.3×10^{-3} using distribution #2.

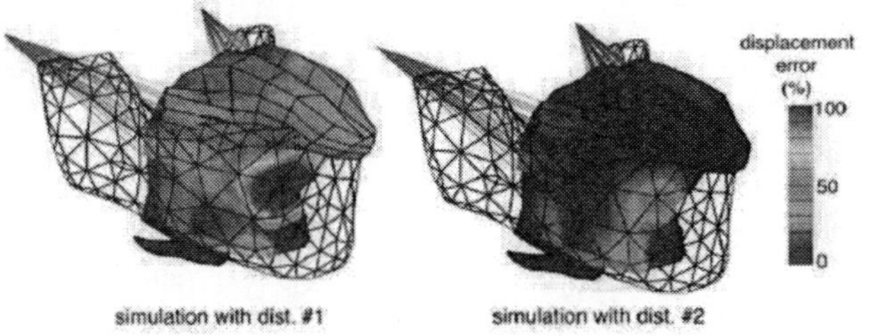

Figure 4: Differences between simulated GG contractions. The color maps shows the displacement error from each of the simulations with approximate fiber distributions (Fig. 3) with respect to the benchmark (DT-MRI solution). Using fiber distribution #1 (left), results in larger error than #2 (right).

DISCUSSION

The magnitude and dispersion of error in Fig. 4 suggests that simulated active contraction is very sensitive to fiber distribution. This result highlights the necessity for adequate modeling of fibers. Figure 4 also shows that if the approximated distribution is too simplistic (as with BC set #1), simulated results can greatly disagree with what would be expected when using experimental data. By incorporating more anatomical details in the boundary conditions of the Laplace's equation solution (as done in BC set #2), it is possible to improve the quality of the estimated fiber distribution. This improves the overall agreement with the experimental data (Fig. 3), and reduces simulation discrepancy (Fig. 4).

Note that DT-MRI data includes some noise. Thus, complete agreement with experimental data may not be necessarily better, and an approximate fiber distribution may be used in conjunction with (rather than in lieu of) experimental measurements. As expected, there was little error associated with passive displacement of the GG via SG contraction, because the material was isotropic.

ACKNOWLEDGEMENTS

The authors would like to acknowledge grants R01DC014717 2R01NS055951 from the National Institutes of Health in the United States.

REFERENCES

[1] Edge SB, et al., *Ann Surg Oncol*, 17(6):1471-4, 2010.
[2] Patel SG, et al., *CA Cancer J Clin.* 55(4):242-58, 2005.
[3] Felton SM, et al., *J Biomech*, 41(8):1782-9, 2008.
[4] Bayer JD, et al., *Ann Biomed Eng*, 40(10):2243-54, 2012.
[5] Takemoto H, *J Speech Lang Hear Res.* 44(1):95-107, 2001.
[6] Stone ML, et al., *CMBBE J.* accepted Apr. 8, 2016
[7] Fujita S, et al., *Oral Sci. Int.* 4(2):97-109, 2007 (via artisynth.org)
[8] Teran JE, et al., *IEEE Trans. Vis. and Comp. Graphics*, 11(3):317-328, 2005.
[9] http://dsi-studio.labsolver.org
[10] Maas, G et al., *J Biomech*, 34:859-871, 2001.

SB^{3}C2017
Summer Biomechanics, Bioengineering and Biotransport Conference
June 21 – 24, Tucson, AZ, USA

MACHINE LEARNING FOR ESTIMATION OF ACTIVATION PATTERNS IN COMPUTATIONAL MODELS OF THE TONGUE

Arnold David Gomez (1), Amod Jog (1), Maureen L. Stone (2), Jerry L. Prince (1)

(1) Electrical and Computer Engineering Dept.
Johns Hopkins University
Baltimore, MD, USA

(2) Department of Neural and Pain Sciences
University of Maryland School of Dentistry
Baltimore, MD, USA

INTRODUCTION

Tissue motion determines many vital processes such as blood pressurization in the heart, locomotion with the limbs, and speech generation with the aid of the tongue. For this reason, the relationship between muscular activation, force generation, and motion is central to biomechanics and biomedicine. The tongue offers a unique example of the complexity of this relationship, because its muscles can activate independently to produce the majority of deformations without direct reliance on joints or other rigid support structures [1]. Despite its complexity, our society has a need for investigating tongue biomechanics given the increasing rates of oral cancer, and the profound detrimental effect that it can inflict [2-3]. An association between activation patterns and different activities can guide the development of surgical tumor removal techniques aimed at reducing functional impact. Thus, a robust method for measuring, or estimating, muscular activation across the tongue's volume would be beneficial.

While non-invasive motion estimation has been successful in different organs, it is currently impossible to perform direct experimental measurements of activation patterns [4-5]. Instead, these can be inferred by optimizing the (activation) input of a computational model so that the simulated output matches measured deformation. This approach, known as inverse modeling, has been demonstrated in the heart and the tongue using finite-element (FE) models [4-5]. Implementation of numerical optimization is challenging because it must include an effective objective function, be numerically stable, and be robust to local minima. For this reason, different strategies make use of regularization. For instance, a common approach consists of grouping contractile elements together and finding each group's activation magnitude [4-5]. However, such grouping, a form of regularization may not be optimal based on observations of the system. This research investigates the use of machine learning without such regularization. Despite being digital in nature, computational models effectively approximate a continuum; thus, the number of possible configurations resulting from all activation patterns is extremely large. One of the main challenges is, then, designing an appropriate training dataset, which is the focus of this research. The goal of this study was to perform inverse estimation via machine learning with a limited training data set and to test the validity of the predictions.

METHODS

The experiments were carried out with an FE model because it can provide training data and test cases (with known activation patterns) outside the training dataset. The strategy (Fig. 1) uses a random forest algorithm to obtain activation predictions and compares them to a known benchmark [6].

The FE model used in this investigation included a representation of the tongue [7], which was constrained using rigid models of the hyoid bone and the jaw. The material was assumed to be isotropic

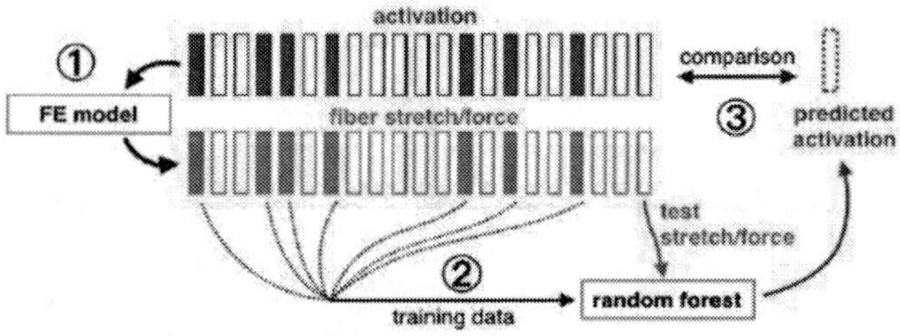

Figure 1: Test strategy. The FE model (1) was used to generate paired training data (2) for random forest regression, which generated predictions (3) to be compared to the known activations.

(neo-Hookean) with coefficients extracted from the literature [8]. To reduce the solution space, the geometry was grouped into the approximate location of 6 tongue muscles (Fig. 2). The training data set consisted of randomly sampling half of the permutations of muscular activation at 4 discrete levels of activation. The largest was 50% (of a maximum contractile stress of 0.35 Mpa [8]) in decrements of 12.5%. The total number of training simulations was 2,048 obtained over a 6-hr period using the FEBio Software suite [9].

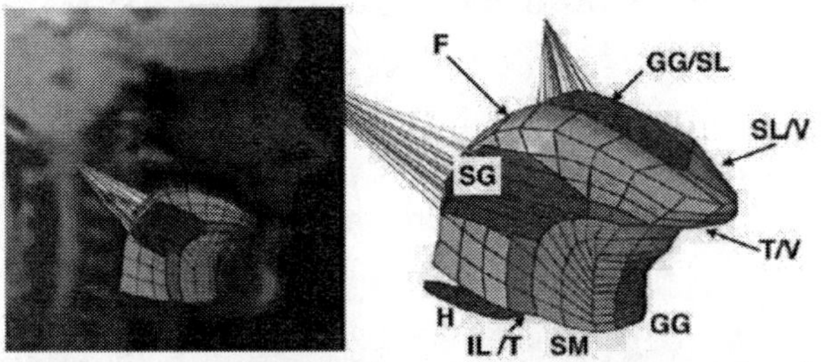

Figure 2: FE model of the tongue. Different colors represent regions shared by different muscles, which included: genioglossus (GG), superior longitudinal (SL), inferior longitudinal (IL), transverse (T), verticalis (V), and styloglossus (SG). The hyoid bone (H), mandible (not shown), and support material (SM) were also included. A flap region (F) was a test modification.

Estimation of activation was framed as a regression problem, but element interactions make it highly nonlinear. The random forest algorithm [6] has been used successfully as a nonlinear regression learning framework in many applications [10]. Given a test simulation k with known fiber strains λ_{ik} and fiber forces f_{ik} (i.e. the strain and stress tensors along the fiber directions, respectively), the trained random forest R_i predicted the activation $\hat{a}_{ik}$ for a given element i ($i = 1, 2, \dots, n$). Borrowing machine learning notation, the set of activations in all elements is $\hat{\boldsymbol{a}}_k$. To train R_i, a feature vector $\boldsymbol{b}_j$ was calculated from each training simulation j ($j = 1, 2, \dots, m$), where $\boldsymbol{b}_j = [\boldsymbol{\lambda}_j \quad \boldsymbol{f}_j]$, such that $\boldsymbol{\lambda}_j \{\lambda_{1j}, \lambda_{1j}, \dots, \lambda_{nj}\}$ was concatenation of fiber strains at all elements for a training simulation j, and $\boldsymbol{f}_j \{f_{1j}, f_{1j}, \dots, f_{nj}\}$ was a concatenation of fiber forces at all n elements. The feature vector $\boldsymbol{b}_j$ was paired with activation of a_{ij} to create a training dataset $\langle \boldsymbol{b}_j, a_{ij} \rangle$ for all m training simulations. A random forest R_i was trained on $\langle \boldsymbol{b}_j, a_{ij} \rangle$. The number of trees in R_i was chosen empirically to be 30. Each tree was trained as described in previous work [6]. The minimum number of samples to be maintained in the leaf of a tree was empirically determined to be one. Given the fiber strains and fiber forces observations from a test case k, a feature vector $\boldsymbol{b}_k = [\boldsymbol{\lambda}_k \quad \boldsymbol{f}_k]$ was generated, and $\hat{a}_{ik}$ resulted from applying the trained random forest to the new feature vector, $R_i(\boldsymbol{b}_k)$.

Test data was constructed using the same muscular groupings, but with random 100 magnitude permutations in the (continuous) range from 0–50%. An additional case used one of these permutations, but with a modification in the spatial grouping: the flap region (labeled F in Fig. 2) was given zero activation. The difference between the approximated activation and the benchmark (for each k) was quantified as a percentage from $\|\boldsymbol{a}_k = \hat{\boldsymbol{a}}_k\| / \|\boldsymbol{a}_k\|$.

RESULTS

Overall, the random forest algorithm provided accurate predictions of activations (pattern and magnitude) across all the magnitude test cases. The mean error across these cases was 10.3% with a standard deviation of 5.8%. Figure 3 (top row) shows at typical solution from these test cases—few differences can be seen visually.

Introducing the flap modification to the mesh (Figure 3, bottom row), affected the prediction by lowering the activation magnitude across the entire muscle region (in this case the SL/V grouping) instead of revealing the expected low-magnitude region. The error in this case was 91.1%.

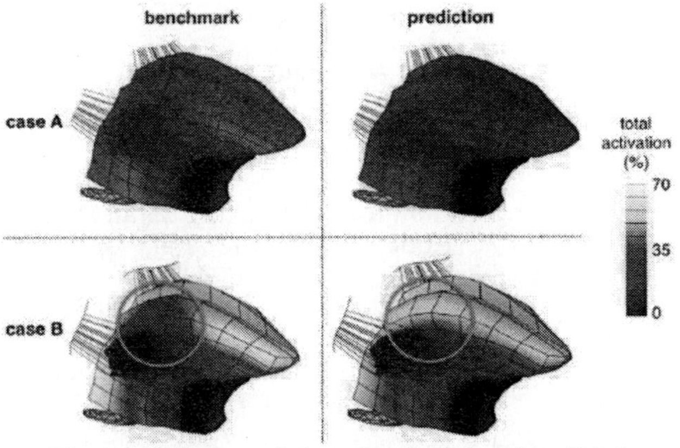

Figure 3: Differences between benchmark and predicted activation patterns. Case A (top row) corresponds to one of the tests simulations differing in magnitude only. Case B (bottom row) resulted from modifying the magnitude and spatial configuration. Both of these were outside the training dataset.

DISCUSSION

Despite using a small training dataset, the algorithm was successful in predicting activation patterns when they differed in magnitude. This can be attributed to the relative smoothness in variation between configurations of varying magnitude. The same was not true when introducing a change in spatial configuration, where the prediction fell back to a type of "best fit" within the spatial groups already in the training dataset. This is a direct consequence of the training approach, which implicitly implemented spatial regularization within the algorithm.

These findings are encouraging because a large number of magnitude and pattern variations can be captured by the regression as is. However, the results also highlight the need for training sets that encompass both magnitude, and spatial variations. Therefore, increasing the size of the training database and improving sampling of the solution space are required to achieve better results.

ACKNOWLEDGEMENTS

The authors would like to acknowledge grants R01DC014717 and 2R01NS055951 from the National Institutes of Health in the United States.

REFERENCES

[1] Takemoto H, *J Speech Lang Hear Res.* 44(1):95-107, 2001.
[2] Edge SB, et al., *Ann Surg Oncol,* 17(6):1471-4, 2010.
[3] Patel SG, et al., *CA Cancer J Clin.* 55(4):242-58, 2005.
[4] Sun K, et al. *J Biomech Eng.*131(11):111001, 2009.
[5] Stavness, et a. *J Biomech.* 15;45(16):2841-8, 2012.
[6] Breiman L, *Machine Learning*, 24(2):123–140, 1996
[7] Fujita S, et al., *Oral Sci. Int.* 4(2):97-109, 2007 (via artisynth.org)
[8] Teran JE, et al., *IEEE Trans. Vis. and Comp. Graphics*, 11(3):317-328, 2005.
[9] Maas, G et al., *J Biomech*, 34:859-871, 2001.
[10] Breiman L, et al. *Classification and Regression Trees*,1984.

BIOMECHANICAL CHARACTERIZATIONS OF THE PORCINE OPTIC NERVE

Sammira Rais-Rohani (2), Sarah Fitzgerald (2), Bryn Brazile (2), Richard L. Summers (3), Robert L. Hester (3), Raj Prabhu (2), Lakiesha N. Williams (2), and Jun Liao (1,2)

(1) Department of Bioengineering, University of Texas at Arlington, Arlington, TX 76019, USA

(2) Department of Biological Engineering, Mississippi State University, Mississippi State, MS 39762, USA

(3) Department of Physiology and Biophysics, University of Mississippi Medical Center, Jackson, MS 39216, USA

INTRODUCTION

In space, astronauts experience several physical abnormalities due to microgravity. Some of these include intraorbital and intracranial pressure changes similar to the effects of idiopathic intracranial hypertension [1]. When exposed to microgravity, many astronauts experience an enlarging of the optic nerve, resulting in the buckling of the optic nerve sheath [2]. These abnormalities may result in blurred vision, which can ultimately lead to permanent blindness. It is believed that the optic nerve may be an integral part of both the cause and the solution to preventing or treating this affliction. Although there have been several studies concerning the correlation between intracranial pressure and the optic nerve, to the best of our knowledge there is no study that investigates the biomechanical properties of the optical nerve itself.

METHODS

Tissue Preparation: Porcine heads (~6-month old) were obtained from a local abattoir, and the optic nerves were dissected. The optic nerves were trimmed of all excess material leaving the sheath intact and placed in phosphate buffered saline (PBS). Biomechanical Testing: Uniaxial tensile tests were performed on the optic nerves by mounting the nerves onto the tissue clamps of a Test Resource uniaxial machine in a PBS bath to obtain the tissue stress relaxation, creep, and failure behavior (N=4).

RESULTS

The stress relaxation test shows that the optic nerve undergoes 57.70 ± 7.46% decrease in stress over a period of 15 minutes (Figure 1A). The creep is found to be negligible in the optic nerve surrounded by the sheath, with a strain increase of 2.50 ± 1.43% over a period of 15 minutes (Figure 1B). Figure 1C shows the average stress-strain curve up to tissue failure. The stress-strain curve has a very short toe region and transitional region, and then reaches a long linear region, followed by yield and tissue failure.

DISCUSSION

Similar to other soft tissues, the sheath-wrapped optical nerve demonstrates large stress relaxation. The tissue complex is able to release a great amount of stress when pulled to a constant strain, which likely provides a protective mechanism for cells and tissues in the optical nerve. It is also found that the sheath-wrapped optical nerve exhibits negligible creep, which means if there is a constant tensile load applied, the sheath-wrapped optical nerve can resist further deformation that might be harmful to the tissues and cells. For the sheath-wrapped optical nerve, the failure stress is ~2.4 MPa and the failure strain is ~17.0%. Both the failure stress and strain appear to provide sufficient protection in an environment insulated by facial bone, ocular muscles, fat, and other extracellular matrix tissues. This study established a strong basis for optical nerve biomechanics. Future work will include histology, bending and buckling tests, and computational simulation of the effects of abnormal pressures.

ACKNOWLEDGEMENTS

This study is supported by CRIS #MIS-351070 and MSU BCoE REU.

REFERENCES

[1] Kramer, Larry A., MD, and Douglas R. Hamilton, MD, PhD. Radiology 263.3 (2012): 819-827.
[2] Soldatos, T., and A. Gouliamos. ''Attikon'' Hospital,Chaidari, Athens, Greece (2008).

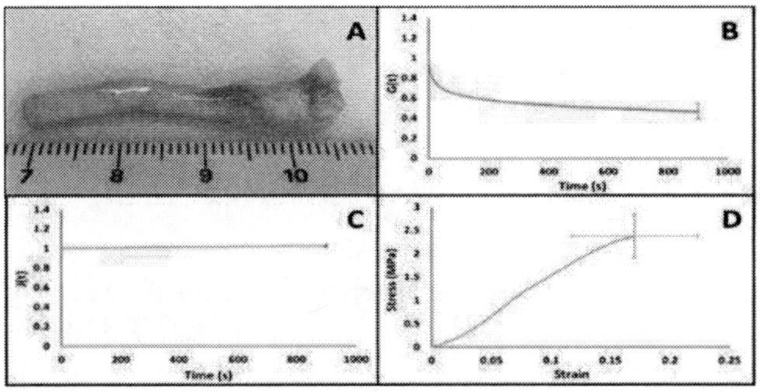

Figure 1: (A) Optic Nerve after Dissection. (B) Stress relaxation behavior - Samples are pulled to 600g and held at the final strain for 15 minutes while monitoring the stress decay. (C) Creep behavior - Samples are pulled to 600g, and the 600g load is maintained for 15 minutes while monitoring the strain changes. (D) Failure behavior - Graph represents the average stress-strain curves up to tissue failure.

SB³C2017
Summer Biomechanics, Bioengineering and Biotransport Conference
June 21 – 24, Tucson, AZ, USA

A NEW GEODESICS MODEL OF COLLAGEN FIBERS IN THE GLOBE; BETTER THAN A REINFORCED SPHERE

Ian A. Sigal (1, 2), Yi Hua (1), Ning-Jiun Jan (1, 2), Andrew P. Voorhees (1)

(1) Department of Ophthalmology
University of Pittsburgh
Pittsburgh, PA, USA

(2) Department of Bioengineering
University of Pittsburgh
Pittsburgh, PA, USA

INTRODUCTION

The fiber architecture of the posterior pole of the eye plays a critical role in globe biomechanics [1]. A central concern in ocular biomechanics is that the posterior pole of the eye has a large opening, the scleral canal, through which pass the nerve fibers that transmit information from the light-sensitive retina to the brain [2, 3]. This opening is often described as a "weak spot" in the globe, suggesting that it can be problematic for the eye as it could lead to concentrations of strain and stress. As an explanation for how the eye could have adapted, several studies have proposed that the scleral canal is surrounded by a circumferential ring of collagen fibers. Thus, the theory is that this ring protects the neural tissues within the lamina cribrosa in the canal from biomechanical insult caused by increases in intraocular pressure [1-4]. The biomechanical models were inspired and are now supported by evidence from imaging studies that have observed what appear to be collagen fibers forming the ring surrounding the scleral canal [5-7]. Thus, the general consensus in the ophthalmology community is that there is a ring of collagen fibers surrounding the scleral canal, and that these exist to provide biomechanical support to the neural tissues within the canal [1-7].

The models proposing the ring of collagen fibers, however, do not explain other aspects of the architecture that have been observed in more recent experiments, such as fibers that are radial from the canal, or the intertwining of fiber bundles further from the canal [5,8]. The ring models also do not explain experimental results showing that increases in IOP can sometimes cause contraction of the scleral canal [9]. Finally, the ring models also typically focus on pressure-induced stretch in the lamina cribrosa, ignoring other mechanical insults to the lamina cribrosa, or to other tissues [1,2].

We propose a new model of posterior pole architecture in which fibers follow near-geodesic curves. In this study we compared the fiber distributions and posterior pole biomechanics arising from our model with previous biomechanical models and experimental data from ours and other laboratories.

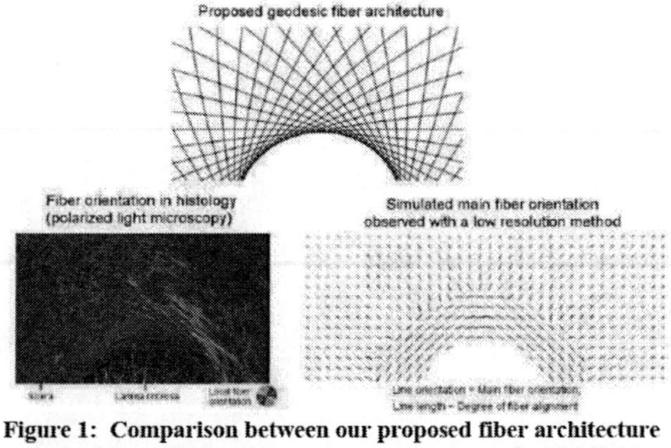

Figure 1: Comparison between our proposed fiber architecture (top) with experimental measurements using polarized light microscopy (bottom left), and a simulated SALS experiment (bottom right). Note how the low resolution of SALS causes the appearance of a ring of highly aligned fibers around the canal, surrounded by a region of low alignment.

METHODS

We constructed virtual posterior poles with our proposed architecture and of other common models (Fig 1). We simulated how these poles would appear in experimental studies done using standard techniques used in ophthalmology to measure collagen fiber orientation, such as small angle light scattering (SALS) and wide angle X-ray scattering (WAXS). We compared the geodesic model with fiber orientations measured in high resolution using our own polarized light microscopy.

We also constructed finite element models of a simplified, generic posterior pole with the sclera collagen fibers organized in five different ways: isotropic (transversely isotropic in the sclera plane), circumferential (the standard ring-reinforcement model), both circumferential and radial fibers (representing recent experimental evidence), radial fibers (representing both a variation of the ring model and the recent experimental evidence), and our geodesic model. The canal tissues were modeled as transversely isotropic in the sclera plane. We then explored the effects of small deviations of the fibers from the geodesic paths.

RESULTS

There was good correspondence between our proposed architecture and experimental observations obtained from SALS, WAXS and PLM (Fig 1). Our geodesic architecture resulted in much lower mechanical insult than all other models (Figs 2 and 3) and reduced concentrations of stress (not shown). Small deviations from the geodesics produced various amounts of canal expansion and even canal contraction

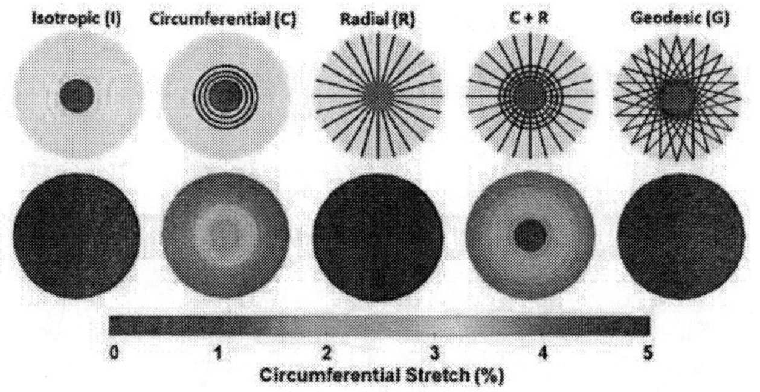

Figure 2: Five models of fiber architecture and the circumferential stretch induced in each by a simulated increase of 50mmHg IOP. Shown in light blue is the posterior sclera, with the scleral canal as a red disc. The black lines represent the collagen fibers in the model. Our model of geodesic fibers (rightmost column) greatly reduced stretch in both the lamina cribrosa and sclera.

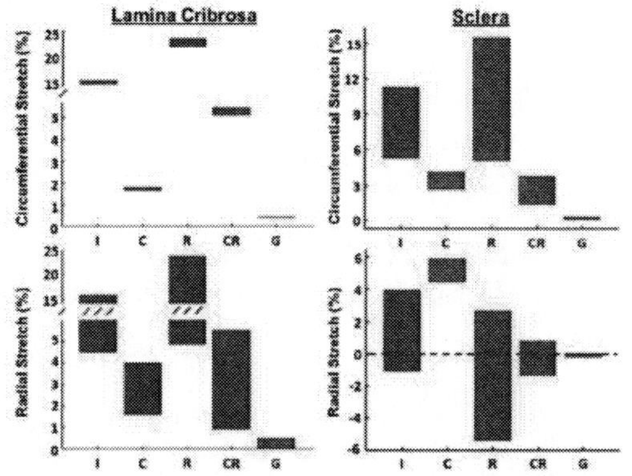

Figure 3: Filled box-plots illustrating the range of circumferential and radial stretch on the five models shown in Figure 2. Our model of geodesic fibers (G) greatly reduced the both types of stretch in both the lamina cribrosa and sclera.

DISCUSSION

The geodesic fiber architecture model we propose was more biomechanically favorable than previous models, reducing biomechanical insult to the canal tissues. In addition, the model also reduced deformations within the sclera. Our model was consistent with experiments, explaining the appearance of a circumferential ring of fibers around the canal in low resolution imaging, as well as the radial and intertwined fibers observed in the more distal sclera at higher resolution. The model architecture is highly adaptable, allowing eye-specific responses that include scleral canal expansion or contraction with relatively small fiber deviations from the geodesic paths.

ACKNOWLEDGEMENTS

NIH R01-EY023966, R01-EY025011, P30-EY008098 and T32-EY017271

REFERENCES

[1] Zhang, L et al., *Investigative Ophthalmology and Visual Science*, 56(3):2031-2042, 2015.
[2] Campbell, IC et al., *Journal of Biomechanical Engineering*, 136(2):021005, 2014.
[3] Sigal, IA et al., *Experimental Eye Research*, 88(4):799-807, 2009.
[4] Grytz, R et al., *Biomechanics and Modeling in Mechanobiology*, 10 (3): 371-382, 2011.
[5] Pijanka, JK et al., *Plos ONE*, 10(2), e0118648, 2015.
[6] Jones, JH et al., *Journal of the Royal Society Interface*, 12(106):20150066, 2015.
[7] Jan, NJ et al., *Biomedical Optics Express*, 6(12), 4705-4718, 2015
[8] Sigal, IA et al., *Investigative Ophthalmology and Visual Science*, 55(13):4257-4257, 2014.
[9] Yang, H et al., *Investigative Ophthalmology and Visual Science*, 50(12):5785-5799, 2009.

ANISOTROPIC AND HETEROGENEOUS FINITE ELEMENT MODELS OF THE HUMAN LAMINA CRIBROSA USING NONLINEAR OPTICAL MICROSCOPY

Reza Behkam (1), Jonathan P. Vande Geest (1,2,3)

(1) Department of Bioengineering, University of Pittsburgh, 300 Technology Drive, Pittsburgh PA 15219

(2) Louis J. Fox Center for Vision Restoration, University of Pittsburgh, Pittsburgh, PA 15219

(3) McGowan Institute for Regenerative Medicine, University of Pittsburgh, Pittsburgh, PA 15219

INTRODUCTION

Visual information is transmitted from the eye to the brain by retinal ganglion cells (RGCs) which are located far from the photoreceptor layer on the surface of the retina and close to the center of the posterior pole. The lamina cribrosa (LC) is a mesh like structure which is located in posterior sclera and is formed by a multilayered network of collagen that attaches to scleral canal wall. The optic nerve LC forms a barrier between the intraocular space (with intraocular pressure) and the retrobulbar space (with retrolaminar pressure). Current literature has shown that the LC is the first site of RGC damage in primary open angle glaucoma [1].

A rigorous understanding of the mechanical behavior of the LC is crucial to better understand the mechanisms of retinal ganglion cell axon damage. Due to several difficulties including inaccessibility and microstructural complexity, there are relatively few studies on the human LC.

Towards this end our research team has devoted significant effort into using nonlinear optical microscopy techniques to generate intraocular pressure (IOP) dependent in-vitro strain maps of the human LC from fresh human donor eyes. This imaging approach utilizes the second harmonic signal of collagen (SHG) as quantified using two photon microscopy. An important limitation of our approach is the lack of signal collected in the pores of the LC that do not contain fibrillar collagenous signal.

The purpose of this abstract is to develop finite element simulations of the LC using geometry from SHG imaging datasets and to work towards identifying the constitutive mechanical properties of the human LC. Our focus will be on developing the workflow that will allow simulation of both collagen and interlamellar nerve tissue in a heterogeneous manner. The simulation work presented here will be further developed in the future and will utilize digital volume correlation displacement data (from pressure dependent SHG datasets) in order to determine the mechanical properties of the collagenous laminar beams of the LC.

METHODS

All sample procurement methods were approved by the University of Arizona's Institutional Review Board. A single human donor eye was acquired from Midwest Eye Bank from which the LC was isolated and underwent a brief mechanical and chemical digestion. The LC was then mounted into a pressure chamber and subjected to four pressures (5, 15, 30, and 45 mmHg) while being imaged from the anterior aspect using a two photon microscope.

Two photon imaging was performed using a Ti:Saph (Mira 900, Coherent, Inc.) tunable 120 fs pulsed light source. The excitation wavelength of the laser was λ= 780 nm and collection was done in the epidirection using a dichroic mirror (405 nm), a 377/50 nm bandpass filter (SHG), and a 460/80 nm bandpass filter. Digital images were acquired using a 4X objective (NA = 0.3) resulting in a pixel size of 2.5μm in the x- and y-direction and 5μm in the z-direction (Figure 1). The SHG images were segmented using an in house multiscale wavelet decomposition method with adaptive scale selection implemented in MATLAB. This approach can efficiently handle segmentation of highly inhomogeneous LC beams.

A subdomain of the entire human LC was chosen for this analysis (Figure 2). Our goal was to investigate how changes in the parameters in an anisotropic hyperelastic constitutive model (Holzapfel model) affect maximum z-displacement in the center of the LC. Based on the aforementioned model, the form of the strain energy potential for a soft tissue with distributed collagen fiber orientation is defined by:

$$U = C_{10}(\bar{I} - 3) + \frac{1}{D}\left(\frac{(J^{el}2)^2 - 1}{2} - \ln(J^{el})\right) + \frac{k_1}{2k_2}\sum_{\alpha=1}^{N}\{exp[k_2\langle\bar{E}_\alpha\rangle^2] - 1\}$$

where $\bar{E}_\alpha \stackrel{\text{def}}{=} \kappa(\bar{I} - 3) + (1 - 3\kappa)(\bar{I}_{4(\alpha\alpha)} - 1)$.

In general κ describes fiber dispersion about the mean fibber orientation at any given location. In this analysis we have chosen $\kappa = \frac{1}{3}$ that shows random distribution. Soft tissues are assumed to be incompressible and as such we have utilized a value of $D = 10^{-6}$ (1/Pa). In our study k_2 was assumed to be a constant value and the computational domain was defined by variation of C_{10} and k_1 (fiber stiffness).

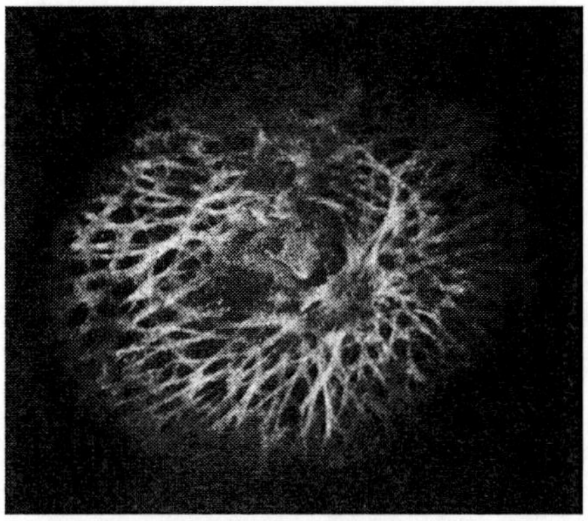

Figure 1. A coronal slice of the LC as imaged using two photon microscopy

Figure 2. Elliptical subdomain segmented from the entire human LC as utilized in this study. Nerve tissue was present in the interlamellar LC spaces but is excluded from this figure for viewing purposes.

The subdomain was located within the body of the LC. We have assumed the trabecular collagen subdomain is embedded in a thin layer of nerve tissue as well as all spaces between the collagen beams. The Gibbon toolset in MATLAB was used to discretize both the nerve and collagen tissue using over 3.9 million linear tetrahedral elements. The average element length of each collagen and nerve element was 1 μm. The boundaries of the domain whose plane normal are orthogonal to the anterior posterior direction were fixed in all three translational degrees of freedom in nasal and temporal directions and the anterior facing surface was exposed to a pressure of 45 mmHg. Both nerve and collagenous tissues were assumed to be hyperelastic and nearly incompressible. Nerve tissue was considered to be isotropic and NeoHookean with values of $C_{10} = 20{,}000$ (Mpa) and $D_1 = 10^{-6}$ (1/Pa).

RESULTS

15 simulations were run and the maximum displacement of the LC center (computational domain) in anterior-posterior direction was captured. Each simulation took approximately one hour on a dual processor computer with 80 cores.

Figure 3. Displacement contour plot of the collagen domain within the LC

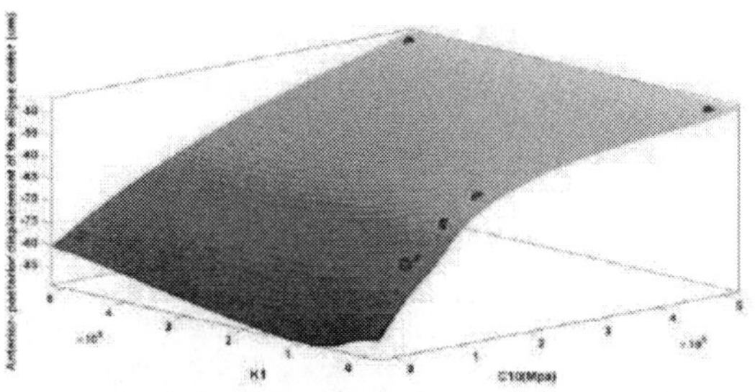

Figure 4. Influence of fiber stiffness (k1) and matrix stiffness (C10) on the anterior-posterior displacement of the human LC.

DISCUSSION

The simulations presented here have several limitations. First and foremost, the fixed boundary condition on the displacement degrees of freedom on the boundary of the LC are not physiological. In future work we will impose digital volume correlation derived strains to the exterior and utilize an optimization within the free LC domain. The approach described will also be modified to include the entire LC volume from SHG images.

ACKNOWLEDGEMENTS

Research reported in this abstract was supported by the National Eye Institute of the National Institutes of Health under award number R01EY020890 to JPVG.

REFERENCES

1. Downs, J.C., M.D. Roberts, and I.A. Sigal, *Glaucomatous cupping of the lamina cribrosa: a review of the evidence for active progressive remodeling as a mechanism.* Experimental eye research, 2011. **93**(2): p. 133-140.

IN-VIVO IRIS STIFFNESS IN PATIENTS WITH OCCLUDABLE ANTERIOR CHAMBER ANGLE FOLLOWING LASER PERIPHERAL IRIDOTOMY

Anup D. Pant (1), Priyanka Gogte (2), Syril K. Dorairaj (3), Vanita Pathak-Ray (2), Rouzbeh Amini (1)

(1) Department of Biomedical Engineering
The University of Akron
Akron, OH, USA

(2) LV Prasad Eye Institute
Hyderabad, India

(3) Department of Ophthalmology
Mayo Clinic
Jacksonville, FL, USA

INTRODUCTION

Glaucoma is a leading cause of blindness worldwide. It is estimated that by 2020, 79.6 million people will suffer from glaucoma and the number of bilaterally blind cases due to glaucoma will surpass 11 million people [1]. Glaucoma is characterized by irreversible optic nerve damage generally coexisting with an increase in the intraocular pressure (IOP). The elevated IOP is generated by abnormally increased resistance to the aqueous humor outflow. For example, in angle closure glaucoma (ACG), a main category of glaucoma, a narrow or closed angle between the iris root and cornea blocks the aqueous humor outflow pathway.

Clinical studies have identified numerous iris biometrics such as concavity, thickness, and chord length as risk factors of ACG. Recent studies have shown that the iris tissue properties could be risk factors for ACG, as well. For instance, irides in ACG eyes were found to be less compressible when compared to normal eyes during dynamic phenomena such as dilation [2]. Also, ex-vivo studies have shown that in patients suffering from ACG, the iris is mechanically stiffer when compared to the normal and primary open angle glaucoma irides [3]. The outcomes of ex-vivo studies should be considered with caution as soft tissue properties are known to be different in vivo. In addition, application of ex-vivo measurements are extremely limited when it comes to diagnostic procedures. Herein, we have developed a method to determine the mechanical properties of human irides using in-vivo imaging and finite element simulation. We have then used our method to determine iris stiffness both in glaucomatous eyes and in healthy control eyes.

METHODS

Anterior segment optical coherence tomography (AS-OCT, Carl Zeiss Meditec, Inc., Dublin, CA) images were acquired from Indian patients with occludable angles (n=8) who had undergone peripheral laser iridotomy (LPI), a procedure for opening the narrow angle via a small hole in the iris, and from healthy volunteers (n=8) at LV Prasad Eye Institute in Hyderabad, India. Images were taken in standard light and following pharmacologically induced dilation. An axisymmetric model of the iris based on our previous finite element model [4] was constructed. A thin layer of the dilator muscle (8.5μm) was identified in the model and the corresponding dilator stress was applied. The iris was modeled as a compressible neo-Hookean solid governed by the stress balance equation:

$$\nabla.\boldsymbol{\sigma} = 0, \tag{1}$$

where $\boldsymbol{\sigma}$ represented the Cauchy stress tensor. The Cauchy stress tensor was defined by neo-Hookean, $\boldsymbol{\sigma}_{NH}$, and active dilator, $\boldsymbol{\sigma}_D$, stress tensors:

$$\boldsymbol{\sigma} = \boldsymbol{\sigma}_{\mathbf{NH}} + \boldsymbol{\sigma}_{\mathbf{D}} \tag{2}$$

The neo-Hookean stress was defined as

$$\boldsymbol{\sigma}_{\mathbf{NH}} = \frac{G}{\det F}(\mathbf{B} - \mathbf{I}) + \frac{2G\nu}{(1-2\nu)\det F}\ln(\det \mathbf{F})\mathbf{I}, \tag{3}$$

where G was the shear modulus, ν was the Poisson's ratio, **I** was the identity tensor, **F** was the deformation gradient tensor, and **B** was the left Cauchy-Green deformation tensor. **F** and **B** were defined as

$$\mathbf{F} = \frac{dx}{dX}, \tag{4}$$

$$\mathbf{B} = \mathbf{F}\mathbf{F}^{\mathbf{T}}, \tag{5}$$

where **x** was the current position of a material point and **X** was its resting position. The dilator stress $\boldsymbol{\sigma}_D$ was applied in the dilator region. The dilator stress was defined by σ_{Act}, the scalar active muscle contraction stress, and $\mathbf{e}_s$, the unit vector representing the direction of the deformed dilator muscle:

$$\boldsymbol{\sigma}_D = \boldsymbol{\sigma}_{Act}\mathbf{e}_s \otimes \mathbf{e}_s \quad (6)$$

As shown in Fig. 1, the commercial software Abaqus (Dassault Systèmes, Velizy-Villacoublay, France) was employed to generate finite element meshes using in-vivo AS-OCT images under standard light. The pharmacologically induced dilated iris images of the same eye were treated as the deformed experimental configuration. We then used an inverse modeling approach [5] to calculate the shear modulus and (G) Poisson's ratio (ν) from the experimental deformation data. The elastic modulus (E) was then calculated from the estimated values of G and ν. The objective function for the inverse modeling was defined as the absolute value of the difference between the chord length of the experimental measurements CL_{exp} and the genetically driven finite element solutionCL_{sim}:

$$SSE = \|CL_{exp} - CL_{sim}\| \quad (7)$$

The simulations were performed on an HP Intel Xeon machine at the Ohio Supercomputing Center (Columbus, OH) for 50 generations.

A
B
C

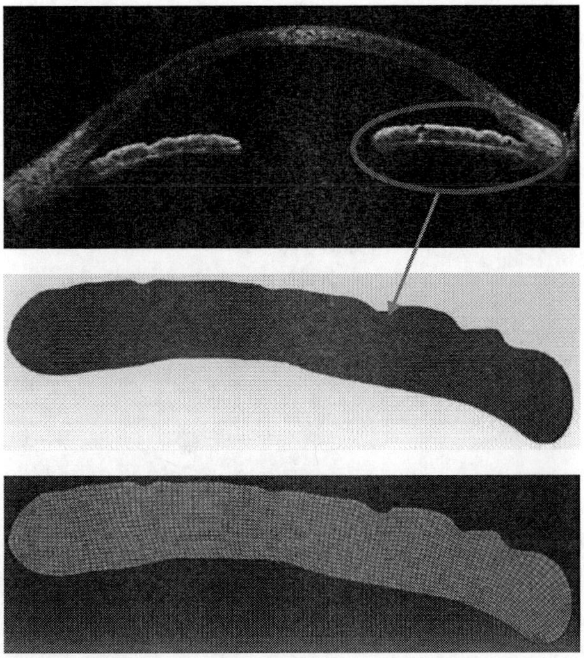

Figure 1: (A) Anterior segment optical coherence tomography image captured under standard light; (B) Segmented iris; (C) Finite element mesh of the iris at the undeformed configuration.

RESULTS

The shear modulus, Poisson's ratio and elastic modulus of the iris in healthy eyes were found to be 0.90 ± 0.32 kPa (mean ± standard error), 0.15 ± 0.02 and 2.05 ± 0.73 kPa respectively. In patients with occludable angles who had undergone LPI for treatment of ACG, the iris shear modulus, Poisson's ratio, and elastic modulus were 4.38 ± 1.74 kPa, 0.18 ± 0.05, and 9.86 ± 3.71 kPa, respectively. The shear modulus (Fig. 2) and the elastic modulus were found to be significantly larger in patients with the history of glaucoma in comparison to the healthy control group ($p < 0.05$, Student t test).

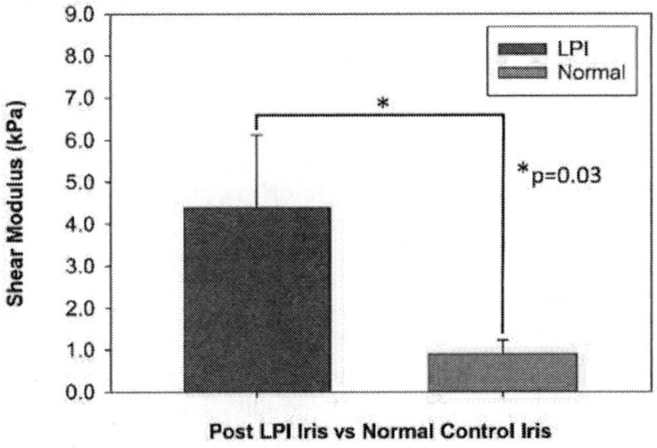

Figure 2: Comparison between the iris shear modulus of patients with occludable anterior chamber angles who had undergone peripheral laser iridotomy (Post-LPI) and those of healthy volunteers (n=8 for both groups, error bars are standard error).

DISCUSSION

Previous studies have suggested that the risk factors that are more in tune with the *dynamic* deformation of the iris (e.g. compressibility) are extremely important in ACG [6, 7]. As tissue stiffness is expected to affects iris *dynamic* deformation, it is not surprising that patients with a history of ACG have stiffer irides in comparison to healthy volunteers. The stiff iris in the patients with occludable anterior chamber angles who had undergone LPI is of particular interest. Since LPI has been performed, the effect of the aqueous humor pressure on the iris deformation has been minimized. However, the anterior chamber angle happens to close following dilation in this patient population. That the stiffer irides lead to the closure of the anterior chamber angle even post LPI cannot be simply concluded from the current study. Future research is needed, especially to identify the in-vivo stiffness of irides in patients with history of ACG who have undergone LPI but do not suffer from occludable anterior chamber angles following this intervention. Since AS-OCT imaging is a non-invasive procedure often performed during the eye exams in the clinical practice, the method developed herein has the potential to become a new diagnostic method following further studies in different patient populations.

ACKNOWLEDGEMENTS

Computations were facilitated by a supercomputing resource grant from the Ohio Supercomputer Center.

REFERENCES

[1] Quigley, H.A. et al., *Br J Opthalmol*, 90:262-267, 2006.
[2] Quigley, H.A., *Ophthalmology*, 117:1–2, 2010.
[3] Narayanaswamy, A.K. et al., *IOVS*, 56:6139, 2015.
[4] Amini, R. et al., *IOVS*, 53:1188-1194, 2012.
[5] Pant, A., Amini R., *IOVS*, 56:6140, 2015.
[6] Quigley, H. A. et al., *J Glaucoma*, 18:173-179, 2009.
[7] Aptel, F. et al., *Opthalmology*, 117: 3-10, 2010.

EVALUATING THE EFFICACY OF CROSSLINKING THE POSTERIOR RAT SCLERA

Bailey G. Hannon (1), Ian C. Campbell (2, 3), A. Thomas Read (2), C. Ross Ethier (1, 2)

(1) George W. Woodruff School of Mechanical Engineering
Georgia Institute of Technology
Atlanta, GA, USA

(2) Coulter Department of Biomedical Engineering, Georgia Institute of Technology/Emory, Atlanta, GA, USA

(3) Rehabilitation Research and Development
Atlanta VA Medical Center, Decatur, GA, USA

INTRODUCTION

Glaucoma is the leading cause of irreversible blindness worldwide and is characterized by a progressive degradation of retinal ganglion cell (RGC) axons in the optic nerve [1]. A major causative risk factor for glaucoma is an elevated intraocular pressure (IOP), and therefore, all current clinical treatments aim to lower IOP. However, these treatments are not always effective: vision in 25-45% of glaucoma patients continues to deteriorate even with treatment [2]. Thus, there is a significant need for development of new therapeutic approaches to prevent this irreversible vision loss.

One promising proposed treatment (extensively reviewed in [3]) is increasing stiffness of the posterior sclera (white of the eye), to counteract biomechanical forces exerted on the optic nerve head by elevated IOP. Since the sclera is composed primarily of collagen fibrils, stiffening can be induced by crosslinking the collagenous matrix. Several stiffening agents have been evaluated in rabbit, porcine, and human sclerae [4-7], however, most *in vivo* glaucoma studies utilize rodents due to their cost effectiveness, similar glaucomatous damage pattern to that in humans, and well-characterized models. Additionally, prior to engaging in *in vivo* studies using these agents in rats, it is necessary to establish a dose-response relationship to determine crosslinker concentrations necessary for a desired stiffening effect. We hypothesize that various concentrations of stiffening agents will produce a desired dose-dependent stiffening response in rat sclera.

METHODS

Eyes were freshly enucleated from a total of 67 male, retired breeder (approximately 9-12 months old) Brown Norway rats, cleaned, and incubated overnight in one of three stiffening agents at various concentrations [Table 1]. Agents (glyceraldehyde (GC), methylglyoxal (MG), and genipin (GP)) were chosen based on biocompatibility and ability to compare to previous studies in other model organisms [4-7].

Table 1: Concentrations tested for all stiffening agents

Stiffening Agent	Concentration (mM)	Concentration (% w/v in PBS)
Glyceraldehyde	10.0, 30.0, 62.5, 125	0.90, 2.7, 5.7, 11
Methylglyoxal	3.5, 7.0, 14	0.25, 0.50, 1.0
Genipin	0.25, 0.50, 1.0, 7.5, 15, 30	0.06, 0.11, 0.23, 1.7, 3.4, 6.8

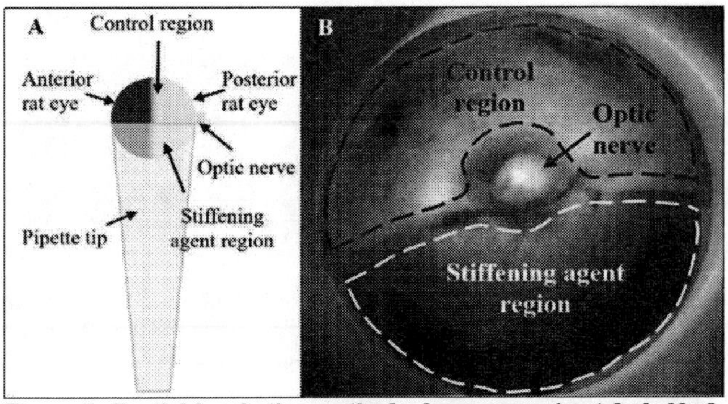

Figure 1: Partial incubation method where approximately half of the eye was immersed in a stiffening agent (GC, MG, or GP), and the control portion was misted with PBS to prevent drying (A). Posterior sclera post-incubation in GP with control and stiffening agent regions outlined (B).

For each concentration of each agent, 4 to 8 eyes were tested. A novel partial incubation technique [Figure 1A] was utilized for each eye, allowing for an internal (within-eye) control region. Eyes were incubated overnight at 37°C to maintain physiological conditions. After incubation, the control and treated regions were marked (to distinguish and compare them during analysis), and inflation testing was carried out to measure the strain response of various spatial regions of the eye. One of the agents tested (GP) is also a blue dye and served as a visual reporter [Figure 1B] to confirm that the agent's distribution was restricted to the stiffening agent region without extensive diffusion into the control portion of the eye.

To evaluate stiffening agent effectiveness, eyes were subjected to whole globe inflation testing, using 3D digital image correlation (DIC) to calculate the strain of the posterior sclera surface. Eyes were mounted on custom-made acrylic blocks cornea-side down, a speckle pattern was applied to the posterior sclera with fine graphite powder, and the corneas were punctured with a 1mm-diameter biopsy punch through a channel in the acrylic block. The end of the channel opposite the eye was then connected to a hydrostatic pressure reservoir. The mounting block and eye assembly was then submerged in a covered, temperature-controlled (37°C ± 2°C) chamber of PBS.

Calibrated stereo cameras (including compensation for refraction through PBS) were used to image the speckle pattern on the surface of the posterior sclera at various pressures throughout the inflation test, and DIC was used to regionally quantify the surface principal strain. Pressure was adjusted by raising the hydrostatic reservoir to heights equivalent to three different IOPs: 3 mmHg (baseline), 13 mmHg (normal IOP), and 28 mmHg (elevated IOP). Eyes were held at each pressure for 30 minutes to ensure that the eye had reached a steady state, and strain was computed in reference to the last image of the baseline (3 mmHg) pressure. The 1st principal Lagrange strain was calculated and used for analysis because it is representative of the main mode of deformation during inflation tests. Averaged strain values in the control and stiffened regions were used to compute the relative stiffness (Equation 1) of the sclera as a percentage.

$$\text{Relative Stiffness} = \frac{E_{exp}-E_{con}}{E_{con}} = \frac{\varepsilon_{exp}-\varepsilon_{con}}{\varepsilon_{exp}} \times 100\% \qquad (1)$$

Here, E_{con} and E_{exp} represent the elastic modulus in the control and stiffened regions, respectively, and ε_{con} and ε_{exp} represent strain in the control and stiffened regions, respectively, where the above formula assumes linearly elastic behavior during scleral inflation. We then compared the relative stiffness of each eye as a function of treatment and inflation pressure using a nested 2-factor ANOVA (relative stiffening as a function of pressure nested within concentration; R 3.3.1).

RESULTS

A significant ($p = 1.03\times10^{-9}$) stiffening effect from application of cross-linkers was observed (relative stiffness of each eye as a function of pressure nested within concentration and pooled over all agents). All of the tested stiffening agents demonstrated a dose-dependent stiffening effect at 13 mmHg [Figure 2A-C] and 28 mmHg (data not shown). Relative stiffness values ranged from 21.8% to 273% for GC, 11.9% to 310% for MG, and 14.7% to 1320% for GP. Concentrations of each agent that resulted in an approximate doubling in stiffness were 30mM for GC, 7mM for MG, and 1mM for GP.

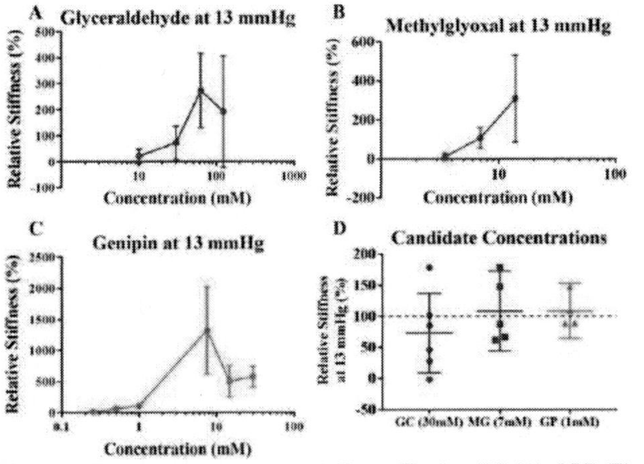

Figure 1: Dose-response curves at 13 mmHg for GC (A), MG (B), and GP (C), and their concentrations that produced approximately 100% relative stiffening (dashed line; D). Data shown as mean ± SD.

DISCUSSION

The results here confirm the efficacy of using collagen crosslinking agents in the rat sclera *ex vivo* and demonstrate the first use of genipin and methylglyoxal in rodent sclera and glyceraldehyde in rat sclera. Each agent induced a relative stiffening of several hundred percent, however, dose-dependent stiffening effects seem to have an upper bound, as can be seen in the glyceraldehyde and genipin plots [Figure 2B and 2C], possibly attributable to cross-linking site saturation.

The novel partial incubation method reported here allows for an internal control from each eye to eliminate inter-eye variability, and it also removes the necessity for preconditioning, as these effects should be uniform in both halves of the sclera. Additionally, inflation testing exerts minimal damage to the integrity of the fibers of this collagenous tissue and allows the eye to maintain its normal physiological shape (in contrast to scleral strip testing).

These results establish concentrations for future *in vivo* studies evaluating these agents as treatments for glaucoma. Future work will characterize potential toxic effects such as damage to the retina or to the scleral fibroblasts responsible for collagen turnover in the eye. We also plan to quantitatively confirm the magnitude of relative stiffening achievable *in vivo* as well as the persistence of the stiffening effect.

ACKNOWLEDGEMENTS

The authors gratefully acknowledge research support from the Department of Veterans Affairs (Career Development Award IK1 RX001791 to ICC), the National Institutes of Health (R01 EY025286 to CRE), and the Georgia Research Alliance (CRE).

REFERENCES

[1] Quigley, H.A. et al., *The Lancet,* 9774:1367-1377, 2011.
[2] Noecker, R.J. et al., *Ther Clin Risk Manag,* 2:193-201, 2006
[3] Campbell, I.C. et al., *J Biomech Eng,* 136: 021005, 2014.
[4] Wollensak, G. et al., *J Cataract Refract Surg,* 34:651-656, 2008.
[5] Wollensak, G. et al., *J Cataract Refract Surg,* 86:887-893, 2008.
[6] Liu, T.X. et al., *Acta Ophthalmol,* 91:e253-257, 2013.
[7] Wong, F.F. et al., *Exp Eye Research,* 103:17-21, 2012.

SB^{3}C2017
Summer Biomechanics, Bioengineering and Biotransport Conference
June 21 – 24, Tucson, AZ, USA

THE EFFECT OF THE REMOVAL OF GLYCOSAMINOGLYCANS ON THE DEFORMATION RESPONSE OF THE HUMAN LAMINA CRIBROSA TO PRESSURE

D.E. Midgett (1), H.A. Quigley (2), T.D. Nguyen (1)

(1) Mechanical Engineering
Johns Hopkins University
Baltimore, MD, United States

(2) Wilmer Eye Institute
Johns Hopkins University
Baltimore, MD, United States

INTRODUCTION

The human lamina cribrosa (LC) is a connective tissue structure in the optic nerve head that mechanically supports the axons of the retinal ganglion cells (RGCs) as they exit the eye. The structure of the LC is formed by approximately 10 perforated plates stacked on top of one another [1]. Albon and coworkers [2,3] showed that the LC is composed mainly of collagen, 24-47%, and elastin, 7-28%, by dry weight depending on age. By comparison, proteoglycans make up a relatively small fraction of the LC, 0.7-2.2% by dry weight. Sulfated proteoglycans (sGAGs) in the LC fall into 3 main categories: chondroitin and dermatan sulfate, which are found near collagen fibrils and are digestible with chondroitenase ABC (ChABC), and heparan sulfate which are found in the basal laminae of astrocytes and blood vessels and are not digestible by ChABC. [4]. Numerous studies have shown that the LC remodels with age and glaucoma. The LC becomes stiffer with age [5] and exhibits an increase in total collagen content [2], a reduction in the amount of sGAGs, and an increase in elastin content [3]. Glaucoma is associated with increases in the amount of sGAGS [6-7]; reduction in the density of type I and III collagen and increases in type IV collagen [8-10]; and morphological changes to elastin, where the fibers appear more curled and disorganized [11].

Our goal is to determine the significance of the chemical alterations in the LC with age and glaucoma. In this paper we examined, for the first time, the change in the pressure-induced strains of the LC caused by the removal of sGAGs. We applied a new *ex vivo* inflation test to measure the three-dimensional (3D) deformation response of the human LC to pressure before and after sGAG digestion [5]. The experiment used a laser-scanning microscope (LSM) to acquire the backscattered second harmonic generation (SHG) signal of the collagen structure and digital volume correlation (DVC) to calculate the displacement field in the LC.

METHODS

Five normal human eyes from 4 donors (ages 42, 49, 63, and 88 years) were obtained from the National Disease Research Interchange (NDRI) and tested within 48 hours post-mortem. Posterior scleral samples were prepared by cutting the optic nerve flush with the sclera to expose the LC and gluing the eye to a polycarbonate ring of custom size, 2-3 mm posterior to the equator of the sclera with the LC centered. Then the cornea, anterior sclera, choroid, and retina were removed and the sample was incubated at 37°C for 4 hours in a Trizma buffer. The posterior sclera was removed from the incubation room and mounted on a custom inflation chamber (Fig 1b). The pressure inside the holder was set using an external water column to 5, 10, and 45 mmHg and the specimen was equilibrated for 30 minutes at each pressure before imaging. A Zeiss LSM 710, tuned to 780nm, was used to acquire Z-stacks of the LC at each pressure (Fig 1a) using SHG. The X and Y directions within each image were aligned with the nasal-temporal (NT) and inferior-superior (IS) axes of the LC respectively, and the out-of-plane direction was designated as Z.

After imaging, the specimen was removed from the inflation chamber, and incubated at 37°C for 4 hours in a Trizma solution containing ChABC to digest the sGAGs within the LC and posterior sclera. The specimen was then remounted and subjected to the same pressurization and imaging protocol. A DVC algorithm [12] was used to analyze the SHG volumes to calculate the 3D displacement fields between the baseline pressure of 5 mmHg and the test pressure. The components of the Green-Lagrange strain were calculated by fitting 6th-order polynomial functions to the components of the displacement field and evaluating the gradient. The in-plane strain components, E_{xx}, E_{yy} and E_{xy}, and the maximum principal strain, E_1, were averaged over the specimen and compared before and after sGAG degradation using a paired t-test (n=5). The DVC errors and uncertainties were evaluated by numerically applying a uniform, triaxial strain field and translation to the SHG volumes and calculating the average absolute error of the DVC displacement and strain components. The strain errors were found to be 0.162±0.216% for E_{xx}, 0.137±0.185% for E_{yy}, and 0.111±0.130% for E_{xy}. Details of the imaging method, image processing, DVC method, and error analysis are described in Midgett et al. [5].

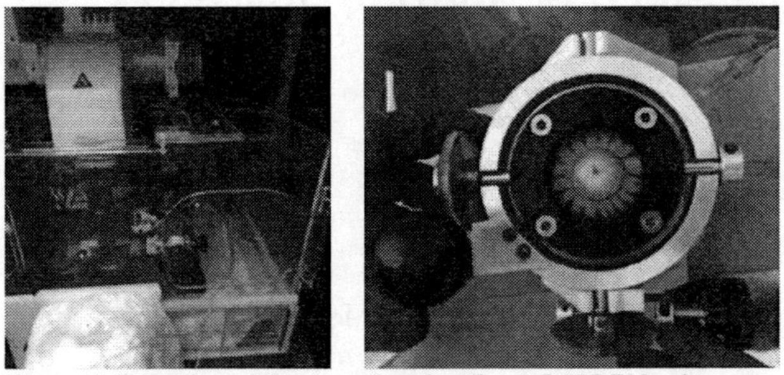

Figure 1, Experimental setup, showing the a) LSM 710 microscope and b) inflation holder with posterior scleral sample

RESULTS

The displacement and strain components in the LC were smaller after 4 hours of sGAG degradation using ChABC at both 10 mmHg and 45 mmHg. E_{yy} showed a statistically significant reduction from 3.08±1.69% to 2.26±1.30% (p=0.0335) at 45 mmHg. E_{xx} showed a smaller reduction from 2.35±0.99% to 1.88±0.78% at 45 mmHg, which was not statistically significant (p=0.097). The maximum principal strain E_1 showed a statistically significant reduction from 4.54±1.98% to 3.92±1.74% (p=0.0212) at 45 mmHg (Fig. 2,3). The eyes from the two youngest donors (age 42 and 49) exhibited a larger average reduction in E_1 of 0.97±0.03% compared to the eyes from the 2 older donors (age 63 and 88), which experienced an average reduction in E_1 of 0.39±0.28%. Preliminary measurements of sGAG levels in the porcine LC and peripapillary sclera (PPS) using a Blyscan assay (Accurate Chemical, Westbury, NY, USA) showed a 90% reduction in sGAGs after 4 hours of incubation in ChABC. These measurements will be repeated for a human eye pair in future experiments.

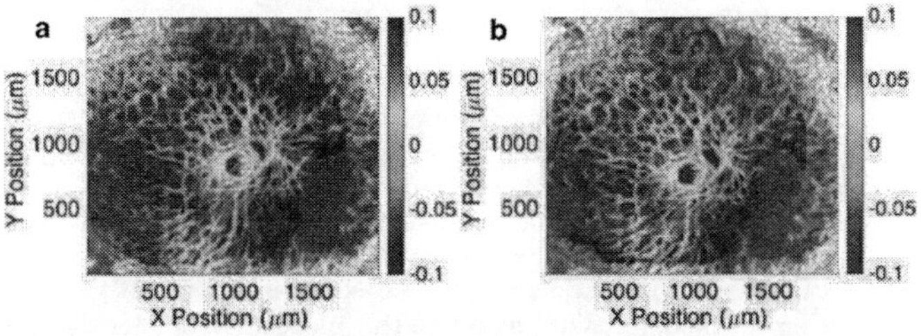

Figure 2, Z-averaged contour plots of the maximum principal strain in the LC of a 49-year Caucasian Female (FC49) a) before degradation in ChABC and b) after degradation in ChABC showing a stiffening effect.

DISCUSSION

This study found that removal of sGAGs resulted in a stiffer LC strain response to inflation. The LC from the 2 younger donors exhibited greater stiffening after sGAG removal than from the 2 older donors. There was also a greater reduction in E_{yy} compared to E_{xx}.

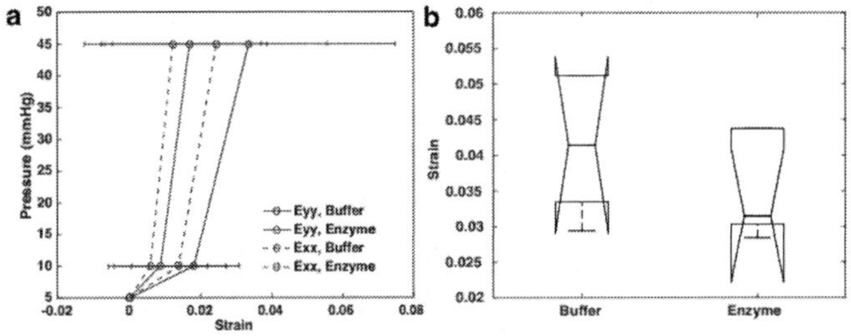

Figure 3, Comparison of a) average pressure-strain response for FC49 before and after enzyme degradation and b) maximum principal strain before and after enzyme degradation (p=0.0212)

The amount of sGAGs in the LC has been shown to decrease with age [3] and increase with glaucoma [6,7]. The results here suggest that the decrease in sGAGs with age may be responsible, in part, for the age-related stiffening in the pressure-strain response with age [10]. Likewise the larger amount of sGAGs in the LC of glaucoma eyes may indicate a more compliant LC, which may affect the susceptibility and progression of glaucomatous axon damage. The difference in the stiffening effect of E_{xx} and E_{yy} measured in this study may be caused by a regional variation in sGAG content, associated with variations in the connective tissue density of the LC [1,13], and by the elliptical shape of the LC. The stiffening effect measured for the human LC is consistent with the findings of Murienne et al. [14], who measured a similar stiffening effect for the human sclera after sGAG removal. It is unclear how much of the reduction in strains measured for the LC is caused by alterations in the inflation response of the surrounding sclera. We will investigate this interaction in the future by measuring the strain in both the PPS and LC simultaneously and also measuring the levels of sGAGs in both tissues before and after degradation. We also plan to increase the sample size for this study to investigate age-related differences and regional variations in the stiffening effect induced by s-GAG removal.

ACKNOWLEDGEMENTS

Grants supporting this project: NSF CAREER Award #1253453, PHSR Grant EY021500, EY02120, EY01765, Brightfocus G2015132.

Equipment, Facilities, and Personnel: Wilmer Eye Institute laboratories and microscopes. Wilmer Eye Institute Personnel Mary Pease, Cathy Nguyen, Lizzie Kimball

REFERENCES

[1] HA Quigley, *Prog in Ret and Eye Res.* 18:39–57 (1998)
[2] J Albon, et al. *Br J Ophthalmol.* 79:368-375(1995)
[3] J Albon, et al. *Br. J. Ophthalmol.* 84:311-317(2000)
[4] S Sawaguchi, et al. *Invest. Ophthalmol.* 33:2388-2398(1992)
[5] DE Midgett, et al. *Acta. Biomat.* (2017),
doi: http://dx.doi.org/10.1016/j.actbio.2016.12.054
[6] T Fukuchi et al. *Exp Eye Res.* 58:231-44(1994)
[7] G Tezel, et al. *Arch Ophthalmol.* 117:917-294(1999)
[8] MR Hernandex, et al. *Am J Ophthalmol.* 109:180-88(1990)
[9] JC Morrison, et al. *Arch Ophthalmol.* 108:1020-24(1990)
[10] HA Quigley, et al. *Curr Eye Res.* 10:877-88(1991)
[11] HA Quigley, et al. *Arch Clin Exp Ophthalmol.* 232:257-61(1994)
[12] E Bar-Kochba, et al. *Exp. Mech.* 55:261–74(2015).
[13] JB Jonas, et al. *Surv. Ophthalmol.* 43:293–320(1999).
[14] B Murienne, et al. *J. R. Soc. Interface.* 13:21060367(2016)

SB^{3}C2017
Summer Biomechanics, Bioengineering and Biotransport Conference
June 21 – 24, Tucson, AZ, USA

LONGITUDINAL REINFORCEMENT IMPROVES POST-INFARCTION FUNCTION BY REDISTRIBUTING TRANSMURAL FIBER STRESS IN THE BORDER ZONE

Ana C. Estrada (1), Samantha A. Clarke (1), Jeffrey W. Holmes (1,2,3)

(1) Department of Biomedical Engineering
University of Virginia
Charlottesville, VA, US

(2) Department of Medicine
University of Virginia
Charlottesville, VA, US

(3) Berne Cardiovascular Research Center
University of Virginia
Charlottesville, VA, US

INTRODUCTION

Myocardial infarction (MI), commonly known as a heart attack, is a prevalent problem, affecting over 1 million Americans each year [1]. The majority of those affected survive the initial attack but continue to live with severely decreased cardiac function and an increased risk of developing heart failure. Myocardial infarctions occur when blood flow to a region of the heart is interrupted, as happens when a coronary artery is blocked. Ischemic cardiomyocytes stop contracting within minutes and die within hours, leaving a mechanically passive region that bulges and stretches during systole rather than contributing to pressure generation. The dead cardiomyocytes are eventually replaced with a stiff collagen scar over several weeks of healing, which results in dramatic changes to the local and global mechanics of the heart. The altered mechanics lead to remodeling in the surviving myocardium, most often resulting in eccentric growth and the further deterioration of pump function. Restoring, or at the very least improving, cardiac function as early as possible is critical to ensuring long term survival and preventing the development of heart failure.

A number of groups have investigated mechanical reinforcement as an approach to improving pump function and preventing the development of heart failure. These interventions are broadly classified into two groups: global restraint and local restraint, which is applied directly to the infarct. Both approaches typically improve systolic function of the heart at the cost of impaired filling, leading to no net change in overall pump function [2]. However, a local restraint approach taken by our lab led to substantial improvements in pump function acutely following infarction: we used a stiff, anisotropic Dacron patch to selectively reinforce large anterior infarcts in the longitudinal direction. The patch was developed based on a finite-element model study of the impact of scar anisotropy on post-infarction stroke volume. The original model predicted that a longitudinally stiff patch would enhance diastolic filling without affecting systolic behavior. Unexpectedly, in animal trials longitudinal reinforcement improved stroke volume by the opposite mechanism: it improved systolic function but had little impact on filling [3-4]. In order to properly understand how the patch restores function, we developed a new finite-element model capable of reproducing the behavior seen in these experiments. The following study aims to understand why longitudinal reinforcement of large, acute myocardial infarcts leads to improvement in stroke volume by comparing the behavior of three patch models (longitudinal, circumferential, and isotropic reinforcement) to a non-patch model of acute ischemia.

METHODS

The effects of longitudinal patch reinforcement were studied using a set of finite-element models of the canine left ventricle (LV) immediately following a large anterior infarction. The models were based on a realistic geometry obtained from MRI scans of a canine heart 48 hours post-MI, before significant remodeling had occurred. Five distinct models were created: a baseline model pre-MI, an acute ischemia model, and three patch models with longitudinal, circumferential, or isotropic reinforcement. The unloaded geometry was the same for all models and was derived by scaling down the end-systolic geometry while preserving wall volume (see below). The models were meshed using MATLAB and the finite element simulations were run using FEBio. Material properties for the models were optimized by matching data from the acute patch experiments previously performed in our laboratory. Three distinct materials were used in these models: normal myocardium, ischemic myocardium, and patch material. All three materials use the transversely isotropic Mooney-Rivlin material implemented in FEBio (equation 1). The parameters in the strain energy function for normal myocardium, as well as the scaling for the unloaded geometry, were obtained by matching the end-diastolic volumes at 5mmHg and 10mmHg, the strains in the circumferential and longitudinal directions between these two pressures, and the slope of the end-systolic pressure-volume curve. The ischemic myocardium and patch material properties were obtained by matching the difference in volumes and the remodeling strains between the new model and the baseline model at and end-diastolic pressure of 10mmHg.

$$W = C_1(I_C - 3) + C_2(II_C - 3) + F\left(\sqrt{IV_C}\right) + \frac{K}{2}[\ln(J)]^2$$
$$\frac{\partial W}{\partial I_C} = C_1, \ \frac{\partial W}{\partial II_C} = C_2, \ \frac{\partial W}{\partial IV_C} = \frac{1}{\lambda_1}C_3\left(e^{C_4(\lambda_1 - 1)} - 1\right) \quad (1)$$

Once the material properties were set, the three patch models were compared to the acute ischemia model at the end of passive inflation to 10mmHg and at peak isovolumic contraction starting from this passive inflation. The fiber stress and strain distributions in the remaining non-infarcted muscle were compared to assess the effect of the different patch orientations on LV contractile function.

RESULTS

The three patch models (longitudinal, circumferential, and isotropic reinforcement) were compared to a non-patch model of acute ischemia in order to understand why the longitudinal patch provides a greater improvement in systolic function following infarction than the other patch configurations. Consistent with the experimental results of improved systolic function without impaired diastolic filling, the differences between the patch models relative to each other and to the ischemia model were seen at the end of the isovolumic contraction but not the end of passive inflation. Table 1 presents the maximum LV elastance calculated from the volumes at the end of passive inflation and the total pressure generated at the end of isovolumic contraction.

Table 1: Maximum elastance (mmHg/mL) for each of the models

Acute Ischemia	Longitudinal Patch	Circumferential Patch	Isotropic Patch
2.53	2.81	2.42	2.64

Fiber stress and strain were calculated at the end of passive inflation and the end of isovolumic contraction. These values were compared between each of the patch models and the acute ischemia model. The most striking difference between the models was that longitudinal reinforcement redistributed stresses transmurally, even in the remote myocardium far from the infarct. As shown in Figure 1 for a representative set of adjacent elements, longitudinal reinforcement reduced diastolic fiber strains and systolic fiber stresses near the endocardium and increased them at the epicardium. The difference in fiber stress for each of the patch models compared to acute ischemia is shown in Figure 2, and shows that a similar redistribution occurred throughout much of the LV. Figure 2 also shows that both isotropic and longitudinal reinforcement reduced systolic fiber stress near the infarct border, while circumferential reinforcement increased it.

DISCUSSION

We created and parameterized a finite-element model capable of reproducing changes in strains and LV function observed during acute ischemia in dogs, and used it to study the mechanism by which selective longitudinal reinforcement following infarction improves systolic function. Our results suggest that the improvement in systolic function produced by the longitudinal patch is due to a transmural redistribution of diastolic fiber stretch and systolic fiber stress. By unloading the endocardium during contraction and increasing stretch and force generation near the epicardium, longitudinal reinforcement transfers load from fibers that are near their peak capacity for force generation to fibers that have excess contractile reserve. These studies offer a physiologically plausible mechanism for the observed increase in global systolic function in the absence of increased filling.

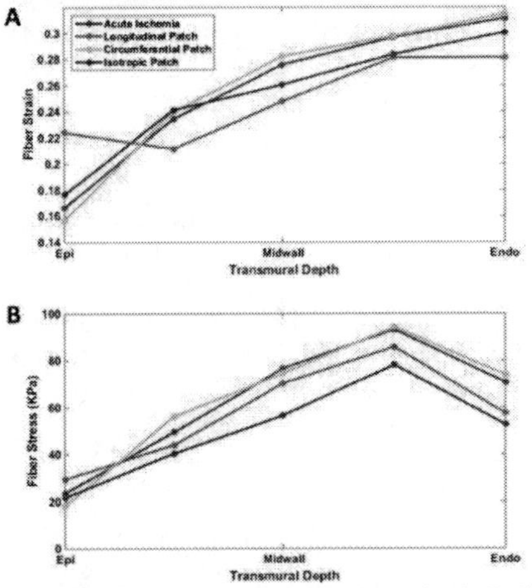

Figure 1: (A) Fiber strains at the end of passive inflation in a representative row of elements running transmurally across the wall. The longitudinal patch model (red) has higher strains in the more epicardial elements but decreased strains in the endocardial elements compared to acute ischemia. (B) Representative fiber stresses at the end of isovolumic contraction in the same elements show a similar trend.

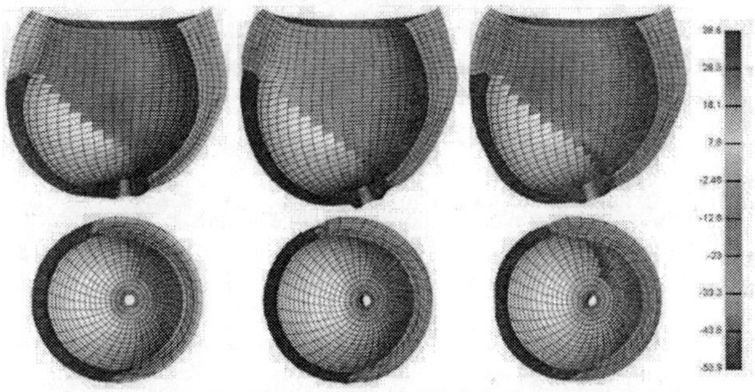

Longitudinal Patch Circumferential Patch Isotropic Patch

Figure 2: Difference in fiber stress between each of the patch models and the acute ischemia model.

ACKNOWLEDGEMENTS

This work was funded by NIH T32 HL-007284 (ACE), an American Heart Association Predoctoral Fellowship (ACE), and NIH R01 HL-116449.

REFERENCES

[1] Mozaffarian, D et al., *Circulation*, 131(4): e29-e322, 2015.
[2] Clarke, SA et al., *In: Cardiovascular and Cardiac Therapeutic Devices*, 2014.
[3] Fomovsky, GM et al., *J Cardiovasc Transl Res*, 1:82-91, 2011.
[4] Fomovsky, GM et al., *Circ Heart Fail*, 5:515-522, 2012.

CHARACTERIZATION OF BIOMECHANICAL PROPERTIES OF HUMAN TRABECULAE CARNEAE

Fatemeh Fatemifar (1), Marc D. Feldman (2), Hai-Chao Han (1)

(1) Department of Mechanical Engineering
The University of Texas at San Antonio
San Antonio, TX, USA

(2) Department of Medicine
The University of Texas Health Science Center at San Antonio
San Antonio, TX, USA

INTRODUCTION

Stiffening of the ventricular wall leads to impaired ventricular relaxation during the left ventricular filling. It has been demonstrated that left ventricular (LV) wall stiffening leads to diastolic dysfunction which is a leading cause of all heart failure hospitalizations [1]. Thickening of the ventricular wall and the trabeculae carneae as well as fibrosis of trabeculae carneae occurs in LV hypertrophy [2, 3]. Our recent results suggested that trabeculae carneae have noticeable effect on the overall stiffness of the left ventricle [1]. However, little is known about the mechanical properties and microstructure of the trabeculae carneae in human heart, though the passive mechanical behavior of the myocardial wall has been well documented.

The aim of this study was to determine the passive mechanical properties and collagen content of the trabeculae carneae in both ventricles of human hearts.

METHODS

Eight human cadaver hearts with average weight of 585.50±97.86g from 7 male and 1 female donors (average age 52±10.40 yrs) were collected within 48 hours post-mortem with permission. The trabeculae carneae and myocardium free wall were isolated from both ventricles (Fig.1) and stored in a 4 °C PBS solution to test within 24 hours of collecting.

Biaxial tensile tests were carried out on specimens isolated from ventricular myocardium and uniaxial tensile tests were performed on the free-running trabeculae carneae isolated from right and left ventricles. Tests were carried out using a CellScale BioTester (Waterloo, ON, Canada). Stretch levels of up to 50% stretch ratio were applied to fully cover the physiological in vivo range [4] and for each case five complete stretch and relaxation cycles were performed for preconditioning since the curves become stable after five cycles. The displacements and the tensile forces were recorded for data analysis. The Cauchy stress and Green strain were calculated based on the tissue incompressibility and neglecting the shear during the uniaxial/biaxial testing.

The Green strains were calculated as

$$E_i = \frac{\lambda_i^2 - 1}{2} \qquad i = x, y \tag{1}$$

where λ is the stretch ratio. The Cauchy stresses were calculated as

$$\sigma_i = \frac{F_i \lambda_i}{A_{0i}} \qquad i = x, y \tag{2}$$

Where F is the loads measured by the load-cell and A_0 is the un-deformed cross-sectional area. The stress-strain data were fitted to the Fung strain energy function:

$$W = \frac{1}{2}c(e^Q\text{-}1); \qquad Q = b_1 E_x^2 + b_2 E_y^2 + 2b_4 E_x E_y \tag{3}$$

where c, b_1, b_2 and b_4 are four material constants determined by fitting experimental data from biaxial testing. In case of uniaxial testing, only c and b_1 are material constants (b_2=b_4=0).

For direct comparison of the nonlinear stress-strain curves, the secant modulus (SM) were calculated at two different stress levels: a) at 5 kPa which represents physiological loading prior to the transitional point in the stress-strain curve (SM_{low}) and b) at 25 kPa which represents the response at higher loads when collagen fibers are lined up with the load direction (SM_{high}).

After mechanical testing, tissues were fixed in 10% formalin overnight and then moved to 70% reagent alcohol. Later, samples were processed and embedded in paraffin and sectioned into 5μm thick sections in fiber direction. Slides were stained with picrosirius red (PSR) and examined using OLYMPUS CX41 microscope at 20x magnification with 10 representative images being taken from each slide. The images were analyzed using ImageJ 1.47v (National Institute of Health) for the collagen density measurement. The collagen content was determined as the percentage of the area occupied by the collagen dividing the total tissue areas.

RESULTS

Both human myocardium and trabeculae carneae isolated from both ventricles demonstrated nonlinear stress-strain curves when subjected to biaxial and uniaxial extension (Fig. 2). For comparison of the stress-strain data from the trabeculae carneae and myocardium, secant modulus (SM) at two stress levels (SM_{low}), (SM_{high}) and maximum Cauchy stress at 20% extension were calculated and summarized in Fig. 3. The reason to choose this stretch ratio is that the myocardium undergoes 12%-22% strain in vivo[4]. The results showed stiffer behavior of tissues from the right ventricle than the corresponding tissues from the left ventricle. The myocardium curves seem stiffer than the trabeculae carneae. The collagen contents were found to be higher in trabeculae carneae than in myocardium (Fig.4).

DISCUSSION

In this study, the mechanical properties and collagen content of human myocardium and trabeculae carneae in both ventricles were measured and compared. We found the stiffness of the right ventricle myocardium and trabeculae carneae to be higher in comparison to the left ventricle myocardium and trabeculae carneae, respectively. Our histological analyses of human cardiac tissues revealed that the collagen content of trabeculae carneae and myocardium from right ventricle were more than those in the left ventricle. This can be a possible explanation of the stiffer behavior of right ventricular tissues. Also, the collagen concentration of the trabeculae carneae were found to be higher than myocardium. However, we found stiffer behavior of the myocardium in comparison to the trabeculae carneae in the left ventricle. Possible explanation is the stiffer behavior of tissues under biaxial loading than under uniaxial loading condition [5]. Future testing needs to be done to compare the trabeculae carneae with the myocardial wall both under uniaxial conditions. The knowledge of the mechanical properties from this study will enable us to build accurate computational models of the heart to explore the underlying biomechanical mechanism and treatment for diastolic heart failure.

ACKNOWLEDGEMENTS

This work was supported by a National Innovation Award (15IRG23320009) from the American Heart Association, grant HL095852 from NIH, and the GREAT program at UTSA.

REFERENCES

[1] Halaney DL, et al. The effect of trabeculae carneae on left ventricular diastolic compliance: improvement in compliance with trabecular cutting. J Biomech Eng. 2016.

[2] Fernandez-Golfin C, et al. Left ventricular trabeculae: quantification in different cardiac diseases and impact on left ventricular morphological and functional parameters assessed with cardiac magnetic resonance. J Cardiovasc Med. 2009;10:827-33.

[3] Gruver EJ, et al. Uniformity of calcium channel number and isometric contraction in human right and left ventricular myocardium. Basic Res Cardiol. 1994;89:139-48.

[4] Urheim S, et al. Myocardial strain by Doppler echocardiography. Validation of a new method to quantify regional myocardial function. Circulation. 2000;102:1158-64.

[5] Demer LL, Yin FC. Passive biaxial mechanical properties of isolated canine myocardium. J Physiol. 1983;339:615-30.

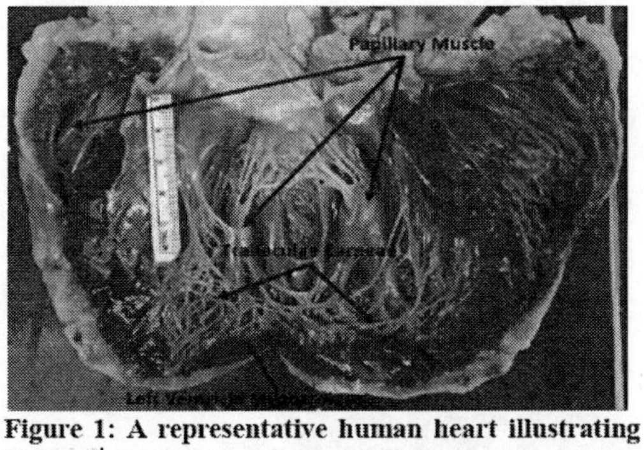

Figure 1: A representative human heart illustrating the trabeculae carneae and myocardium.

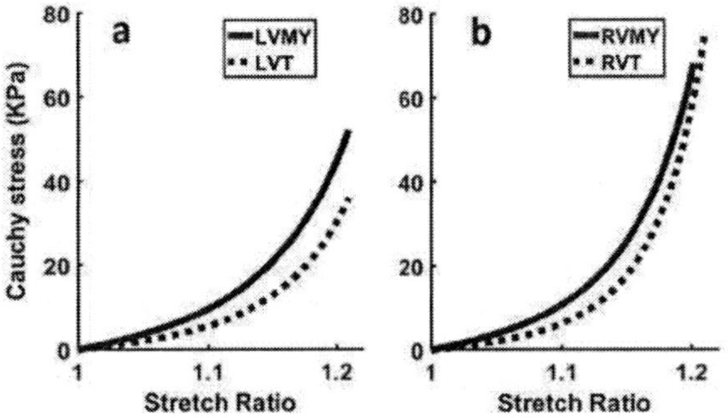

Figure 2: The average Cauchy stress versus stretch ratio curves within the 20% stretch range. a) Left ventricle myocardium (LVMY), Left ventricle trabeculae carneae (LVT); b) Right ventricle myocardium (RVMY), Right ventricle trabeculae carneae (RVT). Sample sizes are given in Figure 3.

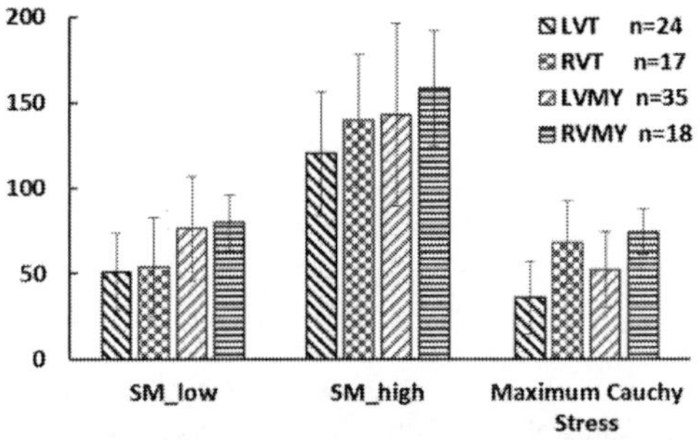

Figure 3: Comparison of the secant modulus SM_{low} at 5 kPa, secant modulus SM_{high} at 25 kPa and Maximum Cauchy stress among left ventricle myocardium (LVMY) and trabeculae carneae (LVT), right ventricle myocardium (RVMY) and trabeculae carneae (RVT) at 20% stretch ratio. Values are mean ±SD.

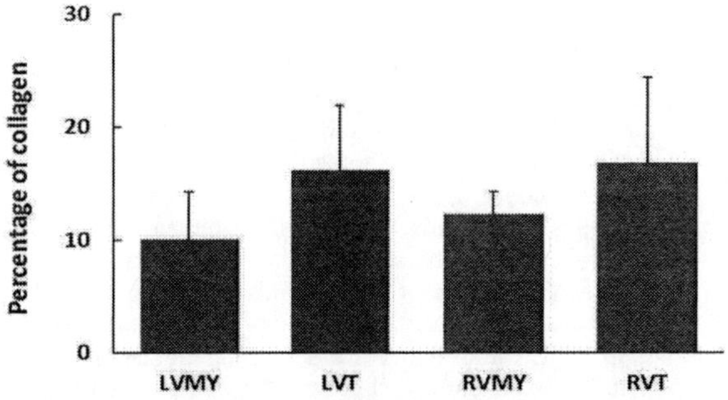

Figure 4: Comparison of Collagen content (% area) in human left ventricle myocardium (LVMY) and trabeculae (LVT), right ventricle myocardium (RVMY) and trabeculae (RVT). Values are mean ±SD (n=8).

CHARACTERIZING THE THREE-DIMENSIONAL MECHANICAL PROPERTIES OF PASSIVE MYOCARDIUM INJECTED WITH HYDROGELS USING A NOVEL NUMERICAL-EXPERIMENTAL INVERSE MODELING APPROACH

**David S. Li (1), Reza Avazmohammadi (1), João S. Soares (1),
Jason A. Burdick (2), Joseph H. Gorman III (3), Robert C. Gorman (3), Michael S. Sacks (1)**

(1) Center for Cardiovascular Simulation
Institute for Computational Engineering and Sciences
The University of Texas at Austin
Austin, Texas, USA

(2) Polymeric Biomaterials Laboratory
Department of Bioengineering
University of Pennsylvania
Philadelphia, Pennsylvania, USA

(3) Gorman Cardiovascular Research Group
Perelman School of Medicine
University of Pennsylvania
Philadelphia, Pennsylvania, USA

INTRODUCTION

The vast majority of heart failure (HF) deaths arise from detrimental remodeling of the left ventricle (LV) after myocardial infarction (MI). Expansion and remodeling of the infarct under the demand of the circulatory system is a major factor in the progression of MI towards HF and mortality [1]. However, recent studies have demonstrated that this process can be mitigated through the injection of biomaterials into the infarct region as bulking agents [2]. These early experiments have generated great interest in this therapeutic methodology due to the potential for developing minimally invasive surgical and percutaneous therapies to limit post-MI LV remodeling.

Despite promising initial success, optimization of the mechanical characteristics and injection patterns of these polymers is limited by a lack of biomechanical and physiological knowledge of the remodeling events that occur as a result of MI and physical support by the injectate. Modeling and simulation of injection therapies in the post-MI heart could guide systematic improvement of these optimal parameters, but require detailed three-dimensional (3D) anatomical, structural, and constitutive relations. Studies on the local structural-mechanical properties of healthy myocardium, as well as how they change in response to the presence of injected material, are needed to obtain critical information for designing simulations and linking their results to the physiological behaviors of the post-MI heart.

In collaboration with the Gorman Cardiovascular Research Group and the Polymeric Biomaterials Laboratory at the University of Pennsylvania, we present a comprehensive investigation on the effect of hydrogel injection on tissue-level mechanical properties of normal myocardium. Our approach utilizes a novel hyaluronic acid (HA)-based hydrogel injectate, combined with state-of-the-art and magnetic resonance imaging (MRI) methodologies and experimentally determined 3D material properties.

METHODS

Mechanical Testing of Viable Myocardium

We employed a previously developed novel triaxial mechanical testing device (TRIAX), designed to apply 3D deformations to single soft tissue specimens, for characterizing the full stress-strain response of non-contracting, healthy, drained LV myocardium in sheep (Fig. 1A).

Specimens of normal myocardium were harvested from a cohort of adult Dorset sheep (N=6). Euthanasia was performed with rapid injection of KCl, and 1x1x1-cm sections were cut out from the LV (Fig. 1B) and immediately placed in cardioplegic solution (Plegisol) with 20 mM 2,3-butanedione monoxime (BDM) in order to prevent hypercontraction due to cutting. Mounted in the TRIAX, each of the samples was subjected to an optimal set of cyclic tensile, compressive, and shear deformations within the span of a few hours. In particular, we applied 3 types of single-axis simple shear and 3 types of biaxial pure shear loading conditions (Fig. 1C), from which stress-strain information was collected. All tests were conducted with the sample immersed in cardioplegic solution heated to 37°C.

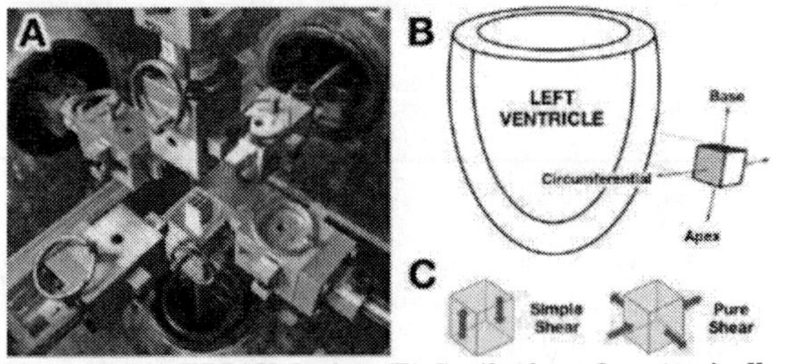

Figure 1: (A) TRIAX device. (B) Sectioning of anatomically oriented samples. (C) Simple/pure shear loading conditions.

Additionally, we monitored the relative change in viability of the healthy myocardium before and after TRIAX testing using modified MTT assays [3]. For each specimen, a Viability Index was computed by scaling the measured absorbance at 570 and 690 nm by the mass of the tissue sample, both immediately after cutting the specimens and upon completion of mechanical testing.

Mechanical Testing After Hydrogel Injection
Concurrently, we performed TRIAX tests on samples of healthy ovine LV injected with dual-crosslinked hyaluronic acid (HA)-based hydrogels, provided by the Polymeric Biomaterials Laboratory [4]. From each sheep of the original cohort, we also cut out adjacent 1x1x1-cm sections from the LV that were then directly injected with either 100 µL of cardioplegic solution (negative control) or 100 µL of hydrogel (treatment condition), equidistant between the epicardial and endocardial surfaces. After injection, the samples were incubated at 37°C and 5% CO_2 for 24 hours, and then underwent the same TRIAX tests as before.

Fiber Structure Analysis Using Diffusion Tensor MRI
Diffusion tensor MRI (DT-MRI) is a non-destructive method for studying the internal structure of soft fibrous tissues. It has recently developed as a means to measure myofiber orientations in the heart, as the diffusion of water molecules in tissue due to magnetic fields tends to be parallel with the orientation of fibers within the sample. DT-MRI produces 3D images of the samples with voxel-based tensors describing the local diffusivity of water. The first eigenvector of the tensor corresponds to the direction of maximum diffusion, and indicates the local orientation of myofibers in cardiac tissue [5].

Following TRIAX testing, each sample was placed into a custom probe and scanned in a 9.4-T imager. The gradient pulses of the DT-MRI were oriented in 7 positions around a unit sphere. The fiber architecture of the samples was obtained through computing and sorting the largest eigenvectors of the tensors in each voxel of the image that contained tissue.

Inverse Modeling with Finite Element Simulations
The combined data were then implemented into an inverse model using iterative finite element (FE) simulations emulating the experimental deformations *in-silico*. We characterized the mechanical behavior of the tissue with the stored energy density function $W(\mathbf{F})$, where $\mathbf{F}$ is the deformation gradient tensor employed by the TRIAX. The orthotropic Fung-type material is given as:

$$W(\mathbf{E}) = \frac{1}{2}c_0\left[e^{Q(\mathbf{E})} - 1\right] \tag{1}$$

where $\mathbf{E} = \frac{1}{2}(\mathbf{F}^{\mathrm{T}}\mathbf{F} - \mathbf{I})$ is the strain tensor, $\mathbf{I}$ is the identity tensor, c_0 is a positive constant, and Q is given as follows:

$$Q(\mathbf{E}) = c_1E_{11}^2 + c_2E_{22}^2 + c_3E_{33}^2 + 4c_4E_{12}^2 + 4c_5E_{13}^2 + 4c_6E_{23}^2 \tag{2}$$

with constants c_{1-6}. The relation for the state of stress of the specimen (the first Piola-Kirchhoff stress tensor $\mathbf{P}$) is written in terms of W as:

$$\mathbf{P} = \frac{\partial W}{\partial \mathbf{F}} - p\mathbf{F}^{-\mathrm{T}} \tag{3}$$

where p is the hydrostatic pressure enforcing material incompressibility. The parameters of the material model were determined through nonlinear least-squares regression.

RESULTS

The stress-strain curves from TRIAX testing showed clearly nonlinear, anisotropic, and slightly hysteretic behavior of the viable myocardium (Fig. 2A). When comparing the viability index of pre- and post-test with a paired t-test (n=15), we found no significant change in viability over the course of the TRIAX experiments, suggesting the tissue indeed remained viable throughout testing (Fig. 2B).

We also confirmed detectable stiffening in mechanical behavior for samples injected with the HA hydrogel (Fig. 2C). Through analysis of the DT-MRI, we were able to clearly distinguish the shape of the injectate within the tissue samples compared to normal un-injected samples (Fig. 3A) by examining intensity maps of the mean diffusivity (Fig. 3B). Examining the largest eigenvector of each voxel also showed a transmural variation in the fiber orientation (Fig. 3C–D), which is consistent with other histological studies of myocardium [6].

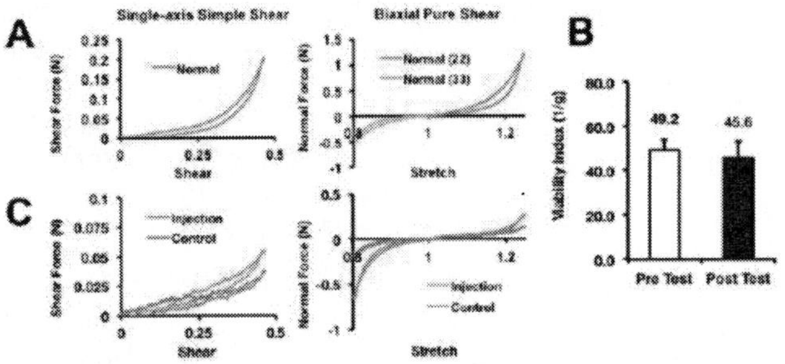

Figure 2: (A) Representative force-strain curves of simple/pure shear loading conditions for normal myocardium. (B) Viability Index before/after testing normal myocardium. (C) Comparison of force-strain curves for control/injection conditions.

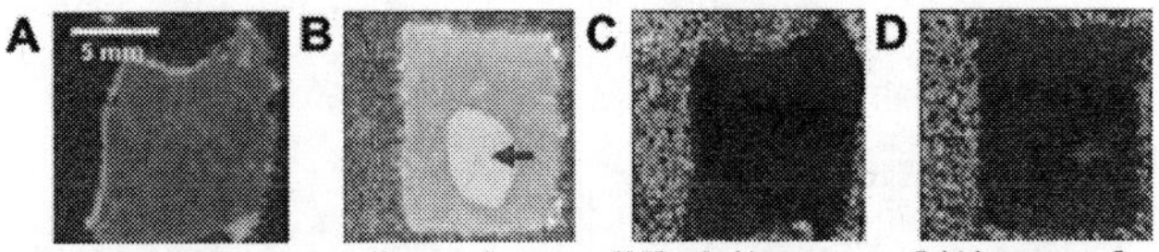

Figure 3: Cross-sections of mean diffusivity maps of (A) normal and (B) injected myocardium, where the injection is clearly visible. Color maps of the largest eigenvector orientation for (C) normal and (D) injected myocardium, showing transmural variation.

DISCUSSION

We have shown an integrated experimental-modeling framework for the characterization of the tissue-level mechanical properties of viable myocardium. Furthermore, we applied this approach toward evaluation of the effect of hydrogel injection on the local behavior of myocardium, confirming stiffening in mechanical properties and remodeling of myofibers around the injectate. In the future, we aim to characterize injected myocardium post-infarction, and the results from these studies can aid the development of clinically relevant organ-level models of injection therapies in the post-MI heart.

ACKNOWLEDGEMENTS

Funding was provided by the NIH (R01 HL068816, HL089750, HL070969, HL108330, HL063954, HL111090, and T32 EB007507).

REFERENCES

[1] Bui, AL et al., *Nat Rev Cardiol*, 8:30-41, 2011. **[2]** Ryan, LP et al., *Ann Thorac Surg*, 87:148-55, 2009. **[3]** Lacerda, CM et al., *Am J Physiol Heart Circ Physiol*, 302(10): H1983-90, 2012. **[4]** Dorsey, SM et al., *Biomaterials*, 69:65-75, 2015. **[5]** Wu, MT et al., *Circulation*, 114(10):1036-45, 2006. **[6]** Scollan, DF et al., *Am J Physiol*, 275(6):H2308-18, 1998.

MESENCHYMAL STEM CELL DELIVERY VIA A NOVEL CARDIAC PATCH IMPROVES RIGHT VENTRICULAR FUNCTION IN PULMONARY ARTERIAL HYPERTENSIVE RATS

Zhijie Wang (1,2), Eric G. Schmuck (3), David A. Schreier (2), Timothy A. Hacker (3), Naomi C. Chesler (2,3)

(1) Department of Mechanical Engineering
Colorado State University
Fort Collins, CO, USA

(3) Department of Medicine
University of Wisconsin – Madison
Madison, WI, USA

(2) Department of Biomedical Engineering
University of Wisconsin – Madison
Madison, WI, USA

INTRODUCTION

Right ventricular failure (RVF) is a common cause of death in patients suffering from chronic pulmonary arterial pressure overload caused by pulmonary arterial hypertension (PAH), an incurable, rapidly developing fatal disease [1]. While current treatments for PAH focus on pulmonary vasculature and only moderately improve symptoms, the gradual deterioration in cardiac function necessitates new treatment strategies that rescue the right heart to reduce mortality.

In the present study, we explored the treatment effects of mesenchymal stem cells (MSCs) for RVF secondary to PAH using an experimental rat model. We hypothesize that local MSCs delivery via a novel cardiac bioscaffold will improve RV performance without affecting PAH development. Our results from RV pressure-volume loops and histology demonstrate a therapeutic effect of the MSC-seeded bioscaffold on PAH RVs despite persistent pressure overload, whereas this therapeutic effect is missing in both intravenous (IV) and intra-myocardium (IM) MSC injection groups. These preliminary findings suggest a new treatment option for patients with failing RVs that could reduce mortality in PAH due to RVF.

METHODS

Animal. All procedures were approved by University of Wisconsin Institutional Animal Care and Use Committee. To induce PAH, 5-6 week old male Sprague Dawley (Harlan, Indianapolis, IN) rats were treated with a single s.c. injection of SU5416 at a dose of 20 mg/Kg and then housed in hypoxia chamber (10% O_2) for 4 weeks and in normal room air for another 2 weeks. These PAH rats then were treated with MSCs via 3 different delivery methods (IV, IM and MSC-seeded patch). Three weeks later, all treated groups were examined by RV function and myocyte morphology as described below. Similar examinations were performed in control rats and untreated PAH rats (age=11-12 week).

Bioscaffold manufacturing and MSCs delivery. A novel bioscaffold was employed for RV MSCs delivery using previously described methods [2]. Briefly, to induce bioscaffold formation, rat cardiac fibroblasts from passage 2-3 were plated at a density of ~1.1×10^5 to 2.2×10^5 per cm^2 in high glucose DMEM + 10% FBS and 1% penicillin/streptomycin for 10-14 days. Then, cardiac fibroblasts and secreted extracellular matrix were incubated with 2 mM EDTA solution at 37°C. To remove remaining cells, the bioscaffold sheet was incubated in alternating hyper/hypotonic tris-buffered saline followed by 24-48 hour incubation with 1% tri-nbutyl-phosphate (TnBp) at 4°C with constant agitation. Contaminating DNA was removed by 1 hour incubation with DNase 1. Finally, the bioscaffold was rinsed in 100% ethanol followed by washes with sterile water and PBS.

EGFP+rMSCs were obtained from Cyagen Biosciences Inc (Santa Clara, CA) and expanded per manufacturers' protocols. 2×10^6 cells were suspended in 50μl of Plasmalyte A and seeded on top of the bioscaffold for 2 hours right before surgery. After the rat was anesthetized and the chest was open, scaffolds were gently rinsed in Plasmalyte A, inverted and placed on the rat right heart. In another group of rats, the MSCs were delivered by intra-myocardial injections in multiple sites on the RV wall. In a third group of rats, the MSCs were injected via jugular veins.

RV in vivo pressure-volume (PV) loop measurements. In vivo RV PV relations were obtained using similar methods established in mice [3]. Briefly, rats were anesthetized with urethane (1.2 g/ kg), intubated and ventilated followed by a ventral midline skin incision. After the chest and lungs were retracted to expose the RV, a 1.9 Fr. admittance PV catheter (Scisense, Ithaca, NY) was introduced into the apex of the

RV. Data were recorded and analyzed using commercially available software (Notocord Systems, Croissy Sur Seine, France).

Echocardiography. In the MSC-seeded bioscaffold treated rats, transthoracic echocardiography was performed to assess cardiac function at baseline, after development of PAH and 3 weeks after MSC-patch treatment, using methods previously described [4].

Histology. After euthanasia, RVs were harvested, fixed in 10% formalin and stained with H&E and Sirius Red. RV cardiomyocyte morphology was measured by the cell area under the cross-section orientation and the cell width under the longitudinal orientation. RV fibrosis was measured by the collagen area percentage in the tissue.

Statistics. An unpaired student t-test was performed between different experimental groups. A paired student t-test was performed for the echocardiography data in the same subjects. All results are mean±SD. $P<0.05$ is considered statistical significant.

RESULTS

MSC-seeded bioscaffold treatment, not IV or IM MSC treatment, improves RV function despite persistent PAH

From the in vivo PV loop measurement, we confirmed pressure overload in the untreated PAH group and the persistence of pressure overload in all 3 MSC-treated PAH groups as evidenced by elevated RV systolic pressure (RVSP) and effective arterial elastance (E_a) (Fig. 1).

We also confirmed reduction in RV stroke volume (SV) and ejection fraction (EF) in the untreated PAH group compared to control, but restoration of SV and EF to control levels in the MSC-patch treated group only (Fig. 2). This trend was observed in cardiac output (CO) as well (data not shown). Importantly, therapeutic effects were missing in the IV and IM MSC injection groups. Moreover, while PAH led to significantly increased EDPVR ($p<0.05$), treatments by the IM MSC injection and MSC-seeded patch both reduced EDPVR to the control levels (data not shown).

MSC-seeded bioscaffold treatment improves RV function and reduces the degree of cardiomyocyte hypertrophy

From the echocardiography measurement (Table 1), we found that RV fractional shortening (FS) was significantly reduced in the untreated PAH group ($p<0.05$), and the MSC-seeded patch treatment limited the reduction. The RV/LV ratio at systole, but not diastole, was increased in untreated PAH group ($p<0.05$), whereas the increase was absent in the MSC-seeded patch treated PAH group. These data suggest that RV enlargement was reduced, at least during systole, by the MSC-patch treatment.

Finally, we examined RV hypertrophy by RV tissue weight and RV cardiomyocyte morphology. As shown in Table 2, PAH development led to an increase in all RV hypertrophy metrics, which remained elevated in the MSC patch treated group ($p<0.05$). However, the degree of RV cardiomyocyte enlargement was less in the treated group as evidenced by cell CSA and width measurements ($p<0.05$). Moreover, RV fibrosis tended to decrease ($p=0.07$, MSC patch vs. PAH) after the MSC patch treatment.

DISCUSSION

In summary, our results suggest potential therapeutic effects of localized MSC delivery to a dysfunctional RV wall via a novel bioscaffold. The restored SV, EF and CO indicates a general improvement in the RV function. FS and RV/LV ratio measured from echocardiography also showed an improved contractility and reduced RV chamber enlargement with the MSC patch treatment. Our initial examination showed that the degree of cardiomyocyte hypertrophy was reduced by the MSC patch treatment. However, how exactly the local MSC delivery via a bioscaffold, not IM injection, led to RV myocardium recovery awaits further investigation. Further refinement of the MSC patch treatment may eventually provide novel treatment for RVF secondary to PAH and reduce mortality in PAH patients.

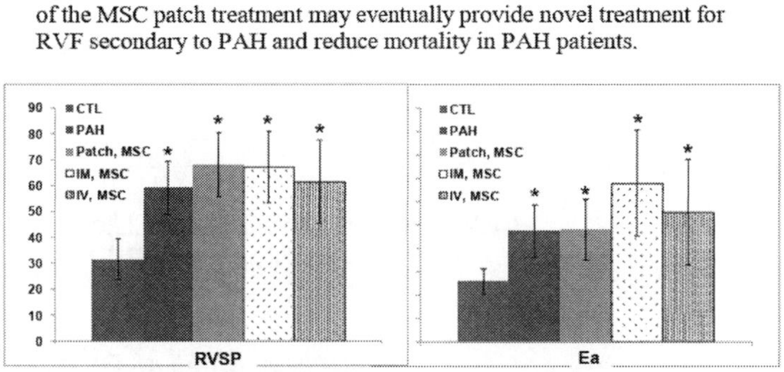

Figure 1: Persistent increases in RVSP (left; mmHg) and Ea (right; mmHg/μL) in all 3 MSC-treated PAH groups. Patch, MSC: MSC-seeded bioscaffold treatment; IM, MSC: IM injection of MSCs; IV, MSC: IV injection of MSCs. N=5-8 per group. *P<0.05 vs. CTL.

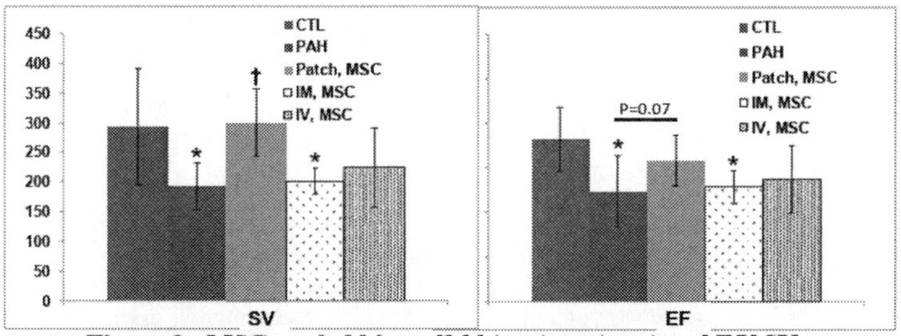

Figure 2: MSC-seeded bioscaffold treatment restored RV SV (left; μL) and EF (right; %) in PAH rats. Patch, MSC: MSC-seeded bioscaffold treatment; IM, MSC: IM injection of MSCs; IV, MSC: IV injection of MSCs. N=5-8 per group. *P<0.05 vs. CTL. †P<0.05 vs. PAH.

Table 1: Echocardiography results of the rats treated with MSCs-seeded bioscaffold. FS: fraction shortening. TAPSE: tricuspid annular plane systolic excursion. For RV/LV ratio: s-systole; d-diastole. N=7-8 per group. *p<0.05 vs. baseline.

	Baseline	PAH	PAH+ MSCs
FS (%)	0.72±0.15	0.42±0.13*	0.50±0.14
TAPSE (mm)	1.87±0.2	1.58±0.36	2.15±0.58
RV ID/LV ID, s	0.31±0.05	0.45±0.08*	0.39±0.1
RV ID/LV ID, d	0.38 ±0.03	0.42±0.03*	0.44±0.05*

Table 2: RV hypertrophy measurements in control, PAH and MSCs-seeded bioscaffold treated PAH rats. BW: body weight. Fulton Index: RV/(LV+S) mass ratio. CSA: cross-section area. N=4-8 per group. *p<0.05 vs. control. †p<0.05 vs. PAH.

	Control	PAH	PAH+ MSCs
Fulton Index (%)	25±2	67±11*	65±8*
RV/BW	0.58±0.03	1.58±0.23*	1.45±0.18*
Myocyte CSA (μm²)	246±45	416±34*	326±35†*
Myocyte width (μm)	11±1	13±1*	12±1†*
RV fibrosis (%)	14±5	29±5*	22±8*

REFERENCES

[1] Voelkel NF et al. *Circulation*, 114:1883-1891, 2006.
[2] Schmuck EG et al. *Cardiovasc Eng Technol*, 5(1): 119–131, 2014.
[3] Wang Z et al. *Physiol Rep*, 1(7): e00184, 2013.
[4] Golob, MJ et al., *J Biomech*, 48:405-412, 2015.

PATIENT-SPECIFIC MODELING OF THE ELECTRO-MECHANO-FLUIDIC FUNCTION OF THE LEFT VENTRICLE AND THE AORTA

Christoph M. Augustin (1), Gernot Plank (2), Shawn C. Shadden (1)

(1) Mechanical Engineering
University of California, Berkeley
Berkeley, CA, USA

(2) Institute of Biophysics
Medical University of Graz
Graz, Austria

INTRODUCTION

Combined electromechanical and hemodynamic computer models of the left ventricle (LV) and the aorta are considered an important tool for analyzing the interplay between LV deformation, valvular anatomy and flow patterns. Typically, image-based kinematic models describing endocardial motion or MRI-based flow measurements in the LV outflow tract are used as an input to blood flow simulations. While such models are suitable for analyzing the hemodynamic status quo, they are limited in predicting the response to interventions that alter afterload conditions, as this approach assumes that the heartbeat itself will remain unchanged. Electro-mechano-fluidic simulations of the LV have the potential to overcome this limitation, but are more difficult to formulate, parameterize and compute.

In this study, we report on our recent advancements in developing an automated workflow for the creation of patient-specific EM models of the LV and the aorta that are suitable to drive blood flow simulations. Models are custom-tailored to individual patients to ensure that the *in-silico* simulation replicates clinical observations for a given patient. These models can be used to probe the effect of different therapeutic options and thus identify the therapy that yields the best post-treatment outcome.

METHODS

Patient Data: Clinical data of patients with aortic valve disease (AVD) and of patients with coarctation of the aorta (CoA) all preceding a cardiac magnetic resonance study were used. This acquired data includes anatomical 3D whole heart MRI scans; LV volume traces, end-diastolic (ED) and end-systolic volume, stroke volume and ejection fraction based on Cine MRI; invasive pressure traces for CoA patients and estimates of LV peak pressure based on combined cuff pressure measurements and ultrasound-based pressure drops across the aortic valve; tagged-MRI based measurements of circumferential and radial strains in basal, mid and apical LV slice. All non-invasively recorded data were acquired also during clinical follow-up after AVD repair or stenting of a coarctation.

Building Anatomical LV and Aortic Root Models: Finite element meshes of the LV anatomy were generated from 3D whole heart MRI acquired at end diastole with 1.458×1.548×2mm resolution. Data were acquired at the German Heart Center Berlin (DHZB). Multi-label segmentation of the LV myocardium and cavity, left atrium and aortic cavities was performed at the DHZB using the ZIB Amira software (https://amira.zib.de/). This segmentation was smoothed and upsampled to a 0.1mm isotropic resolution using a variational smoothing method [1]. The wall of the aorta was automatically generated by dilation of the aortic cavity with a thickness of 1.2mm, and the aorta was manually clipped just before the branch of the brachiocephalic artery. The purpose of this clipped aorta was to provide physiological boundary conditions for the mechanical problem that still enabled a vertical movement of the left ventricular base as it is observed in-vivo. The resulting high resolution multi-label segmentation was meshed using CGAL (https://www.cgal.org/) with a target size of 1.25mm in the LV myocardium and 1mm in the aorta wall. The various stages of the anatomical modeling workflow are illustrated in Figure 1.

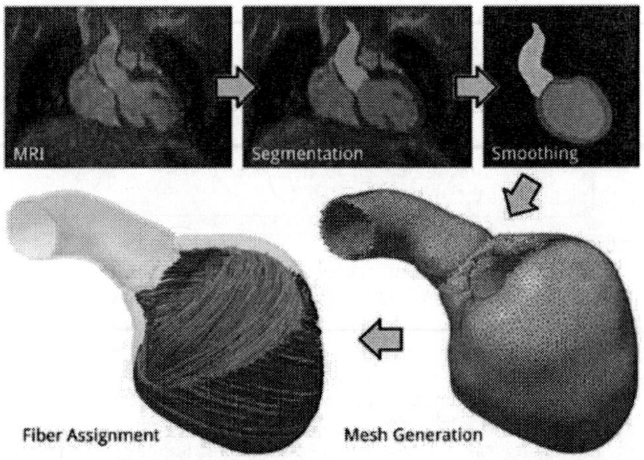

Figure 1: Workflow to build anatomical left ventricle and aortic root models from MRI image data, see [1,3].

Electro-mechanical Model of the LV: The electrophysiology (EP) of the LV was represented by the tenTusscher-Noble-Noble-Panfilov model [2]. Activation sequences and electrical source distribution were computed on the mesh representing the LV in its ED configuration using a coupled reaction-eikonal model. Activation sequences were indirectly parameterized using the QRS complex of a given patient's ECG as guidance [3]. Mechanical deformation was simulated using the equilibrium equations of finite elasticity. The same discrete LV anatomy model with the same fiber architecture was used for both the solid mechanics and the EP problem. The LV myocardial tissue was characterized as a hyperelastic, nearly incompressible, transversely orthotropic material with a non-linear stress-strain relationship [4]. Total stress was additively combined from passive and active stresses and a phenomenological contractile model was used to represent active stress generation [5]. Mechanical boundary conditions were applied to fix the terminal rim of the clipped aorta. The apex of the LV was stabilized by resting the LV on an elastic cushion which was rigidly anchored at its base. A lumped 3-element Windkessel model was used to provide the pressure-flow relationship during ejection (Fig. 2A). Details of the numerical approach were described in [6].

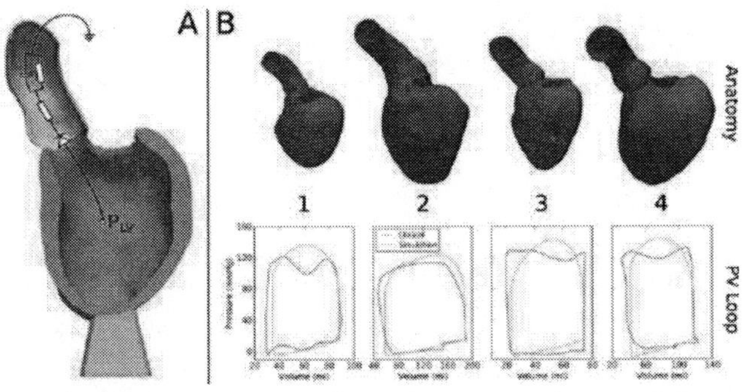

Figure 2: (A) Representation of mechanical boundary conditions. (B) Personalized models and PV loops for four patients.

Fitting of Mechanics and Hemodynamics Model: Using the ED geometry, default material parameters and the recorded ED pressure, an initial guess of the stress-free reference configuration was computed. This unloading procedure was repeated with trial material parameters to fit the ED pressure-volume (PV) relation and stress-free residual volume to the empiric Klotz relation [7]. During the isovolumetric contraction (IVC) phase the cavity volume of the LV was held constant. Rate of contraction and peak stress of the active stress model were adjusted to fit the maximum rate of rise of pressure during the IVC phase. When the LV pressure exceeded the aortic pressure, ejection was initiated by connecting the LV model with the lumped Windkessel model. Recorded volume traces and estimated (AVD) or recorded (CoA) pressure traces of a given patient during ejection were used as input to fit the parameters of the Windkessel model, which represents flow resistance of the aorta as well as resistance and compliance of the systemic circulation.

Model Validation was performed using additional, independent datasets on strains and various kinematic measures such as the motion of the endocardial surfaces as well as rotation and torsion of the LV, evaluated from tagged-MRI based time sequences. These datasets describing the kinematics of the LV were not used in the fitting procedures. Circumferential strain was evaluated in the model and averaged over the same short axis slice used to acquire the 2D tagged MRI sequence in that patient (Fig. 3A). This averaged strain was plotted over time and compared against the circumferential strain determined from analysis of the tagged MRI (Fig. 3B).

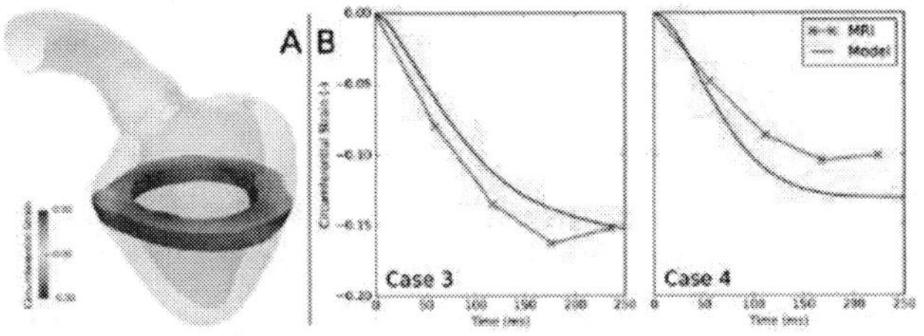

Figure 3: Validation of strains predicted by the in-silico model against MRI data. Strain was averaged over a slice (A) and plotted over systole (B) for the simulation (red) and tagged MRI (blue).

RESULTS AND DISCUSSION

The presented methodology allows the efficient generation and execution of patient-specific EM models of LV and aortic root that are suitable for driving simulations of cardiac and vascular hemodynamics when coupled with a blood flow model. The goodness of fit was verified for all these models under the pre-treatment scenario to ensure that all models replicate the clinical observations with sufficient physiological fidelity. Fitting results are shown for four cases in Fig. 2B and strain validation results are shown in Fig. 3B.

Model fitting efforts were mainly focused on reproducing clinically recorded ECG data during the activation phase and on pressure and LV volume traces during the IVC and ejection phases, as these phases govern cardiac motion that is driving flow in the LV outflow tract across the aortic valve into the aorta. The computational cost of the EP simulation of an entire cardiac cycle including parameterization were minor with an overall expenditure of time of less than 25 min. Fitting of the mechanical parameters was computationally more expensive. The major costs incurred for fitting passive mechanical tissue properties to the Klotz relationship and the fitting of the active stress model summing up to a total of approx. 36 hours. Fitting of the circulatory Windkessel parameter and the synchronization of volume with pressure traces was comparably cheap and could be typically achieved in less than 20 min. Executing the simulation of one heartbeat with the final fitted model lasted around 4 h.

While the modeling performed so far was considering only the pre-treatment scenarios and the quantitative analysis of model predictions has not been completed yet, the efficiency of the developed workflow suggests that such biophysically detailed EM models are becoming a viable option as a driving model for hemodynamic analysis for optimizing therapies. Overall, the developed methodology constitutes an important and nontrivial step towards more holistic fully coupled electro-mechano-fluidic models that represent all the involved physical fields and their bidirectional mechanistic links based on first principles.

ACKNOWLEDGEMENTS

This work was supported by the NSF SSI program, the EU project CARDIOPROOF (ICT-2013.5.2, Grant Agreement 611232) and the Max Kade Foundation.

REFERENCES

[1] Crozier, A et al., *Ann Biomed Eng*, 44(1):58-70, 2015.
[2] ten Tusscher, K et al., *AJP Heart Circ Physiol*, 286:H1573-89, 2004.
[3] Augustin, CM et al., *Europace*, 18(suppl 4):iv121-iv129, 2016.
[4] Guccione, JM et al., *J Biomech*, 28:1167-1177, 1995.
[5] Niederer, SA et al., *Cardiovasc Res*, 89:336-343, 2011.
[6] Augustin CM et al., *J Comput Phys*, 305:622-646, 2016.
[7] Klotz, S et al., *Nat Protoc*, 2:2152-2158, 2007.

ESTABLISHING CREDITABILITY OF THE LIVING HEART PORCINE MODEL

Brian P. Baillargeon (1), Kevin L. Sack (2), Gabriel Acevedo-Bolton (3), Daniel B. Ennis (4), Ghassan S. Kassab (5), Thomas Franz (2,6), Julius M. Guccione (3)

(1) Dassault Systemes SIMULIA Corp
Santa Clara, CA, USA

(2) Department of Human Biology
University of Cape Town
Cape Town, South Africa

(3) Department of Surgery
UCSF
San Francisco, CA, USA

(4) Department of Radiological Sciences Diagnostic Cardiovascular Imaging Section
University of California

(5) California Medical Innovations Institute, Inc.,
San Diego, CA, USA

(6) Faculty of Engineering and the Environment
University of Southampton
Southampton, UK

INTRODUCTION

The SIMULIA *Living Heart Porcine Model* (LHPM) is a dynamic high-fidelity model of a 4-chamber heart corresponding to a healthy, fully-developed pig. The dynamic response of the LHPM is governed by realistic electrical, structural, and fluid (blood) flow physics. As such, it can be used to study the electrical behavior of the heart by itself or to examine the coupled electromechanical behavior wherein the mechanical response is driven by an electrical excitation. Although the model represents a healthy heart, one can also study abnormal (diseased) states by modifying any of its geometry, material properties, and/or loads and boundary conditions. In addition, one can introduce other parts into the model (such as medical devices) to study their influence on cardiac function and explore treatment options. The purpose of the current study is to assess model credibility by comparing the simulation predictions to experimental measurements.

METHODS

The LHPM is an Abaqus 2016 finite element (FE) model using a sequentially coupled electrical-mechanical analysis of a cardiac cycle with a duration of 0.8 seconds, which corresponds to a heart rate of 75 beats per minute [1]. The FE mesh was generated from an explanted heart using Diffusion Tensor MRI (DT-MRI). The images were segmented and the mesh generated by using Simpleware ScanIP. The resulting mesh of 332,146 3D linear tetrahedral elements was used for both the electrical and mechanical simulations. The muscle fiber orientations were calculated at the centroid of each element using the DT-MRI data. The raw DT data at the voxels was filtered to eliminate noise and non-fibrous voxels prior to interpolation to each element centroid. The resultant orientation for the myofibers follow papillary muscle features, are tangent to the epi- and endocardial surfaces, and exhibit the transition from positive to negative helix angles in the left ventricle (LV) from epicardial to endocardial surfaces, respectively. The FE mesh and orientations can be found in Figure 1 below.

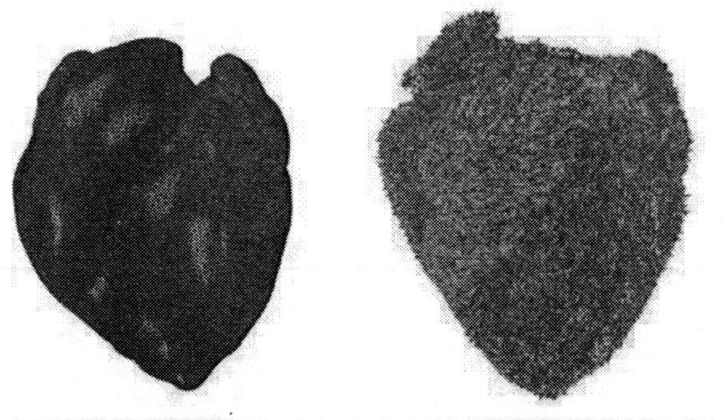

Figure 1: Finite Element Mesh and Myofiber Orientations for the Living Heart Porcine Model.

The constitutive behaviors for both the electrical and mechanical simulations are the same as those described in Baillargeon et al. [2]. The electrical properties were calibrated to provide physiologically observed activation times based upon human physiology, given we did not have experimentally measured activation times or those representative of a pig (see Conduction Systems Tutorial [3]). The passive material properties were calibrated using experimental data from Sommer et al. [4]. The time-varying elastance active tissue response parameters were initially set to be equal to those reported by Guccione et al. [5]; however, several of the values were modified to obtain realistic (i.e. physiologically observed) behavior.

The blood flow is modeled using a combination of surface-based fluid cavities and fluid exchanges [1]. The circuit shown in Figure 2 is a schematic representation of the blood flow model where resistors represent flow resistances and capacitors represent structural compliances.

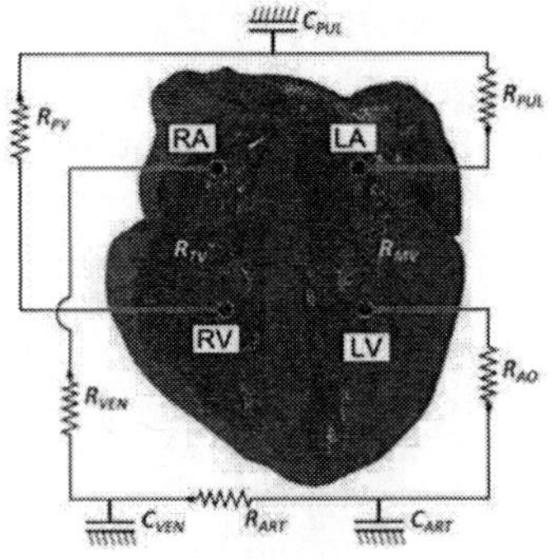

Figure 2: Schematic representation of the blood flow model.

RESULTS

A summary of some key model results obtained during the third cardiac cycle is provided in Table 1 and are compared to those experimentally measured by the California Medical Innovations Institute for the same porcine heart before it was explanted. The LV ejection predicted by the model compares favorably to that of the experimentally measured value, being low by 7.7%. The maximum pressures developed in the LV and RV are over-predicted by 16.6% and 19.2%, respectively. This indicates that while the volume changes (i.e. EF) are well represented, the boundary conditions representing the circulatory system downstream from these structures need refinement.

Table 1: Comparison of key model results to experimentally measured quantities.

Output	Model Result	Experimentally Measured
L. Ventricle (LV) Ejection Fraction	39.8%	43.1%
Max LV Pressure	70.3 mmHg	60.3 mmHg
Min LV Pressure	-1.63 mmHg	-0.57 mmHg
Max RV Pressure	23.5 mmHg	19.7 mmHg
Min RV Pressure	-4.0 mmHg	0.79 mmHg

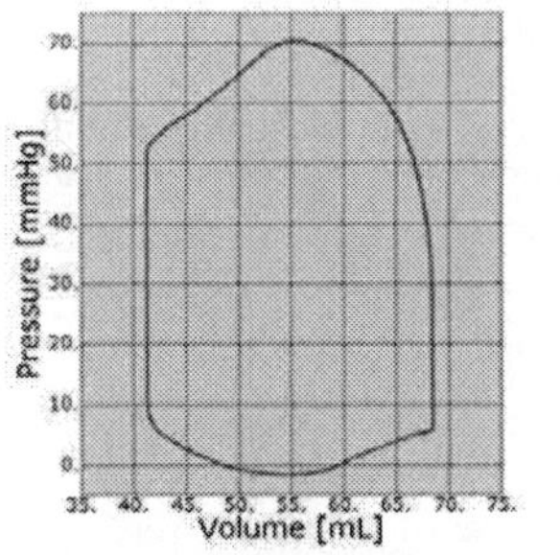

Figure 3: Pressure-Volume Loop of the Left Ventricle for the 3rd Cardiac Cycle.

DISCUSSION

In this study, a high fidelity multi-physics FE simulation of a porcine model was created and the simulation credibility was assessed. The model was constructed using DT-MRI data and methods consistent with those of the SIMULIA *Living Heart Human Model* (LHHM) [1]. Unlike the LHHM model, experimentally measured hemodynamic properties are available for this specific animal, thus allowing for a more rigorous validation of the model. The initial assessment of the model predictions show that the LV volume changes during the cycle are well predicted, but the minimum and maximum pressures predicted by the model are in general over-estimated. This indicates that the downstream boundary conditions imposed by the lump parameter circulation system can be adjusted to improve the model prediction. The computational cost to explore the impact of the various model parameters will be significant, however. Future work will include refinement of the lumped parameter circulation system, as well as a more rigorous validation of the model predictions using measured LV wall strains.

ACKNOWLEDGEMENTS

This work was performed as part of the Living Heart Project.

Financial support: This work was supported by NIH grants R01-HL-077921 and R01-HL-118627 (J.M. Guccione) and U01-HL-119578 (J.M. Guccione); K25-NS058573-05 (G. Acevedo-Bolton). K.L. Sack acknowledges financial support through a scholarship from the Oppenheimer Memorial Trust. K.L. Sack and T. Franz acknowledge financial support from the National Research Foundation of South Africa.

REFERENCES

[1] Abaqus 2016 Analysis User's Guide, 2015. Dassault Systèmes Simulia Corporation.

[2] B. Baillargeon, N. Rebelo, D.D. Fox, R.L. Taylor and E. Kuhl. The living heart project: a robust and integrative simulator for human heart function. Eur. J. Mech. A/Solids. 48:38-47, 2014.

[3] Conduction System Tutorial – Gap Junctions. http://www.vhlab.umn.edu/atlas/conduction-system-tutorial/gap-junctions.shtml.

[4] G. Sommer, A.J. Schriefl, M. Andrä, M. Sacherer, C. Viertler, H. Wolinski, and G.A. Holzapfel, Biomechanical properties and microstructure of human ventricular myocardium. Acta biomaterialia, 24:172-192, 2015.

[5] J.M. Guccione, S.M. Moonly, P. Moustakidis, K.D. Costa, M.J. Moulton, M.B. Ratcliffe, and M.K. Pasque, Mechanism underlying mechanical dysfunction in the border zone of left ventricular aneurysm: a finite element model study. Ann Thorac Surg, 71: 654–662, 2001.

TITLE – BUILDING A BIOMEDICAL IMAGING CLINICAL APPLICATIONS PROGRAM: FLUID DYNAMICS AND DISEASE – ATRIBUTE TO DON GIDDENS

DIEGO R MARTIN (1)

(1) Department of Medical Imaging,
University of Ari□ona, Tucson, A□ USA

INTRODUCTION

This presentation will provide a tribute to Dean Don Giddens, with attention to his contributions in clinical translational sciences while Dean at Georgia Tech. Here I will review the background of individuals, devices, organization and processes that, under Dean Giddens leadership, developed from initiatives to thriving programs with intersections between engineering and medicine.

METHODS

A historical perspective on engineering will be provided in relation to its role in health care. Dean Giddens role in formulating a biomedical imaging group and structure will be reviewed. A specific initiative will be described, related to the use of magnetic resonance imaging for deriving detailed, but limited, vascular flow information, combined with computational fluid dynamics, to gain new insights into important etiological factors of disease.

RESULTS

A sustainable biomedical imaging program has been formulated. The flow initiative has grown has evolved from project to program.

DISCUSSION

The biomedical imaging program has led to dozens of graduate students, post-docs, and MD-PhD graduates who now populate positions in industry and academia. Dean Giddens impact is difficult to grasp quantitatively as it is an exponential function, multiplying through the people trained and their subsequent roles in various areas of focus and in various parts of the country/world, including key leadership positions.

ACKNOWLEDGEMENTS

Thanks to the BBB Conference organizers for arranging this tribute for an important and distinguished contributing member of our community.

REFERENCES

DON GIDDENS AND RESOLUTION OF THE SHEAR STRESS CONUNDRUM IN ATHEROSCLEROSIS

Christopher K. Zarins, MD

Department of Surgery
Stanford University
Stanford, CA, USA

INTRODUCTION

The observation that plaques tend to occur at aortic ostia and bifurcations led to the idea that flow dynamics may play a role in plaque formation. This stimulated a great deal of research in the 1970-80s focused on flow velocity, shear stress, flow disturbances and turbulence and their relation to the endothelium, platelet deposition, cellular proliferation, lipid deposition, inflammation and necrosis. The prevailing hypothesis was that high shear stress induced endothelial injury and that this was a key initiating factor in the cascade of events leading to plaque formation in the arterial tree. While this could be shown in experimental animals, it was unknown exactly where atherosclerotic plaques formed in humans. Was it in areas of high shear or low shear? A young aerospace engineer at Georgia Tech, Don Giddens, addressed this question with a remarkable series of experiments performed in collaboration with a vascular pathologist and vascular surgeon at the University of Chicago [1] [2] [3]. This abstract will review the key evidence demonstrating that human plaques form in areas of low wall shear stress and discuss Don Giddens' seminal contributions in resolving the shear stress conundrum.

METHODS

A glass model carotid bifurcation was constructed based on human angiographic data. Model flow studies were performed under steady and pulsatile flow conditions at Reynolds numbers of 300-800. Flow patterns were visualized using dye-washout and hydrogen bubble techniques. Laser-Doppler anemometry was used to determine flow velocity profiles and shear stresses throughout the bifurcation. The 3D distribution of intimal plaque was quantitated in pressure-perfusion fixed human carotid bifurcations obtained at autopsy. Intimal plaque thickness, measured at 15 degree intervals around the circumference at 5 levels of the human bifurcation, was correlated to shear stress values at corresponding points in the model bifurcation.

RESULTS

Flow streamlines in the bifurcation were skewed towards the inner wall of the carotid sinus with a large area of flow separation along the outer wall (Fig 1). Flow along the inner wall was unidirectional and axially aligned whereas opposite the flow divider in the area of separation, the flow pattern was complex with reversal of axial flow and counter-rotating helical trajectories (Fig 2). Shear stress was high along the inner wall and shear was low along the outer wall where direction of the shear stress vector oscillated within along the vessel wall (Fig 3). Intimal thickness was minimal along the high shear inner wall of the internal carotid whereas significant intimal plaque consistently formed in the low shear outer wall region of the bifurcation (Fig 4). [4] [5]

DISCUSSION

This study was the first to provide quantitative evidence that human atherosclerotic plaques form in regions of low rather than high shear stress, thus providing a resolution of the shear stress conundrum. However, the shear stress story and the contributions of Don Giddens do not end here. Rather, this was just the beginning of a productive 20 year interdisciplinary effort between engineering, pathology and surgery to better understand the complex relationship between hemodynamics and atherosclerosis. While low shear stress induced plaque deposition, high shear stress had the opposite effect, inhibiting intimal thickening. High shear also induced arterial enlargement, thus limiting the effect of stenosing arterial plaques [2]. This could be shown in human coronary, carotid, aorta and peripheral arteries and could be reproduced in experimental animal models [3]. The integration of fluid dynamics engineering, and human cardiovascular pathology with the ability to manipulate flow in both humans and experimental animals was essential for success in this effort. Don Giddens' accomplishments 30 years ago in the productive collaboration between Georgia Tech and the University of Chicago demonstrated the importance of close interaction between engineering, biology and the clinical sciences. Institutions have taken note of this and today we see outstanding GT-Emory and Stanford Bio-X (Engineering-Medicine) programs.

It should be noted that these early experiments were carried out before in-vivo human imaging with CT/MR and effective computational technology was available. However, the findings are still relevant today and have served as the basis for widespread investigations using modern cardiovascular imaging and flow analysis using computational fluid dynamics. These have led to many clinically relevant applications, including the computation of coronary blood flow from patient-specific coronary CT scan imaging data for defining the hemodynamic significance of coronary atherosclerosis [6]. We thank Don Giddens for his pioneering work in introducing quantitative flow assessment to the study of human atherosclerosis.

Figure 1: Hydrogen bubble flow visualization with flow streamlines skewed to inner wall with flow separation along outer wall

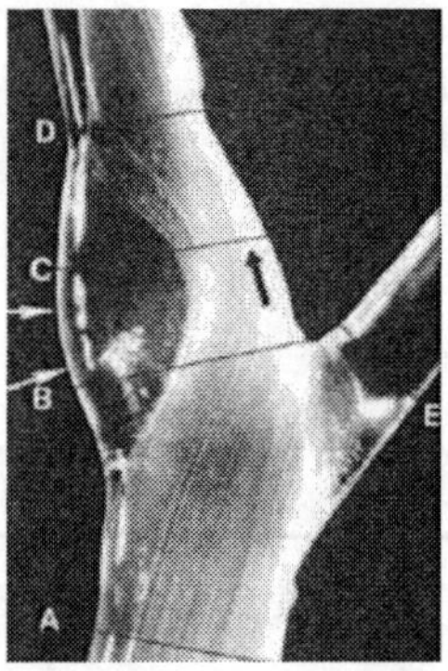

Figure 2: Flow reversal and counter-rotating helical trajectories and delayed dye wash-out along the outer wall of the carotid sinus

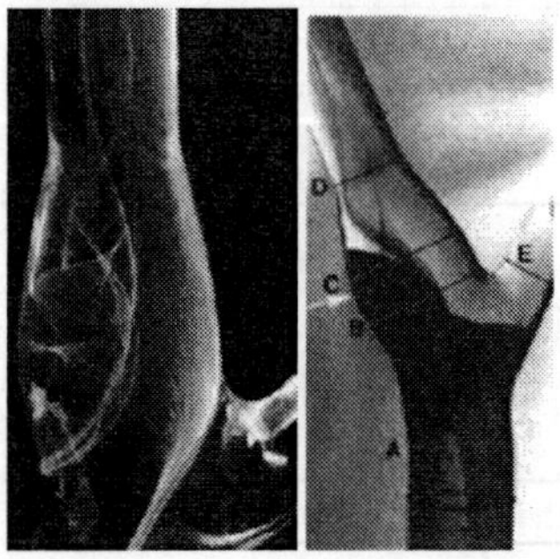

Figure 3: High shear stress and unidirectional shear along the inner wall with low and oscillating shear along the outer wall

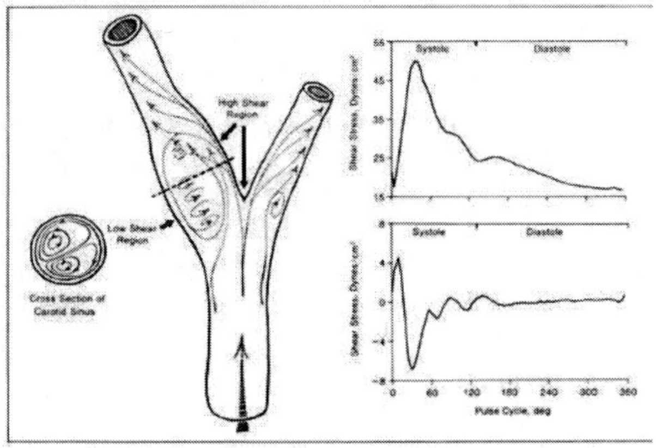

Figure 4: Pressure-perfusion fixed human carotid bifurcation sectioned at 5 levels. Significant intimal plaque is visible along the outer wall of the internal carotid sinus at levels B and C with no intimal thickening along the inner wall

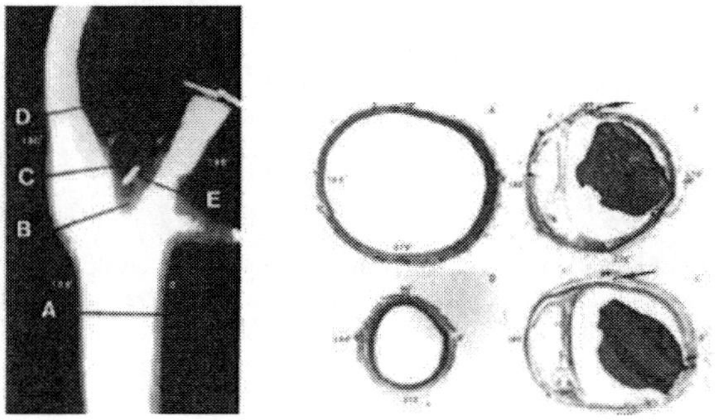

REFERENCES

[1] Giddens DP, Zarins CK, Glagov S, *Appl Mech Rev* 43(5): S98-102, 1990

[2] Giddens DP, Zarins CK, Glagov S, *J Biomech Eng* 116(1): 588-594, 1993

[3] Glagov S, Zarins CK, Giddens DP, Ku DN, *Arch Path Lab Med* 112(10): 1018-1031, 1998

[4] Zarins CK, Giddens DP, Bharadvaj BK, et al, *Circ Res* 53: 502-514, 1983

[5] Ku DN, Giddens DP, Zarins CK, Glagov S, *Arterioscler* 5: 293-302, 1985

[6] Zarins CK, Taylor CA, Min JK, *J CV Transl Res* 6: 708-714, 2013

SB³C2017
Summer Biomechanics, Bioengineering and Biotransport Conference
June 21 – 24, Tucson, AZ, USA

MICROPATTERNING DRIVES ENDOTHELIAL CELL ALIGNMENT AND FUNCTION ON VASCULAR GRAFTS

Matthew W. Hagen, Deirdre E. Anderson, and Monica T. Hinds

Department of Biomedical Engineering
Oregon Health & Science University
Portland, OR, USA

INTRODUCTION

Atherosclerosis, the gradual thickening and hardening of arterial walls is the primary contributor to life-threatening cardiovascular disease in the developed world. Because atherosclerosis is non-reversible, advanced cases are often treated through arterial replacements. Autologous grafting is the current clinical standard for these procedures, however many patients lack suitable donor tissue. Synthetic solutions in the form of ePTFE and Dacron grafts exist only for large diameter vessel replacement. Luminal narrowing resulting from blood reactions to foreign surfaces prevents the application of current synthetics to critical small-diameter applications including coronary and peripheral arteries. While newer materials such as poly(vinyl alcohol) (PVA) have the potential to reduce luminal narrowing, many vascular engineers have turned to the idea of in vitro endothelialization: the growth of a native endothelium on a device prior to implantation.

Native vascular endothelium balances between hemostasis and thromboresistance. The endothelial cell (EC) response to fluid shear stress (FSS) driven by local blood flow is among the key factors determining the atherogenicity and thrombogenicity of an EC. ECs exposed to unidirectional FSS in straight vessels elongate in the direction of that flow and exhibit an atheroprotected phenotype, defined by reductions in immunogenic adhesion molecule expression and increases in the production of antithrombotic factors including nitric oxide. In contrast, ECs exposed to low or oscillatory flow in curved or branching vessel segments maintain a rounded cobblestone shape and exhibit an atheroprone phenotype. Importantly, this effect of morphology has been replicated in static culture, where ECs elongated by growth on a micropatterned substrate exhibit an atheroprotective phenotype analogous to those exposed to unidirectional FSS. This technique has demonstrated effectiveness using both biochemically restrictive patterning to force ECs into narrow lanes, as well as topographical patterning which can drive the elongation of ECs in a confluent monolayer. The ability of micropatterning to drive an atheroprotective phenotype under static and hemodynamic conditions makes it a promising potential tool for synthetic graft development

However, a significant weakness of in vitro graft endothelialization is the invasiveness and risk of harvesting vascular ECs from a hemodynamically compromised patient. Recent attention has turned instead to endothelial outgrowth cells (EOCs), a population derived from circulating progenitors which can differentiate into an endothelial morphology. These cells proliferate rapidly and can be easily isolated through a standard venous blood draw. Several groups including ours have demonstrated that EOCs can function as an endothelium on artificial devices. Cultured EOCs have furthermore shown the same thromboresistant reaction to unidirectional FSS as ECs, however EOC micropatterning has yet to be examined. We hypothesized that micropatterning EOCs using both biochemical and topographical micropatterns will drive elongation and thromboresistance analogous to ECs.

METHODS

Endothelial and EOC isolation: Primary ECs were isolated from the carotid arteries of non-human primates using collagenase and mechanical disruption or purchased HUVECs. EOCs were isolated from a venous blood draw following established methods. [1]

Micropatterned lanes on slides: A vacuum pump pulled collagen Type I through PDMS molds on either glass or polystyrene surfaces to create 25μm wide lanes with 100μm spaces. The collagen was incubated for 1 hr and then rinse extensively and a blocking agent of either BSA or Pluronic® (copolymer of ethylene oxide and propylene oxide) was applied.

Technical Presentation #147

Micropatterned lanes on PVA grafts: An aqueous solution of PVA with sodium trimetaphosphate crosslinker was cast onto 4mm diameter cylindrical molds, which were either smooth or patterned. Patterns consisted of channels.

Cell assessment: ECs and EOCs were imaged at 48hrs. HUVECs were stained for F-actin, VE-cadherin, and nuclei. Cell attachment was quantified at 24hrs.

qtPCR: RNA isolation and reverse transcription were performed on ECs in each condition at 24 hours. Primers for TF, TFPI, vWF, TM, eNOS, and GAPDH were used for qtPCR (Table 4.1). The amplified cDNA was run on 2% agarose gels with SYBR green for UV detection to confirm the amplicon length.

RESULTS

Both primary ECs and EOCs adhered preferentially to the collagen coated lanes compared to the blocking agent. At 48hrs, the ECs were spread and elongated in the direction of the lanes regardless of the blocking agent. In contrast the EOCs, required the blocking agent Pluronic® to maintain cell alignment on the lanes (Figure 1). With bovine serum albumin as a blocking agent, the EOCs rapidly deposited a collagen rich matrix and migrated between the lanes [2].

Topographical micropatterns of 2μm wide lanes on PVA enabled the HUVECs to elongate in the pattern direction (Figure 2). Importantly the elongation occurred as a monolayer, mimicking the in vivo endothelium.

The micropattern induced EC elongation and fluid shear stress induced elongation influenced the thrombogenic gene expression by the ECs. PCR results indicated that micropatterning significantly downregulated vWF and TFPI with fold changes of 0.65 ± 0.25 and 0.72 ± 0.24, respectively (Figure 3). Micropatterning did not significantly affect expression of TF, TM, and eNOS, Fluid shear stress EC elongation also resulted in a significant downregulation of vWF and TFPI with fold changes of 0.15 ± 0.01 and 0.54 ± 0.2, respectively (Figure 3). FSS-induced EC elongation also significantly downregulated TF and significantly upregulated TM, with fold changes of 0.33 ± 0.25 and 1.47 ± 0.35, respectively (Figure 3). FSS-induced EC expression of eNOS was surprisingly variable, with fold change of 1.19 ± 0.62.

DISCUSSION

The continued failure rate of synthetic vascular grafts for the treatment of atherosclerotic lesions is a significant clinical problem. The lack of a functional endothelium on synthetic grafts impedes the ability of the new vessel to properly function. An endothelialized graft that could resist the atheroprone environment has the potential to prevent adverse intimal growth. The micropatterned ECs and EOCs were aligned equivalent to FSS-elongated ECs [3]. The micropattern-elongated ECs also had reduced gene expression of vWF, a prothrombotic molecule. The topographical elongation of cells on PVA indicates that the phenotypic benefits of elongation can now be applied to endothelialize vascular grafts, providing immediate protection against neointima formation.

ACKNOWLEDGEMENTS

We gratefully acknowledge the grant support of NIH R01 HL130274 for funding the PVA work.

REFERENCES

[1] Hinds, MT et al., *J Biomat Res A*, 86:804-12, 2008.
[2] Vartanian, KB et al., *J Biomat Res A*, 91:528-39, 2009.
[3] Vartanian, KB et al., *BBRC*, 371(4):787-92, 2008.

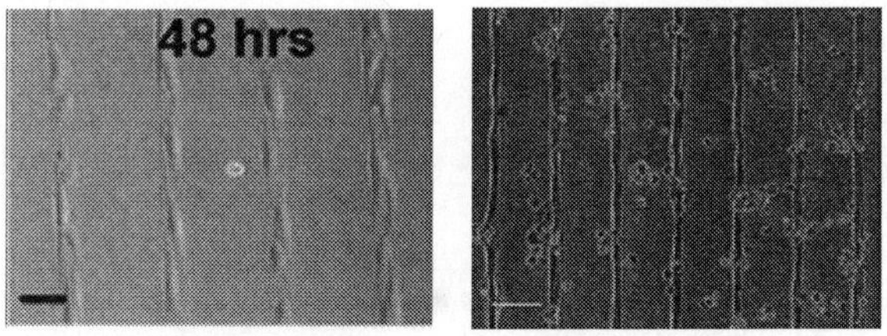

Figure 1: Micropatterned ECs (left) and EOCs (right) after 48 hrs. Both cell types elongated on the lanes. Scale bars are 50μm (left) and 100 μm (right).

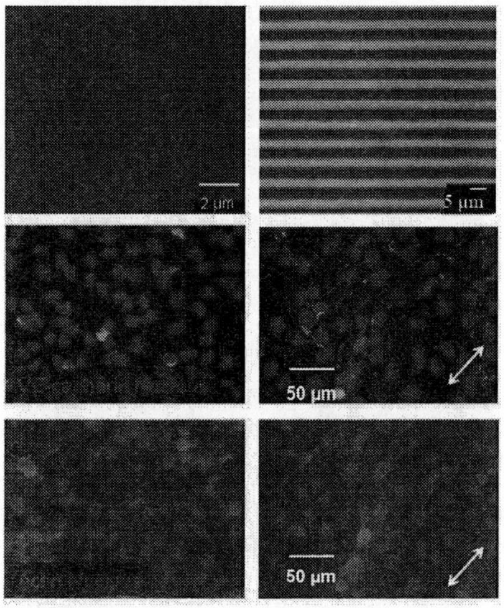

Figure 2: Topographical micropatterns on PVA enabled the alignment of an endothelial monolayer.

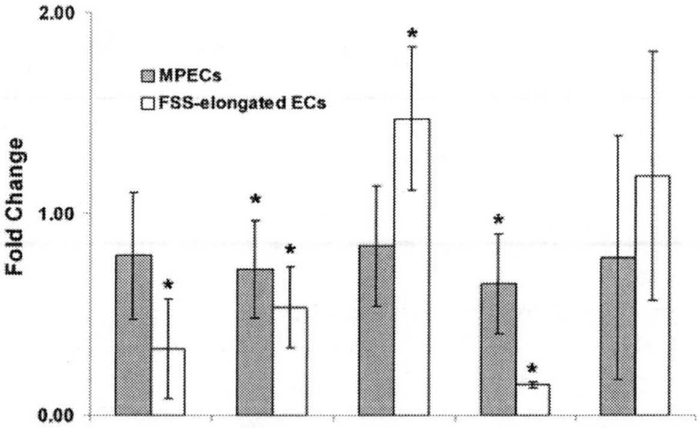

Figure 3: Micropattern and Fluid shear stress alignment regulation of EC thrombogenic gene expression. Micropattern-(MPEC) and FSS-elongated EC fold changes in gene expression were calculated with respect to non-patterned cobblestone ECs (equivalent to 1.0 on graph). Data represented as mean ± SD. *p<0.05, paired Student's t-test.

PARAVALVULAR LEAK IN TRANSCATHETER AORTIC VALVE REPLACEMENT

Ajit P. Yoganathan (1), Ikechukwu U Okafor (1), Prem Midha (1), Vrishank Raghav (1,2), Vasilis Babaliaros (3), Gautam Kumar (1,3)

(1) Biomedical Engineering
Georgia Institute of Technology
Atlanta, Georgia, USA

(2) Aerospace Engineering
Auburn University
Auburn, Alabama, USA

(3) Department of Cardiology
Atlanta VA and Emory University
Atlanta, Georgia, USA

INTRODUCTION

Aortic regurgitation in the form of paravalvular leak (PVL) has been shown to result in worse survival after transcatheter aortic valve replacement (TAVR) (1). PVL results from the improper seal between the outer perimeter of the transcatheter valve and either the aortic root or inner perimeter of the surgical aortic valve in the case of valve-in-valve (VIV) procedures.

The accurate quantification of AR/PVL in the case of TAVR and VIV procedures is necessary as it has been clinically shown that trace to mild PVL leads to worse patient outcomes (1). Current metrics utilized for PVL quantification generally rely on echocardiographic techniques, which have poor precision and accuracy, especially when the leakage is in the trace to moderate range (2,3).

In this work, we evaluated various combinations of transcatheter valve type, size, and deployment height into an Edwards PERIMOUNT surgical heart valve, and their effects on PVL. Furthermore, energy dissipation rate (EDR) was utilized as a metric to stratify PVL/AR.

METHODS

An *in vitro*, programmable piston pump-driven, left heart simulator (LHS) consisting of a flexible-walled anatomical left ventricle (LV) physical model, and an axisymmetric rigid aorta was used for this study (Figure 1a). Flow rates were measured into and out of the LV with transonic flow probes. Pressures in the LV and aorta were measured with strain gage Utah Medical pressure transducers.

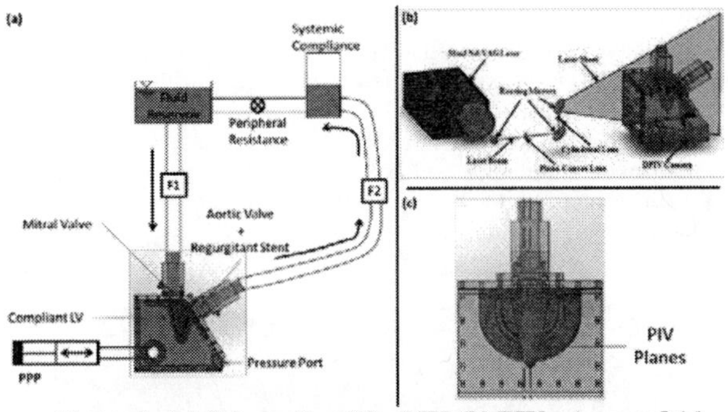

Figure 1: (a) Schematic of the LHS (b) PIV set up and (c) Acquired PIV planes

To examine the effects of THV type, size, and deployment position on PVL, 3 THVs were chosen based on existing recommendations for deployment in a 23mm PERIMOUNT (Edwards Lifesciences, Irvine, CA) surgical bioprosthesis (4). The THVs were a 23mm Edwards SAPIEN, a 23mm CoreValve (Medtronic, Dublin, Ireland), and a 26mm CoreValve. The valves were deployed within a rigid, acrylic chamber designed based on published aortic root dimensions (5,6). Each valve was tested in 5 deployment locations: a nominal position defined by the VIV Aortic app, 2 supra-annular positions, and 2 sub-annular positions.

Two-component, planar, digital particle image velocimetry was used to visualize and characterize LV filling in the presence of AR (Figure 1b and 1c). Out-of-plane vorticity and energy dissipation rate (Equation 1) were used as metrics to characterize the flow fields.

$$\varepsilon = \mu\left[\left(\frac{\partial u}{\partial x}\right)^2 + \left(\frac{\partial v}{\partial y}\right)^2 + 0.5\left(\frac{\partial u}{\partial y} + \frac{\partial v}{\partial x}\right)^2\right] \quad (1)$$

RESULTS

Paravalvular leak (quantified through regurgitant fraction, RF) was shown to be affected by several primary procedural decisions: THV type, size, and deployment positioning. The SAPIEN valve showed reduced RF with increasingly supra-annular deployment, whereas the CoreValves demonstrated the least leakage at a nominal or slightly sub-annular position. In addition, the 26mm CoreValve experienced greater leakage than the 23mm CoreValve in the 23mm PERIMOUNT control. In this configuration, the SAPIEN showed less leakage than either size CoreValve.

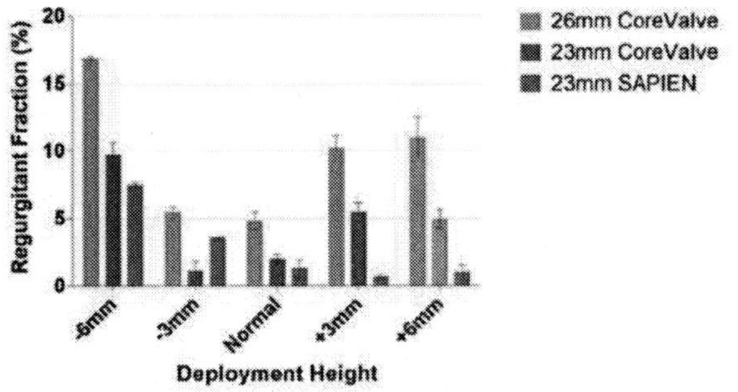

Figure 2: Paravalvular leak can be improved or worsened based on procedural decisions, such as transcatheter valve type, size, and deployment position.

A large vortex that engulfs the whole LV was observed under control conditions. In the presence of AR, the regurgitant jet was observed to disrupt the formation of the large vortex, leading to improper filling. EDR was observed to increase non-linearly with AR severity. EDR for no, trace, mild, and moderate AR were found to be 1.15, 2.26, 3.56, and 5.99 W/m^3, respectively.
It was also observed that as the level of AR increased, the energy dissipation rate per unit volume increased in a non-linear fashion.

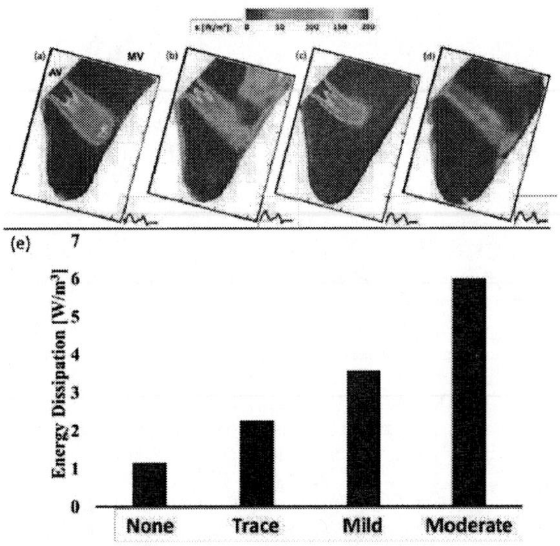

Figure 3: (a-d) EDR field for Central AR at select time points; (e) Spatio-temporal average of EDR for 4 different conditions

DISCUSSION

As PVL is a diastolic phenomenon, it is driven by the large gradient across the closed aortic valve. A supra-annular SAPIEN typically deploys in a somewhat "flower-pot" configuration with the aortic side of the valve flared outward. Increasingly supra-annular deployments allow for further flaring of the outflow. These facts may help explain why PVL in the SAPIEN valve is reduced with supra-annular positioning and increased with sub-annular positioning. The CoreValve, however, interacts not only with the annulus, but also the left ventricular outflow tract (LVOT) and ascending aorta. When in contact with these three regions, the Nitinol stent's complex shape deforms in a less predictable way. In addition, the leaflets of the CoreValve are 12mm from the base of the stent, giving it an already supra-annular leaflet positioning. Thus, PVL around CoreValve does not benefit from additional supra-annular positioning, as does the SAPIEN.

Accurate grading of AR, in the form of PVL, is critical in patients that had undergone TAVR as strong evidence exist showing that patients with ≥ moderate PVL after TAVR have overall worse survival than patients with none or mild PVL (7). Our study suggests that EDR can indeed be used to stratify PVL severity, which could have an impact in the treatment of patients after TAVR. Though PVL is often diagnosed using TTE and TEE, current grading systems rely heavily on a subjective interpretations of the regurgitant jet (8). Thus, EDR could be used to objectively re-stratify those patients presenting after TAVR with worsening heart failure symptoms and borderline PVL (mild vs. moderate) by TTE/TEE. Proper identification of moderate PVL in such cases would result in improvement of resource utilization and selection of early surgical or percutaneous re-intervention.

ACKNOWLEDGEMENTS

We would like to thank Venair for casting the LV models and P&G for providing the glycerin. We would also like thank the machine shop at the School of Chemical and Biomolecular Engineering for building the LHS. Finally, we acknowledge the American Heart Association for partially funding this work.

REFERENCES

[1] Sinning, J. et al., *J. Am. Coll. Cardiol*, 62.11-20, 2013.
[2] Beroukhim R. et al., *Circ. Cardiovasc. Imaging*, 3:542-9, 2010.
[3] Ewe, S. et al., *Am. J. Cardiol.*, 112:560-566, 2013.
[4] Bapat, V. et al., *Cath. Cardiovasc. Int.*, 81:853-61, 2013.
[5] Kinght, J. et al., *Surg. Radiol. Anat.*, 31:597-604, 2009.
[6] Reul, H. et al., *J. Biomech*, 23:181-91, 1990.
[7] Leon, M. et al., N. Engl. J. Med., 374:1609-1620, 2016.
[8] Nishimura, R. et al., J. Am. Coll. Cardiol., 63(22):e57-185, 2014.

TRIBOMECHANICS OF BARE AND a-C:H COATED METALLIC BIOMATERIALS

Konstantinos Kapnisis[1,*], Marios Constantinou[1,2], Maria Kyrkou[1], Petros Nikolaou[1,2], Andreas Anayiotos[1] and Georgios Constantinides[1,2,*]

(1) Department of Mechanical Engineering and Materials Science and Engineering, Cyprus University of Technology, 3036 Lemesos, Cyprus

(2) Research Unit for Nanostructured Materials Systems, Cyprus University of Technology, 3041 Lemesos, Cyprus

INTRODUCTION

Metal implants experience different levels of wear and corrosion due to the mechanical and chemical environment at the implantation site [1,2]. Over the years, there has been growing interest in developing ways to modify the materials' surface properties in an attempt to minimize failures and increase prostheses lifetime. Such interest has motivated the application of protective thin films on load bearing surfaces for increased wear resistance [3]. Amorphous carbon (a-C) and hydrogenated amorphous carbon (a-C:H) coatings are well known for their high hardness, low coefficient of friction, uniform adherence, abrasion resistance, chemical stability and biocompatibility [4]. These characteristics can improve tribological properties at the contact region and allow an increase in prostheses lifetime, avoiding body rejection from the adjacent host tissue.

Although numerous studies have indicated the excellent potential for a-C in biomedical applications, the findings have been variable in terms of material substrate, deposition method and coating characteristics. We here provide a systematic characterization of the morphological, nanomechanical and nanotribological properties of three metallic materials, commonly used for the fabrication of orthopedic, cardiovascular and dental implants: stainless steel (SS), titanium (Ti) and niobium (Nb). The tested materials were subsequently coated with hydrogenated amorphous carbon (a-C:H), using a plasma-enhanced chemical vapor deposition system, in order to evaluate for improved nanotribological response.

METHODS

Hydrogenated amorphous carbon (a-C:H) films were deposited on SS, Ti and Nb substrates (1 cm x 1 cm) using a plasma-enhanced chemical vapor deposition system (PE-CVD). All depositions were performed at room temperature using a chamber working pressure of 10^{-3}-10^{-4} mbar, an RF power of 200W, and a grid voltage of 150V. The density and thickness of the deposited a-C:H film was quantified through X-ray reflectivity measurements in a Rigaku Ultima IV diffractometer, using 3kW X-ray generator with a Cu tube.

SS, Ti and Nb metal plates were pre-and post-coating deposition examined for their topographical and morphological characteristics, using an atomic force microscope (Ntegra Prima, NT-MDT) and a scanning electron microscope (FEI Quanta 200), respectively. Furthermore, the materials were tested for their nanomechanical and nanotribological properties using an instrumented indentation platform (NanoTest, Micromaterials Ltd), from which the hardness, elastic modulus and coefficient of friction (COF) have been extracted.

RESULTS

The bare metal SS specimen was mechanically abraded and polished in successive steps of increasing grit until nanoscale roughness was achieved (12.9 ± 1.4 nm). The Ti and Nb specimens were left in their delivered surface conditions (430.8 ± 227.1 nm and 17.0 ± 7.0 nm respectively) to investigate coating performance on native (unpolished) surfaces. The resulting coating thickness was quantified at ~40 nm with a density value of ~1.7 g/cm³. Evaluation of the a-C:H coated specimens showed an approximately 2-fold decrease in surface roughness for the SS (6.8 ± 0.8 nm) and Ti (250.5 ± 46.5 nm) and a moderate decrease for the Nb (14.5 ± 3.4 nm) sample.

The nanotribological characteristics of the bare metal and a-C:H coated surfaces were studied using nanoscratch testing. In general, SS shows smooth depth and COF responses with scratch distance due to the flat surface finish and nanoscale roughness whereas the native surfaces of Ti and Nb lead to fluctuating responses; nevertheless, the mean response trends of all materials are similar. The steady state friction coefficient of the SS substrate was low, $\mu=0.17\pm0.03$, in contrast to higher values calculated for Ti ($\mu=0.36\pm0.06$) and Nb ($\mu=0.41\pm0.11$).

The effect of a-C:H coating on the nanotribological characteristics of the metallic surfaces are shown in Figures 1-3. Figure 1, in particular, demonstrates the capacity of the nanoscale film to reduce the plastic deformation and wear of the bare metal surfaces; the reduction being more evident in the case of SS. In all cases the deposited film improved the tribological performance of the metal surfaces reducing the residual depth imprint: 28% reduction for SS, 12% for Ti and 35% for Nb. The interatomic bonds developing between the SS / Ti crystals and the a-C:H coating, compose stable structures with minor reduction in friction coefficient (Figure 2). On the contrary, Nb demonstrated significant reduction in COF before the associated critical load (~100mN) that led to film failure and delamination (see Figure 3(f)).

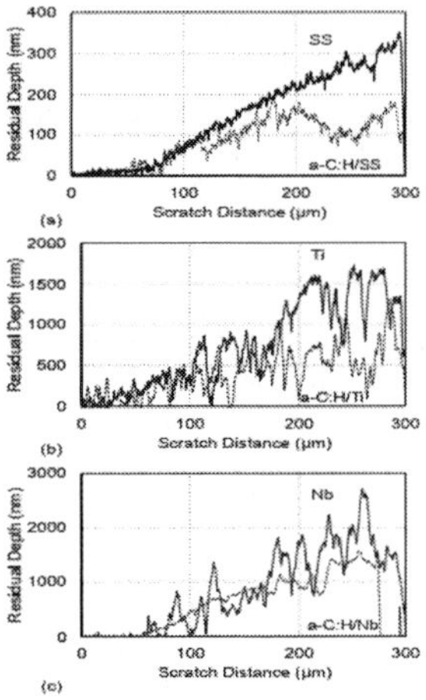

Figure 1: Effect of a-C:H coating on the residual depth response of (a) SS, (b) Ti and (c) Nb metals.

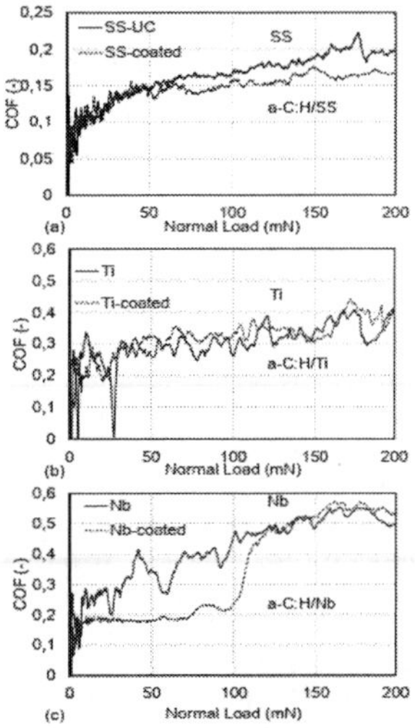

Figure 2: Effect of a-C:H coating on the COF response of (a) SS, (b) Ti and (c) Nb metals.

Figure 3 shows SEM images of the residual imprints left on the surface of the scratched specimens. By comparing the images of the coated (right column) versus the bare metal responses (left column), one can notice the reduction in the size of the residual mark and wear volume, in line with the residual depth results presented in Figure 1. These enhancements are evident in all specimens irrespective of whether the surfaces have been polished or left in their as-synthesized state. The improvements in the tribomechanical response can be attributed in part to the reduction of friction between the indenter and the tested materials and in part to the redistribution of stresses caused by the introduction of a hard, protective coating. The relatively uniform plastic deformations imparted on the a-C:H coated SS and Ti substrates indicate their enhanced tribomechanical response (Figures 3a-b and 3c-d). Surface damage can be observed for Nb substrate (Figure 3(f)) at late stages of the scratch distance which correlates with a critical load for film failure, also evident from a discontinuity in the COF evolution (Figure 2(c)).

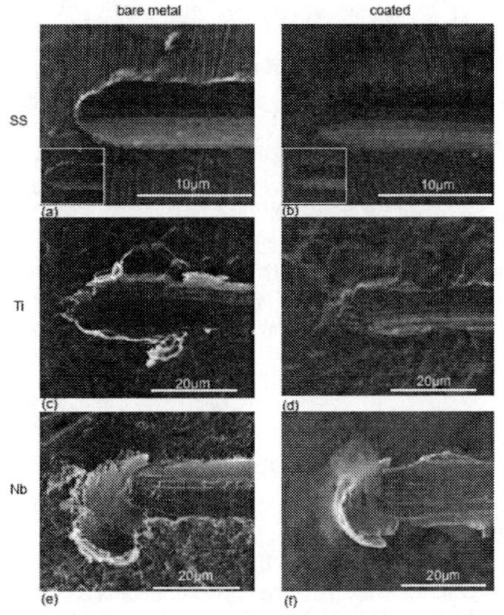

Figure 3: SEM images of residual scratched surfaces for the bare metal ((a), (c) & (e)) and a-C:H coated ((b), (d) & (f)) SS, Ti and Nb.

DISCUSSION

Hydrogenated amorphous carbon films present a promising material for serving as a protective coating on metallic implants. The deposition of even a 40nm thick film improves the tribomechanical response of these metallic surfaces as evidenced through nanotribomechanical tests. Despite the fluctuations of load during the tribomechanical testing of metallic surfaces left in their as delivered state (non-polished), tribomechanical improvements can be observed on all specimens which suggest that the efficiency of a-C:H coating is irrespective of whether the metallic surfaces are mirror polished or left in their native rough state.

ACKNOWLEDGEMENTS

This work was co-funded by the European Regional Development Fund, the European Social Fund, the Cohesion Fund, and the Research Promotion Foundation of the Republic of Cyprus (Strategic Infrastructure Project NEW INFRASTRUCTURE/STRATE/0308/04)

REFERENCES

[1] Wiktor D. M., et al. Coronary Stent Failure: Fracture, Compression, Recoil and Prolapse. *Interv. Cardiol. Clin.* **5**, 405–414 (2016).

[2] Bloch B., et al. Long-term follow-up of a cemented titanium stem. *Acta Orthop. Belg.* **81**, 225–32 (2015).

[3] Alakoski E, et al. Load-bearing biomedical applications of diamond-like carbon coatings-current status. *Open Orthop. J.* **2**, 43–50 (2008).

[4] Joska, Ludek, et al. Properties of titanium-alloyed DLC layers for medical applications. *Biomatter* **4**, (2014).

SB^{3}C2017
Summer Biomechanics, Bioengineering and Biotransport Conference
June 21 – 24, Tucson, AZ, USA

A MULTIPRONGED APPROACH PREDICTS LOW WALL SHEAR STRESS REGIONS THAT CORRELATE WITH THROMBOSIS FORMATION *IN VIVO*

Amanda K. W. Buck (1,2), Joseph J. Groszek (3), Daniel C. Colvin (1), Sara B. Keller (4), Clark Kensinger (5), Rachel Forbes (5), Seth Karp (5), Phillip Williams (5), Shuvo Roy (6), William H. Fissell (3)

(1) Department of Radiology and Radiological Sciences
Vanderbilt University Medical Center
Nashville, TN, USA

(2) Department of Biomedical Engineering
Vanderbilt University
Nashville, TN, USA

(3) Nephrology and Hypertension
Vanderbilt University Medical Center
Nashville, TN, USA

(4) Department of Bioengineering
University of Washington
Seattle, WA, USA

(5) Department of Surgery
Vanderbilt University Medical Center
Nashville, TN, USA

(6) Department of Bioengineering & Therapeutic Sciences
University of California San Francisco
San Francisco, CA, USA

INTRODUCTION

In 2012, the number of patients with end-stage renal disease (ESRD) in the United States exceeded 600,000, and the cost of health care for these patients exceeded $30 billion dollars [1]. For patients with ESRD, renal transplant is the treatment option providing the best survival [2]; however, the number of patients awaiting transplants far exceeds the number of donor organs. To address this organ scarcity, there is a demand for a bioengineered kidney alternative.

Our team is developing an approach to a bioengineered kidney that consists of both a silicon hemofilter and a renal tubule bioreactor [3]. As in any intravascular device, a primary concern in the hemofilter is thrombosis. In this abstract, we test a combined *in silico*, *in vitro*, and *in vivo* approach to assess the flow in a hemofilter prototype and to correlate the observed flow field with thrombosis. Computational fluid dynamics (CFD) simulations were used to simulate the hemodynamic environment; phase contrast magnetic resonance (PC-MR) velocimetry was used for validation of the computational approach; and *in vivo* experiments were used to assess device clotting.

METHODS

Physiologically representative CFD simulations predicted the flow field in a symmetric device (Device A). The simulations were validated *in vitro*, and *in vivo* experiments were conducted to determine thrombogenicity. Upon analysis of results, the device was optimized (Device B) to mitigate potentially thrombogenic flow phenomena; the impact of channel height (500, 750, 1000, 2000 um) on the wall shear stress (WSS) results was investigated; and additional *in vivo* experiments were conducted.

All hemofilter flow path prototypes were manufactured from LEXAN, Resin HPS6, in two pieces that subsequently were vapor polished and thermally bonded together (Hayes Manufacturing Services).

In silico data were validated using an in vitro experiment. Flow through Device A was imaged using a Varian 9.4T MRI scanner (Agilent Technologies, Santa Clara, CA). A pulsatile, peristaltic pump (Cole-Parmer) pumped Magnevist-doped (Bayer HealthCare Pharmaceuticals), citrated bovine blood (Lampire Biological Laboratories) through the prototype. Through-plane velocities were measured across the inlet and outlet portions of the device at 16 time points using PC-MR (TR=700 ms; FOV = 38.4 mm×9.6 mm; slice thickness = 1mm; voxel size = 0.15×0.15×1 mm^3; VENC=150 cm/s). Data were processed in MATLAB (The Mathworks); itk-snap [4]; and tec360 (Tecplot, Inc.). For comparison with *in vitro* data, simulations of blood flow through the device were conducted in Fluent (ANSYS, Inc.), using the PC-MR-determined, pulsatile flow rate as the inlet boundary condition. Blood was considered Newtonian (density = 1053 kg/m^3; viscosity = 0.00368 kg/m-s). Velocity data were extracted from the inlet and outlet planes of the device, and the MR and CFD data were compared at five time points along the cycle.

Thrombogenicity was assessed in vivo. All animal experiments were approved by Vanderbilt University's Institutional Animal Care and Use Committee. Device A prototypes were implanted in five Class-A dogs, and Device B prototypes were implanted in seven Class-A dogs. The hemofilters were grafted between the aorta and the inferior vena cava using end-to-side anastomoses. Ultrasound was used to verify patency, and devices were explanted upon thrombosis or at the end of the experiment (24-31 days)

RESULTS

To validate the simulation approach, the PC-MR-measured velocity values across the inlet and outlet of Device A were compared to the CFD results. The comparisons demonstrated strong qualitative agreement between the *in silico* and *in vitro* experiments, and the mean axial velocity error was 8.9% at the inlet and 10.4% at the outlet.

Due to the relationship between shear and thrombosis, the time-averaged wall shear stress was calculated for Device A (Figure 1). Locally low time-averaged WSS (TAWSS) regions were demonstrated at the outside of the proximal curve and at the inside of the distal curve. Examination of the temporally varying WSS patterns demonstrated a persistent low WSS (<1 Pa) region at the outside of the proximal curve, which was particularly evident when calculating the mean WSS over the cardiac cycle excluding systolic acceleration (Figure 1B). Clots were found in three of five implanted Device A prototypes, with clot formation as early as 11 days post-implantation. Clots in two of the devices were evident in areas of simulated low WSS (Figure 1C).

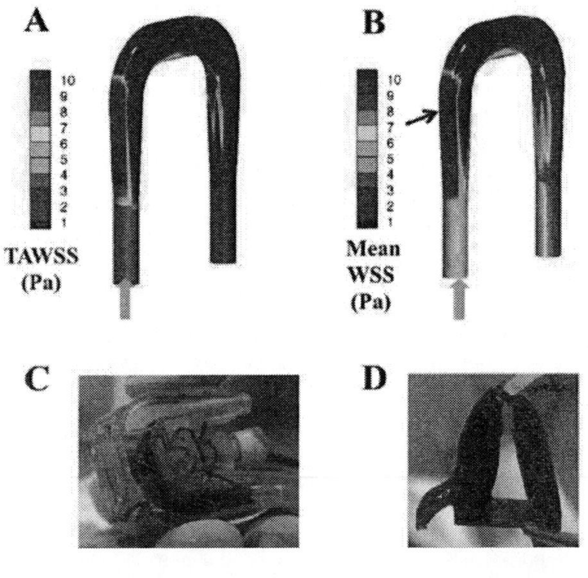

Figure 1: A) Simulated TAWSS map for Device A; and B) the mean WSS averaged over the cardiac cycle excluding systolic acceleration for Device A. Grey arrows indicate bulk flow direction; black arrow highlights low WSS region. C) Explanted device containing clot; D) extracted clot showing platelet-dominated white clot that corresponded with low WSS regions.

Device B demonstrated smaller low TAWSS regions (Figure 2A), and no TAWSS regions <1 Pa were observed. In addition, changing the channel height shifted the location of lower WSS zones, but no TAWSS regions <1 Pa were observed. Of the seven Device B prototypes implanted, no evidence of clotting was observed at the planned explant (28-31 days post-implantation).

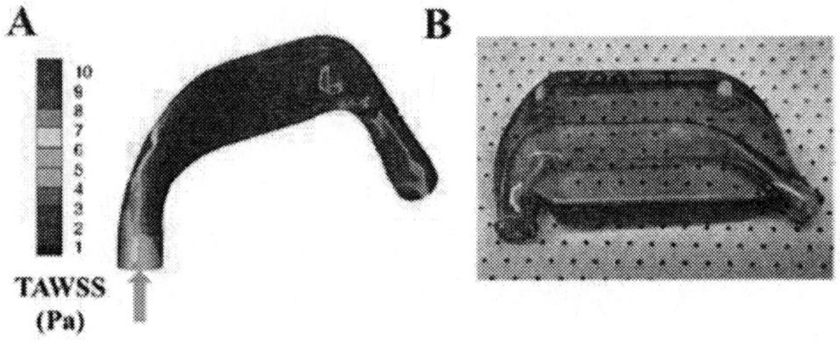

Figure 2: A) Simulated TAWSS map for Device B; and B) explanted Device B devoid of clot.

DISCUSSION

Given the catastrophic risk thrombosis represents in intravascular devices and the relationship between hemodynamics and clot formation in these devices (e.g. [5-7]), it is paramount to produce designs that mitigate thrombogenicity. Using a combined experimental approach we identified low WSS regions that corresponded with device failure, and the locations of these zones corresponded to clot formation in Device A. Subsequently, the flow path of the device was redesigned to minimize persistent low WSS regions, and this revised design showed superior performance *in vivo*.

In this study, *in vivo*, *in vitro*, and *in silico* experiments were used to assess the thrombogenicity of an implantable hemofilter device and to subsequently optimize the design. This combined approach is both feasible and useful for developing bio-artificial, organ-replacement devices.

ACKNOWLEDGEMENTS

This work was supported by grants from the National Institutes of Health (1R01EB014315 and 1U01EB021214), as well as a generous gift from the Wildwood Foundation.

REFERENCES

[1] U.S. Renal Data System, USRDS 2014 Annual Data Report, Volume Two: End-Stage Renal Disease in the United States, National Institutes of Health, National Institute of Diabetes and Digestive and Kidney Diseases, Bethesda, MD, 2014.
[2] Wolfe, RA et al., *N Engl J Med* 341 (23): 1725-30, 1999.
[3] Kensinger, C et al., *ASAIO J* 62 (4): 491-5, 2016.
[4] Yushkevich, PA et al., *Neuroimage* 31 (3): 1116-28, 2006.
[5] Hochareon, P et al., *ASAIO J* 50 (6): 537-42, 2004.
[6] Topper, SR et al., *Cardiovasc Eng Technol* 5 (1): 54-69, 2014.
[7] Yin, W et al., *Ann Biomed Eng* 32 (8): 1058-66, 2004.

SB[3]C2017
Summer Biomechanics, Bioengineering and Biotransport Conference
June 21 – 24, Tucson, AZ, USA

DEVELOPMENT OF LOW COST HUMAN SURROGATES USING ADDITIVE MANUFACTUREING

Travis D. Eliason (1), Art E. Nicholls (1), Daniel P. Nicolella (1)

(1) Musculoskeletal Biomechanics
Southwest Research Institute
San Antonio, Texas, USA

INTRODUCTION

Human surrogates are an important asset to musculoskeletal injury research as the use of live subjects in experiments is strictly limited due to the obvious ethical issues involved. Cadaveric studies are widely used with great success to eliminate these ethical issues but come with their own costs. Procuring, storing, preparing, and disposing of cadaveric specimens is a costly and time consuming process that requires special equipment and procedures, adding significant cost to each experiment. Human surrogates are an attractive alternative as they are reusable, require no special handling, and do not have the safety concerns of working with cadaveric tissue.

However, currently available human surrogates are expensive and have been developed for specific experimental test cases, limiting there usability across multiple research topics. In addition, due to their high cost they must be made sturdy enough to survive multiple experiments which make it difficult for them to accurately represent the reaction of a human body.

Our goal is to develop a low cost human surrogate utilizing additive manufacturing that can be rapidly produced and mimics the material properties and geometry of the structure being represented. By utilizing additive manufacturing these surrogates can be customized to match a target population or even individuals within a population opening up research areas not accessible with the current surrogates that are designed around a population average. Utilizing data from 3D medical imaging, and cadaveric soft tissue experiments we have developed methods to print the hard and soft tissue of the cervical spine that match the structural properties of the native tissues.

METHODS

In order to match the material properties of cervical spine ligaments (ALL, PLL, ISL, LF, and JC) a series of different 3D printing filaments were characterized. The materials characterized included SemiFlex and NinjaFlex, both produced by Ninjatek, as well as PolyFlex made by Polymaker. For each material a series of tension coupons were printed on a LulzBot Taz6 dual- head fusion deposition 3D printers. Each ligament was then tested in tension using a Sintech electromechanical test frame (Figure 1) to determine its tensile elastic modulus.

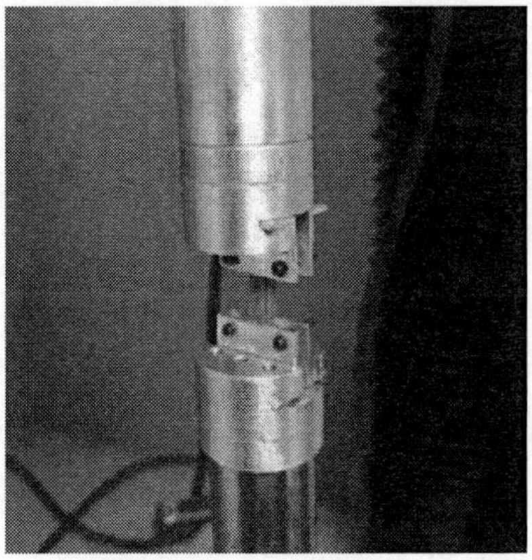

Figure 1: Example coupon loaded into the Sintech load frame for characterization testing.

When producing these coupons with 3D printing, there are several parameters that can be set that modify the structural properties of the end product. For each material tested, these parameters (infill percentage, number of outlines, extrusion width, number of solid layers, infill angle) were varied (Figure 2) and then tested to determine the combination of material and printer parameters that result in a printed ligament structure that mimics the cadaveric ligament and disc properties from literature (1-2). Additionally, multiple materials were printed in combination to match properties that could not be achieved using a single material print.

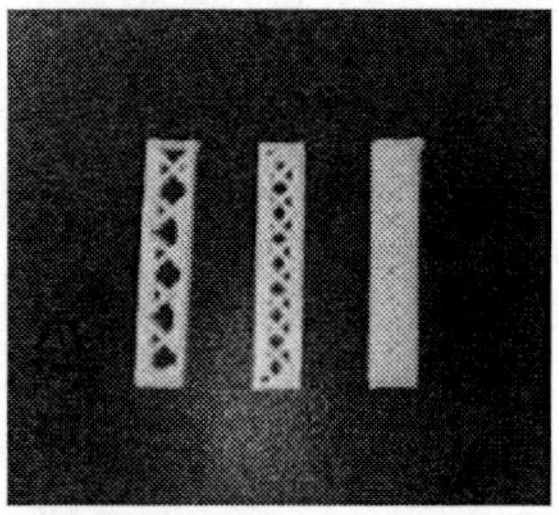

Figure 2: Three example coupons printed with PolyFlex with different infill percentages.

Individual cervical spine motion segments (C4-C5) were printed based on previously created finite element models of the cervical spine developed from 3D medical imaging (3). Due to the limitations of only being able to print two materials at a time, the vertebrae and disc were printed at the same time with the ligaments being printed separately and then attached to complete the motion segment. Additional structures were printed along with the vertebrae to facilitate mounting in our test fixture that replicates the pure moment load testing of cadaveric motion segments (Figure 3).

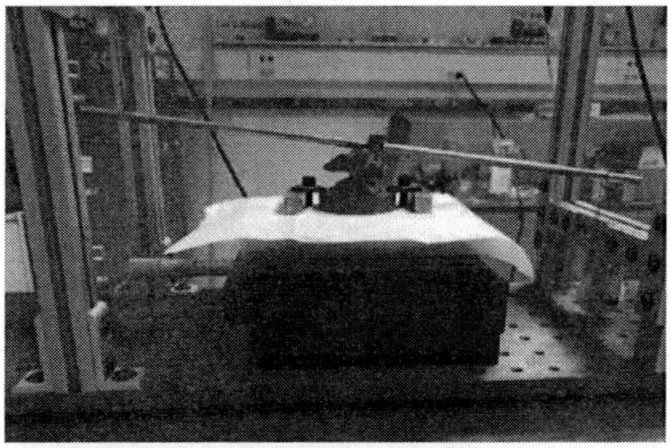

Figure 3: Printed C3-C4 motion segment mounted in test frame with a 2 Nm pure flexion moment applied.

RESULTS

Ligaments properties taken from literature (1) were grouped together and then matched to the material and 3D printing parameters that fit them the best. Ninjaflex was used to fit the JC, ISL, and LF ligaments for both the C2-C5 and C5-T1 levels. For the JC and ISL the print parameters were set to 100% infill with a 45 deg infill angle. The LF is less stiff and therefore only required a 30% infill also with a 45 deg infill angle (Table 1).

Both the ALL and PLL were significantly stiffer than the other ligaments which required moving to Polyflex rather the Ninjaflex used in the other ligaments. At the C2-C5 levels Polyflex was not stiff enough to match so an ABS insert was needed. An insert of 1.5% was found to increase the stiffness up to the literature reported value. At the C5-T1 levels the ALL and PLL were matched with Polyflex alone with 90 deg infill angles and 100% and 75 % infill percentages respectively (Table 2).

Ligament/Surrogate	Modulus (Mpa)	Ligament/Surrogate	Modulus (Mpa)
JC C2-C5	5	LF C2-C5	3.1
JC C5-T1	4.8	LF C5-T1	3.5
ISL C2-C5	4.9	NinjaFlex 30%	3.313
ISL C5-T1	5		
NinjaFlex 100%	5.007		

Table 1: Modulus comparisons between cadaveric data for JC, ISL, and LF ligaments with their corresponding printed surrogates.

Ligament/Surrogate	Modulus (Mpa)	Ligament/Surrogate	Modulus (Mpa)
ALL C2-C5	43.8	ALL C5-T1	28.2
PLL C2-C5	40.9	Polyflex 100%	28.762
Polyflex 1.5% ABS	41.867	PLL C5-T1	23
		Polyflex 75%	22.349

Table 2: Modulus comparisons between cadaveric data for ALL, and PLL ligaments with their corresponding printed surrogates.

Initial comparisons of the pure moment motion segment testing have not matched with the cadaver data in literature, with the surrogates behaving significantly stiffer. At 2 Nm load the measured rotation of the superior vertebrae was only 1/5 what was expected.

DISCUSSION

While the development of a full scale low cost human surrogate produced with additive manufacturing is still in its early stages, this work show initial promise that the concepts holds significant promise. We have shown that by using multiple materials and varying the print parameters and print structure, surrogate ligaments can be printed that closely match the material properties of cadaveric tissues. These ligaments can then be printed along with the boney structures to create functional anatomical structures that closely mimic the morphology and biomechanical behavior of cadaver specimens. The initial motion segment tests did not correlate well with literature data likely resulting from the compressive properties of the surrogate ligaments. These ligaments were initially designed to replicate the tensile behavior of cadaveric tissue and are significantly stiffer in compression compared to natural ligaments.

In addition, these results have been achieved using a low cost 3D printer with commercial off the shelf materials. Future efforts will transition this work to commercial grade equipment and custom developed materials in order to match not only the quasi static properties but the failure and viscoelastic properties as well. With additional work, additive manufacturing has the potential to alter the way human surrogates are used and produced for biomechanics research.

ACKNOWLEDGEMENTS

This work was supported by the Southwest Research Institute Internal Research and Development program (R8692).

REFERENCES

[1] Yoganandan, N et al., *J Biomech Eng*, 122:623-629, 2000.
[2] Stemper, B. et al., *ASME Summer BioEng Conf*, 2011.
[3] Bredbenner et al., *Front Bioeng Biotechnol*, 2:58, 2014.

LONGITUDINAL POSTURE AND ACTIVITY TRACKING IN THE HOME ENABLED BY MACHINE LEARNING AND A CONFORMAL, WEARABLE SENSOR SYSTEM

Ryan S. McGinnis (1,2), Steve DiCristofaro (1), Nikhil Mahadevan (1), Ellora Sen-Gupta (1), Ikaro Silva (1), Elise Jortberg (1), Nirav Sheth (1), John A. Wright, Jr. (1), Brian Murphy (1), Bryan McGrane (1), Milan Raj (1), Melissa Ceruolo (1), Jesus A. Pindado (1), Roozbeh Ghaffari (1), AJ Aranyosi (1), Shyamal Patel (1)

(1) MC10, Inc.
Lexington, MA, USA

(2) Electrical and Biomedical Engineering
University of Vermont
Burlington, VT, USA

INTRODUCTION

Human body posture is an important predictor of human health and as such, is often a target of clinical research, diagnosis, and intervention. For example, upright posture while sitting has been shown to improve stress resilience in asymptomatic patients [1], and to reduce the negative thought bias in depressed patients [2].

Biofeedback has been used to promote improved posture in persons with Parkinson's Diseases [3]. However, to do so effectively requires characterization of human body posture outside of the clinic and in the context of specific activities of daily living (ADLs) such as sitting, standing, and lying down. Researchers have developed a number of methods for objectively characterizing human body posture outside of clinical settings (e.g., body-worn strain gauge [4], instrumented shirt [5]), but these technologies are unable to provide the contextual activity information needed to interpret the posture measurements, and are impractical to wear over multiple days.

To overcome these issues, we present a conformal, wearable sensor system to measure human body posture within the context of activities of daily living and demonstrate its use in the home over 48 hours in healthy subjects. With an eye toward identifying potential behavioral interventions for improving standing posture, we explore relationships between the time spent completing ADLs and posture, and examine whether particular types of activities could be used to predict improved posture during standing.

METHODS

Data from N=28 subjects (sex: 13M, 15F; age: 32 +/- 12 years; height: 1.72 +/- 0.11 m, weight: 74.13 +/- 13.73 kg) over two days (49.34 +/- 5.42 hrs) were collected for analysis. Following collection of informed consent, each subject was instrumented with BioStamp sensors (Fig. 1a, MC10, Inc. Lexington, MA, USA) affixed to the skin of the chest and thigh as shown in Fig. 1b. Subjects performed a simple functional calibration (standing in a neutral posture for 10 seconds), and were then instructed to complete their typical daily activities. After 24 hours, subjects met with study staff to remove the BioStamp devices. This process was repeated on a subsequent day, within 1 week of the initial recording.

The activities completed by each subject are classified according to the machine learning approach described in [6]. Five-second windows of accelerometer data from sensors secured to the chest and thigh of each subject, are used to inform calculation of a set of 62 features, which in turn, are used as input to a hierarchical tree of binary activity classification models. The output of this tree is an activity classification - Recumbent, Sitting, Standing, Walking, or Other - for each five-second window of data.

The accelerometer data recorded by the BioStamp sensor worn on each subject's chest, are used to estimate human body posture. Posture estimation is enabled by the fact that each accelerometer measurement ($\vec{\alpha}_M$) can be modeled as a linear combination of translational acceleration ($\vec{a}_M$), the acceleration due to gravity ($\vec{g}_M$), and white measurement noise ($\vec{w}_M$) as per

$$\vec{\alpha}_M = \vec{a}_M + \vec{g}_M + \vec{w}_M \qquad (1)$$

where accelerometer measurements are reported in a reference frame (F_M) aligned with the measurement axes of the sensor [7] (see Fig. 1b). The orientation of this device-fixed measurement frame is arbitrary relative to the anatomical frame of the body (F_A). The mapping between F_A and F_M, for each subject, follows from the functional calibration activity completed at the beginning of each data collection session [7]. Estimates of human body posture are produced for five-second, non-overlapping windows of accelerometer data, which are averaged to

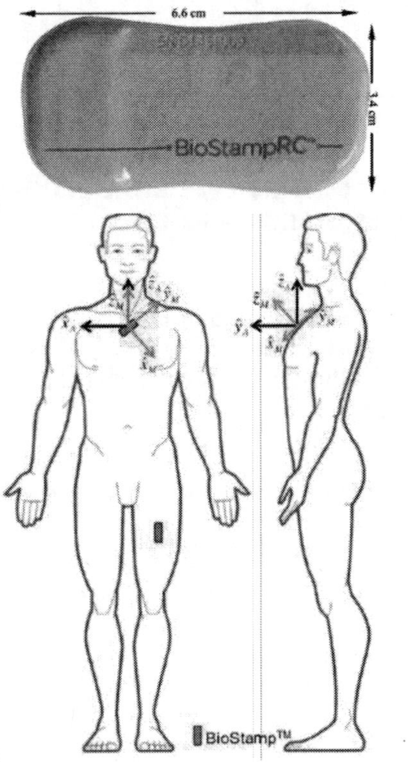

(a)

(b)

Figure 1: BioStamp sensors (a) deployed to the chest and thigh (b) on subjects to characterize posture and ADLs. Posture estimation is enabled by a functional calibration that defines the mapping between reference frames F_A (black) and F_M (orange).

yield the direction of gravity ($\hat{g}_M$), and are transformed into F_A, using this mapping, to yield $\hat{g}_A$. The individual components of $\hat{g}_A$ are used to estimate posture angles in the sagittal (θ_S) and coronal (θ_C) planes following standard techniques. A subject is in a neutral, or upright, posture when $\theta_S = \theta_C = 0$ deg.

We summarize subject activity patterns by computing the percentage of time spent in each ADL, and define posture quality as the average absolute angular deviation from a neutral standing posture in the sagittal plane. Relationships between subject demographics (age, gender, height, weight) and activity patterns are explored using Spearman's rank correlation coefficients. The association between activity patterns and standing posture quality is assessed with linear regression using the isometric log ratio transformation approach for compositional variables [8]. Significance is achieved when $p<0.05$.

RESULTS

Correlation coefficients between subject demographics and activity patterns are reported in Table 1, where significance is indicated with bold text. Significant correlations are observed between age and time spent moving and sitting.

Controlling for age, linear regression reveals an association between standing posture quality and the percentage of time spent moving, sitting, and standing. Model coefficients (β), standard error (std. err), t-scores (t), and p-values (p) for the regression are reported in Table 2.

Table 1: Spearman's rank correlation coefficients between subject demographics and percentage of time spent in each ADL.

	Moving	Sitting	Standing	Recumbent
Age	**-0.39**	**0.44**	-0.22	-0.24
Gender	-0.25	-0.07	-0.17	0.18
Height	-0.15	-0.07	-0.16	0.18
Weight	0.04	0.03	-0.13	-0.08

Table 2: Regression parameters relating average sagittal standing posture angle and percentage of time spent completing activities.

	β (std. err.)	t	p
Intercept	3.79 (3.94)	0.96	0.35
Age	-4.92 (4.07)	-1.21	0.24
Moving %	-11.75 (3.86)	-3.04	0.01*
Sitting %	8.95 (4.70)	1.90	0.07^T
Standing %	7.72 (3.22)	2.40	0.02*
R^2=0.31, * $p < 0.05$, ^{T} $p < 0.10$			

DISCUSSION

Comparing subject demographics to activity patterns in Table 1, we observe that older subjects spend significantly less time moving and more time sitting than younger subjects. This is especially important as we consider data from a relative young sample (oldest participant was 63 yrs. old), and yet these trends are already apparent. These results agree with existing literature where, for example, an increase in inactivity with age has been observed [9]. Future work should extend these techniques to a sample with a wider range of ages to better understand the relationship between aging and activity patterns in the home. This type of analysis could prove useful in identifying early indicators of mobility impairment prior to acute events (e.g. falls).

The regression analysis of Table 2 reveals a significant association between activity patterns and standing posture quality. Specifically, subjects who exhibit better standing posture spend more time moving (β =-11.75), and less time standing (β =8.95) or sitting (β =7.72). These results suggest that movement-based intervention programs could provide a means for improving standing posture. Moreover, they demonstrate that interventions that aim to elicit an increase in the percentage of time spent standing (e.g. standing desks), are unlikely to improve standing posture quality. Future studies should further explore these associations with an interventional study to assess the efficacy of behavioral changes (i.e. increased movement, increased standing, etc.) for improving standing posture. In the future, this type of study could further enable investigation into the efficacy of postural and behavioral intervention for improving mental health.

The results presented herein demonstrate that a new conformal, wearable sensor system is able to track activity patterns and posture noninvasively in the home across multiple days. In so doing, we identify relationships between age and activity patterns and between activity patterns and posture quality. These results point toward future interventional studies aimed at improving posture quality and, more generally, the use of this wearable sensor system for studying human behavior outside of typical clinical environments.

REFERENCES

[1] Nair, S et al., *Health Psychol*, 34:632–641, 2015.
[2] Michalak, J et al., *Clin Psychol Psychother*, 21:519–524, 2014.
[3] Mirelman, A et al., *J NeuroEng Rehab*, 8:35, 2011.
[4] O'Sullivan, K et al., *Manual Therapy*, 17:77–83, 2012.
[5] Sardini, E et al., *I2MTC*, 652–657, 2015.
[6] Patel, S et al., *EMBC*, 5997–6001, 2016.
[7] McGinnis, RS et al., *IEEE TBME*, 62:1503–1513, 2015.
[8] Hron, K et al., *J of App Stat*, 39:1115–1128, 2012.
[9] Caspersen, CJ et al., *Med Sci Sports Exerc*, 32:1601–1609, 2000.

SB^{3}C2017
Summer Biomechanics, Bioengineering and Biotransport Conference
June 21 – 24, Tucson, AZ, USA

IN VIVO TIBIOTALAR AND SUBTALAR KINEMATICS IN CHRONIC ANKLE INSTABITILY PATIENTS AND ASYMPTOMATIC CONTROLS: A HIGH-SPEED DUAL FLUOROSCOPY STUDY

Koren E. Roach (1,2), K. Bo Foreman (1,3), Alexej Barg (1), Andrew E. Anderson (1,2,3)

(1) Department of Orthopaedics
University of Utah
Salt Lake City, UT, US

(2) Department of Bioengineering
University of Utah
Salt Lake City, UT, US

(3) Department of Physical Therapy
University of Utah
Salt Lake City, UT, US

INTRODUCTION

Chronic ankle instability (CAI) often results from trauma to the ankle and is associated with persistent feelings of instability, pain, and frequent ankle sprains [1]. It is clinically hypothesized that deleterious angular and translational (i.e. kinematic) motion at the tibiotalar and subtalar joints causes ankle osteoarthritis (OA) in CAI patients [2, 3]. However, *in vivo* measurements of tibiotalar and subtalar motion in CAI patients are not currently available, and thus, the pathogenesis of OA in CAI patients is unknown.

Joint kinematics are often derived from motion capture techniques that track the positions of reflective markers adhered to the skin over bony landmarks. However, skin-marker motion capture cannot distinguish between motion of the tibiotalar and subtalar joints as there are no reliably palpable landmarks for placement of a skin marker about the talus. Dual fluoroscopy (DF) is an imaging modality that accurately measures three-dimensional *in vivo* bone movement and can be used to identify the independent roles of the tibiotalar and subtalar joints. The purpose of this study was to use DF to evaluate and compare *in vivo* tibiotalar and subtalar joint kinematics between CAI patients and asymptomatic controls during activities of daily living

METHODS

After obtaining Institutional Review Board approval (IRB#65620) and informed consent, ten healthy control subjects (5M/5F; 30.9±7.2 yo; BMI: 23.5±3.5) were enrolled in this study. Additionally, four patients with ankle pain, feelings of instability and symptoms that limited exercise and activities of daily living (1M/3F; 30.8±4.1 yo; BMI: 25.8±3.4) were also enrolled in this study. Standard radiographs were obtained for all subjects including weight-bearing anteroposterior and lateral views of the foot, mortise view of the ankle, and hindfoot alignment view. Radiographs of the control subjects were screened for gross abnormalities and hindfoot OA by a fellowship-trained foot and ankle surgeon (AB). A high-speed DF system previously validated in the ankle to a rotational bias of 0.25±0.81°, and a translational bias of 0.03±0.35 mm was positioned about a dual-belted instrumented treadmill. The DF system acquired images while subjects performed a single-leg balanced heelrise and walked on the treadmill at 0.5 and 1.0 m/s.

Subject-specific CT scans were used in conjunction with the DF data to quantify *in vivo* tibiotalar and subtalar joint kinematics and range of motion (ROM) for all subjects, as reported previously [4]. The mean and 95% confidence intervals (CIs) of the controls were calculated for the tibiotalar and subtalar joint angles throughout each activity. The mean and 95% CIs of the controls were also calculated for the rotational and translational ROM of each joint and activity. Kinematic results from the CAI patients were compared to the mean and 95% CIs of the controls. The portion of each CAI trial that fell outside the 95% CIs of the controls was calculated and expressed as a percentage of the entire trial. We considered CAI joint angles to be clinically different compared to controls if the CAI joint angles fell outside the 95% CIs for more than 50% of an activity.

RESULTS

During balanced heelrise, 70%, 58% and 65% of the measured CAI tibiotalar internal/external rotation (IR/ER), subtalar inversion/eversion (In/Ev) and subtalar IR/ER, respectively, fell outside the 95% confidence interval (CI) of the control subjects. In addition, CAI patients frequently exhibited greater tibiotalar and subtalar IR/ER and subtalar In/Ev with an opposing trend when compared to controls (Figure 1). During 0.5 m/s gait, 50% and 60% of

CAI tibiotalar dorsi/plantarflexion (D/P) and subtalar IR/ER, respectively, fell outside the 95% CIs of the control subjects. During 1.0 m/s gait, 62%, 65%, and 73% of CAI subtalar D/P, In/Ev, and IR/ER, respectively fell outside the 95% CIs of the control subjects

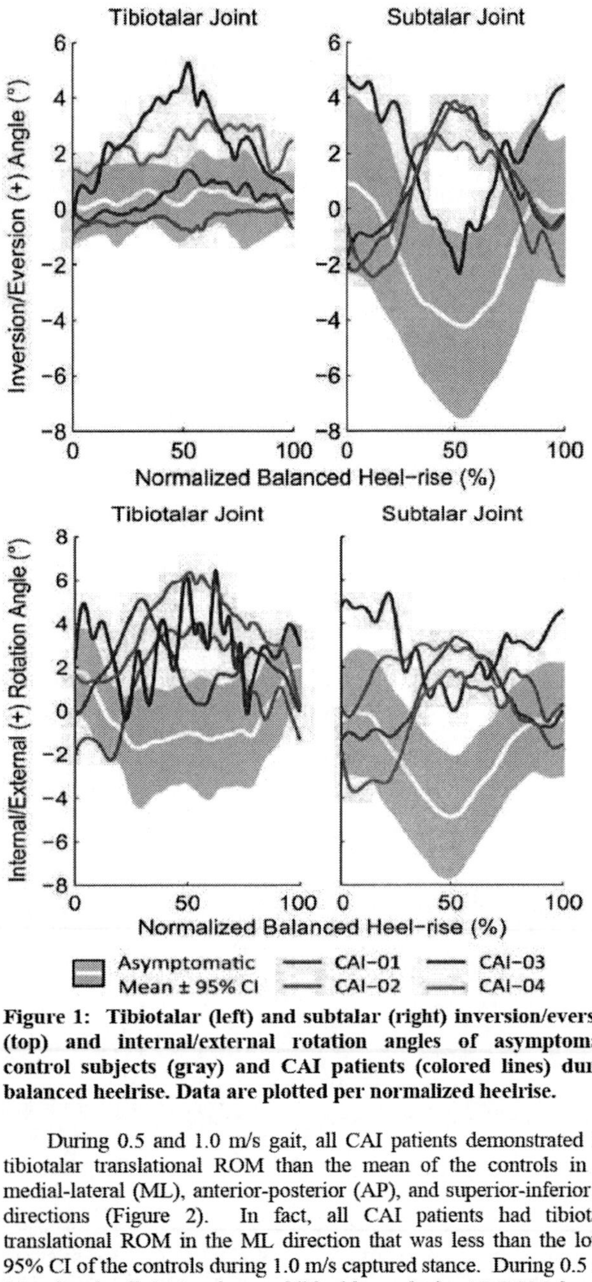

Figure 1: Tibiotalar (left) and subtalar (right) inversion/eversion (top) and internal/external rotation angles of asymptomatic control subjects (gray) and CAI patients (colored lines) during balanced heelrise. Data are plotted per normalized heelrise.

During 0.5 and 1.0 m/s gait, all CAI patients demonstrated less tibiotalar translational ROM than the mean of the controls in the medial-lateral (ML), anterior-posterior (AP), and superior-inferior (SI directions (Figure 2). In fact, all CAI patients had tibiotalar translational ROM in the ML direction that was less than the lower 95% CI of the controls during 1.0 m/s captured stance. During 0.5 and 1.0 m/s gait, all CAI patients exhibited less subtalar AP ROM than the mean of the controls, several falling below the lower 95% CI of the controls. CAI patients appeared to have consistently less In/Ev ROM across all activities and joints, although the individual ROM values were not always less than the lower 95% CI of the controls. CAI-01 and CAI-03 consistently exhibited less subtalar angular ROM than the mean of the controls during 0.5 and 1.0 m/s captured stance, but not during balanced heelrise.

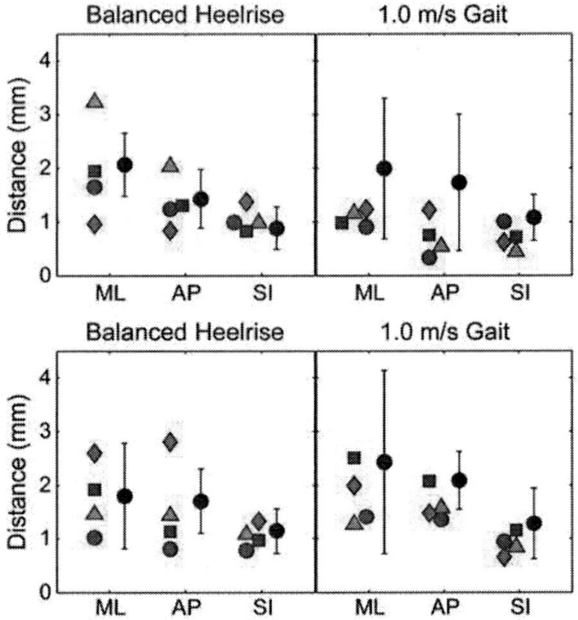

Figure 2: Translational ROM in each direction (ML = medial-lateral, AP = anterior-posterior, SI = superior-inferior) during balanced heelrise (left) and 1.0 m/s gait (right) in the tibiotalar (top) and subtalar (bottom) joints for the CAI patients compared to controls. Each color/symbol represents a different CAI patient.

DISCUSSION

Although a preliminary investigation, this study is the first to quantify tibiotalar and subtalar joint kinematics in CAI patients during dynamic activities and compare them to control subjects. Most notably, our results show that CAI patients exhibit an opposite trend to those of the control subjects during subtalar In/Ev and tibiotalar and subtalar IR/ER. These biomechanical differences may indicate a protective mechanism used by CAI patients to prevent further injury. Our results suggest that increased tibiotalar external rotation, increased subtalar eversion and external rotation, and reduced translational ROM may be mechanical characteristics of CAI. These data could clarify the pathomechanical characteristics of this condition and illuminate the steps required to refine current diagnosis and treatment strategies.

ACKNOWLEDGEMENTS

Funding from the National Institutes of Health (R21AR063844, S10RR026565), American Orthopaedic Foot and Ankle Society, and LS-Peery Discovery Program in Musculoskeletal Restoration is gratefully acknowledged.

REFERENCES

[1] Delahunt, E et al., *Med Sci Sports Exerc*, 42:2106-21, 2010.
[2] Harrington, KD, *J Bone Joint Surg Am*, 61:354-61, 1979.
[3] Valderrabano V et al., *Clin Orthop Relat Res*, 467:1800-6, 2009.
[4] Roach, KE et al., *J Biomech Eng*, 138(9): 091006-1-091006-9, 2016.

IMPACT PERFORMANCE OF BICYCLE HELMETS DURING REAL-WORLD OBLIQUE IMPACTS

Megan L. Bland (1), Craig McNally (1), David S. Zuby (2), Becky C. Mueller (2), and Steven Rowson (1)

(1) Biomedical Engineering and Mechanics
Virginia Tech
Blacksburg, VA, USA

(2) Insurance Institute for Highway Safety
Ruckersville, VA, USA

INTRODUCTION

According to U.S. Consumer Product Safety Commission (CPSC) data, cycling is the number one cause of sport-related head injuries treated in U.S. emergency rooms [1]. Risk of head injury can be reduced by bicycle helmets, which must pass safety standards prior to being sold. In the U.S., the CPSC safety standard for bicycle helmets mandates that helmets reduce linear acceleration to <300 g (~50% skull fracture risk [2]) in impacts from guided drop tests.

Present standards do not reflect common cyclist accidents, which often produce lower head accelerations, averaging around 100 g [3-4] (associated with concussion [5]). Helmets are also not tested at the rim, a very common cyclist impact location [3,6]. Furthermore, standards consider normal impacts and measure linear acceleration, while real-world accidents also involve tangential forces (termed ***oblique***) [6] and rotational acceleration, a major contributor to concussion [7]. It has been suggested that bicycle helmets be evaluated using oblique impacts, but to-date no rig has been standardized, nor has performance of different style helmets under oblique impacts been thoroughly investigated.

The objective of this study was to investigate impact attenuation differences in ten bicycle helmet models under real-world accident conditions using an oblique impact rig. A secondary objective of was to investigate the effects of varying the ratio of normal/tangential forces.

METHODS

Ten bicycle helmet models were selected for comparative testing based on their varied designs (Table 1), and were subjected to oblique impacts using a custom drop tower. The drop mass consisted of a National Operating Committee on Standards for Athletic Equipment head, a Hybrid III 50th-percentile male neck, and a simulated effective torso mass. Impacts were against a flat (0°) anvil or a 30° anvil. The 30° anvil generated both tangential and normal forces, while the 0° anvil generated primarily normal forces. Anvils were coated with 80-grit sandpaper to simulate road surfaces.

Table 1. Helmet models. Road helmets have thinner shells with an aerodynamic shape and more venting, while non-road styles have thicker shells with greater head coverage and less venting. *MIPS: multidirectional impact protection system.

Helmet (Abbreviation)	Cost [$]	Style	Mass [gm]
Bell Solar Flare (BSF)	40	Road	312
Bell Star Pro (BSP)	240	Road	296
Bell Super 2 MIPS (BMIPS)*	155	Mountain	360
Catlike Whisper (CW)	235	Road	283
Giro Sutton MIPS (GMIPS)*	100	Urban	370
Giro Synthe (GS)	250	Road	250
Nutcase Watermelon (N)	70	Urban	465
Smith Optics Overtake (SOO)	250	Road	250
Schwinn Thrasher (ST)	23	Road	340
S-Works Evade (SWE)	225	Road	317

Helmets were impacted at a frontal location at the helmet rim and a temporal location (Fig. 1), as these are common in cyclist accidents [3,6]. Impact sites were mirrored, with each location tested once and impact centers >12 cm apart. Impact velocities were also chosen to reflect cyclist accidents; a normal velocity of 4.4 m/s was found to produce linear accelerations of ~100 g in pilot testing – an average cyclist head impact severity [3-4]. A normal velocity of 5.7 m/s was also chosen for a more severe case. The locations, velocities, and anvils yielded 8 possible impact configurations. Four trials per configuration were conducted for each helmet model, totaling 320 impact tests.

Three linear accelerometers and three angular rate sensors located at the headform center of gravity (CG) measured head kinematics. Data were filtered with a CFC-1000 low-pass filter for linear accelerations (based on SAE J211) and CFC-155 for angular rates. Angular rates were differentiated to determine rotational accelerations, and concussion risk was computed from peak resultant linear and rotational accelerations using a previously published bivariate risk function [6]. Results were compared using ANOVA with Tukey HSD post hoc tests ($\alpha = 0.05$).

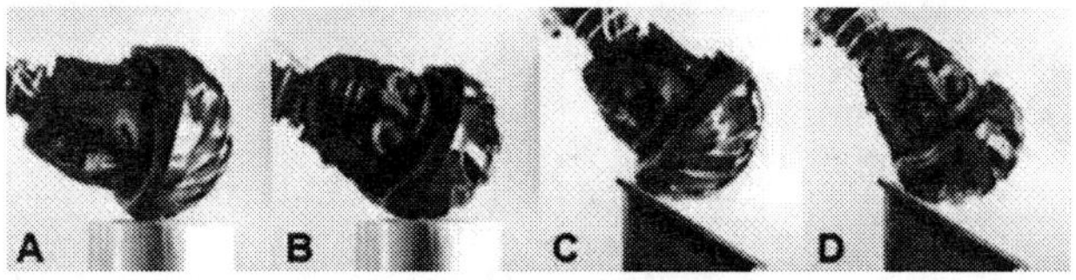

Figure 1. Frontal (A,C) and temporal (B,D) impact locations for the 0° (A,B) and 30° (C,D) anvils.

RESULTS

The ten helmets produced large variations in acceleration, with linear and rotational ranging from 66-234 g and 3.3-12.5 krad/s^2 (Fig. 2). Linear accelerations were unaffected by anvil angle (4.4 m/s averaging 112 g at 0° and 109 g at 30°, and 5.7 m/s averaging 154 g at 0° and 155 g at 30°), while rotational accelerations were significantly higher at 0° ($p<0.0001$). As most impacts in cycling accidents are oblique [6], only the 30° anvil results are further analyzed herein.

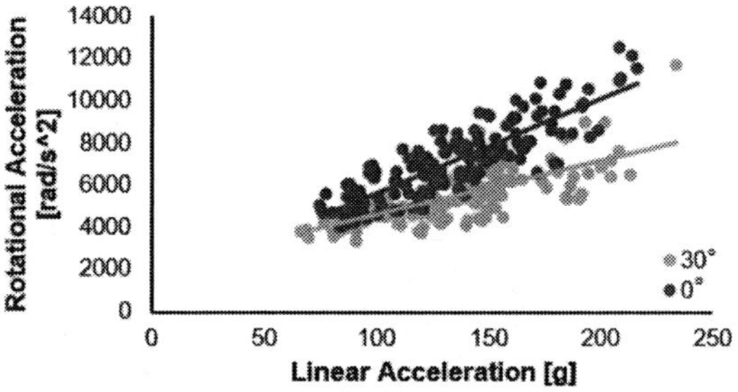

Figure 2. Rotational versus linear accelerations for all tests. Linear did not vary across anvil angles, while rotational was higher at 0°.

Concussion risk ranged from 2% to 99% and varied considerably across helmet model and impact configuration (Fig. 3). The temporal location produced higher risks than the frontal location on average. This is more reflective of linear accelerations than rotational, as linear results were higher for temporal impacts than frontal, while rotational results varied less across location. Significant differences in risk existed between helmet models at each configuration with the exception of frontal-4.4 m/s, which yielded low risks (<15%) for all models. Aside from this case, helmets that produced higher risks in one configuration tended to produce higher risks in the others, and vice versa.

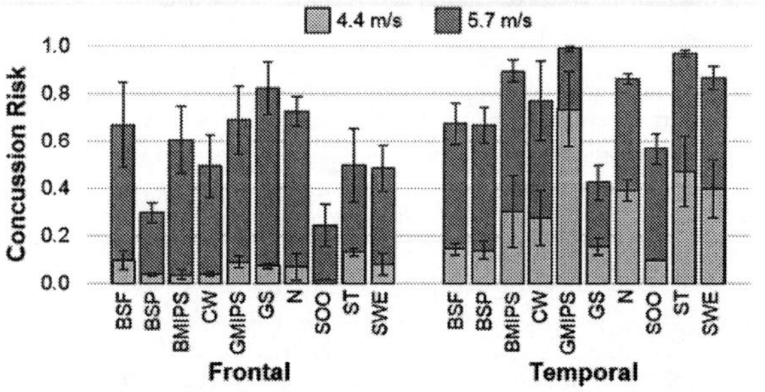

Figure 3. Concussion risk for all helmets (standard deviation error bars). Risk varied significantly by model within most configurations. Temporal impacts generally yielded higher risks.

DISCUSSION

The present results reveal considerable differences in protective capabilities of bicycle helmets in real-world accident scenarios. The anvil angle affected these differences, as rotational accelerations were higher at 0° than at 30°. This can be explained by the forces acting on the head for both anvils: the resultant force is directed closer to the CG of the head-neck assembly for the 30° anvil, producing a smaller moment, and the tangential and normal forces for the 30° anvil create opposing moments, also reducing the resultant moment for this anvil. Most cyclist head impacts are oblique [6], and therefore the 30° anvil provides a more realistic platform on which to evaluate bicycle helmets.

For most helmets, linear accelerations were higher for temporal impacts than frontal, resulting in higher injury risks. This may be partly due to helmets' larger radius of curvature at the sides than the front, resulting in a larger contact area at the temporal location. The larger area reduces pressure for a given force, thus increasing effective liner stiffness and causing it to crush less, dissipating less impact energy. Rotational accelerations, conversely, did not appear to be location-dependent for the 30° anvil setup.

The relative performance of the ten helmets was fairly uniform between impact configurations. BMIPS, GMIPS, and N in particular produced significantly higher risk than a majority of other helmets for nearly every configuration. These were the only non-road helmets, and had thicker shells, larger masses, and generally less venting than the road helmets. These design parameters are likely less conducive to impact protection, but more helmets of this style should be evaluated in further testing. Helmets that produced lower risks across all configurations include SOO and BSP. Although these were among the more expensive helmets, cost was not well correlated to performance. SOO was unique in that it incorporated a plastic honeycomb liner.

Other helmets produced more variance across configurations. GS, for example, yielded the highest risk at frontal-5.7 m/s but the lowest at temporal-5.7 m/s. This may be attributable to local geometry effects, which are exacerbated at higher impact energies. For GS, the frontal location was over a vent, and the surrounding liner material may have been less effective in dissipating the impact energy. This location is commonly impacted in cyclist accidents [3,6], but since the helmet rim is below the testable area defined by standards, its design is unregulated.

Of particular interest were the helmets with MIPS, a technology meant to reduce rotational accelerations upon impact. In this study, GMIPS had significantly higher rotational accelerations at the temporal location than all other helmets, while BMIPS produced average rotational accelerations. No direct comparisons were made between the same helmet with and without MIPS, however, and it is possible that MIPS indeed reduces rotational accelerations within the same helmet.

The bicycle helmets tested herein varied widely in protective capabilities, with concussion risk spanning >60% in single impact configurations. As these configurations reflect real-world cyclist accidents, the present results can be used to inform safer helmet design.

ACKNOWLEDGEMENTS

This work was in collaboration with and funded by the Insurance Institute for Highway Safety.

REFERENCES

[1] CPSC, National Electronic Injury Surveillance System, *US Consumer Product Safety Commission*. Web, 2016.
[2] Mertz, HJ et al., *Stapp Car Crash J*, 47:155-188, 2003.
[3] Williams, M, *Accid Anal Prev*, 23(2-3):119-131, 1991.
[4] Smith, TA et al., *Accid Anal Prev*, 26(6):795-802, 1994.
[5] Rowson, S, Duma, SM, *Ann Biomed Eng*, 41(5):873-882, 2013.
[6] Bourdet, N et al., *J Sports Eng Tech*, 226(3-4):282-289, 2012.
[7] Gennarelli, TA et al., *SAE Tech Paper 720970*, 296-308, 1972.

SB³C2017
Summer Biomechanics, Bioengineering and Biotransport Conference
June 21 – 24, Tucson, AZ, USA

IMPROVING HEAD IMPACT KINEMATICS MEASUREMENT ACCURACY USING SENSOR FUSION OF MULTIPE SENSORS

C. Kuo (1), J. Sganga (2), M. Fanton (1), D. Camarillo (1,2)

(1) Department of Mechanical Engineering
Stanford University
Stanford, CA USA

(2) Department of Bioengineering
Stanford University
Stanford, CA USA

INTRODUCTION

Wearable head impact sensors have gained popularity in the last decade for the study of mild traumatic brain injury (concussions). Concussions have been labeled by the Center for Disease Control as a silent epidemic [1], because of the 1.6-3.8 million concussions occurring in the United States, only an estimated 50% are diagnosed and treated [2]. Diagnosis of concussions is difficult because of the subtlety of neurological symptoms and subjectivity in assessing deficits [3]. Thus, it is imperative that a quantitative be developed for accurate and real-time diagnosis.

Wearable head impact sensors that measure motion kinematics currently represent one solution for diagnosing otherwise unreported injuries. However, current wearable head impact sensor technologies lack the direct skull coupling to accurately measure head motion kinematics [4,5,6]. The instrumented mouthguard shows promise as an accurate measurement device, because it couples directly to the upper dentition, a rigid surface attached to the skull via stiff ligaments [6,7,8].

However, the instrumented mouthguard still has several limitations. First, mouthguards are not ubiquitous in all contact sports. Second, the instrumented mouthguard has been shown to be susceptible to external disturbances, which could falsely trigger the mouthguard during non-impact events [9], and introduce measurement error [7]. Finally, the instrumented mouthguard has limited space, and six degree of freedom measurements effectively necessitate the use of triaxial linear accelerometers and gyroscopes. As a result, angular accelerations are only computable through differentiation of the angular velocity signal, which could amplify noise and is not ideal for unhelmeted impacts [9].

In this work, we present the use of a sensor network to accurately estimate head impact kinematics. This sensor network is a flexible system that addresses limitations of both previous sensor form factors and the instrumented mouthguard. First, the flexibility of this formulation allows the use of any sensor, provided each sensor acts independently to measure head motion kinematics. Second, we implement sensor fusion techniques to mitigate individual sensor errors to obtain an accurate global estimate of head motion kinematics. Finally, the sensor network is inherently well-suited to take advantage of rigid body formulations to directly measure angular acceleration rather than rely on noisy differentiation.

METHODS

To develop and test the head impact kinematics sensor network methodology, we utilized a collection of four X2 Biometrics skin patch sensors placed on the head during mild soccer ball impacts on a single male subject, age 27 (Figure 1). The subject wore an instrumented mouthguard as a ground truth measurement given its published accuracy [7]. We developed a visual localization technique for identifying relative location and orientation of sensors with respect to one another, and a rigid body formulation to combine network sensor data. Finally, we implemented a Kalman filter, an Unscented filter, and a Particle filter to fuse network sensor measurements and obtain a more accurate estimate of the head kinematics.

Ten impacts were delivered using a soccer ball launcher with speeds below 10m/s [6]. Headers were restricted to the forehead, which resulted in head motion primarily in the sagittal plane. Impact sensors were all set to trigger when the linear acceleration magnitude exceeded 3g, recording data for 30ms pre trigger and 70ms post trigger.

Technical Presentation #155

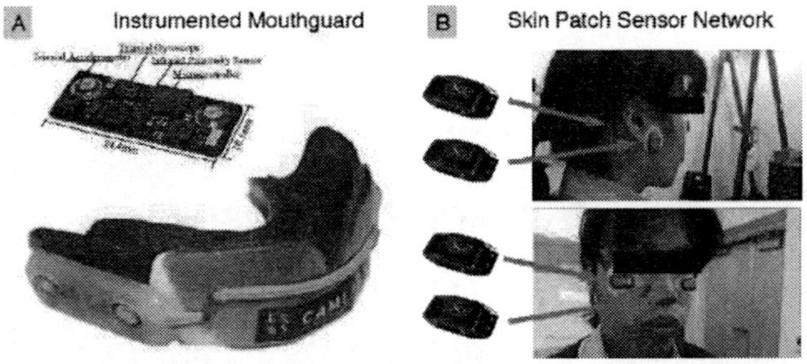

Figure 1: Instrumented Mouthguard Reference and Skin Patch Sensor Network: (A) An instrumented mouthguard was used as the reference sensor to evaluation the kinematic accuracy of B) a network of four X2 skin patch sensors.

We used a visual localization technique to identify the relative position and orientation of sensors with respect to one another. This information was necessary for the rigid body formulation to calculate angular acceleration from the sensor network [6]. We placed a 0.1 inch checkerboard pattern on each of the four skin patch sensors. Photos were then taken of the subject with a calibrated Samsung Galaxy S6 phone camera such that each photo had a pair of skin patch sensors in view. Then, we used Matlab's Vision Toolbox to detect checkerboards and obtain transformations from the checkerboards to the camera.

To combine the data from the sensors in the network, we used sensor fusion techniques to get a best estimate of the kinematics given sensor errors. The state of the system was simply the three axes of angular velocity. The input to the system was the three axes of angular acceleration that propagated the state through time. After state propagation, the state estimate was updated based on measurements taken from sensors in the sensor network. Raw data (angular velocities and linear accelerations) were assumed to have Gaussian noise. The formulation is mostly linear-Gaussian, except for the Coriolis term in the rigid body equations that convert the sensor measurements into an estimate for the angular acceleration input. As a result, the process noise is not necessarily Gaussian. To explore the effects of this non-linearity, we implemented three different filters: Particle Filter, Unscented Filter, and Kalman Filter, which each account for the nonlinearity differently.

For the Kalman Filter, we needed to treat the rigid body equations as a linear process. In the rigid body formulation, we made the assumption that errors only came from the linear Centripetal terms that relate angular acceleration to linear acceleration. In practice, we found that the non-linear Coriolis term contributed less to the rigid body equations than the linear Centripetal terms.

For the Unscented Filter, we sampled 100 particles representing measurements from the skin patch sensors. Each particle could be used to calculate a particle estimate of the angular acceleration using the non-linear rigid body formulation. We then used the particle estimates to obtain an angular acceleration mean and covariance.

Finally, for the Particle Filter, we had a 100 particle sampling of the angular velocity state. We would then take 100 samples representing measurements from the skin patch sensors, and use those samples to propagate each of the 100 state particles. At each time point, we could obtain a state estimate and covariance from the particle samples. This differs from the Unscented Filter in that particle state estimates are not condensed into Gaussian estimates at each time step.

RESULTS

A total of seven of the ten head impacts were collected by all sensor systems (three sensing skin patch sensors and the instrumented mouthguard) and were included in this analysis. In agreement with past work, the skin patch sensors individually exhibited substantial overestimation errors with respect to the reference mouthguard during these head impacts due to skin motion artifacts. The average angular velocity magnitude error for each of the three patch sensors was 117%, 117%, and 83%, while the average angular acceleration magnitude error was 137%, 125%, and 101%.

All filtering approaches were able to improve the estimate of the angular velocity peak magnitude and angular acceleration peak magnitude significantly. Angular velocity peak magnitude estimates were 27%, 25%, and 27% on average for the Kalman, the Unscented, and the Particle Filters respectively. Angular acceleration peak magnitude estimates were 19%, 16%, and 14% on average for the Kalman, the Unscented, and the Particle Filters respectively. While the angular velocity estimates were similar among the filters, it is interesting to note that the angular acceleration estimate improved with increasing fidelity of the non-linear rigid body formulation. Example traces from the mouthguard reference, the three sensing patch sensors, the average of the three patch sensors, and the Particle Filter are shown in Figure 2.

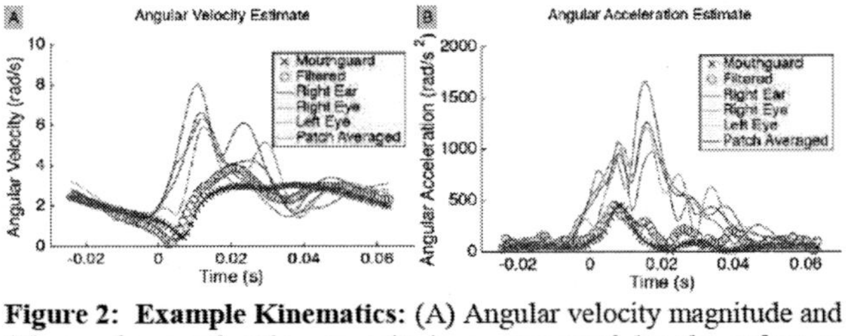

Figure 2: Example Kinematics: (A) Angular velocity magnitude and (B) angular acceleration magnitude as measured by the reference mouthguard, the three measurement X2 skin patch sensors, the average of the skin patch sensors, and the Particle Filter.

DISCUSSION

In this work, we have demonstrated a novel approach to using a sensor network to accurately estimate the angular velocity and angular acceleration during a head impact. The sensor network allows for flexibility in the types of sensors used to measure head kinematics. The sensor fusion techniques employed in this work were effective at smoothing high frequency noise from individual sensors, resulting in superior estimates. Finally, this formulation allows for the calculation of angular acceleration directly using raw measurements from the sensor network, rather than relying on differentiation.

ACKNOWLEDGEMENTS

We would like to acknowledge X2 Biometrics for donating the skin patch sensors. We would also like to thank professor Kochenderfer for assistance on algorithmic development.

REFERENCES

[1] Langlois, J et. al., *J Head Trauma Rehab.* 21(5):375-378, 2006.
[2] Kroshus, E et. al., *J Adolescent Health.* 54(3):269-275, 2014.
[3] Kroshus, E et. al., *Annals Behav Med.* 49(1):95-103, 2014.
[4] Jadischke, R et. al., *J Biomech.* 46(13):2310-2315, 2013.
[5] Campbell, K et. al., *ABME.* 44(4):1246-1256, 2016.
[6] Wu, L et. al., *ABME.* 44(4):1234-1245, 2016.
[7] Camarillo, D et. al., *ABME.* 41(9):1939-1949, 2013.
[8] Kuo, C et. al., *J Biomech.* 49:1845-1853, 2016.
[9] Wu, L et. al., *IEEE Trans Biomed Eng.* 61(11):2659-2668, 2014.

EXPLORING NOVEL OBJECTIVE FUNCTIONS FOR SIMULATING MUSCLE COACTIVATION IN THE NECK

J. Mortensen and A. Merryweather

Department of Mechanical Engineering
University of Utah
Salt Lake City, UT, USA

INTRODUCTION

Previous research suggests that flexing neck muscles in preparation for impacts to the head or trunk, or preflexion, can favorably alter head kinematics by decreasing maximum accelerations [1, 2]. Given adequate preparation, how does the human body protect itself against impacts that could cause traumatic brain injury (TBI)? Our current understanding of muscle preflex and the role it plays in protecting the brain during impacts such as those experienced in popular collision sports is limited. This research aims to add knowledge to our understanding of injury prevention through neck preflexion. This is accomplished through the development of a musculoskeletal model and will be confirmed in the future with human subjects research.

Common muscle control strategies use optimization techniques to minimize an objective function such as muscle activation, metabolic cost, or muscle stress. These are often constrained by muscle activations so the model closely follows kinematics measured experimentally [3, 4]. Unfortunately, these strategies do not work well to calculate coactivation. Coactivation increases joint stiffness, but generates no net moment about a joint [5, 6]. An alternative objective function to model coactivation during preflexion is necessary.

Previous research has used two objective functions to achieve coactivation: maximizing muscle activation and muscle force [5, 7, 8]. Both of these methods include a constraint that muscle activations must not apply a net moment to any joint. Our work adds two new objective functions for computing coactivation with the goal of determining which objective functions best maximize joint stiffness. These models and the acquired knowledge from simulation are a necessary first step prior to conducting experimental studies with human subjects.

METHODS

OpenSim™ was chosen as the musculoskeletal modeling platform, because of its accessibility and open source platform. Vasavada created an OpenSim model of the head and neck, for the purpose of determining the ability of different muscles to cause movement of the head [8]. Her model was used as a starting point for this study. In order to make the model suitable for dynamic simulations, several modifications were necessary. First, the original model only included muscles on the right side of the neck. Muscles were mirrored across the sagittal plane in order to get a more complete muscle set, bringing the total number of muscles in the model to 78. Second, values for mass and inertia were needed for each solid body represented in the model (i.e. head and vertebrae). Third, joint stiffness caused by ligaments and other passive structures was incorporated. Data from Lee et al. (2006) provided mass and inertia values for each solid body and stiffness properties for each joint. These modifications enabled the use of dynamic simulation tools in OpenSim.

Another reason OpenSim was chosen was its ability to interface with MATLAB. Muscle properties from the musculoskeletal model were passed into MATLAB, and a function was created to calculate moments generated about various degrees of freedom given muscle activations. Zero net moment was used as a constraint for the optimization problem to calculate muscle coactivation. The MATLAB function is described as:

$$M_k = \sum_{i=1}^{78} a_i f_i r_i \qquad (1)$$

where M_k is the moment about degree of freedom k, a_i is the activation of a single muscle, f_i is the maximum force a muscle can generate, and r_i is the moment arm of muscle i about degree of freedom k.

Two novel objective functions were constructed to calculate muscle activations. The first was to maximize the total moment applied by each muscle that rotates the head in the sagittal plane (*Maximum Moment*).

The reasoning for this was that muscle activations should be weighted by their ability to resist motion. The second objective function was to maximize the rotational stiffness caused by each muscle in the sagittal plane (*Maximum Stiffness*). This objective function was constructed by determining the contribution of each muscle to rotational stiffness in the sagittal plane. The reason for creating this function was to compare how close other objective functions get to maximizing stiffness in the neck. Using MATLAB, coactivation was calculated using objective functions of maximizing activation, maximizing force, maximizing applied moment, and maximizing stiffness with the "fmincon" optimizer. Below is an example of the setup of one of these four optimization problems, where a is the muscle activation of a particular muscle and M is the net moment about a particular degree of freedom.

Minimize objective function:	$J = -\sum_{i=1}^{36} a_i^2$	(2)
Subject to constraint:	$\sum_{i=1}^{3} M_i = 0$	(3)
With gradient:	$g = -\sum_{i=1}^{36} 2a_i$	(4)
With constraint gradient:	$dc = \sum_{i=1}^{36} M_i/a_i$	(5)

while 78 is the number of muscle actuators; the number of terms for the constraint gradient, and objective functions was reduced to 36 by assuming the solution would be symmetric across the sagittal plane, and ignoring muscles that were unable to apply any moment to the model. The constraint has three terms: one for each degree of freedom.

These muscle activations were then tested for stability and stiffness using OpenSim. Stability simulation required activating muscles according to a solution derived from an objective function, and measuring head rotation. Stability simulation was necessary to verify a solution for coactivation was found. Stiffness simulation involved applying inertial effects from gravity with the body in a prone position (Figure 1), and measuring head rotation and rate of rotation.

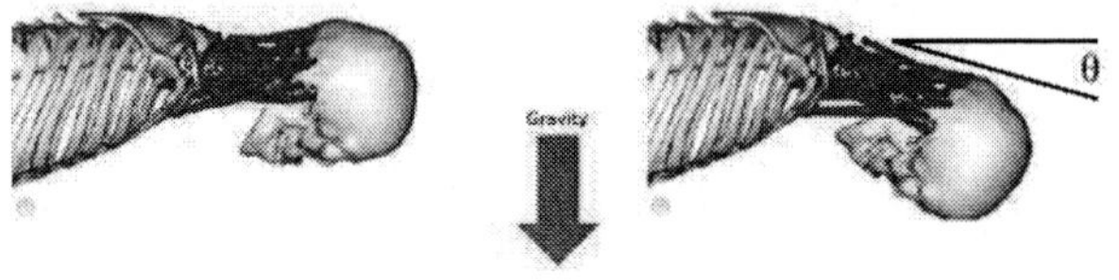

Figure 1: Stiffness test in which simulation of the effect of gravity in a prone position is used to measure neck stiffness

RESULTS

There was a surprisingly high correlation between maximizing stiffness and maximizing moment. Figure 2 shows that maximizing stiffness and maximizing moment have essentially the same levels of stiffness in the neck as measured by the pitch angle. As shown in Figure 3, the maximizing moment approach produces nearly identical muscle activations as the maximizing stiffness approach. Maximizing stiffness produced the highest levels of stiffness in the neck followed by maximizing moment, maximizing activation, and maximizing force.

DISCUSSION

This work highlights the need for additional research to determine how muscle recruitment occurs during unanticipated impacts. We have shown that maximizing the moment applied by each muscle about a particular degree of freedom generates more stiffness in that degree of freedom than maximizing activation or force. This approach to coactivation may prove useful in simulating neck stiffness in particular directions. This work provides a new model and explanation for a coactivation strategy to maximize neck stiffness. With a greater understanding of how different objective functions can affect coactivation, experimental studies aimed at fitting musculoskeletal model parameters with coactivation strategies are feasible. This work also suggests that the musculoskeletal modeling community could benefit from deriving objective functions that not only focus on efficiency, but also on higher objectives such as joint stiffness. Future work is needed to confirm the biological significance of these results.

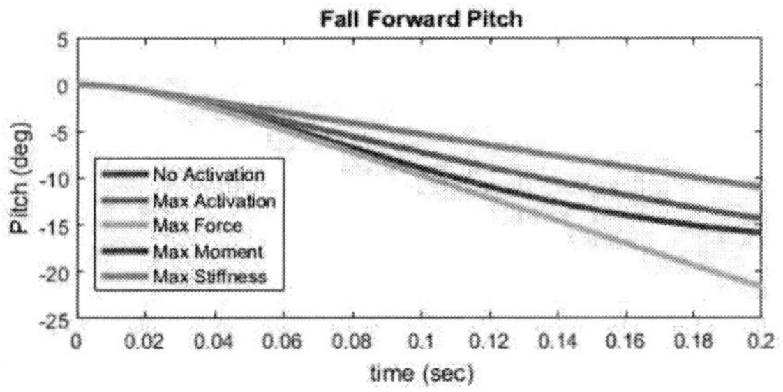

Figure 2: Head rotation for each simulation where smaller angles indicate higher stiffness

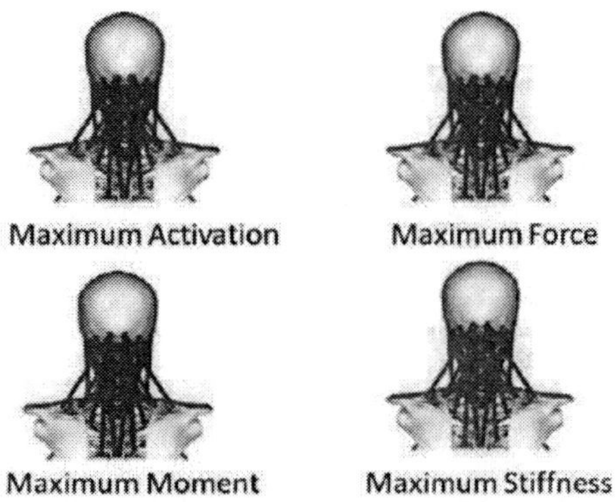

Figure 3: Activation Patterns in OpenSim, where red indicates muscle activity

ACKNOWLEDGEMENTS

This research was funded in part by grants from the National Science Foundation (#1622741) and NIOSH (NIOSH Education and Research Center training grant T42/CCT810426-10).

REFERENCES

[1] Mang, D.W., et al., The Spine Journal, 2015. 15(1): p. 153-161.

[2] Eckner, J.T., et al., The American journal of sports medicine, 2014. 42(3): p. 566-576.

[3] Bottasso, C.L., et al., Multibody System Dynamics, 2006. 16(2): p. 123-154.

[4] McKay, J.L. and L.H. Ting, Plos Computational Biology, 2012. 8(4).

[5] Dibb, A.T., et al., Traffic Injury Prevention, 2013. 14: p. S116-S127.

[6] Lee, S.-H. and D. Terzopoulos, *ACM Transactions on Graphics (TOG)*. 2006. ACM.

[7] Rashedi, E., et al., Proceedings of the Institution of Mechanical Engineers, Part H: Journal of Engineering in Medicine, 2010. 224(3): p. 487-501.

[8] Vasavada, A.N., S. Li, and S.L. Delp, Spine, 1998. 23(4): p. 412-422.

THE TIMING OF MECHANICAL LOADING MODULATES ENDOCHONDRAL OSSIFICATION OF CHONDROGENICALLY PRIMED MSCS

A. McDermott (1), J. Boerckel (1), D. Kelly (2)

(1) Department of Aerospace and Mechanical Engineering
University of Notre Dame
Notre Dame, IN, USA

(2) Trinity Center for Bioengineering
Trinity College Dublin
Dublin, Ireland

INTRODUCTION

Endochondral ossification is a mode of secondary bone formation whereby condensing mesenchymal stem cells form a cartilage template that is remodeled to bone with invading vasculature. In both skeletal development and fracture healing, mechanical loading is essential to induce endochondral ossification. *In utero*, chemical or surgical induction of muscular paralysis inhibits bone growth through reduced chondrocyte proliferation and delayed ossification [1], while, in fracture repair, some compressive motion is necessary for callus formation and endochondral healing [2]. We recently reported that mechanical loading was essential for stem-cell mediated endochondral regeneration of critical-sized defects *in vivo*, dependent on load timing [3]. Mechanical loading initiated in defects after the onset of chondrocyte hypertrophy and osteogenesis (week 4) exhibited the greatest extent of bone regeneration, while immediate loading enhanced bone formation but prolonged chondrogenesis, inhibited early vessel invasion, and caused pseudoarthrosis. Together, these data suggest a significant interaction between mechanical loading and lineage specification of progenitor cells during endochondral ossification that has not fully been explored in tissue engineering, but the influence of the timing of load initiation on endochondral differentiation of mesenchymal stem cells (MSCs) remains unclear. Here we hypothesized that early loading *in vitro* would promote chondrogenesis of MSCs and delay or arrest endochondral ossification, while delayed loading would promote chondrocyte maturation, hypertrophy, and ossification.

METHODS

Bone marrow hMSCs (P4) were encapsulated in fibrin hydrogels at a density of 15×10^6 cells/mL and cultured in chondrogenic media (10ng/mL TGF-β3) in physoxic conditions (5% pO_2). Hydrogels were subjected to displacement-controlled 10% dynamic compression for 2 hours per day, 5 days a week for either 0-2 weeks, 2-4 weeks, 4-6 weeks, or 6-8 weeks, and were analyzed at the end of their loading cycle in comparison to non-loaded, free-swelling controls at the same time point. Samples (n=4 per analysis) were saved at each time point for biochemical analysis, mechanical testing, qPCR, and histology and immunohistochemistry. All statistical analyses were performed as a student's two tailed t-test (*$p<0.05$), where qPCR results were calculated as fold-change over free-swelling controls.

RESULTS

Contrary to our hypothesis, very early mechanical loading (0-2 weeks) in physoxic conditions was not uniformly beneficial for chondrogenesis; loading upregulated collagen-II, but simultaneously downregulated the chondrogenic differentiation factor Sox9 in comparison to free swelling gels, and there were no differences in sGAG and collagen accumulation, or mechanical properties at 2 weeks. However, loading from 2-4 weeks significantly increased chondrogenesis, upregulating collagen-II and aggrecan (Fig. 1A) and increasing sGAG synthesis. A similar trend was observed for total collagen matrix production (Fig. 1B). Alcian blue stained intensely in both groups, but was most intense under dynamic compression (Fig. 1C). Dynamic compression from 4-6 weeks enhanced chondrogenesis from with upregulation of collagen-II and aggrecan, and significantly downregulated the osteogenic markers runx2, osteopontin, and osterix (Fig. 2A). Consistently, while mineralized matrix production was negative for all earlier time points (data not shown), alizarin red staining was observable at 6 weeks in free swelling but not dynamically loaded samples (Fig. 2B).

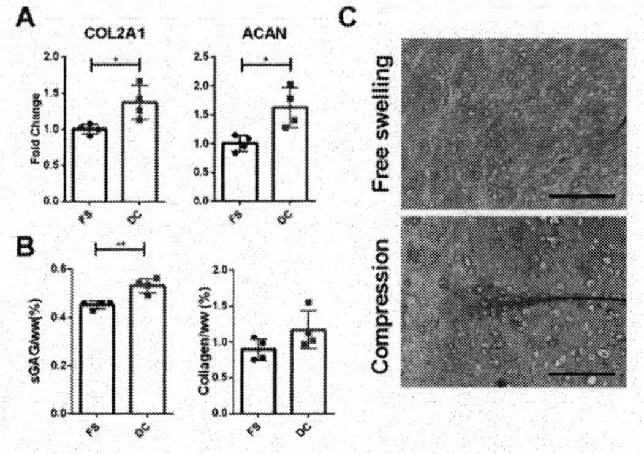

Figure 1 Mechanical loading from 2-4 weeks enhances chondrogenesis. (A) qPCR results after mechanical loading from 2-4 weeks showing upregulation of collagen-II and aggrecan. (B) Biochemical analysis with significant increase in sGAG and a trend towards increasing total collagen, with (C) alcian blue staining (scale bar 250 μm). Bar graphs show mean ± S.D. at 4 weeks with qPCR shown as fold change over free swelling controls.

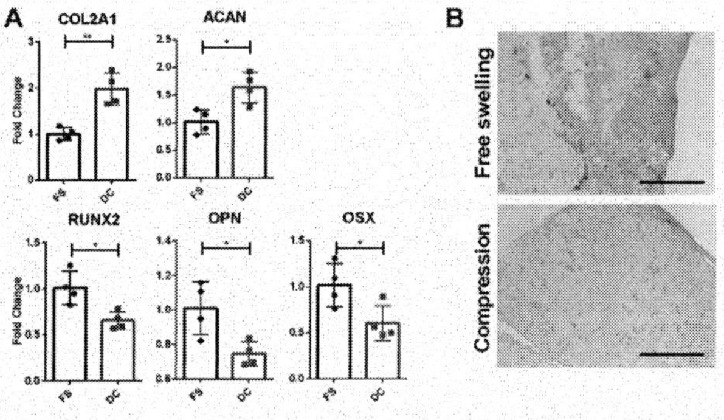

Figure 2 Mechanical loading from 4-6 weeks suppresses endochondral progression. (A) qPCR data after dynamic compression from 4-6 weeks showing an increase in chondrogenesis, and suppression of endochondral ossification. (B) Alizarin Red staining of gels (scale bar 250 μm) showing positive staining in free swelling samples. Bar graphs are mean ± S.D. shown as fold change over free swelling controls at 6 weeks.

Delayed dynamic compression, with loading between 6-8 weeks increased expression of the chondrogenic markers collagen-2 and aggrecan, similar to 4-7 week loading; however, the osteogenic genes osteopontin and runx2 were no longer significantly downregulated, and there was a significant upregulation of collagen-X, indicative of enhanced chondrocyte hypertrophy (Fig. 3A). Consistent with load-induced matrix production, dynamic compression between weeks 6 and 8 significatly increased construct dynamic modulus compared to non-loaded controls at 8 weeks (Fig. 3B). Faint mineralization via alizarin red staining was visible at 8 weeks in both free swelling and dynamic compression (6-8 weeks) groups.

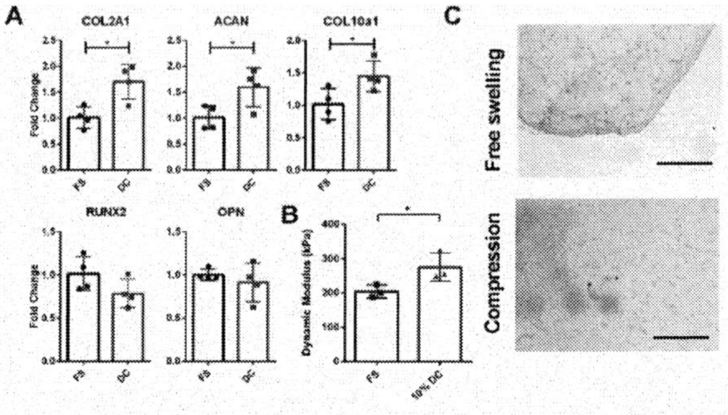

Figure 3 Mechanical loading from 6-8 weeks supports chondrogenesis and osteogeneisis. (A) qPCR results after mechanical loading from 6-8 weeks showing upregulation of collagen-II , aggrecan, and collagen-X, with no downregulation of runx2 and osteopontin. (B) Dynamic compression significantly increased the dynamic modulus compared to free swelling controls, (C) alizarin red staining of gels (scale bar 250 μm) showing mineralization in each group. Bar graphs are mean ± S.D. at 8 weeks with qPCR shown as fold change over free swelling controls.

DISCUSSION

Very early loading of progenitor cells did not measurably alter chondrogenesis, possibly due to lack of time to produce adequate pericellular matrix that may be integral in transducing mechanical load to the cells. However, in all other groups with load initiation delayed to either 2, 4, or 6 weeks, dynamic compression enhanced chondrogenesis with increased matrix production, mechanical properties, and chondrogenic genes, suggesting that mechanical loading *in vitro* may play a role in chondrogenic differentiation. In previous *in vivo* studies, we similarly saw that mechanical loading was necessary to induce endochondral healing of tissue-engineered mesenchymal condensations placed in critically-sized rat femoral defects, exhibiting a mixture of cartilage and bone. Non-loaded limbs failed to functionally regenerate, displaying mainly fibrotic tissue, suggesting a similar role *in vivo* of mechanical-load mediated progenitor differentiation. In addition to inducing chondrogenesis, onset of load at either 2 or 4 weeks also either delayed or arrested endochondral progression through downregulation of osteogenic genes and delay of mineralization. Similarly, early loading *in vivo* exhibited persistent gaps of cartilage and possible pseudoarthrosis, suggesting that persistent early loading may work to maintain a stable cartilage phenotype. However, after 6 weeks, when mineralization is seen in free swelling fibrin hydrogels, dynamic compression supported both chondrogenesis and osteogenesis, and hypertrophic markers such as collagen-X were upregulated, suggesting that late loading may also enhance hypertrophy and endochondral progression. Together, these data demonstrate a significant temporal interaction between mechanical loading and endochondral lineage specification. Future tissue engineering strategies may utilize mechanical priming as a tool to enhance chondrogenesis and stable articular cartilage when applied early, and to support both chondrocyte maturation and osteogenesis when applied late. Ongoing studies are working to elucidate mechanosensitive signaling pathways involved in lineage specification under dynamic compression.

REFERENCES

[1] Hamburger, V. (*et al*), Physiological Zoology, 13: 367-382, 1940.
[2] Kenwright, J. (*et al.*), Lancet, 2:1185-7, 1986.
[3] McDermott, A. (*et al.*), SB3C, National Harbor, MD, 2016

INHIBITION OF RHO KINASE THAT ATTENUATES PAIN ALSO REDUCES EARLY SPINAL GLIAL ACTIVATION & NEUROTRANSMITTER EXPRESSION AFTER MECHANICAL FACET CAPSULE INJURY IN VIVO

Sijia Zhang (1), Christine L. Weisshaar (1), Beth A. Winkelstein (1,2)

(1) Department of Bioengineering
University of Pennsylvania
Philadelphia, PA, USA

(2) Department of Neurosurgery
University of Pennsylvania
Philadelphia, PA, USA

INTRODUCTION

The facet capsular ligament encloses the bilateral spinal facet joints and is a source of pain since it is innervated [1,2]. Excessive stretch of the cervical facet capsular ligament produces morphological and functional changes in its afferents, altered neurotransmitter expression, and neuronal hyperexcitability and inflammation in the spinal cord, all of which are also associated with pain [2-5]. Despite mechanical loading of the facet capsule leading to pain, the molecular pathways involved are poorly understood, hampering pain treatments.

There is growing evidence that Rho-mediated pathways modulate neuronal injury. Mechanical stress on cells can upregulate RhoA via adhesive interactions between the cells and extracellular matrix [6]. Active RhoA binds and activates Rho-associated protein kinase (ROCK), triggering a host of downstream signaling cascades, including those involved in neural injury [7,8]. Inhibiting ROCK has been shown to alleviate inflammatory and neuropathic pain [8] and RhoA activity has been found to increase in the dorsal root ganglia (DRG) soon after painful stretch of the facet capsule [9].

Inhibiting ROCK activity in the facet joint during capsule stretch is hypothesized to prevent pain. To test this, intra-articular ROCK was blocked in a rat model of painful facet capsule stretch that induces pain within 1 day of injury [3-5]. Behavioral sensitivity to mechanical stimuli was assessed to measure pain before and at early times after injury. To probe potential mechanisms by which ROCK may alter nociception, spinal glial and neuronal responses were evaluated using expression of Iba1, a protein associated with microglial activation, and substance P (SP), an excitatory neuropeptide.

METHODS

All procedures were IACUC-approved. Briefly, a controlled bilateral C6/C7 facet joint distraction was imposed using a customized dynamic loading device to stretch the facet capsule [3-5]. Immediately after facet distraction, rats were randomly selected to receive a bilateral C6/C7 intra-articular injection of either the ROCK inhibitor Y27632 (Selleckchem) at a dose of 500μg in 10μL of saline (n=6) or the same volume of only the vehicle saline (n=6). The magnitude of applied stretch was quantified by tracking markers on the laminar bones of the vertebrae and facet capsule during distraction [4]. Vertebral and capsular distractions and the maximum principal strains (MPS) across the capsule were measured and compared between groups using t-tests to ensure that the imposed mechanical injury was not different between the inhibitor and vehicle treatments. A separate group (n=6) received sham surgery and saline vehicle injection but no joint distraction to serve as surgical controls. Pain was assessed before (day 0) and at 8 hours and 1 day after the surgical procedures; the forepaw withdrawal threshold was measured using von Frey filaments [4,5]. A two-way repeated-measures ANOVA with Tukey HSD test determined differences in withdrawal thresholds over time.

After behavioral testing on day 1, C7 spinal cord was harvested after trans-cardiac perfusion and post-fixed. Spinal cord tissue was axially sectioned (at 14μm thickness) and prepared for on-slide immunohistochemical labeling. Slides were blocked using 1% donkey serum (Millipore) in 0.3% Triton X-100 and incubated with rabbit anti-Iba1 (1:1000; Wako) and guinea pig anti-SP (1:500; Neuromics) overnight at 4°C. The next day, sections were incubated with the secondary antibodies donkey anti-rabbit 555 (Fisher) and donkey anti-guinea pig 488 (Jackson ImmunoResearch). The dorsal horns were imaged and uniform-sized regions of interest (ROI) in the superficial dorsal horn (laminae I-II; 750x150 pixels) were cropped from each image for standard analysis [10]. Automated densitometry measured the percent of positively-labeled pixels using a customized MATLAB script [10]. Spinal cord from normal un-operated rats (n=2) was included as negative controls for immunolabeling and analyses. Separate one-way ANOVAs with Tukey HSD tests were used to detect differences in Iba1 and SP expression between the surgical groups.

RESULTS

The severity of the imposed mechanical injury to the facet joint did not differ between the treated and vehicle groups. The mean vertebral distraction was 0.58±0.25mm for vehicle and was 0.43±0.12mm for the ROCK inhibition ($p>0.2$) and the mean capsule distraction was 0.34±0.12mm for vehicle and 0.26±0.09mm for treatment ($p>0.2$). Rats receiving vehicle treatment experienced an average capsular MPS of 13±4% and peak MPS of 28±14%; rats that received ROCK inhibition underwent similar capsule stretch ($p\geq0.1$), with a mean MPS of 9±3% and a peak MPS of 20±12%.

Despite undergoing comparable facet joint distractions, only the group of rats receiving saline vehicle developed pain (Figure 1). The forepaw withdrawal threshold of group receiving vehicle treatment was significantly lowered from baseline (day 0) at both 8 hours and 1 day ($p<0.001$) after facet joint distraction (Figure 1), indicating increased behavioral sensitivity. In contrast, the withdrawal thresholds of the group undergoing injury with ROCK inhibition and the group undergoing sham surgery were not altered from corresponding baseline (day 0) uninjured values (Figure 1).

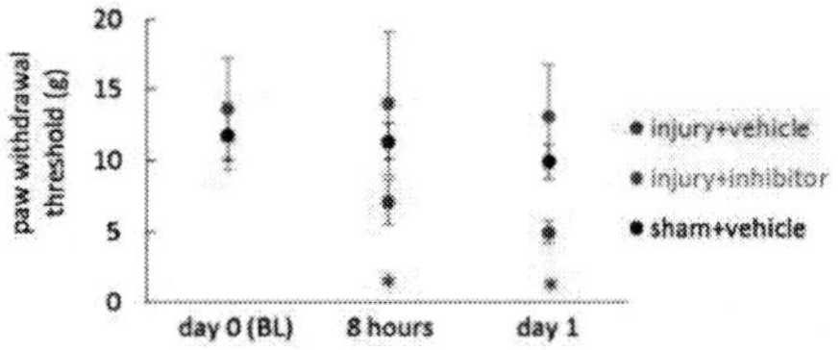

Figure 1: Facet joint distraction decreased the withdrawal threshold at 8 hours and 1 day (*p<0.001) in the injury+vehicle treatment group. Neither the injury+inhibitor nor sham group exhibited altered thresholds from pre-injury baseline (BL) levels.

Paralleling the behavioral responses, spinal microglial activation and SP expression in the injury group with ROCK inhibition was similar to levels in normal un-operated rats and significantly lower than those in the injury group with vehicle injection (Figures 2 & 3). Iba1 expression was increased in the injury group receiving vehicle treatment, and was significantly higher than both the levels in sham ($p=0.003$) and the ROCK inhibition group ($p<0.001$) (Figure 2). In contrast, the Y27632 treatment lowered spinal Iba1 expression below levels in shams ($p<0.001$) (Figure 2). Also, SP expression in the group receiving ROCK inhibition was significantly lower than that in both of the injury with vehicle and sham groups ($p<0.001$) (Figure 3).

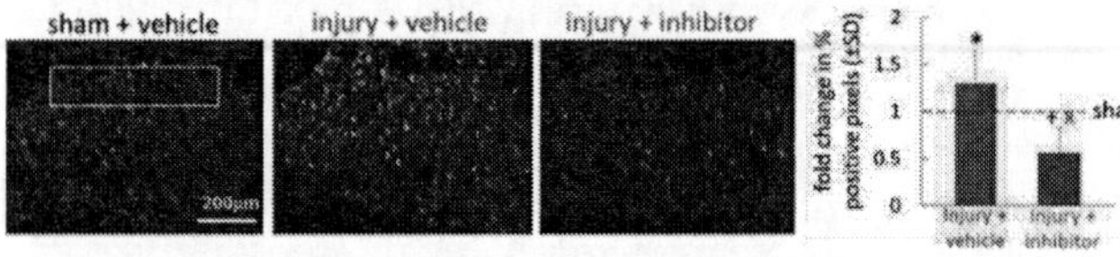

Figure 2: Sample images and quantification of Iba1 in the dorsal horn ROIs (white box). Iba1 expression was lower in the injury+inhibitor group than in both the injury+vehicle (ˣp<0.001) and sham (⁺p<0.001) groups. Iba1 expression in the injury+vehicle group was also greater than expression in sham (*p=0.003).

DISCUSSION

Mechanical facet joint injury can induce spinal inflammation, with increases cytokines at day 1 and glial reactivity at day 7 in association with pain [5,11]. However, this is the first study to evaluate the pain and spinal glial responses within 1 day of injury. In fact, microglial activation in the superficial dorsal horn was prevented, along with pain onset, by blocking intra-articular ROCK (Figures 1 & 2). This effect on microglia is consistent with reports of the Rho/ROCK pathway regulating inflammatory reactions [8,12]. Further, although ROCK inhibition has been found to attenuate pain in neuropathy and inflammation [8,12], our findings suggest, for the first time, that peripheral intra-articular ROCK inhibition can both prevent joint pain from its injury and mediate spinal responses remote from the treatment site. However, only the early responses after injury were investigated here and it is unknown how long the beneficial effects of ROCK inhibition may last. Although the Y27632 dose was selected because it was effective on inflammatory pain [13], a dose response was not evaluated nor were effects on alleviating existing pain.

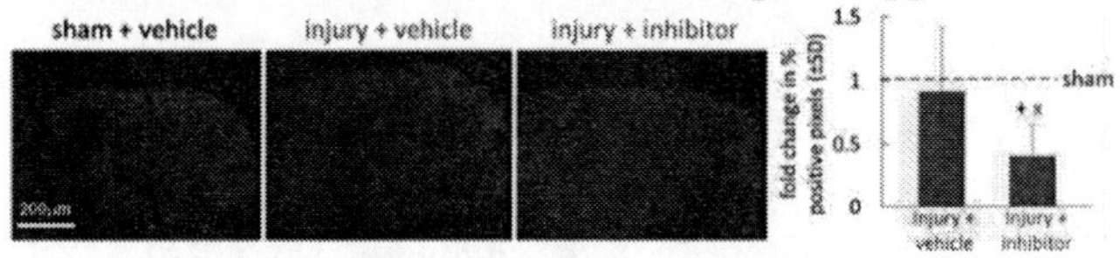

Figure 3: Representative images and quantification of spinal SP. SP expression was significantly lower in the injury+inhibitor group than injury+vehicle (ˣp<0.001) and sham (⁺p<0.001).

In addition to attenuating microglial activation (Figure 2), ROCK inhibition reduced spinal dorsal horn SP levels (Figure 3). Yet, spinal SP expression was not different between the rats undergoing facet injury and those with sham surgery (Figure 3); the comparable levels in both groups is likely due to the surgical effects being still present at day 1. Interestingly, SP increases in axons of an in vitro DRG-gel system simulating the painful facet capsule stretch induced in the current study, and those increases in SP were attenuated by ROCK inhibition [9]. Since NMDA receptors modulate SP release and ROCK regulates afferents' cytoskeletons that affect NMDA receptor activity [14,15], this may be a novel mechanism of ROCK involvement in pain.

Although the current study begins to reveal a possible novel mechanism by which ROCK activation may lead to pain from mechanical joint injury, many other factors contribute to pain, including inflammatory cell infiltration and PKC regulation. In fact, leukocyte infiltration and phosphorylation of a major PKC substrate, MARCKS in the spinal cord are mediated by ROCK [8,11]. As such, further investigations of other Rho/ROCK-mediated changes in the context of pain are needed to fully define if and how ROCK activation is involved in mechanically-induced joint pain. Nonetheless, this study is the first to identify ROCK as a potentially potent mediator of painful mechanical trauma and a potential therapeutic target for joint pain.

ACKNOWLEDGEMENTS

Funding provided by the NIH (U01EB016638), an AMTI Force and Motion Scholarship, and an NSF training grant.

REFERENCES

[1] Kallakuri S et al. *Wold J Orthop*, 3:10-4, 2012. **[2]** Kallakuri S et al. *Eur Spine J*, 17:556-63, 2008. **[3]** Lee K & Winkelstein B. *J Pain*, 10:436-45, 2009. **[4]** Dong L et al. *Brain Res*, 1461:51-63, 2012. **[5]** Weisshaar C & Winkelstein B. *J Pain*, 15:378-86, 2014. **[6]** Marjoram R et al. *Curr Mol Med*, 14:199-208, 2014. **[7]** Mueller B et al. *Rev Drug Discov*, 4:387-98, 2005. **[8]** Tatsumi S et al. *Neuroscience*, 131:491-8, 2005. **[9]** Zhang S & Winkelstein B. *SB³C*, 2016. **[10]** Zhang S et al. *Stapp Car Crash J*, 57:219:42, 2013. **[11]** Kras J et al. *Spine*, 39:207-12, 2014. **[12]** Wu X & Xu X. *Neural Regen Res*, 11: 23-7, 2016. **[13]** Paiva-Lima P et al. *Pain Res Manag*, 19:e172-8, 2014. **[14]** Let S et al. *J Neurosci*, 21:8464-72, 2001. **[15]** Liu H et al. *Nature*, 386:721-4, 1997.

INTERACTION-DRIVEN AGING IN SYNTHETIC TISSUES

Aylin Acun (1), Dervis C. Vural (2), Pinar Zorlutuna (1,3)

(1) Bioengineering Graduate Program
University of Notre Dame
Notre Dame, IN, USA

(2) Department of Physics
University of Notre Dame
Notre Dame, IN, USA

(3) Aerospace and Mechanical Engineering
Department
University of Notre Dame
Notre Dame, IN, USA

INTRODUCTION

Aging is a very complex phenomenon that spans multiple time and length scales. Therefore a fundamental understanding of aging must necessarily take into account factors beyond individual cell damage, and explain how failure cascades from cells to tissues to organs to organisms [1]. In this study we hypothesize that aging in complex organisms is due to the interdependence within the system, rather than the steady accumulation of individual failures. To test this hypothesis we have fabricated biomimetic 3D engineered tissues for which aging damage and interaction parameters can be controlled. Then we used these model tissues to study the failure under various conditions. Our experiments consist of measuring the survival curves of these engineered tissues fabricated from "pre-conditioned" (or "pre-aged") cells under a "post-condition". The pre-condition defines the history of the cells comprising the tissue (prior to tissue fabrication), while the post-condition defines the environment in which we age the engineered tissues and take the measurements.

METHODS

The 3D engineered model tissues were prepared by encapsulating $1x10^3$, $1x10^4$, and $1x10^5$ rat cardiac fibroblasts (CFs) per construct in PEG-RGD hydrogels with pre-aging or no aging (young). The 3 orders of magnitude difference in the encapsulation densities effectively controls the cell-cell interactions within the engineered tissues by controlling the cell-cell distance since the total volume of the tissues are the same for all groups. The CFs were isolated from 2-day old neonates and either used as is ("young" condition) or aged in tissue culture in 3 different ways: 1) Chronological senescence (cells from neonatal animals kept in cell culture without passaging for a month); 2) Oxidative senescence (cells from neonatal animals are kept in cell culture media containing 20 μM hydrogen peroxide (H_2O_2), without passaging, for a month); 3) Replicative senescence (cells from neonatal animals are passaged multiple times until they cannot divide).

RESULTS

First, we tested whether or not our pre-aging conditions mimic the aging "symptoms" of native cells by staining for senescence associated B-galactosidase and by quantifying the DNA damage (**Figure 1).** The young cells did not show positive staining for this marker while cells that went through multiple population doublings as well as the oxidative stress treated cells did. Additionally, we assessed the amount of DNA damage caused by our pre-aging conditions using Comet assay. Our results showed a dose dependent DNA damage with increased H_2O_2 concentration. Then we monitored the survival of the model tissues either without any stress (chronological aging) or with 0.2 mM H_2O_2 exposure, which mimics the oxidative stress in the heart after a myocardial infarct. The model tissues were stained (daily for model tissues with $1x10^4$, and $1x10^5$ CFs; every 2h for tissues with $1x10^3$ CFs) with a nuclear stain that stains only the dead cells and the survival was determined by counting the number of dead cells every day until all the cells in the engineered tissue are dead.

Our results show that although pre-aging affects the cell viability under stress conditions, the cell-cell interactions (defined by the proximity of the cells in the hydrogel) is a more dominant factor and the nature of the pre-aging (i.e. chronological senescence, replicative senescence or oxidative senescence) does not matter. The cells in high encapsulation density tissues survived significantly more compared to cells in low encapsulation density tissues regardless of the pre-aging method (**Figure 2**). In addition, we showed that as the population density decreased, the lifetime of the population was significantly faster in all pre-aging conditions. We also observed that the pre-aging did not affect the death rate if the population size was large enough, emphasizing the importance of intercellular communication in failure of complex systems (**Figure 3**).

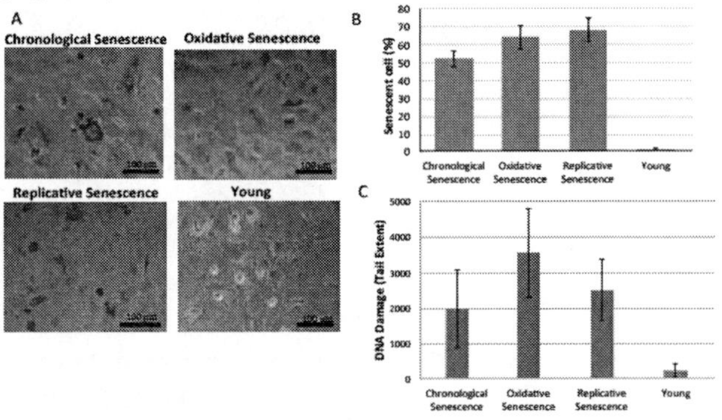

Figure 1: Pre-aging of cardiac fibroblasts. Senescence associated B-galactosidase staining of different pre-treatment conditions (A), and its quantification (B). DNA damage quantification using comet assay in response to oxidative stress (C) (Scale bar 100 μm).

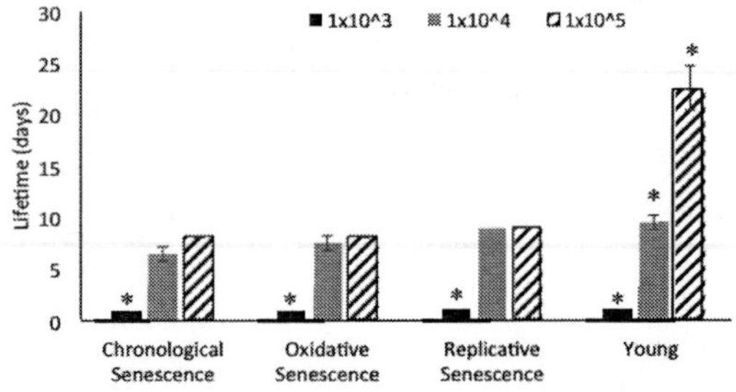

Figure 2: The lifetime of synthetic tissues made from young or pre-aged cells with different population densities (* indicates significance, p<0.05).

DISCUSSION

The 3D engineered model tissues developed in this study allow us to control the population density, intercellular communication and death rate, which are important in understanding how micro-scale failure leads to the eventual collapse of the whole system in complex organisms. We hereby show that the intercellular dependency and not the accumulation of individual failures holds the key to the catastrophic failure in complex systems.

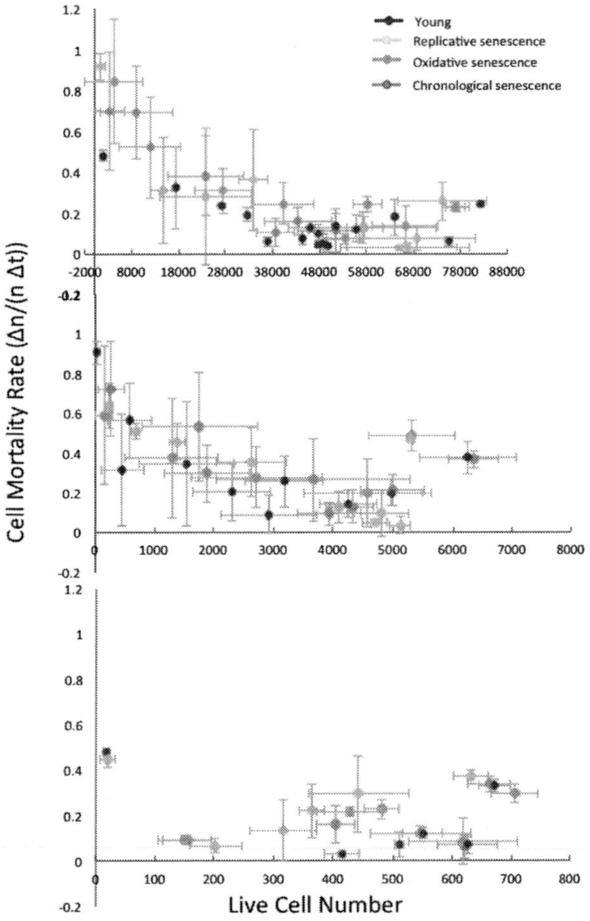

Figure 3: Failure characteristics of synthetic tissues made from young or pre-aged cells as shown by the change of cell mortality rate with respect to live cell number. Denser tissues where cells can interact, display age specific mortality, whereas the mortality rate of sparse tissues is closer to constant.

ACKNOWLEDGEMENTS

This publication was supported by a Project Development Team within the ICTSI NIH/NCRR Grant Number UL1TR001108.

REFERENCES

[1] Vural, D et al., *Phys. Rev. E.* 89(2): 022811-022819, 2014.

A NOVEL BIOREACTOR TO STUDY THE DRIVING MECHANICAL STIMULI OF TISSUE GROWTH AND REMODELING

Mathieu A.J. van Kelle (1,2,*), Pim, J.A. Oomen (1,2,*), Jurgen A. Bulsink (1), Marloes W.J.T. Janssen-van den Broek (1,2), Richard G.P. Lopata (1), Marcel C.M. Rutten (1), Sandra Loerakker (1,2), Carlijn V.C. Bouten (1,2)

** These authors contributed equally to this work*

(1) Department of Biomedical Engineering
Eindhoven University of Technology
Eindhoven, The Netherlands

(2) Institute for Complex Molecular Systems
Eindhoven University of Technology
Eindhoven, The Netherlands

INTRODUCTION

Tissue growth and remodeling are essential processes which ensure long-term functionality of tissue engineered (TE) constructs. Although it is widely recognized that these processes strongly depend on mechanical stimuli, the underlying mechanisms are only partially understood. Nevertheless, it is generally accepted that changes in the mechanical environment trigger a homeostatic response by the cells residing inside these tissues[1]. These cells sense mechanical changes and respond by altering their surroundings, by means of ECM growth and remodeling, to return to their preferred mechanical homeostatic state. However, it remains unclear to which mechanical stimuli these cells respond and thus govern tissue growth and remodeling.

We propose a novel 'versatile tissue growth and remodeling' (Vertigro) bioreactor which can help to determine these driving mechanical stimuli. The bioreactor's unique two-chamber design allows easy, sterile handling of circular 3D TE constructs in a dedicated culture chamber, while a separate pressure chamber facilitates a dynamic loading regime during culture. A custom-made feedback system ensures a constant peak pressure for at least one week of culture. It is hypothesized that this dynamic loading regime will induce mechanical, geometrical and compositional changes in the TE constructs.

Through an extensive series of analysis techniques, we are able to systematically assess these changes. First, non-destructive mechanical testing can be performed, by means of a classic bulge test combined with tracking of the tissue construct's displacement with ultrasound imaging. Second, additional analysis techniques are used to qualitatively and quantitatively asses geometrical and compositional changes due to tissue growth & remodeling.

With this bioreactor and associated analysis techniques, we have developed a powerful toolbox to study and identify the driving mechanical stimuli of tissue growth & remodeling.

METHODS

Bioreactor Tissue culture

Figure 1: A) Vertigro bioreactor. B) Circular TE construct cultured in the bioreactor.

The bioreactor (Figure 1A) consists of two chambers: a pressure chamber and a culture chamber (Figure 2). An insert clamping a TE construct can be mounted inside the culture chamber (Figure 1B). The top culture chamber is detachable from the bottom pressure chamber to enable sterile medium changes. In between the two chambers are two silicon membranes, ensuring proper pressure transmission.

The dynamic pressure during culture is applied through a water-filled pressure chamber, which is connected to a flexible silicone tube. A pneumatic proportional air valve regulates inflow of pressurized air, compressing the silicone tube thus applying pressure in the pressure chamber. The double membrane transmits the pressure into the culture chamber, where a sensor measures the applied pressures. Finally, a custom-made feedback system (NI Labview) maintains these pressures at a constant level by controlling the proportional air valve (Figure 2).

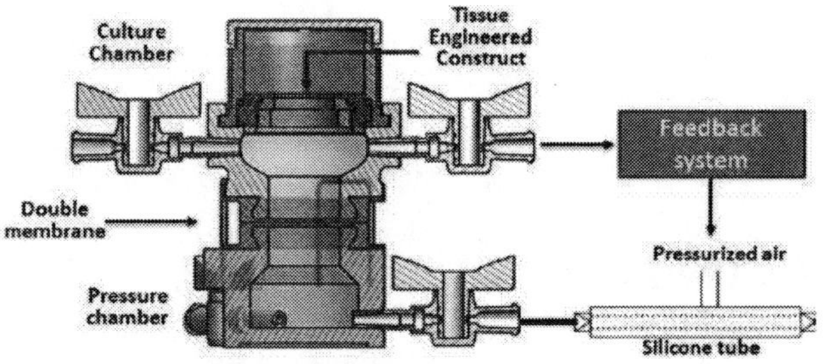

Figure 2: Schematic illustration of the Vertigro bioreactor and the dynamic pressure application system.

Analysis techniques

<u>Mechanical changes</u>
Mechanical testing of the TE constructs is performed by means of a bulge test. The bioreactor hardware to apply the dynamic pressures during culture is used to bulge the constructs, while an ultrasound transducer is positioned directly on top of the bioreactor to measure the deformation profile. In addition, tissue prestretch measurements can be performed by placing markers on the sample, and tracking their displacement upon excision with a standard digital microscope.

<u>Geometrical changes</u>
The ultrasound measurements are also used to measure curvature changes of the TE constructs. Tissue thickness is determined by placing bisected samples under a microscope.

<u>Compositional changes</u>
To determine tissue anisotropy, collagen fibers are stained with CNA[2] followed by imaging using confocal microscopy, making a tile scan of half the tissue sample. Next, gross tissue collagen fiber orientation and dispersity were determined using a custom-made MATLAB script based on the work of Frangi *et al*[3]. Next, a series of established biochemical assays is used to determine the amount of collagen, GAGs and cells. Finally, histological techniques are used to visualize cells and ECM components throughout the tissue constructs.

RESULTS

The tissue constructs were first cultured statically, and could subsequently be cultured under dynamic loading conditions in the bioreactor for at least one week. During this period the maximum pressures were kept at a constant level by the custom-made feedback system (Figure 3).

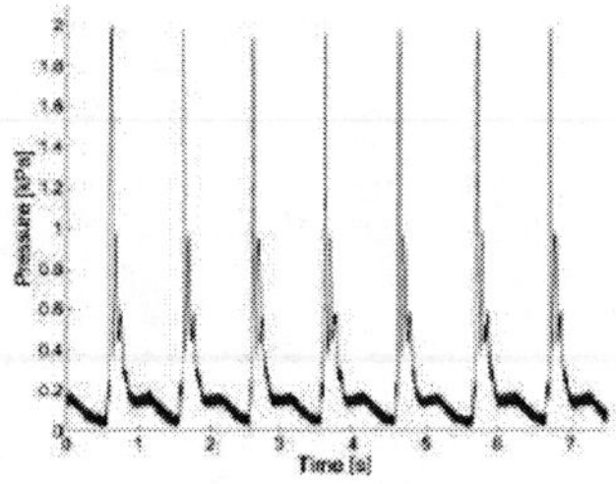

Figure 3: Typical pressure curves during culture, kept constant throughout the culture period by the feedback system.

After culturing the constructs were subjected to the bulge tests. With ultrasound, the displacement of the samples could be accurately visualized (Figure 4A) and tracked (Figure 4C). Next, tissue prestretch (Figure 4B), curvature (Figure 4C) and thickness (Figure 4D) could be determined as described earlier. In Figure 5A a representative image showing the gross collagen architecture is shown. In this particular case, the collagen is oriented in an isotropic fashion (Figure 5B). Figure 5C shows the results for the GAG assay, clearly showing an increase with tissue culture time. Note that during dynamic loading in the final week, the GAG content increases dramatically. Finally, histology shows abundant collagen (red) and alpha-SMA positive cells (green) homogeneously distributed throughout the tissue.

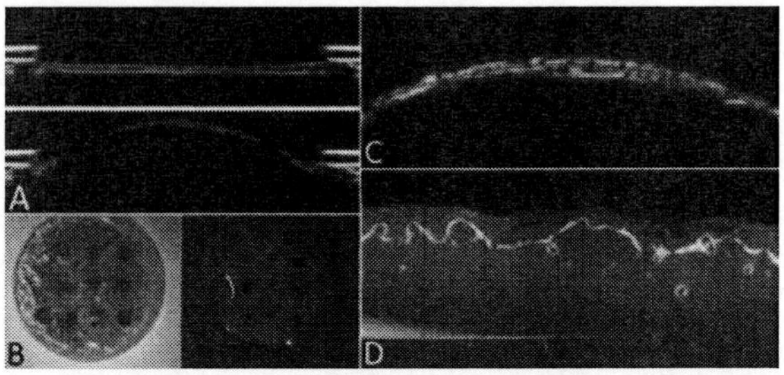

Figure 4: A) Tissue construct imaged during bulge test, flat (top) and during pressurization (bottom). B) Prestretch measurement. C) Tracking of tissue displacement during bulge test, also used to assess geometrical curvature changes. D) Thickness measurement.

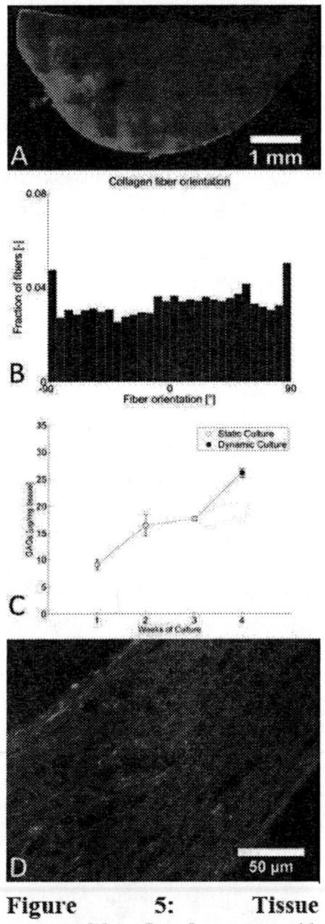

Figure 5: Tissue compositional changes: A) Collagen staining on bisected sample and associated collagen fiber orientation (B). C) Amount of GAGs with culture time and histological image (D).

DISCUSSION

In this study, the potential of the Vertigro bioreactor is demonstrated. First, it is capable of applying a variety of loading conditions to TE constructs that can induce mechanical, geometrical and compositional changes to the tissues. Next, with a large collection of analysis techniques these changes can be assessed. A significant advantage of the used mechanical testing method, is its non-destructive nature, allowing measurements of temporal changes in mechanical properties of a single sample. With all these techniques, we hope to pinpoint the driving mechanical stimuli which dictate tissue growth and remodeling.

ACKNOWLEDGEMENTS

We acknowledge the support from the Netherlands Cardiovascular Research Initiative (CVON 2012-01): The Dutch Heart Foundation, Dutch Federation of University Medical Centers, the Netherlands Organization for Health Research and Development and the Royal Netherlands Academy of Sciences, and the European Union's Seventh Framework Programme (grant agreement No 604514).

REFERENCES

[1] Humphrey, JD *et al.*, *Nat. Rev. Mol. Cell Biol.* 15, 802–12, 2014
[2] Boerboom, RA *et al.*, *J. Struct. Biol.* 159:392–399, 2007
[3] Frangi, AF *et al.*, MICCAI. 1496, 404–410 1998

BIOLOGICAL INTERVENTION TO REDUCE POST-TRAUMATIC JOINT CONTRACTURE: PRELIMINARY EVIDENCE SUPPORTING THE USE OF SIMVASTATIN OR LOSARTAN

Alex J. Reiter (1), Chelsey L. Dunham (2), Ryan M. Castile (1), Aaron M. Chamberlain (3), Leesa M. Galatz (4), Spencer P. Lake (1,2,3)

(1) Department of Mechanical Engineering and Materials Science
Washington University in St. Louis
St. Louis, MO, USA

(2) Department of Biomedical Engineering
Washington University in St. Louis
St. Louis, MO, USA

(3) Department of Orthopaedic Surgery
Washington University in St. Louis
St. Louis, MO, USA

(4) Department of Orthopaedic Surgery
Mt. Sinai Hospital
New York, NY, USA

INTRODUCTION

Post-traumatic joint contracture (PTJC) affects up to 50% of patients following joint dislocation or fracture and often leads to permanent stiffness and debilitating motion loss [1-3]. The elbow is particularly susceptible to contracture following traumatic injury due, in part, to its high degree of natural congruency [4,5]. Nearly, 12-15% of patients with PTJC in the elbow require surgery to release contracted tissue in order to help improve range of motion [6].

Our group recently established the first animal model of a clinically-relevant injury and subsequent PTJC in the elbow [7]. The unilateral injury model (using Long-Evans rats) showed persistent long-term contracture and connective tissue changes following a period of initial joint immobilization and subsequent free mobilization period [8]. With this validated animal model, we are now investigating different treatment and intervention strategies to reduce fibrosis-induced PTJC and preserve joint function.

Statins are a family of FDA-approved drugs used to treat hypercholesterolemia and have been shown to reduce fibrosis in several different tissues including muscle [9], lung [10], and skin [10]. These anti-fibrotic pleiotropic effects are found most consistently for a statin called simvastatin (SV). Losartan (LS), another clinically-available drug approved as an anti-hypertensive medication, has been shown to reduce fibrosis in tissues such as lungs [11] and muscle [12]. However, the potential impact of either drug on connective tissues and joint contracture is unknown. In this preliminary study, we investigated whether biological intervention using SV or LS could prevent mechanical joint contracture in our elbow model of PTJC. We hypothesized that the anti-fibrotic pleiotropic effects of SV and LS (administered separately) would reduce the severity of contracture and help preserve normal joint mechanics.

METHODS

Male Long-Evans rats (300-370g) were anesthetized, the left elbow was exposed laterally, and both an anterior capsulotomy and lateral collateral ligament transection was performed as previously described [7]. Injured limbs were immobilized in flexion, and animals were randomly assigned to receive SV or LS via oral gavage 3x/week for six weeks. Based on previous animal studies and approximations of doses used to treat hypercholesterolemia and hypertension in humans, appropriate doses of 60 and 40 mg/kg of body weight were chosen for SV [13,14] and LS [15,16], respectively. Half the animals (SV: n=3; LS: n=4) were sacrificed after six weeks of immobilization (IM) (Fig. 1). The remaining animals (SV: n=3; LS: n=3) were sacrificed after an additional six weeks of free mobilization (FM) (i.e., unrestricted cage activity after removing immobilization bandages) (Fig. 1).

Both forelimbs were prepared post-mortem for mechanical testing as described previously [7]. A custom mechanical test system was used to evaluate limbs in flexion-extension (F-E) joint testing for five cycles between specified force limits [7]. Force and displacement data from the fifth cycle were converted to torque and angular position, then analyzed for elbow total range of motion (ROM) and neutral zone (NZ) length (i.e., functional range of joint motion). These parameters were chosen as representative measures to determine if the drug treatments helped preserve joint mechanics. Results were compared to previously published data [8] quantifying joint contracture in injured animals without drug treatment (INJ) (IM: n=10; FM: n=10) and uninjured, non-immobilized control animals (C) (IM: n=10; FM: n=12). Groups and study design are shown in Figure 1.

One-way ANOVA tests were used to compare mechanical test parameters between control and injured limbs across groups and control and contralateral (CL) limbs across groups. When significance was found, post-hoc Bonferroni analyses compared each experimental group to control. We also evaluated side-to-side limb differences with paired t-tests to compare injured limbs to CL limbs within each group.

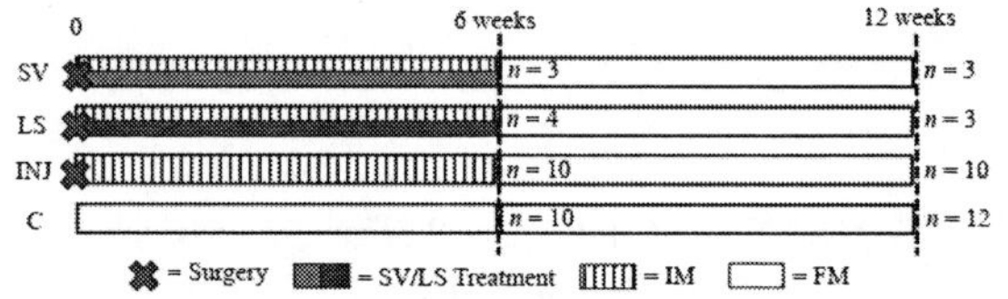

Figure 1: Summary of study groups and treatment plan.

RESULTS

After IM, SV and LS injured limbs showed no improvements in total ROM compared to INJ and remained significantly decreased compared to control limbs (Fig. 2A). Following FM, both groups of drug-treated injured limbs showed an increase in total ROM compared to non-treated injured limbs and were no longer significantly different from control limbs (Fig. 2B). Side-to-side comparisons showed significant differences between injured and uninjured CL limbs within each group at both IM and FM (Fig. 2). However, the magnitude of side-to-side differences were noticeably less for FM animals.

After FM, SV and LS injured limbs also showed no improvements in NZ length compared to INJ and remained significantly decreased compared to control limbs (Fig. 3A). Following FM, injured limbs in both treatment groups showed increased NZ length and were not significantly different from control limbs (Fig. 3B). Side-to-side comparisons showed significant differences between injured and uninjured CL limbs within each group at both IM and FM except for the FM SV group (Fig. 3).

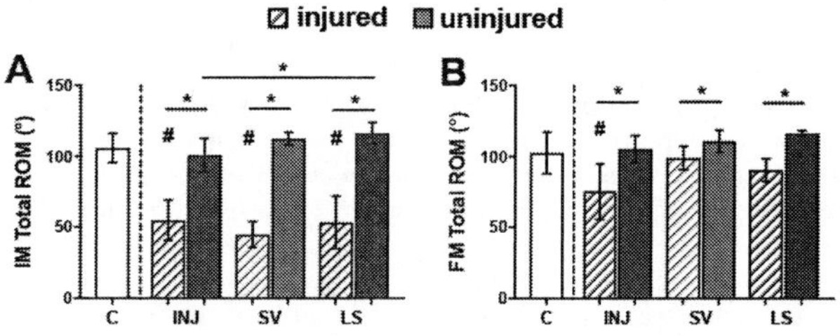

Figure 2: (A) Total ROM in drug treatment groups remained similar to INJ and significantly less than controls following IM. (B) However, following FM, total ROM in drug treatment groups was greater than INJ and not significantly different from controls. (avg ± std, # = sig. diff. from control, * = sig. diff. defined by line endpoints, $p < 0.05$)

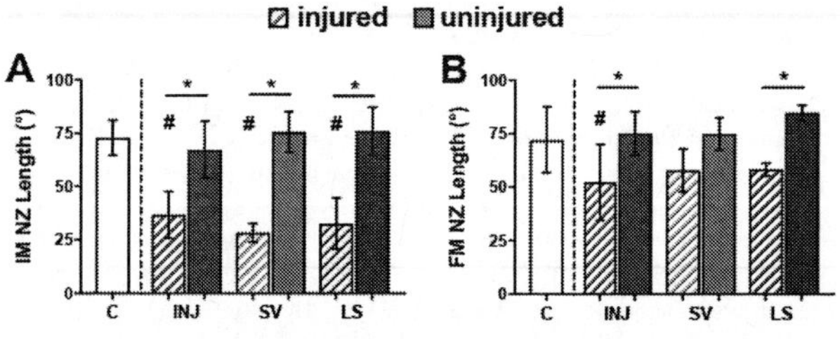

Figure 3: (A) NZ length in drug treatment groups remained similar to INJ and significantly less than controls following IM. (B) However, following FM, NZ length in drug treatment groups was greater than INJ and not significantly different from controls. (avg ± std, # = sig. diff. from control, * = sig. diff. defined by line endpoints, $p < 0.05$)

DISCUSSION

Results of this preliminary study showed that both SV and LS reduced PTJC severity and partially preserved joint mechanics in our animal model as indicated by an improved total ROM (Fig. 1B) and NZ length (Fig. 2B) following the free mobilization period. While no marked improvements were seen following IM, significant improvements were seen following FM. We hypothesize that excessive post-injury ECM production was limited during the IM period, but the FM period was necessary to physically break these weaker and less dense adhesions in the joint for improved function and ROM to return. As previously stated, this is a preliminary study with small drug treatment group sample sizes (n=3-4). If results hold through a more robust study, the anti-fibrotic pleiotropic effects of the two selected drugs may have a wide potential impact on connective tissues following injury.

While the encouraging results from this study need to be confirmed with additional animals, similar anti-fibrotic effects using these drugs have been reported previously in other tissues. Davis et al found that simvastatin-treated rats had reduced collagen accumulation with favorable changes in the expression of fibrosis genes and other markers following a massive rotator cuff tear [9]. Burks et al saw significantly less muscle fibrosis in mice treated with losartan after cardiotoxin-induced injury [12]. We hypothesize that both SV and LS affect the healing and repair response by inhibiting the proliferation of pro-fibrotic mediators transforming growth factor beta (TGFβ) and angiotensin II (Ang II), respectively. Inhibiting these mediators may limit myofibroblast differentiation and decrease collagen production in periarticular tissues and, ultimately, reduce joint fibrosis and preserve joint motion following elbow injury.

Future work will evaluate biological changes and investigate mechanisms responsible for these mechanical changes. Limitations of this preliminary study include small animal numbers in each treatment group and no histological evaluation of tissue changes in the joint, particularly the capsule. Ongoing work is investigating these aspects, as well as the ability of anti-fibrotic drug treatment to preserve pronation-supination (P-S) joint motion, which is also vitally important for full elbow joint function.

ACKNOWLEDGEMENTS

The authors thank the American Shoulder and Elbow Surgeons (ASES) for research funding and the NIH for a training fellowship (T32 EB018266) and research funding (R03 ARR067504).

REFERENCES

[1] Anakwe, R+, *J Bone Joint Surg Am*, 93(13):1220-6, 2011. [2] Gaston, P+, *J Bone Joint Surg Br*, 87(9):1233-6, 2005. [3] Olsen, B+, *Acta Orthop Scand*, S272:67, 1996. [4] Charalambous, C+, *J Bone Joint Surg Am*, 94(15):1428-37, 2012. [5] Monument, M+, *J Orthop Sci*, 18(6):869-77, 2013. [6] Myden, C+, *J Shoulder Elbow Surg*, 20(1):39-44, 2011. [7] Lake, S+, *J Or Res*, 34(2):354-64, 2016. [8] Dunham, C+, *J Shoulder Elbow Surg*, 2017. [9] Davis E+, *J Shoulder Elbow Surg*, 24(2):280-7, 2015. [10] Bagnato B+, *Rhuematology*, 52(8):1377-86, 2013. [11] Guo, F+, *Hum Exp Toxicol*, 34(5):497-505, 2015. [12] Burks, T+, *Sci Trans Med*, 3(82):82ra37, 2011. [13] Aktas, E+, *J Orthop Traumatol*, 12(3):145-51, 2011. [14] Muscaritoli, M+, *Nutrition*, 19(11-12):936-9, 2003. [15] Varo, N+, *J Hypertens*, 17(1):107-14, 1999. [16] Kobayashi, T+, *J Appl Physiol*, 114(2):262-73, 2013.

PROTEOLYTIC BEACON FOR MATRIX METALLOPROTEINASES IMPLICATED IN EXTRACELLULAR MATRIX REMODELLING

Dominic K. Muli (1), J. Oliver McIntyre (2), Jonathan P. Vande Geest (1)

(1) Department of Bioengineering
University of Pittsburgh
Pittsburgh, PA, USA

(2) Departments of Radiology and Radiological Sciences, and Cancer Biology
Vanderbilt University Medical Center
Nashville, TN, USA

INTRODUCTION

Matrix metalloproteinases (MMPs) belong to a large class of proteolytic enzymes whose catalytic activity depends on zinc ion.[1] MMPs are produced in their latent form in the human body. Once activated by free radicals and other enzymes, MMPs participate in a plethora of cell processes which include, but not limited to, activation of growth factors, regulation of cell signaling and physiological processes, tissue repair, angiogenesis and apoptotic process, among others.[2] MMPs contribute to remodeling the extracellular matrix (ECM) and dysregulation of this process can be detrimental, leading to cellular damage and disease. MMPs are ubiquitous during both the initiation and progression of a number of diseases.[3,4]

Several reports describe the detection of individual MMPs and their substrates.[5-7] Diagnostic molecular probes based on fluorogenic protease-targeted substrates have also been developed to assess *in vitro* and *in vivo* MMPs activity.[8-10] We previously reported novel "activatable" proteolytic beacons with peptides substrates engineered for MMP specificity (Figure 1).[11,12] More and highly selective diagnostic tools for MMP detection are still needed. The purpose of this study was to screen our proteolytic beacon (PB) against various MMPs to determine their relative rates of cleavage and selectivity towards the peptide substrate used in the dendritic PB. We used the increased sensor fluorescence and increased sensor/reference ratio, due to loss of FRET (Fluorescence Resonance Energy Transfer), to monitor cleavage of the peptide in the PB by the MMPs.

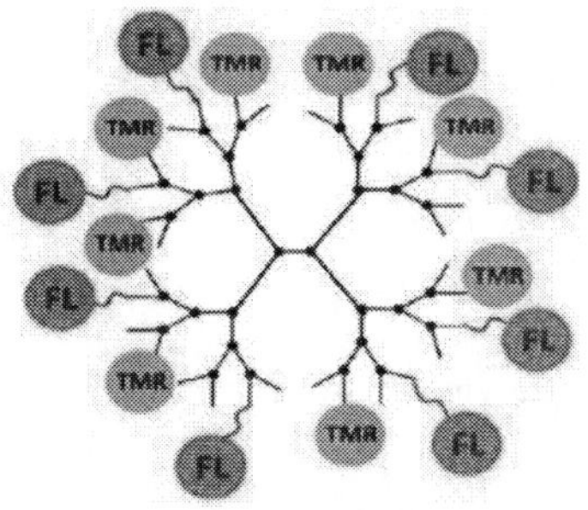

Figure 1: Schematic representation of PB with fluorescein (FL) and tetramethylrhodamine (TMR) FRET pair. FL acts as the sensor that in PB is quenched by energy transfer to TMR reference and becomes highly fluorescent after cleavage. In this case, the peptide substrate specific to certain MMPs is incorporated within the polyamidoamine (PAMAM) dendron scaffold.

METHODS

Synthesis of Fl-TMR PB. The PB was prepared using a linker with RPLAWRS(Ahx)C following our previously reported work.[11] Diafiltration was used to purify the PB and characterization was done using UV/Vis and fluorescence as previously reported.[9, 11-12]

Quantitative analysis of cleavage of PB by MMPs. Active forms of MMPs -1, -2, -3, -7, -8, -9, -12 and -14 were purchased from Enzo Life Sciences, aliquoted and stored at -80 °C. To perform the digestion analysis, ~0.1 μM PB solution was prepared from the stock solution, and DQ-gelatin was used as a positive control. 30 mM EDTA was used as a negative control since it inhibits MMPs. MMPs were quickly thawed and diluted with deionized water to a final concentration of 2 ng/μl. A total volume of 100 μl of either PB alone, PB with MMP or with both EDTA and MMP (n =1) was introduced to a Microfluor fluorescence plate. A plate reader set at 37 °C, 528 nm (sensor), 585 nm (reference) and excitation wavelength at 485 nm was used to kinetically assay the cleavage of the PB by the MMPs. The assay was performed for up to 30 mins, then the plate was taken out of the plate reader, and incubated overnight. Endpoint values were recorded on the plate reader after the overnight incubation. The rate of cleavage of the PB was calculated from the rate of change of the sensor to reference ratio (Fluorescence of Fl@528 nm/Fluorescence of TMR@ 585 nm).

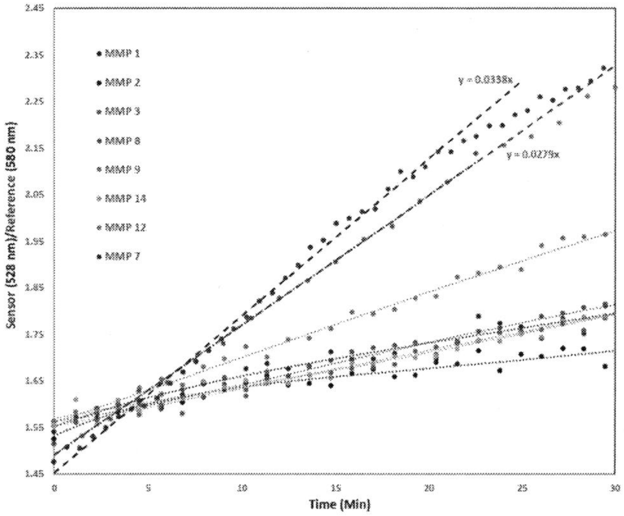

Figure 2: Cleavage of Fl-TMR PB by various 0.1 ng/μl MMPs (n = 1). Fluorescence intensity was monitored using a microplate reader set at 37 °C. Initial rates were used to calculate the relative rates of PB cleavage.

RESULTS

Cleavage rates of the PB by various MMPs were computed using the initial rates for each MMP after plotting the sensor/reference ratio against time as shown in Figure 2. From the graphs, the specific activity (SA) for each MMP was determined using Eqn. 1 and tabulated as shown in Table 1.

Specific Activity (SA) =

$$\{\text{Slope } (\Delta F/\text{min})/\text{MMP conc.}]/\ [(\Delta F_{max})/[\text{PB}])\} \quad \text{(Eqn. 1)}$$

where ΔF_{max} is the difference between the sensor/reference ratio obtained after overnight incubation and the initial ratio.

Table 1: Relative cleavage rates of PB by MMPs as quantified using initial rates.

MMP	Class	Specific Activity* (pmol PB/min/μgMMP)
1	Collagenases	740
8		790
2	Gelatinases	990
9		1300
3	Stromelysin	880
7	Matrilysin	2700
12	Metalloelastase	2100
14	Membrane-type MMP	700

* rates calculated from triplicate measures of each MMP cleavage rate with errors estimated to be ±10%.

DISCUSSION

The Fl-TMR FRET pair was successfully prepared, purified and characterized using UV/Vis and fluorescence spectroscopy. UV/Vis absorption profiles (data not shown) showed two peaks at 490 nm (Fluorescein) and 570 nm (TMR) as expected. Excitation of the sensor (Fl) at 490 nm produced an emission peak at 585 nm suggesting energy transfer to the reference (TMR). Amongst the MMPs tested, the relative cleavage rates (calculated as specific activity) showed that the PB prepared was cleaved fastest by MMP-7 followed by MMP-12, with slowest cleavage by MMP-14. This result is supported by our earlier work[9] where we reported selective cleavage of a similar PB by MMP-7 compared to MMPs -2 and -3. MMPs -1, -3, -8, and -14 were similar in their relative rates to cleave the peptide used in the PB. The MMPs in this study were selected to represent the broader classes of MMPs. From these preliminary results, we find that there are differences in the MMP selectivity of the peptide substrate RPLAWRS(Ahx)C in the PB. We are investigating other substrates that can further distinguish the MMPs because of their roles in various diseases.

ACKNOWLEDGEMENTS

Research reported in this abstract was supported by the National Eye Institute of the National Institutes of Health under award number R01EY020890 to Prof. Jonathan Vande Geest. We also thank members of the McIntyre lab (Vanderbilt University Medical Center) for their support and funding from the Department of Radiology and Radiological Sciences.

REFERENCES

[1] Malemud, C. J., *Front. Biosci.* 11,1696-701, 2006.
[2] Rosenberg, G. A., *Lancet. Neurol.,* 8 (2), 205-216, 2009.
[3] Lu, P. et al., *Cold Spring Harb Perspect Biol.* 3 (12), 2011.
[4] Shay, G. et al., *Matrix Biol.* 44-46, 200-206, 2015
[5] Kaufman, H. E., *Cornea,* 32 (2), 211-216, 2013.
[6] Sambursky, R. et al., *JAMA ophthalmol.* 131 (1), 24-8, 2013.
[7] Seltzer, J. L. et al., *J. Biol. Chem. 264* (33), 19583-19586, 1989.
[8] Zhang, X. X. et al., *J. Control. Release,* 159 (1), 2-13, 2012.
[9] McIntyre, J. O. et al., *Biochem. J.* 377, 617-628, 2004.
[10] Haskett, D. G. et al., *Microsc. microanal.,* 22 (2), 349-60, 2016.
[11] McIntyre, J. O. et al., *Methods Mol. Biol.,* 539, 155-174, 2009.
[12] McIntyre, J. O. et al., *Methods Mol. Biol.,* 622, 279-304, 2010.

SB^{3}C2017
Summer Biomechanics, Bioengineering and Biotransport Conference
June 21 – 24, Tucson, AZ, USA

A NOVEL ANTIOXIDANT POROUS VESICLE TREATMENT PREVENTS THE PAIN & AXONAL DAMAGE THAT DEVELOP WITH NEUROPATHIC INJURY

Sonia Kartha (1), Christine L. Weisshaar (1), Andrew Tsourkas (1) Zhiliang Cheng (1) Beth A. Winkelstein (1,2)

(1) Department of Bioengineering
University of Pennsylvania
Philadelphia, PA, USA

(2) Department of Neurosurgery
University of Pennsylvania
Philadelphia, PA, USA

INTRODUCTION

The spinal nerve root can be injured via compression by disc herniation, spondylosis, or other spine traumas, resulting in many cases in chronic pain [1-3]. Reactive oxygen species (ROS) are associated with persistent pain [4,5]. Elevated levels and accumulation of ROS at the site of neural injury can induce oxidative damage that contributes to pain onset [6]. Although antioxidants, like superoxide dismutase (SOD), show promise in reducing pain after neural injury [6-8], their efficacy is transient since they are prone to degradation [9].

Polymeric vesicles have been used as potential delivery platforms for drugs and/or contrast agents since they have enhanced stability, long blood circulation time, and can be loaded with large quantities of hydrophobic and hydrophilic moieties [10-12]. Yet, the low membrane permeability of polymeric vesicles limits their ability for drug delivery [13]. Porous polymeric vesicles have been developed with increased membrane permeability to molecules <5kDa while retaining molecules >10kDa [14]. These vesicles may provide a delivery platform that protects against degradation while maintaining enzymatic activity.

We hypothesized that by encapsulating SOD in a porous vesicle, ROS would be free to pass into the vesicle interior where it would be degraded by the encapsulated SOD, enabling an effective antioxidant for pain treatment. The toxicity of these polymeric vesicles to neurons was tested in-vitro, and SOD activity within the porous vesicles was measured to characterize the drug delivery platform. Their effectiveness to reduce pain was compared to free antioxidant alone using a rat model of painful nerve root compression. The effect on axonal damage in the injured root was also assessed in-vivo.

METHODS

Studies were IACUC-approved. Nanometer-sized porous vesicles made from triblock copolymers PEG-PBD/PEG-PPO-PEG (75:25 molar ratio) were prepared by a film-hydration technique [14]. Control non-porous vesicles were similarly made using 100%mol PEG-PBD. In-vitro cytotoxicity was tested using cultured dissociated dorsal root ganglia (DRG) isolated from embryonic day 18 Sprague-Dawley rats as described before [15]. Dissociated DRG were plated at a density of 9.25x104 cells/mL; feeding media was changed on days in-vitro 1 and 3. On day in-vitro 4, separate cultures were incubated for 24hrs with vesicle concentrations at 0 (untreated), 1.6, 8, 40, 200, and 1000μg/mL. Average percent cell lysis was measured by a Pierce LDH Cytotoxicity Assay [16] and compared by ANOVA with Tukey's test.

SOD encapsulated porous vesicles were prepared using a Cu/Zn SOD enzyme (MW 32.5kDa) loaded into the aqueous interior of the vesicle through sonication followed by freeze-thaw and extrusion (100nm membrane). SOD activity in both types of vesicles was measured using a ferrictyochrome C assay [17] before and after dissolution with Triton X-100 and separately compared using t-tests.

Porous vesicles and free SOD were administered in separate groups of male Holtzman rats undergoing a painful unilateral C7 dorsal root compression [2,3]. Immediately after compression, either 10mg/mL of SOD porous vesicles (137U/mL; n=6) or empty vesicles (no SOD; n=4) were administered directly to the root in 100μL of saline. A comparable dose of free SOD (137U/mL) or a dose >100-fold higher (17505U/mL) was also applied to the root in separate groups (n=3/group). Sensitivity to mechanical stimuli was measured as the withdrawal threshold of the ipsilateral forepaw before injury at baseline (BL) and for 7 days after [3]. Thresholds were compared between groups by a two-way ANOVA (groupXday) and Tukey's test.

On day 7, the C7 nerve root was harvested and fixed in 4% paraformaldehyde. Roots from naïve and sham surgery (no injury) rats (n=2/group) were also included as controls. The root was cryosectioned at 14μm along its long-axis and mounted on slides. Sections were blocked using 10% normal donkey serum (Vector) with

0.3% Triton-X-100 and incubated in mouse anti-NF200 (1:500; Sigma) and donkey anti-mouse 546 (1:500, Invitrogen). For each rat, 3 sections of the injury root were imaged at 20X (Olympus5X51) and axonal morphology was evaluated by two blinded reviewers using a scale [2], with a (-) assigned for no pathology and normal appearance, a (+) for evidence of axonal thinning, and a (++) for extensive damage. Ratings for each rat were averaged across sections for a single score.

RESULTS

Incubation of empty porous vesicles with neurons does not induce cell lysis different from that measured in untreated cultures nor is there any effect of increasing the SOD vesicle concentration (Fig. 1A). Also, SOD activity within the SOD-encapsulated porous vesicle is not changed after their dissolution with Triton X-100, but does significantly increase ($p<0.02$) in the nonporous vesicles after dissolution with Triton X-100 (Fig. 1B).

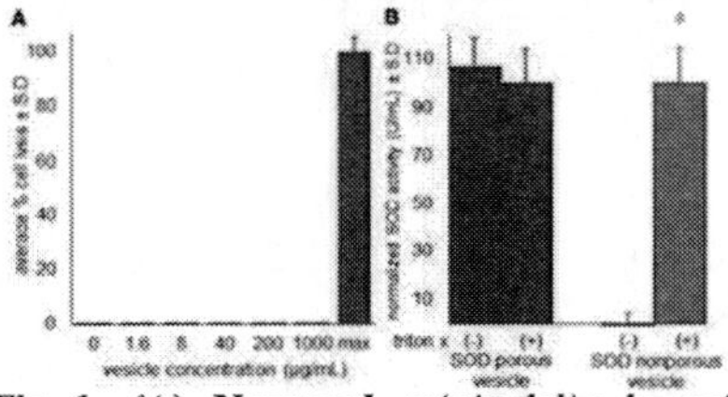

Fig 1. (A) Neuronal cytotoxicity is not evident for any vesicle concentration tested. (B) SOD activity in porous vesicles is unchanged after Triton X-100 but increases (*$p<0.02$) in nonporous vesicles.

SOD-encapsulated porous vesicles prevent the onset of pain that is typically induced by root compression (Fig. 2). The withdrawal thresholds after SOD vesicle treatment are significantly higher ($p<0.01$) than those with either the empty vesicles or the same dose of free SOD (free SOD low) (Fig. 2). The higher dose of free SOD (free SOD high) prevents pain only until day 5, with thresholds at days 6 and 7 significantly below baseline ($p<0.0012$) and not different from injury with empty vesicle treatment (Fig. 2).

Tissue from rats #7 and #11 was damaged during harvest and was not available for rating. SOD porous vesicle treatment partially prevents the axon swelling and thinning of myelinated axons that is typical after root compression (Fig.3; Table 1). Axons exhibit morphology similar to both normal and sham roots (Fig. 3; Table 1). In contrast, axons in the rats with the empty porous vesicles display axonal thinning and discontinuous NF200 labeling (Fig. 3). Treatment with either of the free SOD formulations does seem to prevent the full extent of damage but does not appear to prevent axonal disruption (Fig. 3). The majority of roots from sham rats appear normal (-); only 1 rat exhibits axonal damage with the SOD porous vesicle treatment (Table 1). Roots treated with empty vesicles exhibit the greatest damage, with all but 1 rat receiving the highest score (++) (Table 1). Most rats receiving free SOD display some degree of NF200 discontinuity and axonal thinning.

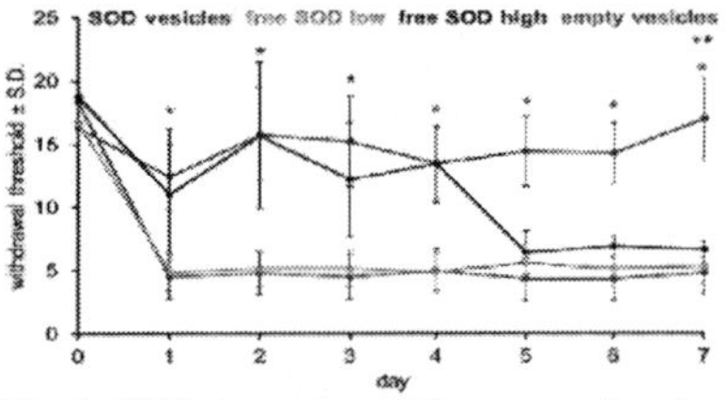

Fig 2. SOD porous vesicles prevent pain, with higher thresholds than both free SOD low and empty vesicles (*$p<0.01$). Pain develops by day 5 (#$p<0.0012$) for free SOD high treatment. Thresholds are lower than for SOD vesicles by day 7 ($p<0.0012$).**

Table 1. Summary of the nerve root NF200 ratings.

Group	SOD vesicle						empty vesicle				free SOD low			free SOD high			sham		normal	
Rat #	1	2	6	7	10	16	9	12	14	15	4	5	13	3	8	11	17	18	N7	N12
Score	-	-/+	-/+	NA	+	-	++	+/++	++	++	-/+	+	+	+	+	NA	-	-/+	-	-/+

DISCUSSION

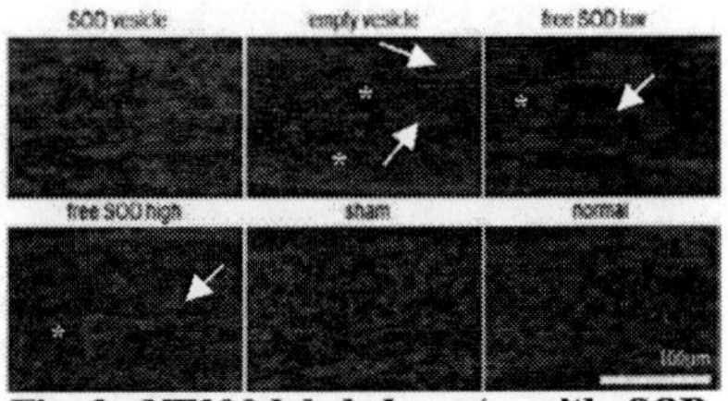

Fig 3. NF200-labeled roots with SOD vesicle treatment have axon morphology similar to sham and normal roots. Roots treated with empty vesicles display the most axonal thinning (*) and disruption in labeling (arrow), while both free SOD treatments show less axonal thinning.

Encapsulating SOD in non-cytotoxic porous polymeric vesicles preserves SOD enzymatic activity (Fig. 1). That formulation also prevents pain onset and axonal damage that are typical after a root compression (Figs. 2 & 3). In fact, SOD vesicles are more effective in preserving axons than comparable or higher doses of free SOD, which also were ineffective in preventing pain (Figs. 2 & 3, Table 1). Together these results suggest porous polymeric vesicles as an effective platform to deliver antioxidants.

Unlike conventional polymeric vesicles, the porous ones maintain SOD activity without Triton X-100 dissolution (Fig. 1), due in part to their permeable membrane that allows ROS to interact with the encapsulated SOD. Similar liposomes that encapsulate antioxidants often require ion transport channels to facilitate ROS interaction and enzymatic activity but are only marginally more effective than free SOD and are not as stable as polymeric vesicles in-vivo [13]. Although both the SOD porous vesicles and the >100-fold higher dose of free SOD given at the injury site immediately after compression prevent pain, the higher dose is only transiently effective, with pain returning on day 5 (Fig. 2). This finding is consistent with a 100mg/kg of a free ROS scavenger not abolishing pain after a nerve injury [6]. While it is possible that even a lower dose of encapsulated SOD could prevent pain, studies are needed to determine its effectiveness.

By reducing elevated ROS, antioxidants preserve neuronal health [4]. SOD porous vesicles partially prevent the axonal thinning and disorganization in the root that occurs after its painful compression (Fig. 3; Table 1) [2]. Although the most extensive axonal damage is evident for treatment with empty vesicles, both treatments of free SOD appear to attenuate axonal damage (Fig.3; Table 1). Free SOD may remain at the injury site long enough to prevent some of the neuronal degeneration cascades that are initiated but not long enough to inhibit those contributing to pain. Since axonal damage is largely prevented by the SOD vesicles, the beneficial long-lasting effects of porous vesicles are supported. Yet, since treatment was given directly after compression, the effects on existing pain are unknown.

ACKNOWLEDGEMENTS

Support was provided by the Sharpe Foundation, the Cervical Spine Research Society and National Institute of Health (CA175480).

REFERENCES

1 Abbed K et al. *Neurosurgery* 60:S28-34, 2007; **2** Nicholson K et al. *J Neurosurg Spine* 20:751-62, 2014; **3** Smith J et al. *Biomaterials* 80:96-105, 2016; **4** Marklund N et al. Acta Neurochir 143:78, 81-2001 131:413-7, 2014; **5** Kim D et al. *PNAS* 107:14851-56, 2010; **6** Gao C et al. *Pain* 131:262-71, 2007; **7** Xie Y et al. *Exp Ther Med* 8:1137-41, 2014; **8** Wang Z et al. *J Pharmacol Exp Ther* 309:869-78, 2004; **9** Muscoli C et al. *Mediators Inflamm* 2013:950947, 2013; **10** Jeong I et al. *Macromol Biosci* 13:946-53 2013; **11** Kim M et al. *Chem Commun* 46:4481-3, 2010; **12** Discher D & Eisenberg A *Science* 297:967-73, 2002; **13** Yang H et al. *Methods Mol Biol* 477:277; **14** Yan L et al. *J Mater Chem B Mater Biol Med* 3:9277-84, 2015; **15** Cullen D et al. *Tissue Eng Part A* 18:2280-9, 2012; **16** Chan F et al. *Methods Mol Bio* 979:65-70, 2013; **17** Shuvaev V et al. *PloSOne* 8:e77002, 2013.

MECHANICAL FUNCTION OF A COMPOSITE NANOFIBROUS BIOMATERIAL ANALOGUE OF THE KNEE MENISCUS INCLUSIVE OF RADIAL TIE FIBER-LIKE ELEMENTS

Sonia Bansal (1,2), Breanna N. Seiber (1,2), Niobra M. Keah (1,2), Robert L. Mauck (1,2), Miltiadis H. Zgonis (1,2)

(1) McKay Orthopaedic Research Laboratory,
University of Pennsylvania
Philadelphia, PA

(2) Translational Musculoskeletal Research Center,
Philadelphia VA Medical Center
Philadelphia, PA

INTRODUCTION

Menisci are semi-lunar shaped fibrocartilaginous wedges located between the femur and the tibial plateau and support the structure and mechanical function of the knee joint [1,2]. Understanding meniscus structure and function is particularly important given the high incidence of meniscal pathology [3,4]. Circumferentially aligned collagen bundles within the meniscus function to convert compressive forces into tensile hoop stresses [5,6]. In addition, the meniscus contains "radial tie fibers" that originate at the meniscus periphery and interdigitate amongst the circumferential fiber population [5-7]. These radial tie fibers (RTFs) vary in size, spatial distribution, and in their degree of arborization [7], and are thought to bind circumferential fibers together and protect against longitudinal splitting [8,9]. Radial tears interject perpendicularly and sever the circumferential bundles and are thought to compromise mechanical function, though recent studies demonstrated that reduction in load transfer occurs only when these tears reach 90% of the meniscus width [10]. This suggests alternative methods for strain transmission in the meniscus. We hypothesize that RTFs play a role in this, particularly in the case of a radial tear. To further this line of inquiry, we quantified RTF density and size in adult menisci as a function of position, and measured the mechanical function of these tissue samples in the context of a radial tear. Additionally, we utilized electrospinning to generate a composite material model of the meniscus, with variable fiber mechanics in each layer, by spinning two polymers with distinct material characteristics.

METHODS

Native Tissue Analysis: Medial menisci (n=14) were harvested from adult (skeletally mature) cows. Six were divided into four equal regions: anterior horn, anterior body (Body-A), posterior body (Body-P), and posterior horn. Each region was cryosectioned in the sagittal plane to 10 μm thickness spanning the cross section and fixed. Three zones were examined: outer, middle, and inner zones. Sections were imaged at 10X via second harmonic generation (SHG, 840 nm excitation). Maximum projections (~5 microns thick) were generated in each zone and the area fraction with positive SHG signal computed as a measure of RTF area fraction **(Fig. 1A)**. Fiber thickness was estimated using the FIJI plugin BoneJ [11-13]. Metrics were compared across regions by ANOVA ($p \leq 0.05$) with Tukey's post-hoc tests; data are presented as the mean value for each region and zone **(Fig. 1B,C)**.

Remaining menisci were trimmed as in **Fig. 2A** to yield Anterior, Posterior, and Body segments. These were sectioned transversely into strips (5 x 15 x 0.3mm) with extended tabs for gripping [14]. Samples were preconditioned for 15 cycles (2%-4%) and then evaluated in tension in a ramp to 15% strain (0.5% strain/s). Post-testing, samples were re-equilibrated in PBS for 1 hr, defected with a 50% radial defect at the inner-zone, and retested as above to 50% strain. Stress versus strain data was curve-fit using a custom bilinear fitting program in MATLAB to define the 'toe' and 'linear' moduli [14,15]. Significance (n=4/group) was analyzed using 1-way ANOVA ($p \leq 0.05$), with Tukey's post hoc; data are presented as mean ± SEM **(Fig 2B,C)**.

Scaffold Fabrication and Mechanical Testing: Nanofibrous 75:25 poly(lactic-co-glycolic acid) (PLGA) and poly(ε-caprolactone) (PCL) scaffolds were electrospun [7]. For aligned (AL) scaffolds, a mandrel rotating at 10 m/s served as the collector. To form non-aligned (NA) scaffolds, the mandrel was slowed to ~3 m/s. To replicate radial tie elements, AL scaffolds with discrete NA layers were produced by lowering the speed of the mandrel for set periods of time. Composite scaffolds had 7 alternating AL and NA layers **(Fig. 3A)** with 33% distributed NA content overall. The composition of each layer was varied such that scaffolds with softer (PCL) or stiffer (PLGA) intervening NA layers were produced. Scaffolds were trimmed to 40 mm length (in the AL direction) x 5 mm width for testing. Two conditions were tested: intact or defect, with the latter having a radial cut spanning 50% of the width. Samples were evaluated in tension (ramp to 100% strain at 0.5% strain/s) **(Fig 3B)**. Stress versus strain data was curve-fit as above and a custom MATLAB program used to determine the yield and failure strain and toughness. Significance (n=6/group) was analyzed by 1-way ANOVA ($p \leq 0.05$), with Tukey's post hoc; data are presented as mean ± SEM **(Fig. 3C-F)**.

RESULTS

SHG revealed differences in RTF fiber density between zones, with the central body lower than the anterior or posterior horns ($p<0.05$). RTF thickness varied as a function of region ($p<0.02$) and zone ($p<0.005$) **(Fig. 1B,C)**. The linear modulus of anterior, body, and posterior sections (35.8±11.8, 24.4±9.5, and 24.0±8.8 MPa) were not different from one another ($p=0.78$). However, we noted an attenuation in the decrease in apparent modulus with defect of posterior samples compared to all other regions **(Fig 2B,C)**.

For scaffolds, the linear region modulus of PLGA-only AL scaffolds was ~4x higher than PLGA NA scaffolds (241±13 vs 53±2 MPa, $p<0.001$) and linear region moduli of AL and NA PLGA scaffolds were ~7 and ~5x higher than AL and NA PCL scaffolds, respectively (34±2 and 11±0 MPa, $p<0.001$) **(Fig. 3C)**. Yield strains for AL and NA PLGA scaffolds (1.5 and 2.4%) were lower than PCL scaffolds (AL: 10% and NA: 6%, $p<0.001$) **(Fig. 3D)**. The greatest toughness of defected scaffolds was seen in AL PLGA scaffolds (1.1±.10 $J \cdot m^{-3} \cdot 10^4$, $p<0.001$). Failure strain ranged from 36% (PLGA, AL) and 46% (PLGA, NA) in single-polymer scaffolds **(Fig. 3E,F)**.

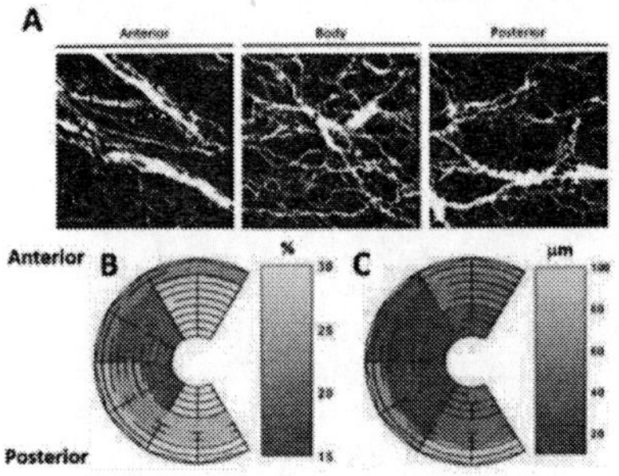

Figure 1: Radial tie fiber (RTF) distribution in adult menisci. (A) SHG images of the middle zone of each region, white shows positive SHG indicative of radial tie fiber presence (SB = 200 μm). (B) Radial tie fiber density, reported as Area Fraction (%) across the meniscus, and (C) Mean fiber thickness (μm).

Composite scaffolds showed distinct mechanical behavior that did not always match expectations, based on a simple weighted mixture model combining the individual components. Expected linear modulus of composites formed from AL PLGA with intervening NA PCL did not match predictions while the reverse configuration exceeded expectations ($p=0.0002$, **Fig. 3C**). Toughness of AL PLGA/NA PCL scaffolds was lower than expected ($p=0.01$) while AL PCL/NA PLGA scaffolds overperformed in this measure ($p=0.008$) **(Fig. 3D)**. Similarly, the yield strain of AL PLGA/NA PCL scaffolds underperformed expected values ($p<0.0001$) while AL PCL/NA PLGA scaffolds overperformed ($p=0.04$) **(Fig. 3E)**. Conversely, strain at failure was underpredicted for both, with this being significant for AL PCL/NA PLGA scaffolds ($p=0.03$, **Fig. 3F**).

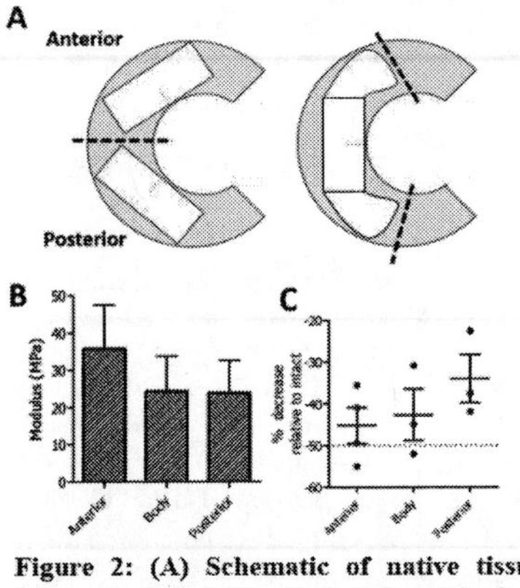

Figure 2: (A) Schematic of native tissue sample preparation (B) Linear moduli (LM) of intact tissues by location. (C) Change in LM of defected native tissue moduli measured prior to defect creation.

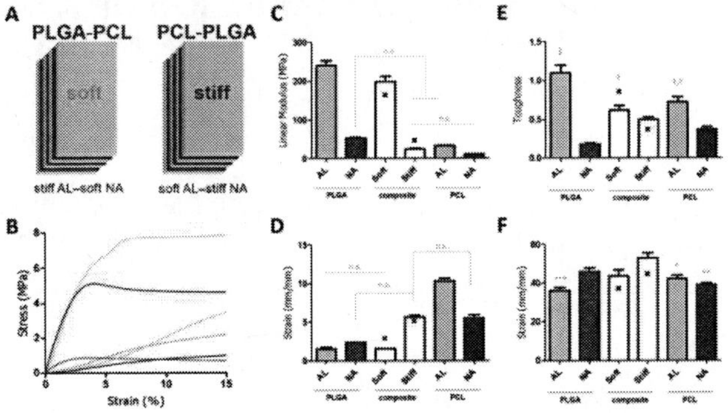

Figure 3: (A) Schematic of composite scaffold preparation (B) Stress-strain plots of each group to 15% strain. (C) Linear moduli and (D) yield strain of intact AL, NA, and composite scaffolds. P>0.01unless otherwise indicated. (E) Toughness and (F) failure strain of defected AL, NA, and composite scaffolds, § = p≤0.05 vs. all other groups, Toughness. ‡ = p≤0.01 vs. PLGA-NA, Toughness. † = p≤0.01 vs. PCL-NA, Toughness. * = p≤0.05 vs. PCL_PLGA, Failure strain. "x" indicates expected outcome from weighted mixture model.

DISCUSSION

Radial tie fibers (RTFs) form a branched fiber network that interdigitates with circumferential fibers in the meniscus, and may function to stiffen the composite and make it more resistant to crack propagation. This effect would likely be contingent upon RTF density, distribution, and thickness. We found the greatest density of RTFs in the posterior horn, potentially reflecting the more demanding in vivo loading experienced in this region. Our finding of a trend towards a higher apparent modulus in the posterior horn (in the context of a radial defect) supports the hypothesis that RTFs play such a functional role. To study these interactions in a more controlled setting, we developed a biomaterial analog containing RTF-like elements. These distinct disorganized fibrous layers (that were interspersed in an otherwise aligned fiber array) altered scaffold mechanics. The finding of increased failure strain (compared to scaffolds of a single composition) supports the beneficial effects of RTF inclusion in native meniscus. This may act by enabling efficient and prolonged load transfer in the context of tears that interrupt the circumferential fibrous architecture. Understanding of this complex composite behavior will inform engineering design towards the fabrication of functional meniscus replacements, and could alter clinical practice with respect to surgical intervention.

ACKNOWLEDGEMENTS: This work was supported by the OREF New Investigator Grant, the NIH, and the VA.

REFERENCES: [1] Kambic+, JOR. 2005. [2] Chevrier+, JOR 2009. [3] Merriam+, AJSM 2015. [4] Mordecai+, World J. Orthop. 2014. [5] Skaggs+, JOR 1994. [6] Makris+, Biomat. 2011. [7] Andrews+, J Anat. 2014. [8] Mauck+, Tissue Eng. 2009. [9] Bullough+, JBJS 1970. [10] Bedi+, JBJS 2010. [11] Saalfeld, FIJI 2012. [12] Dougherty, FIJI 2006. [13] Schindelin+, Nat Meth., 2012. [14] Peloquin+, JBME, 2016. [15] Tanaka+, JBME, 133:074502, 2011.

THE ROLE OF CARBONATE ON PROTEIN–FREE FORMATION OF BONE-LIKE APATITE

Alix C. Deymier (1), Arun K. Nair (2), Baptiste Depalle (3), Zhao Qin (4), Kashyap Arcot (5), Christophe Drouet (6), Claude H. Yoder (7), Markus J. Buehler (4), Stavros Thomopoulos (8), Guy M. Genin (5), Jill D. Pasteris (9)

(1) Dept. of Orthopedic Surgery
Columbia University
New □ork, N□, USA

(2) Dept. of Mechanical Engineering
University of Arkansas
Fayetteville, AR, USA

(3) Dept. of Materials
Imperial College
London, UK

(4) Dept. of Civil and Envir Engr
MIT
Boston, MA USA

(5) Dept. of Mech Engr and MatSci
Washington University
St. Louis, MO, 63130

(6) CIRIMAT
Universit□ de Toulouse
Toulouse France

(7) Dept. of Chemistry
Franklin and Marshall College
Lancaster, PA, USA

(8) Dept. of Orthopedic Surgery
Dept. of Biomedical Engr
Columbia University
New □ork, N□, USA

(9) Dept. of Earth and Planetary Sci
Washington University
St Louis, MO, USA

INTRODUCTION

Hydrated bone is composed of about 50 wt% mineral crystals These crystals provide the bone with its stiffness and strength, allowing it to support and protect the body. In contrast to the prisms typically formed by hexagonal-structured apatite, bone mineral exhibits a plate-like morphology and a nanoscale size. It has been suggested that organics in bone allow and control the unusual crystal shape; however, the chemical complexity of the mineral is rarely accounted for [1]. Chemically, bone mineral is often described as hydroxylapatite; however, the mineral is actually a highly substituted calcium apatite containing 6-7 wt% carbonate. This carbonate level increases with age and is affected by conditions such as acidosis [2-5]. Although the role of carbonate in bioapatite is not well understood, it is believed to help with pH homeostasis, promote bone remodeling, and control mineralized tissue mechanics [3, 6-8]. In order to understand the role that bioapatite mineral plays in calcified tissue, it is necessary to understand how changes in the mineral composition change the physical properties of the crystals and in turn the tissue-level mechanical properties. However, due to the extremely small size of bone mineral crystals (~3 x 50 x 100 nm^3), and the difficulty in extracting the mineral from the organic material without modifying its composition, direct study of bone mineral has been rare. Laboratory precipitated biomimetic apatites provide a unique way to study the precipitation, growth, and mechanics of these crystals [9, 10]. In the current study, we examine biomimetic apatites with varying levels of carbonate substitution in order to identify the role of carbonate in crystal structure and mechanics.

METHODS

Crystals of carbonate-substituted calcium apatite were precipitated from solutions of Na_2CO_3, NaH_2PO_4, and $Ca(NO_3)_2{\cdot}4H_2O$ at 60°C [9, 10]. The levels of Na_2CO_3 were modified to create crystals with a variety of carbonate levels, spanning from 1.0-17.8 wt%. After 2 hrs, the precipitate was filtered and then washed at least four times with water. Carbonate content was then determined by a Carbon-Hydrogen-Nitrogen analyzer and by FTIR and Raman spectroscopy. Crystal size was measured using transmission electron microscopy (TEM) micrographs. *In silico* studies via molecular dynamics (MD) [11] and density functional theory (DFT) [12] simulations were used to determine the effect of carbonate on lattice spacing, surface energy, and modulus. The lattice parameters for the mineral crystals were validated by X-ray diffraction (XRD). Solubility studies were performed using inductively coupled plasma optical emission spectrometry on saturated solutions. A non-standard formulation of Weibull analysis was used to relate the change in lattice strain (ε), as determined by lattice spacing, modulus (E) and surface energy (γ), with respect to crystal size (L) at varying carbonate levels (C) according to:

Figure 1: TEM images of apatite crystals with varying carbonate levels.

$$\frac{L_o(C)}{L_o(C_o)} = \left(\frac{E(C_o)}{E(C)}\frac{\gamma(C)}{\gamma(C_o)}\left(\frac{\epsilon(C_o)}{\epsilon(C)}\right)^2\right)^{\frac{m_o}{2}} \quad (1)$$

where m is the Weibull modulus.

RESULTS

The synthesized samples were all powders of carbonated calcium apatite. FTIR analysis confirmed that the carbonate was present both within the lattice (mostly as B-type CO_3^{2-}) and as labile ions. Although all samples had plate-like morphology, the addition of carbonate to the apatite structure caused a non-linear decrease in crystal size (Fig 1). Experimentally, the addition of carbonate increased the lattice spacing along the c-axis and decreased the lattice spacing in the a-axis, in keeping with B-type substitution. The XRD results agreed with the MD and DFT simulations (Fig. 2). The simulations also indicated a decrease in surface energy and Young's modulus. Addition of carbonate increased the solubility of the powders while decreasing the sample crystallinity. Using all of the above measured factors, the modified Weibull analysis was able to predict a non-linear decrease in size with increasing carbonate content and lattice strain (Fig 3).

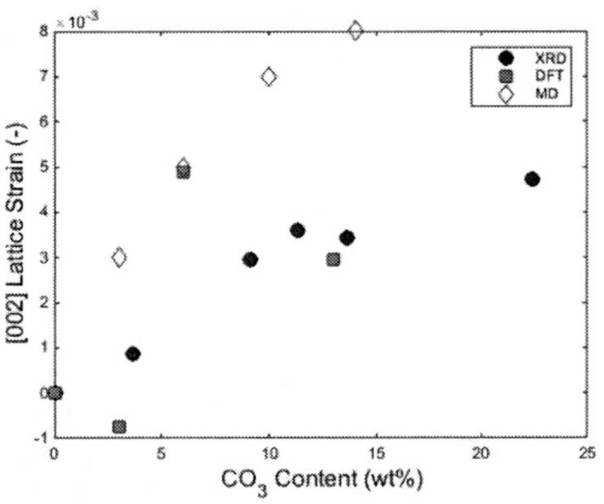

Figure 2: Plot of lattice spacing along the c-axis as a function of CO3 content.

DISCUSSION

Bone-like apatite exhibiting correct composition and morphology for mammalian bones was successfully formed using the protein free in vitro technique presented here. As in bone, this technique produced crystals with primarily B-type substituted carbonate at levels both within and exceeding those measured physiologically and a plate-like morphology. Increases in carbonate content caused a decrease in crystal size (Fig 1, 3) and significant changes in lattice spacings (Fig 2). At physiological carbonate levels between 5-10 wt% CO_3^{2-}, the crystals exhibit sizes commonly measured in bone of ~30 nm by 60 nm (Fig 1 middle). The substitution of carbonate into the lattice structure also resulted in energetic changes to the crystal as seen by the calculated decrease in surface energy, mechanics, and crystallinity along with increased solubility. The modified-Weibull analysis, which correlated the change in crystal energetics to the change in crystal size, was able to accurately predict the experimental results. This indicates that the crystal morphology of bone-like apatite is highly controlled by mineral composition and does not require the intervention of organic molecules. These results are the first to suggest a mechanism for compositional control of size, independent of protein interactions. This work has important implications for the formation of bone-like mineral in tissue engineering and in understanding bone mechanics.

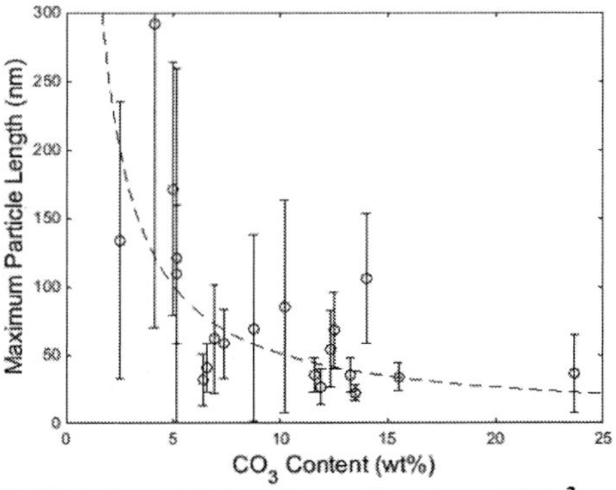

Figure 3: Plot of particle length as a function of CO_3^{2-} content. Points represent experimental data from TEM. The dashed line is the Weibull solution.

ACKNOWLEDGEMENTS

The authors would like to acknowledge Prof. Daniel Giammar at Washington University who performed all of the ICP-MS experiments. This project was funded by the National Institute of Health (NIH) R01 EB016422 and R01 AR057836. Dr. Deymier was funded in part by the National Space Biomedical Research Institute (NSBRI) NSBRI-RFA-13-01 Postdoctoral First Fellow Award. Arun Nair would like to acknowledge support from Department of Mechanical Engineering, University Arkansas. This work was supported in part by the National Science Foundation under the grants ARI#0963249, MRI#0959124 and EPS#0918970, and a grant from Arkansas Science and Technology Authority, managed by Arkansas High Performance Computing Center.

REFERENCES

1.Gu, L.-s., et al., Acta biomaterialia, 2011. **7**(1): p. 268-277
2.Donnelly, E., et al., Journal of Biomedical Materials research Part A, 2010. **92A**(3): p. 1048-1056
3.Akkus, O., F. Adar, and M.B. Schaffler, Bone, 2004. **34**(3): p. 443-453
4.Akkus, O., et al., Journal of Bone and Mineral Research, 2003. **18**(6): p. 1012-1019
5.Rey, C., et al., Connect Tissue Res, 1989. **21**(1-4): p. 267-73
6.Yerramshetty, J.S. and O. Akkus, Bone, 2008. **42**(3): p. 476-482
7.Spence, G., et al., J Biomed Mater Res A, 2010. **92**(4): p. 1292-300
8.Spence, G., et al., Journal of Biomedical Materials research Part A, 2009. **90A**(1): p. 217-224
9.Yoder, C.H., et al., Calcif Tissue Int, 2012. **90**(1): p. 60-7
10.Flora, N.J., et al., Synthesis and Reactivity in Inorganic and Metal-Organic Chemistry, 2004. **34**(3): p. 503-521
11.de Leeuw, N.H., Physical Chemistry Chemical Physics, 2004. **6**(8): p. 1860-1866
12.Menendez-Proupin, E., et al., J Mech Behav Biomed Mater, 2011. **4**(7): p. 1011-20

SB^{3}C2017
Summer Biomechanics, Bioengineering and Biotransport Conference
June 21 – 24, Tucson, AZ, USA

A TUNABLE FLEXIBLE-PLA SCAFFOLD SUITABLE FOR COMPLEX 3D PRINTED TISSUES

Timothy D. Jacobsen (1,2), Andrew Wong (1), Jacob Rigos (1), Nadeen O. Chahine (1,2)

(1) Feinstein Institute for Medical Research, Manhasset, NY

(2) Hofstra Northwell School of Medicine, Hempstead, NY

INTRODUCTION

Biomimetic tissue engineering approaches that mimic the functional mechanical properties of the tissue they are designed to replace are becoming increasingly investigated [1,2]. Constructs which have similar mechanical properties to native tissue help maintain normal mechanical functioning of that tissue, and are necessary for recapitulating the mechanobiology of implanted cells. Additionally, matching the properties of a biomaterial construct to that of its native tissue limits possible complications and negative interactions with neighboring tissues. Matching the properties of complex tissues that have varying mechanical properties along multiple planes of loading still poses significant complications. The intervertebral disc is comprised of three different tissue types each with their own distinct properties [3]. The nucleus pulposus (NP) makes up the center of the disc and is a highly hydrated tissue whose water content and hydraulic pressurization allows it to withstand large compressive loads. The annulus fibrosis (AF) which surrounds the NP has a high tensile strength that helps it contain the NP within the disc. The cartilaginous endplate (CEP) that lines the interfaces between the IVD and the adjacent vertebral bodies must also withstand large compressive loads. The varying mechanical properties of the IVD make designing biomaterial constructs for IVD replacements more complex. A few approaches and studies have recapitulated complex architecture and mechanical properties of IVD. Electrospinning using materials such as PCL and PLGA to create functionally mimetic AF has been used in combination with agarose or hyaluronic acid based NP to create IVD replacement tissues with physiological relevant mechanical properties [4,5]. Disc replacements using a self-assembling gel approach with collagen based AF and alginate based NP has also shown success in recapitulating the structure of the IVD [6,7]. In this study we describe the use of a flexible PLA (FPLA) scaffold which can be 3D printed in varying configurations that allow for the creation of constructs with tunable properties that can vary throughout the construct. Our aim is to tune these FPLA scaffold in a way that allows it to mimic the mechanical properties of AF and NP tissue. The physiological compressive elastic modulus of NP tissue ranges between 5 and 10 MPa, while the physiological tensile elastic modulus of AF varies between 10 and 45 MPa [8,9]. By being able to tune FPLA to achieve both physiological NP compressive modulus and AF tensile modulus FPLA constructs would be well suited for replacement of the mechanically complex IVD tissue.

METHODS

Biomaterial Fabrication FPLA was obtained from a 3D printing material supplier (FormFuture) and its properties were compared to standard 3D printing PLA (Makerbot). Biomaterial constructs were printed with a Makerbot Replicator 2 3D printer; printing speed, temperature and surface were optimized to prevent warping of constructs. Constructs were composed of alternating, perpendicular layers of FPLA fibers. Fiber spacing was varied from 1.5mm to 3mm in printed constructs. Constructs were either printed in sheets from which cylindrical plugs were taken for compressive testing and cytotoxicity testing, or in a dog bone shape for tensile testing. **Mechanical Testing** Constructs were tested on an Instron universal testing machine setup. For compressive testing cylindrical sample were placed under unconfined compression and compressed at a strain rate of 1 mm/min up to a maximum stain of 25% initial thickness. Young's modulus within the elastic deformation range was calculated. For tensile testing constructs were placed under uni-axial tension and stretched until a load of 20N reached, followed by a series of cyclic loading. Young's modulus within the tensile elastic deformation range was also calculated. Hysteresis of the cyclic loading regimens in tensile tests was quantified using MATLAB. **Cytotoxicity Testing**

Technical Presentation #166

NP cells were isolated from young bovine lumbar discs using standard enzymatic digestion and expanded for 1-2 passages in DMEM+10% FBS. bNPs were then seeded in 48 well plates at 20,000 cells per well. Cells were split into two groups: (1) untreated (2) cells cultured in wells containing an FPLA plug, with each group and time point cultured in triplicate. Live/Dead staining assay was performed using 4μM Calcein AM and 2μM Ethidium Homodimer-1 and imaged on an Olympus Fluoview 300 confocal microscope. Images were acquired (three regions of interest per replicate) after 1 and 21 days in culture to determine if the presence of flex PLA affects cell viability over time.

RESULTS

Both compressive (Fig 1) and tensile elastic moduli (Fig 2) of FPLA constructs were found to decrease with increasing fiber spacing. Constructs with a fiber spacing of 1.5mm or 2mm exhibited both compressive and tensile moduli closest to the physiological levels of native NP and AF tissue. Compared to constructs of conventional PLA, FPLA showed significantly lower compressive elastic modulus (Fig 1). During cyclic loading, the hysteresis of any given cycle increased with increased fiber spacing, indicating an increase in energy loss during the loading cycle (Fig 3). No significant changes in cell viability were noted in either group up to 3 weeks in culture, indicating FPLA constructs do not display any cytotoxic effects up to this time (Table 1).

DISCUSSION

In this study we set out to demonstrate the tunability of flexible PLA scaffolds using 3D printing. By varying fiber spacing we were able to adjust both the compressive and tensile properties of the FPLA scaffolds within range of physiological IVD mechanical properties. The 1.5mm spacing scaffold yielded properties both in compression and tension within the normal physiological range. Furthermore, we demonstrate that we are able to modulate the viscoelastic properties of the scaffold by varying fiber spacing. This will allow us to more fully recapitulate the mechanical properties of tissues *in situ*, giving a more suitable scaffold for *in vitro* and *in vivo* testing. Compared to PLA scaffolds with the same specifications, FPLA scaffolds exhibited compressive moduli orders of magnitude lower, making them more suitable for a biomimetic approach than conventional biomaterials like PLA. In addition to varying fiber spacing, FPLA scaffold mechanical properties can be tuned by adjusting fiber diameter and orientation. Adjusting these scaffold parameters in addition to fiber spacing in the future will allow us to further refine the moduli of these scaffolds as well as potentially allow for the replication of other physiological relevant mechanical properties such as ultimate tensile and compressive failure strength, shear modulus, and dynamic and viscoelastic properties. Additionally we have shown FPLA scaffolds do not have cytotoxic effects on NP cells indicating they would be suitable in an IVD environment as well as for incorporation of cellular elements to the bioscaffold design. Embedding cell loaded hydrogels such as alginate in the FPLA scaffold would help promote tissue growth and integration of an implanted scaffold. Lastly the ability of 3D printing FPLA scaffolds allows them to be printed in varying and complex shapes. Creating large complex scaffolds that vary parameters such as fiber size and spacing throughout allows for the creation of complex constructs with varying mechanical properties. This would be well suited for musculoskeletal tissue replacements such as the IVD.

ACKNOWLEDGEMENTS

Funded in part by NSF CAREER Award 1151605 & NIH R01 AR069668.

REFERENCES

[1] Ingber, Tissue Eng, 2006 [2] Murphy, Nature Biotechnol, 2014 [3] Humzah, Anat Rec, 1988 [4] Nerurkar, Spine, 2010 [5] Martin JOR 2016 [6] Bowles, Tissue Eng, 2010 [7] Bowles, PNAS, 2007 [8] Iatridis, Spine J, 2013 [9] Nerurkar, J Boimech, 2010

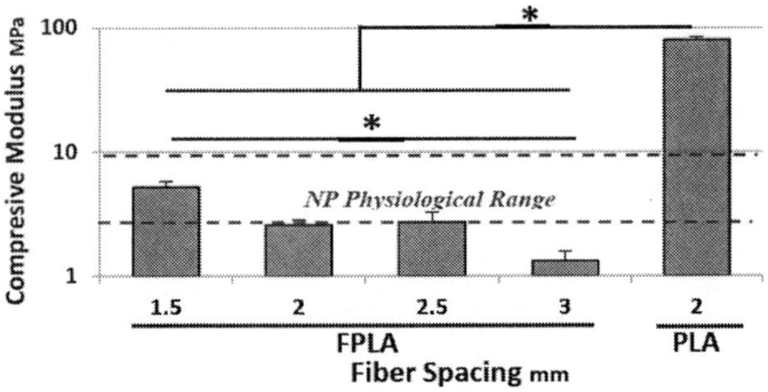

Figure 1: Compressive elastic modulus of FPLA constructs decreases with increasing fiber spacing. 1.5 mm spacing constructs exhibited a modulus within the physiological range of NP tissue. (* $p<0.05$)

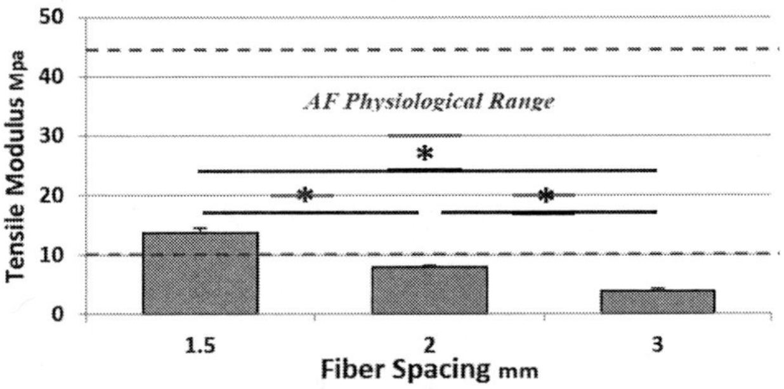

Figure 2: Tensile elastic modulus decreases with increasing fiber spacing. 1.5 mm spacing constructs exhibited a modulus within the physiological range of AF tissue (* $p<0.05$).

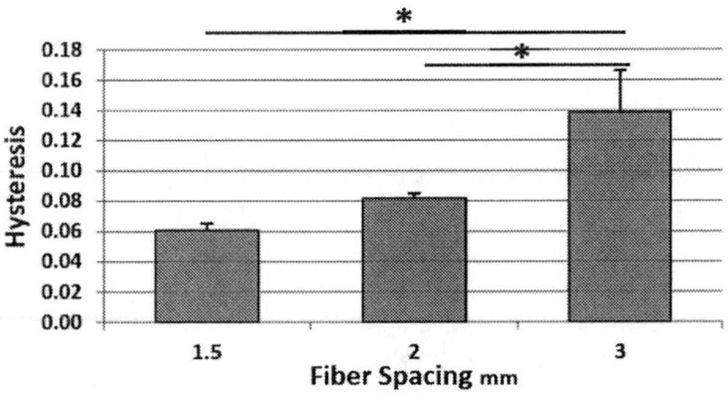

Figure 3: Hysteresis increases w/ increasing fiber spacing. (* $p<0.05$).

Group	Day 1			Day 21		
	Live	Dead	% Live	Live	Dead	% Live
Control	903±220	16±4	98.2±0.1	1068±27	20±1	98.2±0.1
FPLA	1052±261	18±7	98.3±4.1	926±68	18±14	98.2±1.3

Table 1: Average ± standard deviation of live and dead cell count and % live for bNPs cultured with or without FPLA at days 1 and 21.

SB³C2017
Summer Biomechanics, Bioengineering and Biotransport Conference
June 21 – 24, Tucson, AZ, USA

DEVELOPMENT AND TESTING OF A LONG BONE SEGMENT REGENERATING SCAFFOLD FOR PATIENTS

John A. Szivek (1,2), Jacqueline Buchak (1), Andrew M. Wojtanowski (1), David A. Gonzales (1), Adriana Barreda (1), Jordan L. Smith (1), David S. Margolis(1,3)

(1) Department of Orthopaedic Surgery,
University of Arizona,
Tucson, Arizona, USA

(2) University of Arizona Arthritis Center,
University of Arizona,
Tucson, Arizona, USA

(3) Southwest Shoulder, Elbow and Hand Center,
Tucson, Arizona, USA

INTRODUCTION

Critical size long segment bone defects, which do not typically heal without surgical intervention, result from severe trauma or bone resection due to cancer and present significant treatment challenges. Current surgical options have serious drawbacks including treatment failure leading to amputation, resulting in physical and psychological impairment and extensive costs to society. An urgent need exists for a reliable treatment that can induce rapid bone formation to bridge critical size long bone defects.

3D printed scaffolds that support rapid bone growth are being tested in a sheep model providing a potential solution to this problem. Polybutylene terephthalate (PBT) biomimetic scaffolds have demonstrated excellent mechanical properties. They support load bearing and when coated with osteoconductive ceramic powders they induce rapid bone growth. In this study, PBT scaffolds were prepared, mechanically tested, and used in a sheep model to document the extent of bone formation within 3 and 6 months. Preliminary results encouraged the development of allometrically scaled human sized scaffolds which were mechanically tested to establish their suitability to support patient loads for up to 2 years during healing of long segment defects.

METHODS

Constructing and Testing Sheep Long Bone Segment Scaffolds

Sample cylinders of trabecular bone from the femoral head and critical defect-sized cortical bone samples were collected from adult male sheep. They were imaged in a μCT 20 (SCANCO, Bruttisellen, Switzerland) at a 12μm resolution. Files were modified and converted to STL files, then exported to build PBT scaffolds in a Stratsys 1650 FDM (3, 4). Structural accuracy was verified by scanning scaffolds in the μCT and comparing scans to original bone scans.

Rosette strain gauges were attached to scaffolds that were compressed to 294 N at 49 N/s and 294 N/s in an MTS. Data was collected with LabView through a NiDaq board and exported to Excel. Stress vs. principal strain curves were drawn and strength and stiffness determined. A previously published latexing process (3,5) was used to coat scaffolds with 4μm tricalcium phosphate particles and infiltration was verified with microscopy.

Male sheep underwent 2 IACUC approved procedures; the 1st was inguinal fat harvesting. Adipose derived stem cells were extracted using our published procedure (3). Cells were fed, passaged, and perfused into scaffold pores. Scaffolds were implanted 48 hours after perfusion. During surgery, a 42mm femoral segment was removed and a modified lockable IM rod was used to stabilize the scaffold. Sheep were held for 26 weeks. Activity was monitored with 24 hr. video. Monthly x-rays and post-sacrifice histology were used to determine the extent of tissue ingrowth and the rate of bone formation. ANOVA's were used to determine difference between groups.

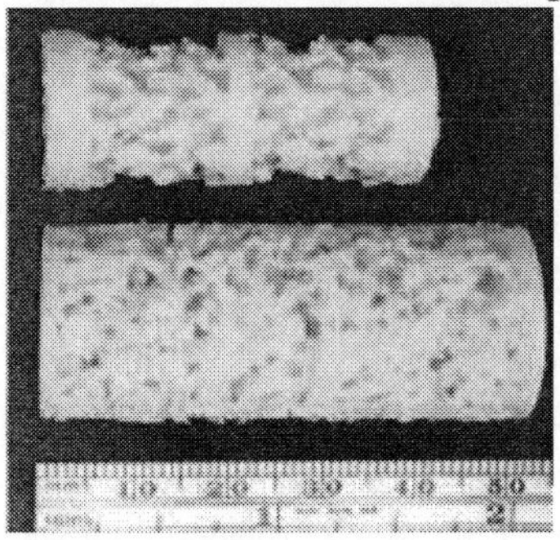

Figure 1: These images show biomimetic PBT scaffolds (top) a sheep scaffold, (bottom) a patient sized scaffold.

Constructing and Testing of Human Scaffolds

A sample of trabecular bone from the femoral head of donated patient bone was collected and scanned with a SCANCO μCT 20 at a 12-micron resolution into a DICOM file. A cylindrical volume of interest was selected from the DICOM file, converted into an STL 3D model and scaled up to improve resolution.

Simultaneously, a 15mm x 28.4mm hollow cylindrical geometric frame was constructed in SolidWorks. The VOI of the bone core and hollow cylindrical frame were combined into one solid 3D model using MeshLab and the resulting STL was uploaded into Quickslice. The model was sliced using a 0.0100 mm slice height and a T16 tip was used for printing in a Stratasys 1650 Fused Deposition Modeler (FDM). Prior to printing, the model was cloned, rotated, and combined to create a cylindrical model measuring 50 mm in height and 28.4 mm in diameter (Figure 1). Base supports were generated in Quickslice and the resulting model was exported as an SML file for printing using polybutylene terephthalate filament.

Single stroke axial compression testing was performed on seven scaffolds and fatigue testing was performed on three

Technical Presentation #167

scaffolds using a servohydraulic materials testing system (MTS). Scaffolds were prepared by removing the base support produced during the printing process and visually inspecting scaffolds for defects. Single stroke compression testing was performed with peak loads of 1471 N (150 kg), 1716 N (175 kg), and 1912 N (195 kg) over successive cycles of 0.5s and 0.3s; each was tested a total of 5 times per scaffold unless scaffold failure occurred. Data was collected with LabView and exported to Excel for analysis.

Cyclic loading was performed on three scaffolds at 441 N (45 Kg) for 250,000 cycles. The scaffolds were examined for failures at the end of each experiment using Spotcheck, a Magnaflux dye penetration crack detection system. These measurements were also collected with LabView and exported to Excel for analysis.

RESULTS

The sheep scaffold configuration was 71.6 ± 0.2% porous. Mechanical testing indicated a linear elastic response and a stiffness of 2.8 ± 0.1 GPa with a compressive strength of 9.44 ± 0.94 MPa (sufficient to support physiological loads). However, scaffolds were 37% as stiff as sheep bone and only 10% the compressive strength. Deposition of TCP was observed deep inside the scaffold pores but cell infiltration was only noted near outer surfaces. Cell counts showed 80% cell depletion following infiltration with 20% adherent to the scaffold surfaces.

Scaffold placement was uneventful and sheep were load bearing within hours. Activity monitoring indicated the sheep spent 23.8 ±6% of their time standing or walking. After 8 weeks, radiographs showed there was 210 sq. mm. of new bone completely bridging the defect. Post sacrifice radiographs and high resolution μCT images showed extensive bone formation at both the 3 and 6-month time points (Figure 2). Cross sectional images showed bone on all scaffold surfaces except where muscle attachment to the linea aspera are present on the sheep's femurs.

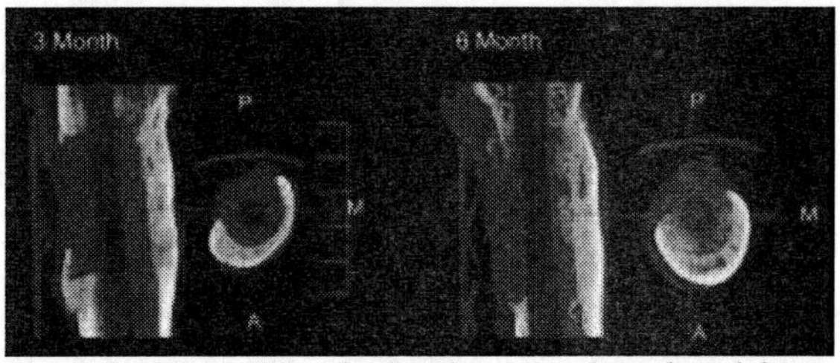

Figure 2: Micro CT's of a 3 and a 6-month explanted femora showing extensive bone formation along the length of the scaffold. There was no bone along the surface where the abductor muscles were attached to the segment that was removed.

Controls, one with a scaffold but no cells, and one with no scaffold showed extensive fibrous tissue formation and some poor-quality bone formation (Figure 3). Both controls had mid regions in the defect where bone was visibly too thin to support the body weight of the sheep and regions which radiographically resembled a non-union.

Human-sized scaffolds tested with cyclic compression for 250,000 cycles showed no evidence of failure. One of seven scaffolds tested simulating gait and impact loading showed evidence of failure at a peak load of 1912 N (195 kg) at a load rate of 6373 N/s. A second failed at a peak load of 1716 N (175 kg) at a load rate of 3432 N/s. These failures were likely due to structural defects that were formed during the 3D printing process which were noted prior to testing of these two specimens.

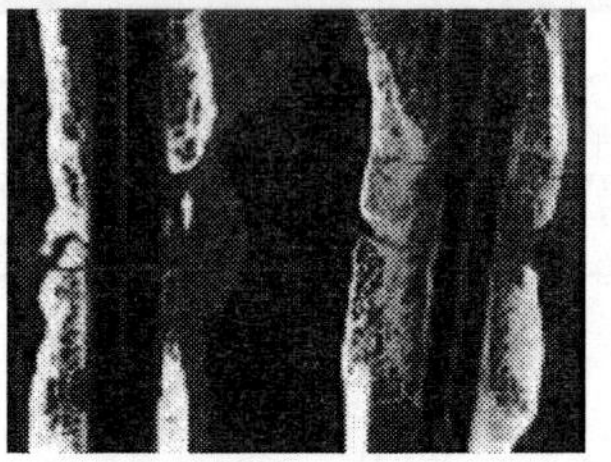

Figure 3: These μCT images show a control limb (from a test animal control) with a scaffold which had not been infiltrated with cells prior to implantation (left) and a control limb (from a test animal control) which received no scaffold but was rodded (right). The central region of the critical sized defects in both these controls did not form adequate bone to support the bodyweight of a sheep once the rod was removed. Although calcified tissue is present, radiographs of both control appear to be similar to the image expected when a non-union occurs in a fracture.

DISCUSSION

A long segment regeneration biomimetic scaffold infiltrated with endogenous adult sheep stem cells facilitated rapid bone growth throughout the length of a critical sized defect in a sheep femur even though sheep were relatively inactive. Results indicate that this design is able to support physiological loads during bone growth as early as 4 weeks. After 3 months the entire length of the scaffold was covered with bone on the anterior and medial surfaces. After 6 months, better bone infiltration into the scaffold was noted and bone remodeling was apparent. Little bone formation was noted on the posterior lateral surface in the region coinciding with the location of the linea aspera.

Preliminary in vivo and bench top test results indicate that patient sized scaffolds having a similar porosity level to sheep scaffolds will be able to withstand peak impact loads of at least 1716 N in patients. Scaffolds experienced no damage when cyclically loaded for the equivalent of 4 months, which will provide sufficient time for bone formation along the length of scaffolds. In the next stage of this study scaffolds instrumented with sensors (strain gauges) will be used to measure loading of the femur during healing using a published procedure (3, 6).

ACKNOWLEDGEMENTS

Preliminary research was supported in part by NSF grant # CMMI 0855493. Student support was through NSF/WAESO and NIH - #T35HL007479 funding, Students in Health Professional School.

REFERENCES

1. Novel Biomimetic Polymer Scaffold Design enhances Bone Ingrowth, CP Geffre, JA Szivek, JT Ruth, DW DeYoung et al., Online Dec 2, DOI: 10.1002/jbm.a.32251 JBMR-A, 2008.
2. Comparison of Bone Ingrowth into Scaffolds with Simple and Novel Biomimetic Porous Architectures, CP Geffre, JA Szivek, DS Margolis, J Invest Med, 56 (1) 242 Suppl., Jan 2008
3. In Vivo Telemetric Determination of Shear and Axial Loads on a Regenerative Cartilage Scaffold following Ligament Disruption, Szivek JA, et al. J Biomed Mater Res–B, DOI: 10.1002/jbm.b.33120, 2014.
4. Trabecular Scaffolds Created Using MicroCT Guided Fused Deposition Modeling, BC Tellis, JA Szivek, CL Bliss, DS Margolis, et al., J Mat Scie & Engineering-C, 28/1, 171-78, 2008.
5. TGF-ß1-enhanced TCP-coated Sensate Scaffolds can detect bone bonding, JA Szivek, D Margolis, BK Garrison, E Nelson, RK, et al., J Biomed Mate Res-B, 73B-1, 43-53, 2005.
6. An Instrumented Scaffold Can Monitor Loading in the Knee Joint, JA Szivek, et al., J Biomed Mater Res, 79B, 218-228, 2006.

SB^3C2017
Summer Biomechanics, Bioengineering and Biotransport Conference
June 21 – 24, Tucson, AZ, USA

DETERMINATION OF THE MECHANICAL PROPERTIES OF DE NOVO ENGINEERED TISSUE IN NEEDLED-NONWOVEN SCAFFOLDS

João S. Soares, Will Zhang, Michael S. Sacks

Center for Cardiovascular Simulation
Institute for Computational Engineering and Sciences, Department of Biomedical Engineering
University of Texas at Austin
Austin, Texas, USA

INTRODUCTION

Regenerative medicine has led to new therapies being used to treat a number of pathologies and has potential to significantly impact medicine as whole by providing the ability to fully restore diseased and injured tissues or entire organs [1]. Such approaches require the development of engineered tissues (ET), and more recently there is growing interest in utilizing ET technologies to perform pharmacological screening [2] or to systematically investigate cellular mechanobiology and interactions between cells and extracellular matrix (ECM) [3]. Scaffold technologies that support these approaches remain an important scientific area of investigation and development.

Multi-filament needled-nonwovens (NNW) scaffolds made of bioresorbable PGA and/or PLA crimped fibers represent one of the earliest and most ubiquitous biomaterials used as ET scaffolds. During manufacture, fibers are crimped and a needle-array punch entangles fibers sufficiently to bind them (although there is no direct adhesion), and yields an interconnected porous structure highly suitable for scaffold-based ET approaches. NNW scaffolds are sufficiently strong and mechanically stable and continue to serve an important role in ET formation, both as a benchmark for evaluating novel biomaterials and for their unique capacity to promote tissue formation, the latter arising from their intricate microstructure and micromechanics.

From a biomechanics perspective, a fundamental issue is the lack of information concerning the mechanics of the NNW scaffold itself and the complex mechanical interactions between nonwoven scaffolds, cells, and especially the de novo forming ECM. Thus, while NNW scaffolds have yielded promising results in diverse ET applications, their use in more fundamental studies is hampered by the lack of: (1) substantial understanding of the complex interactions in between degrading scaffold and forming de novo tissues; and (2) rational models with reliable predictive capabilities describing the evolving mechanical properties of cell-seeded ET constructs based on NNW scaffolds and their response to mechanical stimulation.

Our group has been interested in developing an improved fundamental understanding of the underpinnings of ET growth and development through highly-integrative approaches bridging modeling-simulation-experimentation to address critical barriers in the field of tissue engineering and its over-reliance on pure empiricism [4-6]. Here, we employ an integrative approach involving a systematic experimental program coupled to novel theoretical development to develop a robust mathematical model to predict the mechanical response of the forming ECM by the stiffness amplification of NNW scaffolds [7]. To calibrate our model, we utilize polyacrylamide (PAM) gel of variable known stiffness to mimic the effects of surrounding tissues on the scaffold fiber network. Lastly, we illustrate the impact of our modeling approach by employing it to quantify the stiffness of engineered ECM after in vitro incubation and early stages of in vivo implantation obtained in a concurrent study of ET pulmonary valve conduits in an ovine model [8,9].

METHODS

NNW/ECM composite meso-structural model

A variation of the approach proposed by Garikipati et al. [10] was employed to formulate an analytical model of the mechanical response of a single fiber. This approach is very appropriate for the exploration of the influence of fiber geometry, specifically fiber amplitude and undulation, in its mechanical response; however, the framework requires that fibers remain in a co-planar sinusoidal configuration – while this is valid for extending fibers, this is not suitable to handle fiber compression. Thus, we employ the finite element (FE) method as complement to the analytical study to handle more complex deformations not captured within the elastica framework.

We define the hyperelastic strain energy density ψ of the NNW-ECM composite with contributions of the fiber network and the forming de novo ECM weighted by their volume fractions, i.e. $\psi = \phi_f \psi_f = \phi_m \psi_m$. To model forming ECM surrounding the scaffold fibers, we utilize a single isotropic hyperelastic neo-Hookean model with strain energy density function $\psi_m = (\mu_m/2)(I_1 = 3)$.

The uniaxial response of individual scaffold fibers consists of: (1) the extensional stiffness η_f^{eff} obtained with the elastica framework and corroborated by the FE simulations, but (2) without any stiffness under compression. Normalized fiber distribution $\Gamma(\theta)$ was obtained from SALS data in our previous study [11], and the resulting 2nd Piola Kirchhoff stress in the fiber phase is obtained by integration of the individual fiber responses over the oriented fiber-ensemble.

Fiber-fiber interactions arise due to the diverse ranges of fiber orientations in the NNW scaffolds and the inability for fibers to support compression. The following micro-mechanisms for fiber-fiber interactions are considered: (1) when the material undergoes substantial extension in one direction and shrinkage in the other, fibers aligned with the shrinkage of compression will undergo large-translational motions and will not store significant amounts of elastic energy, and (2) these translational effects of collapsing fibers and their crosslinks with fibers aligned in the extensional direction will indeed contribute towards the extension in the other direction.

The 2nd Piola-Kirchhoff stress of the NNW-ECM composite is

$$\mathbf{S} = \mathbf{S}_f = \mathbf{S}_{\text{int}} = \mathbf{S}_m =$$
$$= \phi_f \left[\eta_f^{\text{eff}} \int_\theta \Gamma(\theta) \left(1 = \frac{1}{\lambda_f(\theta)}\right) (\mathbf{n} \otimes \mathbf{n}) d\theta = \right.$$
$$\left. = \eta_{\text{int}} \iint_{\alpha\,\beta} \Gamma(\alpha)\Gamma(\beta) \left(\frac{\lambda_\alpha - \lambda_\beta}{\lambda_\alpha} \mathbf{n} \otimes \mathbf{n} = \frac{\lambda_\beta - \lambda_\alpha}{\lambda_\beta} \mathbf{m} \otimes \mathbf{m} \right) d\alpha d\beta \right]$$
$$= (1 = \phi_f)\mu_m (\mathbf{I} = C_{33}\mathbf{C}^{-1}),$$

where fiber stretch is $\lambda_f = \sqrt{\mathbf{n} \cdot \mathbf{C}\mathbf{n}}$ in the fiber direction $\mathbf{n}$.

The effective stiffness of scaffold fibers η_f^{eff} and the magnitude of fiber-fiber interactions η_{int} have their effects modulated as functions of the stiffness of the surrounding ECM, i.e.

$$\eta_f^{\text{eff}} = \chi(\mu_m) \frac{\eta_{\text{elastica}}^{\text{PLLA}} - \eta_{\text{elastica}}^{\text{PGA}}}{2},$$
$$\eta_{\text{int}} = \eta_{\text{int}}(\mu_m).$$

Experimental calibration and example application

Mechanical characterization of NNW scaffold-PAM gel tissue-analogs was performed on square samples of tissue-analogs of varying PAM gel stiffness. Load-controlled biaxial testing was conducted (n = 5) to generate data for the determination of the fiber stiffness amplification $\chi(\mu_m)$ and fiber-fiber interaction $\eta_{\text{int}}(\mu_m)$ factors.

We illustrate the impact of our modeling approach by applying it to ETs obtained in a concurrent study of our group on ET pulmonary valve conduits in an ovine model [8,9]. Biaxial tensile mechanical testing was performed on segments of the post-incubation ET samples following the same protocols. Fitting experimental data with fiber stiffness amplification $\chi(\mu_m)$ and fiber-fiber interactions $\eta_{\text{int}}(\mu_m)$ that were previously calibrated with the experimental data collected with NNW scaffolds impregnated with PAM gel of known stiffness allowed the estimation of the stiffness of the engineered ECM (μ_m).

RESULTS

The "apparent fiber stiffness" of undulated fibers when present in the NNW scaffold was found to be 4-fold higher than the stiffness of individual fibers as if acting alone. This reinforcement mechanism was modulated by the environment surrounding the scaffold as the fiber stiffness amplification factor increased with increases in stiffness of the impregnated PAM gel, up to a 20-fold increase with the stiffest PAM gel impregnated. Micro-structurally, the presence of a stiffer environment surrounding the scaffold fibers restricts their deformation, possibly strengthens inter-connecting points in between them and these translate individually to the fiber level as fibers have shorter effective wavelength and amplitude. A similar trend was seen in the interaction term – the energy associated to fiber-fiber interactions increase as the stiffness of the PAM gel increases.

Biaxial testing data on ET constructs were fit simultaneously to find μ_m, now representing the stiffness of the ECM surrounding the scaffold (Figure 1). At post-incubation, the stiffness was rather low ($\mu_m = 5.173 = 0.393$ kPa and $\mu_m = 8.1318 = 1.370$ kPa at t = 0 and 1 day). However, a substantial increase in ECM stiffness was observed at t = 7 day (up to $\mu_m = 18.138 =$ 4.144 kPa).

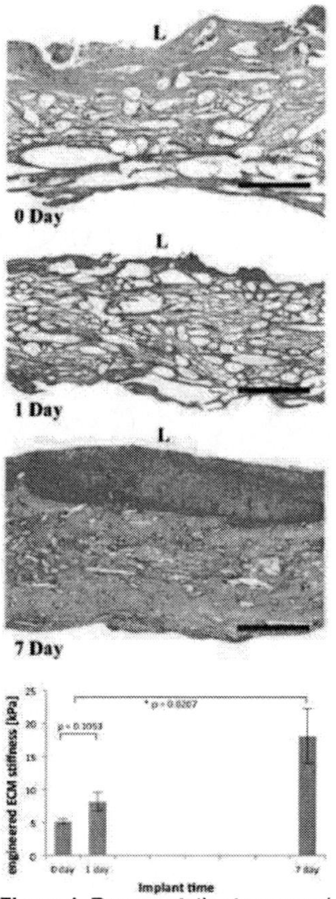

Figure 1. Representative transmural H&E histology of pulmonary valve conduits (top) and determined ECM stiffness (bottom).

DISCUSSION

We have developed a rigorous methodology that complements systematic experimentation with specialized theoretical modeling to expand our understanding of how NNW-ECM composites behave mechanically. The theoretical development yielded critical insight into some particular features of these scaffolds, specifically the underlying mechanisms of fiber buckling, scaffold stiffness amplification, and fiber-fiber interaction behavior that were not possible to conjecture without the insight given by modeling, let alone describe or foresee a priori. As end result, we obtained a robust methodology to infer the stiffness of de novo ECM formed on a NNW scaffold. The ability to obtain reliable and robust measures that quantify de novo ECM would be pivotal in any study that aims to achieve any functional requirement for ET implants.

ACKNOWLEDGEMENTS

NIH grants no. R01 HL068816, HL089750, HL119297, HL63954 and HL108330 (to M.S.S), and all co-authors of References [8,9].

REFERENCES

[1] Mao, AS and DJ Mooney, *Proc Natl Acad Sci U S A*, 112:14452-9, 2015. [2] Hansen, A et al., *Circ Res*, 107:35-44, 2010. [3] Boudou, T et al., *Tissue Eng Part A*, 18:910-9, 2012. [4] Soares, JS and MS Sacks, *Biomech Model Mechanobiol*, 15:293-316, 2016. [5] D'Amore, A et al., *J Mech Behav Biomed Mater*, 62:619-35, 2016. [6] Engelmayr, GC, Jr. et al., *Biomaterials*, 26:175-87, 2005. [7] Soares, JS et al., *Acta Biomater*, 2017. [8] Gottlieb, D et al., *J Thorac Cardiovasc Surg*, 139:723-31, 2010. [9] Eckert, CE et al., *Ann Biomed Eng*, 39:205-22, 2011. [10] Garikipati, K et al., *J Mech Phys Solids*, 56:1693-1713, 2008. [11] Engelmayr, GC, Jr. and MS Sacks, *J Biomech Eng*, 128:610-22, 2006.

THE PREDICTIVE VALUE OF TRANSVERSE SHEAR STRESS ON PLAQUE PROGRESSION IN HUMAN CORONARY ARTERIES

Annette M. Kok (1), David S. Molony (2), Lucas H. Timmins (3), Yi-An Ko (4), Parham Eshtehardi (2), Jolanda J. Wentzel (1), Habib Samady (2)

(1) Department of Biomedical Engineering
Erasmus MC
Rotterdam, The Netherlands

(2) Department of Medicine
Emory University School of Medicine
Atlanta, GA, United States

(3) Department of Bioengineering
University of Utah
Salt Lake City, UT, United States

(4) Department of Biostatistics and Bioinformatics
Emory University Rollins School of Public Health
Atlanta, GA, United States

INTRODUCTION

Decades of cardiovascular research have been devoted to investigate atherosclerotic plaque progression. It is well-known that hemodynamics play an important role in plaque initiation and progression. Low shear stress promotes plaque progression; whereas high shear stress promotes plaque regression [1]. Interestingly, in both the low and high shear stress regions, the necrotic core area increases. Since high shear stress regions also show regression of fibrous tissue, these findings suggest that high shear is involved in the development of a more vulnerable plaque phenotype.

Shear stress is known to influence the endothelial cell morphology and function. Upon exposure to laminar shear stress, endothelial cells align with the mean flow direction, which is associated with activation of atheroprotective pathways. In contrast, oscillatory or multidirectional flow proved to be atheroprone because of the endothelial cell inability to align and thereby induce multiple inflammatory pathways [2]. Different multidirectional flow parameters have been proposed in the literature to investigate these processes. One of these is the transverse wall shear stress (transWSS) which is a measure of the perpendicular flow that an endothelial cell would sense. The transWSS was shown to be positively correlated with plaque initiation in rabbit's aortas [3]. Although the locations of plaque were consistent with high transWSS in the immature rabbits, this was not the case for mature rabbits [4]. Until now, no study has investigated the relationship between transWSS and plaque progression in human coronary arteries.

METHODS

Image acquisition, velocity measurements, and computational modeling of the distal left anterior descending (LAD) were performed in twenty patients. The image acquisition consisted of biplane angiography and virtual histology intravascular ultrasound (VH-IVUS). Doppler derived velocity data was measured in the left main. After 6 months, the image acquisition was repeated. This study was approved by the Emory University Review board, each patients provided a written informed consent.

The IVUS inner and outer contours were manually segmented by an experienced investigator. The serial IVUS images were co-registered by matching anatomical and compositional landmarks. In order to investigate focal effects of shear stress on plaque progressing, each IVUS image was divided into eight sectors as previously described [5]. Change in plaque area (ΔPA) was defined as the follow-up plaque area minus the co-registered plaque area at baseline in each sector.

In order to generate a volume for the computational modelling, the IVUS contours were stacked upon the 3D centerline obtained from angiography. Side branches were attached as straight tubes using directional data from angiography and the diameter from IVUS. A 3D surface was generated using non-uniform rational b-splines, next the volume was discretized. The patient-specific velocity derived measurement was used to impose a plug flow profile at the inlet, furthermore pressure free outlets were assumed. The Navier-Stokes equations were solved with computational fluid dynamics (CFD) as described elsewhere [1].

The time averaged wall shear stress (TAWSS) and transverse wall shear stress (transWSS) were calculated for each node (x):

$$TAWSS(x) = \frac{1}{T}\int_0^T |\tau_w(x,t)|\, dt \quad (1)$$

$$transWSS(x) = \frac{1}{T}\int_0^T |\tau_w(x,t) \cdot P(x)|\, dt \quad (2)$$

$$P(x) = n(x) \times \frac{\int_0^T \tau_w(x,t)\,dt}{\left|\int_0^T \tau_w(x,t)\,dt\right|} \tag{3}$$

where $\tau_w(x,t)$ is the instantaneous shear stress vector and T the time for one cardiac cycle. P is the unit vector perpendicular to the surface normal, $n(x)$, and the mean wall shear stress vector. TransWSS can take any value between 0 and the TAWSS, which makes interpretation difficult. Therefore, Mohamied et al. [6] introduced a normalized version of the transWSS, the cross-flow index (CFI):

$$CFI(x) = \frac{1}{T}\int_0^T \left|\frac{\tau_w(x,t)}{|\tau_w(x,t)|}\cdot P(x)\right| dt \tag{4}$$

For the post-processing, the shear stress parameters were averaged within each sector. The TAWSS was divided into a low (<1 Pa), medium (1–2.5 Pa) or high (>2.5 Pa) group. Since there is no literature on thresholds for the transWSS and CFI, they were divided into three equally distributed groups based on their population average tertiles. A Linear Mixed-Effect (LME) model was used to investigate the association between the shear stress parameter (fixed effect) and ΔPA (dependent variable), which accounted for dependency within patients by subject-specific random intercepts. Moreover, the interaction between CFI and TAWSS was incorporated into the model to determine whether the association between ΔPA and TAWSS depends on CFI. P<0.05 was considered significant.

RESULTS

In total, 13801 sectors were analyzed. The mean and standard deviation were -0.02 ± 0.25 mm^2 for the ΔPA, 2.46 ± 2.01 Pa for the TAWSS, 0.08 ± 0.10 for the CFI, and 0.11 ± 0.14 Pa for the transWSS. Based on tertiles the CFI was divided into low CFI (<0.0285), medium (0.0285–0.0727) and high CFI group (>0.0727). Similarly, the transWSS was divided into low (<0.0509 Pa), medium (0.0509–0.1155 Pa) and high groups (>0.0509 Pa).

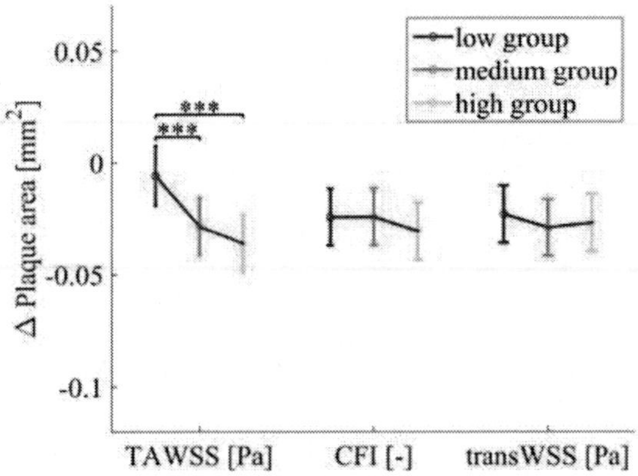

Figure 1: Estimated changes in plaque area with their standard errors among low, medium and high shear stress groups. *** = P < 0.001.

Overall, the TAWSS significantly (p<0.001) contributed to predict ΔPA, whereas CFI and transWSS did not (p>0.05), see Figure 1. Furthermore, the low TAWSS group showed significantly (p<0.001) more ΔPA, less plaque regression, than the medium and high TAWSS groups.

Also, the co-localization of WSS with CFI was investigated, see Figure 2. A significant interaction effect between the co-localization of TAWSS and CFI was found (p<0.001); a higher CFI and a higher TAWSS leads to more plaque regression, and vice versa.

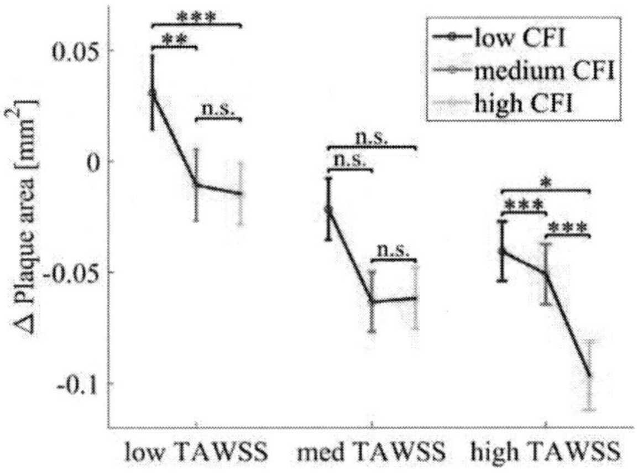

Figure 2: Estimated changes in plaque area with their standard errors for low (<1 Pa), median (1–2.5), and high (>2.5) TAWSS co-localized in three CFI groups. * = P < 0.05, ** = P < 0.01, *** = P < 0.001 and n.s. = not significant

DISCUSSION

Our study showed that plaque progression was not different among the regions exposed to low, medium or high transWSS tertiles. The same results were found for CFI (the normalized transWSS metric), whereas a clear negative relationship was confirmed between plaque progression and TAWSS. Interestingly, co-localization of TAWSS with CFI showed an additional negative effect of CFI on plaque progression.

Our data contradicts earlier findings showing a positive relationship between transWSS and plaque initiation [3,4]. However, a comparison between these studies should be with great caution since the initiation of these diet-induced plaques in rabbits is highly accelerated and might not represent the slow natural development of plaques as found in humans. Moreover, patients in this study received optimal medical treatment, including statins, which slow down and can inhibit the natural development of plaques [7]. In future work we will include the plaque phenotype, in order to investigate the plaque vulnerability.

ACKNOWLEDGEMENTS

The research leading to these results has received funding from the European Research Council under the European Union's Seventh Framework Programme / ERC Grant Agreement n. 310457 and Vereniging Trustfonds Erasmus Universiteit Rotterdam.

REFERENCES

[1] Samady, H et al., *Circulation*, 124(7):779–88, 2011.
[2] Wang, C et al., *Arterioscler Thromb Vasc Biol*, 33(9):2130–6, 2013.
[3] Peiffer, V et al., *J Biomech*, 46:2651–2658, 2013.
[4] Mohamied, Y et al., *Ann Biomed Eng*, 43(1),16–25, 2016.
[5] Timmins, L. H. et al., *Ann Biomed Eng*, 43(1):94–106, 2015.
[6] Mohamied, Y et al., *J Biomech*, in press, 2016.
[7] Nissen, S.E., *J Am Med Assoc*, 295:1556–65, 2006

UNCERTAINTY QUANTIFICATION IN MULTI-SCALE CORONARY SIMULATIONS USING MULTI-RESOLUTION EXPANSION

Justin S. Tran (1), Daniele E. Schiavazzi (3), Abhay B. Ramachandra (2), Andrew M. Kahn (4), Alison L. Marsden (2)

(1) Department of Mechanical Engineering
Stanford University
Stanford, California, USA

(2) Departments of Pediatrics and Bioengineering
Stanford University
Stanford, California, USA

(3) Department of ACMS
University of Notre Dame
Notre Dame, Indiana, USA

(4) Department of Medicine
University of California, San Diego
San Diego, California, USA

INTRODUCTION

Computational simulations of coronary flow can provide non-invasive information on hemodynamics that can aid in surgical planning and research on disease progression. These tools have potential impact in identifying optimal geometries for coronary artery bypass grafts (CABG), in determining risk of re-stenosis in saphenous vein grafts, and in determining thrombotic risk in Kawasaki disease. In this study, patient-specific geometries of the aorta and coronary arteries are constructed from CT imaging data and finite element flow simulations are carried out using the open source software SimVascular to predict quantities of hemodyamic interest that are difficult to measure *in vivo*.

The outputs of these simulations depend on a set of clinically-derived input parameters that define the geometry and boundary conditions. However the values of these input parameters are not known precisely due to limitations in measurement precision and uncertainties arising from population variability. The aim of this study is to assign relative confidence in simulation outputs considering two sources of input uncertainty: uncertainty in the material properties of the vessel wall arising from population variability and uncertainty in the lumped parameter boundary conditions at the inflow and outflows of the model, whose values are estimated by assimilating a set of patient-specific measurements and clinical data on coronary flow.

We use a generalized multi-resolution chaos approach to propagate uncertainties to the outputs of the simulations [1]. The advantages of this approach for the application to multi-scale coronary simulation lie in its ability to support inputs sampled from arbitrary joint distributions and its built-in adaptivity allowing for efficient approximations of stochastic responses characterized by steep gradients.

METHODS

Consider an output quantity of interest u (e.g. wall shear stress at a point or pressure at an outlet) as a function of its random inputs $\boldsymbol{y} = (y_1, y_2, \dots, y_d)$ where d is the number of random inputs. Following the theory of *generalized polynomial chaos*, we seek an expansion of $u(\boldsymbol{y})$ in terms of multi-variate polynomials in the random inputs.

$$u(\boldsymbol{y}) \approx \sum \alpha_i \psi_i(\boldsymbol{y}) \qquad (1)$$

The polynomials $\psi_i(\boldsymbol{y})$ are constructed such that they form an orthonormal set with respect to the underlying joint probability distribution of the input parameters, $\rho(\boldsymbol{y})$.

In the case where all the random inputs can be considered statistically independent (i.e. the joint distribution is product of the individual marginal distributions of each of the parameters), the above expansion yields simple formulas for the statistical moments in terms of the coefficients of the expansion. For example, the mean and variance are given by:

$$\mathrm{E}[u] = \nu_1 = \alpha_0 \qquad (2)$$

$$var[u] = \nu_2 = \sum \alpha_i^2 \qquad (3)$$

Most practical cases, however, cannot assume independence in the random inputs. For this study, we consider a combination of dependent boundary condition parameters [2] and independent wall property parameters. The corresponding joint distribution is then given by:

$$\rho(\boldsymbol{y}) = \rho_{BC}\rho_{WALL} = \rho_{BC}\prod_{i}^{n_{WALL}} \rho_{WALL,i} \tag{4}$$

We then use a change of measure in the calculation of the statistical moments to attain simple expressions for the mean and variance:

$$\begin{aligned} v_p &= \int u(\boldsymbol{y})^p \rho(\boldsymbol{y}) d\boldsymbol{y} \\ &= \int \left[u(\boldsymbol{y})^p \frac{\rho_{BC}\prod_i^{n_{MAT}} \rho_{MAT,i}}{\prod_j^{n_{BC}} \rho_{BC,j} \prod_i^{n_{MAT}} \rho_{MAT,i}} \right] \prod_j^{n_{BC}} \rho_{BC,j} \prod_i^{n_{MAT}} \rho_{MAT,i}\, d\boldsymbol{y} \\ &= \int f(\boldsymbol{y}) \prod \rho_i(y_i)\, d\boldsymbol{y} \end{aligned} \tag{5}$$

The expansion coefficients of $f(\boldsymbol{y})$ will encode the statistics of $u(\boldsymbol{y})$.

The accuracy of a polynomial chaos expansion depends on the number of terms included in the expansion. For a relatively smooth stochastic response, only a few terms are needed to capture essential features. But a stochastic response characterized by steep gradients or discontinuities can require many terms in the expansion to adequately approximate the statistics, which becomes computationally expensive. This is rectified in our study with a multi-element, multi-resolution approach to polynomial chaos. Instead of using global expansion functions which span the entire stochastic domain, we decompose the space into elements. A low order polynomial chaos expansion is computed separately on each element. Instead of increasing the polynomial order of the expansions to increase accuracy, the size of the elements is refined to get a more accurate approximation. The framework can adaptively refine elements only where the stochastic response is poorly estimated by including additional terms that indicate errors at the current resolution level. Expansion coefficients on each element are determined through relevance vector machine regression. Global statistics are then obtained by aggregating contributions from all elements. This framework has been validated against various benchmarks in uncertainty quantification with good performance [1].

RESULTS

We ran 150 multi-scale coronary simulations with inputs sampled from the joint distribution. Using the multi-resolution polynomial chaos framework, we first computed the mean and standard deviations for the time-averaged wall shear stress (TAWSS) on arterial and venous grafts. This quantity has been shown to be significantly different in arterial and venous grafts and could be correlated with venous graft failure [3].

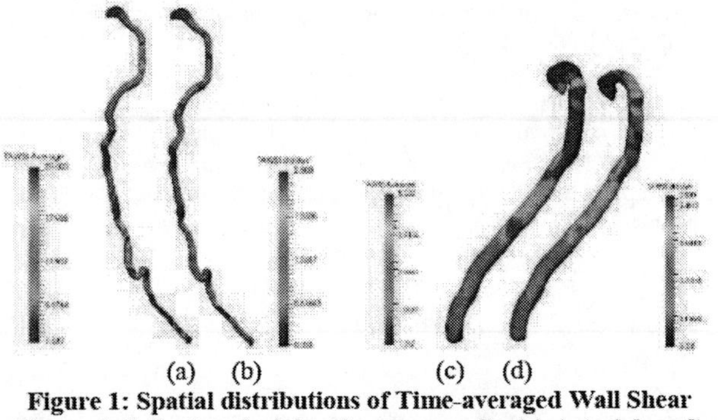

(a) (b) (c) (d)

Figure 1: Spatial distributions of Time-averaged Wall Shear Stress statistics in arterial and venous grafts. (a) Arterial graft mean (b) Arterial graft standard deviation (c) Venous graft mean and (d) Venous graft standard deviation

Figure 1 shows that TAWSS in the grafts is well predicted by our multi-scale model even in the presence of uncertainties. The standard deviation is at least one order of magnitude below the mean values, showing high confidence in the model predictions.

We also computed the area-averaged TAWSS in arterial and venous grafts in the same way as [3] and used the results from the uncertainty propagation to assess confidence in those predictions. To verify the results from the multi-resolution (MR) framework, we also computed a Monte Carlo (MC) estimation of the mean and standard deviation using the same 150 simulations. Standard deviations are reported as error bars in Figure 2.

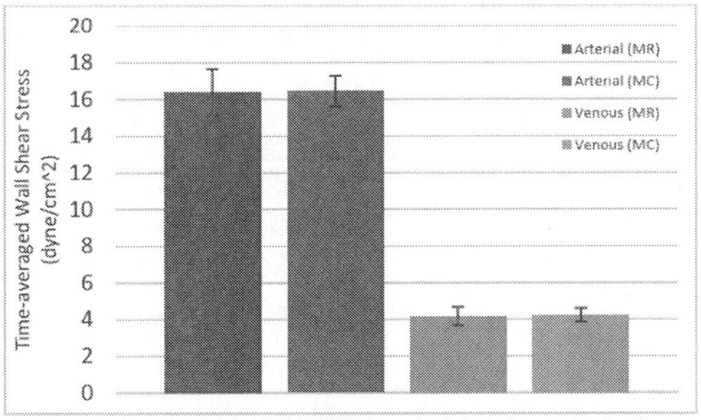

Figure 2: Area-averaged TAWSS in arterial and venous grafts as computed using multi-resolution (MR) and Monte Carlo (MC)

Figure 2 confirms the results of [3] that, even in the presence of uncertainty, the TAWSS is significantly lower in venous grafts than arterial grafts. Also, the means and standard deviations are well correlated between the multi-resolution framework and Monte Carlo.

DISCUSSION

Patient-specific simulations of coronary flow are subjected to many inputs that define the geometry and boundary conditions. Most of these inputs have associated uncertainty due to finite resolution of the measuring instrument, or natural physiologic or population variability. Despite the presence of these input uncertainties, simulation outputs are typically reported as deterministic without any measure of relative confidence in the results. Quantifying these uncertainties is an essential step towards validation of these computational models for clinical use. This study assigns confidence intervals on typical hemodynamic quantities of interest accounting for uncertainty in the boundary conditions and material properties of the wall. Our results show that time-averaged wall shear stress predictions have a high degree of confidence in the presence of these uncertainties.

ACKNOWLEDGEMENTS

This work was supported by NIH RO1-RHL123689A, Burroughs Wellcome Fund CASI Award, and NSF CAREER Award.

REFERENCES

[1] Schiavazzi, D et al., *Computer Methods in Applied Mechanics and Engineering*, 314:196-221, 2016.
[2] Tran, J et al., *Computers & Fluids*, 142:128-138, 2017.
[3] Ramachandra, A et al., *Journal of Cardiovascular Translational Research*, 9.4:279-290, 2016.

AN IMPROVED MICRO-THERMAL SENSOR FOR PLANNING AND GUIDANCE OF PULMONARY VEIN CRYOTHERAPY

Harishankar Natesan (1), Limei Tian (2), Wyatt Hodges (4), Chris Dames (4), John Rogers (2, 3), John Bischof (1)

(1) Department of Mechanical Engineering
University of Minnesota
Minneapolis, Minnesota, USA

(2) Department of Material Science and Engineering
University of Illinois,
Urbana- Champaign, Illinois, USA

(3) Department of Material Science and Engineering
North□estern University,
Evanston, Illinois, USA

(4) Department of Mechanical Engineering
University of California
Ber□eley, California, USA

INTRODUCTION

Atrial fibrillation (AF) affects millions of people every year. If left untreated AF can contribute to the progression of further cardiovascular disease, stroke and even death. While some drugs can help control atrial fibrillation, they have side effects, and their efficacy is not uniform and often diminishes over time. Thus, the use of focal interventions to electrically isolate the pulmonary vein (PV) thereby avoiding drugs is considered a milestone advancement in the treatment of AF. Specifically, thermal and cryoablation for PV isolation has been used for more than 10 years in AF treatment. These treatments require improved monitoring of probe contact, tissue thickness, and freeze or heat completion through the wall for optimal outcomes. Unfortunately, clinical imaging such as Ultrasound, Magnetic Resonance Imaging, and Computed Tomography have resolutions similar in scale (~ 1 - 2 mm) to the PV thickness, and therefore are unable to resolve changes within the PV during treatment. Thus, alternatives are urgently needed to improve monitoring of cryoablation within the PV at the sub-mm level. Here then we present a new, micro-thermal sensor based on the "3ω" technique[1] modified for thin biological systems[2]. We have already generated proof of concept data with the sensor to sense tissue contact vs. fluid contact with and without flow. We have demonstrated using an isolated sensor the ability to measure tissue thickness in the 0.5 to 2 mm regime and the initiation and completion of freezing[3]. In this study, we have improved the sensor design to enable its integration onto a flexible, stretchable substrate such as a cryoballoon. Further, we have tested the improved sensor to measure flow vs tissue contact, thickness and phase change. These new studies confirm the ability of the sensor to monitor thermal therapies for the treatment of atrial fibrillation.

METHODS

The supported 3ω technique is used to measure thermal conductivity (*k*) of thin (< 2 mm) biological tissues. It uses a sensor made up of a thin gold heater line (1 mm X 30 μm X 200 nm) deposited on a glass substrate with a tissue sample on top. An AC current excites the heater line producing temperature oscillations propagating into the sample and the substrate. By measuring the third harmonic voltage ($V_{3\omega,\,op}$) across the heater line and knowing the *k* of the substrate, the *k* of the sample region traversed by the temperature oscillations can be determined. The penetration depth ($PD = (\alpha/2\omega)^{0.5}$) of the temperature oscillations can be controlled by changing the frequency (ω) of the AC current. In this study, the sensors are first modified from a thin straight line to a serpentine shape for integration on a flexible, stretchable balloon. The serpentine shape minimizes local strains during inflation and deflation of the balloon. However, the new shape has a different 3ω voltage response especially at low frequencies. Hence, the modified sensor is tested on a flat substrate to

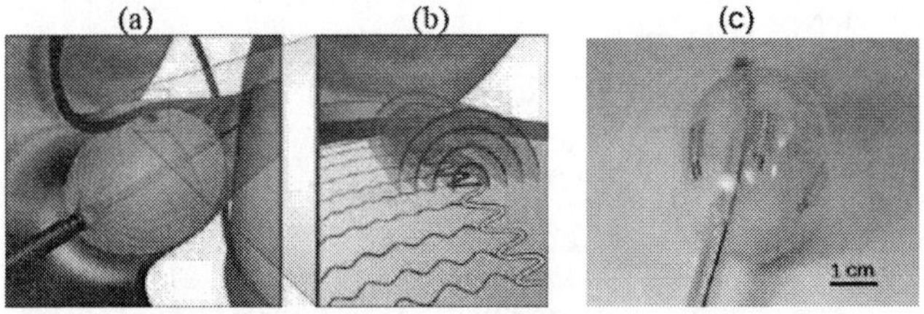

Figure 1. Future vision for the use of micro-thermal 3ω sensors for monitoring balloon cryoablation in pulmonary vein (PV) *(a) Cryoballoon seated inside PV. A sensor array could be microfabricated around the balloon circumference. Each sensor could have 4 wires attached potentially allowing an individual or group of sensors to be activated. (b) The sensor produces temperature oscillations, and measures corresponding 3ω voltage response ($V_{3\omega,\,op}$). (c) Sensors integrated onto a balloon (Rogers group)*

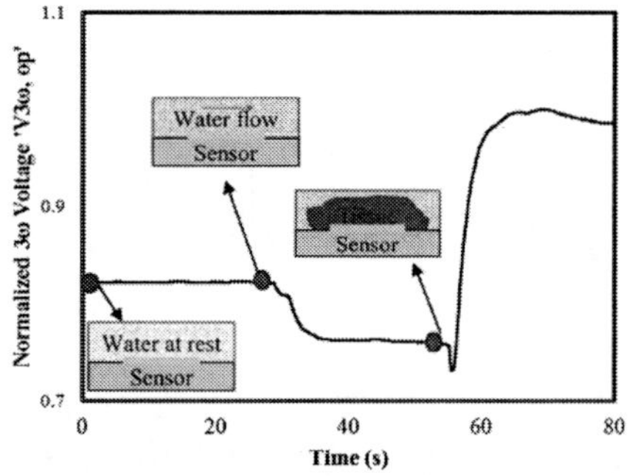

Figure 2. The measurement of flow vs. tissue contact by the serpentine shaped 3ω sensor. *Plot show three cases for the events: (a) water at rest; water flowing; and tissue in contact with 3ω sensor. Red dots represent the beginning of these events.*

sense tissue contact, thickness and phase change. For proof of concept, the modified sensor has been microfabricated onto a flexible, stretchable balloon using transfer printing techniques (Fig 1 (c)) [4].

RESULTS

For the current study, we deposited the modified serpentine shaped sensor on a flat, glass substrate. First, to verify our ability to sense probe contact, we used the sensor to differentiate between tissue contact and flow conditions. Fig. 2 shows 3ω voltage at rest. As water flows in at ~28 s, $V_{3\omega,\,op}$ rises due to convective heat loss. At 50 s, the plot shows that placing a piece of low k tissue (1 mm thick: here represented by mouse liver) on top of the sensor blocks the water flow and thereby increases $V_{3\omega,\,op}$ to the highest value of all cases. Next, to verify our ability to measure thickness, we prepared several samples of thin agargel of varying thickness and recorded the 3ω signal for varying frequency (Fig. 3). At high frequencies, the thermal penetration depth ($(\alpha/2\omega)^{0.5}$) is within the sample as depicted in the inset. At lower frequencies, the penetration depth becomes comparable to the sample thickness as $V_{3\omega,\,op}$ diverge depending on thickness. The frequency at 0.1 Hz decreases with increasing thickness (0.5 to 2 mm). In the end, to verify our ability to sense phase change in real-time, we used the sensor to monitor freezing and thawing of mouse liver, through its effect on the 3ω voltage. The temperature of the stage below the sensor was changed to -25 ^{0}C at ~50 s after which there is a drop in $V_{3\omega,\,op}$. This drop corresponds to the jump in k for biological tissues on freezing. This drop is followed by a stable region where the

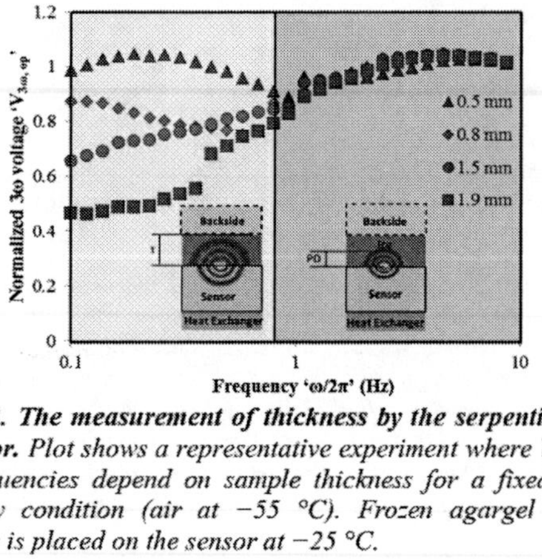

Figure 3. The measurement of thickness by the serpentine shaped 3ω sensor. *Plot shows a representative experiment where V3ω, op at low frequencies depend on sample thickness for a fixed backside boundary condition (air at −55 °C). Frozen agargel of known thickness is placed on the sensor at −25 °C.*

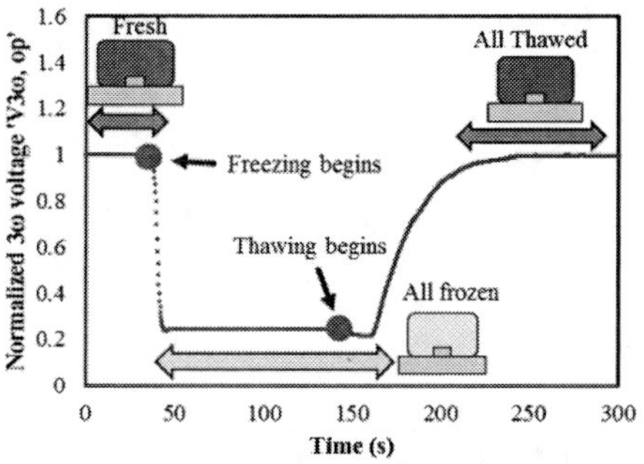

Figure 4. The measurement of phase change onset, completion and thaw by the serepntine shaped 3ω sensor. *Mouse liver is frozen from bottom to top. V3ω,op was measured as a function of time at constant PD (i.e. frequency-10.11Hz). Mouse liver was cooled from below at ~50 s and thawed at ~150 s.*

sample is completely frozen. After ~150 s, the temperature of the stage is set to 15 ^{0}C resulting in increase in $V_{3\omega,\,op}$. The $V_{3\omega,\,op}$ data matches closely before freezing and after thawing showing the reliability of the measurement.

DISCUSSION

We provide proof of principle that the improved 3ω sensor modified for integration onto a balloon can monitor important dynamic aspects of PV cryoablation and other probe based therapies. For instance, we demonstrate the ability to print these sensors onto a flexible, stretchable balloon as shown in Fig 1. Then we show the ability to sense tissue contact vs. fluid in flow in Fig 2. We also demonstrate the ability to measure thickness of thin tissues based on $V_{3\omega,\,op}$ at low frequencies in Fig 3. Finally, we also demonstrate the ability to sense tissue freezing in mouse liver in Fig 4. In summary, this highly innovative work represents several conceptual firsts in the use of a thermal sensor for both static and dynamic measurement within thin biological tissues at the limit of resolution of clinical imaging (i.e. 2 mm) down to 0.1 mm. In future, this sensor will be tested *ex vivo* and *in vivo* to measure thin tissue properties, contact, thickness and the initiation and completion of freezing all with an accuracy of ~0.1 mm. In addition to improving treatment in PV as initially envisioned here, the new thin, flexible sensor could be used in flat or curved geometries with application to thermal therapy (heat or cold) monitoring in peripheral artery disease, renal or hepatic artery treatments (hypertension and diabetes) in the future.

ACKNOWLEDGEMENTS

Funding from Minnesota Biotechnology Partnership grant, Doctoral Dissertation Fellowship (2016/17)

REFERENCES

1. Cahill, D.G. and R.O. Pohl, *Thermal conductivity of amorphous solids above the plateau*. Physical Review B (Condensed Matter), 1987. **35**(8): p. 4067-73.
2. Lubner, S.D., et al., *Reusable bi-directional 3ω sensor to measure thermal conductivity of 100-μm thick biological tissues*. Review of Scientific Instruments, 2015. **86**(1).
3. Natesan, H., et al., *A Micro-Thermal Sensor for Focal Therapy Applications*. Scientific Reports, 2016. **6**: p. 21395.
4. Klinker, L., et al., *Balloon catheters with integrated stretchable electronics for electrical stimulation, ablation and blood flow monitoring*. Extreme Mechanics Letters, 2015. **3**: p. 45-54.

ENHANCED HYPERTHERMIA DUE TO GOLD NANO-PARTICLES DURING MR-GUIDED HIGH INTENSITY FOCUSED ULTRASOUND (HIFU) ABLATION PROCEDURES

S. Devarakonda (1), M. Myers (2), C. Dumoulin (3), M. Lanier (3), R. Banerjee (1)

(1) Department of Mechanical Engineering, University of Cincinnati, Cincinnati, OH, USA

(2) Division of Applied Mechanics, CDRH, US FDA, Silver Spring, MD, USA

(3) Department of Radiology, Cincinnati Children's Hospital Medical Center, Cincinnati, OH, USA

INTRODUCTION

High intensity focused ultrasound (HIFU) therapy has increasingly gained interest as a novel tool for non-invasive ablation of tumors. HIFU temperature measurements can be performed using invasive modalities such as thermocouples (TCs), or by non-invasive modalities such as magnetic resonance (MR) thermometry. Direct measurement of temperature using TCs can lead to errors such as viscous-heating[1] artifacts and positioning errors caused by misalignment between the HIFU beam and the TCs. Alternatively, with the use of MR thermometry, TC artifacts can be avoided. Many researchers have conducted MR-guided HIFU ablation procedures on patients with breast cancer, abdominal tumors and were able to successfully ablate tumors using acoustic powers of greater than 60 W. However, the high acoustic powers used during HIFU procedures have led to collateral damages such as skin burns.

Absorption-enhancing agents such as nanoparticles, can be injected to the tumor site to achieve targeted heating of the tumors. Sun et al.[2], reported enhancement in hyperthermia during HIFU due to magnetic microcapsules (PLGA-coated Fe_3O_4). However, they used non-physiological size of microcapsules (500 – 800 nm) and high powers (180 W - 250 W) which can cause skin burns. Also, the MR thermometry images reported in this study have limited clarity regarding the temperature increase in the tumor.

Dibaji et al.[3] previously conducted HIFU ablations using 1% and 3% volumetric concentrations of magnetic nanoparticles (mNPs) embedded in tissue mimicking material (TMM). They measured the HIFU-induced temperature rise by focusing the beam on embedded TCs. They reported that the peak temperature rise increased upto 2 folds when mNPs were used, for an acoustic power of 14.2 W. However, the concentrations used were non-physiological and magnetic susceptibility of mNPs have been reported to interfere with Magnetic Resonance Imaging (MRI).

In the present study, the effect of gold nanoparticles (gNPs) on the HIFU-induced temperature rise has been assessed using MR thermometry. The TMM phantoms with physiologically relevant concentrations of 0% (control), 0.0625%, and 0.125% gNPs by volume were fabricated. Low acoustic powers - 10 W, 15 W, and 20 W - were used for the HIFU sonications. The temperature rises were calculated from the MR thermometry and were compared for different concentrations of gNPs.

METHODS

Fabrication of Tissue Phantoms. Three cylindrical fixtures with a length of 3 cm and inner diameter of 2.5 cm (volume ~ 15 cm^3) were developed (Fig. 1A). The protocol reported by King et al.[4] and Dibaji et al.[3] were used to fabricate TMM phantoms without and with gNPs, respectively. The size of the gNPs embedded in the TMM was 15 nm (US Research Nanomaterials, Inc., Houston, TX).

MR-HIFU System and Experimental Setup. A clinical MR-HIFU system (Sonalleve™ V2, Philips Medical Systems, Vantaa, Finland), integrated into a 1.5 Tesla (T) whole-body scanner (Philips Ingenia™, Healthcare, Best, The Netherlands) was used for scanning the phantom (Fig. 1B). An optimized resolution of 0.7 mm x 0.7 mm in the radial direction and a slice thickness of 6 mm were achieved using an imaging coil.

MR Imaging. After aligning the phantoms inside the imaging coil, MR images using T_1-weighted and T_2-weighted sequences were obtained. These images were obtained in both sagittal and coronal planes.

HIFU Sonication. Three acoustic powers of 10 W, 15 W, and 20 W were used for the HIFU sonication. A period of 16 sec was chosen for the HIFU sonication of the phantoms.

Temperature Measurements. During the sonication and cooling periods of 16 sec and 15 sec, respectively, MR thermometry was used to acquire the temperature data. The initial temperature of the phantoms and the surrounding medium was fixed at 37 °C. The optimized temporal resolution of the temperature measurements was 6 sec. The maximum temperature values discussed in the manuscript are the volumetric average of all the temperature values inside a pixel. Temperature measurements were recorded from three trials (n=3) for each acoustic power (m=3)

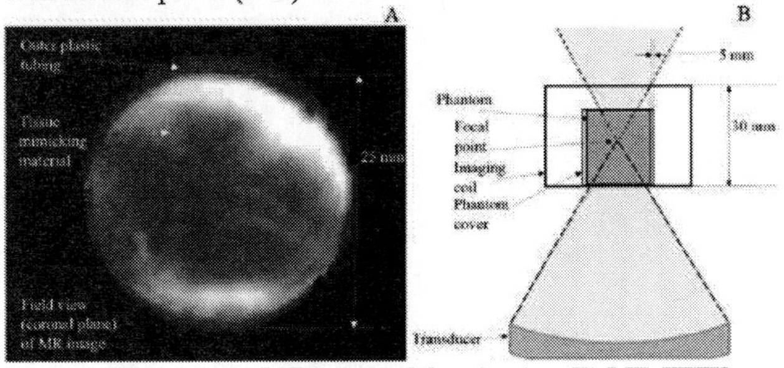

Figure 1: A) MR image of the phantom B) MR-HIFU experimental setup

RESULTS

The temperature maps (Fig. 2) show the variation of temperature for the concentrations of 0%, 0.0625%, and 0.125% gNPs and for the powers of 10 W, 15 W, and 20 W at the plane of maximum pixel temperature. A trending increase in temperature (increase in number of red pixels) can be seen in moving from left to right (increasing gNP) concentration, particularly at the two lower powers.

Figure 3 shows the comparison of maximum temperature rise, which occurs at the end of the 13-second sonication time. For a power of 10 W, the maximum temperature rises of 16±1 °C, 21±3 °C, and 23±3 °C were observed for the concentrations of 0%, 0.0625%, and 0.125% gNPs, respectively. The changes relative to 0% gNP's were 32% and 43% for the two phantoms with gNP's. For a power of 15 W, the maximum temperature rises of 25±0.7 °C, 29±4 °C, and 36±5 °C were observed for the concentrations of 0%, 0.0625%, and 0.125% gNPs. The increases relative to baseline (0% gNP) were 16% and 43% for the 0.0625% and 0.125% gNP values. For a power of 20 W, the maximum temperature rises of 34±3 °C (0% gNP), 43±0.4 °C (0.0625% gNP), and 46±0.7 °C (0.125% gNP). The increase relative to baseline (0% gNP) for the 0.0625% concentration was 24% and for the 0.125% concentration was 34%.

DISCUSSION

Temperature rise in the presence of gNP's is a linear function of power, as it is for no gNP's (Fig. 3). In that sense, the effect of incorporating gNP's in tissue volume can be modeled as a local increase in acoustic absorption.

The temperatures obtained from the MR thermometry are an average of all the temperatures with in a pixel. To estimate the effect of averaging, using the intensity profile provided by Wu and Du[5] (based on a Gaussian-beam assumption) was used. Averaging in the axial and radial directions result in a temperature error of approximately 4.2% and 1%, respectively. For non-Gaussian beams, the error could be higher.

Currently, researchers have used acoustic powers in excess of 60 W - 250 W to achieve cell necrosis during HIFU procedures[2]. Through the use of lower acoustic powers (e.g. 15 W) and shorter sonication times afforded by optimized concentrations of gNP's, undesired damage can potentially be reduced in a clinical setting. The gNPs can be made to accumulate at a desired location in the tumor tissue by injecting the nanoparticles directly into the tumor tissue intravenously. MR-HIFU thermometry can then be performed similar to the work presented in this study to evaluate the influence of gNPs.

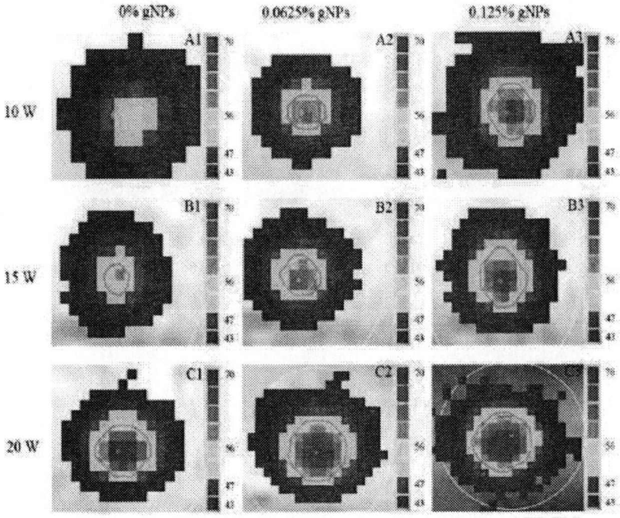

Figure 2: Representative temperature maps of TMM phantoms at maximum temperature with 0%, 0.0625%, and 0.125% gNPs concentration for the acoustic powers of 10 W, 15 W, and 20 W.

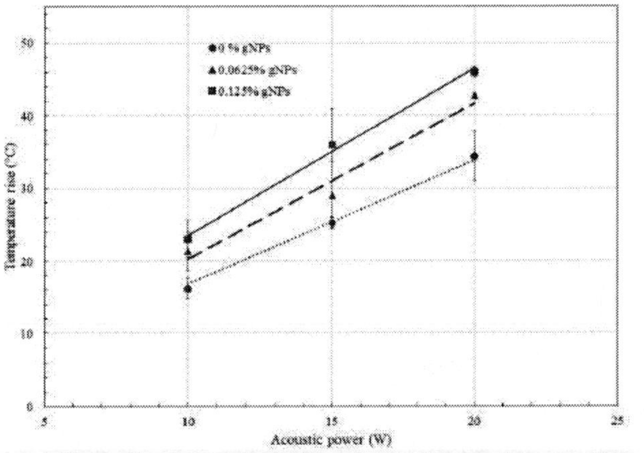

Figure 3: Comparison of maximum temperature rise (°C) for 0%, 0.0625%, and 0.125% gNPs concentration for powers of 10 W, 15 W, and 20 W.

ACKNOWLEDGEMENTS

Financial support from the National Science Foundation (Grant No. 1403356) is gratefully acknowledged.

REFERENCES

[1] Dasgupta, S. et al., *Ultrasonics* **51**, 171-180, 2011.
[2] Sun, Y. *et al. BMC cancer* **14**, 800, 2014.
[3] Dibaji, S. A. R. et al., *Journal of Nanotechnology in Engineering and Medicine* **4**, 2014.
[4] King, R. L. *et al. IEEE transactions on ultrasonics, ferroelectrics, and frequency control* **58**, 1397-1405, 2011.
[5] Wu, J. & Du, G. *Ultrasound in medicine & biology* **19**, 167-176 1993.

SB^{3}C2017
Summer Biomechanics, Bioengineering and Biotransport Conference
June 21 – 24, Tucson, AZ, USA

IN VIVO QUANTIFICATION OF BRAIN TISSUE DISPLACEMENT AND STRAIN USING CINE-DENSE MRI IN A HEALTHY SUBJECT AND A CHIARI MALFORMATION PATIENT

Soroush Heidari Pahlavian (1), Rouzbeh Amini (2), Xiaodong Zhong (3,4), John Oshinski (4), Francis Loth (1)

(1) Conquer Chiari Research Center, Department of Mechanical Engineering The University of Akron, Akron, OH, USA

(2) Department of Biomedical Engineering, The University of Akron, Akron, OH, USA

(3) MR R&D Collaborations, Siemens Healthcare, Atlanta, GA, USA

(4) Radiology & Imaging Sciences and Biomedical Engineering, Emory University, School of Medicine, Atlanta, GA, USA

INTRODUCTION

Chiari malformation (CM) is a neurological disorder that affects more than 300,000 people in the United Sates. This condition is conventionally characterized by cerebellar tonsil herniation (CTH) greater than 5 mm below the foreman magnum, as evident from static morphometric measurements performed on structural magnetic resonance images (MRI). However, several studies showed that a CTH does not necessarily correlate with CM symptoms severity and surgical outcome. As such, developing techniques to evaluate dynamic aspects of the central nervous system (CNS) such as brain deformation and strain can lead to a better understanding of CM etiology and symptomatology.

Cardiac-synchronized pulsatile motion of the brain has been documented in several studies and it is believed to be caused by the transmission of the arterial pulse into the cerebrovascular system [1, 2]. Investigations using various MRI techniques has demonstrated greater CNS tissue motion in CM patients compared to controls [3]. However, previous efforts were mainly focused on quantifying the bulk CNS tissue motion and the distribution of displacements and the resulting strain fields on neural tissues have not been investigated thoroughly.

In this pilot study, we used a cine displacement encoded stimulation echo (DENSE) MRI sequence to quantify tissue deformation and principal strain of the brain tissue in a healthy volunteer and a CM patient. Our goal in the present study, was to evaluate the feasibility of DENSE MRI for measuring deformation and strain on tissues with small motion and to evaluate this technique as a potential tool for obtaining a more dynamic picture of CNS alterations in CM patients.

METHODS

DENSE MRI was carried out, using a work-in-progress sequence, on an adult healthy volunteer and an adult patient diagnosed with CM. For each subject, mid-sagittal brain and cervical spinal cord MR images were localized using a rapid scout scan on a 1.5T whole body scanner equipped with a four-channel head coil and a single channel neck coil (Avanto, Siemens Medical Solutions, Germany). Two multiphase data sets were acquired, one for displacement encoding in the craniocaudal (head-foot) and one in the dorsal-ventral direction. Magnetic field shimming was done to correct the background phase error in DENSE measurements due to field inhomogeneity. The images were acquired for approximately 700 ms during each cardiac cycle after the R wave of ECG with a total of 13–16 cardiac phases. Other imaging parameters were: FOV = 225 – 262 x 300 – 350 mm, matrix = 192 x 256, slice thickness = 7 mm, temporal resolution = 34 ms, echo time = 1.9 ms, flip angle = 15°, and displacement encoding frequency = 1.5 cycle/mm. Subjects were imaged in the supine position with the neck in a neutral position.

An internally-developed MATLAB-based software was used to post-process and analyze the obtained DENSE images. The post-processing steps included: automatic spatial and temporal phase unwrapping [4], manual phase unwrapping (only when necessary), and region of interest (ROI) selection. Furthermore, sagittal 2D Eulerian displacement maps measured by the DENSE sequence were analyzed to quantify the CNS tissue deformation in terms of bulk displacement and motion distribution. Displacement fields were subsequently used to calculate principal Lagrangian Green strain [5], for seven different brain structures outlined in Figure 1.

RESULTS

Figure 1 shows the mean two-dimensional displacements of seven different brain regions as a function of time in the cardiac cycle for the healthy subject and the CM patient. Due to signal-to-noise ratio decay during the acquisition of cine DENSE, only the first two thirds of a cardiac cycle is considered to be reliable data [4]. The discarded portion of the cardiac cycles are shown as shaded regions in Figure 1. In both cases, the CNS tissues constituting the brain stem (midbrain, pons, and medulla) had the largest motion with peak displacement of 0.16 and 0.30 mm for the healthy subject and CM patient, respectively. The smallest intracranial displacement occurred in the occipital lobe with peak displacement of 0.027 and 0.028 mm for the healthy subject and CM patient, respectively. The displacements of intracranial tissues were generally larger in the CM patient compared to the healthy subject. This

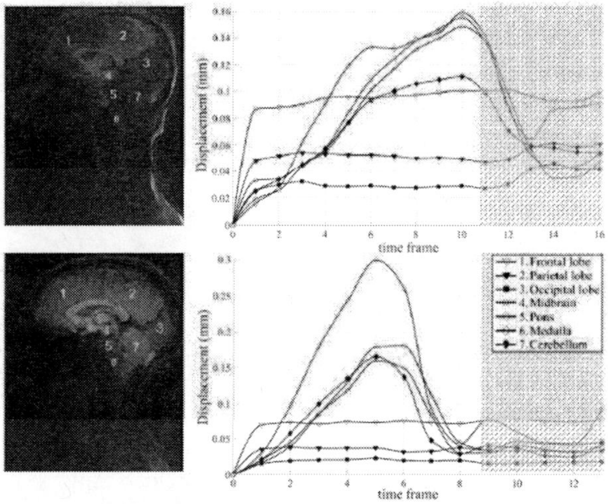

Figure 1: Mean displacement waveforms for different brain structures of the healthy subject (top) and CM patient (bottom).

discrepancy is observed to its greatest degree by examining medulla mean displacement, which was approximately twice as large in the CM patient compared to the healthy subject.

Figure 2 compares the distribution of the craniocaudal displacement and principal strain on the mid-sagittal slices at the time points corresponding to the peak displacement for each case. While the displacement of superior brain structures (frontal, parietal and occipital lobes) were comparable between healthy and CM cases, craniocaudal displacement of midbrain and cerebellum were noticeably greater in the CM patient compared to the healthy subject. The increase in displacement was most prominent on the cerebellar tonsils, as shown with the green arrow in Figure 2.

In both healthy and CM cases, the regions of high principle strain were observed near the boundaries of different brain structures as shown by the yellow arrows in Figure 2. These regions are not real representations of increased strain, as they demonstrate the separation of different structures and not displacement gradient in one structure. Furthermore, a concentrated region of high strain was observed near the cerebellar tonsils in the CM patient and not in the healthy subject (shown by the red arrow in Figure 2). The existence of this region was expected from the increased displacement gradient near the tonsils.

DISCUSSION AND CONCLUSION

In this study, we demonstrated the implementation of a DENSE MRI sequence to quantify CNS tissue displacement and strain and compared these parameters between a healthy subject and a CM patient.

While the measured displacements of brain tissues in the healthy subject agrees well with the other published DENSE measurement [6], previously reported CNS tissue displacements using other MRI techniques (0.17-0.43 mm for healthy and 0.37-0.57 mm for CM patients [2,3]) were markedly larger than the values measured in the current study. This observed discrepancy in CNS tissue motion values could be due to inter-subject variability and/or the difference in various MR modalities

The observed difference between the peak displacement of various brain structures agreed with the trend reported in previous studies on healthy subjects [2, 6]. Also, in accordance with a previous study, the largest motion in the brain stem of the healthy volunteer was observed approximately 330 ms after the R wave [6]. However, the peak

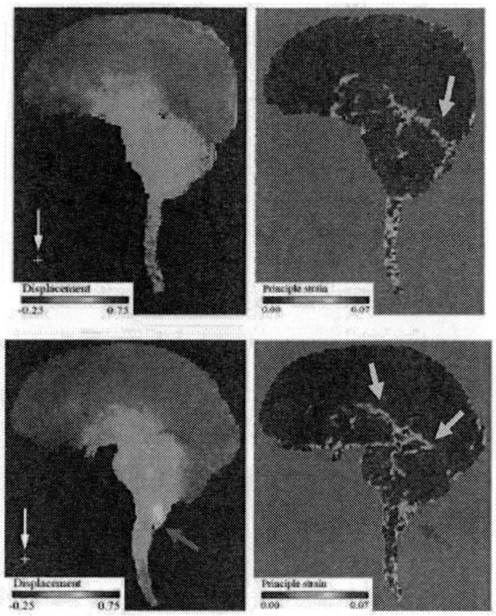

Figure 2: Distribution of peak craniocaudal displacement (left) principal strain (right) in the mid-sagittal slices in the healthy subject (top) and the CMI patient (bottom).

displacement was observed earlier in the CM patient (147 ms after the R wave). Future studies with larger healthy and CM patient populations can help to assess if this timing discrepancy is due to the inter-subject cardiac cycle variabilities or if it is the result of the blockage caused by the tonsillar descent.

While the exact reason for the increased CNS tissue motion in CM patients is not known, there are theories that relates this elevated motion in these patients to brain expansion during systole and the obstruction to the rapid CSF flow across the foramen magnum, caused by the CTH [7]. As the possible significance of mechanotrasduction on the CNS tissue has been investigated [8], the existance of the sustained increased pulsatile motion and strain on the surface of neural tissues can be important in explaining the symptoms experienced by CM patients.

In conclusion, we have demonstrated the feasibility of the cine-DENSE MRI as a tool for quantifying deformation and strain on tissues with small motion. We applied this technique to quantify brain tissue displacement and strain and compared these parameters in a healthy subject and a CM patient. More studies are required to assess if the increased strain observed in the present study is consistent among a larger CM patient population and if its existence correlates with the CM symptomatology before and after surgery.

ACKNOWLEDGEMENTS

The authors acknowledge Conquer Chiari for support of this work.

REFERENCES

[1] Alperin, N, et al., *Neurosurgery*, 75(5): 515-22, 2014.
[2] Enzmann D.R, et al., *Radiology*, 185(3): 653-60, 1992.
[3] Pujol, J, et al., *Neurology*, 45: 1746-1753, 1995.
[4] Spottiswoode B.S, et al., *IEEE Trans Med Img*, 3:7, 2006.
[5] Lake S.P, et al., *Ann Biomed Eng*, 39(7): 1891-1903, 2011.
[6] Zhong X, et al., *Med Phys*, 36(8): 3413-3419, 2009.
[7] Oldfield E.H, et al., *J Neurosurg*, 80(1): 3-15, 1994.
[8] Lee L, et al., *J Neurosci Res*, 91:1117–1132, 2013.

SB³C2017
Summer Biomechanics, Bioengineering and Biotransport Conference
June 21 – 24, Tucson, AZ, USA

CFD MODEL AND MRI MEASUREMENT OF INTRATHECAL CEREBROSPINAL FLUID DYNAMICS IN A CYNOMOLGUS MONKEY

Mohammadreza Khani (1), Tao Xing (2), Christina Gibbs (1), John Oshinski (3), Gregory R. Stewart (4), Jillynne R. Zeller (5), Bryn A. Martin (1)

(1) Department of Biological Engineering
University of Idaho, Moscow, ID, USA

(2) Department of Mechanical Engineering
University of Idaho, Moscow, ID, USA

(3) Department of Radiology
Emory University, Atlanta, GA, USA

(4) Alchemy Neuroscience
Hanover, MA, USA

(5) Northern Biomedical Research
Spring Lake, MI, USA

INTRODUCTION

Cerebrospinal fluid (CSF) plays a vital role in the immunological support, structural protection and metabolic homeostasis of the central nervous system (CNS). CSF is a clear fluid that surrounds the entire brain and spinal cord and has a viscosity approximately that of water. It moves in an oscillatory manner, with zero net flow, in synchrony with intracranial blood pulsations. A detailed understanding of CSF dynamics may improve treatment of several CNS diseases and help to optimize CSF system-based CNS therapeutics. The importance of CSF dynamics has been investigated in several CNS diseases including syringomyelia [1], Alzheimer's disease [2], Chiari malformation [3], and hydrocephalus [4]. Recent studies have examined the possible role of CSF as a conduit for distribution of therapeutic molecules to neuronal and glial cells of CNS tissues [5].

There is a need to develop a CSF hydrodynamic simulator with a realistic geometry and CSF flow distribution. Such a simulator will allow testing and optimization of delivery regimens for CNS therapeutics in development. Since these therapies often require testing on non-human primates (NHPs), our approach was to develop a subject-specific numerical model of CSF hydrodynamics in a cynomolgus monkey, a commonly used species for these studies. The focus of this numerical model was accurate representation of the spinal subarachnoid space (SAS), CSF flow rate, and waveform distribution as intrathecal infusion is primarily conducted within the spine.

METHODS

A healthy four-year-old adult male cynomolgus monkey with a weight of 4.39 kg was selected for the study. Total MRI imaging time was 81 minutes. High-resolution T2-weighted anatomic MR images (375 μm isotopic) were collected on a Philips 3T scanner. Phase-contrast MRI (PCMRI) measurements were collected with retrospective ECG gating and 24 heart phases were reconstructed over the cardiac cycle. Slice location for each scan was oriented perpendicular to the CSF flow direction with slice planes intersecting vertebral discs. These locations included the foramen magnum (FM) and vertebral disks located between the C2-C3, T4-T5, T10-T11 and L2-L3 vertebral levels (Figure 1a). CSF flow was quantified for each of the axial locations shown in Figure 1b. The T2-weighted images were semi-automatically segmented using ITK-snap software. The commercial finite volume computational fluid dynamic (CFD) solver ANSYS FLUENT V17.2 was used to solve the continuity and Navier-Stokes equations. To reproduce the non-uniform distribution of CSF flow along the spine (Figure 1b), a non-uniform deformation of the computational mesh was implemented at each time step. To verify our numerical results, independence studies were carried out to determine the effect of cycle, mesh size and time-step size on velocity results. Results were analyzed based on the second cycle with 11M cells and a 10 ms time-step size.

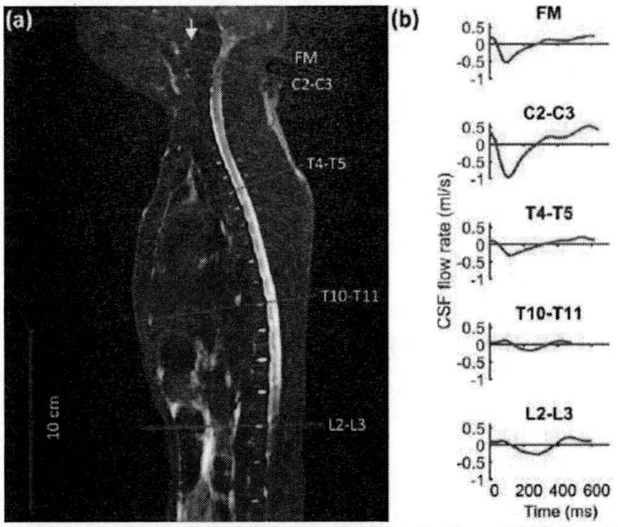

Figure 1: (a) MRI of the NHP spine (b) Distribution of CSF flow at different spinal levels based on the in vivo PCMRI measurements.

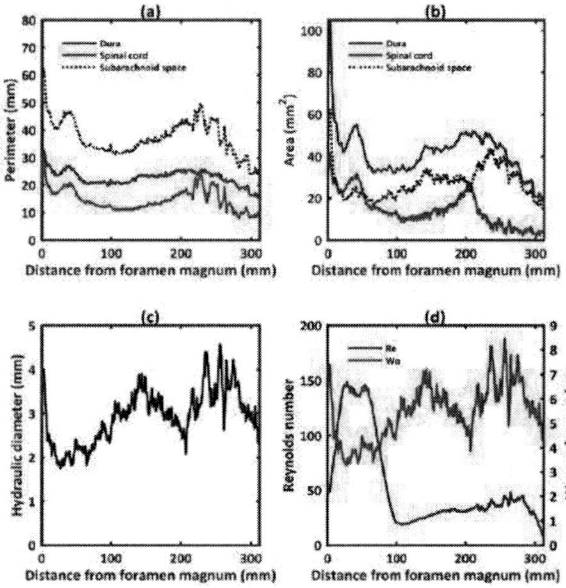

Figure 2: Hydrodynamic parameter distribution for the dura, spinal cord and subarachnoid space computed along the spine.

RESULTS

The following geometric parameters were calculated along the spine at 1 mm intervals (Figure 2) similar to our previous studies [6]: cross-sectional area and perimeter of SAS, spinal cord and dura, and axial distribution of hydraulic diameter, Reynold's number (Re) and Womersley number (Wo). Each of these parameters was calculated within a user defined function (UDF) compiled in ANSYS FLUENT after the computational mesh was formed. Total CSF volume within the SAS from the FM to spinal canal termination was 8.7 ml. For that same region, the spinal cord volume was 4.5 ml. Mean value for the cross-sectional area of the dura and spinal cord was 41.5 and 14.2 mm^2. Mean value of the dura and spinal cord perimeter was 22.7 and 14.5 mm.

Simulation results were compared to the in vivo measurements to help understand how well the simulation reproduced the in vivo CSF flow in terms of distribution and velocity profiles along the spine. Peak flow patterns were also assessed visually to understand any differences in flow fields (Figure 3). Comparison of CSF flow rate at the five MRI slice locations showed that MRI measurements were nearly identical to CFD results (max error < 2.87%).

Maximum peak and mean CSF velocities in the numerical model were present at C4-C5 level (Figure 3). Minimum value of peak and mean CSF velocities occurred in the lower lumbar spine and within the thoracic spine (~ T7-T10, Figure 3).

DISCUSSION

This study provides a methodology for subject-specific reproduction of the non-uniform distribution of CSF flow along the entire spine in a cynomolgus monkey using a non-uniform moving boundary approach. Results were verified by numerical independence studies and validated based on in vivo PCMRI measurements of CSF flow rates along the SAS.

The distribution of CSF flow along the spine is non-uniform and shows local variations in waveform shape and magnitude (Figure 1b and 3). Thus, we implemented a non-uniform moving boundary approach to accurately reproduce the subject-specific PCMRI measurements. This model represents the first validation of numerically modeled subject-specific CSF flow rates along the spine in a cynomolgus monkey.

Importantly, CSF dynamics quantified in the NHP showed similarity with CSF dynamics in humans with parallel changes in Reynold's and Womersley number along the spinal cord. These results serve to support use of the cynomolgus monkey as a preclinical subject for research and therapeutic development. In specific, average Reynolds and Womersley number along the spine were quantified to be ~69 and 10 in a human (unpublished abstract [7]). For the single NHP analyzed in our study, average Reynolds and Womersley number was 53 and 5.4 (Figure 2d). Peak Reynolds number was 150 at C4-C5 vertebral level in the NHP analyzed and 180 at C3-C4 in a human [7]. More research is needed to confirm consistency of these results across NHPs and compare them with humans in health and disease.

To our knowledge, axial variation in spinal SAS geometry in a cynomolgus monkey has not been reported in the literature (Figure 3a, b and c). Our results showed that total CSF volume in the spine was 8.7 ml. We saw similar trends in axial distribution of SAS cross-sectional areas in the cynomolgus monkey as in a human [7].

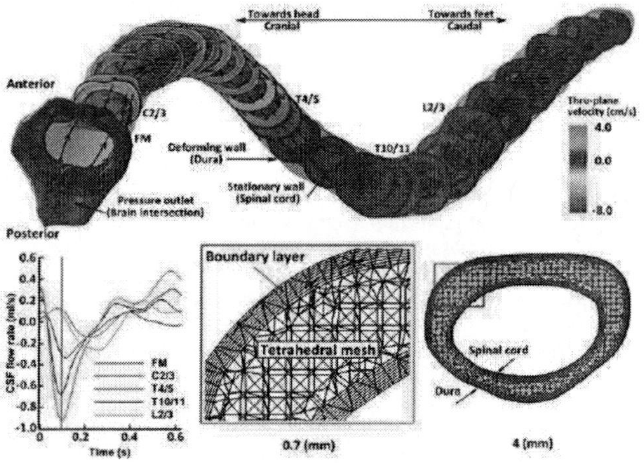

Figure 3: (Top) Peak-systolic thru-plane CSF velocity profiles and (Bottom left) CSF flow waveforms simulated by CFD. (Bottom right) Volumetric mesh visualization in the axial plane within the cervical SAS.

ACKNOWLEDGEMENTS

The authors would like to thank Voyager Therapeutics Corp., University of Idaho Vandal Ideas Project, and an Intuitional Development Award (IDeA) from the National Institute of General Medical Sciences of the National Institutes of health under Grant #P20GM1033408 for funding that supported this study.

REFERENCES

[1] Martin, B. A et al., *Journal of Biomechanical Engineering-Transactions of the Asme,* vol. 132, Nov 2010.

[2] Wostyn, P et al., *Cerebrospinal fluid research,* vol. 6, p. 1, 2009.

[3] Bunck, A. C et al., *European radiology,* vol. 22, pp. 1860-1870, 2012.

[4] Bradley, W. G et al., Jr, *Radiology,* vol. 198, pp. 523-529, 1996.

[5] Papisov, M. I et al., *Mol Pharm,* Feb 12 2013.

[6] Martin, B. A et al., *PLoS One,* vol. 8, p. e75335, 2013.

[7] Sass, L, Khani, M et al. et al., *SB3C Conference Abstract*, 2017.

SHEAR WAVE ELASTOGRAPHY FOR ASSESSING MYOCARDIAL MATERIAL PROPERTIES: AN IN VITRO, EX VIVO AND IN SILICO STUDY

Annette Caenen (1), Mathieu Pernot (2), Abdullah Thabit (1), Darya Shcherbakova (1), Luc Mertens (3), Abigail Swillens (1), Patrick Segers (1)

(1) IBiTech-bioMMeda
Ghent University
Ghent, Belgium

(2) Langevin Institut
Ecole Supérieure de Physique et de Chimie Industrielles
Paris, France

(3) Hospital for Sick Children
University of Toronto
Toronto, Canada

INTRODUCTION

Diastolic (dys)function is a major determinant of surgical decision making and clinical outcome in children with cardiac disorders. However, current diagnostic tools are not ideal as these measure surrogates for ventricular stiffness and compliance, and are either invasive or based on adult-specific guidelines [1].

Shear Wave Elastography (SWE) is a highly promising method for non-invasive assessment of cardiac muscle stiffness [2], as it derives stiffness from the propagation characteristics of the excited shear waves (SWs). We here focus on ultrasonic SWE with impulsive SW excitation, realized through a focused ultrasound beam generated by a conventional transducer. This technique has received wide recognition as a reliable, quantitative imaging tool for evaluating tissue stiffness in unbounded media, such as breast and liver tissue [3, 4]. However, its feasibility for evaluating cardiac stiffness remains to be investigated, as complex SW propagation (e.g. wave guiding, mode conversions, dispersion) is expected to arise due to the myocardium's anisotropic material properties, and ventricular layered thin geometry.

The objective of this work was to investigate the feasibility of SWE in determining cardiac material properties by means of a combined experimental and modeling approach. More specifically, we investigated the effect of ventricular geometry and myocardial anisotropy on SW physics and characterization.

METHODS

Effect of ventricular geometry: We studied the effect of the curved and thin walled ventricular geometry on SW patterns and characterization through a generic left ventricular (LV) model tailored to 10-15 years old adolescents [5]. This model is replicated in (i) polyvinyl alcohol (PVA) for experimental purposes, to investigate the performance of different material characterization techniques and as validation for the numerical results, and (ii) in the finite element (FE) software Abaqus (Abaqus Inc., Providence, USA) for numerical purposes, to gain more profound insights into the complex SW physics (see fig. 1a). PVA's stiffness was obtained via uniaxial mechanical testing, and served as a reference for our material model in Abaqus.

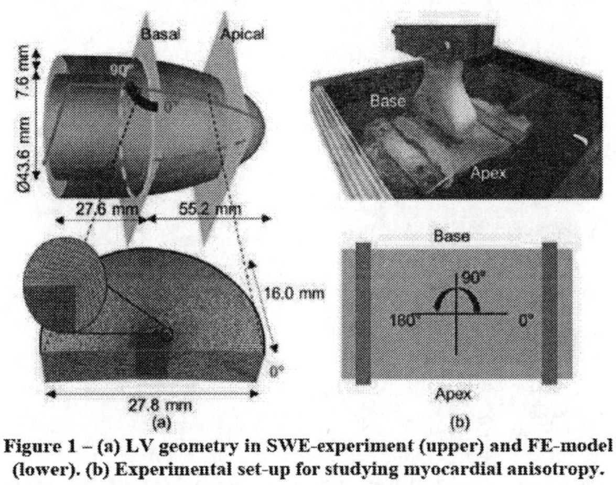

Figure 1 – (a) LV geometry in SWE-experiment (upper) and FE-model (lower). (b) Experimental set-up for studying myocardial anisotropy.

For the SWE experiments, SW's were excited in the submerged PVA phantom via the Aixplorer system (SuperSonic Imagine, Aix-en-Provence, France) using the linear SL15-4 probe. This was done for two different cardiac zones (basal and apical) using four different probe orientations (0°, 30°, 60° and 90°). The obtained axial particle velocities were post-processed with two material characterization methods: (i) a time-of-flight (TOF) method [6], conventionally used for bulk media without dispersion, tracking the SW's position in time to estimate SW speed and subsequently shear stiffness and, (ii) phase velocity analysis [7], standardly applied for media with dispersion, estimating shear modulus through a theoretical fit to the measured frequency spectrum of the SW (via the 2D Fast Fourier Transform).

For the modeling part, we replicated numerically the experiment on the basal zone of the heart and 0° probe orientation. The acoustic pressures of the pushing beam were modeled in the ultrasound simulation software Field II [8]. These pressures are then converted into a time-averaged body force, which is applied on the numerical viscoelastic phantom for the experimental push duration (250 μs). With this model, we studied the effect of tissue surroundings (water) and acoustic loading (interface pressure) on SW physics in the LV.

Effect of myocardial anisotropy: The influence of myocardial anisotropy on SW characteristics was investigated by performing SWE experiments on a pig heart, obtained from a slaughterhouse. A rectangular slab was dissected from the left ventricle by cutting along the interventricular septum. This slab was uniaxially pre-stretched (5%, 10 % and 15%) during SWE acquisitions for 19 different probe directions – 0° to 180° in steps of 10° (see fig. 1b). Assuming the myocardium consists of transversely isotropic layers, the fiber orientation was estimated for each of these layers based on the angle of the tilted major axis of the ellipse fitted to the wave surface [9]. For validation purposes, the myocardial fiber architecture was also derived through a visual image analysis of the sliced myocardium [10].

RESULTS

Effect of ventricular geometry: The modeled SW propagation corresponded well with the experiment (fig. 2a), especially when the surrounding water and the surface radiation force, present at the phantom-water interfaces during SW excitation, were accounted for. In both cases, clear dispersion was present as the SW front has split into two (see arrows in fig. 2a). For stiffness characterization (fig. 2b), the TOF-method underestimated (11.6- 20.8 kPa for all acquisitions) actual stiffness of 24.3 kPa whereas the phase velocity analysis or c_ϕ-method slightly overestimated (21.1-31.2 kPa for all acquisitions).

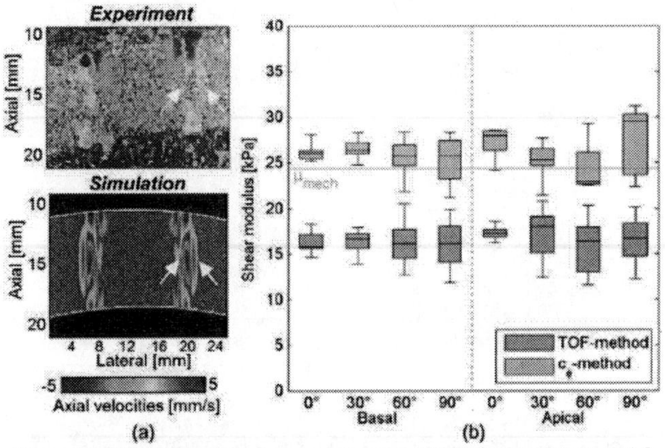

Figure 2 – Effect of ventricular geometry: (a) Axial velocity pattern at 1.8 ms, (b) SWE-derived stiffness vs. mechanically determined stiffness (μ_{mech}).

Effect of myocardial anisotropy: The resulting fiber orientation across myocardial thickness for 5% stretch is illustrated in fig. 3a. The SWE-derived fiber architecture corresponded well to the image based analysis (fig. 3b). Fig. 3b also shows a clear alignment of the myocardial fibers when stretching the LV slab.

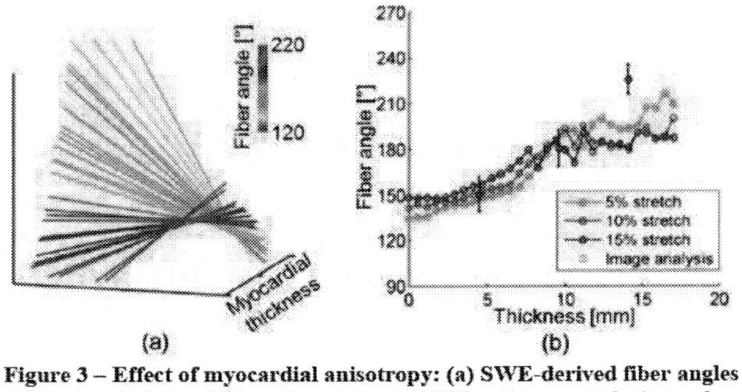

Figure 3 – Effect of myocardial anisotropy: (a) SWE-derived fiber angles for 5% stretch, (b) Fiber orientation via SWE vs. image analysis results.

DISCUSSION

SWE experiments and simulations both improved our understanding of the factors influencing SW propagation and characterization in the LV setting, providing insights in the potential applicability and accuracy of the method in vivo.

In vitro assessment of shear moduli estimators revealed that phase velocity analysis better estimated true stiffness than the TOF-method, independent of the selected cardiac zone and probe orientation (representing LV settings with varying thickness and curvature). This demonstrates the importance of using a material characterization algorithm for dispersive wave regimes, instead of the machine-implemented method for non-dispersive SWs. This conclusion is in line with recent results from Maksuti [11] on models of arteries. Additionally, we presented the potential of SWE in non-invasively extracting myocardial fiber orientation during uniaxial loading, which is corresponding with previous studies [12]. SWE offers the advantages of using low cost ultrasound and acquiring data in a short time, making it an attractive alternative to the well-established Diffusion Tensor MRI [13] for visualizing the myocardial fiber architecture.

We conclude that the ventricular geometry and anisotropic composition of the myocardium pose extra challenges to the use of SWE-based characterization of myocardial stiffness in the clinical setting. Therefore, an important role for FE-modeling is foreseen to ultimately improve these soft tissue characterization techniques.

ACKNOWLEDGEMENTS

This work was supported in part by the Flemish government agency for Innovation and Entrepreneurship (VLAIO), in part by the Research Foundation Flanders (FWO), and in part by the committee for scientific research at Ghent University (CWO).

REFERENCES

[1] Dragulescu, A. et al., *Circ. Cardiovasc. Imaging*, 6:254-261, 2013.
[2] Couade, M. et al., *IEEE Trans. Med. Imaging*, 30 :295-305, 2011.
[3] Athanasiou, A. et al., *Radiology*, 256 :297-303, 2010.
[4] Palmeri, M. et al., *Ultrasound Med. Biol.*, 34 :546-558, 2008.
[5] Caenen, A. et al., *IEEE Trans. Ultrason. Ferroelect. Freq. Control*, accepted for publication.
[6] Couade, M. et al., *Ultrasound Med. Biol.*, 36 :1662-1676, 2010.
[7] Bernal, M. et al., *J. Acoust. Soc. Amer.*, 129 :1344-1354, 2011.
[8] Jensen, J. A., *Proc. IEEE 10th Nordic-Baltic Conf. Biomed. Imaging.*, 34:351-353, 2011.
[9] Wang, M. et al., *IEEE Trans. Med. Imaging*, 32:1671-1684, 2013.
[10] Sommer, G. et al., *Annals of Biomed. Eng.*, 43:2334-2348, 2015.
[11] Maksuti, E. et al., *Ultrasound Med. Biol.*, 42:308-321, 2016.
[12] Lee, W.-N. et al., *IEEE Trans. Med. Imaging*, 31:554-562, 2012.
[13] Tseng, W. Y. I. et al., *J. Mag. Reson. Imaging.*, 17:31-42, 2003.

QUANTIFICATION OF THE EFFECT OF CALCIFICATION ON THE TISSUE – STENT INTERACTION IN A STENOSED AORTIC ROOT

Orla M. McGee(1), Paul S. Gunning (1), Wei Sun (2), Laoise M. McNamara (1)

(1) Biomechanics Research Centre, Biomedical Engineering, College of Engineering & Informatics, National University of Ireland Galway

(2) Wallace H Coulter Department of Biomedical Engineering, Georgia Institute of Technology

INTRODUCTION

Transcatheter Aortic Valve Implantation (TAVI) is a minimally invasive alternative to open heart surgery in the treatment of aortic stenosis. TAVs are mounted onto a collapsible valve stent and delivered to the stenotic valve in a crimped configuration via a catheter. Unlike surgical valves, TAVs are not sutured into place. Thus, the tissue-stent interaction has a great impact on the successful deployment and post-operative device performance. It has previously been shown that excessive radial force can lead to aortic root rupture [1], whereas inadequate radial force can lead to problems such as migration of the stent into the left ventricle [2]. However, the interaction between the TAV and the calcified native valve is not yet fully understood. The objectives of this study were to (1) develop an in vitro experimental model that could represent the in vivo aortic root with calcifications, (2) apply this model with an ultrasound approach to investigate the diameter and determine the force required to pullout the stent, (3) determine the radial expansion force of the stent at a specific diameter, (4) determine the coefficient of friction between the stent and aortic root and finally (5) incorporate this information into a computational model to investigate the stress distribution on the aortic wall following expansion of a TAVI valve into a calcified aortic root.

METHODS

In vitro experimental model representing the in vivo aortic root with calcifications

Artificial calcifications were prepared to mimic the properties of native calcified leaflets using bone glue mixed with water (1:7) with 0.25g hydroxyapatite and 1.75g calcium carbonate added [1]. The calcifications created in this study have Hounsfield units: 535+ which was measured using micro-CT. Calcification fragments were placed on the aortic heart valve of porcine hearts with a 21mm diameter aortic root (Fig.1(c)).

Root – Stent Interaction Experiment

(A) Stent Radial Expansion Experiment

Based on established techniques a stent crimp experiment was performed on the nitinol braided stent [2], to determine the radial expansion force of the stent. A 0.0225" nitinol wire was used to create a 27 mm diameter braided stent. The test was performed by wrapping a Dacron strap around the stent (Fig. 1(a)). The fixture displacement was used to measure the change in the stent circumference and stent diameter. The hoop force exerted from the strap was measured by a load cell that was attached to the uniaxial machine. All experiments were conducted in a 37°C water bath. The hoop force was then used to calculate the radial expansion force using the equation [3]:

$$F_{radial} = 2.\pi F_{hoop} \qquad (1)$$

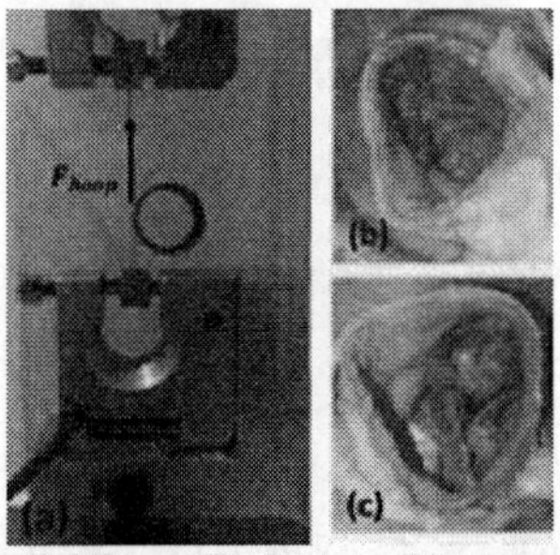

Figure 1 (a) Radial Force Test set up (b) non-calcified and (c) experimentally calcified aortic root.

(B) Root – Stent Interaction Experiment: The stent was then experimentally implanted into the aortic root. The stent was post dilated to ensure that the stent was under the same loading conditions as the experimental test, so as not to assume that the properties of nitinol are symmetric in tension and compression [2]. An ultrasound probe was used to take 2D cross-sectional images and measure the diameter of the stent at the top, middle and bottom (Fig. 2(b)). The radial force was obtained from the experimental results of (A), based on the final diameter of the stent. The pull out force of the stent was measured. This was done by dissecting off the apex of the heart and attaching stiff strings, which were also attached to the end of the stent, to a uniaxial test machine. The force required to pull out the stent was then measured ($F_{pullout}$). Using Amontons Coulombs friction law [2, 3], $F_{pullout}$ can be defined by:

$$F_{pullout} = \mu F_{radial} \quad (3)$$

Using the equation for the *in vivo* force [2]:

$$F_{invivo} = \Delta P_{stent} \pi \left(\frac{D_{max}}{2}\right)^2_{stent} \quad (4)$$

we can determine whether the stent would migrate into the left ventricle at physiological pressure if $F_{pullout}$ is < F_{invivo} [2].

Computational Methods

Aortic Root and Commercial Stent Models: In this study, a patient-specific aortic root geometry was reconstructed from Multi-Slice Computed Tomography images of a 90-year-old female with severe aortic stenosis, with a slice thickness of 0.625 mm, dimensions of 512 x 512 and pixel spacing of 0.5 mm. Mimics 14.1 Imaging Software (Materialise, Leuven, Belgium) was used to threshold the images. The aortic root, leaflets and calcifications were then imported into 3Matic 8.0 (Materialise, Leuven, Belgium). Volume meshes of the aortic root and calcifications were generated using 3D continuum elements (C3D4R). The leaflets were meshed using shell elements with a thickness of 0.384 mm [4]. The geometries were then imported into ABAQUS Explicit 6.13 (SIMULIA, Providence, RI). A braided stent geometry was created using SolidWorks (Dassault Systèmes, Fr) and imported into Abaqus 6.13 as a wire part and meshed using 3-noded quadratic beam elements (B32) and spring connector elements.

Constitutive Models: The braided stent was modelled as a superelastic material using an inbuilt user subroutine (VUMAT). The ascending aorta, the aortic sinus and the leaflets of the aortic valve were modelled as isotropic hyperelastic materials using the first order Ogden model, fitted to uniaxial test data of human tissue [5, 6]. The calcifications were assumed to be isotropic, homogenous and hyperelastic, based on the properties atherosclerotic plaque [7], and thus modelled using a third order Mooney-Rivlin material model.

Boundary and Loading Conditions: The crimping and deployment of the stent were modelled in ABAQUS Explicit 6.13 (SIMULIA, Providence, RI). The stent was crimped inward radially before being deployed by simulating unsheathing of the valve into the aortic root. During stent deployment, a coefficient of friction of 0.1 or 0.15 was applied between the TAV stent (master) and the aortic root (slave) [2] based on the experimental results of Section 2.2. The top and bottom edges of the aortic root were constrained using non-displacement boundary conditions in the longitudinal and circumferential directions.

RESULTS

Experimental: Reducing the crimp diameter from 27 to 21 mm resulted in a 91.1 N increase in the radial force, see Fig. 2(a). The maximum and minimum diameters of the stent deployed into 21 mm porcine hearts were measured (23.5 ± 0.09 mm and 21.7mm ± 0.13 respectively) and the eccentricity of the stent varied from 0.06 to 0.10. The radial force exerted by the stent on the aortic root was 95.54 ± 1.98N. The pullout force of the stent was 14.5 ± 1.2 N. From Eq.3 a coefficient of friction (μ) between the stent and the aortic root of 0.15 ± 0.01 was calculated. At a 100 mmHg pressure gradient the maximum F_{invivo} was calculated as 6.05 N.

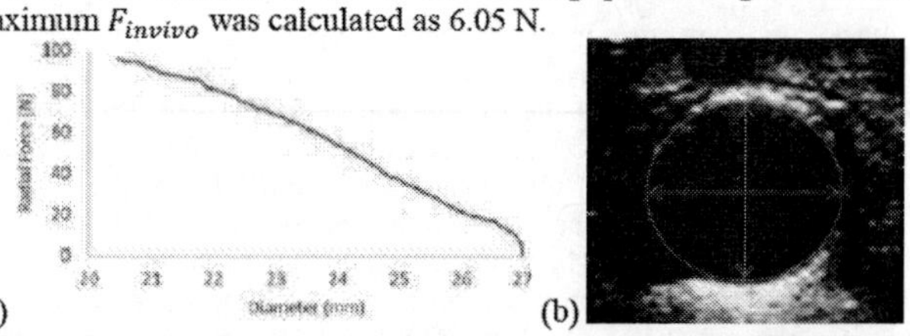

(a) (b)

Figure 2 (a) Radial outward force (N) versus diameter (mm) for the stent (b) Ultrasound measurements of deployed stent.

Computational: The maximum von Mises stresses at the 99th percentile (σ_{99}) were examined to exclude potential outliers due to eschewed elements. For the aortic tissue the σ_{99} predicted was 377kPa for a coefficient of friction of 1.15 versus 326.8kPa for a coefficient of friction of 0.1 (Fig. 3). . The computational models predicted such changes in the interaction properties between the aortic root and the stent can lead to a 15.58% increase in the peak von Mises stresses in the aortic root.

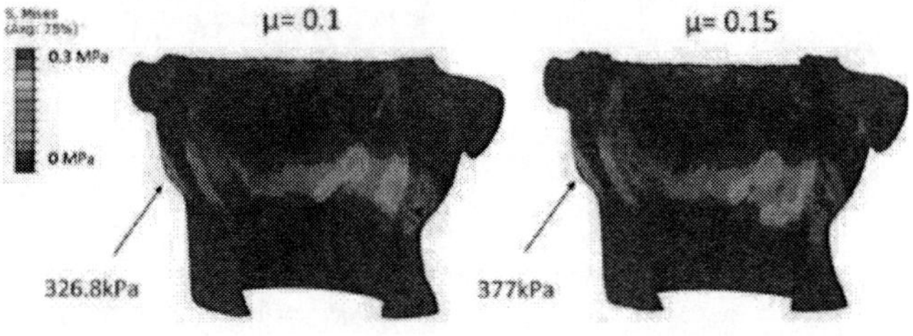

Figure 3: Von Mises stress in (a) non-calcified and (b) calcified aortic root models

DISCUSSION

This study developed an *in vitro* experimental model to represent a calcified aortic root *in vivo*. Using this model, the coefficient of friction in the presence of calcification was determined and was found to be substantially increased (0.151±0.1) compared to that reported for non-calcified valves (0.1±0.01) [2]. Moreover, the range of eccentricity of the stent increased (0.06-0.10), compared to that reported for non-calcified roots (0-0.03) [2]. The pullout force required to dislodge the stent experimentally, $F_{pullout}$ (14.5 ± 1.2 N), was 2.4 times higher than the predicted *in vivo* force, F_{invivo} (6.05 N), which implies that the stent would be unlikely to migrate into the left ventricle *in vivo*. This information was incorporated into a computational model to investigate the stress distribution on the aortic wall following the expansion of a TAVI valve into a calcified aortic root. It was predicted that the change in coefficient of friction resulted in a 15.58% increase in aortic root tissue stress. High stresses have the potential to lead to the compression of the left bundle branch or AV node which has been shown to increase the risk of interferences in the conductance system, leading to the need for pacemaker implantation [8]. In conclusion, we have developed the first in vitro model of aortic calcification and have shown that calcification impacts both the radial stiffness and the friction between the aortic tissue and TAV stent. Computationally we have shown that this can lead to changes in the stress experienced by the aortic tissue and the valve stent. The change in the tissue stent interaction is something that should be considered in future computational models. Furthermore, the impact of calcification and the degree of calcification of a given patient is something that should be further considered in device selection and design.

ACKNOWLEDGEMENTS

This study was funded by an Enterprise Partnership Scheme Postgraduate Scholarship and a NUI travelling studentship.

REFERENCES

[1] D. Wendt, *et al.*, *Eur J Cardiothorac Surg*. 35: 622-7; 27, 2009.
[2] J. Mummert, *et al.*, *Ann Biomed Eng*. 41: 577-86, 2013.
[3] S. Vad, *et al.*, *J Biomech Eng*. 132: 121007-07, 2010.
[4] P. N. Jermihov, *et al.*, *Cardiovasc Eng Technol*. 2: 48-56, 2011.
[5] Q. Wang, *et al.*, *J Biomech*. 45: 1965-71, 2012.
[6] C. Martin, *et al.*, *Eur J Cardiothorac Surg*. 40: 28-34, 2011.
[7] H. M. Loree, *et al.*, *J Biomech*. 27: 195-204, 1994.
[8] C. Fraccaro, *et al.*, *Am J Cardiol*. 107: 747-54, 2011.

INFLUENCE OF OPTIC NERVE HEAD MATERIAL PROPERTIES ON RAT OPTIC NERVE STRAINS DUE TO ELEVATED INTRAOCULAR PRESSURE

Stephen A. Schwaner (1), Marta Pazos (2,3), Hongli Yang (3), Elaine C. Johnson (4), John C. Morrison (4), Claude F. Burgoyne (3), C. Ross Ethier (1,5)

(1) George W. Woodruff School Mechanical Engineering
Georgia Institute of Technology
Atlanta, GA, USA

(2) Ophthalmology Department
Hospital de l'Esperanca-Parc de Salut Mar.,
Institut Mar d'Investigacions Mèdiques
Barcelona, Spain

(3) Optic Nerve Head Research Laboratory
Devers Eye Institute
Portland, OR, USA

(4) Casey Eye Institute
Oregon Health and Science University
Portland, OR, USA

(5) Coulter Department of Biomedical Engineering
Georgia Institute of Technology/Emory
Atlanta, GA, USA

INTRODUCTION

Glaucoma is the second leading cause of blindness worldwide, and is characterized by the death of retinal ganglion cells (RGC) which carry vision information to the brain [1]. Elevated intraocular pressure (IOP) is a key risk factor in glaucoma and is thought to contribute to RGC axon insult by inducing stress and strain in the tissues of the optic nerve head (ONH). The ONH is a region in the posterior eye through which RGC axons pass as they exit the eye and converge to form the optic nerve. Specifically, in humans and primates, these axons pass through a web of connective tissue called the lamina cribrosa (LC) (Fig. 1), which spans a canal in the sclera (connective tissue shell of the eye). Data suggest that the ONH is the initial site of axonal damage in glaucoma and that ONH biomechanics is important to the disease process [2], motivating previous modeling studies of the human ONH [3]. However, there is a need to better understand ONH biomechanics in animal models used in glaucoma research; such models offer a powerful tool for determining the cellular mechanisms by which biomechanical insult leads to RGC death.

The rat is a well-accepted and widely-used animal model for glaucoma research. Studies using the rat are based on IOP elevation and observation of resulting biological changes in the ONH. However, there are significant structural differences between the rat and human ONH which likely affect rat ONH biomechanics (Fig. 1), including: 1) the lack of a collagenous LC spanning the rat neurovascular scleral canal, 2) a perineural vascular plexus (PNVP) surrounding the nerve within the sclera, 3) close apposition of Bruch's membrane (BM) to the superior nerve (creating a BM overhang) and 4) a separate, inferior scleral opening for the central retinal artery (CRA) and posterior ciliary branches. We developed an individual-specific finite element (FE) model of a normal rat ONH to investigate nerve strain levels and patterns under elevated IOP using newly available 3D reconstructions of 8 rat ONHs [4]. We here use this model to perform a sensitivity study, investigating the influence of different tissue stiffnesses on optic nerve strains.

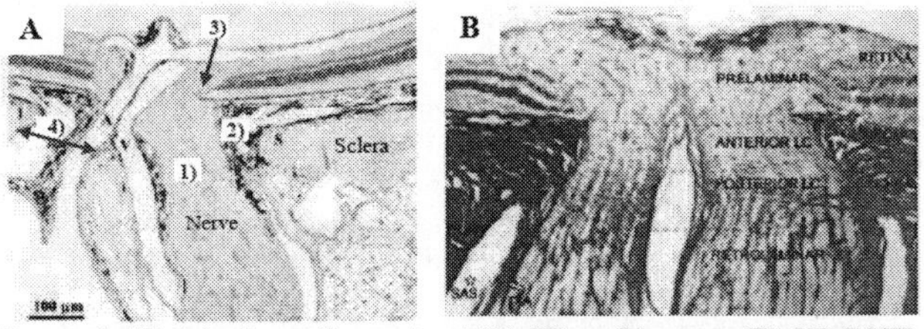

Figure 1: Histologic sections of rat (A) [5] and human (B) [6] ONH illustrating key differences 1)-4) as in text. In (A): superior (S), inferior (I), artery (A), vein (V). In (B): subarachnoid space (SAS).

METHODS

A FE model was constructed from a digital 3D histomorphometric reconstruction of a rat ONH created as follows. After perfusion fixation, the ONH from a normotensive rat eye was embedded in paraffin and serial sectioned. After each cut, the block face was stained and photographed. All images were aligned and stacked into a digital 3D reconstruction [4].

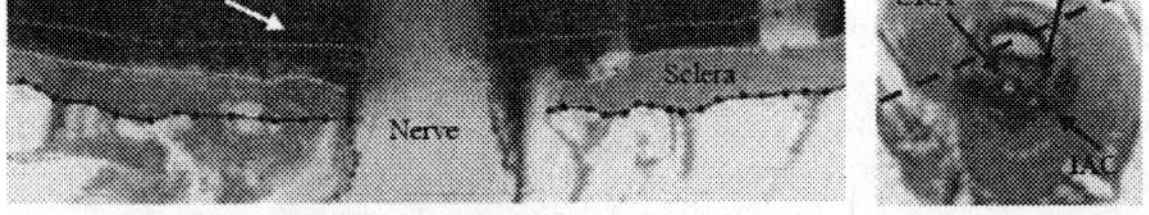

Figure 2: Radial (left) and transverse (right) digital sections of a rat ONH with manual delineations of tissues (colored dots). Dotted line illustrates plane of radial section. See text for abbreviations.

Digital radial and transverse cross sections of the reconstruction were viewed in order to manually place points along borders of key ONH tissues (Fig. 2). Rhino (Robert McNeel & Assoc.) was used to create tissue volumes from the resulting 3D point clouds. Modeled tissues included BM, choroid, sclera, nerve, pia mater, CRA, CRV, PNVP, and IAC.

Tissues were modeled as isotropic, nearly incompressible (υ = 0.49) neo-Hookean solids, and assigned baseline stiffness values based on previous modeling studies in the human ONH [3] and values from the literature [7, 8], as rat-specific tissue properties are not available. The model was meshed using Trelis Pro (csimsoft LLC) with approximately 857,500 linear, hybrid, reduced-integration elements of wedge, hexahedral, and tetrahedral shapes. To simulate IOP, pressures were applied to the model's anterior surface which included anterior surfaces of BM, the nerve, CRV, and CRA. Because the eyes were perfusion-fixed at an IOP of 10 mmHg, we applied a pressure of 20 mmHg in order to simulate ONH conditions at an IOP of 30 mmHg. Arterial and venous blood pressures were applied to the lumen walls of the CRA and CRV. We used a submodeling approach to apply boundary conditions. Briefly, a half-sphere generic posterior eye (GPE) model with simplified ONH anatomy was modeled. Displacements from matching nodes in the GPE were applied as boundary conditions to the eye-specific model edges. Models were analyzed using Abaqus.

We investigated the influence of sclera, pia mater, and PNVP stiffnesses on optic nerve strains by selecting a "baseline", "soft" and "stiff" modulus value for each tissue [3], and individually varying each stiffness, producing one model with baseline values and 6 models with softer or stiffer modulus value for the sclera, pia, or PNVP (Table 1).

Table 1: Soft, baseline, and stiff elastic modulus values (MPa) for tissues considered in sensitivity analysis.

	Baseline	Soft	Stiff
Sclera	3.0	1.0	9.0
Pia	3.0	1.0	9.0
PNVP	0.1	0.03	0.3

RESULTS

1st and 3rd principal strain values were quantified (Table 2) for the anterior nerve region, defined to extend from BM to 150 µm posterior to BM (about even with the posterior scleral surface). Baseline model nerve strain values were within ranges previously seen in modeling studies of the human eye performed at a higher IOP [3]. This difference is expected due to the rat's lack of structural support from a collagenous LC. Strain concentrations were seen in the inferior-posterior region of the anterior nerve, the inferior-anterior region where the CRA and CRV contact the nerve (Fig. 3), and on the temporal and nasal sides of the nerve at the level of BM. Like the human ONH, changes in scleral stiffness had the largest effect on nerve strains. Interestingly, lower PNVP stiffness led to lower optic nerve strains.

DISCUSSION

We created an individual-specific FE model of the rat ONH and investigated strain patterns in the optic nerve under elevated IOP. Our analysis shows that, unlike the human optic nerve, rat optic nerve strains are very asymmetric. Interestingly, small regions of strain concentration were observed near the temporal and nasal edges of the BM overhang on the superior side of the nerve. We were surprised that there were not more pronounced strain concentrations here because we have observed a higher susceptibility to axonal damage in the superior nerve region in previous rat glaucoma studies [5]. This result may change in future modeling studies that include different ONH geometries, more fidelic material properties, and different combinations of material stiffnesses.

We also investigated the sensitivity of optic nerve strains to changes in sclera, pia, and PNVP stiffness values, finding that, like the human ONH, scleral properties are highly influential on optic nerve strains, with a stiffer sclera resulting in lower nerve strains. Lowering PNVP stiffness was shown to decrease nerve strains, suggesting that the PNVP may act as a "buffer tissue" to decrease the amount of load transferred to the pia and optic nerve from the scleral wall.

Table 2: Anterior nerve region strain values (logarithmic).

	1st principal strain		3rd principal strain	
	Mean	95th percentile	Mean	5th percentile
Stiff sclera	2.3%	3.7%	-2.9%	-5.0%
Stiff pia	3.0%	4.7%	-3.4%	-5.7%
Soft PNVP	3.6%	5.4%	-4.1%	-6.6%
Baseline	**4.1%**	**6.0%**	**-4.7%**	**-7.4%**
Stiff PNVP	4.5%	6.5%	-5.3%	-8.2%
Soft pia	5.2%	7.2%	-5.9%	-9.0%
Soft sclera	8.2%	11.8%	-10.0%	-14.5%

In future work, we will address several limitations of the model presented here. 1) Tissue biomechanical properties specific to the rat will be obtained and used. 2) As our analysis suggests that scleral properties are very influential, we will attempt to more closely match true scleral behavior by modeling it as a fiber-reinforced matrix. 3) We will model additional rat ONH geometries to evaluate biomechanical effects of rat-to-rat ONH anatomic variation. After determining if the reported strain patterns and levels hold in different rat ONHs and in models with improved material properties, we will relate strain patterns to regional patterns of biological response in rat glaucoma models, to better understand RGC pathogenesis in glaucoma.

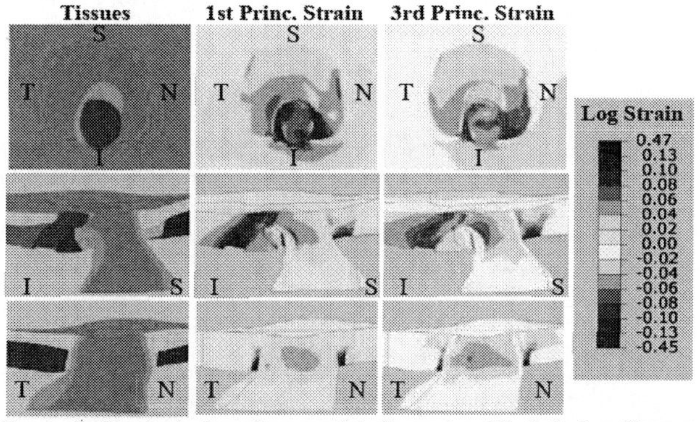

Figure 3: Strain in baseline model. Superior (S), inferior (I), nasal (N), temporal (T) directions are indicated. Top row: en face view. Middle row: S-I cut. Bottom row: T-N cut. Tissue colors: Nerve (green), CRV (pink), CRA (red), BM (orange), choroid (yellow), IAC (magenta), sclera (blue), pia (cyan), PNVP (lavender).

ACKNOWLEDGEMENTS

NIH T32 EY007092 (SAS), Georgia Research Alliance (CRE), NIH RO1 EY010145 (JCM) and P30 EY010572 (JCM), Research to Prevent Blindness (JCM), EY011610 (CFB), Legacy Good Samaritan Foundation (CFB), AGS Mid-Career Award (CFB), Sears Medical Trust (CFB), Alcon Research Institute (CFB).

REFERENCES

[1] Varma, R et al., *Am J Opthalmol*, 152(4): 515-22, 2011.
[2] Burgoyne, CF, *Exp Eye Res*, 93(2): 120-132, 2011.
[3] Sigal, I et al., *Biomech Model Mechanobiol*, 8(2): 85-98, 2009.
[4] Pazos, M and Yang, H et al., *Exp Eye Res*, 139: 1-12, 2015.
[5] Morrison, J et al., *Exp Eye Res*, 93(2): 156-164, 2011.
[6] Kang, M et al., *Neural Regen Res*, 10(8): 1198, 2015.
[7] Chen, K et al., *J of Biomed Mat Res*, 102(6): 2001-2007, 2014.
[8] Curcio, C. et al., *Retina, 5th Ed*, 465-481, 2012.

SB³C2017
Summer Biomechanics, Bioengineering and Biotransport Conference
June 21 – 24, Tucson, AZ, USA

FINITE ELEMENT MODELING OF ACTIVE SKELETAL MUSCLE: MUSCLE FORCE AND INTRAMUSCULAR PRESSURE

Benjamin B. Wheatley (1), Gregory M. Odegard (2), Kenton R. Kaufman (3), Tammy L. Haut Donahue (1,4)

(1) Department of Mechanical Engineering
Colorado State University
Fort Collins, CO, USA

(2) Department of Mechanical Engineering – Engineering Mechanics
Michigan Technological University
Houghton, MI, USA

(3) Department of Orthopedic Surgery
Mayo Clinic
Rochester, MN, USA

(4) School of Biomedical Engineering
Colorado State University
Fort Collins, CO, USA

INTRODUCTION

Intramuscular pressure (IMP) is strongly correlated with muscle force in skeletal muscle and can be measured with a minimally invasive pressure microsensor [1]. However, the link between muscle force and IMP is not fully understood, particularly in regards to inhomogeneous IMP distribution within the tissue. Finite element analysis provides a computational approach to study the behavior of skeletal muscle in lieu of repeated experimental studies.

The goal of this work was thus to develop and validate a finite element model of skeletal muscle under isometric active contractile conditions which can accurately simulate both muscle force and intramuscular pressure. Previous modeling efforts of muscle under passive stretch conditions yielded excellent agreement between experimental force and IMP data and model outputs [3], thus this work is a natural progression in developing a single, robust model.

METHODS

Experimental analysis of muscle force and intramuscular pressure behavior was conducted *in situ* using the New Zealand White Rabbit tibialis anterior. In short, n=8 muscles (one muscle from each rabbit) were passively stretched to and maximally stimulated under isometric conditions at fifteen different muscle lengths. These ranged from -35% to 35% fiber strain where 0% strain was designated as optimal length. Novel pressure microsensors were inserted in two orientations into the muscle midbelly – perpendicular or parallel to muscle fibers to generate two distinct pressure data sets.

A three dimensional finite element model of the New Zealand White Rabbit tibialis anterior was developed using μCT imaging, semi-manual segmentation, and manual hexahedral meshing and implemented in the finite element package FEBio (febio.org). The mesh (18646 elements total) was comprised of three constituents defined by various constitutive modeling approaches and connected directly through the mesh structure by shared nodes. Specifically, these three constituents were "active" muscle (generates contractile force), "passive" muscle (not capable of generating contractile force), and aponeurosis/tendon tissue (Figure 1). The passive response of skeletal muscle was modeled using a previously validated hyper-poro-viscoelastic constitutive formulation [3]. As the decoupling of passive properties of skeletal muscle contractile elements and extracellular matrix is not trivial, the "active" and "passive" constituents were assigned the same constitutive model with the one exception: the "active" constituent did not incorporate poroelasticity.

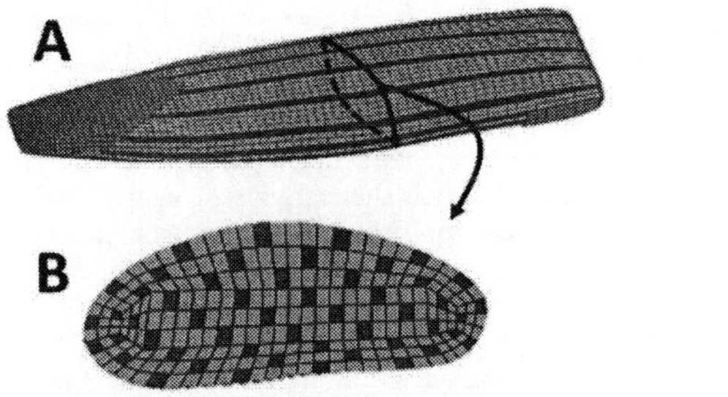

Figure 1. A) Complete finite element model geometry, with skeletal muscle "active" (dark red) and "non-active" (light red) constituents and aponeurosis (gray). B) Cross-sectional image of the same geometry showing distribution of muscle constituents.

The passive response of the model was largely dictated by tension-only reinforcing fibers using a three-dimensional dispersed ellipsoidal fiber distribution that accurately models the unique transversely isotropic tensile behavior of muscle in which the longitudinal orientation is less stiff than the transverse [2]. These were determined by selecting parameters which result in agreement with previously published experimental data on rabbit tibialis anterior muscle [3] and by using a nonlinear optimization approach to fit the model to the passive stretch experimental data. All other passive parameters were identical to previous modeling efforts [2]. Isometric muscle conditions were modeled using uniaxial active contraction (Equation 1, where σ^{active} is active stress, J is the volume ratio, T_0 is the maximum applied stress, $c(t)$ is a time dependent load curve, and $\boldsymbol{n}$ is the direction of muscle fibers).

$$\sigma^{active} = J^{-1} T_0 c(t) \boldsymbol{n} \otimes \boldsymbol{n} \tag{1}$$

Muscle force was converted to fiber stress by dividing by the physiological cross sectional area. Intramuscular pressure was interpreted as the fluid pressure within a small midbelly region of the "passive" model constituent. Following fitting of model stress to experimental data at each of the fifteen lengths by varying T_0, experimental and model pressure values were compared. The normalized root mean square error (NRMSE) was used to evaluate model fits to experimental data.

RESULTS

The ellipsoidal fiber distribution final optimization approach yielded excellent agreement with passive experimental stress and intramuscular pressure for both sensor insertion approaches, similar to previously published data (Figure 2A-C, Table 1) [2]. Model fitting to active stress data was highly successful (Figure 2D and Table 1). Model IMP showed a much stronger correlation to experimental parallel sensor insertion IMP data than perpendicular sensor insertion IMP data both visually (Figure 2B-C) and through statistical analysis (Table 1). The model showed a particularly high level of IMP predictive capability at strains near optimal length, which are most physiologically relevant (Figure 2E inset, NRMSE 26%). While the NRMSE values may seem large for active IMP data, it should be noted that the average standard deviation for each pressure data point was in excess of 50%.

Table 1. NRMSE values for all model-experiment comparisons.

Condition	Stress	Parallel IMP	Perpendicular IMP
Passive	0.765%	12%	5.6%
Active	0.079%	48%	115%

DISCUSSION

This study presents a new approach to studying intramuscular pressure of contracting skeletal muscle using finite element analysis with a hyper-poro-viscoelastic constitutive approach. Previously, this approach was successfully used to model skeletal muscle force and intramuscular pressure (IMP) under passive stretch [2]. This study employed a change in constitutive model parameters and the inclusion of active contraction, resulting in agreement between model and experimental data for both passive and isometric contractile conditions. This work presents the first time a whole muscle finite element model has successfully predicted contractile IMP behavior.

Experimental research has shown IMP readings are highly subject to sensor implementation (Figure 2E-F) [1]. The modeling results from this work suggest that both muscle length and contraction level dictate IMP readings (Figure 2E). Additionally, passive stretch data clearly shows a relationship between tissue deformation and IMP (Figure 2B-C). From a modeling perspective, passive stretch (fiber lengthening) pressurizes the fluid through the Poisson effect. Active contraction (fiber shortening) enacts a contradicting Poisson effect. Passive stretch and active contraction thus enact opposing deformations, thereby reducing the volumetric strain within the model. This results in a decrease of fluid pressure values within the model at longer muscle lengths during isometric activation.

It is unclear exactly what the mechanisms are that lead to measurement variability, although what is known is that anchoring within the tissue is critical to proper sensor use and that sensor insertion location does play a role in IMP measurement [1]. While further work should be completed to identify which measurement approaches are most clinically effective, this work provides for the first time a validated computational approach to studying intramuscular pressure in whole skeletal muscle.

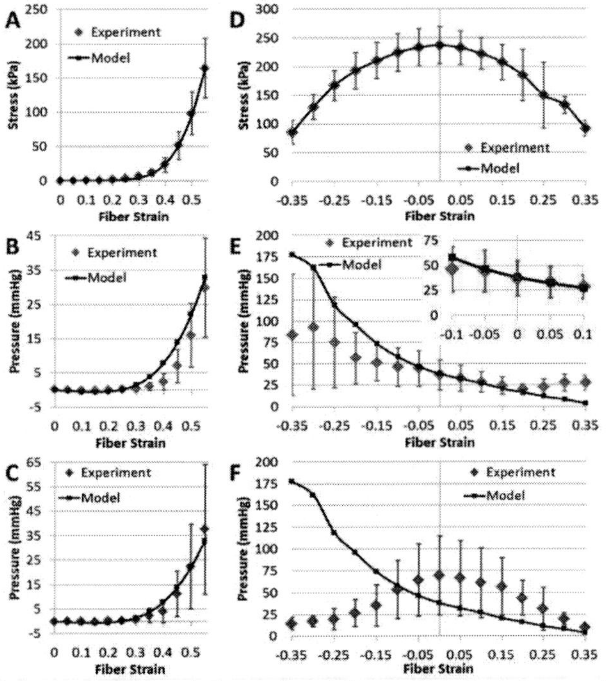

Figure 2. Experimental data and modeling results of passive stretch and active isometric muscle contraction. A) Passive muscle stretch stress data, B) passive stress intramuscular pressure with experimental data using parallel sensor insertion, and C) intramuscular pressure with experimental data using perpendicular insertion. D) Active isometric stress data, E) parallel active intramuscular pressure data (physiological lengths near optimal length shown inset), and F) perpendicular active intramuscular pressure data. All experimental data are shown with standard deviation error bars.

ACKNOWLEDGEMENTS

This work was funded by the NIH: National Institute of Child Health & Human Development [R01HD31476-12]. The authors would like to acknowledge Drs. Richard Lieber, Samuel Ward, and Shawn O'Connor for collecting the reported experimental data.

REFERENCES

[1] Go, S et al., *Ann Biomed Eng*, 2016.
[2] Wheatley, B et al., *Biomech Model Mechanobiol*, 2016.
[3] Wheatley, B et al., *J Biomech*, 49: 3056-3060, 2016.

SB[3]C2017
Summer Biomechanics, Bioengineering and Biotransport Conference
June 21 – 24, Tucson, AZ, USA

A FINITE ELEMENT ALGORITHM FOR LARGE DEFORMATION FRICTIONAL CONTACT OF BIPHASIC MATERIALS WITH APPLICATION TO CONTACT OF ARTICULAR CARTILAGE IN DIARTHRODIAL JOINTS

Brandon K. Zimmerman (1), Krista M. Durney (2), Gerard A. Ateshian (1,2)

(1) Department of Mechanical Engineering
Columbia University
New York, NY, USA

(2) Department of Biomedical Engineering
Columbia University
New York, NY, USA

INTRODUCTION

Contact of biphasic materials is an important topic in biomechanics, particularly in understanding mechanics of diarthrodial joints, where opposing articular cartilage surfaces undergo reciprocal contact loading for thousands of cycles per day. Contact between soft, hydrated, porous-permeable tissues often leads to large deformations and sliding, necessitating a numerical approach such as the finite element method to solve the resulting nonlinear system. To date, finite element algorithms developed for biphasic contact have been restricted to frictionless conditions [1-4]. However, frictional contact of biphasic materials may be critical for accurately capturing the salient mechanics of tissue damage from repetitive loading under physiological joint motion. Consequently, numerical simulation of biphasic friction would allow for hypotheses regarding tissue damage and wear to be evaluated computationally.

FEBio is a free finite element software package designed to meet the computational needs of the biomechanics and biophysics community (febio.org) [5]. This study describes the formulation of a novel finite element algorithm for frictional contact of biphasic materials and its implementation and verification in FEBio. Finite element algorithms for frictional contact of single-phase materials have been well established. These algorithms typically rely on a penalty regularization of the contact constraints, leading to an elastoplastic-type evolution equation for the traction [6]. Extension of these formulations to biphasic contact is nontrivial, as contact of biphasic materials requires special consideration with respect to interstitial fluid pressurization, and has yet to be achieved.

We propose to overcome these challenges by implementing our experimentally validated model of biphasic friction [7] in a large deformation finite element algorithm that relates the frictional traction to interstitial fluid pressurization. In this framework, the frictional traction in a biphasic material is proportional only to the fraction of the normal traction transmitted via solid-to-solid interactions across the contact interface. The frictional traction thus remains low if the pressurized fluid bears a large fraction of the normal load, leading to an effective friction coefficient μ_{eff} that can be orders of magnitude lower than the equilibrium friction coefficient μ_{eq} achieved when the fluid pressure has subsided. With decreasing fluid load support W_P/W, μ_{eff} will approach μ_{eq}, as more of the contact load is transferred to solid-to-solid contact.

METHODS

In a two-body contact analysis between primary and secondary surfaces (denoted by $i = 1,2$), the normal traction transmitted via solid-to-solid contact is

$$t_n^{ss}=t_n+(1-\varphi)p \, , \tag{1}$$

where t_n is the normal traction, φ is the fraction of the contact area in which solid-to-solid contact takes place, and p is the fluid pressure [7]. The effective friction coefficient during slip is evaluated as

$$\mu_{eff}= \mu_{eq}\,|t^n_{ss}|/|t_n| \, . \tag{2}$$

To differentiate between stick and slip, a trial sticking state is assumed, followed by a return mapping based on a slip criterion. During stick the contact interface acts as a tied interface and the traction on the primary surface is calculated as $\mathbf{t}^{(1)}=\varepsilon(\mathbf{x}^{(2)}-\mathbf{x}^{(1)})$, where ε is a penalty parameter and $\mathbf{x}^{(i)}$ is the spatial position of the contacting point on surface i. The slip criterion is $\Psi=|\mathbf{t}^{(1)}-t_n\mathbf{n}^{(1)}| - \mu_{eq}|t_n^{ss}|$, where $\mathbf{n}^{(1)}$ is the unit normal to the primary surface. If $\Psi<0$, stick occurs and $\mathbf{t}^{(1)}$ is calculated as shown above. $\Psi\geq 0$ denotes slip and the return mapping then prescribes the traction as $\mathbf{t}^{(1)}=t_n\mathbf{n}^{(1)} + \mu_{eq}|t_n^{ss}|\mathbf{s}^{(1)}$, where $\mathbf{s}^{(1)}=\mathbf{v}_r/|\mathbf{v}_r|$ is the unit vector in the slip direction and the absolute value for t_n^{ss} arises because t_n^{ss} may occasionally become positive in a biphasic contact analysis, particularly when a sub-ambient fluid pressure causes suction. In accordance with Coulomb's law for kinetic friction, $\mathbf{v}_\tau$ is the tangential component of the relative velocity between the contacting surfaces, formulated as a frame-invariant expression. The contact integral δG_c in our finite element formulation is

$$\delta G_c = \int_{\gamma^{(1)}} \left(\delta \mathbf{v}^{(1)} - \delta \mathbf{v}^{(2)}\right) \cdot \mathbf{t}^{(1)}\, da + \int_{\gamma^{(1)}} \left(\delta p^{(1)} - \delta p^{(2)}\right) w_n^{(1)}\, da \tag{3}$$

where $w_n^{(1)}$ is the normal component of the fluid flux on the primary surface, $\delta \mathbf{v}^{(i)}$ is a virtual velocity and $\delta p^{(i)}$ a virtual pressure on surface i, $\gamma^{(1)}$ is the contact domain on the primary surface, and da is a differential area on $\gamma^{(1)}$. This formulation was implemented in FEBio using the standard approach for linearizing δG_c along increments in displacement $\mathbf{u}$ and pressure p and discretizing the resulting nonlinear system using Newton's method.

RESULTS

There are three characteristic experiments that have been used to validate the friction model implemented here, as reviewed in [7]:

(1) Reciprocal sliding contact between a cylindrical cartilage plug and a flat glass slide, subjected to a step compressive load: This configuration produces a stationary contact area on the articular side of the cartilage plug, producing a monotonic decrease in the fluid load support W_P/W and a corresponding rise in the friction coefficient [8]. The predictions from an equivalent biphasic finite element frictional contact model are presented in Fig. 1 (black lines), faithfully reproducing the prior experimental findings.

(2) Reciprocal sliding contact between a cylindrical cartilage plug and a flat glass slide, subjected to a sinusoidal load: This configuration has been shown to produce a sinusoidally varying friction coefficient μ_{eff} that fluctuates above and below the equivalent response to a static load. In particular, μ_{eff} eventually exceeds μ_{eq} over a portion of the loading cycle, when the interstitial fluid pressure becomes negative (sub-ambient) [9]. The finite element implementation of this configuration reproduces these phenomena, as shown in Fig. 1 (gray lines). The equivalent μ_{eq} response to a constant load is displayed in Fig. 1a as a dashed line. This result demonstrates that the contact algorithm produces valid results even when the interstitial fluid pressure becomes negative.

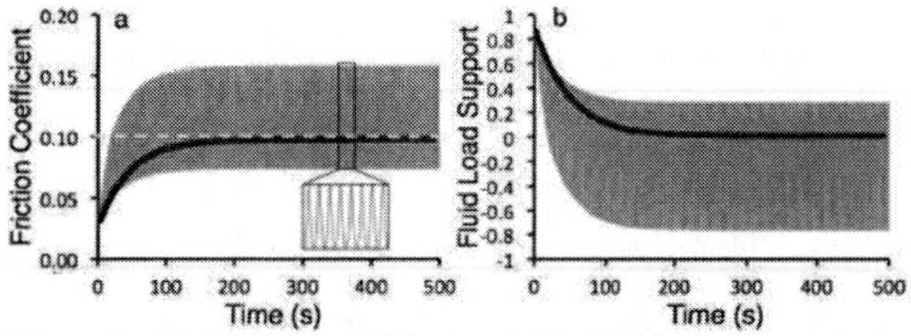

Figure 1. (a) Friction coefficient and (b) fluid load support during stationary contact area sliding between cartilage and a flat glass slide, under a constant (black) and sinusoidal (gray) applied load. The magnitude of the constant load is equal to the mean sinusoidal load. Inset of (a) displays a close-up of the sinusoidal response of the friction coefficient under a sinusoidal applied load. Similar sinusoidal oscillations are present in (b). The dashed line in (a) represents μ_{eq}.

(3) Reciprocal sliding contact between two cartilage layers, subjected to a constant compressive load: This configuration produces a migrating contact area on one or both cartilage surfaces, sustaining elevated interstitial fluid pressure and low friction for a long time [10]. The equivalent finite element predictions are presented in Fig. 2, confirming that the fluid load support remains elevated throughout the loading duration, producing a sustained low value for μ_{eff}.

Each of these three experimentally observed responses is replicated in a finite element model that employs our biphasic frictional contact formulation. The responses are verified by comparing the finite element μ_{eff} to the value predicted analytically from equations (1) and (2), where p and t_n are obtained from the finite element solution. The reader may also compare the responses shown here to the corresponding experimental studies [7-10].

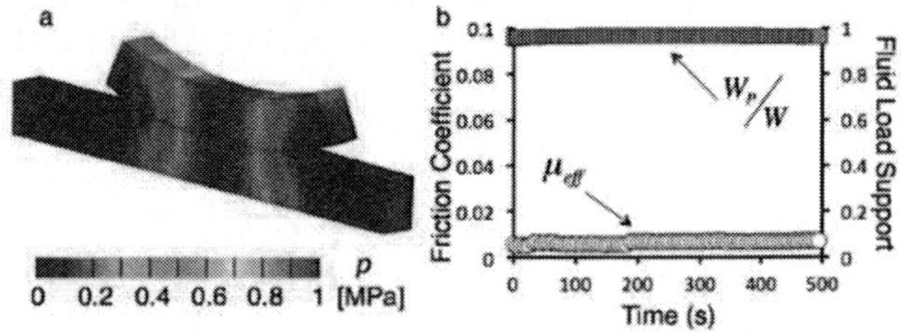

Figure 2. Simulation of migrating contact, reproducing the experiments of [11]. (a) Geometry and mesh utilized in the computational model, with contours displaying fluid pressure. (b) Friction coefficient and fluid load support during sliding, with μ_{eq}=0.1.

DISCUSSION

This study describes the first formulation and implementation of a finite element algorithm for biphasic frictional contact. The friction model adopted for this implementation has been previously validated against characteristic experimental results from the articular cartilage literature [7]. In contrast to the straightforward application of Coulomb's law as used for modeling finite element frictional contact between elastic solids [6], this biphasic friction law accounts for the porous nature of the contacting materials and the fact that interstitial fluid within the biphasic material may contribute to the load support across the contact interface. Consequently, the presented validation cases contain behaviors unique to frictional interactions between biphasic materials, which were previously unobtainable by computational methods. Notably, Fig. 1 demonstrates an effective friction coefficient considerably larger than the equilibrium friction coefficient, due to the effect of suction at the contact interface. It is only through coupling of fluid and solid phases that such effects can now be modeled and further understood.

The ability to model biphasic-biphasic frictional problems opens up new avenues of research; in an earlier study, we demonstrated cartilage fatigue damage under reciprocal sliding contact over thousands of loading cycles [11] and hypothesized such damage was due to reciprocal frictional shear stresses on the cartilage surface. Future work will apply the present framework to evaluate this hypothesis.

ACKNOWLEDGEMENTS

The authors acknowledge support from NSF GRFP DGE-11-44155 and NIH GM083925.

REFERENCES

[1] Ateshian, GA et al., *J Biomech Eng*, 132(6):061006, 2010. [2] Donzelli, PS et al., *Comput. Methods in Appl. Mech. Eng*, 153:63-79, 1998. [3] Guo, H et al., *J Biomech Eng*, 133(11):111001, 2011. [4] Chen, X et al., *JSME INT J C-MECH SY*, 48(4):537-546, 2005. [5] Maas, SA et al., *J Biomech Eng*, 134(1):011005, 2012. [6] Laursen, TA, 2002, Springer-Verlag, Berlin. [7] Ateshian, GA, *J Biomech*, 42:1163-1176, 2009. [8] Krishnan, R et al., *J Orthop Res*, 22:565-570, 2004. [9] Krishnan, R et al., *J Biomech*, 38:1665-1673, 2005. [10] Caligaris, M et al., *Osteoarthritis Cartilage*, 16:1220-1227, 2008. [11] Durney, KM et al., *SB³C*, 2016-1071.

SB³C2017
Summer Biomechanics, Bioengineering and Biotransport Conference
June 21 – 24, Tucson, AZ, USA

RELATIONSHIPS BETWEEN PEAK BONE MICROSTRUCTURE AND RATE OF ESTROGEN-DEFICIENCY-INDUCED BONE LOSS

Yihan Li, Wei-Ju Tseng, Chantal M. J. de Bakker, Hongbo Zhao, X. Sherry Liu

McKay Orthopaedic Research Laboratory,
Department of Orthopaedic Surgery
University of Pennsylvania,
Philadelphia, PA, USA

INTRODUCTION

Postmenopausal osteoporosis affects more than 200 million women worldwide [1]. Reduced estrogen levels post-menopause lead to accelerated bone remodeling, resulting in low bone mass and trabecular bone structural deterioration, which cause bone fragility and increased fracture risk [1-3]. Our previous study in rats suggested that variations in trabecular bone microstructure may impact the degree of bone loss following estrogen deficiency induced by ovariectomy (OVX) surgery [4]. We found that the rats that had undergone pregnancy and lactation had significantly thicker trabeculae than age-matched virgin rats. Moreover, these rats with reproductive history had attenuated OVX bone loss compared to virgins. By pooling the data of reproductive and virgin rats together, we found that the percent reduction in bone volume fraction (BV/TV) was significantly correlated with baseline trabecular thickness (Tb.Th), connectivity density (Conn.D), and trabecular number (Tb.N). However, the study included heterogeneous background of rats. The reproductive history may impact OVX bone loss through other pathways; thus the relationship between trabecular microstructure and OVX bone loss remains unclear. Therefore, the objective of the current study is to longitudinally track the bone microstructural changes before and after OVX surgery in a homogeneous population of virgin rats in order to establish the relationship between peak bone microstructure and bone structural changes induced by estrogen deficiency. We hypothesize that the variations in peak bone microstructure can predict the extent of estrogen-deficiency-induced bone loss.

METHODS

All animal experiments conducted in this study were approved by IACUC. 51 female Sprague Dawley rats underwent OVX surgery at age of 16-17 weeks and developed osteopenia for 4 weeks. *μCT imaging:* μCT scans were performed on right proximal tibiae for all rats at week 0 (before OVX surgery) and week 4 (4 weeks post-OVX) using an *in vivo* μCT scanner (VivaCT 40, Scanco Medical AG, Brüttisellen Switzerland). For each scan, rats were anesthetized (4.0/2.0% isoflourane) and the right tibia was fixed in a customized holder [5]. A 4-mm region of the proximal tibia was scanned at 10.5 μm voxel size. 3D image registration was conducted to detect a constant trabecular volume of interest (VOI) for each rat to evaluate trabecular microstructure at different time points (Fig.1). Standard trabecular bone structural parameters, such as BV/TV, Tb.Th, Tb.N, trabecular spacing (Tb.Sp), structure model index (SMI), and Conn.D were measured. Percent reductions between week 0 and week 4 were calculated for all parameters.

Statistics: Linear correlations and stepwise multiple linear regression analyses were performed to explore the relationship between the baseline (week 0) trabecular microstructural properties and corresponding percent decreases post-OVX. All data were divided into three tertiles, representing groups with Low, Medium, and High relative baseline Tb.Th (adjusted by baseline BV/TV), based on the residuals resulting from the linear correlation of baseline Tb.Th with BV/TV (Fig.2 D). One-way ANOVA with Bonferroni corrections was applied to compare % reductions in trabecular structural parameters among the 3 tertiles.

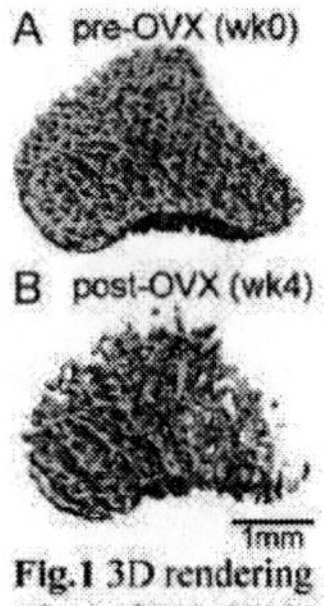

Fig.1 3D rendering of a registered VOI of trabecular bone pre- and post-OVX.

RESULTS

All rats underwent significant bone loss over 4 weeks post-OVX (55.6±7.8% decrease in BV/TV, Fig.1). Correlation coefficients of linear regressions between baseline parameters and % reduction in trabecular bone microstructural properties are shown in Table 1. Baseline BV/TV did not predict the % decrease in BV/TV. In contrast, baseline Conn.D was significantly correlated to % decrease in BV/TV. Baseline Tb.Th was found to be the best predictor for the % reduction in Conn.D. All baseline structural parameters were significantly correlated to % decrease in Tb.Th, while no parameters were significantly correlated to the % decrease in Tb.N or SMI.

As shown in Table 2, stepwise multiple linear regression showed that the combination of baseline Tb.Th and Conn.D was correlated with % reductions in BV/TV, Tb.Th, Tb.Sp and Conn.D, indicating that baseline Tb.Th and Conn.D were the most important predictors to the degree of estrogen-deficiency-induced bone loss.

To further examine the influence of Tb.Th regardless of BV/TV on OVX bone loss, rats were stratified by the relative baseline Tb.Th (adjusted by BV/TV) into 3 groups (Fig.2 A-C). Baseline Tb.Th was significantly correlated to baseline BV/TV (r=0.87, $p<0.001$; Fig.2 D), and the corresponding residuals were applied to determine Low, Medium, and High relative Tb.Th groups. % decrease in BV/TV was 13% lower in the High group compared to the Low group (Fig.2 E), and % decrease in Conn.D in the High group was 15% and 10% lower than Low and Medium groups, respectively (Fig.2 F). Moreover, the % decrease in Tb.N and the % increase in Tb.Sp were 19% and 33% lower, respectively, in the High group than the Low group (Fig.2 GH).

DISCUSSION

This study investigated the relationship between the peak trabecular bone microstructure and the degree of estrogen-deficiency-induced bone loss. Over 4 weeks post-OVX, estrogen deficiency induced substantial trabecular bone loss and microstructural deterioration. Multilinear regression analysis revealed that the extent of bone loss was influenced by the baseline trabecular microarchitecture, most notably the trabecular thickness and connectivity. Moreover, while there was a positive association between baseline Conn.D and the degree of OVX bone loss, Tb.Th correlated inversely, suggesting that a trabecular bone phenotype with thicker trabeculae and fewer connections may be protective against OVX bone loss.

Indeed, a tertile analysis of Tb.Th adjusted for BV/TV suggested that, given the same bone mass (BV/TV), rats with thicker trabeculae had attenuated loss in the volume, number, and connectivity of the trabecular bone network and reduced expansion in the spacing between trabeculae. Our working hypothesis is that the increased bone remodeling in response to estrogen deficiency has variable effects depending on the trabecular thickness. For thick trabeculae, the resorbed bone can be repaired by subsequent osteoblast activities (Fig.3 Left). However, thin trabeculae may be disconnected during remodeling and cannot be repaired (Fig.3 Right). Therefore, trabecular network of low connection but thick trabeculae may be protective against OVX-induced structural deterioration. Future studies with a longer post-OVX duration are necessary to further elucidate the long-term relationship between peak trabecular bone microstructure and the degree of post-OVX bone loss.

ACKNOWLEDGEMENTS

Funding: NIH/NIAMS P30-AR050950 and R03-AR065145.

REFERENCES

[1] Lane, NE et al., *Am J Obstet Gynecol*, 194(2):S3-S11, 2006.
[2] Bouillon, R et al., *Am J Med*, 90:107-110, 1991.
[3] Meunier, PJ et al., *Clin Ther*, 21(6):1025-1044, 1999.
[4] de Bakker, CMJ et al., *ASBMR Annual Meeting*, 2016.
[5] Lan, S et al., *Bone*, 56(1):83-90, 2013.

Table 1: Correlation coefficients (r) between baseline trabecular parameters and % decrease in trabecular microstructure

% decrease	*Baseline Parameters*					
	BV/TV	*Conn.D*	*SMI*	*Tb.N*	*Tb.Th*	*Tb.Sp*
BV/TV	NS	0.28*	NS	NS	NS	NS
Tb.Th	$0.74^{\#}$	$0.59^{\#}$	$-0.72^{\#}$	$0.67^{\#}$	$0.70^{\#}$	$-0.67^{\#}$
Tb.Sp	NS	-0.31*	NS	NS	NS	NS
Tb.N	NS	NS	NS	NS	NS	NS
SMI	NS	NS	NS	NS	NS	NS
Conn.D	-0.41^{+}	NS	0.44^{+}	NS	$-0.60^{\#}$	NS

Minus sign indicates negative correlation. *: $p<0.05$, $^{+}$: $p<0.01$, $^{\#}$: $p<0.001$.

Table 2: Correlation coefficients (r) and independent predictors of stepwise multiple linear regression to predict the degree of bone loss by baseline trabecular structural parameters

% decrease	*r*	*Adjusted r*	*Independent predictors*
BV/TV	0.46	0.41	$Conn.D^{+}$, $-Tb.Th^{+}$
Tb.Th	0.74	0.68	$Conn.D^{+}$, $Tb.Th^{\#}$
Tb.Sp	0.51	0.48	$-Conn.D^{\#}$, $Tb.Th^{+}$
Tb.N	NS		
SMI	NS		
Conn.D	0.68	0.66	$-Tb.Th^{+}$, $Conn.D^{+}$

Minus sign indicates negative correlation. *: $p<0.05$, $^{+}$: $p<0.01$, $^{\#}$: $p<0.001$.

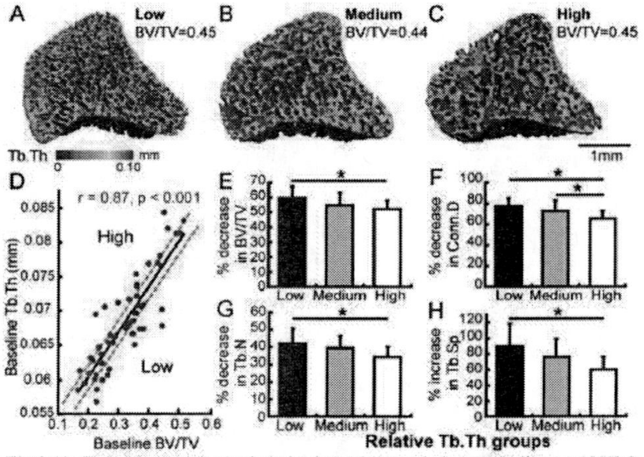

Fig.2 (A-C) Representative trabecular bone images in Low, Medium, and High relative Tb.Th groups. (D) Linear regression of baseline Tb.Th and BV/TV. (E-H) % changes in microstructural parameters in Low, Medium, and High relative Tb.Th groups. * indicates significant difference ($p<0.05$).

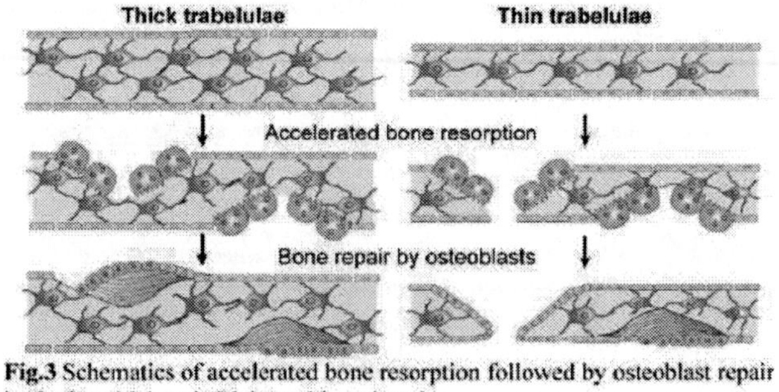

Fig.3 Schematics of accelerated bone resorption followed by osteoblast repair in (Left) a thick and (Right) a thin trabecula.

ANISOTROPIC HYSTERESIS IN VASCULAR SMOOTH MUSCLE CELLS

Zaw Win (1), Justin M. Buksa (1), Patrick W. Alford (1)

(1) Department of Biomedical Engineering
University of Minnesota
Minneapolis, MN, USA

INTRODUCTION

Modern biomechanical models such as constrained mixture models require a firm understanding of the mechanical properties of each component of tissues. Previously, we have developed a method to measure the elastic properties of single vascular smooth muscle cells (VSMCs) [1] and have developed an actin-fiber based strain energy density model that describes active cell mechanics under loading. However, complete mechanical understanding of vascular smooth muscle requires characterization of loading and unloading mechanics of contractile (active) and relaxed (passive) cells. Here, we utilize cellular micro-biaxial stretching (CμBS) microscopy to characterize active and passive mechanics of VSMCs during loading and unloading.

METHODS

Substrate Fabrication. Micropatterned CμBS substrates (Fig 1A) were prepared as previously described [1,2]. Briefly, glass coverslips were micropatterned with fibronectin adhesive islands (32 μm x 128 μm). Pre-polymer solution consisting of acrylamide, bis-acrylamide, acrylic acid N-hydrosuccimide ester, fluorescent beads, tetramethylethylenediamine, and ammonium persulfate was deposited onto a benzophenone-functionalized elastomer. The micropatterned glass coverslip was placed on the pre-polymer solution and exposed to UV to initiate polymerization. CμBS substrates were incubated in PBS for 48-72 h to remove unreacted pre-polymer constituents. Human umbilical artery smooth muscle cells were obtained from Lonza and utilized between passages 4-7. Cells were seeded onto CμBS substrates overnight and cultured at 37°C, 5% CO_2, in growth medium. Cells were serum starved for 24-48 h prior to experiments.

Cell stretching. Stretching experiments (Fig 2A) were conducted in Tyrode's buffer at 37°C on an inverted microscope at 40x magnification. Cell-seeded CμBS substrates were placed into the CμBS device (Fig. 1A) then exposed to either axial or transverse

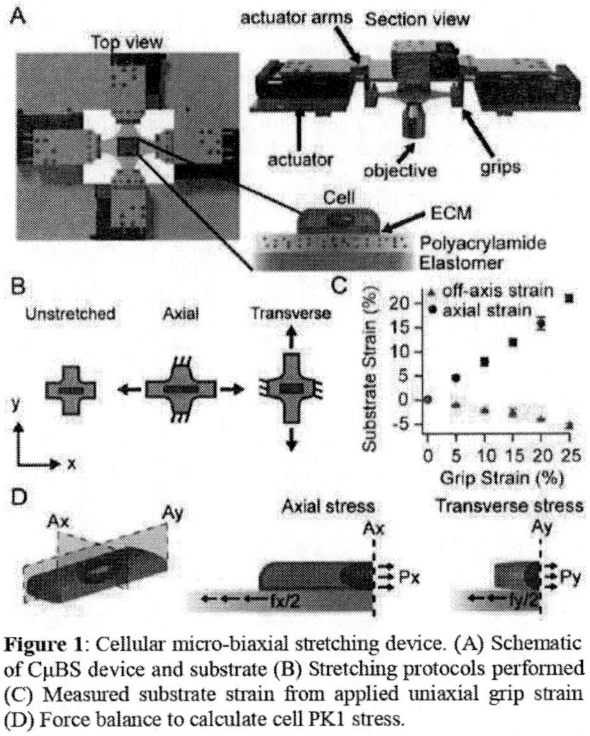

Figure 1: Cellular micro-biaxial stretching device. (A) Schematic of CμBS device and substrate (B) Stretching protocols performed (C) Measured substrate strain from applied uniaxial grip strain (D) Force balance to calculate cell PK1 stress.

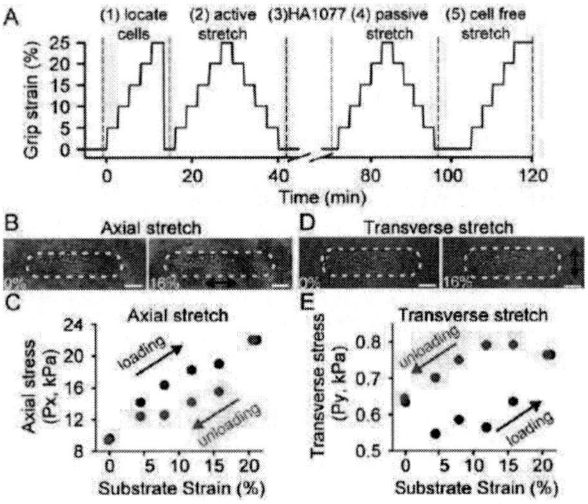

Figure 2: Stretching routine and cell stress. (A) Loading and unloading protocol to determine active and passive mechanics of VSMCs (B,C) Representative bright field images and PK1 stress (P_x) during axial active stretch (D,E) Brightfield images and PK1 stress (P_y) during transverse active stretch. Scale bars: 20 μm.

loading then successive unloading in increments of 5% up to 25% grip strain at 0.1%/s (Fig 2A). Brightfield images of the cells and fluorescent images of the beads in the substrate were acquired at each increment. Cells were treated with 100 μM HA1077 for 30 min. Passive cells were exposed to identical stretch performed on active cells. Cells were then lysed with 0.5% SDS and identical stretch was repeated to obtain cell-free deformation of beads.

Calculation of traction forces and cell stress. A particle image velocimetry algorithm [3] was used to calculate bead displacements at the top surface of the gel by comparing relative bead displacements between cell and cell-free image pairs. Traction stress vectors were determined from bead displacements using an unconstrained Fourier transform traction cytometry algorithm [4] (regularization factor: 1E-9, Poisson's ratio: 0.5), yielding a grid of n substrate traction stress vectors given by $\boldsymbol{T^n} = T^n_x \boldsymbol{e_x} + T^n_y \boldsymbol{e_y}$ where $\boldsymbol{e_i}$ is the unit vector in the i direction. The total traction force components f_x and f_y are given as $f_i = \sum_n - T^n_x A^n \frac{r^n_i}{|r^n_i|}$ where $i = x, y$, A^n is the area of discrete surface n, and $\boldsymbol{r^n} = r^n_x \boldsymbol{e_x} + r^n_y \boldsymbol{e_y}$ is the vector that described the location of surface n with respect to the cell center. The first Piola-Kirchhoff (PK1) stresses (Fig 1D) P_x and P_y were taken as $P_x = \frac{f_x}{2A_x}$ and $P_y = \frac{f_y}{2A_y}$.

RESULTS AND DISCUSSION

We utilize CμBS microscopy to determine the mechanical properties of isolated VSMCs (Fig 1A). The substrate consists of adhesive micropatterned islands patterned on a polyacrylamide gel that is bonded to an elastomer membrane (Fig 1A inset). The CμBS device allows us to perform uniaxial (axial or transverse) stretching on the substrate (Fig 1B). Using this device we applied up 20% uniaxial substrate strain in the direction of stretch (axial) which corresponded compression of up to -5% in the off-axis direction (Fig 1C).

VSMCs micropatterned into a physiological aspect ratio of 4 were then exposed to axial (Fig 2B) or transverse (Fig 2D) strains. Under axial load, P_x (stress in the direction stretch) increased with strain and decreased with decreasing strain during unloading (Fig 2C). Unloading stresses were lower than loading stresses for each increment of strain under axial stretch, suggesting energy loss. When cells were exposed to transverse stretch, P_y increased with load and decreased with unloading (Fig 2E). However, unloading stresses were higher than loading stresses during identical points of strain, suggesting energy gain. This hysteresis behavior was consistent in many cells (n>10) exposed to uniaxial stretching (Fig 3A,B). When cells were stretched in the absence of rho-kinase activity (passive), we found that hysteresis behavior was dissipated compared to active stretch (Fig 3C,D), suggesting contractile mechanisms play a vital role in the regulation of complex cell properties.

Our results show VSMCs exhibit hysteresis behavior during loading and unloading when stretched in the direction of primary fiber alignment, suggesting energy loss during unloading. However, under transverse stretch, slight axial compression occurs during load due to off-axis substrate strain which leads to axial tension during transverse unloading. This could contribute to the apparent energy gain due to engaging of acto-myosin contractile mechanisms during transverse unloading as axial stretch occurs. This study indicates a simple elastic model is not sufficient to describe the anisotropic hysteresis behavior of VSMCs and current efforts are focused towards identifying the mechanism of this behavior and modifying our model to incorporate acto-myosin contractile mechanisms.

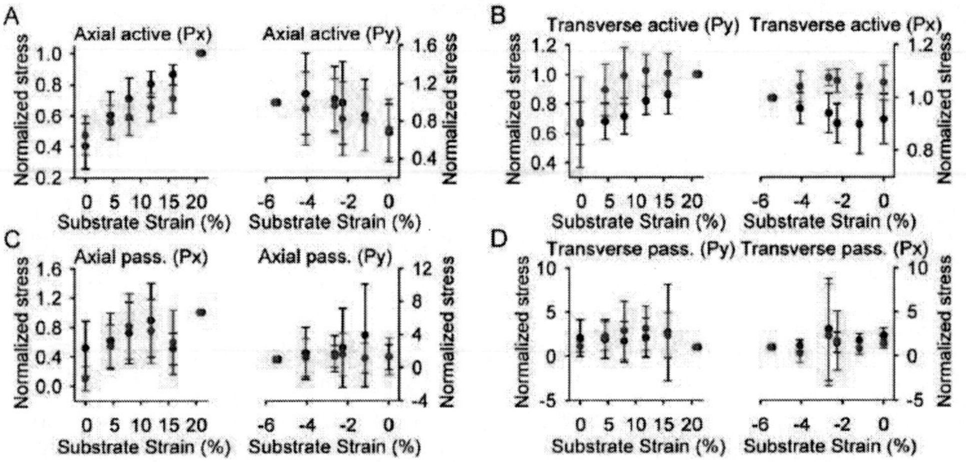

Figure 3: VSMCs display anisotropic hysteresis. (A-B) Active cell loading/unloading stresses during axial (A) or transverse (B) stretch (C-D) Passive cell loading/unloading stresses during axial (C) or transverse (D) stretch. Stresses are normalized to the stress at maximum strain in each cell.

ACKNOWLEDGEMENTS

We acknowledge financial support from the American Heart Association and the National Science Foundation.

REFERENCES

[1] Win, Z et al, *SB3C Conference Proceedings*, 464-465, 2016.
[2] Win, Z et al, *J Biomech Eng*, under review.
[3] Tseng, Q et al, *PNAS*, 109:1506-1511, 2011.
[4] Butler, JP et al, *Am J Phys*, 232:C595-605, 2002.

SB^{3}C2017
Summer Biomechanics, Bioengineering and Biotransport Conference
June 21 – 24, Tucson, AZ, USA

OSTEOCHONDROPROGENITOR PRIMARY CILIA ARE REQUIRED FOR JUVENILE SKELETAL DEVELOPMENT AND ADULT BONE FORMATION

Emily R. Moore, Yuchen Yang, Ya Xing Zhu, Christopher R. Jacobs

Biomedical Engineering
Columbia University
New York, NY, USA

INTRODUCTION

The periosteum is a promising candidate for bone and cartilage regeneration due to the ease with which it can be extracted and the osteochondrogenic commitment of its progenitors [1]. Its inner cambium layer contains a unique population of progenitors that demonstrate osteochondrogenic potential *in vitro* and populate the fracture callus *in vivo* [2,3]; however, investigators have to yet determine how to activate and direct these cell stores to targeted areas. The primary cilium is an extracellular sensory organelle present in most mammalian cell types and is believed to facilitate activation, migration, and differentiation of cells involved in skeletogenesis. For example, disruption of osteochondroprogenitor primary cilia results in dwarfism, polydactyly, and death in embryos [4]. Although these progenitors and their cilia play a critical role in embryonic development [4] and fracture repair [5], their roles in postnatal development and adults are unknown due to a lack of available mouse models [4]. We therefore developed novel models that allow us to temporally disrupt periosteal primary cilia and track the fate of osteochondroprogenitors to study their activation and function during postnatal development and adult bone formation. We hypothesized that cilia are required for migration and differentiation of osteochondroprogenitors to participate in ossification during juvenile skeletal development. This information can be used to recapitulate skeletogenesis in order to regenerate bone and cartilage *in vivo*. Similarly, we hypothesized that adult osteochondroprogenitors are activated to form bone in response to mechanical loading in adults. Together, these studies motivate the use of periosteal progenitors for *in vivo* therapeutic strategies to form cartilage and bone at all stages of life.

METHODS

Mice expressing CreER driven by a periosteum/ perichondrium-specific promoter (Prx1) were crossed with mice containing floxed alleles of an intraflagellar transport protein critical to cilium formation (Ift88). Mutant Prx1CreER-GFP;Ift88$^{fl/fl}$ and control littermates were injected with 33 mg/kg tamoxifen (Sigma) daily from postnatal day seven (P7) until sacrifice on P30. Ulnae were dissected, fixed, and sectioned in 5 um increments. Hematoxylin and eosin (Sigma) stains were performed to quantify length and thickness of the ulnae. Toulidine Blue O (Sigma) stains were carried out to assess the presence of cartilage, calcified cartilage, and bone in tissue sections. IHC was used to determine the number of proliferating cells present in the growth plate (PCNA, abcam) and visualize the hypertrophic zone (Type X Collagen, abcam). Prx1CreER-GFP mice were then bred with fluorescent reporter mice to generate Prx1CreER-GFP;Rosa26^{tdTomato} pups to track the fate of periosteum-derived progenitors. P7 pups received a single dose of tamoxifen to induce tdTomato expression and were sacrificed at P30. Ulnae were dissected, fixed, and cryosectioned. Lastly, Prx1CreER-GFP mice were crossed with a diphtheria toxin ablation model to observe bone formation in mice lacking these osteochondroprogenitors. Prx1CreER-GFP;Rosa26^{DTA} adults were injected with 75 mg/kg tamoxifen for 5 days prior to and concurrent with loading. Right ulnae were exposed to 3 N load (2 Hz for 2 min) for 3 consecutive days. Mice were injected with 10 mg/kg calcein and 70 mg/kg alizarin red (Sigma), as well as tamoxifen, 2 and 6 days after loading, respectively. Upon sacrifice 2 weeks after loading, ulnae were dissected, embedded, and sectioned. Dynamic histomorphometry was performed to quantify bone formation relative to the non-loaded limbs. Statistics were performed using a student t-test with $\alpha = 0.5$ and power of 80%.

RESULTS

Juvenile mice lacking primary cilia in their Prx1-expressing osteochondroprogenitor population had significantly shorter and thinner limbs (Figure 1). These mutants had fewer proliferating chondrocytes and ectopic hypertrophy in the growth plate (Figure 2), a key regulator of endochondral ossification. In general, juvenile mutants had stunted ossification in the trabecular space beneath the growth plate and epiphyseal ossification centers as revealed by Toluidine Blue O stains (not pictured).

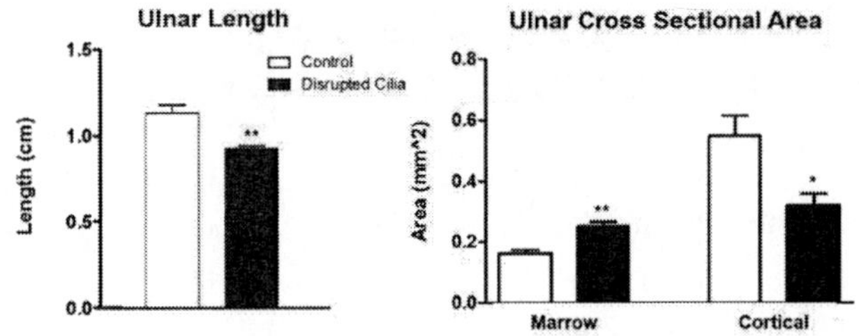

Figure 1: At P30, mutants have shorter limbs and less cortical area than controls. Interestingly, their marrow space is larger.

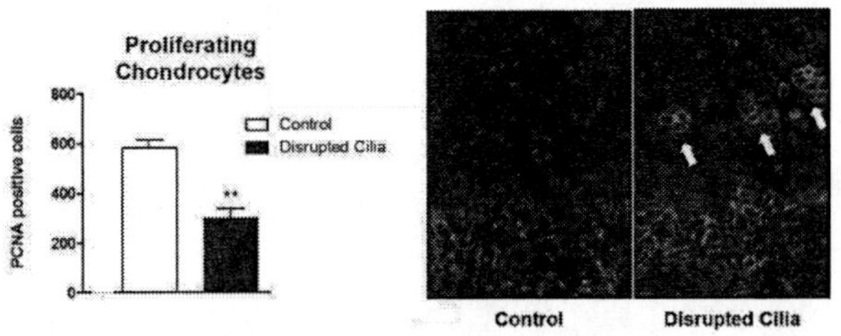

Figure 2: Mutant growth plates contain fewer proliferating chondrocytes and demonstrate ectopic hypertrophy (green) compared to those of control littermates.

Tracking studies indicate that osteochondroprogenitors initially residing in the periosteum/ perichondrium at P7 populate trabecular and cortical bone, the growth plate, secondary ossification centers, and the endocortical surface by P30 (Figure 3), in addition to the cambium layers of the periosteum and perichondrium.

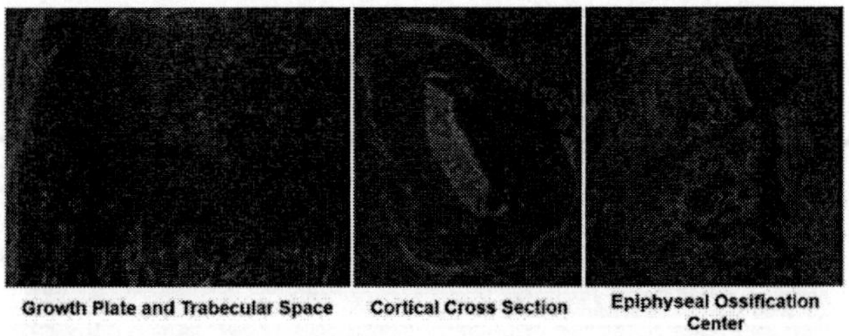

Figure 3: Progenitors (red) located at the periosteal surface at P7 and their progeny populate a variety of skeletal tissues by P30.

Our ulnar loading studies revealed that mice with Prx1-expressing periosteal progenitors that underwent apoptosis prior to loading formed significantly less bone in response to mechanical loading (Figure 4). The mutants had a significantly smaller mineralizing surface as well as reduced mineral apposition rate (not pictured), suggesting a severe lack of osteoblast recruitment to the periosteal surface.

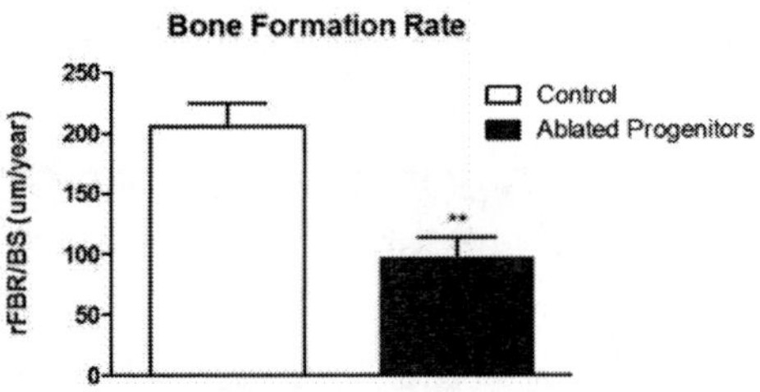

Figure 4: Adults lacking periosteal osteochondroprogenitors form significantly less bone in response to mechanical loading.

DISCUSSION

Our results demonstrate that periosteal progenitors contribute to juvenile skeletogenesis but their role is compromised without functional primary cilia. Our tracking studies indicate these progenitors populate the growth plate and contribute to endochondral ossification, as well as generate bone at the periosteal and endocortical surfaces potentially via intramembranous ossification. Chondrocyte cilia are known to align parallel to the longitudinal axis of the limb and are believed to encourage proper columnar stacking of proliferating cells [6], which facilitates migration within the growth plate and timing of hypertrophy. The lack of proliferating chondrocytes in mutants suggests that cilium disruption perhaps inhibits recruitment of from the perichondrium and impedes columnar alignment for those that do reach the proliferation zone, resulting in premature hypertrophy and stunted endochondral ossification.

Despite being confined to the periosteal surface, these progenitors and their progeny differentiate and populate a surprising number of skeletal tissues during postnatal development. This indicates that progenitors extracted from the periosteum have a great capacity to efficiently regenerate cartilage and bone for large defects without the osteogenic encouragement other progenitor sources require. The presence of tdTomato+ cells on the endocortical surface and larger marrow area observed in ciliary mutants suggests that periosteal progenitors first invade the juvenile cartilaginous structure to form bone, which is later resorbed by osteoclasts to shape the resulting cortical marrow space.

Finally, our ulnar loading studies suggest that these progenitors continue to be active in adulthood and are critical to proper bone formation in response to mechanical loading. Thus, extracting progenitors from the periosteum may prove to be an effective regenerative therapeutic at all stages of life. Overall, we believe these studies may motivate cilium-based therapeutics to trigger activation, migration, and/ or differentiation of periosteal progenitors for bone and cartilage regeneration *in vivo*.

ACKNOWLEDGEMENTS

This work is supported by National Institutes of Health grants R01-AR062177 and T32-AR59038.

REFERENCES

[1] Castro-Silva et al., *Clin Oral Imp Res*. Jan 2012. [2] Kawanami et al., *Biochem Biophys Res Commun*. Aug 2009. [3] Ouyang et al., *Bone*. Oct 2013. [4] Haycraft et al., *Dev*. Jan 2007. [5] Colnot., *J Bone Miner Res*. Feb 2009. [6] Song et al., *Dev Biol*. May 2007.

TARGETING PRIMARY CILIA-MEDIATED MECHANOTRANSDUCTION TO ENHANCE WHOLE BONE ADAPTATION

Milos Spasic, Michael P. Duffy, Christopher R. Jacobs

Biomedical Engineering
Columbia University
New □ork, N□, 10027

INTRODUCTION

Osteoporosis is a conditioned marked by low bone mass that affects nearly 50% of the US population over 50 years old [1]. Decreased bone mass heightens the risk of fracture, which typically results in significant hospitalization, loss of ambulation, and morbidity. Mechanical stimulation is a potent anabolic stimulus of bone formation, yet it is incompletely understood how cells within bone sense and respond to these mechanical cues [2]. The prevailing paradigm is that osteocytes sense mechanical forces, and transduce the input into molecular signals to promote activity of bone-forming osteoblasts. Many molecular mechanotransducers may be in play, but our lab has implicated primary cilia with a critical role in bone mechanotransduction [3].

Primary cilia are single immotile organelles extending from the surface of nearly every mammalian cell and have been identified as mechanosensors in numerous different cell types including kidney, endothelia, cartilage, and bone [4]. Removal of primary cilia impairs cell mechanosensitivity both *in vitro* and *in vivo* [4, 5]. Furthermore, when cilia are removed from osteocytes, mechanically induced paracrine signaling to promote osteogenic differentiation of mesenchymal stem cells is impaired [6]. Primary cilia function as mechanosensing antennae, and when cells are physically stimulated, such as by fluid flow, primary cilia bend and deflect [7]. This generates tension in the ciliary membrane to stimulate stretch-activated ion channels that localize to the cilium, such as polycystin 2 (PC2), or transient receptor potential vanilloid 4 (TRPV4) [7]. Opening of these channels in response to mechanical stimulation initiates a signaling cascade mediated by adenylyl cyclase 6 (AC6), a membrane bound enzyme which catalyzes the conversion of ATP to the second messenger cAMP [8, 9]. These proteins all work in concert to dictate primary cilia-mediated mechanotransduction, and whole bone adaptation.

Recently, several signaling pathways and specific compounds have been identified with the capability of altering cilium structure. Agonists of the cAMP/PKA signaling pathway have been shown to increase cilia length and sensitivity, as well as the dopamine receptor 1 agonist, fenoldopam [10, 11]. Furthermore, treatment with histone deacetylase 6 (HDAC6) inhibitors has been demonstrated to increase primary cilia bending stiffness, and impair mechanosensitivity [12].

The objective of this research is to examine how manipulation of primary cilia structure and mechanotransduction can alter intercellular signaling and whole bone adaptation.

METHODS

<u>*Cell culture:*</u> MLO-Y4 osteocytes were cultured on collagen coated dishes for 72 hours before experimentation. Cells were then placed on a rocker platform to apply oscillatory fluid flow for 12 hours at 1Hz and 0.1 Pa shear stress, or left static as a no flow control [6]. Conditioned media was then collected and used to culture MC3T3 osteoblasts for 24 hours. <u>*mRNA expression:*</u> Osteoblast osteogenic activity was assessed by analysis of osteopontin mRNA expression. Immediately post incubation in conditioned media, osteoblasts were lysed with TriReagent for total RNA isolation. RT and qPCR were performed using *Osteopontin* primer-probes, and *GAPDH* endogenous control. <u>*Drug treatment:*</u> Osteocytes were treated with 10 μM fenoldopam to increase cilia length and mechanosensitivity, or with 5 μM tubastatin (HDAC6 inhibitor) to increase cilium stiffness and decrease mechanosensitivity. <u>*siRNA knockdown:*</u> 48 hours prior to experimentation, osteocyte ciliogenesis was impaired using siRNA against *IFT88*, a transport protein critical for cilia formation [8]. Osteocyte ciliary mechanosensing was also impaired by siRNA knockdown of *AC6, PC2,* and *TRPV4* [7, 8]. <u>*In vivo bone adaptation:*</u> Skeletally mature, 16 week old mice, were subcutaneously injected with

20 mg/kg fenoldopam or vehicle control for 7 consecutive days. On days 5-7, mice were subjected to compressive ulnar loading (3N, 120 cycles, 2Hz sine wave), with contralateral limbs serving as non-loaded control [9]. Calcein and alizarin were injected and standard dynamic histomorphometric analysis was performed to assess bone adaptation. 2-way ANOVA revealed no difference based on gender. *Statistics:* All data reported as mean ± SEM. Analysis performed with one-way ANOVA with Bonferroni post-hoc correction.

RESULTS

Osteocytes with increased cilia length and mechanosensitivity (fenoldopam) displayed increased mechanically-induced paracrine signaling to osteoblasts. When osteocyte primary cilia-mediated mechanotransduction was abrogated by increasing cilia stiffness (tubastatin - HDAC6 inhibitor), this paracrine signaling to osteoblasts was diminished (Fig. 1A).

The role of key components of ciliogenesis and primary cilia-mediated mechanotransduction in osteogenic signaling to osteoblasts was then examined. Primary cilia formation was inhibited by knockdown of IFT88, as well as the key ciliary mechanosensing proteins AC6, PC2, and TRPV4. When either cilia formation, or ciliary mechanosensing was inhibited, osteogenic paracrine signaling to osteoblasts was also inhibited (Fig. 1B, C).

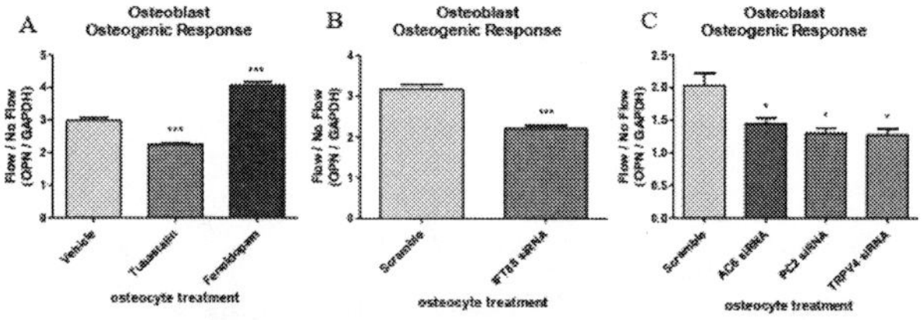

Figure 1: Osteoblast osteogenic activity is diminished with conditioned media from tubastatin treated osteocytes, and enhanced from fenoldopam treated cells (A). Inhibition of osteocyte primary cilia-mediated mechanotransduction also impairs this signaling axis (B, C).

We then sought to translate this work to an *in vivo* model of load-induced bone formation. Mice were treated with the ciliary lengthening agent, fenoldopam, and subjected to ulnar load to assess bone adaptation. Fenoldopam treated mice displayed no change in the amount of relative mineralizing surface (rMS/BS), but significant increases in the relative mineral apposition (rMAR) and bone formation (rBFR/BS) rates (Fig. 2).

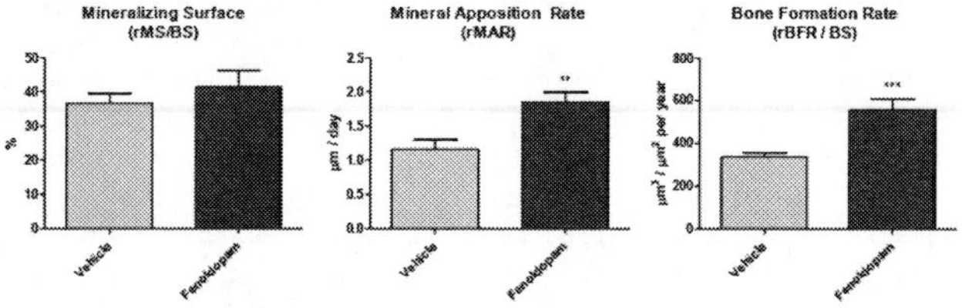

Figure 2: Mice treated with a cilia lengthening agent have increased load-induced bone formation.

µCT analysis was performed, and we found no changes, compared to vehicle control, in normal bone properties (cortical area, cortical thickness, porosity, bone mineral density) as a result of drug treatment, data not shown. Examination of bone ultrastructure also showed no differences in quality of bone formation (Fig. 3A). As an initial examination of toxicity, we also assessed kidney size, weight (Fig. 3B), and morphology, finding no differences with drug treatment.

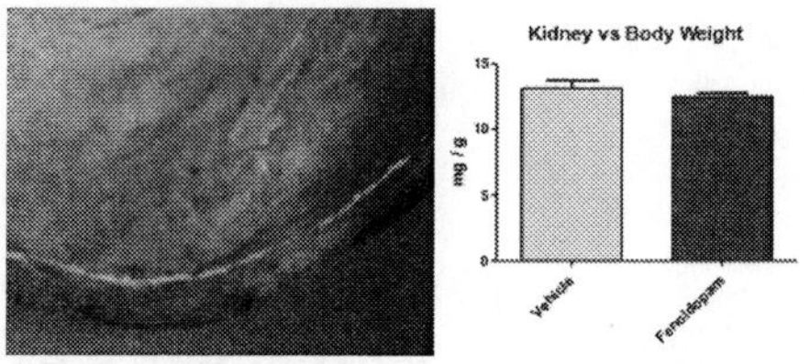

Figure 3: Bone ultrastructure (A) and kidney weight (B) are unchanged in drug treated mice. Calcein, green, and alizarin, red, labels are shown on the bone section.

DISCUSSION

We demonstrate here that the osteocyte primary cilium microdomain plays a critical role in mediating the osteogenic paracrine signaling between osteocytes and osteoblasts. These results do not, however, exclude the possibility for other mechanisms of intercellular communication, such as direct-cell contact. Though, this work does demonstrate that the osteocyte-osteoblast signaling axis can be manipulated by pharmacologically targeting primary cilia-mediated mechanotransduction to augment osteogenesis.

Our *in vivo* results demonstrate, for the first time, that these ciliary targeted therapeutics can sensitize bone to mechanical stimulation. That we did we examined a significant increase in mineral apposition and bone formation rates, but not mineralizing surface, suggests that increasing bone mechanosensitivity did not necessarily enhance osteoblast number, but rather osteoblast activity, consistent with our *in vitro* findings. Fenoldopam is not an osteocyte or bone-specific therapeutic, but our µCT, histology, and kidney measurements all suggest minimal adverse effects of drug treatment. Fenoldopam is a clinically approved treatment for hypertension, but its impact on bone mechanobiology has not been studied. Recently, we have begun a high-throughput drug screen to identify other potential cilia lengthening agents which may prove more suitable for eventual clinical use.

Most current treatments for osteoporosis rely on inhibition of bone resorption (bisphosphonates) or inhibition of anti-osteoblastic signals (sclerostin antibody), both presenting side effects such as atypical fracture or sarcoma. Here, we demonstrate the first potential bone therapeutic strategy based on manipulation of the inherent mechanosensitivity of bone. We utilize a holistic *in vitro* to *in vivo* approach to establish the primary cilium as a promising and viable target for bone disease therapeutics. Moreover, this work warrants the potential of repurposing fenoldopam, which is already clinically approved, for use as a treatment for diseases marked by low bone mass. Targeting primary cilia-mediated mechanotransduction is a propitious method for sensitizing bone to mechanical stimulation and enhancing whole bone formation.

ACKNOWLEDGEMENTS

This work is supported by National Institutes of Health grant (RO1-AR062177).

REFERENCES

[1] US Department of Health and Human Services, 2004. [2] Tatsumi, G et al. *Cell Metab,* 5:464-475, 2007. [3] Temiyasathit, S et al. *PloS ONE,* 7, 2012. [4] Nguyen, A et al. *Bone,* 54:196-204, 2013. [5] Malone, A et al. *PNAS,* 104:13325-30, 2007. [6] Hoey, D et al. *Biochem Biophys Res Com,* 412:182-187, 2011. [7] Lee, KL et al. *Cilia,* 4:7, 2015. [8] Kwon, R et al. *FASEB J,* 24:2859-68, 2010. [9] Lee, KL et al. *FASEB J,* 28:1157-65, 2014. [10] Besschetnova, T et al. *Current Biology,* 20:182-7, 2010. [11] Kathem, S et al. *JGC,* 11:63-73, 2014. [12] Nguyen, A et al. *Biology Open,* 4:1733-8, 2015

SB^{3}C2017
Summer Biomechanics, Bioengineering and Biotransport Conference
June 21 – 24, Tucson, AZ, USA

YAP/TAZ FEEDBACK CONTROL OF CYTOSKELETAL TENSION AND ADHESION REMODELING IS REQUIRED FOR ECFC MOTILITY

Devon E. Mason (1), James H. Dawahare (1), Sherry L. Voytik-Harbin (2), Mervin C. Yoder (3), Joel D. Boerckel (1)

(1) Aerospace and Mechanical Engineering,
University of Notre Dame,
Notre Dame, Indiana, US

(2) Weldon School of Biomedical Engineering and
Dept. of Basic Medical Sciences,
Purdue University,
West Lafayette, Indiana, US

(3) Herman B Wells Center for Pediatric Research,
Indiana University School of Medicine,
Indianapolis, Indiana, US

INTRODUCTION

Endothelial cell migration is essential for vascular morphogenesis in development and for angiogenic sprouting of endothelial cells from existing vessels to ischemic or regenerating tissue [1]. In both early neovascular plexus formation and post-natal angiogenesis, cell-cell and cell-matrix feedback is coordinated to promote dynamic cytoskeletal and cell adhesion remodeling to enable cell motility [1], but the molecular mechanisms that regulate this feedback system remain unclear.

Recent evidence implicates Yes-associated protein (YAP) and transcriptional activator with PDZ-binding domain (TAZ) as key mediators of extracellular matrix mechanosensation [2]. These paralogous transcriptional co-activators translocate to the nucleus to activate the TEA-domain (TEAD) family of transcription factors and have been reported to promote transcription of pro-migratory genes [3]. YAP and TAZ are regulated by two opposing pathways: cytoskeletal tension [2], which activates YAP/TAZ through nuclear translocation, and cell-cell contact, which restricts nuclear localization through the Hippo pathway [4].

We previously found that YAP and TAZ nuclear accumulation in endothelial colony forming cells (ECFCs) is enhanced by extracellular matrix rigidity in both 2D and 3D collagen matrices, and in migrating and vasculogenic ECFCs. YAP/TAZ depletion by RNA interference abrogated 2D ECFC migration and completely abolished 3D vasculogenesis *in vitro* [5], but the molecular mechanisms are unclear. Based on their known regulation of the matricellular growth factors connective tissue growth factor (CTGF) and cysteine rich angiocrine (Cyr61) [2&3], we hypothesized that YAP and TAZ promote cell motility through angiocrine production.

METHODS

ECFCs isolated from umbilical cord blood were cultured in endothelial growth medium (EGM-2) with 10% FBS. YAP and TAZ were depleted using either YAP, TAZ, YAP and TAZ, or non-targeting siRNA, and efficiency of depletion was assessed by Western blot (data not shown). Confluent monolayers of ECFCs depleted of YAP and TAZ were scratched and ECFCs were allowed to migrate for 12 hours into the open space. Cells were then either fixed for immunostaining of Vincluin, GM130 (Golgi), pFAK, α-tubulin, or actin (F- and G-) or lysed for gene expression by qPCR. F-actin polymerization is the ratio of polymerized actin (F-actin, phallodin) and total actin (F- and G- actin, dnase I). Polarization in migrating cells was assessed as the angle between the golgi and the direction of migration (0°) where the nucleus is the center point. Cells are considered polarized if they are within $\pm 60^\circ$ relative to 0°, and was assessed in leading and following cells in a migrating monolayer in comparison to an arbitrary confluent region of cells. Live imaging was performed on ECFCs depleted of YAP and TAZ over several minutes for observing fine morphological changes. ANOVA with Tukey's post-hoc comparisons were performed unless otherwise indicated. Non-repeated letters denote statistically significant differences between groups (α=0.05).

RESULTS

We first tested whether YAP and TAZ regulate migration through angiocrine expression in ECFCs. We found that CTGF, CYR61, and plasminogen activator inhibitor (PAI-1) expression was significantly reduced upon YAP/TAZ depletion; however, in contradiction of our hypothesis, reconstitution of any these proteins was unable to restore cell migration (data not shown). Therefore, to determine whether any YAP/TAZ-induced secreted factors are responsible for ECFC migration, we performed a media exchange experiment between siControl and siYAPTAZ treated cells, but conditioned medium was similarly unable to rescue migration (Figure 1A). Thus, these data

Technical Presentation #184

suggest that YAP and TAZ regulate cell migration in a cell autonomous manner. Dynamic cytoskeletal remodeling is required for cell motility, leading us to examine whether YAP and TAZ can modulate cytoskeletal machinery in migrating cells. YAP-, TAZ-, and YAP/TAZ-depletion significantly enhanced F-actin polymerization from G-actin, resulting in increased numbers of stress fibers (Figure 1B). Consistently, YAP and/or TAZ depletion also increased the number and size of vinculin-stained focal adhesions compared to control cells, indicative of increased focal adhesion maturity (Figure 1C), and abrogated the tightly-controlled relationship between adhesive cell area and adhesion number (data not shown). Taken together, these data demonstrate a cell-autonomous role for YAP and TAZ in modulating cytoskeletal tension and adhesion to enable cell motility. Directional cell movement also requires cell polarization, in which the microtubule organizing center MTOC and the Golgi complex re-orient in the direction of migration. To test whether YAP and TAZ regulate cell polarization, we assessed MTOC and endoplasmic reticulum alignment with the direction of migration in control and YAP/TAZ-depleted cells. While control cells readily polarized in the direction of migration, the ER/Golgi in YAP/TAZ-depleted cells failed to re-orient toward the wound (Figure 1F).

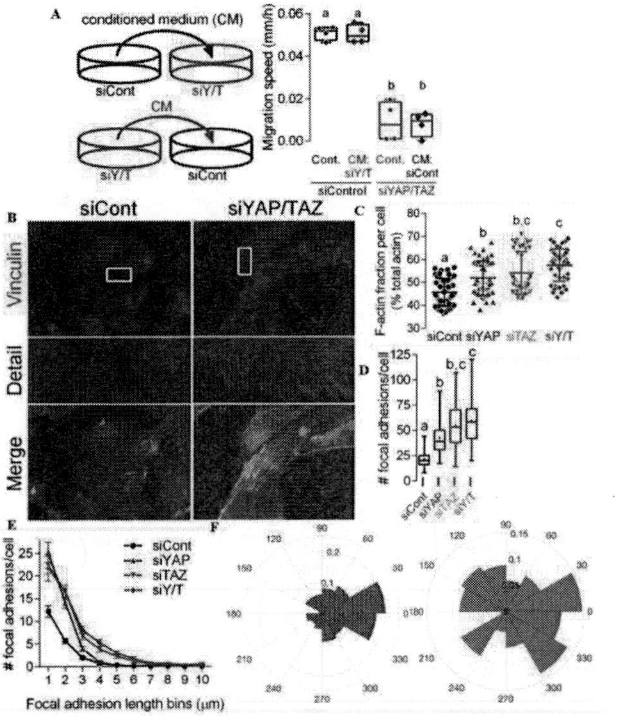

Figure 1: A) Migration speed in cells treated with conditioned medium. B) Focal and the cytoskeleton in siRNA treated ECFCs; vinculin (red), F-actin (green), and nuclei (blue). C) Percentage of polymerized actin per cell. D) Number of Focal adhesions per cell. E) Number of focal adhesions in 1 μm length bins per cell. F) Rose plots of golgi direction relative to the direction of migration in leading cells (siControl is grey and siYAPTAZ is purple).

Because cytoskeletal dynamics and adhesion distribution were impaired in YAP/TAZ-depleted cells, we next examined membrane dynamics and adhesion release during migration by live imaging (Figure 2). Control cells exhibited prominent lamellipodia formation and membrane ruffling, but YAP/TAZ-depleted cells featured minimal lamellipodia at the leading edge and failure of focal adhesion release at the trailing edge, resulting in prominent cytoskeleton-adherent focal adhesions protruding the membrane; suggesting a deficit in focal adhesion remodeling.

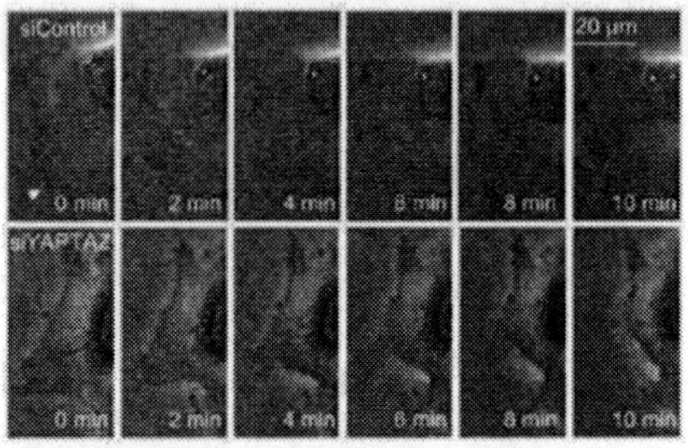

Figure 2: Lamellipodia remodeling in siControl cells (white arrow) and trailing edge protruding adhesions in siYAPTAZ cells (red arrow). Blue arrows indicate the direction of migration.

Intracellular tension generated by myosin and promoted by Rho A can increase focal adhesion formation and maturation. To see if YAP/TAZ-modulated intracellular tension is responsible for this response we treated cells with inhibitors targeting Rho Associated Coiled-Coil Containing Protein Kinase (ROCK) and non-muscle myosin II using Y-27632 and blebbistatin, respectively. Rock and Myosin inhibition normalized the cytoskeleton, reduced focal adhesion length without affecting focal adhesion number, and partially rescued migration in YAP/TAZ-depleted cells. Therefore, we examined gene expression in cells 1 hour after induction of migration. We found that there was a significant increase in NUAK2, cytoskeletal tension, and decrease in AURKA, focal adhesion remodeling, expression in YAPTAZ-depleted cells.

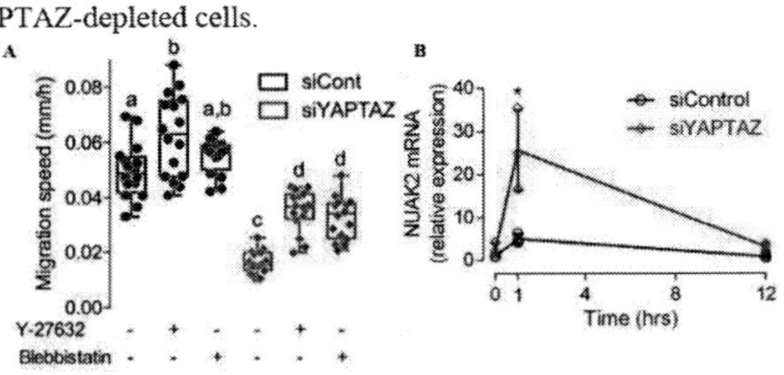

Figure 3: A) Migration speed increases after ROCK (Y-27632) and Myosin (blebbistatin) inhibition. B) NUAK2 gene expression increased in YAP/TAZ-depleted cells 1 hour after induction of migration, *p<.001 using two-way ANOVA.

DISCUSSION

We initially hypothesized that YAP and TAZ regulate ECFC migration and vasculogenesis through upregulation of angiocrine expression. Instead, we found that YAP and TAZ cell autonomously feedback to modulate cytoskeletal polymerization and acto-myosin tension, and to enable cell polarization and trailing edge focal adhesion release. We found this response is partially mediated by controlling myosin-generated tension potentially through the expression of NUAK2, a known promoter of myosin dephosphorylation.

Our experiments reveal a novel mechanism in which YAP and TAZ, activated by acto-myosin tension, serve as cytoskeletal rheostats to provide feedback control of cytoskeletal and focal adhesion remodeling to enable endothelial cell motility. These data implicate YAP and TAZ as potential targets for promoting endogenous angiogenesis or vasculogenic cell therapies. Further, targeted inhibition of YAP/TAZ may enable normalization of tumor vasculature despite tumor-derived growth factors.

ACKNOWLEDGEMENTS

This work was supported by grants from the Advanced Diagnostics and Therapeutics Initiative (to JDB) and the Indiana CTSI (NIH/NCRR UL1TR001108 to JDB).

REFERENCES

[1] Lamalice, L et al. Circ. Research, 100:782-794, 2007
[2] Dupont, S et al. Nature, 474: 179-183, 2011
[3] Liu, X et al. Cell Rep.,14(5): 1169-1180, 2016
[4] Zhao, B et al. Nat. Cell Bio, 13: 877-883, 2011
[5] Mason, D et al. SB3C2016, Abstract: 822, 2016

SB³C2017
Summer Biomechanics, Bioengineering and Biotransport Conference
June 21 – 24, Tucson, AZ, USA

TGFB INDUCES PRIMARY CILIA DISASSEMBLY IN RESPONSE TO CYCLIC LOADING OF TENOCYTES

Daniel T. Rowson (1), Hazel R.C. Screen (1), Martin M.M. Knight (1)

(1) School of Engineering and Materials Science
Institute of Bioengineering
Queen Mary University of London
London, UK

INTRODUCTION

Primary cilia are organelles found singularly on almost every cell type. They are known to regulate various fundamental signalling pathways and have been implicated in a number of disease pathologies[1]. Primary cilia have been shown to respond to cellular mechanical stimulus, with acute cyclic loading in chondrocytes leading to shorter cilia[2]. However, little is known about the role of primary cilia, or their response to mechanical load in tenocytes. It has been shown in tendon explants that stress deprivation leads to increased cilia length and we have shown that this length change correlates with mechanical degradation[3,4]. However, no mechanisms associated with cilia length change have been identified in tendon.

One of the main regulators of cilia length is known to be HDAC6, which deacetylases tubulin allowing for cilia shortening. Previous studies have also identified actin tension as a regulator of primary cilia length and there is suggestion that actin tension may be mechanoresponsive[5]. It has been shown previously that TGFβ is activated in tenocytes by cyclic tensile strain and that it mediates a range of mechanical load induced changes in gene expression[6]. It has also been shown that TGFβ receptors are localised to the cilium[7].

In this study, we investigated the response of the cilia in cultured human tenocytes to cyclic loading, to investigate the mechanisms through which mechanical stimuli may influence cilia length. We hypothesized that mechanical strain would lead to a shortening of cilia length and that this change was TGFβ, HDAC6 and/or actin tension associated.

METHODS

Human hamstring tenocytes (18, m, P4-5) were mechanically loaded using the FlexCell tensile system (Flexcell, Burlington, NC). Cells were seeded onto flexible silicone membranes (50,000 cells per membrane) and subjected to uniaxial cyclic load (3% strain; 1Hz) in serum starved conditions for up to 24h. At the end of the loading period, membranes and cells were fixed with PFA and immunofluorescently stained with anti-arl13b, anti-acetylated α-tubulin and DAPI to enable imaging of the cilium and nucleus. Imaging was done at 63x magnification with confocal microscopy. Cilia length and orientation, as well as cell morphology were calculated for each cell, using ImageJ. For each experiment 3 separate membranes were investigated per loading condition with 20-40 cells imaged on each membrane.

For the first experiment cells were strained for 5, 8, 12 & 24 hours, to provide a time course of cilia response. A further condition where cells were strained for 24h followed by 2 hours unstrained recovery was also performed.

Additional experiments investigating the role actin, HDAC6 or TGFb on tenocyte response, focused on the 24 hour loading time point.

To investigate the role of actin, cells were treated with 20μM Blebbistatin, whilst the effects of inhibiting HDAC6 were investigated by treated cells with 500nM Tubacin (HDAC6 inhibitor) immediately prior to loading.

To investigate whether the addition of TGFβ could replicate the response of cilia to cyclic strain, unstrained serum starved cells were incubated with the addition of 10ng/ml TGFβ for 24 hours. Separately, cells were treated with 2μM TGFβ receptor inhibitor (SB 431542) prior to 24 hours of loading.

RESULTS

3% cyclic loading leads to progressive shortening and disassembly of cilia over time. At 24 hours, mean cilia length had significantly reduced from 3.2±1.7μm to 1.5±0.5μm, and mean cilia prevalence had significantly reduced from 79% to 37%. Notably a 2 hour recovery period after 24 hours of loading fully recovered cilia length and prevalence (fig 1, 2). No significant reorientation of the cell, cell nucleus or the cilia was observed in response to the cyclic strain.

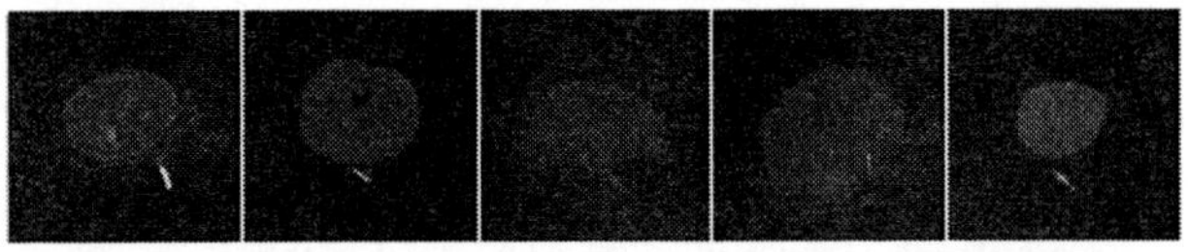

Figure 1: Representative confocal images of primary cilia stained with arl13b, acetylated α tubulin and DAPI after increasing lengths of cyclic loading.

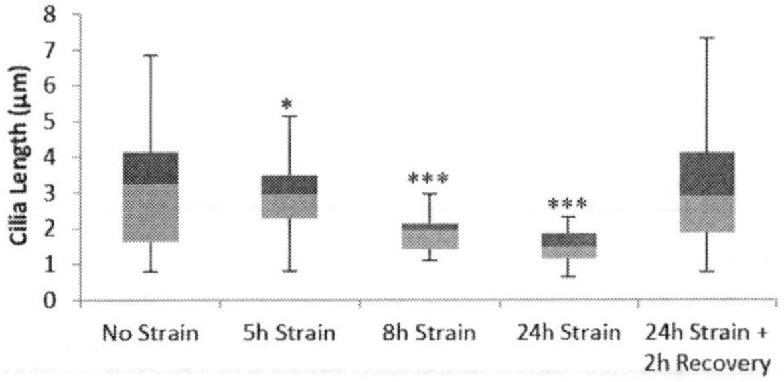

Figure 2: Box and whisker plots showing cilia length versus duration of cyclic strain. * =p<0.05, *** =p<0.001.

Treatment with blebbistatin did not significantly prevent cilia shortening. There was also no significant difference in cilia prevalence between cyclically strained tenocytes with or without blebbistatin. No significant changes in cilia length or prevalence were noted in unstrained cells treated with blebbistatin either.

Treatment with tubacin slightly but still significantly prevented cilia disassembly. By contrast, the significant preservation of cilia prevalence was more striking 55% compared with 28% (fig 3).

Treatment of unstrained tenocytes with TGFβ led to a significant reduction in both mean cilia length (3.2±1.8μm to 1.6±0.5μm) and prevalence (75% to 44%). Indeed, the length and prevalence of cilia in TGFβ treated cells was not significantly different from those of loaded tenocytes. Incorporating a TGFβ receptor inhibitor to tenocytes prior to loading led to a significant reduction in cilia shortening 3.7±1.0μm compared with 2.7±1.3μm (fig 3).

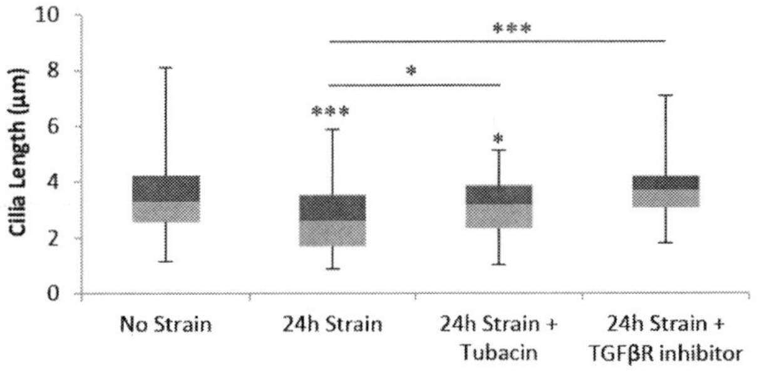

Figure 3: Box and whisker plots showing cilia length after cyclic load and treatment with Tubacin and TGFβR inhibitor. * =p<0.05, *** =p<0.001.

DISCUSSION

This is the first time continuous cyclic loading has been shown to cause near complete cilia shortening and disassembly in tenocytes. We have shown that this shortening can be somewhat reduced by inhibiting HDAC6, in agreement with studies of mechanically induced cilia disassembly in other cell types. However, most striking is the near complete prevention of disassembly through inhibition of TGFβ. This reveals for the first time that the cyclic strain mediated disassembly of tenocyte primary cilia is driven by TGFβ activation.

Interestingly HDAC6 inhibitors have been shown to have anti-inflammatory effects in other tissues and that TGFβR inhibition leads to changes in mechanosensitive gene expression in tenocytes[6]. Further studies will therefore examine whether HDAC6 or TGFβR inhibitors may modulate tendon degradation in response to altered mechanical loading and whether this response is mediated by changes in primary cilia.

ACKNOWLEDGEMENTS

This work was funded by the EPSRC and the Institute of Bioengineering at QMUL.

REFERENCES

[1] Kathem, S et al., *AIMS Mol Sci,* 1:27-46, 2013
[2] Thompson, C et al., *Osteoarthritis Cartilage*, 22:490–8, 2014
[3] Gardner, K et al., *J Orth Res,* 29:582-7, 2011
[4] Rowson, D et al., *J Orth Res,* 34:2146–2153, 2016
[5] Miyoshi, K et al., *Med. Okayama*, 65: 279–85, 2011
[6] Jones, E et al., *Acta - Mol. Cell Res.*, 1833:2596–2607, 2013
[7] Clement, A et al., *Cell Rep.*, 3:1806–14, 2013

INSIGHTS INTO TRIBOLOGICAL REHYDRATION OF ARTICULAR CARTILAGE VIA ANALYSIS OF SOLUTE TRANSPORT *IN SITU*

Brian T Graham (1), Axel C. Moore (2), David L. Burris (1,2), Christopher Price (1,2)

(1) Department of Mechanical Engineering
University of Delaware
Newark, DE, USA

Department of Biomedical Engineering
University of Delaware
Newark, DE, USA

INTRODUCTION

The pressurization of interstitial fluid within articular cartilage permits function with extremely low friction coefficients [1]. However, under compressive loading, fluid pressure subsides as fluid exudes from the tissue, causing strain and friction to increase. Yet, numerous lines of evidence suggest that our joints are somehow able to mitigate this effect, continuously operating in extremely low friction regimes with moderate levels of strain, despite the bouts of compressive loading brought about by daily activity [2].

For decades, *in situ* studies that compressed and slid small, flat-surfaced osteochondral explants against glass platens in a stationary contact area configuration (SCA) have failed to replicate the physiological maintenance of low friction under continuous loading, for even moderate periods of time [3]. Recently, we discovered that performing the same tests on larger, convex-surfaced samples at physiological speeds allows low friction to be maintained indefinitely, and allowed a degree of compressive strain to be actively recovered [4]. This new test setup is termed the convergent stationary contact area (cSCA), because the curved surface geometry of the larger samples creates a wedge-shaped convergence zone at the periphery of the contact, which has a diameter smaller than that of the sample. Although the mechanisms responsible for this response remain unclear, results from *in situ* mechanics experiments conducted thus far are consistent with a tribological rehydration hypothesis, that sliding hydrodynamically entrains and pressurizes bathing fluid, which is subsequently driven into the porous articular surface, thereby reversing the exudation process and restoring interstitial hydration, fluid pressure, and lubrication. In this study, the interstitial flow of fluorescently tagged solutes are observed *in situ*, and in real-time to provide additional insights into the mechanisms underlying tribological rehydration.

METHODS

19-mm diameter osteochondral plugs were harvested from the femoral condyles of adult bovine stifle joints, stained overnight in 5'-DTAF (5-μg/mL), and rinsed prior to testing. Samples were affixed in a custom linear reciprocating materials tester that applied compressive loads and allowed sliding against a 1-mm thick glass platen with simultaneous imaging of the buried cSCA contact via a Zeiss LSM880 confocal microscope. All samples were pre-compressed to 5N for 5 min before adding 1mL of fluorescent solute (AF633, 5μM, MW: ~1200Da in PBS) to the contact periphery. In experiment one, samples were then slid for 30-minutes at speeds of 0-, 1-, or 60-mm/s (n=4/speed). In experiment two, samples experienced the same total duration of sliding at 0-, 1-, or 60-mm/s (n=1 each); however, each was administered in 2-minute bouts with a 5-minute rest/imaging interval between sliding. A third test featured a similar alternating sliding and rest approach, however, the sample was sequentially subject to a different sliding speed (0-, 1-, 5-, 10-, 30-, 60-, 80-mm/s) at each interval. For all experiments, the confocal microscope was used to capture localized, real-time 2D (XZ) images of the contact, in the direction of sliding, during sliding. During the stationary rest periods, 3D volumetric strips (XYZ), in the direction of sliding, were acquired across the contact. Intensity-based quantification of solute accumulation was performed via a custom MATLAB code.

RESULTS

During 30 minutes of continuous of sliding, load-induced deformation at 0- (static) and 1-mm/s were indistinguishable, increasing over time, with friction coefficient for 1-mm/s following a similar upward trend, typical of a SCA. However, 60-mm/s sliding reversed the initial preload deformation and facilitated a rapid drop in the friction coefficient, which was maintained for the duration of the test. (**Fig 1A-B**). Quantification of real-time solute accumulation in the 2D cross sectional time series captured during sliding (**Fig 1C**) and 3D volumes captured after sliding (**Fig 1D-E**) revealed that 60-mm/s sliding allowed rapid and significant accumulation of solute in the center of the contact area. There was little difference between the 0- and 1-mm/s conditions in which the contact was effectively sealed off from solute transport. When the sliding was broken up into 2-minute bouts separated by a 5-minute rest interval, the results were similar to the uninterrupted case. Solute accumulation data from 2D cross-sectional images demonstrated that solute accumulation halted when sliding stopped, regardless of speed, picking up again when the next sliding interval began (**Fig 2A**). 3D volumetric data captured at each rest interval revealed that solute appeared to propagate spatially from the contact periphery to the center (**Fig 2B-D**). In the speed sweep test, significant increases in solute accumulation only occurred at speeds where significant tribological rehydration occurred (>10-mm/s; **Fig 3A**), namely where deformation was recovered (**Fig 3B**).

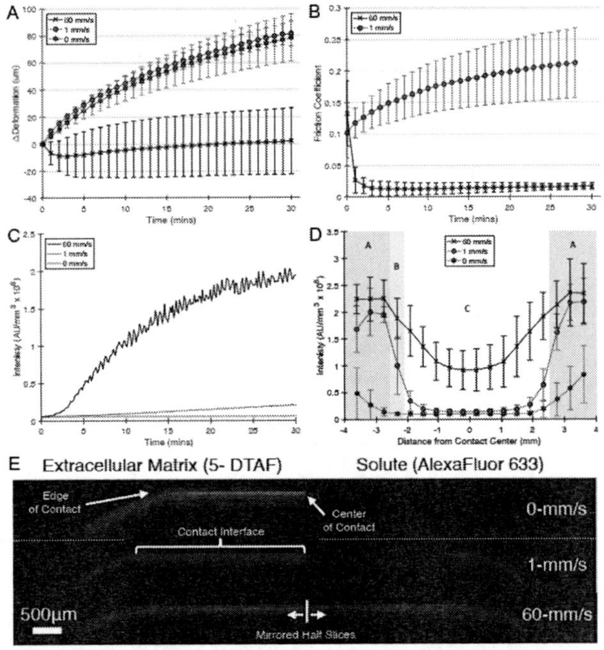

Figure 1: A) Change in deformation and B) friction coefficient over 30 minutes of continuous sliding (n=4) with C) solute intensity for each speed calculated from cross sections (2D) captured at contact center. D) Solute intensity of summed volumetric scans (3D) across contact captured after sliding. 1-way ANOVA, Tukey post-hoc groupings ($p < 0.05$): Zone A = 0- different from 1-, 60-mm/s, Zone B = all different, Zone C = 0-, 1- different from 60-mm/s. E) Single representative slices for each speed from volumetric data in panel D.

DISCUSSION

These data provide strong evidence in support of the tribological rehydration hypothesis, namely that sliding at physiologically relevant speeds (>30mm/s) can force fluid, and thus solute, into the contact, which in turn rehydrates the tissue, re-pressurizes interstitial fluid, and restores lubrication. The 30-minute continuous sliding experiment clearly demonstrated the correlation between low friction, recovered deformation (fluid uptake), and enhanced transport of solute. In static and slow speed sliding, where all mechanical data suggested that fluid exudation is dominant, we saw negligible levels of solute accumulation in the tribological contact. At high speeds, mechanical data showed cartilage thickening (reversal of deformation), which suggested a recovery of interstitial fluid; the significant solute accumulation observed here directly indicated uptake of bath fluid. The speed sweep experiment showed a threshold for tribological rehydration. When solute accumulation was observed, deformation reversal began, and friction rapidly decreased. 10 mm/s appeared to be a transition (threshold) speed for rehydration in our partially equilibrated samples; slower speeds showed no discernable effects. These effects became larger with increasing speed. The present experiments also indicated that as soon as sliding stopped, solute transport is inhibited. This allowed us to perform a detailed 30-minutes single speed intermittent sliding experiment with 3D imaging that allowed us to identify the spatiotemporal profile by which solute penetrates the cartilage. This test suggested that solute migrates from the edges of the contact, through the cartilage, finally reaching the center. While it is clear that high speed sliding has tremendous effect on the rehydration of the cartilage within the otherwise sealed the contact area, we are still investigating whether it penetrates interfacially, via a gap between the glass and cartilage forced open by sliding, or whether pressure is driving fluid and solute through the surface at the wedge, then within the ECM to the contact's center. The present methodology achieved real time observation of the contact by sacrificing the ability to see deeper than 100µm deep into the cartilage, limiting our analysis to the upper-superficial zone alone. However, the data still demonstrated that sliding-induced tribological rehydration of cartilage is a powerful mechanism for promoting cartilage's remarkable ability to provide a low friction contact for very long periods of time.

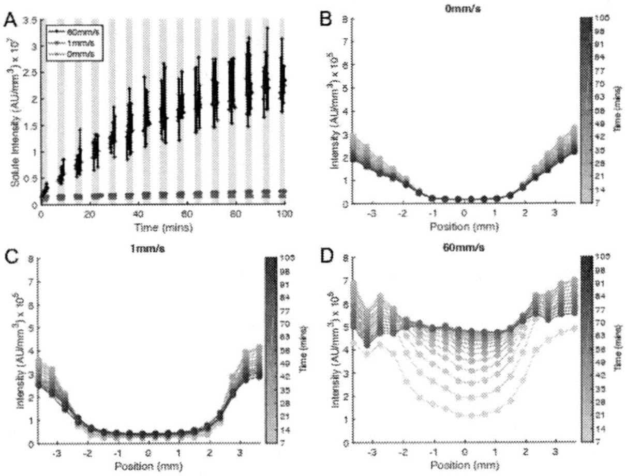

Figure 2 A) Solute intensity during intermittent sliding bouts calculated from cross sections (2D) captured at contact center. Gray shading indicates periods of sliding, white indicates rest periods. B-D) Spatial and temporal evolution of solute intensity of volumetric scans (3D) across the contact captured during the rest intervals.

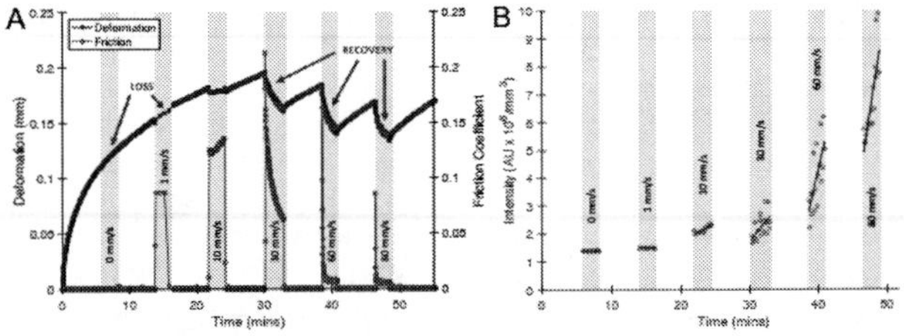

Figure 3 A) Change in deformation, friction coefficient, and B) solute intensity during speed sweep (gray) and rest periods (white).

ACKNOWLEDGEMENTS

This work was funded by the NSF BMMB 1635536, NIH S10 OD016361, and INBRE P20 GM103446.

REFERENCES

[1] Krishnan, R et al., *J Orthopaedic Research*, 22: 565-570, 2004.
[2] Eckstein et al., *Anatomy and Embryology*, 200: 419-424, 1999.
[3] Ateshian, G, *J Biomechanics*, 42:1163-1176, 2009.
[4] Moore, A et al., *Osteoarthritis and Cartilage*, 25.1: 99-107, 2017.

SB^{3}C2017
Summer Biomechanics, Bioengineering and Biotransport Conference
June 21 – 24, Tucson, AZ, USA

REPEATED SUB-THRESHOLD JOINT LOADING PRODUCES PAIN AND ALTERS BIOMECHANICAL & SPINAL GLIAL RESPONSES

Sonia Kartha (1), Ben A. Bulka (1), Nicholas S. Stiansen (1), Harrison R. Troche (1), Beth A. Winkelstein (1,2)

(1) Department of Bioengineering
University of Pennsylvania
Philadelphia, PA, USA

(2) Department of Neurosurgery
University of Pennsylvania
Philadelphia, PA,

INTRODUCTION

The cervical facet joint and its capsular ligament are common sources of spine pain from mechanical injury [1,2]. Even a single sub-failure loading exposure can induce facet capsule laxity and persistent pain [3-7]. Despite maintaining its failure properties, the ACL exhibits laxity and lower stiffness in response to sub-failure loading [8], suggesting altered ligament structure in the absence of its gross injury. In fact, the same macrostrucural changes that suggest injury to the facet capsule and that are large enough to cause pain are accompanied by its collagen realignment and disorganization [5,9,10]. In contrast, a single exposure of the facet capsule to loading below its threshold for initiating pain does not change the microstructure of that ligament [4,5]. Since repeated loading can accumulate microstructural damage and/or alter the local environment of the nerves that innervate the facet capsule [11,12], it is possible that repeated loading, even below the threshold for pain or injury, may be detrimental. It is unknown if repeated sub-threshold loading of the facet capsule can alter its mechanics, which may decrease the threshold for afferent activation, even at loading typically below the pain threshold.

This study evaluated if repeated sub-threshold (ST) loading of the facet ligament induces pain and/or induces glial activation. The failure properties of the isolated facet capsule also were evaluated *after* loading to investigate potential mechanisms of mechanotransduction. Rats underwent either nonpainful ST facet distraction or sham imposed twice, separated by two days. Temporal pain responses were measured, along with capsular biomechanics and spinal glial activation to begin to probe potential pathophysiologic mechanisms.

METHODS

Studies used male Holtzman rats and were IACUC-approved. Rats underwent either a nonpainful ST bilateral C6/C7 facet joint distraction (FJD) (n=12) or a sham surgery procedure (n=12) on days 0 and 2. FJD was applied by a custom loading device at 15mm/sec to joint displacement of 0.2mm [3,5]. Forepaw sensitivity was measured as the withdrawal threshold to stimulation by von Frey filaments and was measured before the 1st surgery (day 0) and on days 1, 2 (before 2nd surgery), 3, and 7. Thresholds were compared between groups by a two-way ANOVA and Tukey's test. For each FJD the vertebral and capsule displacement, force, ligament stiffness, and maximum principal strain (MPS) at 0.2mm were quantified [3-5] and compared between the 1st and 2nd distractions for each rat using paired t-tests.

In half of the rats (n=6/group) on day 7 the C4-T2 spines were excised. The C6/C7 disc and left capsule were transected. Markers were affixed to the right capsule; the facet was set to an interlaminar distance of 2.54mm approximating its in vivo spacing [7]. Pre-conditioning from 0-0.2mm was applied for 60 cycles; tension was applied at 500%/sec until failure. Vertebral and capsule displacement, force, stiffness, and MPS at failure were calculated as with FJDs [3-5]. Failure was confirmed by visible rupture [7,13]. Failure properties were compared between groups by separate t-tests.

The C7 spinal cord was harvested on day 7 from the other rats and fixed in 4% paraformaldehyde. Spinal cord from naïve rats was also included (n=2) as comparison. Tissue was cryosectioned axially (14μm), blocked with 10% normal donkey serum and incubated in mouse anti-GFAP (1:500, Millipore) and rabbit anti-Iba1 (1:1000, Wako). Sections were then incubated in donkey anti-mouse 488 and donkey anti-rabbit 546 (1:500, Invitrogen). The superficial dorsal horns (from 4-6 sections/rat) were imaged at 20X using an Olympus 5X51 microscope. GFAP and Iba1 expression in the superficial laminae of the dorsal horn were quantified with a custom MATLAB script to count the total percent of pixels positive for either GFAP or Iba1 [14]. Expression of each was compared separately between groups using a one-way ANOVA with Tukey test.

Technical Presentation #187

RESULTS

Neither a single ST FJD (FJD1) nor sham (SHAM1) on day 0 altered withdrawal thresholds from baseline on days 1 or 2 (Figure 1). Yet, after the 2nd ST FJD (FJD2) on day 2, thresholds were reduced on day 3 and remained significantly lower on day 7 ($p<0.048$). Withdrawal thresholds in that FJD2 group also were significantly lower than SHAM2 on both days ($p<0.001$) (Figure 1). A 2nd sham did not alter thresholds (Figure 1). Overall, the difference in pain responses between the two groups was also significant ($p<0.001$).

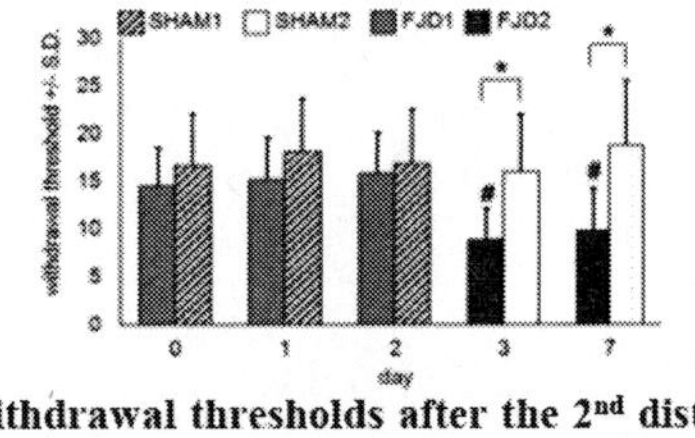

Figure 1: Withdrawal thresholds after the 2nd distraction (FJD2) are lower (#p<0.048) than the 1st distraction (FJD1) & day 0. Thresholds on days 3 & 7 for FJD2 are lower (*p<0.001) than after the 2nd sham (SHAM2). There were no differences for sham.

Although the vertebral displacement was not different between the 1st and 2nd FJDs, the capsule underwent significantly more ($p=0.019$) distraction during the 2nd FJD (Figure 2). Neither peak force nor MPS was different between FJDs. Yet, stiffness significantly decreased ($p=0.008$) during the 2nd FJD (Figure 2). Vertebral displacement at failure was significantly greater ($p=0.014$) for repeated ST loading than for repeated sham ($p=0.014$). But, capsule displacement was not different between groups. Peak force at rupture also was significantly greater ($p=0.041$) after repeated loading. No changes in stiffness or MPS were detected between groups.

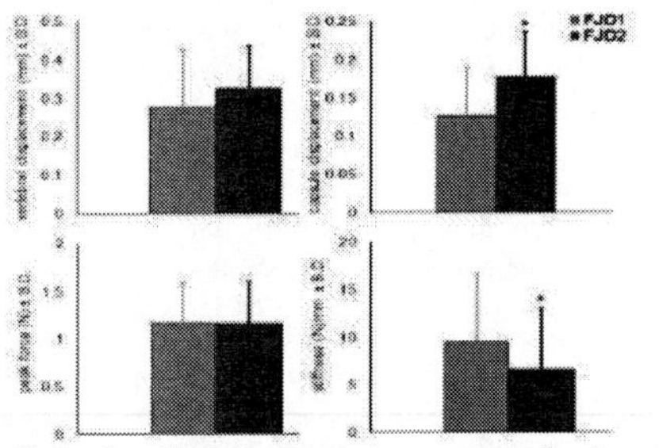

Figure 2: Capsule displacement is greater during the 2nd distraction (*p=0.019) than the 1st, despite the vertebral displacement being unchanged. Peak force is also not different between 1st and 2nd FJD. But, joint stiffness for FJD2 is significantly lower (*p=0.008) than during FJD1.

Spinal glial activation was induced by a repeated ST FJD, with both GFAP and Iba1 expression increased in the superficial dorsal horn (Figure 3). GFAP expression was 40% greater than after a repeated sham ($p=0.007$) and nearly twice that of normal ($p<0.001$) (Figure 3). Iba1 expression was also significantly greater ($p=0.013$) than expression in the repeated sham and normal ($p<0.001$) (Figure 3). A repeated sham did not alter either marker from normal levels.

DISCUSSION

Sub-threshold loading that normally does not induce pain, was painful and altered the ligament's displacement and stiffness during facet joint distraction loading (Figures 1 & 2). Although an initial sub-threshold distraction did not induce pain, when repeated 2 days later, pain and spinal glial activation were induced and remained until day 7 (Figures 1 & 3). The onset of pain and subsequent immune activation

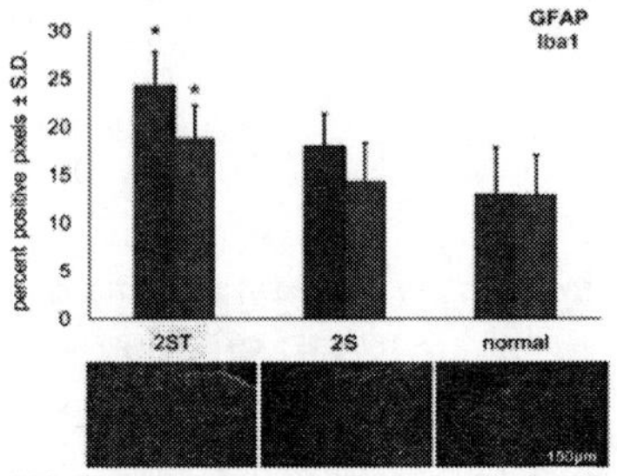

Figure 3: GFAP (green) and Iba1 (red) in the dorsal horn significantly increase at day 7 after repeated FJD (2ST) over levels after repeated sham (2S) and in normal rats (*p≤0.013).

may be due, in part, to a lowered threshold for neuronal activation after the 1st FJD that was activated during 2nd FJD, since the peak force was not different between those distractions (Figure 2). Excessive facet stretch evokes neuronal hyperxcitability in the spinal cord [15]. The deformation during repeated ST loading may induce neuronal activation and pain by cumulative activation. Although these findings suggest that even single ST nonpainful facet loading may lower its injury threshold to subsequent painful loading, studies are needed to define mechanical and physiologic thresholds.

A repeated FJD that is normally non-injurious and nonpainful, did lead to changes in the joint and spinal cord (Figures 2 & 3). The decreased joint displacement and force at failure, along with its decreased stiffness (Figure 2), suggest that a ST exposure alters the capsule's mechanical response. This is in contrast to prior work showing ST tension not changing capsule stiffness or inducing laxity [7]. But, that study calculated the *tangent* stiffness and imposed isolated capsule tension at a rate two orders of magnitude slower than that used here [7]. Since a higher loading rate increases stiffness [16], it is possible that any change is more detectable at the higher strain rate used here. Regardless, a repeated ST trauma seems to have cumulative effects similar to those of a single, more severe exposure [11]. So, *repeated* ST loading may alter the structural response of ligaments similarly to loading that is sufficient to induce pain.

In addition to pain, spinal glial activation increased at day 7 (Figure 3). Immune cell activation is associated with facet joint pain [4,14,17]. This is the first study to demonstrate that repeated tissue loading, even at physiological levels, can induce pain-associated inflammation. Although this work begins to point to a possible mechanism through which non-injurious loading becomes injurious if repeated, more studies are needed to fully define the spatiotemporal biomechanical and physiological mechanisms relating microstructural to macrostructural ligament responses, as well as relative to pain.

ACKNOWLEDGEMENTS

Funded by NIAMS (AR056288) & DOD (W81XWH-10-2-0140).

REFERENCES

[1] Aprill C et al *Spine* 17:744-7, 1992 [2] Bogduk N et al *Spine* 13:610-7, 1988 [3] Lee K et al *J Neurosci Methods* 137:151-9, 2004, [4] Lee K et al *Stapp Car Crash J* 48:373-96, 2004 [5] Dong L et al *Brain Res* 1461:51-63, 2012, [6] Ita M et al,*J Orthop Sports Phys Ther* 201, [7] Quinn K et al *Stapp Car Crash J* 51:169-87, 2007 [8] Panjabi M et al *Clin Biomech* 16:334-40, 1999 [9] Quinn K et al *J Biomech* 43:1870-5, 2010 [10] Quinn K et al *J Appl Physiol* 105:1881-8, 2008 [11] Benoliel R et al *Pain* 14:24-31, 2002 [12] Zhang S et al *J Biomech Eng* 138:021013 2016 [13] Lee D & Winkelstein B *J Biomech* 45:2325-9, 2012 [14] Dong L et al *J Pain* 14(12):1564-72 2013 [15] Quinn, K et al *Pain* 151:414-21, 2010 [16] Zhang G *J Musculoskelet Neuronal Interact* 5:85-90, 2005 [17] Winkelstein B & Santos D *Spine* 33:856-62, 2008.

ANTERIOR CAPSULE IS A LARGER CONTRIBUTOR TO CONTRACTURE THAN MUSCLE IN A RAT ELBOW MODEL OF POST-TRAUMATIC JOINT CONTRACTURE

C. Dunham (1), R. Castile (2), A. Chamberlain (3), S. Lake (1,2,3)

(1) Department of Biomedical Engineering
Washington University in St. Louis
St. Louis, MO, USA

(2) Department of Mechanical Engineering & Materials Science
Washington University in St. Louis
St. Louis, MO, USA

(3) Department of Orthopaedic Surgery
Washington University in St. Louis
St. Louis, MO, USA

INTRODUCTION

Traumatic injury to the elbow leads to post-traumatic joint contracture (PTJC) in as many as 50% of patients [1]. More so than any other joint, the elbow is especially susceptible to contracture due to its highly congruent joint surfaces [2]. The elbow is stabilized by periarticular soft tissues (i.e., capsule, ligaments, tendons, muscles), which are often disrupted when the elbow is injured. However, it is unclear how each of these structures specifically contribute to elbow joint contracture.

In animal models of knee PTJC, arthrogenic tissues (i.e., capsule, ligaments, bones) have been found to contribute approximately 4X more to contracture than myogenic tissues (i.e., tendons, muscles) after six weeks of immobilization only [3,4]. It is unknown if the relative balance of these tissues' contribution to contracture is the same in the elbow. Also, no prior work in any joint model of PTJC has specifically evaluated capsular contribution to contracture. Clinical evidence suggests that the anterior capsule contributes to human PTJC and it has exhibited several biological changes [2,5-7], yet it is still unclear how these biological changes translate to loss in mechanical function.

We previously developed an animal model of PTJC in the rat elbow, which mimicked biological and mechanical symptoms common to human PTJC (i.e., increased adhesions and thickness of the anterior capsule; decreased range of motion) [8-10]. The purpose of this study was to evaluate the myogenic and capsular contributions to elbow contracture to narrow the focus of future research and to determine which tissue(s) should be targeted for therapeutic intervention.

METHODS

In this IACUC approved study, male Long-Evans rats (~300g) underwent unilateral surgery to replicate soft tissue damage seen in elbow dislocation, specifically an anterior capsulotomy and lateral collateral ligament transection. Injured limbs were then immobilized (IM) for six weeks [8]. Control animals were neither injured nor immobilized, but allowed unrestricted cage activity.

Post-mortem, elbow joints were subjected to flexion-extension (F-E) mechanical testing using protocols previously described [8,9] (n = 5-6/group). The limbs were tested a total of three times to determine the contribution of periarticular soft tissues surrounding the elbow to overall joint contracture. F-E testing was completed for each of the following conditions: (1) all soft tissue intact (full), (2) all muscles and tendons removed (no muscle; NM), and (3) all muscles/tendons and anterior capsule removed (no muscle/capsule; NMC). Force and displacement data were converted to torque and angular position, and loading curves were analyzed for maximum extension (Figure 1A). Zero degrees represents full limb extension, therefore the larger the angle of extension, the larger the contracture and range of motion loss (Figure 2A). Analysis focused on extension because previous data using our animal model showed that no significant changes occurred in flexion [8,9]. Post-test analysis also calculated the average curves for each group by averaging the maximum extension and both end points of the neutral zone (average of the loading and unloading curve) to present a visual of the qualitative differences in joint motion between groups (Figure 1B).

Repeated one-way ANOVA tests were used to compare the IM and control limbs separately. When there was significance, post-hoc Bonferroni analyses compared each condition to the full, intact limb. Paired t-tests were used to compare IM to control limbs within each condition. All statistical analysis was performed in GraphPad Prism and significance was defined as $p < 0.05$.

RESULTS

Qualitatively, group-averaged mechanical test data showed differences in the overall F-E loading profiles for IM full and NM limbs

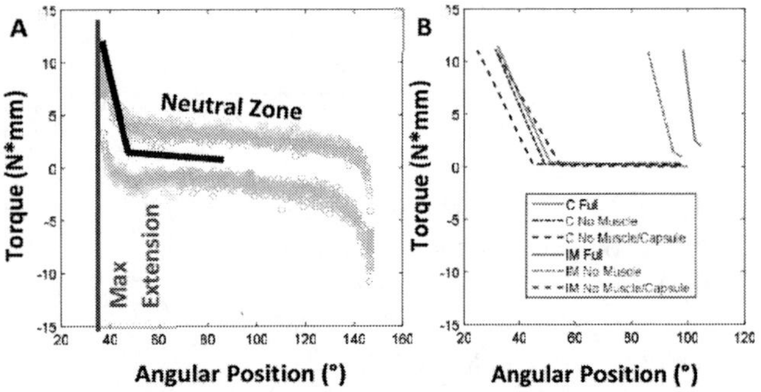

Figure 1: A) Schematic of a torque-angle loading curve for flexion-extension biomechanical joint testing. The light gray circles are data from a representative sample. The neutral zone is the average of the loading and unloading curves. B) Average curves for control (C) and injured (IM) limbs. These average curves represent the mean of all the animals tested per group without standard deviation to make the differences between the curves more apparent. (solid = full limb, dash-dot = no muscle, dash = no muscle/capsule)

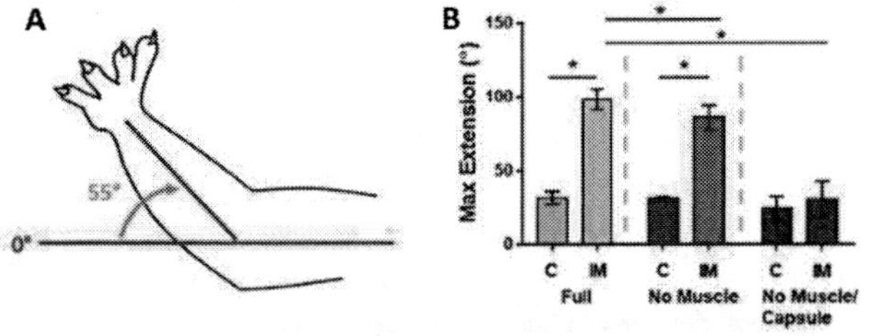

Figure 2: A) Schematic of a rat forelimb angle calculation, zero degrees represents full extension of the limb. B) Maximum extension for injured (IM) limbs were increased compared to control (C) limbs. IM limbs exhibited a significant decrease which was small following removal of muscles/tendons, but large after the subsequent removal of the anterior capsule. (average ± standard deviation; * = significant different, $p < 0.05$)

when compared to all control conditions and IM NMC (Figure 1B). The IM full limbs had a compressed average curve compared to all other groups, representative of extension loss. While removal of the muscles and tendons showed a small change in the maximum extension, the subsequent removal of the capsule showed a much larger change. The direction of these changes demonstrated a larger shift toward increased range of motion, and thus became more similar to the control conditions.

Quantitatively, the maximum extension values of the IM limbs were significantly increased compared to the control limbs for the full and NM conditions (Figure 2B), representative of significant contracture in the IM limbs. The IM NM and NMC conditions were significantly decreased compared to IM full limbs, demonstrating that both the muscle and anterior capsule contributed significantly to elbow contracture.

DISCUSSION

This study showed that the muscle and anterior capsule both contributed significantly to contracture in our rat elbow model of PTJC. After the sequential removal of the muscles/tendons and capsule, there was a 12% and 68% decrease in the maximum extension of the IM limbs, respectively (Figure 1,2). Compared to the maximum extension of non-contracted control limbs (~29°), the muscle/tendons were responsible for ~18% of joint contracture in IM limbs, while the capsule accounted for ~82% of the overall contracture. Thus, while the muscles and tendons play a role, the capsule tissue is the major contributor to elbow PTJC in our animal model.

Human elbow anterior capsule has been described as less compliant following injury and a primary contributor to elbow contracture [6,11]. However, it has never been mechanically evaluated to confirm these observations. In animal models of knee PTJC, arthrogenic tissues were found to be the main contributor to long-lasting contracture, primarily causing loss of motion in extension [4,12], but the capsular contribution was not specifically evaluated. Therefore, our data represents the first time that the capsule was separated from the other arthrogenic tissues to understand its mechanical contribution to PTJC. After the dissection of the muscles/tendons and the subsequent removal of the capsule in our animal model of PTJC, IM limbs were not significantly different from control limbs, confirming that the capsule is a primary component of joint contracture, specifically in the elbow. These data also suggest that the remaining arthrogenic components (i.e., ligaments, bones) do not severely influence contracture in our animal model, although they do limit the ability of the rat forelimb to reach full extension (i.e., to 0 °).

Even though muscle was a smaller contributor to contracture (Figure 1,2), the role of muscle tissue in PTJC remains important because its removal resulted in a significant decrease in maximum extension. In previous work, the biceps brachii from our rat elbow model of PTJC exhibited significantly decreased active and increased passive mechanics compared to control muscles, representative of muscle retraction and fibrosis [13]. These specific changes in muscle properties further support that muscle significantly contributes to elbow contracture following injury and six weeks of immobilization. However, since the decrease in maximum extension is relatively small once the muscle is removed, it will be important in future work to evaluate muscle following a period of free joint mobilization (i.e., free cage activity, no immobilization) to determine if the changes to muscle are permanent and alter joint range of motion long term.

In conclusion, this study showed that muscles/tendons and capsule both contribute to elbow contracture, while the capsule is the primary contributor, suggesting that it should be a primary target for therapeutic intervention. While previous histological evaluation of the capsule has shown increased myofibroblasts, adhesions and thickness [8,9], future work evaluating collagen content and organization will aid understanding of how the capsule should be targeted for treatment to prevent elbow contracture.

ACKNOWLEDGEMENTS

The authors thank the NIH for a training fellowship (NIBIB T32 EB018266) and research funding (NIAMS R03 ARR067504).

REFERENCES

[1] Anakwe, R.E., et al., *JBJS*, 93: 1220-1226, 2011.
[2] Cohen, M.S., et al., *JSES*, 16: 484-490, 2007.
[3] Trudel, G., et al., *Arch Phys Med Rehabil*, 81: 6-13, 2000.
[4] Chimoto, E., et al., *Ups J Med Sci*, 112: 347-355, 2007.
[5] Germscheid, N.M., et al., *Clin Orthop Relat Res*, 450: 219, 2006.
[6] Monument, M.J., et al., *J Orthop Sci*, 18: 869-877, 2013.
[7] Charalambous, C.P., et al., *JBJS*, 94: 1428-1437, 2012.
[8] Lake, S.P., et al., *JOR*, 34: 354-364, 2016.
[9] Dunham, C.L., et al., *JSES*, In press.
[10] Dunham, C.L., et al., *ORS*, 2016.
[11] Lindenhovius, A.L., et al., *The Journal of hand surgery*, 32: 1605-1623, 2007.
[12] Trudel, G., et al., *Journal of Applied Physiology*, 117: 730-737, 2014.
[13] Dunham, C.L., et al., *ORS*, 2017.

SB[3]C2017
Summer Biomechanics, Bioengineering and Biotransport Conference
June 21 – 24, Tucson, AZ, USA

MODELING TENDON VISCOELASTICITY, PLASTICITY, AND DAMAGE USING REACTIVE INELASTICITY

Babak N. Safa (1,2), Michael H. Santare (1,2), Dawn M. Elliott (1,2)

(1) Department of Mechanical Engineering
University of Delaware
Newark, DE, USA

(2) Department of Biomedical Engineering
University of Delaware
Newark, DE, USA

INTRODUCTION

Tendon transmits load from muscle to bone and with large and/or frequent loading a loss in mechanical properties can lead to clinical disorders such as tendinopathy or rupture at loads significantly below the ultimate strength of the tissue. This loss of mechanical properties is a form of *inelastic behavior*. An inelastic behavior is one where, in an arbitrary cycle of deformation, energy is dissipated. The major inelastic behaviors of tendon are viscoelasticity, plasticity, and damage. Despite decades of tendon mechanics studies and several experimental observations of inelastic behaviors (e.g., creep rupture [1] and strain softening [2]), a comprehensive theoretical treatment of inelasticity is missing. The objective of this study was to formulate a comprehensive modeling framework, *reactive inelasticity*, for modeling tendon's inelastic behaviors that is based on molecular bond kinetics.

The adoption of molecular bond kinetics to model inelasticity was pioneered by Green and Tobolsky [3], and is commonly used to study viscoelasticity and plasticity of polymers [4]. This framework was recently reformulated by Ateshian to model viscoelasticity of solid mixtures (reactive viscoelasticity) [5] and has been applied to damage of cartilage constructs [6]. We have recently applied it to describe viscoelastic damage in tendon fascicles [7].

In this study we generalize and extend reactive viscoelasticity [5] to a unifying and comprehensive modeling framework (reactive inelasticity) to explain inelastic behaviors of tendon—viscoelasticity, plasticity, and damage—by defining three types of bonds: (1) Formative (for viscoelasticity), (2) Permanent (for hyperelasticity), and (3) Sliding (for plasticity). These bonds all use the same constitutive equations but at different kinetic rates to account for different mechanical behaviors. Damage is modeled by reducing the number fraction for each type of

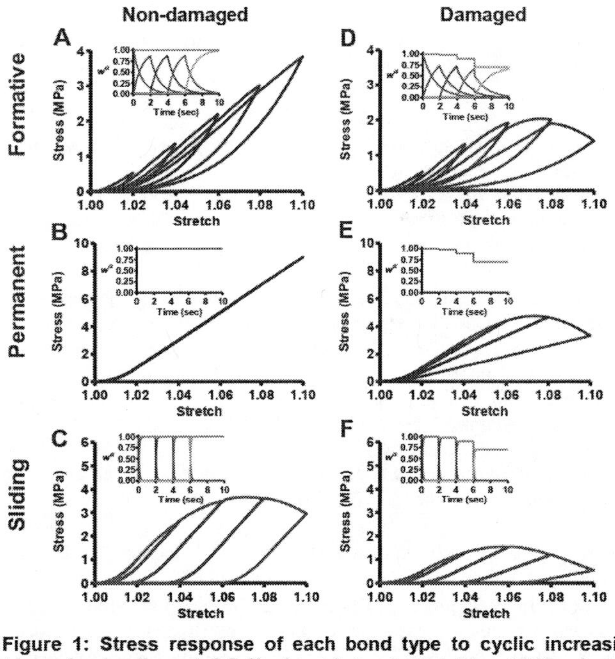

Figure 1: Stress response of each bond type to cyclic increasing triangular loading at 0.5 Hz (stretch, $\lambda = 1.02, 1.04, \ldots 1.10$) . Insets: schematic of evolution of w^{α} for each bond type: formative (*f*), permanent (*p*), and sliding (*s*), with time constants ($\tau_f = 1$ sec, $\tau_p = 10^5$ sec, $\tau_s = 0.1$ sec).

bond. Finally, we provide two examples of the mechanical responses that occur with combinations of these bond types and damage: creep and incremental stress relaxation.

METHODS

Reactive Inelasticity Constitutive Framework: The underlying premise of reactive inelasticity is that the overall free energy of the system is a function of deformation and number fraction of several *bond types* that break and reform (i.e., *react*) in response to external loading. A reforming bond initiates a new *generation*. The bonds are *intrinsically hyperelastic*. That is, there is a unique free energy potential for each type of bond which is a function of the relative deformation gradient tensor ($\boldsymbol{F}_\gamma^\alpha$), defined as $\boldsymbol{F}_\gamma^\alpha \equiv \boldsymbol{F}.(\boldsymbol{\Pi}_\gamma^\alpha)^{-1}$, where $\boldsymbol{F}$ is the overall deformation gradient tensor and $\boldsymbol{\Pi}_\gamma^\alpha$ is the deformation gradient tensor of the reference configuration (subscript γ is the bond type: f for formative, p for permanent and s for sliding; superscript α is the generation of reformed bonds). As a result, the overall free energy density (Ψ) is:

$$\Psi(\boldsymbol{F}_\gamma^\alpha, w_\gamma^\alpha) = \rho \sum_\gamma \sum_\alpha w_\gamma^\alpha(t)\, \psi_\gamma(\boldsymbol{F}_\gamma^\alpha) \tag{1}$$

where ψ_γ is the free energy potential per unit mass and ρ is the overall density of the material. w_γ^α is the number fraction of bonds, which is defined as the number of bonds of a generation divided by the maximum number of possible bonds of that type in the natural state of the material. The Cauchy stress tensor can be derived from (Eq. 1) [5].

Formative bonds break when subjected to loading. Broken bonds simultaneously reform to a new unloaded state at the current configuration (i.e., $\boldsymbol{\Pi}_f^\alpha = \boldsymbol{F}(t^\alpha)$, where t^α is the time of formation of generation α). This change from loaded to unloaded state dissipates energy [3]. A continuous deformation can be discretized into several step deformations that result in new generations. All of the bonds are either in the loaded (breaking) or unloaded (reforming) state (i.e., $w_\gamma^\alpha(t) = 1 - \sum_{t^i<t} w_\gamma^i(t)$), where the number fraction of breaking bonds reduces according to the rate equation:

$$\frac{dw_\gamma^i}{dt} = -\frac{w_\gamma^i}{\tau} \quad (t \geq t^i) \tag{2}$$

Here, τ is the reaction time constant representing a first order kinetics reaction with an exponential decay.

Permanent bonds are the special case of formative bonds, where the rate of breakage of bonds is extremely slow ($\tau \to \infty$). As a result, only one generation of bonds exists ($w_p = 1$, and $\boldsymbol{\Pi}_p^\alpha = \boldsymbol{I}$) and the behavior is hyperelastic.

Sliding bonds are another special case where the rate of breakage and reformation of bonds is extremely fast compared to the characteristic time of the loading ($\tau \to 0$). In this case, if the bonds could reform to the current configuration, the energy would not be stored in the system. However, if the newly formed bonds have a reference configuration other than the current configuration ($\boldsymbol{\Pi}_s^\alpha \neq \boldsymbol{F}(t^\alpha)$), they slide (flow), resulting a plastic deformation.

Damage of bonds can be added to any of these kinetics relations by reducing the number fraction of the bonds:

$$w_\gamma^\alpha = (1 - D_\gamma)(w_\gamma^\alpha)_0 \tag{3}$$

Where, D_γ is a scalar quantity ($0 < D_\gamma < 1$), and $(w_\gamma^\alpha)_0$ is the hypothetical, non-damaged number fraction derived using (Eq. 2).

Constitutive Relations: A 1D exponential-linear constitutive relation was used for intrinsic hyperelastic relation of all of the bond types (linear region modulus (C_3= 100 MPa), toe-region shape parameter ($C_2 = 10$), uncramping stretch ($\lambda_c = 1.02$)), and a Weibull CDF was used for the accumulation of damage ($k = 4, l = 1.10$) [7]. For sliding bonds, the following relation was used to track the change in reference configuration:

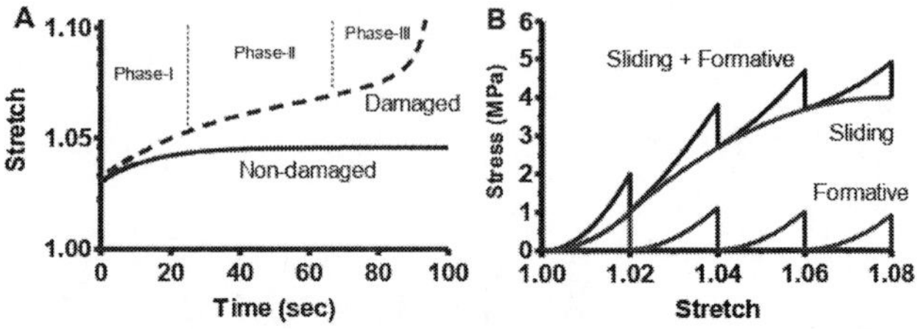

Figure 2: (A) Creep response showing equilibrium response in non-damaged case and three phases of creep damage and rupture when damage is included (B) Incremental stress relaxation response modeled using sliding and formative bonds. ($\tau_f = 10$ sec, $\sigma_0 = 10$ MPa)

$$(\lambda_s^\alpha - 1) = c(\lambda_y - \lambda_y^0)^b \tag{4}$$

where, λ_s^α is the sliding reference stretch, and λ_y is the yield stretch and λ_y^0 is the stretch at the onset of sliding; c and b are positive dimensionless positive numbers ($c = 10$, $b = 2$, $\lambda_y^0 = 1.02$). These model parameters were selected to match typical tendon responses and are illustrative only to show the model response.

Illustrative Examples: Behavior of the basic bond types for increasing cyclic tensile loading was simulated in MATLAB for non-damaged, and damaged cases. Additionally, we simulated the creep response of tendon (formative + permanent bonds with damage) and softening in incremental stress relaxation (formative + sliding bonds).

RESULTS

Each bond type's response is demonstrated in Figure 1. The formative bonds dissipate energy in a cyclic hysteresis (Fig.1 A), the permanent bonds have a hyperelastic response (Fig.1 B), and the sliding bonds undergo plastic elongation (Fig.1 C). By adding damage (decreasing the total number of bonds), softening was captured for all types of bonds (Fig.1 D-E). In particular, for permanent bonds, the Mullin's effect (well-known for elastic rubber materials) was reproduced (Fig.1 E). For sliding bonds, there was no change in modulus without damage, when damage was added the modulus decreased (Fig.1 F).

For the combination of bonds (Figure 2), creep loading showed an equilibrium response in the non-damaged case and the three phases of creep rupture were reproduced when damage was added (Fig.2 A). The softening in incremental stress relaxation was also reproduced when formative and sliding bonds were combined (Fig.2 B).

DISCUSSION

We successfully developed a comprehensive framework to model the inelastic behaviors of tendon (i.e., viscoelasticity, damage, and plasticity). We used the concept of molecular bond kinetics to account for mechanisms of inelasticity, employing the same constitutive equations at different extremes of kinetic rates. Examples using this framework illustrated its potential to model tendon mechanical response. In its current form, this framework is limited because it doesn't include fluid flow effects; these will be added. Future work will solve the model using experimental data to elucidate each bond type's contribution to the mechanisms of inelasticity in tendon.

ACKNOWLEDGEMENTS

National Institutes of Health grants R01EB02425.

REFERENCES

[1] Wang and Ker, *J. Exp. Biol.*, 198: 831–45, 1995.
[2] Szczesny and Elliott, *Acta Biomater.*, 10: 2582–2590, 2014.
[3] Green and Tobolsky, *J. Chem. Phys.*, 14: 80, 1946.
[4] Meng and Terentjev, *Polymers (Basel).*, 8: 108, 2016.
[5] Ateshian, *J. Biomech.*, 48: 941–947, 2015.
[6] Nims *et al.*, *Interface Focus*, 6: 63, 2015.
[7] Safa, *et al.*, *Proceedings of SB3C* 1337–1338, 2016.

ARTERIAL DAMAGE MODEL BASED ON EMPIRICAL STRETCH THRESHOLDS OF COLLAGEN UNFOLDING AND TISSUE YIELDING

Matthew I. Converse (1), Michele Marino (2), Kenneth L. Monson (1, 3)

(1) Department of Mechanical Engineering
University of Utah
Salt Lake City, UT, USA

(2) Institut f□r Kontinuumsmechanik
Leibni□ Universit□t Hannover
Hannover, Germany

(3) Department of Bioengineering
University of Utah
Salt Lake City, UT, USA

INTRODUCTION

Arteries play a critical role in carrying essential nutrients and oxygen throughout the body. However, trauma to the head and neck as well as surgical interventions such as angioplasty can distend arteries beyond their physiological range. Even in the absence of hemorrhaging, such 'subfailure' deformations can damage cells [1-2], cause permanent deformation of the tissue [2], alter vascular mechanics [3-4], and, ultimately, compromise vessel function [1].

A key constituent in all of these processes is collagen. Not only do collagen fibers dominate the passive mechanical behavior at high strains [5], but they also transfer mechanical loads to and from cells. Due to their important role, several structurally-based constitutive models of arterial mechanics have been proposed to incorporate collagen fibers [5]. However, existing models of arterial damage remain purely phenomenological in nature [4, 6]. This may be due to the lack of objectives assays of collagen disruption.

We recently presented the effectiveness of a collagen hybridizing peptide (CHP) in detecting the unfolding of collagen triple helices following arterial overstretch. Not only is this technique the first to directly identify molecular damage to collagen, but the peptide's fluorescent label facilitates straight-forward identification and objective quantification of damage. This progress opens the door to better characterizing and modeling structure-property relationships in arterial damage mechanics.

The objectives of the present work were to:

1. Use the CHP marker to characterize the progression of collagen damage in both axial and circumferential overstretch of cerebral arteries.
2. Demonstrate the relationship between such molecular damage and alterations of the tissue-level mechanical behavior.
3. Develop a constitutive model based on the empirically measured parameters that accurately captures the damaged mechanical response at the tissue level.

METHODS

Mechanical Overstretch: Building off of experiments and methods reported previously [7-8], sheep middle cerebral artery segments were preconditioned and subjected to a single quasi-static overstretch (0.1 mm/s). A total of 21 segments (3-5 mm long) were stretched axially (6 controls) while 28 rings (1 mm long) were distended circumferentially (7 controls). A range of stretch levels up to failure were explored.

Quantification of Collagen Damage: Immediately after overstretch, samples were stained with CHP and imaged with a confocal microscope. Images were taken and quantified within the adventitia for axially loaded samples and within the media for circumferentially loaded samples. Collagen damage was quantified by calculating the percentage of pixels greater than a control-specific intensity threshold. Damage-stretch relationships were graphed and fit with a piece-wise linear regression model to determine the stretch at which the onset of damage occurred.

Quantification of Tissue-Level Yielding: In order to correlate collagen damage with changes in tissue-level mechanical properties, the yield point was calculated for each sample. Yielding was defined at the earliest point in the loading curve where a 30% reduction in the tangent modulus was observed.

Constitutive Damage Model: A constitutive model was developed within the framework of Generalized Standard Materials. Strain energy density was introduced as a function of two damage variables, d_1 and d_2, which evolve from 0 to 1 with increasing damage.

Damage variable d_1 was associated with collagen unfolding (i.e., the loss of molecular structural integrity from the breaking of weak bonds between polypeptide strands). Its evolution was defined by the experimentally derived damage-stretch relationships using CHP. This variable was introduced in the definition of a generalized fourth invariant of deformation which represents the amount of fiber straightening associated with an elastic response. Therefore, evolution of d_1 resulted in a yielding behavior as the reference length of collagen fibers increased and was associated with the onset of residual strain.

Damage variable d_2 was associated with molecular covalent rupture (i.e., the rupture of covalent bonds within polypeptide strands) of collagen molecules. Its evolution was defined by the average failure stretch of the tissue as found in the experiments. This variable scaled the elastic potential energy, leading to a stress drop in the stress-strain response. Therefore, it was associated with final failure of the tissue.

The evolution laws for both variables were defined ensuring the thermodynamic consistency of the developed theoretical framework.

RESULTS

Thresholds and Accumulation of Collagen Damage: The CHP signal was found to increase with overstretch severity for both loading directions (Figure 1). CHP signal was comparable to that of controls for relatively mild stretches, then increased nearly linearly beyond a critical stretch threshold, λ_1. Piece-wise linear regression showed the onset of collagen damage to be $\lambda_1 = 1.17$ and 1.35 and the rate of damage accumulation to be m = 0.53 and 0.67 for circumferential and axial loading, respectively.

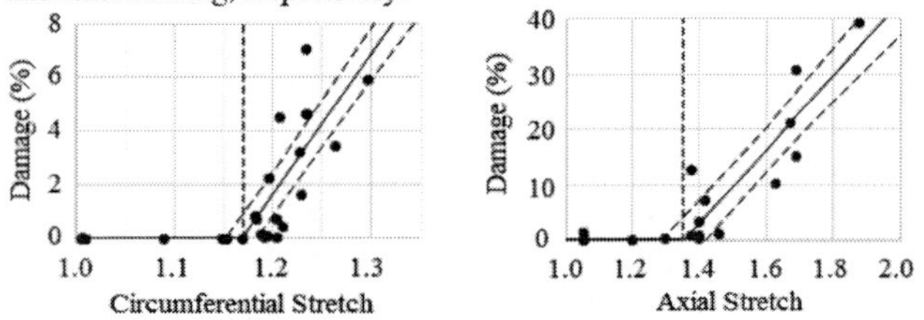

Figure 1: Percentage of pixels above control-specific threshold vs. overstretch (with reference to the in vivo length) for circumferential (left) and axial (right) directions of loading. The intersection of the horizontal and sloped regression lines defines the threshold for collagen damage (red dashed line).

Tissue Yielding: Analysis of tissue-level yielding revealed that stretch thresholds for collagen damage (λ_1) coincided with arterial yield stretches. Yield stretches were found to be 1.19 ± 0.02 (mean ± SD, n=14) for circumferential loading and 1.33 ± 0.06 (n=10) for axial loading. In both cases, the threshold of collagen damage as found by CHP staining fell within one standard deviation of the average yield point (Figure 2).

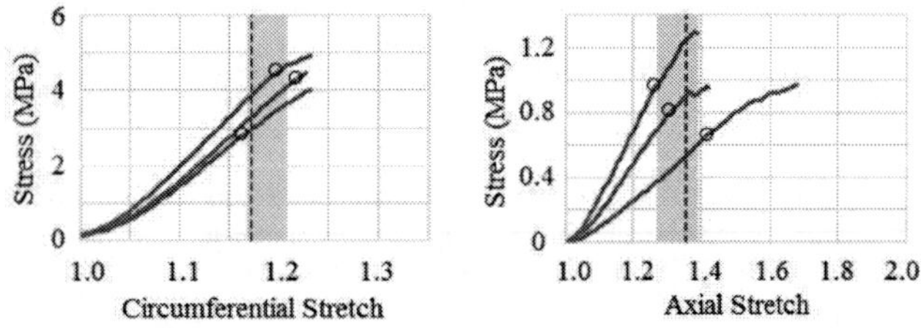

Figure 2: The stretch threshold for collagen damage (red dashed line from Fig. 1) coincided with the yield stretch for both circumferential (left) and axial (right) loading. Representative stress-stretch curves are shown with their yield points circled. The yield stretch for all samples (mean +/- SD) is overlaid in gray.

Constitutive Damage Model: Three experimental parameters were fed into the damage model: the onset of collagen damage (λ_1), the rate of damage accumulation (m), and the mean ultimate failure stretch (λ_2). The model stress response appropriately demonstrated yielding beginning at λ_1 and matched the general behavior of the experimental curve (Figure 3). Only results for the circumferential direction have been included. Furthermore, the damage model captured post-overstretch residual strains comparable to those reported previously (data not shown) [3].

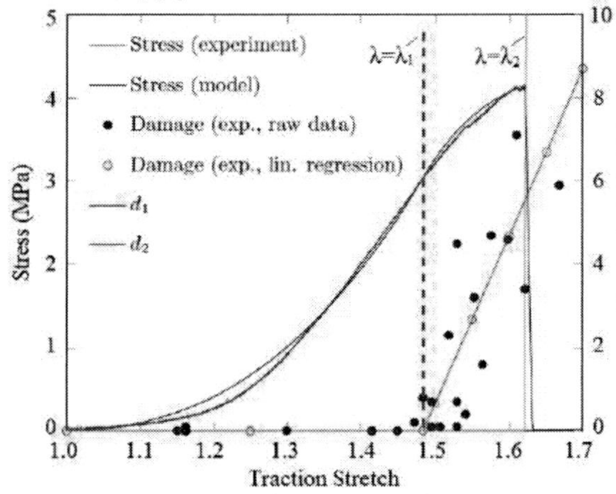

Figure 3: Uniaxial traction (along the fiber direction) of a tissue with one collagen fiber family. Relationship between stress and damage evolution vs. circumferential stretch (with reference to the unloaded length). Comparison between experimental data and model predictions. Red dashed line taken from Fig. 1, left.

DISCUSSION

In addition to the novelty of using CHP to detect molecular level collagen damage in arteries, our most significant experimental finding is the coincidence of the tissue-level yielding with the onset of this damage (inter-strand delamination). To the best of our knowledge, no studies have experimentally demonstrated this correlation before in vascular tissues.

Furthermore, these findings facilitated the development of the first-ever experimentally validated arterial damage model. In classical approaches, internal damage variables are treated as hidden quantities – having no correlation to quantitative measurements [6]. Indeed, the value of damage parameters are classically determined only from fitting the stress-strain constitutive relationship which is ultimately governed by numerous parameters. In contrast, the present work identifies the value of damage parameters from independent measurements on collagen damage and not by means of a phenomenological fit. This approach is a first step in developing experimentally validated multi-scale models of arterial damage.

ACKNOWLEDGEMENTS

3Helix, Inc. for supply of the collagen hybridizing peptide (CHP).

REFERENCES

[1] Jamal, A et al., *Arterioscler Thromb Vasc Biol*, 12:307-317, 1992. [2] Castaneda-Zuniga, W et al., *Radiology*, 135:565-571, 1980. [3] Bell, E et al., *Front Bioeng Biotechnol*, 3, 2015. [4] Weisbecker, H et al., *J Mech Behav Biomed Mater*, 12:93-106, 2012. [5] Gasser, T et al., *J R Soc Interface* 3, 15–35, 2006. [6] Balzani, D et al., *Comput Methods Appl Mech Engrg*, 213:139–151, 2012. [7] Walther, R et al., *SB3C2015*-1020. [8] Converse, M et al., *SB3C2016*-734.

STATIN ATTENUATES THE INFLAMMATORY DAMAGE ON CARTILAGE BY INHIBITING RHO ACTIVITY IN CHONDROCYTES

Mengxi Lv (1), Yilu Zhou (2), Shongshan Fan (2), Olivia Smith (2), X. Lucas Lu (2)

(1) Center for Bioinformatics & Computational Biology
University of Delaware

(2) Department of Mechanical Engineering
University of Delaware

INTRODUCTION

Inflammation after joint injuries is considered a major cause of post-traumatic osteoarthritis. In response to high-level inflammatory cytokines, chondrocytes often undergo a process of swelling, proliferation, hypertrophy, and apoptosis [1]. This process involves the activities of small GTPases (e.g., ras, rho, rac, cdc42) [2]. The prenylation of small GTPases requires geranylgeranyl- and farnesyl-pyrophosphates, which are two down-stream products of the mevalonate pathway [3]. Therefore, inhibition of mevalonate signaling might protect the chondrocytes from inflammatory attack by disabling the activities of small GTPases. As a strong mevalonate pathway inhibitor, statin is a big class of drugs used for the prevention of cardiovascular diseases [4]. In this study, we hypothesized that the interruption of small GTPases in chondrocytes using statin can prevent the cells from responding to the inflammatory attack, and further protect the integrity of cartilage matrix.

METHODS

<u>*RNA-seq of IL-1β Treated Cartilage:*</u> RNA sequencing (RNA-seq) was performed to assess the gene expression changes of small GTPases in chondrocytes under IL-1β attack. Cylindrical cartilage explants (diameter = 3 mm; thickness = 2 mm) were harvested from calf knee joints and assigned into two groups: control and IL-1β-treated group (1 ng/ml) [5]. After 48-hour *in vitro* culture, total RNA was extracted from the explants for RNA sequencing.

<u>*Protective Function of Statin:*</u> Simvastatin (sigma) was added simultaneously with IL-1β into the culture medium of cartilage explants. During the 26 days *in vitro* culture, six groups were included: 1) control, 2) IL-1β, 3) IL-1β + 1 μm statin, 4) IL-1β + 3 μm statin, 5) IL-1β + 10 μm statin, and 6) 10μm statin alone. Statin treatment at 10 μm showed the optimal inhibitory effect on both GAG and collagen loss, and thus 10 μm statin was chosen for all following experiments.

Mechanical properties of cartilage explants were measured via indentation test after 8-day culture. qRT-PCR assessed the expressions of essential chondrogenic genes after 4-day treatment. Histological analysis (H&E) on days 8 and 24 revealed the cell morphology in different groups.

<u>*Role of the Mevalonate Pathway:*</u> To clarify the role of the mevalonate pathway, we further tested another two mevalonate inhibitors, Y27632 (a specific Rho-associated protein kinase inhibitor) and GGTI298 (a selective geranylgeranylation inhibitor), as well as a mevalonate pathway intermediate product, called geranylgeraniol (GGOH) (Fig. 1C). The supplement of GGOH was expected to cancel out the effects of the mevalonate inhibitors. The longitudinal loss of GAG and collagen content was measured during 22 days culture.

<u>*Mechano-transduction of Chondrocytes:*</u> $[Ca^{2+}]_i$ signaling interacts closely with the small GTPases in chondrocytes, especially the activities of rho, ras and cdc42 (Fig. 4A). In this study, cartilage explants cultured with 1) IL-1β, and 2) IL-1β + 10 μm statin were dyed with Fluo-8. The spontaneous calcium signaling of *in situ* chondrocytes was recorded and analyzed.

RESULTS

<u>*Expression of Small GTPases under IL-1β Attack:*</u> IL-1β significantly increased the GAG loss from cartilage explant into culture medium at a dosage- and time-dependent manner (Fig. 1A). After 8-day culture, 1ng/mL IL-1β caused ~50% total GAG loss. RNA-sequencing analysis detected that one of the most significantly changed pathways in IL-1β treated chondrocytes was small GTPases signaling, in which 72 related genes were differentially expressed (Fig. 1B). In the following studies, we aimed to verify that inhibition of the

mevalonate pathway and subsequent small GTPases' activities can protect the chondrocytes from inflammatory attack (Fig. 1C).

Chondro-protective Effect of Statin: Statin at 1, 3, or 10 μm significantly inhibited the IL-1β-induced GAG loss (Fig. 2A). The collagen loss induced by IL-1β was negligible during the first 8-day culture and became substantial from day 8 to 24, which was significantly attenuated by 3 μm statin and almost prevented by 10 μm statin treatment (Fig. 2B). The chondro-protective effects of statin were further reflected on mechanical properties. Statin treatment significantly attenuated the decrease of Young's modulus induced by IL-1β (IL-1β + statin vs. IL-1β alone: 0.16±0.04 vs. 0.09±0.02 MPa, P = 0.045) (Fig. 2C). qRT-PCR showed that statin significantly promoted the expression of anabolic genes (aggrecan and type II collagen) and reduced the expression of catabolic genes (ADAMTS-4, -5 and MMP-1, -9, -13) (Fig. 2D). In the H&E histological image, cell proliferation and swelling were observed in IL-1β-treated cartilage explants, which were inhibited by statin treatment (Fig. 2E).

Role of the Mevalonate Pathway: Similar to statin, another two mevalonate pathway inhibitors, Rho inhibitor and GGTi, significantly reduced GAG and collagen loss from the IL-1β-treated cartilage explants, although the three inhibitors varied in protection effects. The mevalonate pathway intermediate, GGOH, substantially reversed the chondro-protective effects of statin (Fig. 3A-B). Together, the inhibition of the mevalonate pathway can attenuate the matrix degeneration of cartilage explant induced by IL-1β.

Intracellular Calcium Signaling: Spontaneous $[Ca^{2+}]_i$ oscillation of *in situ* chondrocytes was recorded after 4-day drug treatment (Fig. 4B). Rescued by statin, cartilage explants presented more active $[Ca^{2+}]_i$ signaling than the IL-1β alone group, including higher responsive percentages, higher $[Ca^{2+}]_i$ peak magnitudes, and shorter peak recovery times (Fig. 4C).

DISCUSSION

Statin can protect the extracellular matrix integrity of cartilage explants treated with IL-1β via inhibiting the mevalonate pathway in chondrocytes, which further interrupts the activities of small GTPases (ras, Rho, cdc42 etc.). The chondro-protective mechanism of statin was reflected by the spontaneous $[Ca^{2+}]_i$ signaling in chondrocytes. A few clinical trials noticed the beneficial effects of statin in the prevention of osteoarthritis [6]. This study revealed a new chondro-protective mechanism of statin. Over 25 million American people take statin. Inhibition of the mevalonate pathway using statin could be a new therapeutic technique for the prevention of osteoarthritis after joint injuries, such as meniscus tear and ACL rupture.

REFERENCES: [1] Bergmann, G., *J Biomech.* [2] Novakofski, K., *JOR.* [3] Coxon, FP., *Calcif Tissue Int.* [4] Mo, H., *Exp. Biol. Med.* [5] Li, Y., *OAC.* [6] Kadam, U.T., *J Gen Intern Med.*

ACKNOWLEDGEMENTS: DOD-W81XWH-13-1-0148.

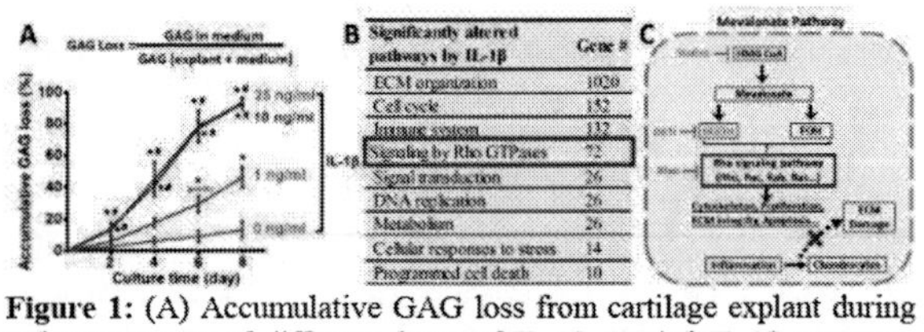

Figure 1: (A) Accumulative GAG loss from cartilage explant during 8-day treatment of different doses of IL-1β. 1ng/ml IL-1β treatment was chosen to mimic a moderately aggressive inflammation injury in the following experiments. N=5 from 3 animals. (B) RNA sequencing showed that IL-1β significantly altered nine OA-related pathways, including Rho GTPases signaling. N=4 from 4 animals. (C) Schematic diagram of the interaction between the mevalonate pathway and small GTPases. The inhibitors of the mevalonate pathway, such as statin, GGTi and Rhoi, might disable the activities of small GTPases and thus protect the chondrocytes from inflammatory attack.

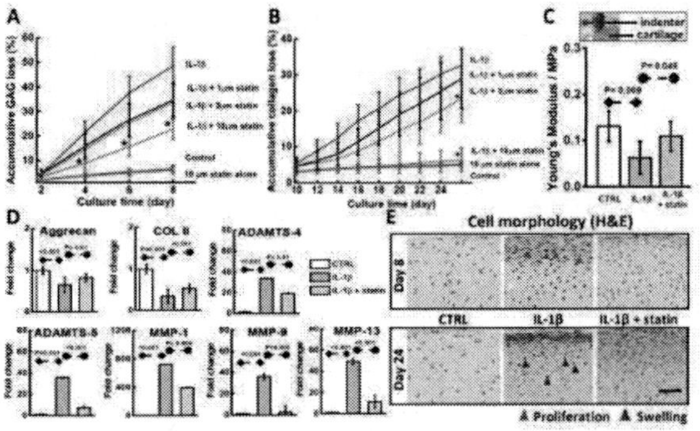

Figure 2: (A-B) Dosage-dependent effect of statin on prevention of GAG and collagen loss from cartilage explant. IL-1β and statin were added into culture medium on Day 0. N=6 from 3 animals. (C) Mechanical properties of cartilage explants after 8-day treatment of statin by indentation test. N=6 from 3 animals. (D) Chondrogenic gene expression of cartilage explants after 4-day culture by qRT-PCR. N=8 from 8 animals. (E) Statin treatment successfully inhibited the chondrocyte proliferation and swelling in cartilage damaged by IL-1β.

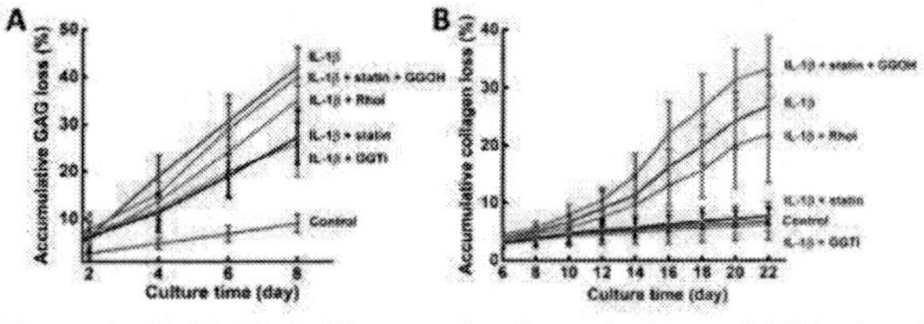

Figure 3: (A-B) Rhoi (Rho-associated protein kinase inhibitor) and GGTi (geranylgeranylation inhibitor) alleviated the IL-1β-induced GAG and collagen loss from cartilage explants. Supplementing a mevalonate derivative, geranylgeraniol (GGOH), cancelled out the protection effect of statin. N=8 from 4 animals.

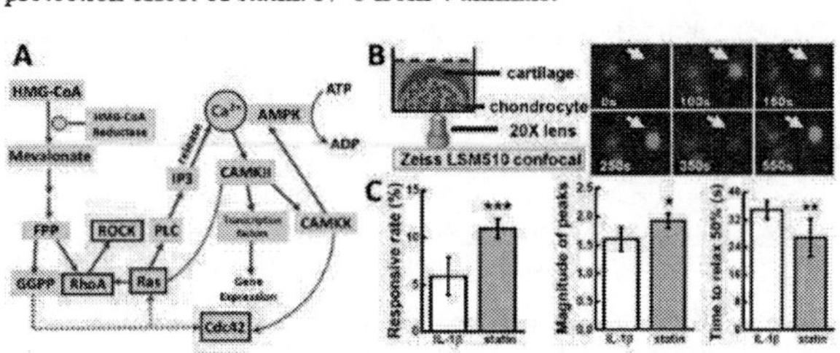

Figure 4: (A) Schematic diagram of the interaction between $[Ca^{2+}]_i$ signaling and small GTPases in chondrocytes, e.g. rho, ras and cdc42. (B) Illustration of the microscopy setup for fluorescent $[Ca^{2+}]_i$ imaging of *in situ* chondrocytes. (C) Percentage of chondrocytes that showed spontaneous $[Ca^{2+}]_i$ peaks, the time to reach a peak, magnitude of peaks, and average number of $[Ca^{2+}]_i$ peaks. N=6 from 3 animals.

SB³C2017
Summer Biomechanics, Bioengineering and Biotransport Conference
June 21 – 24, Tucson, AZ, USA

QUANTIFYING THE RELATIVE IMPORTANCE OF MAXIMUM MYOFILAMENT FORCE AND METABOLITE CONCENTRATION IN RIGHT VENTRICULAR FAILURE: A MULTISCALE COMPUTATIONAL APPROACH

Ryan J. Pewowaruk (1), Shivendra G. Tewari (2), Guanying Wang (3), Diana M. Tabima (1), Anthony J. Baker (3), Daniel A. Beard (2), Naomi C. Chesler (1)

(1) Biomedical Engineering
University of Wisconsin-Madison
Madison, Wisconsin, USA

(2) Molecular & Integrative Physiology
University of Michigan–Ann Arbor
Ann Arbor, Michigan, USA

(3) Medicine
University of California–San Francisco
San Francisco, California, USA

INTRODUCTION

Heart failure is a major cause of morbidity and mortality, with approximately 10% of people over the age of 65 suffering from heart failure in the US [1]. The majority of heart failure literature focuses on the left ventricle (LV), with the right ventricle (RV) assumed to behave similarly. The RV, however, is anatomically, functionally and embryologically different from the LV, and RV failure (RVF) has a pathology distinct from LV failure [2]. At the organ level, RVF is characterized clinically by reduced cardiac output, reduced ventricular ejection, and dilatation [3]. At the cellular level, maximum myofilament forces have been shown experimentally to decrease in RVF [4], and myocardium ATP concentration to decrease while inorganic phosphate (Pi) concentration increases in LV failure [5].

How these changes at the cellular level contribute to RVF at the organ level is a critical knowledge gap. The goal of this study is to adapt the multiscale computational model of Tewari et. al. [6] to the RV, and then quantify the effects of maximum myofilament forces and metabolite concentrations on RV function by simulating RV contraction. Specifically, this study seeks to answer two questions: 1. Can incorporation of decreased myofilament forces in a computational model of the right ventricle simulate RVF? 2. Which alone is a better indicator of RV dysfunction at the organ level: maximum myofilament force or metabolite concentration?

METHODS

The multiscale computational model has been previously reported in detail [6] and will only be described briefly. A system of ordinary differential equations describes the chemical kinetics of actin-myosin interactions, which are dependent upon the concentration of ATP, ADP and Pi. The cross bridge force is calculated from the current state of the actin-myosin kinetic system,

$$F_{XB}(t) = k_{stiff,1}\left(\int_{-\infty}^{\infty} sp_2(t,s)\,ds + \int_{-\infty}^{\infty} sp_3(t,s)ds\right) + k_{stiff,2}\Delta r\int_{-\infty}^{\infty} p_3(t,s)ds, \quad (1)$$

where p_2 is the probability of actin and myosin being bound in the pre power-stroke configuration, p_3 is the probability of actin and myosin being bound in the post power-stroke configuration, s is the sarcomere stretch, and Δr is the length of a power stroke. The mechanical behavior of a myofilament is described by incorporating the active force generated by cross bridges with passive forces of titin and collagen. The myofilament forces are incorporated in the TriSeg heart model [7], which calculates ventricular volumes and pressures from the myofilament forces in the RV wall, LV wall and the interventricular septum (IVS). A lumped parameter model represents the systemic and pulmonary circulations. To simulate RVF, the model was modified so that maximum myofilament forces and metabolite concentrations changes could be isolated to the RV free wall, with the LV free wall and IVS kept at control values.

The multiscale model was first used to determine if incorporating decreased myofilament forces can replicate RVF characteristics. Simulations of myofilament force – pCa^{2+} relationship were conducted over a range of Ca^{2+} concentrations from 0.1 to 100 μM. The relative maximum myofilament force in these simulations was fit to experimental data from a pressure-overload mouse model of RVF [4] by optimizing the parameter $k_{stiff,2}$ in Equation 1. 61 heartbeats were simulated to allow the cardiovascular system to reach a steady-state solution. Simulations were run with control levels of maximum myofilament force, decreased myofilament forces, and decreased forces with heart failure metabolite concentrations. Metabolite concentrations were previously calculated from experimental measurements of total adenine nucleotide, total exchangeable phosphate, and total creatine pool in a canine LV hypertrophy model [6] (Table 1). The simulated RV ejection fractions of the 61st heartbeat were then compared with experimental RV fractional shortening measurements from the RVF mouse model using echocardiography [4]. Given RV anatomy as a free wall attached to the LV, ejection fraction can be assumed equivalent to fractional shortening.

Table 1: Control and failure metabolite concentrations from [6]

Metabolite	ATP	ADP	Pi
Control	8.0 mM	18 μM	0.6 mM
Heart Failure	1.6 mM	4.0 μM	4.0 mM

Simulations were then run to determine if maximum myofilament force or metabolite concentration alone is a better indicator of RV dysfunction, quantified by RV ejection fraction. Five values evenly spaced between healthy and failure values for both maximum myofilament force and metabolite concentrations were used. The resulting 25 combinations of forces and metabolites were then used in the simulation. Simulated RV ejection fractions were correlated with both maximum myofilament force and ATP concentration. The summed squared error (SSE) was used as a metric of predictive capacity.

RESULTS

Experimental measurements of relative maximum myofilament tension vs Ca^{2+}concentration were matched for control (CTL) and RVF conditions with $k_{stiff,2}$ values of 1.08 and 0.69 MPa μm^{-1}, respectively (Fig. 1A). Simulated RV pressure-volume loops (Fig. 1B) with decreased myofilament force show a dilated RV and decreased peak RV systolic pressure compared to control. RV pressure-volume loops with decreased force and heart failure metabolite levels exhibit lower peak pressure, greater dilatation and a markedly decreased stroke volume (Fig. 1B). Simulated RV fractional shortening was much closer to the experimentally measured fractional shortening when the simulation included both the decreased myofilament force and heart failure metabolite levels (Table 2).

Table 2: Experimental and simulated RV fractional shortening

Experimental Condition	Experimental RV Fractional Shortening [4]	Simulation Condition	Simulated RV Fractional Shortening
Control	46.1 ± 4.3 %	Control	46.5 %
RVF	20.2 ± 2.2 %	Decreased Max Force	33.5 %
		Force + Metabolites	25.6 %

RV ejection fraction versus relative myofilament force (Fig. 1C), represents the possible ejection fractions for known maximum myofilament force with varied metabolite concentrations. Similarly, RV ejection fraction versus ATP concentration (Fig. 1D) represents the possible ejection fractions for known metabolite concentrations with varied maximum myofilament force. These results demonstrate that RV ejection fraction correlates positively with both maximum myofilament force as well as ATP concentration. RV ejection fraction has a lower SSE with relative myofilament force than with ATP concentration and is thus more strongly correlated with relative myofilament force than ATP concentration.

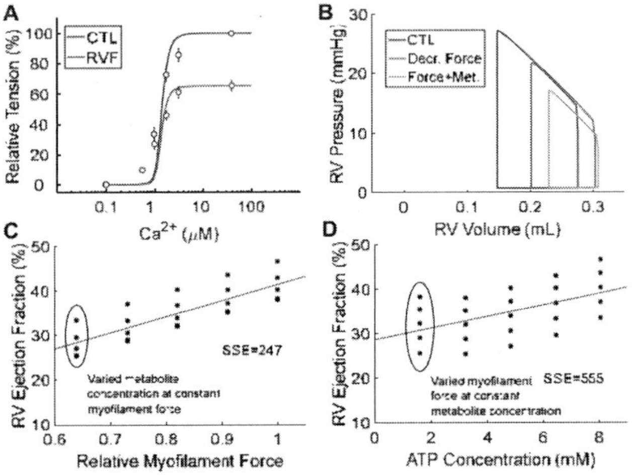

Figure 1: A. Experimental and simulated myofilament force vs pCa, B. Simulated PV loops for control, decreased myofilament force, and decreased force and metabolites, C. RV ejection fraction correlated with relative myofilament tension, D. RV ejection fraction correlated with ATP concentration

DISCUSSION

Our results show that the RV fractional shortening in RVF is better simulated by incorporating decreased maximum myofilament force as well as decreased metabolite concentrations in the RV free wall versus only decreasing the maximum myofilament force. These simulations predict that previously observed decreased maximum myofilament force [4] is not solely responsible for RVF. RV function is dependent upon both maximum myofilament force and metabolite concentrations as well as other variables not considered in this study such as passive mechanical properties of myocytes in the RV and vascular remodeling.

Additionally, simulations in which both maximum myofilament force and metabolite concentrations were varied independently over a pathologically relevant range suggest that maximum myofilament force is a better indicator of RV function than metabolite concentration alone. This suggests that decreases in maximum myofilament force are more responsible for declining RV function in RVF than metabolic changes if these changes are independent. Limitations of this study include that the decrease in myofilament force was not incorporated mechanistically but rather scaled to fit experimental data; the TriSeg model simplifies RV anatomy and structure; and metabolite concentrations measured in the LV might not be appropriate for the RV.

In conclusion, our results predict that decreased RV function in RVF cannot solely be explained by decreased maximum myofilament force. Furthermore, our study demonstrates that the modified Tewari model is a useful tool for exploring mechanisms and predictors of RVF.

ACKNOWLEDGEMENTS

This work was supported by NIH R01 HL086939.

REFERENCES

[1] Hosenpud JD, et. al., *J. Heart Lung Transplant*; 2000. 19:909-31.
[2] Friedberg MK, *Circulation*, 2014. 129(9): 1033-44.
[3] Chen PP, et. al., *Circulation*, 2012. 126:1194-1205.
[4] Baker et. al., Manuscript in preparation.
[5] Stanley WC, *Heart Metab.*, 2005. 49:5–8.
[6] Tewari SG, et. al., *J. Mol. & Cell. Cardiol.*, 2016. 94:162-175.
[7] Lumens J, et. al., *Ann. Biomed. Eng.*, 2009 37:2234–55

MICROMECHANICS OF ELASTIC LAMELLAE IN MOUSE CAROTID ARTERY

Xunjie Yu[1], Raphael Turcotte[2], Francesca Seta[3], Yanhang Zhang[1,4]

[1]Department of Mechanical Engineering
Boston University
Boston, MA, 02215 USA

[2]Wellman Center for Photomedicine,
Massachusetts General Hospital, Harvard Medical
School, Boston, MA, USA

[3]Vascular Biology Section,
Boston University School of Medicine
Boston, MA, 02215 USA

[4]Department of Biomedical Engineering
Boston University
Boston, MA, 02215 USA

INTRODUCTION

Elastin is a major extracellular matrix (ECM) constituent of the arterial wall subjected to billions of stretch cycles during a lifetime. Elastic fibers in the medial layer form concentric layers of elastic lamella, together with smooth muscles cells and collagen fibers, organizing into a lamellar unit that is considered as a functional unit of the arterial wall. The lamellar unit is designed to support and evenly distribute the mechanical load in the arterial wall [1]. However, the micromechanics of elastic lamellae and its response to mechanical loading are not fully understood [2].

Previous studies on the structure of elastic lamellae based on histological images were not able to provide information on the three-dimensional (3D) architecture of the elastic lamella [3]. In this study, we developed a new method to study the micromechanics of elastic lamellae based on 3D reconstructed images. Multiphoton microscopy was used to visualize the structure and mechanical deformation of elastic lamellae in mouse carotid arteries under pressurization and axial stretching. Three-dimensional structure of the elastic lamellae was reconstructed in order to analyze the spatial distribution, waviness and unfolding under biaxial mechanical loading. Combined with a model, which considers thick wall cylindrical deformation, our study provides new understandings of the micromechanics of elastic lamellae and its relationship with tissue-level vascular mechanics.

METHODS

Sample preparation

8-weeks old wild type C57BL/6J mice were used for this study. Carotid arteries were isolated and dissected free of connective tissue. Carotid segments (about 5 mm long) were sectioned (n=7) from the arteries. Four samples were used for imaging and three samples were used for mechanical testing. Samples were maintained at 4°C in 1· phosphate buffered saline and imaged within 24 hours of harvesting.

Multiphoton microscopy and imaging analysis

Multiphoton microscopy was performed to visualize the structure and mechanical deformation of elastic lamellae in mouse carotid arteries under intraluminal pressure from 0 to 120 mmHg and axial stretching from 1 to 1.6. Image z-stacks were acquired with 2μm spacing for a total of about 50 μm from the outer surface of the arterial wall. A custom image processing procedure was developed based on MATLAB Image Processing Toolbox for 3D reconstruction and imaging analysis.

Mechanical testing

Arteries were carefully cannulated on a pressure myograph (DMT, 110P) for pressure-diameter measurements. The arteries were pre-stretched to 1.6× of *ex vivo* length in the longitudinal direction. After preconditioning, the intraluminal pressure was increased from 0 to 120 mmHg with 10 mmHg increments. The change of outer diameter was monitored. The intraluminal pressure and longitudinal force were also recorded.

Constitutive modeling

A structurally motivated three-dimensional strain-energy function was used for the analysis of mechanical behavior of the arterial wall [4].

$$\Psi = \frac{C_e}{2}(I_c - 3) + \sum_{i=1}^{4} \frac{k_1^i}{k_2^i}\left\{exp\left[k_2^i({\lambda_i}^2 - 1)^2\right] - 1\right\} \quad (1)$$

where C_e is a stress-like material parameter associated with elastin , k_1^i and k_2^i are material parameters associated with the kth fiber family. I_c is the first invariant of the right Cauchy-Green tensor $\mathbf{C}$, λ_i is the stretch in the direction of the *i*th fiber family. One circumferential, one axial, and two symmetrically diagonal fiber families were considered. Material parameters were estimated by minimizing the difference between experimentally measured and calculated values of pressure and axial force based on the constitutive model in Equation 1. Transmural variation in stress and stretch was calculated considering thick wall cylindrical deformation.

RESULTS

The 3D view of a reconstructed elastic lamella is shown in Figure 1a. The elastic lamellae appear as continuous concentric layers wavy in both circumferential and longitudinal directions at the unloaded state. Figure 1b shows the cross-sectional view of the lamellar layers in a mouse carotid artery when pressure increases from 0 to 120mmHg and with no axial stretching. During pressurization, the elastic lamellae unfold and the distance between lamellae layers decreases.

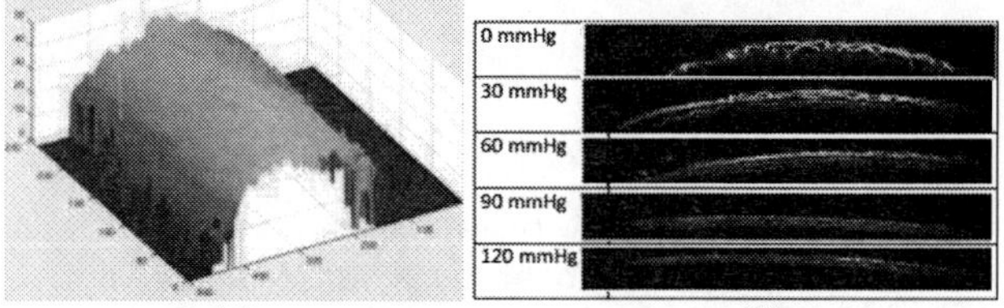

(a) (b)

Figure 1: (a) Representative image of a 3D reconstructed elastic lamella. (b) Unfolding of elastic lamellae (cross-sectional view) in the medial layer when intraluminal pressure increases from 0 to 120 mmHg.

Structural analysis was performed on the 3D reconstructed elastic lamella. Based on the extracted center-lines of the lamellar layers, straightness parameter, defined as the ratio between the length of fitted arcs L_0 and the contour length of the lamella L_f, was calculated and denoted as P_s in Figure 2.

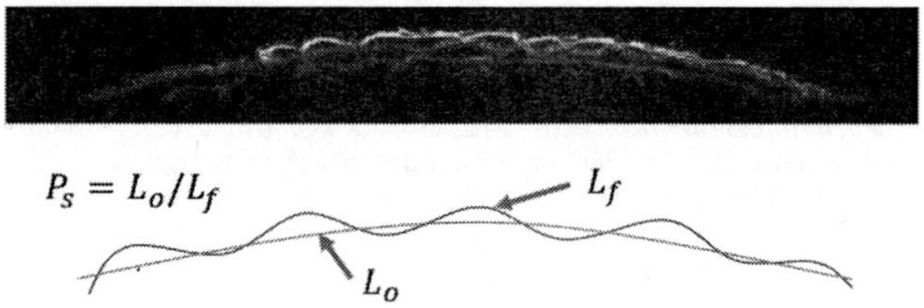

Figure 2: Reconstructed lamellar layers and the calculation of straightness parameter P_s.

As shown in Figure 3, the straightness parameter of each lamella layer increases with pressure. Here the axial stretch was kept at 1.6. The waviness of elastic lamellae varies transmurally with wavier elastic lamella, or smaller straightness parameters, present towards the inner surface of the arterial wall. The unfolding stretch, λ_p, which represents the straightening of the elastic lamellae, was calculated by normalizing the straightness parameters P_s' at each pressure level by the straightness parameter P_s at zero pressure. The tissue level stretch, λ_t, at each lamellar position was calculated from the thick wall model according to their position in the arterial wall (Figure 4b), and its variation with pressure is shown in Figure 5b. We can see that as the pressure increases, lamellar layer closer to the inner surface undergoes higher unfolding stretch, λ_p (Figure 5a), as well as higher tissue level circumferential stretch, λ_t(Figure 5b). However the unfolding stretch in the elastic lamellae is much smaller than the tissue level stretch, indicating that during mechanical deformation, the elastic lamellae layers are not only subjected to unfolding, but also elongation. The elongation stretch, λ_f, which represents the stretching of elastic lamellae during expansion, was thus calculated by dividing the tissue level stretch by the unfolding stretch. Figure 5c shows that the elongation stretch $\lambda_f$remains similar for both lamellar layers.

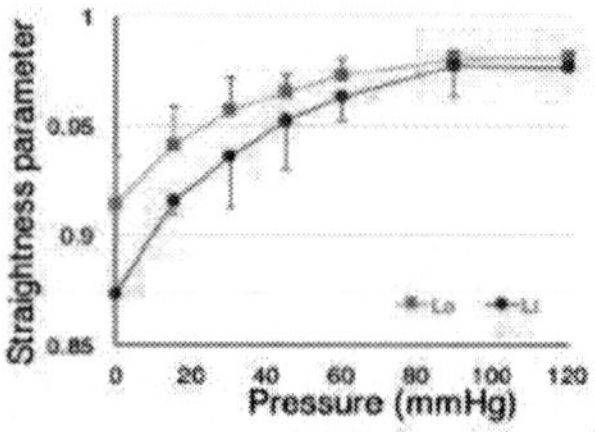

Figure 3: Straightness parameter vs. pressure. L_o, L_i represent outer and inner lamella layers.

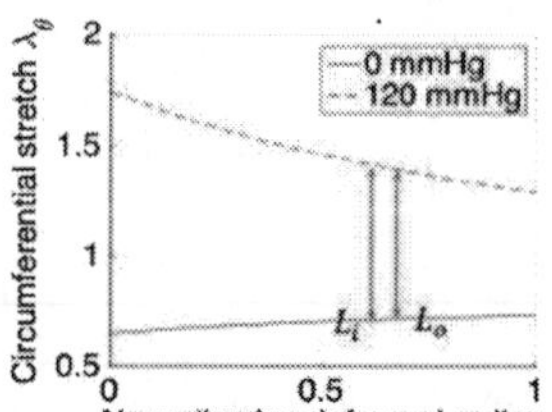

Figure 4: Transmural distribution of circumferential stretch at 0 and 120 mmHg.

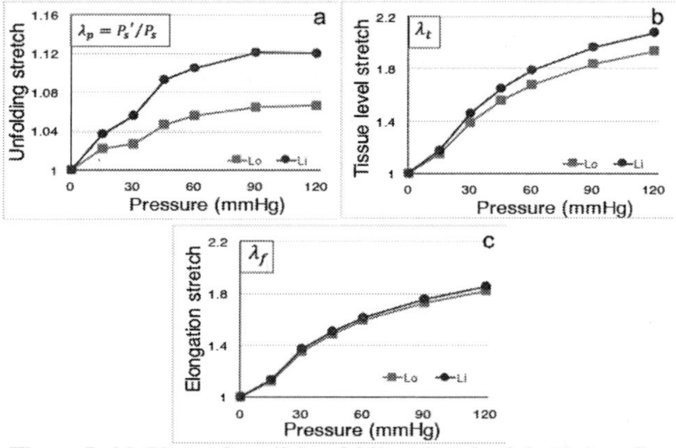

Figure 5: (a) Circumferential unfolding stretch of elastic lamellar layers, (b) tissue level stretch calculated at the location of each lamellar layer, and (c) elongation stretch of elastic lamellar layers vs. pressure.

DISCUSSION

In this study, we studied the micromechanics of elastic lamellae by reconstructing the 3D architecture of elastic lamellae in mouse carotid arteries, and quantified the structural changes in elastic lamellae during biaxial loading. Previous study suggested elastic lamellae unfold at low pressure and straighten at higher pressure [3]. However, our study showed that unfolding/straightening of elastic lamellae, λ_p, only contributes to a small fraction of the circumferential tissue deformation, even at low pressure. Our study suggests that stretching and unfolding of elastic lamellae, occurs simultaneously with deformation, and both contribute to tissue-level deformation, λ_t. The relationship between tissue-level stretch, lamellar unfolding λ_p, and lamellar stretching λ_f can be represented as:

$$\lambda_t = \lambda_p * \lambda_f \tag{2}$$

The higher lamellae unfolding in the inner lamellae layer compensates the larger strain experienced at the inner surface of the arterial wall, and plays an important role in maintaining a more evenly distributed stretching/stress in the lamellar layers through the arterial wall. This study sheds light on the importance of structural inhomogeneity in maintaining tissue homeostasis.

REFERENCES

[1] Wolinsky, H., et al., *Circulation research* 20, no. 1: 99-111., 1967.
[2] Green, E.M., et al., *Interface focus* 4(2), 20130058, 2014.
[3] Sokolis, D.P., et al., *J of biomechanics*, 39(9), pp.1651-1662, 2006.
[4] Holzapfel, G A., et al., *J of Elasticity* 61: 1-48, 2000.

BIOMECHANICAL AND BIOLOGICAL EVALUATION OF ELASTIN STABILIZATION IN RAT ABDOMINAL AORTIC ANEURYSMS USING PENTA-GALLOYL GLUCOSE

Mirunalini Thirugnanasambandam (1), Dan T. Simionescu (2), Eugene Sprague (1, 3), Beth Goins (4), Geoffrey D. Clarke (4), Hai-Chao Han (1, 5), Krysta H. Amezcua (1), Oluwaseun R. Adeyinka (1), Ender A. Finol (1,5)

(1) Department of Biomedical Engineering
University of Texas at San Antonio
San Antonio, Texas, USA

(2) Department of Bioengineering
Clemson University
Clemson, South Carolina, USA

(3) Department of Medicine
UT Health San Antonio
San Antonio, Texas, USA

(4) Department of Radiology,
UT Health San Antonio
San Antonio, Texas, USA

(5) Department of Mechanical Engineering
University of Texas at San Antonio
San Antonio, Texas, USA

INTRODUCTION

Abdominal aortic aneurysm (AAA) is a dilation of the abdominal aorta by 50% or more of its normal diameter, typically occurring below the renal arteries. Upon rupture, AAAs have a 90% mortality rate and this makes it imperative to identify markers that characterize AAA growth and rupture. Since AAAs are a product of the closed-loop feedback mechanism between biological and biomechanical changes of the aortic wall, it is critical to observe aberrant behavior of these parameters with AAA progression. In recent years, alternative AAA treatments have focused on developing drugs that can restrict aneurysm expansion. Typically, these are anti-inflammatory or anti-matrix metalloprotease pharmaceuticals. In this study, we aim to develop a new treatment to change the structural integrity of the aneurysm wall. To this end, we investigated the effects of penta-galloyl glucose (PGG), an elastin and collagen-binding polyphenol, on rat AAA supression. The biomechanical evaluation was based on the first principal stress and the biological evaluation on matrix metalloproteinase-9 (MMP-9) activity.

METHODS

In-vivo study

Six Sprague-Dawley rats were divided into control and PGG treatment groups. All rats were subjected to an initial μMR scan and *in-vivo* molecular imaging. The calcium chloride ($CaCl_2$) model was used to induce an AAA in all subjects during survival surgery. 0.5 M $CaCl_2$ was applied to the infrarenal aorta using a pre-soaked gauze applicator. In the treatment group, PGG was applied peri-adventitially at body temperature (37 °C) before applying $CaCl_2$. The first follow-up phase involved acquiring IVIS *in-vivo* images upon administration of the probe MMPSense 750 FAST intravenously via the tail vein. This probe is an MMP-activatable agent that fluoresces only when it encounters disease-related MMPs. The key element in the structure of the probe is the MMP-cleavable peptide sequence, which induces fluorescence upon cleavage. In the second follow-up phase, all subjects were again subjected to μMR and IVIS imaging, from which fluorescence maps of MMP activity were recorded. At the end of this phase, all rats were sacrificed and the entire aortas harvested for destructive mechanical testing. Specific regions of the aorta were isolated and tested using a Cellscale planar biaxial tensile testing machine to generate force-displacement curves for each specimen.

Constitutive modeling

Cauchy stress-strain curves were generated from the force-displacement graphs for different regions of the aorta for all subjects. The artery was modeled as a nonlinear elastic thick-walled cylinder consisting of two layers – media and adventitia. A Holzapfel-Gasser-Ogden (HGO) multilayer material model was fitted to the experimental Cauchy stress-strain curve [1]. The strain energy density formulation for the 5-parameter HGO model is defined by Eq. (1),

$$\Psi = \frac{c}{2}(I_1 - 3) + \sum_{\substack{layer=media,\\ adventitia}} \sum_{i=4,6} \frac{k_1}{2k_2}\left[\exp\left(k_2\left(I_i^{layer} - 1\right)^2\right)\right] \qquad (1)$$

where c, k_1, k_2 are model parameters and I_i^{layer} are the ith invariants of the Cauchy stress tensor per artery layer. The fourth and sixth invariants implicitly include the fiber directions of the collagen fiber bundles, given by the α_M and α_A angles. An optimization algorithm that utilizes both genetic and Levenberg-Marquardt algorithms was used to identify the best set of parameters that minimize the sum of squares of the differences between experimental and model-predicted

Cauchy stresses. Additional constraints imposing c, k_1, k_2 to be positive were also included to ensure physically reasonable response prediction by the model.

Table 1: HGO material model parameters for rat AAA (C, T indicate control and treatment group subjects, respectively).

Rat #	c (MPa)	k_1 (MPa)	k_2 (-)	α_M (degrees)	α_A (degrees)
C1	0.0351	0.04509	0.57573	96.3	91.4
C2	0.0100	0.00881	2.84458	87.0	87.6
C3	0.0100	0.04145	0.50414	97.3	167.2
T1	0.0100	0.08867	1.57609	86.4	13.0
T2	0.0141	0.00486	2.74487	21.2	172.0
T3	0.0202	0.00363	2.93458	15.8	29.2

Biomechanical evaluation

An in-house segmentation and geometry quantification code (AAAVASC) was used to segment the μMR images of the aorta pre- and post-surgery. The segmentation algorithm of the outer wall and lumen is based on contrast differences. A surface was lofted over the point cloud-derived curves generated from AAAVASC, thus creating an STL file of the outer wall, which was then offset inwards twice using 3-Matic (Materialise, Belgium) such that the layer closest to it is at 1/3 of the wall thickness and the other layer represents the lumen. Thus, a two-layer aortic wall representing the media and adventitia was generated. A finite element mesh was generated in ANSYS ICEM using 3D solid tetrahedral elements, which was later input to FEBio, an open-source FEA solver. Two materials were defined for each domain as a combination of an isotropic Neo-Hookean model and an uncoupled formulation of a single fiber with an exponential-power law using the subject-specific constants presented in Table 1 as derived from the optimization algorithm. The nodes at the ends were fixed in the axial direction and a pressure of 120 mmHg was applied normal to the nodes on the lumen surface. Distribution maps of the first principal stress were obtained using Postview. Maximum, 99th percentile and average first principal stresses were quantified for each subject pre- and post-surgery.

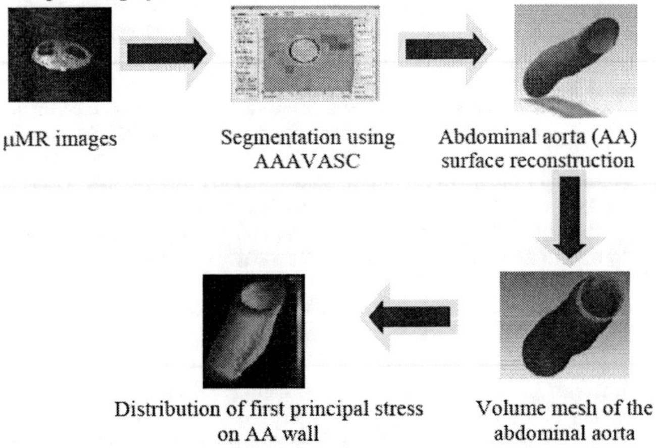

Figure 1: Flowchart for evaluation of wall stress distribution from μMR images of the rat aorta.

RESULTS

Distribution maps of first principal stresses for two exemplary AAAs are presented in Figure 2. In control rats, a mean reduction of 19%, 4% and 9% in maximum, 99th percentile and mean wall stress, respectively, was obtained between pre- and post-AAA induction surgery. Similarly, in PGG-treated rats, a mean reduction of 54%, 38% and 26% was observed between surgeries in these biomechanical parameters. IVIS images showed fluorescence intensity maps representing MMP activity, as shown in Fig. 3.

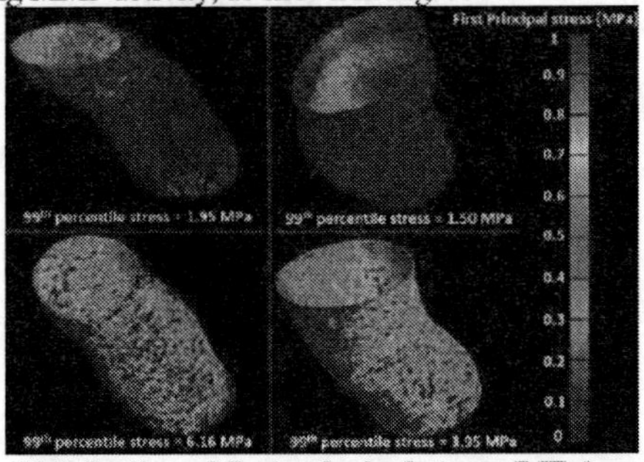

Figure 2: Distribution of first principal stress (MPa) on the AAA wall of a control (top) and PGG-treated (bottom) abdominal aorta pre- (left) and post- (right) AAA induction surgery.

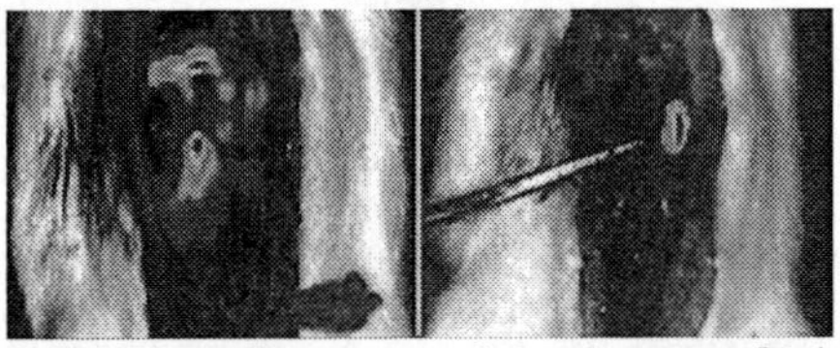

Figure 3: MMPSense fluorescence intensity map obtained from IVIS images for control (left) and PGG-treated (right) rats.

DISCUSSION

PGG functions by preferentially binding to elastin in the arterial wall, thus maintaining its integrity [2]. It was also observed that there was a low percentage of increase in diameter in PGG-treated rats [2]. In addition to concurring with this observation, the present work yielded a considerable reduction in peak, 99th percentile and average first principal stresses in PGG-treated rats, which are known biomechanical metrics of AAA progression and rupture risk. In addition, we observe a smaller area of MMP-activity on the intensity map in PGG-treated rats compared to control rats, thus implying lesser inflammation and proteolytic activity in the treatment group. Hence, from the simultaneous evaluation of biological and biomechanical markers, we can observe a potential AAA stabilization effect of PGG. A major limitation of this study is the small sample size of specimens. Since the same material parameters are used to evaluate stresses pre- and post-surgery, any variation in the distribution map can be considered due to the effect of arterial geometry.

ACKNOWLEDGEMENTS

This work was supported in part by The University of Texas System Board of Regents - Science and Technology Acquisition and Retention (STARS) program and AHA award #15PRE25700288.

REFERENCES

[1] Holzapfel, G. A., et al., *Journal of Elasticity and the Physical Science of Solids*, 61.1-3:1-48, 2000.
[2] Isenburg, J. C., et al., *Circulation*, 115:1729-1737, 2007.

WOVEN COLLAGEN BIOTEXTILES FOR ROTATOR CUFF TENDON REPAIR – AN *IN VIVO* PILOT STUDY

Greg D. Learn (1), Phillip E. McClellan (2), Derrick M. Knapik (3), Jameson L. Cumsky (4), Robert J. Gillespie (3), Ozan Akkus (1, 2, 3)

(1) Dept of Biomedical Engineering
Case Western Reserve University
Cleveland, OH, USA

(2) Dept of Mechanical & Aerospace Engineering
Case Western Reserve University
Cleveland, OH, USA

(3) Dept of Orthopaedics
Case Western Reserve University
Cleveland, OH, USA

(4) School of Medicine
Case Western Reserve University
Cleveland, OH, USA

INTRODUCTION

Tendon injuries are a common cause of pain and disability. Tendons have low intrinsic healing capacity, and surgical repair failures are prevalent, especially for the rotator cuff (RC) tendons [1], with defect size and soft tissue degeneration being risk factors for re-tear. While surgical techniques have vastly improved, the ability to replace missing or defective tendon tissue would benefit repair outcomes. This study explores the use of mechanically robust, biomimetic, woven collagen scaffolds (WCSs) for RC tendon repair and regeneration *in vivo*.

METHODS

Scaffold Fabrication: Acid-soluble Type I collagen (Collagen Solutions) was dialyzed and compacted between 2 stainless steel wire electrodes (30V, 90s) to form electrochemically aligned collagen (ELAC) threads. Threads were combined into 3-ply yarns and crosslinked with 2% w/v genipin in 90% ethanol. Yarns were manually woven and consolidated using a weft fiber [2] to produce a scaffold ~14 x 5 x 2 mm³ (Fig 1). Scaffolds were sterilized prior to use in peracetic acid / ethanol solution.

Fig 1. Woven ELAC scaffold.

Animal Surgeries: Animal procedures were performed in accordance with established protocols approved by the IACUC at CWRU. Fifteen adult New Zealand White Rabbits (Charles River, 3-5kg) underwent surgical creation of a unilateral infraspinatus (IS) tendon defect in the right shoulder, while the left shoulder served as an intact control in every animal (Fig 2a). IS tendons were sharply detached at the enthesis, and immediately either reattached directly with suture (direct repair = 'DR') as a best-case scenario operative control (Fig 2b), or a critical 5mm defect was created and the tendon-bone gap was bridged using a WCS (Fig 2c) either by itself (scaffold-only repair = 'SR'), or pre-seeded with P5 allogeneic marrow-derived CD44+/CD45-/CD90- MSCs (scaffold/cell repair = 'SCR'). The tendon/scaffold were sutured (3-0 Ethibond) using a Krakow technique (3 locking loop pairs held the tendon) and re-anchored to the IS insertion via bone tunnels to the bicipital groove (Fig 2b,c). Rabbits ambulated freely and were euthanized at 3 months post-operatively. Scapulohumeral complexes (SHCs) were harvested, and for a subset of rabbits, were imaged using micro-CT (μCT) to assess for fatty infiltration (FI) of the RC musculature. IS muscles were then dissected away from the rest of the scapula to isolate humerus-tendon-muscle units. 4 animals/group were used for biomechanical testing (BMT) of repair stiffness and max load. Histology was performed on remnant tissues after BMT and shoulders from 1 additional animal/group (3 animals/group total).

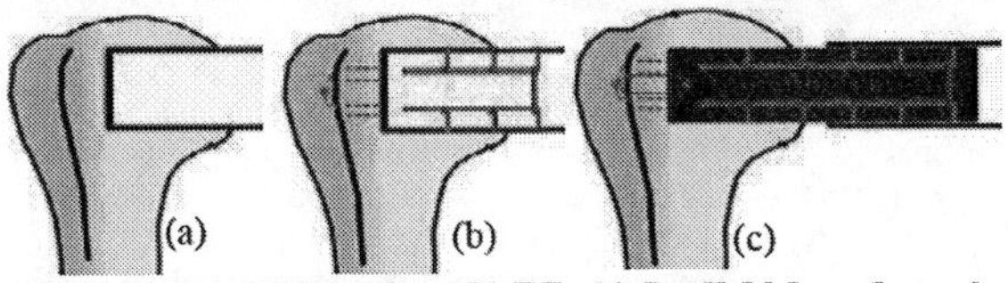

Fig 2. (a) Intact IS tendon. (b) DR. (c) Scaffold-based repair (SR/SCR). Blue = suture. Red dashes indicate bone tunnels.

μCT: SHCs were imaged (70kV, 470μA, 800ms, bin-factor 2, low-med magnification) in a Siemens Inveon PET/CT scanner.

BMT: Humeri were potted and clamped in aluminum blocks, the IS muscle was held using grips cooled with dry ice, and the muscle was pulled in monotonic tension at 10 mm/min until failure using a materials testing instrument (Test Resources) [3]. Stiffness and peak load values were extracted from load-displacement (L-D) data and normalized to intact control values. Linear regression was performed on elastic regions of L-D curves from 40-60% of the peak load value, and slope of the best-fit line was taken to be stiffness (r^2 values were ≥ 0.95). Stiffness and peak load data were assessed for significant differences (SDs) using the Kruskal-Wallis test (significance set at $p \leq 0.05$); where significance was indicated, pairwise comparisons were performed among treatment groups using the Mann-Whitney U test (significance set at $p \leq 0.05$).

Histology: Harvested specimens were embedded, sectioned, stained with Masson's Trichrome (MTc) or Hematoxylin & Eosin (H&E), imaged with a brightfield microscope, and assessed for collagen deposition and cellularity.

RESULTS

The SCR group demonstrated a significant increase in stiffness relative to the SR group (Fig 3a) and a non-significant increase in max load (Fig 3b). No SDs were found between SR and DR, or DR and SCR groups.

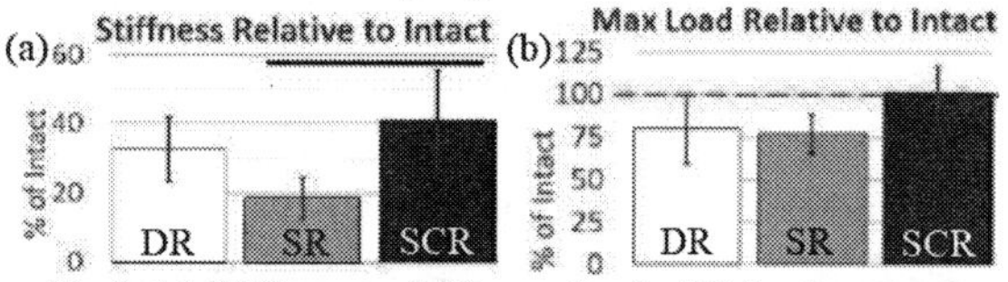

Fig 3. (a) Stiffness and (b) max load of IS tendon repair specimens (n=4/group). Red dashes = intact control value. Black line = SD between groups.

Intact and DR specimens consistently failed at the soft tissue / bone interface (BIF). SCRs consistently failed midsubstance (MS), and roughly half of SRs failed at the BIF while the rest failed MS (Table 1). Note 13 shoulder pairs are included rather than 12, due to one SR rabbit which was repeated because its intact control sample slipped mid-test (this animal's shoulders are excluded from stiffness and max load data).

Table 1. Failure Locations by Group

Group / Site	Intact	DR	SR	SCR
BIF	13	4	3	0
MS	0	0	2	4

µCT revealed moderate FI of the operative (Op) IS muscle belly for DR rabbits (Fig 4b); FI appeared less severe for the Op SR IS muscle belly based on samples from 1 rabbit (Fig 4d). FI was not observed in intact controls (Fig 4a,c) or neighboring RC muscles for any specimen.

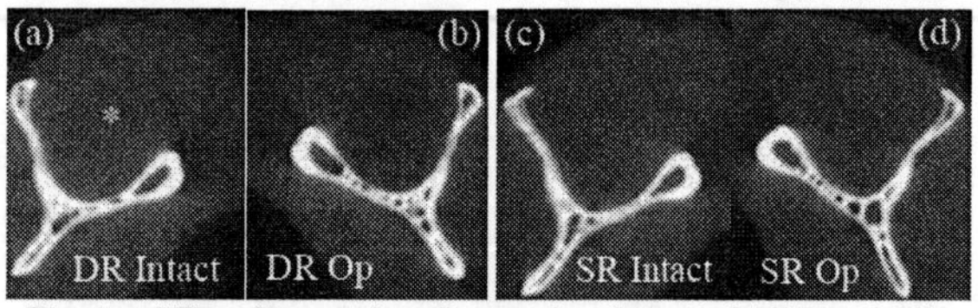

Fig 4. µCT radiographs of scapulae from (a,b) DR rabbits (representative of n=3), and (c,d) SR rabbit (n=1). White = bone, light gray = muscle, dark gray = fat, black = air. Yellow star denotes IS muscle belly.

SCR histological sections demonstrated increased collagen deposition and cell density throughout the scaffold network relative to SR sections (Fig 5). Cell-deposited collagen followed the contour of ELAC threads in both groups.

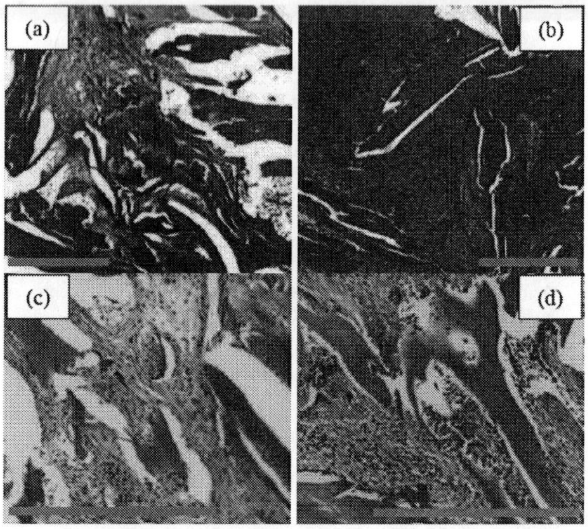

Fig 5. (a,b) MTc stained sections showing scaffold-tissue integration. Navy = *de novo* collagen. Pink = cells. Crimson = ELAC. (c,d) H&E stained sections showing cellularity. Violet = nuclei. (a,c) SR. (b,d) SCR. Scale bar = 500µm.

DISCUSSION

BMT data suggest that addition of MSCs to WCSs enhances repair stiffness and strength, while increasing the chance that the distal scaffold remains fixed to the bone. This could reflect cell-mediated integration between soft and hard tissue. Max load of SR and DR were comparable, suggesting that the WCS is not a detriment to repair strength, considering that DR represents a clinical standard. Histological staining revealed *de novo* collagen between ELAC threads that appears to follow thread contours for both SR and SCR. ELAC may serve as a template which guides cell collagen deposition. SCR sections appear to boast greater cell density within the scaffold continuum than SR sections. These cells may be the implanted MSCs, or they may be host cells drawn to the area. µCT revealed presence of FI in DR Op muscles, suggesting that FI occurs in rabbit IS muscle even after immediate repair of the damaged IS tendon. A qualitative decrease in FI was noted within the SR IS muscle relative to DR group, suggesting FI was prevented by SR. FI is considered irreversible, and is associated with poor functional outcomes [4]. Therefore, FI prevention is desirable. However, as SHCs from only one SR animal have been imaged, this finding could be anomalous. A limitation of this investigation is that sample numbers were low; as such, the findings may be incidental, warranting further study. However, this study presents a novel approach which may hold merit for tendon repair and regeneration. Findings of this study support use of WCSs, possibly in combination with patient-derived MSCs, for repair of RC tendon defects.

ACKNOWLEDGEMENTS

This work was supported through grants from NIH NIAMS: R01 AR063701 to O.A., and Ruth L. Kirschstein National Research Service Award T32 AR007505 to G.D.L. The authors are also grateful for core facility services provided by Case Center for Imaging Research and the Animal Resource Center at CWRU.

REFERENCES

[1] Galatz, L et al. *J Bone Joint Surg Am*. 86-A(2):219-24, 2004.
[2] Younesi, M et al. *Adv Func Mater*. 24(36): 5762–5770, 2014.
[3] Islam, A et al. *Clin Biomech*. 30(7): 669-675. 2015.
[4] Gladstone, J et al. *Am J Sports Med*. 35(5): 719-728. 2007.

SB³C2017
Summer Biomechanics, Bioengineering and Biotransport Conference
June 21 – 24, Tucson, AZ, USA

EFFECT OF OSMOTIC SWELLING IN SOFT TISSUES IS DEPENDENT ON COLLAGEN FIBER ORIENTATION

Bo Yang and Grace D. O'Connell

Mechanical Engineering
University of California, Berkeley
Berkeley, CA, United States

INTRODUCTION

Residual stresses are commonly reported in soft tissues, including the arterial wall and the intervertebral disc, and may act to ensure uniform stress distribution throughout complex fiber-reinforced tissues [1]. Extensive research has shown the important role of collagen fiber composition and architecture on tissue mechanics [2-6]. Collagen fibers are embedded within an extrafibrillar matrix that consists of proteoglycans, water, and fibril-associated collagens that connect collagen fibers, and the role of the extrafibrillar matrix on tissue mechanics is not well understood.

Negatively charged glycosaminoglycans (GAG) act to attract water molecules into soft tissues, resulting in osmotic swelling. Recent studies have shown that osmotic swelling stiffens articular cartilage by engaging the collagen fibers in tension prior to compression testing [7]. In contrast, osmotic swelling in the intervertebral disc caused a decrease in joint stiffness [8]. Moreover, a localized increase in GAG composition in the arterial wall altered stress distribution throughout the wall thickness, which may be a mechanism for thoracic aortic aneurysms [9]. Sophisticated computational models, such as biphasic-swelling and triphasic mixture model theory, have been developed to incorporate tissue swelling in soft tissue simulations [10].

Our previous work showed that tissue shrinking or elongation of rectangular specimens under osmotic swelling conditions was dependent on fiber orientation (with respect to the long axis of the specimen; Fig. 1) [11]. We are particularly interested in the mechanical behavior of the annulus fibrosus (AF) from the intervertebral disc. The AF is a complex fiber-reinforced tissue with collagen fibers oriented at ±45° with respect to the spinal axis in the inner AF and gradually increases to ±60° in the outer AF. Taken together, the mechanical response of fiber-reinforced soft tissues includes contributions from osmotic pressure, elastic deformation of extrafibrillar matrix, and collagen fiber stretch [12]. While it is thought that the extrafibrillar matrix does not play a significant role in tissue tension behavior [13], it may be important for stress distribution between AF lamellae layers. Therefore, the objective of this study was to evaluate the combined effect of osmotic swelling, fiber orientation, and lamellae structure on tissue mechanics. To do this, we evaluated rectangular test specimens used for uniaxial tension testing and an AF ring structure.

METHODS

Rectangular test specimen models were developed with one, two, or three fibrous layers. Similarly, the AF ring model was developed with one, two, or three lamellae rings. Rectangular models were developed and meshed in Preview (FEBio Package) and had initial dimensions of 2 mm X 10 mm X 0.2 mm for width, length, and layer thickness, respectively [14]. Meshes for AF ring models were generated using a custom algorithm to process published disc geometry data [15]. For all models, each layer had a uniform fiber distribution, and the fiber angles alternated between adjacent layers.

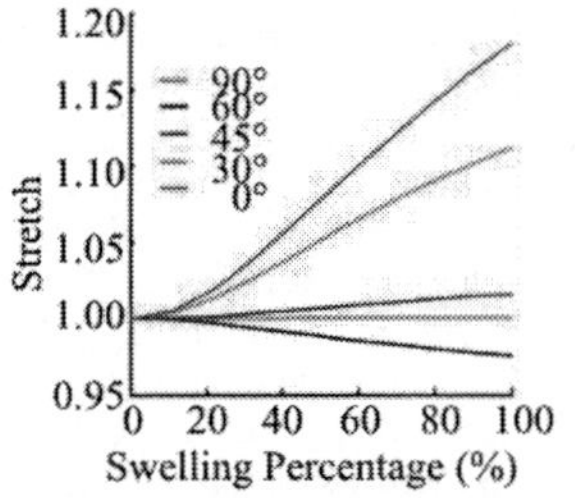

Fig. 1 Tissue shrinking and elongation during swelling [11].

Material coefficients were chosen from our previous model for the AF [11], where the extrafibrillar matrix was described as compressible hyperelastic material: Holmes-Mow (density ρ = 1 g/cm^3, Young's Modulus E = 0.0649 MPa, Poisson's ratio υ = 0.24, and an exponential stiffening coefficient β = 0.95). The initial solid volume fraction was 0.3. Diffusivities of Na^+ and Cl^- ions were 0.00199 mm^2/s and their solubilities were 1 [11]. Fibers were described using an exponential-linear function [14].

For rectangular models, one end was fixed and the other end was free to deform and rotate. The swelling was simulated by increasing the negative fixed charge density from 0 to -100 mmol/L, while the surrounding environment was held fixed to represent 0.15 M PBS. Six cases were simulated, including no fibers (extrafibrillar matrix only) and fiber orientations of 0°, ±30°, ±45°, ±60°, and ±90°. For AF ring models, two cases were considered: one is no fiber case and the other was a consistent fiber orientation (±30° with respect to the transverse plane). The no fiber cases act as negative controls to understand the role of collagen fibers in combination with tissue swelling. Volume ratio was calculated as the volume in the deformed condition divided by the volume in reference (Ref.) configuration. Normalized length (Norm. Length) was calculated as the length after swelling normalized by the length in reference configuration.

RESULTS

For all 'no fiber' cases (both rectangular and AF ring models), osmotic loading resulted in a uniform 42% increase in tissue volume (Fig. 2A & 3A). For single layer tissues, the inclusion of fibers decreased the uniform swelling (36% increase in volume with respect to the reference configuration; Fig. 2A). Furthermore, bulk fiber orientation rotated perpendicular to the direction of the collagen fibers (Fig. 2A). Similar to our previous work, osmotic swelling increased tissue elongation in the 0° case (1.17 normalized length), while tissue elongation along for the 0°<θ<90° cases followed a sigmoidal behavior ($\varepsilon_{xy} = -0.89 * sin(-2\theta)$; Fig. 2C).

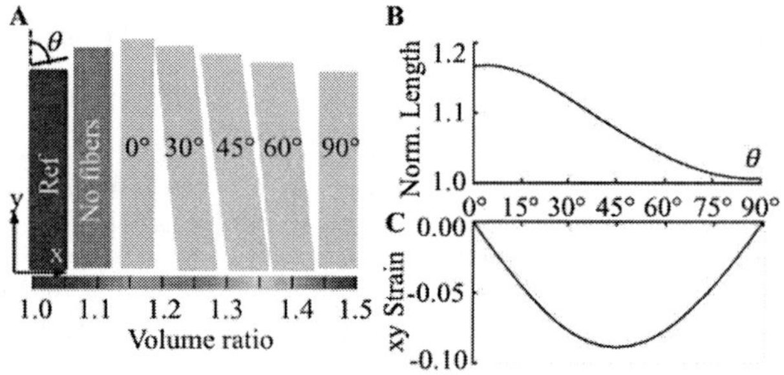

Fig. 2: (A) Volume change due to tissue swelling for one-layer rectangular models. (B) Normalized length vs fiber orientation angle. (C) xy strain vs fiber orientation angle.

Adding a second layer to rectangular specimens did not alter the tissue-swelling ratio when fibers were oriented perpendicular or parallel to long axis of the tissue sample (Fig. 2A *vs.* 3A – 0° and 90° models). In contrast, a cross-ply fiber orientation reduced volume swelling (30% volume increase versus 36%) and caused tissue rotation with a twist angle, α (Fig. 3A). The pitch, P, was defined as $P = l/\alpha * 360°$, where l is the current tissue length. The pitch reached its minimum value when fibers were orientated between ±40° and ±45° (Fig. 3B). Three-layer rectangular models basically repeated observation from our previous study, where tissue length decreased when fiber orientation was between ±45° and ±90° (Fig. 1).

Finally, to understand tissue swelling under *in situ* boundary conditions, we modeled AF rings. The AF ring model experienced a 30% increase in volume with swelling. There was a decrease in the lateral dimension (long axis of the disc) and an increase in the anterior-posterior direction (short axis of the disc; Fig. 4).

DISCUSSION

In this study, we investigated the deformation of fiber-reinforced soft tissue caused by the combined effect of swelling and fiber orientation. Collagen fibers are stiff under tension and buckles under compression, while GAGs absorb water molecules, resulting in

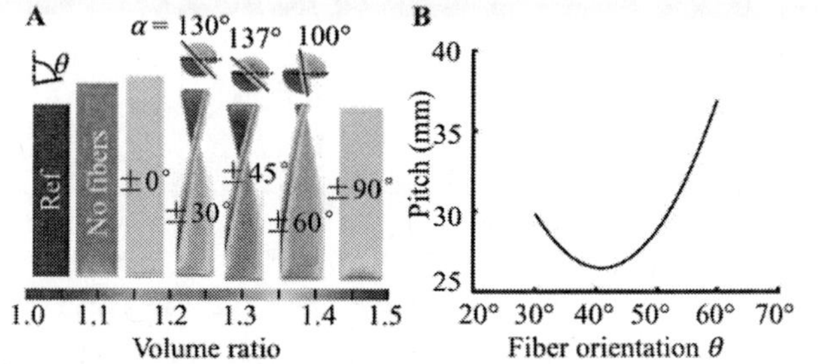

Fig. 3: (A) Volume change and twisting of two-layer rectangular models. (B) Pitch vs fiber orientation (spline fitted).

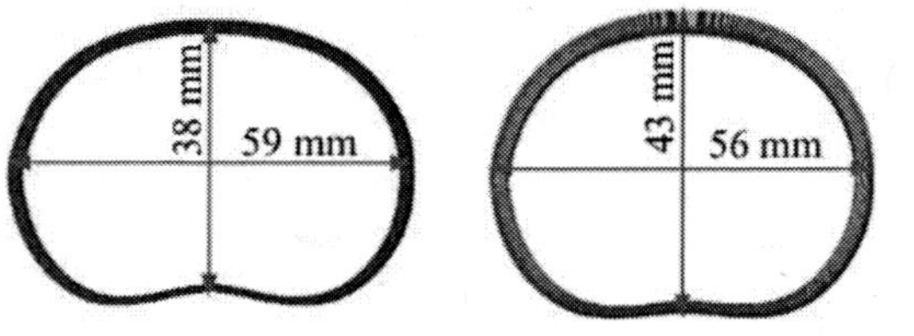

Fig. 4: 3-lamella AF ring model before (left) and after (right) swelling. Color represents volume ratio of 1.3 (same color bar as Fig. 2(A) and 3(A)).

anisotropic tissue swelling. The resistance from stiff collagen fibers combined with anisotropic swelling results in a residual stress state under equilibrium conditions.

While our objective here was to understand AF mechanics with tissue swelling, the fiber architecture of the single fiber control models (0° and 90°) is similar to tendons and ligaments. Interestingly, a single group of fibers resulted in more tissue elongation along the length of the fibers, which would result in higher pre-stress of the collagen fibers (Fig. 2A – 'no fiber' model versus 0° model). However, the volume ratio was lower in model simulations with fibers, suggesting that fibers limited matrix expansion in the transverse direction.

Alternating the fiber orientation between lamellae, as observed in native AF tissues, resulted in a smaller increase in a volume ratio. However, the tissues continued to experience a change in length (Fig. 1 and 3A), suggesting fiber re-alignment during swelling. Previous studies have noted affined fiber realignment during tensile loading, with changes in fiber realignment behavior with degeneration [15].

Taken together, tissue swelling alters stresses transferred to collagen fibers in adjacent lamellae. Understanding how this transfer of stresses alters disc joint mechanics is not trivial; however, the response from the AF ring model suggests that AF fiber architecture with swelling acts create a more circular structure (Fig. 4). In the healthy native disc, this response is counter balanced by the geometry and intradiscal pressure from the nucleus pulposus.

ACKNOWLEDGEMENTS

This study was supported by the Hellman Foundation.

REFERENCE

[1] Azeloglu, EU+, *Am J Physiol Heart Circ Physiol* 294.3: H1197-H1205, 2008; [2] Gasser, TC+, *J. R. Soc. Interface*, 15-35, 2006; [3] Driscoll, TP+, *JMBBM*, 1627-36, 2011; [4] Cassidy JJ+, *J Conn Tiss*, 75-88, 1989; [5] Aspden, RM+. *J. Anat*, 371, 1985; [6] O'Connell, GD+, *BMMB* 493-503, 2012; [7] Soltz, MA+, *J Biomech Eng* 576-586, 2000; [8] Bezci, SE+, *J Biomech Eng*, 101007, 2015; [9] Roccabianca, S+, *BMMB*, 13-25, 2014; [10] Lai, WM+, *J Biomech Eng* 245-258, 1991; [11] Yang, B+; ORS, 2017; [12] Cortes, DH+, *J Biomech*, 2088-2094, 2012; [13] Szczesny, SE+, *J. Orthop Res*, 2017; [14] Yang, B+, SB3C, 2016; [15] Peloquin, JM+, *J Biomech* 2452-2459, 2014; [15] Guerin, HL+, *J. Orthop Res* 25.4: 508-516, 2007.

SB³C2017
Summer Biomechanics, Bioengineering and Biotransport Conference
June 21 – 24, Tucson, AZ, USA

MODELING THE EFFECT OF SPATIALLY-DEPENDENT ECM FIBER DEPOSITION ON CELL TENSIONAL HOMEOSTASIS

Shannen B. Kizilski (1), Rohit Y. Dhume (1), Patrick W. Alford (2), Victor H. Barocas (2)

(1) Department of Mechanical Engineering
University of Minnesota
Minneapolis, MN, USA

(2) Department of Biomedical Engineering
University of Minnesota
Minneapolis, MN, USA

INTRODUCTION

Cells have often been referred to as biological "building blocks", but this term implies that they contribute only passively to tissue mechanics. Though this passive role has been shown to be significant [1], it is known that cells are also responsible for actively mediating changes to the structure of the tissue based on the loading that they experience [2]. Through a phenomenon known as tensional homeostasis, cells grow and modify the surrounding extracellular matrix (ECM) to maintain a target stress state [3]. One of the cell's most powerful tools is collagen, which is employed in nature to provide load-bearing capacity. When a cell is subject to load exceeding its target stress, it deposits collagen fibers in the ECM to shield itself from this loading and return to homeostatic levels.

Models have been developed to study the effect of cell-mediated remodeling on ECM structure [4,5]. In these models, deviation from the target stress σ_0 in the cell is proportionally counteracted by collagen deposition in the ECM. Since models of this type are generally based on the theory of mixtures and treat all components as continuously distributed, they make the implicit assumption that a cell can deposit collagen uniformly throughout the surrounding ECM, independent of distance from the cell. This assumption is reasonable in biological tissues with a high level of cellularity, such that any position within the ECM is adjacent or nearly adjacent to a cell. For tissues with low cell density, however, this assumption may begin to lose its validity.

In this work, we hypothesized that a cell's ability to deposit collagen in the ECM diminishes with distance from the cell. In other words, a cell will deposit most collagen close to its surface and an increasingly smaller amount moving away from the cell surface, in the same manner that when cells are embedded in collagen gels, remodeling is much more dramatic in the region immediately surrounding the cell [6]. Under this hypothesis, if there is a low density of cells within the ECM, then there may be regions where the cells are unable to deposit collagen or can only deposit a small amount. As a result, tissues with low density of cells may end up forming high stress regions due to inhomogeneous collagen deposition [7,8]. One frequently studied tissue to which this condition may apply is aneurysmal tissue. Aneurysms, or weakened dilations in arterial vessels, are characterized by reduced density of smooth muscle cells, when compared to healthy arterial tissue [9]. It is possible that the low density of cells leads to scattered collagen deposition, causing stress concentrations which then contribute to eventual rupture of the aneurysm.

The goals of this work were to model a cell with spatially-dependent collagen deposition, and to see how the resulting stresses in the cell and ECM compare to cases where collagen deposition is assumed to be uniform.

METHODS

Multiscale Model

A multiscale modeling approach was chosen for this problem, to elucidate the effect of microscale fiber deposition on cell and tissue behavior. Cells were modeled as spherical inclusions within an ECM fiber network [1]. Cell density within the matrix was sparse, with each cell one radius away from the next. The model was reduced, through symmetry, to an eighth of a spherical cell in a cube of ECM (Figure 1). For simplicity, the model was loaded through uniaxial extension in the X-direction. The opposing X-oriented face was fixed in X, while the remaining faces (in Y and Z) were given stress-free boundary conditions, enforced through tissue compaction. Up to 8% stretch was imposed for each run, in increments of 1% stretch per step.

Our multiscale code is built upon volume-averaging theory. The macroscopic scale is solved through a three-dimensional Galerkin finite element method. At each Gauss point within an element, a representative volume element (RVE) containing a three-dimensional fiber network is employed to solve the microscale problem. More detail about the general formulation of the code can be found in [10]. In the present model, both the cell and the ECM contained these fibrous RVEs, with the cell's fibrillar network initially set as five times stiffer than the surrounding ECM's. RVEs contained random, isotropic, Voronoi networks of collagen fiber.

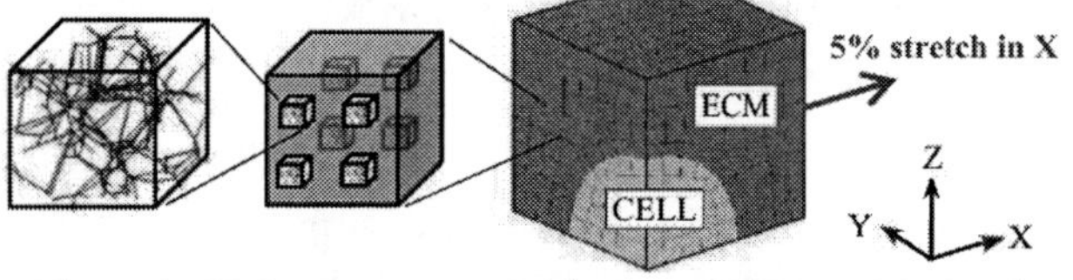

Figure 1: Finite element mesh of cell and ECM model, showing fibrous RVEs within each element.

Collagen Fiber Deposition

ECM remodeling was added to the finite-element code through changes to RVE fiber properties in the microscale calculations. A target stress σ_0 was set for the cell, so that collagen fiber was deposited if the maximum stress in the cell σ_{max} exceeded that value. The total volume of fiber $\Delta V_{fib,tot}$ added to the ECM was given by the proportional relationship:

$$\Delta V_{fib,tot} = \alpha(\sigma_{max} - \sigma_0) \quad (1)$$

In Equation 1, α is the proportionality factor that was set to maintain fiber deposition of under 5% each step. The cell itself was not changed in this study.

Different simulations were run to compare stresses in the cell and ECM when collagen was deposited uniformly throughout the ECM or preferentially closer to the cell. In either case, the total volume of fiber deposited for a given stress state was kept constant. Non-uniform fiber deposition was modeled using an exponential decay that depended on radial distance from the center of the cell.

$$\Delta V(r) = V_0 e^{-(r-r_{cell})/\beta} + C \quad (2)$$

In Equation 2, ΔV is the volume of fiber to be added to an RVE, based on its distance r from the cell's center. The maximum fiber deposition occurs at r_{cell} , which is the surface of the spherical cell. V_0 and C are given by Equations 3 and 4, respectively. The dispersion parameter β dictates how densely the fibers are deposited near the cell surface, with the limit $\beta \to \infty$ causing the equation to become a linear decay.

$$V_0 = \Delta V_{fib,tot}/\beta - (\beta + \Delta R)e^{-\Delta R/\beta} \quad (3)$$

$$C = -V_0 e^{-\Delta R/\beta} \quad (4)$$

The constants for Equation 2 were determined by setting the integral over the domain equal to the total added fiber volume.

For uniform fiber deposition, the total volume of collagen was divided equally among all of the ECM RVEs. The added fiber volume was implemented by increasing the radius of the existing collagen fibers in the RVE networks. The new radius of the fibers at each point was calculated using the current radius and the specified volume increase. Uniform deposition was compared to exponential decay with $\beta = 0.01$ and $\beta = 10$ (Figure 2).

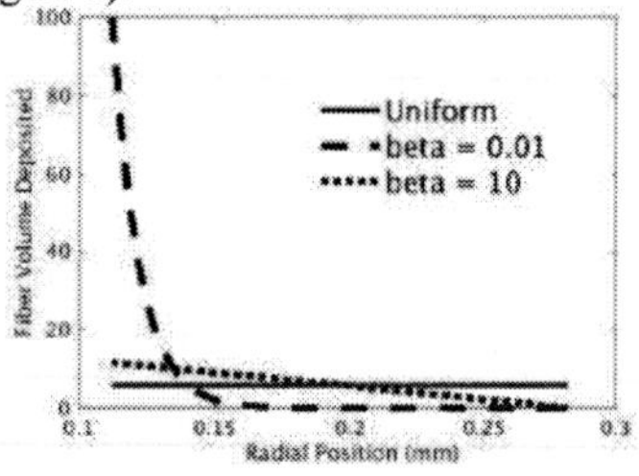

Figure 2: Plot of different fiber deposition distributions. Each curve has the same integral, corresponding to total fiber deposited.

RESULTS

Each simulation was run on 128 processors and ran for an average wall clock time of 70 minutes. Input values for α and σ_0 were tuned to prevent excessive fiber deposition from causing the code to become unstable. For the results reported, $\alpha = 0.5$ and $\sigma_0 = 0.15\ kPa$. As the cube was stretched, collagen was deposited into the ECM per Equation 1. In all cases, stress in both the cell and ECM increased nonlinearly with stretch. Average cell stress is presented here (Figure 3), and average ECM stress exhibited a similar trend, with overall lower stresses due to the lower initial stiffness.

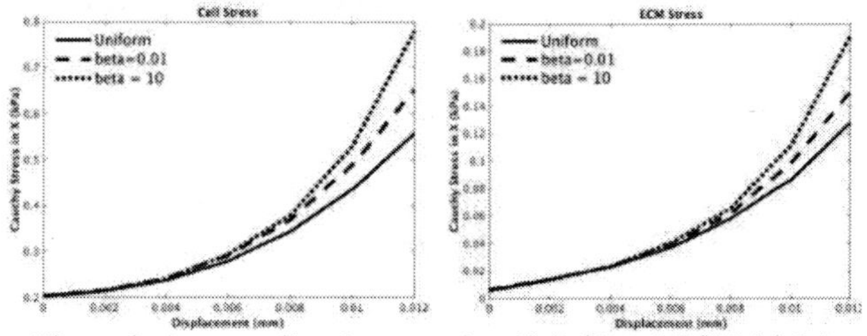

Figure 3: Average Cauchy stress in cell (left) and ECM (right) resulting from uniaxial stretch while collagen is deposited in ECM.

DISCUSSION

Simulation results followed physical expectations, where stress increased as the block was pulled and the ECM fibers became stiffer due to collagen deposition. For stretches above 3%, the three different deposition cases diverged, with uniform deposition maintaining the lowest stresses. By 7% stretch, the $\beta = 10$ case exhibited cell stresses nearly 40% higher than those in the uniform case. The $\beta = 0.01$ case fell in between the other two cases. This was unexpected because when $\beta = 0.01$, most of the fiber is deposited directly at the cell surface. If nonuniform fiber deposition were to cause elevated cell stress, it was expected that the more severe case ($\beta = 0.01$) would cause higher stresses than the more moderate case ($\beta = 10$).

Preliminary results suggest that the behavior of tissues with low density of cells can be affected by how uniformly collagen fiber is deposited during remodeling. In the future, it will be interesting to compare these controlled stretch results to a controlled stress simulation, which may be more physiologically relevant, especially in arteries. Findings from this work show that in cases of low cellularity, distribution of deposited collagen may be as important as amount.

ACKNOWLEDGEMENTS

The authors thank the Minnesota Supercomputing Institute for providing the computational resources required for this project.

REFERENCES

[1] Lai, VK et al., *J Biomech Eng*, 135:071007, 2013.
[2] Humphrey, JD, *J Biomech Eng*, 121:591-7, 1999.
[3] Fung, YC, Biomechanics Motion, Flow, Stress and Growth, 1990.
[4] Gyoneva, L et al., *Interface Focus*, 6:20150069, 2016.
[5] Bellini, C et al., *Ann Biomed Eng*, 42(3):488-502, 2014.
[6] Pizzo, AM et al., *J Appl Physiol*, 98:1909-21, 2005.
[7] Stevenson, MD et al., *Biophys J*, 99:19-28, 2010.
[8] Evans, MC et al., *J Biomech Eng*, 131:101014, 2009.
[9] López-Candales, A et al., *Am J Pathol*, 150(3):993-1007, 1997.
[10] Stylianopoulos, T et al., *Comput Methods Appl Mech Engrg*, 196:2981-90, 2007.

SB³C2017
Summer Biomechanics, Bioengineering and Biotransport Conference
June 21 – 24, Tucson, AZ, USA

PLANAR SHEAR CHARACTERIZATION OF THE FACET CAPSULAR LIGAMENT

Emily A. Bermel (1), Amy A. Claeson (2), Alexander Safonov (1), Victor H. Barocas (1)

(1) Biomedical Engineering
University of Minnesota – Twins Cities
Minneapolis, MN, USA

(2) Biomedical Engineering
University of Delaware
Newark, DE, USA

INTRODUCTION

The facet capsular ligament (FCL) is a spinal ligament that spans the synovial joint formed by the articular facets [1] on the posterior aspect of the spine. The FCL is composed primarily of aligned collagen fibers on the posterior surface and elastin fibers on the anterior surface. The FCLs exist on all levels of the spine, but the lumbar region is of particular interest to us due to its possible role in low back pain (LBP). LBP is a major public health problem, affecting 26.4% of adults in the United States and creating a significant financial burden on the health care system [1]. The presence of nociceptive nerves in the lumbar FCL [2,3] suggest that is may be involved in LBP.

Because of the irregular geometry of the FCL, even routine, simple body motions can cause complex tissue deformations. We have shown previously that the lumbar FCL undergoes a significant amount of in-plane shear during flexion and extension [2]. The mechanical properties of the lumbar FCL during these types of motion, however, have yet to be characterized in detail. Ianuzzi *et al.*, determined the principal strains of the lumbar FCL in situ during flexion and extension, but they didn't report the stresses in situ [3]. The objective of this work was to quantify the mechanical behavior of the lumbar FCL during planar shear testing.

METHODS

Cadaveric right facet joints (n = 2, L4-L5 level), were resected from motion segments (ages 64 – 67 years obtained through the Minnesota Anatomy Bequest program), and the spines were scanned in a 3T MRI system at the University's Center for Magnetic Resonance Research to grade the health of the facet joints and the intervertebral discs. Discs were graded on the Pfirmann scale [4], and the facets were graded on the Fujiwara scale [5]. Spine grading was performed separately by an orthopedic surgeon and a spine research specialist.

In preparation for mechanical testing, all posterior musculature was cleared, and the FCL was stripped of its surface membranes. The curved ligament surface was flattened, and the trabecular bone was removed to facilitate loading of the sample into the testing machine, while still maintaining physiological attachments to the bone.

Mechanical characterization of the FCL was completed on a planar biaxial tester. Two six-degree-of-freedom load cells were used, with each measuring one normal and two shear forces. Testing (figure 1) was completed in two stages. First, uniaxial testing was completed to pre-condition the tissue. Second a shear strain was imposed on the ligament as might occur in flexion and extension (10 cycles each at 0.1 Hz). The ligaments were displaced to 15% strain based on previous research that showed that a 15% strain was a large enough strain to stretch the FCL out of the toe region but not damage the tissue [6].

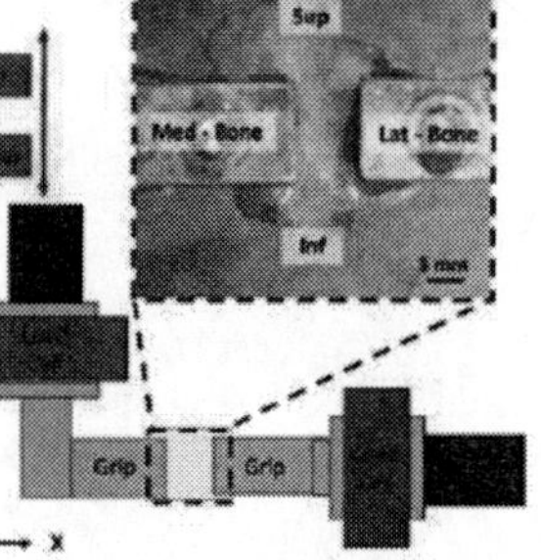

Figure 1: Shear testing setup of right FCL. During extension testing, the actuator on the right (lateral) moved to stretch the sample. Next, for shear testing, the actuator on the left (medial) moved up or down while the lateral actuator remained fixed.

Technical Presentation #198

The force and displacement data from the load cells and actuators were used to calculate the first Piola-Kirchhoff stress and the engineering strain in the ligament. Surface displacement tracking was performed, following a previously described protocol [6]. Briefly, video data from the cycles of interest were converted in to image stacks and the undeformed image was imported into Abaqus CAE to generate a planner mesh of the ligament surface, bordered by the bone on the medial and lateral sides. The images were then run through an image correlation code, and the surface strains were calculated.

RESULTS

Figure 2 shows the shear strain field in the FCL during testing for a representative specimen. For convenience, we use the terms "flexion" and "extension" since the different shear directions correspond roughly to shears that might occur during flexion and extension, but we emphasize that the shear test is not intended to replicate the actual deformations during flexion/extension. The maximum shear strain reached 0.07 during flexion and 0.1 during extension. The shear strain fields were close to homogeneous with the slightly higher strains in the superior medial border. The shear strain values seen here are comparable to those presented previously by Ianuzzi [3]. The average strain in the tissue was 0.15.

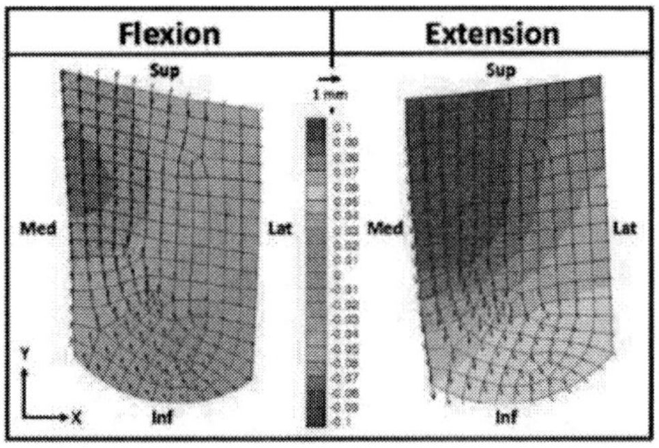

Figure 2: Surface shear strain fields during flexion and extension. The arrows indicate the displacement of the surface.

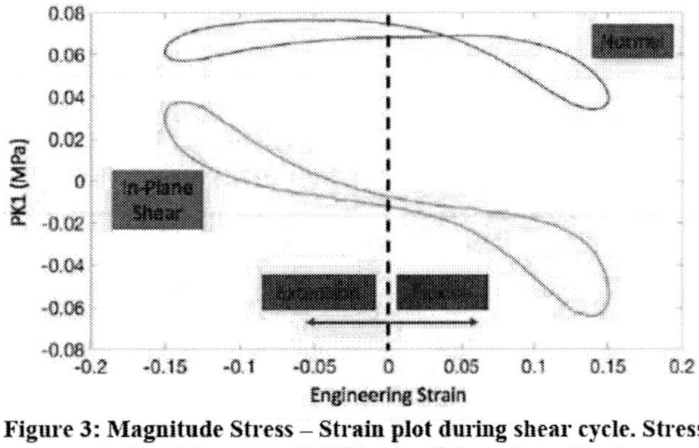

Figure 3: Magnitude Stress – Strain plot during shear cycle. Stress was calculated from the 6DOF load cell force data. Strain was calculated from the displacements of the actuators.

Figure 3 shows the magnitude of First Piola-Kirchhoff stresses in the tissue. The hysteresis region between loading and unloading curves indicates that the tissue is viscoelastic. The measureable hysteresis, in combination with the 0.1 Hz frequency of the test, indicates that the tissue has a relaxation time on the order of a few seconds. Larger stresses were observed during flexion than during extension. This is evident in the in-plane shear stresses (0.03 MPa for extension and 0.07 MPa for flexion). The normal stresses reach values of 0.06 MPa for extension and 0.04 MPa for flexion, indicating that during extension there is more normal stresses present in the tissue during shear.

DISCUSSION

The strain ranges during the flexion and extension shear were compared to ranges seen by Iannuzzi *et al.* in situ [3]. This suggests that the results of the planar shear test are representative of physiologic deformations. The strains, and therefore, stresses are localized in the superior portion of the medial border. It remains to be seen whether this effect is of physiological significance or was merely an artifact of our test geometry and procedures.

In a previous study [6], we estimated fiber orientations (figure 4) from shear forces that arose during biaxial tissue testing. In the current work, we found that the maximum stress attained in the tissue differed for flexion and extension. The higher shear stresses measured during flexion are not in agreement with the orientation of the fibers estimated in [6]. This indicates that there may be more complexity in the fiber alignment than previously estimated, and future work can investigate this difference.

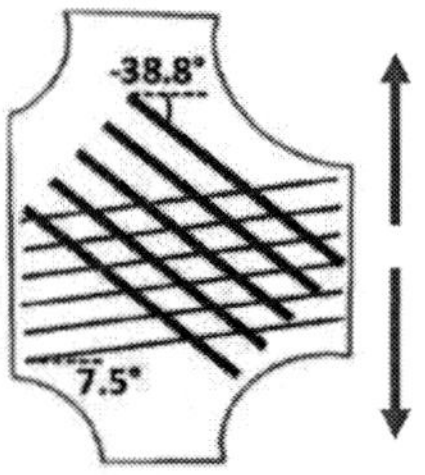

Figure 4: Fiber angels found by Claeson and Barocas when shear was included in the model simulation of the FCL [6]. The blue arrow indicates flexion and the red indicates extension.

ACKNOWLEDGEMENTS

Funding for this work was provided for the research by NIH/NIAMS T32 AR050938 Musculoskeletal Training Grant, and NIH grant U01-EB016638. The authors acknowledge the Minnesota Supercomputing Institute (MSI) at the University of Minnesota for providing resources that contributed to the research results reported within this paper. The authors thank Bjorn Olmanson for his assistance with the experiments.

REFERENCES

1. Deyo, R.A. et al., *Back Pain Prevalence and Visit Rates: Estimates From U.S. National Surveys, 2002.* Spine, 2006. **31**(23): p. 3.
2. Claeson, A.A. and V.H. Barocas, Spine J, 2017. **17**(1): p. 109-119.
3. Ianuzzi, A., et al.,. Spine J, 2004. **4**(2): p. 141-52.
4. Pfirrmann, C.W.A., et al., Spine, 2001. **26**(17): p. 1873-1878.
5. Fujiwara, A., et al., Eur Spine J, 1999. **8**: p. 5.
6. Claeson, A.A. and V.H. Barocas, J Mech Behav Biomed Mater, 2016. **65**: p. 127-136.

SUBSTRATE STIFFNESS DICTATES MACROPHAGE POLARIZATION AND THEIR CROSS-TALK WITH MESENCHYMAL STEM CELLS: IMPLICATIONS FOR BIOMATERIAL DESIGN

Rukmani Sridharan (1,2,3), Andrew. R. Cameron (1,2,3), Daniel. J. Kelly (1,2,3), Fergal J. O'Brien (1,2,3)

(1) Tissue Engineering Research Group, Department of Anatomy, Royal College of Surgeons in Ireland (RCSI), Dublin, Ireland

(2) Trinity Centre for Bioengineering, Trinity College Dublin (TCD), Dublin, Ireland

(3) Advanced Materials and Bioengineering Research (AMBER) Centre, RCSI & TCD

INTRODUCTION

Cells of the innate immune system (eg: macrophages, neutrophils) are the first to respond upon biomaterial implantation. Macrophages direct the tissue remodelling process through pro-inflammatory (M1) and anti-inflammatory (M2) polarization states [1]. Modulation of polarization states can be achieved primarily by providing the right cytokine milieu, but emerging studies implicate biophysical (substrate stiffness, topography, pore size) and biochemical (surface chemistry, ligand presentation) cues in modulating macrophage phenotype, function and polarization state [2]. Another component of the host response to a biomaterial is the interaction with tissue specific stem cells such as Mesenchymal Stem Cells (MSC), which are also influenced by cues provided by the implant environment [3]. Recent evidence suggests that macrophages can direct MSC differentiation and that MSC adapt an immunomodulatory state in the presence of macrophages [4]. While the importance of such cross-talk is established, little is known about the interaction between MSC and macrophages in the context of diverse biophysical/biochemical cues, a situation analogous to the implant environment. The overall aim of this study was thus to elucidate the role of substrate stiffness in macrophage polarization, MSC immunomodulation and the MSC/macrophage cross-talk.

METHODS

Macrophages (derived from THP1 monocytes, a human cell line) were seeded on collagen I coated 2D polyacrylamide gels of three stiffness values (Soft: 11kPa, Medium: 88kpa, Stiff: 323kPa). Cell attachment, cell viability and spread area were measured on different gels. Cells were stimulated with factors that polarized them towards M1 (LPS+IFN-γ) or M2 (IL4+IL13) phenotypes, harvested for gene expression analysis and media was analysed using Enzyme Linked ImmunoSorbent Assay (ELISA). Macrophage phagocytic ability of 1μm latex beads was assessed after 24 hours. Motility of macrophages on gels was also measured over a 24 hour period, using a live cell imaging system. To understand the mechanism of mechanotransduction by macrophages, actin inhibitors (Blebbistatin (Bleb), Cytochalasin D (Cyto-D), ML-9 and Y-27632) were used.

Human MSC were cultured either in regular basal media (BM) or inflammatory media (LPS+IFN-γ) (IM) for 24 hours. Immunomodulatory genes (IDO, PGE2, TGF-β1) and proteins (MCP1, IL6, VEGF) were assessed. For co-culture studies, MSC and macrophages were seeded in direct contact or in transwell inserts (paracrine) and assessed for TNFα and IL10 secretion.

RESULTS

Macrophages assumed a round, aggregated morphology on soft and medium gels and a spread morphology on stiff gels (Fig 1A,B,C). Indeed, the cell spread area was higher on stiff gels compared to soft and medium gels (Fig 1D). Macrophages produced high levels of pro-inflammatory TNFα on soft and stiff gels (Fig 1E) and surprisingly, high levels of anti-inflammatory IL10 on soft and medium gels (Figure 1F).

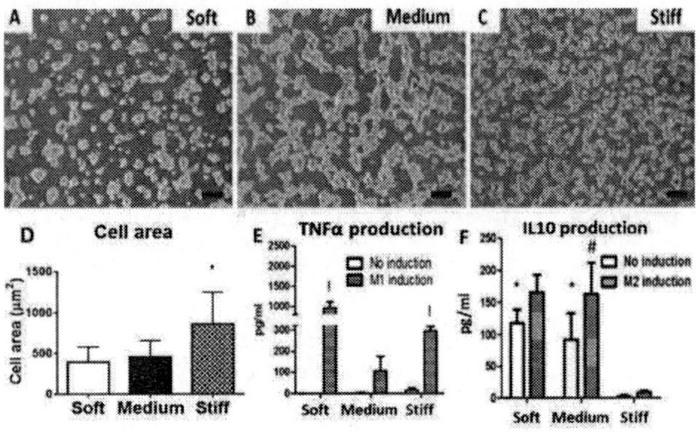

Figure 1 Macrophages behavior on polyacrylamide gels. Cells assume an aggregated, round shape on soft and medium gels (A,B) and a spread phenotype on stiff gels (C). Cell spread area was significantly increased on stiff gels ($p<0.05$) (D). Macrophages secrete high levels of TNFα on soft and stiff gels (E) and high levels of IL10 (F) on soft and medium gels.

Macrophages also had lower capability for phagocytosis (Fig2A) and lower motility on stiff gels (Fig 2B). Together, the data suggests that soft and especially medium stiffness gels promote an anti-inflammatory, highly phagocytic and motile macrophage phenotype with stiff gels consistently producing a pro-inflammatory phenotype. To elucidate the mechanism behind stiffness-sensing by macrophages, we used several cytoskeletal inhibitors. IL10 production by macrophages was only decreased in the presence of Cyto-D and Y-27632 (Fig 2C), suggesting an actin and ROCK mediated mechanotransduction.

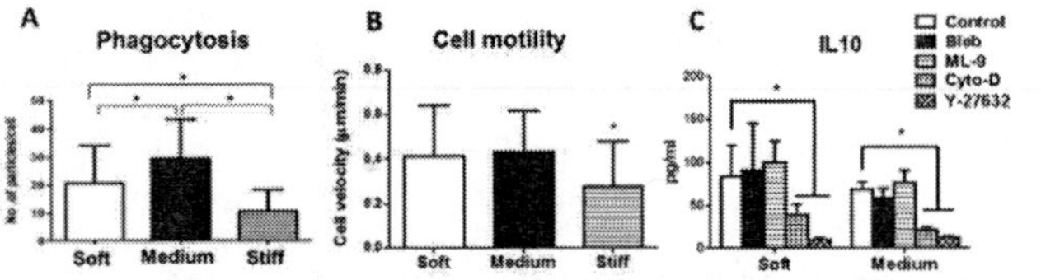

Figure 2 Macrophage functional analysis on polyacrylamide gels. Cells showed decreased phagocytic capability (A) and decreased motility (B) on stiff gels. The use of actin inhibitors Cyto-D and Y-27632 led to significantly lower IL10 production ($p<0.001$) on both soft and medium gels.

In contrast to macrophages, MSC did not display a stiffness dependent gene expression or secretome – cells upregulated immunoregulatory genes and proteins in response to inflammation irrespective of stiffness. However, when cultured with macrophages, MSC decreased TNFα and increased IL10 production from macrophages in both direct contact and paracrine cultures (Figure 3A,B). Interestingly, this immunomodulation was most profound on soft and medium gels and abolished in stiff gels.

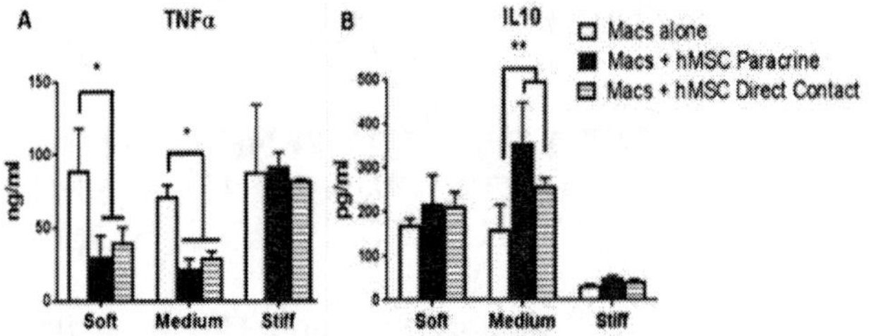

Figure 3 Co-culture with MSC decreases TNFα and increases IL10 production from macrophages on soft and medium gels. Macrophages secrete lower levels of TNFα (A) and higher levels of IL10 (B) when cultured in paracrine or direct contact with MSC only on soft and medium gels. ($p<0.05$)

DISCUSSION

Our results demonstrate that substrate stiffness can be used as a cue to polarize macrophages. Soft (11kPa) and medium (88kPa) stiffness gels promote an anti-inflammatory macrophage phenotype. Additionally, cells on these gels also exhibit increased phagocytic capability and motility, both of which constitute important functions of macrophages to clear and fight infections. We show that this increased IL10 production on soft and medium stiffness gels is mediated by the actin cytoskeleton, suggesting for the first time that macrophage polarization may be an actin-dependent process.

In contrast, MSC produce the same levels of immunoregulatory proteins on different stiffness gels, suggesting stiffness does not change the immunoregulatory properties of MSC. However, stiffness played an important role in facilitating successful cross-talk between MSC and macrophages, with soft and medium gels promoting decreased TNFα and increased IL10 production. Taken together, the data suggests that biomaterial stiffness can be used to control immune and stem cell responses. Such understanding of the cues responsible for directing cell behaviour will eventually lead to better biomaterial design in order to promote successful regeneration after biomaterial implantation.

ACKNOWLEDGEMENTS

This research is supported by a research grant from Science Foundation Ireland (SFI) and AMBER under Grant number SFI/12/RC/2278. The authors acknowledge Brenton Cavanagh, Microscopy Unit, RCSI for his help with obtaining images.

REFERENCES

[1] Anderson,J.M. et al, *Semin Immunol*, 20: 86-100, 2008

[2] Sridharan, R et al., *Materials Today*, 6: 313-325, 2015.

[3] Engler, A.J., et al., *Cell*. 126: 677-89, 2006.

[4] Saldana et al., *Biomaterials*. 37:124-133, 2015.

THE ROLES OF DECORIN IN THE STRUCTURE AND MECHANICS OF CARTILAGE PERICELLUAR MATRIX DURING SKELETAL DEVELOPMENT

D. R. Chery (1), B. Han (1), S. J. Rozans (1), L. Qin (2), D. E. Birk (3), R. V. Iozzo (4), M. Enomoto-Iwamoto (5), L. Han (1)

(1) School of Biomedical Engineering, Science and Health Systems
Drexel University
Philadelphia, PA, United States

(2) McKay Orthopaedic Research Laboratory
Department of Orthopaedic Surgery
University of Pennsylvania
Philadelphia, PA, United States

(3) Department of Molecular Pharmacology and Physiology
University of South Florida
Tampa, FL, United States

(4) Department of Pathology, Anatomy and Cell Biology
Thomas Jefferson University
Philadelphia, PA, United States

(5) Department of Orthopaedics
University of Maryland
Baltimore, MD, United States

INTRODUCTION

The pericellular matrix (PCM) of cartilage is a ≈ 5 μm-thick layer surrounding each chondrocyte, which regulates the biomechanical microenvironment of chondrocytes [1] and protects chondrocytes from overloading [2], while the territorial and interritorial extracellular matrices (T/IT-ECM) between cell lacunae are directly responsible for overall mechanical functions [3]. Thus elucidation of the mechanisms that regulate structuring PCM and T/IT-ECM and maintain their properties is necessary for both basic and translational studies on articular cartilage. The PCM is composed of localized collagen VI in the form of beaded filaments [4] and perlecan [5], as well as other non-fibrillar proteins [6]. Currently, the exact contribution of each molecular constituent to the PCM properties is unclear. Recently, the role of collagen VI in the PCM was studied by the application of immunofluorescence (IF)-guided AFM [7] to cross-sections of collagen VI-deficient (*Col6a1*$^{-/-}$) murine cartilage [8]. In this seminal work, the PCM of *Col6a1*$^{-/-}$ cartilage showed much reduced modulus, despite that the T/IT-ECM appeared to be normal [8]. Our recent studies discovered the indispensable roles of decorin, a class I small leucine rich proteoglycan (SLRP), in the properties of articular cartilage ECM [9]. To this end, we hypothesize that decorin also exerts an important role to the PCM properties. By combining tape-assisted cryo-sectioning that is applicable to mineralized tissues [10] and IF-guided AFM, we developed a new method to directly measure the mechanical properties of murine knee cartilage PCM and examined the impacts of decorin deficiency on the PCM properties.

METHODS

Femoral condyles were harvested from male wild-type (WT) and decorin-null (*Dcn*$^{-/-}$) mice at newborn (3-day old), immature (2-week old) and mature (3-month old) ages (n = 3 from each genotype and age). The condyle was embedded in OCT. Kawatomo's tape-assisted cryo-sectioning [10] was applied to obtain ≈ 5-μm-thick, unfixed sagittal sections containing epiphysis with untreated cartilage and subchondral bone. Following established procedures [7], the cryo-sections were stained by immunofluorescent antibodies specific for collagen VI. Using the Total Internal Reflection Fluorescence (TIRF)-AFM (MFP-3D, Asylum Research), we performed nanoindentation (R ≈ 2.5 μm, k ≈ 5.4 N/m, 15 μm/s rate) in PBS with the guidance of type VI collagen fluorescence imaging to distinguish the PCM versus T/IT-ECM (Fig. 1a). Within each region of interest, nanoindentation was performed as a 40 × 40 grid (1,600 indents) over a 20 × 20 μm^2 region containing both the PCM and the T/IT-ECM of the middle/deep zone uncalcified cartilage. From each indentation curve, effective indentation modulus, E_{ind}, was calculated via the finite thickness-corrected Hertz model [11]. For nanostructural analysis of collagne fibrils, 20-μm-thick cryo-sections were prepared from 3-month-old condyles, treated with Karnovsky's fixative, dehydrated in hexmethyldisilazane and imaged under Zeiss Supra 50VP SEM.

RESULTS

IF images detected distinctive rings of collagen VI-localized PCM surrounding each chondrocyte in both WT and *Dcn*$^{-/-}$ murine cartilage (Fig. 1). Guided by the IF-imaging, from each nanoindentation modulus map, the regions corresponding to the T/IT-ECM versus PCM were registered via our custom Matlab program, and each region was analyzed separately (Fig. 2a).Within each genotype and age, modulus of the PCM was found to be significantly lower than that of the T/IT-ECM. When compared between genotypes, both the PCM and T/IT-ECM of *Dcn*$^{-/-}$ cartilage had significantly lower moduli than WT at all ages (Fig. 2b). For both genotypes, a significant increase in the modulus was observed for both the PCM and the T/IT-ECM between 3 days and 2 weeks of age. Between 2 weeks to 3 months of age, in WT cartilage, the increase in modulus was apparent for both the PCM and the T/IT-ECM. By contrast, in *Dcn*$^{-/-}$ cartilage, a significant modulus increase was observed in the T/IT-ECM, but not for the PCM (Fig. 2b).

DISCUSSION

This study discovers a critical role for decorin in the normal functioning of cartilage PCM. The involvement of decorin in PCM organization is indicated by its binding activities with collagen VI and aggrecan core protein [12, 13]. To this end, this study provides direct evidence supporting that a loss of decorin leads to significantly weakened mechanical properties of the PCM, in addition to its impacts on the T/IT-ECM.

In newborn (3-day-old) mice, the absence of mechanical phenotype in $Dcn^{-/-}$ PCM and T/IT-ECM indicates that decorin mainly influences in the development of cartilage during the post-natal stage, rather than the embryonic stage. By the age of 2 weeks, the salient mechanical phenotype in $Dcn^{-/-}$ PCM illustrates the essential role of decorin during early post-natal development of the PCM. In later stages, from 2 weeks to 3 months of age, for $Dcn^{-/-}$ mice, the lack of an increase in PCM modulus is distinctive from the significant age-dependent increase in WT PCM modulus. This suggests that decorin's influence on PCM is not only at the early stage of post-natal skeletal development, but also throughout maturation. It is possible that decorin can influence the PCM properties through both its binding activities in the matrix, and through regulating chondrocytes synthesis activities. In line with the mechanical phenotype, the presence of highly aligned fibril bundles in $Dcn^{-/-}$ PCM (Fig. 3) suggests that decorin likely influences PCM development by inhibiting fibril lateral growth during cartilage PCM formation. Our ongoing studies aim to reveal the exact mechanism of decorin's roles in the PCM. However, these observations clearly suggest that decorin is an essential component of cartilage matrix, responsible for both direct mechanical function in the cartilage ECMs, as well as specific regional cell biomechanical differences in the PCM.

The combination of tape-assisted cryo-sectioning and IF-guided AFM offers a new method to study cartilage PCM in mice. Importantly, applicability of this test to mature murine knee cartilage enables future studies of PCM not only in uncalcified cartilage, but also in calcified cartilage. As a result, this study is the first to document the development of cartilage PCM from newborn to mature mice. Its further application includes the studies of post-traumatic OA [14]. This new approach enables direct examination of injury-induced changes in the PCM of hypertrophic chondrocytes, which could yield new targets for early OA detection and amelioration.

In conclusion, this study discovered an important role of decorin in cartilage PCM mechanical properties. The new nanomechanical method holds great potential in studying PCM changes in various murine models of cartilage development and OA.

ACKNOWLEDGEMENTS

This work was supported by the National Institutes of Health (AR066824), the U.S. Department of Education iCare for Healthcare Fellowship (DRC), and the Penn Center for Musculoskeletal Diseases (PCMD) (NIH P30AR050950). We also thank the Nano/Bio Interface Center at the University of Pennsylvania for the use of TIRF-AFM.

REFERENCES

[1] Guilak, F et al., *J. Biomech.* 33:1663-1673, 2000.
[2] Poole, CA et al., *J. Orthop. Res.* 5:509-522, 1987.
[3] Maroudas, A. 1979. In *Adult Articular Cartilage*. 215-290.
[4] Soder, S et al., *Osteoarthr. Cartilage* 10:464-470, 2002.
[5] Melrose, J et al., *J. Biol. Chem.* 281:36905-36914, 2006.
[6] Wilusz, RE et al., *Matrix Biol.* 39:25-32, 2014.
[7] Wilusz, RE et al., *J. Royal Soc. Interface* 9:2997-3007, 2012.
[8] Zelenski, NA et al., *Arthritis Rheum.* 67:1286-1294, 2015.
[9] Doyran, B et al., *Trans. ORS* 62:491, 2016.
[10] Kawamoto, T et al., *Methods Mol. Biol.* 1130:149-164, 2014.
[11] Dimitriadis, EK et al., *Biophys. J.* 82:2798-2810, 2002.
[12] Wiberg, C et al., *J. Biol. Chem.* 276:18947-18952, 2001.
[13] Wiberg, C et al., *J. Biol. Chem.* 278:37698-37704, 2003.
[14] Anderson, DD et al., *J. Orthop. Res.* 29:802-809, 2011.

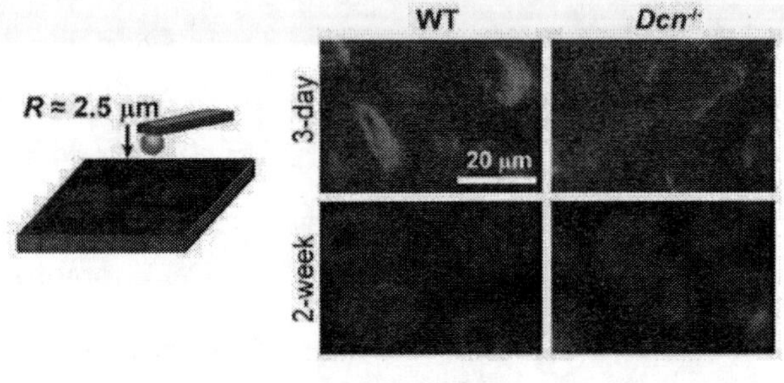

Figure 1: Immunofluorescence Images of type VI collagen identifies cartilage pericellular matrix (PCM) in both WT and $Dcn^{-/-}$ mice (shown for 3-day and 2-week old murine cartilage).

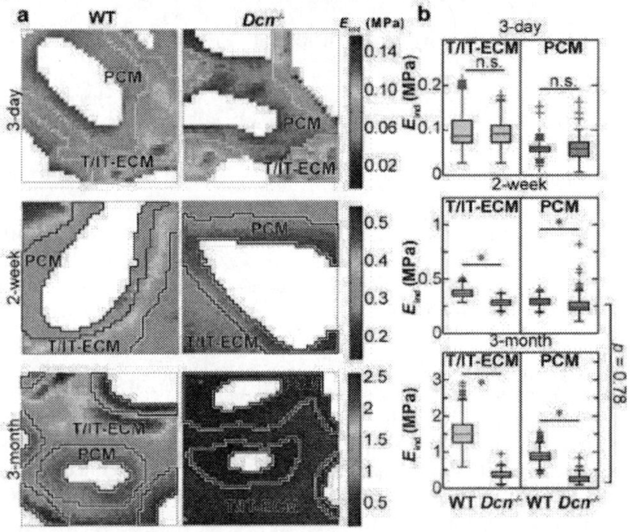

Figure 2: a) Typical indentation modulus maps (20 × 20 µm^2) of 3-day, 2-week and 3-month old WT and $Dcn^{-/-}$ murine cartilage cross-sections. The T/IT-ECM and PCM regions were distinguished by IF-images. b) Box-and-whisker plot of all indentation modulus measured (n = 3 each genotype and age), *:$p < 0.0001$ for WT versus $Dcn^{-/-}$. In addition, modulus of the PCM was significantly lower than T/IT-ECM in all mice, and significant age-dependence was found in all tested regions, except for the PCM in 2-week and 3-month old $Dcn^{-/-}$ mice.

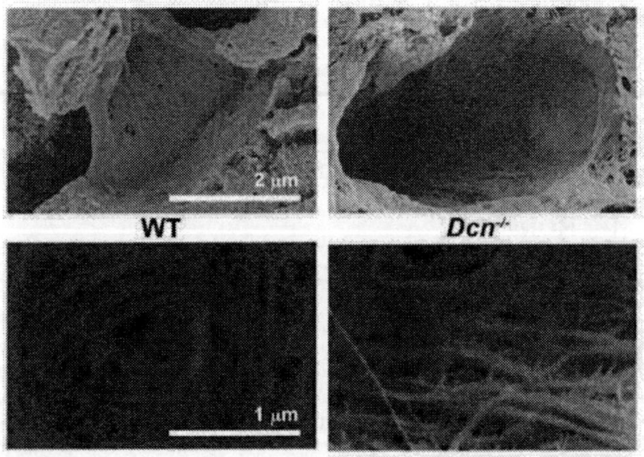

Figure 3: Scanning electron microscope images of 3-month WT and $Dcn^{-/-}$ murine cartilage.

ADHESIVE FILMS FOR ENHANCED TENDON-TO-BONE REPAIR

Stephen W. Linderman (1,2), M. Golman (3), T.R. Gardner (3), D. Yoon (4), V. Birman (5), W.N. Levine (3), G.M. Genin (4), S. Thomopoulos (3)

(1) Orthopaedic Surgery
(2) Biomedical Engineering
(4) Mechanical Engineering and Material Science
Washington University in St. Louis
St. Louis, Missouri, USA

(3) Orthopedic Surgery
Biomedical Engineering
Columbia University
New □ork, New □ork, USA

(5) Missouri S&T Global – St. Louis
Missouri University of Science and Technology
St. Louis, Missouri, USA

INTRODUCTION

Tendon-to-bone repair presents a challenging mechanical problem. Repairs require strong and resilient bonding to accommodate forces from activities of daily living and to avoid repair site elongation or rupture; however, strong bonds between compliant tendon and stiff bone cause high stress concentrations that limit attachment resilience. The healthy tendon enthesis achieves strong and resilient attachment of tendon to bone in several ways, including by distributing force over a relatively large footprint area to reduce local stresses and using a compliant transitional fibro-cartilaginous tissue to optimize stress concentrations and toughen the attachment.[1] Unfortunately, current surgical repair techniques fail to recreate these stress-reduction mechanisms, leading to high failure rates. For example, rotator cuff repairs are notoriously challenging, with post-repair rupture rates ranging from 20% for young, healthy athletes with small tears to as high as 94% for massive tears in elderly patients.[2,3] Due to these high failure rates, operative rotator cuff repair is only indicated for a subset of patients with symptomatic (i.e., painful) shoulders. Repairs are performed to reduce pain but not necessarily to reinstate shoulder function. Approximately half of the US population over 60 years old has a rotator cuff tear, leading to over 500,000 repairs annually. With a growing aging and elderly population, improving on these failure rates is critical to reinstate shoulder function in these patients.

These failure rates are not surprising from a mechanical perspective: single- and double-row rotator cuff repairs transfer almost all of the force from muscle to bone across two to four anchor points, where the suture from a bone anchor punctures through the tendon. The vast majority (86%) of rotator cuff repair ruptures occur by the tendon pulling through the sutures at those anchor points.[4] Our objective is to develop technology to strengthen repairs against these failure mechanisms.

Here, an adhesive-film based approach is proposed to augment standard tendon-to-bone repairs, with a particular focus on supraspinatus tendon rotator cuff repairs for proof-of-concept. This adhesive-based surgical augmentation mimics the natural stress distribution across the repair site to improve repair strength and limit ruptures. We hypothesized that, unlike conventional suture repairs with only a few anchor points, this adhesive repair scheme would reinstate load transfer over the entire tendon-to-bone insertion footprint. This increase in load transfer is expected to improve overall repair construct mechanical properties.

METHODS

We identified adhesive materials that optimize load transfer across the tendon-to-bone interface using a shear lag model, a finite element model, and surgical proof-of-concept experiments. The shear lag model predicted load sharing between repaired tendon and bone by estimating shear stress τ in the adhesive layer, as a function of position x, material properties, and geometry for a human supraspinatus tendon insertion (Fig 1 inset). Equating the peak shear stress (at $x = L$) to the adhesive failure stress, τ_{fail}, and solving for force yielded:

$$\frac{P_{max}}{w} = \tau_{fail} L \frac{\sinh(\beta L)}{\beta L}\left[\frac{\chi}{(\chi-1)\cosh(\beta L)+1}\right] \quad (1)$$

where χ is the dimensionless stiffness and β^2 is the characteristic inverse length scale:

$$\chi = 1 + (E_b t_b / E_t t_t) \quad (2)$$

$$\beta^2 = G_a \chi / t_a t_b E_b \quad (3)$$

in which G, E, and t are the shear and elastic moduli and thickness, respectively, of the tendon (t), bone (b), and adhesive (a), w is the repair width, and L is the repair length.

This model yielded a closed-form law that could be used to identify optimal adhesives. We established the limits of this law for thick and stiff adhesives using a two-dimensional finite element model in the commercial package COMSOL (Fig 1).

To evaluate adhesive properties with relevant adherends, we

developed a new method for lap shear testing using tendon and bone planks. Idealized tendon-to-bone plank repairs were performed using 1/8" thick bovine deep digital flexor tendon planks adhered to 1/4" thick bovine femur cortical bone planks (both planks were 15 mm wide, with 10 mm overlap length). The tendon planks were manufactured using a sliding microtome to plane both sides to a uniform, desired thickness. The bone planks were cut using a diamond wafer blade. The adhesive was a multipartite system consisting of a 1/16" thick bovine tendon to provide a compliant bulk, attached to the adjacent tendon and bone using Loctite 4903 (ethyl cyanoacrylate) to provide the necessary bond strength. The 1/16" thick tendon bulk was first treated with 0 mM, 20 mM, or 100 mM tetrakis (hydroxymethyl) phosphonium chloride (THPC) in saline for 15 minutes to assess the impact of an adhesive bulk material strengthened two-fold by THPC. This cured for 3 hours at room temperature under 100 kPa compressive pressure before lap shear testing. To assess the potential for rotator cuff repairs, human cadaver infraspinatus tendons were dissected away from the humeral head and then adhered to their insertion sites, as described above for lap shear tests, and mechanically tested to failure.

RESULTS

Shear lag and finite element modeling predicted the design space for adhesive mechanical properties that would improve load transfer across tendon-to-bone repairs (Fig 2, band inside blue dotted box, including the elastomer material class). Promising candidate materials were identified by overlaying an Ashby chart on this contour map. Experimental validation yielded 47 N (median, IQR: 34-48 N) of load-carrying capacity over 1.5 cm² in idealized tests, similar to loads measured in cadaver infraspinatus tests, and 66% higher loads with strengthened adhesive bulk interlayers crosslinked with THPC (Fig 3).

DISCUSSION

This work demonstrates an opportunity to dramatically improve tendon-to-bone repair strength using achievable adhesive material properties. The models employed herein predicted that mechanically desirable adhesives would be of millimeter-scale thickness and compliant in shear while maintaining high binding and shear strengths in order to minimize stress concentration and maximize load transfer across the repair (Fig 2, upper left corner). Compliant, thick adhesives allow greater deformation, thereby distributing loads over a larger length than stiff, thin adhesives. Even fairly stiff, off-the-shelf cyanoacrylate-based adhesives bear enough load to improve repair strength by approximately 20% (70 N) over a 2.5 cm² rotator cuff tendon repair footprint, when used in conjunction with current repair techniques.[5] Furthermore, adhesives should prevent micro-motion between repaired tendon and bone, enabling healing. We are actively developing improved adhesive materials to fit within the theoretical optimal material property ranges, which are predicted to yield over 10-fold higher repair strengths, so that the strength of the tissue itself becomes the limiting factor in the repair.

ACKNOWLEDGEMENTS

NIH (U01EB016422, R01AR062947, F30AR069491), WUSTL Musculoskeletal Research Center (NIH P30AR057235), WUSTL Institute of Clinical and Translational Sciences. Prof. Spencer Lake for sliding microtome use.

REFERENCES

[1] Genin, G et al., *Biophys J*, 97:976-985, 2009.
[2] Galatz, L et al., *J Bone Jt Surg Am*, 86-A:219-224, 2004.
[3] Harryman, D. et al., *J Bone Jt Surg Am*, 73:982-989, 1991.
[4] Cummins, C et al., *J Shoulder Elbow Surg*, 12:128-133, 2003.
[5] Gerber, C et al., *J Bone Jt Surg Br*, 76:371-380, 1994.

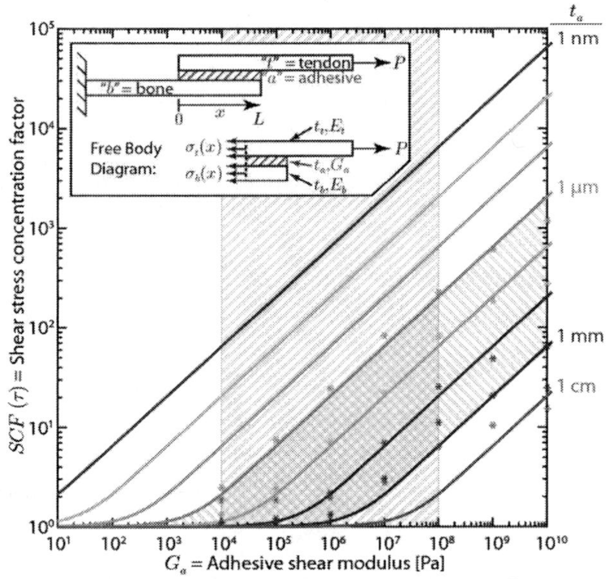

Figure 1: Shear lag (lines) and finite element (*) modeling predicted the lowest stress concentration for compliant, thick adhesives up to 1 mm thickness. Shading indicates relevant ranges for optimal adhesive.

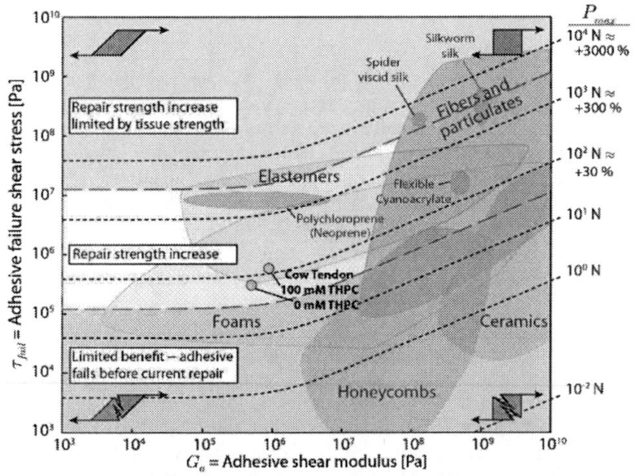

Figure 2: Isoclines of the predictive load capacity of adhesive (0.5 mm thick) repaired tendon-to-bone insertion, overlaid with an Ashby chart of real material properties for several material types. Percent added repair strength approximated from human supraspinatus.[5]

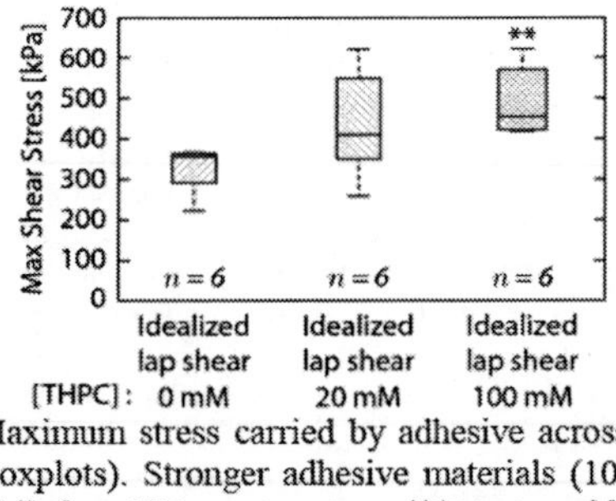

Figure 3: Maximum stress carried by adhesive across idealized lap shear tests (boxplots). Stronger adhesive materials (100 mM THPC-crosslinked) failed at 66% greater stress (**t-test vs. idealized 0 mM: $p = 10^{-3}$).

SELF-ASSEMBLED MICELLE ENABLES ENHANCED DELIVERY OF CRISPR/CAS9 SYSTEM

Yeh-Hsing Lao (1), Mingqiang Li (1), Madeleine A. Gao (1), Kam W. Leong* (1,2)

(1) Department of Biomedical Engineering
Columbia University
New □ork, N□, USA 10027

(2) Department of Systems Biology
Columbia University
New □ork, N□, USA 10032

INTRODUCTION

CRISPR-associated protein-9 nuclease (Cas9) originally is a part of bacterial, adaptive immune system against foreign DNAs. Its corresponding guide RNA (gRNA) specifically directs the Cas9 nuclease to the DNA target and induces a double-stranded break (DSB). This feature enables programmable, precise DNA interference, and thus the CRISPR/Cas9 system has become a powerful tool for DNA editing in a wide spectrum of organisms including mammalian cells.[1] However, delivery of the CRISPR/Cas9 system remains a challenge. Although viral transduction offers better efficiency, its uncontrollable, permanent gene insertions usually raise a safety concern. In contrast, non-viral transfection using nanoparticles provides safer and transient delivery, but the delivery efficiency of Cas9 components, especially for Cas9 plasmid, is unsatisfactory.

To date, only a few delivery systems are optimized for the CRISPR/Cas9 system, and most of them are specialized for the delivery of Cas9/gRNA protein complex.[2-4] Yet, direct Cas9/gRNA complex delivery may result in higher immunogenicity as Cas9 nuclease is a foreign protein, and this method may be limited to gene disruption only as Cas9-based gene correction and insertion need another nucleic acid template, which has a totally different property and may compromise the carrier design. To tackle the aforementioned issues, we propose a self-assembled micellar system, comprising quaternary ammonium-terminated poly(propylene oxide) (PPO-NMe_3) and amphiphilic Pluronic F127, in conjunction with all-in-one gRNA/Cas9-GFP plasmid (pCas9) for delivering the CRISPR/Cas9 system. We evaluated this delivery system in disrupting the human papillomavirus (HPV) E7 oncogene on a human cervical cancer model. The optimized micelle efficiently disrupts the oncogene and inhibits the downstream pathway, suggesting an effective, non-viral approach to advance genome editing-based therapy.

METHODS

PPO-NMe_3 was synthesized through the quaternization on the PPO terminus, and this modification was confirmed by ^1H NMR. The all-in-one gRNA/Cas9-GFP plasmids were constructed by following the protocol established by Ran *et al.*[5]

The micelle was formed by gentle mixing of F127, PPO-NMe_3 and pCas9 at room temperature. Branched polyethylenimine (PEI) or Lipofectamine 2000 was used as a control for transfection. The human cervical cancer cell line, HeLa, was treated with the prepared particles in OptiMEM for 4 h and in complete media afterward.

To confirm the gene disruption, we followed a previous work[6] to sort the cells using the GFP tag and extract the genomic DNAs at 96 h post-transfection. The E7 oncogene disruption was confirmed by T7 endonuclease I (T7EI) assay and Sanger sequencing.

RESULTS

The design of proposed micelle includes three major components, pCas9, PPO-NMe_3 and F127 (**Fig 1A**). The quaternary ammoniums on PPO-NMe_3 terminus provide stronger DNA-binding affinity, and PPO-NMe_3's backbone hydrophobicity could further facilitate the DNA condensation. Yet, the PPO-NMe_3/pCas9 complex was not stable in the transfection condition that contained reduced serum. F127 was therefore introduced to stabilize the complex by forming a F127/PPO-NMe_3/pCas9 micelle via hydrophobic interaction between its PPO segment and PPO-NMe_3.

In this design, the DNA condensation ratio (PPO-NMe_3/pCas9) and F127 blending ratio (PPO-NMe_3/F127) play crucial roles on the micelle formation. We evaluated the pCas9 transfection efficiencies of the micelles prepared with different DNA condensation and F127 blending ratios. Using PEI and Lipofectamine as controls, the micelle with a DNA condensation ratio of 40/1 and a F127 blending ratio of

1/1 showed superior transfection efficiency and lower cytotoxicity (**Fig 1B**). The optimized micelle was nano-sized (z-average diameter = 139 ± 1.37 nm), positively charged (zeta potential = 39 ± 1.04 mV) and narrowly distributed (polydispersity = 0.072 ± 0.05). It was also stable over the 4 h transfection period in the reduced serum condition and transfected HeLa with a transient Cas9 expression profile; the Cas9 expression level achieved the maximum level at 24 h post-transfection with a transfection efficiency of 30.3 ± 2.93%.

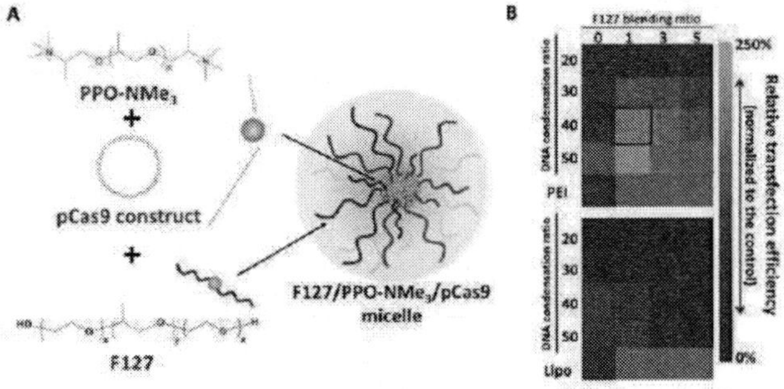

Fig 1. (A) Design and (B) optimization of the proposed micelle.

After optimizing the formulation of our micelle, we next investigated if this approach could provide more efficient gene editing compared with the well-characterized carrier. As Lipofectamine was more potent than PEI, we compared our approach with Lipofectamine side by side on T7EI assay and sequencing (**Fig 2A**). On T7EI assay, the predicted Cas9-induced DSB products are 303 and 248 bp. We detected these products on both micelle- and Lipofectamine-transfected HeLa cells with a comparable efficiency. Nevertheless, sequencing data showed that the proposed micelle induced more disruptions than the Lipofectamine-induction on the transfected cells, and we obtained a similar efficiency when using the micelle with the other E7-targeting gRNA. Furthermore, this Cas9-induced E7 oncogene disruption lowered the downstream proteasome activity by 25.7 ± 10.3 % and thus reduced the proliferation rate on the bulk, F127/PPO-NMe3/pCas9 micelle-treated HeLa cells. Notably, from the sequencing results, we observed a significant population of the micelle-transfected HeLa cells were inserted with a large foreign DNA (8.7% and 4.3%; **Fig 2A**), suggesting that the micelle might facilitate gene insertion during the DNA repair process. To validate this phenomenon, we designed another pCas9 construct targeting chromosome 19 and a RFP-encoded homology-directed repair (HDR) template. Through the co-transfection of pCas9 and the HDR template using our micelle approach, we observed a significant population of the transfected cells expressing RFP without any antibiotics selection (**Fig 2B**).

DISCUSSION

Although non-viral transfection technology has been developed for decades and considerably mature for certain applications, it still encounters difficulty on delivery of the CRISPR/Cas9 system. Especially for Cas9 plasmid, for example, a previously reported work has shown limited efficiencies (<15%) on the easy-to-transfect cells, such as U2OS and A549 cancer cell lines, with one of the most commonly used liposomal system, Lipofectamine.[2] In this work, we have designed a self-assembled micelle, containing amphiphile F127, modified PPO (PPO-NMe3) and all-in-one gRNA/Cas9-GFP plasmid, optimized for Cas9 plasmid delivery (**Fig 1A**). On the *in vitro* HPV model, this approach holds superior transfection efficiency (**Fig 1B**) as well as higher E7 oncogene disruption rate (**Fig 2A**), compared with the well-characterized systems, and therefore it is able to interfere the downstream proteasome pathway. Additionally, as aforementioned, current non-viral CRISPR/Cas9 delivery systems mainly focus on delivering Cas9/gRNA protein complex,[2-4] which may face some challenge on co-delivery with a HDR template. In this work, we have also demonstrated the feasibility of using our micelle to insert a foreign gene (**Fig 2B**). Those findings suggest the promise of the proposed micellar system and its potential on genome editing applications.

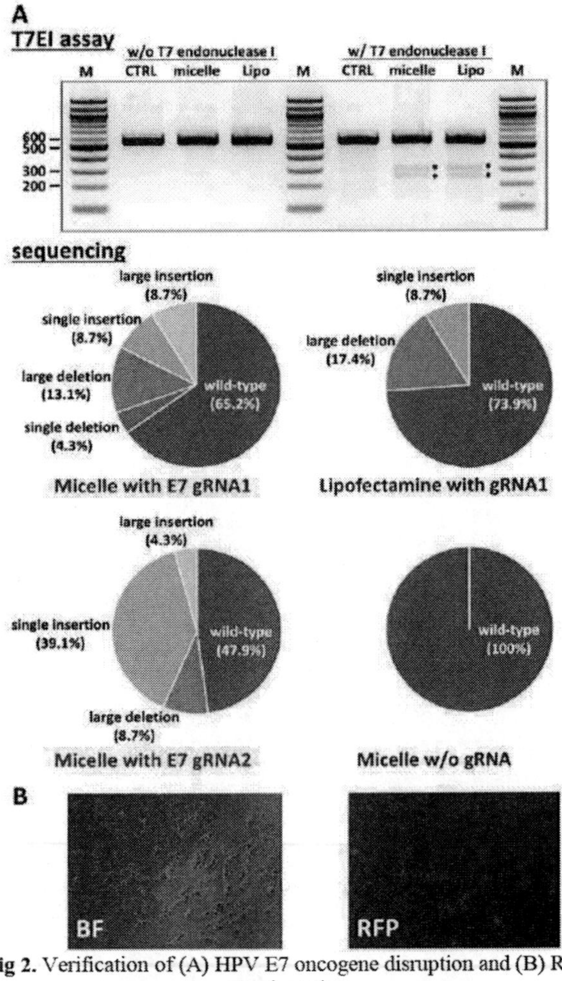

Fig 2. Verification of (A) HPV E7 oncogene disruption and (B) RFP gene insertion.

ACKNOWLEDGEMENTS

This work is supported by NIH (HL109442, AI096305, UH3TR000505, and GM110494). Cell sorting in this research was performed in the CCTI Flow Cytometry Core at Columbia University, supported in part by NIH Office of the Director (S10OD020056).

REFERENCES

[1] Singh, A, et al., *Chem Soc Rev*, 45: 6666-6684, **2016**.
[2] Liang, X. Q., et al., *J Biotechnol*, 208: 44-53, **2015**.
[3] Yu, X., et al., *Biotechnol Lett*, 38: 919-929, **2016**.
[4] Sun, W. J., et al., *Angew Chem Int Edit*, 54: 12029-12033, **2015**.
[5] Ran, F. A., et al., *Nat Protoc*, 8: 2281-2308, **2013**.
[6] Kennedy, E. M., et al., *J Virol*, 88: 11965-11972, **2014**.

A THERMODYNAMIC STATISTICAL MECHANICS MODEL TO INVESTIGATE THE INFLUENCE OF LIGAND DENSITY AND SUBSTRATE STIFFNESS ON CELL SPREADING

Eóin McEvoy (1), Siamak S. Shishvan (2), Patrick McGarry (1), Vikram S. Deshpande (2)

(1) Discipline of Biomedical Engineering,
National University of Ireland Galway,
Galway, Ireland

(2) Department of Engineering,
University of Cambridge,
Cambridge, UK

INTRODUCTION

Several experimental studies have demonstrated that substrate properties have a significant impact on cell behavior. Jacot *et al.* (2008) show that sarcomere development and alignment in cardiomyocytes is dependent on substrate stiffness [1]. A study by Arnold *et al.* (2004) reveals that focal adhesion (FA) and stress-fibre (SF) formation is limited by ligand spacing on the substrate [2]. Finally, Engler *et al.* (2004) have shown that cell spread area is dependent on both substrate rigidity and ligand density [3].

The bio-mechanisms underlying such experimental observations are not fully understood. In the current study we propose a new approach to the analysis of cell spread on ligand coated substrates. We demonstrate that a thermodynamically consistent statistical mechanics model explains several of the key phenomena observed experimentally.

MODEL DEVELOPMENT

Cytoskeletal Remodelling: The thermodynamically consistent kinetic equation for SF formation/dissociation proposed by Vigliotti *et al.* (2015) [4] reduces to:

$$\hat{\eta}(\phi) = \left(\hat{N}_u/\pi\hat{n}(\phi)\right)exp[\hat{n}(\phi)(\mu_u - \mu_b)/kT] \qquad (1)$$

under steady state conditions in a representative volume element (RVE) within the cell. $\hat{\eta}$ is the SF concentration per unit surface area of the RVE, $\hat{n}$ is the number of actin-myosin functional units along the length of the RVE, μ_b is the free energies of the bound contractile units in a stress fibre, μ_u is the free energy of unbound SF proteins, while k and T are the Boltzmann constant and absolute temperature. $\hat{N}_u$ is the number of unbound SF proteins. Here we implement non-local conservation of the total number of SF proteins within the cell, $\hat{N}_T$, under the assumption that the diffusion of unbound proteins through the cytoplasm is infinitely fast relative to the timescale of SF remodelling. The number of unbound proteins in the whole cell is therefore given as:

$$\hat{N}_u = \hat{N}_T - \int_{V_c}\int_{-\pi/2}^{+\pi/2} \hat{\eta}(\phi)\,\hat{n}(\phi)d\phi dV \qquad (2)$$

where the inner integral computes the number of bound proteins in a RVE, and the outer integral provides a summation over all the RVEs in the entire cell volume V_c. At steady state, $\hat{n}$ in a SF at orientation ϕ is given as:

$$\hat{n}(\phi) = \left(1 + \varepsilon_n(\phi)\right)/(1 + \tilde{\varepsilon}_{ss}) \qquad (3)$$

where $\varepsilon_n(\phi)$ is the nominal material strain in the direction of the SF. Extension of a SF results in addition of functional units in series, with the effect that the internal strain in the SF is reduced until a steady state value $\tilde{\varepsilon}_{ss}$ is achieved. Conversely, shortening of a SF results in removal of functional units. Such SF remodelling results in an internal fibre strain $\tilde{\varepsilon}_{ss}$ that is different from the axial material strain in the direction of the fibre. The cytoskeletal free energy density is given as:

$$\bar{G}_c = \rho_c kT\ln\left(\hat{N}_u\right) \qquad (4)$$

where ρ_c is the concentration of cytoskeletal proteins in the cell. The material model is completed by the addition of a passive non-linear Ogden hyperelastic formulation to represent the passive components of the cell. The elastic strain energy density ($\bar{G}_{elas}$) during cell deformation and spreading can therefore be readily computed.

Focal Adhesion Development: The cell-substrate interaction is described by an extension of the focal adhesion (FA) model from Deshpande *et al.* (2008) [5]. The FA complexes are connected to the cell membrane via integrins which have a local area density C_l (μm^{-2}). The number of ways these integrins can be connected to the cell membrane governs the entropy. The chemical potential of the complexes is given by:

$$\chi_c = kT\ln\left(\frac{\bar{C}}{1-\bar{C}}\right) - F_e L_e + \frac{F_e^2}{2\,\kappa_s} \qquad (5)$$

with $\bar{C} = C_l/C_R$, where C_R is the reference integrin concentration (μm^{-2}). The force (F_e) in an integrin-ligand complex is given by $F_e = T/N_H$ where T is the local traction ($kPa\,\mu m^{-2}$), and N_H is the area density of ligands on the substrate surface (μm^{-2}). L_e is the length of the complex, and κ_s is the integrin stiffness. Conservation of the total number of integrins on the cell surface is enforced. As χ_c is spatially uniform at equilibrium, the adhesion energy density is given as:

$$\bar{G}_{adh} = \frac{C_0\, kT\chi_c}{b_0} \tag{6}$$

where C_0 is the initial area density of binding integrins on the cell surface (μm^{-2}), and b_0 is the initial cell thickness. Finally, simulation of cell spreading on a compliant neo-Hookean elastic substrate requires calculation of the substrate energy density ($\bar{G}_{sub}$).

SIMULATIONS AND RESULTS

The Markov chain Monte-Carlo (MCMC) methodology is used to compute the distribution of cell spread states for a given ligand density and substrate stiffness. For each substrate, 2.5 million spread states are considered by imposing a sequence of trial moves on randomly selected control points in the cell (Fig. 1(a)). The random displacement associated with each trial move is given as:

$$u_i = \Delta(Ranf - 0.5) \quad with\ i = x, y \tag{7}$$

where $Ranf$ is a randomly generated variable with $0 \leq Ranf \leq 1$, and Δ is a parameter to control the displacement magnitude. For each spread state, we compute quantities such as the cytoskeletal protein distribution, SF orientation, and focal adhesion traction via a mixed finite element/boundary element method scheme. Results are shown in Fig. 1(b-d) for three different spread states. The total free energy

$$\bar{G}_{tot} = \bar{G}_c + \bar{G}_{elas} + \bar{G}_{sub} + \bar{G}_{adh} \tag{8}$$

is obtained from Eqns. (1-6). A trial move to a new spread state is accepted with a probability:

$$acc(old \rightarrow new) = \exp\{-\beta(\bar{G}_{tot}^{new} - \bar{G}_{tot}^{old})\} < 1 \tag{9}$$

where β, the conjugate to the spread entropy, drives the mean $\bar{G}_{tot}$ of all accepted configurations towards the homeostatic free energy G_0. If the step is accepted, the next trial move is generated. Conversely, if the step is rejected an alternate random trial move is generated from the old spread state. For good convergence, the overall acceptance rate should be between 25% and 45% [6].

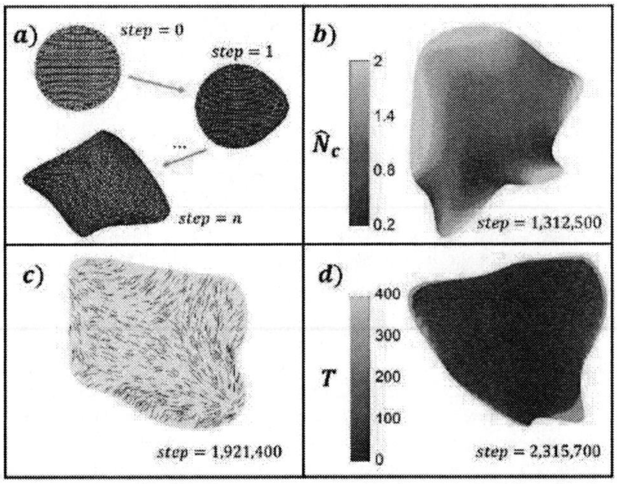

Figure 1: (a) Evolution of cell spread states over steps of Markov chain. (b) Local concentration of cytoskeletal proteins $\hat{N}_c$. (c) Dominant SF orientations. (d) Focal adhesion tractions $T\ (kPa\ \mu m^{-2})$. Different spread states are shown in a-d.

Following completion of the MCMC scheme we can construct the probability distribution for an observable of interest. Fig. 2(a) presents the probability of a spread state having an observed free energy. Additionally, in Fig. 2(b) we show the probability distribution of observed cell spread areas for 3 ligand densities (N_H) on a rigid substrate. The spread area with the highest probability is highlighted in all cases.

Fig 3(c) presents the spread area in terms of mean and standard deviation (SD) as a function of ligand density. For cells on a rigid substrate both the mean spread area and SD increase as the ligand density increases. A peak spread area is observed at $N_H = 750\mu m^{-2}$, with a slight decrease in spread area occurring for higher values of N_H. In the case of cells on a compliant elastic substrates lower spread areas are observed, with a peak spread area occurring at a ligand density of $N_H = 400\mu m^{-2}$. Our computed dependence of spread area (mean and standard deviation) on substrate stiffness and ligand density has been observed experimentally by Engler *et al.* [3] (Fig. 2(d)).

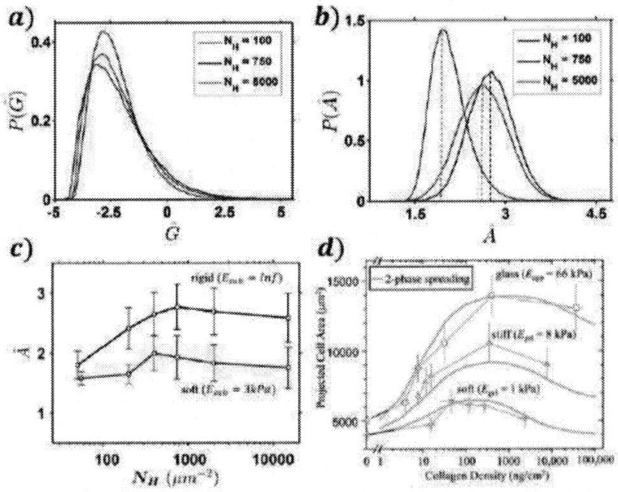

Figure 2: Probability distribution of (a) the total free energy $\hat{G} = G/(\rho_c kT)$ and (b) the spread area $\hat{A} = A/A_0$ for cells on a rigid substrate. (c) Predicted cell spread areas (mean$\pm SD$) at a given ligand density (N_H) and substrate stiffness. (d) Experimental data from Engler *et al.* [3].

DISCUSSION

Incorporation of a thermodynamically consistent SF model into a statistical mechanics framework provides new insight into the role of free energy as a driver of the stochastic process of cell spreading. The framework explores 2.5 million spread states for a given substrate ligand density. The dependence of the spread area on substrate stiffness and ligand density is correctly predicted, both in terms of mean values and standard deviations. This model represents a significant advance on recent deterministic and phenomenological approaches to cellular biomechanics [7,8].

ACKNOWLEDGEMENTS

EMcE acknowledges funding from the Irish Research Council (RCS1434) and the NUI Galway Hardiman Scholarship.

REFERENCES

1. Jacot, JG *et al., Biophysical Journal*, 953479-3487, 2008.
2. Arnold, M *et al., Chemphyschem*, 19;5(3):383-8, 2004.
3. Engler, A *et al., Biophysical Journal*, 86:617-628, 2004.
4. Vigliotti, A *et al., Biomech Model Mechanobiol*, 15:761–789, 2015.
5. Deshpande, VS *et al., J Mech Phys Solids*, 56:1484-1510, 2008.
6. Frenkel, D *et al.* Understanding Molecular Simulation, ISBN: 978-0-12-267351-1, 2001
7. Ronan, W *et al., J Mech Behav Biomed Mater*, 14:143-57, 2012.
8. Reynolds, N *et al., Biomaterials*, 35(13):4015-25, 2014.

INDISPENSABLE ROLES OF DECORIN IN CARTILAGE POROELASTICITY AT THE NANOSCALE

B. Han (1), Q. Li (1), M. Sun (2), H. T. Nia (3), R. Oftadeh (4), L. Qin (5), R. V. Iozzo (6), D. E. Birk (2), L. Han (1)

(1) School of Biomedical Engineering, Science and Health Systems
Drexel University
Philadelphia, PA, United States

(2) Department of Molecular Pharmacology and Physiology
University of South Florida
Tampa, FL, United States

(3) Massachusetts General Hospital
Harvard Medical School
Boston, MA, United States

(4) Department of Orthopaedic Surgery
Beth Israel Deaconess Medical Center
Boston, MA, United States

(5) McKay Orthopaedic Research Laboratory,
Department of Orthopaedic Surgery
University of Pennsylvania
Philadelphia, PA, United States

(6) Sidney Kimmel Cancer Center
Thomas Jefferson University
Philadelphia, PA, United States

INTRODUCTION

Poroelasticity is the dominant mechanism for energy dissipation and impact force protection in articular cartilage during high frequency activities, such as jumping and impactful loading [1,2]. In cartilage, poroelastic mechanics is governed by the interactions between water molecules and glycosaminoglycan (GAG) side chains of aggrecan [3] (mainly chondroitin sulfate (CS)-GAGs) [4]. The collagen fibrillar network serves as a scaffold that holds aggrecan aggregates at ≈ 50% compressive molecular strain that is critical for the hydraulic permeability of the tissue [5]. To this day, while the roles of aggrecan and collagen are well understood, it remains unclear how quantitatively minor molecules, such as small leucine rich proteoglycans (SLRPs), affect cartilage poroelasticity [6]. Such knowledge is critical, as SLRPs can play key roles in directing the assembly of cartilage extracellular matrix (ECM) during the musculoskeletal development stages [7], thereby influencing both elastic and poroviscoelastic characteristics. To this end, this study aims to examine the role of decorin, the most abundant SLRP in cartilage [8], in the nanoscale structure and poroelastic mechanical properties of cartilage. AFM-based nanorheometric test [9] was applied to study the time-dependent nanomechanics of wild-type (WT) and decorin-null (*Dcn*$^{-/-}$) murine cartilage in both intact and CS-GAG-depleted forms.

METHODS

Femoral condyle cartilage was harvested from 3-month old male *Dcn*$^{-/-}$ and WT C57BL/6 mice, and stored in PBS at 4°C with protease inhibitors for < 48 h prior to mechanical testing. *AFM-nanorheometric test*[9] was applied to the surface of medial condyle cartilage using microspherical tips ($R \approx 5$ μm, nominal spring constant 16 N/m) via a Dimension Icon AFM (Bruker) and our custom-built nanorheometer. A 2-3 nm random binary sequence displacement was superimposed onto the ~ 1 μm static indentation depth during a 90-sec ramp-and-hold. The dynamic force, F^*, and indentation depth, D^*, as a function of frequency (1-1000Hz) were extracted via discrete Fourier transform. The complex dynamic modulus magnitude, $|E^*|$, and phase angle, δ, were derived via the Taylor expansion of Hertz model [10]. The self-stiffening ratio was calculated as the ratio of $|E^*|$ at high (800-1000 Hz, E_H) versus low (1-3 Hz, E_L) frequencies. *Finite element modeling:* Previous studies have shown that poroelasticity dominates in the frequency-domain mechanics of murine cartilage under AFM [9]. Fiber-reinforced poroelasticity model [11] was applied to extract the fluid pressure and hydraulic permeability, k. To reveal the roles of CS-GAGs, additional condyles were incubated in 0.1U/ml chondroitinase-ABC for 48hr for CS-GAG removal, and subjected to the same nanorheometric tests. *Collagen nanostructure:* Additional joints ($n = 5$) were used for imaging the collagen network structure via SEM and TEM, following established procedures [12,13]. *CS-GAG and protein content* To quantify the amount of CS-GAGs in cartilage, the femoral head cartilage ($n \geq 3$) from 3-month-old male mice was harvested and weighed. A papain digestion followed by a Dimethylmethylene Blue (DMMB) assay [14] was conducted to quantify the amount of sulfated GAGs. To assess the difference in aggrecan core protein and decorin amount, the harvested femoral head cartilage was subject to western blot.

RESULTS

From the nanorheometric test, we found that WT and *Dcn*$^{-/-}$ cartilage had significant differences in the frequency-dependent spectra of $|E^*|$ and δ, indicating their distinctive poroelastic characteristics (Fig. 2). In comparison to the WT control, besides having a lower elastic modulus [15], *Dcn*$^{-/-}$ cartilage showed a significantly weaker increase in $|E^*|$ with frequency, and a smaller maximum phase angle, δ_m (Fig. 2a). Such results correspond to a 1.8 ± 0.6-fold (mean ± 95% CI) lower self-stiffening ratio, E_H/E_L, as well as a 7.5 ± 2.5-fold higher hydraulic permeability, k, as calculated from the fiber-reinforced FEM (Fig. 2b). In addition, the fiber-reinforced FEM extracted significantly lower pore pressure in $Dcn^{-/-}$ cartilage (Fig. 3). On the other hand, as expected, for each genotype, CS-GAG-depletion resulted in significant decreases in E_H/E_L and δ_m, as well as increases in k. Without CS-GAGs, these properties are similar for WT and $Dcn^{-/-}$ cartilage collagen networks. Despite the similarities in poroelastic mechanical properties, SEM and TEM imaging detected significant variations in the collagen structure, where $Dcn^{-/-}$ cartilage showed larger collagen fibril diameters both on the surface (Fig. 4a) and in middle/deep zones (Fig. 4b). The histology and DMMB assay revealed substantial reduction of CS-GAGs in $Dcn^{-/-}$ cartilage compared to WT (Fig. 1a and 1b). Western blot confirmed that the CS-GAG reduction is due to the decrease in aggrecan core protein content (Fig. 1c).

DISCUSSION

This study highlights indispensable, critical role of decorin in the energy dissipation function of cartilage, as evidenced by the significantly impaired poroelastic mechanical properties of $Dcn^{-/-}$ cartilage (Fig. 2 and 3). While this effect could be due to structural defects of both aggrecan (Fig. 1) and collagen (Fig. 4), the lack of a poroelastic phenotype in $Dcn^{-/-}$ compared to WT CS-GAG-depleted cartilage suggests that the dominating factor is the reduction of aggrecan, instead of changes in collagen. In cartilage, poroelastic mechanics is determined by the 2-3 nm "effective pore size" between GAGs [5]. The regulation of decorin on poroelasticity is thus mainly manifested through its governing of aggrecan content, and therefore, the packing density of GAG chains. The collagen network alone has several orders of magnitude larger pore size (~ 100 nm) [16], and at this length scale, structural changes likely do not have marked impacts on the fluid-solid interactions. Since time-dependent mechanics is a key function of cartilage in load bearing and energy dissipation [17], understanding the role of decorin can provide a new path for documenting osteoarthritis-induced cartilage dysfunction and for evaluating tissue repair products. Our ongoing studies aim to uncover the molecular mechanism whereby decorin influences cartilage ECM structure and mechanics.

ACKNOWLEDGEMENTS

This work was supported by the National Institute of Health (AR066824), and the Penn Center for Musculoskeletal Diseases (PCMD) (NIH P30AR050950).

REFERENCES

[1] Hoshino, A et al., *J. Bone Joint Surg. Br.* 69:807-811, 1987.
[2] Cross, R, *Am. J. Physiol.* 67:304-309, 1999.
[3] Nia, HT et al., *ACS Nano* 9:2614-2625, 2015.
[4] Lee, H-Y et al., *J. Struct. Biol.* 181:264-273, 2013.
[5] Wight, TN. 1991. In *Cell Biology of the Extracellular Matrix*. 45-78.
[6] Heinegård, D, *Int. J. Exp. Pathol.* 90:575-586, 2009.
[7] Chen, S et al., *FEBS J.* 280:2120-2137, 2013.
[8] Poole, AR et al., *J. Orthop. Res.* 14:681-689, 1996.
[9] Nia, HT et al., *Journal of Biomechanics* 48:162-165, 2015.
[10] Mahaffy, RE et al., *Biophys. J.* 86:1777-1793, 2004.
[11] Soulhat, J et al., *J. Biomech. Eng.* 121:340-347, 1999.
[12] Batista, MA et al., *Matrix Biol.* 38:84-90, 2014.
[13] Birk, DE et al., *Dev. Dyn.* 208:291-298, 1997.
[14] Farndale, RW et al., *Biochim. Biophys. Acta* 883:173-177, 1986.
[15] Doyran, B et al., *Trans. Orthop. Res. Soc.* 62:491, 2016.
[16] Clark, IC, *J. Bone Joint Surg. Br.* 53B:732-749, 1971.
[17] Setton, LA et al., *J. Biomech.* 26:581-592, 1993.

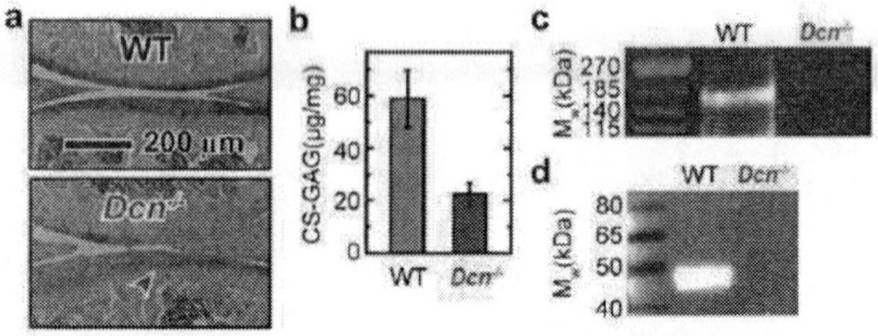

Figure 1: a) Safranin-O/Fast Green histology, and **b)** CS-GAG weight ratio over the weight of the cartilage measured by DMMB assay both showed marked reduction of CS-GAGs in 3-month old $Dcn^{-/-}$ murine cartilage. **c,d)** Western Blot showed c) marked reduction of aggrecan core protein (stained for the core protein G1-globular domain) in 3-month old $Dcn^{-/-}$ murine cartilage, and d) confirmed the absence of decorin in $Dcn^{-/-}$ mice.

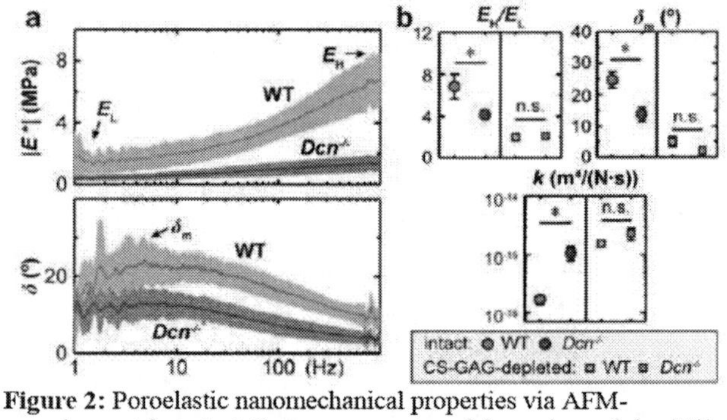

Figure 2: Poroelastic nanomechanical properties via AFM-nanorheometric test. **a)** Frequency spectra of dynamic modulus $|E^*|$ and phase angle δ, from intact WT and $Dcn^{-/-}$ cartilage (mean ± 95% CI of ≥ 5 animals). **b)** Poroelastic mechanical properties of both intact and CS-GAG-depleted cartilage: self-stiffening coefficient E_H/E_L, maximum phase angle δ_m and hydraulic permeability, k (mean ± SEM, $n \geq 4$ animals, *: $p < 0.05$ via Mann-Whitney U test).

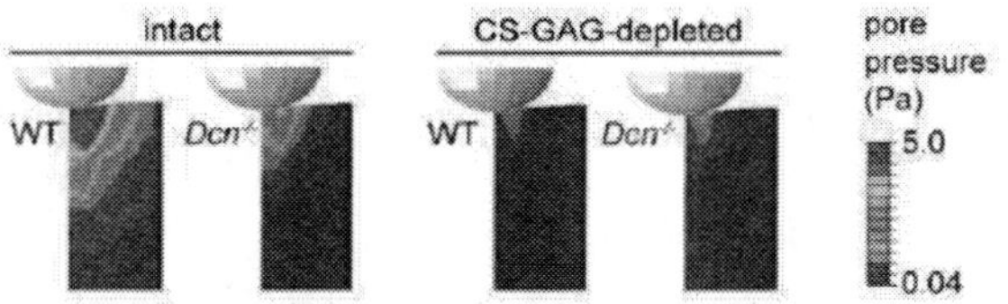

Figure 3: Finite-element simulation of the maximum pore pressure calculated from the fiber-reinforced poroelastic finite element model at frequency of $f = 10$ Hz in intact and CS-GAG depleted cartilages.

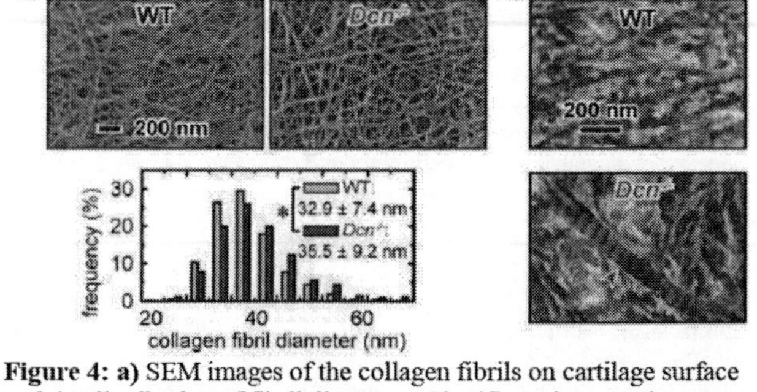

Figure 4: a) SEM images of the collagen fibrils on cartilage surface and the distribution of fibril diameters. Significant increase in average ($p < 0.0001$) and variation ($p < 0.0001$) were found on collagen fibril diameters in $Dcn^{-/-}$ cartilage than the WT control. **b)** TEM on cartilage matrix cross-sections showed the presence of thickened fibrils in the middle/deep zone in $Dcn^{-/-}$ cartilage.

SB³C2017
Summer Biomechanics, Bioengineering and Biotransport Conference
June 21- 24, Tucson, AZ, USA

PROSTHETIC BIKE ATTACHMENT FOR CHILDREN WITH CONGENITAL AMPUTATIONS

Katherine Mavrommati
California Polytechnic State University
San Luis Obispo, CA, USA

Mark R. Oppenheimer
California Polytechnic State University
San Luis Obispo, CA, USA

Marco G. Santini
California Polytechnic State University
San Luis Obispo, CA, USA

Kurt K. Reed
California Polytechnic State University
San Luis Obispo, CA, USA

Caroline E. Skae
California Polytechnic State University
San Luis Obispo, CA, USA

Faculty Advisor

Dr. Lily H. Laiho
California Polytechnic State University
San Luis Obispo, CA, USA

INTRODUCTION

Our team of five undergraduate engineering students from California Polytechnic State University, San Luis Obispo, has been designing and optimizing prosthetic hands and attachments for a seven year old boy born with a partial left hand since November of 2015. Our challenger, Levi, was born with a wrist but no fingers and only a third of his palm. We were connected with Levi through the Quality of Life Plus Student Association, which aims to design and develop prostheses, medical devices and assistive technology for wounded veterans and community members. In June 2016, we finished our first prototype for Levi's prosthetic hand and since then we have been designing a separate Prosthetic Bike Attachment for him to be able to ride his bike safely and effectively. The first prototype of our Prosthetic Bike Attachment was given to Levi in December 2016 and we are now in the final prototype stages.

The first attachment not only allowed Levi to ride his bike successfully for the first time in his life, but also ignited his excitement for using the assistive technology we were creating. We want to spread this excitement to other children that are born with congenital amputations. In the year 2005, it was estimated that approximately 9,860 people were affected in the United States by congenital upper limb loss [1]. The cost and scalability of the design are paramount to keeping up with the rapid growth of a child. Our design can be easily customized to fit a unique variety of upper limb amputations. By using rapid prototyping technology, we were able to keep the manufacturing costs low. This would make the design widely accessible for all children affected by this condition and engage them in the use of assistive technologies to improve their life.

PRODUCT DESIGN

The Prosthetic Bike Attachment contains three main components: the clamp, the socket, and the gauntlet, as seen in the isometric view in Figure 1. The socket and gauntlet connect the device to the user's residual, while the clamp attaches on the bike handlebar, allowing the user to control the bike.

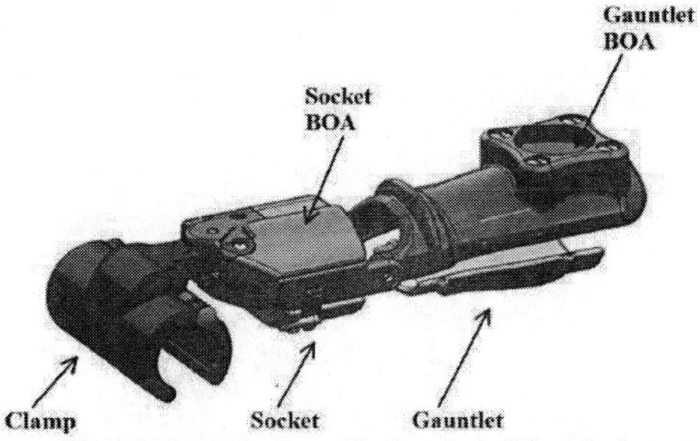

Figure 1: Isometric view of the Prosthetic Bike Attachment.

The clamp's diameter is 30 mm, which reflects the average diameter of a bicycle handlebar. The clamping mechanism is composed of two extruded arcs that are hinged together forming a grip around the handlebar, closed with four 2.5 in-lbs torsion springs along the hinge, resulting in a total of 10 in-lbs of torque. The clamp is released with a pull-tab located on the bottom arc. Since riding a bicycle requires more pulling action than pushing, the pull-tab was placed on the bottom arc so that the top arc, which is necessary for pulling the handlebar, remains rigid. The clamping mechanism is pinned to the socket to allow rotation about the vertical axis. There are two torsion springs located at this pin to provide a resistive torque, regardless of which way the clamping mechanism is rotated. To accomplish this, the torsion springs are oriented opposite one another, hence the clamp is maintained at a straight equilibrium position.

As mentioned earlier, the socket and gauntlet are used to attach the prosthetic device to our challenger's residual limb. Both parts secure the prosthesis using two types of BOA® lacing systems. The BOA® H3

System [2] was used for the gauntlet and a smaller BOA® system, seen in Figure 2, which is found on elbow straps was used for the socket [3].

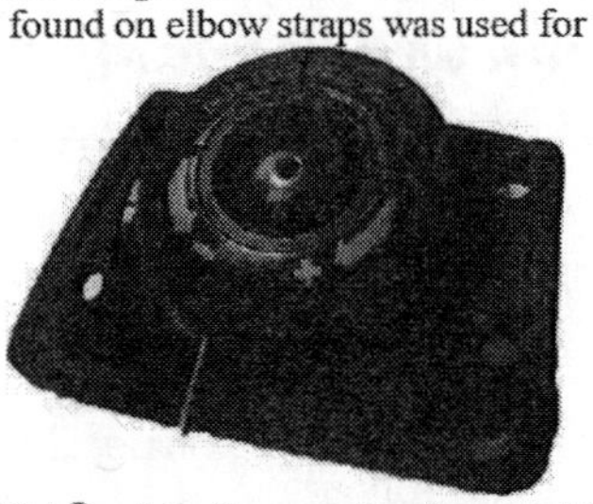

Figure 2: BOA® ratcheting mechanism used for socket.

The socket and gauntlet are each composed of a top half and bottom half that close together around the residual and forearm, respectively. Channels were incorporated into the socket and gauntlet to provide a path for the wires of the BOA®. Once the challenger places the residual limb in the socket and gauntlet, the BOA® mechanisms on both enclosures are simply rotated to tighten the prosthesis to the limb. This provides a distribution of the forces seen by the prosthesis on the challenger's residual limb, resulting in a much more comfortable fit, as opposed to using straps. To provide additional comfort, the socket and gauntlet are designed to accommodate for a liner worn on the challenger's residual limb. The liner was purchased by WillowWood and has a silicone lining on the inside and fabric on the outside. Velcro strips were glued to the inside of the gauntlet and socket to prevent the liner from sliding out of the prosthesis. This secure attachment allows for a greater transfer of energy from the limb to the prosthesis.

To provide a holistic view of how all the parts of the design come together, an exploded view of all parts can be observed in Figure 3.

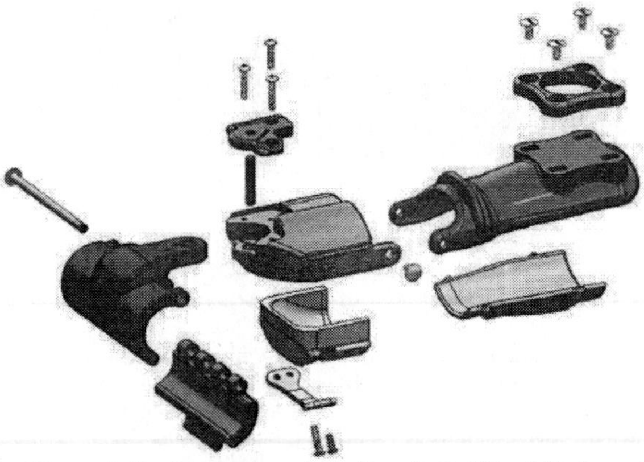

Figure 3. Exploded view of the Prosthetic Bike Attachment.

BUDGET & MARKET ANALYSIS

In the prosthesis industry, if the amputee's family does not have insurance, a basic upper limb prosthesis cost starts at $6,500 [4]. Since the child will grow, the family will have to buy multiple prosthesis, and the cost will double or triple in a matter of years. Hence, it is important to ensure that effective and cheap prosthetics can be produced that can be easily scaled to adjust to the user's growing needs.

The manufacturing cost per unit for the Prosthetic Bike Attachment is derived from the variable costs, fixed costs, and the total amount of units sold. The projected variable costs includes the direct materials at $295.07 per unit. The two most expensive items of the direct materials are the BOA H3 System, and the pediatric liner shown in the bill of materials in Table 1. The pediatric liner is not limited to the Prosthetic Bike Attachment, but may also be used for other prostheses. Since the fixed variable costs are nullified, and the variable costs are confined by price per unit, the estimated manufacturing cost per unit is $295.07. The investments in engineering and tooling is restricted to purchasing the LulzBot Mini 3D Printer at $1250 [5]. 3D printing technology allows for faster production and adaptability of design to conform to customer's needs. The cost of the 3D printer is the total overhead expense, which must be overcome to have a positive return on investment. The cost of the 3D printer could be covered in the first year by selling 250 units. The projected sale price at $350 would overcome the cost of the 3D printer by selling 30 units, and the return on investment for the first year would be 18.31%. The second year selling 250 units would produce an 18.62% return of investment.

As a non-profit business, our prosthesis would be sold at price of $350, which is much more affordable to growing children and would allow them to engage in the use of assistive devices much more rapidly than waiting until they stop growing. The return of investment will be taken and invested into marketing the Prosthetic Bike Attachment, so that more children can have access to the simple task of riding a bicycle.

Table 1. Bill of Materials for Prosthetic Bike Attachment including suggested liner material.

Item	Vendor	Cost/Part Used	Quantity	Cost
Torsion Spring 90 Degree Angle, .375" Spring OD, .048" Wire, Right-Hand	McMaster Carr	6.05	4	$24.20
Torsion Spring 90 Degree Angle, .373" Spring OD, .028" Wire, Right-HandType	McMaster Carr	2.3	2	$4.60
Screws of various sizes	McMaster Carr	0.05	12	$0.60
Aluminum Low-Profile Binding Post for 3/8" to 5/8",Packs of 25	McMaster Carr	0.4624	2	$0.92
Low-Profile Binding Post Aluminum, for 3-1/2" to 3-3/4", Packs of 10	McMaster Carr	0.705	1	$0.71
Low-Stretch Coated Wire - Not for Lifting Ultra-Flexible, Braided, 0.024" Diameter, 5ft Length	McMaster Carr	9.55	1	$9.55
Ace Elbow Strap with Adjustable Custom Dial System	Amazon	21.92	1	$21.92
BOA H3 System	Click Medical	117.5	1	$117.50
Ultimaker Black PLA .75kg, 2.85mm. Black	Dynamism	49.95	0.339 kg	$22.56
Alpha® Classic Pediatric Liners	WillowWood	89.95	1	$89.95
Velcro 2" x 4' - Black	Amazon	9.47	13"	$2.56
			Total	$295.07

ACKNOWLEDGEMENTS

We would like to thank the QL+ Lab for providing us with the opportunity of working with Levi to create this device. We deeply appreciate the space and funding that was provided to us by the QL+ Lab. We would also like to thank our advisor Dr. Lily Laiho for her everlasting support for all of our projects.

REFERENCES

[1] Ziegler-Graham, K et al., Arch. of Physical Medicine and Rehab, 89:422-429.2, 2008.
[2] BOA. "High Power Reel – H3." Theboasystem.com, 2017
[3] Amazon. "Ace Custom Dial Elbow Srap." Amazon.com, 2017
[4] Arms within Reach Foundation."FAQs."Armswithinreach.org, 2017
[5] LulzBot. "LuzBot Mini." Lulzbot.com, 2017

SB³C2017
Summer Biomechanics, Bioengineering and Biotransport Conference
June 21- 24, Tucson, AZ, USA

COUGHING FOR BETTER HEALTH; A PROSTHESIS TO AID IN SPUTUM EXPECTORATION IN LARYNGECTOMEES

Nicole D'Ambrosio
Clarkson University
Potsdam, New York, USA

Matthew Haltermann
Clarkson University
Potsdam, New York, USA

Alden Mitchell
Clarkson University
Potsdam, New York, USA

Kota Tamura
Clarkson University
Potsdam, New York, USA

Kara Van Herwarde
Clarkson University
Potsdam, New York, USA

Faculty Advisor
Byron D. Erath, Ph.D.
Clarkson University
Potsdam, New York, United States

INTRODUCTION

The human larynx contains anatomy essential to the mechanics of a productive cough. During a normal cough, the vocal folds close, allowing the lungs to increase pressure. The sudden opening of the vocal folds releases the pressure, forcing sputum out of the lungs. Cancer of the larynx, or acute trauma, can often necessitate a laryngectomy (removal of the larynx), after which an individual breathes through an opening created in the neck (i.e. stoma). Removal of the larynx renders the normal cough process nearly impossible as the mechanism for building up lung pressure and then suddenly releasing it is lost. In the United States, there are ~ 3,400 new laryngectomies each year, with an estimated 50,000 to 60,000 living laryngectomees [1-2].

The inability of laryngectomees to generate a productive cough has considerable health consequences. Following a laryngectomy, cold, dry air that would normally be heated and humidified as it passes through the nose or mouth, enters the lungs directly. The physiological response is the production of excess mucus, a major problem that all laryngectomized individuals must live with for the rest of their life. In acute care, suctioning must be regularly implemented to remove excess mucus from the trachea in order to prevent bronchial obstructions (i.e. mucus plugs) that are often fatal. In addition to the discomfort associated with mechanical suctioning, it invariably causes bleeding in the airways, greatly increasing the chance of tracheal infection. After discharge from acute care, most individuals do not have access to regular suctioning, increasing infection, mucus plug formation, and mortality rates.

Medical devices exist to aid in mucus extraction, but are designed for non-laryngectomized individuals and must be retrofitted for laryngectomees. These devices are costly ($1000-$6000), and non-portable, requiring the individual to return home to perform the procedure, which is inconvenient as mucus removal is usually necessary ~3-5 times/day [3].

The objective of this study is to develop a cost effective, portable Artificial Cough Prosthesis (ACP) that mimics the mechanics of a cough. The proposed device has the potential to increase patient comfort during mucus clearance, while reducing the incidence of tracheal infections and mortality rates in both acute and long-term care of laryngectomized individuals.

In conjunction with the physiological effects, the psychological effects of a laryngectomy are also quite adverse. Laryngectomees are more susceptible to depression, decreased mental health, social withdrawal, and suicidal thoughts and tendencies [4-5]. The ability of laryngectomees to gain control of their personal care through the use of an ACP will also lead to higher self-esteem as they are able to take responsibility for their personal care, restoring a sense of autonomy.

PRODUCT DESIGN

Design objectives for the ACP were developed in conjunction with Gordon Reitema CCC-SLP, director of the Utica, NY Laryngectomee support group. The objectives include: (1) effective mucus clearance through cough replication, (2) passive actuation, (3) portability, and (4) ease of cleaning.

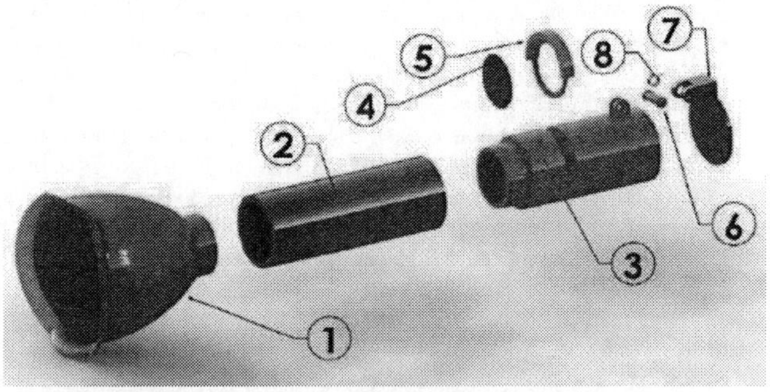

Figure 1: Exploded view of the Mucus Clearance Device

An assembly view of the prototype design is shown in Figure 1, with the respective parts described in Table 1. The stoma interface (part 1) is designed to seal around the stoma by following the contour of a laryngectomee's throat. It will be fabricated from ABS plastic, with the flared edge consisting of soft silicon that will exert less stress on the user as well as create a seal with the neck. The next part of the ACP is the pipe extension (part 2) which enables the user to grasp the device with one hand. It is made from semi-flexible silicone to account for individual ergonomics in grasping and using the device. The main housing (part 3), is manufactured from ABS plastic, and attaches to the

Table 1: ACP part list with prototype and production costs.

Part Name	Prototype Cost	Production Cost
[1] Stoma Interface	$1.17	$0.29
[2] Pipe Extension	$0.45	$0.36
[3] Main Housing	$0.97	$0.48
[4] Filter	$0.59	$0.47
[5] Filter Tray	$0.08	$0.19
[6] Fastener	$0.69	$0.64
[7] Control Valve	$0.12	$0.07
[8] Torsion Spring	$1.03	$0.82
Total:	$5.21	$3.32

pipe extension via hose barbs. A slot in the housing enables the insertion and removal of the filter tray (part 5), which houses a fine nylon mesh filter (part 4) for capturing any expelled mucus. The exit of the housing is covered by a control valve (part 7) which is held in a nominally-open position by a torsion spring (part 8), and which when closed, seals the airway of the device.

The ACP is used by pressing the stoma interface against the neck with one hand, and holding the valve shut with the other. The user generates pressure in their lungs, as during a normal cough, and then releases the valve, expectorating sputum from the lungs.

To achieve the design goal of removing mucus by mimicking a cough, the ACP must not unduly restrict the airflow. Productive cough has been measured to produce maximum flow rates of ~0.005-0.01 m^3/s, when generated by a lung pressure of ~6 kPa [6-8]. The pressure drop across the ACP at the flow rates of normal cough can be calculated using the mechanical energy equation, given by:

$$\left(\frac{\Delta p}{\rho g}+\Delta\alpha\frac{\Delta V^2}{2g}+\Delta z\right)=\left(\Sigma f_i\frac{L}{D}+\Sigma K_L\right)\frac{V^2}{2g} \quad (1)$$

where p is the pressure, V is the velocity, z is the height, α is the kinetic energy coefficient, ρ is the density of the air, and g is the gravitational acceleration. The right-hand-side of the equation is the total head loss that is expressed as the sum of the major and minor losses, respectively, where f is the D'arcy friction factor, L is the length of the device, D is the pipe diameter, and K_L is the minor loss coefficient. Friction factors and minor loss coefficients were computed from the Moody chart, and tabulated data, respectively, and are shown in Table 2 for each component of the ACP. At the desired flow rate of 0.01 m^3/s, the ACP produces additional pressure losses of 0.63 kPa. Therefore, the total lung pressure required for productive cough will be ~6.63 kPa, which is still within the achievable range (~10 kPa) of human lung pressures.

The function of the ACP will be tested using subjects who have been living with a laryngectomy for at least one year to ensure their condition is stable. Initial survey data will focus on the perceived benefit to the user based on single day use. Following any suggested/recommended design changes, comprehensive clinical evaluations will be performed by comparing sputum expectoration between (1) the ACP, (2) active suctioning, and (3) standard manual clearance techniques that most individuals rely upon.

To date, design iterations have included: (1) changing the geometry of the stoma interface to accommodate individuals with larynx/tracheotomy tubes permanently placed in their stoma, (2) configuring the pressure-release valve to be nominally-open (as opposed to nominally-closed) to ensure a quick release of the air pressure when released, and (3) a removable filter for easy cleaning, or operation without the filter, thereby decreasing flow resistance of the

Table 2: Friction factors and minor loss coefficients

K_{Ent}	$K_{1\text{-}2}$	$K_{2\text{-}3}$	K_{Filter}	K_3	f_2	f_3
0.42	0.08	0.08	1.22	0.05	0.091	0.058

device. It is anticipated that with the proposed patient testing, additional modifications will be incorporated based on end-user feedback. Considerations may include ergonomic comfort when using the device, inability or difficulty in holding the exit valve closed, etc.

BUDGET & MARKET ANALYSIS

The projected prototype costs are calculated based on retail pricing. The prototype will consist of parts 1, 3, 5, and 7, to be produced by a 3D printer, and parts 2, 4 and 6, which will be bought from retailers. The total cost for producing a completed device is estimated to be ~$5.21. Clarkson University has resources for 3D printing, so only material costs are being considered. Accounting for prototype iterations and travel to Utica, NY to test the device, $350 is allotted for the entire prototype budget.

Projected revenue growth has been determined in tandem with the Clarkson University Shipley Center for Innovation. It is anticipated that the ACP will result in the development of a new niche market for portable sputum expectoration devices for laryngectomees, thus resulting in rapid market growth. Final production will utilize injection molding for the mass produced parts (1, 3, 5, and 7), resulting in capital start-up costs of $250,000, but decreasing the total production costs to $3.32/unit. Annual labor/overhead costs for two full time employees are budgeted at $150,000/year. The lifespan of the device is anticipated to be 1 year, based on government mandates for respiratory device replacement to ensure adequate health protections. With a total target population of 50,000, a 7% market capture in year one will result in selling 3,500 devices. With a target profit of $50/device, this will produce $175,000 in sales. Assuming an additional 7% market share each year for five years results in a total sustainable market capture of 35%. This will allow capital costs to be paid off by year 3, with a positive annual revenue stream of $725,000/year by year 5.

Due to the challenge of advertising in a niche environment, marketing of the ACP will focus on direct distribution to hospitals (the first point of contact for new laryngectomees) and specialized health care providers (e.g. speech language pathologists that perform vocal and respiratory care for laryngectomees). It is anticipated that partnerships with existing laryngectomy supply companies (e.g. InHealth) may also be established to achieve the desired distributions.

To protect intellectual property, a provisional patent will be filed through the Shipley Center for Innovation at Clarkson University. In addition, resources and contacts within the Shipley center can be leveraged to secure the funding needed to cover the capital costs.

ACKNOWLEDGEMENTS

This work was supported by the National Science Foundation Under Grant No (NSF-CBET 1510367). Thank you to Gordon Reitema, Zachary Daniels and Kristen Land for their help and advice.

REFERENCES

[1] **Maddox, P.,** et al., 2012, OHNS 147(1):85–90
[2] **Benjamin, M.,** et al., 2011, ASHA Conv., 2011, San Diego, CA
[3] **Emerson Cough Assist,** Pat. # US 20120285460 A1
[4] **Eadie, T.** et al., 2005, J Rehab R&D, 42:115-124
[5] **Danker, H.** et al., 2010, Eur Arch ORL, 267:593-600
[6] **Gupta, J.** et al., 2009, Indoor Air 19:517-525.
[7] **Lindsley, W.** et al, 2013, Aerosol Sci Technol. 47(8): 937–944.
[8] **Smith, J.** et al., 2011, J Physiol. 590(3):563-574.

WEARABLE GESTURE RECOGNITION SYSTEM WITH APPLICATIONS TO AMERICAN SIGN LANGUAGE

Isioma R Kasi-Okonye
The University of Texas at Arlington
Arlington, Texas, United States

Simranjit S Ahluwalia
The University of Texas at Arlington
Arlington, Texas, United States

Dinithi S Silva
The University of Texas at Arlington
Arlington, Texas, United States

Arturo Acuna
The University of Texas at Arlington
Arlington, Texas, United States

Faculty Advisor(s)

Oguz Yetkin
The University of Texas at Arlington
Arlington, Texas, United States

George Alexandrakis
The University of Texas at Arlington
Arlington, Texas, United States

INTRODUCTION

American Sign Language (ASL) is the predominant sign language within deaf communities in the United States (US) and most of Anglophone Canada. According to the World Deaf Federation, there are approximately 70 million deaf people in world and about 250,000 to 500,000 native ASL speakers in the US[1].

Although deaf people communicate effectively with each other using ASL, interacting with the rest of the world remains a challenge. This is especially evident in daily activities (e.g. checkout counter at a grocery store, drive-thru at a restaurant, dealing with children or colleagues, or over the phone). And in such settings, writing or typing to communicate with a non-deaf individual may not be appropriate or practical.

The goal of this project is to develop a lightweight, wearable, and cost-effective communication system for native ASL speakers, which can translate the ASL into text or speech. The current prototype of the device is comprised of a set of extremely lightweight ring and fingernail based sensors (Figure 1). This allows input, via hand poses/gestures, to be collected as the user makes various signs. The "sign" detection offered by the current prototype is for the recognition of the ASL manual alphabet, which consist of 26 corresponding letters and signs.

In Human-Computer interaction, there are two main approaches used for gesture recognition and interpretation: vision-based approach and data glove-based approach[2]. In the vision-based approach, hand pose is reconstructed by use of a camera. And through processing techniques, such as gray-scale imaging, relevant features regarding the image can be extracted for interpretation. While, the data glove-approach employs the use of sensors attached to a glove. The sensors generate electrical signals, as a result finger flexions, which are processed to determine hand posture[2].

The appeal of the vision-based approach is the absence of a worn device and the ease of interaction with machines, especially if the user does not have much technical knowledge about the system. Still with this approach, there exists the issue of portability. Although the glove-based approach offers portability, it is not ubiquitous in daily life and so there exists the issue of mass adoption.

The system being developed for this project, is similar to a glove-based approach, but does not require a worn glove and offers more compactness. In addition, since ASL is "spoken" using both hand gesture and movement, the current system can be viewed as a precursor to a fully wearable ASL recognition system which can also take hand motions into account.

PRODUCT DESIGN

Most glove-based ASL recognition systems built are too cumbersome to use. Hence, the present system design scales down the essential portions of a "glove-based" ASL recognition system to miniaturized devices mounted on top of fingernails and rings, leaving the hand free of any haptic encumbrance[3].

At the core of the design is a fingernail mounted light transceiver device instrumented with Infrared Light Emitting Diodes (IR LEDs) and phototransistors[4]. The IR LEDs and phototransistors are placed on both the "nail facing" and "air facing" sides of the device, which is shaped like a plastic fingernail. The device is cast in smooth, transparent resin, and affixed to each fingernail using a temporary nail adhesive. In addition to the fingernail mounted light transceivers, the system also utilizes similar transceivers mounted on transparent rings worn on each finger. Thus, the current system comprises of five dual-sided fingernail sensors/transceivers, along with five transparent rings.

Changes in illumination of the IR LED are observed on each device whenever contact is made between any two fingers. The system exploits the fact that biological tissue is mostly transparent to IR. As such, it can detect contact between two finger pads, a fingernail and a finger pad, two rings, and between a finger pad and a ring. In this manner, the system can detect most hand configurations used in ASL.

Implementation of the device was performed using surface mount technology (SMT) components to build compact fingernail-sized sensors, each driven by an Arduino Pro Mini microcontroller board. This sub-unit was then mounted on a 3D CAD designed ring holder, which was integrated with a battery pack for power supply. In each sub-unit, the sensors serve as transducers, where contact between various parts of the hand are measured as electrical signals. These recorded signals are then processed for letter recognition.

The hardware development of the sensors involved the design of a flexible Printed Circuit Board (PCB), using an electronic design software called KiCad, to fabricate a small dual-sided mounted circuit. The circuit, on both sides, consisted of SMT components: phototransistor, LED, and resistor. To reduce the interference of electrical noise, the sensors were cast in a negative resin mold, the size and shape of a fingernail, using optically clear resin. The software development, for gesture detection and recognition, involved the design of a data acquisition scheme. This entailed, modulation of all 15 LEDS (2 on each fingernail sensor, 1 on each transparent ring) at different frequencies during performance of gestures by the user.

There is a high concentration of vision based systems, specifically Machine or Computer vision-based approaches, to ASL gesture recognition[2]. The system being developed, is unique in that it does not require input collection (gesture recognition) through the aid of camera. In addition, it does not require any extra worn device such as glove and so the user would still be able to use the system with hers/his normally. Instead, the user would wear the lightweight system on the hand for "signing" and the letters will be translated and printed on a graphical user interface (GUI) program. The interface can also identify the words and speak them out loud using existing software systems which run on smartphone and tablet devices.

The technology (hardware and software) and materials used in design of the device provide for optimization and miniaturization that will realize a feasible market-ready product. IR LED eye safety experiments were performed to ensure that the recommended IR radiation intensity, without damaging the eye, of 100W/m^ 2 was not exceeded.

BUDGET & MARKET ANALYSIS

US industry reports conducted by IBISWorld indicate market revenue for translation services at 5.4 billion, with an annual growth rate of 5.2%. As the demand for ASL interpreters continues to increase coupled with a rise in awareness of the language, a growth in ASL technologies is expected to soar as well.

The market demographic for the device consists of deaf or hard of hearing persons, "speakers" of ASL, and those interested in learning ASL. The expected sales is estimated at 100,000 units/year. At this volume, the manufacturing cost per unit is estimated at $10.57. This price reflects the materials required per unit of the device: 15 IR LEDs, phototransistor, and resistors, including a PCB board and wires. Production at this volume allows for a much lower manufacturing cost per unit, versus the expenses associated with a prototype (Table 1).

Investments in engineering will consist of about 200 hours of professional engineering services, including adapting the design for manufacturing and automated assembly. And at about $100/hour, the total cost for engineering services is roughly $20,000. Tooling, which consists primarily of injection molding, is estimated at $13,000. Thus, the total cost of production is estimated at $1,090,000.

The projected sales price is $89.99, which is very affordable compared to competitors such as MotionSavvy ($799). Since the expected revenue is $8,999,000, investment, tooling, and manufacturing costs are expected to be offset after the first year's sales, with a gross profit margin of 87.89%.

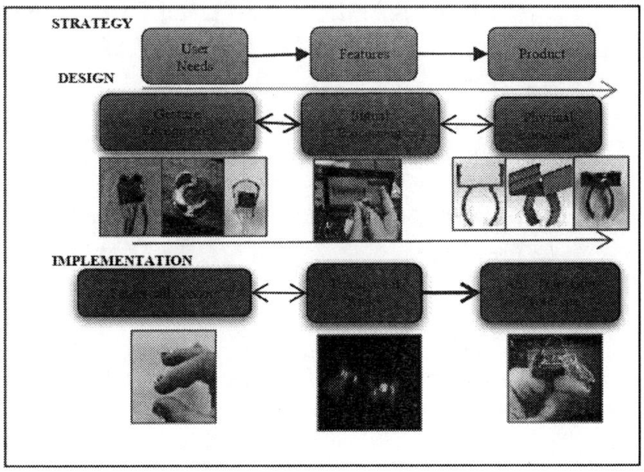

Figure 1: Functional decomposition of the ASL manual alphabet translation system.

ITEM	COST
PCB	
Equipment/parts related to the construction and modification of PCB	**66.82**
Etching	
Related to etching of the PCB (flexible & non-flexible)	**54.83**
Microcontroller assembly	
Related to building the Pro-Mini assembly	**55.48**
Casting	
Material required to make molds and cats for the fingernail sensors and transparent rings	**126.13**
Milling	
Material required fabricate PCB sensors	**26.18**
Soldering	
Material required assembly of PCB sensors	**101.75**
Total cost (Overall)	**431.19**

Table 1: Project budget for designing, building, and testing the prototype.

ACKNOWLEDGEMENTS

We would like to acknowledge the assistance of Dr. Rita Patterson of UNT Health Sciences Center; our design mentors Dr. George Alexandrakis, Dr. Vassilis Athitsos, Dr. Oguz Yetkin of the University of Texas at Arlington Bioengineering Department; Mr. Joshua Baptist and Ms. Rachael Volker in the design and creation of the preliminary device prototypes, and the Dallas Makerspace for general fabrication assistance.

REFERENCES

[1] Ross, E. M et al., *Sign Language Studies*, 6:28, 2006.
[2] Murthy, G. R. S. & Jadon, R.S, *IJITK*, 2:405-410, 2009.
[3] Yetkin, O. Ph.D. in BE, Intuitive Human Robot Interfaces for Upper Limb Prosthetics, 1-214, 2016.
[4] Yetkin, O., Ahluwalia, S., Silva, D., Kasi-Okonye, I., Volker, R., Baptist J. R., Popa, D. O., *SPIE*, 9859:1-8, 2016.

AN ADAPATABLE INTERIM PROSTHETIC SOCKET AS AN ALTERNATIVE FOR BELOW-THE-KNEE AMPUTATION REHABILITATION

Josh Marchese
University of Mount Union
Alliance, OH, USA

Kenneth Muhart
University of Mount Union
Alliance, OH, USA

Edward Cudjoe
University of Mount Union
Alliance, OH, USA

Faculty Advisor(s)

Dr. Joshua Gargac
Department of Engineering
University of Mount Union
Alliance, OH, USA

INTRODUCTION

An interim prosthetic is necessary for proper rehabilitation after a transtibial or below-the-knee (BKA) amputation. As the primary insurance for patients 65 years or older, Medicare requires a patient demonstrate the ability to walk 150 feet unassisted before funding a definitive prosthetic. Recently, the Durable Medical Equipment Medicare Administrative Contractor (DME MAC) for Jurisdiction A, NHIC Corp, has amended Medicare's lower limb prosthesis plan (LCD DL 33787) [1] to eliminate reimbursements for interim prosthetics, creating a barrier to proper physical therapy since these devices can cost as much as $7,000 [2].

During the 2015-16 academic year, a senior capstone design team from the University of Mount Union, attempted to solve this problem by designing an adjustable prosthetic that would be owned by rehab facilities rather than the patient [3]. The design focused on a new articulating ankle, but also included a socket that could adjust to different limb sizes using three inflatable neoprene bladders. Although this prosthetic was novel, it lacked several important aspects that reduced its feasibility as an interim prosthetic. As designed, the socket's air bladders applied pressure uniformly to both the pressure sensitive and tolerant areas of the residual limb. Additionally, the socket could not adjust to differences in residual limb height and lacked a method of suspension to keep the socket on limb.

The overall goal of this year's project was to design a new socket complete with a suspension system, the ability to localize pressures to tolerant areas along the residual limb, and a method to adjust for different residual limb heights. With these improvements, the socket will be incorporated into an adaptable interim prosthetic kit supplied directly to rehab facilities. The socket must adapt to residual limbs ranging in circumference from 36 – 48 cm, and height from 12 – 22 cm, in addition to fitting on a left or right leg. The prosthetic must fit patients between 1.5-1.8 m tall and 100 – 350 lb. To prevent infections, the residual limb must be isolated from the device, and the interface surfaces must be easy to sanitize. The total price of the kits must be less than $3,000.

PRODUCT DESIGN

Adaptable Prosthetic Kits: At the onset of rehab, a physical therapist will assemble a prosthetic for the patient using the components included in the kit: three sockets, 15 aluminum pylons ranging from 7-21 cm, 6 wooden prosthetic feet, and connectors to attach each component (Fig. 1). The three sockets will fit different ranges of the residual limb circumference: small (36-40cm), medium (40-44 cm), and large (44-48 cm) and consist of four design aspects: 1) true fit, 2) height adjustment, 3) interface, and 4) suspension.

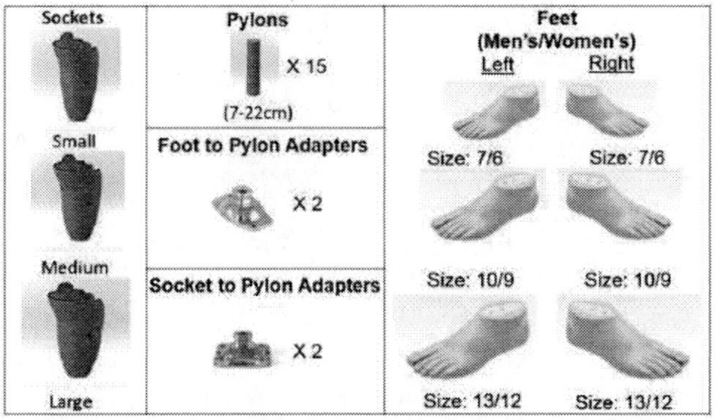

Figure 1: The adaptable prosthetic kit consists of 3 sockets, 15 pylons ranging in sizes of 7-21 cm, two socket and foot connectors, and 6 wooden feet [4].

True fit: The goal of true fit is to eliminate rotational movement of the residual limb in the socket. This can be achieved through triangulation, the focused application of pressure to three tolerant areas (Fig 2). The residual limb will be triangulated using 6 inflatable neoprene air bladders placed around the interior of a thermolyn europlex socket (Fig. 3). These bladders will be inflated using a hand pump and a pressure gauge will provide measurements during inflation. Rigid urethane foam pads shaped to contour with pressure tolerant areas will be inserted into the socket onto the medical bladders (Fig 3). Adjusting the rigid pads allows for triangulation of both left and right legs.

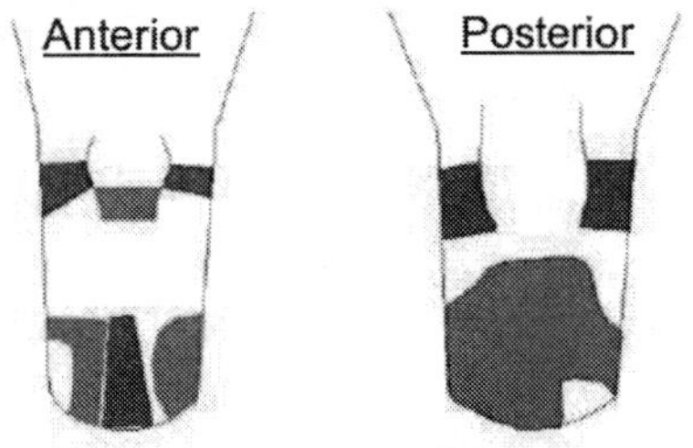

Figure 2: Pressure sensitive (red) and pressure tolerant (green) areas of a BKA.

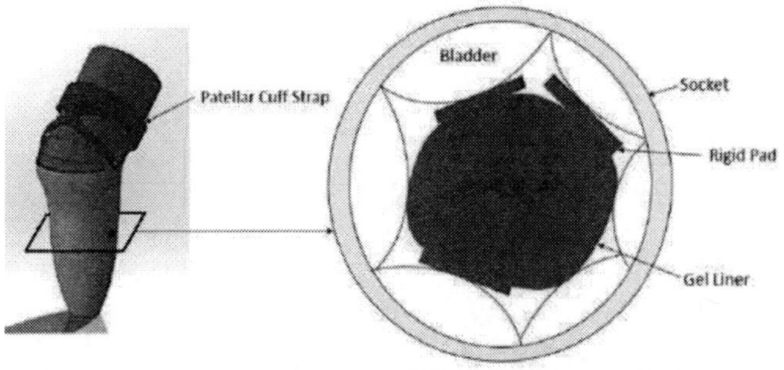

Figure 3: The adjustable socket will include inflatable bladders, rigid pads, a gel liner and a patellar cuff strap.

Height adjustment: By definition, BKA residual limbs range in height from 12 to 22 cm. Memory foam will be inserted into the bottom of the socket to accommodate variations between patients (Fig. 4). At the onset of rehab, each patient will be provided with a personal memory foam to adjust socket height and reduce the risk of cross-patient infection.

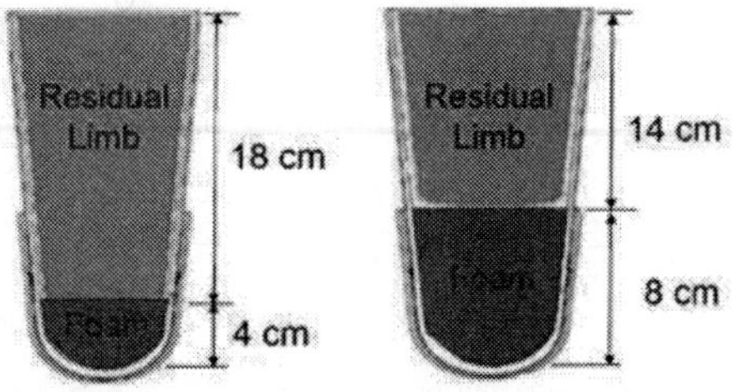

Figure 4: Memory foam inserts will adjust for limb height. All dimensions listed in centimeters.

Interface: Interface describes the materials interacting between the residual limb and the inner socket surface. To reduce the risk of cross-patient infection, the interface must isolate the residual limb from interior surfaces of the socket. Additionally, tight fitting materials must be avoided to prevent the reshaping of the residual limb and rough materials could cause bruising during gait. Therefore, each patient will be provided with a personal silicone-based gel liner for use with the prosthetic. Silicone liners are latex free, hypoallergenic, and provide cushion to prevent abrasion to the residual limb during walking.

Suspension: Suspension anchors the prosthetic to the residual limb, and prevents vertical movement or pistoning within the socket. For suspension, a polyethylene terephthalate patellar cuff strap will connect to both sides of the socket, wrap around the top of the kneecap, and then attach around the back of the knee (Fig. 3). Belt holes on the strap will allow the cuff to tighten around the residual limb. These cuff straps will be available in small, medium, and large sizes to match the socket size. Once again, each patient will be provided with their own cuff strap to reduce infection risk.

BUDGET & MARKET ANALYSIS

More than 100,000 BKAs occur each year in the US [5], and this number is expected to increase by 58,000 by the year 2030 [6]. These amputees will rehab at one of the nearly 1,200 inpatient rehabilitation facilities (IRFs) in the US [7]. Each of these IRFs is a potential customer for the adaptable prosthetic socket kits.

Interim prosthetics are expensive because they require a certified prosthetist to mold a socket specific to each patient. The adjustable interim prosthetic kits shift the paradigm away from many patient-owned to devices to one clinic-owned prosthetic that can be disassembled and reassembled to fit patients with residual limbs of varying sizes. The cost of the one kit will then be spread across many patients reducing the overall cost of rehabilitation.

Based on surveys of physical therapists, the maximum price for prosthetic kit was determined to be $3,000. The cost of materials to create one socket prototype is $200, including the thermolyn europlex exterior, rigid urethane foam pads, and neoprene air bladders. The total cost of materials for the entire kit is $1,000, which includes three sockets ($600), a 3.6 m aluminum tube ($80) that would be cut into smaller pylons, two socket and foot adapters ($200) and six wooden feet ($15 each). This allows for a nearly 3x mark up to account for manufacturing costs and profits. Patients will cover the cost of their personal memory foam insert ($2), patellar cuff strap ($20), and the gel liner ($50).

CONCLUSION

The adaptable interim prosthetic kits allow for a properly-sized prosthetic to be assembled for an amputee at the onset of rehabilitation. As designed, the adjustable socket satisfies the design requirements, and replaces the traditional method of molding a new socket for each patient. As a result, the barriers to proper therapy have been removed, since the need for Medicare reimbursements was eliminated.

ACKNOWLEDGEMENTS

We would like to thank Gabriel Gullia, CP (Yanke Bionics, Akron, OH), Matthew Volansky, DPT, and Megan Salvatore, DPT (Dept. of Physical Therapy, University of Mount Union) for providing expertise on the prosthetics and post-amputation rehab.

REFERENCES

[1] Dunn, S et al., *Lower Limb Prostheses Draft*: 1-6, 2015.
[2] Levengood, N. et al. *SB3C* 2016, National Harbor, MD.
[3] McGimpsey, G et al. Worcester Polytechnic Institution, 2008.
[4] Ottobock SACH Plus Foot, Cascade Orthopedic Supply, 2017.
[5] Gage, BL, et al. 2010. *Report to Congress.* Web: cms.gov
[6] Dillingham TR, et al. *South Med J.* 2002 Aug; 95(8):875-83.
[7] Fletcher DD, et al. *Arch Phys Med Rehabil* 2002 83(10):1389-1393.

TRAINING DEVICE FOR WHEELCHAIR RACING

Sarah C. Peden
University of Delaware
Newark, Delaware, United States

Marjelle F. Scheffers
University of Delaware
Newark, Delaware, United States

Gnanadesikan Somasundaram
University of Delaware
Newark, Delaware, United States

Dylan R. Wergelis-Isaacson
University of Delaware
Newark, Delaware, United States

Faculty Advisor
Dr. Sarah Rooney
University of Delaware
Newark, Delaware, United States

INTRODUCTION

During the year 1960, the first Paralympic games were held in Rome, Italy, where 23 different countries were represented by 400 athletes.[1] This year, 2016, more than 4000 athletes from 160 different countries are participating[2], which makes the Paralympics the second largest sporting event in the world.[3] The extent and growth of the Paralympic games is important because it is quickly becoming a way to impact society's views of disabled people, as well as improve the social wellbeing of disabled people.[4]

Ghana recently became one of the countries participating in the games and are using sports such as wheelchair racing to create more opportunities for disabled people. However, this is a challenge in Ghana because of the lack of training infrastructure and resources, especially in rural areas. For instance, there are only four tracks to train on in Ghana and there is limited availability of training equipment due to a lack of support and budget from the government. As a result, there has been a need for the emergence of nongovernmental organizations (NGOs) such as the Go Get Dem' Racing Club.

The Go Get Dem' Racing Club (GGDRC) was founded by Ghanaian Wheelchair Racing and Paralympic athlete Raphael Botsyo Nkegbe. The club's mission is to improve the lives of disabled persons of Ghana, particularly the youth in rural areas, and empower them to lead fulfilling lives through sport.[5] The club was able to reach their initial goal by securing sponsorships for racing wheelchairs, travel, and accommodations for athletes in various events including Paralympic Qualifiers. The club is working towards its mission to develop Paralympic sports in Ghana. One of the ways they work towards this mission is by providing training for up-and-coming athletes; however, in order for the club to reach its full potential they need proper training facilities and suitable equipment, which is lacking in Ghana.[6] They only have one training device for the 8 members of the club. This need has led to the Go Get Dem' Racing Club's sponsorship of this project in collaboration with the University of Delaware.

The design of an affordable and effective wheelchair racing training device that can be locally made will help the development of the sport in Ghana and help the GGDRC's main goal of empowering the rural disabled youth of Ghana. The end-users, primarily rural Ghanaian athletes, will be able to train at an affordable cost and gain the competitive edge without the limitations of terrain and cumbersome equipment. Moreover, since this device will make training more accessible, it will improve the quality of life of the end-users.[7] The partnership with the International Paralympic Committee allows for the design to potentially be adapted to other developing countries. Since more than 15% of the population of the low-income countries is disabled, this has the potential to impact thousands of people.[7]

This in turn brought us to our main goal which was to design and prototype an affordable training device for wheelchair racing athletes in Ghana, in order to improve the training accessibility.

Through the design process we found that our device had a larger target audience than we originally started with. Our device could be used by any wheelchair user interested in becoming healthier through exercise, any NGO or community organization focused on either persons with disabilities in low income settings, any NGO or community organization focused on introducing youth to wheelchair sports and fitness through sports, or any gym in an urban setting interested in expanding its offerings for cardiovascular fitness to wheelchair users. All over the world there is a need for an affordable and portable wheelchair training device.

PRODUCT DESIGN

The Go Get Dem' Training Roller functions based on the rolling motion of the two drums located at each rear wheel. It is a device consisting of three pieces, two frames with two drums each and a front wheel support, making this a very portable design (see Figure 1). All the framework and the drums are made of aluminum, making it lightweight and allowing for local reproduction in Ghana. The training device is to be set up before use. The three pieces must be placed on the ground based on the dimensions of the wheelchair. The racing wheelchair will then be placed on top of the device and locked in place in two ways. First, the front wheel is locked into supporting frame and secondly, the back wheels sit in between the two drums (see Figure 2).

Overall, this portable design will safely allow for racing wheelchair athletes to train.

The main components of the final design are the drums, the frames, the angled supporters, and the front wheel support. A yoga mat is used to prevent sliding when in use. Solid T-slotted aluminum framing of 1.5" is used to build the frames. The framing is connected using aluminum plates and fasteners, specific to easily attach to the t-slotted aluminum. The frame is 11.5" wide and 24.56" long (see Figure 3). The drums have an outer diameter of 4.5" and are 8" long. The distance between the two drums in a frame is 11". The drums are made by inserting a 1" aluminum rod through the middle and welding caps on the ends of drums. An opening is left at each end of the drum to allow for filling/emptying the drum, and a rubber stopper is used to close it. Low-profile mounted ball bearings made of self-aligning 521000 steel, covered in an aluminum housing, are used to attach the drums to the frame. These bearings are strong enough to hold the applied load and speed, and relatively affordable. Based on a previous iteration of this concept, it was chosen to mount the bearings on t-slotted aluminum that is resting on the ground directly. This is the location where the greatest forces will be acting upon, and by keeping it low to the ground you increase stability and reduce risk of failure of connecting parts.

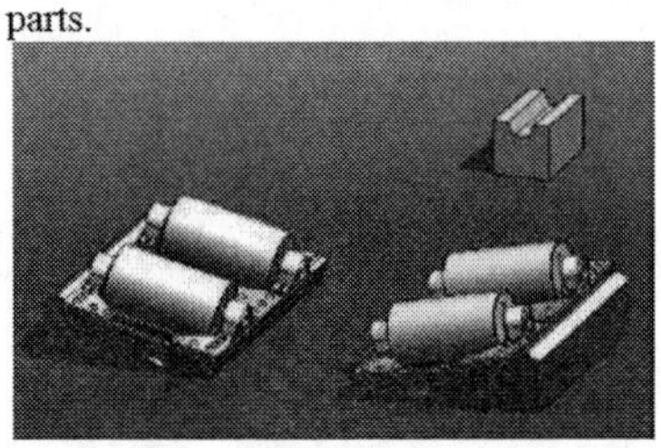

Figure 1: A system level overview of the final design. A base made from t-slotted aluminum, connected by aluminum plates, resting on wooden angled support. Aluminum drums are attached to the frame via low-profile mounted ball bearings. Wooden block for front wheel support.

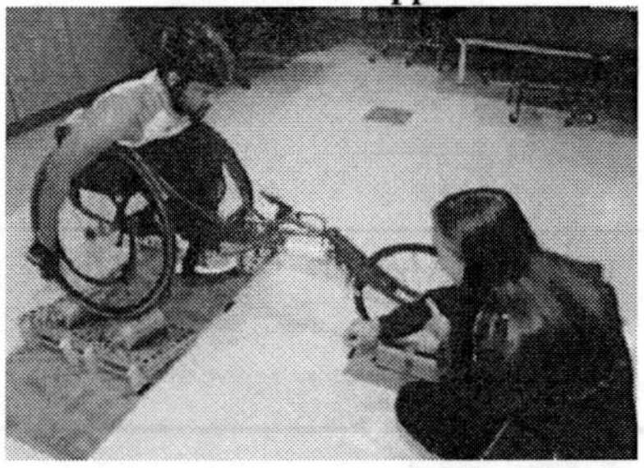

Figure 2: Showing the prototype completely setup, and in use by a racing wheelchair athlete at MOSS Rehab.

Figure 3: A detailed overview of the main component of the final design, including dimensions.

BUDGET & MARKET ANALYSIS

In the beginning, we were allocated $1,000 for our design budget. Of that budget, we used so far $803.61 to design, build, and test the first-generation prototype. We also received more funding recently from National Science Foundation Innovation Corps Sites Program at the University of Delaware. They allocated us $3,000 to use to help develop our design to be more marketable.

We estimate that our total market size would be 9.24 Million with an immediate market size of 20,000 annually. We got the total market size from The World Bank who estimates that the total population of low- and middle-income countries is 6.16 Billion and 15% of those in that category are disabled.[8] Our immediate market size we got from Motivation.org.uk, our partner organization, who conservatively estimates we will reach 1/4th - 1/3rd of their annual market size -- 80,000, with our companion product to their low-cost, high-quality wheelchairs.

The cost per unit volume is currently estimated to be roughly 400 USD. This includes all the material costs. The material cost for the current prototype is 550 USD, but by replacing the t-slotted aluminum with squared aluminum tubing the cost will be reduced. We are currently working on further reducing these costs as far as possible by exploring other material options.

An additional investment to be made is outsourcing the welding process, which will cost 160 USD per device. This is based on the average cost of 20 USD per hour for hiring a certified welder.[9]

Along with reducing material cost, the goal is to bring the total manufacturing cost per unit volume to 460 USD. With a potential sales price of 550 USD, the return on investment is estimated to be 20%.

ACKNOWLEDGEMENTS

- Thank you to:
  - Our Sponsors: Raphael Nkegbe, Maclean Dzidsienyo, Yetsa Tuakli-Wosornu & Adam Bleakney
  - Our Advisor: Dr. Sarah Rooney
  - Biomedical Engineering Department University of Delaware
- Thank you to MossRehab and Global Abilities for helping us facilitate testing of the device
- Thank you to National Science Foundation Innovation Corps Sites Program for additional funding

REFERENCES

[1] History of the Paralympic Movement [Internet]. Bonn (Germany): International Paralympic Committee; [cited 2016Sep9].
[2] Rio 2016 Paralympics - Countries [Internet]. Bonn (Germany): International Paralympic Committee; [cited 2016Sep9].
[3] Rivera D. The Paralympic Games: Changing Perceptions, Popularity Still Challenged [Internet]. Breaking Muscle. [cited 2016Sep9].
[4] Weed M. Developing a physical activity legacy from the London 2012 Olympic and Paralympic Games: a policy-led systematic review. Perspectives in Public Health. 2012Jan;132(2):75–80.
[5] Gogetdem [Internet]. c2016 [cited 2016Sep9].
[6] F.B. Scriven. Sports Facilities for Schools in Developing Countries [Internet]. Paris (France): Educational Studies and Documents; c1973 [cited 2016Sep9].
[7] World Report on Disability [Internet]. Malta: World Health Organisation; c2011 [cited 2016Sep13].
[8] Disability Overview [Internet]. World Bank Group. c2017[cited 2017Feb15].
[9] Hourly Rate for Certification [Internet]. PayScale Human Capital. c2017[cited 2017Feb15].

DESIGN AND ANALYSIS OF A SOFT-ROBOTIC EXOSKELETON FOR RESTORATION OF HAND FUNCTION

Vincent Castonguay-Siu
University of Alberta
Edmonton, Alberta, Canada

Dalen Mimeault
University of Alberta
Edmonton, Alberta, Canada

Pratik Shah
University of Alberta
Edmonton, Alberta, Canada

Craig Trischuk
University of Alberta
Edmonton, Alberta, Canada

Heather Williams
University of Alberta
Edmonton, Alberta, Canada

Faculty Advisor(s)
Michael Lipsett
University of Alberta
Edmonton, Alberta, Canada

INTRODUCTION

One in six people will suffer a stroke in their lifetime, two thirds of which will experience hemiparesis impairing hand and arm function [1]. Through intensive rehabilitation training, including robot-assisted therapy, patients may regain limited control over their arms and hands. However, 66% of hemiplegic stroke patients fail to fully regain motor function in the paretic arm when evaluated six months post-stroke [2]. Beyond stroke numerous neurological and traumatic injuries, such as brachial plexus or spinal cord injury, contribute to a vast population of those affected by impaired hand function. The inability to control the digits to form functional hand conformations can dramatically impact one's independence and ability to interact with their environment. Powered hand exoskeletons are an emerging technology that have demonstrated promising results in alleviating functional challenges associated with hand impairment or weakness [2]. These systems are worn on the hand and actively assist digit flexion and extension.

The majority of exoskeleton gloves currently available are focused on rehabilitation rather than assistance, such as the Festo ExoHand or the Three-layered Sliding Spring Mechanism Exo Glove [3]. As such, they are often limited to laboratory or clinical settings and fail to be useful for day-to-day activities. Many of these systems are high cost, and lack portability as the user is often tethered to a computer and performing tasks in a virtual environment. Consequently, the functional benefits of such hand exoskeleton technology have remained largely inaccessible to clinicians and patients.

Our design team was approached by the Bionic Limbs for Improved Natural Control (BLINC) Lab, to develop of a new prototype hand exoskeleton to overcome the deficiencies of existing devices. Upon examination of the limitations in current designs, the objective became evident: the development of a low-profile, ergonomic, soft-exoskeleton glove that could restore grip and release functions to a wearer suffering from hand motor impairment. Additionally, the client set a design budget of $500 for mechanical components. This value would help ensure a lower cost of a final product facilitating the accessibility to patients.

PRODUCT DESIGN

The design is capable of actively actuating the pointer, middle, and ring fingers individually while the thumb passively opposes the fingers with a thermoformed polylactic acid (PLA) splint. It was developed from an off-the-shelf glove base fitted with 3D-printed PLA components and routed with Bowden cable over each actuated digit (Figure 1). The components are fastened to the glove through Velcro attachments.

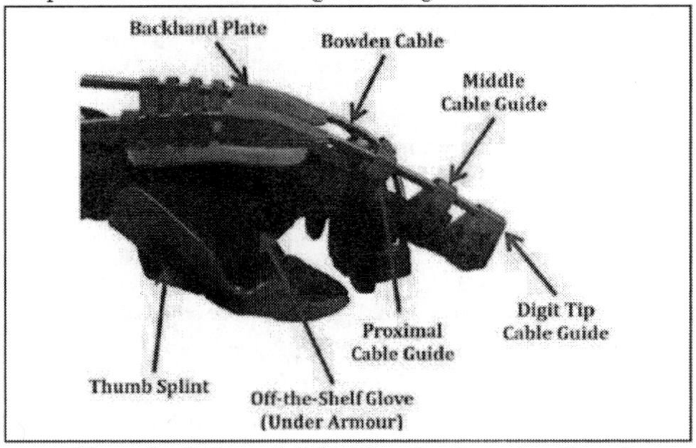

Figure 1: Soft-exoskeleton glove for the restoration of hand function with annotated components

The ends of each of the three Bowden cables are fixed at the digit tip cable guides through an interference fit, whereas the remaining cable is free to slide through the digit cable guides and the backhand plate. As each cable is pushed forward by its respective actuator (not shown), it is forced to bend to accommodate its increased length. The backhand plate and guides ensure that the cable cannot bend anywhere other than above the finger joints. Thus, the cable bending action causes each digit to curl towards the palm of the hand. The stiffness of the Bowden cable ensures the cable arcs, rather than buckles, over the joints, allowing force to be transmitted to the fingertip. In the absence of a push force, the cable returns to its original length over the digit, thereby straightening the finger. One of the key advantages of this design is its ability to pull open the digits, a crucial function for those who suffer from conditions where the hand rests in a closed position.

The thermoformed backhand plate consists of three elevated cable guide structures that act to secure and direct each cable to their associated proximal cable guide. The downward 15 degree bend in the center of the plate minimizes the risk of cable buckling above the knuckle as it reduces the cable's angular rotation when the digits are in flexion. The 3D-printed cable guides are identical in dimensions but can fit most digit diameters due to their half-ring design. The half-rings form complete rings through Velcro attachments. These anchor the cable guides to the glove, prioritize wearer comfort, and provide universal sizing. The digit tip cable guide possesses a similar half-ring design, however it has a closed end and secures the cable through an interference fit and generic cyanoacrylates adhesive ("Krazy Glue").

TECHNICAL FEASIBILITY

The feasibility of the design was validated through the analysis of cable buckling lengths, force requirements, and rigid component fatigue life. However, much of the data required to assess the feasibility of the design was unavailable, and thus empirically determined through experimentation using modular structures, load cells, calipers, and a goniometer.

Experimental results dictated that Bowden cable bending stiffness was 92 ± 10 N·mm and that the exposed cable lengths in the hand exoskeleton should not exceed 60 mm if the risk of buckling was to be mitigated. The anthropometric data of the longest finger was used for geometry, as this digit would require the greatest amount of force to be actuated. From analysis, in order to generate 2.5 N of force at the fingertip, the required input force was estimated to be 8.6 N. The results of the force analysis was validated through force generating experiments using a prototype glove. The testing allowed the maximum generated fingertip force and the ratio of input to output force to be assessed. The empirical data suggested the glove outperformed the original fingertip force estimate by approximately 30% in the closed hand configuration, likely attributed to the conservative assumptions made during analysis.

The microstructure of 3D printed materials is anisotropic because of the linear printing patterns; therefore, the properties of the 3D printed parts will be orthotropic. Commercial 3D printers print each layer as a shell and infill pattern. The latter can be adjusted from 10% to 100% infill, implying the possibility of a degree of porosity in 3D-printed components. To account for the infill pattern, a proportional area deration factor was applied which increased the stress on the component. The positions and values of the maximum stress were determined using static FEM simulations and conservative loading assumptions. The stress values were validated analytically at the locations of maximum stress. Stresses were then recalculated using the area deration factors to obtain corrected maximum stresses. These were all found to fall below PLA's yield strength. Next, the design team conducted a life cycle analysis on the digit tip cable guide, the component subjected to the greatest maximum stress. In order to ensure the success of the device, the perpendicular-to-loading 3D-print orientation and non-stop actuation conditions were assumed. It was found that the digit tip cable guide could meet the desired 100 hour operational time whilst maintaining a safety factor of 5.5 if a 100% infill pattern were applied to the printing process.

BUDGET & MARKET ANALYSIS

The total cost of the prototype was approximately 10% of the 500$ budget. The material cost breakdown is provided in Figure 2.

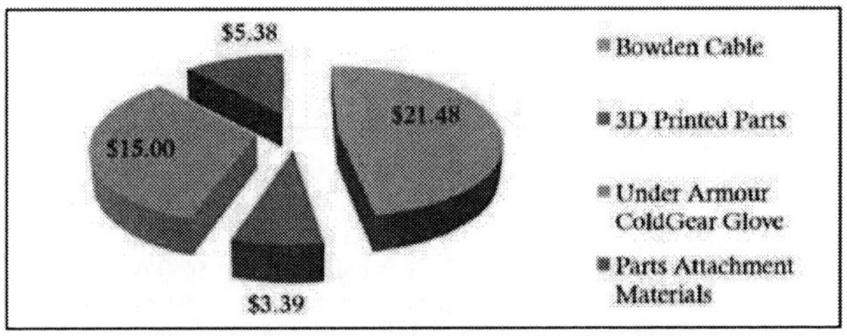

Figure 2: Soft-exoskeleton glove for the restoration for hand function material cost breakdown in CAD$

On a global scale, it is estimated that 15 million people suffer from a stroke annually [1]. Approximately 6.5 million of these stroke victims will fail to fully rehabilitate their paretic arm within six months. As a result, these victims will suffer from prolonged impaired motor function of their upper limb. Thus, it would not be unreasonable to assume that 5% of 6.5 million, or 325,000 stroke victims, would purchase an assistive exoskeleton glove to assist in their day-to-day activities, excluding patients suffering from neurological diseases, recovering from traumatic injuries, or an aging population suffering from muscular atrophy.

Currently, the soft-exoskeleton glove is still in its infancy as it was never intended for immediate commercial use, as such, further laboratory development and user feedback is required. However, if the initial scale of manufacture of 1000 units is assumed, the material costs of the exo-glove amounts to $45. The largest investment would be a commercial high resolution 3D printer incurring an upfront cost of approximately $25,000. The bulk purchase of microcontrollers, myoelectric signal bands, and linear actuators or servo motors create an additional cost of $315, thus resulting in a total material cost of $385 per unit. If Just-In-Time (JIT) manufacturing processes are implemented, the cost of inventory may be neglected. Furthermore, as the target market is global, it is recommended sales be through an international prosthetic distributor, who are likely to charge 10% to 15% of the gross sales revenue.

The current gloves on the market are either rehabilitation gloves, limited to a laboratory or clinical setting, or can only be used to alleviate hand spasticity post-stroke. There are few gloves that have emerged on the global market capable of exerting force to both open and close the hand. Gloves focused only on rectifying spasticity average a retail cost of $300 to $600. The Rapael Exoskeleton Hand is one of the few commercially available gloves that shares the greatest functional similarity to our design and retails for $15,000 [4]. This grants substantial flexibility in our products' final retail cost. However, in order to capture the largest market share, it is estimated that that the device must have a final retail price under $1,500, netting a total return on investment of $890,000 on the first 1000 units sold, assuming the prosthetic distributor charges 15% of gross sales.

ACKNOWLEDGEMENTS

Mr. Jonathon Schofield (client); Professor Mahmoud Salimi; Ms. Quinn Boser; BLINC Laboratory; University of Alberta.

REFERENCES

[1] Seshadri et al. "The Lifetime Risk of Stroke: Estimates From the Framingham Study," *Stroke*, vol. 37, no. 2, pp. 345–350, May 2006.
[2] Heo et al. "Current hand exoskeleton technologies for rehabilitation and assistive engineering," *International Journal of Precision Engineering and Manufacturing*, vol. 13, no. 5, pp. 807–824, 2012.
[3] Arata et al. (2013, May). "A new hand exoskeleton device for rehabilitation using a three-layered sliding spring mechanism," in Robotics and Automation (ICRA), 2013 IEEE International Conference
[3] J. Arata, K. Ohmoto, R. Gassert, O. Lambercy, H. Fujimoto, and I. Wada, "A new hand exoskeleton device for rehabilitation using a three-layered sliding spring mechanism," *2013 IEEE International Conference on Robotics and Automation*, 2013.
[4] D. Furness, "Rapael exoskeleton hand and gaming system is designed for rehabilitation," Digital Trends, 2017. [Online].

SB³C2017
Summer Biomechanics, Bioengineering and Biotransport Conference
June 21 – 24, Tucson, AZ, USA

MUSCLE DRIVEN ELBOW JOINT SIMULATION IN A MULTIBODY FRAMEWORK

Munsur Rahman (1), Mohsen Sharifi Renani (1), Akin Cil (1,2,3), Antonis P. Stylianou (1)

(1) Department of Civil and Mechanical Engineering
University of Missouri-Kansas City
Kansas City, Missouri, USA

(2) Department of Orthopaedic Surgery
University of Missouri-Kansas City
Kansas City, Missouri, USA

(3) Department of Orthopaedics
Truman Medical Centers
Kansas City, Missouri, USA

INTRODUCTION

As the most important joint of upper extremity, the elbow joint serves as a fulcrum of the forearm lever that greatly enhances the spatial positioning of the hand. Because of the centrality of the elbow joint to the upper extremity system, loss or diminished function of the elbow joint results in significant deficits in upper extremity function and loss of independence. A dynamic computational elbow joint model capable of concurrent prediction of muscle and ligament forces, along with cartilage contact forces would be an immensely useful tool in clinical practice. It can enhance our understanding of the interrelationships between joint structures and the musculature facilitating the development of patient specific surgical and no-surgical therapeutic strategies. Computational models of the upper extremity have been employed to study joint biomechanical behavior and analyze musculoskeletal movement simulations [1-2]. However, these models have limited clinical applicability due to simplifications of the joint motion (e.g. hinge joint motions) instead of anatomical based joint motions constrained by ligament forces and cartilage contacts.

Presented here is a forward dynamics multibody model of a subject-specific elbow. The joint was constrained by multiple ligament bundles and deformable articular cartilage contacts. Moreover, the model included the natural oblique wrapping of ligaments. The model was evaluated by comparing bone kinematics of the muscle driven forward dynamics to experimental data. Predicted muscle activation patterns were also compared to collected electromyography (EMG) data.

METHODS

Experimental measures: One healthy volunteer was recruited for the study. High resolution MRI was acquired on the dominant upper extremity. Two localizers made of ABS plastic were attached to the subject's upper arm and the forearm that included two orthogonal tubes. The tubes were filled with mustard which is visible in the MR images and assisted in global coordinate registration later in the modeling process. Optotrack Certus motion capture markers (Northern Digital Inc, Waterloo, Ontario, Canada) were attached to the subject's arm. An Orthopaedic surgeon performed a standard laxity test for the elbow on the day of the test. This test established the kinematic range of motion (KEM) from where ligament zero-load lengths were extracted. After the laxity test, the subject was positioned on a Biodex Multi-Joint Dynamometer system (Biodex Medical Systems, Shirley, NY) where the dominant arm was kept in a supinated rest position. The initial position and orientation of the arm were determined by recording multiple points on the localizers, bony landmarks, and along the surface of the arm by using an Optotrak probe tool. Dynamic testing of the elbow was performed in the dynamometer (Fig. 1a). The subject performed three seated elbow flexion/extension trials in two different protocols: 1) Maximum voluntary contraction (MVC) 2) isokinetic at 10 deg/sec. For each trial, arm segment motions and EMG from the biceps and triceps were recorded.

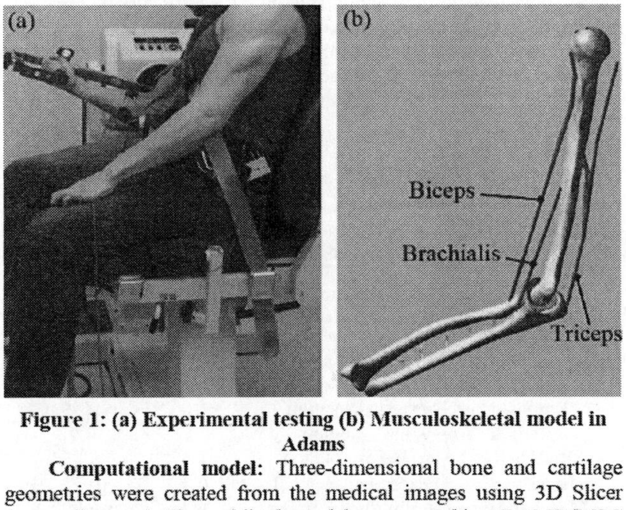

Figure 1: (a) Experimental testing (b) Musculoskeletal model in Adams

Computational model: Three-dimensional bone and cartilage geometries were created from the medical images using 3D Slicer (www.slicer.org). The multibody model was created in ADAMS (MSC Software Corporation, Santa Ana, CA) by aligning these geometries using the initial position and orientation (Fig. 1b). The ligaments and interosseous membranes were modeled as non-linear springs using a piecewise function that includes the "toe" region [3]. The ligament origin and insertion sites were determined from the MRI and published data [4]. Ligaments were wrapped around bones to represent their anatomical physiology. A custom macro was written in ADAMS to automatically divide the humerus cartilage into discrete hexahedral elements. Each element had an approximate 3 x 3 mm cross-sectional area. The macro connected each cartilage element to the humerus bone with a fixed joint located at the center of each element. The radial and ulnar cartilages were attached rigidly to the respective bone using fixed joints. The macro also defined a deformable contact constraint with no friction using modified Hertzian contact law (Eq. 1) between each humerus cartilage element with the radius and ulna cartilage geometry.

$$F_c = k_c\delta^n + B_c(\delta)\dot{\delta} \;\ldots\ldots\ldots\; (1)$$

where k_c is the contact stiffness, δ is the interpenetration of the geometries, and $B_c(\delta)$ is a damping coefficient. An optimization and design of experiment approach was used to determine the contact parameters [5]. The model included three major muscles that cross the elbow joint: triceps (long, lateral, medial), biceps (long, short), and brachialis. Muscle parameters, insertion, origin, and via-point were obtained from previous studies [2]. Point-to-point force elements were added between origin/insertions through via points for all muscles.

Simulation of the computational model was done in two phases. First, the measured motions were used to move the model as constrained by the joint contacts and ligaments. The shortening/ lengthening pattern of each muscle element was recorded during this step. Next the kinematic constraints were removed and muscles served as an actuators to replicate the motions during forward dynamics. The muscle forces were calculated via feedback controllers implemented in Simulink (The MathWorks, Inc., Natick, MA). During forward dynamics simulation, ADAMS and Simulink are linked in co-simulation. In the process, ADAMS sends the current muscle lengths to Simulink and then Simulink sends muscle forces to ADAMS for the next calculation step. The error signal between the current forward dynamics muscle length and muscle length measured during the inverse kinematics simulation is fed into a proportional–integral-derivative (PID) feedback controller. The output of the PID controller is the muscle forces required to track the inverse kinematics muscle length during the forward dynamic simulation. The muscle force is limited such that the muscles can only pull and cannot push. In addition, the PID parameters for each individual muscle are scaled based on the muscle physiological cross sectional area (PCSA). Therefore, muscles with greater PCSA have higher gains than muscles with smaller PCSAs. Furthermore, maximum force generating potential for a given muscle is limited based on PCSA and maximum tissue stress. Local coordinate systems for the each bone segments were created as described by previous study [4] to measure the ulna and radius motions relative to the humerus coordinate system.

RESULTS

RMS errors and correlation coefficients between the inverse kinematics and forward dynamics are presented in Table 1.

Table 1: RMS error (deg, mm) and correlation coefficients for ulna and radius kinematics.

Kinematics description	RMS error	Correlation coefficient
Ulna internal-external rotation	4.4	-0.76
Ulna varus-valgus rotation	7.6	0.61
Ulna flexion-extension rotation	1.2	0.99
Radius internal-external rotation	1.8	0.59
Radius varus-valgus rotation	5.7	-0.72
Radius flexion-extension rotation	1.5	0.99
Ulna superior-inferior displacement	0.4	0.22
Ulna anterior-posterior displacement	0.4	0.99
Ulna medial-lateral displacement	3.2	-0.51
Radius superior-inferior displacement	0.8	0.99
Radius anterior-posterior displacement	1.9	0.98
Radius medial-lateral displacement	0.9	0.85

Kinematics patterns are shown in Figure 2. Normalized muscle forces comparison between experiment and model prediction are shown in Figure 3.

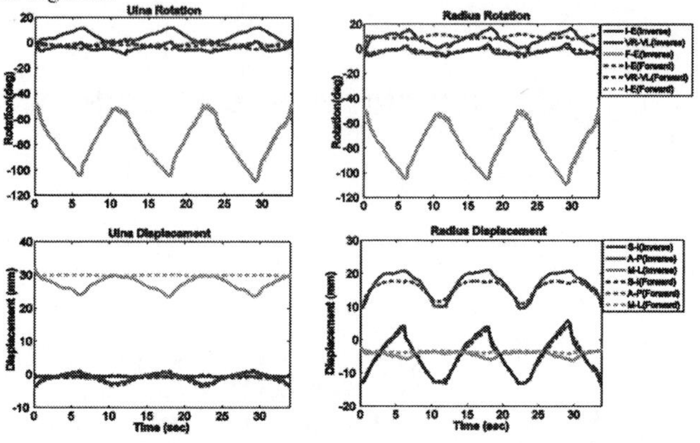

Figure 2: Comparison between inverse kinematics and forward dynamics motion of the ulna and radius relative to humerus.

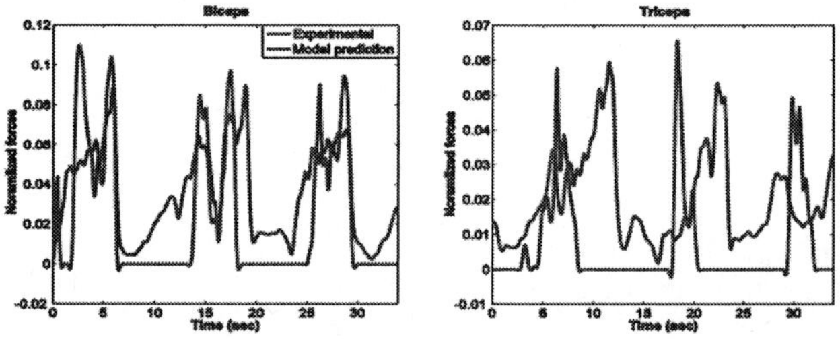

Figure 3: Normalized experimental EMG and normalized muscle forces from forward dynamics

DISCUSSION

This study developed an anatomically correct subject specific musculoskeletal model of the elbow joint in a multibody framework. The model was evaluated by comparing the bone kinematics of a muscle driven forward dynamics with a motion driven inverse kinematics. The normalized muscle forces were also compared to experimental data. The majority of the muscle driven kinematics showed good agreement with the motion driven kinematics. However, some of the forward dynamic predicted motions deviated from the recorded kinematics, especially the varus-valgus directions. Not including the joint capsule in our model may be a contributing factor to this error. Morrey et al. [6] reported that the anterior and posterior capsule provided 32% varus and 33% valgus elbow stability respectively.

A limitation of the muscle modeling scheme is that it does not include the muscle force-length and force-velocity relationship. Future studies will include more muscles with additional optimization of muscle forces and joint capsules around the joint.

ACKNOWLEDGEMENTS

This research was supported in part by funds provided through the University of Missouri-Kansas City, School of Graduate Studies Research Grant program.

REFERENCES

[1] Garner, BA. et al., Comput Methods Biomech Biomed Engin, 4: 93-126, 2001.

[2] Holzbaur, K.R. et al., Ann Biomed Eng, 33: 829-40, 2005

[3] Wismans, J. et al., J Biomech, 13:677-85, 1980

[4] Rahman, M. et al., Med Eng Phys, 38:257-66, 2016

[5] Guess, TM. et al., Comput Methods Biomech Biomed Engin. 16:256-70, 2013.

[6] Morrey, BF. et al., Am J Sports Med.11:315-9, 1983

KINEMATICS OF GLENOHUMERAL JOINT FOLLOWING MULTIPLE ANTERIOR DISLOCATIONS

Masahito Yoshida (1), Tetsuya Takenaga (1), Calvin A. Chan (1), Volker Musahl (1, 2), Albert Lin (1), Richard E. Debski (2, 1)

(1) Department of Orthopaedic Surgery
University of Pittsburgh
Pittsburgh, Pennsylvania, United States

(2) Department of Bioengineering
University of Pittsburgh
Pittsburgh, Pennsylvania, United States

INTRODUCTION

Glenohumeral stability is maintained by contributions from a complex combination of bony contact, muscular and capsular restraints. The majority of glenohumeral joint dislocations occur anteriorly. Following the initial episode, additional dislocations occur in nearly 90% of shoulders in young patients. Patients have been traditionally treated with rehabilitation of the shoulder muscles after a period of rest. However, some clinicians have recommended surgical repair after the first dislocation as the best chance for success in young and active individuals[1]. The kinematics of the glenohumeral joint following multiple dislocations are not well defined for treatment recommendations. Therefore, the purpose of this study was to assess the kinematics of glenohumeral joint, following multiple dislocations in anterior direction, and evaluate the relationship between the number of dislocations and the propensity for recurrence. The magnitude of anterior translation in response to an anterior load was hypothesized to increase and the force for dislocation decrease with increasing numbers of dislocations.

METHODS

Five fresh-frozen cadaveric shoulders (age range 49 to 59 years) were dissected free of all soft tissue except the glenohumeral capsule. The humerus and scapula were fixed in epoxy putty, and each joint was mounted in a robotic testing system that was used to apply external forces and moments to the humerus. **(Figure 1)** To evaluate the function of the intact capsule, a simulated apprehension test was performed at 60 degrees of glenohumeral abduction and 60 degrees of external rotation. At this position, a 50 N anterior force was applied to the humerus, while maintaining a 22 N compressive force, and the resulting joint kinematics were recorded. To assess internal and external rotations at 60 degrees of glenohumeral abduction, a 1.1 Nm torque was applied in each direction. Each joint was then dislocated at 60 degrees of abduction and 60 degrees of external rotation by applying an anterior load to the humerus, while maintaining a 200 N compressive load. The humerus was allowed to move in all three translational degrees of freedom until the anterior translation reached one half the maximum anterior-posterior width of the glenoid plus 10 mm. This definition of dislocation resulted in a Bankart lesion repeatedly in our preliminary study. To evaluate the effect of the dislocation, the same loads were applied to the shoulder and the resulting kinematics were recorded following 1, 2, 3, 4, 5 and 10 dislocations. To assess the propensity for recurrence of the dislocation, the force required to achieve the translation for dislocation was measured. A repeated-measures analysis of variance (ANOVA) was used to compare the kinematics and force to dislocation in the state following each dislocation. Values of $p<0.05$ were considered statistically significant.

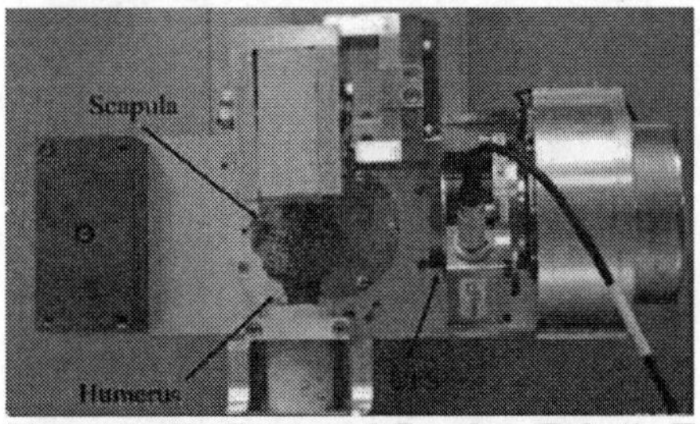

Figure 1. Six Degree of Freedom Robotic Testing System with glenohumeral joint.

RESULTS

Anterior translation in response to an anterior load was statistically different between the intact state, and the states following the 1st and 2nd dislocations ($p < 0.05$). Among the states following the 3rd dislocation, anterior translation did not increase ($p > 0.05$) **(Figure 2)**. Internal and external rotations at 60 degrees of abduction were not significantly different among the intact state and all of the states following multiple dislocations **(Figure 3)**. The anterior force to dislocation was significantly different between the intact shoulder (467 ± 165 N), and the 8th (331 ±139 N), 9th (325 ± 135 N) and 10th dislocations 319 ±134 N) ; ($p < .05$). **(Figure 4)**.

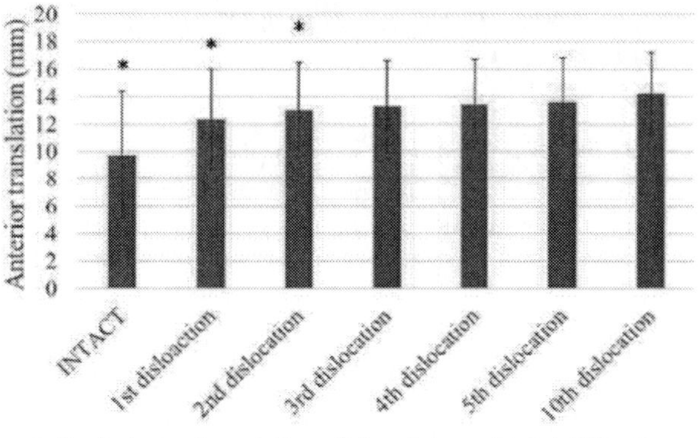

Figure 2: Anterior translation of glenohumeral joint, following anterior dislocation of the glenohumeral joint. (*: $p < 0.05$)

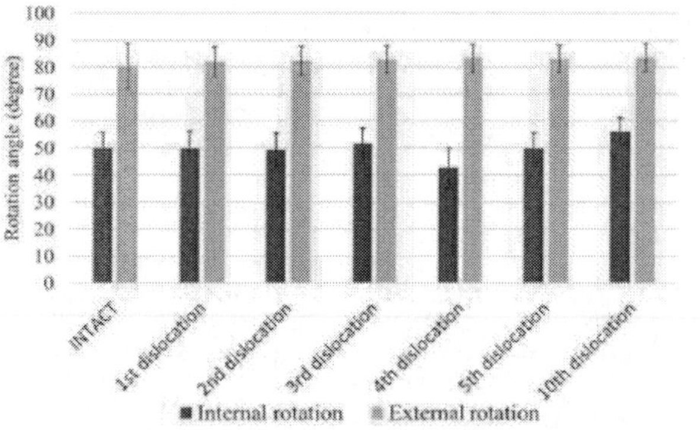

Figure 3: Internal and external rotations at 60 degrees of glenohumeral abduction (*: $p < 0.05$)

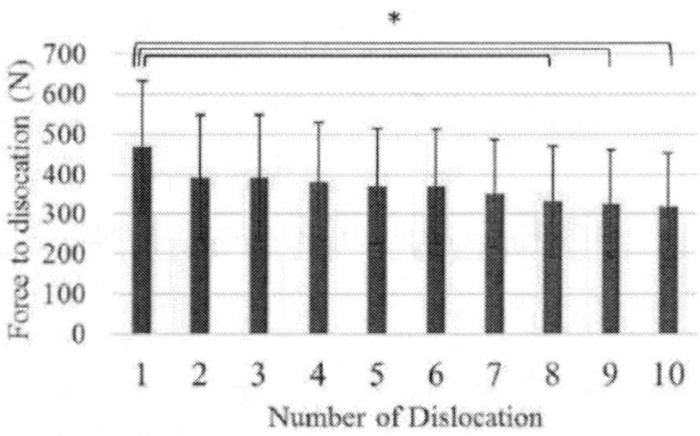

Figure 4. Magnitude of anterior force to anteriorly dislocate the glenohumeral joint. (*: $p < 0.05$)

DISCUSSION

Injury to the glenohumeral capsule was created during application of an excessive load in the anterior direction using a novel 6 degree of freedom robotic testing system. Previous studies have indicated that the second dislocation contributes to increase the propensity for recurrence.[2] Our data suggests that no kinematic and force changes occur between the 3rd to 7th dislocations. In addition, although each dislocation after the 2nd dislocation might not have a significant impact on the kinematics, especially anterior translation in response to an anterior force, accumulation of successive dislocations significantly affects the propensity for recurrence. After the 8th dislocation, the anterior force required for anterior dislocation was decreased in comparison with the anterior force for the 1st dislocation. Thus, the 2nd and 3rd dislocation should be avoided to prevent increasing the anterior instability of the glenohumeral joint. In addition, the repeated dislocations should be avoided to prevent increasing the propensity for recurrence of the dislocation.

As a future direction, the location, direction and magnitude of the injury to the glenohumeral capsule following multiple dislocations will be evaluated as the non-recoverable strain. Based on the non-recoverable strain results, recommendations can be made to perform anatomical capsular plication procedures following multiple dislocations. In addition, the effects of the capsular plication on glenohumeral kinematics following multiple dislocations will be assessed.

ACKNOWLEDGEMENTS

Support from The Albert B. Ferguson, Jr., M.D. Orthopaedic Fund of the Pittsburgh Foundation、the Department of Orthopaedic Surgery, and Nagoya City University Graduate School of Medical Science, Department of Orthopedic Surgery are gratefully acknowledged.

REFERENCES

[1] Bishop JA et al, Operative versus nonoperative treatment after primary traumatic anterior glenohumeral dislocation: expected-value decision analysis. J Shoulder Elbow Surg 2012;21:e17-8.

[2].McMahon PJ et al, Anterior shoulder dislocation increases the propensity for recurrence: a cadaveric study of the number of dislocations and type of capsulolabral lesion. J Shoulder Elbow Surg. 2013 Aug;22(8):1046-52.

INTER-LIMB DIFFERENCES IN KNEE GAIT AND QUANTITATIVE MAGNETIC RESONANCE IMAGING VARIABLES 3 MONTHS AFTER ANTERIOR CRUCIATE LIGAMENT RECONSTRUCTION

Ashutosh Khandha (1), Kurt Manal (2), Jacob J Capin (3), Kevin McGinnis (1), Lynn Snyder-Mackler (3), Thomas S Buchanan (2, 4)

(1) Delaware Rehabilitation Institute
University of Delaware
Newark, DE, USA

(2) Department of Mechanical Engineering
University of Delaware
Newark, DE, USA

(3) Department of Physical Therapy
University of Delaware
Newark, DE, USA

(4) Department of Biomedical Engineering
University of Delaware
Newark, DE, USA

INTRODUCTION

Systematic review and meta-analysis has reported abnormal sagittal plane knee gait biomechanics after anterior cruciate ligament reconstruction (ACLR) [1]. Young subjects who develop medial compartment knee osteoarthritis (OA) 5 years after ACLR demonstrate inter-limb differences in frontal plane moment and medial compartment load during gait, early after ACLR (~ 6 months) [2]. Elevated frontal plane moment and medial compartment load are related to OA progression [3], however, the relation to OA onset is not clear [4]. To detect early OA related changes, cartilage T2 mapping using magnetic resonance imaging (MRI) is used, wherein deterioration of cartilage health is indicated by elevated T2 relaxation time [5]. Currently, it is not known whether inter-limb differences in knee variables during gait are related to inter-limb differences in knee cartilage T2 values as early as 3 months after ACLR. The purpose of the study was to investigate this relationship.

METHODS

11 subjects were enrolled 3 months after unilateral ACLR (6 men, 5 women; mean ± standard deviation (SD): age = 24 ± 7 years, mass = 77 ± 13 kg, height = 1.7 ± 0.1 m). During motion analysis, each subject completed 5 trials at self-selected walking speeds (mean ± SD: 1.5 ± 0.2 m/s). We captured gait data using retroreflective markers, an 8-camera setup (Vicon, Oxford Metrics Limited, London, UK) and a force platform (Bertec Corporation, Worthington, OH). Key variables of interest, peak knee flexion angle (pKFA), peak knee flexion moment (pKFM), peak knee adduction moment (pKAM) and peak medial compartment force (pMCF) during weight acceptance were computed using inverse dynamics (Visual3D, C-Motion, Germantown, MD) and a validated electromyography-informed musculoskeletal model [6].

Each subject underwent supine bilateral knee MRI using sagittal T2 mapping (field of view: 160 x 160 mm, slice thickness = 2 mm, repetition time = 4480 ms, 6 echo times = 12.5 to 75 ms). We calculated T2 maps using exponential fitting (ImageJ, National Institutes of Health, Bethesda, MD). We analyzed the slice corresponding to center of the medial compartment in the frontal plane (primary load-bearing region during gait), which was located based on femoral epicondylar width measurement (figure 1). Key regions of interest (ROI) were defined as the posterior and central load-bearing cartilage layers in the sagittal plane, differentiated by menisci boundaries (figure 2). Each ROI was sub-divided into deep and superficial cartilage sub-layers. We used students t-tests to compare inter-limb differences, and calculated pearson's correlation between inter-limb differences for gait variables vs. inter-limb differences for T2 values.

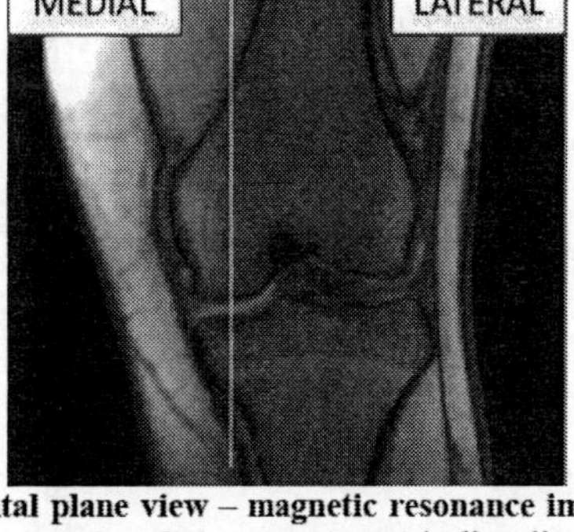

Figure 1: Frontal plane view – magnetic resonance imaging slice located at the center of medial compartment (yellow line in figure is at 25 % of femoral epicondylar width, measured from the medial edge)

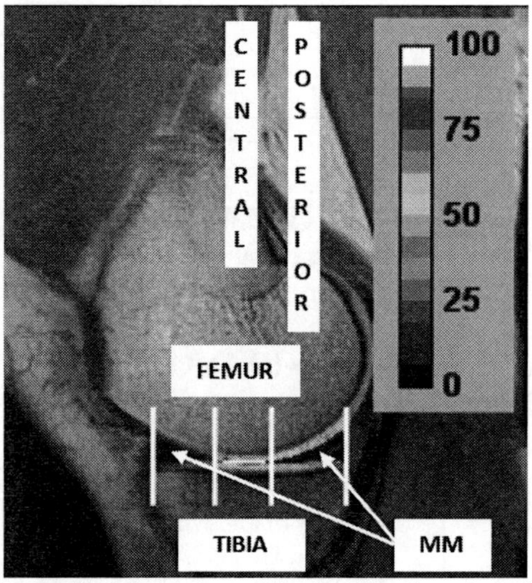

Figure 2: Sagittal plane view – T2 relaxation time map of central and posterior femoral/ tibial cartilage in the medial knee compartment (scale is in milliseconds, MM = medial meniscus) NOTE: Femoral/tibial cartilage layers were further divided into deep and superficial cartilage sub-layers.

RESULTS

In the involved versus uninvolved knee, subjects demonstrated significantly lower values for pKFA and pKFM (table 1). We observed a strong positive correlation for inter-limb difference (involved minus uninvolved, or INV - UN) between pKAM and medial compartment T2 values for posterior tibial cartilage (deep: r = 0.78 and R^2 = 0.6037, superficial: r = 0.67 and R^2 = 0.4435, figure 3). This implies that those with greater pKAM in the INV vs. UN knee also have higher T2 relaxation time values (i.e. potential cartilage deterioration) in the INV vs. UN knee. Inter-limb difference correlations for pKAM versus medial compartment T2 in the deep layers of central tibial cartilage (r = 0.43) and posterior femoral cartilage (r = 0.44) were moderate. A moderate correlation was also observed for pMCF versus medial compartment T2 for the deep layer of posterior tibial cartilage (r = 0.48). Medial compartment T2 values were not significantly different between knees for any cartilage sub-layer.

Table 1: Knee gait variables during weight acceptance. (SD = standard deviation, ACLR = anterior cruciate ligament reconstruction, pKFA = peak knee flexion angle, pKFM = peak knee flexion moment, pKAM = peak knee adduction moment, pMCF = peak medial compartment force, BW = body weight, HT = height)

Variable (units)	Time point: 3 months after ACLR, n = 11		
	INVOLVED Mean ± SD	UNINVOLVED Mean ± SD	p-value
pKFA (degrees)	**16.3** ± 7.7	**21.4** ± 5.9	**0.008***
pKFM (% BW * HT)	**3.2** ± 1.7	**5.7** ± 1.6	**0.003***
pKAM (% BW * HT)	**2.7** ± 1.1	**2.7** ± 0.7	**0.936**
pMCF (BW)	**2.7** ± 0.8	**2.9** ± 0.4	**0.449**

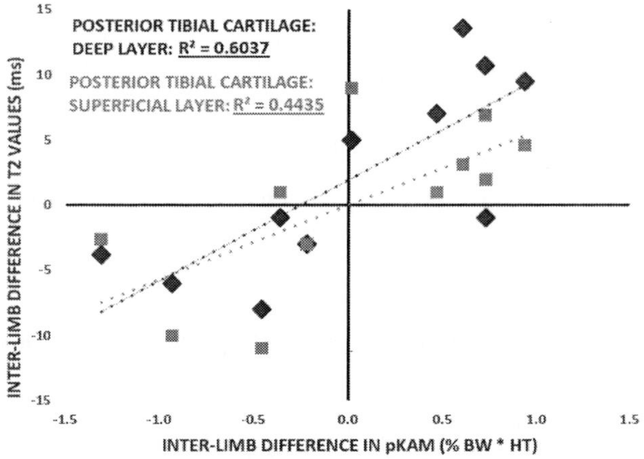

Figure 3: Inter-limb difference (involved minus uninvolved) in medial compartment T2 relaxation times for posterior tibial cartilage (Y-axis, unit = ms) versus inter-limb difference in pKAM (X-axis, unit = % BW * HT). (ms = milliseconds, BW = body weight, HT = height) NOTE: A positive number indicates that the value is greater in the involved versus uninvolved knee, and vice versa.

DISCUSSION

For the involved versus uninvolved knee, subjects demonstrated significant inter-limb differences in sagittal plane knee gait biomechanics, which is in line with previous reports [1]. The strong positive relationship between inter-limb differences in pKAM and medial compartment T2 values for posterior tibial cartilage could be due to excessive anterior tibial translation after ACLR [7], which shifts joint contact posteriorly. This could place those individuals with higher pKAM at increased risk of cartilage degradation in this region (i.e. posterior tibial cartilage of the medial compartment).

Knee gait variables are biomechanical indicators of joint loading, whereas T2 variables are biochemical indicators of cartilage health. The T2 time constant reflects mobility of water protons. High T2 values reflect high proton mobility, which in turn is due to a less-organized matrix of collagen fibers, as is seen during OA onset [8]. Results from the current study warrant further investigation of a potential link between knee gait biomechanics and biochemical changes in knee cartilage after ACLR, with both a larger sample size and long-term follow-up. Evaluating inter-limb differences in knee gait and cartilage MRI variables may aid in early OA detection, as well inform rehabilitation strategies to delay disease progression after ACLR.

ACKNOWLEDGEMENTS

National Institutes of Health: R01-HD087459 & P30-GM103333.

REFERENCES

[1] Hart H, et al. *Br J Sports Med.* 0: 1-17,2015.
[2] Wellsandt E, et al. *Am J Sports Med.* 44: 143-151,2016.
[3] Kutzner I, et al., PLOS ONE. 8: 1-8,2013.
[4] Teichtahl A, et al., *Br J Sports Med.* 37: 289-290,2003.
[5] Baum T, et al., *Osteoarthr Cartil.* 21: 1474-1484,2013.
[6] Manal K, et al., *J Biomech Eng.* 135: 021014,2013.
[7] Abebe E, et al., *J Biomech.* 44: 924-929,2011.
[8] Le J, et al, *Ann N Y Acad Sci.* 1383: 34-42,2016.

FLEXION-PATTERN OF BI-CRUCIATE-RETAINING TOTAL KNEE ARTHROPLASTY-TREATED KNEES

Tetsuya Takagi (1), Yohei Okada (2), Satoshi Yamakawa (1), Atsushi Teramoto (2), Toshihiko Yamashita (2), Hiromichi Fujie (1)

(1) Faculty of System Design
Tokyo Metropolitan University
Tokyo, Japan

(2) Department of Orthopaedic Surgery,
Sapporo Medical University
School of Medicine, Sapporo, Japan

INTRODUCTION

Total knee arthroplasty (TKA) is a successful treatment for severe osteoarthritis (OA) with 8 year's survivorships of 97% [1]. However, certain number of patients are dissatisfied with their TKA [2]. Although standard type of TKA sacrifices the anterior cruciate ligament (ACL) or both the ACL and posterior cruciate ligament (PCL), bi-cruciate-retaining (BCR) TKA preserving ACL and PCL for providing knee stability had been developed. Previous studies reported that the motion of BCR-TKA-treated knee is closed to intact knee motion as compared with conventional TKA (cruciate retaining (CR) TKA)-treated knee [3-4]. However, biomechanical data regarding BCR-TKA were not fully obtained in previous studies, although such data are important for the improvement the TKA. We focused on the determination of flexion axis of BCR-TKA-treated knee based on the helical axis. It is well known that any-3dimensinal motion of a rigid body can be expressed by a translation along and a rotation about the axis. Although physiological knee motion is complex, the helical axis provides easy representation of the motion.

Therefore, the objective of the present study was to determine the flexion axis of BCR-TKA-treated knee using a geometrical representation of the helical axis and to compare them to that of CR-TKA-treated knee.

METHODS

A 6-DOF robotic system consisting of a custom made 6-axis manipulator with a 6-DOF universal force/moment sensor [5] was used. In the original system, a LabView-based control program runs on a windows PC to control the displacement of, and force/moment applied to the cadaveric knee joints with respect to the knee joint coordinate system [6].

Human knee joints (n=3) were dissected down to the joint capsule and fixed to the robotic system. First, passive flexion motion was applied to the intact knee joint at 1.5 degree/s while setting the other 5 DOFs except the flexion-extension DOF under force control with prescribed force/moment at 0. Joint replacement surgery was performed using BCR- and CR-TKA by an orthopaedic surgeon. Then, passive flexion motion was applied to BCR-TKA-treated knees. Finally, passive flexion motion was also applied to CR-TKA-treated knees. The kinematics data collected at every 5 degree from full extension to 90 degree of flexion during the test were used for determining the trajectory of the helical axis of the knees using a calculation scheme developed by us [7].

RESULTS

Results revealed that the flexion axis of intact knee had an inclination at knee extension. The inclination of the flexion axis was decreased with the increase of knee flexion (Fig.1). The flexion axis was located roughly at the line connecting the insertion of the medial collateral ligament (MCL) and lateral collateral ligament (LCL). The flexion axis shifted toward posterior with the increase of knee flexion (Fig.1). The feature of the flexion axis of BCR-TKA-treated knee was similar to that of intact knee (Fig.2), although, the inclination of the flexion axis of BCR-TKA-treated knee was smaller than that of intact knee at knee flexion (Fig.2). The flexion axis of CR-TKA-treated had not an inclination at knee extension (Fig.3). The flexion axis of CR-TKA-treated knee remained inclined at knee flexion locating more distal than intact and BCR-TKA-treated knees. Thus, the flexion axis of CR-TKA-treated knee was quite different from those of intact and BCR-TKA-treated knees (Fig.3).

DISCUSSION

The inclination of the knee flexion axis at low flexion angle is known as screw home movement [8]. The movement was observed in BCR-TKA-treated knee although the inclination was slightly smaller than in intact knee. In contrast, the change of the inclination of the CR-TKA-treated knees flexion axis was different from those of intact and BCR-TKA-treated knees; in which screw home movement was not observed in knee extension and the flexion axis was located posteriorly in knee flexion. It is known that the ACL plays major roles in joint stability and kinematics throughout the range of motion in intact knee [9]. Since there is no significant difference in the geometry of articulating surface between the BCR prosthesis and CR prosthesis, it is considered that the intact knee-like flexion pattern observed in BCR-TKA-treated knees is attributable to the preserved ACL. It should be noted that the kinematics of the CR-TKA-treated knees may be abnormal because the location of flexion axis is far distal. Finally, it is suggested that the flexion pattern of BCR-TKA-treated knees is similar to that of intact knees due to preservation of the ACL.

ACKNOWLEDGEMENTS

The present study was financially supported by the Grant-In-Aid for Scientific Research B (#16H03172) from the MEXT, Japan.

REFERENCES

[1] NJR. National Joint Registry. National Joint Registry for England and Wales: 10th Annual Report; 2013.
[2] Bourne, RB et al., *Clin Orthop Relat Res*, 468:57-63, 2010.
[3] Moro-oka, T et al., *KSSTA*, 15:93-99, 2007.
[4] Cloutier, JM et al., *JBJS Am*, 81:697-702, 1999.
[5] Fujie, H et al., *J Biomech Eng*, 126:54-61, 2004.
[6] Grood ES., Suntay WJ., *JBME*, 105: 136-144, 1983.
[7] Takagi, T et al., *SB3C(ASME)*, 2016.
[8] Blankevoort L. et al., *J Biomech*, 23: 1219-1229, 1990.
[9] Rong, GW et al., *Clin Orthop Relat Res*, 215:65-71, 1987.

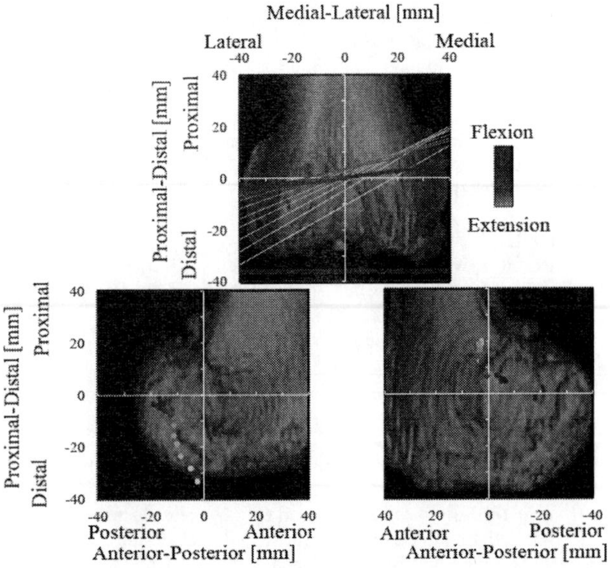

Fig.1 Flexion axis of intact knees. The darker color the deeper flexion, with a resolution of 5 degree of flexion.

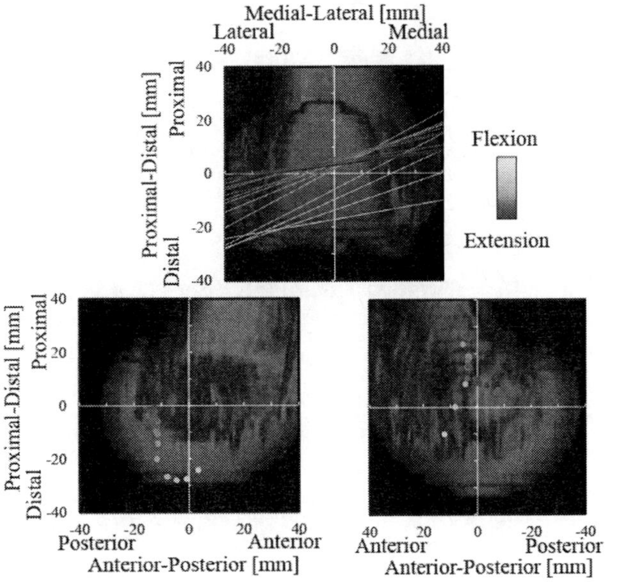

Fig.2 Flexion axis of BCR-TKA-treated knees. The darker color the deeper flexion, with a resolution of 5 degree of flexion.

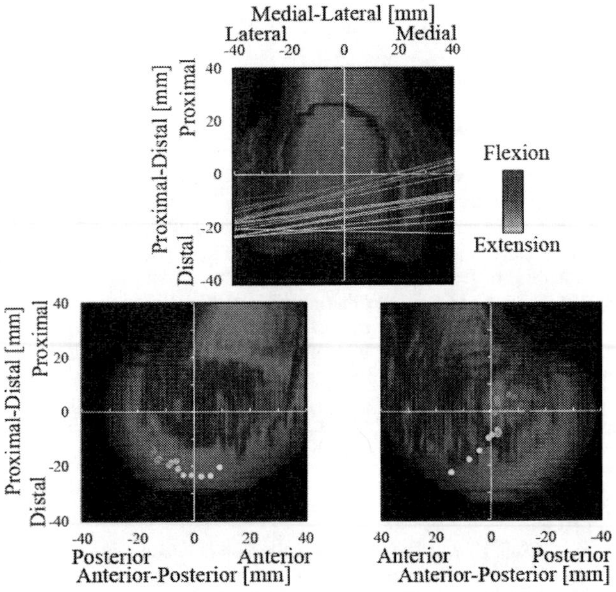

Fig.3 Flexion axis of CR-TKA-treated knees. The darker color the deeper flexion, with a resolution of 5 degree of flexion.

LIGAMENT ENGAGMENT PATTERNS IN A HUMAN CADAVERIC KNEE MODEL: A BASIS FOR PRECISION MEDICINE IN ORTHOPAEDICS

Robert N. Kent (1), James F. Boorman-Padgett (1), Ran Thein (2), Andrew D. Pearle (3), Thomas L. Wickiewicz (3), Carl W. Imhauser (1)

(1) Department of Biomechanics
Hospital for Special Surgery
New York, NY, USA

(2) Department of Orthopaedic Surgery
Chaim Sheba Medical Center
Ramat Gan, Israel

(3) Department of Orthopaedic Surgery
Hospital for Special Surgery
New York, NY, USA

INTRODUCTION

Over 6.5 million knee injuries were presented to U.S. Emergency Departments from 1999-2008, with nearly half (42%) of these injuries being sprains and strains of cruciate and collateral ligaments.[1] Devastatingly, patients who experience such joint injuries are five times as likely to develop post-traumatic osteoarthritis (PTOA) within 15 years, leading to loss of function and a decreased quality of life.[2]

No two knees and no two knee injuries are the same; the presence of large variations in both injury patterns and native physiology suggest that treatment should be personalized.[3] However, surgical reconstruction relies on standardized guidelines, and this discrepancy likely contributes to suboptimal clinical outcomes. As an example, only about half of patients return to full activity following treatment of anterior cruciate ligament (ACL) rupture, and as many as 18% of patients require revision surgery.[4-5]

Variations in the native physiology of the knee also manifest in highly heterogeneous laxity (i.e., the net motion of the tibia relative to the femur in a given direction in response to an applied load) across the population, especially between males and females.[6] These laxity variations are strongly associated with the risk of ligament injury and of joint reinjury following surgical ligament reconstruction.[7-8] In preforming reconstruction procedures, surgeons must pay close attention to joint laxity; improperly tuning laxity to leave the knee under- or over-constrained can accelerate joint degeneration.[9]

Joint laxity is mediated by the way ligaments build force (i.e., engage) to resist motion, but the details of this interaction are poorly understood. Therefore, identifying a relationship between ligament engagement and overall joint laxity would be useful for establishing personalized protocols for ligament reconstruction. To this end, we assessed whether there are distinct patterns relating variations in tibiofemoral laxity to features of how the ACL, posterior cruciate ligament (PCL), and medial collateral ligament (MCL) are engaged.

METHODS

All of the soft tissues save for the ligamentous and capsular structures of the tibiofemoral joint were removed from 20 fresh-frozen human cadaveric knee specimens (mean age: 45 years; range: 20-64 years; 14 male). A six degree-of-freedom robot with a repeatability of ±0.3 mm manipulated the tibia relative to a rigidly grounded femur while recording six orthogonal forces and torques on a universal force-moment sensor. All kinematics and loads were described using an anatomical coordinate system. To identify potential ligament engagement patterns, the function of the ACL, PCL, and MCL was assessed during loading scenarios in which each structure respectively plays a major stabilizing role: (1) an anterior force of 134 N at 30° flexion; (2) a posterior force of 134 N at 90° flexion; and (3) a valgus torque of 8 Nm at 15° flexion.[10]

After preconditioning, neutral tibiofemoral positions were characterized so that ligament engagement could be described relative to a consistently defined starting position across all 20 knees. In the AP and VV directions, neutral positions were respectively defined as the center of the AP load – tibial translation profile and the tibial orientation at which bicondylar contact occurred. While the robot incrementally increased anterior, posterior, or valgus load from 0 N or 0 Nm to the peak applied load, the force carried by each ligament *in situ* was calculated using the principle of superposition.[11]

Ligament engagement was parameterized using three measures: (1) anatomic slack in mm or °; (2) anatomic stiffness in N/mm or N/°; and (3) ligament force at the maximum applied load in N (Fig. 1). The anatomic slack of each ligament was defined as the tibial motion from the neutral position to the engagement point, which was identified using an objective algorithm.[12] The anatomic stiffness was calculated as the slope of the linear portion of the ligament force – tibial motion response following the engagement point (Fig. 1).

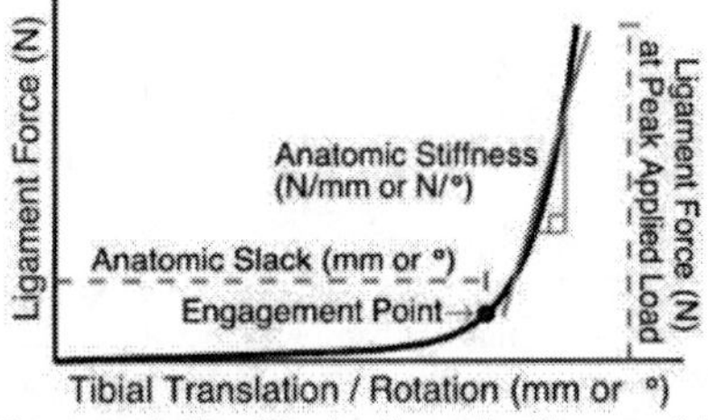

Figure 1: Representative illustration of a ligament force – tibial motion relationship and ligament engagement parameters.

Univariate linear regressions ($p<0.05$) were performed to quantify the ability of each parameter of ligament engagement to explain variations in tibiofemoral laxity. Multiple linear regression models ($p<0.05$) were also developed using stepwise procedures to analyze whether a combination of ligament engagement parameters could explain variations in joint laxity. Normality of each measure was confirmed using Shapiro-Wilk tests (p>0.05).

RESULTS

According to univariate regression models, in response to the anterior load, 83% of the variation in anterior laxity was explained by the variation in the anatomic slack of the ACL (Fig. 2A); anterior laxity was not associated with the anatomic stiffness of the ACL or the force it carried at the peak applied load (both Adj. R^2 = 0.00). In response to the posterior load, 48% of the variation in posterior laxity was explained by variation in the anatomic slack of the PCL (Fig. 2B); posterior laxity was not associated with the anatomic stiffness of the PCL (Adj. R^2 = 0.03) or the force it carried at the peak applied load (Adj. R^2 = 0.13). In response to the valgus torque, variations in neither the anatomic slack of the MCL (Adj. R^2 = 0.17), its anatomic stiffness (Adj. R^2 = 0.22), nor the force it carried at the peak applied load (Adj. R^2 = 0.00) explained variations in valgus laxity (Fig. 2C).

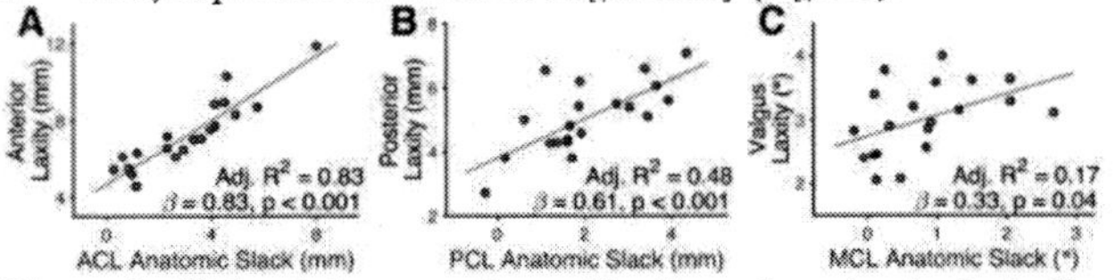

Figure 2: Univariate regressions of laxity and anatomic slack. For each regression, the adjusted coefficient of determination (Adj. R^2) is reported in addition to mean regression coefficients (β) (p<0.05).

According to multiple linear regression models, engagement parameters of the ACL, PCL, and MCL were respectively able to explain 95%, 79%, and 54% of tibiofemoral laxity in their relevant loading directions (Table 1). For each ligament, all three engagement parameters contributed to the model's ability to explain variations in joint laxity (for all β, p < 0.05) (Table 1).

Model/Variable		β	SD	95% CI	p
ACL	Slack	0.84	0.22	(0.74, 0.94)	<0.001
Anterior Load, 30°	Stiff	-0.13	0.10	(-0.17, -0.09)	<0.001
Adj. R² = 0.95	Force	0.03	0.03	(0.02, 0.04)	<0.001
PCL	Slack	0.69	0.44	(0.50, 0.88)	<0.001
Posterior Load, 90°	Stiff	-0.07	0.06	(-0.09, -0.04)	<0.001
Adj. R² = 0.79	Force	0.02	0.04	(0.00, 0.04)	0.044
MCL	Slack	0.43	0.54	(0.19, 0.66)	0.003
Valgus Load, 15°	Stiff	-0.09	0.10	(-0.14, -0.05)	0.001
Adj. R² = 0.54	Force	0.02	0.04	(0.01, 0.04)	0.014

Table 1: 3-Factor multiple linear regressions of laxity and ligament engagement parameters. For each model, the adjusted coefficient of determination (Adj. R^2) is reported in addition to mean regression coefficients (β) (p<0.05), their standard deviations (SD), and their 95% confidence intervals (CI).

DISCUSSION

We have identified unique patterns relating variations in ligament engagement to variations in joint laxity. These 'ligament engagement patterns' can explain a large proportion of the variability in tibiofemoral laxity seen across individuals, which is crucial given that ligament loading behavior, unlike gross joint laxity, cannot be measured *in vivo*. The relationships that we identified could allow clinicians to overcome this limitation by permitting inference of the loading behavior of a primary joint stabilizer based on common clinical laxity exams without direct instrumentation of the tissue.

Univariate regression models indicated that, in the cruciate ligaments, the anatomic slack is the dominant parameter of ligament engagement (Figs. 2A, B). Conversely, the anatomic slack of the MCL could only explain 17% of valgus laxity (Fig. 2C). Moreover, multiple linear regression analyses revealed that variations in knee laxity are best explained by mutually considering all three parameters of ligament engagement (Table 1). Interestingly, the multiple linear regression models left a greater percentage of variability in valgus laxity unexplained compared to variability in anterior and posterior laxity (MCL: 46%, ACL: 5%, PCL: 21% unexplained) (Table 1). This is intuitive given that, while the ACL and PCL are respectively the primary restraints to anterior and posterior loading, the MCL acts more in concert with other stabilizers, such as the ACL, during valgus loading.[10,13]

Previously, investigators have ranked the relative importance of individual ligaments by reporting the force carried by each at the maximum applied load; such work has proven critical for deciding if a ligament should be augmented or reconstructed.[14] However, our results suggest that this commonly-used approach cannot explain variations in knee laxity; thus, it cannot yield guidelines for personalized ligament reconstruction procedures that target a desired level of knee laxity. Only after controlling for both the anatomic slack and the anatomic stiffness of a given ligament did the force carried by that ligament at the peak applied load strengthen the multiple linear regression model (Table 1).

Our findings suggest that, when implanting a graft to reconstruct the cruciates, surgeons must modulate the anatomic slack to achieve a given laxity (Figs. 2A, B). Multiple linear regression analyses indicated that controlling the anatomic stiffness of and force carried by a cruciate graft can also generate higher fidelity to the native kinematics of the joint, but only after the anatomic slack of the graft has been tuned (Table 1). In contrast, when reconstructing the MCL, slack and stiffness must both be considered to achieve a given laxity (Fig. 2C) (Table 1). These findings have direct clinical application: anatomic slack may be adjusted intraoperatively by modulating the flexion angle and pretension at which ligament grafts are fixed and by changing the location of the bone tunnels. Additionally, anatomic stiffness can be adjusted by modulating the graft type, the size of the graft tissue, or the method of bony fixation.

An important limitation of this work is that it does not establish a causal relationship between ligament engagement and joint laxity. However, previous work has shown that variable tightening of an ACL graft, which would likely influence anatomic slack, does affect anterior knee laxity.[15]

In conclusion, we have characterized ligament engagement patterns for both intra- and extra-articular stabilizers, which explain a large portion of the variation in joint laxity. While this work focused on the knee, this framework can be applied to many synovial joints. Furthermore, features of ligament engagement are surgically tunable; thus, they could be adjusted intraoperatively for more precise, patient-specific surgical treatment of ligament injuries. Ultimately, ligament reconstructions that consider factors driving heterogeneity in joint laxity will enable precision medicine in orthopaedics, which holds promise to achieve improved, more predictable clinical outcomes.

ACKNOWLEDGEMENTS

This work was funded by the Clark and Kirby Foundations, the Institute for Sports Medicine Research, the Surgeon in Chief Fund at Hospital for Special Surgery, and the Gosnell Family Foundation.

REFERENCES

[1] Gage, B et al., *Acad Emerg Med*, 19:378-85, 2012.
[2] Gelber, A et al., *Ann Intern Med*, 133:321-8, 2000.
[3] Noyes, F et al., *Orthop Clin North Am*, 16:47-67, 1985.
[4] McCullough, K et al., *Am J Sports Med*, 40:2523-9, 2012.
[5] Webster, K et al., *Am J Sports Med*, [Epub], 2016.
[6] Boguszewski, D et al., *Am J Sports Med*, 43:2982-7, 2015.
[7] Magnussen, R et al., *Am J Sports Med*, [Epub], 2016.
[8] Uhorchak, J et al., *Am J Sports Med*, 31:831-42, 2003.
[9] Schon, J et al., *Am J Sports Med*, 44:2546-56, 2016.
[10] Markolf, K et al., *J Bone Joint Surg Am*, 58:583-94, 1976.
[11] Woo, S et al., *The Knee*, 5:267-88, 1998.
[12] Satopää, V et al., *ICDCSW*, 116-7, 2011.
[13] Schafer, K et al., *Am J Sports Med*, 44:1203-8, 2016.
[14] Debski, R et al., *J Orthop Res*, 17:769-76, 1999.
[15] Amis, A et al., *J Bone Joint Surg Br*, 71:819-24, 1989.

SB³C2017
Summer Biomechanics, Bioengineering and Biotransport Conference
June 21 – 24, Tucson, AZ, USA

ANALYSIS OF UNCERTAINTY IN SUPERPOSITION TESTING: IMPLICATIONS FOR ROBOTICALLY CONTROLLED KNEE JOINT TESTING

Nicholas J. Haas (1), Tara F. Bonner (1), Callan M. Gillespie (1), Robb W. Colbrunn (1)

(1) Bio□obotics and Mechanical Testing Core
Lerner □esearch Institute
Cleveland Clinic
Cleveland, O□ 441□□

INTRODUCTION

The principle of superposition can be applied to in vitro biomechanical testing in order to determine the amount of load carried by specific anatomical structures in comparison to the load carried by the entire joint. This is performed by applying a known overall load to the joint and recording the position of the joint at that load. A ligament within the specimen is then cut, the joint is moved back to the recorded position, and the new load is recorded. The resulting difference between the original and new load accounts for the amount of load that was carried by the cut ligament [1].

Determining the effectiveness of ACL reconstruction techniques by comparing the stability of an intact knee to a knee with a reconstructed ligament is one of the many ways that superposition testing is used in biomedical research [2]. The Joint Stiffness Model, traditionally used in superposition testing, assumes that all the components of the test setup (except the joint ligaments) are infinitely rigid. However, the assumption that the displacement measured by the robot is exactly the same as the displacement of the joint is incorrect. Because each component is compliant, parameters like robot stiffness and bone bending can produce uncertainty in the experimental results. To the best of our knowledge, there has been no quantification or acknowledgment of the effects of this uncertainty in the past two decades in which the superposition principle has been applied in cadaveric knee testing.

Our aim was to investigate the effect that system component stiffness had on the results of superposition testing. We hope that by understanding the impact of this uncertainty on the results we will be able to amend the superposition test method to yield more accurate results in the future.

METHODS

To understand the effect of system component stiffness, the stiffness of all parts of the system were considered. A typical test setup consists of all the following elements, each of which has a unique stiffness that will have an effect on the results.

- Robot
- Bone
- Joint Ligaments
- Fixtures: Load Cells, Clamps, Pots, Base

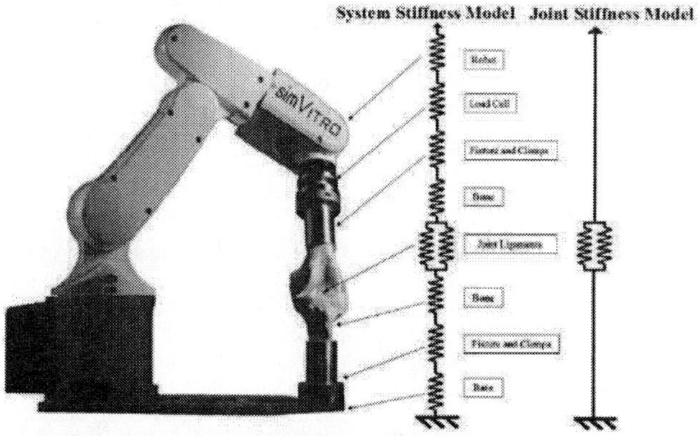

Figure 1: Typical test setup demonstrating the elements analyzed. The Joint Stiffness Model is significantly simplified when compared to the System Stiffness Model.

To simplify the problem, each element can be thought of as a spring with a given multi-dimensional stiffness (Figure 1). We determined the stiffness of several robots in our laboratory. This was done by applying various loads in the three Cartesian axes of each robot while recording the displacement using an optical displacement measurement system (Figure 2). The stiffness was calculated in each axis by dividing the force, applied by weights or ratchet straps, by the measured displacement. The robots tested were: Denso VP-6242G, Mikrolar R2000, Kuka KR16, Fanuc M1-iA, and Kuka KR300 Ultra 2500.

The femur and tibia stiffness were computed by measuring displacement (using a dial test indicator) while applying a cantilever load with an MTS machine (Figure 2). The bones were loaded in five different orientations and average flexural rigidity was calculated. Cantilever stiffness is dependent on the flexural rigidity of the given bone and on the length of the bone extending out of the fixture. The shorter the bone segment, the stiffer the bone becomes.

Data from seven specimens from the Open Knee project (https://simtk.org/projects/openknee) were used to construct the range of average knee stiffness in the anterior-posterior direction.

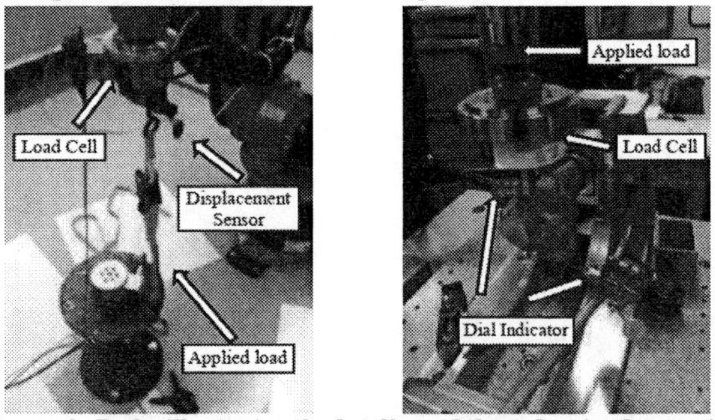

Figure 2: Left- illustrates the loading of the robot in the vertical axis. Right- shows the bone testing setup.

The stiffness of the load cell was specified by the manufacturer. In our analysis the ATI Omega 85 load cell was used, which has a stiffness of 77,000 N/mm.

To allow the analysis to remain general purpose, the fixtures were assumed to be infinitely rigid because of the range of variations from system to system. Most test systems have their own unique fixtures they use for testing and the influence of fixture stiffness should be calculated on a system by system basis.

A custom LabVIEW script was created to calculate the force uncertainty based on the variations in all the stiffness parameters. This script allowed us to compare the effects of cantilever bone length and robot stiffness (Fig 3).

RESULTS

Table 1 illustrates the flexural rigidity and the stiffness of the bone. The experimental values for flexural rigidity were very similar to the numbers found in literature [3]. Table 2 shows the wide range of stiffness that was seen in the robots that were tested. A 3D plot of ligament force uncertainty as a function of bone length and robot stiffness was produced for a reasonable range of these two factors.

Bone	Flexural Rigidity Measured (Nm^2)	Flexural Rigidity Literature[3] (Nm^2)	Stiffness with 10 cm Bone (N/mm)
Tibia	230	217 +/- 95	313
Femur	340	317 +/- 73	275

Table 1: Flexural rigidity and bone stiffness for a 10 cm cantilever segment of bone.

Robot	Average Stiffness (N/mm)
Mikrolar R2000	8100
Mikrolar R2000 with Stage	4300
Kuka KR300 Ultra 2500	1300
Kuka KR16	200
Denso VP-6242G	500
Fanuc M1-iA	300

Table 2: The X, Y and Z axes were tested on all robots. The value displayed in the table is the average stiffness of the three axes for each robot.

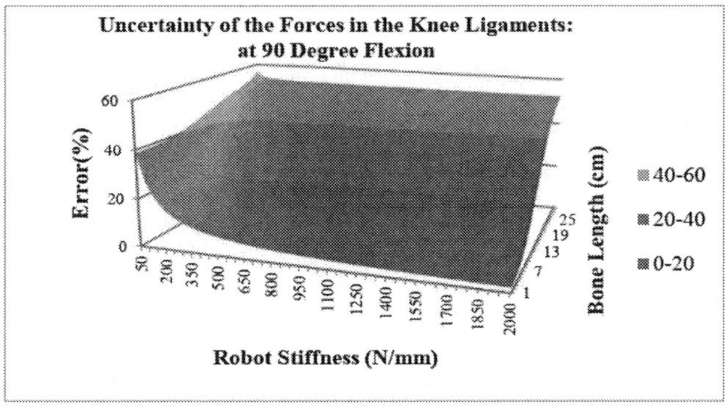

Figure 3: Shows the relationships between force uncertainty, cantilever bone length, and stiffness of robot.

CONCLUTIONS

According to this data, a superposition test performed with a Kuka KR16 robot, and a cadaveric knee with eighteen centimeter cantilever bone length, will result in about 30% uncertainty in the actual load carried by the anatomical structure. In a worst-case scenario, a robot with stiffness less than 100 N/mm, and a bone over twenty five centimeters long, could produce a force uncertainty of over 60%.

These results suggest the superposition testing methodology should be changed to minimize some of the uncertainties. Robots could be programmed to account for the displacement error if the stiffness of the system components were known. It would also be possible to integrate a motion capture system with the robot control system to track actual ligament insertion site motion and ensure the robot moves wherever it is needed in order to create the proper joint displacement. Future work should also include stiffness measurements of more robots, and flexural rigidity measurements of more bones so that this analysis could be applied to joints other than the knee.

ACKNOWLEDGEMENTS

Mike Chaffee of Kuka Robotics for taking stiffness measurements on the KR300 Ultra 2500.

Open Knee(s) Grant: R01EB009643, NIBIB, NIH.

REFERENCES

[1] Livesay G.A., et al., *Ann Biomed Eng* 23(4):467-474, 1995.
[2] Wu, J.L., et al., *Am J Sport Med* 38(3):558-563, 2010.
[3] Gardner M.P., et al., *Ann Biomed Eng* 38(3):613-620, 2010

IN VIVO STRAINS OF BRAIN DEFORMATION WITH MILD ANGULAR AND POSTERIOR HEAD ACCELERATION USING TAGGED MRI

Yuan-Chiao Lu (1), Deva C. Chan (2), Andrew K. Knutsen (1), Sarah H. Yang (1), Philip V. Bayly (3), Wen-Tung Wang (1), John A. Butman (4), Dzung L. Pham (1)

(1) Center for Neuroscience and Regenerative Medicine
Henry M. Jackson Foundation
Bethesda, MD, United States

(2) Department of Biomedical Engineering
Rensselaer Polytechnic Institute
Troy, NY, United States

(3) Mechanical Engineering and Materials Science
Washington University at St. Louis
St. Louis, MO, United States

(4) Radiology and Imaging Sciences
National Institutes of Health Clinical Center
Bethesda, MD, United States

INTRODUCTION

Rapid deformation of brain matter caused by skull acceleration during impact events is the most probable cause of concussion and traumatic brain injury (TBI) [1]. Tagged magnetic resonance imaging (MRI), a noninvasive approach for quantifying motion in vivo, has been previously applied to evaluate regional brain deformation in live human subjects [2,3]; however, the strains generated by different brain motions have yet to be compared. The current study investigated the strains generated in the human brain using an experimental protocol that, for the first time, enables quantification of whole brain deformation during both head rotation and head extension motions.

METHODS

Two custom devices were designed and built to support and constrain the head motion (rotation and extension) of a human subject within the MRI scanner (Fig. 1a & Fig. 2a). For the head rotation device, a rotation of approximately 32 degrees about the inferior-superior axis toward the left shoulder was performed (Fig. 1b), while for the head extension device, the head was rotated in the frontal-occipital direction, constituting a neck extension of about 4 degrees (Fig. 2b). In this study, a male adult (45 yrs, 175.3 cm, and 94.7 kg) with no TBI history performed consistent mild accelerations of the head inside the imaging coil of a clinical MR scanner (3T Siemens mMR Biograph) using each of the two devices. The subject voluntarily released a latch to initiate passive head motion (either rotation and extension) followed by rapid deceleration caused by a padded stop in both devices. The MRI acquisition was similar to those in [3,4], employing a SPAMM tagging sequence optimized for acquisition speed. The angular position for each drop was measured in real-time by an optical sensor [3]. Images with two orthogonal tag line directions for measuring plane displacements were acquired in the axial plane for rotation and in the sagittal plane for extension (tag spacing=8 mm) (Fig. 1c & Fig. 2c). Phase oversampling (14%) was applied to sagittal acquisitions to eliminate wrap artifacts. Using 4 repetitions of the head motion, a time series of a tagged MR slice with 18 ms temporal resolution was acquired. Motion was tracked between image frames using harmonic phase (HARP) analysis with the shortest-path HARP refinement algorithm [5]. Lagrangian strain tensors were computed from the displacement vectors, and the area fractions of the strain field in which the maximum principal stain (E_1) and maximum shear strain (G_{max}) exceed a given threshold λ (λ=2%) were determined for the whole brain and by regions (cortical gray matter (CGM), deep gray matter (DGM), and white matter (WM)).

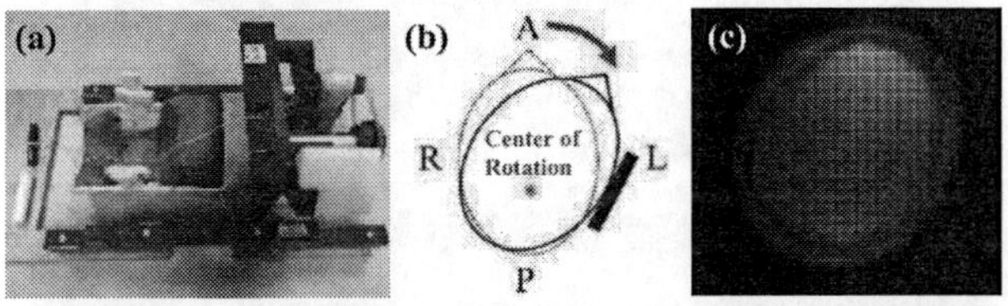

Figure 1: a) The head rotation device; b) Schematic of head at rest (dotted line) and stop (solid line) positions; c) A tagged image.

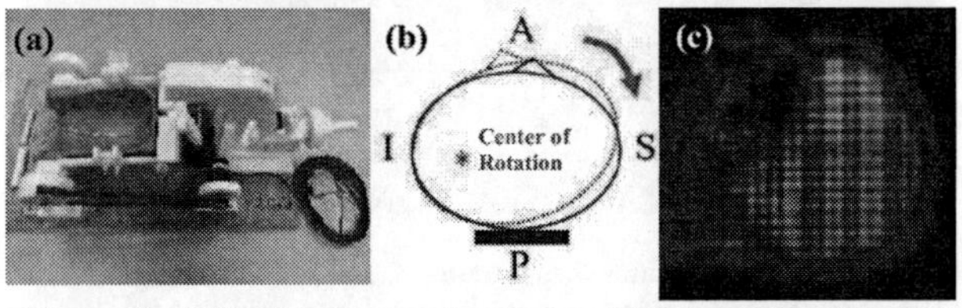

Figure 2: a) The head extension device; b) Schematic of head at rest (dotted line) and stop (solid line) positions; c) A tagged image.

RESULTS

The average peak angular acceleration (rad/s^2) was 186.2±24.3 for head rotation tests (11 axial plane slices with 88 rotations) and 239.5±22.8 for head extension tests (13 sagittal plane slices with 104 rotations). The estimated peak linear acceleration (m/s^2) of the brain was 17.4±2.2 for head rotation tests and 48.4±1.9 for head extension tests. In-plane strains (E_1 and G_{max}) were computed within all brain tissues for the head rotation device (Fig. 3) and for the head extension device (Fig. 4). Strain area fractions at 2% threshold for E_1 and G_{max} of the time series were calculated for the head rotation device (Fig. 5) and for the head extension device (Fig. 6). The maximum area fractions were determined at the time of peak angular acceleration, which coincides with when the peak E_1 and G_{max} of the time series were observed. For head rotation, the maximum area fractions of the brain were 0.25 for E_1 and 0.20 for G_{max}, and the maximum area fraction in CGM was higher than the fractions in WM and DGM (0.31, 0.25, and 0.13 for E_1, and 0.26, 0.19, and 0.05 for G_{max}, correspondingly). For head extension, the maximum area fractions of the brain were 0.03 for E_1 and 0.01 for G_{max}, and the maximum area fraction in CGM was higher than the fractions in WM and DGM (0.05, 0.02, and <0.01 for E_1, and 0.02, 0.01, and <0.01 for G_{max}, correspondingly).

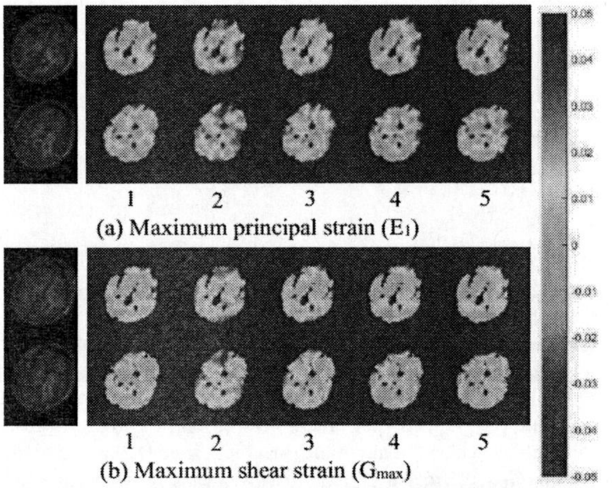

Figure 3: Strain maps (first 5 time frames) of 2 (out of 11) slices of the brain from the time of impact for the head rotation device.

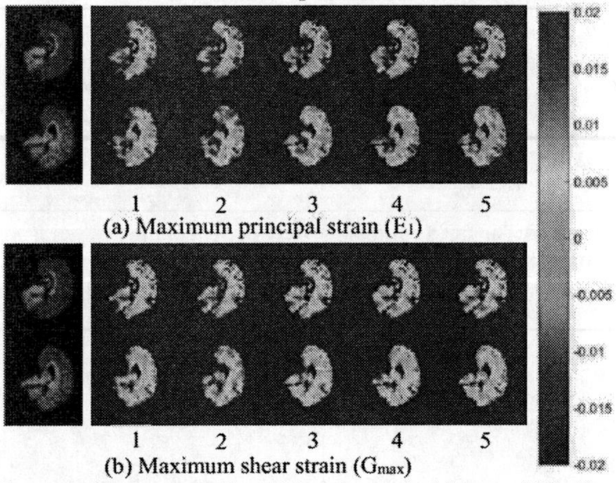

Figure 4: Strain maps (first 5 time frames) of 2 (out of 13) slices of the brain from the time of impact for the head extension device.

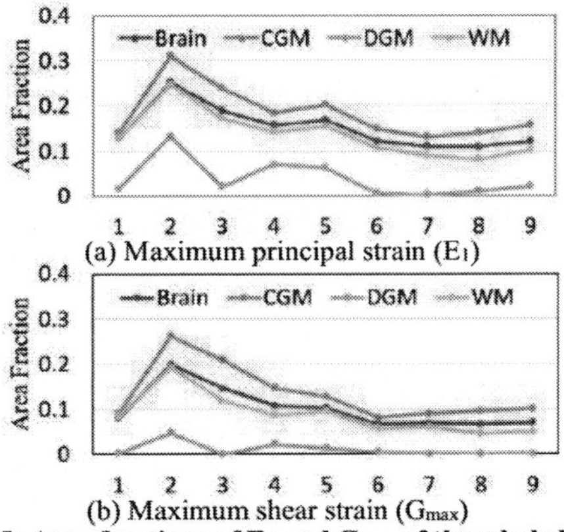

Figure 5: Area fractions of E_1 and G_{max} of the whole brain and by regions (CGM, DGM, and WM) were calculated at a 2% threshold for the first 9 time frames for the head rotation device

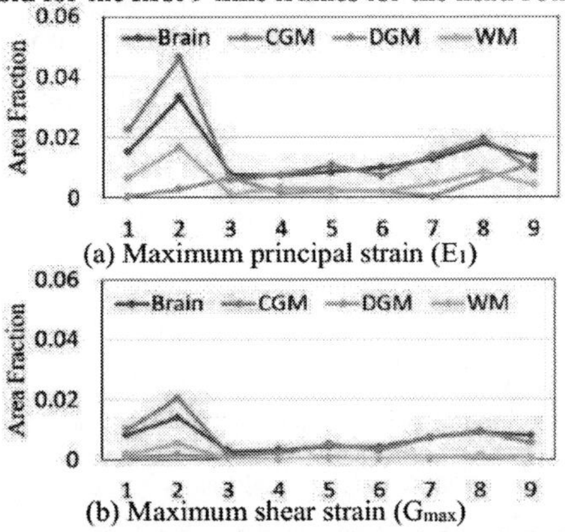

Figure 6: Area fractions of E_1 and G_{max} of the whole brain and by regions (CGM, DGM, and WM) were calculated at a 2% threshold for the first 9 time frames for the head extension device.

DISCUSSION

Strains overall were substantially higher across all regions during the rotation than the extension motion despite a similar level of acceleration. The area fractions of maximum principal strain were higher than the maximum shear strain for both devices during the time of impact. Continuing to better understand how the geometry of the brain and its surrounding anatomy affects the biomechanical response to different accelerations will be invaluable for designing protective equipment and the development of computational injury models [6].

ACKNOWLEDGEMENTS

This work was partially supported by grant NIH NINDS R01 NS055951, the Department of Defense in the Center for Neuroscience and Regenerative Medicine, and NIH Intramural Research Program.

REFERENCES

[1] King, A et al., IRCOBI Conference, Paper #1, 2003
[2] Bayly, P et al., *J Neurotrauma*, 22:845-856, 2005.
[3] Knutsen, A et al., *J Biomech*, 47:3475-3481, 2014
[4] Lu, Y-C et al., *CMBBE Symposium*, Paper #150, 2015.
[5] Liu, X et al., *IEEE Trans Med Imag*, 29:1560-1572, 2010.
[6] Untaroiu, C et al., *Comput Math Methods Med*, Paper #460413, 2013

VALIDATION OF FE HYBRID III, THOR, AND GHBMC M50-OS FOR SPACEFLIGHT CONFIGURATION TESTING

Derek A. Jones (1), James P. Gaewsky (1), Mona Saffarzadeh (1), F. Scott Gayzik (1), Ashley A. Weaver (1), Jacob B. Putnam (2), Jeffrey T. Somers (2), Jessica Wells (3), N. Newby (2), Joel D. Stitzel (1)

(1) Wake Forest School of Medicine
Winston-Salem, NC, USA

(2) KBRwyle
Houston, TX, USA

(3) Leidos,
Houston, TX, USA

INTRODUCTION

Injury assessment reference values (IARVs) used to identify injury risk to occupants are based on post-mortem human subjects positioned in application specific configurations. In automotive research, the occupant is placed in a "super-slouched" position to mimic driving position and is often subjected to either frontal (–X), or side (Y) impact directions. Additionally, the occupant is restrained in a 3-point harness system with airbags. In military research, the occupant is primarily loaded from the bottom (+Z) in a high severity impact while seated upright to simulate under-body blast or ejection seat scenarios. These testing modalities do not directly lend themselves to spaceflight loading directions, postures, or restraint systems. Therefore, the extensibility of IARVs developed for these other scenarios may not translate to spaceflight conditions.

In order to address this concern and to evaluate seat design specifically for spaceflight, this study uses Finite Element (FE) models of Anthropomorphic Test Devices (ATDs) as well as FE Human Body Models (HBMs). The overarching goal is to validate FE models of ATDs and HBMs against spaceflight like conditions, perform sensitivity analysis, and to use these models for analysis of seat design for spaceflight. The first phase of this work and the subject of this paper is validation of FE models against physical test data.

METHODS

A combination of Hybrid III, THOR, and human volunteer tests were selected to form a physical test matrix for simulation comparison. In each test, the subjects were restrained using a 5-point belt system in a flat pan seat with a vertical back [1].

These tests were conducted between 1976 and 2013 and comprise –X, +X, Y, and Z acceleration directions. The acceleration levels range from 3-20 G and the rise times for the acceleration pulses range from 20-110 ms. In full, 49 physical test configurations were selected with a total of 275 tests from multiple studies (Table 1) [1-3].

Table 1: Physical Test Matrix

Model	Direction	# Configurations
Hybrid III	-X, frontal	4
	+X, rear	4
	+Y, lateral	2
	+Z, vertical	5
THOR	-X, frontal	3
	+X, rear	-
	+Y, lateral	4
	+Z, vertical	4
Human/GHBMC	-X, frontal	11
	+X, rear	4
	+Y, lateral	5
	+Z, vertical	3

Simulations were performed using seat and restraint geometries measured from the physical tests. In the case of Y directional impacts, a side guard was added the model to match the physical testing conditions. The Humanetics 50th percentile male Hybrid III, NHTSA THOR 50th Male, and the Global Human Body Models Consortium (GHBMC) 50th male simplified occupant (M50-OS) models were used for FE simulations in LS_DYNA (MPP, Version 971, R6.1.1., LSTC, Livermore, CA) on a computer cluster [4-6] (Figure 1).

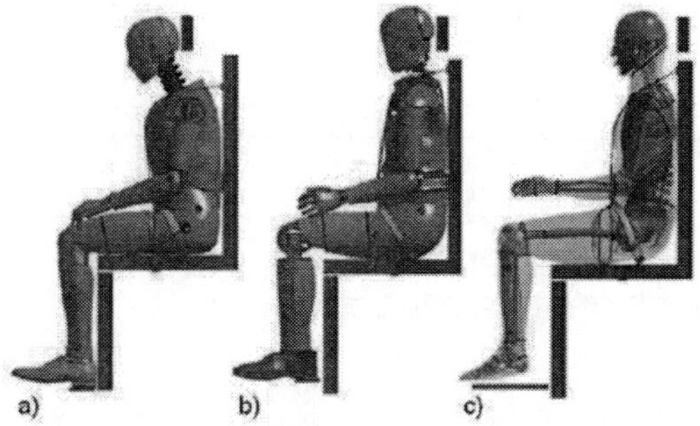

Figure 1: Models used for simulation. A) Hybrid III M50, b) THOR M50, c) GHBMC M50-OS

The models were positioned to match physical testing by applying 150ms of gravity settling, pretensioning belts, and then prescribing acceleration pulses to the seat model. Following simulation, visual comparisons of kinematics were conducted. In addition to qualitative analysis, simulation signals were compared to matched physical signals using the Gehre et al. method (CORrelation and Analysis, or CORA, size phase, and shape) [7]. Regions compared include the head, neck, thorax, and pelvis.

RESULTS

Three models were used in a total of 49 test configurations. In visual inspection of all cases, excursion magnitude and direction of the head and thorax matched the physical test cases. The THOR and Hybrid III FE models were closer to the physical testing than the GHBMC-human comparisons. Figure 2 shows a comparison of the maximum excursion in a matched THOR test scenario. In both the simulation and physical tests, the feet remain in place relative to the chair legs, the hands remain in place relative to the knees, the back comes off of the seat back completely, and the neck flexes forward to a high degree.

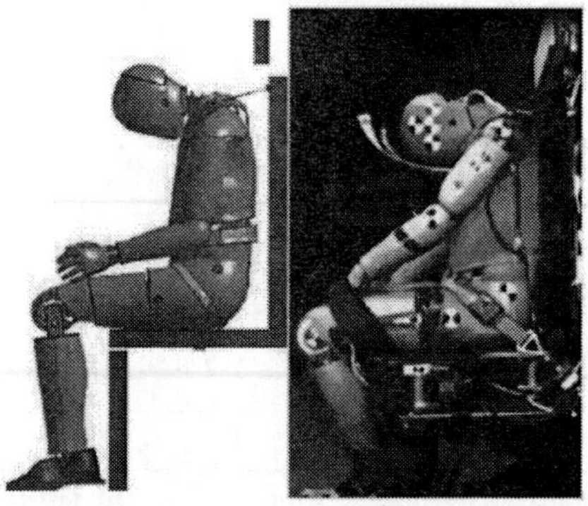

Figure 2: Peak excursion for matched THOR simulation and physical testing. Direction = -X, Peak Acc = 10G, Rise Time = 70ms.

An example comparison between a physical and FE response can be seen in Figure 3. In this example, the resultant head acceleration was compared between the physical and FE THOR model before tuning belt parameters. This served as a baseline test. In this test, the subject was accelerated in the +Z direction to a peak acceleration of 10G with a rise time of 40ms. The average response curve for the experiments is shown in black and reaches a peak of 25 Gs, while the simulation hits 18 Gs. The outer corridors of the physical test data are shown in gray.

Computing the CORA score for each signal results in a set of values that include the shape, size, and phase score as well as the corridor score. For this analysis, the cross correlation score was weighted more heavily. In this example, the size, shape, and phase scores were 0.717, 0.871, and 0.297 respectively on a scale from 0 to 1, with 1 being a perfect score.

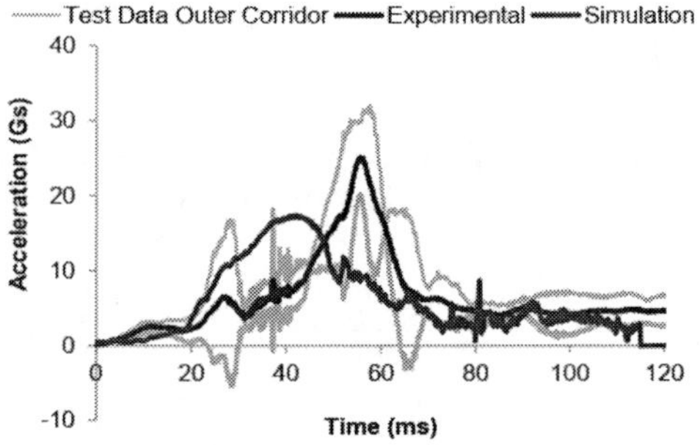

Figure 3: Head resultant acceleration for matched THOR simulation and physical testing. Direction = +Z, Peak Acc = 10G, Rise Time = 40ms.

DISCUSSION

Finite element models of both ATDs and human body models allow for cost and time efficient parametric analysis of seat and restraint design when compared to physical testing. However, it is important to first validate models against physical tests in order to boost confidence in future results. By conducting validation testing on two highly used ATDs and using a well validated HBM in conjunction with human data, we have high confidence in FE model performance in the validated regime. Overall the physical and simulation results were comparable.

Future perturbation analysis will inform physical model selection in physical testing. Comparing ATD results to matched HBM results will allow for comparisons of injury risk and may illuminate potential missed injury risk when using ATDs. The results of this study are highly applicable to both governmental and commercial spaceflight and provide confidence in FE simulation for future design.

ACKNOWLEDGEMENTS

This study was funded by NASA Human Health and Performance Contract (HHPC) through KBRWyle. Views expressed are those of the authors and do not represent the views of NASA or KBRwyle.

REFERENCES

[1] C. Perry, C. Burneka, and C. Albery, "Biodynamic Assessment of the THOR-K Manikin," DTIC Document2013.

[2] B. F. Hearon and J. W. Brinkley, *Aviation, space, and environmental medicine,* vol. 57, pp. 301-12, Apr 1986.

[3] J. R. Buhrman and C. E. Perry, *Aviation, space, and environmental medicine,* vol. 65, pp. 1086-90, Dec 1994.

[4] "H350 Adult Dummy Model LS-Dyna," Humanetics Innovative Solutions, Inc.2012.

[5] J. B. Panzer, S. Giudice, and J. B. Putnam, NHTSA/USDOT2015.

[6] D. Schwartz, B. Guleyupoglu, B. Koya, J. D. Stitzel, and F. S. Gayzik, *Traffic injury prevention,* vol. 16 Suppl 1, pp. S49-56, 2015.

[7] C. Gehre, H. Gades, and P. Wernicke, in *21st International Technical Conference on the Enhanced Safety of Vehicles Conference (ESV), Stuttgart, Germany, June,* 2009, pp. 15-18.

COLLAGEN ALIGNMENT DIFFERENTIALLY INCREASES NEURONAL REGULATION AFTER STRETCH OF NEURON-COLLAGEN CONSTRUCTS

Sagar Singh (1,2), Sijia Zhang (1), Beth A. Winkelstein (1,2)

(1) Department of Bioengineering
University of Pennsylvania
Philadelphia, PA, USA

(2) Department of Neurosurgery
University of Pennsylvania
Philadelphia, PA, USA

INTRODUCTION

Despite having high clinical relevance [1,2], the underlying mechanisms of mechanically-induced pain are not known. The cervical facet capsular ligament, which is a common source of pain from neck trauma, has nerve fibers innervating it [2-5]. Tensile stretch of that ligament exceeding its physiologic range produces collagen disorganization, dysregulation in the neurons embedded in it, and pain [4-7]. Different regions of the facet capsule vary in their collagen organization; collagen fibers in some regions are parallel and have irregular orientations in others [3,8]. Afferents in those regions may experience varied loading and be injured to different degrees during macroscale tissue deformations. Anisotropy of the collagen matrix modulates macroscopic tissue stresses and strains, as well as neuronal responses due to local microscale mechanics [9-11]. Although stretch-induced fiber reorganization occurs at the same strain threshold as that which activates neuron in vitro [11], it is not known if, and how, tissue matrix structure affects the sensory function in the facet capsule.

To better understand the potential interplay between neurons and their matrix environment, an in vitro neuron-collagen construct (NCC) model [11,12] was modified to generate gels with fibers in random or aligned orientation. NCCs underwent uniaxial tension and the response of rat dorsal root ganglia (DRG) embedded in the collagen gels with different fiber orientations were compared to understand effects of matrix orientation on neuronal responses to macroscale stretch. After stretch, neuronal expression of the nociceptive neuropeptide substance P (SP) and phosphorylated ERK (pERK) were evaluated to assess the relative effects on pain signaling and neuronal activation [5,12].

METHODS

NCCs were prepared using a solution of rat tail collagen I (2mg/mL) cast in 12-well plates [11,12]. Randomly oriented gels (n=6) were produced by incubation at 37°C in 5% CO_2 for at least 30 minutes. Aligned gels (n=6) were produced by placing them in a 4.7T small animal horizontal bore MR system (Agilent) at 37°C for 45 minutes [13]. Embryonic Day 18 Sprague Dawley rat DRGs were plated in the center of the gels. Additional collagen was added 3 days later to encapsulate the DRGs. NCCs were cultured for another 4 days.

NCCs were sterilely transferred to 1% penicillin/streptomycin in PBS and cut into vertical strips (21x8mm^2) with markers placed on the surface. Gels were immersed in a 37°C PBS bath, and clamped in a planar testing machine (TestResources). Uniaxial stretch was applied to 4mm (~30% strain) at 3.7mm/s; aligned gels were loaded along the direction of neurite outgrowth. During loading, synchronized force-displacement data and high-speed images were recorded. Unstretched NCCs were included as controls (n=5/alignment group). After testing, gels were cultured for 24 hours and fixed in 4% paraformaldehyde. Marker displacements were measured by ProAnalyst software and input into LS-DYNA software to estimate maximum principal strain (MPS) (Figure 1A). Force-displacement data for each gel were filtered with a 10-point moving average filter and the peak force was extracted (Figure 1B). Average MPS and peak forces were compared between random and aligned gels with separate t-tests.

Gels were labeled and imaged for β-III-tubulin (1:200, Abcam), SP (1:500, Neuromics), and pERK (1:500, Cell Signaling) as before [11]. Images were processed in ImageJ for background subtraction and 3 regions of interest (ROIs) with neurites were isolated. Each ROI was thresholded to quantify SP and pERK intensity (Figure 1C). Intensities were normalized to their respective controls and compared between groups by separate two-way ANOVAs. Images of control NCCs were analyzed with a Fourier Transform method to quantify alignment strength along the major and minor orientations [14]. Differences in axon alignment between NCCs with random and aligned orientation were tested by comparing the minor-to-major-axes ratio using a t-test.

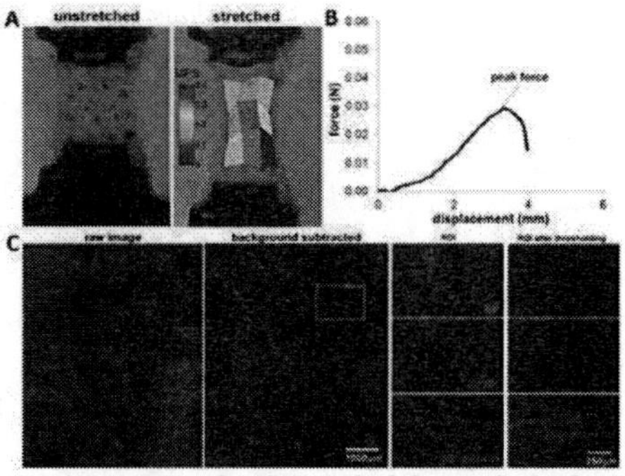

Figure 1: (A) A representative unstretched & stretched gel, with strain map overlay on stretched sample. (B) Corresponding force-displacement data & peak force. (C) Raw image at 10X; ROIs were selected to include neurites. Channels were thresholded to measure pixel intensity.

RESULTS

Significantly more ($p<0.001$) axons were oriented in the same direction in the aligned NCCs (ratio=0.33±0.21) than in random NCCs (ratio=0.80±0.11). Aligned gels experienced greater ($p=0.031$) peak force (0.020±0.007N) than random gels (0.034±0.015N). Strain maps showed similar distributions in both groups and the average MPS was not different ($p=0.462$) between random (0.31±0.13) and aligned (0.30±0.06) gels. Appreciable immunolabeling was present in all stretched gels, regardless of fiber alignment (Figure 2). SP labeling was significantly higher after stretch in random ($p=0.026$) and aligned ($p=0.002$) gels (Figure 3). Labeling of pERK was significantly higher ($p=0.005$) after stretch in aligned gels, which was also greater ($p<0.001$) than levels in random gels after stretch (Figure 3).

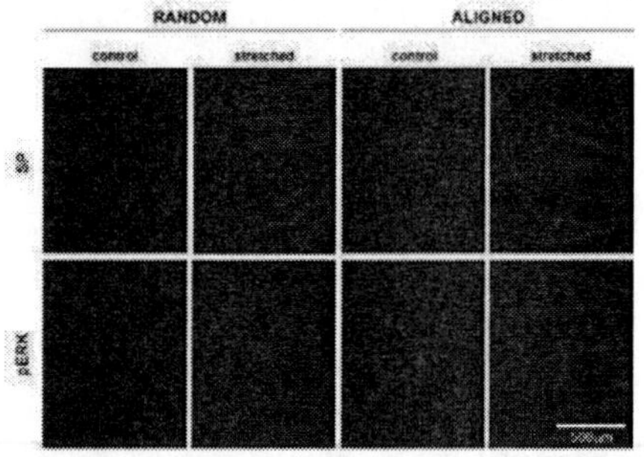

Figure 2: Representative images of neuronal SP & pERK for random & aligned control & stretched gels, showing greater labeling after stretch.

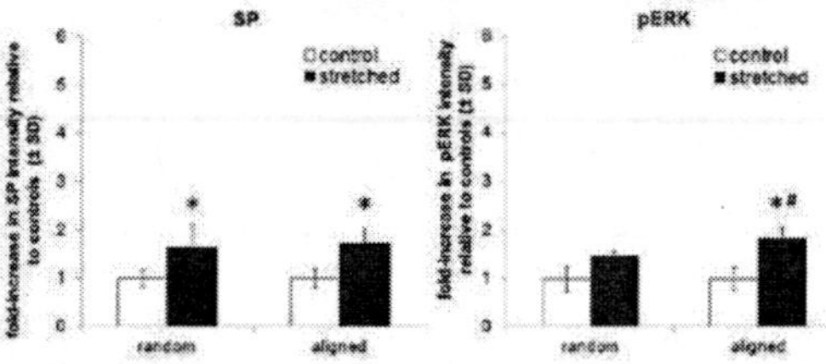

Figure 3: Normalized SP & pERK. SP increases (*p≤0.026) after stretch for both NCCs. pERK only increases (*p=0.005) after stretch in aligned gels, which is greater than pERK in stretched random gels (#p<0.001).

DISCUSSION

This is the first study to show that neuronal nociceptive responses depend on the local tissue structure around them. Although the applied tension did induce similar tissue-level MPS (0.304±0.093) in both NCCs, the corresponding force was greater in the aligned NCCs than those with isotropic fiber orientation. Those differences were also evident in expression of both SP and pERK after stretch (Figure 3). Changes in the macroscopic mechanics likely alter the local fiber and network mechanics, differentially modulating the neuronal responses (Figure 3). Neuronal regulation of SP follows tissue strain, despite gels sustaining significantly different forces. The strain-dependent SP modulation is consistent with its expression being differentially altered in DRG neurons in vivo in response to different magnitudes of facet injury and its levels directly mapping to strain in random gels in this same in vitro NCC system [5,12]. While these findings suggest SP expression may be regulated by deformation and be less sensitive to stress inputs, more work is needed to define the local biomechanics near the neurons since this work made only macroscale measurements.

In contrast to SP, pERK may be regulated by either or both, neuronal stress and strain (Figure 3). ERK phosphorylation can be induced by cellular deformation [15,16]. Stretch simulating painful facet loading in vivo in this same NCC system with random fiber orientation produces greater pERK than physiologic stretch that is "non-painful" [10]. Since pERK mediates neuroplasticity and neuronal excitability [17,18], it is possible that the increased pERK in aligned NCCs (Figure 3) may be due to either greater tissue-level force or local loading to neurons owing to the greater collagen and/or neuronal alignment in those gels. The axon orientations were different between the random and aligned NCCs, with the aligned gels having a nearly 1/3 lower minor-to-major axis ratio (farther from 1 is more aligned). Since axons are oriented with the loading direction in aligned NCCs, they are not only exposed to greater forces under tension, but also likely undergo greater deformations due to higher local collagen fiber strains, which may activate ERK.

Both the higher force and oriented axon growth in the aligned gels (Figure 2) indicate success in aligning collagen by a magnetic field. But, the degree of fiber alignment and its contribution to local forces, deformations and axon orientation were not defined. Although pilot studies quantifying fiber alignment by polarized light confirmed strong alignment (low circular variance of 0.002), studies are needed to rigorously relate fiber alignment to neuron responses. Differences in the micromechanics of the fibers and neurons between aligned and random NCCs would help to better define mechanotransduction. Computational models that predict local mechanics [11] would also provide insight into important regulatory inputs and could evaluate how different loading alters the matrix organization and translates to neuronal responses. Nevertheless, this study suggests that collagen matrix structure differentially modulates neuronal responses likely through varied macro- and micro-scale mechanics, which may relate to heterogeneous pain responses that are observed after injury.

ACKNOWLEDGEMENTS

Funding by the NIH (U01EB016638), Catherine Sharpe Foundation, AMTI Force and Motion, and an NSF training grant.

REFERENCES

[1] Borenstein D *Nat Rev Rheumatol* 9:643-53, 2013. [2] Kirpalani D & Mitra R *Arch Phys Med Rehabil*, 89:770-4. [3] Kallakuri S et al *Wold J Orthop* 3:10-4, 2012. [4] Dong L et al *Brain Res* 1461:51-63, 2012. [5] Lee K & Winkelstein B *J Pain* 10:436-45, 2009. [6] Quinn K & Winkelstein B *J Biomed Opt* 14:054010, 2009. [7] Chen C et al *JBJS Am* 88:1807-16, 2006. [8] Yamashita T et al *Spine* 21:538-43. [9] Lake S et al *J Mech Behav Biomed Mater* 13:25-35, 2012. [10] Khalsa P et al *J Neurophysiol* 76:175-87, 1996. [11] Zhang S et al *J Biomech Eng* 138:021013, 2016. [12] Zhang S & Winkelstein B *SB³C* 2016. [13] Tranquillo R et al *Biomaterials* 17:349-57, 1996. [14] Sander E & Barocas V *J Biomed Mater Res* A 88:322–31, 2009. [15] Samarakoon R & Higgins P *J Cell Physiol* 195:411-20, 2003. [16] Neary J et al *J Neurosci* 23:2348-56, 2003. [17] Gao Y et al *Open Pain J* 2:11-7, 2009. [18] Cheng J & Ji R *Neurochem Res* 22:1970-8, 2008.

DISRUPTION OF CAPILLARY-LIKE STRUCTURE BY IMPULSIVE PRESSURE LOADING

H. Nakadate (1), S. Nakamura (1), S. Aomura (1)

(1) Graduate School of System Design Tokyo Metropolitan University 6-6, Asahigaoka, Hino, Tokyo, 191-0065 Japan

INTRODUCTION

Head impact elicits the generation of a pressure gradient within the brain tissue, resulting in brain injury [1]. Animal studies of head impact have shown that the transient increase in intracranial pressure causes brain haemorrhage and cerebral concussion [2]. Typically, the pressure generated at the impact site (coup pressure) was positive and that generated on the side opposite to the impact site (contrecoup pressure) was negative in experiments using human cadavers [3]. However, the direct influence of impulsive pressure during head impact on the microvasculature has not been well investigated to date. Previously, we developed a pressure loading device to expose cultured endothelial cells to impulsive pressure and thus mimic changes in intracranial pressure after head impact [4].

In the present study, we employed this same pressure loading device to expose the capillary-like structure formed by endothelial cells to impulsive pressure and examined temporal changes in the capillary network.

METHODS

Endothelial cell culture

Human umbilical vein endothelial cells (HUVEC) purchased from Lonza were cultured in Endothelial Basal Medium-2 (EBM-2) supplemented with an EGM-2 SingleQuots containing recombinant human epidermal growth factor (rhEGF), heparin, hydrocortisone, recombinant human fibroblast growth factor-basic (rhFGF-B), ascorbic acid, recombinant human vascular endothelial growth factor (VEGF), recombinant long r insulin-like growth factor-1 (R^3-IGF-1), gentamicin sulphate amphotericin-B (GA-1000), and foetal bovine serum (FBS) under conditions of 5% CO_2 and 100% humidity at 37°C. HUVEC from passages 3–5 were seeded at 50×10^4 cells on BD Matrigel matrix-coated 35-mm culture dishes and were grown to tube formations within 24 h.

Pressure loading experiment

The pressure loading device consisted of a cylinder, piston, pendulum impactor, pressure chamber, and pressure transducer. A pressure chamber equipped with a pressure transducer is connected to the cylinder. The compartment connected between the cylinder and the pressure chamber is filled with water. The lower part of the pressure chamber, which is divided by a silicone membrane, is filled with culture medium. Pressure is generated by striking the piston with the pendulum. A full description of the device configuration and loading mechanism has been previously published [4]. The dishes were placed into the pressure chamber and subjected to impulsive pressure with an amplitude of 200–400 kPa for a duration of approximately 20 ms.

Morphological observation

Capillary-like networks were stained with 2 μM calcein-AM at 24 h post loading and observed by using an inverted fluorescence microscope at 24 h after loading. The capillary length and the number of branching were manually quantified using 5 randomly selected regions per experiment.

Statistical analysis

The density of capillary-like network are expressed as the mean ± standard deviation (SD) of 5 independent experiments. The means were compared by Steel's multiple comparison test. A p value of less than 0.05 was considered significant.

RESULTS & DISSCUSION

The changes in the capillary-like network formed by endothelial cells were observed using an inverted microscope before loading and at 1, 3, 6, and 24 h after loading of the impulsive pressure with amplitude of 400 kPa. The capillary-like structure was broken down after 1 h following impulsive pressure (arrows in figure 1), and the capillary network gradually disappeared until 24 h after loading shown in figure 2. The disappearance of the capillary network was not observed at a pressure amplitude of 100 kPa or less (data not shown).

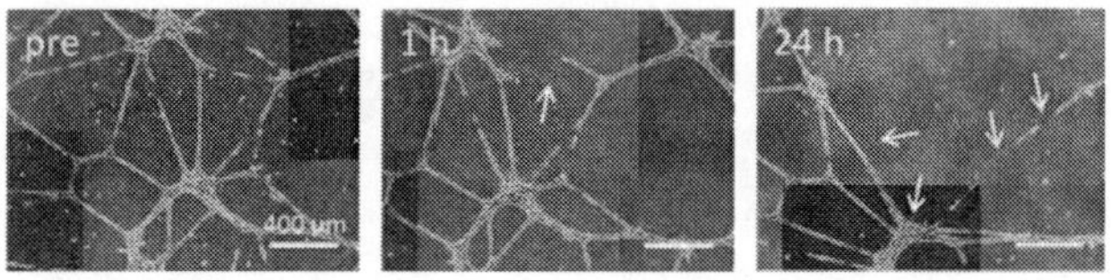

Figure 1: Phase-contrast images in capillary-like network formed by endothelial cells after loading of impulsive pressure.

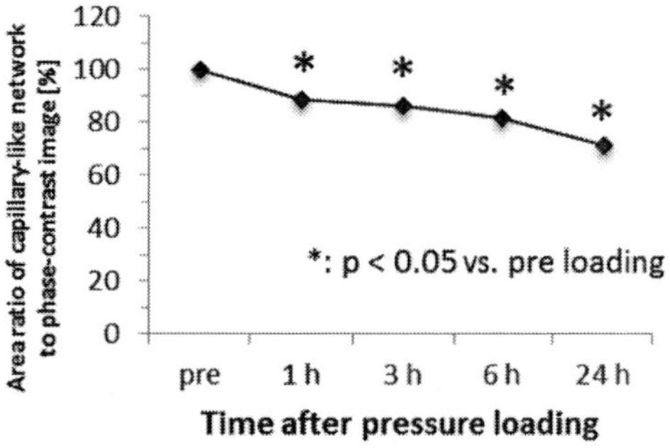

Figure 2: Changes in capillary-like network formed by endothelial cells after loading of impulsive pressure.

The stained networks were observed using an inverted fluorescence microscope at 24 h after pressure loading of 200, 300 or 400 kPa shown in figure 3. There were no significant differences in capillary density of the networks exposed to 200 kPa, but capillary density of the networks exposed to 300 and 400 kPa decreased significantly compared to no-loading (0 kPa) shown in figure 4. In addition, capillary branching of the networks exposed to 400 kPa significantly decreased compared to no-loading, but there were no changes in capillary branching of the networks exposed to 200 and 300 kPa shown in figure 5.

This study demonstrated that impulsive pressure correlated with the disruption of capillary-like structures over time after impact. Recently, some studies have suggested that delayed and progressive haemorrhage during the first several hours after head impact are attributed to molecular events initiated at the time of impact, which lead to later structural failure of microvessels [5]. Simard et al. showed that necrotic death of endothelial cells results in physical disruption of capillaries, leading to the extravasation of blood and formation of petechial haemorrhage in a contusive animal experiment [6]. However, the mechanical factor that influences the molecular abnormality has not been identified. A further detailed investigation of the molecular mechanism of capillary fragmentation using an *in vitro* model is required. In summary, the threshold of the disruption of capillary-like structure formed by cultured endothelial cells was 300 kPa as a surrogate of focal leakage of blood plasma. Moreover, the threshold of 400 kPa would mean pathogenesis of edema and microbleeding.

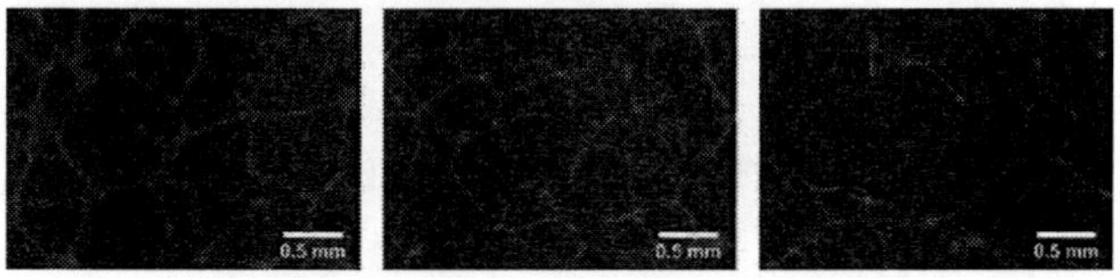

Figure 3: Fluorescent images of capillary-like network in control (left) and exposed to 300 kPa (middle) and 400 kPa (right).

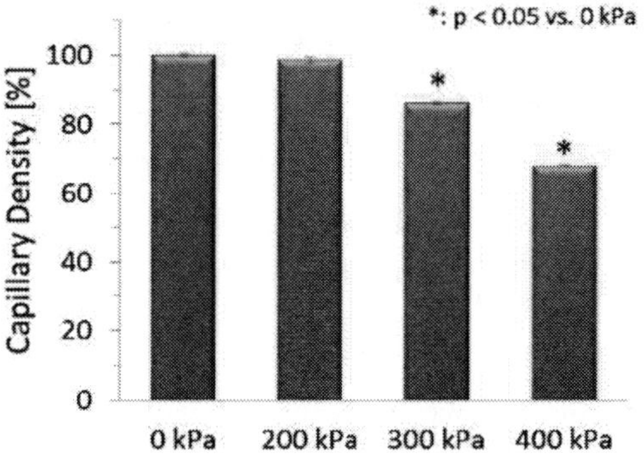

Figure 4: The area ratio of the capillary-like network to fluorescence images.

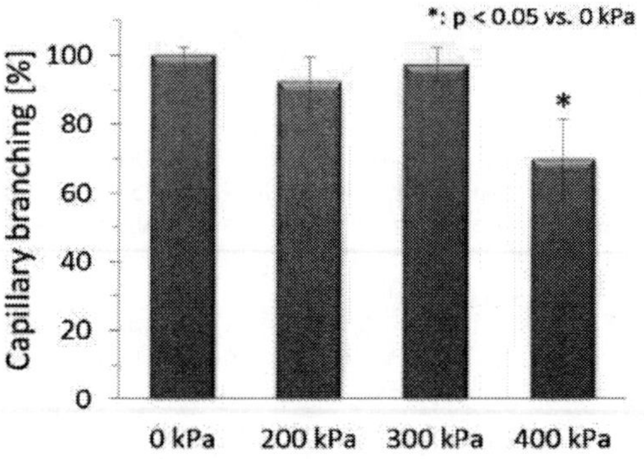

Figure 5: The number of the branching of the capillary-like network.

REFERENCES

[1] Hardy, WN et al., *Int J Impact Eng*, 15:561–586, 1994.
[2] Lindgren, S et al., *Biophysik*, 3:174–180, 1966.
[3] Nahum, A et al, *Proc 21st Stapp Car Crash Conference* :337–366, 1977.
[4] Nakadate, H et al., *Biomed Eng Online*, 13:44, 2014.
[5] Kurland, D et al., *J Neurotrauma*, 29:19–31, 2012.
[6] Simard, JM et al., *J Neurotrauma*, 26:2257–267, 2009.

SB³C2017
Summer Biomechanics, Bioengineering and Biotransport Conference
June 21 – 24, Tucson, AZ, USA

COMPUTATIONAL SIMULATIONS OF LATERAL ANKLE SPRAINS IN TENNIS

Paul D. Heeder (1,2), Feng Wei (1,2,3), Roger C. Haut (1,2,3)

(1) Orthopaedic Biomechanics Laboratories
Michigan State University
East Lansing, MI, USA

(2) Department of Mechanical Engineering
Michigan State University
East Lansing, MI, USA

(3) Department of Radiology
Michigan State University
East Lansing, MI, USA

INTRODUCTION

Ankle sprains are a common injury in sports, and lateral ankle sprains, typically involving the anterior talofibular ligament (ATaFL) and/or calcaneofibular ligament (CaFL), continue to be the leading type of injury sustained by athletes [1]. While foot inversion is involved in a lateral ankle sprain, the complete mechanism of injury remains unclear. Studies have been conducted that describe the kinematics involved in ankle sprain events using motion or video analyses [2,3]. These studies, however, do not provide data on actual ankle ligament strains. Fong et al. [4] document kinematics from five ankle sprain cases in tennis players during competition. While these cases were all diagnosed as lateral ankle sprains, the involved ankle motions were shown to have large variation between cases.

Recently our laboratory has developed a computational ankle model [5] and utilized motion-analysis-based kinematic data from laboratory tests to drive the model for estimations of dynamic ankle ligament strains and joint moments [6]. In addition, the model has been used to study the motions producing a supination ankle sprain [2] that occurred accidentally in a laboratory setting and was diagnosed as a grade I injury to the ATaFL [7].

The purpose of the current study was to use the existing ankle model to analyze the ankle kinematic data documented by Fong et al. [4] from the above-mentioned five ankle sprain cases. It was hypothesized that (1) since the documented ankle kinematic data have such a large variation, the levels of ankle ligament strain obtained from model simulations would also be highly variable, but (2) with inputs of the injury-producing kinematics into the ankle model, the ATaFL and CaFL would have excessively high levels of strain that would explain the clinically diagnosed lateral ankle injuries.

METHODS

A 3D multi-body dynamic foot model [5] was utilized for simulations of the injury cases. The model included 14 bones and 20 ligaments. Details of the model and its validation have been previously described [5]. In brief, the model was constructed from a generic cadaver ankle which was scanned using computed tomography (CT) in a neutral position. CT images were converted into 3D models in MIMICS (Materialise, Ann Arbor, MI) and inserted into dynamic rigid-body motion simulation software (SolidWorks, TriMech Solutions, LLC, Columbia, MD). All ligaments were formulated as linear elastic, tension only springs with their stiffness (N/mm) adapted from the literature [5,7]. The ground was simulated as a rigid platform. The 3D contacts were implemented between adjacent bones as well as between the bones and the ground plate in order to prevent overlaps during simulation. Friction was neglected to simulate cartilage effects. The tibia was only allowed to move vertically. The fibula, talus, and calcaneus, however, were allowed to move with six degrees of freedom, leaving bone motion to be a function of ligament behavior, surface contact, and external perturbations. For simplification purposes, the remaining bones of the foot (tarsal and metatarsal bones) were fused together and moved as a unit, with its motion primarily dependent on motions of the talus and calcaneus.

Approximately two times body weight (1500 N) was applied to the proximal end of the model to simulate dynamic weight bearing. The 3D, temporal kinematic data from the five injury cases reported by Fong et al. [4] were used as input for the simulations. This was accomplished with motor elements, two on the talus for internal rotation and plantarflexion, respectively, and one on the calcaneus for inversion. The three axes of rotation were set as: internal rotation (along the tibial axis, α in Figure 1), plantarflexion (fixed to the talus and initially oriented mediolaterally, β in Figure 1), and inversion (fixed to the calcaneus and initially oriented anteroposteriorly, θ in Figure 1). The SolidWorks Motion package was used to execute these simulations. Continuous rotation-time data were interpolated using the Akima Spline method to drive the talus and calcaneus movements. Ligament strains were determined from the model analysis.

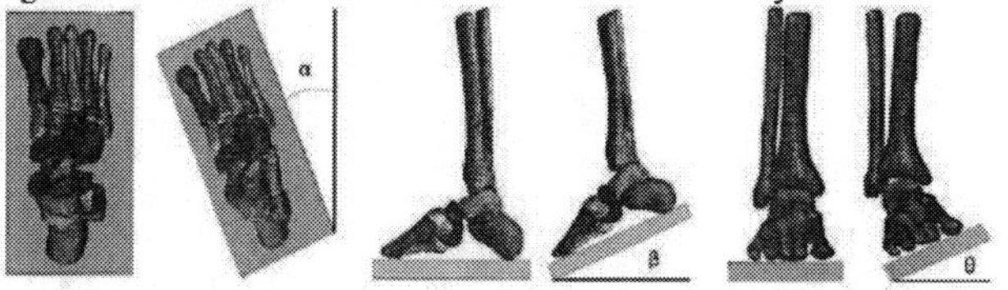

Figure 1: Schematics showing input of internal rotation (α), plantarflexion (β), and inversion (θ) into the ankle model.

RESULTS

For clarity, only strains in the lateral ankle ligaments, namely the anterior tibiofibular ligament (ATiFL), the posterior tibiofibular ligament (PTiFL), the CaFL, the ATaFL, and the posterior talofibular ligament (PTaFL), were reported in the current study.

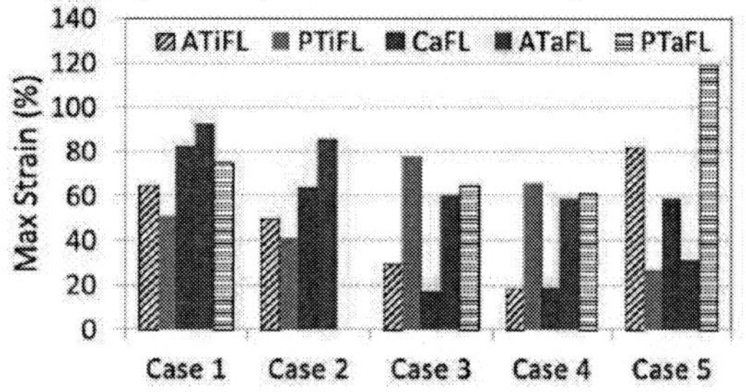

Figure 2: Maximum ligament strains in cases 1-5 with the original kinematic data input into the ankle model.

With inputs of the original injury-producing kinematics into the ankle model, simulations of cases 1 and 2 showed the highest ligament strain in the ATaFL, followed by the CaFL (Figure 2), consistent with a typical lateral ankle sprain. Simulations of cases 3-5, however, demonstrated the highest strain in other ligaments, i.e. the PTiFL in cases 3 and 4, and the PTaFL in case 5 (Figure 2), which was inconsistent with the clinically diagnosed injuries. In general, due to the large variation in the documented kinematic data between cases, ligament strains also varied significantly between cases, being from approx. 60% maximum strain in case 4 to approx. 120% maximum strain in case 5.

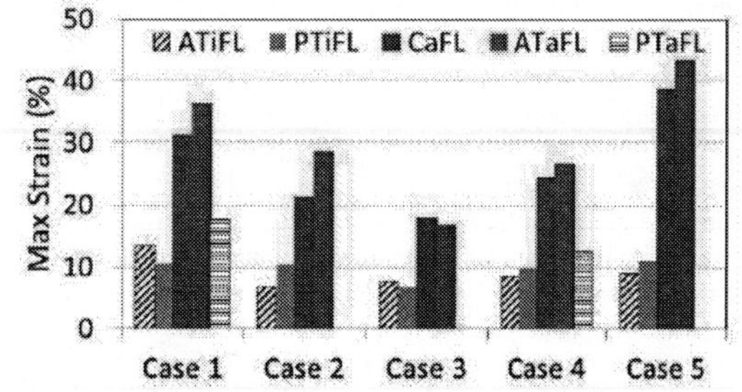

Figure 3: Maximum ligament strains in cases 1-5 with the adjusted kinematic data and the modified axial loads input into the ankle model.

In order to better match ligament strain patterns with those that might produce the clinically diagnosed injuries reported in Fong et al. [4], the ankle kinematic data had to be adjusted to 15% internal rotation for case 3, 30% internal rotation for case 4, and 30% internal rotation, plantarflexion, and inversion for case 5. In addition, the 1500 N axial load had to be modified to 375 N (75% reduction) for cases 1 and 5, and 750 N (50% reduction) for cases 2 and 4 in the new simulations. Results from these new simulations showed maximum ligament strains in all cases within approx. 20-40%, with the ATaFL and CaFL having the largest strains in each case (Figure 3), consistent with the clinically diagnosed injuries reported by Fong et al. [4].

DISCUSSION

With the kinematic data documented in Fong et al. [4] being input into our existing, validated ankle model the simulation results supported the first hypothesis that a large variation in ligament strains would be observed due to the large range of ankle kinematic data reported by Fong et al. [4]. The second hypothesis, however, was not supported by the results in the initial simulations as theoretically inconsistent ankle injuries were potentially generated, as opposed to the lateral ankle sprains involving the ATaFL and/or CaFL (Figure 2). Adjustments to the input kinematics and the input axial loads resulted in injury level strains to the ATaFL and/or CaFL (Figure 3) that were consistent with the clinically diagnosed ankle sprains in all five cases. These results may imply that Fong et al. [4] could have overestimated ankle kinematics in these injury cases, as shoe, rather than ankle, kinematics were monitored in the film analyses [4]. Previous studies in our laboratory, using human cadavers, have shown that during controlled footwear motions the foot and ankle may not follow the same motions, suggesting relative motions may exist between the ankle and footwear [8,9]. The magnitude of such relative motion was shown to depend on footwear design [8]. Future field studies using film analysis should determine potential footwear-ankle interactions in post-injury laboratory investigations. Additionally, levels of ground load may also be essential in a better understanding of the mechanisms of lateral ankle sprain in field studies.

A grade I ankle sprain is characterized as minimal tearing of ligament fibers. Rupture strain of ankle ligaments has been estimated in the range of 30-35% in our model to represent a grade III ankle sprain [7]. Since ligament collagen fibers are thought to tear when half of the rupture strain is reached, a ligament strain in the range of 15-20% has been shown corresponding to a grade I sprain, and a strain in the range of 20-30% has been associated with a grade II ankle sprain [7]. Consequently, the strain results from the current study, after the adjustments to the input kinematics and the input axial loads, may have shown a grade I ankle sprain for case 3, a grade II ankle sprain for cases 2 and 4, and a grade III ankle sprain for cases 1 and 5 (Figure 3). Such clinical data, however, were not previously discussed by Fong et al. [4].

In conclusion this simulation study may show some inconsistency between ankle and footwear kinematics in field studies. This might suggest that future studies may need to involve additional post-injury laboratory investigations before video-derived field data can be used to design future protective devices and/or help in footwear designs.

ACKNOWLEDGEMENTS

The study was supported by the EnSURE 2016 program in the College of Engineering at Michigan State University.

REFERENCES

[1] Gribble, PA et al., *Br J Sports Med*, 50:1496-1505, 2016.
[2] Fong, DT et al., *Am J Sports Med*, 37:822-827, 2009.
[3] Mok, KM et al., *Am J Sports Med*, 39:1548-1552, 2011.
[4] Fong, DT et al., *Am J Sports Med*, 40:2627-2632, 2012.
[5] Wei, F et al., *Ann Biomed Eng*, 39:756-765, 2011.
[6] Wei, F et al., *J Biomech*, 44:2636-2641, 2011.
[7] Wei, F et al., *Comput Methods Biomech Biomed Engin*, 18:243-248, 2015.
[8] Button, KD et al., *J Biomech Eng*, 137:061004, 2015.
[9] Wei, F et al., *J Biomech Eng*, 134:041002, 2012.

PERFORMANCE ASSESSMENT OF A PRE-COMPUTED BRAIN RESPONSE ATLAS IN DUMMY HEAD IMPACTS

Wei Zhao (1), Calvin Kuo (2), Lyndia C. Wu (3), David B. Camarillo (2, 3), Songbai Ji (1,4)

(1) Department of Biomedical Engineering, Worcester Polytechnic Institute, Worcester, MA, USA

(2) Department of Mechanical Engineering Stanford University Stanford, CA, USA

(3) Department of Bioengineering Stanford University Stanford, CA, USA

(4) Thayer School of Engineering Dartmouth College Hanover, NH, USA

INTRODUCTION

Mild traumatic brain injury including sports-related concussion is a major public health problem in the United States [1]. Accurate and reliable prediction of brain injury is important to mitigate this prevailing disease. Model-estimated, response-based injury metrics are desirable, as they can be directly related to tissue injury tolerances. Therefore, finite element models of the human head are playing an increasingly important role in translating external head impact kinematics into tissue mechanical responses. However, their substantial computational cost is a significant barrier to real-world applications, especially for sports-related concussion research that involves simulating a large number of head blows.

To address the computational challenges, recently a pre-computed brain response atlas (pcBRA) was established to enable real-time brain strain estimation [2]. The dramatic improvement in computational efficiency came at the cost of simplifying kinematics profiles. Initially, the pcBRA strain estimation accuracy was validated using two sagittal, resultant rotational impulses [2]. The close match with the directly simulated peak or cumulative strains suggested promise for this technique. However, the prior study was limited to uniaxial, one degree-of-freedom (1-DOF), acceleration-only rotations that were not representative of most field conditions.

In this study, we evaluated the pcBRA estimation performance more systematically. Isolated 3-DOF angular velocity profiles from dummy head drop tests were used as kinematic inputs for direct simulations. Whole-brain strains were then estimated from the pcBRA, which were further compared with the directly simulated counterparts in terms of spatial distribution and magnitude [3]. Results from this study will provide more in-depth insight into the utility of pcBRA for real-world applications.

METHODS

We used angular velocity profiles recorded from a 50th percentile male dummy head in free drop tests to generate kinematic inputs. The drop tests were performed from three heights at six impact locations, each with three trials (total of 54 impacts [4]). The kinematics were originally recorded in a body-fixed coordinate system. They were first transformed into a ground-fixed frame in order to isolate head rotational kinematics for analysis. To simulate arbitrary head rotations, each 3-DOF angular velocity profile was linearly scaled so that its peak velocity magnitude was within the pcBRA parameter sampling range (7.5–37.5 rad/s). The resulting profile was further rotated about a random axis passing through the head center of gravity with a random magnitude (within a range of 0°–90°). This process was repeated twice for each impact, leading to 108 angular velocity profiles (each with a duration of 60 ms). Their corresponding angular acceleration profiles were generated using a 5-point stencil approach [4]. Finally, the kinematics were used as inputs to the Worcester Head Injury Model (WHIM; formerly known as DHIM) [5] for brain strain simulations.

To estimate strains from the pcBRA, the major peak of the resultant angular acceleration was first identified. The corresponding peak magnitude of resultant angular velocity was then determined. At its time of occurrence, the three corresponding orthogonal velocity components were used to define an instantaneous rotational axis, from which the azimuth and elevation angles were readily obtained. This process is illustrated in **Fig. 1**.

The directly simulated brain strains (ε_{actual}) were used as a baseline to evaluate the accuracy of the pcBRA-estimated counterparts (ε_{est}). First, Pearson correlations were conducted between their element-wise, whole-brain strains in order to compare spatial strain distributions. Next, linear regression was conducted between their element-wise ε_{est} and ε_{actual}. The resulting slope indicated an overall over- or under-estimation in magnitude (slope above or below 1.0, respectively, with a zero-intercept constraint in fitting).

RESULTS

Fig. 1 shows results for the case with the highest Pearson correlation between element-wise ε_{est} and ε_{actual} (achieved at the end of impact for the latter). The cumulative strains from the direct simulation continued to evolve after reaching the major peak of resultant angular velocity (15.5 ms after the onset of impact), especially in the central region (**Fig. 1ef**; circles). Despite deceleration (**Fig. 1c**) that led to a decrease in angular velocity (**Fig. 1b**), ε_{est} closely matched with ε_{actual} in terms of both pattern (Pearson correlation coefficient of 0.99) and magnitude (slope of fitted line of 1.06 and R^2 of 0.98).

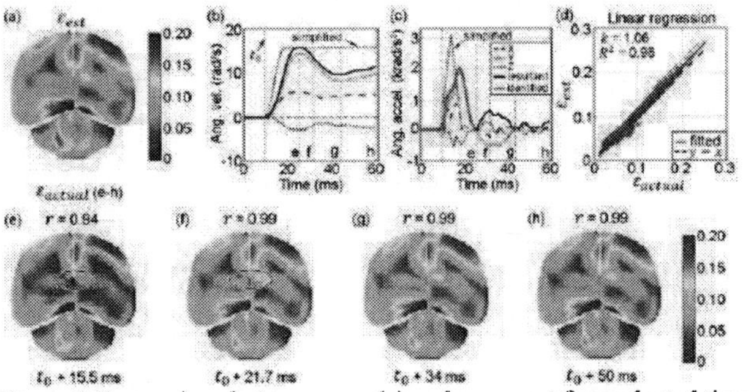

Figure 1 Comparison between ε_{est} (**a**) and ε_{actual} at four selected time points during impact (**e-h** in bottom; i.e., peak strains from onset of impact up to the current time point), as identified in the angular velocity (**b**) and acceleration (**c**) profiles. Comparisons between the two element-wise strains in (**a**) and (**h**) are shown in (**d**).

For the case with the lowest element-wise correlation that also had the greatest under-estimation at the end of impact (**Fig. 2**), it was intriguing to observe that ε_{est} closely matched with ε_{actual} immediately after the first major peak in resultant angular velocity (13.8 ms after onset; correlation coefficient of 0.96, with a slope and R^2 of 1.02 and 0.93 in linear regression, respectively). Apparently, reversing the angular velocity directionality (i.e., the second major peak in **Fig. 2b**) significantly increased brain strains, which substantially degraded the match between ε_{est} and ε_{actual} (correlation coefficient, slope, and R^2 reduced to 0.71, 0.65 and 0.57, respectively, at the end of impact).

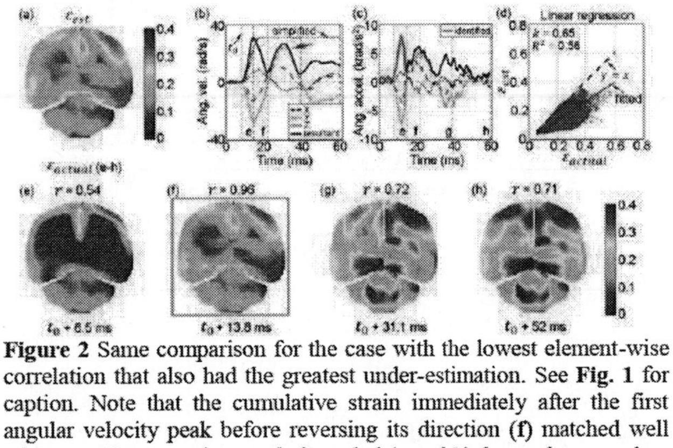

Figure 2 Same comparison for the case with the lowest element-wise correlation that also had the greatest under-estimation. See **Fig. 1** for caption. Note that the cumulative strain immediately after the first angular velocity peak before reversing its direction (**f**) matched well with ε_{est}. However, the match degraded (**g** and **h**) due to the secondary peak in angular velocity (arrow) resulting from substantial deceleration.

With results from all impacts pooled, there was a significant correlation between the element-wise ε_{est}-ε_{actual} Pearson correlation coefficients and their linearly fitted slopes (correlation coefficient of 0.78, p<0.001; R^2 of 0.60; **Fig. 3**). For 57 out of the 108 or 53% cases, the pcBRA accuracy was considered satisfactory (i.e., Pearson correlation coefficient between element-wise strains >0.9 while the fitted slope was between 0.9 and 1.1). Angular velocity profiles corresponding to impacts to the vertex and frontal region yielded poor accuracies, due to the substantial deceleration that led to reversal in angular velocity (**Fig. 3**).

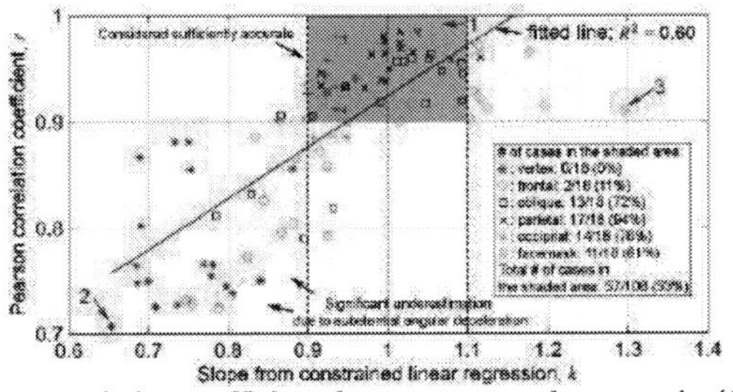

Figure 3 Correlation coefficients between ε_{est} and ε_{actual} significantly correlated with their linearly fitted slopes. For 57 out of the 108 or 53% cases (in the shaded area), ε_{est} was considered sufficiently accurate.

DISCUSSION

Our results showed that the pcBRA could achieve sufficient accuracy in strain estimation relative to the directly simulated baseline, especially for impacts with a single major angular velocity peak without a substantial deceleration. This was not surprising, given that the pcBRA was generated from acceleration-only rotational profiles [2]. For the case with the lowest element-wise ε_{est}-ε_{actual} correlation, there was a substantial head deceleration that led to a reversal in angular velocity directionality. However, the estimation was still accurate up to the point before engaging the secondary angular velocity peak (**Fig. 2a, f**). This indicated that the degraded pcBRA estimation accuracy was a direct result of not accounting for the significant deceleration or the secondary angular velocity peak. On the other hand, all impacts have both acceleration and deceleration (otherwise head motion would become unbounded). The acceleration-only pcBRA estimation yielded sufficient accuracy for some impacts but not others. This indicated a threshold of head deceleration, beyond which cumulative strains could be significantly amplified. For these impacts, the pcBRA estimation accuracy could be further improved, e.g., by parameterizing the secondary angular velocity peak, similarly to parameterizing both acceleration and deceleration in full-stop head rotations [3].

To conclude, the accuracy results reported here were particularly encouraging given that the acceleration-only pcBRA does not yet consider deceleration. These findings support the pcBRA feasibility for accurate real-time strain estimation, and suggest the need to further investigate amplification of cumulative strains due to head deceleration.

ACKNOWLEDGEMENTS

Funding is provided by the NIH Grants R01 NS092853 and R21 NS088781.

REFERENCES

[1] CDC, 2015. [2] Ji and Zhao, *ABME*, 43(8): 1877–1895, 2015. [3] Zhao and Ji, *Biomech. Model. Mechanobiol.*, 2016. DOI: 10.1007/s10237-016-0829-7. [4] Kuo et al., *J. Biomech.*, 49(9): 1845–1853, 2016. [5] Ji et al., *J. Neurotrauma*, 32(7): 441–454, 2015.

PATIENT-SPECIFIC MAPPING OF 2D IN VIVO AORTIC WALL STRAIN IN THE THORACIC AND ABDOMINAL AORTA USING DENSE MRI

J.S. Wilson (1), X. Zhong (2,1), W.R. Taylor (3,4), J.N. Oshinski (1,4)

(1) Dept. of Radiology & Imaging Sciences
Emory University School of Medicine
Atlanta, GA, USA

(2) Magnetic Resonance R&D Collaborations
Siemens Healthcare
Atlanta, GA, USA

(3) Division of Cardiology
Emory University School of Medicine
Atlanta, GA, USA

(4) Department of Biomedical Engineering
Emory University & Georgia Institute of Technology
Atlanta, GA, USA

INTRODUCTION

Patent-specific quantitative medicine is rapidly advancing, particularly in the diagnosis, evaluation, and treatment of cardiovascular disease. With regards to the aorta, significant efforts have been made to measure and model metrics of vascular fluid and soft tissue mechanics in order to understand the development and maintenance of the aorta, as well as the initiation and progression of various aortopathies. For clinical application, a primary challenge is the ability to assess the potentially heterogeneous and evolving mechanical properties of the aorta in vivo for each unique patient. Indeed, the primary clinical concern for many aortopathies can be focal in nature (e.g., aneurysmal rupture, aortic dissection); however, the ability to quantify regional in vivo metrics of strain and stress remains lacking, especially in the thin aortic wall. While computational modeling has attempted to address the need for spatially resolved analyses, the aortic models inherently depend not only on patient-specific geometry but also on the fidelity of the assignment of regional patient-specific material properties – many of which are currently assigned homogeneously from population-averaged in vitro data.

Herein, we hypothesized that magnetic resonance imaging using displacement encoding with stimulated echoes (DENSE MRI) would allow in vivo regional quantification of thoracic and abdominal aortic kinematics, which is a key first step in parameterizing regional material stiffness. Unlike other modalities used to estimate displacement or strain that rely on feature or tissue tracking, DENSE can resolve displacements much smaller than the pixel dimension by utilizing the phase of the image signal [1]. Prior cardiovascular applications of DENSE have focused almost exclusively on cardiac motion. Herein, we attempted to extend DENSE to the aorta by developing novel noise reduction techniques to address the primary technical challenge of acquiring sufficient signal from the 1-2 mm thick aortic wall.

METHODS

A 2D cine DENSE sequence with segmented, spiral k-space sampling was developed with the following parameters: TE 1.21 ms, TR 16 ms, 18 spiral interleaves image, 2 spiral interleaves per heartbeat, slice thickness of 8 mm, and reconstructed pixel size of 1.3 x 1.3 mm. Magnitude and phase images were acquired on one of two 3T MRI systems (Tim Trio or Prisma, Siemens, Erlangen, Germany) using prospective R-wave triggering to capture ~20 equally spaced time frames through mid-diastole. Note that in DENSE, phase is proportional to displacement since a tagging pulse at the r-wave. Phase images were acquired with a displacement encoding value (ω) of 0.25 cycles/mm. For this pilot study, we scanned seven aortas (three thoracic and four abdominal). Aortic locations and orientations for 2D transverse imaging were determined from scout images and targeted the ascending aorta, the descending thoracic aorta at the level of the left atrium, or the infrarenal abdominal aorta above the inferior mesenteric artery.

Custom post-processing algorithms were developed using Matlab (Mathworks, Natick, MA). Briefly, the inner and outer aortic wall was manually segmented, and the displacement of included pixels at each time point relative to their previous position was calculated by $d = \varphi/2\pi\omega$, where φ is the phase value. From this data, the sequentially tracked position of each pixel from the first time point can be interpolated [2]. To reduce noise, the position data was time-smoothed with a 6th order polynomial function. Green strain was calculated using a quadrilateral-based interpolation method [3] with a normalized coordinate system (ζ_1, ζ_2) and the following transformation functions:

$$X_A(\zeta_1, \zeta_2) = \sum_{n=1}^{4} f^n(\zeta_1, \zeta_2) X_A^n \; ; A=1,2 \quad (1)$$

$$f^n(\zeta_1, \zeta_2) = \frac{1}{4}(1+\zeta_1\zeta_1^n)(1+\zeta_2\zeta_2^n) \quad (2)$$

where X_A is the normalized position of any point within a quadrilateral with vertices X_A^n (n=1-4), and (ζ_1^n, ζ_2^n) are the normalized coordinates

(1,1), (1,-1), (-1,-1), and (-1,1). The displacement $u_{A,t}$ is calculated as

$$u_{A,t}(\zeta_1,\zeta_2) = \sum_{n=1}^{4} f^n(\zeta_1,\zeta_2)u_{A,t}^n \; ; \; A=1,2 \qquad (3)$$

from the displacements of the vertices ($u_{A,t}^n$) for all times t. After calculating the referential displacement gradient ($\boldsymbol{H}$), the Green strain ($\boldsymbol{E}$) is calculated by

$$\boldsymbol{E} = \frac{1}{2}(\boldsymbol{H} + \boldsymbol{H}^T + \boldsymbol{H}^T \cdot \boldsymbol{H}), \text{ where } \boldsymbol{H} = \frac{\partial u}{\partial X}. \qquad (4)$$

To reduce noise, 12 equispaced sectors were defined around the aortic circumference, and a single quadrilateral element per sector was defined using data from the nearest pixel to each vertex. In addition, a second overlapping 12-sector map was analyzed that was rotated by 15 degrees such that a weighted mean Green strain could be reported at each of the now 24 sectors by averaging the overlapping strain values in a 1:2:1 ratio. To further reduce noise, reference point averaging was introduced in which a mean strain was calculated for each sector at time $t > R$ by averaging the strain independently calculated in reference to time points 1 through R (where R=2 for the thoracic aorta and R=4 for the abdominal aorta).

RESULTS

Signal-to-noise ratio was adequate for mechanical analysis of the aortic wall in the descending thoracic aorta (DTA) and abdominal aorta (AA), but not in the ascending aorta. The two DTAs demonstrated heterogeneous displacement with peak displacement in the left lateral and inferolateral aortic wall. The four AAs demonstrated heterogeneous displacement with peak displacement in the anterior wall (Figure 1).

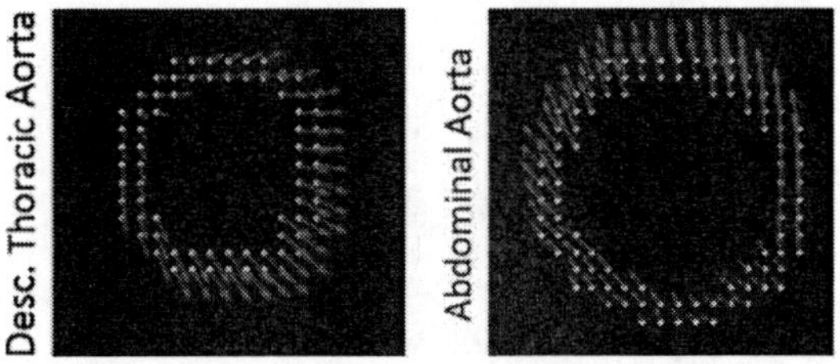

Figure 1: Illustrative vector displacement maps for a descending thoracic aorta (24 yo F) and an abdominal aorta (27 yo M).

In the DTAs, circumferential Green strain was relatively homogeneous (i.e., without large foci of distinct strain) (Figure 2). In contrast, the circumferential Green strain in AAs was regionally heterogeneous, with strain concentrated in the lateral walls. Note that the 32 year-old (Figure 2d) had an accessory left renal artery in the left lateral wall. Mean (homogenized) values of circumferential strain and the range of regional strain at local systole (peak) for each subject are shown in Table 1, which are reasonable compared to prior US data [4].

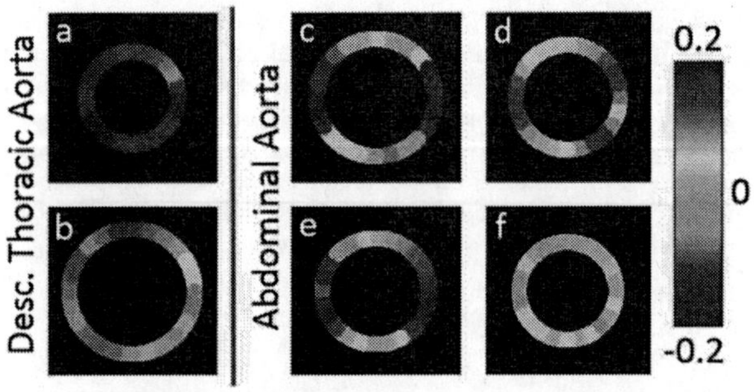

Figure 2: Peak aortic circumferential Green strain maps for (a) 24 yo F, (b) 36 yo M, (c) 27 yo M, (d) 31 yo M, (e) 32 yo M, (f) 36 yo M.

CIRC.	Desc. Thoracic		Abdominal			
STRAIN	24 yo F	36 yo M	27 yo M	31 yo M	32 yo M	36 yo M
Mean	0.16	0.09	0.11	0.10	0.10	0.05
Range	0.08-0.25	0.05-0.14	-0.08-0.38	0.04-0.17	-0.01-0.23	0-0.09

Table 1: Peak mean (homogenized) circumferential Green strain and range of regional strain for each subject and location.

DISCUSSION

A reliable ability to quantify regional patient-specific aortic wall strain in vivo would be immensely useful for both direct clinical correlation to aortic health and disease, as well as improved patient-specific parameterization of aortic growth and remodeling models. This pilot study suggests that 2D cine DENSE MRI is capable of mapping regional circumferential strain in the healthy DTA and AA following implementation of a number of noise reduction techniques including time-smoothing, sectorization, and reference point averaging. Note that the error introduced by reference point averaging is proportional to the strain that occurs between the reference time points used in the averaging and thus is only useful near local end-diastole.

Analysis of the ascending aorta using 2D DENSE may have failed due to the large longitudinal motion of this aortic region and thus movement out of plane; 3D DENSE should be explored in this area. Applying DENSE to the aorta pushes the current limits of resolution. The calculation of 2D strain requires at least two pixels of data in the radial direction; yet the need for sufficiently large pixel dimensions for adequate signal means that peri-aortic tissue is likely included in the analysis of the thin aortic wall (1-2 mm). This tissue is not irrelevant, however, since local peri-aortic tethering may directly affect in vivo kinematics. Notably, we explored alternate post-processing methods to calculate linearized circumferential strain along the single row of luminal pixels and found qualitatively similar results (not shown). These resolution constraints also make the circumferential strain more reliable than the radial and shear strains that can also be calculated.

The pilot results suggest that the circumferential kinematics of the DTA and infrarenal AA are distinct, and a full study is currently underway to compare these differences statistically. In general, both aortic locations demonstrated heterogeneous displacement directed away from the neighboring vertebrae (located along the right lateral wall for the DTA and posterior wall for the AA). Interestingly, while the DTA demonstrated relatively homogeneous circumferential strain, circumferential strain in the abdominal aorta was heterogeneous in this young healthy group. The displacement and strain pattern suggest a bulk displacement of a stiffer anterior wall (which abuts the retroperitoneal duodenum and fascia) with stretching of the lateral walls. Such regional heterogeneities in kinematics could create distinct mechanobiological stimuli to the local tissue and thereby affect regional growth and remodeling and potentially clinical risk. Future studies will seek to explore this relationship in patients at risk for specific aortopathies and in patients with focal aortic lesions (e.g., aneurysms and dissections).

ACKNOWLEDGEMENTS

This work was supported by an NIH-NHLBI T32 Institutional Training Grant (JSW/WRT), a Seed Grant from the Emory Department of Radiology & Imaging Sciences (JSW/JNO), and a Pilot Grant from the Centers for Systems Imaging at Emory University (JSW/JNO).

REFERENCES

[1] Cho, GY et al., *Am J Card*, 97(11): 1661-66, 2006.
[2] Spottiswoode, BS. *IEEE Trans Med Imaging*, 26: 15-30, 2007.
[3] Humphrey, JD. *Cardiovascular Solid Mechanics*, Springer: 2002.
[4] Derwich, W et al., *Eur J Vasc Endovasc Sur*, 51(2):187-93, 2016.

SB³C2017
Summer Biomechanics, Bioengineering and Biotransport Conference
June 21 – 24, Tucson, AZ, USA

TRANSMURAL VARIATION IN FIBER ORIENTATION AND ITS ASSOCIATION WITH THE ANISOTROPIC BEHAVIOR OF ARTERIAL ELASTIN

Xunjie Yu[1], Yunjie Wang[1], and Yanhang Zhang[1, 2]

[1]Department of Mechanical Engineering
Boston University
Boston, MA, 02215 USA

[2]Department of Biomedical Engineering
Boston University
Boston, MA, 02215 USA

INTRODUCTION

Elastin, as one of the major extracellular matrix (ECM) components, imparts elastic property to an artery in order to accommodate the cyclic physiological deformation. The anisotropic behavior of arterial elastin network has been reported in several studies [1], however elastin network is usually modeled as an isotropic material in constitutive models [2]. In a recent study, by assuming an anisotropic strain energy function for elastin, a better prediction of inflation-extension mechanical response was achieved [3]. Our recent multiphoton microscopy study reported a relatively uniform elastin fiber distribution in the arterial wall [4]. However the imaging depth was limited to about 50μm from the intimal surface. To date, the structural basis for the anisotropic mechanical behavior of elastin is not understood. This study focuses on the transmural variation in elastin fiber orientation distribution in the arterial wall, and the incorporation of such structural variations into a structure-based constitutive model for the anisotropic mechanical behavior of elastin network.

MATERIALS AND METHODS

Sample preparation

Descending thoracic aortas were harvested from pigs of 12-24 month old at a local abattoir and transported to laboratory on ice. Purified elastin samples of approximately 20 x 20mm were obtained using a cyanogen bromide (CNBr) treatment method [1]. Elastin samples were kept in 1· phosphate buffered saline (PBS) for mechanical testing and imaging.

Mechanical testing

Equi- and nonequi-biaxial tensile tests were performed using a biaxial tensile testing device to characterize the mechanical properties of elastin network (n=3). Samples were subjected to eight cycles of equi-biaxial tension of 40N/m for preconditioning. Following preconditioning, eight cycles of biaxial tension with f_l:f_c = 1:1, 4:3, and 3:4 were applied to each sample, where f_l:f_c is the ratio of tension applied to each sample in the longitudinal direction to the tension in the circumferential direction.

Multiphoton microscopy and Imaging analysis

A Multi-photon microscope (Carl Zeiss LSM 710 NLO Microscope system) with a tunable femtosecond IR pulse laser (wavelength 810 nm) was used to generate two-photon excited fluorescence (2PEF) from elastin (525/45 nm). The laser scanning system is coupled with an upright microscope with a 20· water immersion objective lens. Elastin samples were imaged from both the adventitial and intima side to a depth of 40 μm to assess the elastin fiber distribution from the outer and inner medial layer, respectively. To assess the transmural variation in elastin fiber distribution, elastin samples were first frozen, and three slices of ~100 um in thickness were cut at different transmural depth of elastin samples and placed on glass slides for imaging.

Two dimensional fast Fourier transform (2D-FFT) analysis using a Directionality plug-in in FIJI was performed to determine fiber orientation and the frequency of fiber angle, following the developer's instructions. Fiber orientation in the spatial frequency domain was determined and a normalized histogram was computed to obtain the amount of fibers at angles from -90° to 90° at 2° increment [4].

Constitutive modeling

A constitutive model of elastin network was developed to incorporate the fiber orientation distribution $R_i(\theta)$ (i = 1, 2, 3) at different transmural depth. The total strain energy function was assumed to be the sum of the strain energy functions at the inner, middle, and outer medial layer, and can be written as:

$$W = \sum_{i=1}^{3} n_i \int_{\pi/2}^{\pi/2} w(\rho) R_i(\theta) d\theta \quad (1)$$

where $w(\rho)$ is the strain energy function at the fiber level, and

$$w(\rho) = k\Theta N \left(\frac{\rho}{N}\beta_\rho + \ln\frac{\beta_\rho}{\sinh \beta_\rho}\right) \quad (2)$$

where n_i is the elastin content, N is the number of rigid links within each chain, ρ is the normalized deformed chain length and is related to the fiber-level Green–Lagrange strain ε by $\rho = P\sqrt{2\varepsilon + 1}$, and $P = \sqrt{N}$ is the normalized undeformed chain length. $\beta_\rho = \mathcal{L}^{-1}(\rho/N)$, where $\mathcal{L}(x) = \coth x - 1/x$ is the Langevin function. $k = 1.38\times10^{-23}$ J/K is Boltzmann's constant, and $\Theta = 298K$ is the absolute temperature. The four material parameters, n_i and N, were determined by minimizing the objective function using the Nelder-Mead direct search method implemented in the *fminsearch* function in Matlab (version R2013b, The MathWorks, Inc.) [5].

RESULTS

Representative 2PEF images from the inner, middle and outer medial layer are shown in Figure 1. Elastin fibers appear to be straight at unloaded state. Differences are observed in elastin fiber distributions in the three transmural positions. FFT analysis results are shown in Figure 2a to quantify the fiber distributions. Elastin fibers in the inner media are relatively uniformly distributed with a slightly preferred distribution in the longitudinal direction. In the middle media, elastin fibers orient with a preferred circumferential direction. Elastin fibers in the outer media are more longitudinally oriented. The fiber orientation ratio of circumferential to longitudinal distributed fibers, defined as the number of circumferential fibers (oriented between 0° ± 20°) divided by the number of longitudinal fibers (oriented between 90° ± 20°), was calculated to compare the fiber alignment at three layers (Figure 2b). The fiber orientation ratio changes from 0.68 ± 0.08 at the inner media, to 2.81 ± 0.32 and the middle media, and to 0.19 ± 0.03 at the outer media.

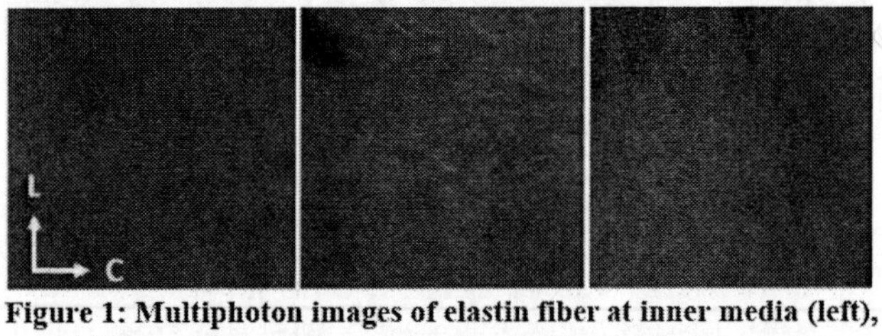

Figure 1: Multiphoton images of elastin fiber at inner media (left), middle media (middle), and outer media (right). Images are $425 \times 425\ \mu m$. L, longitude. C, circumference.

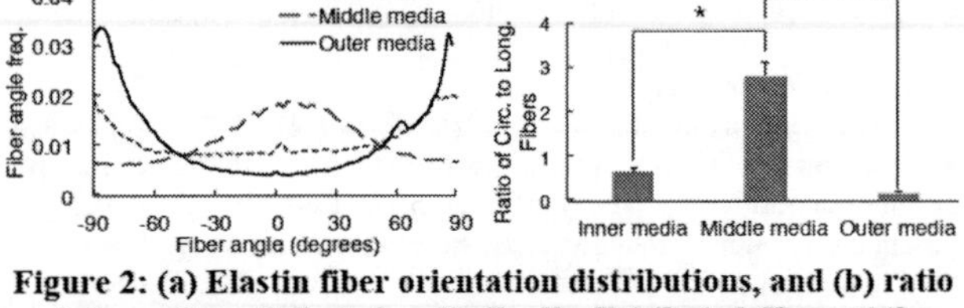

Figure 2: (a) Elastin fiber orientation distributions, and (b) ratio of circumferentially to longitudinally distributed fibers at the inner, middle, and outer transmural positions with 0° being the circumference and 90° being the longitude. (*$p < 0.01$)

Representative stress-stretch responses of elastin under equi- and nonequi-biaxial tension with the corresponding modeling results were shown in Figure 3. The model parameters were obtained by fitting the three sets of stress-stretch responses simultaneously. Overall aortic elastin shows anisotropic mechanical properties with the circumferential direction being stiffer than the longitudinal direction. The model nicely captures the structural variation with material parameter n_2 being much larger than n_1 and n_3, suggesting a higher content in middle elastin.

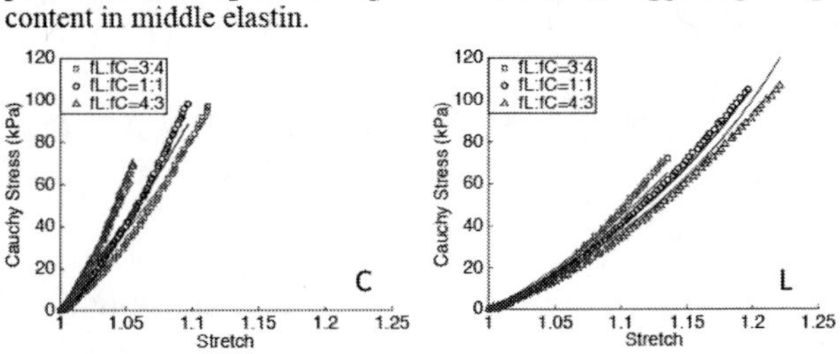

Figure 3 Cauchy stress vs. stretch for elastin network under equi- and nonequi-biaxial tension. Symbols represent experimental measurements and lines represent modeling results. Material parameters in the model: $n_1 = 2.11\times10^{11}, n_2 = 9.61\times10^{24}, n_3 = 1.96\times10^{9}, N = 1.8122$.

DISCUSSION

Elastin is usually modeled as an isotropic material, and such assumption has met challenges in reproducing the anisotropic tissue behavior [3]. The constitutive model developed in this study accounts for the transmural variation in fiber orientation distributions at the inner, middle, and outer medial layer (Figures 1 and 2), and shows promises in capturing the anisotropic behavior of purified elastin.

When studying structure-function relationships, it is important to consider the structural inhomogeneity. Our study shows that there exists a significant transmural variation in fiber orientation distribution in arterial elastin. The much thicker middle media with circumferentially aligned elastin fibers dominate the anisotropic mechanical behavior of the elastin network (Figure 3).

Fiber orientation is regarded as a main source of tissue anisotropy [6]. This study unravels an interesting correlation between mechanical anisotropy of elastin network and its underlying fiber network structure. Chow et al. [4] reported a relative uniform distribution of medial elastin fiber when imaging from the intima side. As the imaging depth is less than 50μm, it is thus questionable whether the fiber distribution obtained from such a small thickness is representative of the tissue structure, especially when developing structure-based constitutive models. Our results show that there is a significant transmural variation in elastin fiber distributions through the media. The slightly longitudinally oriented fibers on the intima side supports the shear stress from the blood flow, while the circumferentially oriented elastin in the majority of the media bears the pulsatile blood flow and the cyclic deformation in the circumferential direction of the arteries. This distinct transmural variation in elastin fiber orientations plays an important role in accommodating the complex loading in the artery.

REFERENCES

1. Zou, Y. and Zhang, Y., *Annals of biomedical engineering*, *37*(8), pp.1572-1583, 2009
2. Holzapfel, G.A., et al., *J. Elast.*, 61(1-3), pp.1-48, 2000
3. Rezakhaniha, R., et al., *Biomechanics and modeling in mechanobiology*, 10(4), pp.599-611, 2011.
4. Chow, M.J., et al., *Biophysical J*, *106*(12), pp.2684-2692, 2014.
5. Wang, Y., et al., *J of biomechanics*, 48(12), pp.2358–2365, 2016
6. Gasser, T.C., et al., *J of the royal society interface*, 3(6), pp.15-35, 2006

ARTERIAL STIFFNESS, AGING, AND ELASTIN DEFICIENCY

Jie Z. Hawes (1), Robert P. Mecham (2), Jessica E. Wagenseil (1)

(1) Mechanical Engineering and Materials Science
Washington University
St. Louis, MO, USA

(2) Cell Biology and Physiology
Washington University
St. Louis, MO, USA

INTRODUCTION

Cardiovascular diseases, like hypertension and diastolic dysfunction, have been linked to increased stiffness or elastic modulus of the large arteries [1, 2]. Increased stiffness or modulus is a common consequence of aging and may be caused by an imbalance in the production of elastin and collagen in the wall. To understand how the accumulation and organization of elastin and collagen uncouple during aging, we used genetically-modified mice in which the elastin and collagen ratio is altered. Wild-type (WT) mice serve as a normal aging model, with increased stiffness and modulus with age. Elastin haploinsufficient (*Eln+/-*) mice have 50% of the normal elastin amounts and may be an adaptive model of aging. Despite reduced elastin amounts, *Eln+/-* aorta remodel to reach a similar physiological elastic modulus to WT during development and maturation [3, 4]. However, it is not known if *Eln+/-* aorta can continue to remodel with aging to adapt the elastic modulus, or if *Eln+/-* aorta loses this ability with aging and shows an increase in stiffness and modulus with age similar to WT mice. The aim of this study is to investigate the structural and mechanical changes with aging in WT and *Eln+/-* mice and quantify how elastin and collagen amount and reorganization contribute to the observed changes. Our goal is to compare WT and *Eln+/-* mice at ages 12, 18, and 24 months, which correspond to human ages of 43, 56, and 69 years. However, at the current time, we only have data on 12 month old mice. In this abstract, we compare our current results with previously published data for 3 month old WT and *Eln+/-* mice [5], equivalent to a 20 year old human.

METHODS

WT and *Eln+/-* mice were sacrificed at 12 months of age. All animal protocols were approved by the IACUC. The left common carotid (LCC) and ascending aorta (ASC) were removed for mechanical testing. Each artery was mounted on stainless steel cannulae in a myograph (Danish Myotechnology), secured with silk surgical suture, and imaged under an inverted microscope connected to a computer. Each artery was preconditioned by 3 cycles from 0 – 175 mmHg in physiological saline solution at 37°C. The in vivo length was determined by the longitudinal length of the artery when it was slightly buckled at 175 mmHg during the preconditioning cycles. Then the mounted artery was tested in the pressure myograph at its in vivo length and pressurized with physiologic saline solution from 0 – 175 mmHg in steps of 25 mmHg. The myograph software recorded the diameter of the artery, axial force, and pressure at a frequency of 1 Hz. After mechanical testing, the artery was cut into narrow rings to measure the unloaded dimensions. Then the rings were cut radially and imaged to measure the residual strain by the opening angle, which was defined as the angle subtended by the lines connecting the midpoint of the inner circumference with the ends of the ring [5]. Image J software (NIH) was used to determine the unloaded diameter and thickness of the artery from the images. Matlab (MathWorks) scripts were used to analyze the mechanical testing data and calculate the opening angle. Compliance was calculated as the local slope of the pressure-diameter curve and is the inverse of stiffness.

RESULTS

The unloaded dimensions in Table 1 show that the outer diameters of the LCC and ASC increase 7-10% for WT and 16-22% for *Eln+/-* mice from ages 3 to 12 months. WT LCC thickness does not vary much during the aging process, while *Eln+/-* LCC thickness decreases slightly. ASC thickness for WT and *Eln+/-* mice increases about the same amount during aging (12-16%). 12 month old *Eln+/-* mice have smaller and thinner arteries than WT, as shown previously for younger mice [5].

Table 1: The unloaded dimensions of the left common carotid (LCC) artery and ascending aorta (ASC) of WT and *Eln+/-* mice at 12 months and 3 months old.

		LCC		ASC		
Age (mo)	Geno-type	Outer diam (µm)	Thick (µm)	Outer diam (µm)	Thick (µm)	N
12	WT	429±22	61±6	1101±50	122±14	3-7
	Eln+/-	438±25	54±7	911±3	113±8	
3	WT	402±10	59±4	1000±12	109±6	6-10
	Eln+/-	358±7	59±4	786±12	97±7	
% diff	WT	7%	3%	10%	12%	
	Eln+/-	22%	-8%	16%	16%	

The pressure-diameter behavior of the LCC for both genotypes does not change much during the aging process from 3 to 12 months. The pressure-diameter curves for the 12 month old mice confirm previous data that the LCC diameter is greater in WT than *Eln+/-* mice at all pressures (Fig. 1a). The pressure-compliance data for the LCC (Fig. 1b) shows that the compliance is decreased in the physiological pressure range for 12 month compared to 3 month old WT mice, which is expected for aging. However, the compliance for *Eln+/-* LCC is similar for 3 and 12 month old mice, which shows the possibility of adaptive remodeling in *Eln+/-* LCC up to 12 months of age. The pressure-compliance curves for 12 month old mice confirm previous data that the *Eln+/-* LCC has reduced compliance at high pressures.

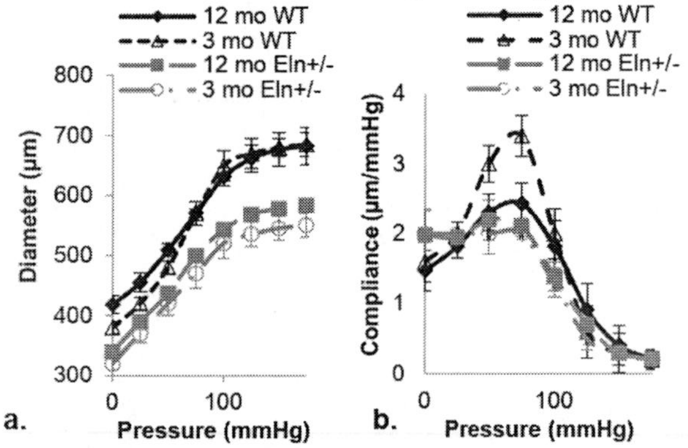

Figure 1: Mechanical testing data for the LCC of WT and *Eln+/-* mice at 3 and 12 months of age. Pressure-diameter data (a) show little change with aging. Pressure-compliance data (b) show a drop in compliance at physiological pressures for WT LCC only.

For the ASC, the pressure-diameter curves do not change much with aging for WT and *Eln+/-* mice (Fig. 2a). The pressure-compliance data for WT ASC shows that the peak compliance shifts to lower pressures during aging (Fig 2b). For the *Eln+/-* ASC, the compliance peaks at about the same pressure for 3 and 12 month old mice, indicating that the aged *Eln+/-* ASC may be able to remodel to maintain the young compliance values.

DISCUSSION

Aging or genetic disease can cause reduced elastin levels in arteries, which leads to increased arterial stiffness and high blood pressure. High blood pressure is associated with an increased risk of adverse cardiovascular events. Previous studies showed that *Eln+/-* mice have decreased compliance and changes in arterial wall structure, but achieved the elastic modulus for optimal cardiovascular function due to remodeling and adaptation of the arteries during development and maturation up to 3 months of age [5].

To explore how the remodeling process continues with aging, we used 12 month old WT and *Eln+/-* mice, which is equivalent to a 43 year old human. We investigated how elastin insufficiency affects older arteries in terms of geometry and mechanical behavior and compared our data to previously published data for 3 month old WT and *Eln+/-* mice. Our results show that the unloaded diameter and thickness increase in both WT and *Eln+/-* ASC from 3 to 12 months of age, as expected with aging. Our results also show that the thickness of the LCC changes much less from 3 to 12 months of age than the dimensions of the ASC in both genotypes, indicating that age-related remodeling may depend on location in the vascular tree. Our pressure-diameter and pressure-compliance data for 12 month old mice confirm previous observations that *Eln+/-* arteries are smaller at all pressures and less compliant at high pressures compared to WT arteries. Comparing 12 to 3 month old mice shows consistent decreases or shifts in the pressure-compliance curves for WT arteries (Figs. 1b and 2b) with aging that are not present in *Eln+/-* arteries, despite similar changes in the unloaded dimensions between 3 and 12 months of age.

Our current results suggest that, unlike WT arteries, *Eln+/-* arteries are able to continually adapt to changing production of elastin and collagen with aging. Future work will increase the statistical power of our results by increasing the sample size for both groups at 12 months of age, determine if similar trends are observed at 18 and 24 months of age, and compare material properties across genotypes and ages. Additionally, we are interested in the mechanism through which the arterial wall adapts in aging. We hypothesize that *Eln+/-* arteries readily adapt collagen amounts, types, and organization with aging through differences in TGF-β signaling. Therefore, future work will also quantify TGF-β signaling and compare microstructural organization of elastin and collagen fibers in the arterial wall.

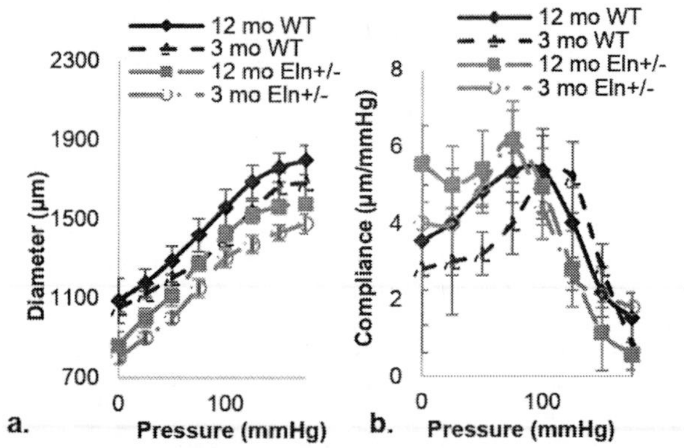

Figure 2: Mechanical testing data for the ASC of WT and *Eln+/-* mice at 3 and 12 months of age. Pressure-diameter data (a) show little change with aging. Pressure-compliance data (b) show a shift in peak compliance to lower pressures for WT ASC only.

ACKNOWLEDGEMENTS

NIH HL-115560 and HL-105314.

REFERENCES

[1] Dernellis J., et al., *Hypertension*. 2005; 45:426-431
[2] Kaess B.M., et al., *JAMA*. 2012; 308:875-881
[3] Wagenseil J.E., et al., *Physiol Rev*. 2009 Jul; 89(3):957-89
[4] Hirano, E., et al., *Circ Res*, 2007. 101(5):523-31.
[5] Wagenseil J.E., et al., *AJP Heart Circ*. 2005; 289: H1209-H1217

SB^{3}C2017
Summer Biomechanics, Bioengineering and Biotransport Conference
June 21 – 24, Tucson, AZ, USA

MODELING THE PULMONARY ARTERIES WITH A FOUR FIBER FAMILY CONSTITUTIVE MODEL

Erica R. Pursell (1), Daniela Vélez-Rendon (1), Daniela Valdez-Jasso (1)

(1) Bioengineering Department
University of Illinois at Chicago
Chicago, IL, USA

INTRODUCTION

Pulmonary arterial hypertension (PAH) is a disease of the heart and pulmonary vessels characterized by elevated blood pressure. The vasculature undergoes remodeling to adjust for these elevated pressures. However, this remodeling also hinders the pulmonary arteries' (PA's) ability to buffer pressure fluctuations in the cardiac cycle, hindering the heart's ability to pump blood with heart failure as the end stage of the disease [1].

Despite the severity of the disease, little is known about the remodeling process of the pulmonary vasculature. One approach to investigating the remodeling process is by understanding how the structure (collagen fiber organization) supports mechanical response. Orientation and content can be directly measured via multiphoton imaging and histology, though these methods can be very time consuming and costly.

To elucidate our understanding of PAH progression, we propose to develop mechanistic models that incorporate both mechanical properties and fiber orientation. Using the modeling efforts on carotid arteries [2], we use biaxial data to predict overall fiber orientation and coefficients related to stress and strain in an animal model of PAH.

METHODS

The experimental protocols used for this study were approved by University of Illinois at Chicago Animal Care and Use Committees. PA segments were harvested from one 12-week-old male Sprague-Dawley rat weighing 200 grams (Charles River Laboratories). At 8 weeks of age, the rat was treated with a subcutaneous monocrotaline (MCT) injection to induce PAH. The animal was then left for four weeks to allow PAH to reach an acute state.

Prior to harvesting the pulmonary arteries, hemodynamic measurements were taken *in vivo* confirm the hypertensive state of the animal. Segments from the left and right pulmonary arteries were harvested and cannulated for tubular biaxial testing in a Bose biodynamic chamber. The proximal portion of the left pulmonary artery (LPA) and distal portion of the right pulmonary artery (RPA) were chosen for mechanical testing. The other portions of the vessels were cannulated and imaged with a multiphoton microscope (MPM) to quantify fiber orientation using the Directionality function in Fiji.

During mechanical testing, the vessel segments underwent two protocols. In the first (circumferential) protocol, the vessels were stretched to 40% of their respective *ex-vivo* length and a pulsatile flow was prescribed to reach pressures matching those measured *in vivo*. For the second (axial) protocol, flow was prescribed to reach a constant pressure matching the average pressure measured *in vivo* while the vessels were cyclically stretched between 40 and 45% of their respective *ex-vivo* length. In both protocols, the diameter was measured using a laser micrometer and axial force was measured with an in-line load cell. More detail on the mechanical protocol can be found in a study by Pursell *et al.* [3].

Data collected from these two protocols were modeled with a four fiber family constitutive model (Equations 1-2).

$$\sigma_z^{th} = c\left(\lambda_z^2 - \frac{1}{\lambda_\theta^2\lambda_z^2}\right) + c_1^1(\lambda_z^2 - 1)\exp[c_2^1(\lambda_z^2-1)^2]\lambda_z^2$$
$$+2c_1^{3,4}\left(I_4^d - 1\right)\exp\left[c_2^{3,4}\left(I_{4,z}^d - 1\right)^2\right]\lambda_\theta^2\cos^2\alpha \quad (1)$$

$$\sigma_\theta^{th} = c\left(\lambda_\theta^2 - \frac{1}{\lambda_\theta^2\lambda_z^2}\right) + c_1^2(\lambda_z^2 - 1)\exp\left[c_2^2\left(\lambda_\theta^2-1\right)^2\right]\lambda_\theta^2$$
$$+2c_1^{3,4}\left(I_{4,\theta}^d - 1\right)\exp\left[c_2^{3,4}\left(I_4^d - 1\right)^2\right]\lambda_\theta^2\sin^2\alpha \quad (2)$$

where the 4th Cauchy invariant (I_4) is

Technical Presentation #226

$$I_4^d = \lambda_z^2 \cos^2 \alpha + \lambda_\theta^2 \sin^2 \alpha \quad (3)$$

c_1^1 and c_1^2 are material parameters with units of kPa, c_2^1 and c_2^2 are dimensionless, and α corresponds to the angle of the diagonally oriented fibers.

For the axial and circumferential stresses (Equations 1 and 2), axial stretch (λ_z) and circumferential stretch (λ_θ) were defined as in Equation 4 where l is the deformed axial length of the vessel, L is the *ex-vivo* axial length of the vessel, and r and r_{min} are the measured and minimum radius for the cycle, respectively.

$$\lambda_z = \frac{l}{L} \; ; \; \lambda_\theta = \frac{r}{r_{min}} \quad (4)$$

Equations 1 and 2 were then fitted to the experimentally derived stresses defined in Equation 5, where f is the measured axial force, h is the vessel thickness, and p is the measured internal pressure.

$$\sigma_z^{ex} = \left(\frac{f}{2\pi h} + \frac{rp}{2h}\right) \; ; \; \sigma_\theta^{ex} = \frac{rp}{h} \quad (5)$$

The parameters c, c_1^1, c_1^2, c_2^1, c_2^2, $c_1^{3,4}$, $c_2^{3,4}$, and α were determined using MATLAB function *fminsearch* with the objective function defined as shown in Equation 6:

$$E = \sum_{i=1}^{n} \left[\left(\frac{(\sigma_z^{th})_i - (\sigma_z^{ex})_i}{(\sigma_z^{ex})_i} \right)^2 + \left(\frac{(\sigma_\theta^{th})_i - (\sigma_\theta^{ex})_i}{(\sigma_\theta^{ex})_i} \right)^2 \right] \quad (6)$$

Here n refers to the number of points, th to the theoretical and ex experimentally obtained stresses.

RESULTS

From visual inspection, the four fiber family constitutive model was able to closely reproduce both axial and circumferential data shown in Figure 1.

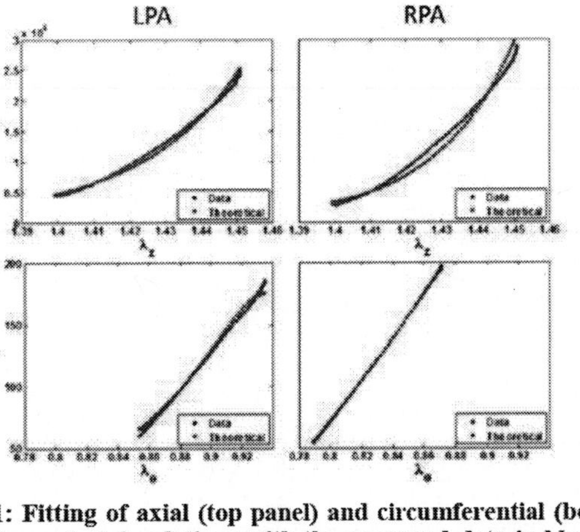

Figure 1: Fitting of axial (top panel) and circumferential (bottom panel) stress-stretch relations with the measured data in blue and theoretical values in green significantly overlapping. Here, the optimized fiber angles for the LPA and RPA were 23.95° and 11.75° respectively. Note how the axial data displayed the typical j-shape whereas the circumferential data was more linear.

For the LPA, the model estimated an off-diagonal fiber orientation of 23.95° while the Directionality function in FIJI gave a histogram of fiber orientation with the peak direction at 25.47°. In the case of the RPA, the model predicted a diagonal orientation of 11.75°, similar to the peak of 9.05° found by FIJI. For our analysis in FIJI, ±90° corresponds to the circumferential direction and 0° to the axial direction. In Figure 2 top panel, the horizontal corresponds to 0°.

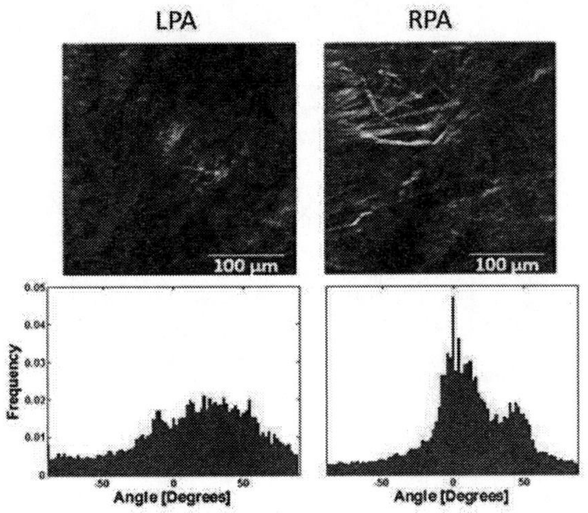

Figure 2: MPM images of the LPA and RPA (top panel) and their fiber angle histograms generated from FIJI (bottom panel). The peak angle was 25.47° for the LPA and 11.75° for the RPA.

DISCUSSION

In this study, the four fiber family model was able to simultaneously fit the axial and circumferential mechanical data from the pulmonary arteries of a hypertensive animal. The close prediction of the angle of preferred direction for the collagen fibers to that measured confirms the relevance of this model for the pulmonary vasculature. However, parameter estimations were highly sensitive to initial conditions. Therefore, it is important to improve our numerical methods for a more robust minimization problem. Future works will also include different stages of PAH to determine whether this model can detect organizational changes during the different disease states of PAH and not only acute.

ACKNOWLEDGEMENTS

Funded by the American Heart Association Scientist Development Grant 16SDG29670010, UIC College of Engineering Seed Fund, and the UIC Lincoln Graduate Student Fellowship. We would also like to acknowledge the experimental assistance from Julie Wagner.

REFERENCES

[1] Stenmark KR., *et al., Circulation Research* 99, 7: 675-691, 2006.
[2] Ferruzzi J, *et al., Cardiovascular Research* 92: 287-295, 2011.
[3] Pursell ER, *et al., J Biomech. Eng.* 138, 2016.

TORTUOSITY AND CURVATURE OF CEREBRAL ARTERIES IN POSTERIOR FOSSA.

D Nakagawa MD PhD(1,2), A Schumacher BS(1), B Berkowitz PhD(1)
D Hasan MD(2), M Raghavan PhD(1)

(1) Department of Biomedical Engineering
University of Iowa
Iowa city, IA, USA

(2) Department of Neurosurgery
University of Iowa
Iowa city, IA, USA

INTRODUCTION

The basilar and vertebral arteries are one of the largest cerebral arteries and known to be affected most commonly by atherosclerosis[1]. Compared to internal carotid artery, that is fixed by skull base bones, these arteries are not fixed in original position. Previous studies have shown some unique methods to assess vascular atherosclerosis by using tortuosity and curvature as parameters[2-5]. In addition, several studies have mentioned the relations between the vascular atherosclerosis and aging[4-8]. Furthermore, tortuosity and curvature secondary to hypertension were found in cerebral arteries[9]. In the present study, we investigated the relationship – if any - between aging and vessel axis morphology – specifically, tortuosity and curvature in basilar and vertebral arteries.

METHODS

Patient population

This study was approved by the Ethics Committee of University of Tokyo (No. 2231-2, approved on 9/30/2013). The patients in this study provided written informed consent. Between January 2016 and June 2016, there were 424 cases with magnetic resonance angiography(MRA). Reviewing the MRA images, we found 18 cases(5 men, 13 women; mean (±SD) age, 49.7 ± 19.7 years) that did not have space-occupying lesions in posterior fossa.

Image acquisition

MRA was performed using a 3.0-T system (Signa 3.0T;GE, Wisc., USA). Recorded images were output in the DICOM(digital imaging communication in medicine) format and imported into an open-source toolkit (Vascular Modelling Toolkit, VMTK)[10].

For segmentation of vascular structures, level set method was performed with colliding fronts algorithm in basilar artery, right vertebral artery, and left vertebral artery. Basilar artery was defined from basilar tip to vertebrobasilar junction. Vertebral artery was defined from vertebrobasilar junction to dura mater. Centerline was extracted using the centerline module in VMTK (Figure1). Finally, tortuosity and curvature of centerline were calculated automatically in basilar artery, right vertebral artery, and left vertebral artery, respectively[10].

Curvature: Curvature of the centerline c(s), is defined as

$$\kappa(s) = \frac{|\mathbf{c}'(s) \times \mathbf{c}''(s)|}{|\mathbf{c}'(s)|^3}$$

Tortuosity: Given the length of the centerline and the Euclidean distance between its endpoints, tortuosity is defined as

$$\chi = \frac{L}{D} - 1.$$

Statistical Analysis

Statistical analysis and graphic display of data were performed by using GraphPad software (version 7.00 for Windows; GraphPad Software, San Diego, CA). All values are reported as mean ± standard error (standard deviation [SD]). Student t test was used for the comparison of two means.

RESULTS

Tortuosity

Tortuosity ranged from 0.02 to 0.14 for basilar artery, 0.08 to 0.46 for right vertebral artery, and 0.03 to 0.41 for left vertebral artery, respectively. Mean (±SD) of tortuosity were as follows, 0.07 ± 0.04 for basilar artery, 0.16 ± 0.1 for right vertebral artery and 0.13 ± 0.09

for left vertebral artery. In patients under 45 years old, mean (±SD) of tortuosity was 0.04 ± 0.006 for basilar artery, 0.17 ± 0.04 for right vertebral artery, and 0.16 ± 0.03 for left vertebral artery. On the other hand, in patient aged 45 and over, mean (±SD) of tortuosity was 0.08 ± 0.01 for basilar artery(P<0.01), 0.16 ± 0.03 for right vertebral artery(P=0.94), and 0.12 ± 0.03 for left vertebral artery(P=0.38) (Figure 2). There was a statistical correlation(P=0.01), and correlation coefficient R^2 was equal to 0.33 ± 0.03, between tortuosity of basilar artery and age. However, there was no statistical correlation(P=0.48), and correlation coefficient R^2 was equal to 0.03, between tortuosity of right vertebral artery and age. Furthermore, there was also no statistical correlation(P=0.72), and correlation coefficient R^2 was equal to 0.01. These results are shown in Figure2.

Curvature

Curvature ranged from 0.02 to 0.13 for basilar artery, 0.03 to 0.14 for right vertebral artery, and 0.03 to 0.18 for left vertebral artery, respectively. Mean (±SD) of curvature were as follows, 0.07 ± 0.03 for basilar artery, 0.08 ± 0.03 for right vertebral artery and 0.07 ± 0.04 for left vertebral artery. In patients under 45 years old, mean (±SD) of curvature was 0.05± 0.01 for basilar artery, 0.07 ± 0.01 for right vertebral artery, and 0.07 ± 0.01 for left vertebral artery. On the other hand, in patient aged 45 and over, mean (±SD) of curvature was 0.08 ± 0.01 for basilar artery(P=0.13), 0.09 ± 0.01 for right vertebral artery(P=0.22), and 0.07 ± 0.01 for left vertebral artery(P=0.94) (Figure 3). There were no statistical correlation(P=0.48, 0.30, 0.59) and correlation coefficient R^2 equal to 0.03, 0.07, 0.02, between curvature of basilar artery or right vertebral artery or left vertebral artery and age. These results are shown in Figure3.

a b

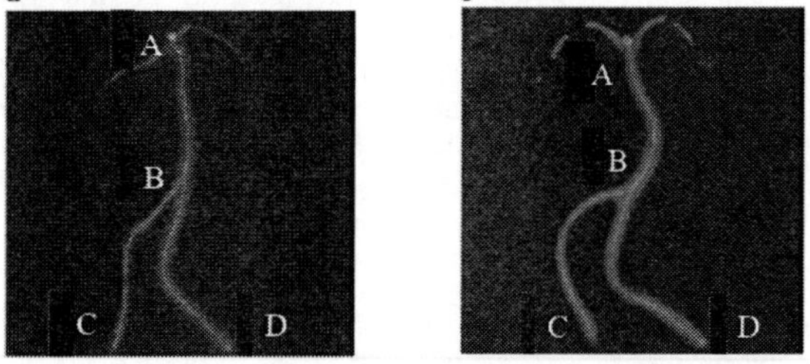

Figure 1: Image of vascular segmentation and centerline
(a) A case of low tortuosity of basilar artery. Patient was 16 years old female. Value of tortuosity of basilar artery was 0.09.
(b) A case of high tortuosity of basilar artery. Patient was 69 years old female. Value of tortuosity of basilar artery was 0.46.
A: Basilar top, B: Vertebrobasilar junction, C: Dura mater of right vertebral artery, D: Dura mater of left vertebral artery

a b c

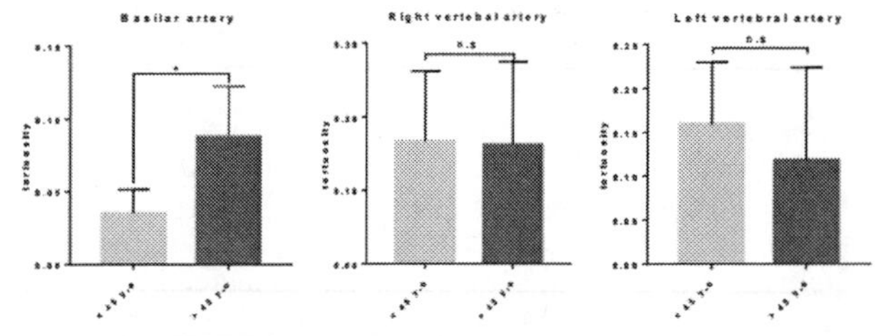

Figure 2: (a)Tortuosity of basilar artery, (b)Tortuosity of right vertebral artery, (c)Tortuosity of left vertebral artery
*:P < 0.05, ns: Not significant, y.o: years old.

a b c

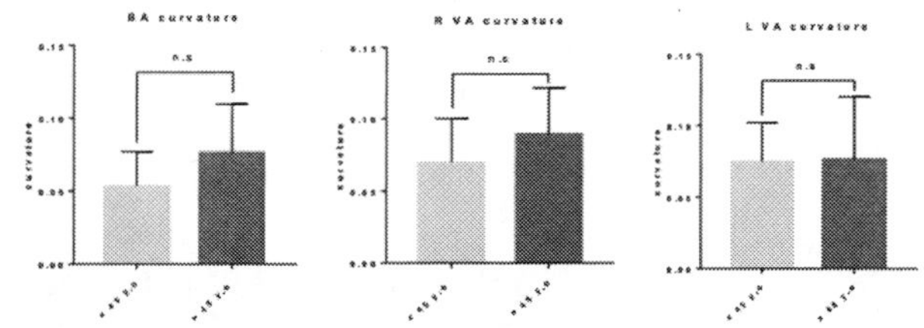

Figure 3: (a)Curvature of basilar artery, (b)Curvature of right vertebral artery, (c)Curvature of left vertebral artery
BA: basilar artery, VA: vertebral artery, R: right, L: left,
ns: Not significant, y.o: years old.

DISCUSSION

Vessel axis morphology may have implications to blood flow (cite and consequently, development of vessel wall pathologies. In the present study, we investigated relations between aging and tortuosity or aging and curvature in basilar and vertebral arteries. In posterior fossa, only the tortuosity of basilar artery was found to increase with aging. On the other hand, we failed to find evidence to support the notion that all the vessel tortuosities in posterior fossa would increase with aging. More detailed analysis would require the morphological information of all the vascular in brain.

ACKNOWLEDGMENTS

The authors are grateful to the VMTK developers and user community for access to the vascular segmentation and analysis software.

SOURCES OF FUNDING

This work was supported by Nakatani Foundation.

DISCLOSURES

The authors report no conflict of interest concerning the materials or methods used in this study or the findings specified in this paper.

REFERENCES

[1] Caplan, LR et al., *Stroke*, 17:648-655, 1986
[2] Cerletti, U et al., *Nissls Histology Histopathology Arbeiten* 4:1-168, 1910
[3] Hassler, O et al., *Acta Neuropathol*, 5:40-53, 1965
[4] Akira, M et al., *Lab Invest*, 55:482-9, 1986
[5] Campbell ACP et al., *Arch Neurol Psychiatr*, 39:1150-1202, 1938
[6] Beskow, J et al., *Acta Psychiatr Scand*, 221:111-19, 1971
[7] Ferszt, R et al., *Raven press*, 133-51, 1983
[8] Ravens, JR et al., *Adv Neurol*, 20:487-501. 1978
[9] Karl, T et al., *BMC Bioinformatics*, 12:515-27, 2011
[10] Marina, P et al., *IEEE transaction on medical imaging* 28:1141-1155, 2009

SB^{3}C2017
Summer Biomechanics, Bioengineering and Biotransport Conference
June 21 – 24, Tucson, AZ, USA

A TRIPHASIC FLUID TRANSPORT MODEL OF THE ARTERIAL WALL

Manuel K. Rausch (1,2), Jay D. Humphrey (2)

(1) Department of Aerospace Engineering & Engineering Mechanics
University of Texas at Austin
Austin, TX, USA

(2) Department of Biomedical Engineering
Yale University
New Haven, CT, USA

INTRODUCTION

Every year approximately fourteen thousand patients suffer from thoracic aortic dissections of which >50% succumb to their affliction [1]. We posit that pathological transmural fluid transport, through its effect on local tissue swelling, may be a contributing factor to the delamination of the thoracic aortic wall [2].

Transmural fluid transport is determined through the nonlinear, multiphysical interplay amongst hemodynamic boundary conditions, wall constituents and microstructure, wall deformation, and wall electrochemistry. In addition, experimental study of transmural fluid transport is complicated for it is difficult to measure locally varying key quantities, particularly pressure. Thus, computational models are needed to shed light on some of the questions about transmural fluid transport that elude us until today.

While transmural fluid transport is primarily governed by the luminal pressure and the permeability of the wall, it also depends on vascularization of the wall (known as vaso vasorum) and the Gibbs-Donnan swelling pressure, which in turn depends on the fixed charge density of the wall as well as the concentration of charged ions in the blood plasma and interstitial fluid. To solve the governing equations for this complex multiphysical problem, we introduce a triphasic finite element model of the arterial wall that takes into account the nonlinear, anisotropic constitutive behavior of a three-layered descending thoracic aorta experiencing large deformations, the spatially varying fixed charge of the arterial wall, charged counter ions in the plasma and interstitial fluid, the spatially varying, mechanically sensitive permeability of the arterial wall, and the vaso vasorum through Starling's law.

Here we establish a baseline transport model of the descending thoracic aorta of an aged, healthy individual. However, our ultimate goal is to understand the role of transmural fluid transport on the local mechanics of the arterial wall as it relates to thoracic aortic dissections. Based on our initial model, we will be able to study the sensitivity of fluid transport on electro-chemo-mechanical wall properties and boundary conditions and thus investigate a number of hypothesized contributors to descending thoracic aortic dissections [2,3], including (i) accumulation of glycosaminoglycans in the aortic wall, which could alter the fixed charge density distribution and thus Donnan swelling, (ii) the effect of alterations in luminal pressure, as seen in hypertensive patients, (iii) increases in the stiffness of the wall, as seen in hypertensive, aged and diabetic patients, (iv) and changes in the microvasculature of the wall (vaso vasorum), as seen in patients with genetic predisposition to aortic dissections.

METHODS

Computational Framework. We model the descending thoracic aorta as an ideal, quarter-symmetric cylinder with three distinct layers: intima, media, and adventitia. Idealizing the behavior of soft tissue, we further assume that each layer is triphasic with an incompressible, Fung-type, anisotropic, hyperelastic solid phase. We also assume that the wall is saturated with an incompressible, inviscid fluid (interstitial fluid), and contains two counter ions, ostensibly Na^+ and Cl^-. Fluid flux and molar flux are modeled to follow a generalized Darcy's law and Fick's law, respectively.

The coupled, nonlinear governing equations for the triphasic model of the aortic wall, namely the balance of linear momentum of the solid phase and the equations of mass balance for the fluid and ions, are solved implicitly and monolithically using the nonlinear finite element solver FEBio (www.febio.org), where the electroneutrality condition is enforced through a penalty term in the expression for the virtual work. For more details see [4].

Technical Presentation #228

Solid Wall Behavior. We model the layers of the arterial wall as constraint mixtures of an isotropic neo-Hookean ground substance with four embedded, Fung-type fiber families. The hyperelastic behavior of the solid portion of the aortic wall is governed by the strain energy function

$$W = \frac{\mu}{2}[I_1 = 3] = \sum_{i=1}^{4} \frac{c_1^i}{4c_2^i}\left[\exp\left(c_2^i[I_4^i = 1]^2\right) = 1\right], \quad (1)$$

where μ, c_1^i, c_2^i are material parameters, I_1, I_4^i are the first and fourth invariants of the right Cauchy-Green tensor $\boldsymbol{C}$, respectively, and where $I_4^i = \boldsymbol{C}:\boldsymbol{M}^i$ with $\boldsymbol{M}^i$ being the structural tensor of each fiber family governed by their respective fiber orientation angles.

Prestress and Residual Wall Stress. We assume that the state of residual stress in the descending thoracic aorta can be model by closing an initially stress-free, opened aortic wall segment. We further include axial prestress by a simple axial extension of the aortic wall segment according to experimentally measured axial prestretch values.

Permeability Model. Under large deformation even an initially isotropic permeability tensor may change its principal values and principal axes depending on the state of deformation, which is governed by residual stresses, the state of swelling, external loads, boundary conditions etc. Here, we consider these effects by implementing a mechanically sensitive permeability tensor for each layer based on recent experimental findings [5], namely

$$\boldsymbol{K} = \sum_{i=1}^{3} K^i \boldsymbol{N}^i \text{, where} \quad (2)$$

$$K^i = k_0 \left[\frac{1-\phi^s}{1-\phi_r^s}\right]^m \left[p_1 = \frac{p_2}{1-\exp(p_3-J)p_4}\right], \quad (3)$$

with $\boldsymbol{K}$ being the referential permeability tensor, K^i the principal values, $\boldsymbol{N}^i$ a second set of structural tensors, J the Jacobian, ϕ^s and ϕ_r^s the solid volume fraction and referential solid volume fraction; k_0, m, p_1, p_2, p_3 are material parameters.

Supply Material. The wall itself is supplied with fluid via the intramural vasculature, the vaso vasorum. Using Starling's law, we include a fluid supply term in the mass balance of the fluid, which depends on the microvasculature hydraulic conductance as well as the luminal pressure of the microvasculature, the local intramural pressure, and the oncotic pressure, viz.

$$\hat{\varphi} = k_p[\tilde{p} = \tilde{p}_m] = c_p[\tilde{c} = \tilde{c}_m], \quad (4)$$

where $\hat{\varphi}$ is the fluid volume fraction supply per unit time, k_p and c_p are material constants, $\tilde{p}$ and $\tilde{p}_m$ are the effective local interstitial pressure and the effective pressure in the vaso vasorum, respectively, and $\tilde{c}$ and $\tilde{c}_m$ are the local effective ion concentration and the effective ion concentration in the vaso vasorum, again, respectively.

Boundary Conditions and Load Steps. To reach the healthy, systolic, in-vivo state of the descending thoracic aorta, we apply mechanical and electrochemical boundary conditions step-wise. First, we close and stretch the initially opened, stress-free wall segment before we apply a physiological intraluminal pressure to the vaso vasorum (see equation (4)). Subsequently, we saturate the wall with Na^+ and Cl^- according to physiological values of blood plasma and the interstitial fluid. Next, we linearly increase the fixed charge density of the intima and media to previously measured values, after which we apply a normal, systolic intraluminal pressure to the wall. All steps are solved assuming steady-state, thus we ignore transient effects.

Finite Element Discretization. We discretize the quarter wall segment with 200 hexahedral finite elements (20 in the radial direction, 10 in the circumferential direction, 1 in the axial direction).

Identification of Stress-free Reference Configuration. Because data for the baseline human descending thoracic aortic geometry are only available in the in vivo state, we employ an inverse finite element approach to estimate the stress-free reference configuration required for the current analysis. To this end, we begin by defining an initial mesh on a unit cube, which we transform into an open wall segment according to opening angle, an initial segment thickness, and initial segment length (results are invariant to the axial length due to symmetry conditions). Since the opening angle has been experimentally measured, we are only required to determine the initial segment thickness and segment length. We employ an Active-Set algorithm to solve a constrained optimization problem in which the in vivo measured wall thickness and radius are the target values. Hence, we run consecutive forward simulations in which we solve the healthy baseline model of the aortic wall for varying initial wall thickness and segment lengths until a set is found that satisfies the error tolerance.

RESULTS

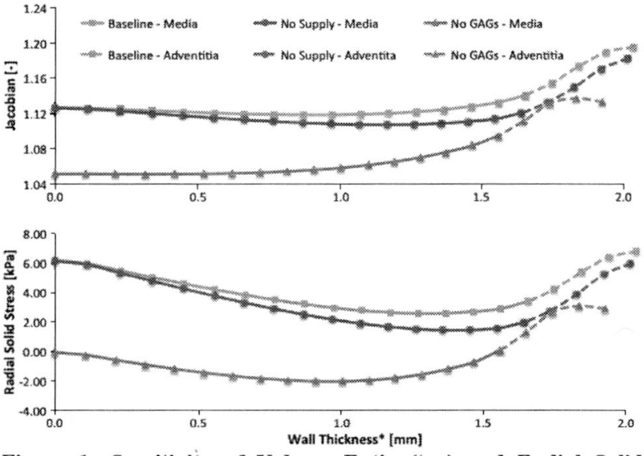

Figure 1: Sensitivity of Volume Ratio (top) and Radial Solid Matrix Stress (bottom) to fluid supply and presence of glycosaminoglycans. *measured relative to the inner wall radius

DISCUSSION

Here we proposed the most detailed transmural fluid transport model of the descending thoracic aorta to date. Notably, our model is the first to consider contributions of both Donnan pressure and vaso vasorum, which strongly affect the state of swelling across the wall and thus the wall stress distribution (Figure 1). Our results therefore demonstrate the complexity of transmural fluid transport as a coupled, bio-electro-mechanical phenomenon. As outlined in our introduction, our next goal is to test key hypotheses about the origin of thoracic aortic dissections and shed additional light on a devastating disease.

ACKNOWLEDGEMENTS

This work was funded, in part, by NIH grants R01 HL086418, U01 HL116323, and T32 HL007974. We also thank Professor Ateshian for his guidance.

REFERENCES

[1] Ollson, C et al., *Circulation*, 114:2611-2618, 2006
[2] Roccabianca, S et al., *J R Soc Interface*, 11:20140397, 2014
[3] Mallat, Z et al., *Arterioscler Thromb Vasc Biol*, 36:00-00, 2016
[4] Ateshian, G et al., *J Biomech Eng*, 135:1111011, 2013.
[5] Chooi, K et al., *J R Soc Interface*, 13:20160234, 2016.

COMPUTATIONAL FLUID DYNAMICS OF AORTIC DISSECTION: 4D FLOW MRI-BASED INLET BOUNDARY CONDITIONS

Sylvana García-Rodríguez (1), Rafael Medero (2), Christopher J. François (1), Alejandro Roldán-Alzate (1, 2, 3)

(1) Department of Radiology
University of Wisconsin-Madison
Madison, Wisconsin, USA

(2) Department of Mechanical Engineering
University of Wisconsin-Madison
Madison, Wisconsin, USA

(3) Department of Biomedical Engineering
University of Wisconsin-Madison
Madison, Wisconsin, USA

INTRODUCTION

Aortic dissection (AD) starts with intimal rupture, allowing blood inflow into the vessel wall and forming a false lumen [1, 2]. There are approximately 5 to 30 AD cases per million people per year [2]. Complications include aneurysm growth, rupture, end-organ malperfusion, and refractory hypertension [2, 3]. Accurate prediction of complications could allow earlier treatment and prevent irreversible or fatal consequences.

Four-dimensional (4D) Flow Magnetic Resonance Imaging (MRI) is useful for in vivo AD evaluation [3, 4]. Using time-resolved, three-directional velocity data, hemodynamics may be non-invasively quantified a posteriori. However, it does not offer predictive studies to foresee treatment or surgical outcomes.

Previous CFD studies [5-7] have provided significant contributions to AD understanding. However, boundary conditions (BCs) are based on literature, two-dimensional (2D) phase contrast MRI, and non-AD subject data [5, 6, 8, 9]. To our knowledge, 4D Flow MRI and CFD have not been combined in AD models. Methods for defining inlet BCs at the high-inertia ascending aorta have not been detailed. The purpose of this study was to investigate the effects of 4D Flow MRI-based inlet BCs on CFD models of in vitro ADs, compared to constant and parabolic velocity derived from volumetric flow rate.

METHODS

Subjects: Following IRB-approved and HIPAA-compliant protocols, routine magnetic resonance angiographies (MRAs) from two type B AD patients (55 y.o. f., Model 1; 54 y.o. m., Model 2) were used for three-dimensional (3D) printing.

3D Printing: Aortic and main arterial branch vessel lumens were segmented from MRA images (Mimics, Materialise; Leuven, Belgium). Surfaces were smoothed (3-matic, Materialise) and models were hollowed (1 mm thickness). Tubing connections were added at inlets and outlets. Both geometries were exported and 3D printed to scale using selective laser sintering (DTM Sinterstation 2500Cl ATC, 3D Systems, Inc.; Rock Hill, SC) with laser power 12 W, scan spacing 0.15 mm, beam speed 5080 mm/s, and nylon 11 powder.

In Vitro Tests: 3D models underwent computed tomography (CT) (Discovery CT750 HD, GE Healthcare; Waukesha, WI) in air with 120 kVp, 345 mA, slice thickness 0.625 mm, slice spacing 0.312 mm and pixel size 0.430x0.430 mm. Each model was connected to a perfusion pump (Stockert SIII Heart-Lung Machine) continuously circulating water plus gadofosveset trisodium contrast agent (Lantheus, N. Billerica, MA) at 3 L/min [10]. Two 4D Flow MRI [11] scans (upper and lower halves) were performed on each model while submerged in water (3T MR750, GE Healthcare) with 1.25 mm isotropic spatial resolution and velocity encoding sensitivity, (V_{enc}) = 120 cm/s. 4D Flow MRI data was processed (EnSight, CEI Inc; Apex, NC) to quantify streamlines, volumetric flow rate, and three-component velocity at the ascending aorta inlet plane.

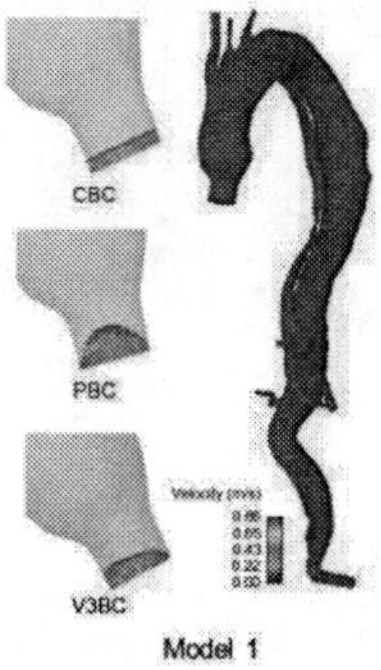

Figure 1: Geometry and BCs.

CFD: The in vitro AD was CFD modeled (Fluent, ANSYS, Inc.; Canonsburg, PA) using CT geometries (**Error! Reference source not found.**1) discretized with approximately 2,000,000 elements (ICEM CFD, ANSYS, Inc.). Continuous flow (density 1000 kg/m³, viscosity 0.001 Pa*s) was defined. No-slip was defined at rigid walls and outlet BCs as zero pressure. Simulations were transient ($\Delta t = 0.05$ s), but BCs were time-constant. Inlet Reynolds numbers were in the order of 700.

CT and 4D Flow MRI geometries were registered (3-matic), to ensure coinciding inlets. For each model, three inlet BCs were compared (Fig. 1): 1) constant BC (CBC): constant, normal velocity using 4D Flow MRI average flow divided by inlet area; 2) parabolic BC (PBC): paraboloid, normal to the inlet, with flow rate equivalent to in vitro average flow; 3) multidirectional velocity BC (V3BC): velocity vectors directly based on 4D Flow MRI, applied by constant interpolation.

RESULTS

Tables 1 and 2 summarize CFD results. Figs. 2 and 3 demonstrate how flow features are highly impacted by inlet BC definition. Pressure gradient was significantly affected as seen in the results. PBC models exhibited higher velocity, pressure gradient and WSS.

Table 1: Maximum CFD velocity, pressure gradient and WSS.

		Velocity (m/s)	Pressure (Pa)	WSS (Pa)
Model 1	CBC	0.57	236.8	8.07
	PBC	0.86	552.4	12.94
	V3BC	0.70	366.4	10.47
Model 2	CBC	0.47	175.4	6.68
	PBC	0.65	354.3	10.13
	V3BC	0.59	298.1	9.16

Table 2: Percent difference (%) comparing inlet BCs.

		Velocity	Pressure Grad.	WWS
Model 1	PBC vs CBC	50.9	133.3	60.3
	V3BC vs CBC	22.8	54.7	29.7
	V3BC vs PBC	-18.6	-33.7	-19.1
Model 2	PBC vs CBC	38.3	102	51.6
	V3BC vs CBC	25.5	70	37.1
	V3BC vs PBC	-9.2	-15.9	-9.6

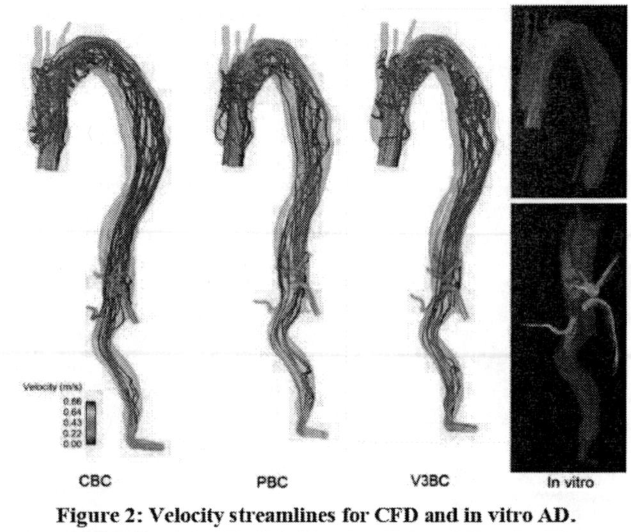

Figure 2: Velocity streamlines for CFD and in vitro AD.

DISCUSSION

Patient-specific CFD models of 3D printed type B ADs were developed, using 4D Flow MRI in vitro velocity data as inlet BCs. Comparisons were made with a constant inlet velocity and a parabolic profile, both derived from in vitro volumetric flow rate. The in vitro system incorporated a patient-specific 3D printed model, allowing controlled repeated scans.

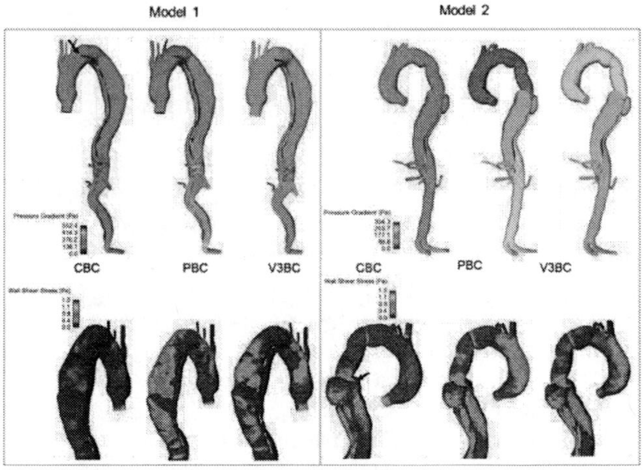

Figure 3: Pressure gradient and WSS. Arrows indicate tears.

Velocity findings give valuable insights in thrombus formation, promoted by recirculation regions, an important AD complication [12]. Maximum velocities were highest for PBC models as expected due to the maximum of the paraboloid velocity for an equivalent in vitro flow. The highest impact of inlet BCs was on pressure gradient, where PBC produced results higher than twice that of CBC. This implies that aneurysm and false lumen growth predictions might be greatly under or overestimated when using flow-derived velocity for inlet BCs [6] rather than velocity-specific conditions. An accurate, patient-specific pressure prediction is relevant for surgical planning or hypotensive medical treatment. This study focuses on inlet effects by defining zero pressure outlets. Vascular resistance and compliance are expected to impact these results. The false lumen exhibited higher WSS with PBC, affecting predictions of vessel wall weakening. The ascending aorta is also highly affected. These results might suggest further risk on the proximal aorta, a greatly complex situation.

In conclusion, this study provides great insight into in vitro AD hemodynamics, showing the importance of inlet BC definition. Velocity, WSS, and especially pressure, are highly impacted. Future steps include flexible 3D printed models, in vitro flow and pressure sensors, and outlet BCs.

ACKNOWLEDGEMENTS

Funding by the UW Dept. of Radiology Research and Development Grant. Authors are grateful to Prof. Tim Osswald, UW Dept. of Mechanical Engineering. GE Healthcare provides MR support at UW-Madison.

REFERENCES

[1] Barbant, S.D. et al., *Am Heart J*, 124: 541-3, 1992.
[2] Khan, I.A. et al., *Chest*, 122:311-328, 2002.
[3] François, C.J. et al., *J Thorac Cardiovasc Surg*, 145:1359-66, 2013.
[4] Clough, R.E. et al., *J Vasc Surg*, 55:914-923, 2012.
[5] Cheng, Z. et al., *J Biomech Eng*, 132:051007-1-051007-9, 2010.
[6] Chen, D. et al., *Biomed Eng Online*, 12:1-16, 2013.
[7] Tse, K.M. et al., *J Biomech*, 44:827-36, 2011.
[8] Alimohammadi, M. et al., *Med Eng Phys*, 36:275-284, 2014.
[9] Karmonik, C. et al., *Technol Health Care*, 16:343-354, 2008.
[10] Roldan-Alzate, A. et al., *J Biomech*, 48:1325-30, 2015.
[11] Johnson, K.M. et al., *Magn Reson Med*, 63:349-55, 2010.
[12] Mohr-Kahaly, S. et al., *Circulation*, 80:24-33, 1989.

PC-MRI DERIVED INLET BOUNDARY CONDITIONS IN CFD MODELS OF HUMAN AORTA: UNCERTAINTY PROPAGATION

S. Bozzi (1), G. De Nisco (2), D. Gallo (2), R. Ponzini (3), G. Rizzo (4), C. Bignardi (2), U. Morbiducci (2), G. Passoni (1)

(1) Department of Electronics, Information Science, and Bioengineering
Politecnico di Milano
Milan, Italy

(2) Department of Mechanical and Aerospace Engineering
Politecnico di Torino
Turin, Italy

(3) HPC and Innovation Unit
CINECA
Segrate, Milan, Italy

(4) IBFM
Research National Council
Segrate, Milan, Italy

INTRODUCTION

In last decades, computational fluid dynamics (CFD) has been extensively applied to study vascular flows, proving to be an effective tool to gain insights into the intricate relationship between hemodynamics and vascular disease. More recently, coupling computational hemodynamics with cardiovascular imaging has allowed to build up even more realistic and personalized CFD models. In general, the reliability of patient-specific CFD results strongly depends on the level of uncertainty introduced in the modeling process, as many sources of uncertainty affect the accuracy of the image-based CFD results. To successfully translate CFD predictions into clinics, all the relevant uncertainty sources should be identified and propagated through the model equations to assess the level of uncertainty of the output variables of interest. This would allow to provide clinicians with model predictions together with the associated uncertainties, thus improving the added value of CFD tools in clinical practice. A number of studies have investigated the sources of uncertainty in computational hemodynamics models: reconstructed vessel geometry [1], boundary conditions (BCs) [2, 3], vessel distensibility and motion [4] and rheological properties of blood [5]. In a recent study, different possible strategies of applying PC-MRI measured flow data as BCs in computational hemodynamic models of human aorta were implemented [2]. The reported findings highlighted that the assumption of idealized velocity profiles as inlet BCs in personalized computational models can lead to misleading representations of the aortic hemodynamics. In this study, we investigate how uncertainty affecting PC-MRI measurements of blood velocity profiles, applied as inflow BCs to a personalized model of human aorta, propagates from the inlet section at the ascending aorta through the aortic territory and results in specific uncertainty of blood flow predictions. By means of Monte Carlo, CFD-based simulations, we provide advice on where, when and how it is important to account for inlet BCs uncertainty affecting PC-MRI measured velocity profiles.

METHODS

PC-MRI was used to obtain the anatomic model of a healthy human thoracic aorta (Fig. 1). Details on in vivo acquisitions and on model reconstruction can be found elsewhere [2]. PC-MRI phase flow measured data were used to prescribe blood flow velocity profiles at the ascending aorta (AAo) inflow section. The finite volume method was applied to solve the fluid motion equations in steady-state conditions. Steady flow analysis was adopted here to limit the overall computational cost of the Monte Carlo procedure. Different flow regimes, corresponding to three different phases of the cardiac cycle were considered: beginning of the systole (T1), peak systole (T2) and halfway of the systolic deceleration phase (T3). The flow regimes were characterized by a Reynolds number, at the AAo section, equal to 608, 5138 and 2497, respectively. Governing equations of motion were solved without turbulence closure, using second-order accuracy. The simulations were performed on a 5 million mesh for flow regimes T1 and T3 and on a finer 18 million mesh for T2. PC-MRI velocity measurements were used to obtain the 3D inlet boundary conditions in terms of velocity profiles at the AAo [2]. An explanatory example of measured velocity profile at the AAo is presented in Fig. 1d. The Monte Carlo method was used to propagate the uncertainty in measured PC-MRI velocity profiles applied as inflow BCs through the CFD model. This technique requires random generation of a large ensemble of inputs from their probability distributions and successive deterministic model simulations to generate many realizations of the output. In this work, at each cell centroid of the AAo inlet section, each PC-MRI velocity component was assumed to have Gaussian distribution, where the mean is equal to the measured PC-MRI value and the standard deviation SD was set by assuming a signal to noise ratio (SNR) of the PC-MRI data equal to 16 [6], corresponding to a coefficient of variation equal to 6.25%. The SD value would imply a maximum deviation of inlet BC velocity data from the mean value of ±18%, when truncating the Gaussian distribution at ±3 SD. Here, the number of Monte Carlo runs was set equal to 100, i.e., the minimum value ensuring the convergence of the probability density functions of the output variables. Hence, for each simulated flow regime and each inlet BC scenario, 100 CFD experiments were performed for a total of 300 numerical simulations. Monte Carlo simulations were used to estimate the empirical probability density functions of hemodynamic quantities of interest at relevant anatomical landmarks. Namely, the uncertainty in the prediction of vessel cross-section averaged flow quantities was estimated at seven aortic cross-sections (Fig. 1). For each cross-section, mean and standard deviation of the empirical cumulative distribution function (ECDFs) of pressure, velocity and vorticity magnitude were estimated in order to estimate the SNR. Wall shear stress (WSS) uncertainty propagation at the luminal surface was also evaluated.

RESULTS

Fig.1 shows the SNR as estimated from the ECDFs of pressure, velocity and vorticity magnitude, at different flow regimes. In general, uncertainty in pressure is the highest and all locations show similar values of SNR. Table 1 reports the SNR of flow variables, averaged over the seven cross-sections selected for the analysis, for flow regimes T1, T2 and T3.

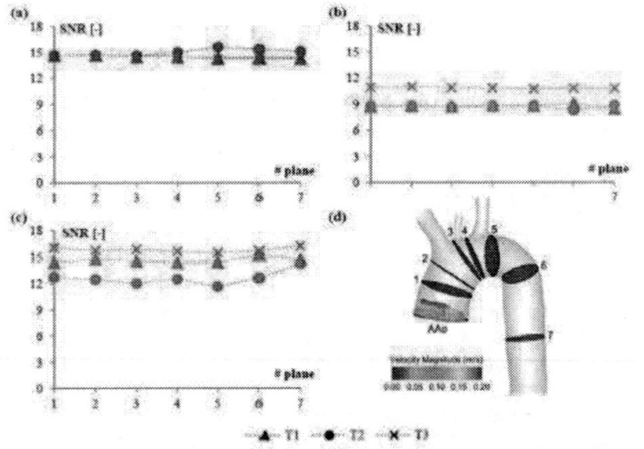

Figure 1: SNR of cross-section averaged blood velocity magnitude (a), pressure (b) and vorticity magnitude (c) at different positions along the aorta (d), at flow regimes T1,T2 and T3.

Data in Table 1 indicate that the predicted uncertainty is always higher than the prescribed input uncertainty (SNR lower than the prescribed SNR=16), with the exception of vorticity magnitude at T3. Noteworthy, blood pressure SNR can be up to 40% lower than SNR source at the inflow section. Regarding uncertainty variability with respect to the flow regime, no common trends can be identified for the three flow variables. In detail, velocity magnitude SNR is almost insensitive to flow regime, while the SNR of vorticity magnitude exhibits the highest variability, with flow regime T2 (peak systole) being the most affected by uncertainty. Finally, pressure uncertainty is higher at flow regimes T1 and T2, characterized by the same SNR, than T3. An additional analysis was performed to investigate the effect of the uncertainty in inlet BCs on WSS magnitude distribution. It was observed that uncertainty affecting WSS markedly differs between flow regime T2 (peak systole), and T1, T3. The value of the coefficient of variation averaged over the whole luminal surface is about 10% at flow regimes T1 and T3, and about 30% at T2. Marked differences were also be observed in the spatial distribution of WSS uncertainty (not shown). In all the cases investigated here, the uncertainty propagation in WSS calculation resulted to be higher than input uncertainty.

Flow regime	mean SNR velocity magnitude	mean SNR pressure	mean SNR vorticity magnitude
T1	14.4	8.8	14.6
T2	15.0	8.7	12.6
T3	14.4	10.9	15.8

Table 1. SNR of the considered flow quantities averaged over the seven cross section in Fig. 1, at flow regimes T1, T2 and T3.

DISCUSSION

This study aimed at investigating the impact of uncertainty in PC-MRI measurements of velocity profiles on patient-specific CFD modeling of aortic hemodynamics. The main findings of the study are that: (1) propagating the inflow BC uncertainty through the Navier-Stokes equations leads to a decrease in the SNR of CFD predictions with respect to the uncertainty source. This result holds for both intravascular flow quantities and WSS distribution, with higher uncertainties for the latter; (2) uncertainty affecting intravascular flow quantities does not present a marked dependence on anatomical location and flow regime. Differently, WSS uncertainty at peak systole is much higher than WSS uncertainty at decelerating/accelerating phases of the systole. One major limit of this work is the steady flow assumption. However, it should be noticed that: in Navier-Stokes equations, uncertainty is mainly propagated by the acceleration operator, where the non-linear advection contribution is expected to be predominant, compared to the local linear term; the computational cost of an unsteady simulation is about two orders of magnitude higher than a steady simulation, hence Monte Carlo unsteady experiments are not affordable unless resorting to HPC platforms. For these reasons, the approach proposed here does make scientific sense and paves the way to further investigations, adopting more efficient stochastic methods to propagate input uncertainty in a more realistic, time-dependent simulation scenario. The approach here adopted emphasizes that PC-MRI flow measurements-derived uncertainty is an important source of uncertainty. This is of utmost importance considering that it is non-linearly related with other uncertainties intrinsic in modeling assumptions. As a consequence, the global effect could not be neglected when looking at model reliability.

ACKNOWLEDGEMENTS

This study was partially funded by the Italian research program (PRIN 2009) "Development and validation of hybrid models for the implementation of patient specific hemodynamics analysis as a tool to support the clinical practice".

REFERENCES

[1] Sankaran, S et al., *J Biomech*, 49(12):2540–2547, 2016.
[2] Morbiducci, U et al. *J Biomech*, 46(1):102–109, 2013.
[3] Schiavazzi, DE et al., *Int J Num Meth Biomed Eng*, in press 2016.
[4] Zhao, SZ et al., *J Biomech*, 33(8):975–984, 2000.
[5] Lee, SW, Steinman DA, *J Biomech Eng* 129(2):273–278, 2007.
[6] Tresoldi, D, et al., *Magn Reson Med*, 71(4):1621–1631, 2014.

NONINVASIVE ESTIMATION OF CORONARY FRACTIONAL FLOW RESERVE (FFR) USING MAGNETIC RESONANCE IMAGING: METHODOLOGY AND PRELIMINARY RESULTS

Jackson B. Hair (1), Lucas H. Timmins (2), John N. Oshinski (1, 3)

(1) Department of Biomedical Engineering
Georgia Institute of Technology and
Emory University
Atlanta, Georgia, United States

(2) Department of Bioengineering
University of Utah
Salt Lake City, Utah, United States

(3) Department of Radiology & Imaging Sciences
Emory University School of Medicine
Atlanta, Georgia, United States

INTRODUCTION

Coronary artery disease (CAD) is one of the leading causes of morbidity and mortality, resulting in over 8 million deaths globally in 2013 and was responsible for 1 in 7 deaths in the United States in 2011 [1, 2]. The underlying cause of CAD is atherosclerotic lesions containing flow-limiting stenoses that develop in the coronary arteries and can cause ischemia or myocardial infarction (MI). Fractional flow reserve (FFR) is the gold standard for determining the functional severity of coronary artery stenosis and is defined as the pressure distal to the stenosis divided by the proximal pressure under hyperemic conditions. Clinically, FFR is measured through percutaneous insertion of a pressure wire.

Recently, coronary computed tomography (CT) has been used with computational fluid dynamics (CFD) modeling techniques to calculate FFR (i.e., FFR_{CT}) noninvasively [3, 4]. However, FFR_{CT} is not optimal as CT requires contrast injection and repeated X-ray imaging, leading to a high radiation exposure. CT also produces imaging artifacts in the presence of calcium, a component frequently present in CAD plaques. Finally, CT cannot measure coronary blood flow, which is of critical importance for appropriate application of boundary condition required for the CFD modelling [4]. Therefore, we developed a methodology using magnetic resonance imaging (MRI) and CFD to estimate FFR (i.e., FFR_{MRI}) and display it as a function of vessel position.

METHODS

We evaluated patients (n = 5) who presented with class III ischemic heart failure and were undergoing a cardiac MRI exam prior to cardiac resynchronization therapy. MRI exams were performed on a 3 T MRI (Trio, Siemens Medical Systems) system using a six-element phased-array cardiac coil. A 3D, whole-heart, navigator and EKG-gated inversion-recovery FLASH sequence with a centric k-space trajectory was used to acquire coronary images in the transverse plane at a resolution of 0.64 x 0.64 x 0.75 mm^3 in diastole during the slow infusion of a Gadolinium-based contrast agent [5].

On the acquired images, a Frangi vessel enhancement filter was applied to extract the lumen contours of the left main (LM), left anterior descending (LAD), and left circumflex coronary arteries in vmtk (the Vascular Modeling Toolkit). A surface was fit to the contours in Geomagic (Geomagic, Inc.), and flow extensions were added to the inlet and each outlet followed by discretization of the geometry in ICEM (Ansys, Inc.). Applied boundary conditions included an inlet velocity of 0.2 and 0.7 m/s for basal and hyperemic flow, respectively, and an inlet pressure of 100 mmHg. Murray's Law was used to approximate the flow split through each branch [6]. CFD simulations were run in Fluent (ANSYS, Inc.) to quantify velocity and pressure values across the computational domain (Figure 1). Following solution convergence, pressure values were extracted within 5 µm of each centerline location and averaged along the LM and LAD. Each averaged centerline pressure value was then divided by the inlet pressure of 100 mmHg to obtain the FFR_{MRI}. FFR_{MRI} was plotted against centerline location, and its value immediately distal to the stenosis was noted to arrive at the final, clinically relevant measurement (Figure 2).

RESULTS

FFR_{MRI} was evaluated immediately distal to the site of stenosis in each vessel under basal and hyperemic conditions (Figure 3). In all five vessels under basal conditions, FFR_{MRI} can be seen to be greater than 0.95, indicating negligible flow obstruction. However, under hyperemic conditions, three of the five exhibit FFR_{MRI} below 0.8, the clinical threshold for intervention [7]. Additionally, in all five vessels, the plotted FFR_{MRI} can be seen to decline very sharply along the site of stenosis but decline negligibly elsewhere (Figure 2).

DISCUSSION

The clinically relevant FFR_{MRI} values obtained align with expectations. As described by Gould *et al*, FFR does not drop significantly under basal conditions until the vessel has been almost entirely obstructed, which was seen here [8]. Conversely, under hyperemic conditions, FFR is expected to drop much more quickly as a function of stenosis anatomic severity. It was also expected that the stenotic regions would exhibit the majority of the drop in FFR_{MRI}, and this can be seen visually (Figures 1C, 2). Despite this trend, however, the anatomic severity was not seen to directly correlate with clinically significant FFR_{MRI} values. While vessel 2 exhibited an anatomically severe stenosis, its FFR_{MRI} was seen to only fall to 0.87 which would indicate a lack of functional severity (Figures 1A, 3). Conversely, vessel 3 appeared to contain only a minor stenosis, but its FFR_{MRI} value fell to 0.79 (Figures 1B, 3). This again was expected as it has been shown previously that anatomic severity is not a strong indication of functional severity [9].

This preliminary study had several limitations. Flow through the coronaries is pulsatile rather than constant as assumed here, and this pulsatility may confer a non-negligible effect to pressure drop. Inlet and boundary conditions were assumed from patient averages obtained with invasive Doppler derived velocity and pressure measurements; it would be more accurate to measure these values through phase contrast magnetic resonance (PCMR) to make the model patient-specific. More data are needed to fully validate the model. Comparisons need to be made between FFR_{MRI} and FFR measured with a pressure wire. These limitations will be addressed in future studies.

In conclusion, these preliminary results obtained using a 3D, contrast-enhanced, whole heart, coronary artery MRI sequence coupled with CFD indicate that this methodology can be used to non-invasively quantify FFR *in vivo*.

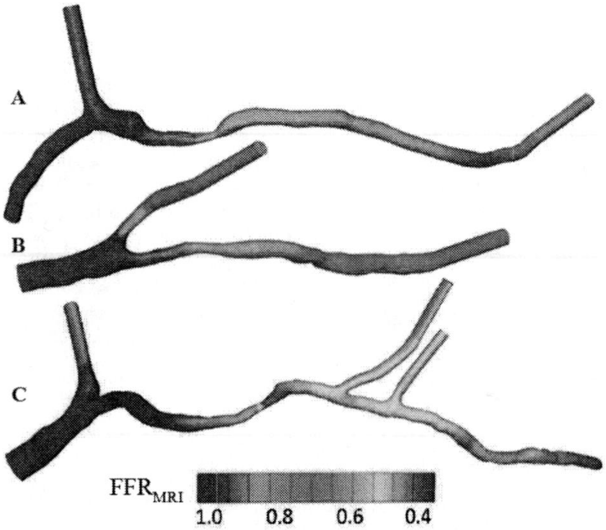

Figure 1: Representative models of vessels 2 (A), 3 (B), and 4 (C) showing estimated FFR_{MRI} values under hyperemic conditions.

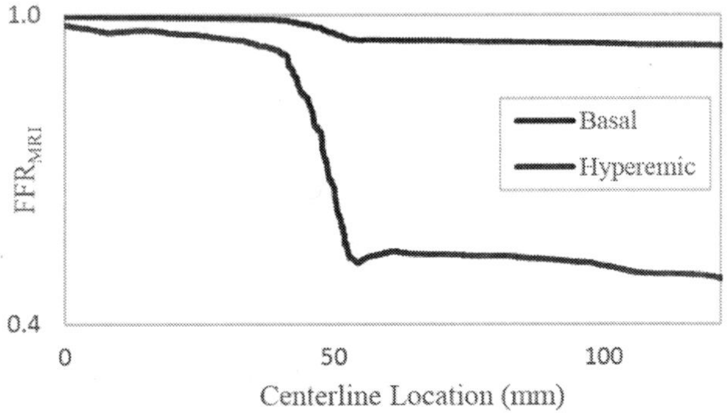

Figure 2: Representative plot of FFR as a function of centerline location for vessel 4 under basal and hyperemic conditions.

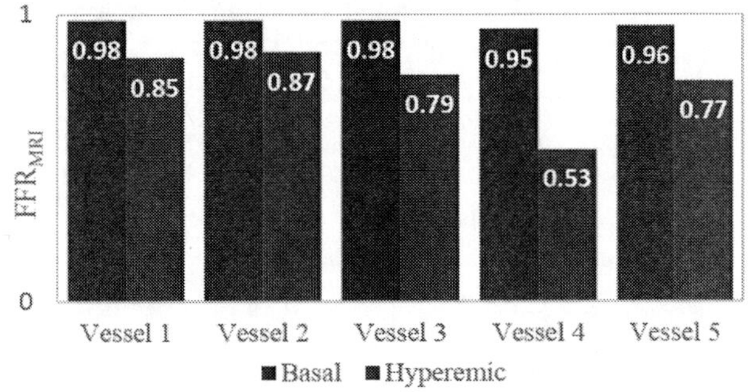

Figure 3: FFR_{MRI} evaluated immediately distal to stenosis in each vessel under basal and hyperemic conditions.

REFERENCES

[1] Naghavi, M., *et al.*, *Lancet,* 385:117-171, 2015.
[2] Mozaffarian, D., *et al., Circulation,* 131:e29-322, 2015.
[3] Koo, B.K., *et al., J Am Coll Cardiol,* 58:1989-97, 2011.
[4] Taylor, C.A., *et al., J Am Coll Cardiol,* 61:2233-41, 2013.
[5] Yang, Q., *et al., J Am Coll Cardiol,* 54:69-76, 2009.
[6] Murray, C.D., *Proc. of the NAS of USA,* 12:207-214, 1926.
[7] Pijls, N.H., *et al., J Am Coll Cardiol,* 49:2105-11, 2007.
[8] Gould, K.L., *et al., Am J Cardiol,* 33:87-94, 1974.
[9] Shaw, L.J., *et al., Circulation,* 117:1283-91, 2008.

FLUID-STRUCTURE-INTERACTION SIMULATIONS OF HEMODYNAMICS IN DATA DRIVEN MODELS OF WILD TYPE AND FIBULIN-5 DEFICIENT MICE

F. Cuomo (1), J. Ferruzzi (2), P. Agarwal (1), C. Li (1), J.D. Humphrey (2), C.A. Figueroa (1,3)

(1) Department of Biomedical Engineering
University of Michigan
Ann Arbor, MI, USA

(2) Department of Biomedical Engineering
□ale University
New Haven, CT, USA

(3) Department of Imaging Sciences and
Biomedical Engineering
King's College London
London, UK

INTRODUCTION

Hypertension affects a large portion of our society; only in the United States 70 million people (1 in 3 adults) suffer from this condition [1]. If untreated, hypertension can lead to heart disease, stroke and kidney failure [2].

Stiffening of the central arteries is recognized as an initiator of hypertension and an indicator of the disease [3, 4]. A stiffer artery propagates the pressure wave faster, which results in an earlier return of reflected waves and an augmentation of central pulse pressure [5, 6]. Furthermore, increased pulse pressure triggers mechanobiological responses with the apparent goal of preserving homeostatic values of circumferential stress and material stiffness [7], often at the expense of increasing structural stiffness (i.e., material stiffness times material stiffness) [8]. This positive feedback loop between local mechanobiology and global hemodynamics can progressively aggravate hypertension and related disease conditions regardless of the underlying cause. There is therefore a clear need to investigate the complex interplay between arterial stiffening, hypertension and hemodynamics.

In this work we combine mice experimental data and computational techniques to assess the effects of regional changes in arterial properties on regional and global hemodynamics. We developed 3-D fluid-solid-interaction mouse-specific models of a wild type mouse and a fibulin-5 deficient mouse (*Fbln5*$^{-/-}$). Fibulin-5 is a protein which is essential in the genesis of effective elastin fiber. First, we assessed anatomical and hemodynamics data in both wild type and *Fbln5*$^{-/-}$. Then we rigorously quantified arterial mechanics, using biaxial testing methods in both mice. Finally, we used the data-driven computational framework CRIMSON [10] which allow to perform fluid-structure simulations of hemodynamics, specifying regionally varying biaxial tissue properties, to perform the simulations of hemodynamics in each mouse.

METHODS

In vivo and in vitro data were collected in wild type and *Fbln5*$^{-/-}$ mice between 20 and 22 weeks of age. Mean blood velocity was acquired with ultrasound at the ascending thoracic aorta and cardiac output was measured with trans-thoracic echocardiography. Central aortic pressure was measured using a SPR-1000 pressure catheter. A micro-CT scan was performed to obtain image data of the mouse aorta and its main branches. The material properties at 4 different locations along the aorta were estimated using biaxial tissue testing and a two-dimensional four fiber family constitutive model. The constitutive model was fitted using pressure-diameter, axial-force and axial stretch data measured in vitro at each location for both wild type and *Fbln5*$^{-/-}$ mice. Specific experimental and theoretical methods for quantifying the biomechanical properties of the vessel wall are described in detail in Ferruzzi et al. [11]. Regional values of linearized biaxial material stiffness over the physiologic range of diastolic-to-systolic pressures were then computed from the constitutive model, using the theory of small on large [12], at in vivo conditions.

The data-driven software CRIMSON was used to simulate the hemodynamics for both mice. The 3D geometrical models were reconstructed from the micro-CT image data; nine pairs of intercostal arteries were subsequently added to each model. A mesh of tetrahedral elements was generated for both models with 2.5 million elements and 500 thousands nodes. A flow waveform, computed from the experimental measurements of blood velocity and cardiac output, was prescribed as inlet boundary condition. Windkessel models were coupled at each outlet, describing the effects of the distal vasculature on flow and pressure. The Windkessel parameters were tuned to match experimentally measured pressure and flow splits. Since a FSI method is necessary to capture pulse wave propagation within an arterial network, a coupled momentum method [13] was employed whereby spatially varying tissue properties were assigned to the vessel wall of the model. The vessel wall was modeled as a linear elastic membrane. Figure 1 reports circumferential and axial material stiffness and thickness assigned at the following locations: ascending thoracic aorta (ATA), descending thoracic aorta (DTA), suprarenal abdominal aorta (SAA), and infrarenal abdominal aorta (IAA). Stiffness and thickness were linearly interpolated to obtain a continuous distribution along the aorta. Finally, an external tissue traction boundary condition was applied on the arterial wall to take into account the effects of perivascular tissues and organs [14].

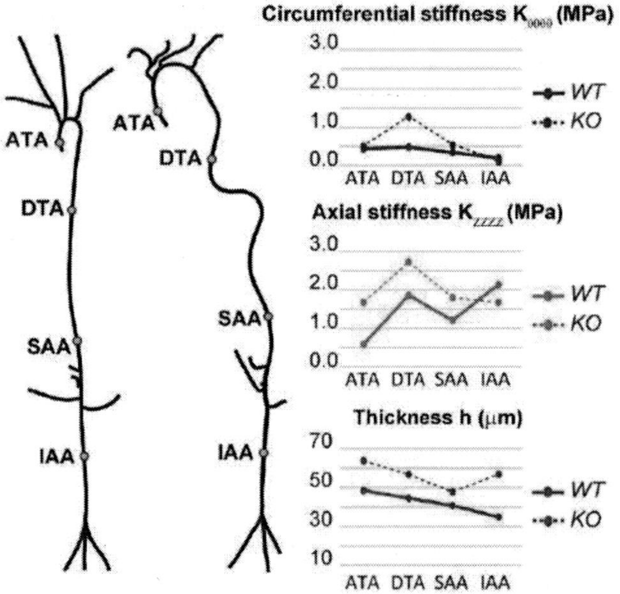

Figure 1: Schematic of wild type (left) and fibulin-5 deficient (right) mice models. Plots show circumferential (dark blue) and axial (light blue) values of material stiffness (MPa) and thickness (green, μm) assigned to four locations: ATA, DTA, SAA and IAA.

RESULTS

Figure 2 shows the experimental and computed local (pressure & flow) and global (pulse wave velocity, PWV) hemodynamics of wild-type (left) and *Fbln5*$^{-/-}$ (right) mice. The *Fbln5*$^{-/-}$ mouse model illustrates how the defective elastic fibers result in longer and tortuous vessels. The computed pressure (red line) reproduces closely the experimental pressure (black line) in the ascending thoracic aorta for both models. In this location, pulse pressure is 43% higher (41 vs. 29 mmHg) in the *Fbln5*$^{-/-}$ model. At the level of the renal arteries, the pulse pressure is 30% higher (25 vs. 19 mmHg) in the *Fbln5*$^{-/-}$ model. PWV, a clinical metric to measure central artery stiffness, was also computed. It revealed, as expected, a 35% larger value (4.6 vs. 3.4 m/s) in the *Fbln5*$^{-/-}$ model.

DISCUSSION

We were able to reproduce the hemodynamics in a wild type and a *Fbln5*$^{-/-}$ mice using a computational framework, which includes: a multi-domain method coupling a 3D vascular model with Windkessel models, a coupled-momentum method for FSI that included regional variation of biaxial stiffness and thickness and an external tissue boundary condition to represent perivascular support.

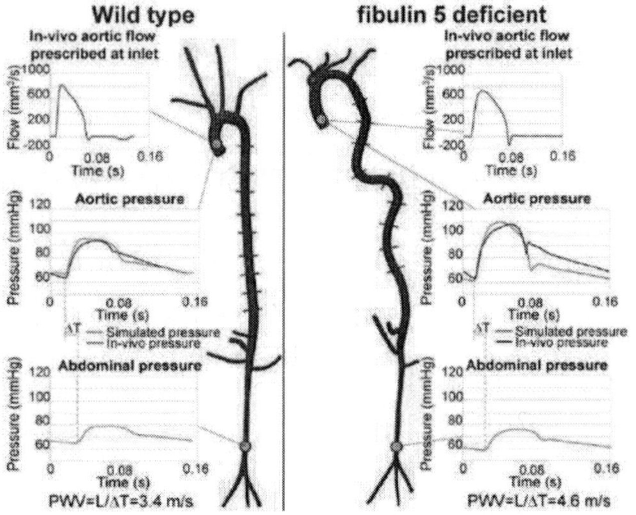

Figure 2: In vivo and computed hemodynamics in both wild type and fibulin-5 deficient mouse models.

We found an increase in pulse pressure and PWV in the *Fbln5*$^{-/-}$ model, due mainly to the increase of thickness. These types of computations can provide valuable insights to investigate the relationship between arterial properties and changes in hemodynamics in mice models of cardiovascular diseases. While hemodynamics metrics such as aortic flow and pressure can be measured experimentally, once calibrated against appropriate data, the computational model provides further information for regions that are not easily accessible experimentally. For example we can compute both flow and pressure in the infrarenal abdominal aorta or in the renal arteries, which are not easily measurable in the mice. This study will be continued to create population average differences in hemodynamics and regional stiffness in wild type and *Fbln5*$^{-/-}$ mice.

ACKNOWLEDGEMENTS

This work was supported by NIH grants (R01 HL086418, R01 HL105297) and by the European Research Council Grant Agreement (n. 307532).

REFERENCES

[1] Heidenreich, PA et al., *Circulation*, 123:933-944, 2011.
[2] Schmieder RE, *Dtsch. Arztebl. Int.*, 107:866-73, 2010.
[3] Dernellis, J. et al., *Hypertension*, 45:426-431, 2005.
[4] Mitchell, GF, *Hypertension*, 64:210-214, 2014.
[5] Najjar, SS et al., *J. Am. Coll. Cardiol.*, 51:1377-1383, 2008.
[6] Wang, KL et al., *Hypertension*, 55:799-805, 2010.
[7] Humphrey, JD, Cardiovascular Solid Mechanics: Cells, Tissues, and Organs, *Springer*, 2002.
[8] Humphrey, JD et al., *Circulation Research*, 118:379-391, 2016.
[9] Kelleher, CM et al., *Curr Top Dev Biol.*, 62:153–188, 2004.
[10] www.crimson.software
[11] Ferruzzi, J et al., *J. Biomech. Eng.*, 137: 031007, 2015.
[12] Baek, S et al., *Comput. Methods Appl. Mech. Eng.* 196:3070–3078, 2007.
[13] Figueroa, CA et al., *Comput. Methods Appl. Mech. Eng.* 195:5685–5706, 2006.
[14] Moireau, P et al., *Biomech. Model. Mechanobiol.* 11:1–18, 2012.

MAPPING LEFT VENTRICLE BLOOD STASIS USING CONVENTIONAL DOPPLER-ECHOCARDIOGRAPHY IN ACUTE MYOCARDIAL INFARCTION

Lorenzo Rossini (1), Pablo Martinez-Legazpi (2), Candelas Pérez del Villar (2), Yolanda Benito (2), Carolina Devesa-Cordero (2), Raquel Yotti (2), Antonia Delgado (2), Ana Gonzalez (2), Andrew M. Kahn (3), Francisco Fernandez-Avilés (2), Javier Bermejo (2), J.C. del Álamo (1)

(1) Mechanical and Aerospace Engineering
UC San Diego
La Jolla, CA, USA

(2) Department of Cardiology
Hospital Universitario Gregorio Marañón
Madrid, Spain

(3) Department of Cardiovascular Medicine
UC San Diego
La Jolla, CA, USA

INTRODUCTION

Cardiac embolism is a major source of mortality and disability worldwide, and intracardiac blood stasis is the major risk factor for cardioembolism [1]. Impaired blood washout in the cardiac chambers is thought to cause the formation of intracardiac clots that may eventually dislodge. However, the mechanistic foundations of intracardiac blood stasis are poorly understood. Clarifying the mechanisms that lead to intracardiac thrombosis therefore warrants visualizing and quantifying chamber stasis. Several methods suitable for assessing intracardiac flows in the clinical setting have been recently implemented [2-5]. Recently, a method for obtaining patient-specific spatio-temporal maps of blood stasis in the LV from conventional color-Doppler ultrasound images has been implemented [6]. We hypothesized that this method is useful for quantifying the abnormalities in cardiac stasis caused by disease, deciphering the mechanisms of increased stasis and, eventually, individualizing therapeutic decisions using tailored indices of cardioembolic risk.

Acute myocardial infarction (AMI) is an excellent clinical scenario to prospectively test these hypotheses. During the first three months after an anterior-AMI, the incidence of LV thrombosis (LVT) can be as high as 15-20% in the presence of systolic dysfunction [7,8]. Thereafter, the prevalence of LVT declines as stunned myocardium recovers after reperfusion. Nevertheless, if systolic dysfunction persists, the risk of cardiac embolism exceeds the general population for years [9]. Hence, sequential studies after AMI allow for addressing the time-evolving three-way interactions between regional myocardial function, intraventricular stasis, and mural thrombosis.

Additionally, a noninvasive method for LVT risk assessment would be of particular clinical value in the setting of AMI because intense antiplatelet therapy narrows the risk-benefit margin of oral anticoagulation. Therefore, the present study was designed to address the clinical potential of an ultrasound-based stasis imaging method in a prospective cohort of patients with an anterior AMI. In particular, we aimed to characterize the stasis abnormalities induced by AMI. We also assessed the biomechanical factors that determine stasis in AMI, and we tested the performance of these stasis metrics to predict LVT.

METHODS

We obtained intraventricular stasis maps from conventional Doppler echocardiography sequences in 73 patients in the early (<72 hours) and follow-up (4-5 months) phases after an anterior-AMI. 37 healthy controls were also studied. We calculated the spatio-temporal maps of blood residence time (T_r) inside the LV as previously described in [6], and characterized the main features of stagnant regions.

The residence time is a metric that quantifies the time spent by a blood particle inside the ventricle before it leaves the chamber. We have shown that spatio-temporal maps of T_r can be efficiently obtained from the 2D+t velocity fields by integrating the equation of advection of a passive scalar with unit forcing. We ran the solver for 8 consecutive cardiac cycles, ensuring temporal convergence. From the T_r maps, we calculated the average residence time of the entire blood volume inside the LV ($T_{r,LV}$). This is a representative metric of global stasis accounting for the full blood pool in the ventricle. To account local stasis, which may be particularly meaningful for mural thrombosis, we also identified and tracked blood regions with a $T_r \geq 2$ s, hereinafter defined as stagnant regions. We measure relative size, mean T_r and perimeter of contact with the endocardium of the stagnant regions. The features of stagnant regions were automatically measured at the instant of mitral valve opening, and regions not spanning a full cardiac cycle or smaller than 2% of LV area were dismissed. Whenever more than one stagnant region was identified, only the apical-most was analyzed. The contact perimeter of the stagnant regions with the endocardium accounts for flow-endocardium interactions that most intensively activate the coagulation cascade in the setting of AMI.

To further account for flow-myocardium and flow-flow interactions ongoing during filling, we classified filling jets as either wall jets whenever most of its trajectory took place attached to the LV wall or free jets in cases in which the filling volume advanced freely in the chamber towards the apex (visual assessment of 4 consecutive beats of the T_r maps).

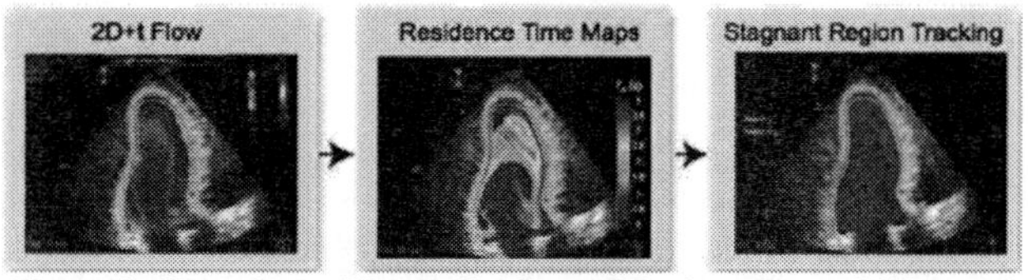

Figure 1: Overview of the methods used for image acquisition and obtaining the T_r maps.

RESULTS

Residence time of intraventricular blood was longer in early than in late phases of AMI (2.2 ± 0.8 *vs* 1.9 ± 0.8 s, $p= 0.009$) and also longer in AMI than in controls (1.6 ± 0.7 s, $p= 0.001$). We identified the largest stagnant regions in early AMI studies, which also showed the longest contact perimeters with the endocardium. Multiple regression predictors of stasis were the propagation of the early-filling jet (std-beta= -0.15), apical myocardial strain (std-beta= 0.21), and impingement of the filling jet to the ventricular walls (std-beta= -0.44; all $p< 0.05$). Stasis metrics performed well to predict LVT (see ROC in Figure 1, C index= 0.80, 95% CI: 0.66 - 0.94) within a selected population of patients on whom the presence or absence of LVT could be unequivocally addressed.

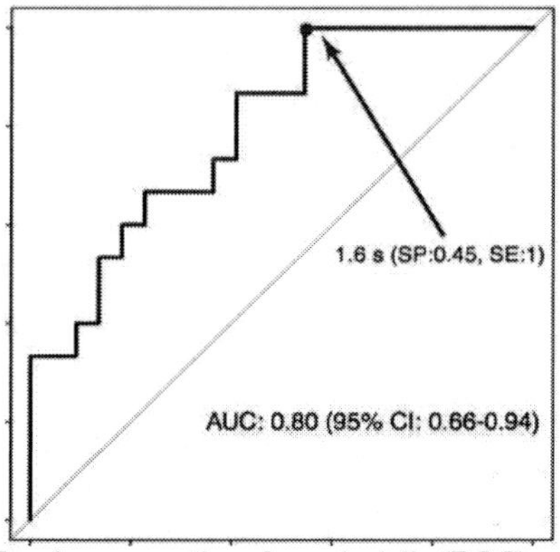

Figure 2: Receiver operating characteristic (ROC) curve for the averaged residence time in the LV after 8 beats in the selected population of patients.

DISCUSSION

We demonstrate that objective indices of ventricular stasis can be derived from bedside Doppler echocardiography. Using this technique, we describe important blood mixing abnormalities in patients after an anterior AMI. We also show that stasis imaging sensitively captures the improvement in blood washout that follows myocardial wall motion recovery over time. Furthermore, we identify several biomechanical factors predisposing to thrombosis. These observations support that stasis imaging techniques not only provide new insight into cardiac physiology but also may be useful to develop risk biomarkers on which to personalize primary prevention anticoagulation strategies.

Two-dimensional LV blood stasis mapping allows quantification of increased intraventricular stasis caused by an anterior AMI and subsequent improvement during follow-up. Anomalous blood stasis is due not only to abnormal apical regional function but also to individual factors of inflow direction. Severe blood transport abnormalities and increased stasis can be identified in patients who develop or are at risk for developing LVT after an anterior AMI. Thus, stasis mapping is a promising method to assess the risk of embolism in cardiac diseases.

REFERENCES

[1] Adams, PC et al., *J Am Coll Cardiol*, 8:76B-87B, 1986.
[2] Garcia, D et al., *IEEE Trans Med Imaging*, 29:1701-13, 2010.
[3] Bermejo, J et al., *Ann Rev Fluid Mech*, 47:315-342, 2015.
[4] Bolger, AF et al., *J Cardiovasc Magn Reson*, 9:741-747, 2007.
[5] Eriksson, J et al., *Am J Physiol Heart Circ Physiol*, 300:H2135-41, 2011.
[6] Rossini, L et al., *J Biomech*, 49:2152-61, 2016.
[7] Vecchio, C et al., *Circulation*, 84:512-9, 1991.
[8] Weinsaft, JW et al., *JACC Cardiovasc Imaging*, 9:505-15, 2016.
[9] Hachet, O al. *Stroke*, 45:3514-20, 2014.

SB³C2017
Summer Biomechanics, Bioengineering and Biotransport Conference
June 21 – 24, Tucson, AZ, USA

COLOR DOPPLER ECHOCARDIOGRAM VELOCIMETRY FLOW RECONSTRUCTION USING STREAMFUNCTION-VORTICITY FORMULATION

Brett A. Meyers (1), Craig J. Goergen (2), Carlo Scalo (1), Pavlos P. Vlachos (1,2)

(1) Department of Mechanical Engineering
Purdue University
West Lafayette, IN, USA

(2) Weldon School of Biomedical Engineering
Purdue University
West Lafayette, IN, USA

INTRODUCTION

Traditional diagnostic measurements of cardiovascular blood flow rely on one-dimensional (1D) Doppler echocardiography. [1, 2] Although these protocols are standard practice, over-simplification of underlying physics coupled with inter- & intra-operator variability can affect overall prognosis. [1] Color Doppler echocardiography provides two-dimensional (2D) blood velocity representations that are less susceptible to the aforementioned constraints; however, velocities are measured perpendicular to the ultrasound probe, making interpretation of complex flow patterns difficult. Estimating the parallel velocity component by vector mapping produces improved measurements providing a foundation for more robust metrics.

The most frequently used vector mapping estimator is Vector Flow Mapping (VFM), wherein the continuity equation is applied to construct 2D two-velocity component (2D-2C) vector maps from 2D one-velocity component (2D-1C) scans. [3-6] This method, however, produces large, non-physical velocity gradients due to over-simplification of the influence of wall motion and surrounding flow on local measurements, impacting accuracy of post-scan measurements.

We propose a new algorithm for vector mapping cardiovascular flow based on the stream function – vorticity (ψ–ω) formulation, a 2D incompressible flow computational method. We believe this is a fundamentally simple yet powerful numerical solver, which we demonstrate as an improved diagnostic algorithm.

METHODS

Solver Formulation

The central term of the ψ–ω formulation relates ψ to the forcing ω through the Poisson equation (Eqn. 1).

$$\nabla^2\psi = \omega \quad (1)$$

The two-dimensional incompressible flow assumption requires the velocity field to be divergence-free. Therefore the velocity, $\boldsymbol{u}$, is related to ψ by partial derivatives (PDE) (Eqn. 2).

$$u = \nabla \times \psi \quad (2)$$

Using the standard equation for ω by taking the curl of $\boldsymbol{u}$, combined with the above equations, constitutes the ψ–ω formulation.

Since Color Doppler scans capture near instantaneous velocity information, we evaluate acquisitions for each velocity frame independent of time. An isotropic approximation model of ω [7] is used to initialize our solver for faster convergence (Eqn. 3).

$$\omega^0 = -2\frac{\partial v_{Doppler}}{\partial x} \quad (3)$$

We employ a Gauss-Siedel algorithm with Successive Over-relaxation to solve the ψ–ω formulation on a collocated Cartesian grid. Dirichlet boundary conditions are implemented for inflow, outflow, and cardiac wall motion. The physical mechanism involved in left ventricle (LV) motion facilitates the assignment of boundary conditions *a priori*. Since ventricular inflow or outflow adds or removes mass from the system, we approximate our boundary conditions as being free-penetration, to mimic motion of the heart wall as either expanding or contracting as reflected in Figure 1. This allows us to provide suitable values for ψ, and for velocities to be approximated and imposed as boundary conditions.

Solver Validation

Validation of our proposed method was compared against results obtained from the VFM algorithm, using a computational LV model. This model contained velocity fields for a single heartbeat based on healthy cardiac conditions. Artificial Color Doppler images for the complete dataset (200 frames) were generated using Field II ultrasound transducer modeling software [8, 9]. Line density and scan noise were parametrically varied to quantify velocity estimate error, with added transducer noise modulated between 0% to 20% of the artificial image dynamic velocity. Relative mean and random error between each algorithm and computational model ground truth were quantified, normalized to the peak inflow velocity.

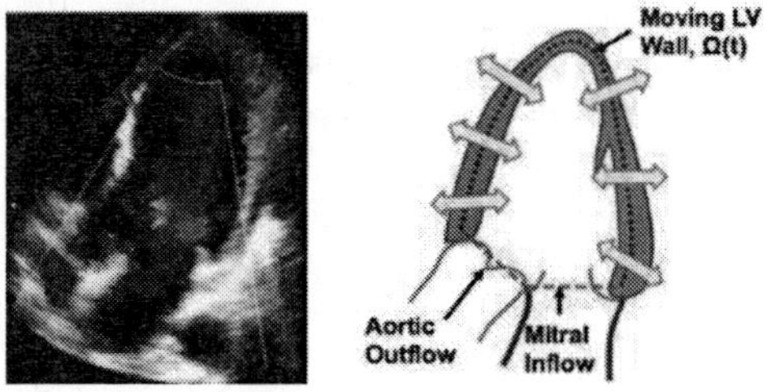

Figure 1: Patient geometry (left) and modeling cardiac function (right) are used to define boundary conditions.

Experimental Data Evaluation

A dataset of mouse two-chamber LV Color Doppler flow was collected to demonstrate performance on experimental data. A Vevo 2100 small animal ultrasound system was used for imaging. Color Doppler echocardiograms were acquired with gating based on ECG and respiration signals. At each gated position along the cardiac cycle, all corresponding frames were averaged to generate the representative frame. In total, 60 equally spaced temporal positions across the cardiac cycle were collected.

RESULTS

A sample set of fields at peak inflow is presented (Figure 2) from the computational model case at 5% added noise and 128 line density. Each field depicts the instantaneous vector field, overlaid onto the colored vorticity field, with closed contours depicting coherent structures. Contours were determined by the λ_{CI} criterion, using a 2% threshold. During peak inflow, a shear layer is expected to develop along the inflow jet, which produces a counter-clockwise vortex near the center of the LV, shown in the CFD field (Figure 2, middle). Our proposed method (Figure 2, left) captures this vortex with similar intensity and position. VFM also captures this vortex (Figure 2, right), although elongated and with additional non-physical structures.

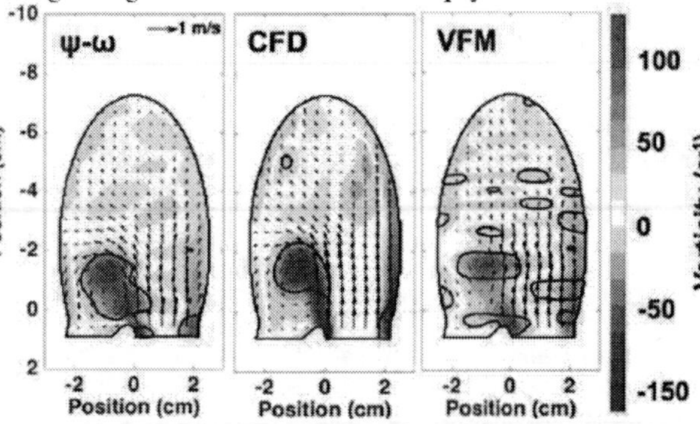

Figure 2: Peak inflow quiver fields for each solver alongside simulation data.

Relative mean error reflects a zero-bias for our proposed method across all test conditions; conversely, VFM mean errors are all negatively biased around 4%. Relative RMS errors reflect a two-fold improvement in accuracy for our proposed method against VFM under all test conditions (minimum error: $\sigma_{\psi\text{-}\omega} = 8\%$, $\sigma_{VFM} = 17\%$).

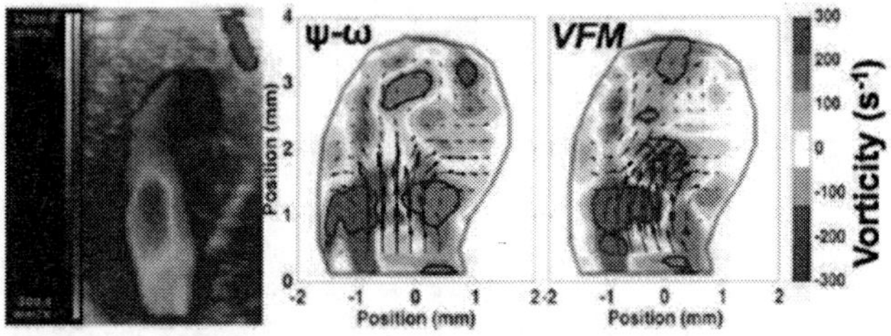

Figure 3: Peak inflow reconstruction for each solver based on mouse two-chamber Color Doppler echocardiography.

Mouse model reconstructions (Figure 3) show similar behavior at peak inflow. Our proposed method captures two vortex structures near the mitral valve (base of the image), suggesting the mitral vortex ring has been captured. The VFM field captures only a portion of the mitral vortex ring and the direction of inflow is biased toward the apex where no flow is present.

DISCUSSION

This work proposes a new methodology for velocity vector field reconstruction from Color Doppler echocardiography using the ψ-ω formulation. We benchmark our method against the most common VFM algorithm by evaluating both simulated and experimental data. Results show a two-fold improvement in the accuracy of lateral velocity estimates, as well as mitigating unfavorable artifacts produced in the VFM reconstruction.

VFM utilizes a 1D line integral; therefore each line is an independent estimate of lateral velocities. This may ensure a smooth velocity field but does not ensure the gradients are smooth, manifesting as "streaks" or "bands" in vorticity measurements. This effect has been noted in recent literature [10-12], but has only been accounted for by excessively smoothing Doppler velocities [11].

The main advantage of our method stems from the integration of the 2D PDE. Our method ensures all components of the velocity vector field gradient are smooth by integrating over the area of the velocity field, not along individual lines, eliminating the need for excessive smoothing operations.

ACKNOWLEDGEMENTS

The authors would like to acknowledge Dr. Patrick Segers for contributing computational data used during validation.

REFERENCES

[1] Nagueh, SF, et al., J Am Coll Cardiol, 30(6) 1527-1533, 1997.
[2] Nishimura, Ra, et al., Depression, 30 8-18, 1997.
[3] Garcia, D, et al., IEEE Trans Med Imaging, 29(10) 1701-1713, 2010.
[4] Hendabadi, S, et al., Ann Biomed Eng, 41 2603-2616, 2013.
[5] Bermejo, J, et al., Am J of Physiol Heart Circ Physiol, 306(5) H718-H729, 2014.
[6] Martínez-Legazpi, P, et al., J Am Coll Cardiol, 64(16) 1711-1721, 2014.
[7] Mehregan, F, et al., Ultrasound Med Biol, 40(1) 210-221, 2014.
[8] Jensen, JA, Proc. Paper presented at the 10th Nordic-Baltic Conference on Biomedical Imaging, 351-353 Year.
[9] Jensen, Ja, The Journal of the Acoustical Society of America, 89 182-190, 1991.
[10] Uejima, T, et al., Ultrasound Med Biol, 36(5) 772-788, 2010.
[11] Itatani, K, et al., Jpn J Appl Phys, 52(7S) 07HF16, 2013.
[12] Pedrizzetti, G, et al., Ann Biomed Eng, 43(1) 26-40, 2015.

ENHANCE DELIVERY OF NANOPARTICLES ACROSS THE BLOOD-BRAIN BARRIER IN BRIAN TUMORS THROUGH AUTOCATALYSIS

Gang Deng (1), Liang Han (1), Sasidhar Murikinati (2), Jaime Grutzendler (2), Joseph M. Piepmeier (1), Jiangbing Zhou (1,3)

Department of Neurosurgery
□ale University
New Haven, CT, USA
(1)

Department of Neurology
□ale University
New Haven, CT, USA
(2)

Department of Biomedical Engineering
□ale University
New Haven, CT, USA
(3)

INTRODUCTION

Improved pharmacological treatment of brain cancer has been limited by the lack of delivery platforms that are able to efficiently overcome the blood-brain barrier (BBB). The BBB in brain tumors is partially disrupted but remains largely impermeable [1]. To overcome this barrier, nanotechnology approaches are promising, as nanoparticles (NPs) can be engineered by a variety of mechanisms for brain-targeted delivery [2]. However, nanotechnology for systemic drug delivery to the brain is still in its infancy. Existing engineering approaches often fail to enhance systemic delivery of NPs to the brain to a degree sufficient for treatment purposes.

To overcome this challenge, we propose an autocatalytic brain tumor-targeting (ABTT) delivery strategy (Fig. 1). With this method, a small fraction of NPs enter the brain tumor microenvironment through a traditional ligand-receptor mediated delivery, in addition to BBB leakage. After reaching tumors, NPs locally release BBB modulators, which in turn transiently enhance BBB permeability to allow additional NPs to enter the same region. Through this secondary autocatalysis mechanism, the delivery procedure creates a positive feedback loop. As a result, the efficiency of NP accumulation in tumors autocatalytically increases with time and subsequent administrations. We tested this delivery strategy by synthesizing ABTT NPs and evaluating them in a variety of mouse models.

METHODS

Polymer III-62% with maleimide functional groups was synthesized according to our recent reports [3,4]. ABTT NPs were fabricated through the standard emulsion procedures [3]. After emulsion and evaporation, NPs were collected by centrifugation and reacted with a selected thiolated ligand.

Tumor bearing mice were established through injection of tumor cells intracranially (U87MG and GL261) or intracardially (MDA-MB-231-BR). NPs were encapsulated with IR780 for imaging, and paclitaxel (PTX) for therapeutic evaluation.

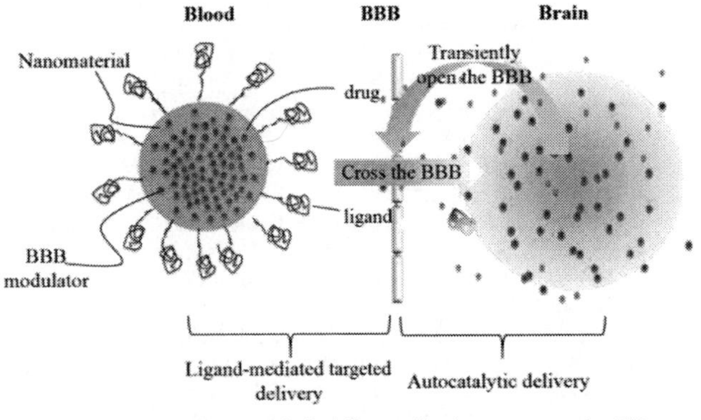

Figure 1. Schematic of autocatalytic delivery of brain tumor-targeting NPs.

RESULTS

Autocatalysis enhanced delivery of NPs to brain tumors ABTT NPs were synthesized using III-62% as the starting materials and were surface conjugated with chlorotoxin (CTX, ~100 per NP), a peptide ligand for brain tumor targeting [4], and internally encapsulated with lexiscan (LEX ~1.0% by weight), a BBB modulator for autocatalysis [5]. The resulting NPs were in spherical morphology and in size of ~160 nm (Fig. 2a). To determine delivery efficiency, ABTT NPs were encapsulated with IR780, a near-infrared dye, and intravenously injected to GL261 glioma-bearing mice. Twenty-four hours later, mice were euthanized. Major organs were isolated and imaged using an IVIS system. We found that ABTT NPs preferentially accumulated in tumors: the concentration of ABTT NPs in brain tumors was 4.3 times greater than that in the liver based on fluorescence intensity (Fig. 2b). The efficiency of ABTT NPs was originated from a synergistic effect of CTX and LEX, as NPs modified with either one alone had significantly lower efficiency (Fig. 2b). By using high-resolution confocal microscopy, we revealed that ABTT NPs accumulated preferentially in tumors but not in normal brain (Fig. 2c-e). Confocal images confirmed that ABTT NPs crossed the BBB, as suggested by the presence of NPs outside of blood vessel (Fig. 2f).

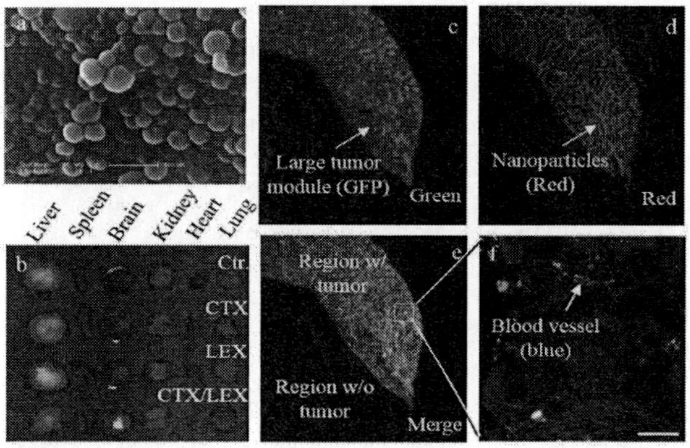

Figure 2. ABTT NPs for systemic drug delivery to brain tumors. (a) Morphology of ABTT NPs, as captured by scanning electron microscope (SEM). Scale bar: 500 nm. (b) Representative fluorescence images of organs isolated from mice received indicated treatments. (c-e) Representative confocal images showed that intravenous administration of ABTT NPs resulted in preferential NP accumulation in tumors. f) A representative confocal image showed that ABTT NPs crossed the BBB and entered tumors.

ABTT as a general mechanism for enhancing brain cancer-targeted NP delivery and treatment ABTT NPs were tested in two additional brain tumor models, including U87MG human glioma model and MDA-MB-231-BRHer2 (231-BR-Her2) human brain metastasis model. Results in Fig. 3a showed that ABTT NPs maintained high efficiency across all tested models. We assessed ABTT NPs for systemic delivery of chemotherapy to brain tumors in the 231-BR-Her2 tumor model. PTX was selected as the model drug, which was encapsulated with an efficiency of 15% by weight. Results in Fig. 3b showed that PTX-loaded ABTT NPs significantly enhanced tumor-bearing mouse survival: median survival time for mice receiving PTX-loaded ABTT NPs, which was 63 days, was significantly longer than those for mice receiving

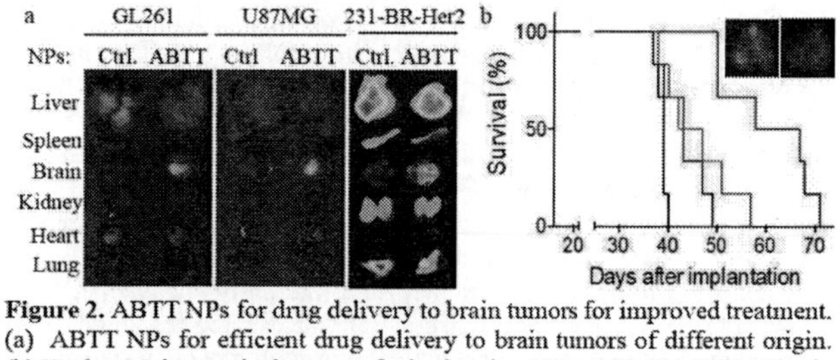

Figure 2. ABTT NPs for drug delivery to brain tumors for improved treatment. (a) ABTT NPs for efficient drug delivery to brain tumors of different origin. (b) Kaplan-Meier survival curves of mice bearing MDA-MB-231-BrHer2 brain metastases with treatment of PTX-loaded NPs (purple), free PTX (brown), blank NPs (blue) and PBS (black). Inserts: Representative fluorescence images of residual tumor after treatment of blank NPs (left) and PTX-loaded NPs (right).

either PBS, blank ABTT NPs or free PTX, which were 39, 43 and 45, respectively, ($p < 0.05$). In accordance with this finding, mice treated with blank ABTT NPs had many large lesions in the brain at day 35, whereas the mice that received treatment of PTX-loaded ABTT NPs had only one single small lesion (Fig. 3b, inserts).

New combinations of ligand and BBB modulator for improved efficiency We explored whether the efficiency of ABTT NPs in the original formulation could be further improved by testing different combinations of ligand and BBB modulator. Results in Fig. 4 showed that several new combinations, particularly the combination of LEX and 4H2, a lupus anti-DNA autoantibody (unpublished), allowed for delivery efficiencies significantly great than that used in the original formulation.

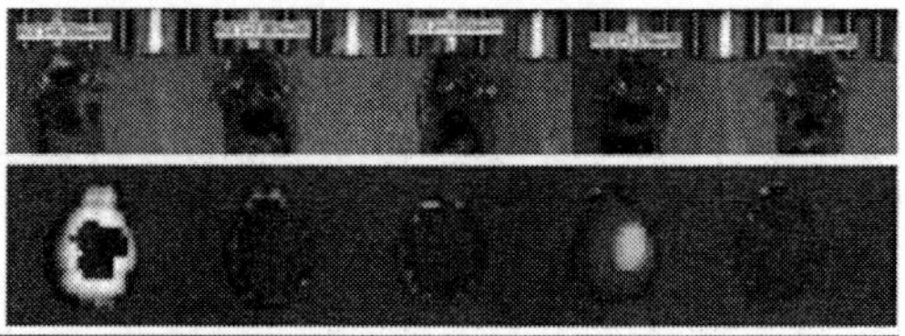

Ligand	CTX	CTX	GA	4H2	3E10
BBB modulator	LEX	Minoxidil	LEX	LEX	LEX

Figure 4. ABTT NPs of different formulation for systemic drug delivery to brain tumors. GA: gambogic acid.

DISCUSSION

We demonstrated ABTT as a novel mechanism for efficiently overcoming the BBB for drug delivery to brain tumors. Due to their unprecedented efficiency in crossing the BBB, their great capacity to accommodate and deliver cargo agents, and their construction from biodegradable materials with limited toxicity, we anticipate that ABTT NPs will serve as a promising approach for clinical management of patients with brain tumors.

ACKNOWLEDGEMENTS

This work was supported by R01NS095817, R21NS095147, American Brain Tumor Association, and the State of Connecticut.

REFERENCES

[1] Blakeley J, Curr Neurol Neurosci Rep, 8, 235-241, 2008.
[2] Patel, T et al., Adv Drug Deliv Rev, 64, 701-705, 2012.
[3] Han L et al., ACS nano, 10, 4209-4218, 2016.
[4] Zhou J et al., Nat Mater, 11, 82-90, 2012.
[5] Deshane, J et al. J Biol Chem, 278, 4135-4144, 2003.
[6] Carman, AJ. et al. J Neurosci, 31, 13272-13280, 2011.

SB³C2017
Summer Biomechanics, Bioengineering and Biotransport Conference
June 21 – 24, Tucson, AZ, USA

ULTRAFAST NEAR-INFRARED LIGHT-TRIGGERED UNCAGING TECHNIQUE FOR PROBING CELLULAR SIGNALING

X. Li (1), Z. Che (2), K. Mazhar (3), T. Price (3), Z. Qin (1,4,5)

(1) Department of Mechanical Engineering
The University of Texas at Dallas
Richardson, Texas, USA

(2) Department of Materials Science and Engineering
The University of Texas at Dallas
Richardson, Texas, USA

(3) School of Behavioral and Brain Sciences
The University of Texas at Dallas
Richardson, Texas, USA

(4) Department of Bioengineering
The University of Texas at Dallas
Richardson, Texas, USA

(5) Department of Surgery
The University of Texas Southwestern Medical Center
Dallas, Texas, USA

INTRODUCTION

Modulation of critical cellular signaling pathways can have profound benefits in restoring cellular function and providing therapeutic treatment [1]. Strategies that can regulate the spatiotemporal dynamics of cell signaling in living systems will open up new avenues to understand cell functions and potentially lead to new disease treatments. So far, the most commonly used approach is by caged compounds. However, the development and validation of a new caged compound is a challenging process to ensure biological inertness, solubility in physiological pH, resistance to aqueous hydrolysis, fast uncaging speed, and high uncaging efficiency [2]. These drawbacks of caged compounds have prompted scientists to develop light-addressable vehicles as a general platform for caging a wide range of biomolecules, especially for on-demand uncaging.

Recently, it has been shown that ultrashort laser pulses activate plasmonic nanoparticles to generate nanoscale cavitation bubbles that are effective to burst release the encapsulated compounds [3]. This is one of the fastest methods to uncage a broad range of molecules and represents an exciting opportunity to probe cell signaling events by accelerating payload uncaging from minutes to sub-milliseconds. This ultrafast speed gives the capability to cross a critical biological threshold to study fast cell signaling processes such as neurotransmission or calcium signaling [4]. In this report, we investigate the use of this nanoscale cavitation phenomena as a new uncaging technique to remotely stimulate intracellular signaling. Specifically, bioactive molecules are packed into plasmonic gold nanoparticles-coated liposomes, or plasmonic liposomes. The nanobubble cavitation generates mechanical forces to eject payloads out of the liposomes and furthermore, for endocytosed plasmonic liposomes, overcoming the endo-lysosomal barrier (Figure 1). Here we demonstrated the capability to activate cell signaling by uncaging inositol 1,4,5-trisphosphate(IP_3), a second messenger for calcium (Ca^{2+}) signaling with single laser pulse exposure.

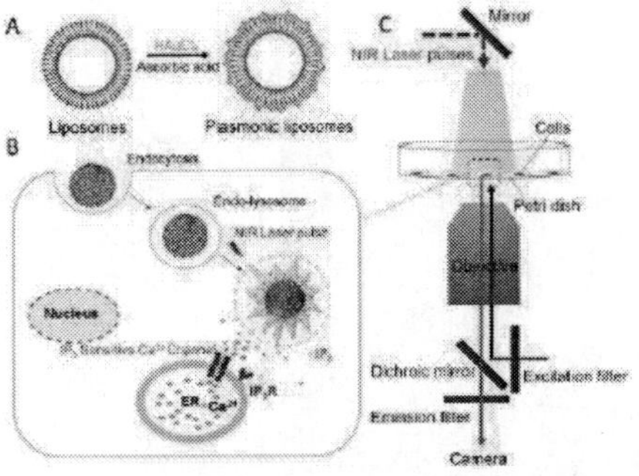

Figure 1 (A) Formation of plasmonic liposomes; (B) Concept of the near-infrared light triggered intracellular uncaging to probe cell signaling; (C) Schematic of the experimental setup.

METHODS

Plasmonic liposomes were prepared by a two-step method. First, liposomes were prepared using the film hydration method, followed by extrusion. Gold nanoparticle was then deposited onto the surface of liposomes via reduction of chloroauric acid ($HAuCl_4$) by ascorbic acid as previously reported [5]. Plasmonic liposomes were characterized by transmission electron microscopy (TEM), dynamic light scattering measurement and UV- VIS spectroscopy.

The ultrafast uncaging behavior of plasmonic liposomes were characterized by imaging real-time calcein uncaging from plasmonic liposomes. Calcein was encapsulated within plasmonic liposomes at 50 mM, a concentration at which its fluorescence is self-quenched.

Technical Presentation #236

For IP_3 uncaging, cancer cells and dorsal root ganglion (DRG) neurons were incubated with plasmonic liposomes containing 730 μM IP_3. After 2 hours incubation to allow endocytosis, IP_3 containing plasmonic liposomes were washed and replaced with fresh cell medium. Single near-infrared laser pulse was applied to trigger IP_3 uncaging. Visualization of IP_3 intracellular uncaging was monitored by loading the cells with a calcium indicator Fluo-4.

RESULTS

The deposition of gold nanoparticles on liposomes surface was confirmed by transmission electron microscopy (TEM). Gold nanoparticles in the range of 2.0-5.3 nm formed discrete gold clusters surrounding the liposome core (Figure 2). The plasmonic coupling between gold nanoparticles leads to a near-infrared absorption at around 750 nm, while non-coated liposomes did not show resonant peak in the near-infrared wavelengths. Upon near-infrared laser pulse irradiation, a new absorption peak appeared around 530 nm and suggests the dissociation of gold nanoparticles from the liposome surface.

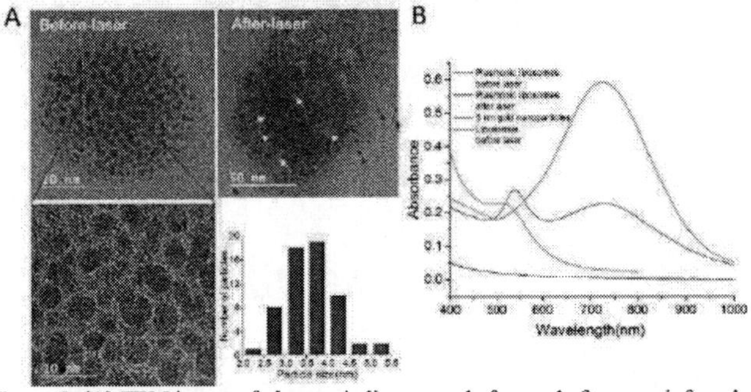

Figure 2. (A) TEM image of plasmonic liposomes before and after near-infrared laser irradiation; (B) UV-Vis spectra of plasmonic liposomes before and after laser irradiation (750 nm, 60 mJ/cm^2), naked 5 nm gold nanoparticles and uncoated liposomes.

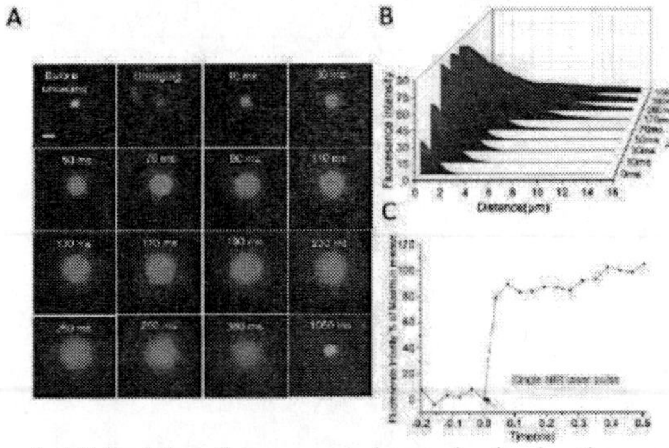

Figure 3. (A) Real-time fluorescence micrographs showing the uncaging of single plasmonic liposome (4 μm) upon irradiation with single picosecond pulse at 750 nm. (B) Fluorescence intensity analysis of individual plasmonic liposome uncaging as a function of time and distance. (C) The uncaging kinetics of 136 nm plasmonic liposomes. Scale bar, 5 μm.

Single near-infrared pulse rapidly release calcein from plasmonic liposomes (Figure 3). The fluorescent intensity immediately increased both within and beyond plasmonic liposomes predominately due to the dilution and thus de-quenching of calcein. The fluorescence intensity inside plasmonic liposomes reaches the maximum at around 70 ms. We also investigated the uncaging kinetics of plasmonic liposomes with diameter around 100 nm and observed an ultrafast calcein uncaging speed within 0.1 ms (Figure 3C).

Calcium imaging showed that single near-infrared laser pulse led to a spontaneous increase in the Fluo-4 fluorescent signal (Figure 4) in the laser pulse energy range 38-127 mJ/cm^2. The fluorescence image sequence in Figure 4B shows rapid efflux of Ca^{2+} from internal stores into the cytosol and then propagation into the nucleus. In control groups, no Fluo-4 fluorescence increase was observed for cells treated with non-coated liposomes, suggesting that IP_3 was trapped inside liposomes and endo-lysosomes and thus unable to bind to the IP_3R to trigger the calcium release. Further, we observed slight Fluo-4 intensity increases in cells that were treated with plasmonic liposome without IP_3, however, the magnitude of the Fluo-4 intensity change is much smaller compared with plasmonic liposomes with IP_3.

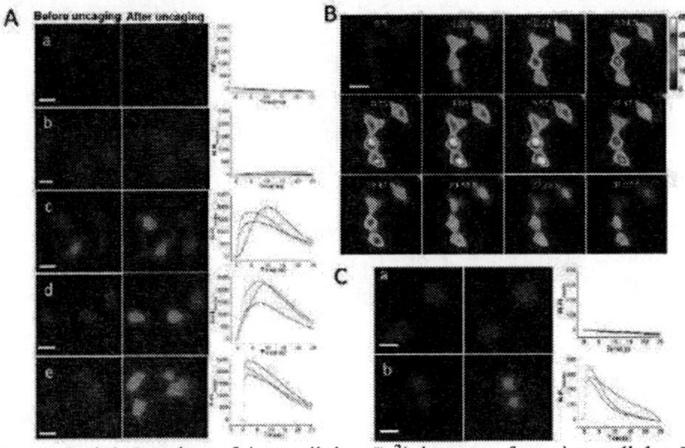

Figure 4. (A) Imaging of intracellular Ca^{2+} increase from intracellular IP_3 uncaging in MDA-MB-231 cells and Fluo-4 fluorescence intensity as a function of time, a) non-coated liposomes with IP_3 ,127 mJ/cm^2; plasmonic liposomes without IP_3, 127 mJ/cm^2; plasmonic liposomes with IP_3, c) 38 mJ/cm^2; d) 63 mJ/cm^2; 127 mJ/cm^2; (B) Fluo-4 fluorescence microscopy image sequence in MDA-MB-231 cells. (C) IP_3 uncaging in primary culture of DRG neurons, a) non-coated liposomes IP_3,127 mJ/cm^2; b) plasmonic liposomes with IP_3, 127 mJ/cm^2. Scale bar, 20 μm.

DISCUSSION

Our approach combines plasmonic liposomes with near-infrared picosecond laser pulses to generate nanoscale cavitation bubbles to uncage the packed biomolecules within 0.1 milliseconds. The gold nanoparticle coating on the liposome surface allows efficient energy conversion from a single laser pulse to nanosecond cavitation events. The ultrafast molecular uncaging technique enabled by nanoscale cavitation provides a general uncaging tool to probe cell signaling events that span a wide timescale using a broad range of bioactive molecules. We anticipate that our technique will help investigate and better understand cell signaling and facilitate the development of improved therapies and diagnostics.

ACKNOWLEDGEMENTS

This work was partially supported by UT Brain seed grant from The University of Texas System--Neuroscience and Neurotechnology Research Institute (UTS--NNRI), a NSF grant (1631910), a postdoc research grant from the Phospholipid Research Center Heidelberg, Germany (1603574), and a NIH grant (NS065926).

REFERENCES

[1] Orsinger GV et al., ACS nano. 8:6151-62, 2014.
[2] Ellis-Davies GC et al., Nature methods. 4:619-28, 2007.
[3] Lukianova-Hleb EY, et al., Nature medicine. 2014;20:778-84.
[4] Sudhof TC et al., Neuron. 80:675-90, 2013.
[5] Troutman TS, et al., Advanced materials. 2009;21:2334-8, 2009.

SB³C2017
Summer Biomechanics, Bioengineering and Biotransport Conference
June 21 – 24, Tucson, AZ, USA

MOTION OF A NANO-SPHEROID IN A CYLINDRICAL VESSEL FLOW: BROWNIAN AND HYDRODYNAMIC INTERACTIONS; IMPLICATIONS FOR TARGETED DRUG DELIVERY

N. Ramakrishnan (1), Y. Wang (2), D. M. Eckmann (1, 3), R. Radhakrishnan (1, 4), P. S. Ayyaswamy (2)

(1) Department of Bioengineering
University of Pennsylvania
Philadelphia, PA 19104, USA

(2) Department of Mechanical Engineering
and Applied Mechanics
University of Pennsylvania
Philadelphia, PA 19104, USA

(3) Department of Anesthesiology and Critical Care
University of Pennsylvania
Philadelphia, PA 19104, USA

(4) Department of Chemical and Biomolecular Engineering
University of Pennsylvania
Philadelphia, PA 19104, USA

INTRODUCTION

In targeted drug delivery (TDD), ligand functionalized nano-sized particles, or nanocarriers (NCs) are commonly used to deliver drugs to specific locations inside the vasculature. The dynamics of these particles in a confined environment, such as in a narrow blood vessel, is governed by a complex interplay between the hydrodynamic forces, Brownian interactions, wall effects, and adhesive interactions of the ligands with specific receptors expressed on the vessel wall. The magnitude of each of these effects is governed by a number of factors including the size and the shape of the NC, the size of the vessel, the flow rate, the hematocrit density, and the expression levels of the receptor molecules on the vascular surface [1].

In TDD, spheroidal shaped nanocarriers are preferred since they afford greater efficacy of binding. Spheroidal shaped nanoparticles are also of interest in microrheology. Here, we numerically determine the motion of a buoyant or a nearly neutrally buoyant nano-sized spheroid in a fluid filled tube without or with an imposed pressure gradient (weak Poiseuille flow) [2]. The Fluctuating hydrodynamics approach and the Deterministic method are both employed. The fluctuation-dissipation relation and the principle of thermal equipartition of energy are both satisfied. The major focus is on the effect of the confining boundary. Results for the velocity and the angular velocity autocorrelations (VACF and AVACF), the diffusivities, and the drag and the lift forces as functions of the shape, the aspect ratio, the inclination angle, and the proximity to the wall are presented. For the parameters considered, the boundary modifies the VACF and AVACF such that three distinct regimes are discernible — an initial exponential decay, followed by an algebraic decay culminating in a second exponential decay. The first is due to the thermal noise, the algebraic regime is due both to the thermal noise and the hydrodynamic correlations, while the second exponential decay shows the effect of momentum reflection from the confining wall. Our predictions display excellent comparison with published results for the algebraic regime (the only regime for which earlier results exist).

METHODS

We consider a spheroidal NC immersed in an incompressible, quiescent or flowing Newtonian fluid contained in a cylindrical tube Σ, as shown in Fig. 1. The inlet and outlet boundaries are denoted by Σi and Σo, respectively, Σw is the wall boundary, and the particle surface is denoted by Γp. The dimensions of the particle are denoted by a, b, and c, and the length and diameter of the tube are L and D, respectively. The position of the particle (i.e., its center of mass) is expressed either in terms of r, the radial distance from the tube axis, or h, the radial distance as measured from the wall boundary. The angular orientation of the particle is measured in terms of the inclination angle θ which denotes an in-plane tilt (in the x-z plane). At $\theta = 0°$, a is the dimension of the NC along the tube axis, while b and c are those along the radial

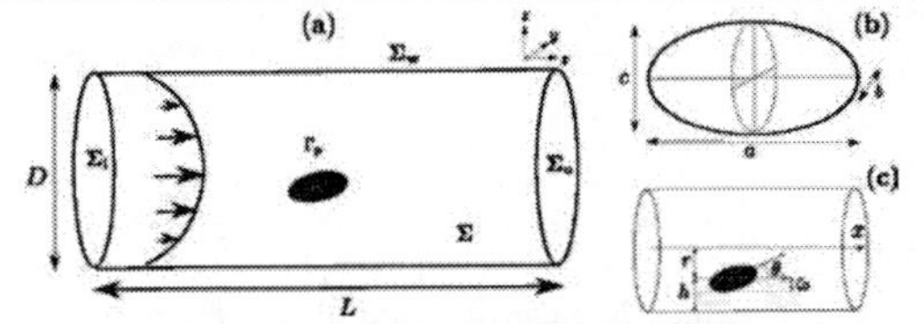

Figure 1: Schematic of the problem.

directions, and this is illustrated in Fig. 1. We describe the shape of the NC with respect to its dimensions at $\theta = 0°$ and define the aspect ratio as $\varepsilon = a/c$. Hence in our notations, $a < b = c$ for an oblate spheroid ($\varepsilon <$ 1.0) and $a > b = c$ for a prolate spheroid ($\varepsilon >$ 1.0).

Fluid equations:

$$\nabla \cdot \boldsymbol{u} = 0;\ \rho \frac{D\boldsymbol{u}}{Dt} = \nabla \cdot \left[-p\underline{I} + \mu(\nabla \boldsymbol{u} + \nabla \boldsymbol{u}^T) + \underline{S}\right], \text{ where}$$

$$\langle S_{ij}(\boldsymbol{x}, t)\rangle = 0, \text{ and,}$$

$$\langle S_{ik}(\boldsymbol{x}, t) S_{lm}(\boldsymbol{x}', t')\rangle = 2k_B T \mu(\delta_{il}\,\delta_{km} + \delta_{im}\delta_{kl})\delta(\boldsymbol{x} - \boldsymbol{x}')\delta(t - t').$$

Particle equations:

$$m\,\frac{d\boldsymbol{U}}{dt} = \int_{\Gamma_p} \underline{\sigma}\,.\,\hat{\boldsymbol{n}}\; dS$$

$$\underline{I}\,\frac{d\boldsymbol{\Omega}}{dt} + \boldsymbol{\Omega} \times (\underline{I}\,\boldsymbol{\Omega}) = -\underline{R}^T \int_{\Gamma_p} (\boldsymbol{x} - \boldsymbol{X}) \times (\underline{\sigma}\,.\,\hat{\boldsymbol{n}})\; dS$$

Initial Conditions:

$$\boldsymbol{U}(t = 0) = 0,\ \boldsymbol{\Omega}(t = 0) = 0, \boldsymbol{u}(t = 0) = 0 \text{ in } \Sigma$$

Boundary Conditions:

$$\boldsymbol{u} = \boldsymbol{u}_{\text{in}} \text{ on } \Sigma_i, \quad \underline{\sigma}\,.\,\hat{\boldsymbol{n}} = 0 \text{ on } \Sigma_o, \boldsymbol{u} = 0 \text{ on } \Sigma_w,$$

$$\boldsymbol{u} = \boldsymbol{U} + \underline{R}\boldsymbol{\Omega} \times\ (\boldsymbol{x} - \boldsymbol{X}) \text{ on } \Gamma_p$$

The above equations subject to IC & BC are solved using an ALE based finite element scheme and weak formulation. Importance is given for the considerations of added masses and added moments of inertia [3]. The VACF and AVACF are calculated using the stochastic approach and the Deterministic method.

RESULTS

We now present the results for $C_{U_\alpha}(t)$, the VACF for $\alpha = x, y$, and $C_{\omega_\alpha}(t)$, the AVACF for $\alpha = 1, 2$, and the direction dependent diffusivities $D_{\alpha\alpha}$ for spheroidal particles. The VACF and AVACF exhibit three distinct temporal regimes and we describe these scaling behavior in units of the viscous relaxation time t_v. Both the VACF and AVACF show an exponential decay for $t < t_v$, which crosses over to

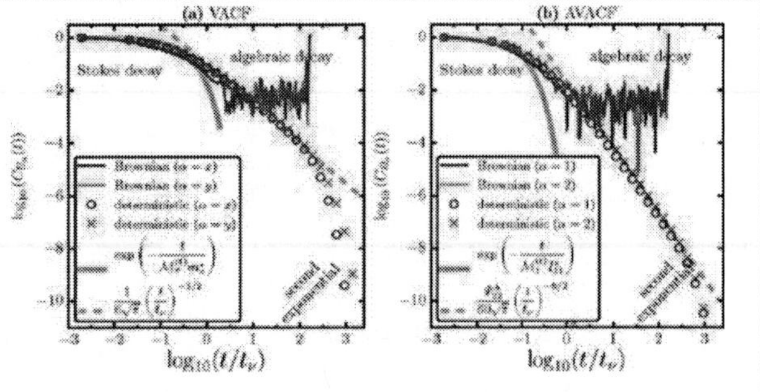

Figure 2: (a) VACF and (b) AVACF for a spheroidal particle.

an algebraic decay for $t > t_v$ and culminates in a second exponential decay for $t \gg t_v$ which is a consequence of the interactions with the confining boundary. In Figure 2, we display these features for an NC (of aspect ratio 1.5 and a = 600 nm) placed at the center of a cylindrical tube. In both the panels, the solid lines correspond to data obtained from stochastic simulations and the symbols denote those obtained using the Deterministic method. The VACF is independent of particle shape

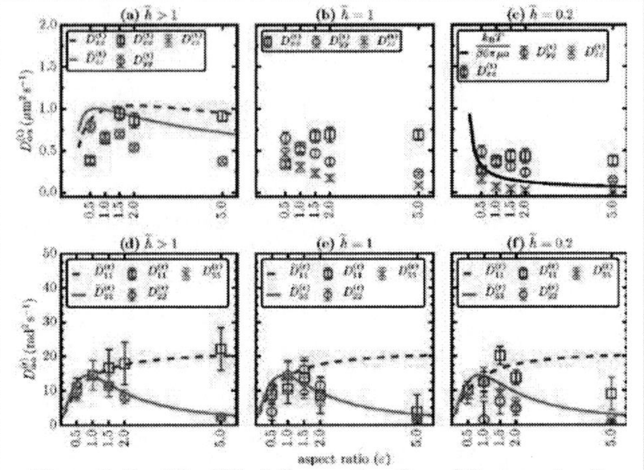

Figure 3: Particle diffusivity as a function of the aspect ratio and the separation from the wall.

while the AVACF shows a strong dependence of the particle aspect ratio.

We estimate the direction dependent particle diffusivities using both the stochastic approach and the Deterministic method and the corresponding values $D_{\alpha\alpha}$ as a function of particle aspect ratio and the separation from the wall are shown in Figure 3. The separation of the particle from the wall are expressed as $\tilde{h} > 1\ (bulk)$, $\tilde{h} = 1\ (near\ wall)$, and $\tilde{h} < 1\ (lubrication)$. The computed values of the direction dependent diffusivities agree well with the Stokes-Einstein long time translational diffusivities for $\tilde{h} > 1$. For $\tilde{h} \leq 1$ we find the diffusivities along all the directions decrease with $\tilde{h}$ and are smaller compared to the bulk values due to increased viscous effects. These observations have profound consequences for predicting particle binding.

DISCUSSION

We have studied the Brownian and flow driven motions of a spheroidal NC in a cylindrical fluid filled vessel as functions of the aspect ratio, the proximity to the bounding wall, and the angle of inclination. The various equilibrium and the transport properties we have developed in this context may be used to understand how hydrodynamic forces influences the binding efficacy of a functionalized nanocarrier. Results obtained also impact microrehology. A concluding observation is that the stochastic hydrodynamics approach presented herein provides a suitable framework for the simulations of biofluids where finite temperature effects should be retained.

ACKNOWLEDGEMENTS

The ALE based DNS methodology was developed in part by support from National Institute of Health (NIH) grant R01-EB006818. The computation of the equilibrium and transport properties was supported in part by NIH grant U01-EB016027.

REFERENCES

[1] Ayyaswamy, P. S et al., J. Nanotechnol. Eng. Med. 4 (1), 011001, 2013.

[2] Ramakrishnan N et al., In Press, Journal of Fluid Mechanics, 2017.

[3] Korotkin A., Added Masses of Three-Dimensional Bodies in Infinite Fluid. Springer, 2009

SYNTHETIC SECOISOLARICIRESINOL DIGLUCOSIDE ATTENUATES MECHANICAL HYPERALGESIA & SPINAL INFLAMMATION IN A RAT MODEL OF PAINFUL RADICULOPATHY

Christine L. Weisshaar (1), Melpo Christofidou-Solomidou (2), Beth A. Winkelstein (1,3)

(1) Department of Bioengineering
University of Pennsylvania
Philadelphia, PA, USA

(2) Department of Medicine
University of Pennsylvania
Philadelphia, PA, USA

(3) Department of Neurosurgery
University of Pennsylvania
Philadelphia, PA, USA

INTRODUCTION

Neck pain is a debilitating disorder that affects up to 70% of adults [1,2]. Radicular pain can develop from many causes including disc herniation and foraminal stenosis which compress the nerve roots [3]. In association with pain, neural injury produces widespread inflammation, including glial activation at the injury site and in the spinal cord [4-7]. Secoisolariciresinol diglucoside (SDG) is the primary bioactive lignan in flaxseed and has both antioxidant and anti-inflammatory properties [8,9]. SDG acts as a free radical scavenger and activates transcription factors that, in turn, regulate enzymes to protect tissue from oxidative damage [10]. For example, SDG reduces pro-inflammatory cytokine levels in peritoneal fluid in mice with asbestos-induced lung inflammation [10]. Although SDG reduces pain, inflammation, and oxidative stress in models of diabetes, peripheral inflammation, hyperplasia, and radiation [10-14], little is known about its effect on pain and inflammation in the context of neural trauma.

Given its robust anti-inflammatory and antioxidant properties, SDG is hypothesized to reduce the inflammation associated with radicular pain from neural trauma. Since other anti-inflammatory drugs have been shown to attenuate pain [15,16], SDG may also provide pain relief. Accordingly, this study administered LGM2605, the synthetic version of SDG [17,18] in an established rat model of painful radiculopathy to investigate its effects on mitigating pain and spinal glial activation. Based on work in other pre-clinical models, SDG was administered after pain was established. Two separate dosing paradigms were selected, with SDG given either as a single dose of 100mg/kg or as multiple doses of 200mg/kg. The timing and amount of dosages were based on efficacy in other rodent inflammatory models [10,19].

METHODS

All experimental procedures were IACUC approved and used male Holtzman rats (275-299g). Rats underwent either a nerve root compression or sham surgery [15,16,20]. The right C7 dorsal nerve root was compressed for 15 minutes using a 10gf vascular clip. Sham surgeries had root exposure but no compression. After compression, rats received either a single dose of *100mg/kg SDG* on 1 day (n=3), repeated doses of *200mg/kg SDG* on days 1-3 (n=8), or repeated *PBS vehicle* injections on days 1-3 (n=5) given subcutaneously. Similarly, *sham* rats received 200mg/kg SDG on days 1, 2, and 3 after surgery (n=6). Forepaw hyperalgesia was assessed by measuring the withdrawal threshold to stimulation by von Frey filaments before surgery (baseline; day 0) and daily until day 7 [7,20]. Treatment was given immediately after behavioral testing. Thresholds were compared between groups and across time using a two-way repeated measures ANOVA with Tukey post-hoc test.

After behavioral testing on day 7, the right C7 spinal cord was harvested from a subset of rats (*200mg/kg SDG* n=6; *100mg/kg SDG* n=3; *PBS vehicle* n=4; *sham* n=4) and sectioned (14μm). Spinal cords from normal naïve rats were also harvested (n=2). Slides were labeled for glial fibrillary acid protein (GFAP; 1:500; Millipore) expression as a measure of astrocytic activation. An ANOVA with post-hoc Tukey test compared the extent of GFAP expression in the superficial dorsal horn on the side of the injury between groups relative to normal.

RESULTS

The withdrawal threshold was reduced immediately by day 1 from baseline for all injury groups ($p<0.0380$) but sham responses were unchanged (Figure 1). The threshold remained reduced in the rats receiving the PBS vehicle through 7 ($p<0.0001$) (Figure 1). As early as

1 day after the first dose of 200mg/kg SDG the thresholds returned to baseline levels and remained above the pre-treatment (day 1) thresholds for the remaining days ($p<0.0021$) (Figure 1). In contrast, a single, lower dose of SDG (100mg/kg) did not alter thresholds from injury responses, with responses remaining significantly reduced from baseline for all test days ($p<0.0382$), except day 3 ($p=0.0852$) (Figure 1). Overall, thresholds for 200mg/kg SDG were not different from sham but were significantly different from both 100mg/kg SDG ($p<0.0008$) and PBS vehicle ($p<0.0003$) responses.

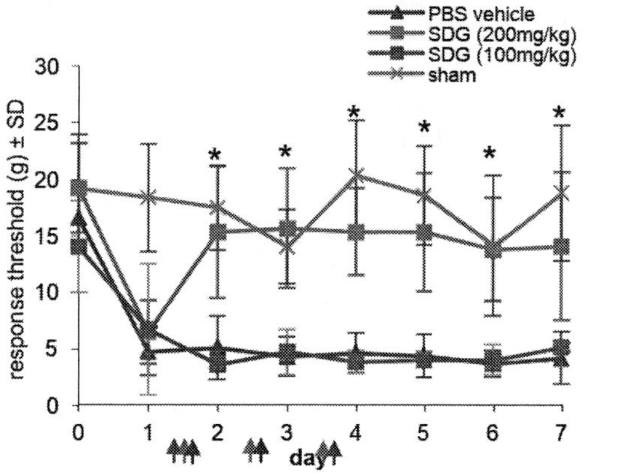

Figure 1: Hyperalgesia is eliminated with repeated doses of 200mg/kg SDG compared to those with PBS vehicle (*p<0.0165). Arrows indicate times of SDG or PBS administration.

Spinal astrocytic expression at day 7 followed pain responses (Figures 2 & 3). After painful nerve compression, GFAP expression in the superficial dorsal horn was reduced only with repeated 200mg/kg SDG administration and similar to sham levels (Figure 2). When quantified, this reduction was significantly lower ($p<0.0063$) than that with the PBS vehicle treatment and was at normal levels (Figure 3).

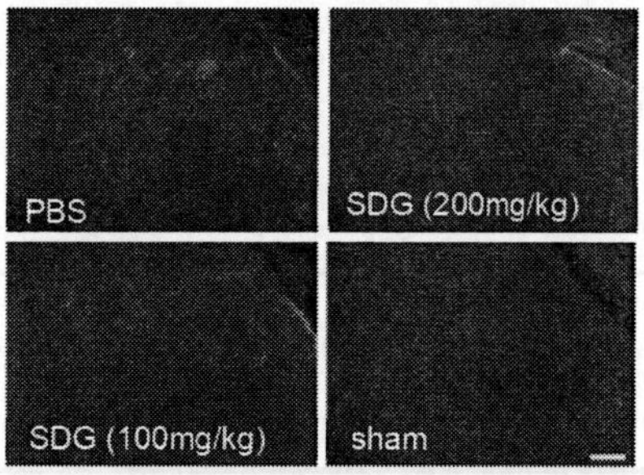

Figure 2: Representative images of GFAP labeling in the ipsilateral spinal dorsal horn at D7. The scale bar is 100μm.

DISCUSSION

This is the first report, to our knowledge, of systemic administration of SDG abolishing radicular pain (Figure 1). However, this effect appears to depend on the magnitude and timing of dosing since only the 200mg/kg SDG given three times was effective (Figure 1). Of note, a single delivery of SDG at 30mg/kg in a mouse model of diabetic neuropathy was ineffective in reducing pain, with a repeated dose being needed to reduce pain [11]. Since our current study varied *both* the dose and the timing of treatments, additional controlled studies are needed to determine whether a single administration of the 200mg/kg dose of SDG may be effective in mediating pain or having long-lasting effects. However, the analgesic effects of SDG do persist for at least three days after treatment was terminated (Figure 1), which is consistent with a prior study showing pain relief for diabetic neuropathy [11]. However, since in that study, the relief was only temporary [11], it is possible that the pain relief is only transient; studies with longer-term follow-up are needed.

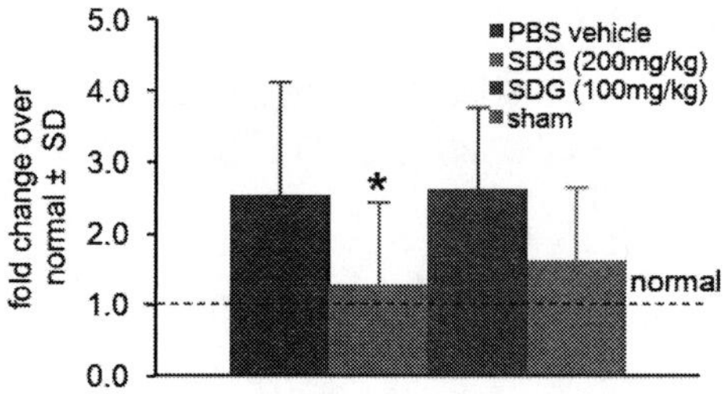

Figure 3: GFAP expression at day 7 decreases (*p<0.006) from injury levels only with the repeated 200mg/kg SDG dosing.

Although the antioxidant and anti-inflammatory effects of SDG are well established [8-11], this is the first study to show the parallel reduction in both pain *and* inflammation using SDG (Figures 1-3). Flaxseed acts via multiple mechanisms to reduce inflammation and oxidative stress, including inhibiting the release of pro-inflammatory cytokines like TNFα [10,21]. Since TNFα has been found to increase in the spinal cord at day 1 after painful nerve root compression [16,22], and its upregulation can activate astrocytes [23], it is possible that SDG is acting, at least partially, through such a mechanism. Indeed, the prevention of the spinal astrocytic activation that is typical with painful radiculopathy (Figures 2 & 3) supports this notion.

Pain reduction may also be attributed to SDG's antioxidant properties. Lipid peroxide which is typically increased in the sciatic nerve and spinal cord in diabetic neuropathy has been shown to be reduced by SDG treatment [11]. Of note, in that study, SDG was administered orally [11], which may have different pharmacokinetic effects than the subcutaneous administration that was used in the current study. Further work is needed to both more-fully characterize the oxidative stress responses after painful nerve root compression and to examine potential optimal delivery approaches for SDG. Nevertheless, these findings suggest SDG or its synthetic counterpart, LGM2605, may be a novel therapeutic for pain.

ACKNOWLEDGEMENTS

This work is funded by an NIH/NCCIH (#AT008291) grant.

REFERENCES

[1] Côte, P et al., *Spine*, 15:S60-74, 2008; [2] Haldemanm, S et al., *Spine*, 15:S5-7, 2008; [3] Abbed, K et al., *Neurosurgery*, 60:S28-34, 2007; [4] Deleo, J, *JBJSurg Am*, 88:58-62, 2006; [5] Chang, Y et al., *J Neurotrauama*, 28:2429-38, 2011; [6] Weisshaar, C et al., *Biomaterials*, 32:9738-46, 2011; [7] Nicholson, K et al., *J Neurosci Res*, 92:116-29, 2014; [8] Herchi, W et al., *J Oleo Sci*, 63:7-14, 2014; [9] Imran, M et al., *Nutr J*, 14:71, 2015; [10] Pietrofesa, R et al., *Carcinogenesis*, 37:177-87, 2016; [11] Hu, P et al., *Eur J Pharmacol*, 767:183-92, 2015; [12] Velalopoulou, A et al., *Int J Mol Sci*, 17:E7, 2015; [13] Prasad, K & Dhar, A, *Curr Pharm Des*, 22:141-4, 2016; [14] Ren, G et al., *Appl Physiol Nutr Metab*, 41:1303-10, 2016; [15] Rothman, S et al., *J Neurosci Res*, 87:2709-17, 2009; [16] Philips, B et al., *Comp Med*, 67:1-9, 2017; [17] Mishra, O et al., *Bioorg Med Chem Lett*, 2013; [18] Mishra, O et al., *Radiat Res*, 2014; [19] Pietrofesa, R et al., *Radiation Research Society Ann Conf*, PS7-22, 2016; [20] Smith, J et al., *Biomaterials*, 80:96-105, 2016; [21] Puukila, S et al., *PloS One*, 30:e0122852, 2015; [22] Rothman, S et al., *J Pain*, 10:90-9, 2009; [23] Zhang, L et al., *Neuroreport*, 11:409-12, 2000.

THE ROLE OF NANOPARTICLE DESIGN IN DETERMINING ANALYTICAL PERFORMANCE OF LATERAL FLOW ASSAYS

Li Zhan (1), Yan Gong (2), David R. Boulware (3), Feng Xu (2), Warren C. W. Chan (4), John C. Bischof (1)

(1) Department of Mechanical Engineering
University of Minnesota
Minneapolis, MN, USA

(2) Bioinspired Engineering and
Biomechanics Center
Xi'an Jiaotong University
Xi'an, Shaanxi, China

(3) Department of Medicine
University of Minnesota
Minneapolis, MN, USA

(4) Institute of Biomaterials and
Biomedical Engineering
University of Toronto
Toronto, ON, Canada

INTRODUCTION

Point-of-care diagnostics are designed to provide fast and simple measurements to facilitate timely medical decision making, thus improving clinical outcomes. However, none are currently able to provide sensitivity and quantitation comparable to laboratory based diagnostics, such as polymerase chain reaction (PCR) or enzyme linked immunoassay (ELISA), which constrains their impact. For instance, lateral flow assays (LFAs), the most widely used point of care diagnostics for over 3 decades, remain roughly 3 $\log_{10}$ lower in sensitivity than alternative laboratory based techniques.[1]

During LFA testing as shown in Fig.1a, the patient sample (i.e. buffer + analyte) is added to the sample pad. The capillary flow convectively drives the analyte and detection antibody labelled gold nanoparticles (GNPs) towards the test and control sections of the LFA. During convective flow, the analyte and labelled GNPs are also subject to diffusion and binding to the capture antibody coated on the test line (or dot) of the membrane. For instance, a "sandwich" complex is formed when the analyte binds with both detection antibody labelled GNP and capture antibody on the test line. Together, this reaction causes the accumulation of GNPs on the test line thereby generating a visually detectable (i.e. red) signal if sufficient GNP label is present. Unfortunately, low GNP capture rate (i.e. <5%) and inefficient label detection approaches (i.e. visual detection by naked eyes) are arguably important issues in LFA performance today.[2]

To improve LFA performance, new labels,[3, 4] pre-concentration steps[2] and signal amplification techniques[5] have been explored. However, this improved performance typically comes at additional cost, complexity, or loss of portability that hinders point-of-care deployment.

Here, we scaled and modeled the impacts of GNP design on the analytical performance of LFAs. We increased LFA sensitivity by using larger sized GNPs with a thermal contrast amplification (TCA) reader.

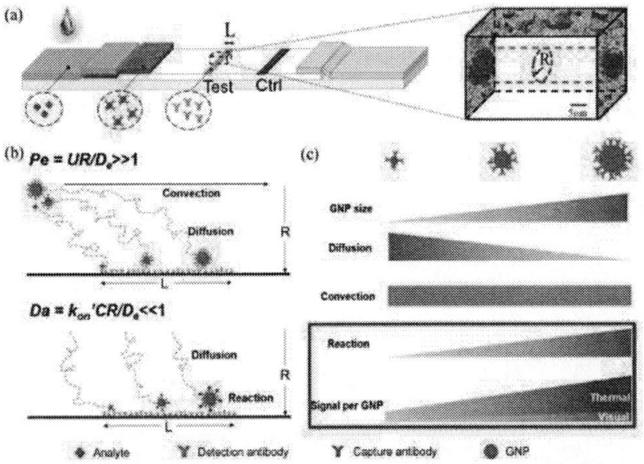

Figure 1. Scaling analysis of the impact of different sized GNPs on LFA sensitivity. (a) Architecture of lateral flow assay with test line width *L*, assuming that nitrocellulose membrane can be simplified as bundles of cylindrical pores with radius *R*. (b) *Pe* is the ratio of diffusion time and convection time, *U* is the convective velocity (0.2mm/s), *R* is the pore radius (5μm), D_e is the effective diffusivity of GNPs in porous membrane. *Da* is the ratio of reaction flux and diffusion flux, *C* is the capture antibody density (~10^{-9} mol/m²), k_{on}' is the effective forward rate constant for antibody labelled GNPs. (c) Comparison of 30nm, 60nm and 100nm GNPs.

Compared to traditional 30nm size, 60 and 100 nm GNPs bear stronger light absorption and scattering abilities, and importantly, exhibit higher reaction affinity by carrying more antibodies (Fig.1c). For better detection of the captured GNPs, a novel TCA reader uses an IR camera to collect the temperature changes of GNPs upon laser irradiation in the test line and provides improved sensitivity and quantitation of the analyte over visual reading.[6] Importantly, the TCA reader already demonstrated an 8-fold improvement in sensitivity on existing commercial GNP-based LFAs for malaria and influenza.[7] Here, by redesigning the LFA with different size GNPs (60 and 100 nm) and using a TCA reader, we achieved a 256-fold improvement over traditional 30nm GNPs with visual detection for C-reactive protein (CRP), a highly useful biomarker to distinguish bacterial from viral infections in order to guide antibiotic use in outpatient settings.[8]

METHODS

The transport and reaction process of different sized GNPs were scaled using Peclet (*Pe*) and Damkohler (*Da*) number (Fig.1b). The effective diffusivity (D_e) of GNPs in porous membrane and effective forward reaction constant (k_{on}') of antibody labelled GNPs are:

$$D_e = \phi D \quad \& \quad k_{on}' = nk_{on} \qquad (1 \& 2)$$

where ϕ is the porosity of membrane (~0.8), D is the diffusivity of GNPs in solution from Stokes-Einstein equation, k_{on} is the forward reaction constant (7.26×10^2 $M^{-1}s^{-1}$) for a single antibody-antigen interaction in LFA membrane,[9] and n can be scaled from 1 to the total number of antibodies per GNP.

A COMSOL model was built to study the diffusion, convection and reaction processes in LFA. First, a special case, where excessive analyte diffuses and binds to the test line without convection, was modeled and validated against a theoretical solution (data not shown).[10] Then, convection was added back to simulate the direct binding LFA where streptavidin labelled GNPs bind a biotin labelled test lines (data not shown). Finally, a full sandwich LFA model was developed for CRP. In the experimental CRP LFAs, test "dots" instead of test "lines" were implemented. Then, the same number of GNPs per LFA (i.e., 6×10^8) were used for comparison. Here each GNP was saturated with detection antibodies. The CRP LFAs were tested with 2-fold serial dilutions of standard human CRP reference (HyTest, Finland). A scanner (Epson XP310) and ImageJ were used to collect the mean grayscale signal of test "dots". A TCA reader equipped with 30nm 532nm laser was employed to record the thermal signal.[7]

Table 1. Parameters, *Pe* and *Da* for different sized GNPs in LFA

	30nm	60nm	100nm
Effective Diffusivity ($D_e = \phi D$) /m^2s^{-1}	1.17×10^{-11}	5.86×10^{-12}	3.52×10^{-12}
Effective rate const. ($k_{on}' = nk_{on}$) /$M^{-1}s^{-1}$	1.25×10^4	5×10^4	1.4×10^5
$Pe = UR/D_e$	85	171	284
$Da = k_{on}'CR/D_e$	7.76×10^{-3}	6.2×10^{-2}	2.8×10^{-1}

RESULTS

The scaling parameters, *Pe* and *Da* for different sized GNPs in LFA are shown in Table 1, where n is calculated by dividing GNP surface area with IgG docking area ($113nm^2$). As $Pe>>1$ in all cases, this implies that the transport process is diffusion-limited. Also as $Da<<1$ in all cases, this implies that reaction (i.e. binding), instead of diffusion, is the rate limit. Together ($Pe >>1$ and $Da <<1$) indicates that captured GNPs on the test line is reaction limited. Thus, we screened and selected antibodies with high binding rate constants to improve LFA performance (Medix, Finland).

For CRP LFA, with visual detection, 100nm GNPs increased the sensitivity by 32-fold versus 30nm GNPs (Fig.2a). Thermal detection is shown to qualitatively and quantitatively improve over visual detection especially with larger size GNPs (Fig.2a&b). Interestingly, using modeling we can predict how the dots fill (i.e. bottom to top) to establish which side of the hook effect we are on. For instance, S from S' (Fig.2b&c). The different patterns of test dots allow to expand analytical range of LFA, i.e., a 5 log_{10} visual detection range from low (0.001 mg/L) to very high (310 mg/L) concentrations of CRP. Importantly, this is currently not possible with traditional test lines and appears to be a unique advantage of test dots. Together, our results demonstrate a 256-fold sensitivity improvement using 100nm GNPs and the TCA reader, versus 30nm GNPs and visual reading alone (Fig.2d).

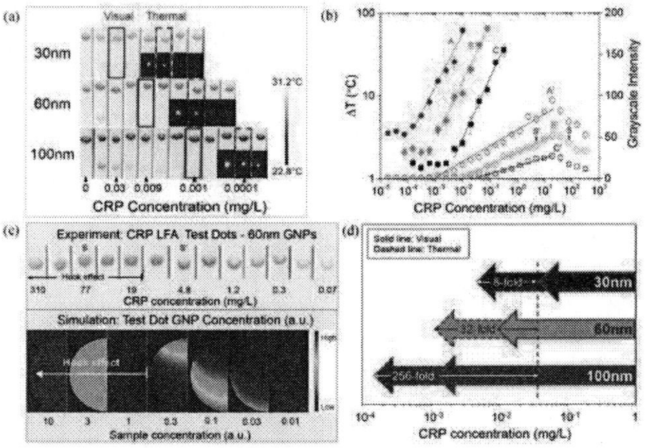

Figure 2. Larger 100nm GNPs and a TCA reader provides 256- fold sensitivity improvement in CRP LFAs. (a) Pictures and IR images of CRP LFAs. (b) Comparison of visual and thermal signals of CRP LFAs. A, A' stand for 30nm thermal and visual, B, B' for 60nm thermal and visual, C, C' for 100nm thermal and visual, respectively. (c) Test dots provide semi-quantitative visual reading and allow detection beyond hook effect. The COMSOL simulation predicted the same attributes. (d) Thermal and visual detection limits of 30/60/100nm GNPs.

DISCUSSION

In conclusion, we studied how GNP design, specifically size, affects LFA analytical performance. The sensitivity of LFAs greatly depends on the reaction process. With more sensitive GNP detection method (i.e. TCA reader), larger size GNPs can improve LFA sensitivity dramatically. Test "dots" expands the analytical range (i.e. >2 log_{10}) beyond "hook" effect and yields semi-quantitative visual readings due to increased test zone area compared to traditional test lines. Thus, modified GNP labels with the TCA reader show promise in creating a first point-of-care diagnostic platform that is competitive in sensitivity and quantitation with lab based PCR or ELISA technologies.

REFERENCES

[1] Eltzov, E., et al., *Electroanalysis,27*:2116-2130,2015.
[2] Moghadam, B.Y., et al., *Analytical Chemistry,87*:1009-1017,2015.
[3] Fu, X., et al., *Biosensors and Bioelectronics,78*:530-537,2016.
[4] Hampl, J., et al., *Analytical Biochemistry,288*:176-187,2001.
[5] Shyu, R.-H., et al., *Toxicon,40*:255-258,2002.
[6] Qin, Z., et al., *Angewandte Chemie,124*:4434-4437,2012.
[7] Wang, Y., et al., *Analytical Chemistry,88*:11774-11782,2016.
[8] Lin, S.-C., et al., *Scientific Reports,6*:38171,2016.
[9] Mosley, G.L., et al., *Lab on a Chip*,2016.
[10] Zimmermann, M., et al., *Biomedical microdevices,7*:99-110,2005.

CONTROLLED ICE NUCLEATION USING FREEZE-DRIED *PSEUDOMONAS SYRINGAE* ENCAPSULATED IN HYDROGEL BEADS

Lindong Weng (1), Shannon N. Tessier (1,2), Anisa Swei (1), Shannon L. Stott (1,3), Mehmet Toner (1,2)

(1) The Center for Engineering in Medicine, Massachusetts General Hospital, Harvard Medical School
Boston, MA, USA

(2) Shriners Hospitals for Children
Boston, MA, USA

(3) Massachusetts General Hospital Cancer Center, Harvard Medical School,
Charlestown, MA, USA

INTRODUCTION

The control of ice nucleation is of fundamental significance in atmospheric science, food processing, pharmaceutical industry, and cryobiology [1,2]. For example, it has been shown that a higher ice nucleation temperature in the extracellular space reduced the probability of detrimental intracellular ice formation (IIF) in slow-freezing cryopreservation of rat hepatocytes [3] and human oocytes [4]. Mechanical perturbation, electromagnetic field and ice-nucleating agents (INAs) have been known to induce ice nucleation with reduced supercooling. But the ice-nucleating methods available so far still suffer from several disadvantages such as the lack of standardization in the case of manual nucleation, the necessity of external fields, the inability to tune the nucleation temperature, and the unknown biocompatibility of INA particles, especially when used in biological systems [1]. Given the autonomous ice-nucleating nature of INAs, a promising solution to overcome the above limitations is to engineer a biocomposite that maintains particle potency but minimizes interactions with biologics, as well as enabling INA recovery [5]. In this study, freeze-dried *Pseudomonas syringae*, a model ice-nucleating agent, was encapsulated into microliter-sized alginate beads. We evaluated the performance of the bacterial hydrogel beads to initiate ice nucleation in aqueous glycerol solutions and investigated variables such as the size and number of the beads and the local concentration of INA particles. In the aqueous sample of a fixed volume, the total mass of the INA particles was found to be the governing parameter that is solely responsible for determining the ice nucleation of the bacterial hydrogel beads.

METHODS

Alginate solutions containing Snomax (Snomax International, Englewood, CO), the freeze-dried form of *P. syringae*, were prepared with a final concentration of 0.1-15 mg/ml Snomax and 2% (w/v) alginate. As illustrated in Figure 1 A, the Snomax-containing alginate solution was loaded into a 3 ml syringe (BD Biosciences, San Jose, CA) which was then mounted onto a syringe pump (Kent Scientific Corporation, Torrington, CT). The 30 and 18 gauge flat needles (SAI Infusion Technologies, Libertyville, IL) were used, which have an inner diameter of 0.159 and 0.838 mm, respectively. The Snomax-containing alginate solution was dropped into a 1 % (w/v) $CaCl_2$ bath. The alginate beads were gelled in the $CaCl_2$ bath for 30 min followed by triple washing with WFI water. A certain number (i.e., 1, 5 or 10) of the prepared hydrogel beads were added into 0.5 ml aqueous 10% glycerol solution contained in a cryovial (Corning Glass Works, Corning, NY).

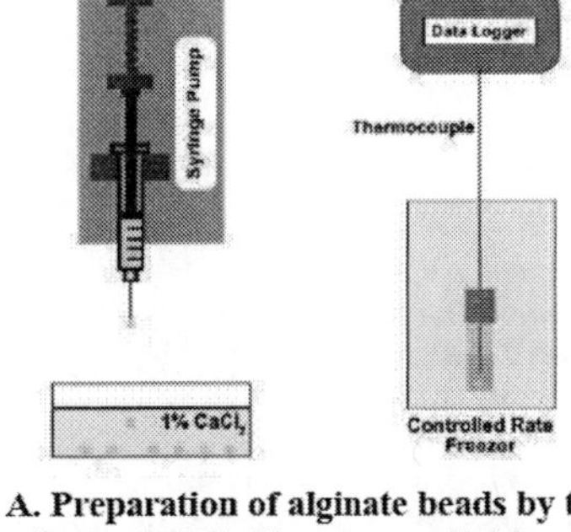

Figure 1: A. Preparation of alginate beads by the ionotropic gelation method using $CaCl_2$ as a crosslinking agent. B. The method for measuring the freezing temperature of the aqueous sample subjected to 1 °C/min cooling in a controlled rate freezer.

As illustrated in Figure 1 B, the cryogenic vial was placed into a Kryo 560-16 controlled rate freezer (Planer, Middlesex, UK) subjected to a cooling ramp at 1 °C/min until freezing. The temperature of each sample was measured by a Type-K thermocouple (Digi-Sense, Cole Parmer Instrument Co., Chicago, IL) that was inserted into the sample. The temperature was recorded every second by a data logger (Omega Engineering, Inc., Stamford, CT). Once ice nucleates in the aqueous sample, the temperature will return to the melting point. Therefore, the freezing temperature can be identified when an abrupt increase appears in the temperature profile.

RESULTS AND DISCUSSION

A range of alginate beads were prepared as summarized in Table 1. With the same Snomax concentration, the beads generated by the 18 gauge needle are around 3.5 times as large as those generated by the 30 gauge needle in terms of volume.

Table 1: The volume of alginate beads containing different concentrations of Snomax generated by the 30 or 18 gauge needle

c (mg/ml)		0	0.1	2.5	15
V (µl)	30-gauge	3.7±0.2	3.8±0.3	3.6±0.2	3.0±0.1
	18-gauge	13.1±0.6	14.6±0.5	12.2±0.3	10.5±0.5

Figure 2 shows the freezing temperature (T_f) of aqueous 10% glycerol solution in the presence of a given number of Snomax-laden alginate beads of different sizes and different Snomax concentrations. Given the stochastic nature of ice nucleation, we presented the data points of all replicates and indicated the median freezing temperature (T_f^{median}) and the interquartile range. It is shown that the freezing temperature is positively related to three parameters under investigation: the number of beads (N), the volume of beads (V), and the local Snomax concentration (c). The median freezing temperature of 0.5 ml aqueous 10% glycerol solution without ice nucleating particles is −11.8 °C. The T_f^{median} of 0.5 ml aqueous 10% glycerol having ten Snomax-free alginate beads prepared by the 18 gauge needle is −11.1 °C. But all the data points shown in Figure 2 are above −11 °C, which demonstrates that the significantly increased freezing temperatures were attributed to the Snomax particles even though they were confined within a hydrogel matrix.

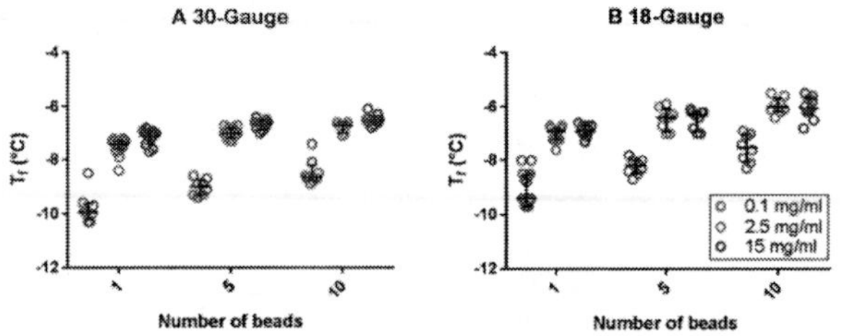

Figure 2: The freezing temperature of aqueous 10% glycerol in the presence of Snomax-laden alginate beads. The beads were generated by a 30 (A) or an 18 gauge (B) needle.

Since multiple variables are able to affect the ice nucleation of the aqueous samples containing Snomax-laden hydrogel beads, it will be highly instructive to provide a universal parameter that is solely responsible for determining the freezing temperature. In this study, we identified the total mass of the Snomax particles contained by the aqueous sample of a fixed volume to be such a parameter. In Figure 3, we plotted the freezing temperatures, as shown in Figure 2, together with those of the Snomax suspensions (0.005-50 mg/ml) prepared in aqueous 10% glycerol solution, as a function of m in a log10 scale. All the data points coincide with the same trend very well. The Pearson's correlation between T_f and $\log_{10}m$ yielded a coefficient (r) of 0.9783, indicating a strong positive linear relationship between these two variables. It is evident that the total mass of Snomax is the sole parameter to dictate the occurrence of ice nucleation in 0.5 ml aqueous glycerol solution regardless of the INA particles being present , either in a suspension or in a confined space.

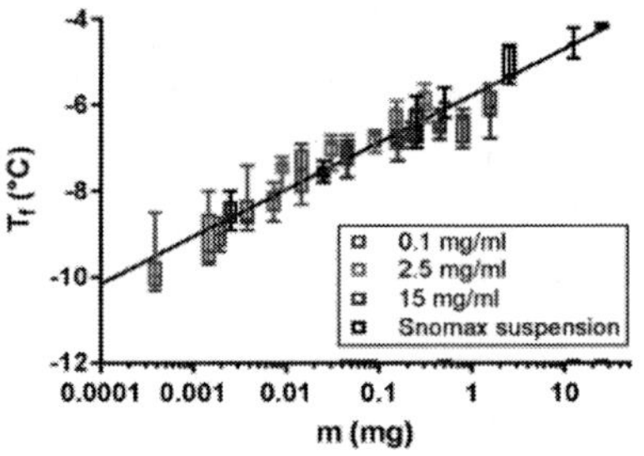

Figure 3: The freezing temperature profile that is plotted as a function of the total mass of Snomax contained in 0.5 ml aqueous 10% glycerol. The black data points represent the freezing temperatures of the Snomax suspensions of 0.005-50 mg/ml prepared in 10% glycerol. The solid line described by T_f=1.093·$\log_{10}m$-5.771 is the best fit to all the data points.

CONCLUSION

Although controlled ice nucleation can greatly benefit many process technologies in food and pharmaceutical industry and cryobiology, the ice-nucleating methods available so far still have several disadvantages. In this study, we encapsulated freeze-dried *P. syringae* into hydrogel beads to develop a tunable, effective ice-nucleating biocomposite that can minimize the interaction between the INA particles and the biological system and enables the recovery of used particles. The proposed approach does not require manual interruption or external fields due to the autonomous ice-nucleating nature of INAs. In the aqueous sample of a fixed volume, the total mass of the INA particles was identified as a universal parameter that dictates the ice nucleation temperature. The findings in this study provide the capacity to tune the ice nucleation temperature within a 6-degree window or even larger by adjusting the size and number of bacterial hydrogel beads and the local INA concentration. The ice-nucleating approach proposed here can improve the outcome and standardization of many ice-assisted process technologies such as the slow-freezing cryopreservation of stem cells for regenerative medicine, cellular therapies, and drug screening.

ACKNOWLEDGEMENTS

This work was supported by grants DoD DHP SBIR H151-013-0141 and NIH P41 EB002503.

REFERENCES

[1] Morris, GJ and Acton, E, *Cryobiology*, 66:85-92, 2013.
[2] Geidobler, R and Winter, G, *Eur J Pharm Biopharm*, 85:214-222, 2013.
[3] Harris, CL et al., *Cryobiology*, 28:436-444, 1991.
[4] Trad, FS, et al., *Hum Reprod*, 14:1569-1577, 1999.
[5] Zamecnik, J et al., *Cryo Letters*, 12:149-154, 1991.

BMP SIGNALING REGULATES DIFFERENTIAL GROWTH TO DRIVE BUCKLING DURING LOOPING MORPHOGENESIS OF THE SMALL INTESTINAL

Nandan L. Nerurkar (1), L. Mahadevan (2), Cliff J. Tabin (1)

(1) Department of Genetics
Harvard Medical School
Boston, MA, USA

(2) Departments of Applied Math, Organismal and Evolutionary Biology, and Physics
Harvard University
Cambridge, MA, USA

INTRODUCTION

Differential growth represents one of the core physical mechanisms driving morphogenesis throughout the vertebrate embryo. Investigations into the mechanics of differential growth have oft illustrated key physical parameters (e.g. tissue stiffness, growth rates, and initial geometry) that determine the resultant tissue shape. Ultimately, however, these parameters must be under genetic control. Yet, at present there are only a limited few examples of how developmental signals are integrated with downstream physical properties to shape the developing embryo. Here we build on our detailed understanding of the physics of loop morphogenesis in the chick small intestine [1] to ask what molecular cues underlie this important physical transformation during gastrointestinal development. We have previously shown that loops arise during embryogenesis through a mechanical instability: the initially straight intestinal tube elongates against the constraint of a thin membranous tissue, the dorsal mesentery. The elongating intestinal tube stretches the mesentery, which in turn compresses the gut tube until it buckles into loops [1]. The present study aims to identify the molecular cues that orchestrate this inherently mechanical morphogenetic event.

METHODS

Fertilized chicken eggs were incubated to E5, at which point they were windowed, and viral particles were injected into the base of the yolk stalk to infect the midgut. Eggs were returned to incubation at 37C until the desired stage. Morphometric analyses of gut loops were measured from brightfield images of dissected small intestines ranging from embryonic day E8 to E16. DAPI/phalloidin staining was performed in whole mount dorsal mesentery using standard protocols. Uniaxial tensile testing was performed with a custom micromechanical tester [2], where a fine tungsten cantilever (of known bending stiffness) was used as the force transducer. Finally, scaling laws were used to relate geometry, stiffness, and differential growth rates to wavelength λ and radius of curvature R of intestinal loops [1]. Briefly, the scaling laws were derived from a static analysis of two structures – tube and mesentery – that are bonded together under differential strain subsequent to differential elongation.

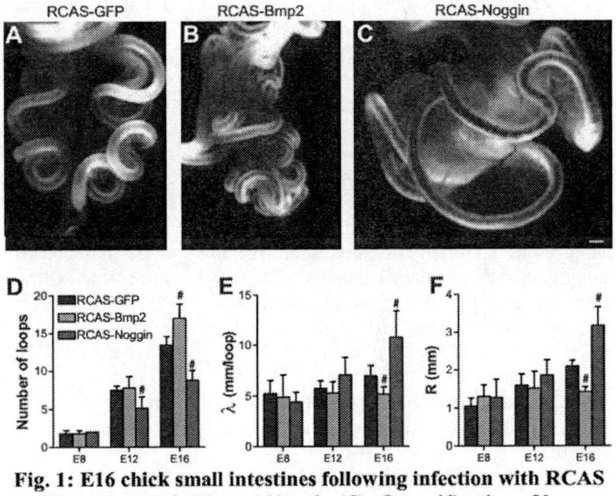

Fig. 1: E16 chick small intestines following infection with RCAS GFP (A), Bmp2 (B), and Noggin (C). Quantification of loop number (D), wavelength λ (E), and radius of curvature R (F); Scale = 1 mm. # = $p < 0.05$ vs. GFP.

RESULTS

Through a candidate *in situ* hybridization screen, we first identified that the gene *Bmp2* is highly expressed in the dorsal mesentery, but absent from the gut tube (not shown). We next developed a strategy using the replication competent retrovirus RCAS to functionally test the role of BMP signaling in looping. Compared to RCAS-GFP (which generates a phenotype indistinguishable from wild-type), overexpression of *Bmp2* using RCAS-Bmp2 resulted in an increase in the number of intestinal loops, while inhibiting BMP activity with RCAS-Noggin (an antagonist of the BMP pathway) conversely caused a decrease in the number of loops (Fig. 1A-D). Activation of the pathway with RCAS-Bmp2 significantly reduced wavelength λ and radius of curvature R, while RCAS-Noggin conversely increased λ and R (Fig. 1E, F). In other words, not only does the number of loops formed vary directly with BMP activity, but so too does the tightness of their curvature.

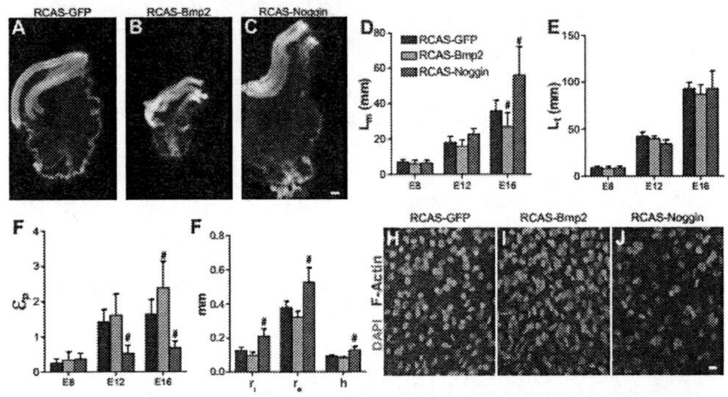

Fig. 2: Dorsal mesentery following removal of intestinal tube for RCAS GFP (A) Bmp2 (B), and Noggin (C) infected embryos at E16. Length of mesentery (L_m, D) and tube (L_t, E), and the differential growth strain ε_o (F). DAPI/Phalloidin staining of dorsal mesentery following infection with RCAS GFP (G), Bmp2 (H), and Noggin (I). Scale = 1 mm (A-C) and 10 um. # = p < 0.05.

Dissection to separate gut tube from dorsal mesentery revealed a dramatic decrease in mesentery length with activation of the pathway by RCAS-Bmp2, and an increase with inhibition of the pathway using RCAS-Noggin (Fig. 2A-D). Importantly, length of the gut tube was unaltered (Fig. 2E), indicating that the differential growth strain ε_p varies directly with BMP activity due to changes in mesentery growth (Fig. 2F). To investigate the cellular mechanism by which BMP signaling negatively regulates mesentery elongation, we next examined cell density and morphology in the mesentery. Cell density in the dorsal mesentery varied directly with BMP activity (Fig. 2H-J), while tissue volume varied inversely. Strikingly, these measurements suggest that the total cell number in the dorsal mesentery is nearly identical for RCAS-GFP, RCAS-Bmp2, and RCAS-Noggin infected embryos, despite clear differences in tissue size (Fig. 2A-C). For RCAS-Noggin, the reduction in cell density was accompanied by a clear increase in cell size (Fig. 2J). Together, these data suggest that Bmp2 functions to establish differential growth by limiting elongation of the mesentery relative to the gut tube, possibly by limiting cellular growth.

Although differential growth is the driving force for looping of the intestine, loop morphology is also subject to stiffness and geometry of the gut tube and mesentery. We therefore quantified each of these as a function of BMP activity, and turned to scaling laws [1] to examine their role in explaining the BMP-mediated looping phenotypes. Uniaxial tensile testing of the gut tube revealed no changes in modulus with either RCAS-Bmp2 or RCAS-Noggin (Fig. 3A, D). In the dorsal mesentery, inhibition of BMP signaling by RCAS-Noggin resulted in a significant decrease in the transition strain, but had no effect on the linear region modulus (Fig. 3B-D). Incorporation of all experimentally measured parameters into the scaling laws resulted in successful prediction of experimentally measured λ and R phenotypes for RCAS Bmp2 and Noggin (Fig. 3F).

DISCUSSION

We show here that buckling of the embryonic gut tube during looping of the small intestine is controlled by Bmp2, which functions to negatively regulate mesentery elongation. By doing so, Bmp2 sets up the differential growth pattern that is responsible for driving looping morphogenesis. The ability to accurately predict loop morphology following BMP activation (which increases looping) and inhibition (which decreases looping) from the scaling laws demonstrates two important points: first, that the physical mechanism of loop formation is the same for both conditions as for wildtype, and second, that while BMP signaling may perform many functions during intestinal development, its role in looping morphogenesis can be almost entirely captured by how BMPs modulate just two geometric properties: tube radius and differential growth strain. In addition to revealing an important role for Bmp2 in intestinal development and organogenesis, these findings provide an important example of how biochemical signals act on tissue-level mechanics to shape the developing embryo. Ultimately, revealing such molecular-mechanical relations in the developing embryo will be indispensable for regenerative approaches aimed at assembling differentiated cells into functional three dimensional organs *in vitro*.

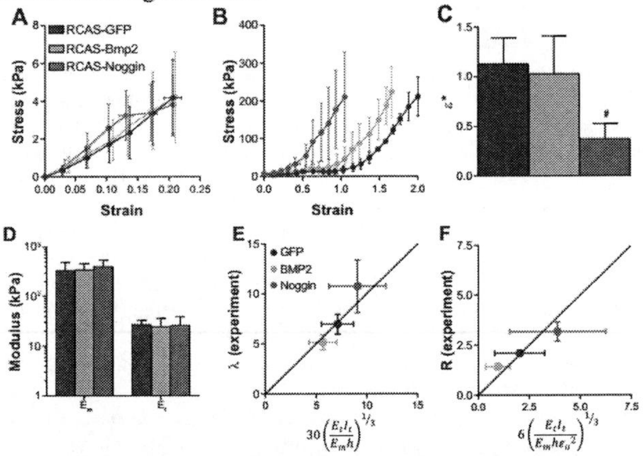

Fig. 3: Uniaxial tensile testing of gut tube (A) and dorsal mesentery (B). Transition strain of dorsal mesentery (C). Linear region modulus of the gut tube E_t and mesentery (E_m, D). Scaling law predictions (x-axis) compared to experimental measurements of loop morphology (y-axis) for wavelength (E) and radius of curvature (F). # = p <0.05 vs GFP.

ACKNOWLEDGEMENTS

Work was supported by the NIH (F32 HD069074, N.L.N; R01 HD087234, C.J.T.) and the MacArthur Foundation (L.M.).

REFERENCES

[1] Savin, T et al., *Nature*, 476(7358):57-62, 2011. [2] Shah R et al. *J Orthop Res*, 33(6):867-73, 2015.

SB³C2017
Summer Biomechanics, Bioengineering and Biotransport Conference
June 21 – 24, Tucson, AZ, USA

CYTOSKELETAL DYNAMICS UNDERLIE GROWTH PLATE CARTILAGE MORPHOGENESIS

Alek G. Erickson (1), Sarah M. Romereim (2), Nicholas H. Conoan (1), Andrew T. Dudley (1)

(1) Genetics, Cell Biology, and Anatomy
University Nebraska Medical Center
Omaha, NE, USA

(2) Reproductive Endocrinology
University Nebraska Lincoln
Lincoln, NE, USA

INTRODUCTION

Growth plate cartilage is the primary determinant of longitudinal bone growth, and is found capping the ends of all growing long bones [1]. In the growth plate, chondrocytes undergo a spatially- and temporally- defined differentiation program that transforms a population of round, randomly oriented "resting zone" cells into highly ordered columns of "proliferative zone" cells [1]. Increased cell volume and extracellular matrix deposition of the columnar chondrocytes promote tissue growth before remodeling of the cartilage matrix into bone [1]. Skeletal dysplasias that result from disruption of growth plate function affect approximately 1 in every 5000 children [2]. Few therapeutic options exist (none restore normal growth) in part as a result of the limited mechanistic understanding of the growth process [2]. For example, **defects in growth factor signaling, cell polarity signaling, cell adhesion, and extracellular matrix structure each result in chondrodysplasias that are characterized by defects in chondrocyte column formation in the proliferative zone, but it is not known how these pathways interact in the signaling network that regulates column formation** and bone growth [1-5]. Understanding the mechanism of column formation is critical for the rational design of growing cartilage constructs. Our previous live cell imaging analysis in cartilage explants showed that that following division daughter chondrocytes rotate 90-degrees to form a column [5]. Rotation is accompanied by expansion of a cadherin cell-cell adhesion surface that connects the daughter chondrocytes [5]. Therefore, **we hypothesized that chondrocyte adhesion surfaces drive rotation through a cytoskeletal force-generating mechanism.**

METHODS

Live cell imaging of cartilage explants: To ask whether cell rotation was associated with the actin cytoskeleton, murine growth plate cartilage explants were cultured in the presence of an ARP2/3 inhibitor and analyzed by our live cell-imaging assay. Fluorescent confocal time series data were analyzed in Volocity software to quantify the resulting effect on daughter cell orientation over time.

Ultrastructural analysis of chondrocyte rotation: ARP/2/3-inhibitor treated and untreated cartilage explants were fixed using a modified Karnovsky's fixative including Safranin O to preserve cell morphology, and were analyzed by transmission electron microscopy.

Immunofluorescence-based quantitative morphometric analysis: To identify domains of cell contractility and adhesion, embryonic murine growth plate cartilage was analyzed by immunofluorescence to detect the cell adhesion protein β-catenin, the motor protein myosin II, and the active form of the regulatory subunit myosin light chain (pMLC). Multi-color 3D images were obtained by confocal microscopy. To deconstruct the process of column formation, daughter cell pairs were classified into dividing, rotating, and separating cells based on the presence and morphology of a β-catenin enriched adhesion surface. The localization patterns of myosin and pMLC were analyzed for each cell and incorporated into the classification based on β-catenin signal. To ask the role of integrins, cartilage explants were cultured overnight in the presence of 1mM RGD or RGE peptide.

RESULTS

Ultrastructural analysis of rearranging chondrocytes revealed nanoscale protrusions located at the daughter cell interface periphery that resembled the actin-based leading edge commonly seen in migrating cells (Figure 1A). We asked whether these protrusions were related to the actin cytoskeleton by analyzing cartilage explants cultured in the presence of a branched actin polymerization inhibitor. ARP2/3 inhibitor treatment blunted protrusions, leading to disrupted cell adhesion (Figure 1B), and blocked chondrocyte rotation in a dose-dependent manner (Figure 1C-F)

This result suggested cytoskeletal forces are an important component to the mechanism of chondrocyte rotation. To investigate the location of cell-generated forces during column formation, we analyzed the relationship between the cell adhesion surfaces and the actomyosin contractile apparatus. In column-forming proliferative chondrocytes (Figure 2C-D), but not in non-rotating resting chondrocytes (Figure 2A-B), pMLC and myosin heavy chain 2b are nearly mutually exclusive with β-catenin. Myosin contractility does not appear to be associated with the cell-cell adhesion surface; instead, myosin and pMLC are enriched at cell-matrix contacts. As daughter cells separate following rotation, β-catenin is replaced by myosin and pMLC at the former cell-cell interface. The appearance of cell-ECM domains of myosin activity suggested interplay between cytoskeleton and integrins in rotation. We tested this by inhibiting integrin binding by culturing cartilage explants with RGD-containing peptides or RGE control. The integrin-blocked, but not control chondrocytes, displayed a disruption of myosin localization patterns (Figure 2, E-F).

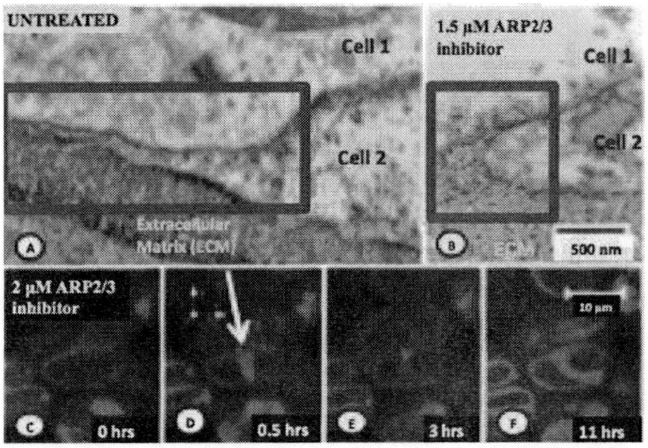

Figure 1: Actin-based protrusions regulate chondrocyte rotation. (A) Electron micrograph of the periphery of a rotating daughter cell interface. Nanoscale protrusion is boxed in red. (B) ARP2/3 inhibitor blunts the protrusion and disrupts cell adhesion. (C) Live imaging analysis of ARP2/3 inhibitor treated explants demonstrating that cells divide but fail to rotate, whereas untreated control explants show 90° cell rotations in 4-5 hours.

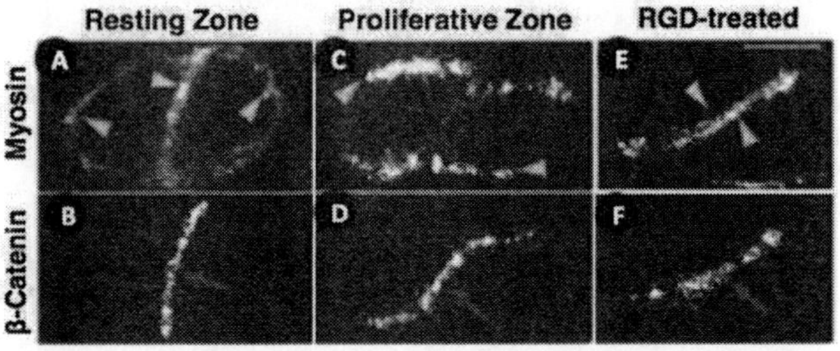

Figure 2: Cortical tension is associated with column formation. Dual immunofluorescence microscopy shows the localization of myosin (A, C, E) and cell-cell adhesion (B-catenin, B, D, F). In non-column forming resting chondrocytes (A-B), myosin is diffusely localized. In actively rotating proliferative zone chondrocytes (C-D), myosin is enriched at cell-ECM border. When cell-matrix adhesion is disrupted using soluble RGD peptides (E-F), myosin is redistributed to the cell-cell surface.

DISCUSSION

Despite genetic characterization of the signaling pathways required for cartilage morphogenesis, the physical and molecular mechanisms used by chondrocytes to form columns are unknown. Our data suggests the driving forces underlying cell rearrangement are produced by cytoskeletal dynamics. These forces are most likely acting at the cell-matrix surface, and potentially the daughter cell interface periphery, specifically through actomyosin contractility and actin-based protrusions. The presence of pMLC at cell-ECM domains during strong cell-cell adhesion seems to be predictive of rotating daughter cells, and can be disrupted by blocking integrin function. This data significantly alters our previously proposed model based on force generated at the cadherin adhesion surface [6]. Instead, it is likely that the integrin surface is the primary driver of cell movement (in agreement with [5]), and that the cadherin adhesion mechanically couples the cortices of daughter cells to ensure coherent rotation. These results indicate a relationship between the signaling networks underlying chondrocyte column formation and cytoskeletal regulation. We are currently exploring this connection using column-disrupting mouse models such as the planar cell polarity deficient Wnt5a and Ror2 mutants. Although we have not yet directly tested the role for myosin function, **the present study identifies domains of apparent force production during column formation, and thus supports innovative scaffold designs that will exploit the cartilage biomechanical environment to create self-organizing cartilage cultures.**

ACKNOWLEDGEMENTS

Thanks to the Advanced Microscopy and Electron Microscopy Core Facilities at UNMC.

REFERENCES

[1] Kronenberg, H, *Nature*, 423:332-336, 2003.
[2] Krakow, D et al., *Genetics in Medicine*, 12:327-341, 2010.
[3] Aszodi, A et al., *Genes Dev*, 17(19):2465-2479, 2003.
[4] Li, Y et al., Development, 136(7):1083-1092, 2009.
[5] Romereim, S et al., Development, 141(10):2085-2095, 2014.

SB[3]C2017
Summer Biomechanics, Bioengineering and Biotransport Conference
June 21 – 24, Tucson, AZ, USA

DIRECT *IN VIVO* QUANTIFICATION OF DIFFERENTIAL MECHANICAL PROPERTIES IN DEVELOPING TISSUES

F. Serwane (1), A. Mongera (1), P. Rowghanian (1), DA Kealhofer (2), AA Lucio (1), ZM Hockenbery (1), O. Campàs (1)

(1) Department of Mechanical Engineering
University of California, Santa Barbara
Santa Barbara, California, USA

(2) Department of Physics
University of California, Santa Barbara
Santa Barbara, California, USA

INTRODUCTION

The sculpting of tissues into their functional morphologies requires a tight spatiotemporal control of their mechanics[1-3]. While cell-generated mechanical forces power morphogenesis, the resulting tissue movements depend on the local tissue mechanical properties, such as its stiffness and viscosity, which govern the system's response to the internally generated forces. Despite their relevance, the role of mechanical forces and mechanical properties in developmental processes remains largely unknown, mainly because of a lack in methodologies enabling direct *in vivo* and *in situ* measurements of cell-generated forces and mechanical properties within developing 3D tissues and organs[4].

METHODS

We use magnetically responsive microdroplets[5] as local micro-rheometers to quantify the mechanical properties of developing tissues in situ, within intact zebrafish embryos.

RESULTS

Using zebrafish as model system, we study the mechanics of the extending body axis. We use ferrofluid microdroplets[5] to measure the local mechanical properties along the anteroposterior axis of developing zebrafish embryos. We show that during tailbud elongation, a model system for vertebrate body axis extension, the tissue behaves as a viscoelastic material with fluid-like behavior at long timescales and displays increasing tissue stiffness and decreasing fluidity toward its most anterior region, where new somites are formed. Our results indicate the tissues are characterized by two stress relaxation timescales, a behavior that can be explained by a two-branch generalized Maxwell model.

The observed gradient in tissue mechanical properties, with lowest viscosity at the posterior end of the embryo, facilitates the extension of the body axis at that location. Moreover, these differential mechanical properties correlate with morphogenetic events and spatial variations in cellular movements.

Altogether, these results indicate that spatiotemporal changes in tissue mechanical properties can play an essential role in the sculpting of embryonic tissues.

ACKNOWLEDGEMENTS

This work was partially supported by NIH grant 1R21HD084285-01 (to O.C.) from the Eunice Kennedy Shriver National Institute of Child Health and Human Development.

REFERENCES

[1] Heisenberg, C-P & Bellaiche, Y, *Cell* 153:948–962, 2013.
[2] Nelson, CM & Gleghorn, JP, *Annu Rev Biomed Eng* 14:129–154, 2012.
[3] Mammoto, T & Ingber, DE, *Development* 137:1407–1420, 2010.
[4] Campàs, O, *Semin Cell Dev Biol* 55:119–130, 2016.
[5] Serwane, F et al., *Nat Methods,* doi:10.1038/nmeth.4101, 2016.

Long-Range Patterning by the Vertebrate Tail Organizer via Mechanical Information

Jamie Schwendinger-Schreck (1), Dörthe Jülich (1), Dipjyoti Das (1), Andrew Lawton (1), Nicolas Dray (1), Corey O'Hern (2,3), Thierry Emonet (1,3) and Scott A. Holley (1)

(1) Department of Molecular, Cellular and Developmental Biology
Yale University
New Haven, CT, USA

(2) Department of Mechanical Engineering and Materials Science
Yale University
New Haven, CT, USA

(3) Department of Physics
Yale University
New Haven, CT, USA

INTRODUCTION

Embryonic organizers are cellular domains that govern the differentiation of adjacent cells by secreting diffusible signaling molecules. The vertebrate tail organizer functions within a flux of tailbud mesodermal progenitors and ectoderm to direct embryonic body elongation. Normal tailbud elongation is necessary for proper spinal column development. Cells transiting the tailbud organizer express the posterior homeobox gene *eve1* as well as the secreted signaling proteins *bmp2b* and *bmp4*. We quantify organizer function by combining pharmacological and localized transgenic perturbations with systems analysis of cell motion. We find that positive feedback between *eve1* and *bmps* locally produces both a stable organizer signaling domain and a transition in cell flux. The long-range effects of localized organizer perturbations reveal that the organizer extends its sphere of influence via mechanical information within the cell flow. Mechanical information, via local cell-cell adhesion and repulsion, can project at the speed of sound through the tissue and thus is more efficient than a diffusible signal. Mechanical information may represent a novel mechanism for long-range orchestration of development by embryonic organizers.

Bmp signaling is required during zebrafish trunk elongation for linear body elongation and appears to function by regulating morphogenesis not cell fate (*1-4*). Bmp signaling, indicated by phosphorylated SMAD 1/5/8 immunolocalization, is restricted to the ventral posterior tailbud by *chordin* and *noggin* expressed in the posterior notochord and PSM (*5, 6*). During trunk elongation, *bmp4* and *bmp2b* are expressed in the posterior tailbud periderm and mesodermal progenitors (*7*). *Even-skipped* (*eve* or *evx*) genes encode transcriptional repressors whose conserved genomic orientation adjacent to the *hox13* end of *hox* clusters suggests a role in posterior vertebrate development (*8-11*). Indeed, *eve1* over-expression produces similar elongation phenotypes as loss of *bmp* signaling, including occasional duplicated tails, and *eve1* is expressed in ectopic tail organizers (*1, 3, 4, 6, 10-15*). The patterns of *bmp* and *eve1/evx1* expression are conserved in *Xenopus* and mouse tailbuds, and *bmp* and *evx1* have tail inducing activity in *Xenopus* (*16-22*). In the mouse and chick, regulation of Bmp signaling from the ventral ectodermal ridge and mesoderm of the tailbud promotes the cessation of gastrulation and is required for subsequent tailbud elongation (*21, 22*). The consistent spatial expression of zebrafish *eve1* in the tail organizer is maintained despite the extensive cell movement (*23-26*). Here, we investigate how the tail organizer sustains itself and regulates both local and long-range cell motion.

METHODS

We use the zebrafish embryo as an experimental system to study a developmental process conserved among all vertebrates including humans. Zebrafish embryos are amenable to genetic manipulation and, given their optical transparency, are the best vertebrate model system for high-resolution live imaging.

We temporally perturbed tail organizer function during trunk elongation using DMH1, a chemical inhibitor of Bmp receptor signaling (*27*). We performed a controlled spatial perturbation of organizer activity by mosaically co-expressing *eve1* and GFP under the control of the *tbx6l* enhancer (*24, 28*). *tbx6l* expression largely overlaps *eve1* expression, so the transgene mosaically disrupts the levels of *eve1* within its normal expression domain in the posterior tailbud.

To determine the mechanical basis of the elongation defect, we performed a systems level analysis of cell motion in the tailbud after perturbation of the tail organizer. We acquired 3D confocal timelapses of embryos expressing a nuclear localized RFP. We segmented the tailbud into four domains plus the periderm and quantified individual cell motion as well as metrics of local and global order and local rotational velocity (*25*).

We used cell velocities relative to motion of the anterior PSM to compensate for growth of the body axis (*25*). We examined the mean square displacement (MSD) and modeled the data using a diffusion coefficient and directional velocity (*24, 29*). Lastly, we examined local and global order of cell motion. We used in silco modeling of tailbud elongation further explore the regulation of cell flow (*30*).

RESULTS

Both DMH1 and transgenic expression of *eve1* lead to similar defects in body elongation. DMH1 treated transgenic embryos had the highest incidence of elongation defects despite having normal levels of Bmp signaling in the tailbud. This observation suggests that the *bmp1-eve1* circuit is required to produce a regulated pattern of Bmp signaling in the tail organizer, rather than a specific level of Bmp signaling, to promote linear body elongation.

Quantitative metrics of cell motion reveal both local and long-range effects of tail organizer perturbation. Cell track straightness was broadly affected, while the coefficient of variation (C.V.) of track mean speed was increased in the ADM and DM in all experimental conditions. The observation that transgenic expression of *eve1* in the PZ is sufficient to affect track straightness and C.V. in the ADM suggests a long-range organizing function for the tail organizer.

We note that transgenic DMH1 embryos exhibit changes in local order, global order and the MSD in the ADM, but neither the decrease in Bmp signaling after DMH1 treatment nor the transgenic misregulation of *eve1* in the organizer is alone sufficient to cause these changes. These observations suggest that the transgenic DMH1 phenotype in the ADM is an indirect long-range effect of the local cell flow alterations within the organizer and not due to direct effects of loss of Bmp signaling in the ADM.

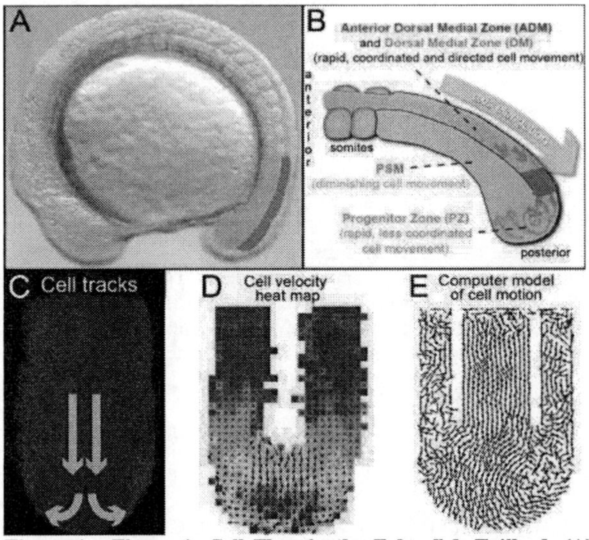

Figure 1: Figure 1. Cell Flow in the Zebrafish Tailbud. (A) An embryo at the end of trunk elongation, i.e. the 16-somite stage. The tailbud domains are colored as in panels B and C. The tail organizer is in the green domain. **(B)** A schematic summary of tailbud cell motion. **(C)** A dorsal view of the cell tracks in the tailbud. Arrows indicate the general direction of cell flow. **(D)** Cell motion visualized by projecting the 3D cell velocities into 2D and averaging over 10 mm^2 sectors. Warmer colors indicate higher average velocities. Note that the ADM has been removed. **(E)** A snapshot of a computer simulation of tailbud elongation. Cells are modeled as self-propelled particles.

DISCUSSION

In close proximity to the tail organizer, Bmp signaling increases directional velocity and promotes local and global order. Computational simulations suggest that changes in cell flow in the organizer (i.e. the PZ) can be propagated mechanically further upstream into the cell flow of the ADM. We call this systems level process "mechanical information." Mechanical information propagates via local, adhesive, and repulsive interactions between neighboring cells and can travel more efficiently than a diffusible signal. Our in vivo data suggest that changes in cell motion in the ADM are indeed propagated from the organizer via mechanical information. Thus, the *eve1-bmp* tail organizer regulates cell signaling and cell motion within both the mesoderm and periderm of the organizer itself, and at longer distances, the organizer projects its influence via mechanical information within the cell flow. Other embryonic organizers may similarly expand their sphere of influence beyond local cell signaling through long-range regulation of tissue mechanics.

ACKNOWLEDGEMENTS

This work was supported by the NIH Common Fund Single Cell Analysis Program (R33GM114257).

REFERENCES

1. S. A. Connors, et al, *Dev Biol* **293**, 191-202 (2006).
2. U. J. Pyati, et al, *Development* **132**, 2333-2343 (2005).
3. H. L. Stickney, et al, *Dev Biol* **310**, 71-84 (2007).
4. Y. Yang, C. Thorpe, *Development* **138**, 2601-2611 (2011).
5. K. O'Neill, C. Thorpe, *Dev Biol* **375**, 117-127 (2013).
6. R. H. Row, D. Kimelman, *Dev Biol* **329**, 55-63 (2009).
7. C. Thisse, B. Thisse, *ZFIN Direct Data Submission*, (2005).
8. C. Li, J. L. Manley, *Mol Cell Biol* **18**, 3771-3781 (1998).
9. L. M. McKay, et al, *Nucleic Acids Res* **27**, 3064-3070 (1999).
10. C. Cruz *et al.*, *Proc Natl Acad Sci U S A* **107**, 3564-3569 (2010).
11. J. L. Seebald, D. P. Szeto, *Dev Biol* **349**, 78-89 (2011).
12. O. Barro *et al.*, *Developmental genetics* **17**, 117-128 (1995).
13. J. S. Joly, et al, *Development* **119**, 1261-1275 (1993).
14. A. Agathon, et al, *Nature* **424**, 448-452 (2003).
15. M. C. Mullins *et al.*, *Development* **123**, 81-93 (1996).
16. C. W. Beck, J. M. Slack, *Development* **126**, 1611-1620 (1999).
17. C. W. Beck, et al, *Dev Biol* **238**, 303-314 (2001).
18. M. K. Dush, G. R. Martin, *Dev Biol* **151**, 273-287 (1992).
19. A. Fainsod, et al, *Embo J* **13**, 5015-5025 (1994).
20. F. Gofflot, et al, *Dev Dyn* **210**, 431-445 (1997).
21. D. C. Goldman, et al, *Development* **127**, 2113-2123 (2000).
22. S. Ohta, et al, *Development* **134**, 4315-4324 (2007).
23. A. Mara, et al, *Nat Cell Biol* **9**, 523-530 (2007).
24. N. Dray *et al.*, *Curr Biol* **23**, 1335-1341 (2013).
25. A. K. Lawton *et al.*, *Development* **140**, 573-582 (2013).
26. R. Fior *et al.*, *Development* **139**, 4656-4665 (2012).
27. J. Hao *et al.*, *ACS chemical biology* **5**, 245-253 (2010).
28. D. P. Szeto, D. Kimelman, *Development* **131**, 3751-3760 (2004).
29. N. Monnier *et al.*, *Biophys J* **103**, 616-626 (2012).
30. D. Das, T. Emonet, S. A. Holley, *in revision*.

BUCKLING DURING MORPHOGENESIS OF THE LUNG

K. Goodwin (1), J. W. Spurlin (1), C. M. Nelson (1,2)

(1) Chemical & Biological Engineering
Princeton University
Princeton, NJ, US

(2) Molecular Biology
Princeton University
Princeton, NJ, US

INTRODUCTION

During development of the mammalian lung, the embryonic airway epithelium folds sequentially to create 23 generations of branches – for a total of approximately 8 million terminal ends – during a period of days to weeks, depending on the species. Many of the biochemical signals that induce this folding have been identified, and include diffusible growth factors and morphogens that are synthesized and secreted by the surrounding mesenchyme. At the same time as the airway epithelium is folding into new branches, the embryonic mesenchyme is differentiating into the different tissues that surround the airways, including the airway smooth muscle. Here, we describe efforts to understand the physical forces that are induced by the mesenchyme to fold the airway epithelium.

METHODS

We used timelapse imaging of embryonic lungs isolated from transgenic reporter mice to uncover the morphodynamics of airway epithelial folding and airway smooth muscle differentiation. To understand the mechanics of the airway epithelium in the absence of smooth muscle, we removed the mesenchyme from the airway epithelium and cultured explanted organoids in extracellular matrix hydrogels.

RESULTS

Timelapse imaging revealed that during terminal bifurcations (where a parent branch splits in half to generate two daughter branches), airway smooth muscle first differentiates at the future cleft site before the epithelium folds [1]. This pattern of airway smooth muscle differentiation is essential for terminal bifurcation events. When smooth muscle differentiation is blocked genetically or pharmacologically, the airway epithelium forms a buckled morphology, which appears to depend on the growth rate of the epithelium. Mesenchyme-free culture experiments support this hypothesis, as the airway epithelium buckles in a growth-rate dependent manner that results from a viscoelastic instability [2].

Interestingly, timelapse imaging revealed a similar role for airway smooth muscle in lateral branching, which is the process through which new branches form off the side of an existing stalk. When smooth muscle is depleted, the airway epithelium forms a similar buckling geometry.

DISCUSSION

Our data suggest that the pattern of branching in the mammalian lung is dictated by the pattern of differentiation of airway smooth muscle, which constrains an intrinsic buckling instability that results from the epithelium growing under confinement. This suggests that the growth factors and morphogens secreted by the mesenchyme are primarily affecting the growth rate of the epithelium and the rate and pattern of differentiation of smooth muscle. Buckling during airway branching joins a growing list of (visco)elastic instabilities during embryonic development [3].

ACKNOWLEDGEMENTS

This work was supported in part by grants from the NIH (GM083997, HL110335, HL118532, HL120142, CA187692), the NSF (CMMI-1435853), the David & Lucile Packard Foundation, the Alfred P. Sloan Foundation, the Camille & Henry Dreyfus Foundation, the

Burroughs Wellcome Fund, and a Faculty Scholars Award from the Howard Hughes Medical Institute.

REFERENCES

[1] Kim, H.Y. et al., *Dev. Cell*, 34: 719-726, 2015

[2] Varner, V.D. et al., *Proc. Natl. Acad. Sci. USA,* 112: 9230-9235, 2015.

[3] Nelson, C.M., *J. Biomech. Eng.,* 138: 021005, 2016.

SB^{3}C2017
Summer Biomechanics, Bioengineering and Biotransport Conference
June 21 – 24, Tucson, AZ, USA

MECHANICAL CONTROL OF CARDIOGENESIS: HOW MECHANICAL CUES GUIDE THE CELL PHENOTYPE OF HEART PRECURSOR CELLS AS THEY FORM A BEATING HEART.

Lance A. Davidson[1, 2, 3]

(1) Bioengineering, University of Pittsburgh, Pittsburgh PA, 15213

(2) Developmental Biology, University of Pittsburgh, Pittsburgh PA, 15260

(3) University of Pittsburgh, Pittsburgh PA, 15213

ABSTRACT: Organogenesis requires the integration of cell mechanics with programs of cell differentiation to generate complex composite structures. The heart is constructed from a diverse set of cells whose origins range from ectoderm, endoderm, and mesodermal precursors. In this presentation I describe a novel pathway that instructs migratory mesenchymal heart precursor cells to switch their phenotype and adopt an epithelial phenotype. This pathway integrates both traditional developmental patterning pathways with cues from the mechanical environment of the heart forming region. Bilateral populations of heart precursor cells (HPCs expressing nkx 2.5) begin their journey to form the heart as mesenchyme cells that associate closely with endoderm and 'ride' endoderm for part of their journey. Along their journey HPCs experience the stresses and strains of ventral convergence and extension and respond by transitioning to a migratory epithelial sheet. HPCs transition to simple epithelial cells via a pathway that first polarizes aPKC and then assembles apical junctions marked by ZO-1. Epithelial cells within the sheet abandon their association with the endoderm but continue collective movements to join with their contralateral mates. We expose the role of mechanics using both advanced biophysical tools such as micro-aspiration and force-traction microscopy to measure mechanical properties and collective behaviors cells in the heart forming region and cell biological tools such as small molecule inhibitors and mutant proteins to modulate mechanical cues in endoderm and HPCs. Live, intravital confocal imaging allows direct interrogation of cell behaviors as HPCs undergo a mesenchymal-to-epithelial transition. Defined methods for controlling MET hold promise for the intelligent design of synthetic organs and dissection of human structural birth defects.

Technical Presentation #246

SB³C2017
Summer Biomechanics, Bioengineering and Biotransport Conference
June 21 – 24, Tucson, AZ, USA

A DEEP LEARNING APPROACH TO PREDICT MILD TRAUMATIC BRAIN INJURY IN CONTACT SPORTS

Yunliang Cai (1), Wei Zhao (1), Zhigang Li (2), Songbai Ji (1,3)

(1) Department of Biomedical Engineering,
Worcester Polytechnic Institute,
Worcester, MA, USA

(2) Department of Biomedical Data Science,
Geisel School of medicine, Dartmouth College,
Hanover, NH 03755, USA

(3) Thayer School of Engineering
Dartmouth College
Hanover, NH, USA

INTRODUCTION

Traumatic brain injury (TBI), including mild TBI and concussion, is a major public health problem in the United States [1]. Reliable injury diagnosis is important to mitigate this prevailing disease. Consequently, developing an accurate injury prediction method is central to the biomechanics studies of TBI. Previous efforts have relied on empirical kinematic or tissue response-based metrics explicitly pre-defined in a specific region of interest (ROI). However, they do not consider other anatomical regions or functionally important neural pathways. A single "training" dataset has also been used to evaluate performance, typically by fitting a logistic regression model to report the area (AUC) under the receiver operating curve (ROC). However, without cross-validation using a separate "testing dataset", there could be uncertainty how the metrics perform when they are, presumably, deployed to predict injury on fresh, unmet impact cases in clinical applications. This is an important issue seemingly under-appreciated.

Injury prediction is a binary classification. Here, we formulated this classical injury prediction problem in TBI biomechanics research into a supervised classification via deep learning. Instead of explicitly defining mechanical response features for injury prediction (e.g., cumulative damage measure (CSDM), peak strain/stress in the corpus callosum), deep learning implicitly utilized response features of the entire white matter (WM). It automatically distilled the most discriminative features for concussion classification. A repeated random subsampling technique was also employed to train and cross-validate the classifier, and report performance metrics more objectively.

Deep learning has dramatically improved the state-of-the-art in numerous research domains [2]. However, its application in TBI biomechanics is non-exist at present. Therefore, findings from this study may provide important fresh insight into this classical research problem: how best to predict brain injury.

METHODS

We used the Worcester Head Injury Model (WHIM; formerly known as the DHIM) [3] to simulate the reconstructed 58 NFL head impacts (including 22 concussions and 35 non-injury cases). Voxel-wise peak WM fiber strains of the entire brain during the head impact, regardless of the time of occurrence, were obtained for each impact. They served as the input for the deep network classifier.

A fully connected, 5-layer deep learning network was developed, whose network structure is illustrated in **Fig. 1**. The structure of the deep network, including the number of layers and the dimensions of each layer, was determined empirically. The deep network training was supervised using labeled subjects (i.e., concussed or non-injury) from the NFL dataset. The training process followed a backpropagation algorithm, which optimized the network parameters independently for each layer to minimize the classification training error [2]. In total, the deep network contained over 1.31×10^8 independent parameters.

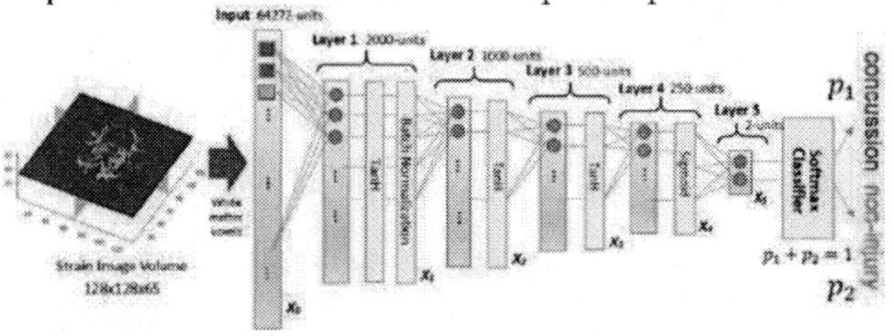

Figure 1 Structure of a full-connected, five-layer deep learning network for concussion classification. The deep network implicitly utilized voxel-wise WM fiber-strain-encoded image features of the entire brain. The network automatically distilled the most discriminative features for concussion classification.

Performance of the deep learning approach was compared with four existing injury metrics: Brain Injury Criterion (BrIC [4]), CSDM for the whole WM (CSDM-WB) and the corpus callosum (CSDM-CC; based on maximum principal strain, at a threshold value of 0.1), and peak fiber strain in the corpus callosum (Peak-CC) [5].

To ensure objective evaluation and comparison of the performances, a repeated random subsampling technique was employed to split the NFL impact cases into training and cross-validation datasets. Four training/validation configurations were used: 19/39, 29/29, 39/19, and 49/9, utilizing all of the 58 impact cases. For each configuration, 20 random samplings were generated to report mean, standard deviation, and range of performances. For each random trial, the deep network launched an independent random initialization for all of the network parameters and then performed supervised training based on the given training dataset. The optimal number of iterations were determined based on the observed convergence behaviors of the error functions of the training and cross-validation datasets. They were determined empirically. After training convergence, the optimized network was used to yield injury predictions based on the cross-validation datasets.

In contrast, the four explicit injury metrics were subjected to conventional logistic regression analysis, using the identical training/validation configurations. We compared the five injury prediction methods using the area under the receiver operating curve (AUC) for the training and cross-validation datasets, as well as cross-validation accuracy, sensitivity, and specificity.

RESULTS

Deep learning achieved the highest AUCs for the training datasets for all configurations (0.95–0.99), significantly higher than all other injury metrics (<0.87; **Fig. 2a**). This was not surprising given its supervised training that iteratively minimized the training errors. However, the high AUCs for the deep learning technique did not translate into the same high levels for the cross-validation datasets (**Fig. 2b**). The AUCs dropped to 0.82–0.86, but were still statistically comparable to BrIC and CSDM-WB ($p>0.5$), and statistically higher than CSDM-CC and Peak-CC ($p<0.01$). Deep learning had the highest mean AUCs when the cross-validation datasets had sizes of $N = 39$ and $N = 19$, and was the second highest for the other two configurations.

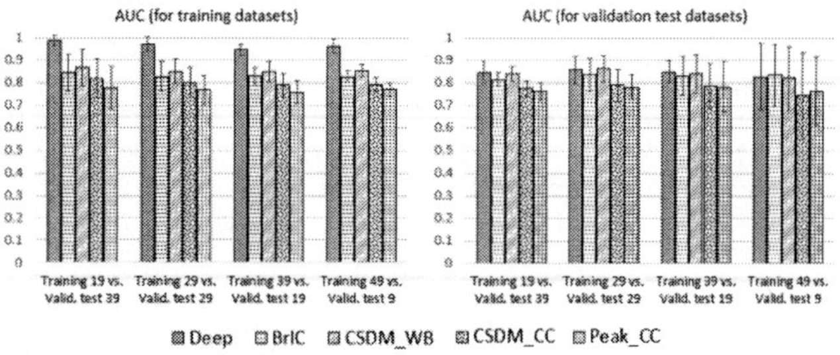

Figure 2 The AUC values over the training and cross-validation datasets. Deep learning outperformed all others in the training datasets while ranked the top 1 or 2 in AUC in the cross-validation datasets.

Deep learning had the highest cross-validation accuracy, consistently outperformed CSDM-WB, CSDM-CC, and Peak-CC, statistically ($p<0.05$; **Fig. 3**). Its performance was also comparable to BrIC (albeit, not statistically better; $p>0.05$), and consistently had a smaller standard deviation (STD), suggesting a more stable performance. Comparisons in sensitivity was inconclusive. BrIC had the best specificity and deep learning ranked the second in this metric (not shown).

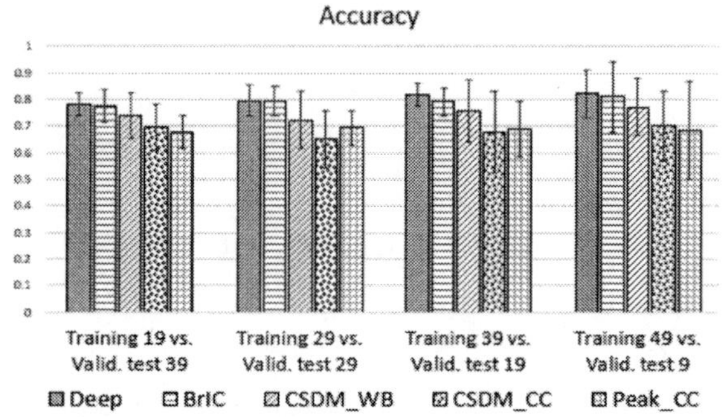

Figure 3 Cross-validation accuracies for the four training/validation configurations. Deep learning had the highest mean accuracy for all configurations and with the smallest standard deviation.

DISCUSSION

Developing an accurate and reliable metric for injury prediction is one of the cornerstones in TBI biomechanics research. Unfortunately, an "optimal" or best injury predictor has yet to be developed/identified. Instead of similarly attempting to pinpoint an explicit response measure pre-defined in a given ROI, here we utilized voxel-wise WM fiber strains from the entire brain. The classical injury prediction problem was formulated into a supervised classification via deep learning. It automatically distilled the most discriminative features implicitly from the strain-encoded images for concussion classification.

Our results indicated that deep learning had the highest AUCs for the training datasets. However, they did not necessarily translate into the same level of AUC or accuracy for the cross-validation datasets (**Fig. 2b** and **Fig. 3**). This observation, while not surprising, suggested the need to cross-validate an injury predictor for more objective evaluation of performance.

Nevertheless, deep learning had statistically comparable and often better cross-validation AUC and accuracy. It consistently outperformed CSDM-CC and Peak-CC. This was expected given that deep learning utilized features of the entire WM, whereas it was essentially "trial-and-error" for the latter empirical metrics pre-defined in the CC.

The main advantages of deep learning included its utilization of estimated brain mechanical responses (vs. kinematics alone) for injury prediction. It directly employed voxel-wise strains of the entire WM, without the need to empirically pinpoint a specific ROI. In addition, no thresholding was necessary, which was otherwise often conducted empirically and without considering inter-regional differences. Further, deep learning is directly applicable to neuroimages for injury prediction as well [6]. Conceivably, this could enable a multi-modal injury prediction framework combining both biomechanical responses and corresponding neuroimages. This is beyond the capabilities of any other kinematic or response-based injury metrics currently in use.

To conclude, we show that deep learning is an attractive and highly competitive alternative approach for brain injury prediction.

ACKNOWLEDGEMENTS

Funding is provided by NIH R01NS092853 and R21NS088781.

REFERENCES

[1] CDC, 2015. [2] Goodfellow et al, *Deep Learning*, MIT Press, 2016. [3] Ji et al., *J. Neurotrauma*, 32(7): 441–454, 2015. [4] Takhounts et al, *Stapp Car Crash J.*, 57: 243–266, 2013. [5] Giordano and Kleiven, *Stapp Car Crash J.*, 58: 29–61, 2014. [6] Cai and Ji, MICCAI mTOP Challenge, 2016.

MULTI-FIDELITY MODELING OF TRAUMATIC HEAD INJURY IN ACCIDENT RECONSTRUCTION

X. Gary Tan (1), Amit Bagchi (1)

(1) U.S. Naval Research Laboratory, Multifunctional Materials Branch
4555 Overlook Ave SW Washington, DC 20375, USA

INTRODUCTION

The traumatic brain injury (TBI) is one of the most common injuries to service members in recent conflicts. The majority of civilian and combat-related TBI cases are categorized as mild (mTBI) which often can result in acute, sub-acute, and/or chronic physiological, neurological, and psychological impairments. Computational models can offer insights in understanding the underlying mechanism of brain injury, which lead to the crucial development of effective personal protective equipment designed to prevent or mitigate mTBI.

Historically many computational models were developed for the brain injury study. However, these models use relatively coarse mesh with a less detailed internal structure (e.g., [1]). In additional these models include the head only while ignoring the rest of body and thus cannot properly model the human head behavior in the real scenario (e.g., accidental fall of pedestrian, blast loading condition etc.). On the other hand, the full body detailed finite element model is very expensive to use in the head injury applications. By combining the high resolution human head model and an articulated body model below the head, we develop the computational multi-fidelity human models which can be used efficiently to investigate the blast- and blunt-related mTBI.

The developed high-fidelity head model includes thirty different organs and accurately reproduces the complex internal and external structures of the head. Without resorting to more expensive high-order tetrahedral elements, we used a novel linear tetrahedral element formulation to remove the parasitic locking problem in modeling the nearly incompressible biological tissues. The simulation of fast running articulated human body model relies on the multi-body dynamics. We use the joint constraints to connect the articulated body with the high-fidelity head model. The resulting models are solved in the DoD open source finite element solver CoBi-FEM, in which the scalability has been verified in the DoD HPC. The bio-fidelity of computational model is validated by the reconstruction of real accident and comparing the numerical results with the medical images of brain injury from the clinical database.

METHODS

We briefly describe the key aspects of our multi-fidelity modeling method for the brain injury applied in the accident reconstruction.

1. High-fidelity head finite element model

The high-fidelity human head finite element model was developed from the high-quality magnetic resonance imaging scans, thus accurately reproduces the complex internal and external structures of the head (Figure 1). The unstructured tetrahedral mesh was used for this highly complicated geometry. The material models and associated material properties for different tissue components in the head are based on the literature findings and from the model calibration using the experimental data. The head model has been validated against the experimental tests for the blunt impact to the head ([2]).

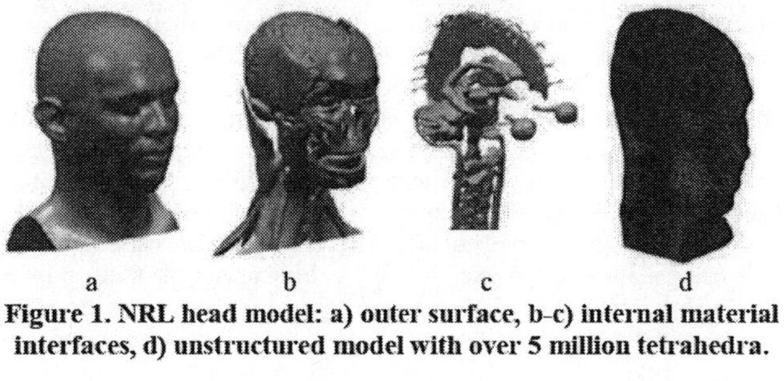

Figure 1. NRL head model: a) outer surface, b-c) internal material interfaces, d) unstructured model with over 5 million tetrahedra.

2. Articulated human body with head finite element model

The articulated human body dynamics has been developed and validated against the experimental data in our previous works ([3]). The fast-running finite element model provides the proper kinematics and the boundary condition for the high-fidelity head model. The human model is based on the Zygote geometry and partitioned into sixteen major body parts such as head, neck, chest, abdomen etc. (Figure 2). The body parts are connected by the fifteen joints to make the articulated human body. The input data of mass, center of mass, and inertia properties for different body parts are computed from the volumetric mesh. The joint parameters was calibrated by comparing the kinematic responses with those obtained from the post mortem human surrogates (PMHS) tests. The articulated human can be subsequently repositioned to different posture by performing a series of rotations on body parts. In this work, we have linked the high resolution head finite element model with the articulated body by installing the joint at the T1 location of head model.

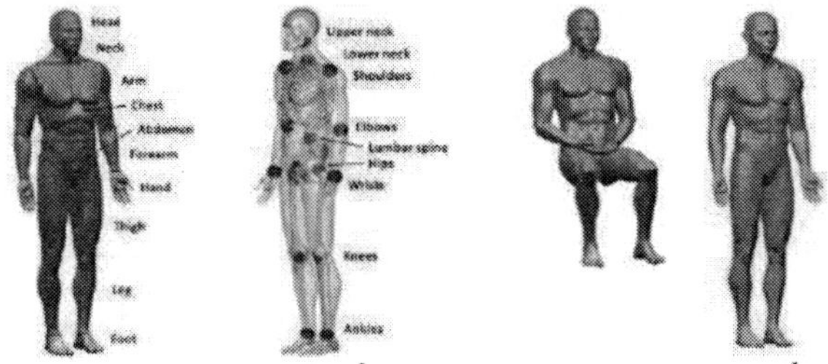

a b c d

Figure 2. a-b) Partition of human body and joint locations, c) repositioning by rotations around joints, d) multi-fidelity model by combining high-fidelity head with articulated body.

3. Scalable Finite Element Solver

The model was simulated by the finite element solver CoBi-FEM, which has been verified in many biomechanics related projects [4]. In terms of element technology, the standard linear tetrahedral element suffers from significant locking effect when nearly incompressible materials such as brain tissues are applied. Conversely we used the averaged nodal pressure (ANP) linear tetrahedral element [5] in which the pressure field is averaged among adjacent elements to overcome volumetric locking. It is quick, convenient and reduces sufficiently the number of constrains in nearly incompressible problems, as compared to the much highly cost in both CPU and memory usage when using the quadratic tetrahedral element. Recently the code was installed in the DOD HPC and the scalability for solving the large problems has been verified by employing the computer cores up to five thousands.

RESULTS

We used the developed multi-fidelity human model for analysis of impact problems to emulate the blunt impact injuries as seen in accidents. Multiple impactor shapes with different impact conditions such as location and velocity on the human head have been chosen to represent the sidewalk impact conditions.

In Figure 3, the human model collides with the rigid plate from different directions with the same velocity of 5 m/s normal to the plate. Initially the feet are in contact with the ground. During the simulation the body parts like extremities may interact with the frictional ground.

We show the simulation results for different loading scenario in Figure 4. The simulation results such as the pressure, principal strain and von Mises stress can be used to compare with medical images from the similar accident event on the injury correlation. It was suggested that taking the cumulative effect of the brain tissue response into account can better correlate the injury location and injury severity than other methods which only evaluated the injury location for one specific time when the maximum value occurred ([6]).

a b c

Figure 3. Model setup for different accidental scenario: a) fall backward, b) fall forward, c) fall sideway.

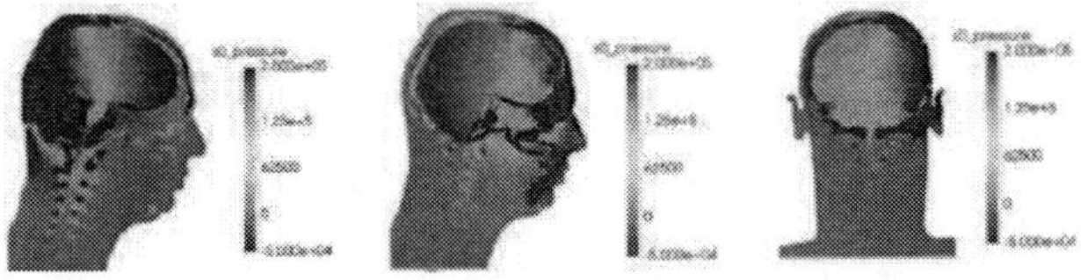

a b c

Figure 4. Biomechanical measures of head injury in different accidents: a) fall backward, b) fall forward, c) fall sideway.

DISCUSSION

In this work we have developed a new multi-fidelity model for the investigation of mild TBI in accident reconstruction. The sophisticated head finite element model with detailed internal structures was connected with the recently developed fast-running articulated human body. By considering the body inertia effect and the body interacting with the surrounding, this new model provides the realistic setting to capture human kinematics in the accidental scenario and meanwhile produces high-resolution biomechanical solution in the head. The model is more efficient in simulating the head injury when compared to the full-body finite element model. Using the highly scalable finite element solver along with the locking-free tetrahedral element, we were able to simulate the detailed head biomechanical response in different pedestrian accidental impact conditions. The resulting mechanical stresses and strains were used to correlate with the medical images obtained from the real accidents. The same human model can also be used for the impact injury caused by the blast event.

ACKNOWLEDGEMENTS

Funding was provided by the Office of Naval Research (ONR) through the Naval Research Laboratory's Basic Research Program. The analysis was supported by the Department of Defense High Performance Computing Modernization Program using the Army and Air Force Research Laboratories and the Engineer Research and Development Center Research Major Shared Resource Centers under project 416, subproject 231. Dr. Ross Cotton from Simpleware® is thanked for creating the head model.

REFERENCES

[1] Kleiven S., *Frontiers in Bioeng. and Biotech.*, 1(15):1-5,2013.

[2] Young P.G., *2015 SIMULIA Community Conference*.

[3] Tan X.G. etc., *Int. J. Human Factors Model.&Simul.*, 2:85–110, 2011.

[4] Tan X.G. etc., *Proceedings of ASME-IMECE* 2012-89072, 2012.

[5] Bonet J. et. al., *Int. J. Numer. Meth. Engng*, 50:119-133,2001.

[6] Doorly M.C., PhD Thesis, University College Dublin, 2007.

SB[3]C2017
Summer Biomechanics, Bioengineering and Biotransport Conference
June 21 – 24, Tucson, AZ, USA

MECHANICAL PROPERTIES OF PORCINE BRAIN TISSUE *IN VIVO* AND *EX VIVO* ESTIMATED BY MR ELASTOGRAPHY

Charlotte A. Guertler (1), Ruth J. Okamoto (1), John L. Schmidt (1), Andrew A. Badachhape (2), Curtis L. Johnson (3), Philip V. Bayly (1,2)

(1) Mechanical Engineering and Materials Science
Washington University in St. Louis
St. Louis, MO, United States

(2) Biomedical Engineering
Washington University in St. Louis
St. Louis, MO, United States

(3) Biomedical Engineering
University of Delaware
Newark, DE, United States

INTRODUCTION

Traumatic brain injury (TBI) causes over 1.7 million new cases of disability and death in the United States each year [1]. TBI occurs when sudden linear or angular head acceleration leads to stretching and shearing of brain tissue [2,3]. In principle, simulations of TBI can be used to develop and evaluate preventative measures, but accurate, validated computer models of TBI are still lacking [2,3]. Such models require accurate mechanical properties of brain tissue. Many mechanical tests of brain tissue have been performed *ex vivo*, but there is limited research on the relationship between tissue properties measured *ex vivo* and behavior *in vivo* [4]. MR elastography (MRE), an imaging modality capable of noninvasively estimating tissue material properties using actuated tissue vibrations, enables characterization of *in vivo* and *ex vivo* brain tissue. There are many factors that differentiate the intact, living brain from *ex vivo* samples [4,5], such as perfusion, residual stress, and metabolic state. In the current study, material properties of porcine brain tissue estimated from MRE were obtained *in vivo* at frequencies of 50, 80, 100, and 125 Hz and *ex vivo* at frequencies of 80, 100, 125, 200, and 300 Hz.

METHODS

In vivo MRE of the Yucatan mini-pig brain (N=6) was performed on a Siemens Prisma® 3T MRI scanner. Shear waves in the brain were induced via skull vibrations at frequencies of 50, 80, 100, and 125 Hz with a commercially available pneumatic system (Resoundant™, Rochester, MN) using a custom end-effector placed between the back molars of the mini-pig jaw (Figure 1A). MRE data with 3D displacements encoded as phase were acquired with a 2D multishot spiral sequence [6] with 1.5 mm isotropic voxels, covering a volume of 180 x 180 x 60 mm^3 (Figure 1B). Data for each mini-pig was collected over the course of 2-3 MRE scans, using one to three frequencies of actuation per scan.

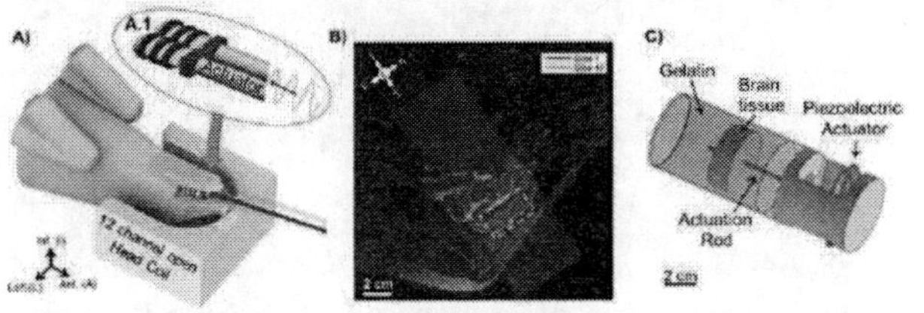

Figure 1: Experimental set-up for *in vivo* (A-B) and *ex vivo* (C) MRE. A) Custom actuator (A.1) driven by Resoundant™ system is placed between the back molars of the supine mini-pig jaw to induce vibrations in the skull and shear waves in brain at 50, 80, 100, and 125 Hz. B) T2-weighted anatomical image (sagittal view) of the supine mini-pig brain with MRE slices denoted. Yellow rectangle shows approximate location of *ex vivo* brain tissue cylindrical sample. C) *Ex vivo* cylindrical brain tissue sample is excited via central actuation rod at 80, 100, 125, 200, and 300 Hz.

Ex vivo MRE of the same Yucatan mini-pig brains was performed on an Agilent/Varian DirectDrive 4.7T small-bore animal MRI scanner. Brains were extracted immediately after euthanasia; 42 mm diameter cylindrical samples, each containing corpus callosum and the superior gray matter, were extracted with a sharp punch and embedded in gelatin/glycerol [7]. MRE studies were performed in brain/gel samples at room temperature (21-23°C) and started within 2 hours post mortem. Shear waves of 80, 100, 125, 200, and 300 Hz were excited by a central actuation rod (Figure 1C) driven by an MR-compatible

Technical Presentation #249

piezoelectric actuator (APA150, Cedrat Technologies, Meylan, France) [8]. Data were acquired with a modified 2D multi-slice spin-echo sequence with 1 mm isotropic voxels, covering a volume of 48 x 48 x 24 mm^3.

Local direct inversion (LDI) was performed on both the *in vivo* and *ex vivo* MRE displacement fields to estimate the mean complex shear modulus of the mini-pig brain sample at each of the measured frequencies using a viscoelastic analog of the Navier equation [7]:

$$(G' + iG'')\nabla^2 U_z(x,y,z,\omega) = -\rho\omega^2 U_z(x,y,z,\omega) \quad (1)$$

where G' is the storage modulus and G'' is the loss modulus. This equation assumes that the material is linear, isotropic, and locally homogenous. Estimates of shear modulus were made for the entire brain *in vivo*. The results were averaged over a region of interest corresponding to the location and dimensions of the *ex vivo* brain samples.

RESULTS

Anatomical T2-weigthed images, brain tissue motion measured by MRE, and estimated storage modulus are shown for both *in vivo* and *ex vivo* data in Figure 2. Estimates of mean storage modulus for *in vivo* and *ex vivo* data are shown in Figure 3. Both *in vivo* and *ex vivo* shear modulus estimates increase with frequency. At the common frequencies of 80, 100, and 125 Hz, estimated shear modulus of *in vivo* brain tissue is higher than that of *ex vivo* tissue.

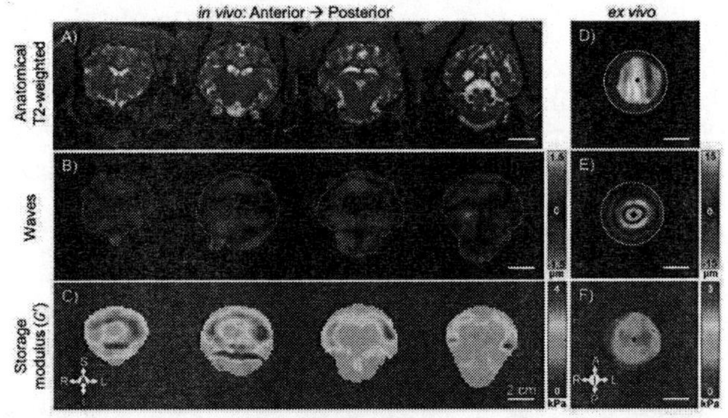

Figure 2: Comparison of experimental data from the porcine brain *in vivo* (A-C) and *ex vivo* (D-F). A) T2-weighted MR images of four non-contiguous (7.5 mm spacing) coronal slices of the brain *in vivo* at 1.5 mm resolution. B) Brain tissue wave motion at 100 Hz measured by MRE *in vivo*. The anterior-posterior (AP) component of motion is shown for the same images slices as in A. C) Shear modulus (G') at 100 Hz estimated using LDI *in vivo*. The AP component of motion is shown for the same images slices as in A and B. G' was only estimated for voxels where >50% of the 7x7x7 fitting kernel was inside the brain. D) T2-weighted image of *ex vivo* cylindrical sample from the same animal at 1 mm resolution. E) Brain tissue wave motion at 100 Hz measured by MRE *ex vivo*. The superior-inferior (SI) component of motion is shown for the same image slice as in D. F) Shear modulus (G') at 100 Hz estimated using LDI *ex vivo*. The SI component of motion is shown for the same image slice as in D and F. G' was only estimated for voxels where >50% of the 7x7x7 fitting kernel was inside the brain. Note: Image scales are the same in each panel (scale bars = 2 cm), but displacement ranges and image slice orientations differ between panels A-C and D-F.

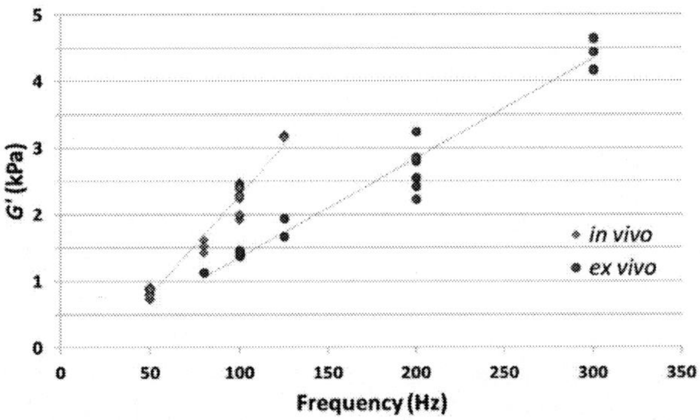

Figure 3: Mean storage modulus (G') of *in vivo* and *ex vivo* mini-pig brains estimated by LDI at frequencies from 50-300 Hz for N=6 animals. Each data point represents the mean storage modulus for one mini-pig scanned at the specified frequency. For *in vivo* data, the markers show the average G' estimates in a cylindrical ROI of dimensions matching that of the *ex vivo* cylindrical sample (Figure 1B). At the common frequencies, 80, 100, and 125 Hz, G' estimates are higher for data obtained *in vivo* than for *ex vivo* data.

DISCUSSION

This study is the first comparison of *in vivo* and *ex vivo* MRE on the same samples of brain tissue. MRE estimates of shear modulus suggest that intact, living brain tissue is stiffer than *ex vivo* samples. Both *in vivo* and *ex vivo* shear modulus estimates increase with frequency, which is expected for viscoelastic tissue. The storage modulus of *in vivo* brain tissue increases more rapidly with increasing frequency than that of *ex vivo* brain tissue. *In vivo* MRE measurements are challenging due to the size of the porcine brain, which limits the number of wavelengths that can be observed, and its position deep inside the skull, which limits the amplitude of shear waves induced in the brain. *Ex vivo* measurements can be performed at a higher resolution and with direct actuation of brain tissue. The dominant direction of shear displacement relative to anatomy differed between *in vivo* and *ex vivo*, but in both cases was perpendicular to corpus callosum fibers. Apart from these methodological differences, the observed differences in tissue stiffness *in vivo* and *ex vivo* may be due to residual stress, perfusion, metabolic state, or other factors.

MRE can detect measurable differences between *in vivo* and *ex vivo* mechanical properties, which are important to our understanding of brain tissue mechanical properties and their use in the development of accurate simulations of TBI. The observation that *in vivo* brain tissue is stiffer than *ex vivo* samples has significant implications for the use in computer models of mechanical properties measured *ex vivo*.

ACKNOWLEDGEMENTS

This project was supported by ONR N00014-15-C-5160, NIH NS055951, NSF CMMI-1332433, and NIH 5T32EB01485505.

REFERENCES

[1] Faul, M et al, *Inj Prev 2010;16:A268*
[2] Shaw, N, Prog Neurobio, 67(4): 281-344, 2002.
[3] Bayly, P et al. Jour Neurotrauma, 22(8): 845-856, 2005.
[4] Rashid, B et al, *J Mech Behav Biomed Mater*, 10:23-38, 2012.
[5] Gefen, A et al. *J Biomech*, 37(9):1339-1352, 2004.
[6] Johnson CL et al., *Magn Reson Med*, 70(2):404-412, 2013.
[7] Okamoto RJ et al., *Phys Med Biol*, 56(19):6379-400, 2011.
[8] Schmidt JL et al. J Biomech 49(7):1042-1049, 2016.

MEASUREMENT OF INTRAOCULAR PRESSURE DURING BLAST WAVE LOADING

Nikolaus A Benko (1), Daniel F. Shedd (1), Brittany Coats (1)

(1) Mechanical Engineering
University of Utah
Salt Lake City, Utah, USA

INTRODUCTION

Ocular primary blast injury has become a subject of great interest due to recent exposures in American and coalition warfighters in the Middle East [1][2]. The two most popular methods of research thus far have been injury replication with small-species animal models using mice, rats, and rabbits, and injury simulation with finite element models [3-6]. To date, there has been little work to relate the two methods of study and no experimental data exists to verify that predicted finite element model pressures are accurate. As such, there is little known about the mechanical forces inside the eye during blast exposure. As part of our lab's recent work on an animal model of ocular blast injury, we conducted the first known *in vivo* measurements of intraocular pressure (IOP) during primary blast injury.

METHODS

All animal studies were reviewed by the Institutional Animal Care and Use Committee of the University of Utah and were performed in adherence to the ARVO Statement for the Use of Animals in Ophthalmic and Vision Research. Long-Evans rats, (n=10, 399±68g) were fit with fiber optic pressure transducers (FISO FOP-LS-2FR-30, Quebec, Canada), surgically placed in each eye to measure the intraocular pressures (IOP) of blast exposed eyes. Animals were anesthetized with an intraperitoneal injection of 64.0 mg/kg of ketamine and 0.25 mg/kg dexmedetomidine. The lateral commissure was clamped with a hemostat for several seconds to limit bleeding, and then cut to gain access to the posterior half of the eye. The IOP transducers were pre-threaded through surgical tubing and placed into the midsection of an 18G hypodermic needle (Fig. 1). Tweezers were used to grasp the conjunctiva near the commissure and rotate the eye in the medial direction. The needle containing the pressure transducer was inserted through the posterior sclera and positioned into the central vitreous chamber. The conjunctiva was released and the eye allowed to gently rotate back into its natural position. Skin sutures and cyanoacrylate adhesive secured the surgical tubing to the back of the rat to limit sensor cable motion during animal positioning and blast wave exposure. The procedure was repeated for the other eye. Rats were secured in the blast tube in a 3D printed holder such that the right eye was directly exposed while the left was indirectly exposed to blast. After blast exposure, animals were euthanized and sensors removed.

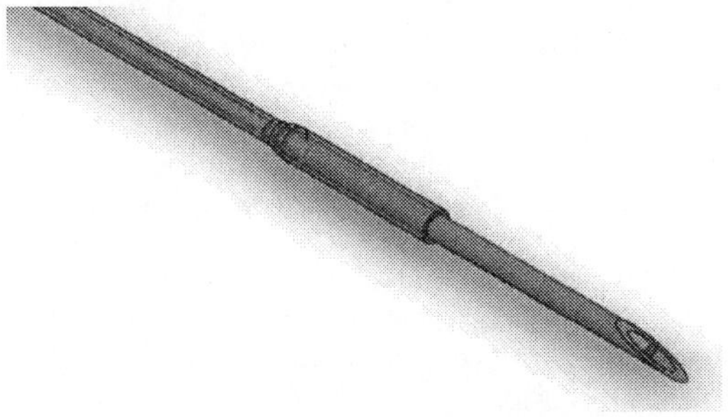

Figure 1: Computer rendering of pressure transducer situated inside an 18 gauge needle and shielded by surgical tubing.

Side-on blasts were generated using a 521 cm long shock tube. The tube has a 64 cm long driver section, 457 cm driven section, and constant internal diameter of 14 cm. The shock tube was tuned to reproduce the Friedlander waveform that is representative of an open field blast. The driver section was pressurized with compressed air until a 0.01" axially-oriented polyethylene terephthalate membrane ruptured. Pressure waveforms within the tube were recorded 2 cm upstream of the right eye for each blast-exposed animal at 1 MHz using a flush-mounted pressure sensor (PCB Piezotronics, NY). The pressure time history profiles were filtered using a custom hardware anti-aliasing filter with fc=180 kHz. Fiber optic transducers were connected to a FISO signal conditioner with a sampling frequency of 15 kHz. Data from the signal conditioner and tube-mounted pressure sensors was recorded using a NI DAQ 9223 with LabView software (National Instruments, Austin, TX). No post-hoc filtering was applied. Pressure data traces were analyzed in MATLAB to record peak pressure, rise time, initial slope, positive, negative, and net impulse for each curve.

RESULTS

Intraocular pressure was successfully captured in both eyes in eight of the ten animals. During initial blast wave contact with the directly and indirectly exposed eyes, IOP closely mimicked the pressure waveform measured in the shock tube. However, after the initial insult, IOP readings became erratic (Fig. 2). This could have been caused by sensor movement within the eye, contact between the ocular lens and the sensor tip, and/or wave reflection in the eye. Based on the initial 5 ms of the pressure waveform it was observed that peak pressure in the directly exposed eye matched the tube pressure measured 2 cm in front of the animal within 0.1% and with an expected 0.6775 ms delay (Fig. 3). Peak intraocular pressure measured in the indirectly exposed eye was lower than the tube pressure by 30%, but this difference was not statistically significant. Pressurization rate decreased by 39% as the blast wave traveled from the tube into the directly exposed right eye and continued to decrease as the wave traveled through the head into the indirectly exposed left eye ($p<0.0001$). No significant difference was observed in net impulse in the tube or either eye.

DISCUSSION

Recent animal models have demonstrated the occurrence of various ocular pathologies as a result of primary blast wave exposure [2]. However, little is known about the mechanical stresses and strains within the eye during exposure. Such knowledge is key for determining injury thresholds, design of protective equipment, and translating findings from animal studies to the human eye. A technique for measuring IOP in rodents during blast exposure has been introduced, and can successfully capture the initial pressurization during blast wave exposure. After peak pressure, however, it is likely there is substantial sensor movement, so identifying reflective waves in the eye is not possible.

In vivo peak ocular pressure during the first 5 ms of blast exposure in directly exposed eyes closely matched external tube

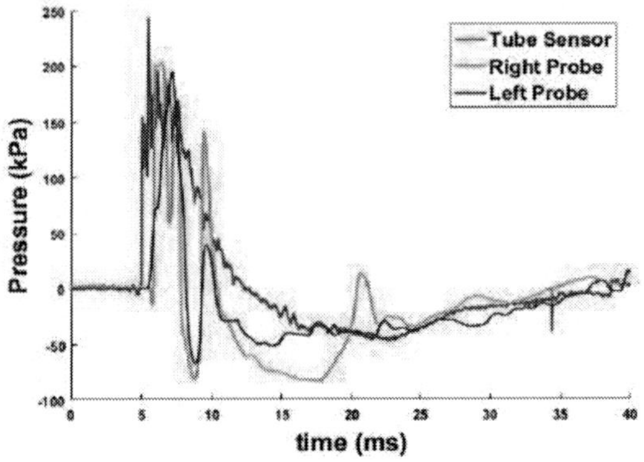

Figure 2: Pressure waveform during a side-on blast insult

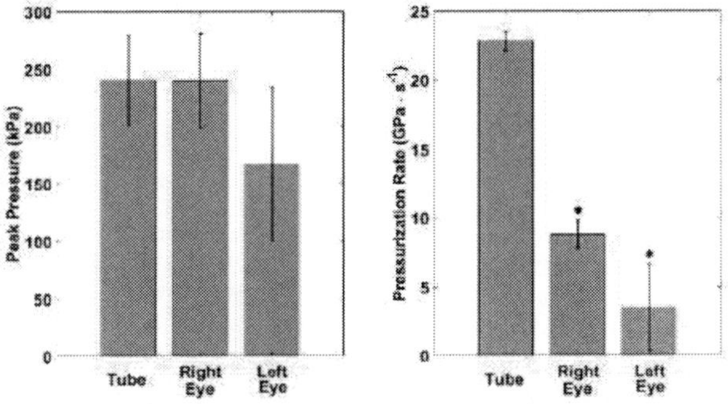

Figure 3:Peak pressure and pressurization rate in the 3 sensor locations. Bars indicate standard deviation (*p<0.0001)

measurements. Indirectly exposed eyes, however, experienced lower mechanical loads than those measured in the tube. Therefore, tube peak pressure measured 2 cm before the animal is a suitable surrogate for peak pressure in directly exposed eyes. Tube pressure is not a good surrogate for peak pressure in indirectly exposed eyes. Furthermore, the tube pressurization rate will likely overestimate the ocular pressurization rate. Decreases in peak pressure and pressurization rate observed in the indirectly exposed eye help account for a lower incidence of ocular injury found in our associated animal studies[7.] These data highlight the important role of facial structures and blast wave direction in determining risk of ocular trauma from blast.

ACKNOWLEDGEMENTS

We are grateful for USAMRAA award #W81XWH-12-1-0243 which financially supported this work.

REFERENCES

[1] Beiran, I et al., *J Opthalmology,* 114:504-505, 1992.
[2] Chaliolias, K et al., *J Army Medical Corps,2007:124-125.*
[3] Bricker-Anthony, C et al., *IOVS,* 55:4853-4862 2014
[4] Petras, J.M. et al, *Toxicology,* 121:41-49. 1997
[5] Esposito, L et al. *Comp Methods in Biomech and Biomed Eng.,* 18:107-115 2015
[6] Watson, R et al. *TVST,* 4:1-11, 2015
[7] Shedd, D., Coats, B., IOVS;56(7):4799. 2015

Technical Presentation #250

SB^{3}C2017
Summer Biomechanics, Bioengineering and Biotransport Conference
June 21 – 24, Tucson, AZ, USA

TWO PHASE THORACIC ORGAN RESPONSE DUE TO BLAST OVERPRESSURE LOADING ON POST-MORTEM HUMAN SURROGATES

Alexander S. Iwaskiw, Constantine K. Demetropoulos, Connor O. Pyles, Timothy P. Harrigan, Edwin B. Gienger, Connor A. Bradfield, Eyal Bar-Kochba, Joseph A. Andrist, Mary E. Luongo, Andrew C. Merkle, Robert S. Armiger

The Johns Hopkins University Applied
Physics Laboratory
Laurel, MD, USA

INTRODUCTION

Injury resulting from blast is a challenging problem that is not limited to the battlefield [1]. Due to the complexities of the event, as well as the increasing number of individuals at risk of blast injuries, the human response to blast is of great interest. Injury causation has long been debated, and experimental animal models as well as computational simulations fall short of understanding the complexities of blast loading and the human response [2,3,4,5]. The need for a post-mortem human surrogate (PMHS) experimental model is apparent, and will ultimately enhance our understanding of reported injuries. In this study, 4 full body PMHS were exposed to simulated blast loads using a shock tube, in order to investigate the mechanisms of blast related injury by evaluating the biomechanical response of vital structures and organs (pressure, strain, motion) in blast exposure to an overpressure insult. This study seeks to understand the differences in loading mechanisms between the vital thoracic organs including the lungs (air-filled) and heart (fluid-filled). Characterization of the difference in pressure transmission through these organs will be used to inform test surrogates and computational models.

METHODS

4 full body male PMHS were tested (ages 30-79) in this study and were selected to minimize the effects of specimen variability on biomechanical response. The specimens were instrumented with pressure sensors, accelerometers, displacement sensors, and strain gauges. The specific focus of this study was to understand the pressure response at the PMHS surface, and within the lungs and heart. The surface pressures were collected using thin profile pressure transducers (PCB 138M156) adhered to the sternum, back and lateral aspects of the body. Care was taken to place these sensors orthogonal to the direction of the shock front. Lung pressure sensors were placed in the left and right lobes (Fiso FOP MIV R3) and in the trachea (PCB 113B21). The fiber optic lung pressure transducers were placed into the lobes of the lung by percutaneous insertion through a cannula. A pressure transducer was placed into the trachea through an endotracheal tube (ETT), inserted through a tracheotomy. The lungs were kept inflated during the simulated blast exposure using positive air pressure through the ETT. Additionally, strain gages (Kyowa single axis strain gages: KFW-2-350-C1-23) were placed on the ribs 3, 5 and 7, both on the anterolateral and posterolateral surface approximately 75 mm from the mid sagittal plane. Rib strain was used to evaluate rib cage motion, as it relates to organ pressure.

Specimens were subjected to shock loads selected to simulate live fire conditions, with a 150 mm diameter shock tube [6,7,8]. A conical diffuser was used to expand the shock wave from the 150 mm driven section to a 460 mm diameter secondary chamber prior to PMHS exposure. Vertical positioning of the PMHS was set with the middle of the sternum (as measured from the midpoint between the manubrium and xiphoid process) in the center of the 460 mm chamber. The PMHS was subjected to three nominal burst pressures per orientation. These burst levels are termed Low (mean = 923.2 kPa), Mid (mean = 1291.8 kPa), and High (mean = 1746.6 kPa) in subsequent analysis. Different orientations of the specimen were used in the study. Front, Rear, Left Lateral and Right Lateral (orthogonal exposure), and Left Oblique and Right Oblique facing (45 degrees) tests were conducted. Specimens were subjected to a repeat Front facing condition to ensure sensor repeatability and sub-injurious specimen conditions.

RESULTS

The overpressure exposure is best characterized by the surface pressure sensors adhered to the surface of the specimen. For Front facing tests across specimen, mean surface pressures were 228.3±37.9 kPa, 278.9 ±32.9 kPa and 320.9±43.6 kPa, for Low, Mid, and High doses, respectively. These pressure sensors were oriented towards the pressure front and have contributions from dynamic and reflected pressures. When comparing those input pressures with mean pressures measured in the lung and heart, large attenuation is seen. Average peak heart pressures in Front facing tests, across increasing dose, were 23.2±9.0 kPa, 33.8±11.4 kPa, 39.6±12.6 kPa. The lungs demonstrated far lower pressures for Front facing tests, with average peak pressures of 4.2±2.2 kPa, 5.6±3.0 kPa, and 7.2±4.0 kPa. For Front tests, pressure response increases with dose (Figure 1). This trend is generally seen in all orientations. While the magnitude sensitivity in pressure response is greater in the heart, where there is a 16.4 kPa increase from Low to High versus a 3 kPa increase for the lungs, the actual percentage increase is equivalent in the heart and the lungs (170% increase). Notionally, repeatability of magnitude and shape within the same specimen was high, with characteristic pressure traces shape remaining across doses and repeats.

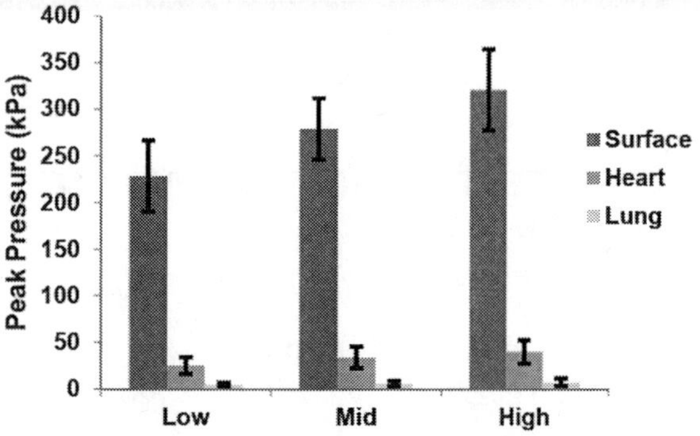

Figure 1: Average Pressure Response and Standard Deviation, Across Doses, at the Surface, in the Heart, and in the Lung for all Front Orientation Tests

Characteristic pressure traces, from a High dose Front facing test, show varying load timing in each response (Figure 2) The surface pressure shows a sharp rise in pressure in <1 ms followed by a triangular positive pulse. The heart then follows with a similarly sharp rise in pressure with its peak occurring approximately 0.5 ms later. The positive pulse duration is longer in the heart than is seen in the surface pressure data. Characteristic lung pressure response (taken from the right lobe of the lung) shows an oscillating pressure rise with a positive peak approximately 5 ms after initial loading. This suggests different pressure transmission mechanics for the heart and lungs.

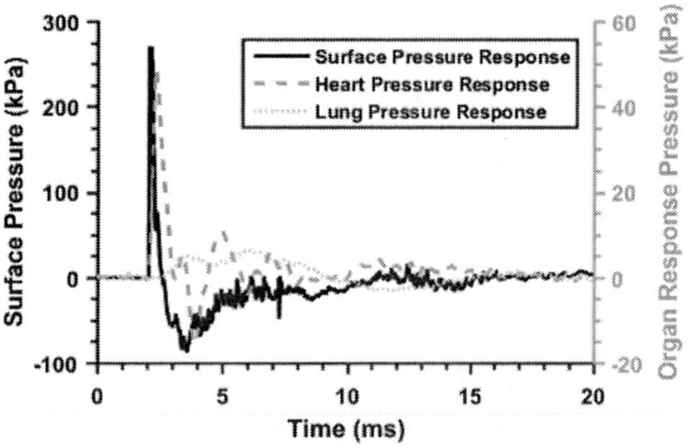

Figure 2: Characteristic Pressure Response Showing Surface Pressure, Heart Response Pressure, and Lung (Left Lobe) Response Pressure for a High (1750 kPa) Dose Test

DISCUSSION

The main observation of this study is that the pressure transmission to the heart and lungs are fundamentally different. The heart demonstrates a greater magnitude pressure response than the lungs, as well as a more rapid response. The heart pressure appears to be more tightly coupled to the surface pressure, where the timing of the positive pressure peak is <1 ms for the heart. Pressures measured in the heart are thought to be generated by pressure wave propagation through the body. The lungs demonstrate a much longer positive pulse duration, without fast wave propagation as observed in the heart. As demonstrated by the rib strain and lung pressure response from a characteristic High dose Front orientation test (Figure 3), the timing of both peaks match favorably. This result supports the thought that the increased pressure in the lungs can be attributed to bulk compression of the thoracic cavity, as opposed to direct pressure wave transmission.

The observed differences in organ pressure transmission are also supported by the fact that lung peak pressure is less sensitive to orientation. The decrease in lung pressure response from Front to Rear is 34%, which is most evident for High doses. The heart sensor demonstrates a much greater sensitivity to orientation, where the decrease in pressure response of 76%, is greater than observed in lung response. It is known that pressure transmission is affected by impedance differences between materials, as a wave propagates. Fewer changes in tissue type (or impedance) exist in Front orientation tests for the heart, as compared to the Rear. For Rear orientation cases, the blast overpressure wave must travel through multiple tissue types (including the air-filled lungs) before reaching the heart. This change in impedance during the wave propagation could lead to the lower pressures seen in the heart for Rear orientation cases. As the lungs are air-filled organs, they are not as sensitive to differences in pressure wave propagation. However, lung pressure response seems to be correlated with bulk torso compression, which may be less sensitive to orientation.

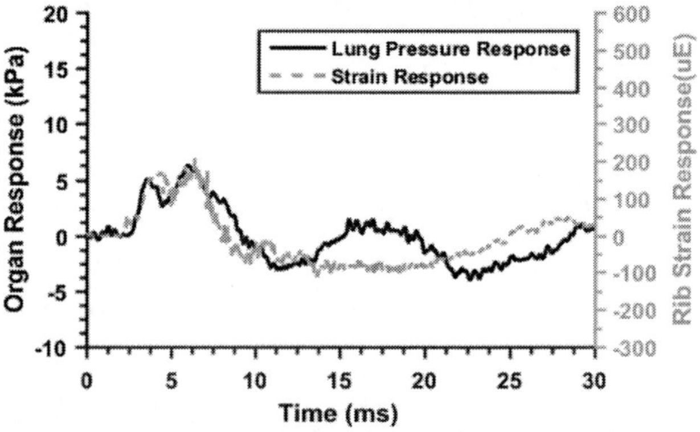

Figure 3. Characteristic Pressure Response Showing and Lung (Left Lobe) Response Pressure with Rib strain (Right Rib 5) for a High (1750 kPa) Dose Test

ACKNOWLEDGEMENTS

This effort was funded by, and in accordance with US Army Medical Research and Materiel Command Office of Research Protections, contract # W81XWH-09-2-0168. The U.S. Army Medical Research Acquisition Activity, 820 Chandler Street, Fort Detrick MD 21702-5014 is the awarding and administering acquisition office. The content included in this work does not necessarily reflect the position or policy of the U.S. government.

REFERENCES

[1] DePalma, R et al *NE J Medicine* 352.13; 1335-1342, 2005.
[2] Bowen, I et al. Lovelace fnd. medical ed & research, 1968
[3] Wightman, J et al *Annals of emergency med.* 37.6:664-678,2001
[4] Stuhmiller, J et al *J Trauma & Acute Surg.*28.1:S132-S139, 1988
[5] Tan, X et al *ASME* 2012.
[6] Merkle, A et al. *ASME* 2009
[7] Merkle, A et al. *NHSTSA Human Subjects Workshop* 2012
[8] Armiger, R et al *Mechanics of Bio.Sys.&Materials* 127-134 2014.

THE INFLUENCE OF COMPRESSIVE STRAIN RATE DEPENDENCY ON THE STRUCTURE-PROPERTY RELATIONS OF FETAL PORCINE BRAIN

Courtney J. White (1), Jun Liao (2), Michaela J. Beasley (3), Michael D. Jones (4), R. Prabhu (1), Lakiesha N. Williams (1)

(1) Department of Agricultural and Biological Engineering
Mississippi State University
Mississippi State, MS, USA

(2) Department of Bioengineering
University of Texas at Arlington
Arlington, TX, USA

(3) College of Veterinary Medicine
Mississippi State University
Mississippi State, MS, USA

(4) Cardiff School of Engineering
Cardiff University
Cardiff, Wales, United Kingdom

INTRODUCTION

Infants are vulnerable due to both their inability to function on their own and the fact that their developing bodies are fragile. This is most notable in respect to weak neck muscles supporting an infant's soft skull and extremely malleable brain. Generally putting them at greater risk for a more severe head injury. This can lead to long term damage in their brain and physical or mental functional disability [1].

Traumatic brain injury (TBI) can affect the functioning of the central nervous system (CNS) and counts for approximately 12% of hospital admissions and brain injury of infants and children [2]. The most common causes of infant TBI are falls, automotive accidents, and assault [3]. TBI research does not usually focus on the first few weeks of life and often involves analyzing brain tissue post trauma. This study will focus on fetal brain tissue from stillborn piglets that have not undergone trauma and were acquired within their first day of life.

When an infant undergoes TBI, it is important to understand the microstructural evolution that occurs during a traumatic event. An understanding of the microstructural evolution would help in developing better injury prevention strategies through the use of animal models, surrogate models, medical databases, or computational models [1,2,4,5]. In this context, it is important to have accurate models of infant biomechanics for determining severity of accidents and trauma. Without an accurate assessment of the amount of injury sustained to an infant's brain, proper steps cannot be taken to intervene and possibly protect from TBI damage. Also more accurate assessments may allow for proper evaluation in a court environment for infant injury cases. Strategies for patient-specific rehabilitation of infants using therapies can also be developed with this proposed research to make infants futures brighter [5].

METHODS

Stillborn fetal piglets were acquired from a local abattoir (Prestage Farms, Crawford, MS, USA) at the morning of death and tested immediately following dissection. The piglets were stored in a cooler of ice at 5°C and transported to the lab of the Agricultural and Biological Engineering (ABE) building of Mississippi State University (MSU) where their brains were extracted. From each of the 45 piglets (weight ~ 680g) two brain samples approximately 22mm in diameter and 11mm in height were dissected from each piglet brain. One brain sample was taken from each hemisphere through the parietal and frontal lobes of the brain. Then, compressive mechanical testing was conducted within an average of three hours after acquisition of stillborn piglets. These tests were carried out at engineering quasi-static strain rates (0.00625-0.10/s). Subsequent to testing, samples were then chemically fixed for histological analysis.

During the compression testing each sample was placed into a steel cylindrical platform with raised sides and then placed onto the Mach-1™ micromechanical testing device (BIOSYNTECH micromechanical systems, Canada). A Phosphate Buffered Saline (PBS) (Sigma-Aldrich, St. Louis, MO, USA) bath was created by submerging the sample in PBS within the stainless steel platform. Samples were then compressed at engineering strain rates of 0.00625/s, 0.025/s, and 0.10/s. For each strain rate the samples were compressed to 15%, 30%, and 45% engineering strains, respectively.

Figure 1: Mach-1™ micromechanical testing device at MSU.

Following compression to the appropriate strain percentage the test was stopped and the PBS solution was removed and replaced by 10% Formalin (Sigma-Aldrich, St. Louis, MO, USA) to chemically fix the sample. The samples were fixed for 24hrs before being removed from the Mach-1™ micromechanical testing device and fixed for an additional 24hrs in a sealed individual container. The fixed samples were then cut into longitudinal and transverse cross-sections and prepped for Haematoxylin & Eosin (H&E) staining.

ImageJ Image Processing and Analysis software [6] will be used to analyze the H&E stained slides from the compression tests (Figure 2). Nearest neighbor distance and cell density will be quantified in each of the (n=45) slides for statistical comparison between the 15%, 30%, and 45% strain levels in compression.

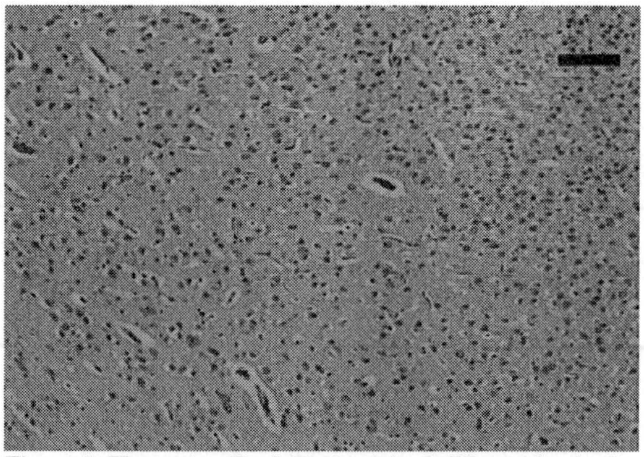

Figure 2: Haematoxylin & Eosin stained slide mechanically tested at a strain rate of 0.1/s to a strain level of 30% (10x magnification with 100µm scale bar).

RESULTS

The mechanical testing shows the response of the porcine infant brain under compression at three different strain rates (Figure 3). The H & E stained slides resulting from compression testing are shown in Figure 2. Nearest neighbor distance and cell density of neuronal cells will be determined for quantifying the microstructural evolution.

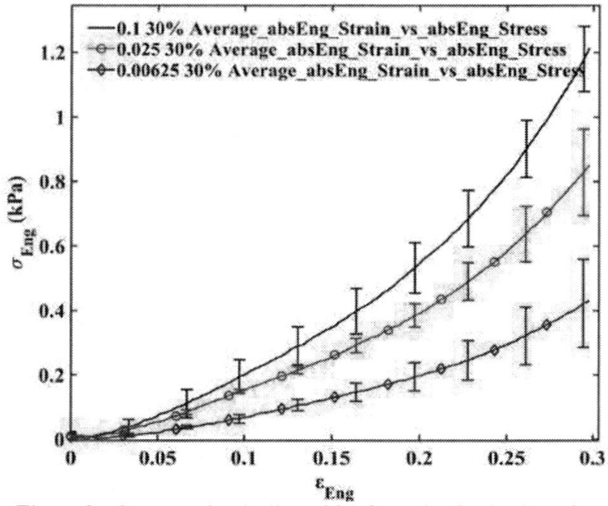

Figure 3: Compression testing of fetal porcine brain tissue by 30% at strain rates of 0.00625/s, 0.025/s, and 0.10/s.

DISCUSSION

The results of this research provided the stress strain response and associated microstructural evolution of fetal porcine brain. As expected, strain rate dependency (shown in Figure 3) was similar to that of the adult pig brain found by Begonia et al. [7]. The stress-strain results from Figure 3 show that the tissue's stress level increases as the strain rates increased. This is due to the cells of the brain tissue being compressed. We propose that as the strain rate and stress increases the cells become more clustered causing the nearest neighbor distance to decrease and cell density to increase.

The findings of this study help to give quantitative data for the stresses of fetal brain from a low rate loading model. There is a lack of information on TBI in children at such a young age. Having this quantitative and histological data will strengthen the ability to develop accurate infant head computational models, physical models, and surrogates. With the histological data processed through ImageJ Imaging and Analysis software [6] we can analyze how much change the tissue undergoes at the different strains and strain rates. This will further help in development of therapies to assist with counteracting and minimizing damage to the neurons and supporting structures suffered subsequent to TBI and in creating better preventative measures for infants.

ACKNOWLEDGEMENTS

This work was supported by the Department of Agricultural and Biological Engineering at Mississippi State University. The histological work was also aided by Dr. Alicia Olivier from the College of Veterinary Medicine at Mississippi State University.

REFERENCES

[1] Meythaler, J.M., et al., *Arch Phys Med Rehabil*, 82:1461-1471, 2001.
[2] Raghupathi, R., et al., *J Neurotrauma*, 21:307-316, 2004.
[3] Coats, B., et al., *Int J Dev Neurosci*, 30:191-200, 2012.
[4] Finnie, J.W., et al., *J Clin Neurosci*, 19:1159-1164, 2012.
[5] Goldsmith, W. and Plunkett, J., *Am J Forensic Med Pathol*, 25:89-100, 2004.
[6] Rasband, W.S., *National Institute of Health*, 1997-2016.
[7] Begonia, M.T., et al., *Ann Biomed Eng*, 38:3043-3057, 2010.

VASCULAR DISTENSIBILTY AND CONSTITUTIVE MODELING OF NORMAL AND PATHOLOGICAL PLACENTAL CHORIONIC ARTERIES

Shier Nee SN. Saw (1), Citra Nurfarah Zaini CNZ. Mattar (2), Arijit A. Biswas (2), Choon Hwai CH. Yap (1)

(1) Department of Biomedical Engineering, National University of Singapore, Singapore

(2) Department of Obstetrics and Gynecology, National University of Singapore, National University Health Systems, Singapore.

INTRODUCTION

Intrauterine growth restriction (IUGR) is a pregnancy complication of high prevalence (3% of pregnancies in developed nations, up to 15% in developing nations [1]). In IUGR, abnormally high placental vascular resistance results in insufficient oxygen and nutrients to the baby, causing it to be abnormally small, and exposing it to medical risks: 5-10x mortality rate, life-long morbidities such as hypertension, diabetes, neuro-maldevelopment etc. [2]. To date, there is no proven method to treat or prevent IUGR [3]. As such, research is urgently needed to further improve our understanding of the disease.

Abnormally high pulsatility in flow velocity waveforms in the umbilical artery had been observed in IUGR cases, as is revealed by abnormally high Doppler indices such as resistance index (RI) and pulsatility index (PI) [4]. In the clinical community, this increased pulsatility is often thought to be caused by the elevated placental vascular resistance due to the reduction of vascular size density [5]. Theoretically, vascular distensibility should also play a role in influencing these flow waveform changes, but to date, there is no direct evidence that IUGR placental chorionic vessels have altered distensibility. In fact, there has been no study on the vascular mechanics of chorionic vessels, either in normal or IUGR pregnancies.

In the current study, we perform mechanical testing of human placental chorionic arteries from normal and IUGR pregnancies, describe measured mechanical properties with constitutive modeling and perform histology to examine microstructures.

METHODS

Specimen Preparation

Post-delivery human placentae were collected and divided into 3 groups based on whether the estimated fetal weight (EFW) and Doppler indices (RI, PI) were normal or abnormal. EFW was normal if above the 10th percentile, while RI, PI were normal if below the 90th percentile. The 3 groups were: (i) Normal - normal EFW and Doppler indices; (ii) IUGR – abnormal EFW with normal Doppler indices; and (iii) severe IUGR – abnormal EFW and Doppler indices. Total 18 normal, 6 IUGR and 9 severe IUGR chorionic arteries were extracted and tested

Mechanical Testing

Pressure-diameter mechanical testing was performed using the setup shown in Fig 1. The artery was mounted onto needles and pulled to it's *in vivo* length. Eight cycles of pressurization and pressure release were performed on the vessel, where the five cycles served as pre-conditioning and the last three cycles were used for analysis. During each pressurization cycle, saline was injected gradually into the blood vessel using the syringe pump with a constant rate of 90ul/min until the luminal pressure reached 45mmHg, at which point a release valve was opened to return the luminal pressure to atmospheric pressure. The valve was then closed in preparation for the next pressurization cycle.

The distensibility was computed by the reciprocal of the gradient of the pressure vs. circumferential stretch graphs (distensibility = change of circumferential stretch /change of pressure). The inner diameter and the thickness of the arteries were measured. After the measurements, the vessels were cut radially and incubated for 4 hours to allowing the relieving of residual stress, and the residual stress was quantified via measurements of the opening angle.

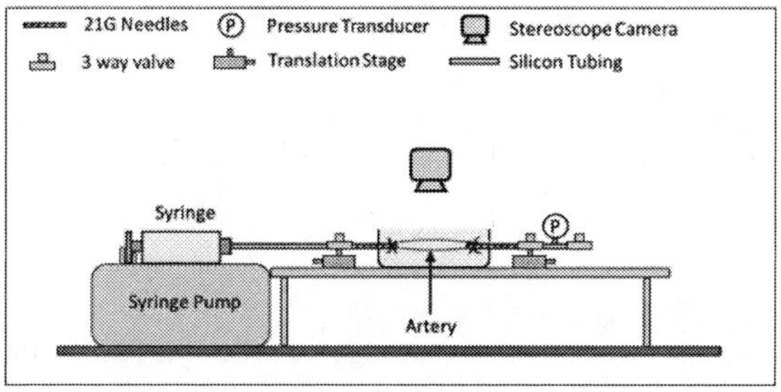

Figure 1. Experiment set up for mechanical testing of chorionic arteries

Constitutive Modeling

The chorionic artery was modeled as a thick wall cylindrical tube that was incompressible, cylindrical orthotropic and homogeneous. A simplified 2-parameters Fung model was adopted to characterize the mechanical properties of the chorionic arteries (Equation 1) since more complex models such as the 4- and 7-parameters Fung model could not provide unique solutions during curve-fitting.

$$W(E) = \frac{c}{2}\left(e^{b_1 E_\theta^2 + b_1 E_z^2 + b_1 E_r^2} - 1\right) \qquad (1)$$

where W is the strain energy function (SEF), c and b_1 are the material parameters and E is the Green Strain.

Histology

After mechanical testing, the arteries were extracted and processed for Van Verhoeff Gieson (VVG) stain to access the collagen and elastin content in the vessels. The collagen to elastin (C/E) ratio was computed by tabulating the number of pink (collagen) and black pixels (elastin).

RESULTS

Mechanical Testing

The mean of the pressure-circumferential stretch curves of the chorionic arteries for three groups are plotted in Fig 2a. The bar charts of the distensibility at different pressures are tabulated in Fig 2b. Results showed that the distensibility of the severe IUGR arteries was significantly higher than the normal when the luminal pressure was higher than 21mmHg.

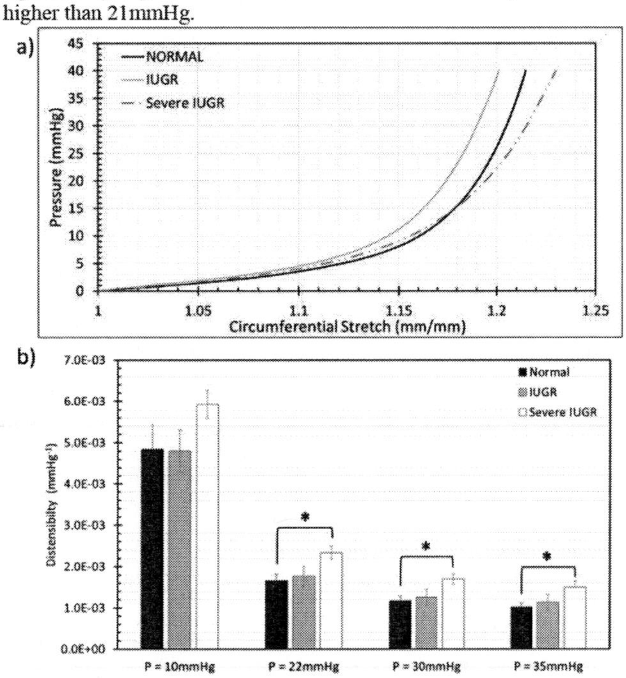

Figure 2. (a) Mechanical responses of the average chorionic arteries for 18 normal, 6 IUGR and 9 severe IUGR chorionic arteries, (b) distensibility of chorionic arteries (mean and standard error) at different pressure for normal, IUGR and severe IUGR. * indicates $p < 0.05$ between two groups as indicated.

Geometrical Measurements

No significant difference was observed among the three groups in term of opening angle, vascular wall thickness and inner diameter. However, the thickness/diameter (T/D) ratio of IUGR group (T/D = 0.75±SD0.30, $p<0.05$) was significantly higher than that of normal (T/D = 0.59±SD 0.13) and severe IUGR (T/D = 0.60±SD 0.14) groups.

Constitutive Modeling

The SEF parameters for the three groups are shown in Table 1. Results showed that the material parameter c was significantly larger in severe IUGR group than that in normal and IUGR groups. No significant difference was observed in parameter b_1 among the three groups.

Table 1. Material constants of the 2-parameters Fung model. Values indicated are mean and standard errors. * represents significance at $p<0.05$ vs. normal and vs. IUGR

	Normal (n=18)	IUGR (n=6)	Severe IUGR (n=9)
c	125.82±50.92	83.46±53.32	436.07±132.23*
b_1	20.63±5.60	13.34±3.75	7.76±1.36

Histology

No significant difference was shown in the C/E ratio among the three groups. From Fig 3b and c, the severe IUGR arteries had lower intima-media/adventitia thickness ratio as compared to normal arteries.

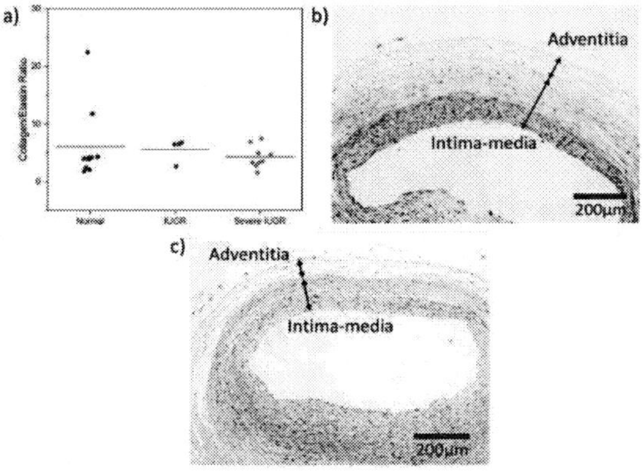

Figure 3. (a) Collagen/Elastin ratio of normal (n=10), IUGR (n=4) and severe IUGR (n=8) chorionic arteries. The lines indicate the mean of the data. Representative images of VVG stain for (b) normal and (c) severe IUGR chorionic arteries.

DISCUSSION

Our mechanical testing results showed that the severe IUGR arteries (abnormal velocity waveform) showed the greatest distensibility compared to normal and IUGR, suggesting that the chorionic arterial vascular distensibility does play a role in affecting the umbilical flow pattern. Since IUGR vessels are most likely hypoxic, our results corroborate with previous studies showing that hypoxic vessels develop increased distensibility [6].

The 2-parameters Fung model was found to be able to describe the mechanical properties and to capture the mechanical difference between severe IUGR vessels and the other two groups. Since more complex models could not give unique solutions, this simplified model may be more useful than more complex models in future modeling analysis.

Finally, histology showed changes to the relative thickness of the various layers of the vascular wall during severe IUGR, which may have played a role in causing the observed changes in mechanical properties.

ACKNOWLEDGEMENTS

Funding was provided by the National University of Singapore Young Investigator Award, entitled "Fluid Mechanics and Mechanobiology of the Congenital Cardiac Outflow Tract" (PI: Yap).

REFERENCES

[1] Barut, F., et.al., *Diagn. Pathol*, 5:24, 2010.
[2] Cosmi, E., et.al., *J.Pregnancy*, 2011:364381, 2011.
[3] Sankaran, S., et.al., *Best Pract Res Clin Obstet Gynaecol*, 23:766-777, 2009.
[4] Lee, S., et.al., *AJUM*, 13:31-36, 2010.
[5] Junaid, T., et.at., *Placenta*, 35: 808-815, 2014.
[6] Mateev, S. N., et.al., *J Appl Physiol (1985)*, 100:1842-50, 2006.

A COMPUTATIONAL STUDY OF THE CONTRIBUTION OF THE COMMONLY IGNORED SUPERFICIAL PERINEAL STRUCTURES DURING VAGINAL DELIVERY

Megan R. Routzong (1), Spandan Maiti (2), Raffaella De Vita (3), Pamela A. Moalli (4), Steven D. Abramowitch (1)

(1) Musculoskeletal Research Center
Department of Bioengineering
Swanson School of Engineering
University of Pittsburgh
Pittsburgh, Pennsylvania, United States

(2) Department of Bioengineering
Swanson School of Engineering
University of Pittsburgh
Pittsburgh, Pennsylvania, United States

(3) STRETCH Lab
Department of Biomedical Engineering and Mechanics
Virginia Tech
Blacksburg, Virginia, United States

(4) Magee-Womens Research Institute
Magee-Womens Hospital
University of Pittsburgh
Pittsburgh, Pennsylvania, United States

INTRODUCTION

The female pelvic floor is a complex network of muscle and connective tissue that work synergistically to hold the pelvic organs in place. During vaginal delivery, the pelvic floor undergoes tremendous strains to accommodate the fetal head as it descends through the vaginal canal, which commonly results in maternal injury. These injuries can set the stage for future pelvic floor disorders, including urinary and fecal incontinence and pelvic organ prolapse [1]. Previous computational simulations have identified the enthesis of the pubovisceral muscle (PVM) as an area that experiences stretch ratios greater that those expected to result in injury [2]. These findings are consistent with magnetic resonance imaging that demonstrates injury at this location for 18% of the parous women observed [3].

Nearly all prior models of vaginal childbirth include the levator ani muscles, which support the pelvic organs and maintain the genital hiatus (the opening in the pelvic floor for the urethra and vagina) [4]. However, they generally ignore superficial perineal structures consisting of the bulbocavernosus (BCM), ischiocavernosus (ICM), and deep and superficial transverse perinei muscles (DTPM and STPM, respectively), which lie directly inferior to the pelvic diaphragm and superior to the perineal body (PB) [5]. Their omission is likely due to the complexity these muscles add to computational simulations and because their contribution appears, based on inspection, to be less significant than that of the levator ani muscles. With the motivation of eventually creating subject specific models of vaginal delivery predictive of maternal injury, the objective of this study was to investigate whether including these additional perineal structures in a simulation of the second stage of labor impacted the stretch ratios measured in the PVM and PB, two common locations for maternal injury. The hypothesis, consistent with the assumptions of previous literature, was that their inclusion would have minimal impact.

METHODS

All muscles and bone were segmented from frozen cryosection images (with a slice thickness of 0.33mm) of the female cadaver from the Visible Human Project (U.S. National Library of Medicine, Bethesda, MD, USA). The geometry was smoothed in 3D-Coat v.4.1.16 (Pilgway, Kiev, Ukraine), the surface meshed using Instant Meshes (Interactive Geometry Lab ETH Zurich, Zurich, Switzerland) and a solid mesh generated in Gmsh v.2.13.2 (C. Geuzaine and J.-F. Remacle, USA). In total, the rigid bones consisted of 43,190 triangular elements and the muscle geometry consisted of 93,857 tetrahedral elements with increased mesh density at regions of high deformations based on a preliminary simulation. The pelvis and sacrum were assigned as rigid bodies and fixed so they could not translate or rotate. The coccyx was deformable only so the desired boundary conditions (rotation about one point limited to the x-axis) could be used, but it was given a stiffness 10^4 times higher than that of the muscles so it essentially behaves as a rigid body. All muscles were continuous and modeled as a nearly incompressible, neo-Hookean 3D continuum. Since representative values of the material parameters are unknown, especially at the time of delivery, and stresses were not considered for this study, values were selected to be the same for all muscles. The fetal head was also a rigid body—free to translate in any direction, but not rotate, with a prescribed downward (z) displacement.

All connective tissue was simulated using nonlinear springs with user-defined force-displacement curves. For all springs, the stiffness was assigned to be 0 N/mm in compression. Tension stiffness values of 123.75 N/mm, 49.5 N/mm, and 66 N/mm were assigned to 62 springs from the coccyx to the sacrum, 37 and 95 springs from the posterior pubococcygeus muscle to the sacrum and coccyx, respectively, and 61 springs from the left and right iliococcygeus to the ischial spines. Finally, 2,146 springs connecting the BCM and ICM to the right and left inferior pubic rami, and the right and left DTPM and STPM to the inferior ischial rami were each assigned a tension stiffness of 16.5 N/mm. For the PVM and pubococcygeus attachments on the pubic bone, the translations of muscle facets were fixed. Contact between all bodies was frictionless sliding.

The analysis was dynamic so temporal evolution could be studied. All tissues were assigned a density of 1 g/mm^2. The fetal head displacement was prescribed as 100% of the head length in the z direction and to occur linearly over 10,000 seconds with automatic time steps ranging from 1e-05 to 1,000s. Numerical damping was employed by adjusting the Newmark time integration parameters beta and gamma to 1 and 2, respectively. All simulations were performed using FEBio Software suite v.2.0 (University of Utah, Salt Lake City, UT, USA).

Two FEM models were created in this way, with the same geometry, material properties, and boundary conditions. For the Without model though, the head could penetrate the superficial perineal muscles and the springs connecting the latter to the pelvis were removed. This assured that the contribution of these structures was the only difference between the two simulations.

1st principal Lagrangian strain vs. time curves were generated for both models sampling data from the PVM to determine when 25%, 50%, 75%, and 100% of the maximum strain occurred. 1st principal Lagrangian strains were also sampled at these time points for all other structures and compared. These strains were then converted to stretch ratio values and normalized for further comparison.

RESULTS

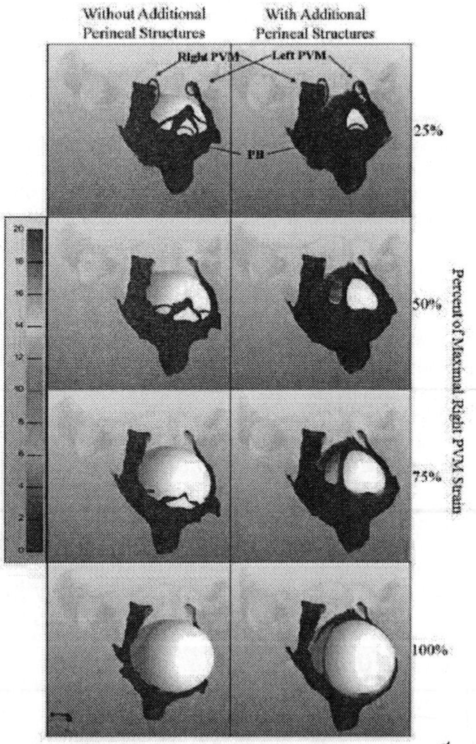

Figure 1: Four images of each model to demonstrate 1st principal Lagrangian strain distributions during fetal head (white) descent.

Figure 1 demonstrates the distribution of strains throughout the second stage of labor to the point of maximal strain experienced near the PVM attachments. For both models, the strains measured at the PVM attachments are higher than and peak before those in the PB. It can also be seen for this geometry that the right PVM experiences higher strains than the left PVM, although they peak at similar times.

Table 1: Stretch ratios normalized to the overall maximal value observed in both models (i.e. right PVM)

Model	Location	Normalized Stretch Ratios
Without	Right PVM	1
With	Right PVM	0.88
Without	Left PVM	0.69
With	Left PVM	0.56
Without	PB	0.35
With	PB	0.86

The Without model had a maximal stretch value 12% and 13% higher than the With model at the right and left PVM, respectively. In the PB, this trend is reversed as the With model predicted values 51% larger than the model without additional superficial perineal structures.

DISCUSSION

The present study demonstrates that excluding the superficial perineal structures results in higher 1st principal strains near the PVM attachment sites and lower strains in the PB. Refuting our hypothesis, these data suggest that it is likely important to consider these structures in simulations of the second stage of labor moving forward, especially if the goal is moving towards patient specific models of maternal injury. It also suggests that the PB is more vulnerable to injury than appreciated by pervious work, which is consistent with the relatively large number of tears to the PB reported in the literature [7].

As this was a first attempt to quantify the importance of including these additional perineal structures, there are several limitations to this study. First, the geometry was based on an older, parous, female cadaver (age 59), which likely does not appropriately represent the geometry of a full-term woman. Second, fetal head molding was not considered and rotation was restricted. These would likely result in lower strains than those observed in this study. Third, the muscles were continuous with each other, hyperelastic, and lacked fiber directions and contractile ability. While these muscles are continuous in-vivo, there are identifiable boundaries connecting one to another and it is not clear how stresses are transferred across these boundaries. Nevertheless, the model with the additional perineal structures in this study provides a better gross impression of the maternal geometry in response to a crowning fetal head and likely warrants more focus than previously given. Future work will focus on a parametric study aimed at better understanding the contribution of these structures by varying relative stiffness values.

ACKNOWLEDGEMENTS

This work was supported by NSF Award #1511504.

REFERENCES

[1] Wu, J. M., *Obstetrics and Gynecology*, *123*(1), 141–8. 2004
[2] Jing, D., *Journal of Biomechanics*, *45*(3), 455–460. 2012
[3] DeLancey, J. O. L., *Obstetrics and Gynecology*, *101*(1), 46–53. 2003
[4] Ashton-Miller, J. A., *Annals of the New York Academy of Sciences*, *1101*, 266–296. 2007
[5] Herschorn, S. *Reviews in Urology*, *6 Suppl 5*(3), S2–S10. 2004
[7] Harkin, R., *European Journal of Obstetrics & Gynecology and Reproductive Biology*, *109*(2), 149–152. 2003

SB[3]C2017
Summer Biomechanics, Bioengineering and Biotransport Conference
June 21 – 24, Tucson, AZ, USA

COMPARING *IN VIVO* ULTRASOUND GEOMETRY AGAINST *IN VITRO* CALCULATIONS FOR BIAXIAL TESTING IN THE NONPREGNANT MURINE CERVIX

Cassandra K. Conway (1), Hamna J. Qureshi (2), Leise Knoepp (3), Laurephile Desrosiers (3), Craig J. Goergen (2), Kristin S. Miller (1)

(1) Biomedical Engineering
Tulane University
New Orleans, LA, USA

(2) Biomedical Engineering
Purdue University
West Lafayette, IN, USA

(3) Urogynecology
Ochsner Clinical School
New Orleans, LA, USA

INTRODUCTION

Improper remodeling of the cervix during pregnancy can lead to detrimental outcomes including preterm birth. Approximately 1 in 10 of all pregnancies result in preterm birth, the leading cause of infant morbidity and mortality worldwide [1]. In order to sustain the fetus and enable safe passage at delivery, the cervix must remain firm and closed during pregnancy and then rapidly shorten and distend during labor. Hence, cervical mechanical properties and geometry play a critical role [2]. The mechanical loading environment of the nonparous and parous cervix, however, is not well understood. Further, cervical structure-function relationships are not fully elucidated. Therefore, there is a need to develop tools to better understand the mechanical environment of the cervix and cervical structure-function relationships [2]. Towards this end, we recently developed a biaxial mechanical testing protocol for the murine cervix that seeks to quantify regional cervical mechanical response to physiologic values of simultaneous extension and inflation at the internal os (uterine cervix) and external os (vaginal cervix) [3]. Herein, we seek to leverage 3D ultrasound imaging techniques [4] in conjunction with biaxial mechanical assessment to provide validation of the calculated *in vivo* conditions during mechanical testing. Prior studies have estimated the *in vivo* load on the utero-cervical complex occurs at 37mmHg of pressure [5]. Therefore, we hypothesize that cervical geometry will be most similar to ultrasound measurements when the *in vitro* pressure is equivalent to 37 mmHg.

METHODS

Reproductive systems from female C57BL/6 mice between 4-6 months of age at estrus (n=7) were imaged using a Vevo2100 ultrasound imaging system. Volumetric ultrasound data was collected from a 40MHz transducer (MS550D) in short axis B-mode with a step size between slices of 0.19mm (Fig. 1). After imaging, the mice were sacrificed and cervical complexes were isolated from the uterus and vagina. The contralateral uterine horn was ligated with 6-0 silk suture for each specimen and then mounted within a custom biaxial myograph testing system. **MECHANICAL TESTING.** Prior to preconditioning,

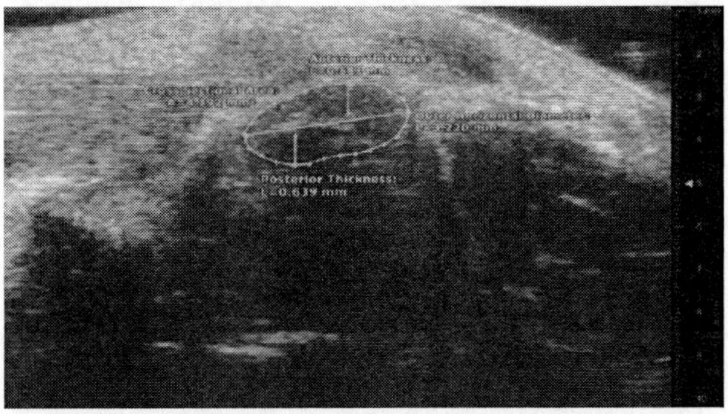

Figure 1: Representative ultrasound image of the external os (region of the cervix closest to the vagina) of a murine cervix measuring diameter, thickness, and horizontal diameter.

the unloaded length of the cervix was measured with a digital micrometer and the unloaded outer diameter was measured optically at the external and internal os. Following preconditioning, the unloaded geometry was re-estimated and then the *in vivo* axial stretch was estimated by identifying the length in which force remained constant over a range of pressures as described previously in arteries (Fig 2) [3,6]. Specimens were subjected to pressure-diameter tests (P= 0 to 200

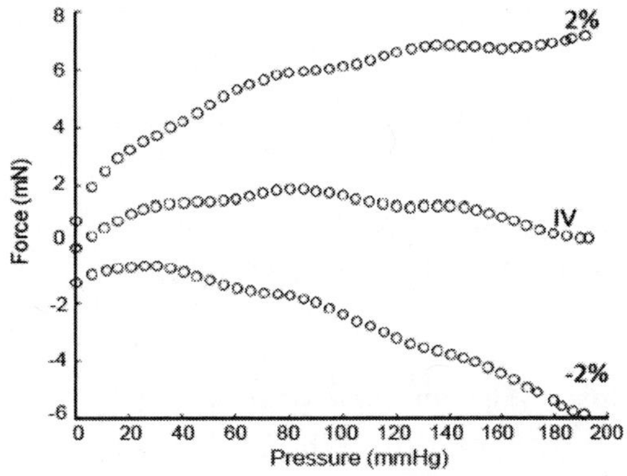

Figure 2: Axial force in response to increasing pressure from a representative sample. The estimated *in vivo* axial stretch (IV) is identified via plot tests where the measured axial force remains nearly constant in response to changes in luminal pressure. The cervixes are tested at 2% above and below this value to ensure that data are collected within the physiologic range of *in vivo* stretches.

mmHg) at three different axial extensions (2% below, 2% above, and at the estimated *in-vivo* axial stretch) and to axial force-length tests at four different pressures (10, 67, 133, 200 mmHg) [6]. During testing, outer diameters of both locations were tracked optically. Following mechanical testing, 0.5mm rings were isolated and imaged from each location. Using a custom MATLAB code, unloaded thickness was determined from the ring images. Assuming the cervix is a thick walled cylinder, unloaded diameter, length, and thickness were utilized to calculate initial cervix volume. The *in vivo* outer diameter was

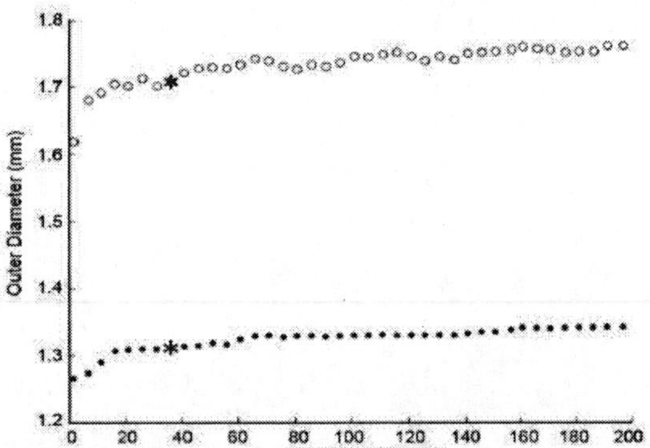

Figure 3. A representative sample demonstrating that the *in vivo* value of outer diameter was calculated from experimental pressure-diameter curves at 37 mmHg (denoted by dark star) for the internal (closed) and external os (open).

calculated from the pressure-diameter curves at the reported nonparous value of physiologic resting pressure 37 mmHg (Fig 3) [5]. The corresponding estimate of *in vivo* thickness was calculated utilizing the unloaded volume, *in vivo* length, and calculated outer diameter assuming the cervix is incompressible (Eq 1).

$$V = \pi * L * (r_o^2 - r_i^2) \quad (1)$$

DATA ANALYSIS. A bilinear curve fit was applied to the local stress-strain data to quantify the moduli in the toe- and linear-regions for each mechanical test [7]. **STATISTICAL ANALYSIS.** To determine if the *in vivo* geometries calculated from the experiments were equivalent to ultrasound measurements, Bland Altman analyses as well as paired t-tests were performed. Mechanical parameters were analyzed using a two-way repeated measure ANOVA (location, axial coupling), followed by paired t-test with Bonferroni corrections when appropriate.

RESULTS

Average calculated *in vivo* thickness values of the internal and external os were 0.410±0.130mm and 0.633±0.247mm respectively. Corresponding values from the ultrasound measurements were 0.527±0.152mm and 0.692±0.175mm. Diameters for the external and internal os from testing were 2.308±0.580mm and 1.956±0.809mm. Ultrasound resulted in diameters of 3.268±0.267mm and 2.012±0.375mm for the external and internal os. Mean experimental and ultrasound lengths were 6.976±0.397mm and 7.025±1.000mm. Paired t-tests performed between the experimental and ultrasound measurements identified a difference between outer diameters at the external os, however, statistically significant differences were not found between the thickness, uterine cervix diameter, or cervical length. Bland Altman analysis for thickness, horizontal diameter, and length between the ultrasound measurements and the estimated *in-vivo* calculations presented data within the confidence limits. Mechanical properties between the internal os and external os yielded statistically significant differences between the strain, toe modulus, and linear modulus.

DISCUSSION

This study demonstrated that the geometry acquired from *in vivo* ultrasound and *in vitro* calculations of physiologic geometry are within the confidence intervals of the Bland Altman analysis and did not yield statistically significant differences with paired t-tests (with exception of the diameter of the external os). The external os, however, is surrounded by a layer of vaginal tissue which was removed prior to mechanical testing possibly contributing to error between the methods. The results from this study suggest that ultrasound can be a valuable tool to inform mechanical testing protocols to ensure that physiologic properties are being quantified for the cervix and other soft tissues. Additionally, this study demonstrated that the mechanical properties of the cervix vary by location. Parameters of strain, toe modulus, and linear modulus were significantly different between the external and internal os. The external os exhibited a higher toe- and linear-region modulus. Prior studies have reported an increase in collagen content near the external os of the cervix in comparison to the internal os [8]. Hence, the higher collagen content at the external os may contribute to the larger moduli identified in this study. Clinically, the internal os is critical for the maintenance of pregnancy. The fetus directly loads the cervix and it is hypothesized that improper cervical remodeling at the internal os may contribute to spontaneous preterm birth [9]. Combining ultrasound imaging and experimental results from mechanical testing can improve the understanding of material properties of the cervix subjected to physiologic loading conditions, and henceforth the cervical mechanical environment.

ACKNOWLEDGEMENTS

The authors would like to acknowledge Katy Robison for assistance with statistical analysis.

REFERENCES

[1] Anum, E et al., Placenta, 30, 207–215, 2009. [2] Myers, K et al., *J of Biomech*, 48:1511-1523, 2015. [3] Morris et al., *SB3C*, 2017. [4] Philips, E et al., *BioMed Research* Int: 1-12, 2015.[5] Bulleti, C et al., *Human Repro*, 15(1):81-89, 2000. [6] Ferruzzi, J et al., Ann Biomed Eng, 41(7): 1311-1330, 2013. [7] Lake et al., JOR, 27(12):1596-602, 2009. [8] Harkness, M et al., J Physiol, 148:524-47, 1959. [9] Myers, K et al., *J of Biomech*, 48:1533-1540, 2015.

MECHANICAL INTEGRITY OF THE CERVIX IS IMPAIRED IN A MOUSE MODEL OF INTRAUTERINE INFLAMMATION AND PRETERM BIRTH

Carrie E Barnum (1), Stephanie N. Weiss (1), Guillermo Barila (2), Amy G. Brown (2), Snehal S. Shetye (1), Michal A. Elovitz (2), and Louis J. Soslowsky (1)

(1) McKay Orthopaedic Research Laboratory, University of Pennsylvania, Philadelphia PA, USA

(2) Maternal and Child Health Research Center, Department OBGYN, University of Pennsylvania, Philadelphia, PA, USA

INTRODUCTION

Preterm birth (PTB) is one of the primary causes of infant morbidity and mortality in the United States affecting 11.2% of live births [1]. There is an emerging concept that the cervix may change or remodel prematurely in many cases of spontaneous preterm birth (sPTB) [2-3]. However, the mechanical role of the cervix in the pathogenesis of sPTB remains unclear [4-5]. Loss of cervical function has been hypothesized to result from the loss of mechanical integrity of the cervix, and has been cited as part of the etiology of PTB [6-7].

Previous research has shown that intrauterine infection and/or inflammation is present in many cases of PTB. When using histological chorioamnionitis as a metric of infection/inflammation, over 85% of infants born at less than 28 weeks have this outcome [9]. To create what is observed clinically, lipopolysaccharide (LPS) has been employed to mimic intrauterine inflammation in mouse models of sPTB [10]. However, the mechanical properties of the cervix in these models have not been investigated. Therefore, the objective of this study was to evaluate the mechanical response of murine cervix after intrauterine LPS compared to saline controls as well as to near term cervices (E18.5). We hypothesized that intrauterine inflammation would result in decreased mechanical properties of the cervix compared to controls but would have similar mechanical properties compared to near term cervices (E18.5).

METHODS

Mouse model of intrauterine inflammation and sPTB: Surgery was performed on embryonic day (E) 15.5 of gestation in CD-1 mice. In both the intrauterine inflammation (IUI) mice and the saline controls (IUS), the right uterine horn was exposed to allow visualization of the first two gestational sacs (those most proximal to the cervix). Each mouse was infused with 250 µg of lipopolysaccharide (LPS) or an equal volume of saline into the uterus between the lower two gestational sacs with care not to enter the amniotic cavity [10]. Control animals (E15.5) received no anesthesia and no intrauterine infusion while IUS animals received anesthesia and intrauterine saline. All animals were harvested for mechanical testing 6 hours after administration of intrauterine infusion of LPS or saline and frozen until mechanical testing. Non-instrumented dams on E15.5 and control dams on E18.5 were harvested in the same manner.

Mechanical Testing: The protocol for cervix mechanical testing has been utilized previously [11]. Briefly, the cervix was dissected free of extra soft tissue and the uterus and vagina were removed. The cervix was then laid flat to expose the lumen. The ends were affixed between two pieces of sandpaper for gripping, such that a uniaxial tensile load on the grips would simulate dilation of the cervical canal. A laser device was used to measure cross sectional area. The cervix was hydrated with PBS during mechanical preparation and the entire test was performed in a saline bath at room temperature. Tensile properties were measured via the following protocol: preload of 0.005N followed by a hold of 5 minutes and then a ramp to failure at 1mm/minute [11].

Collagen Re-Alignment: Collagen fiber alignment maps of the cervix were collected throughout the mechanical testing protocol using our established integrated cross-polarizer system, as described [11]. Collagen alignment was measured at four points during the mechanical test: (1) start of the toe region, defined by the first map after the hold protocol, (2) at the end of the toe region, determined by the last map before the change of slope to the linear region (3) at 45% of the maximum load, and (4) at 90% of the maximum load. Significant re-alignment was defined as a significant change in circular variance between two time points.

Statistics: One-way ANOVA was used to compare between groups for mechanical parameters with Bonferroni-corrected post-hoc tests when appropriate. Comparisons were only evaluated between E15.5, IUS, and IUI, and then separately between IUI and E18.5. The significance was set at $p<0.05$ for all statistical comparisons.

RESULTS

Mechanics: No differences in cross sectional area were observed between any of the groups tested (data not shown). Significant decreases in maximum load were observed between the E15.5 cervix and both the IUS cervix and the IUI cervix. Maximum load of the IUI cervix was significantly larger than the E18.5 group (Fig 1A). No differences in stiffness were found between the WT E15.5 and the IUS cervix. Tissue stiffness was significantly lower for the IUI group when compared with the IUS and E15.5 groups. Stiffness was significantly higher for the IUI group when compared with E18.5 (Fig.1B). Maximum stress was significantly lower when comparing both IUS and IUI to E15.5. The IUI cervices had significantly larger maximum stress when compared to E18.5 (Fig. 1C). No differences were observed in modulus for any of the groups tested (Fig 1D).

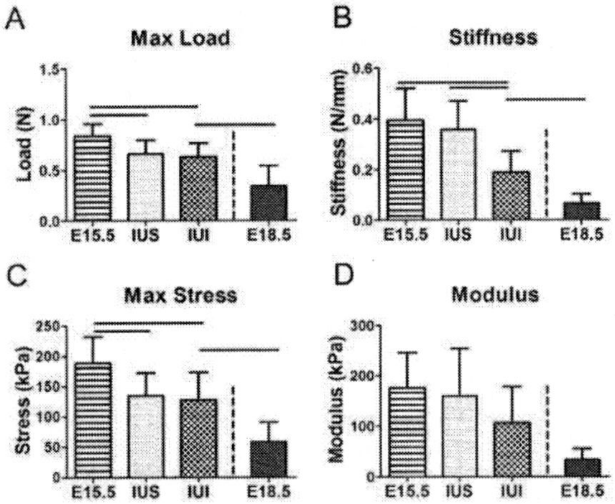

Figure 1: Mechanical properties of LPS mediated PTB in the mouse. (A) Maximum load, (B) stiffness, (C) maximum stress and (D) modulus, for E15.5, IUS, IUI, and E18.5 (n=10-12 for all samples) Lines demonstrate significant changes between groups ($p<0.05$). Data is presented as mean±standard deviation.

Collagen Re-Alignment: Although all groups showed some collagen fiber re-alignment, no differences were found between the E15.5 group and the IUS group. The IUI group did show more collagen fiber re-alignment than the other two groups, but it was not as dramatic as the earlier response to load seen in the E18.5 cervix (Fig.2).

DISCUSSION

In a mouse model of sPTB induced by intrauterine inflammation, cervical mechanical integrity was significantly disrupted. Both the IUI and IUS groups showed a significant decrease in maximum stress and maximum load. However, the IUI cervices had substantially lower stiffness than both the E15.5 control and IUS cervices. The combination of loss of mechanical function through decreased stiffness and increased collagen fiber re-alignment might partially explain the high rate of sPTB observed in the murine IUI model, not observed in the IUS controls [10]. Interestingly, the cervices in the IUI model were still mechanically more robust than near term delivery cervices (E18.5), with the IUI samples having significantly higher stiffness, maximum load, and maximum stress. In contrast, a previous study showed differences in stiffness between IUI and E15.5 cervices, no differences between IUI and E18.5, but made no comparisons to a saline control [12]. This suggests that from a mechanical prospective, the underlying mechanisms governing parturition may differ between the IUI model and term delivery. It has been postulated that molecular mechanisms behind LPS-mediated PTB may differ from term parturition [13].

The decrease in stiffness observed in the IUI cervix as well as the increased collagen fiber re-alignment could be partially attributed to changes in extracellular matrix. These data are consistent with previous studies showing increased spacing between collagen fibers in both the IUI and E18.5 cervix [14]. Furthermore, it has been shown that E18.5 cervices have increased hydration compared to IUI samples, suggesting there could be material compositional differences that partially account for the decreases in mechanical properties only observed in the E18.5 cervix [13].

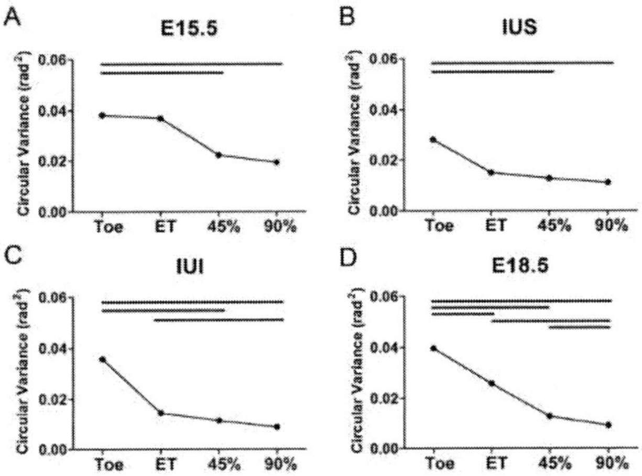

Figure 2: Representative graphs of collagen re-alignment of the mouse model of intrauterine inflammation and PTB during tensile loading. (A) E15.5, (B) IUS, (C) IUI and (D) E18.5 measurements examined at Toe, End of Toe, 45% of maximum load, and 90% of maximum load. Significance, denoted by a bar spanning two comparison groups, represents a p value < 0.05.

This study shows that a trigger such as inflammation could work synergistically with decreases in mechanical properties and changes in collagen re-alignment to cause sPTB. While correlation of these findings to human pregnancy will be necessary, this work lays a critical foundation for investigating cervical biomechanics and role of the cervix in sPTB.

ACKNOWLEDGEMENTS

Supported by MOD Prematurity Research Center at PENN and Penn Center for Musculoskeletal Disorders (NIH/NIAMS, P30 AR050950).

REFERENCES

[1] National Center for Health Statistics, final natality data. Retrieved 2017. [2] Ghartey J et al., Am J Obstet Gynecol. 212:776.e1-776.e12, 2015. [3] Shah, SJ et al., J Proteome Res. 8:2407-17, 2009. [4] Myers KM et al., J Biomech. 48:1511-23, 2015. [5] Peralta L et al., J Biomech. 48:1557-65, 2015. [6] Aspden RM et al., Coll Relat Res. 8:103-112 1988. [7] Mahendroo M, et al. Reproduction 143: 429-438, 2012. [8] Gonçalves LF et al., Ment Retard Dev Disabil Res Rev. 8:3-13, 2002. [9] Mendz,GL et al., Front Cell Infect Microbiol. 3:58, 2013. [10] Elovitz MA et al., Am J Pathol. 163:2103-11, 2003. [11] Barnum CE et al., J Biomech Eng. in press. 2017. [12] Timmons BC et al., Endocrinology. 155:287-98, 2013. [13] Gonzalez JM et al., Am J Pathol. 179:838-49, 2011. [14] Holt R et al., Endocrinology. 152:1036-46, 2011.

BIOMECHANICAL SIMULATIONS OF PREGNANCY: THE INFLUENCE OF FETAL MEMBRANE MECHANICS ON UTERINE AND CERVICAL TISSUE STRETCH

Andrea R. Westervelt (1), Edoardo Mazza (2), Alexander E. Ehret (2), Joy Vink (3), Chia-Ling Nhan-Chang (3), Ronald J. Wapner (3), George Gallos (3), Michael House (4), Kristin M. Myers (1)

(1) Mechanical Engineering
Columbia University
New □ork, N□, USA

(2) Mechanical and Process Engineering
ETH Zurich
Zurich, Swit□erland

(3) Obstetrics and Gynecology
Columbia University Medical Center
New □ork, N□, USA

(4) Obstetrics and Gynecology
Tufts Medical Center
Boston, MA, USA

INTRODUCTION

The mechanical function of the soft tissues supporting the fetus are critical for a healthy pregnancy. These soft tissues that surround the fetus (fetal membranes, uterus, and cervix (Fig. 1)) play a dual-functional role in pregnancy. Throughout gestation, they act to keep the fetus within the womb; the uterus remains mostly passive, the fetal membranes (FM) remain intact to maintain a sterile environment, and the cervix remains closed. Then nearing term, chemical and mechanical triggers signal the onset of labor in a process that is not completely understood. In labor, these soft tissues switch their function; the uterus now becomes contractile, the FM rupture, and the cervix dilates in order to deliver the baby. Early onset of these labor processes is called spontaneous preterm birth (sPTB) and leads to an annual PTB rate of 9.6% in the United States [1]. The goal of this study is to understand how fetal membrane compliance influences the mechanical stretch within the uterine wall and the cervix because we hypothesize that the magnitude of loading on these tissues is one trigger that starts the onset of labor [2]. For example, in twin pregnancies and pregnancies complicated by an excess of amniotic fluid, uterine over-distention has been thought to trigger sPTB because twin pregnancies [3,4,5].

Previous computer modeling of the pregnant abdomen reveals that the mechanical load on the cervix, is influenced by maternal anatomical features and the material properties of the uterus, FM, and cervix [6]. The extent to which each of these structural factors influence tissue loading remains to be determined. This study focuses specifically on the effects of fetal membrane mechanics on uterine and cervical tissue stretch. The fetal membranes consist of the collagen-rich, load-bearing amnion and the cellular chorion [7,8]. The integrity of both layers is essential for the maintenance of pregnancy, where its compliance is necessary to accommodate fetal growth. Yet, membrane rupture and fracture properties are required for normal term delivery [9]. Evidence suggests that the FM grows until the 2nd trimester [10], then stretches over the rest of pregnancy to accommodate the enlarging amniotic cavity. Evidence also suggests that the mechanical properties of the amnion, the load-bearing layer, at term is highly nonlinear [11]. The material properties of the membrane and the extent to which it is stretched at any given gestational age are still unknown.

Here we quantify the amount of cervical stretch as a function of FM material properties using a parametric finite element model of pregnancy [12]. As the stiffness of the FM increases, the amount of cervical stretch decreases.

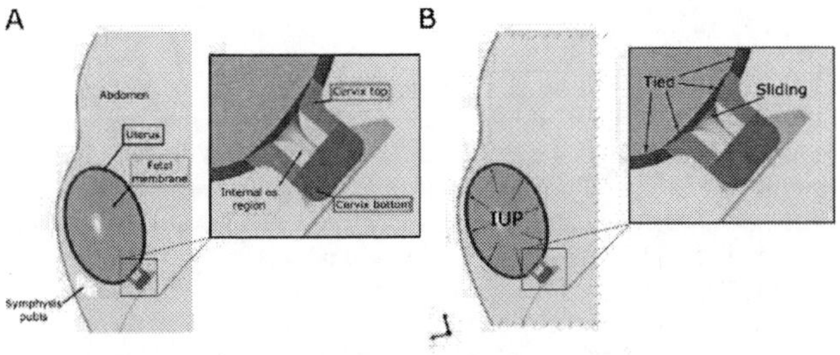

Figure 1: A) Baseline model geometry [12] and B) Prescribed boundary conditions

METHODS

FE models (FEBio 2.3.1) of the pregnant abdomen were built using similar methods published elsewhere [12]. Briefly, a customized computer script (Trelis Pro 15.1.3, csimsoft LLC) generated a parametric solid model of the pregnant abdomen based on a maternal ultrasound scan (Fig. 1). For the baseline model, uterine diameter, uterine thickness, cervical length, cervical diameter, and cervical angle with anterior lower uterine segment (LUS) were measured for a 35 y/o normal P0 patient at 25 weeks.

For FE analysis, the uterus and cervix were modeled as collagenous composite materials meshed using linear tetrahedral elements, based on material fits to passive length-tension curves of pregnant tissue [13,14]. The FM was modeled as an Ogden nonlinear material [15] meshed with hexahedral elements with a thickness of 0.1 mm [11]. The strain energy density for the FM was given by:

$$\Psi^{FM} = \sum_{i=1}^{3} \frac{c_i}{m_i^2}(\lambda_1^{m_i} + \lambda_2^{m_i} + \lambda_3^{m_i} - 3)$$

where for the baseline model c_1=0.859 MPa, c_2=0.004 MPa, c_3=0.756 MPa, m_1=27.21, m_2=27.21, and m_3=-16.64. The baseline model is based on a material parameter fit to bulge inflation data of delivered membranes [11]. To investigate how varying levels of FM compliance effects uterine and cervical loading, the FM stiffness parameters c_i were varied by a factor of 10 from baseline. Such large variation is justified based on the observed variability in mechanical response [11] as well as on the fact that the *in vivo* pre-stretch level is not known.

Boundary conditions were prescribed as in Fig. 1B, where the FM was prescribed a tied contact to the inner uterine wall and a sliding contact to the cervical internal os. Intrauterine pressure (IUP) was applied at both the physiological range for 25 weeks gestation (0.817 kPa) [16] and at contraction magnitude (8.67 kPa) [17], then the magnitude of the principal tissue stretch was calculated. The extent of cervical stretch was evaluated as a percentage of cervical internal os region (yellow in Fig. 1) volume above a 1.05 stretch threshold.

RESULTS

There are visible differences in the magnitude and pattern of tissue stretch at the various levels of fetal membrane stiffness (Fig. 2). When the FM has a membrane stiffness at 10% baseline values, the uterus and cervix see the greatest amount of tissue stretch with contraction-magnitude IUP, where 88.5% of the internal os is stretched above the 1.05 threshold value. As membrane stiffness increases, stretch in both the uterus and cervix decreases. This supports the previous hypothesis that the fetal membranes load share with the surrounding soft tissues throughout pregnancy [6]. The mechanical properties and the nonlinear behavior of the fetal membranes are critical to cervical tissue stretch at the internal os. Softer fetal membranes and larger intrauterine pressures increase cervical tissue stretch.

DISCUSSION

This study represents a primary attempt to model the pregnant abdomen throughout the course of gestation, investigating the contribution of the fetal membrane sharing the load of the fetus with the uterus and cervix. Obtaining longitudinal data on pregnant humans is difficult. Hence, we utilize previous experimental data to inform our FE models of pregnancy at the gestational age of 25 weeks. As we discovered in our previous work [12], various material, anatomical, and contact factors influence the loading of the soft tissues that surround the fetus. The membrane taking on some of the intrauterine load may be protective for the uterus and the cervix, but there's also a limitation to that. If the membrane is too stiff, it may reach its rupture point before term, resulting in premature preterm rupture of membranes. This indicates that there is a delicate balance between the fetal membrane and uterus load sharing that must be understood in subsequent studies. A

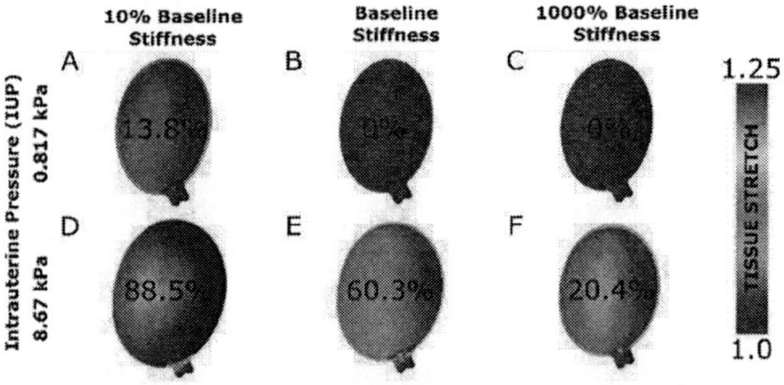

Figure 2: Uterine and cervical stretch patterns for a membrane at 25wk physiological IUP with A) 10% baseline FM stiffness, B) baseline FM stiffness, C) 1000% baseline FM stiffness, and at contraction-magnitude IUP with D) 10% baseline FM stiffness, E) baseline FM stiffness, F) 1000% baseline FM stiffness. Percentages represent the volume percentage of the cervical internal os above a 1.05 stretch threshold.

limitation considered in this study is that the maternal anatomy is idealized as simple geometric shapes, where the bends and folds in the uterine wall are neglected. The influence of these realistic anatomic features is also currently being investigated.

ACKNOWLEDGEMENTS

This work has been funded by the National Science Foundation GRFP DGE-16-44869, and the University of Pennsylvania March of Dimes prematurity research center #22-FY15-002a.

REFERENCES

[1] Martin, J. A., et al., 2015, Statistics (Ber)., **64**(1), pp. 1–104.
[2] Menon, R., et al., 2016, Hum. Reprod. Update, **22**(5), pp. 535–560.
[3] Goldenberg, R. L., et al., 1996, Am J Obs. Gynecol, **175**(4 Pt 1), pp. 1047–1053.
[4] Gardner, M. O., et al., 1995, Obs. Gynecol, **85**(4), pp. 553–557.
[5] Kirkinen, P., and Jouppila, P., 1977, Ann Chir Gynaecol, pp. 117–122.
[6] Fernandez, M., et al., 2015, Comput. Methods Biomech. Biomed. Engin., (May), pp. 1–14.
[7] Bourne, G., 1962, Postgrad. Med. J., **38**, pp. 193–201.
[8] Oyen, M. L., et al., 2006, Am. J. Obstet. Gynecol., **195**(2), pp. 510–515.
[9] Chua, W. K., and Oyen, M. L., 2009, Eur J Obs. Gynecol Reprod Biol, **144 Suppl**, pp. S128--33.
[10] Alger, L. S., and Pupkin, M. J., 1986, Clin Obs. Gynecol, **29**(4), pp. 758–770.
[11] Bürzle, W., et al., 2014, J Biomech Eng, **136**(11), p. 111009.
[12] Westervelt, A. R., et al., J Biomech Eng., Under review
[13] Conrad, J., et al., 1966, Am. J. Obstet. Gynecol.
[14] Myers, K. M., et al., 2015, J Biomech, **48**(9), pp. 1533–1540.
[15] Ogden, R. W., 1972, Proc. R. Soc. London A Math. Phys. Eng. Sci., **326**(1567).
[16] Fisk, N. M.,et al., 1992, Br. J. Obstet. Gynaecol., **99**(1), pp. 18–22.
[17] Buhimschi, et al., 2004, Obs. Gynecol, **103**(2), pp. 225–230.

SB³C2017
Summer Biomechanics, Bioengineering and Biotransport Conference
June 21 – 24, Tucson, AZ, USA

MECHANICAL AND HISTOLOGICAL CHARACTERISATION OF THE HUMAN MALE URETHRA FOR THE PURPOSES OF TISSUE ENGINEERING AN APPROPRIATE REGENERATIVE GRAFT

Eoghan M. Cunnane (1,2,3), Niall F. Davis (4), Alan J. Ryan (1), David A. Vorp (2,5), Fergal J. O'Brien (1), Michael T. Walsh (3)

(1) Tissue Engineering Research Group, Royal College of Surgeons, Dublin, Ireland

(2) McGowan Institute for Regenerative Medicine, Department of Bioengineering, and the Center for Vascular Remodeling and Regeneration, University of Pittsburgh, Pittsburgh, USA

(3) Centre for Applied Biomedical Engineering Research and the Health Research Institute, School of Engineering, Bernal Institute, University of Limerick, Limerick, Ireland

(4) Department of Urology, St. Vincent's University Hospital, Dublin, Ireland

(5) Division of Cardiac Surgery and the Department of Surgery, University of Pittsburgh, Pittsburgh, USA

INTRODUCTION

The human urethra frequently requires reconstruction with a graft to treat damage caused by inflammation, stricture, congenital defects or malignancy. Autologous grafts can be harvested from the mucosa ling of the mouth or the skin of the penis. However, these tissues often have limited availability and their extraction is associated with donor site morbidity. Tissue engineered grafts are not subject to these limitations and have demonstrated varying levels of clinical success by implanting either synthetic polymers or decellularized tissues as tubular implants and patches [1]. However, designing the ideal urethral graft remains challenging as the mechanical properties of native urethral tissue under intraluminal pressure or extension remain largely unknown. Such properties are of critical importance to tissue engineered grafts which must be capable of withstanding in vivo forces and also tailored to provide appropriate biomechanical cues to both seeded and native cells. This study is therefore concerned with characterising the baseline passive mechanical properties of human urethral tissue, and the effect of tissue composition on these properties, to better inform the design of tissue engineered scaffolds intended for use as urethral grafts. Furthermore, this study also develops a bilayered tissue engineered scaffold composed of natural polymers that is intended for use as a vessel-appropriate regenerative urethral graft.

METHODS

Nine human male urethral samples were obtained from living donors undergoing gender reassignment surgery. The overall mechanical response to intraluminal pressure and the regional mechanical response of the tissue to planar tension was characterised. The overall elastic and viscoelastic responses of the tubular samples were characterised by subjecting the samples to dynamic intraluminal pressure and then to a static pressure head of adjustable height. The resulting pressure-diameter data was used to compute tissue compliance over the low (0-3 kPa), mid (4-6 kPa) and high (6-10 kPa) pressure ranges. Beta stiffness, a dimensionless parameter, was determined to describe the full range of the nonlinear response.

The regional elastic and viscous responses of the tissue to circumferential and longitudinal extension were characterised by longitudinally incising the urethras and sectioning them into distal bulbar, proximal bulbar, distal penile and proximal penile regions where available (n=30 in total). The elastic response of these sections was characterised using planar tension testing and the viscous response was characterised using stress relaxation tests.

Masson's trichrome and Verhoeff-Van Gieson stains were employed to histologically examine the collagen and elastin content of the urethral tissue following mechanical characterisation. Ring segments were cut from the middle region of the tubular urethras prior to sectioning and fixed in their traction-free state in 10% formalin, embedded in paraffin, and then sliced into 10 μm thick segments. Immunohistogical staining of smooth muscle actin was also employed to identify the smooth and striated muscle content of the samples.

RESULTS

Figure 1A displays the average mechanical response of the samples to static (blue) and dynamic (red) intraluminal pressure. These results reveal a nonlinear mechanical response typical of biological tissue whereby the urethral tissue becomes less compliant at higher pressures. The significantly higher stretch at each pressure interval during static loading compared to dynamic loading demonstrates the pronounced viscoelastic behaviour of the tissue. There is no significant anisotropy at any stress level for the patient averaged mechanical response to planar tension testing, 1B. Furthermore, there is no significant regional variation in mechanical response to circumferential (1C) or longitudinal

(1D) extension. No significant difference is observed between the stress relaxation response of the urethral tissue to sustained extension in either the circumferential or longitudinal direction, 1E and F. No significant difference is observed between the regional stress relaxation responses (data not shown).

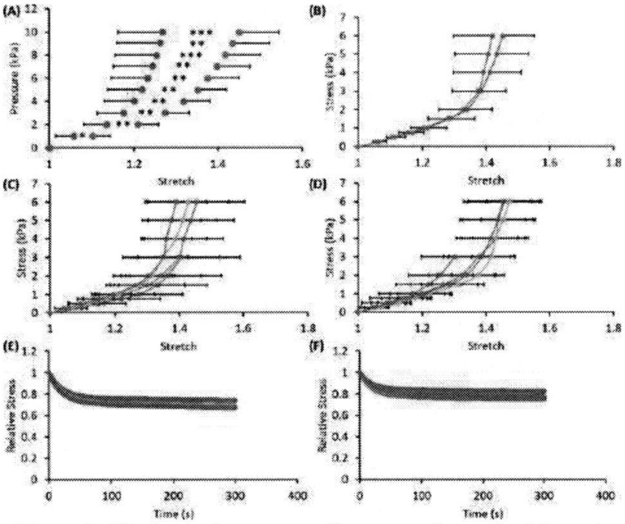

Figure 1: Mechanical response of human male urethral tissue to intraluminal pressure (A), and planar tension testing (B-F).

Figure 2A displays representative stains for elastin, collagen and smooth muscle actin. Elastin is stained black, collagen is stained blue and muscle is stained brown. Staining reveals localisation of elastin and smooth muscle in the inner and middle regions of the urethra, with collagen and striated muscle confined mainly to the outer region. The composition of each sample is analysed using a previously outlined protocol which represents composition as the mean Gray value of the relevant colour intensity with respect to area [2]. Figure 2B-E demonstrates the most meaningful relationships identified between tissue composition and tissue mechanics as characterised by the application of static intraluminal pressure (relationships are considered meaningful at $p<0.1$ and significant at $p<0.05$).

As can be seen, elastin correlates significantly with compliance in the low stretch region whereby increasing elastin content is accompanied by an increase in compliance, 2B. This behaviour is typical of hollow tubular organs as the high distensibility of the elastin fibres increases the compliance of the vessel at low levels of stretch. Collagen displays meaningful correlations with compliance in the mid and high stretch regions which is again typical of such organs whereby decrimping of collagen at high stretch levels induces a stiffening in mechanical response, 2C-D. This is further evidenced by the significant relationship between collagen and overall sample Beta Stiffness, 2E.

DISCUSSION

These findings provide the baseline passive mechanical properties of human urethral tissue and highlight numerous characteristics that tissue engineered grafts should emulate such as a nonlinear mechanical response to loading and an absence of significant differences in regional or directional mechanical responses. Furthermore, histological analysis reveals the significant effect that both elastin and collagen have on the mechanical response of the urethra to intraluminal pressure. This study utilises this data to develop a novel tissue engineered urethral scaffold composed of elastin and collagen.

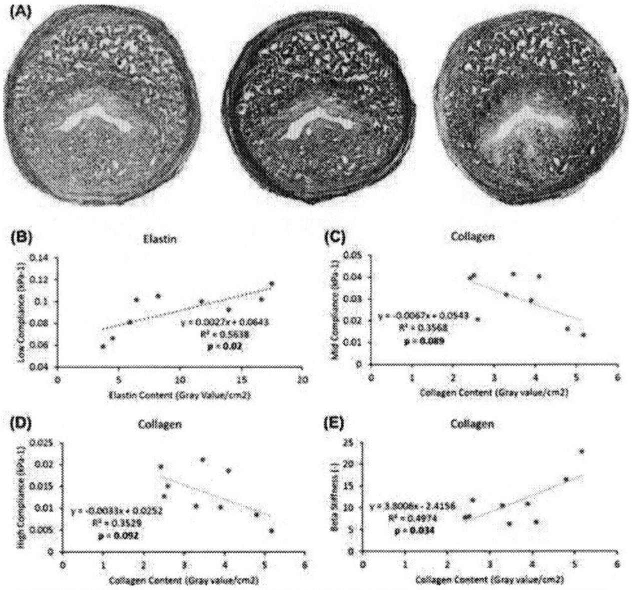

Figure 2: Representative stains for elastin, collagen and smooth muscle actin (A), meaningful relationships identified between urethral tissue composition and mechanical properties (B-E).

Tissue engineered scaffold materials constructed from elastin and collagen have previously demonstrated appropriate viscoelastic behaviour such as exhibited by the native urethral tissue characterised in this study [3]. Bilayered scaffolds were constructed from a combination of these proteins using a dense film as an inner layer and a porous freeze-dried outer layer to act as a suitable environment to encourage cellular ingrowth once implanted. The scaffolds were physically and chemically crosslinked using dehydrothermal and EDAC approaches to improve mechanical stability. These scaffolds will be cultured using an appropriate co-culture of smooth and striated muscle cells to attain a graft that is structurally and mechanically relevant to the native urethral tissue and therefore provides a starting point for the next generation of tissue engineered grafts for regeneration of the urethra.

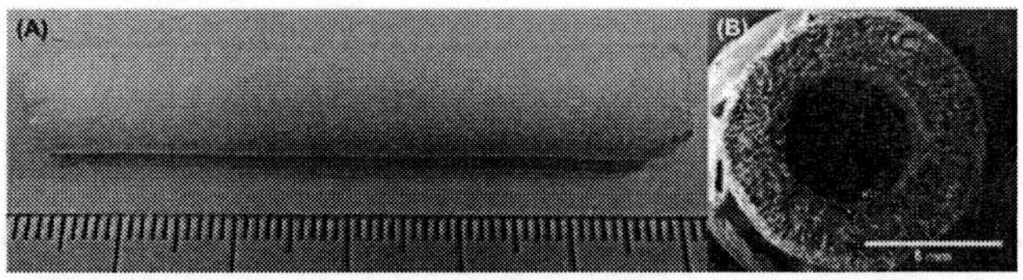

Figure 3: Macro (A) and microscopic (B) images of the bilayed scaffold constructed from elastin and collagen.

ACKNOWLEDGEMENTS

This project has received funding from the European Union's Horizon 2020 research and innovation programme under the Marie Sklodowska-Curie grant agreement No 708867. The findings reported in this abstract reflect only the authors' views and the Agency is not responsible for any use that may be made of the information it contains.

REFERENCES

[1] De Kemp et al 2015. *PLoS One,* 10, e0118653.
[2] Bauman et al 2014. *PLoS One,* 9(10), e109102.
[3] Ryan & O'Brien 2015. *Biomaterials,* 73, 296-307.

COMPARISON OF HUMAN CELL POPULATIONS ON MURINE TENDON REPAIR

Felix Dyrna[1], Leo Pauzenberger[2], Philip Zakko[2], Mary Beth McCarthy[2], David W Rowe[3], Augustus Mazzocca[2], Nathaniel A Dyment[4]

(1) Department of Sports Medicine Klinikum, Rechts der Isar Hospital Technical University of Munich Munich, Germany

(2) Department of Orthopaedic Surgery, UConn Health, Farmington, CT, USA

(3) Department of Reconstructive Sciences, UConn Health, Farmington, CT, USA

(4) Department of Orthopaedic Surgery, University of Pennsylvania, Philadelphia, PA, USA

INTRODUCTION

Traditional tendon and ligament repair strategies are limited and can often result in non-functional repair outcomes due to the poor healing potential of these tissues. This challenge has forced scientists and surgeons to develop novel strategies to improve repair outcomes, including cell augmentation of the repair response. Today bone marrow aspirate (BMA) is a primary source for cell-based therapies with increasing value in the world of orthopedic surgery especially in revision cases of tendon and ligament repairs. However, cells within peritendinous structures, such as the paratenon and adjacent bursa [1,2], contribute to the native tendon healing response and offer promising cell populations for cell-based repair strategies. That being said, few studies have compared the efficacy of these cell populations to the gold standard bone marrow stromal cells (BMSCs) in improving tendon repair, especially from human sources. Therefore, we investigated the repair potential of cells isolated from subacromial bursa tissue vs stromal cells (BMSCs) isolated from BMA. The objectives of the study are to 1) compare the ability of bursa-derived cells and BMSCs to engraft and persist over multiple months following repair and to 2) compare the production of organized matrix by these cells in both in vivo and in vitro models.

METHODS

Institutional IRB or IACUC approved all relevant procedures in this study. Human subacromial bursa and BMA was harvested from the same donor (N=5) during primary arthroscopic rotator cuff repair. Bursa tissue was digested with collagenase and BMA was concentrated by centrifugation in the lab prior to plating. Cells were further expanded to P1-P2 in a 5% O2 incubator and transfected with an ubiquitin-mcherry lentiviral vector to permanently label the cells during the first passage. In Vivo Model. Bilateral, full-length, central-width patellar tendon defects were created and filled with a cell-seeded fibrin clot ($0.15x10^6$ cells per implant, 10mg/ml fibrin gel) in immunocompromised NOD scid gamma (NSG) mice. Bursa cells and BMSCs were implanted into contralateral limbs of the same mouse (n=70). Control mice received a defect filled with the fibrin clot without cells in one limb and defect only in the other (n=40). One day before sacrifice mice were injected with EdU (3mg/kg) to label dividing cells. Mice were harvested at 1, 2, 5, and 8 weeks for histology and two photon microscopy (n=9/group). In Vitro Model. 3D cell-loaded fibrin gel constructs ($0.5x10^6$ cells/ml in 5mg/ml fibrin) were made in the FlexCell Tissue Train system. Gels were cultured either in ambient or low (5%) O2 for 14 days prior to two photon analysis (n=5). Overview images of the gels were taken every other day to document gel compaction. Immunohistochemistry. Limbs were fixed with the knee in 90° flexion, decalcified, and embedded for frozen sectioning. Serial sections (7-8μm) were made at 5 different levels (including intact tendon, transition zone, defect zone) using cryofilm type-2C [1,2]. A staining compilation of 3 different antibody combinations was used. These combinations included 1) collagen I and collagen III, 2) tenascin C (TNC) and EdU, and 3) F4/80 and collagen IV. Imaging. All samples were scanned on the Prairie Ultima IV multiphoton microscope prior to IHC. All IHC images were acquired using the Zeiss Axio Scan.Z1 digital slide scanner. Image Analysis. Cell engraftment and thickness of the healing tissue in the anterior bridge were quantified via two photon microscopy. Coherency, which calculates the level of organization, was measured from the collagen SHG signal using the OrientationJ plugin in Fiji. Antibody staining was also quantified in the tendon healing tissue via Fiji. Biomechanical Testing. At 2, 5, and 8 weeks post-implantation, mice were euthanized and knees were immediately harvested for biomechanical analysis (n=9-12 per group). The harvested knees were dissected down to tibia-tendon-patella units. The distal end of the tibia was embedded within bone cement while the patella was grasped in custom grips and placed into the Bose EnduraTec 3230 testing apparatus. Samples were submerged in physiological saline at 37°C and were preloaded (0.05N), preconditioned (10 cycles, 0-1% strain), followed by a stress relaxation step (ramp to 5% strain, hold for 10 minutes), and loaded to failure at a rate of 1%/sec. Structural and material properties were computed. Statistics. One-way ANOVA with surgical treatment (i.e., Bursa, BMSC, fibrin w/o cells, defect only, and native tendon) as the fixed factor ($\alpha = 0.05$) was used to compare imaging and biomechanical parameters at each time point. One-way ANOVAs with either cell type or oxygen tension as fixed factors were used to compare coherency and collagen thickness in FlexCell gels.

RESULTS

In Vivo Model. Cell implantation had no effect on rupture rate as ~20% of limbs were ruptured in each group. The bursa cells showed greater engraftment into the host tissue (Fig. 1, yellow arrows) as visualized by 3D two photon imaging and 2D cryohistology. The engrafted cells also took on the 3D morphology of the surrounding tissue as cells that infiltrated the adjacent tendon struts arranged in linear arrays similar to native tendon fibroblasts. Comparatively, cells that did not engraft, instead remaining within the fibrin clot, displayed a more rounded morphology and were situated within lacunae with surrounding pericellular matrix consisting of TNC, type I collagen, and type III collagen. Bursa cells displayed greater survivability with 82% of limbs still containing cells at 8 weeks vs. only 50% of BMSC limbs (Fig. 2A). Implantation of human cells also stimulated a greater response from the host cells as the healing bridge that forms over the anterior surface of the tendon was thicker in the cell groups than no-cell controls ($p < 0.05$, Fig. 2B). However, the number of EdU+ cells was not significantly different between the groups and no co-staining of human cells and EdU was found. Fibrin was clearly visible via toluidine blue staining up to the 8-week time point but the fibrin area decreased over time in all groups (BMSC, Bursa, fibrin-only). There was no significant difference in the amount of remaining fibrin between the groups at any time point ($p > 0.05$). In addition, activated macrophages (F4/80) was not significantly different between the groups ($p > 0.05$). Finally, the human cells expressed higher levels of collagen I, collagen III, and TNC within the fibrin implant compared to surrounding host tissue (Fig. 1). Finally, these histology features did not lead to improvements in biomechanical properties as the cell-treated groups were comparable to natural healing at all time points ($p > 0.05$). In Vitro Model. The BMSC gels displayed greater compaction than the Bursa gels (89±4% vs 74±3%). The bursa cells assembled thicker collagen matrix (imaged by two photon SHG) that displayed higher coherency (i.e., greater isotropy) than the BMSC gels. Additionally, gels cultured at 5% O2 yielded thicker and more organized (i.e., higher coherency) matrix than ambient O2 gels.

DISCUSSION

The purpose of this study was to compare the ability of bursa-derived cells to display long term engraftment and to produce organized matrix compared with the gold standard cell source for tendon repair, BMSCs. The advantage of the bursa-derived population is its ease of access from surgical waste tissue vs the additional trauma required for bone marrow aspiration. It was critical for this study to isolate both cell populations from the same patient to minimize the effect of patient differences. In addition, direct comparisons were made via contralateral implantations into patellar tendons of NSG mice to minimize mouse-to-mouse variation. The introduction of stable transfection of the Ub-cherry lentivirus in combination with the 3D two photon imaging of the entire defect volume allowed us to trace the fate of the donor cells reliably over several weeks, which typically is not done in tendon repair studies. Implantation of human cells in the defect space resulted in significantly thicker healing bridge over the defect space (Fig. 2B), suggesting that the human cells stimulate the host cells to produce a more robust healing tissue. The bursa cells displayed superior engraftment and survival in the patellar tendon implants (Fig. 1 & 2A) while also synthesizing greater amounts of organized collagen in the in vitro gels. While this did not result in significant improvements in structural or material properties compared to BMSCs in this study, further study is needed to determine the efficacy of this cell population.

ACKNOWLEDGEMENTS

This work was supported by National Institutes of Health grants K99-AR067283 (ND) and R01-AR063702 (DR).

REFERENCES

1. Dyment et al, PLOS One, 2013; 2. Yoshida et al, Connect Tissue Res, 2016

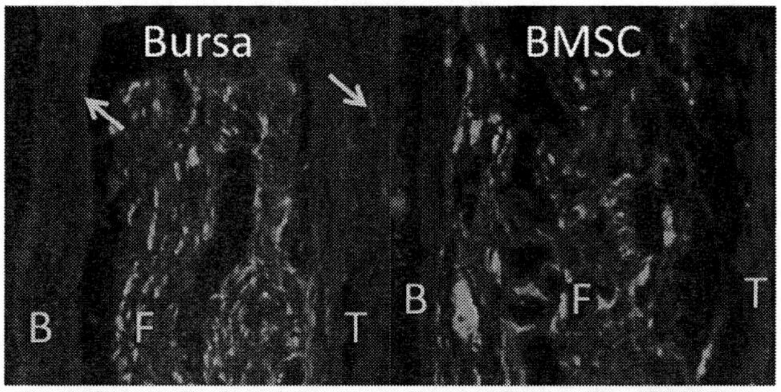

Fig. 1: Bursa displayed greater cell engraftment than BMSCs as denoted by yellow arrows (B = anterior bridge, F = fibrin implant, T = tendon strut). Both cell types produced higher levels of collagen 1 (green) and collagen III (blue) than seen in surrounding host tissue. Ub-cherry cells are in red.

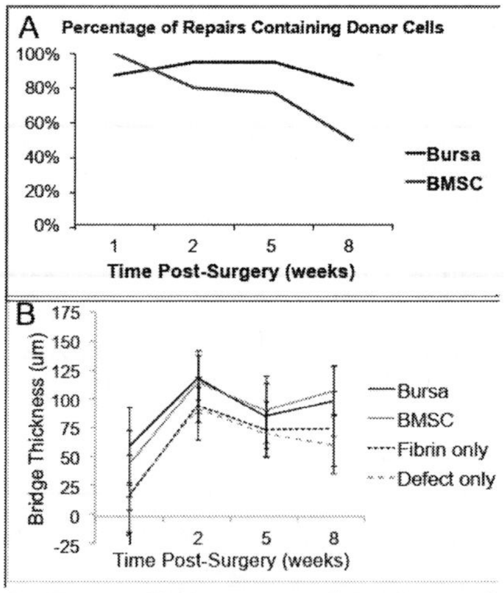

Fig. 2: (A) Bursa cells demonstrated greater survival with over 80% of limbs containing cells at 8 weeks compared to only 50% in BMSC group. (B) Cells also stimulated host cells to produce a thicker healing tissue over the anterior surface of the tendon (i.e., anterior bridge).

LEVERAGING LOCAL BIOMATERIAL PROPERTIES AND MECHANICAL STIMULATION FOR TENDON-BONE-JUNCTION ENGINEERING

William K. Grier (1), Raul A. Sun Han Chang (1), Brendan A.C. Harley (1)

(1) Chemical and Biomolecular Engineering
University of Illinois at Urbana-Champaign
Urbana, Illinois, USA

INTRODUCTION

Tendon and ligament injuries affect a large number and wide range of individuals with more than 32 million occurrences annually in the US[1]. In particular, one of the more common sites of tendon injury is at the tendon-bone-junction (TBJ), such as in the rotator cuff. Rotator cuff injuries account for more than 4.5 million physician visits and 75,000 repair surgeries each year[2]. This high rate of injury at the TBJ is attributed to the mismatch in mechanical properties between highly elastic tendon and much stiffer mineralized bone. This discrepancy in mechanics, when combined with a lack of native healing properties, leads to poor clinical outcomes and increased treatment times and costs to deal with a variety of both acute and chronic injuries. Our lab has previously described the fabrication of both anisotropically-aligned and multi-compartment collagen-GAG (CG) scaffold variants which mimic the natural extracellular matrix (ECM) microstructure of tendon and the biophysical heterogeneities of the TBJ in vitro[3]. Here, we show how biomaterial cues, in combination with biophysically-relevant cyclic tensile strain (CTS), can be used to promote human mesenchymal stem cell (hMSC) differentiation in a spatially-selective manner in order to regenerate the native interface between tendon and bone as a potentially superior repair strategy for acute TBJ injuries. Additionally, we explore incorporation of growth factors to induce region-specific MSC differentiation into relevant cell types of the fibrocartilaginous interface. Here, we employ a hydrogel platform that enables photochemistry based covalent immobilization of biochemical factors within the hydrogel using photoimmobilization approaches previously described by our lab[4]. We further demonstrate techniques to incorporate these hydrogels within the interface of our existing multi-compartment CG scaffolds in an effort to facilitate spatially controlled multi-lineage MSC differentiation to aid healing of osteotendinous insertion injuries.

METHODS

Single and multi-compartment TBJ scaffold variants were fabricated via a previously described directional solidification technique. Non-mineralized (CG) and mineralized (CGCaP) collagen-GAG suspensions were layered into a thermally-mismatched mold prior to freeze-drying to promote the formation of geometrically anisotropic pores. After hydration and carbodiimide crosslinking, scaffolds were then seeded with hMSCs and subsequently subjected to various levels of CTS for up to 6 days in a custom-built bioreactor inspired by the Levenston lab[5]. Western blot, RT-PCR, ELISA, and histological analysis were used to evaluate to evaluate hMSC differentiation and activity. Local scaffold strain in the bioreactor system was monitored by marking discrete locations and using ImageJ to track the distance between each point during cyclic strain.

PEG-SH hydrogels were fabricated via horseradish peroxidase (HRP) polymerization[6]. Hydrogel-CG scaffolds were fabricated using the established liquid phase co-synthesis method by incorporating the hydrogel precursor suspension between collagen-GAG suspensions prior to lyophilization.

RESULTS

When exposed to short-term CTS (1 Hz, 10% ε) on non-mineralized anisotropic scaffolds, hMSCs showed a time-dependent correlation between the activation of the ERK 1/2 and p38 MAPK pathways. Confirming previous findings by Paxton et al in regards to ligament tissue engineering, ERK 1/2 is maximally activated after 10 minutes of CTS with diminishing activation at later time points[7]. Further, reduction in ERK 1/2 activity with increased CTS coincides with an upregulation of the p38 MAPK, which is known to inhibit ERK ½[8]. This led us to proceed to apply a strain paradigm of 10% ε, at 1 Hz for 10 minutes every 6 hours for subsequent long-term studies.

hMSC bioactivity and gene expression in anisotropic scaffold variants was then monitored over 6 days of culture with or without CTS. Increased gene expression levels of ECM components *COL3A1*, *DCN*, and *COMP* along with tendon phenotypic markers *SCX*, *TNC*, and *MKX* were seen in scaffolds in dynamic culture when compared to static conditions. Additionally, cells on scaffolds subjected to CTS showed an increased activation of the Smad 2/3 pathway which is known to be regulated by TGF-β family growth factors and is critical in musculoskeletal development[9].

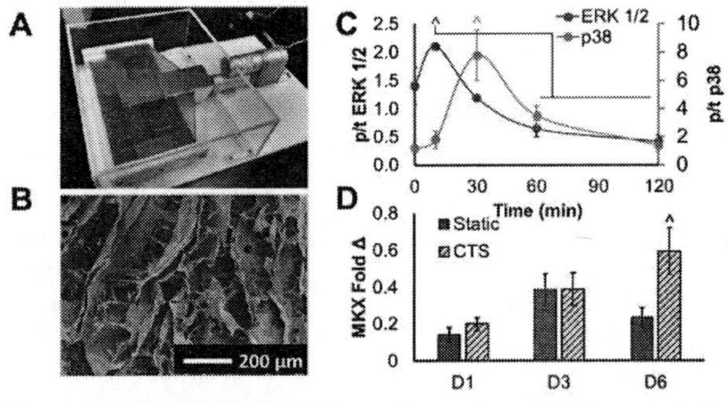

Figure 1: (A) Custom CTS bioreactor system. (B) Representative SEM image of CG scaffold. (C) Time-dependent phosphorylation of ERK1/2 and p38 MAPK with 10% CTS on non-mineralized scaffolds. (D) Upregulation of the tendon marker MKX in non-mineralized scaffolds with CTS compared to static culture.

Uniaxial cyclic stretching of multi-compartment CG scaffolds, with discrete sections of mineral content and pore anisotropy, shows that strain is primarily localized to the softer non-mineralized compartment. This is similar to what is seen in the native TBJ between the two mechanically-mismatched tissue types. This led us to hypothesize that the primary effects of CTS on cell differentiation and bioactivity would be more localized to the non-mineral compartment and interfacial zone.

Indeed, the application of CTS (5% global ε, 1 Hz, 10 min every 6 hrs) to cells on multi-compartment scaffolds shows compartment-specific and interface-specific responses in mechanotransduction pathway activation, integrin subunit expression, cell bioactivity, and gene expression markers.

Additionally, lyophilization allowed a PEG-SH hydrogel to be stably integrated within the construct at the interface between CG scaffolds; ongoing efforts are evaluating fibrocartilaginous differentiation of MSCs within the hydrogel layer as a function of addition of TGF-β and SOX-9.

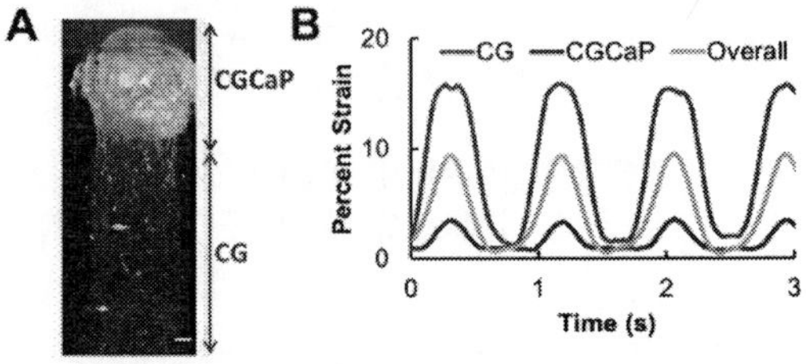

Figure 2: (A) micro-CT image of multi-compartment scaffold (Scale bar: 1 mm)[3]. (B) Compartment-specific strain as a result of 10% global strain in multi-compartment scaffolds.

DISCUSSION

We have described how CG scaffolds, in combination with mechanical stimulation, can be utilized to guide hMSC activity to regenerate tendon-like tissues. When exposed to cyclic tensile strain on anisotropically-aligned scaffolds, hMSCs display upregulation of critical tenogenic markers. Furthermore, when seeded onto multi-compartment CG scaffolds with a discrete mineralized region, hMSCs begin to differentiate based on local material-based cues. When strain is applied to these multi-compartment materials, deformations are primarily localized to the non-mineral compartment, lending to the further augmentation of region-specific hMSC responses.

Ongoing work involves the determination of both upstream effectors and long term studies to evaluate the potential to mechanically reinforce the materials or to produce new tissue suitable for surgical implantation into the native wound environment. Additionally, there is an increased focus on the interfacial zone between tendon and bone and the potential to tailor our materials for the development of this region.

ACKNOWLEDGEMENTS

Research reported in this publication was supported by the National Institutes of Health under Award Numbers R03 AR062811 and R21 AR063331. Addition thanks to undergraduate research assistants Ashley Moy, Matthew Ramsay, and Jordan Meyer for their contributions to this work.

REFERENCES

[1] Butler, D et al., *J. Orthop. Res*, 26:1-9, 2008.
[2] Vitale, M et al., *J. Should. Elbow Surg*, 16:181-187, 2007.
[3] Caliari, S et al., *Adv. Health. Mat*, 4:831-837, 2015.
[4] Mahadik, B et al., *Biomater,* 67:297-307, 2015.
[5] Vanderploeg, E et al., *J. Biomech*, 37:1941-1942, 2004.
[6] Moiyama, K et al., *Chem. Commun.,* 50:5895-5898, 2014.
[7] Paxton, J et al., *Tiss Eng*. 18:277-284, 2012
[8] Zhang, H et al., *J Bio Chem*, 276:6905-6908, 2001
[9] Berthet, E et al., *J Othop Res*, 9:1475-1483, 2013.

SB³C2017
Summer Biomechanics, Bioengineering and Biotransport Conference
June 21 – 24, Tucson, AZ, USA

CYCLIC UNIAXIAL STRAIN INCREASES COLLAGEN III DEPOSITION IN EARLY DEVELOPMENT OF SCAFFOLD-FREE ENGINEERED TENDON FIBERS

Kuwabo Mubyana (1), Connie S. Chamberlain (2), David T. Corr (1)

(1) Biomedical Engineering Department
Rensselaer Polytechnic Institute
Troy, NY, USA

(2) Dept. of Orthopedics and Rehabilitation
University of Wisconsin
Madison, WI, USA

INTRODUCTION

Tendons are highly collagenous structures that transduce the force produced by muscle to skeletal motion, and stabilize joints under high load. Tendons are frequently injured due to overuse and acute trauma (e.g., laceration, strain), and these injuries have significant functional consequence, and a large clinical and financial burden. Due to their poor intrinsic healing capacity and the limited availability of autologous and allogeneic grafts, there is a growing clinical demand to develop surgical replacement grafts for injured tendon. Our laboratory has developed a technique to create individual scaffold-free tendon fibers using directed cellular self-assembly in differentially-adherent growth channels, used this platform to study the effects of biomechanical stimulation on early tendon regeneration [1-3]. Previously, we found that cyclic uniaxial strain improved the tensile and failure properties of these fibers over time [3]. It was hypothesized that these load-dependent increases in fiber modulus, failure stress, and toughness were due to structural and compositional changes in the loaded fibers. The purpose of this study was to examine the influence of cyclic uniaxial strain on fiber composition; namely, the collagen type I (Col-I) and collagen type III (Col-III) composition of our engineered tendon fibers.

METHODS

Tendon fibers were engineered using previously outlined methods [2-3]. In brief, u-shaped growth channels (150 μm width, 17.5 mm length) were molded into agarose, UV-crosslinked, and coated in Human fibronectin (Corning). Human dermal fibroblasts (CCD-1065sk, ATCC) were seeded in growth channels at very high density (1.5×10^7 cells/mL). Cyclic intermittent strain (8 h of 0.1 Hz; 0 - 0.7% sinusoidal elongation, alternating with 8 h of rest) was applied 18 h post-seeding, using a modified FlexCell® Tissue Train® system for 1 ($n = 2$), 3 ($n = 3$), and 7 days ($n = 3$). Age-matched unloaded controls were grown, however due to an inability to maintain fibers for longer than 42h without loading [3], only 1-day control fibers ($n = 3$) were able to be evaluated. **Immunocytochemical Analysis:** Loaded and unloaded fibers were removed from their growth channels and soaked in eosin dye for 5 min., prior to being rinsed in PBS. The fibers were then frozen in OCT (Optimal Cutting Temperature) medium by placement over dry ice and cryosectioned. The purpose of the eosin dye was to increase the contrast between the fiber and OCT medium, for ease of cryosectioning. Sectioned fibers were stained with anti-collagen I antibody and anti-collagen III antibody, and imaged. Collagen levels were then quantified by measuring the intensity of the immunostains in ImageJ. **Statistical analysis** was conducted in R statistical software, using Kruskal-Wallis Rank Sum Tests.

RESULTS

All fibers, even the unloaded, stained strongly for Col-I and Col-III. **Collagen I Immunostains**: Longitudinal sections revealed large (unstained) holes in the fiber structure **(Fig. 1)**. No significant differences in the collagen I stain intensity were found between the unloaded control, 1-day loaded, and 7-day loaded groups. Surprisingly, the 3-day loaded group stained significantly less than others for collagen type I (**Fig. 2**). **Collagen III Immunostains**: Large (unstained) holes were observed in 1-day loaded and unloaded fibers, but were notably diminished in the 3-day loaded and 7-day loaded fibers (**Fig. 1**). Collagen III intensity showed an upward trend with the presence and duration of loading, and was significantly increased by 7 days of loading (**Fig. 3**). In addition, by 7 days of loading, large longitudinal bundles of loosely aligned collagen III could be seen in the fiber sections (**Fig. 1**).

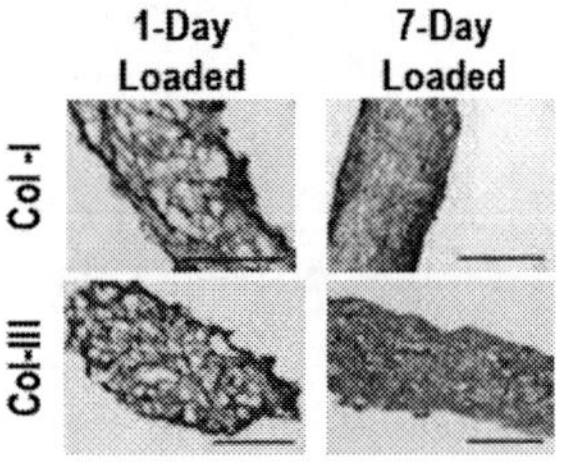

Figure 1: Representative IHC collagen staining (Col-I , Col-III) in 1-day loaded and 7-day loaded engineered tendon fibers (scale bars = 100 μm).

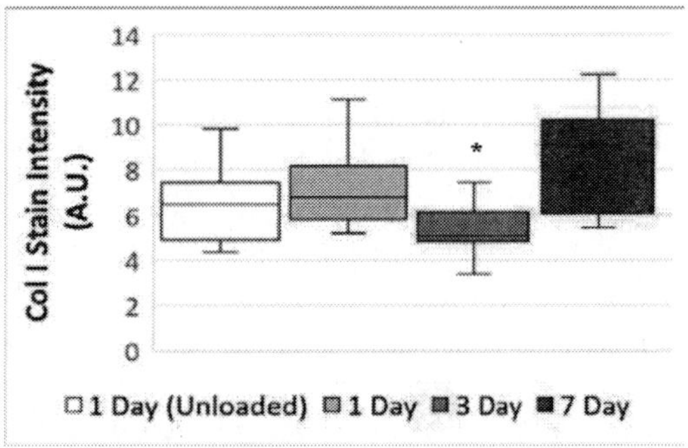

Figure 2: Collagen-I stain intensities of loaded fibers and unloaded controls, indicating no clear influence of cyclic tensile loading on Col-I production (* $p < 0.05$).

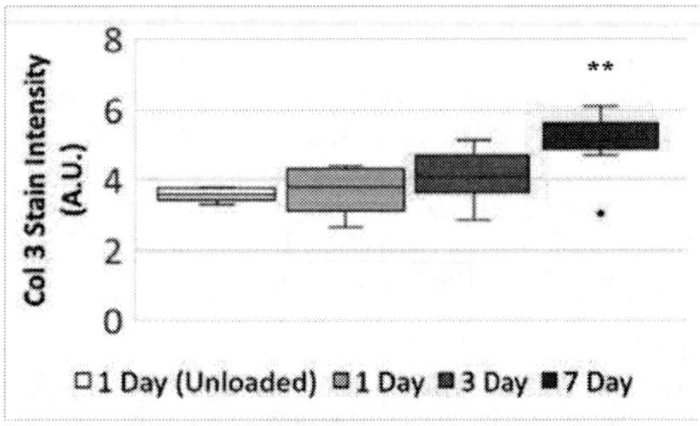

Figure 3: Collagen-III stain intensities of loaded fibers and unloaded controls, showing an increase in Col-III with 7 days of loading ($p < 0.01$).**

DISCUSSION

Collagen I is a major contributor to load-resistance in mature tendon tissue; while, Col-III plays an important role in early wound healing (the formation of fibrotic tissue) and early embryonic Col-I fibrillogenesis [4]. In previous work with these engineered tendon fibers, cyclic uniaxial strain increased modulus, failure stress and toughness over time [3], suggesting that mechanical strain was increasing matrix deposition (specifically fibrillar-forming collagen types I and III). This was supported by the Col-III results herein. While the lower levels of Col-I in the 3-day loaded group could not be readily explained, when considering all other timepoints, it was evident that cyclic loading did not significantly influence Col-I. Given that Col-III is a critical early regulator in collagen I fibrillogenesis[3], it is possible that 7 days of this cyclic uniaxial loading only captures the earliest stages of tendon regeneration, prior to the surge in Col-I. Ongoing studies are investigating the load-dependent changes in collagen fibril alignment, which may also be a contributor to the mechanical properties of these fibers. We believe the large voids observed in the stained fibers to be artifact from the slow freezing in OCT. We are currently replicating these experiments using a more rapid freezing technique, and will employ it for our future collagen alignment studies.

This study provides insight into the influence of mechanical stimulation on fiber development, and the compositional changes that may be contributing to the observed load-dependent increase in tensile strength and failure properties. Such insight can inform current scaffold-based tissue engineering approaches (tendon remodeling), and can also guide a more regenerative, scaffold-free approach to tissue engineering, mimicking key aspect of tenogenesis.

ACKNOWLEDGEMENTS

This work was supported, in part, by NSF CAREER Award CBET-0954990 (DTC).

REFERENCES

[1] Schiele NR, Koppes RA, Chrisey DB, Corr DT. Engineering cellular fibers for musculoskeletal tissues using directed self-assembly. *Tissue Eng Pt A*, 19(9-10): 1223-1232, 2013.
[2] Mubyana K & Corr DT. A micro-mold based technique for the synthesis of scaffold-free, single tendon fibers. *IEEE Proc. of 41st NEBEC*, 2015.
[3] Mubyana K & Corr DT. Cyclic mechanical loading improves tensile and failure properties of scaffold-free engineered tendon fibers. *Proc. of International Symposium on Ligaments & Tendon*, 2016.
[4] Liu *et al.* Type III collagen is crucial in collagen I fibrillogenesis and for normal cardiovascular development. *Proc. Nat. Acad. Sci.*, 94(5): 1852-1856, 1997.

DELETION OF SMAD4 IN ADULT TENOCYTES ENABLES TENDON CELL RECRUITMENT AND FUNCTIONAL RECOVERY AFTER INJURY

Chun Chien (1), Kristen Howell (1), Alice H. Huang (1)

(1) Department of Orthopaedics
Icahn School of Medicine at Mount Sinai
New □ork, N□, USA

INTRODUCTION

Tendon injury is a common clinical problem characterized by slow recovery and high recurrence after treatment. Improving tendon healing to a regenerative, tenogenic state is therefore a crucial research priority. In a previous study, we showed that neonatal mice may harbor a unique regenerative capacity [1]. Lineage tracing further revealed that tenocyte recruitment during healing is a key cellular mechanism of neonatal healing that is absent during adult healing [2].

An additional feature of adult tendon healing is the presence of cartilage and bone deposits within tendon stubs near the muscle and the skeletal insertion [3]. While the etiology is unknown, one theory postulates tenocyte metaplasia into chondrocytes, followed by endochondral ossification. In this study, we test the hypothesis that adult tenocytes may undergo aberrant differentiation toward cartilage at the expense of tenogenic recruitment and differentiation. Since BMP signaling has been strongly implicated in cartilage and bone differentiation [4], ectopic ossification [3], and is a known inhibitor of tendon [5], we also tested whether BMP signaling may regulate tendon cell recruitment and aberrant differentiation after injury via targeted deletion of Smad4 in tenocytes.

METHODS

The following mouse lines were used: ScxGFP [6], Smad4$^{f/f}$ [7], Scx^{CreERT2}, and Rosa26-TdTomato (RosaT) [8]. Achilles tendons were transected at the mid-substance and left unrepaired, in accordance with IACUC, while contralateral tendons were used as non-injured controls. For all studies, tamoxifen was delivered prior to injury to induce gene recombination. Gait analysis was performed on injured mice at d28 after injury using the Digigait Imaging System with treadmill speed of 5 cm/s (n=5-6). Parameters were normalized to Stride length to minimize variations due to mouse size. Histology and fluorescence imaging was performed on cryosections and Western blotting was carried out using standard protocols (Smad 1/5/8 antibody from Cell Signaling). For statistics, gap measurements were analyzed using one way ANOVA with Tukey's posthoc (Systat). Gait data were analyzed using paired t-tests. Significance was determined at $p<0.05$.

RESULTS

To evaluate aberrant differentiation, we used Alcian Blue staining to highlight chondroid deposits in tendons 28 days after injury. While control tendons stained lightly for Alcian Blue, intense, localized staining was observed for the adult injured tendon near the cut site. By d56, ectopic bone formation within the tendon stubs can be detected by faxitron imaging (**Fig 1**). To identify the source of cartilage cells, tamoxifen was given to Scx^{CreERT2} mice to label tenocytes prior to injury. RosaT expression confirmed contribution of labeled Scx^{lin} tenocytes to the ScxGFP-negative chondroid deposits, although non-Scx^{lin} cells were also observed (**Fig 2**). **This data suggests that aberrant differentiation of tenocytes is a feature of adult tendon healing but not neonatal healing, indicating a cell fate switch that may prevent regenerative healing.**

Western blotting for phosphorylated Smads 1/5/8 at d14 after injury showed increased protein after adult injury, suggesting activation of BMP signaling (**Fig 3**). To determine the role of BMP signaling in adult tenocytes after injury, tamoxifen was delivered to adult Smad4$^{f/f}$; Scx^{CreERT2}; RosaT; ScxGFP (mutant) or Scx^{CreERT2}; RosaT; ScxGFP (wild type) mice before tendon transection. Mutant or wild type tenocytes were therefore labeled by RosaT (red) to enable cell fate tracking after injury. Transverse sections through mutant tendons showed comparable contribution of Smad4 mutant tenocytes to cartilage, indicating that Smad4 signaling is not driving this cell fate switch.

Although cartilage differentiation was not affected, at d28, Smad4 mutant cells were observed infiltrating the gap space, which had reduced to ~1mm at this timepoint (down from initial gap of ~3.5 mm). In contrast, the gap space of wild type tendons was still devoid of Scx^{lin} cells at d28 with a gap space not significant different from d3 **(Fig4)**. Gait analysis at d28 showed full recovery of %PropelStride (the parameter most relevant for Achilles tendon function) in Smad4 mutants after injury, compared to continued impaired function in WT. **This data suggests that Smad4 signaling in tenocytes inhibits adult tenocyte cell recruitment and functional recovery after injury.**

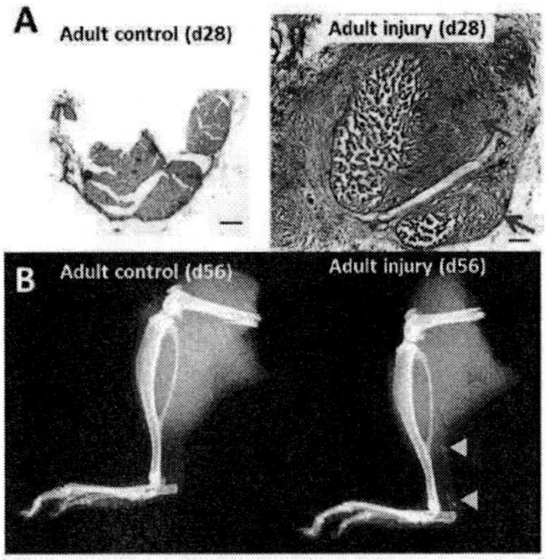

Figure 1: Aberrant bone and cartilage deposits in tendon stubs after adult tendon transection. (A) Radiographs of hindlimbs 56 days after injury reveals bone formation (triangles). (B) Alcian Blue staining of transverse sections shows cartilage formation in injured tendon near transection site (red arrows). Scale: 100 μm.

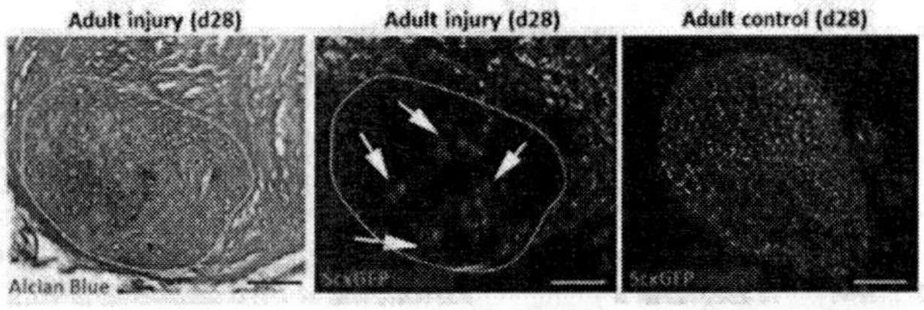

Figure 2: Lineage tracing shows tenocytes contribute to cartilage formation after injury. (Left) Alcian blue staining shows cartilage in tendon. (Right) Labeling of tenocytes prior to injury shows presence of labeled cells (arrows) in cartilage deposits (outline). (Right) Control tendon shows efficient labeling of Scx-lineage cells. Scale: 100 μm.

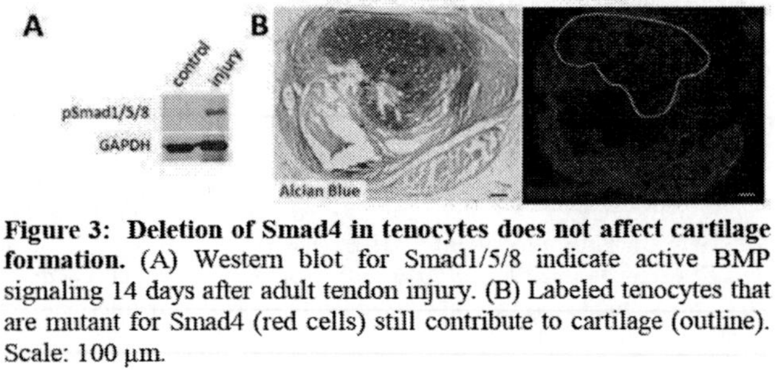

Figure 3: Deletion of Smad4 in tenocytes does not affect cartilage formation. (A) Western blot for Smad1/5/8 indicate active BMP signaling 14 days after adult tendon injury. (B) Labeled tenocytes that are mutant for Smad4 (red cells) still contribute to cartilage (outline). Scale: 100 μm.

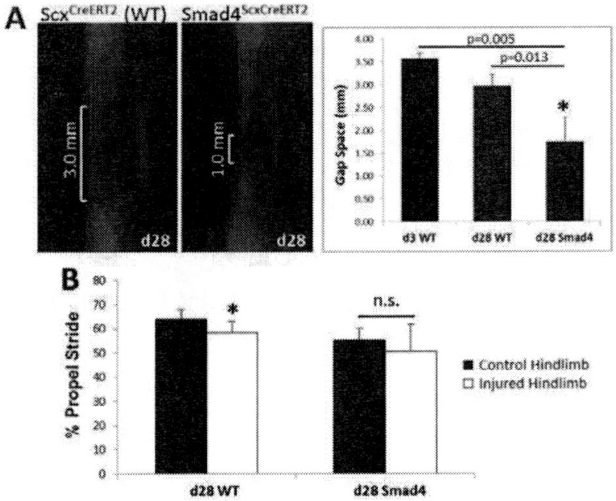

Figure 4: Smad4 mutant tenocytes are recruited into the gap resulting in restored function. (A) Whole mount images and quantification of gap space shows recruitment of Smad4 mutant tenocytes at d28. * indicates $p<0.05$ vs all, n =3. (B) Gait analysis shows recovery of Achilles tendon function in mutant animals after injury. * indicates $p<0.05$ vs control limb (n=5-6).

DISCUSSION

We found that adult tenocytes do contribute to ectopic cartilage formation after tendon injury (although extrinsic cells also contribute) and that aberrant differentiation toward cartilage is uncoupled from tenocyte recruitment. Our loss of function studies showed that Smad4 signaling in adult tenocytes normally represses tenocyte recruitment into the defect after injury. In the absence of Smad4, we showed rescue of recruitment and restored gait function. We propose that our Smad4 results can be attributed to BMP signaling, which we found to be activated after adult injury (shown by elevated pSmad1/5/8). BMP is a known inhibitor of tenogenesis as well as an inducer of chondro/osteogenesis. However, since Smad4 also mediates TGFβ signaling, further studies will be required to distinguish the roles of TGFβ vs BMP signaling in adult tendon healing. Future work will also elucidate the signaling pathways that regulate abnormal heterotopic ossification in the tendon stubs. Although these preliminary studies suggest that Smad4 signaling in tenocytes inhibits regenerative tendon healing in adults, longer-term studies will determine whether fibrotic differentiation is minimized in Smad4 mutants, and whether tendon tensile properties and matrix ultrastructure are also restored.

ACKNOWLEDGEMENTS

This work was supported by NIH/NIAMS R01AR069537 (AHH) and the Department of Orthopaedics at Mount Sinai. We gratefully thank Dr. Ronen Schweitzer for providing the ScxGFP and $Scx^{CreERT2}$ mouse lines.

REFERENCES

[1] Howell+, *Trans ORS*, 2015. [2] Howell+, *Trans ORS*, 2016. [3] Zhang+, *Matrix Biology*, 2016 [4] Oshin+, *Vet Comp*, 2007 [5] Schweitzer+, *Development*, 2001 [6] Pryce+, *Dev Dyn*, 2007 [7] Yang+, *Genesis*, 2002 [8] Madisen+, *Nat Neurosci*, 2010.

ELASTIN IS LOCALISED TO THE INTERFASCICULAR MATRIX OF ENERGY STORING TENDONS AND BECOMES MORE DISORGANISED WITH AGEING

Marta S. Godinho (1), Chavaunne T. Thorpe (2), Steve E. Greenwald (3), Hazel R. Screen (1)

(1) School of Engineering and Materials Science, Queen Mary University of London, London, E1 4NS, United Kingdom

(2) Comparative Biomedical Sciences, The Royal Veterinary College, Royal College Street, London NW1 0TU, United Kingdom

(3) Blizard Institute, Barts and London School of Medicine and Dentistry, Turner Street, London E1 11BB, United Kingdom

INTRODUCTION

Tendons have a highly organised structure, where their subunits (known as fascicles) are oriented in the direction of force application. Fascicles are surrounded by interfascicular matrix (IFM), which facilitates sliding between fascicles. Energy storing tendons, such as the equine superficial digital flexor tendon (SDFT) are more elastic and extensible than positional tendons, such as the equine common digital extensor tendon (CDET). Energy storing tendons have a specialised IFM that facilitates this more elastic and extensible behaviour [1]. However, with ageing, the IFM becomes stiffer and less resilient [2], which may explain why energy storing tendons are more prone to injury then positional tendons. Taken together, these data suggest that the IFM plays a crucial role in tendon mechanics, particularly in energy storing tendons. Interfascicular matrix composition is still poorly defined. However previous data show that the IFM has a distinct composition and greater cell number per unit volume compared to fascicles [3]. Recent studies on bovine flexor tendons reported that the IFM is elastin rich [4], suggesting, elastin may influence IFM and therefore whole tendon mechanics.

Elastin is the core protein of elastic fibres and the major protein in repetitively loaded tissues [5]. It provides extensibility and flexibility to the extracellular matrix [6]. Elastin is extremely insoluble, crosslinked and highly resistant to proteolytic degradation, making it one of the most stable proteins in the body. However, under repetitive loading, it undergoes damage which accumulates over time, thus affecting its function [7]. Overall elastin content in tendon is low and it seems to be sparsely distributed [8]. However, there are reports it is localised between fascicles and around tenocytes [4], which may reflect its role in facilitating fascicle sliding. Therefore, we hypothesise that elastin is more prevalent in energy storing than positional tendons, and is mainly localised to the IFM. Further, we hypothesise that elastin becomes disorganised and decreases in amount with ageing, especially in the energy storing tendons.

METHODS

Forelimbs were collected from horses aged 3-7 years (defined as young group) and 15-19 years (old group) from a commercial abattoir. SDFTs and CDETs were harvested and longitudinally dissected into 4 (SDFT) or 2 (CDET) equal size sections. Small samples, from one of the sections of each tendon, were dissected for biochemical or histological analysis. Samples intended for immunostaining were embedded in optimal cutting temperature compound and snap frozen in hexane cooled on dry ice. Longitudinal cryosections were cut and placed on poly-lysine slides. The remaining samples were used for biochemical analysis. Elastin content was determined using a Fastin Elastin Assay (Biocolour, UK). Elastin concentration was calculated by comparison to a standard curve and expressed as percentage of tissue dry weight.

Tendon elastin organisation was assessed by analysing immunostained sections using confocal microscopy. Cryosections were immunolabelled for elastin (ab9519) and cell nuclei (DAPI). The amount of elastin in the IFM and fascicular matrix (FM) was determined (semi-quantitatively) by calculating the area fraction stained positive for elastin in each region. Confocal images were also analysed using a Fast Fourier Transform (FFT) method, which allows the extraction of quantitative data regarding elastin fibre orientation and alignment [9]. For each image, circular standard deviation was calculated using MATLAB software. Finally, the same images were semi quantitatively scored for elastin organisation by four independent and blinded scorers. Observers scored elastin organisation in each image from 1-5, where 1 denoted highly disorganised elastin and 5 denoted highly organised elastin. Significant differences were determined using ANOVA.

RESULTS

Elastin content in the whole tendon tissue was significantly greater in the energy storing SDFT (3.02% ± 0.60) than in the positional CDET (1.72% ± 0.18, $p<0.001$). With age, the amount of elastin decreased in the SDFT (2.17% ± 0.27, $p<0.01$) (fig. 1).

Immunostaining images showed that most of the elastin (>91.8%) was localised to the IFM in both tendon types (fig. 2). Comparing tendon types, the SDFT IFM showed greater immunostaining for elastin (15.5% ± 4.2) than the CDET IFM (7.6% ± 3.7, $p<0.001$) (fig. 2). With ageing, there was a reduction in the amount of elastin in the SDFT IFM (9.6% ± 2.4, $p<0.001$), however no age-related changes were seen in the CDET IFM (fig. 2). There was also a significantly greater IFM:FM ratio in the SDFT (0.36 ± 0.10) compared to the CDET (0.14 ± 0.05, $p<0.001$), and this ratio increased with age in both tendon types (SDFT: 0.64 ± 0.22, $p<0.01$ and CDET: 0.24 ± 0.09, $p<0.01$).

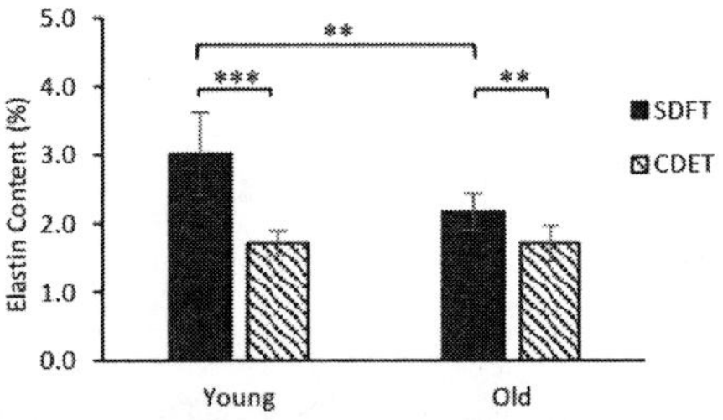

Figure 1: Elastin content expressed as a percentage of dry weight of tendon tissue (n=7). Significant differences are marked as: ** p<0.01 and *** p<0.001. Data are displayed as Mean ± SD

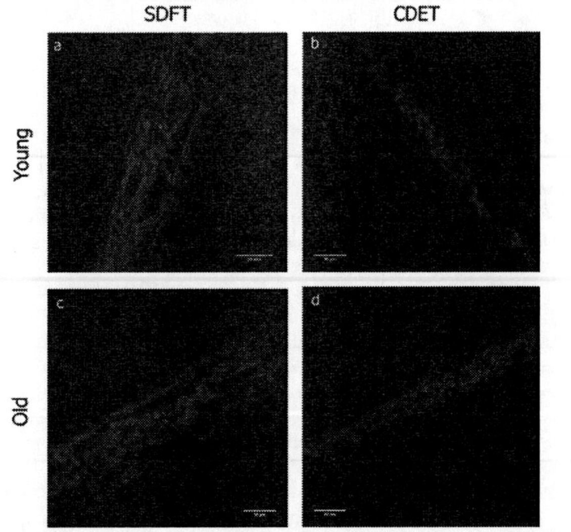

Figure 2: Confocal images showing organisation of elastin in equine energy storing and positional tendons (n=3). Longitudinal sections from young SDFT (a), young CDET (b), old SDFT (c) and old CDET (d) were immunostained for elastin (red). Cell nuclei were also stained (blue). Scale bar: 50µm

The Fourier transformed images showed significant differences between tendon types and age groups. Analyses of the frequency distribution of fibres in each image showed that the old SDFT has a higher circular standard deviation (46.10±3.38) compared to young SDFT (42.81±2.86, $p<0.05$) (fig. 3). Also, in the old group, the SDFT showed a much higher circular standard deviation compared to the old CDET (37.32±5.52, $p<0.001$). These data correlated well with additional image analysis performed by blinded scorers, which indicated a significant decrease in the overall organisation of elastin in old SDFTs (1.75±0.57) compared to young SDFTs (3.02±0.76, $p<0.01$) (average pairwise Cohen's Kappa: 0.454).

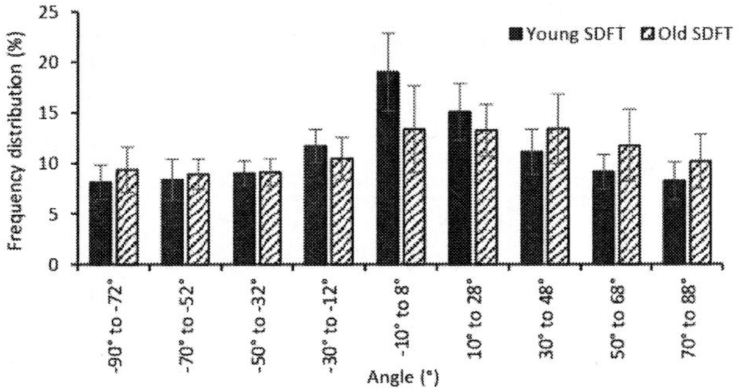

Figure 3: Frequency distribution of elastin fibres for each given direction combining data from 24 images from young and old SDFT (n=3). All data are displayed as Mean ± SD

DISCUSSION

As hypothesised, energy storing tendons were found to have a greater elastin content than positional tendons, with most of the elastin (>91.8%) localised to the IFM. The presence of elastin may provide the IFM with its elastic recoil ability, necessary for efficient energy storage and return.

With ageing, a reduction in the elastin content of the SDFT IFM, may contribute to the previously observed stiffening of the IFM in aged energy storing tendons [2]. Additionally, data showed a significant difference in the distribution of elastin fibres in young and old SDFTs (fig. 3). Circular standard deviation, which is a measure of fibre distribution, was much higher in the old SDFT compared to young SDFT, suggesting a significant decrease in elastin organisation with ageing. This loss of elastin organisation, also confirmed by the blind scoring analysis, may additionally contribute to the loss in extensibility of the IFM in ageing SDFTs, and so, contribute to the increased risk of tendon injuries seen in ageing tendons.

ACKNOWLEDGEMENTS

This work was funded by the BBSRC, UK (BB/K008412/1). Marta Godinho is funded by the Institute of Bioengineering, Queen Mary University of London, through a Bonfield PhD Scholarship.

REFERENCES

[1] Thorpe, C. et al., *J. R. Soc. Interface* 9:3108–17, 2012
[2] Thorpe, C. et al., *Eur. Cell. Mater.* 25:48–60, 2013
[3] Thorpe, C. et al., *Sci Rep.* 6;20455, 2016
[4] Grant, M. et al., *J Anat.* 222(6):573-9, 2013
[5] Braverman, I. & Fonferko, E., *J. Invest. Dermatol.* 78:434-43, 1982
[6] Oxlund, H. et al., *J. Biomechanics* 21(3):213-18, 1988
[7] Robert, L. et al., *Biogerontology* 9:119-33, 2008
[8] Kannus, P. *Scand J. Med & Sci in Sports* 10(6):312-20, 2000
[9] Schindelin, J. et al., *Nat Methods* 9(7):676-82, 2013

SB^{3}C2017
Summer Biomechanics, Bioengineering and Biotransport Conference
June 21 – 24, Tucson, AZ, USA

COMPUTER SIMULATIONS OF BLOOD FLOW IN AORTIC DISSECTIONS WITH FLUID-STRUCTURE-INTERACTION (FSI)

Kathrin Bäumler (1,2), Anna M. Sailer (1,2), Vijay Vedula (3), Alison Marsden (2,3,4), Dominik Fleischmann (1,2)

(1) Department of Radiology
Stanford University
Stanford, CA, USA

(2) Cardiovascular Institute
Stanford University
Stanford, CA, USA

(3) Department of Pediatrics
Stanford University
Stanford, CA, USA

(4) Department of Bioengineering
Stanford University
Stanford, CA, USA

INTRODUCTION

Aortic dissections are a cardiovascular disease enabled by abnormal tissue composition of the arterial wall. Due to a tear in the intimal layer (the primary intimal tear, PIT) and rupture of the media layer blood may enter into the arterial wall, allowing the formation of a new flow channel, the false lumen (Figure 1).

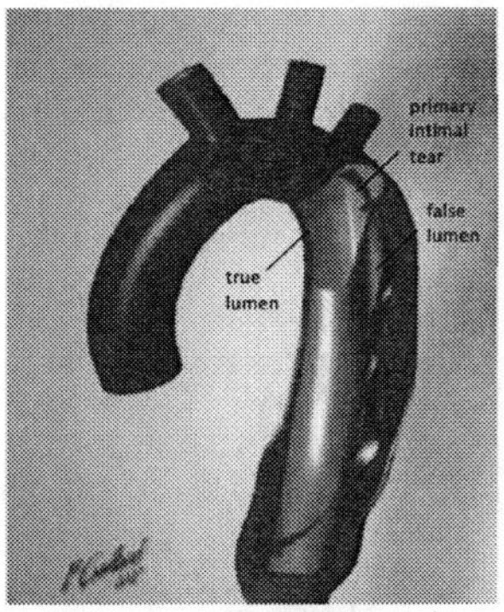

Figure 1: Schematic Type B aortic dissection. Case courtesy of Dr. Gaillard, Radio-paedia.org rID:7640.

Aortic dissections are classified based on the extent of the flap that separates the true from the false lumen: in Stanford type A dissections, the ascending aorta is involved, requiring urgent surgical repair. If the ascending aorta is not involved (defined as Type B), patients are medically managed unless complications occur. Patients face a high risk of late complications, increasing the need for reliable risk prognosis. With improving computational and imaging techniques (e.g. phase-contrast magnetic resonance (MR) imaging) new opportunities have emerged to investigate the role of hemodynamic parameters like false lumen outflow or true and false lumen pressure, [1,2], on the risk of late complications. While CFD simulations have been successfully applied for arterial flows including surgical planning in cardiovascular diseases, [3], simulations of aortic dissections are challenging due to the compliance of the vessel wall. Computational studies of aortic dissections have almost exclusively used rigid wall models, [4,5]. In the present study, we perform fluid-structure-interaction (FSI) simulations in aortic dissections and aim to quantify hemodynamic characteristics including pressure differences between true and false lumen and the deformation of the intimal flap.

METHODS

We perform numerical simulations of blood flow dynamics in both idealized and patient-specific geometries with rigid and deformable walls and with varying elastic modulus of the intimal flap, E_{flap}. We extract the patient-specific anatomic model from computed tomography (CT) scans using SimVascular [6] and Autodesk Meshmixer (www.meshmixer.com). The model includes both true and false lumen, intimal tear and the larger branching arteries (Figure 2). We generate the schematic model with the software Blender (www.blender.com) (see Figure 3).

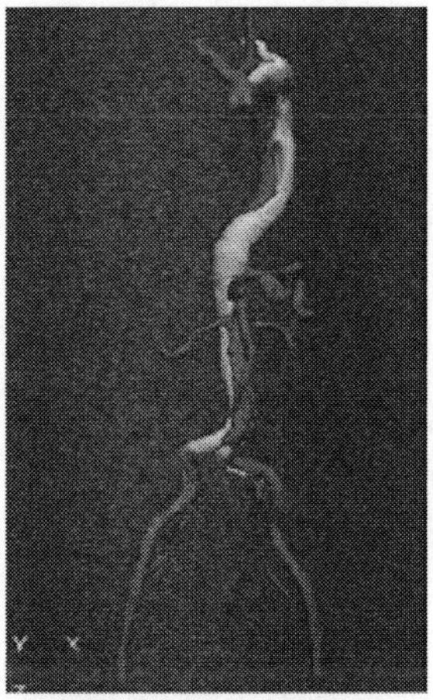

Figure 2: Patient specific model of the true (magenta) and false (white) lumen in a TBAD.

The mathematical model consists of governing equations for blood flow and the arterial wall. The fluid is assumed to be Newtonian and incompressible with constant density ρ (1060 kg/m^3) and viscosity μ (0.004 kg/m s). Flow is governed by the Navier-Stokes equations (1), where u denotes the velocity, and p the pressure.

$$\rho\, \partial_t u = (u \cdot \nabla) u = \mu \Delta u = \nabla p = 0 \quad (1.1)$$

$$\nabla \cdot u = 0 \quad (1.2)$$

The structural momentum equation for the displacement $\hat{x}$ is based on the St. Venant-Kirchhoff model (Eq (2)) in Lagrangian coordinates. The structural density is denoted by $\hat{\rho_S}$ (1200 kg/m^3), and $\hat{P}$ denotes the first Piola-Kirchhoff tensor.

$$\hat{\rho_S}\, \partial_{tt} \hat{x} = \hat{\nabla} \cdot \hat{P} = 0 \quad (2)$$

Fluid and structural equations are coupled via interface conditions. To take mesh deformation into account, we use the arbitrary Eulerian-Lagrangian (ALE) formulation of the above equations.

The numerical discretization of the equations is based on a finite-element formulation (using P1-P1 elements), and a residual based variational multiscale method with a stable second order generalized-α method for time integration. [7], [8] The solver MUPFES [9] is based on a monolithic ALE formulation of the FSI equations. Boundary conditions are incorporated via the coupled multidomain method [10] with resistance outlets. The parameters for the boundary conditions are determined from blood pressure measurements together with quantitative MR imaging that the same patient underwent one day after CT image acquisition, which provides us with patient-specific inflow data.

We evaluate the simulation results by determining the mean pressure p_{tr}, p_{fa} of the true and the false lumen respectively and evaluate the pressure difference $\Delta p = p_{fa} - p_{tr}$. Additionally, we determine the location of the flap midpoint at the PIT.

RESULTS

Idealized geometry

We found an effect of the wall elasticity on the pressure difference Δp in simulations of a schematic geometry, see Figure 3. Δp increases with decreasing flap stiffness E_{Flap}, see Figure 4. A comparison of the flap midpoint location shows a larger deviation from the original location for smaller E_{Flap}, see Figure 5. The flap deforms up to 2.5 mm.

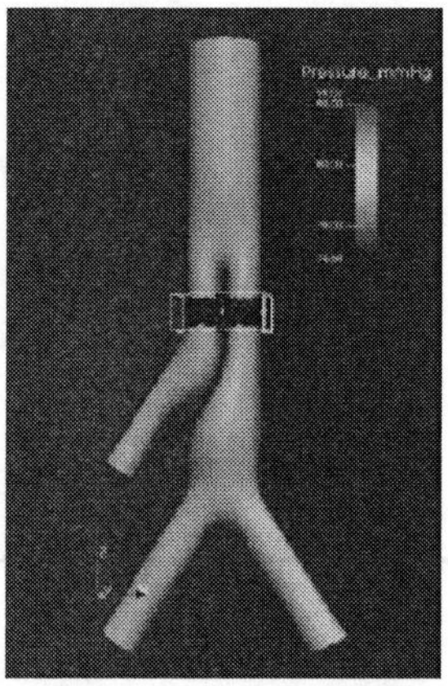

Figure 3: Lumina of schematic geometry. P was measure in the boxes.

Patient-specific geometry

We compared pressure differences between true and false lumen in a patient-specific geometry. We performed a rigid and a deformable wall simulation with E_{Flap} = 280 and E_{Outer} = 700 kPa in Figure 6. We note a pressure difference of 0.11 mmHg in the L1-norm ($\int_0^T |p_{fa} = p_{tr}|$), the flap variation at the intimal tear in the patient specific FSI-simulation is 0.6 mm.

DISCUSSION

Wall deformation clearly affects the pressure difference Δp in simulations in the schematic domain. Additionally decreasing the flap stiffness E_{Flap} increases this effect. Our schematic geometry showed a flap movement of 2.5 mm. From gated CT scans in acute dissections, we know that the flap can move up to several centimeters. In chronic dissections the flap stiffens over time, effectively immobilizing it.

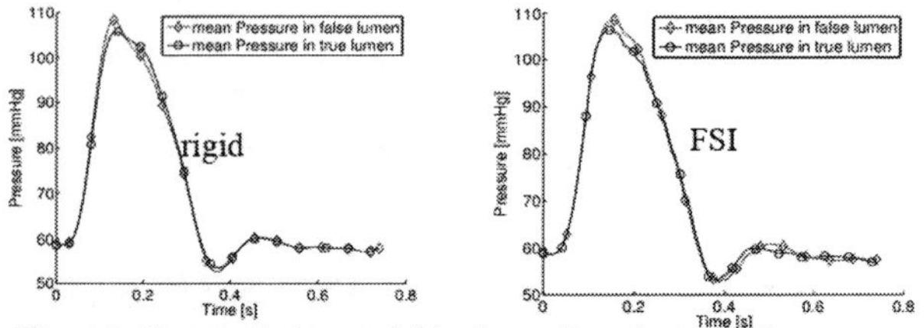

Figure 6: Pressure in true and false lumen in patient-specific geometry. The largest effect can be seen in diastole, where pressure variations are larger in the FSI simulation (right).

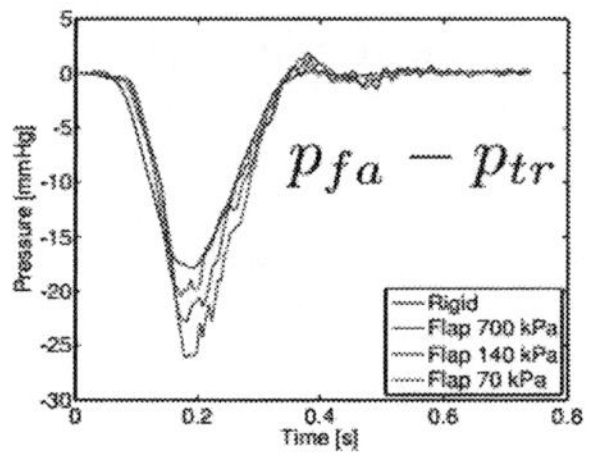

Figure 4: Pressure difference Δp in schematic geometry.

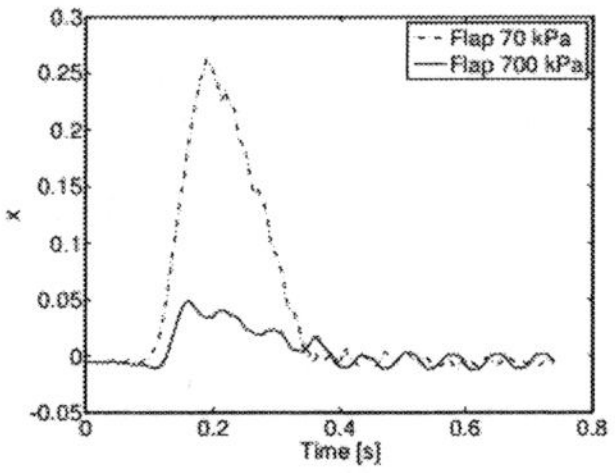

Figure 5: Location of flap midpoint along the x-axis.

For the patient specific geometry, the effects of deformability of the aortic wall and intimal flap were less significant than the schematic geometry. Small pressure differences were captured in the diastolic phase in the FSI simulation which did not appear in the rigid simulation. The overall flap deformation was small, with values of only 0.6 mm.

While the simulation results in the schematic geometry indicate an important influence of wall deformability on Δp, the effect in the patient-specific geometry was less significant. Similarly, the flap movement was larger in the schematic geometry, compared to the patient-specific geometry. Additional flap configurations and patient-specific models should be investigated to corroborate these findings.

Overall, the consideration of wall deformability is necessary to achieve realistic CFD simulations of aortic dissections, but further investigations are needed, e.g. to determine optimal elasticity moduli. Direct comparisons with clinical data will enable validation of patient-specific results, and these are currently underway.

ACKNOWLEDGEMENTS

This research was funded by the Stanford Cardiovascular Institute.

REFERENCES

[1] D. Sailer, AM et al. *Submitt. to Circ. Cardiovasc. Imaging*, 2016.

[2] G. G. Hartnell and J. Gates, *Radiographics*, vol. 25, no. 1, pp. 175–89, 2005.

[3] A. L. Marsden, *Phys. Fluids*, vol. 25, no. 10, 2013.

[4] W. N. Wan Ab Naim et al. *Sci. World J.*, vol. 2014, 2014.

[5] Z. Sun and T. Chaichana, *Int. J. Cardiol.*, vol. 210, pp. 28–31, 2016.

[6] A. Updegrove et al. *Ann. Biomed. Eng.*, 2016.

[7] Y. Bazilevs, et al. *Comput. Methods Appl. Mech. Eng.*, vol. 197, no. 1–4, pp. 173–201, 2007.

[8] T. J. R. Hughes, A. A. Oberai, and L. Mazzei, *Phys. Fluids*, vol. 13, no. 6, pp. 1784–1799, 2001.

[9] A. Marsden and M. Esmaily-Moghadam, *Appl. Mech. Rev.*, vol. 67, no. May, p. 30804, 2015.

[10] I. E. Vignon-Clementel et al. *Comput. Methods Appl. Mech. Eng.*, vol. 195, no. 29–32, pp. 3776–3796, 2006.

UNSTEADY CEREBRAL BLOOD FLOW SIMULATION BASED ON FEEDBACK CONTROL-DATA ASSIMILATION METHOD USING 4D PC-MRI VELOCITY FIELD

Satoshi Ii (1), Yoshiyuki Watanabe (2), Shigeo Wada (1)

(1) Graduate School of Engineering Science
Osaka University
Toyonaka, Osaka, Japan

(2) Graduate School of Medicine
Osaka University
Suita, Osaka, Japan

INTRODUCTION

Image-based simulations using the computational fluid dynamics (CFD) have been developed to reveal a personalized blood flow field [1]. In a general CFD analysis, primitive variables such as a velocity and pressure are obtained by solving the incompressible Navier-Stokes equations. In this regard, a key issue is how to determine the boundary conditions because they are essentially physiological and patient-specific [2]. The phase-contrast magnetic resonance imaging (PC-MRI) has been developed as a noninvasive, observational method for evaluating a patient-specific blood flow field, as given by spatiotemporally discrete velocity vectors. However, the reconstructed flow fields in the PC-MRI data are usually unacceptable because the relatively large measurement error [3], low spatial resolution, and physically inconsistent fields. In these regards, coupled simulations between the CFD and measurement velocity data have been recently proposed by the data assimilation method based on optimal control (e.g. [4]) and feedback control [5] for blood flow analyses. However, it is still under development for practical time-dependent problems.

The purpose of this study is to propose a novel data assimilation approach for the time-dependent flow field using a patient-specific 4D PC-MRI velocity data by extending the method we have proposed, called physically consistent feedback control-based data assimilation (PFC-DA) method [6], and investigate a feasibility of the proposed method for the evaluation of the patient-specific blood flow field and its hemodynamic factors in cerebral aneurysms.

METHODS

Target aneurysm We utilize the patient data acquired from CT and MRI devices, in which an aneurysm appears in an internal carotid artery (ICA). The data was acquired in Osaka University hospital.

Reconstruction of vessel surface A vessel geometry is reconstructed from Digital Subtraction Angiography (DSA) images by the CT measurement, in which the voxel resolution is 0.1 mm. Based on the images, a part of the ICA region including the aneurysm is segmented and the vessel surface is reconstructed with a surface smoothing process using the AMIRA 5.4.2 (Visage Imaging, Berlin, Germany).

Reconstruction of measurement velocity field The measurement velocity vector is obtained from the phase images in each voxel and temporal frame acquired by the 4D PC-MRI measurement. The number of temporal data is 15 in one cardiac cycle, where one cyclic period is 1.47s. First, we subtract the target voxels in the PC-MRI images based on the vessel geometry reconstructed from DSA images. Then, we convert the phase data to the velocity data in each direction and obtain the discrete velocity vectors, which are defined at the mass centers of the target PC-MRI voxels. In this study, the VENC used in the PC-MRI measurement was set to 1 m/s to cover all the blood velocities in one cardiac cycle. The resolution of the PC-MRI voxel is approximately 0.5mm.

In our approach, a temporal change of the measurement velocity is required in a discrete time level of the CFD simulation. In this study, we create a continuous velocity field in time from the PC-MRI temporal data using the Fourier series expansion, where a maximum expansion mode N_L is set to 7.

Flow solution algorithm We apply the incompressible Navier-Stokes equations:

$$\nabla \cdot \mathbf{v} = 0\,, \tag{1}$$

$$\rho(\partial_t \mathbf{v} + \mathbf{v} \cdot \nabla \mathbf{v}) = -\nabla p + \eta \nabla^2 \mathbf{v}\,, \tag{2}$$

where $\mathbf{v}$ is the velocity vector, p is the pressure, ρ is the density and η is the dynamic viscosity. We apply a projection type method based on [7] to couple with the velocity and pressure for equations (1) and (2). We consider the pressure driven system and give the pressure values at the inlet and outlet of the vessel as follows.

$$\begin{cases} p_{\text{in}}(\mathbf{x}) = \phi(\mathbf{x}) & \text{for } \mathbf{x} \in \Gamma_{\text{in}} \\ p_{\text{out}}(\mathbf{x}) = \phi(\mathbf{x}) & \text{for } \mathbf{x} \in \Gamma_{\text{out}} \end{cases} . \quad (3)$$

Here, $\mathbf{x}$ is the Cartesian coordinate, Γ_{in} and Γ_{out} indicate the inlet and outlet region, respectively, and $\phi(\mathbf{x})$ is the scalar potential distributing in a 3D analysis domain Ω that is described later.

We employ the finite difference method for numerically solving the PDEs (1), (2) using a Cartesian mesh system. Here, to deal with the arbitrary vessel geometry, we employ the boundary data immersion method [8], in which the arbitrary geometry is implicitly represented by a phase function in an Eulerian frame. In this study, volume-of-fluid or volume fraction (VOF) function is discretely evaluated in each Cartesian mesh based on the reconstructed vessel surface from the DSA image.

Physically consistent feedback control-based data assimilation (PFC-DA) A key idea of the PFC-DA method [6] is to associate the velocity difference between the CFD and PC-MRI with the pressure boundary conditions (3). In the approach by Funamoto and Hayase [t], to constrain the flow field that satisfies the measurement flow field, the body force vector $\mathbf{f}$ consisting of the velocity difference between the CFD and measurement was incorporated into the momentum equation (2) in a concept of the feedback control theory. However, this force is an artificial and there is no such a force in the actual blood flow field that might offer unphysical flow acceleration and vorticity field. To avoid this problem, we assume that the body force $\mathbf{f}$ is represented as a conservation force including any scalar potential ϕ, i.e. $\mathbf{f}=-\nabla\phi$. Thus, ϕ is associated with the pressure in (2). We also assume that a flow field following the PC-MRI velocity field can be reproduced when the value of scalar potential ϕ is used for pressure values (3) at the inlet and outlet boundaries. According to the vector calculus, the scalar potential field is obtained by solving the following poisson equation.

$$\nabla^2\phi = -\nabla\cdot\mathbf{f}\,, \quad (4)$$

with Neumann boundary conditions $\partial_n\phi=0$.

In the feedback control, it can be regarded that the body force $\mathbf{f}$ is a signal into the system and the difference between the CFD velocity $\mathbf{v}$ and PC-MRI velocity $\mathbf{U}$ is the deviation (or error) as

$$\mathbf{e} = \mathbf{U} - \mathbf{v}\,. \quad (5)$$

In this study, we apply the proportional-derivative (PD) control for the system as follows.

$$\mathbf{f} = K_p\mathbf{e} + K_d\frac{d\mathbf{e}}{dt}\,, \quad (6)$$

where K_p and K_d are the proportional and derivative gains, respectively.

RESULTS AND DISCUSSION

We show temporal changes of L_2 norm for the present PFC-DA and PC-MRI velocities in Figure 1(left) and normalized L_2 norm of velocity difference (5) in Figure 1(right) in the second cardiac cycle. It is seen that the overall velocity change following the PC-MRI velocity field can be obtained. Here, the residual of the velocity difference is approximately 30% except for that of 35 % in the peak systole. In figure 2, we show the PC-MRI velocity field and estimated velocity field interpolated at measurement points in the peak systole and end of the cardiac cycle. Here, we also show the velocity difference vector (5). It is obviously seen that the estimated velocity field is comparable to the PC-MRI one.

The proposed PFC-DA method could reproduce the unsteady PC-MRI velocity field, where the L_2 norm of the velocity difference was approximately 30% in the cardiac cycle. One possible reason to show the level of velocity difference is for measurement errors attributed to the measurement noise and its low spatial resolution. For the future, we will investigate how level of the velocity difference appears and how it affects the exact flow field using a set of noisy data in an ideal flow field, and discuss a reasonability of the estimated flow field by the present analysis.

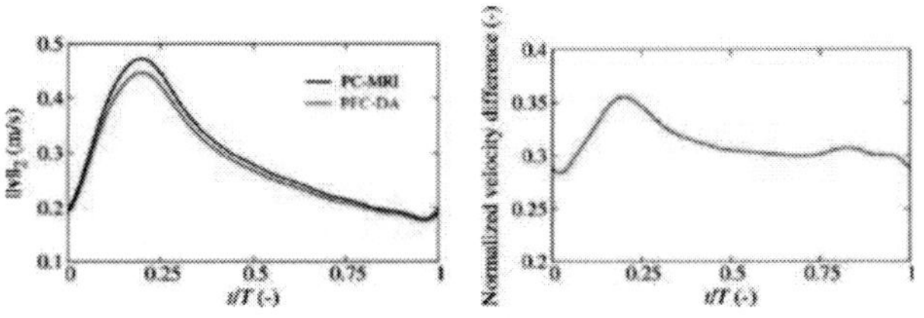

Figure 1: Temporal changes of the L_2 norm for the present PFC-DA and PC-MRI velocities (left) and velocity difference (right) at measurement points in the second cardiac cycle, where the time t is normalized by the cardiac period T.

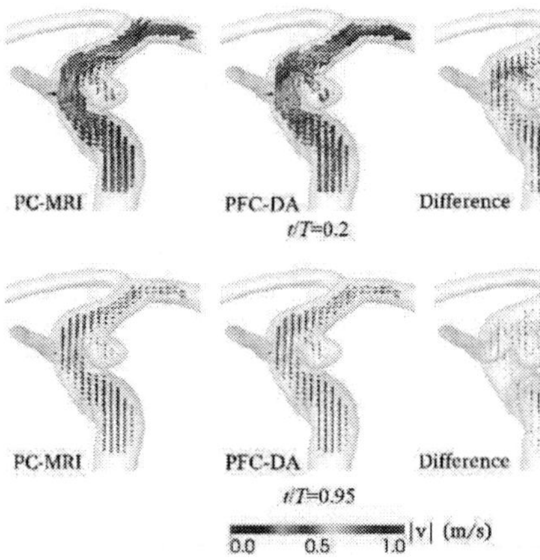

Figure 2: PC-MRI and PFC-DA velocity vectors at measurement points in the peak systole (t/T=0.2) and end of the cardiac cycle (t/T=0.95).

ACKNOWLEDGEMENTS

This research was supported by MEXT as "Priority Issue on Post-K computer" (Integrated Computational Life Science to Support Personalized and Preventive Medicine) (Project ID:hp160218). Y.W. was supported by JSPS KAKENHI Grant Number JP15K09889.

REFERENCES

[1] Taylor, CA et al., *Ann Biomed Eng*, 38:1188-1203, 2010.
[2] Stock, KW et al., *Eur Radiol*, 10:1795–1800, 2000.
[3] Onishi, Y et al., *Int J Numer Meth Biomed Eng*, 29:1089–1103, 2013.
[4] D'Elia, M et al., *J Sci Comput*, 52:340–359, 2012.
[5] Funamoto, K et al., *Ann Biomed Eng*, 33:415–428, 2005.
[6] Mohd Adib, MAH et al., *IEEE BIBE*, 334-337, 2016.
[7] Dukowicz, JK, Dvinsky, AS, *J Comput Phys*, 102:336–347, 1992.
[8] Weymouth, GD, Yue, DKP, *J Comput Phys*, 230:6233–6247, 2011.

IMAGE-BASED COMPUTATIONAL ASSESSMENT OF VASCULAR WALL MECHANICS AND HEMODYNAMICS IN PULMONARY ARTERIAL HYPERTENSION PATIENTS

B. Zambrano (1), N. McLean (1), X. Zhao (2), J.L. Tan (2), L. Zhong (2), L.C. Lee (1), S. Baek (1)

(1) Mechanical Engineering, Michigan State University, Lansing, MI, USA.

(2) National Heart Center, Singapore.

INTRODUCTION

Pulmonary arterial hypertension (PAH) is a complex disorder caused by an increased vascular resistance that leads to an elevated pressure in the pulmonary arterial circulatory system. As the disease progresses, the pulmonary vasculature remodels both structurally and geometrically. A significant part of our current understanding on the vascular mechanical and structural alterations due to PH has been using animal models, which unfortunately, do not fully reproduce all the pathologies found in human PAH. Here, we develop a patient-specific computational modeling pipeline to simultaneously quantify the biomechanical alterations of the deformable vascular wall and hemodynamics.

METHODS

A 3-T Philips scanner with electrocardiography gating was used to acquire cine magnetic resonance (MR) images from one PAH patient as well as one gender- and age-matched normal human subject. Additionally, phase-contrast MR imaging and right heart catheterization (RHC) were performed on the PAH patient to measure blood velocity and pressure in the main pulmonary artery (MPA) trunk, respectively. The patient was a 42-yr-old male who had been diagnosed with severe PAH for 2.5yr at the time of imaging and RHC. All enrolled participants gave written informed consent for the study.

Geometry of the left, right and main pulmonary arteries were reconstructed from the MR images of both normal subject and PAH patient using MeVisLab (http://www.mevislab.de/) (**Fig. 1A**). A finite element (FE) mesh consisting of about 2 million linear tetrahedral elements was generated on each of the reconstructed PA geometries. Fluid-structure interaction between the arterial wall and blood flow was simulated using a coupled momentum formulation that was implemented in CRIMSON (http://www.crimson.software/). In that formulation, blood was assumed to be a Newtonian fluid with a prescribed viscosity η, whereas the arterial wall was assumed to behave as an isotropic linear elastic membrane with prescribed Young's modulus E and Poisson ratio ν. Blood viscosity and Poisson ratio were prescribed as $\eta = 0.004$ Pa s and $\nu = 0.5$, whereas the Young's modulus was calibrated using the measurements. The inlet was prescribed with a velocity field while the outlet was connected to a Windkessel model whose parameters C and R represent the compliance and resistance of the distal vasculature, respectively.

Parameters associated with the mechanical stiffness of the vascular wall (E) and the distal vasculature (C and R) were calibrated in the PAH and normal FE models so that model predictions of inlet pressure waveform and pressure-diameter relationship of the MPA matches quantitatively with the corresponding measurements. A normal PA pressure waveform found in a previous study was assumed for the normal subject who did not undergo RHC. The calibration was performed iteratively using an inner and outer-loop (**Fig. 1B**). In the inner loop, the Windkessel parameters R and C were adjusted based on an initial guess of the Young's modulus E to match the MPA pressure waveform. Thereafter (at the outer loop), E was adjusted (with constant R and C) to match the pressure-diameter relationship. This process was repeated until a good agreement between measurements and model-prediction of these quantities was obtained. Time-averaged wall shear stress (TAWSS) and oscillatory shear index (OSI) were computed using the calibrated parameters.

RESULTS

Compared to the normal subject, the PAH patient has a significantly larger diameter (33.03 vs. 30.25 mm) and pressure (63.62/27.53 vs. 31.04/9.27 mmHg) in

the main PA. Calibrated value of the Young's Modulus E of the proximal vessels is also substantially larger in the PAH patient than the normal subject (1.5 vs. 0.3 Mpa), while distal vasculature of the patient was found to have a higher resistance R and lower compliance C. Using the calibrated parameters, the models were able to predict a pressure waveform and a pressure-

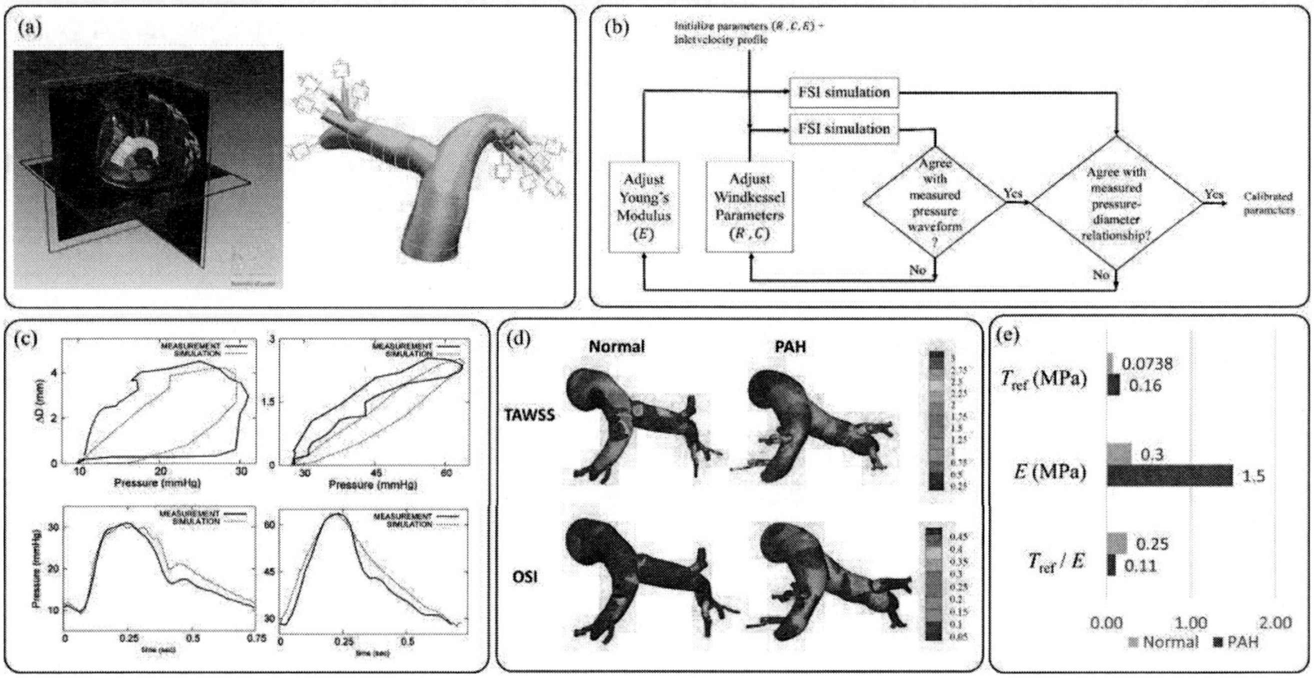

FIGURE 1. (a): Reconstruction of PA geometry from MR images with the outlets each connected to a Windkessel model (b): Schematic of the model calibration. (c): Comparison between model prediction and measurements of the pressure waveform and pressure-diameter relationship at the MPA in the normal subject and PAH patient. (d): TAWSS and OSI distributions computed in the normal subject and PAH patient. (e): Comparison of RV contractility (T_{ref}), PA Young's modulus (E) and their ventricular-vascular coupling (T_{ref}/E) between the PAH patient and normal subject.

diameter relationship that are close to those measured (**Fig. 1C**). Model predictions are within (4.12%) for the pressure waveform and (8.87%) for the pressure-diameter relationship. Comparison of hemodynamics between the patient and normal subject reveals that the latter has a larger TAWSS (1.04 vs. 3.84 Pa, average) and a smaller OSI (0.16 vs. 0.08, average) (**Fig. 1D**).

DISCUSSION

We have described a computational modeling pipeline that simultaneously quantify hemodynamics in the proximal PA as well as the mechanical stiffness of its wall. Our modeling methodology overcomes the limitation from those in previous studies of PAH (e.g., [2]), which assumed that the PA behaves as a rigid wall.

Significant enlargement of the proximal PA was found in the PAH patient compared to that in the normal subject. Our results show that this enlargement was accompanied by material stiffening of the arterial wall as reflected by a substantial increase in E. This result is consistent with animal model studies, which show a higher collagen content in the arterial wall that is exposed to an elevated pressure [3]. The low WSS and high OSI computed in the PAH model in comparison to those in the normal model also suggests endothelial dysfunction in the proximal PA, which may contribute to the accumulation of collagen in the artery.

Because the PA models developed here were derived from the MR images of the same patient and normal subject found in a separate analysis on their ventricular mechanics [1], we also investigated the coupling between the right ventricle (RV) and pulmonary vasculature associated with PAH. In the PAH patient, the ratio between tissue contractility of the RV T_{ref} (found in the other analysis) and E of the PA (found here) was substantially lower than that computed in the normal subject, although RV T_{ref} in the PAH patient was double of that in the normal subject (**Fig. 1E**). This result suggests that the increase in RV contractility is insufficient to compensate for the changes in the stiffness of the pulmonary vessels in this PAH patient, corroborating with the significantly remodeling found in the patient's RV that has comparatively large end-diastolic volume (275 vs. 100 ml). Studies are underway on more patients.

ACKNOWLEDGMENT

This study was supported by NSF CAREER (CMMI-1150376, Baek), Singapore Ministry of Health's National Medical Research Council (NMRC/OFIRG/0018/2016 , Zhong), Goh Cardiovascular Research Award (Duke-NUS-GCR/2013/0009, Zhong) and AHA SDG (17SDG33370110, Lee).

REFERENCES

[1] Xi, C et al. *ASME J. Biomech. Eng.*, 138(11):111001, 2016.
[2] Kheyfets VO et al. *Comput. Methods Programs Biomed.*, 120(2):88-101, 2015
[3] Wang Z and Chesler NC *Biomech. Model Mechanobiol.*, 11(1-2):279-89, 2012

IMPLICATIONS OF SINGULAR INTRACRANIAL ANEURYSM REPAIR IN THE PRESENCE OF CLOSELY-SPACED MULTIPLE ANEURYSMS: A CFD SIMULATION STUDY

Kevin Sunderland[1], Jingfeng Jiang[1], Qinghai Huang[2], Gouthami Chintalapani[3] and Charles Strother[4]

[1]Biomedical Engineering, Michigan Tech. Univ. Houghton, MI, USA

[2]Neurosurgery Department, Changhai Hospital, Shanghai, China

[3]Siemens Medical Solution (USA) Inc., Hoffman estate, IL, USA

[4]Radiology Department Univ. of Wisconsin, Madison WI, USA

INTRODUCTION

Intracranial aneurysms (IAs), ballooning or weakening of the vascular wall, occur in an estimated 3% of the worldwide population [1]. Of those with identified IAs, an estimated 15 to 35 % of patients present with multiple IAs [2]. Serious neurological complications, damage, or death can occur in the event of IA rupture [3]. Current medical practice of IA treatment limits blood flow into an IA through stenting, coiling, or clipping of the IA [4]. Yet depending on factors such as the placement of aneurysms, patient/family medical history, aneurysm characteristics (*i.e.* size), *etc*, all IAs may not be able to be repaired, or require multiple surgical treatments [5].

Individual repair of (multiple) IAs may alter hemodynamic characteristics often associated with IA rupture (Wall Shear Stress (WSS), Oscillatory shear index (OSI), kinetic energy density (KED), swirling flow) within the remaining IAs. **For this study**, the impact that 'repairing' a singular IA has on remaining closely-spaced IAs, all located at the internal carotid artery (ICA), was investigated. Computational fluid dynamic (CFD) simulations of 'patient-specific' vasculature, with either tandem aneurysms (two or more IA in a sequence) or adjacent IAs (two opposite each other on a section of the ICA) were studied. Simulations of all IAs intact and with either the first proximal IA (for tandem) or one IA removed (adjacent) were generated. Hemodynamic conditions were compared between the pre- and post-reconstructions. Understanding the impact that repairing one IA has on remaining IA's could be beneficial in assisting clinical decision making when planning treatment.

METHODS

18 patient vascular structures were studied: 6 with adjacent IAs and 12 with tandem IAs. To model IA repair, removal of either the proximal IA (1st IA in the tandem group) or one of the adjacent IAs, and vessel reconstruction was performed using a computational methodology [6], adapted to the Vascular Modeling Toolkit (VMTK) software package (version 1.2). In the event of vessel shrinkage during editing, the edited vessel was projected back onto the original vessel using the 3-Matic software package (v9, Materialize Inc., Leuven, Belgum) to maintain original vessel characteristics: size and curvature. All vessel geometries were meshed using an open-source tetrahedral mesh generator Tetgen (v1.4.2).

Vascular blood flow was simulated using the ANSYS-FLUENT software (v14.0, ANSYS-FLUENT Inc., Lebanon, NH). Blood was modeled as an incompressible, Newtonian fluid with a dynamic viscosity of 0.004 Pa·s, and mass density of 1050 kg/m^3. Four cardiac cycles were simulated with time steps of 1×10^{-3} seconds (1000 steps per cardiac cycle) with a representative ICA flow waveform based on MR measurements [7]. A Siemens research prototype CDF solver (v4.0, Prototype: not for diagnostic use, Siemens Medical Solution, Inc., IL) was also used to simulate flow, to assure initial findings were not solver dependent. The Siemens CDF solver used the Lattice-Boltzmann Method (LBM) with a Multiple Relaxation Time (MRT-LBM) approximation to simulate blood flow [8].

WSS along the IA wall was calculated by (each) CFD solver. Analysis of flow within IAs was performed by quantifying the spatial-temporal shift (stability) of swirling flow patterns (vortexes) across the cardiac cycle. Using a modified Q-criterion method [9], vortexes were identified by decomposing the velocity gradient field into its rate-of-strain tensor (S) and vorticity (Ω) tensor. Areas of vortex were identified as (1), normalized using the amplitude square of local velocity field (2).

$$Q = \frac{1}{2}[|\Omega|^2 - |S|^2] > 0 \quad (1)$$

$$\overline{Q(x,t)} = \frac{Q(x,t)}{|\vec{u}(x,t)|^2} \quad (2)$$

Vortex areas were mapped to a 3D surface structure, and each structure's degree of volume overlap (DVO) compared to a cycle-averaged vorticity vortex structure was calculated. DVO and variation to the number of structures identified across the cardiac cycle were used to determine (flow) stability. Vortices, and remaining values (OSI, KED) were calculated using in-house VMTK scripts. All parameters were calculated for both pre-and post-reconstruction.

RESULTS

The spatiotemporal averaged OSI decreased in all (remaining) IAs, post-aneurysm repair (Figure 1), whereas a consistent reduction of STA-WSS and STA-KED values (Figure 2; STA stands for spatially-temporally averaged) were only seen when a secondary outlet vessel (offset from the ICA) was in-between the proximal and (1st) distal IA: post-proximal IA removal. This pattern was not seen in either vessels with adjacent aneurysm (regardless of a secondary outlet vessel), or tandem aneurysms with no secondary outlet vessel. Both the Fluent and Siemens (not shown) CFD data exhibited these patterns in WSS and KED.

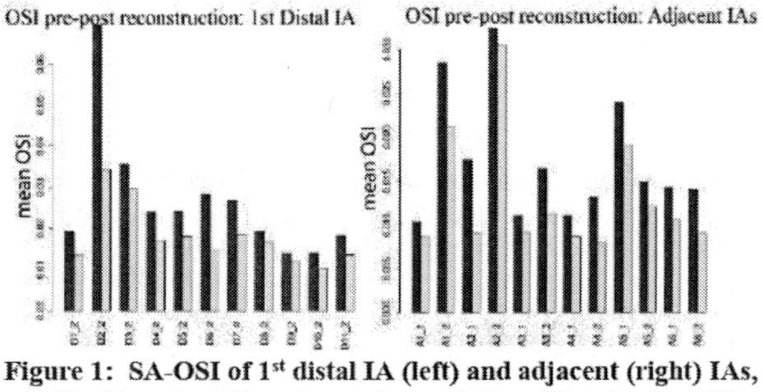

Figure 1: SA-OSI of 1st distal IA (left) and adjacent (right) IAs, pre(black) and post(grey) vessel remodeling: Fluent solver. The OSI decrease post-IA removal was seen across all cases.

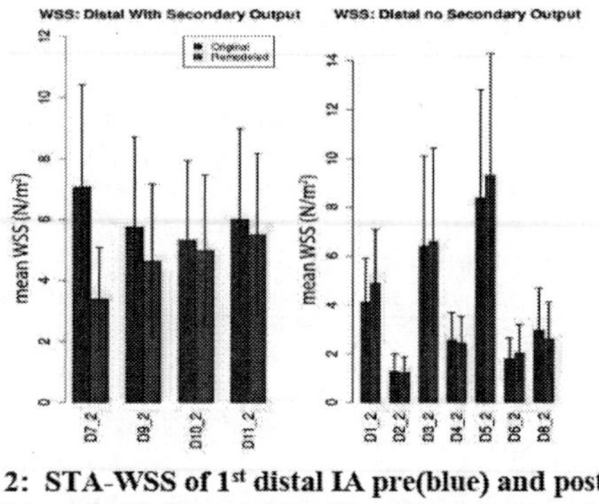

Figure 2: STA-WSS of 1st distal IA pre(blue) and post(red) proximal IA removal: Fluent data. Consistent decrease in values post-IA removal seen with a secondary outlet vessel between proximal and 1st distal IA (left column) versus no secondary outlet (right column).

Vortex DVO and variation to number of identified vortexes showed no distinct pattern between the pre- and post-vessel remodeling in both IA types. Although both the ANSYS-FLUENT and Siemens CFD solver provided consistent results in terms of WSS and KED, we found that vortex characteristics between the Fluent and Siemens CFD solvers were different. In general, the ANSYS-FLUENT data resulted in fragmentary 3D vortex structures, whereas Siemens data resulted in both larger uniform surfaces, and a greater number of identified vortexes. Chosen CFD solver results in different vortex characteristics, yet spatial placement of the largest (main) vortex was similar between solvers (Figure 3). The overall size (volume) of vortex structures post-treatment was reduced (Fluent data only). The mean relative difference (%) between vortex volumes pre-to-post remodeling was -21.56±14.04 for adjacent IAs and -22.03±29.70 for the 1st distal IAs.

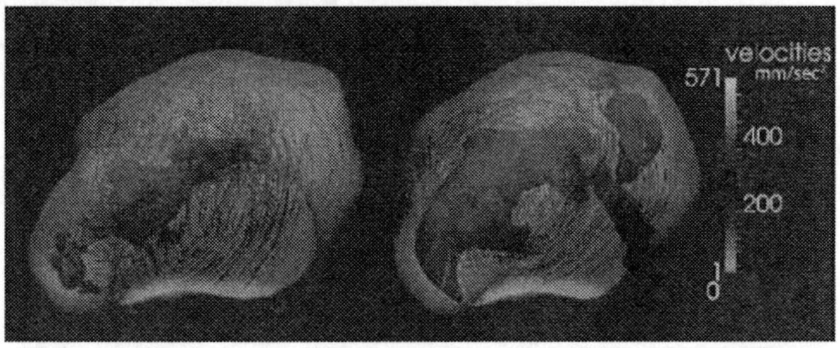

Figure 3: Vortexes (red structure) from two CFD solvers: Fluent (left) and Siemens (right). Colored streamlines show IA flow pattern.

DISCUSSION

Alterations to vasculature in terms of singular aneurysm 'repair' results in variations to hemodynamic characteristics in the remaining IA's. Under conditions (such as secondary outlet vessel between proximal and distal IAs), one IA may alter flow such that faster flow enters subsequent IAs. Removal of said IA may help alter flow away from other IAs, decreasing the stressors and energy within subsequent IAs. Other IAs may act as a 'buffer', altering flow away from (or slowing flow into) subsequent IAs. The repair of said IA may elevate flow into remaining IAs, increasing stressors and energy. This study advocates further assessment and understanding of the impact that repair of IA(s) have on the flow patterns, stressors, and energy within remaining IAs in a system, as it could help impact clinical decision making

ACKNOWLEDGEMENTS

This study was partially supported by a research contract from Siemens to Michigan Tech. University and a Research Excellent Award from the Michigan Tech University.

REFERENCES

[1] Bederson, J.B et al., *Circulation*, 102(18):2300-2308, 2000.
[2] Kaminogo, M et al., *Stroke*, 34(1):16-21, 2003.
[3] Connolly, E.S., et al., *Stroke*, 43(6):1711-1737, 2012.
[4] Steiner, T et al., *Cerebrovascular diseases*, 35(2):93-112, 2013.
[5] Juvela, S et al., *Journal of neurosurgery*, 93(3):379-387, 2000.
[6] Ford, M et al., *The British journal of radiology*, 82:S55–S61, 2014.
[7] Gwilliam, M.N et al., *J Cereb Blood Flow Metab*, 29(12):1975-82 2009.
[8] d'Humières, D et al., *Philosophical Transactions of the Royal Society of London A: Mathematical, Physical and Engineering Sciences*, 360(1792):437-451, 2002.
[9] Sunderland, K et al., *Computational and Mathematical Mehtods in Medicne*, 2016.

PHYSIOLOGICALLY-RELEVANT MEASUREMENTS OF FLOW THROUGH STENTS: TOWARDS IMPROVED MODELS OF ENDOVASCULAR CEREBRAL ANEURYSM TREATMENTS

Michael C. Barbour (1), Michael R. Levitt (2), Luke Johnson (1), Keshav Venkat (1), Christian Geindreau (3), Sabine Rolland du Roscoat (3), Ryan P. Morton (2), Louis J. Kim (2), Alberto Aliseda (1)

(1) Mechanical Engineering
University of Washington
Seattle, Wa, USA

(2) Neurological Surgery
University of Washington
Seattle, WA, USA

(3) 3SR
Université Grenoble Alps
Grenoble, France

INTRODUCTION

In cerebral aneurysms treated with flow diverting stents (FDS), a high-porosity, mesh-like stent is endovascularly deployed in the parent vessel, conforming to the curvature of the vessel and covering the aneurysm inflow neck. These stents are designed to reduce flow into the aneurysm and redirect it into the parent vessel, promoting thrombotic-occlusion and reducing the patient's risk of rupture. Unfortunately, up to 30% of aneurysms treated with a FDS remain patent at mid-term angiographic follow-up[1].

As FDS are designed to alter aneurysmal flow patterns, it is believed the post-treatment hemodynamic environment is linked to treatment outcome. Computational fluid dynamic (CFD) simulations have been used to calculate aneurysm hemodynamic stresses in an attempt to understand treatment mechanics and failure modes[2]. There are two main modeling choices for FDS: use a finite element code to numerically "deploy" a FDS inside the diseased vessel geometry[3], or approximate the stent surface at the aneurysm neck as a porous jump[4]. While the porous media approach is likely an over-simplification, and does not capture the complex hemodynamic environment occurring as flow moves through the stent structure, it requires significantly less computational resources and, as a result, is advantageous in large-cohort, patient-specific studies.

The porous jump condition is applied in a CFD model through the addition of Darcy-Brinkman pressure loss term in the momentum equation. The inertial and viscous loss coefficients are typically derived from published studies of pressure loss across FDS placed perpendicular to the flow in a straight tube or channel[4,5] – an experimental configuration with two shortcomings. First, the flow only has one path and is forced through the stent, while *in vivo*, the aneurysm is often perpendicular to the parent vessel flow where both shear and inertial forces drive flow into the sac. Second, the FDS is kept straight, and does not account for changes in stent porosity as the FDS bends to properly appose to the parent vessel[6].

In this present study, we aim to generate an improved computational model of FDS by experimentally measuring velocity across stents placed in idealized side-branch aneurysm models with varying levels of curvature. Pressure loss measurements across the same stent will be used to determine new model coefficients that will be validated against particle image velocimetry (PIV) data. As a first step, CFD simulations using the standard porous media model of FDS are compared to PIV measurements to assess sources of model error.

METHODS

Three idealized *in vitro,* side-branch aneurysm models are made with a parent vessel diameter of 4mm and an aneurysm diameter of 7mm. The curvature (κ) of the parent vessel at the aneurysm neck for each of the models is: 0.03, 0.07 and 0.11 mm^{-1}. Using a four part wax-casting procedure, 3D printed aneurysm models are converted to clear, silicon phantoms[8]. The *in vitro* models are then "treated" by our neurosurgery collaborators with 4x20mm PipeLine Embolization Devices (Medtronic, Minneapolis, MN).

Stereoscopic-PIV measurements of flow through the treated aneurysm models are made using two Phantom v641 high-speed cameras (Vision Research, Wayne, NJ) and a Terra PIV pulsed laser operating at 527 nm (Continuum, San Jose, CA). The working fluid is a 60:40 water:glycerin mixture that has a viscosity of 3.5 cP and matches the index of refraction of the silicone models. The fluid is seeded with 10 μm diameter RhB fluorescent particles (TSI, Shoreview, NM). Steady flow through the models is driven using a pressure reservoir and measured using an ultrasonic flow-meter. The parent vessel flow rates are: 0.1, 0.2, 0.3, and 0.4 LPM.

CFD simulations are additionally run that mimic the experimental configurations. The inertial and viscous loss coefficients that define the standard porous media model of the FDS are taken from literature[5].

RESULTS

The velocity profile, $\bar{u}$, is extracted from the center-plane of the aneurysm/stent interface from PIV experiments and CFD simulations. The aneurysm inflow velocity is then approximated with the following equation: $U_{Aneurysm} = {}^{1}/_{L} \int \bar{u} \cdot \bar{u}\, dl$, where $\bar{u}$ is the vector normal to the stent surface and L is the length of the 2D integration line. Shown in figure 1, is the aneurysm inflow velocity for each of the three models as a function of parent vessel flow rate. Both CFD and PIV data are shown. Inflow velocity increases with parent vessel flow and with curvature, but large differences are seen between the PIV and CFD calculated values. The degree of error in the porous media model simulations are also shown to increase with increasing curvature, approaching 75% at the highest curvature and flow rate case.

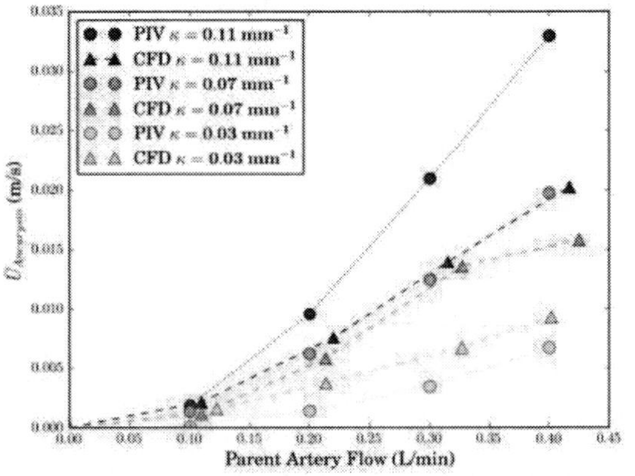

Figure 1: Aneurysm inflow velocity for all three models as a function of parent vessel flow rate. Data points taken from PIV measurements are plotted as circles and from CFD as triangles.

To further explore the error in the porous media models, the velocity vectors at the aneurysm/stent interface are plotted in the center of figure 2 for $\kappa = 0.03mm^{-1}$. The outer panels contain streamlines inside the aneurysm to highlight flow directionality. The CFD calculated flow fields are drastically different from those derived from PIV measurements. At the highest parent-vessel flow rate (bottom row), the bulk of the flow enters the aneurysm at the distal end and creates a clockwise rotating vortex inside the aneurysm. This is seen in both CFD and PIV simulations, but the size of the high velocity zone is much smaller in the CFD solution. The PIV measurements also show that as the parent-vessel flow decreases the rotation direction of the aneurysm vortex shifts and become counter-clockwise at the lowest flow-rate (top row). However, this phenomenon is not captured in the CFD simulations.

DISCUSSION

For the first part of this study, the standard porous media model is compared to PIV measurements of flow through idealized aneurysm models of varying curvature treated with FDS. Significant differences are seen in both aneurysm inflow velocity magnitude and the flow-patterns inside the aneurysm. Errors in the aneurysm inflow velocity are likely due to the fact that standard porous media models do not account for vessel curvature. As the parent vessel curvature increases, the pore size of the stent at the aneurysm inflow location increases as well, increasing porosity[6]. The error in the CFD estimates of aneurysm inflow are greatest for the high curvature case, suggesting that curvature should be included in future models of FDS.

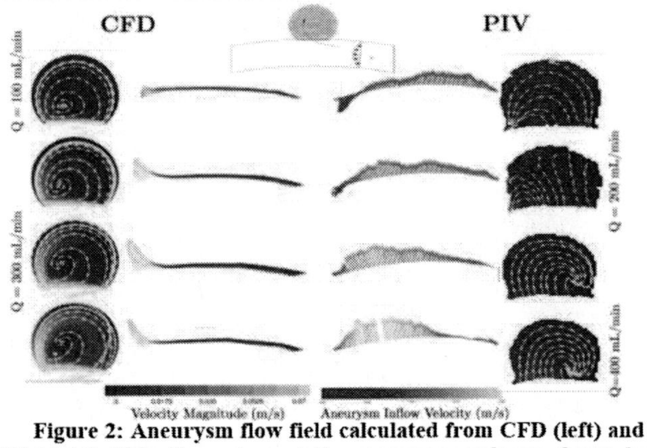

Figure 2: Aneurysm flow field calculated from CFD (left) and PIV (right). Velocity vectors at the aneurysm neck are shown in the center and streamlines inside the dome on the outer panels. Parent vessel flow rate, Q, increases down columns. κ=0.03mm^{-1}

The velocity profile at the aneurysm/stent interface and the characteristics of the aneurysmal flow are significantly different as well. The porous media model is not able to capture the directionality change of the predominant aneurysm vortex that occurs with decreasing parent vessel flow. We hypothesize that the change in rotation at lower flow-rates is due to a backwards-facing step phenomenon. In the lower flow-rate case, the axial momentum at the proximal edge of the aneurysm/stent interface is low enough that the fluid expands into the leading edge of the aneurysm creating a counter clockwise rotation. In the higher flow-rate case, the axial momentum is greater and caries the fluid further downstream before entering the aneurysm at the distal edge. This is likely due to an over-estimation of stent resistance in the CFD models of FDS. Additionally, the parameters of the standard FDS model were derived from experiments that did not accurately represent the *in vivo* flow condition of FDS placed in side-branch aneurysms.

For the second part of this study, pressure measurements will be gathered across the stent with the same models and at the same flow-conditions. These measurements will be used in conjunction with the aneurysm inflow velocity and stent porosity measurements from μCT imaging of the treated *in vitro* models to re-estimate FDS model parameters. CFD simulations using the new model will be compared to PIV measurements for validation.

ACKNOWLEDGEMENTS

This work has been financially supported by the NIH/NINDS (grant 1R03NS078539 and 1R01NS088072). This work was supported by an unrestricted grant in the form of the pipeline devices to our academic institution from Medtronic Inc., which had no role in the experimental design, data analysis or scholarship of this work.

REFERENCES

[1] Molyneux, AJ et al., *Lancet*, 366:809-17, 2005
[2] Levitt, MR et al. *AJNR* 35:143-148 2013
[3] Damiano, R et al. *J of Biomechanics* 48:3332-3340, 2015
[4] Augsberger, L et al. *Annl. of Biomed. Engr.* 39:850-63, 2011
[5] Raschi, M et al. *Int. J.Numer. Meth. Biomed.Engr.* 30:909-25, 2014
[6] Makoyeva, A et al. *AJNR*, 34, 596-602, 2013
[7] Levitt, M el al. *J Neurointervent Surg* 0:1-6, 2016

MORPHOMETRIC AND HEMODYNAMIC IMPACT OF POST ENDOVASCULAR AAA REPAIR: COMPARISON WITH INFRARENAL PHYSIOLOGICAL BLOOD FLOW

P. Tasso (1), A. Raptis (2), M. Xenos (2, 3), D. Gallo (1), M. Matsagkas (2, 4), U. Morbiducci (1)

(1) Dept. of Mechanical and Aerospace Engineering
Politecnico di Torino
Torino, Italy

(2) Institute of Vascular Diseases
Laboratory for Vascular Simulations
Ioannina, Greece

(3) Dept. of Mathematics
University of Ioannina
Ioannina, Greece

(4) Dept. of Vascular Surgery,
Faculty of Medicine,
University of Thessaly,
Larissa, Greece

INTRODUCTION

Abdominal aortic aneurysm (AAA) is a localized expansion of the abdominal aorta. Among treatments for AAA, endovascular aneurysm repair (EVAR) is a reliable technique which is a minimally invasive alternative to classical open-chest surgery. Technically, EVAR is performed by the endovascular delivery and deployment of an endovascular stent graft, to reinforce the wall of the abdominal aorta and to keep the damaged area from rupturing. EVAR of AAAs results in (1) redirection of blood through the deployed endograft, and in (2) reshape of the iliac bifurcation, altering local hemodynamics and adversely affecting its performance [1]. In the last decade post-EVAR effects have been investigated harnessing the power of computational methods [2]. However, the interplay between geometry and hemodynamics in AAAs pre- and post-treatment has been poorly investigated. Here we present a study on the impact of AAA endovascular repair on geometric features and local hemodynamics. The purpose is to define both geometric and hemodynamic features that deviate markedly from their physiological values after EVAR. In detail, local vascular geometric features like curvature, torsion, and cross-sectional area are evaluated in the pre- and post-operative geometries (1 month follow-up after stent graft implantation), and in healthy models. Near-wall flow quantities are calculated with CFD (Computational Fluid Dynamics), and related to the geometric features, to explore if the post interventional reshaping of the vascular territory supports the re-establishment of physiological hemodynamics.

METHODS

Ten subjects suffering from AAA underwent implantation of an Endurant® (Medtronic, Santa Rosa, CA, USA) stent-graft system at the University Hospital of Ioannina, Greece. Computed tomography (CT) scans of the subjects were obtained before and one month after EVAR (as detailed in [1]). Five subjects with no sign of aneurysmal disease underwent CT angiography of the region including the abdominal aorta and common iliac arteries. DICOM images of all subjects were then segmented using Mimics (Materialise, Leuven, Belgium). The postoperative models were restricted to the deployed endovascular graft interior space (no extension to the suprarenal abdominal aorta or the native iliac arteries), while the preoperative and healthy models included the lumen between the renal arteries and the iliac bifurcation [2]. On the reconstructed 3D models, a morphometric analysis based on the geometric centerline was carried out. Centerlines were extracted as the geometrical locus of the centers of the maximum inscribed spheres in the model, as given by the Vascular Modelling Toolkit software (VMTK). Free-knots regression splines were then employed as a basis of representation to provide a noise free, analytical formulation of the centerlines [2]. By differentiation of the free-knots regression spline representations, centerlines were characterized in terms of curvature and torsion, that are known to have a major influence on hemodynamics. Cross-sectional area $A(s)$ along the curvilinear abscissa s were calculated automatically via intersection of a plane normal to the centerline. The reconstructed 3D geometries were meshed with tetrahedral elements [1]. The unsteady incompressible blood flow equations were solved by using the finite volume-based method software package Ansys Fluent (Ansys Inc., Canonsburg, PA). Blood was considered as Newtonian, incompressible fluid. Boundary conditions at inflow and outflow sections were based on flow and pressure measurements by Olufsen et al. [3]. More in detail, a physiological pulse pressure condition was applied as Neumann inflow boundary condition in all the models, while an outflow section area-based flow split strategy was applied at the outflow sections. Near-wall hemodynamics was described in terms of wall shear stress, while intravascular flow structures were described in terms of helical flow.

RESULTS

Local curvature profiles calculated along healthy and post-operative models provide a clear representation of the impact that the stenting procedure had in reshaping the vessel, the final goal being reestablishing healthy geometric features (Fig. 1). The group averaged values of curvature and cross-sectional area diminished from pre- to post-operative geometries. However, post-operative geometric features still differ from healthy models (Fig. 2a-2b).

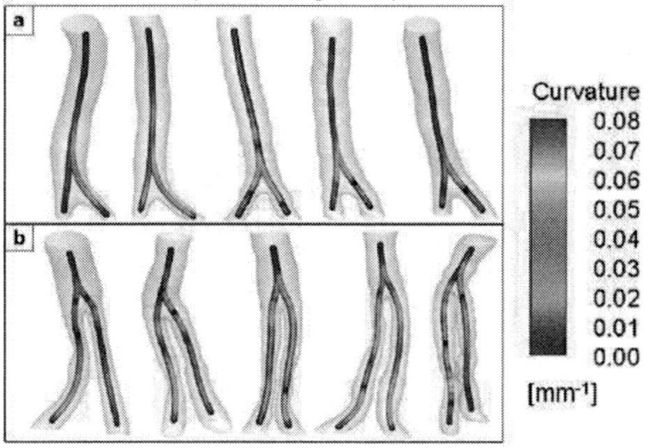

Figure 1: Centerlines with local curvature values: (a) physiological models; (b) post-operative models. Torsion peaks location are also presented (red dots, positive peaks; black dots, negative peaks).

Concerning torsion values, an inversion in sign was observed in post-operative abdominal segments (Fig. 2c). In the abdominal segment, the group averaged rate of cross-sectional area variation in post-operative models was higher than in pre-operative and healthy geometries, while an opposite trend was found in the iliac territory (Fig. 2d).

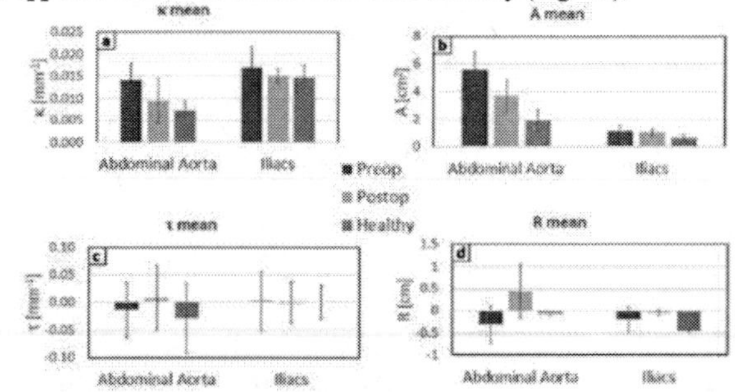

Figure 2: Group averaged values of (a) curvature, (b) cross-sectional area, (c) torsion, (d) cross-sectional area variation-rate.

As for hemodynamics, the observed group-averaged maximum WSS values (Fig. 3a) in the post-operative models were lower than the healthy cases, in the abdominal segments. In the iliac territories, a significant decrease in postoperative maximum WSS, compared to the healthy cases, was observed (8.7±0.5 Pa *vs* 16.8±1.4 Pa, p=0.01). The WSS contour maps (Fig. 4) confirm this finding. In post-operative cases, cycle-averaged helicity intensity in the abdominal segment of the graft was found to be higher than healthy subjects, but the difference was not statistically significant (Fig. 3b). In the iliac territories of the graft, the intensity of helical flow was close to the healthy subjects (Fig. 3b).

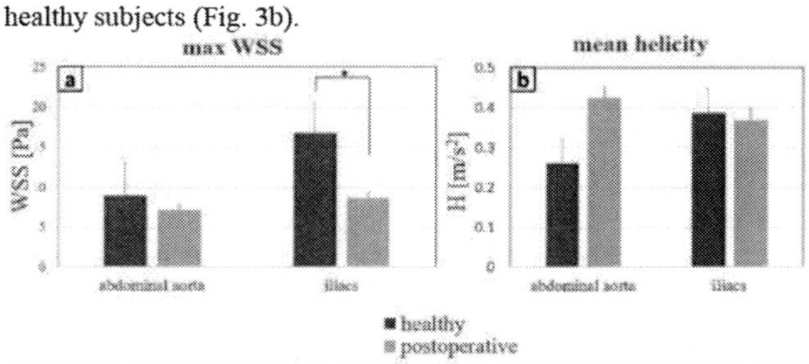

Figure 3: Group averaged (healthy and post-operative) values of (a) maximum WSS, (b) helicity intensity. (*: p<0.05).

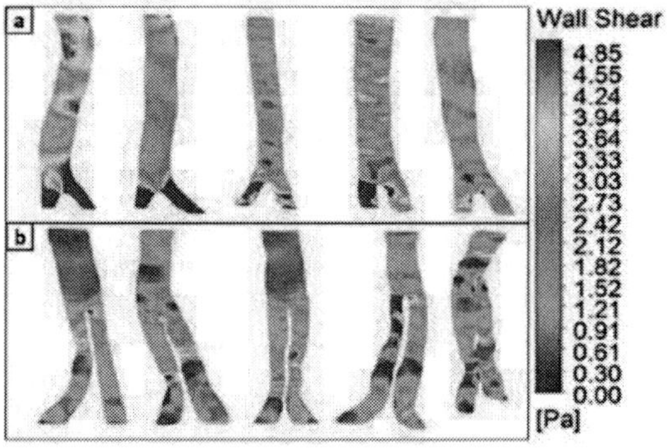

Figure 4: WSS contours at peak systole: (a) healthy cases; (b) post-operative cases (5 of the 10 investigated subjects).

DISCUSSION

The purpose of this study was: (1) to investigate the EVAR post-operative geometric reshaping; (2) to test through numerical simulations whether the geometries of endovascular repaired aortas induce significantly altered hemodynamic flow compared to healthy aortas; (3) to evaluate how vascular reshaping could influence post-operative hemodynamics. The analysis highlighted that the post-operative geometric reshaping, even if has given significant results with respect to the preoperative stage, was still far from restoring the physiological morphometry in the diseased vascular district. This was particularly true in the abdominal aorta territory, where the post-operative higher curvature, torsion sign inversion, cross-sectional area, and area variation-rate could be responsible for the enforcement of helical flow patterns in this territory, thus augmented by factors related to geometry. On the other hand, in iliac territories, where less marked morphometry differences between post-operative and healthy geometries was observed (except for the area variation-rate), comparable values of helicity intensity were observed.

The diminished maximum WSS values observed in post-operative models, with respect to healthy subjects, as well as the overall impact that EVAR has on near wall hemodynamics, and on the long term outcome of the endovascular repair is still subject of investigation, in parallel to the study of the geometric and hemodynamic impact of other commercial stet graft devices.

REFERENCES

[1] Raptis, A et al., *Comput. Methods Biomech. Biomed. Engin.*, 0:1-8,2016.

[2] Gallo, D et al., *Comput. Fluid*, 141:54-61, 2016.

[3] Olufsen, M S et al., et al., *Ann. Biomed. eng.*, 28:1281-1299, 2000.

SB³C2017
Summer Biomechanics, Bioengineering and Biotransport Conference
June 21 – 24, Tucson, AZ, USA

SPRING ASSISTED CRANIOPLASTY: A PARAMETRIC ANALYSIS OF SURGICAL OUTCOMES USING STATISTICAL SHAPE MODELING AND FINITE ELEMENT ANALYSIS

A. Borghi (1,3), K. He (1), J. Bruse (2), N. Rodriguez Florez (1,3), D. Dunaway (1,3), O. Jeelani (1,3), S. Schievano (1,2,3)

(1) GOS-ICH
University College London
London, UK

(2) ICS
University College London
London, UK

(3) Craniofacial Unit
Great Ormond Street Hospital for Children
London, UK

INTRODUCTION

Sagittal craniosynostosis consists of premature ossification of the sagittal suture after birth, with 1 in 1,700 live births affected [1]. The result is an elongated and narrow head shape (scaphocephaly, Figure 1A), with sometimes related functional problems often requiring surgical intervention. Spring assisted cranioplasty (SAC) is a surgical technique to treat scaphocephaly: two osteotomies are performed along the sagittal line of the skull and spring-like metallic distractors are temporarily placed (Figure 1B) to gradually expand the skull and widen it over time. Three spring models are currently available: S10, S12, S14 with the same geometry, but increasing stiffness (Figure 1C). A number of surgical parameters including osteotomy dimensions, spring positioning and choice of distractor stiffness are expected to influence the surgical outcome.

In this work, an average model of scaphocephalic skull was created by means of statistical shape modeling and used to simulate the on-table effect of spring insertion via finite element analysis. Surgical parameters were varied in order to assess the effect of spring positioning and combinations on skull shape and dimensions.

METHODS

Twelve patients (mean age = 3.6±1.2 months [1.6-5.3 months], 10 male) treated for scaphocephaly at Great Ormond Street Hospital for Children, London were chosen for this study. The skull 3D shape of each patient was reconstructed from computed tomography (CT) scans using Mimics (Materialise, Leuven) (Figure 2A) and further processed in Rhino3D (McNeel, WA) and Meshmixer (Autodesk, CA) to: 1) reduce the mesh for ease of processing; 2) create a plane through the Nasion, and the right and left external auditory meatus in order to cut each model consistently; 3) to fill in skull defects and sutures, and obtain a smooth continuous surface; and 4) to separate the inner surface of the skull bone from the outer surface. The outer skull surfaces were aligned and an average model was calculated using statistical shape modeling in Deformetrica (www.deformetrica.org) [2]. A solid skull model was recreated by offsetting internally the average model surface by the same value as the average population skull bone thickness, retrieved for each patient as closest point distance. The average model was subdivided into the following parts: frontal plate, coronal suture, parietal plate, lambdoid suture and occipital plate (Figure 2B). The distance between the osteotomies (*LAT*) measured during surgery from a population of 26 SAC patients was averaged and used to create virtual osteotomies.

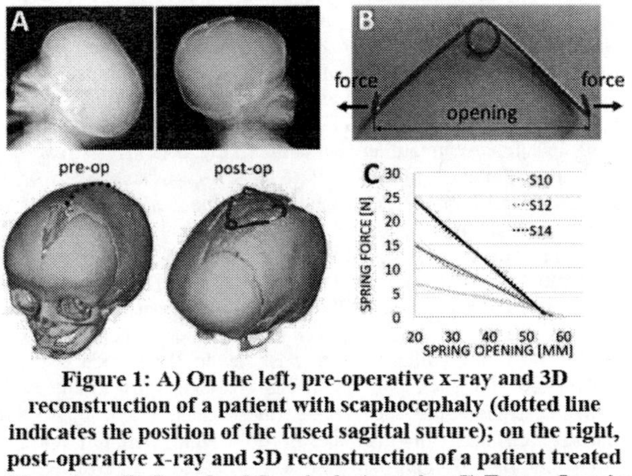

Figure 1: A) On the left, pre-operative x-ray and 3D reconstruction of a patient with scaphocephaly (dotted line indicates the position of the fused sagittal suture); on the right, post-operative x-ray and 3D reconstruction of a patient treated with springs. B) Sample of Cranioplasty spring C) Force- Opening curves for the three spring models used in GOSH.

Spring location from the 26 patient SAC, described by A= distance of anterior spring from coronal suture and B= distance from anterior to posterior spring (Figure 2C), were also averaged and varied on the solid model (min, average and max) in order to create 8 models with same overall geometry, but different spring positions along the sagittal plane.

Each model was imported in ANSYS Mechanical (ANSYS, PA) for finite element analysis: material properties for skull and sutures were retrieved from literature [3] and elastic spring conditions were implemented to simulate the spring implantation. For each model, 9 simulations were carried out, each having a different spring combination. Results were processed to retrieve the skull shape after spring placement assessed through the cephalic index (CI = ratio of maximum breadth of the skull to maximum length), clinically used to evaluate head shape changes achieved by surgery. CI was measured for each model and compared with the pre-operative CI – patients have long and narrow head shapes hence smaller CI before SAC and wider heads after surgery, thus an increased CI. Strains on the sutures were evaluated in the

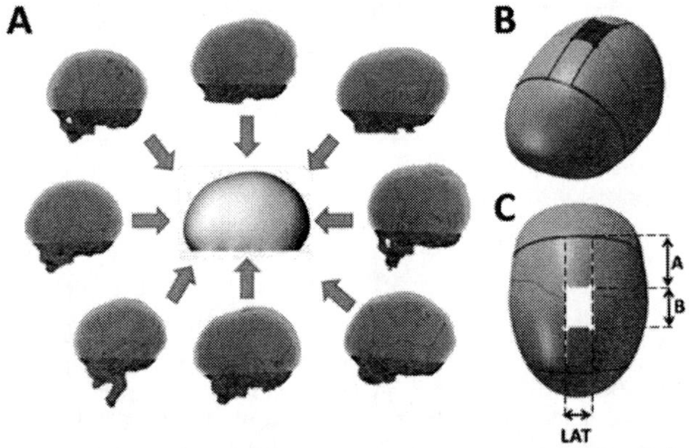

Figure 2: A) Creation of the average model of scaphocephalic skull surface using statistical shape modeling B) Average solid skull including sutures and osteotomies C) Surgical parameters LAT, A, and B.

RESULTS

The average skull thickness was equal to 2.6±0.4mm, while the average solid skull model pre-op CI was equal to 70%. The average values from the surgery data of the 26 patients were LAT = 20.8±2.9mm, A = 41.5±10.5 mm and B = 38.8+12.0 mm. By varying the values of A and B, as well as spring stiffness combinations, post-implantation CI varied from 71.2% to 74.2% (Figure 3). Change in spring locations maintaining the same spring combination caused a variation of final CI from 0.46% (S10-S10 combination) to 0.92% (S14-S14 combination). For each spring combination, A= 41.5mm (mean distance of anterior spring from coronal suture) and B = 50.8mm (max distance between springs) resulted in the lowest final CI, while A = 41.5mm and B = 26.8mm (min distance between springs) had the highest final CI, thus suggesting that with an anterior spring in average position, a closer posterior spring results in wider impact. The most evident effect was due to the choice of spring combination: for each pair of spring positions, post-expansion CI varied linearly with the sum of the stiffness of the two springs (Figure 4). Maximum and minimum values in suture strain were found for A = 52mm and B = 38.8mm; lowest when S10-S10 implantation was simulated (16.2%) and highest when S14-S14 was considered (91.4%).

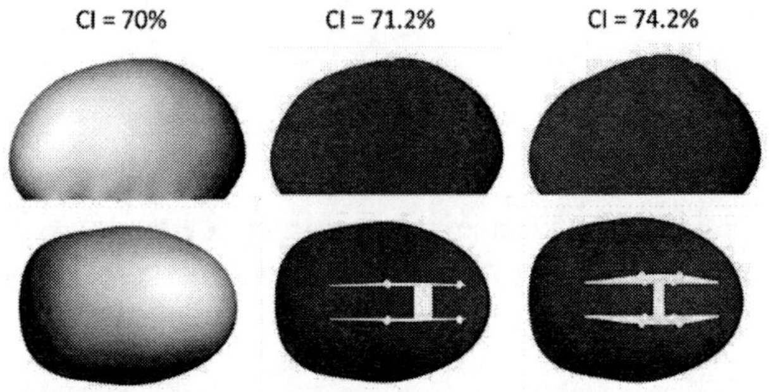

Figure 3: Skull shape with cephalic index (CI) pre-operatively (grey); at lowest post-op CI (blue); and at highest post-op CI (red)

DISCUSSION

SAC is a successful minimally invasive technique to treat craniosynostosis, but it allows the surgeon only minimal control over the final head shape outcome. In this work, we have investigated the effect of surgical parameters on a generic, average model of scaphocephalic skull, which serves as a representative of a sagittal craniosyostosis population, in order to better understand the current procedure and provide some indications to the surgeon to optimize results. The results show that the choice of spring combination has the highest impact on the final remodeling of the skull in terms of changes in CI but also on sutural strains. The positioning of the posterior spring in relation to the anterior spring can also affect results. Future work will include analysis of the viscoelastic behavior of the skull to assess the effect of surgical parameters on long-term changes in cranial shape.

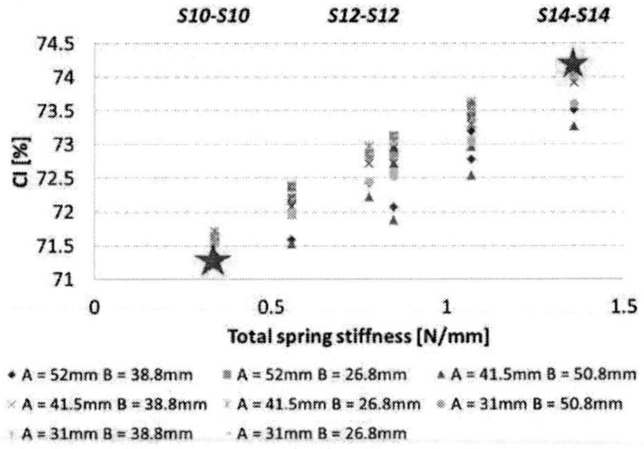

Figure 4: A) Change in post-op cephalic index according to spring combination and implantation distances: in red the configuration with highest CI, in blue the configuration of lowest CI

ACKNOWLEDGEMENTS

The work has been funded by the Great Ormond Street Hospital for Children Charity (Grant n. 12SG15).

REFERENCES

[1] Fearon, J et al., *Plastic and Reconstructive Surgery*, 133:1261-75, 2014.

[2] Bruse J et al. *BMC Med Imaging*, 16(1):40, 2016.

[3] Zhigang Li et al. *Computer methods and Programs in Biomedicine*,, 112:309-319, 2013.

SB³C2017
Summer Biomechanics, Bioengineering and Biotransport Conference
June 21 – 24, Tucson, AZ, USA

NEW APPROACH FOR WORST CASE DETERMINATION OF HIP STEM USING FEA AND ABAQUS GUI

Mohsen S. Renani (1,2), Jeffrey E. Bischoff (2)

(1) Department of Mechanical Engineering
University of Missouri – Kansas City
Kansas City, MO, USA

(2) Biomechanics Research
Zimmer Biomet, Inc.
Warsaw, IN, USA

INTRODUCTION

Numerous test standards exist for pre-clinical evaluation of implants for total joint arthroplasty, based on failure modes seen during early generation devices or on features common to implant systems that could introduce a new failure mode into the field. Historically, these test standards prescribe physical testing to ensure product performance [1]. However, as computational modeling and simulation (CM&S), including finite element analysis (FEA), has become both more common and more important to the industry in recent years for design development, verification, and validation, standards are emerging to officially recognize the utility of and to govern the use of CM&S. These standards identify CM&S for use in support of determining appropriate boundary conditions for the corresponding physical testing method [4], for determination of worst case configuration within a given test method [2,3,6], or if appropriately validated, in place of a physical test [5].

Standards-based physical testing and associated CM&S are intended to be prescriptive in nature, so as to ensure uniform evaluation across a class of devices. As such, an unavoidable characteristic when using CM&S to support standards-based device evaluation is the replication of a common model setup across many device sizes and/or modular assembly combinations. Advancements in design and FEA tools are increasingly supportive of automation of model assembly and evaluation for such systems. In particular, contemporary FEA software typically provides an application program interface (API) to enable customization of either the software graphical user interface (GUI) and/or the underlying model setup directives. With such APIs, efficient and robust customizable applications can be developed that both facilitate quick evaluation of many designs, as well as removing repetitive manual steps that are prone to human error. Additionally, such applications can then be deployed among analysts with a range of experience with computational methods, but nevertheless with confidence in the quality of the results.

Evaluation of the strength of hip stems in total hip arthroplasty is a classical, standardized test method [1] that has an adjacent standard for use of FEA for worst case determination [2]. The objective of this study was to develop an automated method for evaluation of hip stems per ASTM F2996-13 [2].

METHODS

A GUI was developed in ABAQUS CAE 2016 (Dassault Systemes) for automation of the test method (**Figure 1**). In the GUI, the user can select a stem from an identified stem library or can manually enter the key features of the stem including neck angle, taper length, and neck length. Once the stem has been selected, the user identifies key parameters of the test method (configuration properties, boundary conditions, and loading conditions), as well as of the FEA (e.g. material, mesh). Finally, the GUI enables automated execution of the fully established boundary value problem and post-processing of the results once complete (**Figure 2**).

Underlying the GUI is a Python script that drives all aspects of the model preprocessing, execution, and post-processing. Specific essential algorithm details for model setup include the following:
1. The stem and neck axis of each stem is identified based on the neck angle and the distal tip point of the selected stem(s).
2. The head center location is projected onto the neck axis according to the offset specified by the user and the range of head centers supported by the implant system. Alternatively, the location of the head center can be directly specified by the user. A reference point is created at the head center location and tied to the stem taper for application of the load.

Technical Presentation #272

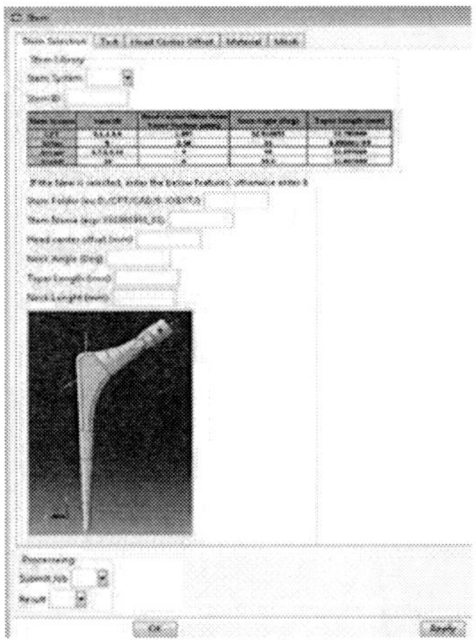

Figure 1: Front page to the ABAQUS GUI for hip stem evaluation.

3. Parameters of the test setup are utilized to virtually pot the stem in bone cement [1], including the distance from the head center to the potting level, the orientation of the stem in frontal and lateral planes, and the applied load. The user may also select default parameters based on a chosen standard [1-2], thus ensuring rigorous adherence to the standard.
4. The solid model of the stem is trimmed below the target potting level, and fully constrained boundary conditions are applied on the trimmed surface of the distal stem.
4. A library of common materials is available for drop-down selection by the user. Alternatively, the user may directly enter the essential material properties (elastic modulus; Poisson's ratio).
5. The stem is meshed based on specified global and local mesh seeds.
6. A complete input file is generated and submitted to the solver.
7. Once complete, peak stress values in the stem are extracted from the output database, and representative contour plots are saved to file.

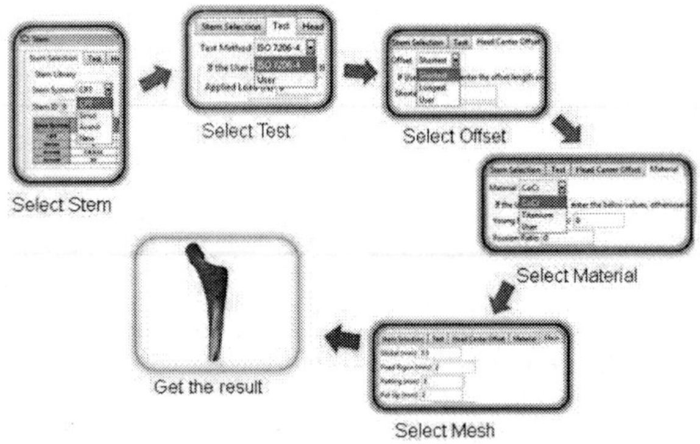

Figure 2: Schematic of the automated work flow.

RESULTS

The stress distribution for exemplars of two different classes of hip stems, generated completely through automation, is shown in **Figure 3**. For both cases, as is expected with this test method, the peak stress occurs around the potting region.

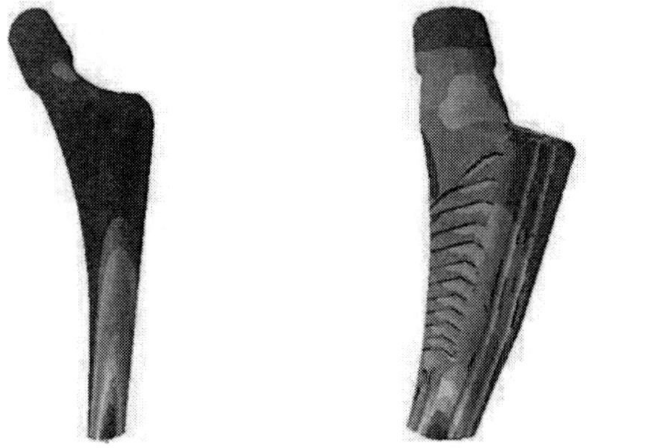

Figure 3: Stress distributions for two unique but representative hip stems.

DISCUSSION

The main objective of this project was to develop and utilize an automated tool for FE analysis for worst case determination of hip stem strength, to make worst case determination more efficient and consistent. The tool was successfully deployed to evaluate stems of varying complexity. With tools such as these, the laborious and error-prone manual process of model development is replaced by a quick and robust tool that enables engineering focus to reside on specification of parameters and interpretation of results, rather than model development and execution. With the associated GUI, the tool may be deployed to engineers across the enterprise with minimal dependency on pre-existing practical and technical expertise with FEA.

With the existing algorithmic base, further work may extend the application to automated design of experiments, parametric studies, mesh convergence analyses, and other types of analyses that rely on a parametric FE model design. Additionally, the approach taken here may be similarly followed to automate other standards-based FEA methods that are expected for pre-clinical component evaluation.

REFERENCES

[1] ISO 7206-4:2010 (E), Implants for surgery — Partial and total hip joint prostheses— Part 4: Determination of endurance properties and performance of stemmed femoral components," ISO, 2010.
[2] ASTM_F2996-13, Standard Practice for Finite Element Analysis (FEA) of Non-Modular Metallic Orthopaedic Hip Femoral Stems. ASTM International, West Conshohocken, PA.
[3] ASTM F3161-16, Standard Test Method for Finite Element Analysis (FEA) of Metallic Orthopaedic Total Knee Femoral Components under Closing Conditions. ASTM International, West Conshohocken, PA.
[4] ASTM F1800-07, Standard Test Method for Cyclic Fatigue Testing of Metal Tibial Tray Components of Total Knee Joint Replacements. ASTM International, West Conshohocken, PA, 2007.
[5] Class II Special Controls Guidance Document: Knee Joint Patellofemorotibial and Femorotibial Metal/Polymer Porous-Coated Uncemented Prostheses; Guidance for Industry and FDA. US Food and Drug Administration, Washington, DC, 2003.
[6] Assessment of Radiofrequency-Induced Heating in the Magnetic Resonance (MR) Environment for Multi-Configuration Passive Medical Devices; Guidance for Industry and Food and Drug Administration Staff. US Food and Drug Administration, Washington, DC, 2016.

FINITE ELEMENT SIMULATION AND EXPERIMENTAL CHARACTERIZATION OF SURGICAL KNOT PERFORMANCE

Arz Y. Qwam Alden (1), **Peter A. Gustafson** (1,2)

(1) Mechanical and Aeronautical Engineering
Western Michigan University
Kalama□oo, MI, USA

(2) Western Michigan University
Homer Stryker M.D. School of Medicine
Kalama□oo, MI, USA

INTRODUCTION

Suture materials are used clinically to hold the tissue together after an injury or surgery. Sutures are manufactured from a variety of different absorbable and non-absorbable materials, and may be a single filament or braided with or without a coating. The integrity of the suture and tissue is of great importance. When selecting a surgical suture and stitch (knot), the surgeon should consider many properties of the suture material and knot, including the suture size, the tensile performance (non-knotted and knotted), the length of the cut ends, the number of throws, and holding capacity of that knot [1,2]. Measurement and comparison of the strength of knotted suture materials is a problematic because there is an absence of a consensus test method. In general, there are few investigations reporting the influence of suture properties performance (e.g. gauge length and knot tensioning force) [3].

The objective of this study is to develop a finite element (FE) model of a surgical knot and to examine the mechanical performance of surgical sutures. Additionally, a series of experiments are conducted on multifilament No.2 FiberWire suture to determine the tensile performance of non-knotted suture for input into the model. The outcomes of this work will be an improved understanding of factors influencing knot performance, which can assist surgeons in selecting knot configurations.

METHODS

Tensile Test on braided suture: in order to obtain material properties and validation data for the FE model, an experimental study of braided non-absorbable suture materials No.2 FiberWire (Arthrex, Naples, FL) was performed. FiberWire suture is composed of a biocompatible polyethylene core of several small individual strands covered with a multi-strand braided jacket of polyester [4]. In addition, the core has been separated from the jacket to test each individually to determine the mechanical properties more precisely.

The experimental setup consists of two pulleys made from aluminum with a diameter of 63 mm, each pulley has a groove and compression plates to grip the suture. The compression plates were slotted where depth of the slot varied from the suture diameter at the gauge section end to zero at the terminal end. This slot provides support for the specimen and theoretically makes the load transition from the free end to the clamped end more gradual. Therefore, these slots minimize the stress concentration and prevent breakage of the specimen in the clamping area. Compression of these plates was accomplished using two screws. The upper and lower pulleys include a plate, which connected to the MTS Bionix servo hydraulic load frame (Model 370.02, Eden Prairie, Minnesota) and alignment of the specimen was maintained along testing axis of the load frame.

To prepare the specimen for testing, the suture was wrapped around each pulley and passed between the compression plates. The end of the specimen was then wrapped around one of the screws at each pulley side to maintain position and to further decrease the likelihood of suture slippage. The screws were tightened to fix the suture in place. The suture material was marked close to the plates edges to allow the visual detection of any slippage in the suture during the test. The removable plates of each side also have a slot.

A 10 N tensile preload was applied to each specimen to set a reference displacement. Subsequently, the specimen was pulled in tension to failure. Displacement was applied at a rate of 0.1 mm/sec. Displacement and force were recorded at a rate of 102.4 Hz. Images were captured by a high-resolution camera. Strain was computed from measured displacement (referencing the gauge length of the specimen) and also based on the imaging where the centers of two marked dots were tracked. The two strain calculation methods were found to be in agreement.

FEA Model of Knot: ABAQUS (Version 6.13-1, Dassault Systemes Simulia Corp., Johnston, RI, USA) was used to create FE model of the knot. The model consists of one throw that was created by using reduced integration hexahedral elements (C3D8R). The material was assumed to be homogenous and isotropic elastic. Knots involve complex contact conditions and large deformations. Therefore, an explicit dynamic FE solution was used to overcome these severe nonlinearities.

Displacement control of the boundary was used to apply 0.1 mm/sec. Mass scaling was used to increase the stable time increment; thus, the knot was modeled assuming a quasi-static load in an explicit dynamic model. The ratio of kinetic to internal energy was monitored to ensure the mass scaling and a quasi-static assumption was appropriate.

RESULTS

Material Characterization: for all sutures, the modulus of elasticity, maximum stress at failure, the peak load to failure, and suture dimensions were calculated and compared to published literature and were found to be consistent, Table 1. The cross-section of the suture was assumed to be a circular. The suture diameter was calculated along the suture length utilizing an image of undeformed suture. Moreover, the diameter was monitored using subsequent images with the suture under strain. Poisson's ratio values were calculated at various levels of strain and found to be approximately 0.38. This value is approximate as it required an invalid assumption of homogeneity of the suture.

Table 1: Selected literature for comparison of suture properties.

Study type	Elastic modulus [Gpa]	Max Stress [Mpa]	Peak Load [N]	Diameter [mm]
Present study	8.5±0.92	646±29	258±12	0.703± 0.002
Najibi et al [5]	4.76±0.5	762±81	282±30	0.69
Wright et al [6]	7.2±3.52	803±185	255.3±10.37	0.705
Arthrex website[4]	-	-	271	-
Barber et al [7]	-	-	259.7±85.81	-
Barber et al [8]	-	-	187.9±12.2	-
Abbi et al [9]	-	-	276±24	-

Figure 1a shows the load-displacement behavior of the whole suture. The maximum load of the core and jacket are approximately 65 N and 210 N respectively. While the maximum strain is 2.6% for the core and 9% for the jacket. Representative load-elongation behavior of the core and jacket are shown in figure 1b. Stress is not shown in the figure because it is difficult to establish the cross sectional area of the individual constituents of the suture.

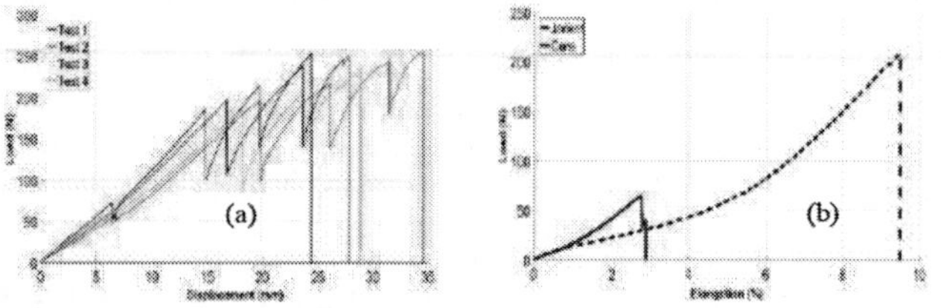

Figure 1: (a) Load versus displacement curves of the suture, and (b) Load versus elongation curves of core and jacket.

FEA Results: the material properties of FE model were determined through experimentation. Figure 3a shows an example result of the FE models. Loads were small until the knot tightened, and the subsequently the load increased dramatically with additional applied displacement. The predicted stresses and strains were in locations that are qualitatively consistent with expectation.

The total kinetic energy of the model was found to be small relative to the strain energy over the whole loading cycle, excluding the initialization where strain was very small. The kinetic energy was less than 1% of the internal energy in critical phases, as depicted in figure 3b. Therefore, it was deemed reasonable to treat the results as quasi-static.

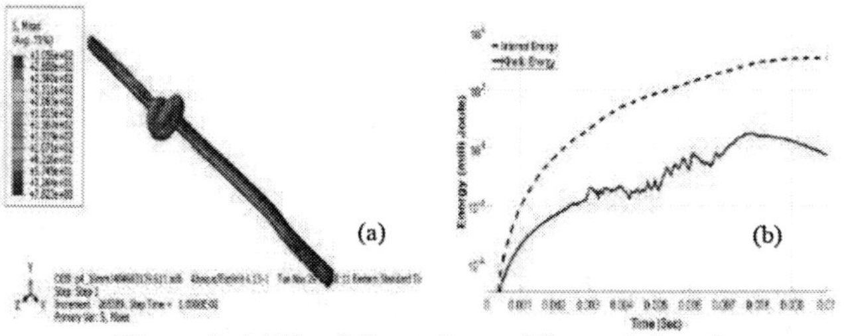

Figure 3: (a) Von Misses stress of the model, and (b) Internal energy and kinetic energy in logarithmic scale.

DISCUSSION

Several prior studies assess the mechanical properties of FiberWire suture. Most of these studies evaluate material properties of specific suture knots [1,7,10], or comparison of one type of suture versus another [4,5,8]. The current data was found to be consistent with these studies, however, no prior study reported the core and jacket separately.

The pulley fixtures ensured that the suture failure occurred between the two pulleys for all tested suture samples. Furthermore, there was no significant slippage observed in any of the samples.

The jacket is braided and thus undergoes a complex strain behavior. Under tension, the braided jacket lengthens and narrows while the angle between the warp and weft threads changes. Therefore, the circumference shrinks with increases in tension. Consequently, the jacket compresses on the core. In doing so, additional shear load is transferred between the core and the jacket. Therefore, the jacket was observed during experiments to "reload" the core even after core failure. This is a shear lag phenomenon familiar in long fiber composite materials. In one example, the core failed 3 times prior to jacket failure. Thus, core failure does not immediately lead to complete load removal. No load reversals were applied in the current experiments, however, load reversals would exhibit complex behavior that would likely decrease the load transfer to the core. The behavior of the suture is similar and can be understood when compared to the action of a "Chinese finger trap" (a common toy for children).

The FE analysis results of the knot are qualitatively correct and are based the experimental results. Thus, the FE model can be used in additional investigations. The experiments with knots are ongoing as well as further validation studies for the FE model of the knot.

The limitations of the current study are several. First, the experiments focused on the suture while the FE investigated the knot. Second, the FE model of the knot assumed a homogenous isotropic elastic material, however, the suture consists of two strand materials. Work is ongoing to overcome these limitations.

REFERENCES

[1] Heward, A et al., *Textile Research Journal*, 74(1), 83-90, 2004.
[2] Hockenberger, A et al., *IJFTR*, 29(3), 271–277, 2004.
[3] Carr, D et al., *The Journal of The Textile Institute*, 100(1), 51-56, 2009
[4] http://www.arthrexvetsystems.com/en/mediacenter/whitepapers.
[5] Najibi, S et al., *Iowa Orthop. J.*, 30, 84–88, 2010.
[6] Wright, P et al., *Arthroscopy*, 22(12), 1270–1275, 2006.
[7] Barber, F et al., *Arthroscopy*, 25(2), 192–199, 2009.
[8] Barber, F et al., *Arthroscopy*, 22(10), 1063–1069, 2006.
[9] Abbi, G et al., *Arthroscopy*, 22(1), 38–43, 2006.
[10] Ilahi O., et al., *Am J Sports Med*, 36, 2407–2414, 2008.

SB³C2017
Summer Biomechanics, Bioengineering and Biotransport Conference
June 21 – 24, Tucson, AZ, USA

EVALUATION OF METAPHYSEAL RECONSTRUCTIVE KNEE REVISION IMPLANT IMPACTION WITH SURGICAL CADAVERIC OPERATION

Gregg Schmidig (1), Mayur Thakore (1), Damon Servidio (1)

(1) Stryker Orthopaedics
Mahwah, NJ USA

INTRODUCTION

A bone defect may become evident during primary or revision orthopaedic knee implant surgery. A 3D printed metal device has been designed to replace the defect and to reconstruct the metaphyseal area in a cemented or cementless application. The component is placed within the bone to fill the gap from the defect and can be used with the tibial or femoral implant. The process of implanting this device has not been studied as much as primary orthopaedic devices [1, 2].

The purpose of this study was to use surgeon evaluation of the device in a cadaver setting to capture the impaction forces and speeds applied to help quantify the energy input to this device.

METHODS

A commercial pure Titanium material was utilized with a new manufacturing process for this particular design. The implant was sized to the cadaver bone as it would typically be performed for the standard orthopaedic implant. The reconstructive device was a truncated cone shape (Stryker Orthopaedics®, Mahwah, NJ, USA) that would fit into a matching cavity preparation in the bone, have cement placed in the interior, and a standard orthopaedic femur or tibia implanted, followed by final seating in the patient. The instrumentation for the system was utilized by experienced orthopaedic surgeons to implant the device into the bone without cement between the cone implant and bone.

Fourteen (14) surgeons, familiar with the product and proposed implantation technique, had their impaction force and speed recorded during implantation of the device into twenty (20) cadaveric bones. The cadaver bones were ranked for quality by the surgeons to use bones considered "good" candidates for this operation simulation. During this task, instrumented hammer data was recorded with a camera system to capture the impaction force and speed of the hammer swing.

The force data collection was completed with a custom fabricated instrumented hammer mounted with a 50KNewton range load washer (9031A, Kistler, Instrument Corp., Amherst, NY) and an acquisition rate of 100KHz. The hammer speed was recorded with a high speed camera system (Del Imaging Systems, LLC, Cheshire, CT, USA) and analyzed with ProAnalyst 3-D Professional motion analysis software (Xcitex, Inc., Cambridge, MA). The system has been validated for speed to an accuracy better than 98% [3].

The energy was calculated using the impulse-velocity method with the force-time curve impulse data and speed data from the video collection. The reported values were the maximum impaction force occurring at impact for each impaction swing, the maximum speed during the swing phase of the hammer that occurred immediately before impact, and the calculated energy of these two results.

RESULTS

The development of a new device requires various data collection tools to quantify the system's integrity. The collection of impact force is vital to understanding the energy applied to the instrumentation system and the product to minimize the potential for intraoperative fractures that others have seen in competitive products [4, 5]. The testing conducted, using good quality cadaver bone and experienced surgeons, did not result in any bone fracture during implantation. The data shown in Table 1 represents essentially equivalent averages for the tibia and femur on good quality bone at 11,035N and 11,701N, respectively for force. The speed and energy followed suit with a 3.8m/s and 2.4J for the tibia and 3.9m/s and 2.6J for the femur. This data demonstrates that the surgeons applied similar implantation techniques to the tibia and femur for these devices when considering the average of all the peak values. The median values are also very similar indicating the impact data is not skewed from the average. The use of surgeons helps capture the inherent variability between these users as noted by the higher standard deviation.

Average values of surgeon impactions into cadaver bones					
		Average max force, N	Average max speed, m/s	Average max energy, J	Average impacts
Tibia	Average	11035	3.82	2.36	12.1
	std dev	5565	1.88	1.55	5.7
	Median	11306	3.79	2.68	
Femur	Average	11701	3.86	2.62	12.4
	std dev	5386	1.20	1.58	2.1
	Median	12378	3.85	2.20	

Table 1: Data collection results from a cadaver lab. The table shows the product type, force, speed, and energy with standard deviation.

DISCUSSION

A reconstructive orthopaedic device may be a useful option for a surgeon when bone loss is evident during surgery. The study of the impact mechanics associated with the implantation of the devices may help guide mechanical testing of the instrumentation and fixation of the device. The use of multiple surgeons and cadaver specimens provides a clinical perspective of the products' use.

Significance-Surgeon testing of products with cadaver specimens assists in the evaluation of performance and simulates patient application. Surgeon impaction data values can assist manufacturers in designing testing. The knee revision implant testing provides support for performance and fit in a simulated patient application

ACKNOWLEDGEMENTS

N/A

REFERENCES

[1] Scholl, L et al "Comparison of Evaluation Methods to Determine Intraoperative Impaction Energy Associated with Total Hip Arthroplasty", EFORT 2013, EFORT 13-2514.

[2] Schmidig, G. et al, "Impaction Test Method Replicating Simulated Use for Cancellous Bone Punch Arthroplasty", EFORT 2013, EFORT 13-2805.

[3] Schmidig, G. et al, "Validation Of A Novel Technique To Measure Surgical Strike Speed Using A High Speed Camera System", EFORT 2014, EFORT 14-419.

[4] Villanueva-Martinez, M. et al, "Tantalum Cones in Revision Total Knee Arthroplasty. A Promising Short-Term Result with 29 Cones in 21 Patients", Journal of Arthroplasty, 28 (2013) 988-993.

[5] Derome, P. et al, "Treatment of Large Bone Defects with Trabecular Metal Cones in Revision Total Arthroplasty, Short Term Clinical and Radiographic Outcomes", Journal of Arthroplasty, 29 (2014) 122-126.

SB³C2017
Summer Biomechanics, Bioengineering and Biotransport Conference
June 21 – 24, Tucson, AZ, USA

PERFORMANCE TESTING OF TISSUE CONTAINMENT BAGS FOR POWER MORCELLATION

Alexander Herman(1), Nandini Duraiswamy(1), Thomas E. Claiborne(2), George J. Gibeily(3), Veronica A. Price(4), Prasanna Hariharan(1)

(1) Division of Applied Mechanics, Office of Science and Engineering Laboratories, Center for Devices and Radiological Health, US Food and Drug Administration, Silver Spring, MD, United States

(2) General Surgery Devices Branch II, Division of Surgical Devices, Office of Device Evaluation, Center for Devices and Radiological Health, US Food and Drug Administration, Silver Spring, MD, United States

(3) Plastic and Reconstructive Surgery Branch I, Division of Surgical Devices, Office of Device Evaluation, Center for Devices and Radiological Health, US Food and Drug Administration, Silver Spring, MD, United States

(4) Obstetrics and Gynecology Devices Branch, Division of Reproductive, Gastro-Renal and Urology Devices, Office of Device Evaluation, Center for Devices and Radiological Health, US Food and Drug Administration, Silver Spring, MD, United States

INTRODUCTION

Laparoscopic power morcellators are medical devices used to divide tissue into smaller fragments to facilitate removal using minimally invasive techniques that rely on small incisions. In some cases, tissue fragments can be left behind leading to significant complications. A recently published FDA safety communication [1] highlighted the risk to patients undergoing treatment of uterine fibroids with these devices during surgery to remove the uterus (hysterectomy) or remove uterine fibroids (myomectomy). Uterine fibroids are noncancerous growths; however, in some cases women undergoing treatment for fibroids may have an unsuspected uterine sarcoma, a type of uterine cancer. Use of laparoscopic power morcellators in these patients carries the risk of spreading cancer cells within the abdomen and pelvic region which may worsen survival. To minimize the risk of spreading cancerous cells, some investigators have recommended the use of tissue containment bags deployed inside the body to isolate the extirpated tissue and morcellator from surrounding tissue/organs [2]. However, they have been using commercially available specimen bags that are not indicated for this type of use. FDA has recently permitted marketing of the first tissue containment system for use with laparoscopic power morcellators (ASC PneumoLiner currently marketed in the U.S. by Olympus, Inc., DEN150028). Currently marketed tissue containment bags are made out of biocompatible polymers designed to prevent penetration of cellular material through the pores present in the membrane. Device manufacturers evaluate the performance of these devices using standard tensile, burst, dye penetration, and puncture tests. However, these tests are not adequate to evaluate the physiological forces experienced by the tissue containment bags during power morcellation. In addition to the force exerted on the bag as a result of the high speed rotation of the morcellator blade, investigators have advocated insufflating the bag with CO_2 gas to create a working space for the morcellator and a laparoscope for visualization. As a part of this study, we are developing new performance test methods that evaluate the propensity of tissue containment bags to leak when subjected to the forces imparted during a power morcellation procedure (including the insufflation pressure). We present preliminary results from our study, and hope that the performance test methodologies will help FDA reviewers and device manufacturers evaluate the safety and effectiveness of tissue containment bags when used with power morcellators.

METHODS

The integrity of different tissue containment bags were tested by subjecting new and pre-used sample bags to a series of dye and bacteriophage penetration tests, as well as tensile strength testing.

Test Bags

Legally marketed tissue containment bags from three different manufacturers were used for our preliminary testing. These bags were not indicated for use with the power morcellators. The three bags were anonymized and labeled as TC#1, TC#2, and TC#3 in this study. TC#1 was made of an ethylene-methacrylate copolymer. TC#2 and TC#3 were both made from a composite of nylon fibers overlaid with a polyurethane coating.

Dye penetration testing (at 2 psi)

American Society for Testing and Materials (ASTM) F1670 standard was used to test the penetration of artificial blood through the test bags when subjected to a standard upstream pressure of 2 psi. The test set-up is detailed in Figure 1. According to the standard, the artificial blood [3] was introduced to the dye cell and allowed to rest for 5 minutes. It was then subjected to a pressure of 2 psi for 1 minute, then vented and allowed to stand for another 54 minutes. The sample was considered to have passed the test if no artificial blood diffused to the receiver side of the sample. Three samples for each bag were tested to determine statistical significance.

Dye penetration testing (up to 50 psi)

If the tissue containment bag passed the ASTM F1670 dye penetration testing, the bag was subjected to higher pressure loads (up to 50 psi which is the limit for our experimental setup) for about 20 minutes while letting the artificial blood sit in the cell for 5 minutes before and after applying the pressure.

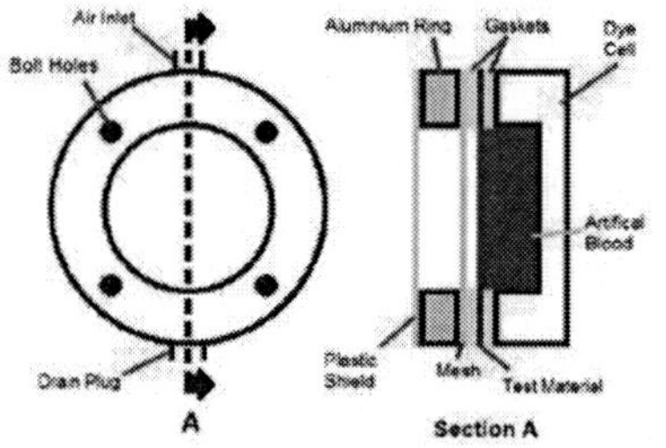

Figure 1: Diagram of the dye penetration setup

Tensile strength testing

Three ASTM test methods (D412, D882, D6324) were used to develop a tensile test protocol for the bag material. The ASTM D6324 test method is used for measuring the tensile properties of male polyurethane condoms which requires the containment of cellular pathogens similar to the tissue containment bags used during uterine fibroid power morcellation. A strain rate of 50 mm/min was used for uniaxially pulling our samples to failure using a uniaxial tensile test machine (Instron Tabletop). The test samples were cut according to ASTM D412. The average thickness of the three bags was measured to be 0.04 mm, 0.14 mm, and 0.13 mm, respectively.

RESULTS

Dye penetration testing

All three tissue containment bags (TC#1, TC#2, and TC#3) passed the 2 psi dye penetration test, and through a cotton swab test, no blood was detected on the outside surface of the sample.

Subsequently, the pressure acting on the bag was increased from 2 psi to 50 psi and the modified F1670 test was repeated again. Similar to the 2 psi tests, no blood was detected on the outside surface of the bags suggesting that the bags passed the high pressure test.

Tensile strength testing

Figure 2 shows the stress-strain curve for TC#3; note that the strain obtained here is calculated from the crosshead displacement. The yield and ultimate tensile strength (UTS) for this material was estimated to be 7.3±0.875 MPa and 63±5 MPa, respectively. The failure strain (@ UTS) for this material was 0.25±0.003.

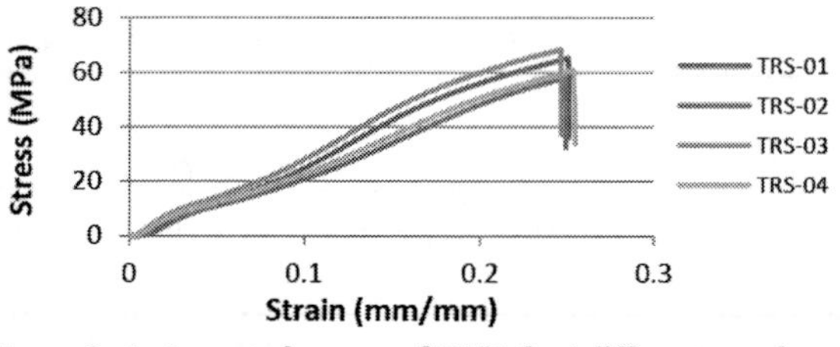

Figure 2: A stress strain curve of TS#3 for 4 different samples

Figure 3 compares the UTS for the three bag materials tested. Since the material construct was different for each of these bags, the UTS and the failure strain @ UTS varied widely between the bags. TC#1 was purely a homogenous polymer and hence could withstand large amount of strain (400%) before failing. TC#2 and TC#3 were a composite of both polymer and fabric, and had a lower ultimate strain but a higher ultimate strength value (75 MPa and 63 MPa, respectively).

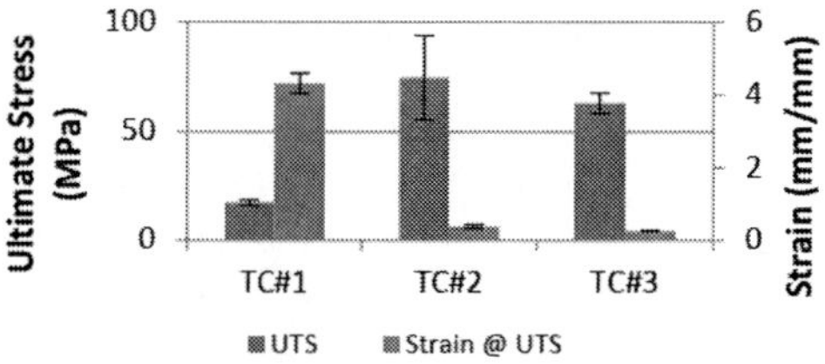

Figure 3: Bar plot of the UTS and the strain at the UTS.

DISCUSSION

As a part of this ongoing study, we presented the preliminary results of our efforts to develop test standards that can be used to evaluate the safety and effectiveness of tissue containment bags used with power morcellators. Failure of these devices may lead to spread of undetected cancerous cells within the body during hysterectomy or myomectomy procedures. We tested three different legally marketed tissue containment bag materials. This experiment enabled us to estimate the safety factor for each bag by comparing the threshold pressure at failure to the insufflation pressure exerted on the bag during the morcellation procedure. The standard pressure recommended by investigators for insufflating these bags in the abdominal cavity is between 10 and 15 mmHg [2]. We demonstrated that each material can withstand 166x times expected insufflation pressure and the pressure load was not high enough to cause penetration of blood analog fluid through the bag layers. Additional experiments are currently being planned to estimate the forces imparted by the power morcellator on the tissue bags.

Ongoing studies will examine potential failure modes for these bags when used with power morcellators. Our study will be expanded to include more types of tissue containment bags to represent the market variety. The results from this study may aid in the development of FDA guidance documents and new testing standards for pre-clinical testing of tissue containment bags used for power morcellation.

ACKNOWLEDGEMENTS

This work was funded by Office of Women's Health, Office of the Commissioner, US Food and Drug Administration and the Oak Ridge Institute for Science and Education (ORISE)

REFERENCES

[1] US Food and Drug Administration, 2014, FDA Safety Communication Laparoscopic Uterine Power Morcellation in Hysterectomy and Myomectomy http://www.fda.gov/MedicalDevices/Safety/AlertsandNotices/ucm424443.htm

[2] Cohen, S et al., J. Minimally Invasive Gynecology, 21(5):935-39, 2014

[3] Inter. Org. of Standardization, Standard # 16603:2004(E), Annex A Synthetic Blood Formula, 1(2004)

ELASTIC FIBER NETWORK STRUCTURE AFFECTS MASS TRANSPORT INTO THE ARTERIAL WALL

Austin J. Cocciolone (1), Jessica E. Wagenseil (2)

(1) Biomedical Engineering
Washington University
St. Louis, MO, U.S.A

(2) Mechanical Engineering and Materials Science
Washington University
St. Louis, MO, U.S.A

INTRODUCTION

The native structure of the arterial wall directly affects mass transport through the wall. The endothelial cell (EC) layer operates as a barrier to transport for biological macromolecules and drugs. The media is usually considered homogeneous to mass transport and is often modeled as a porous material where the porosity is the space between smooth muscle cells (SMCs). Several studies indicate, however, that the wall is non-homogeneous to transport.[1,2] In the present study we aim to test whether the elastic fiber network within the media is responsible for the non-homogenous mass transport. We have a mouse deficient in fibulin 5 (*Fbln5-/-*), a protein critical to the production of mature elastic fibers, in which the large arteries have fragmented elastic lamellae within the media (Fig. 1).[3] Additionally, we have developed a method for testing permeability *ex vivo* with the artery stretched and pressurized in an *in vivo* conformation. Through a series of permeability experiments we have characterized mass transport in the wild-type (WT) carotid artery in the *in vivo* conformation. We have also demonstrated a 2-fold decrease in the uptake of 70 kDa FITC-dextran in *Fbln5-/-* arteries compared to WT. We are continuing to characterize the transport properties in arteries with fragmented elastic fibers.

METHODS

WT and *Fbln5-/-* mice were sacrificed at 3-4 months of age. All animal protocols were approved by the IACUC. Left and right carotid arteries were dissected, mounted on a pressure myograph, and submerged in phosphate-buffered saline (PBS). The lumen was flushed with 10 ml of air to remove the EC layer. The unloaded length of the artery was determined visually and recorded. 2.5 mg/mL FITC-dextran solution of different molecular weights from 4 – 70 kDa was introduced into the lumen of the artery. A hydrostatic pressure column was generated to pressurize the artery to 100 mmHg. The artery was then stretched longitudinally to 1.4 times the unloaded length, which is considered the *in vivo* stretched conformation.[5] The loaded length and diameter were recorded. The testing apparatus was protected from light to prevent bleaching of the FITC-dextran solution.

After the desired incubation time the artery was flushed once with PBS. The artery section exposed to the FITC-dextran solution was removed and the artery was opened with a lateral cut. A series of three elution steps were performed by submerging the artery section in a known volume of fresh PBS and incubating it for 24 hours at 37°C with gentle shaking. Fluorescence from each elution fraction was measured and summed to obtain the total amount of FITC-dextran that moved into the artery wall during the experiment. This FITC-dextran concentration was normalized to the stretched length of the artery.

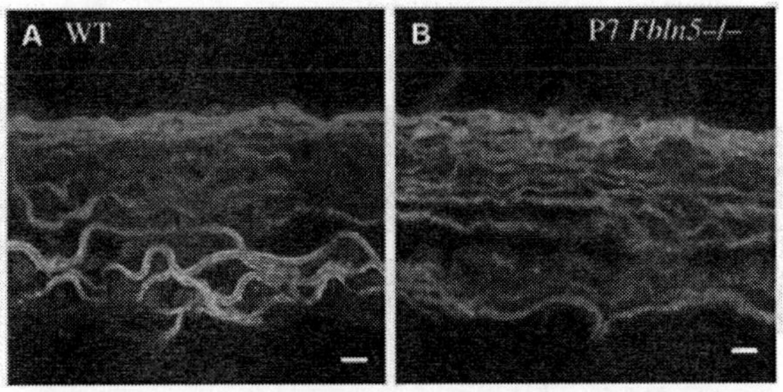

Figure 1. Arterial wall fluorescent staining of elastin (red), collagen (green), and SMC nuclei (blue) in WT (A) and *Fbln5-/-* (B) mice. Note the spaces within the elastin lamellae in *Fbln5-/-* arteries that are filled in with collagen.[4]

RESULTS

We performed a series of experiments investigating mass transfer into the arterial wall with our novel experimental set-up. Our method involves testing arterial wall permeability in the *in vivo* structural conformation by pressurizing and longitudinally stretching the arteries.

Comparing the uptake of FITC-dextran of different sizes into WT carotid arteries shows that larger molecules trend toward reduced permeability through the wall (Fig. 2). This result is as expected and validates our experimental method.

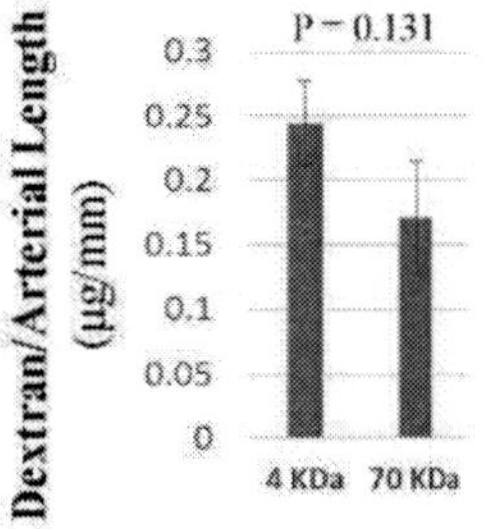

Figure 2. Total uptake of 4 and 70 kDa FITC-dextran into WT carotid arteries after 4 hours.

Further, we show that with increasing time there is increased FITC-dextran uptake into the arterial wall (Fig. 3). Unexpectedly, even after 8 hours the FITC-dextran uptake into the wall has not reached a saturated level, whereas several results from the literature indicate that saturation should occur after a few hours.[6] It is possible that our set-up in the native structural conformation is responsible for this difference. Further testing is in progress to confirm the time course of FITC-dextran uptake in the WT carotid artery.

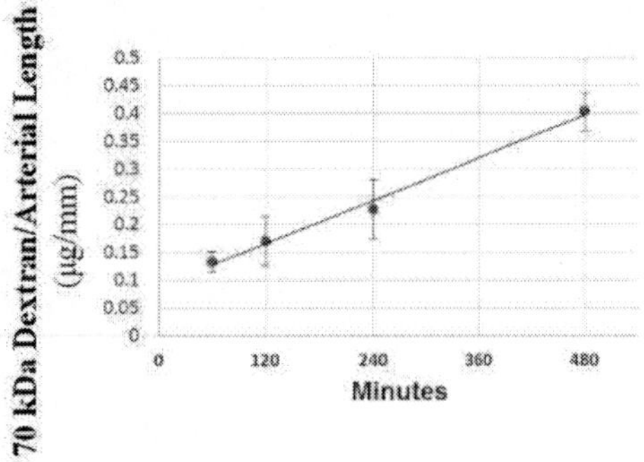

Figure 3. Total uptake of 70 kDa FITC-Dextran into WT carotid arteries at different time points.

We have also demonstrated the capacity of our testing method to detect differences in mass transfer between different artery types. The wall of the carotid artery is less permeable than the mesenteric artery (Fig. 4A). It is possible that the reduced permeability of the carotid compared to the mesenteric artery wall is a result of the difference in medial ultrastructure. Specifically, the carotid has more elastic lamellae than the mesenteric and so it is likely the elastic fiber network is important for determining mass transport in the arterial wall.

To further test the importance of the elastic fiber network to arterial wall permeability, we compared FITC-dextran uptake in WT and *Fbln5-/-* carotid arteries (Fig. 4B). Since *Fbln5-/-* arteries have a disrupted elastic fiber network, any difference in permeability may be a result of this disruption. Our results indicate a reduced uptake of dextran in the *Fbln5-/-* artery compared to WT after 2 and 4 hours. This difference indicates that the elastic fiber network is important for determining mass transfer properties of the arterial wall.

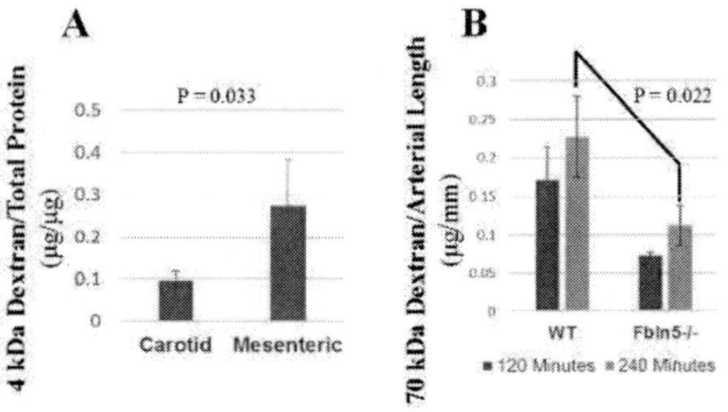

Figure 4. A) Total uptake of 4 kDa FITC-dextran into WT carotid and mesenteric arteries. B) Total uptake of 70 kDa FITC-dextran into WT and *Fbln5-/-* carotid arteries after 2 and 4 hours.

DISCUSSION

Our goal was to design and test methods to investigate how arterial wall structure determines mass transport properties. We developed a novel experimental set-up to investigate transport with the artery in its *in vivo* configuration. We confirmed that transport depends on molecular size and incubation time. We also found differences in transport depending on wall structure. We found increased uptake in mesenteric arteries that have fewer layers of elastic lamellae compared to carotid arteries and decreased uptake in *Fbln5-/-* arteries that have fragmented elastic lamellae compared to WT. We are continuing this study by performing permeability experiments with various molecular sizes and incubation times in WT and *Fbln5-/-* arteries. The results will allow us to characterize how fragmented elastic fibers changes mass transport properties of the arterial wall. The results could have implications for how arteries with fragmented elastic lamellae due to aging or disease handle transport of signaling molecules important for communication of different cell types, cholesterol and lipids in atherosclerosis development, and drug dispersion from drug-eluting stents.

ACKNOWLEDGEMENTS

NIH HL-115560 and HL-105314

REFERENCES

[1] Hwang, C. et al., *Circulation Research,* 90(7):826-832, 2002.

[2] Tada, S. et al., *Am. J. Physiol. Heart Circ. Physiol.*, 287(H):905-913, 2004.

[3] Le, V.P. et al., *J. R. Soc. Interface*, 12:20141250, 2015.

[4] Borel, A. et al., *J. Biol. Chem.*, 284:24553-67, 2009.

[5] Cardomone, L. et al., *Biomech Model Mechanobiol.*, 8(6):431-446, 2009.

[6] Sasaki, K. et al., *J of Surgical Research*, 43:565-570, 1987.

EFFECT OF VASCULAR HETEROGENEITY ON FLUID FLOW AND TRANSPORT IN TUMORS

M. Alamer (1), X.Y. Xu (2)

(1) Department of Chemical Engineering
Imperial College London
London, United Kingdom

(2) Department of Chemical Engineering
Imperial College London
London, United Kingdom

INTRODUCTION

Cancer remains one of the world's leading causes of mortality and morbidity. Whilst much work has focused on the molecular mechanisms of drug resistance at cellular level, numerous studies have shown that the in solid tumors, the physiology and microenvironment limits the ability of anticancer drugs to transport and distribute uniformly and hence they're unable to reach some cancer cells at lethal concentrations. Aberrant tumor vasculature, dense extracellular matrix (ECM), high interstitial fluid pressure (IFP) and hypoxia are all common properties of tumors that contribute to growth and therapy resistance. Angiogenesis taking place as the tumor grows creates vessels that are non-uniformly distributed featuring excessive loops and branching, with a lack of vessel hierarchy and no clear distinction between arterioles, venules and capillaries. Perfusion in tumors is poorer as a result, which can limit the supply of nutrients, oxygen and any drug molecules. In addition to disorganized structure, individual vessels tend to be immature lacking in pericyte and basement membrane coverage and employing defective endothelial cells thus causing vessels to exhibit higher permeability. Consequently, fluid leakage from vessels in tumors occurs at a high rate raising the pressure in the interstitium. High IFP can prevent therapeutic macromolecules from penetrating through the vessel walls into the interstitium. Additionally, drugs able to penetrate the vessel walls will be limited in their ability to reach all cancer cells due to increased ECM density and steep pressure gradient formed in the interstitum[1]. The tumor microenvironment is incredibly complex and dynamic where it varies spatially and temporally, from patient to patient with properties that are tightly coupled in which the alteration of one property can influence the other. Computational and mathematical modelling methods can allow for a wide range of parameters to be tested and varied simultaneously in a way that is not viable through experimental means. These methods may hold some promise in understanding the complex interplay between the intrinsic properties of tumors and the influence of these properties on drug uptake and distribution within tumor tissue. Several studies have attempted to investigate the transport of fluid and drugs in tumors using computational modelling, however the models were usually oversimplified where the heterogeneous morphological properties of the vessels were rarely represented explicitly. Here we attempt to investigate the influence of different properties of the tumor vasculature on fluid flow and pressure distribution by coupling a tumor angiogenesis model with a fluid mechanics model.

METHODS

An angiogenesis model based on a mathematical model for tumor induced angiogenesis developed by Anderson and Chaplain is used to construct the tumor vasculature and geometry. The model describes vascular geometry growth as function of endothelial cell migrate from the parent vessel through random motility, chemotaxis in response to tumor angiogenic factors (TAFs) and haptotaxis in response to fibronectin [2]. Matlab was used in this study to implement the angiogenesis and fluid flow model. A 3D simulation region for a tumor 268 mm^3 in volume was used and discretized into a mesh of grid points in each orthogonal direction. Governing PDEs in Anderson and Chaplains work were solved and discretized to obtain spatial and temporal distributions of endothelial cells, TAF and fibronectin. Given this vessel networks could be generated. The interstitum is modelled a homogenous porous medium composed of uniformly distributed ECM and cancer cells.

Pozrikidis, developed a mathematical model for fluid flow at a single capillary level coupling capillary and interstitial flow for highly permeable vascular walls considering the discrete nature of the vasculature [3]. Vascular, transvascular and interstitial flow are coupled and described by Poiseuilles, starlings and Darcy's equation respectively. Greens function formulation is used to describe fluid flux as a set of source points along the vessel coupling vascular and interstitial pressure. This allows for interstitial pressure to be represented as function of vascular distribution and capillary pressures. Pozrikidis's method was coupled with the angiogenesis model to compute flow and pressure distributions within vascular networks generated. To validate the model morphological and hemodynamic parameters were calculated and compared to literature. A sensitivity analysis is performed to understand how different parameters influence interstitial pressure.

RESULTS

The tumor geometry and vasculature are shown for one case in figure 1. Morphological and hemodynamic parameters calculated showed good agreement with those found in the literature as seen in table 1. Interstitial pressure was calculated for different tumor geometries and different parameter values. Figure 2 shows the average interstitial pressure across the length of the tumor for different vessel permeability with interstitial pressure being highest at the center of the tumor and lowest at the periphery. Simulations performed with varying interstitial hydraulic conductivity showed a significant pressure drop when conductivity was increased. Analysis of the effect of vascular density showed a trend of increasing overall interstitial pressure as the vascular density increased. The influence of necrosis on overall pressure was investigated and its effect was found to be negligible.

Table 1: Morphological and hemodynamic parameters for generated vascular network model and literature data

Property	Model	Literature [4]
Vascular density (%)	0.23	0.15-1.25
Mean diameter (micrometers)	21.1	5-225
Mean length (mm)	0.277	0.06-0.3
Maximum extravascular diffusion distance *R*(micrometers)	242	30-250
Mean Volumetric flowrate (m^3 s^{-1})	2.39×10^{-11}	$<3\times10^{-8}$
Mean velocity $\times10^{-4}$ (m s^{-1})	85.21	1-250
Mean shear stress (pa)	5.98	1-10

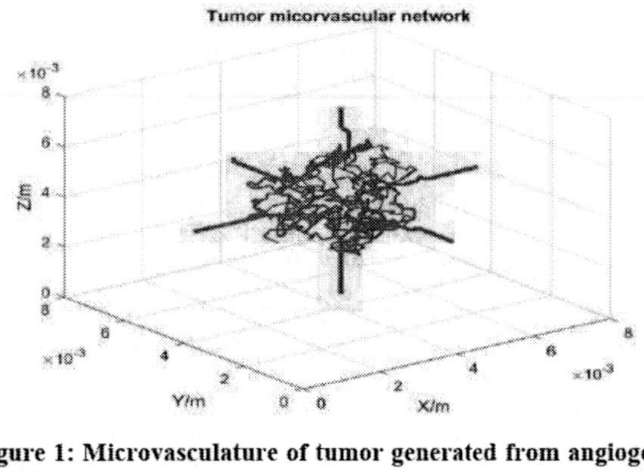

Figure 1: Microvasculature of tumor generated from angiogenesis model

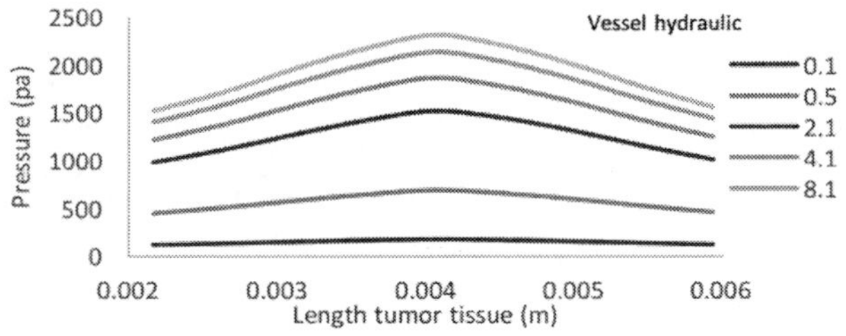

Figure 2: Interstitial pressure along the Length of the tumor tissue for different vessel hydraulic conductivities

DISCUSSION

The model predicted elevated interstitial pressure values that fit well with experimentally determine values (1000-4000pa) [1]. The effect of lowering vessel permeability on interstitial pressure highlights the need normalize the vessels. Spatial distributions of interstitial pressure within the tumor (Fig 3) showed pressure spikes in some regions and can be attributed to the effect of the discrete nature of the vasculature included in this model. The will limit the ability of drugs in reaching cancer cells in this region and hence they can escape treatment cycles. Additionally, based on the results from this model the assumption of uniformly elevated interstitial pressure that has been made in other simplified tumor models does not appear to be completely valid. Incorporating drug transport into this model is the next step which can show how different drugs, identified by their molecular weight, distribute within the tissue and what parameters influence drug distribution most and what tumor properties to target to enhance distribution. Work on applying the fluid flow model to realistic tumor vasculature images to obtain clinical results is currently ongoing.

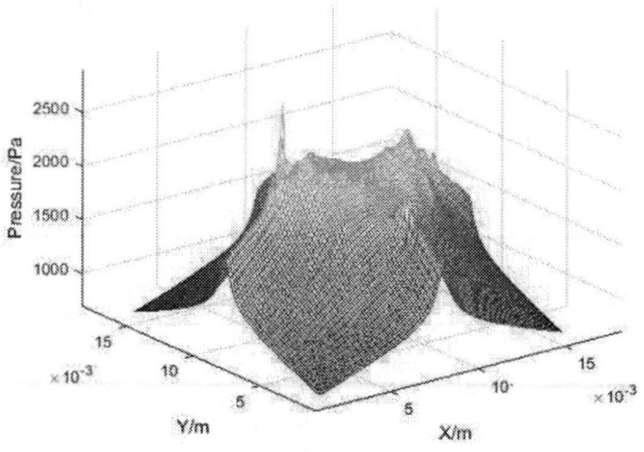

Figure 3: Spatial distribution of interstitial pressure at mid height cut section of the tumor

REFERENCES

[1] Heldin, C-H. et al., *Nature Reviews Cancer,* 4:806-813, 2004
[2] Anderson, A.R.A et al., *Bulletin of Math. Biol.,* 60:857-900, 1998
[3] Pozrikidis, C., *J. Math. Biol.,* 60:75-94, 2010.
[4] Stamatelos, S.K et al., *Microvascular Research,* 91:8-21,2014

SB[3]C2017
Summer Biomechanics, Bioengineering and Biotransport Conference
June 21 – 24, Tucson, AZ, USA

A PARALLEL FLUID SOLID COUPLING TOOL WITH APPLICATIONS IN PARTICLE TRANSPORT IN BLOOD CELL SUSPENSIONS

Jifu Tan (1,2), Talid Sinno (1), Scott Diamond (1,2)

(1) Department of Chemical & Biomolecular engineering
University of Pennsylvania
Philadelphia, PA, USA

(2) Institute for Medicine & Engineering
University of Pennsylvania
Philadelphia, PA, USA

INTRODUCTION

Particle transport in blood flow has many applications in biology and medicine. For example, in drug carrier design, size and shape has been found to influence their transport in blood. Long flexible filaments persisted in the circulation up to one week after the intravenous injection in rodents. This is about ten times longer than their spherical counterparts [1]. Lateral migration was observed for platelets in blood cell suspensions [2,3], as the blood cells tend to migrate toward the center of the blood vessel. In these applications, the hydrodynamics and the deformable cells played a major role in influencing the motion of the small particles. To model such coupled problems, it requires a computational tool that can handle both fluid flow and large solid deformation, as these motions are coupled together. In this paper, we present a partitioned approach to solve the coupled Multiphysics problem. The fluid field is solved by the Lattice Boltzmann method [4] using open source code Palabos (parallel lattice Boltzmann solver), while the solid deformation is simulated by Lammps (Large-scale Atomic/Molecular Massively Parallel Simulator) [5]. The coupling is achieved through the immersed boundary method so that different physics can exchange information with each other. The reason to use these open source codes are that both codes are highly parallelized, with a lot of documentations and a large online support community. Open source codes also encourage the numerical model sharing and verification of the computational results.

METHODS

The Lattice Boltzmann method (LBM) is a mesoscale approach to modeling fluid dynamics that has been used extensively in blood flow modeling [6–8]. LBM is usually considered as a second order accurate method in space and time [9]. Reviews of the underlying theory for the LBM can be found in the literature [4,10]. Here we use the open source Palabos as the fluid solver due to its parallel performance. In blood flow modeling, a strain energy based continuum cell model is usually used [11]. Recently, a particle based cell model has emerged as a popular alternative due to its simplicity in mathematical description [12,13], which will be adopted here. The potential energy of the cell membrane includes stretching (U_s), bending (U_b), and area/volume conservation terms (U_A and U_V). The stretching and bending energy are:

$$U_s = \sum_j \left[\frac{k_B T l_m}{4p}\frac{3x_j^2-2x_j^3}{1-x_j} + \frac{k_p}{l_j}\right],\quad U_b = \sum_j k_b(1-\cos(\theta_j-\theta_0)) \qquad (1)$$

where l_j, l_m are the instantaneous and maximum bond length, p is the persistence length, $k_B T$ is the thermal energy, k_p is the repulsive coefficient, $x_j = l_j/l_m$, k_b is the bending constant, θ_j, θ_0 are the instantaneous and maximum angle formed by two triangle surfaces.

The area and volume conservation energy is given by

$$U_A = \frac{k_g(A-A_0)^2}{2A_0} + \sum_j \frac{k_l(A_j-A_{j0})^2}{2A_{j0}},\quad U_V = \frac{k_v(V-V_0)^2}{2V_0} \qquad (2)$$

where k_g, k_l are global and local area constant, k_v is the volume potential constant, A is the area, V is the volume, subscript 0 means their equilibrium value. Interested readers can refer [12] to for more details. All these potential energy was implemented through extension of Lammps.

The Immersed Boundary method (IBM) was selected to model the interaction between the fluid and the cell due to the algorithm's efficiency [14,15]. The coupling scheme enforces velocity continuity at the fluid cell boundary, and transfers forces from the cell back into the fluid through an effective force density, that is

$$\boldsymbol{f}(\boldsymbol{x},t) = \int \boldsymbol{F}(s,t)\,\varphi(\boldsymbol{x}-\boldsymbol{X}(s,t))ds \qquad (3)$$

$$\boldsymbol{u}(\boldsymbol{X},t) = \int \boldsymbol{u}(\boldsymbol{x},t)\,\varphi(\boldsymbol{x}-\boldsymbol{X}(s,t))d\boldsymbol{x} \qquad (4)$$

Technical Presentation #279

where $\boldsymbol{f}(\boldsymbol{x},t)$ is the fluid force density, $\boldsymbol{F}(s,t)$ is the solid force, $\boldsymbol{u}$ is the velocity, $\mathbf{X}(s,t)$ is the parametric cell membrane description. $\boldsymbol{x}$ is the fluid Cartesian coordinates. $\varphi(\boldsymbol{x}) = \delta(x)\delta(y)\delta(z)$ is the interopolation function in 3D Cartesian coordinate system with $\delta(r)$ as a zero function except when $-2 \leq r \leq 2$, $\delta(r) = (1 + \cos(\pi r/2))/4$.

In our implementation, IBM is realized as an interface code where Palabos and Lammps can exchange information. Palabos is the driving code calling LAMMPS as an external library. The same processor grid is used to partition both the solid and fluid domain, an example of simulation grid layout running with 16 processors is shown Figure 1(a).

RESULTS

I. Performance of the coupling code

In the performance test, the fluid domain consists of rectangular box ($50 \times 50 \times 4096$) with 6 nonslip boundary walls, where it has 204 red blood cells, consisting of 130968 particles forming 391680 bonds, 261120 angles, and 391680 dihedrals. In the test, each computer has 2xAMD Opteron 8-core 6128 processor with 2.0 GHz. The CPU time for the simulation running up to 100 time steps were recorded. All the processors are equally distributed in the axis aligned with 4096 lattices to achieve the best performance. The simulation time, averaged over 3 independent runs, for the fluid solid coupling, the fluid solver (Palabos) and the solid solver (Lammps) were plotted in Figure 1(b). It shows an almost linear relation between the computational time and the number of processors. The computation time decreases from 911.4 s to 74.0 s as the number of processor increases from 1 to 16 processors. The same linear relation were observed for both Palabos and Lammps, which demonstrates that both packages are well designed for parallel computing.

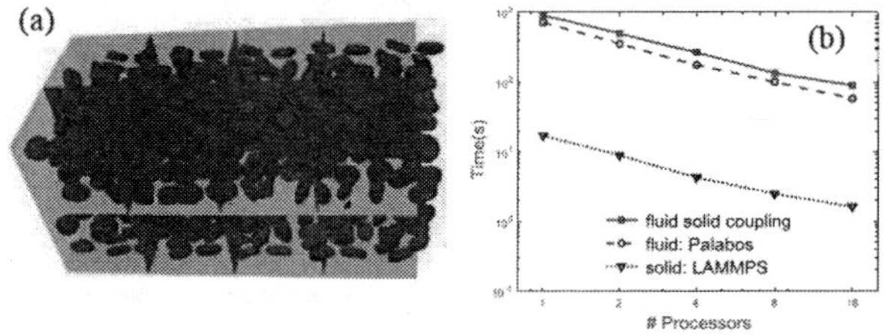

Figure 1. (a) A 16 processor grid layout example for blood cells transport in a rectangular box. (b) Comparison of the performance between the fluid solid coupling, the fluid solver (Palabos) and the solid solver (Lammps).

II. Rod particle transport in blood flow

To investigate the rod particle transport in blood flow, a blood vessel model with a diameter of 30um and length of 50 um was created. 100 cells were put in the vessel, resulting a hematocrit of 25%. 156 rod particles were modeled as polymer chains with relatively large bending stiffness. The cell is a biconcave shape with a diameter of 8 um, with shear modulus 6 uN/m, bending stiffness 70 k_BT. The rod is 2 um in length. Initially cells were randomly distributed in the vessel, while the rods were uniformed placed on the cross section. A pressure gradient was applied to drive the flow. The shear rate generated by the pressure gradient at the wall is 1000/s. The actual wall shear rate is less than that due to the cells and rods. A snapshot of the simulation was shown in Figure 2(a). An excess of rods were aggregated near the vessel wall, as shown in Figure 2(b). To quantify the aggregation of the rods, the averaged rod center position in radial direction <*r*> was plotted with time, as shown in Figure 3(a). The mean radial position increases very fast and then reached to an equilibrium positon. The fraction of the rods within 3 um cell free layer is also shown in the figure. The tread is consistent with the mean radial position. The mean square displacement (MSD) is also shown in Figure 3(b). The dispersion coefficient is $8.76 \times 10^{-11} m^2/s$, which is much faster than thermal diffusion coefficient $1.11 \times 10^{-13} m^2/s$.

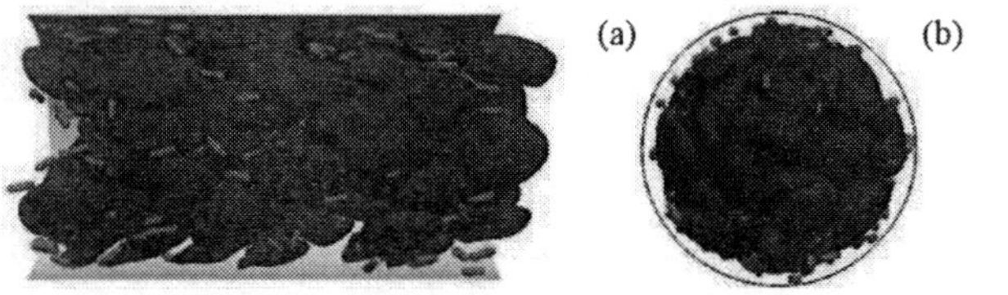

Figure 2. (a) The snapshot of the simulation with rod particles (green) mixing with blood cells (red). (b) The cross section view of the simulation showing rods aggregate near the vessel wall.

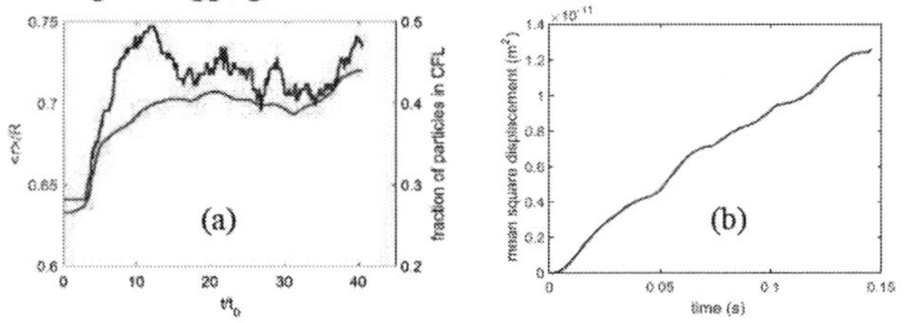

Figure 3. (a) the time history of the average center of mass for the rods <*r*> normalized with vessel diameter R, and the time history of the number of rods in the cell free layer (CFL) defined as 3 um distance within the vessel wall. The time is normalized with t_0 within which the average flow rate would travel distance R. (b) The time history of the mean square displacement of the rods in the blood flow.

DISCUSSION

We showed that the IBM coupling with Palabos and Lammps can achieve high performance, demonstrating almost linear scaling with 16 processors used. An excess of aggregation of rods was observed near the vessel wall, which is consistent with observations for platelets in literature [2,16]. In summary, the paper presented a fluid solid coupling tool based on widely used open source code Palabos and Lammps. It is useful in a lot of applications where both hydrodynamics and large solid deformation are important, for example, white blood cell migration, cancer cell capture, drug delivery and binding, blood clotting and bleeding.

REFERENCES

[1] Y. Geng, et. al., Nat. Nanotechnol. **2**, 249 (2007).
[2] L. Crowl et.al., Int. J. Numer. Method. Biomed. Eng. **26**, 471 (2010).
[3] K. Vahidkhah, et.al. J. Biomech. Eng. **135**, 51002 (2013).
[4] S. Chen and G. D. Doolen, Annu. Rev. Fluid Mech. **30**, 329 (1998).
[5] S. Plimpton, J. Comput. Phys. **117**, 1 (1995).
[6] J. Wu and C. K. Aidun, Int. J. Numer. Methods Fluids n/a (2009).
[7] T. Krüger, et.al., Soft Matter **9**, 9008 (2013).
[8] J. Tan, et.al., Nanomaterials **6**, 30 (2016).
[9] T. Inamuro, et.al., http://dx.doi.org/10.1063/1.869426 (1998).
[10] S. Succi, *The Lattice Boltzmann Equation for Fluid Dynamics and beyond* (Clarendon Press, 2001).
[11] R. Skalak, et.al., Biophys. J. **13**, 245 (1973).
[12] D. A. Fedosov, et.al., Biophys. J. **98**, 2215 (2010).
[13] M. Dao, J. Li, and S. Suresh, Mater. Sci. Eng. C **26**, 1232 (2006).
[14] C. S. Peskin, Acta Numer. **11**, 479 (2002).
[15] R. Mittal and G. Iaccarino, Annu. Rev. Fluid Mech. **37**, 239 (2005).
[16] T. Krüger, Rheol. Acta **55**, 511 (2016).

'GRAYSCALE' LITHOGRAPHY TO CREATE 3-D CHANNELS: APPLICATION TO HIGH SHEAR THROMBOSIS ASSAYS

Michael T. Griffin (1,2), David N. Ku (1,2)

(1) G.W. Woodruff School of Mechanical Engineering
Georgia Institute of Technology
Atlanta, GA, USA

(2) P.H. Petit Institute for Bioengineering & Biosciences
Georgia Institute of Technology
Atlanta, GA, USA

INTRODUCTION

High shear thrombosis assays are utilized in both clinical and research settings to analyze and assess platelet function and thrombus formation characteristics [1]. However, these assays are susceptible to large amounts of variability due to factors such as individual blood biochemistry, anti-coagulation methods, experimental setup, and device channel geometry. It is worthwhile to reduce all sources of variability in order to acquire the most accurate metric that is desired in the assay. While all four areas require attention, this abstract details a method by which to reduce variability due to the geometry of the device.

Previous microfluidic applications in the field of high shear thrombosis have largely utilized standard 2-dimensional (2D) lithography approaches in device fabrication [2]. However, 2D lithography, by definition, is limited to geometric variations in the x- and y-directions of the wafer substrate. The height (z-direction) is therefore conserved at typically 100 μm or less in both the nominal and constricted or stenotic test sections of the flow channel [3-5]. This leads to a drastic increase in total resistance along the flow path, making it difficult to achieve physiological shear rates (500-1000 s^{-1} in the nominal and > 5000 s^{-1} in the test sections) under constant pressure due to gravity driven flow. Therefore, these assays must make use of pumps to drive flow, which can lead to large undesirable increases in pressure within the flow channels as an occlusive thrombus forms.

3-dimensional (3D) lithography approaches have the potential to alleviate the issues presented within standard methods of high shear platelet assays. The introduction of 'grayscale lithography' provided a drastic change in microfabrication technology; however, it remains an expensive option that is limited mainly to industry [6]. A few simulated grayscale approaches have been devised, such as with the use of a liquid photomask with axial variations in dye concentration along the channel, thus altering the amount of UV exposure to the substrate [7]. However, the control of dye concentration along the channel is not highly regulated and can thus lead to variations arising from channel to channel. CNC machining is an alternative for 3D geometries, but is not as desirable as lithography in terms of surface roughness and geometric repeatability [8]. An automated lithography practice is therefore preferred in the development of 3D geometries in microfluidics [6]. This abstract introduces a method for the creation of a microfluidic chip for physiologically relevant high shear thrombosis assays through the use of an automated 'grayscale' 3D lithography approach.

METHODS

Positive photoresist Microposit SC 1827 (MicroChem, USA) was uniformly coated on a 100 mm silicon wafer with an SCS G3 Spin Coater (SCS Equipment, USA). In order to develop a defined slope gradient on a silicon wafer, a simulated grayscale approach was taken with the use of an optical ultra-violet method. This method was specifically designed for this application through the characterization of parametric effects on photoresist thickness, based on measurements made with a Dektak 150 contact profilometer (Veeco, USA).

Appropriate parameters were selected from the parametric investigation and were utilized to create the microfluidic mold shown in Figure 1, based on a similar design by Li et al [8]. The converging, stenotic, and diverging regions of this geometry were created with the novel method, while the remaining device geometry was created with a secondary standard soft lithography step via mask aligning. Trimethylchlorosilane was utilized to create a hydrophobic surface on the wafer prior to pouring polydimethylsiloxane (PDMS) to create usable microfluidic chips. After inlet and outlet holes were punched in the PDMS, the chips were plasma bonded to 25x75 mm glass slides for use in flow studies.

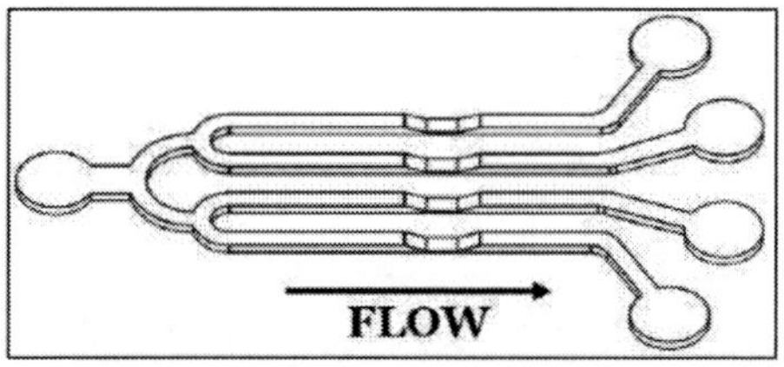

Figure 1: 3D lithography geometry designed for high shear thrombosis assays.

RESULTS

First, the parametric analysis showed that the photoresist thickness begins to converge to 0 μm thickness as the gain of the optical method is increased. In addition, nominal photoresist thickness converges to 4.5 μm with a decreasing bias parameter. The gain and bias difference were adjusted to ensure a full change from nominal to zero photoresist thickness. A parametric analysis under these settings was performed with varying intensities, and is detailed in Figure 2. While 45, 47, and 50% mostly cleared the photoresist, some noticeable roughness was observed at 45%, so a secondary analysis was selected.

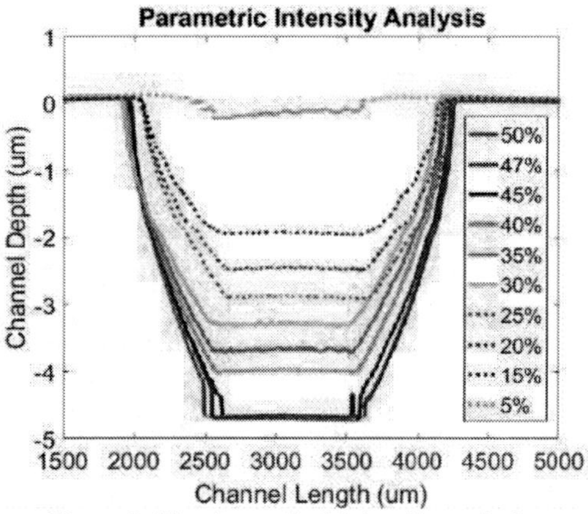

Figure 2: Varying optical intensity analysis.

A full range analysis of the grayscale effects with the selected gain and bias settings, shown in Figure 3, indicated that a 2.0% exposure was necessary to initiate a noticeable change in photoresist thickness. The thickness continued to decrease until an intensity value of 50%, which is when all of the photoresist was completely removed from the wafer. Measurements beyond 50% in Figure 3 suggests that there is still photoresist on the wafer; however, this was due to instrument error at the step-up the end of the channel. Visual inspection of this region confirmed that no photoresist remained on the wafer beyond 50% laser intensity.

The 3D device from Figure 1 was created using the newly developed optical method. The relevant geometric measurements are detailed in Table 1. These values represent averages over the 4 channels within the device. In a similar design with the same stenotic test section parameters, with the exception of respective target angles of 20 and 15 degrees for the grayscale and CNC approaches, a CNC machined device was measured for the same dimensions. The CNC variation in stenotic height was about 3 compared to 0.1 μm for the grayscale. Through the use of an F-test to determine the statistical relationship between CNC machining and 3D 'grayscale' lithography, the proposed 'grayscale' method was found to significantly decrease channel to channel height variability (p-value < 0.001). No statistical difference was found between the variance of the angle measurements of the two groups.

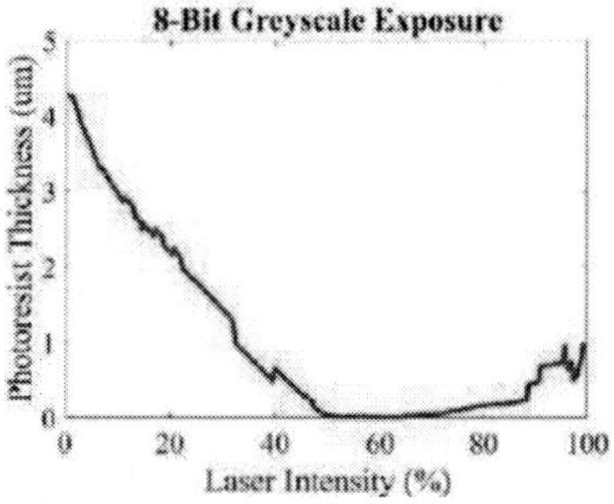

Figure 3: Full gain-bias range impact on photoresist thickness.

Table 1: Intra-device 3D profilometer measurements for grayscale (n=4) and CNC (n=8) methods. *p<0.001

Method	Stenotic Height (μm)	Nominal Height (μm)	Angle (degrees)
Grayscale	73.7 ± 0.1	256.0 ± 0.4	20.0 ± 0.6
CNC	82.4 ± 2.9*	265.1 ± 3.4*	13.5 ± 0.8

DISCUSSION

This newly developed method provides a means to fabricate complex 3D microfluidic geometries. 3D changes in geometry, such as the 20 degree converging and diverging regions within this proposed design, are now made possible via lithography. Controlled geometries in all three of the x-, y-, and z-planes will allow researchers to better design devices with more physiologically relevant flow patterns.

In addition, it was found the proposed 3D lithography method was significantly more repeatable in geometry than 3D CNC machining. By reducing the variability associated with the geometry of the device, overall variability associated with thrombosis assays will decrease. While the application presented here is for the creation of high shear rate regions in microchannels for occlusive thrombus studies, this may be seen as a general microfluidic technique that may extend to many other future applications.

ACKNOWLEDGEMENTS

The authors would like to acknowledge the work of staff engineers from the Georgia Institute of Technology's Institute for Electronics and Nanotechnology, especially that of Mr. Ben Hollerbach. Funding was provided by L.P. Huang Chair Professorship at Georgia Tech.

REFERENCES

[1] Paniccia, R et al., *Vasc Health Risk Manag*, 11:133-148, 2015.
[2] Branchford, B et al., *Thrombosis Res*, 136:13-19, 2015.
[3] Hosokawa, K et al., *Microvascular Res*, 83:154-161, 2012.
[4] Jain, A et al., *Nature Comm*, 7:1-10, 2016.
[5] Colace, T et al., *Arterioscler Thromb Vasc Biol*, 33:105-113, 2013.
[6] Guo, C et al., *Optics Express*, 17:19981-19987, 2009.
[7] Chen, C et al., *PNAS*, 100:1499-1504, 2003.
[8] Li, M et al., *Lab Chip*, 12:1355-1362, 2012.

SB^{3}C2017
Summer Biomechanics, Bioengineering and Biotransport Conference
June 21 – 24, Tucson, AZ, USA

USING CFD TO QUANTIFY CHANGES IN WALL SHEAR STRESS BETWEEN COMMON AND INNOVATIVE CELL SEEDING TECHNIQUES

Jake E. Rabidou (1), Andrew W. Holt (2), William E. Howard (1), Elizabeth T. Ables (3), David A. Tulis (2), Stephanie M. George (1)

(1) Department of Engineering
East Carolina University
Greenville, NC, USA

(2) Department of Physiology
Brody School of Medicine, East Carolina University
Greenville, NC, USA

(3) Department of Biology
East Carolina University
Greenville, NC, USA

INTRODUCTION

Parallel-plate flow chambers (PPFCs) are used for flow simulations of various cell types (endothelial cells, smooth muscle cells (SMCs)) [1]; where the in vitro model, the PPFC, is supposed to mimic and simulate the in vivo fluid shear stresses that the cells would be exposed to in their natural environment. A single layer of cells is attached to one of the internal surfaces of the PPFC and is subjected to fluid flow by creating a pressure gradient throughout the chamber [1]. Because adherent cells primarily rely on reorganization of their actin cytoskeleton to translocate and migrate [2, 3], and the application of fluid shear stress to adherent cells influences their cytoskeletal makeup and regulates their migratory phenotype; the development of tools for studying and understanding cell migration in the context of fluid shear stress while minimizing disturbances of the seeded cells is crucial.

Two common seeding techniques that avoid disturbing the PPFC surface during seeding are; first, a pipette tip method, in which a standard P1000 pipette is modified by removing the coned portion of the tip, allowing the now flat end of the tip to be inverted and used as a tool to scratch an area onto the seeding surface where cells can be seeded successfully, and two a surface tension technique that limits cell adherence to the cell suspension-liquid interface. A novel method of seeding cells being investigated here uses a 3D printed insert that fits inside the PPFC and allows a confluent layer of cells to be seeded, because the height of the cells doesn't change, the wall shear stress is not affected in the same ways the previously mentioned seeding methods are.

The objective of this study was to investigate and quantify the impact the pipette tipped method and 3D printed insert seeding method had on the applied levels of WSS of the cells.

METHODS

First, to compare the WSS imposed on adherent cells, the cell heights of the pipette tipped method were measured at both 1 and 24 hours. These cell heights were measured from SMCs that were seeded then fixed and stained, and then the Z-stacks were imaged using a confocal microscope to accurately measure adherent cell height. To quantify cell height on the substrate, cell height was measured at the center of the nucleus across the y-z plane using Zeiss ZEN software. The cell heights were taken from the highest point of the cells and averaged to get the overall height of the cell area.

Next, three CFD models were developed based on substrate surface and cell profile for the 3D-printed insert and pipette tip method at both 1 and 24 hours. A model of the flow chamber (Glycotech circular flow chamber) was developed using SolidWorks (Dassault Systemes, Waltham, MA) with dimensions 2 x 1 x 0.0254 cm. Inlet and outlet diameters were measured using microcalipers (0.159 cm for the inner diameter of the connecting tubing). The inlet and outlet were extended to limit end-effects in the computational region of interest. To mimic the pipette seeding technique, a cylinder, representing the cells, was included in the model using the cell heights measured at 1 hour and 24 hours post attachment.

The solid model was imported in ANSYS Workbench (ANSYS, Inc., Canonsburg, PA) and a 500,000 element mesh was generated using tetrahedral/hybrid shaped elements. A mesh independence study was completed which showed a less than 1% change in average WSS after 400,000 elements. The properties of the media used in the lab experiments were calculated using an electromagnetic viscometer (VISCOlab 4000, Medford, MA) with temperature control jacket set to 37°C. A viscosity value of 0.75 ± 0.01 cP was measured at 37.04 ± 0.08 °C then imported into FLUENT v17.1 (ANSYS, Inc., Canonsburg, PA). A standard flow rate of 8.6 ml/min was used to

Technical Presentation #281

determine the inlet velocity which was applied as a blunt profile. The model extension allowed the flow to develop as it reached the computational region of interest. The outlet boundary condition was set as a pressure-outlet, while the solver settings were defined as a second order pressure based solver with steady state flow. The solution converged when the residuals reached a value of 10^{-4} and post-processing was completed in CFD-Post (ANSYS, Inc., Canonsburg, PA).

RESULTS

Figure 1 below represents the cells heights being measured, where the blue dyed areas represent the nucleus and the red areas represent the plasma membranes. The average height after fixing the cells for 1 hour and 24 hours were 10 and 8.541 μm respectively.

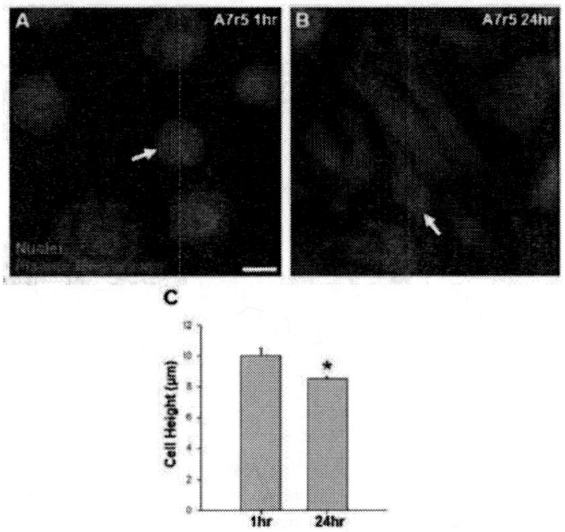

Figure 1. (A–C) (A-B) cells plated for 1 hour (A) or 24 hours (B). (C) Average cell height at the highest point of the cells.

CFD models were used to illustrate the impact the cell seeding methods had on the applied level of WSS. Figure 2 represents the average wall shear stress along the bottom face of the models (A-C), along with the shear stress across the extruded cell heights at 1 and 24 hours (B-C). Figure 2 below shows that seeding methods which do not maximize the perfusion area (pipette tip method) (B-C) showed more than a 5% increase in average WSS compared to the 3D printed insert cell seeding method (A).

DISCUSSION

Most adherent cells attach and spread on the substrate as a function of time, which could influence the magnitude of fluid shear stress or flow profile that is being reported. Results demonstrate that the longer time cells have for attachment, the more flattened and spread out they become. Because the common pipette tipped, cell seeding methods are preferred to obstruction-based methods but lack the ability to fully maximize the perfusion area, we performed CFD to illustrate and quantify this theory. Using various approaches (3D insert and pipette tip) and the cell heights measured at both 1 and 24 hours, three CFD models were tested (Figure 1. A–D). Because the cell area in the 3D insert occupies the entire perfusion area, cell height does not change the applied fluid shear stress. However, the smaller cell areas defined by the standard P1000 pipette tip method showed >6% (1 h attachment; $P < 0.05$) and 5% (24 h attachment; $P < 0.05$) difference in the mean WSS compared to the 3D printed insert model (Figure 2. D). The data demonstrates the importance of maximizing perfusion area and complete cell spreading prior to perfusion onset.

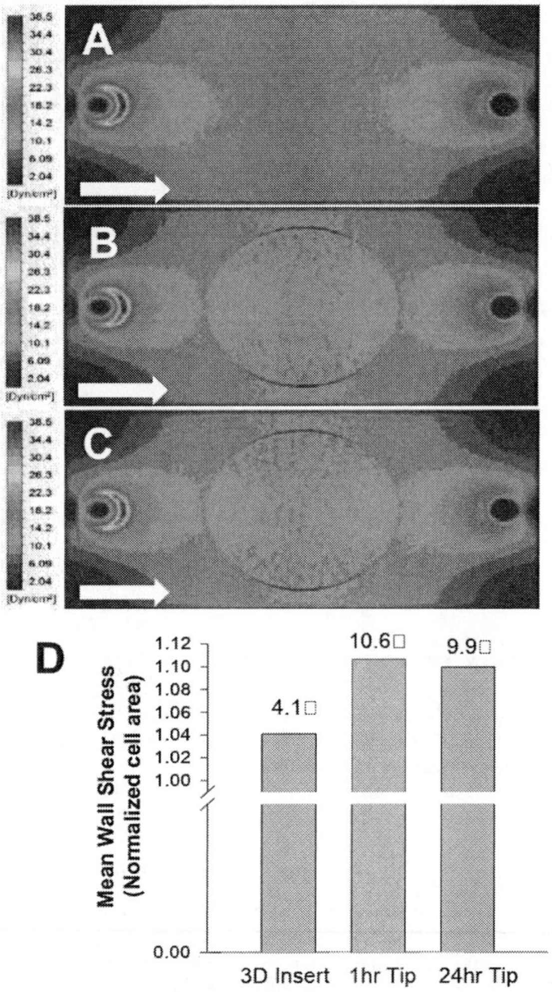

Figure 2. (A–D) Simulations of (SMCs) seeded within 3D printed inserts (A) and within modified polypropylene pipette tips 1 h (B) and 24 h (C). (D) Represents the mean wall shear stress between the models. In (A-C) the white arrows dictate the direction of flow.

ACKNOWLEDGEMENTS

We would like to acknowledge Cindy Kukoly, Jake Francisco, and Angel Chukwu for their contributions to this work. The authors declare no conflicts of interest for this work.

REFERENCES

[1] Bacabac, R et al., *J Biomech,* 38:159-167, 2005.
[2] Carlier, B et al., *Annu Rev Biophys,* 39:449-470, *2010.*
[3] Gerthoffer, W. *Circ Res,* 100:607-621, 2007.

COEFFICIENT OF FRICTION BETWEEN CARBOXYMETHYLATED HYALURONIC ACID (CMHA-S) FILMS AND THE OCULAR SURFACE

Jourdan Colter (1), Hee-Kyoung Lee (2), Brenda Mann (2), Barbara Wirostko (2,3), Brittany Coats (1)

(1) Department of Mechanical Engineering
University of Utah
Salt Lake City, UT, USA

(2) EyeGate Pharmaceuticals, Inc.
Waltham, MA, USA

(3) Department of Ophthalmology and Visual Sciences
University of Utah
Salt Lake City, UT, USA

INTRODUCTION

Novel carboxymethylated hyaluronic acid-based polymers (CMHA-S, EyeGate Pharma) are currently being developed as ocular drug-delivery vehicles for the treatment of various ophthalmic conditions. One existing challenge with the development of the CMHA-S film for clinical use is retention in the eye for extended time periods. Previous work by our group using finite element analysis has shown that retention of CMHA-S in the eye is largely dependent on the frictional interactions between the CMHA-S film and the eye [1]. That is, the coefficient of friction on the eyelid side of the film must be greater than the coefficient of friction on the globe side for the device to be retained in the inferior fornix. In order to achieve this friction differential, a potential modification to the eyelid-side of the CMHA-S film is to layer the polymer with methylcellulose [2, 3].

In this experimental study, we designed a custom tribometric system to measure the coefficient of friction of CMHA-S on the eye. Specifically, the research objectives were to (1) quantify the coefficient of friction of CMHA-S with the sclera, and to (2) compare the effects of added methylcellulose on the friction of CMHA-S.

METHODS

CMHA-S films with (n=5) and without (n=15) methylcellulose were tested against ovine and human sclera. A custom tribometric system adapted from an AR-G2 rheometer (TA Instruments) was used to measure the coefficient of friction of the samples (Fig. 1). A scleral annulus (8 mm outer diameter) was attached to the upper rheometer head, and the CMHA-S film was sandwiched between stationary rubber grips concentrically below the annulus. Blink Tears lubricating eye drops (Abbott Laboratories) were applied to a surrounding well to maintain lubrication throughout testing. The test protocol was carried out at axial loads of 0.3, 0.5, and 0.7 N, which are commensurate with the reported range of physiological forces experienced by the eye during blinking [4, 5]. For each axial load (starting at 0.3 N and incrementing upward), four clockwise and four counterclockwise revolutions were applied at increasing effective linear velocities of 0.3, 1.0, 10.0, and 30.0 mm/s, with a 12 second relaxation time between each set of four revolutions. Sliding velocities were selected based off of eyelid blink speeds [6]. Static coefficient of friction was extracted from within the first 10 degrees of revolution, and kinetic coefficient of friction was averaged over the last two revolutions. Two-way ANOVAs with repeated measures were used to evaluate the effect of sliding velocity and formulation (CMHA-S+/-methylcellulose) and sliding velocity and species (ovine, human) on friction.

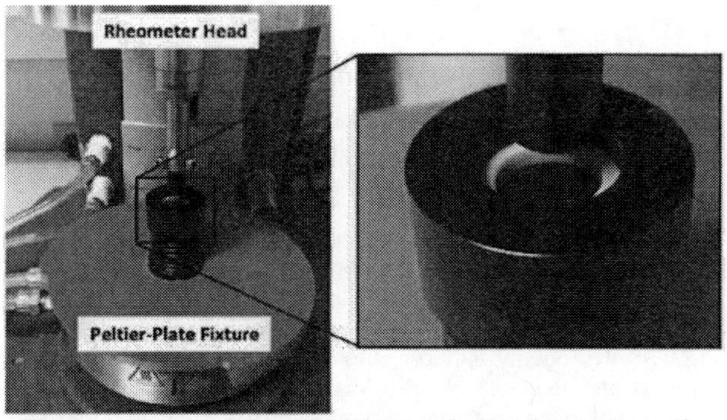

Figure 1: Custom tribometric fixtures designed to measure the friction between the CMHA-S film and sclera.

RESULTS

Static and kinetic coefficients of friction of the CMHA-S film relative to ovine sclera were found to be 0.29±0.1 and 0.15±0.1, respectively (Fig. 2). Static coefficient of friction was significantly dependent on rate ($p<0.0001$), especially for CMHA-S with methylcellulose ($p=0.0002$). Kinetic coefficient of friction was not affected by rate. CMHA-S with methylcellulose had 1.6 and 1.8 times higher static and kinetic coefficients of friction, respectively, than CMHA-S, but began to wear starting at loads of 0.5 N. CMHA-S tested against human sclera was not statistically different from CMHA-S tested against ovine sclera.

A

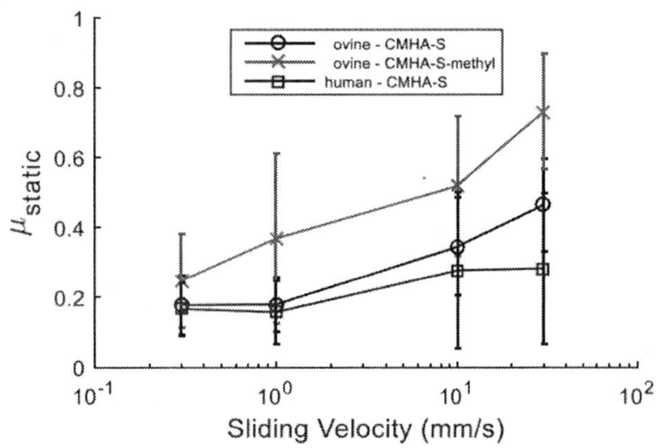

B

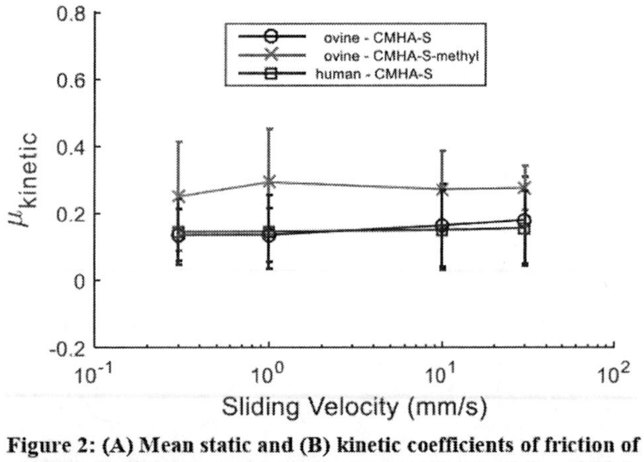

Figure 2: (A) Mean static and (B) kinetic coefficients of friction of CMHA-S with ovine and human sclera at 0.3, 1.0, 10.0, and 30.0 mm/s. Error bars indicate standard deviation from the mean.

DISCUSSION AND CONCLUSION

While the addition of methylcellulose to the CMHA-S film increased the friction, it also increased wearing of the films. The CMHA-S with methylcellulose substantially degraded at all rates for forces greater than 0.5 N. This suggests that the formulation with methylcellulose may wear rapidly following placement into the eye. If a layering approach is used, the CMHA-S with methylcellulose will be placed on the side of the film in contact with the inner eyelid. There may not be as much movement on this side of the film as compared to the globe side, so the degradation of CMHA-S with methylcellulose may be a slower process *in vivo* as compared to the experimental tests. In summary, the addition of methylcellulose will achieve the requisite friction differential for improved retention. There is an increased risk of degradation of methylcellulose with use, but this degradation may be negligible or delayed giving the limited movement of the lower eyelid with the film. This suggests methylcellulose can be used to tailor CMHA-S frictional interaction with the globe surface and eyelid, but careful consideration of potential increased degradation over time must be considered.

ACKNOWLEDGEMENTS

This material is based upon work supported by the National Science Foundation under award number IIP-1430921, and the Department of Defense under award number W81XWH-14-C-0025. Any opinions, findings, and conclusions or recommendations expressed in this abstract are those of the authors and do not necessarily reflect the views of the National Science Foundation or Department of Defense.

REFERENCES

[1] Colter, J, MS Thesis, University of Utah, 2016.
[2] Nasatto, PL et al., *Polymers*, 7:777-803, 2015.
[3] Ramesha, R et al., *Drug Dev Ind Pharm*, 25:685, 1999.
[4] Hung, G et al., *Am J Optom Physiol Opt*, 54(10):678-690, 1977.
[5] Miller, D, Arch. *Ophthalmol*, 78(3):328-330, 1967.
[6] Jones, MB et al., *Bull Math Biol*, 70:323-343, 2008.

SB[3]C2017
Summer Biomechanics, Bioengineering and Biotransport Conference
June 21 – 24, Tucson, AZ, USA

FINITE ELEMENT FORMULATION OF MULTIPHASIC SHELL ELEMENTS FOR CELL MEMBRANE ANALYSES IN FEBIO

Chieh (Jay) Hou (1), Steve A. Mass (2), Jeffrey A. Weiss (2), Gerard A. Ateshian (1)

(1) Mechanical Engineering
Columbia University
New York, NY, USA

(2) Bioengineering
University of Utah
Salt Lake City, UT, USA

INTRODUCTION

Cells such as chondrocytes are embedded in a relatively stiff, electrically charged extracellular matrix and subjected to large deformations. The regulation of cell homeostasis in such environments remains a challenging area of investigation [1]. Mathematical models for cell physiological regulation are usually limited to 1D analyses of isolated cells, often precluding cell deformations. With the recent implementation of multiphasic materials in the open-source finite element (FE) software FEBio [2-4], 3D models of cells embedded within tissue may now be analyzed, accounting for solid matrix deformation, transport of fluid and solutes, membrane potential and reactions. For example, models for cell volume, pH, and Ca^{2+} regulation using FEBio have been recently reported [5,6].

The cell membrane is a critical component in cell models, which regulates transport of fluid and solutes in the presence of large concentration and electric potential gradients, while also facilitating transport of various proteins. The cell membrane is much thinner (~10 nm) than the cell (~10 μm) [7]; therefore, in an FE environment, shell elements would be preferred for modeling the cell membrane, for the convenience of mesh generation from image-based data, and the application of boundary conditions [8]. However, porous or multiphasic shell elements are yet to be developed in the FE literature and commercial FE software. This study demonstrates the formulations and implementation of multiphasic shell elements in FEBio.

FINITE ELEMENT IMPLEMENTATION

In this study, a degenerated solid approach is used [8], where the shell formulation is equivalent to a thin-layered solid, with nodes located on the front face. The nodal degree of freedoms of the shell include the displacements of the front and back faces of the solid, shown in Fig. 1.

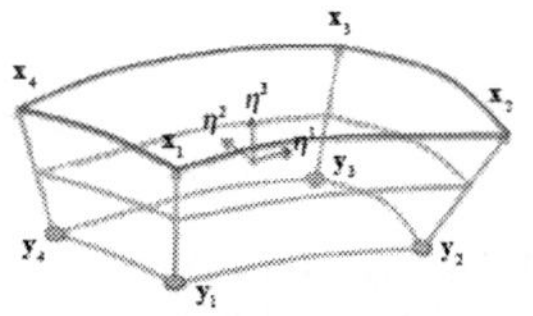

Figure 1. Degenerated solid approach. The shell surface represents the front face of the thin solid.

The position $\mathbf{x}$ of a point in the shell element is interpolated with front and back face nodal positions $\mathbf{x}_a$ and $\mathbf{y}_a$, respectively,

$$\mathbf{x} = \sum_{a=1}^{n} \frac{1}{2} M_a(\eta^1,\eta^2)\left[(1+\eta^3)\mathbf{x}_a + (1-\eta^3)\mathbf{y}_a\right]. \qquad (1)$$

Here, n is the number of shell element nodes, $M_a(\eta^1,\eta^2)$ are the in-plane interpolation functions and η^3 is the out-of-plane coordinate ($-1 \le \eta^i \le 1$). Four and 8-node shell elements have been implemented in FEBio. The degenerated-solid approach for the 4-node shell is known to lead to significant thickness and transverse-shear locking [8], which can be reduced significantly using the 8-node shell. In multiphasic elements, a similar interpolation is adopted for the effective fluid pressure $\tilde{p}$ and solute concentrations $\tilde{c}^\alpha$, e.g.,

$$\tilde{c}^\alpha = \sum_{a=1}^{n} \frac{1}{2} M_a(\eta^1,\eta^2)\left[(1+\eta^3)\tilde{c}_a + (1-\eta^3)\tilde{d}_a\right], \qquad (2)$$

where $\tilde{c}_a$ and $\tilde{d}_a$ represent front and back face nodal effective concentrations. Following the same procedure as in [3,4], the virtual work δW is expressed in terms of the nodal variables (front and back face displacements $\mathbf{u}_a$ and $\mathbf{w}_a$, fluid pressures $\tilde{p}_a$ and $\tilde{q}_a$, and concentrations $\tilde{c}_a$ and $\tilde{d}_a$). This expression is linearized along increments in these variables for use in an iterative Newton scheme.

Technical Presentation #283

VERIFICATIONS

All verification problems compare the shell formulation to an equivalent thin 3D element and are reported with user-defined units. The verifications aim to demonstrate that multiphasic shells may form stand-alone thin structures, or may serve as an outer porous membrane around a 3D porous material, or may represent membranes sandwiched between 3D porous materials. In particular, multiphasic shells may be ascribed significantly different transport properties (hydraulic permeability, solute diffusivity) than adjoining materials.

A cantilever elastic shell beam is subjected to a torque of 10^{-5} along its free end. The deformed shell beam shows the bending response and the free end deflection increases with the torque in agreement with a thin 3D model (Fig. 2).

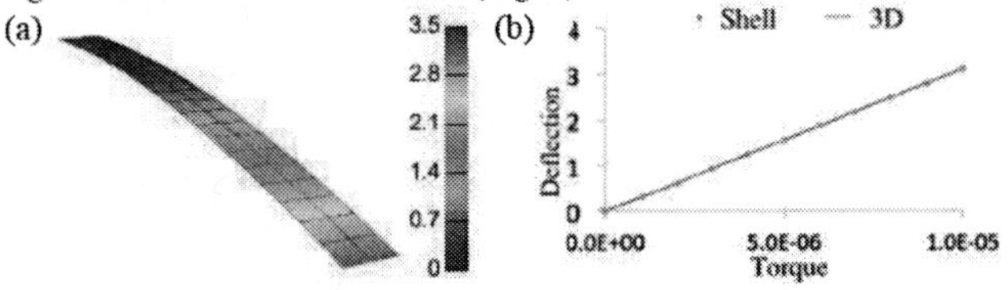

Figure 2: (a) Deflection of elastic shell beam under a prescribed torque at the free end. (b) Deflection vs. torque, compared to thin 3D layer. Properties: dimensions $20\times 2\times 0.01$, neo-Hookean solid, Young's modulus ($E_Y=1$), Poisson's ratio ($\nu=0$).

A biphasic shell disk is subjected to a fluid pressure difference of 10^{-3} across the thickness, and its edge is fixed in position. The total fluid flux at the center increases with the fluid pressure, in agreement with a thin 3D biphasic model (Fig. 3).

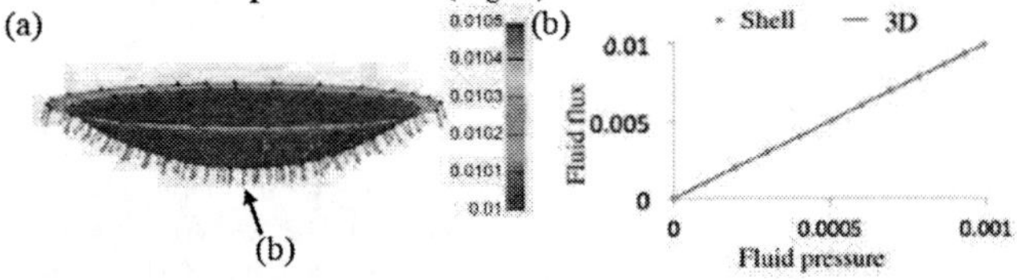

Figure 3: (a) Total fluid flux of biphasic shell disk with applied fluid pressure difference across the thickness. (b) Total fluid flux at the center vs. fluid pressure, compared to thin 3D biphasic model. Properties: dimension $\varnothing 5\times 0.1$, solid volume fraction ($\varphi_0=0.3$), neo-Hookean solid ($E_Y=1$, $\nu=0$), isotropic hydraulic permeability ($k=1$).

A spherical biphasic cell is compressed with a flat impermeable platen by 40% of its radius over the time interval $0\le t\le 10$. The cytoplasm uses 3D biphasic elements and the membrane uses biphasic shells. The fluid flux at the rightmost node increases and then decreases back with time, in agreement with a model using a single layer of thin 3D biphasic elements to model the membrane (Fig. 4).

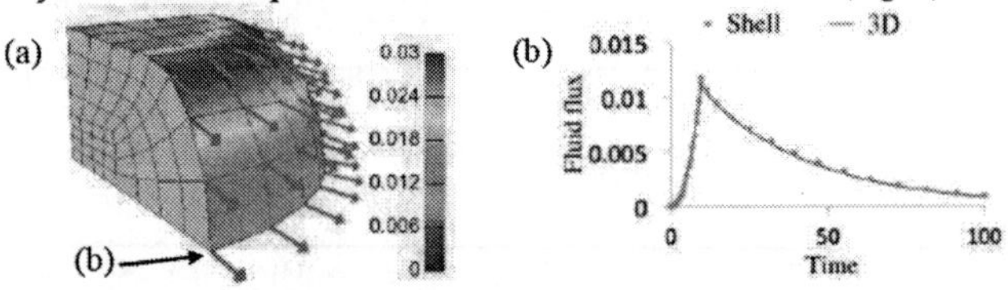

Figure 4: (a) Fluid flux on membrane of biphasic cell under compression ($t=10$). (b) Fluid flux at rightmost node vs. time, compared to 3D membrane. Properties: cytoplasm, $\varnothing 5$, $\varphi_0=0.3$, neo-Hookean solid ($E_Y=10^{-3}, \nu=0$), $k=10^3$; membrane, thickness 0.1, $\varphi_0=0.3$, Mooney-Rivlin solid ($c_1=10^{-3}$, bulk modulus $K=1$), $k=10$.

A biphasic cell (3D biphasic cytoplasm, biphasic shell membrane) embedded within a 3D biphasic matrix is compressed on the top by 10% of the matrix height ($0\le t\le 10$). The fluid flux at the rightmost node increases and decreases with time, in agreement with a model using a thin 3D biphasic membrane (Fig. 5).

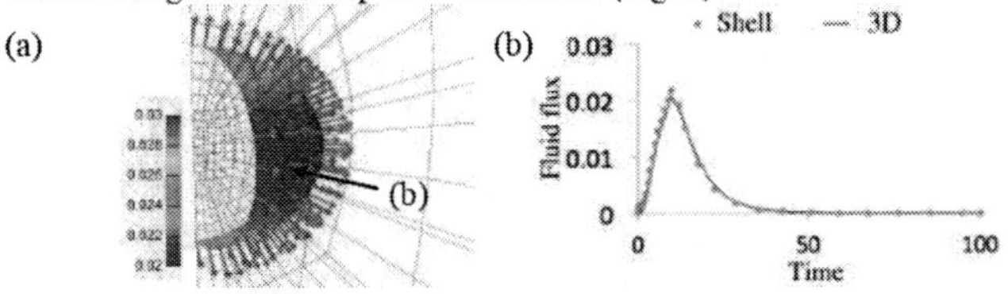

Figure 5: (a) Fluid flux on membrane of biphasic cell embedded in tissue under compression ($t=10$). (b) Fluid flux at rightmost node vs. time, compared to thin 3D biphasic membrane. Properties: cytoplasm and membrane have same properties as previous example, except for membrane $K=10^5$ and $k=10^{-1}$; tissue, dimension $100\times 100\times 100$, $\varphi_0=0.3$, neo-Hookean solid ($E_Y=0.64$, $\nu=0$), $k=600$.

A biphasic-solute cell (3D cytoplasm and shell membrane) is subjected to osmotic loading by diluting the ambient solute concentration to 50% of its initial value of 0.1 over $0\le t\le 10$. The solute flux at the rightmost node increases and then decreases back with time, in agreement with a model using a thin 3D biphasic-solute membrane (Fig. 6).

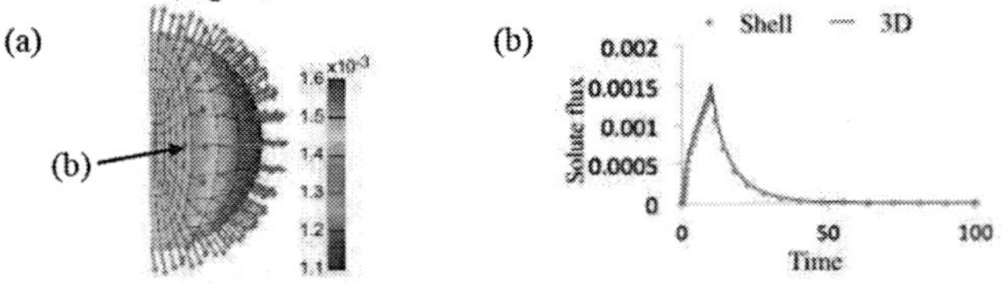

Figure 6: (a) Solute flux on membrane of biphasic-solute cell under osmotic loading at time ($t=10$). (b) Solute flux at rightmost node vs. time, compared to thin 3D biphasic-solute membrane. Properties: cytoplasm and membrane have same dimensions and materials for solid matrix as previous example; cytoplasm, $k=10^3$, solute diffusivity ($d=10^3$); membrane, $k=10^{-1}$, $d=10^{-2}$.

DISCUSSION

This study presents a novel finite element formulation for porous (biphasic and multiphasic) shell elements, which has been implemented in FEBio. This implementation addresses an important need in the study of cell mechanics and transport using 3D models that may capture the geometric complexity of cells and intracellular membrane-bound compartments, and their surrounding environments. Future efforts will address active transport processes as chemical reactions occurring across multiphasic shell elements.

ACKNOWLEDGEMENTS. NIH R01 GM083925

REFERENCES.

[1] Wilkins, R. et al., *J Membrane Biol*, 177(2):95-108, 2000.
[2] Maas, S. et al., *J Biomech Eng*, 134(1), 2012.
[3] Ateshian, G. et al., *J Biomech Eng*, 135(11):111001, 2013.
[4] Ateshian, G. et al., *J Biomech Eng*, 133(8), 2011.
[5] Hou, C. et al., *SB3C2015*, 356, 2015.
[6] Hou, C. et al., *SB3C2016*, MD, 2016.
[7] Keener, J. et al., *Mathematical Physiology*, Springe, 2009.
[8] Bischoff, M. et al., *Encyclopedia of computational mechanics*, Wiley, 2004.

A MULTI-SCALE MODEL PREDICTS INCREASING FOCAL ADHESION SIZE WITH DECREASING STIFFNESS IN FIBROUS MATRICES

Xuan Cao (1), Ehsan Ban (1), Brendon M. Baker (2), Yuan Lin (3), Jason A. Burdick (4), Christopher S. Chen (5), Vivek B. Shenoy (1, 4)

(1) Department of Materials Science and Engineering
University of Pennsylvania
Philadelphia, PA 19104, USA

(2) Department of Biomedical Engineering
University of Michigan
Ann Arbor, MI 48109, USA

(3) Department of Mechanical Engineering
University of Hong Kong
Hong Kong SAR, China

(4) Department of Bioengineering
University of Pennsylvania
Philadelphia, PA 19104, USA

(5) Department of Biomedical Engineering
Boston University
Boston, MA 02215, USA

INTRODUCTION

Focal adhesions (FAs) play important roles in many cellular behaviors, including proliferation, differentiation and locomotion, and pathological processes like tumorigenesis and wound healing (1). Numerous observations have demonstrated that cells form larger FAs as well as develop higher intracellular traction forces on stiffer extracellular matrices (ECMs) (2). In most investigations, the substrates considered were flat (2D) and linear elastic. However, *in vivo*, many cells reside within three-dimensional (3D) fibrous scaffolds where the density and diameter of fibers can vary depending on the nature of the tissue (3). The local architecture of these fibrous networks may change significantly when cells exert forces on them, leading to phenomena such as non-linear stiffening, reorientation and physical remodeling of the ECM (4, 5). Interestingly, our recent study on cells in synthetic fibrous matrices showed that increasing fiber stiffness suppresses spreading in contrast to hydrogels where increased stiffness always promotes cell spreading (6). Clearly, a quantitative description of how cells are able to physical remodel matrices in order to mature FAs, which in turn can lead to greater spreading is currently lacking.

Here, we propose a multiscale chemo-mechanical model to describe the evolution of FAs in crosslinked fibrous networks that native ECMs. By combining the mechanics of fiber recruitment with stress dependent growth kinetics of FA plaques, we predict a phase diagram for the stable size of focal adhesions as a function of ECM stiffness and a new parameter we introduce, namely the recruitment index of the ECM that characterizes how easily fibers can be recruited by contractile cells. Our model explains how cell-driven fiber recruitment can lead to a departure from the monotonic stiffness versus cell spreading relationship observed in hydrogels.

METHODS

The rate-limiting step in the growth of the FAs is the incorporation of new plaque units, a process influenced by the stress level in the plaque exerted by the actomyosin network. The contractility of the network, in turn depends on the effective stiffness of the adhesion complex determined by the size of the plaque, the stiffness of the ECM and the density of integrin links between the ECM and the plaque. Thus, predicting the growth kinetics and size of focal adhesions requires us to consider the two way cross-talk between matrix reorganization and cell contractility, achieved with the multiscale procedures shown in Fig. 1.

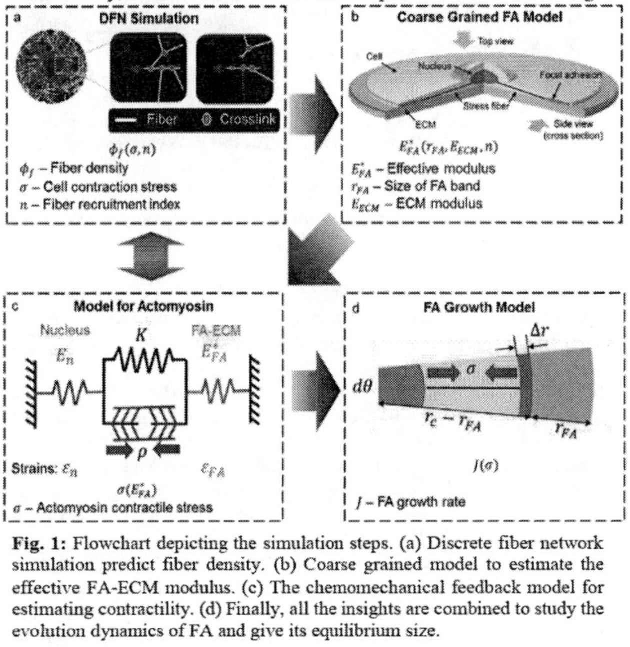

Fig. 1: Flowchart depicting the simulation steps. (a) Discrete fiber network simulation predict fiber density. (b) Coarse grained model to estimate the effective FA-ECM modulus. (c) The chemomechanical feedback model for estimating contractility. (d) Finally, all the insights are combined to study the evolution dynamics of FA and give its equilibrium size.

RESULTS

Crosslink breakage enables ligand recruitment in fibrous networks

Our DFN simulations showed that the fiber strains decay gradually away from the periphery of the cell where contractile forces are applied. Furthermore, fibers oriented in the radial direction are stretched while strains in the fibers aligned circumferentially were predominantly compressive. As the level of contractile force increases, the compressed fibers buckle while the crosslinks between the radial fibers could undergo higher stretching. Rupture of crosslinks takes places once the forces the crosslinks transmitted exceed a critical level, eventually allowing the fibers to be pulled into the cell area. As expected, significant fiber recruitment was observed in networks with weak crosslinks (Fig. 2a) that could rupture easily, while no recruitment was observed when the fibers were "welded" together, in agreement with experimental observations (Fig. 2c). The density of fibers (underneath the cell) under different levels of contractile stresses and crosslink strength are shown in Fig. 2b. Interestingly, increase in the density of recruited fibers with force can be well fitted by the phenomenological relationship that is characterized by two parameters, namely the fiber recruitment index (n) and the threshold stress for the crosslinks to rupture (σ_c). We found that ECM stiffness indeed had a significant impact on fiber recruitment. Specifically, the cell recruits significantly less fibers on stiffer ECMs as shown in Fig. 3c, a phenomenon that is well captured by our model.

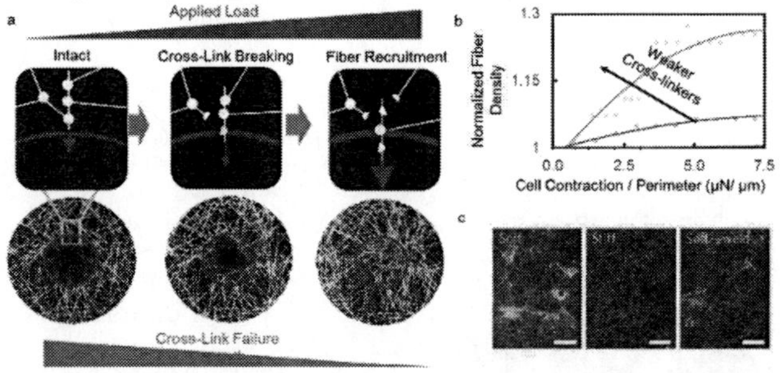

Fig. 2: (a) Tensile forces generated by cellular contraction leads to the breakage of the fiber crosslinks, which allows the cell to recruit more fibers. (b) Normalized fiber density as a function of cell contraction for networks with crosslinks of different breaking strengths. (c) The cell induces large deformations to the soft network, while induced deformations are much smaller for the stiff and welded soft network (stronger crosslinks).

FA size-ECM modulus relation becomes non-monotonic when cells can recruit fibers

We first examined the correlation between FA size and ECM modulus without fiber recruitment, a scenario relevant to most elastic hydrogels and crosslinked ECMs that cannot be physically remodeled by the cells. As shown in Fig. 3a, we found that cells cannot form stable FAs on very soft ECMs. Furthermore, FA size increases monotonically with the ECM modulus. This can be explained by the fact that higher contractile stress will be developed on stiffer ECMs, which eventually leads to larger FAs. When cells can recruit fibers, the stable FA size increases compared to the cases without fiber recruitment (solid red curve in Fig. 3a) for a given level of ECM stiffness. This is because more ligands for the integrins are available with the recruitment of fibers, leading to a stiffer FA-ECM complex and therefore to higher levels of contractility. More interestingly, the FA size will reach its maximum at a certain intermediate ECM modulus when fiber recruitment is possible (Fig. 3a), in direct contrast to the monotonic trend observed on substrates that cannot be remodeled. The reason is that crosslinks are ruptured more easily in a softer ECM due to the large deformation caused by cell contraction. This competition between increase in the fiber/integrin density and the ECM modulus (both promoting the formation of larger FAs) leads to a peak in the FA size at intermediate levels of the ECM modulus (Fig. 3a). These findings are consistent with our recent experimental observations. By varying the values of n and E_{ECM}, the stable FA band size as a function of the ECM modulus and the fiber recruitment index is shown in Fig. 3b.

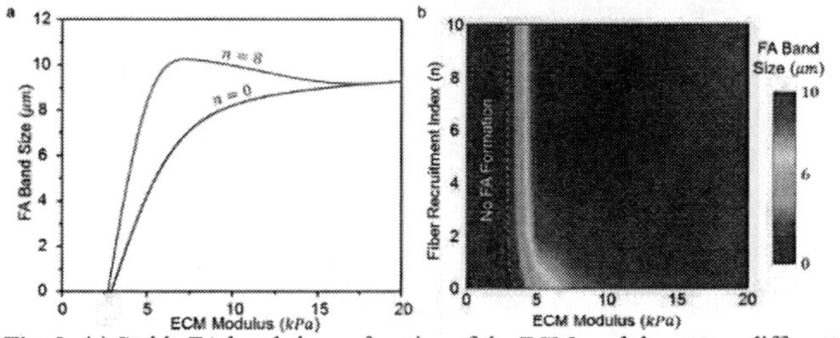

Fig. 3: (a) Stable FA band size as function of the ECM modulus at two different recruitment indexes. (b) Heat map of the stable FA band size as a function of the ECM modulus and fiber recruitment index.

DISCUSSION

The behavior of cells on different ECM substrates is best summarized by Fig. 4. In particular, for ECMs that cannot be remodeled (i.e. hydrogels), the ECM modulus and ligand density are decoupled from each other. Therefore, as the ECM becomes stiffer the response of cells follows a linear path with no variation in ligand density, resulting in a positive correlation between ECM modulus and FA size. In contrast, for fibrous ECMs that can be physically remodeled, cells are able to recruit more fibers from the soft microenvironment and engage more integrins. This results in a more complicated path where ligand density decreases with increasing ECM modulus, leading to the non-monotonic FA size – ECM modulus relationship.

We believe that in addition to providing explanations to a variety of experimental observations, this study can also serve as a theoretical framework for accessing the intracellular – extracellular communication in various non-linear extracellular environments.

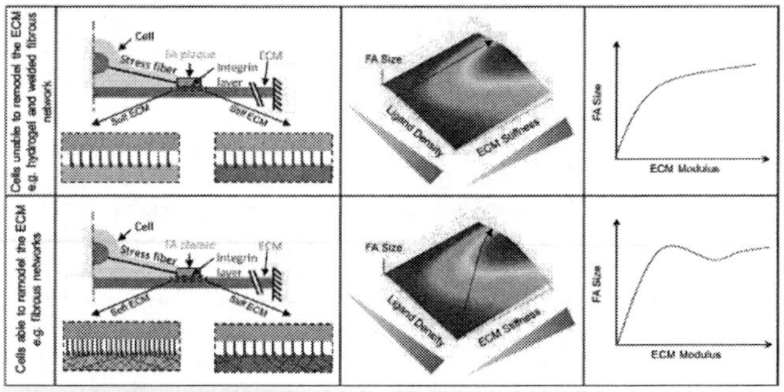

Fig. 4: Summary on the results.

ACKNOWLEDGEMENTS

This work was supported by the NCI (U01CA202177 and U54CA193417) and the NIH (R01EB017753, HL124322 and R01AR056624).

REFERENCES

[1] Discher, D.E., et al., *Science*. 310: 1139-43, 2005.
[2] Goffin, J.M., et al., *J. Cell Biol.* 172: 259-68, 2006.
[3] Baker, B.M., et al., *J. Cell Sci.* 125: 3015-24, 2012.
[4] Abhilash, A.S., et al., *Biophys. J.* 107: 1829-40, 2014.
[5] Wang. H.., et al., *Biophys. J.* 107: 2592-2603, 2014.
[6] Baker, B. M., et al., *Nat. Mater.* 14: 1262-68, 2015.

SB^{3}C2017
Summer Biomechanics, Bioengineering and Biotransport Conference
June 21 – 24, Tucson, AZ, USA

EFFECTS OF INFLAMMATION ON CELLULAR DEFORMATION OF NUCLEUS PULPOSUS CELLS: A BIPHASIC FINITE ELEMENT MODEL

Quynhhoa T. Nguyen (1), Nadeen O. Chahine (1,2)

(1) Feinstein Institute for Medical Research, Manhasset, NY USA

(2) Hofstra Northwell School of Medicine Hempstead, NY, USA

INTRODUCTION

Nucleus pulposus is the central region of the disc that is comprised of sparsely distributed cells embedded in a matrix that is rich in proteoglycans and has high water content [1,2]. Disc degeneration is characterized by changes in extracellular matrix (ECM) properties, including loss of proteoglycans and collagens, degenerative fibrillation, and decreased water content, as well as by elevated levels of inflammatory cytokines that have been implicated in matrix degradation [3,4]. While inflammatory signaling has long been recognized to alter cell metabolism, our recent studies have shown that cellular biomechanical properties and cell morphology are also altered in inflammatory environment [5,6]. Although disc cells are known to be mechanosensitive, the mechanism of cellular mechano-transduction in pathological conditions (e.g. in inflammatory environment) is poorly understood. In the current study, we hypothesized that changes in NP cell biophysical properties induced by inflammatory stimuli would alter the deformational behavior of cells in response to applied deformational loading. To address this, we modeled the multiscale cell deformation behavior using biphasic finite element analysis.

METHODS

To investigate microscopic deformation behavior of nucleus pulposus cells, a multiscale finite element analysis was performed using FEBio software suite, whereby an NP cell was embedded in the center of an agarose block (0.5 x 0.5 x 0.5 mm^3). Agarose was used as a model system because NP cells maintain a rounded morphology in this biomaterial and because its compressive properties are on same order of magnitude as native NP ECM. A constant 10% displacement was applied along the 3-direction (or z-direction) in unconfined configuration between 2 impermeable platens on top and bottom surfaces of the agarose disk. The simulation was done over a course of 1000 seconds to ensure that equilibrium was reached. The agarose disk was modeled as a biphasic matrix with isotropic elastic solid (Young's modulus, E_y = 13.1 kPa, Poisson's ratio, ν = 0.15) and constant hydraulic permeability (k_p = 5.14 x 10^{-14} $m^4/(N.s)$) [7]. The cell assumed a spherical geometry with 10 µm radius and was modeled as a homogeneous gel, representing the protoplasm (cytoplasm, cytoskeleton, and all enclosed organelles). The cell protoplasm was also modeled as a biphasic material with isotropic elastic solid and constant hydraulic permeability.

Deformation behavior of healthy NP cells (E_y = 2kPa, ν=0.3, and k_p = 5 x 10^{-15} $m^4/(N.s)$) as well as cells exposed to inflammatory stimulation (E_y = 1kPa, ν = 0.3, and k_p = 10×10^{-15} $m^4/(N.s)$) were analyzed based on prior experimental measures [5]. A hexahedral mesh was generated in PreView, nonlinear finite element analysis was performed in FEBio 2.4, and post-processing was done in PostView (**Fig. 1**). Time course of principal stresses was reported. Equilibrium principal strains, equilibrium fluid pressure, and total fluid flux were calculated and compared between healthy NP cell, inflammatory treated NP cell, and agarose.

Technical Presentation #285

RESULTS

Both at peak ramp (data not shown) and equilibrium, the cells were deformed more than the surrounding agarose material as indicated by higher magnitude of principal strains (**Fig. 2A**). In general, inflammatory treated cell experienced slightly higher deformation than healthy untreated cells (**Fig. 2A**). Fluid pressure at equilibrium was comparable between healthy and treated cells (**Fig. 2B**); however total fluid flux was higher in inflammatory treated cell compared to control (**Fig. 2C**).

Both non-treated and treated cells experienced compression in all 3 principal directions as indicated by negative principal stresses whereas agarose was only compressed in principal directions 2 and 3 (**Fig. 3**). Inflammatory treated NP cell had higher compressive stresses at peak, but lower stresses at equilibrium compared to untreated cell.

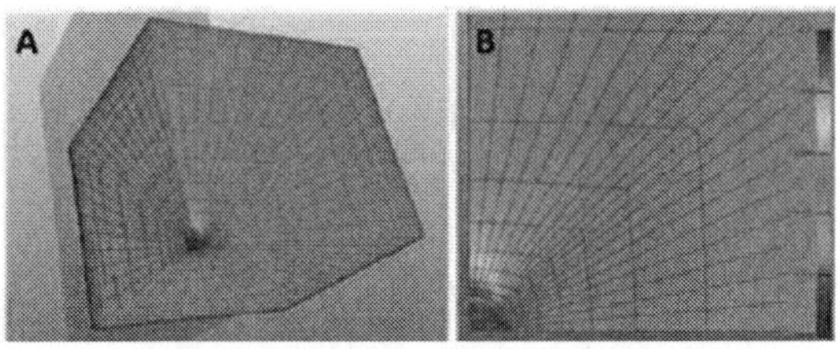

Figure 1: (A) A cell embedded at the center of an agarose block, and (B) a color map showing principal strain 3 of the cell and surrounding matrix.

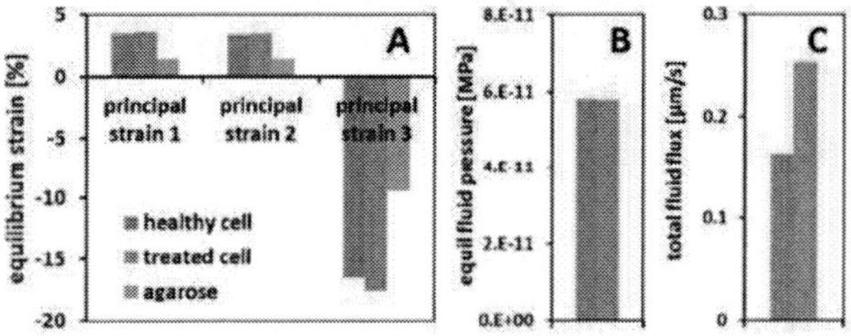

Figure 2: (A) Equilibrium principal strains, (B) fluid pressure at equilibrium, and (C) total fluid flux of healthy cell, inflammatory treated cell, and agarose.

DISCUSSION

The results of this model show that cells respond differently to an applied compression in an inflammatory environment compared to control condition, mediated by altered biomechanical properties of the cells in inflammation. NP cells that are exposed to inflammatory stimuli have lower stiffness, and hence experience larger compressive deformation (principal strain 3) in response to the same macroscopic displacement when compared to untreated healthy cells. Interestingly, we observed that all 3 principal stress levels at equilibrium were lower in inflammatory treated cell. This may be mediated by the increase in hydraulic permeability due to inflammation. Indeed, preliminary findings suggest that increased fluid flux may contribute to decreased principal stresses on the cell.

Future studies will include analysis of cell deformation in response to macroscopic dynamic loading and validating the model with empirical measurements.

ACKNOWLEDGEMENTS

NSF CAREER 1151605, NIH R01AR069668.

REFERENCES

[1]. Humzah MD et al, *Anat Rec,* 220(4):337-56, 1988

[2]. Urban JP, *Biochem Soc Trans,* 30(Pt 6):858-64, 2002

[3]. Setton LA *et al, J Bone Joint Surg Am,* 88 Suppl 2:52-7, 2006

[4]. Antoniou J *et al, J Clin Invest,* 98(4):996-1003, 1996

[5]. Maidhof R *et al, PLos One,* 9(6):e99621, 2014

[6]. Rajan NE *et al, Spine,* 38(16):1343-51, 2013

[7]. Mauck RL *et al, J Biomech Eng,* 125(5):602-14, 2003

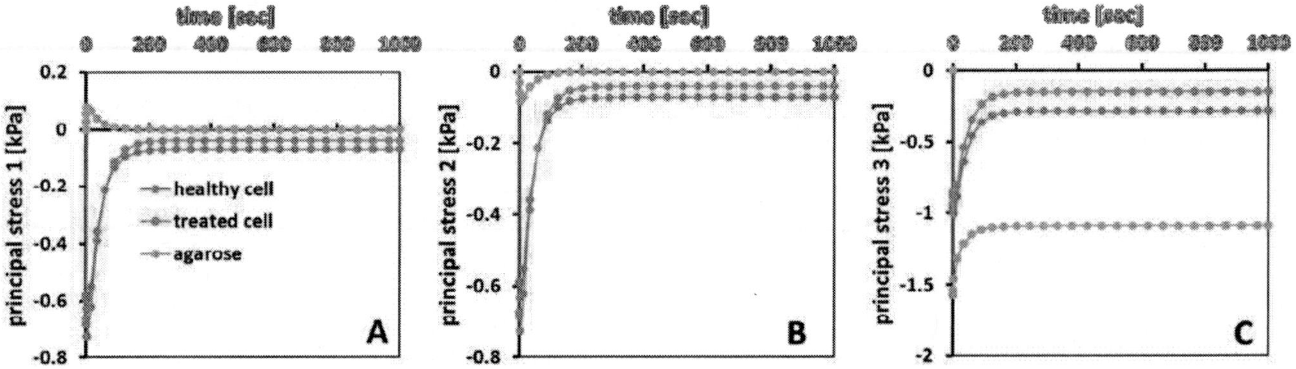

Figure 3: Principal stresses of healthy cell, inflammatory treated cell, and agarose as functions of time.

CELL FORCE GENERATION IN BIAXIALLY AND UNIAXIALLY LOADED TISSUES

Noel Reynolds (1), Eoin McEvoy (1), Vikram Deshpande (2), Patrick McGarry (1)

(1) College of Engineering and Informatics
National University of Ireland Galway
Galway, Ireland

(2) Department of Engineering,
University of Cambridge,
Cambridge, United Kingdom

INTRODUCTION

Mechanical priming strategies have been developed in an ongoing drive to engineer tissues with increased functional viability [1, 2]. Previous studies demonstrate that the application of substrate stretching can up-regulate gene expression, promote differentiation, cause cell reorientation, increase proliferation, and increase extracellular matrix synthesis [3, 4]. Given that cells actively respond to loading in the 3D microenvironment, the development of a fundamental mechanistic understanding of the effects of mechanical conditioning on cell biomechanics is critical for the advancement of the field of tissue engineering. In this study we develop an experimental methodology to measure actively generated cell forces in tissues subjected to uniaxial and biaxial dynamic loading conditions. We then use a thermodynamically motivated cell contractility model to provide insight into our experimental observations.

EXPERIMENTAL METHODS

Synthetic tissues were fabricated using human cardiomyocytes and rat tail type I collagen, as shown in Figure 1. Using customized teflon molds both uniaxial and biaxial tissues were created, with stainless steel gripping systems embedded in the tissues. Tissues were then connected to a biaxial test machine fitted with custom built load cell transducers which provide accurate force measurement in the sub-mN range. Tissues were subjected to cyclic deformation at a frequency of 1Hz and an applied nominal strain of 20% in one (uniaxial) or both (biaxial) loading direction for a period of one hour. Tests were also performed on collagen hydrogels without cells, and on tissues in which cell contractility was impaired using CytoD. Tissues were fixed and stained with phalloidin and confocal microscopy imaging was performed to determine cell distributions and orientations following testing.

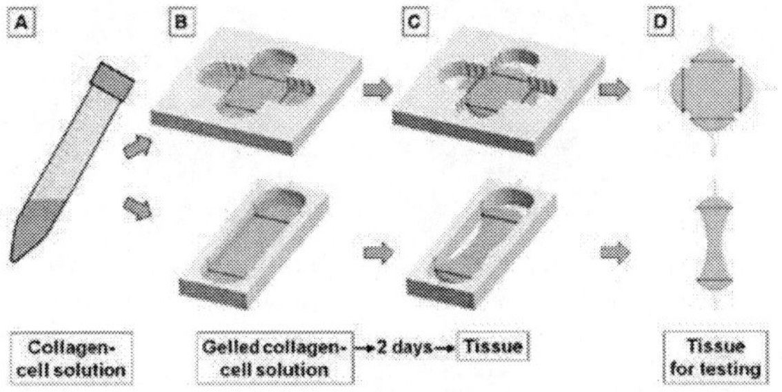

Figure 1: Illustration of sample fabrication from collagen-cell solution using customized molds for biaxial tissues (top) and uniaxial tissues (bottom).

EXPERIMENTAL RESULTS

Measured forces for contractile tissues containing cells were significantly higher than for corresponding forces measured on cell-free hydrogels and CytoD treated tissues, as shown in Figure 2. This illustrates the significant contribution of active traction generation by cardiomyocytes, both in a uniaxially and biaxially tissues. In the case of passive hydrogels, biaxial forces are 100% higher than uniaxial forces as a result of the geometric constraints. For tissues with contractile cells the biaxial forces are only 36% higher than uniaxial forces, suggesting that cell contractility contributes more strongly in uniaxial experiments.

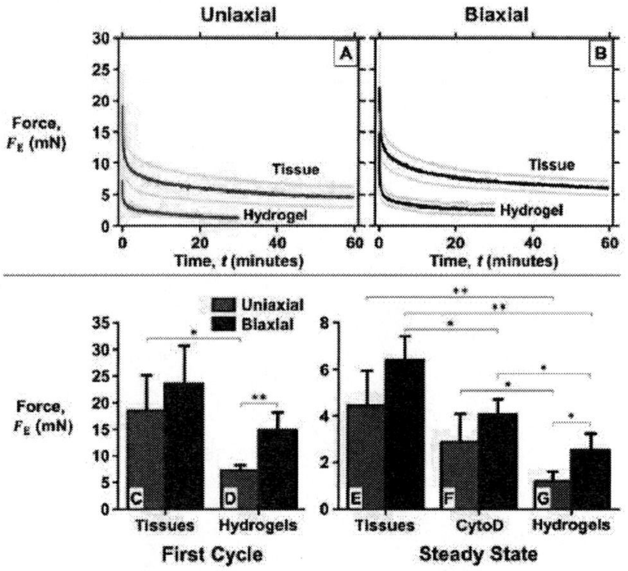

Figure 2: Time-dependent force measurements during cyclic deformation of contractile tissues containing cardiomyocytes and hydrogels (without cells). Statistically significant differences between uniaxial forces and biaxial forces are also indicated. Results are also shown for tissues treated with CytoD.

Confocal microscopy images (Figure 3) reveal that cells in the center of uniaxial tissues are highly aligned, whereas in biaxial tissues cells are randomly orientated.

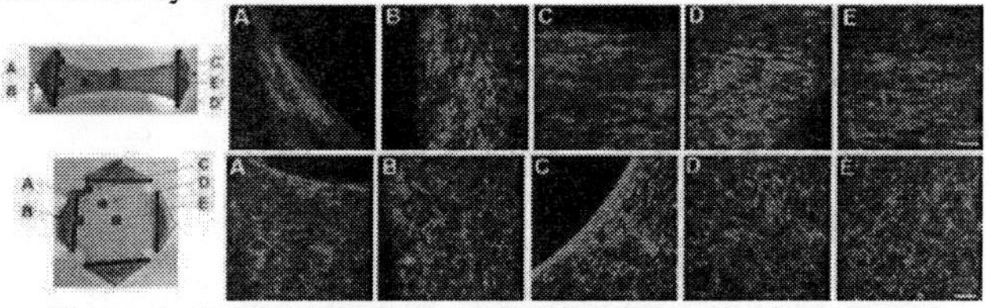

Figure 3: Spatially dependent distribution and orientation of cardiomyocytes in uniaxial (top) and biaxial (bottom) tests.

THERMODYNAMICALLY CONSISTENT CELL MODEL

Uniaxial and biaxial experiments are simulated using an active thermodynamically motivated model for cell contractility [5] in which the rate of sarcomere formation in a direction ϕ is given as

$$\dot{\eta}(\phi) = \frac{\hat{N}_u}{\pi\hat{n}}\omega_n \exp\left[-\hat{n}\frac{\mu_a-\mu_u}{k_BT}\right] - \hat{\eta}(\phi)\omega_n \exp\left[-\hat{n}\frac{\mu_a-\mu_b(\phi)}{k_BT}\right]$$

The dynamic force generation due to cross bridge cycling is based on a fading memory formulation whereby the internal strain rate in a sarcomere is given as

$$h(\phi) = \sum_{i=1}^{n} A_i \int_{-\infty}^{t} e^{-\alpha_i(t-\tau)}\dot{\varepsilon}_m(\phi)\,d\tau$$

following from the single cell dynamic models of Reynolds and McGarry [6]. Model results shown in Figure 4(A,B) predict that highly aligned high concentration sarcomeres are formed when the stress state is uniaxial $\hat{S} \approx 1$, whereas a biaxial stress state $\hat{S} \approx 0$ leads to the computation of unaligned low concentration sarcomeres. Highly accurate predictions of experimental tissue deformations (Figure 4(C,D)) and tissue forces (Figure 4(E)) are provided by this active modelling framework. Model results confirm that the small experimentally measured difference (36%) in uniaxial and biaxial force for contractile tissue (compared to the large difference (100%) for a passive hydrogel) is due to stronger sarcomere formation and alignment induced by a uniaxial stress state.

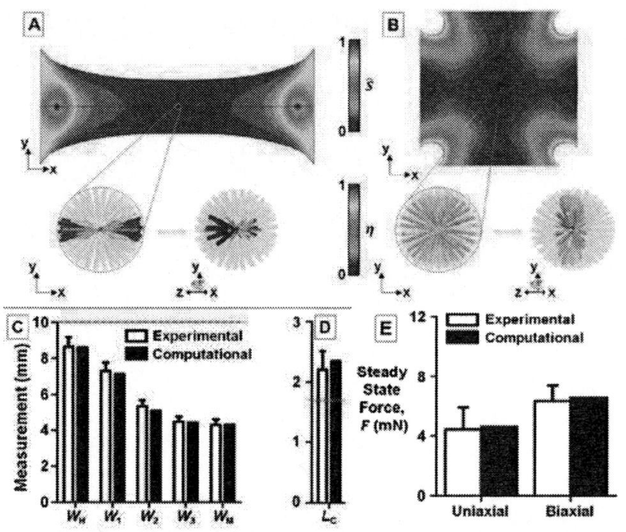

Figure 4: Deformed tissue geometry, Contours of stress uniaxiality $\hat{S}$ plotted on the deformed tissue geometries for (A) uniaxial tissue, and (B) biaxial tissue. Inserts show predicted sarcomere alignment. (C) Comparison of predicted deformations with experiments. (D) Comparison with predicted tissue forces with experiments.

DISCUSSION

This study provides for the first time an experimental comparison of active cell force in biaxial and uniaxial tissues constructs. A previous study by Thavandiran et al. [7] observed that sarcomere formation is more pronounced in uniaxial tissues, but measurement of contractile force was not implemented so the relationship between mechanical constraint and active cell force generation was not established. The current study also demonstrates the ability of a dynamic thermodynamically consistent active modelling formulation to accurately predict a complex series of experimental observations. This modelling approach, which has previously been validated for single cell experiments [6, 8] has the potential to guide tissue engineering strategies so that required levels of cell alignment and active force generation can be achieved for a range of applications.

ACKNOWLEDGEMENTS

This study is funded by Science Foundation Ireland grant SFI-12/IP/1723.

REFERENCES

[1] Billiar *et al.*, *J Biomed Mater Res A,* 73, 182-91, 2005.
[2] Mauck *et al.*, *J Biomech Eng*, 122, 252-60, 2000.
[3] Campbell *et al.*, *Biochem Biophys Res Comm,*361,329-334, 2007.
[4] Foolean *et al.*, *Biomaterials,* 33, 7508-18, 2012.
[5] Vigliotti *et al.*, *Biomech Model Mechanobiol,*15(4),761–789, 2016.
[6] Reynolds and McGarry, *Acta Biomaterialia,* 27, 251-263, 2015.
[7] Thavandira *et al.*, *PNAS,* 110(49), 4698-4707, 2013.
[8] Reynolds *et al.*, *Biomaterials*, 35(13),4015-4025, 2014.

SB3C2017
Summer Biomechanics, Bioengineering and Biotransport Conference
June 21 – 24, Tucson, AZ, USA

MODELING THE TWO-WAY FEEDBACK BETWEEN CONTRACTILITY AND MATRIX REALIGNMENT REVEALS A NON-LINEAR MODE OF CANCER CELL INVASION

H. Ahmadzadeh[1], M. R. Webster[2], R. Behera[2], A. T. Weeraratna[2], V. Shenoy[1]

(1) Department of Materials Science and Engineering, School of Engineering and Applied Science, University of Pennsylvania, Philadelphia, Pennsylvania 19104, USA.

(2) Tumor Microenvironment and Metastasis Program, The Wistar Institute, Philadelphia, Pennsylvania 19104, USA.

INTRODUCTION

Cell invasion from primary tumors is mediated by an intricate intercoupling between intracellular forces (such as cell contractility) and extracellular forces (adhesions and protrusions) that depend on the stiffness of the surrounding stroma and the alignment of matrix fibers. Here, we incorporate a mechano-chemical free energy-based approach to elucidate how the two-way feedback loop between cell contractility (induced by the activity of chemo-mechanical interactions such as Rho pathway) and matrix fiber realignment and strain-stiffening enables the cells to polarize and develop contractile forces to break free from the tumor spheroids and invade into the ECM. The predictions of the model are also validated by analyzing the invasion of melanoma cells in collagen matrices of varying concentration. Interestingly, through this computational model we are able to identify a critical stiffness which is required by the matrix to break intercellular adhesions and initiate cell invasion. The model also shows that there is a positive correlation between the elongated morphology of the invading cells and the alignment of fibers in the matrix.

METHODS

Invasion of Melanoma Cells in Collagen Matrix: Isolated melanoma cell spheroids of radius 200 μm are cultured inside collagen matrix with 0.1 to 2 mg/mL collagen concentration, and the invasiveness of the clusters and the polarization of the invading cells are investigated. In order to eliminate the effect of cell proliferation and matrix proteases, the cells were treated with 1 μg/mL Mitomycin C and MMP inhibitors.

Chemo-Mechanical Model for Cell Invasion from Tumor Clusters: The model consists of a closely-packed spheroidal cluster of contractile cells embedded in a non-linear fibrous material representing the ECM (Fig. 1). The cells are treated using a combination of passive mechanical (representing cytoskeleton) and active chemo-mechanical (representing intracellular contractility induced by myosins) elements. Cell contractility varies depending on the stiffness of the surrounding matrix, via the activation of mechanosensory molecular pathways such as Rho-pathway. Specifically, when the cell is subject to tensile forces, the activation of Rho-pathway leads to enhanced phosphorylation of myosins and ultimately, increased cell contractility. The ECM is also modeled by using a non-linear element which accounts for the alignment of the fibers and strain-stiffening. Collagen fibers in the ECM are oriented randomly in the absence of external stresses. When the local strain in the ECM is tensile and is above a critical threshold (~5%), the fibrous matrix becomes considerably aligned and undergoes strain-stiffening. Our computational model employs an energy-based approach, where the total free energy (U_{total}) of the system with contributions from the contractile cells, non-linear matrix and cellular adhesions is calculated to determine the dynamics of the cell invasion.

RESULTS:

The Driving Force of Cell Invasion Increases with the Stiffness of the Matrix: By incorporating the interplay between matrix fiber realignment and strain-stiffening with the activation of Rho-pathways, our model predicts the driving force for cell invasion. Specifically, in response to the fiber realignment due to the intrinsic contractility of the spheroid, large tensile stresses are exerted on the cells. Consequently, the Rho-pathway is activated and the cell contractility is increased. In turn, in the presence of highly contractile cells, the matrix is further stretched and more fibers become realigned. This process initiates a two-way feedback loop between the cell contractility and the matrix realignment and strain-stiffening. As this process progresses, the matrix fibers become highly aligned and the cells become sufficiently contractile to break free from intercellular adhesions and move within the matrix.

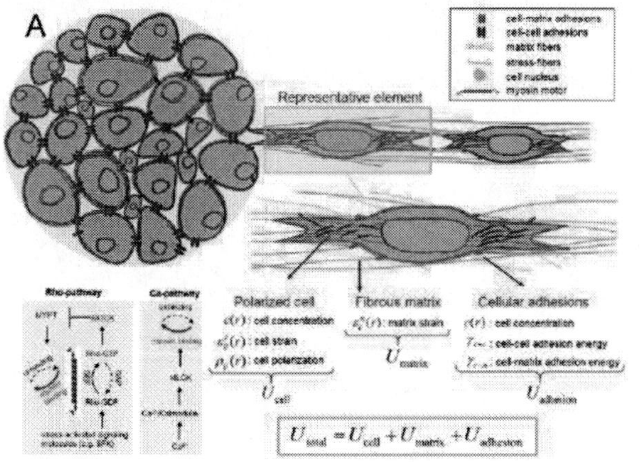

Fig. 1. (A) Model of cell invasion from a tumor spheroid into a fibrous matrix. The total free energy (U_{total}) of cells within a tumor spheroid embedded in a matrix is a function of the chemo-mechanical energy of the cells (U_{cell}), the mechanical energy of the matrix (U_{matrix}), and the adhesion energy associated with cellular adhesions ($U_{adhesion}$).

Our model shows that, by increasing the stiffness of the matrix, the driving force of invasion increases (Fig. 2). At low stiffness of the matrix, this driving force cannot overcome the cell adhesions and therefore the spheroids remain stable. While beyond a critical stiffness, the driving force can break the adhesions and induce invasion. In agreement with the model, invasion of melanoma cells in collagen matrix also shows that at low concentrations of collagen, the melanoma spheroids are stable, while by increasing the concentration, cell invasion increases (Fig. 2). Similarly, a critical collagen concentration can be identified, which demarcates the spheroids into stable vs. invasive phenotypes.

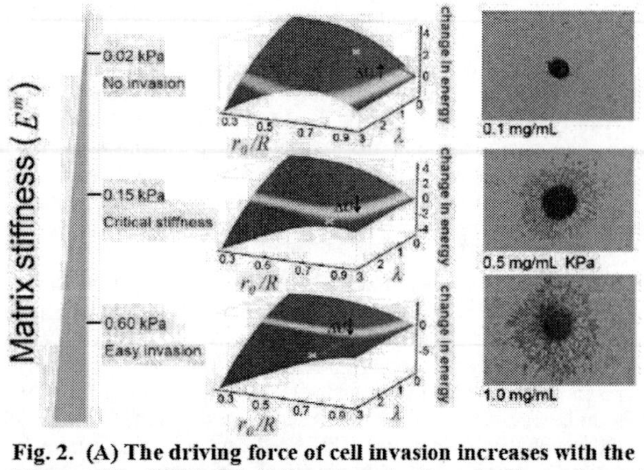

Fig. 2. (A) The driving force of cell invasion increases with the stiffness of the ECM. (B) At the concentration of collagen beyond the critical value (0.5 mg/mL), melanoma cells leave the cluster and invade into the 3D collagen matrix. (scale bar = 500 µm).

The Interplay Between Cell Contractility and Matrix Fiber Realignment Induces Polarization in the Invading Cells: Measurements of the cell polarization (defined as the ratio of the major axis to the minor axis of the cells) shows that the cells invading the matrix are elongated and this elongation increases with the concentration of collagen (Fig. 3A). Similarly, our model also shows that the cells inside the cluster are not polarized while the cells located at the periphery of the spheroid (in the invaded region) are highly polarized. Furthermore, by increasing the elastic modulus of the matrix from, the model shows an increase in cell polarization (Fig. 3B). We also find that once the cells invade into the matrix, those located close to the spheroid are more elongated than the cells that have invaded further into the matrix. This agrees with the spatial variation of the fiber alignment and that the fibers are highly aligned near the spheroid and become random with increasing distance from the spheroid; therefore, losing the ability to polarize the cells in regions far from the spheroid (Fig. 3C).

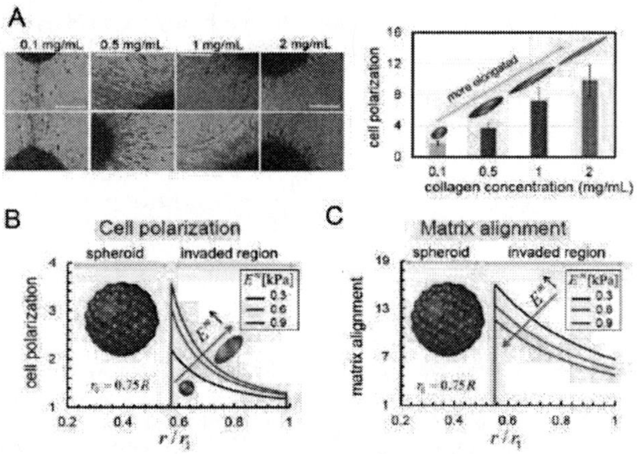

Fig. 3. (A) Cell polarization increases with collagen concentration (scale bars = 100 µm) (B) Our model predicts that the cells inside the cluster are not polarized while the cells in the invaded region are highly polarized. (C) Cell polarization is directly proportional to the alignment of the matrix fibers.

Discussion

In summary, our model presents a mathematical explanation for the role of the interplay between intracellular and extracellular forces in the initiation and progression of cell invasion from tumor clusters. Importantly, the model shows that mechanical principles mediated by the contractility of the cells and the non-linearity of the ECM behavior play a crucial role in determining the phenotype of the cell invasion[1].

ACKNOWLEDGEMENTS

This work is supported by National Cancer Institute Grants U01CA202177 and U54CA193417, NIH Grant R01EB017753 and the National Science Foundation Grant CMMI-1548571 (to V.B.S.). M.R.W is supported by K99 CA208012-01. A.T.W and R.B. are supported by R01CA174746.

REFERENCES

Ahmadzadeh et al., Modeling the Two-way Feedback Between Contractility and Matrix Realignment Reveals a Non-linear Mode of Cancer Cell Invasion. In press.

IN SITU CHARACTERIZATION OF NATIVE EXTRACELLULAR MATRIX FIBRIL DEFORMATION

Andrea Acuna (1), Michael A. Drakopoulos (1), Benjamin J. Sather (1), Craig J. Goergen (1), Sarah Calve (1)

(1) Weldon School of Biomedical Engineering
Purdue University
West Lafayette, IN, USA

INTRODUCTION

The extracellular matrix (ECM) is a dynamic meshwork of active biopolymers that controls cell behavior by defining the stiffness of their environment and activating intracellular signaling. Cells directly interact with ECM fibrils in their microenvironment at a mesoscale between the single molecule and the bulk network scales [1,2]. The mechanical behavior of the matrix at this mesoscopic level is not well defined, due to the insolubility of ECM networks that prevents the extraction of individual fibrils for mechanical testing. Previous efforts to characterize the ECM at this scale (0.5-2 µm in diameter) used fibrils generated in vitro or in silico and have a number of limitations. Without cells to orchestrate proper assembly, it is unlikely that fibrils synthesized in vitro have the same organization and material properties than those synthesized in vivo. Fibrils created in silico likely do not recreate the complex morphology and organization of native fibrils and the networks they form in vivo. Furthermore, the in situ characterization of fibril mechanics is hindered by the presence of cells and a dense ECM in adult tissues, where it is difficult to resolve individual fibrils.

To address the aforementioned problems, our laboratory has developed a technique to visualize the 3D organization of ECM fibrils by decellularizing embryonic tissues. At this stage of development, the ECM architecture is less dense and easier to visualize and manipulate experimentally. The objective of this work is to mechanically characterize native ECM fibrils in situ by quantifying fibril displacements after load in order to develop constitutive equations that describe the ECM mechanical behavior at the mesoscale. Better understanding of the ECM building blocks responsible for the initial tissue assembly is needed in order to engineer scaffolds that closely mimic in vivo conditions. This information will better inform tissue engineering efforts seeking to replicate the native cellular microenvironment for cell culture and regenerative medicine.

METHODS

C57Bl/6 murine embryos were harvested at E14.5 and incubated in an acrylamide-based hydrogel mixture that creates an interpenetrating framework to maintain the 3D architecture of the ECM during decellularization with 0.05% sodium dodecyl sulfate in PBS. Decellularized embryos were sectioned with a vibratome to create 400 µm-thick samples with homogenous thickness, facilitating the application of normal loads. Sections were immunohistochemically stained to visualize fibrillin-2 (FBN2, white) and fibronectin (FN, red) as well as a subset of proteoglycans by staining with AF488-conjugated wheat germ agglutinin (WGA, green; Figure 1A).

Samples were imaged before and after the application of static loads (0.3-2 grams) normal to the horizontal plane at 63x on a Zeiss LSM710 confocal. Image stacks were processed and rendered in 3D using FIJI (NIH). Intensities were normalized so the undeformed and deformed z-stacks have the same intensity distributions. To ensure the same region of interest in both configurations was correlated, FIJI was used to create sub-stacks that correspond to the same location in x, y, and z directions (Figure 1B).

Fast iterative digital volume correlation (FIDVC) was used to quantify the displacements of fibrils in the network, using a MATLAB algorithm that is able to evaluate large, non-linear deformations [3]. We developed an algorithm to remove displacements corresponding to dark regions (not fibrils) in both the undeformed and deformed intensity arrays, based on the intensity profile of each pair of images correlated. The magnitude and direction of fibril displacement was plotted in 3D using MATLAB. The distributions of displacements in x, y, z, and displacement magnitude of each ECM protein were plotted with MATLAB for comparison.

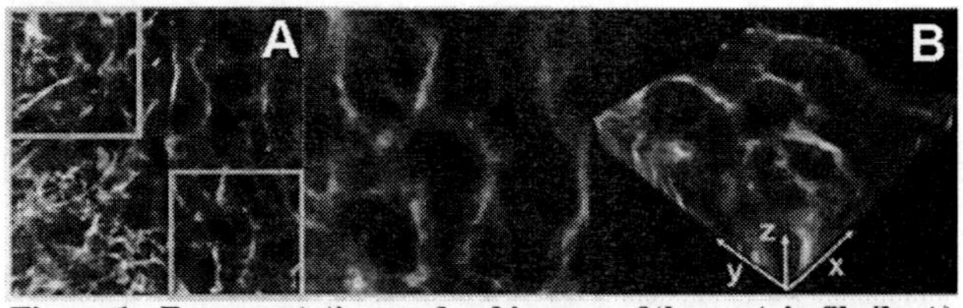

Figure 1: Representative confocal images of the protein fibrils. A) Composite of the undeformed fibrils (FBN2 = white, WGA = green, FN = red) showing the location of the two investigated sub-regions (boxed). B) Composite of the undeformed (blue) and deformed (pink) configurations of FBN2 from the bottom right sub-region of A analyzed using FIDVC (25x25x5 μm).

RESULTS

The displacement fields for the two sub-regions in Figure 1A were analyzed using FIDVC [3]. Displacements were unevenly distributed across the sample sub-regions, correlating positively with fibril density (Figure 2). There were differences in the localization of high displacement regions across FN, FBN2, and WGA. In sub-region A, FBN2 appeared to experience a lower displacement than FN and WGA, as shown by the predominantly blue and light blue regions in the displacement field (Figure 2A). This observation was confirmed by plotting overall displacement magnitude frequencies, showing that the displacement distribution curve of FBN2 has a peak at 0.6 μm whereas FN and WGA have a peak at 1 μm displacement magnitude. In sub-region B, FN experienced greater displacements than FBN2 and WGA; however, all displacement distributions had a peak at around 1.25 μm. In both sub-regions, FN experienced the greatest displacement magnitude (sub-region A: 2.55 μm; sub-region B: 3.25 μm).

The overall displacement magnitudes and the displacement magnitude in the xy plane had very similar distributions in all three proteins. This indicated that the applied load normal to the horizontal plane results in significantly greater displacements in the xy plane than in the direction of the load (-z).

DISCUSSION

The obtained displacements in sub-region A indicate FBN2 is undergoing lower deformations than FN when subjected to the same load. Since the proteins have different fibril organizations, the observed stiffness, or resistance to deformation, is likely related to the different geometries as well as differences in material behavior. Alternatively, the differences in displacement distributions between the two sub-regions could be due to the heterogeneous fibril organization in the sample and across sub-regions, leading to region-dependent displacements. In order to validate the observed trends in displacements, additional regions with a homogenous fibril distribution will need to be imaged and correlated.

Quantification of these displacements is a first step towards developing constitutive equations that describe the response of the ECM to mechanical loading. Information gathered from the proposed work has the potential to facilitate the fabrication of physiologically relevant scaffolds for regenerative medicine by establishing mechanical guidelines for microenvironments that support the assembly of tissues.

ACKNOWLEDGEMENTS

This work was supported in part by NIH grant R03 AR065201 and the Frederick N. Andrews Fellowship.

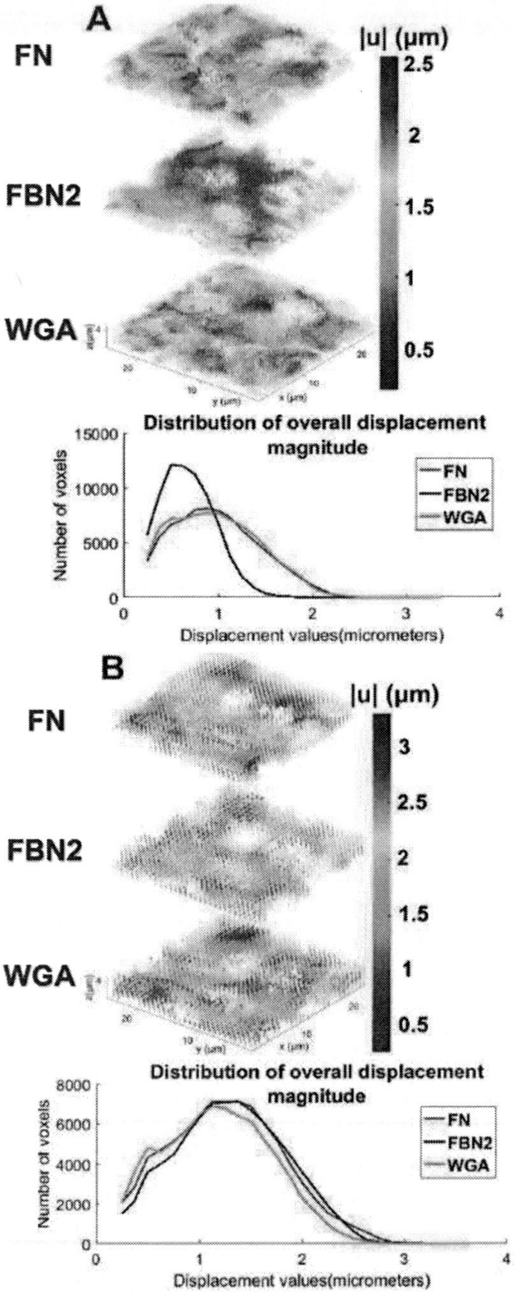

Figure 2. Displacement fields obtained from FIDVC of the unloaded and loaded z-stacks and the distributions of the overall displacement magnitude. A) Resulting displacement fields in sub-region A (bottom right corner of Figure 1A) and the magnitude distributions in this sub-region. B) Resulting displacement fields in sub-region B (top left corner of Figure 1A) and the magnitude distributions in this sub-region.

REFERENCES

[1] Bradshaw, MJ., et al., *Acta Biomater.* 10:1524–1531, 2014.
[2] Vairo, G, *J. Biomed. Sci. Eng.* 6:1–5, 2013.
[3] Bar-Kochba, E. et al., *Exp. Mech.* C:261–274, 2015.

INJURY PREDICTION USING STRAIN AND SUSCEPTIBILITY MEASURES OF THE DEEP WHITE MATTER VIA REPEATED RANDOM SUBSAMPLING

Wei Zhao (1), Yunliang Cai (1), Zhigang Li (2), Songbai Ji (1,3)

(1) Department of Biomedical Engineering
Worcester Polytechnic Institute
Worcester, MA, USA

(2) Department of Biomedical Data Science
Geisel School of medicine
Dartmouth College
Lebanon, NH, USA

(3) Thayer School of Engineering
Dartmouth College
Hanover, NH, USA

INTRODUCTION

Traumatic brain injury is a leading cause of morbidity and mortality in the United States [1]. Reliable prediction and diagnosis of TBI, including concussion, are important for effective management of this prevailing disease. Model-estimated, response-based injury metrics are desirable, as they can be directly related to tissue injury tolerances. Consequently, finite element models of the human head play an important role in brain injury studies. Model-estimated tissue responses have been shown to be more effective in injury prediction than empirically derived, kinematics-based brain injury metrics alone [2,3].

A number of tissue response-based injury metrics have been proposed; however, it remains unclear which one is the most effective, or the "best". For example, maximum shear stress in the brainstem [2], strain in the gray matter, and CSDM at strain threshold of 0.1 in the white matter (WM) [4] were separately reported as having the best accuracy in injury prediction. After incorporating WM material property anisotropy, peak axonal strain in the brainstem performed the best [3]. These conflicting reports based on the same injury dataset suggest challenges in brain injury studies.

One of the challenges is that a single training dataset has been typically used to evaluate and compare injury prediction performance. A separate cross-validation, while considered important and necessary [5], has not been typically utilized for objective performance verification. Further, previous studies attempted to empirically pinpoint a specific brain region of interest; however, they did not systematically consider other anatomically important regions or neural tracts.

In this study, we employed a repeated random subsampling technique to generate training and testing datasets and to evaluate the injury prediction performances of the entire deep WM. Findings from this study may provide important insight into how best to predict injury for better clinical concussion management in the future.

METHODS

Brain responses from the 58 reconstructed NFL head impacts [6] were simulated using the Worcester Head Injury Model (WHIM) [7]. The 50 deep white matter regions of interest (ROIs) were localized in the WHIM by registering the JHU-ICBM-DTI 81 WM atlas [8] with the T1-weight MRI used to develop the baseline WHIM. Using these 50 deep WM ROIs as anatomical constraints, their corresponding WM fiber tracts were identified [9]. A total of eight injury predictor variables were utilized, including the peak magnitude and regional average of maximum principal strain (ε_p) and WM fiber strain (ε_n) as well as ROI- and tract-wise injury susceptibility indices of the 50 WM ROIs/tracts.

To objectively evaluate their injury prediction performances, the 58 impact cases were randomly sampled into training (39 cases, or approximately two-thirds) and testing (the remaining 19 cases) datasets. A univariate logistic regression was conducted using the training data, while the testing dataset was used to assess the injury prediction performances. This process was repeated 100 times to calculate average performance metrics, including the area under receiver operating characteristic curve (AUC), accuracy, sensitivity, and specificity. For each ROI-/tract-wise injury susceptibility index, a Wald χ^2 test was conducted based on logistic regression in each subsampling trial. An optimal strain threshold was then determined when it resulted in the largest statistical significance of risk-response relationship. Each random subsampling trial generated an 8-by-50 matrix, representing the performance metrics of the 8 variables in the 50 ROIs.

To assess the relative performances among these injury predictor variables, pair-wise permutation tests based on t-statistics were performed using the performance metrics averaged from 100 random subsampling trials. Finally, the predictor variables were ranked by the number of times that a given candidate had a significantly larger average value than others (as determined by the permutation p-values).

RESULTS

Figures 1–4 summarize the average AUC, accuracy, sensitivity, and specificity for the eight injury predictors across the 50 ROIs/neural tracts. The two tract-wise injury susceptibility indices consistently had the highest AUC, accuracy, and sensitivity (averaged from the 50 neural tracts). This was confirmed by the pair-wise permutation tests (**Figure. 5**). Among the 50 neural tracts for the two tract-wise injury susceptibility indices, superior longitudinal fasciculus right (SLF-R) yielded the best AUC, accuracy, and specificity.

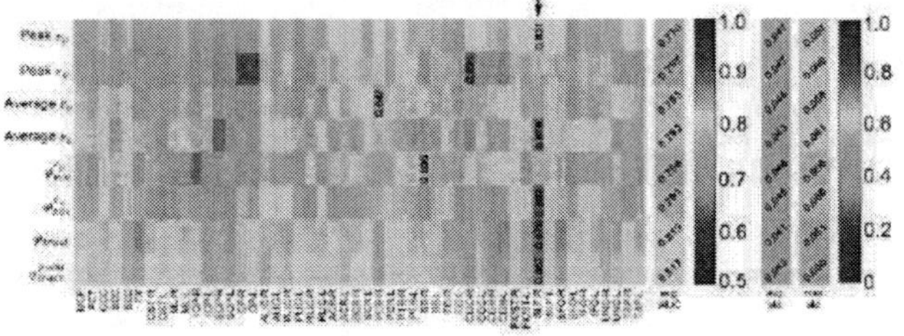

Figure 1 Summary of AUCs averaged from 100 random subsampling trials based on the training datasets, shown as an 8×50 AUC-encoded image. Each row corresponds to an injury predictor variable while each column represents a WM ROI/neural tract. The average AUC value for each injury predictor variable, regardless of the region, is also shown, along with the average and maximum standard deviations of the AUC samples across the trials. Five of the eight injury predictor variables achieved the largest AUCs (averaged across trials) in the SLF-R (arrow). $\varphi_{ROI}^{\varepsilon_p}$ and $\varphi_{ROI}^{\varepsilon_n}$: ROI-wise injury susceptibility index based on ε_p and ε_n, respectively. φ_{tract}: tract-wise injury susceptibility index based on the fraction of injured fibers for a given neural tract, weighted by the portion of injured sampling points for each given fiber. φ_{tract}^{point}: tract-wise injury susceptibility index based on the fraction of injured fiber sampling points.

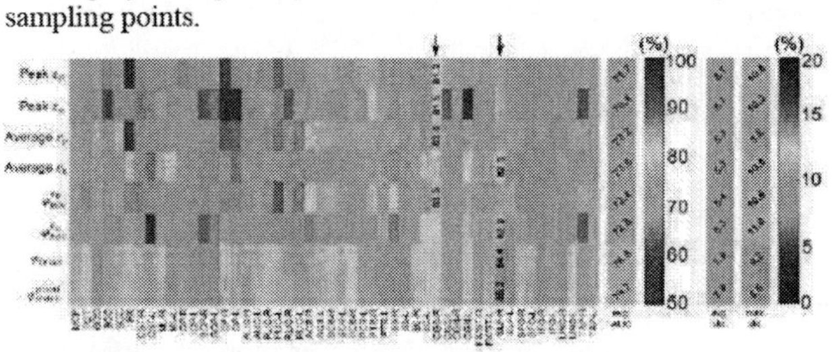

Figure 2 Summary of average accuracy measures based on the testing datasets. Four of the eight injury predictor variables achieved the best accuracy in the cingulate gyrus right; (CGC-R) and SLF-R (four for each; arrows).

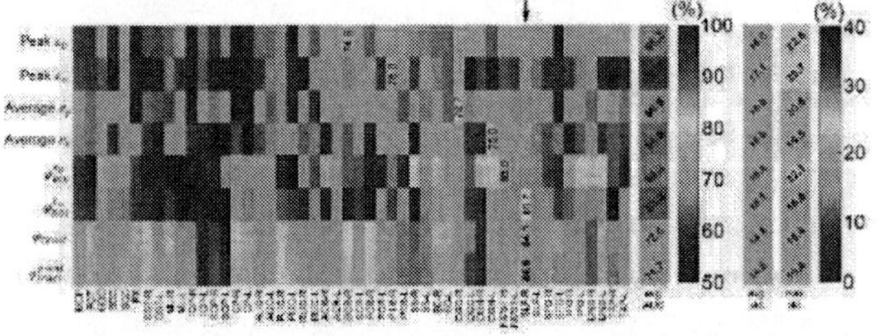

Figure. 3 Summary of average sensitivity measures based on the testing datasets. Three of the eight injury predictor variables achieved the best sensitivity in the SLF-R (arrow).

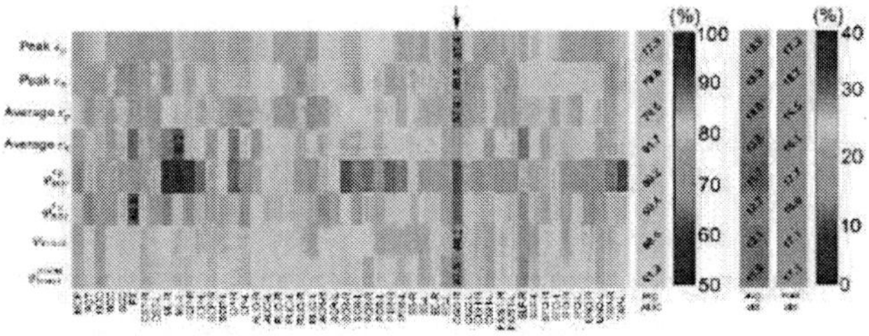

Figure 4 Summary of average specificity based on the testing datasets. Five of the eight injury predictor variables achieved the best specificity in the CGC-R (arrow).

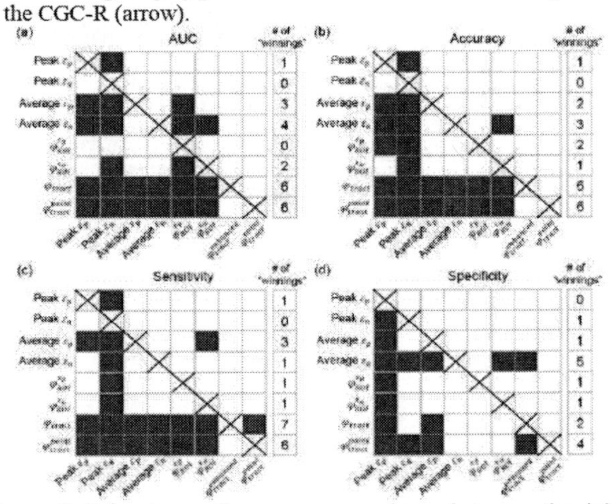

Figure 5 Pair-wise performance comparisons between the injury predictor variables in terms of AUC (a), accuracy (b), sensitivity (c), and specificity (d). Each square in a row represents whether the performance measure was significantly larger (dark gray; otherwise, white) than that in a column (self-comparisons along the diagonal excluded).

DISCUSSION AND CONCLUSIONS

In this study, we employed a repeated random subsampling technique to objectively evaluate brain injury prediction performances of the entire deep WM ROIs/neural tracts. This technique minimized data "over-fitting" concerns that may present in previous studies where a single training dataset has been typically used. Results indicated that injury susceptibility indices based on fiber strain along WM neural tracts had the best overall injury predictive power. SLF appeared to be among the most injury discriminative neutral tracts. These findings may offer a better understanding of brain injury prediction in the future.

ACKNOWLEDGEMENTS

Funding is provided by the NIH grants R01 NS092853 and R21 NS088781.

REFERENCES

[1] CDC, 2015. [2] Zhang et al., *J. Biomech. Eng.*, 126(2): 226–336, 2004. [3] Giordano et al., *Stapp Car Crash J.*, 58: 29–61, 2014 [4] Kleiven, 2007, *Stapp Car Crash J.*, 51: 81–114, 20007. [5] Anderson et al., *Comput. Methods Biomech. Biomed. Engin.*, 10(3): 171–184, 2007. [6] Newman et al., *IRCOBI Conf.*, 223–233, 2000. [7] Ji et al., *J. Neurotrauma*, 32(7): 441–454, 2015. [8] Mori et al., *Neuroimage*, 40(2): 570–582, 2008. [9] Zhao et al., *J. Neurotrauma*, 33:1834–1847, 2016.

PROS AND CONS OF ARBITRARY LAGRANGIAN EULERIAN METHOD FOR FLESH SIMULATION IN A WHOLE BODY FINITE ELEMENT MODEL FOR ACCELERATIVE VERTICAL LOADING

JiangYue Zhang, Timothy P. Harrigan, Connor O. Pyles, Connor A. Bradfield, Edna S. Wong Emily Crane, Drew P. Seker, Robert S. Armiger, Andrew C. Merkle

Research and Exploratory Development Department
The Johns Hopkins University - Applied Physics Lab
Laurel, MD, USA

INTRODUCTION

Whole body finite element models of the human body have been increasingly used to predict injury outcome from trauma producing loading events [1,2]. However, most of these models typically have been validated for impacts in the horizontal plane such as automotive crashes. In contrast, the current study focused on the dynamic accelerative loadings to the human body in the vertical direction, such as injuries sustained from underbody blast (UBB), pilot seat ejection, or helicopter crash landing. In these events, vertical accelerations often range in a few hundred Gs [3,4]. Inertial forces from the flesh to the skeleton can significantly modulate vertical loadings transferred through the skeletal system and affect skeletal injury outcome. Therefore, the interaction between the skeleton and the flesh needs to be accurately simulated. Studies have developed continuum Lagrangian finite element meshes or use simplified methods to account for the cushion effect of flesh in impact loading scenarios. However, these methods often face the challenge of element distortion and failure with severe loading. Consequently, the current study investigated the possibility of using Arbitrary Lagrangian Eulerian (ALE) method for flesh/skeleton interaction, and the pros and cons of this method in simulated UBB loading scenario.

METHODS

A 50th percentile whole body Lagrangian finite element model of the human body (Figure 1a) has been recently developed at JHU/APL. The geometry of the model was obtained from the Visible Human Project, scaled to simulate 50th percentile male subjects, and meshed using hexahedral discretization [5,6]. The model was sub-divided into head/neck, thorax, lumbar, pelvis and lower extremity body regions, and each region hierarchically validated at tissue, component and sub-system levels individually with experimental data internally produced at JHU/APL. The sub-system models were integrated into a whole body skeleton model with tied surfaces to allow force/moment transmission. A continuous inner-bone-surface was generated to encase the underlying skeletal model (Figure 1b). Tied contacts were defined between the skeletal model and the wrapping surface. The outer-skin-surface was produced by repositioning the supine skin surfaces from the visible human into 90-90-90 seated position (Figure 1c).

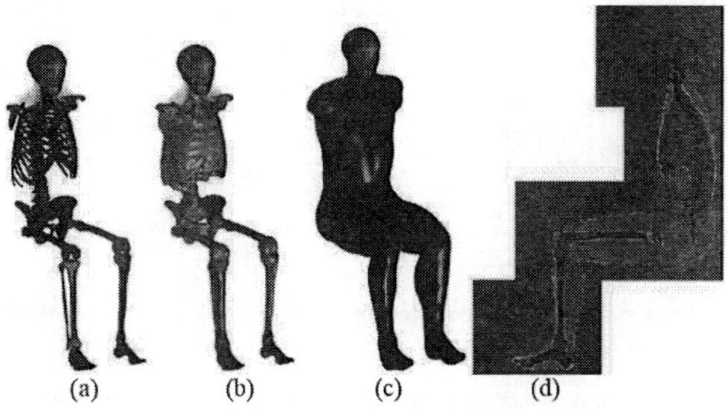

Figure 1: The skeletal models (a) were wrapped using a continuous surface to create the inner surface on the ALE flesh (b) and outer-skin-surface of the whole body model (c). Mid-sagittal cross-cut showing flesh material were filled between the inner-bone-surface and outer-skin surface (d: red colored region) and the space outside of the body and inside the inner-bone surface were treated as vacuum (d: blue regions)

An Eulerian space was defined to encase the whole body model (Figure 1d, blue). Volumes outside of the inner skeleton shrink wrap surface and inside the outer skin surface were filled with flesh materials using the Eulerian filling method as shown in the red area in Figure 1d (red). A finite element model of the Vertically Accelerated

Load Transfer System (VALTS) developed at JHU/APL was used to apply simulated UBB seat and floor acceleration profiles to the model (Figure 2). The final model consisted of approximately a total of 3.2M elements with approximately 1.2M elements for the human body, 1.2M elements for the VALTS and 800k elements for the Eulerian space. The model was simulated with MPP LS-DYNA on a cluster system with 40-80 CPUs and 256GB memory.

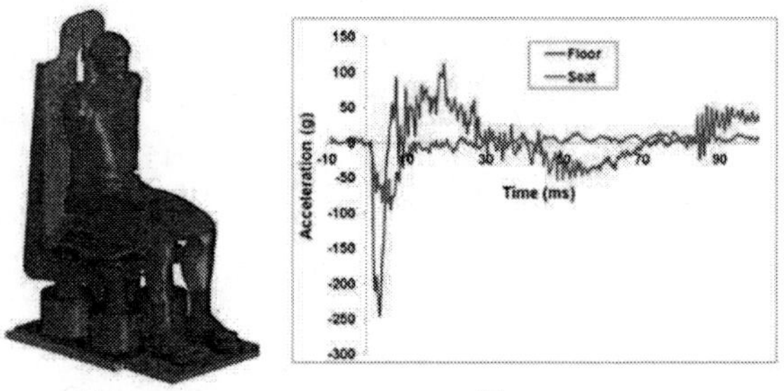

(a) (b)

Figure 2: Fully integrated model in the VALTS system (a) and acceleration history of seat and floor from a simulated UBB loading (b).

RESULTS

The ALE method was successfully implemented in the integrated with the whole body Lagrangian finite element skeleton model. The fill operation of the Eulerian method automatically generated the flesh volume between the skin surface and the skeleton wrap surface. As the model was loaded, the contact interaction of the outer skin to the VALTS seat and floor produced skin motion, which then drove the filled Eulerian flesh material that acts on the surfaces tied to the skeleton. The inertial effect of the flesh material was simulated as the Eulerian flesh was accelerated due to the coupling between the outer skin and inner skeleton wrap surfaces.

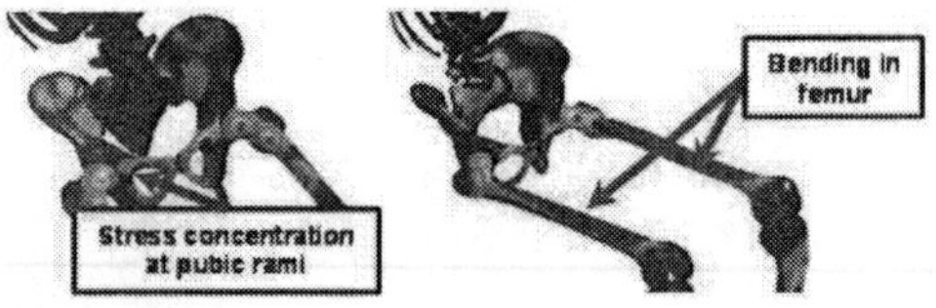

Figure 3: Accentuated stress in pubic rami and femur bending from inertial loading of pelvic content and thigh flesh

The simulation successfully completed well beyond the initial kinetic loading phase acceleration. Peak force and accelerations have already been reached during this initial kinetic phase. Accentuated stress in pubic rami and femur bending from inertial loading of pelvic content and thigh flesh (Figure 3) coincided well with reported injuries from field data and PMHS testing [7].

DISCUSSION

Modeling the flesh of the human body is a challenging task through the traditional Lagrangian approach due to a few obvious obstacles. First, the inner surface of the flesh that is coupled to the skeleton is complex and irregular. It is a challenging task that requires a major effort to produce a high quality Lagrangian mesh for such a complex geometry, not to mention it would be very time consuming as well. Second, the flesh of the human body undergoes large deformations during severe trauma-producing loads such as those occurring due to UBB. The severe distortion and large deformation of Lagrangian elements often lead to numerical instabilities and premature termination of the model. To overcome these challenges, an Eulerian/Lagrangian coupled approach was used to allow the simulation of flesh interactions with skeletal and internal structures, without the challenges of meshing and risk of numerical instability [8]. The automatic filling operation to generate the flesh volume eliminated the challenging and time consuming task of generating a finite element mesh for the irregular volume of the flesh. Using a stationary Eulerian mesh and allowing the flesh materials to move through the mesh allowed the simulation of the severe distortion and deformation of the flesh under UBB loading without the numerical instability typically associated with severely distorted/deformed Lagrangian meshes. The inertial force transmitted from the flesh to the skeleton (Figure 3) offered an explanation to the observed field injuries in otherwise hard to explain pubic rami and femur injuries.

However the convenience of the ALE method also comes with additional costs. A large volume of Eulerian space is needed to encase the whole volume into which the Eulerian flesh will likely move. This often leads to significant increases in total number of elements of the model. In the current study, approximately 1/4 of the elements are Eulerian elements. Each Eulerian element needs to be filled with appropriate materials (flesh or void) at the start of a simulation. This fill operation takes significant computational hours to complete. The volume fraction information may be reloaded if the Eulerian space and volume to be filled did not change between simulations. In addition, the material within each Eulerian element needs to be individually tracked for mass flow in/out for the proper simulation of the flesh deformation and the coupling between Eulerian and Lagrangian parts. This process consumes significantly more memory than traditional Lagrangian methods with a similar number of elements. Approximately 10x more memory is required in the current study with the ALE method. In summary, the ALE approach is still a resource extensive method. However, with the advancements in computational hardware and MPP scalability in finite element software, it has now become a viable solution for complex computation problems such as flesh/skeleton interactions in UBB loading events.

ACKNOWLEDGEMENTS

This effort was funded by contract # W81XWH-09-2-0168. The U.S. Army Medical Research Acquisition Activity, 820 Chandler Street, Fort Detrick MD 21702-5014 is the awarding and administering acquisition office. The content included in this work does not necessarily reflect the position or policy of the U.S government.

REFERENCES

[1] Schwarz, D. et al., Traffic Inj. Prev., 16:Sup1, S49-56, 2015.
[2] Maeno, T. et al., 17th ESV, Paper# 494, 2001
[3] Foster, C. et al., 2nd Workshop on numerical analysis of human and surrogate response to accelerative loading, 2016
[4] Pope, D. et al., 2nd Workshop on numerical analysis of human and surrogate response to accelerative loading, 2016
[5] Armiger, R.S. et al., Military Health System Research Symp., 2014
[6] Zhang, J. et al., RTO-MP-HFM-207-A Survey of Blast Injury across the Full Landscape of Military Sci. NATO: Halifax, CA, 2011
[7] Possley, D. et al., 2nd Workshop on numerical analysis of human and surrogate response to accelerative loading, 2016
[8] Harrigan, T.P. et al., ARL Numerical Workshop: July 7-10, 2014

SB^{3}C2017
Summer Biomechanics, Bioengineering and Biotransport Conference
June 21 – 24, Tucson, AZ, USA

MODULAR USE OF HUMAN BODY MODELS OF VARYING COMPLEXITY FOR THORACIC ORGANS

William B. Decker (1,2), Bharath Koya (2), F. Scott Gayzik (1,2)

(1) Biomedical Engineering
Wake Forest University School of Medicine
Winston-Salem, NC, USA

(2) Virginia Tech – Wake Forest School of
Biomedical Engineering and Sciences
Winston-Salem, NC, USA

INTRODUCTION

Finite element human body models (HBMs) can be used to investigate injury mechanisms and tolerance of the human body. However, the complexity required to achieve reasonable biofidelity in HBMs can lead to long compute times. The significant computational resources required to simulate detailed human body finite element models has motivated the development of faster running simplified models [1].

The current study focuses on models developed as part of the Global Human Body Model Consortium (GHBMC) project. The aim of the GHBMC is to consolidate development of computational human models for the blunt injury environment. As part of this project, simplified occupant (OS) models have been developed (e.g. GHBMC M50-OS). This simplified model retains the same body habitus and bony structures as the detailed model (GHBMC M50-O), but has simplified soft tissue structures, contact schemes, materials and reduced mesh density. The organs in the M50-OS are not individually represented but rather modeled with a simplified throaco-abdominal cavity with a lung, mediastinal, and lower abdomen component. All structures inferior to the diaphragm in the peritoneal and retroperitoneal cavity are represented by a homogenized volume [1].

Previous studies have shown the ability to modularly incorporate high detail components, such as the GHBMC M50-O brain model, into the simplified model. This technique allows for localized analysis of the region of interest in a fraction of the computational time required for the detailed model, while retaining biofidelity in the region of interest [2, 3]. Regional analysis such as this is not restricted to the brain, which has been shown in prior studies, but can be used in regions such as the organ cavity or lower extremity.

This study focuses on the modular capabilities of incorporating detailed thoracic organs in the simplified model (M50-OS+TO). There are a number of motivating factors for developing a modular capability in the thorax of the model. Chest deflection is commonly used to determine injury probability for HBM chest models, and was used during the validation of the GHBMC models. The modular approach will reduce runtime for researchers focusing only on chest injury by enabling local analysis of thoracic region, including thoracic viscera, in less time. Finally, the modular thorax will enable greater throughput. Focused analysis of organ response in HBMs has yet to be explored, but may prove to be a valuable tool in the future. Injury criteria for thoracic organs are lacking, and a more rapidly running model will reduce the computational burden of parametric studies that could lead to improved criteria.

METHODS

A detailed heart, lungs, major blood vessels, fat, and pleural space, defined by 10 parts and 6318 elements, replaced the simplified thoracic cavity of 2884 elements. Materials for the modular organs were retained during this process. The existing simplified abdominal cavity was modified to fit against the detailed organs without creating new penetrations or intersections of the mesh. No further modifications were required as the M50-OS retains the same bony structures as the M50-O. The new detailed organs were added to the existing part set for the M50-OS organ contact. This contact definition is the same for both the M50-OS and M50-O.

The M50-O, M50-OS, and M50-OS+TO were each simulated through a 6.7 m/s frontal chest impact with a rigid cylindrical impactor. The resulting chest force-displacement relationship and the runtime of each model were reported. All models were simulated using

LS-DYNA and run on the DEAC high performance computational cluster; 48 processors were used for all simulations.

RESULTS

Runtime of the simplified model approximately doubled with inclusion of the detailed organs, yet the M50-OS+TO finished the simulation in approximately one-tenth of the time of the M50-O. Normalized run times were determined as real-time minutes for every millisecond (ms) of simulation, shown in Table 1 as simulation rate.

Table 1. Runtime, simulation rate, and element density comparison for the three models.

Model	Runtime	Simulation Rate	Total # of Elements
M50-O	386 min	6.43 min/ms	2,190,806
M50-OS	19 min	0.32 min/ms	358,649
M50-OS+TO	41 min	0.68 min/ms	394,438

Maximum chest deflection was measured for each of the models as a percentage of the distance from the anterior to posterior flesh (Figure 1). This measurement was accomplished using the methodology provided by [4, 5]. Additional measures such as the max chest deflection, force, and time of max force can be seen in Table 2. Figure 1 shows a sagittal-cut view of each model at maximum deflection. This figure provides visual confirmation that the contact for the detailed organs are functioning properly in the M50-OS+TO.

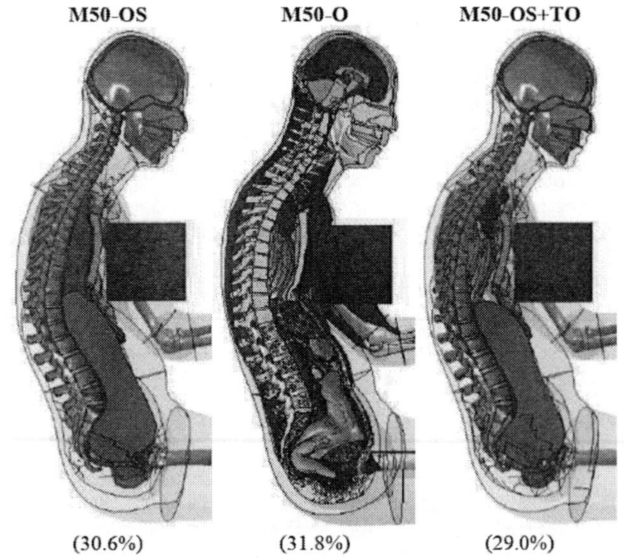

Figure 1. Sagittal cross-section view of each model at (maximum deflection).

Chest impact force was plotted against percent chest deflection for the three models in Figure 2, along with the experimental mean and one standard deviation response corridors from [5]. The force-deflection data shows an improvement from the M50-OS to the M50-OS+TO when compared to the M50-O. The M50-OS+MO match the corridors better than the M50-OS, but shows less chest deflection.

Table 2. Chest impact measurements for each model.

Model	Max Deflection	Max Force	Time of Max Force
M50-O	76.7 mm	4.36 kN	16.1 ms
M50-OS	73.6 mm	5.09 kN	12.2 ms
M50-OS+TO	69.7 mm	4.97 kN	15.6 ms

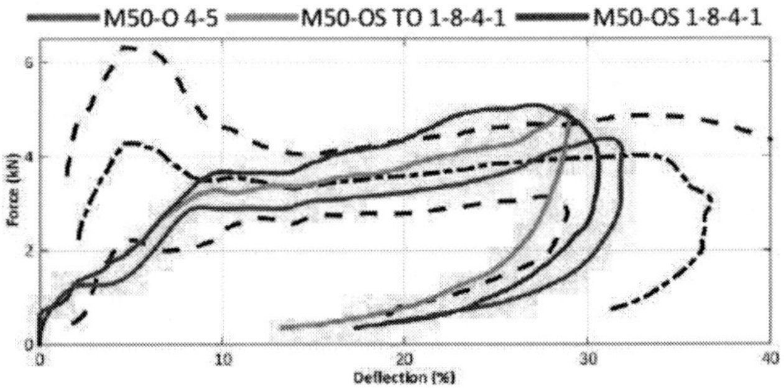

Figure 2. Chest force vs percent chest deflection of each model.

DISCUSSION

This preliminary study of the M50-OS+TO demonstrates that for the thoracic components, the model can provide comparable chest deflection data compared to the M50-O using one-tenth the compute time. With the addition of the organs in the chest, the M50-OS+TO brings the force response within the corridors, yet exhibits less deflection than the M50-O and M50-OS models. This is thought to be due to the combined material model differences and contact definitions required to integrate the thoracic viscera. Further efforts will be made to bring the results more in line with the M50-O model. Normalized runtimes for the M50-OS+TO were observed at 0.68 min/ms compared to 6.43 min/ms for the M50-O. This reduction in computational time can reduce the burden of high computational requirements when using detailed HBMs.

ACKNOWLEDGEMENTS

Funding was provided by the Global Human Body Models Consortium, LLC (GHBMC) which is supported by its member companies, participants and the National Highway Traffic Safety Administration (NHTSA).

REFERENCES

[1] D. Schwartz, *et al.*, "Development of a computationally efficient full human body finite element model," *Traffic injury prevention*, vol. 16, pp. S49-S56, 2015.

[2] W. B. Decker, *et al.*, "Quantitative Evaluation of Head Motion Kinematics Between Human Body Models of Varying Complexity," *Biomed Sci Instrum*, vol. 52, p. 9, 2016.

[3] W. Decker, "Modular use of Human Body Models of Varying Levels of Complexity: Validation of Head Kinematics," *In Review*.

[4] C. Kroell, *et al.*, "Impact Tolerance and Response of the Human Thorax," *Stapp Car Crash J*, vol. 15, 1971.

[5] M. Lebarbé and P. Petit, "New biofidelity targets for the thorax of a 50th percentile adult male in frontal impact," in *2012 International IRCOBI Conference on the Biomechanics of Impact. Dublin, Ireland*, 2012, pp. 856-870.

BRAIN MORPHOMETRICS THAT PROVIDE A BETTER UNDERSTANDING OF CHIARI TYPE I MALFORMATION

Maggie S. Eppelheimer (1), James R. Houston (2), Soroush H. Pahlavian (3), Audrey M. Braun (3), Dipankar Biswas (3), Dorothy M. Loth (2), Aintzane Urbizu (3,4), Richard Labuda (5), Philip A. Allen (2), Francis Loth (1,3)

(1) Conquer Chiari Research Center, Department of Biomedical Engineering The University of Akron Akron, OH, U.S.

(2) Conquer Chiari Research Center, Department of Psychology The University of Akron Akron, OH, U.S.

(3) Conquer Chiari Research Center, Department of Mechanical Engineering The University of Akron

(4) Duke Molecular Physiology Duke University Medical Center Durham, NC

(5) Conquer Chiari Wexford, PA

INTRODUCTION

Chiari Malformation Type I (CMI) is typically diagnosed through radiographic characterization of a tonsillar position (TP) of three to five millimeters below the foramen magnum, and the presence of symptoms such as occipital headaches, neck pain, balance problems, and many other diverse symptoms[1]. Magnetic Resonance Imaging (MRI) provides physicians and radiologists a tool for diagnosis and treatment of CMI patients. A recent study demonstrated that out of a large sample of participants, 1-2% had TP of at least five millimeters [2]. This demonstrates the importance of determining other morphometrics related to CMI which may better differentiate between asymptomatic and symptomatic individuals. Through examination of MR images, additional brain and cranial structures have been shown to differ significantly between CMI patients and healthy individuals. In previous studies, a reduced posterior cranial fossa (PCF) is the principal attribute of CMI [3]. Yet, other studies did not find a relationship between PCF area and CMI [4].

The main objective for this study was to determine additional parameters aside from TP that differentiate between CMI patients and healthy controls, with the potential to improve diagnosis and prognosis of this disorder. A subset of these parameters located within the PCF compartment will be discussed within this abstract.

METHODS

Participants and MRI Protocol. 162 CMI and 140 healthy control adult female T1 or T2-weighted pre-surgery MR images were used for this study. Demographic information and anonymized medical image data were provided voluntarily from participants through the Chiari1000 project, a web-accessed database established at The University of Akron. Healthy control participants' demographic information and MR images were obtained through Washington University, University of Minnesota, the Akron General Medical Center, and Oxford University Human Connectome Project consortium [5]. Measurements were taken on midsagittal images, where specific image quality criteria must be met for each midsagittal image to be analyzed.

Measurement Software and Parameters. Twenty-nine linear, angular, and area parameters within the PCF compartment, craniocervical junction, oral cavity, and intracranial structures were measured using an internally-developed MATLAB (Mathworks, Natick, MA) software, *MorphPro*, at The University of Akron. Motivation for the development of this software was the need for a more efficient program compared to commercially available DICOM viewers. *MorphPro* has an image viewer-based interface that allowed each operator to mark specific parameter locations through point-by-point selection. Additionally, this program provided automated storage for parameter pixel locations and geometric parameter calculations. Implementation of *MorphPro* decreased measurement acquisition time from 2-3 hours using conventional measuring techniques to approximately twenty minutes per image/participant. Of the twenty-nine parameters within this study, ten located within the PCF compartment will be discussed in this abstract. The ten parameters located within the PCF compartment consist of McRae line length (1), TP (2), fastigium height (3), clivus length (4), PCF height (5), PCF diameter (6), PCF area (7), basal angle (8), Wackenheim angle (9), and Boogard angle (10), as shown in Figure 1.

Statistical Analysis. Group differences were obtained through two-tailed independent samples t-tests for each morphometric measurement. Bivariate Pearson correlations were evaluated to find structural associations. All analysis was conducted on SPSS (SPSS Inc., Chicago, IL) and Microsoft Excel (Microsoft, Redmond, WA). A restricted dataset which controlled for age, BMI, and racial differences between CMI participants and healthy controls was used for the analysis.

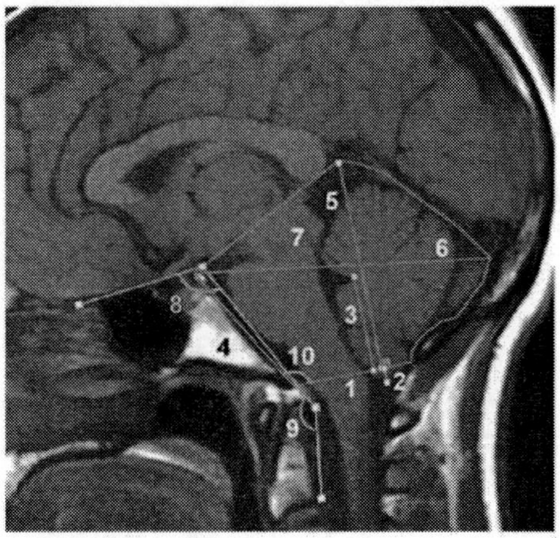

Figure 1: 10 Morphometric measurements taken on T1 midsagittal MR image of CMI participant. Numbers correspond to measurement numbers in methods.

RESULTS

PCF areas (7) between CMI participants and controls were not found to be significantly different (p = 0.006). In addition, PCF diameters (6) did not demonstrate a reliable group difference (p = 0.416), however PCF height (5) ($p < 0.001$) and fastigium heights (3) ($p < 0.001$) in CMI patients were found to be significantly shorter compared to controls. CMI patients also presented wider basal angles (8) ($p < 0.001$) and Boogard angles (10) ($p < 0.001$) with narrower Wackenheim angles (9) ($p < 0.001$) compared to controls. McRae line lengths (1) were found to be wider and TP (2) were found to be significantly lower in CM patients vs. controls.

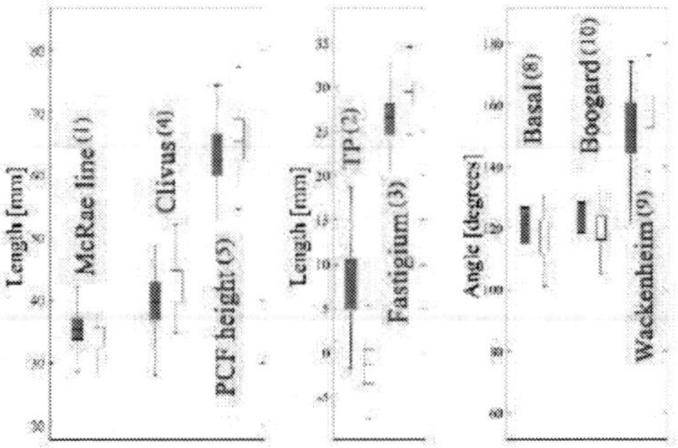

Figure 2: Boxplots of morphometric measures within the PCF compartment demonstrating significant differences between CMI patients and controls.

Measurement correlations showed interesting relationships between structures within the PCF compartment in CMI patients. Wider Boogard angles (10) and narrower Wackenheim angles (9) were significantly correlated with larger TP (2) (R = 0.32, $p < 0.001$; R = -0.29, $p < 0.001$) and reduced fastigium heights (3) (R = -0.26, $p < 0.001$; R = 0.30, $p < 0.001$). Additionally, clivus lengths (4) were positively correlated with Wackenheim angles (9) (R = 0.22, $p < 0.001$).

DISCUSSION

In this study, we developed and implemented a morphometric image processing program that enabled the analysis of a large sample of CMI patients and healthy controls through automatic storage of parameter locations and geometric calculations.

The results of this study demonstrated the descent of hard and soft tissue structures in CMI patients which was evident by a 2-3 millimeter reduction of multiple height parameters within the PCF compartment. Many studies also supported a reduction in PCF height measurements demonstrated in additional parameters including PCF height [6, 7]. The absence of PCF diameter differences between CMI patients and controls differed from previous studies which found narrower PCF diameters [6, 8]. These findings, in addition to a wider McRae line, demonstrated a flattening of the PCF compartment rather than a significant change in PCF area in CMI patients compared to controls. Since PCF area was found to be similar between CMI patients and controls, this study does not support PCF area reductions acting as a clinical marker for CMI. The changes in PCF form are also demonstrated by a horizontally-angled clivus bone. This is evident with a wider basal and Boogard angle, and a narrower Wackenheim angle. In addition, measurement correlations supported a flatter PCF compartment and demonstrated a lower cerebellum in CMI patients. Further studies are necessary to assess the possible impact of the observed morphometric differences on the strain/stress environment around the neural tissues in the PCF compartment.

While this study shows insight into CMI morphology, many limitations were present. Such limitations included the use of self-report data for diagnostic conditions and demographic information. Since MR images were voluntarily provided, MR machine manufacturers and image clarity varied widely. Our sample was also restricted to adult while females due to a lack of other ages, genders, and race.

In conclusion, group comparisons and measurement correlations not only suggested an overall change in PCF form in CMI patients, but also supported the impact of the clivus bone in the abnormal development of PCF structures indicative of CMI. With this finding, our future goal is to determine the impact of these structural differences on symptom severity and surgical outcome to better help and evaluate CMI patients.

ACKNOWLEDGEMENTS

We would like to thank all CMI patients for participating in this study and for Conquer Chiari for financial support.

REFERENCES

[1] Fischbein R, Saling JR, Marty P, Kropp D, Meeker J, Amerine J, et al. *Neurological Sciences,* 36:1617-24, 2015.

[2] Smith BW, Strahle J, Bapuraj JR, Muraszko KM, Garton HJ, Maher CO. *J Neurosurg*, 119:812-9, 2013.

[3] Milhorat TH, Chou MW, Trinidad EM, Kula RW, Mandell M, Wolpert C, et al. *Neurosurgery*, 44:1005-17, 1999.

[4] Sekula RF, Jr., Jannetta PJ, Casey KF, Marchan EM, Sekula LK, McCrady CS. *Cerebrospinal Fluid Res*, 2:11, 2005.

[5] Van Essen DC, Ugurbil K, Auerbach E, Barch D, Behrens TE, Bucholz R, et al. *Neuroimage*, 62:2222-31, 2012.

[6] Aydin S, Hanimoglu H, Tanriverdi T, Yentur E, Kaynar MY. *Surg Neurol*, 64:237-41; discussion 41, 2005.

[7] Urbizu A, Poca MA, Vidal X, Rovira A, Sahuquillo J, Macaya A. *J Neuroimaging*, 24:250-6, 2014.

[8] Yan H, Han X, Jin M, Liu Z, Xie D, Sha S, et al. *Eur Spine J*, 25:2202-9, 2016.

SB^{3}C2017
Summer Biomechanics, Bioengineering and Biotransport Conference
June 21 – 24, Tucson, AZ, USA

MECHANICAL PROPERTIES OF INJURED MOUSE BRAIN TISSUE

Y. Feng (1, 2), Y. Gao (3), T. Wang (3), L. Tao (3), S. Qiu (1), X. Zhao (1)

(1) Center for Molecular Imaging and Nuclear Medicine, School of Radiological and Interdisciplinary Sciences (RAD-X) Soochow University, Suzhou, Jiangsu, China

(2) School of Mechanical and Electronic Engineering, Soochow University, Suzhou, Jiangsu, China

(3) Department of Forensic Science, Soochow University Suzhou, Jiangsu, China

INTRODUCTION

Traumatic brain injury (TBI) affects a large portion of population in both developed and developing countries [1-5]. Understand the mechanical behavior of brain tissue is crucial to uncover the mechanism of TBI. Although many studies have been dedicated to study the mechanical properties of the brain tissue, many of them investigate only the healthy brain tissue [1, 4, 6, 7]. However, limited studies have been done to investigate the mechanical properties of the injured tissue.

For TBI studies using animal models, typical experiments such as weight-drop, controlled cortical impact and fluid percussion are among the mostly used. For the animal species, it is shown that only in transgenic mice the genetic factors and be readily explored [8]. Therefore, in this study, we adopted the weight-drop method to investigate the injured brain tissue in a mouse model.

Many different methods have been used to characterize the mechanical properties of brain tissue. Shear testing and indentation tests are among the mostly used methods [7, 9, 10]. Although shear test provides the desirable homogenous deformation of the soft tissue, small size of the mouse brain tissue posed challenges to fix the tissue. However, indentation test provides a relatively easy way to test the small samples and has the advantage to test specific regions within a small sample.

In this study, we used indentation method with a circular shaped punch to measure the mechanical properties of the injured brain tissue sing a mouse model. The injury was imposed by a weight-drop device. Regions of the injury site and the surrounding tissues were tested and compared. The results provided information to understanding the injury mechanism and to guide in vivo imaging methods such as MRE.

METHODS

A total of 6 adult male ICR (Institute of Cancer Research) mice were used in this study. All mice were raised and obtained from the Laboratory Animal Center of Soochow University. All of the animal procedures and protocols were approved by the Institutional Animal Care and Use Committee of the Soochow University and conducted in accordance with the National Institutes of Health Guide for the Care and Use of Laboratory Animals.

Each mouse was anesthetized before imposing injury with the weight-drop device. A midline incision was created to expose the skull (Figure 1a). The impact was produced by a free-fall of a 40-gram weight from 20 cm height onto a 4-mm diameter footplate resting on the Dura (Figure 1b). The impact position was centered at the bregma with a 2.7 mm to the lateral left (Figure 1c). The controlled depth of the impact was 1 mm. The brain injury was imposed on the left hemisphere of the mouse

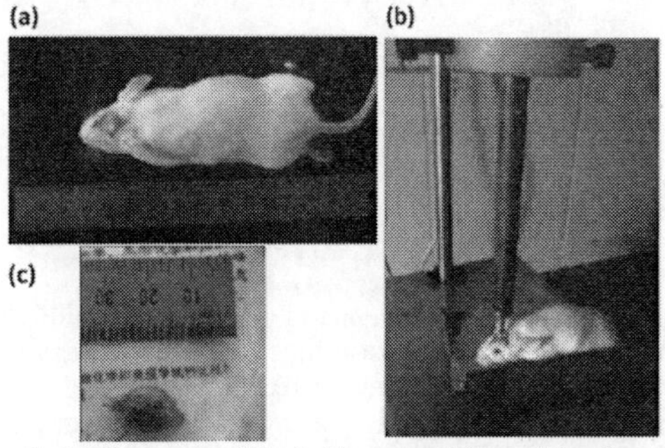

Figure 1. (a) A mouse with skull exposed for getting impact injury. The skin wrapping the skull was peeled off. (b) The brain injury was induced by impacting the left hemisphere of the brain by a custom built impacting device. (c) Sample brain tissue harvested after impact injury.

The tissue properties were measure 1 hour after the injury. A custom-built asymmetric indentation device was used to characterize white matter [11]. A circular flat indenter head with a diameter of 2 mm was used to indent samples. The indenter was first lowered to get into contact with the sample surface. Then each sample was indented 8% of its sample thickness to minimize the boundary effect [12]. The indentation stiffness k (mN/mm) of the sample were calculated by linearly fitting the indentation force-displacement curve, and k is the slope of the fitted line. With the assumption of the incompressibility, the Poisson's ratio was taken as 0.5 and the Young's modulus of the sample was calculated by [12].

$$E = \frac{3}{4}\frac{k}{D} \quad (1)$$

where k is the indentation stiffness and D is the diameter of the circular indenter head.

To compare the injury site and it surrounding tissues, and the injured hemisphere with its contralateral non-injured hemisphere, we selected eight regions of interests for measurements (Figure 2).

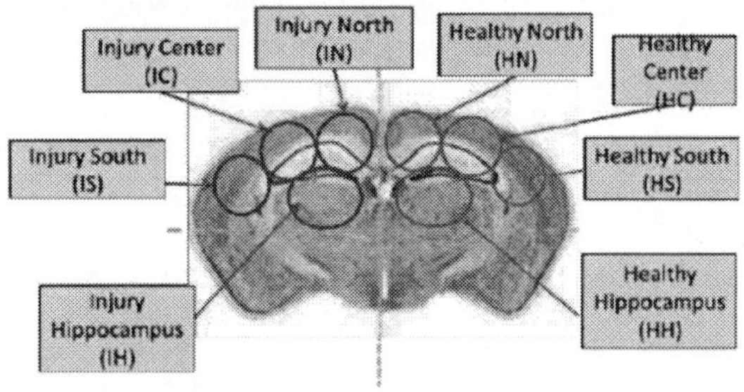

Figure 2. Regions of interests tested for a mouse brain.

RESULTS

Typical indentation force-displacement curves of both healthy and injured ROIs are shown in Figure 3. We fitted the largely linear force-displacement curves and calculate the E values. Comparisons of the E values between each ROI pair are shown in Figure 4. Significant differences of the E values were observed between the injured and healthy ROIs one hour post-injury.

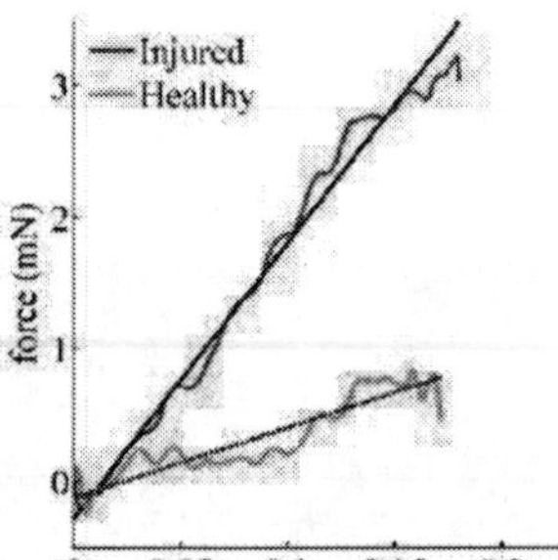

Figure 3. Sample force-displacement curves from the indentation of the mouse brain 1 hour post-injury. The injured region (IC) and its contralateral healthy region (HC) are compared. The solid and dotted black lines are linear fittings of the experiment data.

We also compared the modulus differences among different regions. For the injured hemisphere, the hippocampus region (IH) appeared to have the lowest mean *E* value (3.34 kPa) while the highest *E* value was observed at IS (13.52 kPa). For the uninjured contralateral hemisphere, HN had the highest mean *E* value (7.13 kPa), white IH had the lowest mean *E* value (1.25 kPa). Between the injured and uninjured hemispheres, the IC had significantly larger *E* value than the HC (student t-test, $p<0.05$). No significant differences were observed at the other pairs. The largest differences between the mean values of the injured and its contralateral regions also happened between HC and IC (2.86 kPa).

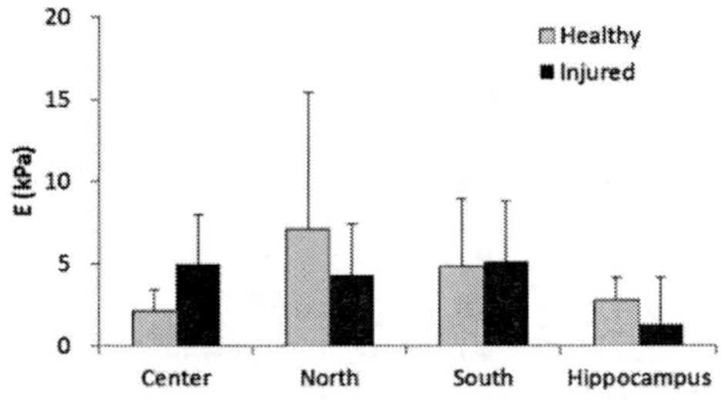

Figure 4. (a) Sample indentation force-displacement curves of the injured and healthy tissue. (b) Comparisons of the healthy and injured regions at the four tested locations.

DISCUSSION

In this study, we investigated the regional mechanical properties of the brain tissue post-injury. Using indentation techniques, we observed that, compared with its contralateral region, the elastic modulus at the center of the injured region was significantly higher one-hour post-injury. However, no significant differences were found at adjacent regions. Results indicated that the most significant changes of the mechanical properties of the brain tissue after injury concentrated at the injured site. Increased modulus could be due to the flux of fluid due to hematoma and formation of the astrocyte. Although brain tissue has viscoelastic properties, we considered only elastic properties as one perspective to illustrate the property change of the tissue after the injury. Future studies will include more accurate modeling and take longitudinal effect into considerations.

ACKNOWLEDGEMENTS

Funding is provided by grant 61503267 (YF) and 81530062 (LT) from National Natural Science Foundation, grant BK20140356 (YF) and 16KJB460018 (YF) from Jiangsu Province, and by grant K511701515 (YF) from Scientific Research Foundation for the Returned Overseas Chinese Scholars, State Education Ministry. We thank Prof. Philip V. Bayly for the helpful discussion and suggestions.

REFERENCES

[1] Bayly, P.V., et al., *Ann Rev BME*, 2012. **14**: p. 369-396.
[2] Crisco, J.J., et al., *J Biomech*, 2011. **44**(15): p. 2673-2678.
[3] Meaney, D.F., et al., *J Biomech Eng*, 2014. **136**(2).
[4] Prange, M.T. and S.S. Margulies, *J Biomech Eng*, 2002. **124**(2): p. 244-52.
[5] Goriely, A., et al., *BMMB*, 2015. **14**(5): p. 931-965.
[6] Chatelin, S.,et al., *Biorheology*, 2010. **47**(5-6): p. 255-276.
[7] Destrade, M., et al., *I J Non Mech*, 2015. **75**: p. 54-58.
[8] Brody, D.L., et al., *J Neurotrauma*, 2007. **24**(4): p. 657-673.
[9] Rashid, B., et al., *JMBBM*, 2013. **28**: p. 71-85.
[10] Feng, Y., et al., *JMBBM*,, 2013. **23**: p. 117-132.
[11] Feng, Y., et al., *JMBBM*,, 2017. **65**: p. 490-501.
[12] Budday, S., et al., *JMBBM*,, 2015. **46**: p. 318-330.

VISCOELASTIC BEHAVIOR OF ISOLATED CERVICAL SPINAL CORD AND PIA MATER TISSUES

Nicole L. Ramo (1), Kevin L. Troyer (2), Christian M. Puttlitz (1,3)

(1) School of Biomedical Engineering
Colorado State University
Fort Collins, CO, USA

(2) Component Science and Mechanics
Sandia National Laboratories
Albuquerque, NM, USA

(3) Department of Mechanical Engineering
Colorado State University
Fort Collins, CO, USA

INTRODUCTION

The pia mater is the innermost of the meningeal layers and closely envelops the spinal cord. Due to the difficulty in isolating spinal pia mater samples for experimental testing, the majority of conclusions about its mechanical importance have been inferred from comparisons between data obtained from constructs of the intact pia and cord complex to data obtained with the pia excised. While these comparative studies have shown that the thin membrane has important functions related to overall cord-pia complex stiffness [1,2] and recovery after compression [1], the independent mechanical properties of pia mater are required in order to fully model all of the relevant spinal cord complex components and ultimately understand the role it plays in spinal mechanics. To the authors' knowledge, only one study has examined the mechanical properties of isolated spinal pia mater, reporting the elastic moduli of the tissue at three strain-rates [3]. Since the spinal cord is subjected to time-varying loads *in-vivo*, there exists a significant need to characterize the non-linear viscoelastic behavior of the pia mater. This study represents a first step in addressing this need by comparing the non-linear viscoelastic properties of: (i) the spinal cord with the pia mater attached, (ii) the isolated cord, and (iii) the isolated pia mater.

METHODS

Following a method previously established by our laboratory [4], a section of freshly harvested (unfrozen) ovine cervical spinal cord was loaded into a custom built uniaxial test stand (Figure 1A). Following a 100-cycle preconditioning protocol, 4 cyclic tests to 5% strain and two stress-relaxation tests (one to 5% strain and the other to 3% strain) were performed in a randomized order with 4 minutes of recovery between each test. Three of the dynamic cyclic test frequencies (0.0035, 0.035, 0.35 Hz) correspond to the three largest decadal values of the Prony series time constants used to approximate the relaxation moduli ($\tau = 100$ s, 10 s, 1 s; see RESULTS section). The highest frequency was dictated by the speed of the actuator, and therefore, depended on the gauge-length of the test condition (0.75 Hz for spinal cord with pia attached, 0.61 Hz for the isolated cord, and 1.22 Hz for the isolated pia mater). The ramping rate for both stress-relaxation tests was 6 mm/s. Following data collection from the cord with pia mater attached, a superficial longitudinal incision was made along the cord and the pia mater was carefully removed. The isolated pia was laid flat, wrapped in saline-soaked gauze, and stored in an airtight bag while the aforementioned testing was repeated on the isolated cord. Using a second uniaxial test stand (Figure 1B), the isolated pia mater was also tested using the same protocol as the previous two test conditions.

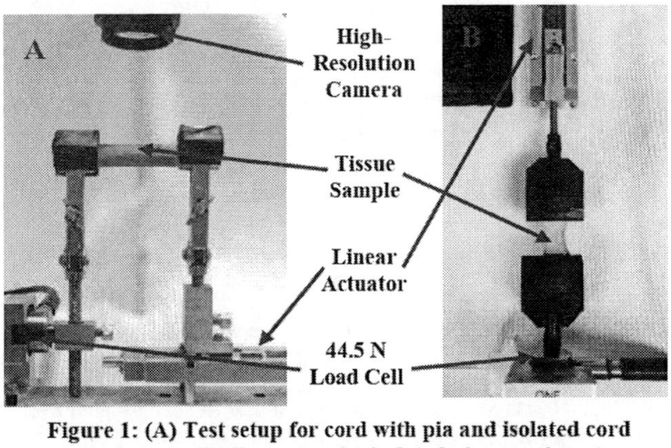

Figure 1: (A) Test setup for cord with pia and isolated cord experiments. (B) Test setup for isolated pia experiments.

Uniaxial force data were converted to engineering stress (σ) using measured cord cross-sectional area and literature reported pia thickness [1]. Displacement data were converted to engineering strain (ε) using the grip-to-grip distance as the gauge length.

Using a novel data fitting method (presented in a separate abstract submission), data from the four cyclic tests were simultaneously fit to the following non-linear viscoelastic model:

$$\sigma(t,\varepsilon) = \int_0^t E(t-\tau,\varepsilon)\frac{d\varepsilon(\tau)}{d\tau}d\tau \qquad (1)$$

where the strain-dependent relaxation modulus was represented as:

$$E(t,\varepsilon) = E_\infty(\varepsilon) + \sum_{i=1}^{4} E_i(\varepsilon)e^{\frac{-t}{\tau_i}} \qquad (2)$$

with time constants fixed at $\tau_1 = 0.1$ s, $\tau_2 = 1$ s, $\tau_3 = 10$ s, $\tau_4 = 100$ s. The strain-dependent Prony weights were cast as:

$$E_\infty(\varepsilon) = C_1^\infty\varepsilon + C_2^\infty\varepsilon^2 \qquad (3)$$

$$E_i(\varepsilon) = C_1^{\tau_i}\varepsilon + C_2^{\tau_i}\varepsilon^2 \qquad (4)$$

resulting in a total of 10 fitted parameters.

RESULTS

Despite the preconditioning protocol, relaxation was still apparent over the course of the five cycles of each dynamic test (Figure 2). The magnitude of the force response for each test condition (cord with pia attached, isolated pia mater, isolated cord) was consistent across the cyclic frequencies tested. Considering the significantly smaller cross-sectional area of the pia mater, this finding is consistent with past studies that have shown the pia mater to be significantly stiffer than the cord with the pia mater, which itself is stiffer than the isolated spinal cord [1,2].

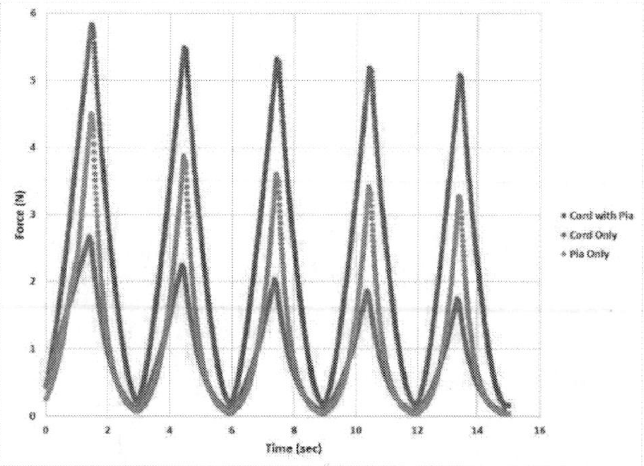

Figure 2: Force-time results for the 0.35 Hz tests showing intra-test relaxation and varying tissue stiffness.

The resulting fitted parameters are shown in Table 1. These fitted parameters were then used to predict the stress-relaxation response (Figure 3). The average RMSE of the predictions were as follows: Cord with Pia – 0.006 MPa (average of 12% of peak stress); Isolated Cord – 0.001 MPa (average of 7% of peak stress); Isolated Pia – 0.38 MPa (average of 6% of peak stress).

Table 1: Fitting results for dynamic test data (units MPa).

	Cord with Pia	Isolated Cord	Isolated Pia
$C_1^{\tau_1}$	52.29	-27.05	-200.31
$C_2^{\tau_1}$	-301.37	1204.85	68125.74
$C_1^{\tau_2}$	117.45	8.97	283.01
$C_2^{\tau_2}$	-4030.31	82.63	131214.11
$C_1^{\tau_3}$	-92.00	-1.29	-3559.55
$C_2^{\tau_3}$	4849.56	440.73	131982.42
$C_1^{\tau_4}$	107.24	71.94	-361.07
$C_2^{\tau_4}$	-2617.90	-2723.36	93141.86
C_1^∞	52.75	-16.49	4810.81
C_2^∞	-1377.30	496.41	-78100.19
Avg. RMSE	**0.005**	**0.002**	**0.473**

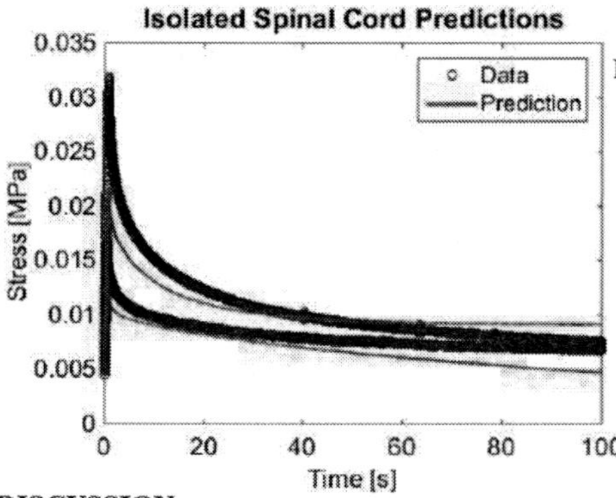

Figure 3: Non-linear viscoelastic stress-relaxation prediction of isolated cord using coefficients obtained from cyclic data.

DISCUSSION

In order to address a critical need in understanding spinal cord viscoelasticity, this study presented preliminary results of non-linear viscoelastic modeling of the cervical spinal cord with pia mater attached, isolated spinal cord, and isolated pia mater tissues. Our novel experimental procedure demonstrates the feasibility of performing experimental tests on isolated pia mater to elucidate its contribution to spinal mechanics.

Qualitative results on the relative stiffnesses obtained in the current study are consistent with the limited previous research available [1,2]. Although the RMSE of the predicted stress-relaxation response is relatively high compared to the peak stress, these predictions represent a first attempt to predict the behavior of these tissues under loading conditions not used to fit the data.

In the future, the methodology described above will be applied to additional samples to confirm that the predicted results fall within experimental variability. The results may then be used as material behavior input data for computational models of the spinal cord to better predict the spinal mechanics related to spinal cord injuries.

ACKNOWLEDGEMENTS

Sandia National Laboratories is a multi-mission laboratory managed and operated by Sandia Corporation, a wholly owned subsidiary of Lockheed Martin Corporation, for the U.S. Department of Energy's National Nuclear Security Administration under contract DE-AC04-94AL85000.

REFERENCES: [1] Ozawa H et al. *J Neurosurg Spine* 2004; 1:122–7
[2] Mazuchowski E, Thibault L. *ASME Summer Bioeng Conf.* June 2003
[3] Kimpara H et al. *Stapp Car Crash J* 2006; 50:509–44
[4] Shetye S et al. *Acta Biomaterialia*, 2014 10:792-197

SB3C 2017
Summer Biomechanics, Bioengineering and Biotransport Conference
June 21-24, 2017, Tucson, AZ, USA

PLANAR BIAXIAL MECHANICAL PROPERTIES OF SWINE VAGINAL TISSUE

Jeffrey McGuire[1], Steven Abramowitch[2], Spandan Maiti[2], and Raffaella De Vita[1]
[1]STRETCH Lab
Department of Biomedical Engineering and Mechanics
Virginia Tech
Blacksburg, Virginia 24061

[2]Department of Bioengineering
University of Pittsburgh
Pittsburgh, Pennsylvania 15260

INTRODUCTION

Pelvic floor disorders (PFDs), such as urinary incontinence, fecal incontinence, and prolapse, affect one in three adult women in the United States. One of the major risks for developing a PFD is injury to the pelvic floor during pregnancy and vaginal delivery. Particularly, mechanical damage and tears of the vagina are quite common and can often propagate to the surrounding organs. These tears initiate at the vagina but extend to the perineum, sphincter, and rectum, ultimately compromising the structure and function of the entire pelvic floor. Unfortunately, up to 65% of women incur an obstetric laceration to the vagina during vaginal delivery [1].

The vagina is a multilayered organ comprised of smooth muscle and dense connective tissue. To date, mechanical testing of the vagina has been performed primarily through uniaxial testing methods and mechanical properties of vaginal tissue have been found to be influenced by factors such as pregnancy, PFDs, and age [2]. Little has been done to study the biaxial properties of the vagina, despite the importance in examining the vaginal tissue in a biaxial state of stress. Biaxial tests are more physiologically relevant than uniaxial tests and can better portray the anisotropic behavior of the tissue. In this preliminary study we characterize, for the first time, the biaxial mechanical properties of swine vaginal tissue. Our long-term goal is to mechanically investigate how lacerations form and propagate within the vaginal wall.

METHODS

The swine was used as the animal model due to the established similarity between the physiology of swine and human vaginas. The animal protocols were approved by the Virginia Tech Institutional Animal Care and Use Committee. Vaginal tissue was dissected from four adult female domestic swine reproductive tracts, and the thickness of each specimen was measured with calipers (accuracy ± 0.05 mm, Mitutoyo Absolute Low Force Calipers Series 573, Japan). The resulting average thickness of the specimens was 2.6 ± 0.71 mm. Specimens were cut to have a 15 x 15 mm square central region and then dyed, speckled, and loaded equi-biaxially into an Instron planar biaxial machine (Instron, UK). Non-contact strain measurements were performed using high-resolution cameras and a digital image correlation (DIC) system (Correlated Solutions, Columbia, SC). Loads were recorded with four 20 N load cells simultaneously. Specimens were pre-loaded to 0.1 N, preconditioned for 10 cycles from 0.2 N to 1.0 N at a load rate of 0.05 N/s, unloaded for five minutes, and stretched at 0.05 N/s until an equi-biaxial load of 18 N was obtained in both directions. The DIC software was used to calculate the average Lagrangian strain in the longitudinal direction (LD) and circumferential direction (CD) over a specific region of interest, while stress was calculated by dividing the recorded loads by the initial cross-sectional area. Figure 1 displays Lagrangian strain maps, specifically in the CD, computed at different equi-biaxial loads for one specimen.

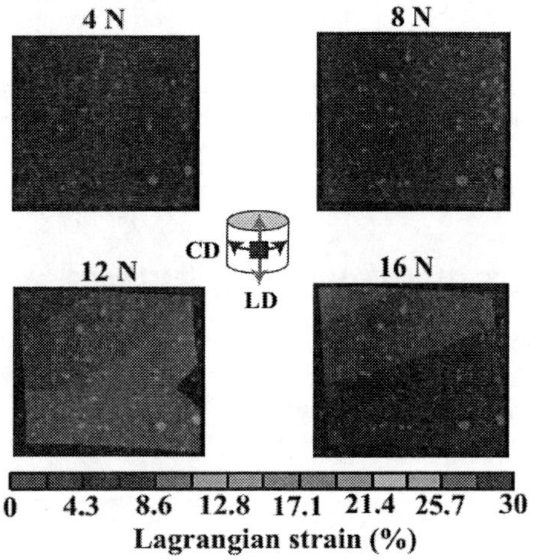

Figure 1. Lagrangian strain map in the CD for a representative specimen subjected to equi-biaxial loads of 4, 8, 12, and 18 N.

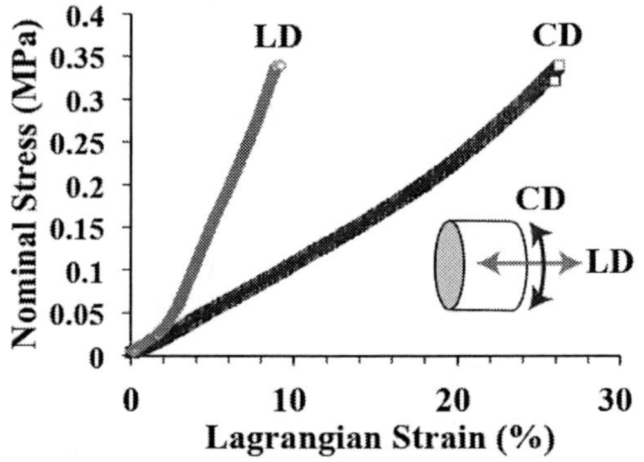

Figure 2. Stress-strain results of a representative specimen.

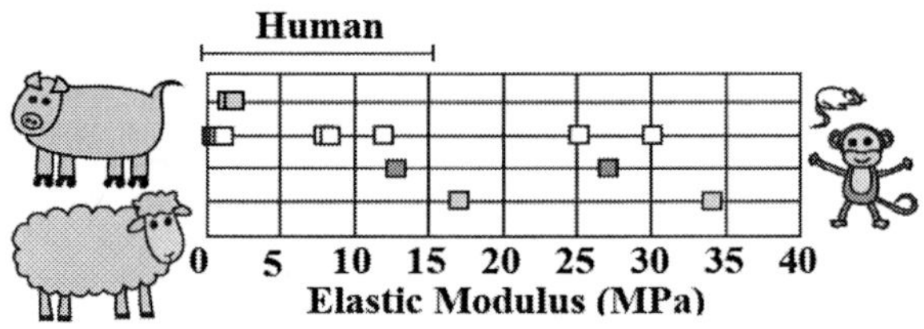

Figure 3. Ranges of elastic moduli reported in literature compared with our results from swine.

RESULTS

Equi-biaxial test results from a representative specimen are presented in Figure 2. The swine vaginal tissue exhibited a nonlinear and anisotropic response, most likely due to the presence of collagen fibers preferentially oriented in the longitudinal direction. Stresses in the longitudinal and circumferential directions were found to be similar at small strains, but significantly larger in the longitudinal direction compared with the circumferential direction at large strains, further highlighting the tissue's anisotropy. These differences in the mechanical properties of specimens should be studied further with scanning electron microscopy (SEM) to determine fiber orientation as a function of location along the length of the swine vaginal tissue. Indeed variability in initial fiber orientation could significantly alter the biaxial response of the tissue.

Figure 3 displays our data for swine elastic moduli compared with those of other animal models. Tangent moduli for the longitudinal and circumferential directions were found to be 2.78 MPa and 1.4 MPa, respectively. These results were on the lower end of the ranges reported in the literature [2]. We expect the average value of the elastic modulus of the swine specimens tested here to be lower since such value was obtained through biaxial testing rather than uniaxial testing, as done in other studies, and thus the specimens were not pulled to failure [2].

DISCUSSION

This is the first study to present planar biaxial data on the swine vaginal wall. The tissue clearly presents a nonlinear anisotropic behavior. SEM and histology studies will be performed to examine the tissue micro-structure along the length of the vaginal wall for development of constitutive models. In addition, planar biaxial tests will be performed on swine specimens to examine tear propagation and characterize the pre-failure response of the tissue. Computational models will be developed and evaluated from these data. This work will aid in patient specific recommendations regarding term delivery, as well as surgical and physical therapy treatment for PFDs.

ACKNOWLEDGMENT

This research was supported by NSF award #1511603.

REFERENCES

[1] K. J. Stepp, et al., "Textbook recommendations for preventing and treating perineal injury at vaginal delivery." Obstetrics & Gynecology, 2006. 107(2, Part 1): 361-366.

[2] A. Baah-Dwomoh, et al., "Mechanical Properties of Female Reproductive Organs and Supporting Connective Tissues: A Review of the Current State of Knowledge." Applied Mechanics Reviews, 2016. 68(6): 060801.

SB^{3}C2017
Summer Biomechanics, Bioengineering and Biotransport Conference
June 21 – 24, Tucson, AZ, USA

CHANGES IN THE TIME-DEPEDENT MECHANICAL BEHAVIOR OF THE CERVIX IN A NORMAL MOUSE PREGNANCY

Kyoko Yoshida (1), Mala Mahendroo (2), Kristin Myers (1)

(1) Department of Mechanical Engineering
Columbia University
New York, New York, USA

(2) Department of Obstetrics and Gynecology
UT Southwestern Medical Center
Dallas, Texas, USA

INTRODUCTION

During pregnancy, the female reproductive system undergoes a dramatic remodeling process in which the uterus and cervix must synchronously grow and remodel for the successful birth of a healthy baby. In particular, the remodeling process of the uterine cervix has implications for a successful development and delivery of the fetus. For the majority of gestation, the cervix remains rigid and closed. Towards the end of a normal pregnancy, the cervix dramatically softens and dilates to allow for the safe delivery of a healthy, term baby. Preterm birth (PTB) is a current clinical dilemma, which annually affects 15 million babies worldwide [1]. Since an acceleration of the cervical remodeling process can lead to PTB, our goal is to better understand this process for normal pregnancy in an effort to reduce the PTB rate.

The cervix is a collagenous structure made up of Type I collagen fibers. Cervical collagen fibers undergo tremendous remodeling during pregnancy as stronger intermolecular collagen crosslinks are replaced by weaker crosslinks [2]. This remodeling process is reflected in second harmonic images as straight, aligned fibers in nonpregnant (NP) cervical tissue and wavy fibers that are less organized in pregnant tissue. Based on this evidence, we concluded that the breakdown of collagen is in part responsible for the cervical remodeling process [3]. In previous publications, we demonstrated that a porous fiber composite constitutive model captures the *equilibrium* mechanical behavior of NP and pregnant cervices from humans [4] and mice [5]. Using this equilibrium material model, we characterized the cervical remodeling process in a normal mouse pregnancy as a decrease in fiber stiffness by 4 orders of magnitude [5]. As a collagenous tissue, however, the cervix exhibits a time-dependent response to loading. Therefore, the objective of this study is to characterize the time-dependent response of cervical tissue towards a more complete material model for the cervix.

Here, we provide further analysis of the mechanical data from our previous stress-relaxation tests [5]. We also present additional mechanical data from instantaneous load-to-break tests and compare this data to our previously published equilibrium response. Based on our results, we discuss the changes in the time-dependent response of the cervix as gestation progresses and highlight the importance of evaluating both equilibrium and instantaneous tissue response.

METHODS

Mechanical testing

Mechanical testing methods were previously published [2]. Briefly, whole, intact cervical tissue were dissected out from NP and nulliparous pregnant female mice at gestation days (d) 6, 12, 15, and 18. Each cervical sample was tested using a ring test configuration with a universal testing machine (Instron 5948 MicroTester, 10 N load cell) while two CCD cameras (Point Grey, GRAS-50S5M-C 75 mm, f/4 lens) tracked the geometry of the tissue throughout testing.

Samples were allowed to swell to equilibrium in physiological saline (PBS) prior to loading and kept in the solution bath throughout testing. To determine the equilibrium properties of the tissue, cervical samples were subjected to stress-relaxation testing (n=3-5 per gestation group). Each cervix sample was displaced in increments of 0.5 mm grip-to-grip displacement. NP and d6 samples were allowed to relax for 30 mins, and d12, d15, and d18 were allowed to relax for 60 mins at each displacement hold. To determine the instantaneous response of the tissue, additional NP, d15, and d18 cervix samples were subjected to load-to-break testing (n=4-5 per gestation group). For these tests, tensile grips were continuously displaced until the sample broke. For all mechanical tests, samples were loaded at a 0.1mm/s grip-to-grip displacement rate.

Technical Presentation #296

Load-to-break analysis

In our previous study, we conducted a full inverse finite element analysis (IFEA) to determine the equilibrium material properties of the tissue [5]. The IFEA was necessary due to the complex loading conditions of the mechanical test conducted here. For the analysis of the instantaneous load-to-break tests, we conducted a simpler analysis by calculating the *structural* properties of the cervix. This structural analysis is commonly used in the literature to analyze load-to-break tests on mouse cervical tissue [2,6].

RESULTS

Stress-relaxation behavior

All samples exhibited a force-relaxation in response to a displacement hold (Fig. 1A,B). To evaluate the ability of the cervix to stretch, we compared the maximum fiber stretch in our previous IFEA analysis at the highest sustained deformation for each sample [5]. We also calculated the peak to equilibrium force ratios at this highest level of deformation to determine the degree of relaxation for each sample (Table 1). In NP and early pregnant d6 samples, the cervix exhibited relatively small amounts of relaxation, leading to similar peak and equilibrium forces. These samples stretched by a relative small amount before breaking. By gestation d12, significant relaxation response was observed as indicated by higher F_{peak}/F_{eq} ratios at higher sustained stretches. Neither average values for d12, however, were significantly different compared to NP and d6. By gestation d15, tissue samples sustained significantly higher stretch and larger relaxation behavior compared to NP and d6 samples. These results indicate that with progressing gestation, the cervix is able to stretch significantly more and exhibit more force-relaxation in response to a displacement hold.

Table 1: Pregnant samples sustain higher stretches and exhibit more force-relaxation at the maximum sustained stretch. * indicates $p<0.05$ vs. NP and d6, † indicates $p<0.05$ vs. d12 (one-way ANOVA).

Gestation day	Max fiber stretch	F_{peak}/F_{eq}
NP (n=5)	1.27 +/- 0.16	1.10 +/- 0.03
d6 (n=4)	1.18 +/- 0.07	1.16 +/- 0.09
d12 (n=3)	1.56 +/- 0.20	1.98 +/- 0.80
d15 (n=4)	2.27 +/- 0.61*	3.33 +/- 0.96*
d18 (n=4)	2.70 +/- 0.45*†	3.65 +/- 1.41*

Instantaneous vs. Equilibrium tensile response

Equilibrium stiffness curves for NP cervix were similar to instantaneous load-to-break curves (Fig.1C). In late pregnant samples, however, the instantaneous load-to-break response was much stiffer compared to the equilibrium response (Fig.1D). To compare instantaneous and equilibrium tissue structural stiffness, the maximum slope was calculated for both tests based on the loading curves, averaged for each test for NP, d15, and d18. For NP cervix, the maximum stiffness was not significantly different compared to the instantaneous stiffness. For the pregnant cervix, the instantaneous stiffness was 15 times stiffer in d15 tissue and 16 times stiffer in d18 tissue compared to the equilibrium stiffness. These results indicate that the NP cervix is less sensitive to time scales of loading, whereas the pregnant cervices have very different responses to quick and sustained mechanical loading.

DISCUSSION

The objective of this study was to characterize the time-dependent mechanical response of NP and pregnant mouse cervices. Our analysis demonstrates that the viscoelastic behavior of the cervix changes dramatically with progressing gestation. In our stress-relaxation tests, pregnant cervices relaxed significantly more at higher sustained stretches compared to NP and early pregnant d6 cervices. These late pregnant samples also exhibited much stiffer responses to instantaneous loading compared to equilibrium loading, whereas NP tissue exhibited similar stiffness responses to both. The relaxation response during a stress-relaxation test can be attributed to the combination of collagen fibers rearranging into a preferred equilibrium state and fluid flow into and out of the tissue. Based on the known collagen fiber changes during pregnancy, we postulate that the straight, aligned fibers in the NP cervix do not rearrange significantly and are less porous, leading to small amounts of relaxation. On the other hand, we postulate that the loosely arranged fiber network in the pregnant cervix rearrange significantly when allowed to relax and the higher porosity allows more free movement of the fluid. Since *in vivo* loading conditions of the cervix during pregnancy likely include both instantaneous (i.e. Braxton Hicks contractions, fetus kicking) as well as equilibrium loading conditions, we suggest that both responses should be considered to fully characterize cervical mechanical response, particularly for pregnant samples.

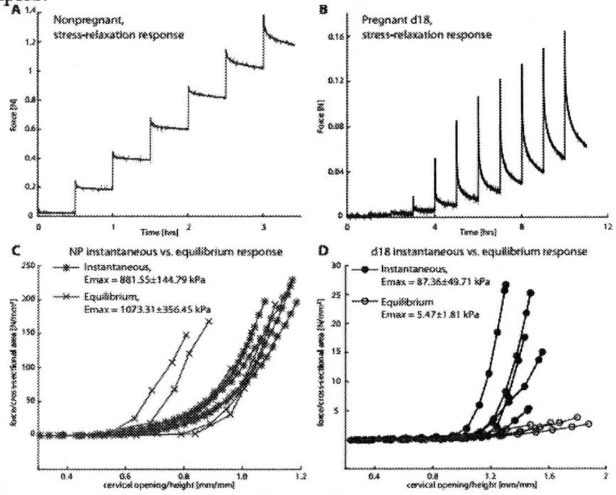

Figure 1: Representative stress-relaxation behavior of NP (A) and pregnant d18 (B) cervix. Instantaneous vs. equilibrium loading response in (C) NP and d18 (D) cervices.

The main limitation in our study is the calculation of tissue structural properties rather than material properties for the instantaneous response. We attempted to fit our material model to the load-to-break data using our previous IFEA framework to quantify the material changes in the fiber stiffness. In this attempt, however, we were unable to fit our model to the load-to-break data because of the sudden stiffening at high levels of deformation seen in our pregnant samples (Fig. 1D). Low fiber stiffness values were required to capture the initial long linear region, but these values were unable to capture the sudden stiffening at higher deformations. Therefore, we are currently formulating a full viscoelastic material model to capture both instantaneous and equilibrium responses of the cervix.

ACKNOWLEDGEMENTS

This work was funded by the NSF Graduate Research Fellowship, the NSF CAREER grant, and the March of Dimes Prematurity Initiative.

REFERENCES

[1] WHO, http://www.who.int/mediacentre/factsheets/fs363/en/.
[2] Yoshida, K. et al, PLOS One, 9(11): e112391, 2016.
[3] Akins, M. et al, J Biomed Opt, 15(2): 026020, 2010.
[4] Myers, K. et al, J Biomech, 48(9): 1533-40, 2015.
[5] Yoshida, K. et al, Acta Biomat, 36: 195-209, 2016.
[6] Timmons, B. et al, Endocrinology, 155(1): 298-98, 2013.

FINITE ELEMENT (FE) MODELING OF MONKEY OPTIC NERVE HEAD (ONH) BIOMECHANICS: METHODS AND PRELIMINARY RESULTS

Fanwei Kong (1), Andrew Feola (1), Stephen A. Schwaner (2), Hongli Yang (3), Howard Lockwood (3), Juan Reynaud (3), Claude F. Burgoyne (3), C. Ross Ethier (1,2)

(1) Coulter Department of Biomedical Engineering, Georgia Institute of Technology/Emory, Atlanta, GA, USA

(2) George W. Woodruff School of Mechanical Engineering, Georgia Institute of Technology, Atlanta, GA, USA

(3) Devers Eye Institute, Optic Nerve Head Research Laboratory, Portland, OR, USA

INTRODUCTION

Glaucoma is the second most common cause of blindness worldwide [1]. Intraocular pressure (IOP) is the primary risk factor for its onset and progression, but the disease can occur at all levels of IOP. A common hypothesis is that optic nerve head (ONH) deformation drives connective tissue remodeling and ultimately results in retinal ganglion cell (RGC) loss. Axonal damage likely begins within the lamina cribrosa (LC), a porous connective tissue spanning the scleral canal through which RGC axons pass as they exit the eye [2]. However, the sensitivity to IOP is eye-specific and differs among individuals. Here, we model the biomechanical environment of the ONH in response to elevated IOP in monkeys to identify and understand the eye-specific determinants of ONH susceptibility to IOP.

Finite element (FE) models have been used to study the complex biomechanical environment of the ONH. However, most previous FE studies have used idealized or simplified geometries that did not incorporate complex ONH geometry[3], and none compare their results with *in vivo* OCT image sets to refine the models behavior. We present a method for building eye-specific FE models to compare stresses and strains between normal and early glaucomatous (EG) monkey eyes using ONH geometries obtained from 3D histomorphometric reconstruction. We compared the acute ONH deformations predicted by our eye-specific models to *in vivo* OCT measurements obtained prior to sacrifice. We then adjusted the material properties to determine if we could match those OCT targets.

METHODS

Early glaucoma (EG) was unilaterally induced in a monkey (26 years old), with the contralateral eye serving as a control. Prior to sacrifice, both eyes (control and EG) underwent OCT imaging at IOPs of 10 and 30 mmHg. At each pressure the eyes were allowed to equilibrate for 30 minutes. Eyes were then perfusion fixed at an IOP of 10 mmHg and a mean arterial pressure (MAP) of 55 mmHg [4]. The ONHs were embedded in paraffin, serial sectioned and 3D reconstructed [4]. The ONH anatomy was manually delineated within 40 radial digital section images of both the 3D histomorphometric and OCT images including the LC, pre-laminar neural tissue (PLNT), optic nerve (ON), retinal vessel (RV), pia mater, and sclera (SC). Surfaces were constructed (Rhino V5.0, Seattle, WA) from these delineations and meshed (Hypermesh V13.0, Troy, MI). Boundary conditions were determined by solving a simple posterior eye model and applying the local displacements to a refined ONH (Figure 1) in the FE solver FEBio (V2.5).

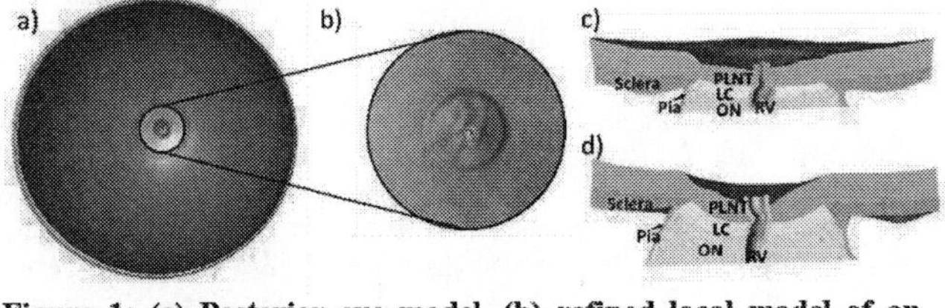

Figure 1: (a) Posterior eye model, (b) refined local model of an ONH, and (c) and (d) cross-sectional views of the control and EG eye local models (as perfusion fixed, i.e. at IOP 10 mmHg).

Tissues were considered to be linear-elastic and nearly incompressible (ν=0.49). Baseline Young's modulus values were taken from earlier FE work: sclera and pia: 5 MPa; ON and PLNT: 0.05 MPa; RV: 0.3 MPa, LC: 0.5 MPa [5,6]. The scleral Young's modulus falls within the range reported for aged monkey eyes [6]. We compared the computed 1st and 3rd principal strains between control and EG eyes at baseline material properties at an IOP of 30 mmHg.

To directly compare deformations predicted from our FE simulations and those measured from OCT images we assessed the following parameters: 1) LC deformation, defined as the change in LC position relative to the anterior scleral canal opening (ASCO; Figure 2a). 2) & 3) ASCO radius and ASCO area expansion; and 4) sclera bowing, defined by the depth of ASCO relative to a peripheral scleral reference plane (ASCO depth-reference plane; Figure 2b&2c). All FE and OCT outcomes parameters were obtained as IOP was elevated from 10 to 30 mmHg. After simulations with our baseline material properties, we increased and decreased the modulus of the sclera and LC by up to 4 fold to investigate their influence on our outcomes.

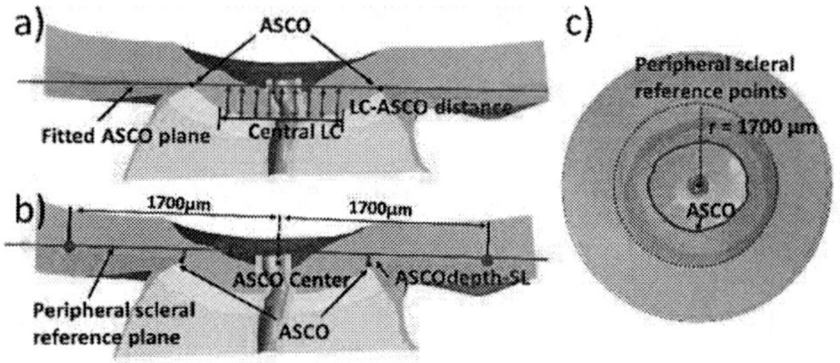

Figure 2: ONH deformation parameters within the FE model geometry for the EG eye. Cross-sectional views showing: (a) LC-ASCO distance and (b) ASCO depth-SL measurements. (c) En face view of the model with peripheral scleral reference points.

RESULTS

FE Computed Strains. There were differences in the ONH strain distributions between control and EG eyes (Figure 3). Relative to the control eye, the EG eye had a 24% lower peak (95th percentile) 1st principal strain, and a 17% higher peak (5th percentile) 3rd principal strain.

Figure

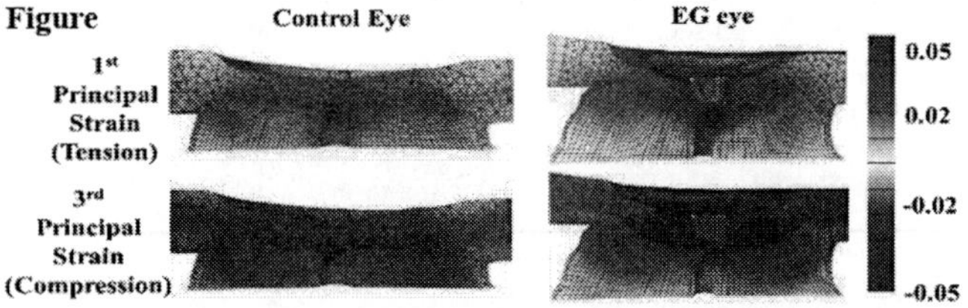

3: Computed contour maps of the 1st and 3rd principal strain distributions of the control (left) and EG (right) eyes.

FE Comparison to OCT Measurements. At baseline material properties, the FE-predicted LC deformation and scleral bowing were less than those measured from OCT images in both control and EG eyes (Figures 4a&4b). However, the ASCO radius and area expansion in the EG eye were near the OCT measurements at baseline properties (Figure 4c&4d). We then considered more compliant or stiff sclera properties (1.43MPa, 10MPa, 15MPa and 20MPa) and a more compliant LC (0.143MPa). As scleral stiffness increased, the LC deformation increased while ASCO radius and area expansion decreased for both the control and EG eyes. Sclera bowing plateaued when scleral stiffness was above 10 MPa in the control eye. However, there was a lack of agreement of LC deformations and scleral bowing between FE models and OCT measurements for both eyes for all simulations. In the EG eye, a compliant LC increased sclera bowing and LC deformation while ASCO expansion showed little change (Figure 4c&4d). In the control eye, a compliant LC increased LC deformation and ASCO expansion, leading to a better agreement between the predicted ASCO radius and expansion from our FE model vs. OCT measurements (Figure 4c&4d).

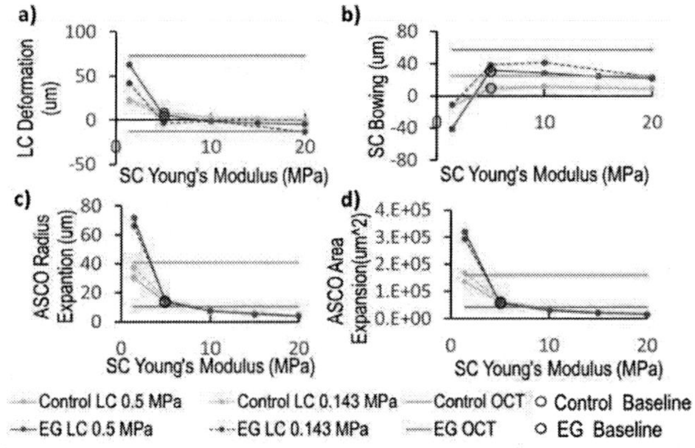

Figure 4: Comparison between FE results and OCT measurements in control and EG eyes for various scleral (SC) and LC material properties.

DISCUSSION

This approach provides a robust framework to investigate the differences in ONH biomechanics between control and EG eyes. We are specifically working towards model validation by comparing FE simulation outputs to OCT imaging data to refine tissue material properties of control and EG eyes. A major advantage of our approach is the use of eye-specific geometries; however, attempting to refine our material properties failed to capture the complex behavior observed by OCT measurements, specifically the LC deformation and scleral bowing.

The lack of agreement between our model results and OCT data may be related to several factors. One possibility is tissue shrinkage that occurs during the perfusion fixation process, and we are developing methods of adjusting OCT data to account for this confounder. We also intend to use different material properties for the peripapillary sclera and annular scleral ring vs. the remaining posterior sclera, which will potentially allow for more complex and realistic scleral and ONH deformations. Lastly, our present model uses linear-elastic material models. In future, we will incorporate more representative material properties that include collagen microarchitectural information for the sclera and LC, collected on aged animals. We will also regionalize the results to correlate LC strains to longitudinal OCT structural alterations by foveal-Bruch's membrane opening sectors, and changes in axon counts to improve our clinical significance.

ACKNOWLEDGEMENTS

Georgia Research Alliance (CRE), NIH T32 EY007092 (SAS), EY011610 (CFB), Legacy Good Samaritan Foundation (CFB), AGS Mid-Career Award (CFB), Sears Medical Trust (CFB), Alcon Research Institute (CFB)

REFERENCES

[1] Varma, R et al., *Am J Opthalmol*, 152(4): 515-22, 2011.
[2] Burgoyne, CF, *Exp Eye Res*, 93(2): 120-132, 2011.
[3] Sigal, I et al., *Biomech Model Mechanobiol*, 8(2): 85-98, 2009.
[4] Burgoyne, CF et al., *IOVS*, 45, 4388-4399 , 2004
[5] Feola, A et al. IOVS, 57(4) 1901-11, 2016.
[6] Girard, M et al, Invest Ophthalmol Vis Sci 50(11): 5226–5237

SB^{3}C2017
Summer Biomechanics, Bioengineering and Biotransport Conference
June 21 – 24, Tucson, AZ, USA

POSTERIOR SCLERA AND OPTIC NERVE DEFORMATION COMPARISON BETWEEN GLAUCOMATOUS AND NORMAL HUMAN EYES

Ehab A. Tamimi (1), Jeffery D. Pyne (2), Stephen J. Howerton (3) Jonathan P. Vande Geest (1,4,5)

(1) Department of Bioengineering
University of Pittsburgh
Pittsburgh, PA, United States

(2) Department of Mechanical Engineering
University of California Berkeley
Berkeley CA, United States

(3) Department of Aerospace and Mechanical Engineering
University of Ari□ona
Tucson, AZ, United States

(4) McGowan Institute for Regenerative Medicine
University of Pittsburgh
Pittsburgh, PA, United States

(5) Louis J. Fox Center for Vision Restoration
University of Pittsburgh
Pittsburgh, PA, United States

INTRODUCTION

Glaucoma is the second leading cause of blindness and is expected to affect 80 million by 2020 worldwide [1]. The biomechanical properties of the lamina cribrosa and its surrounding regions are hypothesized to play an important role in the mechanism of ocular disease. Many studies have measured deformations in normal pressure-inflated human posterior poles using various types of imaging modalities including 3D digital image correlation (DIC) [2], electronic speckle pattern interferometry [3], and 3D ultrasound speckle tracking[4]. A study by Coudrillier et al. is one of studies to compare strains between glaucomatous (G) and non-glaucomatous (NG) human eyes using DIC [2]. The study focused on measuring deformations in the sclera, which excluded the optic nerve (ON). To the authors' knowledge, no study has investigated regional ON strain differences between G and NG eyes using deformation measuring imaging protocols, most likely due to current measurement limitations and the aspherical geometry of the ON region.

Our research group has recently developed a sequential digital image correlation (S-DIC) method with advanced 3D capabilities to map posterior sclera and ON deformations with a reconstruction accuracy of 0.17% and 8 μm uncertainty in the out-of-plane direction [5]. The purpose of this study is to use S-DIC to compare strains between NG and G scleral shells across three regions (ON, peripapillary (PP) sclera and non-PP sclera). This comparison was done using data from four inflation pressures: 5, 15, 30 and 45 mmHg.

METHODS

Posterior scleral shells from human donors older than 50 years old were received from the Banner Sun Health Research Institute in Sun City, AZ, the Donor Network of Arizona in Phoenix and Tucson, AZ, USA, the Illinois Eye Bank in Chicago and Bloomington, Illinois, USA, the Michigan Eye Bank in Ann Arbor, Michigan, and the San Diego Eye Bank in San Diego, CA. The eyes were categorized as NG (n=9) and G (n=3) based on next of kin questionnaires and confirmed via ON axon counts (data not shown). Scleral deformation measurements were collected following the protocol described in our previous study [5]. Briefly, the measured geometry and deformation data was used to create a mesh for each eye. The Green-Lagrange strain, E, was calculated for each mesh element, which was used to calculate in-plane principal strains, E1 and E2. Each geometry was divided into the ON and sclera region using the ON saddle ring points determined using Hessian curvature. The scleral region 2 mm away from the ON saddle ring was designated as PP sclera while the remainder of the sclera was designated as non-PP sclera. The principal strains were determined for each element with in each zone and region for inflation pressure 15, 30 and 45 mmHg, assuming the undeformed state was at 5 mmHg for each pressure.

For our statistical approach, a linear mixed model with random effects (lmer) was performed in R assuming region, pressure and health status as fixed variables. The model assumed individual subjects as a random intercept variable using all elemental E1 and E2 data for each eye. Post hoc pairwise least square mean E1 and E2 comparisons were performed for all factors and factor interactions and the p-values were presented and analyzed. Additionally, pressure was considered a continuous variable to evaluate strain effects as a function of pressure for each fixed variable term.

RESULTS

The E1 and E2 values for the 15, 30 and 45 mmHg were evaluated for all regions and disease status. The data showed that the majority of the E1 and E2 data were positive (tensile) and negative (compressive), respectively (as expected). E1 and E2 strains projected on the undeformed geometry for one NG sample and one G sample are shown in Figure 1.

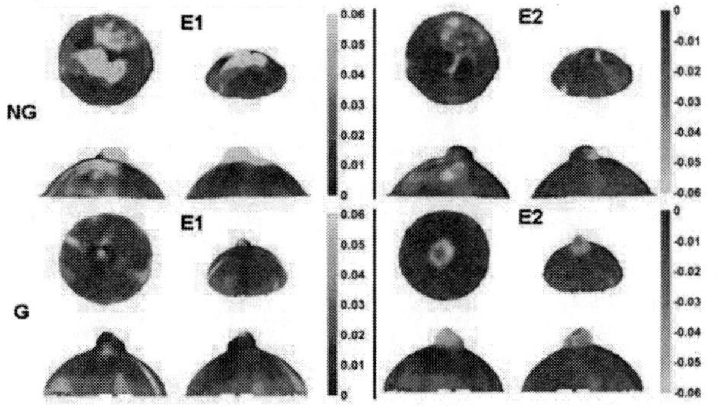

Figure 1. E1 and E2 strain maps at 45 mmHg for a NG sample (top) and a G sample (bottom). Four views are included: top view (upper left), isoparametric view (upper right), back side view (lower left) and front side view (lower right). Note: the color map for E2 strain maps has been reversed to emphasize absolute value.

The statistical model revealed that the E1 strain-pressure slope for NG eyes was statistically higher than G eyes (p-value<0.001), while the E2 strain-pressure slope for NG was significantly lower than G eyes (p-value < 0.001). Furthermore, the piecewise least square mean comparison showed that E1 values for NG was significantly higher than G eyes for the ON region only (p-value=0.004). Conversely, the absolute values of E2 was higher for G eyes compared to NG eyes for both the ON and PP scleral region (p-value<0.001 and p-value=0.0486, respectively). The model regional estimates for E1 and E2 across health status is shown in Figure 2.

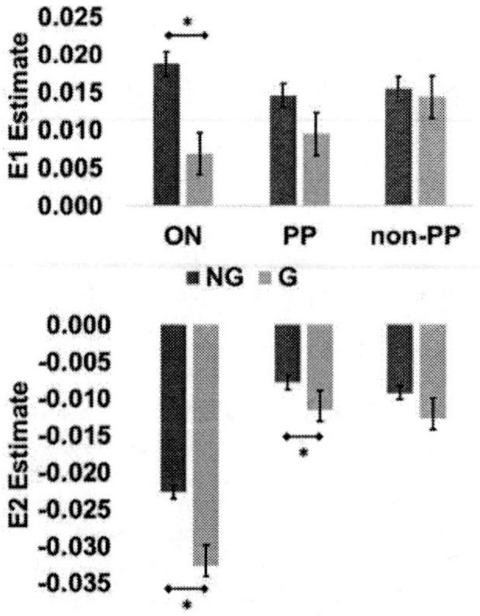

Figure 2. E1 and E2 estimate comparisons between NG ad G eyes for each zone, single asterisk indicates p-value <0.05, error bar denotes estimate standard error

DISCUSSION

A regional in-plane strain comparison was conducted across three inflation pressures for NG and G human scleral shells using displacement measurements collected via S-DIC. The NG tensile strains were found to have a higher increase with pressure as compared to the G, while the G compressive strains were found to be higher compared to the NG. The most significant differences in strain were observed in the ON region for both compressive and tensile strains, while the PP scleral region showed significant differences for compressive strains only. Health status appeared to have little effect on strains in the non-PP scleral region.

Studying the strains of the ON, which serves as the sheath for nerves, is an important consideration in biomechanical evaluation of the posterior pole. Our results suggest that the ON and the surrounding regions of glaucoma eyes different significantly than normal eyes. This may be correlated to the loss in axon count exhibited in the ON of G eyes. Future studies will focus on quadrant evaluation of strains in human eyes and relating it to glaucoma that is localized to one or more meridional direction. Furthermore, evaluating regional strain across racioethnic groups could provide more information on the mechanisms that drive ocular disease in groups that are more susceptible to glaucoma.

ACKNOWLEDGEMENTS

Funding for this work was provided by the National Institute of Health (NIH) grant to JPVG (5R01EY020890). The authors would like to thank Dr. Katia Genovese for her help in setting up the original S-DIC device.

REFERENCES

[1] Quigley, H. A., and Broman, A. T., 2006, "The number of people with glaucoma worldwide in 2010 and 2020," Br J Ophthalmol, 90(3), pp. 262-267.

[2] Coudrillier, B., Tian, J., Alexander, S., Myers, K. M., Quigley, H. A., and Nguyen, T. D., 2012, "Biomechanics of the human posterior sclera: age- and glaucoma-related changes measured using inflation testing," Invest Ophthalmol Vis Sci, 53(4), pp. 1714-1728.

[3] Fazio, M. A., Grytz, R., Morris, J. S., Bruno, L., Gardiner, S. K., Girkin, C. A., and Downs, J. C., 2014, "Age-related changes in human peripapillary scleral strain," Biomech Model Mechanobiol, 13(3), pp. 551-563.

[4] Cruz Perez, B., Tang, J., Morris, H. J., Palko, J. R., Pan, X., Hart, R. T., and Liu, J., 2014, "Biaxial mechanical testing of posterior sclera using high-resolution ultrasound speckle tracking for strain measurements," J Biomech, 47(5), pp. 1151-1156.

[5] Pyne, J. D., Genovese, K., Casaletto, L., and Vande Geest, J. P., 2014, "Sequential-digital image correlation for mapping human posterior sclera and optic nerve head deformation," J Biomech Eng, 136(2), p. 021002.

SB³C2017
Summer Biomechanics, Bioengineering and Biotransport Conference
June 21 – 24, Tucson, AZ, USA

BIAXIAL MECHANICAL RESPONSE OF SMALL BOWEL MESENTERY: EXPERIMENTAL MEASUREMENTS AND CONSTITUTIVE MODELING

Keyvan Amini Khoiy (1), Sophia Abdulhai (2), Ian C. Glenn (2), Todd A. Ponsky (2), Rouzbeh Amini (1)

(1) Biomedical Engineering Department
The University of Akron
Akron, Ohio, USA

(2) Division of Pediatric Surgery
Akron Children's Hospital
Akron, Ohio, USA

INTRODUCTION

Intestinal malrotation, an incomplete rotation and fixation of the gastrointestinal track during fetal development, occurs approximately 1 in 200-500 of live births [1]. In pediatric patients with malrotated small bowel, the risk of midgut volvulus is significantly increased. Midgut volvulus is one of the most critical abdominal emergencies in the pediatric population. It is characterized by torsion of small bowel and mesentery that could lead to blockage of mesenteric blood vessels and ischemia of the small bowel. Midgut volvulus may lead to intestinal necrosis within a few hours and it, thus, requires immediate surgical intervention [2]. Considering the risks involved when midgut volvulus is manifested, upon detection of intestinal malrotation, many surgeons are in favor of aggressive preventative surgical approaches.

Although malrotation is a predisposing risk factor to the development of midgut volvulus, there are different degrees of malrotation, and it is unknown which type of malrotation is at higher risk. Development of computational models of the native and malrotated small bowel is useful to investigate the propensity of the small bowel to deform at different levels of malrotation. An important step in developing such models is the characterization of the mechanical properties of the small bowel mesentery. In this study, we used a biaxial mechanical testing equipment to quantify the stress-strain response of the tissue. We further used a phenomenological constitutive model to specify tissue material properties.

METHODS

A custom-made biaxial tensile testing equipment were used [3]. Porcine mesenteries were obtained from a local slaughterhouse (Duma Meats Inc., Mogadore, Ohio) immediately after the animals were slaughtered. As shown in Figure 1, 11 mm by 11 mm square-shaped specimens were removed from three different regions of the mesentery in the proximal jejunum. The specimens were cut in such a way that the edges were aligned in the radial and tangential directions. The average thickness of each specimen was measured as described in a previous publication [3]. In total, twenty four specimens (n=8 for each region) were tested. All tests were performed within 2 to 6 hours postmortem to eliminate probable erroneous results due to freezing and thawing of the tissues.

After preconditioning, each specimen experienced five different tension-controlled loading protocols with the tension ratio of $T_c\!:T_r = 1\!:1, 1\!:0.75, 0.75\!:1, 1\!:0.5, 0.5\!:1$, similar to our previous work [3]. As the strength of the tissue was not the same at the different regions, three different maximum loads of 50, 100, and 130 N/m were selected based on the tissue strength for the distal avascular region, the distal vascular region, and the root of mesentery, respectively.

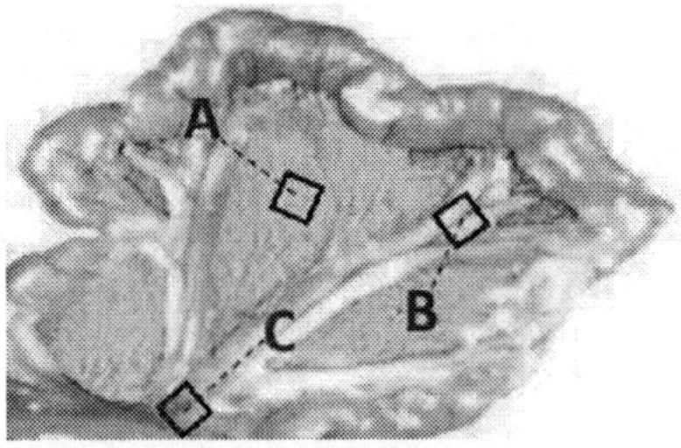

Figure 1: The specimens were excised from (A) the distal avascular region, (B) the distal vascular region and (C) the root of porcine mesenteries.

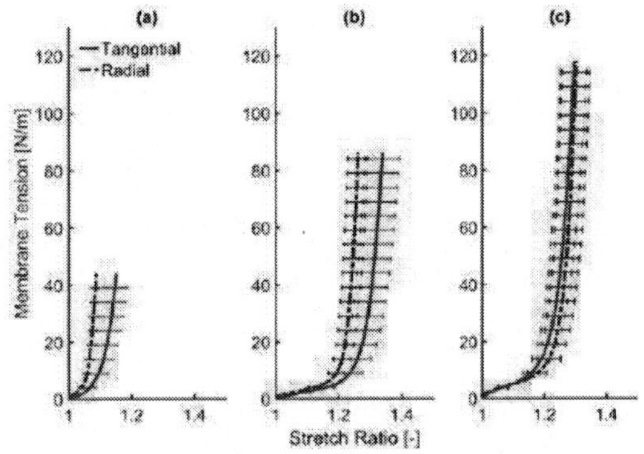

Figure 2: The average membrane tension versus stretch ratio for the equbiaxial loading protocol for (a) the distal avascular region, (b) the distal vascular region, and (c) the root of mesenteries (n=8, the bars are the standard errors).

For each testing protocol, the deformation gradient tensor **F**, the Green strain tensor **E**, and the second Piola–Kirchhoff stress tensor **S** were calculated. To evaluate the amount of shear deformation, the shear angle θ was also calculated:

$$\theta = \cos^{-1}(\frac{C_{12}}{C_{11}C_{22}}) \quad (1)$$

where C_{ij} were the components of right Cauchy–Green deformation tensor **C** defined by

$$\mathbf{C} = \mathbf{F}^{\mathrm{T}}\mathbf{F} \quad (2)$$

After careful examination of the stress-strain data using methods described by Vande Geest et al. [4], a Fung-type strain energy function [5] was chosen to model the mechanical behavior of the mesentery:

$$W = \frac{c}{2}(e^{a_1E_{tt}^2+a_2E_{rr}^2+2a_3E_{tt}E_{rr}} - 1) \quad (3)$$

where c and a_i were material constants and E_{tt} and E_{rr} were the component of the Green strain tensor in the tangential and radial directions, respectively. In this model, the tissue was assumed to be incompressible, homogenous, and hyperelastic undergoing finite deformations. The elements of the second Piola-Kirchhoff stress tensor were thus defined by

$$S_{ij} = \frac{\partial W}{\partial E_{ij}} \quad (4)$$

RESULTS

Figure 2 shows the average membrane tension versus the stretch ratio in the different regions of the porcine mesentery. Both in the avascular region and in the vascular region of the distal mesentery, the deformation was larger in the tangential direction in comparison to that of the radial direction, indicating an anisotropic tissue behavior. In the root region, however, no notable difference between the stretch values were observed indicating a relatively isotropic response.

The maximum shear angles for the equibiaxial protocols were 1.29 ± 0.24 (mean $\pm$ standard error), 3.05 ± 0.68, and 4.07 ± 0.65 degrees for the distal avascular region, the distal vascular region, and the root of mesentery, respectively. The small values of the shear angle indicated that the assumption of shear-free biaxial extension was reasonable.

As demonstrated by a typical five-protocol fit in Fig. 3, the Fung-type strain energy function represented the responses of the porcine mesentery relatively well. Table 1 also shows the material constants for representative samples of each region of the tissue. The experimental data fitted the model with average R-squared values of 0.97, 0.95, and 0.92 for the distal avascular region, the distal vascular region, and the root of mesentery, respectively.

Table 1: Material parameters of Fung-type model for typical samples in different regions of the porcine mesentery.

Region	$c[kpa]$	a_1	a_2	a_3	R^2
Distal Avascular	11.70	46.90	54.97	22.64	0.991
Distal Vascular	0.45	2.45	6.11	1.59	0.981
Root	0.27	4.88	10.04	2.95	0.972

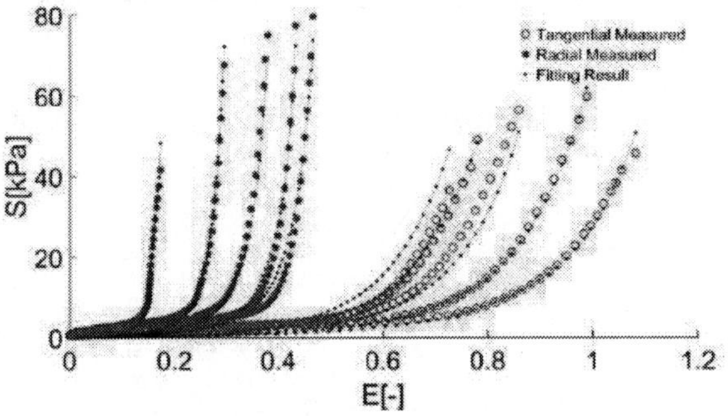

Figure 3: The constitutive model fit (dot-line) plotted over the experimental data of all five loading protocols for the tangential (circles) and radial (stars) directions.

DISCUSSION

Our biaxial mechanical testing experiment showed that, similar to many other soft tissues, the porcine mesentery responded in a nonlinear manner. The classic toe and hill regions were observed in all stress-strain graphs of the tested samples. The stiffening transition was most likely due to straightening of the collagen fibers from their crimped configurations in the unloaded state of the tissues.

The mechanical response of the three selected regions were notably different. The root of the mesentery responded in an isotropic manner, however, the distal region was more anisotropic. In addition, although thinner, the avascular distal region of the porcine mesentery was much stiffer than the vascular distal region.

The presented study is the first step in the development of computational models of the small bowel to investigate the biomechanics of intestinal malrotation and midgut volvulus. Currently, no reliable criteria are used to identify appropriate candidates with malrotated intestine for preventative surgical operations. Since this preventative operation is a major abdominal surgery almost always performed in pediatric patients, more research is critical to identify the biomechanical factors contributing to midgut volvulus if malrotation is present.

ACKNOWLEDGEMENTS

Funding for this work was provided in part by an Akron Children's Hospital research grant. Provision of porcine tissues by Duma Meats Inc. (Mogadore, Ohio) is also acknowledged.

REFERENCES

[1] Torres, A. M., *World J Surg*, 17:326-331, 1993.
[2] Pracros, J., *Pediatr Radiol*, 22:18-20, 1992.
[3] Amini Khoiy, K., *J Biomech Eng*, 138:104504, 2016.
[4] Vande Geest, J. P., *J Biomech Eng*, 126:815-822, 2004.
[5] Tong, P., *J Biomech*, 9:649-657, 1976.

STENTING THE PATIENT-SPECIFIC, ACTIVELY CONTRACTING AND BUCKLING ESOPHAGUS: A FINITE ELEMENT ANALYSIS

M. Peirlinck[1], N. Debusschere[1], F. Iannaccone[1], P. Siersema[2], B. Verhegghe[1], P. Segers[1], M. De Beule[1]

[1]Biofluid, Tissue and Solid Mechanics for Medical Applications Lab (IBiTech, bioMMeda), Ghent University, Ghent, Belgium

[2]Department of Gastroenterology and Hepatology, University Medical Center Utrecht, Utrecht, The Netherlands

INTRODUCTION

The esophagus is responsible for the transport of a food or fluid bolus down to the stomach. Any malfunctioning of this multi-layer muscular tube organ easily leads to considerable discomfort and problems. Knowledge of the large deformations and transmural stress and strain distributions in the esophageal wall has been proven crucial in the understanding of esophageal physiology and pathophysiology [1]. A thorough understanding of the biomechanics of the (diseased) esophagus is also important in case of esophageal stenting, an effective treatment strategy for non-resectable tumors or benign obstructions. Indeed, reported migration problems and complications such as perforations, bleeding and fistula formation demonstrate the importance of a full biomechanical understanding of this gastro-intestinal organ and the interaction between stent and esophageal wall. This work set forth to develop a full mechanical framework, incorporating highly detailed numerical esophagi and stent models enabling an enhanced understanding of esophageal physiology on the one hand and symptoms or complications following stent implantation on the other.

METHODS

Esophagus model: The esophageal tissue shows, similar to other biological (e.g. cardiovascular) tissues, highly non-linear, pseudo-elastic and anisotropic behavior. Previous constitutive models focused on characterizing the esophagus's passive behavior, ignoring its active function. For that reason, we propose a new constitutive material model based on Hill's muscle model where a passive and contractile unit in parallel determine the muscle's mechanical behavior. The proposed strain-energy function (SEF) was additively decomposed in a volumetric Ψ^{vol} and deviatoric part Ψ^{dev}, where the deviatoric part constitutes the non-collagenous ground matrix, the collagen fibers and the muscle cells in the esophageal wall. The corresponding constitutive parameters were fitted based on the mechanical response of ovine tissue in biaxial tensile tests [2].

$$\Psi = \Psi^{vol} + \Psi^{dev} = \Psi^{vol} + \Psi^{mat} + \Psi^{fib1} + \Psi^{fib2} + \Psi^{smc}$$

Based on esophageal histology, the human esophagus was modeled as a five-layered system consisting of the mucosa, the collagen-rich submucosa, an interfacial layer, and the inner and outer muscular layer. As the output of numerical models depends heavily on geometry [3], a CT acquisition of a 69-year old male's esophagus suffering from an esophageal carcinoma was segmented in *Mimics* (Materialise; Leuven, Belgium) and meshed into a patient-specific finite element model using our in-house developed finite element pre- and postprocessor *pyFormex* (Ghent University; Gent, Belgium) (see Fig. 1).

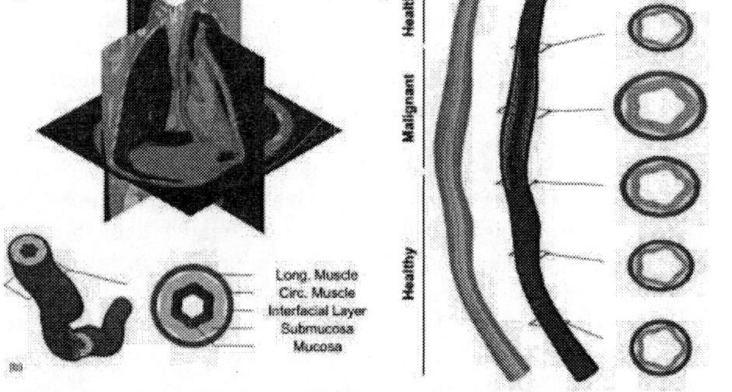

Figure 1: Patient-specific pathologic esophagus geometry – (a) CT segmentation, (b) 3D model with a cross-section depicting the 5 esophageal layers and (c) representation of the 8cm long malignant submucosal proliferation caused by the esophageal carcinoma.

As this CT acquisition was made in an in-vivo stressed organ state inside the human body, the axial residual strain and in vivo pressure-induced stress and strain state were incorporated in the model using a Backward Incremental Method [4].

Virtual Stent Implantation Procedure: A commercially available new biodegradable esophageal stent design (Ella-BD stent) was subjected to in-vitro stent compression and tensile tests to determine the radial strength of the stent and the elasticity of its braided wires. An in-vitro degradation study was performed, in which the full stent and separate stent wires were conserved in a phosphate buffered solution (pH=7.2) at 37°C for 10 weeks. These tests served to validate a numerical stent model, which was geometrically described and meshed in *pyFormex* and virtually bench tested in the finite element analysis (FEA) software *Abaqus* (Simulia, Dassault Systemes; Providence, Rhode Island, USA). A correct numerical implementation of the friction between the thick wires of the polymeric braided stent design was found in the incorporation of connector elements with rotary friction in between the braided stent wires. A tuning of the friction coefficients between the wires proved adequate to successfully capture the stent's temporally varying mechanical behavior [5]. The esophageal stent was virtually implanted in the patient-specific esophagus model. A user subroutine was programmed in which the stent got crimped onto a catheter, bended as if it were endoscopically inserted in the esophagus and implanted by withdrawing the crimping sheath.

RESULTS

The in vivo stress state of the patient-specific esophagus is depicted in Fig. 2 and shows clear stress differences between the (sub)mucosal and the muscular layers presenting stresses ranging between 0.001 kPa and 5.0 kPa and up to 0.001 kPa respectively.

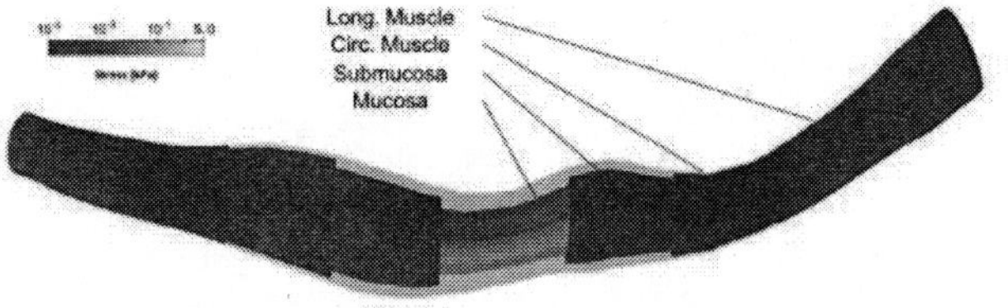

Figure 2: Patient-specific esophageal in vivo von Mises stress state (peeled muscular and (sub)mucosal layer representation)

A subsequent spatiotemporal activation of the muscular layers' myosin heads leads to a relative sliding between the actin and myosin filaments which results in the peristaltic contraction of the patient's esophagus. The mucosa and submucosa both buckle as a response to the compressive forces exerted by the outer muscularis externa, which leads to the full occlusion of the esophageal lumen needed to push the food or fluid bolus down (peak contact pressures on the inner mucosal layers of 3.496 ± 0.7873 kPa at 100% lumen occlusion). Local compression and extension, as well as local stress states can be found in Fig. 3. The stent model was virtually implanted following the procedure described in the stent's Instructions For Use and is depicted in Fig. 4. The combination of the stent model's validated mechanical dilating behavior and the complex (patho)physiological patient-specific model leads to interesting insights on the stent's capability to open up the constricted lumen (see Fig. 4), the contact pressures between stent and esophagus (varying locally from 0.522 to 2.089 kPa) and the stress state the device induces in the esophageal wall.

DISCUSSION

Finite element analysis (FEA) has the potential to expand our knowledge on the esophagus' functioning and disease states [1], however some shortcoming, such as the absence of active constitutive models, three-dimensional geometries and the incorporation of in-vivo stresses were still impeding the use of FEA to study the relationship between stress, remodeling and mechanosensory function. To our knowledge, this computational framework is the first of its kind to overcome these deficiencies. The framework proved capable to simulate the active peristaltic contraction of the esophageal muscle tissue and the buckling of the (sub)mucosal layers in a pathophysiological state. The computed stress and strain states of the esophageal wall allow us to assess the remodeling processes of esophageal tissue in diseased states [1], to get more insight in the physiological process of food swallowing and to perform virtual device (e.g. stent) implantation or surgery, as shown during the virtual stent implantation.

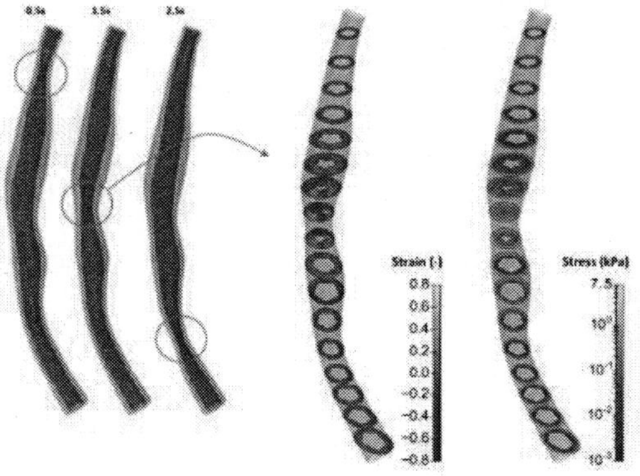

Figure 3: Peristaltic contraction

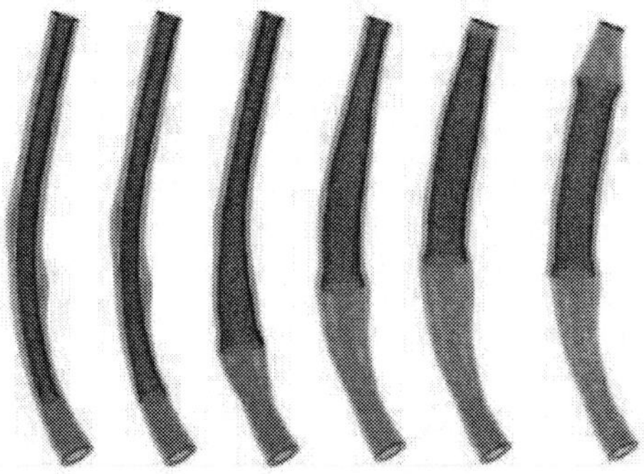

Figure 4: Virtual Stent Implantation procedure

ACKNOWLEDGEMENTS

This work was supported by a Strategic Research grant from the Flemish government's Department for Innovation and Entrepreneurship (VLAIO).

REFERENCES

1. Gregersen, H., *Biomechanics of the gastrointestinal tract: new perspectives in motility research and diagnostics*. 2003: Springer Science & Business Media.
2. Sommer, G., et al., *Multiaxial mechanical response and constitutive modeling of esophageal tissues: Impact on esophageal tissue engineering*. Acta Biomater, 2013. **9**(12): p. 9379-91.
3. Liao, D., et al., *Two-layered quasi-3D finite element model of the oesophagus*. Med Eng Phys, 2004. **26**(7): p. 535-43.
4. De Putter, S., et al., *Patient-specific initial wall stress in abdominal aortic aneurysms with a backward incremental method*. Journal of biomechanics, 2007. **40**(5): p. 1081-1090.
5. Peirlinck, M., et al. *An in silico biomechanical analysis of the stent-esophagus interaction*. in *2015 Summer Biomechanics, Bioengineering and Biotransport Conference*.

A STRUCTURE-BASED CONSTITUTIVE MODEL OF ARTERIAL TISSUE

Tarek Shazly (1), Alexander I. Rachev (1, 2)

(1) Department of Mechanical Engineering
University of South Carolina
Columbia, South Carolina, USA

(2) Institute of Mechanics
Bulgarian Academy of Sciences
Sofia, Bulgaria

INTRODUCTION

One of the basic tasks of vascular biomechanics is quantification of the mechanical properties of arterial tissue in terms of continuum mechanics-based constitutive equations. Completion of this task enables the formulation and solution of boundary value problems that provide predictive results for the mechanical response of arteries and enable understanding of both their normal functions and the genesis/ progression of arterial pathologies, such as aneurism formation. The constitutive modeling of arterial tissue is based on the notion that the major load-bearing constituents are elastin and families of collagen fibers that are oriented in one or more preferred directions and are undulated in the natural configuration. Collagen is mechanically involved in producing stress in a gradual manner as fibers become unfolded and stretched as the tissue undergoes extension. All known structure-based models based on constrained mixture theory use as a weighting factor of the contribution of collagen to the total strain energy function (SEF) and total stress the mass fraction of the *total* collagen of a given family, which is in contradiction to the assumption of *gradual* engagement of collagen [e.g. 1, 2]. The aim of this study is to propose a novel approach for constitutive modeling of arterial tissue considered as a constrained mixture. It explicitly accounts for the gradual involvement of collagen fibers, stretch-induced increase in collagen mass fraction, and mechanical properties of collagen. The performed parametric study allows prediction of the stress-strain response after independent variation of model parameters.

CONSTITUTIVE MODEL

The arterial tissue is considered as a constrained mixture of elastin, collagen and constituents that do not bear load, including the relaxed smooth muscle. Under loads the tissue undergoes a finite elastic biaxial deformation, which is described by principle stretches, λ_1 and λ_2. For simplicity we consider the case when collagen fibers are aligned only in one direction, say λ_1. The mechanical response of collagen fibers is governed by their configuration in the state of no load, their mass, and mechanical properties. In contrast to the existing models, we separately incorporate each of these factors in the constitutive equations. We assume that collagen exists as a collection of an infinite number of fiber families with identical orientation but continuously varying waviness in the state when the tissue is not stretched. The tissue stretch $\lambda_{r(i)}$ that is needed to straighten the i-th fiber family is called the recruitment stretch; $\lambda_{r(i)} = \lambda_{r(min)} + \alpha_{(i)}\Delta\lambda_r$, where $\Delta\lambda_r = (\lambda_{r(max)} - \lambda_{r(\min)})$, and $\alpha_{(i)}$ is a dimensionless parameter. For given tissue stretch λ_1 collagen fibers that are straightened and stretched to $\lambda_{c(i)}$ satisfy the condition $\lambda_{c(i)} = \frac{\lambda_1}{\lambda_{r(i)}} > 1$.

Strains produce stress. We adopt the assumption that the mechanical properties of collagen fibers are described by a quadratic strain energy function

$$W_{c(i)} = \frac{1}{2}c_{col}E_{c(i)}^2, \qquad E_{c(i)} = \frac{1}{2}(\lambda_{c(i)}^2 - 1).$$

c_{col} is collagen Young's modulus.

We follow the conclusions in previous studies, [2-5], that only a portion of total collagen mass, denoted here by M_{col_r} , produces passive stress, and assume that it is distributed continuously among families with different waviness labeled by recruitment stretch. To be specific, we describe the distribution of mass M_{col_r} via mass intensity selected

$$m_{col(i)} = k\left[\frac{1}{1+exp\left(-b(\lambda_{r(i)}-\lambda_{r(min)})\right)} - \frac{1}{1+exp\left(-b(\lambda_{r(i)}-\lambda_{r(max)})\right)} - a\right],$$

here k , a, and b are the model parameters. Elastin is considered as a neo-Hookean solid. Within the framework of the theory of the constrained mixture the total tissue stresses are

$$t_{T(1)} = 2\varphi_{el}c_{el}\left(\lambda_1^2 - \frac{1}{\lambda_1^2\lambda_2^2}\right) + c_{col}\Delta\lambda_r \int_0^{\widehat{\alpha}} \lambda_{c(i)}^2 E_{c(i)}\widehat{m}_{col(i)}d\alpha_{(i)},$$

$$t_{T(2)} = 2\varphi_{el}c_{el}\left(\lambda_2^2 - \frac{1}{\lambda_1^2\lambda_2^2}\right)$$

$$\hat{\alpha} = \begin{cases} 0 & if \quad \lambda_1 < \lambda_{r(\min)} \\ \dfrac{\lambda_1 - \lambda_{r(min)}}{\Delta\lambda_r} & if \ \lambda_{r(\min)} < \lambda_1 < \lambda_{r(\max)} \\ 1 & if \quad \lambda_1 > \lambda_{r(\max)} \end{cases}$$

c_{el} and φ_{el} are the elastin material constant and mass fraction, respectively.

RESULTS

To illustrate and validate the proposed constitutive formulation of arterial tissue some model parameters pertaining to a specific arterial vessel were taken from the literature [2,3]. The remaining parameters were then adjusted in a manner such that the model yields reasonable stress-strain curves. Some of the results obtained are shown in Fig.1. The tissue is stiffer if collagen Young's modulus is higher, the amount of collagen mass that is capable to bear load is increased, the recruitment stretch at which the first collagen fibers are engaged is smaller, and the intensity of distribution of collagen mass more rapidly tends to the maximal value

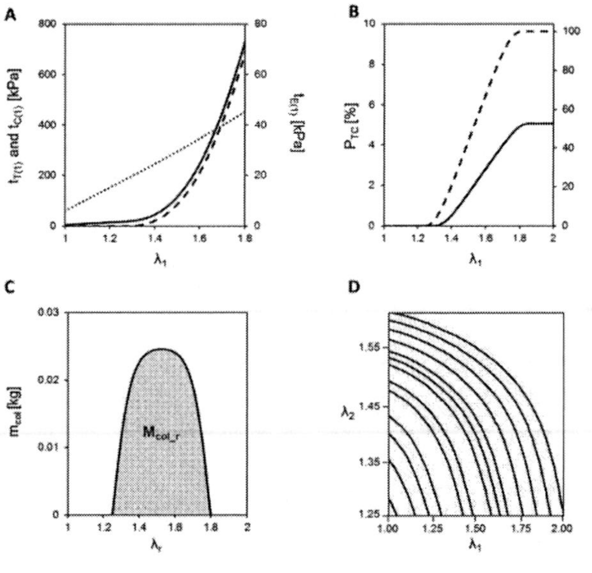

Figure 1. (A) Stress vs. tissue stretch (dot cure refers to elastin, dashed curve - to collagen, and continues curve - to total stress); (B) Percentage of engaged collagen with respect to the total collagen (continuous curve) and total recruitable collagen (dashed curve) vs. tissue stretch; (C) Intensity of recruitable collagen mass vs. recruitment stretch; (D) Representative iso-SEF contour plots in the plane $\lambda_1 - \lambda_2$ proving material stability.

DISCUSSION

The major novelty of proposed model is that in each deformed state the contribution of collagen to the stored SEF or total tissue stress are integrals of SEF, or respectively produced effective stress, weighted by the corresponding differential mass fraction of the fibers that are *actually* involved in load bearing. Thus the model corrects the overestimation of the collagen contribution to the passive mechanical properties included in previous constitutive models, though this flaw is in part compensated for in previous models when parameters are determined to fit experimental data.

Our model assumes three pathways for collagen to affect the passive mechanical properties of the arterial tissue, namely the individual mechanical properties of collagen, the variable wavy configuration of fibers embedded in the tissue in the state of no load, and the amount and distribution of collagen among fibers with different undulation. Each factor is governed by distinct biological processes. An advantage of the proposed model, provided these factors can be quantified from histological and morphometric investigations, is that the estimation of their effects on arterial mechanics is a well posed direct mathematical problem. However, identification of factors that cause experimentally recorded changes in mechanical properties of the tissue due for instance to aging or progression of vascular diseases, is an inverse mathematical problem. Because of nonlinearity it does not exhibit a unique solution. For given tissue strain identical stress response might be obtained due to different combinations of factors that considered individually have qualitatively similar effects. To avoid discrepancy in interpretation of experimental data, specification of model parameters has to be done with regard to analysis of supplementary biological information.

Like some previous structure-based constitutive models of the arterial tissue [2, 3] or entire arterial wall [4, 5], our model predicts that a relatively small part of collagen is involved in load bearing under physiological stretches (Fig.1 B). The amount of the engaged collagen is modulated by setting as a input parameter the value of the collagen modulus. It could be speculated that the remaining collagen serves together with the adventitia as a supporting structure under extreme high pressures. On the other hand, authors in [5] assumed that the passive mechanical response of the tissue is governed by the deformation of elastin and a small portion of collagen arranged "in parallel". It is speculated that most of collagen is arranged "in series" with the smooth muscle fibers and is not engaged in bearing load when the muscle is relaxed.

In conclusion, the model proposed in this study captures the basic characteristics of collagen architecture with focus on the progressive involvement of collagen fibers. The constitutive formulation is performed in the framework of constrained mixture theory with due account for waviness-specific collagen mass fractions.

ACKNOWLEDGEMENTS

This research was supported by the NIH INBRE Grant for South Carolina (P20GM103499).

REFERENCES

[1] Zhou, B et al., *Acta Biomaterialia* 45: 286-295, 2016
[2] Zulliger, M, et al., *J Biomech* 37: 989-1000, 2004
[3] Roy, S et al., *J Biorheol* 24:84-93, 2010
[4[Wuytsts, FL et al., *Phys. Med. Bid:* 40 1577-1597, 1995
[5] Bank, AJ et al., *Circulation* 94 (12): 3263-3270, 1996

COLLAGEN NETWORK MICROSTRUCTURE OF THE ASCENDING THORACIC AORTIC MEDIA PREDICTS EXPERIMENTAL UNIAXIAL FAILURE BEHAVIOR

James R. Thunes (1), Julie A. Phillippi (1,2,3,4), Thomas G. Gleason (1,2,3,4), David A. Vorp (1,2,3,4), Spandan Maiti (1)

(1) Department of Bioengineering
University of Pittsburgh
Pittsburgh, PA, USA

(2) Department of Cardiothoracic Surgery
University of Pittsburgh
Pittsburgh, PA, USA

(3) McGowan Institute for Regenerative Medicine
University of Pittsburgh
Pittsburgh, PA, USA

(4) Center for Vascular Remodeling and Regeneration
University of Pittsburgh
Pittsburgh, PA, USA

INTRODUCTION

Understanding of pathophysiology of the ascending thoracic aorta (ATA) requires the knowledge of the biomechanical response of all its constituents. We are particularly interested in understanding the mechanics acute aortic dissection, the delamination of the media of the ATA wall. Recent experimental works in the literature have reported uniaxial [1] as well as biaxial failure properties of the aortic wall. However, how wall tissue strength is related to the underlying collagen fiber architecture, the primary load bearing constituent of the aortic wall, is currently not known. Continuum models of the aorta predominate in the literature but recent work by us and others have proposed structurally based models that explicitly model the fiber microstructure. However, almost all of these reports only study the pre-failure response of the tissue and cannot predict how the material properties and geometrical features of the collagen network influence tissue strength. In this abstract, we present a finite element based modeling approach where image based architecture of the collagen fibers was considered, and tissue failure was predicted from the failure behavior of individual collagen fibers. We focused only on aortic media, as dissection primarily propagates in this layer. Using published collagen network parameters for human aneurysmal ATA media, we were able to create tissue models that captured the failure stress and stretch of the tissue. Further, after fitting the material parameters of the collagen fibers from experimental stress-stretch curve in one direction, the tissue pre-failure and failure response in the orthogonal direction was predicted.

METHODS

The ATA media is a structure composed of repeating lamellar units (LU) (Fig 1a). The LU consists of two elastic lamellae (dense sheets of elastin fibers) separated by an interlamellar space (GAGs, vascular smooth muscle cells, etc. Running parallel to the lamellae are collagen fiber networks. A portion of the LU complex was used as a representative volume element (RVE) of the aortic media. To model the LU, collagen networks were embedded within a 3D continuum (Fig 1c). 1800 special 8-noded fiber-embedded hexahedral elements were used to represent the solid while 140 fiber segments were required to model the collagen network. Of the two collagen networks in the model, we assumed that one was oriented with a mean angle of $=\gamma$ and the other of $-\gamma$ degrees. Multiphoton imaging from [2] reported collagen networks with $\gamma = =35°$ and $OI = 0.60 = 0.03$ (mean=s.d.), where orientation index, OI, is a measure of the fiber dispersion with 0.5 denoting an isotropic network and 1.0 a perfectly aligned network.

Along with γ and OI for the collagen fiber network, the areal density (AD) was used to create 20 networks for aneurysmal tissue using a custom Matlab script (Matlab 2014b, Mathworks, Natick, MA). Fibers were added to the network until the required areal density was reached. The fiber angle was normally distributed about γ using OI. After network creation, normalcy was checked using a Lilliefors test ($p \leq 0.25$). Non-normal networks were recreated, and tested again. Networks were also recreated if the OI of the simulated network was more than one standard deviation from the experimentally reported data. A representative fiber network is shown in Fig 1b.

The non-collagenous matrix was modeled with a 1-parameter isotropic incompressible neoHookean material: $\Psi = \mu/2(I_1 = 3)$ with μ the shear modulus and I_1 the first invariant of the right Cauchy-Green deformation tensor. The collagen fibers were modeled with the following stress-strain relationship

$$\sigma = \begin{cases} 0 & \text{if } \lambda \leq \lambda_r \\ E_f(\lambda = \lambda_r) & \text{if } \lambda_r \leq \lambda \leq \lambda_p \\ E_f(\lambda_p = \lambda_r)\left(1 = \dfrac{\lambda = \lambda_p}{\lambda_{max} = \lambda_p}\right) & \text{if } \lambda_p \leq \lambda \leq \lambda_{max} \end{cases}$$

where E_f is the fiber modulus and λ is the fiber stretch. $\lambda_r, \lambda_p,$ and λ_{max} are the recruitment stretch, peak stretch, and max stretch of the fiber, respectively. The initial zero stress portion of the response represents the initial crimp in the fiber. A gradual decrease in stress after the max stress, σ_{max}, is reached (at λ_p) was included for numerical stability.

The domain was stretched in the circumferential (CIRC) direction to an applied stretch of $\lambda_{appl} = 1.75$ and the ensuing stress-stretch curve was fitted against experimental stress-stretch curve [2] to determine the material parameters ($E_f, \lambda_r, \lambda_{max}, \sigma_{max},$ and μ). Using the same parameters, the domain was then loaded to the same stretch in the longitudinal (LONG) direction, and the computational stress-stretch response was recorded. The load was applied over 1000 load steps and force at the loaded face was collected to construct the stress-stretch curve.

RESULTS

Separate uniaxial stretching simulations were performed for each of 20 different instances of the simulated collagen network. The differences in the simulation networks lead to variation in the stress-stretch response shown by the shaded region in Fig 1d,e. The average of the CIRC stretching simulation derived stress-stretch curves was regressed against CIRC experimental data [2]. The material parameters matching the experimental data were: $E_f = 70$ MPa, $\sigma_{max} = 12.5$ MPa, $\lambda_r = 1.25$, $\lambda_{max} = 1.5$, and $\mu = 335$ kPa. Average difference between the failure stress and stretch (Table 1) for the simulations and experiment were 0% and 2%, respectively. When stretched in the LONG direction using these material properties, the average differences for the failure stress and stretch obtained from experiments [2] and computation were 1.4% and 6.5%, respectively.

Table 1: Failure stress and stretch (mean=s.d.) for CIRC and LONG loading. Experimental data taken from [2].

	CIRC		LONG	
	λ_p	σ_{max} (MPa)	λ_p	σ_{max} (MPa)
Exper	1.49 = 0.09	1.61 = 0.49	1.46 = 0.25	1.25 = 0.35
Sim	1.46 = 0.02	1.61 = 0.12	1.48 = 0.02	1.33 = 0.09

DISCUSSION

Minimal material models for the non-collagenous matrix and collagen fibers in conjunction with image-derived collagen fiber microstructure were able to recreate the experimentally observed non-linear pre-failure and failure response for the aortic media tissue. As the collagen networks were constructed using fiber network properties ($AD, \gamma,$ and OD) gathered from multiphoton images of the aortic media, the presented structural model is physically based. In addition, regressed material parameters are also within reported ranges [2]. With the introduction of a failure mechanism for the collagen fibers, the failure stress and stretch derived experimentally could be simulated. The predictive capability of our model was demonstrated through the simulation of LONG response. The material parameters were fitted in the CIRC direction and kept constant for the simulations in the LONG direction. The close agreement with experimental failure data in LONG direction indicates that failure response of the ATA media depends on the architecture of the fiber network, and mechanical properties of collagen and non-collageneous matrix. Our modeling will be useful for predicting wall tissue failure response for multi-axial in-vivo loading conditions, for which experimental data may not be available.

The current study has some limitations. First, we neglected any active contraction of the tissue due to the vascular smooth muscle cells in the interlamellar region. While we feel this is a reasonable assumption, forces from this mechanism may affect the model – especially locally due to the collagen network microarchitecture. Secondly, we allow only fiber failure. Failure in the matrix may play a role in the failure response of the diseased tissue where, for example, elastic lamellae are fragmented. We will include these additional scenarios in our future work.

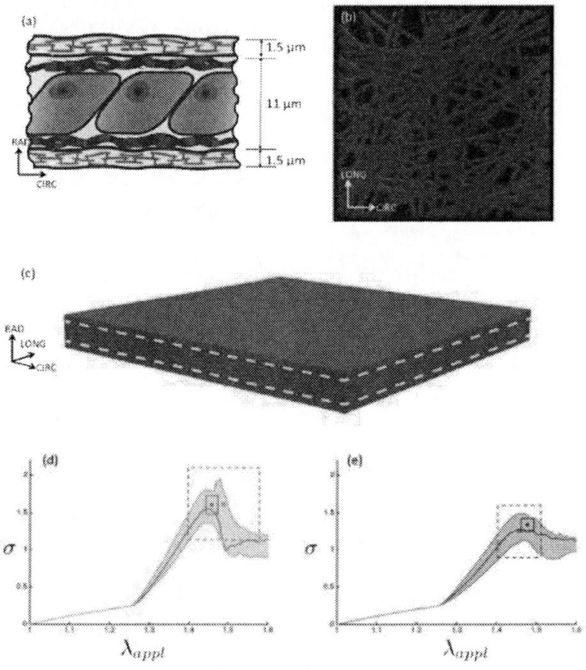

Figure 1: (a) The lamellar unit of the aortic media. (b) Representative collagen fiber network. (c) Finite element model of the RVE. Yellow lines denote the location of the collagen networks. (d,e) Stress-stretch response of the tissue under CIRC and LONG loading. Solid curve is the response of a representative network while the shaded region is the extents of the response for the 20 networks. Filled and empty circles are the average failure points for the simulations and experimental tests, respectively. Boxes denote the standard deviation: solid (simulation), dashed (experimental).

ACKNOWLEDGEMENTS

This work was funded in part by the National Heart, Lung and Blood Institute of the National Institutes of Health under Award Number R01HL109132, and the National Science Foundation under Award Number CBET 1511504.

REFERENCES

[1] Duprey, A et al.. Biaxial rupture properties of ascending thoracic aortic aneurysms. Acta Biomater 2016;42:273-85.

[2] Pasta, S et al. Constitutive modeling of ascending thoracic aortic aneurysms using microstructural parameters. Med Eng Phys 2016;38:121-30.

AUTHOR INDEX

AUTHOR INDEX

AUTHOR INDEX

AUTHOR INDEX

AUTHOR INDEX

AUTHOR INDEX

AUTHOR INDEX

AUTHOR INDEX

AUTHOR INDEX

AUTHOR INDEX

AUTHOR INDEX

AUTHOR INDEX

AUTHOR INDEX

AUTHOR INDEX

AUTHOR INDEX

AUTHOR INDEX

AUTHOR INDEX

AUTHOR INDEX

AUTHOR INDEX

Summer Biomechanics, Bioengineering and Biotransport Organizing Committee
201 Waterfront St
National Harbor, MD 20745
USA

ISBN 978-1-5108-5993-7
9 781510 859937
CURRAN ASSOCIATES INC.
proceedings.com

Summer Biomechanics, Bioengineering and Biotransport Conference 2017

Tucson, Arizona, USA
21 - 24 June 2017

Volume 2 of 2

SB3C
biomechanics.
bioengineering.
biotransport.
SB3C.org

Summer Biomechanics, Bioengineering and Biotransport Conference 2017

Tucson, Arizona, USA
21 - 24 June 2017

Volume 2 of 2

ISBN: 978-1-5108-5993-7

Printed from e-media with permission by:

Curran Associates, Inc.
57 Morehouse Lane
Red Hook, NY 12571

Some format issues inherent in the e-media version may also appear in this print version.

Printed by Curran Associates, Inc. (2018)

Summer Biomechanics, Bioengineering and Biotransport Organizing Committee
201 Waterfront St
National Harbor, MD 20745
USA

info@sb3c.org

Additional copies of this publication are available from:

Curran Associates, Inc.
57 Morehouse Lane
Red Hook, NY 12571 USA
Phone: 845-758-0400
Fax: 845-758-2633
Email: curran@proceedings.com
Web: www.proceedings.com

TABLE OF CONTENTS

VOLUME 1

THROMBOSIS (FLUIDS)

BIOTRANSPORT AND MICROFLUIDICS (BTR)

INSTRINSIC AND EXTRINSIC REGULATION OF CELLULAR MECHANOTRANSDUCTION (CTE)

SOFT TISSUE CHARACTERIZATION AND MODELING (SOLIDS)

CELEBRATION FOR LARY TABER: GROWTH AND REMODELING IN DEVELOPMENT AND DISEASE (SOLIDS/CTE)

ATHEROSCLEROSIS (FLUIDS)

BIOFLUIDS (FLUIDS)

STRATEGIES TO IMPROVE REHABILITATION TREATMENTS (DDR/IAB)

MICRO-ENGINEERED PHYSIOLOGIC SYSTEMS (CTE)

MICROENVIRONMENTAL CONTROL OF TISSUE FORMATION AND CELL FUNCTION (CTE)

MULTISCALE ANALYSIS OF CARTILAGE AND INTERVERTEBRAL DISC (SOLIDS)

SOFT TISSUE MECHANOBIOLOGY (SOLIDS)

HEART VALVE FLOW AND FUNCTION (FLUIDS/SOLIDS)

CARDIOVASCULAR DEVICE (FLUIDS/SOLIDS)

ACTIVE LEARNING IN BIOMECHANICAL ENGINEERING EDUCATION (EDU)

HYPERTHERMIA, CRYOTHERAPY, AND CRYOPRESERVATION (BTR)

DISEASE MODELS AND ENGINEERED THERAPIES (CTE)

MECHANICS AND MODELING OF MUSCULOSKELETAL SOFT TISSUES (SOLIDS)

OCULAR BIOMECHANICS (SOLIDS)

CARDIAC MECHANICS (SOLIDS)

DON GIDDENS' IMPACT ON CARDIOVASCULAR FLUID DYNAMICS AND ATHEROSCLEROSIS (FLUIDS)

MEASUREMENT IN MOVEMENT AND TRAUMA (DDR/IAB)

MECHANICAL REGULATION OF REMODELING AND REPAIR (CTE)

THERAPEAUTIC MATERIALS FOR REPAIR AND REGENERATION (CTE)

PHD PAPER COMPETITION: IMAGING, BIOFLUID MECHANICS, AND BIOTRANSPORT

PHD PAPER COMPETITION: TISSUE MECHANICS AND CHARACTERIZATION

PHD PAPER COMPETITION: CELLULAR MECHANICS AND MECHANOBIOLOGY

PHD PAPER COMPETITION: DISEASES, INJURY, AND REMODELING

PHD PAPER COMPETITION: EXTRACELLULAR MATRIX BIOMECHANICS

PHD PAPER COMPETITION: BIOMATERIALS AND MATERIAL-CELLULAR INTERACTION

UNDERGRADUATE DESIGN COMPETITION

UPPER AND LOWER EXTREMITY JOINT MECHANICS (SOLIDS)

HEAD INJURY & INJURY BIOMECHANICS 1 (SOLIDS)

VASCULAR MECHANICS (SOLIDS)

IMAGING AND DIAGNOSTICS (FLUIDS)

NANO AND MIRCROTHERAPEUTICS (BTR)

MECHANICAL REGULATION OF MORPHOGENESIS (CTE)

HEAD INJURY & INJURY BIOMECHANICS 2 (SOLIDS)

REPRODUCTIVE BIOMECHANICS (SOLIDS)

TENDON TISSUE ENGINEERING AND REGENERATION (CTE/SOLIDS)

ANEURYSM (FLUIDS)

EXPERIMENTAL MODELING FOR CLINICAL SURGICAL APPLICATIONS (DDR/IAB)

VASCULAR, LYMPHATIC, AND OCULAR TRANSPORT (BTR)

MEASURING AND MODELING CELL MECHANCIS AND THE MICROENVIRONMENT (CTE)

HEAD INJURY & INJURY BIOMECHANICS 3 (SOLIDS)

REPRODUCTIVE, OCULAR, AND GASTROINTESTINAL BIOMECHANICS (SOLIDS)

ANEURYSM MECHANICS (SOLIDS)

VOLUME 2

TENDON MECHANICS AND STRUCTURE (SOLIDS/CTE)

PEDIATRIC FLOW (FLUIDS)

SURGICAL DEVICE DESIGN APPLICATIONS (DDR/IAB)

MULTI-SCALE MEASURES AND MODELS OF ENGINEERED MATERIALS AND TISSUES (CTE)

POSTER SESSIONS

BACHELORS LEVEL STUDENT PAPER COMPETITION I – DYNAMICS & INJURY, DEVICES, AND IMAGING

BACHELORS LEVEL STUDENT PAPER COMPETITION II – FLUIDS & MICROFLUIDICS, CELLULAR & TISSUE MECHANICS, PHYSIOLOGY & DISEASES

MASTERS LEVEL STUDENT PAPER COMPETITION I – PHYSIOLOGY & DISEASES, CELLULAR & TISSUE MECHANICS, DEVICES

MASTERS LEVEL STUDENT PAPER COMPETITION II – DYNAMICS & INJURY, FLUIDS & MICROFLUIDICS, BIOTRANSPORT & HEAT TRANSFER

BIOTRANSPORT POSTERS

EDUCATION POSTERS

DESIGN, DYNAMICS, AND REHABILITATION POSTERS

CELL AND TISSUE ENGINEERING POSTERS – MECHANOBIOLOGY AND THE MICROENVIRONMENT

CELL AND TISSUE ENGINEERING POSTERS – TISSUE ENGINEERING AND DISEASE MODELS

CELL AND TISSUE ENGINEERING POSTERS – MEASUREMENTS AND MODELING IN CELL AND TISSUE ENGINEERING

SOLIDS POSTERS – BONE AND CARTILAGE

SOLIDS POSTERS - MUSCULOSKELETAL

SOLIDS POSTERS – TISSUE MECHANICS

SOLIDS POSTERS - CARDIOVASCULAR

FLUIDS POSTERS – CARDIOVASCULAR DEVICES, VALVES, AND FLOWS

FLUIDS POSTERS – CARDIOVASCULAR DIAGNOSTICS AND FLOW ANEURYSMS

FLUIDS POSTERS – BIOLOGICAL FLOWS

UNDERGRADUATE DESIGN COMPETITION POSTERS

CORRELATIONS OF WALL STRESS AND GEOMETRY IN SYMPTOMATIC AND RUPTURED ABDOMINAL AORTIC ANEURYSMS

Sathyajeeth S. Chauhan (1), Carlos A. Gutierrez (1), Mirunalini Thirugnanasambandam (1), Victor De Oliveira (2), Satish C. Muluk (3), Mark K. Eskandari (4), Ender A. Finol (1)

(1) Joint Graduate Program in Biomedical Engineering
University of Texas at San Antonio
San Antonio, Texas, USA

(2) Department of Management Science and Statistics
University of Texas at San Antonio
San Antonio, Texas, USA

(3) Department of Thoracic & Cardiovascular Surgery
Alleghany Health Network
Pittsburgh, Pennsylvania, USA

(4) Feinberg School of Medicine
Northwestern University
Chicago, Illinois, USA

INTRODUCTION

Abdominal aortic aneurysm (AAA) is a prevalent cardiovascular disease characterized by the focal dilation of the aorta. With the AAA increasing in diameter over time, the risk of aneurysm rupture is generally associated with the size of the aneurysm. The criterion to decide on surgical intervention is determined by measuring the maximum diameter of the aneurysm relative to the critical value of 5.5 cm, which is an evidence-based criteria and not individual predictor of aneurysm rupture risk. However, a more reliable approach could be based on understanding the biomechanical behavior of the aneurysmal wall. In addition, geometric features that are proven to be significant predictors of the AAA wall mechanics could be used as surrogates of the AAA biomechanical behavior and, subsequently, of the aneurysm's risk of rupture. The purpose of the present work is to perform quantitative wall mechanics and geometric analyses for patient-specific AAAs that were either symptomatic or ruptured, with the objective of assessing whether a substantial correlation exists between specific geometric indices and wall stress.

METHODS

The abdominal computed tomography angiography (CTA) scans of 75 patients who underwent emergency repair of their AAA were obtained retrospectively from existing databases in the Departments of Radiology at Allegheny General Hospital (Pittsburgh, PA) and Northwestern Memorial Hospital (Chicago, IL). The Standard Digital Imaging and Communications in Medicine (DICOM) images are input into the MATLAB (Mathworks Inc., Natick, MA) code AAAVasc, which utilizes the contrast gradient of the image to identify the lumen, inner wall and outer wall contour. AAAVasc generates a point cloud representing the AAA surface and volumetric binary masks that represent the three regions of interest of the AAA (namely, the lumen, ILT and wall) [1]. The masks are used to generate surface and volume meshes using the in-house meshing code AAAMesh.

Finite element analysis (FEA) was performed using the aforementioned meshes with the solver ADINA (Adina R&D Inc., Watertown, MA). The FEA simulations were run subjecting the AAA to an intraluminal pressure of 120 mmHg. The outcome of the simulations was used to assess the first principal stress (PS1) distributions using the post-processing software EnSight (Computational Engineering International, Apex, NC). The spatially averaged PS1 was recorded as the biomechanical parameter representative of the AAA wall mechanics for subsequent analyses.

In-house MATLAB scripts were used to calculate 52 patient-specific geometric indices that quantify the shape, size, curvature, and wall thickness of each aneurysm. The point clouds generated from the image segmentation process were used to calculate the 1D and 2D indices, while the 3D indices are quantified from the surface and volume meshes. The biquintic Hermite finite element method (BQFE) was used to evaluate the curvature-based indices. The complete definition and mathematical formulation of the geometric indices are found in Shum et al [2].

With the goal of identifying the geometric indices that have substantial correlation with wall stress, we carried out a series of test of hypotheses. The null hypothesis $H_0: \rho = 0$ was tested against the alternate hypothesis$H_A: \rho \neq 0$, where ρ is the correlation coefficient between a geometric index and the wall stress. A geometric index may be identified as having a substantial correlation with wall stress if the p-value of its corresponding test is smaller than a chosen level of significance. The Bonferroni correction was applied to overcome the possibility of type 1 error, resulting in a corrected significance level of 0.00096. Hence, obtaining $p < 0.00096$ was considered to be a significantly correlated geometric index with wall stress.

RESULTS

The results of the FEA simulations, as presented in Fig. 1, yielded the spatially averaged PS1 with a mean = 29.3 ± 10.8 N/cm^2 and a maximum = 55.1 N/cm^2 among the 75 AAAs. In addition, the average maximum diameter among the 75 AAAs was 6.9 ± 1.7 cm with the largest aneurysm having a maximum diameter of 11.0 cm.

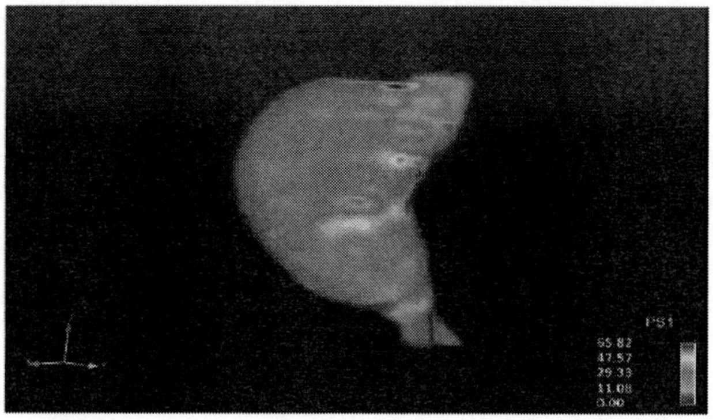

Figure 1: Exemplary spatial distribution of PS1 - first principal stress [N/cm^2].

The geometric indices that showed positive correlation with the spatially averaged PS1, as indicated in Table 1 were: 1D size-indices maximum diameter (D_{max}), AAA length (L), sac length (L_{sac}), and bulge height (H_b), and 3D size indices vessel surface area (S) and vessel volume (V). The geometric indices that showed negative correlation with the spatially averaged PS1 were average wall thickness ($t_{w,ave}$), average wall thickness at D_{max} ($t_{w,Dmax}$), mean wall thickness variance ($t_{w,meanVar}$), and median wall thickness variance ($t_{w,MedianVar}$). Amongst the curvature indices, area averaged mean curvature (MAA) and area averaged major principal curvature ($K1AA$) were significantly correlated with wall stress.

Table 1: Pearson's correlation coefficients and p-values for the relationships of spatially averaged wall stress with each of the significantly correlated geometric indices.

Geometric Index	Correlation coefficient (r)	p-value
D_{max} (mm)	0.48	0.000018
L (mm)	0.53	<< 0.00096
L_{sac} (mm)	0.49	<< 0.00096
H_b (mm)	0.38	0.00086
V (cm^3)	0.47	<< 0.00096
S (cm^2)	0.52	<< 0.00096
$t_{w,ave}$ (mm)	-0.63	<< 0.00096
$t_{w,Dmax}$ (mm)	-0.62	<< 0.00096
$t_{w,meanVar}$ (mm)	-0.45	<< 0.00096
$t_{w,MedianVar}$ (mm)	-0.48	<< 0.00096
MAA (mm^{-1})	-0.61	<< 0.00096
$K1AA$ (mm^{-1})	-0.44	0.00011

A stepwise regression analysis was performed starting from the 12 geometric indices identified above. It resulted in the smaller number of geometric indices listed in Table 2, with coefficients and standard errors also described. The coefficient of determination for this model is 0.76, so the four geometric indices that were ultimately selected explain 76% of the variability of wall stress.

Table 2: Coefficients of the stepwise regression predictive model with their respective standard errors for the 4 significantly correlated geometric indices with spatially averaged wall stress. The coefficient of determination of this model was 0.76.

Geometric Index	Coefficient of linear regression	Standard error
D_{max} (mm)	0.29	0.037
L (mm)	0.068	0.012
$t_{w,Dmax}$ (mm)	-7.68	1.23
$t_{w,MedianVar}$ (mm)	-17.55	9.53

DISCUSSION

The largest spatially averaged first principal stress was 55.13 N/cm^2 while the largest maximum diameter was 10.98 cm; evidently large aneurysms subject to high average mechanical stresses. This led to 6 of the 12 highly correlated indices with wall stress being size-related geometric indices. Among the geometric indices that exhibited high correlations with wall stress were wall thickness type indices. The average wall thickness and average wall thickness at D_{max} bore the highest correlations with the wall stress, followed by the mean and median variances in the wall thickness. This underscores the importance of accurately estimating individual AAA wall thickness distributions from clinical images. In an autopsy study conducted by Raghavan et al. [3], it was revealed that all aneurysms had a significant reduction in wall thickness near the site of rupture.

The area averaged mean curvature and area averaged major principal curvature were also highly correlated with wall stress. Shape and curvature based indices have been identified as predictors of rupture risk, due to the high correlations obtained between the mean and maximum centerline curvatures with peak wall stress, as reported by Giannoglou et al. [4] The stepwise regression analysis further reduced the number of geometric indices by eliminating redundancy in their predictability of wall stress. Such analysis suggests that maximum aneurysm diameter (D_{max}), AAA sac length (L), average wall thickness at D_{max} ($t_{w,Dmax}$), and the median of the wall thickness variance ($t_{w,MedianVar}$) can be used to predict the spatially averaged wall stress with an accuracy of 76%. We infer from this outcome that the two size- and two wall thickness-related indices may play the role of reliable surrogates of wall stress for symptomatic and ruptured AAAs. To this end, the outcome of this work supports the use of global measures of wall thickness and size indices as the geometric surrogates of wall stress for emergently repaired AAAs.

ACKNOWLEDGEMENTS

This work was supported in part by research funding from National Institutes of Health award R01HL121293 and American Heart Association award 15PRE25700288.

REFERENCES

1. Shum, J. et al., *Medical Physics*, 37(2):638-648, 2010.
2. Shum, J. et al., *Ann Biomed Eng*, 39(1):277-286, 2011.
3. Raghavan, M.L. et al.., *J. Biomech*, 39(16):3010-3016, 2006.
4. Giannoglou, G. et al., *Angiology*, 57(4):487-494, 2006.

SB^{3}C2017
Summer Biomechanics, Bioengineering and Biotransport Conference
June 21 – 24, Tucson, AZ, USA

CROSSLINKED ELASTIC FIBERS ARE NECESSARY FOR RESISTANCE TO STRETCH AT LOW PRESSURE AND FOR LOW ENERGY LOSS IN THE ASCENDING AORTA

Jungsil Kim (1), Marius C. Staiculescu (1), Hiromi Yanagisawa (2), Robert P. Mecham (3), Jessica E. Wagenseil (1)

(1) Mechanical Engineering and Materials Science
Washington University
St. Louis, MO, USA

(2) Life Science Center
University of Tsukuba
Japan

(3) Cell Biology and Physiology
Washington University
St. Louis, MO, USA

INTRODUCTION

In the large arteries, it is believed that elastin provides the resistance to stretch at low pressure, while collagen provides the resistance to stretch at high pressure [1]. It is also thought that elastin is responsible for the low energy loss observed with cyclic loading and unloading. These tenets have been supported through experiments that alter component amounts through protease digestion, vessel remodeling, normal growth, or in different artery types. Genetic engineering provides an opportunity to revisit these tenets through the loss of expression of specific wall components. We used mice lacking expression of elastin (*Eln-/-*) or lacking expression of two key proteins (lysyl oxidase, *Lox-/-*, or fibulin-4, *Fbln4-/-*) that are necessary for assembly of functional elastic fibers to investigate the contributions of elastic fibers to large artery mechanics. We imaged the wall structure and quantified the nonlinear, viscoelastic mechanical behavior of newborn *Eln-/-*, *Lox-/-*, and *Fbln4-/-* aorta and wild-type controls.

METHODS

Eln+/-, *Lox+/-*, and *Fbln4+/-* mice were bred to produce the associated wild-type (WT) (+/+) and knockout (KO) (-/-) pups. Pups were used within 24 hours of birth and euthanized by CO_2 inhalation. Animal protocols were approved by the IACUC. The ascending aorta (AA) was removed for fluorescent staining to visualize microstructure or for mounting in a pressure myograph to quantify mechanical behavior. For mechanical testing, the AA was stretched to its in vivo length, preconditioned, then inflated cyclically with physiologic saline from 0 – 60 mmHg. The energy loss was quantified by numerically integrating the pressure-outer diameter data to determine the area under the loading and unloading curves and the percent hysteresis (Fig. 1A). The resistance to stretch was quantified by fitting linear functions to the low and high pressure regions [2] (Fig. 1B). The best fit was determined by systematically changing the transition point between the two linear functions until the minimum error was obtained. KO AAs were compared to WT AAs using a two-tailed t-test with unequal variance. $P < 0.05$ was considered significant.

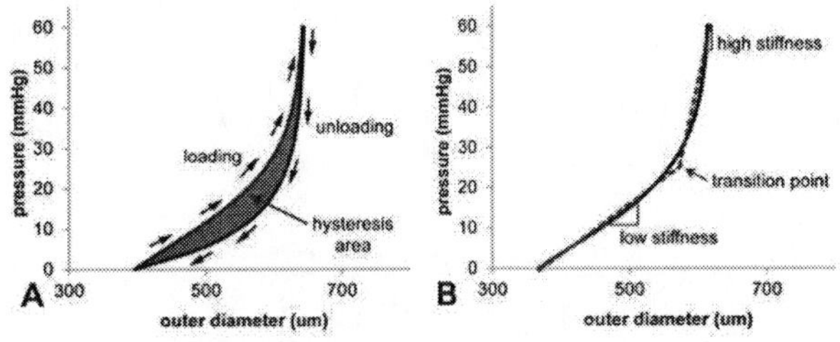

Figure 1: Example calculations for the percent hysteresis (A), low and high stiffnesses, and transition point (B).

RESULTS

Fluorescent microscopy images confirm that the lack of elastin, lysyl oxidase, or fibulin-4 results in absent or severely fragmented elastic fibers in the aortic wall (Fig. 2A-D). There are no obvious changes in collagen fiber organization (Fig. 2E-H). There are an increased number of smooth muscle cells (SMCs) in the KO AAs, and the cells appear disorganized, especially near the lumen (Fig. 2I-L).

Pressure-outer diameter curves for loading and unloading were compared to examine the contribution of elastic fibers to the structural behavior of the aortic wall. WT AAs show similar average pressure-diameter behavior, with slight shifts in the absolute diameter values. (Fig. 3A). The loading and unloading curves follow similar paths for the WT AAs and the percent hysteresis is 11 – 31% (Fig. 4). Average

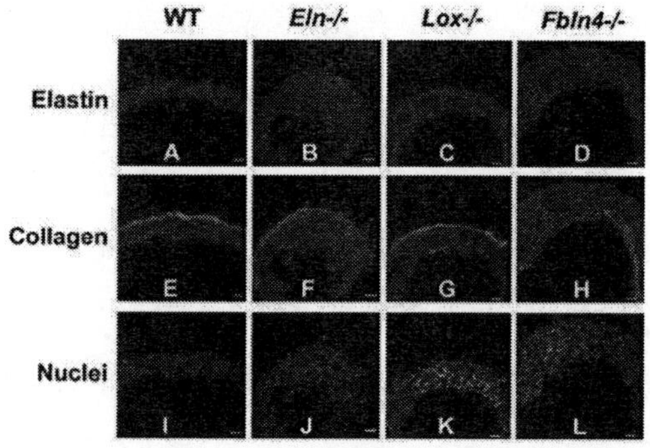

Figure 2: AA sections stained for elastin (red), collagen (green), and cell nuclei (blue). Lumen at bottom. Scale = 10 µm.

pressure-diameter curves for *Eln-/-* AA are shifted to lower diameters than *Eln+/+*, while pressure-diameter curves for *Lox-/-* and *Fbln4-/-* AA are shifted to higher diameters than *Lox+/+* and *Fbln4+/+*, respectively (Fig. 3B). The loading and unloading curves follow very different paths for the KO AAs (Fig. 3B). The percent hysteresis is 48 – 72%, with *Eln-/-* AA having the highest values (Fig. 4).

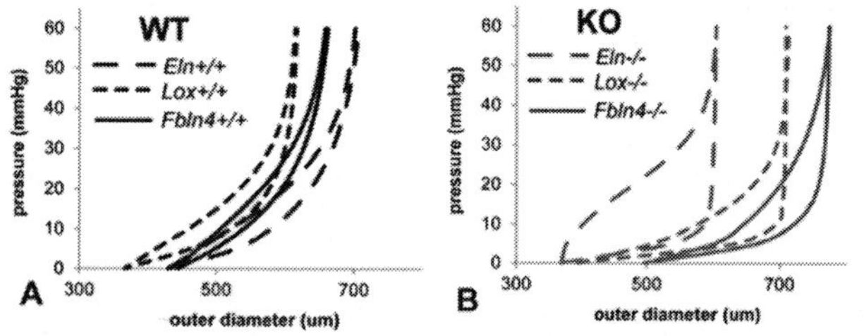

Figure 3: Average pressure-outer diameter curves for loading and unloading cycles of WT and KO AA.

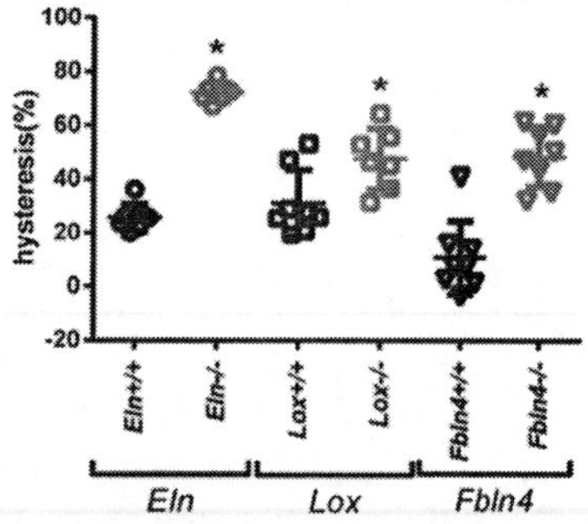

Figure 4: Percent hysteresis for WT and KO AA. * = P < 0.05.

The diameter shifts are supported by the lower transition diameters for *Eln-/-* compared to *Eln+/+* AA and the higher transition diameters for *Fbln4-/-* and *Lox-/-* compared to *Lox+/+* and *Fbln4+/+* AA, respectively (Fig. 5A). When examining the loading curves, there are surprisingly few significant differences between the transition pressure, low stiffness, or high stiffness values for WT versus KO AAs (Fig. 5B – D). Only *Lox-/-* and *Lox+/+* AA have significantly different transition pressures (Fig. 5B) and low stiffness values (Fig. 5C). In contrast, the unloading curves show significant differences in all of the quantified parameters for all KO groups. The transition pressure is lower (Fig. 5B), the low stiffness is smaller (Fig. 5C), and the high stiffness is larger (Fig. 5D) for the unloading curves for all KO AAs compared to WT controls.

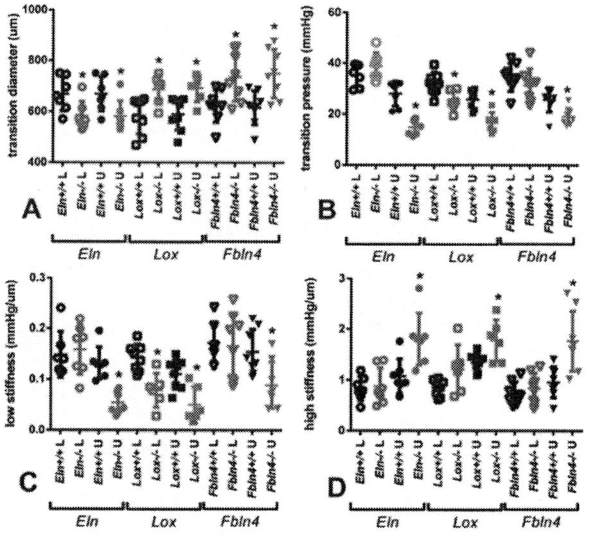

Figure 5: Quantification of loading (L) and unloading (U) behavior. Transition diameter (A), transition pressure (B), low stiffness (C), and high stiffness (D) are shown. * = P < 0.05.

DISCUSSION

Eln-/- AA has 0%, *Lox-/-* AA has 39% [3], and *Fbln4-/-* AA has 6% [4] of the normal level of elastin-specific crosslinks. Presuming that elastin crosslinking is necessary for elastic fibers to resist stretch, one would predict that the resistance to stretch at low pressures is decreased in *Eln-/-*, *Lox-/-*, and *Fbln4-/-* AA. This is true for the low stiffness for all of the unloading data for KO AAs, but it is not always true for the loading data (Fig. 5C). Therefore, our results support the tenet that elastic fibers are responsible for the resistance to stretch at low pressure [1], with the additional caveats that the elastic fibers must be properly assembled and crosslinked and that both loading and unloading cycles should be examined for arteries with highly disrupted elastic fibers. Our results extend previous results where arterial elastin is digested with specific proteases.

KO AAs have significantly increased dissipated energy compared to WT, as measured by the percent hysteresis. The large increase in energy loss is consistent with the tenet that elastic fibers are responsible for low energy loss in the large, elastic arteries. The major component of elastic fibers, elastin, alone is not enough, however, correctly assembled and crosslinked elastic fibers are necessary for low energy loss. Increased arterial viscoelasticity can affect pressure wave dissipation in the cardiovascular tree [5] and altered substrate viscoelasticity can alter cell phenotype [6]. Hence, altered viscoelastic behavior, in addition to altered elastic behavior of the arterial wall, may be an important factor in cardiovascular diseases associated with elastic fiber reduction or fragmentation.

ACKNOWLEDGEMENTS

NIH HL-115560 and HL-105314.

REFERENCES

1. Roach, M.R., et al., *Can J Med Sci*, 35(8): 681-90. 1957.
2. Tabima, D.M., et al., *J Biomech*, 43(10): 1864-9. 2010.
3. Hornstra, I.K., et al., *J Biol Chem*, 278(16): 14387-93. 2003.
4. McLaughlin, P.J., et al., *Mol Cell Biol*, 26(5): 1700-9. 2006.
5. Maxwell, J.A., et al., *Biophys J*, 8(8): 920-50. 1968.
6. Chaudhuri, O., et al., *Nat Mater*, 15(3): 326-34. 2016.

PATIENT-SPECIFIC MECHANICAL CHARACTERIZATION OF ABDOMINAL AORTIC ANEURYSMS AND HEALTHY AORTAS USING 4D ULTRASOUND: AN IN VIVO COMPARISON STUDY

Emiel M.J. van Disseldorp (1,2), Niels J. Petterson (1), Frans N. van de Vosse (1), Marc R.H.M. van Sambeek (2), Richard G.P. Lopata (1)

(1) Department of Biomedical Engineering
Eindhoven University of Technology
Eindhoven, the Netherlands

(2) Department of Vascular Surgery
Catharina Hospital Eindhoven
Eindhoven, the Netherlands

INTRODUCTION

An abdominal aortic aneurysm (AAA) is a localized enlargement of the abdominal aorta, which is generally the case when the maximal aortic diameter is larger than 3 cm. These aneurysms can rupture, which lead to a fatal hemorrhage in roughly 80% of all cases [1]. Therefore current clinical guidelines suggest to perform aortic repair when the risk of surgery outweighs the risk of rupture. In clinical practice this is the case when the AAA has grown more than 1 cm a year or when the diameter is larger than 5.5 cm [2]. However, this is an inadequate criteria since some aneurysms rupture before this threshold while others remain stable above, sometimes up to 10 cm. Therefore a more patient-specific rupture risk assessment is needed.

Wall stress analysis has been widely proposed as a new technique for patient-specific rupture risk prediction, usually based on computed tomography (CT). However, CT uses nephrotoxic contrast agents and radiation exposure which makes it not applicable for small aneurysms and follow-up. As an alternative, 4D (3D+t) ultrasound (US) overcomes these disadvantages, has a high temporal resolution and is relatively cheap. Moreover, the vessel's motion can be captured and used to calibrate the wall stress analysis to the 4D US data to determine patient-specific wall stresses, but also the material properties. The disadvantages of 4D US are the limited field-of-view and reduced image contrast. Recent studies showed that 4D US is applicable for patient-specific wall stress analysis and provides similar results as shown CT, but now also for small aneurysms and with patient-specific material properties [3, 4].

In this study, 4D US-based wall stress analysis is used to estimate the mechanical properties and wall stresses of healthy volunteers and AAA patients. Thereby, this characterization is performed using an iterative forward finite element approach, and in a more direct manner using elastometry. Both methods are mutually compared for both healthy volunteers and AAA patients.

METHODS

4D US data were acquired for 15 volunteers and 40 AAA patients using a Philips IU22 system (Philips, Bothwell, WA, USA) equipped with a 3D matrix probe (type: X6-1, center frequency: 3.5 MHz). Data were acquired during breath hold for 5 seconds, and the brachial pressure was measured using an arm cuff during the US acquisition.

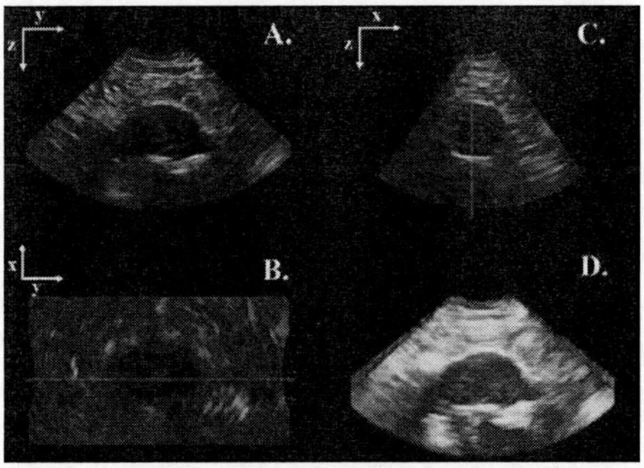

Figure 1: Example of a 4D US acquisition of an AAA patient. With a longitudinal (A), coronal (B) and transversal (C) view of the 3D render (D).

In the US data, the aortic wall was manually segmented and tracked over time to estimate the 3D displacement field of the vessel. The diastolic geometry was extracted from the 4D US data and converted into a quadratic tetrahedral mesh with a wall thickness of 2 mm. An example of the generated mesh is visualized in Figure 2.

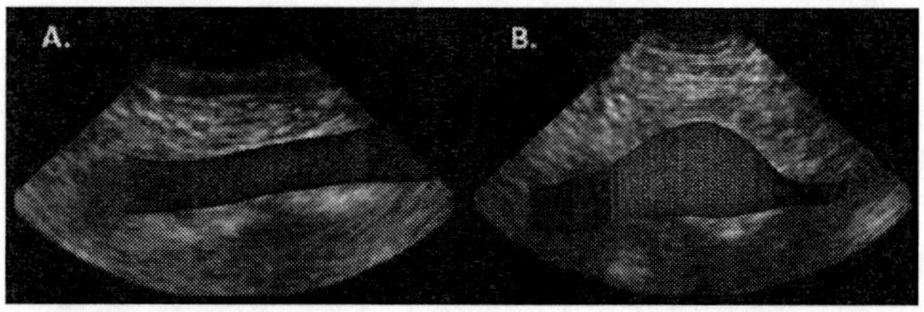

Figure 2: Example of the mesh of a normal aorta (A) and abdominal aneurysm (B). The blue mesh is the segmented aorta and the gray mesh is added to avoid boundary effects.

The mesh was elongated by 3cm in the proximal and distal direction to avoid boundary effects in the finite element analysis. Subsequently, wall stress analysis was performed using a neo-Hookean material model. The initial diastolic wall stress was reconstructed with a backward incremental method after which the model was inflated until the patient-specific systolic pressure was reached. The patient-specific incremental shear modulus (G_{FEM}) was obtained by iteratively adapting the modulus until the model displacement output matched the *in vivo* US displacements found by the 3D tracking algorithm.

In the direct elastometry approach, the diastolic and systolic circumferential wall stress is calculated using the LaPlace's law (Equation 1 – 2):

$$\sigma_{\theta\theta}(p_{dia}) = \frac{p_{dia} \cdot d_{dia}}{2h_{dia}} \quad (1)$$

$$\sigma_{\theta\theta}(p_{sys}) = \frac{p_{sys} \cdot d_{sys}}{2\frac{h_{dia}}{\lambda_{\theta\theta}}} \quad (2)$$

with p_{dia}, d_{dia}, and h_{dia} the diastolic pressure, diameter and wall thickness respectively, and $\lambda_{\theta\theta}$ the measured circumferential stretch between diastole and systole. Finally the patient-specific incremental shear modulus (G_{ELASTO}) was calculated by Equation 3:

$$G_{ELASTO} = \frac{\sigma_{\theta\theta}(p_{sys}) - \sigma_{\theta\theta}(p_{dia})}{\lambda_{\theta\theta}^2 - \frac{1}{\lambda_{\theta\theta}^2}} \quad (3)$$

All AAA patients were divided in three groups based on their diameter and together with the volunteers the material properties G_{FEM} and G_{ELASTO} were calculated. Also the patient-specific 99th percentile wall stress, known as potential rupture risk estimator, was derived from the finite element model (FEM).

RESULTS

Results reveal that the patient-specific material property of volunteers is significantly smaller for healthy volunteers compared to AAA patients ($p << 0.01$). Figure 3 also suggest a possible increase in arterial stiffness between the small, middle, and large AAAs. However, only a significant difference between the small and large AAAs was seen using the FEM method ($p = 0.03$).

In Figure 4 the 4D-US based 99th percentile wall stress is shown for the volunteers and AAA patients. A clear significant difference in peak wall stress is shown between the healthy volunteers and the AAA patients. In addition, within the AAA population a large spread in peak wall stress is observed.

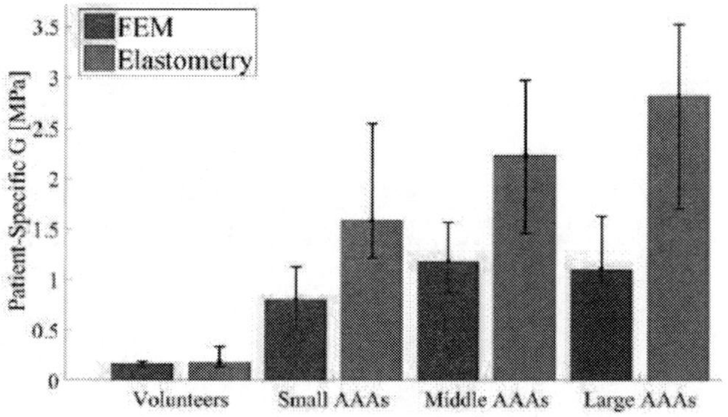

Figure 3: Patient-specific shear modulus (G) derived using FEM (red) and elastometry (gray) is shown for healthy volunteers and small, middle and large AAAs with its interquartile range.

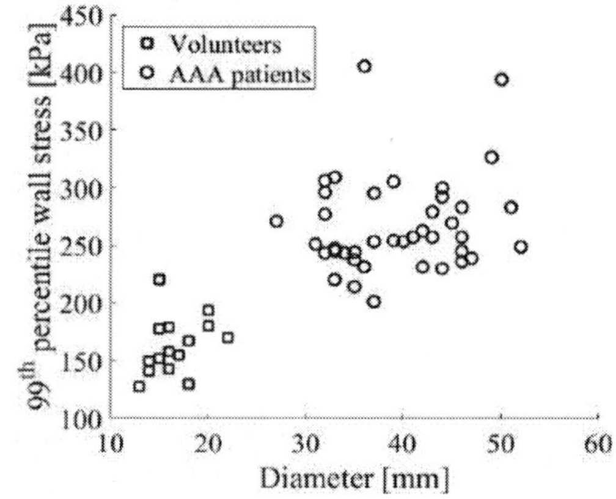

Figure 4: The 99th percentile wall stress as function of the aortic diameter for the healthy volunteers (squares) and AAA patients (circles).

DISCUSSION

This study shows that 4D-based wall stress analysis is feasible for AAA patients and healthy volunteers, providing significant differences in material properties and peak wall stresses between the two groups. Furthermore, current study suggest an increase in arterial stiffness within the AAA population with respect to the diameter. However, inclusion of more patients is needed to strengthen this finding.

This study elicits the opportunity to follow the progression of disease in a longitudinal study and monitor changes in peak wall stress and material properties over time, which could eventually improve AAA rupture risk assessment.

ACKNOWLEDGEMENTS

This study was supported with an unrestricted educational grant from the Lijf en Leven Foundation.

REFERENCES

[1] Reimerink, JJ et al., *Br J Surg*, 100:1405-13, 2013.
[2] Moll, FL et al., *Eur J Vasc Endo Surg*, 41:S1-58, 2011.
[3] Kok, AM et al., *J Vasc Surg*, 61:1175-84, 2015.
[4] van Disseldorp, EMJ et al., *Eur J Vasc Endo. Surg*, 52:635-42, 2016.

FAILURE BEHAVIOR OF HUMAN ASCENDING THORACIC AORTIC ANEURYSMS IN SHEAR LAP VS. UNIAXIAL LOADING

C. Korenczuk (1), R. Dhume (2), C. Witzenburg (2), V. Barocas (1)

(1) Department of Biomedical Engineering
University of Minnesota – Twin Cities
Minneapolis, MN, USA

(2) Department of Mechanical Engineering
University of Minnesota – Twin Cities
Minneapolis, MN, USA

INTRODUCTION

Ascending thoracic aortic aneurysms (ATAAs) are a high-risk pathology characterized by abnormal dilatation of the aorta between the aortic root and aortic arch. The dilatation is typically accompanied by disorganization or loss of elastin and vascular smooth muscle cells within the tunica media [1]. Experimental testing of ATAA tissue in uniaxial [2], biaxial [3], and peel [4] testing regimes has shown significant anisotropy, with greater stiffness in the circumferential direction, as well as increased stiffness in all directions when compared to non-aneurysmal tissue. Variations in mechanical strength between the lesser and greater curvature have also been reported [5]. The mechanisms of aneurysm rupture and dissection, however, are still largely unknown. It has been proposed [6] that laminal strength between wall layers play a significant role in the failure process. Experimental geometries, such as the peel geometry which quantifies radial strength, explore the strength between the laminal layers.

Despite the value of current experimental results, there remains a compelling need to characterize the shear strength in ATAAs, as the contribution of shear stresses in aneurysm failure must not be overlooked. Furthermore, the regional mechanical strength in all directions and possible loading configurations is not well documented. Understanding the mechanical strength of ATAA tissue in all directions and loading configurations is the first step in properly characterizing its behavior and is essential to predict failure using computational models.

Here, we study the mechanical behavior of ATAA tissue in multiple loading configurations and orientations, and compare results to previously studied healthy porcine ascending aortic tissue [7].

METHODS

Resected human ATAA tissue (n=16, 5 male, 4 female, ages 53-81 years old) was obtained from patient surgeries at the University of Minnesota Medical Center. Following surgery, tissue was stored in 1x PBS at 4° C. Samples were cleaned of excess connective tissue and cut open axially. Uniaxial dogbones, peel, and lap sample geometries (Fig. 1) were cut in the circumferential (C) and axial (A) directions on both the greater and lesser curvature. Biaxial samples (Fig. 1) were cut from the greater curvature due to size constraints on the lesser curvature. Samples were photographed to obtain the undeformed sample geometry.

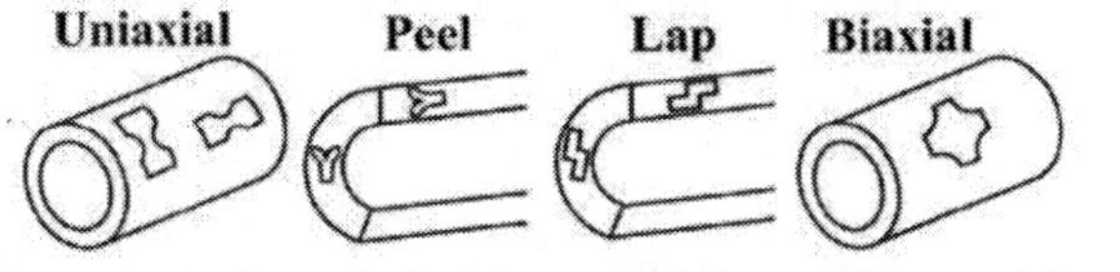

Figure 1: A schematic showing each of the sample geometries.

Uniaxial, peel, and lap samples were prepared by initially using a rectangular tissue piece (approximately 10mm x 5mm). Uniaxial samples had the adventitia removed, and were created by cutting half circles (r = 2.5mm) out of the rectangle using a biopsy punch. Peel samples were prepared from the rectangle by making an incision in the media parallel to the vessel wall to initiate peel propagation. Lap samples were cut by removing half of the thickness on each side of the sample, leaving an overlap region in the medial layer center. Biaxial samples had the adventitia removed and were cut from a square (approximately 20mm x 20 mm) into a cruciform shape using biopsy punches (r = 5mm).

Uniaxial, peel, and lap samples were clamped with custom grips, placed in a 1x PBS bath at room temperature, and pulled at 3mm/min in strain-to-failure experiments on a uniaxial testing machine (MTS, Eden Prairie, MN). A static 10N load cell recorded the forces and grip stretch was calculated using the actuator displacement. Normal and shear first Piola-Kirchhoff stresses (σ and τ, respectively) for uniaxial and lap samples, respectively, were computed by dividing the force by the undeformed area. Peel tension (T^{peel}) was calculated by dividing the force by the undeformed sample width.

Biaxial samples were tested in equibiaxial displacement controlled experiments (Instron 8800 Microtester) to 15% strain for each arm. Forces were recorded by 5N load cells, and PK1 stresses were calculated based on the undeformed geometry. All statistical analysis (Tukey's Multiple Comparisons) was performed in GraphPad Prism 6.

RESULTS

Uniaxial samples showed a statistically significant ($p < 0.01$) difference in failure stress in favor of the circumferential direction in both the greater and lesser curvatures (σ_C = 1219.2 ± 252.9 kPa, σ_A = 482.6 ± 119.1 kPa (mean ± 95%CI) for the greater curvature and σ_C = 1044.2 ± 307.6 kPa, σ_A = 458.5 ± 145.3 kPa for the lesser curvature). There was no significant difference in failure stress between the greater and lesser curvature for either orientation. Uniaxial samples on the greater curvature exhibited an increased stiffness in both the circumferential and axial directions compared to porcine uniaxial samples, while the lesser curvature exhibited a decreased stiffness in the circumferential direction and no change in stiffness in the axial direction when compared to the porcine uniaxial samples. All uniaxial locations and directions exhibited lower failure stretches than porcine data.

Peel samples showed no significant difference in peel tension between the circumferential and axial direction on the greater curvature (T_C^{peel} = 27.69 ± 6.19 N/m, T_A^{peel} = 38.24 ± 6.8 N/m), but did show a statistically significant (p = 0.02) difference in favor of the axial direction on the lesser curvature (T_C^{peel} = 27.9 ± 4.63 N/m, T_A^{peel} = 42.79 ± 10.04 N/m). No significant difference was found between the greater and lesser curvatures for either orientation.

Lap samples exhibited anisotropy ($p < 0.05$) and were stronger in the circumferential direction on both the greater and lesser curvatures (τ_C = 81.72 ± 7.69 kPa, τ_A = 55.34 ± 11.42 kPa for the greater curvature and τ_C = 87.9 ± 8.4 kPa, τ_A = 54.89 ± 11.85 kPa for the lesser curvature). No significant difference was found between the greater and lesser curvature for either orientation. Lap samples on both the greater and lesser curvatures showed increased stiffness in both directions compared to the porcine data, with the lesser curvature exhibiting a slightly higher stiffness. All orientations and locations exhibited a lower failure stretch compared to porcine data.

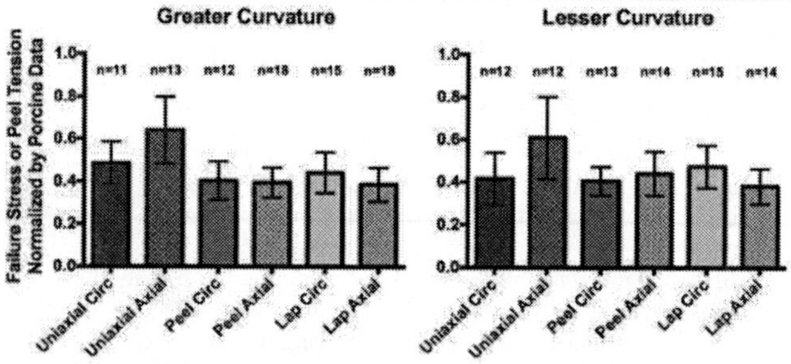

Figure 2: Human ATAA data normalized by porcine data for each experimental geometry. Error bars indicate 95%CI.

When normalized by the porcine data for each one of the experimental tests, there were no significant differences between orientation on either the greater or lesser curvature (Fig. 2). There were also no significant differences for each geometry on the greater or lesser curvature. Each sample corresponded to roughly half of the porcine tissue strength in every testing geometry and location.

Biaxial samples showed an increase in stiffness compared to porcine data, as an increased failure stretch. Human ATAA biaxial samples only reached a stretch of 1.15, while porcine biaxial samples were stretched up to 1.4.

DISCUSSION

Human ATAA results for uniaxial and biaxial samples were in agreement with previous studies [2,3]. Peel samples exhibited lower average peel tension (~65 N/m) for both the circumferential and axial directions compared to [4] but showed a similar trend in anisotropy (axial > circumferential). To our knowledge, shear lap results for ATAA tissue have not been published. Lap samples exhibited significantly lower failure shear stresses compared to uniaxial samples, highlighting the importance of studying the shear stress contribution to the failure of ATAAs.

Uniaxial, lap, and biaxial samples all exhibited higher stiffness compared to porcine ascending aorta, as expected in the ATAA pathology. It is interesting to note, however, that increased stiffness was only seen on the greater curvature for the uniaxial samples, while both curvatures saw increased stiffness for the lap samples, suggesting a regionally heterogeneous change in the tissue microstructure due to the pathology.

In comparing ATAA failure stress and peel tension to porcine ascending aorta, the tissue was weaker in every direction and location by a similar ratio, which suggests that the weakening may involve multiple components. These results emphasize the importance of fully understanding the structural and mechanical changes that occur in ATAAs. Tissue composition is clearly affected in different ways, meriting further exploration of the constituent's orientation and composition.

ACKNOWLEDGEMENTS

This material is based upon work supported by the National Science Foundation Graduate Research Fellowship Program under Grant No. 00039202 (CEK). This work was supported by the National Institutes of Health (R01EB005813), and CEK is a recipient of an ARCS Foundation Scholar Award. Porcine tissue was provided by the Visible Heart Laboratory at the University of Minnesota.

REFERENCES

[1] Humphrey, J, Cardiovascular solid mechanics: cells, tissues, and organs. 2013.

[2] Vorp, D, et al., *Ann. Thorac. Surg.*, 75(4), pp. 1210–1214, 2003.

[3] Okamoto, R, et al., *Ann. Biomed. Eng.*, 30(5), pp. 624–635, 2002.

[4] Pasta, S, et al., *J. Thorac. Cardiovasc. Surg.*, 143(2), pp. 460–467, 2012.

[5] Duprey, A, et al., *Eur. J. Vasc. Endovasc. Surg.*, 39(6), pp. 700–707, 2010.

[6] Pal, S, et al., *J. Biomech.*, 47(5), pp. 981–988, 2014.

[7] Witzenburg, C, et al., *J. Biomech. Eng*, Accepted, 2016.

AGED SUPRASPINATUS TENDONS HAVE ALTERED DYNAMIC COMPRESSIVE AND POROELASTIC PROPERTIES

Brianne K. Connizzo (1), Alan J. Grodzinsky (1,2,3)

(1) Department of Biological Engineering
Massachusetts Institute of Technology
Cambridge, MA, USA

(2) Department of Mechanical Engineering
Massachusetts Institute of Technology
Cambridge, MA, USA

(3) Department of Electrical Engineering and Computer Science
Massachusetts Institute of Technology
Cambridge, MA, USA

INTRODUCTION

Rotator cuff disorders are one of the most common causes of shoulder pain and disability, with an incidence reported to exceed 50% in the population over the age of 60 [1]. One factor thought to contribute to the progression of disease is compression of the rotator cuff tendons [2]. While some compression occurs during normal motion, several factors associated with age, such as rotator cuff weakness and poor posture, can also cause excessive compression of the tendon, which would be further exacerbated by repetitive use and normal age-related tendon degeneration. However, few studies have investigated the baseline compressive response of tendons and how that response is altered with aging. The compressive response is highly dependent on fluid exudation from the tissue and no studies to date have measured the poroelastic behavior of supraspinatus tendon directly. We recently adapted our high-frequency AFM-based rheology technique to characterize the regional poroelastic behavior of tendon, and found that tendons exhibit a complex dynamic response with both a viscoelastic and poroelastic peak in the phase angle [3]. The objective of the present study was to use this technique to investigate how regional poroelastic properties of mouse supraspinatus tendons are altered with age.

METHODS

Supraspinatus tendons, along with muscle and bone attached, were harvested from the limbs of 4 month (mature) and 18 month (aged) male C57BL/6 mice (n=7-8/group). Excess soft tissue was gently removed from the tendons, but the bony insertion and myotendinous junction were kept intact. Samples were then affixed to a custom stage for wide bandwidth rheological testing using cyanoacrylate glue, and kept hydrated with PBS throughout experiments (Fig. 1A,B).

Indentations were performed at fourteen regions along the length of the tendon from bone to muscle. The first region was defined as the closest indentation site to the humeral head, typically less than 100μm from the bone. Samples were then indented every 50μm for the first 200μm and then every 100μm until the junction with muscle. Indentations at each region were performed in the middle of the tissue (away from the glued edges) at 9 locations within each region. Built-in top-view optics and translational stage were used to visualize positions of locations.

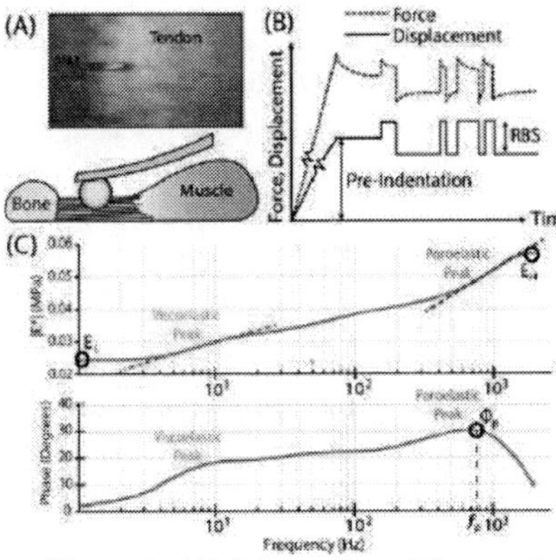

Figure 1. (A) Schematic and image of indentation of supraspinatus tendon (B) Indentation loading profile (C) Sample dynamic response of tendon

We indented samples with ~25 μm diameter polystyrene colloidal probe tips attached to tipless cantilevers having nominal spring constant $k \sim 7.4$ N/m. Data were obtained over a high frequency range (1 Hz to 10 kHz) using our custom rheology system coupled to the MFP-3D AFM (Asylum Res, CA), as described [4]. The indentation loading profile (Fig. 1B) consisted of an initial load-controlled ramp-and-hold pre-indentation of approximately ~3.5 μm followed by random binary sequence (RBS) displacements of 8-12 nm, which were applied using previously established methods. A discrete Fourier transform was used to obtain the fundamental frequency component of the force F_{osc} and displacement δ signals. The magnitude and phase of the dynamic complex modulus at each frequency was thereby obtained from the measured force and applied displacement [4].

Tendons exhibit a dynamic phase response having two characteristic peaks in the frequency range measured here, but only the 2nd peak, which occurs at higher frequencies, is associated with fluid-solid poroelastic energy dissipation [3]. From the measured dynamic modulus at low and high frequency limits, we quantified the equilibrium (low-frequency) and instantaneous (high-frequency) moduli, the self-stiffening ratio (ratio of high to low frequency moduli [5]), and the peak phase angle and frequency of the peak phase angle (Fig. 1C). For pooled comparisons, the insertion site was defined as the first 300μm from the first indentation and the midsubstance was defined as the next 400-1100μm. Pooled comparisons were made using two-way ANOVAs, and with Bonferroni corrected post-hoc t-tests where applicable ($p<0.05$ significant (solid bar); $p<0.10$ trend (dashed bar)).

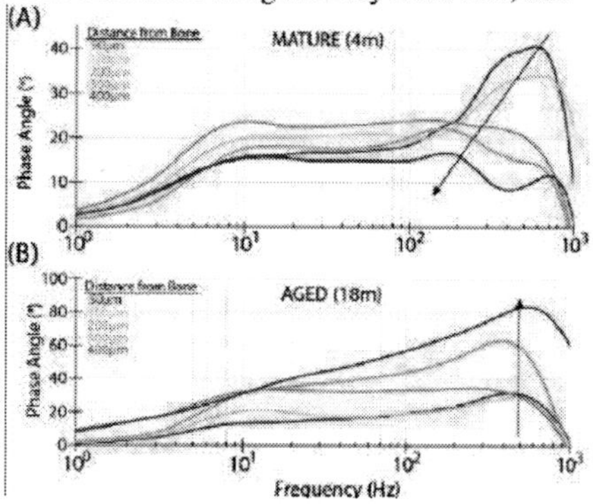

Figure 2. Representative phase responses of (A) mature and (B) aged tendons.

RESULTS

All 2-way ANOVAs for pooled data were significant for age, region, and the interaction. We therefore discuss here the Bonferroni-corrected post-hoc tests. As in our previous study, all tendons exhibited a two-peak phase response (Fig. 2). We now report that mature tendons have decreasing hydraulic permeability from the insertion site to midsubstance, as evidenced by the downward shift of the poroelastic peak phase angle (Fig. 2A). Aged tendons did not exhibit a similar regional response, as indicated by little to no shift in the poroelastic peak (Fig. 2B). Equilibrium modulus tended to increase in the aged group at the insertion site (Fig. 3A). There were no differences in high frequency modulus at either region (Fig. 3B). The self-stiffening ratio tended to increase in the aged group at the insertion site (Fig. 3C). Finally, the peak phase angle was significantly increased at the insertion site (Fig. 4A).

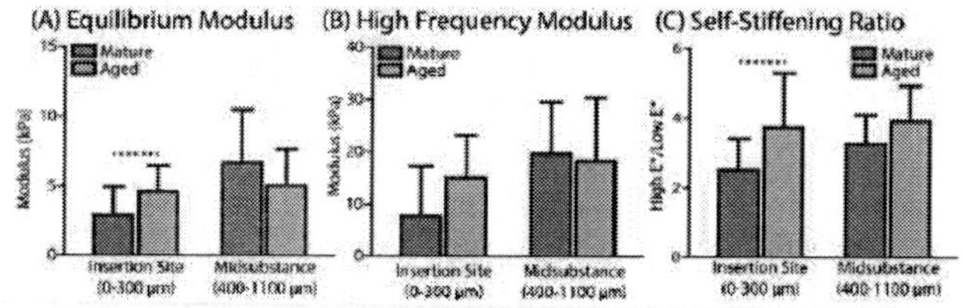

Figure 3. (A) Equilibrium, modulus, (B) High frequency modulus, and (C) Self-stiffening ratio for mature and aged tendons

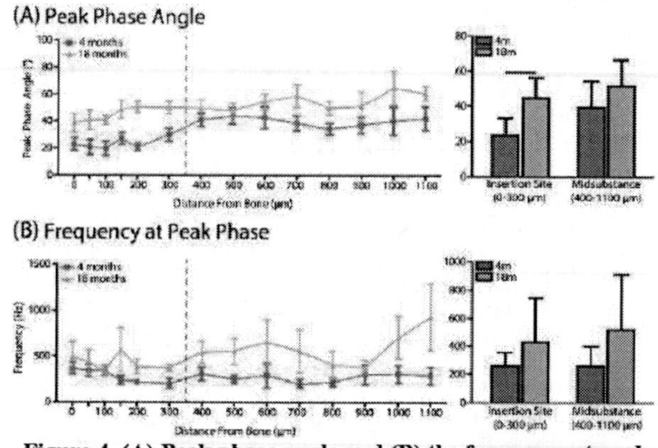

Figure 4. (A) Peak phase angle and (B) the frequency at peak phase angle [(Left) Per region, (Right) Pooled by tendon]

DISCUSSION

Aged tendons had increased equilibrium modulus at the nanoscale at the insertion site, which is consistent with previous studies at the macroscale [6]. Interestingly, self-stiffening at the insertion site was also increased in aged tendons, suggesting an increase in high frequency modulus as well, although we did not reach significance with the current sample size. This could be a feature of overall joint stiffening that occurs with age or suggest an accumulation of glycosaminoglycans associated with tendon degeneration.

Interestingly, the dynamic response of the tissue was significantly altered in the aged tendons as well. In mature tendons, the insertion site typically had a higher hydraulic permeability than the midsubstance. We hypothesize that the poroelastic response may be more heavily impacted by the organization and density of the collagen fibrils, both of which vary by region, than by the smaller proteoglycans present in tendon when compared to other tissues studied [7]. This regional variation, which could be crucial for directed fluid flow along the tendon length, is lost in aged tendons.

With aging, the phase angle and peak frequency of the phase angle were both increased, suggesting increased energy dissipation and hydraulic permeability. Organization, density and diameter of collagen fibrils have all been shown to decreased with age in several tendons [8], which could be responsible for the increases found here. Increased hydraulic permeability could be detrimental as the resulting decrease in fluid pressurization within the tendon makes it unable to resist compressive loading, resulting in direct loading of the collagen fibers and cells, particularly during high impact loading. Furthermore, this mechanism could promote transdifferentiation of tenocytes to a more cartilage-like phenotype, consistent with observations in tendinopathy samples. Finally, the increase in hydraulic permeability could also allude to an increased ability of cytokines to reach cells in the interior of the tendon, thereby spreading injury from one tendon to the rest of the rotator cuff more easily.

This study is the first to directly investigate the poroelastic properties of mature and aged supraspinatus tendons at the nanoscale. While we continue to add samples to this dataset, we have already found that many of the properties governing fluid flow and energy dissipation are altered. We hypothesize that this could be a contributing factor to the development of age-related tendinopathy. Ongoing studies aim to further investigate the findings here by increasing sample size, adding a 'young' group at 1 month old, and studying the bursal side of the tendon.

ACKNOWLEDGEMENTS

This study was supported by an NIH/NIA NRSA Postdoctoral Fellowship (F32-AG052284) and NSF Grant CMMI-1536233.

REFERENCES

[1] Buckwalter JA et al., J Bone Joint Surg Am, 85-A(4):748-758, 2003. [2] Thornton GM et al., Br J Sports Med, 44(10):698-703, 2010. [3] Connizzo BK et al., in press at J Biomech, 2016. [4] Nia HT et al., J Biomech, 48(1):162-165, 2015. [5] Nia HT et al., ACS Nano, 9(3):2614–2625, 2015. [6] Connizzo BK et al. Trans Orthop Res Soc 42:586, 2017. [7] Connizzo BK et al., J Orthop Res, 34(12):2154-2161, 2016. [8] Strocchi R et al., Foot Ankle, 12(2):100-104, 1991.

SB3C2017
Summer Biomechanics, Bioengineering and Biotransport Conference
June 21 – 24, Tucson, AZ, USA

THE HUMAN ACHILLES TENDON SHOWS SPECIALISATION TOWARDS ENERGY STORAGE THAT IS AFFECTED BY AGEING

Dharmesh Patel (1), Ewa M. Spiesz (2), Chavaunne T. Thorpe (3), Helen L. Birch (4), Graham P. Riley (5), Peter D. Clegg (6), Hazel R.C. Screen (1)

(1) School of Engineering and Materials Science, Queen Mary University of London, London, United Kingdom

(2) Department of Bionanoscience, Delft University of Technology, Delft, Netherlands

(3) Comparative Biomedical Sciences, Royal Veterinary College, London, United Kingdom

(4) Institute of Orthopaedics and Musculoskeletal Science, University College London, London, United Kingdom

(5) School of Biological Sciences, University of East Anglia, Norwich, United Kingdom

(6) Department of Musculoskeletal Biology, University of Liverpool, Liverpool, United Kingdom

INTRODUCTION

Tendons primarily function to transmit forces from muscle to bone. However, some tendons, such as the human Achilles tendon and equine superficial digital flexor tendon (SDFT), also function to store and release energy during locomotion. These tendons are known as energy storing tendons, and have specialised mechanical properties. However they are also prone to injury (1,2).

The specialised mechanical properties of energy storing tendons have been primarily studied in equine tendons. The SDFT has been shown to exhibit higher failure strains than the common digital extensor tendon (CDET; a positional tendon) during *in vitro* testing, indicating specialisation of the tendon to meet its secondary function (3). Subsequent micromechanical studies have found that the matrix between fascicles, the inter-fascicular matrix (IFM), exhibits lower stiffness in the SDFT than the CDET, which potentially facilitates the additional extension and energy storing capacity of the SDFT (3,4). Ageing studies in the equine SDFT have also identified age related stiffening of the IFM and reduced functionality, which is likely to compromise tendon function *in vivo* (4).

Limited data are currently available on tendon microstructure and the related micromechanics in human tissues. Human tendons are generally smaller than those in the equine model, and are also likely to have different functional demands, potentially leading to differences in the properties and significance of the IFM. This study investigates the mechanical properties of the IFM and fascicular matrix (FM) of two functionally distinct human tendons (Achilles and anterior tibialis) and explores the effect of ageing on the mechanics of each tendon component as well as the mechanics at the quarter tendon level. It is hypothesised that the IFM of the Achilles tendon exhibits less hysteresis and behaves more elastically to facilitate energy storage, but that this specialisation is lost with ageing.

METHODS

The mechanical properties of two lower leg tendons were compared; the Achilles tendon (an energy storing tendon) and the anterior tibialis tendon (a positional tendon). All tendons were obtained post-mortem with ethical approval (14/NE/0154). Matched samples for each test and from each tendon type were compared, and samples categorised into two age groups: middle-aged tendons (9 donors; 31-58 years old) and old-aged tendons (9 donors; 72-94 years old).

Quarter tendon, FM and IFM samples were prepared for mechanical characterisation from each tendon, from all 18 donors. Quarter tendons were prepared for quasi-static mechanical testing by cutting tendons along their long axis into quarters. The cross-sectional area of each specimen was measured using an alginate mould (5), before specimens were secured in cryogrips, preloaded to either 20N (Achilles) or 10N (anterior tibialis), pre-conditioned for 10 loading cycles from 0 to 2.5% strain at 1Hz, and then pulled to failure at a strain rate of 5%/second. Fascicles, approximately 40mm length, were dissected from a second tendon quarter, for FM testing. The diameter of each fascicle was measured using a laser micrometer, and the fascicle loaded into pneumatic grips. A pre-load of 0.02N was applied, then samples pre-conditioned with 10 loading cycles between 0 and 2.5% strain at 1Hz, before pulling to failure at a strain rate of 5%/second. For IFM testing, fascicle pairs (i.e. two adjoining parallel fascicles) were isolated from the same tendon quarter used to isolate FM specimens. The opposing ends of the fascicle pairs were dissected such that 10mm of IFM was left intact between the two fascicles, enabling the IFM to be tested in shear by pulling the opposing fascicle ends in a uniaxial

Technical Presentation #308

manner (3). The IFM specimens were pre-loaded to 0.02N, pre-conditioned with 10 loading cycles between 0 and 0.25mm at 1Hz, and then pulled to failure at a rate of 1mm/second. The force and extension data from all mechanical tests were used to calculate failure properties, hysteresis and relaxation.

For immunohistochemistry, transverse and longitudinal pieces from a third quarter of the tendons of 5 of the donors in each age group were embedded in paraffin wax and sectioned. The sections were stained for decorin, fibromodulin, lubricin and versican, and histochemical stains for elastin and mucins/proteoglycans were performed as described previously (6). Images for each stain were acquired at x20 magnification using a digital slide scanner and scored from 0 to 5 for abundance of each stain in FM and IFM regions by two independent assessors, blinded to tendon type and stain.

RESULTS

The failure properties of FM, IFM and quarter tendon samples from middle-aged donors showed no significant differences between tendon types. However, FM, IFM and quarter tendon samples from the Achilles all exhibited significantly less hysteresis (Fig. 1), and the Achilles FM and IFM samples exhibited less relaxation (4% less for FM, and 13% less for IFM) than the anterior tibialis, suggesting the compartments of the Achilles tendon have a specialisation towards energy storage.

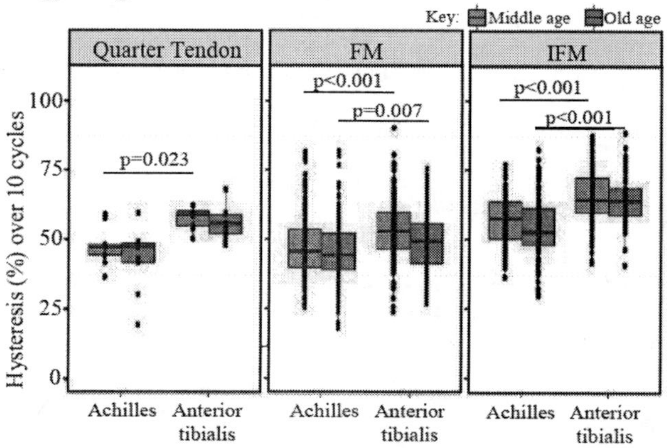

Figure 1: The hysteresis for Achilles quarter tendon, FM and IFM, over 10 cycles of 0 to 2.5% strain, are all significantly less than that of the anterior tibialis tendon.

Changes with ageing were identified by comparing middle-aged and old-aged tendons. Ageing predominately affected the mechanical properties of the Achilles tendon. However, fascicle stiffness and ultimate tensile strength were reduced with age in both the Achilles (p=0.011 and p=0.049 respectively) and anterior tibialis tendons (p=0.033 and p=0.017 respectively). The fascicles from old Achilles tendons also showed a reduction in relaxation (p=0.035). Further, the IFM of the Achilles tendon also showed a decrease in maximum stiffness with age (p<0.001). At the quarter tendon level, a decrease in Achilles tendon ultimate tensile stress (p=0.025) was observed.

Immunohistochemical analysis showed that IFM and FM composition differed. Overall proteoglycan content was significantly higher in the IFM than the FM in both tendon types (p=0.015 and p=0.022 respectively). More specifically, decorin was higher in the IFM than FM of the Achilles tendon (p=0.037) while versican was higher in the IFM than FM of the anterior tibialis tendon (p=0.029). Elastin was also observed with higher abundance in the IFM compared to the FM in both tendon types (p=0.006). Ageing was accompanied by a decrease in fibromodulin in the FM of Achilles tendon specifically (p=0.045), while lubricin and versican both showed a trend towards decreased levels in the IFM of both tendon types.

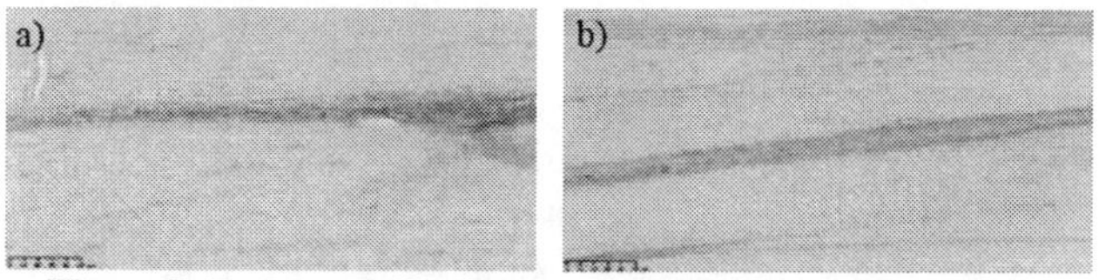

Figure 2: Representative images of a) Achilles and b) anterior tibialis tendons stained with alcian blue/periodic acid-Schiff. Note the stronger staining of the IFM compared to the FM indicating a higher abundance of proteoglycans. Scale bar = 100μm.

DISCUSSION

This is the first study to assess the mechanical properties and composition of different structures across functionally distinct human tendons. The IFM in equine tendons has previously demonstrated an ability to resist cyclic loading, particularly in the energy storing SDFT, where the IFM has a greater capacity for recovery than in the positional CDET (3,4). The current data provide confirmation that similar trends are present in functionally different human tendons. While there were no differences in failure properties between tendon types at the quarter tendon, FM or IFM level, less hysteresis and reduced viscoelasticity was observed in the Achilles tendon and its components as hypothesised. Reduced viscoelasticity is beneficial for energy storage, thus the more elastic behaviour of the Achilles tendon substructures emphasises specialisation towards this tendon's function.

Further, changes in mechanical properties during ageing were also investigated in both tendon types. Studies of the equine SDFT have found ageing reduced elasticity of the IFM (4). In the current study ageing predominately affected the Achilles tendon; the maximum stiffness of the FM and IFM both decreased, while the FM also showed increased relaxation. These changes suggest ageing potentially affects tendon function, and indicate structural or compositional changes occur with age in energy storing tendons.

Further characterisation of the tendons identified compositional differences between the two tendon regions; the IFM contained a higher overall proteoglycan content than the FM, whilst also exhibiting a higher elastin content. Regions within the equine SDFT show similar compositional specialisations, i.e. higher elastin content in the IFM (6). Together with the mechanical properties observed, it is apparent that tendons exhibit mechanical properties specialised towards function as hypothesised and that these properties change with age. Furthermore, proteoglycans and elastin are likely to play a key role in facilitating tendon energy storing capabilities and alteration of these components with ageing could lead to increased injury risks.

ACKNOWLEDGEMENTS

Funded by Arthritis Research UK (Grant No. 20262).

REFERENCES

[1] Knobloch, K et al., *Foot Ankle Int*, 29:671-6, 2008.
[2] Kasashima, Y et al., *Equine Vet J*, 36:346-50, 1999.
[3] Thorpe, C et al., *J R Soc Interface*, 9:3108-17, 2012.
[4] Thorpe, C et al., *J Mech Behav Biomed Mater*, 52:85-94, 2015.
[5] Goodship AE et al., *J Biomech*, 38:605-8, 2005.
[6] Thorpe, C et al., *J Anat*, 229(3):450-8, 2016.

STRUCTURAL REMODELING OF FATIGUE DAMAGED TENDONS BY EXERCISE IS ASSOCIATED WITH INTEGRIN SUBUNITS αV AND α5

Rebecca Bell (1), Remi Gendron (2), Jack Brenneman (1), Evan L. Flatow (2), Nelly Andarawis-Puri (1,3,4)

(1) Sibley School of Mechanical and Aerospace Engineering
Cornell University
Ithaca, NY, USA

(2) Department of Orthopaedics
Icahn School of Medicine at Mount Sinai
New York, NY, USA

(3) Nancy E. and Peter C. Meinig School of Biomedical Engineering
Cornell University
Ithaca, NY, USA

(4) Hospital for Special Surgery
New York, NY, USA

INTRODUCTION

Tendinopathy is a common debilitating musculoskeletal disease. Using our established in vivo rat model of fatigue loading, we have shown that fatigue damaged tendons exhibit decreased structural alignment and decreased mechanical properties out to at least 6-weeks after injury [1-3]. Since physiological loading is a powerful modulator that is often used as a therapeutic, we have been evaluating the utility of exercise to promote remodeling of fatigue damaged tendons. We have found that exercise that is initiated 2-weeks after onset of injury results in remodeling of the tendon leading to improved structural alignment [3,4]. However, the mechanisms by which exercise promotes remodeling in fatigue damaged tendons are unknown.

Integrins are an essential component to mechano-sensing ability of the cells and are expected to be important in an outcome of an effective response to exercise. In particular, integrin subunits αV and α5 play distinct roles in the adhesion to the extracellular matrix [5]. αV initiates the contact with the ECM, however fails to maintain the connection under high loads. The failure of the αV connections causes recruitment of α5 that has adherence strength that can withstand higher loads [5]. The increase of α5 increases the capacity of the cell to contract [5,6]. Therefore, the objective of this study is to quantify the effect of delayed exercise which has been shown to promote remodeling, on the percentage of positively stained cells with integrin αV and/or α5. Our hypothesis is that exercise will decrease the percentage of cells with αV and increase percentage of cells with α5 to allow greater adhesion strength to the ECM and greater contractile force from the cell, ultimately promoting structural remodeling.

METHODS

Under IACUC approval, 64 Sprague-Dawley female retired breeders were fatigue loaded as previously described [1-4]. Briefly the tibia and patella were exposed and clamped. The patella clamp was connected to an actuator and a material testing system. The tendon was loaded from 1 to 40 N for either 500 or 7200 cycles to induce a sub-rupture fatigue injury. At 2-weeks post fatigue injury, rats initiated treadmill running exercise at 17 m/min for 30 min/day for 5 days/week (500 Exercise or 7200 Exercise) or remained cage active (500 Cage or 7200 Cage). Animals were sacrificed at either 4- or 10-weeks post-fatigue injury. The patella-patellar tendon-tibia complex was harvested, fixed, embedded in paraffin and sectioned. Sections were stained for α5 (abcam ab150361) and αV (abcam ab179475). Alexa Fluor 488 (Thermo Fisher) was used as the secondary antibody to indicate positively tagged cells and NucRed (Thermo Fisher) stained all cells. Using confocal microscopy, images were acquired under 20x magnification at the insertion, midsubstance, and the origin. The percent of positively stained cells was calculated using a custom Matlab program. As previously described, second harmonic generation (SHG) images were analyzed using a custom Matlab program to determine the damage area fraction (DAF) and identify areas of damage and no damage [7]. The percent of positively stained cells was then also reported for the damage and undamage areas for each of those regions [3]. Comparisons between exercise and cage within their respective cycle group were made with Student's t-test. Significance (*) and trend (#) were set at $p<0.05$ and $p<0.10$, respectively.

RESULTS

Early timepoint: As expected, in the total area, exercise decreased the percentage of positively stained cells for αV at the insertion for both 500 and 7200 cycle groups (Figure 1). In the midsubstance, exercise decreased the percentage of αV for the 7200 cycle group only. There was no effect from exercise on αV at the origin. Further analysis of the damaged and undamaged areas showed that exercise significantly decreased the percentage of αV in the damaged area for both 500 and 7200 cycle group in the insertion (Figure 1). However, there were no changes due to exercise in the undamaged areas.

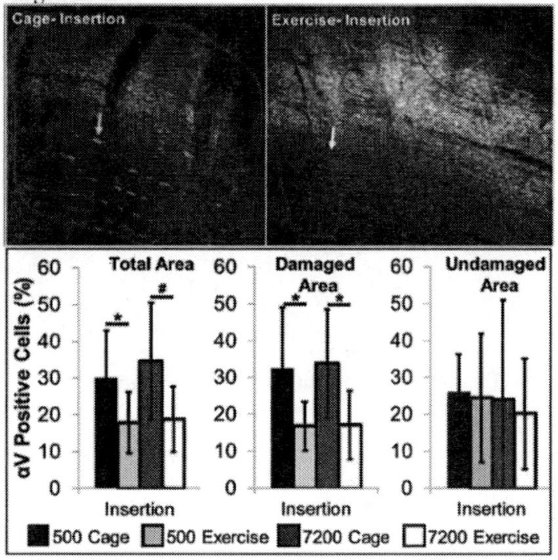

Figure 1: Exercise decreased integrin subunit αV for both 500 and 7200 cycle groups in the total area and damaged areas but not the undamaged areas. Red outline indicates damage area. Yellow and white arrows indicate positive and negative cells, respectively.

As hypothesized in the total area, exercise increased the percentage of α5 positive cells in the midsubstance for the 7200 cycle group, and in the origin, for both the 500 and 7200 cycle groups (Figure 2). There was no effect of exercise at the insertion site on the percentage of α5 positive cells. The analysis of the damaged and undamaged areas showed that the observed increase of α5 in the midsubstance was in both the damaged and undamaged areas. As for the origin, exercised increased the percentage of α5 only in the undamaged areas for both 500 and 7200 cycle groups (Figure 2).

Late timepoint: Surprisingly, in the total area, exercise significantly increased the percentage of αV in the insertion site for the 500 cycle group. There was no effect of exercise on αV for midsubstance or origin. There was also no effect of exercise for any of the regions for the 7200 cycle group. Analysis of the damage and undamaged areas showed that the increase in αV for the 500 cycle group due to exercise was influenced by its increase in the damaged areas of the insertion.

Surprisingly, in the total area, exercise decreased the percentage of α5 positive cells in the origin for both 500 and 7200 cycle group. Exercise had no effect for any of the other regions. The decrease of α5 by exercise at the origin occurred in the damaged areas.

DISCUSSION

In this study, we have found that exercise which promotes remodeling also modulates integrins αV and α5. Integrins have been shown to be key players in mechano-sensing by providing a physical link between cells and the ECM. αV and α5 are of a particular interest due to their suggested role in tendon and ligament remodeling [8,9]. Even though these integrins share the same ligand (fibronectin), they each have distinct yet collaborative roles [5,6]. The role of αV is to initiate the initial bond to the ECM under low forces. The breaking of that bond under high forces triggers recruitment of α5. The cycle of attachment and breaking of the bond between the cell and ECM through αV continues until enough α5 adheres to the ECM. At this point, αV is no longer needed and the increased cluster of α5 creates sevenfold increase of contractile force the cell can apply to the matrix [5,6]. Our finding that exercise that leads to remodeling also leads to a decrease in the percentage of αV and an increase in percentage of α5 at the early timepoint suggests that this could be one mechanism by which exercise improves the structure of fatigue damaged tendons [3].

The effect of exercise on αV and α5 at the late timepoint was surprising. We expect that damaged regions at this timepoint could be exhibiting an increase of αV and decrease of α5 because they were more severely damaged and were less capable of remodeling. This idea is consistent with the fact that in severe osteoarthritis, there is an increase of αV [10]. It is suspected that αV is increased in diseased states to attempt to re-establish the connection between the cell and ECM that would eventually lead to a coordinated attempt at remodeling. Further investigation into these areas to determine the cause of the surprising result is ongoing.

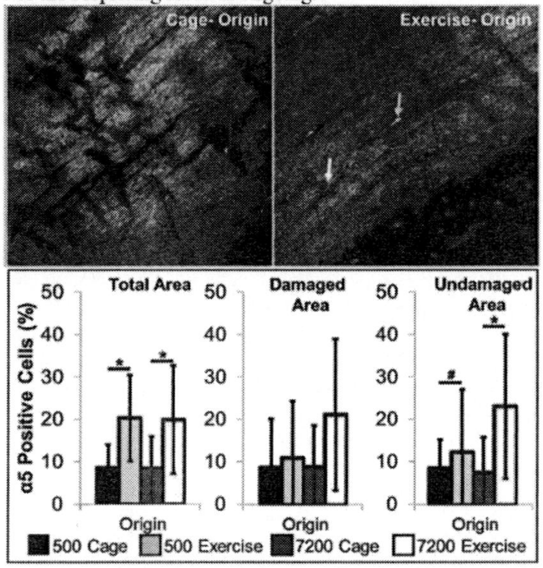

Figure 2: Exercise increased integrin subunit α5 for both 500 and 7200 cycle groups in the total area and undamaged areas but not the damaged areas. Red outline indicates damage area. Yellow and white arrows indicate positive and negative cells, respectively.

ACKNOWLEDGEMENTS

Authors would like to acknowledge Matt Anderson for his help on this project. Funded by NIH AR52743. Imaging data was acquired through the Cornell University Biotechnology Resource Center, with NIH S10OD018516 funding.

REFERENCES

[1] Andarawis-Puri, N et al, *J Orthop Res*, 32(9):1097-103, 2014. [2] Fung, DT et al, *Ann Biomed Eng*, 38(5):1741-51, 2010. [3] Bell, R, et al, *J Orthop Res*, 33(6):919-25, 2015. [4] Bell, R et al, *Trans of ORS*. 2015. [5] Roca-Cusachs, P et al, *Proc Natl Acad Sci*, 106(38):16245-50, 2009. [6] Mierke, CT et al, *J Cell Sci*, 124(Pt 3):369-83. [7] Sereysky, J et al, *J Biomech*, 17;43(13):2641-4, 2010. [8] Harwood, FL et al, *Connect Tissue Res*, 39(4):309-16,1998. [9] AbiEzzi, SS et al,

J Ortho Res, 13(4):594-601, 1995. [10] Garciadiego-Cazares, D et al, 10(5):e0127166, 2015.

SB³C2017
Summer Biomechanics, Bioengineering and Biotransport Conference
June 21 – 24, Tucson, AZ, USA

MULTI-SCALE STRUCTURE AND FUNCTION OF RAT ACHILLES TENDON

Andrea H. Lee (1), Dawn M. Elliott (1)

(1) Department of Biomedical Engineering
University of Delaware
Newark, DE, USA

INTRODUCTION

Although Achilles tendon is the largest and strongest tendon in the body, it is susceptible to damage, which can lead to pathologies including pain and often rupture [1]. Damage is defined as an irreversible change in the microstructure that alters the macroscopic mechanical parameters [2]. Previously, we used rat tail tendon fascicle to investigate loading and damage mechanisms at fascicle and fibril-levels [2,3], and similar technique and rigor need to apply to study mechanisms of more complicated tendon structures like the Achilles tendon. The Achilles tendon is a fusion of sub-tendons from the gastrocnemius and soleus muscles that inserts into the calcaneus bone. Previous studies have demonstrated that sub-tendon fusion results in non-uniform deformations in Achilles tendon [4,5]. Further, many studies hypothesize that these non-uniform deformations are caused by sub-tendon sliding, in addition to interfascicular and interfibrillar sliding within each sub-tendon [4]. However, there is a critical lack of experimental data to support these hypotheses, and the multiple length scales mechanisms for load transfer and tendon damage at the tendon-level are still unknown. The structural complexity of Achilles tendon is underappreciated in previous studies, and experimental evaluation at multi-scale is needed. Thus, the objective of the study was to establish the structure and mechanics of rat Achilles tendon at multi-scale.

METHODS

Sample Preparation: Achilles tendons (n=9) from five 4-6 month Long-Evans female rats were dissected with the medial gastrocnemius (MG), lateral gastrocnemius (LG), soleus (SOL) muscle complex, and calcaneal bone attached. Using a fine dissection microscope, the sheath that surrounds the Achilles tendon was removed. These samples were divided into two groups: structure (n=3) and mechanical testing (n=6).

Structure of Achilles Tendon: To study the structure and organization of the Achilles tendon, additional dissection was performed after sample preparation. We performed blunt dissection until the end of tendon-muscle junction to separate each sub-tendon. At the end of the junction, we performed sharp dissection by using micro spring scissors to carefully cut the sub-tendon interfaces. The macro-scale structure of sub-tendons and their fusion were observed.

Multi-Scale Mechanical Testing: Tendon was stained with 20 μg/ml 5-DTAF (Life Technologies), and tested on a custom-made uniaxial device mounted on an inverted confocal microscope (LSM 5 LIVE) as previously described [2,3]. Tendon deformations were measured using ink markers and a CCD camera. Inter-tendon deformations, including inter-tendon strain and sliding, were measured using four photobleached lines separated by 125 μm. Previously, we referred to "inter-tendon" deformation as "interfibrillar" deformation

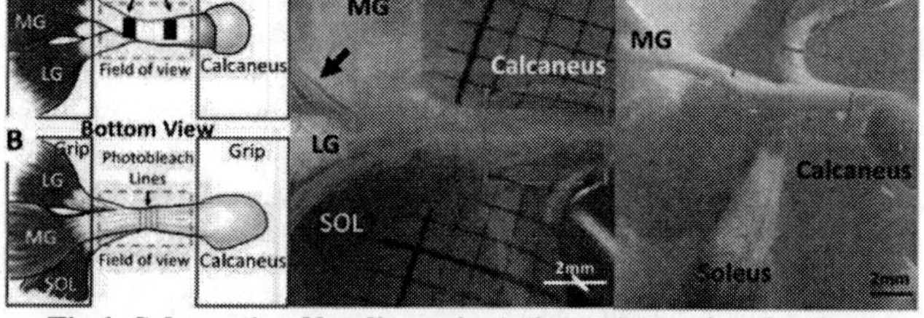

Fig 1. Schematic of loading orientation: A) top view for tissue strain calculation, and B) bottom view for inter-tendon deformations. C) Mechanical testing sample with small fascicles bridging sub-tendons (arrow). D) After sharp dissection. Blue dashed boxes show the same field of view.

[2]. However in the Achilles, there is more than one fascicle within the confocal field of view, and we cannot distinguish fascicle boundaries with the DTAF stain. Thus, the deformation was called "inter-tendon" deformation to encompass both fascicle and fibril deformations. We gripped the muscle complex on one end and calcaneus bones on the other end (Fig 1). We oriented the tendon with the SOL sub-tendon facing the CCD camera for tissue strain measurements (Fig 1A), while the MG sub-tendon faced the confocal objective (Fig 1B).

Testing (Fig 2) included a preload to 0.1N to define the reference length and preconditioning (5 cycles of 3% grip strains). The sample was then loaded to 14% grip strain, defined as BASELINE (Fig 2). Tendon was held at the loaded strain level for a 15 min relaxation period and unloaded to the reference length, where it rested for 40 minutes. Finally, tendon was loaded to failure, defined as DIAGNOSTIC (Fig 2). All loading and unload rates were 1%/s.

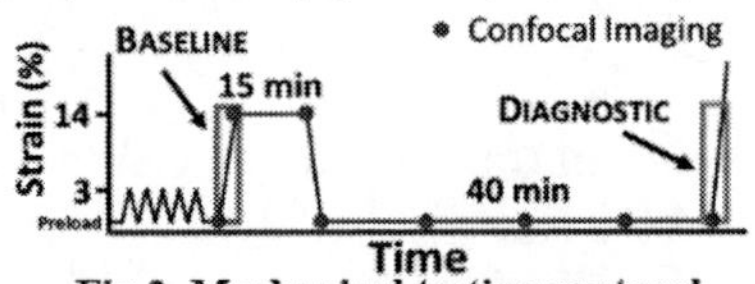

Fig 2. Mechanical testing protocol

Data Analysis: The tendon-level stress-strain (σ-ε) data from zero to the inflection point, defined as a point where the σ-ε curve shifts from strain-stiffening to strain-softening [2], was fit with a nonlinear constitutive model that optimized the transition stress and strain (the end of toe-region) and linear region modulus [6]. These parameters were calculated for in the BASELINE and DIAGNOSTIC ramps (Fig 2). To determine the effect of loading on tendon-level parameters, a two-tailed t-test was performed between BASELINE and DIAGNOSTIC with significance set at p =0.05.

Inter-tendon material parameters (e.g., inter-tendon strain and sliding) were determined from confocal microscope images. Inter-tendon strain was the change in distance between pairs of photobleached lines, and sliding was the averaged tortuosity of these lines [3]. Non-recoverable sliding was defined as the permanent sliding that remained after the 40-minute rest period.

RESULTS

Structure of Achilles Tendon: Rat Achilles tendon is composed of three tendons: LG, MG, and SOL (Fig 1). Near the tendon-muscle junction, the sub-tendon can be separated through blunt dissection (Fig 1C). In addition, we observed fascicle boundaries within each sub-tendon. However, the boundaries among sub-tendons became unclear near the tendon-calcaneus junction, and the crimp pattern became less identifiable, showing traits of fibrocartilage (Fig 1D). Between sub-tendons, we observed structures (~150 μm in diameter) that bridge sub-tendons together with crimp pattern (Fig 1C, arrow). The diameter and crimp pattern suggest that these structures are collagen fascicle.

Tendon-Level Mechanics: All tendon-level parameters increased from BASELINE to DIAGNOSTIC (Fig 3). In particular, the linear region modulus, inflection point strain, and stress showed a significant increase between BASELINE and DIAGNOSTIC. In addition, the tissue strain averaged 13.68 ± 0.039%, which demonstrates that there was minimal slip.

Inter-Tendon Mechanics: Inter-tendon strain fully recovered (Fig 4A). However, inter-tendon strain magnitude only averaged 0.5% at the end of loading. Inter-tendon sliding did not fully recover and non-recoverable sliding averaged 61.96 ± 20.34% (Fig 4B).

DISCUSSION

This study highlighted the complexity of Achilles tendon and provided insights into the underappreciated structure and mechanics of rat Achilles. Importantly, rat Achilles tendon has sub-tendon and fascicle structures, similar to human Achilles tendon [7]. While the tissue strain was close to the applied grip strain, the magnitude of inter-tendon deformations were much smaller than expected. For example, in rat tail tendon fascicles, fibril strain averaged 3% and sliding averaged 12 degrees when loaded to 8% (80% ultimate strain) [2]. In this study, since the tissue strain and inter-tendon deformation were quantified on different sub-tendons (Fig 1A, B), the small inter-tendon deformation suggests non-uniform deformation of tendon. We will quantify other sub-tendon deformations in future studies.

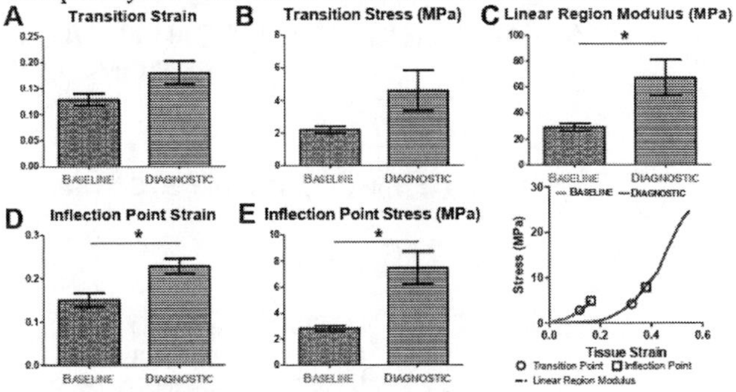

Fig 3. Tendon-level mechanics, comparing BASELINE and DIAGNOSTIC. Error bars in SEM and * represents p<0.05

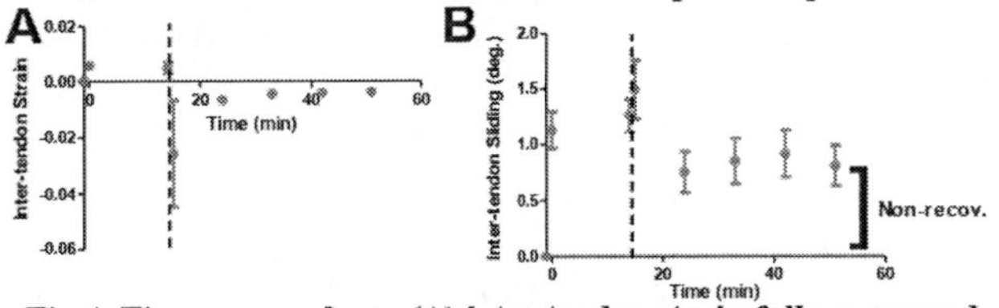

Fig 4. Time course shows (A) inter-tendon strain fully recovered, but (B) inter-tendon sliding did not. Dotted line represents when samples were unloaded. Error bars: SEM

In addition, the small inter-tendon deformation suggests some loads may be transferred between sub-tendons as suggested by other studies [4,5]. Previously, we suggested that smaller diameter fibrils may transfer shear loads at the fibril-level [8]. Here, we propose a similar mechanism at a larger scale, where the smaller diameter fascicles that bridge sub-tendons (Fig 1C) may be the load transferring structures at the tendon-level.

Similar to our previous study in rat tail tendon fascicles [2], the effect of damage was observed in both tendon and inter-tendon levels. We observed loading to 14% grip strain induces damage, permanently changing material parameters at the tendon-level and non-recoverable inter-tendon sliding at inter-tendon level.

We demonstrated that rat Achilles tendon preserves the structural complexity that is often observed in human tendon. Although Achilles tendon is often considered as a single tendon, careful experimental design is needed to highlight the physiological non-uniform deformations. Future studies will use this model to investigate loading and damage mechanisms at multi-scale. Establishing the loading and damage mechanisms at multi-scale can improve prevention and rehabilitation strategies for tendon pathology.

ACKNOWLEDGEMENTS

The National Institutes of Health grant R01 EB02425 and NSF IGERT program at the University of Delaware.

REFERENCES

[1] Maffulli, N et al., *J R Soc Med*, 2004. [2] Lee, A et al., *in review*, 2017. [3] Szczesny, S et al., *Acta Biomat*, 2014. [4] Slane, L et al., *J Biomech*, 2014. [5] Handsfield, G et al., *J Biomech*, 2016. [6] Tanaka, M et al., J *Biomech Eng*, 2011. [7] Szaro, P et al., *Ann. Anat*, 2009. [8] Szczesny, S et al., *J Orthop Res*, 2014

STRUCTURAL AND MECHANICAL CONSEQUENCES OF UNLOADING ON THE TENDON-TO-BONE ATTACHMENT

Alix C. Deymier (1), Andrea G. Schwartz (2), Zhonghou Cai (3), Guy M. Genin (4), Stavros Thomopoulos (1)

(1) Orthopedic Surgery
Columbia University
New □ork, N□, USA

(2) Orthopedic Surgery
Biomedical Engineering
Washington University
St Louis, MO, USA

(3) Advanced Photon Source
Argonne National Laboratory
Argonne, IL, USA

(4) Mechanical Engr and Materials Science
Washington University
St Louis, MO, USA

INTRODUCTION

The musculoskeletal system is sensitive to its loading environment. Unloading leads to loss of tissue mass and compromised mechanical properties for tendon, muscle, and bone [1, 2]. However, it remains unclear how the adult tendon enthesis, a complex transitional tissue between tendon and bone with gradients in composition and mechanics [3], is affected by unloading. The mechanical function of the tendon enthesis relies on a number of features across many length scales to dissipate stresses at the tendon-bone interface and decrease the risk of failure. Modification of this complex structure with unloading from paralysis, space flight, or bedrest could lead to increased risk of injury. We previously studied the effects of muscle loading on the development of the tendon enthesis [4, 5]. In the current study, the effects of unloading on the adult tendon enthesis were investigated via local paralysis due to botulinum toxin A (Btx) injections [4]. We hypothesized that unloading would lead to structural defects and compromised mechanics.

METHODS

To induce local muscular unloading, 5-6-month old CD-1 mice were injected with 1μL/g of body mass of 0.02 U/μL Btx in the left supraspinatus muscle of the shoulder (N=20). An equal volume of saline was injected into the contralateral supraspinatus muscle as an internal control. Animals were sacrificed after 21 days of paralysis. Age-matched mice were allowed cage activity and served as normal controls (N=20). Btx (B), saline (S), and control (C) samples were mechanically tested in tension to obtain stress-strain curves from which mechanical properties could be determined (N=15). Microcomputed tomography (μCT) was performed prior to mechanical testing to determine bone density and architecture as well as the tendon minimum cross-section. Post-testing μCT analysis was performed to measure the tendon enthesis area of failure. Nanometer-scale high-energy X-ray diffraction (XRD) was used to investigate the size and orientation of mineral crystals across the supraspinatus enthesis (N=5). Supraspinatus-humeral head complexes from the Btx and control groups were frozen at -80 °C in optimal cutting solution and sectioned into 20 μm-thick slices along coronal oblique plane. Sections were placed on amorphous silicon nitride windows and examined at the Advanced Photon Source (APS, Argonne National Laboratory), station 2-ID-D. The 10.1 keV X-ray beam was 200 nm x 200 nm in cross-section. For each sample, 2-3 X-ray diffraction scans were taken along a straight line across the tendon-to-bone attachment (resolution ~200 nm, line length of ~30 μm). Diffraction rings were analyzed to obtain crystal orientation and size as a function of position across the attachment. The full width half maximum (FWHM) of the 00.2 peak intensity as a function of azimuth was used to determine the level of alignment of the mineral crystals. The c-axis crystal sizes were determined from the 00.2/00.4 peak broadening using the Scherrer equation [6]. Raman analysis was also performed on these sections to determine the composition of the bone mineral.

RESULTS

After 21 days of Btx-induced unloading, unloaded samples exhibited a significant decrease in strength (B: 11.9 ± 6.7, S: 33.2 ± 34.8 C: 32.2 ± 21.9 GPa) and toughness (B: 3.6 ± 2.4, S: 11.7 ±10.1, C: 13.1 ± 10.4 GPa) as compared to the control samples (Fig 1). Stiffness and resilience were unaffected by unloading. Tendon cross-sectional area (B: 0.44 ± 0.23, S: 0.56 ± 0.33 C: 0.68 ± 0.23 mm^2) was also decreased between the Btx and control samples. Structurally, unloading led to a 39% decrease in the ratio of bone volume to total volume compared to control samples. There was also a 21% decrease in trabecular thickness and an 18% increase in trabecular spacing due to unloading (Fig 2). In all cases, failure occurred via bone avulsion from the humeral head. The area of the resultant bone crater, projected onto the bone surface (referred to here as the "failure area"), was significantly larger in the Btx group compared to both saline and control (B: 0.44 ± 0.30, S: 0.22 ± 0.13, C: 0.19 ± 0.11 mm^2).

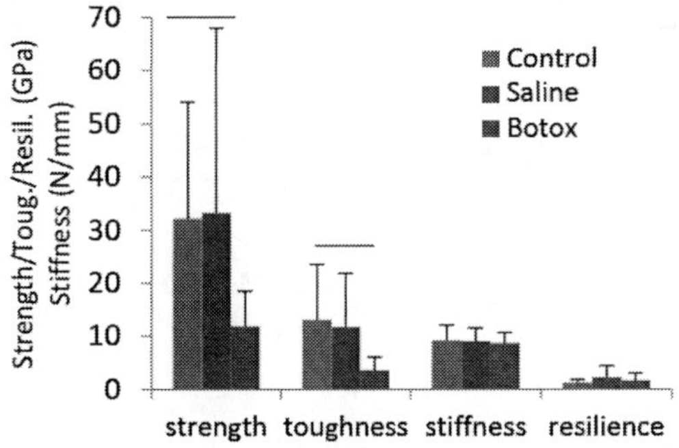

Figure 1: Mechanical properties for B, S, and C groups.

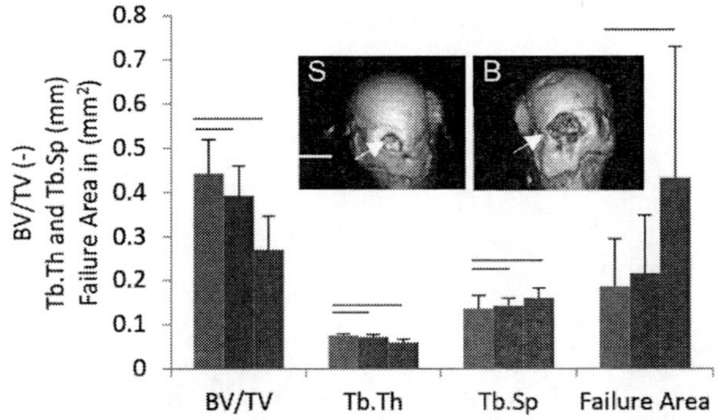

Figure 2: BV/TV, trabecular thickness (Tb.Th) and spacing (Tb.Sp), and failure area for the B, S, and C groups. Inset shows the failure areas in S and B samples.

The XRD results indicated that both the Btx and control samples had a 20% decrease in the FWHM of the 00.2 intensity peak, which is inversely proportional to the alignment of the crystals, with increased mineralization across the enthesis. Crystal size as a function of mineral content was found to increase with increased mineralization for both Btx and control groups. On average, Btx samples exhibited larger FWHM values and smaller crystals size compared to controls, indicating that unloading decreases crystal alignment and crystal size. Raman results indicated that the carbonate content in the bone crystals was decreased with unloading.

DISCUSSION

The failure properties of the tendon-to-bone attachment were significantly compromised with unloading. In addition, the failure area, which results from avulsion of tendon from the bone, was increased in Btx-treated samples. As discussed below, the increased risk and extent of failure seen in the unloaded samples likely occurred due to: (1) the presence of a compromised compliant zone and (2) loss of enthesial bone.

Previous work has demonstrated that the tendon-to-bone attachment contains a compliant zone in the region between the tendon and the bone [3, 7]. The ability of this region to deform is due to collagen realignment and low mineral content, and this allows it to absorb mechanical energy while avoiding catastrophic failure of the attachment [8]. In the unloaded samples, the mineral crystals exhibited decreased size, greater disorder, and a change in composition. Based on a rotation model, it was determined that these changes in crystal size and orientation decreased the total energy absorption in the compliant zone (Fig 3). In addition, the bone mineral in unloaded samples exhibited decreased carbonate content, which was associated with an increase in mineral stiffness. These changes led to a decrease in the attachment's ability to absorb mechanical energy, thereby increasing the failure risk.

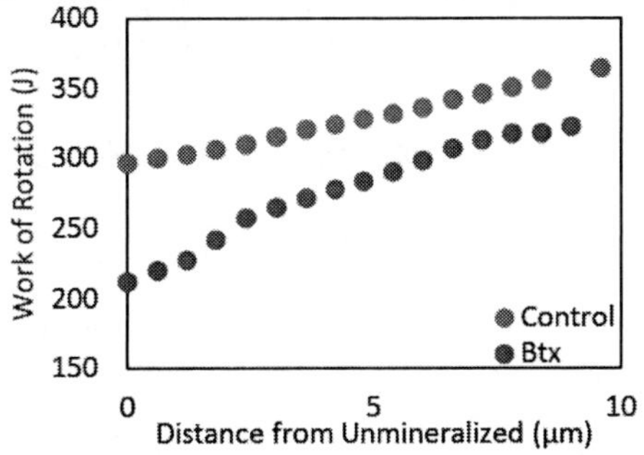

Figure 3: The rotation model showed that work of rotation increased with mineralization and decreased with unloading at all positions along the tendon-to-bone attachment.

In addition to modification of the compliant zone, there was significant loss of bone (~40%), in the region below the attachment. If we consider the mineralized fibrocartilage at the attachment as a circular plate that is clamped at its edges by the trabeculae, the maximum deflection of the plate can be calculated from the equation $\omega_{max} = -\frac{qa^4}{64D}$, where ω_{max} is the maximum deflection of the plate, q is the applied force, D is the flexural rigidity, and a is distance between clamps [9]. Thus, as bone is lost and the trabecular spacing increases, the distance between clamps, a, will increase, resulting in an increase in deflection proportional to the 4th power. This dramatic increase in deflection will lead to greater stress concentrations and failure. The combined effect of these changes to the attachment structure and composition explain the increase in failure risk and extent seen experimentally.

ACKNOWLEDGEMENTS

This project was funded by National Institute of Health (NIH) R01 EB016422 and R01 AR057836. Dr. Deymier was funded in part by the National Space Biomedical Research Institute (NSBRI) NSBRI-RFA-13-01 Postdoctoral First Fellow Award.

REFERENCES

1.Cann, C.E., *Adaptations of the skeletal system to spaceflight*, in *Introduction to space life science*, S. Churchill, Editor. 1993.
2.Reeves, N.D., et al., Journal of Applied Physiology, 2005. **98**(6): p. 2278-2286
3.Thomopoulos, S., et al., J Orthop Res, 2003. **21**(3): p. 413-9
4.Schwartz, A.G., et al., Bone, 2013. **55**(1): p. 44-51
5.Thomopoulos, S., et al., J Orthop Res, 2007. **25**(9): p. 1154-63
6.Stock, S.R. and B.D. Cullity, *Elements of x-ray diffraction*. 2001, Upper Saddle River, NJ: Prentice Hall.
7.Deymier, A.C., et al., Acta Biomaterialia, 2016. **Under Review**
8.Genin, G.M., et al., Biophys J, 2009. **97**(4): p. 976-85
9.Birman, V., *Plate structures*. 2011, New York, NY: Springer Science and Business media.

ABSENCE OF ESTROGEN DURING MATURATION UNIQUELY AFFECTS PROGESTERONE RECEPTOR IN EXTRA-ARTICULAR LIGAMENT AND TENDON: POTENTIAL MECHANISM FOR MECHANICAL CHANGES

Devin B. Lemmex, Natalie C. Rollick, Yohei Ono, David A. Hart, Ian K.Y. Lo, Gail M. Thornton

McCaig Institute for Bone and Joint Health
University of Calgary
Calgary, Alberta, Canada

INTRODUCTION

Progesterone receptor (PR) and estrogen receptor (ER) have been found in ligament and tendon [1,2,3]. As a consequence, the mechanical properties of ligament and tendon could be affected by hormonal changes. Hormonal influences have been suggested as a possible factor in the sex-related disparity in ligament injury, and tendon pathology and injury [4]. Hormonal changes include those that occur during the menstrual cycle, pregnancy and menopause. Also, amenorrhea occurs in 2%-5% of the general population [5] and 20%-50% of the athletic population [6]. Estrogen deficiency during maturation can occur due to cancer treatments or surgical intervention following trauma, as well as due to vigorous athletic training. Correspondingly, the current study examines the effect of ovariohysterectomy (OVH) in adolescence on the expression of PR and ER in ligament and tendon of the adult rabbit knee: medial collateral ligament (MCL), lateral collateral ligament (LCL), anterior cruciate ligament (ACL), posterior cruciate ligament (PCL) and patellar tendon (PT).

We have previously documented the increased creep strain (cyclic, static, total) in the MCLs of adult rabbits that underwent OVH surgery as adolescents when compared with the MCLs of intact normal adult rabbits [7]. Interestingly, the only molecular change that was detected in the MCL with adolescent OVH was a decrease in PR [7]. While there is limited information on the effect of PR in knee ligaments, decreased PR has been implicated in the increased stretching of the uterosacral ligament in pelvic organ prolapse [8] and the pubic symphysis ligament in pregnancy [9]. Taken together, these suggest a possible relationship between decreased PR and increased ligament stretch or creep.

We had yet to evaluate whether PR and ER expression changed in other knee ligament and tendon when adult rabbits had matured in the absence of estrogen. Our purpose in this current study was to examine the effect of adolescent OVH on the molecular expression of hormone receptors (ER, PR), matrix metalloproteinases (MMP-1, MMP-3, MMP-13) and tissue inhibitors of metalloproteinases (TIMP-1, TIMP-2, TIMP-3) in extra-articular ligament and tendon (MCL, LCL, PT) and intra-articular ligament (ACL, PCL). Evaluation of MMPs and TIMPs was included as MMP/TIMP imbalance has been implicated in the degenerative changes in tendinopathy [10]. Consistent with our observation in the MCL [7], our hypothesis was that PR would be decreased in ligament and tendon of the knee due to adolescent OVH.

METHODS

Knee joint tissues from adult female New Zealand White rabbits underwent molecular assessment where two groups were evaluated: intact normal (n=8) and adolescent OVH (n=9). For adolescent OVH, the OVH surgery was performed at 15 weeks of age and assessments were performed at a minimum of 48 weeks of age.

Tissues (PT, MCL, LCL, ACL, PCL) underwent RT-qPCR with rabbit specific primers for hormone receptors (ER, PR), MMPs (MMP-1, MMP-3, MMP-13), TIMPs (TIMP-1, TIMP-2, TIMP-3), and the housekeeping gene 18S [7]. Data were compared using Mann-Whitney *U* tests.

RESULTS

Progesterone receptor (PR) was decreased when estrogen was absent during maturation for the PT, MCL, and LCL which are extra-articular tendon and ligament (Fig. 1; $p \leq 0.05$). For intra-articular ligament, PR was not affected by adolescent OVH in ACL and PCL but ER was decreased in the ACL with adolescent OVH (Fig. 2; $p \leq 0.05$).

In extra-articular PT, TIMP-2 and TIMP-3 were decreased and MMP-3 was increased comparing adolescent OVH and intact normal groups (Fig. 3,5,6; $p \leq 0.05$). In extra-articular LCL, MMP-13 was increased with adolescent OVH (data not shown; $p \leq 0.05$), similar to the increased MMP-3 in extra-articular PT. In intra-articular ACL, TIMP-1 was increased with adolescent OVH (Fig. 4; p≤0.05), opposite to the decreased TIMP-2 and TIMP-3 in extra-articular PT.

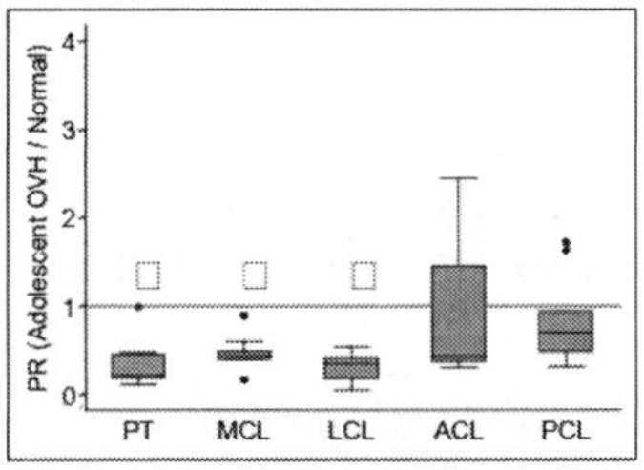

Figure 1: Progesterone Receptor (PR) mRNA levels. *p≤0.05

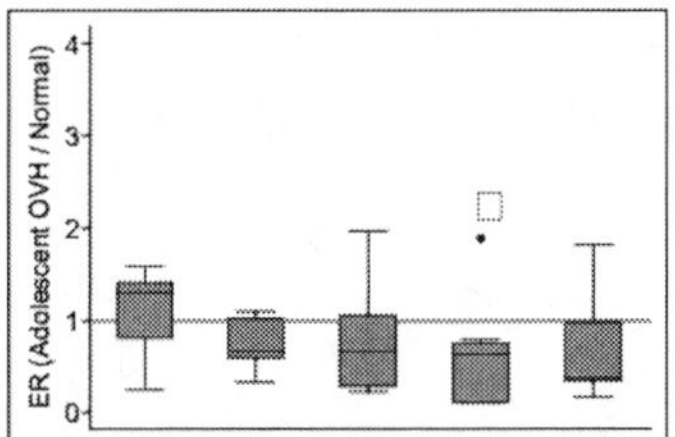

Figure 2: Estrogen Receptor (ER) mRNA levels. *p≤0.05

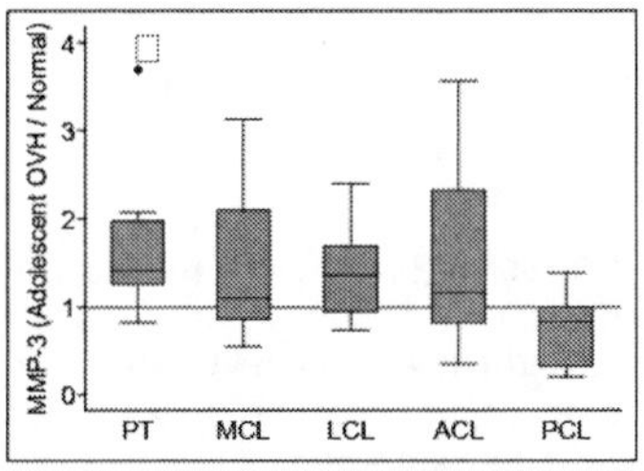

Figure 3: MMP-3 mRNA levels. *p≤0.05

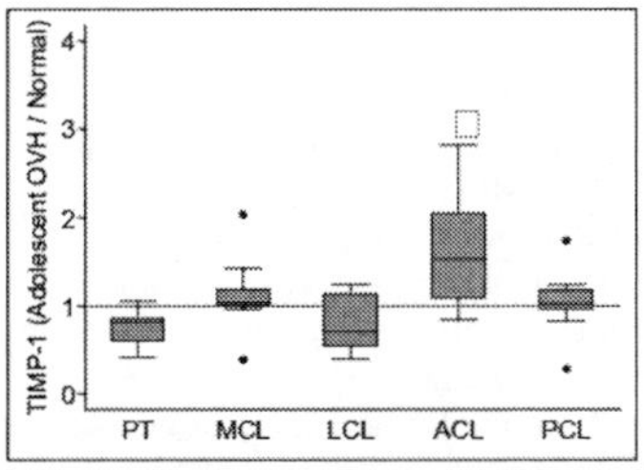

Figure 4: TIMP-1 mRNA levels. *p≤0.05

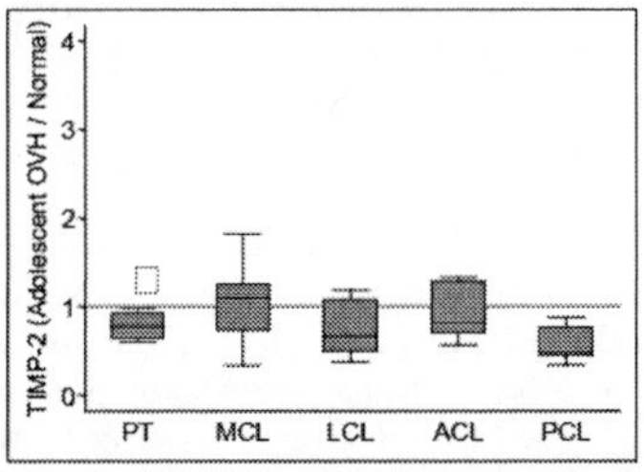

Figure 5: TIMP-2 mRNA levels. *p≤0.05

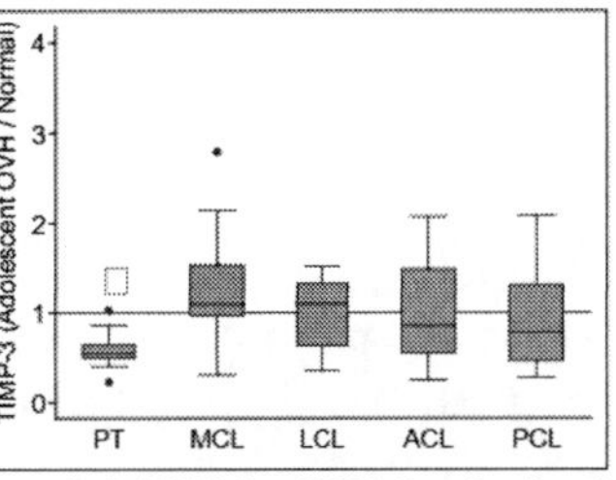

Figure 6: TIMP-3 mRNA levels. *p≤0.05

DISCUSSION

Progesterone receptor (PR) was decreased in extra-articular tendon and ligament of the knee (PT, MCL, LCL) when adult rabbits had matured in the absence of estrogen. We had previously found in the MCL an abnormally increased creep strain concurrent with decreased PR with adolescent OVH [7]. Decreased PR has been implicated in increasing the ability of ligaments to stretch or creep [7,8,9]. If the decreased PR was concurrent with increased creep in the PT and LCL as we observed in the MCL [7], then this could have important implications for the possible relationship between hormonal changes and increased potential for tendon and ligament injury.

An MMP/TIMP imbalance may contribute to degenerative changes in tendinopathy [10]. Comparing the adolescent OVH group to the intact normal group, the PT exhibited increased MMP-3 and decreased TIMP-2 and TIMP-3, suggesting that adolescent OVH resulted in an MMP/TIMP imbalance towards degradation in PT. This may also have important implications for hormonal influences on increased potential for tendon pathology and injury.

In summary, absence of estrogen during maturation led to a unique decrease in progesterone receptor (PR) in the extra-articular ligament and tendon of the knee. If these decreases in PR in the PT and LCL are associated with increases in creep consistent with our previous observations for the MCL [7], then these results suggest that the relationship between decreased PR and increased creep should be further investigated as a potential mechanism for mechanical changes that may contribute to the sex-based discrepancy in injury risk in ligament and tendon.

ACKNOWLEDGEMENTS

Natural Sciences and Engineering Research Council of Canada (NSERC); Canadian Institutes of Health Research (CIHR) Institute of Musculoskeletal Health and Arthritis (IMHA).

REFERENCES

[1] Liu, SH et al., *J Orthop Res*, 14:526-533, 1996.
[2] Sciore, P et al., *J Orthop Res*, 16:604-610, 1998.
[3] Hart, DA et al., *Clin Orthop Relat Res*, 351:44-56, 1998.
[4] Hansen, M, Kjaer, M, *Adv Exp Med Biol*, 920:139-149, 2016.
[5] Yeager KK et al., *Med Sci Sports Exerc*, 25:775-777, 1993.
[6] Teitz, DA et al., *J Am Acad Orthop Surg*, 5:87-96, 1997.
[7] Lemmex, DB et al., *J Biomech*, 49:382-387, 2016.
[8] Bai, SW et al., *Int Urogynecol*, 16:492-496, 2005.
[9] Rodriguez, HA et al., *J Steriod Biochem Mol Biol*, 111:247-254, 2008.
[10] Jones, GC et al., *Arthritis Rheum*, 54:832-842, 2006.

EFFECTS OF AORTIC COARCTATION ON VENTRICULAR ENERGETICS IN HYPOPLASTIC LEFT HEART SYNDROME

Lauren Carter (1), Tianqi Hang (1), Giovanni Biglino (2), Chad Smith (1), Tain Yen Hsia (3), Richard S Figliola (1)

(1) Departments of Mechanical Engineering and Bioengineering
Clemson University
Clemson, SC 2963-0921, USA

(2) Bristol Heart Institute
University of Bristol
Bristol, UK

(3) Cardiorespiratory Unit
Great Ormond Street Hospital for Children,
NHS Trust, London, UK

INTRODUCTION

Aortic coarctation (CoA) is a congenital heart defect and often presents itself as a co-morbidity for patients born with hypoplastic left heart syndrome (HLHS), and thus relying on a single functional systemic right ventricle (RV). The presence of CoA in HLHS patients can be an additional significant problem when a Norwood procedure is performed. Moderate to severe CoA is known to deteriorate RV function. Re-coarctation after the Norwood remains a cause of morbidity and mortality and occurs in 17% of Norwood patients. Reconstruction of the neoaorta during the Norwood procedure creates a patient-specific morphology impacting the circulation impedance for the RV. CoA presents an increased impedance, which likely contributes to unfavorable ventricular energetics. Yet a full appreciation of the effects of the severity of CoA on single RV mechanics remains lacking. While such information is difficult to gather *in vivo*, validated engineering modeling techniques can be used to study the aspects of such complex physiologies. Multi-scale *in vitro* benchtop models have proven capable of incorporating patient-specific HLHS circulation characteristics and recapitulating features of HLHS physiology [1-2], with the appealing feature that they allow for performing parametric studies in a stable and reproducible environment. The purpose of this study is to investigate the combined effects of aortic morphology and CoA severity on the single RV after the Norwood procedure.

METHODS

Patient-specific anatomical models for four patients (mean age 4±0.8 months) were derived from HLHS patients using cardiovascular magnetic resonance (CMR) imaging, by processing the 3D CMR datasets with appropriate commercial software (Mimics, Materialise NV, Leuven, Belgium). The neo-aortic arch morphologies presented by these patients show a range from a tubular (uniform) transverse arch and a moderate shaped arch to two with more dilated and bulging transverse arches. CoA was quantified with the traditionally accepted coarctation index (CI), the ratio of aortic diameter at the isthmus over aortic diameter in the descending aorta at the diaphragm.

For the purpose of testing a range of CoA severities, each patient model was then modified in the virtual domain to simulate progressively severe CI. Each virtual reconstruction was then created as an accurate physical three-dimensional (3D) aortic test phantom by stereolithographic printing using a transparent rigid resin. Each 3D test phantom was then inserted into an experimental mock circulatory system (MCS) for individual testing.

The mock circulatory system (MCS) of the Norwood circulation used here has been well described and validated previously [2]. The MCS coupled a lumped parameter network (LP) model of the Norwood circulation [3] with an anatomically accurate, three-dimensional (3D) test phantom of the neoaorta. The model includes three major circulation branches: upper body, lower body and pulmonary branch. The single ventricle was simulated by a ventricular assist device (Excor® 25cc, Berlin Heart, Berlin, Germany). Mean aortic pressure, heart rate (HR) and stroke volume (SV) were set to patient-specific values derived from the patient CRM report. To compensate for the rigidity of the aortic model, an aortic compliance chamber was placed in-line just prior to the test phantom.

As described in Hang et al [2], pressures were measured using liquid-filled transducers. Flow rate was measured in all districts of the MCS by means of electromagnetic flow probes. Flow measurements included flow in the descending aorta (i.e. lower body flow, Q_{lb}), flow in the brachiocephalic vessels (i.e. upper body flow, Q_{up}), and flow in the mBT-shunt equivalent conduit (i.e. pulmonary flow, Q_{pul}). System operation was controlled and all data recorded using a data acquisition board. A saline-glycerine blood analog was used for the working fluid.

For these tests, cardiac output (CO) was set to the patient clinical condition and then maintained constant for each test across the range of simulated CoA severities. Other tests in which upper body perfusion was held constant instead provided similar outcomes. The atrial pressure was maintained constant at the patient clinical condition.

Coarctation resistance (R_{CoA}) was calculated as the resulting impedance on the lower body as:

$$R_{CoA} = (P_{pre} - P_{post})/Q_{lb} \qquad (1)$$

where P_{pre} and P_{post} are the mean pressures before the CoA and 10 diameters downstream of the CoA, respectively. Ventricular power (VP) was calculated as an indicator of the effect of progressively severe CI on the single ventricle. Ventricular power was determined as:

$$Power = VP = \frac{1}{T_{cycle}}\left(\int Q_{out}(t)P_{out}(t)dt - \int Q_{in}(t)P_{in}(t)dt\right) \qquad (2)$$

where T_{cycle} is the duration of the cardiac cycle; Q_{out} ,Q_{in} , P_{out} , and P_{in} are the instantaneous flow rates and pressures, respectively, at the outlet and inlet of the VAD.

Regression analysis using the coefficient of determination (R^2) and a root-mean-square-error (RMSE) were used to assess the impact of R_{CoA} and CI on pressure gradient and ventricular power. VP per unit stroke volume (SV) and Q_{lb} versus CI were assessed for qualitative trends based on morphology. Comparative mean values were examined using a *t* test with a value of $p < 0.05$ judged to indicate a statistically significant difference.

RESULTS

Ventricular power (Figure 2 and presented as power per unit stroke volume, [VP/SV]) is notably higher for the most highly dilated morphology (HLHS-5, $p < 0.001$) and differences between morphologies is apparent. However, the differences between VP/SV with CI is not significant between HLHS-1 and HLHS-2 ($p > 0.15$) whereas differences are significant between HLHS-3 and HLHS-4 and when compared to HLHS-1 and HLHS-2 ($p < 0.05$). Regardless of morphology, less severe coarctation (CI ~ 0.6 - 0.9) requires a nearly constant ventricular power. However, the relative change in power becomes more pronounced for severe CoA (CI < 0.4), significantly increasing ventricular energy expenditure per cycle.

Overall, coarctation resistance R_{CoA} increases as CI deceases (Figure 3), but the largest change is concomitant with severe coarctation. Comparing R_{CoA} to the baseline total vascular resistance of the circulation (TVR) (Figure 3), the amount of coarctation resistance is nearly independent of morphology ($p > 0.05$), an important finding. The R_{CoA} increases at a rate defined by a negative fourth power function

$$R_{CoA} = 0.8\ CI^{-4} \qquad (3)$$

consistent with Poiseuille and obstruction meter theory. Only as $R_{CoA} \rightarrow$ TVR does VP begin to increase significantly, indicating that the total impedance seen by the ventricle is the determining factor in power expended, an important finding. The data for all morphologies fits equation (3) with high correlation and low error (R^2=0.90, RMSE=19.5 WU). With each morphology, the VP measurably increases as R_{CoA} achieves about 20% of total TVR and VP substantially increases as R_{CoA} nears parity or exceeds the TVR of the circulation.

DISCUSSION

The severity of CoA correlated significantly with progressively increasing pressure gradient across the narrowing, as expected. Differences in single ventricle power exertion was pronounced between extreme morphologies. For a given morphology, CoA adds to the power burden of the single ventricle as CI < 0.6 and particularly as CI < 0.4 and only more specifically when R_{CoA} > TVR. Total impedance determines ventricular power.

The data suggest that the geometry of the neo-aorta plays a role in determining ventricular power through the baseline vascular resistance. Even after CoA repair, ventricular hypertrophy is known to occur, due to factors like increased afterload from small residual gradients and possible on-going mild hypertension. These observations are also in keeping with previous CMR work that discusses how a more dilated transverse aortic arch might be hemodynamically unfavourable in palliated HLHS patients and detrimental to VA coupling. Aortic coarctation and aortic arch morphology both influence energy expenditure of the single ventricle in their own ways.

ACKNOWLEDGEMENTS

This work was supported by a Leducq Foundation Network grant.

REFERENCES

[1] Biglino G, et al., *ASAIO J*, 2012;58(1):25-31.
[2] Hang et al., *ASAIO J*, 2016;62(3):317-324.
[3] Hsia et al., *Circulation*, 2011;124(S11):S204-1010.1161

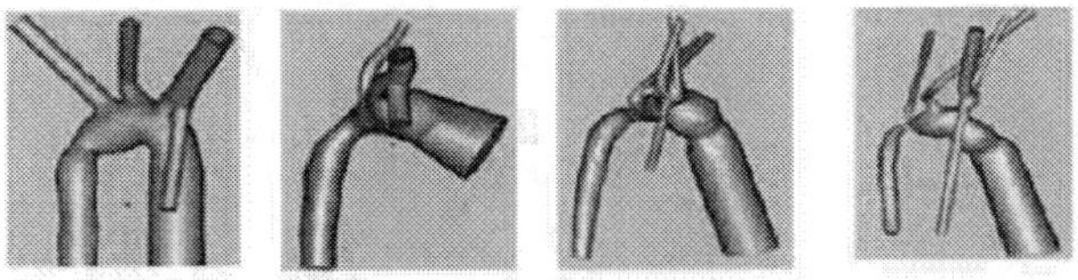

HLHS-1 HLHS-2 HLHS-3 HLHS-4

Figure 1: Patient-specific morphologies

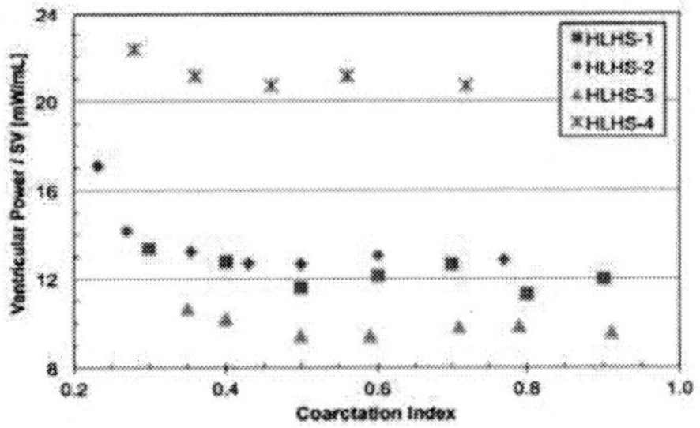

Figure 2: Single ventricle power versus coarctation

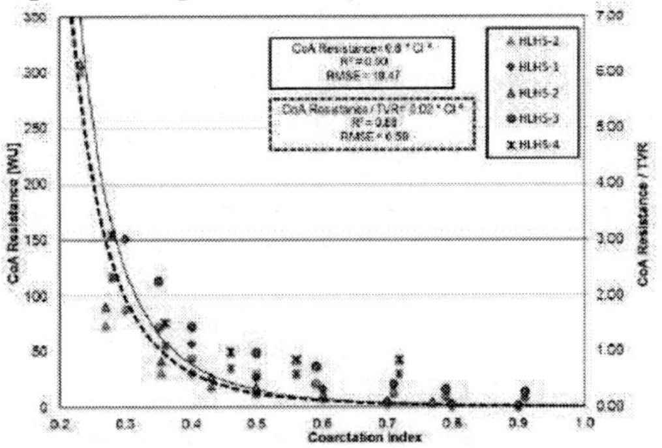

Figure 3. Coarctation resistance versus coarctation

EFFECT OF PERISTALSIS LIKE MOTION OF THE RIGHT VENTRICLE ON THE FLUID DYNAMICS IN 20 WEEKS OLD HUMAN FETAL RIGHT VENTRICLE

Hadi Wiputra(1,2), Khong Chun Chua(2), Nivetha Raju(2), Hwa Liang Leo(2), Choon Hwai Yap(2)

(1) NUS Graduate School for Integrated Sciences and Engineering, National University of Singapore, Singapore.

(2) Department of Biomedical Engineering, National University of Singapore, Singapore.

INTRODUCTION

Congenital Cardiovascular Malformation (CCM) such as tetralogy of fallot and hypoplastic left heart syndrome affects 0.6-1.9% of the population [1] and is the leading cause of birth-defect related death. Recent work provided evidence that abnormal hemodynamics condition can affect the development of animal embryonic hearts to cause CCM [2]. It is thus important to characterize blood flow mechanical forces in fetuses and understand how they affect cardiovascular development.

The Right Ventricle (RV) plays a central role in fetal circulation, since it pumps to most parts of the body and RV failure significantly affect patients' outcome [3]. In both adult human and canine hearts, it is known that the RV exhibits a peristalsis-like motion during contraction [4,5], where muscles near the infundibulum (inflow tract) first contract and the ones at sinus (outflow tract) followed soon after. In the human adult heart, this contraction delay between the inflow and outflow tracts averaged at 15% of the cardiac cycle [4]. In canine RV, the contraction delay was estimated to produce approximately 15mmHg of intraventricular pressure difference [5].

Currently, whether the presitalsis-like motion occurs in human fetal heart is unknown and the mechanics of the fetal heart is poorly understood. Further, the function of this peristalsis-like motion and its impact of RV mechanics is unknown. These factors motivate our current study. We hypothesize that this peristalsis-like motion exist in the human fetus, and is an adaptation to reduce energy required for the ejection of blood from RV. We test this hypothesis in the current study.

METHODS

This study is approved by the Domain Specific Review Board (DSRB) at the National Healthcare Group, Singapore, and informed consent was obtained from all patients. 4D ultrasound images of six normal 20 weeks old human fetal hearts were obtained using ultrasound machine under STIC (Spatio-Temporal Image Correlation) mode.

We first determine whether the peristalsis-like motion existed in human fetuses and quantify the contraction delay. The luminal cross-sectional areas parallel to the apical four chamber view was extracted over all time points, at multiple slice locations between the RV inflow tract and the outflow tract (Figure 1a). At each cross-section, the luminal area versus time was quantified, and the time point where luminal area peaked was recorded. This marked the onset of contraction at that particular slice. Linear regression was performed on the plot of time of contraction versus slice location to determine the overall contraction delay from the inflow tract to the outflow tract (Figure 1b).

Next, dynamic mesh Computational Fluid Dynamics (CFD) of human fetal RV was performed based on clinical ultrasound images, using methodologies previously established [6]. Modifications were made to include modelling of the RV peristalsis-like motion. The motion of the ventricle wall was described as the change in radial location (R) of all surface coordinates from the centroid of the right ventricle in spherical coordinate system:

$$R(\theta,\phi,t) = \alpha(\theta,\phi)\Omega\big(t+\Delta t(\theta,\phi)\big) + R_0(\theta,\phi) \qquad (1)$$

where Ω is the characteristic motion waveform, formulated based on the pulsed wave Doppler ultrasound reading, R_0 is the initial radius, α is the amplitude of radius motion over one cycle, curve-fitted to ultrasound measurements using spherical harmonic functions, θ and ϕ are spherical coordinates angles and t is time. To model the peristalsis motion, an additional term, $\Delta t(\theta,\phi)$, was added to describes the distribution of the contraction delay at various wall locations. Simulations were performed with ANSYS Fluent with a UDF to describe wall motions

From the simulation results, cumulative work done (Wd) by the RV for ejection was calculated by:

$$Wd(t) = \int_{systole} \int_{RV\ wall} P.(v.n).dA.dt \quad (2)$$

Where P is the RV wall pressure, v is wall velocity and A is wall area. Simulation were performed on 3 hearts at various peristalsis contraction delays to investigate effects of the peristalsis motion.

RESULTS

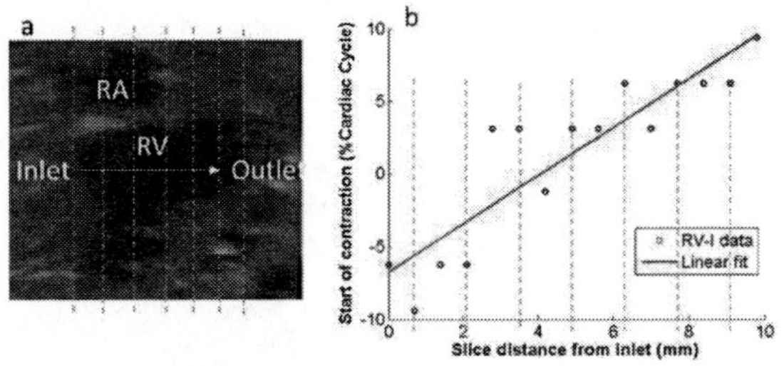

Figure 1a) Slices taken parallel to the four chamber view as illustrated by dashed line b) Time of initiation of contraction at various slices from inflow tract to outflow tract, demonstrating a linear delay.

Figure 1 shows how slices were taken to quantify the peristalsis-like motion from human fetal ultrasound images. The quantification results are shown in table 1, peristalsis-like motion was found to have varying delay in contraction between inflow and outflow tract and peaked at around 16% of the cardiac cycle. Surprisingly, some two hearts demonstrated peristalsis delay in the reverse direction (contraction at the outflow tract before that at inflow tract).

Table 1 Contraction delay between inflow tract and outflow tract for six RVs studied. * indicates reversed peristalsis motion.

Case	Contraction Delay (% Cardiac Cycle)
RV I	16.36404
RV II	-3.12494*
RV III	-16.56396*
RV IV	0.54964
RV V	15.7164
RV VI	2.26184

CFD was conducted with varying degree of peristalsis delay (15% to -15%). The wall shear distribution and vortex dynamics were not significantly altered with different peristaltic motion (Figure 2a), neither was the overall kinetic energy dynamics of the RV (Figure 2b).

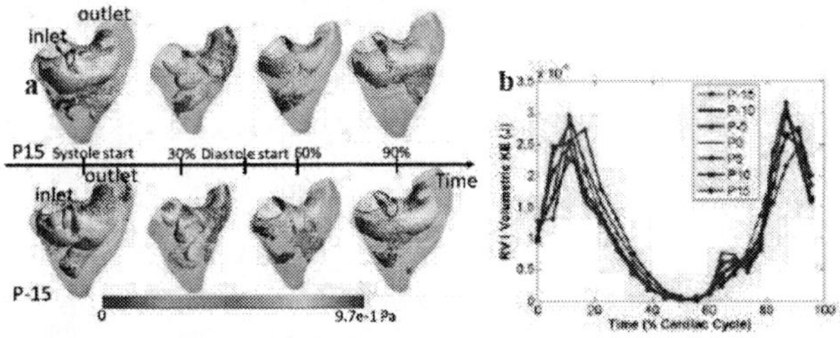

Figure 2a) CFD results: wall shear stress contour and lamda 2 vorticity iso-surfaces. b) Total kinetic energy in the RV over time, plotted for various peristalsis contraction delays.

Based on the CFD result, the work done to eject blood during systole was calculated which was plotted in **Error! Reference source not found.**. For fair comparison, we ensured that the stroke volumes among the different peristalsis cases differ by less than 1%. Results showed that systolic work done for ejection decreases with increasing prominence of the forward peristalsis. Moreover, when the direction of peristalsis were reversed, the work done required for ejection increases with increasing peristaltic contraction delay. Compared to the no peristalsis motion case, the +15% contraction delay (forward peristalsis) required 14.4% less energy for ejection. Compared to the no peristalsis case, the -15% contraction delay (reversed peristalsis) required 13.4% more energy for ejection.

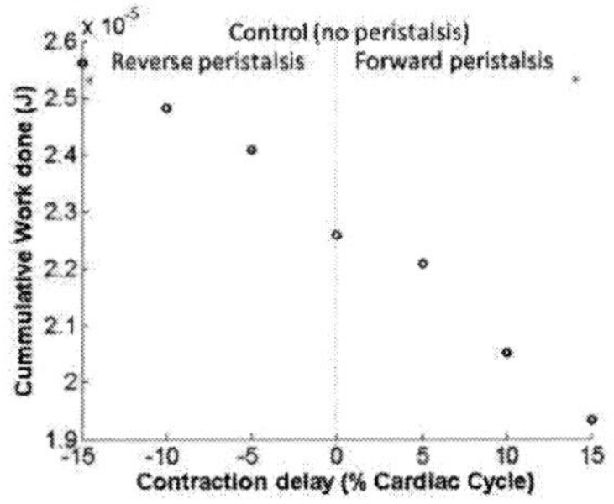

Figure 3 Cumulative work done at different peristaltic values for one of the RV investigated.

DISCUSSION

We elucidated for the first time, the presence of peristaltic-like contraction motions in the human fetal RV, where the RV wall contracts in a wave motion between the inflow tract and the outflow tract. Upon quantification of systolic work done for ejection by the fetal RV, forward peristaltic motion was found to provide favorable energy reduction required for ejection, while reversed peristalsis was found to require additional work for ejection. Collectively, these demonstrate that the peristalsis-like motion in the human fetal heart could be an adaptation to reduce energy expenditure of the heart.

On the other hand, this peristalsis-like motion may be an artefact of the fetal heart tube development. The infundibulum first present 20 days post fertilization, and the sinus formed later from proximal bulbus cordis few days later [7]. Part of the motion may also be accounted from RV electrical activation. Electrical map of the adult RV also shows a slight delay in the activation of the muscles near the right ventricle outflow tract in human adult RV [8], which might produce the peristaltic-like motion observed. Further studies into this interesting phenomenon is definitely warranted.

ACKNOWLEDGEMENTS

Funding is provided by the Singapore Ministry of Health, NMRC Grant Number NMRC/BNIG/2020/2014 (PI: Yap), and the National University Young Investigator Award (PI: Yap, Choon Hwai).

REFERENCES

[1] Hoffman, J.I et al., *J. Am Coll Cardiol*, 39(12):1890-900, 2002.
[2] Yang, Q., et al., *Birth Defects Res A Clin Mol Tetratol,* 76(10):706-713, 2006.
[3] Davlouros, P., et al., *Heart* 92: i27-i38, 2006.
[4] Geva, Tal, et al., *Circulation* 98.4, 1998.
[5] Raines, R. A., et al., *Am J Physiol,* 231(5):1395-1400, 1976.
[6] Wiputra, H., et al. *Am J Physiol Heart and Circ,* 311.6 (2016): H1498-H1508.0
[7]Clark, E.B., et al., Heart Disease in Infants, Children, and Adolescents. 4th ed. Baltimore, Md: Williams & Wilkins; 1989:2–15.
[8]Ramanathan, C., et al., *Proc. Natl. Acad. Sci* 103.16 (2006): 6309-6314.

PORCINE SMALL INTESTINAL SUBMUCOSA MITRAL VALVE FUNCTIONALITY UNDER PEDIATRIC CONDITIONS

Omkar V. Mankame[1], Sharan Ramaswamy[1], Lilliam Valdes-Cruz[2], Steven Bibevski[2], Frank Scholl[2], Ivan Baez[2]

(1) Department of Biomedical Engineering, Florida International University, Miami, Florida, United States

(2) The Heart Institute, Joe DiMaggio Children's Hospital, Hollywood, Florida, United States

INTRODUCTION

Congenital heart disease has been for years a fairly common ailment among infants and small children and also the recurrent cause of death among these young populace [1]. In the United States, there are an estimated 650,000 to 1.3 million children and adults living with congenital heart disease [2]. Majority of cardiac abnormalities occur due to cardiac valves defect resulting in 30% of all congenital heart problems [3]. Treatment alternatives for pediatric patients presenting with critical valve disease are exceptionally limited. Use of available valve substitutes in infants and children have restrictions in terms of somatic growth, durability and susceptibility to infection. It has been shown that acellular porcine small intestinal submucosa (PSIS) scaffolds have potential capacity to support hemodynamic functions and permit tissue growth and remodeling after implantation [4, 5]. Our group has implanted custom-made PSIS valves in 4 infants with critical valve disease unable to receive standard valves, with promising follow-ups. In our previous study, we evaluated acute hydrodynamic functionality of PSIS bioscaffold constructs in the aortic location under normal human physiological conditions [6]. In this study, we specifically focused on obtaining PSIS mitral valve characteristics under pediatric physiological conditions.

METHODS

PSIS (n=2) were tested in an in-house pulse duplicator system (Vivitro Laboratories, Victoria, Canada). Bi-leaflet PSIS valves assembled from sheets of PSIS bioscaffold (Cormatrix, Roswell, GA) were manually sutured in Dacron conduits by referring to the procedure previously utilized in heart valve repairs [7]. These constructs were subsequently immersed in a protease inhibitor cocktail (Sigma Aldrich, St. Louis, MO), with phosphate buffer solution (PBS) as the solvent. Valve holders were designed (Solidworks, Waltham, MA) and customized for each PSIS valve to secure the valve during *in vitro* hydrodynamic testing. Precise measurements of the dimensions of each valve were recorded and the computer-aided drawing (CAD) of the valve holder was custom-made to accurately accommodate the valves. The valve holders were then fabricated using white-colored, Poly-L-lactic acid raw filament material (Dynamism, Chicago, IL) via 3-dimensional printing (Makerbot replicator series, Makerbot Industries LLC, Brooklyn, NY). Once the holders were printed, the valves were secured compactly to the holders by suturing the valve annulus around the circumference of the holders. The secure valve was then press-fitted into the mitral location within the pulse duplicator system.

We have utilized the pulse duplicator unit in our previous experiments to evaluate flexible-leaflet valve implants and to assess functionality of PSIS tri-leaflet aortic valves [6, 8]. In the current study, a PSIS valve was mounted in the mitral position within the test unit, whereas a bi-leaflet mechanical valve was mounted in the aortic position. The mechanical component of the pulse duplicator system comprises of atrial, ventricular and aortic units, which mimic those to left side of the heart, a piston pump, and an integrated software system. To enable the measurement of flow and pressure data, a flow probe was positioned at the inlet of the mitral valve and two pressure transducers at atrial and ventricular location in the system. Finally, a 0.9% saline solution was introduced through the atrium chamber, to fill the loop with fluid. Before commencing the experiment, we calibrated the entire system at heart rate of 70 bpm and stroke volume of 80 ml/beat. Calibration was performed to warrant for accurate pressure and flow output readings. After the system was checked for any observable leaks, testing was initiated. The established parameters incorporated two sets of flow profiles depicting left circulatory conditions in the pediatric (Table 1(a) & (b)). A physiological waveform representative of the pumping action of the left ventricle was selected (in the case of the Vivitro system the S50 waveform was chosen for all hydrodynamic tests) to drive flow through the loop [6]. At the end of each run, data was recorded over 10 continuous cycles

Table 1(a): Varying heart rates at constant stroke volume = 40 ml/beat

Pediatric Hemodynamic Conditions at Constant Stroke Volume	Heart Rate (bpm)
	110
	125
	140

Table 1(b): Varying stroke volumes at constant heart rate = 145 bpm

Pediatric Hemodynamic Conditions at Constant Heart Rate	Stroke Volume (ml/beat)
	20
	25
	30

The effective orifice area (EOA) and closing energy loss for the mitral valves was defined as follows:

$$EOA = \frac{Qrms}{51.6 * \Delta P} \quad (1)$$

$$Closing\ Energy\ Loss = \int_{t0}^{t1} \Delta p(t) * q(t)dt \quad (2)$$

where t0 = beginning and t1 = end respectively of the systolic phase of the cardiac cycle, Δp(t) is the pressure gradient between atrial chamber and ventricular chamber over the cardiac cycle and q(t) is the pulsatile flow rate of the fluid.

RESULTS

Table 2(a): Mean ± SEM of Hydrodynamic Metrics of PSIS Mitral Valves at constant stroke volume of 40 ml/beat.

Heart Rate (bpm)	ΔP (mmHg)	Q_{rms} (ml/s)	EOA (cm^2)	Closing Energy Loss (mJ)
110	3.83 ± 0.07	128.6 ± 22	1.28 ± 0.2	21.6 ± 5.13
125	6.16 ± 0.37	150.6 ± 7.85	1.26 ± 0.03	28.3 ± 7.94
140	6.37 ± 0.6	152 ± 4.42	1.17 ± 0.09	41.1 ± 3.26

Table 2(b): Mean ± SEM of Hydrodynamic Metrics of PSIS Mitral Valves at constant heart rate of 145 bpm.

Stroke Volume (ml/beat)	ΔP (mmHg)	Q_{rms} (ml/s)	EOA (cm^2)	Closing Energy Loss (mJ)
20	4.94 ± 0.18	93.4 ± 1.80	0.82 ± 0	7.88 ± 1.57
25	5.37 ± 0.14	114.4 ± 1.49	0.96 ± 0.02	19.7 ± 4.45
30	5.74 ± 0.17	133.7 ± 0.95	1.08 ± 0.01	31.6 ± 2.87

Results demonstrated robust, physiological pressure and flow waveforms. Higher energy losses were observed with increasing cardiac output (heart rate * stroke volume) in both the cases. Energy loss for the mitral valves to close reached a maximum at a cardiac output of 5.6 L/min (Heart rate = 140 bpm, stroke volume = 40 ml/beat).

DISCUSSION

PSIS valves may offer the potential of somatic growth which make them appealing for the treatment of critical congenital valve diseases in the young [9]. We observed that a 27% increase in heart rate augmented energy losses by ~ 90 %. Meanwhile, a 25% increase in the stroke volume led to a 150% increase in the energy losses. These early findings suggest that energy losses across the mitral PSIS valves are more sensitive to the changes in pediatric stroke volume. As energy losses can be a concern for abnormal workload on the heart, it may be important to monitor the stroke volume clinically, as an indirect measure of this increased workload, with the growth of the pediatric patient. On the other hand, the heart rate will decrease with the patient's age and is thus less likely to be of concern. Nonetheless, a larger sample size is needed to confirm these findings.

REFERENCES

[1] Shuler, C.O., Black, G.B. & Jerrell, J.M., Population-Based Treated Prevalence of Congenital Heart Disease in a Pediatric Cohort. Pediatr Cardiol (2013) 34: 606. doi:10.1007/s00246-012-0505-3.

[2] Mozaffarian, D., Benjamin, E.J., Go, A.S. et al., American Heart Association Statistics Committee; Stroke Statistics Subcommittee, heart disease and stroke statistics-2016 update: a report from the american heart association. Circulation. 2016 Jan 26;133(4):447-54.

[3] Hinton, R.B. & Yutzey, K.E., Heart Valve Structure and Function in Development and Disease. Annu Rev Physiol. 2011; 73(1): 29–46.

[4] Bibevski, S. & Scholl, F.G., Feasibility and early effectiveness of a custom, hand-made systemic atrioventricular valve using porcine extracellular matrix (CorMatrix) in a 4-month-old infant. Ann Thorac Surg, 2015. **99**(2): p. 710-2.

[5] Kalfa, D. & Bacha, E., New technologies for surgery of the congenital cardiac defect. Rambam Maimonides Med J, 2013. **4**(3): p. e0019.

[6] Ramaswamy, S., Lordeus, M., Mankame, O.V. et al., Hydrodynamic Assessment of Aortic Valves Prepared from Porcine Small Intestinal Submucosa. Cardiovasc Eng Tech (2016). doi:10.1007/s13239-016-0290-x

[7] Slachman, F.N., Constructive remodeling of CorMatrix extracellular matrix after aortic root repair in a 90-year-old woman. Ann Thorac Surg, 2014. 97(5): p. e129-31.

[8] Ramaswamy, S., Salinas, M. et al., Protocol for relative hydrodynamic assessment of tri-leaflet polymer valves. J Vis Exp 2013(80):e50335.

[9] Zafar, F., MD, Physiological Growth, Remodeling Potential, and Preserved Function of a Novel Bioprosthetic Tricuspid Valve Tubular Bioprosthesis Made of Small Intestinal Submucosa-Derived Extracellular Matrix. Journal of the American College of Cardiology Vol.66, No.8, 2015 Issn 0735-1097

A 4-D COMPUTATIONAL STUDY OF DEVELOPMENTAL CARDIAC MECHANICS IN ZEBRAFISH EMBRYOS

Vijay Vedula (1), Juhyun Lee (2), Hao Xu (3), C.-C. Jay Kuo (3), Tzung K. Hsiai (2), Alison L. Marsden (1)

(1) Departments of Pediatrics (Cardiology), and Bioengineering
Stanford University
Stanford, CA, USA

(2) Department of Bioengineering, and Division of Medicine (Cardiology)
University of California Los Angeles
Los Angeles, CA, USA

(3) Department of Electrical Engineering
University of Southern California
Los Angeles, CA, USA

INTRODUCTION

Ventricular formation during cardiac development is tightly regulated by both genetic and mechanical factors. [1] Ventricular trabeculation is an essential process that involves a complex network of endocardial protrusions (trabeculae) into the ventricle and form ridges and grooves. Disruptions in normal cardiac morphogenesis can lead to either excessive trabeculation, a congenital heart disease known as non-compaction cardiomyopathy, or a significant reduction in trabeculation that is usually associated with ventricular compact zone deficiencies. Both these conditions can lead to heart failure and high mortality during embryonic development. While the genetics and molecular mechanisms underlying cardiac morphogenesis has been extensively studied, the impact of hemodynamic shear on the mechano-transduction remains less understood, partly attributed to the difficulty in measuring hemodynamic forces during the rapid developmental stages. [2, 3]

We developed a computational framework to quantify biomechanical forces in terms of endocardial wall shear stress (WSS) and oscillatory shear index (OSI) in zebrafish embryos (*danio rerio*). To provide new insights into the shear modulation of cardiac developmental mechanics, we genetically manipulated the zebrafish embryos. An ErbB2 signaling inhibitor, AG1478, was used to inhibit differentiation and proliferation of trabeculation during cardiac morphogenesis. The use of *gata1a morpholino oligomer* (*gata1a* MO) inhibited red blood cell production, thereby reducing blood viscosity and WSS. Also, *weak atrium* (wea) mutant was used to inhibit atrial contraction, leading to reduction in WSS and attenuation in trabeculation. Next, we performed 4-D (3-D space + time) light-sheet fluorescent microscopy (LSFM) [4] to image the contracting hearts in response to these genetic manipulations at 4 days post fertilization (dpf).

We applied our computational framework to the 4-D LSFM images in response to control, ErbB2 inhibitor (AG1478), *gata1a* MO, and *wea* mutation. While the control fish developed endocardial trabeculation, genetic manipulations attenuated trabeculation and cardiac contractile function. We compared hemodynamic parameters, WSS and OSI, in the presence and absence of trabeculation, and we investigated the implications of hemodynamic shear on the initiation of trabeculation. Our goal is to analyze the spatial and temporal variations in the computed shear profiles at the trabecular ridges and grooves, where the developmental Notch signaling mediates differentiation and proliferation of trabecular myocytes. Insights developed in the current study will pave way for improved understanding of the role of embryonic hemodynamics on the onset of congenital heart disease.

METHODS

We developed an efficient image to flow pipeline to compute hemodynamic shear in zebrafish embryos (Figure 1). Steps include:

- The morphology and motion of the zebrafish hearts were extracted using 4-D LSFM at high spatial and temporal resolution ($65X65X200 \mu m^3$, 100 frames per cycle). [4]
- Segmentation was performed at one selected cardiac phase to extract the ventricular morphology using 3-D level set segmentation tool in Simvascular. [5]
- The endocardial surface in direct contact with the blood, is subsequently extracted as a triangulated surface. The extracted surface is further processed for segmentation related errors such as hole filling, smoothing, extrusion and surface remeshing in Meshmixer (Autodesk Research INC.).

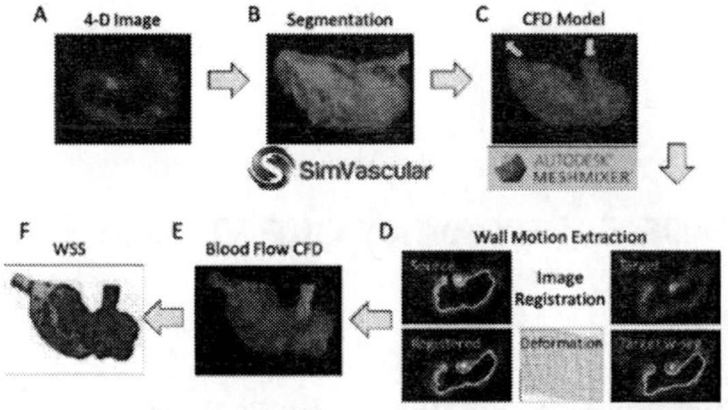

Figure 1: Computational framework for computing ventricular hemodynamic shear from 4-D image data in embryonic zebrafish.

- Registration was performed on the 4-D image data using B-spline based deformable registration methods, [6] and the computed deformation field was used to morph the segmented endocardial surface to extract the ventricular motion.
- Intraventricular hemodynamics was simulated by solving the Navier-Stokes equations governing incompressible blood flow with imposed ventricular wall motion using an arbitrary Lagrangian Eulerian (ALE) framework. [7]
- We use our in-house stabilized finite element method based flow solver [7] to conduct blood flow simulations on largescale multicore high-performance computing clusters.
- The computed velocity field is then used in conjunction with the ventricular wall motion and wall orientation to extract hemodynamic metrics such as WSS and OSI.

Since we perform segmentation only at one cardiac phase and use deformable registration methods to extract the cardiac motion, [6] we significantly reduce preprocessing effort avoiding the need to segment all the cardiac phases. This makes the workflow highly efficient and improves modeling accuracy as it reduces user intervention and associated errors while performing multiple segmentations. Zebrafish ventricles execute large deformations and pose significant challenge for simulating blood flow confined by the moving walls. We employ well-established variational multiscale methods [7] which are stable and second order accurate, and efficiently implemented dynamic remeshing to prevent run-time mesh distortion owing to the moving walls.

RESULTS

Hemodynamic shear characteristics including instantaneous WSS, area-averaged WSS (AWSS) and OSI are compared between control and genetically treated zebrafish variants in Figures 2 and 3.

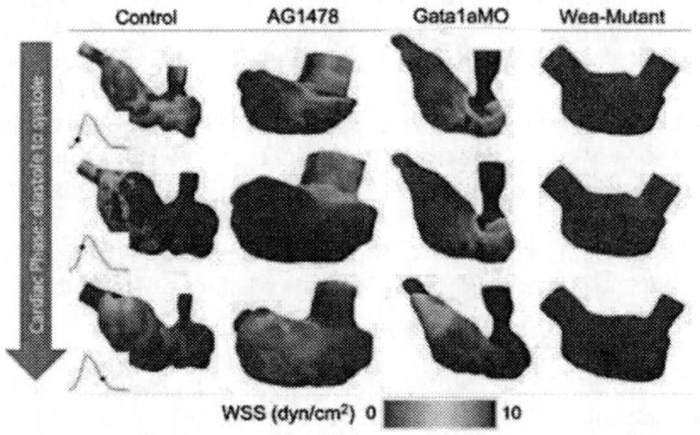

Figure 2: Comparison of wall shear stress (WSS) profiles between control and genetically manipulated zebrafish embryos.

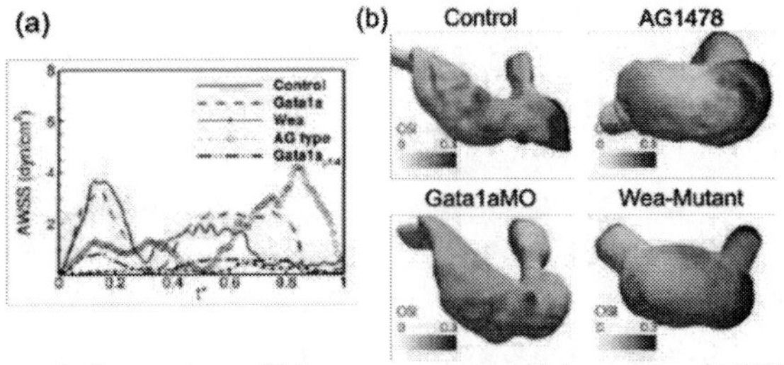

Figure 3: Comparison of (a) area-averaged wall shear stress (AWSS), (b) oscillatory shear index (OSI), in response to genetic manipulations. Data was phase-averaged over 4 cycles.

In Figure 4, we compare AWSS and OSI at the trabeculae grooves (crests) and ridges (troughs) for controls at 5dpf when the trabeculations are nearly developed. A high value of AWSS developed throughout the cardiac cycle accompanied with a higher OSI at the trabeculae grooves (crests) than at the ridges (troughs).

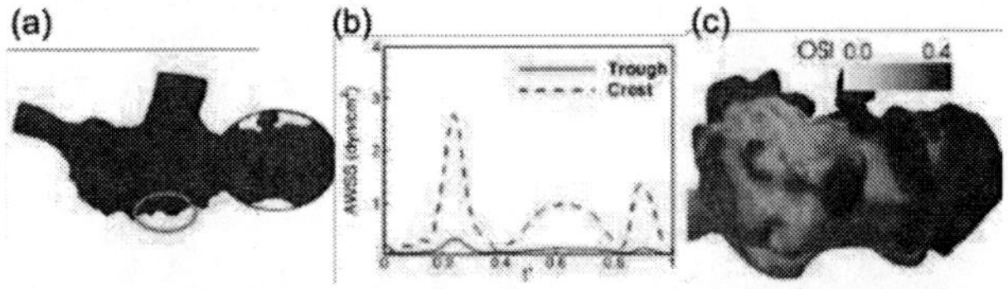

Figure 4: (a) Sites of trabeculae grooves (crests) and ridges (troughs). (b) AWSS plotted as function of time. (c) OSI distribution.

DISCUSSION

We have presented a novel and efficient computational framework for accurately computing hemodynamic shear from 4-D image data in zebrafish embryos. We have demonstrated the utility of the framework by comparing hemodynamic shear characteristics in four different zebrafish models: one control and three untrabeculated variants (AG1478, gata1aMO, and wea mutant). Our results indicate that high WSS and OSI are strongly linked with the initiation of cardiac trabeculation. We have further shown that the ridges of the trabeculae are subjected to high AWSS and OSI in a control fish acquired at 5dpf. Although the present analysis is performed on a single fish of each type, we plan to conduct a future study to demonstrate statistical significance with N>1 fish. We also aim to establish local correlations between blood shear and Notch signaling to quantitatively analyze shear-modulated mechano-transduction during cardiac morphogenesis.

ACKNOWLEDGEMENTS

The authors would like to acknowledge financial support through NIH NHLBI R01 grant 5R01HL129727-02 for the current study. VV & AM would also like to acknowledge computing resources received through NSF XSEDE grant TG-CTS30034 and software resources provided by the SimVascular project (www.simvascular.org).

REFERENCES

[1] Lindsey, S.E., et al., *Front. Physiol.*, 5:1-15, 2014.
[2] Jamison, R. A., et al., *PLOS One*, 8(10):e75722, 2013.
[3] Hove, J. R., et al., *Nature*, 429:172-177, 2003.
[4] Lee, J., et al., *J. Clin. Invest.*126(5):1679-1690, 2016.
[5] Updegrove, A., et al., *Annals Biomed. Eng.*, 2016.
[6] Myronenko et al., *IEEE Trans. Med. Imag.*, 29(11):1882-91, 2010.
[7] Esmaily-Moghadam, M., et al., *Comput. Meth. Appl. Mech.Eng.*, 286:40-62, 2015.

POPULATION BASED CHARACTERIZATION OF EARLY AVIAN GREAT VESSEL MORPHOGENESIS

Stephanie E. Lindsey (1,2), Irene E. Vignon-Clementel (2), Jonathan T. Butcher (1)

(1) Department of Biomedical Engineering
Cornell University
Ithaca, NY, USA

(2) REO group
INRIA-Paris
Paris, France

INTRODUCTION

The embryonic heart begins as a singular tube connected to bilateral pairs of symmetric arch artery outlets. At the common ventricle/common atria stage, before the heart is divided into four chambers, the outflow tract consists of the vessels flow must pass through in order to exit the heart. The uniform tube and 3rd, 4th and 6th pharyngeal arch artery (PAA) pair mature to form the aorta, pulmonary artery, and their branches. In the event outflow vessel maturation does not proceed as planned, congenital heart defects (CHDs) emerge. More than 20% of CHDs in the United States arise from malformation of the pulmonary artery and aorta, the precursors to which are PAAs. While the formation of these abnormalities is poorly understood, most can be traced back to early PAA perturbations.

Remodeling of the primitive paired arch arteries into the great vessels of the heart involves a delicate sequence of events yet to be fully elucidated. The PAA network experiences substantial local growth, which requires quantitative analysis and correlation to changing hemodynamic and biological parameters. While early studies examined the cellular composition of the arches, examination of their local hemodynamics is relatively recent. Here, we work to establish natural variations in PAA morphology and local hemodynamics during a critical window of development. We quantify day 3, day 4, and day 5 PAA morphogenesis and the relationship between geometrical and hemodynamic changes across days and arches. Our population based approach allows us to establish the robustness of hemodynamic and geometric norms. Our findings show distinct population based shape characteristics associated with stage of development that affect functional parameters and correlate with movement of the outflow tract relative to the arch arteries. Ultimately, this work will facilitate the identification of CHD origins and provide insights into the optimal timing for restorative interventions.

METHODS

Experimental preparation/measurements: Fertile white Leghorn chicken eggs were incubated for three, four, and five days at 37.5°C. Outflow tract (OFT) velocity and that of the three paired pharyngeal arch arteries were measured using B-mode guided Doppler Ultrasound. Embryos were then preserved, soaked in iodine solution and sent to undergo nano-computed tomography scans (nano-CT).

Reconstruction of 3D geometries: Embryo-specific 3D geometries of the day3, day 4, and day 5 PAAs were generated by importing nano-CT images into MIMICS (Materialise, Louvain, Belgium) and 3MATICs (Materialise, Louvain, Belgium). For 3D mesh generation and adaptation, INRIA's *ghs3d* and *feflo* were used.

Generation of computational parameters: A Poiseuille profile was assumed for day 3 and day 4 OFT flow. Plug flow was assumed the for the day 5 embryos. 0D circuit representations of arch artery morphology, in the form of a system of ordinary differential equations, were created for day 3, day 4, and day 5 geometries. This 0D model was used to tune the RCR or Windkessel outlet boundary conditions. 0D circuitry was adapted from Yoshigi et al's single compartment 0D model [1].

3D Numerical Simulation & Post Processing: Multiscale CFD simulations were conducted at INRIA-Paris. A population of five embryos were examined for each day. Blood was treated as a Newtonian fluid with constant hemodynamic properties (ρ = 1060 kg/m^3, μ = 3.71 × 10^{-3} Pa.s). Rigid, impermeable vessel walls were assumed with no slip boundary conditions. Flow was simulated with finite-element numerical treatment through FELiScE (http://felisce.gforge.inria.fr). The Vascular Modeling Toolkit (vmtk),supported by Orobix, sri, was used to obtain morphological parameters along the centerline of each geometry.

RESULTS

Overall Pressure increased dramatically between days (Fig. 1, Fig. 3), while wall shear stress maintained the same global level. Vessel geometries were shown to differ from the presumed circular phenotype. Vessel cross-sectional area (CSA) and shape were shown to play a large role in the functionality of a vessel. CSA has been shown to correlate with flow distribution [2,3], with the "dominant" arch possessing the largest diameter and greatest percentage of flow over one cardiac cycle. When arch diameters vary largely over the course of an arch, it can be uncertain which diameters or CSAs hold the most meaning. Figure 2 illustrates CSA and shape parameter variation across a population of five day 3 embryos.

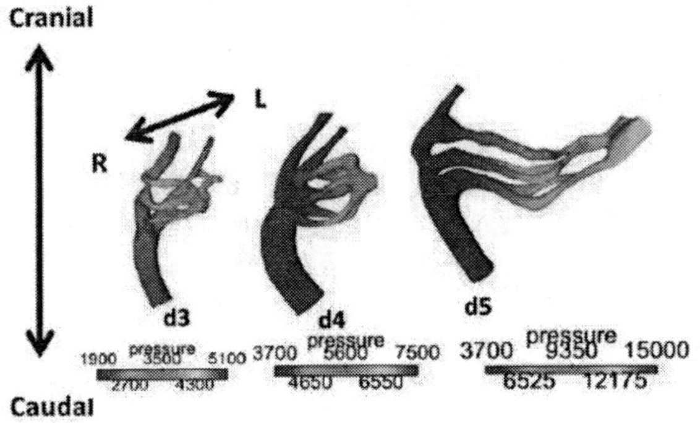

Figure 1: Pressure distributions [dynes/cm^2] for a day 3 day 4 and day 5 geometry with arrows to indicate notations

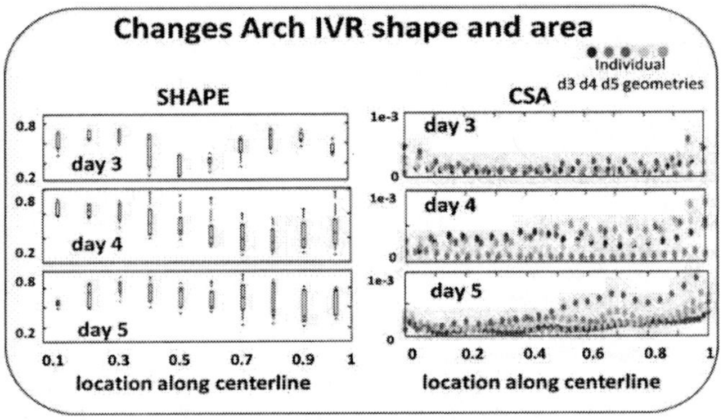

Figure 2: Box plot and CSA [cm^2] for PAA IVR across days.

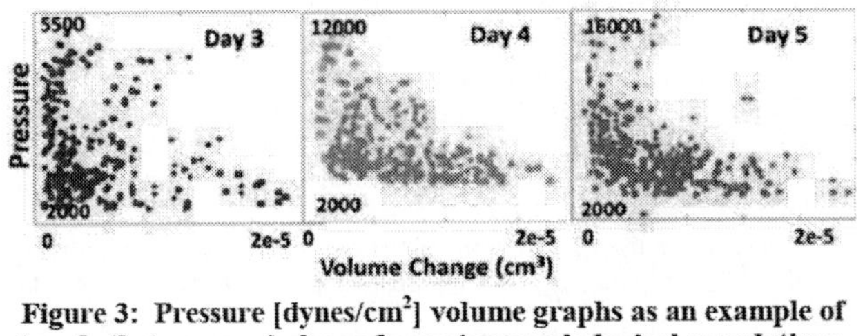

Figure 3: Pressure [dynes/cm^2] volume graphs as an example of trends that emerge in hemodynamic-morphological correlations.

The lumped parameter modeling provided an additional quantitative assessment of individual vascular segment growth and shape change with flow (Fig 4). Though resistance here is a hemodynamic parameter that represents a pressure drop to flow ratio, it is also a functional parameter. A 3D vessel is reduced to a single resistor in our 0D model and is able to capture its flow curve as seen in the 3D simulation. As expected, the regressing PAA IIR and PAA IIL have the highest resistance and smallest mean CSA. PAA IVL, day 3, and PAA VIL day 4 have almost identical resistance area ratios. These caudal arches are both growing in on their respective days. For the established arches (PAA IIIR and PAA IIIL) day 4 resistance values increase from day 3 to day 4, consistent with the narrowing of the vessel diameters and decrease in CSA. Day 3 and day 4 vessel resistance is roughly proportional to A^{-2}, while day 5 follows a trend of its own.

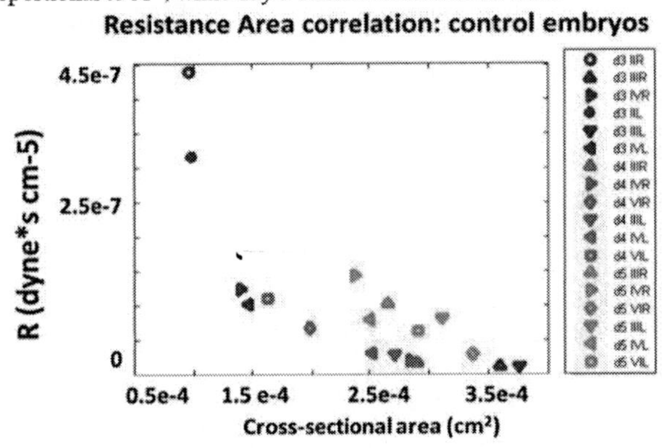

Figure 4: RA graph highlights emergence and regression of PAAs.

DISCUSSION

A lack of population based local hemodynamic quantitative analysis prohibits the establishment of regions of high variability and low tolerance in the delicate yet rapidly expanding PAA network. In using this population based approach with multiscale CFD modeling, we begin to quantify growth and underlying vessel trends in a way that fosters the development of accurate growth laws. Across days, the pressure magnitude was consistently greatest at the outflow tract junction before being dissipated both cranially through the aortic sac and dorsal -laterally over the arch arteries as they diverge from the aortic sac and reach their dorsal connection. In day 3 embryos, peak pressure extends farthest along the developing IV arches. Though the diameter of the IV arch approaches that of the second arch, the fourth arches' proximity to the ventricle and short vessel length render it more susceptible to pressure gradients over the course of the arch. Similarly to day 3, day 4 caudal arches experience the highest pressure magnitude over the length of the arch artery, as they have privileged access to outlet diameters. These caudal arches will experience the largest increases in vessel diameter within their subsets from one day to the next. We identify a strong correlation between segment morphogenesis and flow that appears to be independent of PAA origin. Information obtained from a detailed subject specific population study can prove to be particularly useful when assessing the effects of abnormal flow patterning and arch artery morphology on downstream circulation. Lumped parameter representations of vessel morphology provide another means of quantifying the hemodynamic properties of the vessels themselves and understanding growth across the days.

ACKNOWLEDGEMENTS

NSF GRFP, NSF GROW, NIH R01 HL110328.

REFERENCES

[1] Yoshigi, M et al., *Comput Methods Programs Biomed, 63:29-41, 2000.* [2]Wang, Y et al, *Ann Biomed Eng*, 37: 1069–81, 2009 [3] Lindsey, S et al., *Biomech Model Mechanobiol*,14:735-751, 2014.

RESPIRATORY CHANGES IN PULMONARY FLOW DISTRIBUTION IN FONTAN CIRCULATION: A COMPARISON BETWEEN "5-D" MRI AND CFD SIMULATION

David R. Rutkowski (1,2), Christopher J. Francois (2), Oliver Wieben (2,3,4), Alejandro Roldán-Alzate (1,2,4)

(1) Mechanical Engineering
University of Wisconsin
Madison, WI, United States

(2) Radiology
University of Wisconsin
Madison, WI, United States

(3) Medical Physics
University of Wisconsin
Madison, WI, United States

(4) Biomedical Engineering
University of Wisconsin
Madison, WI, United States

INTRODUCTION

Complex congenital heart diseases (CHD) with single ventricle physiology are often treated through a series of surgical procedures, the last of which is known as the Fontan procedure, to create a total cavopulmonary connection (TCPC). Altered hemodynamics in the TCPC can result in long-term complications.[1],[2] In particular, artero-venous malformations may form in the lungs, which are thought to arise from unbalanced flow distribution to the pulmonary arteries from the hepatic veins through the Fontan.[3] Before treatment for such malformations, the flow through the TCPC can be quantified to determine the severity of flow imbalance. However, hemodynamics of thoracic vasculature have been shown to vary between stages of the respiratory cycle, which can skew the results and interpretation of clinical analyses.[2],[4] The effects of respiratory variation on TCPC specific hemodynamics have not yet been thoroughly examined. Evaluation of these effects can provide insight into the changes in pulmonary flow distribution and flow imbalance among stages of respiration.

The influence of respiratory dynamics on TCPC flow can be studied with 4D flow MRI using radial projections (PC-VIPR[5]), which allows for flexible retrospective sorting of cardiac and respiratory data, adding a "fifth" dimension. However, imaging alone cannot be used for analysis of prospective geometric or physiologic variations in the TCPC. Computational simulation offers this capability. Nonetheless, the accuracy of computational simulation in complex flow situations must be validated and adequately represent in vivo flows before clinical implementation. The purpose of this study was to examine flow distribution variations between respiratory plateaus in TCPC subjects, and evaluate the ability of computational fluid dynamics (CFD) simulation to replicate *in vivo* results.

METHODS

In this IRB-approved and HIPAA-compliant study, six (6) patients with TCPC (two atriopulmonary and four extracardiac) were imaged on either a 1.5T or 3T MRI scanner (Discovery MR750, GE Healthcare) using PC-VIPR prescribed over a large chest imaging volume (FOV = 32 x 32 x 32 cm3, TR/TE = 5.5/2.3 ms, $\alpha = 15°$, Venc = 150 cm/s, projection number ≈ 22000, 16 reconstructed cardiac time frames, scan time: 11 minutes 30 seconds). A scheme was employed that allows for double gating to the ECG and respiratory cycles based on the bellows signal to provide a cardiac series of flow data for separate respiratory phases (inspiration and expiration). [6],[7]

The TCPC anatomy of each subject was segmented using semi-automatic segmentation software (MIMICS, Materialise, Leuven, Belgium), through which a three-dimensional geometry was generated from the phase contrast angiograms obtained from 4D MRI imaging. Subsequently, the two image sets (inspiration and expiration) from 4D MRI imaging were exported to an advanced visualization software package (EnSight, CEI). Cut planes were placed perpendicular to the vessel flow axis in the superior vena cava (SVC), inferior vena cava (IVC), left (LPA), and right pulmonary artery (RPA) of each TCPC model, and velocity vectors were generated. Total flow (mL/min) and kinetic energy were computed from the velocity vectors at each plane for both respiration plateaus.

To analyze the distribution of flow through each TCPC branch, and its variation over respiratory plateaus, particle traces were emitted from the IVC and SVC cut planes to produce streamlines throughout the model. Note that data were averaged over each respiratory plateau, therefore flow was analyzed as "steady," and streamlines are representative of the fluid particle paths. A streamline analyzer tool was then used to count the number of particles that flowed through each plane of the model. The number of particles passing through each pulmonary branch were recorded and categorized based on their origin (IVC or SVC), and flow percentages were obtained. This process was performed for both inspiration and expiration data sets for each patient. Flow distribution differences between respiratory plateaus were analyzed using a statistical t-test for significance ($p<0.05$).

To examine the ability of CFD to represent in vivo particle path distribution between respiratory plateaus, simulations were performed on each TCPC. To do this, the TCPC geometries from the segmented expiration MR image data of each of the six patients were meshed and prepared for simulation in commercial software ICEM and FLUENT (Ansys, inc Cannonburg, PA, USA). Two simulations were conducted on each TCPC, one for inspiration and one for expiration. Inlet flow rates were set at the IVC and SVC boundaries, using flow rates from the in vivo MR data. Outflow conditions, representing downstream vasculature resistances, were set at the LPA and RPA boundaries, using the MR data. Steady state simulations were run using no-slip conditions at vessel walls, and blood density and viscosity of 1060 kg/m^3 and 0.0035 kg/m-s, respectively. Using simulation results, streamlines, representing the fluid particle paths through each model, were generated and quantified at each outlet plane as was done in the in vivo analysis.

RESULTS

In vivo analysis showed that between-plateau differences in pulmonary streamline distribution were significant ($p=0.046$ (IVC), $p=.037$ (SVC)) in this data set. CFD pulmonary streamline distributions were also significantly different between inspiration and expiration plateaus ($p=0.0452$(IVC), $p=.046$(SVC)). The flow distributions from the in vivo analysis were not significantly different from CFD analysis for flow from the SVC ($p=0.0838$) and IVC ($p=.0804$) in inspiration. However, the streamline distributions obtained through CFD simulation were significantly different from those obtained from in vivo analysis for both the IVC ($p=.0052$) and the SVC ($p=0.05$) in expiration. Visual representations of the streamlines of a select number of subjects for both in vivo and CFD analyses are shown in Figure 1. Additionally, the streamline distributions in each plateau for both analysis methods are shown in Table 1. Lastly, the flow rates through the pulmonary artery branches were not significantly different between CFD and in vivo analysis in expiration. However, these rates were significantly different between methods in inspiration.

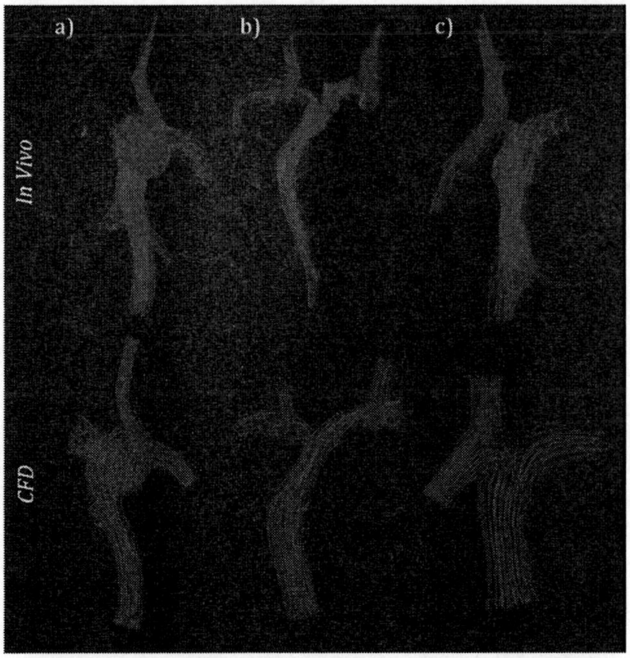

Figure 1: Streamlines through TCPC geometries of patient number a) 2, b) 4, and c) 5, obtained through in vivo MR imaging and computational fluid dynamics simulations. Flow from the SVC and IVC is shown in blue and red, respectively.

Table 1: *In vivo* and CFD flow distribution from the SVC and IVC to RPA and LPA branches during inspiration and expiration.

	TCPC1		TCPC2		TCPC3		TCPC4		TCPC5		TCPC6	
	Ins	Exp	Ins	Exp	Ins	Exp	Ins	Exp	Ins	Exp	Ins	Exp
	Distribution of Flow from the IVC (RPA%/LPA%)											
In Vivo	49/51	19/81	71/29	85/15	58/42	66/34	37/63	37/63	44/56	34/66	95/5	100/0
CFD	36/64	28/72	65/35	68/32	58/42	55/45	32/68	30/70	10/90	33/67	100/0	90/10
	Distribution of Flow from the SVC (RPA%/LPA%)											
In Vivo	86/14	74/26	35/65	59/41	18/82	20/80	1/1	1/1	59/41	67/33	30/70	13/87
CFD	100/0	89/11	33/67	56/44	26/74	27/73	1/1	1/1	94/6	1/1	24/76	0/100

DISCUSSION

Unbalanced flow distribution through the TCPC in single ventricle patients can lead to negative consequences. The severity of flow imbalance can be estimated with medical imaging techniques and computational simulation. However, if not accounted for, the dynamics of the respiratory cycle may lead to skewed clinical hemodynamic quantification results. To examine the effects of these dynamics in single ventricle patients, the flow paths through six TCPC geometries were analyzed during inspiration and expiration plateaus with both "5D" MR imaging and computational fluid dynamics simulations.

Both in vivo analysis and CFD simulation showed that the streamline (representative of flow path in steady flow) distributions were significantly different between inspiration and expiration plateaus. However, there were some discrepancies in the magnitude of these differences between methods. These discrepancies may have arisen from a number of factors. First, a uniform mass flow rate was imposed at the inlets in the CFD simulations. In reality, this profile may not be uniform and the varying velocity trajectories could have an impact on downstream flow. Second, the error associated from obtaining flow at cut planes from "5D" flow MRI data is propagated to the CFD simulation boundary conditions. If this flow is not completely accurate in CFD simulations the streamline distributions will vary from those obtained from in vivo data. Lastly, rigid walls were assumed in simulations, when in reality the walls may experience motion due to cardiac and respiratory dynamics.

In conclusion, this study showed that significant differences is pulmonary flow path distribution can occur due to dynamics of the respiratory cycle. These changes can be observed through use of "5D" MRI and computational simulation; however, assumptions and inaccuracies in simulation input data can inhibit complete replication of observed *in vivo* behavior.

ACKNOWLEDGEMENTS

We gratefully acknowledge funding by AHA (14SDG19690010 (AR)), NIH (UL1TR000427, TL1TR000429NIH, and 2R01HL072260), and GE Healthcare for their assistance and support.

REFERENCES

[1] Khairy, P., et al., *Circulation*, 117.1: 85-92, 2008
[2] Korperich, et al., *Eur Heart J*, 16.2: 198-209, 2014
[3] Imoto, Yet al., *Pediatric Cardiology*, 27.4: 490-92, 2006
[4] Hsia, T.-Yet al., *Circulation* 102.Supp 3, 2000
[5] Johnson, K M, et al., JMRI 2008
[6] Roldán-Alzate, et al., ISMRM, Singapore, May 7-13, 2016.
[7] Schrauben EM, et al. ISMRM 2011

USING ARTIFICIAL MUSCLE TO FABRICATE ARTIFICIAL HEARTS – HARNESSING GIGANTIC DEFORMATION OF DIELECTRIC ELASTOMERS FOR LARGE VOLUME FLUID PUMPING

Zhe Li (1), Yingxi Wang (2), Choon Chiang Foo (3), Jian Zhu (2), Choon Hwai Yap (1)

(1) Department of Biomedical Engineering, National University of Singapore, Singapore

(2) Department of Mechanical Engineering, National University of Singapore, Singapore

(3) Institute of High Performance Computing, 1 Fusionopolis Way, #16-16 Connexis, Singapore

INTRODUCTION

A common problem with conventional centrifugal blood pumps or artificial hearts is the severe blood damage caused by the high speed rotating impeller (2,000-15,000 rpm) [1]. High speed rotation causes blood damage and severe hemolysis [2], leading to severe outcomes. For example, 18% of patients with left ventricular assist devices were reported to suffer from hemolysis, reducing the one-year survival rate from 89% to 39% [3]. Developing a soft actuator fluid pump capable of mimicking the natural heart would be a solution to this problem.

Dielectric elastomer (DE), commonly known as artificial muscle, is an electroactive polymer under electrical stimulation. Due to its high energy density (capable of providing large actuation power with small mass) and large deformation capability, it is promising to use DE for the development of next generation artificial blood pumps of minimized blood damage. A typical DE actuator is composed of a DE membrane sandwiched between two compliant electrodes. Actuated by voltage applied onto the electrodes, DE membrane would be deformed by the electrostatic compressive force generated in DE. Development of fluid pump using DE was previously reported, but to date, the achieved actuation volume and flow rate were very small [4,5].

In this study, we demonstrated that large volume and high flow rate fluid pumping can be realized by harnessing DE's giant deformation properties and we also investigated the optimal conditions for achieving large volume and high flow rate fluid pumping.

METHODS

Quasi-static pressure-volume (PV) curves of the DE pump at various voltages were first characterized by infusing water into the pump using a syringe pump at a constant flow rate of 10 ml/s. Based on PV curves, work loop for large volume fluid pumping was designed. Feasibility of pumping large volume fluid from low pressure to high pressure was then experimentally confirmed.

The DE fluid pump experimented with composed of one layer of 2×2 pre-stretched DE membrane (VHB™ 4910 membrane with carbon grease electrodes) mounted onto a chamber (diameter: 56 mm; height: 60 mm, see Figure 1). Inlet and outlet were connected to large water tanks, and inlet/outlet pressures were controlled by adjusting the water level in the water tanks. Voltage was applied at a high slew rate by a high voltage amplifier (10/40A, Trek Inc., USA), controlled by a custom Labview program (National Instrument Inc., USA). Pressure change in the pump was recorded by a pressure sensor (MLT844, ADInstruments Inc., Australia). A camera was used to capture pump's front view at 50 Hz. The captured images were processed by a customized Matlab algorithm to calculate the pump volumes.

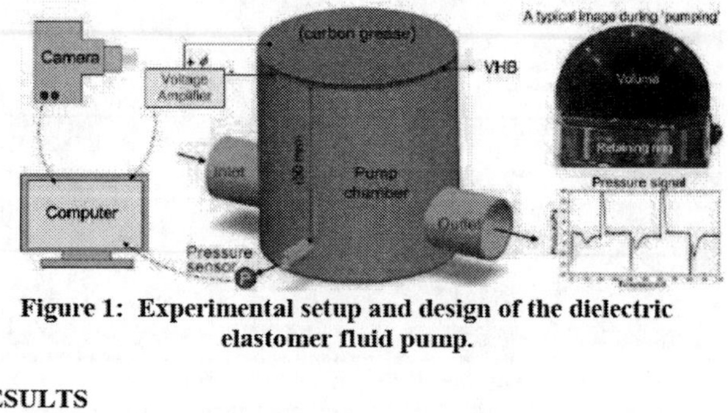

Figure 1: Experimental setup and design of the dielectric elastomer fluid pump.

RESULTS

The measured quasi-static PV curves revealed the basis of the giant deformation or snap-through phenomenon. Under specific conditions, the DE membrane would experience electro-mechanical instability, caused by a combination of pump pressure and electrostatic Maxwell stress, leading it to expand rapidly. This can be observed in the PV curve in Figure 2(a). Under 5 kV, at point C, the membrane expands to point E due to the characteristic dipping shape of the PV curve. Image of the membrane before and after the snap-through expansion is shown in Figure 2(b). This phenomenon has been harnessed for large volume fluid pumping in this study.

Based on quasi-static PV curves, work loop for large volume pumping was proposed in Figure 2(a). A→B→C→D→E→F→A is the work loop representing fluid pumping at $P_{OUT} = P_{IN}$ (P_{OUT}: outlet pressure; P_{IN}: inlet pressure). At point A, voltage application would cause pump pressure drop to B and initiate inflow. Inflow would occur subsequently through states C and D to E. At point E, pump pressure equals to P_{IN}, and inflow finishes. Upon voltage removal, pump pressure would increase to P_F at point F. P_F being large than P_{OUT}, fluid would be pumped out (F→A). Similarly, the work loop for pumping fluid from a low P_{IN} to a high P_{OUT} is designed as A"→B"→E→F→A" (Figure 2(a)).

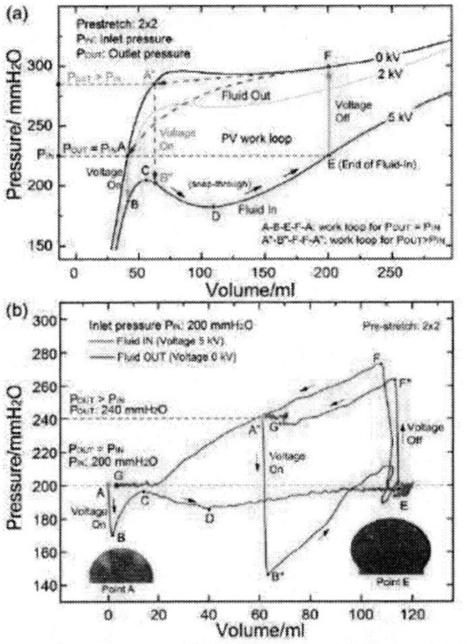

Figure 2: (a) Design of work loop for large volume fluid pumping based on quasi-static measured PV curves; (b) PV loops for dynamic fluid pumping experiments respectively for $P_{IN} = P_{OUT} =$ 200 mmH_2O, and for $P_{IN} = 200$ mmH_2O with $P_{OUT} = 240$ mmH_2O.

We verified the feasibility of large volume fluid pumping via dynamic pumping experiments, performed with 5 kV square wave voltage with 50% duty cycle. Typical work loops are shown in Figure 2(b) for the case of $P_{OUT} = P_{IN}$ (work loop A→B→E→F→G) and $P_{OUT} > P_{IN}$ (work loop A"→B"→E→F"→A"). The realized pumping volume per cycle was large, 110 ml for the former, and 50 ml for the latter.

We further investigated the optimal conditions of inducing snap-through deformation for large volume pumping. Results showed that firstly, a minimum of 4kV is necessary for the PV curve to have a N-shape snap-through curve. Secondly, an optimal inlet pressure is necessary, as shown in Figure 3(a). If the inlet pressure were too small, the DE membrane would not have giant deformation, and the achieved pumping volume was very small (less than 5 ml per cycle). From 180 mmH_2O to 210 mmH_2O, the achieved pumping volume significantly increased from about 10 ml/cycle up to about 110 ml/cycle. However, if the pressure is too high, above 230 mmH_2O, pumping volume decreases.

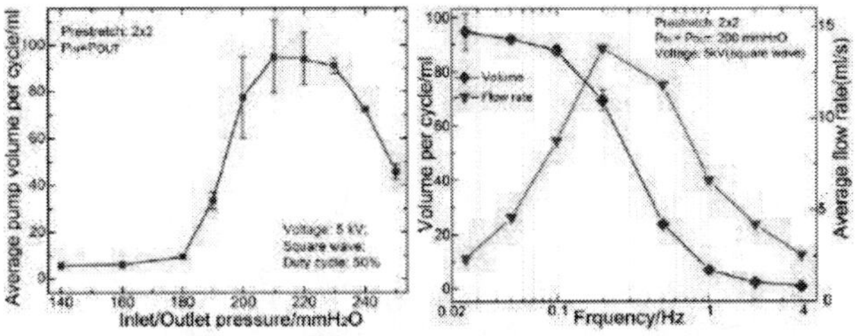

Figure 3: (a) Effects of inlet pressure on volume pumped per cycle by the DE fluid pump; and (b) Effects of pumping frequency on volume pumped per cycle and average flow rate.

Thirdly, there is an optimal pumping frequency, as shown in Figure 3(b). Large pumping volume (>80 ml/cycle) could be maintained at an actuation frequency ≤ 0.1 Hz, due to the occurrence of giant deformation. Above 0.5 Hz, the viscoelasticity of the DE membrane prevented the membrane to achieve large deformations within the short cycle durations. In terms of flow rate (volume pumped per cycle multiplied by frequency), the optimal frequency was observed to be 0.2 Hz, where flow rate peaked, due to an optimal compromise between number of pumping cycles per unit time and pumping volume per cycle.

DISCUSSION

In this study, large volume and high flow rate fluid pumping by DE actuator has been realized experimentally for the first time. The achieved pumping volume in this study is up to 110 ml per cycle; previous studies only realized a pumping volume of less than 20 ml per cycle [5]. Moreover, the developed DE fluid pump is capable of pumping fluid from low pressure to high pressure, which has also not been reported before with the artificial muscle, dielectric elastomer.

DE actuation voltage, inlet pressure and the actuation frequency were found important in inducing giant deformation of DE for large volume and high flow rate fluid pumping. In this study, we confirmed the feasibility of pumping fluid from low pressure to high pressure using a soft DE pump. In our experiment, only a single layer of 2×2 pre-stretched DE membrane was used. Multiple layers of DE membrane of different pre-stretches may help further improve its performance. In conclusion, we have demonstrated preliminary feasibility of using artificial muscle, dielectric elastomer, for fluid pumping, which could lead to an artificial heart that mimics the native human heart.

ACKNOWLEDGEMENTS

Fund for this work was provided by the Singapore Millenium Foundation, Grant Entitled "Minimally-invasive, Thoracoscopically Implantable Cardiac Assist Device for Palliative Care of the Elderly with Failing Hearts" (PI: Yap Choon Hwai).

REFERENCES

[1] Song X, et al., *J Fluid Eng*, 126:410-418, 2004.
[2] Lou S, et al., *Crit Care Med*, 42:1213-1220, 2014.
[3] Ravichandran AK, et al., *J Heart Lung Transplant*, 33:44-50, 2014.
[4] Tews AM, et al., *Smart Structures and Materials*, 159-169, 2003.
[5] Bar-Cohen Y, et al., *Smart Structures and Materials*,122-133, 2004.

SB[3]C2017
Summer Biomechanics, Bioengineering and Biotransport Conference
June 21 – 24, Tucson, AZ, USA

VALIDATION OF EXPERIMENTAL SETUP TO SIMULATE AND MODEL NON-VALVED GLAUCOMA DRAINAGE DEVICES

Tabitha HT. Teo (1), Paul M. Munden, MD. (2), Sara E. Wilson, Ph.D. (1), Ronald L. Dougherty, Ph.D. (1)

(1) Department of Mechanical Engineering
University of Kansas
Lawrence, Kansas, USA

(2) Department of Ophthalmology
KU Eye Center
University of Kansas Medical Center
Kansas City, Kansas, USA

INTRODUCTION

Glaucoma is an optic neuropathy associated with an increased intraocular pressure (IOP) and is the leading cause of irreversible blindness globally.[1] IOP control is essential in order to prevent ongoing vision loss and eventual blindness. Surgery is required when topical medications or laser therapy don't control IOP. A common and effective surgical procedure is the implantation of a glaucoma drainage device (GDD) which lowers IOP by shunting aqueous humor (AH) from the anterior chamber of the eye through a small silastic tube to a silicone plate placed on the posterior globe in the subconjunctival space. There are two types of GDDs: valved (VGDD) and non-valved (NVGDD). The most commonly used VGDD is the Ahmed implant which controls IOP with a butterfly valve placed at the termination of the silastic tube.[2] NVGDDs cannot be implanted without some type of modification because the AH-tube offers little resistance to flow, resulting in a low IOP (hypotony) in the immediate post-operative period. Ad hoc modifications to NVGDD are inconsistent and unpredictable for preventing early post-operative hypotony. . A fibrous capsule forms around the NVGDD in the 2 to 6 weeks following implant surgery, becoming the source of resistance to AH outflow, and thus providing post-surgical IOP control. Randomized clinical trials have demonstrated the superiority of the Baerveldt (NVGDD) implant over the Ahmed (VGDD) in terms of lower long-term IOP control.[3] However, post-surgical hypotony related complications are more common for the Baerveldt (NVGDD) implant.[3]

We propose to investigate a new method to consistently and effectively modify NVGDD surgery in order to predictably control IOP, and significantly reduce the risk of early post-operative hypotony. We hypothesize that a degradable annular insert of appropriate diameter and length, placed in the tubing of a NVGDD, will provide sufficient resistance to AH outflow to maintain an acceptable IOP in the early post-operative period.[4] The insert would degrade over a clinically relevant time period until AH outflow resistance following formation of the fibrous capsule around the implant plate becomes the controlling factor in determining final post-operative IOP.

A reliable test apparatus is needed to evaluate current and proposed solutions for NVGDD IOP issues. This study is focused on validating and obtaining preliminary pressure/flow data from such an experimental setup for tube flow and annular flow. Our proposed setup uses hypodermic tubes, microtubes, and stiff metal inserts to characterize the pressure-flow characteristics of the proposed NVGDD annular insert device over a clinically relevant range of IOP (0-30 mmHg) and AH production (2.5 μl/min). Experimental measurements of pressure drop are validated using theoretical calculations.

METHODS

Experimental Setup: Two duplicate microflow pressure-measuring setups were developed. Each setup consists of a precision pneumatic pump, a bottle containing deionized (DI) water, microfluidic tubing and connections, a pressure transducer, a data logger, and a test section (Fig. 1)[4]. The test sample is connected in a temperature-controlled incubator (± 2 °F).

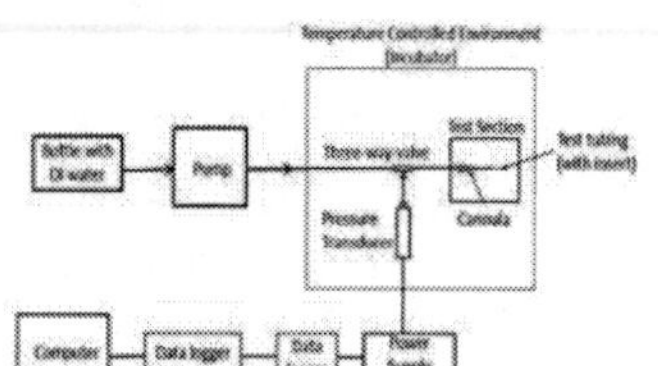

Figure 1: Schematic of microflow pressure-measuring test setup

Preliminary Trials: Long-term (2-4 hours) tube flow trials were performed with 50 μm and 75 μm inner diameter (ID) 35 mm long fused silica microtubes to validate the ability of the setup to measure intraluminal microtube pressures. Short-term (30 minutes) preliminary annular trials were performed using varying ID stainless steel hypodermic tubes with stiff metal inserts (kanthal, nichrome, stainless steel), and flexible NVGDD tubes with flexible inserts (nylon, polypropylene, and poliglecaprone sutures) in order to determine the ability of the setups to provide accurate initial annular flow pressure data. All inserts were 8 mm long. The pumps were primed at 50 µl/min for approximately 10 minutes before the trials in order to ensure that undetected air bubbles were removed from the setup..

Data Analysis: The intraluminal pressures for the long-term microtube trials were measured and the average pressure was calculated after the pressure remained relatively constant (Figure 2). The fluid was assumed to be incompressible and non-Newtonian, and the flow was assumed to be fully developed and laminar (calculated Reynold's numbers were less than 2100).[5] The average measured pressures were compared with the theoretical pressure ranges calculated using Hagen-Poiseuille (H-P) theory:[5]

$$\Delta P = \frac{8\,\mu L_t Q}{\pi r_2^4} \qquad (1)$$

The initial intraluminal pressure data for short-term annular trials were compared with the theoretical pressure curve calculated from the modified H-P theory for annular flow:[5]

$$\Delta P = \frac{8\mu L_t Q}{\pi}\left(r_2^4 - r_1^4 - \frac{(r_2^2-r_1^2)^2}{\ln\left(\frac{r_2}{r_1}\right)}\right)^{-1} \qquad (2)$$

The results were plotted as a function of effective diameter, D_{eff}:[5]

$$D_{eff} = 2\left[r_2^4 - r_1^4 - \frac{(r_2^2-r_1^2)^2}{\ln\left(\frac{r_2}{r_1}\right)}\right]^{\frac{1}{4}} \qquad (3)$$

Here, r_1 and r_2 are the inner and outer radii of the annulus between the insert and the tube, L_t is the insert length, Q is the flow rate, and ΔP is the intraluminal pressure differential across the length of the insert.

RESULTS

Long-Term Microtube Trials: The experimental pressures measured for flow through the fused silica microtubes are shown in Fig. 2. The average measured pressures after the pressures had reached steady state were within the theoretical pressure ranges (Table 1). The standard deviation of the average pressures in the 50 µm and 75 µm microtubes were 9.01 mm Hg and 0.08 mm Hg, respectively, at steady-state. The initial high pressures were due to the priming of the pumps.

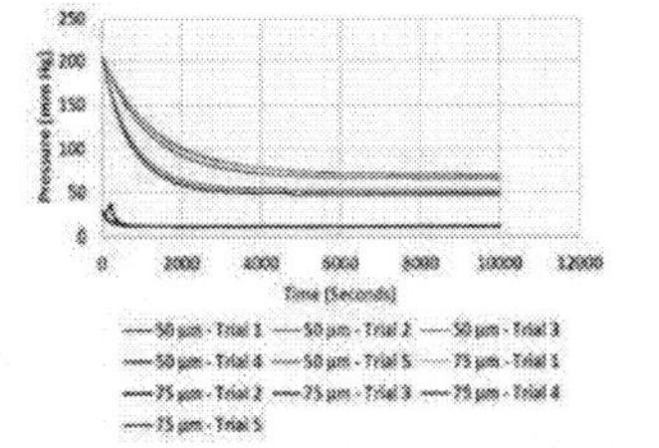

Figure 2: Experimental intraluminal pressure data in 50 µm and 75 µm microtubes as a function of time

Table 1: Theoretical and experimental intraluminal pressure ranges for microtube flow trials (includes modified table)[6]

Flow rate (µl/min)	ID (µm)	Length (mm)	Theoretical ΔP (mm Hg)	Experimental ΔP (mm Hg)
2.5 ± 0.125	50 ± 5	35 ± 2	46.0 – 115.1	47.9 – 68.3
2.5 ± 0.125	75 ± 5	35 ± 2	10.3 – 19.7	10.6 – 10.8

Short-Term Annular Trials: The intraluminal annular flow pressures were found to be higher than the theoretical pressure curve (Fig. 3). The average difference between the theoretical pressures and the experimentally measured pressures appears to be independent of tube and insert material flexibility.

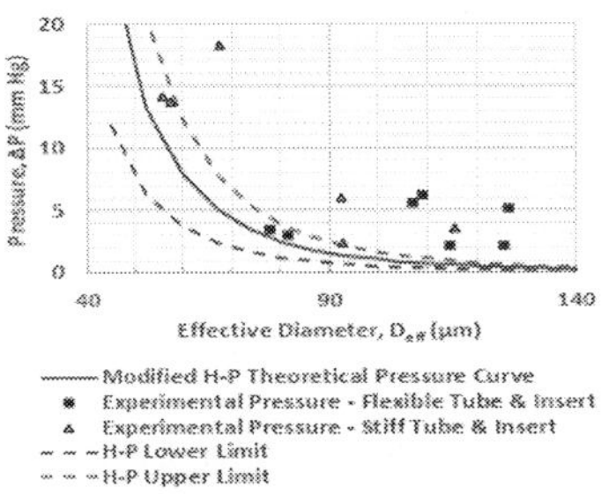

Figure 3: Experimental & theoretical annular pressure as a function of effective diameter

DISCUSSION

Long-Term Microtube Trials: The intraluminal pressures measured for the 50 µm and 75 µm tubes conform well to the predicted pressure ranges, indicating that the experimental setups developed can produce pressures within the predicted pressure ranges. The deviation in pressures obtained with the 50 µm tubes is greater than that of the 75 µm tubes because the resulting intraluminal pressure becomes more sensitive to differences in tube ID at smaller IDs, agreeing with the findings of Jiang *et al.*, who studied DI water flow through microtubes.[7]

Short-Term Annular Trials: For flows at the microscopic level, flow characteristics such as surface roughness, tube and insert surface wettability, surrounding pressures and temperatures can affect fluid flow.[8] The deviation of the experimental annular flow pressures from H-P theory may be due to the surface roughness of the tubes and inserts. An experiment by Tang *et al.* showed that roughness is the main culprit for higher friction factors which then leads to higher intraluminal pressures, especially in stainless steel microtubes with IDs between 50 µm and 530 µm.[9] These effects may be further amplified since the flow is through an annular orifice, involving an insert with additional roughness considerations. It is important to note that there are two major phenomena at the microfluidic level that are not accounted for with conventional theory, which are the electro-viscous effect and the capillary effect.[8,10] Another factor that was not accounted for in this experiment was concentricity of the insert in the tube. These effects may have influenced the pressure data as well.

Based on the results, the setups prove to provide sufficiently accurate pressure results which correspond with previous work. More experiments are necessary to validate the ability of the setup to reliably provide 2+ week pressure results for the testing and simulation of AH flow through NVGDD tubes with degradable inserts.

ACKNOWLEDGEMENTS

This work was supported in part by the University of Kansas.

REFERENCES

[1] Tham, Y. C, et al., *Ophthalmology* **2014,** *121* (11), 2081-90.
[2] Christakis, P.G, et al., *Ophthalmology* **2013,** *120* (11), 2232-40.
[3] Budenz, D. L., et al, *Ophthalmology* **2015,** *122* (2), 308-16.
[4] Munden, P., et al, 2015 Assoc for Research in Vision and Ophthalmology (ARVO) Annual Meeting, **2015** May 3-7; Denver, Colorado.
[5] Munson, B. R., et al., *Fundamentals of Fluid Mechanics*. 6th ed.; J. Wiley & Sons: Hoboken, NJ, **2009**.
[6] Ramani, A. J.. MS Thesis, Mechanical Engineering Department, University of Kansas, Lawrence, KS, **2015**.
[7] Jiang, R. J., et al, *China Physics Letter*. **2006**, *23* (12), 3305-8.
[8] Wang, F., et al, *Chinese Science Bulletin*. **2009**, 54 (13), 2256-2262.
[9] Tang, G. H., et al. *Journal of Micromechanics and Microengineering*. **2007**, 17, 539-550.

SB³C2017
Summer Biomechanics, Bioengineering and Biotransport Conference
June 21 – 24, Tucson, AZ, USA

CHARACTERIZATION OF ALIPHATIC URETHANE SHAPE MEMORY POLYMERS FOR BIOMEDICAL DEVICE DESIGN

Jingyu Wang[1], Shoieb Chowdhury[1], Yingtao Liu[1], Bradley Bohnstedt[2], Chung-Hao Lee[1]

(1) School of Aerospace and Mechanical Engineering
The University of Oklahoma
Norman, OK, USA

(2) Department of Neurosurgery
The University of Oklahoma Health Sciences Center
Oklahoma City, OK, USA

INTRODUCTION

Shape memory polymers (SMPs) have been developed as a critical technology platform for biomedical applications in the past decade, such as implants, surgical instruments, wound covers and controlled drug delivery systems [1-3]. In particular, SMPs are clinically essential for the development of novel medical devices to significantly improve long-term surgical outcomes.

In this study we proposed experimental procedures to synthesize and characterize aliphatic urethane SMPs. The fully characterized SMPs will be fabricated as thin wires for the development of novel device for the endovascular embolization of intracranial (or cerebral) aneurysms [4-6]. It has been hypothesized that the critical polymer properties, such as the glass transition temperature and shape memory function, can be tailored, by adjusting the polymer formula, to obtain the optimal endovascular embolization for treating cerebral aneurysms.

In this work, we presented our preliminary results, which demonstrate that the synthesized materials possess excellent shape memory function and suitable functional properties. In addition, the critical properties can be optimized and controlled by adjusting the weight ratio of the used monomers during material preparation.

METHODS

Three monomers were used to synthesize the aliphatic urethane SMPs. All the materials were purchased from Sigma Aldrich and used without any additional modification. The monomers used were: (i) Hexamethylene diisocyanate (HDI), (ii) N,N,N0,N0-tetrakis (hydroxypropyl) ethylenediamine (HPED), and (iii) Triethanolamine (TEA). Five different SMPs using various molar ratios of each monomer were prepared, as summarized in Table 1.

Table 1. Molar compositions of five synthesized SMPs

Polymer Designation	HDI (mole)	HPED (mole)	TEA (mole)
A1	1.000	0.500	0.000
A2	1.000	0.450	0.067
A3	1.000	0.400	0.133
A4	1.000	0.350	0.200
A5	1.000	0.300	0.267

Dynamic mechanical analysis (DMA) was employed to determine the thermal transition properties of the synthesized SMPs. Experiments were conducted using a TAQ800 dynamic mechanical analyzer (TA Instruments, New Castle, DE). Beam specimens (17.5 × 15 × 2 mm) were tested using three-point bending mode from 30°C to 130°C. All the synthesized SMPs (A1-A5) were tested and repeated for four times.

Cyclic uniaxial tensile tests were carried out using dog-bone samples (length $L = 9.5$mm, gauge width $w_c = 3.18$mm, thickness $t = 4$ mm) at the temperature above their glass transition temperature to determine the suitability for device

Technical Presentation #322

applications. The specimen were tested in a dual-column Instron mechanical testing system (Instron, Norwood, MA) with an environmental chamber. The specimen were clamped in the chamber and kept at the required temperature for 10 minutes before conducting the mechanical testing. Due to the low Young's modulus of such SMP materials, a load cell with the maximum load capacity of 100N was used.

RESULTS

The shape memory function of the synthesized SMPs were first investigated. Each straight beam sample was first immersed into hot water (5-10°C above the expected Tg), bended up to 180°, and cooled back to room temperature while maintaining the bended shape. Then the SMP specimen was placed back to hot water and the material restored to the initial straight form autonomously. SMP specimen at different shape is illustrated in Figure 1.

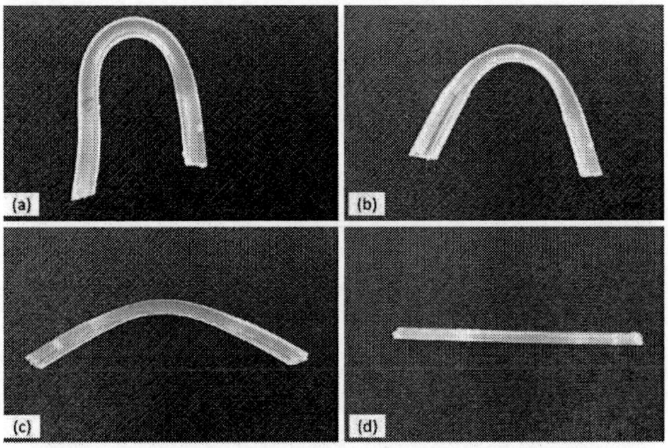

Figure 1. SMP specimen changes shapes as heated up to Tg.

DMA tests were applied to evaluate the glass transition temperature of polymers with various molar ratios and monomer formulations. As shown in Figure 2, the glass transition temperature of each polymer is obtained as the temperature where tan(delta) reached the peak value. As the molar ratio changed, the glass transition temperature of SMPs varied in the range of 65°C-80°C. More experiments will be conducted to fully investigate the wide glass transition temperature range for all the potential SMP formula using the selected three monomers.

Cyclic uniaxial tensile testing were used to evaluate the level of strain recovery for shape memory applications. Representative results of A3 formulation was shown in Figure 3. The experiment was conducted under displacement control and the strain were applied linearly for the total of ten cycles. It has been observed that in the first four cycles there was initial hysteresis, which may be due to the residual stress, material relaxation, or the re-arrangement of dangling chains and side-groups in the polymer. Once the SMP have been fully stretched and relaxed for four cycles, the hysteresis became relatively insignificant.

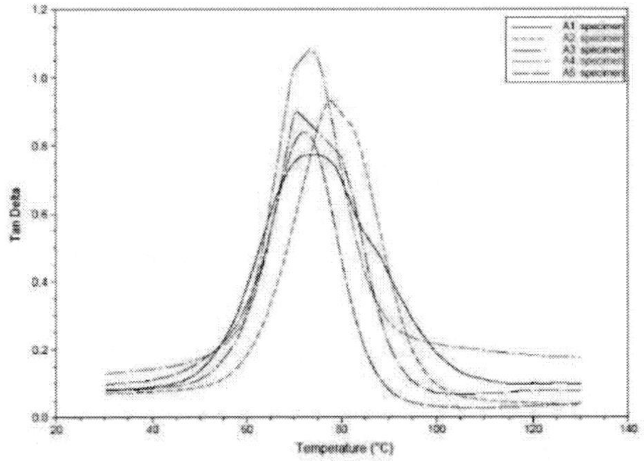

Figure 2. tan(delta) of DMA temperature sweeps, showing the Tg of synthesized SMPs

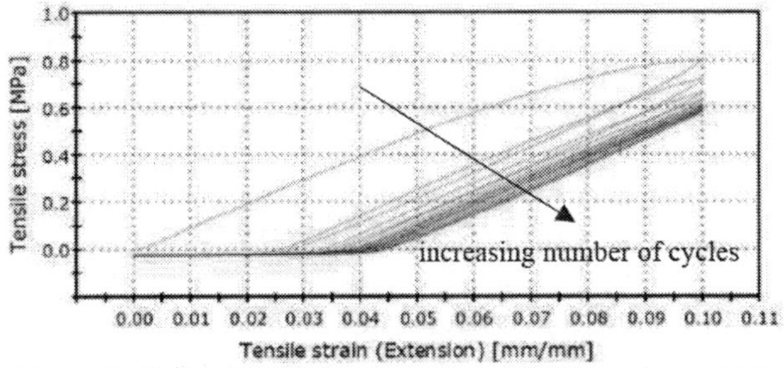

Figure 3. Cyclic uniaxial tensile testing of SMP specimen A3 to 10% engineering strain up to first 10 cycles

DISCUSSION

The objective of this work was to develop aliphatic urethane SMPs with optimal shape memory function and properties for potential applications in the endovascular embolization of intracranial aneurysms. Our preliminary results indicated that the synthesized polymer had good shape memory function when heated up to their glass transition temperature. The critical properties, such as their glass transition temperature, can controlled and optimized by adjusting the molar ratio of the used monomers during polymer synthesis. Such a design procedure could be tailored based on predictive finite element simulations, which is currently under investigation. Moreover, cyclic mechanical tensile tests showed that the SMPs had initial hysteresis when operated above their glass transition temperature. After four cyclic conditions, the synthesized SMPs performed stable with negligible variations.

REFERENCES

[1]Small, W. et al., *J Mater Chem*, 20:3356-66, 2010.
[2]Lendlein, A. et al., *Expert Rev Med Devices*, 3:357-79, 2010.
[3]Zhao, Q. et al., *Progres Poly Sci*, 49-50:79-120, 2015.
[4]Rodriguez, J. et al. *J Biomed Mater Res*, 103A:1577-94, 2015.
[5]Horn, J. et al., *J Biomed Mater Res*, doi: 10.1002/jbm.b.33725, 2016.
[6]Boyle, A. et al., *J Biomed Mater Res*, 104B:1407-15, 2016.

IMPROVING TISSUE MANIPULATION IN LAPAROSCOPIC RESECTION TRAINING USING VISUAL FORCE FEEDBACK

R. Hernandez (1), A. Onar-Thomas (2), F. Travascio (1), S. Asfour (1)

(1) Department of Industrial Engineering
University of Miami
Coral Gables, FL, USA

(2) Department of Biostatistics
St. Jude Children's Research Hospital
Memphis, TN, USA

INTRODUCTION

Laparoscopy is a type of minimally invasive surgery that relies on thin elongated instruments inserted through an inflated abdomen to perform operations. It is a challenging type of surgery that requires extensive training. The application of visual force feedback during training allows trainees to improve their tissue manipulation. Improvements due to force feedback are based on the successful moderation of applied forces toward an idealized goal, which is usually the average performance of expert surgeons. Research in this area is due in large part to the work of Horeman and Delft University, with notable improvements found when training in laparoscopic tasks such as needle driving [1] and tissue manipulation with graspers [2,3]. However, it is unclear if the benefits found in previous studies with this kind of training will transfer over when applied to other important laparoscopic tasks, such as organ resection. To this end, a simplified laparoscopic cholecystectomy (gallbladder removal) task was designed, due to the popularity of this procedure and since it requires both precision cutting and tissue manipulation. The objective of this research is to perform a randomized longitudinal study using the cholecystectomy task in order to determine whether a group with visual force feedback training will have improved tissue manipulation compared to a control cohort who does not have feedback. Consequently, the differences in applied force between the two groups will be estimated throughout the training period.

METHODS

An FLS Trainer System (VTI Medical, Waltham, MA) was used for the study as it is the official device used for training and evaluation in the Fundamentals of Laparoscopic Surgery certification exam. The trainer was retrofitted to measure forces exerted in both pulling and pushing directions using a load cell. The laparoscopic task was modeled after the latter part of gallbladder removal surgery (Figure 1). Six expert laparoscopic surgeons performed the task in order to acquire ideal performance values. A visual force feedback display was designed with *Processing 3* software [4] to provide warnings to the trainee should they approach forces above those of the expert surgeon group (Figure 2). The procedure, methods, and consent forms used in this study were approved by the Internal Review Board of the University of Miami. Twenty-two university students with no previous experience in laparoscopy first underwent basic proficiency training and then trained on the task sixteen times within two weeks. Participants were randomly assigned to either a treatment group that had visual force feedback during training or a control group which did not. An initial baseline and post-training trial without feedback were also included. Outcomes measured include task completion time and maximum pulling and pushing forces. A Longitudinal analysis was performed on the data with linear mixed-effect regression models run with the *R* program [5] and the *nlme* library [6].

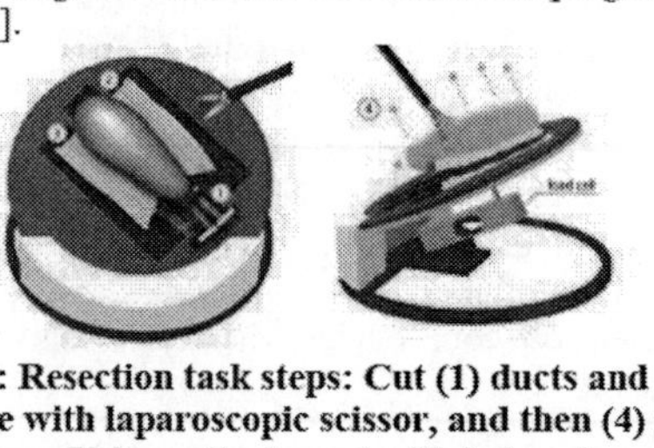

Figure 1: Resection task steps: Cut (1) ducts and (2-3) connective tissue with laparoscopic scissor, and then (4) remove gallbladder from Velcro attachment with babcock grasper

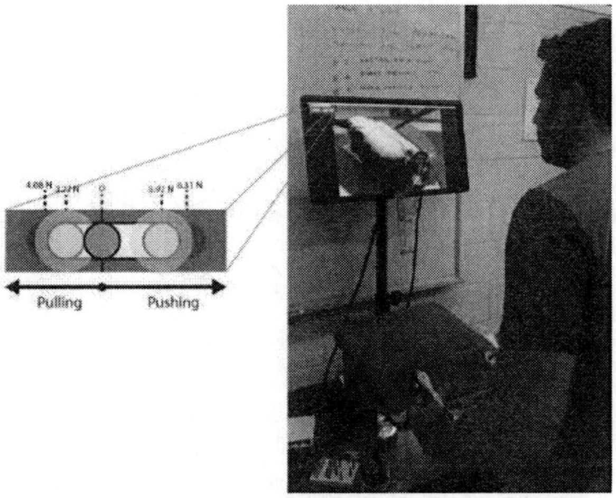

Figure 2: Visual force feedback system design: ball slides left/right when pulling/pushing, and warnings appear when specific force values are reached

RESULTS

The results showed that visual force feedback improved on certain aspects of force moderation (Figure 3). Extreme maximum pulling force values were tapered throughout training. Average maximum pushing forces were significantly lowered towards the end of training. On the other hand, task completion time was higher throughout the training period, yet lower times were found for both groups compared to the surgeon mean line. Significant differences in aptitude across participants were found.

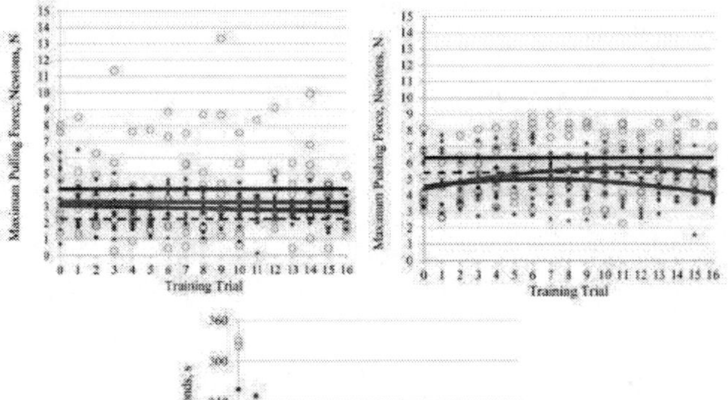

Figure 3: Maximum Pulling Force, Maximum Pushing Force, and Task Completion Time scatterplots with fitted mixed-model curves (red = control group, blue = treatment group, black = feedback max value, dashed = surgeon mean)

DISCUSSION

The significant improvements in force moderation found in the study are supported by the literature, but on different laparoscopic tasks [1-4] as well as those done without laparoscopy [7, 8]. Specifically, the average maximum force is known to decrease significantly after training with visual force feedback, as was the case for the applied pushing force. However, no significant improvement in the average maximum pulling force was found. Although extreme values were successfully tapered, lower mean values were expected. Therefore, there is indication that the feedback maximum values were set too high or that a longer training period is required to observe significant changes for the resection task. On the other hand, the completion time for both groups was significantly lower than the surgeon mean value, indicating unnecessary rushing to complete the task. The results of this study show the benefits of visual force feedback training when applied to a laparoscopic resection task. Trainees were gentler when manipulating tissue, reflected in either lower extreme force values and/or overall average values. Furthermore, accounting for the difference in aptitude between individuals is essential when assessing the benefit of learning with feedback training. As demonstrated by this study, it is possible to incorporate force feedback capabilities into existing laparoscopic trainers. Surgical instructors can create tasks that are pertinent to their specialty and establish objective performance standards of tissue handling which students can later train to replicate.

ACKNOWLEDGEMENTS

The study was supported by funds donated to the Biomechanics Research Group of the University of Miami.

REFERENCES

[1] Horeman, T et al., *Surg Endosc*, 26:242-248, 2012.
[2] Horeman, T et al., *Surg Endosc*, 28:1961-70, 2014.
[3] Smit, D, Spruit, E, et al., *Surg Endosc*, 31:299-308, 2017.
[4] Fry, B, Reas, C, *Processing 3*, https://www.processing.org/, 2016.
[5] R Core Team, *R*, https://www.R-project.org/, 2016.
[6] Pinheiro, J et al., *nlme*, http://CRAN.R-project.org/package=nlme, 2016.
[7] Rafiq, A et al., *J Surg Educ*, 65:270-4, 2008.
[8] Rodrigues, SP et al., *Brit J Surg*, 101:1766-73, 2014.

IMPROVED SUCTION DEVICE FOR AIRWAY MANAGEMENT IN EMERGENCY AND MILITARY CLINICAL SCENARIOS

F. Akhter (1), M. Lasch (1), E. Liu (1), R. Pescador (1), R. A. DeLorenzo (2), B. D. Adams (2), R. L. Hood (1), Y. Feng (1)

(1) Department of Mechanical Engineering
University of Texas at San Antonio
San Antonio, Texas, United States

(2) Department of Emergency Medicine
University of Texas Health Science Center
at San Antonio
San Antonio, Texas, United States

INTRODUCTION

In the event of medical emergencies, such as trauma or cardiac arrest, securement of the airway is one of the first priorities [1]. This life-saving intervention involves clearing any obstruction and, if deemed clinically necessary, placing an endotracheal tube into the patient's airway to maintain a pathway for ventilation. Airway compromise is the second leading cause of potentially survivable battlefield deaths [2]. In combat casualty care environments, the survival rate for wounded soldiers depends significantly on the skill of the combat medics, who often must face particularly chaotic situations with limited time. Consequently, the efficiency of the equipment used for the procedure has significant impact on the relative success of procedural outcomes [3]. Process improvements directed toward pre-hospital care can improve survival rate from combat injury as 90% of all battle field deaths occur prior to the casualty reaching a medical triage facility [4] [5].

The current method of addressing a compromised airway is first to perform a rapid evaluation of the patient's condition followed promptly by establishing and maintaining access to a patient's airway. In civilian emergency settings, the provider often utilizes a medical suction device to remove fluids, vomit, teeth, or other debris that have accumulated in the patient's airway. Unfortunately, this resource is often not readily available to prehospital or combat critical care providers who are deployed to the location of injury [6]. Consequently, the development of a powerful, lightweight, hand held suction device would greatly assist both military and civilian responders by enhancing the ability for those providers to increase visibility in the airway, clear obstructions, and maintain the airway. While commercially available suction devices do exist and are routinely used in hospital environments, none of the currently available units were specifically developed for the combat environment [6]. Devices for military medical scenarios are required to be compact, rugged, and portable while capable of maintaining clinical reliability and performance standards. Desirable performance characteristics for such an airway device include a suction flow rate of 3 L/min with vacuum pressure up to 550 mmHg. The present work approaches this gap in battlefield care by employing these design constraints to develop a suction device that can be readily accessible and practical for the military environment.

METHODS

The current prototype was designed to utilize an air diaphragm pump, an inline hydrophobic filter, and a collection canister. The diaphragm pump was chosen as the suction mechanism, while the inline hydrophobic filter served to protect the pump from any contaminants and to prevent the passage of any fluid to the air pump. These design choices allowed the pump to remain reusable while the filter, canister, and tubing require replacement after usage. In addition to an on/off switch and a low-battery LED notification, the prototype was outfitted with supplementary power options including an overdrive switch, which accommodates higher viscous fluids or obstructions, as well as a potentiometer to regulate the pumping power. The control system of the device was composed of an Arduino Leonardo microcontroller mounted on a printed circuit board (PCB) and a 12V DC brushless motor (Parker BTC-IIS Series). The device was designed with dual power sources: one 12V Lithium ion battery (MG Electronics) dedicated to the pump and one 9V battery to power the PCB. A series of tests were conducted to evaluate the performance of the device with simulated bodily fluids. The simulated vomitus solution was prepared in accordance with Section 59.9 of ISO standard no. 10079-1:2015 for electrically powered medical suction equipment. The formula comprised 10g of food grade xanthan gum dissolved in 1L of distilled water, with 100g of 1 mm diameter glass beads with specific gravity of 2.55 added to the dissolved xanthan gum solution. The glass beads were stirred into the solution and dispersed as evenly as possible.

RESULTS

The described prototype (Figure 1) is 4.4" x 4.3" x 10.7" and weighs 3.8 lbs with a 500 ml waste canister and 5/8" (inner diameter) suction tube. Von Mises stress analysis and failure simulation (Figure 2) showed that the device can withstand maximum 650 N (484 lbs) of compressive force before yielding. When tested with simulated bodily fluids, the device generated a maximum vacuum pressure of 305 mmHg while evacuating 200 ml vomitus solution in 3.9 second (flow rate 3.07 L/min). While running, noise level remained beneath 70 db.

Figure 1: Combat airway suction device (a) housing for pump/motor, (b) Canister (c) suction tube (d) On/Off switch, (e) Potentiometer, (f) overdrive switch, (g) low battery LED

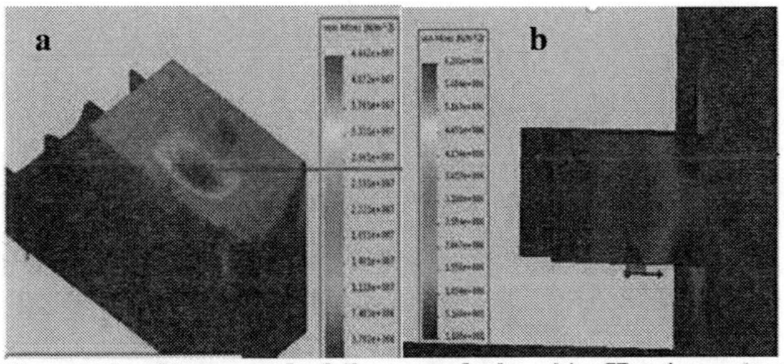

Figure 2: Stress and failure analysis: (a) Housing stress concentration, (b) Latch stress concentration

DISCUSSION

Structurally, the present airway suction device achieves a smaller footprint than commercially available devices, such as Laerdal Compact Suction Unit® 4 (LCSU4, 800 ml) that measures 9.3" x 7.5" x 9.3" and weighs 4.3 lbs [7]. The simulated vomitus performance and stress analysis tests also support its achievement of clinical reliability and durability, despite this decreased footprint. The drop in maximum vacuum pressure observed during the vomitus testing was likely due to the hydrophobic filter used in the device which can be eliminated by a secondary pump. On the other hand, such design modifications would increase footprint and weight, which is not desirable for military applications. Taking into account the feedback from military partners, ongoing prototype iterations will focus on optimizing these design tradeoffs to maintain a small device footprint and weight, while enhancing the maximum vacuum pressure.

ACKNOWLEDGEMENTS

The funding from NSF I-Corps Program, the Office of Commercialization at the University of Texas at San Antonio, San Antonio Life Science Institute, and the University of Texas Health Science Center are greatly appreciated.

REFERENCES

[1] C. Prokakis, et al. *"Airway trauma: A review on epidemiology, mechanisms of injury, diagnosis and treatment"* Journal of Cardiothoracic Surgery, 9(1):117, 2014.

[2] Eastridge, B.J., et al, *"Death on the battlefield (2001-2011): Implications for the future of combat casualty care"* J Trauma & Acute Care Surg, 73(6), S431-S437, 2012.

[3] Peake JB. *"Beyond the Purple Heart: continuity of care for the wounded in Iraq"* N Engl J Med, 352(3):219-222, 2005.

[4] Champion HR, et al, *"A profile of combat injury"*. J Trauma, 54(5):S13-S19, 2003.

[5] A.T. Simpson. *"Transporting lazarus: Physicians, the state, and the creation of the modern paramedic and ambulance 1955-73."* J History of Medicine & Allied Sciences, 2013.

[6] Calkins, M. D. *"Evaluation of possible battlefield suction pumps for the far-forward setting."* Military medicine, 167(10): 803, 2002.

[7] http://www.laerdal.com/us/doc/2115/Laerdal-Compact-Suction-Unit-4-LCSU-4#/Specifications

FIBROUS DOUBLE NETWORK MODEL TO MATCH OBSERVED FAILURE BEHAVIOR OF COLLAGEN-FIBRIN CO-GELS

David S. Nedrelow (1), Danesh Bankwala (1), Jeffrey D. Hyypio (1), Victor K. Lai (2), Victor H. Barocas (1)

(1) Biomedical Engineering
University of Minnesota
Minneapolis, MN, USA

(2) Chemical Engineering
University of Minnesota
Duluth, MN, USA

INTRODUCTION

The mechanical behavior of collagen-fibrin (Col-Fib) Co-gels is both clinically applicable and scientifically interesting. Col-fib networks are a staple of tissue engineering research, but, the mechanical consequences of changes in co-gel composition, specifically failure, have remained difficult to predict or even explain. We previously observed fundamental differences in behavior between collagen-rich and fibrin-rich co-gels, suggesting an essential change in how the two components interact as the co-gel's composition changes.

In this work, we explored the hypothesis that the co-gel behavior is due to a lack of percolation by the less concentrated component. We generated a series of computational models based on interpenetrating fiber networks in which the major network component percolated the model space but the minor component did not, instead occupying a small island embedded within the larger network.

METHODS

Network Parameters: Once generated, the individual Voronoi networks for collagen or fibrin were scaled such that the total protein concentration was constrained to 2 mg/mL, and both collagen and fibrin fibers were set to a diameter of 100 nm. The size of the embedded island was determined by the protein volume fraction. Thus, for example, the 40/60 col/fib network model contained a collagen island with 2/3 of the fibrin network cube's side length. Because the fiber diameters remained constant, and only the fiber lengths decreased when the networks for the islands were scaled down, total fiber *length*, not total fiber *volume* scaled with island size. Links between the two fiber types were generated when fibers lay within a user-specified distance.

Mechanical Model: The mechanical response of the collagen and fibrin fibers was determined by the following constitutive relation:

$$F = \frac{A}{B}\exp(BE_f) - 1) \tag{1}$$

where F is force on the fiber, A is a measure of fiber stiffness, and B is the degree of nonlinearity. Based on model fits to the pure-component experiments, the values (A, B) were set to (24.24 x 10^6 N, 3.0) for collagen and (2.88 x 10^6 N, 0.6) for fibrin; thus, the collagen was stiffer (larger A) and more nonlinear (larger B) than the fibrin.

Uniaxial mechanical simulations of double networks were completed by displacing boundary nodes on one side (the $+x$ face) of the Residual Volume Element (RVE) at 5% strain increments while holding the opposite ($-x$) face fixed. The four transverse surfaces (+/- y, and +/- z) of the RVE were allowed to move inward so as to maintain zero net force on each surface. Interior nodes in the network were permitted to rearrange so as to remain at mechanical equilibrium (i.e., no net force on any interior node) before iterating the next strain step.

Following Chandran et al. 2008[1], the average Cauchy stress σ_{ij} for the network was calculated as follows:

$$\sigma_{ij} = \frac{1}{V} \sum_{\substack{boundary\\nodes}} x_i F_j \tag{2}$$

where V is the volume of the RVE, x is the position of a boundary node, and F is the force on that node. The RVE was rescaled to the physical domain as described by the following equation.

$$\xi = \sqrt{\frac{LA_f}{\theta_o}} \tag{3}$$

where ξ is the RVE dimension in physical units, L is the total fiber length in the RVE (normalized to the unit cube), A_f is the fiber cross-sectional area, and θ_o is the initial fiber volume fraction. The Green strain of a fiber, E_f, was computed as

$$E_f = 0.5(\lambda_f^2 - 1) \quad (4)$$

where λ_f is the fiber stretch.

Simulated Stretch to Failure. The stretch at which fibers failed ($\lambda_{col} = 1.36$, $\lambda_{fib} = 3.4$) was determined by fitting individual networks to previously determined ultimate Green strains of 5.1 for pure fibrin and 0.7 for pure collagen[2]. After each strain step, a check was performed on each fiber. Upon a fiber reaching its critical stretch in our simulations, the prefactor A in Eq. 1 was reduced by a billionfold for that fiber, effectively removing it from mechanical significance without requiring a change to the network topology. After the fibers were so modified, the network calculation was repeated, and additional fibers were failed as needed. This process was repeated until no more fibers failed. Networks were stretched until network failure, defined where a negative slope in the stress-strain curve, arose. In spite of it simplicity, this model has proven successful in describing the failure of collagen gels[3,4]. The number of fibers that failed, by type, was tallied at each strain step, and fiber stretch histograms were generated for both inter- and intra-connected fibers.

Study Design. Simulated co-gel double networks were formed over 5 Col island scales (1, 10, 20, 30, 40) and 4 Fib island scales (40, 30, 20, 10) resulting in networks of the compositional range ~10, 30, 38, 43, 46, 55, 58, 63, 71% collagen. For each composition, five network simulations were performed, each with the same inter-fiber linking distance.

RESULTS

After equilibrium was reached for each strain step, illustrations of the networks were generated and First Piola-Kirchhoff stresses were plotted against Green strains (Fig. 1). Ultimate failure strains were plotted for each replicate and the previously published experimental failure strains were overlaid onto the plot (Fig. 2).

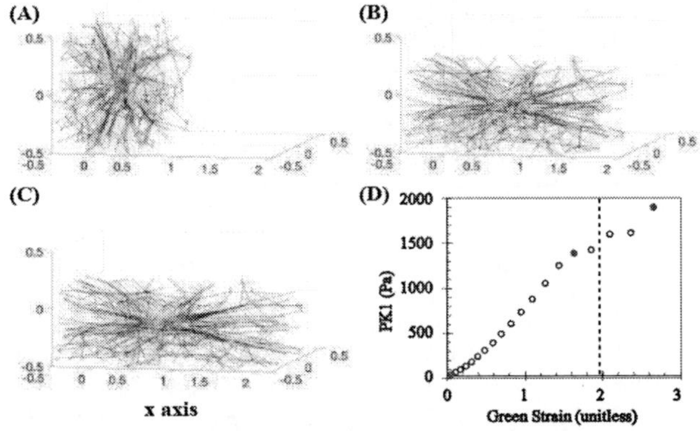

Figure 1: (A) Undeformed 40% Col/Fib double network, fibrin outer fibers (maroon), and embedded collagen island (gold). (B) Double network at stretch step where failure occurred. (C) Final stretch step on plot. The gray lines in the illustrations indicate failed fibers. (D) 1st Piola-Kirchhoff Stress vs. Green Strain with filled in solid maroon data points at strains shown in illustrations A, B, and C. The vertical dashed line shows experimental failure strain.

For double networks at 50% Collagen concentration (and 50% Fibrin), the lesser "island" domain was equal to the greater double network domain, so the smaller islands lie at the lower and higher extremes of the x-axis in figure 2.

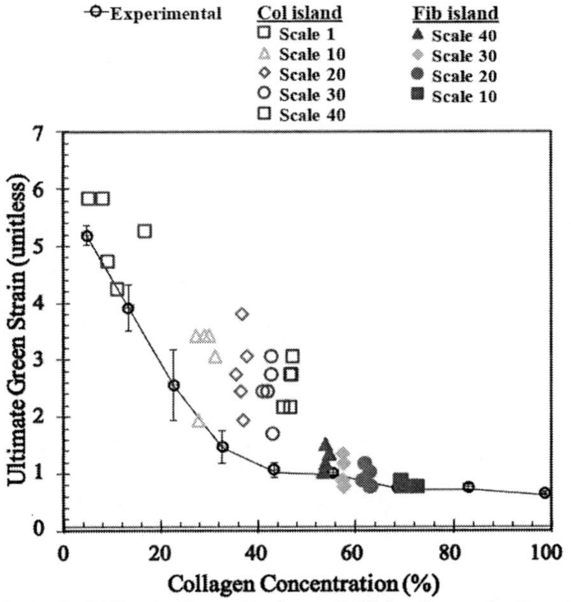

Figure 2: Ultimate Green Strain vs. Collagen Concentration of Col/Fib double network. The "Scales" in the legend indicate the %length of the embedded island network relative to the larger network.

DISCUSSION

From previous *in vitro* experiments, it was suggested that a transition from series to parallel mechanical behavior occurred as the concentration of collagen approached 50%[2]. Our double network model simulates this series/parallel transition behavior by exploiting two important features in the double network structure. Firstly, the lower-density minority component was limited, by relative concentration, from spanning across the domain, attributing a series character to the double network. As the embedded island was scaled up in size (as collagen concentration approached 50%), the double network made a transition to more parallel behavior resulting in lower failure strains approximating those of the experiment. Secondly, the extent of interconnectivity between the two fiber types affected the overall stiffness of the network.

The "island" double network hypothesis for the failure behavior of Col/Fib Co-gels was consistent with microscopy of Co-gels, and with a series-parallel transition in mechanical behavior, but there was a tendency to overestimate failure strains as the double network concentration approached 50/50 Col/Fib, especially in the Col island case. Accurate simulations of Col/Fib Co-gel failure behavior with a computer model are important to expanding multiscale model functionality and increasing accuracy for modeling failure in soft living tissues.

ACKNOWLEDGEMENTS

This work was supported by NIH grant [RO1 EB005813]. We extend our gratitude to the University of Minnesota Supercomputing Institute for making the necessary resources available to complete this study's simulations. ±

REFERENCES

[1] Chandran, P et al., *SIAM*, 36(2):828–845, 2014. [2] Lake, S. P. et al., *J. Biomech. Eng*, 134 (1):11004, 2012. [3] Vanderheiden, S. M. et al. *J. Biomech. Eng*, 137 (12):121002, 2015. [4] Hadi, M. F. et al., *J. Biomech. Eng*, 134(9):91005, 2012.

PLASTICITY OF FIBROUS COLLAGEN TRACTS FORMED BY CONTRACTILE CELL CLUSTERS

Ehsan Ban (1), Matthew Franklin (2), Hailong Wang (1), Lucas Smith (3), Rebecca G. Wells (3), Jan T. Liphardt (2) and Vivek B. Shenoy (1,4)

(1) Department of Materials Science and Engineering, School of Engineering and Applied Science, University of Pennsylvania
Philadelphia, PA 19104, USA

(2) Department of Bioengineering, Stanford University
Stanford, CA 94305, USA

(3) Department of Medicine, Perelman School of Medicine, The University of Pennsylvania
Philadelphia, PA 19104, USA

(4) Department of Bioengineering, School of Engineering and Applied Science, University of Pennsylvania
Philadelphia, PA 19104, USA

INTRODUCTION

Cells are affected by both elastic and dissipative (inelastic) mechanical properties of their environment. Inelastic extracellular matrix (ECM) deformation may be viscoelastic or plastic, the latter leading to long-term irreversible deformations. Since the architecture and the elastic modulus of the ECM influence cell spreading, motility [1] and differentiation [2], matrix plasticity after mechanical deformation is likely to be an important determinant of cellular behavior. Interestingly, plastic behaviors have been observed in several previous cell-ECM experiments [3, 4]. Despite the importance of ECM plasticity, its underlying mechanisms are not well understood and current models for ECM mechanics fail to account for it.

Here, we used mammary acini embedded in collagen type I gels as models to study plastic deformation in fibrous ECMs. Matrix deformation and fiber alignment were studied in a system of contractile acini. We observed that the tracts persist at least 24 hours after eliminating cell contractility. These observations were repeated using a system with fibroblast cells where fiber alignment was quantified using second harmonic generation. A reactive fiber network model was developed that allows for crosslink formation and breaking between aligned fibers.

This model illustrates the possibility of plastic deformation by the formation of new crosslinks during the course of network deformation and provides insights about its dependence on the rate and magnitude of the applied strain and the yield strain of plastic deformation of individual fibers. Additionally, a coarse-grained constitutive model is proposed that reproduces the viscoplastic behaviors of collagen networks by accounting for crosslink formation and breaking. It also captures the ECM densification and plasticity observed in the fibrous tracts.

METHODS

Mammary acini were seeded atop 2 (mg/mL) collagen I gels (Fig. 1a). Gels were stained with a dye-labeled collagen binding protein (CNA35-SNAP-Alexa647) to provide estimates of local collagen density. We tested whether collagen shows mechanical plasticity by inhibiting cell contractility in the gels with established collagen tracts. The culture was treated using a drug cocktail (latrunculin, cytochalasin D, and focal adhesion kinase inhibitor 14) to inhibit cellular traction forces.

A Mikado network of interpenetrating fibers was generated by randomly and placing fiber segments in a planar domain (inset of Fig. 1e). The fibers are attached by two types of crosslinks: the pre-existing crosslinks formed at gelation and the crosslinks that form when the network is highly stretched. The distance-dependent formation and breaking of crosslinks is modeled by rate constants, k_{on}, and k_{off}, which are exponentially related to the distance between the crosslink sites of the fibers. The probability of the existence of each bond, p_b, evolves as $\dot{p}_b = k_{on} - k_{off} p_b$.

A coarse-grained model was developed to enable modeling ECM plasticity at the continuum scale. This model builds on the elastic fibrous constitutive law we previously developed for modeling long range force transmission in fibrous matrices [5]. It takes into account the kinetics and energy associated with the crosslinks preexisting in the matrix prior to loading and the newly incorporated crosslinks formed at stretched states of deformation. Similar to the network case, the crosslink formation and breaking in the coarse-grained model also follow exponential relations against distance. Strain energy of the deformed crosslinks is calculated as a function of their relative deformation with respect to the configuration where they were formed.

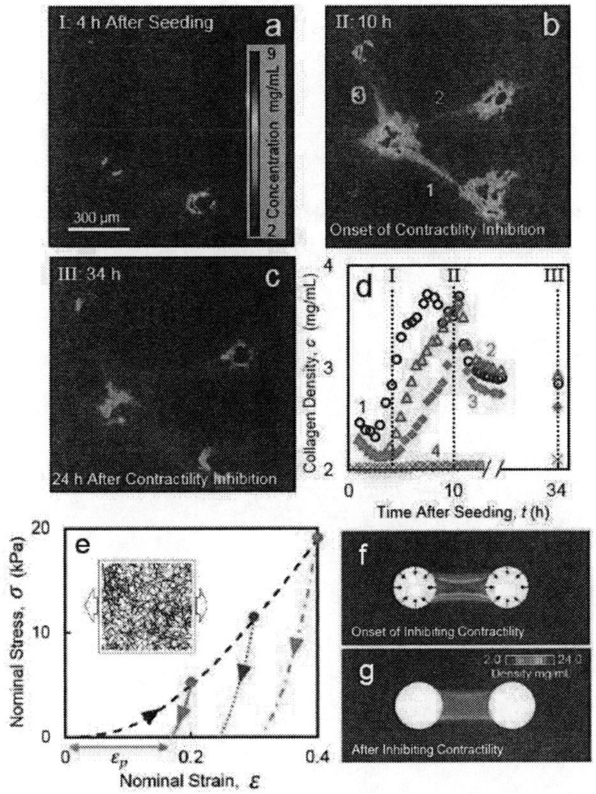

Figure 1: (a-c) Formation of collagen tracts formed between mammary acini and their persistence after inhibiting cell contractility. (d) Collagen density at the tracts and in the background against time. (e) Stress-strain response of the reactive fiber network. (f and g) The coarse-grained model reproduces collagen plasticity at the tracts by crosslink formation.

RESULTS

As previously reported, the acini are contractile and began pulling the matrix towards themselves after adhering to the fibrous matrix [6]. Within a few hours, the accumulated strains lead to large-scale fiber alignment and matrix densification and high-density collagen tracts formed along the axes connecting neighboring acini (Fig. 1b). In order to test matrix plasticity, we added a drug cocktail to inhibit cellular contractions. We observed that after the drug cocktail was added ECM reorganization stopped, strains partly relaxed and collagen density slightly decreased along the tracts. Interestingly, significant residual strains and collagen density remained in the tracts even 24 h after inhibiting contractility (Fig. 1c and d). Similar results were obtained in the experiments using fibroblast cell spheroids.

We used the reactive fiber network model to study the possibility of plastic network deformation by the formation of new crosslinks. Uniaxial loading and unloading using six timesteps were performed by stretching the network. As expected the network stress-stain curve was initially linear and then transitioned to a nonlinear strain-stiffening regime, characteristic of collagen networks (Fig. 1e). As previously reported, stretch of the network was accompanied by fiber alignment and large contractions in the transverse direction [7]. Therefore, reactive sites on neighboring aligned fibers come in close proximity, facilitating the formation of new crosslinks. The new crosslinks are stress-free when they first form, but are deformed when the network is unloaded and resist the complete recovery of network deformation, leading to plastic axial strains. The plastic axial strain, ε_p, is the amount of residual strain after external forces are removed (Fig. 1e). Up to 30% plastic strain is observed in the fiber network loading-unloading tests, which is in agreement with the amount plastic strain observed along the tracts in our experiments.

Next, the coarse-grained constitutive law was used to model matrix densification and plasticity by the contractile acini. A three-dimensional domain modeled the contraction of the tethered fibrous collagen gel by the cell clusters. The acini diameter and spacing are chosen to replicate a pair of acini in our experiments. The experimentally measured strains at the periphery of the acini were applied to the matrix as external loads. Upon the gradual application of external strains, the matrix model develops strain fields similar to the experimental results. The matrix was stretched along the tract bridging the acini and contracted in the transverse direction. ECM deformation by a pair of acini is accompanied by a large increase in matrix concentration along the tract axis (Fig.1 f). The coarse-grained model indicated that in addition to matrix contraction transverse to the tract axis, contraction in the out-of-plane direction also significantly contributes to the densification at the tract. The matrix out-of-plane deformations were highly localized at the tracts and form a narrow trench-like patterns in the matrix along the fibrous tracts. Next, to model plastic deformation at the tracts we eliminated cellular forces; we gradually reduced the applied cellular contraction until the average external forces vanished at the acini periphery. The resulting deformations indicated that about half of fiber density persists at the tracts, which was in agreement with our experimental results (Fig. 1g).

DISCUSSION

We used insights from our experiments and fiber network model to develop a coarse-grained model to explain our experiments of collagen plasticity and to enable studying the mechanical plasticity of ECMs by cellular forces. On one hand, the existing models for cellular force transmission within fibrous matrices reproduce the elastic matrix deformation, but lack the capability to account for plastic deformations. On the other hand, the existing lumped viscoplastic constitutive models (e.g. the Bingham plastic model) do not reproduce the initial matrix densification that leads to the formation of tracts between communicating cell clusters. These shortcomings prevent theoretical modeling of long-term mechanical ECM remodeling by cellular forces. Our constitutive model addresses these problems by accounting for the underlying microstructural phenomena that cause plasticity. In addition to explaining our experimental results, the developed constitutive model enables the study of mechanical remodeling in processes such as fibrosis, morphogenesis and cancer cell metastasis.

ACKNOWLEDGEMENTS

This work was supported by National Cancer Institute awards U01CA202177 and U54CA193417 (to V.B.S.) and National Institute of Biomedical Imaging and Bioengineering award R01EB017753 (to V.B.S. and R.G.W.).

REFERENCES

[1] Discher, D et al., *Science*, 310: 1139–1143, 2005.
[2] Engler, A et. al., *Cell*, 126, 2006.
[3] Sawhney, R and Howard, J, *J. Cell Biol.*, 157: 1083–1092, 2002.
[4] Guidry C and Grinnell, F, *Coll. Relat. Res.*, 6: 515–529, 1987.
[5] Wang, H et al., *Biophys. J.*, 107: 2592-2603, 2014.
[6] Shi, Q et al., *Proc. Natl. Acad. Sci.*, 111: 658–663, 2014.
[7] Vader, D et al. *PLOS ONE*, 4: e5902, 2009.

TYPE III COLLAGEN IS CRITICAL TO THE PROPER FUNCTIONING OF KNEE CARTILAGE AND MENISCUS DURING SKELETAL DEVELOPMENT

C. Wang (1), B. K. Brisson (2), Q. Li (1), K. Hoxha (1), M. Enomoto-Iwamoto (3), S. W. Volk (2), L. Han (1)

(1) School of Biomedical Engineering, Science and Health Systems
Drexel University
Philadelphia, PA, United States.

(2) School of Veterinary Medicine
University of Pennsylvania
Philadelphia, PA, United States.

(3) Department of Orthopedics
University of Maryland
Baltimore, MD, United States.

INTRODUCTION

It has long been recognized that the collagen network of cartilage extracellular matrix (ECM) is dominated by the heterotypic fibrils of type II, IX and XI collagens [1]. However, recent studies suggested that type III collagen is also present, and can make up for ~ 10% total collagen in mature human cartilage [2] and osteoarthritis (OA) cartilage [3]. In fact, in human genetic diseases associated with type III collagen (COL3A1) deficiency, such as vascular Ehlers-Danlos syndrome [4], premature OA is a common symptom [5]. Additionally, recent biochemical analyses indicated that type III collagen has the potential to bind to type II collagen fibril surface and facilitate inter-fibril cross-linking (Fig. 1a) [6]. These evidences suggest that type III collagen could be an important structural constituent of cartilage ECM. However, the roles of type III collagen in cartilage structure and mechanical properties are unclear. To this end, to test whether type III collagen has essential roles from developmental stage to maturation in joint functioning, this study examined the structural and mechanical phenotype of articular cartilage, as well as its direct loading counterpart, meniscus, in Col3a1 heterozygous (Het) mice (*Col3a1*$^{+/-}$).

METHODS

Sample preparation. Articular cartilage and meniscus were harvested from 2-week and 2-month old wild-type (WT) and *Col3a1*$^{+/-}$ BALB/c male mice. The null mice (*Col3a1*$^{-/-}$) were not included because they are perinatal lethal [7].

Histology and immunohistochemistry (IHC). Whole joints of both knees were harvested, fixed in 4% paraformaldehyde, decalcified in 10% EDTA, and embedded in paraffin. Serial 6-μm-thick sagittal sections were cut, and one every consecutive six sections was used for staining of type III collagen by its primary IHC antibody (ab7778, Abcam), joint and cell morphology by Hematoxylin and Eosin (H&E), as well as sulfated glycoaminoglycans (sGAGs) by Safranin-O/Fast Green.

AFM-nanoindentation was performed in PBS using a microspherical tip ($R \approx 5$ μm, $k \approx 5.4$ N/m, μMasch) and Dimension Icon AFM, following established procedures [8, 9]. For each condyle and meniscus, at least 20 locations were tested on the intact surface to account for spatial heterogeneity. The effective indentation modulus, E_{ind}, was calculated for each location via Hertz model [8].

Structural analysis. Samples were treated with hyaluronidase and trypsin for 24h to remove proteoglycans, fixed with Karnovsky's fixative for 3h in 25C, then dehydrated in graded ethanol and hexamethyldisilazane [10]. Collagen fibrils on cartilage and meniscus surfaces were then visualized by SEM (Zeiss Supra 50 VP).

Statistical test. Two sample student's t-test was used to detect the structural and mechanical differences between WT and *Col3a1*$^{+/-}$ mice. The significance level was set as $\alpha = 0.05$.

RESULTS

In comparison of the *Col3a1*$^{+/-}$ mice against the WT control at age of 2-month, IHC showed the deposition of type III collagen in cartilage and meniscus was greatly reduced, while histology did not detect appreciable differences in articular cartilage organization or proteoglycan content (Fig. 1b). In WT, both cartilage and meniscus at the age of 2-week showed a significant lower E_{ind} compared to which at 2-month. At the age of 2-week, significant decrease of E_{ind} was detected in *Col3a1*$^{+/-}$ mice cartilage compared to WT, while no difference in meniscus. Also, both cartilage and meniscus in 2-month *Col3a1*$^{+/-}$ mice showed significantly lower E_{ind} compared to WT (Fig. 2b). Furthermore, in both the cartilage and meniscus of 2-month-old mice, the collagen fibril diameters of *Col3a1*$^{+/-}$ were significantly larger than WT (Fig. 3).

DISCUSSION

Immunohistochemistry (Fig. 1b) confirmed that type III collagen is not only present in murine cartilage, but also is substantially reduced in *Col3a1*$^{+/-}$ mice, ascertaining these mice as an *in vivo* model for examining the roles of type III collagen. In *Col3a1*$^{+/-}$ mice, the thickening of collagen fibrils in cartilage (Fig. 3) highlights the active roles of type III collagen in the fibrillar network structure. Such observation is consistent with previous hypothesis that type III collagen co-polymerizes with type II collagen on fibril surfaces (Fig. 1) [6], in which, it may contribute to limiting the lateral growth of type II/IX/XI collagen fibrils and increasing the overall inter-fibril cross-linking. On the other hand, the absence of phenotype in sGAG staining or gross-level histology (Fig. 1b) suggests that the roles of type III collagen could be limited to the collagen content, at least at the tissue-level. The reduction in cartilage E_{ind} (Fig. 3) can thus be mainly attributed to the altered collagen structure. In cartilage ECM, the collagen network functions as the scaffold that holds aggrecan aggregates up to ~ 50% compressive molecular strain, even in unloaded cartilage [11], changes in collagen structure can affect not only the modulus of collagen network itself, but also the molecular strain and apparent stiffness of aggrecan, which cannot be detected by histology. In addition, it is also possible that given type III collagen's potential roles in inter-fibrillar cross-linking [6], its reduction can result in decreased collagen network cross-linking density.

Besides cartilage, we also observed the modulus reduction in type I collagen-dominated meniscus (Fig. 2b). This observation suggests that type III collagen is also involved in the formation of type I collagen-based synovial tissues. Therefore, its impact in joint functioning is likely not limited to articular cartilage, but other fibrous synovial tissues, such as the meniscus, anterior cruciate ligament (ACL) and synovium. It is therefore necessary to examine multiple synovial tissues simultaneously to fully understand the roles of type III collagen in the knee function and OA.

During early skeletal maturation stage (2-week), type III collagen presents significant effects on cartilage mechanical properties in WT compared to *Col3a1*$^{+/-}$ mice, but less so in meniscus, indicating that the roles of type III collagen is not only tissue-specific, but also age-dependent. Our ongoing studies are developing joint-specific targeted inducible knockout of Col3a1 using the aggrecan-CreER mice [12], which will allow for uncovering the specific roles of type III collagen in cartilage and meniscus with temporal specificity.

ACKNOWLEDGEMENTS

This work was supported by the National Institutes of Health (R21AR066824) and the Penn Center for Musculoskeletal Diseases (PCMD) (NIH P30AR050950).

REFERENCES

[1] Eyre, D, *Arthritis Res.*, 4:30-35, 2002.
[2] Eyre, DR et al., *Eur. Cell. Mater.*, 12:57-63, 2006.
[3] Hosseininia, S et al., *Osteoarthr. Cartil*, 24:1029-35, 2016.
[4] Pepin, M et al., *N. Engl. J. Med.*, 342:673-80, 2000.
[5] Castori, M, *ISRN Dermatol.*, 2012:751768, 2012.
[6] Wu, JJ et al., *J. Biol. Chem.*, 285:18537-44, 2010.
[7] Brisson, BK et al., *Am. J. Pathol*, 185:1471-86, 2015.
[8] Batista, MA et al., *Matrix Biol.*, 38:84-90, 2014.
[9] Li, Q et al., *J. Biomech.*, 48:1364-1370, 2015.
[10] Bray, DF et al., *Microsc. Res. Tech.*, 26:489-495, 1993.
[11] Wight, TN et al. 1991. *In Cell Biology of the ExtracellularMatrix*. 45-78.
[12] Henry, SP et al., *Genesis*, 47:805-14, 2009.

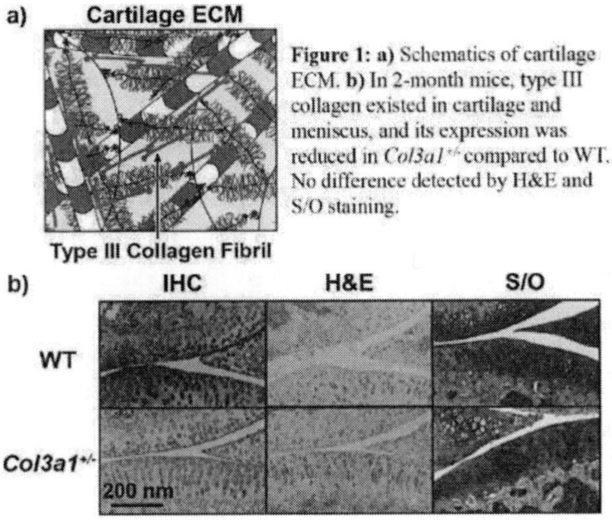

Figure 1: **a)** Schematics of cartilage ECM. **b)** In 2-month mice, type III collagen existed in cartilage and meniscus, and its expression was reduced in *Col3a1*$^{+/-}$ compared to WT. No difference detected by H&E and S/O staining.

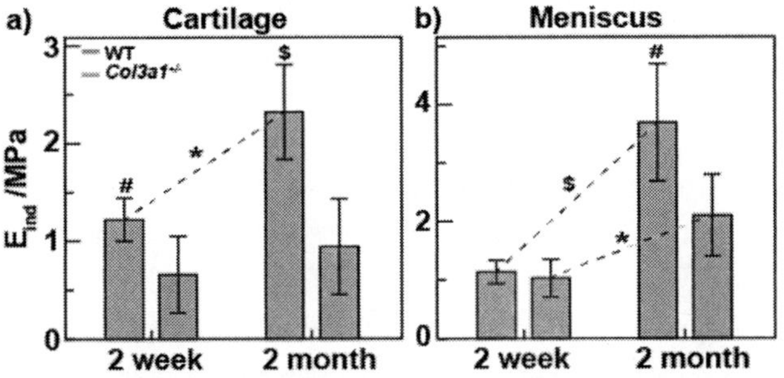

Figure 2: Nanoindentation modulus on **a)** articular cartilage and **b)** meniscus in WT and *Col3a1*$^{+/-}$ at age of 2-week and 2-month (mean ± 95% CI, 2-week cartilage, WT: n = 5, Het: n = 4; 2-week meniscus, WT: n = 4, Het: n = 4; 2-month cartilage, WT: n = 7, Het: n = 8; 2-month meniscus, WT: n = 8, Het: n = 8). #: $p < 0.05$, *: $p < 0.01$, $: $p < 0.001$.

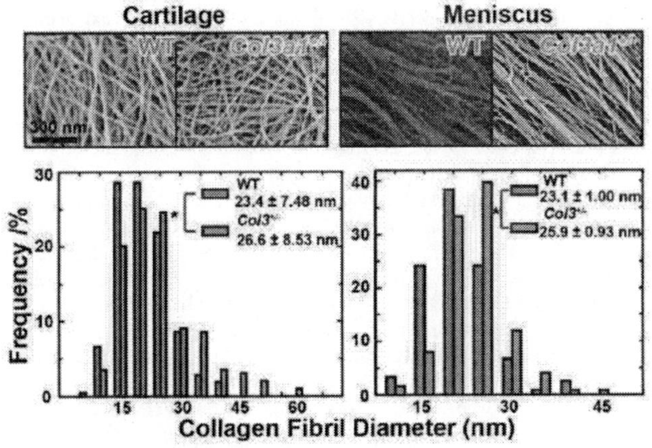

Figure 3: Nanostructure images and histograms of 2-month cartilage and meniscus collagen fibril diameter in *Col3a1*$^{+/-}$ and WT, mean ± 95% CI, ≥ 120 fibrils from n = 3 animals). *: $p < 0.0001$.

SB^{3}C2017
Summer Biomechanics, Bioengineering and Biotransport Conference
June 21 – 24, Tucson, AZ, USA

ON THE 3D MICROENVIRONMENT OF VALVE INTERSTITIAL CELLS UNDER PHYSIOLOGICAL LOAD

Salma Ayoub, Karen C. Tsai, Amir H. Khalighi, Michael S. Sacks

Center for Cardiovascular Simulation
Institute for Computational Engineering and Science
Department of Biomedical Engineering
The University of Texas at Austin
Austin, Texas, USA

INTRODUCTION

Of the four heart valves, the mitral valve (MV) is subjected to the greatest hemodynamic forces. Structurally, leaflets consist of four histologically-distinct layers: the *atrialis* and the *ventricularis*, both of which are thin elastin-rich layers, the *spongiosa*, which consists mainly of proteoglycans (PGs) and glycosaminoglycans (GAGs), and the *fibrosa*, the main load-bearing layer, composed of circumferentially-oriented collagen fibers. The structure-function relationship between mitral valve interstitial cells (VICs) and the surrounding extracellular matrix (ECM) enables the valvular tissue to maintain a homeostatic state in different hemodynamic and biomechanical environments.

While the ECM provides the tissue with its structural integrity and influences cellular processes through matricellular, matricrine, and mechanical processes, it is the mitral VICs (MVICs) residing within the valve's four layers that maintain the structural integrity of the leaflet tissue through protein synthesis and enzymatic degradation [1]. These cells are thus critical to the remodeling demands of the valve and tissue homeostasis. The interaction of MVICs with the surrounding ECM, particularly the collagen fibrils, is fundamental in modulating cellular response. The ECM can transmit external mechanical stimuli, such as stretch, shear stress, and pressure, to the cell via integrins and other membrane-bound protein receptors. We have shown that while MVICs have the same stiffness and phenotypic state across all layers, their deformations vary considerably during loading due differences in layer-specific ECM structure [2], highlighting the importance of the VIC/ECM coupling in the mechanoregulation of these cells.

Yet, despite this phenomenal level of reliability, more than five million people are diagnosed with heart valve disease in the United States each year, with approximately 95,000 annular valve replacement surgeries and 20,000 mortalities per year [3]. Though ring annuloplasty for mitral regurgitation is beneficial in the short-term, it has been shown to be less promising in the long term, with repair failure as high as 60% [1-2]. Mechanical stress is a strong etiological factor: alterations in mechanical loading caused by surgical repair lead to stress-induced changes in mitral valve interstitial cell (MVIC) function that affect both tissue structure and composition, ultimately leading to repair failure. While cell phenotype and extracellular matrix (ECM) regulation under physiological stress have been previously studied, little attention has been paid to the three-dimensional microstructure and microenvironment of VICs. Our goal is first to develop a biaxial bioreactor system that will allow us to fix tissue at physiological loads and second, employ state-of-the-art electron microcopy combined with advanced image analysis techniques to characterize the three-dimensional ultrastructure of VICs.

METHODS

Tissue Preparation: Porcine hearts were acquired from a local abattoir. The MV anterior leaflet was isolated, mounted in the biaxial bioreactor specimen chamber, and fixed in 2.5% EM grade glutaraldehyde for 4 hours (n=3 per tension level). Fixed tissue was cut into small (1cm x 0.5cm) blocks, stained with osmium tetroxide, dehydrated with alcohol and acetone (microwave-enhanced method). Resulting tissue blocks were infiltrated in epoxy resin overnight and cured for 48 h in a 60°C oven. Samples were sectioned with a diamond knife (70nm), imaged with an FEI Tecnai transmission electron microscope to acquire a sample map, then placed in the focused ion beam scanning electron microscope for continuous milling and imaging.

Focused Ion Beam Scanning Electron Microscopy: Resin-embedded samples were adhered to a 45° aluminum stub, coated with carbon paint, sputter coated with 15 nm of Pt/Pd and placed in the FEI

Helios NanoLab 660 DualBeam system for block-face milling and imaging. Two datasets were acquired for each sample: the "Microenvironment Only" group focused on the collagen fibrils at a higher resolution and thus, had a slice thickness of 50nm and a horizontal field width (HWF) of 20.7 µm. The "Whole Cell" group captured the MVIC as well as its surrounding microenvironment: the slice thickness was equal to 200nm and the HFW was 53.5 µm. All samples underwent cross-section milling to prepare a flat surface. FEI's *Auto-Slice-and-View* software was used to collect serial SEM images from the freshly-milled surface without moving the sample.

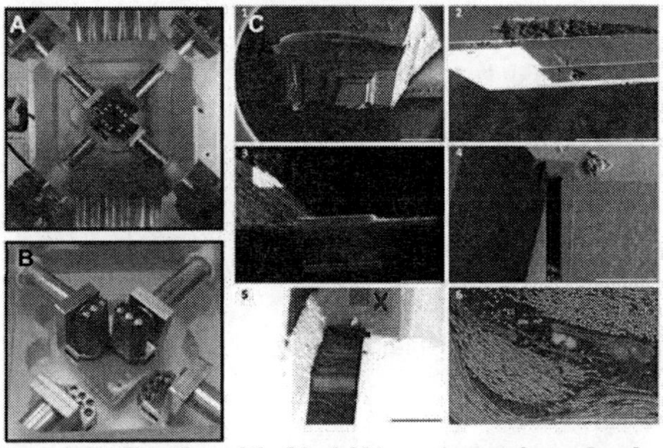

Figure 1. (A) Top view of the biaxial bioreactor specimen chamber. (B) Pin attachment system. (C) FIB/SEM workflow: 1) SEM view, 2) Side view, 3) Ion beam side view, 4) SEM image after cross-section milling, 5 and 6) SEM images of the samples after milling.

Image Processing: Resulting images went through a series of image processing steps before final volume rendering and analysis: (1) post-processing, (2) registration, and (3) segmentation. A custom python script with *Contrast Limited Adaptive Histogram Equalization* (CLAHE) and *Denoise* algorithms were applied. Images were registered using *Image J's StackReg* and *TrackEM* functionalities with a least square translational cost function [4]. Images were then loaded into *Simpleware ScanIP* as a 3D image for segmentation. Collagen, elastin, and VIC cytoplasm and nuclei were extracted using a combination of segmentation tools, such as flood fill and threshold.

3D Volume Rendering and Analysis: 3D volume rendering and subsequent analyses were performed in *Simpleware ScanIP*. The "Microenvironment Only" dataset was used to characterize the microenvironment. As such, the analysis focused on volume and surface areas of collagen and elastin, as well as the packing density, average number of fibrils, and diameter of collagen fibers and fibrils. The "Whole Cell" dataset was used for analysis of cell dimensions and interaction of the cell with the surrounding microenvironment.

RESULTS

3D reconstructions from FIB/SEM images (Fig. 2 and 3) allowed us to quantify the collagenous VIC microenvironment. The average collagen fibrillar diameter is 61.2 ± 2.01 nm with an average of 605 ± 113 fibrils per fiber. Collagen fiber bundles have an average surface area of 227 ± 47 μm^2 in the unloaded configuration.

Resulting data from 3D reconstructions was then used to quantify cellular morphology and subsequent deformation under physiological loads. Improving the osmium staining protocol will result in enhanced staining of collagen fibrils, which will allow the quantification of the cellular microenvironment—particularly, the circumferentially oriented collagen fibrils surrounding the VICs: (1) transmural-radial representation of fibrils from which the diameter of each can be determined as well as the distance from the cell membrane, (2) transmural-circumferential representations of fibrils are used to quantify the collagen fibril length and degree of crimp in both the loaded and unloaded configurations.

Figure 2. Volume Rendering of the MVIC microenvironment only from serial FIB/SEM data with 50nm sections (Grey: elastin, Blue: collagen fiber bundles, with individual fibrils identified).

Figure 3. Volume Rendering of an MVIC and its microenvironment.

DISCUSSION

The technique outlined above allows us to capture the previously unknown VIC geometry and surrounding microenvironment with a high level of detail. An advantage of serial TEM over single section TEM and other microscopy techniques is that it allows us to track individual collagen fibrils, which have an average diameter of 300 nm. This study highlights the previously unrecognized complexity of the interconnection between VICs and ECM fiber networks and enables us to develop more structurally accurate computational models that incorporate the heterogeneous complexity of the mechanobiologically-stimulated microenvironment.

ACKNOWLEDGEMENTS

National Institute of Health grant R01HL119297 to MSS and American Heart Association Pre-Doctoral Fellowship PRE33420135 to SA.

REFERENCES

[1] Taylor et al. *Int J Bioc & Cell Bio*, 113-8, 2003.
[2] Lee et al. *J Theor Biol,* 373: 26-39, 2015.
[3] Go et al. *Circulation*, 129:28-292, 2014.
[4] Schindelin et al. *Nature Methods*, 9(7): 676-82, 2012.

BIOLOGICAL TISSUES SHOW POROELASTIC AND VISCOELASTIC BEHAIVIOUR AT DIFFERENT FREQUENCY SPECTRUMS

Ramin Oftadeh (1), Alan J. Grodzinsky (1,2,3)

(1) Department of Biological Engineering
Massachusetts Institute of Technology
Cambridge, MA, USA

(2) Department of Mechanical Engineering
Massachusetts Institute of Technology
Cambridge, MA, USA

(3) Department of Electrical Engineering and Computer Science
Massachusetts Institute of Technology
Cambridge, MA, USA

INTRODUCTION

Soft connective tissues experience a wide range of loading rates and frequencies, from physiologic up to high rate impact injury. Frequency-dependent tissue behavior may generally be governed by fluid-solid frictional dissipation and fluid pressurization associated with poroelasticity, and by flow-independent viscoelastic mechanisms. Viscoelastic rate processes are often associated with stretching and sliding of collagen fibrils (and other macromolecules) within tissue ECM, while poroelastic processes are the result of deformations leading to local compressive loading and f
luid flow [1]. However, the mechanical behavior of many tissues has not been explored over a full frequency spectrum relevant to functional loading. Here we hypothesize that multi-scale analysis associated with AFM-based nanoindentation may enable identification and separation of viscoelastic and poroelastic rate processes that can occur within the same tissue at different loading frequencies. We first develop a finite element poro-viscoelastic model for wide bandwidth nanoindentation, and then test the ability of the model to predict poroelastic and viscoelastic tissue responses of mouse skin subjected to nanoindentation over a 4-decade frequency range.

MATERIALS AND METHODS

Theoretical Model: A fibril-reinforced poro-viscoelastic finite element model was implemented to determine the tissue mechanical properties measured via AFM nanoindentation using a spherical probe tip of radius R with tip-tissue contact length d (Fig. 1A) . This model is an extension of that used previously for poroelasticity of cartilage [2] by adding viscoelastic capability. The model (Fig. 1B) consists of an isotropic nonfibrillar matrix (representing non-collagenous ECM) and a fibrillar network (e.g., collagen). The model is created in ABAQUS software using its soil mechanics capability. The tissue poroelastic

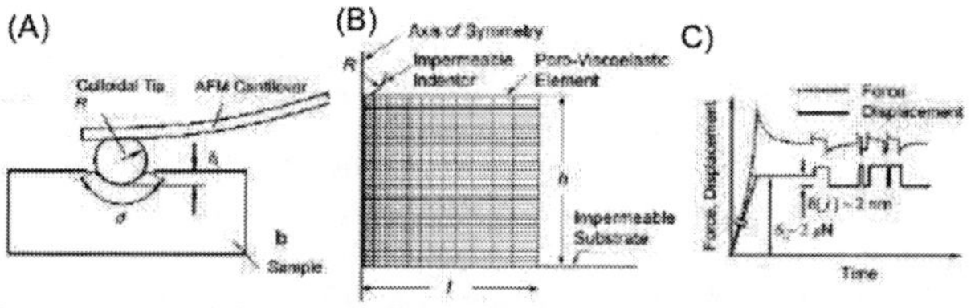

Figure 1. (A) Schematic of compressive deformations (B) Finite element model (C) Indentation loading profile

mechanical properties include the nonfibrillar elastic modulus E_m, hydraulic permeability k, fibrillar modulus E_f and Poisson's ratio ν. Viscoelastic properties are represented by a standard three element linear solid consisting of the same modulus E_m in parallel with a Maxwell solid (i.e., a modulus E_v in series with dashpot η, such that the viscoelastic relaxation time is $T = [\eta/E_v]$. Parameter values are obtained from a best fit of the model to the magnitude $|E^*|$ and phase angle ϕ of the dynamic complex modulus measured in frequency domain (see Results below). **Experiments:** Cylindrical disks of dermal skin, 9mm diameter, were harvested from mature (4 months old) mice and cut to 0.5 mm thickness using a tissue slicer. Dynamic nanoindentation tests were performed using a wide bandwidth rheology system coupled to a commercial AFM (MFP-3D, Asylum , CA) [3]. Polystyrene colloidal probe tips with end radii either 3 μm or 25 μm were attached to tipless cantilevers with nominal spring constant k~7.4 N/m (Fig. 1A). A second piezo increased the system's high frequency response to ~10 kHz (compared to the commercial AFM's intrinsic high frequency limit of ~100 Hz). A preindentaion of $\delta_0 \sim 1 - 2\ \mu m$ was first applied (Fig. 1C). After stress relaxation, random binary sequence (RBS) displacements (~2 nm amplitude) were then applied (Fig. 1C). The magnitude and phase of the dynamic complex indentation modulus versus frequency were computed from the measured force and applied displacement using the fourier transform built in function "efte" in MATLAB.

RESULTS

Theory: The FE model exhibits both poroelastic and viscoelastic relaxation times, $[d^2/E_m k]$ and T, respectively. In the frequency domain, these relaxation times are manifested as peaks in the phase angle of the complex modulus (e.g., Fig. 2, which exemplifies data from mouse skin experiments). Both poroelastic (blue) and viscoelastic (red) components of the phase superpose to give the total phase (violet) in Fig 2. With the viscoelastic relaxation time fixed at 0.08 sec and the hydraulic permeability decreased from 7e-13 to 7e-15 [m^4/N.s.] (Fig. 2A,C,E), the peak of the poroelastic component of the phase (ϕ_P) shifts sequentially to lower frequencies (blue), while the peak in phase of the viscoelastic component (ϕ_v) is unchanged (red). (These parameter values were in the range obtained from mouse skin.) Decreasing hydraulic permeability can make the two peaks merge into one broader peak (Figs. 2 C,E). In contrast, with k held constant [7e-13 m^4/N.s.] and the viscoelastic relaxation time increased from 0.01 to 5 sec (Figs. 2B,D,F), the viscoelastic phase peak shifts to lower frequencies but the poroelastic peak is unchanged. Thus, the model correctly demonstrates that poroelastic viscoelastic behavior are independent of one another and can superpose. The overall poro-viscoelastic response (both magnitude and phase) also vary with preindentation depth δ_0 and tip radius R (Fig. 3). Increasing both parameters results in shifting the poroelastic phase peak (blue) to lower frequency: this further confirms that the poroelastic peak frequency is geometry-dependent and is related to k, E_m and d^2 by $f_{peak} \propto kE_m/d^2$ [4]. However, viscoelastic behavior is not geometry-dependent and the location of the viscoelastic peak (red) does not change.

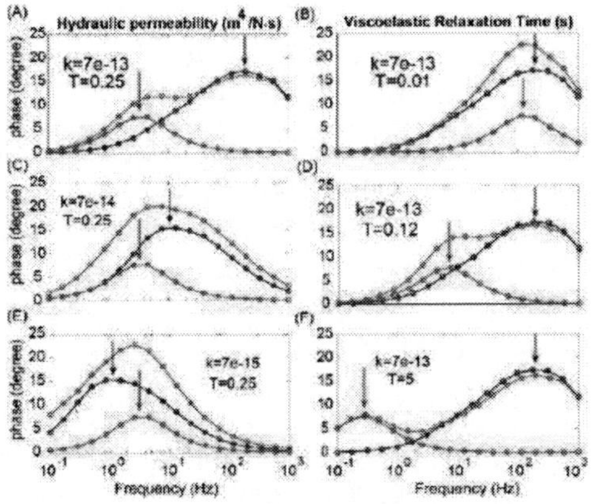

Figure 2. FEM model predictions (Left) Hydraulic permeability (Right) viscoelastic relaxation time

Experiments (Fig. 4): To obtain model parameters from the data, the low frequency asymptote of the measured $|E^*|$ was taken to be the equilibrium modulus E_m (varying other model parameters did not affect this low frequency asymptote). After fixing E_m, the value of $|E^*|$ corresponding to viscoelastic peak was taken to be E_v, the transient elastic modulus.

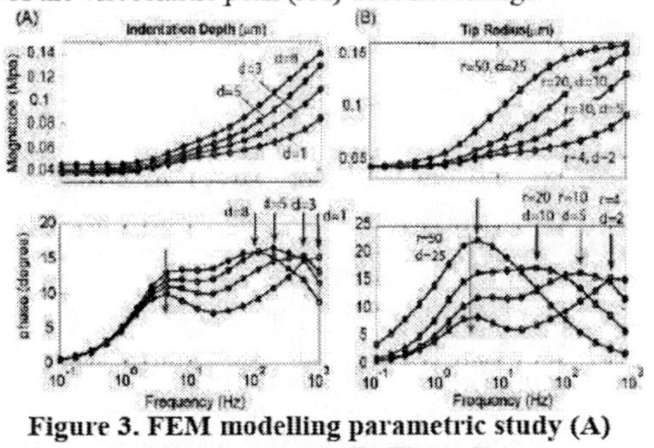

Figure 3. FEM modelling parametric study (A) indentation depth (B) Tip radius

Then, with both E_m and E_v fixed, the value of E_f, the fibrillar modulus, was varied so that the high frequency asymptote of the model matched that of $|E^*|$. Values of k and T were obtained by matching the frequencies of poroelastic and viscoelastic peak angles of the FE model to those of experiments, respectively. (A parametric study showed that varying ν in the range $0.1 < \nu < 0.4$ did not significantly affect the frequency response; we therefore used $\nu = 0.1$). Two different probe tips were used ($R = 3\ \mu m$ and $25\ \mu m$) and the indentation depth was held at ($\delta_0 \sim 1\ \mu m$). More than 10 tests were performed at each location. The finite element model was fit to the experimental results as described above, and presented in Fig. 4 as only poro (blue), only visco (red) and total poro-viscoelastic (violet) response. (k was found to be 3.1 e-13$\pm$ 0.23 e-13 m^4/N.s; T was 0.67 $\pm$ 0.11 s.) Two distinct phase peaks were observed with tip radius R=3μm; the poroelastic peak shifted to lower frequency with R=25 μm, while the viscoelastic peak did not change.

DISCUSSION

The poroelastic and viscoelastic behavior of connective tissues have been analyzed and reported extensively at both macro and micro scales including studies of cartilage [3, 5], tendon [6, 7] and skin [8, 9]. It has not been obvious how to quantitatively identify and separate poroelasticity versus viscoelasticity within tissues when both are present simultaneously. In this study, AFM-based dynamic nanoindentation was used in conjunction with poro-viscoelastic finite element modeling to measure and predict the frequency dependent nanomechanical behavior of murine skin samples. We found a peak in the phase angle of the complex modulus that shifted with frequency when AFM tip radius was altered, in a manner consistent with the specific length scaling of poroelastic behavior. Thus, at the nanoscale, poroelastic behavior is dominant at higher frequencies (>100 Hz) whereas viscoelastic behavior appears dominant in the lower frequency spectrum. These results exemplify tissues in which poro and viscoelastic responses are independent and can jointly govern the overall mechanical behavior. However, poroelastic behavior scales in a unique manner with the square of the characteristic length for fluid motion, while viscoelastic behavior does not.

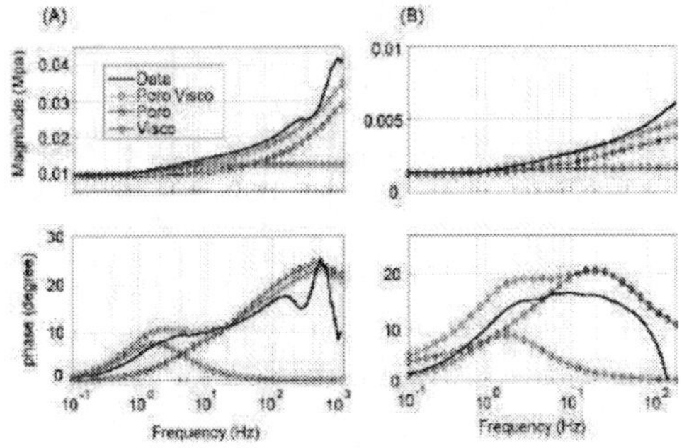

Figure 4. Mouse skin oscillatory nano indentation (A) R=3 μm (B) R=25 μm

ACKNOWLEDGEMENTS: Supported by NSF Grant CMMI-1536233. The authors thank Drs. Brianne Connizzo and Hadi Nia for their comments and technical expertise.

REFERENCES

[1] Grodzinsky, A., Fields, Forces and Flows in Biological Systems, Garland Science, 2011. [2] Nia, H.T et al.. Biophys J, 101:2304-2313, 2011. [3] Nia, H.T et al., Biophys. J, 104:1529-1537, 2013. [4] Mahaffy, R., et al., Biophys. J., 86:1777-1793, 2004. [5] Mak, A, J Biomech Eng, 1986. 108:123-130. [6] Atkinson, T S et al, J Biomech, 32:907-914, 1999. [7] LaCroix, A.S., et al., Ann Biomed Eng, 41:1120-1128, 2013 [8] Crichton, M.L et al, Biomaterials 34:2087-2097, 2013. [9] Grant, C.A et al, Acta Biomater, 2012. 8(11): p. 4123-4129.

SB^{3}C2017
Summer Biomechanics, Bioengineering and Biotransport Conference
June 21 – 24, Tucson, AZ, USA

A NOVEL SMALL-SPECIMEN PLANAR BIAXIAL TESTING DEVICE FOR INVERSE MODEL VALIDATION OF SOFT TISSUES

S. Potter (1), J. Graves (2), B. Drach (3), T. Woodard (2), T. Leahy (2), C. Hammel (2), A. Feng (2), A. Baker (2), M. Sacks (1,2)

(1) Center for Cardiovascular Simulation
Institute for Computational Engineering and Sciences
Department of Mechanical Engineering
University of Texas at Austin
Austin, TX, USA

(2) Center for Cardiovascular Simulation
Institute for Computational Engineering and Sciences
Department of Biomedical Engineering
University of Texas at Austin
Austin, TX, USA

(3) Department of Mechanical and Aerospace Engineering
New Mexico State University
Las Cruces, NM, USA

INTRODUCTION

A fundamental goal in constitutive modeling is to predict the mechanical behavior of a material under a generalized loading state. For both natural biological tissues and tissue-derived soft biomaterials, there exist many physiological, surgical, and medical device applications where rigorous constitutive models are required. Typically, constitutive models have been developed by direct mechanical testing through *in vitro* experiments. In contrast, inverse models are generated from *in vivo* measures of tissue behavior and represent an exciting tool for developing constitutive models without the invasiveness and inaccuracy inherent in excising tissue for *in vitro* testing. As with any class of model, however, inverse models must be verified against experimental data. Planar biaxial testing remains a common method for characterizing the complex mechanical behavior of soft tissues. As new inverse models are developed they must be evaluated against careful, rigorous biaxial experimentation. These experiments must cover all relevant deformation modes and collect data for all mechanically distinct regions of a tissue specimen.

Towards this end, we have recently developed a new biaxial testing system that is able to control full in plane deformation, test small specimens (3-4 mm per side) in order to perform regional verification of inverse models, and apply deformation homogeneously. The data that this device can provide will be instrumental in the verification of constitutive inverse models of biological tissues.

METHODS

Device requirements. For the device to achieve its goal of collecting experimental data for the validation of inverse models for soft tissue, its performance must meet the following minimum metrics. First, the device must be able to fully prescribe and control in-plane deformation for all deformation modes, including shear, in real time. Consequently, the device must be able to measure specimen strain in real time and have a control scheme that is flexible enough to account for unpredicted response to induced displacement at the specimen boundary. The second performance requirement for the device is the capability to generate a homogenous strain field within the inner region of interest of the sample. Without this homogenous strain field, material parameters determined for constitutive modeling are unreliable. This need for homogenous strain field places emphasis on the need for symmetric, well placed specimen attachment points[1]. Beyond applying fully controlled, homogenous strain fields, the device must also be able to test small specimens (3-4 mm a side). This requirement is driven by the need for gathering data on regional properties for inverse model verification. Finally, the device must be able to subject the sample to the full range of physiological loading. Stress values during biaxial tests of biological tissue generally range up to 1.4 MPa for normal stress and between -600 kPa and 600 kPa for shear. Correspondingly, the Green's strain in the normal direction range up to 0.25 along the axis aligned with the tissue fibers and 0.4 in the direction perpendicular to the tissue fiber axis. Green's strains for shear range between -0.2 and 0.2[2].

Device Design. An overview of the test system as seen from above is shown in Figure 1. The device used the following solutions to the design specifications described above. Twelve individually actuated needle attachment points are used to apply strain to the sample. The individual actuation allows the device to prescribe all relevant in-plane deformation modes, while the use of needles allows for small samples to be tested. To improve strain field homogeneity, a custom aluminum jig is used to guarantee the attachment arms are properly positioned when mounting the specimen. To measure strain in real time, four fiducial markers, defining a 1 mm^2 region, are glued to the sample. These markers are then imaged at a rate of 100 Hz during testing. The images are passed to a custom LabView (National Instruments

Corporation, Austin, TX) program that runs a sliding mode control (SMC) algorithm[3] to apply the user prescribed deformation. The SMC scheme was chosen for its ability to handle coupling between specimen axes and adapte to unknown tissue material properties.

Device Evaluation. Fresh porcine hearts were acquired from an abattoir. The hearts were brought to the lab no more than 1 hour after excision. Upon arrival, the aortic valve leaflet (AVL) was isolated, submerged in PBS and stored at 4° C. No more than 1 hour passed before square specimens from the basilar and nodular areas of the belly region were stamped out with razor blades. The locations from which the specimens were obtained are shown in Figure 3. The square specimens' circumferential and radial dimensions (L1, L2) were measured with digital calipers under a microscope. Thickness (L3) measurements were made using a micrometer under no optical assistance. The specimen from the basilar region was measured to be 5.25 mm x 4.58 mm x 0.674 mm along the circumferential, radial and longitudinal directions respectively. The specimen from the nodular region was measured to be to be 4.83 mm x 4.23 mm x 0.807 mm, likewise. The specimens were marked with fiducial markers as described above.

Samples were mounted on the device so that the tissue preferred fiber direction was aligned along one of the test axis. For both regions, a load control procedure was used to determine strain ranges which would produce a 60-90 N/m tension response along the axes. The strain ranges from the load tests was then used to determine initial strain ranges for subsequent strain control tests. The actual and prescribed fiducial markers were tracked during testing and stored for later analysis, including strain field analysis with a four node, bilinear finite element.

RESULTS

Device Performance. The positions of the actual vs. prescribed locations of the fiducial markers were plotted together. The results can be seen in Figure 2A and Figure 2C. Note that in both the case of biaxial stretch (Figure 2A) and more complex deformation cases like pure shear (Figure 2C), the device enforces the requested deformation mode with minimal tracking oscillation through the duration of the 40 second test. The results of the strain field analysis with the four node, bilinear finite element formed from the fiducial marker position data show that the device is capable of achieving homogenous in-plane strain. Figure 2B demonstrates this capability in a biaxial stretch load condition, while Figure 2D shows a pure shear scenario. In each figure, the corners of the plots represent the nodes of the bilinear finite element and the color map is an interpolated calculation of the deformation. Solid colored plots indicate highly homogenous strain.

AVL Results. As seen in Figure 3, the basilar and nodular regions of the AVL have markedly different mechanical responses. Moreover, the response of the tissue along the fiber preferred direction (circumferential direction) is significantly stiffer than in the perpendicular, radial direction. These results highlight the broad range of tissue responses in the AVL as a function of fiber orientation and location within the leaflet.

DISCUSSION

The device described above has been shown to meet the requirements needed to obtain experimental data for verification of inverse models of soft, biological tissues. The device can handle specimens of small size and apply fully controlled in-plane strain with a high degree of homogeneity. Further, the device can apply and control all relevant deformation modes, e.g. shear, needed for inverse model refinement. The device's performance was verified with tests on porcine aortic valve leaflet and the results are within required specifications.

While the results are promising, the device has been only verified against a small set of tissue types. Future work will continue verifying the device's performance and then begin to use if for collection of data for inverse model verification. Further improvements on the strain field measurement technique will be implemented by integrating digital image correlation (DIC) techniques into the existing test system.

In summary, this device represents an novel tool with exciting potential to build the data set necessary for developing robust inverse and constitutive models of soft, biological tissues.

ACKNOWLEDGEMENTS

This work was funded by the National Institute of Health (NIH) grant R01HL119297.

Figure 1. Top view of testing device.

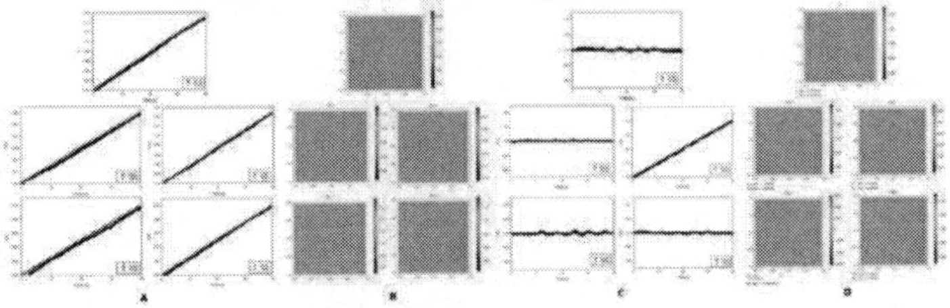

Figure 2. (A) Ability of device to enforce prescribed deformation for biaxial stretch. (B) Plots of the interpolated strain field of the specimen during biaxial stretch. (C) Ability of device to enforce prescribed deformation for pure shear. (D) Plots of the interpolated strain field of the specimen during pure shear.

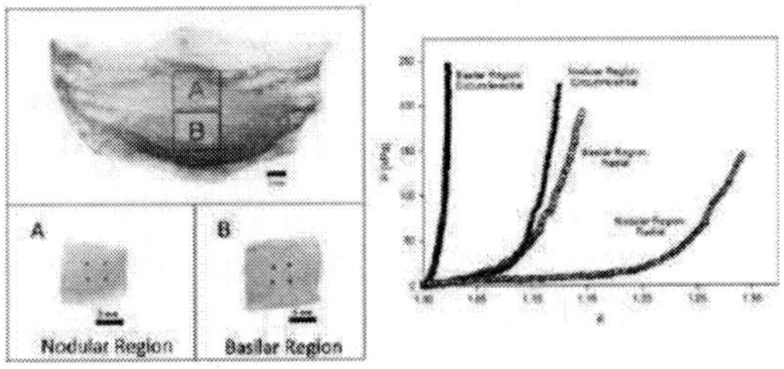

Figure 3. Regional and directional differences of the mechanical response of a porcine aortic valve leaflet.

REFERENCES

1. Eilaghi, A. et al. *J. Biomech. Eng.* **131,** 91003 (2009).
2. Sacks, M. S. *J. Elast.* **61,** 199–246 (2000).
3. Drakunov, S. V. & Utkin, V. I. *Int. J. Control* **55,** 1029–1037 (1992).

THE EFFECT OF FLOOR STIFFNESS ON STANDING POSTURE AND SWAY

Daiane Aizen Grill (1), Sara E Wilson (1)

(1) Department of Mechanical Engineering
University of Kansas
Lawrence, KS, United States

INTRODUCTION

LBP and lower extremities' discomfort have been associated to occupations requiring standing for long periods of time [1]. Several studies have investigated the effect long-term standing has on a person's perceived discomfort, on measurable biomechanical variables, and on the development of LBP [2]. However, few studies have been performed in order to analyze the effect material properties of the floor have on posture and sway. In a study by Cham and Redfern (2001), different flooring conditions (mostly viscoelastic foams) were examined [3]. In that study, increases in center of pressure (COP) weight shifts were associated with increased discomfort as subjects. COP lateral weight were found to have a nonlinear relationship with floor stiffness in which an ideal flooring condition appearing to minimize this biomechanical parameter. Another study suggested that the probability of developing LBP while performing long-term standing is not a function of the standing time, rather it is dependent on the posture adopted by the subject on the first 15 minutes of the standing period [4]. The objective of the current study is to investigate whether standing posture or COP parameters vary as a function of axial floor stiffness.

METHODS

Twenty-eight (28) healthy individuals (13F, 15M) participated in this experiment. All subjects voluntarily signed a consent form prior to the commencement of the study, as well as filling a health questionnaire. Subject were recruited from University settings and their ages ranged between 18 and 30 years old. Subjects' height ranged between 1.52m and 1.93m (mean 1.65 ± 0.09m). Weights ranged between 52 kg and 104 kg (mean 70 ± 13kg). 8 subjects reported LBP in the past year. The experimental procedure was IRB approved.

Subjects were asked to stand in an upright position in 5 different flooring conditions of differing stiffnesses including a high stiffness control condition. Data was collected in five 1-minute intervals per flooring condition. Between intervals, subjects were asked bend their knees and bow forward. During the minute intervals, subjects were asked to stand in a comfortable stance and avoid movement. Subjects performed the study bare-footed. The order of the flooring conditions was randomized and subjects were asked to sit between the testing of each floor.

A variable-stiffness platform was constructed consisting of a double-layered surface with four identical springs between the layers. The springs were held in place by steel guiding rods. The platform's stiffness was modified through the exchange of these springs. The overall stiffness of each flooring condition was 75.59 kN m^{-1}, 100.79 kN m^{-1}, 123.18 kN m^{-1}, and 139.98 kN m^{-1}. The control flooring condition was achieved by placing a wooden block (assumed to be relatively inelastic) between the platform's layers to prevent deformation of springs. Stiffnesses were chosen to span stiffnesses seen in elastic flooring such running tracks and gymnastics/tumbling flooring that have stiffnesses of 50-400 kN m^{-1} [5].

The measuring equipment used consisted of electromagnetic motion sensors (MotionStar, Ascension Technologies, VT) and a force plate (Bertech, Columbus, OH). The variable stiffness platform was placed on top of the force plate and the former was zeroed for every flooring condition. A total of 7 magnetic sensors were placed on each subject: one over the seventh cervical vertebrae, one over the twelfth thoracic vertebrae, one over the first sacral vertebrae, and one half-way through each thigh and lower leg. Data was collected for 60s at a rate of 100 Hz.

The biomechanical parameters collected by the electromagnetic sensors were of position and rotation of each sensor while the force plate collected COP. Additionally, Motion Monitor software was used to digitize the knee and hip in order to obtain knee flexion angles and hip flexion, rotation and abduction angles. The postural variables measured include thoracic flexion, lumbar flexion, hip flexion, lateral rotation, hip torsion, and knee flexion. Additionally the range and standard deviation (SD) of the COP in the anterior-posterior (AP) and lateral (LAT) directions were assessed. A repeated measures ANOVA was performed. A significance level of 0.05 was used. Within-subjects contrasts were assessed for the results of each flooring condition compared to the control for those conditions that were found significant in the ANOVA.

RESULTS

No statistically significant differences were found for the following postural parameters: thoracic flexion, lumbar flexion, hip flexion, hip lateral rotation, hip torsion and knee flexion. The minimum p-value was of 0.096 and the maximum p-value was of 0.978. This suggests that posture did not generally change significantly as a function of floor stiffness.

Statistically significant patterns were found for the results in the range and standard deviation of the COP positions in the AP direction; and for the COP position range in the LAT direction. The results for the ANOVA performed on these variables is shown in Table 1. The variation in the position of the COPs is expressed in terms of the standard deviation exhibited by each sample.

Table 1: ANOVA Results for COP in x and y directions

Variable	p	Within-Subjects Contrast of Floors to Control			
		75.59 kNm^{-1} vs control	100.79 kNm^{-1} vs control	123.18 kNm^{-1} vs control	139.98 kNm^{-1} vs control
COP AP SD	**0.025**	**0.036**	**0.020**	**0.040**	0.064
COP AP range	**0.022**	0.064	**0.003**	**0.013**	0.075
COP LAT SD	0.068	**0.002**	**0.006**	0.200	**0.006**
COP LAT range	**0.009**	**0.002**	**0.002**	0.054	**0.017**

The results for the COP range and SD in the AP direction are shown in Figure 1. The points represented by a red diamond showed a statistically significant difference to the control flooring condition. The range in AP COP increased with floor stiffness to the third flooring condition, above which the COP range dropped (p=0.022). Similarly, the variation in the COP in the AP direction showed an increase in magnitude with stiffness up until the second flooring condition, after which it dropped in the third and fourth conditions (p=0.025).

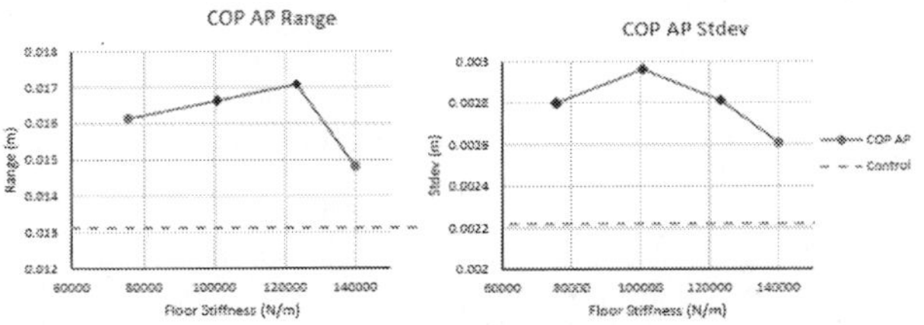

Figure 1: Position Range and Standard Deviation for the COP in the AP Direction

Figure 2 shows a decreasing trend in LAT COP with increasing floor stiffness (p=0.009). The points represented by a red diamond were the results that showed a statistically significant difference to the Control flooring condition.

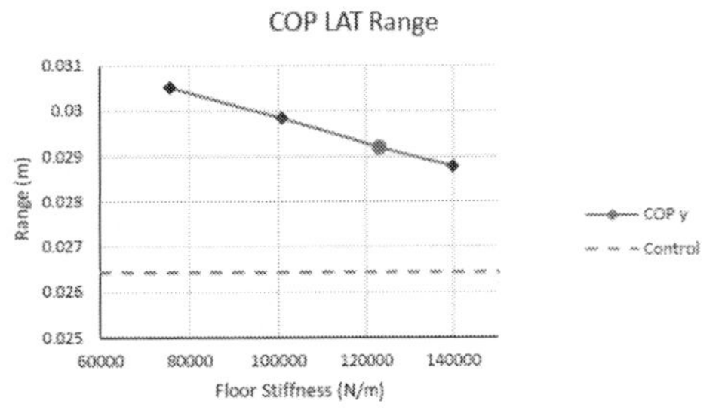

Figure 2: Variation in the COP Range in the LAT Direction

DISCUSSION

The results obtained showed that there are no significant changes in the postural parameters due to floor stiffness but did show changes in the COP sway. In particularly, we observed greater sway with decreasing floor stiffness. In the AP direction, a peak sway was observed between the 100 and 123 kNm^{-1} floor stiffnesses. This range of stiffness is around that currently used in gymnastics floors and running tracks to improve performance and reduce injury. Greater sway may be useful in varying posture and reducing discomfort [5].

Unlike this study, Cham and Redfern found that greater COP weight shifts were exhibited by harder floors which the authors attributed to an increase in discomfort and fatigue [3]. Weight shifts are different than sway. In this study the measurement period was short and subjects were asked to maintain the same posture, so weight shifts are less likely to have occurred.

In this study, subjects stood for only 5 minutes on each floor. Future studies should examine extended standing. Such studies could also examine perceived levels of discomfort and its association with COP sway.

Additionally, Rys and Konz found that posture is dependent of the position of the feet during standing [6]. If the movement of the COP in the AP direction is discovered to be associated with perceived discomfort, the relationship between floor stiffness and feet placement during standing should be studied.

ACKNOWLEDGEMENTS

We would like to acknowledge the generosity of the Undergraduate Research Award Program at the University of Kansas for their financial and logistic support.

REFERENCES

1. National Institute for Occupational Safety and Health, 1997 "Musculoskeletal Disorders and Workplace Factors - A Critical Review of Epidemiologic Evidence for Work-Related Musculoskeletal Disorders of the Neck, Upper Extremity, and Low Back" DHHS 97-141.
2. Redfern, M. and Cham, R., 2000, AIHAJ, 61 (5), pp. 700-708.
3. Redfern, M. et al 2001, Human Factors, 43 (3), pp. 381-391.
4. Gregory, D. et al 2008, Gait & Posture, 28, pp. 86-92.
5. Nigg, B.M., et al, 2000, Biomechanics and Biology of Movement, Human Kinetics. Champaign, IL.
6. Rys, M. et al, 1994, Ergonomics, 37 (4), pp. 677-687.

SB^{3}C2017
Summer Biomechanics, Bioengineering and Biotransport Conference
June 21 – 24, Tucson, AZ, USA

PRINCIPAL COMPONENT ANALYSIS OF GAIT AND CYCLING EXPERIMENTS: CROSSTALK ERROR REDUCTION AND CORRECTED KNEE AXES

Jordan M. Skaro (1), Harsh Goel (1), Scott J. Hazelwood (1,2), Stephen M. Klisch (1,2)

(1) Mechanical Engineering Department
California Polytechnic State University
San Luis Obispo, CA, USA

(2) Biomedical Engineering Department
California Polytechnic State University
San Luis Obispo, CA, USA

INTRODUCTION

Crosstalk is a leading source of error in motion analysis [1-2]. Due to incorrect flexion axis direction that develops from marker placement error, crosstalk results in a strong, anatomically incorrect correlation between flexion-extension (FE) and adduction-abduction (AA) motions [1-2]. Thus, crosstalk limits the ability of biomechanical models to reflect the "true" motion of the knee. Principal Component Analysis (PCA) has been proposed as a post-hoc correction for crosstalk in prior gait studies [1-2]; however, previous studies have not proposed a method to determine PCA corrected knee axes. Further, it is not clear how PCA should be implemented in motion analysis studies that involve several exercises, on the same subjects, involving a relatively high range of flexion angles.

The long-term goal of this study is to determine accurate knee kinematics in a variety of exercises performed by the same subjects. This study tests two hypotheses: (1) PCA corrects for crosstalk between FE and AA angles in gait and cycling and (2) PCA corrected knee axes are similar for gait and cycling. The aims are to (1) determine PCA corrected knee angles in gait and cycling for the same subjects and their corresponding FE-AA correlations, (2) develop and implement an algorithm for determining PCA corrected knee FE and AA axes, and (3) compare the PCA corrected FE and AA axes for the same subjects to determine if they are similar in gait and cycling.

METHODS

Subject Selection. Subjects were male (n=5) and female (n=1), 21-26 years of age, and non-obese. Subjects were screened for prior leg injuries or malalignment that could bias results. Subjects 1, 2, 4 and 5 were right leg dominant while subjects 3 and 6 were left leg dominant. Protocols were approved by Cal Poly's Human Subjects Committee to minimize risk to human subjects.

Experimental Procedure. An enhanced Helen Hayes marker set with retroreflective markers was used to determine kinematics. A ten-camera motion capture system and Cortex software (Motion Analysis, Santa Rosa, CA, USA) were used to record marker position and process kinematic data. Subjects stood motionless for a static trial to create virtual axes for body segments. Subjects walked across the load cell walkway leading with their dominant leg to capture a full gait cycle. Subjects then pedaled a stationary bicycle (LifeFitness LifeCycle GX, Rosemont, IL, USA) at 70 rpm for 15 seconds.

PCA Analysis. PCA was implemented to reduce crosstalk by conducting a coordinate system transformation of calculated knee angles that minimizes FE-AA correlations [3]. A covariance matrix $[\boldsymbol{S}]$ of the knee angle data was calculated as

$$[\boldsymbol{S}] = \frac{1}{n-1}[\boldsymbol{X}_{centered}]^T[\boldsymbol{X}_{centered}] \quad (1)$$

where $[\boldsymbol{X}_{centered}]$ is the original knee angles, $[\boldsymbol{X}]$, with the means of each knee angle subtracted. An eigendecomposition of matrix $[\boldsymbol{S}]$ was calculated to produce a matrix of column eigenvectors, $[\boldsymbol{P}]$, according to

$$[\boldsymbol{S}] = [\boldsymbol{P}]^T[\boldsymbol{X}][\boldsymbol{P}]. \quad (2)$$

Finally, the original knee angles, $[\boldsymbol{X}]$, were projected onto a new set of axes, as described by the eigenvectors in matrix $[\boldsymbol{P}]$. This results in the calculation of an $nx3$ matrix $[\boldsymbol{Z}]$ which contains PCA corrected FE, internal-external rotation (IR), and AA angles:

$$[\boldsymbol{Z}] = [\boldsymbol{X}][\boldsymbol{P}]. \quad (3)$$

The coefficient of determination (R^2) between FE and AA angles was used to quantify crosstalk both before and after PCA. Larger R^2 values indicate the presence of more crosstalk.

Calculating PCA Corrected Knee Axes. PCA corrected knee axes were determined by finding the axes that, when used with PCA corrected knee angles, resulted in thigh, shank and ankle positions that

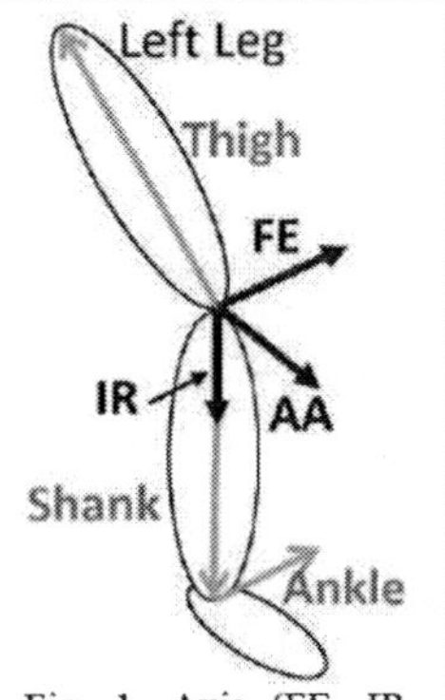

Fig. 1. Axis (FE, IR, AA) and body (Thigh, Shank, Ankle) vectors used in analysis.

were most similar to the corresponding positions determined by Cortex (which used the experimental marker data). A floating axis (i.e. AA axis) coordinate system was used [4]. In a local coordinate system, different for left and right leg dominant subjects, anterior and lateral directions were aligned with positive x- and y-directions, respectively, and inferior directions were defined as positive and negative for left and right leg dominant subjects, respectively (Fig. 1).

Statistics. Regression analyses were performed on FE vs. AA angles pre- and post-PCA treatment to assess FE-AA correlations in gait and cycling. Spherical directional statistics (i.e. Watson-Williams tests) [5-6] were used to assess for significant differences in the directions of the PCA corrected FE and AA axes between gait and cycling across all subjects. A one sample t-test was performed on the calculated angles between the corrected FE axes for gait and cycling to test if these angles were statistically similar to zero. For all statistical analysis tests, $p<0.05$ denotes statistical significance.

RESULTS

R^2 correlation values (Table 1) between FE and AA knee angles (Fig. 2) were reduced by 3 and 4 orders of magnitude for gait and cycling, respectively. Regression analyses found reduced correlations for gait FE-AA knee angles ($p=0.000$ for pre-PCA [strongly correlated] and $p=0.857$ for post-PCA [not correlated]) and for cycling FE-AA knee angles ($p = 0.000$ for pre-PCA and $p=0.956$ for post-PCA). The spherical directional statistical tests found FE (Table 2) and AA (Table 3) axes to be similar among subjects for gait and cycling ($p=0.289$ for FE and $p=0.259$ for AA). The one sample t-test on the angles between the corrected FE axis for gait and cycling showed significant differences from zero ($p=0.022$).

Table 1: R^2 values (mean ± 1 standard deviation) for FE-AA angles pre- and post-PCA correction for gait and cycling.

	Gait	Cycling
Pre	.66 ± .28	.21 ± .28
Post	2.5E-04 ± 2.7E-04	6.7E-05 ± 1.5E-04

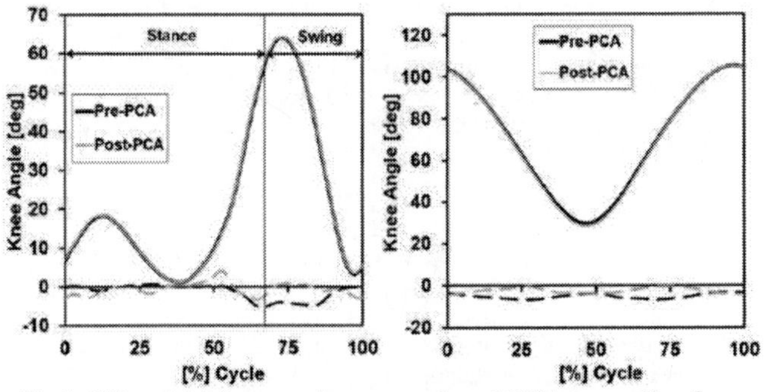

Fig 2: FE and AA knee angles pre- and post-PCA correction for gait and cycling for one subject. Solid/dashed lines = FE/AA.

Table 2: PCA corrected FE axes (in [x,y,z] format) and angle between axes (in degrees) for gait and cycling.

Subject	Gait	Cycling	Angle
1	[0.28,0.96,0]	[0,1,0]	16
2	[0.28,0.96,0]	[0.31,0.95,0]	2
3	[0.10,0.99,0]	[0.62,0.79,0]	32
4	[0.28,0.96,0]	[0.10,0.99,0]	10
5	[0.50,0.87,0]	[0.10,0.99,0]	24
6	[-0.17,0.98,0]	[0.62,0.79,0]	48

Table 3: PCA corrected AA axes (in [x,y,z] format) and angle between axes (in degrees) for gait and cycling.

Subject	Gait	Cycling	Angle
1	[0.96,-0.28,0.02]	[1.00,0,0.02]	16
2	[0.95,-0.27,-0.04]	[0.94,-0.31,-0.04]	2
3	[0.99,-0.10,0.08]	[0.79,-0.61,0.07]	32
4	[0.96,-0.27,0.05]	[0.99,-0.10,0.05]	10
5	[0.86,-0.50,-0.05]	[0.99,-0.10,-0.03]	24
6	[0.97,0.17,-0.13]	[0.78,-0.61,-0.05]	48

DISCUSSION

The statistical analyses demonstrated that there is substantial crosstalk between knee axes for gait and that PCA can correct for it. It is unsure if the corrected knee axes are similar between gait and cycling due to the large angles between corrected and uncorrected FE and AA axes. Similarities between those axes were found using the spherical statistical analysis while the one sample t-test indicates the FE axis between gait and cycling may be different. However, the spherical statistical test may find significant differences with more subjects. Other studies reported that a correlation exists between FE and AA angles at high flexion angles (> 60 deg.) [7]; thus, the predicted knee axes may be incorrect for cycling due to the high flexion angles measured in cycling (maximum flexion angles were 106 deg. in cycling and 62 deg. in gait). Thus, when using PCA to correct for crosstalk error in subjects performing exercises with high-flexion motions, these results suggest that PCA corrected axes from gait analysis may be used as a standard set of knee axes for other motions.

This study has several limitations. First, the number of subjects was relatively low; inclusion of additional subjects may lead to detected differences in the corrected knee axes for gait and cycling using the spherical directional statistics test. Second, methods were not used to reduce errors induced by soft tissue artifact, which is considered another leading source of error in motion analysis. Despite these limitations, this study has shown that PCA can be used to correct for crosstalk in both gait and cycling experiments and may be used to motivate further studies to determine the optimal method for reducing crosstalk when analyzing knee motion for subjects performing multiple exercises or high-flexion motions.

ACKNOWLEDGEMENTS

This work was supported by the Donald E. Bently Center (SK), W.M. Keck Foundation (JS), and the Defense Health Program (JS, SH, SK) through the Department of Defense Broad Agency Announcement for Extramural Medical Research Program #W81XWH-BAA-14-1 under Award No. W81XWH-16-1-0051. Opinions, interpretations, conclusions and recommendations are those of the authors.

REFERENCES

[1] Baudet, A. et al,. *PLOS ONE*,, 9:1-7, 2014. [2] Jensen, E. et al., *J Biomech*, 49:1698-1704, 2016. [3] Shlens, J., *arXiv*:1404.1100, 2014. [4] E. S. Grood et al., *J Biomech*, 105:136-144, 1983. [5] Watson, G. S et al., *Biometrika*, 43:344-52, 1956. [6] Watson, G. S et al., *Geophysical J International*, 7:153-59, 1956. [7] D. R. Wilson et al., *J Biomech*, 33:465-473, 2000.

DEVELOPMENT OF A HEAD IMPACT DEVICE FOR THE STUDY OF INDIRECT TRAUMATIC OPTIC NEUROPATHY

Elizabeth M. Konopacki, Yik Tung Tracy Ling, Thao D. Nguyen, Kaliat T. Ramesh

Hopkins Extreme Materials Institute
Mechanical Engineering Department
Johns Hopkins University
Baltimore, Maryland, United States

INTRODUCTION

The optic nerve is vulnerable to damage due to blunt trauma to the head as a result of road accidents, falls, military ballistic blast damage, etc. A patient with Indirect Traumatic Optic Neuropathy (ITON) reports reduced vision and a pupillary defect, regardless of a normal slit lamp examination, computer tomography (CT), and acute magnetic resonance imaging (MRI) of the optic nerve and canal [1]. This is in contrast to Direct Traumatic Optic Neuropathy, where the optic nerve axons are directly damaged and show symptoms of nerve avulsion, transection, and sheath hemorrhage [2]. Degree of blindness as a result of ITON is variable; however, it is still considered a significant cause of permanent visual loss, because approximately 50% of patients diagnosed have "light perception" or "no light perception" vision [3]. As of today, the cause of ITON is still unclear and there are no proven methods of treatment for this condition [4].

There are currently a number of devices developed for the investigation of traumatic brain injuries, including studies of blast-impact using a shock tube [5][6], the study of direct head impact using a piston arm [7], and studies of skull fracture due to ballistic impact [8][9]. However, there are a limited number of models that study head impacts with direct contact, and none that focus on Indirect Traumatic Optic Neuropathy.

It is hypothesized that frontal impacts on the head are amplified in the eye orbit such that the optic nerve is damaged more than other parts of the brain [4]. In addition, the displacement of the brain relative to the skull may increase the forces experienced in the visual pathway. Therefore, the objective of this study is to create an experimental setup that is able to measure mechanical quantities such as strain, pressure and acceleration in the optic canal and brain during impact. Then this data will be used to develop a computational model with the ability to simulate biomechanical loads on the optic nerve.

METHODS

Pendulum Impactor: To determine how impacts on the head affect the optic nerve, an experimental setup was designed with the ability to replicate impacts to the head of different velocities and energies. It utilizes a large, hinged pendulum arm that can be released from different angles to swing down and impact a cadaver head (Figure 1).

It is important to note that in order to study ITON, the head cannot be damaged during impact. If the skull fractures, then the desired indirect trauma has been compromised due to transection of the optic nerve. This was taken into account when determining the dimensions of the pendulum arm and frame. Previous studies have shown that the average energy of a boxing punch is 17.2 J [10], while the energy needed to cause a skull fracture ranges from 14.1-68.5J [11]. In addition, it was found that a velocity of 5m/s with a mass of 5.59lg induces a severe brain injury [12]. These values set the desired range for the effective energy of the impactor. Equations (1) and (2) were used to determine the corresponding range of velocity, v, and release height, h, of the arm.

$$\Delta E = \frac{1}{2}mv^2 \tag{1}$$

$$v^2 = 2gh \tag{2}$$

The vast majority of the device is composed of 8020 aluminum beams (80/20 Inc., Indiana). The T-slot design allows connections to be made anywhere along the axis of the profile, thus creating flexibility to modify aspects of the design such as the position of the impact point on the pendulum arm.

The neck is held in place with an iron stand that is bracketed to the aluminum frame at its four corners. The stand has a circular collar with four setscrews that have rubber disks at the end. These screws are tightened simultaneously from all four sides until the disks press firmly into the neck. This method secures the head in a complete upright position, but prevents shearing of the neck tissue around the setscrew once the head is impacted. A semi-circular inflatable cushion is placed around the upper neck to provide support for the jaw, while still allowing for natural motion of the neck upon impact (Figure 1).

Figure 1: Experimental set-up

Sensors: Four types of sensors were selected to measure mechanical quantities during impact to the cranium: triaxial accelerometers, accelerometers, pressure sensors, and strain gauges. The experimental configuration allows for one triaxial accelerometer, two regular accelerometers, two pressure sensors, and five strain gauges.

The triaxial accelerometer (356A01 PCB Piezotronics, NY) were placed on the tragus of ear to measure linear acceleration of the head and corresponding rotational acceleration at the head's center of gravity using equation (3), where C refers to the center of gravity and P to the point where the sensor was attached.

$$\vec{a}_C = \vec{a}_P = \vec{\alpha} \times \vec{r}_{PC} = \vec{\omega} \times (\vec{\omega} \times \vec{r}_{PC}) \quad (3)$$

The three uniaxial accelerometers (352A73 PCB Piezotronics, NY) interface with a simple constant-current signal conditioner (482C05 PCB Piezotronics, NY). They will be placed on the visual cortex and the impactor arm. The pressure sensors (138M156 PCB Piezotronics, NY) will be placed on the optic nerve to measure intracranial pressure. Finally, the strain gauges (c2a-06-015lw-120 Vishay Precision Group, PA) will be placed on the optic nerve and chiasm. The specifications are in Table 1 and the layout is shown in Figure 2. All data was acquired through LABVIEW program (National Instruments, TX).

Table 1: Sensor Information

Sensor	Measurement Range	Sensitivity	Width (mm)	Height (mm)
Tri-Axial Accel.	± 1000 g pk	5 mV/g	6.35	6.35
Accel.	± 1000 g pk	5 mV/g	4.1	8.6
Pressure	200 psi	25 mV/psi	1.78	2.54
Strain	± 3%	0.6 Ω	1.34	1.99

The strain gauges will be attached using cyanoacrylate adhesive (M-Bond 200, Vishay Precision Group, PA) bond after applying GC-6 Isopropyl alcohol on the optic nerve sheath. The pressure sensors and accelerometers will be embedded in orbital fat surrounding the optic nerve.

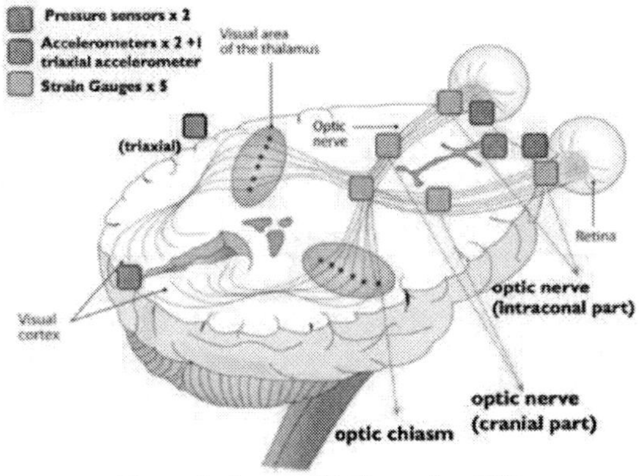

Figure 2: Sensor Configuration [13]

RESULTS

A pendulum head impact device was developed along with a data acquisition program that synchronizes data collected from a total of ten sensors. The impactor was made to allow adjustable impact velocity (0-6.6m/s) and impact energy (0-107.9J) by varying the launch height of the swing arm (0-2.2m). A neck-securing stand was also developed to control the starting position of each head and remain adjustable to accommodate variations in size. The device can study impacts from different directions (360 degrees) with variable magnitude of impact energy and acceleration.

Initial tests were conducted on a plastic head (ERP#1345, Sawbones, WA) with a surrogate neck using a frontal impact. The swing arm was released at a height of 0.91m, and recorded an acceleration of 5.67g's, which is within the theoretical range of 5.55-5.74g for a swing time between 0.55-0.59s.

DISCUSSION

The initial experiment on the plastic head was able to confirm the functionality of the pendulum impactor and data acquisition program; however, the plastic head model did not contain an interactive brain and optic nerve preventing accurate data measurement. Proceeding forward, a porcine head will be used in order to refine the methods of data collection and ensure functionality of the device. Then the impactor will be used on three cadaver heads, and the final data sets will be recorded and analyzed to verify the hypothesis and examine the causes of ITON. Future applications of the project include development of a computational model with the ability to simulate biomechanical loads on the optic nerve.

ACKNOWLEDGEMENTS

This study is funded by The Wilmer Pooled Professor Fund for the project title "The Biomechanics of Indirect Traumatic Optic Neuropathy: A Multimodal Approach" at Johns Hopkins Medical Institute. Elizabeth Konopacki would like to thank Dr. Eric Singman for the collaboration opportunity.

REFERENCES

[1] Singman, E et al., *Military Medical Research,* 3.1 2016, [2] Anderson, R et al., *Ophthalmology,* 89: 445-455, 1982, [3] Levin, L et al., *Ophthalmology* 106.7:1268-277, 1999, [4] Carta, A et al., *J Neurol Neurosurg Psychiatry,* 74:246-248, 2003, [5] Awad, N et al., *Arabian Journal for Science and Engineering,* 40.1:945-958, 2015, [6] Wang, H et al., *J Neuropathol Exp Neurol,* 75.2: 183-196, 2016, [7] Hardy, W et al., *Stapp Car Crash J.,* 51:17-80, 2007, [8] Rafaels, K et al., *Journal of Forensic Science,* 60.1:219-225, 2015, [9] Alexandropoulou, C et al., Health Science Journal, 4:225-36, 2010, [10] Walilko, T et al., *British Journal of Sports Medicine,* 39:710-19, 2005, [11] Yoganandan, N et al., *Neurotrauma,* 12(4):659-668, 2009 [12] Rezaei, A et al., *Int. Brain Injury Association,* 2014, [13] University of Minnesota, *Intro to Psychology*, 2010

EMG-DRIVEN INVERSE DYNAMIC ANALYSIS OF KNEE CONTACT FORCES DURING GAIT AND CYCLING USING OPENSIM

Megan V. Pottinger (1), Katherine Mavrommati (1), Scott J. Hazelwood (1,2), Stephen M. Klisch (1,2)

(1) Biomedical Engineering Department
California Polytechnic State University
San Luis Obispo, CA, USA

(2) Mechanical Engineering Department
California Polytechnic State University
San Luis Obispo, CA, USA

INTRODUCTION

Joint contact forces determine the loading experienced by cartilage tissue and, thus, may be used to predict risk of cartilage tissue damage and osteoarthritis (OA). Participating in low impact and/or non-weight bearing activities such as cycling may help reduce knee OA risk by limiting forces exerted during exercise [1]. Cycling is a common recommendation for rehabilitative or fitness sustainment exercise for select patients [1]. Although knee joint contact forces have been directly measured in gait and cycling using instrumented knee implants [2,3] and calculated in gait using EMG-driven analysis [4]; they have not been calculated in cycling using EMG-driven analysis.

The long-term goal of this study is to identify weight control exercises for overweight (OW) and obese (OB) subjects that minimize OA risk. This current study tests the hypothesis that knee joint contact forces are significantly lower during cycling than gait. The objectives are to: (1) conduct motion analysis experiments and EMG-driven OpenSim analyses for gait and cycling, (2) compare predicted tibiofemoral (TF) contact forces to published values, and (3) test for significant differences in maximum TF compressive forces in gait and cycling.

METHODS

Equipment. Kinematic data was collected using a 10-camera motion analysis system with Cortex software (Motion Analysis Corp., Santa Rosa, CA, USA). Gait experiments were conducted using 3 ground force plates (Accugait, AMTI, Watertown, MA, USA). Cycling experiments used a stationary bike (Lifecycle GX, Life Fitness, Schiller Park, IL, USA) retrofitted with custom pedals containing 6-axis load cells (AMTI, Watertown, MA, USA). EMG data was collected using 4 wireless EMG sensors (Trigno, Delsys, Natick, MA, USA).

Experimental Studies. Six subjects (average body mass index (BMI) = 25.0) aged 18-26 years and with no previous knee injury history participated. Protocols were approved by Cal Poly's Human Subjects Committee and were designed to minimize risk to human subjects. EMG sensors were placed on the following dominant leg muscles: lateral gastrocnemius, vastus lateralis, vastus medialis and semimembranosus. An enhanced Helen Hayes marker set with 32 retroreflective markers was used to track kinematics. First, the subjects performed 10 gait trials, 5 analyzing each leg, at self-selected walking speeds. Subsequent analyses used the middle three trials of the dominant leg. Then, each subject performed 3 cycling trials at a moderate machine resistance level (10) and 70 RPM. A static trial was captured at the end of the experiments for determining reference knee angles and scaling procedures in OpenSim (Stanford University, Palo Alto, CA, USA). Finally, body mass, height and Q-angle measurements were recorded.

Data Processing and Analysis. Trials were processed using Cortex to identify markers and create virtual markers to generate vectors of body segments. Kinematic and kinetic data were filtered (4th order Butterworth filter, cutoff frequency = 6 Hz) and exported to Matlab (MathWorks, Natick, MA, USA). EMG data were also exported to Matlab and time synced with the kinematic data. A bandpass filter of 20Hz to 450Hz [5] was applied to the time synced EMG data to create a Control Constraints file for muscle activation input to OpenSim [4].

OpenSim analyses proceeded with the following steps. (1) A full body musculoskeletal model was scaled to each subject [6]. (2) The Inverse Kinematics (IK) and Residual Reduction Algorithm (RRA) tools were used to output joint kinetic data and a corrected kinematic file. (3) An EMG-driven Computed Muscle Control (CMC) analysis was used to obtain muscle activations. (4) CMC results were used in Joint Reaction (JR) analysis to calculate knee joint contact forces. Results were normalized by body weight (BW) and trimmed to one full gait cycle of the dominant leg (0% = 1st heel strike, 100% = 2nd heel strike), and one full crank revolution (0 deg. = 1st top dead center, 360 deg. = 2nd top dead center).

Statistics. A paired t-test was used to compare the knee joint compressive contact forces for the two types of exercises (gait vs. cycling). Significance levels were defined by $p = 0.05$.

RESULTS

Results for TF compressive forces were more similar to publications for gait [4] than for cycling [2, 3]. For gait, average TF compressive contact forces (Fig. 1) closely resembled published results when subjects walked at 1.25 m/s [4]; the 6 subjects of this current study had an average walking speed of 1.32 m/s. For cycling, average TF compressive contact forces (Fig. 1) resembled published results [3], which reported knee forces during cycling at select power outputs. In particular, in [3] the reported maximum TF contact force at a power of 25 W was 0.6xBW, whereas in this study the maximum TF compressive contact force at an average power of 26 W was 0.75xBW. Also for cycling, our maximum TF contact forces were 27% lower compared to those reported in [2]. Peak compressive forces experienced during gait were 2.9xBW while for cycling they were 0.75xBW, a difference that was found to be significant (Fig. 2).

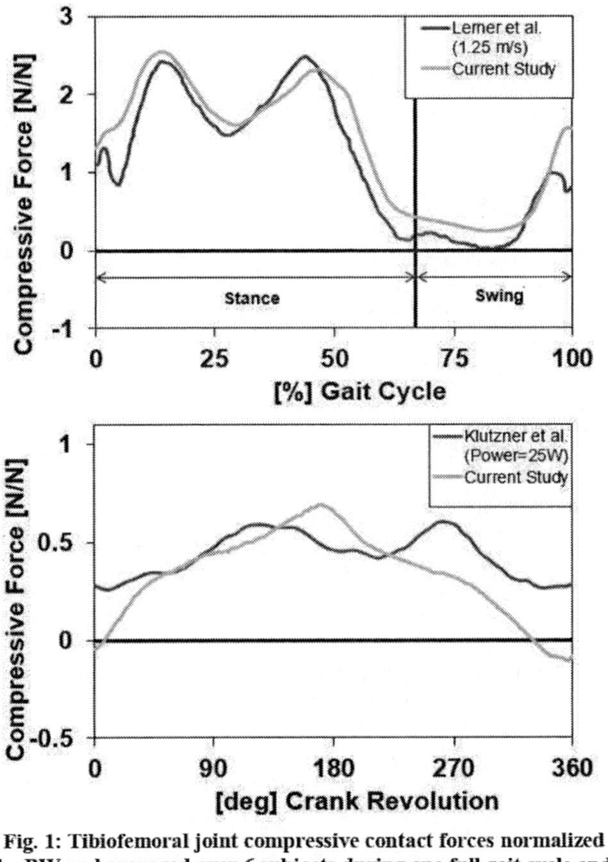

Fig. 1: Tibiofemoral joint compressive contact forces normalized by BW and averaged over 6 subjects during one full gait cycle and one full crank revolution.

DISCUSSION

These findings reinforce the hypothesis that cycling results in relatively low TF joint compressive contact forces. A novel finding of this study is that OpenSim may be used to calculate knee contact forces during cycling as the results generally agreed with published results.

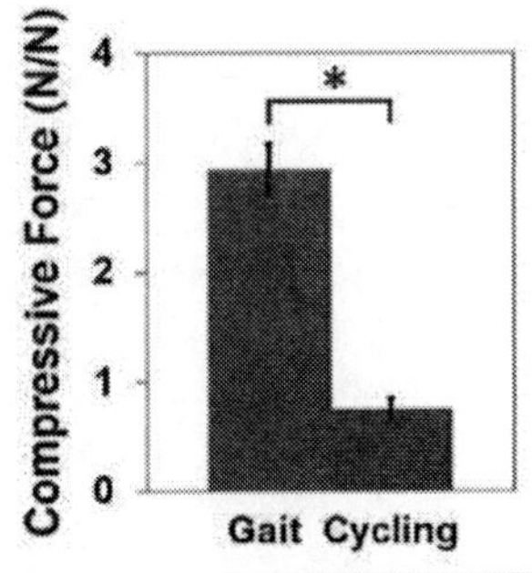

Fig. 2: Maximum TF compressive force for gait and cycling (n=6). Mean ±1 S.D. shown. *Significant difference (p<0.0001).

The difference in the graphs for compressive forces during cycling, observed in Fig. 1, might be due to the difference in setup of the subject on the stationary bike, the type of stationary bike, and the pedal design (i.e. clipped vs strapped). For this study, an upright stationary bike was used and the seat height was positioned such that at 180° crank angle the subject's leg was almost straight. It should be noted that the published results in Fig 1. were recorded at a cadence of 40rpm while our subjects kept a 70rpm constant cadence. Also, the subjects used in [4] were older and were suffering from OA.

It is emphasized that in OpenSim's RRA, a pelvic residual is introduced for dynamic consistency. For gait, that pelvic residual is minimized to a value close to zero, but for cycling the pelvic residual is minimized to a relatively large value that should represent the effect of seat and handlebar forces. A future study should directly measure the seat and handlebar forces and, thus, provide an experimental target for the pelvic residual.

This study calculated TF joint contact forces for cycling and gait non-invasively using an EMG-driven inverse dynamics OpenSim analysis. The results suggest that cycling may be the preferred exercise for limiting OA risk in populations at high risk for knee OA. In order to better identify exercises that minimize the OA risk, future studies may include other potential weight control and fitness sustainment exercises and populations that are at high risk for knee OA such as obese, amputee, and ACL reconstructive surgery subjects.

ACKNOWLEDGEMENTS

This work was supported by the Donald E. Bently Center (SK), W.M. Keck Foundation (MVP, KM), and the Defense Health Program (SK, SH) through the Department of Defense Broad Agency Announcement for Extramural Medical Research Program #W81XWH-BAA-14-1 under Award No. W81XWH-16-1-0051. Opinions, interpretations, conclusions and recommendations are those of the authors.

REFERENCES

[1] McLeod, W et al., *Am J Sports Med*, 8:175-180, 1980.
[2] D'Lima, D et al., *Clin Orth Rel Res* 469:2953-2970, 2011.
[3] Kutzner, I et al., *J Orth Sports Phys Ther.* 42:1032–1038, 2012.
[4] Lerner, Z. F et al., *Gait & Posture.* 39:978–84, 2014.
[5] De Luca, C et al., *J Biomech*, 43:1573-1579, 2010.
[6] Anderson F et al., *J Biomech Eng*, 123:381-390, 2001.

DIFFERENCES IN MATERIAL PROPERTIES OF THIGH AND GLUTEAL SOFT TISSUE BETWEEN MALES AND FEMALES

Zachary, J. Sadler, Joshua, P. Drost, Wu Pan, Tamara Reid Bush

Mechanical Engineering
Michigan State University
East Lansing, MI, USA

INTRODUCTION

Human soft tissue is composed of different components that together yield unique mechanical characteristics. The modeling and analysis of soft tissue can be aided by an improved understanding of these characteristics [1], [2]. Limited work has been conducted on *in vivo* material properties; existing studies have a single participant in either the supine [3] or prone position [4], [5]. There is no work on the properties of bulk soft tissue in the seated position. Further, to be able to input tissue characteristics into a model, data from multiple individuals must be collected and analyzed. The medical, automotive, and ergonomic seating industries are especially interested in these material properties. With a better understanding of the variance in material properties between males and females, seating in all three environments can be improved to aid in comfort, support, and in the case of medical seating avoid soft tissue injuries (e.g., pressure ulcers). The purpose of this study is to characterize the bulk soft tissue properties in the buttocks and thighs and examine the differences in mechanical properties between males and females.

METHODS

A total of twenty-nine individuals participated in the study (IRB # 15-889) and were comprised of both males and females (average ages 20.6 and 20.9, respectively). Force-deflection data were gathered through a load cell paired with a motion capture system. A tissue indenter was constructed with a specialized contact surface and was attached to the top of a calibrated multi-axis load cell (AMTI, Watertown, MA). Positioning of the load cell and participant was tracked using reflective markers. The femur was identified by markers placed at the greater trochanter and lateral epicondyle.

Six different locations along the buttocks and thigh were tested (Fig. 1(a)). Participants were seated in a custom chair. This chair contained removable sections to allow the tissue to be exposed in the different testing areas. The test administrator applied load by hand from either behind the participant (upper and lower buttocks) or below the chair (ischial tuberosities, proximal thigh, medial thigh, and distal thigh). The load was applied at a constant rate until a physical resistance was reached, and then the indenter was removed at the same rate. Force-deflection data were then analyzed using Qualysis Track Manager and MATLAB.

The tissue deflection was calculated as the displacement of the indenter with respect to the greater trochanter for the upper and lower buttocks regions. For all other regions, the displacement of the indenter was compared to a vector created between the greater trochanter and lateral epicondyle. Physical measurements were gathered from testing as well as cadavers to approximate the original tissue lengths of the different body regions. Measurements were also extracted from magnetic resonance images to help determine a range of original tissue lengths [6].

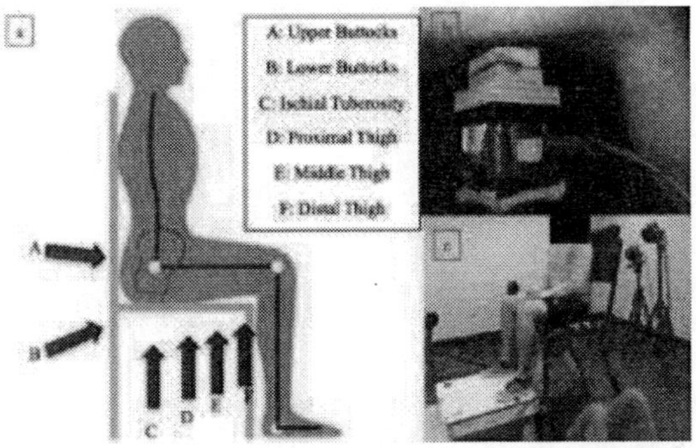

Figure 1: Soft Tissue Testing: (a) Testing locations (b) Tissue indenter composed of load cell and reflective markers (c) Full testing setup with tester below chair applying load

These lengths were measured from the point of contact on the soft tissue to the bony structure and were then used as the original lengths to compute the resulting strain in the soft tissue. Stress was calculated from the force and contact area of the indenter, and the stress-strain relationships were plotted. It is important to note that these are compressive stresses.

Initial work was also conducted to fit a participant's data with the Mooney-Rivlin strain energy:

$$\Psi = \frac{\mu}{2}[\alpha(I_1 - 3) + (1 - \alpha)(I_2 - 3)] \quad (1)$$

Where μ represents the shear modulus, α is a material constant, and I_x are the invariants of the deformation tensors. The material was assumed to be under a uniaxial compressive load with only normal stress and strain considered. The soft tissue was also assumed to be an isotropic, incompressible, and hyperelastic material.

RESULTS

Representative male and female stress-strain plots are presented (Fig. 2). Upon initial analysis of the data, the soft tissue in the buttocks appears to be stiffer than that of the thighs. The upper buttocks is generally the stiffest, and this may be due to a low amount of adipose tissue in this region. The male tissue also appears to be stiffer than the female data. One exception is the proximal thigh region. This region was a difficult region to test due to interference from a support structure on the chair and showed high variability across participants.

Preliminary μ and α values from a single participant demonstrate the same trend (Table 1). The μ value is related to stiffness and is thus higher in the regions of the buttocks.

Table 1: Calculated μ and α of representative male

Region	Up. Butt.	Low. Butt	I.T.	Prox. Thigh	Med. Thigh	Dist Thigh
μ (MPa)	0.0446	0.0305	0.0214	0.044	0.0143	0.0091
α	3.02E-09	1.14E-07	5.76E-07	2.93E-07	2.20E-07	0.786

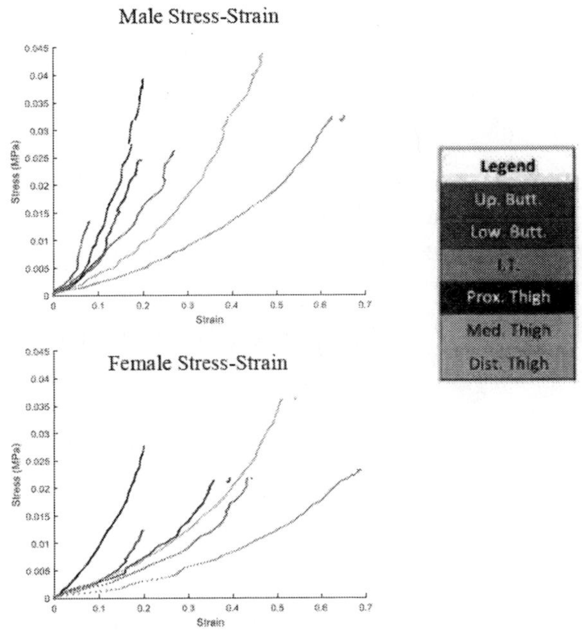

Figure 2: The stress-strain data for a representative male (top) and female (bottom). Strain is plotted as a positive value, but it is compressive.

DISCUSSION

The thigh and gluteal bulk soft tissue display differing material properties with respect to body region and sex. For body region, stiffness is generally higher in the buttocks than the thighs and male data is generally stiffer than female data across all regions.

An understanding of the differences between the mechanical properties of male and female soft tissue will aid in modeling the human body and improve design of devices. This information is particularly important for devices that interact with the body, such as seats. Further, this study is particularly unique in that these data are *in-vivo*, across a range of individuals, for a range of body regions and was conducted in the seated position.

One potential limitation to this work is that the original tissue lengths used for each body region were the averages of data compiled between measurements taken on our participants and data available in the literature. Individual-specific measurements will be obtained in future studies to improve data sets.

ACKNOWLEDGEMENTS

The authors would like to thank L Nault, K Patterson, R Geary, Q Gao, and M Jakeway for their help in this study. This work is supported by NSF CBET, grant number 1603646.

REFERENCES

[1] Setyabudhy, R et al., *SAE Tech.Pap.Ser.*, 1997.
[2] Kauer, M et al., *Med. Image Anal.*, 6:275-287, 2002.
[3] Sun, Q et al., *RESNA 28th Annual Conference*, 2005.
[4] Vogl, T et al., *Acad Radiol*, 17:1486-1491, 2010.
[5] Then, C et al., *Technol. Heal. Care*, 15:385-398, 2007.
[6] Sonenblum, S et al., *J. Tissue Viability*, 24:51-61, 2015.

MEASUREMENT OF RETINAL BLOOD VESSEL STRAIN DURING CYCLIC ROTATION

Kendall R. McMillan (1), Brittany Coats (1)

(1) Department of Mechanical Engineering
University of Utah
Salt La□e City, Utah, USA

INTRODUCTION

Widespread and multi-layered retinal hemorrhages (RH) are well-documented in cases of abusive head trauma (AHT). [1] Several causative mechanisms have been proposed, but no conclusive evidence for the correct mechanism exists. One favored mechanism is that during abusive shaking the eye experiences inertial forces on the vitreous that pull on the retina and result in retinal detachment and hemorrhage. [2] To begin evaluating the feasibility of vitreoretinal traction as a potential mechanism of RH, we sought to learn more about the biomechanics at the vitreoretinal interface during rapid rotation. Specifically, the objective of this study was to design a device and image analysis system to measure deformation of retinal blood vessels during cyclic rotation. Movement of retinal vasculature at increasing velocities was compared to baseline error measurements.

METHODS

System Design. The synthetic socket was designed in SolidWorks based on publically available MRI images of the adult eye and scaled to the well-characterized dimensions of adult sheep provided in literature. [3] An adult sheep eye was used for initial development and evaluation of the device. A pediatric eye socket has also been created, but is not utilized in this study. The adult model was 3D printed with ABS plastic. The 3D printed synthetic socket was attached to a rocker and four bar assembly driven with a 1/7 HP, 4200 RPM DC motor. A 13/84 step down was created to lower the speed from 440 rad/s to 68 rad/s. The system has variable degrees of rotation depending on an adjustable base length, and was programmed using an Arduino motor controller to follow the velocity trace for abusive shaking reported in the literature. [4]

Synthetic EOT was created using a formula of 2.3g Agar and 2g Bovine serum per 50 mL of water. This was determined experimentally to follow the shear storage modulus of strain sweep tests performed on fresh ovine EOT with an AR-G2 rheometer (**Fig. 1**).

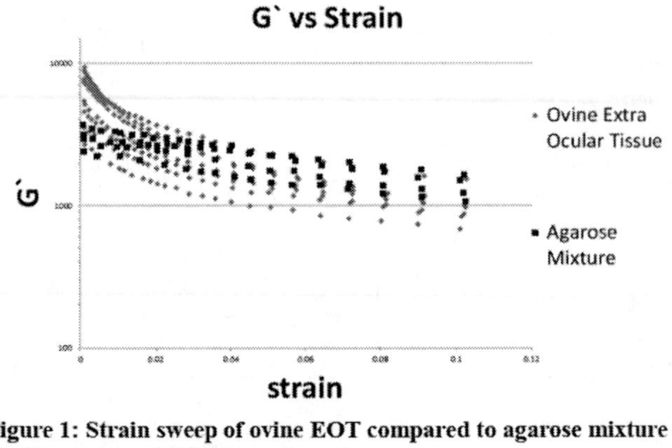

Figure 1: Strain sweep of ovine EOT compared to agarose mixture used to represent EOT in cyclic tests.

Experimental Setup. Fresh adult ovine eyes (n-3) were collected from existing studies on campus. The cornea was removed, and the globe was placed into the socket already prepared with the agarose mixture. A glass cover slip was glued over the eye and socket in a manner such that no air bubbles were trapped within the globe. The camera system was mounted onto the socket, and the cyclic device initiated to produce 13, 16, 21, or 28 rad/s rotations for 1.7 sec.

Data Analysis. Angular velocity was captured using an Adafruit 9-DOF IMU mounted vertically to the back of the socket. A micro USB camera was mounted to the front. Camera and sensor data was transmitted to a laptop computer at 250 Hz using an Arduino Uno microcontroller. Video and velocity data were synced in MATLAB and a custom image processing algorithm was developed to track blood vessel branch points and calculate the geodesic lengths between them (Fig. 2). Vessel segment strain was calculated at each frame by dividing the difference in segment length from baseline by the baseline length (engineering strain). The maximum and average frame strains for each video were extracted. Tensile maximum strain for each vessel was statistically correlated with the peak to peak angular velocity. The analysis error was assessed by running the code on video of an eye with no rotation.

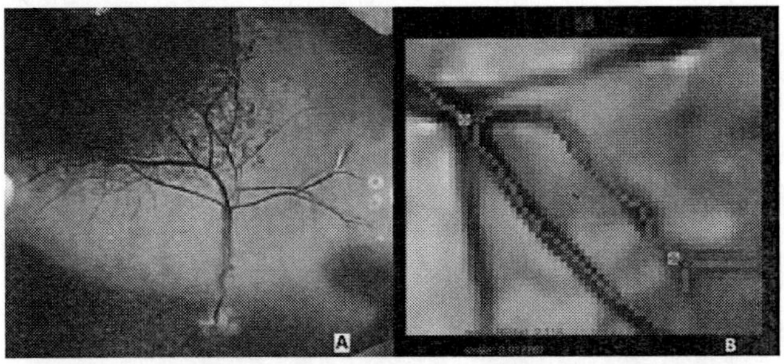

Figure 2: (Left) Image of retinal vessels prior to testing. (Right) Output of one frame of the vessel tracking code. A red trace shows the original location of the selected vessels, and the green trace shows the current location. Strain is determined by branch points, designated in the image as blue pentagrams.

RESULTS

Strain measurements from the video during shaking oscillated about zero in a rhythm similar to the measured velocity (Fig. 3). Maximum positive strains ranged from 2.2% to 15%. Compressive strains were also during the cyclic motion and ranged 4.3% to 13.2%. An average error of 2% was found from analysis of the static video. The strain had a moderate positive correlation of 0.627 with angular velocity.

DISCUSSION

In this study we designed a device to capture retinal blood vessel strain during rapid cyclic rotation. Measured strains were above the imaging noise threshold and reached up to 15% strain. The maximum velocities explored in this study are approximately 30 rad/s lower than those reported in shaking [4]. Therefore, it is likely that strains greater than 15% are

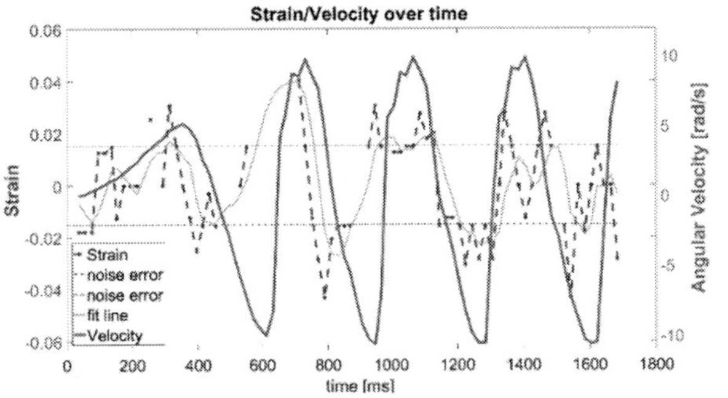

Figure 3: Strain and angular velocity from a single vessel segment. Error measured from static images was ±0.018.

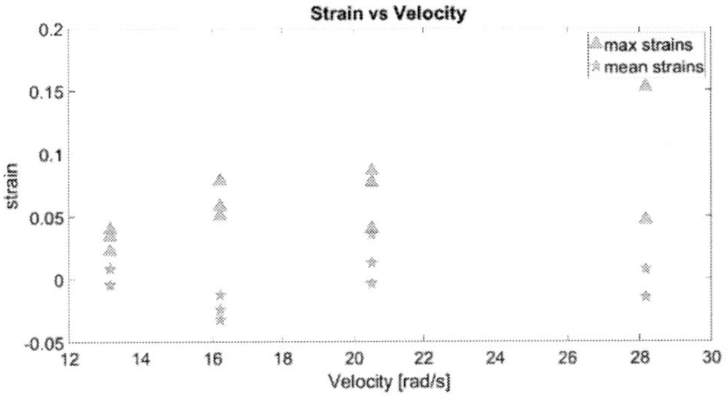

Figure 4: Correlation between maximum strains (triangles) and velocity. Average strains for each vessel also provided (stars).

experienced by retinal vasculature during shaking as we saw an increase in strain with velocity. The design system is capable of these high speeds and will be explored in future studies

Maximum strains occurred just after spikes in angular acceleration. In order to replicate the velocity and acceleration traces of an actual AHT shaking event, the four bar system was designed to have a low transition angle which causes knocking points in the rocker. This simulated the rapid deceleration that might be seen as a child's chin hits their sternum. At these time points, the sudden change in velocity of the rocker, and resulting spike in acceleration, caused the eye to rapidly change direction and resulted in a larger measurable strain value in the vessel.

One limitation of this study is that the curvature of the retina was not accounted for in the measurements of strain. Future work will explore the error produced by this limitation, as well as determine the effect of age (i.e., pediatric vs. adult) on vessel strain.

ACKNOWLEDGEMENTS

Funding was provided by the University of Utah Undergraduate Research Opportunities Program.

REFERENCES

[1] Binenbaum et al, *JAAPOS*, 2009
[2] Waney Squier, *Acta Neuropathol*, 2011

[3] Timothy A. Turvey, *Oral Maxillofac Surg Clin North Am.*, 2012
[4] Michael T. Prange, *J Neurosurg,* 2003

IDENTIFICATION OF HYSTERESIS BEHAVIOR OF PRESSURE-MEASURING INSOLES

A. Ghanem (1), J. DeBerardinis (1), M. Trabia (1), J. Dufek (2), D. Lidstone (2)

(1) Department of Mechanical Engineering
University of Nevada Las Vegas
Las Vegas, Nevada, USA

(2) Department of Kinesiology and Nutrition Sciences
University of Nevada Las Vegas
Las Vegas, Nevada, USA

INTRODUCTION

Extensive research has been conducted on commercial pressure-measuring insoles and their material characteristics[1–7]. Earlier research indicated that these insoles experienced variability between sensors [1,2]. They typically performed best within a certain range of pressures with substantial increases in error observed outside that range[4]. Furthermore, creep and hysteresis were observed[3,5–7]. It was found that the accuracy of pressure-measuring insoles depended on the calibration methods[3]. The purpose of the current experiment was to understand the behavior of a commercial insole (Medilogic®, Shönefield, Germany). The loading and unloading calibration curve produced in this study may provide a more accurate representation of the pressures measured by the insole.

METHOD

Equipment: Medilogic® pressure-measuring insoles (Schönefeld, Germany, 60Hz) were used for in these experiments. As shown in **Figure 1(A),** each insole has a grid of sensors. Six sizes were tested with the number of sensors ranging between 93 and 162 sensors. The sensors, which are 0.75x1.5 cm rectangles, measure the change in electrical resistance, which is proportional to the pressure applied to these sensors. A data acquisition system collected the bit output of each sensor on the insole. The insoles were loaded using a Tinius Olsen® (Horsham, USA, 1Hz) axial loading machine to induce a dynamic loading. The machine is equipped with a 22,250 N load cell. An extensometer, attached to the upper (moving) crosshead was used to determine the displacement of the crosshead of the machine. Load and displacement data were recorded during the experiments.

Experimental Protocol: The insoles were placed under an aluminum plate (302x230x12.8 mm). Aluminum shims (0.6 mm thick) were placed around the insoles to ensure that the insole would not be damaged during the experiments. Loading was applied to the center of the plate to create an even distribution of the force onto the insole, **Figure 1 (B).**

For each experiment, the upper crosshead was placed 0.6 mm above the plate. The insoles were loaded up to a force of approximately 6700 N at a rate of 3.81 mm/min for 30 seconds. The direction of the load was then reversed for 30 seconds. Each experiment was repeated ten times.

Data Processing: The insole data were down-sampled to 1 Hz to match the testing machine force data. Loading and unloading curves of average pressure versus the average number of bit per sensor were created. These curves were qualitatively analyzed for consistency.

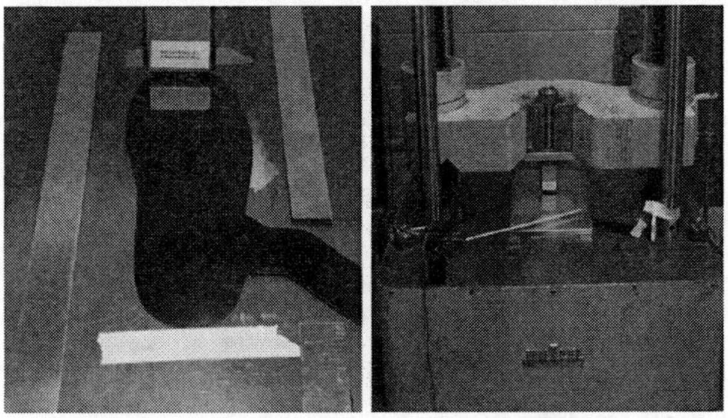

(A) (B)

Figure 1: Experimental setup: (A) An insole and shims on the base of the testing machine. (B) The crosshead applying the load the insoles through the aluminum plate with the insole under it

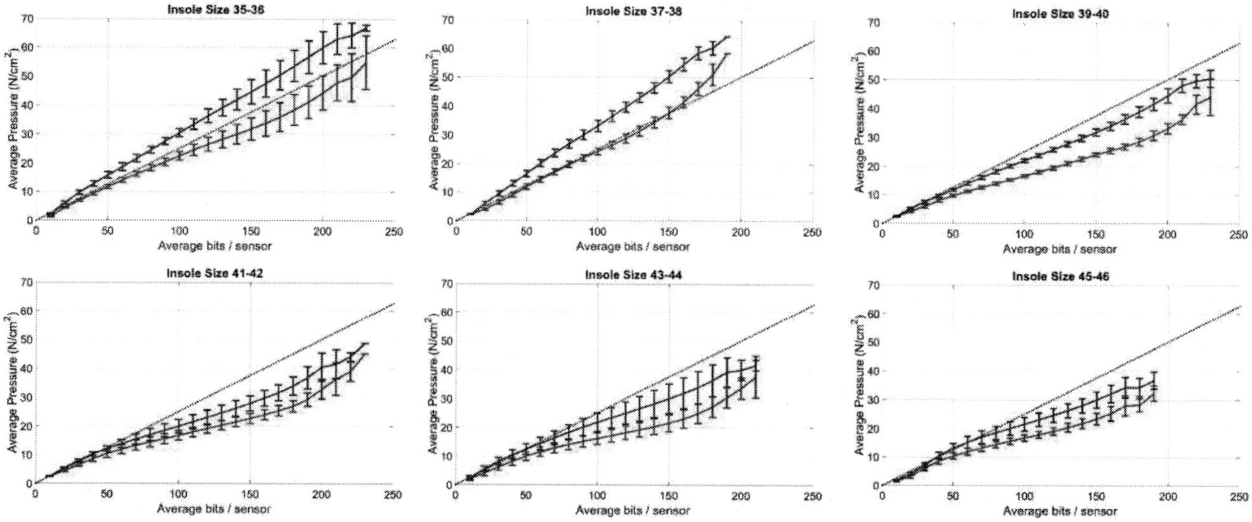

Figure 2: The loading (red) and unloading (blue) curves of the insole in comparison to the manufacture's recommended linear calibration (black)

RESULTS

Qualitative analysis consistently showed that the first loading/unloading experiment for any insole resulted in the loading curve being above the unloading curve. This feature was reversed in subsequent testing. It was therefore decided to combine tests 2 through 10 to create the average hysteresis curve of each insole. The left and right curves were then combined to create a curve for each insole size. Associated error bars were also generated. **Figure 2** shows the resulting loading and unloading curves along with the linear calibration suggested by the manufacturer (64 N/m^2 for 255 bits) for comparison purposes.

DISCUSSION

The goal of these experiments was to assess the validity of using the calibration method suggested by the manufacture and to understand the hysteresis behavior of the insoles. The results showed that the left and right insoles of any size were not significantly different, as could be seen by observing the relatively small error bars of **Figure 2**. The figure also shows that while the behavior of the insole sizes were different all of them exhibited hysteresis behaviors. It was of interest to observe that the insole curves gradually produced lower average pressure for the same average bit values as the insole size increased. The results also showed that as the load increased, more deviation from the linear calibration suggested by the manufacturer was observed.

The experiments indicated that the Medilogic® insoles experienced hysteresis, which confirmed the finding of other researchers[1,3,5,6]. The results showed preconditioning is important for accurate use of the Medilogic® insoles, which confirms the findings of Hall et al[7]. Different hysteresis behavior of different sizes required the calibration of each size independently. The results also showed that the use of suggested manufacturer calibration curve should be limited to relatively light loads.

We plan to apply the results of this research to analyzing ground reaction force during walking to increase the measurement accuracy of these portable insoles. A limitation to this work was that the rate of load application, which was slower than what is observed in walking or running. Future experiments may be needed to investigate the effects of the rate of loading on the hysteresis of these insoles. We also plan to consider the combined effects of hysteresis, and bending and shear loading on the insoles calibration.

ACKNOWLEDGEMENTS

This publication was made possible by a grant from the National Institute of General Medical Sciences (GM103440) from the National Institutes of Health. This contents of this work are solely the responsibility of the authors and do not necessarily represent the official views of NIH.

REFERENCES

1. Woodburn J, Helliwell PS. Observations on the F-Scan in-shoe pressure measuring system. *Clin Biomech*. 1996;11(5):301-304. doi:10.1016/0268-0033(95)00071-2.
2. Luo ZP, Berglund LJ, An KN. Validation of F-Scan pressure sensor system: a technical note. *J Rehabil Res Dev*. 1998;35(2):186-191.
3. Nicolopoulos C., Anderson E., Solomonidis S., Giannoudis P. Evaluation of the gait analysis FSCAN pressure system: clinical tool or toy? *Foot*. 2000;10(3):124-130. doi:10.1054/foot.1999.0536.
4. Hsiao H, Guan J, Weatherly M. Accuracy and precision of two in-shoe pressure measurement systems. *Ergonomics*. 2002;45(8):537-555. doi:10.1080/00140130210136963.
5. Giacomozzi C. Appropriateness of plantar pressure measurement devices: a comparative technical assessment. *Gait Posture*. 2010;32(1):141-144. doi:10.1016/j.gaitpost.2010.03.014.
6. Price C, Parker D, Nester C. Gait & Posture Validity and repeatability of three in-shoe pressure measurement systems. 2016;46:69-74.
7. Hall RS, Desmoulin GT, Milner TE. A technique for conditioning and calibrating force-sensing resistors for repeatable and reliable measurement of compressive force. *J Biomech*. 2008;41(16):3492-3495. doi:10.1016/j.jbiomech.2008.09.031.

HEAT AND MASS TRENDS WITHIN A REBUILDABLE DRIP ATOMIZER ELECTRONIC CIGARETTE

Phoebe C. Belser (1), Timothy M. Raymond (1), Dabrina D Dutcher (1), James W. Baish (2)

(1) Chemical Engineering
Bucknell University
Lewisburg, PA, USA

(2) Biomedical Engineering
Bucknell University
Lewisburg, PA, USA

INTRODUCTION

Electronic cigarettes and vaporizers have become one of the fastest growing industries in the United States. Even though these technologies are widely used, they may be hazardous and are not well understood. This past August, the FDA implemented stricter regulations[1] for electronic cigarettes. With so much still unknown about these devices, research is still needed such that appropriate and reasonable regulations can be put forth.

The market of electronic cigarettes is evolving toward rebuildable drip atomizers (RDA) in which the user wraps a heating coil, to the complexity and design of their own choosing. These devices also offer some level of power (and thus temperature) control.

One concern with such user-customized devices is when a so-called 'dry hit' occurs. A dry hit is the condition under which the wick that supplies fluid, known as e-liquid, to the vaporizer coil is not fully saturated. A burnt taste and excessively high temperatures may result, releasing unwanted toxins in the effluent. For example, previous studies have found the presence of formaldehyde[2] in the effluent released during the dry hit. To better understand dry hits, it is imperative that the wicking and evaporation trends are understood.

The hypothesis of this project is that a dry hit occurs when the evaporation of the e-liquid occurs more quickly than wicking can replace the e-liquid, resulting in the burning of the cotton wick and e-liquid. In this experiment, the wicking and evaporation trends in a dual coil RDA were investigated. The dual coil is considered to be the most basic configuration, and therefore was used throughout the entire project, such that more complicated configurations can be understood in the future. Before the wicking and evaporation trends can be understood, an understanding of the temperature variation throughout the coil and wick must be established. The evaporation trends were investigated with respect to the composition of the e-liquid, the applied voltage from the controller, and saturation of the wick. Likewise, the wicking rates were researched in relation to the density of wick, the composition of e-liquid and the evaporation rates.

METHODS

First, it was imperative that a procedure was developed such that the e-cigarette was set up identically for each trial. A commercial power supply and atomizer were used. The power supply was chosen for its capability of measuring the resistance in the coil, which allows the user to control the applied power. The set up of the atomizer is shown in Figure 1.

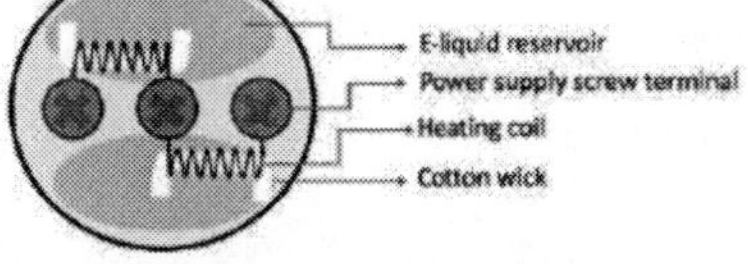

Figure 1: Diagram of the atomizer set up from above, using two coils of 30 gauge Ni200 wire.

Due to the uncertainty of the composition of commercially available e-liquids and flavoring agents, all e-liquids used in this experiment were composed solely of propylene glycol (PG) and vegetable glycerin (VG).

The coil temperature and voltage were measured simultaneously using a thermocouple resting just inside the coil and voltage leads wired through the atomizer, respectively.

In order to see wicking as a result of different compositions and densities, wicks of varying fiber densities were hung vertically adjacent to one another above a beaker of desired e-liquid composition and the rate of wicking was recorded.

The wicking and evaporation trends were then further investigated with respect to one another within the atomizer. Using e-liquid dyed with green food dye, the saturation of the cotton could be observed based on the shade of green of the cotton. Digital images of the wick were taken at 12 second intervals beginning with a fully saturated wick. A script was written to relate the gray scale images to the saturation profile in the wick.

RESULTS

The temperature of the coil when evaporating different compositions, taken with identical voltages and set up, are shown below (Figure 2).

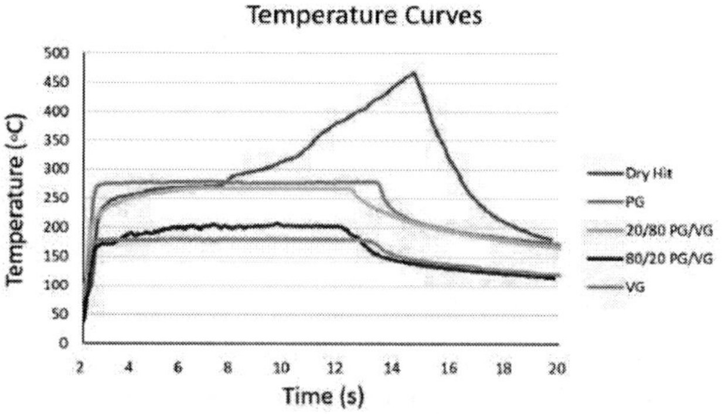

Figure 2: Temperature of the coil with different ratios of propylene glycol and vegetable glycerin. The temperature of a 20/80 PG/VG dry hit is also present. Power was applied to the coil from approximately time 2.5 s to time 13 s.

This data shows that the temperature of the coil closely matches the boiling point of the e-liquid present. When a solution of only pure VG or pure PG is used the temperature data shows that they trend towards the known boiling points, 290 °C[3] and 187 °C[4], respectively. For compositions of the two liquids, the PG, the component with the lower boiling point, evaporates first. Then the temperature increases to the boiling point of the less volatile component. This is seen in the shape of the curves in Figure 2. The temperature within the coil during a dry hit also shown in the above figure, the e-liquid evaporates as expected with the 20/80 curve, but once the wick runs dry, the temperature of the coil increases dramatically. With these dry hits, as this temperature increase occurs, the wick sometimes ignited.

Regardless of the voltage, when e-liquid is present at the coil, the temperature will always be the boiling point. The voltage and resulting power level, however, determine the rate of evaporation. The high temperature conditions of dry hit arise when the evaporation rate exceeds the wicking rate regardless of the temperature or voltage selected by the user.

Figure 3 shows the gradient in the saturation levels along the wick as determined from the digital images. This gradient shows the drastic difference in wicking and evaporation rates, as all the e-liquid adjacent to the coil has evaporated as the e-liquid present in the reservoir has not yet been wicked up. From the other wicking experiments, it was found that the denser wick and PG have significantly faster wicking rates. The effect of composition on wicking speeds stems from the difference in properties such as viscosity, surface tension and heat of vaporization. We found that viscosity is the dominant property with PG wicking faster than VG because of its lower viscosity[3,4].

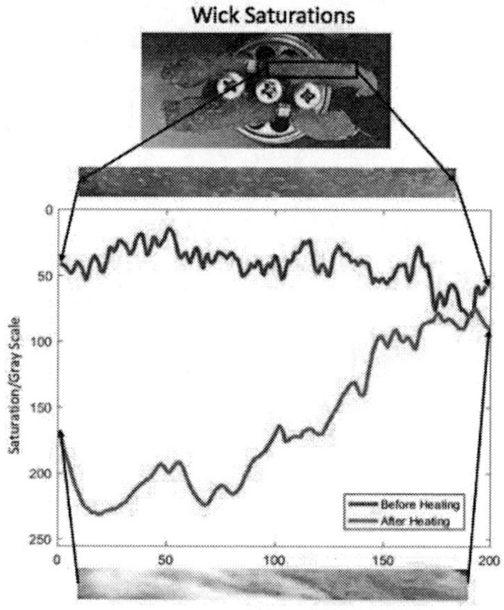

Figure 3: A cropped image of the wick before and after 4 V hit with designated color graph showing the saturation of e-liquid across the wick.

DISCUSSION

From the above results the wicking and evaporation trends, along with hazards associated with e-cigarettes are represented clearly. Evaporation rates are higher with larger voltages, while resulting temperatures remain constant as long as wicking can keep up with evaporation. These temperature and evaporation trends demonstrate that automatic temperature control by a power source has little effect on the temperature inside the coil. When coupled with relatively slow wicking rates, the wick proximal to the coil will become dry quicker with higher voltages. Therefore, at higher voltages the frequency of "dry hits" increases, along with likelihood of ignition. In the user community, a large amount of vapor is desired with each hit, encouraging the use of a higher voltage. It is also more popular for there to be more VG than PG in an e-liquid, resulting in a higher temperature, again increasing the likelihood of hazards occurring. In conclusion, a vaperizer operated at a lower voltage, using predominantly PG, is less likely to produce hazards from high temperatures than the alternatives.

ACKNOWLEDGEMENTS

This research was sponsored and funded by the Chemical and Biomedical Engineering Departments at Bucknell University.

REFERENCES

[1] FDA. *FDA's New Regulations for E-Cigarettes, Cigars, and All Other Tobacco Products, 2016.*

[2] Jenser, R et al., *New Engl J Med,* 374:392-394, 2015.

[3] Thermo Fisher Scientific. *Glycol SDS*, 2016.

[4] Thermo Fisher Scientific. *Propylene Glycol SDS*, 2016.

A CLINICAL STUDY: THERMAL CONTRAST AMPLIFICATION READER IMPROVES THE DETECTION OF STREP THROAT FOR LATERAL FLOW ASSAYS

Erin M. Louwagie (1), Yiru Wang (1), Daniel Larkin (2), David R. Boulware (3), and John C. Bischof (1)

(1) Department of Mechanical Engineering
University of Minnesota – Twin Cities
Minneapolis, MN, US

(2) HealthEast Clinic
Saint Paul, MN, US

(3) Department of Medicine
Infectious Diseases and International Medicine
University of Minnesota – Twin Cities
Minneapolis, Minnesota, USA

INTRODUCTION

Strep throat is estimated to infect millions of people every year in the US [1]. For instance, in 15%-30% of pediatric cases and 5%-20% of adult cases of acute sore throat, Strep throat, or group A beta-hemolytic streptococcus (GABHS), is the cause [2]. Strep throat is highly contagious, due to GABHS bacteria's ability to be spread through both the air and physical contact. Though strep throat itself is not a serious illness, prolonged infection of GABHS can lead to more hazardous inflammatory illnesses, such as rheumatic fever or kidney inflammation [3]. Therefore, it is imperative that diagnosis of Strep be timely and accurate to allow prompt treatment and reduced transmission.

Diagnosis of Strep is initially based on running a rapid diagnostic test (RDT) (Fig. 1) based off a throat swab in the clinic. RDTs require only 10-15 minutes to run, and easy to use, and can be performed directly at the point of care. When the RDT is visually positive (Fig. 1 - S2), the health care provider can then quickly and correctly provide the patient with antibiotics or other treatment. Unfortunately, due to low RDT sensitivity (only 70%-90%), there are often patients who test visually negative with RDTs but are confirmed positive by culture later [5]. These RDTs are "false negative" (Fig. 1 - S1) and therefore lead to delays in treatment and unnecessary exposure of others to potential infection. Due to this hazard, visually negative RDTs lead to a further confirmatory blood agar plate (BAP) culture performed on a second throat swab at the time of initial testing [4]. BAP cultures are considered the "gold standard" for diagnosing Strep throat with diagnostic sensitivity ≥ 90% [5]. Nevertheless, confirmatory cultures require an 18 – 48 hour turn around, resulting in extra cost and labor from a skilled professional. It also delays appropriate treatment. Thus, improving the sensitivity of the Strep throat RDT to catch visually negative RDTs would reduce the need for confirmatory culture and improve access to treatment and reduction of transmission.

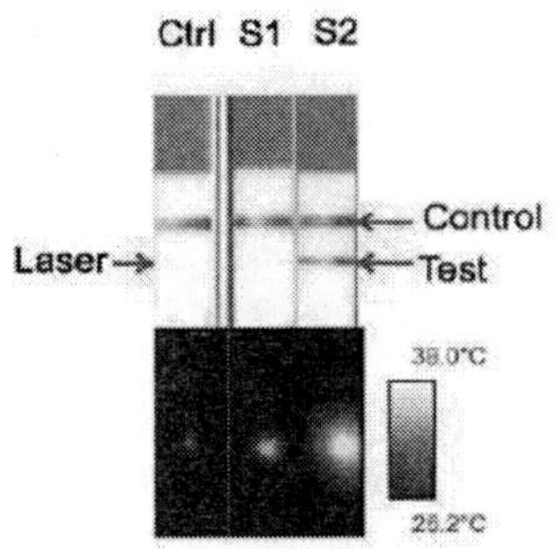

Figure 1. Visual and Thermal Contrast. S2 is visually positive. S1 is visually negative, but thermally positive thereby improving sensitivity of the RDT [9].

In this collaborative work with an urgent care clinic in St. Paul, MN, we show that we can identify close to 50% of the clinical "false negative" GABHS RDTs in the form of lateral flow immunoassays (LFAs). LFAs are commonly used in rapid detection tests because they are fast, inexpensive, and easy-to-use. Antibody-labeled gold nanoparticles (GNPs) are typically used in LFAs to provide visual results, providing a red test line to indicate a positive test result [6,7]. We achieved this significant improvement in the sensitivity by using a novel thermal contrast amplification (TCA) reader (Fig. 2). These results suggest a possible path forward to improve RDT testing with TCA, thereby reducing or even eliminating confirmatory testing for Strep and perhaps other diseases (i.e. flu).

Poster Presentation #P9

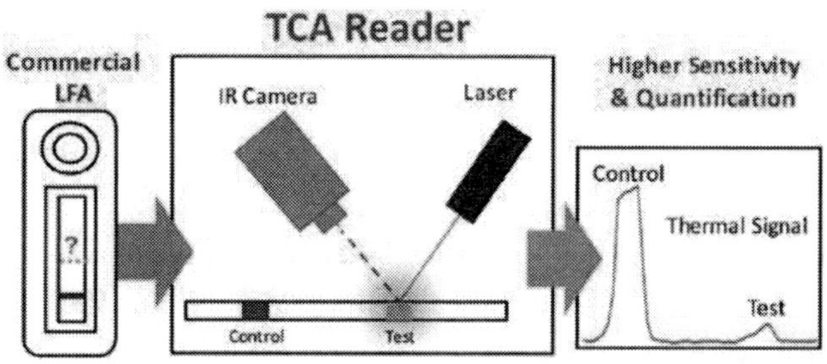

Figure 2: The principle of thermal contrast amplification reader. A false negative LFA is detected as TCA positive [8].

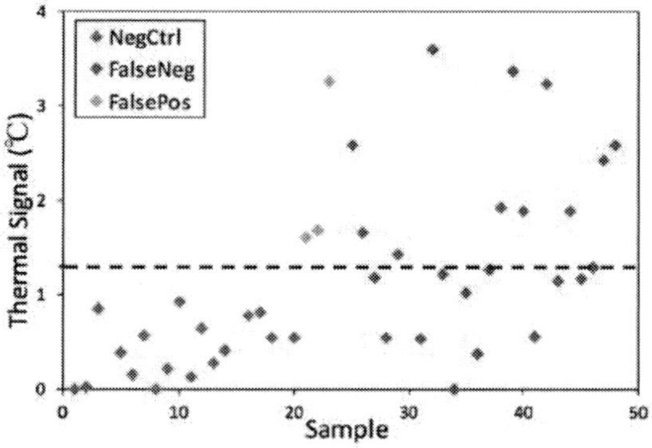

Figure 3: TCA reader caught about 50% of the false negatives for Strep LFAs in the study. 20 true negatives and 23 false negatives as shown here. More will be shown in the final poster.

METHODS

When a clinical GABHS LFA (Quidel QuickVue) showed a negative result, a subsequent BAP culture (GenProbe DNA/RNA) was performed on the same patient to find if the LFA result was a visually false negative. Then all patient specific information was removed from the LFA samples, leaving only a numerical identifier. The false negative LFAs were collected by Dr. Daniel Larkin at HealthEast Clinic and transferred to the Bischof lab in University of Minnesota roughly once a week in a rolling fashion until the total number reached 100. All LFAs were read with the TCA reader. 60 true clinically negative LFAs were tested beforehand to set up the detection criteria. The protocol and limit of detection were described in detail in our published work [8].

The TCA reader is a single step add-on to traditional LFAs. When irradiated by a green laser (532 nm), the GNPs in an LFA will generate heat due to surface plasmon resonance [7,9]. This heat is detected by the IR camera and analyzed in order to give a thermal signal [7]. In its use to date, the TCA reader has shown 8-fold improvement for disease detection with LFAs for malaria, influenza, and *C. difficile* GDH [8].

Each sample was tested once in the TCA reader. The output of the IR camera in the reader was then analyzed in MATLAB, which gave an area under the curve (AUC) based thermal signal for the test line. The thermal signals that fell outside the limit of detection established by the true negatives were defined as TCA positive by the TCA reader [8].

RESULTS

The thermal signals of all LFA samples (including true negatives and false negatives) are plotted in Figure 3. The TCA reader was able to detect ~50% of the known false negatives thus far.

In Figure 3, each point marks an individual sample. The dashed line shows the detection criteria determined by the 60 negative controls and our previous work [8], and the false negatives above this line were those that were detected by the TCA reader. The x-axis denotes the numerical identifier for each sample, and the AUC thermal signal of the sample.

A dilution study with Strep positive control is in process. Data will be shown on the final poster.

DISCUSSION

We have shown that TCA reader improves the diagnostic sensitivity of GABHS LFAs by detecting ~50% of the clinically false negatives. This was done by setting up a limit of detection with true negative samples and applying it to visually false negative samples. TCA reader has the potential to save time and budget and avoid secondary infections in Strep throat diagnosis. However, we did observe some false positives during the test of true negative samples (Figure 3). This is attributed to sample preparation. For instance, if the absorbent pads are not removed at the end of the test, GNPs can wick back to the test line thereby artificially enriching the unbound GNP concentration during drying. A solution we are exploring is the removal of all extra pads (sample pad, conjugate pad and absorption pad) from the LFA and leaving the membrane alone after visual evaluation.

As this study is not finished yet, more LFAs are coming to the lab. In the future, a dilution study of positive Strep sample is planned to find the analytical sensitivity improvement for the TCA reader with Strep LFAs. Also, we would like to field-test the TCA reader in a clinical setting to access its performance with fresh LFA samples.

AKNOWLEDGEMENT

COI (Bischof): CTO Vigilant Diagnostics, LLC translating TCA technology.

REFERENCES

[1] https://www.cdc.gov/groupastrep/
[2] Choby, BA, *Am Fam Physician*, 79(5):383-390, 2009.
[3] http://www.mayoclinic.org/diseases-conditions/strep-throat/home/ovc-20165963
[4] Bisno, AL et al, *Clin Infect Dis*, 35(2):113-125, 2002.
[5] Alcaide, ML et al, *Infect Dis Clin of NA*, 21(2):449-469, 2007.
[6] WHO. *Summary Results of WHO Product Testing of Malaria RDTs: Rounds 1-5*; World Health Organization: Geneva, 2014.
[7] Posthuma-Trumpie, GA et al, *Bioanal Chem*, 393(2):569-582, 2009.
[8] Wang, Y et al, *Anal Chem*, 88(23):11774-11782, 2016.
[9] Qin, Z., Angew. Chem. Int. Ed. 2012. 51: p. 4358–4361

EXPERIMENTAL MOTION TRACKING OF THE MEMBRANE IN THE PENN STATE PEDIATRIC VENTRICULAR ASSIST DEVICE

P. Crompton (1), B. Good (1), K. Manning (1,2)

(1) Department of Biomedical Engineering
Pennsylvania State University
University Park, PA, USA

(2) Department of Surgery
Penn State Hershey Medical Center
Hershey, PA, USA

INTRODUCTION

While there are several ventricular assist devices available for adults, counterparts in pediatric care have been left behind, despite an ever increasing need for circulatory support in infants and children. This difficulty in producing a viable pediatric version can be attributed to factors relating to the variability in patient size and smaller number of patients in need (which can also hinder clinical trials) [1]. For example, the hemodynamics in the Penn State pediatric ventricular assist device (PVAD) vary significantly from the adult version, thus making further analysis of its behavior necessary [2].

At this time, little work has been done to characterize the behavior of the inner components of the PVAD, most notably the segmented poly-(ether polyurethane urea) (SPEUU) membrane that separates its air and blood chambers. In order to understand and predict the hemodynamics involved in the operation of this device, it is necessary to experimentally observe and characterize the complex motion of this membrane between two different fluids while under physiological conditions. Ultimately, these data will be used to validate a fluid-structure interaction (FSI) computational model that accurately predicts the hemodynamics within the PVAD.

METHODS

An experimental mock PVAD setup was designed consisting of two chambers placed together as shown in Figure 1. The smaller compartment on the left is designed to imitate the air chamber within the PVAD, and the larger compartment on the right holds a blood analog fluid (Figure 1).

Two CCD cameras (Allied Vision, Exton, PA) were placed on the far side of the blood analog chamber at a 15° angle so that the images from each camera could be cross referenced to create a three-dimensional picture. Care was taken to place the cameras exactly perpendicular to the side walls of the model (Figure 1). A PVAD membrane is placed between the two chambers, convex to the air side (left chamber). The membrane is marked with a uniform circular array of black dots drawn on it. In post processing, the location of these points will be cross-referenced between two cameras to generate a set of three dimensional coordinates corresponding to the entire membrane.

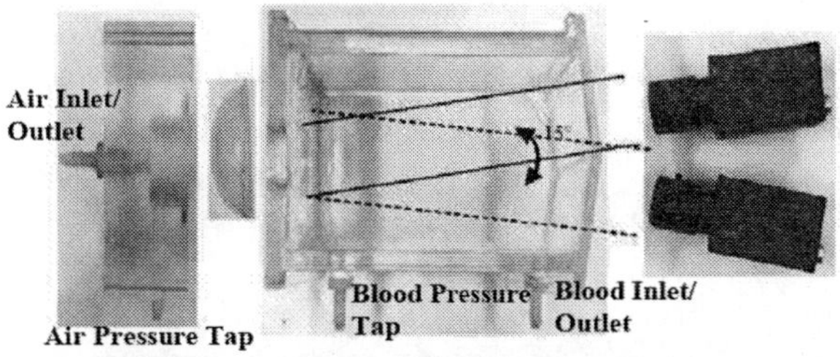

Figure 1: Mock PVAD Experimental Setup

Images of the membrane were acquired at steady pressures of 30, 118, 934, and 3730 Pa, and also during two pulsatile waveforms: a purely sinusoidal waveform at 60 bpm with a 0.5 seconds systolic duration, and a physiological waveform at 75 bpm with a 0.34 seconds systolic duration. The pressures on both sides of the membrane were measured and recorded for validation.

The blood analog solution used was designed to mimic the viscoelastic nature of blood and consisted of 55% water, 45% glycerin, and 0.01% Xanthan gum by weight. The viscosity of this fluid was measured and fit to a modified Oldroyd-B viscoelastic model [3]. The coefficients of fit were found to be $\eta = 8.54$ cP, $\eta_1 = 10.6$ cP, and $\mu_E = 1.621$ cP/s with density $\rho = 1.13$ g/mL.

Before each trial, the cameras were calibrated using two specially designed calibration volumes (Figure 2) with easily distinguishable points spaced in the positive and negative spaces of three orthogonal directions, encompassing the entire domain through which the membrane moves during its full extension.

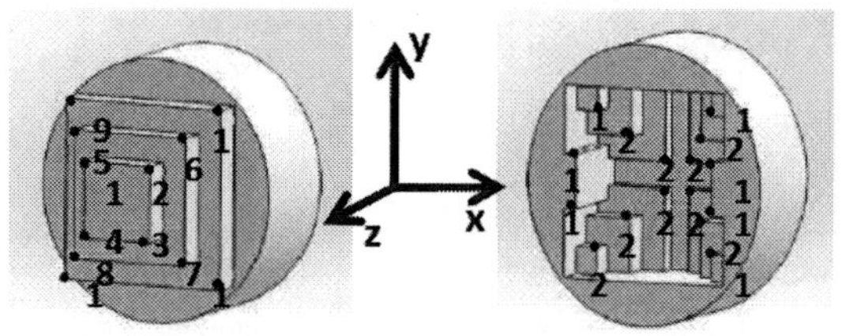

Figure 2: Positive/Negative Calibration Volumes of Known Points

Comparison of the known points of calibration with the image space coordinates from each camera was used to calculate the coefficients needed to perform a direct linear transformation (DLT) to three-dimensional coordinates using a MATLAB script. This process maps two image coordinates (u_1,v_1) and (u_2,v_2) to one real coordinate in physical space (x,y,z) (Figure 3).

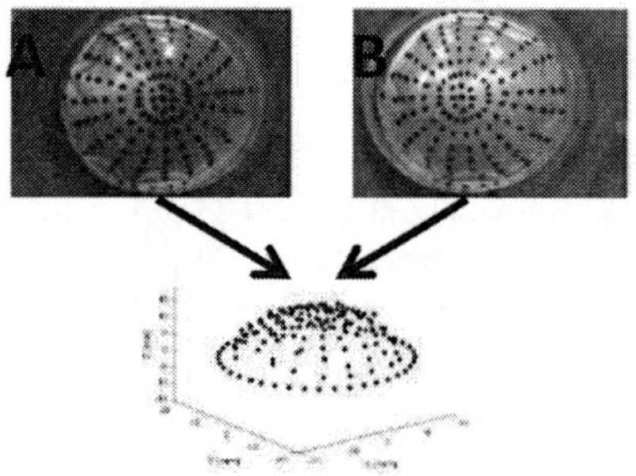

Figure 3. DLT of Image Pair to 3D Coordinates

Equation 1 are the relationships between these coordinate sets, where A-K are coefficients generated in the calibration process.

$$u = \frac{Ax+By+Cz+D}{Ix+Jy+Kz+1} \qquad v = \frac{Ex+Fy+Gz+H}{Ix+Jy+Kz+1} \tag{1}$$

RESULTS

Figure 4 illustrates the membrane deformation at the four steady pressures demonstrating that the increase in pressure causes the membrane to invert completely. The time dependent results were averaged over three cycles in order to reduce their statistical variation, yielding results with standard deviations on the order of 1 mm, depicted by the red markers. Figure 5 shows a full cycle at 60 bpm in 200 ms time steps illustrating the dynamic changes occurring temporally similar to the actual pump operation.

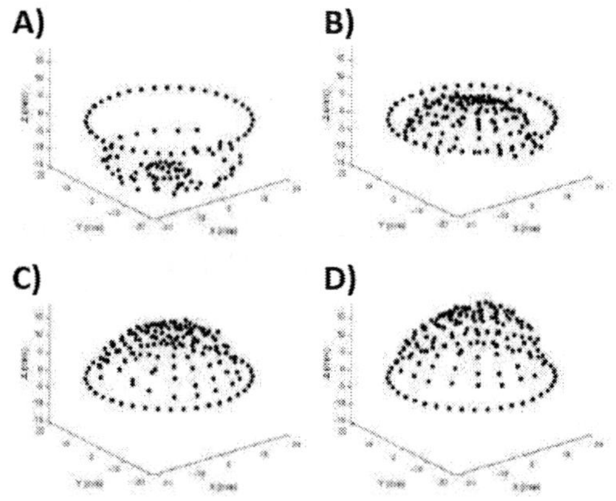

Figure 4: Steady Results A) 30 Pa B) 118 Pa C) 934 Pa D) 3730 Pa

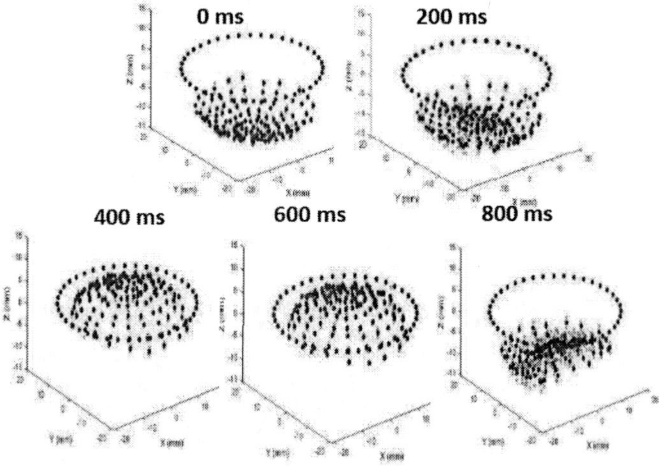

Figure 5: Full Cycle Results at 60 bpm with Air

DISCUSSION

These data provide an accurate and repeatable experimental design for tracking the PVAD's membrane. Figures 4 and 5 provide insight into the full motion of the membrane under a variety of situations, both steady and physiological, and demonstrates the structural changes of the PVAD membrane. As a result, the membrane motion of the FSI model can be validated using the results of this experiment.

ACKNOWLEDGEMENTS

This work was supported by NIH NHLBI HL108123.

REFERENCES

[1] VanderPluym, C et al., *Circulation* 129:1520-1537, 2014.
[2] Bachmann, C et al. *Artificial Organs*. 24(5):362-372, 2000.
[3] Good B, et al. *Annals of Biomedical Engineering*. 44(4):1019-1035, 2016.

DESIGN AND CHARACTERIZATION OF A HELMHOLTZ RESONATOR FOR BRAIN MAGNETIC RESONANCE ELASTOGRAPHY

Rachel E. Mickelson (1), Charlotte A. Guertler (1), Dennis J. Tweten (1), Ruth J. Okamoto (1), Philip V. Bayly (1)

(1) Mechanical Engineering and Materials Science
Washington University in St. Louis
St. Louis, MO, USA

INTRODUCTION

Magnetic resonance elastography (MRE) is a non-invasive imaging technique in which external vibrations are used to induce shear waves in tissue [1]; the resulting wave field is imaged by MRI and used to estimate mechanical properties. MRE can be used to characterize brain tissue in vivo in humans or animals to study traumatic brain injury (TBI). In mini-pig MRE experiments, external vibrations are applied by a pneumatic actuator clamped between the animal's jaws at frequencies of 50 to 125 Hz [2]. At higher frequencies, the induced waves are attenuated and must be amplified to elicit accurate measurements of tissue mechanical properties. Figure 1 is an image displaying the current amplitude of the shear waves induced at 100 Hz during MRE.

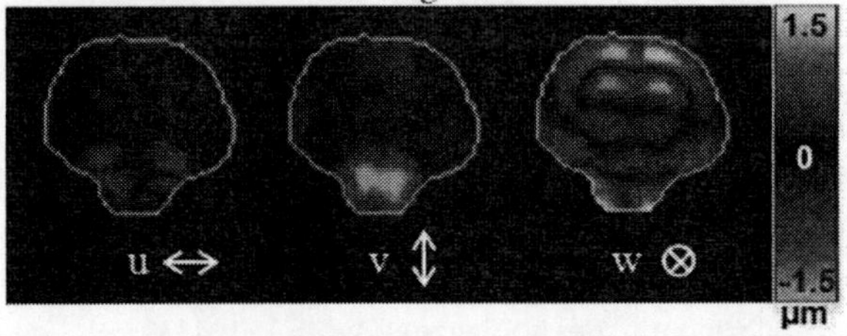

Figure 1: Example MRE shear waves at 100 Hz are shown in a mini-pig brain for the three directions of motion

A Helmholtz resonator can be used to amplify a single frequency during MRE. The following equation shows the relationship between the different parameters and corresponding resonant frequency:

$$f_{res} = \frac{c}{2\pi}\sqrt{\frac{S_p}{V_p l_p}}$$

Where f_{res} is the resonant frequency (Hz), c is the speed of sound in air (m/s), S_p is the cross sectional area of the neck (m^2), V_p is the volume of the cavity (m^3), and l_p is the length of the neck (m) [3]. The Helmholtz resonator consists of a cavity volume connected to an acoustic system by a narrow neck. The equation is valid only when V_p is much greater than the volume of the neck, V_n. When V_p is zero, V_n behaves like a side branch with a closed end and determines the resonance of the system. By altering the dimensions of the resonator, it can be tuned to a specific resonant frequency. When this resonant frequency matches the input frequency of an acoustic system, the amplitude of the external vibration is amplified, allowing more accurate and robust characterization of the mini-pig brain during MRE.

METHODS

The Helmholtz resonator was constructed from a 0.025 m dia. PVC pipe ("neck") connected by a 3D printed adapter to a section of 0.15 m dia. PVC pipe (cavity). The cavity was sealed with an adjustable expansion plug, providing a maximum volume of 9 L. The resonator was connected to an acoustic experimental system, consisting of a subwoofer that generated pressure waves through flexible PVC tubing (0.01905 m inside dia., 10.75 m in length) to a custom multi-directional jaw actuator. In the MRE setup, a minimum PVC tubing length of 9.2 m was required to reach from the Resoundant (Resoundant; Rochester, MN), which supplied the pressure waves, to the jaw actuator. Figure 2 displays a schematic of the acoustic experimental system setup.

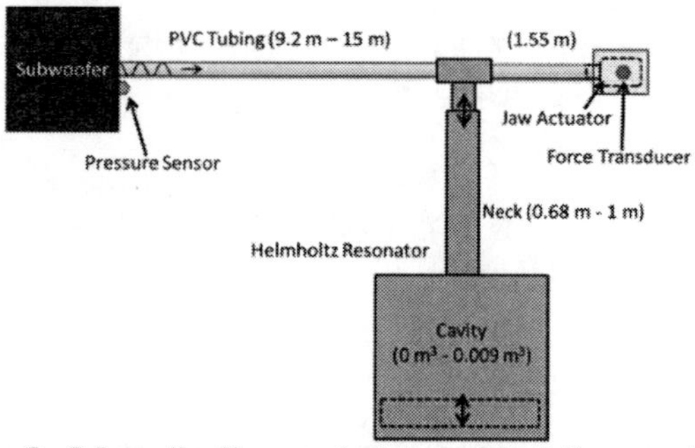

Figure 2: Schematic diagram (not to scale) of the experimental system with subwoofer (black), pressure sensor (green), PVC tubing (gray), Helmholtz resonator (red), and force transducer (blue). The Helmholtz resonator can be adjusted by changing the height of the neck and/or cavity.

A dynamic force transducer (Model SN 14368, PCB Piezotronics) was used to measure the amplitude of the vibrations that would be delivered to the mini-pig's skull through the jaw actuator. The subwoofer (Model SA-W2500, Sony) provided multi-frequency excitation (chirp) at frequencies from 10 Hz to 120 Hz. The resonator was tuned by testing different neck lengths and cavity volumes as well as altering its placement along the tubing. These experimental results were compared to theoretical results from a finite element (FE) acoustic model (COMSOL Multiphysics, v5.1).

RESULTS

In absence of the resonator and side branch, the total length of PVC tubing determined the position of the systems resonant peaks; increasing the total length of tubing resulted in smaller resonant frequency values. By altering the dimensions of the resonator, it was possible to optimize the resonator to amplify actuator vibrations and shear waves at 100 Hz.

Table 1: Theoretical (FE) and experimental results for resonator with neck length of 0.85 m, neck inner diameter of 0.026m, resonator volume of 0.009 m^3, total tube length of 10.75 m, pre-resonator length of 9.2 m, post-resonator length of 1.55 m

FE Model Eigenfrequencies (Hz)	Experimental Resonant Frequencies (Hz)
22.4	17.25
33.8	33.5
44.5	37
-	52.5
64.8	69
79.3	-
85.5	85
101.2	99.75
116.7	113.5

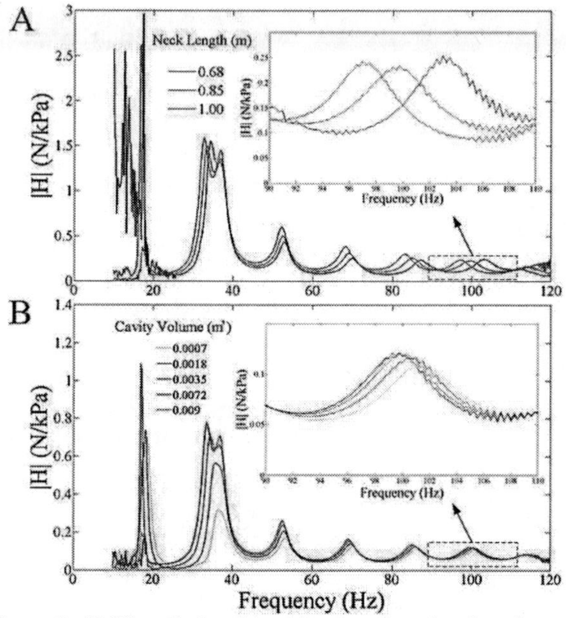

Figure 3: A) Magnitude of transfer function showing changes in resonant frequencies for varying neck lengths and cavity volume of 0 m^3. B) Magnitude of transfer function showing changes in peak amplitudes at resonant frequencies for cavity volumes ranging from 0.0007 m^3 to 0.009 m^3 and neck length of 0.85 m. Inset panels enlarge region close to 100 Hz.

Keeping other parameters constant, it was found that the neck length controlled the frequencies of resonant peaks in the transfer function, H, from input pressure to output force (Figure 3a); increasing the neck length decreased all resonant frequencies. The cavity volume controlled the magnitudes of the peaks in the transfer function (Figure 3b); increasing the cavity volume increased the magnitude of the peaks. Furthermore, experimentally observed resonant frequencies in the experiment were consistent with the eigenfrequencies of the corresponding FE model (Table 1).

DISCUSSION

A Helmholtz resonator was designed and used to alter the performance of an acoustic system for MRE. By changing its parameters, the resonator can be tuned such that its resonant frequency matches and thus amplifies a desired tone. When applied to MRE, this could lead to more accurate characterization of brain tissue *in vivo*. Furthermore, a FE model of the experimental setup was able to produce theoretical results that closely matched the experimental results. Thus, this theoretical model is potentially useful for designing MRE systems with desired resonant frequencies.

ACKNOWLEDGEMENTS
Financial support from NSF grant CMMI-1332433.

REFERENCES

[1] Venkatesh S.K., et al., JMRI, 2013, 37(3): 544-555.
[2] Guertler, C.A. et al., Abstract 1821, BMES Annual Meeting, Minneapolis, MN, Oct 5-8, 2016.
[3] Han, Myonghyon, Theses and Dissertations, paper 1015, Lehigh Univ, 2008.

NEUROCHI™ VIRTUAL REALITY SIMULATOR OF THE CEREBROSPINAL FLUID SYSTEM

G. Conley Natividad (1), Brian Cleveley (2), Lucas R. Sass (1), Tao Xing (3), Olivier Baledent (4), Vartan Kurtcuoglu (5), Bryn A. Martin (1)

(1) Biological Engineering
University of Idaho
Moscow, ID, USA

(2) Virtual Technology and Design
University of Idaho
Moscow, ID, USA

(3)Mechanical Engineering
University of Idaho
Moscow, ID, USA

(4) BioFlow Image
University of Picardy Jules Verne
Amiens, France

(5) University of Zurich
Institute of Physiology
Zurich, Switzerland

INTRODUCTION

Cerebrospinal fluid (CSF) is a clear, water-like fluid that surrounds the brain and spinal cord and is located in interconnected reservoir-like spaces within, and around, the brain called ventricles and cisterns (**Figure 1**). Total CSF system fluid volume in an adult is ~150 ml with an average pressure of ~10-15 mmHg when lying in the supine position [1]. CSF moves in an oscillatory manner in synchrony with the cardiac cycle due to cerebral blood flow pulsations and is also impacted by the respiratory cycle [2]. CSF plays a mechanistic role to cushion the delicate central nervous system tissue by the Archimedes principle and also plays a functional role to remove waste products and deliver nutrients to the CNS tissues. Recent research has shown that tracers introduced to the CSF rapidly move into the brain tissue via the perivascular spaces [3].

Many medical applications are rapidly developing that incorporate virtual and/or augmented reality (VR / AR). At present, VR is used widely as a rehabilitation tool as it can safely create an augmented reality using realistic scenarios [4]. The presence of VR in the medical field is an accessible cost cutting tool; however, less than 20% of medical VR software uses immersive, VR technology [1].

Visualization of the complex CSF system is helpful to understand its geometry and functional importance. The CSF system contains micro-anatomical features, such as nerve rootlets, the cerebellar and cerebral fissures, as well as the ventricular volumes and passages that are difficult to visualize without physical or VR representation. A VR simulator of the CSF system could be helpful as a teaching tool and as robotic simulator for neurosurgical procedures. Our aim was to create a CSF system VR simulator that allows a) endoscopic-like navigation within the CSF system and b) colocation of the endoscope location and medical imaging used to create the model.

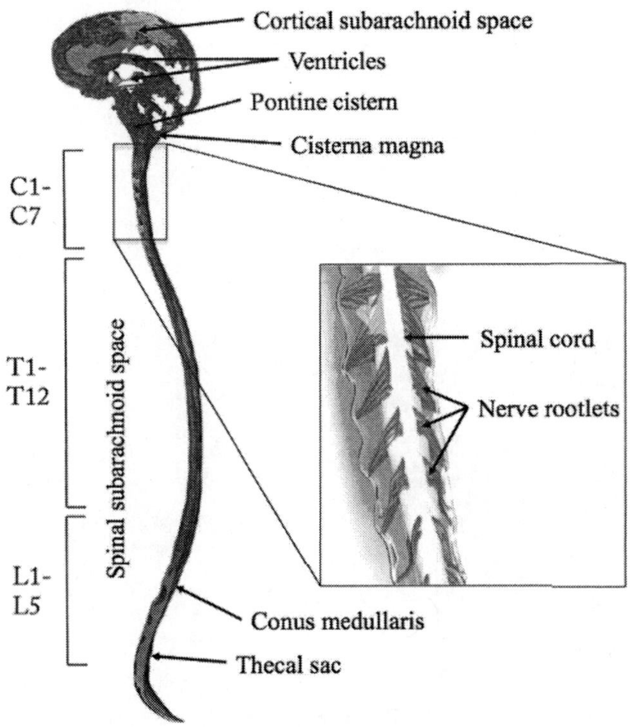

Figure 1: 3D representation of the CSF system surrounding the brain and spinal cord and within the ventricles.

METHODS

Gross Anatomy: MR images were collected within three high-resolution, T2-weighted, MRI volumes (30 x 30 x 7 cm). Each volume was segmented using ITK-SNAP (Version 3.4.0, University of Pennsylvania, U.S.A.), exported as a .STL and then imported into Blender (Version 2.77a, Amsterdam, Netherlands). The spinal geometry was then converted to a low-poly quadrilateral mesh from the native triangulated .STL format. The process, in brief, consisted of axially slicing the spinal geometry into a series of profiles that were then decimated and rejoined to reconstruct a low-resolution surface for both the dura and spinal cord. Manual organic sculpting was utilized to improve the original and reconstructed fit before final shrink-wrap.

Fine Anatomy: Due to limitations in MR image quality, only some of the spinal nerve rootlets were discernable. Each nerve rootlet was idealized as a simple cylinder; the descending angle, radicular line, and thickness were determined through a combination of best available segmentation and tabular data at each location [5]. The complex CSF volume surrounding the brain stem and cerebellum was manually segmented left of the sagittal mid-plane. In addition, the third and fourth ventricle including the Sylvian aqueduct, the foramen of Magendie and the foramina of Luschka were segmented.

VR Creation: The complete mesh was imported into 3D Studio Max (Version 2017, Autodesk, Inc.) for final mesh preparations and UVW mapping development. Prior to virtual world deployment, surface materials were developed using Mudbox (Version 2017, Autodesk, Inc.) and Adobe Photoshop CC (Version 2017.0.1, Adobe Inc.). Unity 3D (Version 5.5.0f3, Unity Technologies) was used as the target 'real-time' platform for virtual reality delivery. Once the textured model was imported into Unity, a 3D environment and navigational controls were added to navigate and explore the CSF space from a first person perspective. Finally, an anatomical map and MRI viewport were integrated into the VR environment that can be implemented in Oculus, Vive or Google VR systems.

RESULTS

The final VR model (**Figure 2**) consists of the ventricular and cortical regions of the brain, dura matter, spinal cord, coccygeal nerve and 30 dorsal and ventral nerve rootlet pairs. The primary component of the VR simulator is a navigable spinal canal with six degrees of freedom (DOF). The secondary component of the simulator is a real time anatomical index with a 3D orientation gimbal and a MRI viewport with 100% and 150% window size levels. The results of the segmentation process showed that the total volume of the spinal segmentation below the foramen magnum was 100.5 cm^3. After remeshing, the spine model had a total volume of 102.6 cm^3. The volumetric error was calculated at 2.1%.

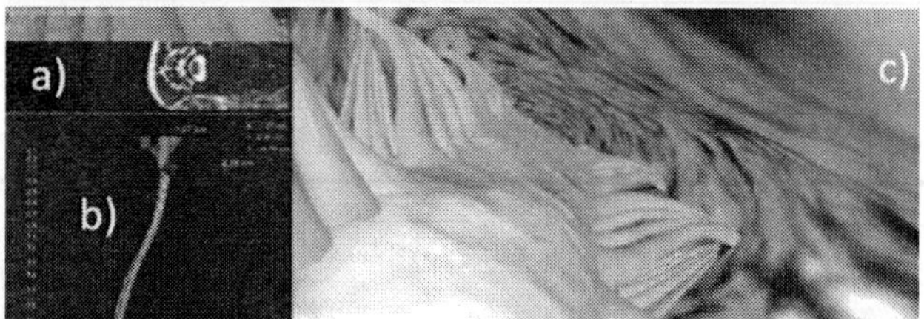

Figure 2: Neurochi™ VR simulator of the CSF system: (a) MRI data slice where user is located, (b) 3D anatomical index and locator, and (c) rendering within the spinal subarachnoid space showing the spinal cord, nerve rootlets and dura.

3D Model creation: A 3D model (**Figure 1**) was successfully created of the entire adult CSF system for a healthy adult including the spinal subarachnoid space, cortical subarachnoid spaces, lateral ventricles, 3rd ventricle, 4th ventricle, pontine cistern, Sylvian cisterns, cisterna magna and other cisterns of the brain. This model included the 30 spinal cord nerve root pairs on the dorsal and ventral side of the spinal cord with anatomic reference to subject specific MR imaging.

VR Neurochi™ software model: The VR model included three major aspects a) visualization of MR image at the viewer location within the model, b) orientation of the viewer within the 3D model and c) VR rendering of the user within the model (**Figure 2**). This system allowed the user to orient within the model in terms of space and within the medical imaging (MRI) used to create the model. This visualization is akin to the "Google street view" feature where the street map helps with overall orientation of the user.

DISCUSSION

This project represents the first VR simulator of the CSF system, Neurochi™, that is equipped with navigational viewports and indices allowing an immersive tool for learning and medical applications. Due to the spinal canal's cylindrical shape, creating navigation from a first person perspective presented the following challenges:

1) User disorientation. It was difficult for users to orient themselves with the default 6 DOF controller. We introduced an Xbox game controller that mitigated this issue. We also explored the use of fixed pathways for positional navigation while maintaining a user defined camera orientation. The path navigation option supports multiple paths, allowing the user to dynamically switch between paths. Ultimately we employed both the 6 DOF and path controllers and allowed the user to choose.

2) Surface textures and lighting solutions. Given the cylindrical and organic nature of the model, creating acceptable texture solutions involved an extensive UVW mapping process combined with development of acceptable organic texture maps. The lighting solution also presented challenges from the shape of the model and the confined nature of the navigation space. The current solution addresses global lighting affecting the entire model and local lighting affecting the space at the user's present location. The local lighting is patterned after an endoscope.

3) Movement in a confined space. While moving through CSF space, the user encounters collisions with the dura, the spinal cord and the nerve roots that can make it difficult to navigate. Our solution was to provide collision with the dura and spinal cord (inner and outer structures) and make the nerve roots invisible to collision.

ACKNOWLEDGEMENTS

This work was supported by an Intuitional Development Award (IDeA) from the National Institute of General Medical Sciences (NIGMS) of the National Institutes of health (NIH) under Grant #P20GM1033408, NIH NIGMS Grant #4U54GM104944-04TBD, University of Idaho Vandal Ideas Project, and Idaho Academy of Engineers (Engineering Grand Challenges Scholar Program).

REFERENCES

1. Martin, B.A., et al.,. Am J Physiol Heart Circ Physiol, 2012.
2. Elliott, N.S.J., et al., Journal of Fluids and Structures, 2013.
3. Simon, M.J. and J.J. Iliff, Biochim Biophys Acta, 2016.
4. Riva, G., Methods Inf Med, 2003.
5. Heidari Pahlavian, S., et al., PLoS One, 2014.

MODELING THE SKULL-BRAIN INTERFACE USING SYLGARD 527 PHANTOMS

Jake A. Ireland (1), Andrew A. Badachhape (2), Ruth J. Okamoto (1), Ramona S. Durham (2), Philip V. Bayly (1,2)

(1) Mechanical Engineering and Materials Science
Washington University in St. Louis
St. Louis, Missouri, USA

(2) Biomedical Engineering
Washington University in St. Louis
St. Louis, Missouri, USA

INTRODUCTION

Traumatic brain injury (TBI) occurs when an external force causes skull acceleration, which in turn leads to brain tissue deformation and damage. Computational models of TBI can provide valuable data for injury prevention; however these models often lack accurate boundary conditions at the skull-brain interface. Using phantoms to model the transmission of motion is useful for understanding TBI mechanics. Magnetic resonance elastography (MRE) is a noninvasive imaging method that has been used to estimate the mechanical properties of soft biological tissue from measured displacements during applied harmonic vibration[1]. In this study, two cylindrical phantoms were fabricated to model the mechanical behavior of the skull-brain interface during MRE. A homogeneous phantom was constructed to model only the skull and brain and an "interface phantom" was constructed to simulate the skull, brain, and a soft interface. Brain tissue is simulated using a two-part dielectric gel Sylgard 527, with tunable properties[2]. To assess how different material interface affected motion transmission; both phantoms underwent MRE testing at 50 Hz. Finite-element modeling of the MRE experiments was also performed to characterize interfacial effects.

METHODS

The phantoms were fabricated by mixing Sylgard 527 base (Part A) and catalyst (Part B) in their designated ratio, degassing the mixture, and then setting it aside to cure at room temperature. Sylgard cylinders were enclosed in acrylic casing with an inner diameter of 152.4 mm and a height of 127 mm. Different ratios of Sylgard base and catalyst respectively, were tested to determine an appropriate mixture approximating the storage modulus of brain tissue at 50 Hz (~2-3 kPa)[3]. Using dynamic shear testing (DST), the storage modulus of a sample was approximated as a function of frequency (Figure 1)[4].

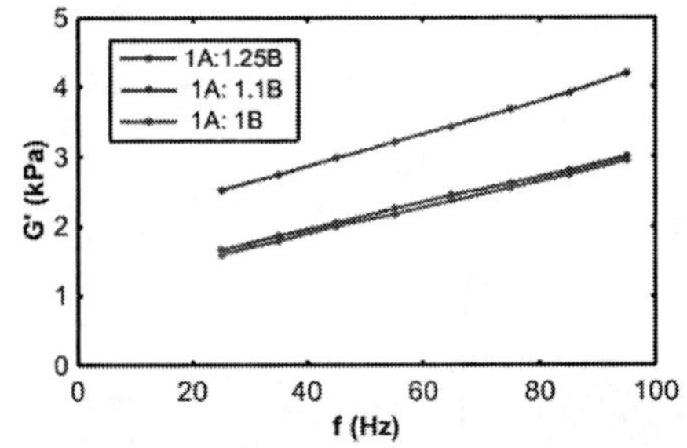

Figure 1: Storage modulus of different ratios of Sylgard 527 was estimated from DST. Ratios of base (A) and catalyst (B) included 1:1 (red), 1:1.1 (blue), and 1:1.25 (green).

Lower concentration of catalyst produced Sylgard samples with low stiffness. As a result, a ratio of 1:1.25 Part A to Part B was chosen to construct the phantoms. Sylgard 527 Part A was used to simulate a liquid interface between the cured Sylgard cylinder and the case. Due to the softness of the Sylgard cylinder, flexible plastic tubes (drinking straws) were used in the interface phantom to maintain even spacing between the cured Sylgard and case (Figure 2).

Three accelerometers were attached firmly to the case of the phantom, which was placed in the bore of a Siemens Trio® 3T MRI scanner. An echo planar imaging (EPI) MRE sequence was then conducted to assess rigid-body motion and wave motion in the phantom. Imaging parameters included motion encoding gradient (MEG) strength of 25.9 mT/m with 3 MEG orientations, 8 phase offsets, and 24 axial

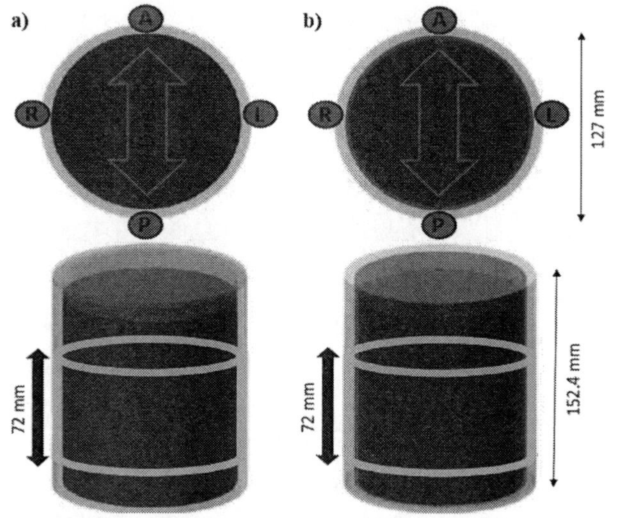

Figure 2: Schematic illustration of top and isometric views of homogenous phantom (a) and interface phantom (b). During MRE, the case undergoes harmonic motion in the AP direction. In (b), flexible supports (green cylinders) act as spacers between the Sylgard cylinder and the case. Case motion was reconstructed at Anterior (A), Right (R), Left (L), and Posterior (P) case points from accelerometer data. MRE acquisition planes (orange cylinder) encompass 24 slices at 3 mm voxel resolution.

slices with 3 mm isotropic voxels, actuator (Resoundant™, Rochester, MN) amplitude of 5%, and actuator frequency of 50 Hz. Rigid-body motion (RBM) was reconstructed from the accelerometer recordings at four points around the case center: anterior (A), right (R), left (L), and posterior (P). Motion is described by components in three directions: right-left (RL), anterior-posterior (AP), and superior-inferior (SI). MRE images were processed to estimate the three components of harmonic motion at each voxel in the image volume[5], which were separated into RBM and wave motion. Finite element (FE) simulations (COMSOL™, Burlington, MA) of the MRE experiments were performed, using a 3D model of the Sylgard cylinder excited by AP harmonic motion (15 µm, 50 Hz) of the rigid container. Cured Sylgard 527 was modeled with a storage modulus (G') of 3 kPa and loss factor (η) of 0.6. Sylgard 527 part A was modeled with $G' = 1$ Pa and $\eta = 100$.

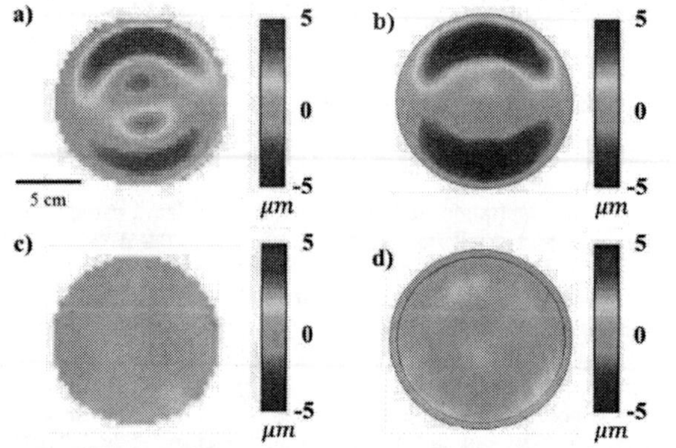

Figure 3: Experimentally collected values for SI wave displacement in the homogenous phantom (a) and interface phantom (c) are compared with simulated SI wave displacement fields in the homogeneous phantom (b) and interface phantom (d).

RESULTS

Experimental and simulated displacement fields due to wave motion were compared between the homogenous and interface phantom (Figure 3). In the experiments, the homogenous phantom exhibited higher amplitude wave motion, particularly in the SI direction. Simulations confirmed that the presence of a soft interface led to lower wave motion in the interface phantom. Quantitative comparison of case RBM to Sylgard RBM and wave motion is provided in Figure 4. Case RBM was estimated by averaging displacements estimated at the four case reconstruction points. Sylgard RBM and wave motion were estimated from the full 24 slice MRE volume. RBM amplitudes in Sylgard were similar to those in the acrylic case. Though both phantoms experienced similar excitation, the interface phantom had substantially lower AP wave motion.

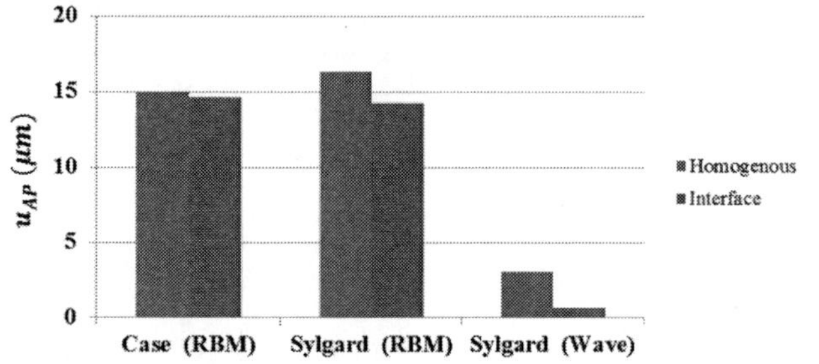

Figure 4: Comparison of average AP displacement for case RBM, Sylgard 527 RBM, and Sylgard 527 wave motion for the homogeneous phantom (blue) and interface phantom (red).

DISCUSSION

Experimental data indicate that the presence of a soft interface attenuates wave motion transmitted to the internal Sylgard gel cylinder. Unlike the homogeneous phantom, the interface phantom does not have a free surface at its end, which may further limit the amount of wave motion seen in the interface phantom. Comparisons of experimental data with simulation results confirm this interfacial attenuation. Some limitations of this study are acknowledged: the fluid interface was simulated using a solid mechanics model, for efficient simulation of harmonic motion. The Sylgard 527 samples are also slightly stiffer than brain tissue. Despite these limitations, we conclude that the presence of a soft interface provides substantial, quantifiable, attenuation of wave motion. Future work will involve the creation of more advanced physical models that will better represent the skull-brain interface, including some of the attachments points between the skull and brain (such as the stiff membranes, the falx cerebri and tentorium cerebelli).

ACKNOWLEDGEMENTS

Funding provided by NIH Grant RO1 NS-055951

REFERENCES

[1] Manduca, A., et al., *Med. Imag. Analysis*, 2001, 5(4): 237-254.
[2] Arbogast, K. et al., *J. of Biomech.*, 1997, 30(7): 757-759.
[3] Green, M., et al., *NMR in Biomed.,* 2008, 21(7): 655-764.
[4] Okamoto, R. et al., *Physics in Med. and Biol.,* 2011. 56(19):6379.
[5] Clayton E. et al., *J. Roy. Soc. Int.,* 2012, 9(76): 2899-2910.

MR ELASTORAPHY AS TECHNIQUE FOR INVESTIGATION OF BLAST INDUCED TRAUMATIC BRAIN INJURY

S. Ingram (1), G. Burnett (1), J. Vancura (2), D. Tighe (2), A. Robbins (1), M. Moreno (1,2)

(1) Biomedical Engineering Department
Texas A&M University
College Station, TX, USA

(2) Mechanical Engineering Department
Texas A&M University
College Station, TX, USA

INTRODUCTION

Over the course of the military conflicts in Iraq and Afghanistan, improvements in protective gear have reduced casualties and injuries overall, especially those due to explosive blast. This prevention of fatalities and life-threatening wounds has allowed focus to fall on injuries that, while mild in comparison, impact patients' lives in severe ways. One of these injuries is blast induced traumatic brain injury, or bTBI. BTBI is caused by the direct interaction of a shock wave with the brain, and has been called the "signature injury of America's 21st century wars" [1]. This injury has an especially high prevalence in military conflicts due to the use of improvised explosive devices (IEDs) and other explosives, and often goes undiagnosed and untreated [1].

Studies investigating bTBI may be performed using a compressed gas shock tube to simulate the injury in animal models, but methods of evaluating damage due to blast exposure vary between imaging, necropsy, behavioral studies, and others. Traditional imaging techniques such as MRI and CT provide limited information in the case of mild to moderate bTBI. Elastography is a modality of medical imaging which maps the stiffness of tissues, and could be useful in studying mechanical changes within the brain due to blast exposure. In this study, an elastographic imaging tool for studying bTBI in animal models was developed and evaluated with brain phantoms of varying complexities.

METHODS

MR elastography uses mechanical sheer waves induced in the subject to generate a displacement as a result of the sheer wave. Inverse finite element methods can then be used to calculate a map of the mechanical properties of the subject A system (**Figure 1**) was designed and developed which can produce this sheer disturbance and allows animal models to be situated in the ideal region of the MR machine with repeatable precision.

The system includes a restraint where animal models are encased for the duration of testing, and can be quickly and smoothly transitioned from the site of blast exposure to the imaging system. Within the imaging system, the restraint interfaces with aVarian RF coil (6 cm diameter), used to receive signals from the subject to produce images, so the head of the animal model is located in center of the coil along its longitudinal axis to allow for maximum image quality. The mechanical sheer wave required for elastography was produced by applying an oscillating current to a solenoid, producing an oscillating magnetic field within the solenoid perpendicular to that of the MRI. As a result, the solenoid vibrates according to the oscillation of the applied current. The solenoid is fixed to a rod which passes through a pivot point restraining it to translation on the horizontal plane, and has a clamp on the opposite end. The clamp is fixed to the head of the animal model inside the Varian RF coil.

Figure 1: Elastography System which will be used with an MRI machine to perform elastography.

In order to test the ability of this device to produce elastographic images, several MR stiffness gradient phantoms (**Figure 2**) were designed.

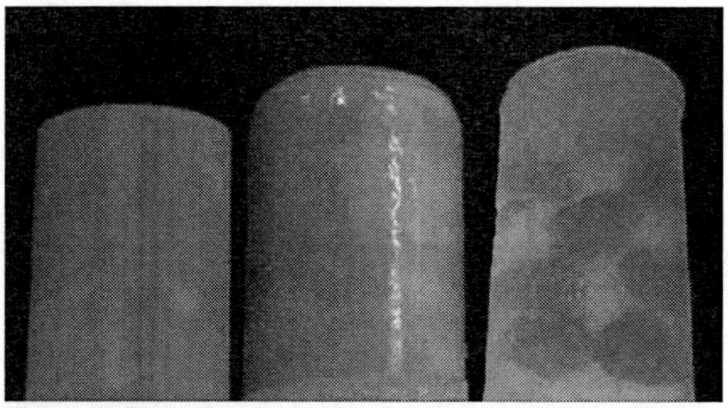

Figure 2: Elastography brain phantoms with open ABS casing with silicone end cap (left), solid ABS casing (middle), and silicone casing (right).

The most basic phantom was made of a soft silicone shell (SORTA-Clear 40 translucent silicone, Shore 40A hardness, 100A:10B). The intermediate phantom had a shell of 3D printed ABS that was open on both ends, then sealed with soft silicone caps. The shell of the third phantom was made completely of 3D printed ABS with a completely closed end. This hierarchy of shell stiffnesses allowed testing of the imaging system under a progression of phantom complexities, first to ensure the imaging equipment was capable of performing elastography, and ultimately allowing simulation of the skull of an animal model to determine whether the mechanical sheer wave was strong enough to pass into the phantom cores.

The core of each phantom consisted of water based ultrasound gel mixed with pellets of general purpose agar of ranging concentrations. The agar was made with a copper (II) sulfate solution to decrease the relaxation time of the pellets and simplify imaging.

Each of these brain phantoms was placed in the elastography system, which was then inserted into the bore of an MR machine for imaging while a 400 Hz sinusoidal sheer disturbance was applied.

RESULTS

The elastography system was able to produce displacement maps of each phantom (**Figure 3**).

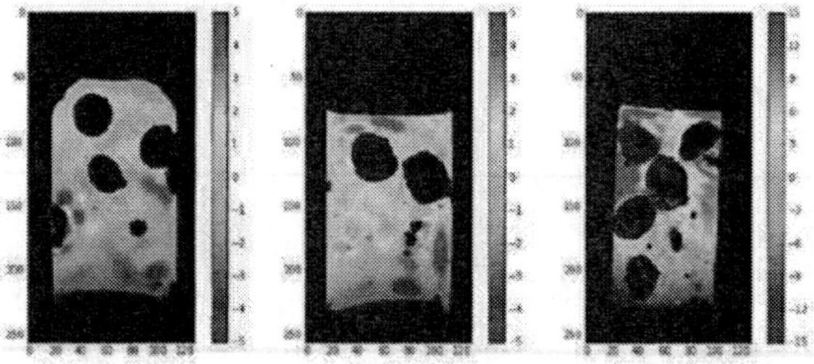

Figure 3: Displacement maps of phantoms with solid ABS casing (left), open ABS casing with silicone end cap (middle), and silicone casing (right).

These maps show the displacement of the ultrasound gel in each phantom due to the mechanical sheer disturbance. The magnitude of these displacements can be used to measure the interior stiffnesses of the phantoms. Overall, these images indicate that the system is capable of reliably producing displacement maps in a range of samples.

DISCUSSION

The images shown in **Figure 3** indicate that the imaging system can successfully perform MR elastography. The results of this study show that the mechanical sheer wave produced displaced the ultrasound gel throughout the core of each phantom, and allowed the production of displacement maps showing these changes.

It is clear from the visible differences in displacement in the ultrasound gel that the mechanical disturbance successfully penetrated even the closed ABS shell, indicating that similar results could be produced when imaging through the skull of an animal model.

The black portions shown within the ultrasound gel medium are the agar pellets of ranging concentrations. However, too much copper (II) sulfate was added during mixing, reducing their relaxation times so much that they were not imaged successfully. In the future, the pellets will be remade with a smaller concentration of copper (II) sulfate (1 g/L of water). Once this change is made, imaging can be repeated to ensure that the mechanical disturbance can travel through the phantom cores and produce displacement within the stiffer portions of agar, indicating that it is also sufficient enough to have similar effects when applied to an animal model.

It is also likely that the ultrasound gel used in the phantoms is much less stiff than what would be observed in the brain of an animal model, causing the phantoms to produce excessively promising results. To determine whether the system can be used with animal models, the phantoms will be improved to model physiological conditions more accurately by replacing the gel with a more viscous solution of agar and copper (II) sulfate.

As can be observed in **Figures 2 and 3**, none of the phantoms were designed to mimic the shape of a rat's head, which was used to design the animal restraint in the elastography system. This disparity caused deformation of the restraint during imaging that would not occur when using an animal model, so the phantom casings will be redesigned with a more realistic shape, allowing more accurate images.

After the changes mentioned above have been implemented, the updated phantoms will be imaged again. If this is successful, a euthanized animal model will be imaged in the system to thoroughly evaluate its capabilities to perform MR elastography for the investigation of blast induced traumatic brain injury.

ACKNOWLEDGEMENTS

This research was supported by the Magnetic Resonance Systems Laboratory in the Department of Electrical Engineering at Texas A&M University.

REFERENCES

[1] Prine, Carl. "U.S. combat veterans not alone with brain injuries." Pittsburgh Tribune Review (PA) 28 Dec. 2015: Points of View Reference Center. Web. 20 Dec. 2016.

HIGH FREQUENCY MAGNETIC RESONANCE ELASTOGRAPHY IN-VIVO OF THE SPINE

Sean M. Rothenberger (1), Dooman Akbarian (1), Daniel H. Cortes (1, 2), Thomas Neuberger (2), Corina S. Drapaca (3)

(1) Department of Mechanical and Nuclear Engineering
Pennsylvania State University
University Park, PA, USA

(3) Department of Biomedical Engineering
Pennsylvania State University
University Park, PA, USA

(2) Department of Engineering Science and Mechanics
Pennsylvania State University
University Park, PA, USA

INTRODUCTION

The degeneration of intervertebral discs (IVD) is closely correlated to lower back pain, which affects nearly 85% of people at some point in their lives [1]. The late stage effects of this degeneration are more easily recognizable with current diagnosing methods through the use of disc morphology, proposed by Pfirrmann et al. [2]. However, to more closely study the underlying mechanisms behind disc degeneration and the effects of possible treatments, it is necessary to have the ability to quantify the degree of degeneration in the affliction's earlier stages.

Previous studies have shown that T2 relaxation time of the IVD correlates to its degeneration. However, our previous study showed that this degeneration more strongly correlates to a decrease in the shear modulus of the nucleus pulposus (NP) [3]. Magnetic Resonance Elastography (MRE) was adapted within this previous study to measure the NP shear modulus in intact human cadaveric disc segments. MRE is a technique where small-amplitude mechanical waves are excited and mechanical properties are calculated from measurements of wave propagation. Previous tests on ex-vivo spine segments indicate an optimum frequency range of 500 to 1200 Hz [3]. To date, no MRE tests have been conducted in-vivo at higher frequencies. The objective of this study is to more closely study the hypothesized optimal frequency range of MRE in-vivo, along with study the relationship between IVD stiffness and T2 relaxation time.

METHODS

MRE Setup: An electro-mechanical actuator was designed to make contact with the subject's lumbar spine to induce shear waves in the IVD. The actuator was tuned to the proper frequency of operation by an external function generator. The design of this actuator was explained in detail within a previous abstract [4]. This actuator was then positioned inside the core of an MRI machine with the subject laying on top. Displacement data was measured using a 2D MRI sequence similar to that used in our previous study [3]. Based on preliminary data, it does not seem feasible to effectively utilize 1000 Hz, however 500 Hz provided reliable results [4]. The frequencies used in this study were 100, 200, 250 and 500 Hz. MR parameters include TR = 100 ms, slice thickness 5mm, field of view 300x300 mm, and a matrix size of 128x128. The TE values were reduced as much as possible for each frequency in order to achieve the highest signal from the tissue. The number of motion encoding gradients for 100, 200, 250 and 500 Hz were 1, 2, 2 and 5, respectively. For each of the frequency, 4 time offsets were acquired. The displacement data was analyzed using MREWave software to generate elastograms.

T2 Maps: An apparent T2 map was obtained in the sagittal plane using a turbo spin-echo sequence with 31 echoes equally spaced from 16 to 256 ms. TR was set to 2000 ms. The field of view was 300x300 mm with a matrix size of 256x256.

Subjects: Seven asymptomatic subjects were selected for the study (5 males and 2 females). The individuals tested ranged between 21 and 37 years of age with an average age of 25.57 years old and a standard deviation of 5.83 years. All MRE measurements were applied to healthy L4-L5 discs. Additionally, MRE was applied to the L3-L4 disc of one participant with signs of asymptomatic degeneration (Pfirrmann grade 3). Subjects were imaged using MRE and T2 during the same session in order to reduce any variation of IVD properties which may depend on the time of day.

Statistical Test: The possible relationship between MRE shear modulus and T2 relaxation time was evaluated using a Pearson correlation test.

Poster Presentation #P15

RESULTS

Frequency Study: Figure 1 shows the stiffness distribution of the L4-L5 IVDs at the four tested frequencies. The average stiffness values of 100, 200, 250 and 500 Hz were 2.37, 9.55, 14.50 and 50.53 kPa, respectively, along with standard deviations of 0.50, 2.24, 3.86 and 11.32, respectively as well. The stiffness values for the L3-L4 disc showing signs of degeneration were 2.96, and 50.18 for 100 and 500 Hz, respectively. The average value of apparent T2 relaxation time for the L4-L5 was 153.44 ms.

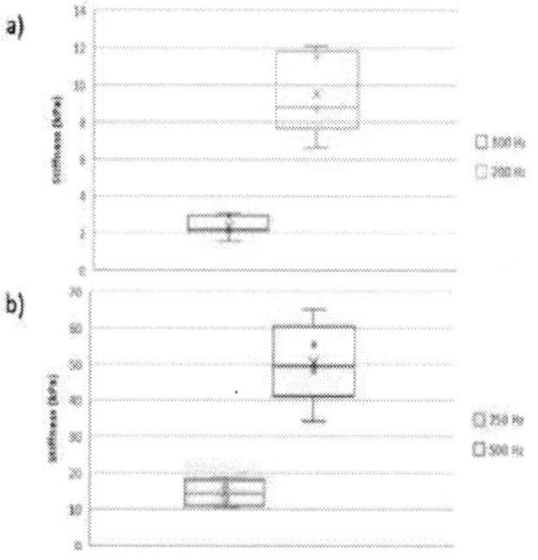

Figure 1: IVD Stiffness Values

Stiffness vs. T2 Relaxation: The stiffness values for each participant was plotted against T2 relaxation time. This correlation was separated into each frequency used to collect the stiffness values (Figure 2a and b). A best-fit line was added to the data sets in order to show a potenial relationship which exists between stiffness and T2 relaxation time. It was found that for 100, 200, 250 and 500 Hz correlation coefficients of 0.21, 0.34, 0.50 and 0.22 were present, respectively.

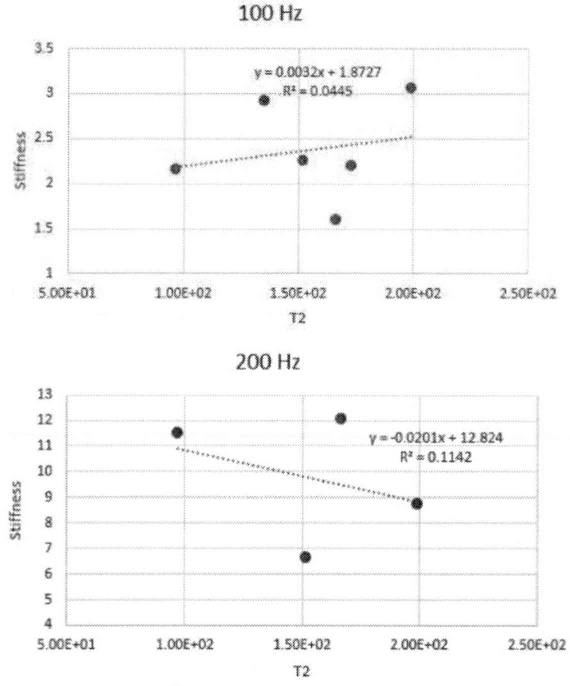

Figure 2a: Stiffness vs T2 (100 and 200 Hz)

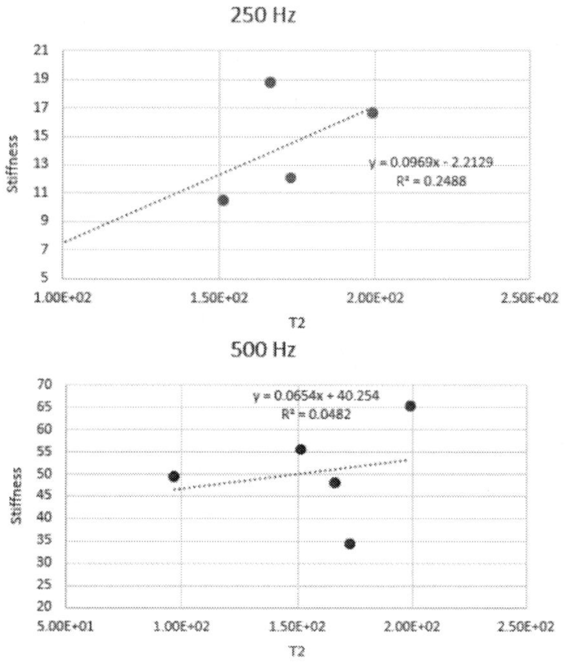

Figure 2b: Stiffness vs. T2 (250 and 500 Hz)

DISCUSSION

Frequency Study: The results gathered indicate an increase in perceived stiffness with the increase of the operational frequency. This finding is consistent with the findings in our previous study [3]. The average values and standard deviations derived from this study will serve as a range of stiffness values for healthy IVDs corresponding to the operational frequency chosen. The stiffness value of the L3-L4 degenerated disc was far above the average for asymptomatic discs at 100 Hz and close to the average value for asymptomatic discs at 500 Hz. Further studies will closely analyze this abnormal 100 Hz stiffness value to see if it is significant or not and whether this may signify an optimal frequency range at even higher frequencies.

This study has successfully demonstrated the application of high frequency MRE in-vivo. High frequency MRE could be used to analyze a wide range of impact or workplace-related injuries in the future.

Stiffness vs. T2 Relaxation: The frequency tests exhibited low correlation coefficients between the stiffness and the T2 of the discs. This shows that stiffness may be an independent means to measure IVD degeneration. Future studies will work towards collecting more data to generate a more substantial conclusion.

ACKNOWLEDGEMENTS

This study financed using start-up funds provided by the Pennsylvania State University.

REFERENCES

[1] Ben-Abraham, E et al., Magnetic Resonance Imaging 2015.
[2] Pfirrmann, CW et al., *Spine* 26:1873–1878 2001.
[3] Cortes, D et al., *Mag. Resonance in Medicine*, 00:4–8 2013.
[4] Rothenberger, S et al., *Bioeng. and Biotrans. Conference*, 2017.

AUTOMATED OPTICAL THICKNESS MEASUREMENT SYSTEM

Raghav Malik (1,2,3), Ahmet Erdemir (1)

(1) Department of Biomedical Engineering and Computational Biomodeling (CoBi) Core Lerner Research Institute, Cleveland Clinic, Cleveland, OH, USA

(2) Electrical and Computer Engineering, Purdue University, West Lafayette, IN, USA

(3) Mentor High School Mentor, OH, USA

INTRODUCTION

The ability to measure the thickness of a tissue is a requirement for many biomechanical analyses. Thickness measurements on soft tissue such as cartilage or tendons are needed to calculate stresses and strains from tensile and compression tests. Several methods already exist for determining the thickness of a tissue such as calipers, Linear Variable Differential Transformers (LVDT), and thickness gauges. Many of those rely on contact with the tissue of interest and measurement is commonly a manual process. Manual measurements necessarily come with several drawbacks. A manual thickness measurement system relies on a human to take the measurement and there is some error inherently associated with the human factor. Furthermore, for high-throughput analysis that requires hundreds or thousands of measurements the repeated process of measuring tissue after tissue quickly becomes very tedious for a human to perform. Finally, most manual thickness measurement systems directly interact with the tissue they are measuring, often deforming the specimen and resulting in inaccuracies in the results. Optical thickness measurement systems, when automatically identifying tissue boundaries, may resolve many of the issues associated with each of these methods [1,2].

The goal of this study is therefore to develop and evaluate such a system, the AutoOTMS. The system needs to operate on high-resolution medical images of the tissues eliminating the need for physical contact and therefore the chance for any kind of deformation. Furthermore, the precision of the system is bounded only by the resolution of the image, and the marginal cost of obtaining an extremely high-resolution image is relatively low when compared to other high-precision gauges such as the LVDT (a high-resolution camera can be found for under \$100 while LVDT probes can cost upwards of \$800). In addition, the identification of tissue boundary and computation of the thickness are carried out by software, resulting in minimal human input required. Thus, not only the data collection process can be both streamlined and made inexpensive, but the collected data can be consistent. The AutoOTMS deals with physical thickness units instead of voltages or resistances, necessitating very little calibration apart from the presence of a gauge of known thickness. Finally, the tissue is only present during photography, meaning that the entire system can operate in a separate environment, making it completely insensitive to environmental conditions such as humidity or temperature.

METHODS

The hardware necessary for the AutoOTMS is minimal by design, requiring no specialized components and making it universally usable for any lab. It consists of only a camera to take the images of the tissue, a frame for the tissue to be placed on (for instance, a small table with a solid dark color), a computer with AutoOTMS installed, and a single caliber or gauge. If the camera is connected to the computer, AutoOTMS can interface with it and get tissue images directly; otherwise, the images must be taken separately and then transferred to the computer with the AutoOTMS installation.

The software requirements for AutoOTMS are also very minimal. As the entirety of AutoOTMS is written in Python, it requires an interpreter for Python 3.0 or above to run. AutoOTMS does not use many external libraries to be more portable but does require NumPy for numerical manipulation, the Python Imaging Library (PIL) for image manipulation, and Tkinter for its graphical user interface.

The AutoOTMS takes as its input a single image consisting of the tissue placed in the same plane as a physical gauge of known thickness on top of a high-contrast table (to assist with detection of the bottom edge) (Figure 1). A single run of the system can determine the average and standard deviation of pixel heights of both the tissue and the caliber. These are used to proportionally calculate the actual thickness of the tissue, and the results can be stored in an XML file for future reference.

An edge detection algorithm runs on individual columns of the image to obtain thickness measurements at various horizontal locations. The three-dimensional color signal is reduced to a single dimensional grayscale value by computing a weighted average designed to mimic the eye's relative sensitivity to each of colors [3] (Figure 1). The spatial derivative of this single dimensional signal is then calculated. To account for the noise in the derivative, the peaks are clustered based on height. The spikes belonging to the "upper" cluster are labeled as edges. The process is repeated for each relevant column to get an approximation of an edge.

Errors are often unavoidable, due to discolorations or other noise introduced to the image. The error removal happens after an edge has been defined for an object. The detected edge is once again transformed into the differential domain. Peaks and valleys are identified and eliminated automatically, effectively cleaning out the edge.

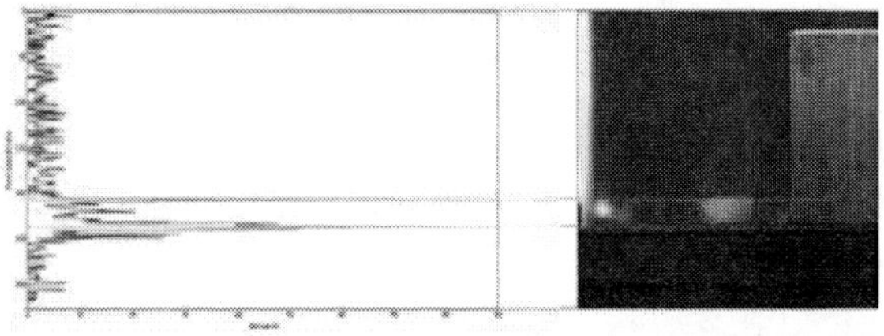

Figure 1: Spike detecting algorithm to identify tissue boundary.

Several trials were conducted to determine both the consistency of AutoOTMS measurements as well as to compare AutoOTMS measurements with those obtained using the benchmark Manual OTMS, which relies on the user picking the tissue boundary for measurements. The trials were run using two tissue samples: a femoral cartilage sample labeled FMC-MCXX (Figure 2), and a lateral meniscus sample labeled MNS_LCuX. For each trial, the tissue sample was placed on the table next to the caliber, and an image was taken of the setup and loaded in to the AutoOTMS for analysis.

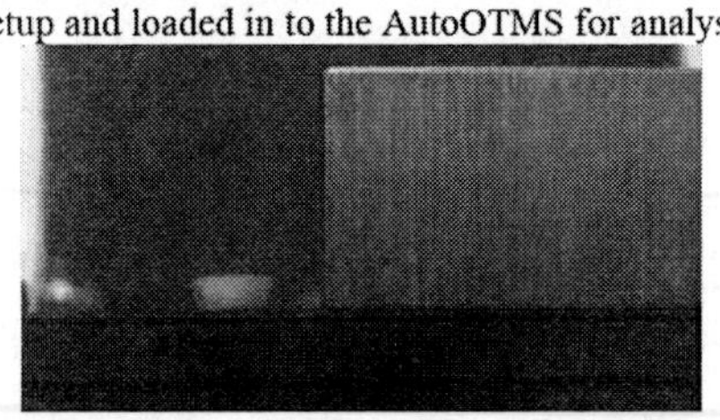

Figure 2: Setup for imaging femoral cartilage with gauge.

Measurements from the Manual OTMS were used as a benchmark because the Manual OTMS does not suffer from any of the issues associated with physical gauges as described above. Furthermore, as the Manual OTMS is currently in use by another project in the laboratory – Open Knee(s) – the sample thickness measurements were readily available for the tested tissue specimens.

RESULTS

Operation of the AutoOTMS is shown in Figure 3. Repeated measurements of femoral cartilage and lateral meniscus samples are provided in Table 1 along with the measurements of the Manual OTMS.

Table 1: AutoOTMS vs Manual OTMS Results

Sample/Trial	FMC-MCXX	MNS_LCuX
Trial 1	2.030 mm	1.013 mm
Trial 2	2.027 mm	0.996 mm
Trial 3	2.028 mm	1.088 mm
Average	2.0276 mm	1.0323 mm
ManualOTMS	1.952 mm	0.854 mm

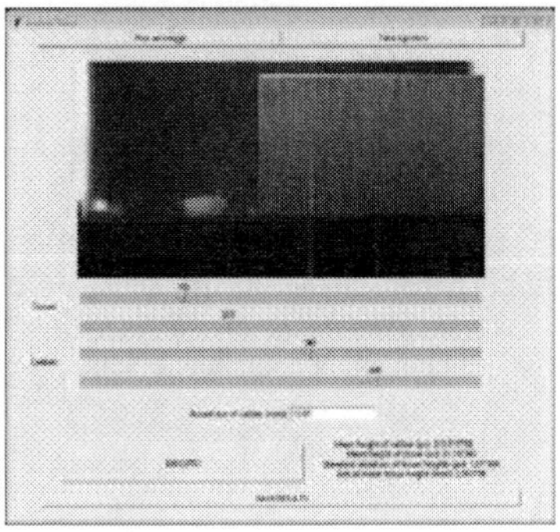

Figure 3: AutoOTMS Output

DISCUSSION

For the test samples, the measurements of the AutoOTMS analysis appeared to be consistent depending on tissue type. The cartilage thickness measurements were within 0.003 mm of each other (<0.2% uncertainty), and the meniscal thickness measurements were within 0.092 mm of each other (<9% uncertainty). The mean thickness for the femoral cartilage was close to the benchmark Manual OTMS thickness, with a difference of approximately 3.9%. However, the meniscal sample was much further from its benchmark value of 0.854 mm, with approximately a 21% difference from the Manual OTMS number. The irregular shape (which may affect point measurements) and color of the meniscal specimen (which may influence boundary detection) are likely the cause of the much higher error. However, this is not prohibitive for the use of the AutoOTMS as the measurements appear to be consistent. The boundary detection algorithm can be enhanced to minimize uncertainties in measurements or an additional calibration factor can be used if necessary. Further steps to be taken with this project include implementing better image processing techniques to help account for the irregularities in the meniscal specimen and hopefully reduce the error. In addition, two tissue samples are not sufficient to prove reliability or reproducibility, therefore more data needs to be collected to be able to make stronger statistical claims about the performance of AutoOTMS.

ACKNOWLEDGEMENTS

This project was funded through the Open Knee(s), which is supported by NIGMS, NIH (R01GM104139). The authors would like to appreciate the advice and guidance from Snehal Chokhandre, Craig Bennetts, and Robb Colbrunn.

REFERENCES

[1] O'Leary, S.A. et al., *J Biomech*, 46:1955-1960, 2013.
[2] Tomasetti, J.J. et al., *The Angle Orthodontist*, 74:351-357, 2001.
[3] Rec. ITU-R BT.7

DYNAMIC CHANGES IN IRIS BIOMETRICS IN NORMAL AND GLAUCOMATOUS EYES FOLLOWING PHYSIOLOGICAL DILATION

Matthew Wojcik (1), Anup D. Pant (1), Priyanka Gogte (2), Chidiebere Aninweze (1), Allie Stanley (1), Syril K. Dorairaj (3), Vanita Pathak-Ray (2), Rouzbeh Amini (1)

(1) Department of Biomedical Engineering
The University of Akron
Akron, OH, USA

(2) LV Prasad Eye Institute
Hyderabad, India

(3) Department of Ophthalmology
Mayo Clinic
Jacksonville, FL, USA

INTRODUCTION

The iris makes up the colored portion of the eye and functions to control the amount of light that can be transmitted to the lens. In low light environments, the iris constituent muscles contract and enlarge the pupil area, increasing the amount of light that can enter the eye. The angle that is formed between the anterior surface of the iris and the posterior surface of the sclera, i.e. the anterior chamber angle (ACA), is important in the pathophysiology of glaucoma. For example, in individuals with angle closure glaucoma (ACG), a narrow or closed ACA leads to blockage of the aqueous humor drainage canals and consequently to an increase in the intraocular pressure (IOP). If not managed properly, ACG can lead to blindness [1].

A laser peripheral iridotomy (LPI) procedure is typically performed in ACG patients as a way to move the iris so that the drainage canals are exposed again. During the LPI procedure, a hole is made in the tissue to equalize the pressure between the two sides of the iris. In most cases, following LPI, the IOP drops because the blockage of the drainage area is removed.

In certain patients, even after an LPI is performed the angle remains narrowed or closed. These occludable angles could still block the outflow pathway and contribute to the sustained IOP increase following LPI. While the exact reason behind the existence of post-LPI occludable angles is not known, it is postulated that the mechanics of the iris tissue and the dynamic changes that occur during dilation can provide some explanations [2]. In addition, previous examination of the eyes in ACG patients has shown that their irides are less compressible than those of healthy individuals. This incompressibility results in the iris responding differently to changes in the lighting environment. Specifically, it has been shown that the iris volume of individuals with narrow angles do not change significantly as a result of dilation [3, 4]. Our study focused on patients with occludable angles that had received an LPI procedure. We hypothesized that the irides of these individuals would react differently when dilated compared to the irides of a healthy control group.

METHODS

Images of the anterior segment cross section were taken under dim and standard lighting conditions (Fig. 1) at the LV Prasad Eye Institute in Hyderabad, India. The imaging was performed on two different populations: patients that have occludable angles and had undergone an LPI procedure to lower their intraocular pressure (n=20) and a control group of healthy individuals (n=20). The images were obtained using a Visante anterior segment optical coherence tomography device (Carl Zeiss Meditec, Inc., Dublin, CA). Each image was then analyzed using an in-house software to quantify fifteen different parameters of the cross section of the anterior segment. Image analysis was performed by two different individuals in order to minimize any subjectivity that could have affected the results of a single technician.

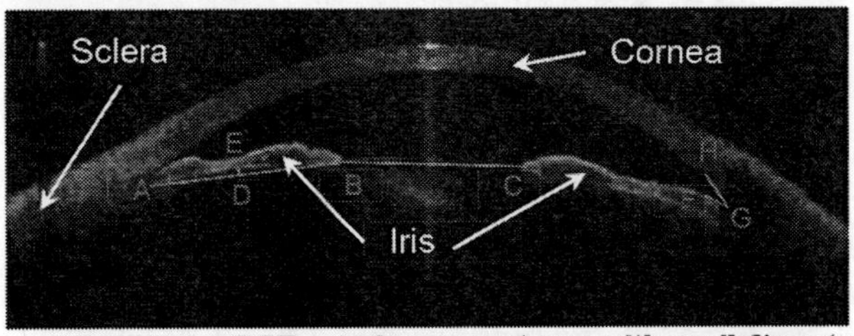

Figure 1: An OCT anterior segment scan with pupil diameter (BC), iris chord length (AB), iris concavity (DE), and anterior chamber angle (FGH) displayed.

A total of four scans were taken at each lighting condition for each subject, representing a view of the anterior segment at 0, 45, 90, and 135 degrees with respect to nasal-temporal (i.e. vertical) axis. Some of the parameters (e.g. iris cross-sectional area) were measured on the temporal and nasal sides of the scans to account for the clock hour positions corresponding to the other four angles. Therefore, a total of eight clock hour images were measured for each lighting condition and each subject. An average value of the eight different positions was calculated.

In the current study, only a number of parameters involved in the dilation-induced dynamic deformation of the iris were used in the statistical analysis. In particular, we examined the iris volume, pupil diameter, iris chord length, and iris concavity (Fig. 1). Iris volume was calculated as a function of the cross-sectional area of the iris. Pupil diameter was identified as the distance between the two tips of the iris on each scan. The iris concavity was defined as the distance between the posterior surface of the iris and the iris chord length (Fig.1). A single tailed student t-test was used to check for any significant differences in the normal and post-LPI groups between the standard and dim lighting environments. A p-value less than 0.05 was considered significant for this study.

RESULTS

For the normal group there was a significant increase ($p<0.05$) in iris volume as the average volume under standard light was 63.63 ± 5.76 mm^3 (mean ± standard deviation), up from 61.39 ± 6.25 mm^3 in dim light. The same increase was not observed in the post LPI group as there was no significant change in the iris volume of these individuals ($p>0.05$, Fig. 2). Similarly, there was a significant difference ($P<0.05$) in the iris concavity of normal patients increasing from 0.214 ± .059 mm under dim lighting, to 0.231 ± .058 mm under standard light. However, no significant increase was seen in the concavity of the post-LPI group.

Both groups experienced a significant change in pupil diameter due to a change in the lighting conditions. The pupil diameter was significantly larger under dim light dilation compared to a standard lighting environment (Fig. 3). This observation confirmed the expected behavior of the iris constituent muscles in healthy and post-LPI individuals. A related parameter, iris chord length, also showed significant changes in both groups. Chord length was significantly larger ($p<0.05$) under standard light than it was in dim light as expected.

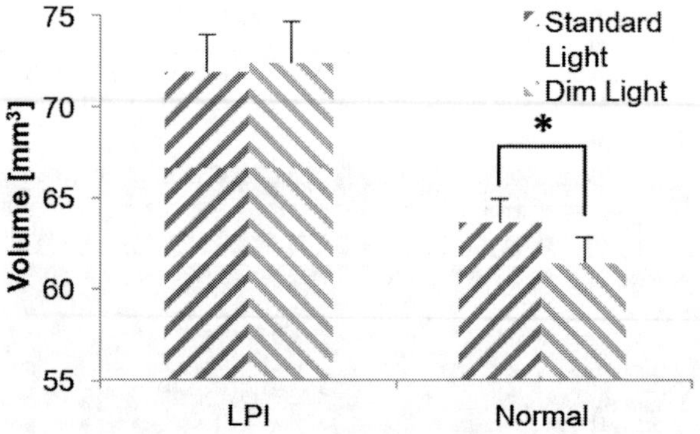

Figure 2: Changes in the iris volume for normal healthy subjects and post LPI patients under different lighting conditions (n=20 for each group). Error bars show standard error, asterisk denotes a significant difference (p<0.05).

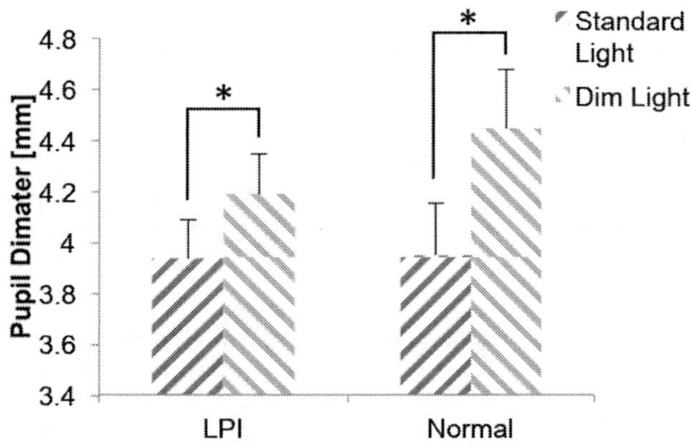

Figure 3: Changes in the pupil diameter under standard and dim light scenarios for the post-LPI group and the normal group (n=20 for each group). Error bars show standard error, asterisk denotes a significant difference (p<0.05).

DISCUSSION

The irides of individuals with occludable angles that had undergone an LPI procedure responded differently to changes in lighting conditions compared to the irides of healthy subjects. Our analysis showed that the post-LPI group retained more iris volume when dilated than the control group. We also found that iris concavity was relatively unchanged due to dilation in the post-LPI group however healthy individuals underwent a significant decrease in concavity when dilated. It has been previously shown that the patients suffering from angle-closure glaucoma have less compressible irides [4]. We observed that even after the performance of an LPI in certain population of patients, the iris maintains its incompressibility.

The LPI procedure eliminates pupillary block and equalizes the aqueous pressure in the anterior and posterior chambers of the eye. As such, the effect of aqueous humor pressure differences on the behavior of the iris is minimized in the post-LPI eyes. Therefore, the narrowing of the anterior chamber angle in occludable post-LPI eyes is most probably due to crowding of the iris periphery into the surrounding tissues following dilation as described in previous theoretical studies [5].

The differences between the dynamic behaviors of the iris in healthy individual and those with the history of angle-closure glaucoma further indicates that irides could be mechanically different in these two group. Quantification of such mechanical parameters (e.g. elastic moduli and Poisson ratio) could provide further insights into the mechanical behavior of the iris in relation to angle-closure glaucoma.

ACKNOWLEDGEMENTS

The authors would like to thank James Houston for the helpful discussions.

REFERENCES

[1] Quigley, H., *Br J Ophthalmol*, 80: 389-393, 1996.
[2] Amini, R. et al., *Invest Ophthalmol Vis Sci*, 53: 1188-1194, 2012.
[3] Quigley, H., *Am J Ophthalmol*, 148: 657-669, 2009.
[24 Aptel, F., Denis, P., *Ophthalmology*, 117: 3-10, 2010.
[5] Jouzdani, S. et al., *Invest Ophthalmol Vis Sci*, 54: 2977-2984, 2013.

ERROR ANALYSIS AND OPTIMIZATION OF NONINVASIVE ULTRASOUND ELASTICITY IMAGING FOR ESTIMATING MECHANICAL PROPERTIES OF HUMAN TENDON

Hannah Schmitz
University of Arizona
Tucson, AZ, US

Liang Gao
University of Washington
Seattle, WA, US

Andres Nuncio Zuniga
University of Arizona
Tucson, AZ, US

Cindy Fastje
University of Arizona
Tucson, AZ, US

Mihra Talijanovic
University of Arizona
Tucson, AZ, US

Faculty Advisor(s)

Russell Witte
University of Arizona
Tucson, AZ, US

Daniel Latt
University of Arizona
Tucson, AZ, US

INTRODUCTION

Posterior tibial tendon dysfunction (PTTD) is a common tendinopathy that is most prevalent in women over 40 [2]. The PTT runs along the inside of the ankle and is particularly vulnerable to degeneration and overuse injury with aging. Initial treatment for advanced PTTD is often conservative (e.g., months of bracing and physical therapy), which is only successful in approximately 70% of patients [2]. Others depend on aggressive surgery for repair of the tendon; however, surgery also has some failure rate and thus risk. However, there is currently no established technique that reliably predicts which treatment option is optimal. This unmet need is a particular concern because prolonged physical therapy can be an unnecessary burden of time and expense to the patient; similarly, surgery can lead to a long recovery time with the potential for chronic weakness or damage to the tendon and surrounding tissue.

Previous studies have suggested that the mechanical properties and material composition of the tendon are altered with advanced tendinopathy or injury [2]. Moreover, these properties might be quantified using a noninvasive imaging technique that assesses the elastic properties of a tendon. We propose Ultrasound Elasticity Imaging (UEI), which combines mechanical force sensing, ultrasound imaging, and signal processing to calculate the elastic properties of human tendon. This technique has been validated in an *ex vivo* preparation of the human PTT [2]. Further work is needed to optimize UEI *in vivo* and assess sources of error (e.g., tendon size/structure, out of plane motion) for estimating tendon displacement, strain and elastic properties.

To further analyze factors that affect UEI and examine their tradeoff, we have developed a dynamic 3D ultrasound model of the human PTT. Because the simulation software allows for precise control of the tendon's deformation, error associated with key model parameters can be directly calculated. Optimization of the model parameters can then be applied to *in vivo* imaging for guiding treatment decisions and improving outcomes for patients with advanced PTTD. For this study, we analyze the effects of out-of-plane motion (OPM) and tendon strain (magnitude) on tracking error for calculating lateral strain and Young's modulus of the PTT using a simple elastic model.

METHODS

We have developed a dynamic 3D ultrasound model of the PTT for testing key parameters of UEI and sources of error. Ultrasound simulation were executed in FOCUS software [1] and Matlab™ [2]. The PTT was modelled as a long cylinder with properties shown in **Table 1**. The tendon was fixed at one end and free to stretch from the other. Ultrasound data was simulated using a 14 MHz hockey stick linear ultrasound array having similar parameters to an *in vivo* setup for imaging the PTT [2]. The deformation of the tendon was simulated according to a predetermined "programmed" strain profile. Two-dimensional displacement and strain between ultrasound frames were calculated using a 2D cross correlation phase sensitive speckle tracking program [2]. The strain is measured by linear fitting of averaged displacement values across each column for identifying the slope, i.e. the overall strain. For this study, a simple linear elastic model was implemented. Young's Modulus is calculated from stress divided by strain.

RESULTS

OPM exceeding 1.5 wavelengths (231 μm) creates decorrelation below 0.70 (see **Figure 1 Left**). The correlation map corresponds to some percent error which is shown in **Figure 1 Right**. Both images were calculated using accumulated strain data for five different programmed sets with a constant defined magnitude of strain between frames. From this, the optimal strain seems to be between 2-3% between frames. **Figure 2** further indicates that, without OPM, UEI accurately estimates the lateral strain of the tendon. Error in estimating lateral strain directly affects calculations of Young's Modulus, which is computed from the slope of the stress-strain curve (**Figure 3**). For the model, stress was determined from an average Young's modulus of the PTT (450 MPa) multiplied by the programmed strain. (**Figure 4**). Strain is the primary source for error in the model because stress is typically estimated from the load and cross section area of the tendon.

DISCUSSION

The optimal strain between frames is an important parameter because this can be controlled during *in vivo* UEI experiments. Given a maximum frame rate of UEI (e.g., 50 Hz) and expected strain rates of the PTT during an inversion task, an ideal step size for speckle tracking can be calculated. The variability for estimating strain across the tendon can be reduced by add adding spatial filtering during or after speckle tracking.

Future work need be done to further compare this model to *in vivo* data which has been previously collected. Additional analyses of the elastic properties could also incorporate effects of nonlinear properties.

Table 1

PTT Parameter	Programmed Value
Diameter	4.5 mm
Scatterer density	50 scatterers/mm^3
Width of alternating bands	300 μm-wide bands

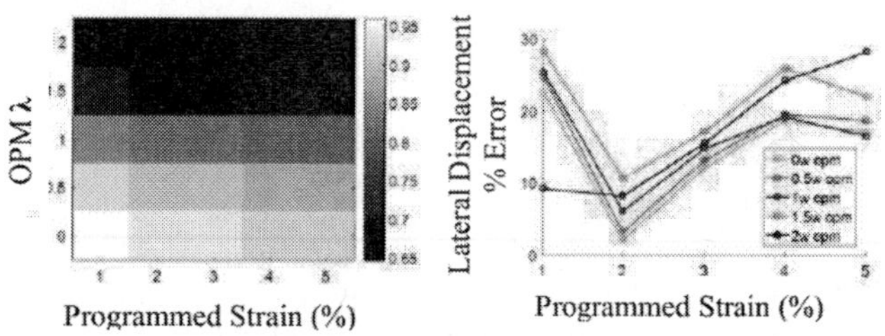

Figure 1: (Left) Correlation map demonstrates how OPM as a function of strain magnitude causes decorrelation in speckle tracking and is a primary factor that affects error in estimating tendon mechanical properties. (Right) Effect of strain magnitude on tracking error for estimating lateral displacement.

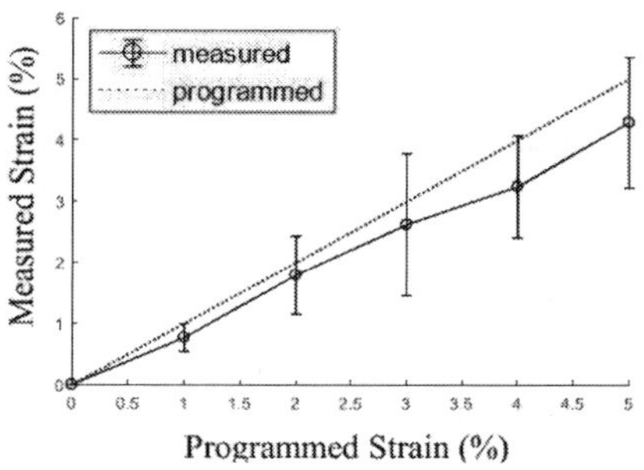

Figure 2: Demonstrates how measured data compares to expected programmed values as well as the spatial standard deviation of these measurements.

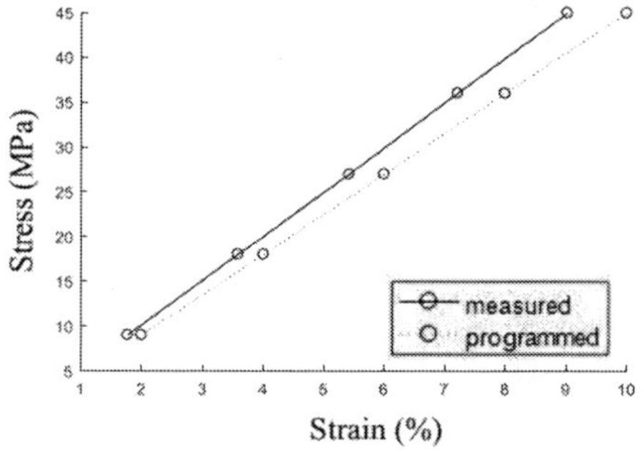

Figure 3: These mechanical properties yield Young's Modulus though analysis of the slope of a line.

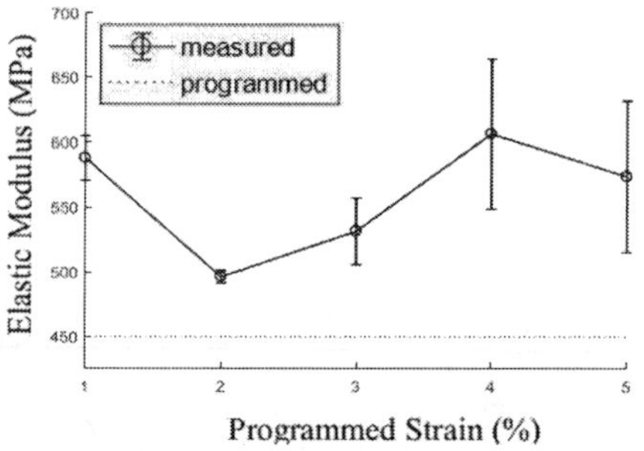

Figure 4: Measured elastic modulus for each programmed strain value relative to expected modulus with standard deviation.

ACKNOWLEDGEMENTS

Funding provided by NIH grant # NIH R21AR065732 and the Undergraduate Biology Research Program (UBRP).

REFERENCES

[1] FOCUS web page: http://www.egr.msu.edu/~fultras-web/
[2] Gao, L et al,. *IEEE Trans on Biom Engineering*, 62.4 (2015): 1179-1184.

IMPACT OF SHEAR RATE ON VON WILLEBRAND FACTOR UNFOLDING

Joshua M. Riley (1), Xavier J. Candela (1) William O. Hancock (1), Peter J. Butler (1), Keefe B. Manning (1,2)

(1) Department of Biomedical Engineering
The Pennsylvania State University
University Park, PA, USA

(2) Department of Surgery
Penn State Hershey Medical Center
Hershey, PA, USA

INTRODUCTION

When a patient suffering from heart failure is waiting for a viable heart donor for transplant, a left ventricular assist device (LVAD) is often used to maintain blood circulation. Acquired von Willebrand Syndrome (AvWS) is a frequent problem for patients with implanted centrifugal or axial continuous flow LVADs [1]. Similar to von Willebrand Disease (vWD), the most common inherited bleeding disorder, AvWS results in spontaneous bleeding, prolonged bleeding time, and increased blood loss in the event of injury [2]. This syndrome is caused by abnormalities in the von Willebrand factor (vWF), which helps recruit platelets to the site of blood vessel tear or rupture. The implanted ventricular assist devices can cause shear rates exceeding 10 times that of regular blood flow, and is thought to be the underlying cause of AvWS. The wall shear rates of normal, healthy blood flow range from 300-1600 s^{-1} in arteries and arterioles [3]. In both axial and centrifugal LVADs, blood experiences shear rates of over 10,000 s^{-1} [4] and sometimes reaching over 100,000 s^{-1}. This increased shear induces conformational changes in vWF and the cleavage of heavy weight multimers responsible for normal clotting.

Whether AvWS results from the mechanical cleavage or increased proteolytic cleavage by ADAMTS13, an enzyme that maintains vWF length homeostasis, is yet to be resolved. The duration and magnitude of shear that causes the opening of the vWF A2 domain to degradation by ADAMTS13 is currently unknown. To design a ventricular assist device that does not result in AvWS, a minimum shear rate threshold must be established to preserve the function of vWF.

METHODS

To measure the shear induced vWF conformational change, the protein was adhered to 1-µm radius polystyrene beads. The beads were suspended in a DPBS/BSA/glycerol/NaN_3 solution and isolated using an optical trap. The displacements of the beads, induced by stage motion, were measured using a quadrant photo diode system. The bead was trapped using an IX71 Olympus epifluorescence inverted microscope with computer-controlled piezo-z-form, a Mad City Labs piezo tilt actuator, and an Arroyo 200W 635nm picosecond pulsed laser.

Uncoated, serving as a control, and vWF-coated beads were sheared at rates of 1253, 2010, and 3012 s^{-1} for 30 seconds prior to data collection. The purpose of this pre-shearing is to unfurl the vWF for testing conformational change and dislodge any free-floating vWF. Using in-house LabView programs, the displacement of the bead and phase difference between microscope stage and bead movement were measured during data collection. Each bead's displacement and phase are collected over ten consecutive 20-second intervals. The ten values are averaged and taken as the displacement for that bead. The displacement due to velocity can be related to the radius of the bead. Hooke's Law and Stokes' Law can be set equal to one another resulting in Equation 1:

$$k^*x = v * \beta \;\; or \;\; k^*x = v * 6\pi\eta r \qquad (1)$$

The "effective radius" of the bead can be calculated by rearranging Equation 1 to form Equation 2.

$$r_{eff} = \frac{k*x}{v*6\pi\eta} \quad (2)$$

The control and experimental bead solutions were placed on a microscope slide in distinct chambers (Figure 1) separated by a sealed boundary, underneath a single coverslip. Measuring control and vWF-coated beads with the same slide and coverslip ensures consistency in optical trap measurements and creates a closer comparison between the values measured for effective radius and phase.

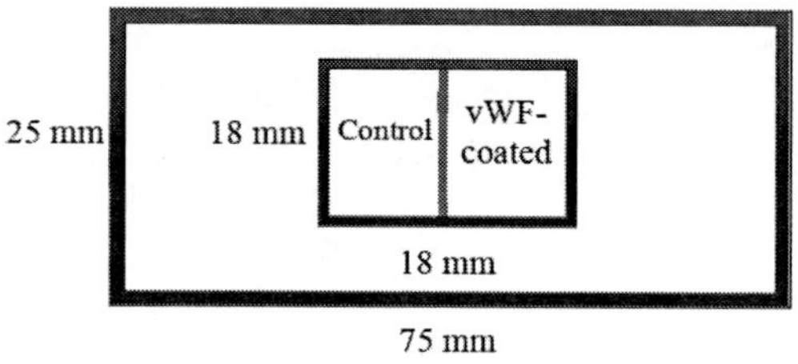

Slide thickness- 0.96-1.04 mm Coverslip thickness- 0.15-0.19 mm

Figure 1: Dual-chamber microscope slide setup for data collection (not to scale)

RESULTS

The effective radii and phase were measured following application of shear (n=10), as shown in Table 1. Control data were also collected at the same shear rates. The effective radius of the vWF-coated beads increased with increasing shear rates. The vWF-coated beads have significantly larger effective radii than the control as shown in Figure 2.

Table 1. Effective radius and phase averages

Shear Rate	Effective Radius (μm)	Standard Error (μm)	Phase (degrees)
vWF – 1253 s^{-1}	1.905	0.164	72.4
vWF – 2010 s^{-1}	2.143	0.190	73.2
vWF – 3012 s^{-1}	2.461	0.101	73.9
Control	1.194	0.046	69.9

The average radius of the control beads was measured at 1.194 μm. The control beads maintained a constant radius value when experiencing different shear rates prior to data collection. There was neither significant change in phase between the vWF-coated beads at the tested shear ratess nor a difference between the phases of the vWF-coated and control beads.

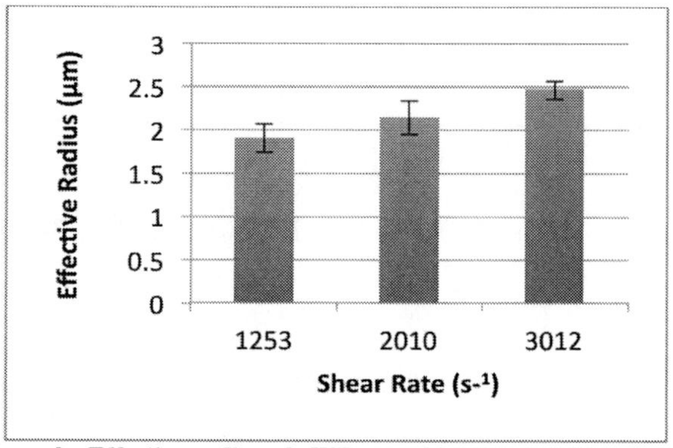

Figure 2: Effective radius of vWF-coated beads at each shear rate with error bars

DISCUSSION

The significance of these results is that an increase in shear causes an increase in the effective radius of the vWF-coated beads. This indicates that the shear rates present in LVADs unfurl the vWF more than normal flow. As the control bead radius was not affected by the increase in shear, the increase in radius of the vWF-coated beads can be contributed solely to conformational changes in the vWF protein (Figure 3). Knowledge of the shear rate at which vWF-coated bead radius plateaus at a peak value will determine the maximum shear for the opening of A2 domain to possible ADAMTS13 cleavage.

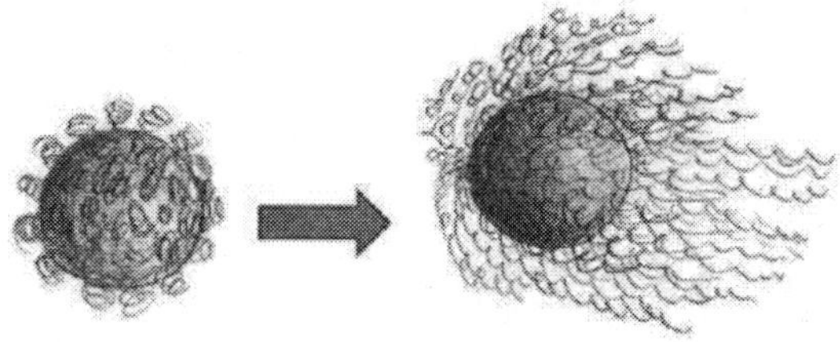

Figure 3: Indicated conformational change in vWF by shear application [5]

ACKNOWLEDGEMENTS

Funding was received from the Penn State Grace Woodward Grant and the Rodney A. Erickson Discovery Grant Program.

REFERENCES

[1] Meyer, A. et al *JACC Heart Fail*, 2(2): 141-145, 2014.
[2] Uriel N. et al. *J Am Coll Card.*, 56:1207-13, 2010
[3] Kroll, M. et al. *Blood*, 88(5):1525-1541, 1996.
[4] Selgrade, B. et al. *Artif. Organs*, 36(4):89-96, 2012.
[5]Corsetti, M. BS Thesis at The Pennsylvania State University. 2014.

SB³C2017
Summer Biomechanics, Bioengineering and Biotransport Conference
June 21 – 24, Tucson, AZ, USA

Left Coronary Artery Thermal Modeling During Targeted Hypothermic Cooling

Tyler C. Diorio (1), Nesrine Bouhrira (2,3), Jennifer E. Mitchell (2), Thomas L. Merrill (2,3),

(1) Department of Chemical Engineering
Rowan University
Glassboro, NJ, USA

(2) FocalCool, LLC
Mullica Hill, NJ, USA

(3) Department of Mechanical Engineering
Rowan University
Glassboro, NJ, USA

INTRODUCTION

Heart disease is the leading cause of death in America. Each year about 735,000 Americans experience a heart attack and 610,000 die from heart disease [1]. Following heart attacks, as well as strokes, large areas of tissue death follow oxygen delivery deficiencies during an ischemic event. Research has offered a way of reducing ischemic related damage by providing cooling to the afflicted region to induce a state of hypothermia.

Current cooling methods target the entire body. However, these approaches have limitations including low cooling rates and side effects such as shivering [2]. Localized intracoronary cooling provides rapid and direct cooling without systemic side effects. One direct approach is a cooling catheter that can be inserted directly into the blood pathways that are nearest to the ischemic region [3]. This allows for immediate delivery of cooling to the specific region at risk.

The efficacy of this method naturally depends heavily upon the vascular structure of each patient. Since this structure can vary greatly between each patient, to provide optimal therapeutic hypothermia treatment care providers need to understand the impact of these structural differences. Current models cannot accurately predict discrete temperatures at specific areas prone to an ischemic event [4].

The focus of this work is to build and explore thermal fluid models to accurately predict the cooling efficacy inside two different structures of a patient's vascular. Future work will be aimed at developing a localized therapeutic hypothermia metric for potential patient tissue salvage benefit based upon model validation through in vivo experimentation.

METHODS

To accurately model targeted hypothermic cooling in a patient, we (1) analyzed an MRI file of an actual patient using the DICOM imaging software Mimics to construct the Left coronary artery (LCA) and, (2) developed thermal fluid models of the LCA using a finite volume method to model pulsatile infusate flow through a patient's vasculature during the induction of therapeutic hypothermia.

Two LCA structures were modeled (1) a healthy coronary artery and (2) an occluded artery. The MRI file for patient HH55 was obtained through cooperation with the Visual Heart Project [5]. Figure 1 shows the 3D model of the LCA of patient HH55 as a volume rendering in Mimics. The 3D vasculature was constructed using a dynamic region growing function. This is a region-based image segmentation method that allows users to select portions of the scan based upon the value of a specified seed point. This 3D structure was then smoothed and repaired using Materialise 3Matic software.

The occluded region was removed from the initial healthy model by severing the model at the onset of the occlusion point. This was achieved using a basic cell deletion method in the surface repair tab of StarCCM+. It was then sealed by creating a flat layer of cells which was attached to the surface of the vasculature. Therefore, the blood supply was cut off entirely to the occluded portion of the model, shutting down the chilled infusate flow.

Boundary conditions for both models were set to simulate environments within a patient's heart. Therefore, they involved, (1) fixing the wall temperature of the artery to body temperature and assuming the no-slip condition, (2) fixing the inlet temperature to 28 ℃ setting the mass flow rate of the chilled blood and (3) fixing the outlet pressure to zero for the twenty-five outlets. The non-occluded model included thirteen individual outlets. Pulsatile flow was accomplished using an unsteady state solver that pulsed according to a 120 beats per minute function. The wall at the occluded region and the rest of the vasculature walls were modeled assuming a no-slip condition and a static temperature of 37℃.

Poster Presentation #P20

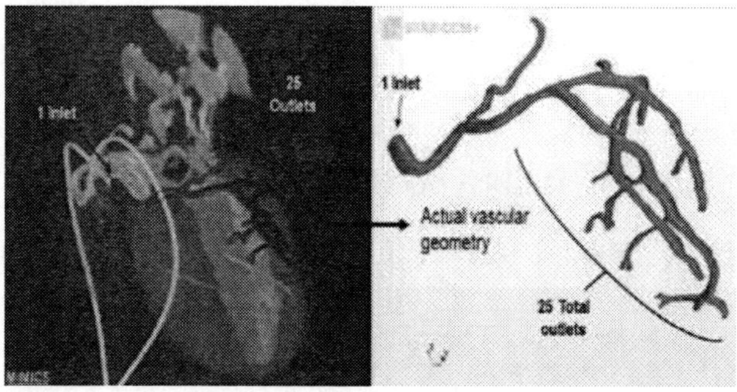

Figure 1: Left: 3D model patient HH55's LCA as a volume rendering in Mimics. Right: Geometry scene of LCA in StarCCM+. The vasculature was constructed in Mimics then imported into StarCCM+ after refinement in SolidWorks and 3Matic.

The mesh was generated using a polyhedral mesher with a total number of 318,825 cells for the non-occluded model and 272,131 cells for the occluded one. Figure 2 shows the volume mesh generated for the two models of the artery.

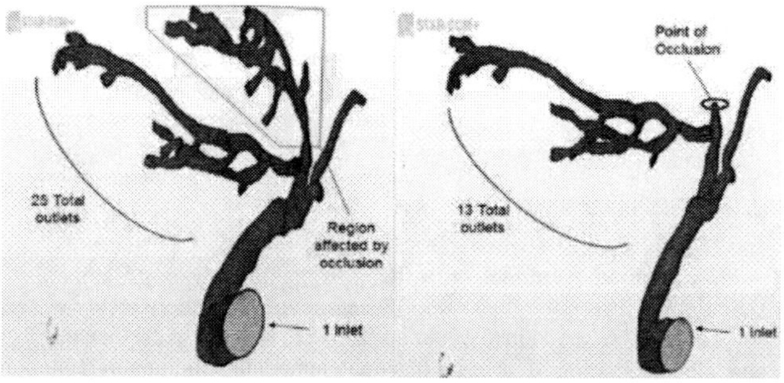

Figures 2: Volume mesh generated for the non-occluded structure on the left and the occluded structure on the right. These models include one large inlet as well as 25 and 13 individual outlets, respectively. The non-occluded mesh consisted of 318,825 cells, 5 prism layers, a base size of 8.0e-4mm, and a polyhedral mesh. The occluded model consisted of 272,131 cells with the same mesh set-up.

RESULTS

The temperature profiles of the pulsatile blood flow along the occluded and non-occluded vasculatures are shown in Figure 3. The non-occluded model showed cooling of 3-5℃ along the main branch and 0-4℃ at the most distant outlets. Given the same inputs, the occluded model showed cooling of 4-6℃ along the main branch and 1-5℃ near the farthest outlets. Overall, temperature reductions of 1-2℃ were predicted throughout the coronary vasculature. Keep in mind, the constant wall temperature boundary condition of 37°C applied in our model provides a conservative estimate of expected chilled blood infusion since thermal energy will be extracted from tissue surrounding the vasculature, lowering the actual vasculature wall temperature.

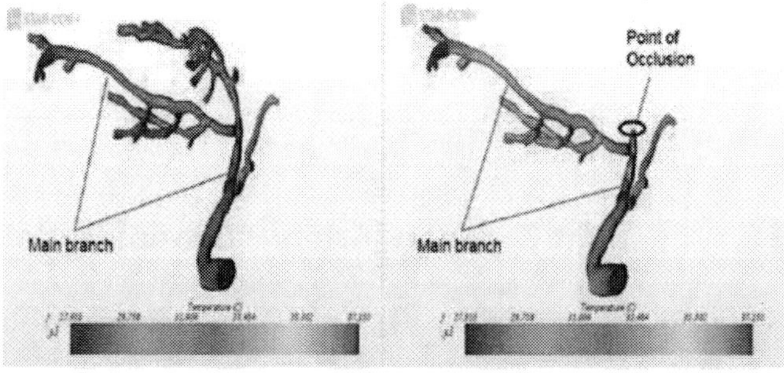

Figures 3: Temperature scenes the non-occluded (left) and occluded (right) structures in StarCCM+. There were 25 outlets with 1 inlet in the non-occluded vasculature and 13 outlets with 1 inlet in the occluded vasculature.

DISCUSSION

Based on our CFD models, the non-occluded vascular structure showed large regions that were cooled to 34℃ and lower. This temperature approximately marks the onset of useful hypothermic cooling. Regions within this range will experience reduced infarct size and increased tissue recovery [6]. With twenty-five outlets, our non-occluded simulation showed that the vascular structure of patient HH55 has the potential to distribute useful cooling over a large area to surrounding tissue.

Vascular occlusions are common in patients. The occluded model demonstrated the impact of an occlusion on vascular cooling. This simulation showed that an occlusion will force chilled blood flow to redirect from the point of occlusion and flow downstream. Occlusions also reduce the available flow area which forces coolant to flow along a more direct path to distant outlets. Compared to the non-occluded vasculature, greater cooling was observed along the main branch and at the farthest outlets.

Limitations of our work were related to (1) the high level of complexity in modeling blockages in the vascular structure of a patient (2) unavailability of experimental data to validate the models with and (3) MRI files are available only in moderate quality which hinders our ability to capture the entire vasculature of the LCA.

Future work will include performing in-vitro testing to validate our CFD modeling to help achieve a successful in vivo testing. Investigating higher quality imaging of different patients will also be considered to allow for more accurate CFD predictions as well as different vasculatures comparison.

ACKNOWLEDGEMENTS

"Our work was supported by NIH's National Heart, Lung and Blood Institute (NHLBI). Grant Number: 1R43HL117403-01. The content is solely the responsibility of the authors and does not necessarily represent the official view of NHLBI or NIH."

REFERENCES

[1] "Heart Disease Facts," Heart Disease in the United States, 10-Aug-2015. [Online] [Accessed: 12-Feb-2017].
[2] Kurtz A., et al, J. Clin. Anesth., 7 5 , pp. 359–366.
[3] Merrill T.L. et al, *J. Med Devices*. 4(3): 035001(8pgs), 2010.
[4] Konstas A.A. et al. *J Appl Physiol*. 102(4):1329–40, 2007.
[5] Tinen Iles, Private Comunication, Visual Heart Lab, 13-Sept-2016
[6] Dae, M.W., Am. J. Physiol. Heart Circ. Physiol., 282(5):H1584-91, 2002.

SB^{3}C2017
Summer Biomechanics, Bioengineering and Biotransport Conference
June 21 – 24, Tucson, AZ, USA

IN VIVO BIOMECHANICS OF TRAPEZIOMETACARPAL JOINT

Ryan Downing (1), Ken Fischer (1), Lance Frazer (1), Nolan Norton (1), E. Bruce Toby (2), Phil Lee (2), Terrence E. McIff (2)

(1) University of Kansas
Lawrence, KS, USA

(2) University of Kansas Medical Center
Kansas City, KS, USA

INTRODUCTION

Osteoarthritis (OA) in the basal joint of the thumb (trapeziometacarpal joint) is much more common in women than men, and the joint is the second most common location for OA in the hand [1]. Basal joint OA will cause severe pain and numbness at the thumb base and loss of the ability to grip and move the thumb [2, 3]. Patients with thumb basal joint OA may find it difficult to perform everyday tasks such as turning keys or writing. Preferred treatments for early thumb basal joint OA (Stage 1) are nonsurgical, but these treatments fail for more than 60% of patients with intermediate OA (Stage 2) or worse [4]. Surgery to excise the trapezium (wrist bone articulating with the thumb metacarpal) will be performed in cases of severe thumb basal joint OA for pain relief, though the recovery time is lengthy and it results in inhibited range of motion and permanent reduction of strength [5].The purpose of this study is to compare joint mechanics in the thumb basal joint between younger (ages 20-30) and older (50+) asymptomatic females. Other factors analyzed were estrogen levels and radial subluxation of the basal joint. The study aims to find a relationship between the contact mechanics, estrogen levels, and radial subluxation with the increasing prevalence of OA by age. This may improve the ability to diagnose OA in earlier stages and allow for more conservative treatments to maintain the integrity of the basal joint thumb.

METHODS

Subjects

Six consenting female adults in accordance with the local institutional review board participated in this study. Three of the subjects were in the younger age group (20-30) and the other three were in the older age group (50+). All subjects were asymptomatic of osteoarthritis.

Measurement of Estrogen Levels

Saliva samples were taken from all the subjects and analyzed for estrogen levels.

Measurement of Dynamic Instability

Dynamic instability of the thumb basal joint was measured. Four fingers of each hand were placed on a flat surface. The thumbs were pressed together with a thin-film force sensor between them. The subjects were required to reach a force of 20 N, and an X-ray image was obtained with the hands in this position. Dynamic instability was measured as the percentage of radial subluxation seen at the basal joint.

Imaging

Each subject's non-dominant hand underwent MRI scans in both a relaxed posture and active loading tasks. A high resolution scan was taken during the relaxed position (Figure 1A), and lower resolution scans were taken for the loading tasks. The selected loading task was tip pinch. The task was performed using a pneumatic bulb with real-time visual feedback of applied pressure to ensure each subject maintained a constant pressure on the bulb. Subjects' wrists were braced to assure consistent positioning.

Segmentation and Meshing

Each scan was manually segmented in ScanIP (Simpleware). Unloaded scans were segmented to include the metacarpal, trapezium, and their accompanying cartilages, while loaded scans only included the subjects' bones (Figures 1B-C). Loaded bones were used for image registration only, whereas the unloaded models were meshed into piece-wise quadratic tetrahedral elements (C3D10) for future finite element analysis.

Registration

We first performed image registration of the loaded metacarpal to the unloaded metacarpal for each subject (Figure 1D). The respective kinematic transformations for these registrations were applied to the loaded trapezium. We then registered the unloaded trapezium to the loaded trapezium in order to obtain the kinematic transformation that puts the unloaded trapezium into its loaded configuration.

Finite Element Analysis

Using the unloaded mesh, the kinematic transformations found during the registration process were applied to each of the subject's thumb model using either ABAQUS 6.14 (SIMULIA) or FEBio 2.5. Cartilage was modeled as linear, isotropic with a Young's Modulus of 4 MPa and a Poisson's ratio of 0.45. Bone was modeled as linear, isotropic with a Young's Modulus of 12,000 MPa and a Poisson's ratio of 0.3. Contact was defined using the general contact algorithm in ABAQUS. Contact area, average contact pressure, and the average of the top 5% contact pressure were recorded for each subject.

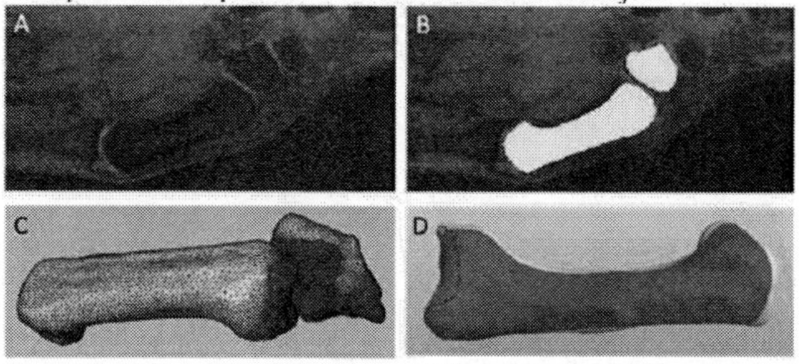

Figure 1. A) Unloaded MRI image. B) Segmentation of the unloaded image to include the metacarpal, trapezium, and cartilage (red and blue). C) Finite element mesh using the unloaded segmentation. D) Example of the registration process with the loaded bone highlighted in blue and the unloaded bone as beige.

RESULTS

Younger subjects had similar values for quantitative contact mechanics measures for tip pinch (Table 1). Peak and average pressure for all subjects are shown in Figure 2. Contact areas for two of the three older subjects were larger than the younger subjects, and the contact patterns for the older tip pinch subjects were quite different from the younger subjects (Figure 3). Neither estrogen nor subluxation appeared to correspond to joint contact measurements.

Table 1. Quantitative results for each subject for tip pinch.

Subject	20-30	20-30	20-30	50+	50+	50+
Top 5% (MPa)	2.9	2.8	2.9	3.9	3.0	2.2
Average (MPa)	1.3	1.3	1.6	1.6	0.97	1.1
Area (mm)	28.5	29.5	28.5	27.5	36.6	36.2
Force (N)	37.0	38.8	44.6	44.4	35.4	39.8
Sublux. %	45	39	68	54	33	63
E2 (pg/mL)	2.9	3.6	0.9	1.3	1.0	0.7

DISCUSSION

The hypothesis of this study was partially verified due to the fact that two of the older subjects had different contact patterns than the younger subjects. The change in the placement of contact pressure may be related to an increased risk of OA. All of the older subjects showed contact on the edges of the articular surface while the contact areas for two of the three younger subjects were closer to the center. The contact areas and forces were fairly consistent across both groups and did not present any clear trends that would indicate an increased OA risk with increasing age. Subluxation on stress X-rays was found to not have any clear association with the contact pattern or force. Thus, this may not be a useful marker for OA risk in the thumb carpometacarpal joint, but more data is needed before it should be ruled out. Lower estrogen levels were measured in all of the older subjects. One of the younger subjects had lower estrogen, but there was no deviation in the force or contact pattern. Thus, estrogen level may not be a definite marker for OA risk in the thumb basal joint.

This is limited by the low number of current subjects. Testing of additional subjects and analysis of additional functional tasks will reinforce the findings of this pilot study.

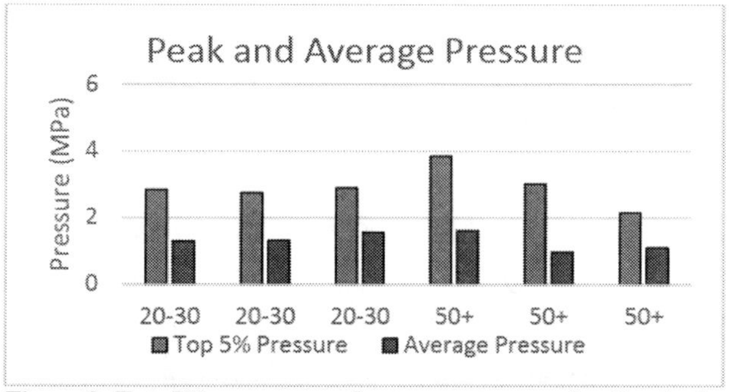

Figure 2. Top 5% pressure and average pressure for tip pinch of each subject.

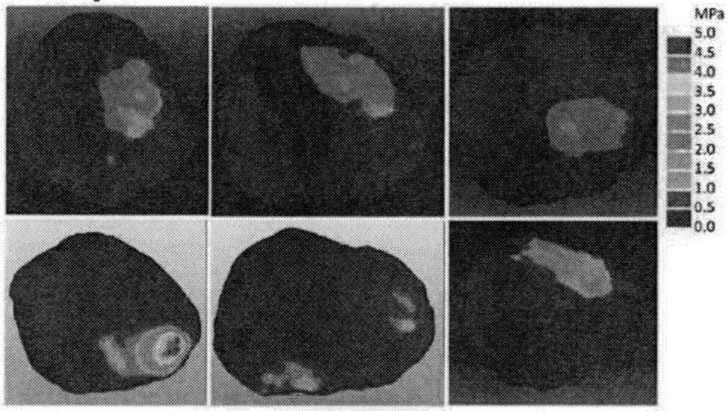

Figure 3. Contact pressures for tip pinch on the proximal metacarpal for each subject. Top row) younger subjects. Bottom row) older subjects.

ACKNOWLEDGEMENTS

This project was supported by an Institutional Development Award (IDeA) from the National Institute of General Medical Sciences of the National Institutes of Health under grant number P20 GM103418. The content is solely the responsibility of the authors and does not necessarily represent the official views of the National Institute of General Medical Sciences or the National Institutes of Health. This study was also funded by the Madison and Lila Self Graduate Fellowship at the University of Kansas.

REFERENCES

[1] Dahaghin S, et al. Ann Rheum Dis. 2005; 64:682-7.
[2] Dias R, et al. Postgrad Med J. 2007; 83:40-3.
[3] Acheson RM, Chan YK, Clemett AR. 1970; 29:275-86.
[4] Day CS, et al. Journal of hand surgery. 2004; 29:247-51.
[5] Martou G, Veltri K, Thoma A. Plast Recon Surg. 2004; 114:421-32.

SB³C2017
Summer Biomechanics, Bioengineering and Biotransport Conference
June 21 – 24, Tucson, AZ, USA

AN EXAMINATION OF STRESS CONCENTRATIONS DUE TO MYOCARDIAL INFARCTION IN THE WALL OF THE HUMAN LEFT VENTRICLE

Arlynn C. Baker (1), Sudhir Kaul (1), Heather B. Coan (2), Martin L. Tanaka (1)

(1) School of Engineering and Technology
Western Carolina University
Cullowhee, NC, US

(2) Department of Biology
Western Carolina University
Cullowhee, NC, US

INTRODUCTION

The American Heart Association estimates that about 790,000 people suffer from heart attack annually in the United States alone. During a heart attack or myocardial infarction, blood supply is cut off to portions of the heart, starving the tissue of oxygen, and often resulting in cell necrosis. Myocardiocytes, the primary cell type in heart muscle, have limited regenerative potential. Although resident stem cell populations exist in the heart, ischemia/reperfusion injuries sustained from an infarction limit regrowth of functional cardiac muscle. Instead, remodeling of damaged heart muscle often results in the formation of scar tissue by fibroblasts. These fibroblasts not only fail to function as contractile elements but also secrete fibrous matrices that possess very different material properties than the surrounding healthy muscle tissue. Both the loss of function and the differing material properties may have a significant impact on the heart as time passes, influencing remolding, function, and cell growth.

One procedure proposed to reduce the adverse effects of heart attack is cell therapy. In this procedure, stem or progenitor cells are introduced directly into the damaged tissue either through the bloodstream or by direct injection. The newly introduced cells should assist in regrowth/repair of the damaged myocardium; however, recent studies suggest that the microenvironment of damaged myocardium is not conducive to stem/progenitor repair. To adequately promote repair through cell therapy, the damaged tissue should be similar in nature to a healthy cardiac environment - supplying adequate mechanical loading, electrical stimulation, and other "environmental" factors.

In this study, we investigate the material properties of the infarcted zone. To date, very little research has focused on understanding how these material properties affect stress in the wall of the heart. The purpose of this research is to determine how different values of the elastic modulus within the infarcted zone affect the stress levels within and around the infarcted zone. This research will enable us to better understand how abnormal stresses may affect the heart during recovery, or influence stem or progenitor cell-mediated repair.

METHODS

In order to determine how stresses in the damaged heart are distributed, a three-dimensional (3D) model of the human left ventricle was created using the modeling software CREO 2.0. For this model, the ventricle was assumed to be a truncated ellipsoid with a wall thickness of 10mm, a total length of 70mm, and a major diameter of 50mm. The infarcted zone was defined as a half circle section of the ventricle with a radius of 5mm located 40mm from the bottom of the ventricle. To reduce solve time, symmetry was used by defining only a quarter section of the geometry (figure 1).

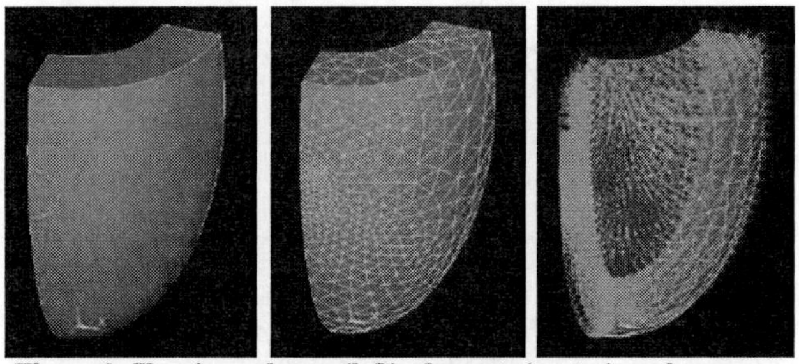

Figure 1: Showing volumes (left), elements (center), and pressure and displacement loads (right).

The 3D model was then imported, meshed, and solved in the ANSYS 17.2 ADPL environment for finite element analysis. The apex of the section was constrained in the Y direction to represent how the ventricle would naturally be constrained by the attached atrium. To implement boundary conditions resulting from symmetry, the X=0 wall was constrained in the Z direction and the Z=0 wall was constrained in the X direction. A Tet-10 element was chosen for meshing. This is a tetrahedral element with 10 nodes, each node has three degrees-of-freedom. The ventricular pressure was set at 140 mm Hg and the pericardial pressure was considered to be negligible.

The elastic modulus of both the infarcted zone and the surrounding healthy tissue was assumed to be linear, homogeneous, and isotropic. The elastic modulus of the healthy heart tissue was set at 5.88 kPa [1]. Several different elastic moduli for the infarcted zone were evaluated (160 kPa, 40 kPa, 10 kPa, 2 kPa, 0.5 kPa, and 0.125 kPa). Both tissue types were considered incompressible and assigned a Poisson's ratio of 0.49. Convergence analysis was performed to check the accuracy of the results of the finite element model.

RESULTS

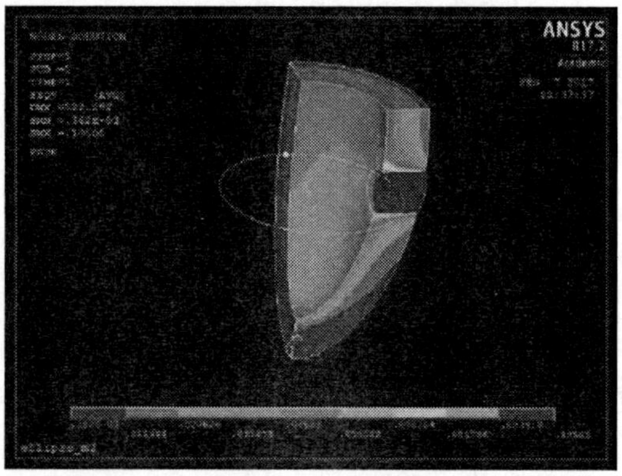

Figure 2: Element von Mises stress contour plot with circular path ($E_{infarct}$=0.125kPa).

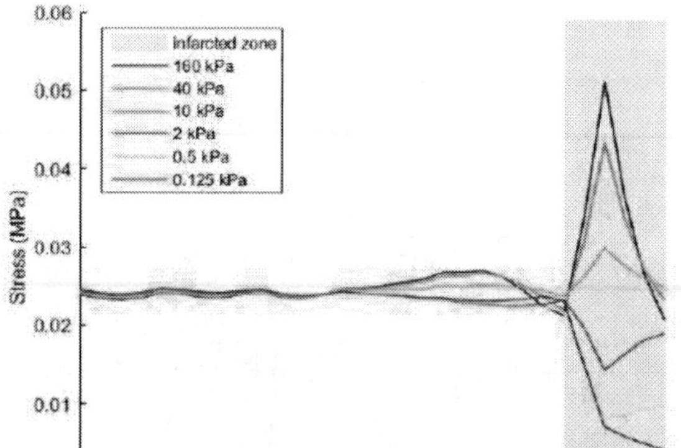

Figure 3: Von Mises stress along the circular path where s is the distance from the XY plane.

To evaluate stresses, a circular path was defined parallel to the XZ plane with the center located at Y=40mm, X=0mm, and Z=0mm and a radius of 20mm (figure 2). The model was solved multiple times, each with a different infarcted zone modulus. The von Mises stress for each different modulus was mapped onto the path and is shown in figure 3.

The results indicate that ventricular wall stress was affected by regional differences in material properties. For infarcted zones with lower elastic modulus than the surrounding tissue, lower stresses were observed within the infarcted zone. Conversely, infarcted zones with higher elastic modulus showed exacerbated stresses, especially along the border between the infarcted and healthy tissue. As the material properties move further away from the healthy tissue properties, the effect intensified. An additional observation was that elements closer to the interior surface had a higher stress than those located on the outer surface of the ventricle.

DISCUSSION

The elevated stresses seen in infarcted zones with a higher elastic modulus are likely due to added shearing, compressive, and tensile forces, which develop when the more deformable healthy tissue stretches under pressure next to the less deformable infarcted tissue. This happens all around the boundary of the infarcted zone. Towards the center of the infarcted area, these added stresses are dissipated by interfacing elements with the same material properties. In infarcted zones with reduced elastic modulus, stresses decrease through the infarcted zone for precisely the same reason. In these cases, the surrounding material is stiffer than the infarcted zone which produces a protective effect similar to stress shielding.

While many sources disagree on the magnitude of the elastic modulus of both the infarcted and non-infarcted tissues, they do agree that infarcted tissue is stiffer than healthy tissue after fibroblasts have moved in [2]. Based on our results, it is easy to see how this stiffness may affect the heart in many ways. The resulting increase in wall stresses could be a motivating factor in cardiac remolding following a myocardial infarction. Increased stresses may also affect the efficacy of cell therapy. It is known that the mechanical environment experienced by stem cells is a major determining factor in their differentiation. Increased stresses may provide a poor environment for implanted cells.

The results indicate that a softer infarcted region, achieved through early treatment or pharmacological methods, may increase the likelihood of successful cell therapy. The use of biomaterial scaffolds may also change cell therapy outcomes. Many biomaterials, including keratin, can be altered to exhibit different material properties [3]. These biomaterials may then be injected and used to reduce stresses on stem cells while also providing a scaffold on which cells can differentiate and mature.

More research is needed to fully understand how these stresses affect the heart on a long term basis. Important steps moving forward include determining an accurate range of material properties for both healthy and infarcted cardiac tissue and; taking into account the time elapsed since infarction. This may change treatment timelines and help us to understand remodeling. Future models should include various infarcted zone sizes and shapes, nonlinear and anisotropic material properties, and the inclusion of a biomaterial zone.

ACKNOWLEDGEMENTS

We would like to thank Mr. Michael Clare from the Rapid Center at Western Carolina University for his advice on how to improve our 3D models. Funding was provide by NSF Grant No. 1355872.

REFERENCES

[1] Janz, R and Grimm, F, *Biophysical J*, 13(7):689-704, 1973.
[2] Pislaru, C et al., *Ultrasound Med Biol*, 40(8):1785-1795, 2014.
[3] Rouse, J and Van Dyke, M, *Materials*, 3: 999-1014, 2010.

SB³C2017
Summer Biomechanics, Bioengineering and Biotransport Conference
June 21 – 24, Tucson, AZ, USA

DEVELOPING AND EVALUATING A MECHANICAL BIOREACTOR SYSTEM TO INVESTIGATE TENDON MECHANICS AND MECHANOBIOLOGY

Abigail R. Raveling and Nathan R. Schiele

Department of Biological Engineering
University of Idaho
Moscow, Idaho, USA

INTRODUCTION

Tendon is a strong, collagen-rich tissue that transfers mechanical forces from muscle to bone. Due to this force transfer, tendons are exposed to mechanical loads. Unfortunately, tendons are frequently injured and have poor healing ability (1). Mechanical stimuli may be an important factor in understanding tendon injury (2) and developing regenerative therapies as well as engineered tissues using stem cells (3, 4). There is a growing need to understand how mechanical forces influence cell behavior (e.g., mechanobiology) in a controlled *in vitro* environment. There is also a need to evaluate the mechanical properties of engineered tendon replacement tissues as well as native tendons to assess functional tissue development. To conduct these studies requires a system that applies mechanical loads to cell-seeded engineered tissue constructs in culture and measures their mechanical properties. Previous studies have utilized independent systems such as mechanical bioreactors or tensile load frames. A few recent studies demonstrated a bioreactor that combines these capabilities into a single system (5, 6). However, these combined bioreactor/tensile load frame systems tend to be expensive and hard to modify while off-axis forces and friction may impact force measurements. Therefore, there remains a need for a low cost, customizable, and multifunctional mechanical bioreactor system. The objective of this project was to design, build and evaluate a multifunctional mechanical bioreactor system that can be used to mechanically stimulate cells *in vitro* and to evaluate the mechanical properties of soft tissues.

METHODS

Mechanical bioreactor design: The mechanical bioreactor consists of an aluminum frame, linear actuator, load cell, and removable culture chambers (**Figure 1**). The entire system has been designed to fit into a standard cell culture incubator. The culture chambers, soft tissue grips, upper moving grip and actuator rod were drawn in a 3-dimensional (3D) CAD design software package (SolidWorks, Waltham, MA). The chambers and grips were printed in acrylonitrile butadiene styrene (ABS) plastic using a FlashForgeUSA (Los Angeles, CA) 3D printer. The culture chamber was printed with a 10% infill and the grips with a 30% infill. The soft tissue grips are fitted with stainless steel cap screws and backed with stainless steel nuts for securely gripping soft tissues. The ABS plastic culture chambers were made leak-proof by treating them with an acetone vapor bath for 1 h, and then sealing the inner portion of the chamber with cyanoacrylate. The chamber was designed to be sealed with a clear polycarbonate lid that compresses against an o-ring using stainless steel wing-nuts and bolts. The culture chambers are mounted into the custom aluminum stand and clamped in place.

Data acquisition and control: A size 14, linear actuator (Haydon Kerk, Waterbury, CT) is controlled by a motor controller (Peter Norberg, Ferguson, MO). A 1000g capacity load cell (Honeywell Sensotec, Columbus, OH) collects force data through a NI data acquisition device (DAQ) (NI, Austin, TX). Custom LabView™ programs and graphical user interface were developed to control the movements of the actuator, providing user-control over strain magnitude, strain rate, repetitions, and dwell time between stretches. Using these custom LabView™ controls, we can apply a continuous ramp for measuring mechanical properties or cyclic tensile strain for long-term cell culture studies. Calibration of the load cell was conducted using calibrated weights and actuator displacement from digital images using ImageJ (NIH, Bethesda, MD).

Mechanical analysis of tendon fascicles and collagen sponges: To evaluate the mechanical bioreactor as a tensile load frame, the mechanical properties of mouse tail tendon fascicles and collagen sponges (DSM Biomaterials, Exton, PA) were measured. The tendon

Poster Presentation #P23

fascicles (n=5) were isolated in phosphate buffered saline (PBS), clamped into the grips with sandpaper to limit slipping, and placed into the bioreactor chamber. The chamber was then filled with PBS. Cross-sectional areas and initial lengths were measured from calibrated digital images. The samples were preloaded to 0.02 N to remove the slack, and then preconditioned for 10 cycles to 5% strain, at a strain rate of 1%/s. The samples were then pulled in tension to failure at a strain rate of 1%/s while recording the force and displacement. The collagen sponges (n=3) were first cut into dumbbell-shaped specimens with a consistent gage length of 12 mm and width of 4 mm, and evaluated using the same mechanical testing protocol.

Figure 1. Custom soft tissue mechanical bioreactor and culture chamber.

Use as *in vitro* bioreactor: To evaluate the bioreactor for *in vitro* cell culture, collagen sponges were prepared as previously described (7). Briefly, adult mesenchymal stem cells (MSCs) were cultured in standard culture conditions until 70% confluence, and then seeded at 80,000 cells per construct into sterile collagen sponges (n=3) and incubated for 24 h in custom culture wells to allow initial cell attachment. The constructs were then mounted in the grips and culture chamber of the bioreactor, which had been previously sterilized in an ethanol bath. Cell-seeded constructs were preloaded to 0.02 N, and cyclically loaded to a peak tensile strain of 10% at a strain rate of 1%/s at 720 cycles/day for four days. Static loading (e.g., 0 cycles) of cell-seeded constructs was used as a control. After 4 days of loading, the constructs were fixed in 4% paraformaldehyde, stained with Hoechst to visualize the cell nuclei, and imaged using a 40x objective on a fluorescent microscope (Lecia).

RESULTS

The mechanical bioreactor system (**Figure 1**) was successfully used for both mechanical evaluation of small, soft tissues and mechanical stimulation of cells in culture. The bioreactor chamber, grips, and actuator rod were fabricated in ABS plastic using 3D printing to allow for rapid production and customization. This appeared to provide the mechanical integrity and sterile cell culture environment required for an *in vitro* mechanical bioreactor.

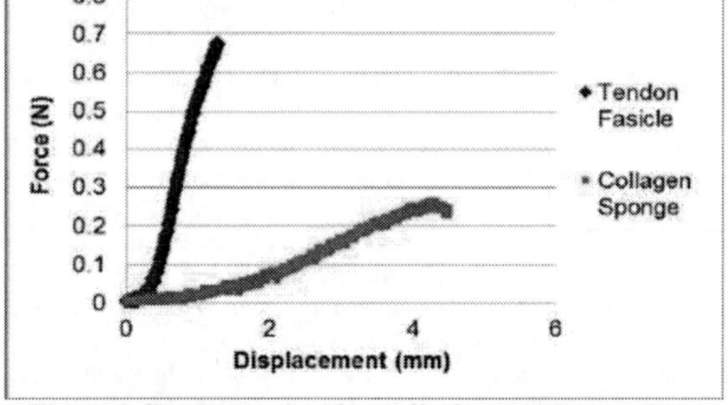

Figure 2. Representative force-displacement curve for a single tail tendon fascicle and collagen sponge.

The mechanical properties of mouse tail tendon fascicles and collagen sponges were evaluated using the bioreactor (**Figure 2**). The custom grips were capable of gripping the fascicles (average diameter of 216 ± 67 μm) without slipping. The fascicles had an average maximum (max) force of 0.59 ± 0.34 N and displacement of 1.3 ± 0.3 mm, stiffness of 0.70 ± 0.4 N/mm, max stress of 15.7 ± 4.0 MPa and strain of 9.2 ± 2.4%, and elastic modulus of 266 ± 72 MPa. The collagen sponges had an average max force of 0.24 ± 0.03 N and displacement of 4.1 ± 0.4 mm, stiffness of 0.09 ± 0.007 N/mm, max stress of 18.7 ± 2.3 kPa and strain of 20.6 ± 2.1%, and elastic modulus of 143 ± 7 kPa.

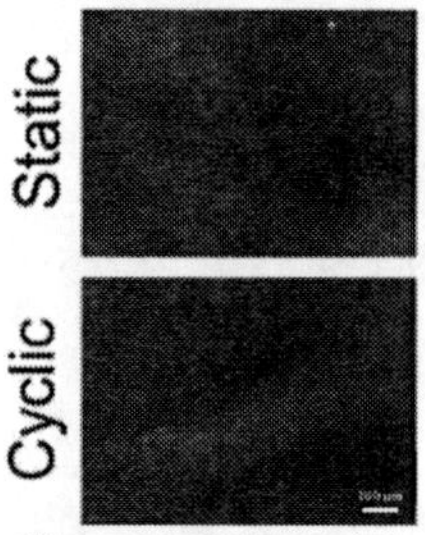

Figure 3. Cell nuclei in sponges following cyclic and static loads.

Cyclic tensile strain was applied to the MSC-seeded collagen constructs for 4 days, demonstrating the bioreactor's performance for mechanically stimulating cells in culture. A 10% cyclic strain appeared to increase the number of MSCs present in the collagen sponges, compared to static controls (**Figure 3**).

DISCUSSION

The need for a multifunctional mechanical bioreactor motivated this project. Our results demonstrate the design, fabrication, and evaluation of a custom bioreactor system that can be used to measure the mechanical properties of soft tissues as well as apply controlled mechanical stimuli to cells in culture. We also showed that 3D printing technology is feasible as a platform for developed a multifunctional bioreactor system. 3D printing is rapid, easily customizable, and dramatically lower cost, especially in comparison to commercially available systems or customized machined parts.

The mechanical properties of the tendon fascicles we measured in this study are similar to those in other studies (8, 9). Our initial *in vitro* cell culture study appeared to show an increase in the number of cells with cyclic loading when compared to static controls, consistent with other studies (5). Future studies aim to quantify this effect as well as its impact on functional tendon tissue formation by stem cells. Although more studies are needed, current results are promising and indicate that we can indeed perform long-term *in vitro* cell culture. This system does have some limitations. It is not appropriate for whole tendons from larger animals. Small imperfections in the 3D printed chamber need to be addressed through an acetone vapor bath and cyanoacrylate to prevent leaking. Future studies will use this bioreactor system to investigate the impact of mechanical loading on stem cell-seeded tissue constructs for tendon tissue engineering.

ACKNOWLEDGEMENTS

The authors thank Hee Jun Um and Dr. Jonathan Kluge for their assistance with developing the LabView™ programs. This work was supported by the University of Idaho Office of Undergraduate Research - Summer Undergraduate Research Fellowship (to A.R.R.).

REFERENCES

[1] Lin, TW et al., *J Biomech,* 37(6):865-77, 2004.
[2] Arnoczky, SP et al., *J Exp Pathology,* 88(4):217-26, 2007.
[3] Kuo, CK et al., *Tissue Eng Part A,* 14(10):217-26, 2008.
[4] Chokalingam, K et al., *Tissue Eng Part A,* 15(9):2561-70, 2009.
[5] Kluge, JA et al., *Biomed Eng,* 39(5):1390-402, 2011.
[6] Youngstrom, DW et al., *J Orthop Res.,* 33(6):911-18, 2015.
[7] Butler, DL et al., *Nat Protoc.* 5(5):849-63, 2010.
[8] Reuvers, J et al., *J Struct Biol.,* 176(1):41-5 2011.
[9] Robinson, PS et al., *J Biomech Eng,* 126(2):252-7, 2004.

SB^{3}C2017
Summer Biomechanics, Bioengineering and Biotransport Conference
June 21 – 24, Tucson, AZ, USA

THE EFFECT OF FIBER ORIENTATION ON FAILURE PATTERNS IN THE BOVINE MENISCUS DURING TENSILE LOADING

Derek Q. Nesbitt (1), Madison E. Krentz (1), Trevor J. Lujan (1)

(1) Department of Mechanical & Biomedical Engineering
Boise State University
Boise, ID USA

INTRODUCTION

Knee meniscal tears are one of the most common injuries in the U.S. with over 500,000 acute incidents in the U.S. each year [1]. Tearing of the meniscus can cause pain, swelling, joint instability, and increased risk of developing osteoarthritis [2-3]. Unfortunately, for most meniscus injuries, there is no effective treatment to fully restore meniscus function. Without effective treatment options, prevention of tears is of paramount importance.

A current barrier in preventing meniscus injuries, is a lack of knowledge on how fibrous soft tissue fails. The prevention of static failure in conventional non-biological materials has relied upon the development and validation of mathematical models that define physical limits of failure based upon stress or strain. For ductile materials, cracks propagate along the plane of maximum shear stress; while for brittle materials, cracks propagate along the plane of maximum normal stress. These experimental observations have led to the common use of distortion energy and Modified Mohr theories to predict failure risk in ductile and brittle materials, respectively [4-5]. Measuring the planes of crack propagation in the meniscus would similarly be helpful in developing models to describe failure, but to our knowledge, this information is unknown. Moreover, the meniscus is a fiber reinforced composite that consists of collagen fibers embedded in a ground substance, where the majority of collagen fibers are aligned longitudinal to the meniscus circumference. Radial and horizontal meniscus tears occur when stresses longitudinal and transverse to the circumferential fiber direction cause tensile strains to exceed a failure limit, respectively. Due to the composite architecture, failure may originate in the meniscus ground substance or fiber network, and the principal fiber alignment may influence crack propagation.

Objective. The objective of this study was to determine the failure plane of the meniscus with respect to fiber orientation. To identify the failure plane, digital image correlation (DIC) was used in combination with a high speed camera to create strain color maps during tensile testing.

METHODS

Materials. Four medial menisci were harvested from the hind knees of four unpaired bovine knees, aged 16 to 30 months. Each meniscus was vertically layered into 1 mm slices using established methods [6]. Dog-bone shaped specimens were then punched longitudinal (length = 14 mm) and transverse (length = 7 mm) to the principal fiber direction.

Testing Procedure. Longitudinal and transverse specimens were individually loaded onto a mechanical test system (Instron, Norwood MA; E10000). Preconditioning was applied by a 20-cycle triangular tension-tension wave at 2Hz and 8%-gauge strain. After preconditioning the sample was again preloaded to 0.10 N. Samples were then speckled for DIC analysis using black india ink applied using an airbrush tool with a pressure of 35 psi at a distance of 18 inches. Prior to spraying, the specimen surface was lightly dabbed with a dry cloth to ensure proper adhesion of the ink and to prevent running. After speckling, the sample was pulled in tension to failure at a rate of 1% strain/second. A total of 28 specimens were tested, but eight were not analyzed due to clamp-site ruptures, and four were discarded due to misapplication of surface ink. This resulted in 16 tested specimen (eight longitudinal and eight transverse).

Imaging. Image data during testing was collected by a high speed camera system (Photron, San Diego CA; FASTCAM Mini UX50) at 500 frames per second and a shutter speed of 1/2000 seconds. Specimens were illuminated using an LED floodlight. In order to eliminate glare on the specimen, the floodlight was placed at a 45-degree angle to the illuminated surface, and a circular polarizing lens was placed on the high speed camera.

Digital Image Correlation. For each specimen, 300 images with a resolution of 1280x240 were captured. DIC post-processing was done using Ncorr, an open source 2D-DIC MATLAB program (MathWorks Inc., Natick MA). The DIC program was validated using a rubber sample punched with the same dog-bone shaped geometry as the meniscus specimen, with a grid of black applied to the surface. Displacement and Green-Lagrange strains were calculated from this grid using ImageJ (U.S. National Institutes of Health, Bethesda, MD) and then compared to values given by the DIC program. All values between the DIC program and optical measurement were within three percent across six trials.

Analysis. Failure angle was determined by using ImageJ to take measurements of the angle of crack propagation identified by the DIC E_{yy} colormap at the time of failure. The time of failure was identified as the DIC image corresponding to the time of maximum stress on the stress-strain curve. Three measurements were averaged for each trial. A failure angle of 0 degrees signified that the failure plane was normal to the loading direction. A one-way ANOVA test was performed to determine the statistical differences between the failure parameters measured between the longitudinal and transverse specimens.

RESULTS

Transverse tensile specimens had nearly 4-fold less UTS and linear modulus than longitudinal specimens (**Table 1**). The longitudinal specimens, with the principal fibers parallel to the loading direction, had a two-fold greater failure angle than specimens loaded transverse to the principal fiber direction (**Fig. 1**, **Table 1**).

TABLE 1. Failure Parameters Relative to Fiber Orientation

Fiber Orientation	UTS (MPa)	Linear Modulus	Failure Strain (%)	Failure Angle (°)
Longitudinal	16.9±5.1	80.6±23.3	26.9±4.1	37.4±8.5
Transverse	4.2±2.1*	21.8±12.3*	25.1±5.3	18.7±13.5*

*Significantly different than longitudinal ($p \leq 0.05$)

DISCUSSION

This study measured the failure patterns that the meniscus experiences with respect to the preferred fiber orientation. The experimental data shows that failures in the longitudinal fiber alignment most closely followed the plane of maximum shear strain, which would occur in a tensile test at 45 degrees to the loading direction. This would indicate that failure during longitudinal loading initiates in the ground substance, since failure of the collagen fibers would have resulted in a failure pattern with an angle close to zero degrees. Therefore, distortion-energy theory may be appropriate for modeling ground substance failure, where the failure threshold of the ground substance governs the failure response of the fiber composite. At first glance, the results from the transverse test seem to contradict this finding, as the ground substance failed along a plane that would experience a greater normal stress and less of a shear stress. However, the transverse fiber orientation could have reduced the stress and strain experienced by the ground substance at oblique angles. The tissue also exhibited a noticeable reduction in cross sectional area prior to failure, indicating a ductile failure type.

To our knowledge, this is the first paper to measure the angle of fracture propagation in meniscus tissue using DIC techniques. However, other studies on non-biological materials have measured crack propagation using DIC and high speed photography [7-8]. For example, DIC techniques have been used to identify areas of high strain in carbon composites prior to their failure along regions of high strain [9]. In addition, studies looking at glass fiber-reinforced composites were able to use DIC techniques to visualize if failures were initiating in the fibers or in the matrix [8].

A limitation of this study is that tensile tests were performed on bovine meniscus, and results may not translate directly to human tissue. However, bovine meniscus has comparable structure and function to human meniscus, and UTS results in this study were within 12-53% of properties reported for human meniscus [6, 10, 11]. This study also did not examine the fibrous microstructure after failure, which could provide information on where failure originates. It's important to note, that on a large scale, when global failure of a composite material occurs, the crack propagation can be treated similar to that of a traditional isotropic material [12].

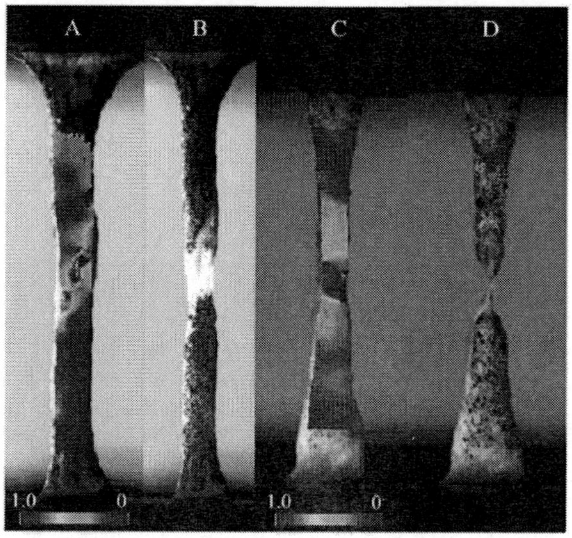

Figure 1: Representative failure images. A) Vertical strain in a longitudinal sample before B) rupture. C) Vertical strain in a transverse specimen before D) rupture.

In conclusion, this study suggests that failure in the meniscus during tensile loading first occurs in the ground substance. These experimental findings may help inform numerical and computational models that predict failure in meniscus and other soft fibrous tissue. The validation of predictive models with experimental data can eventually help clinicians determine risk factors for meniscal tears and preventative strategies to reduce this common injury.

ACKNOWLEDGEMENTS

This project was supported by the National Science Foundation under grant no. 1554353 and NIH NIGMS P20GM109095.

REFERENCES

[1] Kim, S et al, *J Bone Joint Surg Am*. 93 (11) 994-1000[2] White, L et al., *Radiology*, 222(2):421-429. 2002[3] Papalia,R et al,. *Br Med Bull* 99: 89-106. 2011[4] Shigley, J et al., *McGraw Hill*. 230-264. 2015[5] Callister, W et al., *Wiley*. 253-289. 2012 [6] Creechley, J et al., *J Mech Behav Biomed Mater*, 69:185-192, 2017.[7] Roux, S et al., *J Physics: Applied Physics*, 42:21. 2009[8] Lee, D et all., *J Composite Mater*, 2009[9]Qin, L et all., *Applied Science and Manufacturing* 43(2):310-316, 2012[10] Tanaka, M et al., *Int J Polym Mater*, 63:891-897, 2014.[11] Peloquin, J et al., *J Biomech Eng*, 138:021002, 2016.[12] Knauss, W et al, *NASA grant NAG 1-1975*, 2001.

VISCOELASTIC HEATING OF BOVINE INTERVERTEBRAL DISC

Harrah Newman (1), Robby D. Bowles (2), Mark R. Buckley (1)

(1) Department of Biomedical Engineering
University of Rochester
Rochester, New York, USA

(2) Department of Bioengineering,
Department of Orthopaedics
University of Utah
Salt Lake City, Utah, USA

INTRODUCTION

Intervertebral discs (IVD) are composite fibrocartilaginous structures that act as cushions between vertebrae in the spine. IVD are viscoelastic tissues; when cyclically loaded, viscoelastic materials can dissipate energy in the form of heat, causing an increase in the tissue temperature. This effect has been previously observed in other collagenous tissues, including equine tendons[1]. IVD undergo compression and other multiaxial deformations in the body as a result of regular movements and over time, discs can degrade, resulting in significant pain. Previous studies have shown that degenerative IVD are particularly sensitive to temperature such that even minimal heating causes increased neural firing[2]. Moreover, viscoelastic heating has been shown to adversely affect cell health[3].

Our objective in this study was to investigate whether IVD experience viscoelastic heating due to cyclic compressive loads applied *in vitro*. We also sought to characterize regional differences in IVD viscoelastic heating (outer annulus fibrosus versus inner nucleus pulposus) and how IVD viscoelastic heating depends on loading frequency. We hypothesized that IVD would exhibit measurable viscoelastic heating that increases with frequency and is highest in the more fluid-like nucleus fibrosus.

METHODS

Cyclic Strain Experiments

IVD were obtained from healthy bovine tails and frozen until testing. Disc diameters were 23.07 ± 2.55 mm (mean ± SD) and disc thicknesses were 7.23 ± 1.14 mm. Prior to testing, IVD were defrosted and kept hydrated with phosphate buffered saline (PBS). IVD were then placed between two flat, custom fabricated platens in an Instron Electropuls E10000 materials testing device. The top platen was attached to the actuator, while the bottom platen was attached to a 1 kN load cell. Thus, movement of the actuator compressed the IVD in an unconfined geometry.

After thermal equilibration (~20 minutes), the temperature of the insulated IVD annulus and nucleus were recorded with a sharp k-type thermocouple prior to testing. After applying a compressive preload of 1 N, the specimen was compressed with a triangular waveform to 20% strain for a period of 10 minutes at a frequency f of 0.1 Hz, 1 Hz or 15 Hz, depending on the trial. After the test, the temperatures of the annulus and nucleus were recorded.

Specimen Insulation

In the body, IVD are surrounded by bone and soft tissue; several methods of insulation were attempted in order to mimic the insulation that IVD experience in the body. In a set of pilot experiments, IVD were tested inside a PBS bath. However, we found that even without a specimen between the platens, the surrounding PBS exhibited a substantial temperature increase due to viscous dissipation caused by movement of the actuator in the bath. Thus, it was not possible to separate out viscous heating in the disc from viscous heating in the PBS bath and this testing method was abandoned. Instead, to avoid thermal effects from a surrounding bath, wet discs were tested wrapped in insulating sheets of polyethylene surrounded by woven fabric.

Control Experiments

Cyclic displacement control tests were performed to ensure that the materials testing system itself did not transmit heat to the disc. Specifically, discs (n=2) were adhered to the top platen of the mechanical testing equipment and the test was run independent of the bottom platen such that the IVD did not experience any compressive loads. In addition, *static contact control tests* were performed by comparing disc temperature before and after applying a nominal (0.2 N) static load for ten minutes (n=7).

Statistical Analysis

For each frequency, temperature change was compared across regions using a paired Student's t-test. Since temperature did not vary across regions (see below), temperature change was averaged across the region for each specimen and compared across frequency using a one-way ANOVA with Bonferroni post-hoc analysis.

RESULTS

Control Experiments

Both cyclic displacement control tests (data not shown) and static contact control tests (Figure 1) induced negligible temperature change in both the nucleus and annulus of IVD.

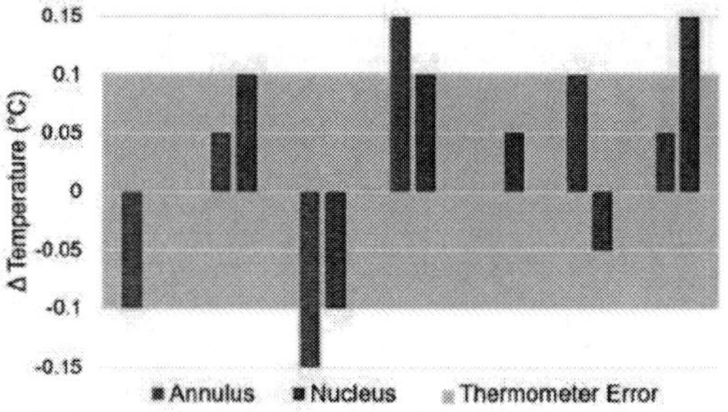

Figure 1: Temperature change for static contact control tests. The temperature change for each static contact control test (blue and red bars) was generally within the error of the k-type thermocouple (± 0.1)

Cyclic Strain Experiments

In all cyclic strain experiments, there were no significant differences in heating between regions (Table 1). Therefore, the average temperature change between the nucleus and annulus was used to represent each specimen.

Table 1: Heating between Annulus & Nucleus

Test Type	n	p
Static	7	0.692
0.1 Hz	7	0.567
1 Hz	11	0.791
15 Hz	11	0.512

Temperature change ΔT in the IVD for cyclic strain experiments was significantly higher than the static control test at f = 1 Hz and f = 15 Hz, but not at f = 0.1 Hz (Figure 2). ΔT was also higher at 1 Hz versus 0.1 Hz and at 15 Hz versus 0.1 and 1 Hz.

DISCUSSION

To our knowledge, this study is the first to investigate viscoelastic heating in IVD. As hypothesized, IVD do experience measurable viscoelastic heating *in vitro* that increases with loading frequency; however contrary to our expectation, there was no significant difference between heating of the annulus and nucleus. Importantly, all control experiments yielded negligible heating, strongly suggesting that the observed heating was due to viscoelastic dissipation.

The absence of regional differences in viscoelastic heating between the nucleus and annulus could be due to thermal equilibration during the timescale of the experiment. However, the observed increase in viscoelastic heating with frequency is

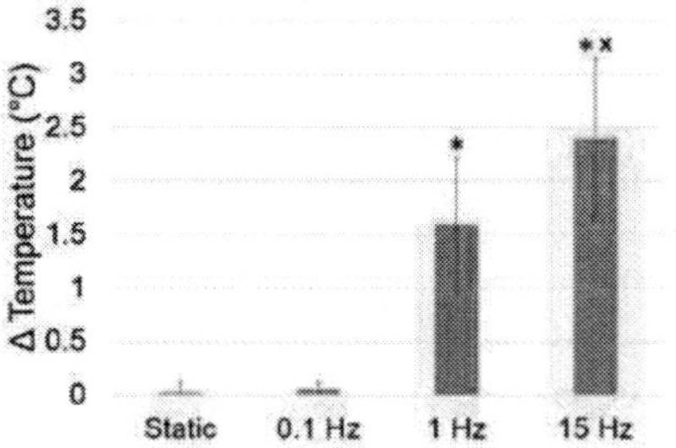

Figure 2: Temperature increase in IVD after cyclic strain experiments. Data are mean ± SD (n = 7-11 per group). * indicates $p<0.05$ vs static and 0.1 Hz and x indicates $p<0.05$ vs 1 Hz

expected since more loading cycles are performed in the same time span for higher loading frequencies, leading to greater energy dissipation. The loss tangent – a parameter that governs the capacity of a material to dissipate energy – does decrease slightly with frequency for compressed IVD, but only by a factor of ~30%[4].

This study has several limitations. All experiments were performed at room temperature and not at body temperature, though the *changes* in temperature that were assessed should not depend strongly on the initial temperature. In addition, the insulating capacity (i.e., the thermal resistivity) of the polyethylene/woven fabric used to wrap the tested specimens was not evaluated and compared to *in vivo* conditions. In the body, surrounding tissues may insulate the disc; however, nearby vasculature may also help prevent viscoelastic heating by regulating the temperature around the disc.

Future directions for this project include implementing a recently fabricated external temperature chamber to test viscoelastic heating in discs where the initial and surrounding temperature are set to 37° C. In addition, we will test healthy and degenerative human IVD, which are expected to exhibit greater viscoelastic heating due to their larger dimensions. Finally, we will correlate the observed temperature increase with viscous energy loss computed from stress/strain data. Collectively, the data acquired in this study suggest that IVD has the capacity to heat as a result of viscoelastic dissipation, and the resulting increase in temperature may be clinically significant due to its potential impact on pain sensation[2] and cell health[3] in degenerative discs.

ACKNOWLEDGEMENTS

We would like to thank Dr. Hani Awad, Emma Gira and the University of Rochester Center for Musculoskeletal Research for use of mechanical testing equipment. We would also like to thank the entire Buckley lab for assistance and support.

REFERENCES

[1]Goodship AE and Wilson AM, *J Biomech* (1994)
[2]Stover J et. al., *SB³C2016* (2016)
[3]Birch HL et. al., *J Exp Biol* (1997)
[4]Costi J et. al., *Spine* (2008)

SB³C2017
Summer Biomechanics, Bioengineering and Biotransport Conference
June 21 – 24, Tucson, AZ, USA

USING ASTM STANDARDS TO REDUCE CLAMPSITE FAILURES IN TENSILE TESTS OF SOFT FIBROUS TISSUE

Madison E. Krentz, Derek Q. Nesbitt, Jaremy J. Creechley, Trevor J. Lujan

Department of Mechanical & Biomedical Engineering
Boise State University
Boise, ID, USA

INTRODUCTION

Tensile testing is frequently used to quantify the mechanical properties of many soft fibrous tissues in the musculoskeletal system, including ligament, tendon and meniscus. The primary function of these tissues is mechanical in nature, and therefore measuring accurate mechanical properties is critical to understand injury mechanisms, wound healing, normal and pathological phenotypes, and developing meaningful computational models. In order to acquire mechanical properties that are truly representative of the material being tested, it is necessary to cut specimens to a specified geometry, called a coupon. The geometries of these coupons are selected to promote failure within the gauge length or mid-substance, where cross-sectional area is consistent and load can be reliably estimated. For conventional non-biological materials, the American Society for Testing and Materials (ASTM) has published standard geometries for tensile testing, but ASTM standards have not been developed for soft fibrous tissue. This lack of international standards has resulted in research groups using a wide range of preparation methods and coupon geometries when performing tensile tests of the meniscus and other soft fibrous tissues.[1,2,3,4] As a consequence, a high incidence of failures continue to occur at the clamps,[5] which either increases the number of experiments needed to acquire a sufficient amount of midsubstance failures, or may result in inaccurate calculations of mechanical properties if clamp failures are not excluded from analysis.

The objective of this research project is to determine the effect of coupon geometry on midsubstance failures and mechanical properties during quasi-static tensile testing of bovine meniscus. Coupon geometries were selected from specimen dimensions that have been used in the literature to test soft fibrous tissue, and from ASTM standards for materials that are most comparable to soft fibrous tissue, including elastomers and fiber reinforced composites.

METHODS

The medial menisci from six unpaired hind bovine knees were harvested, and the central region was vertically layered to a 1mm thickness using previously validated methods.[6] Test coupons were then cut using 3D printed punches with bendable razors (**Fig. 1A**). Four groups of coupon geometries were tested, which used either a rectangular or dog-bone (DB) shaped geometry. The wide rectangular coupon design was based upon recommended geometry for fiber-reinforced composites (**Fig. 1B**, ASTM D3039), while the narrow rectangular coupon increased the aspect ratio of the ASTM standard (**Fig. 1C**). The dog-bone specimens used either a wide tab design based on previous research on ligament and meniscus (**Fig. 1D**),[2,6] or a narrow tab design based on ASTM recommended geometry for elastomers (**Fig. 1E**, ASTM D638). A total of 40 coupons were acquired for testing, with ten in each of the four groups.

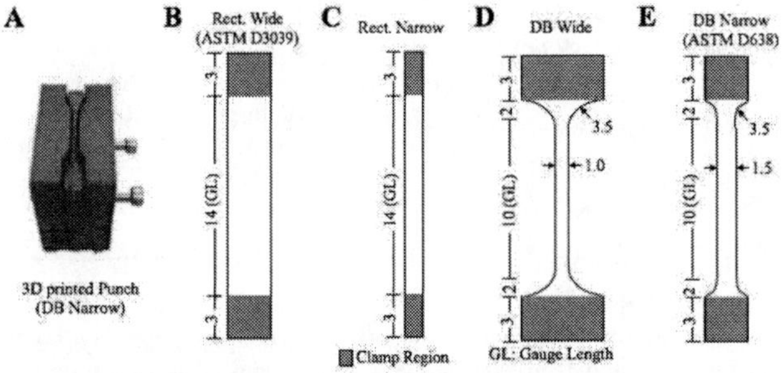

FIGURE 1. COUPON PUNCHES. A) PUNCH DEVICE EXAMPLE. DIMENSIONS FOR B) WIDE RECTANGULAR C) NARROW RECTANGULAR D) WIDE DOG-BONE, AND E) NARROW DOG-BONE COUPON. ALL DIMENSIONS IN MM.

Poster Presentation #P26

Coupons were covered in saline-soaked gauze and placed in the freezer for 24 hours at −4°C. Prior to testing, coupons were removed from the freezer and allowed to thaw for 1-2 hours. Coupons were then mounted in adjustable clamps positioned 14 mm apart, and preloaded to 0.1 N. After preconditioning coupons to 8% strain for 20 triangular waveform cycles, a secondary preload of 0.1 N was applied prior to pulling to failure at a rate of 1% strain/s. The zero strain position was based from the second preload. A one-way MANOVA was performed on the ultimate tensile strength (UTS), failure strain, and linear modulus obtained from stress-strain curves during ramp to failure.

Failure mode was separated into three categories: clampsite, mixed rupture,[1] and midsubstance (**Fig. 2A-C**). Midsubstance ruptures occurred in the uniform gauge region away from the clamps and tabs. Clampsite ruptures originated at the clampsite. Mixed ruptures signified a failure in the tabbed region, or near the clamps.

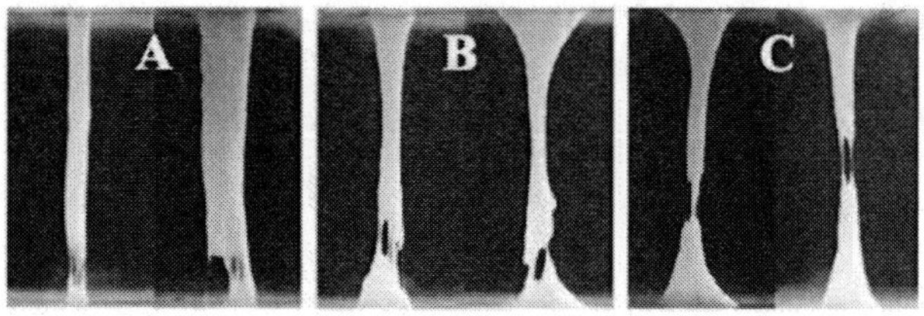

FIGURE 2. CHARACTERISTIC FAILURES. A) CLAMPSITE, B) MIXED, AND C) MIDSUBSTANCE RUPTURES.

RESULTS

The average UTS and failure strain for both groups of rectangular coupons was on average 55% and 25% less than the wide dog-bone coupons ($p<0.05$, **table 1**). All coupons that had clampsite failures had on average 60% less UTS and 25% less failure strain than midsubstance failures ($p<0.05$), but there was no significant difference in mechanical properties between coupons with midsubstance and mixed failures. Both dog-bone groups had more midsubstance ruptures than the rectangular groups, as over half the rectangular coupons had clampsite failures. The inter- and intra-variation in thickness and width for all test coupons was less than 6% and 12%, respectively.

Table 1. Quasi-Static Mechanical Properties for each Coupon Shape

Coupon Shape	Linear Modulus (MPa)	UTS (MPa)	Failure Strain[a] (%)
Rect. Narrow	60.0 ± 39.0	4.9 ± 3.5	12.0 ± 3.4
Rect. Wide	42.7 ± 19.0	4.9 ± 3.3	14.9 ± 3.6
Dog-bone Narrow	62.7 ± 28.9	8.3 ± 3.3	17.9 ± 3.9
Dog-bone Wide	76.5 ± 27.9	11.0 ± 4.3[*]	18.2 ± 2.3[*]

*Significantly greater than both wide and narrow rect. coupons ($p<0.05$)
[a]Clamp-to-clamp strain

Table 2. Failure Modes for Each Coupon Shape

Coupon Shape	Midsubstance Ruptures	Mixed Ruptures	Clampsite Ruptures
Rect. Narrow	10%	30%	60%
Rect. Wide	10%	20%	70%
Dog-bone Narrow	40%	30%	30%
Dog-bone Wide	50%	20%	30%

DISCUSSION

This study investigated the effect of coupon geometry on the tensile mechanical properties and failure modes of bovine meniscus. Rectangular coupons had twice the amount of clampsite failures than dog-bone coupons, and coupons that failed at the clampsite had UTS values that were 60% less than coupons that failed at midsubstance. Collectively, these results suggest that rectangular strips should be avoided when measuring mechanical properties of soft fibrous tissue, and clampsite failures should always be excluded from mechanical analysis. Another important finding was that the measured mechanical properties were not significantly different between mixed or midsubstance failures (**FIG. 2B-C**), regardless of coupon geometry. This supports the inclusion of mixed failure modes when measuring mechanical properties of soft fibrous tissue.

A novel aspect of this study was the comparison of coupon geometries based on ASTM standards. While numerous international test standards have been developed for tensile testing of conventional materials, no accepted standards exist for fibrous soft tissue. Researchers have therefore had to create their own custom specimen geometries, which may contribute to the variability observed in mechanical properties between research groups. ASTM states that a key factor in geometry choice comes from 'acceptable' failure modes in a 'reasonable' number of tests. This study found that dog-bone specimens based on ASTM standards for elastomers resulted in a 70% rate of midsubstance and mixed failures. Using rectangular specimens based on ASTM standards for fiber-reinforced composites resulted in only a 30% rate of midsubstance and mixed failures. These results suggest that fibrous soft tissue has failure behavior better represented by ASTM standards for elastomers. Furthermore, dog-bone geometries from prior studies (**FIG. 1D**) had failure behavior that was equivalent or better than coupons using ASTM recommendations. Although these results suggest that a nearly perfect midsubstance failure rate may not be achievable for fibrous soft tissue, midsubstance failure rates could potentially be improved by using different coupon geometries or by using synthetic inserts between the specimen and clamps.

Another important contribution of this study is the use of 3D printing and bendable razors to fabricate tissue punches. Researchers commonly face the problem of creating a punch device to excise the specimen at the correct dimensions for testing. Established methods include custom machined punches[2,3] and die cutting devices.[7] Custom machined punches and die cutting devices are expensive and require sharpening. Conversely, 3D printed punches can be assembled for less than $1 each, and the use of replaceable razors eliminate the need for sharpening blades. These 3D printed punches resulted in a width variation of 12% for all coupon shapes, which was comparable to the width variability of 10% and 32% when using custom machined punches and die cutters, respectively.[8,9] The novel development of these 3D printed punches enabled us to economically design and test multiple coupon geometries that closely matched ASTM standards.

This study has limitations. Tensile tests were performed on bovine meniscus, and results may not translate directly to human tissue. However, bovine meniscus has comparable structure and function to human meniscus, and mechanical properties in this study were within 5-50% of properties reported for human meniscus.[2,10] In addition, the fiber network in the meniscus is less aligned than ligament and tendon, and therefore tendon and ligament may exhibit different failure behavior.

In conclusion, this study found that using ASTM standards for elastomers provides reasonable results for tensile testing of meniscus, while rectangular specimens, with dimensions based on ASTM standards for fiber composites, are unsuitable. Research groups can use these results to inform decisions on specimen geometry when conducting mechanical tests on soft fibrous tissue.

ACKNOWLEDGEMENTS

This project was supported by the NSF under grant no. 1554353 and by NIGMS under grant no. P20GM109095.

REFERENCES

[1]Peloquin, J et al., *J Biomech Eng*, 138:021002, 2016.[2]Lechner, K et al., *J Orthop Res*, 18:945-951, 2000.[3]Lakes, E et al., *J Biomech*, 48:4333-4338, 2015.[4]Danso, E et al., *J Biomech*, 27:200-206, 2014. [5]Abdelgaied, A et al., *J Biomech*, 48:1389-1396, 2015.[6]Creechley, J et al., *J Mech Behav Biomed Mater*, 69:185-192, 2016. [7] Leroux and Setton, *J Biomech*, 124:315-321, 2002.[8] Henninger, H et al., *Acta Biomater*, 25:304-312, 2015.[9]Criscenti, G et al., *J Mech Behav Biomed Mater*, 54:141-148, 2016. [10]Tanaka, M et al., *Int J Polym Mater*, 63:891-897, 2014

CHANGE IN SKELETAL MUSCLE STIFFNESS AFTER RUNNING COMPETITION IS DEPENDENT ON BOTH RUNNING DISTANCE AND RECOVERY TIME

Cassidy Newman (1), Seyedali Sadeghi (1), Daniel H. Cortes (1,2)

(1) Mechanical and Nuclear Engineering Department
Penn State University
State College, PA, USA

(2) Biomedical Engineering Department
Penn State University
State College, PA, USA

INTRODUCTION

Quantifying muscle adaptation and remodeling in runners is important to design proper training strategies and avoid injury. Delayed onset muscle soreness (DOMS) is characterized by muscle tenderness, stiffness or reduced performance and is typically observed following extreme muscle stimulation, such as a workout [1]. Previous studies have been unsuccessful at determining the specific cause of the delayed soreness, but several theories have been developed. DOMS has been hypothesized to be caused by lactic acid buildup, muscle spasms, inflammation, or most commonly: muscle or connective tissue damage [2]. The intensity of DOMS typically peaks between 24 and 72 hours after exercise and subsides within 5 to 7 days [3]. If, however, the muscle damage is more extreme, the subject may have a muscle injury resulting in an extended recovery period. Muscle mechanical properties change during DOMS [4], and mechanical properties of quadriceps muscles change during extreme running activity [5]. Therefore, mechanical properties could potentially be used as biomarkers for muscle health. The objective of our study is to use ultrasound shear wave elastography (SWE) to measure mechanical properties of leg muscles before and after running races of various distances. Knowledge of the relationship between muscle exertion and mechanical properties can help us to better understand muscle injuries and their onset, as well as to develop modes of prevention and treatment.

METHODS

For this study, 22 subjects (age: 23 ± 5 years) were tested, 15 males and 7 females. Subjects were recruited through the Penn State Club Cross Country Team and the Nittany Valley Running Club. The subjects were split into three race categories: short distance (3 – 5 miles), middle distance (10 – 13 miles), and long distance (26+ miles). For each subject, SWE measurements were taken in both legs and included the Rectus Femoris, Vastus Lateralis, Vastus Medialis, Soleus, Lateral Gastrocnemius, Medial Gastrocnemius, Biceps Femoris, and Semitendinosus muscles.

The shear wave speed for each muscle of each subject was measured 5 times to calculate a robust average. Of the five measurements, the median value was used for data analysis. For the middle and long distance races, three measurements were taken: one a day before the race, one a day after the race, and one a week after the race. For the short distance races, only two measurements were taken (one before the race and one a day after the race) because of the inability to ensure that results seen one week after the race were not affected by training of equal or greater distance during the post-race week.

The analysis of the data included a correlation of shear wave speed values between right and left leg muscles to determine if those can be treated as independent data points. A two-way ANOVA analysis was performed for each muscle with recovery time and race distance as factors.

RESULTS

A significant correlation, with a very low r^2, was obtained from the correlation of wave speed between right and left leg muscles (Figure 1). The low r^2 value suggests that the stiffnesses of bilateral muscles are weakly related, and that other factors may have a stronger effect on these measurements.

From the two-way ANOVA analysis, time was found to have a significant effect on the Soleus, Rectus Femoris and Semitendinosus muscles ($p = 0.0242$, 0.0444 and 0.0490, respectively). Distance was found to also have a significant effect for the Biceps Femoris, Rectus Femoris and Semitendinosus muscles ($p = 0.0178$, 0.0124 and 0.0463, respectively). Figures 2 and 3 show the change in shear wave speed through two muscles that exhibit a significant change with both distance and time (Semitendinosus and Rectus Femoris muscles). The Semitendinosus muscle shows a consistent decreasing trend with time.

From t-tests of the individual muscles, separated by distance, there were 3 instances that showed a significant difference in the means of the before and after measurements. One instance was the Soleus muscle in the short distance group. The test returned a p value of 0.0136 and the graph shows a decrease in velocity from a mean value of 3.5732 m/s before the race, to a mean value of 2.8138 m/s one day after the race. Another instance was the Rectus Femoris muscle, also in the short distance group. The test returned a p value of < 0.0001 and the graph shows a decrease in velocity from a mean value of 3.7163 m/s before the race, to a mean value of 2.8963 m/s one day after the race. The last instance was the Biceps Femoris muscle in the long distance group. The test returned a p value of 0.0431 and the graph shows an increase in velocity from a mean value of 1.8607 m/s before the race, to a mean value of 1.9350 m/s one day after the race, to a mean value of 2.1397 m/s a week after the race.

DISCUSSION

The results from this study suggest that running distance has an effect on the change of mechanical properties of skeletal muscles. This effect, however, seems to vary between different muscles or different people. This discrepancy is most likely due to the many factors that affect the forces applied to individual muscles. Forces applied to muscles can depend on a variety of things including running form, topography of training courses and intensity of training.

Running form, according to Olympic coach, Roy Benson, is like an electric current, "Your body follows the path of least resistance" [6]. That path varies from person to person, depending on one's specific height, weight and body type. Different running forms cause different forces to be applied to muscles. For example, a person that runs with an uneven arm swing may exert more forces on the anterior side of left leg and the posterior side of the right leg. This is also the reason that the two legs of one subjects may exhibit unrelated behavior (Figure 1).

The same can be said for the topography of training courses and intensity of training. A person who spends more time running up hills will apply greater forces to their posterior muscles, while a person who trains on mainly flat courses may have a more even distribution of forces to each muscle. Intensity of training can also affect the forces applied to muscles. Top running speeds are accomplished by applying greater forces to the ground [7]. Running speeds however, depend on the distance of the course. For example, a person running a 3-mile race would have an average speed higher than a person running a 50-mile race, and therefore greater forces applied to each muscle.

Limitations of this study include limited sample size, especially for the long distance group. Additionally, subject pre-race mileage, recovery time, topography of training courses and intensity of training were not controlled, which may have had an influence on the variability of the data. In conclusion, SWE effectively measured the mechanical properties of skeletal muscles before and after running races of various distances. Both time and distance were found to have a significant effect on muscle stiffness values, although the effects varied between muscles and subjects.

ACKNOWLEDGEMENTS

Penn State Start-Up Funds

REFERENCES

[1] Pearcey, David J et al., *Journal of Athletic Training*. 50(1): 5-13, 2015. [2] Cheung, Karoline et al., *Sports Med.* 33:145, 2003. [3] Braun, William et al., *Delayed Onset Muscle Soreness*. American College of Sports Medicine, 2011. [4] Agten, CA et al., *American Journal of Roentgenology*. 208(2): 402-412, 2016. [5] Andonian, P et al., *Shear-Wave Elastography of Quadriceps Stiffness during MUM.* PLoS One 11(11), 2016. [6] Douglas, Scott. *Why Running Form Matters*. Runner's World, 2010. [7] Weyand, Peter G et al., *Journal of Applied Physiology*, 89(5): 1991-1999, 2000.

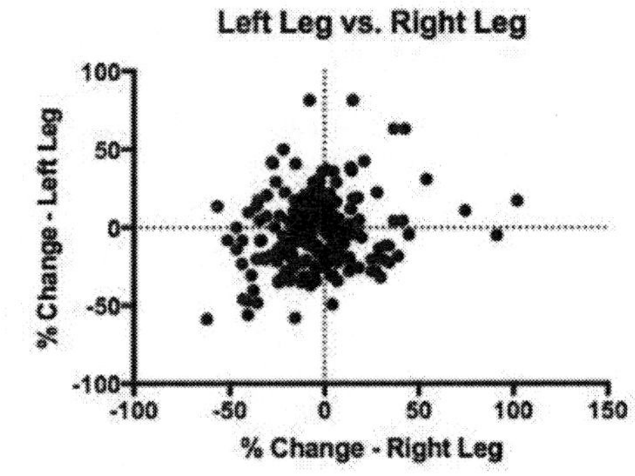

Figure 1. Percent change in shear wave speed through different muscles before and after competition. Shows the relationship between the right and left leg.

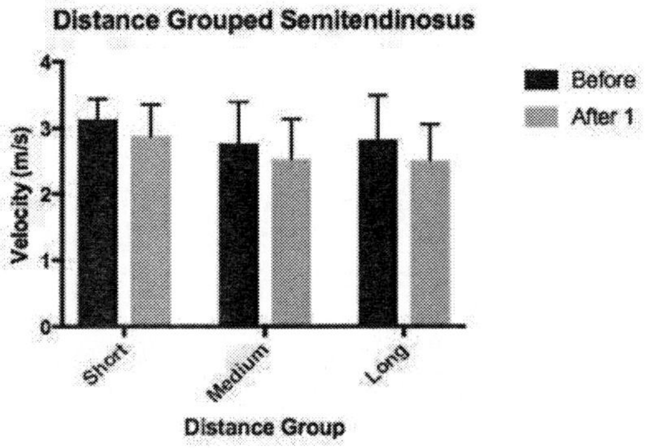

Figure 2. Shear wave speeds through the Semitendinosus muscle before and after competition. This muscle's stiffness is affected significantly by both distance and time and shows a consistent decreasing trend with time.

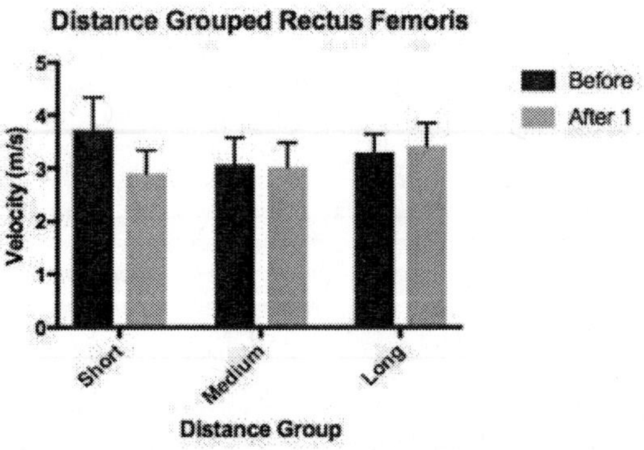

Figure 3. Shear wave speeds through the Rectus Femoris muscle before and after competition. This muscle's stiffness is affected significantly by both distance and time, but shows inconsistent behavior with time.

STEADY-STATE CHARACTERIZATION OF THE MECHANICAL PROPERTIES OF THE PACINIAN CORPUSCLE

Ellen T. Bloom[1], Julia C. Quindlen[1], Amy A. Claeson[1], Laura E. Ortega[1], Amy Moeller[2], Victor H. Barocas[1]

[1]Department of Biomedical Engineering, University of Minnesota, Minneapolis, MN

[2]Department of Orthopaedic Surgery, University of Minnesota, Minneapolis, MN

INTRODUCTION

Of the five basic senses that enable us to interact with our surrounding world, the sense of touch is the least understood. Different mechanoreceptors vary in number, location, and specific function to cooperatively construct our sensation of touch. Pacinian corpuscles (PCs), one of four types of mechanoreceptors located in skin, are onion-shaped structures of concentric collagenous layers separated by viscous fluid [1], and they respond to vibrations between 10 and 1000 Hz [2]. PCs generate a neural response when stress applied to the corpuscle deforms the nerve fiber and alters the permeability of the ion channels in the neuron's cell membrane [3].

Currently, the mechanical characteristics of the PC are unknown. Previous modeling studies used only estimated properties for the PC: Biswas et al. asserted that the PC would be similar to the basement membrane with a Young's modulus of 1kPa [4], while Loewenstein and Skalak used a modulus of 500 kPa, similar to that of the arterial wall [5]. Obtaining the value of the modulus for the PC experimentally would provide for more accurate modeling of the PC and its neural encoding.

In this study, we applied two different techniques to measure PC properties. First, micropipette aspiration (MPA), was used, in which a pressure-deformation curve was generated as a PC was sucked into a micropipette. After measuring a Young's modulus for the cadaveric PC by MPA, we embedded a PC in a polyacrylamide gel and performed uniaxial tensile tests to asses the validity of the estimate.

METHODS

PCs were isolated from the palm and index finger of cadaveric human hand specimens acquired through the University of Minnesota Anatomy Bequest Program. Suction was applied to a PC over regular pressure and time increments using a pipette connected to an adjustable-height water reservoir [6]. Images were acquired at each pressure step using a Leica S6D microscope with a MotiCam 1000 digital microscopy camera. The protrusion length of the sample into the pipette was measured at each pressure increment using ImageJ, and the slope of the resulting plot of protrusion length vs. applied suction pressure was used to calculate the Young's modulus of each sample. Protrusion length data were collected for both increasing and decreasing pressure in 5 mm H_2O (48.75 Pa) increments.

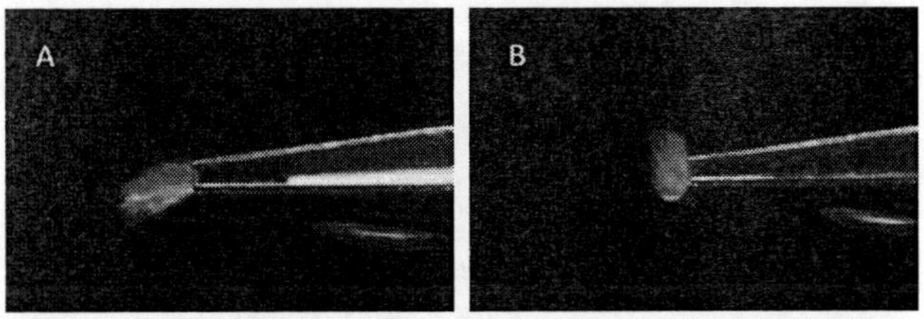

Figure 1: MPA was performed on PCs in the (A) tip and (B) side orientations.

PCs for tensile testing experiments were harvested from the index, middle, ring, and small fingers of a cadaveric human hand. The PCs were then embedded into polyacrylamide dogbones. [7]. The tensile tests were performed using a uniaxial testing machine (TestResources) with a 10N load cell and a Canon Rebel T3i with a 100mm macro lens. In order to approximate the slow deformation of our MPA experiments, a strain rate of 0.002%/second was applied to a total strain of 30%. The surface of the gel was speckled with dried Verhoeff's stain to allow for displacement tracking (Fig. 2A). The video of the tensile test was converted to an image stack. The first (undeformed) image was used to create a mesh of the tissue boundary in ABAQUS with quadrilateral elements. The surface of the 3D mesh was input into a displacement tracking algorithm [8], then successive pairs of images were correlated to calculate the displacement of each node throughout the sequence. The 3D mesh from ABAQUS was imported into FEBio (Fig. 2B). An example of a speckled, PC-embedded dogbone and the corresponding model is shown in Figure 2.

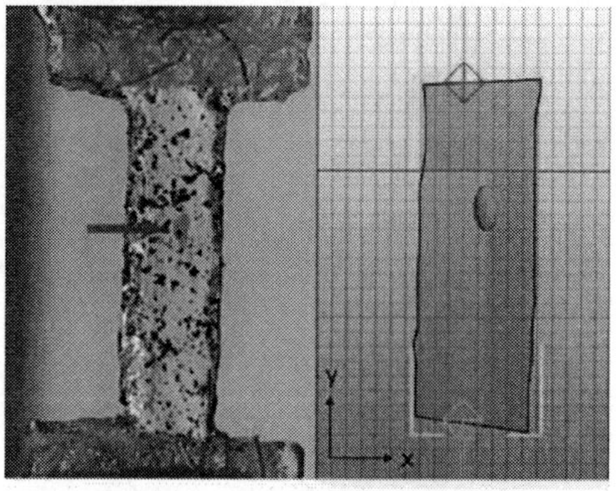

Figure 2: (A) A PC within the embedded gel for a uniaxial experiment. (B) The corresponding 3D mesh in FEBio.

Green strains from the strain-tracking code were compared to those calculated from the FEBio simulation. The Young's modulus of the polyacrylamide gel was calculated using an optimization code utilizing an adapted *fminsearch* to the match experimental and simulated force-length curves. The ratio of strain along the sample's long-axis measured directly above the PC to average strain along the long-axis over the entire gel surface was computed for the FEBio simulations with various PC modulus values. The same strain ratio was calculated from the speckled experimental images.

RESULTS

Fig. 3A shows the protrusion length vs. suction pressure for a typical MPA experiment; a slight hysteresis was observed in all cases. The average moduli calculated for all cases are shown in Fig. 3B; the side and tip cases had no significant difference in modulus ($p=0.54$), but the decreasing pressure modulus was significantly higher than the increasing pressure modulus ($p=0.02$) The overall average modulus for the increasing pressure experiments was 2.75 ± 0.68 kPa. The increasing modulus was deemed a more accurate measure since the slight plateau before the subsequent decline in protrusion length during the decreasing pressure experiments suggested that friction between the sample and the pipette may have limited the sample's motion.

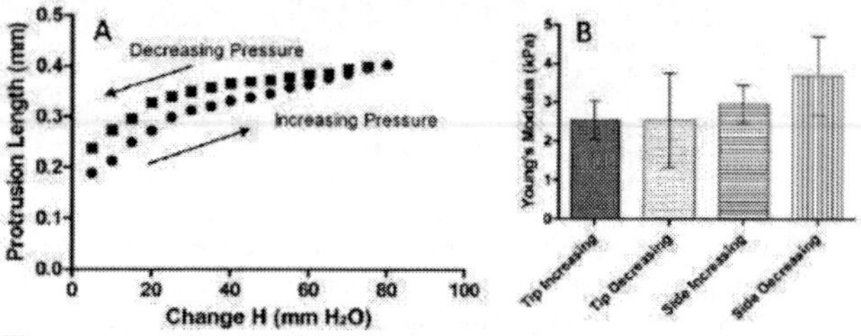

Figure 3: (A) Results from a single MPA experiment. (B) Young's modulus estimated for PC (n=10 for each case, mean ± 95% CI).

Before the PC modulus could be estimated in the dogbone experiments, the modulus of the polyacrylamide gel had to be verified. The results for the optimization of two different gel samples gave a Young's modulus of 15 kPa for the polyacrylamide gels, which was used in subsequent FEBio simulations.

In order to estimate a Young's modulus for the PC from the uniaxial experiments, a ratio of the strain in a finite element directly over the PC to the average strain in the sample ($\varepsilon_{YY,PC}/\varepsilon_{YY,AVG}$) was calculated from the FEBio simulation and was parameterized over a range of possible modulus ratios (E_{PC}/E_{Gel}), as plotted in Fig. 4. Using the output from the strain-tracking code, the same strain ratio was calculated for the experiment, and a modulus ratio was extrapolated from the graph to yield E_{PC} of 1.056 kPa.

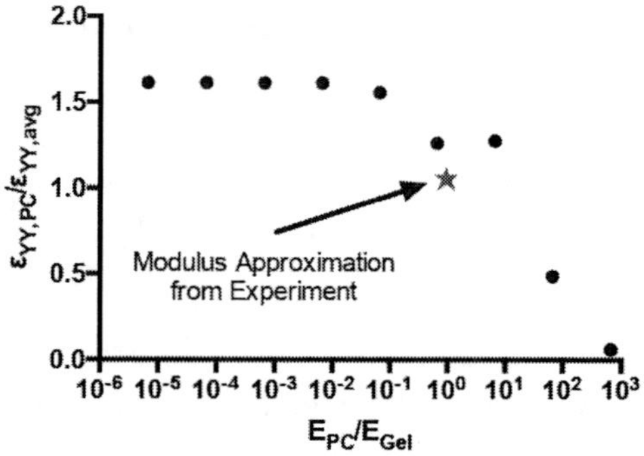

Figure 4: The ratio of PC strain and average strain vs. the ratio of moduli for one experimental model.

DISCUSSION

The PC modulus from the MPA experiments was determined to be 2.75 kPa, while the uniaxial results for one sample provided an estimate of 1.056 kPa. This range of 1-2 kPa is much smaller than the value used by Loewenstein and Skalak [5], but is comparable to Biswas et al.'s approximation of 1·kPa [4]. Also, it must be recognized that the experiment and analysis in this study consider the PC as a homogenous material, but in fact it is a complex, multilayered structure; the models of Biswas and Loewenstein used a layer modulus, and not an effective overall tissue modulus, so caution must be taken in comparing values.

Some of the differences between the MPA and the uniaxial test results may be a result of the different donors from which the PCs used in the two different experiments were obtained. The structure of the PC may be altered with aging due to the addition of further lamellar layers [9]. The modulus may be also affected by disease; diabetes may induce glycosylation of the tissue, thus altering the mechanical properties of the lamellar layers and the PC as a whole [10]. Future work could explore the mechanical properties of the PC under dynamic conditions, or least on a time scale that is more physiologically relevant. While steady-state characterization is an important metric, the properties of the PC during its normal deformation would be a natural next step. However, these experiments are still acceptable because they provide the first experimental approximation of a PC's properties.

ACKNOWLEDGEMENTS

The authors thank Patrick Alford, Zaw Win, and Kerianne Steucke for their assistance with the polyacrylamide gels. This research was supported by a University of Minnesota undergraduate research scholarship, URS #11143. Computational resources were provided by the Minnesota Supercomputing Institute.

REFERENCES

[1] Güçlü, B et al., *Somatosensory,* 23:119–26, 2006. [2] Hubbard, S. *J Physiol*, 141:198–218, 1958. [3] Johnson, K. O. *C O neurobiology*, 11:455–61, 2001. [4] Biswas, A et al., [5] Loewenstein, W et al., *J Physiol*, 182:346–378, 1966. [6] Shojaei-Baghini, E et al., *Annals of Biomed Eng,* 41:1208–16, 2013. [7] Adhikari, N et al., *J of Cardiovascular Translational Research*, 8:158-163, 2015. [8] Raghupathy,R et al., *J of Bio Eng*, 133: 1011-7, 2011. [9] Verrillo, R. T., *J. Gerontol.*, 35:185–193, 1980. [10] Alsunousi, S. et al., *Front. Neuroanat.*, 8:79, 2014.

Poster Presentation #P28

AN EXPERIMENTAL SETUP TO QUANTIFY PRESSURE-INDUCED MICROSTRUCTURAL CHANGES IN TRICUSPID VALVE ANTERIOR LEAFLETS

Anthony L. Black (1), Anup D. Pant (1), Vineet S. Thomas (1), Taylor Verba (1), Rouzbeh Amini (1)

(1) Department of Biomedical Engineering
The University of Akron
Akron, Ohio, USA

INTRODUCTION

The tricuspid valve is a three-leaflet valve that allows blood to flow in one direction from the right atrium to the right ventricle during atrial contraction. Functional tricuspid valve regurgitation (FTR) is a common disorder associated with this valve. Elevated right ventricular pressure due to pulmonary hypertension (PH) has been associated with FTR. Previous studies have shown that the tricuspid valve annulus dilates and the geometry and normal function of the valve leaflets alter in patients suffering from PH-induced FTR [1-3]. Such changes are believed to have been caused by chronic pressure overload [1].

Both in healthy and in diseased heart valves, analysis of the underlying microstructural components of the extracellular matrix (ECM) is particularly important as they determine the overall mechanical properties of the tissues and subsequently its function. For example, collagen fibers are one the basic structural elements of the valve ECM, providing mechanical strength to the tissue [4]. Since PH exposes the valve leaflets to chronic pressure overload, one can postulate that the changes in the valve geometry and function are caused by the remolding response of the ECM structural components. Although, such a response takes place on longer time scales, in our study, we hypothesized that immediately after increasing ventricular pressure, a significant change in the leaflet microstructural architecture can be detected. If such modification occurs immediately after increases in pressure, it is expected that a more permanent remodeling response is also expected to take place chronically.

METHODS

Fresh porcine hearts were obtained from a local slaughterhouse (3-D Meats, Dalton, Ohio) within 2-3 hours after the animals were slaughtered. The hearts were thoroughly rinsed with phosphate buffer solution (PBS) to remove blood and blood clots. For the pressurized groups (n=8), the hearts were set up in a fixation apparatus with a hose connected to a separatory funnel entering through the pulmonary artery into the right ventricle (Fig. 1). The funnel was filled with 0.5% v/v glutaraldehyde and kept at a height of 35 cm from the valve to ensure a hydrostatic pressure of approximately 26 mmHg being applied on the tricuspid valve. An additional pump system was incorporated with the experimental setup to deliver the leakage from the vena cava back into the reservoir and to maintain the constant hydrostatic pressure for 6-8 hours. The hearts were then dissected and each leaflet was separated. For the non-pressurized group (n=8), the hearts were simply dissected and the three leaflets were fixed for 6-8 hours. The dissected valves were subsequently rinsed with PBS and were dehydrated in 100% glycerol for an hour to ensure optical clarity necessary for microstructural characterization (Fig. 2).

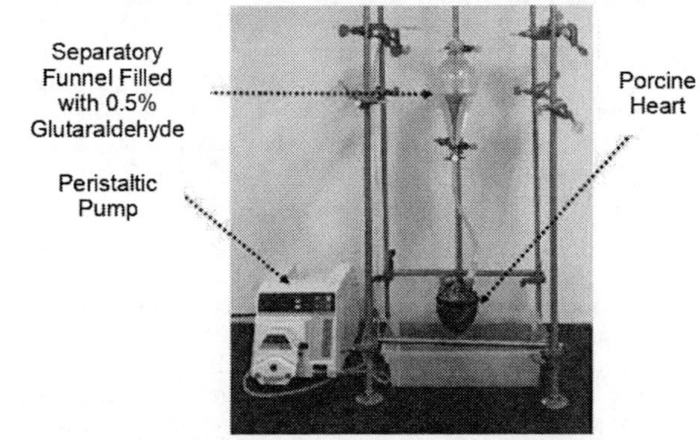

Figure 1: Experimental set-up to fix tricuspid valve leaflets exposed to hydrostatic pressurize

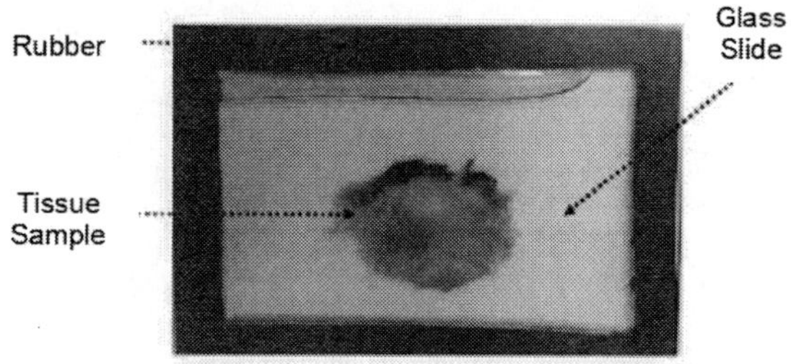

Figure 2: Tissue sample preparation

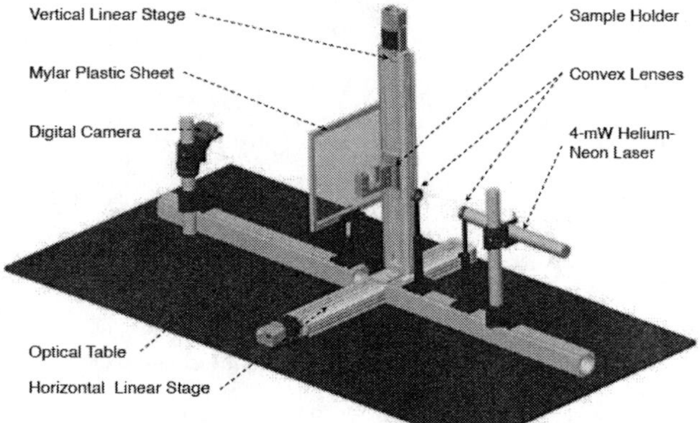

Figure 3: Small Angle Light Scattering (SALS) equipment

The microstructural architecture of the leaflets was obtained using custom-made small angle light scattering (SALS) equipment (Fig 3). [5-6]. This equipment consisted of a light source (Lumentum HeNe Laser of a wavelength 632 nm, Model 1125, 5 mW Random Polarization), and two plano convex lenses (Edmund Optics 25.0 mm Diameter x 125.0 mm FL, 25.0 mm Diameter x 35.0 mm FL, 633 nm V-Coat, PCX Lens), which when combined acted as a beam expander. The equipment also contained two linear actuators (Zaber A-LST0500B: Motorized linear stage, 500 mm travel, medium resolution, medium speed) perpendicular to each other for the movement of the sample holder in horizontal and vertical direction, followed by a projection screen (Mylar sheet) and a camera (Grasshopper3 2.3 MP Mono USB3 Vision, GS3-U3-23S6M-C).

The Normalized Orientation Index (NOI), a quantitative measure of the ECM fiber distribution was calculated from the SALS images. A higher value of NOI corresponded to a more organized ECM fiber network [7]. Since the geometry of each leaflet was slightly different, both pressurized and non-pressurized leaflets were divided into 9 different regions using a curvilinear coordinate system that followed the free edge and annulus of the valve (Fig. 4B). The NOI values were then averaged for each region.

RESULTS

Continuous monitoring of the pressurization setup and the necessary manual adjustment of the pump flow enabled us to maintain a constant pressure and successfully fix the hearts. Figure 4 shows a typical anterior leaflet of a pressurized valve and its corresponding SALS scan. Although the regional NOI average for the pressurized leaflets has a relatively large variability, the values in all nine regions were significantly higher in the pressurized valves compared to the non-pressurized valves (p<0.05, Student t test, Fig. 5).

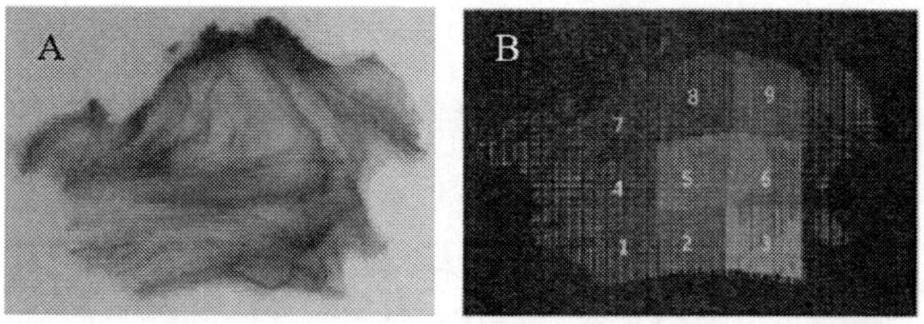

Figure 4: (A) A typical anterior leaflet and (B) the corresponding SALS image divided into 9 different regions.

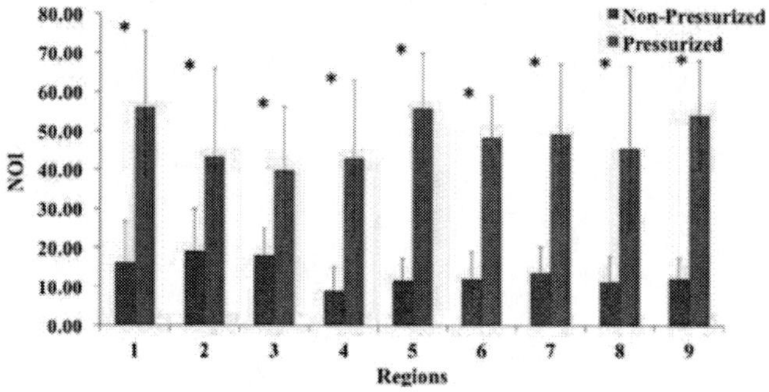

Figure 5: Comparison between the average Normalized Orientation Index (NOI) of the pressurized and non-pressurized tricuspid valve anterior leaflets for regions 1 to 9 (n = 8 for both groups, error bars are standard errors). Asterisks indicate significant difference between (P< 0.05).

DISCUSSION

Consistent with our proposed hypothesis, we observed that the ECM fibers were realigned and the network became more organized in response to increased ventricular pressure. Similar to the other cardiac valves [6], pressure-induced changes in the valve microstructure were expected in the tricuspid valve leaflets. Such rotations and realignments of the ECM fibers are reversible during the normal a cardiac cycle, when the right ventricular pressure changes from systolic to diastolic values. However, one would except that excessive increases in the right ventricular pressure due to the pulmonary hypertension could intensify such network responses beyond its homeostatic state. As such, the tricuspid valve cells and ECM are expected to undergo a remodeling response that could potentially lead to the mechanical, geometric, and functional alteration of the valve.

ACKNOWLEDGMENT

This project was funded in part by the American Heart Association (Grant No. 16SDG27770088). Assistance from John Lesicko in the development of the SALS equipment is acknowledged.

REFERENCES

[1] Sukmawan, R. et al., *JASE*, 20: 470-476, 2007.
[2] Sukmawan, R. et al., *Am. J. Cardiol*, 94:1096-1099, 2004.
[3] Praagh, R. et al., *Am. Heart J*, 135:772-785, 1998.
[4] Fung, Y., *Biomechanics: Mechanical Properties of Living Tissues*. Springer Science & Business Media, 2013.
[5] Sacks, M. et al., *Ann. Biomed. Eng.*, 25:678-689, 1997.
[6] Joyce, E. et al., *Ann. Thorac. Surg.*, 87:1240-1248, 2009.
[7] Amini, R. et al., *J Biomech Eng*, 136.3:031003, 2014.

CONTRIBUTION OF COLLAGEN FIBERS AND MYOCYTES TO RESIDUAL STRESS IN THE LEFT VENTRICULAR WALL

Marissa R. Grobbel (1), Sheikh M. Shavik (1), Emma Darios (2), Stephanie W. Watts (2), Lik Chuan Lee (1), and Sara Roccabianca (1)

(1) Mechanical Engineering
Michigan State University
East Lansing, MI, United States

(2) Pharmacology and Toxicology
Michigan State University
East Lansing, MI, United States

INTRODUCTION

Residual stress is a prominent feature of cardiac tissue. We hypothesize that the mechanical interaction of two key constituents of cardiac tissue, collagen fibers and myocytes, contributes to the strength of residual stress in the left ventricular wall. In order to quantify the contribution of each of these constituents to residual stress, we have isolated them and performed opening angle tests on three groups of adult rat hearts: control (intact tissue), collagenase-treated tissue (isolated myocytes), and decellularized tissue (isolated collagen fiber network). While past studies have performed opening angle experiments to quantify residual stress in heart tissue, all of these studies have been performed on whole, intact tissue. No previous studies have quantified the contribution of individual tissue constituents to residual stress. Additionally, we have included both male and female samples in order to determine differences between sexes, not only in overall residual stress of the left ventricular wall, but in residual stress fields of isolated collagen fibers and myocytes.

METHODS

An opening angle experiment was performed on hearts explanted from adult male ($n = 15$) and female ($n = 15$) Sprague Dawley rats. In this experiment, two lateral slices were cut from the intact heart, creating "ring-like" samples that expose the left and right ventricles. These rings were then divided equally into three test groups, each with an equal number of males and females. These test groups were: (1) the control (intact tissue) group, (2) the collagenase-treated (isolated myocyte) group, and (3) the decellularized (isolated collagen) group. The samples are subjected to different treatments based on their group.

The samples in the control group were soaked in a buffer containing a myosin-inhibitor in order to prevent active muscle contraction during testing. After 30 minutes of soaking, opening angle tests were performed on these samples. In the isolated myocyte group, samples were treated with collagenase, a collagen-disrupting enzyme that breaks down the collagen fiber network. Based on a previous protocol [1], each sample was soaked in the same buffer as the control group for 15 minutes, then in a buffer containing the enzyme for 96 hours before being subjected to the opening angle tests. Finally, the extracellular matrices of the decellularized group were isolated through the removal of cells and cellular residua. Based on a previous protocol for whole-heart decellularization [2], samples were first soaked (for 15 minutes) in a heparinized saline solution, and then they were put into a 10% sodium-dodecyl sulfate solution for 96 hours. Thereafter, they soaked first in deionized water (15 minutes), then in a 10% Triton X-100 solution (30 minutes) before the opening angle tests.

Opening angle tests were performed on all samples following constituent isolation. The test consists of making a radial cut across the left ventricular wall directly across from the right ventricle. This cut results in the ring-like samples opening up as they approach a stress-free configuration [3].

Pictures of the samples were taken both immediately before and after the radial cut, and then every 3 minutes for the first 15 minutes. Thereafter, pictures were taken and at 30, 60, 75, and 90 minutes following the radial cut. These pictures were analyzed using Imagej to calculate the opening angle. This was done by first establishing a centerline through the middle of the opening that extends through the right ventricle. Then, this line is used, along with the inner radius of the left ventricle- measured from the pre-cut pictures- to locate the center of the left ventricle. From this center point, two lines are created that connect the center to either side of the cut. The angle between these two lines, then, is the opening angle.

RESULTS

Our results show that the samples in which the collagen fiber network was isolated (decellularized group, $\phi = 106.45°$) had a significantly higher opening angle when compared to the intact tissue (control group, $\phi = 57.88°$) containing both constituents ($p < 0.05$). Conversely, the samples in which the myocytes were isolated (collagenase-treated group, $\phi = 21.00°$) had a significantly lower opening angle when compared to the intact tissue ($p < 0.05$). These differences were visible immediately after the cut and persisted throughout the test (**Figure 1**). This behavior was found in both the female and the male groups. A comparison of the male and female opening angles showed only a significant change in the collagenase-treated group (*$p < 0.05$) (**Figure 2**). From a histological analysis, samples stained with Picrosirius red show a decrease in the collagen content (red) in the collagenase-treated group, and a decrease in myocytes (yellow) in the decellularized group when compared to the control (**Figure 3**).

DISCUSSION

Our results suggest that, although the collagen fiber network accounts for such a small portion of the mass fraction of cardiac tissue, it is the main contributor to residual stress in the left ventricular wall. These results are further confirmed by the histological analyses, as it can be seen that the constituent isolation treatments produced their expected effects. Additionally, the same results were achieved in both the male and female test groups, only varying significantly from one another in the collagenase-treated group. These results could potentially lead to a better understanding of the mechanical changes in the heart associated with pathological or physiological growth and remodeling processes, during which the relative amounts of tissue constituents have been known to change.

FIGURES

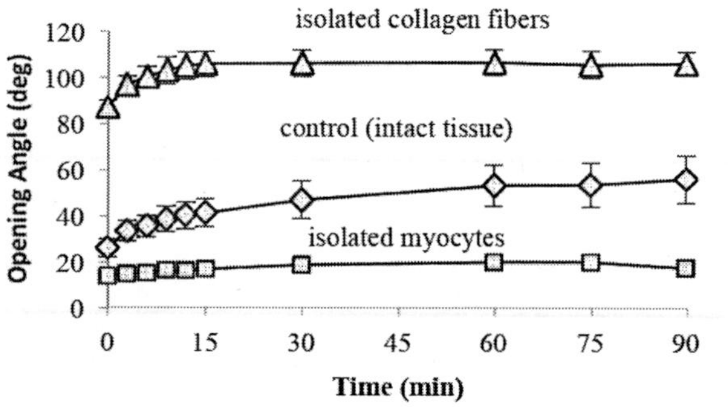

Figure 1 Plot of the opening angle measure in the male heart samples at different time points after the radial cut.

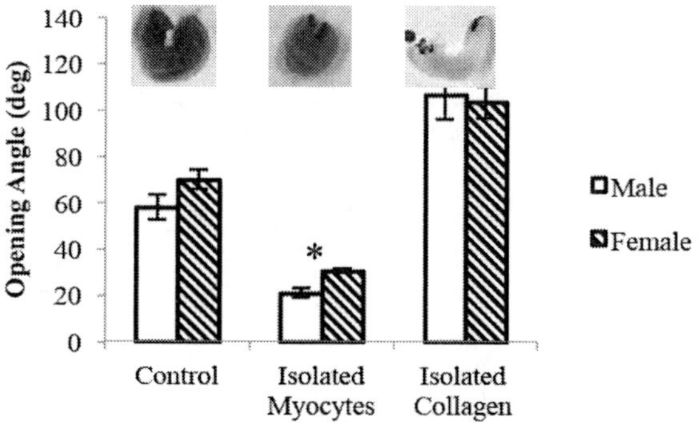

Figure 2 Final opening angle of all groups. Pictures of the resultant configurations can be seen above their respective columns (*$p < 0.05$).

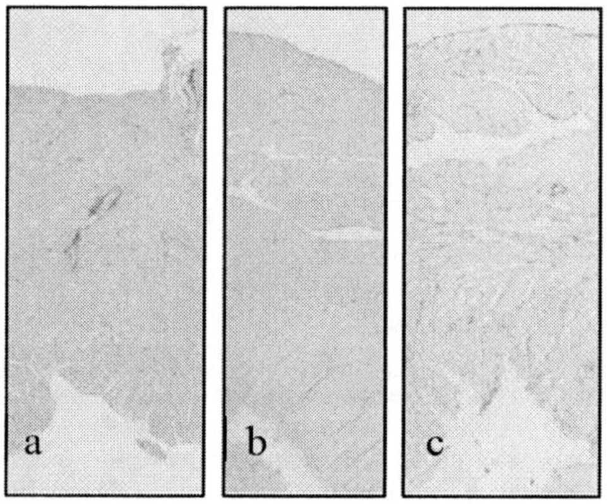

Figure 3 PSR stains of (a) intact tissue, (b) isolated myocytes, and (c) isolated collagen fibers. Cells are in yellow and collagen fibers are in red.

ACKNOWLEDGEMENTS

We would like to thank Professor Stephanie Watts for kindly providing us with the rat heart samples. We also thank Ms. Emma Darios for helping us to dissect the hearts.

REFERENCES

[1] D. A. MacKenna et al, *Am J Physiol Heart Circ Physiol*, 266, H1007-H1018, 1994

[2] H. C. Ott et al, *Nature Medicine*, 14, 213-221, 2008

[3] J. H. Omens, and Y.C. Fung, *Circulation Research*, 66, 37-45, 1990

OPTIC NERVE AXON COUNT AND STRAIN COMPARISONS BETWEEN NORMAL AND GLAUCOMATOUS HUMAN EYES

Kelsey T. Sadlek (1), Katelyn F. Axman (2), Ehab A. Tamimi (2), Jonathan P. Vande Geest (2,3,4)

(1) Department of Chemical Engineering
University of Pittsburgh
Pittsburgh, PA, USA

(2) Department of Bioengineering
University of Pittsburgh
Pittsburgh, PA, USA

(3) McGowan Institute for Regenerative Medicine
University of Pittsburgh
Pittsburgh, PA, USA

(4) Louis J. Fox Center for Vision Restoration
University of Pittsburgh
Pittsburgh, PA, USA

INTRODUCTION

Glaucoma is the second leading cause of permanent blindness in developed countries, expected to affect over 79 million people world-wide by the year 2020 [1]. The most common age-related optic neuropathy, glaucoma results in vision loss due to retinal ganglion cell (RGC) death [1, 2].

Axon counts have been used to better understand the pathogenesis and permanent vision loss occurring in glaucoma. Axons that traverse the optic nerve become considerably degraded and are eventually lost all together in patients with glaucoma [2, 3]. Comparing the axon count in glaucomatous (G) and non-glaucomatous (NG) optic nerve (ON) samples is one method that has been shown to distinguish normal from diseased tissues [4]. Previous studies have used semi-automated axon counting methods predominantly to examine non-human tissue[4-6].

It has been hypothesized that both RGC loss and glaucoma are associated with biomechanical stress within the ON. ON strain is quantified as a measure of the local deformation in glaucomatous tissue induced by applied stress [7]. The ON and surrounding regions of the posterior eye are the most susceptible to biomechanical effects such as these.

Though glaucoma is associated with the loss of axon count, it is unclear whether ON biomechanics are correlated with ON axon counts for glaucomatous eyes. There are few literature studies that compare measured strains to axon counts. The objective of this study is to compare axon counts between G and NG eyes calculated using a semi-automated axon count extrapolation method. A statistical analysis between strain measurements and axon counts for all eyes was also performed, which could shed more light on the relationship between RGC loss, ocular strain, and glaucoma, ultimately leading to a better understanding of the disease.

METHODS

Optic nerves were received from Midwest Eye Bank (n=5), San Diego Eye Bank (n=1), and Banner Health (n=1), containing nerves from both the left (n=4) and right (n=3) eyes. All eyes were from donors of European descent over the age of 50. The ON was cut, fixed in Poly/LEM, transferred to vials containing 2.5% glutaraldehyde for 24 hours, and finally transferred to PBS and stored at 4 °C.

The samples were then cross-sectioned, stained with OsO_4, and embedded into slides that were visualized under a microscope (Nikon Eclipse 90i). Each cross section was individually imaged at 60x magnification using Nikon NIS-Elements software; 15% overlap as well as autofocus capabilities were used for each pixel, allowing for successful and detailed image montaging. Each montaged image was then read individually through a Matlab program written to perform a semi-automated axon count across 100% of the given optic nerve cross section 9 (Figure 1). The user was prompted to manually count the axons in each in 10 randomly chosen sections. These manual counts were used to extrapolate across the entire cross-sectional area. Each image was run through the program three to four times for repeated measures.

A one-way repeated measure ANOVA test was performed to evaluate statistical differences in axon count across health status. A statistical linear mixed model was used to predict axon count number with the peak values of principle strains, E1 and E2, as fixed effect continuous variables. Region (superior, inferior, nasal and temporal) and pressure (15, 30 and 45 mmHg) were considered random effect variables within each eye. This model utilized strain data calculated by measuring ON deformations using an in-house sequential digital image correlation protocol [8].

RESULTS

For the NG and G samples counted, mean semi-automated axon counts ranged between 710,276-1,1143,456 and 95,093-387,029 axons, respectively. The statistical results showed that the NG eyes had significantly higher average axon count compared to the G eyes(p-value<0.001) (Figure 2). The mixed model statistical analysis showed that there was no significant relationship between E1 ON strains and axon count (p-value = 0.21). However, there was a significant relationship between E2 strain values and axon count (p-value<0.001) as shown in Figure 3.

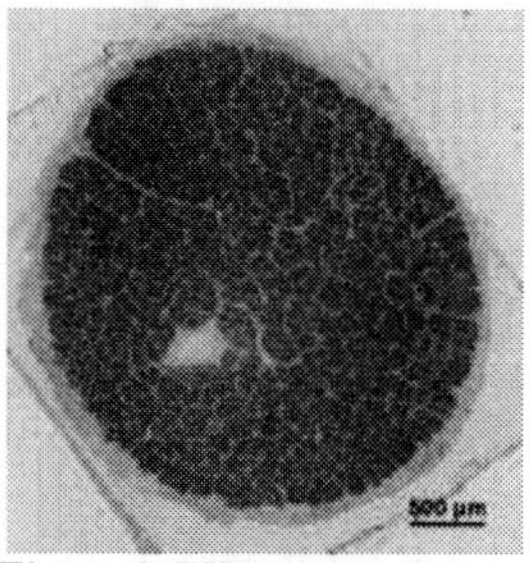

Figure 1. Microscope image of full optic nerve cross section used for semi-automated axon counting. Dark sections indicate bundles of axons separated by lighter connective tissue.

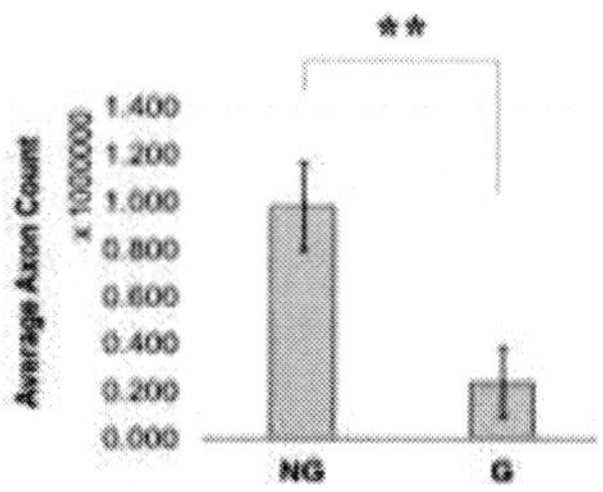

Figure 2. Mean semi-automated axon counts in NG and G eyes, where n=4 and n=3 for NG and G, respectively. Double asterisk indicates p-value<0.001 Error bars indicate one standard deviation.

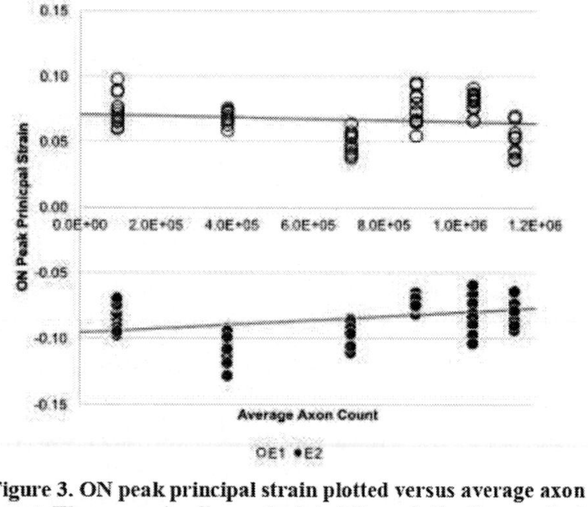

Figure 3. ON peak principal strain plotted versus average axon count. The regression lines calculated through the linear mixed model are also shown.

DISCUSSION

The difference between NG and G mean axon counts was statistically significant, suggesting NG eyes have more axons than that of G samples. NG axon counts also exhibited axon counts within the normal range for healthy human eyes, a verification that these samples are a true representation of normal tissue supporting previous findings [5].

Based on the statistical analysis results, only the compressive strain component increased with axon count, while tensile strain component had no relationship to axon count. This may suggest that increased compressive strains in the ON could be linked to the loss of axons, as a possible mechanism for the onslaught of glaucoma.

There are several limitations in the present study. First and foremost, our axon counting was performed by one observer in a semi-automated manner which also required extrapolation. Our research team is currently working towards a more automated and user independent procedure for generating axon counts. Furth more, in the future, we will also utilize a statistical model that estimates axon counts as a function of both disease status and ethnic background, which will allow a more thorough correlation analysis. We will also explore how the regional differences in ON biomechanical response correlates with the region-specific axon counts within the ON.

ACKNOWLEDGEMENTS

Funding for this work was provided by the NIH (NEI 5R01EY020890-05 to JPVG). The authors would also like to thank the Swanson School of Engineering, the Department of Bioengineering, and the Office of the Provost at the University of Pittsburgh for contributing to the funding that facilitated this research.

REFERENCES

1. Alqawlaq, S., et al., *Challenges in neuroprotective nanomedicine development: progress towards noninvasive gene therapy of glaucoma.* Nanomedicine (Lond), 2012. **7**(7): p. 1067-83.
2. Quigley, H.A. and A.T. Broman, *The number of people with glaucoma worldwide in 2010 and 2020.* Br J Ophthalmol, 2006. **90**(3): p. 262-7.
3. Nickells, R.W., *The cell and molecular biology of glaucoma: mechanisms of retinal ganglion cell death.* Invest Ophthalmol Vis Sci, 2012. **53**(5): p. 2476-81.
4. Johnson, B.M., M. Miao, and A.A. Sadun, *Age-related decline of human optic nerve axon populations.* AGE, 1987. **10**(1): p. 5-9.
5. Mikelberg, F.S., et al., *The normal human optic nerve. Axon count and axon diameter distribution.* Ophthalmology, 1989. **96**(9): p. 1325-8.
6. Ebneter, A., et al., *Estimation of axon counts in a rat model of glaucoma: comparison of fixed-pattern sampling with targeted sampling.* Clin Exp Ophthalmol, 2012. **40**(6): p. 626-33.
7. Downs, J.C., M.D. Roberts, and C.F. Burgoyne, *The Mechanical Environment of the Optic Nerve Head in Glaucoma.* Optometry and vision science : official publication of the American Academy of Optometry, 2008. **85**(6): p. 425-435.
8. Pyne, J.D., et al., *Sequential-Digital Image Correlation for Mapping Human Posterior Sclera and Optic Nerve Head Deformation.* Journal of Biomechanical Engineering, 2014. **136**(2): p. 0210021-02100212.

SB³C2017
Summer Biomechanics, Bioengineering and Biotransport Conference
June 21 – 24, Tucson, AZ, USA

FINITE ELEMENT BASED SIMULATION OF GROWTH MORPHOMECHANICS OF THE PHARYNGEAL ARCH ARTERIES

Mark A. Lantieri (1), Jonathan T. Butcher (1)

(1) Biomedical Engineering
Cornell University
Ithaca NY, USA

INTRODUCTION

Computer simulation software is a valuable alternative to traditional laboratory work. It allows for the examination of certain phenomena where experimental investigation would be difficult if not impossible. This includes the use of mechanical models to study the interactions of animal and human tissue with the surrounding environment. One such modeling technique is fluid-structure interaction studies (FSI), which examine the link between deformable bodies and their surrounding fluid flow. Past studies have examined tissue elastic deformations and stress states that result from fluid flow. Few have attempted to quantify the tissue growth that can occur as a result of these physiological loads.

This study utilizes a stress-based growth law, which states that the growth a tissue undergoes is related to its current stress state, to determine the inelastic deformation (growth) of several tissue models. Tissue geometries studied include a simplified, arbitrary vessel geometry and a micro-CT scanned pharyngeal arch artery (PAA). The specific PAA considered is the right fourth PAA of a chick, which through growth and remodeling evolves into the aorta [1]. The basic models discussed in this abstract can be developed and used as a tool in the formation of testable hypotheses regarding the mechanobiology of embryonic growth and malformation.

METHODS

Simulations were performed using ANSYS Parametric Design Language (APDL) software. Fluid and solid domain meshes of varying quality were linked at their interfaces via common nodal points. A pressure drop was applied to simulate in-vivo conditions, in addition to no-slip walls [2]. Entry and exit regions were added for flow development. The fluid was modeled as incompressible and Newtonian, with properties approximating those of blood [3].

After solving the fluid flow problem, resulting pressure and wall shear stress loads were applied to the solid domain via surface effect elements. The tissue was then allowed to grow for a set, non-physiological time step and unloaded.

Tissue growth was assumed to be inelastic deformation, which was calculated via the stress-based growth law of equation (1) [4]. It is derived using the Principle of Virtual Work and the Entropy Inequality.

$$\boldsymbol{L} = \dot{\boldsymbol{F}}\boldsymbol{F}^{-1} = a\boldsymbol{f}^{T}(\boldsymbol{\sigma} - \boldsymbol{\sigma}^{*})\boldsymbol{f}^{-T} \tag{1}$$

$\boldsymbol{L}$ is the deformation rate of the inelastic tensor, $\boldsymbol{F}$, $\boldsymbol{f}$ is the elastic tensor, $\boldsymbol{\sigma}$ is the current stress state, $\boldsymbol{\sigma}^{*}$ is the homeostatic stress state, and a is a growth rate parameter. Equation (1) states that the growth is related to the difference between the current stress state and the homeostatic stress state. The concept of a homeostat has been used in other growth models, and represents a stress state at which no tissue growth would occur. These simulations assume it to be isotropic to eliminate directional bias [4]. It is calculated as

$$\sigma^{*} = p\boldsymbol{I} \tag{2}$$

This growth law was implemented via the UserMat user-programmable function capabilities of APDL.

Two geometries were considered. The first, a hollow, cylindrical straight tube, represents a simplified vessel geometry. The second is a micro-CT scanned Hamburger-Hamilton stage 21, right fourth pharyngeal arch artery of a chick embryo (HH21 PAA-IVR). Material properties of developing HH27 chick atrioventricular valves were used as an approximation due to limited data availability [5].

Multiple simulations were run on both simplified and native geometries, with the goals of observing/validating growth and examining the potential roles of the growth rate parameter and homeostat.

RESULTS

Initial and deformed cylindrical geometries are show in Figure 1. Two simulations were conducted. The first assumed a constant growth rate parameter. As such, the simulation was essentially axisymmetric. Cylinder length and diameter increased, as expected. The second assumed a linearly decreasing growth rate parameter with increasing x-coordinate. This biases the growth, as one side would undergo faster growth under identical fluid loading and homeostat. Cylinder length and diameter increased, as expected. Increasing curvature was also observed due to the biased growth.

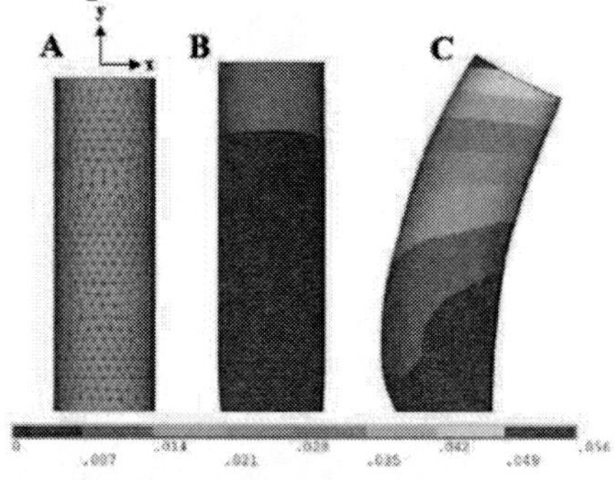

Figure 1: A: hollow cylindrical geometry with mesh. B: constant growth rate parameter growth. C: linearly decreasing growth rate parameter growth. Color plots are displacement vector sums.

For HH21 PAA-IVR simulations, the growth rate parameter was assumed constant. Instead, homeostat was varied, with both tensile and compressive stresses considered. An assortment of homeostats were defined as multiples of the pressure drop ΔP across the geometry.

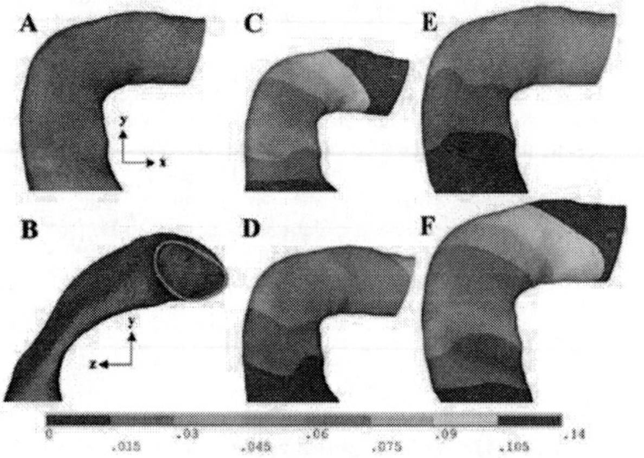

Figure 2: A/B: front and side views of undeformed HH21 PAA-IVR geometry with mesh. C: $\sigma^* = 2\Delta P$. D: $\sigma^* = \Delta P$. E: $\sigma^* = -\Delta P$. F: $\sigma^* = -2\Delta P$. Color plots are displacement vector sums.

DISCUSSION

The simplified cylindrical geometry simulations demonstrate the efficacy of the stress-based growth law. As intuitively expected, fluid flow results in an increasing vessel diameter and length. Furthermore, a position-dependent growth rate parameter biases growth, as expected. This implies that if individual cells were to acquire locally-specific growth rate parameters, perhaps from chemical or biological signaling, a wide assortment of growth paths could occur.

The HH21 PAA-IVR simulations represent a practical application of stress-based growth law simulations. They provide preliminary support that stress states can potentially drive the maturation of these geometries. Of critical importance appears to be the sign of the homeostat. As demonstrated in Figure 2, a tensile homeostat results in overall geometry shrinkage, while a compressive homeostat results in growth. In all cases, growth and shrinkage were associated with changes in arch lengths, diameters, and wall thicknesses. Similar results have been found in developing chick atrioventricular valve leaflets [4]. Similar to the growth rate parameter, it seems that a locally specific homeostat can lead to many growth paths, including locally-specific shrinkage and growth.

It also appears that the homeostat cannot exceed a certain magnitude in relation to the pressure drop across the vessel. In this case, a large homeostat magnitude would easily render the stress state from fluid flow negligible. Hemodynamic loading then loses the ability to influence growth and remodeling, with a more uniform growth occurring as in Figure 3.

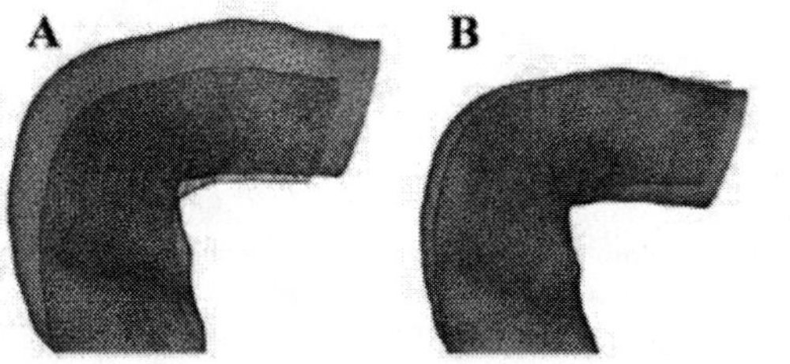

Figure 3: Large homeostats render fluid stresses negligible, producing more uniform growth. A: $\sigma^* = -2\Delta P$. B: $\sigma^* = -\Delta P/2$. Deformed and undeformed geometries are shown.

Future work will include more detailed flow parameters and material properties. It is likely that with correct parameters, growth between stages can be modeled. Critical tissue regions and those of lesser importance could be identified from a growth perspective. The genesis and growth of malformations can also be examined.

ACKNOWLEDGEMENTS

The authors would like to acknowledge Phil Buskohl and Stephanie Lindsey for their intellectual support, as well as grants NIH HL110328 and NSF-CMMI-1635712 for funding.

REFERENCES

[1] Hiruma, T., *Anat Embryol*, 191:415-423, 1995
[2] Yalcin, H.T. et al., *Dev Dynam*, 240:23-35, 2011
[3] Al-Roubaie, S. et al., *Am J Physiol Heart Circ Physiol*, 301:2473-2481, 2011
[4] Buskohl, P.R. et al., *Biomech Model Mechanobiol*, 11:1205-1217, 2012
[5] Buskohl, P.R. et al., *J Biomech*, 45:895-902, 2012

CHARACTERIZTION OF TRANSMURAL MORPHOLOGICAL PROPERTIES IN PORCINE THORACIC DESCENDING AORTA USING MULTIPHOTON FLUORESCENT MICROSCOPY AND IMAGE PROCESSING

T. Gillin (2), A. Hemmasizadeh (1), B. Gligorijevic (2), K. Darvish (1)

(1) Department of Mechanical Engineering
Temple University
1947 N Broad St, Philadelphia, PA 19122

(2) Department of Bioengineering
Temple University
1947 N Broad St, Philadelphia, PA 19122

INTRODUCTION

Characterization of major proteins across the aortic wall can illuminate the baseline functionality of the aorta microarchitecture. If it can be characterized how the normal structural integrity of these proteins is compromised during a vascular disease, then new diagnosing techniques can be conceived from understanding the architectural composition and mechanical properties of aortic proteins. A multidisciplinary approach was used to characterize the properties of two proteins, collagen and elastin, using multiphoton fluorescent microscopy to illuminate the auto-fluorescence of collagen and elastin to acquire detailed images of the fibers across the aortic wall. Then, extensive image processing techniques (including MATLAB, FIJI, CTFIRE, EXCEL) were implemented to determine the orientation of each specific protein fiber. Previous research suggests that the structural strength of the proteins is correlated with the fibers "straightness" [1]. For straighter fibers, the mechanical properties showed a higher Young's Modulus, with variation from the intima layer to the adventitia layer. The image processing characterized the straightness of the proteins based on fiber angle relative to a fix coordinate plane. Histograms of angle orientation were produced, making it easy to determine fiber straightness. These results confirmed previous mechanical testing of the transmural mechanical properties of aorta and give more details on the orientation of aortic proteins.

METHODS

Six porcine thoracic descending aortas were sectioned at three different locations in the longitudinal direction of the aorta. 60 micrometer sections were made to optimize the multiphoton fluorescent images. Second harmonic generation and specific laser wavelengths were used to image the proteins of interest (elastin and collagen). Images were taken at specific locations across the wall, from the intima layer to the outer layer (Figure 1), to analyze how the protein composition changes through the wall. The first step of image processing began in FIJI. Fiji functionalities allowed for increasing the resolution of the images to help delineate proteins to distinguish them. The images were then converted to a 4-bit image and further processed in MATLAB. MATLAB has even finer resolution processing power for delineation of specific fibers. The bwareaopen command was utilized to exclude any unwanted fragmentations from the fibers to allow for neater processing in the CTFIRE program. The CTFIRE program is tailored towards determining orientation of collagen proteins (Figures 2 and 3). This extensive image processing software produced histograms of angle orientation with angle count. These results were processes in excel using Boolean functions to determine

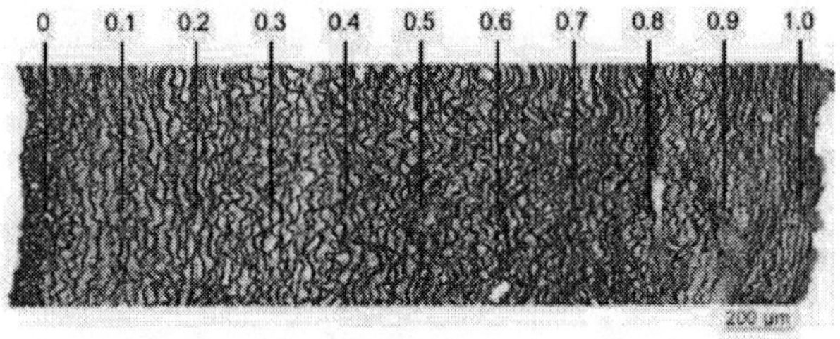

Figure 1 Illustration of transmural direction (Intima to Adventitia)

straightness based on angle. For horizontal photos, 90 degrees +/- 2.5 degrees was considered to be a straight fiber (i.e., in the circumferential direction), then the percentage of straight fibers were calculated for different layers in each slice.

Figure 2: Multi-Fluorescent Microscopy Image of Collagen

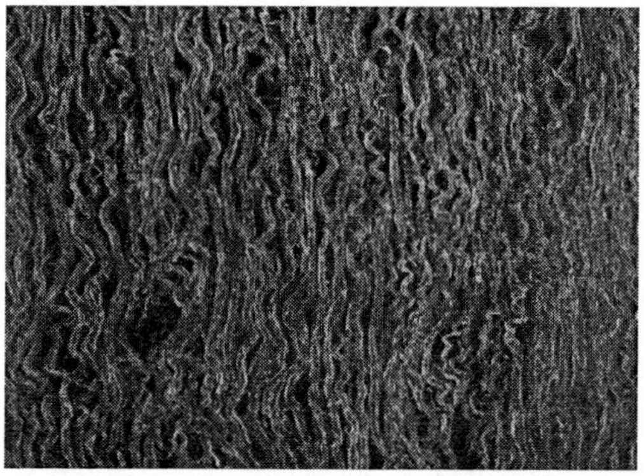

Figure 3: Processed image of Figure 2

RESULTS

The results showed a distinct difference in morphological properties of collagen and elastin density percentage across the aortic wall. The density percentage of the straight proteins increased from the intima layer to the adventitia layer. This increase is as between 3 percent to 7 percent across adjacent layers. Particularly, between the inner and outer halves, there was a significant difference in protein densities. For example, a medial slice was imaged at four different areas across the wall, one image for the intima, two images across the medial layer, and a final image near the adventitia layer (from 0-.2,.2-.5, .5-.8-.8-1.0 in Figure 1). The percentage of straight collagen was 6.52 %, 8.56 %, 10.85%, and 13.15% respectively. These percentages in different slices were within a range between 5% and 15%. A similar increasing pattern was identified among all imaged slices for both protein density percentages, which substantiates the hypothesis that a strong correlation exists between protein orientations and the relative transmural location in the aorta. A rigorous statistical analysis of the results is under way.

DISCUSSION

The results confirmed the preconceived evidence that the transmural fiber orientations change across the wall. These results support a correlation between the morphological properties with transmural location. It can be suggested that the orientation of major aortic proteins is aligned more in the circumferential direction as we move from intima to adventitia. These results were obtained from healthy porcine aortas. Future research can determine how vascular diseases alter, if any, the normal microarchitecture outlined in this report. A data reservoir of these protein compositions can be collected, with both normal aortic function and aorta undergoing vascular diseases. If a diagnosing imaging device can be conceived to acquire such protein composition of patients, it can potentially make diagnosing cardiac diseases more efficient and less evasive.

ACKNOWLEDGEMENTS

The work reported was partially supported by the American Heart Association.

REFERENCES

[1] Hemmasizadeh, A., Autieri, M., & Darvish, K. (2012). Multilayer material properties of aorta determined from nanoindentation tests. Journal of the mechanical behavior of biomedical materials, 15, 199-207.

TRPV1 ION CHANNEL MEDIATED THERMAL RESPONSE OF CA3 HIPPOCAMPAL PYRAMIDAL NEURON – A SIMULATION STUDY

R. Renato (1), J. Xu (2)

(1) Department of Biology
Tarleton State University
Stephenville, TX, USA

(2) Department of Engineering Technology
Tarleton State University
Stephenville, TX, USA

INTRODUCTION

Traditional electrical stimulation such as deep brain stimulation in which a stimulating electrode is implanted in the brain has been used to treat neuronal disorders such as epilepsy and Parkinson's disease. Because the electrodes are placed in the brain itself, deep brain stimulation involves a small chance of bleeding in the brain. Recently it has been shown that pulsed infrared lasers can be used to evoke and detect neuronal responses. Comparing with traditional electrical stimulations, the lasers are free of physical contact with target tissues, have better spatial resolution, and less stimulation interference. The plausible mechanism that underlies the laser-induced neuronal responses is photo-thermal effects mediated by TRPV ion channels that are activated by heat (1). For example, Shibasaki et al. (2) has shown that hippocampal neurons, especially CA3 hippocampal neurons, heavily express functional transient receptor potential vanilloid (TRPV), one of the thermosensitive TRP ion channels that can be activated by heat, and that activation of TRPV depolarized the resting membrane potential in hippocampal neurons by allowing cation influx, indicating that TRPV is a key regulator for hippocampal neural excitabilities as shown in Figure 1. In order to utilize the infrared lasers to stimulate the neurons that are safe and effective under physiological conditions, we need to first explore the thermal response of heated neurons.

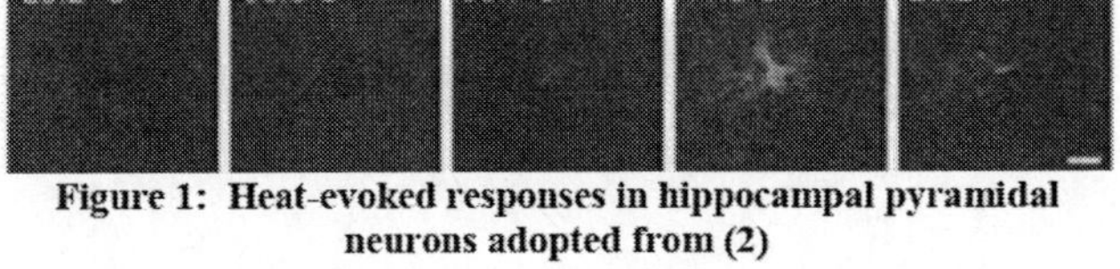

Figure 1: Heat-evoked responses in hippocampal pyramidal neurons adopted from (2)

The purpose of this study is to examine the effect of temperature variation on the response of a Hippocampal Pyramidal Neuron that features Transient Receptor Potential cation channel subfamily V member 1 (TRPV1). Currently a hippocampal pyramidal neuron model that encompasses the TRPV1 ion channel does not exist. Such a model can be useful in creating a more accurate representation of neurons for future use in research.

METHODS

The CA3 neuron model of Migliore et al. (3) was used as shown in Figure 2, modified to include a two-state ionic channel model developed by Voets et al (4).

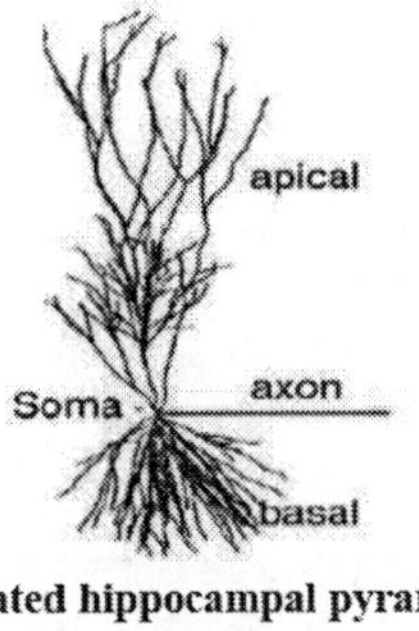

Figure 2: Simulated hippocampal pyramidal CA3 neuron

Explicitly, the propagation of the electrical signal through dendrites and axons is modeled by cable theory:

$$C_k \frac{dV_k}{dt} = \sum_l \gamma_{l,k}(V_l - V_k) - I_{ion} \quad (1)$$

Where V_k is the membrane voltage, C_k is the membrane capacitance, γ_{lk} is the conductance between the two adjacent compartments, and I_{ion} is the total ionic current

$$I_{ion} = I_{Na} + I_{CaT} + I_{KDR} + I_A + I_M + I_{AHP} + I_{KC} + I_{Ca} + I_{Leak} + I_{TRPV1} \quad (2)$$

The model includes the delayed-rectifier potassium current I_{KDR}, the transient potassium current I_A, the M current I_M, the fast sodium I_{Na}, the calcium-dependent potassium current I_{AHP}, the calcium-dependent potassium current I_{KC}, calcium currents: low-threshold I_{CaT}, I_{CaL}, and I_{CaN}, leak current I_{leak}, and TRPV1 current I_{TRPV1}.

All the channels with Hodgkin-Huxley-like kinetics were taken from ref (3) except TRPV1 channel kinetics, which was modeled as a two-state model and was taken from ref (4). The two-state model for the temperature sensitive opening (α) and closing (β) rates, which are both voltage and temperature dependent, of the TRPV1 voltage gated channel are shown here:

$$Close \underset{\beta(V,T)}{\overset{\alpha(V,T)}{\rightleftharpoons}} Open \quad (3)$$

$$\alpha = Ae^{\frac{E_{a,open}}{RT}} e^{\frac{\delta zFV}{RT}} \; and \; \beta = Ae^{\frac{E_{a,close}}{RT}} e^{\frac{-(1-\delta)zFV}{RT}} \quad (4)$$

Based on experimental data (4), α is steeply temperature-dependent and β shows a shallow temperature dependence. The parameters used were experimentally deduced to reproduce the temperature dependence of TRPV1 ion channels. See original references for definitions of parameters and their values. All the simulations were performed using NEURON, a simulation environment for modeling neurons (5).

RESULTS

The CA3 Hippocampal Pyramidal Neuron model fitted with TRPV1 responded to temperature increases as expected. The heat invoked response increased as the temperature exposure increased at the dendritic locations as shown in Figure 3. The largest depolarization was observed in the somatic membrane as shown in Figure 4. The depolarization events were heat invoke rather that initiated by non-thermal stimuli such as electrical stimuli or short-duration depolarizing current. Additionally, both of the dendritic and somatic membrane potentials of the neuron gradually increased as the temperature increased from 20 °C to 35 °C. At 35 °C, there was sufficient increase in the membrane potential to elicit depolarization as bursts of action potentials. It is important to note that at higher temperatures than that of the internal human of 37.2 °C the neuron began to exhibit sustained depolarizations with this modeled simulation (not shown).

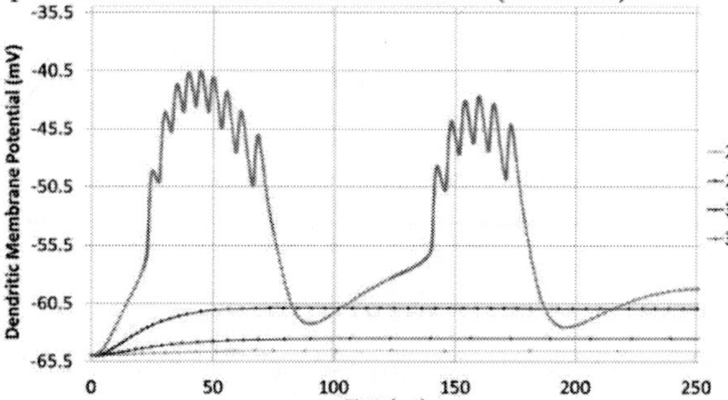

Figure 3: Dendritic membrane temperature at varying temperature.

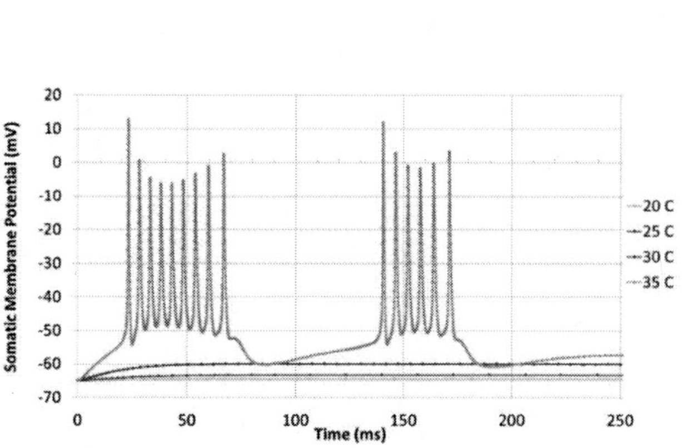

Figure 4: Somatic membrane temperature at varying temperature.

DISCUSSION

TRPV1 is a heat-sensitive channel contributes to temperature detection due to its presence in sensory nerve endings. This temperature sensitivity can be attributed to sufficiently different activation energies of the opening and closing probabilities of the channel. Our results show that the introduction of TRPV1 voltage gated ion channel to the CA3 Hippocampal Pyramidal Neuron model exhibited an expected heat invoked response that resulted in increased membrane potentials and the presence of depolarizations as temperature increased.

ACKNOWLEDGEMENTS

This study was supported in part by the Faculty Development Grant from Tarleton State University.

REFERENCES

1. Izzo AD, Richter CP, Jansen ED, Walsh JT, Jr. Laser stimulation of the auditory nerve. Lasers Surg Med. 2006;38(8):745-53. Epub 2006/07/28. doi: 10.1002/lsm.20358. PubMed PMID: 16871623.
2. Shibasaki K, Suzuki M, Mizuno A, Tominaga M. Effects of body temperature on neural activity in the hippocampus: regulation of resting membrane potentials by transient receptor potential vanilloid 4. The Journal of neuroscience : the official journal of the Society for Neuroscience. 2007;27(7):1566-75. Epub 2007/02/16. doi: 10.1523/JNEUROSCI.4284-06.2007. PubMed PMID: 17301165.
3. Migliore M, Cook EP, Jaffe DB, Turner DA, Johnston D. Computer simulations of morphologically reconstructed CA3 hippocampal neurons. Journal of neurophysiology. 1995;73(3):1157-68. Epub 1995/03/01. PubMed PMID: 7608762.
4. Voets T, Droogmans G, Wissenbach U, Janssens A, Flockerzi V, Nilius B. The principle of temperature-dependent gating in cold- and heat-sensitive TRP channels. Nature. 2004;430(7001):748-54. Epub 2004/08/13. doi: 10.1038/nature02732. PubMed PMID: 15306801.
5. Carnevale NT, Hines ML. The NEURON Book. Cambridge, UK: Cambridge University Press; 2006.

SB^{3}C2017
Summer Biomechanics, Bioengineering and Biotransport Conference
June 21 – 24, Tucson, AZ, USA

BONE PROPERTIES SURROUNDING SURFACE MODIFIED DENTAL IMPLANTS: A NANOINDENTATION STUDY

Ryan Doud (1), Ramzi V. Abou-Arraj (2), Jack E. Lemons (2), Alan W. Eberhardt (1)

(1) Department of Biomedical Engineering
University of Alabama at Birmingham
Birmingham, AL, USA

(2) School of Dentistry
University of Alabama at Birmingham
Birmingham, AL, USA

INTRODUCTION

Endosteal dental implants have a long history of success; however, research continues to further improve osseointegration. Surface roughening techniques such as plasma spraying, grit blasting, acid etching, and anodization have been shown to enhance tissue response, improving healing times [1-3]. Porous coating increases friction between the implant and bone allowing ongrowth/ingrowth to strengthen the bone-implant binding [4]. Chemical surface modifications using calcium phosphates (CaP), such as hydroxyapatite (HA), are achieved by immersion or plasma spraying. CaP slowly releases from the implant during the healing process leading to increased apatite formation on the surface of the implant and faster fixation of the implant to the bone [1,2,5].

The lack of sufficient bone quantity, or the presence of poor bone quality due to local or systemic conditions, has led to the introduction of osteoinductive growth factors in implant site preparation. Demand for enhanced bone formation and improved integration rates for dental implants resulted in the use of bone morphogenic proteins (BMPs), which play key roles in the development and growth of new bone [6]. Specifically, the role of BMP-2 in stimulating bone growth led to the development of a synthetic recombinant, rhBMP-2, which has become widely available for clinical research and application. CaP-coated surfaces adsorb BMP-2, allowing its release during the healing process after implantation. Research has shown that application of rhBMP-2 to dental implants also improves the healing response [5,7].

Studies of a polyglutamate domain (E7) support consideration for use in dental implants as a way to improve binding of the bone to the implant. E7 enables the attachment of peptide sequences to HA-coated implants. Culpepper et al [8] found that HA disks treated with E7 bound to the collagen mimetic peptide DGEA exhibited greater new bone growth than disks treated with DGEA alone or untreated disks.

In the present study, nanoindentation was used to characterize bone mechanical properties surrounding CaP-coated titanium dental implants with surface modifications to compare untreated samples with those treated with E7-rBMP-2 peptides, rBMP-2 peptides (no E7), or BMP-2 full-length proteins.

METHODS

Specimens were supplied by the UAB School of Dentistry and non-decalcified preparation was completed by the UAB Histology Laboratory. Institutional IACUC approval was obtained for all activities involving animals. Threaded CaP-coated titanium dental implants in the shape of 3.25 x 6.5mm screws received one of four different surface modifications prior to implantation: BMP-2 full-length protein, BMP-2 peptide ± calcium-binding peptide, or no treatment (control). They were surgically implanted in eight New Zealand white rabbits, each rabbit receiving four implants (two in each proximal tibia). After a recovery period of 2 weeks, the rabbits were euthanized for samples. Each sample was assigned to one of four groups (A, B, C, D) based on its surface modification, while the investigators were kept blind to their identities.

Preparation of nanoindentation samples began by fixing in formalin for 24 hours, followed by a series of dehydrations in ethanol, all done under vacuum. Three xylene baths were then used to flush out excess ethanol, also under vacuum. The samples were embedded in poly-methyl methacrylate (PMMA), allowing the methyl methacrylate liquid monomer to infiltrate the bone samples for 4 weeks before adding a polymerizing catalyst. Sections were then cut and affixed to slides, which were cut and polished down to the appropriate depth with 600, 800, 1200, and P4000 grit sandpaper, and final polishing using a 0.05 μm alumina paste. The specimens were then stained with Sanderson's rapid bone stain with a Van Gieson counterstain.

Poster Presentation #P36

Nanoindentation testing was performed with a Nano Indenter® G200 testing system (Keysight, Santa Rosa, CA), following the methods of Oliver and Pharr [9]. Indents were made with a Berkovich diamond indenter tip to a prescribed depth of 2500 nm, with a peak load hold time of 100 seconds at max displacement. Drift calculation was performed at 90% unloading. Mature bone was identified as pale pink, while newly formed bone was a dark pink (Fig. 1). A region of mature bone (Zone 5) was indented to provide a baseline against which newly formed bone properties were compared. Additionally, indents were made in the PMMA in each sample to provide lower limits of hardness and modulus for development of exclusion criteria. Five zones of interest were defined - Zones 1 and 3 being adjacent to mature bone and extending to the center of the next thread and zone 2 and 4 encompassing the bone between the adjacent pair of threads. A minimum of 9 indents were made in each zone.

Upon completion of the indentation testing, samples were inspected with the nanoindenter microscope. If indents were not visible, Scanning Electron Microscopy (SEM) on an FEI Quanta FEG™ 650 (FEI Company, Hillsboro, OR) was performed to confirm that the indents did not stray into the resin or implant, or into bone lacunae.

The data were analyzed with two-way ANOVA via regression followed by an F test and t-test of individual pairings to determine which treatment groups showed significant differences in modulus and hardness of new bone growth. The regression method for the two-way ANOVA was required since all sample sizes were not equal (unbalanced). Statistical significance was determined based on whether or not the calculated p-value was below the threshold value, $\alpha=0.05$, for a 95% confidence level.

RESULTS

At the time of this abstract, tests had been completed on only one specimen per group, the results for which are illustrated in Figure 2. Two-way ANOVA revealed that the modulus of bone was significantly different between different implant treatment groups ($p<0.001$). It also showed significant differences between zones of immature bone (zones 1-4) and zones of mature bone (zone 5) ($p<0.001$). Additionally, the analysis found no significant effects due to interaction between the treatment type and the region of bone that was indented – that is to say, the type of treatment received by the implant did not affect the relations of immature bone properties to mature ($p=0.070$).

After the ANOVA, samples were paired and analyzed with an F test followed by a t-test. The F test determined whether variances should be treated as equal or unequal for the t-test, which determined whether or not the sample means were significantly different from each other. The t-tests revealed that, for the new bone growth (zones 1-4, combined), the elastic modulus for treatment group D (8.8 ± 1.5 GPa) was significantly higher than the moduli of groups B (7.58 ± 2.02 GPa, $p=0.001$) and C (7.7 ± 1.9 GPa, $p=0.004$). The moduli for groups B and C were not significantly different from each other ($p=0.8$). The modulus for group A (7.8 ± 2.3 GPa) was not significantly different from any other group ($p=0.7$ for A-B comparison, $p=0.8$ for A-C, and $p=0.0502$ for A-D).

Similar to the modulus, the two-way ANOVA showed that both treatment and the region indented had a significant effect on the hardness of the bone ($p<0.001$ for both), while there was no interaction between treatment type and the region indented ($p=0.083$). The subsequent F tests and t-tests revealed that, in the zones of new bone growth (zones 1-4), the hardness of group D (0.35 ± 0.11 GPa) was significantly higher than that of group B (0.31 ± 0.11 GPa, $p=0.044$), but was not significantly different from either group A (0.30 ± 0.13 GPa, $p=0.076$) or group C (0.31 ± 0.11 GPa, $p=0.060$). Additionally, no other significant differences were observed ($p=0.9$ for A-B comparison, $p=0.8$ for A-C, and $p=0.97$ for B-C).

DISCUSSION

This preliminary work demonstrated the utility of the nanoindentation testing; however, additional testing is needed to determine differences between the various surface treatments. Due to the inconclusive and proprietary nature of the results, the individual treatments may not be revealed. Ongoing work includes the testing of additional specimens to increase sample sizes, and the de-coupling of Zones 1-4 to determine if the location along the implant further affects the mechanical properties of the healing bone. It should also be noted that the present specimens were harvested at 2 weeks, post-surgery. Additional time points and longer healing times may also be a focus of future work.

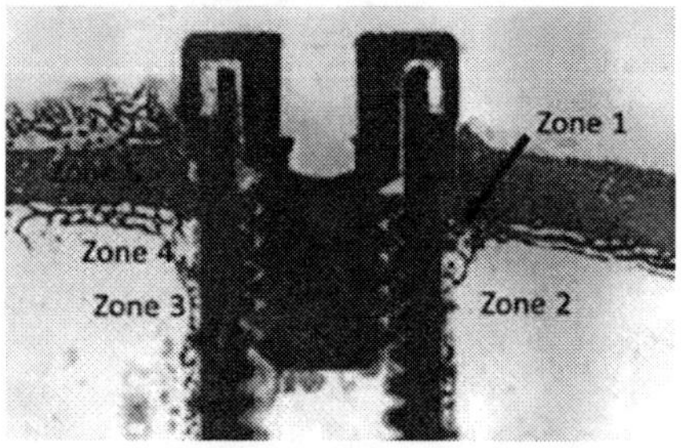

Figure 1: Zones 1-5 established for nanoindentation, with Zone 5 representing mature bone

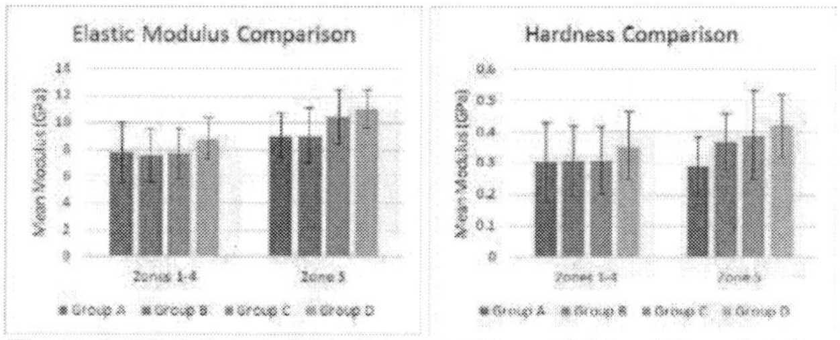

Figure 2: Modulus and hardness results combining Zones 1-4 for comparison with Zone 5.

ACKNOWLEDGEMENTS

The authors gratefully acknowledge financial support from the Zimmer-Biomet.

REFERENCES

[1] Buser, D., et al., *Periodontol. 2000*, 73, 1, 7–21, 2017.
[2] van Oirschot, B.A., et al., *Clin. Oral Implants Res.*, 24, 4, 355–362, 2013.
[3] Adell, R., et al., *Int. J. Oral Maxillofac. Implants*, 5, 4, 347–59, 1990.
[4] Luo, G., et al., *J. Biomech.*, 32, 3, 275–284, 1999.
[5] Liu, Y., et al., *J. Dent. Res.*, 86, 1, 84–9, 2007.
[6] Wozney, J. M., *Prog. Growth Factor Res.*, 1, 4, 267–280, 1989.
[7] Sykaras, N., *Int. J. Oral Maxillofac. Implants*, 19, 5, 667–678, 2004.
[8] Culpepper, B. K., et al., *Biomaterials*, 31, 36, 9586–9594, 2010.
[9] Oliver W. C., and Pharr, G. M., *J. Mater. Res.*, 7, 6, 1564–1583, 1992.

SB³C2017
Summer Biomechanics, Bioengineering and Biotransport Conference
June 21 – 24, Tucson, AZ, USA

TISSUE CORING THROUGH UN-RETRACTED CANNULA INSERTION

Alexandro A. Gonzalez and Malisa Sarntinoranont

Mechanical & Aerospace Engineering
University of Florida
Gainesville, Florida, United States

INTRODUCTION

In the medical field, there are generally two kinds of tissue coring; the intentional coring obtained through tissue biopsies [1], and the unintentional coring that is believed to occur during drug delivery [2-8]. Tissue coring is best described as the volume of tissue that can displace into a cannula during insertion. Tissue biopsies include, but are not limited to the skin, breasts and brain, and are typically used to investigate the health of an organ in the human body. In regard to drug delivery, the most common types of injections include: intravenous, intradermal, intramuscular, and subcutaneous. For this study, we focused on convection-enhanced delivery (CED). CED is a drug delivery method developed to circumvent the blood-brain barrier by delivering therapeutics directly into brain tissue. Therapeutic agents are pumped through a cannula and then into the brain at a specified volumetric flow rate. Unfortunately, CED has some common areas of concern which include tissue clogging (plugging, coring, blocking, etc.) [2-6, 8] and backflow of therapeutics [2-7].

One of the earliest and more formal recordings of tissue coring during in-vivo CED was documented by Bobo [2]. To explain an initial pressure spike in the infusion line, the idea of a "tissue plug" was created. Tissue coring during CED has been referred to interchangeably as "clogging" by tissue, "coring of tissue" and "large aggregations" in scientific literature presented in [8]. Tissue coring is also thought to produce time delays in tracer appearance and spread [4], and contribute to backflow [7]. Proposed methods to reduce incidences of coring include infusion before insertion to prevent catheter blockages [8], and inserting a stepped stylet before cannula insertion to prevent tissue coring in the porcine brain [6].

To our knowledge, there are no studies that attempt to characterize this clogging or tissue coring, in brain tissue during insertion. Therefore, to determine if unintentional tissue coring could be observed, we chose to isolate the insertion step that is part of the CED process by using simple, straight-through insertion tests. The results from the tests are then used to examine the effect of varying inner and outer diameter ratio on porcine brain tissue coring.

METHODS

Cannula Preparation A series of blunt tip, polyimide-coated cannulae (Polymicro Technologies Capillary Tubing, Molex) with different inner diameters (IDs) were tested. The cannulae had IDs of 100, 200, 300, 400 and 500 μm, while the outer diameter (OD) for all cannulae was maintained constant (~ 21 gage). Only one cannula was used per insertion test. After cutting each cannula to 32 mm in length, each sample was viewed with a CMOS camera (OptixCam Summit Series OCX-5) to inspect uniformity and perpendicularity of end surfaces compared to an underlying grid. All cannula tips were then sanded using a 320 fine grit sanding block until blunt tips were achieved. A blunt tip geometry was used due to the ease of its reproducibility.

Tissue Sample Preparation Brain tissue was acquired over the course of a month from five adult-sized white pigs at the University of Florida's Meat Processing Center. Brain tissue was used within 2-3 hours after slaughter. The whole brain was removed, stored in phosphate-buffered saline [9] and transported on ice to our laboratory. Each brain was cut into smaller samples approximately 10 mm thick and positioned in an acrylic tissue container. A total of 60 samples were removed from the cortex to the end of the frontal lobe.

Poster Presentation #P37

Un-retracted Insertion Test Having properly positioned a brain slice into its container (Figure 1), a cannula was inserted straight into the tissue using a stereotactic guide (Graduated Knob UniSlide, Velmex) at a speed of 1-2 mm/s. Velocity was determined by using the duration of each test and the total distance each cannula traveled. After a cannula had fully penetrated through a brain slice, it was unfastened and pulled by the penetrating end through the drilled hole the tissue container had in one of its walls. Then, through visual and camera inspection, the incidence of tissue coring was recorded for each cannula. If coring was present, the tissue core height was also recorded so that it could be used to calculate the volume of the core, under the assumption that the core was cylindrical. There were 12 straight-through insertion tests (n = 12) conducted for each cannula.

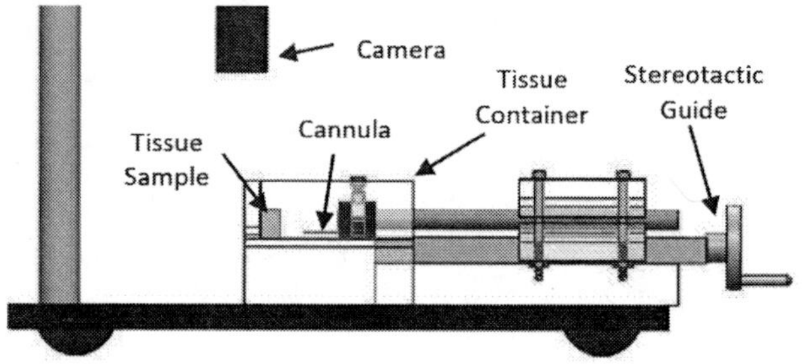

Figure 1: Un-retracted, straight-through insertion test apparatus and key components.

RESULTS

The lowest coring frequency (Figure 2) occurs at the smallest ID/OD ratio (100:795), while the highest coring frequency occurs at an intermediary ID/OD ratio (200:795). It is also noted that after the second to smallest ID/OD ratio, there is a decreasing trend in coring frequency. A normalized measure of tissue coring (Figure 3) was created to compare the tissue core volume (V_c) to the total possible displaceable volume (V_t). Both volumes were treated as cylinders. Tissue core volume was calculated using the measured height of each tissue core and the known ID of each cannula. The total possible displaceable volume was calculated using the constant thickness of each tissue sample and the constant OD of each cannula. An upward trend in V_c/V_t was found as cannula ID/OD ratio increased. Standard deviation also increased for larger cannula IDs.

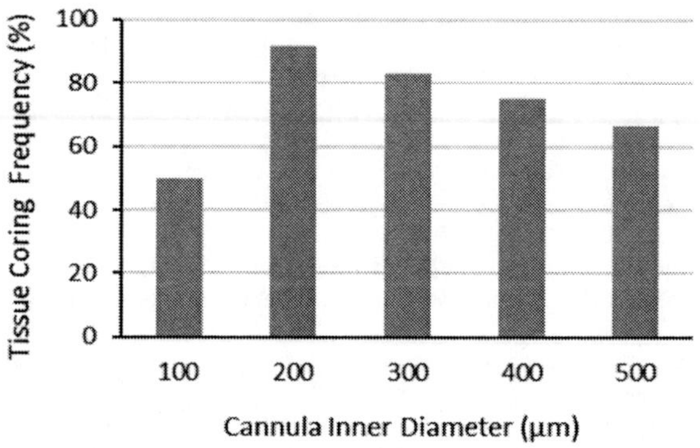

Figure 2: Frequency of tissue coring in brain tissue for cannulae with varying ID (100-500 µm) and constant OD (795 µm).

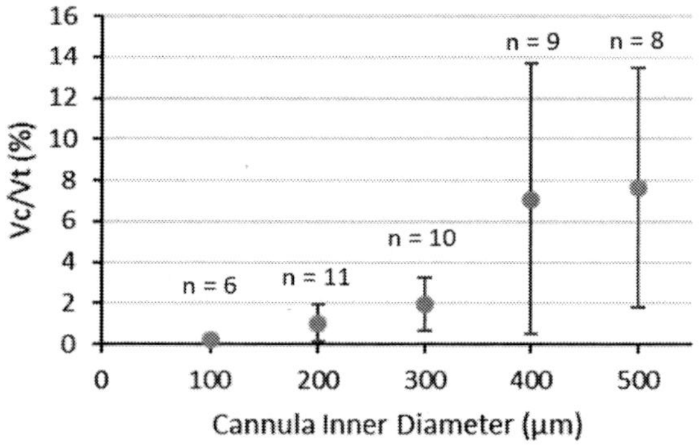

Figure 3: Ratio of V_c/V_t for each tested cannula. Error bars represent standard deviation and *n* is incidents of coring.

DISCUSSION

Our study complements the observations of tissue coring in previous studies by measuring coring frequency and quantifying core volume. Both coring frequency and core volume were dependent on cannula ID. Cannulae with an ID/OD ratio of 200:795 cored the most often compared to the remaining four cannulae with the same OD Examining tissue coring frequency and normalized volume data, higher coring frequency did not necessarily correlate with greater core volume. Although the largest ID/OD ratio cannulae produced less instances of coring, the instances when a tissue core was recorded, produced the largest and most variable tissue core volumes compared to those of the smaller ratio cannulae.

In our study, it cannot be assumed that brain tissue properties were the same in each tissue sample tested due to the manner in which the brain slices were cut and the range of brain regions tested. This variability in tissue properties and that in the range of insertion velocity could contribute to the variability in tissue core volumes.

This study provides new data to characterize tissue coring and simple, reproducible experimental methods for ex-vivo testing. Results of this study may be helpful when considering new cannula tip designs. Proper selection of specific ID/ODs ratio in blunt tip cannulae could reduce the frequency of tissue coring in brain tissue.

REFERENCES

[1] Martin, H. E. & Ellis, E. B., *Ann. of Surgery*, 92:169–181, 1930.
[2] Bobo, R. H. et al., *Proc. of the National Academy of Sci.*, 91: 2076–2080, 1994.
[3] Prabhu, S. S. et al., *Surgical Neurology*, 50: 367–375, 1998.
[4] Kim, J. H. et al., *Ann. of Biomed. Eng.*, 40: 2043–2058, 2012.
[5] Casanova, F. et al., *PLoS ONE*, 9:e94919, 2014.
[6] Pleticha, J. et al., *J. of Neurosurgery*, 121:851–858, 2014.
[7] Kim, J. H. et al., *J. of Neuroscience Methods*, 209:62–73, 2012.
[8] Jain, K. K., *Drug Delivery to the Central Nervous System*. Humana Press, 2016.
[9] D'Arceuil, H. & de Crespigny, A., *NeuroImage*, 36:64–68, 2007.

CONTRIBUTION OF REPETITIVE STRETCHING TO NEURITE INJURY IN CORTEX PRIMARY NEURONAL CELLS

S. Shirasaki (1), H. Nakadate (1), S. Aomura (1), A. Kakuta (2)

(1) Graduate School of System Design Tokyo Metropolitan University 6-6, Asahigaoka, Hino, Tokyo, 191-0065 Japan

(2) Advanced Course of Mechanical and Computer Systems Engineering Tokyo National College of Technology

INTRODUCTION

Repetitive mild traumatic brain injury (r-mTBI) has been gaining increasing attention from the researchers since several studies have reported that the cognitive dysfunctions after single mTBI become measurably long-term deficits, such as delayed speed of processing and memory dysfunction, after r-mTBI, contributing to the emerging hypothesis that r-mTBI may cause cumulative damage to the brain, and in the absence of cell death, could result in cognitive deficits which may ultimately progress to memory and learning dysfunction [1]. Studies also associated r-mTBI with "second-impact syndrome" and chronic traumatic encephalopathy (CTE) as possible consequences of r-mTBI [2]. However, the potential injury mechanisms involved in r-mTBI remain unclear and research on r-mTBI is still in early stages. This study introduces an *in vitro* model in which subsequent injuries were induced by using a uniaxial stretching device for realization of the post-injury sequelae of r-mTBI. Proposed *in vitro* uniaxial stretching model provides a reliable environment to study the mechanisms underlying cellular dysfunction and for a better characterization of the cellular degradation and dysfunction following both single and repeated injuries.

METHODS

Uniaxial stretching device

The uniaxial stretching device mainly consists of a servo actuator, a linear sensor for measuring tensile displacement, a load cell for measuring tensile loading and a polydimethylsiloxane (PDMS) chamber. A full description of the device configuration and loading mechanism has been published [3]. The Green-Lagrange strains of the culture substrate in the PDMS chamber were calculated by microscope images before and after the stretching. In this study, the history of the displacement shown in Figure 1 was applied to the chamber with a strain of 0.10 at a strain rate of 5 s^{-1}. The strain rate was obtained by dividing the maximum strain by the time to maximum strain.

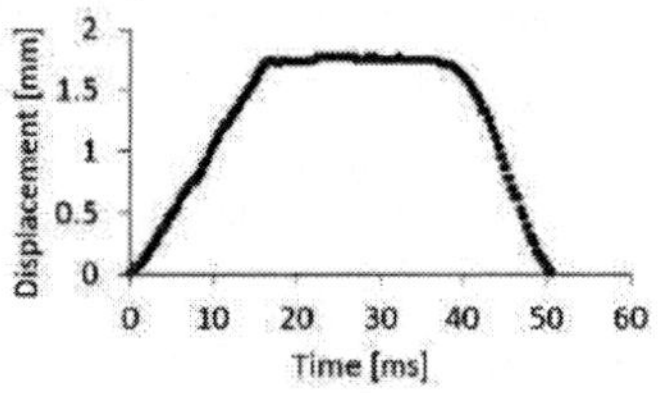

Figure 1: PDMS chamber displacement. The displacement corresponds to a strain of 0.1 and a strain rate of 5 s^{-1}

Culturing the rat brain cortex neuronal cells

Rat brain cortex neuronal cells were cultured in Primary Neuron Basal Medium supplemented with 2mM L-glutamine, 50 ug/ml Gentamicin, 37 ng/ml Amphotericin, and 2% NSF-1 under conditions of 5% CO_2 and 100% humidity at 37°C. The cells were seeded in PDMS chamber at 15 × 10^4 cells/cm^2 for 7–9 days. The direction of axonal growth is controlled by using micro stamp of Matrigel. The stump wide was matched with the size of the cell.

Single stretching experiment

2 types of stretching experiments were conducted to decide interval between initial and second stretchings for repetitive stretching experiment. The single loading groups were subjected to stretching with a strain of 0.1 at a strain rate of 5 s^{-1}. Strain and strain rate were based on previous studies [3]. The sham control groups were set in and removed from the uniaxial stretching device without receiving mechanical loading.

Cell migration

It is reported that cell migration is activated by mild stimulated [4]. In this study, cell migration was used as an indicator of the recovery process after mild stretching. Phase-contrast images were obtained immediately before and after stretching every 6 h until 48 h. Cell migration was observed from the phase-contrast images at each time. The number of cells and cell migration distance were manually measured with ImageJ. Cell migration was defined as more than 10 μm movement of a cell body and rate of cell migration was calculated.

Repetitive stretching experiment

3types of stretching experiments were conducted. The single loading groups were subjected to stretching with a strain of 0.1 at a strain rate of 5 s^{-1}, the repetitive loading groups were subjected again to the same stretching 24 h after initial stretching with a strain of 0.1 at a strain rate of 5 s^{-1}. The sham control groups were set in and removed from the uniaxial stretching device without receiving mechanical loading.

Damage rate of neuronal cells

Fluorescence images were obtained at 48 h after initial injury. Microtubule-associated protein 2 (MAP2) was stained to label dendrites and glial fibrillary acidic protein (GFAP) was stained to label astrocytes. Cell nuclei were stained with DAPI. Dysfunction of the axonal transport was defined as the rate of neurons that have swellings. Disruption of the axonal transport was defined as the rate of neurons that have bulb. Injured neuron was defined as the rate of neurons that have neurite swellings and bulb.

Statistical analysis

Results were expressed as the mean ± standard deviation (SD) of 3 independent experiments. 100–200 neurons per a PDMS substrate were analyzed totally. Means were compared by Steel's multiple comparison test. A p value of less than 0.05 was considered significant.

RESULTS & DISSCUSION

Cell migration were observed in neuronal cells all conditions. Rate of cell migration in the single loading groups maximized immediately after stretching and decreased gradually. There was significant difference until 24 h compared to the sham control group (Figure 2). Therefore, cells were stretched at an interval oh 24 h in repetitive experiment.

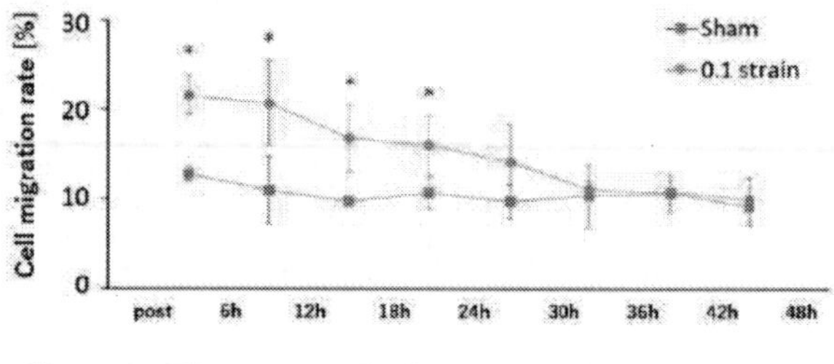

Figure 2: Time course of cell migration rate. The * symbol represents a statistically significant difference ($p < 0.05$) versus sham controls at each time point.

MAP2 was used to identify swellings and bulbs. After stretching we quantitatively observed and counted them from fluorescent images as shown in Figure 3. Swellings formation in the repetitive loading groups increased significantly at 24 h after second stretching but there was no significant difference in bulb formation at 24 h compared to the sham control group (Figure 4). Although bulb formation tends to increase in the repetitive group, much of neuronal damage would not progress to bulb at a strain of 0.1. These data show that even though initial insult induced some level of swelling formation, swelling formation increased by second stretching confirming that r-mTBI causes increased amounts of cellular damage when compared with single insults of the same magnitude and indicate that if injured neurons are subsequently subjected to low strain at the same level, the rate of injury significantly increases confirming the emerging hypothesis on repetitive mTBI exacerbating traumatic axonal injury.

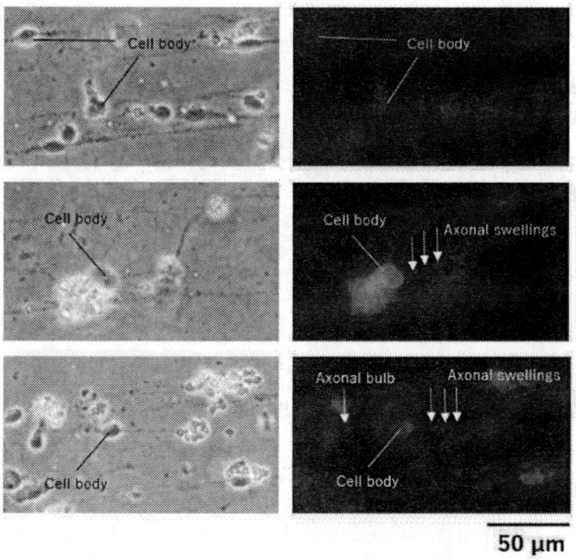

Figure 3: Phase-contrast image (left) and fluorescence image (right) of primary rat cortex neurons (green) and astrocytes (red) and nucleus (blue) in sham control (upper), single stretching (middle) and repetitive stretching (lower).

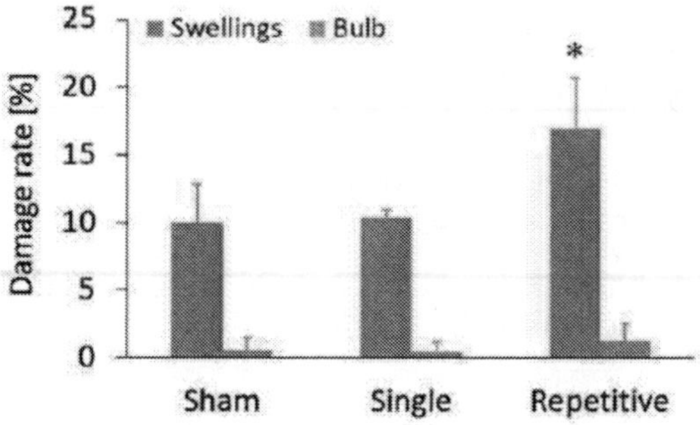

Figure 4: Damage rate of neuronal cells. Rate of neurons that have axonal swellings (left) and axonal bulbs (right). The * symbol represents a statistically significant difference ($p < 0.05$) versus sham controls.

REFERENCES

[1] Kamnaksh A, et al., J Neurotrauma 28: 2145-2153, 2011.
[2] Smith DH, et al., Nat Rev Neurol 9: 211-221, 2013.
[3] Aomura, S et al., Inte Mol Med 3: 654–660, 2016.
[4] Zito, A et al., Brain Structure and Function 219.1: 105-118, 2014.

LOCAL DISCONTINUITIES IN ALIGNED FIBROUS NETWORKS ATTENUATE TISSUE-TO-NUCLEAR STRAIN TRANSMISSION

Tonia K. Tsinman (1), John M. Peloquin (2), Spencer E. Szczesny (1), Su-Jin Heo (1), Dawn M. Elliott (2), Robert L. Mauck (1)

(1) Department of Bioengineering
University of Pennsylvania
Philadelphia, PA, USA

(2) Department of Biomedical Engineering
University of Delaware
Newark, DE, USA

INTRODUCTION

Tensile load bearing soft tissues, such as the knee meniscus, tendon, and the annulus fibrosus of the intervertebral disc, have an anisotropic and hierarchical organization. The extracellular matrix (ECM) of these tissues is comprised of aligned type I collagen whose organization is crucial for proper transmission of loads and strains across the ECM-to-cell length scales and subsequent normal mechanical signaling and homeostasis. During aging and pathologic remodeling, however, the ordered structure of these tissues is often lost. For instance, we recently reported that microscopic proteoglycan-rich domains emerge within the otherwise ordered collagenous ECM of the knee meniscus [1]. These micro-domains, with differing composition and mechanical properties from surrounding fibrous tissue, altered local strain transmission and created heterogeneous mechano-signaling patterns [1]. While the origin of these micro-domains remains to be determined, one hypothesis is that local interruptions of the fibrous architecture (i.e., local micro-damage) change the mechano-biologic stimuli received by resident cell populations, resulting in altered differentiation and matrix remodeling. To begin testing this hypothesis, this study developed an aligned polymer scaffold inclusive of local interruptions in the fibrous architecture in order to computationally and experimentally probe their impact on macroscopic and cell scale strain transmission.

METHODS

Aligned poly (ε-caprolactone) (PCL) scaffolds were fabricated by electrospinning 14.3% w/v PCL solution onto a rotating mandrel (as in [2]). Generated scaffolds were trimmed into 6.5cm long strips with either a 1 or 2cm width and with the prevailing fiber direction oriented in the long axis of the scaffold. Interruptions in the fibrous architecture were created by introducing two incisions (oriented perpendicular to the fibers) longitudinally spaced 2cm apart. The width of the cuts in 1cm and 2cm wide scaffolds was 0.4cm and 1cm, respectively (**Fig. 1**). To map strain at the macro-scale, strips were speckle coated, and grip-to-grip strain was applied in the fiber direction at a rate of 0.5%/s to 15%. Images were acquired during the test (at 15fps) and analyzed using Vic-2D (Correlated Solutions) to determine E_{11} Lagrangian strain across the sample. At least 150 points per sample were tracked. To model the potentially altered strain fields for these varying scaffold geometries, FE models were implemented for each scaffold geometry. Meshes were constructed from 6-node brick elements as in [3] with a 0.25 mm element size. The material was modeled as a neo-Hookean and exponential fiber mixture using previously reported material properties [4]. Grip-to-grip tensile strain was simulated from 0% to 15% using FEBio. For analysis, the mesh was partitioned into edge and middle regions (**Fig 2**, left) as in the Vic-2D strain fields. The Lagrange strain (E_{11}) for each region was evaluated at the element centroids, and summarized using median and interquartile range.

To determine Lagrangian strain at the cell scale (as well as to measure subcellular deformations), mesenchymal stem cells (MSCs) isolated from juvenile bovine bone marrow (expanded through passage 1) were seeded onto 1cm wide scaffolds at a density of 10^5 cells/scaffold. Prior to seeding, scaffolds were rehydrated through an ethanol gradient and coated overnight with 20μg/mL fibronectin. Seeded constructs were cultured for two days in a chemically defined medium to allow for cell attachment. To measure nuclear deformation (denoted as the change in nuclear aspect ratio, NAR, relative to baseline), cell-seeded scaffolds (with and without local incisions, 0.4cm in width) were stained with Hoechst 33342. Next, uniaxial strain was applied in 3% grip-to-grip strain increments using a custom micromechanical test device mounted on top of an inverted fluorescent microscope [5]. Nuclei of cells were imaged at 3, 6, 9, 12, and 15%

strain steps using a 20X objective. The NAR was defined as the ratio of the long to short axis of the nucleus at each strain step, and was quantified using a custom MATLAB code (n>50 for each group) [5]. The same software was used to calculate Lagrangian strains at the cellular level, based on the original and deformed positions of nuclear triads. Statistical differences were established using a two-way ANOVA with Bonferroni post-hoc testing to make comparisons between groups (significance threshold: $p<0.05$).

RESULTS

Strain mapping of intact speckle coated PCL scaffolds demonstrated relatively uniform 'global' E_{11} strain distributions that largely matched the magnitude of the applied grip-to-grip strain (**Fig. 1a**). As expected, introduction of discontinuities in the aligned fiber network resulted in non-uniform strain fields across the length and width of the scaffold, where measured E_{11} strains were substantially diminished in the region located between the two incisions as compared to the outside edges of the scaffold (**Fig. 1b,c**). Increasing the width of the scaffold exaggerated the reduction in strain transmission between regions with intact versus severed fibers, allowing for an area with virtually no measured E_{11} strain, even at physiologically relevant grip-to-grip strains of 3 or 6% (**Fig. 1c**).

The response of scaffolds with and without fiber interruption of varied geometries to applied strain was also successfully predicted computationally. The FE model correctly showed the lesser effects of fiber discontinuities on strain attenuation in narrower (1cm wide) scaffolds (**Fig. 2**). Determination of 'local' Lagrangian strains in cell-seeded scaffolds through cell nuclei tracking showed similarity between local and grip-to-grip strains of control and cut scaffolds (**Fig. 3a**), validating nuclear tracking as a method for strain mapping in this experimental system. Significant strain attenuation at the cellular level in the E_{11} direction was observed in the middle of the cut region compared to all other locations, and at all applied grip-to-grip strains (**Fig. 3a**). Importantly, perturbations in the local strain field were reflected in differences in nuclear-scale deformation. Cells positioned along fully continuous fibers (uncut scaffolds, edge of cut scaffolds) showed a monotonic increase in NAR (up to 23%) with increased applied strain. In contrast, the NAR of cells located in areas containing a fiber discontinuity (middle of cut scaffolds) did not change with the applied strain (**Fig. 3b**).

DISCUSSION

This study shows that local interruptions in scaffolds composed of aligned nanofibers result in localized strain attenuation at the macroscopic (**Fig. 1**) and microscopic (**Fig. 3**) level. Additionally, a Neo-Hookean matrix/exponential fiber mixture model was able to predict the variation in strain transmission for two cases of cut scaffold geometries, highlighting its potential use as a tool for assessing strain fields for other defected fibrous scaffold geometries that may have biologic relevance. The diminished strain transfer was particularly evident at length scales relevant to the cellular mechano-response. Similar trends in NAR and local strain fields confirmed that strains in the principle directions influence the magnitude of nuclear deformation. Importantly, the local fiber discontinuities affected nuclear deformation *only* in regions within the defect, while cells residing within regions of uninterrupted fibers experienced similar deformations in the direction of applied strain as those of uncut control samples. These data illustrate how local fiber disruption can generate distinct mechanical microenvironments within the same scaffold, with neighboring cells experiencing markedly different mechanical inputs based on local boundary conditions. Future work will use this system to query how perturbed local mechanics leads to a differential mechano-response in resident cell populations and ultimately drives the development of heterogeneous cell-mediated ECM deposition and remodeling. Together, this work will provide insight into the cellular response to tissue micro-damage and the mechanism of formation of micro-domains in the meniscus and other fibrous connective tissues.

ACKNOWLEDGEMENTS

This work was supported by the National Institutes of Health (R01 EB02425) and the Penn Center for Musculoskeletal Disorders.

REFERENCES

[1] Han, W., et al, *Nat Mater*, 2016. **[2]** Baker, B et al., *Proc Natl Acad Sci*, 2012. **[3]** Peloquin, J et al., *J Mech Behav Biomed Mater*, 2015. **[4]** Baker, B et al., *J Biomech Eng*, 2009. **[5]** Driscoll, T., et al, *Biophys J*, 2015.

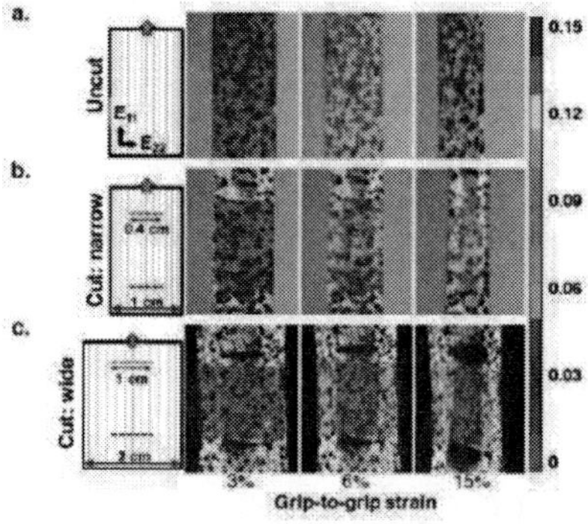

Figure 1. Vic-2D calculated strain maps using particle tracking for (a) uncut, (b) 'narrow' cut scaffolds, and (c) 'wide' cut scaffolds, at 3, 6, and 15% grip-to-grip strains. Cuts in both scaffold geometries were positioned 2cm apart.

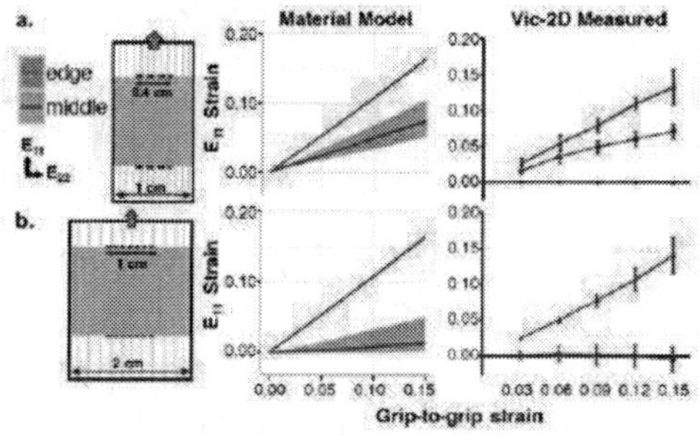

Figure 2. Calculated scaffold strains ('middle': area between cuts, 'edge': area flanking the cut region) using FE modeling compared to strain mapping. (a) 'narrow cut scaffold' geometry and (b) 'wide cut scaffold' geometry, as described in Fig. 1. For FE graphs, dark line represents the median and light shading the IQR. Vic-2D measurements reported as the mean and standard deviation.

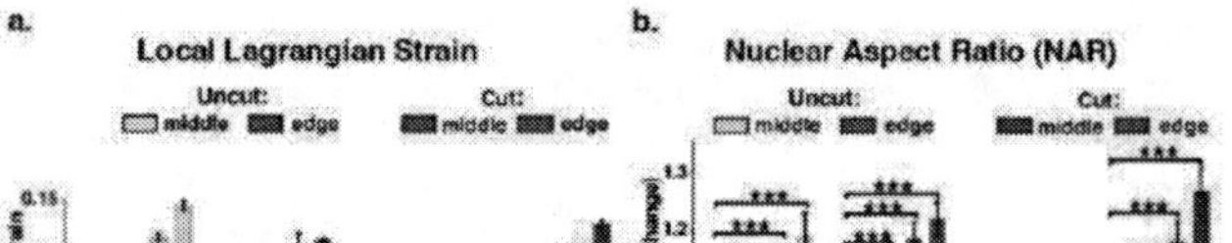

Figure 3. (a) E_{11} measured through nuclear tracking ('local strain') at denoted applied strains. 'Cut: middle' cohort compared to all other groups at each strain. (#:p<0.001) (b) NAR (normalized to 0% strain) as a function of applied strain. ns: not significant, *p<0.05, **p<0.01, ***p<0.001.

A COMPUTATIONAL ANALYSIS OF AORTIC PULSATILE FLOW CONDITIONS FOR VALVE TISSUE FORMATION

Alex Williams, Manuel Perez, Arash Moshkforoush, Omkar Mankame, Manuel Salinas, Nikolaos Tsoukias, Sharan Ramaswamy

Department of Biomedical Engineering
Florida International University
Miami, Florida, United States of America

INTRODUCTION

Fluid induced shear stress has been shown to influence collagen formation in the growth of engineered tissue [1, 2, 3]. Previous studies have examined collagen formation in engineered tissue, under zero flow (static), steady-state flow, pulsatile square-wave flow, cyclic flexure as well as combined cyclic flexure and steady-state flow simulations (Flex-Flow) [1]. These conditions however, do not fully represent cardiac pulsation experienced by native cardiovascular tissues.

This study examined flow physics that more closely mimic physiologically-relevant conditions. A bioreactor was used to simulate in-vivo conditions experienced by engineered tissue using rectangular strips of specimens immersed in flow media. Computational fluid dynamics (CFD) software was used to analyze flow and flow-based parameters around the specimens. Computational findings were compared to previous tissue engineering studies conducted in our laboratory [1].

METHODS

ANSYS computational fluid dynamics software (ANSYS-CFX, ANSYS Inc., Canonsburg, Pennsylvania, USA) was used to perform the CFD simulations and model the flow parameters and flow physics under viscous laminar flow within the chamber.

The simulations were performed using a bioreactor mesh consisting of a U-shaped tubular conditioning chamber that houses the tissue samples and through which flow media was pumped across the face of the tissue samples (Fig. 1).

A physiologically relevant pulsatile waveform from the human aorta was obtained from literature [4]. Subsequently, the aortic velocity profile (Fig. 2) was prescribed as the inlet boundary condition.

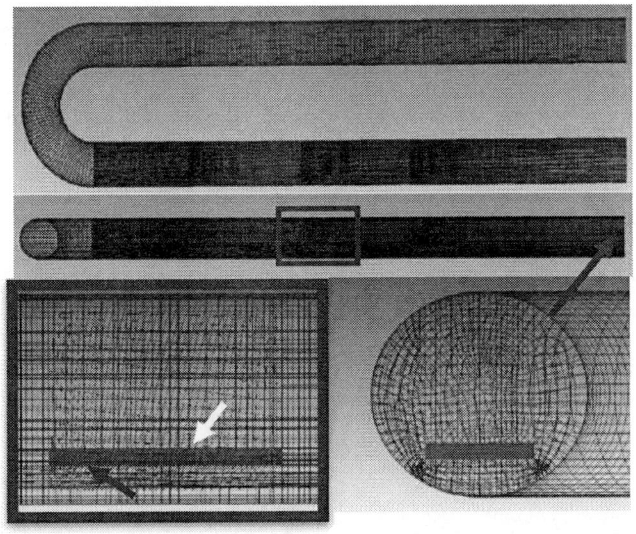

Figure 1 – Bioreactor and Sample Mesh Visualization with purple arrow referring to distal sample wall and white arrow referring to proximal sample wall.

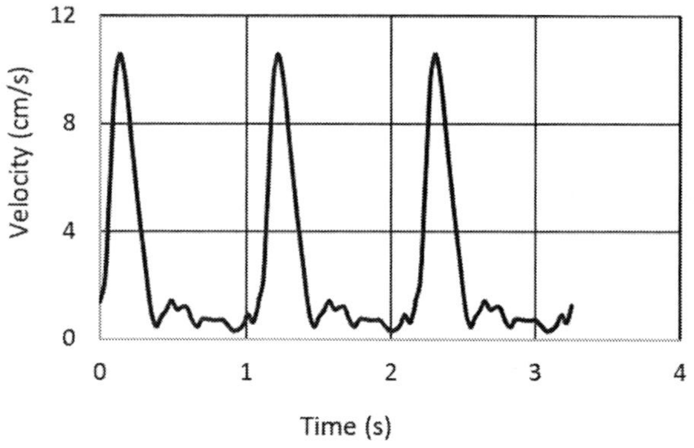

Figure 2 –Aortic velocity waveforms, taken from literature [4], were utilized to replicate flow dimensions, such as wall shear stress, of cardiac valve tissue.

A zero pressure outlet boundary condition was prescribed. Material properties, including density and viscosity, were included in the setup of the analysis. Convergence criteria was set at a 10^{-9} residual for both continuity and momentum equations. Shear stress data was subsequently computed over a discrete set of points across the face of the specimens, after which the data was loaded into post-processing software (Tecplot Inc., Bellevue, WA, USA) for application as contour representations of time-averaged wall shear and oscillatory shear index (OSI) at the proximal and distal walls of the specimens.

RESULTS

While square-wave pulsatile conditions gave rise to heightened ranges of OSI (0.37 ± 0.021), Flex-Flow and aortic pulsatile conditions comparatively gave significantly lower OSI ranges (Fig. 3). The aortic pulsatile conditions displayed higher uniformity in terms of specimen shear stress distribution. The overall OSI averages, considering both the proximal and distal walls of the specimens, were 0.37 ± 0.021, 0.1055 ± 0.052 and 0.175 ± 0.002 for square pulsatile, Flex-Flow and aortic pulsatile conditions respectively.

DISCUSSION

Our recent evidence suggests that an optimal range of oscillatory flow may elicit augmented engineered heart valve tissue formation, gene expression and supporting valvular phenotype from bone marrow stem cells [1, 5]. Here we showed, through computational analysis, that a physiological flow regime could lead to an altered range of OSI. We interpret that these marked differences may promote enhanced biological responses from bone marrow stem cells. Hence in conclusion, physiologically-relevant pulsatile flow waveforms may provide enhanced *in vitro* bioreactor culture conditions, thus warranting consideration in the dynamic culture of engineered structures.

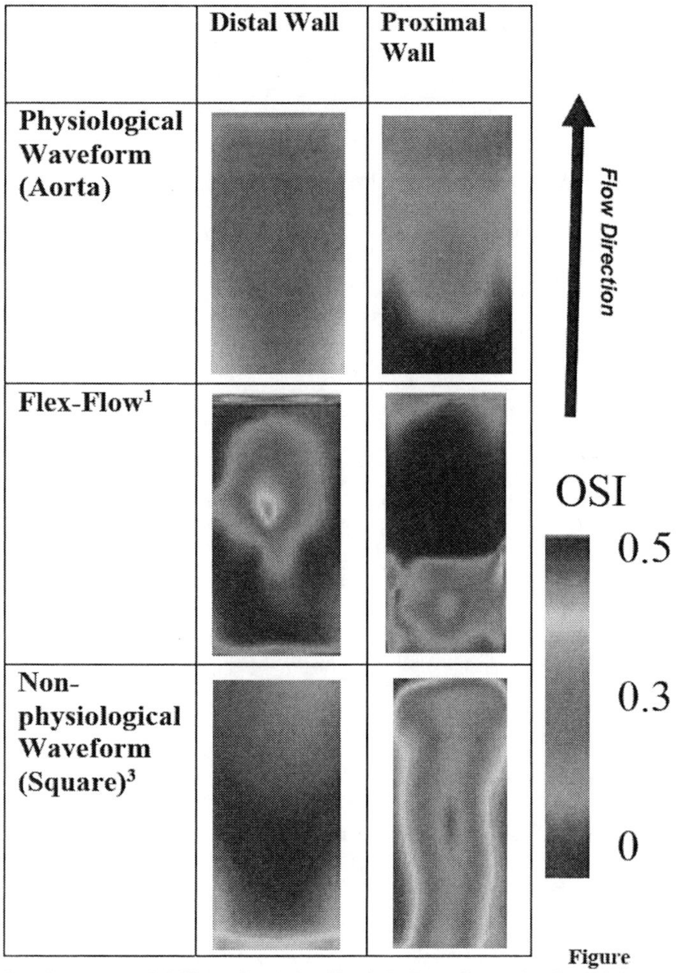

Figure 3 – Contours of OSI (n=3) at the distal (left) and proximal (right) specimen walls of the bioreactor during Flex-Flow, square-wave and physiological pulsatile conditions

REFERENCES

[1] Rath, S et. al. *PLoS One*. 2015

[2] Engelmayr GC Jr. *Biomaterials*. 2006

[3] Salinas, M., et al. *Cardiovascular Engineering and Technology. (2016)*

[4] Lotz, Joachim, et al. *Radiographics* (2002)

[5] Salinas, M, Ramaswamy, S. *Journal of Biomechanics*, 2014 Nov

[6] Salinas, Manuel, et al. *Computer Methods in Biomechanics and Biomedical Engineering* (2014)

[7] Ramaswamy S .et. al. *J Biomech Eng*. 2014 Dec

VALIDITY OF DYNAMIC MECHANICAL ANALYSIS FOR SHAPED MENISCUS

Reo Tanabe (1), Seido Yarimitsu (2), Hiromichi Fujie (2)

(1) Division of Human Mechatronics Systems,
Graduate School of System Design,
Tokyo Metropolitan University
Tokyo, Japan

(2) Department of Intelligent Mechanical Systems,
Faculty of System Design,
Tokyo Metropolitan University
Tokyo, Japan

INTRODUCTION

Dynamic mechanical analysis (DMA) under physiological conditions is necessary for deeper understanding of biomechanical properties of knee meniscus. Although biomechanical evaluations of meniscus have been conducted in previous studies [1-5], the specimens shaped by cutting tools were used due to ease of handling and testing. However, shaping treatment damages the original tissue structure and may change the biomechanical properties of meniscus. Therefore, the objective of the present study is to assess the validity of DMA for shaped specimen through the evaluation of site- and frequency-dependent viscoelastic properties of intact and cylindrically-shaped sheep meniscus.

METHODS

Specimen preparation

Medial menisci (n=8) were harvested from the left and right knee joints of mature sheep (female, average age 93.8 ± 25.6 months, average weight 69.8 ± 19.1 kg) and stored at -35 °C until experiment. Prior to mechanical testing, menisci were completely thawed and then immersed in saline solution at 37 °C for 30 min for equilibration. All tests were completed within 4 hrs after thawing, while specimens remained in saline solution at 37 °C throughout. The measurement sites at the anterior, mid body and posterior segments of meniscus were marked with a surgical marker according to International Society of Arthroscopy, Knee Surgery and Orthopedic Sports Medicine (ISAKOS) classification [6]. Subsequently, the thickness of specimens of intact meniscus group (n=4) at each measurement sites was measured along the direction of indentation using a digital point jaw neck caliper (Niigata Seiki Co., Ltd., Niigata, Japan). The meniscus was then glued on the bottom of liquid bath with cyanoacrylate adhesive (Aron Alpha, Toagosei Co., Ltd., Tokyo, Japan). The specimens of shaped meniscus group (n=4) were punched at measurement sites using a biopsy punch (ϕ 4mm, Kai Industries Co., Ltd, Tokyo, Japan) along the direction of indentation, and their tibial side were cut to be formed in cylindrical shape (Fig.1). Thereafter, the thickness was measured using a light microscope (VHX-1000, KEYENCE Corporation, Osaka, Japan) and glued on the bottom of liquid bath with cyanoacrylate adhesive.

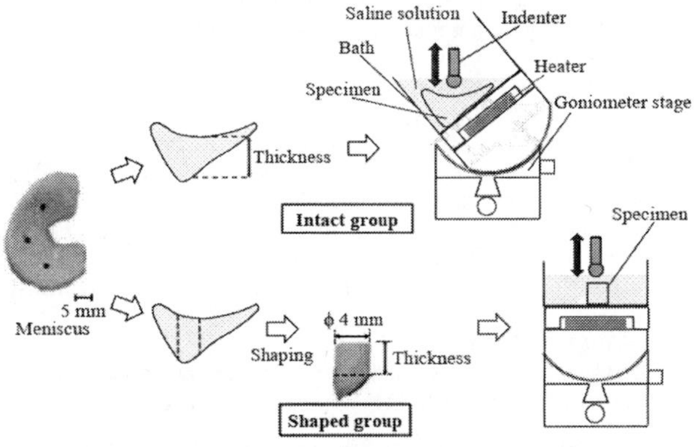

Fig.1 Specimen shaping and schematics of experimental setup.

Dynamic mechanical analysis (DMA)

DMA was performed using a mechanical testing system (ElectroForce 5500, TA Instruments, Minnesota, USA) equipped with a 22.2 N capacity load cell. The liquid bath was aligned to obtain perpendicular meniscus-indenter contact using goniometer stage (Fig.1). A spherical probe (ϕ 3 mm, SUS440C) was used as an indenter. After equilibration at 37 °C for 10 min in saline solution, a reference of contact deformation of specimen was determined by applying pre-load of 0.005 N with a compression speed of 3.25 mm/s to eliminate bias from fluid surface tension. And then, pre-strain of 2% of each specimen thickness with a compression speed of 3.25 mm/s was applied to prevent specimen detaching from the indenter during the test. After 3 min of relaxation, DMA was conducted using sinusoidal compression strain with amplitude of 2.5% of each specimen thickness at a frequency rate of 0.1, 0.3, 0.6, 1, 3, 6 and 10 Hz. The storage modulus (E′), loss modulus (E″) and loss factor (tan δ) were determined from DMA. The influences of shaping treatment and compression frequency on E′, E″ and tan δ were assessed by two-way analysis of variance (ANOVA) with Tukey's honestly significant difference (HSD) post hoc analysis for multiple comparisons.

RESULTS

Frequency had a significant influence on E′ at all segments ($p < 0.01$ in two-way ANOVA) regardless of the treatment (Fig. 2 (a)). However, shaping treatment had a significant influence on E′ ($p < 0.05$ in two-way ANOVA) and the shaping group show significantly lower E′ than intact group at 6 and 10 Hz ($p < 0.05$ in Tukey's HSD) (Fig.2 (a)) at posterior segment. In contrast, shaping treatment had no significant influence on E′ at anterior and mid body segments in both groups. No significant influence of shaping treatment was detected in E″ at all segments (Fig.2 (b)). Shaping treatment had a significant influence on tan δ at anterior and posterior segments ($p < 0.01$ in two-way ANOVA) and significant differences between two groups were detected at lower frequency at all segments (Fig. 2 (c)). Frequency had a significant influence on E′, E″ and tan δ at almost all the segments.

DISCUSSION AND CONCLUSION

In viscoelastic tissues, it is well known that a higher indentation rate or frequency leads to generate a higher stiffness because of increasing of interstitial fluid pressure. Results reveal that shaping treatment had a significant influence on elastic property and frequency dependency of E′ at posterior segment at least. It is considered that the shaping treatment damage the collagen structure in meniscus and this leads to the decrease of collagen modulus in meniscus. In addition, squeezing out of water from tissue easily occurs and this leads to the decrease of fluid pressure under compressive condition. For E″ and tan δ that represent the capacity of dissipating energy and damping property, frequency dependencies were observed with a similarity to previous study confirmed [3]. The influence of shaping treatment was observed in tan δ at identical frequencies although detail mechanisms remained unknown.

The results indicated that the shaping treatment affects the viscoelastic property of meniscus, especially in posterior segment. For further investigation, we have to focus on the effects of structure and constituents in each segment on the biomechanical properties of menisci.

ACKNOWLEDGEMENTS

The present study was financially supported by the Grant-In-Aid for Scientific Research B (#16H03172) from the MEXT, Japan.

REFERENCES

[1] Baro, V. J., et al., *Bone,* 51:232-240, 2012.
[2] Gaugler, M., et al., *Knee Surg Sports Traumatol Arthrosc*, 2015.
[3] Pereira, H., et al., *Osteoarth Cart*, 22:1271-1281, 2014.
[4] Danso, E. K., et al., *J Biomech*, 48:1499-1507, 2015.
[5] Lee, C. H., et al., *Sci Transl Med*, 6:266ra171, 2014.
[6] Anderson, A. F., et al., *Am J Sports Med,* 39:926-932, 2011.

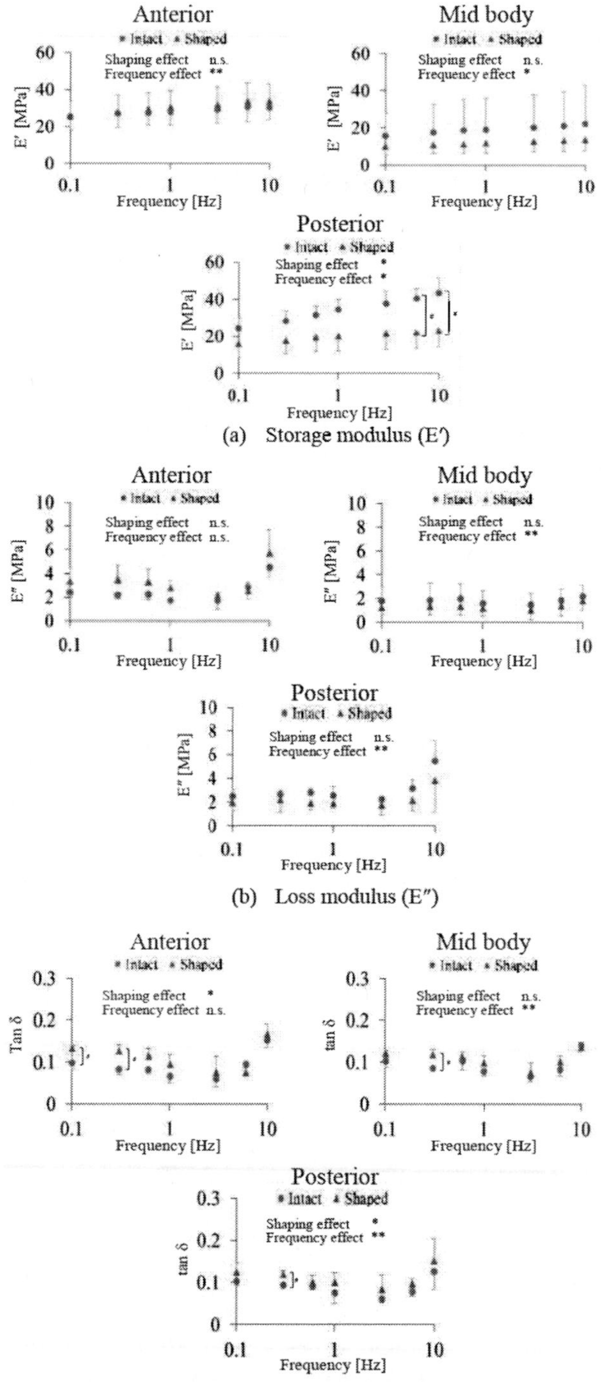

Fig.2 DMA of medial meniscus at several frequencies. Values are mean ± SD (n=4). (*p < 0.05, **p < 0.01 in two-way ANOVA, #p < 0.05 in Tukey's HSD)

SB³C2017
Summer Biomechanics, Bioengineering and Biotransport Conference
June 21 – 24, Tucson, AZ, USA

EFFECT OF FIBER ARCHITECTURE ON TISSUE FAILURE DYNAMICS: A FINITE ELEMENT STUDY

Minhao Zhou*, Benjamin Werbner*, Grace O'Connell

Mechanical Engineering
University of California, Berkeley
Berkeley, CA, United States

INTRODUCTION

Fiber-reinforced tissues experience large, complex loads during daily activities. Failure of these tissues often results in debilitating pain and reduced mobility. Understanding the failure behavior of tissues with limited self-healing capabilities, such as the meniscus in the knee and annulus fibrosus (AF) in the intervertebral disc, is of particular importance, as their failure alters the stress distribution throughout the tissue and adjacent tissues. Increased stresses result in tissue remodeling and may lead to a degenerative cascade [1].

Unfortunately, the failure mechanics of fiber-reinforced tissues, particularly those with fibers oriented off-axis from the applied load, is not well understood. Previous studies have suggested that this lack of understanding is due to limited specimen sizes and high strains at the testing grips, which may cause unpredictable tissue failure [2, 3]. Recent work by Peloquin *et al.* identified five different failure modes in meniscus specimens, with less than 25% of specimens failing at the mid-substance [3]. Unpredictable failure behavior may be responsible for the large variation of failure properties reported in the literature.

The fiber architectures of load-bearing, fiber-reinforced soft tissues vary greatly. For example, tendon is considered a unidirectional composite with fibers aligned primarily along the loading direction (Fig. 1a). Meniscus (Men.) is considered a cross-ply laminate composed of alternating layers in which the fibers are oriented at 0° and 90° relative to each other (circumferential and radial fibers, Fig. 1b, c). The AF is considered an angle-ply laminate containing alternating layers of fibers oriented between ±30° and ±45° to the transverse plane (Fig. 1d) [4].

While fiber orientation is known to affect sub-failure mechanical properties [3, 5], few studies have quantified the effects of fiber architecture on failure behavior [6]. Therefore, the objective of this study was to investigate the effects of various fiber architectures on the predicted failure location of fiber-reinforced tissues in uniaxial tension using finite element analysis. Our analysis also considered the effect of applied compression at the grips. The results from this model help explain large variations in failure behavior reported in the literature.

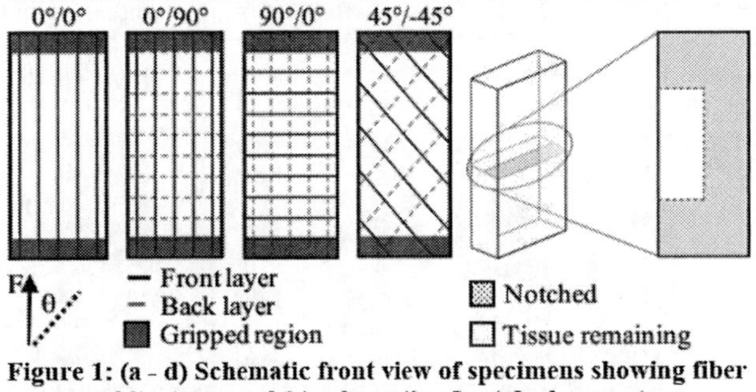

Figure 1: (a - d) Schematic front view of specimens showing fiber architectures and (e) schematic of notched geometry.

METHODS

Finite element models (FEMs) were developed representing fiber-reinforced tissue specimens (PreView 1.19.0; FEBio 2.5.2, ~180k nodes, ~170k hexahedral elements). Bulk tissue dimensions were defined to be 4.8 mm x 2 mm x 0.8 mm, which is proportional to experimental specimens from our previous work [7]. Four 200 µm-thick, welded-interface lamellae were included [4]. Full-width sandpaper grips were simulated to cover 0.4 mm of the tissue at both ends (Fig. 1a-d – grey region). A homogeneous, hyperelastic material description was used [8], with material coefficients based on data obtained from preliminary experiments on AF specimens in tension (ρ= 1 g/cm^3, C_1= C_2= 0.1 MPa, C_3= 0.02 MPa, C_4= 30, C_5= 25 MPa,

Poster Presentation #P42

k= 50 MPa, λ= 1.15). While FEMs were developed based on AF material properties, these methods can be applied to other fiber-reinforced tissues as well.

Tendon specimens were modeled with layers of fibers aligned to the loading direction (Fig. 1a). Meniscus specimens were modeled with layers of fibers alternating between 0° and 90° relative to loading. Meniscus A refers to a sample with front layer fibers aligned with loading ('0°/90°', Fig. 1b), while Meniscus B refers to a sample with front layer fibers oriented at 90° with respect to loading ('90°/0°', Fig. 1c). Inner AF specimens were modeled with alternating layers of fibers oriented at ±45° from loading (Fig. 1d), while outer AF specimens were modeled with alternating layers of fibers at ±30°.

In addition to the intact models described above, a notched specimen geometry, which increased the likelihood of mid-substance failure, was included [7]. In these cases, elements were removed at the mid-substance (MS) such that one quarter of the original cross-sectional area remained (Fig. 1e). Loading was simulated by first applying 10% compressive strain at the grips, followed by uniaxial tension to 30% global engineering strain along the length. Grip boundary conditions were fixed to represent no slipping between the grips and the sample.

Local strains for each node in the mid-substance and at the grip-line were sorted in descending order, and the average strain for each region was calculated as the average of the top 35 nodes. The MS:grip strain ratio was calculated as the average local mid-substance strain divided by the average local grip strain (i.e. MS:grip ratio greater than 1.0 indicates higher local strain at mid-substance). An average local strain of 50% served as an estimate of bulk failure initiation: failure was predicted at the location reaching this strain threshold earliest in the simulation.

RESULTS

Average local strains reached 50% at either the mid-substance or at the grip-line for all simulations. For all intact simulations, local strains were always higher at the grip-line than at the mid-substance, so these specimens are predicted to fail at the grips. For all notched simulations except Meniscus A, the MS:grip ratio at predicted failure was at least 1.10 (Fig. 2a, 2c-e – front, white arrows; Fig. 3), so mid-substance failure is predicted in these samples. Notched Meniscus A was predicted to fail at the grips because the maximum local strain at predicted failure was at the grip-line (Fig. 2b – back, white arrows).

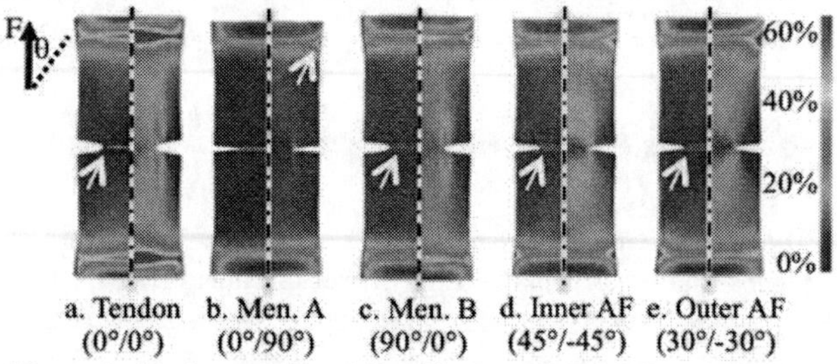

Figure 2: Strain map (front | back) at predicted failure initiation. Arrows indicate maximum local strains.

DISCUSSION

In this study, we used FEMs to investigate the effect of fiber architecture on the predicted failure location of various fiber-reinforced soft tissues. The results of this study indicate that the fiber angle relative to the loading direction and the relative fiber orientation between layers both have a significant effect on the predicted failure location under uniaxial tension. Our results also reinforce the importance of modeling compressive strains caused by gripping when investigating failure properties. Additionally, this study further validates the effectiveness of the described notched geometry for inducing mid-substance failure of fiber-reinforced soft tissues [7].

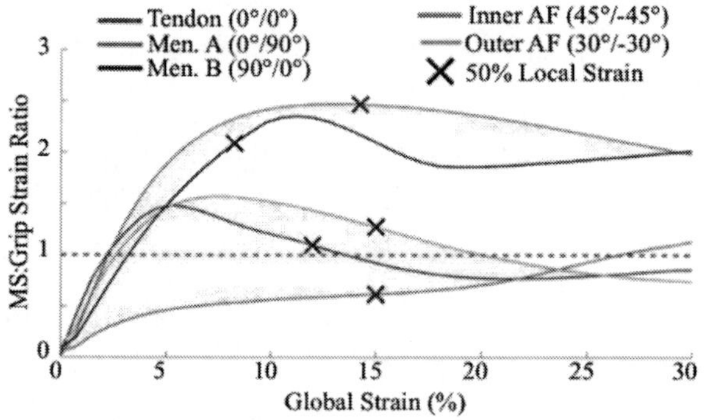

Figure 3: MS:grip ratio versus global strain for four fiber architectures (all Notched). X denotes predicted failure.

We found that increasing fiber alignment towards the loading direction increased the likelihood of failure at the grip-line. For example, the MS:grip strain ratio of notched tendon specimens was close to 1.0 when the maximum local strain reached 50% (Fig. 3 – blue line). The simulations also suggest that mid-substance failure was more likely in notched specimens taken from the inner AF, where fibers are less aligned with loading (Fig. 3 – pink line), than the outer AF, where fibers are more aligned with loading (Fig. 3 – green line).

Predicted failure location differed greatly between the two Notched meniscus cases (Fig. 3– red and black lines), where the models differed only by the fiber orientation of the layer adjacent to the notch. The model strongly predicted mid-substance failure for Meniscus B (90° layer at front of notch), but suggested grip-line failure for Meniscus A (0° layer at front of notch). The 0° fiber layer is stiffer, so it develops higher local stresses, whereas the 90° layer develops higher local strains. This dependency on specimen alignment helps explain the large variation in reported meniscus failure properties [3].

While it is not well understood whether tissue failure is a stress- or strain-based phenomenon, we chose 50% local strain as a threshold to predict tissue failure. Furthermore, our model did not include a description for damage, which may cause differences in global failure strain predictions. Finally, it is likely that local failure behavior depends on tissue composition; therefore, future work will utilize a heterogeneous model to describe the extrafibrillar matrix separate from the collagen fibers in order to better investigate tissue failure mechanisms [8].

This study expands upon our previous work that aimed to ensure repeatable mid-substance failure of fiber-reinforced tissues in uniaxial tension [7]. We found that both the notch geometry and fiber architecture have a significant impact on predicted failure location for such specimens. These observations are valuable for guiding time-intensive experimental studies. Moreover, as electrospun scaffolds are explored for biological repair strategies of fiber-reinforced materials, the findings of this study emphasize the importance of replicating the native fiber architecture to maintain native stress distribution throughout the tissue [9].

ACKNOWLEDGEMENTS * Both authors contributed equally.

REFERENCES [1] Adams, M.+, *Spine*, 2006; [2] Jacobs, N.+, *J Biomech Eng*, 2013; [3] Peloquin, J.+, *J Biomech Eng*, 2016; [4] Cassidy, J.+, *Conn Tiss*, 1989; [5] Adams, M.+, *Eur Spine J*, 1993; [6] Isaacs, J.+, *JMBBM*, 2014; [7] Werbner, B.+, *ORS*, 2017; [8] Yang, B., et al., *SB3C*, 2016; [9] Li, W.+, *J Biomech*, 2007.

Mechanical Analysis of Heterogeneous Pulmonary Acinus Structure Using Image-based and Mathematical Models

Keisuke Nishimoto (1), Kenichiro Koshiyama (1), Satoshi Ii (1), Shigeo Wada (1)

(1) Graduate School of Engineering Science
Osaka University
Toyonaka, Osaka, Japan

INTRODUCTION

A pulmonary acinus is the peripheral airway region which forms complicated branching structure composed of airways and alveoli. The acinus is the largest functional unit for gas exchange. The tissue of the acinus deforms during breathing and cells in the tissue express and adjust their function in response to such mechanical stimulations [1]. Furthermore, in mechanical ventilation, excess tissue deformation occasionally occurs and it damages the tissue, known as ventilator-induced lung injury [2]. Hence, it is important to know the details of mechanical fields at the tissue and cell levels to understand the cellular response to mechanical stimulations in the acinus and to develop better mechanical ventilation protocols.

The tissue of the acinus is composed of elastin, collagen, ground substance, and cells. A liquid film, water and lung surfactant, covers the interface between air and the tissue. Because of the complicated material compositions and the complicated structure, it is difficult to investigate the mechanical fields by experimental approaches. Accordingly, computational mechanical analysis of the acinus has been conducted [3], [4]. The simulations have used the acinus structure of the image-based model constructed from a part of whole acinus [3] or of the mathematical model, which is an assemblage of simple polyhedra [4]. We recently proposed the heterogeneous acinus structure model composed of alveoli of irregular sizes, shapes, and locations [5]. Also, we obtained the image-based structure model from the synchrotron micro-CT images of the whole acinus [6]. In this context, we here perform mechanical analysis of the acinus using image-based and mathematical models of the structure and analyze three dimensional mechanical fields of the whole acinus for expansion.

METHODS

Acinus structure model The structure models are one image-based structure model (Model A) extracted from the mouse lung CT image [6] and two mathematical structure models (Model B, ModelC) [5] (Figure 1). The mathematical model is an assemblage of polyhedra and the model regards each polyhedron as an airspace region and the faces as tissue region. The ductal tree is generated by eliminating the common walls between these polyhedra based on their connectivity. The modeling conditions for the mathematical model are average acinar and alveolar volume characteristics based on the Model A (Table 1) . Note that since the mathematical models are represented by an assemblage of polyhedra, we do not consider large alveolar ducts found in the Model A. The Model B is a heterogeneous structure model constructed by randomly arranged polyhedra and the Model C is a homogeneous structure model constructed by assemblage of polyhedra arranged in body centered cubic lattice. The difference between the Model B and the Model C is that the variations of the alveolar volume and shape in the Model B are larger than in the Model C.

Mechanical model The lung tissue is modeled by a hyperelastic material and the strain energy density function [7] is

$$W = C_1(\hat{I}_c - 3)^2 + C_2(\hat{I}_c - 3)^3 + C_3(-\ln III_c + III_c - 1), \qquad (1)$$

where $\hat{I}_c$ is the reduced first invariant of the right Cauchy-Green deformation tensor, and III_c is the third invariant. C_1, C_2 and C_3 are material parameters, which are 4.1, 20.7, 4.1 kPa, respectively.

In the acinus, the surface tension coefficient changes with the pulmonary surfactant concentration. The model of the surface tension coefficient γ [8], which is a function of the inner surface area of the acinus, is

$$\gamma(A) = \gamma_{\max}[1 - d_1 \exp(-d_2\, A/A_0)], \qquad (2)$$

where $\gamma_{\max}$ is the maximum value of the surface tension coefficient, d_1 and d_2 are constants, which are 0.004 N/m, 9.9924, 2.35, respectively, A is the current surface area inside the acinus, and A_0 is its initial value.

Expansion simulation of pulmonary acini Expansion simulation of the acinus is performed by adding transpulmonary pressure to the inner surface of the acinus structure model. Specifically, equilibrium state is obtained by solving weakly formulated motion equation

$$\int_V \rho\ddot{\mathbf{u}} \cdot \delta\mathbf{u}\, dV + \int_V \mu\dot{\mathbf{u}} \cdot \delta\mathbf{u}\, dV + \int_V \mathbf{T} : \nabla\delta\mathbf{u}\, dV + \int_S \mathbf{t} \cdot \delta\mathbf{u}\, dS = 0\,, \qquad (3)$$

where V and S are the volumes and surface areas of arbitrary regions, ρ is the density, μ is the artificial viscosity coefficient, $\mathbf{u}$ is the displacement vector, and $\mathbf{T}$ is the Cauchy stress tensor. $\delta\mathbf{u}$ represents a virtual displacement vector, and $\mathbf{t}$ represents a surface force vector. The $\mathbf{t}$ in Eq. (3) is decomposed into the force from the transpulmonary pressure $\mathbf{t}_p$ and from the surface tension $\mathbf{t}_s$. The $\mathbf{t}_p$ is given by

$$\mathbf{t}_p = p\mathbf{n}, \quad (4)$$

where p is the given transpulmonary pressure and $\mathbf{n}$ is the normal vector on the inner surface. The $\mathbf{t}_s$ is given by

$$\mathbf{t}_s = \nabla_s \cdot \boldsymbol{\tau}_s, \quad (5)$$

where ∇_s is the surface gradient, which is expressed as $\nabla_s = \mathbf{P} \cdot \nabla$. The τ_s is the surface tension tensor per unit length and is defined as $\boldsymbol{\tau}_s = \gamma\mathbf{P}$, where $\mathbf{P}$ is the surface projection tensor, which is expressed as $\mathbf{P} = \mathbf{I} - \mathbf{nn}$ by the normal vector $\mathbf{n}$ and the unit tensor $\mathbf{I}$.

RESULTS AND DISCUSSION

Figure 2 shows the closeup view of the spatial distribution of the 1st principal stress of the Model A with and without surface tension for $p = 0.5$ kPa. The Model A has an uneven stress distribution throughout the tissue and the stress is relatively high in the edge region of the opening of the alveoli. Additionally, when considering the effect of surface tension, the stress in the tissue decreases, while the stress in the edge region maintains relatively high. These trends are also found in the Models B and C (data not shown).

Figure 3 shows the frequency distribution of the 1st principal stress with and without surface tension. The distributions are generated for the stress in the center region of the acinus (see Figure: 1) to eliminate the influence of the boundary effect. In all models, the mean and the standard deviation become small when considering surface tension effect. In the Model A, the mean and variation are larger than those of the Models B and C, which might be due to the lack of large alveolar ducts in Models B and C. Interestingly, despite of the heterogeneity of the Model B, there are no significant differences between the Model B and the Model C.

In summary, we performed mechanical analysis of acinus using the image-based and mathematical structure models. All models had uneven 1st principal stress distributions throughout the tissue, and the stress was relatively high in the edge region. By considering the effect of surface tension, the mean value and the standard deviation of the frequency distribution of the 1st principal stress decreased. The mean value and the variation of the stress for the mathematical models were smaller than those for the image-based model. In the mathematical models, at least for the frequency distribution of the 1st principal stress, there were no significant differences between the current heterogeneous and homogeneous models. It might be important for mathematical models to include more heterogeneous structure, e.g. large alveolar ducts, to analyze the effect of heterogeneity on mechanical fields.

REFERENCES

[1] Roan, E. et al., *Am J Physiol Lung Cell Mol Physiol.*301:L625–35,2011.

[2] Vlahakis NE. et al., *Am J Respir Crit Care Med.*171:1328-1342,2005.

[3] Denny, E., Schroter, R.C., *J. Biomech.*39:652–663,2006.

[4] Rausch S.M.K. et al., J. Mech. Behav. Biomed. Mater.4:583–592,2011.

[5] Koshiyama K and Wada S, *Comput. Biol. Med.*62:25–32,2015.

[6] Xiao, L. et al., *Comput Math Methods Med,* 1-10,2013.

[7] Rausch S.M.K,et al., *Annals of Biomedical Engineering,39*:2835–43,2011.

[8] R. Kowe et al., *J. Biomech.*,19:541–549,1986.

Table 1: The acinar volume and alveolar volume of each model.

	Acinar volume [mm³]	Alveolar volume [mm³] (mean value ± standard deviation)
Model A	5.91×10^{-2}	3.55×10^{-5}
Model B	5.71×10^{-2}	$3.61 \times 10^{-5} \pm 1.07 \times 10^{-5}$
Model C	6.05×10^{-2}	$3.41 \times 10^{-5} \pm 0.10 \times 10^{-5}$

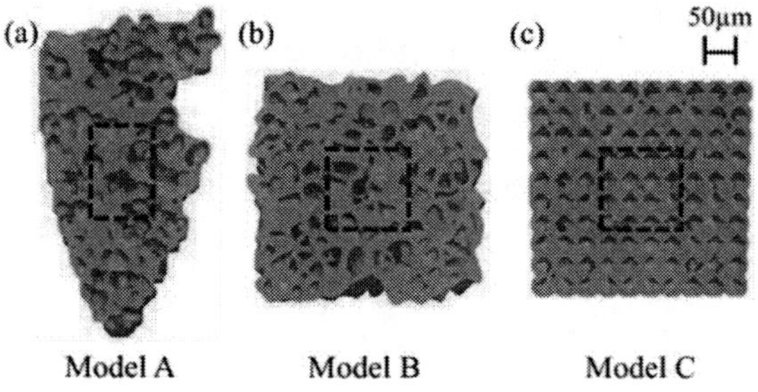

Figure 1: The cutaway of 3D acinus structure models of the image-based structure model (a), the heterogeneous model (b) and of the homogeneous model (c). The region surrounded by a broken line indicate the center region of the model.

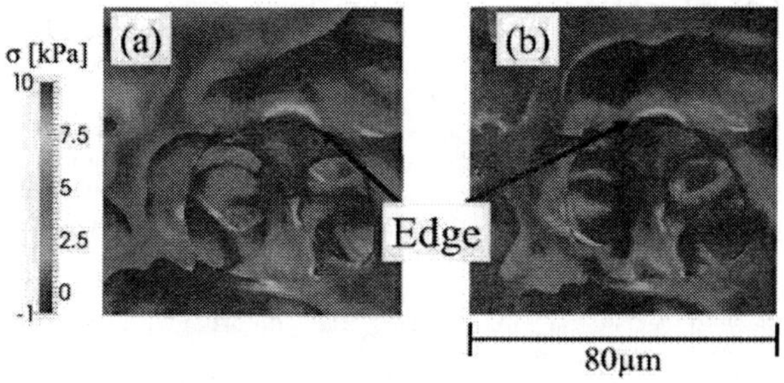

Figure 2: The representative spatial distribution of 1st principal stress for the Model A. Panel (a) is the result without surface tension effect, and Panel (b) is the result with surface tension effect.

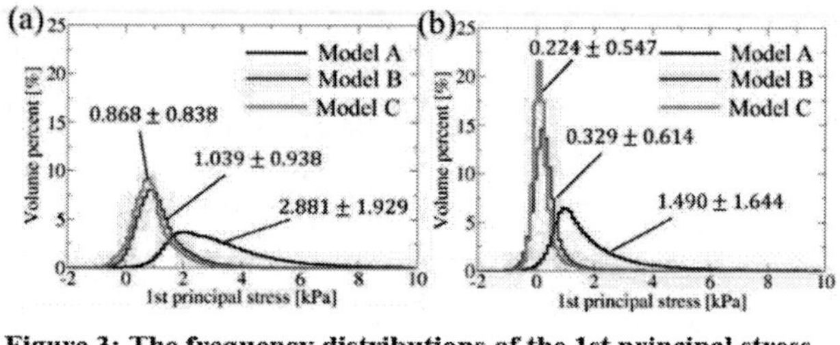

Figure 3: The frequency distributions of the 1st principal stress. Fig. 3(a) is the result without surface tension effect, and Fig. 3(b) is the result with surface tension effect. The mean value and standard deviation of each model are shown.

SB^{3}C2017
Summer Biomechanics, Bioengineering and Biotransport Conference
June 21 –June 24, Tucson, AZ, USA

GEOMETRIC MODELING OF ABDOMINAL AORTIC ANEURYSMS UNDER SURVEILLANCE: A RETROSPECTIVE STUDY

Shalin A. Parikh (1), Aura Teasley (1), Mirunalini Thirugnanasambandam (1), Victor De Oliveira (2), Satish C. Muluk (3), Ender A. Finol (4)

(1) Department of Biomedical Engineering
University of Texas at San Antonio
San Antonio, Texas, USA

(2) Department of Management and Statistics
University of Texas at San Antonio
San Antonio, Texas, USA

(3) Division of Vascular Surgery
Allegheny General Hospital
Allegheny Health Network
Pittsburgh, Pennsylvania, USA

(4) Department of Mechanical Engineering
University of Texas at San Antonio
San Antonio, Texas, USA

INTRODUCTION

Abdominal aortic aneurysm (AAA) is a vascular disease that carries a cumulative mortality rate of 90% [1]. The current standard for elective repair is a maximum diameter of 5.5 cm or an expansion rate of 1 cm/year. Studies have emphasized that maximum diameter and expansion rate should not be the only criteria that determine rupture risk clinically. It has been observed from autopsy data that 60% of large aneurysms (diameter > 5.5 cm) remain stable for a lifetime whereas 13% of small aneurysms (diameter < 5.5 cm) rupture [2]. Geometric features like tortuosity, centerline curvature and wall thickness have been found to have high correlation with peak wall stress in AAAs [3-5]. Aneurysm morphology plays an important role in wall mechanics and AAAs under surveillance are not recommended elective repair due to their small size relative to the maximum diameter criterion. Thus, the objective of this study is to quantify the patient specific geometry of surveillance AAAs and correlate it with wall stress.

METHODS

Geometry Quantification

Using an in-house image segmentation and geometry quantification tool (dubbed AAAVasc), computer models were generated for 22 patient specific AAA from patients under surveillance. The models were generated for three domains of the aorta: lumen, inner wall and outer wall. A MATLAB (Mathworks, Natick, MA) script was used to calculate geometric indices with respect to the z-axis of the coordinate system. The input for the script is the point cloud obtained from AAAVasc for each of the aforementioned domains and the DICOM information file generated for each image data set. Forty-eight shape and size-based geometric indices were quantified for each AAA. These include 11 (1-D) size indices such as length of the AAA centerline, height of the neck, distal and proximal neck diameter, etc. Nine (2-D) shape indices, some of which include tortuosity, diameter-height ratio and average lumen compactness. Four (3-D) size indices (AAA volume, surface area, thrombus volume, and ratio of AAA sac to ILT volume) were also calculated from the volumes enclosed by each domain surface. Ten (3-D) shape-based curvatures were also quantified using a BQFE (biquintic hermite finite element) algorithm. The BQFE script makes use of the segmented images obtained from AAAVasc as input. Thrombus and wall thickness were also calculated. The complete definition and mathematical formulation of the geometric indices are found in Shum et al [6].

Biomechanical analysis

The first CT scan available during the surveillance period was acquired retrospectively for each AAA and used as input to AAAVasc. The output of the segmentation was binary masks comprising the volumetric domains wall, lumen and thrombus. The masks were then used as input to AAAMesh, an in-house mesh generation script to generate volume (quadratic hexahedral) and surface (linear triangular) meshes of the AAA. Quasi-static finite element analysis (FEA) was performed using these meshes in the solver ADINA (Adina R&D Inc., Watertown, MA), while applying an endoluminal pressure of 120 mmHg. The results were post-processed using Ensight (CEI Inc., Apex, NC) to obtain the first principal stress distributions on the AAA wall. Using these, the peak wall stress (PWS), the spatially averaged wall stress (AWS) and the 99th percentile wall stress (99th–WS) were calculated for each AAA model.

Statistical analysis

The 48 geometric indices were regressed on each of the 3 biomechanical parameters (PWS, AWS, 99th–WS) by means of a feature selection technique and stepwise regression using SPSS (IBM Corp., Armonk, NY). This yielded a subset of geometric indices that were highly correlated with the 3 biomechanical parameters based on their respective Pearson's correlation coefficients.

RESULTS

The sample size had a mean maximum diameter of 44.8±4.7 mm. The mean values of PWS, AWS and 99th–WS were 94.9 ± 61.9 N/cm^2, 16.7 ± 5.0 N/cm^2, and 39.1 ± 14.0 N/cm^2, respectively. Three independent regression analyses were performed. A stepwise regression with significance level of entry, $p = 0.05$, and exit, $p = 0.1$, was used to obtain the statistically significant predictors. Table 1 describes the geometric indices that were highly correlated with the biomechanical parameters while Table 1 indicates the correlation coefficients of these geometric indices. The R^2 value in each of the models indicates the goodness of fit of the regression line.

The results of the stepwise regression indicated that wall thickness bore a negative correlation with all the biomechanical parameters. Surface curvatures MLN and MAA (the L2-norm and area-averaged Mean curvatures, respectively) were also negatively correlated with PWS and AWS. The 1-D geometric indices bulge height and proximal neck diameter were positively correlated with PWS and 99th–WS, respectively.

Table 1: Pearson's correlation coefficient for the relationships of wall stress with the significantly correlated geometric indices.

Biomechanical Parameter	Geometric Index	Pearson's Correlation Coefficient
PWS	Bulge height (mm)	0.672
	MLN (mm^{-1})	-0.053
	Mode of wall thickness (mm)	-0.628
	Vessel volume (cm^3)	0.043
	Non-fusiform index	0.449
AWS	Average wall thickness (mm)	-0.789
	Vessel volume (cm^3)	0.567
	MAA (mm^{-1})	-0.311
99th–WS	Mode of wall thickness (mm)	-0.730
	Minimum wall thickness (mm)	-0.646
	Proximal neck diameter (mm)	0.126

Table 2: Goodness of fit (R^2) of the regression model for each of the biomechanical parameters.

Biomechanical Parameter	R^2
PWS	0.898
AWS	0.897
99th–WS	0.708

DISCUSSION

Wall stress computed from FEA has been used to assess rupture risk of AAAs. Fillinger et al. [4] reported that wall stress has a better accuracy in differentiating ruptured and unruptured AAAs compared to maximum diameter. It can be observed in Table 2 that the Pearson's correlation coefficient is negative for the average wall thickness and higher than the other significant geometric indices. These results are corroborated by Venkatasubramaniam et al. [3], who demonstrated that decreasing the wall thickness by 25% increased the maximum wall stress by 20%. Furthermore, an experimental study conducted by Raghavan and co-workers [7] indicated a significant difference in wall thickness between ruptured site and calcified sites. Our results support the need for quantifying wall thickness in patient specific AAAs as it may be used as an early marker of rupture risk in surveillance AAA.

Giannoglou et al. [5] reported that centerline curvatures are highly correlated with PWS. Our results indicate that MLN and MAA are both negatively correlated with all three biomechanical parameters. This suggests that surface curvatures, in addition to wall thickness and maximum diameter, can be used as predictors of wall stress in surveillance AAAs. This outcome is supported by the study of Tang et al., which reported that MAA was strongly associated with rupture risk of AAAs [8]. The positive correlations of bulge height and proximal neck diameter are consistent with the fact that higher stresses are concomitant with aneurysm size. This outcome is supported by [9], in which it was demonstrated that aneurysm shape affects PWS more effectively than maximum diameter.

Although PWS is an established marker of AAA wall mechanics, our correlation analyses were also performed with AWS as the geometric parameters are calculated globally taking into account the entire AAA wall surface. To mitigate the possibility of erroneous stresses due to mesh distortions, the 99th–WS was calculated, yielding regressions similar to those of PWS, with the mode of wall thickness being the commonly regressed geometric index.

The limitation of this study is the relatively small sample size of patient-specific data. The primary outcome of this work supports the quantification of AAA shape, volume-based features and wall thickness indices for having a stronger correlation with wall stress than maximum diameter. These indices may be utilized in a clinical setting as surrogates for wall stress and, in the future, for prediction of rupture risk in surveillance AAAs to determine the need for elective repair.

ACKNOWLEDGEMENTS

This work was supported in part by an American Heart Association Student Scholarship in cardiovascular disease, National Institutes of Health award #1R01HL121293, and American Heart Association award #15PRE25700288.

REFERENCES

1. Karkos, C. et al., European Journal of Vascular and Endovascular Surgery, 19(3):299-303, 2000.
2. Greenhalgh, R.M., The Lancet, 364(9437):843-848, 2004.
3. Venkatasubramaniam, A. et al., European Journal of Vascular and Endovascular Surgery, 28(2):168-176, 2004.
4. Fillinger, M.F. et al., Journal of Vascular Surgery, 37(4):724-732, 2003.
5. Giannoglou, G. et al., Angiology, 57(4):487-494, 2006.
6. Shum, J. et al., Annals of Biomedical Engineering, 39(1):277-286, 2011.
7. Raghavan, M.L. et al., Journal of Biomechanics, 39(16):3010-3016, 2006.
8. Tang, A. et al., Journal of Vascular Surgery, 59(4):894-902, 2014.
9. Georgakarakos, E. and C.V. Ioannou, Medical Hypotheses, 79(1):71-73, 2012.

SB^3C2017
Summer Biomechanics, Bioengineering and Biotransport Conference
June 21 – 24, Tucson, AZ, USA

PULMONARY ARTERY AND SOMATIC GROWTH IN FONTAN PATIENTS

Akash Gupta (1), Ethan Kung (1) (2)

(1) Department of Mechanical Engineering
Clemson University
Clemson, SC, USA

(2) Department of Bioengineering
Clemson University
Clemson, SC, USA

INTRODUCTION

The multiphase surgical palliation of the single ventricle physiology involves routing the entire systemic venous return directly into the pulmonary arteries. Consequently, the flow to the pulmonary arteries is devoid of a pumping source and is driven by the central venous pressure. Exercise intolerance is one of many sequelae that diminish the quality of life of such patients. It has been hypothesized that the size and geometry of the Fontan junction and pulmonary arteries have significant effects on the patient's cardiovascular performance[1]. Understanding the growth of the pulmonary arteries in Fontan patients is a critical step in attempting to improve our insight into the Fontan physiology.

Scattered attempts have been made to study pulmonary artery size in single ventricle patients[2]. A majority of the relevant literature focuses either on changes in the size of the pulmonary arteries between the various stages of the surgical palliation[3], or linking the pre-operative pulmonary artery size to functional outcomes post-surgery. Studies that constitute the former category of literature lack comprehensive data tracking pulmonary artery growth through to adulthood. The majority of the investigations that constitute the latter category of literature presented normalized metrics of pulmonary artery size such as the Pulmonary Artery Index (PAI) and their utility in the optimal selection of patients for the Fontan procedure [4]. No model has been created to describe the temporal evolution pulmonary artery growth in Fontan patients from infancy to adulthood. In this study, we address this gap by consolidating existing literature through weighted regression analysis.

METHODS

An extensive literature review revealed several studies that have measured the diameter of the Left Pulmonary Artery (LPA) and the Right Pulmonary Artery (RPA) in Fontan patients. Most studies have presented data in normalized metrics such as PAI or z-scores. Additionally, a majority of these studies have reported the mean Body Surface Area (BSA) of their patient samples. When the assumption that the LPA and the RPA sizes are equal[5][6] is applied to the definition of PAI, the pulmonary artery size can be extracted using the following equation:

$$PAI = \frac{\pi d^2}{2\,BSA} \tag{1}$$

Where d is the mean of the systolic and diastolic diameter of the LPA or the RPA.

Some investigators presented their findings in the form of z-scores. The findings of Rammos et al.[7] combined with data from the CDC growth charts[8] allowed us to de-normalize the z-scores and calculate the mean measured LPA and RPA diameters.

In some studies, investigators had measured systolic diameters rather than the mean diameter. In such cases we have employed the data on distensibility as reported by Bossers et al[2] to calculate the mean diameter.

The mean pulmonary artery diameter data extracted from the aforementioned studies was then plotted against the mean age of the patient sample in each study. Regression analysis was employed to derive the best-fit trend line for the data using a physiologically realistic function. In addition, the mean BSA of the patient samples was plotted against the mean age of the patient sample and a similar weighted regression analysis was performed to derive the best-fit trend line for the data. Using these two mathematical relationships, we derived the PAI vs age function. Finally, we assume that the growth of the pulmonary arteries is negligible after the age of 18 and thus present our fitted model up to that time point.

RESULTS & DISCUSSION

Our fitted model for Fontan pulmonary artery size shows retarded pulmonary artery growth in Fontan patients compared to normal children (Figure 1). However, the somatic growth of Fontan patients closely follows that of normal children (Figure 2). Our model for somatic growth matches well (maximum difference of 9%) with the observations made by Hasan et al[9]. The combination of retarded pulmonary artery growth and close-to-normal somatic growth leads to a precipitous drop in PAI over time (Figure 3). This trend closely follows the observations made by Adachi et al[10]. The substantial difference in the temporal evolution of PAI in Fontan patients as compared to normal children indicates a severe insufficiency of the growth of pulmonary arteries relative to the somatic growth, indicating a potential to restrict blood flow and adversely affect cardiac output.

The predictions made by our model for somatic growth and temporal evolution of PAI match well with existing literature both qualitatively and quantitatively. This model for pulmonary artery growth should be verified and expanded by longitudinal and individual studies. We anticipate that these results could eventually serve as a useful reference for clinicians. The phenomenon of the late failure of the Fontan is poorly understood and requires investigation of the adult Fontan physiology. While the utilization of patient specific geometries in computational fluid dynamics simulations can provide insightful information, acquiring patient imaging data at multiple time points through to adulthood is difficult in practice. By utilizing our model, patient-specific geometries of the pulmonary vasculature can be scaled to create evolving patient models without the necessity of acquiring patient data at multiple time points.

ACKNOWLEDGEMENTS

This work was supported by an Award from the American Heart Association, The Children's Heart Foundation, and the Department of Mechanical Engineering at Clemson University.

REFERENCES

[1] R. H. Khiabani et al., *Heart*, vol. 101, no. 2, pp. 139–143, 2015.
[2] S. S. M. Bossers et al., *Pediatr. Cardiol.*, Jan. 2016.
[3] G. Buheitel et al., *Heart*, vol. 78, no. 5, pp. 488–92, 1997.
[4] H. Senzaki et al., *Circulation*, vol. 89, no. 3, pp. 1196–202, 1994.
[5] J. S. Baek et al., *Ann. Thorac. Surg.*, vol. 91, no. 4, pp. 1240–1246, 2011.
[6] M. Restrepo et al., *Ann. Thorac. Surg.*, vol. 99, no. 1, pp. 163–170, 2015.
[7] M. Rammos et al., *Herz*, vol. 14, no. 6, pp. 348–357, 1989.
[8] R. J. Kuczmarski et al., *2000 CDC growth charts for the United States: Methods and development*, vol. 11, no. 246. 2002.
[9] B. S. Hasan et al., *Congenit. Heart Dis.*, vol. 3, no. 5, pp. 330–335, 2008.
[10] I. Adachi et al., *Eur. J. Cardio-thoracic Surg.*, vol. 32, no. 1, pp. 156–162, 2007.

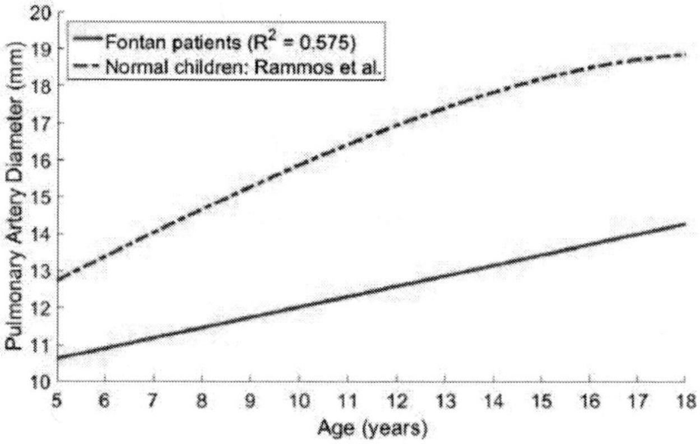

Figure 1: Pulmonary artery diameter vs age

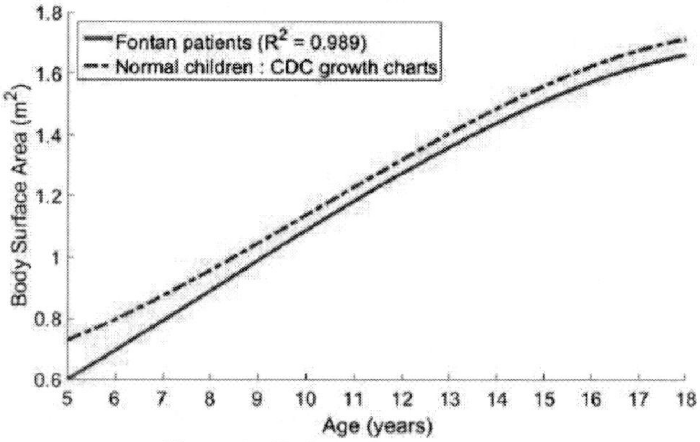

Figure 2: Body surface area vs age

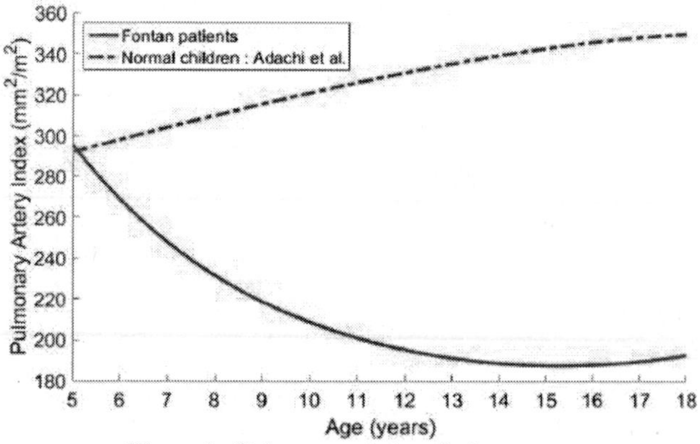

Figure 3: Pulmonary Artery Index vs age

CLINICAL OUTCOMES IN MICROVASCULAR DISEASE PATIENT-SUBGROUP WITH EPICARDIAL STENOSIS: A PILOT STUDY TO ASSESS A NEWLY DEVELOPED PRESURE-FLOW DIAGNOSITC ENDPOINT

Ullhas U. Hebbar (1), Mohamed A. Effat (2), Srikara V. Peelukhana (1), Imran Arif (2), Rupak K. Banerjee (1)

(1) Department of Mechanical and Materials Engineering
University of Cincinnati
Cincinnati, OH, USA

(2) Division of Cardiovascular Diseases
University of Cincinnati Medical Center
Cincinnati, OH, USA

INTRODUCTION

Accurate assessment of intermediate epicardial stenosis (ES) is a clinical challenge for interventional cardiologists. Functional diagnostic indices such as Fractional Flow Reserve (FFR) and Coronary Flow Reserve (CFR) show good agreement with stress testing [1], although significant microvascular dysfunction (MVD) limits their efficacy as both indices rely solely on either pressure or flow measurements [2, 3].

For the purpose of simultaneous detection of ES or MVD using a single parameter, we recently introduced the pressure drop coefficient (CDP), a functional diagnostic endpoint based on fundamental fluid dynamics which combines pressure and flow measurements [4]. Evaluation of CDP via *in-vitro* [5], *in-vivo* animal experiments [6, 7] and patient level studies [8] demonstrated its ability to differentiate between ES and MVD. An equivalent cut-off to the well-established FFR<0.75 was determined for CDP (CDP>27.9) in order to assess interventional decisions for single-vessel ES [9].

The proposed cut-off value for CDP was determined using an earlier clinical study [10] where positive clinical outcomes were observed for the CDP>27.9 group when compared with the FFR<0.75 group. To assess the efficacy of CDP in delineating MVD in the presence of ES, the *objective* of this *follow-up* study is to compare the clinical outcomes between CDP>27.9 and FFR<0.75 for a subgroup of patients with possible MVD (*CFR<2.0* or *diabetes*).

METHODS

The study protocol was approved by the institutional review board at the University of Cincinnati (UC) and Cincinnati Veteran Affairs Medical Center, and informed consent was obtained from all 86 patients [10].

CDP can be defined as the ratio of trans-stenotic pressure drop to distal dynamic pressure.

$$CDP = \frac{\Delta p}{0.5 \times \rho \times APV^2} \quad (1)$$

where, Δp is the pressure drop across the lesion (Pa); the distal dynamic pressure being the product of blood density (ρ) (kg/m³), the square of average peak flow velocity (APV) (m/s) and a constant (0.5).

All consented patients were followed up for a minimum period of 1-year. The primary outcomes consisting of *major adverse cardiac events (MACE)* were determined over this period. MACE was computed as the composite of all-cause mortality, myocardial infarction (MI), and repeat revascularization.

The subgroup analysis for patients suffering from MVD was performed by extracting two subgroups from the complete patient group data: one composed of patients exhibiting an abnormal CFR value (CFR<2.0); the other composed of patients suffering from diabetes.

The patient data for each subgroup was divided according to cut-off values for FFR (FFR<0.75) and CDP (CDP>27.9). The %MACE in the FFR<0.75 group was compared with the corresponding %MACE in the CDP>27.9 group. Additionally, comparisons performed for the complete patient group [10] have also been presented. The comparisons were performed using the two-tailed chi-squared test and further evaluated using the Fischer's exact test.

Further, Kaplan-Meier *survival curves* were used to compare the event-free survival of the FFR<0.75 group and the CDP>27.9 group for each subgroup using the log-rank test (MedCalc V10.2, Mariakerke, Belgium). Patients lost to follow-up were counted as censored data. All results were considered statistically significant if $p<0.05$.

RESULTS

For the CFR<2.0 subgroup, the %MACE outcomes in the FFR<0.75 group (3/14, 21.4%) were higher than the outcomes seen in the CDP>27.9 group (2/26, 7.7%). However, the difference was not statistically significant, evidenced by p=0.21 for the chi-squared test and p=0.32 for the Fischer's exact test. The results for the diabetic subgroup and the complete patient group [10] followed a similar trend to the CFR<2.0 subgroup (Table 1).

Table 1: Comparison of %MACE between FFR < 0.75 and CDP > 27.9 groups for the CFR < 2.0 subgroup, diabetic subgroup and the complete patient group [9]. The p-values are provided for the chi-squared test, and the p-values for the Fischer's exact test are provided in parentheses.

	CFR<2.0	
	FFR<0.75	CDP>27.9
%MACE	3/14 (21.4%)	2/26 (7.7%)
p-value	0.21 (0.32)	
	Diabetic	
	FFR<0.75	CDP>27.9
%MACE	2/11 (18.2%)	2/16 (12.5%)
p-value	0.69 (1.00)	
	Complete patient group [10]	
	FFR<0.75	CDP>27.9
%MACE	4/20 (20%)	2/35 (8.6%)
p-value	0.24 (0.24)	

The Kaplan-Meier survival analysis performed for the CFR<2.0 subgroup (Figure 1) resulted in a hazard ratio of 0.26 (95%CI: 0.04-1.82) indicating that the survival probability for the CDP>27.9 group is 3.85 times the corresponding probability in the FFR<0.75 group. The difference in survival time was borderline significant (p=0.10). The results for the diabetic subgroup and the complete patient group [10] were again similar to the CFR<2.0 subgroup with higher long-term survival rates for the CDP>27.9 group.

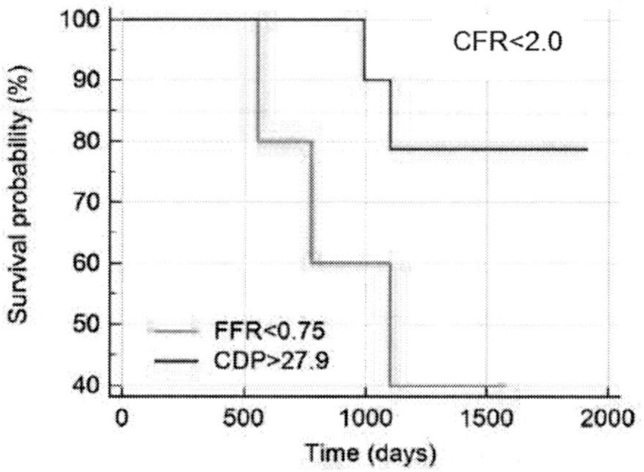

Figure 1: Comparison of Kaplan Meier survival curves between FFR<0.75 and CP>27.9 for CFR<2.0 subgroup

DISCUSSION

A significant contributor to improved quality of life is reduced incidence of MACE. For the primary outcome comparison, the first subgroup methodology in the study was based on CFR values. Abnormally low CFR values (*CFR<2.0*) are associated with MVD and an inability to achieve peak hyperemia [11]. Under this scenario, constant minimum microvascular resistance may not be achieved, leading to an underestimation of pressure drop across the lesion and a consequent over-estimation of FFR values. The reduction in hyperemic flow due to MVD typically dominates over the corresponding reduction due to ES, and the formulation of CDP accounts for this through the square of flow velocity seen in the denominator. This gives CDP an improved resolving power compared to FFR for accurate evaluation of the stenosis, and the results of this study clearly indicate this improvement.

Similar to patients exhibiting abnormal CFR values, patients suffering from *diabetes* are also associated with potential MVD [12]. The second subgroup analysis consisting of diabetic patients was performed as an additional confirmation to our hypothesis. The results obtained displayed a similar trend to the CFR<2.0 subgroup analysis. Further, the complete patient group analysis performed in our earlier study reported reduced %MACE outcomes for the CDP group, thus clearly highlighting the improved resolving power of CDP for ES, particularly in the presence of MVD.

Another significant outcome that affects the quality of life for cardiac patients is long-term event-free survival. Comparison of survival curves between the CDP-based and FFR-based strategies was performed for both subgroup methodologies. Positive results were seen with the CDP groups exhibiting higher long-term survival rates. Furthermore, the survival analysis conducted for the complete patient group again indicated a statistically significant improvement in long-term event free survival for the CDP>27.9 group.

Therefore, this study proposes CDP as an improved functional diagnostic endpoint compared to FFR for accurate prediction of major ischemic events, particularly in the presence of confounding scenarios such as MVD. Despite not reaching statistical significance, consistent improved measures were achieved for all measures.

All clinical decisions in the study were made based on FFR values alone. A larger sample size, and a prospective randomized clinical trial is required to evaluate the performance of CDP and assess as well as confirm the outcomes of this exploratory cohort study.

ACKNOWLEDGEMENTS

This study was supported by VA Merit Grant (I01CX000342-01), Department of Veteran Affairs (PI: Dr. Rupak Banerjee).

REFERENCES

[1] Kern MJ et al., *Circulation*, 114; 12:859-871, 2006.
[2] Hoffman JI, *Ann Biomed Engg*, 28; 8:884-896, 2000.
[3] van de Hoef TP et al., J Mol Cell Cardiol, 52; 4:786-793, 2012.
[4] Sinha Roy A et al., MJ et al., *The Journal of Surgical Research*, 150; 1:24-33, 2008.
[5] Peelukhana SV et al., *J Biomech*, 42; 16:2753-2759, 2009.
[6] Kolli KK et al., *J Invasive Cardiol*, 24; 1:6-12, 2012.
[7] Peelukhana SV et al., *Am J Physiol Heart Circ Physiol*, 302; 8:1563-1573, 2012.
[8] Kolli KK et al., *Cath Cardiovasc Interv*, 83; 3:377-385, 2014.
[9] Kolli KK et al., *Cath Cardiovasc Interv*, 87:273-282, 2016.
[9] Effat MA et al., *World J Cardiol*, 8; 3:283-292, 2016.
[11] Feigl EO, *Phsiol Rev*, 63; 1:1-205, 1983.
[12] Picchi A, *World J Cardiol*, 2:377-390, 2010.

DEVICE TO APPLY LOADS AT TARGETED MAGNITUDES AND STROKE FREQUENCIES DURING INSTRUMENT ASSISTED SOFT-TISSUE MOBILIZATION

John B. Everingham (1), Peter T. Martin (1), Trevor J. Lujan, Ph.D. (1)

(1) Department of Mechanical & Biomedical Engineering
Boise State University
Boise, ID, USA

INTRODUCTION

Ligaments are bands of dense fibrous tissue that provide joint stability by binding bone to bone. Ligament tears account for 50% of sporting injuries [1], are painful, restrict physical activity and can lead to chronic impairment [2]. Instrument assisted soft tissue mobilization (IASTM) is a manual therapy technique used by over 25,000 clinicians for treating dysfunctions in ligaments and other soft tissues [3]. Several IASTM techniques are practiced [4], and while each technique has minor differences in tools and treatment protocols, they all involve the manual application of dynamic compressive loads by cyclic stroking of the damaged tissue through the skin with an instrument. Notably, none of the prevalent IASTM techniques specify the loading magnitudes and stroke frequencies that are recommended during treatment. A need therefore exists to identify loading parameters that result in optimal patient outcomes; however, a practical research device has not yet been validated to accurately apply IASTM treatments at targeted compressive loads and stroke frequencies.

Current techniques to apply IASTM in a research setting include instrumented hand held devices and robotic manipulators. Hand held devices are practical, but have not yet been designed or validated to apply targeted loading parameters [5-8]. Robotic manipulators have been used to control the compressive force and stroke frequency during soft tissue mobilization of muscle along a single axis. However, these devices can be cumbersome, expensive and time intensive to operate, making them impractical for large animal experiments and human clinical studies. The objective of this study is to develop and validate an instrumented hand held device that enables users to accurately apply targeted forces and stroke frequencies during IASTM treatments. This device was validated by measuring accuracy at variable force magnitudes and stroke frequencies, and by using this device to apply IASTM treatments to injured ligaments in a rat model.

METHODS

The device (**FIGURE 1A**) is made up of a custom machined aluminum tip, uniaxial force sensor (Omegadyne, Sunbury, OH; max load = 10 N, error = 0.3%), stainless steel shaft for holding the device, and data acquisition system (National Instruments, Austin TX; cDAQ-9171 USB chassis & NI 9237 Module). The aluminum tip has a beveled edge similar to tools used in physical therapy clinics for treating finger joints (**FIGURE 1B**). A custom LabVIEW program was used to view and record force data at 20 Hz from the force sensor. The program allows the user to input a targeted treatment duration, peak compressive force, and stroke frequency. The program displays a live waveform of the applied load overlaid with a square waveform that shows the targeted loading profile and creates an auditory signal that indicates when to start and end a stroke; the signal acts as a metronome to guide stroke frequency. The peak compressive force is determined by taking an average of the data points measured during tissue contact. These data points are selected using a graded threshold algorithm that identifies clustered regions of the sample data (**FIGURE 1C**). The stroke frequency is determined in post-processing by using a Fourier series summation to fit the raw force measurement data using a non-linear least squares fitting method (**FIGURE 1D**). The measured stroke frequency is the frequency of the fitted curve.

Two tests were executed to evaluate device accuracy. The first test analyzed the device's accuracy in measuring compressive load when varying the angle between the device's shaft and the surface being treated (i.e. shaft angle). This test was performed using 3D printed blocks to anchor the device at fixed angles between 50° to 90°. At each fixed angle, the device was manually loaded three times from 0 - 5 N, and data was collected for three minutes. To determine the device's accuracy in measuring compressive force, the force normal to the surface was simultaneously measured with an auxiliary uniaxial force sensor (**FIGURE 1E,** Instron, Norwood, MA; 1kN Dynacell).

The second test determined the device's accuracy in applying loads and stroke frequencies over a range of targeted values. This test was performed by having three individuals dynamically load a foam block at eight targeted loads (0.1-5.0N) and three targeted frequencies (0.5, 1.0, 2.0Hz). Each test was performed for one minute, and every test condition was performed three times by each operator. Each

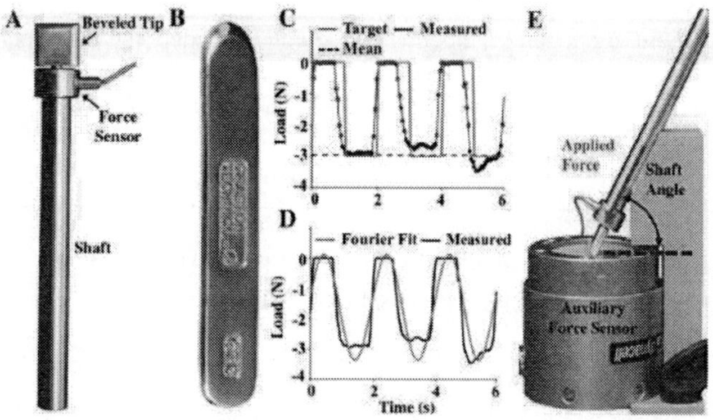

Figure 1. Device Overview (A) Device Components (B) Graston tool GT3 (C) Sample of measured load data overlaid with the target load profile. Filled circles indicate the data used to calculate the mean. (D) Fourier series fit for measured load data, target frequency of 1Hz, R^2 = .851 (E) Device fixed at a 60° shaft angle for accuracy testing

operator was allowed 30 minutes of time to practice using the device before live testing. A 1-minute practice run was performed at each test condition prior to the live test.

In order to test the repeatability of the device over time in a research setting, an animal study was conducted utilizing six Sprague-Dawley rats (body mass = 450-500g, age = 6 months). The device was used to perform IASTM to treat medial collateral ligament (MCL) injuries induced through surgical transection. The experiment followed an established protocol [8], and IASTM treatments were based on guidelines for the Graston Technique [9]. Three of the animals received IASTM and the other three served as controls. For the treated group, a 2 N force was applied along the length of the MCL, between the tibial and femoral insertion, at a rate of 1 Hz for 1 minute **(FIGURE 3A)**. Treatments were repeated 2x per week for 3 weeks.

RESULTS

For device shaft angles between 70° - 90°, the device measurement error was less than 5%. Below 70°, the device measurement error was greater than 15% **(FIGURE 2A)**. Load was applied with less than 10% error for target magnitudes of 0.3 - 5.0N. For loads below 0.3N the error was greater than 15% **(FIGURE 2B)**. The percent error in applied load was significantly higher for target loads less than 0.2N ($p = .005$). For target loads greater than 0.3N the percent error was not significantly different ($p = 0.1$). There was no impact of target frequency on the percent error in load across all load magnitudes ($p = 0.9$). The stroke frequency was accurately applied with less than 0.2% error for all target values. Additionally, changing operators had no effect on device accuracy ($p = 0.9$).

The average percent error in load and frequency across IASTM rat treatments was 5.5% ± 22.8% and 0.06% ± 0.12% respectively **(FIGURE 3B)**. There was no effect of time of IASTM rat treatment on the accuracy of the device ($p = 0.3$) **(FIGURE 3C, D)**.

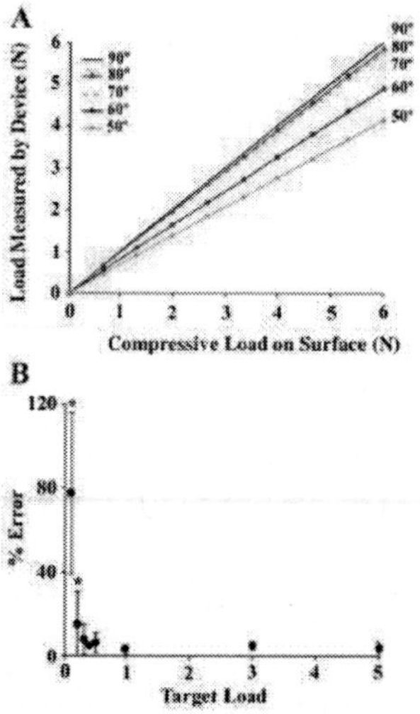

Figure 2. Device accuracy. (A) Load measured by the device vs. the surface normal load measured by the auxiliary force sensor. (B) Percent error in applied load at multiple target loads, averaged for three operators. Note that the S.D. shown is between each operator not the pooled SD of each 1-minute test.

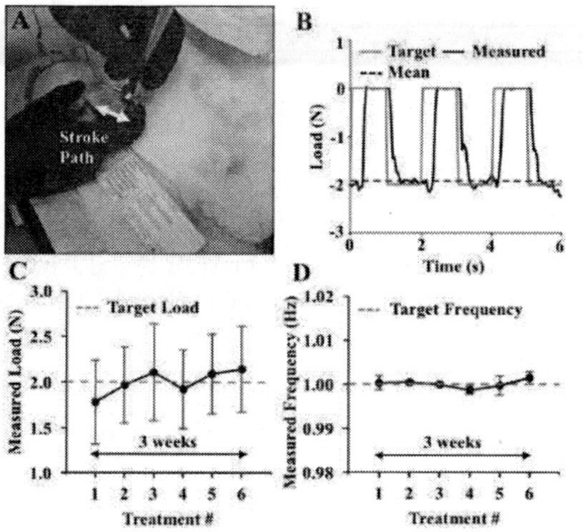

Figure 3. Experimental validation. (A) Device stimulating a rat. (B) Sample of targeted load profile, overlaid with measured load. (C, D) Average applied load and frequency during treatments.

DISCUSSION

The described device provides a practical method to accurately apply targeted forces and stroke frequencies during IASTM treatments. Based on the results of the performance tests, it is recommended that the device be held at shaft angles greater than 70° during use and only be utilized for targeted loads between 0.3 – 5.0 N, where % error in applied load is less than 10% **(FIGURE 2B)**. These force magnitudes replicate parameters used during animal research focused on IASTM [8]. By designing a device and software that are intuitive to users, this level of accuracy was achieved for new operators with only 30 minutes of practice.

To our knowledge this is the only hand held device that has been validated to apply loads at targeted magnitudes and frequencies during IASTM treatments. Several researchers have attempted to prescribe forces during soft tissue mobilization through various techniques including: using an IASTM tool on a force plate to acquire a kinesthetic feel for the amount of force applied prior to treatment [6,8], attaching pressure transducers to a soft tissue mobilization instrument [5], or to the fingers [7]. Yet none of these methods were validated to apply a targeted force or stroke frequency. A robotic IASTM device could potentially provide the highest degree of force and frequency control. However, a robotic device built to apply IASTM, produced only a 4% better coefficient of variation in applied force than our handheld device and the robot was constrained to one axis of motion [10].

This study has limitations. While the rat sample size for the experimental validation was large enough to measure the device's ability to apply a targeted load magnitude and stroke frequency across multiple weeks of treatment, this sample size was too small to measure an effect of IASTM treatments on ligament mechanical properties. Also, the device only measures the normal component of the applied load. The device could be modified to use a tri-axial load cell to capture both normal and transverse components of the applied load and allow accurate load measurement from any shaft angle.

In conclusion, a hand held device was developed to accurately apply targeted loading parameters. This validated device can now be used to investigate the effects of varying IASTM loading parameters on soft tissue healing. This type of research can support optimization of IASTM protocols and ultimately improve patient outcomes.

REFERENCES

[1]Fernandez WG. et al. *Acad. Emg. Med.* 14:641-645, 2007 **[2]**Van Rijn, RM. et al. *Am J Med.* 121:324-331, 2008 **[3]**Ploski, M. Director of Strategic Planning-PT, Graston Technique, LLC, Indianapolis, Ind. (personal communication) **[4]**Cheatham, SW. et al. *J Canadian Chiro. Assoc.* 60(3):200-211, 2016 **[5]**Gehlsen et al. *Med.& Sci. in S& E. 31*(4):531-535, 1999 **[6]**Imai, K. et al. *J Manip. & Physio. Therap.* 38(2):112-118 2015 **[7]**Lee, H. et al. *Manual Therapy.* 14(5):501-507, 2009 **[8]**Loghmani, M. T. et al. *J Orthop Sports Phys. Ther.* 39(7):506-514 2009 **[9]**Hyde *Cons. Manag. of Sports Injuries.* 2007. Print. **[10]**Zeng, H. et al. *J. Med. Devices* 2(4):041003, 2008

DESIGN, TESTING, AND IMPLEMENTATION OF CONTROLS AND INTERFACE FOR AN ADAPTABLE EXERCISE DEVICE FOR PEOPLE WITH PHYSICAL DISABILITIES

John M. Hoyle (1), Alan W. Eberhardt (2)

(1) Department of Biomedical Engineering
University of Alabama at Birmingham
Birmingham, AL, USA

(2) Department of Biomedical Engineering
University of Alabama at Birmingham
Birmingham, AL, USA

INTRODUCTION

In a recent study, people living with physical disabilities made up less than 30% of the total population, but 40% of the obese population. They suffered more than their able-bodied peers from a wide array of secondary disabilities including pain, fatigue, depression, and pressure sores [1]. Although exercise could help to improve their quality of life [2], those with physical disabilities consistently have not achieved the recommended levels of exercise [3].

Recent studies have investigated using commercial devices to increase energy expenditure [4]. Unfortunately, commercial products have not been designed to be inclusive technologies, which has necessitated the creation of customized test platforms [5]. The described project has undertaken to design an Advance Virtual Exercise Environment Device (AVE2D), which combined a custom physical device with a virtual environment.

METHODS

The physical device was designed and fabricated by PUSH Product Design, Birmingham, AL. This device has a variety features to accommodate users with disparate functional abilities. The exercise is currently derived through applying energy into crank inputs on both the arm and the leg axes. The physical device has been instrumented and paired with a virtual environment display to increase enjoyment and immersion during exercise.

The virtual display loads and shows a video and modulates the play rate to mimic the exercise input speed. Whenever the displayed path forks, the user is able to select the direction they want to take. Choices can be made through mouse clicking in designated areas, voice commands, and directional leaning mediated through a custom Velostat system. The display can also connect to Bluetooth enabled heart rate monitors and display target ranges for the user in real time. Figure 1 shows the display.

Figure 1: Virtual display with current position along trail, and continuation choices in the top right corner; balance information in the top left corner; option bar along the bottom; and heart rate information displayed (green and orange bars).

In order to verify the utility of the path choice mechanisms, the voice recognition and balance sensing components were tested.

Voice Recognition The voice recognition system (Microsoft Speech 11.0 in Visual Studio 2015) was tested by having a user attempt to say a displayed phrase and recording the number of attempts required for recognition to occur. One hundred samples were taken at a variety of background noise levels with the phrases being randomly selected from the database of recognized phrases.

Balance Sensing The Velostat (3M, St. Paul, MN) system was tested for functionality using three different filtering schemes, both while standing and while seated in a wheelchair. The schemes used were unfiltered, low pass filtered at 10 Hz, or by a 20 point running average. The display would generate a target at a random position and record the time and path taken to move to the target area. The path taken divided by the shortest possible path is know as hand path ratio (HPR) and was used as a measure of movement efficiency. One hundred movements were recorded for each filtering condition.

RESULTS

Voice Recognition Levene's test demonstrated ($p<0.05$) that the variances were unequal ($p = 3.18*10-28$) for the various background interference levels, so Welch's Test determined that the means were unequal [$F(4,216.12) = 17.48$, $p = 1.90*10-12$]. A Tukey HSD showed that the lowest noise level (43.5 dBs) took significantly fewer attempts for each recognized phrase compared to the two highest noise levels, and that the highest noise level (65 dBs) took significantly more attempts for each recognition compared to the all other conditions. Table 1 shows the breakdown of the results.

Table 1: Results for voice recognition with attempts/recognition, standard deviation, the number of completed samples, and the number of phrases completed on the first attempt.

dBs	Attempts/ Recognition	Standard Deviation	Completed	First Attempt
43.5	1.06	0.28	100	95
50	1.36	0.85	100	79
55	1.42	0.85	100	74
60	1.60	1.41	99	73
65	2.35	1.93	94	47

The most noticeable connectetion between phrases that showed multiple attempts to complete was the presence of difficult consonant combinations.

Balance Sensing Based on Levene's test, time and HPR while standing had unequal variances ($p = 9.39*10-5$, $p = 0.019$ respectively), so Welch's Test was used to test the sample means with a Tukey HSD post hoc test. The means for all three filters while standing were not the same [$F(2,193.12) = 3.14$, $p = 0.045$], because the low pass filter took significantly less time than the unfiltered condition ($p = 0.020$). For HPR while standing, the means were significantly different [$F(2,187.27) = 3.92$, $p = 0.012$], with both filtered and averaged values being significantly smaller than unfiltered values ($p = 0.046$ and $p = 0.022$, respectively). Levene's test did not find the variances to be different for time and HPR while seated, so a one way ANOVA with Tukey HSD was used in each case to determine mean differences. Mean time values were statistically different [$F(2,297) = 4.32$, $p = 0.014$], with the averaged mean meeting significance relative to the unfiltered mean ($p = 0.014$). The HPR for the wheelchair trials also showed statistical differences in means [$F(2,297) = 5.40$, $p = 0.005$], as both filtered and unfiltered data were smaller than the unfiltered values ($p = 0.0366$ and $p = 0.0052$, respectively). Table 2 shows the average and standard deviation for all conditions.

DISCUSSION

Recognition began to plummet once background noise intensity became close to 65 dBs. This should only be an issue for users who attempt to operate with loud conversations in the background.[6] Improving recognition rates should be improved, but it is possible that the phrase selection could be the key to improving the recognition. This idea is strengthened by the fact that nearly three quarters of the phrases were recognized on the first attempt even at the 60 dB noise level.

Table 2: Averages and standard deviations of the time and HPR results for all conditions.

	Time (s)	HPR
Standing Unfiltered	3.44 ± 1.16	1.81 ± 0.85
Standing Filtered	3.09 ± 0.79	1.58 ± 0.71
Standing Averaged	3.14 ± 0.77	1.54 ± 0.49
WC Unfiltered	4.26 ± 1.04	3.71 ± 1.01
WC Filtered	4.18 ± 0.98	3.35 ± 0.87
WC Averaged	3.88 ± 0.82	3.25 ± 1.24

There were two main implications of the balance protocol. First, this system was functional for both standing and seated users. Second, both the traditional low pass filtering and the averaging techniques provided a consistent advantage over the unfiltered operation mode.. The balance system was used during the workout session in which heart rate was recorded, and the user was able to successfully make 10 path selections while operating the device. Taken together, these results indicate that this balance system added profound functionality to the overall experience by allowing users to make path choices through leaning, whether or not they could stand on their own.

While the device was primarily designed for those with physical disabilities, the one participant involved in the current testing was an able bodied individual, which was the main study limitation. The next step will be to recruit a full cohort of individuals with physical disabilities to provide feedback. Regardless, the device as currently constructed shows full functionality, with an engaging exercise experience and potential to empower people with physical disabilities.

ACKNOWLEDGEMENTS

Project supported by RERC RecTech funded through grant 90RE5009-01-00 from Department of Health and Human Services, Administration for Community Living, National Institute on Disability, Independent Living, and Rehabilitation Research (NIDILRR). NIDILRR is a Center within the Administration for Community Living (ACL), Department of Health and Human Services (HHS). The contents of this report do not necessarily represent the policy of NIDILRR, ACL, HHS, and you should not assume endorsement by the Federal Government. Thanks to Lloyd Cooper from PUSH Product Design for his continuing help.

REFERENCES

[1] CDC, Related Conditions, 2016, from http://www.cdc.gov/ncbddd/disabilityandhealth/relatedconditions.html
[2] Heath, GN, & Fentem, PH, *Exer & Sport Sci Reviews*, 25: 195-234, 1997.
[3] Boslaugh, S, & Andresen, E, *Preventing Chronic Disease*, 3:1-14, 2006.
[4] Sato, K et al., *Games for Health J*, 4:161-167, 2015.
[5] Widman, LM et al., *J of Spinal Cord Med*, 29:363-370, 2006.
[6] Daniel, E, *J of School Health*, 77:25-231, 2007.

SB^{3}C2017
Summer Biomechanics, Bioengineering and Biotransport Conference
June 21 – 24, Tucson, AZ, USA

HIGH MAGNITUDE HEAD IMPACT EXPOSURE IN YOUTH FOOTBALL GAMES

Eamon T. Campolettano, Ryan A. Gellner, and Steven Rowson

Biomedical Engineering and Mechanics
Virginia Tech
Blacksburg, VA 24060

INTRODUCTION

Despite youth football players representing 70% of all players in the United States, most head impact exposure research has focused on older populations [1]. Effective teaching and rule modification are considered to be the preferred methods for limiting head impact exposure and mitigating concussions in football [2].

Using helmet-mounted accelerometer arrays, previous research with youth football players has shown that most head impacts occur in practice [1, 3-4]. Head impact exposure can be reduced by as much as 40% by limiting contact in practice [3]. In 2016, a classification system developed to assess the number of high magnitude impacts produced by specific practice drills found that most tackling drills were associated with a greater impact severity than game impacts [4].

With a valuable framework already in place to assess head impact exposure in football practices, the primary objective of this study was to analyze youth football games and determine what aspects were associated with high magnitude impacts.

METHODS

Two youth football teams composed of 9-12 year old players were included in this study approved by the Virginia Tech Institutional Review Board. A total of 45 players participated and received helmets instrumented with accelerometer arrays (Head Impact Telemetry (HIT) System). The players had a mean age of 10.7 ± 1.1 years and a mean body mass of 43.8 ± 12.1 kg.

All players were given a Riddell Revolution or Speed helmet with a six accelerometer array mounted inside. To ensure measurements of head acceleration, and not helmet acceleration, the accelerometers are spring-mounted to maintain contact with the head throughout the impact [5]. Players wore the instrumented helmets at each practice and game throughout the season. Data acquisition for impacts was automatically triggered when an individual channel exceeded a 14.4 g threshold. Impact data, which were transmitted wirelessly from the helmets to a sideline computer, were processed to computer linear and rotational resultant accelerations based on previous research [6, 7].

Games and practices were filmed to facilitate video verification of head impacts. All impacts exceeding 40 g, which are considered high magnitude, were visually verified in order to identify the game scenario associated with the head impact. Activity logs for each session were used to parse out spurious impacts most often associated with players throwing their helmets on the ground during breaks in action.

Game video was used to develop offensive and defensive play counts for each player. With these data, a normalized metric for a player's game activity level was developed called the Player Usage Rate (PUR). It is defined as the ratio of the number of plays in which a particular player participated and the total number of plays that player could have played and is expressed as a percent. Similarly, the average number of impacts per practice and game were determined for each player by noting the number of sessions in which each player was active.

High magnitude game impacts were classified as either offensive or defensive impacts. Further distinction was made between impacts occurring in the open field and at the line of scrimmage. Open field impacts were defined as those that occurred outside of the zone where the offensive linemen were positioned at the snap while line of scrimmage impacts occurred within this zone. Beyond this, impacts were classified based on the impacted player's role on the play in which the high magnitude impact occurred.

Each player was also assigned to one of three position groups, corresponding to their role on the field (Table 1). Youth players generally play both offense and defense with their roles mirrored. Using traditional football positions, the mirror position for an offensive lineman would be a defensive lineman.

Table 1. Each player was assigned a position classification according to their role on offense and defense.

Classification	Positions
Backs	Quarterback, running back, and linebacker
Line	Offensive and defensive line
Perimeter	Wide receiver, cornerback, and safety

ANOVA with a post-hoc Tukey HSD Test was used to compare differences in impact severity among the three position groups, while a Student's t test was used to compare differences between open field and line of scrimmage impact severity. Statistical significance was determined using a threshold of $P < 0.05$ for all tests.

RESULTS

A total of 7590 impacts were recorded and verified, 584 of which exceeded 40 g. These high magnitude impacts were comprised of 386 (66%) practice impacts and 198 (34%) game impacts. Among game impacts, 122 were classified as open field impacts, with the remaining 76 as line of scrimmage impacts. Open field impacts [57.9±16.8 g] were associated with a higher average impact severity ($P = 0.001$) than line of scrimmage impacts [50.7±13.0 g] (Figure 1).

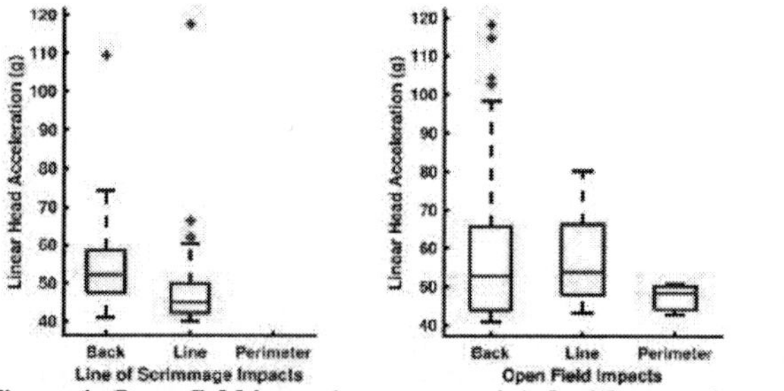

Figure 1. Open field impacts were associated with a greater severity than line of scrimmage impacts as players were able to achieve greater speeds prior to impact. No perimeter players experienced a Line of Scrimmage impact.

Table 2. On average, backs were utilized most frequently and experienced the most high magnitude head impacts. Non-count values are reported as average (standard deviation).

	Backs	Line	Perimeter
# of Players	14	18	13
Impacts > 40g	127	67	4
Average Severity [g]	23.1 (3.1)	19.8 (5.5)	18.6 (2.9)
95% Impact Severity [g]	49.1 (14.5)	41.5 (12.5)	29.0 (10.2)
Impacts Per 100 Plays	24.1 (14.1)	24.1 (13.5)	19.7 (12.4)
Impacts Over 40 g Per 100 Plays	2.0 (2.2)	1.7 (1.8)	0.2 (0.6)
Average PUR [%]	68.3 (25.0)	58.6 (30.3)	17.8 (16.4)

Most high magnitude head impacts occurred for players in the Backs group (Table 2). Impacts per 100 plays did not vary among the three groups ($P > 0.85$). Backs experienced high magnitude impacts in all five impact categories, with impacts while tackling or carrying the ball representing the majority (Figure 2). Impact categories consisted of blocking (Blocker) or being blocked (Blocked), tackling (Tackle) or being tackled (Carrier), and Loose Ball, where players pursued a fumble.

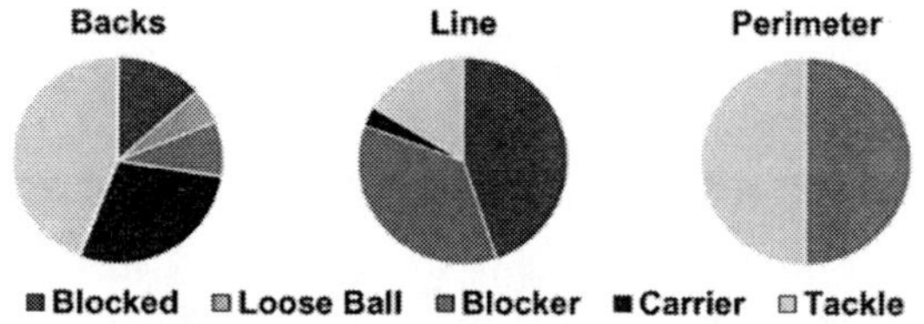

Figure 2. Line players were most frequently blocked or blocking during their high magnitude impacts. Perimeter players experienced the fewest high magnitude impacts.

DISCUSSION

While the proportion for high magnitude practice impacts (66%) is low compared to previous studies, it should be noted that the average age in this study is older than previous youth teams[1, 3-4]. As players age, more high magnitude impacts occur during games than in practices. [3] For high magnitude game impacts, a greater proportion occurred in open field environments than at the line of scrimmage. Players were able to reach greater speeds and produce higher magnitude head impacts there, leading to the greater impact severity observed (Figure 1). Open field impacts usually involved players classified as Backs. These players had higher severity impacts than either the Line ($P < 0.0001$) or Perimeter ($P < 0.0001$) groups despite not experiencing more impacts when normalizing for play count (Table 2). Backs, who have the highest PUR, are the most active and athletic among the player groups, while the Line (52.8±12.4 kg) tends to be larger and the Perimeter (33.5±5.8 kg) the smallest. These players were the most likely to be involved in the tackle for a given play. Backs were found to have not only the most high magnitude impacts, but also the greatest variety among impact scenarios (Figure 2).

It should be noted that head impact exposure varied among individuals and between the teams in this study. Further, for individual measurements, the HIT system is associated with random error up to 15.7%, though the average error is only 1% [8].

Head impact kinematic data were collected from two youth football teams of players 9-12 years old in order to determine where high magnitude head impacts occur in games. Impacts in the open field usually involved players grouped in the Backs classification and were observed to result in more severe head impacts than other configurations. Coupled with the practice framework previously developed, this analysis may prove useful to researchers and policy makers aiming to reduce head impact exposure in youth football.

ACKNOWLEDGEMENTS

The authors gratefully acknowledge the NIH-NINDS under award number R01NS094410 for supporting this work and the Blacksburg youth football teams for their participation in this study.

REFERENCES

[1] Daniel, R. et al., *Ann. Biomed. Eng.*, 40(4):976–981, 2012.
[2] Crisco, J. J., *Curr. Sports Med. Rep.*, 10:7–9, 2011.
[3] Cobb, B. et al., *Ann. Biomed. Eng.*, 41(12):2463-2473, 2013.
[4] Campolettano, E. et al., *J Neurosurg Pediatr*, 18(5):536-541, 2016.
[5] Manoogian, S. et al., *Biomed. Sci. Instrum.*, 42:383–388, 2006.
[6] Rowson, S. et al., *Ann. Biomed. Eng.*, 40(1):1–13, 2012.
[7] Crisco, J. J. et al., *J. Biomech.*, 126(6):849–854, 2004.
[8] Beckwith, J. G. et al., *Ann. Biomed. Eng.*, 40(1):237–248, 2012.

CHARACTERIZATION OF ELEVATED HEAD IMPACT EXPOSURE BETWEEN INDIVIDUAL YOUTH FOOTBALL PLAYERS

Ryan A. Gellner, Eamon T. Campolettano, and Steven Rowson

Biomedical Engineering and Mechanics
Virginia Tech
Blacksburg, VA, United States

INTRODUCTION

Football is among the most popular sports in the United States at all ages, with the majority of its participants being at the youth level. This team sport is also associated with the highest incidence of concussion among participants [1]. As research has suggested that sports-related concussion, even at the youth level, can have adverse long-term effects [2], preventing injury through interventions aimed at reducing high magnitude impacts and head impact exposure has become a key interest. Examples of these interventions include teaching proper tackling technique and eliminating drills that expose players to high magnitude head impacts not representative of game situations [3].

Head impact exposure varies in magnitude and frequency between individual players because of differences in position, amount of playtime, athleticism, and level of play. Tolerance to these head impacts also likely varies by individual, due to variance in age, gender, past exposure to head injury, size, and a range of other factors [4]. Previous studies have used aggregate analyses to characterize head impact exposure in youth football players [5]. The objective of this study was to characterize differences in head impact exposure between individuals and identify characteristics associated with elevated head impact exposure.

METHODS

Head impact data from 45 youth football participants (age 10.7 ± 1.1) from two recreational league teams were collected during a single season. Participants were recruited by obtaining written consent from parents of each youth athlete and verbal consent from each athlete. This study was approved by the Virginia Tech Institutional Review Board.

Helmets of subjects were instrumented with the Head Impact Telemetry (HIT) System (Simbex, Lebanon, NH; Sideline Response System, Riddell, Chicago, IL). This system consists of a six-accelerometer array inserted into either a Riddell Revolution or Speed model helmet and communicates with a sideline computer during all games and practices. Resultant linear and rotational head acceleration values are measured for each impact players experience during games or practices.

Video recordings and activity logs of each session (game or practice) were used to verify all individual head impacts resulting in head accelerations exceeding 40 g, defined as high magnitude impacts. Practice drills were divided into general categories based on previous analyses [3]. Of these, two practice drill types were considered: "Tackling" and "Offense vs Defense." Additionally, participants completed assessments measuring parent perception of child behavior. The portion of the Behavior Rating Inventory of Executive Function (BRIEF) survey dealing with a player's aggressive behavior was used to compare participants.

Players typically played one offensive and one defensive position during competition and were grouped accordingly into position groups. The three position groups used were Backs (n = 14 - quarterbacks, running backs, linebackers), Line (n = 18 - offensive/defensive line, tight end), and Perimeter (n = 13 - wide receiver, safety, cornerbacks).

For each player, high magnitude impacts per session, game, and practice were calculated for the season. Position groups were compared using a one-way ANOVA and Tukey HSD post-hoc comparisons. A subset of players with elevated head impact exposure relative to comparable players was identified. These High Impact Players (HIP) were defined as players who received a greater number of high magnitude head impacts than the mean of all players plus one standard deviation in both practices and games. Statistical comparison between the Backs position group and these HIP was made using a two-sample independent pooled t-test. Results were deemed significant for p <

0.05. Minitab 17 (Minitab Inc., State College, PA) was used for all statistical calculations.

RESULTS

A total of 584 high magnitude head impacts were experienced by all players during the season. Players in the Backs position group experienced the majority of high magnitude impacts with 349, followed by Line players with 198 and Perimeter players with 37.

The average player experienced 0.43 ± 0.55 high magnitude impacts per session. Backs (0.80 ± 0.77) experienced more high magnitude impacts per session than Line (0.36 ± 0.33) or Perimeter (0.13 ± 0.14) players (Figure 1). Backs were found to have significantly greater impacts per session than Perimeter and Line players ($p=0.003$).

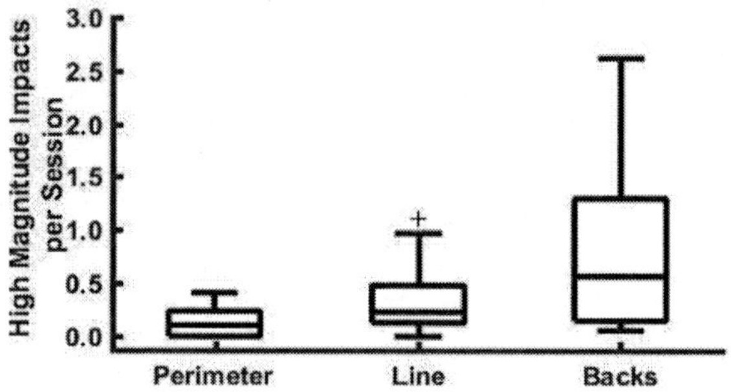

Figure 1: High magnitude impacts per session for each of three player position groups. Backs had the greatest number of high magnitude impacts per session at 0.80 ± 0.77.

The average player experienced 0.37 ± 0.46 high magnitude impacts per practice and 0.69 ± 1.09 impacts per game. Three players had high magnitude impact rates more than one standard deviation greater than the mean for both games and practices (Figure 2). Overall, these three players (7%) accounted for 213 (36%) of the high magnitude impacts for the season.

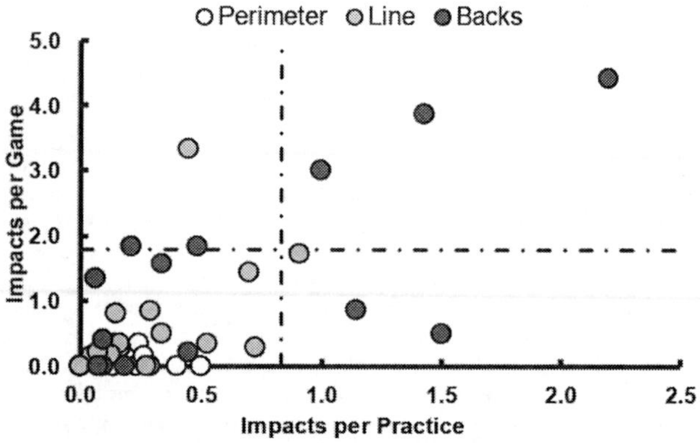

Figure 2: Three Backs were identified as High Impact Players because they exceeded one standard deviation greater than the mean for all players in both impacts per game and impacts per practice. HIP thresholds were 0.83 impacts per practice and 1.78 impacts per game (dash-dotted lines).

The HIP group did not differ significantly from all other players in age, weight, or shortened BRIEF score, but did differ in their impacts per activity type across all scenarios. The HIP were significantly greater than other Backs in their weight ($p = 0.003$) and showed a greater occurrence of high magnitude impacts in Offense vs Defense drills ($p = 0.003$), Tackling drills ($p = 0.018$), and Games ($p < 0.001$) than other Backs (Figure 3).

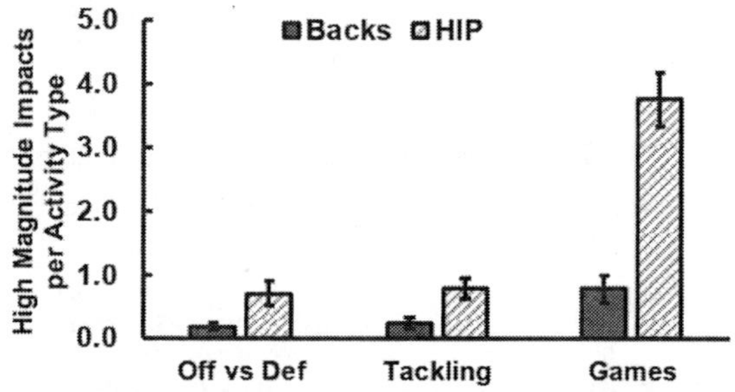

Figure 3: High magnitude impact rates by activity type. The HIP group, all Backs themselves, had a greater number of impacts per session in all activity types and games than other Backs. Error bars shown are standard error for the mean.

DISCUSSION

Previous studies have presented individual-specific head acceleration data which reveals the existence of players on youth football teams who experience greater head impact exposure than their peers [5]. Head impact exposure varied in this cohort of 45 youth football players. This variance can be partially explained by player position, as Backs had more impacts than Line or Perimeter players. However, even within the Backs position group, head impact exposure varied greatly. Three players were identified as HIP and experienced a greater rate of high magnitude head impacts in practices and games than other Backs. These three players were unarguably among the most athletic and competitive players on their team and therefore received the most playing time, ball carrying time, and were involved with the most plays during games and offense vs defense drills during practice. Additionally, these players would have faced other teams' most athletic players during games, increasing the chances of high magnitude collisions. Players with a greater number of impacts per game were not necessarily the same players with a greater number of impacts per practice. Although all players participated similarly in tackling drills, the Backs' ball-carrying aptitude resulted in more repetition with this scenario and likely greater confidence, leading to a larger number of high magnitude head impacts when these players were involved in a play.

Limitations of this study include the error associated with the HITS system – subject to up to 15.7% error for a given individual impact, but less than 1% error on average [3, 6]. The behavioral survey used in this study was not designed solely to capture aggressiveness information. Therefore, future studies should incorporate a more sensitive behavior scale to determine if aggressive behavior differs between HIP and other players.

ACKNOWLEDGEMENTS

The authors would like to thank the NIH-NINDS for supporting this work under award number R01NS094410, and the Blacksburg Recreational Football League players and parents for their participation.

REFERENCES

[1] Lincoln, A.E. et al., *Amer Jrl of Sports Med*, 39:958-963, 2011.
[2] Gavett, B.E. et al., *Clin Sports Med*, 30:179-188, 2011.
[3] Campolettano, E.T. et al., *J Neurosurg Pediatr*, 18:536-541, 2016.
[4] Guskiewicz, K. M. et al., *Exerc. Sport. Sci. Rev.*, 39:4–11, 2011.
[5] Cobb, B. R. et al., *Ann Biomed Eng*, 41:2463-2473, 2013.
[6] Beckwith, J.G. et al, *Ann Biomed Eng*, 40:237-248, 2012

SB^{3}C2017
Summer Biomechanics, Bioengineering and Biotransport Conference
June 21 – 24, Tucson, AZ, USA

MORPHOMETRIC ANALYSIS OF THE HUMAN ANKLE JOINT

Tia Arvaneh (1,3), William E. Lee (1), Roy Sanders (2), Peter Simon (1,2,3)

(1) Department of Chemical & Biomedical Engineering
University of South Florida
Tampa, FL, USA

(2) Department of Orthopaedics and Sports Medicine
University of South Florida
Tampa, FL, USA

(3) Department of Biomechanics
Foundation for Orthopaedic Research and Education
Tampa, FL, USA

INTRODUCTION

In general, osteoarthritis (OA) can be defined as a non-inflammatory degenerative disorder of hyaline cartilage and subchondral bone with subsequent hypertrophy of surrounding tissue[1]. Despite the fact that OA and its sequelae are the most common cause of disability and work limitations in adults in the US, its etiology is unknown. Its pathogenesis, on the other hand, has been described as a complex system of mechanical, biological, biochemical and enzymatic processes and changes. Regardless of ever-closer understanding of the disease, its societal impact, and wide spread prevalence, little has been done to reverse its progression. Instead, focus was directed towards the treatment options of the final stage of the disease.

Degenerative changes of the ankle can cause pain and alter joint function, which can affect the quality of life and daily activities, including walking mechanics. The basic patterns of motion are described by the geometric features of the articulating surfaces of the talus and of the bones of the lower leg. The presence of osteoarthritis can disrupt those patterns and affect ankle kinematics.

Commonly, ankle OA is diagnosed and graded radiographically, identifying joint specific morphometric and morphologic features such as joint space narrowing, osteophyte formation and subchondral bone sclerosis. Additionally, effects of talus hypertrophy are noted as they disturb the ankle soft tissue balance. Subsequent treatment options are limited to ankle fusion or arthroplasty, both with limited success.

Previous works on ankle joint morphometry primarily focused on 2D x-ray based evaluation of talar size and shape. Subjectivity of such measurements can be questioned, as their accuracy may be highly dependent on operator's skill[2] and feet position during scanning. 3D approaches (CT, laser scanner based) previously reported[3] offer more adequate description of ankle joint morphometry in both healthy[4] and diseased state[5].

The goal of this study is to assess the geometric parameters of the talus. Additionally to gender and side, effects of OA and its progression will be evaluated. We hypothesize that changes of the talar size and shape are a function of OA progression (reflected by KL grade).

METHODS

The pilot study was performed using a dataset of 68 CT scans comprised of 45 OA and 23 normal ankles. Inclusion criteria were no fracture/previous talus surgery and no rheumatoid arthritis. The scans were previously graded for OA using the Kellgren and Lawrence classification (KL). 3D models were created using MIMICS. Volumetric point-cloud data was then imported into MATLAB. A custom-written, fully automatic algorithm was created to measure and calculate the following parameters: medial and lateral height (MH, LH), and medial and lateral radius (MR, LR).

Prior to measurements, the subject-specific coordinate system was normalized for all subjects using principal component analysis. Upon normalization, the talus was separated into its respective medial and lateral regions. For each side, the apex was identified by calculating the most superior point, creating the medial and lateral apexes. The apexes were used as references points for parameter measurements.

Height. Starting at the apex point of the talus, a vertical line was drawn downwards toward the most inferior point of the body, perpendicular to the most plantar aspect. The distance between these two points was calculated. This was done for both medial and lateral sides of the talus body. In order to account for deformities that may have affected the inferior portion of the talus, height was normalized to a transverse plane located at the center of the talus.

Radius. Three sagittal planes were defined using the medial apex, lateral apex, and center point of the talus as references. The points of the talus that overlapped with each plane were extracted. The area of each surface was calculated and individually equated to the area of circle in order to find the radii (Figure 1).

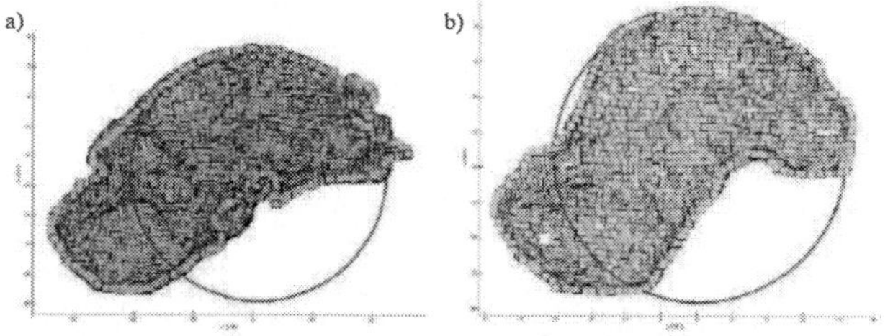

Figure 1: a) MR measurement for subject with OA; b) MR measurement for normal subject

Mean and standard deviation was calculated for every parameter. Two-way ANOVA (p=0.05) and LSD post-hoc test were used to evaluate gender, side, group (healthy, OA) and KL grade differences.

RESULTS

The study population consists of 68 ankles (23 healthy, 45 OA; 29 females, 39 males; 28 left, 40 right). Average age was 59.2±17.3 years (44.8±17.2 in healthy group; and 66.6±12.1 in OA group), and OA was grouped as follows: 22 subjects graded KL 1; 4 subjects graded KL 2; 27 subjects graded KL 3; and 14 subjects graded KL 4.

Average LH was 31.5±5.0mm. LH of male individuals was significantly larger than LH of females (33.2±4.1mm, vs. 29.2±5.2mm; p=.0007). There were no significant differences between right and left ankles (p=.0807), between healthy and OA groups (p=.47), and KL grades (p=.572).

Average MH was 29.8±4.5mm. Male ankles had significantly larger MH than female ankles (31.4±3.8mm, vs. 27.6±4.4mm; p=.0003). There were no significant differences between the left and right sides (p=.904). MH in the OA group was significantly different from MH in the healthy group (p=.003), and there were significant differences between KL grades (overall: p=.048, Fig. 2a).

Average LR was 17.5±2.6mm. Males had significantly larger LR than females (18.6±2.4, vs. 16.2±2.3; p<.0001). There were no significant side differences (p=.5222). LR in the OA group was significantly different from healthy individuals (p=0.0006, Fig. 2d), and there were significant differences between KL grades (overall: p=.0134, Fig. 2b).

Average MR was 22.5±2.0mm. Male ankles had significantly larger MR than female ankles (23.9±1.2, vs. 20.7±1.3; p<.0001). There were no significant side differences (p=.1861). OA group had significantly larger MR than the normal group (p=.408, Fig. 2c), however, there was no significance between KL grades (p=.0594).

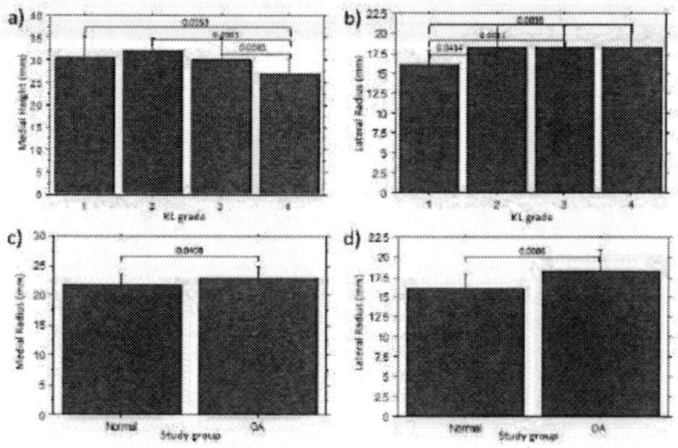

Figure 2: a) MH±SD by KL grade; b) LR±SD by KL grade; c) MR±SD by study group; d) LR±SD by study group (significant p-values)

DISCUSSION

Ankle joint OA is a degenerative disease that affects morphology and morphometry of the talus and the mortise formed by distal tibia and fibula. Disputable signs of OA include changes in joint space, bone density and bone size (hypertrophy). Available surgical treatment of end-stage OA includes total ankle joint replacement or arthrodesis, both of which result in severe limitations. Despite the fact that OA is a slow, insidious process that is likely amenable to remission early in the process, little has been done to identify an early signs of OA progression.

This current work is, to the best of our knowledge, the first to report on 3D morphometric parameters of the talus at all 4 stages of OA progression (KL grades). Results showed that talar morphometry is gender specific and is independent of anatomical side. Medial talar height was significantly different in the OA group and at differing KL grades, while the measure of the lateral talar height did not show significant differences between healthy and OA group and between individual KL grades. This may be an indicator of the uneven flattening of the talus. On the other hand, both lateral and medial radii of curvature of the talar dome were significantly larger in the OA group, confirming previously published work[5].

Being a preliminary study, our work has limitations. First, using a total of 68 subjects made it difficult to match a cohort to every evaluated group (KL 2 had only 4 subjects). Second, only small number of talar parameters (height and radius) were included. We aim to include more parameters in the future.

In conclusion, talar morphometry is gender-dependent and the changes of the talar bone (flattening) is a function of OA progression. While our findings increase basic understanding of the OA process in the ankle, further research is necessary to fully understand etiology of OA in order to optimize diagnostics and treatment.

ACKNOWLEDGEMENTS

This project was internally funded by Foundation for Orthopaedic Research and Education.

REFERENCES

[1] Kuettner, KE, Cole, AA, *Osteoarthritis Cartilage,* 13:93-103, 2005
[2] Stagni, R et al., *Clin Biomech*, 20:307-311, 2005.
[3] Hayes, A et al., *Iowa Orthop J,* 26:1-4, 2006.
[4] Cher, WL et al., *J Biomech,* 49:3026-3030, 2016.
[5] Wiewiorski, M et al., *Foot & Ankle Int*, 37:644-651, 2016.

SB³C2017
Summer Biomechanics, Bioengineering and Biotransport Conference
June 21 – 24, Tucson, AZ, USA

MORPHOLOGICAL ANALYSIS OF OVINE RETINA AS A FUNCTION OF AGE

Matthew P. Byrne and Brittany Coats

Mechanical Engineering
University of Utah
Salt Lake City, UT, USA

INTRODUCTION

Retinal hemorrhage in children can be an indicator of abusive head trauma (AHT) [1]. However, being able to discern whether the trauma originated from abuse or from an accident is of critical importance to the primary care provider, as catching abuse early is important for the survival of the young patient. Our long-term research goal is to understand the microstructural and mechanical characteristics of the immature retina to better identify retinal injury thresholds in infants. To achieve this goal, an understanding of retinal vessel morphology, and how it changes with age, will be critical for defining potential mechanisms of failure.

Existing studies of retinal vascular morphology have used 2D fundus images taken from human patients in an effort to characterize "normal" vessel characteristics [2] or study prediction markers for maladies such as hypertension [3]. Several descriptive studies have evaluated retinal vascular morphology through the depth of the retina [4-6], but very little quantitative characterization has been performed. Furthermore, no studies have quantitatively evaluated changes in retinal vascular morphology with age. The objective of this study was to quantify changes in the morphological features of retinal vasculatures with retinal depth and age (neonate, adolescent, adult). These data will be useful in identifying age-specific vasculature characteristics that may associate with higher risks of retinal hemorrhage from trauma.

METHODS

Adult, 6-month, and preterm ovine eyes (n=1) from existing non-ocular related studies on campus were enucleated with 4 hours post-mortem and fixed in 4% paraformaldehyde for a period of no less than 36 hours at 4°C. The retina was extracted from the eye in a PBS solution and placed immediately in a block buffer (5% BSA and 0.5% Triton X-100, Sigma Aldrich) for 24 hours at 4°C. The buffered retinas were washed with PBlec to provide calcium for the lectin antibody stain. The samples were then labeled with Alexa Fluor® 488 (Thermo Fisher Scientific, Waltham, MA) and washed once more in PBS.

Labeled specimens were imaged on an A1R+ confocal microscopy system (Nikon Instruments Inc, Melville, NY) beginning at the vitreous and increasing in depth towards the retinal pigment epithelium of the eye retina. Fig. 1 illustrates an overview of the layers of the retina as well as the primary region of retinal vascularization.

Custom image processing software developed in MATLAB was used to quantify 6 morphological features at 3 depths (25, 60, and 80 μm) in the image stack. These depths roughly represent the top, middle, and bottom of the retinal vasculature region. Some small adjustment to

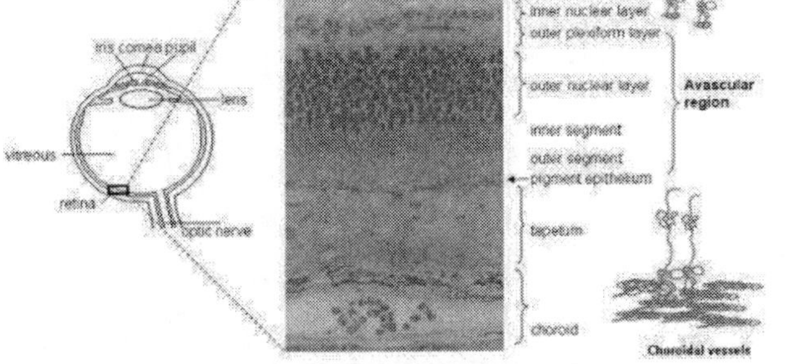

Figure 1: Layers of the retina and their vascularization. [5]

these depths was made to find a layer with good staining and imaging quality. Fig. 2 depicts an example image of retinal vasculature obtained from a depth of 29 μm.

The morphological features captured by the image processing software were (1) length of each vessel segment, (2) diameter of each vessel segment, (3) ratio of length to diameter, (4) number of branch points, (5) tortuosity, and (6) angular asymmetry. Vessel segment length was measured from the vertex of one branch point to its terminal point. Three diameters along the segment length were averaged for the vessel diameter. Tortuosity was defined as the ratio of the true length of the vessel and the Euclidean distance between its two end points. To define angular asymmetry, three angles are measured at each branch point, and the absolute difference is taken from the angles between the trunk and the two branches.

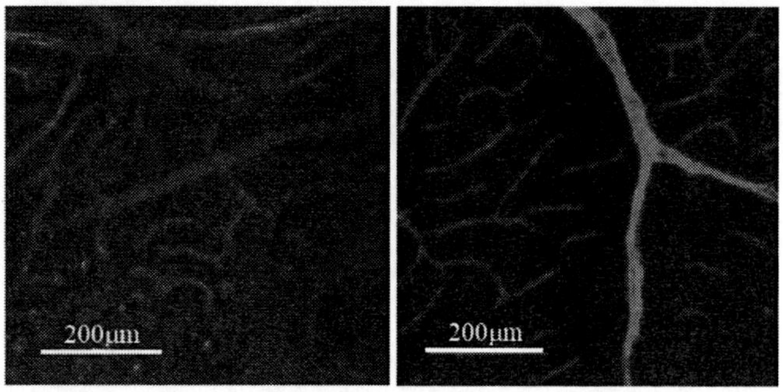

Figure 2: Image of a preterm (left) and 6-month-old (right) retinal vasculature at a depth of 26 μm.

RESULTS

Vessel segments in adult and 6-month-old ovine eyes were longer than vessel segments in preterm eyes and decreased with image depth (Fig. 3A). Preterm vessel segment length did not change with depth. Vessel diameters followed a similar trend (Fig. 3B). The length to diameter ratio (L/D, Fig. 3C) appeared fairly constant across age and depth. The number of branch points for all ages were highest at the deepest vascular region, followed by the shallowest region (Fig. 3D). The middle of the vasculature had the least number of vessels. Tortuosity ranged from 1.09 to 1.15 and did not appear to have any relationship with age or depth (Fig. 3E). Angular asymmetry also did not vary with age or depth (Fig. 3F).

DISCUSSION AND CONCLUSION

The objective of this study was to identify changes in retinal vasculature morphology with age and depth as this may provide insight into locations with higher risk of failure. As might be expected, vessel segment length and diameter increased with age, suggesting developed eyes can withstand more retinal force than premature eyes. The more superficial vessels of the adult and adolescent retina had larger diameters and segment lengths than deeper structures, and are likely more protected from injury. No such distinction was made in the preterm eyes, suggesting equal vulnerability throughout the vasculature.

The finding of increased branching in the superficial and deeper regions of the vasculature is interesting, and agrees with how vascular differentiation has been presented previously in adults [5]. Branch points create stress concentrations in vessels. Therefore, there is likely a higher risk of vessel failure in the deeper structures of the retinal vasculature. The tortuosity and angle of the branches is an important consideration when studying stress concentration progenitors, but they do not appear to be dependent on age or depth.

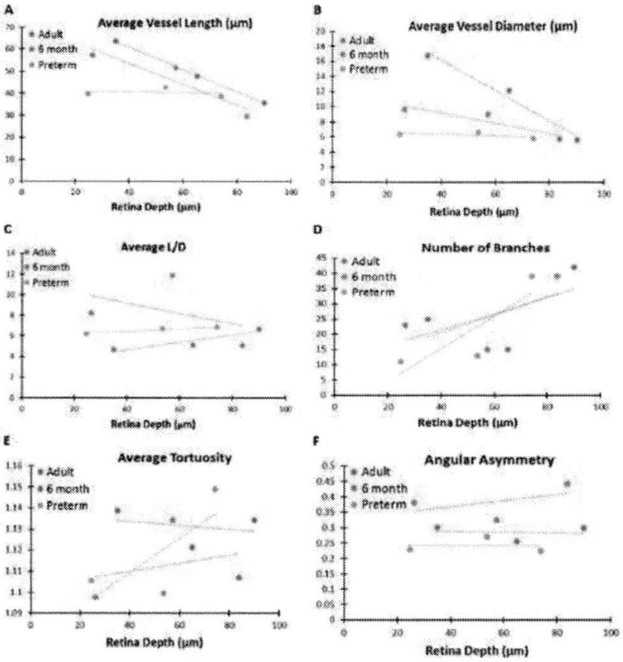

Figure 3: (A) Average vessel length, (B) average vessel diameter, (C) average length/diameter, L/D, (D) number of branch points, (E) average tortuosity, and (F) angular asymmetry of ovine retinal vasculature across three age groups and at three depths within the retina.

The current study is limited to a single specimen at each age. Additional specimens are currently being processed to allow a more in depth statistical comparison between developmental ages and vasculature depth. Even with the size limitation, however, we found unique age and depth-dependent patterns for vessel segment length, diameter, and branching that could be useful for identifying differences in failure mechanisms at different stages of development. Future computational studies will investigate the effect of these features on stress and strain distribution during vitreoretinal traction.

ACKNOWLEDGEMENTS

We would like to thank Kendall McMillan for his expertise in image processing and the Kurt Albertine lab at the University of Utah for their donation of ovine ocular tissues. We'd also like to thank Dr. Chris Rodesch at the Fluorescence Microscopy Core Facility at the University of Utah for his expertise. Microscopy equipment at the core was obtained using a NCRR Shared Equipment Grant #1S10RR024761-01.

REFERENCES

[1] S. A. Maguire, et al., *Eye*, 2013.
[2] A. Hellstrom, et al., *Acta Opthalmol.* 76:260-267, 1998.
[3] A. Hughes, et al., *J Hypertens*, 24:889-894, 2006.
[4] G. Chan et al., *IOVS*, Vol. 53, No.9, 2012.
[5] T. Chang et al., *J Magn Reson Imaging*, 23(4):465-472, 2006.
[6] P. Ern Zhi Tan, et al., *IOVS*, 53: 9, 2012.

KNEE BIOMECHANICS DURING CYCLING ARE SIMILAR FOR NORMAL WEIGHT AND OBESE SUBJECTS

Juan D. Gutierrez-Franco (1), Jordan M. Skaro (1), Scott Hazelwood (1,2), Stephen M. Klisch (1,2)

(1) Mechanical Engineering Department
California Polytechnic State University
San Luis Obispo, CA, USA

(2) Biomedical Engineering Department
California Polytechnic State University
San Luis Obispo, CA, USA

INTRODUCTION

Osteoarthritis (OA) is a degenerative disease of cartilage and bone tissue, and is linked to more than 70% of total hip and knee replacements [1]. In 1994 the direct and indirect costs of OA in the United States were $155 billion [2] and in 2006 OA resulted in approximately $10.5 billion in hospital charges [3]. Obesity is a risk factor for OA [1, 3, 4], likely due to increased knee loading [5, 6] and varus malalignment [7] in gait. Seated cycling has been recommended as a weight-loss exercise with lower knee loads than walking or jogging [8]. However, lack of biomechanical studies for obese subjects in exercises, other than gait, impedes selection of exercises that may best prevent knee OA development in the obese population.

This study tests the hypothesis that cycling knee kinematics and kinetics are not different for normal weight (NW) and obese (OB) subjects. The long-term goal of our research group is to calculate knee joint loading and kinematics during select exercises to aid in selection of weight-loss exercises that minimize risk of OA development. The objectives of this study are to (1) conduct cycling experiments with a motion capture system to calculate internal knee kinematics and kinetics and (2) compare knee kinematics and kinetics for normal weight and obese subjects during cycling.

METHODS

Bicycle Development. The pedals in an upright stationary bicycle (LifeFitness LifeCycle GX, Rosemont, IL, USA) were modified to include 6-channel load cells (GEN5, AMTI, Watertown, MA, USA) [9] (Fig. 1). The pedals included a marker set for use with a motion capture system to track crank angle.

Subject Selection. Subjects were separated into two populations, NW (n=4) and OB (n=4), determined by body mass index (BMI). Protocols were approved by Cal Poly's Human Subjects Committee to minimize risks to human subjects.

Experimental Procedure. Retroreflective markers were placed on subjects using a lower body Helen Hayes marker set. An eight-camera motion capture system and Cortex software (Motion Analysis, Santa Rosa, CA, USA) were used to record marker position and process kinematic data. Subjects stood motionless for a static trial to create virtual axes for body segments. The dominant and non-dominant legs for each subject were identified. Subjects pedaled the modified bicycle with a cadence of 70 RPM at low (C1) and moderate (C2) intensities, measured using the bicycle's resistance levels, for 2 minutes after reaching a steady cadence. Kinematic and kinetic data collected were processed in Cortex. Custom MATLAB (MathWorks, Natick, MA, USA) codes were used to format and average data for three crank cycles. Knee angles were corrected for crosstalk error using custom code with Principal Component Analysis (PCA) [9]; briefly, PCA minimizes the flexion-adduction correlation (R^2) value that is considered a quantitative measure of crosstalk caused by error in flexion axis direction. Only absolute magnitude values are reported here.

Statistical Analysis. Three-way repeated measures ANOVAs, followed by Tukey pairwise comparisons, were performed to determine differences in knee forces, moments, and angles using BMI, cycling intensity, and leg

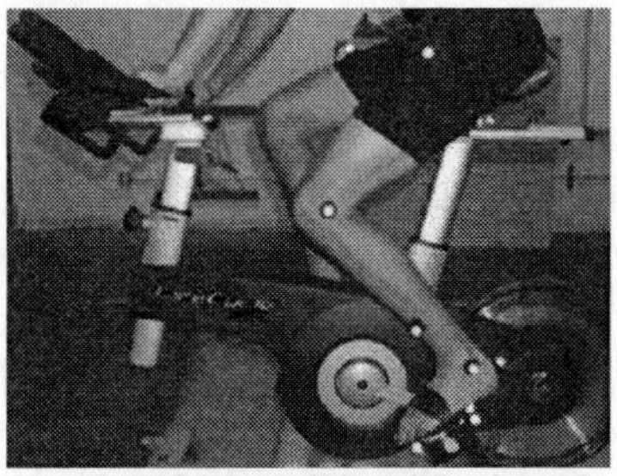

Fig. 1: Subject pedaling bicycle with custom instrumented pedals.

dominance as factors. Statistical significance was defined by $p<0.05$. Interactions between BMI, intensity, and leg dominance were considered, as well as differences within each factor alone.

RESULTS

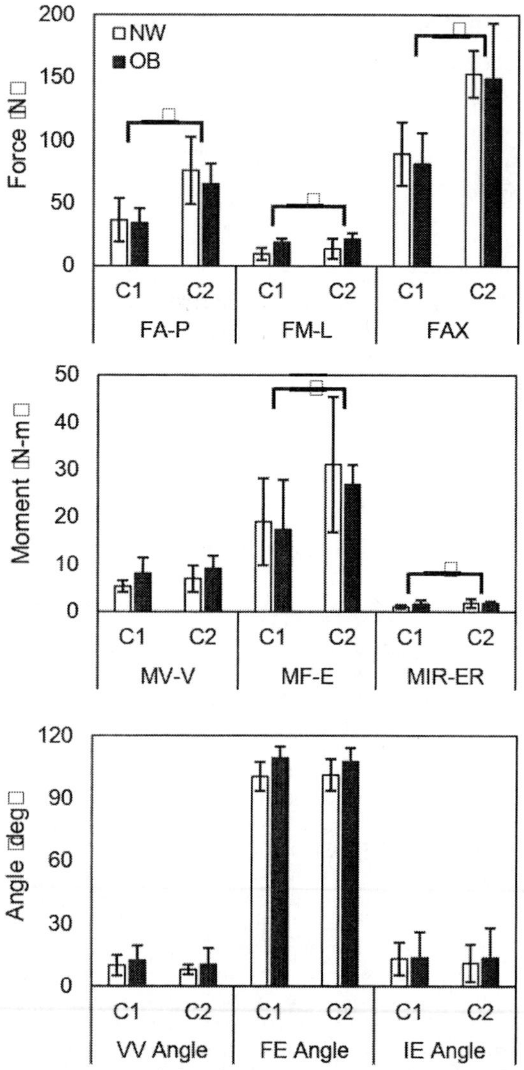

Fig. 2: Knee biomechanics during low (C1) and moderate (C2) cycling intensities for NW and OB subjects. Dominant leg shown. FA-P, FM-L, and FA-X represent anterior-posterior, medial-lateral, and axial knee forces. MV-V, MF-E, and MIR-ER represent varus-valgus, flexion-extension, and internal-external rotation moments. V-V, F-E, and IR-ER Angles represent varus-valgus, flexion-extension, and internal-external rotation knee angles. *Significant difference due to intensity ($p<0.05$).

Knee loads did not differ for NW and OB subjects, expect when comparing cycling intensity levels (C1 and C2) (Fig 2). For all statistically significant cases, C2 had higher loads than C1 ($p=0.006$ for FA-P, $p=0.034$ for FM-L, $p<0.001$ for FAX, $p=0.016$ for MF-E, $p=0.005$ for MIR-ER). All other results showed statistical similarities for knee kinematics and kinetics in cycling between NW and OB subjects. PCA reduced the knee flexion-adduction angle correlations measured using R^2 values which were decreased by three orders of magnitude, thus showing a decrease in knee angle cross-talk.

DISCUSSION

All knee forces and the axial and flexion-extension moments for moderate cycling intensity (C2) were higher than low intensity cycling (C1). This is expected as the higher intensity with constant cadence causes the cycling effort to increase. Knee loading and kinematics were similar for BMI, leg dominance, and their interaction. This is beneficial as similar knee loads are seen in OB and NW subjects during cycling, which could translate to substantially lower OB knee loads in cycling as compared to gait.

The varus-valgus moment does not show statistical significance for any of the tests performed. This result suggests that cycling could minimize the effects of varus misalignment linked to gait in OB subjects. Thus, these results suggest that cycling, likely due to its status as a non-weight bearing exercise, may be a preferred weight-loss exercise as knee loads are not increased due to BMI as occurs in full weight-bearing exercises such as gait [4, 6].

This study has several limitations. First, the sample size was relatively low. Although a power study performed indicates as few as 14 subjects per subject population (NW and OB) could highlight more significant differences due to BMI, the measured knee loads for cycling are substantially lower than previous results for gait [4-6]. Second, soft tissue artifact (STA) (skin and adipose tissue moving around bone tissue causing marker position to differ from bone position) likely produced errors in knee angles. Third, this study reported resultant loads, which differ from the joint contact force that is the true load seen by articular cartilage tissue. Our ongoing cycling studies are using algorithms to minimize STA and employing EMG-driven inverse dynamics to calculate knee contact loads. Regardless, this study produced novel comparisons of knee biomechanics during cycling for NW and OB subjects that suggest that non-weight bearing exercises, such as cycling, should be recommended in weight-loss programs.

ACKNOWLEDGEMENTS

This work was supported by the Donald E. Bently Center (SK), W.M. Keck Foundation (JDGF), and the Defense Health Program (JS, SH, SK) through the Department of Defense Broad Agency Announcement for Extramural Medical Research Program #W81XWH-BAA-14-1 under Award No. W81XWH-16-1-0051. Opinions, interpretations, conclusions and recommendations are those of the authors.

REFERENCES

[1] Felson, DT., *Am J Clin Nutr*, 63.3:430S-432S, 1996.
[2] Sun, J et al., *Ann Epidemiol*, 17.1:51-56, 2007.
[3] Sowes, MR et al., *Curr Opin Rheumatol*, 22.5:533-537, 2010.
[4] Messier, SP et al., *Arthritis Rheum*, 52.7:2026-2032, 2005.
[5] Lerner, ZF et al., *J Appl Biomech*, 30.2:197-205, 2014.
[6] Browning, R et al., *Med Sci Sports Exerc*, 39.9:1632-1641, 2007.
[7] Sharma, L et al., *Arthritis Rheum*, 43.3:568-575, 2000.
[8] Ericson, M, *Scand J Rehabil Med Suppl*, 16:1-43, 1986.
[9] Kraemer, L et al., *SB3C Proceedings*, 1033, 2015.
[10] Baudet, A et al., *PLOS ONE*, 2014.

DEVELOPMENT OF A NUMERICAL METHOD FOR ASSESSMENT OF CEREBROVASCULAR RESERVE USING 1D-0D HEMODYNAMIC SIMULATION WITH CEREBRAL AUTOREGULATION MODEL

Changyoung Yuhn (1), Marie Oshima (2)

(1) Department of Mechanical Engineering
The University of Tokyo
Tokyo, Japan

(2) Interfaculty Initiative in Information Studies
The University of Tokyo
Tokyo, Japan

INTRODUCTION

In order to lower the risk of stroke associated with a severe carotid stenosis, surgical treatment such as carotid endarterectomy (CAE) or carotid artery stenting (CAS) is utilized. However, those treatments could cause postoperative complications such as cerebral hyperperfusion syndrome (CHS) for patients with low capacity of cerebrovascular reserve (CVR) [1]. Therefore, evaluation of CVR is necessary for the surgical planning in the preoperative stage. One of the useful methods for assessment of CVR is utilization of single photon emission computed tomography (SPECT) in conjunction with acetazolamide (ACZ)-challenge [2]. CVR is evaluated by a difference between cerebral blood flow before and after ACZ-challenge. However, the measurement technique requires an invasive procedure that takes radionuclides into a patient.

In this study, a numerical method for assessment of CVR using one-dimensional and zero-dimensional (1D-0D) hemodynamic simulation has been developed. In our method, patient-specific arterial geometry of the circle of Willis (CoW) was used. An autoregulation model was coupled to find the patient-specific cerebral peripheral resistances. Then, those resistances were set to lower limits to simulate the condition after ACZ-challenge virtually. As a result of this procedure, flow distributions over efferent arteries of the CoW before and after ACZ-challenge were obtained. The present method was applied to three patients with unilateral carotid stenosis and the results were compared with the SPECT measurements for validation.

METHODS

The present method considers the cardiovascular model presented in [3,4]. This model describes the entire cardiovascular system as the 1D wave propagation model for large arteries coupled with 0D closed loop lumped parameter model for the remainder such as capillaries, veins, and heart. The equations governing the 1D simulation are the continuity, Navier-Stokes equations integrated over the arterial cross-section, and the pressure-area relationship. The governing equations for the 0D simulation are the sets of ordinary differential equations (ODEs) which are obtained by mass and momentum conservation on electric circuit analogy.

The 1D simulation consists of a total of 83 arteries. The geometries of arteries containing the CoW are obtained from medical images of a patient, while those of the other arteries are assigned based on the literature data summarized in [3,4]. Since the patient-specific arterial geometries are introduced, it is required to find appropriate cerebral peripheral resistances for 0D simulation [5]. Thus, we consider oxygen transport from heart to brain and cerebral autoregulation based on oxygen concentration in brain tissue [6-8] in order to obtain the patient-specific cerebral peripheral resistances.

In the 1D simulation, oxygen transport is described by [7,8]:

$$\frac{\partial(AC)}{\partial t}+\frac{\partial(QC)}{\partial z}=-\pi h D K_w \tag{1}$$

where A, Q, and C denote the cross-sectional area, volumetric flow rate, and averaged concentration, respectively. In addition, h, D, and K_w represent the wall thickness, arterial diameter, average oxygen consumption rate of arterial wall, respectively. The source term on the right hand side describes the oxygen consumption of an arterial wall, assuming arterial blood to be the only oxygen source for the arterial wall tissues. Based on Q and C from the 1D simulation, the oxygen concentration in brain tissue C_T is computed from the following equation [7,8] at the terminal of each efferent artery of the CoW:

$$V\frac{dC_T}{dt}=Q(C-\sigma C_T)-2K_0V\frac{C_T}{\overline{C_T}+C_T} \tag{2}$$

where V denotes the volume of brain tissue perfused by each efferent artery. Note that V is distributed with the same distribution of flow obtained from the simulation using the original model presented in [3,4], summing up the total brain volume V_{brain}. K_0 and $\overline{C_T}$ are the average consumption rate and average concentration of oxygen in brain tissue, respectively. The last term on the right hand side is the rate of oxygen consumption described by the Michaelis-Menten equation. σ, the partition coefficient, can be determined from the steady-state solution of Eq. (2):

$$\bar{Q}\left(C_{O_2}-\sigma\overline{C_T}\right)-K_0V=0 \tag{3}$$

where $\bar{Q}$ and C_{O_2} denote the mean flow rate over a cardiac cycle and mean oxygen concentration in arterial blood, respectively. Finally, cerebral peripheral resistance R is determined from the sigmoidal function as presented in [6]:

$$\frac{dA_R}{dt} = G_R(C_T - \overline{C_T}) \tag{4}$$

$$R = \frac{R_L + R_U e^{A_R - \hat{C}}}{1 + e^{A_R - \hat{C}}}, \quad \hat{C} = -\log\left[\frac{R_{SP} - R_L}{R_U - R_{sp}}\right] \tag{5}$$

where A_R is an activation for R to change, R_{SP} is the set point of R. In general, a proportional gain G_R needs to be set to match the transient dynamics. However, since our aim is to obtain the converged resistances which are assumed as a long-term consequence of cerebral circulation, G_R can be chosen as large as possible to fasten convergence. R_U and R_L are the upper and lower limits of R and are set to $1.30R_{SP}$, $0.74R_{SP}$, respectively [6].

In this study, Lax-Wendroff scheme with minmod flux limiter was used to solve the Eq. (1) and the other equations in 1D simulation were solved by two-step Lax-Wendroff scheme. For the ODEs, 4th order Runge-Kutta scheme was used. The simulation started with constant boundary condition of C_{O_2} at the aorta inlet and peripheral resistances tuned to match patient's cuff blood pressure. The cerebral autoregulation described above was conducted in every time step until convergence. After obtaining mean flow rates over a cardiac cycle in each artery, the cerebral resistances were set to their lower limits i.e., $0.74R_{SP}$ to mimic the condition after ACZ-challenge, keeping the other peripheral resistances fixed. Finally, mean flow rates were obtained again and the simulation was stopped.

RESULTS

The present method was applied to three patients with unilateral carotid stenosis (Table 1). The data were collected before treatment. Note that the patient in Case 3 had undergone two-stage CAS because the measurement data indicated significantly diminished CVR on the stenosis side. Physical and physiological parameters used in the simulation are listed in Table 2. In order to validate the present method, the results from the simulation were compared with those obtained from the SPECT measurements (Figure 1).

DISCUSSION

In general, cerebral blood flow measured after ACZ-challenge shows an uneven distribution if there exists a significant unilateral stenosis. This is because the cerebral arteries on the stenosis side undergo compensatory vasodilatation up to a maximal level, resulting in blunted response to ACZ-challenge [2].

As shown in Figure 1, the results from the present method are in good agreement with the measurement data qualitatively in terms of change in flow distribution after ACZ-challenge. Also, the results showed pathologically and physiologically reasonable tendency such that flow distribution after ACZ-challenge gets more unbalanced if a stenosis is more severe.

On the other hand, the present method has a limitation that individual differences on compensatory autoregulation are not taken into account. Therefore, further work should be focused on developing a methodology for better estimation of the patient-specific cerebral peripheral resistances, which is necessary for quantitative assessment of CVR.

Table 1: Patient-specific data in three cases.

Case No.	Age	Cuff [mmHg]	Stenosis (%NASCET)	Treatment
1	63	97.3	Rt. ICA (50)	CEA
2	70	82	Lt. ICA (73)	CAS
3	63	112	Lt. ICA (74)	CAS (two-stage)

Table 2: Physical and physiological parameters [8,9].

Parameter	Value	Unit
V_{brain}	1.071	$10^{-3} \cdot m^3$
K_w	2.0	$10^{-5} \cdot ml\ O_2/(ml\ tissue \cdot s)$
K_0	6.067	$10^{-4} \cdot ml\ O_2/(ml\ brain \cdot s)$
$\overline{C_T}$	7.235	$10^{-4} \cdot ml\ O_2/ml\ brain$
C_{O_2}	1.959	$10^{-1} \cdot ml\ O_2/ml\ blood$
G_R	2.0	$10^{4} \cdot ml\ brain/ml\ O_2$

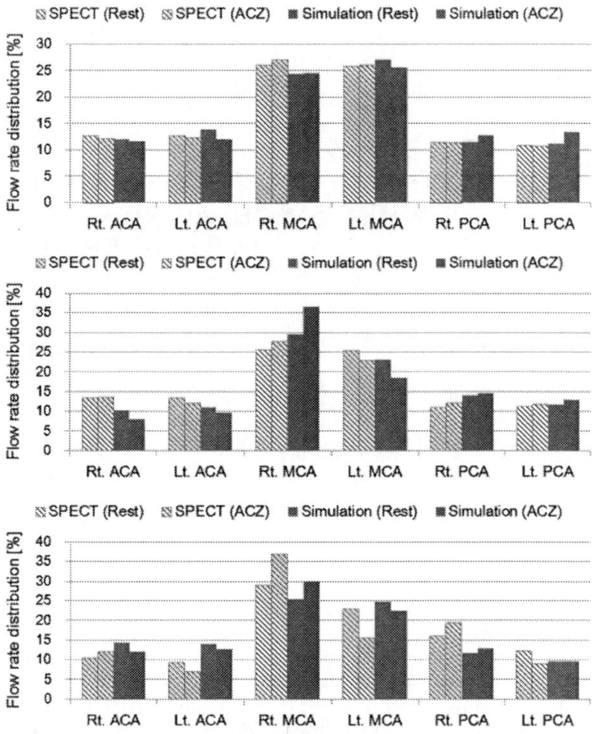

Figure 1: Comparison of flow distribution over efferent arteries of the CoW between SPECT and simulation; Case 1 (top), Case 2 (middle), Case 3 (bottom).

ACKNOWLEDGEMENTS

This work was supported by JSPS KAKENHI Grant Number JP16H04264. The authors acknowledge Shigeki Yamada, MD, PhD and Motoharu Hayakawa, MD, PhD for providing the patient-specific data used in this work.

REFERENCES

[1] van Mook, WN et al., *Lancet Neurol*, 4:877-88, 2005.
[2] Vagal, AS et al., *Am J Neuroradiol*, 30:876-84, 2009.
[3] Liang, F et al., *Med Biol Eng Comput*, 47:743-55, 2009.
[4] Liang, F et al., *Biomed Eng Online*, 10:1-22, 2011.
[5] Zhang, H et al., *Ann Biomed Eng*, 44:2351-63, 2016.
[6] Alastruey, J et al., *Int J Numer Methods Fluids*, 56:1061-7, 2008.
[7] D'Angelo, C, *PhD thesis*, EPFL, Lausanne, Switzerland, 2007.
[8] Köppl, T., *Med Eng Phys*, 36:905-14, 2014.
[9] Crawford, DW et al., *J Clin Invest*, 65:1498-508, 1980.

IMPROVEMENT OF SIMULATED ARTERIAL WAVEFORMS USING MEASURED PARAMETERS BY ULTRASONOGRAPHY

K. Hirayama (1), K. Niki (1), M. Oshima (2), M. Sugawara (3)

(1) Department of Biomedical Engineering
Tokyo City University
Tokyo, Japan

(2) Interfaculty Initiative in Information Studies
The University of Tokyo
Tokyo, Japan

(3) Department of Medical Engineering
Himeji Dokkyo University
Himeji, Japan

INTRODUCTION

Arterial stiffness and flow velocity can be evaluated by ultrasonography, but measurable arteries are limited. Computational fluid dynamics (CFD) has potential for providing the information from various arteries. The cardiovascular system model developed by Liang [1] was used for CFD. However, there were discrepancies between simulated and measured blood flow and pressure waveforms from the lower extremities. The aim of this study was to improve the simulated waveforms by comparing these with measured personal data.

METHODS

We used Liang's computational model of the entire cardiovascular system (Figure 1). The arterial tree is consisted of 55 systemic vessels and 28 cerebral vessels. The arterial system was described as a one-dimensional (1D) model coupled with lumped parameter (0D) models (Figure1) for remaining parts, which contain peripheral vessels and capillaries and the heart. The 0D and 1D model were proposed by Olufsen [2], and the 0D model yielded a closed loop multi-scale model of the entire cardiovascular system. 1D governing equations are given as follow,

$$\frac{\partial A}{\partial t}+\frac{\partial Q}{\partial z}=0 \qquad (1)$$

$$\frac{\partial Q}{\partial t}+\frac{\partial}{\partial z}\left(\frac{Q^2}{A}\right)+\frac{A}{\rho}\frac{\partial P}{\partial z}+8\pi\nu\frac{Q}{A}=0 \qquad (2)$$

$$P-P_0=\frac{4}{3}\frac{Eh_0}{r_0}\left(\sqrt{\frac{A}{A_0}}-1\right) \qquad (3)$$

where A, Q and P are the cross-sectional area, the flow rate and the pressure, respectively. ρ, ν, E, h and r are the density, the dynamic viscosity, Young's modulus, the vessel wall thickness, and the radius of blood vessel. The subscript 0 indicates a value at the reference state. The Lax-Wendroff method was used to solve equations (1) - (3).

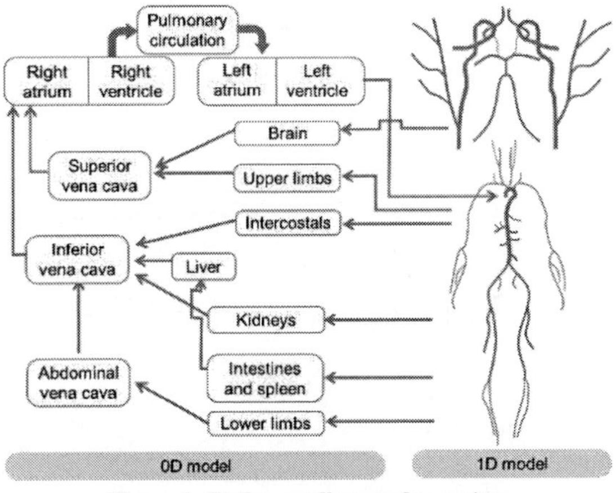

Figure 1: Entire cardiovascular system

We also applied wave intensity (WI) analysis to interpreting hemodynamic changes. WI is a hemodynamic index to evaluate the interaction between the heart and the arterial system [3]. WI is calculated as the product of the time derivatives of pressure and velocity, which is defined as

$$\mathrm{WI} = \frac{dP}{dt}\frac{dU}{dt} \tag{4}$$

where U is the blood flow velocity. The positive values of WI mean forward waves generated by the heart are dominant, and negative values mean the reflected waves from the peripheries are dominant [3]. To obtain WI noninvasively, we measured arterial diameter change waveform by ultrasonography, which can be used as a surrogate for a pressure waveform after being calibrated by the systolic blood pressure and the diastolic blood pressure [4].

Since the pressure and velocity waveforms is composed of forward (P_+, U_+) and reflected waves (P_-, U_-), it can be separated by the following equations [5].

$$P_\pm = \frac{(P - P_0) \pm \rho c U}{2}$$

$$U_\pm = \pm\frac{P_\pm}{\rho c}$$

where c is the pulse wave velocity and P_0 is end-diastolic pressure in the artery. We used these equations to compare forward and reflected waves, respectively.

We made two modifications to the entire cardiovascular system model. Firstly, systemic arterial diameter was modified by a value measured by ultrasonography. Secondly, we correlated the equation for the change in arterial radius with age by using values reported by Niki et al [6].

The measurement by echocardiography was performed in a 55 year old man and the simulation was conducted for a 55 year old model.

RESULTS

The shape of the waveform of blood flow velocity was improved (Figure 2). By correcting arterial diameter, negative waves in late systole appeared in the blood flow velocity waveforms of the model popliteal artery, which were characteristic waveforms in the lower extremity arteries. On the other hand, the waveform of blood pressure was not improved. Wave intensity analysis showed that the forward wave peak was improved but the reflected wave was not (Figure 3).

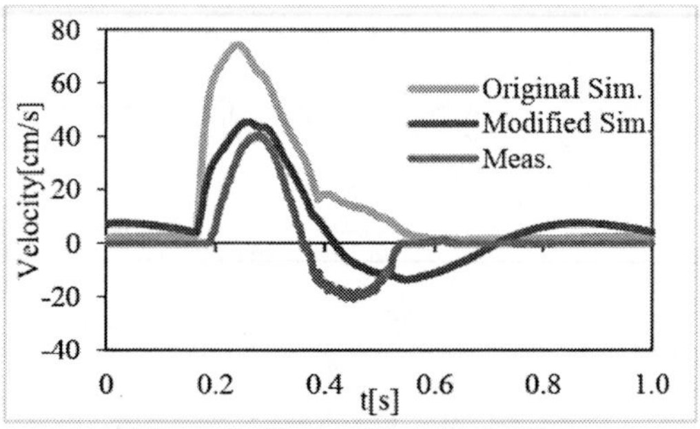

Figure 2: Comparison of velocities

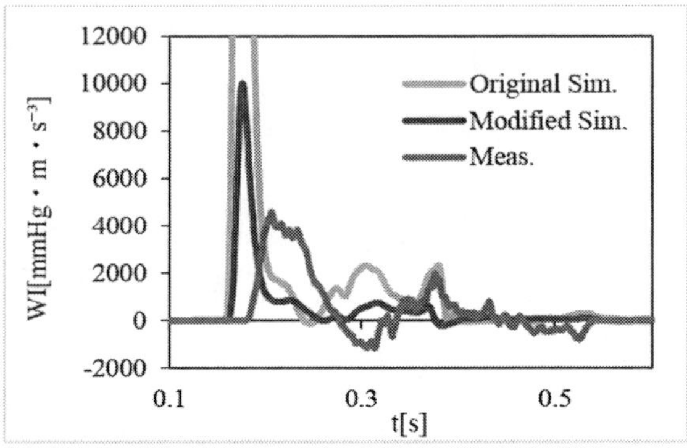

Figure 3: Comparison of wave intensities

DISCUSSION

The shape of the waveform of blood flow velocity was improved. However, the pressure waveform was not improved. The analysis of wave intensity separating pressure waveforms into forward and reflected waves showed delayed return of reflected wave in the measured wave, which caused pressure wave forms inadequate. Moreover, the stiffness parameter in simulation model was lower than the measured value.

CONCLUSIONS

By using measured velocity and diameter change waveforms obtained by echocardiography, the simulated waveforms were improved. Other parameter is required to improve the pressure waveform.

REFERENCES

[1] Liang, F et al., Med Bio Eng Comput, 47(7):743-755, 2009.
[2] Olufsen, MS et al., Ann Biomed Eng, 28(11):1281-1299, 2000.
[3] Sugawara, M et al., Med Biol Eng Comput. 47:197-206, 2009
[4] Niki, K et al., Heart Vessels, 17(1):12-21, 2002
[5] Westerhof, N et al., Cardiovasc Res 6:648-656, 1972
[6] Niki, K et al., J Am Soc Echocardiogr, 16:539, 2003

STEREO AND TOMOGRAPHIC PARTICLE IMAGE VEOCIMETRY - 4D FLOW MRI VALIDATION

Rafael Medero (1,2), Alejandro Roldán-Alzate (1,2,3)

(1) Department of Mechanical Engineering
University of Wisconsin
Madison, WI, United States

(2) Department of Radiology
University of Wisconsin
Madison, WI, United States

(3) Department of Biomedical Engineering
University of Wisconsin
Madison, WI, United States

INTRODUCTION

4D flow MRI has shown promising results assessing in vivo hemodynamics. To keep expanding its clinical use, reliable validation is needed for improved accuracy and precision. Velocity and flow measurements from 4D flow MRI have been previously validated with numerical simulations [1,2] and in vitro systems using 2D PC-MR [3] and Doppler ultrasound. However, these techniques are limited due to temporal resolution. On the other hand, validation of 4D Flow MRI using particle image velocimetry (PIV), which uses a set of cameras to capture the movement of particles illuminated by a laser and optically measure flow velocities, has yet to be thoroughly investigated. PIV has been commonly used for flow visualization and quantification in numerous engineering applications [4], with an increase on biomedical fluids applications in the past decade [5,6,7].

The purpose of this study was to compare stereoscopic PIV (Stereo-PIV), where the three velocity components (3C) are measure on a plane, and tomographic PIV (Tomo-PIV), that uses three-dimensional cross-correlation to determine the corresponding 3D-3C velocity field [8], in an in vitro carotid artery bifurcation model. Additionally, PIV results were compared to velocity measurements obtained from 4D Flow MRI as the first step for advanced validation of 4D Flow MRI measurements.

METHODS

PIV Measurements: A silicone model of a carotid artery bifurcation was connected to a continuous pump and PIV was performed using a Flowmaster PIV system (LaVision, Göttingen, Germany). The system consisted on a dual-pulse 527 nm Nd:YLF laser perpendicular to two high speed cameras Phantom v341. The light sheet thickness was 1 mm for Stereo-PIV, and 4 mm for Tomo-PIV. A solution, seeded with polymer particles (diameter = 10 μm), of 36.9% glycerol, 47.4% water and 15.7% sodium iodide was continuously pumped at 1.4 L/min through the model. This solution matched the refractive index of the silicone model (1.41), while keeping an appropriate dynamic viscosity at room temperature (4.31 ± 0.03 cP) [9].

The system acquired 100 sets of images (25 μm pixel size) at a frame rate of 402 Hz. The time separation between images on each set was 300 μs yielding a maximum particle displacement of 125 μm in the center of the vessel. Data was analyzed in Ensight (CEI Inc., Apex, NC), where velocity vectors were calculated from measurements with each PIV method. A longitudinal plane was extracted from Tomo-PIV at the location of the stereo experiment for comparison with Stereo-PIV. Vorticity at the plane was calculated from velocities.

MR Imaging: The silicone model was scanned on a clinical 3T scanner (Discovery MR 750, GE Healthcare, Waukesha, WI) using a wrist coil. 4D flow MRI was performed with a 5-pt radial-undersampled technique, PC-VIPR [10]. Imaging parameters were: imaging volume: 24 x 24 x 24 cm; 0.625 mm isotropic spatial resolution; TR/TE = 6.4/1.8 ms; VENC = 75 cm/s. MRI was performed while a solution of water and glycerol circulated through the model at 1.4 L/min. Gadofosveset trisodium (Ablavar, Lantheus Medical Imaging, N. Billerica, MA) was added to the solution to improve the signal to noise ratio.

4D Flow MRI Quantification: Vessel segmentation was performed in Mimics (Materialise, Leuven, Belgium), where a three-dimensional (3D) geometry was generated from the PC angiogram. Visualization and quantification of 4D Flow MRI data was done in Ensight at the analysis plane as described above, where velocity and vorticity were obtained.

RESULTS

Qualitative and quantitative agreement in the flow distributions was observed with the three methods as shown by the velocity color-coded vectors in Figure 1. Similarly, the vorticity distribution at the assigned plane showed good agreement between techniques as shown in Figure 2. Finally, Figure 3 shows the whole volume reconstruction of Tomo-PIV revealing good agreement with 4D flow MRI.

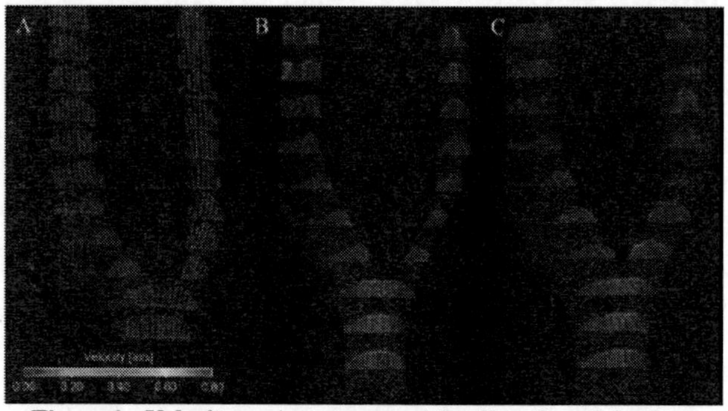

Figure 1: Velocity vectors represent the direction of the flow obtained with 4D flow MRI (A), Stereo-PIV (B), Tomo-PIV (C).

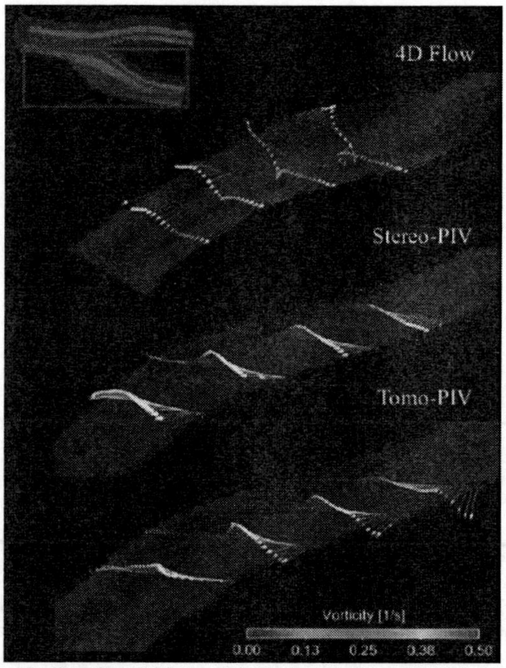

Figure 2: Results within a central plane (red rectangle). The 3D vectors indicate the flow pattern at the left branch of the bifurcation, overimposed on vorticity color-coded contours.

DISCUSSION

Peak velocities as well as a recirculation region were found downstream of the bifurcation using three methods. Better agreement was found between Tomo-PIV and 4D flow MRI when comparing vorticity, suggesting the presence of higher out-of-plane velocities, not detected by Stereo-PIV. Tomo-PIV was able to measure velocity vectors going in both z-directions at the created plane, similar to 4D flow MRI (Figure 2). This was not obtained in the Stereo-PIV, which could be improved if measurements are taken in multiple planes to later reconstruct a 3D flow field. A non-illuminated area was obtained in the right branch of the bifurcation, affecting the comparison between techniques. However, left branch and main vessel results allowed the comparison of the three techniques.

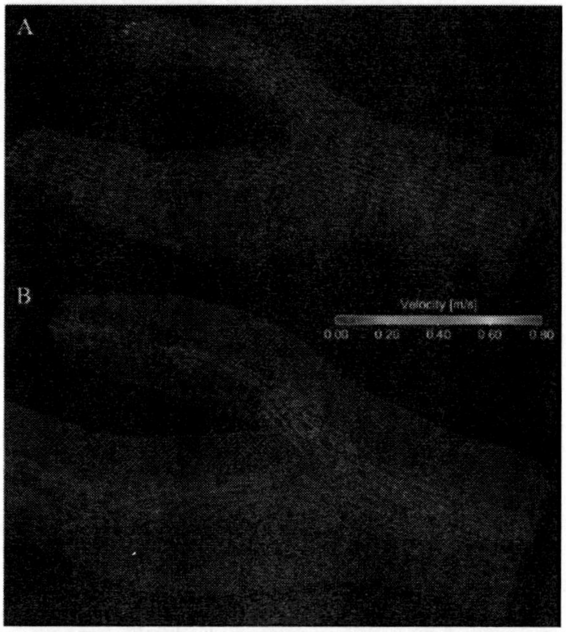

Figure 3: Three-dimensional representation of the velocity field obtained by 4D flow MRI (A) and Tomo-PIV (B).

The PIV system provides great insight into the velocity field using an excellent acquisition quality. Stereo-PIV results demonstrated to be a faster and easier technique to measure the 3D velocity field. On the other hand, Tomo-PIV can take longer to calibrate and post-process but results reveal the complexity of the flow, demonstrating a better agreement with 4D flow MRI. Future work will include visualization of the wall movement and wall shear stress measurements. Further analysis of PIV results will allow evaluation of different flow regimes, such as turbulent flow and boundary layer, not possible with MRI due to the limited temporal and spatial resolution.

ACKNOWLEDGEMENTS

We are grateful to the Scientist Development Grant (AR), American Heart Association and to GE Healthcare for its support.

REFERENCES

[1] Roldán-Alzate, A., et al. *J Biomech*, 48:1325–1330, 2015.
[2] Jiang, J., et al. Med. Phys, 38:6300–12, 2011.
[3] Carlsson, M., et al. J. Cardi MR, 13:55, 2011.
[4] Day, S.W., et al. J. Biomech. Eng. 127:244–253, 2005.
[5] Töger, J., et al. Magn. Reson. Med. 75:1064–1075, 2016.
[6] Kitajima, H.D., et al. J. Biomech. Eng. 130:41004, 2008.
[7] Khodarahmi, I., et al. J. Magn. 1485:1477–1485, 2014.
[8] Buchmann, N.A., et al. Exp. Fluids50, 1131–1151, 2011.
[9] Yousif, M.Y., et al. Exp. Fluids 50:769–774, 2011.
[10] Johnson, K.M., et al. Magn. Reson. Med. 63:349–355, 2010.

SB³C2017
Summer Biomechanics, Bioengineering and Biotransport Conference
June 21 – 24, Tucson, AZ, USA

A NOVEL RIGHT–SIDE ASSIST IMPLEMENTATION COULD BRING POTENTIAL HEMODYNAMIC IMPROVEMENTS IN FONTAN PATIENTS

Ehsan Mirzaei[1], Minoo Kavarana[3], Dimitrios Georgakopoulos[4], Ethan Kung[1,2]

(1) Department of Mechanical Engineering
Clemson University
Clemson, SC, USA

(2) Department of Bioengineering
Clemson University
Clemson, SC, USA

(3) Division of Cardiothoracic Surgery
Medical University of South Carolina
Charleston, SC, USA

(4) Sunshine Heart, Inc.
Eden Prairie, MN, USA

INTRODUCTION

Infants born with only one functional ventricle need a 3-staged surgical intervention to survive. The majority of these children might suffer from heart failure at later ages. We believe that implanting a Ventricular Assist Device (VAD) at the junction of a patient's inferior vena cava (IVC) and pulmonary artery during the third stage of the operations known as Fontan procedure could improve their cardiac output and reduce their inferior vena cava pressure. The C-Pulse heart assist device manufactured by Sunshine Heart was selected for this purpose. The C-Pulse system which consists of a cuff and an external driver is typically used in patients with heart failure; the cuff is wrapped around the outside of the aorta, and is inflated and deflated in counterpulsation with the heart. In the proposed Fontan right-side assist application, the C-pulse device would be wrapped around the Fontan extra-cardiac graft and implanted together with the graft during the Fontan surgery. In order to investigate the physiologic impacts of such a device application, a computational physiology model was utilized. The aim of this study is to determine whether the current C-Pulse system can generate physiologic benefits, and to investigate device modifications to further improve patient hemodynamics.

METHODS

A previously published[1] lumped parameter network (LPN) describing a Fontan patient with weight=50kg and height=150cm at rest is utilized to model the systemic physiology (Figure 1). To model the Fontan right-side assist scenario, the pressure waveform generated by the C-Pulse system is prescribed to the Fontan graft and two one-way valves are added at the inlet and outlet of the graft to facilitate uni-directional flow. Another valve is added in the Superior Vena Cava (SVC) to protect the upper body from elevated pressure while the C-Pulse is pumping. To model these valves, non-linear resistors are used; the pressure drop across which is a quadratic function of flow, i.e.

$$\Delta P = kQ^2 \qquad (1)$$

A value of 1.0797×10^{-3} mmHg/(mL/s)^2 is used for the coefficient k which was obtained from literature[2]. The Fontan graft is modeled as a fluid reservoir with a volume of 16 mL. We assumed that the blood pressure inside the graft is equal to the pressure generated by the C-Pulse device except when the graft is completely filled. When the non-elastic graft is completely filled during device diastole, the C-pulse device has no effect on the flow and as a result the blood pressure in the graft is affected only by the upstream pressure.

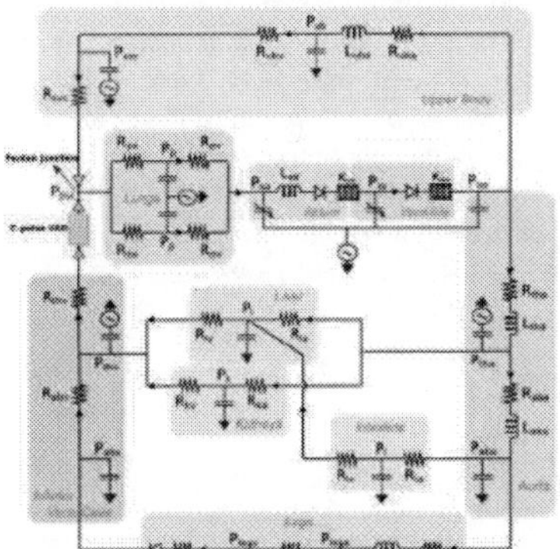

Figure 1: LPN model of the Fontan circulation with the C-Pulse device implemented for right-side support

RESULTS

The actual pressure waveform measured from the C-Pulse device is shown in Figure 2. Using this pressure waveform in our model and setting the VAD heart rate to the maximum of the operation limit of the current C-Pulse device (120BPM), we observed that no significant improvement in cardiac output could be obtained compared to the no-VAD reference case.

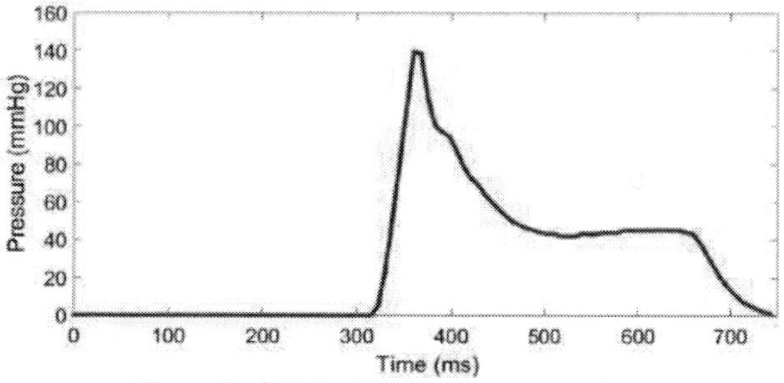

Figure 2: C-Pulse VAD pressure waveform

Therefore, we investigated the following hypothetical modifications to the VAD operation to identify potential device improvements that could result in hemodynamic benefits:

1) Increasing VAD heart rate:

The C-pulse heart rate was increased from 80 to 180BPM. We observed that the cardiac output increased with heart rate until 150BPM and remained constant thereafter (Figure 3). An 11.2% increase in cardiac output (from the reference no-VAD case) was observed at 150BPM.

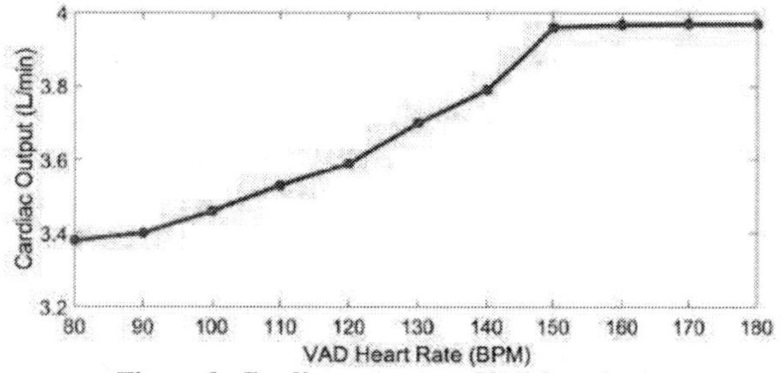

Figure 3: Cardiac output vs. VAD heart rate

2) Decreasing systolic-diastolic time ratio

Using a new pressure waveform generated by compressing the VAD systole from 57% of the VAD cycle down to 34%, we ran simulations for a range of VAD heart rates from 80 to 180BPM (Figure 4). Cardiac output continued to increase with VAD heart rate until 170BPM, where a 25.3% improvement (compared to the no-VAD reference case) was attained.

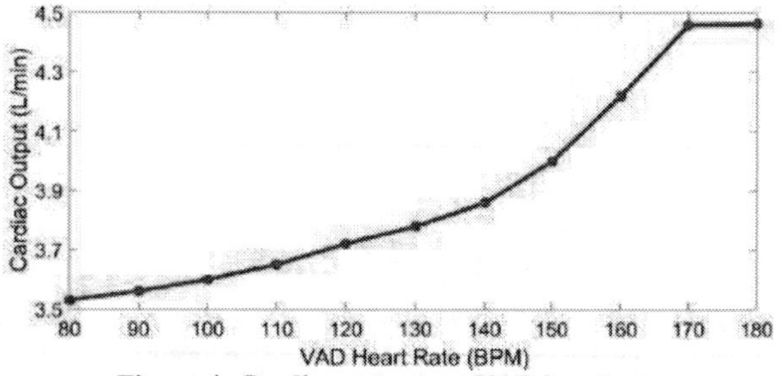

Figure 4: Cardiac output vs. VAD heart rate

3) Creating negative cuff pressure during diastole

We observed that the Fontan graft does not fill completely at high VAD heart rates (e.g. 160BPM), and thus investigated using a negative pressure of 5 mmHg during the diastolic period to help alleviate this issue. We obtained improved cardiac output by 18.5% at 160BPM compared to the no-VAD reference case.

	CO (L/min)	P_{IVC} (mmHg)
Reference (no-VAD)	3.56	13.33
VAD (160BPM)	3.97	10.77
VAD with negative diastole pressure (160BPM)	4.22	9.01

Table 1: Simulation results with negative filling pressure

DISCUSSION

Simulation results indicate that the current C-pulse device, even at maximum drive setting (120BPM), is unlikely to produce hemodynamic benefits for Fontan right-side support. Three different hypothetical modifications to device operation for improving cardiac output of a Fontan patient were examined. Increasing VAD heart rate up to 150BPM, decreasing systolic fraction to 34%, and generating 5 mmHg of negative pressure during diastole all produced clinically significant improvements in cardiac output and reductions in inferior vena cava pressure. In all cases examined there was a negative correlation between inferior vena cava pressure and cardiac output. Increased VAD heart rate improves cardiac output until incomplete graft filling limits further improvement. Extending VAD diastole leads to prolonged period of graft filling and enables further increased cardiac output. Creating a negative pressure in the cuff during diastole also aids filling by sucking blood into the graft. Operating C-Pulse device at 80BPM results in a cardiac output of 3.38 L/min (Figure 3) which is smaller than the cardiac output in the no-VAD reference case (3.56 L/min). This demonstrates that certain device settings could be counter-productive to Fontan support and a careful analysis should be performed for each patient if this application were to be adopted clinically.

While a number of assumptions were made in implementing the computational model of the C-pulse device, we have performed preliminary experimental study to validate the computational model. The results of this computational study suggest that by making proposed modifications to the C-Pulse driver, improvement in the cardiovascular performance of Fontan patients are attainable and the application of the C-pulse device for Fontan right-side assist could be a potential clinical option for single-ventricle palliation.

ACKNOWLEDGEMENTS

This work was supported by the Saving tiny Hearts Society, the American Heart Association, and The Children's Heart Foundation. We also acknowledge Sunshine Heart for providing C-pulse device data required to perform simulations.

REFERENCES

[1] Kung, E et al., *J Biomech Eng*, 136:081007-1-13, 2014.
[2] Vandervoort, P.M et al., *J Circulation*, 92:3464-3472, 1995.

SB^{3}C2017
Summer Biomechanics, Bioengineering and Biotransport Conference
June 21 – 24, Tucson, AZ, USA

NETWORK MODEL OF EXTRACELLULAR FLUID FLOW THROUGH RAT CEREBRAL CORTEX PARENCHYMA AND PERIVASCULAR SPACES

Julian Rey and Malisa Sarntinoranont

Mechanical Engineering
University of □lorida
□ainesville, □L, USA

INTRODUCTION

In animal models, dissolved compounds in the subarachnoid space (SAS) have been found to preferentially transport through the perivascular spaces (PVS) of vessels in the cortex at rates faster than would be expected from diffusion alone [1][2][3]. The PVS are extensions of the subpial space formed by cylindrical arrangements of glial cells that surround descending arterioles (DA) and ascending veins (AV) in the cortex [4]. Compounds injected into the parenchyma (PCY) move into the PVS improving clearance [4]. These findings suggest that a network of extravascular channels may serve as a means for facilitated transport of dissolved compounds and exchange between interstitial fluid and cerebrospinal fluid [5]. As such, it may substitute for an absent lymphatic vessel network in the parenchyma by collecting excess interstitial fluid and metabolic wastes. Though the PVS may play an important role in solute movement through the brain, the hydrodynamics and transport phenomena involved are unclear. These are essential aspects of our understanding of brain physiology and, more importantly, how changes in anatomy caused by disease, therapies, or environmental changes affect brain function.

The rate of imaging tracer transport from the SAS into the PVS of penetrating arterioles has been positively correlated with arterial pulsatility [2][3]. Pulsatility refers to the periodic changes in blood vessel volume caused by heart contractions. This pulsatility effect and slow rates of diffusion through tissue suggest a mechanism of convective transport generated by pulsatility may be involved.

In this study, a hydraulic network model of a representative section of cortical brain tissue was used to explore potential drivers of convective transport. Flow within tissue regions of the SAS, PVS, and parenchyma was modeled assuming idealized geometries with corresponding resistances to flow. The goal of this study was to model the effect of pulsatility on flow patterns in a cortical slice. Pulsatility derived flow drivers included (i) a difference in pulse amplitude between DAs and AVs, (ii) differences in pulse timing between DAs and AVs, and (iii) vessel wall peristalsis. The relationship between simulated flows and solute transport is also discussed.

METHODS

The hydraulic network model of a representative cortical slice included the PVS of a surface arteriole and vein, as well as the local SAS and parenchyma (PCY) (Figure 1). The rectangular parenchyma section was 300 µm tall, 200 µm wide and 200 µm deep. The local SAS was 100 µm tall and 200 µm deep. Both arteriole and vein outer diameters were 10 µm and the PVS formed a 1 µm thick layer around the vessels. Hydraulic resistances for the PVS and SAS were derived from the Navier Stokes solutions for viscous flow through a long, rigid annulus and extended parallel plates respectively. The parenchyma was assumed a rigid porous matrix governed by Darcy's law with hydraulic conductivity $2.05\text{x}10^{-12}$ m^2/Pa*s [6]. To account for the pulsatility of the surface arteriole and vein, sinusoidal flow sources were introduced along the perivascular channels (Figure 1). The flow rate of each arterial source was determined by first defining vessel radius variation with time [2] and proceeding to find the volume displacement rate as a function of time. To explore the effect of (i) a difference in pulse amplitude between descending arterioles and ascending veins, the venous flow sources produced a sinusoidal flow rate function with a smaller amplitude. To model (ii) a phase difference in arterial and venous pulse timing, time delays ranging from 0 s to 0.2 s were applied to venous PVS sources. The flow patterns produced by (iii) peristaltic vessel wall movement in the brain tissue section were simulated by orienting flow sources in series with the perivascular resistances and setting the flow rate according to the analytical solution in [7]. Simulations were performed in OrCAD 16.6 Lite (Cadence Design Systems, San Jose CA, 2015).

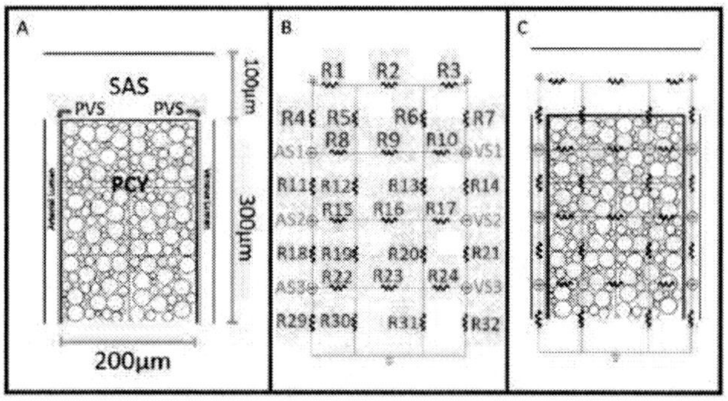

Figure 1 – A) Cross section of a representative brain tissue slice. The outermost boundary of the arterial and venous lumen are the centerlines of the arteriole and vein respectively. Subarachnoid space (SAS), perivascular space (PVS), and parenchyma (PCY) are indicated. B) Hydraulic network model of brain tissue section. Resistances are labeled R1-R32. Sinusoidal flow sources (circle with arrow) account for pulsatile expansion of the arteriole and vein (arterial sources labeled AS1-AS3; venous sources labeled VS1-VS3). Grounds where flow may enter or exit the network are found at either end of the local SAS, the bottom of the PCY, and the bottom of the PVS. C) Superposition of the tissue cross section and network model.

RESULTS

For the case when (i) sinusoidal flow sources in the arterial PVS were set to a larger amplitude than flow sources in the venous PVS, flow was observed through all resistances in the network. The flow rate in every region fluctuated with time like a sinusoid centered at zero (no net flow) with no phase difference between regions (Figure 2A). Peak flow rates in the perivascular spaces were greater than peak flow rates in the parenchyma because resistance in the former was lower than in the latter. Peak flow rates through regions near the artery were greater than regions near the vein because arterial pulsatility was greater than venous pulsatility.

When a phase difference was introduced between arterial flow sources and venous flow sources (ii), flow rate values through all regions remained sinusoidal with time, but the waveform amplitudes changed. Peak flow rate through horizontal resistances was maximized when the arterial and venous flow sources were 180 degrees out of phase, that is, when arterial expansion coincided with venous contraction. This trend was more pronounced at lower flow rates in regions further from the artery and vein, such as in R9 and R16.

Peristaltic wall movement (iii) produced flow that was mostly limited to the PVS in the direction of pulse propagation: away from the SAS in arterial PVS and toward the SAS in venous PVS. Flow rate into the parenchyma was low.

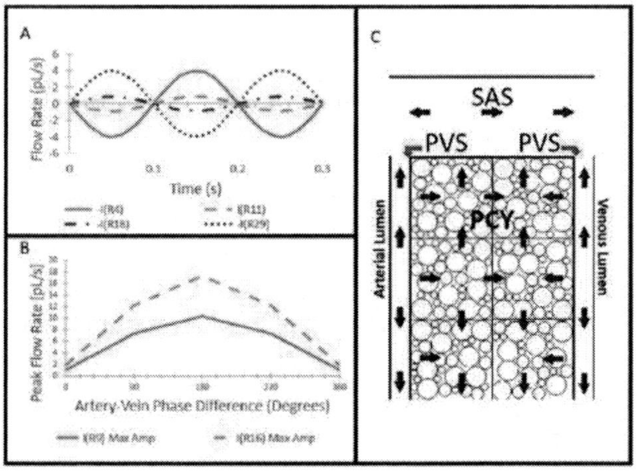

Figure 2 – A) Plot of flow rate versus time for the arterial perivascular resistances. Positive and negative flow rate values indicate flow away and toward the SAS, respectively. B) Peak flow rate in horizontal parenchyma resistances versus flow source phase difference in degrees. Peak flow rate was maximized when the arterial and venous flow source sources are perfectly out of phase. C) Arrow diagram depicting flow direction in the representative tissue slice during in-phase vessel expansion (0s-0.17s).

DISCUSSION

The uptake rate of dissolved compounds from the SAS into the perivascular spaces of penetrating cortical arteries and clearance rates from the cortical parenchyma cannot be explained by diffusion alone [4]. Convective transport (i.e. bulk flow) through the PVS related to vascular pulsatility has been forwarded as a rationale for observed transport rates.

In this study, a difference in pulsatility magnitude produced convective currents in the PVS and parenchyma but did not produce a net flow for convective solute transport. Varying pulse timing affected peak flow rate, but did not generate net flow over time. When flow through PVS corresponding to vessel wall peristalsis was introduced, very little penetrated the parenchyma because of its high hydraulic resistance relative to PVS. Under these flow conditions, tracer uptake could possibly be enhanced by the dispersive effect of the transient convection currents formed [8]. In short, spatial and temporal variations in velocity could improve mixing in cortical tissue giving rise to observed transport patterns despite low net flow in parenchyma.

REFERENCES

[1] Iliff, J., Nedergaard, M., *Science Translational Medicine*, 4(147), 2012

[2] Iliff, J., Nedergaard, M., *J. Neuroscience*, 33(46):18190-18199, 2013

[3] Rennels, M., Grady, P., *Brain Research*, 326:47-63, 1985

[4] Ichimura, T., Cserr, H., *Brain Research*, 545:103-113, 1991

[5] Abbott N.J., *Neurochemistry International*, 545-552, 2004

[6] Neeves K., Olbricht W., *J. Controlled Release*, 111:252-262, 2006

[7] Wang P., Olbricht W., *J. Theoretical Biology*, 274:52-57, 2011

[8] Deen W., *Analysis of Transport Phenomena, 2nd Ed.*, Oxford University Press, 2012

STOCHASTIC MODELING OF BIOTRANSPORT IN A TUMOR WITH UNCERTAIN MATERIAL PROPERTIES

M. Lu (1), A. Alexanderian (2), M. Salloum (3), L. Zhu (1), R. Ma (1), and M. L. Yu (1)

(1) Department of Mechanical Engineering,
University of Maryland Baltimore County,
Baltimore, MD 212□0, USA

(2) Department of Mathematics,
North Carolina State University,
□aleigh, NC 27□□□, USA

(3) E□treme Scale Data Science □ Analytics Department,
Sandia National Labs,
Livermore, CA □4□□0, USA

INTRODUCTION

Cancer treatments usually cannot kill or remove all cancerous tissues due to the irregular structures of tumors. Those remaining tumor cells have contributed to tumor recurrence and/or cancer metastasis. As a result, one grand challenge in cancer treatment is how to effectively deliver therapeutic agents to the entire tumor region with uncertain material properties [1]. However, most bio-transport models based on the Darcy's or Brinkman's law rely on the assumption that the tissue properties (e.g., permeability) are uniform inside tumors. Micro CT imaging following nanofluid infusion often reveals highly irregular distributions due to the spatial heterogeneity of the tissue structure [2].

It is of clinical significance to model enhanced convection drug delivery to tissues with heterogeneous material properties, and account for the uncertainty resulting from the structural randomness in the tissue. In the current study, Monte Carlo (MC) methods are used in our uncertainty quantification process. We apply this approach to determine the solution of the fluid pressure, velocity, and concentration distribution in a two-dimensional (2D) domain during a positive pressure infusion process. Quantifying the relationship between the tissue heterogeneity and the irregularity of the concentration contour would contribute to the clinical field, such as to obtain a predictive fidelity of the simulation of the drug delivery process in the porous tissue with uncertain material properties.

METHODS

Governing Equations

Consider the 2D Darcy's law constrained by mass conservation in the spherical coordinate system with a fixed polar angle $\theta = \pi/2$,

$$\frac{\partial}{\partial r}\left(\frac{\kappa r^2}{\mu}\frac{\partial p}{\partial r}\right) + \frac{\partial}{\partial \varphi}\left(\frac{\kappa}{\mu}\frac{\partial p}{\partial \varphi}\right) = 0, \qquad (1)$$

where p is the gauge pressure, κ is the permeability, and μ is the fluid viscosity.

During an injection process, the concentration field is governed by the convection-diffusion equation expressed as follows,

$$\frac{\partial C}{\partial t} + \nabla \cdot (\boldsymbol{v}C) = \nabla \cdot (D\nabla C) + R, \qquad with\, \boldsymbol{v} = -\frac{\kappa}{\mu\phi}\nabla p \qquad (2)$$

where C is the concentration, D is the diffusivity, R describes source or sink of concentration C, $\boldsymbol{v}$ is the interstitial velocity, and ϕ is the porosity. More details on the problem statement, including parameter setup, can be found in Ref. [3]. Eq. (1) is discretized with the 2nd-order central finite difference scheme, and solved using the implicit Gauss-Seidel iteration. The convection part of Eq. (2) is discretized with the 2nd-order upwind scheme, and marched in the temporal direction using the explicit two stage TVD Runge-Kutta scheme.

Statistical Models

The permeability κ is modeled as a log-normal random field with mode (i.e., the value that appears most often in a set of data) 0.5 md. A truncated Karhunen–Lòeve (KL) expansion is used to represent the random field $log(\kappa(\boldsymbol{x},\omega)) = a(\boldsymbol{x},\omega) = a_0(\boldsymbol{x}) + z(\boldsymbol{x},\boldsymbol{\xi}(\omega))$, where $\boldsymbol{x}$ is the coordinate of the spatial domain, ω is the random event, $\boldsymbol{\xi}$ is the multi-dimensional random variable, a_0 is the pointwise mean of the stochastic process, and z is a centered Gaussian process with pointwise variance $\sigma_a^2 = 0.25$. The correlation function for z is assumed as $c(\boldsymbol{x},\boldsymbol{y}) = \sigma_a^2 exp\left\{-\frac{1}{l}\|\boldsymbol{x}-\boldsymbol{y}\|_2\right\}$, where l is the correlation length. The porosity ϕ is modeled as a log-normal random variable with mode 0.2. A linear function is used to approximate $log(\phi(\omega)) = b(\eta(\omega))$ as $b(\eta) = b_0 + \sigma_b\eta$, where η is a standard normal random variable, and $\sigma_b = 0.5$ is the standard deviation.

Monte Carlo Methods

The MC methods are used to obtain realizations of pressure, velocity, and concentration. The solution mean can be estimated as

$$E[Y(\xi)] \approx \frac{1}{M}\sum_{i=1}^{M} Y(\xi_i), \quad (3)$$

where Y is the uncertain model observable, ξ_i are i.i.d. realizations of uncertain parameters, and M is the size of random samples.

Irregularity Measurement

Once the concentration contour is obtained, we will be able to locate the coordinates of the inner and outer boundary points of that irregular shape. Furthermore, we can calculate the area of that irregular shape by using the Heron's formula. Therefore, we can get an equivalent circle with the same area. The irregularity indicator is defined as the ratio between the circumference of the true boundary and the circumference of the equivalent circle. Clearly, a ratio of one suggests that the shape is perfectly round; a ratio larger than one implies that the shape is irregular. The larger the ratio is, the large the shape irregularity is.

RESULTS AND DISCUSSIONS

The permeability and pressure contours for the cases with correlation lengths of 0.5 mm and 10 mm are displayed in Figure 1(a)-(d), respectively. It is observed that small correlation length (e.g., $l = 0.5\ mm$) generates large randomness in permeability. This can further result in large pressure shape asymmetry. Comparisons of the irregular shape of the concentration contour with the equivalent circle are shown in Figure 2.

The objective of this study is to quantify the relationship between the tissue heterogeneity and the irregularity of the concentration contour. Note that averaged irregularity indicators evaluated from MC simulations are reported in Figure 3 and Figure 4. From Figure 2 and Figure 3, it is observed that when the correlation length decreases (the permeability field will become more fluctuated), the concentration contour will have a more irregular shape. From Figure 4, it is found that when the correlation length is fixed, as the number of KL modes increase, the irregularity would also increase. This is because that when the number of KL modes increased, highly fluctuated modes will be incorporated into the permeability field. As observed from Eq. (2), if convection is dominated in the flow, there is no damping effect on these highly fluctuated modes. More studies will be performed to study this phenomenon.

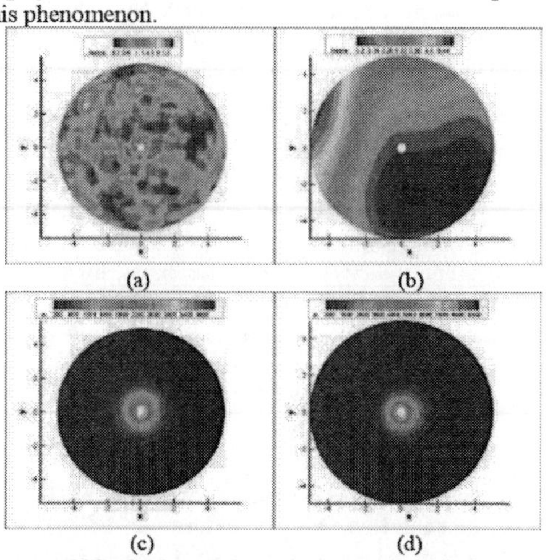

Figure 1: Permeability for the correlation length of (a) 0.5 (i.e., $l = 0.5\ mm$) and (b) $l = 10\ mm$, and the corresponding pressure fields for (c) $l = 0.5\ mm$ and (d) $l = 10\ mm$.

To summarize, the proposed statistical models are capable of modeling the heterogeneity of the tissue structure, and the irregularity indicators can quantitatively measure the material uncertainties in the tumor.

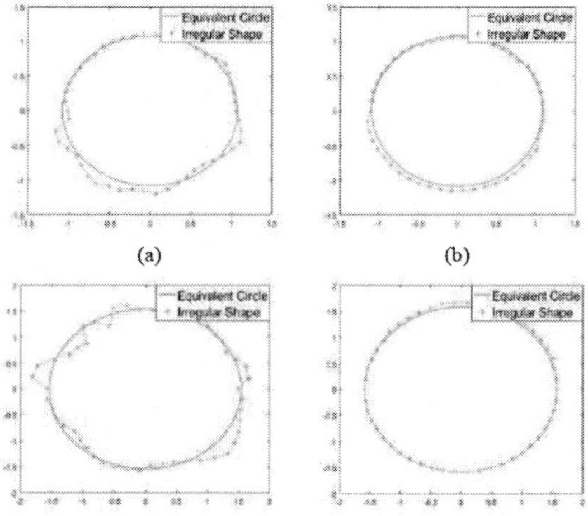

Figure 2: The simulated concentration contours vs. equivalent circle contour at 5 mins with (a) $l = 0.5\ mm$ and (b) $l = 10\ mm$, and at 15 mins with (c) $l = 0.5\ mm$ and (d) $l = 10\ mm$.

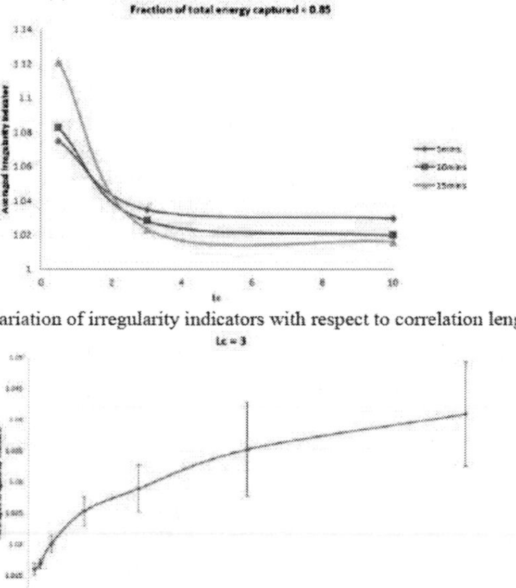

Figure 3: Variation of irregularity indicators with respect to correlation lengths.

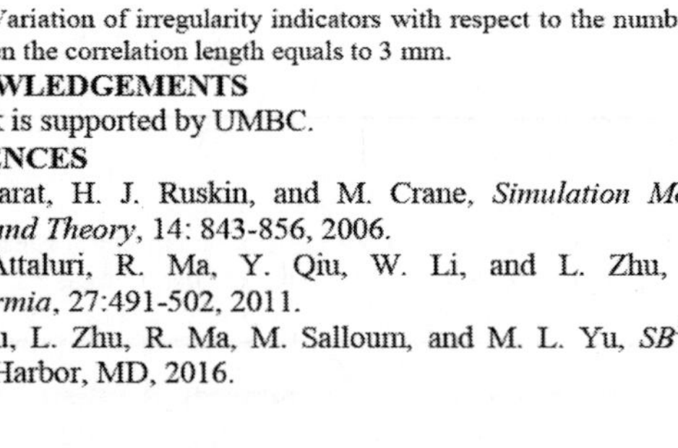

Figure 4: Variation of irregularity indicators with respect to the number of KL modes when the correlation length equals to 3 mm.

ACKNOWLEDGEMENTS

This work is supported by UMBC.

REFERENCES

[1] A. Barat, H. J. Ruskin, and M. Crane, *Simulation Modelling Practice and Theory*, 14: 843-856, 2006.

[2] A. Attaluri, R. Ma, Y. Qiu, W. Li, and L. Zhu, *Int. J. Hyperthermia*, 27:491-502, 2011.

[3] M. Lu, L. Zhu, R. Ma, M. Salloum, and M. L. Yu, *SB^{3}C2016*, National Harbor, MD, 2016.

SB³C2017
Summer Biomechanics, Bioengineering and Biotransport Conference
June 21 – 24, Tucson, AZ, USA

EXTRACELLULAR MATRIX COMPOSITION MODULATES THE MIGRATORY RESPONSE OF BREAST CANCER CELLS IN A 3D MICROFLUIDIC CULTURE

Karina M. Lugo- Cintrón (1), Lucas Tomko (2), Patrick N. Ingram (1), Patricia Keely (3), David J. Beebe (1)

(1) Department of Biomedical Engineering
University of Wisconsin
Madison, Wisconsin, USA

(2) Molecular and Cellular Pharmacology
University of Wisconsin
Madison, Wisconsin, USA

(3) Cell and Regenerative Biology
University of Wisconsin
Madison, Wisconsin, USA

INTRODUCTION

The composition of the breast tumor microenvironment (Fig. 1A) greatly influences cancer progression, leading to differential invasion, migration, and metastatic potential [1,2]. During their interaction with the microenvironment, cancer cells start a crucial crosstalk with stromal cells (such as fibroblasts) that will define tumor outcomes. In addition, biophysical and biochemical cues, such as those from the extracellular matrix (ECM), greatly alter cell invasion [3,4]. However, these complexities are often overlooked in traditional *in vitro* migration models [5]. Creating reproducible models that incorporate organotypic structure, ECM, and stromal support have the potential to improve migration and invasion assays. Here, we have developed a microfluidic breast cancer migration platform, which utilizes a luminal structure filled with a mixture of cancer cells and collagen, and enables the testing of heterogeneous ECMs with and without stromal components (Fig. 1A). Incorporation of breast cancer cells (MDA-MB-231) into different stromal cultures (human mammary fibroblasts (HMF) and cancer associated fibroblasts (CAF)) in a 3D microenvironment with various ECM proteins (Laminin, Fibronectin, Trombospondin-1) is easily achievable and more closely mimics *in vivo* invasion conditions. Furthermore, preliminary screening of ECM and stromal effect on breast cancer migration have been guided and informed by patient ECM proteomic data that identifies matrix components upregulated in tumor microenvironments. Future work will focus on high throughput screens of cancer cell, ECM, and stromal influences on breast cancer metastasis.

METHODS

Microfluidic Device Fabrication & Experimental Set Up: Fabrication of LumeNEXT devices was performed as has been previously published [6] and LumeNEXT set up is shown in Fig. 1B. Here, the luminal structure was filled with a solution of cancer cells embedded in a collagen hydrogel and is surrounded by an extracellular matrix as seen in Fig. 1A.

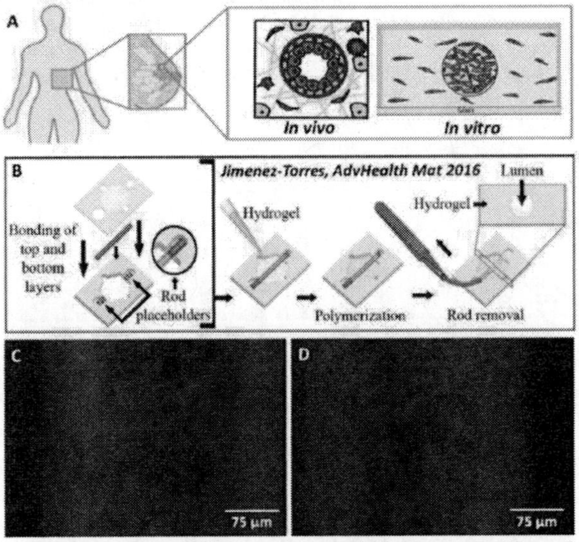

Figure 1: Breast Microenvironment, Device Assembly and Stromal Influence in Organotypic Model. (A) Breast microenvironment composition found in vivo can be incorporated *in vitro*. (B) LumeNEXT Assembly. Lung endothelial lumen with fibroblasts in the matrix (C) and lung endothelial lumen (D) with no stromal cells present in the matrix.

Poster Presentation #P60

Hydrogel Preparation: High-density rat-tail collagen (Corning) was diluted with 5X PBS and neutralized with 0.5 M NaOH to a final concentration of 3 mg/ml collagen and a pH of 7.4. For ECM studies, Laminin (LN), Fibronectin (FN), and Trombospondin-1 (TSP-1) were added into the collagen mixture before polymerization or as a lumen coating for 30 minutes before cancer cells were added. LN was added at a final concentration of 100ug/ml, FN at a final concentration of 100ug/ml, and TSP-1 at a final concentration of 1.8ug/ml. The 3 mg/ml collagen solution was diluted 1:4 in media (for the monoculture control) or with 5,000 cell/µl human mammary fibroblast cell or cancer associated fibroblast cell solution diluted in media (for the co-culture).
Cell Seeding: High-density rat-tail collagen was diluted with 5X PBS and neutralized with 0.5 M NaOH to a final concentration of 1.5 mg/ml collagen and a pH of 7.4. Then, a 1.5 mg/ml collagen solution was diluted to a 1:2 ratio with 50,000 cell/ul MDA-MB-231 cell solution. 2 ul of the collagen solution with cancer cells was pipetted into the LumeNext port to create a cancer-filled luminal structure.
Imaging and Cell Migration Analysis: MDA-MB-231 invasion was imaged with a Nikon Titanium for 5 days. The migration distance was obtained from bright-field and fluorescent images using ImageJ-Fiji on Day 2. Distance traveled was calculated measuring the total cell distance from the lumen border to the final cell location and was averaged across all migrating cells per condition.

RESULTS

As cancer cell migration potential is heavily influenced by the tumor microenvironment, the creation of *in vitro* assays capable of more accurate recapitulation of the microenvironment have the potential to advance our understanding of these critical progression events. In previous investigations of endothelial biology, we found structural elements (lumen) and the addition of stromal signaling were both critical to create *in vitro* vessels with proper endothelial morphology (Fig. 1B, C). Similarly, we observed that ECM composition and stromal signaling worked in tandem to more greatly alter cancer cell invasion distance (Fig. 2, Fig. 3). Furthermore, stromal cells enhance the migration of cancer cells when ECM proteins such as Laminin, Fibronectin and TSP-1 are present. Cancer cells were found to migrate farther when fibronectin and cancer associated fibroblasts were present in the matrix as shown in Fig. 2 and Fig. 3.

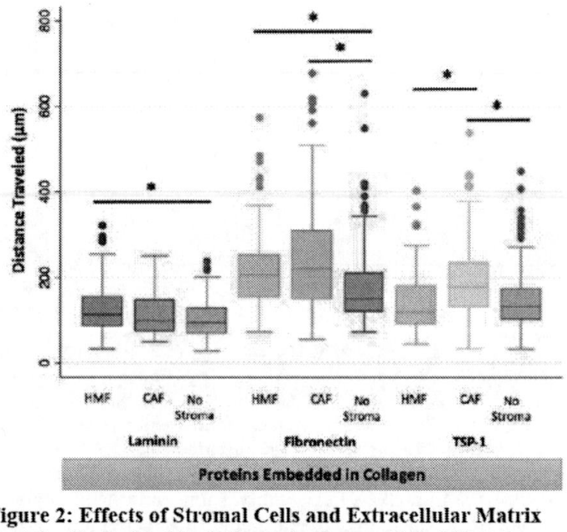

Figure 2: Effects of Stromal Cells and Extracellular Matrix Proteins Embedded in Collagen in Cancer Cells Migration.

The distance traveled by cancer cells were analyzed by comparing the stromal cell component when a specific protein was present in the matrix. Significance differences ($P<0.05$) were found in between different stromal cells as shown in Figure 2 and Figure 3

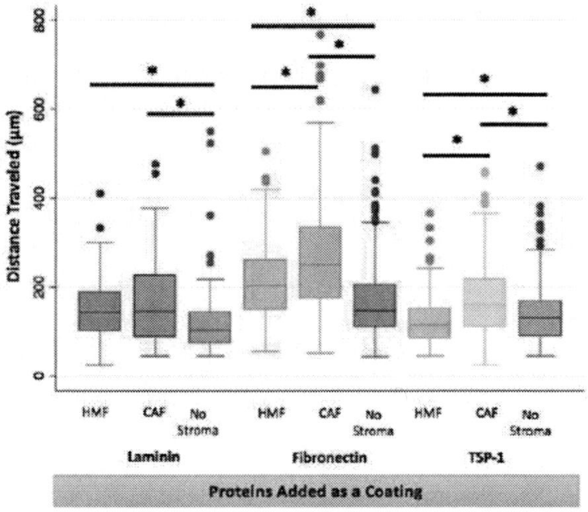

Figure 3: Effects of Stromal Cells and Extracellular Matrix Proteins Added as a Coating in Cancer Cells Migration.

DISCUSSION

The creation of a model of breast cancer invasion capable of modulating ECM protein composition (LN, FN, TSP-1) and incorporating stromal cells (HMFs and CAFs) in a 3D microenvironment was successful. Stroma cells exposed to different ECM proteins differentially regulated the migration of cancer cells. In all ECM conditions in which when no stromal cells are present in the extracellular matrix, cancer cells migrate the least distance. However, when HMFs are present in the extracellular matrix, cancer cell migration distance increases. Following this trend, CAF stromal signaling further enhanced migration. This trend is statistically significant in most conditions. Patient proteomic data showed increased presence of FN and TSP-1 and decreased LN in tumor microenvironments, and our model, following this expectation, showed TSP-1 and fibronectin resulting in greater increases in invasion when compared to laminin. Moreover, when comparing embedded proteins with the proteins added as a coating, proteins added as a coating promoted cancer cells to migrate farther, except when TSP-1 was present. Further experiments need to be done to understand the crosstalk between cancer and stromal cells and to optimize ECM component concentrations based on the proteomic estimates. In summary, this model provides a promising method for examining cancer cells-stromal interactions.

ACKNOWLEDGEMENTS

Thanks to Molly Morgan for Fig. 1A. This work was supported by NIH R01EB010039 BRG and the University of Wisconsin-GERS.

REFERENCES

[1] Multhaupt, H et al., *Adv. Drug Deliv. Rev.*, 97:28-40, 2016.
[2] Oskarsson, T et al., *Breast J.*, 22:S66-S72, 2013.
[3] Conklin, M et al.,*Cell Adh Migr*, 6(3):259-260, 2012.
[4] Mao, Y et al., *Cancer Metastasis. Rev.*, 32:303-315, 2013.
[5] Barney, L et al., *Integr Biol (Camb).*, 7(2):198-212, 2015.
[6] Jiménez, J et al., *Adv Healthc Mater.*, 5(2):198-204, 2016.

SB³C2017
Summer Biomechanics, Bioengineering and Biotransport Conference
June 21 – 24, Tucson, AZ, USA

CANCER ASSOCIATED FIBROBLAST-INDUCED SPATIOTEMPORAL CONTRACTION IN PANCREATIC DUCTAL ADENOCARCINOMA

Michael J. Bradney (1), Yi Yang (2), Stephen F. Konieczny (2), Bumsoo Han (1,3)

(1) School of Mechanical Engineering
Purdue University
West Lafayette, IN, USA

(2) Department of Biological Sciences
Purdue University
West Lafayette, IN, USA

(3) Weldon School of Biomedical Engineering
Purdue University
West Lafayette, IN, USA

INTRODUCTION

Pancreatic ductal adenocarcinoma (PDAC) is the fourth most common cause of death from cancer, with a fatality rate of 95%. Early detection of this disease is difficult, and patients with metastatic pancreatic cancer have a median life expectancy of just 3-6 months[1,2]. Even when a tumor is identified, treatment is notoriously difficult due to unsuccessful delivery of chemotherapeutic drugs and resistance to these agents[3].One explanation for this challenge in treatment of PDAC is thought to be caused by dense stromal tissue surrounding the tumor, characterized by an abundance of cancer-associated fibroblasts (CAFs), excess extra-cellular matrix (ECM) proteins, immune cells, and inflammatory cells[4]. More recent research suggests that not only does the dense stroma of PDAC make it more difficult to treat, but it may also play a pivotal role in disease progression. In response to small mechanical loads from their surrounding tumor microenvironment, CAFs tend to secrete additional ECM proteins such as collagen and fibronectin and directly remodel the ECM[5]. While this heavily remodeled ECM has become a key component of PDAC pathology, there is still debate as to how exactly it affects tumor development. Some suggest that the presence of CAFs, through both solid stress and paracrine signaling, limits tumor growth and cancer cell metastasis[6–8]. Others report that the presence of CAFs promotes tumor malignancy through both ECM remodeling and crosstalk with cancer cells[9–11].

These conflicting viewpoints suggest that current models are not sufficient for determining the role of CAFs in PDAC. Two-dimensional cultures oversimplify the tumor microenvironment, and *in vitro* mouse models are time-consuming, expensive, and difficult to control[12]. To span this gap, we have developed an *in vitro* tumor model, the tumor-microenvironment-on-a-chip (TMOC), which allows us to study the complex tumor environment of PDAC in a controlled fashion. This platform allows us to replicate PDAC *in vitro* by recreating the 3D histological architecture of the pancreatic duct surrounded by stromal tissue where CAFs are embedded in a collagen matrix. In this study we utilize the TMOC to show the correlation between CAF contractility and cancer cell behavior in PDAC. Based on our experimental results, we hypothesize that the mechanical presence of CAFs in the tumor microenvironment limits the ability of pancreatic cancer cells to invade the adjacent extracellular matrix and metastasize.

METHODS

Mouse PDAC cells (KPC-2) and CAFs were cultured in RPMI-1640 culture medium supplemented with 5% fetal bovine serum and 1% penicillin streptomycin. For device loading, cells were trypsinized using 0.05% Gibco Trypsin-EDTA. A cell suspension of 20,000 cells/μL was used to seed cancer cells on the devices. Microfluidic devices were created using polydimethylsiloxane (PDMS) soft lithography. Two mating PDMS pieces were surface-treated with a corona discharge and bonded to form a complete device (**Figure 1A**).

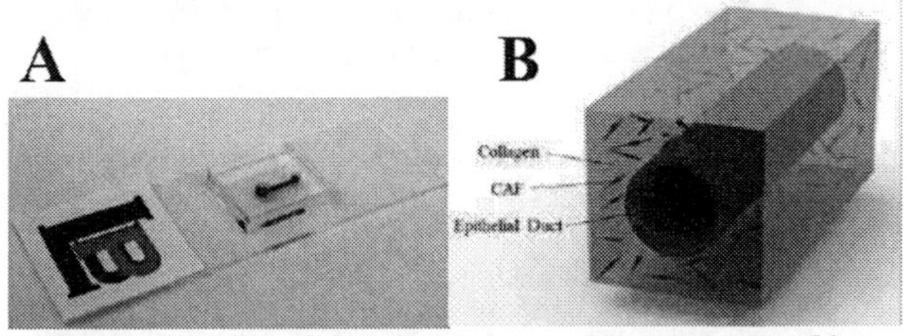

Figure 1: A) Complete tumor-microenvironment-on-a-chip. B) 3D rendering of the artificial tumor microenvironment.

6.0 mg/ml Type I collagen was created from a stock solution of rat tail collagen. CAFs were seeded in collagen at concentrations of 0, 1500 or 2500 cells/µL. 20 µL of collagen was loaded into the device channel, and a single 20 µL droplet of culture medium was added to the collagen, creating a patent lumen via viscous fingering[13]. The devices were allowed to incubate at 37C for 15-20 minutes to allow the collagen gel to polymerize. Devices were perfused with a cancer cell suspension to coat the inner lumen, forming a 3D ductal structure **(Figure 1B)**

Strain was calculated by comparing the difference between the cross-sectional width of the collagen and the cross-section of the lumen at a given time t (δ_t) to the initial difference at $t = 0$ (δ_0) (Equation 1). Measurements for each device were taken at a minimum of three cross-sections along the channel. Lumens were assumed to be radially uniform at each cross-section.

$$Strain\,(\%) = 100 \times \frac{(\delta_t - \delta_0)}{\delta_0} \qquad (1)$$

RESULTS AND DISCUSSION

Cells were cultured on these devices for four days with bright field images taken every day to visualize the temporal change in the tumor microenvironment. **Figure 2** shows a KPC-2 monoculture compared to a co-culture with RFP-fluorescent CAFs. The most apparent change with added CAFs is the collapse of the collagen matrix due to contractile forces **(Figure 2, bottom right)**.

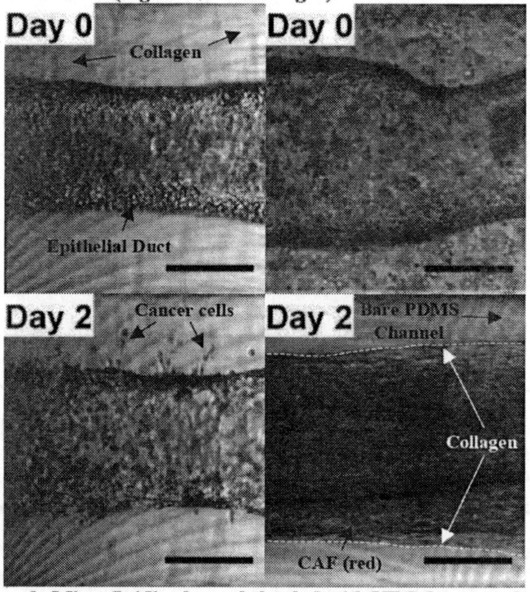

Figure 2: Microfluidic channels loaded with KPC-2 monocultures (left) and KPC-2 co-cultures with RFP-CAF (right). CAFs shown at 2500 cells/µL. Scale bars represent 250 µm.

These images show how the CAFs change morphology and tend to align along the direction of the lumen, suggesting the direction of minimal stress. Concomitantly, this implies a significant stress occurring in the radial direction due to the compaction of the matrix. This radial stress may play a role in preventing cancer cell invasion into the adjacent ECM. Observable changes in the microenvironment were quantified by calculating the radial strain imparted by the cells on the collagen **(Figure 3)**. For samples with high CAF concentration, collagen strain is initially positive, indicating that the collagen is stretching as the lumen is contracting but the collagen is still attached to the PDMS chamber walls. By Day 2, the collagen detaches and the strain becomes negative, indicating that the collagen is now compacting due to the contractile forces of both the cancer cells and CAFs. For KPC-2 monocultures and samples with low CAF concentrations, the strain is increasingly positive as the lumen contracts while the collagen remains attached to the PDMS.

Another key difference in the samples was the invasiveness of the cancer cells. In KPC-2 monocultures and samples with low CAF concentrations, it was common to see cells invading into the adjacent ECM **(Figure 2, left)**, while there was no such invasiveness seen in samples with high CAF concentrations.

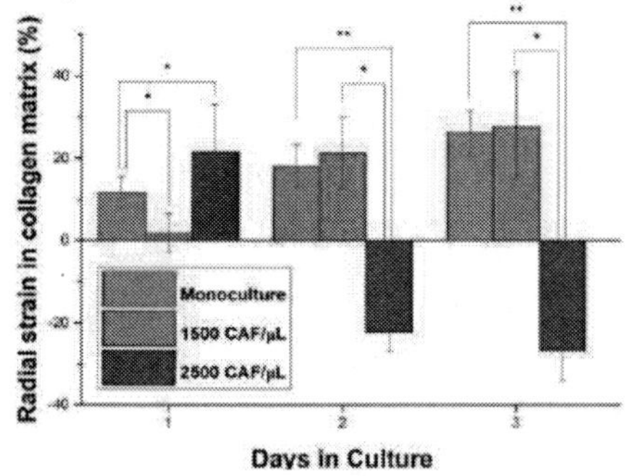

Figure 3: Strain imposed on collagen matrix due to cellular contraction. *=p<0.05, **=p<0.001

By varying the CAF concentration within the collagen matrix, we were able to study how the cells altered the mechanical environment. Additional CAFs caused a compacted matrix within 2 days of culture and limited cancer cell invasion into the collagen matrix. These higher CAF concentrations, more relevant to the physiological PDAC conditions[4], suggest the significance of stromal-induced stress *in vivo*. By mimicking the 3D structure and biological makeup of the PDAC microenvironment, we believe our TMOC serves as a valuable tool for the study of the role of CAFs in tumor development.

ACKNOWLEDGEMENTS

This research was partially support by the Purdue University Center for Cancer Research, NIH grant P30 CA023168.

REFERENCES

[1] Yadav, D. & Lowenfels, A. B. *Gastroenterology* **144,** 1252–1261 (2013). [2] Wachsmann, M. B., Pop, L. M. & Vitetta, E. S. *J. Investig. Med. Off. Publ. Am. Fed. Clin. Res.* **60,** 643–663 (2012). [3] Longo, V. *et al.*. *Oncotarget* **7,** 58649–58658 (2016). [4] Carr, R. M. & Fernandez-Zapico, M. E. *EMBO Mol. Med.* **8,** 80–82 (2016). [5] Butcher, D. T., Alliston, T. & Weaver, V. M. *Nat. Rev. Cancer* **9,** 108–122 (2009). [6] Helmlinger, G., Netti, P. A., Lichtenbeld, H. C., Melder, R. J. & Jain, R. K. *Nat. Biotechnol.* **15,** 778–783 (1997). [7] Özdemir, B. C. *et al. Cancer Cell* **25,** 719–734 (2014). [8] Rhim, A. D. *et al. Cancer Cell* **25,** 735–747 (2014). [9] Paszek, M. J. *et al. Cancer Cell* **8,** 241–254 (2005). [10] Karagiannis, G. S. *et al. Mol. Cancer Res.* **10,** 1403–1418 (2012). [11] O, D. W., P, D., M, M. & M, B. *Int. J. Cancer* **123,** 2229–2238 (2008). [12] Yamada, K. M. & Cukierman, E. *Cell* **130,** 601–610 (2007). [13] Bischel, L. L., Lee, S.-H. & Beebe, D. J. *J. Lab. Autom.* **17,** 96–103 (2012).

SB³C2017
Summer Biomechanics, Bioengineering and Biotransport Conference
June 21 – 24, Tucson, AZ, USA

DISTRIBUTION OF ENCAPSULATED CELLS IN A PHASE-SEPERATED ORMOSIL GEL TO OPTIMIZE BIODEGRADATION

Joey J. Benson (1), Lawrence P. Wackett (2,3), Alptekin Aksan (1,3)

(1) Department of Mechanical Engineering
University of Minnesota
Minneapolis MN, USA

(2) Department of Biochemistry, Molecular Biology, and Biophysics
University of Minnesota
Minneapolis, MN, USA

(3) BioTechnology Institute
University of Minnesota
St. Paul, MN, USA

INTRODUCTION

Bioremediation has emerged as a cost-effective and environmentally friendly method for the removal of toxic chemicals from the environment. Encapsulation of the biodegrading cells in a three-dimensional matrix allows the cells to retain their activity, but isolates them from the environment and protects them from predation and environmental stressors. Silica gel is an ideal candidate for bioencapsulation due to its favorable mechanical properties, chemical stability, and biocompatibility [1]. Silica gels can further be organically modified to introduce a wide range of chemical groups into the gel, allowing the chemical and physical properties of the gel to be tuned.

Organically modified silica (ORMOSIL) gel has been shown to be superior to inorganic silica gels with regard to transport properties, biocompatibility, and adsorption of organic molecules [2]. Enhanced adsorption in ORMOSIL gels can provide pre-concentration of substrate within the gel, leading to enhanced biodegradation rates. Our lab has shown that recombinant *E. coli* encapsulated in ORMOSIL gels containing phenyltriethoxysilane (PTES) or methyltriethoxysilane (MTES) have greater atrazine biodegradation rates than *E. coli* encapsulated in inorganic silica gels. However, a direct correlation between organic precursor content in the gel, and biodegradation was not observed, and the lack of correlation was attributed to changes in the localization and distribution of organic groups in the gel.

The distribution of organic groups in the ORMOSIL gels was visualized using the hydrophobic probe Nile red. Nile red revealed that the ORMOSIL gels were heterogeneous, and contained hydrophobic aggregates rich in organic groups that formed due to phase separation. The concentration and size of these aggregates could be controlled by changing organic content in the gel. Confocal images of *E. coli* expressing green fluorescence protein (GFP) encapsulated in ORMOSIL gels showed that *E. coli* adhered to the surface of the hydrophobic aggregates, likely due to a hydrophobic interaction.

The hydrophobic aggregates are expected to adsorb greater concentrations of the hydrophobic substrate than the surrounding gel, locally increasing the concentration of substrate close to the *E. coli* surface. We hypothesize that the local increase in the concentration of the substrate is the main reason for the enhanced biodegradation by cells that are in contact with the aggregates. Therefore, maximizing the contact between the encapsulated cells and the aggregates is expected to maximize the biodegradation rate of the gel. This "watering hole" hypothesis could explain enhanced biodegradation in ORMOSIL gels discussed previously

Because *E. coli* is approximately 1.5-2μm in length and 0.5 μm in diameter, we hypothesized that there should exist an optimum aggregate size that causes the most bacteria per gel volume to be in contact with the hydrophobic aggregates. Therefore, in this study, a model was developed to determine the optimum aggregate size, so that gels with enhanced biodegradation rates can be designed.

METHODS

The model developed here calculates the number of bacteria (approximated as cylinders with spherical caps) that can be packed onto the surface of a spherical hydrophobic aggregate. The input parameters are the non-dimensionalized bacteria length (L^*) and width (w^*), the angle of contact between the bacteria and the sphere (Ψ), and the fraction of the bacteria cell that sticks to the contour of the sphere (s):

$$L^* = L/R \qquad w^* = w/L \tag{1}$$

where L and w are the bacteria length, and width, respectively and R is the aggregate radius.

The model divided the sphere into n cross sections occupying the same surface angle (Φ) as w^* as:

$$\Phi = 2\sin^{-1} {w^*}/{2} \qquad n = {\pi}/{\Phi} \tag{2}$$

The model then calculated each cross-sectional radii r_{ci} and determined the number of bacteria (B_{ci}) that fit around each cross section by determining the angle (θ) occupied by L^* as follows:

$$\theta = 2\tan^{-1} {L^*}/{2r_{ci}^*} \qquad B_{ci} = \theta/2\pi \qquad r_{ci}^* = r_{ci}/R \tag{3}$$

The total number of bacteria that can fit on the sphere is calculated by summing the number of bacteria on all cross sections as follows:

$$\mathrm{B} = \sum_{i=1}^{n} B_{ci} \tag{4}$$

The aggregate size that allowed maximum contact area between bacteria and the aggregate surface was determined by the merit function, M. The merit function breaks a total amount of aggregate volume V, into z spheres of equal radius r and calculates the total number of bacteria on all spheres. It then normalized this value by the number of bacteria that could be packed on one large sphere of volume V and radius R as follows:

$$M = {\mathrm{zB(r)}}/{\mathrm{B(R)}} \tag{5}$$

RESULTS

Bacteria packed onto a sphere as described by the developed model are shown in **Figure 1** in green. The sphere appears to be completely and uniformly covered by cells, and there does not appear to be room for more bacteria.

Figure 1: Graphic representation of the maximum number of bacteria than can be in contact with a spherical hydrophobic aggregate (L^*=0.5, w^*=0.25, B=180).

The maximum number of bacteria that can be in contact with a spherical aggregate is shown in **Figure 2** as a function of L^*. At low values of L^*, B was very large, reaching 1,200 bacteria at $L^* = 0.2$. B decreased rapidly with L^* initially down to 46 at $L^* = 1$. B decreased more gradually with L^* for $L^* > 1$ and reached a value of 14 at $L^* = 2$.

A contour plot of the merit function is given in **Figure 3** as a function of L^* and r/R. The plot contains two bands with large merit functions, moving from the lower left to upper center of the plot. Choosing an aggregate size in one of these bands would be ideal for maximizing the contact between bacteria and the spherical aggregates. For example, the location of crosshairs on **Figure 3** has a merit function of 7.92 ($L^* = 1.10$, $w^* = 0.25$, and $r/R = 0.28$), which would be an ideal aggregate size to maximize biodegradation rates in a gel.

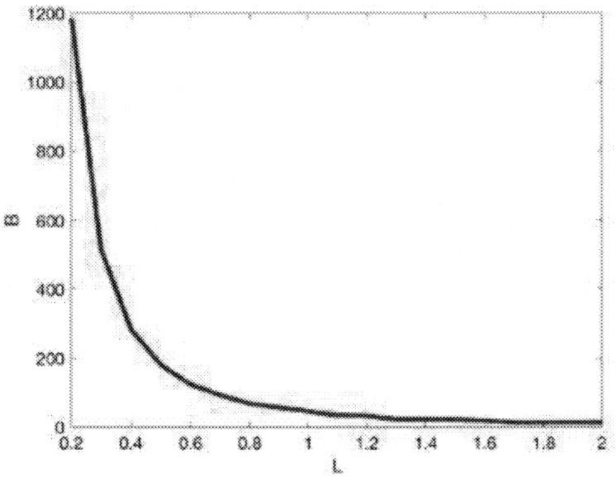

Figure 2: Number of bacteria that can stack on a sphere as a function of L^*

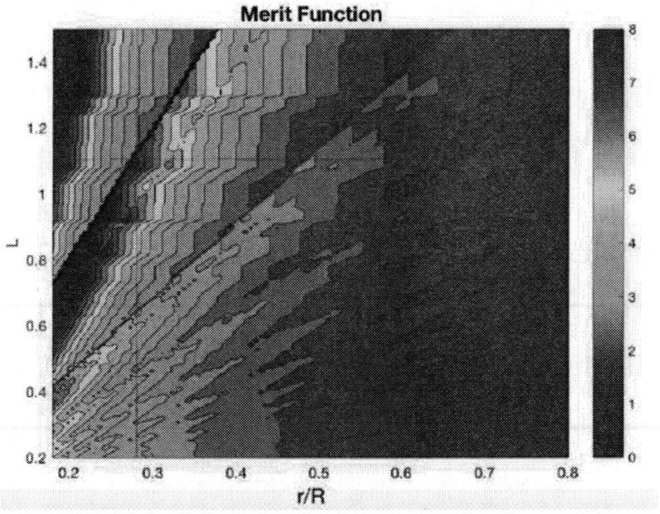

Figure 3: Contour plot of the merit function

DISCUSSION

The model has calculated the optimum aggregate size that creates maximum contact between bacteria and aggregates. Future studies will determine if designing gels with this aggregates size enhances the biodegradation rate in the gel.

REFERENCES

1. Mutlu et. al., J Mater Chem A 1,11051, 2014
2. Ciriminna R., et al., Soft Matter 1(3):231-237, 2005

Poster Presentation #P62

SB[3]C2017
Summer Biomechanics, Bioengineering and Biotransport Conference
June 21 – 24, Tucson, AZ, USA

CHARACTERING INTRACELLULAR ICE FORMATION DURING FREEZING AND THAWING OF LYMPHOBLASTS USING LOW TEMPERATURE RAMAN SPECTROSCOPY

Guanglin Yu (1)

Allison Hubel (1)

(1) Department of Mechanical Engineering
University of Minnesota
Minneapolis, MN, USA

INTRODUCTION

Cells used to treat disease have the potential to treat up to 100 million Americans per year[1]. Poor or inadequate methods of preservation have limited the ability to use cells to treat disease. Current theories of cryopreservation were developed in the 1960's[2]. Most cell types exhibit an 'inverted U-shaped' variation in survival with cooling rates. At high cooling rates, it has been hypothesized that the formation of ice inside the cell (i.e. intracellular ice formation, IIF), causes loss of viability. At low cooling rates, it is postulated that high concentration of electrolytes or exceeding of a minimum tolerable cell volume results in damage to cells[3,4]. Thawing is just as critical as cooling for cell survival. The warming rate that results in highest cell survival depends on the cooling rate that is used to freeze cells. Cells frozen at high cooling rates have small ice crystals with high surface energy and tend to grow when cells are thawed[5]. Growth of ice crystals can result in mechanical damage to the cells[6].

Freezing is a stochastic process, and individual cells in the population may respond to freezing differently. Confocal Raman spectroscopy have been an important tool in single cells analysis of freezing response. The objective of this investigation is to use low temperature Raman spectroscopy to analyze single cells and correlate their freezing and thawing response to the distribution of ice inside each cell. This analysis can advance our understanding of freezing and thawing damage of individual cells within population.

METHODS

Jurkat Cell Culture and Freezing Studies

The studies involved the use of Jurkat cells (ATCC, TIB-1522), which were incubated at 37C with 5% CO2 in media composed of high glucose RPMI 1640, and 10% fetal bovine serum. Cells were grown in suspension and maintained at a concentration of 1-2 x 10^6 cells/ml.

Cells samples were prepared by washing cells twice in Dulbecco's Phosphate Buffered Saline (DPBS). The cells were then suspended in each experimental solution of interest and frozen on the thermally controlled stage described previously[7]. For thawing studies, cells were first frozen on the stage and thawed afterwards to different holding temperatures (-50, -40, -30°C).

Confocal Raman System

Confocal Raman spectroscopy measurements were conducted using Witec Confocal Raman Microscope System Alpha 300R (WITec, Ulm, Germany) with a UHTS300 spectrometer and DV401 CCD detector. A wavelength of 532 nm Nd:YAG laser powered at 10 mw was used as excitation source. A 100X air objective (NA 0.90; Nikon Instruments, Melville, NY) was used for focusing the 532 nm excitation laser to the sample. Data collection and analysis were performed by Windows-based WitecControl_1.38 software.

Raman images were assembled by integrating the spectrum at each pixel based on characteristic wavenumbers of common intracellular and extracellular materials (ex. water, organics and ice). Images of ice were generated by integrating the peak centered at 3125 cm^{-1} [7]. The release of cytochrome c from mitochondria will be used in this investigation to quantify cell viability. For these experiments, the Raman peak centered at 1127 cm^{-1} was used to generate Raman images of cytochrome c.

RESULTS

Figure 1 showed confocal Raman images of ice for cells frozen at 1°C/min, 10°C/min and 50°C/min. Intracellular ice is formed in all of the cells. The amount of intracellular ice increased with increasing cooling rate. At the same cooling rate of 10°C/min, all cells cryopreserved in different DMSO concentration also showed IIF. The amount of IIF was clearly decreased as DMSO concentration increased. Cells cryopreserved in 0% DMSO (DPBS) solution showed large chunks of IIF, which were different from all the other cells that showed small ice crystals.

Figure 1: Raman images of ice for cells cryopreserved in 10% DMSO at different cooing rates (1, 10, 50°C/min) as well as cells cryopreserved at 10°C/min in different concentration of DMSO (0, 5, 10%). Ice formation in the extracellular solution was seeded at -6°C and the sample was cooled down to -50°C.

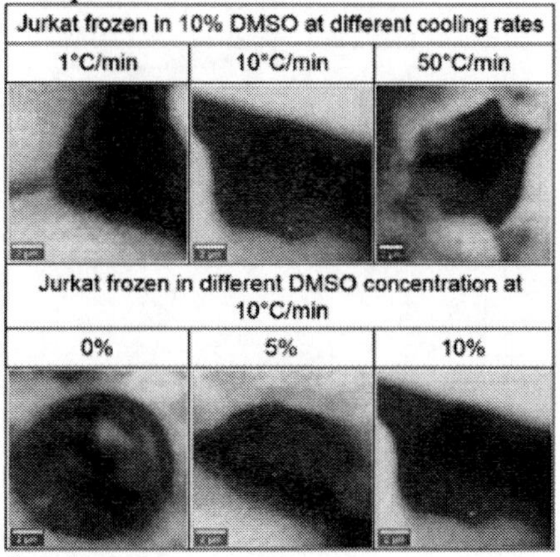

Scale bar 2 μm

The next phase of investigation involved using cytochrome c as a measure of cell viability. Figure 2 showed Raman images of ice and cytochrome c for two cells that both had IIF. For the first cell, cytochrome c was distributed widely in the cell with low intensity, while for the second cell; cytochrome c was localized in a small region with strong intensity. In other studies, we established that release of cytochrome c is associated with a loss of viability when compared to conventional trypan blue measures (data not shown).

Figure 2: Raman images of ice and cytochrome c for cells cryopreserved in 10% DMSO. Seeding of the extracellular ice was at -6°C and the sample was cooled at 10°C/min down to -50°C.

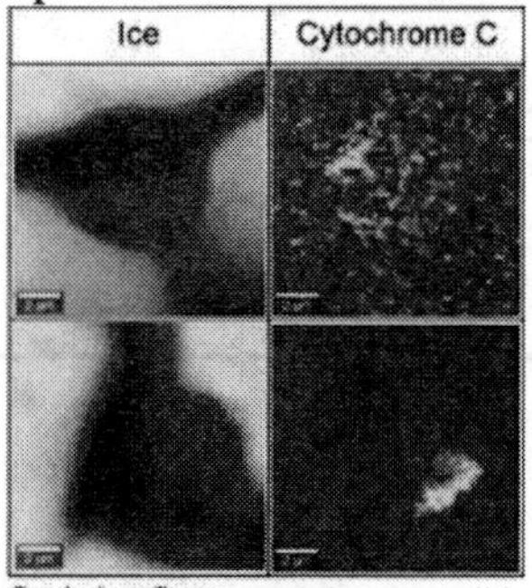

Scale bar 2μm

Cell viability does not merely depend on the cooling process. Thawing of cells also affects cell viability. Growth of intracellular ice crystals causes cell damage during thawing[6]. However, different phenomenon during thawing process was observed in this study. Figure 3 showed Raman images of ice for cells cryopreserved in 1% DMSO or DPBS solution and thawed afterwards at 120°C/min to -40°C and -30°C. Cell cryopreserved in 1% DMSO showed less of IIF as temperature increased. Cell cryopreserved in DPBS showed no change of IIF as temperature increased.

Figure 3: Raman images of ice for cells cryopreserved in 1% DMSO or DPBS solution and thawed afterwards at 120°C/min to different temperature. Seeding of the extracellular ice was at -6°C and the sample was cooled at 10°C/min down to -50°C.

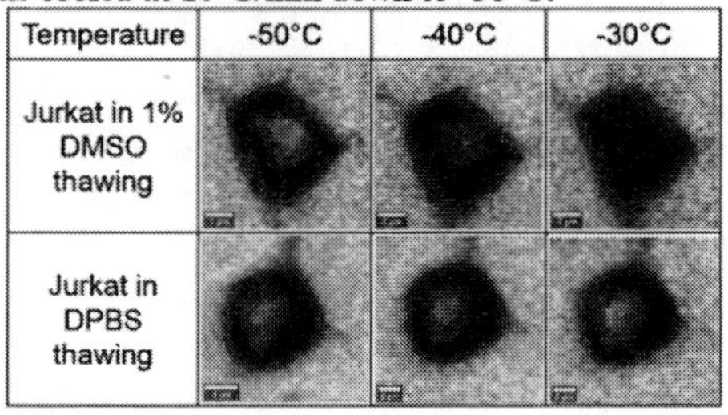

Scale bar 2μm

DISCUSSION

Intracellular ice formation was quite common. The general trend observed is that the amount of IIF increased with increasing cooling rate, as would be expected. DMSO concentration also influences IIF greatly. Most cells had only small ice crystals present (<0.5μm). The presence of large ice crystals (2-3μm) was observed when cells were cryopreserved in DPBS solution. Cells cryopreserved in DPBS solution had much lower cell viability than cells cryopreserved in DMSO solution based on vial freeing experiments (data not shown), which indicated large chunks of ice crystals were more damaging than small ones. Furthermore, intracellular ice crystals were found co-located to the sections of cell membrane in close proximity to extracellular ice, supporting Toner's hypothesis that extracellular ice affects the cell membrane making it a favorable location of seeding IIF.

Earlier investigations have shown that IIF is a lethal event[8]. This investigation showed that cells with IIF could have entirely different distribution of cytochrome c, suggesting that IIF alone might not be a lethal event. The amount of IIF as well as the size of ice crystals might also contribute to cell viability. There could be a threshold value of the amount of IIF, only above which cells would be dead.

For thawing studies, the growth of intracellular ice crystals during thawing was not observed for all of the experimental conditions tested. On the contrary, melting of ice crystals present was observed as the sample thawed and melting was observed at very low temperatures. Growth of extracellular ice was however observed and resulted in changes in cell shape and may represent a potential mechanism of damage during thawing.

ACKNOWLEDGEMENTS

This work was funded in part by NIH grant R21EB016247. Parts of this work were carried out in the Characterization Facility, University of Minnesota, a member of the NSF-funded Materials Research Facilities Network (www.mrfn.org) via the MRSEC program.

REFERENCES

[1] Mason, C et al., *Regen Med*, 6(3): 265-262, 2011.
[2] Mazur, P., *J.Gen.Physiol*, 47:347-369, 1963
[3] Meryman, H., In *Ciba Foundation Symposium-The Frozen Cell*, 51-67, 1970.
[4] Lovelock, J. E., *Biochim. Biophys. Acta*, 10: 414-426, 1953.
[5] Fuller, B. J et al., CRC Press, 53, 2004.
[6] Mazur, P., *Am. J. Physiol*, 247(3 Pt 1): C124-42, 1984.
[7] Dong, J et al., *Biophys. J*, 99(8):2453-2459, 2010.
[8] Mazur, P et al., *Biol.Reprod*, 66(5): 1485-1490, 2002.

Thermal Fluid Models of a Temperature Controlled Sheath Used to Deliver Thermosensitive Hydrogel Inside Pancreatic Cancer Lesions

Nesrine Bouhrira (1, 2), Thomas L. Merrill (1,2)

(1) Department of Mechanical Engineering
Rowan University
Glassboro, NJ, USA

(2) FocalCool, LLC
Mullica Hill, NJ, USA

INTRODUCTION

Pancreatic cancer is a devastating cancer. This disease, in which malignant tumors form in the organ responsible for creating digestive enzymes, has a five year survival rate of as little as a 14% for stage IA, and only 1% for stage IV [1]. The survival rate is low partly due to the pancreas's location deep within the body making its detection difficult. Chemotherapy is a common treatment option, but surgery is often necessary for tumor removal. However, surgical tissue removal is only a viable option for about 20% of patients [1]. There is a need for a localized minimally invasive method to effectively deliver anti-cancer agents into these lesions.

It has been hypothesized that direct drug injection into the tumor would enhance treatment effectiveness[2]. Consequently, we examined the use of endoscopic ultrasound (EUS) combined with a fine needle injection (FNI) to deliver a drug-eluding thermosensitive hydrogel directly into the tumor. However, normal body temperatures surrounding the EUS can warm the hydrogel drug combination beyond its phase transition temperature or lower critical solution temperature (LCST) before its destination inside the tumor. This transition leads sharp viscosity increases, requiring unrealistic delivery pressures[3]. To address this problem, a modified version of FocalCool's technology CoolGuide™ catheter [4], now called the CoolGuide™ sheath, was used to provide temperature control along the injection pathway.

The objective of this work is to build and explore thermal fluid models that can investigate the hydrogel delivery temperature and pressures using two different sheath configurations. Using experimental results for validation we intend to demonstrate that only the temperature controlled device makes the hydrogel delivery possible under body temperature conditions.

METHODS

To explore hydrogel delivery, we (1) developed a mock Endoscopic Ultrasound system (EUS) that mimics the procedure accessing the pancreas under body temperature conditions (without the needle) and, (2) developed thermal fluid models using a finite volume method to predict hydrogel delivered temperatures that we validated with experiments.

Two sheath configurations were explored 1) a standard circular sheath and 2) a temperature controlled sheath called CoolGuide™. The CoolGuide™ is tri-lumen sheath with a central lumen for hydrogel flow as well as wing-shaped lumens for coolant flow.

A schematic of the in-vitro experimental set-up is shown is Figure 1. 30% (w/v) Pluronic F127 hydrogel was injected at different flow rates through the CoolGuide™ using a syringe pump. The hydrogel has a liquid to-gel transition temperature of 14°C. Viscosity values transition from <1 Pa-s at 4°C to 3200 Pa-s at 37°C.

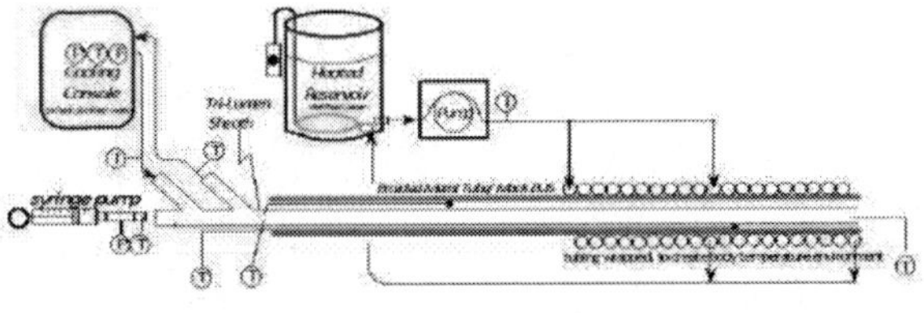

Figure 1: Schematic of the test setup without the FNI-Needle. The working channel was created with braided stainless steel tubing partially surrounded with body-temperature fluid from a heated reservoir. A cooling console was used to provide coolant to the CoolGuide™ sheath. Our numerical analysis was directly compared to experimental results. Boundary conditions were fixed in a manner suitable to reproduce the experimental conditions. The model consisted of different domains, the

fluid domains for hydrogel, and coolant flow, as well as the solid domains for the CoolGuide™ sheath and the mock EUS. Boundary conditions for all models involved a) fixed inlet temperatures and flow rates for hydrogel and coolant, b) fixed outlet pressure to zero gage and c) fixed external wall temperatures and assuming the no-slip condition. The mesh was generated with polyhedral cells and prism layers were added to the fluid side of the fluid-solid interface. The final model contains 19792075 cells. Figure 2 shows the boundary conditions imposed on the model as well as the mesh generated for the CoolGuide™ sheath.

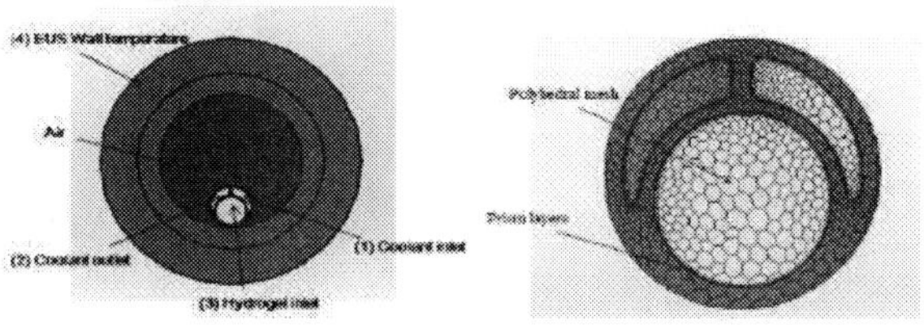

Figure2: Schematic of the Mock EUS model with the boundary conditions as well as the mesh generated for the CoolGuide sheath.

RESULTS

Hydrogel delivered temperatures using the CoolGuide™ sheath were first conducted and validated with experiments. Figure 3 shows the hydrogel delivered temperatures as a function of the flow rate. The temperature behavior showed that all model predictions were within 2 % of experimental data. Using the temperature controlled sheath, the delivered temperature was about 9.65°C for a flow rate of 2 ml/min with a reasonable delivery pressure of 9.5 Psi.

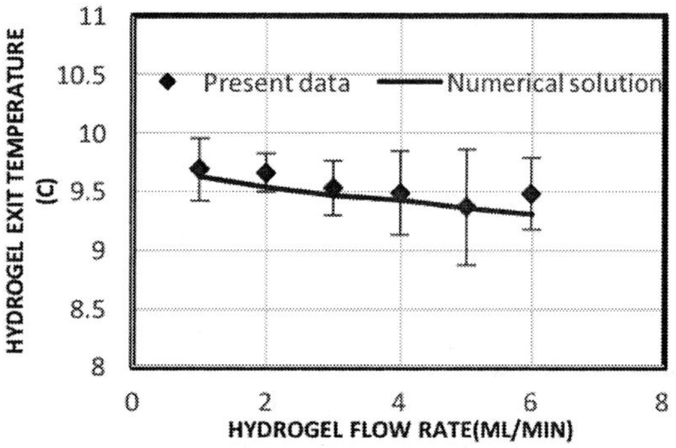

Figure 3: Hydrogel delivery temperatures using the temperature controlled sheath as a function of the flow rates. The error bars describe the standard deviation for three different data sets.

Given the same inputs as provided in the experiments, modeling results of the standard sheath showed that hydrogel delivered temperature exceeded the LCST (14°C). For a flow rate of 2 ml/min, hydrogel delivered temperature was about 27°C with an average viscosity of 3010 Pa-s. The pressure recorded with numerical solution to deliver hydrogel through the sheath exceeded the 100 psig limit provided with a standard syringe pump. Figure 4 shows the temperature and viscosity distributions of the hydrogel through the CoolGuide™ and the standard sheath.

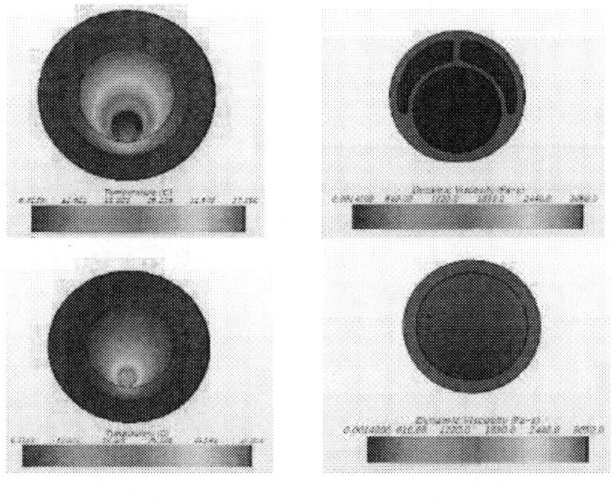

Figure 4: Hydrogel delivery temperatures and viscosity profiles with the two sheath configurations at a flow rate of 2 ml/min.

DISCUSSION

Based on both experimental testing and numerical modeling, the CoolGuide™ sheath showed the ability to maintain thermosensitive hydrogel in liquid form and below the LCST along the injection pathway for a range of tested flow rates. The device has therefore the ability to help avoid clogging along the injection pathway and unrealistic pumping pressures compared to the standard sheath configuration. Without temperature control, hydrogel delivery is not possible with standard syringe pumps.

Limitations of our application were related to the hydrogel thermal fluid properties. Also, the fidelity of the measured exit temperature was limited due to the probe orientation repeatability at the exit location.

Future work will include testing and modeling a mock EUS system using an FNI- Needle. While our experiments were instrumental in the validation of an initial thermal fluid model, additional modeling will explore a boarder range of possible designs for the CoolGuide™ sheath to enhance temperature controlled delivery of drug-eluding thermosensitive hydrogel deep inside the body.

ACKNOWLEDGEMENTS

Please acknowledge any grant or other funding support or the assistance of others as appropriate.

REFERENCES

[1] A. C. Society, "Pancreatic cancer survival by stage." [Online]. Available: http://www.cancer.org/cancer/pancreaticcancer/detailedguide/pancreatic-cancer-survival-rates. [Accessed: 01-Dec-2015].

[2] "Endoscopic Ultrasound," *Johns Hopkins Medicine*. [Online]. Available: http://www.hopkinsmedicine.org/medart/gallery/student/esiegel_10.jpg. [Accessed: 01-Dec-2015].

[3] M. J. Levy, et al "716 EUS Guided Intra-Tumoral Gemcitabine Therapy for Locally Advanced and Metastatic Pancreatic Cancer," *Gastrointest. Endosc.*, vol. 73, no. 4, p. AB144-AB145, 2011.

[4] T. L. Merrill, J. E. Mitchell, and D. R. Merrill, "Heat transfer analysis of catheters used for localized tissue cooling to attenuate reperfusion injury," *Med. Eng. Phys.*, vol. 38, no. 8, pp. 758–766, 2016.

TUNING THE GOLD NANOPARTICLE COLORIMETRIC ASSAY BY NANOPARTICLE SIZE AND CONCENTRATION

Varsha Sanjay Godakhindi[1 *], Peiyuan Kang[2 *], Maud Serre[3 *], Naga Aravind Revuru[2], Michael Roner[4], Jeffrey Kahn[5], Jaona Randrianalisoa[6] and Zhenpeng Qin[1,2,7]

[1] Department of Bioengineering, The University of Texas at Dallas, Richardson, TX 75080, USA

[2] Department of Mechanical Engineering, The University of Texas at Dallas, Richardson, TX 75080

[3] Ecole Supérieure d'Ingénieurs de Reims (ESIReims), University of Reims Champagne - Ardenne, 3 Espl. Roland Garros, 51100 Reims, France

[4] Department of Biology, University of Texas at Arlington, 701 S Nedderman Dr, Arlington, TX 76019

[5] Departments of Pediatrics & Microbiology, University of Texas Southwestern Medical Center, 5323 Harry Hines Blvd, Dallas, TX 75390; Children's Medical Center, 1935 Medical District Dr, Dallas, TX 75235

[6] Groupe de Recherche en Sciences pour l'Ingénieur (GRESPI) - EA 4694, University of Reims Champagne - Ardenne, 51687 Reims Cedex 2, France

[7] Department of Surgery, The University of Texas at Southwestern Medical Center, 5323 Harry Hines Blvd, Dallas, TX 75390

*Co- first authors

INTRODUCTION

Infectious disease associated mortality has emphasized the significant need to develop point-of-care (POC) diagnostics to allow for early and rapid detection and screening.[1] Gold nanoparticles (GNP) have proven to be essential candidates for the development of fast screening methods. In particular, GNP colorimetric assay is a simple and fast detection method and is based on the aggregation of GNPs that results in a color change due to the plasmonic coupling.[2] Due to its simplicity, GNP colorimetric assay has found broad applications including the detection of proteins, nucleic acids, viruses, heavy metals, aptamers, and bacteria. Recent innovations have been attempted to improve the sensitivity and to simplify experimental conditions for nucleic acid detection using GNP aggregation assay. However, it is unclear how other GNP design parameters impact the colorimetric assay performance. In this report, we systematically investigated the effects of GNP design parameters, including GNP size and concentration, with the goal of further improving the assay performance.

METHODS

Small GNPs (15nm) were synthesized using Fren's method with slight modifications[3, 4] and larger GNPs (30 and 50nm) were synthesized by hydroquinone reduction.[5] The synthesized particles were characterized by dynamic light scattering (Malvern, Zetasizer) and transmission electron microscopy (JEOL JEM 2100). The experiment was designed considering malaria as the clinically relevant target and GNPs were conjugated with oligonucleotide sequences that recognize malaria DNA to produce probes (A and B). The probes were mixed with the target in the presence of hybridization buffer.[6] UV-Vis spectrum of the samples were then measured using a microplate reader (Biotek Synergy 2). The effect of GNP size and concentration for 15, 30 and 50nm particles were studied.

Discrete dipole approximation was employed to calculate the optical properties of the GNPs. Briefly, the target structure is discretized into a finite array of dipoles (N). After solving a large number (3N) of complex linear equations with unknown dipole moments, the extinction, absorption and scattering cross-sections can be predicted. In this study, the DDA package DDSCAT 7.3 was implemented to solve the scattering and absorption properties of individual and aggregated GNPs [7]. The dipole structures were visualized in an open-source software package ParaView. The extinction coefficient was calculated by using the following formula:

$$Q_{ext} = \frac{C_{ext}}{\sum_{i=1}^{M} \pi r_i^2} \tag{1}$$

where Q_{ext} is the extinction efficiency factor, C_{ext} is the extinction cross-section, and r_i is the radius of individual GNP, and M is the number of GNPs in the aggregate ($M \geq 1$). The denominator of the above equation refers to the summation of geometric cross-sectional areas for nanoparticles that form the aggregate, and is used here for convenient comparison with individual GNP. The data was plotted using OriginLab.

RESULTS

As shown in Figure 1, GNP aggregation due to DNA hybridization was confirmed by the plasmonic peak shift in the UV-Vis spectrum and a visible color change. Figure 1 (b) describes the effect of the GNP size on the assay performance while keeping the absorbance at the peak wavelength constant (i.e., abs=0.75). Figure 1(c) compares the assay performance for GNP concentrations indicated in terms of absorbance 0.15, 0.75 and 1.85, while keeping the GNP size constant (i.e. diameter =30 nm).

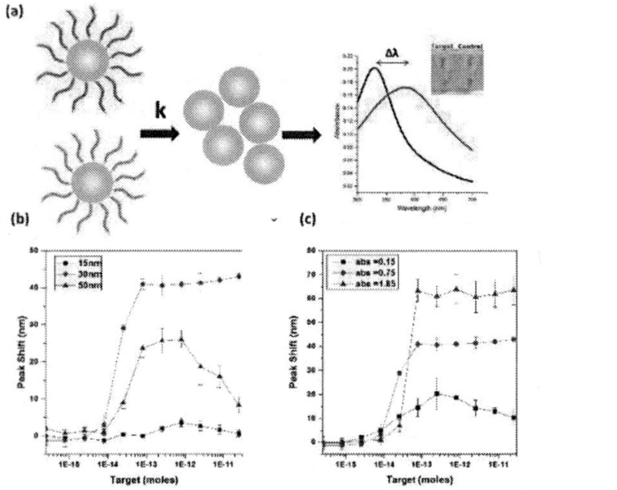

Figure 1(a) Schematic of target induced GNP aggregation and subsequent plasmonic shift and color change (inlet). (b) Effect of GNP size on assay performance. (c) Effect of GNP concentration on assay performance.

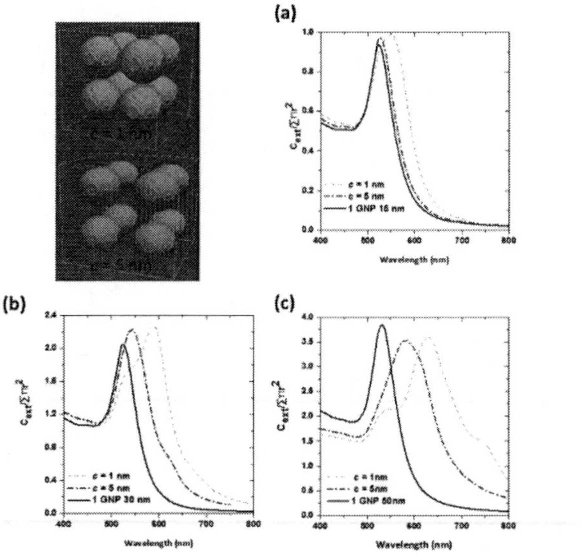

Figure 2 Eight GNPs positioned in a simple cubic lattice representing aggregate with clearance distance c=1 and 5nm (a) 15nm (b) 30nm and (c) 50nm.

To illustrate the underlying physics and to keep a reasonable simulation domain, we studied the plasmonic coupling eight GNPs. Figure 2 (a), (b), and (c) show the aggregate optical properties with different clearance distances (1 and 5 nm), for GNP sizes including 15nm, 30nm and 50nm respectively. It can be observed that larger GNP leads to a more significant peak shift under the same clearance distance, suggesting a stronger plasmonic coupling.

DISCUSSION

Our study focuses on effects of GNP size and concentration on the colorimetric assay performance. We systematically examined the effect of GNP size and found that 30 nm GNP gives the best performance in signal strength and sensitivity. In terms of GNP concentration, lower GNP probe concentrations lead to better sensitivity but weaker overall signal. This can be attributed to the lower level of background signal (i.e., non-aggregated GNP). We further analyzed the plasmonic coupling strength by numerical simulation. When placed in close proximity, larger GNPs demonstrate stronger plasmonic coupling with a more significant plasmonic peak shift towards longer wavelengths. The final performance of the GNP aggregation is likely a result of several factors including the plasmonic coupling strength and the speed of nanoparticle aggregation. In summary, our work provides new guidelines to design GNP aggregation-based POC devices to meet the sensitivity needs to diagnose different infectious diseases.

ACKNOWLEDGEMENTS

This study is supported by a Texas Medical Research Collaborative (TxMRC) grant and startup fund from The University of Texas at Dallas (UTD). The authors acknowledge the Texas Advanced Computing Center (TACC) at The University of Texas at Austin and the ROMEO HPC Center hosted by the University of Reims Champagne-Ardenne for providing HPC resources that have contributed to the research results reported within this work.

REFERENCES

1) Sun, J.; Xianyu, Y.; Jiang, X. *Chem.Soc.Rev.* **2014**, *43*, 6239-6253.
2) Storhoff, J.; Elghanian, R.; Mucic, R.; Mirkin, C.; Letsinger, R. *J.Am.Chem.Soc.* **1998**, *120*, 1959-1964.
3) Frens, G. *Nature-Phys. Sci.* **1973**, *241*, 20-22.
4) Kimling, J.; Maier, M.; Okenve, B.; Kotaidis, V.; Ballot, H.; Plech, A. *J Phys Chem B.* **2006**, *110*, 15700-15707.
5) Perrault, S. D.; Chan, W. C. *J.Am.Chem.Soc.* **2009**, *131*, 17042-17043.
6) Cordray, M. S.; Amdahl, M.; Richards-Kortum, R. R. *Anal.Biochem.* **2012**, *431*, 99-105.
7) Draine, B.; Flatau, P. *J. Opt. Soc. Am. A.* **1994**, *11*, 1491-1499.

SB³C2017
Summer Biomechanics, Bioengineering and Biotransport Conference
June 21 – 24, 2017, Tucson, AZ, USA

Shear-Augmented Dispersion Affects Cerebrospinal Fluid Solute Transport within the Subarachnoid Space but not within the Basement Membranes of the Brain

M. Keith Sharp (1), Roxana O. Carare (2), Bryn Martin (3)

(1) Department of Mechanical Engineering
University of Louisville
Louisville, KY, USA

(2) Institute for Life Sciences
Southampton General Hospital
University of Southampton
Southampton SO16 6YD, UK

(3) Department of Biological Engineering
University of Idaho
Moscow, ID, USA

INTRODUCTION

The influence of shear-augmented dispersion on solute transport in the cerebrospinal circulation was evaluated in two locations with very different geometric and kinematic characteristics. First, amyloid-β and other tracers have been observed to be cleared from the brain along basement membranes surrounding cerebral arteries at rates that cannot be explained by molecular diffusion alone. However, solvent flow in these small channels (~100 nm thick) has not been documented due to the difficulty of measurements in such a small channel in a living animal. Nonetheless, several models of mechanisms by which flow might carry solutes along the perivascular spaces in the direction opposite that of blood flow have been developed [Schley, *et al.* 2006, Colomba, *et al.* 2016, Sharp, *et al.* 2016].

Second, the spinal subarachnoid space (SSAS) has been proposed as a channel for drug delivery to the brain. The SSAS is considerably larger (~3 mm thickness), which has allowed estimations and measurements of flow during the cardiac cycle [Martin, *et al.* 2012]. The measured flow is oscillatory with near zero mean, thus drugs, which are too large to be delivered to the brain effectively by molecular diffusion alone, cannot be delivered by advection alone either. Therefore, shear-augmented dispersion is clearly a key to transporting the drugs to the brain as intended.

These two examples bracket (in terms of size) other regimes of CSF flow and transport. Thus, determining the potential influence of shear-augmented dispersion in these locations may provide insight that can be applied more broadly throughout the CSF circulation.

METHODS

1. *Mathematical model*

Watson [1983] derived an expression for dispersion enhancement in a 2D channel. Fully-developed oscillatory flow is assumed between rigid plates that provide no-slip boundary conditions, and also are impervious to transport. A linear concentration gradient is also assumed (similar to Taylor [1953]), which makes the transport also fully-developed and oscillatory. For these conditions, the enhancement of axial transport relative to molecular diffusion is given by

$$R=\frac{1}{2}\frac{\cosh\alpha\sqrt{2}-\cos\alpha\sqrt{2}}{\cosh\alpha\sqrt{2}+\cos\alpha\sqrt{2}}\frac{C\left(\alpha\sqrt{2}\right)-C\left(\beta\sqrt{2}\right)}{1-\sigma^{-2}}P^2\alpha^{-2} \quad (1)$$

where

$$C\left(\alpha\sqrt{2}\right)=\frac{1}{\alpha\sqrt{2}}\frac{\sinh\alpha\sqrt{2}-\sin\alpha\sqrt{2}}{\cosh\alpha\sqrt{2}-\cos\alpha\sqrt{2}} \quad (2)$$

and $\alpha=h\sqrt{\omega/\nu}$ is the Womersley number, $\beta=h\sqrt{\omega/\kappa}$ will be called the Watson number, $\sigma=\nu/\kappa$ is the Schmidt number, $P=(dp/dx)/(\rho\omega\nu/h)$ is the nondimensional pressure gradient magnitude, h is the channel half height, ω is angular frequency, ν is kinematic viscosity, ρ is density, κ is molecular diffusivity, p is pressure and x is the axial coordinate.

Because of the behavior of the hyperbolic functions with increasing values of their arguments, limits of equation 1 are useful. These limits also represent the three possible regimes of flow and transport. For $\alpha,\beta\rightarrow 0$, which represents viscous-dominated flow (for which the velocity profile is Poiseuille throughout the cycle), and diffusive transport (diffusion dominates advective transport), the enhancement becomes

$$R=\frac{\sigma^2}{945}\left(1-\frac{5\sigma^2+82}{1980}4\alpha^4\right)P^2\alpha^4 \quad (3)$$

For $\alpha\rightarrow 0$ and $\beta\rightarrow\infty$, which represents viscous flow and unsteady, advective transport, the enhancement becomes

$$R = \frac{1}{6}\left(1 - \frac{3}{\beta\sqrt{2}}\right)P^2 \quad (4)$$

Finally, for $\alpha, \beta \to \infty$, which represents unsteady flow (with Womersley velocity profiles) and unsteady transport, the enhancement becomes

$$R = \frac{2^{-3/2}}{\left(1+\sigma^{-1}\right)\left(1+\sigma^{-1/2}\right)} P^2 \alpha^{-3} \quad (5)$$

2. Parameter values

In both example locations, the frequency of oscillatory flow is assumed to be that of the heart beat, $\omega = 2\pi$ rad/s and CSF density and viscosity can be approximated as that of water, $\rho = 993$ kg/m^3 and $\nu = 7 \times 10^{-7}$ m^2/s, respectively. Amyloid-β in the perivascular space has a diffusivity of $\kappa = 5 \times 10^{-11}$ m^2/s [Goodhill 1997], for a Schmidt number of 14000. In the SSAS, Methotrexate, $\kappa = 5.26 \times 10^{-10}$ m^2/s [Blasberg 1975], gives a Schmidt number of 1330.

The pressure gradient amplitude in the SSAS is estimated as 45 Pa/m [Martin, *et al.* 2012] for $P = 155.7$. The pressure fluctuations driving flow in the perivascular space are not known, so the largest possible fluctuation is adopted here in order to test the ultimate feasibility of shear-augmented dispersion in this space. A recent model suggests that high pressure may persist to the capillaries [Gould, *et al.* 2016], so a 100 mmHg pressure amplitude over an estimated 0.1 m length of cerebral vessel gives $P = 1.526$.

RESULTS

Three regions of dispersion enhancement differentiated by the different slopes of each curve correspond to the limits $\alpha, \beta \to 0$ on the left, $\alpha \to 0$ and $\beta \to \infty$ in the middle and $\alpha, \beta \to \infty$ on the right (Fig. 1). Solid dots indicate the operating points for the heart beat frequency and the dimensions assumed for each. For the perivascular space, $R = 2.42\text{E-}10$ and for the SSAS, $R = 91.8$.

DISCUSSION

For clearance of amyloid-β along periarterial basement membranes, shear-augmented dispersion as predicted by the Watson model is insignificant, and would be so for any operating conditions on the red curve in Fig.1, given that substantially increased transport compared to molecular diffusion is needed to explain experimental observations. If the pressure amplitude is less than the extreme value assumed, the enhancement would be even lower. However, a 1 cm artery length would lead to 10X larger P and 100X greater, but still insignificant, enhancement (see equation 3). R scales approximately with σ^2 if α is held constant, *e.g.*, 3 kDa dextran ($\kappa = 1.28 \times 10^{-10}$ m^2/s) gives smaller σ and β, and reduces R to 3.71E-11. On the other hand, the enhancement in the SSAS is by two orders of magnitude, which is useful in facilitating intrathecal drug transport. In the SSAS, R is relatively insensitive to σ, but scales with P^2.

The Watson model includes simplifications that affect the enhancement of transport. First, the ultrastructure in both spaces blunts the superficial velocity profile, which decreases the fraction of the cross section with shear, in which dispersion is enhanced. On the other hand, the ultrastructure also causes local shear, as well as transverse mixing, that both increase axial transport. For instance, flow through a hexagonal array of cylinders enhances longitudinal transport by about four orders of magnitude [Edwards, *et al.* 1991]. More aggressive transverse mixing, particularly when the mixing time scale is matched to the oscillatory period, can enhance transport by an additional two orders of magnitude [Pedley & Kamm 1988, Sharp, *et al.* 1991].

Second, channel curvature (which applies more to the perivascular compartment) [Sharp, *et al.* 1991] and varying cross section (which applies to both examples) also leads to further enhancement of transport compared to the simple parallel plate geometry. Whether more accurate modeling of the perivascular space will result in significant enhancement is questionable, since the Watson result would need to be exceeded by at least ten orders of magnitude. However, for the SSAS, it appears that further work to more faithfully represent the actual geometry would be useful for understanding intrathecal drug transport and designing optimal drug delivery protocols.

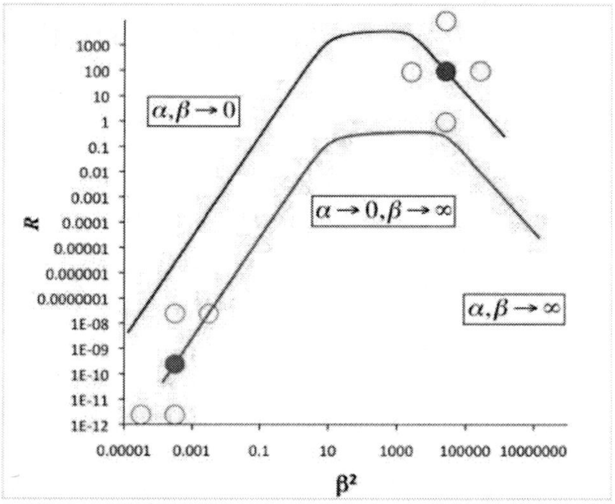

Figure 1. Enhancement of dispersion (compared to molecular diffusion) for the σ and P in the perivascular space (red) and SSAS (blue). Solid dots indicate the estimated operating condition in each example. Open dots are for σ and P varied by an order of magnitude higher and lower.

REFERENCES

Blasberg RG, Patlak C, Fenstermacher JD. *J Pharmacol Exp Ther* 195, 73–83, 1975.

Coloma M, Schaffer JD, Carare RO, Chiarot PR, Huang P. *J Math Biol* 73:469–490, 2016.

Edwards DA, Shapiro M, Brenner H, Shapria M. *Transport Porous Media* 6:337-58, 1991.

Goodhill GJ. *Eur J Neurosci* 9:7:1414-21, 1997.

Martin BA, Reymond P, Novy J, Balédent O, Stergiopulos N. *Am J Physiol Heart Circ Physiol* 302:H1492–509, 2012.

Pedley TJ, Kamm RD. *J Fluid Mech* 193:347-67, 1988.

Schley D, Carare-Nnadi R, Please CP, Perry VH, Weller RO. *J Theoretical Biol* 238:4:962-74, 2006.

Sharp MK, Kamm RD, Shapiro AH, Kimmel E, Karniadakis GE. *J Fluid Mech* 223:537-63, 1991.

Sharp MK, Diem AK, Weller RO, Carare RO. *Ann Biomed Eng* 44:5:1553-65, 2016.

Taylor GI. *Proc R Soc Lond A* 219:186-203, 1953.

SB³C2017
Summer Biomechanics, Bioengineering and Biotransport Conference
June 21-24, 2017,Tucson, Ari□ona, USA

USING MICRO-CT TO INVESTIGATE NANOPARTICLE DISTRIBUTION IN SOLID TUMORS AFTER INTRATUMORAL INFUSION

Myo Min Zaw (1), Timothy Munuhe (1), Jeffrey Li (2), Liang Zhu (1), Ronghui Ma (1)

[1] Department of Mechanical Engineering
University of Maryland Baltimore County
Baltimore, MD, USA

[2] Centennial □igh School,
Ellicott City, MD, USA

INTRODUCTION

Magnetic nanoparticles have been used in applications of hyperthermia treatments for cancer due to their abilities to generate impressive levels of heat when excited by an external magnetic field [1-3]. One commonly used method to deliver nanoparticles in solid tumors is intratumoral injection where nanoparticles dispersed in aqueous solution are injected into a tumor under a positive pressure gradient. As nanoparticles are the heating source, their distribution in a tumor is a critical factor that defines the temperature elevation of the tissue and the consequent treatment outcome. One leading issue in this treatment approach is the limited knowledge of nanoparticle distribution in tumors. Salloum et al. [3] studied the injection of a ferrofluid in semi-transparent agarose gels, and the study suggests a non-uniform nanoparticle distribution with a high nanoparticle concentration near the needle tip. Recently, micro-CT has been used to quantify the magnetic nanoparticles distributions in tumors and to optimize treatment protocols based on the micro-T scans [1]. Due to inhomogeneous tumor properties, the nanoparticle distribution is randomly irregular. These studies have found that a lower inject rate tends to generate a more controllable and repeatable nanoparticle distribution in the tumor. However, detailed information on three-dimensional nanoparticle distribution in a tumor is still missing. Numerical simulation has also been performed to investigate convection and diffusion of the nanofluid in deformable tumor tissue [3]. However, it remains unclear to what extent the tumor heterogeneity affects the final distribution of the nanoparticles. The objective of this study is to construct three-dimensional nanofluid distribution volume in a tumor based on the previously obtained micro-CT scans, to extract detailed information on heterogeneous tumor structure, and to reevaluate nanoparticle transport mechanisms in solid tumors during intratumoral infusion.

METHODS

The tumors used in this study were PC3 prostatic tumors implanted on the flanks of BALB/c Nu/Nu male mice. Water based ferrofluids (EMG705 series, Ferrotec (USA) Corporation, Nashua, NH) with a concentration of 5.8% by volume and a nominal particle size of 10 nm were loaded on a syringe pump (Genie Plus, Kent Scientific, Inc. Torrington, CT) to control the flow rate (3-5 μl/min). A 26-gauge Hamilton needle (Fischer Scientific, Springfield, NJ) was used for injecting ferrofluid of 0.1 cc into the tumors. Three injection rates (3, 4 and 5 μl/min) were used on three groups mice (five in each group) respectively to investigate the effect of injection rate on nanoparticle distribution.

After the experiments, the mouse was euthanized and the tumors were resected and scanned in a microCT imaging system (skyscan1173, MicroPhotonics, PA), resulting in 523 individual images and each image has 1052x1944 pixels. The vast data were processed using the software SAS (SAS Institute, Cary, NC) to generate a single file that gives the pixel index values in the form of a three-dimensional matrix. In order to reduce the size of the data, we averaged the pixel index values over a 3-D cluster of voxels (7x7x7). The cluster size can be adjusted to yield a reasonably sized file without sacrificing the accuracy of the imaging quality. The resulting pixel index value distributions were plotted with MATLAB R2014a (Mathworks, Natick, MA). The data post-processing was aided by the use of the iso2mesh MATLAB toolbox, a set of functions used for processing and analysis of grayscale data and images from MRI and CT scanners.

The iron-based nanoparticles have a significantly higher density (5240 kg/m^3) than the tumor tissue. Our previous study has established a linear relationship between the pixel index value and nanoparticle concentration [1]. Thereby, voxels with higher pixel index values on a micro-CT scan represent tumor tissue containing a higher concentration of nanoparticles. We expect that the pixel index values in the micro-CT scans will not only provide information on the detailed three-dimensional nanoparticle distribution, but also reveal the heterogeneous tumor structure and the nanofluid flow pattern during an infusion.

RESULTS

In this study, the micro-CT scans of 15 tumors have been processed, and the three-dimensional shapes of isosurfaces with cutoff values of 40, 75, and 100 are constructed. Figure 1 shows the 3D shapes of the isosurfaces in a tumor with an injection rate of 3 μl/min. The cutoff value of 40 has been used to distinguish nanoparticle region from the surrounding tumor tissue, yielding a tumor of a nearly spherical shape with a diameter around 1 cm as shown in Fig. 1(a). Shown in Figures 1(b) and 1(c) are the isosurfaces with higher cutoff values of 75 and 125, respectively. The volumes enclosed by these isosurfaces are considered to have a high nanoparticle concentration. The shapes of the 3-D volumes are highly irregular and it appears that the nanofluid accumulates in one side of the tumor despite injection at the center of the tumor. In addition, the two volumes shown in Fig. 1(b) and 1(c) are similar in shape in spite of their different volume sizes. Figure 1(d) show the pixel index value distributions on three slices of the tumor. They confirm that the majority of the nanofluid accumulates on one side of the tumor instead of the central region. In addition, there is a hollow region in the tumor core, contradictory to our previous speculation that the nanoparticle distribution follows a Gaussian distribution, i.e., high at the injection site, and low in the periphery region.

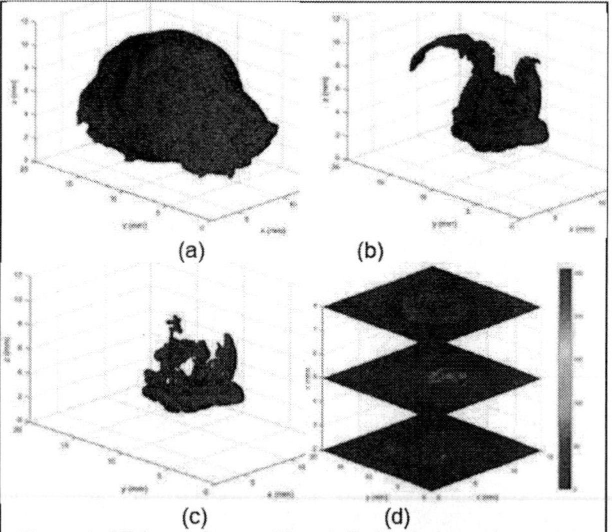

Figure 1□3D isosurfaces with cutoff pixel index values of (a) 40□(b) 75 and (c) 125. Figure 1 (d) displays the pixel index value distributions on three slices.

The processed micro-CT scans of 15 tumors exhibit the same pattern of nanoparticle distribution, such as irregular distribution and similarity in 3D isosurface shapes with cut-off values of 75 and 100. Another interesting observation in the tumor scans is the lack of nanoparticle accumulation in the central region of the tumors. In this study, effort has been made to ensure that the needle tip was placed in the tumor center by visual inspection and control of the insertion depth of the needle. However, the volumes containing the highest concentration of nanoparticles in 15 tumors all exhibit a scattered pattern.

The irregularity and similarity of the isosurfaces observed in all tumors confirms the dominant role of tumor heterogeneity on the nanofluid transport during an infusion process. Driven by the elevated pressure at the injection site, the nanofluid will follow paths with least hydraulic resistance towards the tumor periphery, such as tumor tissue with higher porosity, micro-cracks induced by injection, or the annular space along the needle track. The nanofluid then spreads into the regions of higher porosities along a path before they disperse into those of lower porosity. After injection, nanoparticles continue to spread in the tumor by diffusion. Diffusion plays a less important than convection due to the short period of injection time and low diffusivity of nanoparticles in the extracellular matrix. We speculate that the volume shown in Figure 3(c) contains cracks and high porosity tumor tissue.

In an ideal situation where a nanofluid is infused into a homogenous porous medium, the nanoparticles are supposed to accumulate near the injection site, and the isoconcentration surfaces should have a spherical shape. Even in the presence of backflow, there is still a significant amount of nanoparticles concentrated at the injection site [2, 3]. One explanation for the lack of nanoparticle deposition at the infusion site is the high interstitial pressure (IFP) at the tumor core. As illustrated in Figure 2, it is possible that the nanofluid is pushed back along the needle track by the high IFP at the tumor core. The backflow of the nanofluid then seeps through small cracks along the path to form a scattered distribution pattern. Another possible scenario is that nanoparticles may migrate towards the periphery driven by the IFP gradient after the injection. More experimental and theoretical studies are needed to investigate the effect of IFP on the nanoparticle distribution during and after intratumoral infusion.

In summary, micro-CT imaging method has been used to study the pattern of nanoparticle distribution in tumors after injection of 0.1 cc ferrofluid into 15 tumors planted on mice at three different injection rates. The results show that irregular shell-like nanoparticle deposition pattern in all the tumors, despite that a single nanofluid infusion was performed at the tumor center. A lack of nanoparticle accumulation in the central region of the tumors has been observed, suggesting that high interstitial fluid pressure might affect the nanofluid flow and diffusion during and after an injection process.

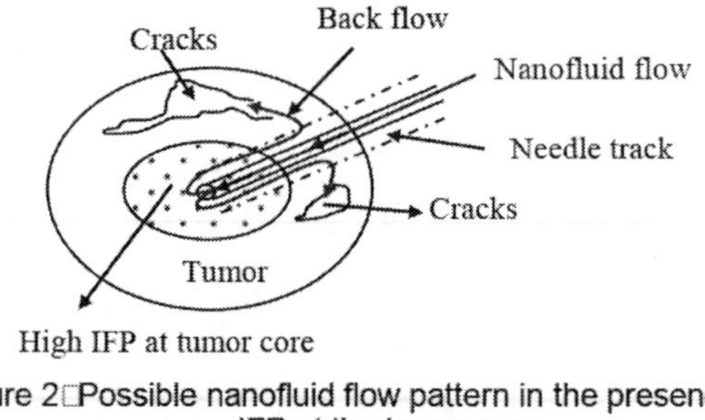

Figure 2□Possible nanofluid flow pattern in the presence of high IFP at the tumor core.

ACKNOWLEDGEMENTS

This study was supported in part by an NSF research grant CBET-1335958 in partial fulfilment of the requirements for the PhD degree from UMBC by Myo Min Zaw.

REFERENCES

1. A. LeBrun, T. Joglekar, C. Bieberich, R. Ma and L. Zhu. *International Journal of Hyperthermia*, 32(2):132-143, 2016.
2. D. Su, R. Ma, and L. Zhu, *Medical and Biological Engineering & Computing*, vol. 49(11), pp.1233-40, 2011
3. M. Salloum, R.-H. Ma, L. Zhu, *International Journal of Hyperthermia*, vol. 24(4), pp. 337-345, 2008.

NANO-BIO-THERMAL INTERFACE: NANOSECOND PLASMONIC HEATING INDUCED SELECTIVE PROTEIN INACTIVATION

Peiyuan Kang (1), Zhuo Chen (2), Steven O. Nielsen (2), Kenneth Hoyt(3,4), Sheena D'Arcy(2), Jeremiah J. Gassensmith(2), Zhenpeng Qin (1,3,5)

(1) Department of Mechanical Engineering
The University of Texas at Dallas
Dallas, Texas, USA

(2) Department of Chemistry and Biochemistry
The University of Texas at Dallas
Dallas, Texas, USA

(3) Department of Bioengineering
The University of Texas at Dallas
Dallas, Texas, USA

(4) Department of Radiology
The University of Texas at Southwestern Medical Center
Dallas, Texas, USA

(5) Department of Surgery
The University of Texas at Southwestern Medical Center
Dallas, Texas, USA

INTRODUCTION

Proteins are the functional machinery of biological systems. The development of techniques to change a protein's conformation and control its function with both high spatial and temporal resolutions is a crucial pursuit in the development of next-generation medicine. This capability is being rigorously pursued as it offers significant and broad implications for uncovering a protein's folding and unfolding mechanisms, elucidating a protein's biological function *in situ*,[1] non-invasively manipulating biological activity [2], and developing novel therapeutics for disease treatment. This is partially evidenced by the development and application of optogenetic approaches that have utilized light-responsive, activity-modulating proteins in the last decade[2,3]. While the genetic modification represents a challenge for clinical translations, such novel approaches to spatiotemporally modify protein conformation and function promise to lead to paradigm shifting advances.

Nanoparticles offer a direct interface with proteins and there have been significant advances in using nanoparticles for the long-term labeling and tracking of proteins in living on account of their excellent photostability[4]. While nanoparticle labeling has led to new understanding of protein localization and dynamics in live cells, the ability to further manipulate protein structure and activity with a high spatiotemporal resolution represents a significant advancement with a wide range of new possibilities and applications. Pulsed laser heating of plasmonic gold nanoparticle (GNP) is known to create a highly localized heating [5–8]. Here we utilize GNP and pulsed laser heating to generate highly localized ultrahigh temperatures in a few nanoseconds to cause ultrafast selective protein inactivation. The heating can be precisely controlled within a few nanometers around the nanoparticle without changing the temperature of the bulk medium. Both the high spatial (nanometers) and temporal (nanoseconds) capabilities offered by this new technique will find broad applications in the inactivation of protein activity *in situ* to elucidate protein function, and remote manipulation of biological activity.

METHODS

We covalently conjugate enzyme protein (α-chymotrypsin, Cht) onto gold nanoparticles by heterofunctional polyethylene glycol (PEG) molecules (1kDa, thiol and carboxyl functionalization on the two ends) using EDC/NHS chemical reaction. Another enzyme, alkaline phosphatase (ALP), solution was mixed with Cht coated GNPs colloidal solution. After one single nanosecond laser pulse, the activity of ALP and Cht in solution were examined seperately though enzyme catalysis reaction. For Cht, the activity signal was read by the rate of absorbance change at 410 nm which was due to the end product of substrate (Suc-Ala-Ala-Pro-Phe-NA) digestion. Similarly, for ALP, the signal was read at 405 nm where the absorbance was yielded by substrate (pNPP) decomposition. The mechanism of this selective inactivation phenomenon was futher studied by computational simulation combining the heat conductive equation and the Arrhenius model.

RESULTS

We experimentally demonstrated protein inactivation is effective and specific to the protein that is tagged or linked to the GNP in solution. By linking GNP to Cht, but not ALP, the Cht activity was selectively reduced without significantly affecting the function of ALP. As shown in Fig. 1B, one laser pulse decreased the Cht activity to 9% but ALP activity still remained high at 79%. The slight decrease in the ALP activity may be due to some passive absorption of ALP to GNP. Through the modeling and simulation, the temperature of GNPs rises rapidly due to plasmonic absorption. The heat diffuses from GNP to the surrounding water and causes a significant temperature increase and steep temperature gradient within several nanometers around GNP. Importantly, the temperature rise, which is insufficient to heat up the entire solution, is limited to the nanoparticle and its immediate surrounding medium. Thus, highly localized plasmonic heating accounts for the selective protein inactivation phenomenon.

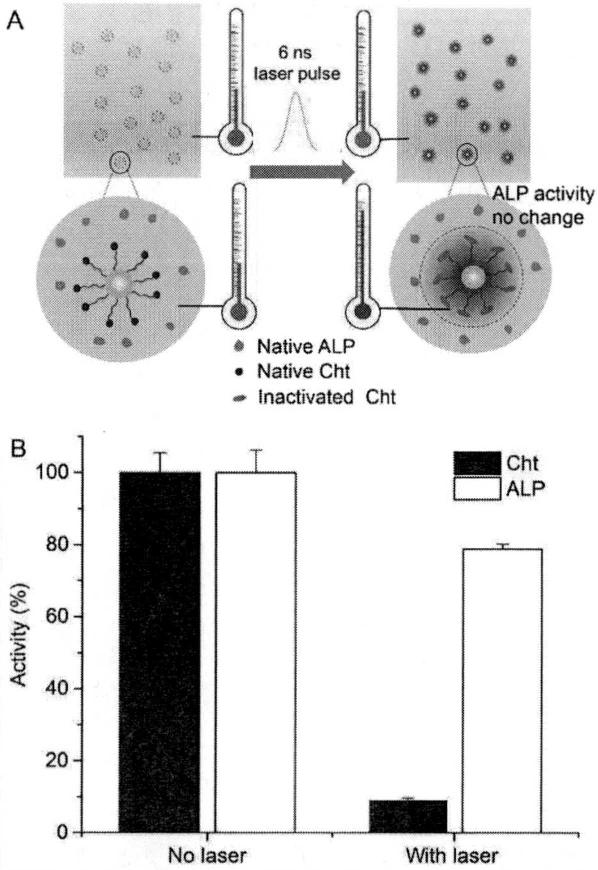

Figure 1: Selective protein inactivation due to highly localized plasmonic heating in 6 nanoseconds. (A) The nanosecond laser pulse heating up GNPs without increasing the solution temperature causes selective protein inactivation. (B) Specificity of the nanosecond protein inactivation. Laser energy density is 510.06 ± 24.60 mJ/cm^2, GNP size is 15 nm.

DISCUSSION

We have proved the feasibility of selective protein inactivation due to localized nanoseconds plasmonic laser heating, which is generated by GNPs. This novel method can be potentially used as an easier approach to control the protein behavior in cells, which offers a platform to remotely manipulate protein structure and activity without genetic modification. This will allow for protein function study *in vitro* as well as *vivo*. Further work is warranted to investigate the ultrafast kinetics of the protein unfolding by time-resolved X-ray scattering[9] and detailed structural change using hydrogen/deuterium exchange and mass spectrometry (H/DX-MS)[10]. Also, we would like to see its effective utilization in cell study. The selective protein inactivation combined with nanomaterial labeling may prove useful for basic protein functional research,[11,12] remote control of biological activity and behavior, and in therapeutic applications.

ACKNOWLEDGEMENTS

The authors acknowledge the startup support, Zachary Campbell for use of q-PCR, and useful discussions with Dr. Stephen Levene from the Department of Bioengineering, from The University of Texas at Dallas.

REFERENCES

[1] J. Kubelka, *Photochem. Photobiol. Sci.* **2009**, *8*, 499–512.
[2] L. Fenno, O. Yizhar, K. Deisseroth, *Annu. Rev. Neurosci.* **2011**, *34*, 389–412.
[3] A. Berndt, S. Y. Lee, J. Wietek, C. Ramakrishnan, E. E. Steinberg, A. J. Rashid, H. Kim, S. Park, A. Santoro, P. W. Frankland, et al., *Proc. Natl. Acad. Sci. U. S. A.* **2016**, *113*, 822–9.
[4] D. J. Nieves, Y. Li, D. G. Fernig, R. Lévy, *R. Soc. Open Sci.* **2015**, *2*.
[5] Z. Qin, J. C. Bischof, *Chem. Soc. Rev.* **2012**, *41*, 1191–1217.
[6] P. Berto, M. S. A. Mohamed, H. Rigneault, G. Baffou, *Phys. Rev. B - Condens. Matter Mater. Phys.* **2014**, *90*, 1–12.
[7] A. O. Govorov, W. Zhang, T. Skeini, H. Richardson, J. Lee, N. A. Kotov, *Nanoscale Res. Lett.* **2006**, *1*, 84–90.
[8] G. Hüttmann, B. Radt, J. Serbin, R. Birngruber, *Ther. Laser Appl. Laser-Tissue Interact. Proc. SPIE* **2003**, *5142*, 88–95.
[9] D. Arnlund, L. C. Johansson, C. Wickstrand, A. Barty, G. J. Williams, E. Malmerberg, J. Davidsson, D. Milathianaki, D. P. DePonte, R. L. Shoeman, et al., *Nat. Methods* **2014**, *11*, 923–6.
[10] G. Wang, I. A. Kaltashov, *Anal. Chem.* **2014**, *86*, 7293–7298.
[11] M. S. Robinson, D. A. Sahlender, S. D. Foster, *Dev. Cell* **2010**, *18*, 324–331.
[12] J. Lin, S. Sann, K. Zhou, S. Nabavi, C. Proulx, R. Malinow, Y. Jin, R. Tsien, *Neuron* **2013**, *79*, 241–253.

FEASIBILITY STUDY OF A NEW THERMAL PLASTY BALLOON

S Zhao (1),Jincheng Zou(1), Yuntao Ma(1), A. Zhang* (1), Lisa X. Xu (1, 2)

(1) School of Biomedical Engineering
Shanghai Jiao Tong University
Shanghai, China

(2) Med-□ Institute
Shanghai Jiao Tong University
Shanghai, China

INTRODUCTION

Atherosclerosis, the narrowing of the blood vessels is a life threatening disease, as it is the major cause of stroke, cardiovascular disease[1]. Percutaneous Transluminal Coronary Angioplasty (PTCA) is the major treatment in clinic. It uses mechanical force to open up the occluded passage of the vessel by dilation of the angioplasty balloon. Stents would also be positioned in the diseased region if necessary. However the forces also caused the endothelium injury, which triggers the upcoming inflammatory process leading to restenosis. The proliferation and migration of the vascular smooth muscle cells (VSMCs) also plays a significant role in restenosis[2]. Recently Tang *et al* [3] have identified the multipotent vascular stem cells (MVSCs) as the source of the VSMC proliferation.

Thermal methods have also been used to treat the atherosclerosis. McCaslin et al combined PTCA with cryoplasty and used mild hypothermia (-15°C to -10◦C), to induce apoptosis of smooth muscle cells and prevent the narrowing of affected arteries[4]. However, Fossaceca et al obtained a contrary result, that cryo methods can't reach a promising therapeutic effect, and the vessels re-stenosed after 6 months [5]. Becker et al designed a gold-electrode-embedded balloon and found that radio frequency can fuse tunica media and intima, without damaging the integrity of the vascular structure[6]. Other studies also found that by combining the RF ablation with PTCA has a much better therapeutic outcome[7]. However, there are still high rates of restenosis after treatment[8].

In order to treat the unwanted region of the plaque, suppress the proliferation and differentiation of MVSCs, while protecting the endothelium cells, a novel thermal balloon is proposed. As shown in Figure 1, the micro-electrodes manufactured on the surface of balloon are connected to the Radiofrequency generator with individual control on each electrode. Meanwhile the cooling agent is delivered into the balloon which would decrease the temperature on the inner surface of the blood vessel due to RF heating and dilate the balloon at the same time. This paper investigates the feasibility of this design which includes the heating pattern of the microelectrodes and correspondent influencing factors of the heating.

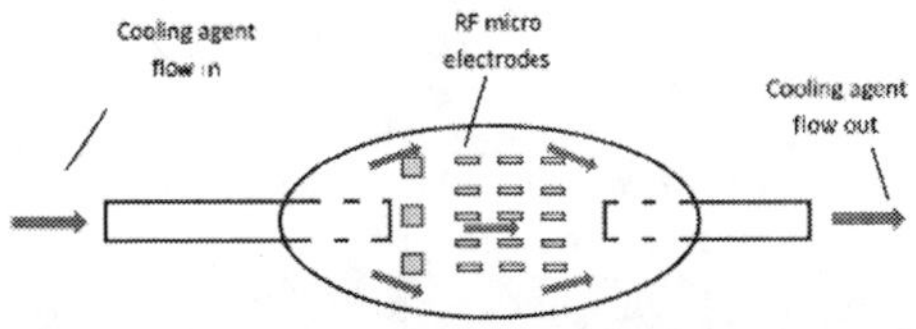

Figure 1 Schematic illustration of the designed RF angioplasty balloon

METHODS

Microelectrodes thermal balloon manufacture: A microelectrodes patch is made with the traditional manufacture technology of flexible printed cable board (FPCB). On the patch, different sizes of the electrodes and distances between the electrodes can be made.

Phantom preparing: To mimic the electrical property of human body repeatedly, a phantom was applied in our study. Our phantom mixture contains 88.5% deionized water, 5.5% gelatin, 0.46% NaCl powder and 5.54% formaldehyde solution (volume concentration:37%). To mimic the lipid pool inside the atherosclerosis plaque, a small piece of pig fat is placed in the middle of the phantom gel as shown in Figure 2a. To investigate the influence of the electrodes spacing distance on the heating pattern, a block of gel with 1mm in height is used.

Temperature measurement: For evaluation of the heating ability of the design for the plaque, the experiment setup is shown in Figure 2a.

While for study of the influence of the electrodes distance on RF heating pattern, the setup shown in Figure 2b is used. The heating ability of the design. The infrared camera A310 (FLIR, Oregon, USA) with a micro lens (the resolution is 50 μm) is used for temperature measurement. The distance between the electrodes is 1.00, 1.25, 1.50, 1.75, 2.00, 2.25 mm, respectively. The heating lasts for 60 seconds.

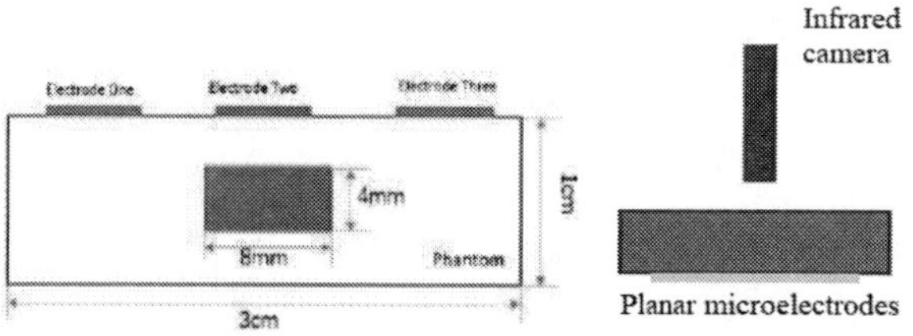

Figure 2 Experimental Configuration. a. for mmimicking of the real plaque. b. for investigation of the influence of electrodes spacing distance.

RESULTS AND DISCUSSION

The temperature map on the surface of the gel mimicking the plaque during different electrodes on/off mode is shown in Figure 3. The "+" means the electrode is used as positive, while "-"means the opposite. Obviously found from Figure 3, when the 3 adjacent electrodes are all used as positive electrodes, the heating range is the least and only a thin layer near the electrodes are heated. While when the center electrode is set to be the opposite of the two electrodes on its two sides, the penetration is the deepest, and the pig fat (as labeled by the dotted line) in the middle could be effectively heated. This result suggest that for the lipid pool in the plaque, a centerized heating is necessary, and it can be realized by control of the small electrodes. 21

Figure 3 Heating pattern of three electrodes with different on/off mode.

The results indicating the effect of the electrode spacing interval are shown in Figure 4. It could be seen that the shape of the high temperature region (the red region) changed from the circle to the ellipse with the electrodes spacing increased. The area with a temperature increase larger than 13°C is the largest when the spacing between two electrodes is 1mm. This area also decreases with the electrodes spacing.

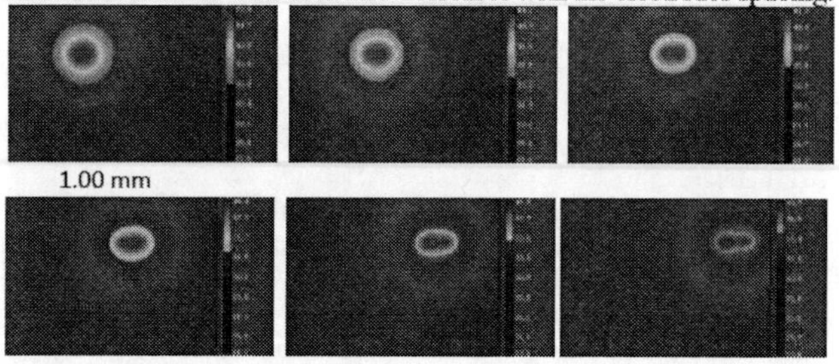

Figure 4 Temperature Distribution on the surface of the gel when heated with two electrodes (one positive and the other one negative). The distance between the two electrodes are 1mm, 1.25, 1.5mm, 1.75, 2.0mm, 2.25mm respectively

The increasing of the maximum temperature on the gel surface with time is shown in Figure 5. The maximum temperature decreases with the increase of the electrodes spacing. It increases very fast at the beginning of the heating, and gradually slows down until equilibrium is reached.

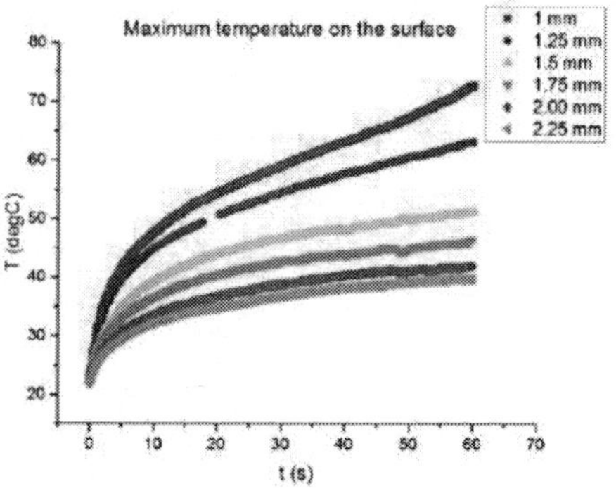

Figure 5 The Surface Temperature Varies by the Time

CONCLUSION

This study demonstrates the heating ability of microelectrodes array and its feasibility for treatment of the atherosclerosis plaque. Through control of the electrodes, a centralized heating can be realized. The distance between two electrodes remarkably affects the heating ability. The temperature on the surface which is 1 mm away from the electrode plane could increase by 50 degC. But with the distance increase, the high temperature region is larger, at the same time, the maximum temperature is lower. The heating is more uniform.

ACKNOWLEDGEMENTS

This work was supported by National Natural Science Foundation of China (NSFC 51476101) and Ministry of Science and Technology of China (2016YFC0106200).

REFERENCES

1. Mozaffarian. Heart Disease and Stroke Statistics-2015 Update: A Report From the American Heart Association (vol 131, pg e29, 2015). *Circulation* 2016,**133**:E417-E417.
2. Ross R. The pathogenesis of atherosclerosis: a perspective for the 1990s. *Nature* 1993,**362**:801-809.
3. Tang ZY, Wang AJ, Yuan FL, Yan ZQ, Liu B, Chu JS, *et al.* Differentiation of multipotent vascular stem cells contributes to vascular diseases. *Nature Communications* 2012,**3**:13.
4. Fava M, Loyola S, Polydorou A, Papapavlou P, Polydorou A, Mendiz O, *et al.* Cryoplasty for femoropopliteal arterial disease: late angiographic results of initial human experience. *Journal of vascular and interventional radiology* 2004,**15**:1239-1243.
5. Fossaceca R, Guzzardi G, Di Terlizzi M, Divenuto I, Malatesta E, Cerini P, *et al.* Comparison of cryoplasty and conventional angioplasty for treating stenotic-occlusive lesions of the femoropopliteal arteries in diabetic patients: immediate, mid-term and long-term results. *La radiologia medica* 2012,**117**:1176-1189.
6. BECKER GJ, LEE BI, WALLER BF, BARRY KJ, KAPLAN J, CONNOLLY R, *et al.* Radiofrequency Balloon Angioplasty Rationale and Proof of Principle. *Investigative radiology* 1988,**23**:810-817.
7. Fram DB, Gillam LD, Aretz TA, Tangco RV, Mitchel JF, Fisher JP, *et al.* Low pressure radiofrequency balloon angioplasty: evaluation in porcine peripheral arteries. *Journal of the American College of Cardiology* 1993,**21**:1512-1521.
8. Sreeram N, Townsend P, Morton D. Radiofrequency thermal balloon angioplasty in an experimental model of peripheral arterial stenosis. *International journal of cardiology* 2000,**74**:25-32.

PRO-ANGIOGENIC HEMATOPOIETIC CELLS MEDIATE PATHOLOGIC REMODELING DURING PULMONARY HYPERTENSION THROUGH SEROTONIN 2B RECEPTOR SIGNALING

Nathaniel C. Bloodworth (1), James D. West (2), Christa Gaskill (2), Santhi Gladson (2), Sheila Shay (2), Susan Majka (2), and W. David Merryman (1)

(1) Biomedical Engineering
Vanderbilt University
Nashville, TN, United States

(2) Allergy, Pulmonary, and Critical Care Medicine
Vanderbilt University
Nashville, TN, United States

INTRODUCTION

Pulmonary arterial hypertension (PAH) is a fatal illness of the pulmonary vasculature with limited options for management and no disease-modifying therapies. The pathogenesis of PAH is characterized by small vessel stiffening and remodeling. Evidence from both humans and animal models suggests bone marrow (BM) derived pro-angiogenic cells may contribute to this remodeling process, though to date no study has conclusively demonstrated their necessity or role in disease pathogenesis[1]. Additionally, expression of the serotonin 2B receptor (HTR2B) in the myeloid compartment is necessary for mice to develop hypoxic pulmonary hypertension, with HTR2B ablation resulting in lower percentages of BM derived cells expressing endothelial markers[2]. This study evaluates the hypothesis that pro-angiogenic hematopoietic cells contribute directly to PAH by mediating arteriole remodeling. This study further seeks to determine if HTR2B signaling mediates pro-angiogenic hematopoietic cell recruitment to the lung vasculature during PAH.

METHODS

BM Transplant. All animal protocols were approved by the Vanderbilt IACUC prior to the study. Adult C57BL/6 cogenic recipients (CD45.1) were irradiated with a split 10 Gy dose from a Cs^{137} source. Following irradiation, a 100μL solution of 2.5×10^5 cells from age- and sex-matched donors was administered retro-orbitally. Mice were maintained on acidified water with neomycin and polymyxin B for 2 weeks, followed by an additional 8 weeks to allow for full reconstitution of the hematopoietic system.

Transgenic Animals. For controlled ablation of BM-derived pro-angiogenic cells, donor cells were collected from a transgenic animal with a tamoxifen-inducible endothelial-specific Cre and a floxed-STOP diphtheria toxin driven by the ROSA26 promoter (5'-SCL-Cre^{ERT2}/$DTa^{fl/-}$). For lineage tracing studies, donor marrow cells were collected from animals with a Tie2 promoter driven Cre and a floxed-STOP YFP driven by the ROSA26 promoter (Tie2-Cre/$YFP^{fl/-}$).

Induction of PAH. To induce PAH, SU5416 was administered at 20mg/kg/week while mice were maintained in hypoxia (10% O_2) for 3 weeks. Control animals were maintained on room air while receiving vehicle injections. Mice receiving 5'-SCL-Cre^{ERT2}/$DTa^{fl/-}$ BM were simultaneously given intraperitoneal tamoxifen at 2mg every other day in order to destroy BM-derived pro-angiogenic cells, while mice receiving Tie2-Cre/$YFP^{fl/-}$ BM were implanted with subcutaneous Alzet pumps delivering the HTR2B antagonist SB204741 (1mg/kg/day). Control animals received vehicle injections or pumps. Animals were hemodynamically phenotyped following insertion of a catheter into the right ventricle while under anesthesia (Avertin) to measure right ventricular systolic pressures (RVSP).

Flow Cytometry. Lung cells were harvested and stained with CD31 PE-Cy7 and CD11b AF700. Peripheral blood mononucleocytes were harvested and stained with CD45.1 PE and CD45.2 PerCP-Cy5.5 to assess engraftment. The BD LSRFortessa was used for data acquisition and FlowJo for data analysis.

Immunohistochemistry (IHC). Paraformaldehyde-fixed OCT-embedded lung sections were stained for α-smooth muscle actin (αSMA) to assess vessel remodeling, αSMA, CD31, and GFP to assess BM-cell localization, and imaged at 20x on an Olympus BX53.

RESULTS

BM-Derived Pro-angiogenic Cells are Necessary for PAH Pathogenesis. In order to determine if BM-derived pro-angiogenic cells contribute directly to PAH pathogenesis, we bred transgenic animals that express a floxed-STOP diphtheria toxin activated by Cre recombinase specific to endothelial cells (5'-SCL-Cre^{ERT2}/$DTa^{fl/-}$) to use as BM donors. Administration of tamoxifen activates Cre in BM-derived pro-angiogenic cells[3] and subsequently allows for the expression of diphtheria toxin in these cells, efficiently ablating them. After simultaneous administration of tamoxifen and induction of PAH, we found that animals receiving tamoxifen have significantly lower RVSP than those receiving vehicle (Figure 1A) as well as significantly fewer remodeled arterioles (Figure 1B), confirming their necessity for disease pathogenesis.

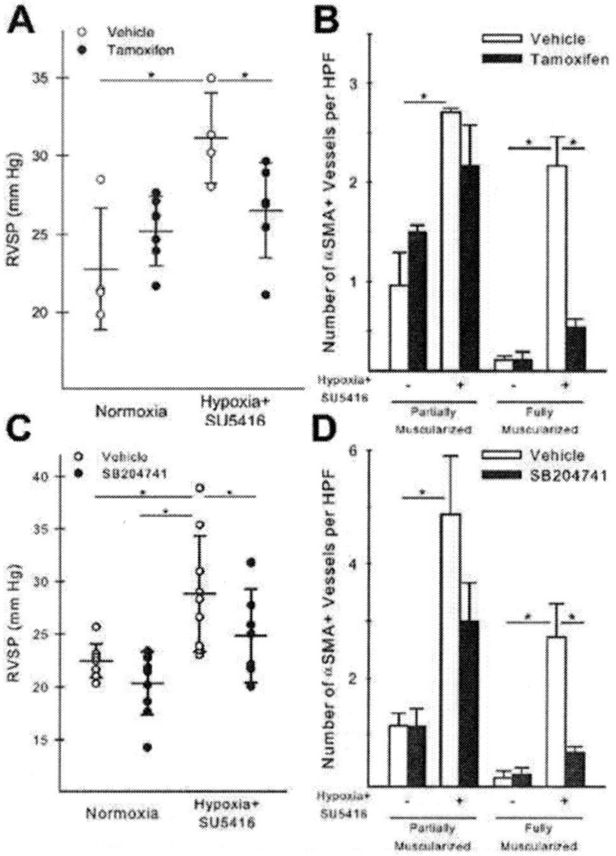

Figure 1: (A-B) Both BM-derived pro-agngiogenic cell depletion and (B-C) HTR2B antagonism reduce RVSP and vessel remodeling in a mouse model of PAH (Mean +/- S.E., n=4-8, *p<0.05 2-way ANOVA followed by Holm-Sidak post-hoc).

HTR2B Antagonism Reduces the Fraction of BM-Derived Pro-angiogenic Cells in the Lungs During PAH. To determine if HTR2B antagonism would reduce BM-derived pro-angiogenic cell recruitment, we bred transgenic animals expressing YFP in all BM cells (Tie2 is expressed on common hematopoietic progenitors[4]) to use as BM donors. Following induction of PAH, we found that SB204741 treatment lowered RVSP and reduced the number of remodeled arterioles (Figure 1C-D). Additionally, we found that these animals also had a smaller fraction of YFP+CD31+CD11b- cells in their lungs (Figure 2A-B), with fewer BM-derived CD31+ cells in the walls of their pulmonary arterioles than vehicle treated controls.

DISCUSSION

This is the first study to conclusively demonstrate that BM-derived pro-angiogenic cells contribute directly to small vessel

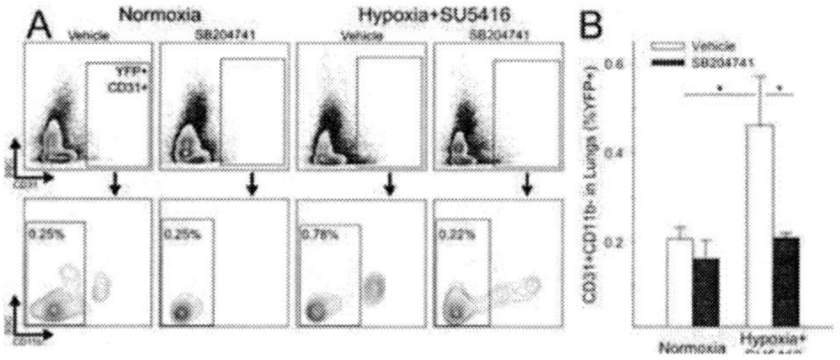

Figure 2: (A-B) HTR2B antagonism reduces the fraction of BM-derived pro-angiogenic cells in the lungs during PAH (n=3-4). (C) HTR2B antagonism normalizes the stiffness of pulmonary arterioles when compared to vehicle treated controls. (n=3) (Mean +/- S.E. *p<0.05, 2-way ANOVA followed by Holm-Sidak post-hoc)

remodeling and PAH development, and that their recruitment to the lungs is predicated on intact HTR2B signaling. Furthermore, evidence from IHC and BMPR2 mutant mice (published previously[5]) suggests that HTR2B antagonism may alter pathologic vessel stiffening, an early but important consequence of vessel remodeling. Given our results it is quite possible that the previously observed changes in vessel elastic modulus occur secondary to changes in BM-derived pro-angiogenic cell recruitment and activation. While it is clear from our results that this cell population contributes to the PAH phenotype, it is also quite probable that other independent pathologic mechanisms also contribute alongside as evidenced from the partial (but significant) reduction in RVSP and non-significant changes in the number of partially muscularized pulmonary arterioles. Further work remains to determine the precise mechanism whereby HTR2B antagonism reduces the number of these cells and exactly how they mediate small vessel remodeling.

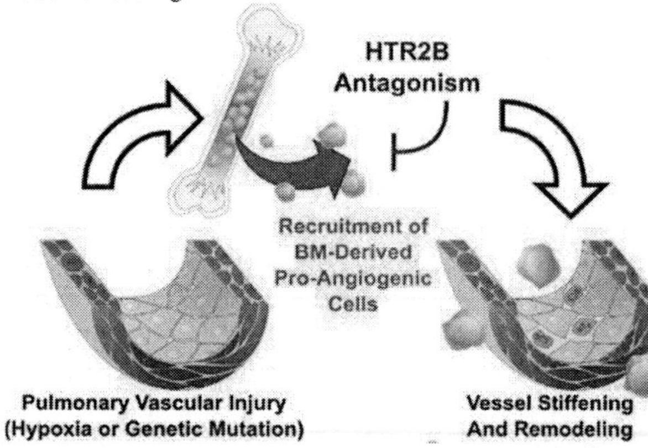

Figure 3: BM-derived pro-angiogenic cells are recruited to the pulmonary vasculature in a manner dependent on HTR2B signaling where they help mediate remodeling.

ACKNOWLEDGEMENTS

This work was supported by the NIH (R01 HL115103, R35 HL135790, and F30 HL126280) and the AHA (15PRE23260021).

REFERENCES

[1] Lanzola, E et al. *Pulm Circ.* 4:781-791, 2013.
[2] Launay, JM et al. *Blood.* 7:1772-1800, 2012.
[3] Gothert, JR et al. *Blood.* 6:1769-1777, 2004.
[4] Tang, Y et al. *Genesis.* 9:563-567, 2010.
[5] West, JD et al. *PLoS One.* 11:e0148657, 2016.

SB³C2017
Summer Biomechanics, Bioengineering and Biotransport Conference
June 21 – 24, Tucson, AZ, USA

CONSTRUCTING ANALYSIS NURBS FROM DISCRETE IMAGE-BASED MODELS

Adam R. Updegrove (1), Nathan M. Wilson (2), Shawn C. Shadden (1)

(1) Department of Mechanical Engineering
University of California, Berkeley
Berkeley, CA, USA

(2) Open Source Medical Software Corporation
Santa Monica, CA, USA

INTRODUCTION

Isogeometric analysis (IGA) is an emerging technique in finite element analysis in which aspects of an analytically defined geometry can be used directly for analysis; thus, skipping the explicit meshing of the geometry. In order for the geometry to be used directly for analysis, the geometry must be defined using basis functions that satisfy the requirements of a finite element basis and contain a globally valid parametric space. One such definition is a non-uniform rational B-spline (NURBS) surface or volume, which is the industry standard for computer aided design (CAD) and the original definition utilized in isogeometric analysis [1]. Isogeometric analysis is gaining momentum in numerical analysis due to improved accuracies, quicker simulation times, and the lack of an explicit finite element mesh [2]. This makes it superior for a wide range of problems including fluid-structure interaction and design optimization. Image-based modeling and simulation-based treatment planning are applications for which isogeometric analysis could have substantial impact. Thus, methods to form analysis suitable NURBS of image-based geometries are required.

It remains a challenge to represent complex geometries in an analysis suitable NURBS format for isogeometric analysis [3]. In addition, most image segmentation methods create a discrete (triangulated) boundary representation surfaces. In this work, we discuss a method to create analysis suitable NURBS from input triangulated surfaces of image-based vascular geometries, which can be particularly challenging given their branched and multiscale structure. This work is carried out using the open-source image-based cardiovascular modeling software, SimVascular [4].

METHODS

Currently, there are few methods to create analysis suitable NURBS for vascular geometries. Zhang et al. [3] outlined methods to construct NURBS from isosurfaces of medical image data in cases of smooth, idealized geometry. Al Akhras et al. [5] focused on forming analysis suitable NURBS from triangulated surfaces of varying geometries, but unsuitable for vascular geometries containing branching structures, long tubular regions, and substantial changes in scale.

Our process to create an analysis suitable vascular geometry follows four steps: (1) Decomposing the geometry into patches, (2) forming a polycube structure for the surface, (3) parameterizing each individual patch, and (4) forming the analysis suitable NURBS geometry. We describe each of these steps below.

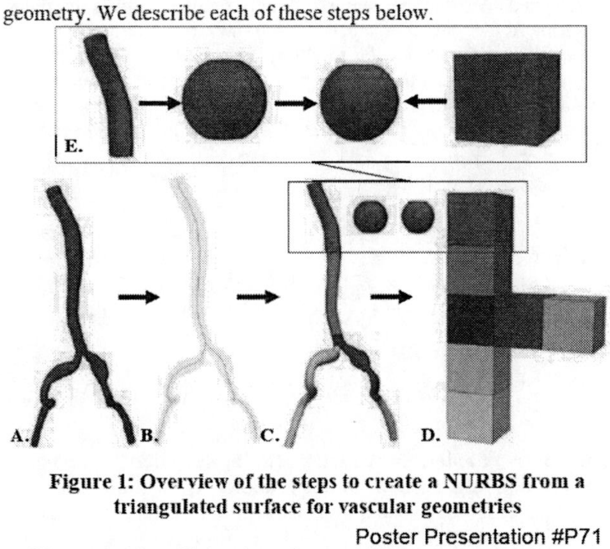

Figure 1: Overview of the steps to create a NURBS from a triangulated surface for vascular geometries

Poster Presentation #P71

(1) Decomposing the geometry into patches:

We use the VMTK (http://vmtk.org) to generate the centerlines of the input triangulated surface (Figure 1). The output is a series of centerlines containing field data at each point indicating the maximum inscribed sphere radius of the geometry at that point. We then separate the geometry into patches based on the centerlines.

(2) Forming the polycube structure for the surface

NURBS are mutli-dimensional splines where the splines span a parametric space, which in 2D is topologically equivalent to a rectangle and in 3D is topologically equivalent to a cuboid. Thus, the input geometry needs to be decomposed into a series of rectangles (in 2D) and cuboids (in 3D). To achieve this, we add an additional patch at each bifurcation that acts as a link between the branching patches (Figure 1C). The polycube structure is then formed based on the decomposed geometry (Figure 1D).

(3) Parameterizing each individual patch

We form the rectangular parameter lines that act as the guidelines for the multi-dimensional splines of NURBS; thus, the parameter lines need to be as uniform as possible. The typical method for forming parameterization lines on a complex surface is to map the surface or portions of the surface to another domain where the uniform parameterization lines are more easily formed. In this case, we conformally map (preserving angle magnitude and direction) each individual patch to the spherical domain (Figure 1E). To reduce distortions, we map long branches to multiple spheres. It is necessary that parameterization lines match at patch boundaries, so constraints are set on the mapping to ensure we obtain matching parameter lines and a globally valid NURBS geometry.

(4) Forming the analysis suitable NURBS geometry

A NURBS surface has the following definition:

$$S(u,v) = \sum_{i=0}^{n} \sum_{j=0}^{m} \frac{N_{i,p}(u)\, N_{j,q}(v)\, w_{i,j}}{\sum_{k=0}^{n} \sum_{l=0}^{m} N_{k,p}(u)\, N_{l,q}(v)\, w_{k,l}} \boldsymbol{P}_{i,j}, \quad (1)$$

where $N(u)$ and $N(v)$ are the basis functions in the respective u and v parameter directions. The $\boldsymbol{P}_{i,j}$ are the control points of the surface, and $w_{i,j}$ are the corresponding weights for the control points. We solve for the control points of each individual patch using global surface fitting techniques. At patch boundaries, we apply derivative information to insure continuity across patch boundaries.

RESULTS

We show step-by-step results of the algorithms for a bifurcation geometry. In many cases, as is displayed in Figure 2D, individual vessels may need to be segmented into smaller patches to reduce distortion in the mappings. Because each piece is conformally mapped to the sphere (Figure 3), the resulting parameter lines of each piece (Figure 3E) are smooth and conform to the original geometry. The resolution of the parameter lines is variable and dependent on the size of the geometry being parameterized. The final full parameterization

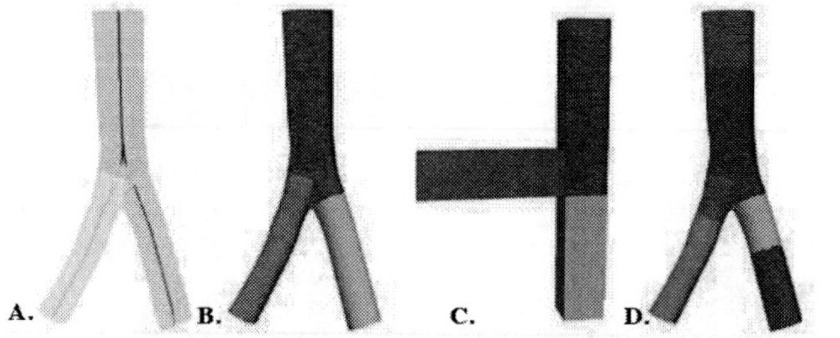

Figure 4: (A.) Centerline extraction, (B.) initial patch decomposition, (C.) polycube construction, and (D.) smaller segmentations to reduce mapping distortion.

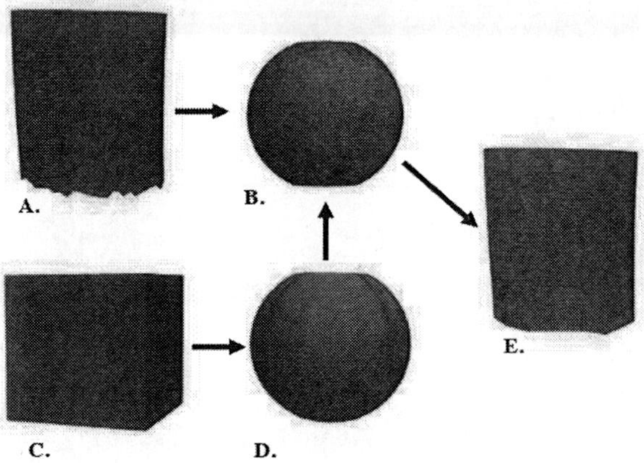

Figure 2: Parameterizing an individual patch. (A.) A portion of the surface, (B.) the corresponding conformal mapping to the sphere, (C.) the corresponding cube, (D.) the cube mapped to the sphere, and (E.) the final parameterization of the surface portion.

matches at the boundaries between the individual pieces, which allows for a smooth NURBS surface to be fitted to the geometry. Derivative constraints on the boundaries of the NURBS patches provide a continuous NURBS surface (Figure 4).

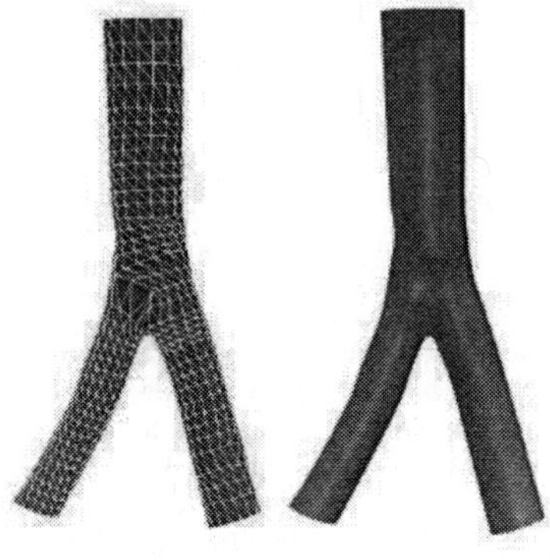

Figure 3: The full parameterization of the surface and the corresponding NURBS lofted surface of the geometry.

DISCUSSION

This work demonstrates a novel framework to create an analysis suitable NURBS representation from a discrete image-based segmentation. While we do not discuss forming a final volumetric representation, this is a straightforward extension on the presented framework. The algorithm computation time is reasonable; for a dense triangulation of a single bifurcation, the algorithm operates in under a minute. In addition, the algorithm currently handles bifurcations well, but trifurcations and more general 3D geometries, including vascular geometries with aneurysms, can require user intervention in specifying patch decompositions. The algorithms are currently being expanded to robustly handle these more complicated vascular scenarios.

ACKNOWLEDGEMENTS

Support was provided by an NSF Graduate Fellowship (NSFGF) and NSF SSI Award 1407834. SimVascular is hosted by simtk.org.

REFERENCES

[1] Hughes, T et al., *Comput Methods Appl Mech Eng, 194*:4135-4195, 2005.
[2] Nguyen, V et al., *Math Comput Simul, 117*:89-116, 2015.
[3] Zhang, Y et al., *Comput Methods Appl Mech Eng, 196*:2943-2959, 2007.
[4] Updegrove, A et al., *Ann Biomed Eng*, 1-17, 2016
[5] Al Akhras, H et al., *Comput Methods Appl Mech Eng*, 2016.

SB[3]C2017
Summer Biomechanics, Bioengineering and Biotransport Conference
June 21 – 24, Tucson, AZ, USA

NANOPARTICLE RE-DISTRIBUTION IN TISSUE-EQUIVALENT GELS INDUCED BY MAGNETIC NANOPARTICLE HYPERTHERMIA

Qimei Gu, Myo Min Zaw, Timothy Munuhe, Ronghui Ma, Liang Zhu

Department of Mechanical Engineering
University of Maryland Baltimore County
Baltimore, MD, USA

INTRODUCTION

Magnetic nanoparticles have gained prominence in recent years for use in clinical applications such as imaging, drug delivery, and hyperthermia. Magnetic nanoparticle hyperthermia is a minimally invasive and effective approach for confining heat in tumors with collateral damage below acceptable levels. Our previous *in vivo* animal experiments have suggested possible change in nanoparticle distribution in tumors after heating [1]. Although not being directly visualized in most previous studies, our observation is consistent with a phenomenon known as "thermal by-stander effect" [2]. Previous experimental studies have shown a much more uniform temperature elevation in tumors after repeated heating treatments using nanoparticles. Repeated heating has been used as a technique to kill tumor cells completely in magnetic nanoparticle hyperthermia [2]. We speculate that nanoparticle hyperthermia increases the tumor porosity due to heat-induced tumor cell death. It further results in changes in nanoparticle transport properties in tumors, such as nanoparticle diffusivity in the extracellular matrix. The migration of nanoparticles in a porous medium is possible once the diffusion force overcomes other forces trapping them in their initial locations.

In our previous experiments there was a significant amount of nanofluid leaking out through the needle track, as well as seeping into mouse tissue [1]. Thus, our study was limited by having a very large standard deviation in the obtained nanoparticle distribution volume. In addition, heterogeneous tumor structure and high pressure induced crack formation in the tumors also added uncertainty on the calculated results to draw concrete conclusions. In this study, we plan to use agarose gels injected with nanofluids with a very slow infusion rate to minimize crack formation, therefore, to isolate the effect of heating on nanoparticle redistribution. Agarose gel is a widely used alternative to animal tissue in tests of drug delivery in biological systems. It has similar convection/diffusion properties and extracellular space to that of tumors. Gel concentration and solidification temperature can be adjusted to mimic different types of tissues. It should be noted that gels are homogeneous in comparison to the heterogeneous tumor morphology, allowing evaluation of 1-D nanofluid advection and diffusion with an obtained spherical deposition pattern.

In this study, a commercially available ferrofluid containing magnetic nanoparticles was injected into tissue equivalent agarose gels, to study nanoparticle spreading in the gel. A high-resolution microCT imaging system was used to investigate nanoparticle concentration distribution before and after the gel specimen was subjected to heating of 15 minutes induced by an alternating magnetic field. The high resolution three-dimensional microCT images of the density distribution were used to quantitatively evaluate the role played by heating on nanoparticle redistribution.

METHODS

The agarose gel was prepared by dispersing agarose powder (Sigma-Aldrich, Saint-Louis, MO) in a 10% buffer solution (Tris-Borate EDTA, Gibco BRL, Rockville, MD). The mixture was then heated and later cooled down in room temperature until solidification [3]. The gel concentration is 0.1% in this study.

Water based ferrofluids (EMG705 series, Ferrotec (USA) Corporation, Nashua, NH) with a concentration of 5.8% by volume and a particle size of 10 nm were injected in the agarose gel via a syringe pump (Norm-ject, Fischer Scientific, Springfield, NJ), which enables a precise control of the amount and infusion rate. The injection amount was selected as 0.01 cc and the infusion rate was adjusted to be as low as 0.5 μL/min, so that the resulted nanoparticle distribution was almost spherical. Figure 1 illustrates a semi-transparent gel with or without injected nanofluid. The nanoparticle region is represented by the dark color. Needle track was barely seen.

Our previous experience on nanoparticle spreading has suggested that there is very little change in the nanoparticle spreading within 12 hours after the injection ends [3]. A high-resolution microCT imaging system (Skyscan 1172, Microphotonics, PA) was used to image the density of the agarose gel with nanoparticles. The gel block was placed in a low-density Styrofoam, which would absorb little X-ray during the scan. A medium resolution scan of 34.6 μm (pixel size) was done at 100 kV and 100 μA without a filter. The total CT scan time was approximately 20 minutes. The final microCT image gives gray scale values ranging from 0 to 255. The images acquired from the microCT scan were reconstructed using the NRecon® software package provided by Microphotonics. Scan parameters and reconstruction parameters were kept the same for all the gel blocks in the study, therefore, the same final voxel gray scale value of the images is correlated to the same local density. The average value of the gel without nanofluid infusion is used as the cutoff to determine locations of voxels containing nanoparticles. The nanoparticle distribution volume is defined as the total volume in the gel with nanoparticle presence, as a voxel location with a gray scale value higher than the cutoff value.

After the nanofluid injection, the agarose gel specimen was first scanned by the microCT system. The same gel specimen was then placed in an alternating magnetic field (5 kA/m, 190 kHz) and heated for 15 minutes. Only the surface temperature of the gel block was recorded using an infrared camera and temperature elevation is approximately 10°C from its initial value. No thermocouples were inserted into the gel due to concern of causing small cracks in the gel and nanofluid backflow from the needle track. After the heating, the specimen was immediately scanned again to evaluate whether heating induced any further particle spreading from the infusion site.

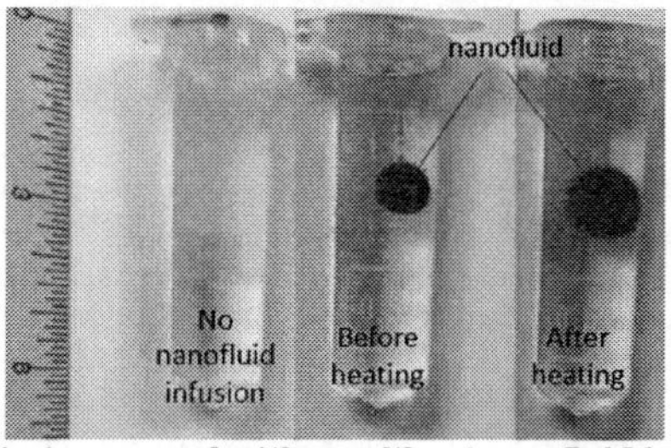

Figure 1: Agarose gel with or without nanofluid injection.

RESULTS

Visual examination of the same gel specimen before and after heating (Figure 1) shows that the dark black region becomes bigger in volume, implying that the nanofluid has migrated outwards, possibly caused by local heating in the gel. However, since nanoparticles may not be uniformly distributed in the nanofluid/gel after infusion, the dark color region alone is not sufficient to quantify nanoparticle spreading in gels. Figure 2 gives the microCT slices of a gel without infusion, and nanoparticle distribution in a gel before or after heating. Although microCT (mm) does not allow direct visualization of individual nanoparticles (nm), the accumulation of nanoparticles results in density variations in gels that can be detected by the microCT system. The average gray scale value in the gel sample without nanofluid infusion is 65, which is the cutoff value for later calculation of the nanoparticle distribution volume. Before heating, the nanoparticles occupy a circular region with elevated gray scale values, shown as the brighter region in the middle panel in Figure 2. After heating, the occupied region by nanoparticles is getting bigger, however, less bright with a more blurry boundary.

Figure 3 illustrates the 3-D color contours of the CT gray value distribution (top row), and gray value distribution on the cross-sectional plane A-A (bottom row). 3-D visualization of the region with nanoparticle presence deviates from a spherical shape. The nanoparticle distribution inside the gel before heating is not uniform, varying from 220 at the center near the infusion site to 135 at the periphery. After heating, the region is getting bigger, however, the nanoparticles are less concentrated than that before heating with a more uniform gray scale value of approximately 190. After heating the nanoparticle penetration front is also less defined than that before heating.

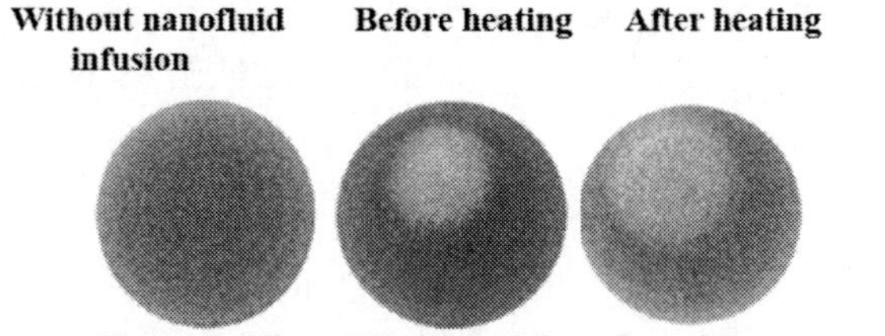

Figure 2: MicoroCT slices of the gel specimens.

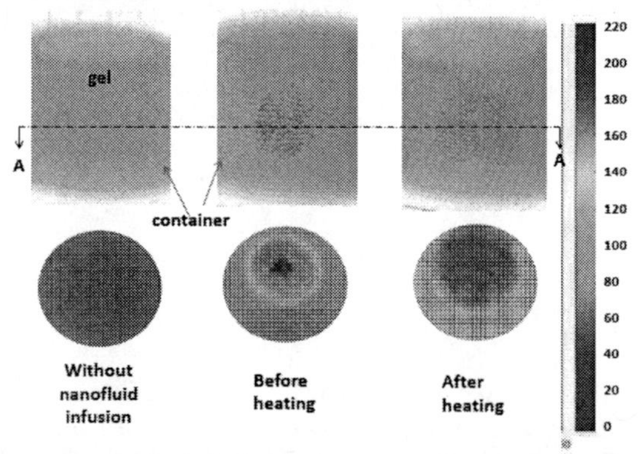

Figure 3: 3-D color contours of nanoparticle distribution in gels before and after heating.

Experiments were repeated ten times to show the variability of the data. The nanoparticle distribution volume is calculated as 0.58±0.15 cm^3 (mean±SD, n=10) using a cutoff gray scale value of 65. After heating the nanoparticle distribution volume increases by 26% to 0.73±0.14 cm^3 (n=10). Student t-test was performed to show that the p value is 0.03, therefore, the difference between the average values of the two groups is statistically significant.

In summary, we performe experiments to evaluate nanoparticle distribution and re-distribution after heating in a very diluted gel, with a controlled infusion rate. The obtained nanoparticle region is almost spherical, manifesting minimized crack formation and back flow during the infusion. Nanoparticle presence in the gel can be visualized in microCT images and the nanoparticle concentration in the gel is not uniform. Magnetic nanoparticle heating induced by an alternating magnetic field for 15 minutes alters nanoparticle distribution in the gel via pushing nanoparticles further away from the infusion site, resulting in a 26% increase in nanoparticle distribution volume.

ACKNOWLEDGEMENTS

This study was supported by an NSF research grant (CBET-1335958). The research was performed in partial fulfilment of the requirements for the PhD degree from UMBC by Qimei Gu.

REFERENCES

[1] Attaluri, A et al., *Int J Hyperthermia,* 27(5):491–502, 2011.
[2] Jordan, A et al. *Int J Hyperthermia,* 13(6):587-605, 1997.
[3] Attaluri, A et al., *Journal of Heat Transfer,* 133:011003(1-5), 2011.

THERMAL EXPANSION OF THE CRYOPROTECTIVE AGENT COCKTAIL DP6 IN COMBINATION WITH VARIOUS SYNTHETIC ICE MODULATORS

Prem K. Solanki and Yoed Rabin

Biothermal Technology Laboratory
Department of Mechanical Engineering
Carnegie Mellon University
Pittsburgh, PA, USA, 15213

INTRODUCTION

Differential thermal expansion is the driving mechanism of thermo-mechanical stress during cryopreservation by vitrification (*vitreous* means *glassy* in Latin). Thermo-mechanical stress can be harmful in cryopreservation, causing structural damage such as fractures and plastic deformations. Its effect only intensifies with the size of the specimen, with wide implications on organ banking and transplant medicine. Unfortunately, the correlation between structural damage and the development of thermo-mechanical stress is largely unexplored in the context of cryopreservation.

Cryoprotective agents (CPAs) are commonly used in order to suppress crystallization in cryopreservation processes, where ice formation is the cornerstone of cryoinjury. Unfortunately, relatively high cooling rates are required in order to facilitate vitrification, which may also intensify the resulting thermo-mechanical stress. One alternative to reduce the cooling rate while controlling the rate of ice growth is by introducing synthetic ice modulators (SIMs) into the CPA cocktail [1]. Recent studies on thermal expansion of the CPA cocktail DP6 demonstrated that SIMs indeed hamper crystal growth without affecting the thermal expansion coefficient of the mixture [2].

As part of an ongoing effort to investigate thermo-mechanical effects in cryopreservation, the current study expands on previous efforts [2,3] to study thermal expansion of SIM+CPA cocktails. The unique contribution of this study is in measuring the thermal expansion of M22 [4], which has shown very promising results in kidney cryopreservation [5], in measuring various ingredients of M22 when mixed with DP6, which has been studied as a base solution [2,3,6], and in measuring the thermal expansion of DP6 mixed with sucrose, which has recently shown promising results in blood vessels cryopreservation in large samples [7].

METHODS

<u>*Method of Analysis:*</u> The experimental apparatus used in the current study has been reported previously [3,8]. In brief for the completeness of presentation, it consists of a passively controlled closed system, where expansion of the specimen and its holding

Table 1: Coefficients of best-fit polynomial approximation for thermal strain, $\varepsilon = a_2T^2+a_1T+a_0$, and the coefficient of thermal expansion, $\beta=2a_2T+a_1$, where n is the number of data sets for corresponding material and R_ε is the thermal strain ratio measured over the temperature range of -80°C and 20°C with reference to DP6+UCV

Material	Temperature Range, °C	$a_0 \times 10^3$	$a_1 \times 10^4$	$a_2 \times 10^7$	n	$\sigma \times 10^4$	R_ε
DP6	20…-41.9	-3.62	2.14	9.50	7	1.45	N/A
DP6+UCV	20...-87.8	-7.44	2.28	6.84	5	3.67	1
DP6+UCV+1% X1000	20...-84.4	-7.33	2.32	6.57	7	1.15	1.037
DP6+UCV+1% Z1000	20…-84.3	-7.37	2.49	6.55	5	1.87	1.134
DP6+UCV+1% X1000+1%Z1000	20…-84.4	-5.75	2.43	6.72	6	2.68	1.094
DP6+UCV+0.5M Sucrose	20…-81.7	-5.07	2.27	5.16	7	2.82	1.075
M22	20…-91.1	-4.71	2.52	6.24	6	5.61	1.167

chamber drive pressure changes, which is the measured parameter. Simultaneous measurements of temperature by means of thermocouples, facilitate the correlation between pressure and temperature. First, pressure calibration provides the relationship between pressure and displacement in the system. Next, the linear thermal strain in the system is calculated by:

$$\varepsilon = \frac{1}{3}\frac{\Delta V_{\text{CPA}} + \Delta V_{chamber}}{V_0} \quad (1)$$

where ΔV_{CPA} is the volume change in the specimen, $\Delta V_{chamber}$ is the calculated volume change of the CPA vessel, and V_0 is the initial volume of the cooling chamber containing the specimen. The volume change in the cooling chamber is calculated by:

$$\Delta V_{chamber} = 3V_0 \int_{T_0}^{T_i} \beta_{chamber} dT \quad (2)$$

where T_0 and T_i are the initial and final temperatures of the cooling chamber, respectively, $\beta_{chamber}$ is the thermal expansion coefficient of the chamber (brass). Finally, the thermal expansion of the sample is calculated by:

$$\beta = \frac{d\varepsilon}{dT} \quad (3)$$

The uncertainty in the thermal strain is within 4%, when calculated over a temperature range of -90°C and 20°C [8].

Materials: The primary ingredients in DP6 are 3M dimethyl sulfoxide (DMSO) and 3M propylene glycol in Unisol (UCV) as a vehicle solution in the current study. M22 is a more complex cocktail containing 5 permeating cryoprotectants (pCPAs) and 3 non-permeating cryoprotectants (npCPAs). The pCPAs in M22 are DMSO (22.3%), formamide (12.9%), ethylene glycol (16.8%), N-methylformamide (3%), and 3-methoxy-1,2-propanediol (4%), while the npCPAs are polyvinyl pyrrolidone K12 (2.8%), X-1000 ice blocker (1%) and Z-1000 ice blocker (2%). Table 1 lists the tested materials in the current study, where some of the ingredients of M22 where further mixed with DP6 in effort to improve its vitrification qualities. Additionally, DP6 mixed with sucrose has also been tested, following recent promising results of blood vessels vitrification in large volumes [7].

Thermal Protocol: When coated with a layer of a thermal insulator, the cooling chamber is immersed in liquid nitrogen. The boiling of liquid nitrogen on the surface of thermal insulator facilitates cooling. The typical cooling rate ranges from 10°C/min at 0°C and 5°C/min at -94°C, with an average value of 7.3°C/min. These rates are above the critical cooling rate for vitrification of the respective solutions, which range below 3.5°C/min. The solutions become so viscous around -95°C that thermal expansion measurements become impractical at lower temperatures. Passive rewarming is performed by free heat convection to room temperature. The typical rewarming rates range between 10°C/min at -92°C to 2°C/min at 0°C, with an average value of 4.9°C/min. The thermal strain measurements reported here were taken during cooling. In order to study crystallization events during rewarming, differences between thermal strain measurements during rewarming and cooling were further compared [3].

RESULTS AND DISCUSSISON

Table 1 lists the coefficients of polynomial approximation for the measured thermal strain, while Fig. 1 displays the results graphically. One can see from Fig. 1 that the vehicle solution Unisol enables thermal expansion measurements of DP6 down to a lower temperature than that for pure DP6 in water. This is due to the fact that Unisol contains 0.194M glucose, which is a glass promoting agent—it creates

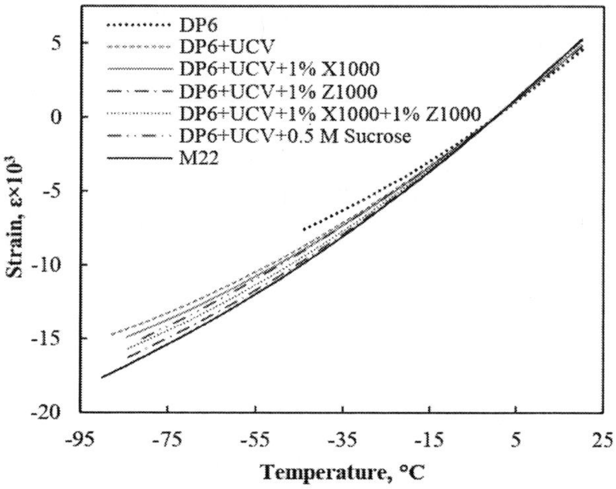

Figure 1 Polynomial approximations of thermal Strain for all the materials used in the study.

more favorable conditions for vitrification. Note that the lower temperature limit for measurements in the current system is around -95°C, when the vitrified material becomes too viscous to flow freely.

DP6 in water begins to crystallize around -35°C at the cooling rates achieved in this study. The addition of SIMs like X1000 and Z1000 to DP6 further suppresses crystallization and, hence, permits thermal expansion measurements down to lower temperatures when compared with DP6 in water.

The addition of SIMs showed an increase in the magnitude of thermal strain, where R_ε in Table 1 is the ratio of thermal strain for a particular DP6+SIM cocktail with respect to that of DP6 in Unisol, when calculated over the temperature range of -80°C and 20°C.

In conclusion, the current study shows that CPA cocktails of DP6 mixed with selected SIMs promote vitrification, which is evident when the SIM cocktail behaves as liquid down to lower temperatures. Interestingly, Unisol also displays vitrification promotion qualities, although it is more generally considered for its biocompatible vehicle solution characteristics. With DP6 in Unisol solution as a reference, linear thermal strain may increase by up to 17% over the temperature range of -80°C and 20°C for the other cocktails tested, which may increase thermo-mechanical stresses when the liquid-like CPA is trapped in a solid-like surroundings. Out of all the cocktails tested, M22 displayed the highest thermal strain, which is consistent with its highest solute concentration.

ACKNOWLEDGEMENTS

This project has been supported by in parts by award R01HL127618 from the National Heart Lung and Blood Institute (NHLBI). The authors would also like to thank Dr. Gregory Fahy, 21st Century Medicine, Inc., and Dr. Michael J. Taylor, Sylvatica, Inc., for their invaluable discussions about the nature of CPAs.

REFERENCES

[1] Taylor, M. J., et al., *Life in the Frozen State*, 603–641, 2004.
[2] Eisenberg, D. P., et al., *Cryobiology*, 68(3):318–326, 2014.
[3] Eisenberg, D. P., et al., *Cryobiology*, 65(2):117–125, 2012.
[4] Fahy, G. M., et al., *Cryobiology*, 48(2):157–178, 2004.
[5] Fahy, G. M., et al., *Organogenesis*, 5(3):167–75, 2009.
[6] Ehrlich, L. E., et al., *Cryobiology*, 73(2):196–202, 2016.
[7] Taylor, M., et al., *Cryobiology*, 73(3), 2016.
[8] Rabin, Y., and Bell, E.,*Cryobiology*, 46(3):254–263, 2003.

SB³C2017
Summer Biomechanics, Bioengineering and Biotransport Conference
June 21 – 24, Tucson, AZ, USA

PREFERENTIAL ENTRAPMENT OF SOLUTES IN ICE PHASE DURING FREEZING OF PROTEIN-CRYOPROTECTANT SOLUTIONS

Sampreeti Jena (1), Raj Suryanarayanan (2), Alptekin Aksan (1)

(1) Department of Mechanical Engineering
University of Minnesota
Minneapolis, Minnesota, USA

(3) Department of Pharmaceutics
University of Minnesota
Minneapolis, Minnesota, USA

INTRODUCTION

Freezing is often utilized for the long term preservation of pharmaceutical and biological specimens for longer retention of their bio-activity [1]. But freezing introduces extreme shifts in pH, osmotic pressure and temperature [1]. When subjected to these harsh conditions, proteins tend to transition, sometimes irreversibly into the thermodynamically favourable, unfolded state. To avoid this, cryo-protectants are added to stabilize the protein molecules against these stresses by the preferential exclusion/binding mechanisms [2].

During freezing, as ice crystals grow, it has been shown that there may not be uniform exclusion of both protein and cryo-protectant by it. Sometimes, either the protein or cryo-protectant may be preferentially entrapped in the ice leading to micro-heterogeneity in the freeze concentrate. Preferential entrapment of the cryo-protectant in ice phase would result in increasing relative concentration of the protein in the freeze concentrate and thus reduced protection from the cryo-protectant. The reverse would suggest the protein has an affinity for the ice phase and binds to the ice interface. This is detrimental as it could lead to unfolding and aggregation of the protein in the ice phase [3]. This behavior has been previously observed in case of antifreeze proteins [4].

In this study, the objective was to investigate the nature and extent of preferential exclusion of solutes from the ice phase as well as its effects on the secondary structure of protein during near-equilibrium freezing in Bovine Serum Albumin (BSA)-Dimethyl Sulfoxide (DMSO) solutions. Albumin is the most abundant protein in human plasma and readily available. DMSO is a widely used cyo-protectant with a strong tendency to depress the freezing point of solutions [5].

METHODS

Fourier Transform Infra-red (FTIR) was used to study the composition of the freeze concentrate liquid (FCL) of the partially frozen samples. For analysis, 30% DMSO (w/w) solutions containing 10% (w/w) albumin were prepared gravimetrically with DMSO (99.9% purity, Sigma-Aldrich) and bovine serum albumin (MW 66.5 kDa, ≥99% Purity, Sigma-Aldrich) in ultrapure water.

An FTIR spectrometer (Thermo-Nicolet Continuum, Thermo Electron) equipped with a freeze-drying cryostage was used with a resolution of 4 cm^{-1} and a total of 128 scans being averaged per spectrum. In order to avoid supercooling, a method of equilibrium freezing was used to introduce nucleation sites for the ice crystals to form and grow very slowly. To achieve this, the solution was rapidly cooled at 30^0/min till homogenous nucleation and then slowly warmed at 1^0/min to its equilibrium melting temperature. When a few small crystals could be seen dispersed in the solution through the microscope, the sample was cooled at a very slow rate (0.3^0/min), to avoid concentration variation due to diffusional limitations. During cooling at 0.3°/min, spectra were collected from 5 different regions of the freeze concentrated phase while holding temperature at intervals of 5^0 until -70^0C.

A differential scanning calorimeter (Q2000, TA Instruments) equipped with a refrigerated cooling accessory was used to determine glass transition temperature of the freeze concentrate. The instrument was calibrated using tin and dry nitrogen was used as the purge gas at a flow rate of 50 ml/min. The sample was subjected to equilibrium freezing up to -80^0C (identical to the IR analysis) and the frozen solution was heated at 10°C/min from -70^0C to room temperature.

A powder X-ray diffractometer (D8 ADVANCE; Bruker AXS) equipped with a variable temperature stage and a Si strip one-dimensional detector was used to detect crystallizing phases. The sample was subjected to identical thermal history as during IR analysis. During the slow equilibrium freezing up to -70^0C, XRD patterns were collected continuously with a dwell time of 900 s. Solutions were exposed to Cu Kα radiation (40 kV × 40 mA) over an angular range of 5–30° 2θ with a step size of 0.05°.

RESULTS

Relative concentrations of albumin and DMSO were measured from the base line corrected areas of the albumin Amide II band (1525–1570 cm^{-1}) and the DMSO ν-HCH band (1367–1485 cm^{-1}) [6]. From the ratio of the integrated areas of the albumin Amide II and DMSO ν-HCH bands, the albumin to DMSO mass ratio (R) in the freeze concentrate was calculated using calibration curves and plotted as a function of temperature (Figure 1). The equilibrium freezing temperature of the 30% DMSO sample was -11°C. Spectra acquisition from the freeze concentrate during equilibrium freezing was started at -13°C. While cooling up to -33^0C, as the ice phase volume increased to 49%, the relative albumin concentration w.r.t. DMSO (value of R) in the FCL increased, suggesting preferential entrapment of DMSO in ice. On the other hand while cooling between -33°C to -59°C, as the ice phase volume increased from 49% to 56%, the relative albumin concentration w.r.t. DMSO in the FCL was found to decrease, suggesting preferential entrapment of albumin in ice. The amount of DMSO entrapped in ice between -13°C and -59°C was calculated (from the calibration curves and the DMSO-water phase diagram) to be 3.28 mg per 1g of water in the FCL. Similarly the amount of albumin entrapped in ice between -13°C and -59°C was 2.88 mg per 1g of water in the FCL. Upon cooling below -59°C, the albumin to DMSO mass ratio remained constant.

The ($\nu_2 + \nu_3$) combination band of water, known to be sensitive to the hydrogen bonding structure of water, split into 2 peaks in presence of DMSO (Figure 2). The higher frequency peak S_1 located at 5116 cm^{-1} was attributed to water molecules engaged in self-association and a second peak S_2* at 4800 cm^{-1} was attributed to the formation of clusters of water and DMSO molecules engaged in hydrogen bonding [7]. The relative peak intensity of S_2* (w.r.t. S_1) was essentially unchanged until -28°C, below which it started to decrease drastically until -32°C, when the S_2* peak disappeared and did not reappear upon further cooling. This indicated that the water-DMSO complexes started to break apart between -28°C to -32°C, below which water molecules were mostly engaged in self-association. The decrease in the population of water-DMSO complexes may have been a consequence of the decreasing concentration of water in the FCL during freezing. The gradual shift of S_2* towards lower frequencies occurred due to replacement of water by DMSO in the complexes.

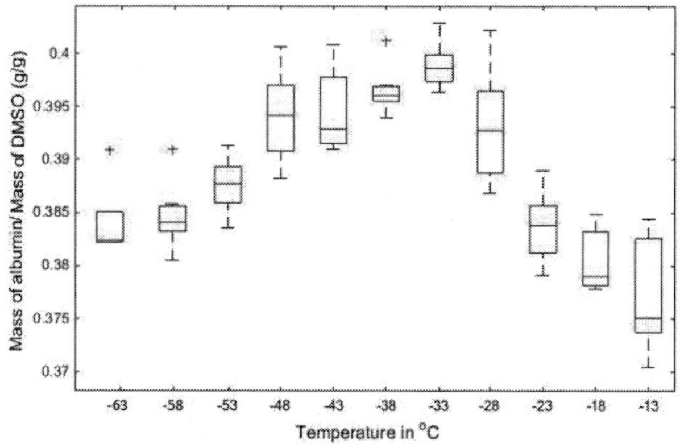

Figure 1: Albumin/DMSO mass ratio (R) at 5 different regions in the Freeze Concentrated Liquid during equilibrium freezing of 30%DMSO + 10% (w/w) albumin solution.

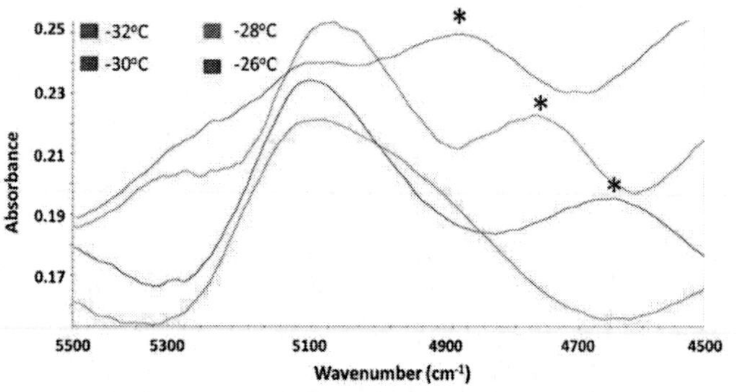

Figure 2: ($v_2 + v_3$) band of water in the NIR region during equilibrium freezing of 30%DMSO +10% (w/w) BSA solution. * indicates the S_2* peak arising due to DMSO-water clusters

From the XRD patterns collected during freezing, hexagonal ice was the only crystalline phase detected. Although formation of the water-DMSO trihydrate eutectic has been reported to occur at -63±1°C [5], crystallization of DMSO could not be detected in the XRD patterns (as peaks) or in the DSC scans (as crystallization exotherms) during the slow freezing. An amorphous (non-crystallizing) phase usually transforms into a viscous glass on cooling below a characteristic temperature (T_g). This reversible transition called glass transition or vitrification appears as a baseline shift in DSC curves.

When the frozen 30%DMSO +10% (w/w) BSA solution was heated back to room temperature at 10°C/min, a glass transition could be detected at -63°C (T_g) in the DSC scan (Figure 3), when the previously vitrified freeze concentrate underwent devitrification. The T_g could not be detected during the slow freezing due a negligible baseline shift.

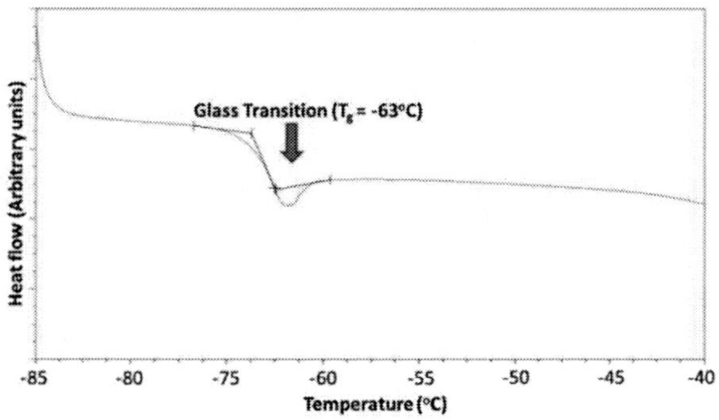

Figure 3: DSC scan during heating of frozen 30%DMSO+10%BSA (w/w) solution.

DISCUSSION

During near equilibrium freezing of DMSO-albumin solutions, it was observed that the relative concentration of DMSO w.r.t. BSA in the FCL decreased up to -32°C. The preferential entrapment of DMSO in the ice phase was attributed to its reduced mobility in DMSO-water complexes. These complexes subsequently break apart and disappear at -32°C, below which albumin was seen to be preferentially entrapped in ice, suggesting an affinity for ice. Crystallization of DMSO did not occur on cooling below its eutectic temperature and the freeze concentrate vitrified instead at -58°C. The increase in viscosity of the FCL upon glass transition arrested any further compositional changes below that temperature.

CONCLUSION

Preferential exclusion of albumin and then DMSO from a growing ice phase was observed during equilibrium freezing, causing changes in their relative concentrations in the freeze concentrate. This phenomenon was governed by DMSO-water interactions and albumin's affinity for ice. Compositional changes in the freeze concentrate halted below -58°C due to glass transition. Addition of excipients such as surfactant may disrupt DMSO-water complexes and prevent accumulation of protein at the ice surfaces, yielding a relatively homogenous freeze concentrate.

ACKNOWLEDGEMENTS

This research was funded by an NSF grant (CBET-1335936) to A.A. Parts of this work were carried out in the Characterization Facility, University of Minnesota.

REFERENCES

[1] Bhatnagar, BS et al., *Pharm Develop Tech,* 12(5):505-523, 2007.
[2] Carpenter, JF et al., *Cryobio*, 25(3):244-255, 1988.
[3] Eckhardt, BM et al., *Pharm Res*, 8(11):1360-1364, 1991.
[4] Raymond, JA et al., *Proc Nat Acad Sci*, *74*(6):2589-2593, 1977.
[5] Rasmussen, DH et al., *Nature*, 220(5174):1315-1317, 1968.
[6] Twomey, A et al., *J Phys Chem B*, 117(26):7889–7897, 2013.
[7] Malsam, J et al., *J Phys Chem B*, 113(19):6792-6799, 2009.

BIOHEAT TRANSFER IN LACTATING HUMAN BREAST

Mohammad Aliakbari Miyanmahaleh (1), S. Negin Mortazavi (2), Fatemeh Hassanipour (1)

(1) Mechanical Engineering Department
University of Texas at Dallas
Dallas, TX, USA

(2) Department of Integrative Biology,
University of California
Berkeley, California, USA

INTRODUCTION

The female human lactating breast is a biological tissue of great importance to both mother and infant lifelong health. Many researchers have attempted to study the female human breast tissue from different points of view. In the area of thermal behavior of the breast, most of the studies are limited to clinical investigations. The effect of warming the breast tissue on milk ejection has been studied by Hill et al. [1], Yigit et al. [2] and Kent et al. [3] among others. The above studies reported a significant enhancement of milk flow when heat is applied on the surface of the breast. Betzold [4] studied locational breast inflammation and effect of different factors, such as therapeutic ultrasound and heat treatment, on reversing inflammation. Kimura et al. [5] measured skin temperature during breastfeeding in eleven mothers. They isolated 5 stages of breastfeeding in regards to temperature changes in the breast. They found that the breast core temperature is affected by arterial blood temperature and flow. Based on their results, there is an increase in the breast skin temperature after the first or second day postpartum and before 4 or 5 days postpartum, and the skin temperature had a higher value during suckling and 5 minutes later.

This study is a mathematical approach to model the lactating human breast tissue temperature distribution using Wulff [6] equations while a heat flux (Q) on the outer surface of the breast is applied. The effect of milk flow on the thermal distribution inside the breast has been investigated. Also the temperature distribution is obtained inside the breast versus time, and also temperature rise over

METHODS

A 1D modeling of bio-heat transfer in female human lactating breast has been done by using Wulff continuum model. The effect of a constant heat flux, Q, applying on the outer surface while there is always a metabolic heat generation, g, within the biological tissue has been investigated. Also, effect of milk is considered by adding a new extra term to this equation:

$$\rho_t c_t \frac{\partial T}{\partial t} = k_t \nabla^2 T - (\rho_b c_b v_b + \rho_m c_m v_m)\nabla T + g \tag{1}$$

Initial and boundary conditions are:

$$\begin{aligned} t = 0 &\rightarrow \quad T(x,0) = T_0 = 37°C \\ x = 0 &\rightarrow \quad -k_t \frac{\partial T}{\partial x} = Q \\ x = D &\rightarrow \quad \frac{\partial T}{\partial x} = 0 \end{aligned} \tag{2}$$

Using the non-dimensional parameters (3),

$$X = \frac{x}{D},\ \eta = \frac{k_t t}{\rho_t c_t D^2},\ \theta = \frac{k_t}{D^2 g}(T - T_0) \tag{3}$$

the above equations can be rewritten as:

$$\frac{\partial^2 \theta}{\partial X^2} - A\frac{\partial \theta}{\partial X} + 1 = \frac{\partial \theta}{\partial \eta} \tag{4}$$

$$\begin{aligned} \eta = 0 &\rightarrow \quad \theta(X,t) = 0 \\ X = 0 &\rightarrow \quad \frac{\partial \theta}{\partial X} = B \\ X = 1 &\rightarrow \quad \frac{\partial \theta}{\partial X} = 0 \end{aligned} \tag{5}$$

where

$$A = \frac{\rho_b c_b v_b + \rho_m c_m v_m}{k_t},\ B = -\frac{Q}{Dg} \tag{6}$$

By solving the bio-heat equation analytically and applying the boundary condition and dimensionless parameters, the open form of the T can be derived as:

$$T(x,t) = -B\left[\frac{Ae^{\frac{A}{2}}}{2\sinh\left(\frac{A}{2}\right)}\frac{k_t t}{\rho_t c_t D^2} + \frac{e^{\frac{Ax}{2D}}}{A\sinh\left(\frac{A}{2}\right)} cosh\left(\frac{A}{2}\left(\frac{x}{D}-1\right)\right) + \frac{\left(1-\frac{x}{D}\right)e^{\frac{A}{2}}}{2\sinh\left(\frac{A}{2}\right)}\right] + \tag{7}$$

$$B.\sum_{n=1}^{\infty} e^{s\frac{k_t t}{\rho_t c_t D^2}}\, e^{\frac{Ax}{2D}}\left[\frac{\left(A-\sqrt{A^2+4s}\right)e^{\frac{\sqrt{A^2-4s}}{2}\left(\left(\frac{x}{D}\right)-1\right)}-\left(A+\sqrt{A^2+4s}\right)e^{\frac{\sqrt{A^2-4s}}{2}\left(1-\left(\frac{x}{D}\right)\right)}}{-4s^2\left[\frac{1}{\sqrt{A^2+4s}}\cosh\left(\frac{\sqrt{A^2+4s}}{2}\right)\right]}\right] +$$

$$\frac{k_t t}{\rho_t c_t D^2} + T_0$$

where D is the breast tissue thickness and s=$-(n^2\pi^2 + A^2/4)$. Also, subscripts t, b and m indicate tissue, blood and milk respectively.

RESULTS

Without considering the milk flow, the temperature profiles from the current study and Shih et al. [7] have been compared. The applied constant heat flux on the surface is 1000 W/m^2, k_t=0.5 $W/m°K$, $\rho_b = 1050\ kg/m$, c_b= 3770 $J/kg°K$, $\rho_m = 1030\ kg/m$, c_m= 3078 $J/kg°K$, [8] $\rho_t = 1058\ kg/m$, c_t=3396 $J/kg°K$ blood perfusion rate W_b is 0.5 kg/m^3s, $0.5 < v_b < 630\left(\frac{mm}{s}\right)$, $5\times10^{-5} < v_m < 5\times10^{-3}\left(\frac{mm}{s}\right)$ and the metabolic heat generation g=368$\left(\frac{W}{m^3}\right)$. As shown in Figure 1, the temperature profile trends in different depth of the tissue are in good agreement with a previous study by Shih et al [7].

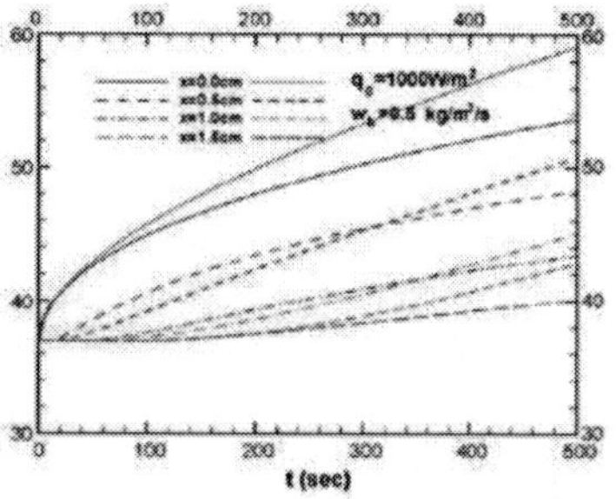

Figure 1: Wulff continuum model (red lines), analytical solution in comparison with Shih et al. [7] using PBHE (black lines)

The temperature distributions for different heat fluxes versus position and time are shown in Figure 2 (a) and (b) respectively. As expected, for various Q, tissue temperature increased at all depths by increasing the heat flux. When moving away from the surface, the tissue temperature decreases until the temperature of the breast tissue, at x=0.03 (m) (maximum depth of glandular tissue), reaches the body core temperature (37°C). In Figure 2(b), the temperature of the outer surface is shown from t=0 to 600s and an increase in temperature can be seen for all heat fluxes versus time. Choosing the maximum value for Q is confined by the maximum temperature that skin tissue can tolerate without any damage within and after the time period of breastfeeding.

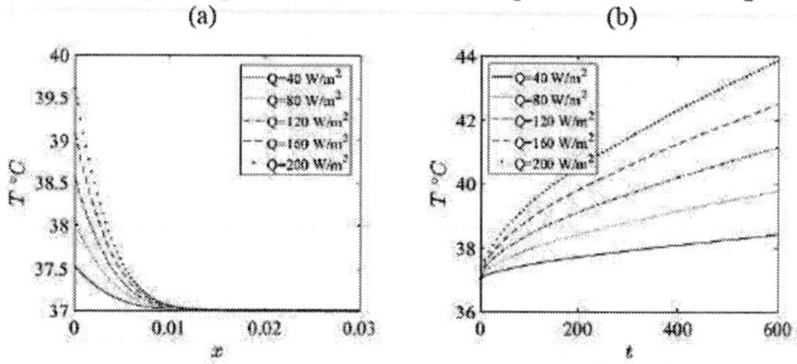

Figure 2: Effect of heat flux on tissue temperature distribution, a) along the depth of the tissue at t=120 s, and b)on the surface versus time, $v_b = 3.5\times10^{-6}, v_m = 4.8\times10^{-6}\left(\frac{m}{s}\right)$

Figures 3(a) and (b) represent the effect of milk velocity variation on temperature in lactating breast tissue with respect to position and time, respectively, by considering a constant heat flux (Q = 50 W/m^2) on the outer surface. Mentioning that the milk expression can change in different mothers and different stages of lactation is really important. Therefore, we have tried some values for milk velocity which are close to the corresponding values of different milk expression rates. As seen in Figure 3, the milk velocity affects the tissue temperature. Figure 3(a) demonstrates an increased tissue temperature as milk flow velocity increases. The slope of the line with higher milk velocity in 3(b) is greater than the others. In fact, the higher milk velocity can lead to greater heat transfer at a faster rate through the tissue.

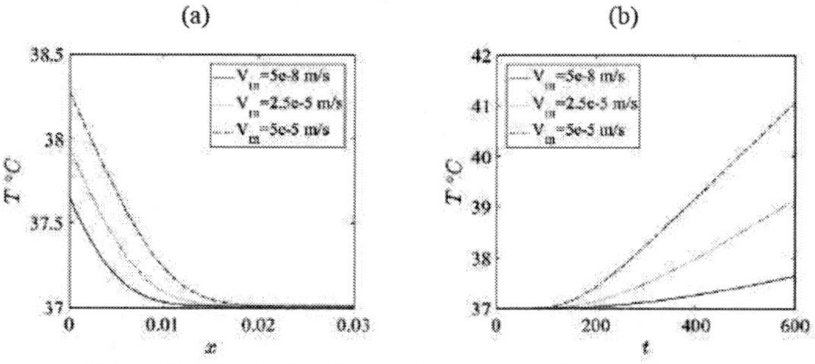

Figure 3: Effect of milk velocity on tissue temperature distribution a) along the depth of the tissue at t= 120s, and b) at x=D/2 versus time, $Q = 40\frac{W}{m^2}, v_b = 3.5\times10^{-6}\left(\frac{m}{s}\right)$

DISCUSSION

This paper creates a mathematical thermal analysis in human breast tissue. Using the Wulff model assumptions and adding the transport factor of milk flow, a new bioheat equation (energy equation in biological organ) is derived for human lactating breast. We validated the results of this analytical approach by comparing with Shih et al. [7]. The effects of milk flow and heat flux have been studied. The results show that both the application of heat flux and the velocity of milk have a direct effect on the temperature of the tissue. This modelling can be used to evaluate the amount and duration of heat flux application on the breast to determine safe protocols to avoid tissue damage.

ACKNOWLEDGEMENTS

The authors would like to acknowledge the financial support by National Science Foundation, Grant #1454334.

REFERENCES

[1] P. Hill et al, *Human Lactation,* 17(1):9-13, 2001.
[2] F. Yigit et al, *Breastfeeding Med.,* 7(6):487-488, 2012.
[3] J. Kent et al, *Human Lactation,* 27(4):331-338, 2011.
[4] C. Betzold, *Midwifery & Woman Health,* 52(6):595-605, 2007.
[5] C. Kimura, *Human Lactation,* 23(1):60-69, 2007.
[6] W. Wulff, *IEEE Trans.Bio-Med. Eng.,* 21:494-495, 1974.
[7] T. C. Shih et al, *Med. Eng.Phys.,* 29(9): 946-953, 2007.
[8] M.C. Neville et al, American J. Clinic. Nut.,48(6):1375-1386,1988

SB³C2017
Summer Biomechanics, Bioengineering and Biotransport Conference
June 21 – 24, Tucson, AZ, USA

INDUSTRIAL DESIGN FOR A MASTER OF ENGINEERING PROJECT COURSE IN MEDICAL DEVICE DEVELOPMENT

Shea Tillman (1), Alan W. Eberhardt (2)

(1) Department of Biomedical Engineering
University of Alabama at Birmingham
Birmingham, AL, USA

(2) School of Industrial & Graphic Design
Auburn University
Auburn, AL, USA

INTRODUCTION

Healthcare spending has grown substantially in the past 30 years. In fact, the medical and healthcare industries constituted 17.5% of the gross domestic product of the U.S. in 2014 [1], and are estimated to be 20% by 2020 [2]. Many important medical innovations of the past 25 years represent engineered solutions, such as hip and knee replacement, mammography, MRI and CT, balloon angioplasty, and coronary artery bypass grafts [3]. With an aging population that is more obese, medical device needs continue to present real opportunities for engineering product development.

The need for engineering innovation of medical devices also provides an outstanding opportunity for multi-disciplinary student training. As such, curricula are needed where students master hands-on methods while applying their knowledge to real world engineering design challenges. The Master of Engineering (MEng) in Design and Commercialization at the University of Alabama at Birmingham (UAB) was created to deliver hands-on training that involve interdisciplinary teams and innovative environments [4]. The MEng curriculum consists of 24 credit hours in engineering and 12 credit hours from the Master of Business Administration (MBA) program. Student teams engage in hands-on design and prototyping of Innovation-Commercialization (IC) projects and partner with MBA students to perform customer discovery and develop business models. The present manuscript describes the implementation and assessment results from a first offering of an industrial design course established to introduce students to human-factors and product use in order to generate marketable innovations.

METHODS

EGR 695 Project III represents the final project course of a three-semester sequence in product development focused on developing and applying hands-on skills. The primary learning outcomes for this course were for students to integrate human-centered design approaches into their current IC project prototypes, consider product use context, incorporate human-factors design refinements, and visualize product concepts as marketable innovations. The key aims were to improve the overall ease-of-use and perception of prototyped devices, while increasing their potential marketability through more compelling visual communication. Students revisited three IC projects: a wheelchair scale, a hemostasis tool, and a cervical pessary in order to refine and better communicate the user benefits of their designs.

The first step involved a complete reassessment of the existing prototype based on the primary user's perspective, which involved a neurosurgeon in the case of the hemostasis tool. Students sought to gain insights into how well the prototype worked, fit the body, and was visually perceived by the user. This led to a detailed sequence-of-use activity analysis to better understand user behavior and use patterns, and to identify potential problems and opportunities to refine the design. Students investigated anthropometric considerations and the role of the surrounding context in how a product is used, transported and stored.

Students were next tasked with redesigning physical touchpoints and interfaces of their prototype, based on their use reassessments. This included expanding and smoothing grip profiles to enable more flexible handle orientations. Students were also challenged to simplify, integrate, and unify features of their prototype designs, in order to bring perceptual clarity. This included the review of case studies that illustrate over-complexity due to feature-creep, a common pitfall that occurs when arbitrary functions are added to a product as a means of "out featuring" the competition. Students were introduced to "emotional value propositions" [5] and deliberate decisions about what the product should do and, just as importantly, what it should not do. They studied and crafted visual semantics of their product and provided visual cues to its use. For the wheelchair scale and the pessary, perceptual improvements were emphasized such as reducing fear, increasing confidence, or minimizing a negative stigma.

The final focus of the course was to enhance students' abilities in effective visual communication using KeyShot 3D rendering software (Luxion USA, Tustin, CA). This activity allowed them to create photorealistic images of their refined 3D CAD models (color, finish, transparency, etc.) and develop a visual narrative that communicated the benefits of the design and a complete production product vision.

To examine the levels of student satisfaction at the end of the course, IDEA surveys (IDEA, Manhattan, KS) were given at the course conclusion. For the present manuscript, the categories "Progress on Relevant Objectives" and Overall Ratings – Excellent Course were used in the following.

RESULTS

The deliverables for the course included visual posters that realistically demonstrated how the designs worked and their core innovations. These visual narrative posters enabled the students to "pitch" their designs for commercialization opportunities, as well as to advance their own professional work portfolios. Figure 1 shows representative examples of the hemostasis hand piece, the wheelchair scale turned Wii-fit board, and the cervical cerclage pessary. This last design debuted colors, clear materials, and forms refined to be more friendly to anxious, expectant mothers. These renderings are far more realistic than those created in most contemporary 3D CAD packages.

The IDEA surveys yielded high student scores for EGR 695 Project III, averaging 4.7/5 for Progress on Relevant Objectives and 4.9/5 for Excellent Course. Student comments included, "… I really enjoyed the industrial design … which provided excellent perspective on aesthetics …really enjoyed learning about product perception and display."

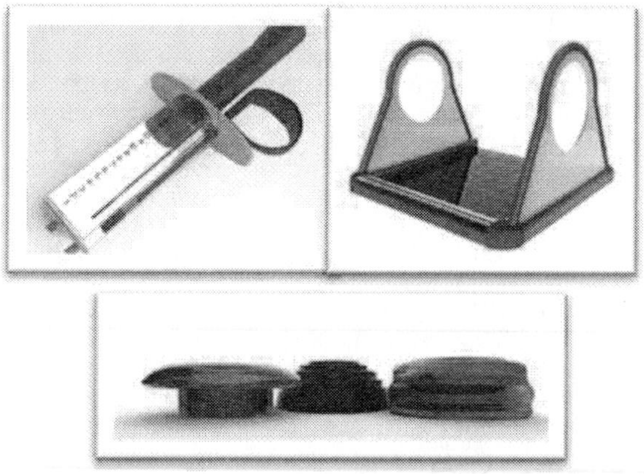

Figure 1: KeyShot renderings of (top) hemostasis tool and wheelchair scale/game board; and bottom (pessary)

DISCUSSION

Overall, the introduction of industrial design (ID) for the third and final project course surpassed all expectations. The students welcomed the opportunity to study the artistic elements of ID and were happy to spend time re-evaluating their original design concepts with a fresh eye on ease and logic of use. The students were concurrently enrolled in another course that emphasized design for manufacture and assembly, which provided additional incentive for re-evaluation of their designs. The inclusion of their final Keyshot renderings, designed for commercial advertising of their user friendly devices, made for an ideal closure to the course and served to enhance the students' design portfolios.

In the future, we hope to repeat these activities with each group of MEng students. Discussions centered around when exactly is best to bring in ID are ongoing. While the third semester offered the opportunity to reflect and refine the designs, another potential option would be to bring in ID during the original brainstorming and evaluation of prototypes. This remains a topic for further study.

ACKNOWLEDGEMENTS

The authors gratefully acknowledge financial support from the National Science Foundation GARDE Program, the National Institutes of Health Research Education Program (R25), and VentureWell.

REFERENCES

[1] https://www.cms.gov/research-statistics-data-and-systems/
[2] Keehan, S.P., et al., *Health Affairs*, 30(8), 1594–1605, 2011.
[3] Lee, J.-S., Biomedical Engineering Entrepreneurship, 2010.
[4] Eberhardt, A. W., et al., *J Biomech Eng,* 138, pp. 070803-1-5, 2016
[5] Kolko, J., *Harvard Business Review*, 66-71, 2015.

'EXPLORING "ME"CHANICS: THE MULTISCALE MECHANICS OF ME!' SUMMARY OF OUTREACH LESSONS LEARNED

Stephany Santos (1), Hannah Kackley (1), David M. Pierce (1,2)

(1) Department of Biomedical Engineering
University of Connecticut
Storrs, CT, USA

(2) Department of Mechanical Engineering
University of Connecticut
Storrs, CT, USA

INTRODUCTION

Eighty-two percent of 16 and 17 year-olds in the UK believe engineering is a vital key for the future of technology innovation, yet only 21% are interested in engineering careers [1]. Many factors could contribute to such statistics, for example: perception of need for natural ability in math and science, reputation of engineering as a difficult subject, notions that engineering is not for women, and more. Perceptions like these do not take hold in high school, but rather at much earlier, even as young as pre-primary education age.

During previous outreach activities we asked a group of K-1 girls, "What do engineers do?" One kindergartner immediately raised her hand and exclaimed, "They drive trains!" A partial list of the students' ideas regarding engineering jobs includes: drive trains, build trains, make cars, build planes, build bridges, and fix computers. Nearly 85% of the responses indicated that engineers create transportation related machines or infrastructure. The last response connects to engineers and technology, but the keyword "fix" implies that engineers do not create but only find solutions when something goes wrong. This representative handful of responses misses the breadth of engineering tasks, and furthermore the breadth of fields, e.g. biomedical engineering.

To successfully respond to increasing global demand, the US must increase domestic enrollment and retention in STEM areas currently in decline [2]. Our broad objective as educators is to positively impact the attitudes, behavioral intentions [3], and learning of a wide range of students – from underrepresented elementary school students to active researchers – by integrating multiscale biomechanics research with education and outreach activities.

In this effort we have developed an outreach event targeted specifically at female middle-school students as this is a crucial age to foster STEM interest [4] and challenge stereotype threats [5].

METHODS

We designed an event titled, "Exploring MEchanics: The Multiscale Mechanics of Me!" to introduce students to biomedical engineering, biomechanical analysis, and our research in cartilage biomechanics. We developed four interactive stations for students to visit at their own pace. The content translates our research on structure-function relationships across length scales in biomechanics using simple models that middle school students can make, deform and take home.

Station 1: Your body is a machine. At this station, we introduce students to some of the 360 joints in the body and the common joint motions of flexion, extension, adduction, abduction, and rotation. Students explore their own joint motion using XBox® Kinect Skeletal View, and a biomechanics-inspired 'Simon Says' game (Fig. 1).

Figure 1: At Station 1 students explore their own joint motion using XBox® Kinect

Poster Presentation #P77

Station 2: You kneed tension to function. At this station we introduce students to muscle tension, and how muscles work in pairs to create movement in the knee. Students build a joint and use balloons to represent the quadriceps and hamstring muscles. With movement they observe muscle tension/compression during knee flexion/extension.

Station 3: Cartilage helps you move freely. At this station students learn that cartilage facilitates smooth joint motion and naturally absorbs shock in our joints. We introduce students to the terms *viscoelastic* and *biphasic*, and explore the general properties of cartilage using "oobleck," a non-Newtonian fluid made of cornstarch and water.

Station 4: Looking from the inside out. At this station, students look at a slide of human cartilage through a microscope, and begin to understand its microstructure and function (Fig. 2). Students create special drawings that, when stained, reveal their own "histology" image.

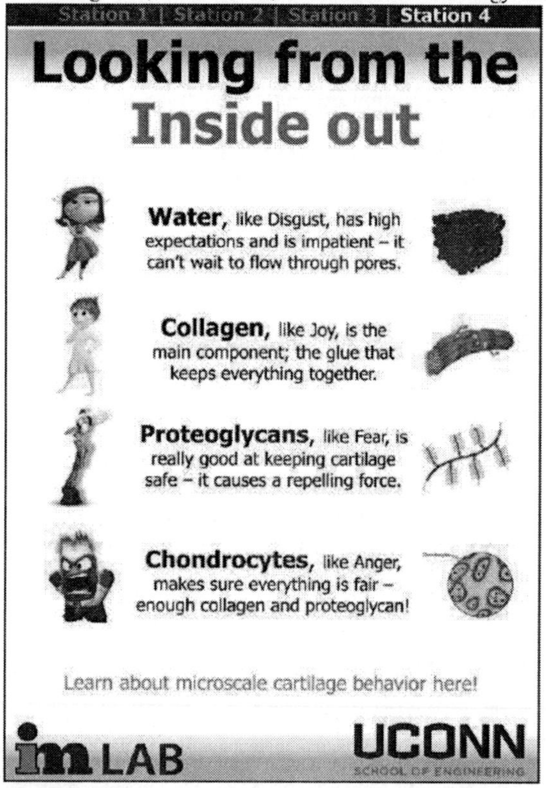

Figure 2: Sample poster. This poster helps students understand cartilage constituents, and microscale structure and function.

To deliver this outreach event we actively partner with the Connecticut Science Center, particularly the Women in Science Initiative, as well as Danbury Public Library (both in Connecticut). We hosted events run by Stephany Santos, a Ph.D. candidate in the PI's lab, and supported by undergraduate women Hannah Kackley, Phoebe Szarek, and other women from the PI's course BME 3600W "Biomechanics." We completed an initial test event at the Connecticut Science Center on Dec. 12th, 2015, and a follow-up at Danbury Public Library on Dec. 7th, 2016. The latter event served PK-6 grade students in a community with more than 40% Hispanic population.

In a post-survey, we asked students participating in the event: (Q1) Have you previously heard of biomedical engineering?; (Q2) What was your opinion of science and engineering BEFORE attending Exploring MEchanics? (rank on a scale of 1-6); (Q3) What was your opinion of science and engineering AFTER attending Exploring MEchanics? (rank on a scale of 1-6); and (Q4) What is one thing you learned today?

We flagged the following keywords from Q4: Cartilage, Force/Compression/Tension, Muscle/Contraction, Knee, and Cells. In addition, we attributed responses as being most related to station 1 and 2, 3, or 4. We ran a Wilcoxon Rank Sum test to probe connections among age, race, gender, and Q1-4.

RESULTS

Our preliminary data is only from the event at Danbury Library. Unfortunately there were no statistically significant correlations. Instead we summarize the general trends. From the survey responses, Q1 revealed that only white male participants recognized the term "biomedical engineering" prior to attending the event. Looking at improvements in score from Q2 to Q3, every surveyed female ranked their opinion of science and engineering higher, whereas males either remained equal or improved. Specifically, half of the females reported a score of less than 3 for Q2, compared to one quarter of the males.

In the open-ended Q4, all student responses were macroscale, relating to stations 1 or 2. None of the students chose to include "cells" or the microscale in their responses. Female students tended to include the keyword "cartilage," while males often included the keyword "knee." Responses mentioning the structure-function relationship of cartilage and the knee were only given by 4th and 5th grade students.

DISCUSSION

Our event exposed students to biomedical engineering, many for the first time. Q1, in our study of PK-6 students, seems to align with the common stereotype that engineering is only for white males. Female affinities to science and engineering improved through participation in our event. This is promising, and demonstrates that outreach events like these can improve the perceptions of female students. Our survey showed that the females had a lower initial liking for STEM, but the interactions with female role models appeared to have a positive effect.

Students responded well and retained information about macroscale biomechanics. Responses such as, "I learned that there's a cushion on your joint so when you bend it, it doesn't get damaged," indicate a direct impact from our event, and an initial understanding of biomechanics of the knee. Many students visited the large-scale mechanics station last, so Q4 may be influenced by the lapse in time.

For many of the students, this event provided their first view through a microscope. Further, several students had no prior knowledge of cells, or even their existence. Thus, while this first exposure is important, it was often hard for them to conceptualize.

Future improvements will include multiple microscale stations to help students better understand what they cannot see with the naked eye alone. Utilizing techniques such as macroscale metaphors, repetition, and more experiments may improve outcomes on the microscale. We hope to expand this outreach by bringing the event (min. 2× per year) directly to local middle schools.

ACKNOWLEDGEMENTS

Cailah Carroll, Kristen Rose, Phoebe Szarek, and Alyssa Wilczynski. The Connecticut Science Center and Danbury Public Library.

REFERENCES

[1] McDonald C, *ComputerWeekly.com*, March 16, 2016.
[2] National Science Board, *Science and Engineering Indicators*, 2014.
[3] Ajzen I and Fishbein M, *The Handbook of Attitudes*, 173-221, 2005.
[4] Lemming J, *DecodedParenting.com*, March 11, 2014.
[5] Spencer SJ et al, *J Exp Soc Psychol*, 35:4-28, 1999.

SB³C2017
Summer Biomechanics, Bioengineering and Biotransport Conference
June 21 – 24, Tucson, AZ, USA

CREATING VIRTUAL LABORATORIES IN BIOMECHANICS

Sara E Wilson (1)

(1) Department of Mechanical Engineering
University of Kansas
Lawrence, Kansas, United States

INTRODUCTION

It has been well documented that active learning practices can improve learning outcomes and student performance in engineering [1]. At the undergraduate level, students in traditional lecture courses are 1.5 times more likely to fail than in active learning settings [1]. Laboratories can often provide valuable active learning opportunities. However, with larger class sizes and dwindling budgets, the resources for such labs can be limited. In these situations, virtual labs can provide a method to allow students to get hands on practice with concepts at little costs outside of the cost to initially create the virtual lab. Additionally, students can often do such labs in their own time, without the need for dedicated time and laboratory space. Such virtual labs are also useful for online and hybrid courses where students may be offsite.

At the University of Kansas, I am in the process of developing virtual labs for several senior/graduate level courses including Biomechanical Systems, Physiological Systems, and Control Systems. In this abstract, I will describe two virtual labs that have been created for the course, Biomechanical Systems. This course starts with muscle mechanics, muscle energetics, and muscle mechanical models. It also covers nervous system physiology, modeling of neuromuscular dynamics, and human/animal motion mechanics (including locomotion, effects of scale, stability, human motion modeling). The two labs described in this abstract come from the first portion of the course focused on muscle.

HUXLEY & SIMMONS VIRTUAL LAB

The first virtual lab created for this course is based on Huxley and Simmons 1971 model of early tension transients in muscle [2,3]. In this model, the tension as a result of a quick shortening of a sarcomere is modeled on the millisecond time scale. The model was created by Huxley and Simmons to model the experimental results presented by Huxley in 1974 (Figure 1) [4].

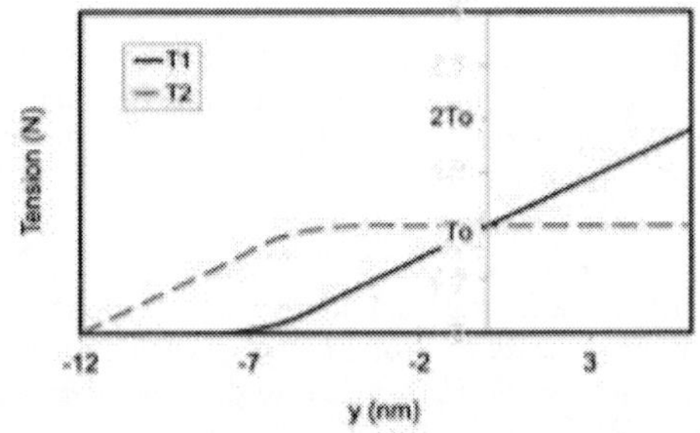

Figure 1 Instantaneous Tension (T1) and early tension recovery (T2) observed experimentally by Huxley (1974) and modeled by Huxley and Simmons (1971) [2,4].

In this model, the crossbridge attachments (S-1) were modeled as having two stable states and a spring element modeled the S-2 fragment of the myosin head.

Using the rate and tension equations developed by Huxley and Simmons, a virtual lab instrument was created to allow the students to do virtual experiments with the model [2,3]. The underlying rate equations are:

$$n_1 + n_2 = 1 \qquad 1$$

$$n_2 = n_2^{\infty} + (\tfrac{1}{2} - n_2^{\infty})e^{-\frac{t}{\tau}} \qquad 2$$

$$n_2^{\infty} = \tfrac{1}{2}(1 - \tanh\left(\tfrac{Khy}{2\kappa\theta}\right)) \qquad 3$$

where n_1 and n_2 are the fraction of crossbridges in each of the two states, K is the spring stiffness, h is the distance between the states, y is the shortening length, k is Boltzmann's constant and θ is the absolute temperature. The tension equation is:

$$T = K(y_o + y + \left(n_2 - \tfrac{1}{2}\right)h) \qquad 4$$

where Ky_o is the initial tension.

The virtual laboratory experiment was created in Labview. This software allows the student to have a graphic user interface (GUI) that shows only the pertinent information (Figure 2). In the code (block diagram), a while loop takes the model through 0.5 ms timesteps that are slowed down 200 times to allow easy viewing.

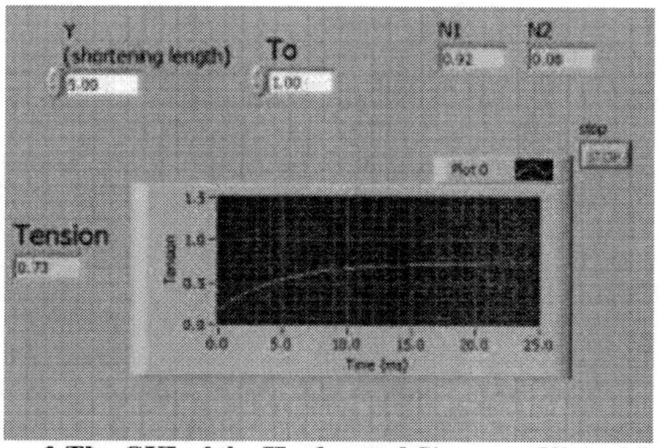

Figure 2 The GUI of the Huxley and Simmons Virtual Lab

This virtual lab has been used as an experimental problem to allow the students to explore the Huxley and Simmons model. The students are asked to use the virtual lab to run the experiments required to create figure 1.

MUSCLE HEAT & FUEL VIRTUAL LAB

The muscle heat and fuel lab is based on McMahon's description of energetics within the muscle as an ATP carburetor. This lab allows a student to do several experiments including:

1. Oxygen deprivation and iodoacetic acid poisoning (no glycolysis) – to observe a quick drop in ATP
2. Oxygen deprivation – to observe alactic oxygen debt followed by anaerobic glycolysis and lactic acid build up
3. High ATP usage, breathing rate – to observe the transition from aerobic to anaerobic glycolysis and lactic acid build up

The numbers used in this virtual lab are qualitative rather than representing actual amounts of given substances. The GUI displays bars representing the amount of a substance (ATP, glucose, oxygen, lactic acid, PCr) present in the muscle. Switches allow one to perform the virtual experiments.

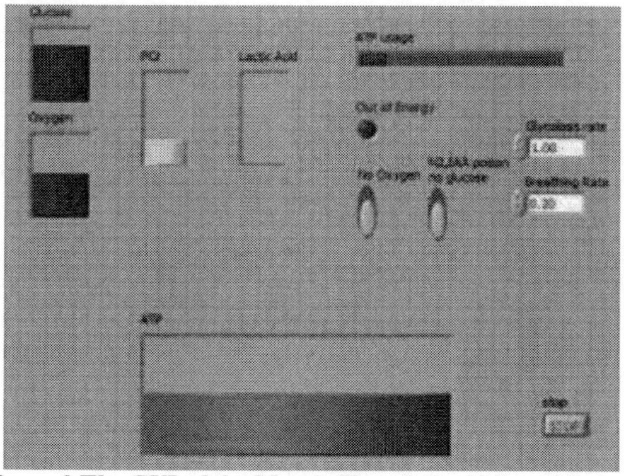

Figure 3 The GUI of the Muscle Heat & Fuel Virtual Lab

Underlying this GUI, the code contains four simultaneously running while loops, one for glycolysis, one for breathing, one for ATP usage, and one for the PCr buffer reaction.

DISCUSSION

These virtual labs are relatively simple, but allow students to experience concepts normally presented as equations and text. They can be used as demonstrations in class, but are most effective when students are able to play with them and to perform virtual experiments. These two virtual labs were created in National Instruments' Labview software. However, other software that allows the creation of a GUI interface could be used including Matlab. In the future, I plan to continue to develop similar virtual laboratories, particularly for concepts of feedback control.

REFERENCES

1. Freeman S, Eddy SL, McDonough M, et al. Active learning increases student performance in science, engineering, and mathematics. Proceedings of the National Academy of Sciences of the United States of America 2014;111:8410-5.
2. Huxley AF, Simmons RM. Proposed mechanism of force generation in striated muscle. Nature 1971;233:533-8.
3. McMahon T. Muscles, reflexes, and locomotion. Princeton, NJ: Princeton University Press; 1984.
4. Huxley AF. Muscular contraction. J Physiol 1974;243:1-43.

INTRODUCING REHABILITATIVE DESIGN TO MECHANICAL ENGINEERING STUDENTS USING A PROBLEM-BASED LEARNING APPROACH

Joshua Gargac (1)

(1) Department of Engineering
University of Mount Union
Alliance, OH 44617

INTRODUCTION

Although engineering departments at small, undergraduate-focused institutions often do not offer a biomedical engineering major, many faculty and students are interested in applying mechanical engineering (ME) principles to biomechanics-related projects. At Mount Union, the ME curriculum provides ample opportunities for hands-on experience from automotive, manufacturing, and robotics projects, but there are few opportunities to engage in biomechanics work. As an introduction to the field, biomechanics projects can be introduced into existing ME courses. One sub-discipline of biomechanics that relies heavily on ME principles is the design of rehabilitative devices.

One area of particular interest involves designing devices for children with tetra- or paraplegia. Tetraplegia is the partial or complete paralysis of both the arms and legs, while paraplegia affects only the legs. The primary cause of tetra- and paraplegia is injury to the spinal cord, which can result from motor vehicle crashes or other traumatic accidents [1]. Tetra- and paraplegic children have difficulty completing everyday activities and can miss out on recreation and athletics. Therefore, a project was created to design devices that improve access to recreational activities for tetra- and paraplegic children. The project was integrated into a pre-existing ME course using a problem-based learning (PBL) approach to provide a real-world application of the course material. Since non-profit agencies and government institutions often offer grants aimed to improve accessibility for the physically handicapped, the project was modeled as a funding grant proposal.

METHODS

A PBL project was introduced as part of a junior-level mechanical engineering course called Kinematics and Dynamics of Machinery (Theory of Machines). This course applied rigid-body dynamics and computer aided engineering to design and analyze one degree-of-freedom (DOF) mechanisms. Four bar linkages are the focus of the course since they are the most common 1-DOF mechanism. The course was worth four credit hours, had an enrollment of 17 students, and met for 65-minute lectures three times each week. The projects were completed in four teams with 4-5 students..

Student groups were provided with a *Call for Proposals* explaining that an imaginary rehabilitative design association was soliciting grant proposals for the design of machines to aid tetra- and paraplegic children. While the primary goal of the grants was to fund devices that would improve a child's access to recreational activities (i.e. kicking a soccer ball into a net), designs that helped with everyday functions would also be considered (i.e. reaching a toy on the ground). To reinforce course material, the designs were required to primarily consist of a four bar linkage. Student teams could design for either (or both) gender, any age group between 3 and 16 years old, and any degree of paralysis. Overall device size, weight, and cost were taken into account when selecting the winning design.

To start, students generated a list of activities in which tetra- and paraplegic children could not participate. After selecting an activity, a detailed list of constraints and performance criteria was created. The project followed the paradigm (Fig. 1) from Norton's *Design of Machinery* text [2]. The motion of the mechanism necessary to complete the activity was defined and linkage dimensions were calculated to provide that motion (linkage synthesis). A vector loop equation was used to calculate the spatial orientations of the links with respect to the input link (position analysis). Given an input motion (i.e. motor input), the first and second derivatives were taken to calculate the velocity and acceleration of each link (velocity and acceleration analysis). Joint and inertial forces were then calculated through inverse dynamics (force analysis). Finally, the project teams were required to apply the knowledge gained from their Mechanics of Materials course in order to perform a stress analysis with their selected material.

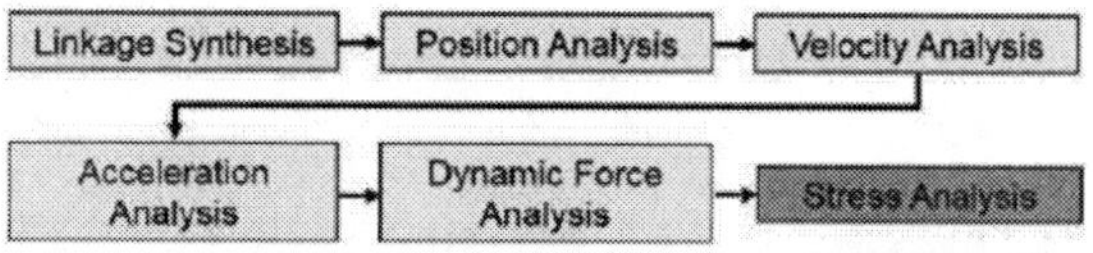

Figure 1: The steps for mechanism design and analysis. Stress analysis was included for the project but not covered in lecture.

Approximately 25% of the course was devoted to the project. During the second week of the semester, all three course meetings were used to introduce the project, work with teams to develop project constraints and performance criteria, and complete a linkage synthesis using MATLAB (MathWorks, Natick, MA). Three course meeting were devoted to adapting the MATLAB scripts to perform the (1) position, (2) velocity, and (3) acceleration analyses. During the final two weeks of the course, three course meetings were designated as project work time. Teams used this time to identify suitable materials, complete the force and stress analyses, create 3-D solid models in SolidWorks (Dassault Systèmes, Waltham, MA) or Autodesk Inventor (Autodesk, San Rafael, CA), and consider other design aspects including user interaction, safety features, and drive motors, if needed.

Project teams were required to submit progress updates to the instructor after completing each step (Fig. 1). Written proposals were collected at the last course meeting of the semester, during which groups presented their designs. Proposals were scored in four categories:

- *Motivation*: Why is a mechanism needed to complete the activity?
- *Technical Analysis*: Was theory properly applied to prove the mechanism will work?
- *Overall Design*: Were all design parameters considered? How will this mechanism interact with a wheel chair?
- *Feasibility*: Is it reasonable to make the device based on cost, size, weight, and torque requirements. Are there safety concerns?

RESULTS

Student Progress: Each team selected an appropriate activity that could be completed using a four-bar linkage. At the onset of the project, students needed more guidance to synthesize the linkages and establish performance criteria and constraints. By the velocity analysis step, the groups became comfortable with the process and were capable of working independently to implement the successive steps.

Grant Proposal Evaluations: The grant proposals were completed with minimal guidance from the instructor. Proposals were submitted for mechanisms that could putt a golf ball into a hole (Fig. 2a), cast a fishing pole (Fig 2b), hit a baseball off a tee (Fig. 2c) and kick a soccer ball into a goal (Fig. 2d). The progress updates ensured that each group appropriately applied the course theory to prove the efficacy of their design. Scores varied significantly for motivation, overall design, and feasibility. The putt-putt mechanism was the highest scoring proposal and featured designs for mounting onto a wheel chair (Fig. 3) and a description of how a child would use the device.

DISCUSSION

The assistive four-bar mechanism project provided a practical application for vector-based analysis while reinforcing the engineering design process, rigid body dynamics, mechanics of materials, programming in MATLAB, and 3-D solid modeling. As such, the project prepared the students for future coursework, including the senior capstone design course. The project also introduced biomechanics to the students without establishing an entirely new course.

As implemented, the project involved an imaginary scenario but could be improved by integrating real-life connections. For example, a person with paraplegia or a representative from an organization that serves the physically handicapped could serve as a client for the project and provide feedback to the students. If more resources were available, prototypes could be fabricated to demonstrate the mechanisms. Finally, the instructor could work with the each group to write an abstract and submit it to the SB^{3}C undergraduate design competition.

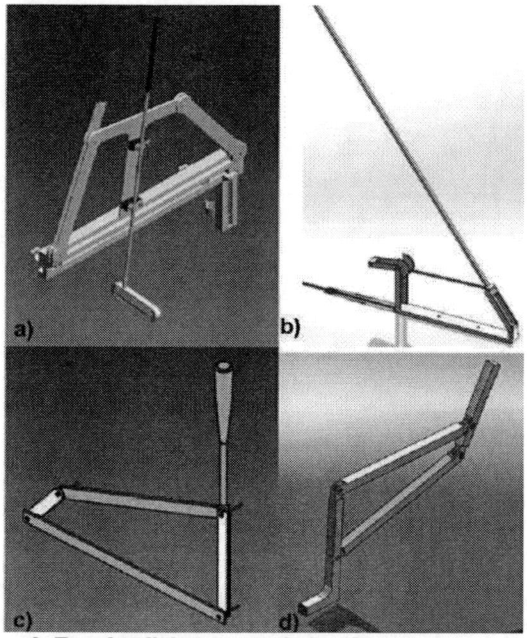

Figure 2: Four-bar linkages were designed to (a) putt a golf ball, (b) cast a fishing pole, (c), hit a baseball of a tee, and (d) kick a soccer ball.

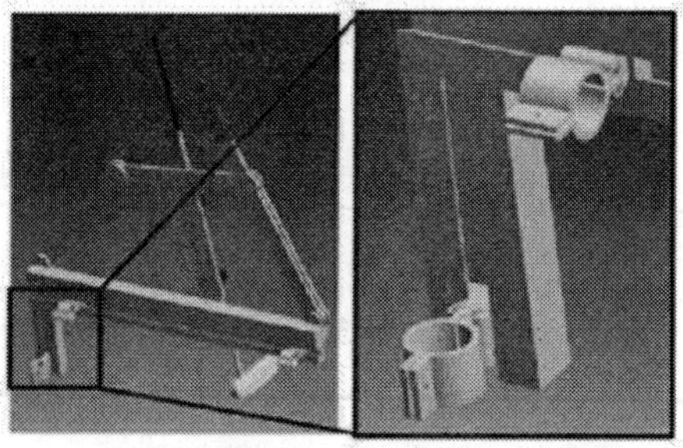

Figure 3: Clamp design and positioning to connect putting mechanism to a wheel chair.

ACKNOWLEDGEMENTS

The author would like to acknowledge the students enrolled in the EME 310: Kinematics and Dynamics of Machinery at the University of Mount Union during the Fall 2016 semester.

REFERENCES

[1] Baker, C et al., *Am J Emerg Med*, 17(3):230-234, 1999.
[2] Norton. R. *Design of Machinery 5th edition*. McGraw Hill, 2012.

SB^{3}C2017
Summer Biomechanics, Bioengineering and Biotransport Conference
June 21 – 24, Tucson, AZ, USA

THE USE OF JOURNALS CAN EXPOSE STUDENT LEARNING METHODS IN CAPSTONE DESIGN

Ferris M. Pfeiffer (1,2), Suzanne Burgoyne (3), Rachel E. Bauer (3), Jennie J. Pardoe (3)

(1) Department of Bioengineering
University of Missouri
Columbia, MO, USA

(2) Department of Orthopaedic Surgery
University of Missouri
Columbia, MO, USA

(3) Department of Theatre
University of Missouri
Columbia, MO, USA

INTRODUCTION

As engineering educators, it is imperative to understand what students are learning and what they are not learning. It is just as important to understand why students learn. Understanding why and how students learn is extremely difficult and we are not provided with many measurement tools to quantify such measures. Engineers typically thrive on data. We acquire date, analyze data, and make educated hypotheses based on data. However, when it comes to evaluating what and how engineering students are learning, the data is limited. Typically, we acquire such data objectively using traditional measures such as exams, quizzes, and student response systems. We sometimes acquire data in less formal subjective measures including classroom discussions, 1 on 1 meetings with students, and even non-verbal communication before, during, or after class. All of these measures have limitations which prevent forming accurate conclusions and optimal interventions. Tests, quizzes, and student response systems give a measure of what students know at a particular time, but these methods usually do not reveal how or why students acquired the knowledge, or how long they will retain what they have learned. Subjective methods often require active initiation and/or participation on the student's part. Any instructor who has asked a question in a lecture class and then waited patiently while no one raised their hands knows just how ineffective such subjective measures can be. With this in mind, we set out to evaluate the potential for using reflective journaling, a method borrowed from our theatre department) as a new method for understanding how students learn and implement the design process in a senior capstone design course.

METHODS

Senior Bioengineering (BE) capstone design students (n=60) were allowed to self-select into groups (n=5 per group) based upon capstone design projects pre-identified by the faculty mentors facilitating the class. A sub-set of these students (6 groups; n=30 total students) were randomly assigned to be in the experimental (E) group. All other students were assigned to be in the control (C) group. Prior to the beginning of coursework, all students completed a validated survey measuring engineering design self-efficacy [1]. The control and experimental groups both received standard instruction, but in addition the experimental group received one hour per week of creativity training developed by a theatre professor. This active learning process integrated techniques drawn from actor training, improvisation, and theatre of the oppressed [2] with creative problem-solving methods drawn from multiple research-based sources [3]. For instance, a short lecture comparing convergent and divergent thinking was immediately followed by a theatre exercise in which the class, divided into two groups, put divergent thinking into action by quickly inventing forty ways to cross a room. Theatre games helped the students develop a safe, supportive environment in which students could begin to venture outside their comfort zones. Unless one is willing to risk trying something new—and making mistakes--one can't be creative. Theatre exercises also enabled students to open their minds, question assumptions, and see things differently, since creativity is more about seeing things differently than seeing different things. Other exercises involved active listening, framing and reframing their creative problems, using analogies to brainstorm solutions, and evaluating and refining solutions. During the course of the semester, students reflected on and recorded in journals various aspects of the creativity curriculum that assisted their process of learning. For some, active learning pedagogies (learning by doing) seemed effective.

RESULTS

Throughout the semester, we identified two general ways in which journaling seemed helpful to the students.

1. As instructors, we got some insight into what students are and are not understanding about concepts and skills presented in the class. So if students were having problems, we find out about them and could address those problems quickly and personally. We also learn whether the activities we were doing were effective in teaching what we wanted the students to learn.

2. For the students, reflecting on their learning helped them understand what they were learning and why it was important. The approach we took to this class was active learning, in which students learn things by doing activities and then reflecting on what they've learned from those activities. The assignment of keeping a journal encouraged them to take the time to do that reflecting, which is as much a tool for learning as were the activities.

The following are some sample excerpts from the student journals this semester:

"So you want me to make goofy noises and motions in a group of serious, non-smiling engineering students? First thoughts? Ugh. I'm silly and very open at home and amongst people who really know me, but when it comes to strangers, classmates and coursework, it's all business. **This exercise showed me I'm afraid to step out and let go around people who aren't in my inner circle of trust.** This exercise also showed me that there's no reason I can't be myself. There's no rule I must be straight faced and closed off at all times. Everyone else seemed to be just as worried about the exercise as I. **There is no rule that states you cannot be an excellent engineer if you step outside the traditional engineering box.** "

"During the last class period we did a number of activities. First, we were given a piece of paper with a grid of circles printed on it and were told to draw using the circles as part of the drawing. Even though everyone started with the same base template, drawings varied drastically between each person. To me this showed that everyone sees different things and has different ideas even when looking at the same template that is objectively just a grid of circles. **In terms of the group project this tells me that it's important to listen to everyone's ideas even when we think we already have the best solution, because it's possible someone has thought of something we hadn't even considered.**"

"This assignment really highlights a major strength of tackling a project as a group. When given the same constraints, people will come up with drastically different solutions. **The strength of any group, be it a college senior project, a board of executives, or a legislative body, comes from its diversity.** People of different identities, interests, and experiences approach problems with a unique set of assumptions. Considering those differences and giving special weight to those ideas that are unlike your own is key to innovation and to success in our group project."

"During this week, I tried to use the concepts we discussed about active listening to pay more attention to the things people were discussing with me. **It was almost scary how much I find myself not actually absorbing half the information the person who is talking to me is actually saying.** Even when I was aware of it, I still found it hard to really listen to the person to understand what they were talking about. More often than not, I discovered that I would pay enough attention to the words they were saying to latch on to the general concept of what they were talking about, and then say something from my own experience that was somewhat related. Actually, I feel somewhat embarrassed to say how many times I caught myself interrupting a story or something similar that someone was telling me just to share an experience of mine that I thought fit into the subject matter we were talking about."

"After having a couple days to think about the points that stuck with me the most from this week's reading, I have concluded that the author's description of creativity as being a skill that is possessed by all, but requires regular exercises to fully acquire is the most significant point thus far. **I had never thought of creativity as being something that took practice and discipline to acquire, but instead fell into the category of people who believed it to be a quality that some possessed and some didn't.** I have seen the effects of thinking like this in my life. There are many times where I get frustrated when I cannot accomplish a task, and instead of sticking with it and trying to find different ways to improve upon what I currently have, I usually ditch the task at hand and blame it on myself for not being good enough. I think if anything, I hope that I can help to reverse some of this and apply this eight step method to something and see if it helps me to accomplish some of my goals."

DISCUSSION

These selected excerpts from student journals highlight the significant concepts that students learned in the respective sections of the course. In addition, one student began using journals as a way to communicate personally how he was doing with his project. He wrote about his own hang-ups and difficulty connecting and how it influenced his work with a group as an engineer. The journal became a place to let me know when he found new ideas on his project, and when he needed help communicating. He was pretty quiet in class and only spoke when working with his group all semester (and even then, he sometimes stayed quiet). His is a pretty strong example of why journaling is so important, as the instructor was able to have a weekly conversation with him about applying what we were learning to his work as an engineer. This student's journals were very open because of the trust that was built through this form of communication. The use of student reflective journaling became an invaluable tool for rapid and thorough evaluation of student progress.

REFERENCES

[1] Carberry, A.R., H.-S. Lee, and M.W. Ohland, *Measuring Engineering Design Self-Efficacy.* Journal of Engineering Education, 2010. **99**(1): p. 71-79.

[2] Boal, A. *Games for Actors and Non-actors,* 2nd ed. 2002. Trans. A. Jackson. London: Routledge.

[3] Sawyer, K. *Zig-Zag: The Surprising Path to Greater Creativity.* 2013. San Francisco: Jossey-Bass.

SB[3]C2017
Summer Biomechanics, Bioengineering and Biotransport Conference
June 21 – 24, Tucson, AZ, USA

USE OF AN EDUCATIONAL TOOL KIT TO TEACH MECHANICS OF MATERIALS

Rita M. Patterson (1), Robin Bartoletti (2), Dennis P. Chou (3), John J. Dignam (3), and Vijay Vaidyanathan (4)

(1) Family Medicine
University of North Texas Health Science Center
Fort Worth, TX USA

(2) Center for Innovative Learning
University of North Texas Health Science Center
Fort Worth, TX USA

(3) Mentis Sciences, Inc.
Manchester, NH USA

(4) Biomedical Engineering
University of North Texas
Denton, TX USA

INTRODUCTION

Mechanics of materials is a fundamental mechanical engineering course that explores the topic of solid objects subjected to stress and strain. The aim is to predict the response of engineering structures to various types of loading, design new more efficient structures, and to analyze the vulnerability of these structures to various failure modes. These same principles are applied to tissues in the human body to predict fracture and to design synthetic materials to replace or augment tissue and organ function in bioengineering curricula.

In Biomedical engineering education students typically learn these concepts in biomaterial or biomechanics courses. Lab demonstrations using a Materials testing machine are extremely useful to teach these core concepts and the importance of viscoelasticity in human tissue. However, access to these large expensive machines may not be available to undergraduate students as many courses are dropping the lab component of the course. In addition, these concepts are also taught in kinesiology or exercise science curricula or at smaller Universities that that do not have money to buy this equipment or dedicated lab space.

The Mentis Sciences Educational Toolkit (MSET) is a unique, portable, and affordable educational tool that is being introduced into the STEM (science technology, engineering and mathematics) community (http://www.mentissciences.com/mset-stem/index). It is designed to provide students with a glimpse into the world of material testing and physical science. The system provides experiments covering topics in physical sciences, simple machines, material science and data acquisition. There are 42 experiments that use a common platform shown in figure 1.

This device offers an opportunity to use active learning and interactive engagement techniques in biomechanics. Engineering students can benefit from using the device to apply biomaterial concepts to real world problems. In addition, because of the portability of this

Figure 1 MSET device ready for a 3 point bending experiment.

device, it can be taken to junior and high schools for demonstrations or hands on learning with the goal of increasing student interest in STEM fields. This paper describes the MSET device and outlines its use in an undergraduate bioengineering biomechanics class and during National Biomechanics day.

METHODS

The MSET device will be used in an undergraduate biomechanics course. Students will use the device in class to evaluate material properties of various man made and biologic materials. Student engagement and confidence in learning the material will be assessed using a subset of the Student Course engagement questionnaire (SCEQ)[1].

The MSET device will also be used during National Biomechanics day April 6, 2017. National Biomechanics Day (NBD) is a world-wide celebration of Biomechanics in its many forms for high school students and teachers sponsored by the American Society of Biomechanics. Its goal is to advance Biomechanics science and education by increasing the awareness and appreciation of Biomechanics among the high school community around the world. 2017 will be the second NBD celebration following a hugely successful NBD 2016 which had over 2,000 participating high school students and teachers in Biomechanics Labs across the U.S learning that "Biomechanics is one cool science and a fabulous, exciting professional career". Students attending the fair will be asked their level of interest in science before and after using the device.

REFERENCES

[1] Handelsman, M.M., Briggs, S.L., Sullivan N., and Towler A. A Measure of College Student Course Engagement, J Educational Research; 98(3) Jan-Feb 2005.

BIOMECHANICAL CHANGES PRECEDE RADIOGRAPHIC EVIDENCE OF NONTRAUMATIC VERTEBRAL FRACTURE UNDER CYCLIC LOADING: AN EX-VIVO STUDY

Nicole C. Corbiere-Gale (1), Stacey L. Zeigler (2), Christopher Towler (2), Kathleen A. Issen (1), Arthur J. Michalek (1), and Laurel Kuxhaus (1)

(1) Mechanical and Aeronautical Engineering Department
Clarkson University
Potsdam, N□, USA

(2) Physical Therapy Department
Clarkson University
Potsdam, N□, USA

INTRODUCTION

Vertebral fractures are a painful injury. While they are often a result of trauma, fractures can also occur during the repetitive loading of daily life. For example, ring apophysis fractures (RAFs) occur most frequently in healthy young athletes [1-2] without memorable trauma, and osteoporotic fractures can also occur without memorable trauma in older adults. Our previous *ex-vivo* work showed that the small, repetitive movements (≤ 15° of flexion, the most common movement [3]) of activities of daily living (ADLs) can contribute to these fractures [4] despite the low loads and low amplitudes of motion.

We hypothesized that application of high cycle non-traumatic motion to an *ex-vivo* model (absent of bone remodeling) would induce fractures in vertebral motion segments resulting in biomechanical symptoms. This hypothesis was tested by applying low-load, low-angle cyclic flexion to cervine and cadaver 5-vertebra motion segments. Cervine (white tailed deer, *O. virginianus*) specimens had a similar degree of skeletal maturity and have a similar range of motion [5] to adolescent humans. The cadaver specimens were older, with some degree of osteoporotic degeneration; we expected similar mechanical behavior to the cervine specimens.

METHODS

Seven cervine (2 male, 5 female; 2.4 ± 0.2 years) and five human cadaver (1 male, 4 female; 64.2 ± 4.8 years) spines were dissected into 5-vertebra motion segments (L1-L5) and potted using a previously described method [6]. Specimens were cyclically loaded in eccentric compression with pinned end conditions [4] from a neutral position to 15° of flexion, as measured by the regional centroid angle. X-ray images were obtained before and after cyclic loading on all specimens. Strains on the surface of the cortical bone were measured using three-dimensional digital image correlation (3D-DIC) before and periodically during loading, as specified below. A motion capture system (4 Oqus 500 cameras, Qualisys AB, Sweden) was used to track the regional centroid angle (α) and the intervertebral angles (IVAs) periodically throughout the cyclic loading. Load-displacement, kinematic, and deformation data were collected for each specimen.

Specimens were cyclically loaded (Instron 1331, Norwood, MA) to 20,000 cycles at 0.5Hz, cycling between displacements corresponding to 0° and 15° of flexion to simulate a low-flexion ADL such as bending slightly forward. The initial displacement required to achieve 15° of flexion was determined for each specimen using the regional centroid angle as measured with the motion capture system, and this displacement was used for subsequent cycles. The specimens were preconditioned (four cycles); a fifth cycle was used for analysis as described below. Periodically (after cycles 0, 1,000, 3,000, 6,000, 10,000, 15,000 and 19,995), 3D-DIC and motion capture imaging occurred during five witness cycles (slower, at 0.01Hz), the fifth of which were analyzed. Tests were performed at ambient temperature (approximately 22°C) and specimens were kept moist with physiologic saline.

Load-displacement data were post-processed to apply a 4th order Butterworth filter with a cutoff frequency of 0.01Hz using a custom MATLAB code (The Mathworks Inc., Natick, MA). The areas of the hysteresis loops were computed. During the witness cycles, kinematic data were collected at 120Hz using 4.0 mm hemisphere markers, with an average calibration residual of 0.40 mm. These data were processed with QTM (Qualisys Track Manager Version 2.10, Qualisys AB, Sweden) to ensure continuity of markers and minimize the effects of marker occlusion, and the relative IVAs were computed. Kinematic data from the analyzed cycles were aligned at the peak angle of each cycle to permit comparison across cycles and then filtered using a 4th order Butterworth filter with a cutoff frequency of 0.0075Hz using custom MATLAB code. Surface strain results captured at 4Hz from two-camera (Scorpion IEEE-1394, Point Grey Research Inc., Richmond, BC, Canada) 3D-DIC were enabled by a white-on-black speckle pattern on the surface of the specimen. During the witness cycles, regions of high localized cortical surface strain were identified using VIC3D 7 (Correlated Solutions Inc., Columbia, SC).

RESULTS

All tests were completed without obvious evidence of catastrophic fracture and without radiographic evidence of fracture on plain x-rays. The hysteresis loop area trends are similar between the cervine and cadaver specimens (Figure 1), and agree with our previous similar work that created fractures in cervine specimens. One cervine specimen test was halted after cycle 10,005 due to equipment failure. One cadaver specimen's cyclic loading was interrupted by unexpected failure of the potting material; this specimen was thus cyclically loaded on two separate days and appears as an outlier in cycle 10,005 (indicated by the arrow in Figure 1). Note that after cyclic loading resumed, the original trend continued.

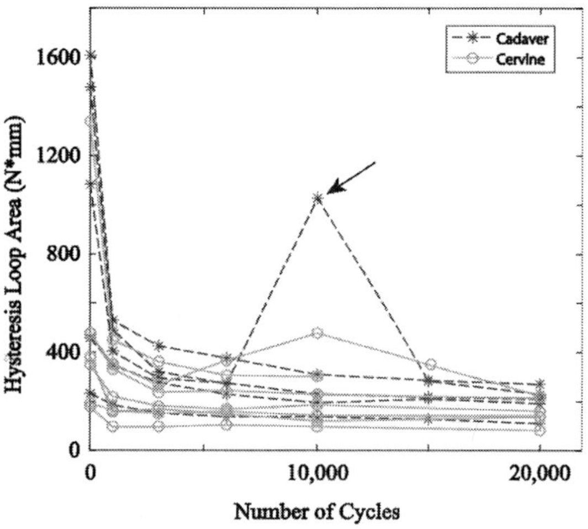

Figure 1: Hysteresis loop area for all specimens. The arrow indicates the specimen with interrupted loading.

There is evidence of fracture formation from both 3D-DIC and IVA changes during the witness cycles, despite the absence of radiographic findings. 3D-DIC provides evidence of cracks opening and closing on the cortical shell. Comparing images from these specimens (Figure 2, top) to our previous research (Figure 2, bottom) [4] that created RAFs under a different loading protocol suggests that the current cervine specimens also have RAFs. This evidence is also echoed in the observed kinematic changes. The shape of the L2/L3 IVA changes from the beginning of the test to the time of observable crack opening (cycle 10,005), an effect of damage accumulation (Figure 3). The increase in apparent compliance (increase in angle) near extension from cycle 5 to 10,005 indicates damage. The IVAs during each witness cycles show observable change in kinematics throughout the test. When comparing the 5th and $10{,}005^{th}$ cycles there is a clear change in change of IVA, the angle increases, and the behavior throughout the application of the same displacement condition is markedly different. This change suggests that damage modified the kinematics of the motion segment.

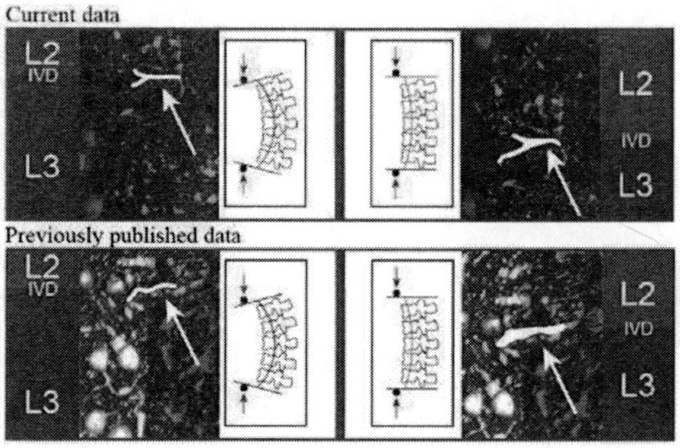

Figure 2: Top: Evidence of fracture during this study. Bottom: Evidence of fracture from previous work [4].

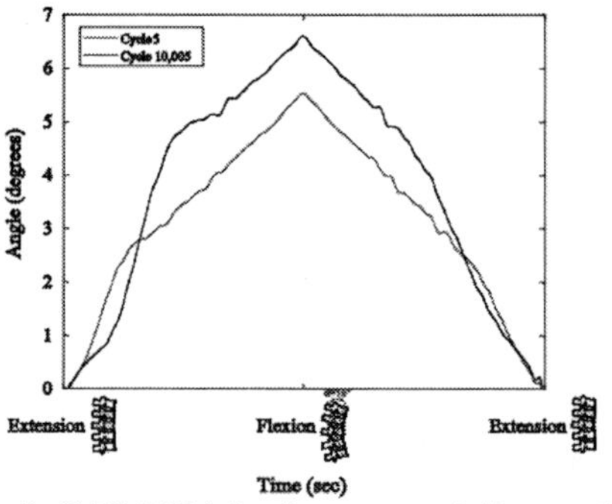

Figure 3: L2/L3 IVA for the representative specimen.

DISCUSSION

Vertebral fractures can occur during low-load, low-angle cyclic loading as confirmed by surface strain and kinematic changes during loading. This loading protocol leads to injuries that would likely be symptomatic, but would most likely not be diagnosed as fractures. Our work has demonstrated that vertebral fractures can occur in the absence of radiographical findings, which may alter recommendations for fracture diagnosis and prevention.

Future laboratory work will explore additional imaging modalities to assess type and severity of fractures in our specimens. Future clinical work will identify optimal radiologic imaging to detect these fractures. Efforts to prevent fracture formation in both healthy, physically active adolescents and older adults should consider the role that small movement ADLs may play in fracture risk.

ACKNOWLEDGEMENTS

We thank Mark Hedgeland and Mario Ciani for technical assistance; Sandra Thitten, Arlene Roach, and Ashley Fish, at Canton-Potsdam Hospital; and Glenn Seymour and Chad Green at Donaldson-Seymour Funeral Home. The authors have no conflicts of interest.

REFERENCES

[1] Takata K, et al. *JBJS [AM]*, 70A:589-594, 1988.
[2] Singhal A, et al. *Pediatric Neurosurgery*, 49:16-20, 2013.
[3] Cobian D, et al. *J of ASTM Int*, 9:1-10, 2011.
[4] Corbiere N, et al. *J Biomech*, 49:1477-1481, 2016.
[5] Lingle S. *Canada Journal of Zoology*, 71:708-724, 1993.
[6] Corbiere N, et al. *JBME*, 126:064504-1-4, 2014.

THIS HAND IS MY HAND, THIS HAND IS YOUR HAND

Joshua P. Drost, Tamara Reid Bush

Mechanical Engineering
Michigan State University
East Lansing, MI, 48823

INTRODUCTION

Most daily tasks require use of the hand, so that when hand function is lessened due to arthritis or injury the function of the entire body decreases. Thus, when hand function is diminished it is critical that we understand objectively how much function was lost, and where, so that the best possible treatment can be provided.

Currently, changes in hand function are measured using patient surveys, visual based pain scales and radiological exams, all of which are subjective [1, 2]. To understand the complete changes in hand function, objective methods to quantify changes in function are needed. A model that includes the total force and motion abilities will allow health care experts to easily compare changes in hand function, and better rehabilitate and treat the individual.

Previously, we modeled changes in the motion abilities of the hand caused by arthritis [3]. This model included the full range of motions of the hand; however, force data for each finger were not incorporated into that model. Both motion and forces are necessary to generate a comprehensive hand model for clinical use. The goal of this work was to develop a model that could predict the forces over the kinematic workspace of participants with and without reduced hand functionality.

METHODS

Sixteen participants (7 female and 9 male, average age 25.6 years, SD 6.1 years) without any reported injury or arthritis, termed "Healthy", and fifteen participants (13 female and 2 male, average age 73.5, SD 4.8 years) with doctor diagnosed arthritis, termed "Arthritic", were included in this study.

Forces due to changes in flexion/extension of the index finger (no adduction or abduction) were measured using a "U-shaped" metal bracket placed in seven positions along a line, each 15mm apart. The participant was asked to push on the bracket, and then pull on the bracket with maximum force using the pad of the index finger (Figure 1 A&B).

Maximum abduction/adduction forces were also measured in six positions. Three of the positions were at maximum extension (called "Adduction/Abduction Push"—Figure 1C) and the other three were at a mid-range flexion of the interphalangeal joints (called "Adduction/Abduction Pull"—Figure 1D). During data collection participants were asked to continually grip a cylindrical handle with their other fingers to isolate the finger forces and to maintain a consistent orientation of the wrist.

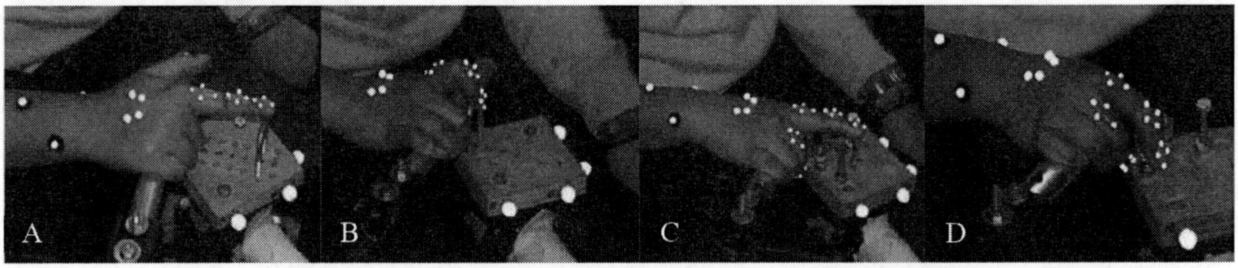

Figure 1: Experimental setup and force data for (A) flexion/extension push, (B) flexion/extension pull, (C) adduction/abduction push and (D) adduction/abduction pull trials

After collection, the data were analyzed in terms of the 1) fingertip posture (y,z coordinate i.e. distal and palmar displacement) 2) the direction of the force applied (i.e. angle within the sagittal plane from the positive distal axis), and 3) the magnitude of the force applied. To compare between different subjects the 3-D position of the fingertip for each test was normalized using the participant's finger length.

A linear mixed effect statistical model was then created for each health and gender population (healthy males, arthritic females, etc.). The model included the covariates of normalized distal displacement, normalized palmer displacement, force direction angle and the force direction angle squared along with their interactions. Subject number was used as a random variable to remove error between subjects.

RESULTS AND DISCUSSION

Sagittal plane views of the force data are presented in Figure 2 (Left). Force vectors were plotted as colored arrows based on the finger posture in 3D space and normalized based on the length of the individual's finger. Arrow color is related to the magnitude of the force in Newtons. The origin for all force vectors was at the MCP (metacarpal phalange) joint of the index finger. The plots are oriented so that a fully extended index finger would sit along the horizontal axis with the pad of the index finger pointing in the positive vertical direction. Shown are the Healthy Male data, but similar data exist for the other gender and population groups.

Predicted forces are also shown in a sagittal plane view (Figure 2 Right). The covariates (3D position and force application angle) are the same as the observed data, but the force magnitudes have been predicted using the model. This model can closely predict the forces over the range of motion. Of the covariates, the most accurate are the three-way interaction between the displacements and the angle and the three-way interaction between the displacements and the angle squared ($p=0.01$). The least accurate is the palmar displacement ($p>0.05$) but it is included due to highly significant interaction terms.

The purpose for a prediction model is to facilitate translation into the clinical environment. It is not practical (i.e. increase in time) to add numerous measurements to a therapy session, however, if only a few measurements could be gathered and used to map the entire functional space of the hand, then it becomes clinically viable. Using this model clinicians could measure a small number of forces over the range of motion and from these understand the entire force profile of the individual. This profile could be compared to the healthy and arthritic populations to determine the amount of function that has been lost and determine the best course of action for rehabilitation. Throughout treatment the patient can be retested to generate new force profiles and determine objectively how treatment is impacting had function.

Currently the model has only been developed for the index finger. Future work will include forces and associated models for all fingers. Also, measures of motion and force will be gathered prior to intervention; mid-way through rehabilitation and after rehabilitation to test its usefulness in a clinical situation. Other issues that cause loss of hand function, such as rheumatoid arthritis, can also be investigated and added to the model.

There is a need for an objective method for diagnosing loss in hand function. We are working to model the force and motion abilities of the hand and how they change with arthritis. This model is highly innovative and useful: comparisons with these models will be able to determine what level of function was lost and how much was restored due to treatment.

ACKNOWLEDGEMENTS
The authors would like to thank the Pearl J. Aldrich Endowment in Aging Related Research for their funding as well as Sam Lietkam, Wu Pan, and Jessica Buschman for assistance in data collection.

REFERENCES
[1] Chung K C., et al The Journal of Hand Surgery, 575-587. 1998.
[2] Katz S, et al. Journal of the American Medical Association, 914-919. 1963.
[3] Leitkam S, Bush TR. ASME Journal of Biomechanical Engineering. 2015.

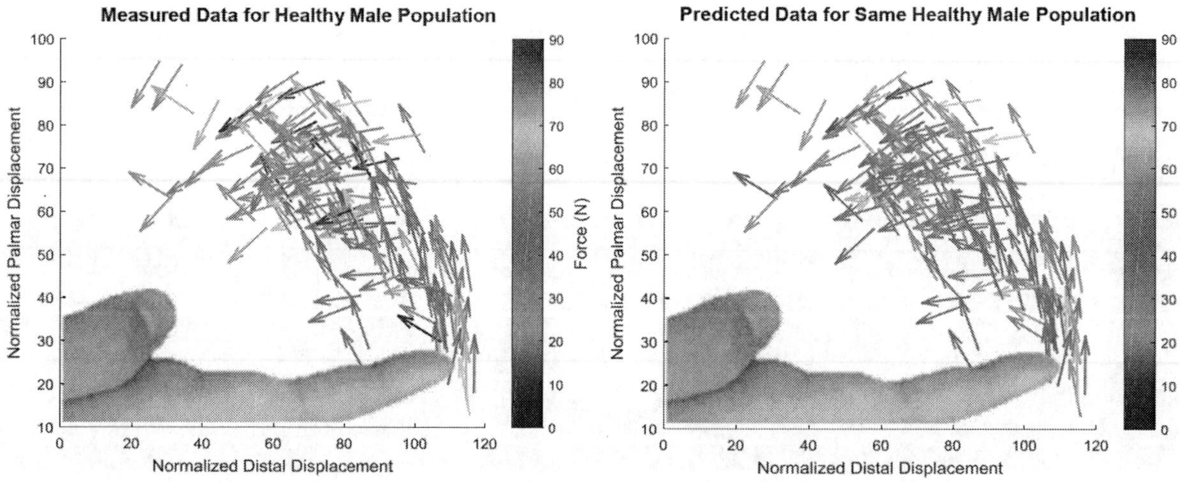

Figure 2: Sagittal projection of force data for healthy male subjects. A finger has been superimposed over the image to help represent position and direction of forces. Left: observed forces. Right: model prediction of force magnitudes using the same positions and directions.

DOES PATHOLOGICAL HUMAN TENDON ADAPT TO LOAD AND IS THIS RELATED TO CLINICAL OUTCOME? A SYSTEMATIC REVIEW

K. Färnqvist (1), P. Malliaras (2), S. Pearson (3)

(1) Haninge rehab
Handens vårdcentral
Haninge, Sweden

(2) Department of physiotherapy
Monash University
Melbourne, Frankston, Australia

(3) Centre for Sport, health and rehabilitation
Salford University
Manchester, United Kingdom

INTRODUCTION

Tendinopathy is common among athletes and sedentary individuals and can result in significant morbidity. Modulation of loading via exercise is the most evidence based treatment for managing tendinopathy (1). There is lack of strong evidence to support a certain type of load type (eccentric or concentric) in terms of beneficial tendon structure adaptation (2, 3). A systematic review by Drew et al (4) found that anteroposterior diameter and imaging measures of tendon structures did not consistently improve or change in parallel to improved self-reported pain and function. Many of the studies in this review included changes in tendon hypoechoic regions on ultrasound or intratendinous signal on MRI that are subjectively rated from baseline to follow-up. The reliability of these measures, when reported, has been shown to be poor (5- 10), bringing into question the validity of the conclusions. The review by Drew et al. (4) also did not include studies that investigate change in mechanical properties among pathological tendons.

It is important to understand whether pathological tendon adapts to load and whether this is associated with clinical outcome. This will provide information about whether the tendon should be a target of loading interventions and lead to future studies investigating optimal loading strategies. Therefore, the aim of this review is to synthesis the current evidence examining changes in tendon size (AP diameter, CSA), tendon mechanical properties and structure (UTC) in response to chronic load (>4 weeks), and relate this to changes in clinical outcome (e.g. symptoms, function, or patient satisfaction).

METHODS

The PRISMA statement protocol (11) guidelines were followed in this systematic review. A systematic search of the databases PubMed, CINAHL, EBSCO and Google Scholar was undertaken in June 2016. Fifteen different cohorts met the inclusion criteria. A total of 607 patients underwent a period of tendon loading. Structural, mechanical and clinical outcome measures were collected after intervention. Due to the heterogeneity among the studies and lack of comparative studies a meta-analysis of treatment effect was not possible (12, 13) and a qualitative evaluation was undertaken. Included studies were evaluated for risk of bias using the Pedro scale. Guidelines regarding level of evidence were taken from van Tulder et al (13).

RESULTS

This review identified fifteen cohorts investigating changes in tendon size (AP diameter or cross-sectional area) following HSR loading in the Achilles, patellar and plantar fascia (3 categories), eccentric loading in the Achilles, patellar and lateral elbow tendons (3 categories), and medical exercise in the rotator cuff (1 category). There was either no change in tendon size or conflicting findings in five of the seven categories. One study actually found lateral elbow tendon thickness increased (became more abnormal) following eccentric loading, in parallel to improved clinical outcome (14). Positive tendon adaptation (reduced tendon thickness, more towards normal) was only seen for HSR loading in the Achilles, but this was based on a single study (15).

This review found conflicting evidence that change in structure on UTC is associated improvements in clinical outcomes. A cohort with a high risk of bias and longer follow up time (52 weeks) found that improved echotypes I and II and decreased echotypes III and IV were associated with improved clinical outcomes at 52 weeks (16-18). In

contrast, in a cohort with a high risk of bias and shorter follow-up time (26 weeks) a non-significant change in UTC was not associated with improved clinical outcomes (19).
Overall, there is conflicting evidence that patellar tendon mechanical properties improve in parallel to improvement in clinical outcomes following 12 weeks of HSR loading one study utilising HSR and eccentric loading in tendinopathic subjects with low risk of bias showed no change in tendon stiffness or modulus following either loading program despite improvement in pain function and patient satisfaction at 12 weeks (20). Whereas, a study with high risk of bias found tendon stiffness decreased, and no change in tendon strain or stress, was associated with improvement in pain and function at 12 weeks (21).
Regarding loading type there is conflicting evidence that HSR loading is superior to eccentric loading for reducing tendon thickness (only when considering the patellar and not the Achilles tendon). One study with low risk of bias among people with patellar tendinopathy found HSR but not eccentric loading changed tendon thickness at 12 weeks whilst neither loading program bought about a change in mechanical properties despite improvements in pain, function and satisfaction (20). In contrast to the earlier study by Kongsgaard et al. (20), a later study examining HSR loading with high risk of bias found patellar tendon thickness did not change (21) in parallel to improvements in pain and function. One study with low risk of bias comparing HSR and eccentric loading in Achilles tendinopathy showed no superiority for HSR or eccentric loading based on change in tendon thickness but HSR was superior regarding patient satisfaction at 12 weeks but not at 52 weeks (15).

DISCUSSION

The main finding from this systematic review is that there does not appear to be a relationship between objective measures of pathological tendon adaptation and clinical outcome following exercise interventions (>4 weeks). Objective tendon adaptation outcomes (change in tendon AP diameter or cross sectional area, tendon structure or mechanical properties) generally do not show change in parallel to improvement in clinical outcome (pain, function and treatment satisfaction). The current literature clearly demonstrates that pathological tendon adaptation, as determined by currently available objective measures, is not necessary for improvement in clinical outcomes following exercise interventions applied over a number of weeks. Although this does not mean that tendon adaptation is not occurring, if it is currently applied methods are not able to consistently detect tendon adaptation and the adaptations is clearly different to the normal tendon.
This review did not find any evidence for adaptation in pathological tendon when considering objective outcomes of tendon size, structure (UTC) and mechanical properties. This is consistent with evidence the tendon core has almost no turnover regarding collagen synthesis in adult life (23). In contrast, Langberg et al. (24) found evidence for increased collagen production in the peritendon following loading of pathological Achilles tendons.
Recently HSR was considered in a prior review (4), to be potentially superior for tendon turnover in patellar tendinopathy when compared to eccentric loading. The current review included a more recent evidence comparing HSR and eccentric loading in Achilles tendinopathy where there is no evidence of the superiority of HSR (based on change in tendon thickness only), therefore, the literature overall is conflicting.
We judged studies against the high standards of an RCT because as argued by Drew et al (4) the best design to answer our systematic review question would be an RCT where participants are randomised into an exercise and sham exercise group. It can be argued that this review is limited because we have included data from different tendons demographic groups (e.g. athletes versus sedentary people) and various exercise protocols. There is also very limited description of load intensity and compliance so it is not possible to draw any firm conclusions regarding the relationship between load and pathological tendon adaptation. There are also differences in follow up time with 12 weeks being the most common but some studies following participants for a year or longer. Another key limitation of the literature is that very few studies investigate structure with UTC and mechanical properties. There are also new technologies being used to assess tensile stiffness (e.g. elastography) but until now, as far as we could find, these have not been applied to investigating change in stiffness with loading in pathological tendons.
To conclude, the evidence is limited with regard to the relationship between objective measures of pathological tendon adaptation and clinical outcome following exercise interventions and a major limitation is poor reporting of load interventions. So although we can say that tendon adaptation does not appear necessary for improved clinical outcomes, no conclusions can be made about whether pathological tendon adapts to load.

ACKNOWLEDGEMENTS

The authors have no conflict of interest to declare that are relevant to the content of this review. No funding was received to assist in the preparation or completion of this review.

REFERENCES

1. Scott A, et al., Br J Sports Med. 47 :536-44. 2013
2. O'Neill S et al., Int J Sports Phys Ther. 10 :552-62. 2015
3. Couppé C et al., J Orthop Sports Phys Ther. 45:1-25. 2015
4. Drew BT et al ., Br J Sports Med. 48: 966-72. 2014
5. Sunding K et al., Knee Surg Sports Traumatol Arthrosc. 24: 1988-96. 2016
6. Collinger JL et.,. Acad Radiol. 16: 1424-32. 2009
7. Weinreb JH et al., Muscl Lig Tend J. 4: 66-73. 2014
8. O'Connor PJ et al., Eur Radiol. 2004: 14: 1968-73. 2004
9. Ingwersen KG et al., BMJ Open. 6: 011746. 2016
10. Sein ML et al., Br J Sports Med. 41: 9. 2007
11. Liberati A et al., BMJ. 339: b2700. 2009
12. Candy B, et al., BMC Med Res Methodol.:11:124. 2011
13. van Tulder M et al., Spine (Phila Pa 1976). 28: 1290-9. 2003
14. Krogh TP., et al Am J Sports Med 41: 625-35. 2013
15. Beyer R et al., Am J Sports Med. 43: 1704-11. 2015
16. de Jonge Set al., Am J Sports Med. 39: 1623-9. 2011
17. de Jonge Set al., Am J Sports Med. 43 :2950-8. 2015
18. de Vos RJ et al., Br J Sports Med. 45: 387-92. 2011
19. de Vos RJ et al., J Sport Rehabil. 21: 34-43. 2012
20. Kongsgaard M et al., Scand J Med Sci. 19: 790-802. 2009
21. Kongsgaard M et al., Am J Sports Med. 38: 749-56. 2010
22. Ohberg L et al., Br J Sports Med. 38: 8-11. 2004
23. Heinemeier KM et al., FASEB J. 5: 2074-9. 2013
24. Langberg Het al., Scand J Med Sci Sports. 17: 61-6. 2007

SB[3]C2017
Summer Biomechanics, Bioengineering and Biotransport Conference
June 21 – 24, Tucson, AZ, USA

DESIGN AND TESTING OF A 3D PRINTED LOWER LIMB PROSTHESIS

McKenzie C. Evans, Cooper H. Welch, Hunter T. Dender, Nathaniel A. Godwin, Connor L. Martin, Elizabeth M. Scheig, S. Nima Mahmoodi, Beth A. Todd

Department of Mechanical Engineering
University of Alabama
Tuscaloosa, AL, USA

INTRODUCTION

As 3D printing has become more popular and less expensive, there have been a number of instances where hobbyists have created upper limb prosthetics, particularly for children. These devices are often associated with open-source data. [1] [2] While a major role of the arm is to position the hand where it can function and do useful work, the leg must be able to support the total body weight when walking as it passes through the single stance phase. [3] Thus the amount of force for the upper limb to withstand may be an order of magnitude smaller than the lower limb. [4] This led to the question of whether a lower limb prosthesis could be 3D printed that would support the weight of an adult.

In further refining the design problem, constraints including those related to the available facilities were developed. As this was an exploratory problem, the design team was given an initial spending limit of $500. SolidWorks was the most readily available CAD software, and the available 3D printers could produce parts 11.55 inches by 7.55 inches by 5.85 inches. Printer resolution varied from 0.0039 inches to 0.01 inches. The design team also considered questions of how much to rely on ABS plastic for strength, how to incorporate other materials, and how to include a damping system.

For an initial design, the weight of 300 lb was used as an approximate force that could account for both static and impact loads. Other considerations were compatibility with standard prosthetic ankle and knee joints, weight of the prosthesis, aesthetics, manufacturability, and cost of production.

METHODS

The first question addressed was materials. Because of the weight of ABS plastic and the cost of 3D printing, it was determined that the 3D printed component would be utilized for aesthetics only. Due to high strength to weight ratio, durability, availability, and affordability, 6061 aluminum alloy was selected for the primary load bearing component of the prosthesis. The ABS plastic shell was designed to fit the contour of a human calf muscle and the shell was left hollow with a 1" through hole for the aluminum support rod. It was determined that a traditional damping system would exceed the desired weight and cost leading to the use of silicone rubber fixed to each end of the aluminum for shock absorption. Circular aluminum plates were connected to either end of the shank with countersunk 6-32 screws leaving room for the rubber to be compressed. Figure 1 shows a SolidWorks rendering of the final design.

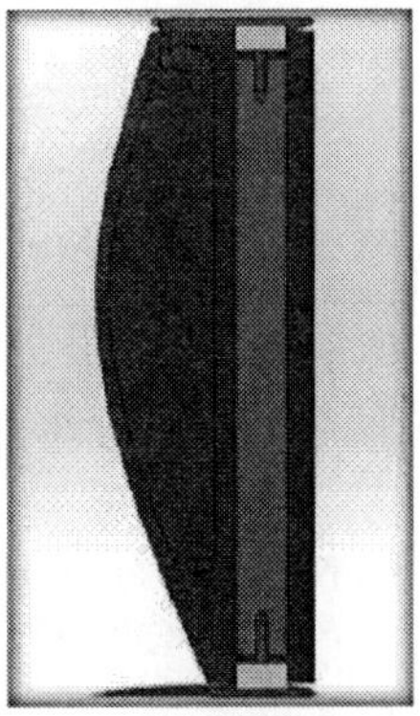

Figure 1: Final Prototype Design

There was concern that the total deflection would exceed the allowable 0.25 inches resulting in unwanted load being transferred to the ABS plastic shell. In order to calculate the theoretical displacement, the axial stress was calculated to be 382 psi. Next, the theoretical displacements of the aluminum and rubber parts were calculated as to be 4.2 x 10^{-4} in and 5.27 x 10^{-2} in, respectively. The resulting total theoretical deflection was 0.0531 inches which is well below the allowable 0.25 inches.

Static compression was performed on the prototype using an MTS uniaxial load frame. Since the target load was 275 pounds, compression testing was performed to 300 pounds and deflection was measured using MTS TestSuite TW Software considering zero displacement at 100 pounds. The integrity of the part was preserved under 300 pounds static loading.

In order to perform dynamic testing, a new set of custom end plates was designed allowing the prototype to be tested using an existing testing apparatus designed to allow an able-bodied person to walk using a lower limb prosthesis. Due to time and budget constraints, a rubber peg was used in place of a prosthetic foot as shown in Figure 2.

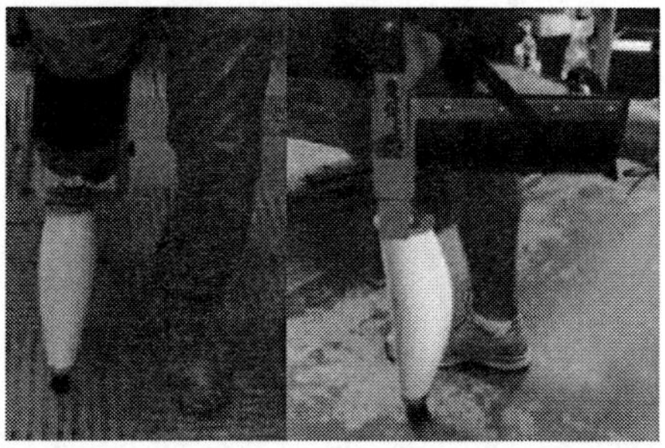

Figure 2: Front and side views of prosthesis attached to dynamic testing apparatus

With practice, various test subjects were able to successfully use these devices to ambulate without the use of a functioning lower leg.

RESULTS

The experimental displacement measured by the MTS uniaxial load frame was 0.0468 inches which is 11.9% lower than the theoretical displacement calculated. Using the axial stress and assuming yield strengths of 40,000 psi for aluminum and 4351 psi for rubber [5], the factor of safety of the aluminum was calculated to be 105 and the factor of safety of the rubber was found to be 11 giving the combined factor of safety of 11.

Dynamic testing was successful for various users of different heights and weights demonstrating the versatility of the design. Additionally, after repetitive use by many test subjects, the prosthesis showed no visible signs of wear.

DISCUSSION

The design, development, and testing of a functioning 3D printed lower limb prosthesis was completed for under $500. The parts and labor for a single prototype cost approximately $180. With more efficient mass manufacturing techniques, this cost could be further decreased, allowing easier access to amputees who may not otherwise have the means to purchase a prosthesis.

The difference between theoretical and experimental displacement of the prototype under a 300-pound load could have been caused by different factors. The most significant cause of the difference was that the change in displacement was not measured from zero; instead the change in displacement was measured from 100 pounds to 300 pounds in 50-pound increments. Another potential cause of the difference is the steel screws embedded in the rubber dampers, which could increase the rigidity of the rubber.

The overall factor of safety well exceeds the desired factor of safety. The weight of the finished prototype was only 3.26 pounds and could be increased as desired by adding more material in the hollow shell or using denser materials. The versatility was demonstrated in the redesign of the top and bottom plates to seamlessly connect with the existing dynamic testing device. If time and budget had allowed, further testing could have been conducted using a standard prosthetic foot and knee. This design also leaves room for customization of the size and appearance of the shell.

Limitations of the design are its susceptibility to aesthetic or in severe cases structural damage to the shell through accidental application of forces by surrounding objects. Additionally, the study did not include analyses of torsional or bending stresses and yielded no quantifiable dynamic results through repetitive loading.

Despite these limitations, this design provides a platform for further research into the use of alternate methods to create customizable and affordable prosthetic devices.

ACKNOWLEDGEMENTS

The authors express their appreciation to Mr. Jim Edmonds of the College of Engineering Machine Shop. Dr. Mark E. Barkey provided facilities and guidance for compression testing. The senior design team "Pirate Prosthetics" allowed us to use their dynamic testing device.

This project was funded through the James and Mildred Hire Fund in the Department of Mechanical Engineering at the University of Alabama.

REFERENCES

[1] Krausz, N. E., et al., *IEEE Transactions on Neural Systems and Rehabilitation Engineering*, v 24, n 5, p 562-572, May 2016.

[2] Zuniga, J., et al., *BMC research notes* 8.1 (2015): 10.

[3] Webster, J. G., *Encyclopedia of Medical Devices and Instrumentation* (2nd ed.). 2006.

[4] Enderle, J., S. Blanchard, J. Bronzino, *Introduction to Biomedical Engineering* (2nd ed.). 2005.

[5] Budynas, R. Nisbett, K., *Shigley's Mechanical Engineering Design* (8th ed.). 2006.

DEVELOPMENT OF A HEAD SUPPORT DEVICE FOR PEOPLE WITH HYPERMOBILE-TYPE EHLER-DANLOS SYNDROME

Robert S. Pierce (1), Candace Ireton (2), Martin L. Tanaka (1), David Hudson (3)

(1) Engineering and Technology
Western Carolina University
Cullowhee, NC, USA

(2) Candace Ireton, MD
Asheville, NC, USA

(3) Health and Human Sciences
Western Carolina University
Cullowhee, NC, USA

INTRODUCTION

Ehler-Danlos Syndrome (EDS) is a group of heritable disorders that are characterized by abnormal collagen synthesis, leading to multisystemic connective tissue disorder. Six major types of EDS are recognized, the most common of which is the hypermobile type. EDS hypermobile-type (EDS-HT) is characterized by connective tissue laxity resulting in excessive joint mobility due to weak and less elastic ligaments and tendons. Patients with EDS-HT suffer from frequent joint dislocations and tears in their connective tissue that are slow to heal. Cumulative damage to joint connective tissue often results in frequent, intense pain in several different joints. This pain, combined with joint fragility, has the effect of severely limiting the activity levels of many EDS-HT patients, with a corresponding decrease their quality of life. [1]

A particularly common symptom amongst EDS-HT patients is pain and instability in the neck. Laxity of the ligaments of the neck may result in multilevel cervical instability, including craniocervical (CCI) and atlanto-axial instability (AAI). CCI and AAI may cause repetitive brainstem and spinal cord compression and stretch (medullary kyphosis), producing neurologic symptoms and longstanding damage. The vertebral arteries pass through the vertebrae in the cervical region, so instability at these levels may also result in repetitive decrease in blood flow through the vertebral arteries. In EDS-HT, the extensive complex of cervical ligaments may fail to provide the normal support and endpoints for range of motion, thus the muscles of the neck and upper back must fire almost continuously to stabilize the head and neck. This leads to muscle fatigue and pain from muscle spasm. EDS-HT patients describe the feeling of balancing a bowling ball on top of their necks, constantly working to support and stabilize their heads and unable to rest the muscles in their necks and shoulders.

In cases in which there is CCI and/or AAI, many EDS_HT patients will choose to undergo surgical fusion of occiput to C1, C1-C2, or more extensive multilevel fusion. These procedures have been shown to significantly decrease head instability, however the procedures carry a high risk, are extremely expensive, are somewhat controversial in the neurosurgical community with regard to the EDS population, and there are few neurosurgeons willing to do them. Furthermore, the long-term effects of this surgery on the spines of EDS-HT patients is uncertain. The procedures may transfer loads to vertebrae located below the fusion site, raising the possibility of need for subsequent fusions.

An alternative to spinal fusion surgery for EDS-HT patients is long-term use of a cervical collar in order to provide support for the head. It is common for EDS-HT patients to own one or more different types of cervical collar and to use these collars when they need to unload their neck. However, commercially-available collars were not designed with the needs of EDS-HT patients in mind. Many cervical collars are designed to be used after surgery or after an injury. They are stiff and completely immobilize the head, making it impossible to participate in many common activities. Furthermore, by completely immobilizing the head the post-surgical type of collars may contribute to atrophy and weakening of the muscles in the neck that are needed to support the head. An alternative to the post-surgical collars is the soft collar, usually made from foam wrapped in a cloth cover. These collars allow some movement of the head, however they are bulky and hot and they completely enclose the neck. This makes them very uncomfortable for many EDS-HT patients who are intolerant of pressure surrounding their necks.

In this abstract we describe work that has been performed to design a head support specifically for EDS-HT patients. We applied the methods of "human-centered design" (HCD), a method that has its origins in the field of information technology and was developed in response to the need for better human-machine interfaces [2]. A key to the method is to develop an understanding of the physical and psychological needs of the human user. As it is applied to product development, HCD treats the end-user of a product as a member of the design team. The designer works to build an understanding of the user community at a human level. Product design becomes a highly iterative process in which customer interviews are quickly followed by a round of brainstorming, design, and prototyping. In this manner, designers and customers work together to converge on a design solution.

METHODS

HCD dictates that the designer become familiar with the community of intended users. Towards this end, a survey of websites that provide support for people with EDS was performed. This provided insight into the type of head support devices that people with EDS-HT currently find useful. Next, we contacted the leaders of the Asheville area EDS Support Group who invited us to attend a group meeting. At this meeting we met several EDS patients who described their experiences with neck pain and confirmed the need for an improved head support device.

Seven people with EDS-HT (six female and one male) were recruited to participate in interviews. The goal of these interviews was to uncover desirable characteristics of a head support. The interview script and process were approved by the Institutional Review Board. Transcripts of these interviews were analyzed to determine the frequency with which interviewees gave particular responses. Table 1 gives a summary of the most frequent interview responses.

Using Table 1 as a starting point, a "voice of the customer" statement was formulated: "I want a device that will support and stabilize the weight of my head while allowing limited motion. It should fit well, be well-ventilated, and be unobtrusive."

Interview Question	Most frequent responses
What causes pain?	Sitting upright with head unsupported Pressure on occipital region of head Looking up or down Physical activity Staying in any one position for long
How is the pain manifested?	Sharp pain in back of neck Fatigue and tightness in neck muscles Headache centered on occipital region Sensation of upper vertebrae rubbing together Looseness and pain in the jaw
Thoughts on existing "hard" cervical collar designs	Too confining Fear that immobilizing head will lead to atrophy of neck muscles Fit is often poor Draws unwanted attention in public
Thoughts on existing "soft" cervical collar designs	Some allowed motion is a plus Contact around entire neck is confining Doesn't provide enough support of head weight Draws unwanted attention in public
Desirable characteristics of a new head support design	Allows some motion Stabilizes the head on the neck Supports the weight of the head Well ventilated Adjustable or custom fit Unobtrusive

Table 1: Most frequent responses given by EDS-HT patients during interviews

Starting with this statement, a functional analysis of the head support was performed [3]. Primary functions were identified and broken down into sub-functions in order to construct a function tree. Several solution methods were then formulated for each sub-function. Sub-function solutions were then combined in order to look for promising solution variants. Two of these variants were selected for detailed design and prototyping.

RESULTS

Figure 1 shows solid models of the two approaches that were selected for the first round of design. The "headband" design, shown in Figure 1a is intended to stabilize the head while allowing limited motion in all three rotation axes. The "space-frame" design is intended to support the weight of the head while also resisting motion. Both designs are constructed from molded ribs that can be engineered to flex in a prescribed manner. This allows limited motion of the head, while providing support and preventing excessive motion. The range of motion in different axes can be controlled by changing the stiffness and geometry of the ribs. Both designs contact the user's body at points that are not painful for people with EDS-HT. The designs are open, well-ventilated and unobtrusive. They leave the user's face and the front of the neck open and they follow the contours of the user's body so that they can be worn under clothing.

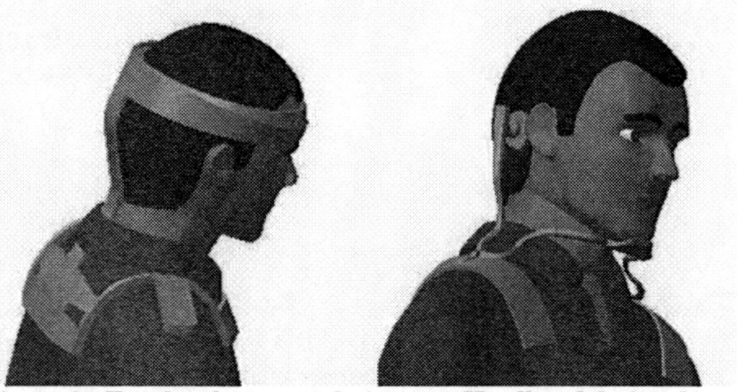

Figure 1: Two head support designs. a) Headband design, b) Space frame design.

DISCUSSION

Using the methods of HCD, two head supports have been designed to meet the needs and desires of people with EDS-HT. The supports stabilize and support the head while allowing limited, prescribed motion. They contact the body in areas that people with EDS-HT do not find painful. They are open, well-ventilated, and can be worn under clothes, increasing the probability that they will be used in public.

This work is the first iteration in a highly-iterative process. Prototypes of the two designs are currently under construction. These prototypes will be taken to EDS-HT support group meetings in for feedback that will be used in the next design iteration. Over time, we hope to converge on designs that provide highly-usable, affordable devices for head stabilization and pain relief for EDS-HT patients.

ACKNOWLEDGEMENTS

We would like to thank the members of the Asheville Area EDS Support Group for their participation in this work.

REFERENCES

[1] Rombaut, L. et al., *Disability and Rehabilitation*, February, 2010.
[2] Maguire, M., *Methods to Support Human-Centred Design*, Int. J. Human-Computer Studies, (2001) **55**, 587-634.
[3] Cross, N., Engineering Design Methods, *Strategies for Product Design, 4th Ed.*, Wiley, 2000.

DESIGN OF A NOVEL MULTIDIRECTIONAL FLUID SHEAR STRESS BIOREACTOR FOR AORTIC TISSUE

Janet Liu, Philippe Sucosky

Department of Mechanical and Material Engineering, Wright State University, Dayton, OH, U.S.A.

INTRODUCTION

The normal aortic valve consists of three leaflets and is also known as the tricuspid aortic valve (TAV). The bicuspid aortic valve (BAV) is one of the most common congenital cardiac defects in which two of the leaflets are fused together, and has been associated with secondary aortopathy, such as aortic dilation, aortic aneurysm and aortic dissection. Although both TAV and BAV patients can develop aortic dilation, BAV patients are at higher risk of developing the disease[1]. Furthermore, while the dilation patterns are symmetric in TAV patients, they only affect the convexity of the aortic wall in BAV patients. Interestingly, the expression of dilation seems to correlate with the local blood flow patterns in TAV and BAV aortas. Blood flow in the TAV aorta is essentially streamlined and aligned along the axis of the aorta wall; it is much more helical in BAV aortas[2] (**Figure 1**). As a result, the TAV and BAV subject the convexity of the ascending aorta to substantially different wall shear stress (WSS) vectors, characterized by different time-varying magnitudes and directionalities. The elucidation of the possible cause-and-effect relationships between BAV hemodynamics and aortic dilation requires a device capable of subjecting native aortic tissue to this local WSS environment.

Two different bioreactors, the parallel plate flow chamber and the cone-and-plate, have been designed previously to study the effects of shear stress on cardiovascular tissue. The parallel plate bioreactor has been typically used in applications requiring exposure of a cell monolayer to steady and unidirectional WSS[3]. The cone-and-plate bioreactor has added the capability to subject cells or tissue to a time-varying WSS, pulsatile or oscillatory WSS[4]. While these devices have provided critical insights into the response of cardiovascular cells and tissue to idealized WSS environments, they are not able to replicate the multidirectional nature of the native WSS signal experienced by the aortic wall. To address this limitation, this study aimed at designing a new bioreactor capable of replicating the magnitude and directionality of the local WSS on native aortic tissue.

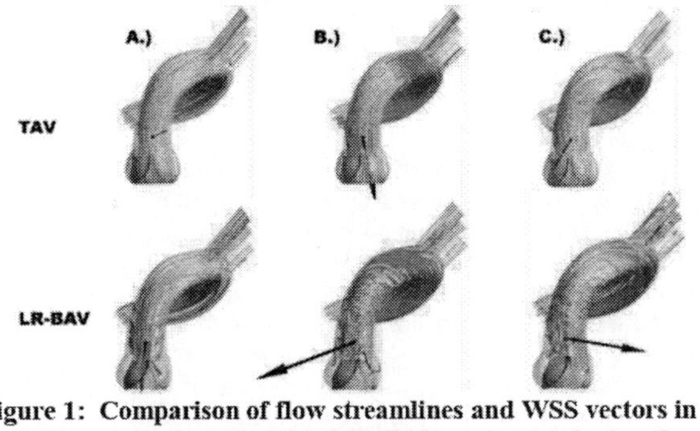

Figure 1: Comparison of flow streamlines and WSS vectors in a TAV aorta and a left-right (LR)-BAV aorta: A.) during the acceleration phase; B.) at peak systole; and C.) during the deceleration phase.

METHODS

Geometrical and Mechanical Requirements

The requirements of the design were to subject multiple aortic tissues simultaneously to desired pulsatile WSS magnitude and directionality, while maintaining them in a sterile culture environment. The target WSS waveform consisted of a time-varying signal with a period of 0.86 s (heart rate: 70 beats per minutes), a magnitude τ varying between 0 and 4 Pa (**Equation 1**), and a direction angle θ varying between +30 and -30° (**Equation 2**):

$$\tau(t) = |4\cos(\tfrac{2\pi t}{0.86})| \quad (1)$$

Poster Presentation #P87

$$\theta(t) = \frac{\pi}{6}\sin\left(\frac{2\pi t}{0.86}\right) \quad (2)$$

This target WSS waveform was created based on computational predictions of the native WSS waveform on the BAV aorta convexity, which was shown to vary between 36 and 54 dyn/cm^2 over one cardiac cycle[5].

Design Principle

The bioreactor design was adapted from the cone-and-plate viscometer principle, in which a rotating inverted cone generates a fluid flow on top of a stationary plate, resulting in a uniform shear stress environment (τ) directly proportional to the cone velocity (ω):

$$\tau = \mu\omega r/(h + r\alpha)\,, \quad (3)$$

where h is the distance between the cone apex and the bottom plate, μ is the dynamic viscosity of the working fluid, and α is the angle between the cone and the plate. The angular orientation of the WSS vector was achieved by rotation of four mounts supporting the tissue specimens.

Validation Methodology

The device was validated computationally using the computational fluid dynamics (CFD) software ANSYS CFX. Due to the device symmetry, only a 90° section was modeled. The cone surface and the tissue surface were modeled as rotating walls. The cone angular velocity was derived from **Equation 1** in order to achieve the desired WSS magnitude. The tissue surface was rotated to satisfy the instantaneous angular orientation of the target WSS. The working fluid (Dulbecco's Modified Eagle's Medium – high glucose, Sigma-Aldrich Co., St Louis, MO) was modeled as a Newtonian fluid. The simulations were performed over two cardiac cycles, and the flow data were extracted from the second cardiac cycle. Validation consisted on comparing the target instantaneous WSS magnitude and directionality to those predicted by the model at the same instant of time. The WSS predictions were also compared to the target WSS in terms of temporal shear magnitude (TSM),

$$\text{TSM} = \frac{1}{T}\int_0^T |\tau|dt, \quad (4)$$

where T is the cardiac period.

RESULTS

Design Solution

The cone-and-plate device uses culture medium (viscosity: 0.00095 kg/m.s) as the working fluid and the stationary bottom plate is designed to accommodate four circular tissue samples (**Figure 2**). Changes in WSS magnitude are achieved by the rotation of the inverted cone with a prescribed angular velocity derived from **Equation 3**. The rotation is achieved by a servo motor connected to a programmable servo drive. Changes in WSS directionality on each sample are achieved by rotating the four tissue mounts using a second servo motor coupled to a driving belt and connected to a second servo drive. Perfusion pumps are used to recirculate culture medium through the system.

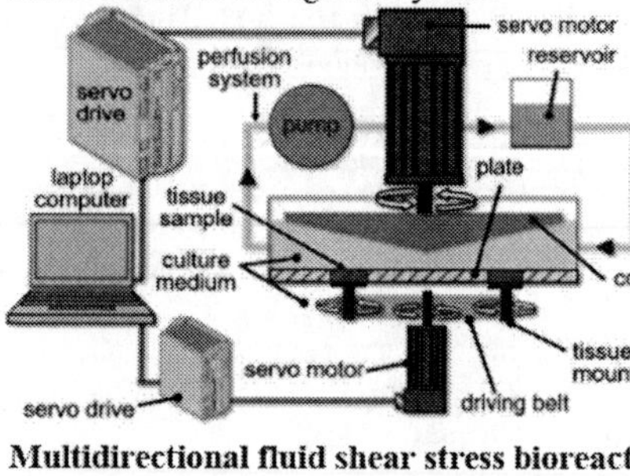

Figure 2: Multidirectional fluid shear stress bioreactor schematic

Mechanical Validation

The WSS contours predicted by the CFD model on the surface of one tissue sample were captured at each instant of time throughout a cycle (**Figure 3A**). The WSS magnitude was extracted at the center of the tissue and compared to the area-average WSS captured on the entire tissue surface. The percent difference at each instant of time was less than 5%. The computed TSM value predicted on the tissue surface was 2.6 Pa, which represented a 7% error relative to the target TSM (**Figure 3B**).

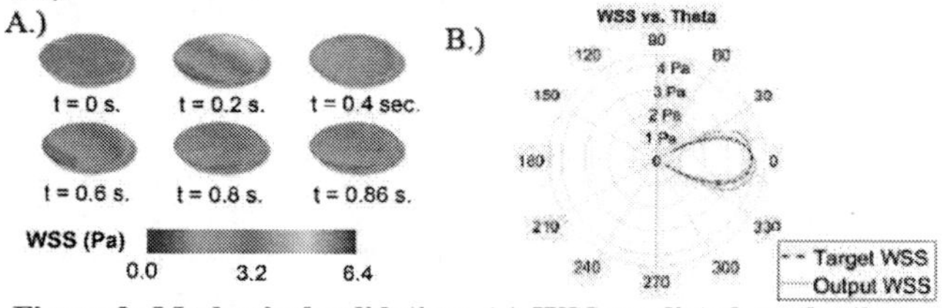

Figure 3: Mechanical validation: A.) WSS predicted on the tissue surface; B.) Comparison between the target and predicted (output) WSS in terms of magnitude and directionality

Table 1. Comparison of WSS magnitude captured A.) at the tissue center, and B.) over the entire tissue surface

Time	0 s	0.2 s	0.4 s	0.6 s	0.8 s
Tissue-center WSS (Pa)	3.13	1.17	3.77	2.02	2.62
Tissue-average WSS (Pa)	3.10	1.18	3.96	2.10	2.59
% difference	1.0%	1.4%	3.5%	4.8%	1.2%

DISCUSSION

A new multidirectional WSS bioreactor was designed. The device was validated computationally with respect to its capability to subject tissue samples to a desired time varying WSS magnitude and angular orientation. Ongoing work is being performed to verify the capability of the device to replicate a more realistic WSS environment, more similar to that present on the surface of the TAV and BAV ascending aortic wall. The next step will consist of validating the device biologically in terms of its capability to maintain tissue under sterile conditions. This device will be instrumental in elucidating aortic tissue mechanobiology and more broadly the role played by WSS in pathologies (e.g., atherosclerosis, aortopathy, aneurysms), which may open new perspectives toward the management of cardiovascular disease.

ACKNOWLEDGEMENTS

1. National Science Foundation (CAREER grant CMMI-1148558)
2. College of Engineering and Computer Science at Wright State University

REFERENCES

1. Braverman, A. C. *et al.* The Bicuspid Aortic Valve. *Curr. Probl. Cardiol.* **30,** 470–522 (2005).
2. Mahadevia, R. *et al.* Bicuspid Aortic Cusp Fusion Morphology Alters Aortic 3D Outflow Patterns, Wall Shear Stress and Expression of Aortopathy. *Circulation* **129,** 673–682 (2014).
3. Frangos Eskin, S. G., McIntire, L. V., Ives, C. L., J. A. Flow Effects on Prostacyclin Production by Cultured Human Endothelial Cells. *Science (80-.).* **227,** 1477–1479 (1985).
4. Sucosky, P. *et al.* Design of an ex vivo culture system to investigate the effects of shear stress on cardiovascular tissue. *J Biomech Eng* **130,** 35001 (2008).
5. Atkins, S. K., Cao, K., Rajamannan, N. M. & Sucosky, P. Bicuspid aortic valve hemodynamics induces abnormal medial remodeling in the convexity of porcine ascending aortas. *Biomech. Model. Mechanobiol.* **13,** 1209–1225 (2014).

HIP CUP HICCUPS: VALIDATING A COMPUTATIONAL MODEL FOR HIP CUP STABILITY

Mohsen S. Renani (1,3), Philippe Favre (2), Jeffrey E. Bischoff (3)

(1) Department of Mechanical Engineering
University of Missouri – Kansas City
Kansas City, MO, USA

(2) Research
Zimmer Biomet GmbH
Winterthur, Swit□erland

(3) Biomechanics Research
Zimmer Biomet, Inc.
Warsaw, IN, USA

INTRODUCTION

Elimination of aseptic loosening associated with insufficient primary fixation is an unrealized clinical goal of cementless acetabular cups in total hip arthroplasty. Achieving good fixation relies on surgical (e.g., press fit, component positioning, surgeon experience), implant (e.g., surface coating, cup stiffness), and patient (e.g., bone quality, activity) factors [1]. Preclinical evaluation and optimization of these factors can lead to arthroplasty solutions with reduced failures and improved patient outcomes.

Computational modeling is a proven method for evaluating the primary stability of orthopedic implants [2-4], enabling the effective and efficient understanding of the influence of variations in design, technique, and biomechanical factors on expected performance. However, there is no established consensus perspective on the essential aspects of such models that enable comparisons to be made with confidence, including the appropriate way to model insertion of the cup into the pelvis [2,5-7]. The goal of this study is therefore to examine the influence of numerous reasonable but different approaches for modeling cup stability on the ability of the model to predict differences in stability due to different arthroplasty configurations.

METHODS

A series of steps was utilized, both experimentally and computationally, to evaluate the resistance of a pressfit (noncemented) hip cup to torque-induced micromotion. A single cup design (23mm diameter) was evaluated, characterized essentially by a cobalt chrome alloy substrate with a thin plasma spray coating to facilitate bony ingrowth. The cup was inserted via impaction into a polyurethane foam block (10 or 15 pounds per cubic foot (pcf) density, for 1 and 2mm pressfit respectively) intended to represent low or normal quality bone. Subsequently, a rotational torque was applied to the cup leading to micromotion.

Cup insertion. Figure 1 shows the experimental and numerical approach for modeling insertion. Experimentally, the foam was prepared to allow for either a 1mm or 2mm diametric pressfit with the cup. A static load was applied until the application platen was flush with the surface of the foam (Fig. 1a). The platen was then released and the cup was allowed to recoil based on the pressfit in the foam. Two modeling approaches were used to simulate this process: (1) Direct placement of the cup in a flush condition with the foam, with reliance on the finite element solver to resolve the interference due to the diametric pressfit; and (2) Quasi-static insertion of the cup to the flush condition (Fig. 1b).

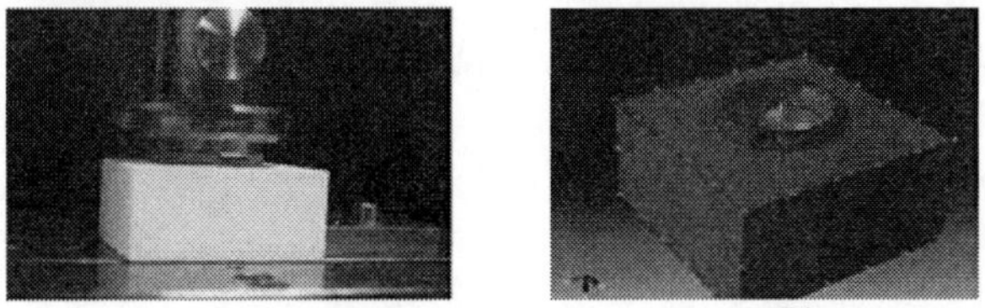

(a) (b)

Figure 1. Cup insertion a) experimental set up and b) model setup

Torque out. Torque was applied experimentally by grasping the cup in three locations and applying a rotational moment, while preserving a small (50N) normal force to keep the cup seated (Fig 2a). This step was modeled with two sets of boundary conditions: (1) A fixed constraint on all external surfaces of the foam; and (2) Inclusion of rigid blocks to more closely mimic the experimental setup (Fig. 2b).

(a) (b)

Figure 2. Torque out a) experimental set up and b) model setup

A pressure-based friction law was utilized to model the interaction between the porous coating and the foam [8]. The cobalt chrome cup was modeled using linear elastic material properties; foam was modeled assuming isotropic plasticity. All simulations were performed using Abaqus 2016 (Dassault Systemes).

RESULTS

Insertion force and maximum torque for 1mm and 2mm diametric pressfit values, with results normalized by the corresponding 2mm pressfit value, are shown in Figure 3. Experiments showed that a reduction in pressfit from 2mm to 1mm decreased insertion force and maximum torque by approximately a factor of 2, whereas the model predicted a 3-4x decrease due to pressfit reduction and reduced density foam.

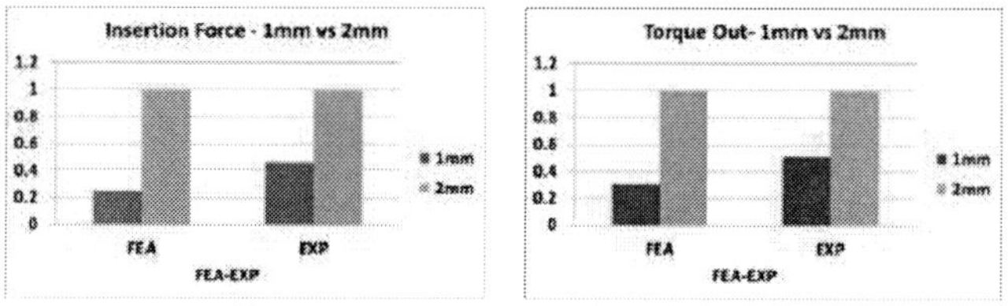

(a) (b)

Figure 3: Normalized finite element (FEA) and experimental (EXP) results for a) insertion force and b) torque out for 1 and 2 mm pressfit

The force required to maintain a cup initially placed flush with the bone foam during resolution of the interference was significantly less than the force required to insert the cup into the foam via quasi-static simulation, even though final diametric pressfit values were equivalent. Similarly, resolution of pressfit interference resulted in significantly less bounce back of the cup once load was removed. The extent to which the rim of the cup was constrained during insertion (e.g. constrained to move as a rigid body during insertion, versus allowing radial compression / rotation as pressfit occurred) had minimal impact on insertion force, attributed to the low stiffness of the foam as compared to the shell.

The predicted response of the system to both insertion and torque out was significantly impacted by the far-field boundary conditions on the foam. Using fixed constraints on the sides of the foam resulted in a stiffer overall system, which translated to higher insertion forces and higher torque out values.

DISCUSSION

The main objective of this study was to develop a framework for modeling pressfit and torque out of acetabular shells in bone foam, for subsequent use to predict the effects of different implant designs (geometries and/or materials) or techniques (e.g. amount of pressfit) on system performance. A consistent trend was observed between FE predictions and experimental measurements for two different values of pressfit, though the effect of pressfit was greater in the experiment as compared to the model.

Numerous factors were explored in the model development process, highlighting a couple key modeling assumptions that should be maintained in future work. Most significantly, approximating cup insertion by resolving shell/foam interference for a nominally placed cup impacts the predicted insertion force, cup bounce back once the force is removed, and subsequent maximum torque out of the cup, as compared to quasi-static insertion of the cup. Thus, the potential benefit of interference resolution in terms of total solution time (insertion required approximately 1.5x longer solver time compared to interference resolution) results in compromised simulation accuracy, and should be avoided. On the other hand, treating the cup rim essentially as a rigid entity has negligible effects on the predictions, and thus can be used for a more efficient model. Finally, the extent to which pressfit impacted insertion force and torque out was impacted by the boundary conditions on the foam (e.g. Fig. 1b), and thus careful attention to those modeling assumptions is important.

Several other aspects of model complexity could be more fully explored to further close gaps between experimental work and the model predictions. Most notably, both polyurethane foam [9] and bone [10] are known to be highly inelastic materials, and more complex material modeling could elucidate the effect of low-strain elastic nonlinearities or damage [11] on stability.

The current model may be expanded in numerous ways to improve the clinical relevance of the predictions of primary stability. Currently available liners (metal, ceramic, or polyethylene) could be added to the shell to more accurately reflect the stiffness of the complete reconstruction. More physiological bone support conditions may be modeled through development of patient specific virtual implantations of the shell/liner into the acetabulum. Physiologic loads that may occur in vivo and which challenge the primary stability of the reconstruction may be incorporated. Finally, quasi-static insertion of the shell may be replaced by simulation of the dynamic impaction of the shell into the pelvis as occurs during surgery.

REFERENCES

[1] Miettinen SSA, et al., Int Orthop 2016.
[2] Amirouche F, et al., Clin Biomech 2014;29:1177-1185.
[3] Favre P, et al., Clin Biomech 2016;36:46-51.
[4] Fitzpatrick CK, et al., J Biomech 2014;47:1718-1726.
[5] Udofia I, et al., J Bone Joint Surg 2007;89-B:549-556.
[6] Yew A, et al., Proc Inst Mech Eng H 2006;220(2):311-319.
[7] Spears IR, et al., J Biomech 1999;32:1183-1189.
[8] Grant JA, et al., J Biomech 2007;40:1158-1164.
[9] Patel PSD, et al., BMC Musc Disord 2008;9:137-144.
[10] Prot M, et al., J Biomech 2016;49:1050-1057.
[11] Bishop NE, et al., J Biomech 2014;47:1472-1478.

SB^{3}C2017
Summer Biomechanics, Bioengineering and Biotransport Conference
June 21 – 24, Tucson, AZ, USA

THE EFFECT OF FLOOR STIFFNESS ON STANDING POSTURE AND SWAY

Daiane Aizen Grill (1), Sara E Wilson (1)

(1) Department of Mechanical Engineering
University of Kansas
Lawrence, KS, United States

INTRODUCTION

LBP and lower extremities' discomfort have been associated to occupations requiring standing for long periods of time [1]. Several studies have investigated the effect long-term standing has on a person's perceived discomfort, on measurable biomechanical variables, and on the development of LBP [2]. However, few studies have been performed in order to analyze the effect material properties of the floor have on posture and sway. In a study by Cham and Redfern (2001), different flooring conditions (mostly viscoelastic foams) were examined [3]. In that study, increases in center of pressure (COP) weight shifts were associated with increased discomfort as subjects. COP lateral weight were found to have a nonlinear relationship with floor stiffness in which an ideal flooring condition appearing to minimize this biomechanical parameter. Another study suggested that the probability of developing LBP while performing long-term standing is not a function of the standing time, rather it is dependent on the posture adopted by the subject on the first 15 minutes of the standing period [4]. The objective of the current study is to investigate whether standing posture or COP parameters vary as a function of axial floor stiffness.

METHODS

Twenty-eight (28) healthy individuals (13F, 15M) participated in this experiment. All subjects voluntarily signed a consent form prior to the commencement of the study, as well as filling a health questionnaire. Subject were recruited from University settings and their ages ranged between 18 and 30 years old. Subjects' height ranged between 1.52m and 1.93m (mean 1.65 ± 0.09m). Weights ranged between 52 kg and 104 kg (mean 70 ± 13kg). 8 subjects reported LBP in the past year. The experimental procedure was IRB approved.

Subjects were asked to stand in an upright position in 5 different flooring conditions of differing stiffnesses including a high stiffness control condition. Data was collected in five 1-minute intervals per flooring condition. Between intervals, subjects were asked bend their knees and bow forward. During the minute intervals, subjects were asked to stand in a comfortable stance and avoid movement. Subjects performed the study bare-footed. The order of the flooring conditions was randomized and subjects were asked to sit between the testing of each floor.

A variable-stiffness platform was constructed consisting of a double-layered surface with four identical springs between the layers. The springs were held in place by steel guiding rods. The platform's stiffness was modified through the exchange of these springs. The overall stiffness of each flooring condition was 75.59 kN m^{-1}, 100.79 kN m^{-1}, 123.18 kN m^{-1}, and 139.98 kN m^{-1}. The control flooring condition was achieved by placing a wooden block (assumed to be relatively inelastic) between the platform's layers to prevent deformation of springs. Stiffnesses were chosen to span stiffnesses seen in elastic flooring such running tracks and gymnastics/tumbling flooring that have stiffnesses of 50-400 kN m^{-1} [5].

The measuring equipment used consisted of electromagnetic motion sensors (MotionStar, Ascension Technologies, VT) and a force plate (Bertech, Columbus, OH). The variable stiffness platform was placed on top of the force plate and the former was zeroed for every flooring condition. A total of 7 magnetic sensors were placed on each subject: one over the seventh cervical vertebrae, one over the twelfth thoracic vertebrae, one over the first sacral vertebrae, and one half-way through each thigh and lower leg. Data was collected for 60s at a rate of 100 Hz.

The biomechanical parameters collected by the electromagnetic sensors were of position and rotation of each sensor while the force plate collected COP. Additionally, Motion Monitor software was used to digitize the knee and hip in order to obtain knee flexion angles and hip flexion, rotation and abduction angles. The postural variables measured include thoracic flexion, lumbar flexion, hip flexion, lateral rotation, hip torsion, and knee flexion. Additionally the range and standard deviation (SD) of the COP in the anterior-posterior (AP) and lateral (LAT) directions were assessed. A repeated measures ANOVA was performed. A significance level of 0.05 was used. Within-subjects contrasts were assessed for the results of each flooring condition compared to the control for those conditions that were found significant in the ANOVA.

RESULTS

No statistically significant differences were found for the following postural parameters: thoracic flexion, lumbar flexion, hip flexion, hip lateral rotation, hip torsion and knee flexion. The minimum p-value was of 0.096 and the maximum p-value was of 0.978. This suggests that posture did not generally change significantly as a function of floor stiffness.

Statistically significant patterns were found for the results in the range and standard deviation of the COP positions in the AP direction; and for the COP position range in the LAT direction. The results for the ANOVA performed on these variables is shown in Table 1. The variation in the position of the COPs is expressed in terms of the standard deviation exhibited by each sample.

Table 1: ANOVA Results for COP in x and y directions

Variable	p	Within-Subjects Contrast of Floors to Control			
		75.59 kNm^{-1} vs control	100.79 kNm^{-1} vs control	123.18 kNm^{-1} vs control	139.98 kNm^{-1} vs control
COP AP SD	**0.025**	**0.036**	**0.020**	**0.040**	0.064
COP AP range	**0.022**	0.064	**0.003**	**0.013**	0.075
COP LAT SD	0.068	**0.002**	**0.006**	0.200	**0.006**
COP LAT range	**0.009**	**0.002**	**0.002**	0.054	**0.017**

The results for the COP range and SD in the AP direction are shown in Figure 1. The points represented by a red diamond showed a statistically significant difference to the control flooring condition. The range in AP COP increased with floor stiffness to the third flooring condition, above which the COP range dropped (p=0.022). Similarly, the variation in the COP in the AP direction showed an increase in magnitude with stiffness up until the second flooring condition, after which it dropped in the third and fourth conditions (p=0.025).

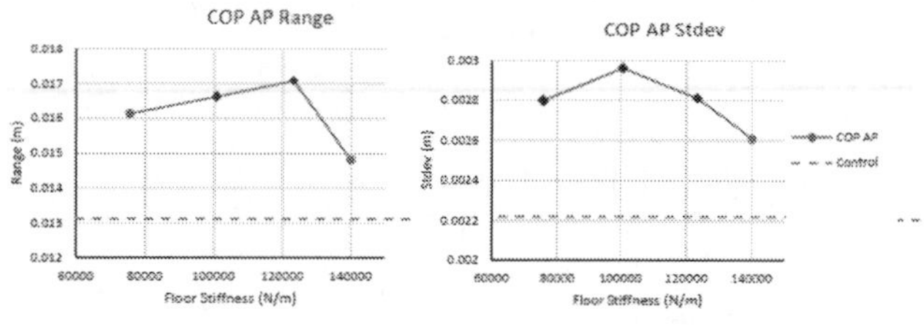

Figure 1: Position Range and Standard Deviation for the COP in the AP Direction

Figure 2 shows a decreasing trend in LAT COP with increasing floor stiffness (p=0.009). The points represented by a red diamond were the results that showed a statistically significant difference to the Control flooring condition.

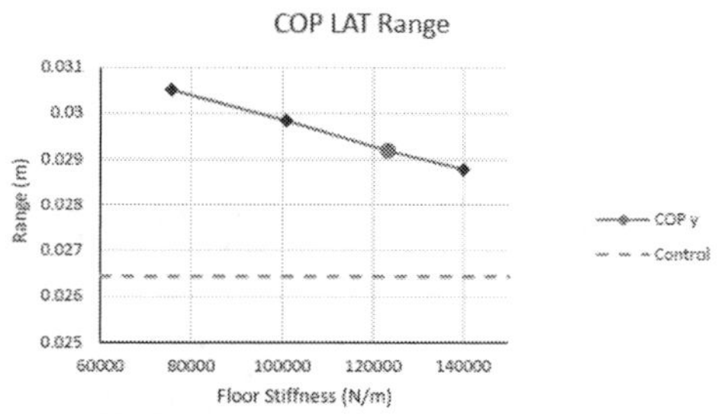

Figure 2: Variation in the COP Range in the LAT Direction

DISCUSSION

The results obtained showed that there are no significant changes in the postural parameters due to floor stiffness but did show changes in the COP sway. In particularly, we observed greater sway with decreasing floor stiffness. In the AP direction, a peak sway was observed between the 100 and 123 kNm^{-1} floor stiffnesses. This range of stiffness is around that currently used in gymnastics floors and running tracks to improve performance and reduce injury. Greater sway may be useful in varying posture and reducing discomfort [5].

Unlike this study, Cham and Redfern found that greater COP weight shifts were exhibited by harder floors which the authors attributed to an increase in discomfort and fatigue [3]. Weight shifts are different than sway. In this study the measurement period was short and subjects were asked to maintain the same posture, so weight shifts are less likely to have occurred.

In this study, subjects stood for only 5 minutes on each floor. Future studies should examine extended standing. Such studies could also examine perceived levels of discomfort and its association with COP sway.

Additionally, Rys and Konz found that posture is dependent of the position of the feet during standing [6]. If the movement of the COP in the AP direction is discovered to be associated with perceived discomfort, the relationship between floor stiffness and feet placement during standing should be studied.

ACKNOWLEDGEMENTS

We would like to acknowledge the generosity of the Undergraduate Research Award Program at the University of Kansas for their financial and logistic support.

REFERENCES

1. National Institute for Occupational Safety and Health, 1997 "Musculoskeletal Disorders and Workplace Factors - A Critical Review of Epidemiologic Evidence for Work-Related Musculoskeletal Disorders of the Neck, Upper Extremity, and Low Back" DHHS 97-141.
2. Redfern, M. and Cham, R., 2000, AIHAJ, 61 (5), pp. 700-708.
3. Redfern, M. et al 2001, Human Factors, 43 (3), pp. 381-391.
4. Gregory, D. et al 2008, Gait & Posture, 28, pp. 86-92.
5. Nigg, B.M., et al, 2000, Biomechanics and Biology of Movement, Human Kinetics. Champaign, IL.
6. Rys, M. et al, 1994, Ergonomics, 37 (4), pp. 677-687.

TIME DOMAIN ANALYSIS OF LOCAL DYNAMIC STABILITY MAY BE USEFUL IN PREDICTING A CRITICAL EVENT BEFORE IT OCCURS

Martin L. Tanaka and Chaoke Dong

School of Engineering and Technology
Western Carolina University
Cullowhee, North Carolina, USA

INTRODUCTION

Low back pain is a common medical problem that affects 4 out of every 5 people some time during their life. Idiopathic low back pain has been associated with torso instability. One method used to quantify local dynamic stability is to calculate the maximum finite time Lyapunov exponent. The Lyapunov exponent describes how quickly two initially close points diverge in state space. Previous researchers have calculated the Lyapunov exponents and averaged them to obtain a single scaler value [1]. This value was used it to evaluate if deterministic chaos existed in a system. Others have plotted the Lyapunov exponent in state space to find regions of stability and located boundaries between stable and unstable regions [2]. However, we are not aware of any researchers who have calculated the Lyapunov exponent and examined how it changes within the time domain. The focus of this research was to determine how the Lyapunov exponent changes in the time domain as it approaches a critical transition (i.e. falling, in our case).

METHODS

The initial investigation consisted of thorough testing of two participants. The study was approved by the Institutional Review Board (IRB) and participants signed informed consent prior to beginning the study. Each participant balanced on an unstable sitting apparatus capable of attaining large deflection angles (Figure 1). Participants maintained balance until falling into a foam padded safety frame. Roll, pitch and yaw angles were collected using a gyroscopic sensor and a custom LabView program during the balancing and falling portions of the trail. Data was imported in to Matlab to generate plots and calculate the Lyapunov exponent. Participants repeated the trial multiple times (55 for participant 1) so that a large data set could be establish for analysis. Trials were performed in multiple sets over several days in order to avoid fatigue.

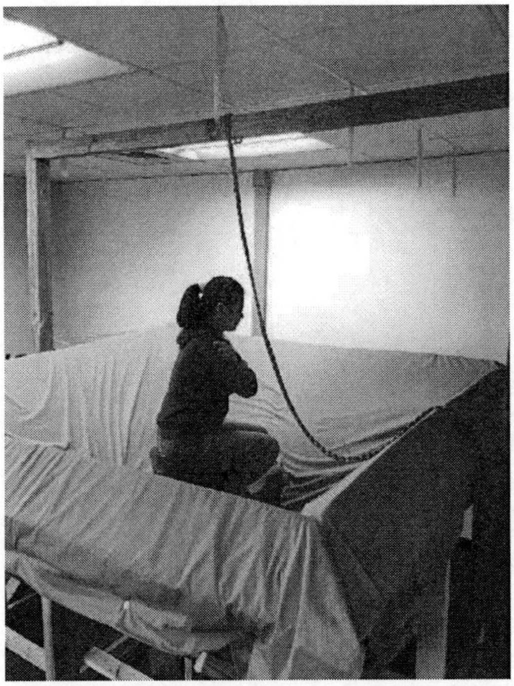

Figure 1: Experimental configuration. Participant balances on a kneeling chair in the center of a foam padded safety frame.

RESULTS

Figure 2 shows the results of a typical set of torso balancing trials. The participant balances around zero degrees until falling backwards (negative angle) or forwards. The time of zero indicates the beginning of data collection after initial balance is attained.

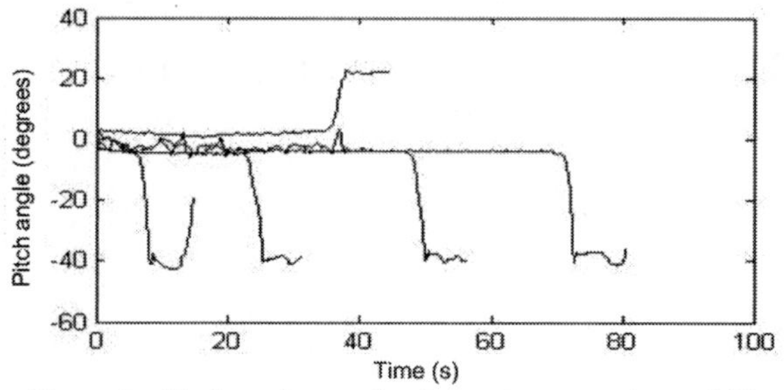

Figure 2: Pitch angle remained close to zero prior to falling.

In order to compare the behavior over multiple trials, the zero reference time was changed to the midpoint of the fall. Using this reference negative time indicates the time prior to falling and positive time indicates the time elapsed after the fall. All trials were aligned based on this new time reference and the results are shown below (Figure 3).

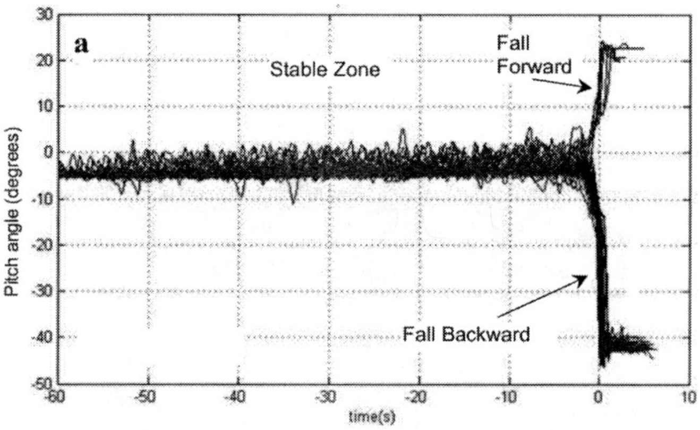

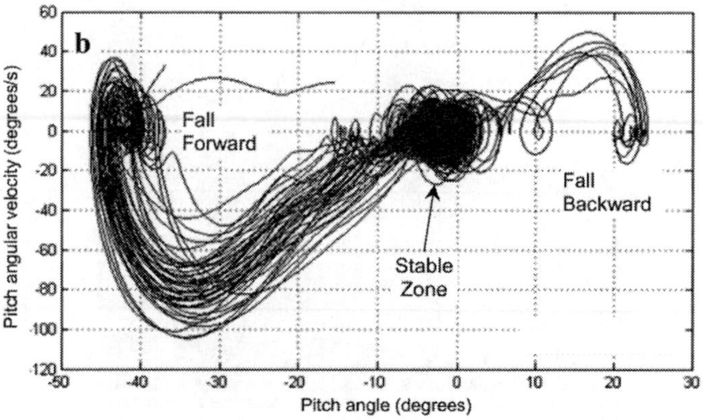

Figure 3: Time domain (a) and state space (b) representation of 55 trials by a single study participant.

The Lyapunov exponent for each data point in the time domain was calculated and the average value for 55 trials is shown (Figure 4). The central line (blue) is the mean value and the upper (green) and lower (red) lines indicate +/- 1 standard deviation, respectively. Several zones are noticeable in this time domain plot of the Lyapunov exponent. The first is the stable zone where the Lyapunov exponent undulates around 105 while the participant maintains balance. The second is the transition zones, where a small peak is observed. This peak may be an indicator that the test subject is beginning to lose stability. Next is the falling zones where the Lyapunov exponent sharply drops from about 95 to 47. Motion during falling is more consistent behavior and results in a lower Lyapunov exponent. Within the post fall zone the participant impacts the safety frame, bounces off, and then comes to rest. This causes a spike in the Lyapunov exponent which then decreases.

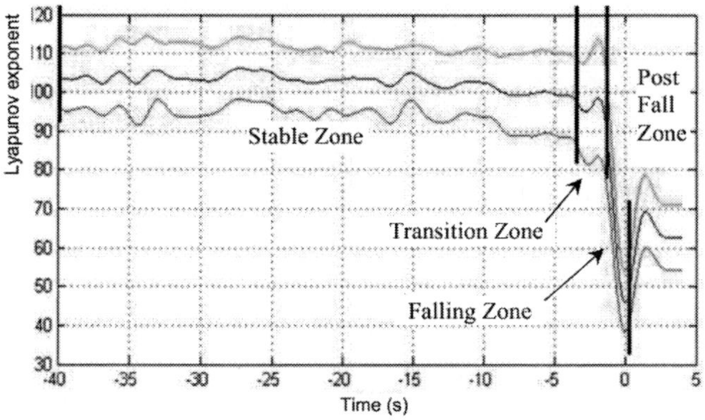

Figure 4: Average Lyapunov exponent (blue) for 55 trials. Note the 4 regions, stable, transition, falling, and post fall.

DISCUSSION

A key discovery of this research was that the Lyapunov exponent changed over time as it approached the critical event (the fall). The standard deviation bands show the trends to be consistent over multiple trials. The small peak observed in the transition region was observed just prior to the fall. Further investigation is needed to determine if this is a real effect or an artifact.

Tracking the Lyapunov exponent in the time domain may be useful as an indicator to predict a future event. This approach not only can be used for this research, but is also generalizable to other dynamic systems that have critical events. The ability to predict a future event *prior to its occurrence* may be used as a warning so that preventative action may be taken to avoid the occurrence or minimize potential negative effects. This is similar to an air bag system that detects high accelerations and uses this information to inflate the bag to protect the occupants during a crash.

ACKNOWLEDGEMENTS

This research was funded by Western Carolina University Graduate School and the Rapid Center at WCU.

REFERENCES

[1] Rosenstein, M. T., Collins, J. J., and Deluca, C. J., 1993, "A Practical Method for Calculating Largest Lyapunov Exponents from Small Data Sets," Physica D, 65(1-2), pp. 117-134.

[2] Ross, S. D., Tanaka, M. L., and Senatore, C., 2010, "Detecting dynamical boundaries from kinematic data in biomechanics," Chaos, 20(1), p. 017507.

QUANTIFING LOCOMOTION STABILITY BY MEASURING THE DEVIATION OF THE EXTRAPOLATED CENTER OF MASS FROM THE CENTROID OF BASE OF SUPPORT

M. Alamoudi, F. Travascio, S. Asfour

Department of Industrial Engineering
University of Miami
Miami, FL, USA

INTRODUCTION

In all human movements, control of locomotion stability is a complex task that requires the interaction between sensory, nervous, and motor systems to regulate the body center of mass (CoM) over the base of support (BoS) [1]. In a full gait cycle, the body is in a continuous state of imbalance, since approximately 80% of the time the CoM is outside the BoS with each subsequent step preventing a fall [2]. Recently, dynamic stability has received research attention. Hof et al. [3] presented Margin of Stability (MoS), which is an inverted pendulum model that considers the state of the CoM expressed as the extrapolated CoM (XCoM). Two components are required to generate the XCoM: the vertical projection of the CoM trajectory, and its velocity. The reason that CoM velocity should be accounted for is because if the velocity is directed outward the BoS, balance may be impossible to achieve, and the reverse is also possible [4]. Then MoS is described as the distance between the XCoM and the boundary of the BoS in the anterior-posterior (A/P), and medial-lateral (M/L) directions. However, this method does not differentiate between single and double limb support phases in measuring locomotion stability. In a full gait cycle, 20% of the time the body will be supported by double limb, which occurs twice, at the beginning, and end of the stance phase. In the remaining 80%, the body will be supported by a single limb. From biomechanical perspective the stability while standing on a single limb is worse than standing on double limb support. Therefore, the goal of this study was to introduce a new measure of gait stability that incorporates the concept of the XCoM, BoS area during each phase (i.e. single and double limb support phases), and the percentage of time of each phase, and comparing it with MoS in terms of reliability.

METHODS

Proposed Gait Stability Measures

When the CoM of a human body lies at or near the center of the BoS (CBoS), it is more stable than when it lies far from it or near the edge of the BoS [5]. Therefore, CBoS was accounted for in the proposed measure, which will act as a reference point at each phase of the gait cycle. The BoS was determined by the trajectories of feet markers. Namely, toe, lateral malleolus, and heel markers that follow the classical Golem model of Vicon®. The model was modified to add an extra marker on the 5th metatarsal, in order to allow the motion capturing system to capture all the area under the feet (i.e. BoS). Several calculations should be done in order to find the CBoS. The BoS will create a convex polygon with n sides, which consists of n-2 triangles, as shown in Figure 1. The area of each triangle will be calculated using equation (1).

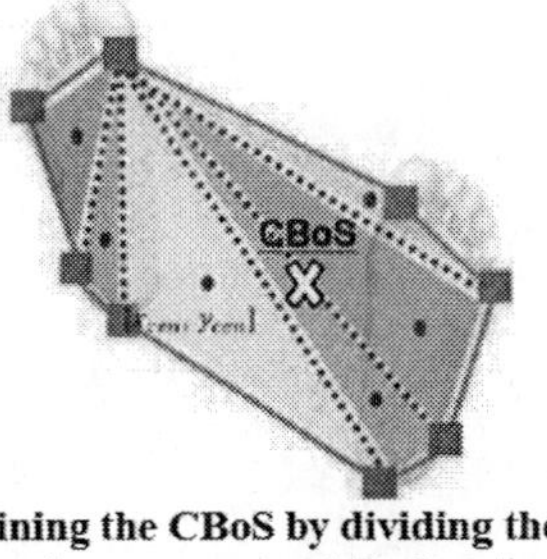

Figure 1: Determining the CBoS by dividing the BoS into set of triangles. The green boxes represent the location of the markers used in the experiment. The purple circles represent the center of each triangle.

$$Area_i = \sqrt{\left(\frac{a+b+c}{2}\right)\left(\frac{b+c-a}{2}\right)\left(\frac{a+b-c}{2}\right)\left(\frac{a+c-a}{2}\right)} \quad (1)$$

Where, a, b, and c are the triangle sides.

After calculating the area of each triangle, the center of each one was calculated using equation (2), and (3).

$$x_{cen} = \frac{x_1+x_2+x_3}{3} \quad (2)$$

$$y_{cen} = \frac{y_1+y_2+y_3}{3} \quad (3)$$

Where, $[x_1, y_1]$, $[x_2, y_2]$, and $[x_3, y_3]$ are a triangle corners.

Then, using equations (4), and (5), the center of a polygon (i.e. CBoS) was found.

$$x = \frac{\sum(x_{cen_i} * Area_i)}{\sum Area_i} \quad (4)$$

$$y = \frac{\sum(y_{cen_i} * Area_i)}{\sum Area_i} \quad (5)$$

Walking stability is calculated by finding weighted average of the normalized root mean square (RMS) deviation of the XCoM from the CBoS. The RMS deviation is normalized by the square root of the BoS area of that phase (i.e. either single or double limb support) to have a unitless number that quantifies gait stability. The normalized value is assigned a weight, which is the duration of limb support (in %) in a full gait cycle, as shown in equation (6). The proposed measure was used to quantify stability in the A/P, and M/L directions.

$$Gait\ Stability = \frac{\sum_{i=1}^{4} Support\ duration_i\% \times \sqrt{\frac{\frac{\sum_{j=1}^{k}(CBoS-XCoM)^2}{k}}{BoS\ Area_i}}}{4} \quad (6)$$

Equipment and Tools

The experiment was conducted at the Biomechanics laboratory at the University of Miami using the Vicon® Motion Capturing System (Oxford Metrics, United Kingdom) Nexus software with its 12 cameras at a sampling rate of 120 Hz.

Experimental Protocol

Eight healthy male subjects participated in the study (age: 29.25 ± 2.38 years; height 173.13 ± 8.09 cm; body mass: 72.5 ± 7.25 kg). The protocol of the experiment was approved by the University of Miami's Institutional Review Board (IRB). Subjects who had fallen within the past 6 months, with history of dizziness, tremor, alcoholism, neurological disorders, diabetic symptoms, vestibular disorders, or back pain or injuries were excluded from the study. After attaching the 41 reflective markers to the subject's body, they walked with their self-selected speed. Each participant repeated the trial for 3 times.

Statistical Analysis

Measurements are almost always exposed to various types of errors, which cause the measured value to differ in each trial of an experiment. If reliability is high, measurement errors are small in comparison to the true differences between subjects. The parameter that is used to measure reliability is called intraclass correlation coefficient (ICC), with a value close to one corresponding to small amount of variability, and vice versa. Reliability refers to the extent by which the measurement is reproducible, or the degree to which the measurement influenced by measurement errors [6, 7].

RESULTS

The proposed stability measure produced smaller standard deviation than the MoS, as shown in Table 1.

Table 1: Mean (and standard deviation) of dynamic stability measures for healthy subjects performing normal gait.

Measure	Gait stability	Margin of Stability (cm)
A/P	1.828 (0.219)	17.88 (3.269)
M/L	0.165 (0.019)	7.12 (1.401)

In terms of ICC values, the proposed measure was higher than the MoS in the A/P by 6.23%, and the M/L by 8.76%, as shown in Figure 2.

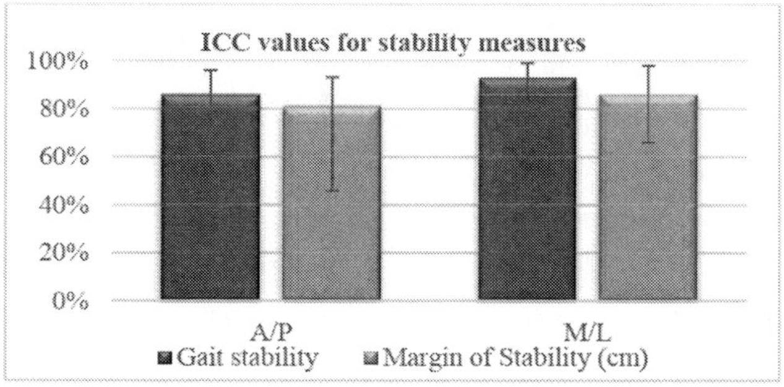

Figure 2: ICC values for stability measures with its 95% CI.

DISCUSSION

Although MoS showed a good reliability, three indications depicts that the proposed gait stability measures were more reliable. First, smaller standard deviation value, which means the data are clustered closely around the mean. Second, higher ICC values, which indicates higher reliability. Third, narrower 95% CI, which corresponds to a small amount of variability due to error. According to Bartlett [8], in comparing between different measures, using the one with higher reliability will give greater statistical power to find differences between groups. However, there were several limitations of this study. First, small sample size participated in the study. Second, all the participants were young and healthy adults. One might investigate the locomotion stability of elderly adult, who are physically limited. In conclusion, this study proposed a new method that will evolve gait stability research in the laboratory settings.

ACKNOWLEDGEMENTS

The authors would like to thank the Biomechanics Research Group at the University of Miami for funding this study.

REFERENCES

[1] Bruijn, S et al., *J R Soc Interface*, 10: 1-23, 2013.
[2] Patla, A, Adaptability of Human Gait: Implications for the Control of Locomotion. Amsterdam: North-Holland. 1991
[3] Hof, A et al., *J Biomech*, *38*(1), 1-8, 2005.
[4] Hof, A, *Hum Mov Sci*, *27*(1), 112-125, 2008.
[5] Whiting, W et al,. Dynatomy. Champaign, IL: Human Kinetics; 2006.
[6] Bruton, A et al., *J Physiother*, 86:94-9, 2000.
[7] Pinsault, N et al., *Med Eng Phys*, 31(2):276-286, 2009.
[8] Bartlett, J et al., *Ultrasound Obstet Gynecol*, 31:466 – 475, 2008.

ELBOW AND SHOULDER JOINT TORQUES ARE CORRELATED WITH BODY MASS INDEX BUT NOT GAME PITCH COUNT IN YOUTH BASEBALL PITCHERS

Jim D. Darke (1), Eshan M. Dandekar (3), Arnel Aguinaldo (4), Scott Hazelwood (1,2), Stephen M. Klisch (1,2)

(1) Biomedical Engineering Department
California Polytechnic State University
San Luis Obispo, CA, USA

(2) Mechanical Engineering Department
California Polytechnic State University
San Luis Obispo, CA, USA

(3) Kinesiology Department
California Polytechnic State University
San Luis Obispo, CA, USA

(4) Kinesiology Department
Point Loma Nazarene University
San Diego, CA, USA

INTRODUCTION
The number of ulnar collateral ligament (UCL) reconstructive (i.e. "Tommy John") surgeries performed on youth baseball pitchers have more than doubled since 2000 [1]. Routinely pitching while fatigued is considered a leading factor associated with UCL injuries; adolescent pitchers who had elbow or shoulder surgery were 36 times more likely to have routinely pitched with arm fatigue [1]. MLB/USA Baseball Pitch Smart guidelines limit 9-10 yr. old pitchers to a maximum 75 pitches per game, a figure based on long-term studies related to injury prevention [2]. Several studies have shown that pitching kinematics (e.g. elbow flexion/extension and pronation/supination, scapulothoracic internal-external rotation) may change as adult pitchers reach muscular fatigue [3], and such kinematic changes could result in higher elbow and shoulder rotational torques that may increase injury risk [4]. Several biomechanical studies have been done on ~12 yr. old youth pitchers [5,6] but none have been reported at the 9-10 yr. old level. This study aims to predict elbow and shoulder joint torques throughout a simulated game of 75 pitches for 9-10 yr. old youth pitchers and investigate joint torque correlations with pitch count, pitch speed, and body mass index (BMI; kg/m^2).

METHODS
Experiments
Protocols were approved by Cal Poly's Human Subjects Committee to minimize risks to human subjects. We conducted motion analysis experiments on 9 experienced baseball pitchers whose ages qualified them as 9-10 year olds during the 2016 season at Cal Poly's Human Motion Biomechanics (HMB) Lab. The HMB Lab includes a motion analysis system (Motion Analysis Corp., Santa Rosa, CA, USA) with 8 digital cameras used to track retroreflective markers and characterize motion kinematics. Height and weight measurements were recorded for BMI calculations. Subjects completed a pre-game warm-up including stretching, jogging, and 20-25 non-pitching throws. Then, subjects changed into compression clothing and retroreflective markers were placed on anatomical landmarks based on the UETrak (Motion Analysis Corp.) pitching marker set [5]. Subjects pitched a simulated game of 75 fastball pitches consisting of 2 innings. The pitchers threw 38 and 37 pitches in the 1st and 2nd innings, respectively. Each inning was separated by a 15-minute break to simulate the team on offense.

Subjects pitched off of a mound in the room's center and into a net 23 feet away with a scaled strike zone (Fig. 1). Every 5 pitches the pitcher was asked to take a 30 second break to simulate a change in batters and every 10 pitches was asked if any soreness or pain was felt (all pitchers were able to throw 75 fastballs with no pain). Marker trajectory was recorded in Cortex analysis software (Motion Analysis Corp.) at 200 Hz and filtered using a 4th order Butterworth filter with a cutoff frequency of 13.4Hz. A radar gun was

Fig. 1: Youth pitcher with retroreflective markers.

used to track ball speed, which was not disclosed in order to prevent any mechanical changes based on having that information available.

Analysis

Biomechanical outputs were processed using UETrak software. Marker and pitch speed data were collected for 3 "pitch periods" defined by pitches 1-5, 34-38, and 71-75. For each subject, 3 pitches were selected from each pitch period for analysis. Maximum values of elbow varus torque and shoulder internal rotation torque between foot contact and ball release were averaged to obtain one value at each pitch period for each subject. Also, pitch speeds were averaged to obtain one value at each pitch period.

Statistics

Repeated measures ANOVAs were performed to determine significant differences in joint torque and pitch speed between pitch periods at a significance level of 0.05. Both absolute and normalized (by body weight times height) torques were analyzed. Also, regression analyses were performed to determine significant correlations between joint torque and pitch speed or BMI at each pitch period ($p<0.05$ significant).

RESULTS

Subjects' age, height, weight, and BMI values (mean ± 1 S.D.) were 10.2 ± 0.4 yrs., 143.4 ± 8.2 cm, 35.7 ± 5.8 kg, and 17.3 ± 1.6 kg/ m^2, respectively. Absolute (Fig. 2) and normalized elbow varus and internal rotation torques did not change by pitch period. While most studies reported peak torque values occurring during the late cocking

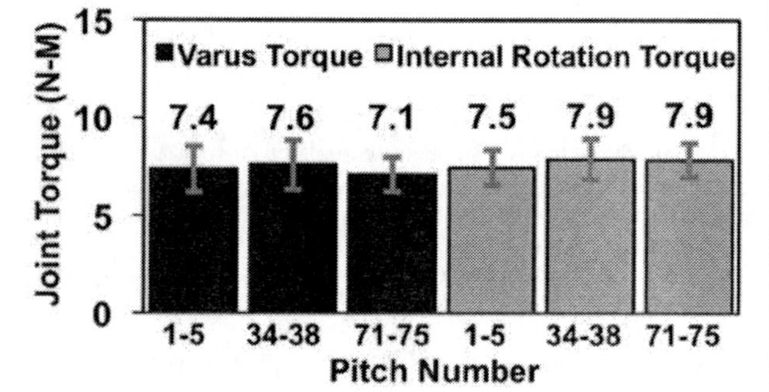

Fig. 2: Absolute elbow varus and shoulder internal rotation torques at each pitch period. Mean ±1 S.D. values shown.

phase, at or before when the shoulder is at maximum external rotation (MER) [5,6], peak values in this study continually occurred just after MER during the acceleration phase. Pitch speed did not change throughout the simulated game. Pitchers averaged 37.0±4.2 mph for pitches 1-5, 38.3±4.1 mph for pitches 34-38, and 35.8±2.2 mph for pitches 71-75. While absolute joint torques were not correlated to pitch period or speed, significant positive correlations existed between joint torques and BMI at each pitch period ($p<0.05$) (Fig. 3), with the exception of varus torque at pitches 71-75 ($p=0.07$).

DISCUSSION

While no significant changes in joint torques were found throughout the simulated game, there were some notable discoveries for an age group that has not been included in previous biomechanical studies. A pitching study with 12 yr. olds revealed that a maximum elbow varus torque of ~ 18 N-m occurs just before MER [6]. A different pitching study with 12 yr. olds revealed that a maximum shoulder internal rotation torque of ~ 33 N-m also occurs just before MER [5]. Our study revealed that for 9-10 yr. old pitchers, these peak values don't

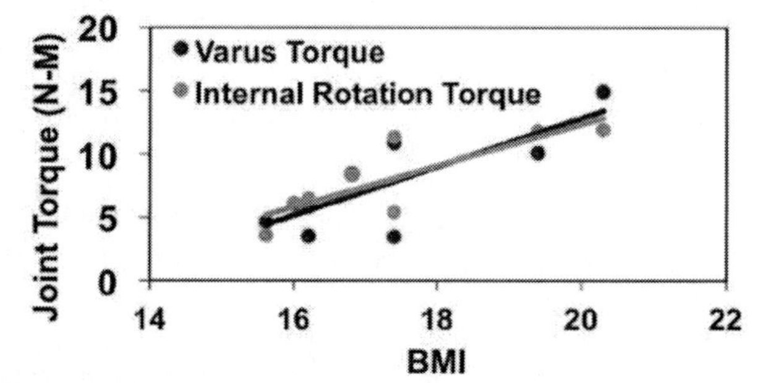

Fig. 3: Correlations between absolute joint torques (pitches 34-38) and BMI. R^2 values were 0.64 and 0.70 for elbow and shoulder torques.

occur until just after MER, right before ball release, with the exception of one pitcher whose elbow varus torques occurred just before MER for 67% of his pitches. This could be the result of less experience and development of pitching mechanics for this age group. The significantly smaller peak values are most likely the result of the smaller mass and pitch speed of 9-10 yr. olds as compared to older youths of previous studies [5,6]. Also, this study revealed positive correlations between joint torques and BMI; that finding is similar to previous results linking joint torque and body mass with older (12-16 yr. old) pitchers [9].

Limitations include the use of an adult anthropometric model in the UETrak calculations. A study on anthropometric properties in children [7] showed that 10 yr. olds have a larger mass % in their hand, lower arm, and upper arm of 50%, 3%, and 7% when compared to adults [8]. Also, the ball mass used (145 g) is less than the MLB ball mass (170 g) assumed in UETrak. Although these mass errors should only slightly over-estimate joint torques, a future study should correct these errors. Another limitation is that the size of the lab did not allow for the regulation pitching distance to be achieved.

Although elbow and shoulder torques did not change throughout the simulated game, it is important to note that muscle groups around the elbow provide dynamic valgus stability [10] and if they become fatigued, a player may be putting higher loads on their UCL. Major findings of this study include the positive correlations between joint torques and BMI, suggesting that pitch limits may be altered based on weight and height as young overweight pitchers may be at higher risk for overuse injuries. These novel findings suggest that ongoing studies should aim to produce a more comprehensive, biomechanical understanding of injury risk factors in youth baseball pitchers.

ACKNOWLEDGEMENTS

This work was supported by the W.M. Keck Foundation (JDD, EMD) and the Donald E. Bently Center (SK).

REFERENCES

[1] Fleisig, GS and Andrews JR. *Sports Health,* 4.5: 419-24, 2012.
[2] Lyman, S et al. *Med Sci Sports Exerc*, 33(11):1803-10, 2001.
[3] Escamilla, RF. *AM J Sports Med*, 23-33, 2016.
[4] Fleisig, GS et al. *Am J Sports Med,* 23.2: 233-39, 1995.
[5] Aguinaldo, AL et al. *J Appl Biomech*,23: 42-51, 2007.
[6] Sabick, MB et al. *J Shoulder and Elbow Surg,* 13.3: 349-55, 2004.
[7] Jensen, Robert K. et al. *J Biomech,* 19.5: 359-68, 1986.
[8] Winter, DA. New York: Wiley, 1990.
[9] Garner, JC et al. Ped Exer Sci, 23: 379-87, 2011.
[10] Lin, Fang et al. *J Shoulder and Elbow Surg,*16.6: 795-802, 2007.

SB^{3}C2017
Summer Biomechanics, Bioengineering and Biotransport Conference
June 21 – 24, Tucson, AZ, USA

THE EFFECTS OF DIFFERENT CARRYING METHODS ON SPATIO-TEMPORAL GAIT PARAMETERS

M. Alamoudi, F. Travascio, S. Asfour

Department of Industrial Engineering
University of Miami
Miami, FL, USA

INTRODUCTION

Over the years, lifting and loading activities have been shown to be potentially hazardous occupational tasks. Therefore, several studies investigated the effect of these activities on physical, physiological, and psychophysical stresses [1]. Yet, one of the routine manual material handling activities that has less often been studied is load carrying, which occurs daily in the industrial and construction fields in different ways. Load carrying is considered a perturbation factor that alters the biomechanics of the gait [2]. However, to our best knowledge, none of the previous studies investigated and studied different carrying postures on locomotion biomechanics. Therefore, the main goal of this study was to determine the optimal carrying method by analyzing and comparing different carrying methods in terms of gait characteristics, such as walking speed, stride length, double support duration, and cadence, by finding which carrying method has less deviation from the normal gait.

METHODS

Equipment and Tools

The experiments were all conducted at the Biomechanics laboratory at the University of Miami using the Vicon® Motion Capturing System (Oxford Metrics, United Kingdom) Nexus software version 2.5 and 10 cameras at a sampling rate of 120 Hz. Moreover, to simulate different carrying methods, a 20 lb. weight was held in a 15"× 6" × 6" box or in a 12.5"× 6" × 15.3" backpack.

Experimental Protocol

Eight male subjects participated in the study (age: 29.25 ± 2.38 years; height 173.13 ± 8.09 cm; body mass: 72.5 ± 7.25 kg). The protocol of the experiment was approved by the University of Miami's Institutional Review Board (IRB). Subjects who had clinically conditions that might affect their gait patterns were excluded from the study. After attaching the 39 reflective markers to the subject's body, subjects walked with their normal speed for 3 times per each condition. Normal gait (i.e. unloaded condition) served as a baseline for the comparison. Participants were assigned randomly to each condition of the experiment.

Independent Variables

Four different carrying postures that occur in the daily activities were the independent variables, as shown in Figure 1.

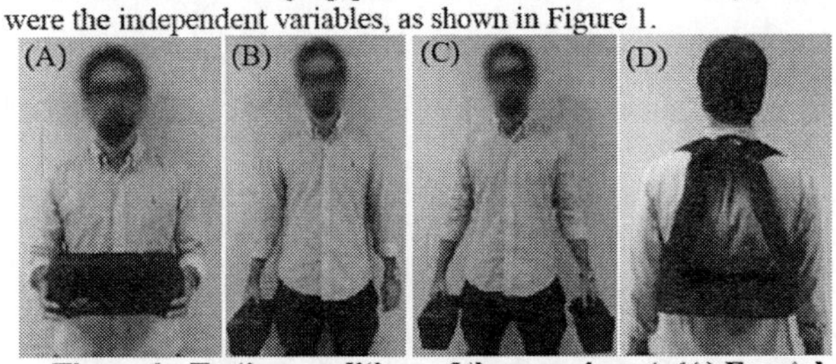

Figure 1: Testing conditions of the experiment. (A) Frontal carriage. (B) Unilateral Carriage with the dominant hand. (C) Bilateral carriage. (D) Posterior carriage. Note that a hole was made in the backpack to allow the motion capturing system to capture the T10 reflective marker.

Dependent Variables

Vicon® Motion Capturing System was used to measure the spatio-temporal gait parameters. Namely, double support duration, expressed as percentage of gait cycle, stride length, measured in meters, walking speed expressed in meters per second (m/s), and cadence, expressed in steps per minute (steps/min).

Statistical Analysis

Since each participant received the same 4 carrying strategy treatment conditions, a repeated measures ANOVA was used. Moreover, Tukey's post-hoc analysis was used to check the significant difference between the levels of carrying methods. For all analyses, a significance level of $\alpha = 0.05$ was set.

RESULTS

Higher mean values for cadence, velocity, stride length, and lower double support duration (i.e. which are indicators of normal gait patterns [3]) were observed for the unloaded condition than all the loaded ones. The ANOVA showed that there was a statistical significant effect of carrying methods in terms of cadence, and double support duration. Tukey's post-hoc results revealed that bilateral, and posterior carriage showed insignificant difference with normal gait except in the double support duration, as shown in Figure 2.

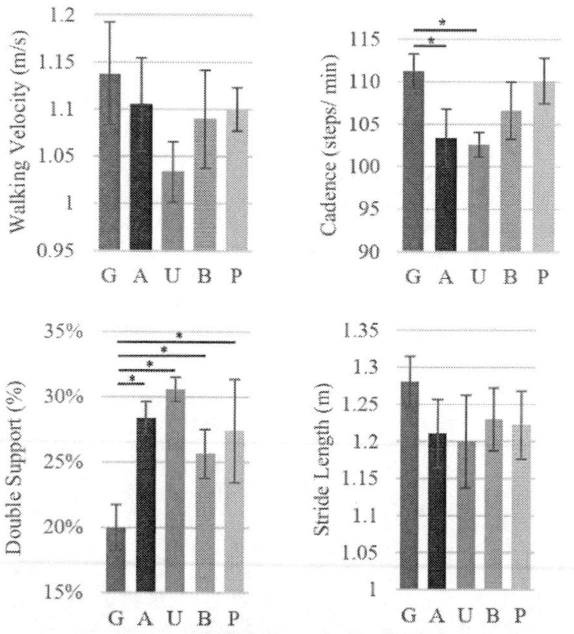

Figure 1: Mean with 2 standard error ranges of spatio-temporal gait parameters. (G) Normal gait (unloaded condition), (A) Anterior, (U) Unilateral, (B) Bilateral, (P) Posterior carrying methods. (*) indicates significant difference $p < 0.05$

DISCUSSION

Previous studies have associated double support time, walking velocity, and cadence as a measure of dynamic postural instability that can precipitate a fall [3]. It has been suggested that a lower walking velocity, cadence, and higher double support time are consequences of an altered gait patterns [4, 6]. All the carrying method described earlier were significantly different from the unloaded condition in terms of double support duration. This finding was consistent with previous studies that found significant effect on double support time from posterior [6], anterior [7], and lateral carriages [8]. As shown earlier, the highest duration of double support was observed in the unilateral carriage, followed by anterior, posterior, and finally bilateral carriages. This increase might be a strategy to minimize either the induced gait instability [9] or to minimize the strain on the musculoskeletal system in terms of possible higher lower limb joint moments for the lower configuration [10]. In the present study, the effects of all carrying methods on walking velocity, and stride length were consistent with the previous studies as they were found lower but insignificant [11, 12]. Lower walking velocity, and cadence could be a minimization of the higher energy cost for load being carried [13]. The results showed that anterior, and unilateral carriage were significantly lower than unloaded condition. Human body adopts lower cadence as a compensation strategy to maintain gait stability [14]. In conclusion, comparing different carrying methods to normal gait, unilateral and anterior carriages were the most deviated in terms of spatio-temporal parameters. The results of the study will help in understanding the effect of carrying methods, and can be harnessed in the formulation of recommendations and legislations on safe carrying methods.

ACKNOWLEDGEMENTS

The authors would like to thank the Biomechanics Research Group at the University of Miami for funding this study.

REFERENCES

[1] Gabbett, T et al., *Sports Med*, 44:989-1003, 2014.
[2] Attwells, R et al., *Ergonomics*, 46: 1527-1537, 2006.
[3] Hong, Y et al., *Gait Posture*, 22:63-68, 2005.
[4] England, S et al., *Gait Posture*, 25:172-178, 2007.
[5] Dingwell, J et al., *J Biomech*, 39:444-452, 2006.
[6] Xingda, Q et al., *J Biomech*, 44:1529-1263, 2011
[7] Perry, C et al., *Gait Posture*, 32:57-61, 2010.
[8] Zhang, X et al., *Comput Methods Biomech Biomed Engin*, 13:339-344, 2009.
[9] Hong, Y et al., *Gait Posture*, 11:254-259, 2000.
[10] McGraw, B et al., *Arch Phys Med Rehabil*, 81:484-489, 2000.
[11] Wang, Y et al., *Ergonomics*, 44:858-569, 2001
[12] Knapik, J et al., *Mil Med*, 169:45-56, 2004.
[13] Singh, T et al., *Gait Posture*, 29:49-53, 2009.
[14] Rogers, H et al., *J Aging Phys Act*, 16:85-96, 2008.

SB^{3}C2017
Summer Biomechanics, Bioengineering and Biotransport Conference
June 21 – 24, Tucson, AZ, USA

SIMULATING INGRESS FOR CAB DESIGN

Hyun-Jung Kwon (1), Yujiang Xiang (2)

(1) Transportation Research Center Inc., East Liberty, OH 43319 USA

(2) Department of Mechanical Engineering, University of Alaska Fairbanks, Fairbanks, AK 99775 USA

INTRODUCTION

Design of cabs for heavy earthmoving equipment has been based primarily on the expertise and prior experience of the designer. Among the few design studies reported in literature on ingress, most focus on studying the motion itself and rely heavily on the experimental data [1, 2]. Very little attention has been given to simulating ingress motion for heavy earthmoving equipment. Availability of such a simulation capability can be useful to understand the risks associated with the ingress process and for the proper design of the cab for these equipment. This work describes the development of a virtual model to perform ingress motion for heavy equipment, and subsequently, use the simulation model to perform virtual studies to rank seven different cab designs with focus on ergonomics and safety factors.

The ingress motion is simulated using predictive dynamics approach [3]. A general two-objective optimization of the problem is formulated where dynamic effort and joint discomfort are minimized subject to physical and environmental constraints. The cab geometry is modeled with finite primitive elements. Collisions between body segments themselves, and cab elements and body segments are avoided by imposing explicit constraints.

The developed methodology is used to study different cab designs for three digital human models with different anthropometries, namely, tall-heavy, medium, and short-lean. Then different cab design metrics for propensity for injury, comfort, and accessibility are proposed to rank each design case based on the simulated ingress motion. Finally, each cab design is ranked based on the design metrics to identify the best design for a range of anthropometries. These results help designers make decisions and plan further design changes.

METHODS

In this study, a 55-DOF (degrees of freedom) three-dimensional digital human skeletal model [4] is used to simulate ingress motion. The design variables are the joint angle profiles $q_i(t)$ for ingress motion where $i = 1$ to 55. Two objective functions are optimized simultaneously: the dynamic effort (the integral of the squares of all joint torques) and the joint discomfort (the integral of the squares of all joint discomforts), as shown in Equation (1):

$$f = c_{de} \sum_1^{ndof} \int_0^T \left(\frac{\tau_i(q,t)}{\tau_i^{up} - \tau_i^{low}}\right)^2 dt + c_{jd} \sum_1^{ndof} \int_0^T \left(\frac{q_i(t) - q_i^N(t)}{q_i^{up} - q_i^{low}}\right)^2 dt \quad (1)$$

where τ_i^{up} is the maximum value and τ_i^{low} is the minimum value for the torque τ_i, and q_i^{up} is the maximum value and q_i^{low} is the minimum value for the angle q_i. T is the total time of simulation. *ndof* = 55 is the number of DOF. c_{de} and c_{jd} are coefficients for dynamic effort and joint discomfort respectively.

Several constraints are proposed and implemented in this work to satisfy the laws of physics and boundary conditions throughout the ingress process. These constraints include joint angle limits, joint torque limits, ground penetration, foot-contacting positions, zero moment point location, hand position, obstacle avoidance, and continuity conditions [5].

The three-step ingress motion is planned from standing position outside the cab to sitting position inside the cab (Fig. 1). The ingress motion is simulated for seven different cab designs with different seat locations and ceiling height (Fig. 2). Then the cab design metrics for propensity for injury, comfort, and accessibility are calculated from the simulation kinematic and kinetic data. These design metrics include [5] overall torque, maximum torque, stability, joint discomfort due to deviation from neutral angles, joint discomfort due to reaching the limits, door accessibility, and seat accessibility (Table 1). Each

Poster Presentation #P94

design metric is scaled between 0 and 10, and the smaller the number, the better the design.

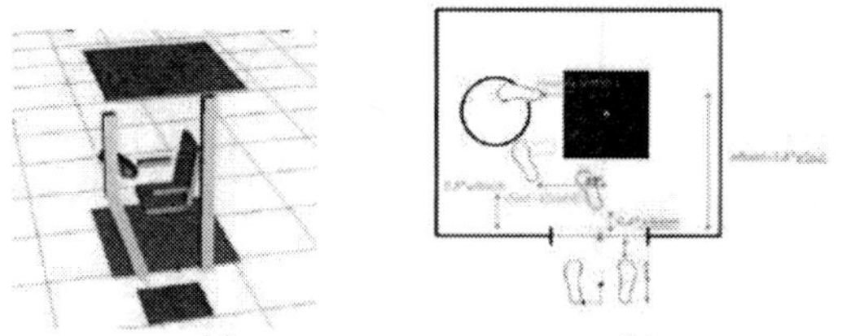

(a) (b)

Figure 1: CAD model of the cab (a), foot positions for ingress (b)

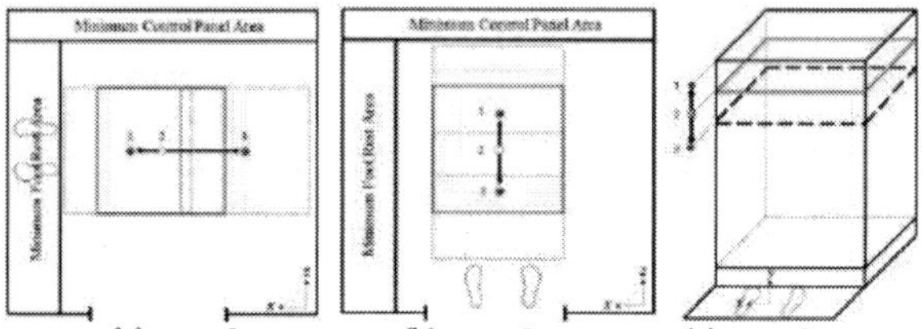

(a) case 1 (b) case 2 (c) case 3

Figure 2: Cab design for ingress, (a) seat change fore-aft, (b) seat change sideways, (c) ceiling height change, point 2 in the middle for each case is the default design

Table 1: Metrics for the cab design

Attribute	Selected Metric
Propensity to get injured	1) Overall torque 2) Maximum torque at key joints 3) Stability at the deck
Comfort (Feeling)	1) Difference from neutral position for key joints 2) Joint that reaches the limits
Accessible	1) Door – ladder position 2) Seat – door position

RESULTS

A computer with an Intel(R) dual core 3.16 GHz CPU and 8 GB RAM is used to solve the optimization problem. The optimal solution is obtained in about 30 seconds of CPU time starting from a feasible solution. The knee joint torque is presented in Fig. 3 (other joint torques are also available from the simulation). The motion is depicted in Fig. 4. Next, all the metrics are averaged for each avatar after obtaining the individual magnitude of each metric. All the metrics can be presented as one number per avatar (Fig. 5). Then, the ranks of the three different avatars are averaged again to make one number for each design case (Fig 6).

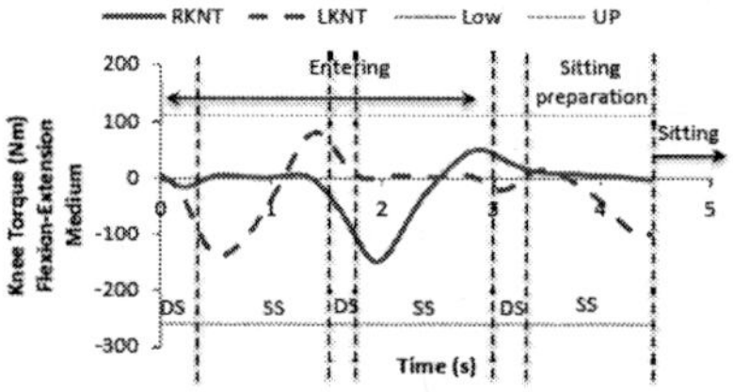

Figure 3: Knee torque for ingress with medium avatar

Figure 4: Predicted motion for ingress

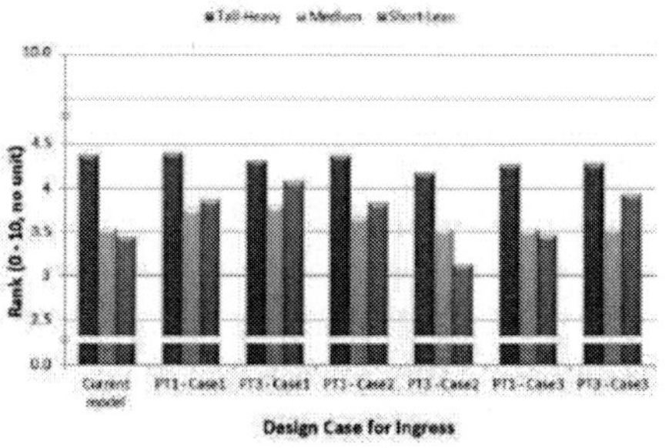

Figure 5: Avatar-wise cab design ranking for ingress

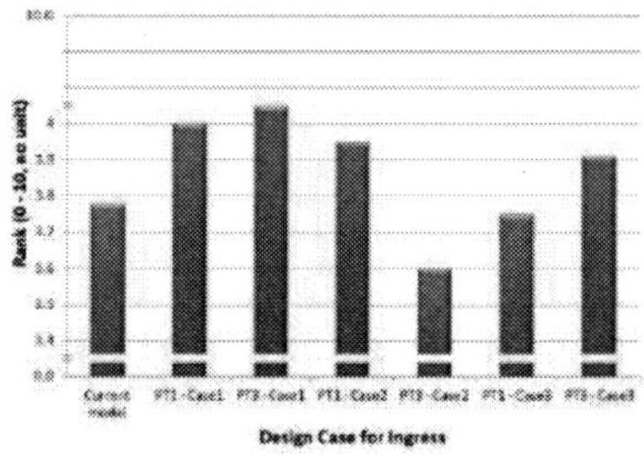

Figure 6: Overall cab design ranks for ingress

DISCUSSION

In this study, several different heavy earthmoving equipment cab designs were presented to study the effect of cab layout on human ingress motion. Two more digital human models were introduced to study the effect of anthropometric changes. We obtained simulation results and evaluated metrics for different cab designs. Additional metrics for the cab design evaluation can be developed by designers. However, with the presented design metrics, Figure 6 showed that point 3 of case 2 was marked as the lowest rank. Therefore, we could suggest that the designer place the seat position close to the door like the design case (point 3 of case 2). This would make a better design than the current model for ingress for different-sized operators from the points of view of injury, accessibility, and comfort.

ACKNOWLEDGEMENTS

This research was partially supported by Caterpillar Inc. and partially by the Virtual Soldier Research Program at the Center for Computer-Aided Design.

REFERENCES

[1] Lin, LJ and Cohen, HH, *Int J Ind Ergonom*, 20:287-300, 1997.
[2] Moore, SM et al., *J Safety Res*, 40:455-460, 2009.
[3] Xiang, Y et al., *Struct Multidisc Optim*, 41:465–479, 2010.
[4] Xiang, Y et al., *J Biomech*, 44: 683–693, 2011.
[5] Kwon, HJ, *PhD Thesis*, University of Iowa, USA, 2011.

VALIDATION OF A PATELLOFEMORAL JOINT MODEL DRIVEN BY KNEE JOINT KINEMATICS

Jonathan A. Gustafson (1), Kyle A. Berkow (1), John J. Elias (2), Richard E. Debski (1), Shawn Farrokhi (3)

(1) Department of Bioengineering
University of Pittsburgh
Pittsburgh, PA, USA

(2) Akron General Medical Center
Akron, OH, USA

(3) Naval Medical Center San Diego
San Diego, CA, USA

INTRODUCTION

Knee osteoarthritis (OA) is a joint disease that affects 25% of the population at the patellofemoral joint [1]. Patellofemoral OA has been associated with significant functional limitations and pain [2], with the cause commonly attributed to increased joint contact pressure at the articular surfaces due to maltracking. A prior study by Farrokhi et al. [3] found significant differences in patellofemoral kinematics between patients with knee OA and age-matched controls; however, no method exists to evaluate how altered kinematics affect underlying articular joint contact pressure. It is necessary to understand the effect of altered kinematics on joint contact mechanics in order to better prescribe treatment solutions for individuals with knee joint dysfunction.

Computational modeling is an attractive solution for studying the effect of joint injuries and diseases on joint mechanics. Specifically, the use of subject-specific kinematics as inputs to drive models would allow for researchers and clinicians to better understand joint contact mechanics during functional tasks and provide a novel tool for prescribing and evaluating subject-specific rehabilitative treatment solutions.

The objective of this work was to validate a modeling framework for estimating patellofemoral joint contact mechanics employing the discrete element analysis (DEA) method driven by accurate knee joint kinematics. Specifically, patellofemoral average contact pressure and peak contact pressure were validated across multiple knee flexion positions.

METHODS

Two fresh-frozen cadaveric knees (mean age = 49 ± 4 years) were utilized for this study and prepared by: 1) isolation of quadriceps tendons, 2) Potting of distal tibia and proximal femur ends, and 3) affixing rigid registration blocks to the femur, patella, and tibia for digitization. The framework for the computational models in this validation study can be described in three parts: 1) creation of subject-specific model geometry; 2) collection of experimental kinematics and contact mechanics as input and validation data; and 3) performing computational simulations and validating contact variables.

Subject-Specific Model Geometry High resolution magnetic resonance (MR) images (slice thickness = 0.7mm) were acquired for each knee in an extended, unloaded position. Specimen-specific geometry of bones and cartilage were manually segmented (Mimics®, Materialise, Leuven, Belgium) and imported into a meshing software (Hypermesh, Altair Engineering Inc., Troy, MI). Tetrahedral, shell elements with an average element size of 1 mm were used to model the bones and articular cartilage.

Experimental Kinematics and Contact Mechanics Each specimen was placed in a custom-designed, six degree of freedom testing jig, where physiologic compressive loading (35 kg) was applied to the tibiofemoral joint and tensile loading to the quadriceps tendons [4] at the following tibiofemoral flexion positions: Full extension and 15° of flexion with 5° variations in internal/external femoral rotation. Kinematic data was collected at each position using a mechanical digitizer (Faro Arm®, Lake Mary, FL). Patellofemoral pressure distributions were recorded at each position using a calibrated thin, flexible pressure sensor (Tekscan Inc., South Boston, MA).

Computational Simulations and Validation DEA was utilized for computational analysis at each position by creating compression-only springs (E=1MPa, ν=0.42 [5]) at the centroid of each element of the segmented femoral and patellar cartilage surfaces [6]. The knee joint was loaded in each position based on the experimental kinematics using a local to local coordinate system registration technique [7]. Contact pressure distributions were estimated using the following linear elastic modeling equation:

$$P = \frac{(1-\nu)E}{(1+\nu)(1-2\nu)h}d \quad (1)$$

where h is the undeformed surface to surface cartilage thickness and d is the spring deformation upon positioning the joint in its loaded position. Spring deformation was determined through a penetration analysis between the loaded patellar and femoral articular cartilage surfaces [5, 6]. Contact pressure was calculated for each element and divided into medial and lateral patellar facet regions, defined by a line drawn through the inferior apex to the superior-most point on the base of the patella. The DEA algorithm was implemented using MATLAB (2015a, The Mathworks, Natick, MA). For validation, lateral to medial facet ratios of average and peak pressures were compared between the experimental and computational model results.

RESULTS

The contact pressure distribution between the experimental data and computational model followed a similar pattern when an internal rotation was applied to the knee (**Figure 1**). For both knee specimens, the computational model replicated the pressure increases on the lateral facet with an increase in internal femoral rotation.

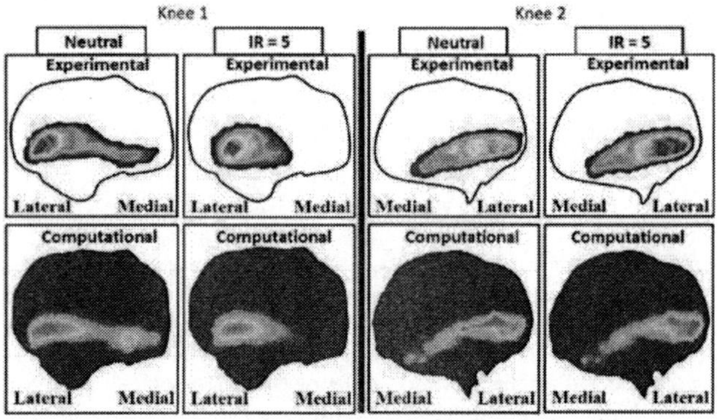

Figure 1: Experimental and computational patellofemoral pressures of two knees at 15° of flexion with 5° of femoral rotation (IR = internal rotation)

The lateral to medial facet ratio between the computational model and the experimental data were in general agreement across the flexion positions (**Figure 2 & 3**). A value greater than 1 indicates greater loading on the lateral facet. There was greater loading on the lateral facet (ratio > 1) in almost every position tested for both knees except when applying 5° of external femoral rotation at full extension (**Figure 2 & 3**: Knee 1 & 2) and 15° of flexion (**Figure 2**: Knee 2, **Figure 3**: Knee 2). When averaged across all of the knee joint positions tested for both knees, the computational model was able to predict the experimental data for average contact pressure (mean error = 14.4%) and peak contact pressure (mean error = 8.0%).

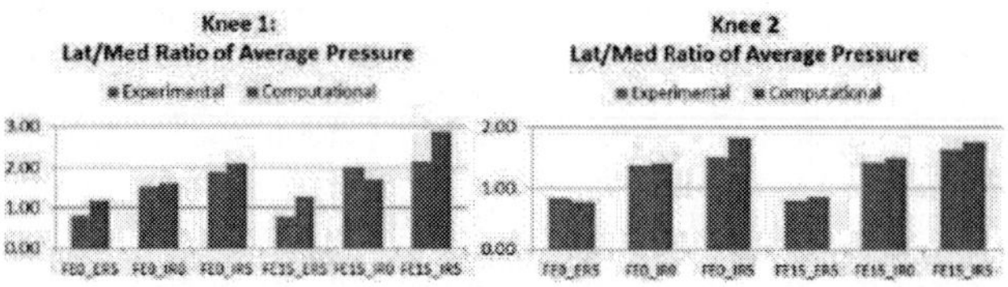

Figure 2: Lateral to medial facet ratio of average pressure across the tested positions for both knees (FE = flexion; IR/ER = internal/external rotation)

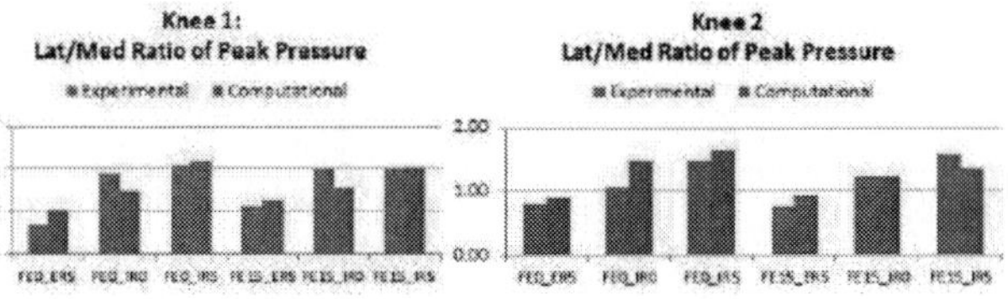

Figure 3: Lateral to medial facet ratio of peak pressures across the tested positions for both knees (FE = flexion; IR/ER = internal/external rotation)

DISCUSSION

This study has validated a framework for employing displacement-driven computational models to estimate patellofemoral joint contact mechanics with experimental data. In general, the contact pressure distributions from the computational model exhibited similar patterns to the experimental contact pressures plots. Specifically, our model was able to predict the lateral to medial facet ratio average contact pressure within 15% error and, more clinically relevant, the peak contact pressure ratio within 8% error across the tested positions. The intended use of this model is to observe changes in contact pressures (both average and peak) due to changes in kinematics.

Lee et al. [8] reported greater than 22% changes in pressure due to internal femoral rotation compared to average changes of 47% experimentally and 38% computationally in our study. However, the prior study only observed the effect of femoral rotation at flexion angles of 30° and greater. The flexion angles tested in our study simulate the loading response phase of gait (0° - 15°), which is a vulnerable position for the patella as it has not become fully engaged with the trochlea and represents conditions where patients may experience high contact pressures that cause cartilage damage.

Limitations of the current model include representing cartilage as a linear elastic material, though this assumption can be suitable under small durations of loading. Additionally, this model is not intended to determine differences in absolute magnitudes of pressure, but rather examine the effect of changes in contact pressures due to changes in kinematics, such as the case of improved kinematic motion after rehabilitation/surgical interventions. Future studies will employ these validated models with accurate, in vivo 3D knee joint kinematics during functional tasks to determine differences in contact patterns for patients with and without patellofemoral OA.

ACKNOWLEDGEMENTS

The project described was supported by the National Institutes of Health (K12 HD055931) and the Pittsburgh Claude D. Pepper Older Americans Independence Center (P30 AG024827).

REFERENCES

[1] Arden, N et al., *Best Pract Res Cl RH*, 20:3-25, 2006.
[2] Duncan, R et al., *Osteoarthr Cart*, 17.9:1151-1155, 2009.
[3] Farrokhi, S et al., *J Biomech*, 48:1707-1712, 2015.
[4] Hofer, JK et al., *J Arthroplasty*, 27.4:620-624, 2012.
[5] Blankevoort, L et al., *J Biomech*, 24:1019-1031, 1991.
[6] Elias, JJ et al., *J Biomech Eng*, 135:081011, 2013.
[7] Fischer, KJ et al., *J Biomech*, 34:377-383, 2001.
[8] Lee, TQ et al., *J Orthop Sports Phys*, 33: 686-693, 2003.

Poster Presentation #P95

CHARACTERIZING BRAIN INJURY CRITERIA FOR CONCUSSION THROUGH RECONSTRUCTIONS OF COLLEGIATE FOOTBALL HEAD IMPACTS

Bethany Rowson, Steven Rowson, and Stefan M. Duma

Department of Biomedical Engineering and Mechanics
Virginia Tech
Blacksburg, Virginia, United States

INTRODUCTION

Head injury criteria have been used in a variety of industries including sports protective equipment and motor vehicles. The criteria used today are generally based on the Wayne State Tolerance Curve (WSTC), which was derived from experimental data collected during the 1950's and 1960's [1]. Although the WSTC and resulting injury criteria have been criticized, they have been largely successful at preventing severe and fatal head injuries because they limit the amount of energy delivered to the head during an impact. In an effort to continually improve head protection and decrease the number of less severe head injuries like concussions, a number of new injury criteria have been introduced. Existing criteria are based on data from cadavers, anthropomorphic test devices (ATDs), and animal models. Functional injuries like concussion cannot be evaluated with cadavers or ATDs, and data from animal models must be scaled to be applied to humans, making it difficult to relate injury metric values to these types of injuries. The objective of this study was to determine ranges of existing brain injury criteria for concussion by reconstructing impacts from instrumented collegiate football players. These impacts were reconstructed to have the full 6-degree-of-freedom linear and rotational kinematics necessary to calculate different brain injury criteria.

METHODS

Eleven impacts resulting in a diagnosed concussion recorded from instrumented Virginia Tech football players between 2004 and 2015 were used for this study. Players were instrumented with the Head Impact Telemetry (HIT) System (Simbex, Lebanon NH). Impacts were reconstructed in the lab with a pneumatic linear impactor (Biokinetics, Ottawa, Canada). The impactor struck a modified medium NOCSAE headform mounted on a Hybrid III 50th percentile male neck. The headform was instrumented with 3 linear accelerometers (7264B-2000, Endevco, San Juan Capistrano, CA) and 3 angular rate sensors (ARS3 PRO-18K, DTS, Seal Beach, CA). The head and neck were mounted on a 5-degree-of-freedom slide table that simulated the effective mass of the torso and allowed for adjustment of impact location.

Helmet models were matched for each reconstructed impact. The location reported by the HIT System was used as a starting point for each reconstruction (Figure 1). The location and impact speed were varied until the peak resultant linear and rotational accelerations matched the HIT System data within +/- 5%. Linear acceleration data were filtered to channel frequency class (CFC) 1000, and angular rate data were filtered to CFC 155 using a 4-pole phaseless Butterworth low-pass filter. Angular rate data were differentiated to determine rotational acceleration for each impact.

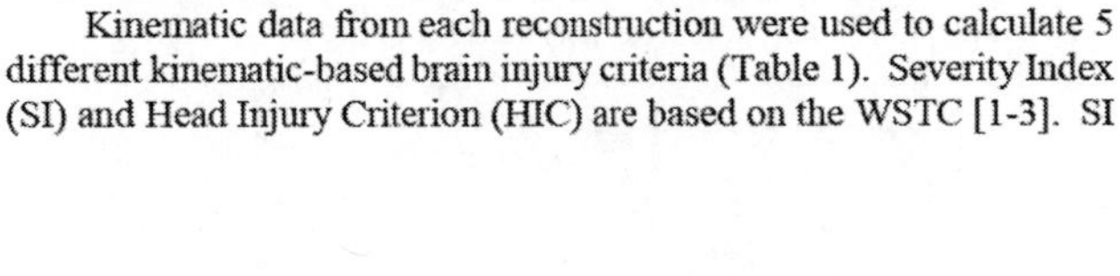

Figure 1: Concussive impacts recorded from players were reconstructed with a pneumatic linear impactor. Information from the helmet sensor was used to set boundary conditions for reconstructions in the lab.

Kinematic data from each reconstruction were used to calculate 5 different kinematic-based brain injury criteria (Table 1). Severity Index (SI) and Head Injury Criterion (HIC) are based on the WSTC [1-3]. SI is commonly used as a metric for helmet safety standards, while HIC is used most notably in automotive safety standards and performance ratings. HIP represents the rate of change of kinetic energy of the head, and was developed using reconstructions of concussive impacts in professional football players [4]. BrIC was proposed by NHTSA as a rotational brain injury metric to supplement HIC in automotive and other safety standards [5]. CC is the linear portion of a bivariate risk function that was developed using on-field data from instrumented football players [6, 7]. Linear regressions were used to quantify correlations between all criteria and kinematic parameters evaluated.

Table 1: Kinematic-based brain injury criteria evaluated with reconstructions of head impacts from instrumented football players. a = linear acceleration; t = time; α = rotational acceleration; ω = rotational velocity; x, y, and z indicate axis direction; m indicates maximum value

Brain Injury Criteria Equations
$SI = \int a^{2.5}\, dt$
$HIC_{15} = max\left[\frac{1}{t_2 - t_1}\int_{t_1}^{t_2} a\, dt\right]^{2.5} (t_2 - t_1)$
$HIP = 4.5a_x \int a_x dt + 4.5a_y \int a_y dt + 4.5a_z \int a_z dt + 0.016\alpha_x \int \alpha_x dt + 0.024\alpha_y \int \alpha_y dt + 0.022\alpha_z \int \alpha_z dt$
$BrIC = \sqrt{\left(\frac{\omega_{xm}}{66.25}\right)^2 + \left(\frac{\omega_{ym}}{56.45}\right)^2 + \left(\frac{\omega_{zm}}{42.87}\right)^2}$
$CC = -10.2 + 0.0433a_m + 0.000873\alpha_m - 0.000000920a_m\alpha_m$

RESULTS

For the 11 concussive impacts that were reconstructed, the median [25th 75th percentiles] peak linear and rotational accelerations of the HIT System data were 118 g [77 138] and 4871 rad/s^2 [3767 7235]. The laboratory reconstruction accelerations were 121 g [77 136] and 4841 rad/s^2 [3765 7167]. Median values for reconstruction peak linear acceleration (PLA), peak rotational acceleration (PRA), peak rotational velocity (PRV), SI, HIC, HIP, BrIC, and CC are reported in Table 2.

Regression analyses showed that all criteria with the exception of BrIC were highly correlated with one another, and with peak linear acceleration ($R^2 > 0.8$). HIP and CC were the only criteria highly correlated with both peak linear and rotational acceleration. BrIC was only highly correlated with peak rotational velocity ($R^2 = 0.97$). All other criteria had low correlations with rotational velocity ($R^2 < 0.2$).

DISCUSSION

Kinematics from the reconstructed impacts in this study can be compared to previously reported concussive distributions. For reconstructions of NFL impacts resulting in concussion, the mean kinematics were 98 g for peak linear acceleration, 6432 rad/s^2 for peak rotational acceleration, and 34.8 rad/s for peak rotational velocity [8]. A large dataset of head accelerations from football players wearing the HIT System had mean accelerations of 105 g and 5022 rad/s^2, and mean rotational velocity of 22.3 rad/s for concussive impacts [9, 10]. The impacts reconstructed in the current study had mean accelerations of 115 g and 5433 rad/s^2, and mean rotational velocity of 24.6 rad/s. Based on these comparisons, the concussive impacts reconstructed in this study are similar in severity to those previously reported.

Table 2: Median kinematics measured from reconstructed concussive impacts and calculated brain injury criteria.

Parameter	Median [25th 75th percentiles]
PLA (g)	121 [77 136]
PRA (rad/s^2)	4841 [3765 7167]
PRV (rad/s)	27 [17 30]
SI	351 [180 519]
HIC	277 [148 426]
HIP (kW)	16.4 [8.5 23.0]
BrIC	0.47 [0.35 0.53]
CC	-1.8 [-3.4 1.4]

The median values for injury criteria from these reconstructions can also be compared to current thresholds for context. NOCSAE standards for football helmets limit SI to 1200, while the median value for SI in this study was 351. FMVSS 208 limits HIC_{15} for the Hybrid III 50th percentile male to 700, and the median value here was 277. It is expected that the impacts in this study would result in substantially lower SI and HIC values, as the values used in safety standards are meant to reduce severe and fatal head injuries rather than concussions. Thresholds have not yet been suggested for the other criteria evaluated here as they are not used for safety standards. However, the risk function associated with CC has been used to evaluate relative performance of hockey helmets, and the median CC value from this study is equivalent to a 14% risk of concussion. Although the impacts in this study were similar to the average concussion kinematics from previous studies, there are still many more impacts that occur at that severity that do not result in concussion, making the risk relatively low.

The correlations between all criteria and kinematic parameters evaluated show which values provide similar information regarding impact severity. These correlations are intuitive, as criteria are highly correlated with the parameters they are dependent on. For example, HIC and SI had the highest correlation ($R^2 = 0.99$), because they are both weighted functions of linear acceleration. These correlations do not differentiate the predictive capability of different criteria, only how related different metrics are.

The data presented here provide unique insight into concussive ranges of existing brain injury criteria. These values could be used to adjust existing standards or protective equipment evaluation to better protect against concussive injuries.

ACKNOWLEDGEMENTS

The authors would like to acknowledge the UVa-VTC Neuroscience Seed Fund Program for their support of this research.

REFERENCES

1. Gurdjian, ES et al., *J Trauma*, 6:600-604, 1966.
2. Gadd, CW, SAE Technical Paper 660793. 1966.
3. Versace, J, SAE Technical Paper 710881. 1971.
4. Newman, JA et al., *Stapp Car Crash J*, 44:215-247, 2000.
5. Takhounts, EG et al., *Stapp Car Crash J*, 57:243-266, 2013.
6. Rowson, S and SM Duma, *Ann Biomed Eng*, 41(5):873-882, 2013.
7. Rowson, B et al., *Ann Biomed Eng*, 43(10):2429-2443, 2015.
8. Pellman, EJ et al., Neurosurgery, 53(6):1328-41, 2003.
9. Rowson, S and SM Duma, *Ann Biomed Eng*, 39(8):2130-40, 2011.
10. Rowson, S et al., *Ann Biomed Eng*, 40(1):1-13, 2012.

SB^{3}C2017
Summer Biomechanics, Bioengineering and Biotransport Conference
June 21 – 24, Tucson, AZ, USA

A METHOD FOR EXAMINING THE ROLE OF MECHANICS IN APOPTOSIS

Zachary E. Goldblatt, Heather A. Cirka, Kristen L. Billiar

Biomedical Engineering
Worcester Polytechnic Institute
Worcester, MA, USA

INTRODUCTION

A balance between cell death and cell proliferation is crucial for the proper progression of various biological processes, such as wound healing, normal tissue maintenance, and tissue development. Dysregulation of this process can lead to various disease states. In particular, apoptosis is thought to play a critical role in many cardiovascular diseases such as calcific aortic valvular disease (CAVD), atherosclerotic plaque rupture, and aortic aneurysm formation.

The two main mechanistic pathways that induce apoptosis are the intrinsic pathway and the extrinsic pathway. The intrinsic pathway is also known as the mitochondrial pathway, and is activated through signals generated inside the cell due to stress and is dependent on cytochrome c released from the mitochondria. The extrinsic pathway is activated by external signals that bind to cell membrane ligands. Apoptosis can be initiated by a multitude of factors such as inflammatory cytokines, oxidative stress, DNA damage, radiation, or infection. These factors are implicated in causing apoptosis through chemical and biological signals. More recently, there is increasing evidence that the mechanical environment plays a critical role in apoptosis.

Many mechanical signals have been shown to initiate apoptosis. Previous studies have shown that apoptosis is correlated with the rate and wave form of both dynamic and static stretch (1). Both increasing and decreasing shear stress via fluid flow has also been demonstrated to change apoptotic behavior in endothelial cells (2). Additionally, apoptosis rates have increased due to a lack of stress, such as in cases of anoikis, low substrate modulus, or absence of shear flow (3). Restricting cell spread area by culturing cells on small micropatterned features also results in an increase in apoptosis in cells (4). Cells on these small protein "islands" have been shown to generate smaller traction forces as spread area is restricted and aspect ratio decreases (5). Further, cells cultured on low modulus (soft) substrates have higher instances of apoptosis and generate lower traction forces compared to cells on stiffer substrates (6). Thus, both reduced spread area and low traction forces are correlated with cell apoptosis, yet these mechanical parameters are convoluted in that they occur concomitantly in the aforementioned studies.

The goal of this work is to develop a system capable of decoupling mechanical factors such as cell spread area, shape, and cell-generated traction force to elucidate the relationship of each of these single physical parameters on regulating apoptotic signaling in individual valvular interstitial cells. Our approach is to micro-contact print protein islands onto polyacrylamide (PA) gels attached to compliant polydimethylsiloxane (PDMS) wells able to be cyclically stretched.

METHODS

To control and quantify cell size, shape, and tension with cyclic stretch and correlate these to cytoskeleton organization and the amount of apoptosis, we developed a system involving micro-contact printed protein patterns on PA gels (Young's Modulus of 0.6–40kPa) attached to flexible custom-made PDMS wells.

PDMS wells were created using a custom-made device. These wells were treated with benzophenone and rinsed with methanol. PA was prepared, pipetted into the center of the PDMS well, allowed to polymerize under UV light, and rinsed for 72hrs in PBS before treatment for seeding cells (7). These PA gels were treated with hydrazine hydrate to make it reactive to monomeric collagen that is transferred from PDMS stamps. Micro-contact stamps were created from negative imprints of PDMS on patterned silicon wafers. Stamps were soaked in collagen and then placed onto the PA gel to "stamp" the collagen. This leaves collagen residue on the PA gel, which cells can attach. Cells are unable to bind to uncoated PA, hence controlling attachment.

As a proof of concept, valvular interstitial cells were seeded onto PA gels and allowed to adhere overnight. Cells were then subjected to cyclic stretch from a custom made stretching device (Fig1A). Cells underwent cyclic uniaxial stretch at 1Hz with 10% strain. This stretch was parallel or perpendicular to the long axis of the cell, making the cell's intracellular stress more anisotropic or isotropic, while keeping cell area and substrate modulus constant. Dynamic stretch was used since it is known that cells adapt to static stretch (8).

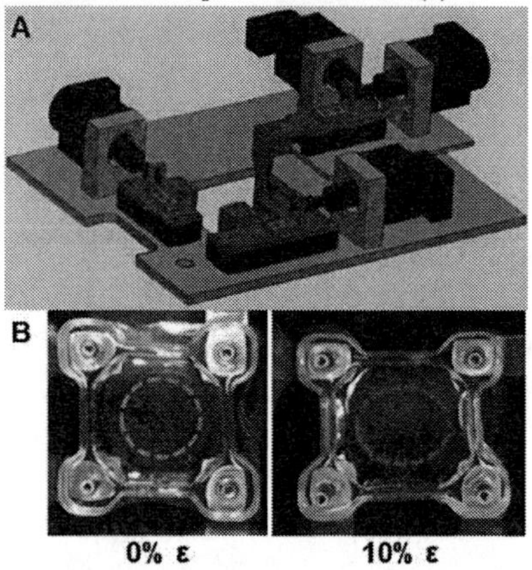

Figure 1: A) CAD image of PDMS stretching device. B) PMDS well inserted into mechanical stretching device under 0% strain (left) and 10% strain (right). PA gel is outlined in blue and red for 0% and 10% respectively to show stretching of the gel.

Apoptosis was quantified via Caspase 3/7 expression (CellEvent Caspase 3/7 Detecting Reagent, Invitrogen) as previously described (9) and then the images were analyzed in MATLAB. F-actin and αSMA stress fibers will be stained for using Phalloidin (Alexa Fluor 488, Life Technologies) and anti–αSMA respectively. Fiber orientation will be analyzed using 2D FFTs to create an alignment index. In order to determine traction forces generated by cells, fluorescent microbeads will be dispersed onto PA gels, and imaged over time as previously described (10). These images will be analyzed in MATLAB and ANSYS in order to generate force maps.

RESULTS

Under static conditions, using this system in larger cell aggregates (200µm diameter), there is a high density of rounded cells in the center of each aggregate, coupled with large, elongated cells along the edges (not shown). These central cells transfer lower traction forces to the substrate than peripheral cells (Fig 2A); these cells also exhibit fewer αSMA stress fibers (not shown), and higher intensity of caspase 3/7 activity (Fig 2B). These results indicate that traction forces and cell spread area affect apoptosis. However we have not yet determined a relation between these two factors and how they mechanistically initiate apoptosis, thus single cell studies are required to test each mechanical parameter individually.

Preliminary single cell studies have revealed collagen deposits on PA gels small enough to permit only individual VIC adherence (Fig 2C,2D). Cells are restricted to areas of 300-4000µm^2 (20-70µm diameter circles). These areas are of varying shape, ranging from circles to ovals with aspect ratio of 8:1, demonstrating the ability to use this system for controlled single-cell experiments.

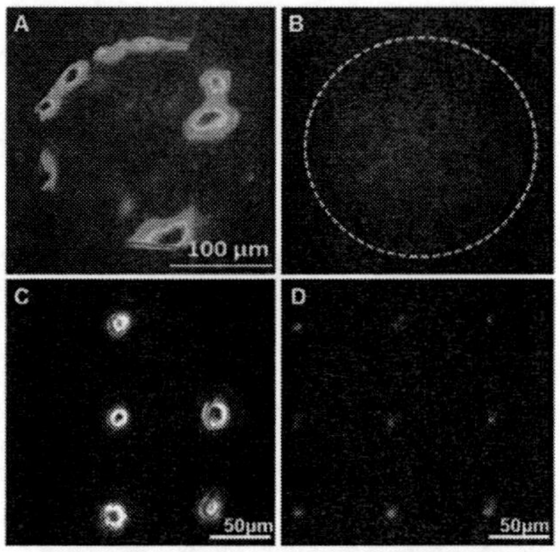

Figure 2: A) Traction force maps generated after 24hrs in multicellular aggregates indicate low transfer of traction from the cells at center of aggregate to the substrate. B) Corresponding caspase 3/7 activity localized to central regions of aggregates. Single VICs in phase-contrast (C) and corresponding fluorescent labeled collagen (green) (D) on 25µm diameter micro-contact printed patterns on 40kPa PA gels.

DISCUSSION

Comprehending how specific physical parameters (stretch, intracellular tension) interconnect with cytoskeletal organization and ultimately cell fate, will have profound impact on our knowledge of how mechanical stimuli contribute to cardiovascular disease. Understanding the mechanical mechanisms that control cell apoptosis is crucial to facilitating control of cell fate in diseases. Using the system described herein, we will be able to study the inter-relationship between intracellular stress, actin organization, and apoptosis.

We have previously shown that introducing cyclic stretch to cells on low substrates can enable them to recover traction forces and elongate (10). Cells with higher traction forces and larger aspect ratios have been seen to have fewer instances of apoptosis. Thus we hypothesize that by introducing mechanical cyclic stretch to cells, we can reduce the amount of apoptosis that occurs.

The system developed promises to offer new insight in how mechanical stress regulates apoptosis, and subsequently affects cell survival. Determining the specific mechanical cues and their associated intracellular pathways that regulate apoptosis has the potential to aid in identifying targets for therapeutic intervention.

ACKNOWLEDGEMENTS

This work was funded in part by a National Science Foundation IGERT (DGE 1144804) and the WPI-University of Massachusetts Medical School Seed Grant Program.

REFERENCES

[1] Chen, WG et al, *IEEE*, 2-5, 2009
[2] Wu, CC et al, *Proc. Natl. Acad. Sci.*, 104:1254-1259, 2007
[3] Kaiser, D et al, *Biochem Biophys Res Commun*, 231:586-590, 1997
[4] Chen, CS et al., *Science*, 276:1425-1428, 1997.
[5] Rape, AD et al., *Biomaterials*, 32:2043-2051, 2011.
[6] Zhang, YH et al., *Matrix Biol.*, 30:135-144, 2011.
[7] Simmons, CS et al., *Lab Chip*, 13:646-649, 2013.
[8] Cui, Y et al., *Nat. Commun*, 6:6333-6340, 2015
[9] Khalili, JS et al., *Clin. Cancer Res.*, 18:4345-4355, 2012.
[10] Cirka, H et al., *Biophys. J.*, 110:1845-1857, 2016

SB[3]C2017
Summer Biomechanics, Bioengineering and Biotransport Conference
June 21 – 24, Tucson, AZ, USA

MYOSIN MEDIATES ANISOTROPIC MECHANOSENSING

Shin-Min Wen and Pen-Hsiu Grace Chao

Institute of Biomedical Engineering
Nation Taiwan University
Taipei, Taiwan

INTRODUCTION

The native cellular microenvironment is often heterogeneous and anisotropic. Cell elongation and alignment in response to the environmental cues correspond with changed cytoskeletal structure and behaviors [1]. Furthermore, anisotropic stimulation leads to changes in cell structure, orientation, and phenotype [2]. Focal adhesion kinase and actomyosin architecture have been proposed as regulators of the orinetatio-dependent mechanosensing, yet much of the underlying mechanism remains unclear [3, 4]. In the current study, we investigate the initial actin cytoskeleton response to interactions of anisotropic cell shape and uniaxial loading.

METHODS

Microfabticaton PDMS membranes were treated with 2 mg/ml dopamine hydrochloride and microcontact printed with 25 μg/ml fibronectin. Unprinted regions were blocked with 1% BSA. The elliptical patterns were ~1250 μm^2 in area with an aspect ratio of 5.

Cell Culture Human bone marrow-derived mesenchymal stem cells (MSCs) were kindly provided by Dr. Shih-Chieh Hung [5]. Cells were seeded on the patterned PDMS membrane for four hours. Myosin activity was inhibited with 50 μM blebbistatin in serum-free medium for 30 minutes.

Mechanical Stimulation Static 10% uniaxial strain was applied for 30 and 60 seconds (strain rate: 0.1/sec) either parallel or perpendicular to the long axis of the pattern. Cells were fixed in formalin on the bioreactor prior to release.

Imaging and Analysis Cytoskeleton organization was examined by fluorescently labeling f-actin with Alexa Flour 488-conjugated phalloidin (Invitrogen) and the nuclei were labeled with by DAPI. F-actin length and number were quantified tracking the actin fibers Imgae J plugin NeuronJ (short: 0-32.4 μm, medium: 32.4-64.8 μm, long: 64.8-97.2 μm). Zyxin was visualized by immunofluorescence with anti-zyxin antibody (Abcam). Zyxin on actin ratio was determined by the ratio of cells exhibiting fibrous zyxin structures.

Statistical Analysis SPSS was used to perform ANOVA analysis with post-hot test (LSD, α=0.05). Chi-square tests with Bonferroni correction were used to compare proportions. Error bars represent standard deviations or standard errors of a proportion.

RESULTS

Static uniaxial loading parallel to cell orientation significantly reduced actin length after 30 seconds, which recovered and further increased after 60 seconds of stretch (Figure 1AB). These length changes corresponded with the initial increase in the number of actin fibers at 30 seconds, and subsequent decrease at 60 seconds (Figure 1C). Perpendicular loading also led to initial reduction of actin length and recovery after 60 seconds, but did not surpass the original actin length. The number of actin fibers did not change with perpendicular stretch. When the number of actin fibers in each cell was divided into short, medium, and long groups by length, stretch significantly decreased the long fibers and increased the short fibers at 30 seconds, suggesting that the long fibers broke into shorter segments (Figure 1D). As zyxin, the focal adhesion protein, have been proposed as an actin-remodeling agent, we examined zyxin localization in response to stretch [6]. Interestingly, zyxin displayed a significant actin localization response only to parallel, but not to perpendicular, loading (Figure 2). When myosin activity was inhibited with blebbistatin treatment, actin length reduced initially with parallel stretch, and the recovery was reduced (Figure 3B). Myosin inhibition also abolished the zyxin relocalization response in parallel stretch (Figure 3). In perpendicular stretch, myosin inhibition suppressed the initial shortening response without changes to zyxin distribution.

DISCUSSION

We investigated the initial cytoskeletal and zyxin response to uniaxial mechanical stretch and explored the role of myosin activity on actin and zyxin interactions. Our results demonstrate that parallel stretch initially breaks actin fibers, in a myosin-independent manner, and the fibers are repaired by myosin-dependent zyxin recruitment. In perpendicular stretch, while myosin-mediated actin shortening is observed, zyxin is not involved in the repair process. Our result is consistent with reports of zyxin recruitment of α-actinin or Ena/VASP to repair actin fibers in response to mechanical loading [6, 7], and extended the previous understanding with direction-dependent behaviors. As myosin regulates zyxin-mediated actin polymerization [8], future studies will investigate the role of cell tension on anisotropic mechanosensing and zyxin activation

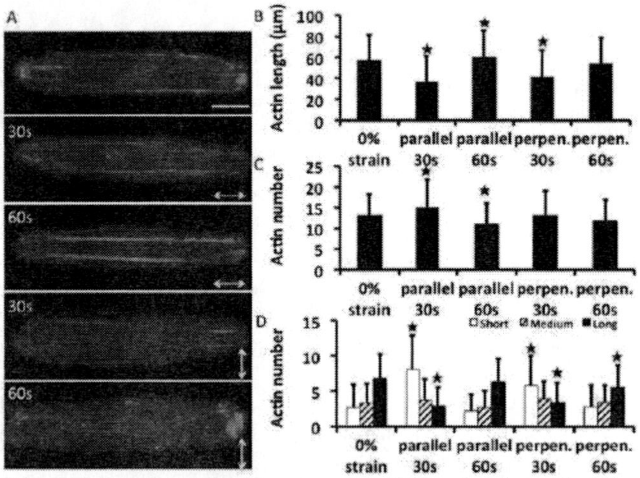

Figure 1: F-actin structure in response to 10% uniaxial stretch. (A) Representative actin structure; scale bar = 10 μm. (B-D) Actin length and number (n= 42-143 ★$p<0.05$ vs 0% strain)

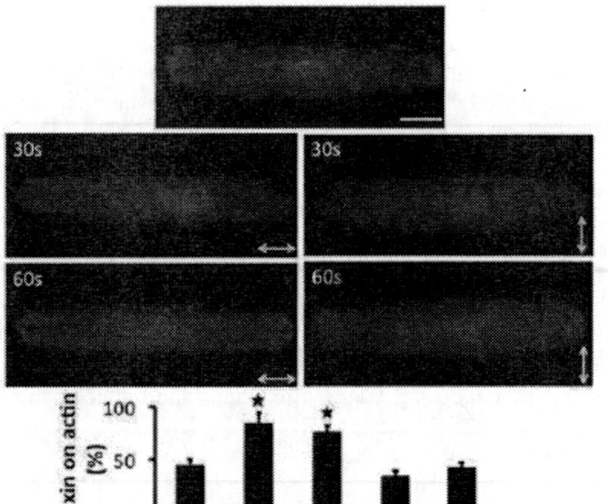

Figure 2: Representative images of zyxin distribution and ratio of zyxin colocalization with actin in response to mechanical stretch. Scale bar = 10 μm. (n=54-143 ★$p<0.05$ vs 0% strain)

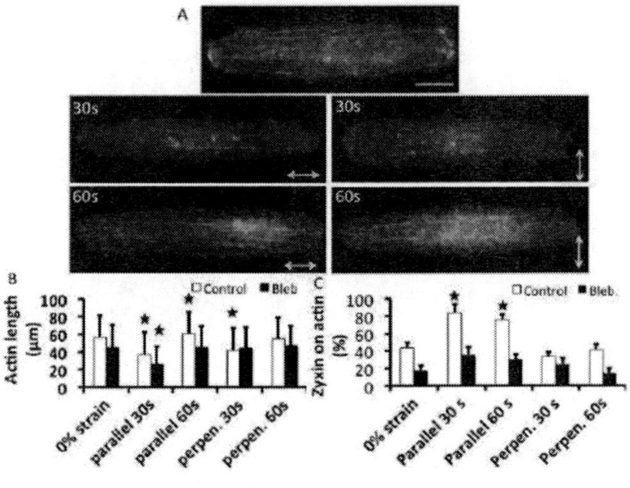

Figure 3: Actin and zyxin response to mechanical stretch with blebbistatin treatment. (A) Representative images of actin structure and zyxin distribution, scale bar = 10 μm. (B) Actin length (n=31-55 ★$p<0.05$ vs 0% strain). (C) Ratio of zyxin on actin (n=30-143 ★ $p<0.005$ vs 0% strain).

ACKNOWLEDGEMENTS

The work was supported by the Taiwanese. MOST (105-2221-E-002-006) and NHRI (EX106-10411EI)

REFERENCES

[1] Teixeira AI, Abrams GA, Bertics PJ, Murphy CJ, and Nealey PF. Epithelial contact guidance on well-defined micro- and nanostructured substrates. Journal of Cell Science 2003;116:1881-92.
[2] Kurpinski K, Chu J, Hashi C, and Li S. Anisotropic mechanosensing by mesenchymal stem cells. Proceedings of the National Academy of Sciences of the United States of America 2006;103:16095-100.
[3] Shao Y, Mann JM, Chen W, and Fu J. Global architecture of the F-actin cytoskeleton regulates cell shape-dependent endothelial mechanotransduction. Integrative Biology : quantitative biosciences from nano to macro 2014;6:300-11.
[4] Chen Y, Pasapera AM, Koretsky AP, and Waterman CM. Orientation-specific responses to sustained uniaxial stretching in focal adhesion growth and turnover. Proceedings of the National Academy of Sciences of the United States of America 2013;110:E2352-61.
[5] Hung SC, Pochampally RR, Hsu SC, Sanchez C, Chen SC, Spees J, et al. Short-term exposure of multipotent stromal cells to low oxygen increases their expression of CX3CR1 and CXCR4 and their engraftment in vivo. PloS one 2007;2:e416.
[6] Hoffman LM, Jensen CC, Chaturvedi A, Yoshigi M, and Beckerle MC. Stretch-induced actin remodeling requires targeting of zyxin to stress fibers and recruitment of actin regulators. Molecular Biology of the Cell 2012;23:1846-59.
[7] Smith MA, Blankman E, Gardel ML, Luettjohann L, Waterman CM, and Beckerle MC. A zyxin-mediated mechanism for actin stress fiber maintenance and repair. Developmental Cell 2010;19:365-76.
[8] Hirata H, Tatsumi H, and Sokabe M. Mechanical forces facilitate actin polymerization at focal adhesions in a zyxin-dependent manner. Journal of Cell Science 2008;121:2795-804.

SB³C2017
Summer Biomechanics, Bioengineering and Biotransport Conference
June 21 – 24, Tucson, AZ, USA

SUBSTRATE DISPLACEMENTS INDUCE DIRECTED KERATINOCYTE MIGRATION

H. Zarkoob (1), S. Chinnathambi (1), J.C. Selby (2), E.A. Sander (1)

(1) Department of Biomedical Engineering
College of Engineering
University of Iowa
Iowa City, Iowa, USA

(2) Department of Dermatology
Carver College of Medicine
University of Iowa
Iowa City, Iowa, USA

INTRODUCTION

Cell migration is essential to many physiological and pathological processes in the body, including wound healing. During wound healing, chemical and physical gradients direct cell migration, control re-epithelialization, and guide tissue remodeling [1]. Until recently, the regulation of these processes were largely understood in terms of biochemical signals, but recent interest in the mechanobiology of tissues suggests that physical forces and mechanical cues from the wound bed may also play an important role in the healing process.

In a previous study, we explored this possibility and found that mechanical cues did affect human epidermal keratinocyte (HEKn) behaviors related to the process of re-epithelialization [2]. Specifically, HEKn cultured at low density on soft (1.2 kPa) polyacrylamide (PA) gels versus stiff (24 kPa) PA gels had significantly higher migration velocities, increased rates of aggregate formation, and more cells per aggregate. In addition, the HEKn on soft PA gels appeared to migrate in a directed manner towards the forming aggregates in response to mechanical communication through the deformable substrate. Directed migration was not apparent on the stiff PA gels. However, we were unable to clearly demonstrate that the HEKn were responding to these mechanical signals, possibly because multiple cells produced a spatially and temporally heterogeneous milieu of deformations in the substrate.

The goal of this study was to determine whether HEKn sense and respond to mechanical signals from deformations in the substrate induced by other HEKn, and if so, what are the underlying mechanisms. To test this hypothesis, we imaged and analyzed the response of single HEKn to local substrate deformations induced by the micromanipulation of a needle inserted into the gel. We also examined the effect of adding either Y-27632, a rho kinase inhibitor, or blebbistatin, a myosin II inhibitor, on keratinocyte behavior. Our results indicate that keratinocytes do migrate in the direction of continuous mechanical deformations in the substrate, and that inhibiting different elements of the mechanosensing pathway disrupted this behavior. The results of this study give new insights on the mechanism by which HEKn sense and response to mechanical signals from their environment and could have important implications for devising new wound healing strategies.

METHODS

Type I collagen coated PA gels embedded with fluorescent microspheres were prepared with nominal stiffnesses of 1.2 kPa (soft) and 24 kPa (stiff), as previously reported [2]. HEKn in KSFM medium were cultured on the gels at a density of 200 cells/cm^2. The Ca^{2+} of medium was elevated from 0.09 mM to 1.2 mM by adding $CaCl_2$. The HEKn were allowed to attach and stabilize for 3 hours, before the gels were moved to an environmentally-controlled chamber for time-lapse imaging with a Nikon Eclipse Ti microscope operating in both DIC and epi-fluorescence mode.

Continuous local mechanical deformations in the PA gel substrate were produced by inserting a sterile 0.25 mm x 75mm needle into the gel. The needle was secured to a high-resolution linear actuator (PI), controlled with MTESTQuattro software (ADMET), and mounted to a high-precision, manual linear stage (Newport). The needle was inserted either 100 μm or 200 μm away from a single HEKn at least 500 μm away from any neighboring cells. The needle was then displaced at a constant rate of 1 μm/minute to mimic displacement rates observed previously [2]. Time-lapse images were taken at a five minutes interval for two hours. As many as three sequential experiments were conducted on each gel, such that all experiments were conducted within 10 hours from the start of the experiment.

Poster Presentation #P99

To assess how inhibiting components of the mechanosensing and force generating machinery of the cytoskeleton affected HEKn behavior, in some experiments, 50 μM of either Y-27632 or blebbistatin were added to the culture medium one hour after culturing the cells (*i.e.*, 2 hours before imaging started). The needle was then inserted into the gel 100 μm away from a single HEKn and displaced as described above. Additional control experiments without a needle were conducted up to 24 hours on soft and stiff PA gels with and without drug in order to characterize HEKn migration in the absence of any symmetry breaking gradients.

Standard metrics for characterizing cell migration were assessed, including the confinement ratio (CR), defined as the net HEKn distance traveled over the total distance traveled, and the directionality ratio (DR), defined as the ratio of the sum of the projected cell velocity in the direction of needle movement to the sum of the cell velocities over the duration of the experiment. For the experiments involving the needle, the adjusted migratory cell path was determined by subtracting out displacement of the underlying substrate. The substrate displacement was measured by using an algorithm based on cross-correlation of images of the fluorescent beads embedded in the PA gels, as was done previously [2].

RESULTS

Our results indicate that HEKn on soft PA gels tended to follow the local mechanical deformations made by the needle (Fig.1, Fig. 2A,B). This directed migration towards the needle happened more often when the needle was inserted 100 μm away from a single HEKn than when it was inserted 200 μm away. The addition of Y-27632, which disrupts the Rho/ROCK pathway, essentially blocked mechanical cues from substrate displacement, resulting in cell migration patterns that resemble a persistent random walk (Fig. 2C) The addition of blebbistatin, which reversibly blocks non-muscle myosin II engagement with actin and the ability to generate traction forces, severely impaired cell motility. CR and DR values agreed with these observations (Table 1). Note that the high CR value for Soft+Blebbistatin likely reflects the fact that HEKn were basically stationary.

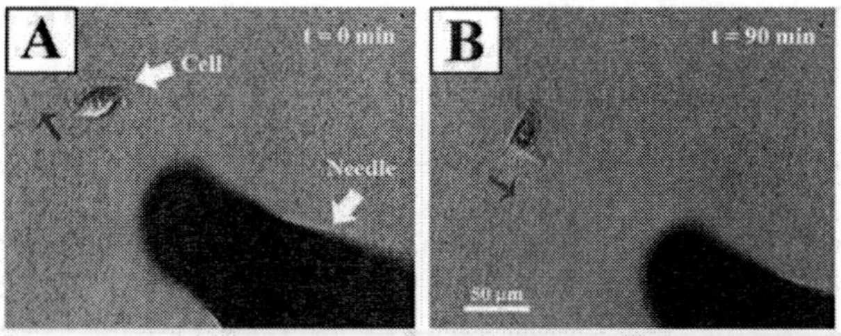

Figure 1: DIC Images (A) t = 0 min and (B) t = 90 min showing the direction of keratinocyte migration (red arrow) in response to a constant rate of substrate displacement induced by a needle.

DISCUSSION

We have found tantalizing evidence that single HEKn subjected to the calcium switch, which transforms the cell from a proliferative state to one of cell aggregation, move in a directed fashion towards mechanical deformations in the PA substrate. We hypothesize that such mechanical signals could be important regulators of the re-epithelialization process in wound healing. These results agree with previous studies by Lo *et al.* [3] and Reinhart-King *et al.* [4] that reported that cells respond to local mechanical deformations made by either a microneedle or an adjacent cell, respectively. We are further analyzing these results to help further elucidate this signaling process and its role in healing wounds.

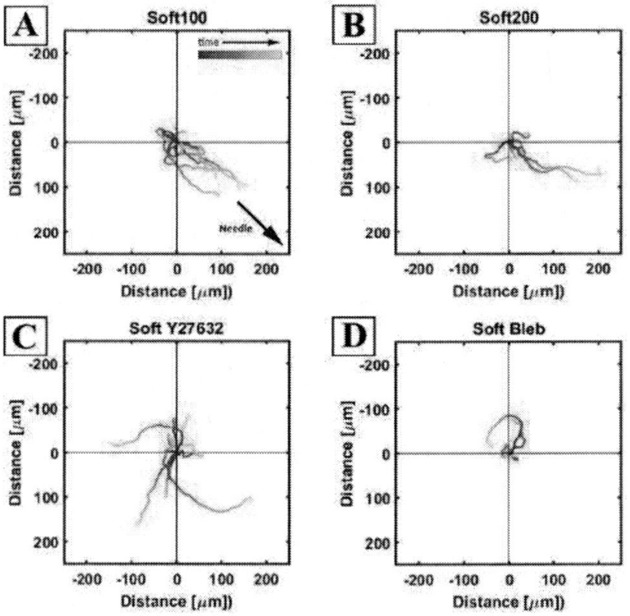

Figure 2: Cell Migration in Response to Substrate Displacement (n=9 or 10 for each condition). Line color shows the temporal progression of cell path from t=0 (blue) to t = 2hrs (yellow).

Sample Name	CR	DR
Soft-100 μm	0.49 ± 0.26	0.46 ± 0.28
Soft-200 μm	0.49 ± 0.25	0.43 ± 0.40
Soft+Y27632	0.29 ± 0.17	0.05 ± 0.40
Soft +Blebbistatin	0.53 ± 0.28	− 0.09 ± 0.31

Table 1: Confinement ratio (CR) and directionality ratio (DR) for Different Experimental Conditions.

ACKNOWLEDGEMENTS

Support for this work was provided by the National Science Foundation (CAREER 1452728) to E.A.S. In addition, J.C.S. acknowledges the Dermatology Foundation for their support of this work through a career development award.

REFERENCES

[1] Martin, P., Science, 1997. **276**(5309): p. 75-81.
[2] Zarkoob, H., et al., Cell Mol Bioeng, 2015. **8**(1): p. 32-50.
[3] Lo, C.M., et al., Biophysical Journal, 2000. 79(1): p. 144-152.
[4] Reinhart-King, C.A., et al., Biophys J, 2008. 95(12): p. 6044-51.

ESTROGEN DEFICIENCY CHANGES MECHANOBIOLOGICAL RESPONSES OF OSTEOBLASTS TO FLUID FLOW EFFECTING OSTEOBLAST INDUCED OSTEOCLAST DIFFERENTIATION

[1]Allison, H., [1]Deepak, V., [1]McNamara, LM

(1) Biomedical Engineering, College of Engineering and Informatics
National university of Ireland, Galway
Galway, Ireland

INTRODUCTION

Osteoporosis is associated with bone loss, which leads to bone fractures, pain and immobility. Reduced levels of circulating estrogen during osteoporosis are associated with an increase in the number and activity of osteoclasts as well as an increase in osteoblastogenesis [1]. The differentiation of osteoclasts is dependent on a member of the tumour necrosis factor (TNF) cytokine, RANKL, which is produced by osteoblasts. Osteoblasts also express osteoprotegerin (OPG), a decoy receptor for RANKL, which binds RANKL and prevents binding to RANK and osteoclastogenesis (Figure 1). Thus osteoblasts play an important role in inducing osteoclast maturation and bone resorption [2]. Osteoblasts have been shown to act as mechanosensors responding to shear stress created by interstitial fluid flowing through the canalicular network in bone [3]. When osteoblasts sense this mechanical stimulation this leads to intracellular calcium signalling, which is later transduced into biochemical signals and changes in gene expression [4]. One pathway effected is the RANKL-OPG pathway (Figure 1); following mechanical stimulation, an increase in OPG mRNA and a decrease in RANKL mRNA has been shown. The decrease in RANKL/OPG ratio has been shown to cause a decrease in osteoclast formation when mechanically stimulated osteoblasts are co-cultured with osteoclast pre-cursors [5]. Osteoblasts treated with estrogen show attenuated RANKL induced osteoclast formation [6]. Osteoblastic cells deprived of estrogen display deficient osteogenic responses to mechanical stimuli in vitro [7]. However, the precise role of osteoblastic cells in regulating osteoclast differentiation under mechanical loading and during estrogen deficiency is not known.

OBJECTIVE

To determine how osteoblasts regulate osteoclast differentiation under mechanical loading and estrogen deficiency.

METHODS

Cell culture: MC3T3-E1 osteoblastic cells were maintained in expansion media (α-MEM supplemented with 10 % FBS, 2 mM L-glutamine, 100 U/mL penicillin, 100 μg/mL streptomycin). RAW264.7, a macrophage cell line, were maintained in DMEM supplemented with 10 % heat inactivated FBS, 2 mM L-glutamine, 100 U/mL penicillin, 100 μg/mL streptomycin. **Estrogen deficiency:** For 6 days, MC3T3-E1 cells were treated with estrogen (E: 10 nM 17β-estradiol) or under simulated menopausal conditions using a selective estrogen receptor degrader Fulvestrant (FE: 10 nM 17β-estradiol plus 100 nM) or by withdrawing the estrogen from the culture media, after cells had become accustomed to estrogen for 3 days (EW). **Mechanical stimulation:** MC3T3-E1 cells were seeded onto slides (150,000 cells per slide) and subjected to either laminar oscillatory fluid flow (±1 Pa shear stress applied at a frequency of 0.5 Hz, for 1 h using a NE-1600 syringe pump and 50 mL syringe) or static conditions. After 24 hours conditioned media (CM) was collected from the MC3T3-E1 cells. **Co-culture and conditioned media experiments:** RAW cells were seeded 5,000 per well in a 96 well plate before being treated with 50% CM from MC3T3-E1 cells and cultured for 7 days. The MC3T3 cells, which were used to collect CM, were trypsinized and seeded in a 48 well plate (15,000 cells per well). RAW264.7 cells were seeded directly on top (10,000 cells per well) and co-cultured for 8 days. TRAP activity assays and TRAP staining were performed on both co-cultures and CM experiments. One way ANOVA analysis was performed on GraphPad Prism $P < 0.05$.

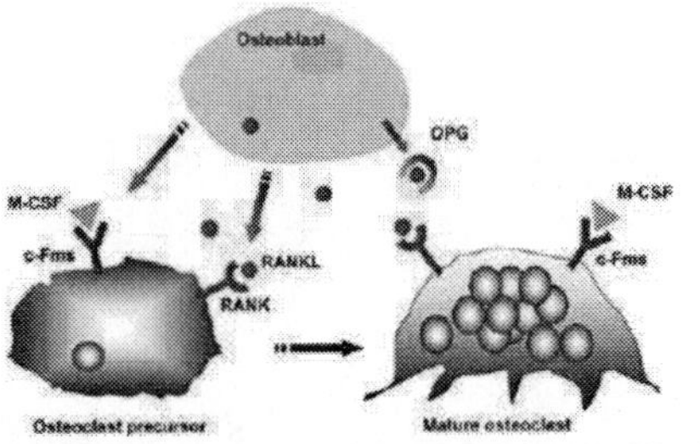

Figure 1: Shows the communication between osteoblasts and osteoclasts. RANKL and OPG are both produced by osteoblasts, RANKL binds to RANK receptors on osteoclast progenitors and leads to their differentiation. And OPG acts as a decoy receptor for RANKL [8]

RESULTS

When RAW cells were co-cultured with flow stimulated MC3T3 cells (Fig. 2E, I), there were less multinucleated TRAP positive (+) cells compared to static control (Fig. 2A, I). There were also less TRAP+ cells when RAW cells were co-cultured with MC3T3-E1 cells treated with estrogen under both flow and static conditions (Fig. 2B, F). A higher number of TRAP+ cells were seen when RAW264.7 cells were co-cultured with MC3T3-E1 cells under simulated menopausal conditions static + EW (Fig. 2C), flow + EW (Fig. 2G), static + FE (Fig. 2D) or flow + FE (Fig. 2H) compared to Static or flow + Estrogen (Fig. 2D, I). CM experiments showed similar changes (Fig. 3).

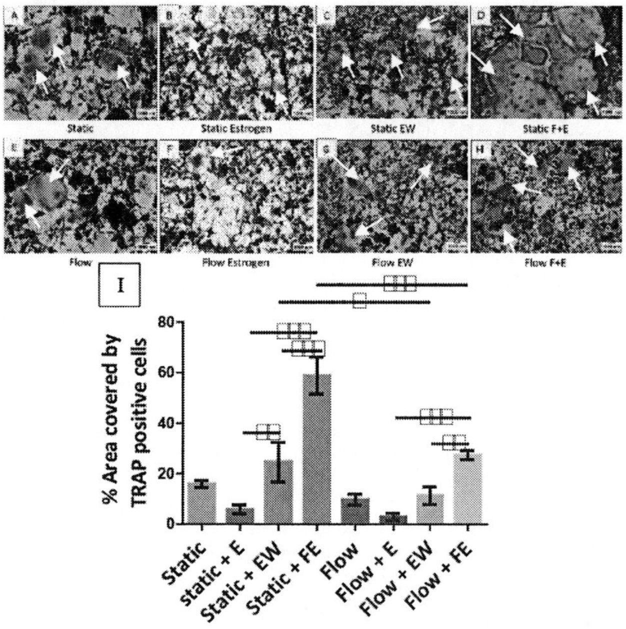

Figure 2: A to H-Multinucleated TRAP+ RAW264.7 cells (white arrows) after co-culture with static/mechanical stimulated MC3T3-E1 cells for 8 days. I- Show percentage area covered by multinucleated TRAP+ cells (P < 0.05).

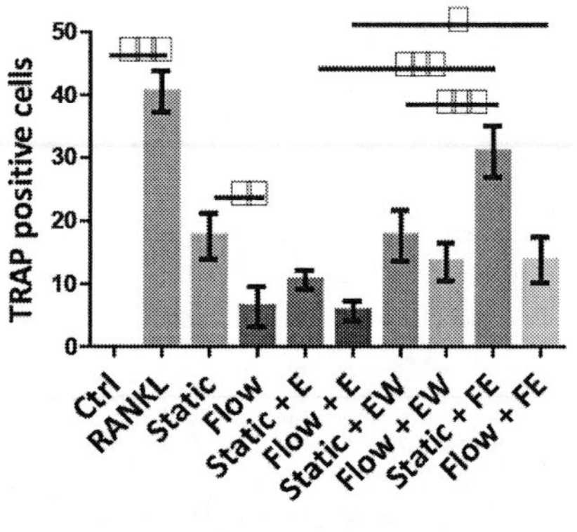

Figure 3: Amount of TRAP positive cells present when RAW 264.7 cells were treated with 50% CM from MC3T3-E1 cells subjected to various conditions (P < 0.05).

DISCUSSION

The results of this study reveal that when osteoblasts are subjected to fluid flow, their ability to induce osteoclast differentiation is reduced (Fig 2 and 3). This supports previous research, which has found that mechanical stimulation leads to an increase in OPG mRNA, which is important in osteoclast differentiation (Fig1) [5]. Pre-treatment of estrogen also reduces the capacity of the osteoblasts to induce differentiation and the combination of estrogen and fluid flow seems to have an additive effect in reducing the amount of osteoclast differentiation seen (Fig 2I). We have shown for the first time that under postmenopausal conditions, stimulated by either the withdrawal of estrogen or inhibiting its receptor, a significant increase in osteoblast induced osteoclast formation occurs (Fig 2I and 3). This might explain, in part why in postmenopausal osteoporosis, where low levels of circulating estrogen are seen, bone resorption exceeds that of bone formation. This imbalance is due to the fact estrogen has a protective role against bone resorption by the modulation of cytokines such as IL-1 and TNF expressed by osteoblasts. The elevated levels of these cytokines due to the lack of estrogen, effects both the mature osteoclasts and their precursors leading to an increase in differentiation and resorption [11]. The antagonist Fulvestrant had a more potent effect on osteoblast-induced osteoclast differentiation compared to EW in both static and flow conditions (Fig 1I and 3). This could be due to the fact that the agonist blocks but also down regulates the estrogen receptor [12], whereas in EW conditions the receptors are still present and there could be residual estrogen present. Under combined postmenopausal conditions and mechanical loading osteoclast differentiation occurred (Fig 2G, H), but not to the same extent as static conditions (Fig 2C, D). One explanation is that estrogen deficiency alters the mechanoresponsiveness of osteoblasts by reducing biochemical signaling, resulting in deficient osteogenic responses [13]. The defect in mechanoresponsiveness could mean that a down regulation of OPG is seen in response to mechanical loading as opposed to the up regulation of OPG which has previously been observed [5]. The down regulation of OPG and potential up-regulation of RANKL, could lead to a decrease in osteoclast apoptosis and an increase in osteoclast differentiation and function. Current findings implicate that osteoblast induced osteoclast differentiation might contribute to bone loss in post- menopausal osteoporosis.

ACKNOWLEDGEMENTS

This publication has emanated from research conducted with financial support of Science Foundation Ireland (SFI) and is co-funded under the European Regional Development fund under grant number 14/IA/2884.

REFERENCES

1. Rosen, C. J. 2000. Clinics in Laboratory Medicine. 20, 591 –602.
2. Asagiri, M. & Takayanagi, H., 2007. Bone, 40(2), pp.251–64.
3. Lee, D. *et al.*, 2008. Bone. 23, 1140-9
4. Huo, B. *et al.*, 2008. Cellular and molecular bioengineering, 1, 58-66
5. Kim, C.H. *et al.*, 2006. Bone, 39(5), pp.1043–1047
6. Qu, Q. et al., 1999. Bone, 25(2), pp.211–215.
7. Sterck, J. G, *et al.*, 1998. Am. J. Physiol. 274, 1113–1120.
8. Lee, N. K. 2010. Endocrinology and Metabolism, 25(4), 264.
9.Birmingham, E. *et al.*, 2012. Eur Cell Mater, 23(353), pp.13–27.
10. Jaalouk, D & Lammerding, J., 2009. Nat Rev Mol Cell Biol. 10, 63-73
11.Feng, X and McDonald, J, M. 2011. Annu Rev Pathol. 6, 121-145
12. Oakman, C. *et al.*, 2011. Future Oncol. 7, 173-186
13. Brennan, M,A. *et al.*, 2014. Horm Metab Res. 46, 537-545

PREDICTING CELLULAR (RE)ORIENTATION IN CYCLICALLY STRETCHED COLLAGEN GELS DUE TO MECHANICAL AND TOPOGRAPHICAL CUES

Tommaso Ristori (1,2), Thomas M.W. Notermans (1), Frank P.T. Baaijens (1,2), Sandra Loerakker (1,2)

(1) Department of Biomedical Engineering
Eindhoven University of Technology
Eindhoven, The Netherlands

(2) Institute for Complex Molecular Systems
Eindhoven University of Technology
Eindhoven, The Netherlands

INTRODUCTION

Cellular orientation is one of the most important determinants of the mechanical behavior of soft tissues. Along their main direction, cells exert traction forces[1] and produce collagen[2], which is the main load-bearing component of most soft tissues. Therefore, to analyze the mechanical behavior of these constructs and ultimately optimize their design, predicting and understanding cellular orientation is essential.

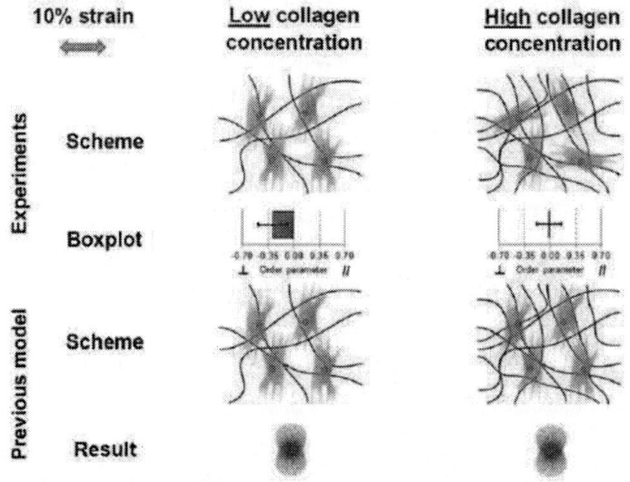

Figure 1: Scheme of experimentally observed cellular orientations[4,5] and of computational results obtained with a previous mechanical model[6].

Contractile cells usually align along the direction of actin stress fibers[3], the main contractile component of the cytoskeleton. Cells are generally able to remodel the direction of these fibers in response to mechanical stimuli[3]. However, collagen fibers can influence this process of realignment, a phenomenon which seems to depend on collagen concentration. In particular, by analyzing cells in biaxially constrained collagen gels that were statically cultured for 3 days and uniaxially cyclically stretched for other 3 days, it has been observed that high collagen concentrations can hinder stress fiber remodeling[4,5] (Fig. 1). This was not the case when the analyzed collagen gels contained a relatively low concentration of collagen or when the cyclic strain was applied immediately after seeding[4,5].

Recently, a mechanical model able to predict stress fiber alignment and stress fiber stress in response to a range of mechanical stimuli has been developed[6]. However, this model cannot capture the impact that collagen concentration has on stress fiber remodeling, and thus on cellular (re)orientation (Fig. 1). During this project, we aimed at overcoming this limitation.

METHODS

The previously proposed mechanical model was extended and coupled with a model for collagen prestretch and rotation. In particular we hypothesized that, in the first 3 days of culture, the weakly stable collagen network cannot influence cellular (re)orientation. On the contrary, in this period collagen fibers are prestretched and rotated by cellular forces. A previously published numerical algorithm was used to model the collagen stress and prestretch[7], while collagen rotation was simulated with a newly developed mathematical model based on the assumption that collagen fibers tend to assume the same distribution as stress fiber stresses.

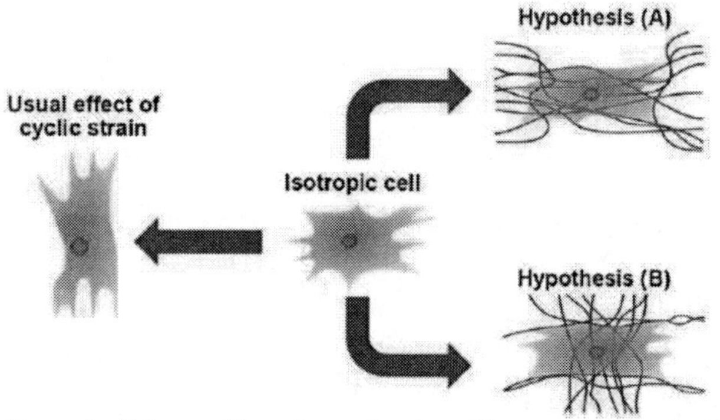

Figure 2: Scheme of hypotheses for stress fiber remodeling. Cyclic strain usually causes cell realignment (left), but in case of high collagen concentrations (right) cells do not react to strain because they (A) locally follow the direction of collagen fibers or (B) do not have space to reorient due to collagen fibers.

For the remaining 3 days of culture, we assumed that the collagen network is stabilized by non-covalent interactions[8] and crosslinks provided by cells and that, under these conditions, collagen fibers can influence stress fiber remodeling with two possible mechanisms (Fig. 2):

A. collagen fibers provide cells with topographical cues, such that stress fibers tend to align along the directions with higher collagen densities;
B. collagen fibers constitute a spatial obstruction for cellular realignment, such that cells can reorient only when the collagen density in their surrounding is below a certain threshold.

The previously proposed mechanical model for stress fiber remodeling was then adapted using the two hypotheses separately. The resulting numerical algorithms were tested by simulating collagen and stress fiber remodeling occurring in biaxially constrained collagen gels that were uniaxially cyclically stretched immediately, or after 3 days of static culture, varying the concentration of collagen. The computational results were then compared with experimental results present in the literature[4,5].

RESULTS

In accordance with experimental results, the evolution law for collagen fibers predicted that immediate stretching of collagen gels caused collagen rotation perpendicular to the direction of stretching (Fig. 3, rows 1 and 2, red lines). On the other hand, a delayed stretching resulted in an isotropic collagen distribution (Fig. 3, rows 3 and 4, red lines).

Concerning cellular (re)orientation, the extended mechanical model for stress fiber remodeling predicted that cells are able to reorient their cytoskeleton in response to cyclic stretch when this mechanical stimulus is applied to the tissue constructs right after seeding, or when the engineered constructs contain a relatively low concentration of collagen (Fig. 3, rows 1 and 2, blue lines, and Fig. 3, rows 1 and 2, first column). Conversely, the model extended with both the hypotheses (separately) predicted that stress fibers cannot realign perpendicular to cyclic strain when gels contain relatively high collagen concentrations and the cyclic strain is applied after the stabilization of the collagen network by non-covalent interactions and crosslinks (Fig. 3, rows 3 and 4, third column). This is in line with experimental observations, and an improvement compared to the original model.

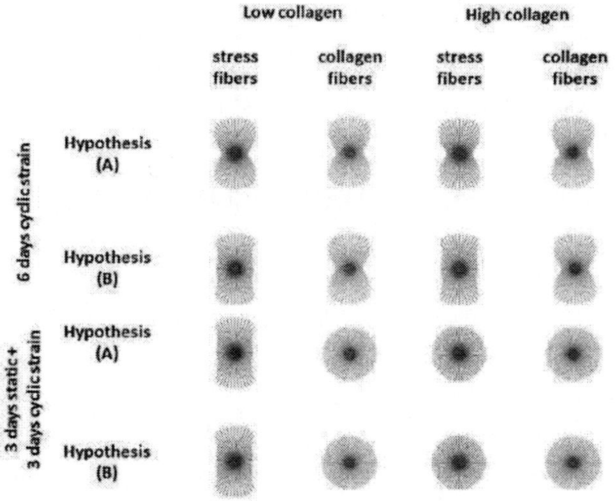

Figure 3: Results of the computational simulations of the extended model. The length of the lines is proportional to the stress fiber (blue lines) and collagen (red lines) volume fractions along that particular direction.

DISCUSSION

In this study, a previously proposed mechanical model for stress fiber remodeling was extended to capture the effects that different collagen concentrations have on cellular (re)orientation. Moreover, to predict the correct direction of the topographical cues provided by collagen, a mathematical model for collagen rotation in response to cellular forces was developed. The coupled evolutions equations for collagen and stress fiber remodeling were tested by simulating biaxially constrained collagen gels that were uniaxially cyclically stretched. The obtained results were qualitatively in agreement with experiments, which demonstrated that predicting cellular (re)orientation in collagen gels with different collagen concentrations is possible. However, since the different proposed hypotheses did not show significant differences in results, future studies are necessary to identify the exact processes that link collagen concentration with cell remodeling. Nevertheless, the developed numerical algorithm could be used in future works to further optimize engineered soft tissues or to understand the remodeling of native tissues.

ACKNOWLEDGEMENTS

This study was supported by the People Programme (Marie Curie Actions) of the European Union's Seventh Framework Programme [FP7/2007-2013/]; REA [grant agreement number 317512].

REFERENCES

[1] Ghibaudo, M et al., *Soft Matter*, 4:1836–1843, 2008.
[2] Wang, JHC et al., *J Biomech*, 36:97–102, 2003.
[3] Faust, U et al., *Plos One*, 6:e28963, 2011.
[4] Foolen, J et al., *Biomaterials*, 33:7508–7518, 2012.
[5] Foolen, J et al., *Acta Biomater*, 10:1876–85, 2014.
[6] Obbink-Huizer, C et al., *Biomech Model Mechanobiol*, 13:227-236, 2014.
[7] Loerakker, S et al., *Biomech Model Mechanobiol*, 13:985–1001, 2014.
[8] Guidry, C and Grinnel, F, *Collagen Rel. Res.*, 6: 515-529, 1987.

SB³C2017
Summer Biomechanics, Bioengineering and Biotransport Conference
June 21 – 24, Tucson, AZ, USA

DOSE-DEPENDENT EFFECTS OF BETA-AMINOPROPIONITRILE ON OSTEOBLAST GENE EXPRESSION AND COLLAGEN PRODUCTION

Silvia P. Canelón (1), Joseph M. Wallace (2,3)

(1) Weldon School of Biomedical Engineering
Purdue University
West Lafayette, IN, USA

(2) Department of Biomedical Engineering
Indiana University-Purdue University at Indianapolis
Indianapolis, IN, USA

(3) Department of Orthopaedic Surgery
Indiana University School of Medicine
Indianapolis, IN, USA

INTRODUCTION

The fibrillar structure of type I collagen is stabilized by enzymatic crosslinks catalyzed by lysyl oxidase (LOX), a step which can be interrupted using the lathyritic agent beta-aminopropionitrile (BAPN). Murine *in vivo* bone studies have confirmed effects of BAPN on collagen nanostructure[1] and the objective of this study was to evaluate the mechanism of these effects on type I collagen produced *in vitro* by MC3T3-E1 osteoblasts. Effects of BAPN were assessed by quantifying gene expression of type I collagen, LOX, and bone morphogenic protein-1 (BMP-1), a LOX activator, measuring collagen morphology (D-spacing), and evaluating the ratio of mature to immature crosslinks.

METHODS

Quantitative Reverse Transcription Polymerase Chain Reaction (qRT-PCR): To investigate a possible dosage-dependent effect on expression of collagen-related genes, an experiment was conducted for qRT-PCR analysis, using 5 different BAPN concentrations (0.125mM, 0.25mM, 0.5mM, 1.0mM and 2.0mM). Osteoblastic cells were cultured and differentiated in 60mm dishes in the presence or absence of BAPN (n=5/group). Cells proliferated in complete α-MEM media for 7 days then differentiated for 7 days using ascorbic acid. Following RNA isolation and RT, qRT-PCR was performed using SYBR Green primers and master mix (COL1A1, COL1A2, LOX, BMP-1 and β-actin as reference) and gene expression levels analyzed using an efficiency-calibrated mathematical model (Eq 1) implemented in the REST® program[6].

$$Ratio = \frac{(E_{target})^{\Delta CT_{target}(control-sample)}}{(E_{ref})^{\Delta CT_{ref}(control-sample)}} \quad (1)$$

Atomic force microscopy (AFM): The experiment was repeated using 0.14mM BAPN (n=4/group) to explore potential effects of a low dosage on type I collagen morphology using fibril D-spacing, a metric sensitive to differences in collagen structure across tissue types and disease states[2,8]. After 7 days of differentiation, media was removed from the dishes and cells were dissociated from the matrix with an EDTA buffer. The matrix was allowed to dry and imaged in air using AFM in 5 locations/dish. A 2D Fast Fourier transform was performed on 10 fibrils/location to extract D-spacing (50 fibrils/dish, ~200 fibrils/group).

Fourier Transform Infrared Spectroscopy (FTIR): The low-dose experiment (0.14mM BAPN) was repeated for FTIR analysis (n=5). Cells were seeded into 12 dishes (6 control and 6 treated). After 14 days of differentiation, the dishes were rinsed with PBS and water and transferred directly to a barium fluoride window to air-dry. Spectroscopic data were collected from the samples, peaks under the amide I region were located using 2nd derivative analysis, and the underlying peaks at ~1660cm^{-1} and ~1690cm^{-1} were fit using Gaussian curves. These peaks have been shown to represent mature (HP, hydroxylysylpyridinoline) and immature crosslinks, respectively[5,7].

RESULTS

A downregulation of both LOX and BMP-1 was observed with exposure to 0.25mM BAPN or greater, with statistically significant fold changes at 0.25mM, 0.5mM, and 1.0mM BAPN concentrations relative to 0mM BAPN controls (Table 1). No significant difference was detected in genes coding for type I collagen.

Poster Presentation #P103

Table 1: Fold changes in mRNA expression of BAPN-treated samples relative to controls (n=5).

Fold changes	0.125 mM	0.25 mM	0.5 mM	1.0 mM	2.0 mM
LOX	0.783	0.58*	0.54*	0.727	0.912
BMP-1	1.044	0.675*	0.71	0.666*	0.945
COL1A1	1.236	0.918	0.872	0.941	1.001
COL1A2	1.215	0.878	0.93	1.123	1.396

*Indicates statistically significant changes ($p<0.05$).

After sample processing and AFM imaging, 50+ fibrils per dish were analyzed (217 total for control and 251 for BAPN-treated). The mean D-spacing measurement for each dish ranged from 65.8nm to 66.8nm for the control group and 66.6nm to 67.6nm for the BAPN-treated group with a mean from each group of 66.4nm ± 0.4nm (control) and 67.1nm ± 0.4nm (BAPN, $p=0.060$). The analysis revealed a distribution of D-spacing that was shifted towards higher values in the BAPN-treated group (A-D test, $p<0.0001$) (Fig 1).

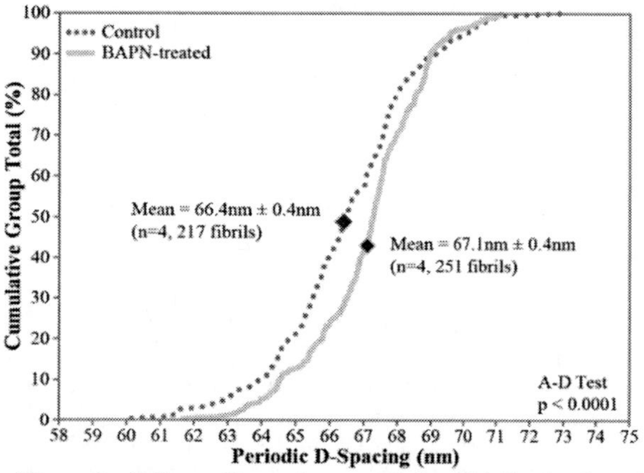

Figure 1: Collagen D-spacing cumulative distribution function (CDF) created from D-spacing measurements in each group (n=4).

Peak fitting of FTIR data resulted in consistent peaks in the 1654cm^{-1} and 1680cm^{-1} regions (Fig 2) representing mature HP and immature crosslinks, respectively. The area under these peaks was used to obtain the ratio of mature (~1660cm^{-1}) to immature (~1690cm^{-1}) crosslinks. BAPN treatment resulted in a decrease in the collagen crosslink ratio driven by a significant reduction in the HP crosslink peak percent area ($p<0.05$). There was no difference in the percent area of the peak corresponding to immature crosslinks (Table 2).

Table 2: Information regarding underlying peak fittings at ~1660cm^{-1} and ~1690cm^{-1} (n=5).

		Mean Peak Percent Area	Mean Area Ratio
Control	~1660 cm^{-1}	16.2868 ± 4.1089	3.9068 ± 1.6353
	~1690 cm^{-1}	4.7963 ± 2.2037	
BAPN	~1660 cm^{-1}	8.2149 ± 3.4959	1.9865 ± 0.6145
	~1690 cm^{-1}	4.4880 ± 2.3100	
p-value	~1660 cm^{-1}	0.0048	0.0338
	~1690 cm^{-1}	0.8177	

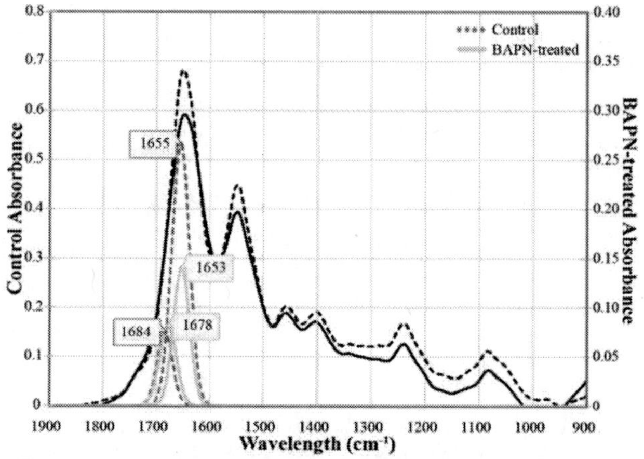

Figure 2: Representative mature and immature crosslink peak fittings underneath the FTIR spectral.

DISCUSSION

This study demonstrates significant effects of BAPN treatment on gene expression, as well as the morphology and enzymatic crosslinking in type I collagen produced *in vitro* by osteoblasts. qRT-PCR confirmed a dose-dependent response of LOX and BMP-1 to BAPN treatment. Both were downregulated with 0.25mM BAPN, as seen in similar studies[7]. This coupled downregulation at 0.25mM indicates that the irreversible binding of LOX by BAPN drives a decrease in LOX expression, potentially driven by decreased BMP-1. The lack of this coupled effect with higher dosage suggests BAPN binding alone may be responsible for the downregulation of LOX and BMP-1, or that other contributing factors exist. BMP-1 interacts with periostin in dense connective tissues such as the periosteum, aorta, and tendons[4], and BMP-1 is necessary to activate LOX[3]. Thus, future research addressing BAPN effects on other mechanisms would provide critical information. The lack of an effect on any target gene at a low concentration encouraged an exploration into potential low dosage effects on matrix output. AFM analysis revealed a distribution of D-spacing values similar to that seen with treatment *in vivo*[1] and an upward shift in D-spacing with treatment. Similarly, a low dosage of BAPN decreased the ratio of mature to immature crosslinks, seen elsewere[7], driven by a reduction in the mature crosslink HP. Together, these results highlight that BAPN is able to produce post-translational nanoscale structural changes to the collagen matrix in absence of a response in expression of genes relating to collagen synthesis or enzymatic crosslinking. The study also shows that AFM and FTIR analysis can be used to characterize collagen produced *in vitro* in its native state.

ACKNOWLEDGEMENTS

This work was supported by NIH K25AR067221-01A1, the IUPUI Office of the Vice Chancellor for Research and the IUPUI Biomechanics and Biomaterials Research Center.

REFERENCES

1. Hammond, M. A. *Bonekey Rep.* 4:645, 2015.
2. Kemp, A. D. *J. Struct. Biol.* 180:428–38, 2012.
3. Maruhashi, T. *J. Biol. Chem.* 285:13294–13303, 2010.
4. Norris, R. A. *J. Cell. Biochem.* 101:695–711, 2007.
5. Paschalis, E. P. *J. Bone Miner. Res.* 16:1821–8, 2001.
6. Pfaffl, M. W. *Nucleic Acids Res.* 30:36e–36, 2002.
7. Turecek, C. *Calcif. Tissue Int.* 82:392–400, 2008.
8. Wallace, J. M. *Langmuir* 26:7349–54, 2010.

EFFECTS OF LOW-INTENSITY ULTRASOUND WITH NANOPARTICLE CONCENTRATION ON STEM CELL OSTEOGENESIS AND CHONDROGENESIS

A. Qin (1), M. Hu (1), Y-X. Qin (1)

(1) Department of Biomedical Engineering
Stony Brook University
Stony Brook, N□ 11794, USA

INTRODUCTION

Musculoskeletal disorders and diseases are major health problems affecting millions of people. Approximately 1.5 million individuals suffer a fracture caused by bone disease annually [1,2]. Osteoporosis and related fractures reduce the lifespan of patients. An estimate of 50% of patients who suffer osteoporosis fracture die within two years due to associated complications. Osteoarthritis (OA) is one of highly prevalent joint disorders that lead to joint dysfunction and physical disability in US. It is predicted to affect 67 million people by 2030 [3,4] Low-intensity pulsed ultrasound (LIPUS), as mechanical stimuli, has been observed to promote osteogenesis in mesenchymal stem cells (MSCs) and effects on chondrogenesis [5]. Combined with nanocomposite scaffolds it is further promoted. However, the application of nanoparticle is still largely unknown. The objective of this study was to evaluate whether there is a positive synergic effect between nanoparticles and ultrasound in osteogenesis and chondrogenesis of MSCs. The combined use of a high and low ultrasound and nanoparticle dosages were tested.

METHODS

Mouse mesenchymal stem cells (C3H) were incubated in either a solution of 0, 10, or 50 ug/ml of HGC (hydrophobic ally modified glycol chitosan) nanoparticles in media. The media used was either a collagenic differentiation media, osteogenic differentiation media, or media made out of a mixture of 89% alpha-MEM (Gibco, Life Technologies, Grand Island, NY, USA), 10% fetal bovine serum (Gibco, Life Technologies) and 1% Penicillin-Streptomycin (Gibco, Life Technologies). The Sonicator 740® (Mettler Electronics Corporation, CA, USA) with gel-coupled plane wave US applicator (ME7410) was used to apply ultrasound stimulation to the cultured dishes (n=4) (Table 1). Groups included a control with no treatment, experimental group that only had an ultrasound of 0.2 or 0.4W/cm^2, experimental group that only had a nanoparticle concentration of 10 or 50 ug/ml, and an experimental group that had a combination of both ultrasound and nanoparticle treatments. The experiment group received treatment for 20 minutes per day for 5 days per week. Low intensity group received a sinusoidal ultrasonic pulse at 1 ms/MHz with a repetition frequency of 100 Hz and spatial-average and temporal-averaged (SATA) intensity of 20 mW/cm^2, with 20% duty cycle. Medium intensity group received a sinusoidal ultrasonic intensity of 40 mW/cm^2, with 20% duty cycle. An ALP assay test was used to test the expression of osteocytes that have differentiated in the osteogenic differentiation media using the SpectraMax i3 plate reader machine set at a wavelength of 405 nm. An alcian blue staining was done to image the collagen activity under a Zeiss microscope at a magnification of 10x. The Graph Pad Prism 6 software was used to perform one-way ANOVA and Tukey's pairwise multiple comparisons test on all groups to calculate significance, and create graphs. Graphs are presented in a mean ± standard deviation format, and P-values < 0.05 were considered significant.

Table 1. Combination of LIPUS magnitude and nanoparticle.

Control	0.2 W/cm^2	0.4 W/cm^2
10 ug/ml	0.2 W/cm^2+10 ug/ml	0.4 W/cm^2+10 ug/ml
50 ug/ml	0.2 W/cm^2+50 ug/ml	0.4 W/cm^2+50 ug/ml

RESULTS

ALP Assay Analysis: Osteogenesis was enhanced by the treatment of the 0.4W/cm^2 ultrasound and 50ug/ml nanoparticle either single or combined treatment. The 0.4W/cm^2 of ultrasound, 50 ug/ml nanoparticle, and combination of the two showed an increase of 1%, 3%, and 2% ALP enzyme activity, respectively (Fig. 1). *Collagen Analysis:* For the chondrogenesis analysis, the representative collagen images showed that combination of 50ug/ml of nanoparticles and 0.4W/cm^2 had significant increase in collagen content compared to the control and other groups (Fig. 2a, 2b). The result showed that combination of 50ug/ml of nanoparticles and either 0.2W/cm^2 and 0.4W/cm^2 had significant increase in collagen content compared to other groups.

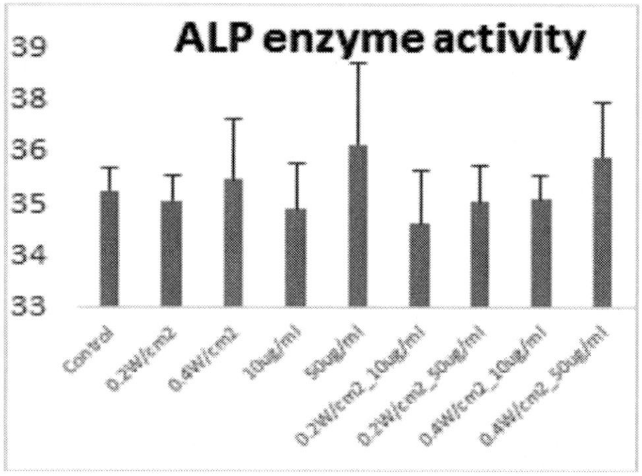

Figure 1. ALP enzyme activity. The 0.4W/cm^2 of ultrasound, 50 ug/ml nanoparticle, and combination of the two showed an increase of 1%, 3%, and 2%, respectively.

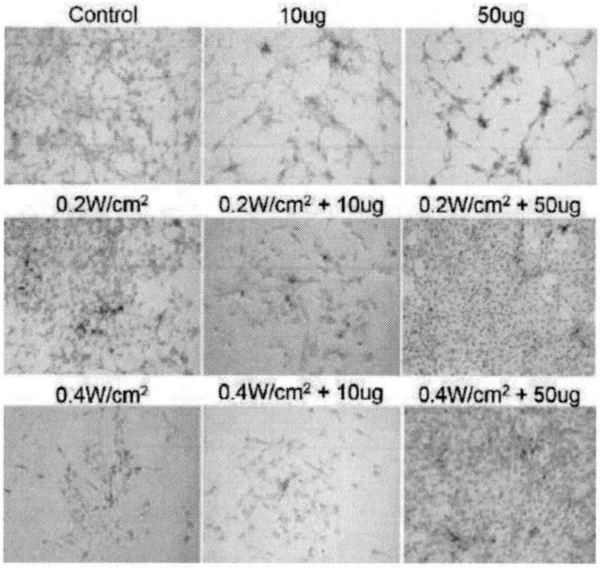

Figure 2a. Alcian blue staining. The combination of 50ug/ml of nanoparticles and either 0.2W/cm^2 and 0.4W/cm^2 had significant increase in collagen content compared to others.

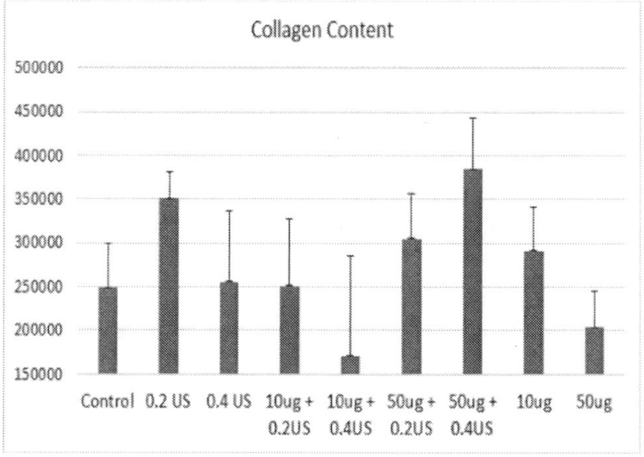

Fig. 2b. Collagen expression showed that combination of 50ug/ml of nanoparticles and 0.4W/cm^2 had significant increase in comparison to control and ultrasound or nanoparticle alone groups.

DISCUSSION

While the ultrasound and nanoparticle treatments appeared to be ineffective in the normal growth of cells, the differentiation of stem cells towards osteogenesis showed promising response to either ultrasound and nanoparticles specific dosages. The combination of ultrasound of 0.4 W/cm^2 and 50 ug/ml was outcompeted by the 50ug/ml alone nanoparticle indicating optimized ultrasound or nanoparticle dosage alone would enhance MSC osteogenesis. However, the chondrogenesis proliferation image under the combination of 50ug/ml of nanoparticles and either 0.2W/cm^2 and 0.4W/cm^2 have shown promising in chondro-differentiation. In this study, we have developed a novel combination treatment using both ultrasound and nanoparticles for MSC osteogenesis and chondrogenesis. In this innovative approach, combined effect seems to promote stem cell growth and differentiation. Because ultrasound can provide treatment noninvasively, to develop an injectable nanoparticle along with LIPUS treatment may create a unique approach for potential clinical application.

Significance: This work provides better understanding of ultrasound and nanoparticle treatment on osteogenesis and chondrogenesis of stem cells. It demonstrated that integrated nanoparticle with low intensity acoustic radiation force could enhance both osteogenesis and chondrogenesis of stem cells if optimized amplitude of ultrasound such as 0.4W/cm^2 with 50 ug/ml HGC nanoparticles.

ACKNOWLEDGEMENTS

This work is kindly supported by the National Institute of Health (R01 AR52379 and R01AR61821), and the National Space Biomedical Research Institute through NASA contract NCC 9-58.

REFERENCES

[1] Zhang D, Qin YX. et al., Bone. Jun;87:161-8, 2016.
[2] Chan CK et al., Br Med Bull. Aug 24, 2016.
[3] Joseph A et al., Advanced Drug Delivery Rev, 58: 150-167, 2006.
[4] Mow VC et al., Biomaterials, 13: 67–97, 1992.
[5] Uddin SM et al., Osteoarthritis Cartilage. Jun 27, 2016

SB³C2017
Summer Biomechanics, Bioengineering and Blotransport Conference
June 21–24, 2017, Tucson, AZ, USA

EFFECT OF EXTRACELLULAR MATRIX ON SMOOTH MUSCLE CELL MIGRATION BEHAVIOUR

Toshiro Ohashi (1), Yasufumi Hagiwara (2)

(1)Faculty of Engineering
Hokkaido University
Sapporo, Hokkaido, Japan

(2) Graduate School of Engineering
Hokkaido University
Sapporo, Hokkaido, Japan

INTRODUCTION

Adherent cells are known to adhere to their extracellular matrix at focal adhesions, where actin filaments are indirectly connected to integrins via multiple proteins. Integrins are in contact with the extracellular matrix mediating various cellular signals, possibly leading to modulations of a variety of physiological functions in cell proliferation, migration, etc. Integrins are obligate heterodimers containing two distinct chains, called the α and β subunits, very specific to the types of extracellular matrix such as fibronectin, vitronectin, type I collagen, etc. In previous studies [1, 2], it has been found that fibronectin and vitronectin might direct smooth muscle cell phenotype into synthetic type while type I collagen contractile type. It is also known that inhibition of α-SMA expression, a marker of contractile phenotype, could enhance cell migration [3]. Therefore, it is speculated that extracellular matrix-mediated integrin expression may modulate physiological functions of smooth muscle cells through alterations in phenotypes.

In this study, cell migration assays was performed on smooth muscle cells cultured on three different types of extracellular matrix such as fibronectin, vitronectin, and type I collagen. The total cell migration length is evaluated for 24 h under a microscope.

METHODS

Cell preparation – Smooth muscle cells from bovine thoracic aortas were purchased (Cell Applications, USA). Cells were seeded in tissue culture flasks with DMEM + 10% FBS. Cell populations from the 4th to 9th generation were studied.

Migration assays – The experimental protocol is shown in Fig. 1. For experiments, the tissue culture dishes were coated with fibronectin, vitronectin, and type I collagen at 50 μg/ml. In order to create a space that allows cells to migrate, a block was made from polydimethylsiloxane (PDMS, Dow Corn-ing, Japan) and was placed in the center of the dishes. Cells were then cultured in the dishes to induce the phenotypic modulation. After reaching confluent, the dishes were transferred into a CO_2 chamber maintained at 37°C and 95%Air/5%CO_2 under an inverted microscope (IX71, Olympus, Japan). After removing the PDMS block, cell migration into the space was observed.

Images were captured with a digital CCD camera every 10 min up to 24 h. The total cell migration length was evaluated with the ImageJ software (National Institute of Health, USA).

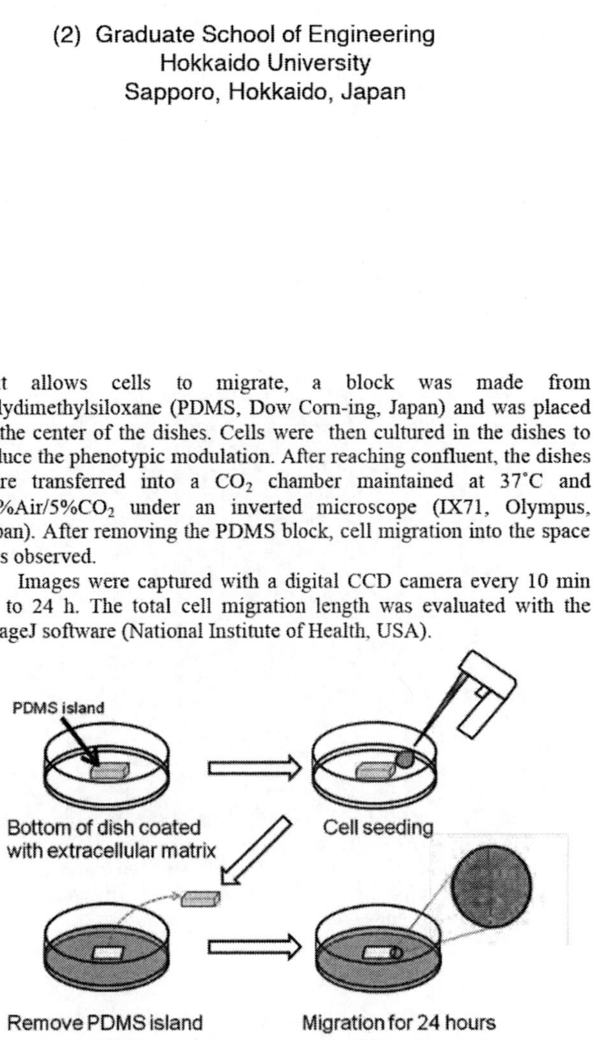

Fig. 1 Experimental protocol. Cells are cultured on a dish coated with the three different extracellular matrix. When they reach confluent, the PDMS block is removed to allow cells to migrate.

RESULTS AND DISCUSSION

Figure 2 shows cell migration on the three different extracellular matrix at 0 h, 12 h and 24 h. The dashed lines indicate the edges of the PDMS blocks that were placed before cell migration while the solid lines indicate the front lines of cells after migration. Under all the three conditions, cells randomly migrated but, on the whole, migrated towards the space where the PDMS block had been placed. Cell trajectory was then manually measured, as shown in Fig. 3. The total length of cell migration is summarized in Fig. 4. For fibronectin, cells migrate for 810 ± 332 μm, while for vitronectin and type I collagen, cells migrate for 658 ± 255 μm and 573 ± 198 μm, respectively. Cell migration on fibronectin was significantly enhanced compared to the other two extracellular matrices.

In a previous study [1], fibronectin-coated substrate directed smooth muscle cell phenotype into synthetic phenotype. Moreover, cellular traction force measurements revealed that traction forces were higher on type I collagen than those on fibronectin and vitronectin. This previous result is not in consistent with the present migration behaviour because high traction forces do not produce high migration rate. This indicates that not only the magnitude of traction forces but also the direction of traction forces may contribute to the driving forces of cell migration to a specific direction. It is also known that α5β1 integrin specifically binding to fibronectin may enhance cell proliferation. In fact, in a separate study, we measured the density of cells from region to region in Fig. 2 and found that the rate of cell proliferation was higher for fibronectin than vitronectin and type I collagen. Taken together, fibronectin could modulate smooth muscle cell phenotype into synthetic type, enhancing cell migration as well as proliferation.

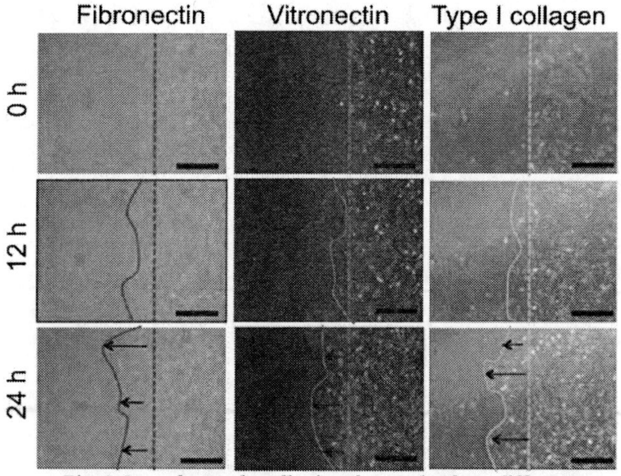

Fig. 2 Smooth muscle cell migration on the three different extracellular matrix-coated substrate. Cells cultured on fibronectin more migrate into the space than those on vitronectin and type I collagen. Scale bars: 500 μm.

CONCLUSIONS

This study performs cell migration assays with different types of extracellular matrix on smooth muscle cells. The result indicates that fibronectin-mediated smooth muscle cell phenotype, synthetic type, might enhance cell migration as well as proliferation, possibly via modulation in integrin formation.

REFERENCES

[1] Ohashi T, Ichihara H, Sakamoto N, Sato M (2008) Specificity of traction forces to extracellular matrix in smooth muscle cells. 13th In-ternational Conference on BioMedical Engineering, Singapore (in CD-ROM).

[2] Hedin U, Bottger BA, Forsberg E, Johansson S, Thyberg J (1988) Diverse effects of fibronectin and laminin on phenotypic properties of cultured arterial smooth muscle cells. J Cell Biol 107: 307-319.

[3] onnov-Jessen L, Petersen OW (1996) ADP-ribosylation of actins in fibroblasts and myofibroblasts by botulinum C2 toxin: influence on microfilament morphology and migratory behavior. Electrophoresis 17: 1776-1780.

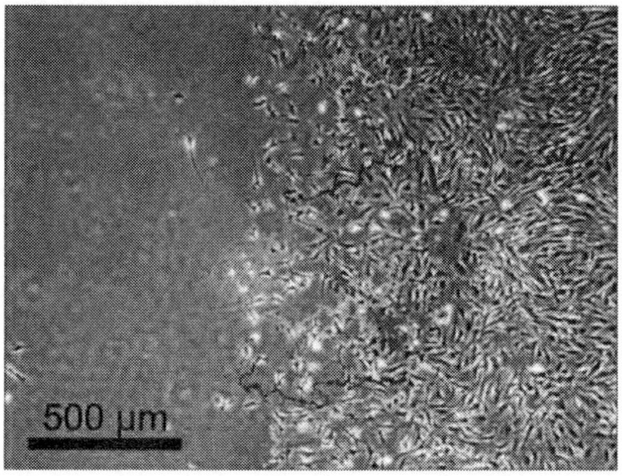

Fig. 3 Cell trajectory was manually measured to determine the total length of cell migration.

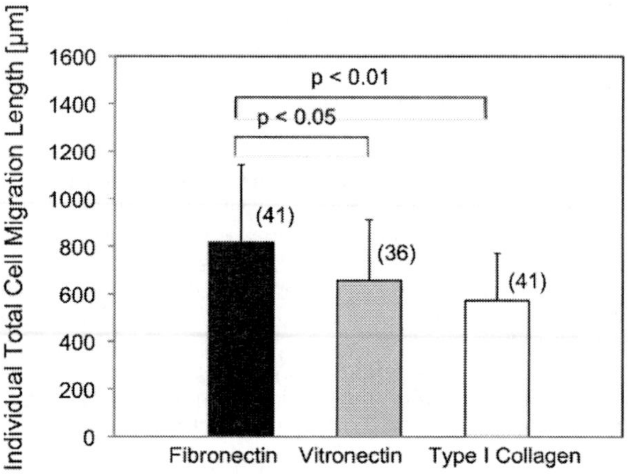

Fig. 4 The total length of smooth muscle cell migration on the three different extracellular matrix-coated substrate. Data are represented by mean + SD. ( ) represents the number of cells.

PREDICTING INDIVIDUAL CARDIOMYOCYTE FIBER ORGANIZATION IN SPATIALLY CONSTRAINED CELLS

William F. Sherman (1,2), Anna Grosberg (1,2,3,4)

(1) Center for Complex Biological Systems
University of California, Irvine
Irvine, CA, USA

(2) The Edwards Lifesciences Center for Advanced Cardiovascular Technology
University of California, Irvine
Irvine, CA, USA

(3) Department of Chemical Engineering and Materials Science
University of California, Irvine
Irvine, CA, USA

(4) Department of Biomedical Engineering
University of California, Irvine
Irvine, CA, USA

INTRODUCTION

A functional, healthy heart relies on coordination across multiple length scales [1, 2]. Since the use of stem-cell derived cardiomyocytes is an essential component in cardiac tissue engineering, it is important to be able to predict how the intracellular architecture of individual cardiomyocytes responds to changes in boundary conditions.

Cells operate by converting chemical energy into mechanical forces, which are then transmitted through sarcomeres, the force-producing units of the myofibrils. As a result, force is compounded as numerous sarcomeres synchronize along each myofibril [3, 4]. The net force can then be transmitted to the cellular membrane through focal adhesions, which anchor the myofibrils to the extracellular matrix. The coordination continues at larger length scales with synchronized contractions of cells within the laminar sheets of the heart allowing it to pump blood throughout the body [3, 4]. Conversely, structural changes in the heart at the millimeter and centimeter scale result in dramatic changes to individual cardiomyoctes at the micrometer scale. One such change is the elongation and deformation of individual cells in dilated cardiomyopathy [4]. In turn, this change in cellular geometry influences the organization of myofibrils (Figure 1A-C). To understand heart disease or to engineer heart tissue, it is important to be able to predict how this structural remodeling affects the internal mechanisms of the individual cells.

Several predictive models for self-assembly have been proposed (review [5]). However, predictive models for whole cardiomyocyte myofibrilogenesis have relied on a key assumption: they assumed the dimensions of the myofibrils are significantly smaller than the dimensions of the cell. This assumption allowed for models to be created which did not consider the size of individual fibers. We hypothesize that this is the reason model predictions of the fiber distribution in highly elongated cells fail. Experimentally, it has been observed that in cases where a competition for space exists, some fibers bend to accommodate spatial constraints. This problem is demonstrated in Figure 1A, where an elongated cell has a vertical width that is on the same scale as the nucleus. The other limitation to the models is their inability to predict the placement and length of z-lines. To simulate this, we must be able to model a larger range of length scales.

To collectively address these model limitations, our aim is to include these spatial scales and explore why their inclusion is essential in predicting self-assembly in elongated cells. Moreover, this will provide a deeper understanding of the physical meaning of the parameters used and how they influence cellular dynamics.

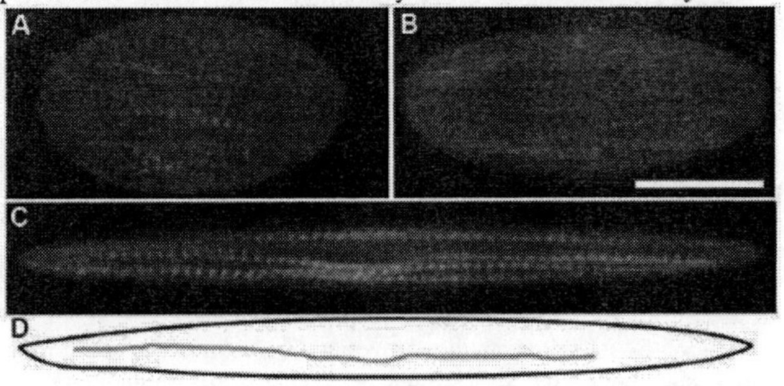

Figure 1: Immunostained actin (green) demonstrating fibers reorganized in a cell with aspect ratio 1.65 (A), 2.38 (B), 8.36 (C), and an outline of the cell in (C) with the trace of a bent fiber (D). Scale bar is 20 μm.

METHODS

We began by incorporating scaling terms into the Grosberg et al. model of myofibrillogenesis [6]. It was possible to use many of the previously fitted parameters by appropriately scaling them based on the properties of the experiments originally used to fit them. New parameters were included to maintain dimensional homogeneity, and their values were determined using previous experimental data or obtained from the literature. It was then necessary to consider the physical meaning of each term in the model to determine which parameters should be allowed to vary based on the scaling equations and which should be held constant. The new model was tested and verified by checking for consistency in the results when subjected to the same conditions as the previously validated model. Simulations were run by discretizing the equations and solving using MatLab. Resulting image outputs were compared to experimental data, which was collected using previously described methods [1]. Limiting cases were constructed to test the validity of the scaling properties of the model.

RESULTS

Spatial considerations were incorporated into the Grosberg et al. model through rescaling:

$$F(\boldsymbol{x},t) = f_p \rho_{sat} R_p(\boldsymbol{x}) \int_{\Delta(\boldsymbol{x})} \rho_{sat} R_p(\boldsymbol{x}')[\boldsymbol{x}' = \boldsymbol{x}] d^2\boldsymbol{x}'$$
$$= f_n \rho_{sat} R_n(\boldsymbol{x}) \int_{\Delta(\boldsymbol{x})} \rho_{sat} R_n(\boldsymbol{x}')[\boldsymbol{x}' = \boldsymbol{x}] d^2\boldsymbol{x}' \quad (1)$$

$$R(\boldsymbol{x}) = R_p(\boldsymbol{x}) = R_n(\boldsymbol{x})$$
$$= \frac{\rho_p(x,t)}{\rho_0 = \rho_p(x,t) = \rho_n(x,t)} = \frac{\rho_n(x,t)}{\rho_0 = \rho_p(x,t) = \rho_n(x,t)}. \quad (2)$$

This new force equation describes the net force acting on a focal adhesion and incorporates an integrin-saturation term, ρ_{sat}, which is related to the total available connections in a unit area. The quantities f_p and f_n denote the force per number of fibers per unit length for pre-myofibrils and nascent myofibrils, respectively, where the density of integins bound to pre-myofibrils and nascent myofibrils are denoted ρ_p and ρ_n. Using these two distinctions, we are able to write the fraction of force bearing myofibril connections, described by the function R, as the sum of the fraction of force bearing pre-myofibril connections and nascent myofibril connections. By rewriting the force in this way, we are able to identify key parameter relationships and study how fiber dimensions influence the focal adhesion distributions. For instance, ρ_0, which controls when a unit area is considered saturated with integrins, can be reinterpreted to reveal a connection between the saturation mechanism and the size of a connector complex anchoring the fiber into the focal adhesions.

Using this saturation control parameter, we are able to derive a first estimate for the size of a single fiber connector, A_{con}:

$$A_{con} = \frac{1}{2\rho_0}. \quad (3)$$

This allows us to estimate the size of individual connectors using a known parameter. We can also test the limits of our model by considering large and small cells, which was not possible with the previous model. As one extreme case, we simulated the bound integrin density for an unrealistically small cell (Figure 2Aiii) that is on the same order as A_{con} and compared the result to simulations for a normal sized cell (Figure 2Aii) and a large sized cell (Figure 2Ai).

Since ρ_0 is linked to the size of a fiber connector, we further tested our model by simulating the scenario of an increased or decreased fiber connector size (Figure 2B). As expected, increasing or decreasing the individual fiber connector size by manipulating ρ_0 changes the concentration of bound integrins in regions of large curvature (compare Figure 2B to Figure 2Aii).

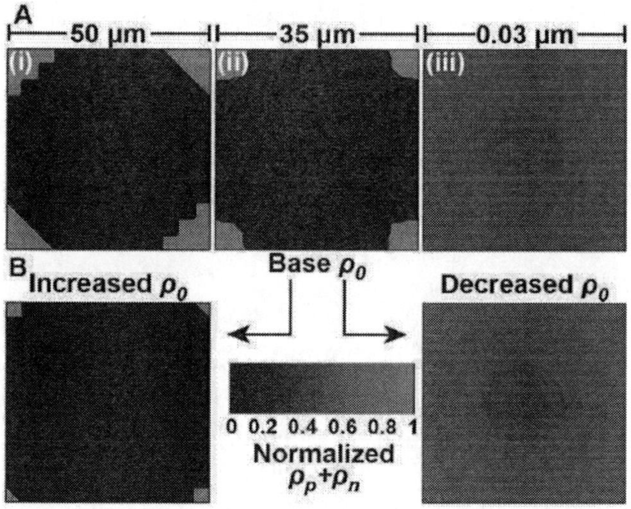

Figure 2: (A) Simulated normalized bound integrin density at steady state for (i) 2500 μm^2, (ii) 1250 μm^2, and (iii) 0.001 μm^2 cells. (B) For the 1250 μm^2 cell, ρ_0 was increased (left) and decreased (right) by a factor of 10.

DISCUSSION

By incorporating spatial dimensions into a previously validated model, we are able to test the influence of cell size and geometry on focal adhesion and fiber distribution. Rescaling previously fitted parameters allows for the exploration of important parameter relationships. In particular, we altered the force equation to allow for integrin saturation as well as control over the saturation mechanism. This allows us to explore when saturation occurs and how changing the saturation control parameter influences bound integrin distribution.

Increasing ρ_0 corresponds to increasing the density of integrins that can accumulate in a unit area of the cell. This results in an increased saturation density threshold and decreased fiber connector size as described in equation 3. Smaller sized connectors coupled with increased saturation threshold yields a reduction in the number of fiber connectors that accumulate in regions of large curvature. On the other hand, a lowered threshold and increased fiber connector area leads to bound integrins being distributed more uniformly across the cell.

Previous models were not designed to explore these types of parameter relations. Our reformulation gives new insights into how the integrin saturation mechanism influences focal adhesion dynamics for different sized cells. We also now have the ability to include other spatial dimensions, vary the aspect ratio of the cell and include the nuclei.

ACKNOWLEDGEMENTS

We thank all members of the Cardiovascular Modeling Laboratory for contributing useful discussion and the use of experimental data from [1].

REFERENCES

[1] Drew, N. et al., *Journal of Biomechanical Engineering* 138, 2016
[2] Kuo, P. et al., *The American Journal of Pathology* 181(6), 2012
[3] Knight, M. et al, *Biophysical Journal* 110(7), 2016
[4] Gerdes, A. et al., J Mol Cell Cardiol 27, 1995
[5] Drew, N. et al., *Cardiomyocytes: Methods and Protocols*, Methods in Molecular Biology 1299, 2015.
[6] Grosberg, A. et al., *PLoS Comput Biol* 7(2), 2011.

CELLS ALIGN ALONG TOPOGRAPHICAL CUES AS A RESULT OF FREE ENERGY MINIMIZATION AND HOMEOSTASIS

Tommaso Ristori (1,2), Siamak S. Shishvan (3), Gitta A.B.C. Buskermolen (1), Frank P.T. Baaijens (1,2), Sandra Loerakker (1,2), Vikram S. Deshpande (3)

(1) Department of Biomedical Engineering
Eindhoven University of Technology
Eindhoven, The Netherlands

(2) Institute for Complex Molecular Systems
Eindhoven University of Technology
Eindhoven, The Netherlands

(3) Department of Engineering
University of Cambridge
Cambridge, United Kingdom

INTRODUCTION

Experimental studies have shown that cells can remodel their cytoskeleton and orientation in response to topographical stimuli. For example, in vitro, when seeded on substrates with micro/nanogrooves[1-9] or linear patterns made with microcontact printing[10-13], cells tend to realign their stress fibers and main orientation along the direction of these patterns. This phenomenon, known as contact guidance, is often observed also in vivo, where cells align along topographical cues provided by collagen, the most abundant protein in our bodies. Thus, a clear understanding of these phenomena is of great importance.

Recently, Vigliotti et al.[14] proposed a computational model able to predict the in vitro experimental results. In their study, they proposed that cells align along topographical patterns as a result of a complex interplay between focal adhesion intracellular signaling and stress fiber traction forces and remodeling. Their biomechanical model can potentially explain the mechanisms that determine the realignment of cells along topographical cues, but it cannot explain the preference of cells for these directions and the variability present in experimental results.

To overcome this limitation, in this study we hypothesized that cells reorient along the anisotropic cues provided by micro/nanogrooves or printed lines as a result of free energy minimization and homeostasis. Our hypothesis was tested by computationally simulating cellular shape and stress fiber remodeling for cells seeded on a substrate with anisotropic patterns (lines). The computational results were compared with experimental observations.

METHODS

A computational framework was developed based on the following hypotheses:

- the free energy of a cell can be computed as a summation
$$G_{tot} = G_{elas} + G_{cyto} + G_{adh}, \quad (1)$$
where G_{tot} is the total free energy, G_{elas} the elastic energy, G_{cyto} the energy associated with the cytoskeleton, and G_{adh} is the energy associated with the adhesions between cell and substrate;
- cells are more likely to assume configurations with low free energies, such that
$$P^{(j)} = \frac{\exp(-\beta G_{tot}^{(j)})}{\sum_j \exp(-\beta G_{tot}^{(j)})}, \quad (2)$$
where $P^{(j)}$ is the probability for a cell to acquire a configuration (j), $G_{tot}^{(j)}$ is the total energy associated with that configuration, and β is a parameter characterizing the substrate where cells are spreading;
- cells remodel their shape and cytoskeleton while keeping the average of the free energies associated with all their possible configurations (labeled G_{avg}) to a constant homeostatic value, which we indicate with G_0.

For a cell having a certain shape and internal deformations, the associated total energy G_{tot} was computed as follows. G_{elas} was obtained as the elastic energy of the cell resulting from a two-dimensional material model obtained by adapting the Ogden model. G_{cyto} was given by a previously proposed thermodynamically motivated model for stress fiber remodeling[15]. G_{adh} was computed by assuming that the cell-surface adhesions can be approximated as springs and that $k_{adh} \gg k_{noa}$, where k_{adh} and k_{noa} are the elasticity constants associated to substrates with adhesive and non-adhesive patterns respectively.

For each substrate, G_{avg} was computed with a Monte Carlo method, varying the shape and internal deformations of a cell and computing for each configuration the associated $G_{tot}^{(j)}$. The parameter β, characterizing the substrate in exam, was fitted in such a way that $G_{avg} = G_0$, where G_0 was calculated as the energy of a free standing cell.

Each simulation resulted in a collection of accepted configurations of cellular shapes and deformations. Each cellular shape was then fitted with an ellipse, from which the main cellular orientation was extracted. These extracted data were then compared with results from experiments performed in our lab with the microcontact printing technique[16].

RESULTS

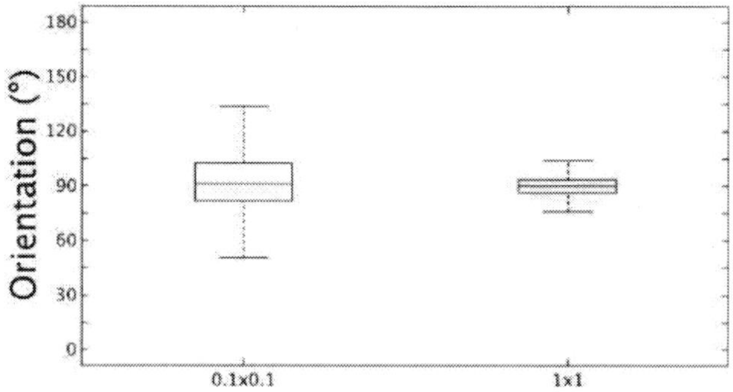

Figure 1: Boxplot of cellular orientations resulting from computational simulations of cells on substrates with anisotropic patterns (lines). 90° orientation corresponds to the direction of the lines. The width of adhesive lines was equal to the one of non-adhesive lines. Their non-dimensional values are reported along the x-axis.

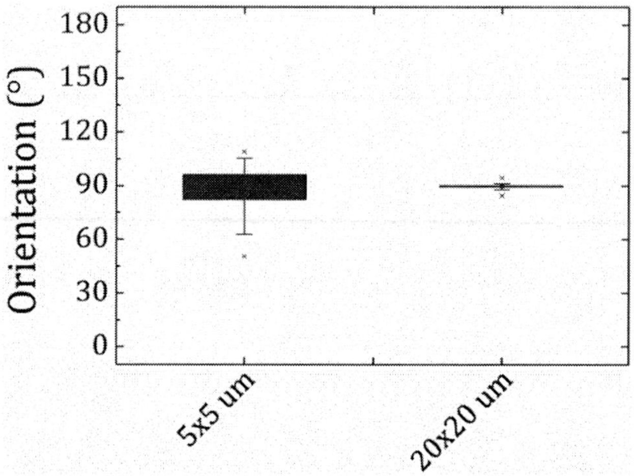

Figure 2: Boxplot of cellular orientations resulting from experiments of cells on substrates microcontact printed with lines. 90° orientation corresponds to the direction of the lines. The width of adhesive lines was equal to the one of non-adhesive lines. Their values are reported along the x-axis.

Computational simulations of cellular remodeling on substrates with linear adhesive patterns were performed for various widths of adhesive lines and spacing. These simulations predicted that cells prefer to align along the direction of topographical patterns for both relatively small and large widths of lines (Fig. 1), in agreement with experimental observations (Fig. 2).

Increasing the width of adhesive lines and spacing induced a greater degree of alignment (Fig. 1). For smaller lines the 99.3% of cells was oriented between 50° and 134°, where 90° is the direction of adhesive lines. This interval of cellular orientations decreased for larger lines, for which the 99.3% of cells aligned between 75° and 104°. This trend was qualitatively in agreement with the experimental results, although in that case the observed intervals were smaller compared to the computational results (Fig. 2).

DISCUSSION

In this study, a possible explanation for the biophysical reasons for cells to align along linear patterns on substrates was proposed. We hypothesized that cells orient along topographical cues to minimize their total free energy and maintain a homeostatic value for this energy averaged over all the possible cellular configurations. Our hypothesis was tested by developing a computational framework that could compute the average of the free energy via a Monte Carlo approach, taking into account the energies associated with stress fibers, cellular adhesions, and elasticity of cells.

In accordance with experiments, computational simulations predicted that cells seeded on patterned substrates align along the direction of topographical cues, with the degree of alignment proportional with the width of adhesive and non-adhesive lines.

In a recent study, Livne et al.[17] computationally showed that cells (re)orient in response to mechanical stimuli to minimize their free energy. In the present study, we demonstrated that the same happens when cells are subjected to topographical stimuli, and that the variability present in the experimental results is due to the tendency of cells to conserve a homeostatic value for their free energy. In future research, a similar approach can be used to verify the validity of these hypotheses when both mechanical and topographical cues determine cellular remodeling.

ACKNOWLEDGEMENTS

This study was supported by the People Programme (Marie Curie Actions) of the European Union's Seventh Framework Programme [FP7/2007-2013/]; REA [grant agreement number 317512].

REFERENCES

[1] Den Braber, ET et al., *Biomaterials*, 17:1093–1099, 1996.
[2] Lamers, E et al., *Biomaterials*, 31:3307–3316, 2010.
[3] Walboomers, XF et al., *J Biomed Mater Res*, 46:212–220, 1999.
[4] Walboomers, XF et al., *J Biomed Mater Res*, 51:529–534, 2000.
[5] Sarkar, S et al., *Acta Biomater*, 1:93-100, 2000.
[6] Saito, AC et al., *Exp Cell Res*, 327:1-11, 2014.
[7] Dreier, B et al., *Am J Physiol Cell Physiol*, 305: C290-C298, 2013.
[8] Loesberg, WA et al., *J Biomech*, 28:3944–3951, 2007.
[9] Pilipchuk, SP et al., *Adv Healthc Mater*, 5:676-687, 2016.
[10] Alford, PW et al., *Integr Biol*, 3:1063-1070, 2011.
[11] Zimerman, B et al., *IEEE Proc Nanobiotechnol*, 151:62-66, 2004.
[12] Kalinina, S et al., *Biomaterials*, 29:3004-3013, 2008.
[13] Williams, C et al., *Biomaterials*, 32:5625-5632, 2011.
[14] Vigliotti, A et al., *J R Interface*, 12: 20141320, 2015.
[15] Vigliotti, A et al., *Biomech Model Mechanobiol*, 15:761-89, 2015.
[16] Chen, CS et al., *Science*, 276:1425-1428, 1997.
[17] Livne, A et al., *Nat Commun*, 5:3938, 2015.

SB^{3}C2017
Summer Biomechanics, Bioengineering and Biotransport Conference
June 21 – 24, Tucson, AZ, USA

DORMANCY-CAPABLE CANCER CELL ISOLATION VIA PHYSICAL PROLIFERATION INHIBITION

Julian A. Preciado[1], Samira Azarin[2], Emil Lou[3], Alptekin Aksan[1]
[1]Department of Mechanical Engineering, [2]Department of Chemical Engineering and Materials Science, [3]Hematology, Oncology and Transplant, Department of Medicine, University of Minnesota, Twin Cities
Minneapolis, USA

INTRODUCTION

Metastatic cancers account for approximately 90% of cancer-related deaths[1]. Certain metastatic cells can undergo dormancy, a stage in tumor progression when cancer cells cease proliferating but remain alive in a quiescent state. Recent studies suggest that metastasis-initiating cells enter dormancy in response to signals within their new microenvironment and can remain dormant for years; after which they regain a tumorigenic phenotype due to various poorly understood reasons[1]. Therapeutic targeting of dormant cancer cells (DCCs) is difficult because most treatments target only rapidly proliferating cells. Therefore, understanding dormancy and reawakening is crucial in developing therapeutics to kill DCCs, awaken them in a controlled fashion so that they can be targeted by chemotherapeutic agents, or maintain them in a dormant state to convert cancer into a manageable chronic disease.

Current methods to study dormancy include serum starvation in cell culture and animal models involving injections with cancer cell lines that exhibit dormant behavior *in vivo*. Serum starvation has been shown to induce quiescence in cancer cell lines that do not exhibit a dormant phase *in vivo*, showing no discernment between dormancy capable and incapable cells[2]. Current *in vivo* dormancy models are limiting since it can take several months for enough cells to metastasize for tissue examination. More recently, encapsulation techniques have attempted to bridge the gap between cell culture and *in vivo* conditions using a basement membrane extract system[2]. This was a successful model in differentiating dormancy-capable from dormancy-incapable cell lines. However, biologically based gels contain bioactive molecules such as fibrin, which have been shown to awaken dormant cells[18]. Animal-derived products also have an inherent batch-to-batch variability and lack the chemical and physical control to study specific aspects of dormancy. These limitations have restricted the isolation of DCCs to further understanding of dormancy and screen for therapeutic targets.

We have developed silica-PEG hydrogels in which cell migration and proliferation are inhibited. We hypothesize that inhibiting cell proliferation and movement creates a biphasic, stress-inducing stimulus where dormancy-capable cancer cell lines can enter quiescence whereas highly invasive, dormancy-incapable cells undergo cell death. To our knowledge, we are the first group to use physical proliferation inhibition to induce cancer cell dormancy.

METHODS

Silica-PEG (SPEG) gel synthesis: Gelling solution was made by mixing cell media, 4 arm PEG (Creative PEGWorks, Chapel Hill, NC) and NS125 silica nanoparticles (Nyacol, Ashland, MA) at a ratio of (0.9:0.1:0.001). Gelling solution was then mixed with tetrakis(2-hydroxyethyl) orthosilicates (Gelest, Morrisvalle, PA) at a 5:1 ratio. Cell solution was quickly added at the same volume ratio as gelling solution, poured into a multiwell plate, allowed to gel for 2 min and covered with media.

Alamar blue metabolic assay: 10 μL of Alamar Blue solution (ThermoFischer, Massachusetts, MA) was added to 44 μL of gel with 54 μL media. Fluorescence was measured 12 hrs. later using a spectrophotometer (560nm/590nm filter). Cells cultured cells on TCPS were used as positive controls. Cells killed prior to metabolic activity assay were used as negative controls.

Live/dead staining: Calcein AM and propidium iodide were used as a live and dead cell stains (ThermoScientific, Massachusetts, MA), respectively, at optimized concentrations.

Ki-67 staining: Cells were stained for Ki-67 (CellSignaling, Dancers, MA), a marker of actively cycling cells. Live cells were stained with calcein AM prior to fixation. Cells were fixed daily and permeabilized with 4% paraformaldehyde and 0.3% triton x-100, respectively. Cells cultured in TCPS under serum and serum-free conditions were used as positive and negative controls, respectively. The effect of glycemic media levels (4.5 and 1 g/L) on the transition into quiescence was also assessed. Gel staining control involved staining cells plated on SPEG to confirm no antibody diffusion limitations.

RESULTS

Metabolic Activity: Metabolic activity (MA) data indicated a significant distinction between MCF-7, a dormancy-capable cell line and MDA-MB-468, a dormancy-incapable cell lines as shown in **Fig. 2**. MCF-7 MA decreased to 30% of its initial value after 3 days of encapsulation and remained constant up to day 7. MDA-MB-468 MA had a sharp decrease in MA and within 2 days of encapsulation it reached the same levels as the dead controls.

Live/Dead staining: Viability staining confirmed MA data as seen in **Fig. 1**. 3T3 fibroblasts, MDA-MB-231, MDA-MB-468 cells are all dead after 2 days of encapsulation. A significant number of MCF-7 cells also die, however a subpopulation survives prolonged encapsulation.

Ki-67 expression: Ki-67 staining of live MCF-7 cells shows a steady decrease in Ki-67$^+$ cells during encapsulation **(Fig. 3),** indicating a transition into quiescence. This transition was independent of media glycemic levels. Cells cultured on TCPS under serum-free conditions had a significant decrease in Ki-67$^+$ cells as expected. However, the decrease was most pronounced with cells cultured in low glucose media, indicating serum starvation is not as a robust method to induce quiescence as physical proliferation inhibition.

DISCUSSION

Subpopulation isolation: Metabolic activity data showed a differential response to encapsulation between MCF-7 (dormancy-capable) and MDA-MB-468 (dormancy-incapable) cell lines. MDA-MB-468 cells lost all MA after 1 day of encapsulation, while MCF-7 MA levels off at ~30% of its initial value. Live/Dead staining reaffirmed MA results, showing that 3T3, MCF-7, MDA-MB-231 and MDA-MB-468 cells experience a significant drop in cell number. However, a subpopulation of MCF-7 cells survives prolonged encapsulation, whereas no 3T3, MDA-MB-231 and MDA-MB-468 cells are viable after 1 day of encapsulation. Elimination of 3T3 illustrates the ability of physical proliferation to eliminate normal cells in addition to dormancy-incapable cells. This isolation behavior can be attributed to a cell response to physical proliferation inhibition of the gel, where cells that cannot enter quiescence eventually undergoing cell death.

Induction of quiescence: Ki-67 staining confirmed that viable MCF-7 cells enter quiescence as a response to physical proliferation inhibition (**Fig. 3**). Viable MCF-7 cells had a gradual decline in Ki-67$^+$ cells and reach similar expression levels as the serum starved TCPS control. Serum starvation of cells grown on TCPS showed the most significant decrease in Ki67$^+$ cells when cultured under low glucose media. Interestingly, physical proliferation inhibition showed no dependence on glucose levels; thereby indicating physical proliferation inhibition as a more widely applicable method to induce quiescence than serum starvation.

These findings conclude that physical inhibition of proliferation can be used to **isolate** dormancy-capable cell lines in a **quiescent state**. The SPEG hydrogel's biocompatibility, high permeability to nutrients, and specificity for dormancy-capable cell lines make it an improved method to induce cancer cell dormancy compared to current approaches. Further studies will focus on the clinical applications of this platform to isolate and study patient-derived DCCs.

REFERENCES

1 Sosa MS, Bragado P, Aguirre-Ghiso JA. Mechanisms of disseminated cancer cell dormancy: an awakening field. *Nature reviews Cancer*. 2014;14(9):611-622.

2 Barkan D, Kleinman H, Simmons JL, et al. Inhibition of Metastatic Outgrowth From Single Dormant Tumor Cells by Targeting the Cytoskeleton. *Cancer research*. 2008;68(15):6241-6250.

3 Reátegui, E., Kasinkas, L., Kniesz, K., Lefebvre, M. A. & Aksan, A. Silica–PEG gel immobilization of mammalian cells. *J. Mater. Chem. B* 2, 7440–7448 (2014).

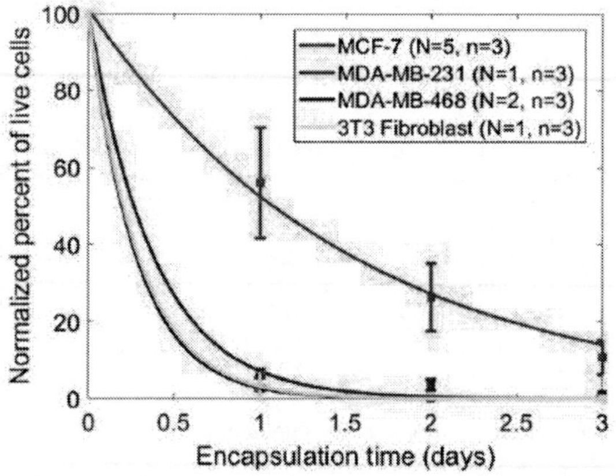

Fig. 1 Encapsulated cell viability confirmed metabolic activity data with MCF-7 cells (red) showing a slower decrease in viable cells and reaching a steady-state point compared to MDA-MB-231 (red) and MDA-MB-468 (black), which have no viable cells after 3 days of encapsulation.

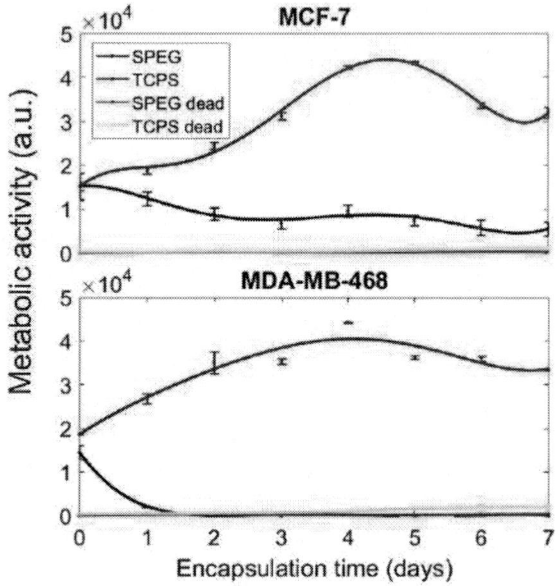

Fig. 2 Metabolic activity showed a variant response between MCF-7 (top) and MDA-MB-468 (bottom) when encapsulated in silica-PEG hydrogel (blue).

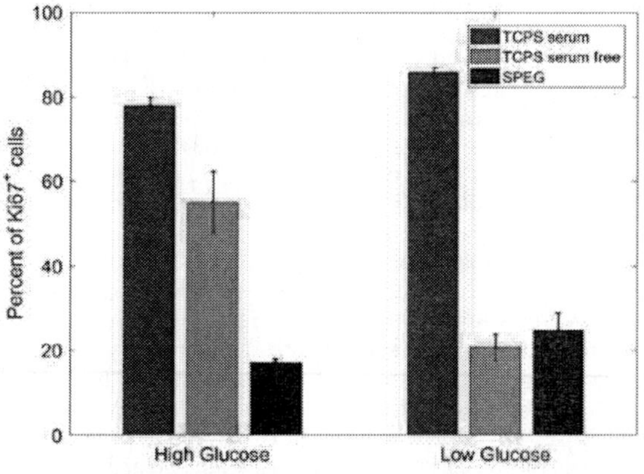

Fig. 3 Ki-67 staining showed an increase in the number of quiescent cells (Ki67$^-$) when encapsulated (black), under both high and low glucose conditions, reaching same levels as cells grown in TCPS under serum free conditions (green, right)

ACTOMYOSIN CONTRACTILITY REGULATES NUCLEUS PULPOSUS CELL BIOPHYSCIAL AND BIOMEHCANICAL PROPERTIES

Timothy D. Jacobsen (1,2), Paula Hernandez (1), Nadeen O. Chahine (1,2)

(1) Feinstein Institute for Medical Research, Manhasset, NY

(2) Hofstra Northwell School of Medicine, Hempstead, NY

INTRODUCTION

As part of its normal physiological functioning the intervertebral disc (IVD) and its central nucleus pulposus (NP) region are exposed to a number of different mechanical and biophysical cues. In addition to compressive forces, loading and unloading of the highly hydrated NP induces hydrostatic and osmotic forces on NP cells [1]. The regulation of cell volume and morphology in response to osmotic changes is important to NP function, and alteration of volume regulation parameters may play a crucial role in the loss of cellular function in IVD degeneration (DD) [2]. DD is characterized by elevated levels of pro-inflammatory cytokines [3]. Inflammatory stimulation of NP cells within the disc can lead to a catabolic breakdown of disc ECM and consequentially to a positive feedback of further loss of tissue integrity and further inflammation [4]. We have previously shown that inflammatory stimulation of isolated NP cells in 2D leads to significant and irreversible changes in cellular biomechanical properties which consequently effects the response of cells to environmental cues and may perpetuate degenerative changes. These changes include increased hydraulic permeability (Lp), increased cell size, and disruption of the actin cytoskeleton [5]. The actomyosin network and actomyosin contractility is regulated by Rho/ROCK pathway and is known to regulate cell morphology [6]. Indeed we have shown that inhibition of actomyosin contractility with the ROCK inhibitor (Y27632) and a non-muscle myosin-II inhibitor (blebbistatin) mimics inflammatory induced changes to cell biophysical properties [7] In this study in order to further investigate the role of actomyosin contractility on NP cell biophysical properties, we tested the effect of increasing contractility on cell biophysical properties. We hypothesize that increased actomyosin contractility will have an inverse effect on cell biomechanical properties compared to those seen with inflammatory stimulation, and will mitigate the effects of inflammatory stimulation on cell mechanobiology. To test this hypothesis, cells were treated with a Rho pathway activator (CN03) in the presence or absence of pro-inflammatory cytokines. We evaluated the effect of CN03 on Myosin light chain (MLC) phosphorylation as an indicator of actomyosin contractility [8], as well as on cell biophysical properties including hydraulic permeability, cell size, and cell elastic modulus.

METHODS

NP Cell Isolation: NP tissue was isolated from young bovine lumbar discs. Cells were expanded for 1-2 passages in DMEM+10% FBS, and cultured in 3D culture in 1.2% alginate beads to promote round morphology. **Treatments:** Cells were exposed to the following treatments for 24h: (1) untreated (2) 10ng/ml TNF-α (3) 1ug/ml Rho Activator CN03 (4) 1ug/ml CN03 + 10ng/ml TNF-α. **Immunofluorescence:** Beads were fixed, permeabilized, and immunostained with α-pMLC primary antibodies and counterstained with Draq5 nuclear stain. Cells were released from alginate beads and run through an Amnis Imagestream X Mark II Imaging Flow Cytometer that images individual cells in a high throughput manner (N~40,000). pMLC staining intensity was determined for individual cells, averaged for each group, and normalized to untreated group. **Osmotic properties:** Cells were seeded into a custom osmotic loading microfluidic chamber [5, 9] where they were allowed to adhere to a Poly-D-Lysine coated slide for 30 minutes and then placed under an inverted microscope for observation. Step hyper- (333 to 466 mOsm/L NaCl) and hypo- (466 to 333mOsm/L NaCl) osmotic loads were applied at room temperature. Volume response was analyzed using multiphasic mixture theory [10,11] to compute material properties: hydraulic permeability (L_p) and osmotically active intracellular water content (φ_w). Cell equilibrium was also observed in cells seeded within the microfluidic changes as 333mOsm/l NaCl.

Atomic Force Microscopy: Force indentation curves were obtained from individual, rounded, live NP Cells using an Asylum MFP-3D BIO atomic force microscope with spherical tip probes. Obtained curves were fitted to a Hertzian model in order to determine cell elastic modulus. Data was analyzed with 1-way ANOVA and Fisher LSD post-hoc test, with $p<0.05$ significant.

RESULTS

Stimulation with TNFα significantly reduced pMLC staining in NP cells confirming that actomyosin contractility decreases with inflammatory stimulation (Fig 1). CN03 significantly increased actomyosin contractility as indicated by increased pMLC staining, and was able to mitigate the effects of TNF-α (Fig 1). TNFα stimulation caused an increase in L_p under hyper-osmotic loading (Fig 2), in agreement with our previous findings [5]. Furthermore treatment with CN03 increased L_p and mitigated the effects of TNF-α. Additionally CN03 increased circularity observed by actin staining and mitigated the effect of TNFα. CN03 was not found to have a direct effect on cell radius or cell elastic modulus but did mitigate the effect of TNF-α on these properties (Fig 3). Lastly the relationship between actomyosin contractility and biophysical properties we observed by comparing the values of all groups. A significant correlation was observed between pMLC and L_p (Fig 4) as well as between pMLC and circularity and between L_p and circularity (data not shown).

DISCUSSION

The goal of this study was to examine the effects of Rho activity in regulating single cell biophysical properties. Our findings confirm that inflammatory stimulation with TNF-α decreases actomyosin contractility increases hydraulic permeability, decreases cell elastic modulus, decreased cell circularity, and increases cell size. Increasing actomyosin contractility through activation of the Rho pathway with CN03 decreased hydraulic permeability and increased circularity and was able to mitigate changes induced by TNF-α in all observed measures. Furthermore by comparing actomyosin contractility vs biomechanical properties measured among the various groups we observed significant correlation between actomyosin contractility and biomechanical properties including L_p and circularity. We postulate that inflammatory stimulation disrupts the biomechanical properties of NP cells in a mechanism that is dependent on actomyosin contractility and have furthermore shown increasing actomyosin contractility is sufficient to prevent inflammatory induced changes in NP cells. This is the first report that directly links actomyosin contractility with the regulation of cellular biomechanical and osmotic properties in NP or other chondrocyte-like cell. Modulating Rho signaling by using a Rho activator fully restores cell stiffness, cell permeability, and cell morphology suggesting a potential intervention strategy to maintain the cellular biomechanical homeostasis in the threat of inflammatory stimulation

ACKNOWLEDGEMENTS

Funded in part by NSF CAREER Award 1151605 & NIH R01 AR069668.

REFERENCES

[1] Urban, Biochem Soc Trans, 2002 [2] Urban, Spine, 1989 [3] Le Maitre, Arth Res Ther, 2005 [4] Millward-Sadler, Arth Res Ther, 2009 [5] Maidhof, PLoS ONE, 2014 [6] Henson Microsc Res Tech, 1999 [7] Jacobsen, ORS Annual Conf 2017 [8] Kolodney, J Biol Chem, 1993 [9] Chao, J Biomech, 2005 [10] Ateshian, J Biomech, 2006 [11] Albro, Biophys J, 2009.

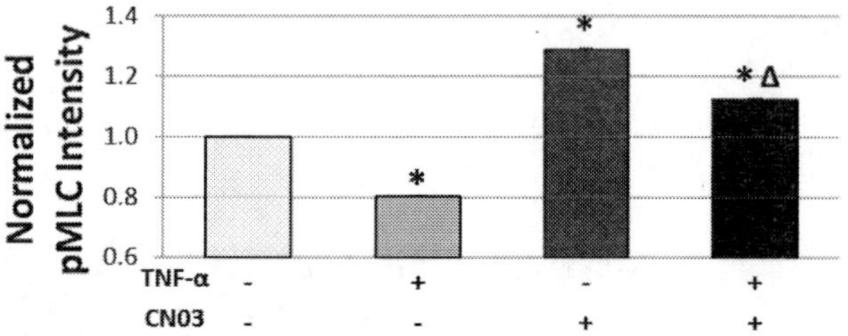

Figure 1: TNF- α decreases actomyosin contractility as indicated by pMLC staining. CN03 Increases contractility as well as mitigates the effect of TNF-α (* $p<0.05$ vs Control) (Δ $p<0.05$ vs TNF-α alone)

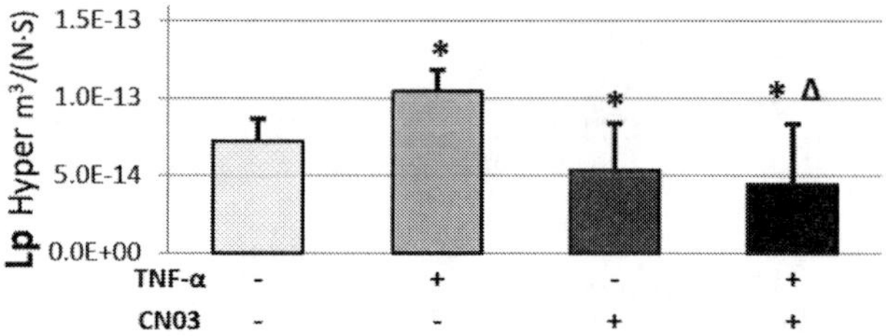

Figure 2: Increasing actomyosin contractility with CN03 is protective of TNF-α induced changes to hydraulic permeability (Hyperosmotic step) (* $p<0.05$ vs Control) (Δ $p<0.05$ vs TNF-α alone)

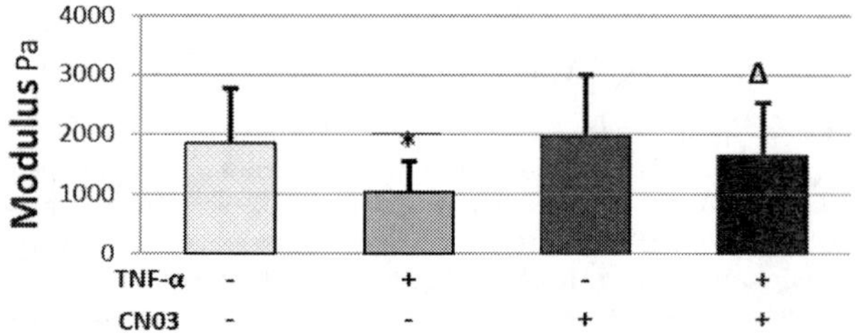

Figure 3: Increasing actomyosin contractility with CN03 is protective of TNF-α induced changes to elastic modulus (* $p<0.05$ vs Control) (Δ $p<0.05$ vs TNF-α alone)

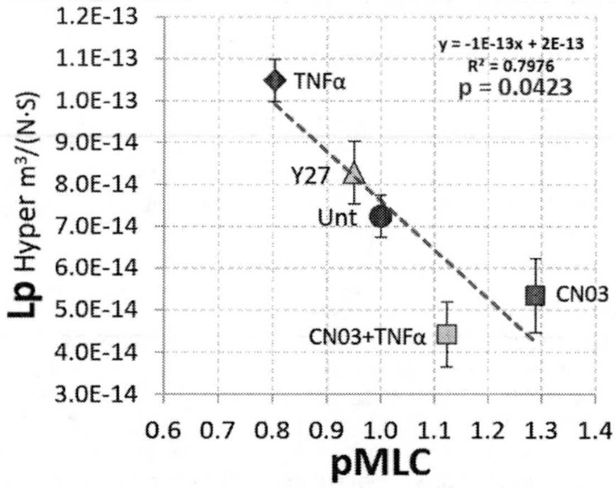

Figure 4: Normalized pMLC staining intensity significantly correlates with hydraulic permeability (hyper-osmotic step)

SB^{3}C2017
Summer Biomechanics, Bioengineering and Biotransport Conference
June 21 – 24, Tucson, AZ, USA

ENGINEERING TENDON THROUGH A MULTISCALE APPROACH AND CONDITIONING IN A BIOREACTOR

Brittany L. Banik (1) and Justin L. Brown (1)

(1) Department of Biomedical Engineering
The Pennsylvania State University
University Park, PA USA

INTRODUCTION

More than half of the 30 million musculoskeletal injuries worldwide involve tendons and ligaments, with 50% of sports injuries involving tendons.[1] Damage to these soft tissues is painful, debilitating, and often times healing is incomplete. Current therapeutic solutions and replacements remain ineffective with repeated mechanical failure, poor integration, and instability. There is an inherent and significant need for a better synthetic tendon replacement option.

Tendons are dynamic, highly organized, fibrous tissues that function in the musculoskeletal system to transfer forces between muscle and bone, store energy, stabilize joints, aid in mobility, and absorb shock. The structure of tendons is hierarchical; this complex architecture stems from a single collagen unit to a collagen bundle intricately organized to form a tendon. This complex architecture is difficult to design and synthesize; added complications include recapitulating the viscoelastic response and high load-to-failure capability of tendons.

We propose a (poly)caprolactone (PCL) scaffold design incorporating bi-axially aligned, bioinstructive nanofibers to create a microenvironment capable of organizing and directing stem cell growth and differentiation while maintaining appropriate mechanics for tendon tissue engineering applications. A bioreactor has been developed to apply physiological cyclic stretch while providing a real-time readout of resistance measurements.

Bioreactors are apparatuses utilized in tissue engineering to begin cultivating tissue replacements in a more physiological environment. Bioreactors are critical in tissue morphogenesis because cells respond to subtle changes in their microenvironment, which can be regulated by various bioreactor inputs. With sensitivity to temperature, mechanical loads, growth factors, fluid flow, and other environmental changes, the cell niche is critical in tissue development. Bioreactors assist in creating a more natural environment for cells to mature and develop; they are clinically relevant by providing a means to reduce significant donor site morbidity associated with current reconstructive procedures.[2] Previous research has demonstrated increases in ultimate tensile strength and elastic moduli values following mechanical treatment in a bioreactor.[3] This project evaluates bioreactor culture as a means to determine an ideal scaffold structure based on remodeling and degradation events. Furthermore, real-time monitoring of scaffold remodeling is introduced. It is proposed that following *in vitro* uniaxial stretch conditioning, scaffolds will be positively altered to reflect improved biomechanical function, enhanced extracellular matrix alignment, and increased mechanical properties of ultimate tensile strength and elastic moduli.

METHODS

A bioreactor was constructed with the following criteria established: real-time readout of resistance, ability to be easily clean and sterilize the device, control of strain and time parameters, adjustable clips to load constructs, and capability to withstand 3 weeks of continuous load with minimal maintenance. The tissue chambers were designed to hold up to six scaffolds for simultaneous testing; the containers allow for easy media changes and sterilization. An IDEA Stepper Motor Controller and Drive (Haydon Kerk PCM4826E) and linear actuator (Haydon Kerk 43HP-2.33.815, 43000 Series, size 17) worked in concert with the IDEA^{TM} drive interface program to provide easy programming through Graphic User Interface (GUI). The program allows the ability to modify the following features: speed, distance (i.e. strain), and delay time. A conductive rubber cord stretch sensor (Adafruit, Product ID: 519) was utilized as a real-time readout of resistance to monitor scaffold remodeling. As the sensor cord is stretched, the resistance increases. Likewise, at a relaxed state, the cord returns to the original resistance output. The base of the bioreactor is made out of poly(methyl methacrylate) with additional parts 3D-printed from t-glase (Rostock Max™ V2 Desktop 3D Printer).

The bioreactor was verified for functionality by investigating the time required for the resistance readings to have minimal fluctuations (i.e. delay time between loading and unloading) and testing various strains to understand the deviations between readings when a force is applied or removed.

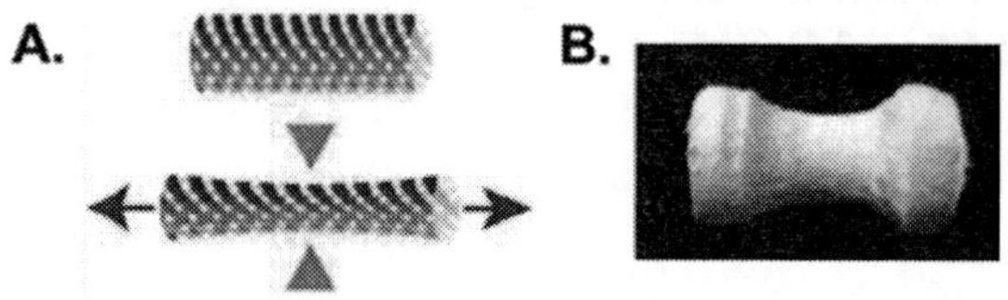

Figure 1: A) Chinese fingertrap tendon scaffold designed to improve mechanics and integration, B) Synthesized polymeric tendon scaffold

RESULTS

We have previously demonstrated the feasibility of using a multiscale scaffold design (**Figure 1**) for tendon tissue engineering.[4] The approach analyzes the problems in tendon tissue engineering from a hierarchical perspective through the design ideas behind a Chinese fingertrap toy. As such, the nano- level is described by the nanofibers, which act as a bioinstructive surface to direct cells down the tendon lineage. The nano- level represents the cellular level and thus is critical as a defining factor in the cell niche. The micro- or tissue level is defined by the scaffold mechanics and response as a whole construct. The bioreactor completes the macro- or body level investigation by providing information as to how the scaffold may perform in a more physiological environment with mechanical stimulation.

The main considerations for the bioreactor can be broken down into three categories—tissue culture chambers, motor system, and platform design. A novel aspect of the bioreactor design (**Figure 2**) is the output of real-time readout of the resistance. The conductive rubber cord (**2), Figure 2**) is directly connected to an ohmmeter to provide instead feedback on scaffold performance. The programmability of the stepper motor controller (**4), Figure 2**) and associated computer interface allows for variability in programs to determine the best inputs for tendon tissue engineering applications. Small deviations of resistance readings following extension and retraction are demonstrated for the bioreactor (**Figure 3**).

DISCUSSION

According to literature, cyclic strain can increase cell proliferation and alignment, promote collagen I production, increase secretion of growth factors, and maintain tenocyte morphology *in vitro*.[5,6] This proposal seeks to utilize a uniaxial cyclic strain bioreactor to improve scaffold performance at mechanical and cellular levels. We have created a device that meets specifications to function as a bioreactor for tendon tissue engineering applications and demonstrated the feasibility of using the bioreactor to test our scaffolds.

While we have already proven mechanical suitability based on scaffold design, the incorporation of the bioreactor will enhance the scaffold biomechanics and functionality towards natural tendon performance. Cells will be cultured on the scaffolds while in the bioreactor; as such, the bioreactor design provides the ability to monitor scaffold mechanics in real-time. Additionally, we anticipate that the application of cyclic uniaxial strain will encourage cell and extracellular matrix alignment and serve as a bioinstructive mechanism towards the tendon lineage. The scaffold design and construct is a promising avenue for not only tendon tissue engineering applications but also areas of ligament and vascular engineering.

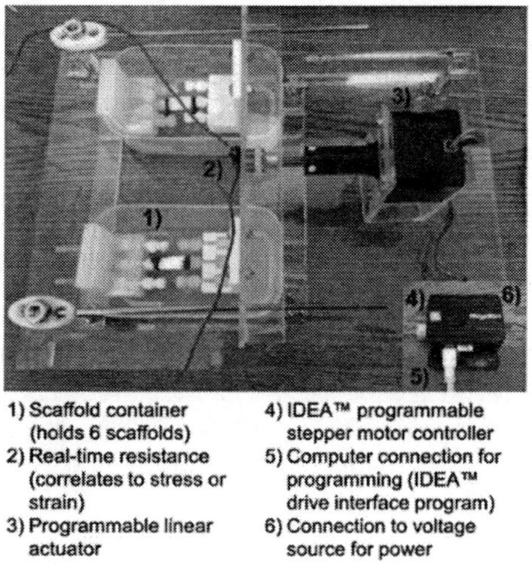

Figure 2: Bioreactor design to apply physiological cyclic loading.

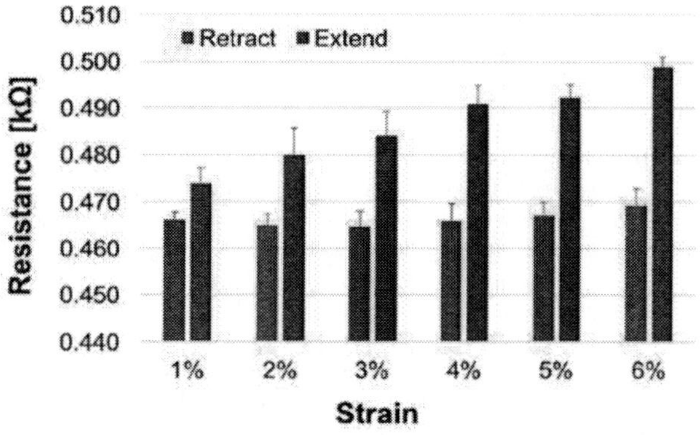

Figure 3: Evaluation of a real-time stress gauge incorporated into the bioreactor in response to a fixed strain applied to two scaffolds in parallel.

ACKNOWLEDGEMENTS

Special thanks to Gene Gerber and Tony Banik for bioreactor construction. This material is based upon work supported by the National Science Foundation (NSF) under Grant No. DGE1255832. Any opinions, findings, and conclusions or recommendations expressed in this material are those of the authors and do not necessarily reflect the views of the NSF.

REFERENCES

[1] Walden, G. *et al. Tissue Eng. Part B*. 23(1):1-15, 2017.
[2] Mace, J. *et al. Curr Stem Cell Res Ther*. 11(1):35-40, 2016.
[3] Saber S. *et al. Tissue Eng. Part A*. 16(6):2085–90, 2010.
[4] Banik, B.L. *et al. Regen. Eng. Transl. Med.* 2(1):1-9, 2016.
[5] Riboh, J. *et al. J. Hand Surg. Am*. 33:1388–96, 2008.
[6] Nirmalanandhan, V.S. *et al. J. Biomech. Eng.* 129:919–23, 2007.

SB³C2017
Summer Biomechanics, Bioengineering and Biotransport Conference
June 21 – 24, Tucson, AZ, USA

AN IN-VITRO PLATFORM TO INVESTIGATE VASCULAR ACCESS GRAFTS FOR IN-SITU TISSUE ENGINEERING UNDER HEMODYNAMIC LOADING

Eline E. van Haaften (1, 2), Marcel C.M. Rutten (3), Jurgen A. Bulsink (3), Nicholas A. Kurniawan (1, 2), Carlijn V.C. Bouten (1, 2)

(1) Soft Tissue Biomechanics and Tissue Engineering
Eindhoven University of Technology
Eindhoven, The Netherlands

(2) Institute for Complex Molecular Systems
Eindhoven University of Technology
Eindhoven, The Netherlands

(3) Cardiovascular Biomechanics
Eindhoven University of Technology
Eindhoven, The Netherlands

INTRODUCTION

The leading cause of vascular access dysfunction in hemodialysis patients is occlusion of the graft due to thrombosis. *In-situ* tissue-engineered (TE) vascular access grafts are envisioned to improve long-term functionality. Pursuing this approach requires the design of instructive scaffolds with optimized parameters such as bio-functionalization, structural organization, and degradation rate.

In-situ TE relies on the principles of growth and remodeling (G&R) to preserve functionality throughout the regeneration process, i.e. to reach and maintain a mechanical homeostasis [1]. Testing vascular grafts on their potential to facilitate tissue G&R towards homeostasis is ideally tested *in-vitro* under systematic, challenging hemodynamic conditions. Several bioreactor systems to culture vascular grafts *in-vitro* exist, such as pulsatile perfusion systems applying both circumferential stress and shear stress in a TE blood vessel. However, since pressure, stretch, and shear stress are coupled to each other in such geometries, systematically investigating the contribution of the different mechanical cues on G&R remains challenging.

The current study presents the design and validation of an *in-vitro* platform that allows for independent control of circumferential stretch and fluid shear stress in a 3D tubular scaffold. The platform enables systematic assessment of the impact of hemodynamic conditions on cell-produced matrix and mechanical properties of developing vascular constructs.

METHODS

The bioreactor is capable of providing four types of physiologically relevant mechanical loading conditions during *in-vitro* growth of eight tubular scaffolds: (1) static culture, (2) circumferential straining (ε_θ) only, (3) perfusion at predefined wall shear stress (WSS) only, and (4) ε_θ and WSS combined.

The test platform consists of a custom-made tubular 3D-scaffold culture chamber (Figure 1a, b). The scaffolds, centered in a glass tube, are pressurized from the inside, generating ε_θ, and perfused through the annular gap between the scaffold and glass wall, generating WSS. Using this approach, ε_θ (0-10 %, 0-1 Hz) and WSS (0-5 Pa) can be controlled independently.

Computationally driven design: The effect of changing channel dimensions, as a result of ε_θ, on the WSS distribution under pressure driven flow conditions was numerically investigated in an axisymmetric 2D problem in TFEM [2]. The fluid flow is governed by the unsteady Navier-Stokes equations for Newtonian and incompressible fluids.

With zero pressure at the output, a pressure between 0.5-2.5mmHg was prescribed at the inlet to generate low WSS (1.5Pa) and high WSS conditions (4.5Pa). The time-dependent deformation and velocity were prescribed for the scaffold wall and fluid using a sinusoidal waveform (1Hz, 5-10%). Velocity and WSS distribution at the scaffold wall were analyzed to tune the graft length and slit height towards more homogenized loads.

Poster Presentation #P111

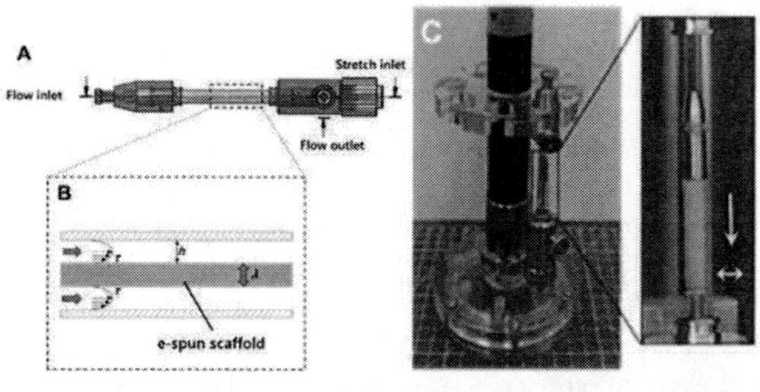

Figure 1: Schematic (A, B) and picture (C) of the platform, consisting of a tubular scaffold centered in a glass tube. The flow in- and outlets are connected to the annular ring, for WSS application. The stretch inlet is connected to the silicone-mounted scaffold for ε_θ application.

Bench-top experiment: The culture chamber is connected to an Ibidi pump system to generate flow. The scaffolds are mounted around a PBS-filled silicone tubing, which can be compressed with a custom-made displacement-controlled pump to apply cyclic ε_θ in sync with WSS (Figure 1c). As a proof-of-concept, tubular electrospun polycarbonate-bisurea (PC-BU) scaffolds (ID 3mm, fiber ø5 µm) are seeded with human vena saphena cells (HVSCs). The cell-seeded constructs are exposed to the four types of mechanical loading for 7 days.

Collagen architecture and cellular distribution in the cultured constructs are labeled with CNA35-OG488 and propidium iodide, respectively, for visualization with confocal laser scanning microscopy. Based on Hessian's 2nd derivative, principal directions of the green pixels were calculated to obtain a measure of the collagen distribution, which is compared against the scaffold fiber distribution and preferred orientation of the cell nuclei.

RESULTS

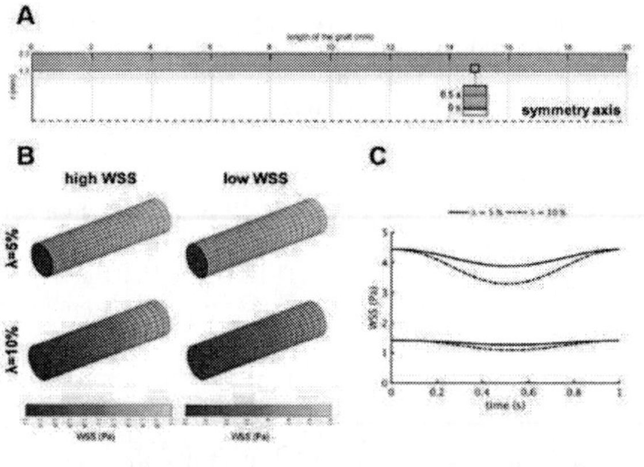

Figure 2: A pressure-driven axisymmetric annular flow simulation with a 1Hz sinusoidal moving membrane (A, at ε = 10%) was used to tune the graft length and slit height towards more homogenized loads. Due to changing channel dimensions as a result of cyclic stretch, the wall shear stress shows a spatially (B, at t = 0.25 s) and temporally (C, at the graft center) inhomogeneous distribution.

The optimized chamber dimensions enables us to systematically distinguish different types of loading (shear stress, stretch) at different magnitudes (0-5 Pa, 0-10 %) without inducing flow disturbances. The required pressures are within the limits of the 2 pump systems and allow for long-term culture experiments (>2 weeks, Figure 2).

After a culture period of 7 days, the bench-top experiment at predefined WSS (1.5Pa, 80mmHg) showed that perfusion alone resulted in a preferential direction of collagen matrix deposition in the flow direction, whereas the collagen organization under static conditions appeared to be mainly contact-guided by the scaffold's microfibers (Figure 3, n=1).

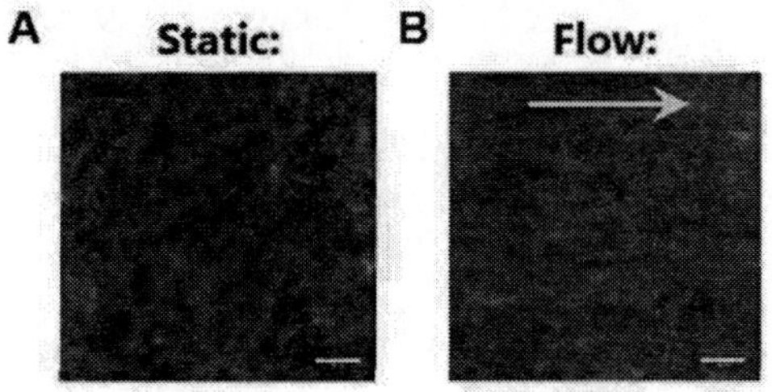

Figure 3: Collagen (green) and nuclei (blue) in an isotropic scaffold after 7 days of perfusion (B) and static culture (A). Arrow indicates flow direction, bar = 50 µm.

DISCUSSION

The developed platform relies on the versatility of microfluidic devices to apply 'active cues' (hemodynamic loading) and includes the 3D environment of biodegradable scaffolds to apply 'passive cues'. This way, the system can be used to gain fundamental insight in the interaction between cells, their environment, and mechanical cues, and also to precondition tissue engineered vessels. Such insights contribute to the rational design of new scaffolds for vascular (access) applications.

The scaffold's initial structural organization is an important factor for engineered tissue architecture, via mechanisms referred to as contact-guidance [3]. Relating structure to function, a rational design of vascular scaffolds contains a micro-structure with similar fiber distributions as native-like collagen. In the long-term, the relative rate of tissue growth/maturation versus scaffold degradation under hemodynamic loading will also contribute to the tissue architecture. To this end, we plan to evaluate the effect of scaffold anisotropy on tissue composition and architecture under degradation conditions using this platform.

ACKNOWLEDGEMENTS

We are grateful to A.I.P.M. Smits for fruitful discussions. This study (436001003) is financially supported by ZonMw and the Dutch Kidney Foundation.

REFERENCES

[1] Humphrey, J. *Cell Biochem Biophys*, 50:53-78, 2008.
[2] Hulsen, M.A., *User guide*, 2007.
[3] de Jonge, N. et al., *Tissue Eng Part A*, 20:1747-1757, 2013.

SB³C2017
Summer Biomechanics, Bioengineering and Biotransport Conference
June 21 – 24, Tucson, AZ, USA

DESIGN FEATURES TO ENABLE PHYSIOLOGICAL-RELEVANCE IN FLOW FOR OPTIMIZING ENGINEERED VALVE TISSUES

Manuel Perez-Nevarez, Omkar Mankame, Elnaz Pour Issa, Alex Williams, Alejandro Piñero and Sharan Ramaswamy

(1) Tissue Engineering, Mechanics, Imaging and Materials Laboratory,
Department of Biomedical Engineering, College of Engineering and Computing,
Florida International University, 10555 W. Flagler Street, EC 2614, Miami, FL 33174, USA

INTRODUCTION

Despite advancements in prosthetic heart valve technologies to treat heart valve disease, no current options offer permanent solutions for this debilitating disease. Tissue engineered heart valves (TEHV) offer new options in the treatment of this disease. Regenerative medicine researchers have determined that flow, stretch and flex are important mechanical modes of stimuli during for tissue engineered heart valve construct preparation. To achieve this conditioning, flow-stretch-flexure bioreactors have been designed to aid researchers in developing TEHV constructs[1,2,3].

Our previous studies with bioreactors indicate that flow induced shear stress can influence vascular phenotype development and tissue formation. These studies have examined vascular tissue under either zero flow (static), steady state flex-flow, or under square-wave flow simulations. Although these flow conditions did influence vascular tissue formation, they however do not fully represent the flow and pressure profiles produced by cardiac pulses and experienced by native vascular tissue. Additional findings of these studies indicated that collagen formation in square-wave flow alone was comparable to flex-flow (steady state) and exhibited similarities in gene expression. The studies also found that although traditionally time-varying flow has been replicated through specimen movement, physiologically relevant OSS can be maximized with the use of true pulsed flow conditioning environment.

This study reports on the development of a bioreactor system that can be used to examine how physiologically relevant oscillatory flow shear stress influences vascular tissue formation. Tissues in the vascular system are subjected to sinusoidal flow and pressure variations produced by heart pulsation during its heartbeat. This pulsatile flow subjects vascular tissue to Oscillatory Shear Stress (OSS) regimes not present in either static or steady-state flow conditions.

We have developed a robust flow perfusion bioreactor system for in-vivo simulation of these physiologically relevant flow environments experienced by vascular tissue with the use of tissue specimens immersed in flow media that simulates hydrodynamic cardiac output (Fig.1). The novel aspect of our design is the ability to fully recreate physiologically relevant pulsatile waveforms. This reduces a substantial layer of complexity as it isolates the OSS variable from other environmental factors affecting tissue development. In studying how OSS affects vascular tissue development, our overall approach is to generate implantable tissues by seeding cells in biodegradable scaffolds, culturing the cell-seeded constructs in our bioreactor to simulate native OSS conditions that promote tissue formation, and implanting the treated constructs in vivo. A successful vascular tissue developed with native OSS conditioning must be composed of cell-secreted ECM that remodels in response to local mechanical forces (pulsed flow), allowing it to adequately function long term.

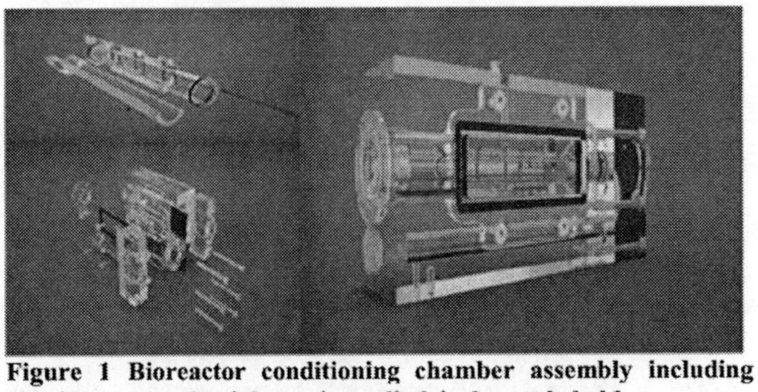

Figure 1 Bioreactor conditioning chamber assembly including viewing port and axial-opening cylindrical sample holder.

Poster Presentation #P112

METHODS

Our flow perfusion bioreactor was used to subject vascular tissue samples to physiologically relevant pulsatile flow conditions. The bioreactor system consisted of a U-shaped tubular conditioning chamber that houses the tissue samples and is connected in a flow loop with the pulsatile pump through which flow media was pumped across to drive flow across the face of the tissue samples (Fig.2). Components of the system include the bioreactor conditioning chamber, the pulsatile pump, conical adaptors to connect the chamber to the pump outflow and inflow, and flow as well as the pressure sensors (provided by Vivitro Labs, Victoria, BC, Canada). The bioreactor chamber, which was produced using commercially available CNC equipment, also includes a viewing port used to image the surface of the samples without having to physically remove them from the chamber. A pulse duplicator pump system (Vivitro Labs, Victoria, BC, Canada) was used to generate pulsatile flow through the bioreactor. The pump system's flow and pressure probes provide the data used by the testing system software to obtain relevant hydrodynamic analysis.

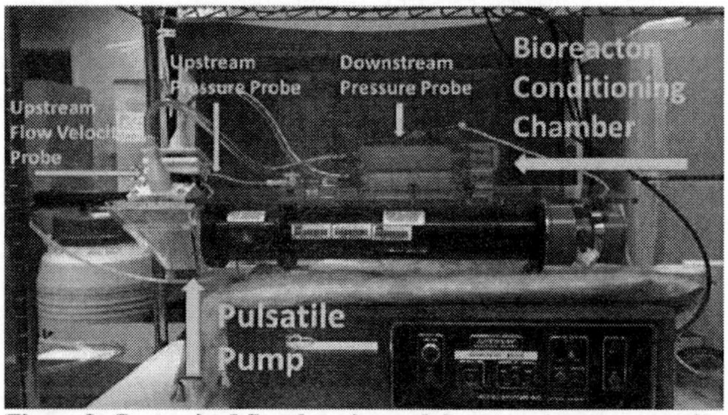

Figure 2: Customized flow loop in our laboratory to measure point flow rates and pressures at the pump exit point (upstream probe) as well as within the FSF bioreactor's conditioning chamber (downstream probe).

RESULTS

Using the flow and pressure sensors embedded in the flow loop, the downstream flow profile for a full cardiac cycle was determined based on the sensor data. Bernoulli's theorem along a streamline and conservation of mass were subsequently invoked to compute the downstream flow waveform within the bioreactor conditioning chamber (Fig.3).

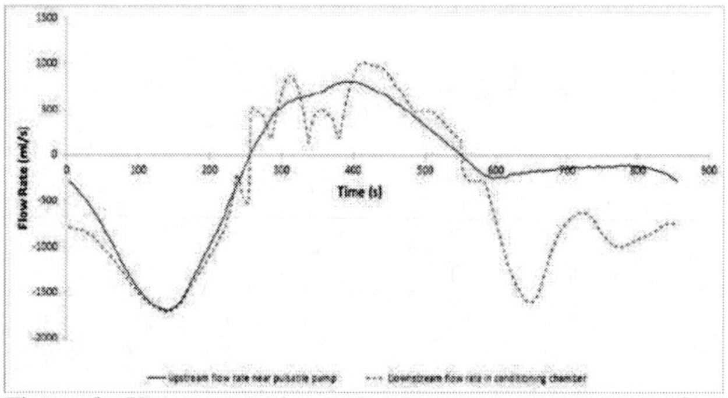

Figure 3: Upstream and calculated downstream pulsatile flow waveforms.

Waveform results for the first 3rd of the cardiac cycle showed close correlation between the measured flow output at the pump exit and downstream flow rates at the conditioning chamber. For the second 3rd portion of the cardiac cycle, there was moderate discrepancy between upstream flow rate vs. chamber flow rate. In the last 3rd of the cardiac cycle, pump flow rates and chamber flow values diverged considerably, possibly due to backflow effects in this late stage of the cardiac cycle. Future development of the bioreactor system will be required in order to reduce error rates and divergence of values.

DISCUSSION

Efforts in heart valve tissue engineering are in general agreement that bioreactor mechanical pre-conditioning augments important biological properties of the constructs, including enhanced tissue formation and gene expression[1,2,4]. Previously we, as well others, have extensively used a bioreactor system for predominantly running steady flow experiments in combination with cyclic flexure[2,3]. However, native circulatory flow conditions are highly pulsatile and have very specific waveform shapes. Our recent findings suggest that the oscillating component of the fluid is important in enhancing tissue formation and gene expression[1,2]. Therefore, here we have updated our previous system[5] to incorporate the physiological components of the flow, so that a wide-array of circulatory pulsatile flow waveforms could be introduced during the engineered tissue conditioning process. We verified that the flow rates at the vicinity of the specimens, i.e., downstream within the centerline axis of the conditioning chamber, were similar to the upstream flow rates (at the pump exit) imparted during pump operation. However, there were some major differences: we found that in the early part of cardiac cycle there was close agreement between pump output and flow rate within the chamber, some divergence in the mid part of the cardiac cycle, and major discrepancies in the latter part of the cycle. This discrepancy may be due to backflow conditions present within the flow loop and/or possible air entrapment within the chamber that disrupts pressure measurements in that segment of the cycle. Future development work remains and will be focused on determining the sources of these discrepancies.

Ultimately, a high-fidelity pulsatile pump and further fine-tuning of the set-up of our flow loop will enable additional improvements for maintaining the integrity and accuracy of the flow waveform downstream, where the engineered specimens will be located. Along these lines, from a practical viewpoint we have already updated the design of the system to be easier to assemble and use (Fig. 1 – new bioreactor). This in turn will minimize time-to-assembly of the system as well as reduction in contamination during tissue engineering experiments.

REFERENCES

[1] Salinas, M et al., "Relative Effects of Fluid Oscillations and Nutrient Transport in the In Vitro Growth of Valvular Tissues," *Cardiovascular Engineering and Technology*, Feb 2016.
[2] Rath, S et al., "Differentiation and Distribution of Marrow Stem Cells in Flex-Flow Environments Demonstrate Support of the Valvular Phenotype," *PLOS One*, Nov 2015.
[3] Engelmayr, G et al., *Biomaterials*, 37:6083-6096, Dec 2006.
[4] Ramaswamy, S et al., "The role of organ level conditioning on the promotion of engineered heart valve tissue development *in-vitro* using mesenchymal stem cells," *Biomaterials*, 31:1114-1125, 2010.
[5] Ramaswamy, S et al., "A Novel Bioreactor for Mechanobiological Studies of Engineered Heart Valve Tissue Formation Under Pulmonary Arterial Physiological Flow Conditions," *Journal of Biomedical Engineering*, Vol.136, 2014.

SB³C2017
Summer Biomechanics, Bioengineering and Biotransport Conference
June 21 – 24, Tucson, AZ, USA

MECHANICAL ANALYSIS OF PULMONARY HYPERTENSION VIA ADJOINT BASED DATA ASSIMILATION OF A FINITE ELEMENT MODEL

H. Finsberg (1), C. Xi (2), J.L Tan(3), L. Zhong (3), LC. Lee(2), S.Wall(1)

(1) Scientific Computing
Simula Research Laboratory
Oslo, Norway

(2) Mechanical Engineering
Michigan State University
East Lansing, MI, USA

(3) National Heart Centre
Singapore

INTRODUCTION

Pulmonary hypertension (PH) is associated with an elevated pressure in the pulmonary arterial system. Without treatment, PH can quickly lead to decompensated RV failure and death. Our current understanding of PH has largely been obtained through animal models, which do not fully reproduce the etiology of human PH. To overcome this limitation, we describe an inverse computational analysis framework that uses patient data consisting of MR images and pressure measurements to quantify the biomechanical alteration in ventricular mechanics associated with PH. The framework was developed based on an adjoint-gradient based data assimilation method that can efficiently handle optimization in high dimensional parameter spaces.

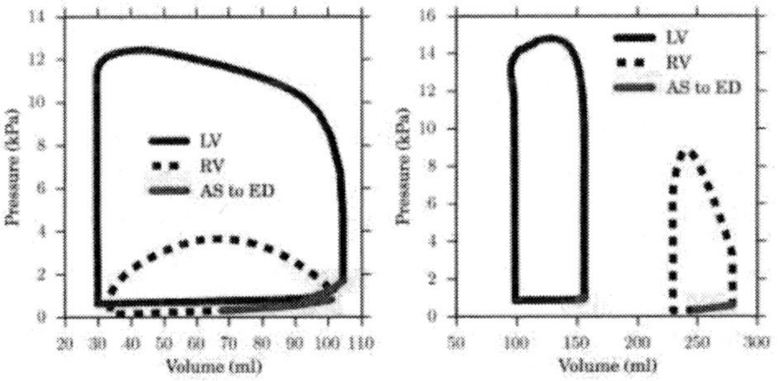

Figure 1: Pressure volume loops for patient diagnose with pulmonary hypertension (left) and healthy control (right). Red line indicates part of the loop that belongs to atrial systole.

METHODS

We consider two subjects, where one is diagnosed with PH and one is an age- and gender- matched healthy control subject. Cine magnetic resonance (MR) images were obtained from both subjects. Ventricular, atrial and artery pressure were also measured in the PH patient by right heart catheterization. On the other hand, ventricular pressure was estimated for the normal human subjects based on previous studies[7]. Left and right ventricular (LV and RV) volumes were measured at different time frames in a cardiac cycle from the MR images by segmentation of the biventricular geometry. These measurements (as well as the pressure estimation for normal human subjects) were used to construct pressure-volume loops of the RV and LV for the PH patient and the normal subject (Figure 1). Biventricular finite element (FE) geometries for the PH patient and normal subject were reconstructed from the corresponding MR images at atrial systole (when the heart is not contracting).

We model the myocardium as an incompressible hyperelastic material, with a transversally isotropic version of the strain-energy density function presented in [1]:

$$\Psi = \frac{a}{2b}\left(e^{b(I_1=3)} = 1\right) = \frac{a_f}{2b_f}\left(e^{b_f(I_{4f}=1)^2_=} = 1\right) \quad (1)$$

There are two main approaches in modeling of the active response of the tissue[2], namely the active stress approach where you decouple the stress tensor into passive and active stress, and the active strain approach where you decouple the deformation gradient in into an active and passive deformation. Here we employ the active strain approach, which assumes a multiplicative decomposition of the deformation gradient into an active and an elastic part:

$$\boldsymbol{F} = \boldsymbol{F}_e\boldsymbol{F}_a, \quad (2)$$

with

$$F_a = (1 = \gamma) f_0 \otimes f_0 = (1 = \gamma)^{-1/2}(I = f_0 \otimes f_0). \quad (3)$$

Fiber directions are assigned using the ruled-based algorithm in [3], with a helix fiber angle of 60 and -60 degrees on the endo- and epicardium respectively. Measured LV and RV pressures are assigned as a Neumann boundary condition on the LV and RV endocardium respectively, and the ventricular base is fixed in the longitudinal direction and limited in the other directions subjected to a linear spring with constant $k = 1$. The displacement field $\boldsymbol{u}$ and hydrostatic pressure p is computed at each time point by minimizing the total elastic energy in the system, that is solving $\delta\Pi(\boldsymbol{u}, p) = 0$, where Π represented the total elastic energy. To avoid numerical instabilities such as locking we use Taylor-Hood finite elements with $(\boldsymbol{u}, p) \in P_2 \times P_1$. The data assimilation is formulated as a PDE-constrained optimization problem, where the objective functional represents the L_2 difference between simulated and measured LV and RV volume and the constraints is given by the force-balance equation $\delta\Pi(\boldsymbol{u}, p) = 0$. Formally, for each time point i, we want to solve the following problem

$$\min_m \mathcal{J}(\boldsymbol{u}^i, p^i, m), \text{ subject to: } \delta\Pi(\boldsymbol{u}, p) = 0, \quad (4)$$

with

$$\mathcal{J}(\boldsymbol{u}^i, p^i, m) = \sum_{k=LV,RV} \alpha_k \left(\frac{V_k^i = \tilde{V}_k^i}{V_k^i}\right)^2 \quad (5)$$

The end-diastolic volumes are used to estimate the linear isotropic martial parameter a in (1) by inflating the mesh from 0 kPa to end diastolic pressure and solving (4) with $m = a$, and assuming $\gamma = 0$. The remaining parameters in (1) are fixed according to biaxial test results in [1]. For the remaining points in the cardiac cycle, we fix the optimized material parameter and let $m = \gamma$ (2,3) be the new control parameter. The control parameter m is modeled as a piecewise linear function over the vertices during the optimization of the material parameter and a constant function with different parameters on the LV and RV during the active phase. More details on the data assimilation method is found in [4].

The solver is implemented in FEniCS[5], and uses a gradient-based optimization method to solve (4), where the gradient is computed by solving an automatically generated adjoint equation[6].

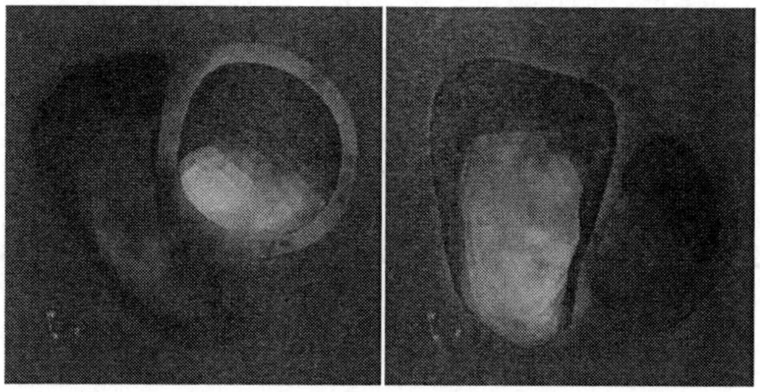

Figure 2: Showing optimized material parameter for the healthy (left) and PH (right) subject. Red/blue color represents stiffer/softer material.

RESULTS

The described method is applied to two subjects; one healthy and one diagnosed with PH. In Figure 2 we see the resulting material parameters optimized in order to fit the LV and RV end-diastolic volumes. In Figure 3 we show the peak value of γ from (3), where γ is an index of the amount of relative active fiber shortening.

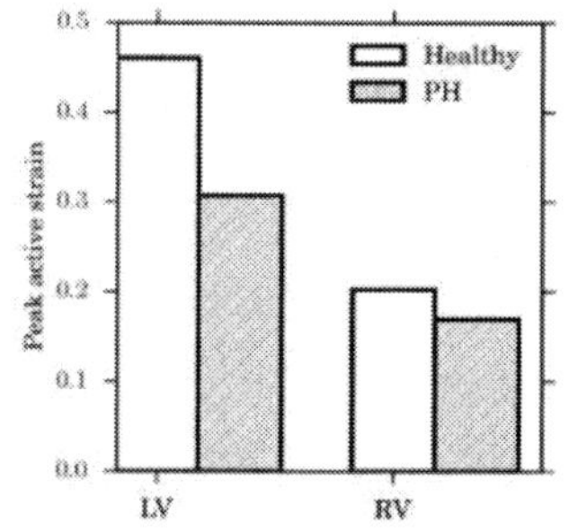

Figure 3: Showing the peak active strain (γ in (3)) for the LV and RV for the two subjects considered.

DISCUSSION

From Figure 1 is clear that the PH patient has an increased pressure in the right ventricle. Moreover, in the same figure we see that the healthy subject has an ejection fraction of about 80% which is challenging to recapitulate due to numerical instabilities when the amount of active contraction is high. For that reason we were not able to assimilate the whole cycle for this patient but terminated at the end of the ejection phase.

The material parameters where chosen to be spatially resolved at the level of the vertices. From Figure 2 we see that the optimized material parameter indicate that the healthy subject has a slightly stiffer LV, while the opposite case is indicated for the PH patient.

It is also evident from Figure 3 that the amount of active force generation is much higher for the healthy patient with peak γ being 150% and 120% higher for the healthy patient in the LV and RV respectively.

The accuracy of data assimilation technics is limited by their ability to fit a large amount of control parameters. By utilizing tools, which automatically compute the functional gradient[6] we are able to fit a large number of parameters without sacrificing too much computational time.

ACKNOWLEDGEMENTS

This study was funded by Research Council of Norway: Center for Biomedical Computing at Simula Research Laboratory and Center for Cardiological Innovation at Oslo University Hospital. Computations were performed on the Abel supercomputing cluster at the University of Oslo via a Notur project. This study was also partially supported by Singapore Ministry of Health's National Medical Research Council (NMRC/OFIRG/0018/2016, Zhong), Goh Cardiovascular Research Award (Duke-NUS-GCR/2013/0009, Zhong) and AHA SDG (17SDG33370110, Lee).

REFERENCES

[1] Holzapfel, G et al., *Philos Trans A Math Phys Eng Sci*, 1902:3445-3475, 2009.
[2] Ambrosi, D et al., *J. Elasticity*, 2:199-212, 2012
[3] Bayer, J.D. et al., *Ann Biomed Eng*, 40:2243-2254, 2012.
[4] Balaban, G et al., *Int. J. Numer. Meth. Biomed. Engng.*, e02863. doi: 10.1002/cnm.2863
[5] Logg, A. et al. The FEniCS book (Vol. 84), 2012
[6] Farrell, P et al. *SISC*, 4:369-393, 2013
[7] Redington, A.N, *Brit Heart J*, 59.1:23-30, 1988

DETERMINATION OF OSTEOGENEIC MARKERS USING RNA SEQUENCING IN HUMAN ADIPOSE TISSUE DERIVED ADULT STEM CELLS

S. Shaik (1), E. Martin (2), D. Hayes(3), R. Devireddy (1)

(1) Mechanical Engineering
Louisiana State University,
Baton Rouge, LA, USA

(2) Biological & Agricultural Engineering,
Louisiana State University,
Baton Rouge, LA, USA

(3)Biomedical Engineering,
Pennsylvania State University
University Park, PA, USA

INTRODUCTION

Spatio-temporal modulation of osteogenesis in a 3D stromal/stem cell model adipose tissue derived stem cells (ASCs), when grown on collagen coated methylcellulose based thermally reversible system, form in to cell sheets that are self-detachable at room temperature [1-3]. The novelty of the format allows to stack the cell sheets and co-differentiate into different pathways within close proximity of each other using selective light activation of transfected miRNA in ASCs [4,5]. For example, we could study the changes within an ASC undergoing an osteogenic differentiation pathway that is some fixed length away from another ASC undergoing angiogenesis. Sequencing of mRNA from the isolated single cells from specific locations in cell sheets undergoing osteogenesis will provide insights on the expression of known and unknown genes and further lead to elucidation of intricate and elusive pathways/mechanisms involved in cell differentiation and establish baseline (control) data of mRNA sequencing for undifferentiated and osteogenically differentiated (chemical media induction) ASCs. This baseline data will be used in subsequent studies to compare and contrast the spatio-temporal effects of ASC differentiation.

METHODS

Cell culture: Frozen/thawed ASCs at passage P0 from 3 different donors were pooled together and plated. The cells were cultured till the passage 1 and replated in 12 well plates at a density of 3x104 cells/cm2 to reach confluence.

Osteogenic induction: To induce osteogenesis the cells were fed media containing 10mM β-glycerophosphate, 50ug/ml L ascorbic acid 2 phosphate sesquimagnesium salt hydrate, 10nm of dexamethasone for every 3 days till 21 days.

Alizarin red S staining: After 21 days the cells were washed with PBS and fixed with 70% ethanol for 30 minutes. The ethanol was removed and wells were washed with PBS and stained with 2% alizarin red s (pH adjusted to 4.2) for 30 minutes. Later, the stain was removed and the wells were washed till the loosely bound stain is removed.

RNA isolation: The purelink RNA isolation kit was used to isolate RNA and the quality and quantity was assessed by using nanodrop spectophotometer. Briefly, the cells were lysed with lysis buffer containing 10% beta mercaptoethanol and homogenized. The lysate is added to the silica membrane spin columns and treated with buffers supplied by manufacturer containing guanidine thiocyanate and ethanol. After removing other biomolecules by centrifugation the bound RNA to silica membrane is eluted in nuclease-free water. The RNA with A260/280 ratio equivalent to 2 is considered for further processing.

Library preparation: The mRNA isolated in the above steps was used to generate cDNA using the Smart seq v4 ultralow input RNA sequencing kit as described in the manual. The synthesized cDNA was amplified and purified using AMPure XP magnetic beads as per the manufacturer's instructions. The quality assessment of cDNA was performed using fragment analyzer. After the assessment the purified cDNA was fragmented using AfaI enzyme to generate library and the quality of library was assessed using fragment analyzer to determine the library size.

Sequencing and data analysis: The libraries were sequenced using the ion PGM Chip on ion proton sequencer. The data obtained from the sequencer was analyzed using various bioinformatics tools such as STAR, RSEM, and David bioinformatics on high computing clusters at LSU.

RESULTS

Osteogenic differentiation of ASCs after 21 days induced using chemical media. The image below depicts the staining of ASCs that underwent osteogenesis whereas the undifferentiated ASCs remain unstained after treating with alizarin red S stain.

Osteogenesis

Undifferentiated control

Table 1 lists the osteogenic markers that were found to be upregulated. Other genes such as Jun, FOS, and MAP kinases that are known to modulate the osteogenic genes and markers were also found to be upregulated. Further analysis is currently being performed on the sequencing data to elucidate the pathways/mechanisms that are involved with ASC osteogenesis.

Table 1: Upregulated osteogenic markers

Gene Name	Control Count	Control TPM	Control FPKM	Osteo Count	Osteo TPM	Osteol FPKM	Fold Change
ALPL	6860.62	119.98	71.52	20461.99	309.95	207.68	2.983
BGLAP	1.02	0.05	0.03	3.15	0.14	0.1	3.088
BMP4	91	2.84	1.69	144	4.29	2.88	1.582
BMPR1B	214	2.09	1.24	386	4.04	2.71	1.803
COL10A1	5	0.08	0.05	281	5.44	3.65	56.2
CSF1	5126.15	67.05	39.96	17666.67	246.9	165.43	3.446
CTSK	3045.95	84.94	50.63	71657.01	1738.42	1164.82	23.525
EGFR	5323	37.16	22.15	9910.34	61.08	40.93	1.862
FN1	1222356	8119.97	4839.91	3391541	18631.62	12484.05	2.775
ICAM1	1376	22.72	13.54	6718.76	92.86	62.22	4.883
IGF1	50.92	1.21	0.72	137.92	4.92	3.3	2.7086
ITGA3	1370.39	14.54	8.67	2189.73	26.09	17.48	1.598
MMP2	84805.54	1950.48	1162.58	146515.2	2871.53	1924.06	1.728
NOG	2	0.05	0.03	91.04	1.87	1.25	45.52
TGFBR2	14775.1	134.17	79.97	41096.61	329.49	220.77	2.782
TNF	2	0.05	0.03	5	0.12	0.08	2.5
TWIST1	347.99	17.37	10.36	753.89	33.57	22.5	2.166
VCAM1	75	1.03	0.62	801.89	9.83	6.58	10.692
VEGFB	1231	66.86	39.85	3432.78	153.42	102.8	2.789

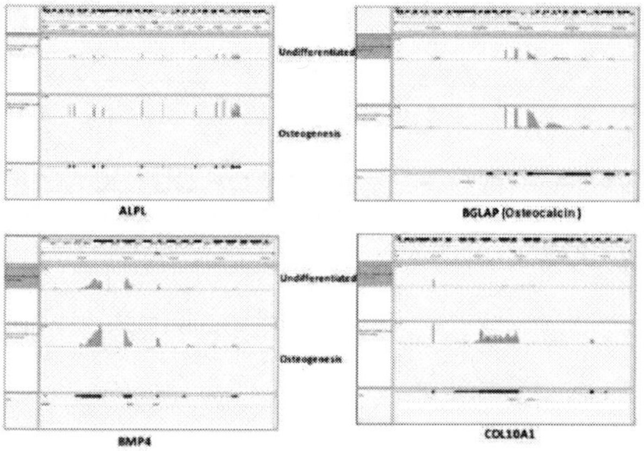

Fig. 1: Representative osteogenic genes visualized in *igv* viewer

DISCUSSION

Sequencing of mRNA from ASCs undergoing osteogenic differentiation in comparison to undifferentiated ASCs indicated upregulation of several osteogenic markers (shown in Table 1). Further studies, including pathway analysis by david bioinformatics tools and validation of data using QPCR has to be performed to establish osteogenic control which can be used as a reference for the spatio-temporal modulation of osteogenesis.

ACKNOWLEDGEMENTS

Generous financial support for this work was provided by the Department of Mechanical Engineering at the Louisiana State University and by the National Institute Dental and Craniofacial Research of National Institutes of Health under award R01 DE024790. The content is solely the responsibility of the authors and does not necessarily represent the official views of the National Institutes of Health.

REFERENCES

[1] Thirumala, S et al., *Cells* 2.3 (2013): 460-475.
[2] Altomare, L et al., *J Mater Sci Mater Med.* 2016 May;27(5):95.
[3] Lee, W et al., *Sci Rep.* 2016 Jul 6;6:29408.
[4] Qureshi, A et al., *Biomaterials* 34.31 (2013): 7799-7810.
[5] Qureshi, A et al., *Acta Biomater.* 2015 Jan;12:166-73.

JUNCTION PROTEIN AND TRANSPORT CHARACTERIZATION OF RECONSTRUCTED ENDOTHELIUM IN A MICROFLUIDIC CELL ARRAY WITH MIMICKED TUMOR MICROENVIRONMENT

Chun-Wei Chi, Chenghai Li, A.H. R. Ahmed, Elizabeth Benoy, Zeynep Dereli-Korkut and Sihong Wang

Department of Biomedical Engineering,
CUN□- City College of New □ork
New □ork, N□, USA

INTRODUCTION

According to a FDA study, more than 90% of cancer drug fails in clinical trials because the patient outcome differs from the *in vitro* results [1]. *In vivo* model is considered as the most reliable study, which has conditions close to human's physiological microenvironment, but animal studies are time-consuming and costly. Therefore, there is a demand of novel approaches that can replicate *in vivo* microenvironment for pilot, prescreening studies. Although conventional 3D culture technology enables a more realistic extracellular microenvironment, they are still limited to static culture [2]. In contrast, a microfluidic platform offers a spatial control of cell distribution and reconstruction of microenvironment, which allows a precise control of flow to mimic real, *in vivo* conditions, such as nutrient supply and waste removal.

We previously developed a three-layers microfluidic cell array to reconstruct the physiologically spatial structure of tumor microenvironment by co-culturing cancer cells with microvascular endothelial cells in two compartments, separated by a thin layer with clustered pores [3]. Compared with static co-culture of lung cancer cells and endothelial cells, this structurally co-culture platform showed attenuated and delayed drug responses. We speculate that the monolayer of endothelial cell in the top channel of the device acts as a barrier for drug delivery, which may closely represent *in vivo* tumor microenvironment. Therefore, in this study, we characterize the structure of endothelial cell monolayer by immunostaining of junction proteins in the microfluidic device. Furthermore, the flow pattern in this microfluidic cell platform and diffusive permeability from the mimicked microvessel channel to the chambers with cancer cells and mesenchymal cells are being subsequently characterized.

METHODS

The 3D microfluidic cell array was constructed by three layers of polydimethylsiloxane (PDMS) as shown in **Figure 1**. Each layer was fabricated by soft lithography techniques. After oxygen plasma treatment, layers were aligned and permanently bonded to one and other. The microfluidic device was sterilized by an autoclave before use. In order to improve cell attachment, fibronectin was coated on the surface of channels after changing surface property to hydrophilic by oxygen plasma treatment. Human non-small cell lung cancer cell (PC9) was mixed and encapsulated in 0.25% PuraMatrix in the bottom chamber. On the second day, the human dermal blood microvascular endothelial cell (HMVEC) was introduced to the top channel. Continuous flow of medium was provided after HMVEC attached on the middle layer.

For cell characterization, GFP-expressed PC9 cell was used, and CD31 was immunostained on the endothelial cell layer to outline the structure. PC9 and HMVEC were seeded in the device. On the third day of experiment, cells were fixed by 4% paraformaldehyde for 60 minutes. After washing, 0.2% Triton X-100 was used to increase cell permeability for further staining. In order to reduce the background signal of immunostaining, 3% BSA was used for blocking. Sequentially, anti-CD31 antibody was introduced and incubated with cells at 4°C overnight after washing with PBS buffer. Secondary antibody (Alexa Fluor 594 Goat anti-mouse IgG) was subsequently stained, and the nucleus was stained with DAPI.

In order to investigate the flow of solute in the device, FITC-labeled dextran (10kDa) was used to trace the characteristic of transport. In the study, the solute transport was monitored in the side view of the three-layer construction. Before testing, the top channel and bottom chamber were filled with water. Dextran-FITC was injected into top channel by syringe pump while the bottom channel was closed. Based on the fluorescence signal, the diffusive permeability can be estimated by following equation:

$$P = \frac{dI/dt}{\Delta I_0} \times \frac{vol.}{area} = \alpha \times \frac{vol.}{area} \qquad (1)$$

I and ΔI_0 represent the fluorescent signal and initial intensity in the filled top channel, respectively. *vol.* is the volume in the top channel and *area* is the area between top channel and bottom chamber.

RESULTS

As shown in **Figure 2A**, a monolayer of HMVEC was formed on the middle layer of the device, which contained clustered pores allowing interaction with the PC9 cells encapsulated in peptide hydrogel in the bottom chamber. We further used immunostaining of CD31 to depict the outline of endothelial cells with red fluorescence and observed the green fluorescent signal from GFP-tagged PC9 cells (**Figure 2B**). CD31, known as platelet endothelial cell adhesion molecule-1, is mostly expressed at junctions of adjacent endothelial cells. Therefore, the image revealed the integrity of endothelial cell monolayer formed under continuous flow in the device.

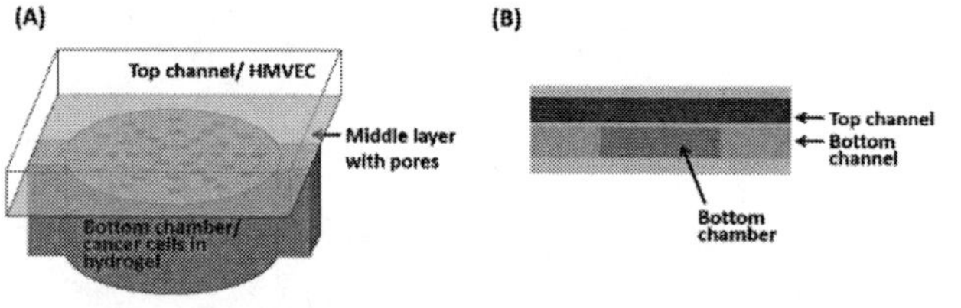

Figure 1. Illustration of three-layers microfluidic device which mimicking the tumor microenvironment with microvessels (A) 3D schematic diagram (B) Side view of the device

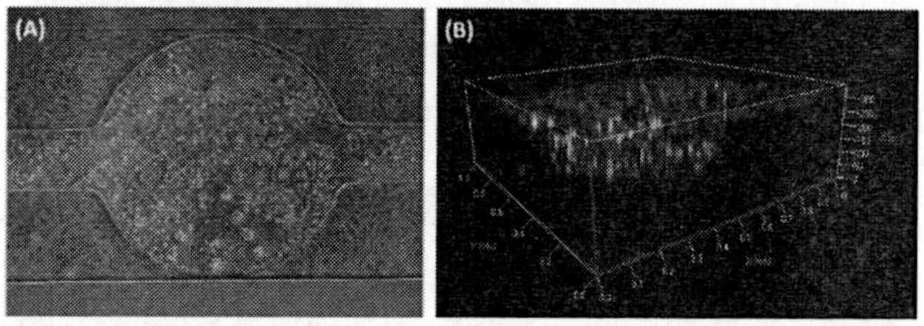

Figure 2. Images of cells seeded in the microfluidic device. (A) phase contract image (B) 3D fluorescence image by confocal microscope (green: GFP-fused PC9 cells; red: CD31 immunostaining of HMVEC; blue: DAPI nuclear staining)

Figure 3 shows the transport of 10kDa dextran-FITC from top channel to the bottom chamber observed from the side view of the device at different time points. In addition, two different perfusion rates (50 μL/min and 1ml/min) of 0.1mg/mL dextran-FITC were monitored, and the calculated solute permeability was quite similar.

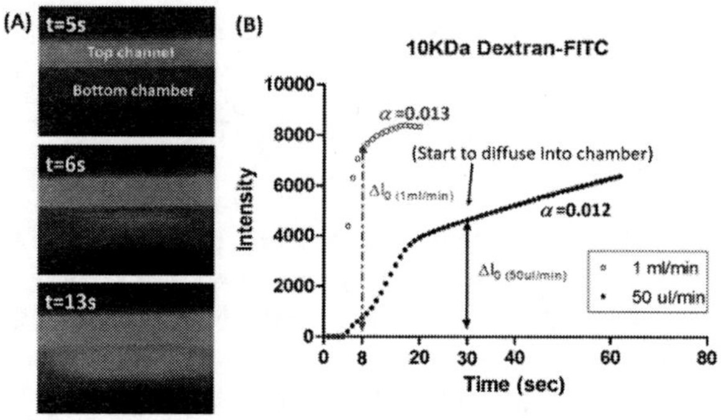

Figure 3. Dextran-FITC transport in the microfluidic device. (A) transport of 10kDa dextran-FITC at different time points (B) Fluorescence intensity was monitored with time

DISCUSSION

In this study, we have demonstrated the integrity of the monolayer of endothelial cells formed in the device by immunostaining of the junction proteins CD31. Endothelial cells, which are joined laterally by cell-cell junction, are known to function as a barrier for the movement of water, proteins, blood cells and solutes [4]. We previously found that the drug response of cancer cells in our *in vivo*-mimicking microfluidic device was slower and reduced. Herein, the demonstrated intact monolayer of endothelial cell can be an evidence to explain the attenuated effect. Additionally, the transport of low molecular weight particles in this device was traced by 10kDa dextran-FITC. A method to estimate solute permeability in this device was also established in this study. A further study is ongoing, regarding measuring solute permeability to cancer cells plus mesenchymal cells encapsulated in hydrogel in the bottom chamber of the device across or not across endothelial cells to verify the role of the monolayer of endothelial cell in solute transport. Higher molecular weight dextran (e.g. 40kDa) will be used in these experiments too. Furthermore, comparing the *in vivo* permeability of microvessels would help us optimize the seeding density of endothelial cells in our device to recreate a more physiological relevant tumor microenvironment.

ACKNOWLEDGEMENTS

This work was supported by NIH/NCI 1U54CA137788 (CCNY-MSKCC Partnership), NSF CBET-1055608, and Pershing Square Sohn Cancer Research Alliance.

REFERENCES

[1] *FDA Press Announcements*, **2006**.
[2] Walker, G. M., et al., *Lab Chip*, 4: 91-97, **2004**.
[3] Dereli-Korkut, Z., et al., *Anal Chem*, 86: 2997-3004, **2014**.
[4] Rodrigues, S. F., et al., *Tissue Barriers*, 3: e978720, **2015**.

USING MULTICELLULAR BUILDING BLOCKS TO ADVANCE BIOPRINTING OF 3D TISSUES

Swathi Swaminathan (1), Mi Thant Mon Soe (1), Qudus Hamid (1), Wei Sun (1), Alisa Morss Clyne (1)

(1) Mechanical Engineering and Mechanics
Drexel University
Philadelphia, PA, USA

INTRODUCTION

3D human tissue biomanufacturing would enable mechanistic study of healthy and diseased tissues, deliver a powerful platform for drug screening, and potentially provide a replacement for diseased tissues. The standard 3D tissue biofabrication process is layer-by-layer printing of a bioink composed of cells within a matrix material. While this process allows cells to be printed in a specific pattern to guide them towards the desired 3D structure, this process relies primarily on cellular self-assembly into a 3D architecture that hopefully recapitulates the *in vivo* tissue. Unfortunately, cellular self-assembly may take days or weeks, may require complex spatial and temporal environmental cues, or may not occur when two cell types are co-cultured together. Because of these challenges, critical features of the tissue microenvironment cannot be recapitulated, and therefore cell-cell interactions cannot be studied in a physiologically relevant in vitro model.

3D tissue models have already demonstrated that the tissue microenvironment, including different cell types, extracellular matrix, growth factors, and cytokines, determines how cells perceive and interpret both biochemical and biomechanical cues [1]. Thus physiologically relevant 3D cancer models are a critical intermediary between conventional 2D *in vitro* culture and human disease [2].

Our long term goal is to biomanufacture vascularized cancer tissue models to understand tumor cell interactions with the vasculature *in vitro*. The objective of this study was to develop an entirely different approach to 3D tissue biomanufacturing. Rather than bioprint individual cells and wait for these cells to form their 3D architecture, we bioprinted multicellular building blocks (e.g., breast epithelial cell spheroids) to create a model tissue that could be used almost immediately (Figure 1). The multi-cellular building blocks were either grown *in vitro* prior to printing, or they were derived from tissues (e.g., breast organoids).

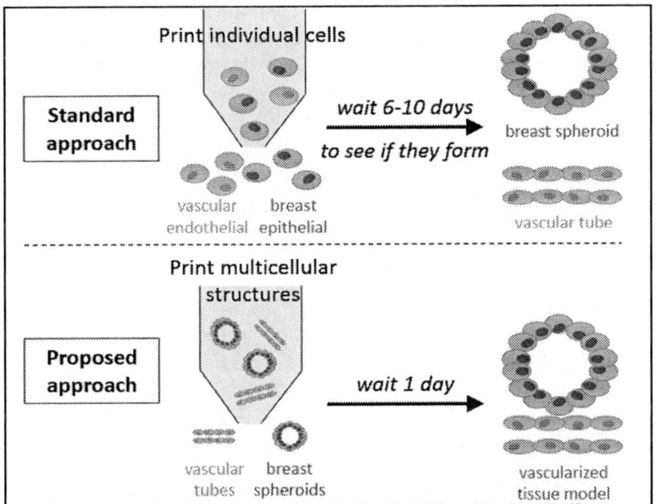

Figure 1: Standard vs. multicellular building block printing.

METHODS

Breast epithelial spheroids were created *in vitro* from the breast cancer cell lines MCF-10A and MCF10A-NeuN, which stably express human ErbB2. Breast cells were cultured in DMEM/F12 with 5% horse serum, 20 ng/ml epidermal growth factor (EGF), 10 µg/ml bovine insulin, 10 ng/ml cholera toxin, 500 ng/ml hydrocortisone, and 1% penicillin-streptomycin. Breast cell lines were used to passage 25.

To grow 3D breast spheroids, 30 µL growth factor reduced Matrigel was added to each well of an 8 chamber BioCoat Falcon

Culture slide and incubated at 37°C for 20 minutes to solidify the Matrigel. MCF10A or MCF10A-NeuN cells were trypsinized and resuspended in the appropriate medium with 20% Matrigel (3D medium). 4,000 cells were added to each chamber well. MCF10A cells formed spheroids in 8-10 days, whereas MCF10A-NeuN cells formed spheroids in 5-6 days. Medium was replaced every 4 days. Breast spheroids were then carefully pipetted out of their wells using a 1 ml pipette with 0.5 cm cut off the pipette end.

To create primary mammary organoids, normal breast tissue was removed from the mouse and minced to a fine paste. The tissue paste was then digested with 200 mg/ml collagenase and 200 mg/ml hyaluronidase in RPMI 1640 for 2 hours at 37°C. Samples were centrifuged and washed four times with additional digestion buffer. The final tissue pellet was resuspended, filtered through a 40 μm filter mesh, and stored in a low-attachment dish until printing.

A multi-nozzle solid freeform fabrication-based computer controlled direct cell writing system was used to create 3D tissue constructs as previously described [3]. The system consists of three motion arms that allow micron-scale spatial control of material deposition as well as two screw driven motors that deposit biological material from 10 mL syringes. This 3D printing system is functionalized with HEPA filtration system as well as UV light to keep a sterile environment during the bioprinting process. An independent syringe heating enclosure provided temperature control of the bioink within the syringe. The bioink was composed of MCF10A and MCF10A-NeuN cells as either individual cells (200,000 cells/ml) or as spheroids and suspended in either Matrigel or gelatin/alginate. Bioink was drawn into each syringe, capped with a nozzle tip of the appropriate gauge, and stored at the appropriate temperature to prevent gelation prior to printing. Bioink was then microextruded using a screw-driven motor (1 mL/min, 25 gauge nozzle).

After 1 to 8 days, samples were fixed with 4% paraformaldehyde, blocked with immunofluorescence (IF) wash with 10% goat serum for 90 minutes, followed by IF wash with 10% goat serum and F(ab')2 fragment goat anti-mouse IgG (1:100). Samples were incubated with a primary antibody for integrin α6 (1:100, breast spheroids) or VE-cadherin (1:500, endothelium) overnight at 4°C, followed by an Alexa Fluor 488 secondary antibody (1:200) and Hoescht 33342 (10 μg/ml, nuclei) for one hour at room temperature protected from light. After thorough washing, samples were detached from the slides and mounted on coverslips in PBS with 0.75% w/v glycine. Samples were imaged using a Zeiss LSM 700 confocal microscope.

RESULTS

We first tested whether individual MCF10A and MCF10A-NeuN cells would self-assemble into spheroids after printing in Matrigel or a gelatin/alginate hydrogel. Both cell types maintained their viability 24 hours after printing in both matrix materials (Figure 2). However, breast epithelial cells only formed spheroids after 8 days of culture in Matrigel but not in the gelatin/alginate.

We then printed preformed breast spheroids in either Matrigel or gelatin/alginate using the same printing parameters. After 48 hours, breast spheroids printed in either Matrigel or gelatin/alginate maintained their 3D architecture, including the hollow lumen characteristic of MCF10A spheroids (Figure 3, inset in bottom left shows spheroid center plane). In addition, we bioprinted primary mammary organoids in Matrigel and showed that they maintain their viability and 3D architecture after 48 hours. The organoids also showed endothelial cells, highlighting the benefit of bringing tumor stromal cells along with the mammary organoids.

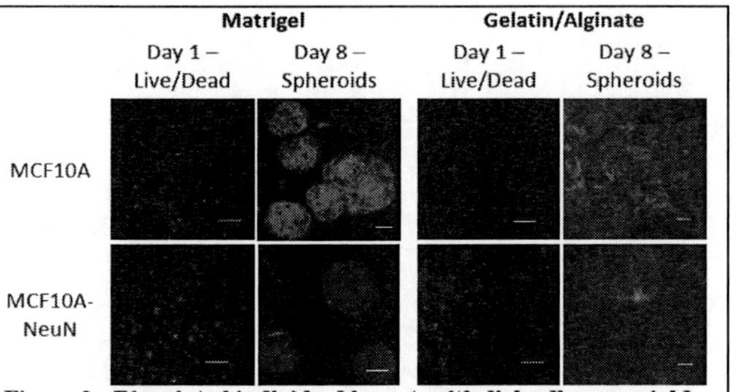

Figure 2: Bioprinted individual breast epithelial cells were viable, but only formed spheroids in Matrigel. MCF10A or –NeuN cells were trypsinized, mixed in either a Matrigel or gelatin/alginate hydrogel, and bioprinted. A Live/Dead assay was performed after 1 day, and spheroid architecture assessed by confocal microscopy (blue = nuclei, green = integrin α6). Scale bar = 20 μm.

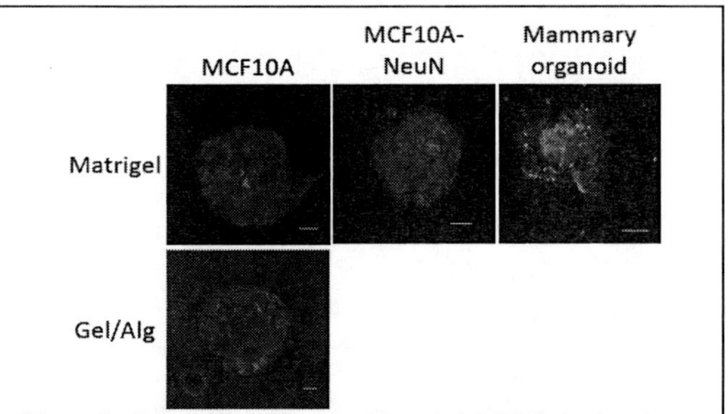

Figure 3: Spheroids and organoids maintained their structure after bioprinting in Matrigel or gelatin/ alginate. Confocal microscopy (blue = nuclei, green = integrin α6). Scale bar = 20 μm.

DISCUSSION

In this research, we created a new bioprinting paradigm in which multicellular building blocks that recapitulate *in vivo* architecture were used to create a 3D breast cancer tissue model. This is fundamentally different from bioprinting individual cells or cell aggregates without tissue-specific architecture and then providing them with biochemical and biomechanical cues to self-assemble into a tissue over time. Our method decreases the time between bioprinting and experimental assay from weeks to days; increases physiological relevance by allowing the investigator to more precisely control tissue architecture; and enables the use of primary human tissues together with stromal cells and local extracellular matrix. In the future, we will bioprint breast spheroids together with microvascular fragments to create a vascularized breast cancer model.

ACKNOWLEDGEMENTS

We thank Mauricio Reginato for providing the breast cell lines, mammary organoids, and spheroid culture techniques.

REFERENCES

[1] Radisky, D *et al. Cancer Invest,* 20(1):139-53, 2002.
[2] Bissell, MJ. *Int Rev Cytol,* 70:27-100, 1981.
[3] Khalil S, Sun W. *J Biomech Eng,* 131(11): 111002, 2009.

SB^{3}C2017
Summer Biomechanics, Bioengineering and Biotransport Conference
June 21 – 24, Tucson, AZ, USA

Characterization of 3D Bioprinted Tissue Functionality

Likitha Somasekhar (1), Cameron Hume (2), Carlos Martino (1), Kenia Nunes Bruhn (3), Kunal Mitra (1)

(1) Department of Biomedical Engineering
Florida Institute of Technology
Melbourne, Florida, United States

(2) Department of Mechanical Aerospace Engineering
Florida Institute of Technology
Melbourne, Florida, United States

(3) Department of Biological Sciences
Florida Institute of Technology
Melbourne, Florida, United States

INTRODUCTION

Engineered 3D constructs have gained attention as in vitro tools for the study of cell-cell and cell-matrix interactions and are being explored for potential use as experimental models or therapeutic replacements of human tissues. The ability to print and design all the components that constitute a tissue comprising of cells and matrix materials in three-dimension (3D) to generate structures similar to tissues is an exciting prospect of bioprinting. Vascularization of large tissue constructs is one of the primary challenge that needs to be addressed. Very few types of cells can tolerate being more than 200 µm away from a blood vessel because of the limited oxygen diffusion rate. Without a vasculature system, engineered thick tissues or organs cannot get sufficient nutrients, gas exchange or waste removal, so nonhomogeneous cell distribution and limited cell activities are observed. Systems must be developed to transport nutrients, growth factors and oxygen to cells while extracting metabolic waste products such as lactic acid, carbon dioxide and hydrogen ions so the cells can grow, proliferate and form the extracellular matrix (ECM), forming large-scale tissues and organs. However, available biomanufacturing technologies encounter difficulties in manufacturing and integrating vasculature networks into engineered constructs. In this research a 3D culture model is developed to produce tissue constructs within the surrounding extracellular matrix (ECM). Developments in understanding the many roles that the ECM plays as a key regulator in tissue architecture, cell-cell communication, proliferation, differentiation, migration, apoptosis, metastasis, etc. further highlight the need for 3D tissue culture models that more accurately represent the native ECM environment. In order to create a similarity between the human body and 3D cell culture models, a unique bioink must be created for the cells to survive. This bioink should include the extracellular matrix proteins, growth factors, and enzymes that are particular to the environment that the cells usually interact with in vivo. The current bioink being used to print the cells is an alginate-gelatin solution. This alginate-gelatin solution is enough to suspend the cells in a 3D formation, but it does not provide the nutrients or directions for the cells to grow properly. By creating a unique bioink with all the essential biological ingredients, the cells can be printed, directed, and grown into living tissue samples. A syringe based extrusion (SBE) deposition method employed in this research is based on a fluid dispensing system for extrusion and an automated three-axis system for printing. The extrusion method of bioprinting at the microscale level is currently one of the most common methods used as it is rapid and provides the best structural integrity to the material. It also allows for a wider selection of biomaterials since high viscosity biomaterial can be printed through the micro-nozzle which results in better fabrication and scale-up of 3D structures using viscous biomaterials. Additionally, cell viability in the fabricated structure by extrusion based bioprinter is reportedly higher than 90% [1]. Thus, the custom SBE system will provide a reliable, inexpensive and efficient platform for manufacturing cell-embedded 3D ECM environments with enhanced accuracy and reproducibility. The architectural design of the tissue construct is optimized to improve the diffusion limitation required for the alginate beads to diffuse into the tubular matrix in the 3D constructs and test for mechanical stabilization. Goal of this research is to perform a parametric study to optimize the cell-laden ECM matrix for enhancing tissue functionality.

EXPERIMENTAL METHODS

Bioink Formulation

Hydrogel is prepared using sodium alginate (Fischer Scientific) at different concentrations of 3.5, 2.5, 1.5 (w/v %) and gelatin (Sigma-Aldrich) at 1.5, 2.5, 3.5 (w/v %) that is dissolved respectively in DI water and 10X PBS solution under constant stirring to make a 5% (w/v

%) solution. Then 100µl of 0.5M calcium chloride (Fischer Scientific) is added dropwise to the mixture. The hydrogel is then placed in a 3ml syringe based extruder while calcium chloride is placed in another syringe controlled dispenser as shown in Figure 1. Rheology of the hydrogel was characterized to evaluate shear-dependent viscosity using a rheometer.

Human Umbilical Vein Endothelial Cells (HUVECs Promo Cell) are maintained in fresh endothelial cell growth medium and are then used to bioprint tissue constructs using the proposed SBE system which functions by dispensing filaments of alginate crosslinked with 0.5M $CaCl_2$ in a layer-by-layer fashion to create a vessel-like microfluidic channels wherein the cellular spheroids made of alginate-gelatin mixture containing cells are extruded between these channels which allow transport and diffusion of oxygenated perfused media to the cellular environment to maintain cell viability.

Live-Dead Assay

To demonstrate the viability of the cells live-dead assay is performed using Calcein-AM measured at 488-515 nm and ethidiumhomodimer-1 (Fischer Scientific) measured at 570-602 nm. Calcein-AM labels living cells with bright green fluorescent. Ethidium homodimer is a red fluorophore that stains nonviable cells but cannot penetrate living cells. Each sample is washed with PBS before live/dead staining. After 15-min incubation, imaging was performed on a ZEISS Axio Observer A.1 fluorescent microscope (Carl Zeiss Microscopy, Thornwood, NY) and a CoolSnap MYO CCD camera (Photometrics, Tucson, AZ) for image acquisition. Three different fields were counted for each sample. ImagePro software is then used for automated merging of red- and green-stained cell images to evaluate cell viability.

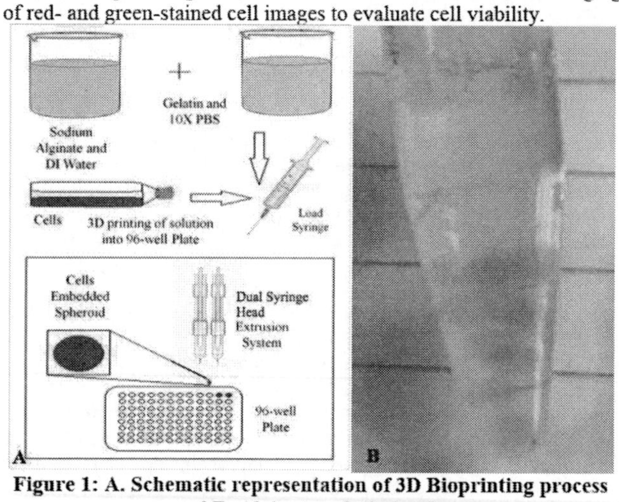

Figure 1: A. Schematic representation of 3D Bioprinting process and B. Alginate-gelatin beads.

RESULTS

To understand the viscosity of the gel and its shear thinning properties, the hydrogel was characterized using a rheometer. With increasing amounts of gelatin, the solution tends towards a more viscous behavior. The viscosity will affect the amount of gel extruded and the overall printing fidelity of the construct. The thickness of the scaffolds can be precisely controlled by regulating the thickness of one layer or printing different layers. The printing bioink (alginate/gelatin materials) has suitable mechanical properties to self-support for layer-by-layer fabrication and is suitable to be used in extrusion bioprinting.

To print the cell-laden gelatin/alginate gel an important criterion is that the material should come out of the nozzle with minimum applied shear force. Otherwise, the applied shear force may damage the cells and reduce the cell viability in the printed constructs [2]. The cell viability using the live/dead assay is imaged in the printed construct after 3D printing. As shown in Figure 2, there is minimal cell death caused by the printing process. Thus, the printing cell-laden materials and parameters used in this experiment did not have any harmful effect on the encapsulated cells (HUVECs) and can support cell viability without disruption of the constructs.

Figure 2: A. Live dead assay images of HUVECs culture and B. Bioprinted constructs at 10X magnification (scale bar, 500 µm).

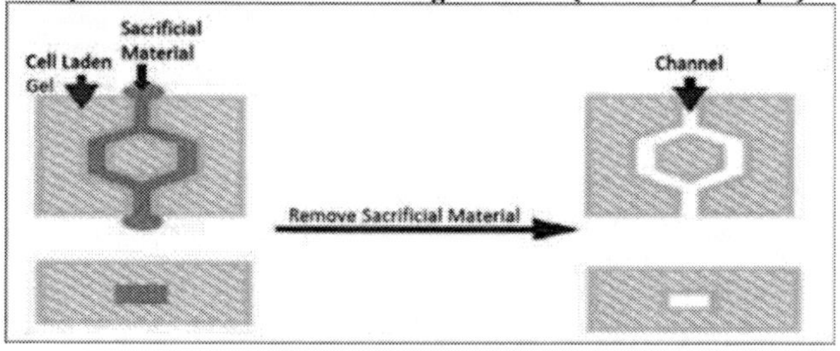

Figure 3: 3D sacrificial printing technique.

DISCUSSIONS

The proposed bioink will provide an accurate representation of the natural ECM to support the growth and function of the chosen cell lines, and the SBE system will allow for modeling of the bioink scaffold with the precise geometries and concentration gradients that are observed within mammalian tissues. Although the long-term stability of the constructs during the culture time in the medium is a requirement for potential applications in 3D tissue regeneration, alginate's non-degradable property also negatively affects bioprinted cell behavior and tissue formation [3]. Future work in this research includes sacrificial bioprinting wherein, a network of channels is fabricated through this system by creating a rigid 3D matrix of filaments by printing using alginate-gelatin mixture and then sacrificing the matrix to reveal a microfluidic architecture in the bulk material as represented in Figure 3. This extrusion system allows for a wider selection of biomaterials since high viscosity biomaterial can be printed through the micro-nozzle which results in better fabrication and scale-up of 3D structures using viscous biomaterials. A major functional requirement of an engineered vasculature is its efficacy to sustain cellular activity in metabolically demanding settings such as high cell densities. This method rapidly generates/reproduces multiscale microvascular architectures in aqueous-based matrix containing live cells.

REFERENCES

[1] Seol, Y. J. et al., *European Journal of Cardio-Thoracic Surgery*, 2014.

[2] Derby, B., *Science*, 338(6109):921-926, 2012.

[3] Rutz, A. L. et al., *Advanced. Materials*, 27.9:1607–1614, 2015.

ENGINEERING EXTRACELLULAR MATRIX BIOFIBERS BY HOLLOW FIBER CELL CULTURE

K. Roberts (1), J Schluns (2), J Jones (2), K Quinn (2), J. Hestekin (3), J. Wolchok (2)

(1) Cell & Molecular Biology Program
University of Arkansas
Fayetteville, AR, USA

(2) Department of Biomedical Engineering
University of Arkansas
Fayetteville, AR, USA

(3) Department of Chemical Engineering
University of Arkansas
Fayetteville, AR, USA

INTRODUCTION

Over 1,000,000 surgical meshes are implanted each year for the repair of hernia and other soft-tissue injuries. While synthetic materials such as polypropylene have been used to great success, the foreign body response to such materials precludes their application as regenerative biomaterials. Extracellular matrix (ECM) biomaterials are seeing increased adoption due to their preferable wound-healing and biocompatibility properties. To create ECM biomaterials our group has previously explored methods to farm the ECM secreted by populations of mammalian cells during growth in culture [1,2].

Toward this end, we have developed a technique to capture and collect cell-secreted molecules, using hollow fiber membranes (HFMs) as a cell culture platform. When cells are seeded into the HFM and grown in culture, they secrete a multitude of ECM molecules that accumulate within the lumen. The innovative step is that once the ECM has accumulated, the HFM can be thoroughly dissolved using a solvent, stripping away the synthetic HFM, and leaving behind only the accumulated ECM. The result is an engineered ECM bio-fiber comprised of the molecules secreted by living cells. In this study, we describe this novel scheme, and present the characterization results for several key ECM biofiber physical and chemical properties.

METHODS

Mesoporous polysulfone (35 kD, Sigma-Aldrich) HFMs were manufactured using a nonsolvent-induced phase inversion process known as dry-jet wet spinning. HFM pore diameter distribution was evaluated by evapoporometry. Polysulfone HFMs were rinsed thrice in deionized water to remove residual solvent and sterilized in 70% ethanol for 24 hours, and incubated in fibronectin (20ug/ml) in preparation for culture. Low passage 3T3/NIH fibroblasts, murine astrocytes, or rat skeletal muscle fibroblasts were seeded directly into the lumina of prepared HFMs. HFMs were cultured for three weeks in DMEM-F12 with 10% ascorbic acid and 5 ng/mL human TGF-β1 (Peprotech). Cultured HFMs were dissolved in n-methyl-2-pyrrolidone (NMP) and rinsed three times in deionized water to yield whole biofibers. Biofibers were decellularized using a common detergent-based method [3].

Biofiber porosity (% open space) was measured from histological sections using image analysis software (ImageJ), while the surface structure was investigated using scanning electron microscopy. Biofiber composition was investigated via fibronectin immunofluorescence, second harmonic generation microscopy, and tandem mass spectrometry with filter-assisted sample preparation. ATR-FTIR spectroscopy was used to validate the effectiveness of the NMP-based dissolution process. To evaluate cytotoxicity, decellularized biofibers were seeded with NIH/3T3 cells and stained with Calcein-AM and EthD-1 at 72 hours post-seeding. HFMs and biofiber mechanical properties were evaluated with a uniaxial tensile test.

RESULTS

HFMs exhibited an asymmetric structure, with outer and inner skin layers supported by a highly porous intermediate layer. Evapoporometry analysis indicated a mesoporous pore size distribution with an average pore diameter of 39.8 nm ± 3.9 nm SD.

Upon dissolution of cultured HFMs with NMP and subsequent water rinsing, continuous filaments of cell-derived material were collected (Figure 1B, right). For the highly productive 3T3 culture, approximately 1 mg of biofiber was collected for each mm of HFM cultured. H&E staining of cryosectioned biofibers and subsequent image analysis revealed an irregular distribution of interconnected pores, with an average porosity of 32.7% ± 5.8%. SEM revealed a fibrous external appearance consistent with that of native tissue ECM.

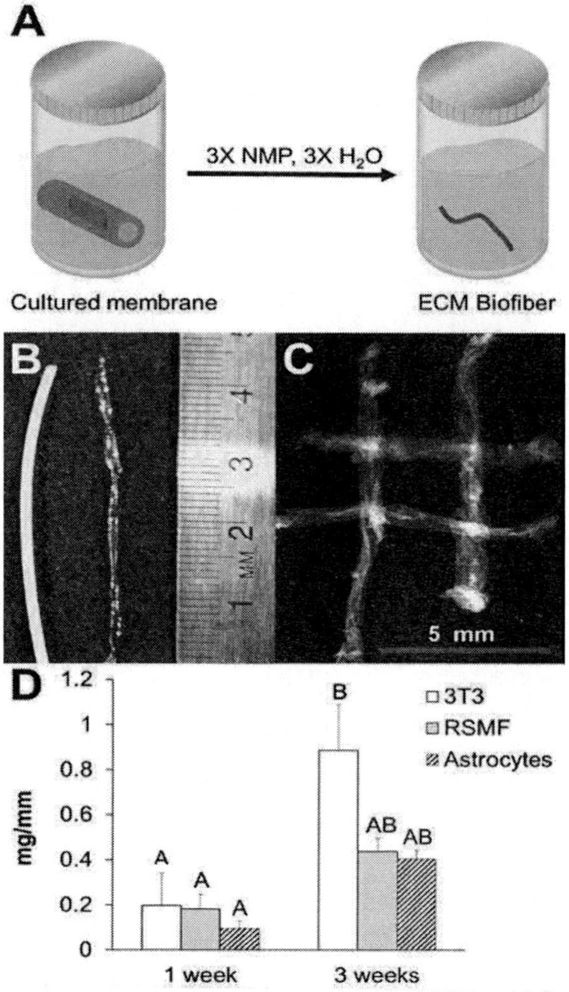

Figure 1: (A) Following 3 weeks in culture, cell seeded HFMs were (A) dissolved and the accumulated ECM was collected as a fiber (green). ECM biofibers (B and C) were amenable to handling, allowing for formation into a simple weaves. ECM biofibers were collected (D) for each cell line tested, suggesting robustness between cell lines.

DAPI and phalloidin staining revealed an absence of cellular components, suggesting thorough decellularization. Decellularized biofibers remained strongly reactive to antibodies for fibronectin. Second harmonic generation microscopy indicated the presence of fibrillary collagen. Tandem mass spectrometry and subsequent MASCOT queries revealed an array of extracellular matrix components (Table 1), including several collagens, proteoglycans, growth factors, and metalloproteinases, suggesting bio-complexity

Biofibers remained amenable to handling following decellularization and could be used to create simple mesh geometries. Average elastic modulus and ultimate tensile strength for biofibers were 34.6 ± 15.3 kPa and 5.2 ± kPa, respectively. Calcein-AM staining of cultured biofibers indicated a confluence of viable fibroblasts by 72 hours post seeding. A lack of significant EthD-1 fluorescence showed minimal cell death and was suggestive of good material cytocompatibility.

Table 1: ECM constituents identified by MS/MS.

Structural and Proteoglycans	Other
Biglycan	EFEMP 1, 2
Collagen I, III, VI, XVIII	EMILIN 1
Decorin	Fibrillin 1
Fibronectin	Fibulin 2, 5
Glypican 1, 6	HDGF
Perlecan	IGF-II
Prolargin	MMP2
Versican	MMP14
	PEDF
	Periostin
	Tenascin
	Thrombospondin 1, 2
	TIMP1
	TIMP3

DISCUSSION

In addition to their use in woven implants, ECM biofibers could potentially find utility as the print material in 3-D bio-printing systems The most common type of printers, fused deposition, utilizes long filaments to build up an object of interest. We could envision a system in which the synthetic fibers are replaced by ECM biofibers, which are then used to print organs and tissues. Furthermore, similar in concept to the various synthetic polymer chemistries available with traditional 3-D printing systems, it may be possible to print using an assortment of ECM biofibers synthesized using vascular, nerve, and cardiac, cells. Using such a system, it may be possible to bio-print tissues and organs with heterogeneous structures and compositions (ex. vascular fibers embedded within dermal fibers).

The elongated architecture of the biofibers would suggest usage as implants targeting the repair the of highly aligned muscle tissues. Of particular interest to our group is the repair of damaged skeletal muscle. Biofiber bundles fabricated using the ECM secreted by muscle derived fibroblasts like those described in this study, could be used to recapitulate the parallel alignment of native muscle ECM. We, as well as others, anticipate that scaffolds with muscle mimetic network alignment could enhance myogenesis by providing the appropriate topographical cues during healing. In whole muscle, ECM surrounds and supports multinucleated myofibers which are highly aligned in the direction of muscle contraction. The individual fibers range from 5 to 100 um in diameter, and can be as long as several centimeters, values that are comparable to biofiber synthesis capabilities.

ACKNOWLEDGEMENTS

Research reported in this publication was supported by the National Institute of Arthritis and Musculoskeletal and Skin Diseases of the NIH under award number (R15AR064481) as well as the Arkansas Biosciences Institute.

REFERENCES

[1] Wolchok JC, Tresco PA. The isolation of cell derived extracellular matrix constructs using sacrificial open-cell foams. Biomaterials. 2010;31:9595-603.
[2] Hurd S, Bhati N, Walker A, Kasukonis B, Wolchok JC. Development of a Biological Scaffold Engineered Using the Extracellular Matrix Secreted by Skeletal Muscle Cells. Biomaterials. 2015;49:9-17.
[3] Mirsadraee S, *et al.* Development and characterization of an acellular human pericardial matrix for tissue engineering. Tissue engineering. 2006;12:763-73.

SB³C2017
Summer Biomechanics, Bioengineering and Biotransport Conference
June 21 – 24, Tucson, AZ, USA

IMPROVED CHARACTERIZATION OF SPATIALLY-GRADED MECHANICAL PROPERTIES OF NANOFIBROUS SCAFFOLDS VIA INVERSE PROBLEM TECHNIQUES

Nicholas R. Hugenberg (1), David T. Corr (2), Assad A. Oberai (1)

(1) Department of Mechanical Engineering
Rensselaer Polytechnic Institute
Troy, N□, United States

(2) Department of Biomedical Engineering
Rensselaer Polytechnic Institute
Troy, N□, United States

INTRODUCTION

Due to high rates of injury and poor intrinsic healing of ligaments and tendons there is significant interest in engineering their functional replacements. In order to successfully mimic the performance of their natural counterparts, these replacements must have very specific properties. They must have a nonlinear stiffening response to loading that ensures easy movement at low strains (i.e., appropriate toe-in region behavior) and must be spatially-graded; in particular they must be stiff near bony attachment sites and more compliant in the middle.

Traditional tensile tests are sufficient to determine bulk structural and material properties and to assess whether an artificial candidate exhibits the needed low-load strain-stiffening behavior. However, these tests alone cannot determine the spatial distribution of a sample's material properties. Moreover, grafts for tendon and ligament replacement are too large for practical examination by methods such as atomic force microscopy, presenting a serious obstacle to verifying that potential tendon or ligament replacements are constructed with appropriate spatially-graded properties.

Our group has previously developed a technique to circumvent this difficulty, using Digital Image Correlation (DIC) to collect the full-field displacement data within samples during tensile tests, discretizing these to nodal displacements, and using these data to solve an inverse elasticity problem, generating a reconstruction of the spatial distribution of the shear modulus (SM) within a given sample [1]. However, the reconstructed results, while promising, tended to have artifacts and had poor user-to-user reproducibility. To overcome these limitations, this research aims to refine the reconstruction process by reducing noise in the displacement data and modifying the parameters of the inverse problem.

METHODS

Trials were conducted using eight 0.6 x 3.6 cm graded fiber samples; these samples were prepared as previously described [1] and displacement data was collected during tensile testing using a two-camera DIC system (Correlated Solutions, Inc.). This data was then used as the input for the inverse problem solver, NLACE, which determined the shear modulus distribution [2]. In order to improve the reconstruction process, we made adjustments to NLACE and the input data by: developing a more objective approach to regularize solutions; changing the boundary conditions used; eliminating assumptions about the incompressibility of the samples and the quality of the data in the axial and lateral directions; and using Principal Component Analysis (PCA) to remove noise from our data.

We first addressed the selection of regularization parameters for NLACE. The solver minimizes a sum of two terms: a displacement-matching term, which is the difference between predicted and measured displacements, u and u^m, and a Tikhonov regularization term which helps to account for noise in the data by smoothing the reconstruction [2]. The influence of the smoothing term is controlled by a regularization parameter α whose value was simply determined by the operator in previous tests. In order to make this process more accurate and less subjective, we chose α using an L-curve. A separate L-curve was generated for each sample by creating a series of reconstructions using different α values and plotting the log of the displacement matching term against the log of α. We compared reconstructions to identify characteristic regions in these curves that represented the optimal compromise between smoothing and displacement matching. Using these regions allowed us to choose α in the same way for each sample.

Poster Presentation #P119

Next, we proceeded to change the boundary conditions of the problem. The conditions needed to conform to the conditions of the tensile tests, so we imposed Dirichlet conditions on combinations of both the axial and lateral displacements on all sides of the sample, primarily investigating conditions that were fixed on all sides of the sample and those that were fixed on the top and bottom surfaces (i.e., zero relative displacement between sample and grip) and free on the sides.

Following this we inspected two of the basic assumptions made in previous tests: first, that the accuracy of the lateral and axial displacement data was the same; second, that the samples were perfectly incompressible. This first required examining the effect of diminishing the weights given to the lateral and axial data in NLACE, while the second involved dropping the enforced incompressibility condition. The change to a compressible model also allowed us to get more information about each sample: we reconstructed both the shear modulus and the first Lamé parameter λ simultaneously, allowing us to calculate the spatial variation of the Poisson's Ratio υ.

Finally, we used PCA to remove noise from the displacement data. We kept the maximum average strain in the sample below 5% in order to work in the small-strain regime and used incremental displacement data, treating each loading step as an independent realization of the "same" measurement. Applying PCA allowed us to separate the data into a number of modes equal to the number of loading steps used. Each of the modes holds data accounting for a fixed percentage of the total variance in the original data; weighted combinations of the modes accounting for the highest percentages were treated as the de-noised signal while the other modes were discarded as noise. The de-noised data were then used in NLACE to produce our final reconstructions.

RESULTS

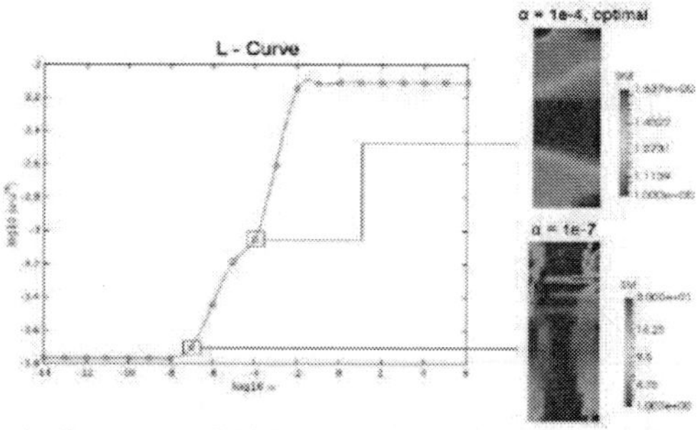

Figure 1: L-curve and SM reconstructions, incompressible case

We found that L-curves across all samples had a characteristic short plateau in the middle of the curve, as seen in the α = 1e-4 to 1e-5 range in Fig. 1. The α values producing the best results were consistently the largest of the values spanning this plateau. Once the assumption of incompressibility was dropped, α = 1e-3 became the final value used for all tests.

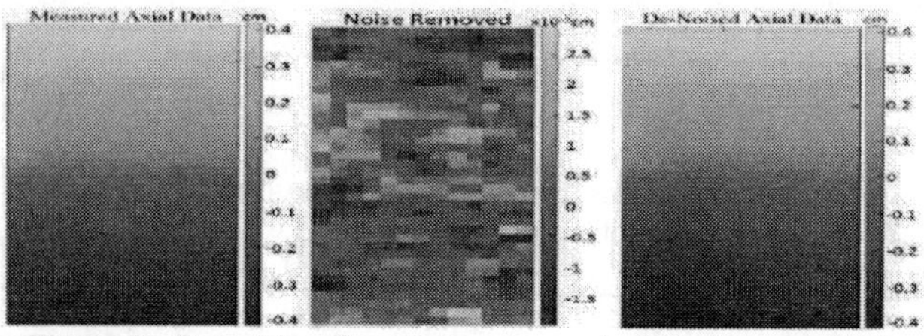

Figure 2: De-noising the axial component of displacement data

PCA allowed us to effectively remove noise from our data and easily visualize that noise (see Fig. 2), which allowed us to qualitatively verify that we were not removing the systematic parts of the signal. The difference between the de-noised and raw data, though it seemed small, was enough to sharpen our reconstructions and make them less sensitive to changes in α.

The following additional conditions consistently yielded good reconstructions: enforcing Dirichlet conditions on the bottom edge of the samples for both axial and lateral data and for the axial data on the top edge; assuming compressibility; and scaling the weight of the lateral data from 1 down to approximately 0.35. Fig. 3 demonstrates that we were able to significantly improve the quality of reconstructions by achieving higher contrast, better symmetry, and more realistic estimates of υ than in our previous work. In the final reconstruction we can observe the graded material properties (large SM values on the ends and smaller values in the center) that are desired for our application.

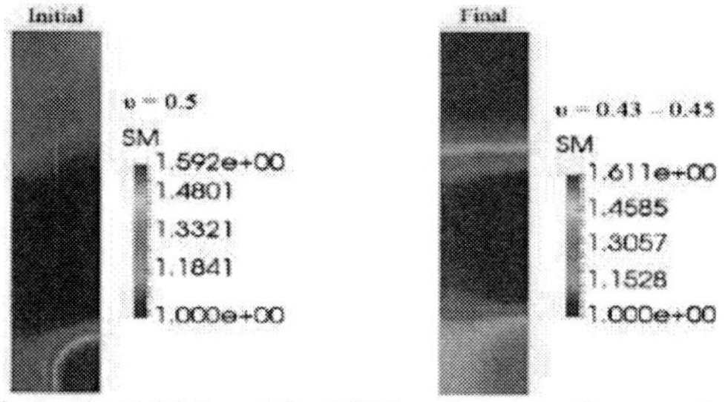

Figure 3: Initial and final SM and υ range for sample 8

DISCUSSION

The shear modulus reconstructions are much better, but are still limited by the presence of artifacts: notably, almost all samples showed spots of high or low modulus in one or more corners, which in some cases could extend through significant portions of the domain. The source of these artifacts is difficult to determine: they could be results of the optimizer, the data itself, or the boundary conditions. However, the impact of the artifacts is lessened by their tendency to remain small in magnitude. Their modulus is typically offset from the modulus of the surrounding area by no more than ± 10% in the final reconstructions, as opposed to much larger variations seen in older results. This, along with the increase in robustness afforded by using PCA and L-curves, gives us confidence that our improved method can be used to effectively characterize properties in spatially-graded samples, allowing for verification of the composition of tendon and ligament replacements and similar heterogeneous structures. Future work in this direction involves determining the spatial variation of the nonlinear elastic properties of samples.

ACKNOWLEDGEMENTS

This project was supported, in part, by the RPI Presidential Graduate Fellowship and NSF CAREER (#0954990, DTC)

REFERENCES

[1]Oberai A.A. et. al. "Engineering and Characterizing Nanofibrous Composite Scaffolds with Spatially-Varying Architecture", *ISL&T-XIV*

[2]Oberai A.A. et. al, Solution of inverse problems in elasticity imaging using the adjoint method. *Inverse Probl,* 19(2),297, 2003.

SB^{3}C2017
Summer Biomechanics, Bioengineering and Biotransport Conference
June 21 – 24, Tucson, AZ, USA

HUMAN ADIPOSE DERIVED STEM CELLS CULTURED ON POROUS POLY (L-LACTIC ACID) SCAFFOLDS PREPARED BY THERMALLY INDUCED PHASE SEPARATION METHOD

Harish Chinnasami and Ram Devireddy

Department of Mechanical Engineering,
Louisiana State University,
Baton Rouge, Louisiana 70803, USA

INTRODUCTION

A tissue-engineered scaffold structurally mimics the intended tissue to be developed with adequate pore sizes for the stem cells to proliferate and differentiate. The initial scaffolds developed were primarily composed of ceramic compounds like hydroxyapatite, akermanite and tricalcium phosphate that are structurally rigid and resemble the mineral composition in bones. In recent years, many research works stressed on the advantages of using bio-compatible polymers such as poly (l-lactic acid) (PLLA), Polycaprolactone (PCL), poly (l-glycolic acid) (PLGA) as a scaffold material. Some of the prominent methods used to make polymer scaffolds are solvent casting, fiber bonding, membrane lamination and particulate leaching [1]. Absolute control over the 3D micro-structure of the scaffolds has been achieved by advanced methods such as computational topology design (CTD) and solid free-form fabrication (SFF) [2]. But these methods are very expensive when compared to TIPS, which could be easier to implement in large scale production.

The objective of this study was to (i) select suitable PLLA scaffolds for cell culture, (ii) to observe the viability of human adipose stem cells (hASCs) when cultured on these scaffold and (iii) to quantify the cellular growth inside the PLLA scaffolds selected. The future scope of this project is to make cryo-preservable bone-grafts for *in vivo* implantation into patients when needed.

MATERIALS AND METHODS

In thermally induced phase separation (TIPS) method, crystalline PLLA was dissolved in 1,4 Dioxane at 3, 7 and 10 (wt/vol)% to form a homogenous clear solution. These solutions were each frozen to a low temperature (-60°C), at 1, 10 and 40°C/min cooling rates (C.R.). The frozen solutions were lyophilized (freeze-dried) to remove the frozen solvent resulting in the final scaffold, which was 3D porous PLLA structure. The porous morphology of the scaffold was partly determined by the crystal formation of dioxane in the solution which was influenced by the applied freezing profile.

Porosity (%) of a scaffold was calculated at two positions (0-5 mm and 5-10 mm from bottom) by weighing a known volume of the scaffold and from the following formula:

$$Porosity\ (\%) = \left(1 - \frac{Scaffold\ Density}{Polymer\ Density}\right) \times 100 \quad (1)$$

Table 1: Porosities of the scaffolds made using the specified PLLA:Dioxane (wt/v) composition and the C.R.

PLLA: Dioxane	-1°C/min		-10°C/min		-40°C/min	
	(0-5) mm	(5-10) mm	(0-5) mm	(5-10) mm	(0-5) mm	(5-10) mm
3% (wt/v)	96.96%	96.89%	96.34%	96.75%	95.56%	95.50%
7% (wt/v)	94.16%	93.98%	93.92%	93.96%	91.81%	93.28%
10% (wt/v)	90.47%	89.46%	89.25%	89.69%	88.82%	88.39%

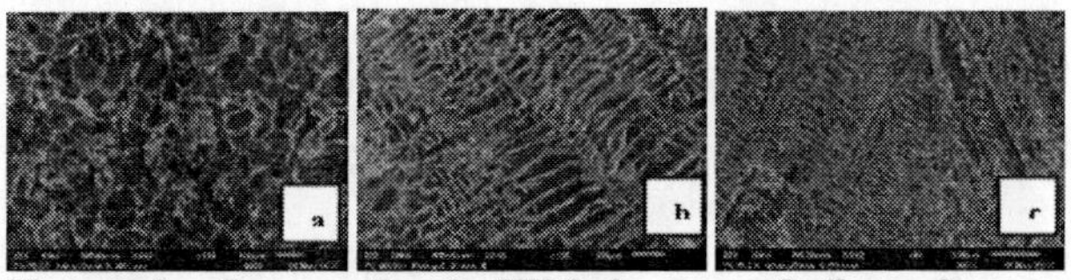

Figure 1: Cross-sectional SEM images of the micro-pore structures of the 7% (wt/vol) scaffolds cooled at (a) 1°C/min, (b) 10°C/min and (c) 40°C/min

Scaffold sections of 12mm diameter and 5mm height were coated with Platinum (15nm thick) using EMXS550X Sputter Coater. JEOL JSM-6610LV Scanning Electron Microscope was used to scan the samples and capture SEM images to qualitatively observe the pore structure and measure the pore sizes.

INSTRON 5900 Series – Advanced Mechanical Testing Systems Compression tester was used to compress scaffolds of 12mm in diameter and 10mm in height to 20% of its initial height, while simultaneously recording the applied load and strain. Compressive modulus was calculated as follows:

$$Compressive\ Modulus = \frac{\left(\frac{Applied\ Load}{Cross\ sectional\ area\ of\ scaffold}\right)}{Compressive\ strain} \qquad (2)$$

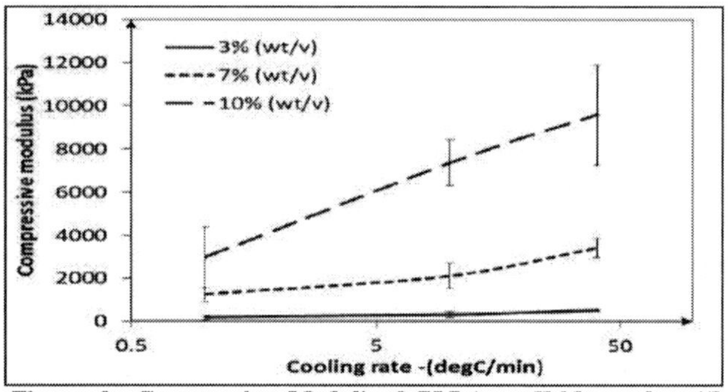

Figure 2: Compressive Moduli of PLLA scaffolds made with PLLA in Dioxane in 3, 7 and 10% (wt/vol) initially and also their variation with increasing cooling rate.

From the compression test results shows that 3% (wt/vol) scaffolds have very low modulus resulting in a weak scaffold that is difficult to manipulate during cell culture. Scaffolds cooled at 1°C/min showed irregular microporous structure. Therefore 7 & 10% (wt/vol) frozen at 10°C/min was selected for cell culture.

Cell Culture: Cryo-preserved hASCs were thawed and cultured in traditional Stromal Medium (SM) containing 90%(v/v) DMEM/F-12 (Gibco), 10%(v/v) Fetal Bovine Serum (Gibco), Penicillin - 90 IU/mL , Streptomycin – 90 µg/mL and Amphotericin B – 225 ng/mL. The PLLA Scaffolds were attached to the bottom of the 6-well plates using 1% Agarose (Sigma) as an adhesive. Equal number of hASCs was cultured simultaneously on both 3D porous scaffolds and 2D wells without scaffold as a control.

DNA Quantification: Number of cells on each sample at time point 0: ~2×10^5. The PLLA Scaffolds, homogenized with Dounce tissue grinder pestle (7 mL) (Sigma) as well as the control samples were incubated at 37°C with Lysis buffer (0.2 mg/mL Sodium Dodecyl Sulfate and 0.2 mg/mL Proteinase K) for 20 mins. The DNA present in the Lysis buffer was quantified using Quant-iT™ PicoGreen® dsDNA Assay Kit (Invitrogen) according to the manufacturer's protocol. The total DNA in each sample was normalized with the surface area exposed to SM in 1mm (150.80 mm^2), 5mm (263.89 mm^2) thick scaffolds and compared to 2D cell growth on control (962 mm^2).

Live/Dead Viability Assay: Number of cells on each sample at time point 0: ~10^4. Samples were incubated in staining solution of 2 µM Calcein AM and 4 µM Ethidium homodimer-1 (EthD-1) was prepared using LIVE/DEAD® Viability/Cytotoxicity Kit (Molecular Probes) stock solutions in DPBS for 20 mins at 37°C and were taken for fluorescence imaging using SteREO Lumar.V12 (Zeiss).

RESULTS

The proliferation of hASCs in scaffolds was measured by quantifying the amount of DNA in scaffolds and the viability was checked by fluorescence imaging after staining with live/dead dyes.

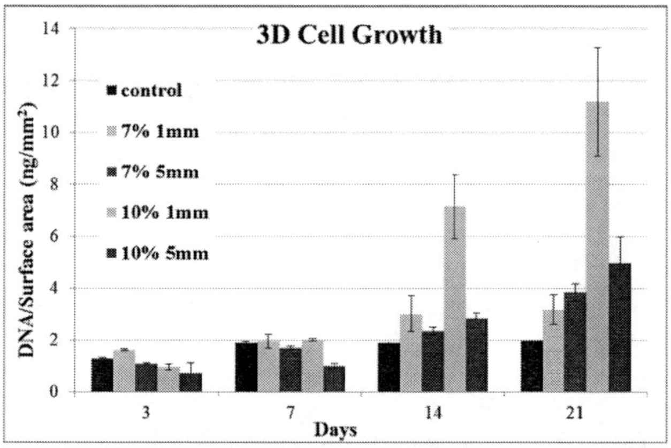

Figure 3: 3D cell growth measured by relative increase of DNA surface density (ng/mm^2) of scaffolds over control with time progression. Sample size, n=6 for each data point.

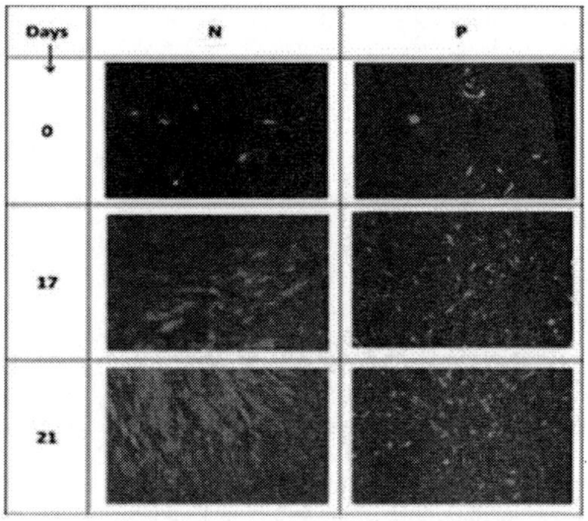

Figure 4: Live Cells (green) and dead cells (red) in Control samples (N) and PLLA Scaffolds (P) during time points 0, 17 and 21 days.

DISCUSSION

Fig.3 shows DNA surface density increases with time in 3D scaffolds compared to 2D control where the cell growth stops at ~1.95 ng/mm^2 at day 7. This implies that cells are permeating through the scaffolds' micro-porous structure. 10% scaffolds show better cell growth than 7% even though the pore-sizes were comparatively low. This maybe because PLLA indeed promotes cell growth as seen in live/dead stain showing very little dead cell (similar to control). Therefore, PLLA is also bio-friendly and shows no cytotoxicity to hASCs. The abundant growth in 10% 1mm scaffolds maybe because of better exposure to SM. Hence, either regular microstructure in scaffolds or simple increase in PLLA quantity could be favorable for cell growth, which will be investigated in future.

REFERENCES

1. Qingpu Hou et al., *J.Biomed Mater Res. Part B: Applied biomaterials*, 67(2):732-740, 2005
2. Scott J. Hollister, *Nature Materials*, 4:518-524, 2005.

IMPACT OF CELLULAR CHOLESTEROL ON MONOCYTE CHEMOTAXIS

Amit K. Saha (1,2), Shatha F. Dallo (1), Anand K. Ramasubramanian (1,2)

(1) Department of Biomedical Engineering
The University of Texas at San Antonio
San Antonio, Texas, USA

(2) Department of Biomedical, Chemical & Materials Engineering
San Jose State University
San Jose, California, USA

INTRODUCTION

Cholesterol plays an important role in maintaining the integrity and functionality of plasma membrane [1]. Cholesterol has been shown to modulate cell stiffness [2] as well as affect various cellular processes such as protein sorting and cell signaling [3]. Apart from the changes in plasma membrane, cellular cholesterol plays also significantly affects the cytoskeleton. This in turn influences the biophysical response of cells. For example, decrease of membrane cholesterol content causes enhanced cell spreading and formation of stress fibers in fibroblasts . Cholesterol has also been shown to dictate the cytoskeletal properties in other cell types such as endothelial cells and human embryonic kidney cells . One important cellular process where the membrane and cytoskeleton act in tandem is chemotaxis, which is the directed movement of cells in response to chemical signals. Chemotaxis plays an extremely vital role in inflammation [4]. Cholesterol has been shown to dictate the chemotactic response in neutrophils and T cells among others.

The plasma membrane is segregated into lipid rafts, which are dynamic, heterogeneous regions, rich in gangliosides, cholesterol and glycosphingolipids. Lipid rafts regulate a variety of cellular processes such as protein trafficking and signal transduction . We have already observed in our lab that modification of cellular cholesterol content alters the membrane lipid raft organization and the overall membrane fluidity in human monocytes. Consequently, various receptors also exhibit varied expression and distribution levels.

The effect of cholesterol on the chemotactic and spreading response as well as the underlying causes remain understudied despite the interplay between cholesterol and monocytes in inflammatory diseases such as atherosclerosis. We depleted and enriched the baseline cholesterol content in human monocytic cell line THP-1 to understand the role of cholesterol in THP-1 biomechanics. Cholesterol depletion resulted in a decreased chemotactic potential, and enrichment caused enhanced chemotaxis. In order to probe the probable causes for these changes, we assessed the alterations in both the membrane and cytoskeletal properties.

METHODS

Cell Culture. Human monocyte cell line, THP-1 (ATCC) was cultured in RPMI 1640 (ATCC) supplemented with 10% FBS and .05 mM mercaptoethanol (Sigma-Aldrich), at 37°C and 5% CO_2. The cells were passaged into fresh media when the cells reached a density of 10^6/ml. The cell viability was measured by trypan blue exclusion assay using the Countess automated cell counter (Life Technologies).

Cholesterol treatment. Cholesterol depletion was performed using Methyl-β-cyclodextrin (MβCD; Sigma-Aldrich) and enrichment using MβCD-cholesterol complex, respectively. Cells were washed once with RPMI and centrifuged at 130xG for 7 minutes at room temperature (RT), and resuspended at a concentration of $1x10^6$/ml in RPMI. Desired concentration of either MβCD or MβCD-cholesterol solution was mixed with the cell suspension, and incubated for 30 mins. at 37° C and 5% CO_2. After 30 minutes, cell viability was measured using the trypan blue exclusion assay. The cell suspension was centrifuged at 130xG for 7 minutes at RT, and then resuspended at a concentration of $1x10^6$/ml in RPMI. This was repeated thrice, and cell viability was measured again. The cell suspension was centrifuged one more time. The supernatant was discarded and the cells were resuspended in suitable media and concentration as required by a particular assay. Before utilizing the cells for any assay, the cell viability was assessed one more time.

Cell spreading. 8-well chamber slides (ThermoFisher Scientific) were coated with 500 μL 10μg/cm² type 1 collagen per well (Corning Inc.). 50000 cells were seeded per well and the slides were left undisturbed at 37°C for 3 hours. 100 nM PMA were added to activate the THP-1 cells. Upon reaching the specific time point, the cells were washed once with PBS and fixed with 2% Formaldehyde, followed by processing for scanning electron microscopy. The cells were fixed with a solution of glutaraldehyde (2.5% w/v) in 0.1 M sodium

Poster Presentation #P121

cacodylate buffer at pH 7.4 for 2 h at 37°C. Following fixation, they were treated with a solution of osmium tetroxide (1% w/v) in 0.1 M sodium cacodylate buffer at pH 7.4 for 2 h at room temperature. The samples were rinsed with water and soaked in a series of ethanol solutions, ending with 100% ethanol. After dehydration, the samples were dried overnight in a vacuum dryer and subsequently coated with a 60:40 gold-palladium alloy; approximately 10 nm thick using a Cressington Sputter coater for a duration of 30 sec. The cells were imaged using a Hitachi 1510 Scanning Electron Microscope, followed by subsequent image analysis.

Chemotaxis. Monocyte chemotactic response to MCP-1 and IL-8 were assessed using 48 well plate transwell inserts. The cells were allowed to undergo chemotaxis for 1.5, 3, and 6 hours. Following that comparative cell analysis between the different conditions were done from fluorescence readings using a plate reader (Biotek Synergy 2). To investigate the relative importance of the various receptors, they were blocked with the relevant antibodies and the experiment was repeated.

Receptor distribution and expression levels. The changes in the relevant receptors' expression and distribution levels were investigated using multispectral imaging flow cytometry.

Overall cellular deformability. A high throughput microfluidic device was fabricated for investigating single cell deformability. The designed pattern was transferred to silicon wafers using standard photolithographic techniques. The test channels in the device had a cross-section area of 5μmx5μm. The silicon mold was used to make polymeric micro-channels using Polydimethysiloxane (PDMS; Dow Corning). The PDMS blocks with the embedded micro-channels were attached to clean glass slides (Fisher Scientific) using oxygen plasma treatment (Harrick Plasma). Silicon connector kits (Cole-Parmer) were used so as to get better adhesion with PDMS. The channels were passivated by perfusing 5% HSA and washed with PBS.

A custom nitrogen based delivery system was used to deliver the cell suspensions at a concentration of 0.75 x 10^6/ml. High speed videos were taken at 20X and 40X magnifications (Leica Microsystems) using a high speed sCMOS camera (PCO-Tech) and subsequent image processing using ImagePRO (Media Cybernetics Inc.) was done.

Statistics. All experiments were performed in triplicate, and each experiment was repeated at least thrice under independent conditions. The results are represented as mean ± SEM from one representative experiment. Statistical differences between treatments were evaluated using one way ANOVA with Tukey's post-hoc test, and significance was reported at α=0.05.

RESULTS

Monocyte chemotaxis. THP-1 cells did not show much difference in chemotaxis behavior in response to IL-8. However, significant changes were observed towards IL-8. Cholesterol enriched cells exhibited the highest degree of chemotaxis (Figure 1). Blocking the receptors eliminated the differences for shorter periods of duration. Over longer times some difference was observed with cholesterol enriched cells having the largest number of monocytes passing through the transwell inserts. This was plausibly due to the differences in cellular deformability.

Spreading behavior of monocytes. Alteration of cellular cholesterol significantly impacted the spreading properties of monocytes. Cholesterol depletion led to more elongated cells as compared to the untreated control or the cholesterol enriched cells. The phenomenon was consistent for shorter (1 and 3 hours) as well as longer (8, 24 and 48 hours) time periods

Changes in receptor expression and distribution. We probed the changes in expression and distribution levels of CCR2, which is the most well-known receptor on THP-1 towards MCP-1. We found that cholesterol enriched cells had highest expression levels (Figure 2).

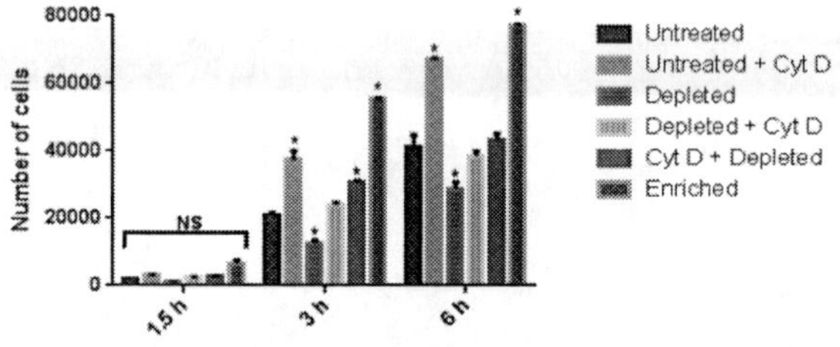

Figure 1. THP-1 Chemotaxis in response to MCP-1

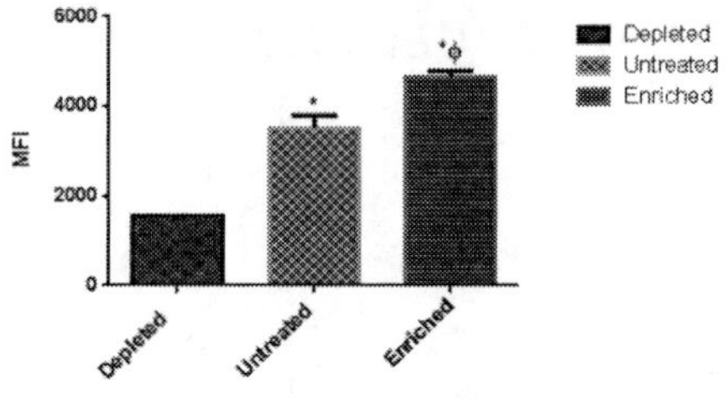

Figure 2. Expression levels of CCR2

Changes in cellular deformability. We probed the changes in cellular deformability as a function of cholesterol load in order to understand possible links with the changes in monocyte biophysical response. Cholesterol depletion was observed to decrease the cellular deformability and cholesterol enrichment resulted in the highest cellular deformability.

DISCUSSION

This work establishes a link between cellular cholesterol load and chemotactic response in monocytes. We observed significant changes in chemotaxis of monocytes due to altered cholesterol loads. And as probable reasons for the observed changes, we investigated the changes in membrane properties and cellular deformability.

Since cholesterol is responsible in maintaining the integrity of the plasma membrane, disturbance in its natural balance alters the membrane fluidity and the lipid raft distribution and expression. Consequently, the receptors on the membrane also show altered expression levels and distribution. These can greatly influence the biophysical response of monocytes. Differences in cellular deformability might also have a big role to play. We are currently developing computational models to assess the individual contributions from the different receptors as well as the deformability. Our results can provide the foundation for an enhanced understanding of mechanistic features of diseases like atherosclerosis.

ACKNOWLEDGEMENTS

We gratefully acknowledge the help of Dr. Shatha Dallo, Dr. Chiungyu Hung, Dr. Marylene Palard (Microelectronics Research Center, UT Austin) and Dr. Shankar J. Evani. This work was supported by NIH HL112629. This work received technical and instrument support from The Immunology Defense Core at UTSA, funded by the National Institute on Minority Health and Health Disparities from the National Institutes of Health.

REFERENCES

1. Roduit, C., et al., *Elastic membrane heterogeneity of living cells revealed by stiff nanoscale membrane domains.* Biophys J, 2008. **94**(4): p. 1521-32.
2. Saha, A.K., et al., *Cellular cholesterol regulates monocyte deformation.* Journal of Biomechanics, 2016.
3. Pande, G., *The role of membrane lipids in regulation of integrin functions.* Current Opinion in Cell Biology, 2000. **12**(5): p. 569-574.
4. Wong, C.H.Y., B. Heit, and P. Kubes, *Molecular regulators of leucocyte chemotaxis during inflammation.* Cardiovascular Research, 2010. **86**(2): p. 183-191.

NUMERICAL INVESTIGATION OF THE ROLE OF INTERCELLULAR FORCES ON COLLECTIVE CELL MIGRATORY BEHAVIORS

Liqiang Lin (1), Xiaowei Zeng (2)

(1) Department of Mechanical Engineering
University of Texas at San Antonio
San Antonio, Texas, US

(2) Department of Mechanical Engineering
University of Texas at San Antonio
San Antonio, Texas, US

INTRODUCTION

Cell migration is a fundamental biological process throughout all the stage of animal life, from its commencement to its end. When cells migrate in loosely or closely associated groups, it can be referred to as collective cell migration. A variety of fundamental processes in biological system, such as embryonic morphogenesis, wound healing and tumor progression, depend on the coordinated motion of cell groups. Although factors affecting migratory behavior of single cell are beginning to be understood, still little is known about directed migration when cells are in collective groups.

Coordinated movement of cell group could be regulated through dynamic interaction with both neighboring cells and the extracellular matrix (ECM). The adhesion of cell-cell junction would regulate cell migration from random to collective migration [1]. On the other hand, the integrity of cell-cell junctions plays an important role on collective cell movement in morphogenesis, tissue repair and cancer[2]. The cell-cell adhesion combined with contractility would largely determine the rearrangement of epithelial cell layers. In addition, a study on collective cell migration through cooperative intercellular forces demonstrated that the mechanical stresses exerted at cell-cell junctions will regulate the monolayer coordinated movements by navigating cells motion along orientations of minimal intercellular shear stress [3].

To further study the effects of cell-cell interaction property, including the interaction forces in different directions in cell-cell junction, we adopted an interfacial zone model to model the cell-cell interaction. Via the proposed interfacial zone model, we can describe different behaviors and properties of cell-cell interaction. Through this study, the simulation results were consistent with the experimental observation that the orientation of local principal stress was alignment with that of migration velocity. It was also found that the normal interfacial strength has dominant effects on coordinated migration of epithelial monolayer sheet.

METHODS

In the collective epithelial cell migration study, the forces control the collective cell migration can be modeled as adhesion force between cell and ECM, protrusion force at the cell leading edge due to actin polymerization, the drag force arising from viscous resistance to cell movement, and the cell-cell adhesion force for collective cell migrations:

$$f = f_{adhesion} + f_{protrusion} + f_{resistance} + f_{cell-cell}$$

The normal or shear traction from intercellular adhesion in cell-cell junction is shown in Fig. 1.

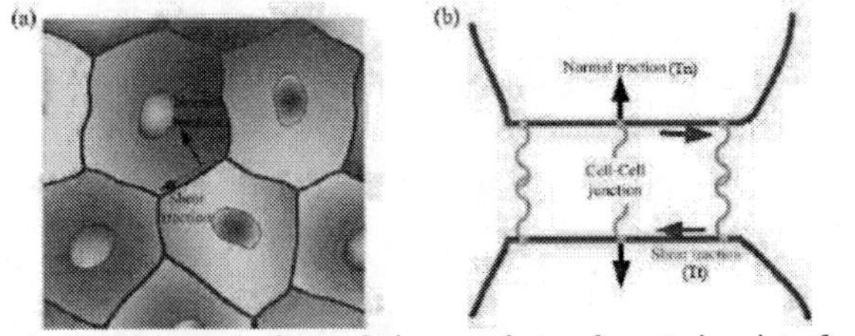

Fig. 1 Traction-separation relation used to characterize interface behavior:(a) normal direction, (b)tangential direction

The mechanical intercellular interactions in cell-cell junction were governed by an interfacial zone model[4] and it takes the forms as:

$$T_n = \begin{cases} \sigma_c \left(\frac{\Delta_n - \delta_0}{\delta_{dn} - \delta_0}\right) \left[e^{1 - \frac{\Delta_n - \delta_0}{\delta_{dn} - \delta_0}} \right]^{q_n} & \Delta_n \le \delta_{dn} \\ \sigma_c \left(\frac{\delta_{fn} - \Delta_n}{\delta_{fn} - \delta_n}\right)^{p_n} & \delta_{dn} < \Delta_n < \delta_{fn} \\ 0 & \Delta_n \ge \delta_{fn} \end{cases} \quad (1)$$

$$T_t = \begin{cases} \tau_c \left(\frac{\Delta_t}{\delta_{dt}}\right) \left[e^{\frac{1}{2} - \frac{\Delta_t^2}{2\delta_{dt}^2}} \right]^{q_t} & 0 \le |\Delta_t| \le \delta_{dt} \\ \tau_c \frac{\Delta_t}{|\Delta_t|} \left(\frac{\delta_{ft} - |\Delta_t|}{\delta_{ft} - \delta_{dt}}\right)^{p_t} & \delta_{dt} < |\Delta_t| < \delta_{ft} \\ 0 & \Delta_n \ge \delta_{ft} \end{cases} \quad (2)$$

In the proposed interfacial zone model, there are six independent parameters (σ_c, δ_{dn}, δ_0, δ_{fn}, q_n and p_n) in normal direction and five independent parameters (τ_c, δ_{dt}, δ_{ft}, q_t and p_t). Here, σ_c and τ_c are intercellular adhesion strength in normal and tangential direction, respectively; the δ_0 is the original gap of cell-cell junction and $e = 2.71828$; δ_{dn} and δ_{dt} are initial cutoff interactions between cells in normal and tangential directions; the shape parameters (q_n, q_t) are introduced to describe different intercellular behaviors before initial cell-cell surface separation and the shape parameters (p_n, p_t) are used to describe different intercellular behaviors after initial cell-cell surface separation; δ_{fn} and δ_{ft} are normal and tangential cutoff interaction distance beyond which no interaction occurs between cells. The Δ_n and Δ_t are normal and tangential cell-cell surface separation, respectively.

From the experimental observation [3], it was found that the local normal stress is severely heterogeneous and the value of that is exceeding $300Pa$ in tension. To compare with the experimental observation, the distribution of average normal stress and maximum shear stress was first plotted with respect to the time. Then, we studied the effects of cell-cell interaction adhesion on the collective behavior.

RESULTS

As shown in Fig. **2**, we observed that the distribution of average normal stress is heterogeneous compared to maximum shear stress. The fluctuations of average normal stress occur over the whole monolayer sheet. In addition, the value of average normal stress is greater than that of maximum shear stress and the dominant maximum shear stresses in the monolayer were close to zero (Fig. **2**). These observations were consistent to the experimental investigation [3].

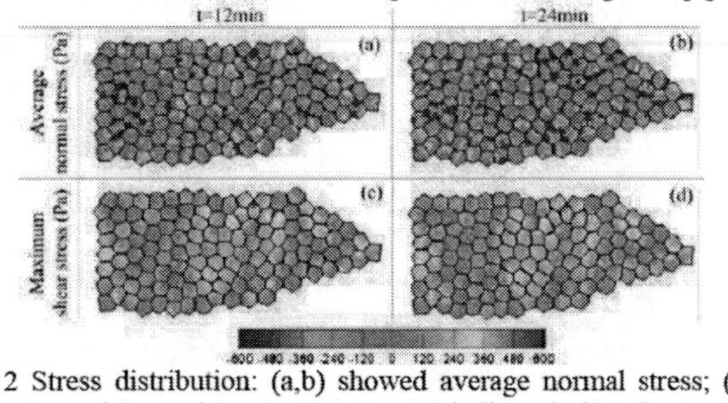

Fig. 2 Stress distribution: (a,b) showed average normal stress; (c,d) showed maximum shear stress. It was indicated that the average normal stress $(\boldsymbol{\sigma_{max}} + \boldsymbol{\sigma_{min}})/\mathbf{2}$ is greater than shear stress $(\boldsymbol{\sigma_{max}} - \boldsymbol{\sigma_{min}})/\mathbf{2}$ and the dominant maximum shear stresses were close to zero.

By plotting the migration direction versus time, we observed that the tangent of migration direction decreased with the increase of normal intercellular adhesion strength (Fig. **3**(a)). However, the tangent of migration direction was almost the same with different shear intercellular adhesion strength (Fig. **3**(b)), which means the shear intercellular adhesion strength has very minor effects on the migration direction. This result demonstrated that the normal intercellular adhesion strength has the significant effects on the migration direction as well.

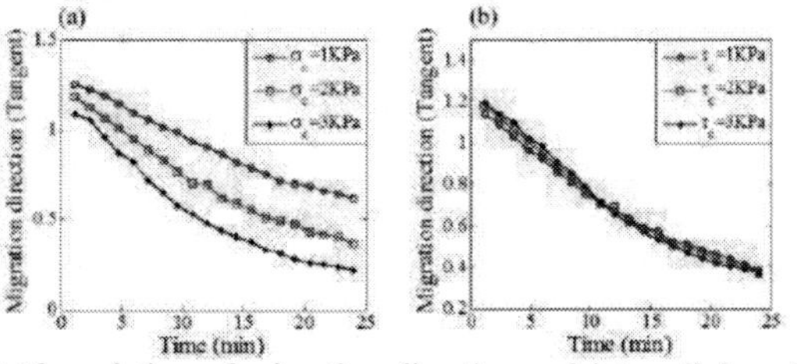

Fig. 3 The relation of migration direction and intercellular adhesion strength: (a) normal direction (b) shear direction

DISCUSSION

Study on the influence of cooperative intercellular forces on cell group movement inferred that E-cadherin was engaging in regulating local intercellular forces throughout a monolayer [3]. Other study indicated that E-cadherin clusters concentrate in the cell-cell junction region and form an "adhesive belt" to stitch cells together [5]. In fact, from a mechanical point of view, the "adhesive belt" formed by the concentration of E-cadherin in the cell-cell junction will balance the cell-cell interactions mainly in the direction perpendicular to the contact surface of cell edge. From our simulation results, it showed that the normal intercellular adhesion strength had dominated effects on the coordinated movements of monolayer sheet (Fig. **3**). This might the reason that when the cell-cell junction could not carry the tangential loading, then the loading in the cell-cell junction will be transferred to normal direction and the cell-cell junction will carry most of the loads in normal direction. This will stretch the cell-cell junction in the direction perpendicular to edge surface and cause the result that the average normal stress is greater than the average shear stress (Fig. **2**). In return, the cell will sense the normal intercellular adhesion forces from the neighboring cells and tune it to migration along the direction of maximum principal stress.

ACKNOWLEDGEMENTS

This work is supported by a grant from National Institutes of Health (Grant No. SC2GM112575) and a grant from the University of Texas at San Antonio, Office of the Vice President for Research.

REFERENCES

[1] Vitorino, P., Meyer, T., Genes & development 22: 3268-3281, 2008.
[2] Friedl, P. et al., International Journal of Developmental Biology 48: 441-449, 2004.
[3] Tambe, D. T. et al., Nature materials 10: 469-475, 2011.
[4] Lin, L. et al., Journal of the Mechanical Behavior of Biomedical Materials 65: 224-235, 2017.
[5] Guillot, C., Lecuit, T., Science 340: 1185-1189, 2013.

COLLAGENASE EXPOSURE ALTERS NEURONAL ACTIVITY & BIOCHEMICAL REGULATORS WITH IMPLICATIONS FOR DEGENERATIVE PAIN

Meagan E. Ita (1), Modupe F. Adegoke (1), Beth A. Winkelstein (1,2)

(1) Department of Bioengineering
University of Pennsylvania
Philadelphia, PA, United States

(2) Department of Neurosurgery
University of Pennsylvania
Philadelphia, PA, United States

INTRODUCTION

Joint degeneration results from aging or trauma and is often associated with pain [1]. Tissue degeneration is driven by degradation by enzymes like collagenase, with both mRNA and protein increases in degenerated joints [2]. Degradation can alter the biomechanics of the tissues in the joints [3,4]. Although degradative enzymes increase and alter joint mechanics, the mechanisms of pain from such changes are unknown. Since the local biomechanical environment of neurons mediates their signaling, any changes in joint biomechanics that result from degradation can affect the innervating joint afferent neurons [3-5]. Yet, degeneration-induced pain exists in the absence of mechanical injury [4]; there is no correlation between the magnitude of structural joint damage and pain in knee degeneration patients [6]. Pro-inflammatory cytokines and chemokines that are implicated in degeneration can also sensitize neurons [7]. Although intra-articular injection of collagenase induces degenerative structural changes in the lumbar facet joint [8] and movement-induced pain [9], it is unknown whether collagenase can directly sensitize neurons.

We investigated if collagenase induces a nociceptive response in neurons by exposing dorsal root ganglia (DRG) to it. Because neuronal hyperactivity contributes to pain [10], we hypothesized increases in neuronal activity after collagenase exposure; calcium fluorescence was used as a proxy for action potentials and neuronal firing. Gene expression of several nociceptive molecules, signaling kinases, and inflammatory cytokines was also measured since they are all implicated in pain [7,11]. The effect of collagenase on the immediate and delayed expression of the signaling protein phosphorylated extracellular signal-regulated kinase (pERK) and the neuropeptide substance P (SP) was assessed since their rapid and delayed post-translational regulation mediates nociceptive signaling after noxious afferent stimulation [7,10,11].

METHODS

All DRGs were isolated from embryonic day 18 Sprague-Dawley rats and plated in 200μL of feeding medium (FM) on dishes treated with PDL and laminin [12]. Dissociated DRGs were plated at 3.0×10^5 cells/mL. The FM was changed starting on day-in-vitro (DIV) 1 and every other day. DRG cultures were exposed to 60 units of purified collagenase (CLSPANK) in FM and incubated for 1 hour at 37°C. Matched untreated cultures were included as controls in each study.

To test if collagenase alters neuronal viability, cytotoxicity was measured by a Pierce Cytotoxicity Assay for LDH [13]. LDH levels were measured in FM samples at 0, 3, 6, 12, and 24 hours after exposure on DIV4. The percent cell lysis in treated and untreated dissociated DRGs (n=6/group) was compared by ANOVA.

Neuronal firing was evaluated in dissociated DRGs (n=4) after transducing them on DIV1 with an adeno-associated virus expressing GCaMP6f (#AAV1.Syn.GCaMP6f.WPRE.SV40; 1:6,000) [14] that fluoresces with rapid changes in intracellular calcium. Time-lapse images were acquired (20Hz; 2mins) on DIV11 [15] immediately before (PRE) and 1 hour after (POST) collagenase exposure. The same field of view (Figure 1A) was evaluated in each culture for both image sets to identify the same neurons; neuronal firing was quantified as the number of Ca^{2+} events per neuron using FluoroSNNAP in MATLAB [15]. Images were segmented by defining 42 neurons (Figure 1B), and normalized fluorescence traces ($\Delta F/F$) were calculated for each neuron (n=168) (Figure 1C) [15]. The number of Ca^{2+} events was counted for each trace using a template-matching algorithm with known Ca^{2+} waveforms [15,16]. Neurons with no events in both image sets were not analyzed (n=47). The difference in the number of Ca^{2+} events per neuron between PRE and POST was compared by a paired t-test.

To test if collagenase treatment has an immediate effect on gene expression, RNA was extracted from DRGs by Trizol homogenization immediately after collagenase exposure on DIV4. Samples were reverse-transcribed and real-time PCR was performed [17] with primer sequences specific for neuropeptides, kinases, and inflammatory cytokines (Figure 2). Gene levels were calculated and represented as fold-changes over the housekeeping gene cyclophilin-A (CyA) (Figure 2). Separate t-tests compared levels for each gene between the collagenase treated (n=5) and untreated cultures (n=4).

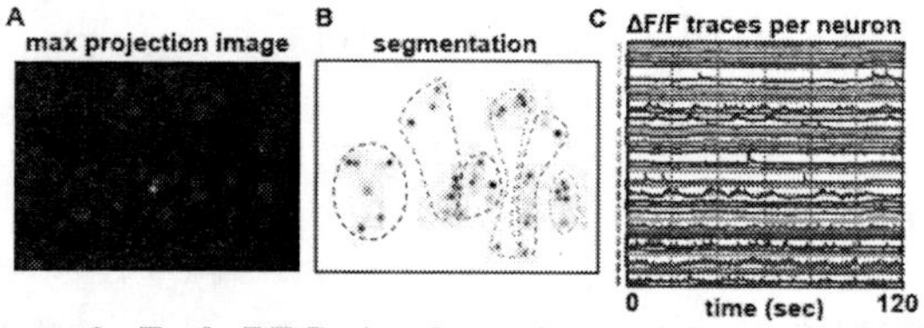

Figure 1: Each DRG time-lapse image (A) was manually segmented to delineate neurons (B). The normalized fluorescence traces (ΔF/F) were computed for each (C). Colored regions and neuron colors in (B) correspond to the y-axis colors and neuronal traces in (C), respectively.

Gene	Primer Sequence (5' – 3')
SP	Fwd: GACCAAATCAAGGAGGCAATG Rev: TCTTCGGGCGATTCTCTGA
CGRP	Fwd: GACTCCAATATGCCCGAAGA Rev: GTGGCTTGAATAAAATCCATG
p38	Fwd: CCGAGCGATACCAGAACCTG Rev: CAGACCAATCACATTCTCGTGC
ERK	Fwd: GTGGCACAACAGACGACAAC Rev: GGCATCCGAGACATCCTCAG
TNFα	Fwd: ATCATCTTCTCAAAACTCGAGTGACAA Rev: CTGCTCCTCTGCTTGGT
IL1β	Fwd: GCAATGGTCGGGACATAGTT Rev: GAGACCTGACTTGGCAGAGG
IL1α	Fwd: AAGACAAGCCTGTGTTGCTGAAGG Rev: TCCCAGAAGAAAATGAGGTCGGTC
CyA	Fwd: TATCTGCACTGCCAAGACTGAGTG Rev: CTTCTTGCTGGTCTTGCCATTCC

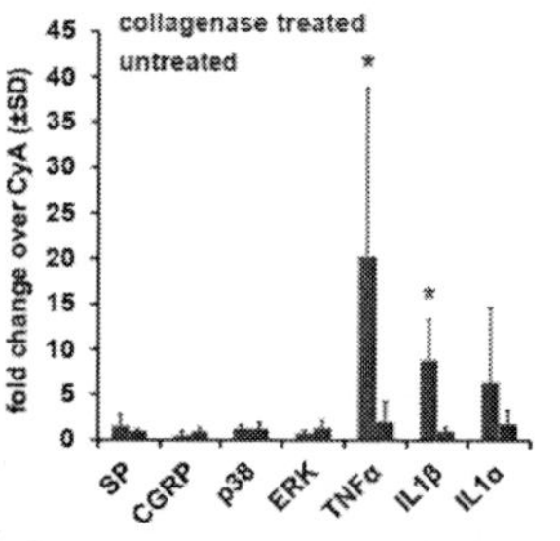

Figure 2: Primer sequences for target and housekeeping (CyA) genes. Collagenase immediately increases TNFα (*p=0.046) and IL1β (*p=0.016) mRNA after exposure.

Neuronal expression of pERK and SP was measured by immunolabeling in dissociated DRGs immediately after 1 hour of collagenase incubation on DIV4 (n=2/group) and 24 hours later (n=4 untreated; n=5 treated). Cultures were fixed, blocked, and incubated overnight at 4°C with antibodies to neuronal marker β-III tubulin (anti-mouse 1:200; Biolegend), pERK (anti-rabbit 1:200; Cell Signaling), and SP (anti-guinea pig 1:500; Neuromics), and then for 2 hours at room temperature with secondary antibodies. Randomized images (n=5/dish) were taken at 20X using a confocal microscope. For each image, the co-localization of β-III tubulin and pERK or SP labeling was quantified by densitometry and normalized by β-III tubulin. Separate t-tests compared each protein in groups at each time point.

RESULTS

Although collagenase exposure did not alter the neuron viability in DRGs compared to untreated responses at any time after treatment, it did significantly change ($p<0.001$) the number of Ca^{2+} events by 2.08±3.22. Collagenase *decreased* Ca^{2+} events by 2.81±1.93 ($p<0.001$) in 105 neurons, but *increased* them by 2.82±2.67 ($p<0.001$) in only 16 neurons. There was no difference between cultures dishes. Only mRNA for TNFα ($p=0.046$) and IL1β ($p=0.016$) was increased (Figure 2). Collagenase increased protein expression of pERK ($p=0.034$) and SP ($p=0.026$) only at 24 hours after exposure (Figure 3).

DISCUSSION

Although collagenase exposure for 1 hour was not toxic it did decrease activity in the majority (~87%) of neurons. Here, changes in Ca^{2+} events were measured relative to each neuron's baseline and did not directly compare changes to those at comparable times in untreated DRGs. Since neurites become more active and connected with longer culture times [15], and collagenase disrupts connective tissue [18], it is possible that incubation with collagenase for even 1 hour may disrupt neuron connectivity and decrease activity. Collagenase treatment alters receptor binding and kinetics of cortical neurons [19], suggesting collagenase may similarly affect receptors in DRGs and alter their activity. Of note, 13% of the collagenase-treated neurons did increase their activity. DRGs contain many types of sensory neurons with distinct electrophysiological properties and responses to stimuli [10]. Given the different phenotypic responses to collagenase observed here, it is possible that the two types of responses reflect neurons from different DRG populations. Indeed, different populations of sensory neurons respond differently to biochemical molecules that are present in degeneration [7]. Certainly, the specific population(s) of nociceptors that mediate degeneration-associated pain remains unknown.

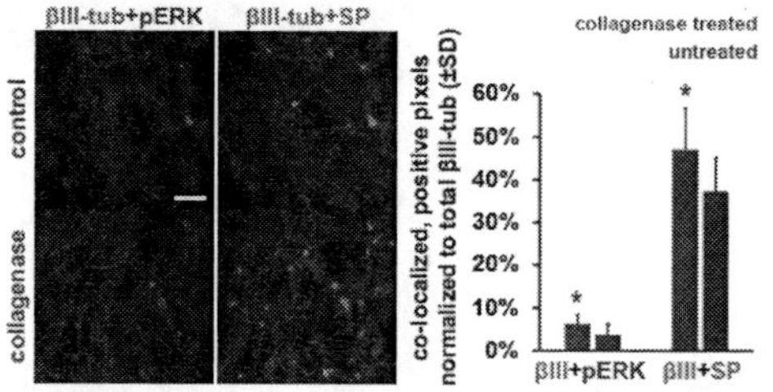

Figure 3: Collagenase increases neuronal pERK (*p=0.034) and SP (*p=0.026) at 24 hours. Scale=100μm; applies to all panels.

Collagenase exposure immediately increased TNFα and IL1β mRNA and neuronal pERK and SP 24 hours later (Figures 2 & 3). These results are consistent with in vivo findings of joint degeneration that support a role of inflammation early and later nociceptive signaling in degenerative pain [7]. Collagenase-induced degeneration in the knee induces late-onset altered expression of nociceptive proteins in the DRG [9], which may be explained by the in vitro findings here. By isolating sensory neurons from their physiological biomechanical environment, these studies provide the first evidence that collagenase alone may be sufficient to initiate cascades involved in pain. Integrating this work with neuronal responses to collagenase in vivo will help elucidate the biochemical, biomechanical, and cellular responses involved in degenerative pain.

ACKNOWLEDGEMENTS

Support provided by the NIBIB (U01EB016638). We thank Dr. Dave Meaney for the virus and use of spinning disk confocal microscope.

REFERENCES

[1] Neogi et al *Osteoarthr Cartilage* 21:1145-53, 2013. [2] Kim et al *Osteoarthr Cartilage* 23:2242-51, 2015. [3] Varady & Grodzinsky *Osteoarthr Cartilage* 24:27-35, 2016. [4] Hill et al *Arthritis Rheum* 52:794-9, 2005. [5] Zhang et al *J Biomech Eng* 138:021013, 2016. [6] Torres et al *Osteoarthr Cartilage* 14:1033-40, 2006. [7] Miller et al *Curr Osteoporos Rep* 13:318-26, 2015. [8] Yeh T et al *Eur Spine J* 17:734-42, 2008. [9] Adaes et al *Osteoarthr Cartilage* 23:914-24, 2015. [10] Basbaum et al *Cell* 139:267-84, 2009. [11] Cheng & Ji *Neurochem Res* 33:1970-8, 2008. [12] Cullen et al *Tissue Eng Part A* 18:2280-9, 2012.[13] Chan et al *Methods Mol Bio* 979:65-70, 2013. [14] Chen et al *Nature* 499:295-300, 2013. [15] Patel et al *J Neurosci Methods* 243:26-38, 2015. [16] Schultz et al *Soc Neurosci* 29:8005-15, 2009. [17] Smith et al *PLoS One* 8:e80006, 2013. [18] Hayar et al *J Neurosci Methods* 173:74-82, 2008. [19] Guarnieri et al *Brain Res* 93:337-42, 1975.

SB³C2017
Summer Biomechanics, Bioengineering and Biotransport Conference
June 21 – 24, Tucson, AZ, USA

INDUCTIVE ELECTRIC FIELDS HINDER EGF GRADIENT PROMOTED BREAST CANCER CELL MOTILITY

Ayush A. Garg (1), Travis Jones (1), Sarah M. Bushman (2), Jessica Shuman (1), Jacob Enders (2), Vish Subramaniam (1), Jonathan W. Song (1,3)

(1) Department of Mechanical and Aerospace Engineering, The Ohio State University, Columbus, Ohio, USA

(2) Department of Biomedical Engineering, The Ohio State University, Columbus, Ohio, USA

(3) The Comprehensive Cancer Center, The Ohio State University, Columbus, Ohio, USA

INTRODUCTION

In a normal breast duct, epithelial cells are separated from the extracellular matrix by the basement membrane. This microarchitecture helps maintain a trans-epithelial potential (30mV) across the barrier resulting in an endogenous electric field (0.01–1 V/cm) from the luminal side (anode) to the basal side (cathode). During invasion, breast cancer cells first migrate into the lumen/anode side [1,2]. Subsequently the cancer cells breach the basement membrane resulting in an electric discharge and disruption of this endogenous E-field. Based on this observation and previous findings [3], we hypothesized that weak inductive electric fields (iEF) applied with the direction of migration can potently hinder breast cancer cell motility. Moreover, while it has been shown that there is a direct correlation between EGFR expression and electrotactic migration to DC E-fields *in-vitro* [4], we wish to establish if EGFR stimulation by EGF sensitizes cancer cells to E-fields.

Actin is an important cell protein whose dynamic properties provide the driving force for cell motility [5]. Actin polymerization at the leading and/or the trailing edge is a hallmark of a polarized cell [6] and this actin polymerization is regulated by small Rho GTPases which get recruited by activated EGFR [7]. Thus, we are also interested in measuring changes to the intracellular actin under the influence of E-fields to elucidate possible mechanisms that regulate the cell motility response to iEFs.

METHODS

We used three cell lines for these studies: 1) MCF10A – a control epithelial breast cell line 2) MCF10IV – a Ras transformed variant of MCF10A cells, and 3) MDA-MB-231 – a triple negative metastatic breast cancer cell line. Control experiments for all of the cell lines were conducted in 0.1% serum media without any exogenously applied growth factors or iEF. iEFs (~10μV/cm) were applied with custom fabricated air-cored electrical coils, which produced E-fields using principles of electrical induction. The iEFs were applied with or without of EGF stimulation (Fig 1D).

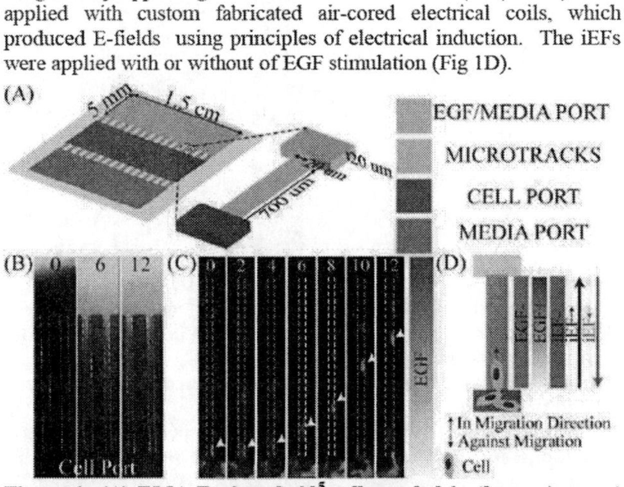

Figure 1: (A) BMA Design, 2x10⁵ cells seeded in the center port and tracked as they migrate along these tracks; (B) EGF gradient characterization with FITC-Dextran (10kDa) over 12 hrs; (C) MDA-MB-231 cells tracked as they migrate under EGF gradients (EGF+) and no E-Field (iEF-), arrow head follows the tracked cell, dotted lines represent microtracks and time in hours; (D) Schematic of all treatments w.r.t. direction of migration.

We used two approaches to monitor the migration of each cell line: 1) modified transmembrane or Transwell Assay to quantify the number of cells migrating over short distances (~10μm) and 2) Custom Bi-directional Microtracks Assay (BMA) fabricated out of polydimethylsiloxane (PDMS), which mimicked the topological cues encountered by migrating tumor cells during metastatic invasion.

For Transwell Assay (8 μm pore size), 1.5x10^5 cells were seeded in each fibronectin (10 μg/mL) coated well and run for 8 hours. For the BMA, the microtracks were aligned in parallel (Fig 1A) and were fabricated out of PDMS using standard photo- and soft-lithography techniques [8]. EGF gradients were characterized using a fluorescent FITC-Dextran (Fig 1B). In BMA, microtracks were flushed with fibronectin (10μg/mL) for 90 minutes following which 2x10^5 were seeded in the central port and incubated for 48 hours following which cells were tracked over 12 hours in intervals of 5 min (Fig 1C).

To quantify actin distribution, we devised a new metric called the Actin Aggregation Ratio (AAR), which is a dimensionless number that quantifies the top 20% actin intensity distribution (red spokes on rose plots – Fig 2B) in a cell. AAR equals 1 if this top 20% is confined to leading (60°-120° sector) and/or trailing edges (240°-300° sector) and 0 if none of it is in that region of interest. We used standard immunofluorescence staining for actin and images were processed using a custom MATLAB script for single cell analysis that generated rose plots depicting the spatial actin distribution for each cell.

RESULTS

Hindering effects conferred by E-fields is sensitized by EGF stimulation in Transwell Assay: MDA-MB-231 cells stimulated with EGF showed suppressed migration by ~60% when iEFs were applied compared to no applied E-fields. The same trend was observed for epithelial MCF10A cells which showed 250% suppression under similar conditions in Transwell assay. Furthermore, migration numbers show no significant change in response to iEF without EGF.

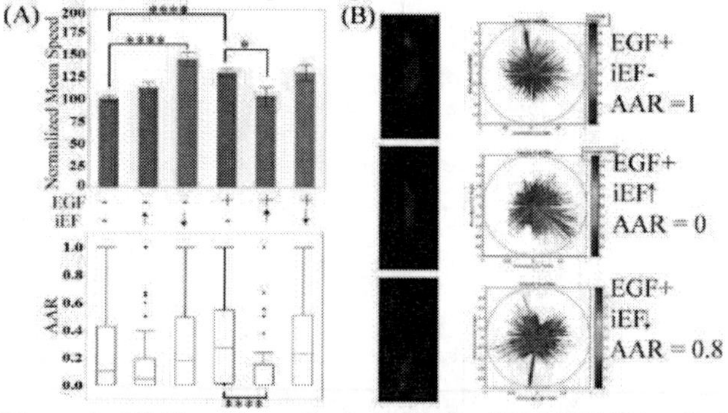

Figure 2: (A) Top, Normalized speeds for MDA-MB-231 cells in BMA (standard student t-test); Bottom, Quantitative actin distribution for MDA-MB-231 cells (Wilcoxon test); (B) Representative images for MDA-MB-231 cells for actin distribution under 3 conditions, Red-F-actin, Blue-nucleus; *p<0.05, ****p<0.0001.

iEF directionally increased migration speeds without EGF and hindered motility in presence of EGF in BMA: iEF when applied against the direction of migration for MDA-MB-231 (Fig 2A Top), MCF10IV, significantly increases migration speeds in absence of EGF whereas, under stable EGF gradients, MDA-MB-231 show suppressed migration speeds when iEF is applied along these gradients. MCF10IV cells in the latter case show bi-directional hindering under EGF gradients. The normal epithelial MCF10A cells, do not show any negative migratory response to iEFs in presence of EGF, moreover, their speeds increase significantly when iEFs are applied against the EGF gradients. The AAR results (Fig 2A Bottom) also captures the significant suppression in speeds for MDA-MB-231 when iEFs are along a stable EGF gradient. Moreover, the differences in the genetic make-up of the cells results in subtle differences in their response to E-fields but the major difference is in response between metastatic and non-metastatic cell lines. Only the metastatic cell lines (MCF10IV and MDA-MB-231) in the BMA showed a suppressed response to iEFs in presence of EGF.

DISCUSSION

Here we report effects of iEFs on cell migration and intracellular actin distribution. We note that these iEFs, which potently modulate the migration of malignant and non-malignant epithelial cells, are also comparatively weak being ~5 orders of magnitude smaller than all previous studies conducted [2,4]. In the Transwell Assay, cells had to physically squeeze to navigate sub-cellular (8μm) pores while only traversing short distances (~10μm). This physical squeezing of cells can alter the mode of migration (e.g. amoeboid versus mesenchymal) compared to mode of migration exhibited in the BMA assay. These differences in modes of migration may explain in part the different responses on cell migration in response to iEFs between the two assays.

We observe uni-directional suppression due to iEFs in MDA-MB-231 cells and a bi-directional suppression in MCF10IV cells. The difference between MDA-MB-231 and MCF10IV cells may be attributed to the expression of the EGF receptor, ErbB2, on MDA-MB-231 cells. Interestingly, the outcome that both metastatic cell lines exhibit significantly higher migration speeds against the E-fields might explain the *in-vivo* outcome of proliferating malignant cells filling up the positive lumen prior to breaching the basement membrane during invasion. Moreover, in a normal, intact breast duct, E-fields are directed from the lumen to basal side. However, upon invasion, this trans-epithelial potential is disrupted once the basement membrane is breached leading to metastasis. Here we show that iEFs exhibit a hindering effect of iEFs in the direction of migration. Therefore, we speculate that iEFs can be applied to restore the disrupted endogenous fields to suppress metastatic invasion. Furthermore, our actin aggregation results suggest that it can predict in part the migration speeds; it is able to capture the suppressive effects but not the increased effects on migration in response to iEFs. Therefore, it is possible that iEFs affect cell migration by directly affecting actin as it has been shown that EGRF binds to F-actin directly in certain conditions [9]. As part of future studies, we plan on probing the molecular mechanism of EGF-EGFR pathway for a better understanding on progression and metastasis in breast cancer as well as possibly better therapeutics and prognosis using non-invasive E-fields.

ACKNOWLEDGEMENTS

We thank the Luker lab for GFP labeled MDA-MB-231 cells, the Ostrowski Lab for the dsRed labeled MCF10A cells, and the Ganju lab for the Ras transformed MCF10A cells (MCF10IV). This work was supported by funding from the American Cancer Society, the OSU Institute for Materials Research, and the OSU Pelotonia.

REFERENCES

[1] Nuccitelli, R. *Current Topics in Dev Bio*, 58:1-26, 2003
[2] McCaig, C.D. et. al., *J Cell Sci*, 122:4267-4276, 2009
[3] Ahirwar, D.K. et. al., *Scientific Reports*, 5, 2015
[4] Pu, J. et. al., *J Cell Sci*, 120:3395-3403, 2007
[5] Hall, A, *Science*, 279:509-514, 1998
[6] Lauffenburger, D.A. et.al., *Cell*, 84:359-369, 1996
[7] Wells, A., *The Intern. J of Biochem & Cell Bio*, 31:637-643, 1999
[8] Whitesides, G.M. et. al., *Nat Rev Microbiology*, 5:209-218, 2007
[9] Hartigh J.C. den et. al., *J of Cell Bio*, 119 (2): 349, 1992

SB3C2017
Summer Biomechanics, Bioengineering and Biotransport Conference
June 21 – 24, Tucson, AZ, USA

A THERMODYNAMICALLY-MOTIVATED MODEL FOR STRESS FIBRE REORGANIZATION

William Ronan (1,2), Andrea Vigliotti (2,3), Vikram S. Deshpande (2)

(1) Biomedical Engineering
National University of Ireland Galway
Galway, Ireland

(2) Department of Engineering
University of Cambridge
Cambridge, UK

(3) Innovative Materials Laboratory
Italian Aerospace Research Centre
Capua, Italy

INTRODUCTION

The response of cells to their mechanical environment has been explored for a variety of different loading environments: Discher et al. [1] report increased cytoskeletal activity for different substrate stiffness; Tan et al. [2] show that changing the arrangement or stiffness of micropillar substrates also affects the intensity and distribution of the cytoskeleton and surface tractions. Similarly, micropipette aspiration, cell shearing, and cell compression are all affected by the configuration of the cytoskeleton and the mechanical environment.

In turn, the response of the cytoskeleton to the mechanical environment also affects the regulation of cellular functions: MSCs on different substrates are directed towards neurogenesis, myogenesis, or osteogenesis by increasing the substrate stiffness [3]. Cyclic loading of bone-marrow MSCs promotes chondrogenesis [4].

Cyclic stretching of cells on substrates which are relatively stiff compared to the cells produces a cytoskeletal network which is oriented perpendicular to the imposed strain [5]. This observed avoidance of strain increased with increasing imposed strain amplitude at a fixed frequency of loading, suggesting that the response was sensitive to the strain rate [5]. Other studies, however report increased strain avoidance with increasing strain amplitude for a fixed strain rate [6]. Investigations on the shape of the strain waveform by Tondon et al. [7] suggest that the cytoskeletal arrangements are more sensitive to the lengthening rate compared to the shortening rate.

In contrast to the strain avoidance in two-dimensional (2D) environments, cells cultured in a three-dimensional (3D) matrix or gel do not demonstrate the same strain avoidance. If the matrix/gel is uniaxially constrained, the cytoskeleton of the cells aligns in the constrained direction [8] and this preference remains even when cyclic strains are applied [8].

Understanding such phenomena requires accurate and rich knowledge of the conditions in the cell during mechanical loading. Much progress has been made understanding the role of biochemical factors in cell function; however, mechanical regulation is less well understood and there is no consensus on the best approach.

Early models used to investigate these phenomena assumed linear elastic or viscoelastic properties which were useful to interpret data, but offered limited insight into the underlying. Tensegrity approaches highlighted the importance of the evolving cytoskeleton [9]; however, tensegrity requires a predefined cytoskeleton and it has been shown experimentally that the disruption of microtubules results in an increase in the traction force generated by cells [10].

A recent family of models have sought to explain these phenomena by including stress fibre kinetics and contractility relationships. The models of Deshpande et al. [11] and Vernerey and Farsad [12] accurately predict the strain avoidance for the cyclic response of cells on 2D substrates but cannot model the alignment of the stress-fibers with the imposed strain in 3D. On the other hand, the model proposed by Obbink-Huizer et al. [13] accounts for the cyclic response of cells in both the 2D and 3D settings by including a strain dependence in the stress-fiber kinetics; however, such strain dependency is hard to justify over longer time scales as the cell will remodel and lose this strain memory.

In the current study, we develop a micro-mechanical approach for the constitutive law for stress-fiber contractility within cells. The model employs thermodynamic considerations to motivate the dependence of stress-fiber kinetics on stress, strain, and strain-rate.

METHOD

In order to develop a robust model for stress fibre kinetics and contractility, we first accurately quantify fibres at a point in the cell. In summary, η quantifies the number of stress-fibers in parallel at a particular orientation while n gives the number of functional units in series within each of those stress-fibers (Figure 1A,B).

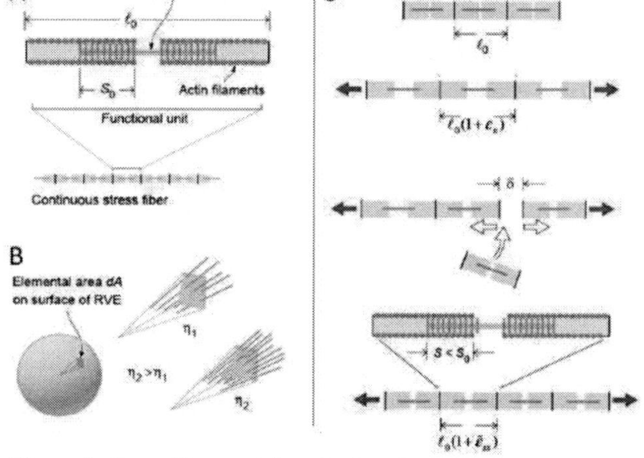

Figure 1: Stress fibres are described by the number of functional units (A) in a given representative volume element (RVE) at each point (B) . Stress fibres remodel to reach a steady state strain $\tilde{\varepsilon}_{ss}$ via addition of units (C).

The kinetics of fibre growth and dissociation are driven by the enthalpies of the bound state μ_b (which is different in each fibre orientation) and the unbound state μ_u relative to the enthalpy of an activation barrier μ_a:

$$\dot{\eta} = \frac{N_u}{2\pi\, n} \exp\left[-n\frac{\mu_a - \mu_u}{k_B T}\right] - \eta \exp\left[-n\frac{\mu_a - \mu_b}{k_B T}\right]$$

where N_u is the number of unbound units available [14]. The enthalpy of the bound state increases as a functional unit is strained away from its ground state. We note that when a stress fibre remodels in response to a macroscopic strain ε via the addition a unit, the strain in each unit $\tilde{\varepsilon}$ may return to its ground state. Thus, the response of the fibres will have a transient dependence on the macroscopic strain. The enthalpy of the unbound state μ_u depends on the intensity of a representative cellular signaling cascade.

Hill-like contractility relates the stress in the fibres to the strain rate, similar to Deshpande et al [11]. It is crucial to note that the fibre tension depends on the strain in the unit $\tilde{\varepsilon}$, not the macroscopic strain.

The material stress is determined by summing the contribution of the tension in each fibre (a series of units) in a given direction (by multiplying by η) and subsequently summing over all possible directions within the RVE. We also note that changes in the size of the RVE caused by the addition of functional units is accounted for. An elastic contribution representing all other cellular components is included in parallel to this active model. It is also necessary to solve for mechanical equilibrium, depending on the boundary conditions. Solutions to these equations were obtained using an Euler-backward integration scheme in the commercial package MATLAB.

RESULTS

Results are presented for two mechanical environments: a cell on a comparatively stiff substrate which is subjected to cyclic strain in one direction and constrained from contraction in the orthogonal direction (hereafter referred to the 2D case) and for a cell seeded in matrix and then subjected to stretch in one direction and free to contract in the other two directions (Figure 2A,B). Cells are allowed to equilibrate in response to an activation signal for 6 hours from the initial condition of no formed stress fibres; everything is in the unbound configuration. The cell is held fixed with $E_{11} = 0$ during this 6 hours, following which the cell is cyclically stretched for a further 6 hours at 1 Hz to 10% strain.

In order to quantify the results and readily compare them to experiment observations of stress fibre alignment, we define the quantity $\xi = \eta n$ for each possible fibre direction. Circular histograms (normalized such that the maximum possible uniform and isotropic distribution of fibres would have $\xi = 1$) are presented.

Figure 2 shows that an isotropic distribution of fibres is predicted for the 2D substrate with no strain in either direction. Application of the cyclic stretch leads to strain avoidance, with approximately twice as many fibres perpendicular to the stretch. In contrast, the 3D substrate shows fibres aligned with the stretch/constraint direction both before and after stretch.

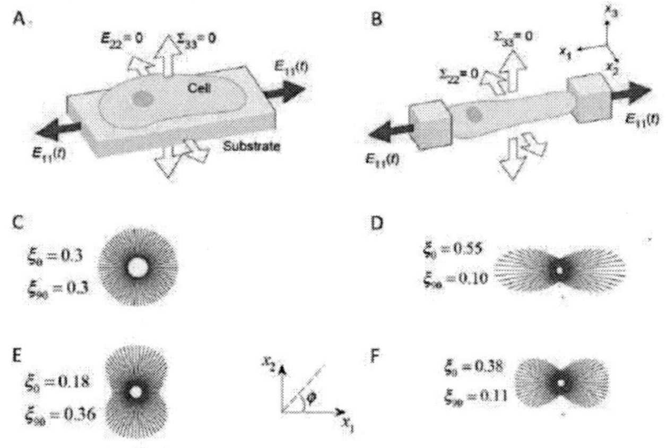

Figure 2: Boundary conditions (A), and stress fibre distributions for 2D case before (C) and after (E) cyclic stretching. Boundary conditions (B), and stress fibre distributions for 2D case before (D) and after (F) cyclic stretching.

DISCUSSION

The material model presented was motivated by the energy states of stress fibres. The thorough description of the stress fibre arrangement within an RVE with the thermodynamically driven kinetics accounts for dissipation and changes in energy during remodeling. This approach naturally establishes the stress, strain, and strain rate dependence of stress-fibre dynamics.

In contrast to previous work by the authors and others, it is not necessary to alter the material formulation or material properties in order to capture the contrasting experimental phenomena of strain avoidance in 2D and strain alignment in 3D. The framework presented here captures the cellular response of a wide range of boundary conditions (including others not discussed here).

ACKNOWLEDGEMENTS

AV: the Royal Society Newton International Fellowship. Discussions with R.M. McMeeking (UCSB) and F.P.T. Baaijens (TU/e) are gratefully acknowledged.

REFERENCES

[1] Discher DE et al. *Science*, **310**, 1139, 2005. [2] Tan JL et al. *PNAS* **100**,1484, 2003 . [3] Engler AJ et al. *Cell*, 126, 677, 2006. [4] Huang C et al. *Stem Cells* **22**, 313, 2004. [5] Kaunas et al. *PNAS*, 102, 15895, 2005. [6] Faust et al. *PLoS One* 6(12):28963, 2011. [7] Tondon et al. *J Biomech*, **45**, 728. 2012. [8] Nieponice et al. *J Biomed Mater Res A*, **81**, 523, 2007. [9] Ingber, DE, *Ann Rev Physiol*, **59**, 575, 1997. [10] Kolodney and Elson. *PNAS.*, **92** 10252, 1995. [11] Deshpande et al. *PNAS.*, **103**, 14015, 2006. [12] Vernerey FJ, Farsad M. *JMBBM*, **4**, 1683-1699, 2011. [13] Obbink-Huizer et al. *BMMB*, **13**, 227-236, 2014. [14] Vigliotti et al. *BMMB*, **15** 761, 2016

SB[3]C2017
Summer Biomechanics, Bioengineering and Biotransport Conference
June 21 – 24, Tucson, AZ, USA

IMPLEMENTATION OF A RIGOROUS LINEAR VISCOELASTIC MODEL FOR MEASURING CELL MECHANICAL PROPERTIES USING A MICROFLUIDIC EXTENIONAL FLOW DEVICE

Joanna B. Dahl

Engineering Department
University of Massachusetts Boston
Boston, MA, USA

INTRODUCTION

The quantification of cellular mechanical properties is of tremendous importance in biology and medicine. As a result, there is much interest in developing platforms to quickly and accurately quantify cellular mechanical properties. Recently we reported a simple microfluidic platform that uses hydrodynamic forces at low Reynolds number and low confinement to elongate single cells near the stagnation point of a planar extensional flow (Ref. 1). In tandem, we presented a novel analytical framework that enables determination of cellular viscoelastic properties (stiffness and fluidity) from these deformation measurements. However, in a simplistic approach, we considered the cell to be an elastic sphere with a constant 'effective' shear elastic modulus for the timescale of the particular experiment. Thus we did not explicitly consider the time-dependent shear modulus in the governing balance of momentum equations. While our final results provided informative scaling information and we successfully detected perturbations to cytoskeletal assembly and mechanics, the reported stiffness values were not consistent with the balance of linear momentum and therefore not the true stiffness values of the assumed stress-strain relationship.

Here I revise the analytic equation for the deformation of a viscoelastic sphere in extensional flow through rigorous derivation using the viscoelastic correspondence principle. This new model is applicable to any viscoelastic sphere with a time-dependent shear modulus. With this new equation, I revisit the cell deformation data from Ref. 1. Then I show preliminary results of a new analysis that extracts the mechanical properties of a single cell from its deformation history as is passes through the uniform extensional flow region.

METHODS

A microfluidic cross-slot, consisting of two channels that intersect at 90 degrees, generates planar extensional flow. Cells flowing in both cross-slot inlet channels were elongated at constant strain rate in extensional flow and observed passing through the stagnation point region (Figure 1). Cell strain was defined as $\varepsilon = (a - b)/(a + b)$, where a and b are the long and short axes, respectively, of an ellipse fitted manually to the outer edge of the cell membrane. We studied 3T3 fibroblasts and primary glioblastoma tumor initiating cells (GBM TICs) under control conditions and in the presence cytochalasin D (CytoD), which depolymerizes the cytoskeleton, and paraformaldehyde (PFA), which irreversibly cross-links cellular proteins.

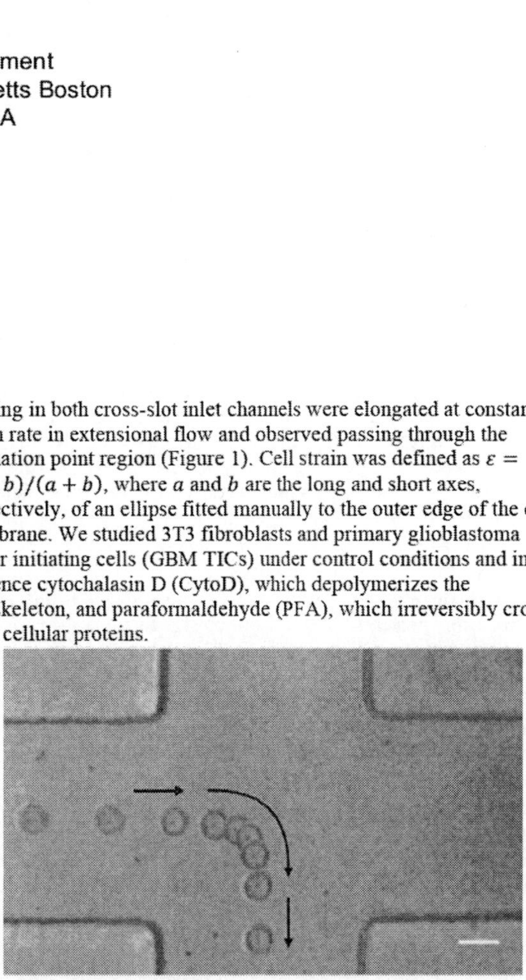

Figure 1: Time-lapse images of a 3T3 fibroblast stretching in extensional flow as it passes through the stagnation point region. Reproduced from Ref. 1 with permission of Elsevier.

Population-averaged mechanical property measurements: A single image of cell deformation near the stagnation point due to the known viscous forces is analyzed. At a given flow rate, cell type, and drug treatment, the reported deformation under those experimental conditions was computed as the average deformation of $10 \leq n \leq 30$ cells. Mechanical property parameters were obtained through linear regression with a least-squares fit of the observed strain ε to cross-slot extensional strain rate $\xi U/D$ on a log-log plot based on the theory described in the Results Section where ξ, U, and D are defined. *Single cell mechanical property measurements:* A sequence of images capturing the time evolution of strain for a single cell is analyzed. Mechanical property parameters were obtained through linear regression with a least-squares fit of the observed strain ε with time t on a log-log plot. For both procedures. the uncertainties in the linear fits are evaluated using analysis of covariance (ANOCOVA).

RESULTS

Using the viscoelastic correspondence principle (Ref. 2), the Laplace transformed viscoelastic solution is obtained directly from the corresponding elastic solution, and the final result is obtained upon inversion. From Murata's solution of an incompressible elastic sphere deforming in an arbitrary linear, low Reynolds number flow field (Ref. 3). We derived a strain rate $\xi U/D$, where $\xi \sim 1.5$ is expressed analytically as a function of the aspect ratio of the device, U is the average flow velocity, and D the channel half-width (see Ref. 1 for details). The simple two-parameter power law captures single cell viscoelastic behavior (Ref. 4):

$$G(t) = G_0 \left(\frac{t}{t_0}\right)^{-\alpha} \quad (1)$$

t_0 is an arbitrary reference time, G_0 is the value of the shear modulus at time t_0, and the fluidity parameter α describes the dependence of the shear modulus on time. To arrive at the solution of our viscoelastic problem, reinterpret the elastic field variables in the elastic solution as their Laplace transform. The viscoelastic response function $G(t)$ is replaced by $s\bar{G}(s)$. Time-dependent boundary conditions must also be transformed. When the cell enters the cross-slot region, the cell suddenly feels the extensional flow field. So Ω is interpreted as $\Omega h(t)$ in the viscoelastic problem utilizing the Heaviside step function $h(t)$ and is replaced in by its Laplace transform Ω/s. Making these replacements into the elastic solution and then inverting yields the desired final solution for $\varepsilon(t)$:

$$\varepsilon(t) = \frac{5}{2}\frac{\xi U}{D}\frac{\mu}{G_0}\left(\frac{t}{t_0}\right)^{\alpha}\frac{\sin \pi\alpha}{\pi\alpha} \quad (2)$$

This new analytic model that satisfies the governing equations differs from the theory presented in Ref. 1 only by the factor $\sin(\pi\alpha)/\pi\alpha$.

The new model measures lower stiffness parameters than the original phenomenological model due to the $\sin(\pi\alpha)/\pi\alpha$ factor, which for the measured values of α mean new stiffness values that are 40-50% less than the values reported in Ref. 1 (Table 1). The fluidity parameter measurements are unaffected by the updated model because the fitted slope only depends on α.

Table 1: Comparison of population-averaged 3T3 fibroblast and GBM TIC stiffness parameters for the simplistic model (Ref. 1) and the mechanically rigorous model from this work. Specified reference time $t_0 = 5$ ms, the average time of deformation across all cell experiments.

	Control		10 µM CytD		4% PFA	
	Ref. [1]	*This work*	*Ref. [1]*	*This work*	*Ref. [1]*	*This work*
3T3 $G_0(t_0)$ (kPa)	0.59 ± 0.05	0.39 ± 0.03	0.40 ± 0.05	0.23 ± 0.03	0.93 ± 0.08	0.61 ± 0.05
TIC $G_0(t_0)$ (kPa)	0.44 ± 0.03	0.28 ± 0.02	0.22 ± 0.04	0.11 ± 0.02	0.73 ± 0.05	0.47 ± 0.03

Preliminary viscoelastic measurements of a single CytoD-softened 3T3 cell continually deforming as it dwells in extensional flow yield fluidity and stiffness values are close to the population-averaged values (Figure 2).

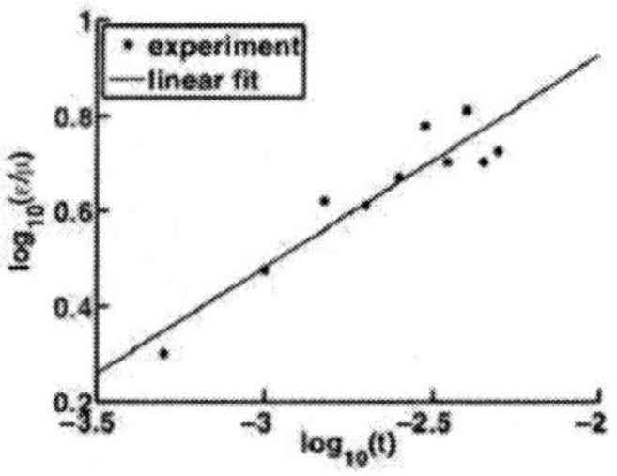

Figure 2: Measuring viscoelastic properties of a single cell (3T3, CytoD) from strain history and application of new rigorous model. Measured values: $\alpha = 0.44 \pm 0.03$, $G_0(t_0) = 0.29 \pm .07$ kPa

DISCUSSION

The model is derived for steady state conditions with the deforming sphere located at the stagnation point, but here we are applying it to cells observed passing through the extensional flow region. We could achieve better adherence to model conditions—though at the cost of throughput—by hydrodynamically trapping cell at stagnation point and observing $\varepsilon(t)$. More images would also yield more certainty in the measured viscoelastic parameters. and a more accurate measurement.

Previous measurements of suspended 3T3 cells using optical stretchers yielded shear moduli of $\sim 70 - 80$ Pa (extracted from Figs. 2A and 3B from Ref. 5) and 100 ± 10 Pa (extracted from Fig. 2 from Ref. 6) for a deformation timescale of $t = 200$ ms. To compare with these earlier results, we set the timescale $t_0 = 200$ ms and find that $G_0(200\text{ ms}) = 66 \pm 5$ kPa. The corresponding Ref. 1 comparison was $G_0(200\text{ ms}) = 101 \pm 8$ Pa. So the improved model does not change the conclusion that our method compares favorably with other reported measurements, and in this instance our new result is almost identical to the recently published values in Ref. 5. Realizing only an order of magnitude agreement with previously published microscale mechanical property measurements is typical. There is variation among previously published results, is likely due to differences in measurement modalities across these reports (e.g. forces applied at different timescales, whole-cell vs localized deforming forces) that are exacerbated by the small microscale dimensions of the specimens. Finally while the absolute values of the viscoelastic parameters reported here are not too much different from those in Ref. 1, the theory is more sound and can be applied to any isotropic, incompressible viscoelastic sphere with a shear modulus time-dependent behavior.

ACKNOWLEDGEMENTS

The author wishes to thank Dr. Vivek Narsimhan for discussions on modeling linear viscoelastic behavior.

REFERENCES

[1] Guillou L., Dahl J.B., Lin J.-M. G., Barakat A.I., Husson J., Muller S.J., and Kumar S., *Biophys. J.*, 111(9):2039–2050, 2016.
[2] Christensen, R.M., *Theory of Viscoelasticity: An Introduction*, 2nd ed., New York: Academic Press, 1982.
[3] Murata T., *J. Phys. Soc. Jpn.*, 50(3):1009–1016, 1981.
[4] Fabry B., et al., J.J., *Phys. Rev. Lett.*, 87:148102, 2001.
[5] Chan C.J., et al., *Biophys. J.*, 108:1856–1869, 2015.
[6] Wottawah, F., et al., *Phys. Rev. Lett.*, 94:098103, 2005.

SB³C2017
Summer Biomechanics, Bioengineering and Biotransport Conference
June 21 – 24, Tucson, AZ, USA

INVESTIGATION OF FIBER ARCHITECTURE EFFECTS ON AXONAL DEFORMATION DURING TRANSVERSE AND AXIAL LOADING VIA A COUPLED NETWORK-AXON MODEL

Vahhab Zarei (1), Sijia Zhang (2), Beth A. Winkelstein (2), Victor H. Barocas (3)

(1) Department of Mechanical Engineering
University of Minnesota
Minneapolis, MN, US

(2) Department of Bioengineering
University of Pennsylvania
Philadelphia, PA, US

(3) Department of Biomedical Engineering
University of Minnesota
Minneapolis, MN, US

INTRODUCTION

The sensory function of a mechanoreceptor is based on its deformation, which itself depends on its interaction with surrounding extracellular matrix (ECM). The nerve fibers' (axons) strains arise from the microscopic matrix constituents (e.g. collagen fibers), which deform based on the heterogeneous macroscopic structure of the innervated tissue. Although axonal stretch has been considered in isolated neurons and in neurons embedded in model ECMs [1,2], we know of no previous theoretical work on local deformation of sensory neurons during loadings. Also, because of the complex, heterogeneous tissue architecture that can exist (e.g., in the facet capsular ligament of the spine [3,4]) and the complicated motions that some tissues undergo [5], the magnitude and direction of the local load relative to the axon orientation can vary greatly. Hereby, we developed a model to study strain transmission from collagen fiber networks to an axon, and we studied the effect of collagen fiber network structure on the axon's mechanical behavior during transverse (to axon's long axis) and axial loading.

METHODS

Delaunay networks of collagen fibers were generated within a rectangular 3-D domain in MATLAB and contained nonlinear fibers following the constitutive equation $F=AS\,[exp(BE)-1]/B$, where, F, A, S, B and E represented fiber force, stiffness, cross section area (radius=50nm), nonlinearity, and Green strain, respectively. The fiber properties A and B were set to 43.3MPa and 2.5, respectively [6]. A cylindrical region along the Z direction (diameter=1μm, length=3 μm) was removed from the networks, to make space for embedding a simplified axon. Clipped fibers were attached to the axon at pin-joint insertion points. The network size was 5x5x3 and 5x5x6 (μm^3), for transverse and axial loading cases, respectively. To ensure complete wrapping of the axon by collagen fibers, in addition to the Delaunay network fibers, circumferential fibers were constructed along the length of the cylinder. The axon was modeled as a linear elastic cylinder (ν=0.48) with longitudinal fibers (Figure 1). The linear elastic component represented the axoplasm and microfilaments. Three values – 20, 100, and 500 - were considered for the stiffness ratio between the collagen fibers of the ECM and the axoplasm. The longitudinal fibers modeled the microtubules and were implemented as linear elastic fibers (k=9.4n/m [7]). Networks were stretched by 20% while the lateral-to-load faces were kept fixed. To allow lateral contraction in Y in transverse loading, the fibers close (distance<0.5μm) to the Y-facing surfaces were made much more compliant compared to other fibers in the network (with stiffness 10 times lower). For axial loading, fibers close to X-facing and Y-facing surfaces were treated similarly. To eliminate edge effects, nodes located on the Z-facing surfaces were free to move in X and Y. These boundary conditions were specified to represent loading of a small segment of a long axon and the surrounding ECM.

Axonal strains were correlated with collagen fiber *network alignment* and *volume concentration*. The *network alignment* was characterized by calculating the difference between the eigenvalues of the orientation tensor (δ), as described in [3,8]. Networks can be either isotropic or transversely isotropic with preferential alignment in the Z-direction (axon axis). Three cases (n=8 for each) of δ≈0 (isotropic), δ≈0.4 (moderately aligned) and δ≈0.8 (strongly aligned) were simulated and compared. To assess the influence of collagen *volume concentration* on the axon deformation, isotropic networks with volume fractions of approximately 10%, 30%, and 50% were compared (n=8). Simulations were run on Abaqus/Standard. Brick elements (C3D8, 13600 elements) and connective elements (CONN3D2) were used for the axon and the fibers, respectively.

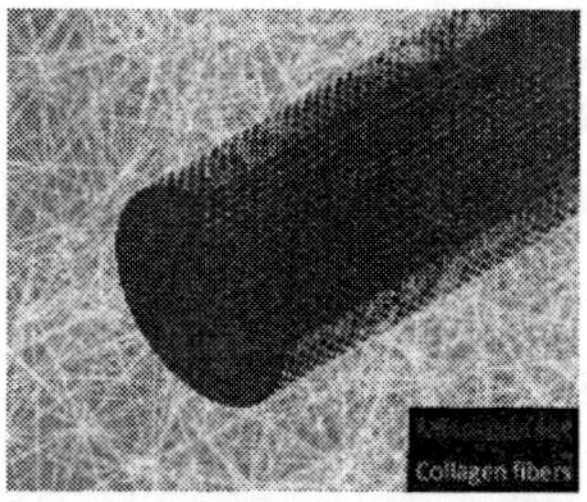

Figure 1. Axon (blue elements) reinforced by microtubules (red lines) and surrounded by collagen fibers (white lines)

RESULTS

Our model predicted localized axonal strains for a variety of network organization, for transverse and axial loading conditions. Figure 2 shows the strain field (maximum principal Green strain) on a representative axon embedded in a fiber network stretched laterally.

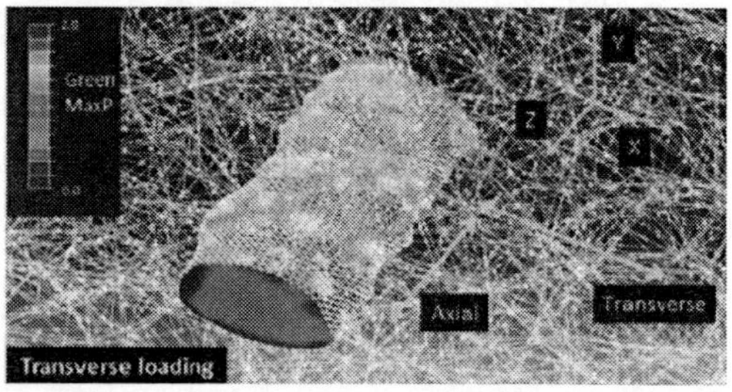

Figure 2. Strain field on the axon during transverse loading

The impact of network structure on the axonal deformation is shown in Figure 3, where the average value of maximum principal Green strain (MaxPS) over the axon is plotted with respect to anisotropy (δ) and volume concentration, for each loading. For each case, different stiffness ratios showed similar trends, with the maximum principal Green strain increasing with increasing fiber stiffness.

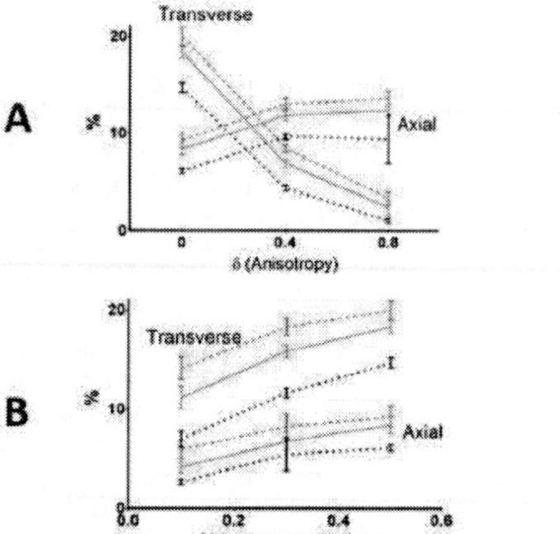

Figure 3. Axon strains correlated with network anisotropy (δ) and collagen fiber volume concentration (n=8). Error bars are 95% confidence intervals.

Increasing network alignment along the long-axis of the axon showed opposite effects for each loading case. Networks with more strongly aligned fibers have higher stiffness in the axial direction, while due to fiber reorientation in the loading process, the transverse stiffness decreases. Therefore, axonal strains increased and decreased respectively, during axial and transverse loading, with increase in fiber alignment in axon's long axis. As Figure 3A suggests, as anisotropy changes from δ=0 (isotropic) to δ=0.8 (strongly aligned), the average of MaxPS drops from ≈20% to ≈5% for transverse loading. Although these values follow the opposite trend for axial loading, the effect is less pronounced.

For both loading scenarios, the average values of MaxPS (Figure 3B) increased with collagen volume concentration. The greater number of insertion points (due to the higher volume concentration) transmitted strain more effectively to the axon.

The contribution of axial microtubules in the axon resulted in roughly 2 times lower strains during axial loading compared to the transverse case (same network stretch) for isotropic networks. The microtubules might have contributed slightly to the overall stiffness during transverse loading as well, due to their local reorientation along the loading direction.

DISCUSSION

The current study used a computational model to quantify how structural variations can lead to changes in the deformation of an embedded nerve fiber for two specific loading scenarios. Based on our observations in spinal facet capsular ligaments [3,4], fiber networks with different architectures were investigated. Moreover, since tissues undergo complex deformations during physiological loadings [5], two different loading cases (axial and transverse to the long-axis of axon) were studied and compared.

It is important to note that the predictions apply only for displacement-control loadings, and force-control loadings (e.g. pressurization of spine facet capsules) might result in a different response.

Studying the mechanical response of the axon is important both in understanding its *sensory function* and in predicting its *injury*. In terms of *sensory functions*, axonal deformation might lead to an electrical signal, if strains are large enough to open stretch-gated ion channels [9]. Axonal deformation has also been reported as a useful *injury* predictor [9]. Thus, in attempting to understand how macroscopic stretch leads to neuronal responses, it is necessary to consider the tissue architecture because the same macroscopic stretch is not always transmitted to the axon at the same level.

ACKNOWLEDGEMENTS

This work was supported by the National Institutes of Health (U01EB016638). Computational resources were provided by the Minnesota Supercomputing Institute (MSI) at the University of Minnesota.

REFERENCES

[1] Bernal, R et al., *Phys. Rev. Lett.*, 99(1), 018301, 2007.
[2] Spedden, E. et al., *Int. J. Mol. Sci.*, 14(8), 16124-16140, 2013.
[3] Zarei, V et al., *Biomech Model Mechanobiol.*, under review
[4] Quinn, K et al., *Appl Physiol.* 105.6: 1881-1888, 2008.
[5] Siegmund, G et al., Spine, 26(19), 2095-2101, 2001.
[6] Lai, V et al., *J Biomech Eng.*, 134(1), 011004, 2012.
[7] Pampaloni, F et al., *Proc. Natl. Acad. Sci.*, 103(27), 10248-10253, 2006.
[8] Chandran, P et al., *J Biomech Eng.*, 129(2), pp.137-147, 2007.
[9] Tang-Schomer, M et al., *FASEB J*, 24(5), 1401-1410, 2010.

SB³C2017
Summer Biomechanics, Bioengineering and Biotransport Conference
June 21 – 24, Tucson, AZ, USA

DETECTING ENVIRONMENTAL PH USING MECHANICAL PROPERTIES OF MICROORGANISM

W. Zheng (1, 2), H. Yang (3), G. Xuan (1, 2), L. Dai (3), Y. Hu (3), S. Hu (3,4), S. Zhong (5), Z. Li (3), M. Gao (1), S. Wang (3), Y. Feng (1)

(1) Center for Molecular Imaging and Nuclear Medicine, School of Radiological and Interdisciplinary Sciences (RAD-X) Soochow University, Suzhou, Jiangsu, China

(2) School of Mechanical and Electronic Engineering, Soochow University, Suzhou, Jiangsu, China

(3) College of Resources and Environmental Sciences, Nanjing Agricultural University, Nanjing, Jiangsu, China

(4) Department of Plant Pathology, North Carolina State University, Raleigh, NC, USA

(5) School of Iron and Steel, Soochow University, Suzhou, Jiangsu, China

INTRODUCTION

The metabolism, growth, and development of cells and tissues are influenced by many different factors. Among them, environmental pH plays an important role [1-4]. Although challenging, it is of great interest to detect the historical time course of environmental pH. Many different methods have been proposed such as magnetic resonance imaging [5], fluorescent dyes [6, 7], and particle-based nano-sensors [8, 9]. However, none of them can trace back the history of the environmental pH.

The very first AFM-like instruments were invented for the purpose of measuring cell mechanical properties [10]. Although certain assumptions were needed for estimating the mechanical properties, AFM has been used to obtain much useful information about the mechanics of cells. Phosphorous solubilizing fungi (PSF) are common microorganisms living in the environment. They have a strong adaptability to the environment with a relatively higher acidity. Therefore, possible relationships exist between the mechanical properties of PSF and the environmental pH.

In this study, we explore the potential of using AFM nano indentation to correlate the mechanical properties of the fungi with the culturing environment. For the first time, we investigated the change of the mechanical properties of *Aspergillus niger* (*A. niger*) over time in environments with different pH values. Our goal was to explore the potential of using mechanical properties as biomarkers for uncovering a history of environmental pH.

METHODS

A. niger strain (NJDL-12) [11] was cultured in acidic medium with an initial pH value of 1.5 (pH=1.5). Potato dextrose agar (PDA) for culturing the strain was composed. The liquid PDA was prepared without agar. We used 1 mol/L HCL and 1 mol /L NaOH to modulate the pH value of the solution. To compare with the mechanical responses of the *A. niger* in the acidic environment, we also cultured the strain with a pH value of 6.5 (pH=6.5). A drop of the medium was deposited on a silicon wafer and was then air-dried for the AFM analysis. All of the sample preparations were in a clean bench.

A commercial AFM device (Dimension Icon, Bruker) was used for the nanoindentation tests at room temperature. We used an aluminum reflex coated silicon cantilever probe with a cone tip (Tap 150AI-G, Budget Sensors, Innovative Solutions Bulgaria Ltd., Sofia, Bulgaria) for the nanoindentation. The half cone angle of the tip is 10 degrees. The tip radius is 10 nm, and the nominal force constant is 5 N/m. We used the tapping mode (PeakForce Tapping) to measure the indentation force-displacement curve with the depths ranged between 100 nm to 300 nm (Figure 1).

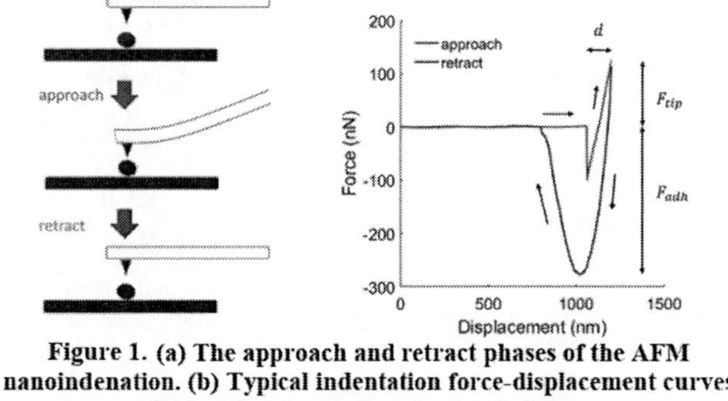

Figure 1. (a) The approach and retract phases of the AFM nanoindenation. (b) Typical indentation force-displacement curves during the approach and retract phases.

Recognizing the limitations of indentation, we calculated only an effective modulus. The DMT model was adopted to calculate the effective modulus [12]:

$$E = \frac{3}{4}\frac{\left(F_{tip} - F_{adh}\right)}{4\sqrt{Rd^3}}(1 - \nu^2), \quad (1)$$

where F_{tip} is peak force of the AFM probe in the approach phase (Figure 1b), F_{adh} is the adhesive force between the AFM tip and the sample during the retract phase, R is the tip radius, d is the indentation depth, and ν is Poisson's ratio. We note that we treated the organism as incompressible; because isotropy was also assumed, this was achieved by choosing ν=0.5.

RESULTS

Sample AFM images and the estimated moduli of spore and hypha are shown in Figure 2. To demonstrate the inter-sample differences, we selected two different spores and two different hyphae under the same culturing condition (pH=1.5) for a comparison (Figure 2). The numbers of testing points for the two spore samples were 4 and 5. For the two hyphae samples, the numbers were 12 and 5. The *p*-value of the measured E values between the two spores was 0.76, and for the two hyphae was 0.67. Therefore, because no significant intersample and intrasample difference were observed, we used the all the measured E values from each sample for analysis.

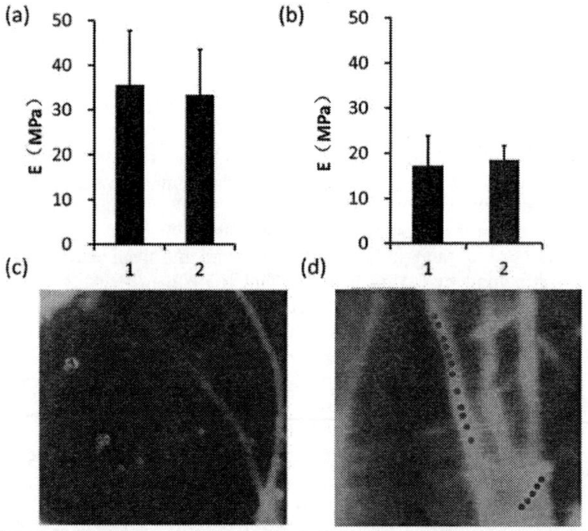

Figure 2. Elastic moduli from the indentation of the (a) spore and (b) hypha. The red and blue dots represent the indentation locations for the two different (c) spore and (d) hypha samples.

After 3 days of culturing, the estimated effective moduli for spore and hyphae at pH 1.5 were 34.96 ± 21.73 MPa and 29.48 ± 17.14 MPa, respectively. At culturing environment with a pH of 6.5, the estimated moduli for spores and hyphae were 85.75 ± 43.69 MPa and 235.49 ± 93.26 MPa, respectively. By comparing the E values measured in a culturing environment with pH=1.5 and pH=6.5, we observed significant differences at each incubation time points for both spores and hyphae (Figure 3).

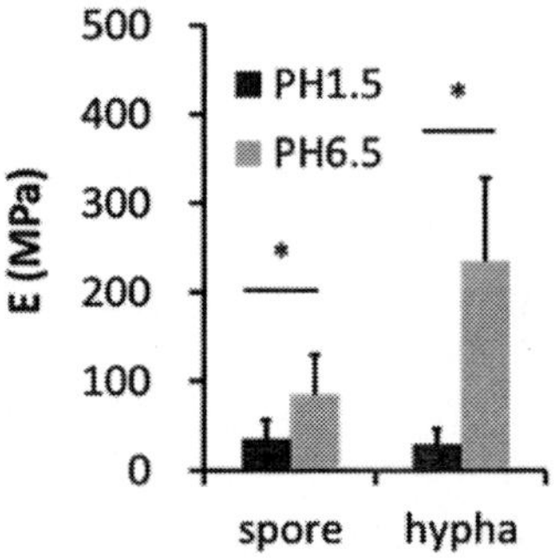

Figure 3. Comparisons of the E values of the spores and hyphae cultured in an environment with pH=1.5 and 6.5 after 3 days of incubation. The * symbol indicates a significant difference between the two groups (p<0.05).

DISCUSSION

Mechanical properties of cells and tissue are closely related to their developing conditions. In this study, we developed a method using nanoindenation to detect the environmental pH values. We observed that in the acidic environment, the elastic moduli of spore and hypha were softer than that cultured in the neutral environment. By correlating with the change of mechanical properties, we aim to use mechanical property as a biomarker to track the pH history of the environment. Future studies involve testing the mechanical properties of the microorganism cultured in environment with different pH values with more sampling time points.

ACKNOWLEDGEMENTS

This work was supported by grant 61503267 (YF) from National Natural Science Foundation of China, National Program on Key Basic Research Project No. 2015CB150504, grant BK20140356 (YF), grant 16KJB460018 (YF) from Jiangsu Province, and by grant K5115701515 (YF) from Scientific Research Foundation for the Returned Overseas Chinese Scholars, State Education Ministry. Support from Priority Academic Program Development of Jiangsu Higher Education Institutions (PAPD) is also acknowledged. We greatly thank the technical assistance of Hui Zhang from the Analytic and Experiment Center (AFM division) at Soochow University.

REFERENCES

[1] Casey, J.R., S. Grinstein, and J. Orlowski, Nat Rev Mol Cell Biol, 11:50-61, 2010.
[2] Webb, B.A., et al., Nat Rev Cancer, 11:671-677, 2011.
[3] Neri, D. and C.T. Supuran, Nat Rev Drug Discov, 10:767-777, 2011.
[4] Uria, N., et al., Talanta, 147:364-369, 2016.
[5] Gallagher, F.A., et al., Nature, 453: 940-943, 2008.
[6] Ni, Y. and J. Wu, Organic & biomolecular chemistry, 12:3774-3791, 2014.
[7] Guo, Z., et al., Chemical Society Reviews, 43:16-29, 2014.
[8] Xiong, H., et al., Biosensors and Bioelectronics, 86:164-168, 2016.
[9] Chen, P.-C., et al., Nanoscale, 6:3503-3507, 2014.
[10] Petersen, N.O., W.B. McConnaughey, and E.L. Elson, PNAS, 79:5327-5331,1982.
[11] Li, Z., et al., Scientific Reports, 6:25313. 2016.
[12] Young, T.J., et al., Measurement Science and Technology, 22:125703, 2011.

SB^{3}C2017
Summer Biomechanics, Bioengineering and Biotransport Conference
June 21 – 24, Tucson, AZ, USA

BIOINSPIRED POLYMER INFILTRATED HYDROXYAPATITE NANOCOMPOSITE HYBRIDS

Rohit Khanna (1), Xiaodu Wang (1)

(1) Department of Mechanical Engineering
University of Texas at San Antonio
San Antonio, TX 78249, USA

INTRODUCTION

Recent advances in materials research are focused on fabricating advanced materials with high strength and toughness, inspired from natural materials such as nacre, bone or teeth, all of which resemble a nanocomposite structure composed of hard nanocrystals embedded in a soft organic matrix. For instance, remarkable mechanical properties of nanocomposites have been achieved by mimicking brick and mortar structure of nacre using alumina as the hard material and PMMA as the soft matrix [1,2]. Energy dissipation as a result of sliding between PMMA and alumina platelets was registered as the toughening mechanism in these nanocomposites. However, these toughening mechanisms were not exploited at the nanoscale because thickness of organic interfaces between alumina platelets was in the microscale in these artificial hybrid composites as compared to that of only ~20 nm between mineral platelets in natural materials. In addition, it is technically very challenging to fabricate a nanocomposite with thin organic interface between the hard nanoparticles.

As a promising alternative, we propose here to fabricate a ceramic template with nano-porosity, followed by infiltration of a hydrophilic biopolymer to fill up the pores to make compliant interfaces between the ceramic constituents. Our research work is particularly inspired from nano-architecture of natural materials, such as bone, nacre, and tooth. We believe that bone is significantly toughened by energy dissipation through interface sliding between intertwining compliant collagen fibrils and hard mineral. To prove this concept, we fabricated hydroxyapatite (HA) templates with interconnected nano-pores with a size close to those of collagen by the compaction of HA nano-powders followed by partial sintering. Then, we infiltrated a hydrophilic polymer (*e.g.* Poly Vinyl Alcohol, PVA) into the nano-porous HA template. Finally, we tested the new nanocomposites mechanically and observed microscopically. The results of this pilot study clearly indicated that infiltration of hydrophilic polymers into nano-pores in the HA templates significantly improved the *in situ* mechanical properties of the nanocomposite.

METHODS

Commercial nano-powders of HA were cold pressed to circular disc samples of 20 mm by 1 mm at a pressure of 40MPa and subsequently, sintered at a temperature of 1000°C for 2 hours in a standard furnace in air to form nano-porous HA templates. The density of the porous templates was determined by Archimedes Principle. PVA (molecular weight 25 kDa) powder was dissolved in deionized water by continuous stirring at 75°C for 30 minutes and subsequently, infiltrated into synthesized HA templates by drop-wise addition onto the surface of sample, which was kept inside a desiccator connected to a low suction pump. The sample was weighed at the end of each cycle to measure the weight change in order to monitor the infiltration process. The polymer infiltration was repeated until the sample weight became saturated. Phase analysis of powders before and after sintering was done by x-ray diffraction to confirm that the HA phase was not changed during the sintering. TEM was performed to characterize the size and morphology of the HA nano-powders. High resolution images of synthesized HA template before and after infiltration were acquired by high resolution Field-emission SEM (FE-

SEM). Mechanical properties of the samples were evaluated using Nano Indenter XP system with a cube corner tip to determine the *in situ* toughness of the synthesized samples using the nano-scratch approach formulated by our group [3].

RESULTS

Figure 1a shows TEM image of agglomerates of several HA particles on the Cu grid. High magnification TEM image, Fig. 1b revealed the nano-plate morphology of HA powder with sizes in the length 90-130 nm and width 20-40 nm.

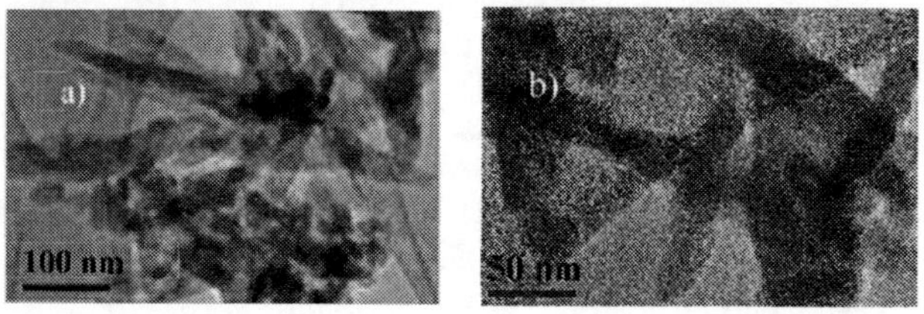

Figure 1 TEM images of agglomerates of HA powders (a) and plate-like morphology (b).

Figure 2a shows FE-SEM image of fracture surface of synthesized HA templates. After sintering, the size of HA crystals increased to about 200-300 nm and formed bridges with neighboring HA particles leading to interconnected pores in the range of 200-400 nm. Relative density of nano-porous HA template was ~94%. After addition of 4 wt % PVA solution into nano-porous HA templates, the weight of the sample increased by ~18%, indicating the infiltration of PVA into nano-porous HA, as shown by dense microstructure of infiltrated PVA-HA hybrid in FE-SEM image, Fig. 2b.

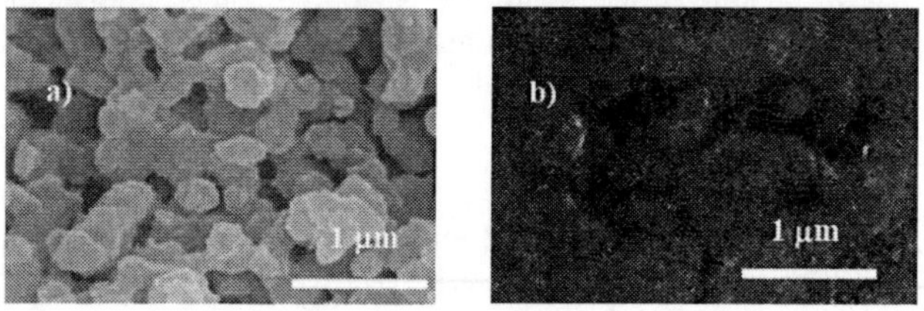

Figure 2 FE-SEM images of synthesized HA template before (a) and after (b) infiltration of PVA.

Figure 3a shows the typical line profiles across the nanoscratch. Nanoscratch test on HA template without PVA infiltration revealed the wider cross-section of scratch ~30 μm with a maximum penetration depth ~7500 nm. Whereas after PVA infiltration, narrow cross-section of scratch ~8 μm wide with relatively smaller penetration depth ~1800 nm clearly revealed the increase in abrasion resistance after polymer infiltration. This result was further corroborated by toughness calculations adopting our previous nanoscratch formulations which revealed that toughness of nano-porous HA was dramatically increased from mean value of 0.5 to 5.7×10^8 J/m^3 as shown in Fig. 3b. Overall, these results suggests significance improvement in mechanical performance of designed HA templates after polymer infiltration.

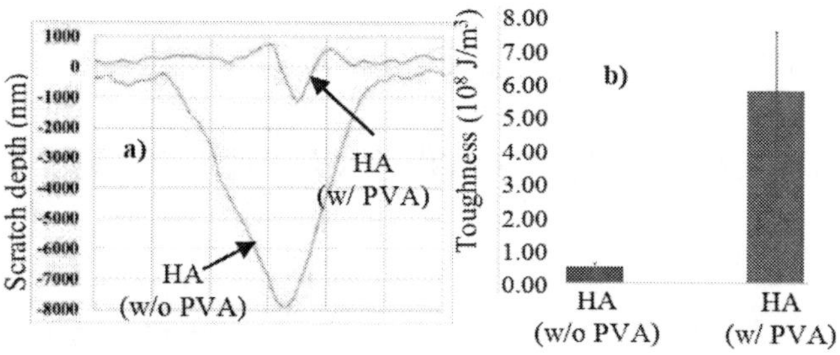

Figure 3 Line profiles across the scratch (a) and toughness (b) of the HA template with and without PVA infiltration

DISCUSSION

Results of this pilot study revealed that scratch groove is much wider and deeper in nano-porous HA template without infiltration, suggesting poor mechanical response, whereas nanoscratch measurement after polymer infiltration revealed a narrow groove with shallow penetration depth, suggesting polymer infiltration makes the material more resistant to scratch. Furthermore, scratch toughness is one order of magnitude higher in polymer infiltrated HA template which strongly supports our hypothesis. These results suggest that compliant polymer is able to transmit load between hard particles and dissipate energy by deformation as a result of sliding between polymer and HA materials. These results are significant in our opinion, since formulation of this approach to fabricate thin compliant polymer interfaces by nano-porous HA fabrication, followed by polymer infiltration represents a trajectory from conventional methods of polymer-ceramic nanocomposite fabrication by simple mixing the constituents. Further studies of efficacy of toughening in these hybrid systems and optimization of HA templates and polymer infiltration are underway to fabricate tough polymer-HA hybrids by closely mimicking bio-inspired natural materials.

CONCLUSION

These material design concepts can be potentially used in the fabrication of orthopedic medical devices that require enhanced mechanical performance.

REFERENCES

[1] Munch et al, *Science*, 322: 1516-1520, 2008.
[2] Wegst et al., *Nature Mater., 14:23-36, 2015.*
[3] Islam et al., *J. Mech. Behav. Biomed. Mater.*, 5:156-164, 2012

SB³C2017
Summer Biomechanics, Bioengineering and Biotransport Conference
June 21-24, 2017, Tucson, AZ, USA

MYOCYTE-COLLAGEN INTERACTION IN THE HEART: AN EXPERIMENTALLY-GUIDED MODELING STUDY

Sheikh Mohammad Shavik
Michigan State University
Lansing, MI, USA
shaviksh@egr.msu.edu

Marissa Grobbel
Michigan State University
Lansing, MI, USA

Lik Chuan, Lee
Michigan State University
Lansing, MI, USA

Sara Roccabianca
Michigan State University
Lansing, MI, USA

INTRODUCTION

Myocytes and the collagen fibers are key constituents of the heart tissue and they contribute collectively to the overall heart's mechanical and functional behaviors. Although it is generally known that myocytes and the collagen fiber network both affects ventricular mechanics and can interact mechanically, it is not known, however, how they interact and how that interaction affects the tissue's mechanics. Consequently, current descriptors of cardiac tissue mechanics do not take into account the direct mechanical interactions between collagen and myocytes. Here, we describe a experimentally-guided modeling approach to elucidate the mechanical interactions between myocytes and collagen in the normal rat heart.

METHODS

Results from our experiments conducted on normal rats have showed that the opening angle arising from a radial cut in short-axis slices of the left ventricle (LV) is $58 \pm 13°$ ($n = 5$, adult male rats weighing 250 - 400g). When collagen or cells were removed (by collagenese treatment or decellularization, respectively), however, the opening angle increased (to $106 \pm 23°$, $n = 5$) and decreased (to $21 \pm 4°$, $n = 5$), respectively (**Figure 1a**). Constrained mixture (CM) modeling was applied here to interpret these experimental results.

The CM model, which describes the overall tissue mechanical behavior as the sum of its constituents mechanical behavior weighted by their corresponding mass fraction, was used to infer the existence of any mechanical interactions between the myocytes and collagen. We note that the CM model does not take into account any explicit inter-constituent mechanical interactions between the myocytes and collagen. Specifically, the CM model describes the overall tissue mechanical behavior by a strain energy function that is given mathematically by

$$W_{\text{heart}} = \phi_{\text{cell}}\, W_{\text{cell}} + \phi_{\text{coll}}\, W_{\text{coll}} - p(J-1)\,, \tag{1a}$$

where

$$W_{\text{cell}} = c_1 \left(\exp^{k_1 (I_1((\mathbf{F}\mathbf{F}_{\text{p, cell}})^T \cdot (\mathbf{F}\mathbf{F}_{\text{p, cell}})) - 3)} - 1 \right)\,, \tag{1b}$$

$$W_{\text{coll}} = c_2 \left(\exp^{k_2 (I_1((\mathbf{F}\mathbf{F}_{\text{p, coll}})^T \cdot (\mathbf{F}\mathbf{F}_{\text{p, coll}})) - 3)} - 1 \right)\,, \tag{1c}$$

In Eq. (1), p is the Lagrange multiplier (for enforcing incompressibility), J is the Jacobian of the deformation gradient $\mathbf{F}$, while ϕ_{cell} and ϕ_{coll} denote the mass fraction of the myocyte and collagen fibers in the tissue, respectively. The strain energy functions of the myocyte (W_{cell}) and collagen fibers (W_{coll}) were prescribed based on uniaxial mechanical tests conducted on single myocytes [1] and collagen fibril [2], in which the corresponding material parameters c_1, c_2, k_1 and k_2 were fitted from the test results. Because collagen fibers were observed to be largely wavy in the unloaded LV, we offset the collagen fibril stress-strain data horizontally before fitting. The offset corresponds to a straightening stretch of about 1.2, a value prescribed based on a previous study showing that perimysial collagen fibers in the heart straightened considerably as sarcomere length is increased from 1.85μm to $2.3\ \mu$m [3]. Both W_{cell} and W_{coll} depend on the deformation map from the corresponding natural stress-free configuration of the myocytes and collagen, respectively. These mappings were decomposed sequentially into (1) a map between the cut ring of the "cell-only" and "collagen-only" samples (assumed to be stress-free) to the intact ring (i.e., $\mathbf{F}_{\text{p, cell}}$ and $\mathbf{F}_{\text{p, coll}}$) and (2) a map between the intact mixture to its deformed configuration $\mathbf{F}$ (**See Figure 1b**). We note that $\mathbf{F}_{\text{p, cell}}$ and $\mathbf{F}_{\text{p, coll}}$ can be calculated respectively from the opening angle measured in the "cell-only" and "collagen-only" samples.

Parameterized using quantities measured experimentally in the intact tissue as well as tissue isolated with collagen and myocytes (i.e., ϕ_{cell}, ϕ_{coll}, c_1, k, c_2, $\mathbf{F}_{\text{p, coll}}$ and $\mathbf{F}_{\text{p, cell}}$), the CM model was used to predict the opening angle of the mixture to test the hypothesis on the existence of a inter-constituent mechanical interaction between the myocytes and collagen. Specifically, should the CM model be able to reproduce the opening angle measured in the intact tissue using measured quantities associated with the constituents will suggest that any explicit mechanical interaction between the constituents is negligble. Conversely, should the model fail to reproduce the measured opening angle would suggest the existence of an inter-constituent mechanical interaction that cannot be ignored.

RESULTS

Using quantities obtained from the experiments, the CM model predicted an opening angle that is far off from that measured in the "control" samples containing both myocytes and collagen. Instead, the model predicted an opening angle of 104.0°, a value that is close to that found in the "collagen-only" samples. The CM model can only reproduce the measured opening angle found in the "control" samples using a prescribed mass fraction of $\phi_{\text{cell}} = 0.9995$ and $\phi_{\text{coll}} = 0.0005$ (i.e., myocytes and collagen accounting for 99.95% and 0.05% of the mass in the mixture).

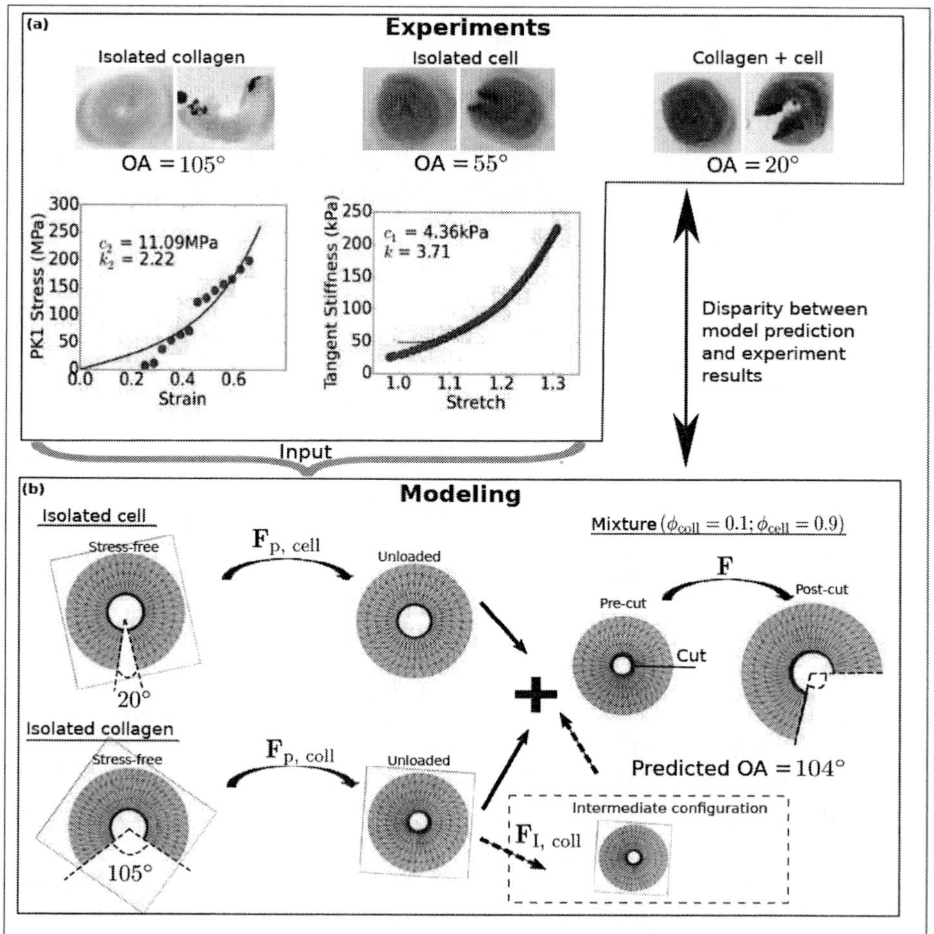

FIGURE 1. (a): Opening angle test results and mechanical tests on isolated single collagen fibril [1] and myocyte [2]. Inset: fitted material parameters. (b) Constrained-mixture model using experimental data of measurements of constituent prestress ($\mathbf{F}_{\text{p, coll}}$ and $\mathbf{F}_{\text{p, cell}}$), mass fraction (i.e., ϕ_{cell}, ϕ_{coll}) and material parameters (c_1, k, c_2, k_2). Disparity between model prediction and experiment measurements of the opening angle in the mixture suggest the presence of myocyte-collagen interaction via an intermediate configuration

DISCUSSION

We have used a CM model to intepret the experimental findings that the opening angle is substantially different when collagen or cells are removed in the LV of normal rat heart. Because the CM model takes into account the contribution of collagen and myocytes to the overall mechanical behavior by their mere presence, and does not consider any explicit mechanical interaction between them, the failure of the model to reproduce the opening angle measured in the "control" samples (mixture) would suggest the existence of an inter-constituent mechanical interaction in the cardiac tissue. A mass fraction of $\phi_{\text{cell}} = 0.9995$ and $\phi_{\text{coll}} = 0.0005$ necessary to reproduce the opening angle in the CM model is far beyond the values measured in this study as well as in previous studies. This disparity could be resolved by the presence of some direct myocyte-collagen physical interaction in the cardiac tissue, which is difficult to ignore as scanning electron microscopy images show that the myocytes and collagen fibers are tightly connected one another in the tissue ultrastructure.

One way of accounting for the mechanical interaction between myocyte and collagen when they are co-existing in the tissue is by the introduction of an intermediate configuration in the deformation maps. Associated with this configuration is a corresponding deformation map that allows the constituents to "fit" together in the tissue. In other words, cell and collagen fibers undergo additional physical deformation when co-existing (or "put together") in the intact tissue. To make our case here, we introduced an intermediate configuration for the collagen fibers that is characterized by a deformation gradient tensor $\mathbf{F}_{\text{I, coll}} = \alpha\mathbf{I}$ with $\mathbf{I}$ (**Figure 1b**) denoting an identity tensor. Using the measured mass fractions $\phi_{\text{cell}} = 0.9$ and $\phi_{\text{coll}} = 0.1$, we find the measured opening angle can be reproduced with a value of $\alpha = 0.6$, which suggests that the collagen fibers are, on average, compressed by the myocytes when co-existing in the mixture. In summary, our study suggests the presence of physical interactions between the cells and the collagen fiber network that cannot be ignored, and is likely to play a significant role in tissue mechanics under physiological and pathophysiological conditions.

REFERENCES

[1] Shen, ZL et al. *Biophys. J.*, 95(8):3956-3963, 2008.
[2] Brady, AJ *ASME J. Biomech. Eng.*, 106(1):25–30, 1984.
[3] Hanley, PH et al. *J. Physiol.*, 517(3):831–837, 1999.

SB³C2017
Summer Biomechanics, Bioengineering and Biotransport Conference
June 21 – 24, Tucson, AZ, USA

INFLUENCE OF THE DIVALENT CATION CROSSLINKER AND ITS CONCENTRATION ON THE ELASTIC MODULUS AND PERMEABILITY OF ALGINATE HYDROGELS

David M. Kingsley (1) and David T. Corr (1)

(1) Biomedical Engineering Department
Rensselaer Polytechnic Institute
Troy, NY, USA

INTRODUCTION

Hydrogels are three-dimensional hydrated polymer matrices that in recent years have become widely used for biomedical applications. In tissue engineering related fields, the popularity of hydrogels is largely due to its structural and mechanical similarities to native biological extracellular matrix (ECM) [1]. Furthermore, some hydrogels have been shown to have widely tunable properties, which can be leveraged to mimic those of the native ECM. Controlling the mechanical features of a cellular environment, specifically within 3D matrix materials (*e.g.* elastic modulus and porosity), is essential for mimicking native tissue mechanical cues.

The importance and implications of ECM structural properties on cellular behavior have only begun to be fully realized. It has been clearly demonstrated that cells have machinery to sense and respond to mechanical cues from the surrounding matrix [2]. These cues are a key factor in guiding stem cell differentiation, cancer metastasis, and countless other biological phenomena [3], [4]. It is therefore essential that researchers have the tools to manipulate the mechanical properties of the extracellular matrix for next generation tissue engineering, applications in regenerative medicine, and *in vitro* diagnostics.

Alginate, a polysaccharide composed of covalently linked mannuronic acid (M) and guluronic acid (G) subunits, is an ideal material for such applications because of its tunable properties and biocompatibility [5]. Soluble alginate gels through ionic interactions with divalent cations (*e.g.* Ca^{2+}, Sr^{2+}, Ba^{2+}) shared with two G blocks of adjacent chains. When seeking to control the mechanical properties of alginate gels, the divalent cation, the relative cation concentration, alginate gel concentration, and alginate M:G ratio have all been identified as factors directly influencing the gel's resultant mechanical properties. In this study, we examine how the crosslinking divalent cation, and its concentration, influence the mechanical properties of alginate gels. Specifically, we characterize the elastic modulus and permeability of alginate gels, and further look at the stability of these properties over 24h in culture.

METHODS

Alginate hydrogel fabrication:

Alginate gels mechanical test specimens were prepared as cylindrical disks within PDMS-punched wells, using 2% alginate/0.9% NaCl (w/v) dissolved in deionized water and subsequently run through 0.8/0.2μm filters. Alginate was then pipetted as 250-μl volumes into 10mm-diameter cylindrical PDMS molds, and subsequently covered with crosslinker ($CaCl_2$ or $BaCl_2$) for a 15-minute duration. Samples were then removed and stored in Dulbecco's Modified Eagle's Medium (DMEM). Mechanical testing was performed at either 1 hour or 24 hours after fabrication.

Alginate gel characterization:

To determine the elastic modulus of each alginate gel, cylindrical test specimens were mechanically characterized by parallel-plate compression using a Mach-1 V500cs micromechanical system (Biomomentum, Laval, QC), equipped with a 70.0 N load-cell (±3.5mN resolution). To test, each specimen was placed in the center of a chamber filled with DMEM to remain hydrated throughout testing, and the initial height was determined using the "find-contact" mode, until a 0.5gF load was detected. Based on its initial height value, each specimen was then subjected to unconfined compression to 15% strain, at a constant rate of 1% strain/second. The resulting force-displacement data were converted to stress-strain using measured specimen geometries. Elastic modulus was determined as the slope of the stress-strain curve over a defined linear range, approximately 8-10% strain (unless otherwise stated).

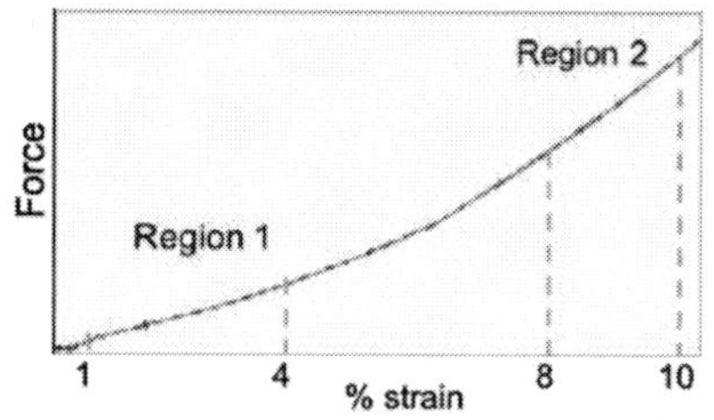

Figure 1. Representative force-displacement curve for alginate compression data.

Permeability of alginate gels was estimated by characterizing the diffusion of a tracer molecule (70 kDa FITC-labeled dextran) from alginate gels. Briefly, alginate gels were fabricated as previously described, and subsequently soaked in 10 mg/mL tracer molecule for 24 hours. Samples were then moved to chambers containing deionized water within a 12-well plate, and media was drawn from independent wells at (15min, 30min, 1hr, and 4hr), and the tracer concentration was estimated through spectrophotometry. Last, tracer concentration was plotted vs. time (i.e., square root of time in media) for the different gel samples. Using **equation (1)** and the slope of these plots, a diffusion coefficient parameter could then be estimated for each gel.

$$C_{media} = C_{matrix} \frac{A_{matrix}}{V_{bath}} 2 \sqrt{t \frac{D_{matrix}}{\pi}} \qquad (1)$$

RESULTS

Compression of alginate hydrogels produced force-displacement curves with two distinct linear regions within the first 10% strain: 1-4% strain (region 1), and 8-10% strain (region 2) (**Figure 1**). Elastic moduli determined from region 2 were consistently ~1.7-2.1x greater than those of region 1, for all gels. For clarity, and comparison to prior results, all elastic moduli reported hereafter are from region 2.

When characterized 1-hour post fabrication, the average elastic moduli of alginate gels ranged from ~5-20kPa (**Figure 2**). For both alginate-Ba^{2+} and -Ca^{2+} gels, elastic modulus increased with crosslinker concentration. Further, hydrogels crosslinked with Ba^{2+} had a greater modulus at the same concentration of crosslinker than those crosslinked with Ca^{2+}. After 24 hours in culture, gels crosslinked with either 2% Ba^{2+} or 2% Ca^{2+} exhibited a 12.6% and 62.7% reduction in elastic modulus, respectively (**Figure 3**).

Plots of tracer vs time and the associated diffusion coefficients, shown in **Figure 4,** indicate that the 2% alginate gels crosslinked with 2% Ca^{2+} exhibited significantly greater (faster) diffusion than all other gels tested.

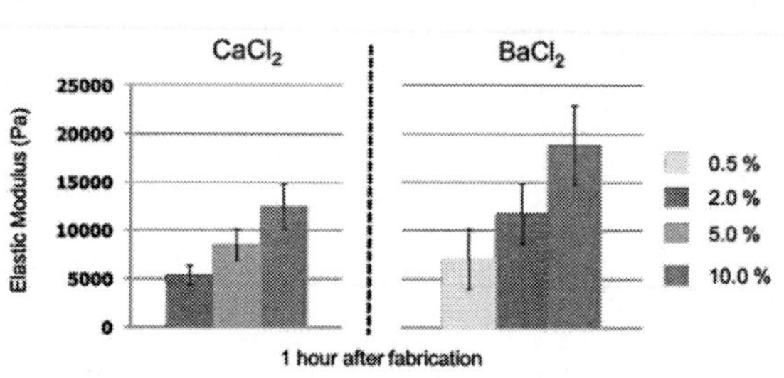

Figure 2. Elastic moduli of 2% alginate hydrogels, at 1 hour in media, crosslinked with $BaCl_2$ or $CaCl_2$, illustrating that moduli increase with crosslinker concentrations. (n=6, all groups).

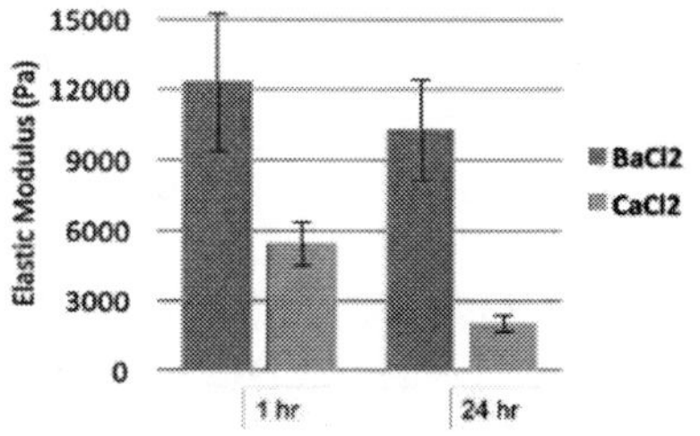

Figure 3. Elastic moduli of 2% alginate hydrogels, crosslinked with either 2% $BaCl_2$ and 2% $CaCl_2$, at 1 hour and 24 hours in media, suggesting that Ba^{2+}-crosslinked gels exhibit greater mechanical stability in culture. (n=6, all groups).

DISCUSSION

We showed that a variety of alginate gel stiffnesses could be generated by manipulating the divalent cation and its relative concentration. Barium produced gels with greater elastic modulus than those crosslinked by calcium, and barium gels appeared more stable with time in culture. It is possible that barium, the larger divalent cation, leaches more slowly from alginate gels due to changes in permeability or electrostatic interactions. To examine this, we explored the permeability of alginate disks made with different divalent cations, divalent cation concentrations, and alginate concentrations. However, there was no apparent change in the permeability of any of the gels, except for the low concentration calcium-crosslinked gels which exhibited faster diffusion. The fact that these gels were the softest (lowest elastic modulus) of the gels tested, suggests a possible link between the gel permeability and elastic modulus, however this needs further work to substantiate.

In our ongoing work, we are currently assessing how the mechanical properties of alginate bulk gels compare to those of alginate microbeads. It is possible that microbeads have different gelling behavior than large bulk gels due to their substantial difference in the ratio of surface area/volume. Characterization of microbeads is a more complex procedure, in part, because these spherical structures are micro-scale, and require sensitive testing equipment. Overall, we seek to utilize these alginate structures (bulk gels and microbeads) as synthetic ECM for cell-based applications, where the mechanical properties of the ECM could be tuned for specific applications, or to mimic the stiffness of a desired biological tissue.

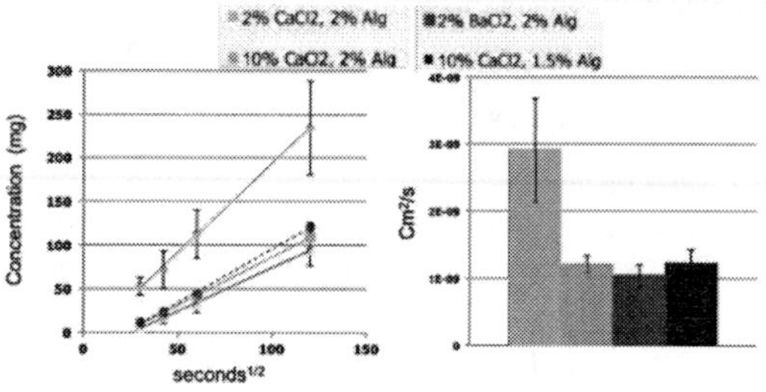

Figure 4. *(left)* Plot of tracer in media vs time for alginate gels, and *(right)* the corresponding diffusion coefficients (n=3/group).

REFERENCES

[1] Hoffman, A et al., *Adv. Drug Deliv. Rev.*, 64: 18–23, 2012.
[2] Discher, D et al., *Science*, 310: 1139–43, 2005.
[3] Engler, A et al., *Cell*, 126: 677–89, 2006.
[4] Swaminathan, V et al., *Cancer Res.*, 71: 5075–5080, 2011.
[5] Rowley, J et al., *Biomaterials*, 20: 45–53, 1999.

MEASUREMENT OF THERMAL CONDUCTIVITY OF CORTICAL BONE

Seon Jeong Huh (1), Hee Joon Lee (1), JuEun Lee (2)

(1) Department of Mechanical Engineering
Kookmin University
Seoul, Korea

(2) Department of Mechanical Engineering
University of the Pacific
Stockton, CA, USA

INTRODUCTION

It is important to avoid thermal necrosis during orthopaedic surgery since it can lead to tissue damage and infection. Hence, it is essential to predict temperature distribution at cortical bone under bone drilling [1-2]. However, thermal conductivity of a cortical bone is not a fixed value. It is dependent on the species, the age, the living environment, etc. [1-3] In this study, a thermal conductivity of cortical bone was experimentally measured. There are several experimental values, but the value varies in the literatures. Recent study was conducted by Zhang *et al.* [4] and they use Raman thermometry for measurement. However, concerning reproducibility, Parallel plate method was adopted to measure the thermal conductivity of a cortical bone. This method is based on the steady-state measurement, which maintains temperature to get the value of temperature drops between samples. From the temperature drop, thermal conductivity could be calculated using heat equation.

METHODS

Two different specimens were used in this experiment. One is bone as a subject for the experiment and the other is a reference sample. Davidson and James [5] reported that bovine bone is easy to purchase and substitutable for human tissue. Therefore, bovine bone was used for bone sample specimen. Three-year-old united states bovine tibia was obtained from local butcher shop. It was processed to a cylinder with 6 mm in diameter and height, respectively. First, core bore drill was used to make cylindrical shape and lathe was used to adjust height. After this work, the specimens were soaked in saline solution and stored in -21℃ in a freezer for the experiment.

Quartz was used for a reference sample. It has thermal conductivity of 1.4 W/mK. It was processed to a cylinder with 6 mm of diameter and height respectively as well.

The experimental setup is illustrated schematically in Fig. 1. Figure 2 shows the photo of experimental setup.

Parallel plate consists of three PMMA plates with cartridge heater and copper heat sink. It was designed to create one-dimensional heat flow through a bone specimen.

Quartz sample (A in Fig. 1) and a bone specimen (B) were placed into closed-cell PE foam (C) insulator. Polycarbonate (D) was used for insulation as well, but its main purpose was to fix the specimens and the closed-cell PE foam insulator. They were located between the top and middle PMMA plates (E). Cartridge heater (F) was put into the remaining hole of closed-cell PE foam insulator and held with two PMMA plates and bolts (I). Copper was positioned underneath the setting of bone and submerged in water bath to make steady temperature. Height of apparatus could be adjusted with bolts (G). To measure temperature between each specimen, three thermocouples (H) were located properly.

Bone specimen, quartz sample and the cartridge heater were placed in a row in the insulator. They were fixed with bolts. Three thermocouples were located between the cartridge heater and the bone specimen, the bone specimen and the quartz sample and quartz sample and copper. Thermal paste was used to fill gap between each material to avoid contact resistance. Height of whole setup was adjusted using the longer bolts. After adjustment, it was put into water bath. Half of copper was submerged in the water. Temperature of the heater on the top of the bone specimen was set to ambient temperature plus 15℃ with using temperature controller (OMEGA CN7500) which was connected

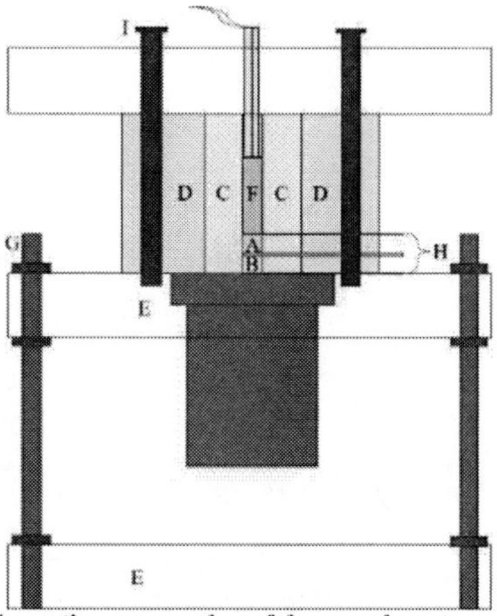

Figure 1: Schematic cross-section of the experiment apparatus used to measure thermal conductivity of cortical bone. A, quartz; B, bone specimen; C, closed-cell PE foam; D, Polycarbonate; E, PMMA plate; F, cartridge heater; G, bolts; H, thermocouples, I, bolts.

to the heater. Temperature of bottom part was set using water bath which maintain ambient temperature minus 15℃.

When temperature was in steady state after setting, temperature was recorded from the DAQ (National Instrument cDAQ-9178 with 9219 temperature module) for 1,000 seconds.

Both the heat loss to surroundings and the effect of contact resistance between samples were calibrated by using two quartz rods before measuring thermal conductivity of cortical bone.

To get thermal conductivity of bovine cortical bone, heat equation was used for calculation. It was described in Davidson and James [2] literature. The equation is

$$k_b = k_0 \frac{t_b}{t_0} \frac{T_0}{T_b} \qquad (1)$$

where k_0 is thermal conductivity of quartz [W/mK], k_b thermal conductivity of bone [W/mK], t_0 quartz sample thickness [m], $t_b$bone thickness, T_0measured temperature drop across the quartz sample [°C], and T_b measured temperature drop across the bone specimen [°C].

To validate current experimental setup and procedure, two quartz samples were put into the testbed instead of bone specimen. The thermal conductivity of two quartz samples is 1.4 W/mK, so that the temperature between two quartz samples should be average value of both top and bottom temperature. Figure 2 shows temperature measurement at top, bottom and middle (between quartz samples). The average top and bottom temperature were 33.6°C and 3.2°C, respectively. The middle temperature was measured as 18.9°C. Temperature difference and the corresponding thermal conductivity are is 0.5°C and 1.31 W/mK compared to the exact temperature (18.4°C) and thermal conductivity (1.4 W/mK), respectively. The relative error of temperature and thermal conductivity is 0.03% and 6.4% in this experimental setup and procedure, respectively.

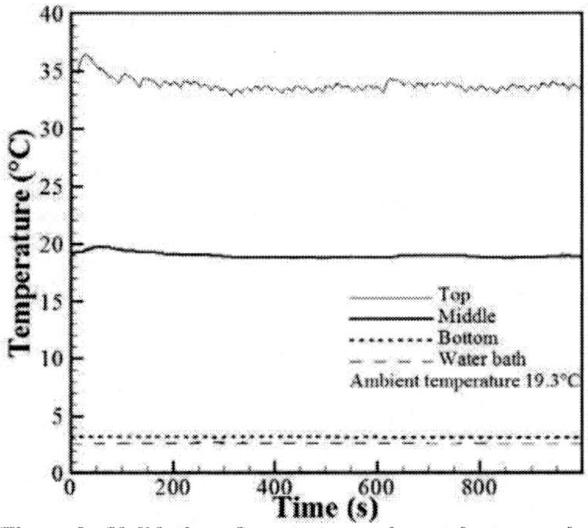

Figure 2: Validation of current experimental setup and procedure using two quartz samples.

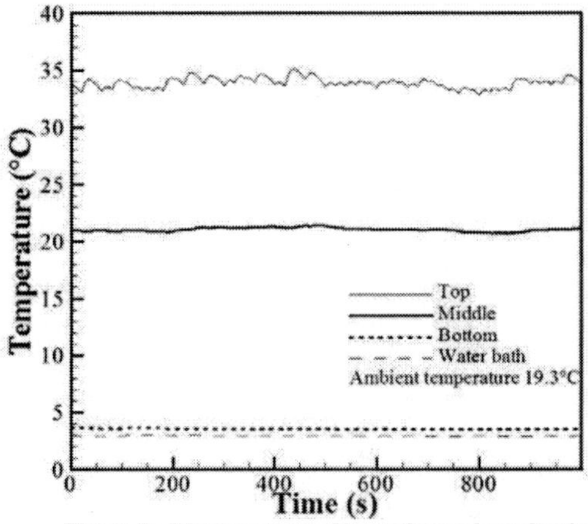

Figure 3: Measurement of bone thermal conductivity.

RESULTS

This study presents experimental measurement of thermal conductivity of cortical bone using parallel plate method. Thermal conductivity from the experiment was 1.03 ± 0.03 W/m K shown in Fig. 3. In comparison to the existing values, the thermal conductivity is slightly greater than expected value based on very recent literature. This value varies widely depending on species of bone and dry/fresh condition. Hence, the thermal conductivity obtained could be affected by those factors.

REFERENCES

[1] Lee, JE et al, *Med Eng Phys*, 33:1234-1244, 2011.
[2] Lee, JE et al, *Med Eng Phys,* 34:1510-1520, 2012.
[3] Lee, JE et al, *J Biomech*, 45:1076-1083, 2012.
[4] Zhang, Y et al., *J Nanotechnol Eng Med*, 5:21003, 2014.
[5] Davidson, SR, and James, DF, *Med Eng Phys*, 22:741-747, 2000.

SB³C2017
Summer Biomechanics, Bioengineering and Biotransport Conference
June 21 – 24, Tucson, AZ, USA

DETERMINING THE OPTIMAL SCREW CONFIGURATION FOR TIBIA PLATE FIXATION OF COMPOUND FRACTURES: A FINITE ELEMENT STUDY

Andrew L. Sori, Shihab Asfour, Francesco Travascio

Department of Industrial Engineering
University of Miami
Coral Gables, Florida, United States

INTRODUCTION

Tibial fractures often require the support of plate fixation for proper healing to occur, which can have as few as 6 holes or as many as 18 holes where screws can be applied. However, there is no consensus among manufacturers as to how many screws are necessary for proper fixation or where optimal placement should be, leaving it to the surgeon's discretion.[1-2] It is common practice to apply as many screws to the plate as the fracture allows to ensure mechanical stability, however it is suggested that applying more than a certain number of screws may not only be mechanically redundant, but detrimental to recovery due to soft tissue stripping and inducing non-optimal cell differentiation patterns.[3-4] Ideally, a surgeon would want to implant the least amount of screw while maintaining the structural integrity of the plate and inducing chondroblast differentiation at the callus. There is a lack of literature investigating this issue. Through computational analysis, this study aims to investigate how varying screw number and configuration affects the stress at the fracture site, callus, and throughout the plate, as well as its effects on the regeneration of the bone during reparative phase of healing. This is the critical phase of bone healing which will determine the long term health and integrity of the reformed bone.

METHODS

For this study, 3D geometry of a generic bone model was reconstructed using Mimics and 3-Matic software (Materialise, Belgium) (see figure 1). The mechanical properties of the cortical bone, cancellous bone, and callus were taken from literature. This study is modeled during the reparative phase of bone healing, where granulation tissue forms into hyaline cartilage and woven bone is formed. A complete fracture was simulated at the center of the tibial shaft with a gap of 2mm and a callus was modeled and placed at the gap. A tibial plate with different screw configurations (see figure 2 and table 1) was applied to the fracture on the anterior medial surface. A mechanical load of 750 N was applied to the condyles. Using FEBio Suite (University of Utah), a computational analysis was conducted to yield stresses in the bone at the fracture site (see figure 1), callus and throughout the plate. The shear strain was also yielded at the fracture.

A biphasic analysis was conducted to yield the fluid flow at the fracture. This information was used to determine the cell differentiation pattern based on the mechanoregulation theory:

$$S = \frac{\gamma}{a} + \frac{v}{b} , \qquad (1)$$

where S is the mechanical stimuli index, a = 0.0375, b = 3 $\frac{\mu m}{s}$, γ is the shear strain, and v is the interstitial fluid flow. The value of S predicts the cell differentiation pattern at the callus.[5-8] If S is less than 1, osteoblast differentiation occurs which could inhibit callus formation and delay healing. If S is greater than 1 and less than 3, chondroblast differentiation occurs, promoting healing through increased growth of the soft callus and cartilage. If S is greater than 3, then fibroblast differentiation occurs, causing stiff fibrous tissue to form. If S is greater than 6, there is excessive strain and fluid flow, and a high amount of fibrous tissue will form, delaying healing and potentially causing non-union of the bone.

RESULTS

Peak effective stress at all observed sites differed among all configurations. The highest observed peak effective stress at the plate of 275 MPa was seen in configuration 1 (see figure 3 and table 1). This was far greater than the next highest force of 142 MPa as seen in

Poster Presentation #P133

configurations 3 and 5. The lowest peak effective stress in the plate was seen in configuration 2 at 76 MPa. The peak effective stress seen on the tibia plate was observed at the center, between screws 4 and 5, closest to the fracture site.

The peak effective stress at the fracture site was observed in configuration 3 with 28.6 MPa. Configuration 1 had a similar stress of 22.6 MPa. Configurations 2, 4, and 5 saw similar stresses in the fracture site speak stresses around 13 MPa.

The highest peak effective stress at the callus was observed in configuration 1 of 342 MPa. The next highest was 72 MPa and was observed in configuration 4. Configuration 5 had the lowest effective stress at the callus of 17 MPa.

Configurations 1 and 5's mechanical stimuli index (S) fell in the same range between 3 and 6, indicating that fibroblast differentiation would occur (see figure 4). Configurations 2 and 4 fell between 1 and 3, indicating that chondroblast differentiation would occur. Configuration 3 had the highest S of 8.7, indicating that there is excessive strain and fluid flow, leading to very high fibrous tissue formation.

DISCUSSION

Although it is standard practice for surgeons to apply the maximum number of screws the plate and fracture site allows as in configuration 1, this may induce far greater stresses on the plate and callus than other configurations, as well as inducing non-ideal differentiation patterns, contributing to the possibility of implant failure and reduced healing. Alternatively, implanting fewer screws would compromise less bone and muscle, leading to better recovery. However, implanting too few screws or choosing improper placement may lead to increased stress at the fracture site, as demonstrated by configurations 1 and 3, increasing the risk of malunion, nonunion, and infection.[3,5] The mechanical stimuli index also differs among different configurations; ideally a value between 1 and 3 is preferred to induce chondroblast differentiation. Configuration 1, 3, and 5 may inhibit wound healing by promoting a high level of stiff fibrous tissue formation.

The results of this study suggest that using four screws in configuration 2 is the optimal method for implantation, rather than the commonly used 9 screw configuration. By using fewer screws, this configuration compromises less soft tissue, while also yielding lower stresses on the plate, fracture site and callus, as well as inducing chondroblast differentiation to promote wound healing.

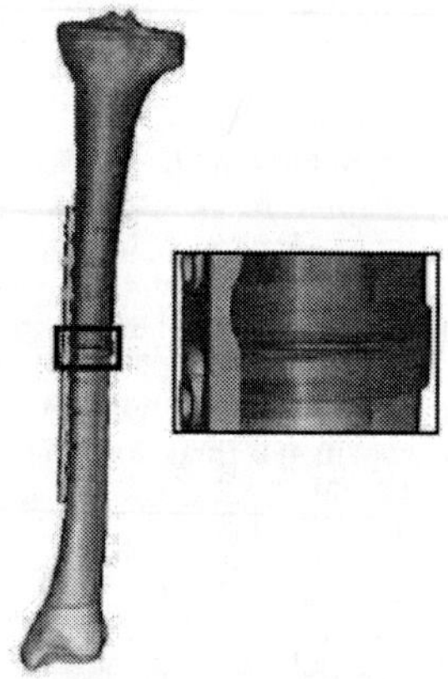

Figure 1: The fractured tibia model with a callus and a fixated plate. The observed fracture site is highlighted yellow.

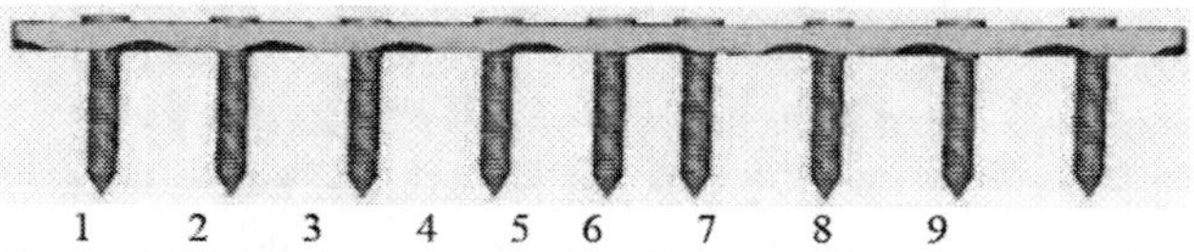

Figure 2: Tibial plate model. Screw 1 is proximal; screw 9 is distal.

Table 1: Configurations used. Screw numbers correspond to figure 2.

Configuration		Screws Used
1		1-9
2		1,4,5,9
3		1,2,8,9
4		1-3,7-9
5		3-6

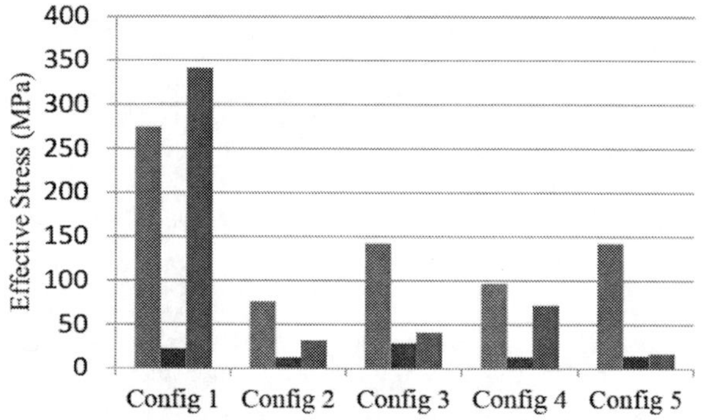

Figure 3: Peak effective stresses on the models. (Blue) Peak effective stress on the plate. (Red) Peak effective stress at the fracture site. (Green) Peak effective stress of the callus.

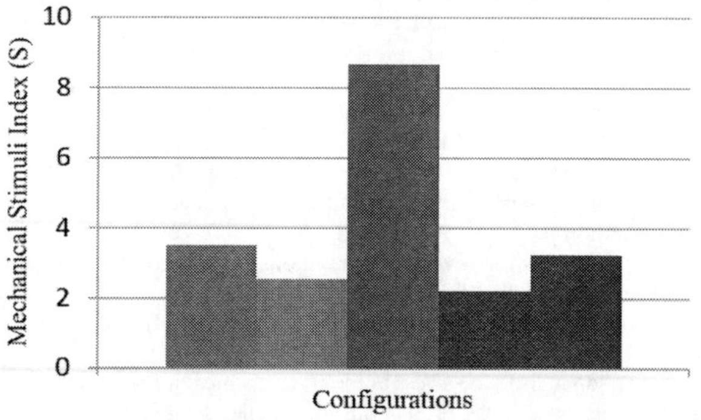

Figure 4: Calculated mechanical stimuli index (S). (Blue) Configuration 1. (Orange) Configuration 2. (Green) Configuration 3. (Red) Configuration 4. (Black) Configuration 5.

REFERENCES

[1] SYNTHES Inst & Imp Tech Guide
[2] Stryker AxSOS Op Techiques
[3] Greiwe, et al. Locking Plate Tech, 2007
[4] Muller, et al. Man of Int Fixation, 1995
[5] Miramini, et al. Medical Eng and Physics, 000 1-9, 2016
[6] Morgan, et al. J Biomech 43(12): 2418-2424, 2010
[7] Qin, Y, Minyi, H. Biomed Res Int. 2014: 863421, 2014
[8] Fritton, Weinbaum. Annu Rev Fluid Mech. 41: 347-374. 2009

ALTERATIONS IN EQUINE TIBIAL CONTACT PRESSURE AND BONE STRESS DUE TO FEMORAL CYSTS ARE INDEPENDENT OF KINEMATIC CONSTRAINTS

Lance L. Frazer (1), Elizabeth M. Santschi (2), Kenneth J. Fischer (3)

(1) Bioengineering Graduate Program
University of Kansas
Lawrence, KS, USA

(2) College of Veterinary Medicine
Department of Clinical Sciences
Kansas State University
Manhattan, KS, USA

(3) Mechanical Engineering
University of Kansas
Lawrence, KS, USA

INTRODUCTION

Subchondral bone cysts occur in the medial femoral condyle of the young horse and in many human osteoarthritic joints.[1,2] The most common treatment of equine subchondral lesions is cyst debridement which can be followed by the addition of various biological constituents. Successful treatment occurs in about 70% of horses and rarely results in radiographic healing, questioning if debridement is the ideal therapy. Part of the uncertainty is due to a lack of understanding of the biomechanics of the joint affected by subchondral bone cysts (SBCs). In order to develop and implement rational treatment strategies, a deeper insight into SBCs is required.

Finite element modeling provides a tool to investigate the mechanics of SBCs. However, kinematic data of an equine stifle joint is scarce making it difficult to validate the absolute results from such a model. By performing a sensitivity analysis of equine stifle joint finite element models we can better understand the stifle through comparative studies. The primary purpose of this study is determine the changes in contact pressure and bone stresses that occur when a medial femoral condylar bone cyst is present and how changing constraints on the joint affects these data. The secondary purpose is to compare results from various joint constraints to clinical evidence and infer the most appropriate kinematic constraints.

METHODS

Segmentation and Meshing – Computerized Tomography (CT) images of an extended stifle joint of a yearling Thoroughbred were imported into ScanIP (Simpleware). The image set was manually segmented to include the femur, tibia, patella, cartilage, and menisci. The 3D rendering of the segmentation was meshed with 4-node tetrahedral elements (C4D10) in ScanIP and subsequently imported into ABAQUS 6.14 (Simulia). Cartilage and menisci elements were converted to hybrid elements (C4D10H) for better contact accuracy.

Material Properties – Linear, elastic material formulations were used for each 3D structure (Table 1), and ligaments (MCL, LCL, PCL, ACL, meniscal-femoral ligaments, and meniscal-tibial ligaments) were modeled using nonlinear 1D spring elements. The cyst was given very low properties relative to the surrounding tissue. Bone tissue moduli were related to their corresponding CT Hounsfield unit using the following equations:

$$Mass\ Density = (1.067 * 10^{-3}) * GrayScale \quad (1)$$
$$Young's\ Modulus = (6.0 * 10^{3}) * MassDensity^{2.5} \quad (2)$$

Boundary Conditions – Frictionless, non-linear surface-to-surface contact was defined for all articulating surfaces using the general contact function in ABAQUS. Twenty springs were connected between each outer meniscal rim and ground to simulate the constraining effects of the joint capsule. Rotations were constrained at various levels (see sensitivity analysis), but all three translations were unconstrained. The distal tibia was encastred. A uniform surface pressure resulting in a 7200N compressive force, which is the peak ground reaction force during gallop, was placed on the proximal femur.

Contact Pressure and Stress Analysis – Contact pressure maps of the medial and lateral plateau and compressive bone stresses in the frontal plane at the articulation site of the femur and tibia were compared between the normal model and a model containing a 1 cm³ cyst in the medial femoral condyle. Also measured was the tensile strain in the medial meniscus with and without a cyst.

Sensitivity Analysis – Intact and cystic models were tested at 7200N femoral compression and 1500N patellar tension under 3 conditions: 1) no rotation allowed, 2) valgus/varus rotation only allowed, and 3) all rotations except flexion allowed. A fourth condition was no rotation and 3000N patellar tension. Contact pressure and stresses were analyzed for each set of conditions

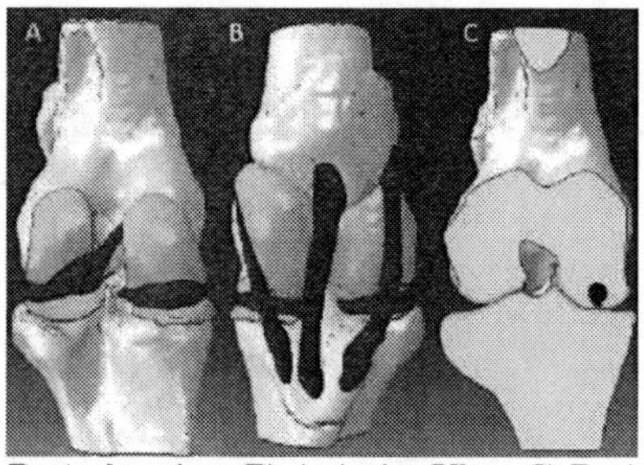

Figure 1. A) Posterior view. B) Anterior View. C) Posterior planar cut to reveal the 1 cm^3 cyst.

Table 1. Material properties used in this study.[3-5]

Structure	Type	E (MPa)	v
Bone	Iso.	50-21,000	0.3
Menisci	Anis.	Radial, Axial = 20, Circumferential = 120	Radial, Axial = 0.3, Circumferential = 0.45
Cartilage	Iso.	50	0.45
Ligaments	Iso.	Nonlinear Springs	N/A
Cyst	Iso.	1	0.45

RESULTS

When comparing a cystic femur vs. a normal femur, our models predicted the same relative change in tibial contact pressure, regardless of the selected rotational constraint (Figure 2). The cystic models demonstrated a slight decrease in average contact pressure on the medial tibial cartilage and a slight increase on the medial meniscus. Average compressive principal stresses in the medial femoral condyle increased with the addition of the cyst, most notably at the lateral, distal portion of the cyst. Lastly, we found that tensile strain in the meniscus increases in the cystic models compared to models of the normal stifle. Each rotational constraint that was lifted in the model resulted in an overall increase in lateral tibial contact pressure, decrease in the medial tibial contact pressure, and the kinematics of the model were altered by each additional rotational degree of freedom. Increasing the quadriceps force also increased lateral contact pressure. Stiffening and softening our cartilage material properties resulted in no relative difference in our results when comparing normal vs. cyst.

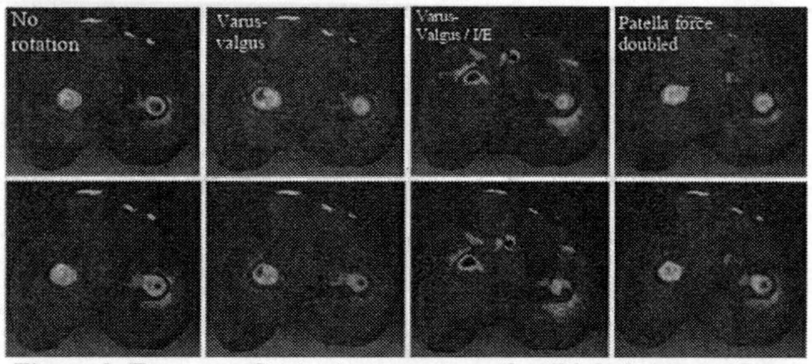

Figure 2. Top row: Contact pressure on the tibial plateau without a cyst present for the various model changes (see upper left corner of each image). Bottom row: Contact pressure on the tibial plateau with a cyst in the medial femoral condyle. The same color/pressure scale is used for each contour map.

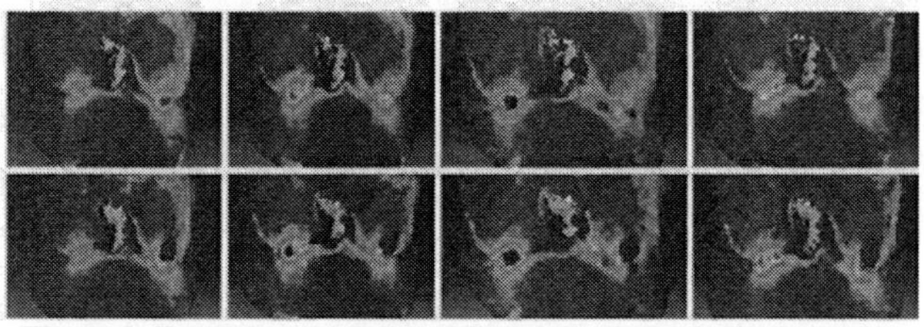

Figure 3. Frontal cross-section of compressive principal stresses for same states/conditions as Figure 2.

DISCUSSION

The primary purpose of this study was to determine how a bone cyst would alter tibial contact pressures, principal stresses in the medial femoral condyle, and/or medial meniscal tension, compared to the normal femur. Because kinematic and kinetic data of an equine stifle joint are unavailable, we tested the sensitivity of our comparitive findings by changing our rotational constraints and the quadriceps force. We found that a cyst increases principal stresses in the medial femoral condyle, as well as tibial contact pressures independent of our selected rotational constraint or quadriceps force. Also, meniscal strain increases with a cyst, consistent with common secondary injury in the meniscus after cyst formation.

Because the equine cruciate ligaments very tightly connect the femur and tibia, and are stout and infrequently injured, we believe that rotation (valgus/varus and internal/external) in extension in the equine stifle joint is minimal. In the model, as rotational constraints were removed and quadriceps force increased, load shifted to the lateral tibial plateau, which is a very uncommon location of equine injury. These anatomic and clinical observations suggest that lifting rotational constraint does not reflect in vivo mechanics/kinematics. Constraining the rotation in 3 planes appears to result in a reasonable estimation of contact pressures and compressive stresses.

Our findings are preliminary, and there is much work to be done in obtaining experimental data of an equine stifle joint. However, the use of comparative models allow reasonable inferences about the effects of a bone cyst on surrounding tissues. The relative pressure/stress changes were independent of the kinematic constraints. Because the cyst alters tissue mechanics, it is logical that treatment strategies for subchondral bone cysts should also address the altered mechanics that arise with cystic development.

ACKNOWLEDGEMENTS

The authors of this study acknowledge the funding support from the Self Graduate Fellowship that made this research possible.

REFERENCES

[1] Jeffcott, LB et al., *Equine Vet J.* 15(4):304-11, 1983.
[2] Audrey, H.X et al., *Open Orthop J.* 8:7-10, 2014.
[3] Donahue, T et al., *J Biomech.* 36(1), 36(1):19-34, 2003.
[4] Malekipour, F et al., *J Mech Behav Biom Mat.* 26:127-135. 2013.
[5] Kiapour, A et al. *J Biomech Eng*, 136(1):011002, 2014.

SB³C2017
Summer Biomechanics, Bioengineering and Biotransport Conference
June 21 – 24, Tucson, AZ, USA

A NEW REACTION-DIFFUSION-STRAIN MODEL FOR PREDICTING THE PROCESS OF SKULL GROWTH AND DEFECT FORMATION

Chanyoung Lee (1), Reuben H. Kraft (1,2)

(1) Department of Mechanical and Nuclear Engineering
The Pennsylvania State University
University Park, PA, USA

(2) Department of Biomedical Engineering
The Pennsylvania State University
University Park, PA, USA

INTRODUCTION

Bones of the mouse cranial vault form through intramembranous ossification that includes direct differentiation of multi-potent undifferentiated mesenchymal cells to osteoblasts, which would be mineralized and become bone. The process is controlled by a cascade of reactions between extracellular molecules and cells. Alan Turing [1] proposed a mathematical model (referred to as reaction-diffusion model) for the regulation between interacting molecules for pattern formation. The model has been used for various biological pattern formation problems including pattern of bone formation however it considers only molecular interactions without any mechanical influences, which are revealed to play important roles in bone formation. In this study, we propose an integrated mechanobiological model for bone formation in the cranial vault by coupling the reaction-diffusion model with structural analysis in the growing domain.

METHODS

From related biological information we derived several assumptions about behaviors of key molecules for bone formation and mechanical effects on bone formation as follows:

- The differentiation of mesenchymal cells is activated by a key molecule (the activator) and the behavior of the activator is regulated by another molecule (the inhibitor) so that the regulatory loop of the two molecules can be modeled using a Turing's reaction-diffusion model.
- Differentiated osteoblasts express the activator molecule to lead to bone growth and the inhibitor molecule inhibits cell differentiation to form sutures, which are spaces of undifferentiated cells between bones in the cranial vault.
- Mechanical stimuli from the underlying brain growth affect bone formation in a way that moderate tensile hydrostatic strain promotes the expression of the activator while excessive magnitude of hydrostatic strain inhibits the cell differentiation.

Based on the assumptions, we modified the activator-inhibitor model, which is a kind of reaction-diffusion model, by adding mechanical input and suggest a fully coupled reaction-diffusion-strain model for the cranial growth. The mathematical expressions of the reaction-diffusion-strain model are as follows.

$$\frac{\partial a}{\partial t} = \frac{o_s^n}{o^n = o_s^n}\left((E_{hyd} = E_0)\alpha_a = \gamma_a \frac{a^2}{h} = D_a \nabla^2 a\right) = \beta_a a = \alpha_o o \quad (1)$$

$$\frac{\partial h}{\partial t} = \frac{o_s^n}{o^n = o_s^n}(\alpha_h = \gamma_h a^2 = D_h \nabla^2 h) = \beta_h h \quad (2)$$

$$\frac{\partial o}{\partial t} = \eta \frac{o_s^n}{o^n = o_s^n} \frac{a^n}{a^n = a_T^n} \frac{h_l^n}{h^n = h_l^n} \frac{E_l^n}{{E_{hyd}}^n = E_l^n} \quad (3)$$

The term a, h and o represent the concentration of activator, inhibitor and differentiated osteoblasts, respectively. The production ($(E_{hyd} = E_0)\alpha_a$ and α_h), reaction ($\gamma_a \frac{a^2}{h}$ and $\gamma_h a^2$) and diffusion ($D_a \nabla^2 a$ and $D_h \nabla^2 h$) of activator and inhibitor are assumed to take place only in mesenchymal cells, not in differentiated osteoblasts and the term $\frac{o_s^n}{o^n = o_s^n}$ represents that component of the model. The production of activator is assumed to be proportional to the local hydrostatic strain $(E_{hyd} = E_0)$, where E_0 is a value for making the production term to be positive, so that strain field can initially drive the distribution of activator. Hydrostatic strain (E_{hyd}) can be defined by ($E_{xx} = E_{yy} = E_{zz}$)/3. o_s represents saturation concentration of

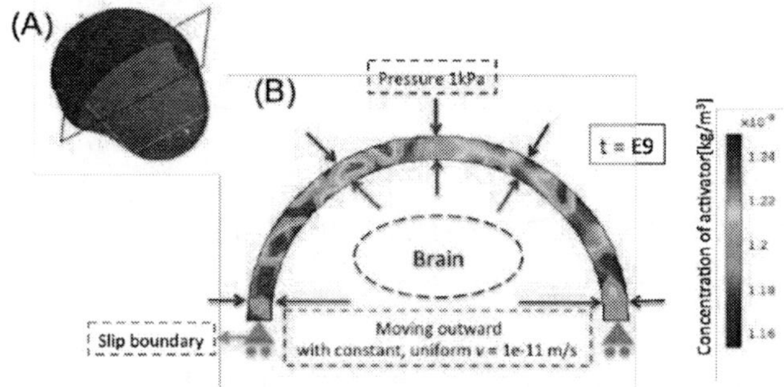

Figure 1: (A) Geometry of the computational domain (B) initial condition for reaction-diffusion model and boundary condition for structural analysis in a cross sectional view.

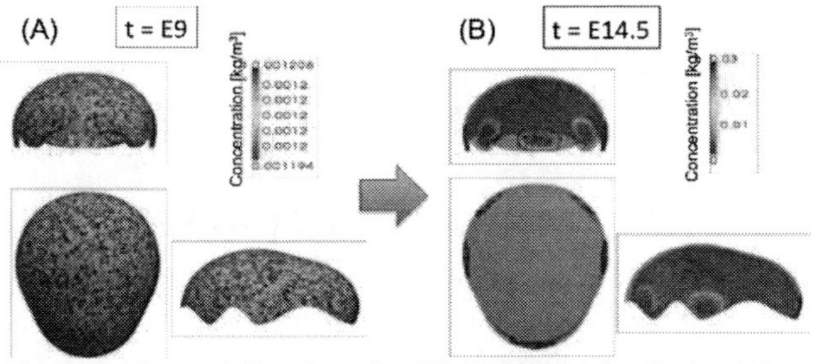

Figure 2: Computational results. Distributions of activator (A) at initial time (B) at E14.5 Random perturbations form an inhomogeneous spatial pattern.

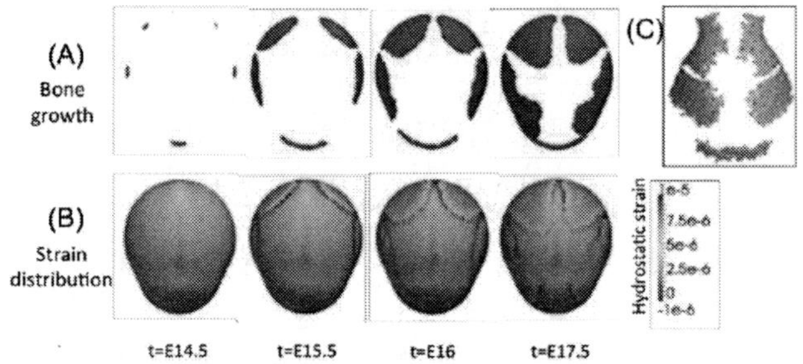

Figure 3: Computational results. Distributions of (A) osteoblasts and (B) hydrostatic strain over time. Bone grows from the primary centers of ossification and sutures form. The distribution of strain affects the pattern of bone formation. The prediction agrees well with (C) experimental observation at P0

osteoblast that sets the upper limit of generation of osteoblast. The terms $\mu_a a$ and $\mu_h h$ quantify degradation of the molecules. The rate of differentiation of mesenchymal cells to osteoblasts $(\frac{\partial o}{\partial t})$ is promoted by activator which has higher concentration than a threshold value (a_T) and is limited by inhibitor which has a higher concentration than a limitation value (h_l). The rate is also limited by the strain which has a higher magnitude than the limitation value (E_l). The value n in the terms $\frac{o_s^n}{o^n=o_s^n}, \frac{a^n}{a^n=a_T^n}, \frac{h_l^n}{h^n=h_l^n}$ and $\frac{E_l^n}{E_{hyd}{}^n=E_l^n}$ is a constant characterizing how sharply the terms increase or decrease according to the change in variables o, a, h and E_{hyd}. Osteoblasts and activator are coupled in the way that osteoblasts enhances activator by the term $\alpha_o o$ and the activator in turn activates differentiation of cells near the osteoblasts, which models the growth of bone. Because material properties change as cell differentiation proceeds, the strain field $(\boldsymbol{E})$ is estimated at every time step from structural analysis with updated material properties of the domain according to the concentration of osteoblasts.

The computational domain is constructed as a dome-shaped thin three-dimensional volume (Figure 1(A)) that represents a layer of cells around the brain. Figure 1(B) shows boundary and initial conditions in a cross sectional view of the domain. The inner surface moves outward to represent underlying brain growth and the outer surface is exposed to pressure. The bottom surfaces slips in horizontal direction. The initial distribution of activator is random with small perturbations of concentration (±0.5%)

An open source code OpenFOAM was used to solve the mathematical model and to conduct a structural analysis using the finite volume method.

RESULTS

Figure 2 shows distribution of the activator. At initial time (E9, embryonic days 9) the distribution is nearly uniform with small random perturbations (Fig.2(A)). At E14.5, through the reaction-diffusion process which is guided by mechanical strain the concentration of the activator forms an inhomogeneous spatial pattern showing five locations of high concentration of activator (Fig. 2(B)). From our assumption, cells at the locations will differentiate to osteoblasts so the locations become primary centers of ossification, which consist of the first differentiated bone cells.

Figure 3 shows distributions of osteoblasts and hydrostatic strain over time. Region of differentiated osteoblasts expands from the primary centers of ossification with different speed in different direction leaving gaps between them, which can be considered as sutures. Distribution of strain shows that global distribution of strain due to the global shape of the domain and local strain change due to the change of material properties at the front of growing bones guide relative growing speed of bones and the final subtle pattern of bone formation.

DISCUSSION

The results show that our mechanoiological model predicts some key features of morphology of cranial vault, which can be found in experimental observation (Fig. 3(C)), including relative position and size of five bones and position and angle of sutures between bones. The model provides a possible explanation for the positioning of primary ossification centers, cranial vault bone growth and the positioning of cranial vault sutures by suggesting important roles of mechanical strain in the process of bone formation.

The model is expected to be improved by applying real shape and growth pattern of brain as computational domain and boundary conditions.

ACKNOWLEDGEMENTS

This work was supported in part through instrumentation funded by the National Science Foundation through grant OCI0821527, a Burroughs-Welcome Fund 2013 Collaborative Research Travel Grant, and by National Institutes of Health grants R01 DE022988 and P01HD078233. The content is solely the responsibility of the authors and does not necessarily represent the official views of the National Institutes of Health.

REFERENCES

[1] Turing, AM, Philosophical transactions of the royal society of LondonSeries B, Biological sciences 237(641):37–72, 1952

SB[3]C2017
Summer Biomechanics, Bioengineering and Biotransport Conference
June 21 – 24, Tucson, AZ, USA

A NOVEL METHOD FOR IMAGING WHOLE BONE 3D FRACTURE DURING MECHANICAL TESTING

Kyle A. Bodnyk (1), Michael J. Heyden (1) Richard T. Hart (1)

(1) Department of Biomedical Engineering
The Ohio State University
Columbus, OH, USA

INTRODUCTION

Studying the fracture mechanics of bone may help us understand how osteoporotic fractures occur and possibly to evaluate osteoporosis treatment therapies. The current study uses a novel bending device to visualize murine femur fractures in 3D. A bone bending device was designed to be compatible with the material testing stage inside the Bruker Skyscan Material Testing Stage compatible with the 1172d micro-CT system (Kontich, Belgium) (provisional patent number 62/419,682, 2016). The device was constructed out of PEEK, a strong yet radiolucent material allows the loaded bone to be visualized in the scanner. The objective of this feasibility study is to see if we could use μCT to visualize whole bone 3D crack propagation during bending loads.

METHODS

For this initial study, six femurs were obtained from female mice sacrificed at 10-12 weeks of age, wrapped in saline soaked gauze and stored at -40°. Femurs were thawed for 1 hour before mechanical testing. Specimens were placed in a custom saline filled notch jig that guided a razor blade to form a consistent notch transverse to the long axis of the bone, 55% of the bone length from the proximal epiphysis on the anterior surface. Femurs were wrapped in saline soaked gauze and mounted in the custom bending jig, positioned so the load is placed on the posterior diaphysis directly opposite the notch. Loaded in this way, the crack will open with applied load similar to mode I (tensile opening) fracture, shown in Figure 1. The custom bending jig was mounted in the Bruker Skyscan Material Testing Stage (Kontich, Belgium). Bones were first scanned while unloaded to obtain notch geometry with the Bruker Skyscan 1172d (Kontich, Belgium). Scan settings were 2.98 μm pixel size, 70kv, 142 μA, Al filter. After initial scan, bones were loaded with displacement control until crack propagation occurred. Displacement was then stopped to make another scan. The process was repeated until the bone fractured unstably. Load and displacement data was recorded during testing, shown in Figure 2. Scans were first reconstructed using NRecon (Kontich, Belgium), then imported to Synopsys Simpleware ScanIP (Mountain View, California). The bones were segmented from the background making sure the notch

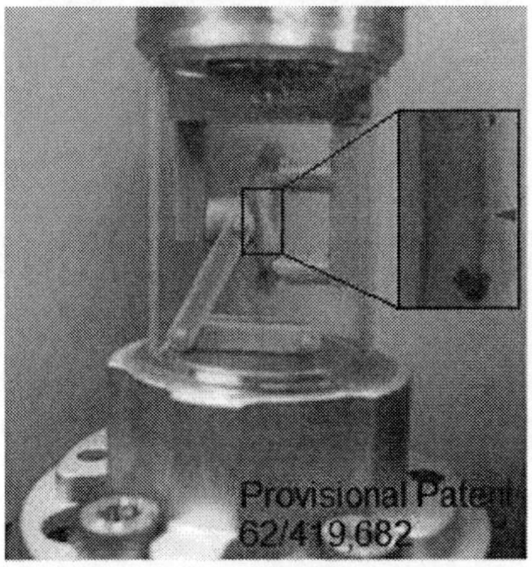

Figure 1: Femur mounted in custom bending jig [2]. Insert: Raw μCT image of bone under bending load with stable crack. Red arrow marks notch.

and/or crack was clearly outlined. Using the segmented bone, the stable crack was visualized in 3D.

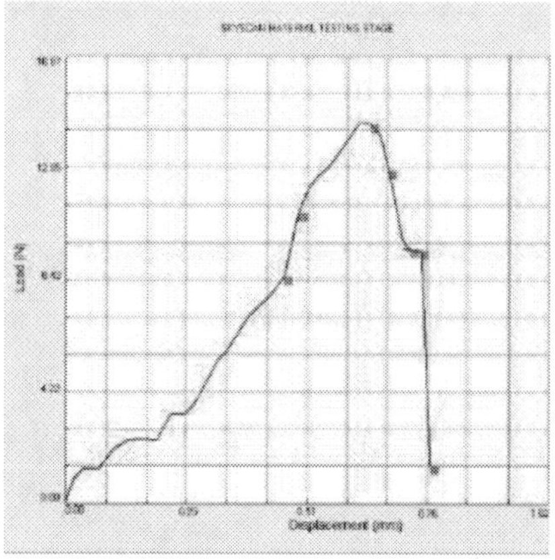

Figure 2: Bending load vs. Displacement during bone bending. Blue squares are when displacement was stopped to check for crack or to take μCT scan

RESULTS

From the 6 femurs tested, only one had a stable crack propagate. The remaining femurs fractured unstably starting at the crack tip and traveled at an angle through the entire bone. For the one test that arrested the crack, displacement was stopped for a scan when the crack began to propagate (14.94 N). The crack propagated further, now at 12.55 N, as shown by the arrows in Figures 3 and 4.

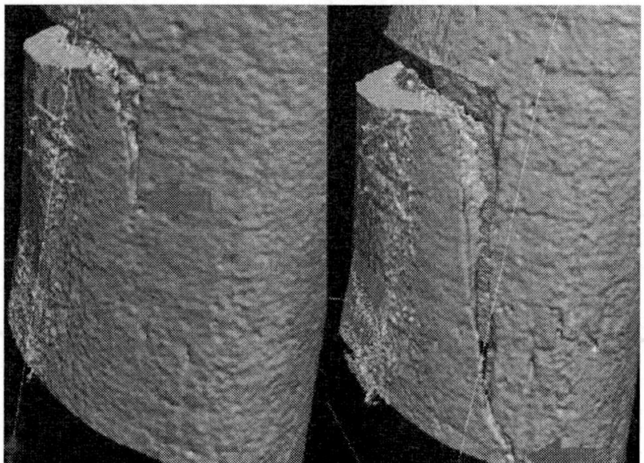

Figure 3: Regions of interest from figure 1 rendered in Simpleware. left: initial stable crack. Right: propagated stable crack. Crack tips marked by blue arrow

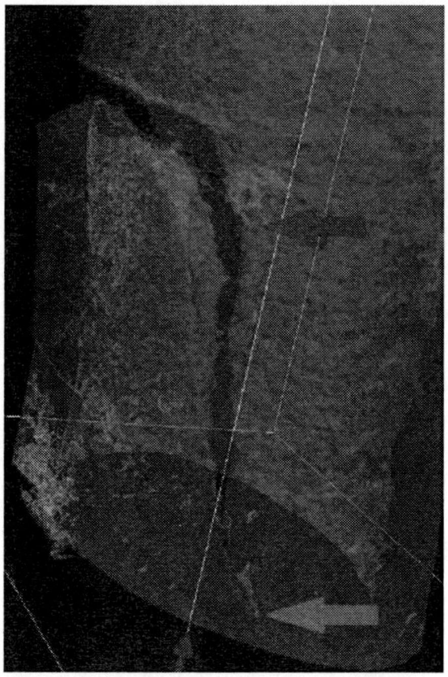

Figure 4: Bone rendered translucent to show stable crack through both cortices. Blue arrow marks crack tip from figure 3: Right. Red arrow marks crack tip on far side of bone.

DISCUSSION

This feasibility study demonstrates that it is possible, but not common, to use our device to visualize whole bone 3D stable fracture propagation with a bending load using μCT. More testing and further methods need to be developed to accurately quantify crack surface area in order to calculate work to fracture as well as the work per unit surface area to propagate a stable crack. These preliminary results expose the difficulty of stopping the displacement at the fine line between stable crack propagation and fracture instability in order to take μCT images of the crack. The authors' believe this is the first time stable crack propagation in a bone specimen has been captured in 3D.

ACKNOWLEDGEMENTS

Partial support comes from the Edgar C. Hendrickson funds. Special thanks to the Breuer lab and James Reinhardt at Nationwide Children's Hospital for the gift of specimens.

REFERENCES

[1] Ritchie, R.O, et al. Measurement of the toughness of bone: a tutorial with special reference to small animal studies. Bone. 43 (2008) 798–812

[2] K.A. Bodnyk, R.T. Hart, A.S. Litsky, Bending Apparatus for Material Testing and Micro-CT Imaging, Provisional patent 62/419,682, 2016.

ANALYSIS OF MINERAL DISTRIBUTION IN THE TRABECULAR BONE OF NORMAL AND ESTROGEN DEFICIENT RAT ULNAE AND RADII USING MICRO CT AND NANOINDENTATION

Laura M. O'Sullivan (1), Eoin P. Parle (1), Laoise M. McNamara (1)

(1) National University of Ireland,
Galway,
Ireland

INTRODUCTION

Osteoporosis results in severe bone loss, significantly reduced strength and altered bone tissue porosities. Recent studies have shown that bone tissue-level mineral distribution is altered in a sheep model of osteoporosis [1], and changes in mineralized crystal maturity, mineral-to-matrix ratio, and collagen cross-linking occur [2]. In particular quantitative backscattered imaging (qBEI) of trabeculae of ovine femurs was applied to analyse the mineral heterogeneity of bone tissue. This study showed that heterogeneity of mineral distribution in estrogen deficient animals compared to controls, was dependent on the anatomical region and the age of the animal [3]. Although fracture of the distal forearm is one of the most frequent osteoporotic fractures, the mineral changes in the trabecular bone of ulnae and radii during estrogen deficiency have not been investigated. The objective of this study was to use µCT and nanoindentation to conduct a comprehensive analysis on the effect of estrogen deficiency on the mineral distribution and mechanical integrity of trabecular bone in the distal forearm of an ovariectomised (OVX) rat model. **Micro CT (µCT)** analysis was conducted at the **Sub-Macro level** (volumes of the trabecular network) to evaluate the extent of bone loss and overall mineral content changes due to estrogen deficiency. **Nanoindentation** was carried out on individual trabeculae from these bone samples to obtain localised material properties at the tissue level. **Micro CT** was used for **Tissue level** analysis of these same individual trabeculae to evaluate the mineral distribution at precise locations in the trabecular network and to gain understanding of the material level mineral changes which occur in the distal forearm in estrogen deficiency.

METHODS

The bone tissue used in this study originated from the skeletally mature rats of a European Research Council funded project, which were sacrificed 34 weeks post-ovariectomy (OVX). The distal end of the ulna and radius were studied in the OVX (n=4) and age-matched control (n=4) animals.

Micro-CT, Sub-Macro level: Trabecular regions of approximately 0.005mm³, were evaluated using µCT (µCT100, Scanco) at a resolution of 4.9µm. A threshold of 770 mgHA/ccm was determined based on calibration of the grey scale images from the first ulna and radius samples scanned. This thresholded scan data was used to evaluate the microarchitecture and mineral content, in particular the bone volume fraction (BV/TV), trabecular connectivity, trabecular number and bone mineral density distribution (BMDD). The full width at half max (FWHM) was calculated for the BMDD histogram of each sample in order to evaluate the mineral heterogeneity, with a higher FWHM indicating a more heterogeneous mineral density distribution (see Fig.1).

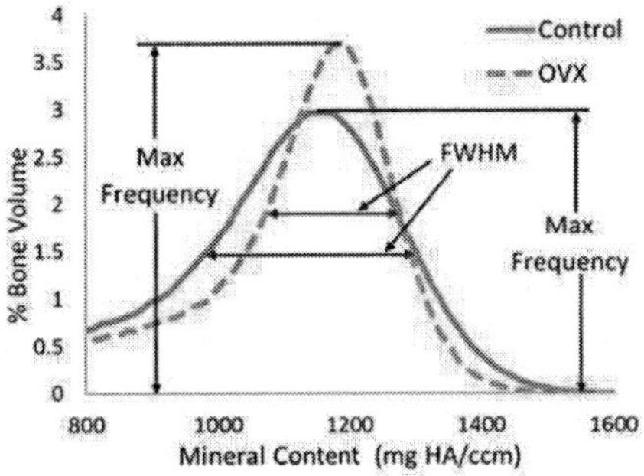

Figure 1: Representative Bone Mineral Density Distribution plot (BMDD) of one OVX sample and one control sample.

Nanoindentation, Tissue level: Ulnar bone specimens were sectioned along the longitudinal axis using a precision diamond blade saw (IsoMet™ LS Saw, Buehler) and then embedded in epoxy (Epo-thin low viscosity epoxy, Buehler) and polished to achieve a high quality test surface. Six indents were made using a Berkovich diamond tip indenter on a CSM Nano-Hardness Tester across the width of a trabecular strut on each sample. Each indent was made by driving the indenter into the sample for 60 seconds to a max load of 20mN, holding for 120 seconds and unloading. The equations (1), (2) and (3), developed by Oliver and Pharr [4], were implemented to calculate the tissue elastic modulus and tissue hardness from the unloading segment of the indentation load-displacement curves (see Fig. 2).

$$S = \beta \frac{2}{\sqrt{\pi}} E_{\text{eff}} \sqrt{A} \quad (1)$$

$$\frac{1}{E_{\text{eff}}} = \frac{1 - \nu^2}{E} + \frac{1 - \nu_i^2}{E_i} \quad (2)$$

$$H = \frac{P_{\max}}{A} \quad (3)$$

Where, S is the initial unloading stiffness, β is the indenter geometry constant, E_{eff} is the effective elastic modulus, A is the Indenter-Sample contact area, E and ν are the Elastic constants of the sample, E_i *and* ν_i are the elastic constants of the indenter, H is the sample hardness and P_{max} is the max indentation load applied.

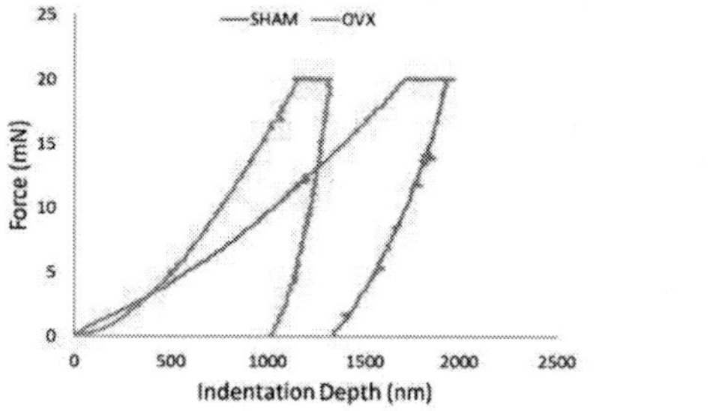

Figure 2: Force-displacement curves from nanoindentation of trabecular tissue from estrogen deficient and control ulnar trabeculae samples.

Micro-CT, Tissue level: A subset of the nanoindented ulnar trabeculae were µCT scanned again at 6.6µm resolution (with previously mentioned scan parameters) to assess the mineral distribution at the tissue level. This thresholded scan data was used to evaluate the mineral content changes in estrogen deficiency through analysis of the bone mineral density distribution (BMDD).

Statistics: Student t-tests were used to analyse the statistical significance of the difference between the OVX and age-matched control groups, with significance of $p<0.05$.

RESULTS

Micro-CT, Sub-Macro level: Significant differences were seen in the microarchitecture of the OVX ulna specimens compared to the controls; BV/TV (-40%, p=0.04), trabecular connectivity density (-60%, p=0.03) and trabecular number (-30%, p=0.026), see Fig. 3B inset. Trabecular thickness was higher in OVX animals (+11%, p=0.09) There was no significant difference in average mineral content (1195 mg Ha/ccm vs. 1192 mg Ha/ccm, p=0.85). The mineral distribution in OVX bone (FWHM: 241.18 mg HA/ccm) was significantly (p=0.04) more homogeneous than controls (FWHM: 298.78 mg HA/ccm).

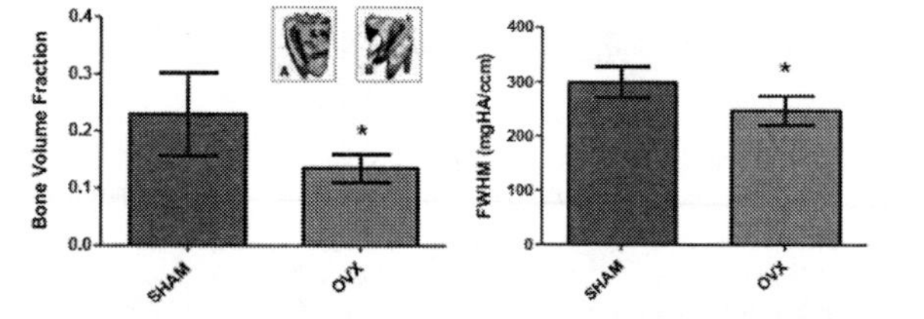

Figure 3: Bone Volume Fraction and Full Width at Half Max values for sub-macro level Micro-CT analysis of control and OVX samples. Micro-CT 3D representations are also shown inset (A, B).

There was no significant difference in the microarchitecture or mineral distribution of OVX radius specimens compared to control; BV/TV (-11%, p=0.2), trabecular connectivity density (-6%, p=0.42), trabecular number (-19%, p=0.057) and average FWHM (240.38 mg HA/ccm vs. 271.86 mg HA/ccm, p=0.23).

Nanoindentation: The average elastic modulus by nanoindentation was 9.04 GPa for OVX (n=2) and 3.28 GPa for control samples (n=2) and the hardness values were 535.33 MPa and 171 MPa respectively, p=0.35 (see Fig.2 for representative test curves).

Micro-CT, Tissue level: Preliminary results from analysis of individual ulnar trabeculae (OVX, n=2 and Control n=1) showed increased Peak, Mean and FWHM values from the BMDD curves of OVX samples compared to controls.

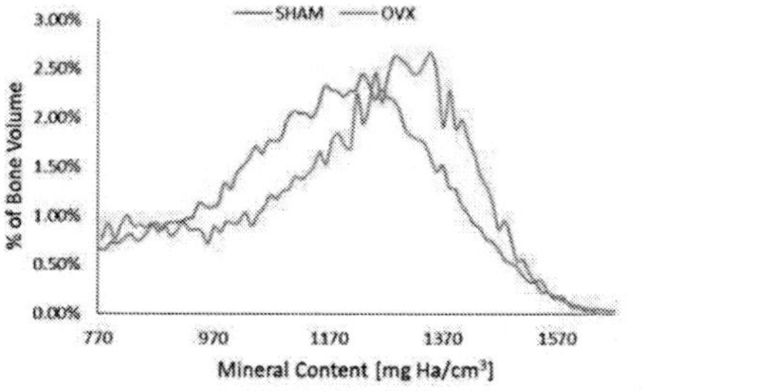

Figure 4: Bone Mineral Density Distribution plot for individual trabeculae of estrogen deficient (OVX) and Control ulnar trabeculae samples

DISCUSSION

Micro-CT, Sub-Macro level: The results from analysis of trabecular volumes suggest that, although the bone mass is lower in OVX animals, there is a decrease in the mineral heterogeneity of trabecular tissue in the distal ulna and distal radius. However, there seems to be no difference between OVX and healthy tissue in the peak and average mineral content. This suggests that overall range of mineral content densities seen in OVX bone has reduced, but the peak and average mineral densities observed in the bone have not changed.

Nanoindentation, Tissue level: The preliminary nanoindentation results suggest that OVX ulnar trabecular tissue is stiffer than healthy tissue. These material properties did not match our observations in the sub-macro level bone mineral analysis. Therefore we sought to conduct a localized analysis of mineral distribution at the tissue level (individual trabeculae).

Micro-CT, Tissue level: This localized µCT analysis of mineral distribution indicated more heterogeneous mineral distribution and higher mineral content in the OVX ulnar trabeculae that had been indented. These finding suggest that the mineral distribution at the tissue level of estrogen deficient rat ulnae is similar to that seen previously in sheep femurs, but further study is required to confirm this observation.

The increased elastic properties and higher mineral content at the tissue level, might render the ulnar tissue more brittle and explain distal forearm fracture, but further research is required to explore this.

ACKNOWLEDGEMENTS

This publication has emanated from research conducted with the financial support of Science Foundation Ireland (SFI) and is co-funded under the European Regional Development fund under Grant number 14/IA/2884.

REFERENCES

[1] Brennan, M (et al.), Eur Cell Mater, 21:396-406, 2011.
[2] Brennan, O (et al.), Calcif Tissue Int, 91: 440-449, 2012.
[3] Brennan, M (et al.), J Mech Behav Biomed Mater 29: 161-170, 2014.
[4] Oliver, W.C. and Pharr, G.M., J Mater Res, 19: 3-20, 2004.

ULTRASTRUCTURAL ORIGIN OF BRITTLENESS OF BONE USING A FINITE ELEMENT APPROACH

Abu-Saleh Ahsan, Mohammad Maghsoudi-Ganjeh, Xiaowei Zeng, and Xiaodu Wang*

Mechanical Engineering, University of Texas at San Antonio, Texas

INTRODUCTION

Bone is a natural composite of mineralized collagen fibrils and hydroxyapatite mineral crystals with a complex hierarchical organization at multiple length scales. In the hierarchy, lamella is the basic structural unit for both cortical and trabecular bones, bridging between the nano and mesoscales of bone. From the perspective of composite mechanics, the mechanical behavior of lamella depends on the material property and spatial arrangement of the collagen fibrils, mineral crystals, non-collagenous proteins, and matrix water. Such ultrastructural property directly affects the bulk mechanical behavior of bone. In fact, previous evidence shows that bone toughness is rooted at its ultrastructure level.

Age or disease related bone brittleness is a major health concern and has been extensively investigated. However, the relationship between the ultrastructure and brittleness of bone is still poorly implicit. It was recently reported that the degree of intrafibrillar mineralization may play a significant role in the fragility of bone in osteogenesis imperfecta (OI), a genetic bone disorder that leads to severe brittleness of bone tissues. To elucidate the underlying mechanism, a 2-D plane strain multiscale cohesive finite element model was built to mimic the structure of bone lamella and simulated under both compressive and tensile load to capture effect of intrafibrillar mineralization on the fragility of OI bones.

METHODS

A novel 2-D structural model of the bone lamella was proposed to establish a multiscale computational framework that preserves all the major ultrastructural features of bone. In this model, mineralized collagen fibrils were considered to be embedded in an extrafibrillar matrix that is composed of hydroxyapatite poly crystals separated by a thin organic interface layer. First, sub models of the mineralized collagen fibril and the extrafibrillar matrix were built. Then, the two sub-models were assembled together *via* an interface in between them to represent the bone lamella (Figure 1). The organic interface was modeled using cohesive zone element. Thickness of the interface was selected as 1nm. The width of the mineralized collagen fibril and extrafibrillar matrix was selected as 90nm and 70nm respectively based on a calculation from the estimated volume fraction of the mineral phase (42-50% of total volume of bone). The overall dimension of the finite element model in this study was 340 nm long in the direction of mineralized collagen fibril and 576 nm wide in transverse direction.

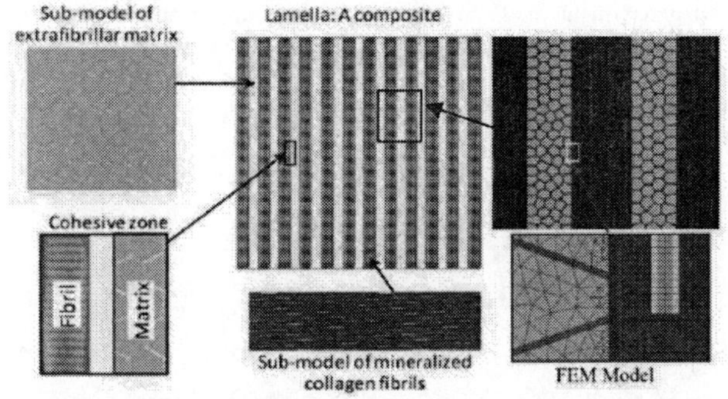

Figure 1: Multiscale modeling of bone lamella using sub-models of mineralized collagen fibrils and extrafibrillar matrix *via* cohesive zone

A staggered structure of mineralized collagen fibrils was adapted from the geometrical model proposed by Olszta et al. [1]. The main components of mineralized collagen fibril are collagen molecules, hydroxyapatite mineral crystals, and interfaces between the collagen and mineral phases. In this model the length of each collagen molecule was 300nm. The gap and the overlap regions were 40nm and 27nm, respectively. The overall D-spacing to continue the staggered periodic manner was 67nm. The interface between the collagen and mineral crystals were modeled as a cohesive zone.

Structurally, the extrafibrillar matrix was considered as a hybrid composite of granular mineral crystals bounded through thin organic interface [2]. The geometric model of mineral crystals and the organic interface were created by Voronoi generated grains and cohesive interfacial zones respectively. Centroidal Voronoi Tessellation (CVT) approach was used to mimic the polygon shape of the mineral crystals. The approximate grain size was set 22-25nm and the thickness of the organic interface was set to 2nm, which was selected based on the calculation from the estimated volume fraction of non-collagen proteins (~10% by volume, in the extrafibrillar matrix) [2-3].

In this study, the organic interface was modeled as a cohesive zone. The constitutive response of the cohesive zone was defined by the bi-linear traction-separation law using the following parameters: the critical stress as 38MPa and 20MPa, representing the damage initiation criterion, and the energy release rate as 0.08J/m^2 and 0.02J/m^2 for opening and sliding mode, respectively, to mimic the water surface tension and van der Waals bonding at the interface.

ABAQUS CAE/Explicit was used for modeling and finite element simulations (Figure 1). The cohesive finite element (FE) model was subjected to compressive and tensile load, respectively. Isotropic linear elastic behavior were chosen as E_m = 125GPa and v_m = 0.3 for mineral crystals and E_c=3.0GPa and v_c = 0.4 for collagen. Here, E is the elastic modulus and ν is the Poisson's ratio. Velocity boundary condition along fibril direction was applied in both compressive and tensile loading simulations.

RESULTS

The simulation results indicated that the stress-strain relationship of lamella in compression and tension changed with different degree of intrafibrillar mineralization (Figure 2 and Table 1). The compressive mechanical properties of lamella estimated from the stress-strain curves revealed that the elastic modulus (E), strength (σ_u), and toughness (U_T) of lamella decreased with decreasing intrafibrillar mineralization. Similarly, the elastic modulus, strength, and toughness of lamella in tension also decreased with diminishing intrafibrillar mineralization (Figure 2).

The snapshots of the model indicated that failure occurred first at the cohesive interface between extrafibrillar mineral crystals and propagated in an inclined direction by bulking during compression (Figure 3) or in an opening mode fracture during tension. After failure of the extrafibrillar matrix, load was transferred to the mineralized collagen fibrils, with stretching in tension and an excessive buckling in compression (Figure 3).

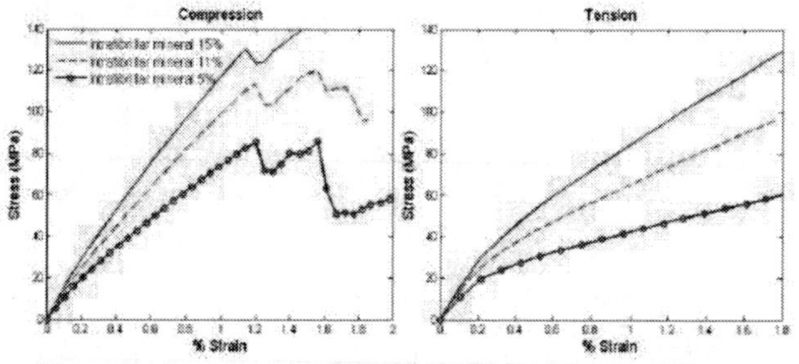

Figure 2: Comparison of the stress-strain curve for different degree of intrafibrillar mineralization in different loading mode

Table1: Mechanical Behavior of bone Lamella from simulation

Mechanical Property	Intrafibrillar Mineralization					
	Compression			Tension		
	5.0 %	11.0 %	15.0 %	5.0 %	11.0 %	15.0 %
E (GPa)	7.0	9.5	11.5	7.5	10.0	13.0
σ_u (MPa)	85.3	120	138	60	97	130
$U_T \times 10^6$ (J/m^3)	0.76	1.02	1.21	0.51	0.78	0.98

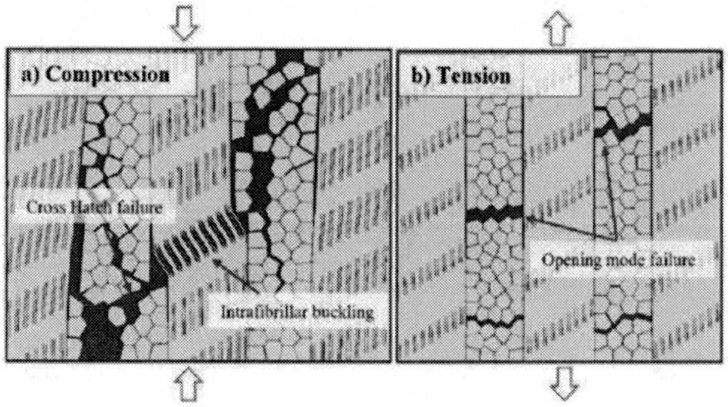

Figure 3: Damage progression in bone lamella, a) Cross-hatch and buckling failure pattern in compression, b) Opening mode failure pattern in tension

DISCUSSION

The results of this study indicate that irrespective of the loading mode, the bulk mechanical behavior of bone lamellae is dependent on the degree of intrafibrillar mineralization. With decreasing amount of intrafibrillar mineralization, showing decreased elastic modulus, strength and toughness of bone as the intrafibrillar mineralization decreases. This suggests that diminishing intrafibrillar mineralization may be an origin of bone brittleness caused by OI.

Buckling of mineralized collagen fibrils in compression may lead to the cross-hatch failure in bone lamellae. Since collagen molecules are much less rigid compared to the mineral crystals, the non-mineralized portion (overlap region) would be relatively less resistant to buckling than the mineralized portion (gap region). More mineralization would make collagen fibrils more resistant to buckling. However, buckling could occur much more easily in bone as intrafibrillar mineralization in OI bone diminishes.

In tension, however, the mineralized collagen fibrils play a pivotal role in load bearing after opening mode failure of extrafibrillar matrix. In addition, the load carried by the mineralized collagen fibrils would decrease with decreasing amount of intrafibrillar mineralization, thus leading to a reduction in the elastic modulus, strength, and toughness of bone in tension.

In summary, bone fragility may largely depend on intrafibrillar mineralization. Irrespective of the loading mode (compression or tension), bones become more fragile as the degree of intrafibrillar mineralization diminishes.

ACKNOWLEDGEMENTS

This work was partially supported by an NIH/NIAMS (AR065641) and a NSF grant (CMMI-1538448).

REFERENCES

[1] Olszta et al., j.mser.2007.05.001
[2] Lin, Samuel et al., jmbbm.2016.08.027
[3] Sroga, Vashishth, Curr. Osteoporosis Rep. 10, 141-150

DEVELOPMENT OF STOCHASTIC STRUCTURAL FINITE ELEMENT MODEL FOR TRABECULAR BONE

Saif Alrafeek (1), Peter A. Gustafson (1), James R. Jastifer (2,3)

(1) Mechanical and Aerospace Engineering
Western Michigan University
Kalamazoo, MI, USA

(2) Homer Stryker M.D. School of Medicine
Western Michigan University
Kalamazoo, MI, USA

(3) Borgess Orthopedics
Kalamazooo, MI, USA

INTRODUCTION

Bone can be described as a multi-level composite arranged in a hierarchical form. At the macrostructure scale, bone is grouped into two types: cortical bone (also known as compact bone or dense bone) and trabecular bone (known as cancellous bone or spongy bone) [1]. Generally, cortical bone surrounds trabecular bone. While cortical bone has a porosity of five percent to fifteen percent, trabecular bone has a porosity that varies from forty per cent to more than ninety percent [1]. Because cortical bone has low porosity, the continuum based finite element methods are likely for most current modelling purposes.

In contrast, trabecular bone consists of a three-dimensional network structure mainly composed of rod-shaped and plate-shaped fundamental units named "trabeculae". Most published finite element models of trabecular bone incorporate the use of geometrically homogenized continuum elements (i.e., elements which ignore the porous structure of the tissue) [2]. However, the assumption of continuity may not be appropriate for high fidelity models or for models which quantifying factors that impact the fundamental mechanical behaviour of trabecular bone. Thus, some investigators approach finite element modelling of trabecular bone with the intent of capturing its porous nature. Some models use continuum element methods where porosity is explicitly included in the geometry which itself is based on high resolution CT, MRI, or other advanced imaging [3].

FE models based upon a stochastic geometric approach have the advantage of being less dependent on detailed CT and MRI scans. This work investigates the use of beam elements to represent the trabecular micro-architecture. This method is hypothesized to capture the dominant geometric factors of the trabecular architecture while providing a mechanism for modelling the stochastic nature of the tissue in a cost efficient manner (both in model creation and in computational cost). Beam properties (such as length, orientation, cross section and connectivity) were set stochastically. The FE models in this work will be compared to published biomechanical experiments and cadaver studies [4].

METHODS

Trabecular bone consists of a three-dimensional network structure mainly composed of rod-shaped and plate-shaped fundamental units named "trabeculae." In this work, the trabeculae were modelled as beam elements. Two methods were investigated using differing assumptions. The first method assumed repetition of beam elements arrange as hexogonal unit-cells over a cubic trabecular bone specimen with 4 mm sides. The second method assumed randomly oriented beams also within a 4 mm cube. The cross section of the beams for method one was defined as a circular shape with a 0.1 mm radius. Hence, hexogonal unit-cell is the first of many planned unit cell architectures to be investigated.

For the second method, beams were created algorithmically so that beam properties could be applied stochastically [Figure (1)]. The algorithm is general to apply a stochastic approach to all beam properties. However, in this article, stochastic properties included length and orientation while the cross section was assumed to be a circular shape with a 0.025 mm radius. For both methods, the beam element type was a 2–node linear beam element (B31). Abaqus (V. 6.14, Simulia Ltd) was used as an FE solver.

Strain field was enforced on the cube structure by constraining the nodes on one face and applying load on the other opposite parallel face of the cube. Material constants were extracted from constitutive materials relationships. Apparent densities were calculated by obtaining the ratio of the actual beams mass to the cubic bulk volume. The models' materials were assumed orthotropic for the purpose of material property extraction and three modulii (representing the normal-normal stress-strain relationships) were extracted.

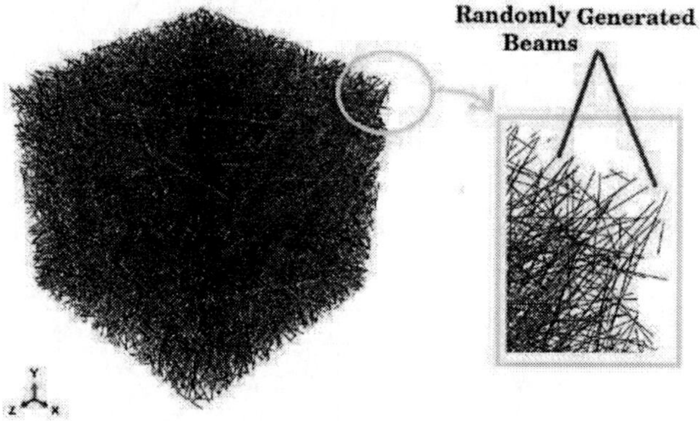

Figure 1: Trabecular bone specimen (structured from randomly generated beams).

RESULTS

The Individual FE models created in this study exhibited the mechanical properties listed in Table (1). Each model demonstrated effective modulii within published ranges of trabecular bone [1]. For the model generated from unit-cells, the three effective moduli varied by hundreds of megapascals as is consistent with the preferred directionality of the unit cell beam orientations. Effective Exx and Eyy in the stochastic model were similar in value as is consistent with the random nature of the beam orientations. The apparent densities of the current models are also within the published ranges [1, 5]. In the present work, the apparent density of the unit-cell model is slightly greater than that of the randomly generated beams model.

Table 1: Independent Elastic Constants, Apparent Density and beams circular section radii.

		FE Models		Literature	
		FE Model (unit-cells)	FE Model (random beams)	Bartel, D.L. [1]	Dalstra, M. [5]
Elastic Constants	E1 (Exx)	858.5	40.98	237	96.4
	E2 (Eyy)	246.06	43.7	309	49.1
	E3 (Ezz)	1199.04	67.22	823	25.2
Parameters Considered	Apparent Density (g/cm3)	1.625	1.18	0.09 ~ 1	0.1 ~ 0.959
	Beams circular section radius (mm)	0.1	0.025	-	-

E's are in (MPa).

DISCUSSION

In a fully random stochastic model with a sufficient number of randomly oriented beams, the effective modulii should produce approximately isotropic effective properties. In models with a finite number of beam elements randomly oriented, the effective modulii will not be isotropic but will not have predictable preferred directions. Trabecular bone has been reported to be orthotropic material with preferred directions in line with mechanical loads [5]. Such orthotropy can be introduced into the stochastic model by controlling the parameters such that they are not fully random. Rather, they should follow some stochastic process with a distribution of properties that resulting in orthotropy. For example, the beam elements can have a distribution of orientations centered around a preferred orientation.

The apparent densities of the FE models are dependent on the number of beams and on their cross sections which were dissimilar between the models. Future models will emphasize cross sections and lengths that are consistent with the nominal (and stochastic) geometry of physical trabecular architecture.

The methods described in this article using beam elements to model trabecular bone provide a foundation for future parametric study and research into efficient bone FE models. Such models may be used to establish a better understanding of the relationships between trabecular bone architecture (i.e apparent density, beam cross section, beam shape) and the macroscopic properties. Parametric study may be conducted employing trabecular bone apparent density and trabeculae radii (beam radii and shape). Other parameters may be included. Furthermore, if the efficacy and efficiency of these models are established, then they may be useful for patient specific modeling while not requiring invasive and costly imaging of each patient.

One of the advantages of utilizing beam elements in creating trabecular bone FE models is that porosity is preserved in the geometry without need for detailed imaging and costly mesh generation. The current beam-based structural FE models incorporate trabecular length, mean radii, and apparent density in quantification of the gross mechanical properties of the trabecular bone tissue. However, additional parameters may be accounted and might enhance the utility of the quantification or improve accuracy.

ACKNOWLEDGEMENTS

This work was supported by MoHESR (Iraqi Ministry of Higher Education and Scientific Research). Additional support was provided by: Western Michigan University (WMU) [College of Engineering and Applied Sciences (CEAS) and Homer Stryker M.D School of Medicine (WMed)]. And Borgess Orthopedics.

REFERENCES

[1] Bartel, D.L., Davy, D.T. et al Orthopedic Biomechanics, Pearson Education, Inc., 2006.

[2] Evans, S.P., Parr, W.C.H. et al., J Biomech, 45: 2702 -2705, 2012.

[3] Depalle, B., Chapurlat, R. et al. J Mechanical behavior of Biomedical Materials, 18:200 – 212, 2013.

[4] Jastifer, J., Alrafeek, S. et al., American Orthopaedic Foot and Ankle Society, 37(4): 419-426, 2015.

[5] Dalstra, M., Huiskes, R. et al., J Biomech, 26: 523 – 535, 1993.

[6] Goda, I., Assidi, M., et al., J Biomechanics and Modeling in Mechanobiology, 13: 53-83, 2014.

SB[3]C2017
Summer Biomechanics, Bioengineering and Biotransport Conference
June 21 – 24, Tucson, AZ, USA

POST-YIELD ANISOTROPIC HARDENING BEHAVIOR OF TRABECULAR BONE

David Nolan (1, 2), Patrick McGarry (1)

(1) College of Engineering and Informatics
National University of Ireland Galway
Galway, Ireland

(2) Trinity Centre for Bioengineering,
Trinity College Dublin
Dublin, Ireland

INTRODUCTION

Trabecular bone (TB) is a complex highly porous material consisting of a structural system of rods and plates [1]. When subjected to large mechanical uniaxial compressive loads it exhibits three distinct phases: a stiff elastic phase, yielding followed by an approximately constant stress, and finally densification where the material stiffness increases as the pores fully collapse. It has previously been shown by McGarry and Kelly that yielding of TB should be modelled using a pressure-dependent yield function which is ellipsoidal in principal stress space [2].

The current study demonstrates that TB exhibits differential hardening where the post-yield hardening rate is dependent on the loading mode. Microstructural finite element models are used to elucidate the anisotropic inelastic behavior of TB, and the suitability of a class of anisotropic yield functions to model this behavior is investigated.

MODEL DEVELOPMENT

Isotropic Mode-Dependent Strain Hardening

Cylindrical specimens of trabecular bone were extracted from ovine vertebrae. In order to investigate macroscopic strain hardening a series of experimental tests were performed for three modes of deformation: unconfined uniaxial compression, confined compression, and torsion. In each test specimens were loaded beyond initial yield stress and post-yield strain hardening was measured.

Figure 1 shows the stress-strain curves for the three deformation modes. Note that the post-yield hardening rate increases as the induced ratio of von Mises stress (q) to hydrostatic stress (p) decreases. Confined compression loading induces high levels of hydrostatic pressure, and consequently a low p/q ratio, exhibits the highest hardening rate. To simulate the dependence of trabecular bone strain hardening on the mode of deformation we propose a new hardening function:

$$\kappa = \kappa_{vol}\left(\varepsilon^{p}_{vol}\right) + \kappa_{dev}\left(\varepsilon^{p}_{dev}\right) \qquad (1)$$

where κ_{vol} and κ_{dev} are volumetric and deviatoric hardening contributions which are functions of the volumetric and deviatoric plastic strains ε^{p}_{vol} and ε^{p}_{dev} respectively ($d\varepsilon^{p}_{\mathrm{dev}} = \sqrt{2/3\, d\varepsilon^{p\prime} : d\varepsilon^{p\prime}}$; $d\varepsilon^{p}_{\mathrm{vol}} = \mathrm{tr}[d\varepsilon^{p}]$).

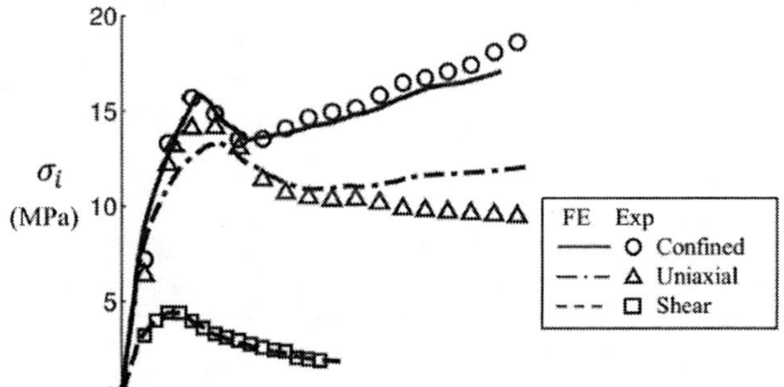

Figure 1: Experimental stress-strain relationship for confined compression, uniaxial compression and torsion (shear) for ovine vertebral TB. n=12 for each test group. Only mean data points are included for comparison with FE predictions using the hardening function proposed in equation (1).

The hardening function is incorporated into an isotropic plasticity constitutive law which incorporates pressure dependent yielding [3]. A fortran user defined material subroutine is used to implement the new formulation into the commercial finite element (FE) code Abaqus. Simulation of the large deformation multi-mode experiments reveal that proposed hardening function provides a reasonably accurate prediction of the post-yield behavior observed experimentally, as shown in Figure 1. This represents a significant improvement on non-differential isotropic hardening simulations reported in a recent study by Kelly and McGarry [2].

Anisotropic Mode-Dependent Strain Hardening

We next consider trabecular bone anisotropy (simulations presented in Figure 1 assume isotropic material behavior). In order to establish the influence of TB microstructure on post-yield anisotropic material behavior we construct microstructural FE models of 8 mm TB cubes using micro-CT scans of ovine vertebrae, as shown in Figure 2. The trabecular material within these microstructural models is assumed to be isotropic elastic-perfectly plastic, so that macroscopic anisotropy and strain hardening emerges solely die to the complex microstructural morphology. The TB cubes are subjected to both positive and negative loads in each of the three principal directions. Uniaxial stress, confined compression, and shear loading are considered. The macro level inelastic behavior is determined from these μFE models, as shown in Figure 3 for the case of confined compression. The on-axis 3-direction exhibits a higher elastic stiffness and a higher initial yield stress compared to the 1- and 2-directions. However, hardening in the 3-direction is lower than other two directions.

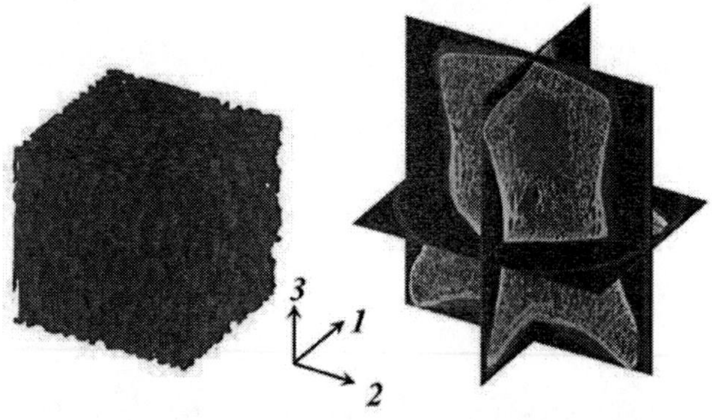

Figure 2: Schematic indicating the location in the L6 ovine vertebra from which the 8 mm cube of ovine trabecular bone is taken from the mCT scan for construction of a mFE model.

We next propose a normalized anisotropic yield function f_y , as outlined in equation (2) below. This new formulation builds upon previous work by Xue and Hutchinson [4] to explicitly include a hydrostatic stress term.

$$\left(\frac{f_y}{\sigma_0}\right)^2=\left(\frac{\sigma_{11}}{\kappa_{11}}\right)^2+\left(\frac{\sigma_{22}}{\kappa_{22}}\right)^2+\left(\frac{\sigma_{33}}{\kappa_{33}}\right)^2-2\nu_{32}^p\left(\frac{\sigma_{22}\sigma_{33}}{\kappa_{33}^2}\right)$$
$$-2\nu_{31}^p\left(\frac{\sigma_{11}\sigma_{33}}{\kappa_{33}^2}\right)+\left(\frac{1}{\kappa_{hyd}^2}-\frac{1}{\kappa_{11}^2}-\frac{1}{\kappa_{22}^2}+\frac{2\nu_{32}^p+2\nu_{31}^p-1}{\kappa_{33}^2}\right)\sigma_{11}\sigma_{22} \qquad (2)$$

This yield function incorporates four inelastic hardening functions, κ_{ij}, in addition to three plastic Poisson's ratio functions, ν_{ij}^p. Again this constitutive formulation is implemented in a user defined material subroutine. Large deformation FE analyses using continuum cubes are implemented with the goal of replicating the macroscopic behavior that emerged from our microstructural models (Figure 2). Model parameters are calibrated using uniaxial compression data. Confined compression behavior is then predicted and compared to microstructural TB confined compression hardening curves.

As shown in Figure 3, the proposed anisotropic pressure dependent yield function (equation 2) accurately predicts initial yield and hardening of TB under confined compression loading in the three principal directions. This constitutive law, incorporating anisotropy, mode dependent hardening and pressure dependent yielding represents a significant advance on previous TB continuum models,

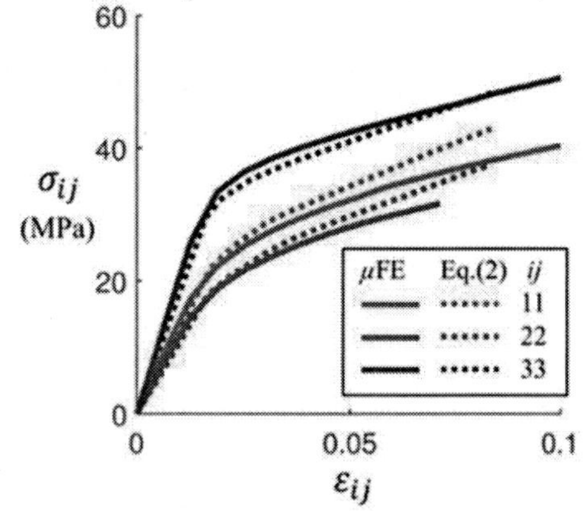

Figure 3: Stress-strain behavior for confined compression in the three principal material directions determined from the μFE models (solid lines) and predicted curves using the proposed continuum anisotropic constitutive law outlined in equation (2) (dashed lines).

DISCUSSION

The mode dependent differential hardening function and the anisotropic plastic yield function proposed in this study represent significant contributions to the current state-of-the-art in continuum modelling of TB. Continuum level models allow for the macroscopic simulation of orthopedic device implantation and the computational efficiency of such models provides a significant advantage over discrete microstructure based models. The constitutive laws developed here should be used for multi-axial large deformation of trabecular bone, as encountered during implantation of press-fit devices.

ACKNOWLEDGEMENTS

This study was funded by the Irish Research Council

REFERENCES

[1] Keaveny, T.M. et al. Ann rev biomed eng (2001).
[2] Kelly and McGarry J Mech Behav Biomed Mat (2012)
[3] Deshpande, V. and Fleck, N. J Mech Phys Sol, (2000)
[4] Xue and Hutchinson CMES (2005).

PROBABILISTIC COMMONALITY OF TRABECULAR BONE STRUCTURES: IS IT A RESULT OF NATURE'S DESIGN?

Matthew L. Kirby (1), Anuradha Roy (2), Feng Zhao (3), and Xiaodu Wang (1)

(1) Mechanical Engineering
University of Texas at San Antonio
San Antonio, TX, USA

(2) Management Science and Statistics
University of Texas at San Antonio
San Antonio, TX, USA

(3) Biological and Medical Engineering
Beihang University
Beijing, China

INTRODUCTION

Fragility fractures of trabecular bone are not only dependent on bone mineral density (BMD), but also on changes to the trabecular architecture [1]. However, the heterogeneity and individual randomness of trabecular architecture have made it difficult for researchers to quantitatively determine the structure-function relationship of this structurally complex tissue.

Recently, image processing techniques, such as volumetric decomposition and skeletonization, have made it possible to obtain local morphological information about the architecture in terms of features of individual trabeculae, which has shown that the geometry of trabecular plates and rods follows certain probabilistic distributions [2]. This result raises an interesting question: Are the underlying distributions of structural features of trabecular bone similar, independent of individual differences? Presumably, nature's design follows a set of optimized design principles that result in some commonality among trabecular structures in individuals and at a variety of anatomical locations. Could it be that this commonality is manifested in the underlying probabilistic distributions of structural features in trabecular bone? If so, quantifying this commonality would be a significant step toward elucidating the structure-function relationship and may provide insight into Nature's design principles, optimization principles that many believe exist and govern the structural design of trabecular bone.

Our hypothesis is that trabecular commonality is manifested in the underlying probabilistic distributions of structural features of trabecular bone. We will test this by analyzing the underlying probabilistic distributions of the following structural features: geometry, spatial arrangement, and orientation of trabecular plates and rods.

METHODS

Twenty-three (N = 23) human trabecular bone samples (twelve samples from the femoral neck and eleven samples from the L1 vertebral body) of a mixture of male and female cadaveric donors, ranging from 40 to 90 years of age, were used in this study. On-axis cylindrical (Φ8 mm×10 mm) trabecular bone samples were cored out, according to an established protocol in the literature [3]. Next, μ-CT scan images of the cylindrical specimens were taken, and a volumetric decomposition and skeletonization software, Individual Trabecular Segmentation (ITS) [2], was used to decompose the structure into individual trabecular plates and rods and provide information pertaining to the geometry, spatial arrangement, and orientation of the plates and rods.

The geometry, spatial arrangement, and orientation of plates and rods are described using eight parameters. Plate and rod geometry is described by univariate distributions of (1) plate area and (2) plate thickness and (3) rod length and (4) rod diameter, respectively. Spatial arrangement was described by the univariate distributions of plate and rod centroid's (5 and 6) nearest neighbor distance [4]. Plate and rod orientation is described by (7 and 8) bivariate angular distributions of the colatitudinal (γ) and azimuthal angles (ϕ). Briefly, the orientation of each plate and rod is represented as a unit surface normal vector and can be uniquely expressed as a function of two angles: γ is the angle between the point and the z-axis and ϕ is the angle between the projection of the point in the x-y plane and the x-axis. For ease of probabilistic comparison, univariate marginal projections of γ and ϕ were obtained.

In order to compare the underlying probabilistic distributions, all univariate parameters were normalized against the mean of the distribution, and the bivariate orientation distributions were normalized to align with the principal axes using principal component analysis. The probabilistic approaches used in this study were pairwise quantile-quantile linear regression analysis, to determine if underlying distributions are correlated, and pairwise hypothesis testing, a more rigorous method to determine if the populations of two distributions are the same. The hypothesis test used in this study was the nonparametric Mann-Whitney U-test with a pairwise significance $\alpha = 0.05$. However, given that a total of 253 independent comparisons are performed for each parameter, the significance was adjusted to $\alpha = 0.0002$ via the Šidák correction to account for the familywise error rate (known as the multiple comparisons problem) [5].

RESULTS

The relative frequency histogram line-plots (Figure 1) illustrate that there is a trend in the shape of the underlying distributions of the parameters, and it seems that the parameters do follow a similar distribution across all the samples in our study group. This qualitative analysis is confirmed by the quantitative results of the pairwise quantile-quantile linear regression analysis and the pairwise Mann-Whitney U-test.

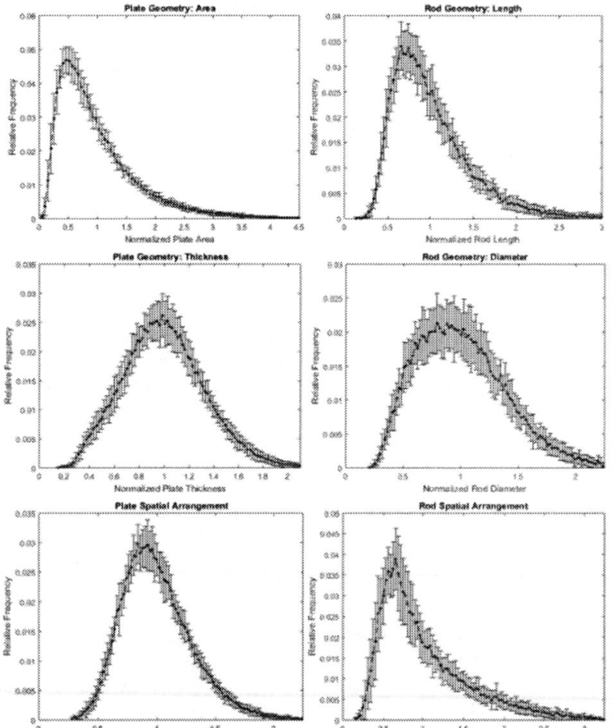

Figure 1: Relative frequency histogram line-plots of the 8 structural parameters (aggregate of all 23 samples)

The quantile-quantile linear regression analysis revealed that the underlying distribution of each parameter is strongly correlated across all samples, as the R^2 value of every pairwise linear regression was greater than 0.94.

The more rigorous Mann-Whitney U-test further indicates that the underlying distributions of each parameter are similar across all samples. All parameters had greater than a 93% acceptance rate for all of the pairwise hypothesis tests, excluding plate area and rod spatial arrangement with a greater than 75% acceptance rate, at the adjusted significance level of $\alpha = 0.0002$ (Table 1). Interestingly, it seems that the acceptance rate may have been negatively affected by outliers at the tails of the distributions. The following abbreviations are used to refer to the structural parameters and their distributions: Plate Area (PA), Plate Thickness (PT), Plate Spatial (PS), Plate Orientation colatitudinal projection (PO (γ)), Plate Orientation azimuthal projection (PO (ϕ)), Rod Length (RL), Rod Diameter (RD), Rod Spatial (RS), Rod Orientation colatitudinal projection (RO (γ)), and Rod Orientation azimuthal projection (RO (ϕ)).

Table1: Results of pairwise Mann-Whitney U-test

Sig. Level (α)	Null Hypothesis (H_0) Acceptance Rate by Parameter (%)									
	PA	PT	PS	PO (γ)	PO (ϕ)	RL	RD	RS	RO (γ)	RO (ϕ)
0.05	47.0	77.5	85.0	69.6	86.2	67.6	90.1	54.5	67.2	60.9
0.0002	75.9	98.0	100	96.8	97.6	96.0	99.6	83.8	93.3	91.3

DISCUSSION

Based on the relative frequency histogram line-plots, the quantile-quantile regression analysis, and probabilistic hypothesis testing, our results demonstrate that the underlying distributions of each structural parameter are similar across all samples, irrespective of individual differences that were present in our sample group. This finding supports our hypothesis that trabecular commonality is manifested in the underlying probabilistic distributions of structural features of trabecular bone. Although this study was limited in sample size, the individual differences (anatomical location, age, gender, and bone mineral density) appear to have a very limited effect on the underlying distributions. Nonetheless, we may need to increase our sample size to confirm this important finding.

The outcome of this study may have several significant contributions to the field. First, by elucidating the probabilistic model that describes each one of the 8 structural parameters, a representative mathematical model of all trabecular bone architectures could be readily created and used to illuminate its structure-function relationship. Additionally, knowing what is common among all trabecular structures could help clinicians identify the primary individual differences that influence structural integrity, which would lead to improved fragility fracture prediction. Finally, the discovery and mathematical representation of trabecular commonalty provides some evidence that the structure is a by-product of Nature's design and a quantitative framework for further investigation of governing optimization principles.

REFERENCES

[1] Parkinson, I. H. and Fazzalari, N. L., *Stud Mechanobiol Tissue Eng Biomater*, 5:31-51, 2013.

[2] Liu, X. S. et al., *J Bone Miner Res*, 23:223–235, 2008.

[3] Keaveny, T. M. et al., *J Biomech*, 27:1127–1136, 1994.

[4] Illian, J. et al., *Statistical Analysis and Modelling of Spatial Point Patterns*, 10-12, 2008.

[5] Šidák, Z. K., *J Am Stat Assoc.*, 62:626-633, 1967.

Poster Presentation #P142

SB^{3}C2017
Summer Biomechanics, Bioengineering and Biotransport Conference
June 21 – 24, Tucson, AZ, USA

MATERIAL SENSITIVITY ANALYSIS OF ELBOW JOINT CARTILAGE PARAMETERS IN A FINITE ELEMENT MODEL

Mohsen Sharifi Renani (1), Munsur Rahman (1), Akin Cil (1, 2, 3), Antonis Stylianou (1)

(1) Department of Mechanical Engineering
University of Missouri - Kansas City
Kansas City, MO, USA

(2) Department of Orthopaedics Surgery
University of Missouri-Kansas City
Kansas City, Missouri, USA

(2) Department of Orthopaedics
Truman Medical Centers
Kansas City, Missouri, USA

INTRODUCTION

Only few millimeter thick, articular cartilage withstands very high compressive and shear forces while protecting the bone from excessive loading, and provides a smooth articulation for the joint. Better understanding of elbow cartilage mechanics can provide a valuable insight into cartilage degeneration mechanisms and osteoarthritis development. Computational methods, such as the finite element (FE) method can be very useful in predicting cartilage contact mechanics of the joint [1, 2]. However, finite element models should be validated against experimental data and extensively tested for robustness. When performing contact analysis, some specific model parameters, such as the cartilage modulus and incompressibility of the material, govern the contact characteristics (e.g. contact area and pressure) of the joint soft tissues [2].Therefore, before applying these models in any clinical investigation, it is necessary to understand how changes in these input parameters impacts the simulated output results.

The objective of this project was to develop a finite element elbow joint model, evaluate it against experimental data, and use this model to perform a sensitivity analysis to investigate which contact parameters have greatest impact on the model simulation results.

METHODS

Two fresh-frozen cadaveric elbow specimens (right and left arm) from a 69 years old male were used in this study. The I-Scan System (Tekscan Inc.) was used to measure the joint contact pressure experimentally. The specimens were cemented in custom fixtures at 20 degree flexion, and placed in a bi-axial Instron mechanical testing machine. Before applying any axial loading, a point cloud at the initial position was collected using an Optotrak rigid probe (NDI, Waterloo, Ontario, Canada) for alignment of the models. An 110N axial load was applied to the proximal humerus and maintained for 10 min to allow the viscoelastic deformation of cartilage to become constant (Fig.1). Contact pressure data was collected during the entire experiment, but only the last frame of 10min was used for comparison with finite element solutions.

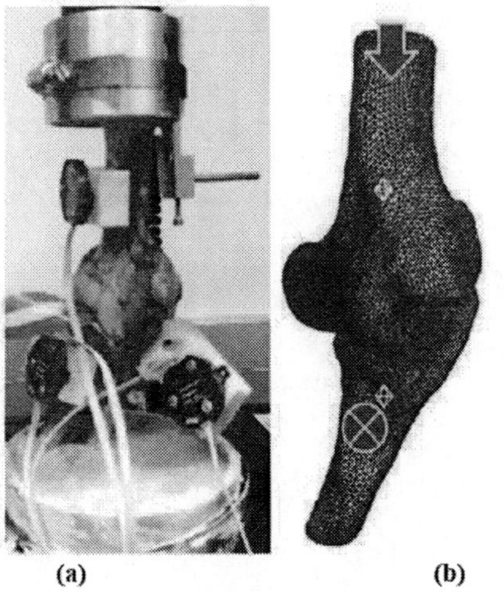

Figure 1: a) Physical experiment and b) Finite Element model

The specimens were imaged using Magnetic resonance imaging (MRI). Auto thresholding, manual segmentation, and subtraction of cartilage geometries from bones were employed to create the bony and cartilage geometries in MIMICS (Materialise, Ann Arbor, MI). For developing the FE models, the geometries were converted into solid 4-node tetrahedral elements using Hypermesh (Altair Engineering, Inc., Troy, MI). The element size for cartilages and bone were assumed to be 1mm and 2mm respectively. Bones were modeled as rigid bodies and the cartilages were modeled as Moony-Rivlin hyper elastic materials (E=0.7 MPa, v=0.47). Cartilage to bone and cartilage to cartilage contact surfaces were modeled using rigid tie and frictionless surface to surface contacts respectively [2, 3]. Boundary conditions identical to the experiment were applied to the model. FE computations were done in FEBio (version 2.4; mrl.sci.utah.edu/software.php).

One-at-a-time parametric analysis were performed to investigate the sensitivity of the model's contact mechanics to cartilage material properties, including Young's modulus (E) and Poisson ratio (v). First, the baseline cartilage young's modulus was altered by ±20% while the Poisson ratio remained constant. Then, while the Young's modulus was kept constant, the Poisson ratio was varied from 0.43 (bulk to shear modulus of 10:1) as the minimum value to 0.49 (bulk to shear modulus of 100:1) as the maximum values with increment of 0.01. The contact area, average, and peak contact pressure were computed for each case as an output for the analysis [1, 4-5]. In order to guarantee both accuracy and low computational cost, a mesh convergence study was also conducted using the cartilage with two smaller element size 0.7mm and 0.5mm

RESULTS

The FE models and the experiments are in good agreement in terms of contact characteristics (Fig.2). The sensitivity analysis results show that decreasing the Poisson's ratio from 0.49 to 0.43 caused a significant decrease in peak pressures, while changes in average pressure and contact area were less than 10% (Fig. 3). Changing the young's modulus did not have a remarkable effect (less than 6%) on the contact mechanics (Fig. 3).

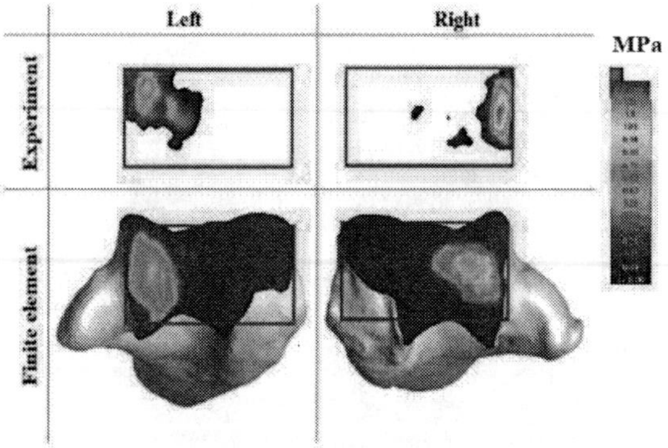

Figure 2: Contact pressure distribution

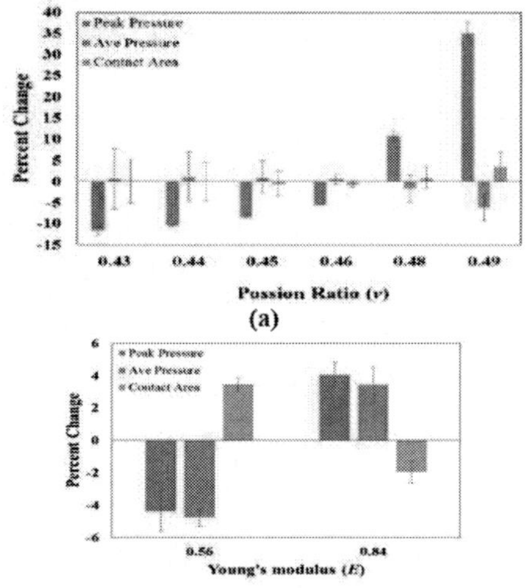

(b)

Figure 3: Percent changes in peak pressure, average pressure and contact area due to alterations in Poisson ratio (a) and Young's modulus (b). Error bars specify standard deviations over the two finite element models.

DISCUSSION

The primary goal of this project was to develop finite element models and use them to improve current modeling techniques for parameter selection. This study found that the Poison ratio is a more significant parameter on contact mechanics, especially peak contact pressure than Young's modulus. This suggests that selection of this parameter is critical for accurate prediction of contact area and pressure. This investigation also shows that prediction of joint contact characteristics under static conditions is not very sensitive to material modulus. These result is in good agreement with a finite element study performed on hip joint [1]. Although the models are in agreement with the experimental data and literature, future refinements could improve the contact pressure predictions [2, 6] .The limitations of the current study include: contact mechanics analysis on the radius bone and its cartilage was not performed in this study, the ligaments and capsule around the joint were not included in these computational models, and the investigation of the contact mechanics was performed at one angle and in a static position. The validity of the current developed models in dynamic conditions and the contact mechanics on the radial cartilages will be explored in the future studies by incorporating the ligaments and joint capsule.

ACKNOWLEDGEMENTS

Please acknowledge any grant or other funding support or the.

REFERENCES

[1] Anderson, A.E., et al., J Biomech Eng, 2008. 130(5): p. 051008.
[2] Willing, R.T., et al., J Biomech, 2013. 46(10): p. 1767-71.
[3] Schenck Jr,R.C. et al. Clinical Orthop. Related Research, 299: 305–312, 1994.
[4] Hamby, D.M., Health Phys, 1995. 68(2): p. 195-204.
[5] Kiapour, A., et al. J Biomech Eng, 2014. 136(1): p. 011002.
[6] Chantelot, C., et al. Surg Radiol Anat, 2008. 30(2): p. 103-11.

SUSTAINING LOW FRICTION BY LOAD SHARING MECHANISM IN HYDROGELS FOR CARTILAGE IMPLANTS

E. Porte, P. Cann, M. Masen

Tribology group
Department of Mechanical Engineering
Imperial College London
London, United Kingdom

INTRODUCTION

As the population ages the development of materials for articular cartilage repair is becoming increasingly important. Such biomedical materials would reduce the need for intensive and costly joint replacement surgery. Hydrogels seem a promising possibility, as they are bi-phasic, porous and have appropriate mechanical stiffness. However, for these to be successfully implemented as cartilage implants it is important to understand their tribology performance under physiological loading and shear conditions. Friction and lubrication of the rubbing implant contact will have a significant effect on the wear and fatigue of the materials and thus their durability.

One of the main contributions to the very low friction of cartilage is its very efficient load sharing mechanism [1]: during compression the interstitial fluid in cartilage pressurizes because of the deformation of the solid matrix. Because of this pressurisation, part of the applied load is carried by the interstitial fluid rather than the solid matrix [2]. In combination with the lubricating action of the fluid phase, the load reduction on the solid phase significantly reduces the friction force in the sliding contact. The aim of this study is to show the ability of hydrogels to have a load sharing mechanism and to gain insight into the principles of this mechanism to optimise the performance.

Friction tests with different sliding conditions were performed on poly(vinyl alcohol) (PVA) hydrogels to show a low friction coefficient can be sustained due to fluid load support. A photoelastic PVA hydrogel is successfully developed to explore the use of photoelasticity in the contact behaviour of hydrogels. Photoelastic materials show fringe patterns based on the stress state of the material. The higher the stress in the material, the more fringes will show [3]. This property can be used to study the stress state in the hydrogel to find direct evidence of the load sharing mechanism and to gain a fundamental understanding.

METHODS

The PVA hydrogels were tested using a newly developed physiological tribometer (PCS Instruments London UK) which is shown in Figure 1. The device has an upper specimen which is loaded and reciprocates (frequency range 0.01-5 Hz) against a lower stationary specimen. Independent actuation of the upper specimen in x-, y- and z-direction allows for complex loading cycles, comparable to those in joints. The advantage of this small scale test equipment for relatively short tests compared to joint simulators is that it allows for a more detailed study to gain insight into the load sharing mechanism.

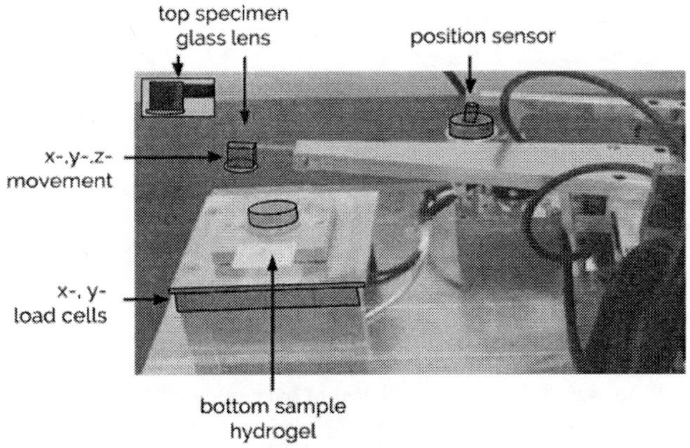

Figure 1: Physiological biotribometer test setup

Tests were performed with different sliding conditions between a PVA hydrogel sample (lower-) and a glass lens (upper specimen). Two different motion patterns were used: stationary contact (SC) and migrating contact (MC). In SC motion the relative motion between the specimens is smaller than the contact area, as visualised in the top left corner of Figure 3. As a result, the overlapping area is a continuously loaded area on the hydrogel. In a MC the contact area moves over a larger distance to ensure unloading the area after contact.

Before hydrogels are tested, they are characterised for their properties such as compressive modulus. In addition, the photoelastic properties of the hydrogels are explored in order to relate the stress in the hydrogel to its friction coefficient in sliding tests. A special camera setup, as shown in Figure 2, is required to record the photoelastic response of the hydrogel. The setup comprises a light source, a red filter, two polarisation filters and a camera. The colour filter is used to limit the bandwidth of the light spectrum to simplify the analysis of the obtained fringe pattern. Polarisers are necessary to obtain the fringe pattern in the material.

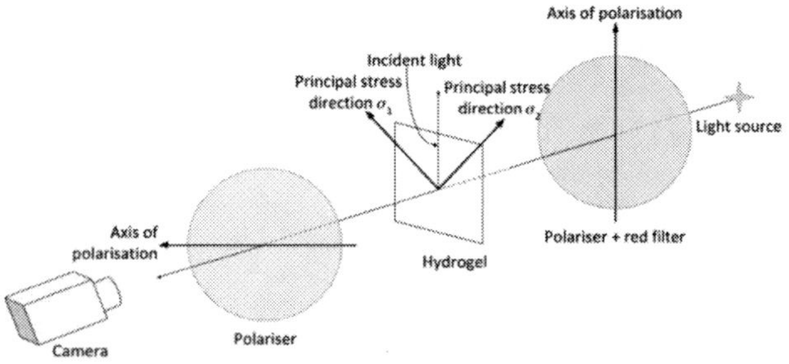

Figure 2: Camera setup for photoelastic characterisation

RESULTS

Tests on the biotribometer show that the friction coefficient of the PVA hydrogel depends on the sliding conditions. Figure 3 displays the average measured friction coefficient for the two different sliding regimes. The friction coefficient significantly increases over time for a stationary contact. In a migrating contact a low friction coefficient is sustained over a long period of time.

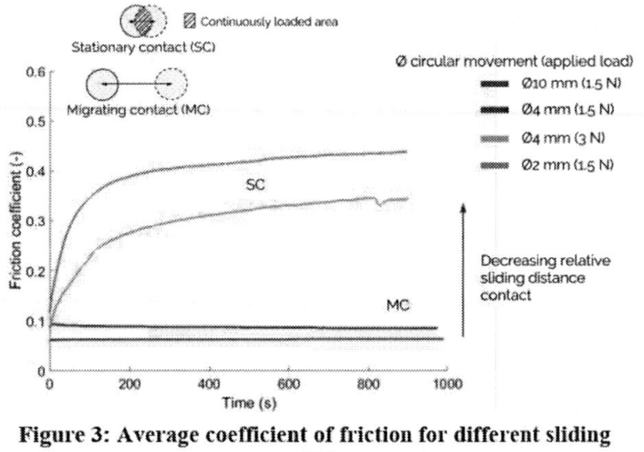

Figure 3: Average coefficient of friction for different sliding conditions

The maintained low MC friction is attributed to the ability of the hydrogel to sustain fluid load support when the loaded area moves over a sufficiently large distance, because of rehydration of the contact. To find direct evidence for this load sharing mechanism between the solid and fluid phases of the hydrogel, a photoelastic hydrogel was developed.

Figure 4 shows the photoelastic PVA hydrogel under different compressive loads with a cylindrical indenter. The change in the shape and number of fringes indicates the change in stress state in the material. The sample with a 1.5N load shows the maximum stress most clearly, pointed out in Figure 4. The fringe pattern in the material is in accordance with a Hertzian contact stress distribution. The results show a proof of principle that PVA hydrogels have photoelastic properties and that these can be used to study the stress response in the solid matrix.

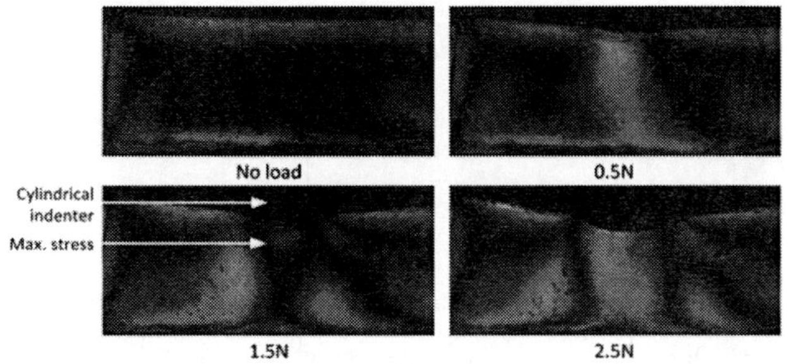

Figure 4: Photoelastic response in compressive loading

The next step will be to visualise the stresses in the hydrogel while performing friction tests. Stationary and migrating contact conditions will be compared. It is expected to see the stresses in the material increase under stationary contact conditions and a constant stress in migrating contact tests.

DISCUSSION

It can be concluded that the load sharing mechanism is an important feature for frictionally optimised hydrogels. The friction in the material strongly depends on the contact conditions. Preliminary experiments using the photoelastic technique show promising results to relate the stress state in the material to its friction coefficient. This technique has not been used on this type of experiments before and is therefore still in development. The friction results in this study show qualitative resemblance with tests performed on articular cartilage by [1], which proves the ability of hydrogels to mimic behaviour similar to cartilage. Photoelasticity is a promising technique to carry out an in depth study to gain a thorough understanding of the mechanics of the fluid load support.

ACKNOWLEDGEMENTS

We would like to thank Clive Hamer and Tom Welham of PCS Instruments for the collaboration in the development of the biotribometer.

REFERENCES

[1] Caligaris, M et al, *Osteoarthritis & Cartilage*, 16: 1220-1227, 2008

[2] Ateshian, G, *J of Biomechanics*, 42:1163-1176, 2009

[3] Shukla, A et al, *Experimental Solid Mechanics 2nd ed*, Part III: 259-509, 2014

EVALUATION OF THE "MEMBRANE" EFFECT OF THE LAMINA SPLENDENS OF ARTICULAR CARTILAGE: IMPLICATIONS FOR OA

Ferris M. Pfeiffer, PhD (1,2), Joe T. Rexwinkle (3), Andrew Polk (1), Aaron M. Stoker, PhD (4), Nikki Werner (4), Sydney Timmerman (1)

(1) Department of Bioengineering
University of Missouri
Columbia, MO, USA

(2) Department of Orthopaedic Surgery
University of Missouri
Columbia, MO, USA

(3) Department of Mechanical Engineering
University of Missouri
Columbia, MO, USA

(4) Veterinary Medicine
University of Missouri
Columbia, MO, USA

INTRODUCTION

The role of articular cartilage is to provide for smooth articulation of joints against each other and to distribute loads during locomotion. In order to optimally serve this purpose, the surface of the tissue, called the lamina splendens, must be healthy in order to provide both a proper interface and to preserve the health of the tissue. The lamina splendens is not well characterized. Published data varies as to thickness, composition, and function. It has been shown in several studies that a symptom and possible cause of osteoarthritis is disruption of the cartilage surface. One function of the lamina splendens appears to be a "membrane" like structure which imparts some mechanical purpose to the underlying tissue as well as provides a biochemical boundary which promotes homeostasis in the joint. In this study, the integrity of the lamina splendens was tested via needle puncture in order to characterize the strength of the "membrane" against the degree of degradation as quantified via India ink staining. We hypothesize that damaged tissues will require a lower amount of force to rupture the lamina splendens layer and that the difference in force before and after breaking will be larger for damaged (i.e. osteoarthritic) tissues as compared to normal controls.

METHODS

Following IRB approval (IRB# 1208392), 8mm diameter osteochondral plugs (n=72) were collected from two human patients who had undergone total knee replacements. Samples were taken from regions with varying levels of tissue damage. Tissue from a total of three knees (two from one patient and one from a second patient) was collected for evaluation. Of the 72 human samples, 20 were excluded for excessive degradation or mishandling. ***Biomechanical testing***: All explants were tested for the biomechanical properties using an Instron 8821S servo-hydraulic testing machine. A needle probe was used to determine tissue puncture force (TPF) of the cartilage and cartilage thickness. A stress relaxation test (3.9mm diameter flat punch) was then used to determine aggregate modulus (AG) and permeability (PERM) of the cartilage tissue.The lamina splendens of these samples was evaluated using a puncture test with a blunted 22 g needle (F igure 1). This was performed using an Instron 8821s machine with the cartilage samples held in place within a well. The blunted needle was advanced at 0.1mm/s while the force was monitored. An initial increase in force was noted as the needle contacted the cartilage surface. A precipitous drop in the force was noted as the lamina splendens "membrane" was punctured (Figure 2). Finally, an increase in the force was noted when the subchondral bone was contacted. A measure of membrane integrity was calculated as the difference between the measured force at initial contact and the maximum force prior to membrane rupture. Severely osteoarthritic tissues did not demonstrate a clear membrane breaking force and thus were given a measure of zero. Osteochondral plugs were stained using India ink to determine the severity of cartilage lesion present prior to testing (Figure 3). ***Media biomarker analysis***: Media from all patients were assessed for MMP-1, MMP-2, MMP-9 and MMP-13; general MMP activity; prostaglandinE2 (PGE2); nitric oxide (NO); GRO-α, IL-6, IL-8, MCP-1; C2C; PIIANP; and GAG content.

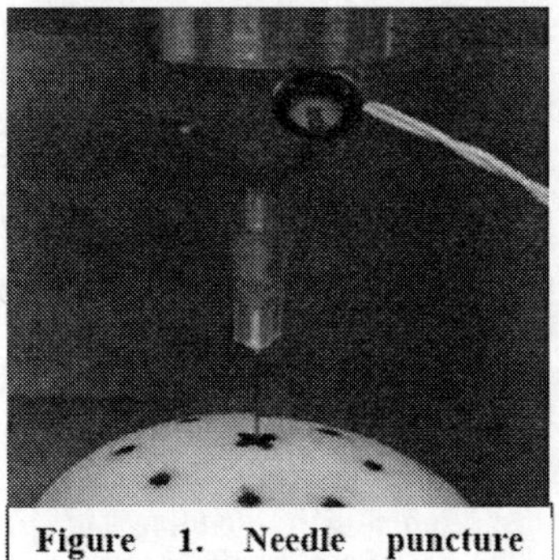

Figure 1. Needle puncture testing of lamina splendens

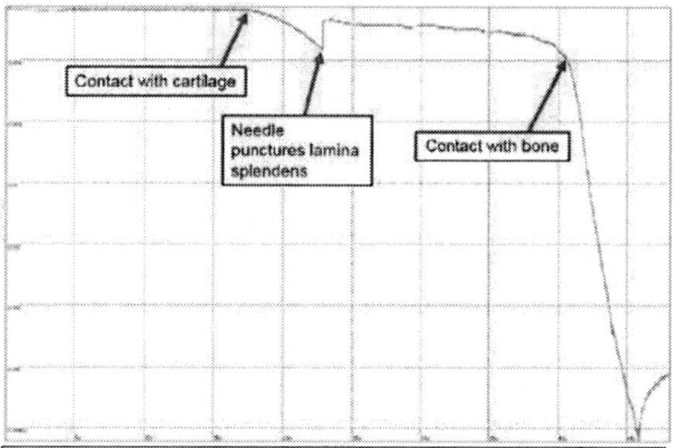

Figure 2. Needle puncture representative test data indicating contact with the cartilage surface, as wells as puncture of the lamina splendens and contact with the subchondral bone.

Media from patients 1 and 2 were further assessed for CASPASE-3, CYTOCHROME C, PARP; TNF-RI, TNF-RII, DR5; and hyaluronic acid using commercially available assays. Additionally, media from patients 3-6 were assessed for IL-1β, RANTES, TNF-α; and CS846.

RESULTS

OCA plugs evaluated ranged between no lesion (n=12), mild lesion (n=17), and severe lesion (n=23). There was a strong negative correlation (R^2>0.93) between severity of the observed lesion and the membrane breaking strength (Figure 4). Permeability had a negligible to weak negative correlation (r=-0.189-[-0.295], p≤0.001-0.034) to Tissue GAG, Media GAG, NO, PGE_2, IL-6, IL-8, and MCP-1, a weak positive correlation (r=0.251-0.387, p≤0.001-0.045) to MMP-1, MMP-13 general MMP activity, AG, TNF-RI, DR5, and RANTES, and a moderate positive correlation (r=0.421, p<0.001) to MMP-9. Aggregate Modulus had a weak to moderate negative correlation (r=-0.213-[-0.402], p≤0.001-0.024) to Tissue HP, MMP-2, IL-6, IL-8, and MCP-1, HISTO score, and CS846, and a weak positive correlation (r=0.252-0.304, p≤0.001-0.003) to MMP-1, MMP-9, MMP-13, and TPF. Tissue Puncture Force had a weak negative correlation (r=-0.289-[-0.337], p≤0.001-0.013) to GRO-α, NO, IL-1β, RANTES, TNF-α, and VCD, a moderate negative correlation (r=-0.425-[-0.511], p<0.001) to MMP-2,

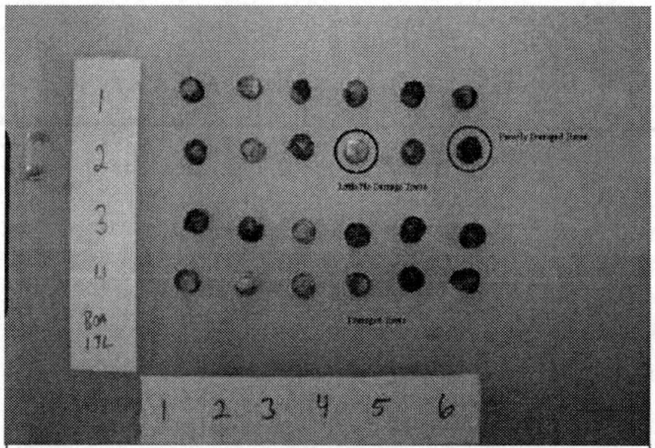

Figure 3. Representative India ink staining of osteochondral plugs indicating severity of cartilage lesion damage from none to severe.

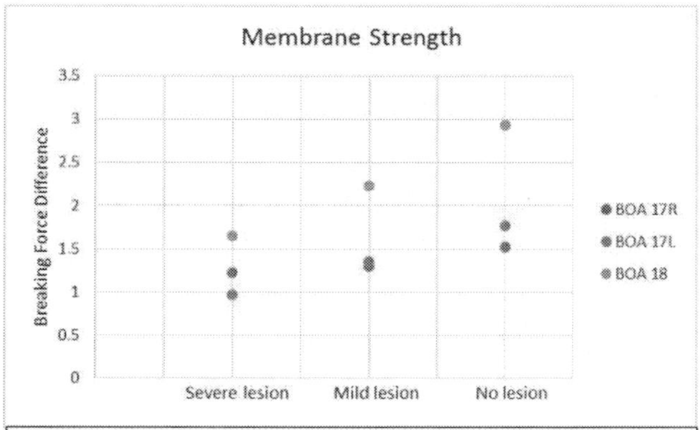

Figure 4. Measured membrane strength as compared to severity of cartilage lesion as determined using India ink staining.

IL-6, IL-8, and MCP-1, a weak to moderate positive correlation (r=0.215-0.549, p≤0.001-0.017) to general MMP activity, Media GAG, MMP-9, and CS846, and a strong positive correlation (r=0.716, p<0.001) to MMP-13.

DISCUSSION

Testing of OCA plugs in this study using a blunted 22g needle was able to determine the presence of the lamina splendens as confirmed using India ink staining and biomarker analysis in a representative sample set that included normal and pathologic tissues.

ACKNOWLEDGEMENTS

This work was supported by the Comparative Orthopaedic Lab

SB³C2017
Summer Biomechanics, Bioengineering and Biotransport Conference
June 21 – 24, Tucson, AZ, USA

EFFECTS OF FREEZING ON MECHANICAL PROPERTIES OF BOVINE, OVINE, AND PORCINE ARTICULAR CARTILAGE

Kelly J. Vazquez (1) and Corinne R. Henak (1,2)

(1) Department of Mechanical Engineering
University of Wisconsin-Madison
Madison, Wisconsin, United States

(2) Department of Biomedical Engineering
University of Wisconsin-Madison
Madison, Wisconsin, United States

INTRODUCTION

Articular cartilage is a connective tissue that provides a low-friction, wear-resistant surface in synovial joints [1]. Cartilage is comprised of a type II collagen network surrounded by a hydrated proteoglycan matrix [2]. Cartilage hydration provides resistance to compression, as well as apparent viscoelasticity as fluid flows out of the collagen when articular cartilage is loaded, and then is reabsorbed when cartilage is unloaded [3].

Due to limited availability of fresh tissue, cartilage is frequently stored prior to mechanical testing [2]. If articular cartilage is damaged during freezing, mechanical properties may be altered and therefore not representative of the *in vivo* tissue. The effects of storage on the mechanical properties of articular cartilage are not fully known. Previous studies have found that short term storage of cartilage at -20°C has no significant effect on fibril modulus, matrix modulus, permeability, dynamic modulus, or damping [2,4]. These studies have focused on a single animal model (bovine), and have not compared the effects of freezing across species.

Single animal models are often used to evaluated cartilage properties due to the limited availability of human or other tissue. The selection of a single animal model potentially limits the applicability of findings across species. Common animal choices are bovine [2-4] and porcine [5], partially due to the availability of tissue from abattoirs. Previous studies have shown inter-species variation in biphasic material constants, as well as in dynamic moduli [6,7]. However, variations between porcine, ovine, and bovine tissue, coupled with storage effects, have not yet been evaluated.

Thus, the aim of this study was to assess both inter-species variation and the effects of freezing on bovine, ovine, and porcine articular cartilage.

METHODS

Sample Preparation

Fresh patellae were harvested from 4 bovine, 6 ovine, and 5 porcine animals from a local abattoir within 3 hours of slaughter. Patellae were kept hydrated in a Phosphate-buffered saline (PBS) plus protease inhibitor (PI) solution and patellae were separated for immediate testing (fresh) or storage (frozen). Frozen samples were wrapped in PBS/PI-soaked gauze at -20° C until mechanical testing (3-10 weeks). Prior to mechanical testing, frozen tissue was thawed in PBS/PI solution at room temperature. Fresh samples were tested within 3 hours of harvest. For both fresh and frozen (thawed) samples, osteochondral cores were removed from patellae using a 6 mm diameter coring tool and a scalpel. Full thickness cartilage samples were then made by removing the thin layer of subchondral bone. Cartilage thickness was measured using calipers. Following measurement, samples were speckled with India ink prior to mechanical loading and were returned to the PBS/PI solution.

Mechanical Testing

All samples remained hydrated in PBS/PI throughout testing. Samples were loaded into a custom well on a tabletop test machine (Fig. 1) and a 10.30 ± 0.26 g tare load was applied until equilibrium was achieved. Samples were then loaded for 10 cycles to 10% strain at 1 Hz. Force

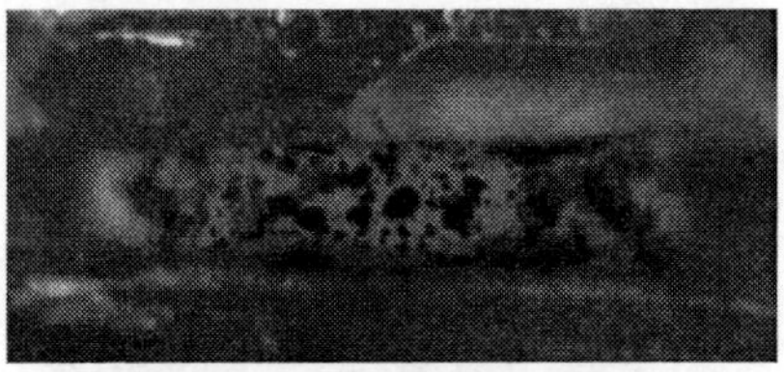

Figure 1: Representative cartilage sample during testing.

and displacement data were collected to calculate full-thickness stress and strain. Data were analyzed using custom MATLAB code. Averages of the last 5 cycles were used to calculate results for each sample, since the mechanical testing was repeatable over these cycles. The loading portion of each cycle was fit to a line to determine the linear modulus:

$$\sigma = E \cdot \varepsilon \tag{1}$$

Where E is the linear modulus, ε is strain, σ is the stress. Loading curves were also fit to an exponential expression [8], from which the tangent moduli could be calculated:

$$\sigma = A \cdot (e^{B\varepsilon} - 1) \tag{2}$$

$$\frac{d\sigma}{d\varepsilon} = AB \cdot (e^{B\varepsilon}) \tag{3}$$

Where A and B are coefficients, and $\frac{d\sigma}{d\varepsilon}$ is the tangent modulus at a given strain value. The energy dissipated per cycle was calculated using the area between the loading and unloading curves using trapezoidal numerical integration.

Statistical Analysis

Statistical significance was determined between measured parameters and interaction terms using a two-way analysis of variance of the mean (ANOVA) to compare the effect of cartilage storage and species on moduli and dissipated energy. Values of $p < 0.05$ were considered to be statistically significant. All measured data are reported with mean ± standard deviations for five samples per group.

RESULTS

Freezing had no effect on linear moduli (Fig. 2(a)), tangential moduli at zero strain (Fig. 2(b)), tangential moduli at 10% strain (Fig. 2(c)), or energy dissipation (Fig. 2(d)) ($p > 0.2$).

Conversely, species significantly affected all results ($p < 0.0001$). Porcine cartilage had the highest moduli and energy dissipation. Ovine cartilage had the lowest moduli and energy dissipation.

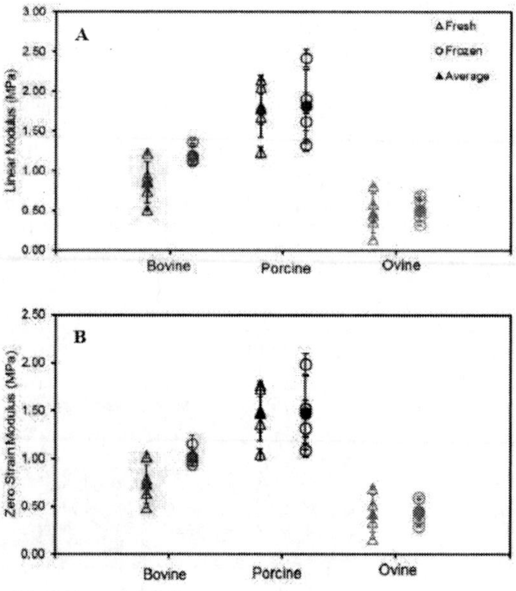

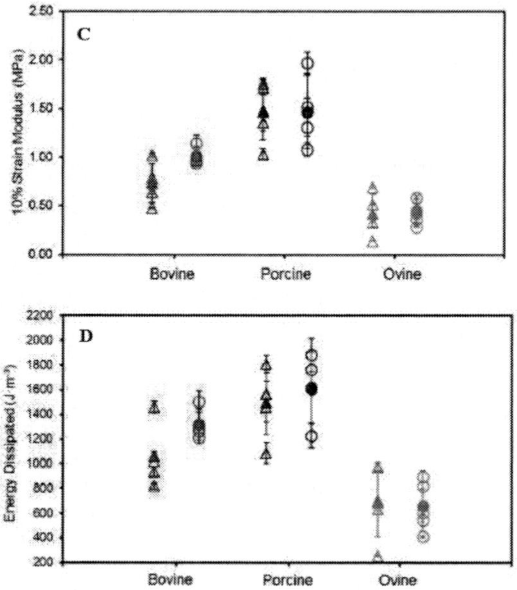

Figure 2: Calculated (a) linear moduli, (b) tangential moduli at zero strain, (c) tangential moduli at 10% strain, and (d) energy dissipation for each animal and sample type.

DISCUSSION

Understanding the effect of freezing on cartilage mechanical properties ensures that testing stored samples accurately reflects the tissue's properties. The current results are consistent with previous findings that a single freeze-thaw cycle does not significantly change the mechanical properties of articular cartilage [2-4]. These results imply that no substantial damage to cartilage integrity occurs during a single freeze-thaw cycle at -20°C regardless of the animal model. Whether a single freeze-thaw cycle will have an effect on other properties, such as depth-dependent moduli, has yet to be determined.

Significant differences in mechanical properties across species suggests caution in selecting an animal model. These results are consistent with previous studies [6,7], but evaluate a different set of material properties. The aggregate modulus of porcine cartilage was significantly lower than that of bovine and ovine cartilage in a previous study [6]. In the present study, porcine cartilage had the largest moduli. This discrepancy emphasizes the potential differences in applicability of an animal model depending on the specific mechanical output of interest. Together, the present and previous studies suggest that inter-species comparison of mechanical properties is necessary prior to assuming that results are consistent across species.

In conclusion, this study achieved the aim of assessing both inter-species variation and the effects of freezing on bovine, ovine, and porcine articular cartilage. While inter-species variation was significant, freezing did not affect mechanical properties.

REFERENCES

[1] Mow, Basic Orthopaedic Biomechanics, 1997;[2] Changoor+, *J Biomech Engg*, 2010; [3] Kiefer+, *J Orthop Res*, 1989; [4] Szarko+, *BMC Muscoloskel Dis*, 2010; [5] Gannon+, *Osteoarth Cart*, 2010; [6] Taylor+, *Proc IMechE Part H*, 2011; [7] Temple+, *BMC Musculoskel Dis*, 2016; [8] Park+, *Ann Biomed Engg*, 2008.

SB^{3}C2017
Summer Biomechanics, Bioengineering and Biotransport Conference
June 21 – 24, Tucson, AZ, USA

SENSITIVITY OF CARTILAGE CONTACT MECHANICS PREDICTIONS TO SUBJECT SPECIFIC LOADING CONDITIONS

Penny R. Atkins (1-2), Niccolo M. Fiorentino (2), Samuel A. Colby (1), Andrew E. Anderson (1-3)

(1) Department of Bioengineering
University of Utah
Salt Lake City, UT, USA

(2) Department of Orthopaedics
University of Utah
Salt Lake City, UT, USA

(3) Scientific Computing Institute
University of Utah
Salt Lake City, UT, USA

INTRODUCTION

Cam femoroacetabular impingement (FAI), a condition marked by asphericity of the femoral head and reduced femoral head-neck offset, is proposed to alter hip joint kinematics (rotations and translations) and kinetics (muscle and joint reaction forces (JRFs)). Due to reduced congruency and early abutment between the aspherical portion of the femoral head and the acetabulum of the pelvis, the biomechanics of the hip joint with cam FAI are likely altered. Strong evidence supports the idea that increases in cartilage stresses hasten the degeneration of the joint resulting in the early onset of osteoarthritis (OA), which correlated well with clinical findings of cam FAI morphology in patients with end-stage OA [1].

The evaluation of in-vivo contact mechanics in the native hip is not currently possible. As such, we have developed and validated a finite element (FE) pipeline to predict cartilage stresses from subject-specific surface reconstructions of bone and cartilage [4].

Historically, FE analyses of the hip joint have been predominantly driven using generalized kinematics and JRFs from the literature (e.g. patients with instrumented implants [5]). Recently, subject-specific kinematics, defined using skin marker (SM) motion analysis using reflective skin markers and infrared cameras, have been incorporated into FE analysis of the hip joint [6]. However, the evaluation of hip joint kinematics in patients with cam FAI using SM motion analysis has to date been somewhat inconclusive. The inaccuracies in measuring hip kinematics with SMs, including errors in the estimate of the hip joint center and the effects of soft-tissue artifact, may make this technology ill-suited to study kinematics in patients with cam FAI [2]. To resolve this issue, we have developed and validated a dual fluoroscopy (DF) system to quantify hip kinematics with sub-millimeter and sub-degree errors [3]. Improved accuracy and precision provides the capability to identify the more subtle variations in movement patterns that may result from the aspherical morphology of the femoral head and affect the mechanics of the hip joint.

Given that cam FAI is defined as a condition for which both hip anatomy and joint articulation are altered, patient-specific boundary and loading conditions may be necessary to accurately predict hip contact mechanics. The objectives of this study were to determine the sensitivity of FE predictions to changes in kinematics and kinetics, using data from SM, DF, and generalized values obtained from the literature [5]. Our results from these sensitivity studies may illuminate aspects of the modeling protocol that could be streamlined and identify the effect of subject-specific kinematics and kinetics on hip biomechanics.

METHODS

One female subject with cam FAI signed informed consent to participate in this University of Utah IRB approved study (21 years of age, weight: 65.4 kg, BMI: 24.8 kg/m2). The finite element pipeline followed our previously validated methods [4]. Volumetric computed tomography (CT) arthrography images of the proximal femur and pelvis were acquired (Siemens SOMATOM 128 Definition, Munich, Germany). Bone and cartilage surfaces were segmented and reconstructed to provide subject specific anatomy using Amira (v5.6, Visage Imaging, San Diego, CA, USA). Preprocessing for the models was completed in PreView (v1.19, University of Utah, Salt Lake City, UT, USA).The volume of the cartilage and labrum were meshed with volumetric hexahedral elements using TrueGrid (XYZ Scientific, Livermore, CA, USA). For the proximal femur and pelvis, the distance between surfaces of the inner and outer cortex provided the cortical thickness at each surface vertex which served as the computational thickness for the triangular shell elements of the outer cortex (Figure 1). Trabecular bone was not included as it has little effect on cartilage contact stress predictions.. For all models, cortical bone was modeled as isotropic, hyperelastic (E=17 GPa, v=0.29), the cartilage as isotropic, neo-Hookean hyperelastic (G=13.6 MPa, K=1359 MPa), and the labrum as a transversely isotropic hyperelastic (C_1=1.4 MPa, C_3=0.05 MPa, C_4=36, C_5=66 MPra, λ=1.103). FE analysis was completed in Nike3D (v3.4.2, Lawrence Livermore National Laboratory, Livermore, CA, USA).

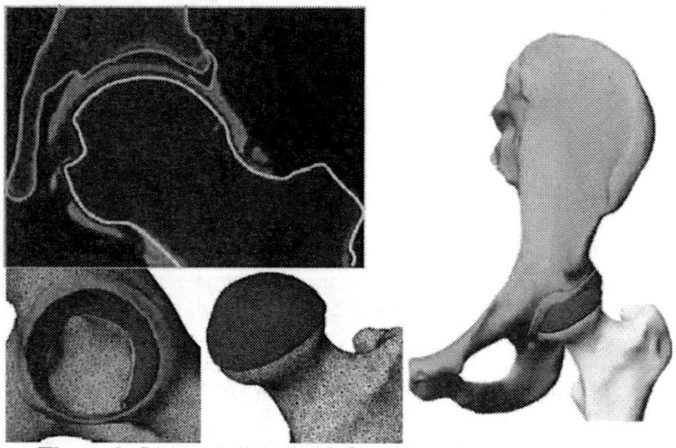

Figure 1: Segmentation (top left) and surface reconstruction (right) of model surfaces. Acetabulum of the pelvis (bottom left) and proximal femur (bottom middle) with hexahedral cartilage (purple) and labrum (blue) volumes and triangular shell elements.

To obtain boundary and loading conditions, the subject walked at a self-selected speed (1.08 m/s) on an instrumented treadmill as hip kinematics were measured simultaneously using DF and a 10-camera Vicon motion capture system. Spherical markers were placed on bony landmarks of the lower limbs [2]. The corresponding bony landmark positions were determined from the DF data using model-based tracking [3]. Virtual DF marker and SM positions were combined with the ground reaction force (GRF) data from the instrumented treadmill in OpenSim (https://simtk.org) to calculate kinematics and JRFs at the hip (Figure 2).

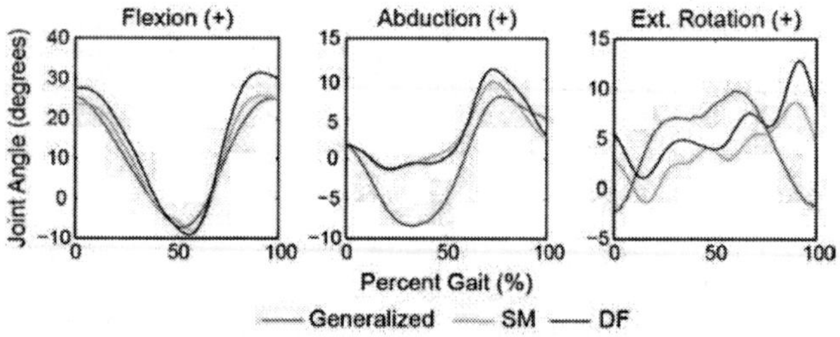

Figure 2: Kinematics from generalized data [5] and subject specific kinematics from skin marker (SM) and dual-fluoroscopy (DF) motion capture data.

Two loading scenarios, based on the impact and active peaks of the GRF data, were evaluated for each method of defining kinematics and kinetics (generalized, SM and DF). Joint angles were applied to the femur and pelvis of the subject in Preview (https://febio.org). The sacroiliac and pubis symphysis joints were held rigid, while motion of the femur was applied in the direction of loading until the JRF was reached. Translation of the femur perpendicular to the direction of applied loading was allowed so as to reduce unrealistic stress risers caused by improper seating into the acetabulum. Cartilage and labrum contact stresses and contact areas were evaluated for the two loading scenarios using kinematic and kinetic inputs from generalized data [5], and subject-specific SM and DF kinematics and calculated JRFs. For the impact peak, using DF boundary and loading conditions, contact stresses and contact areas were compared to evaluate the result of using hexahedral and tetrahedral elements to represent the cartilage and labrum.

RESULTS

For this subject with cam FAI, impact and active GRF peaks occurred at 15% and 47% gait, compared to 15% and 46% for the generalized data [5]. Joint angles at the impact and active peaks varied by less than 4.8° in flexion, 7.0° in abduction, and 8.8° in rotation between the three modes of data collection (Figure 2). For the generalized data, the JRF at the first (impact) peak was larger than the second (active) peak (1479 vs. 1295 N), while the active peak was smaller than the impact peak for both SM and DF (SM: 1115 vs. 1482 N and DF: 1172 vs. 1579 N).

The highest stresses were observed in the anterior acetabulum during the impact peak using DF model inputs (14.1 MPa). Compared to the generalized data, contact occurred more anteriorly in both the SM and DF models (Figure 3). For all three modes of kinematic and kinetic model inputs, more of the labrum was in contact during the active peak (14-26% of the overall contact area) than the impact peak (4-7% of the overall contact area). The difference between labrum contact for the loading scenarios was larger for the subject-specific boundary conditions (16% difference for SM and 22% for DF) than for the generalized boundary conditions (7% difference).

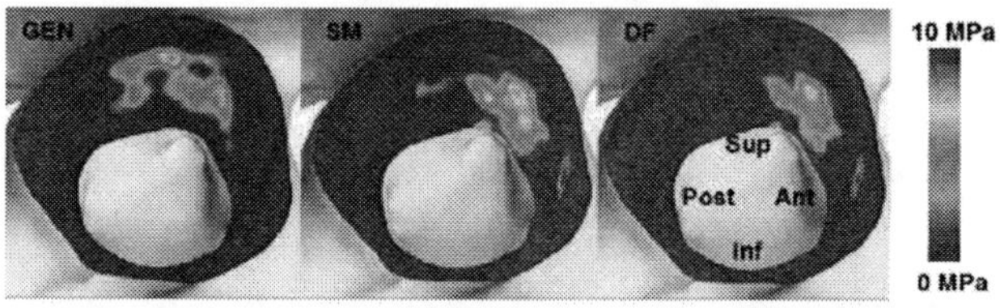

Figure 3: Contact stress during active peak for models driven using kinematics and joint reaction forces from generalized data (GEN), skin markers (SM), and dual-fluoroscopy (DF).

DISCUSSION

From this case study, subject-specific kinematics and calculated JRF model inputs from SM and DF resulted in altered magnitude and location of maximum cartilage contact stresses. This data may be crucial to our understanding the cause of OA of the hip. Future efforts will be focused on expanding the analysis to include additional subjects and activities to determine whether the differences between the subject-specificity of SM and DF model inputs are clinically relevant and whether these results are applicable to other patient populations.

ACKNOWLEDGEMENTS

Financial support was provided by the National Institutes of Health (R21-AR063844, S10-RR026565) and LS Peery Discovery Program in Musculoskeletal Restoration.

REFERENCES

[1] Buckwalter, J.A. et al. Cartilage. 4(4):286-294, 2013.
[2] Fiorentino, N.M. et al. Gait Posture, 50:246-251, 2016.
[3] Kapron, A.L. et al. J Appl Biomech, 30(3):461-470, 2014.
[4] Anderson, A.E. et al. J Biomech, 43(7):1351-1357, 2010.
[5] Bergmann, G. et al. J Biomech, 34(7): 859-871, 2001.
[6] Ng, K.C. et al. Clin Orthop Relat Res. 2016. [In press]

THREE DIMENSIONAL MEASUREMENT OF METATARSAL PRONATION IN PATIENTS WITH HALLUX VALGUS

Bradley C. Campbell (1), Stephen F. Conti (2), Mark Carl Miller (1, 3, 4)

(1) Dept. of Mechanical Eng. & Materials Science
University of Pittsburgh
Pittsburgh, PA, USA

(2) Orthopaedic Practices
UPMC
Pittsburgh, PA, USA

(3) Department of Bioengineering
University of Pittsburgh
Pittsburgh, PA, USA

(4) Allegheny General Hospital
Pittsburgh, PA, USA

INTRODUCTION

Hallux valgus (HV), the most common forefoot deformity, is a painful protrusion of the first metatarsophalangeal joint. As the deformity progresses, the metatarsal dorsiflexes, abducts and pronates[1]. The standard 2D radiographs typically employed in clinical settings cannot accurately capture this pronation. CT scans feature 3D data, but the images are usually presented as planar projections and associated solid model representations do not quantify the bony structure. Decomposition of rotation matrices based on bodily landmarks readily provides the three angles needed to describe HV deformities. However, the common Grood-Suntay[2] decomposition is based on the relationship between two articulating long bones, a situation that does not apply to the intermetatarsal angle used to compare HV deformities. In order to assess the 3D HV deformity with a 3D technique and still provide clinically relevant description with varus/valgus (abduction/adduction) and internal/external rotation (supination/pronation), we sought a rotation sequence and a reference position that fit with standard clinical practice and led to a standard mathematical decomposition. Fortunately, standard clinical practice considers the second ray as a relatively rigid keel in the forefoot. In that HV is primarily a forefoot deformity and that the second metatarsal forms the foundation for the forefoot, the metatarsals offer the landmarks for the establishment of the axes needed for a local HV coordinate system. Here, we posit our methodology is sufficient to create axes for the description of HV and this description encompasses standard clinical measurements and fits standard clinical terminology. We sought to find good agreement between the calculated angular orientations and traditional clinical measures. The current work sought to quantify pronation of the first metatarsal relative to the second metatarsal and of the first phalanx of the first toe relative to the first metatarsal.

METHODS

Three-dimensional models were reconstructed from loaded and unloaded CT images (10 HV, 10 normal) patients enrolled in an IRB approved study. The orientation of specific bones was determined by selecting landmarks on the surface of the bones (Fig. 1).

First, the coordinate data from the selected landmarks were used to calculate the dot product of the long axes of the bones. The resulting angles, the first metatarsal relative to the second and the first phalanx relative to the first metatarsal, were compared against traditional (X-ray and CT Plane) hallux valgus (HVA) and intermetatarsal (IMA) angles taken in clinical setting of the same patient groups (Figure 2).

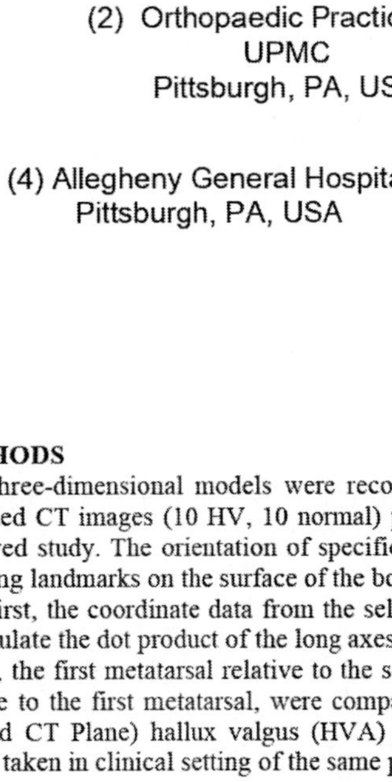

Figure 1: The most medial and lateral points on the head and base of the first metatarsal, the second metatarsal and the proximal first phalanx were selected and located in the solid models.

Second, coordinate system triads were computed from the landmark selection data and the rotation matrix representations of the first metatarsal relative to the second, and the first phalanx relative to the first metatarsal were created (Figure 3).

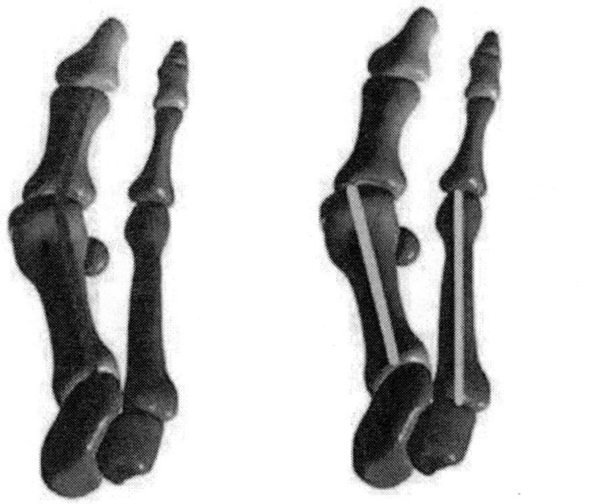

Figure 2: HVA (left) and IMA (right) angles. In X-ray and CT, lines are drawn along the axial direction and the angle between them. Our DP angle was compared against them measures.

The resulting decomposition calculations, resolved in the X-Z-Y sequence, output a set of angles to determine the pronation/supination of the first metatarsal relative to the second (IMA Pronation), and of the first phalanx relative to the first metatarsal (HVA Pronation). The loaded and unloaded models were compared to determine the changes in angulation due to weightbearing.

A regression analysis was conducted to examine the relationship between adduction and pronation (known here as intermetatarsal and pronation).

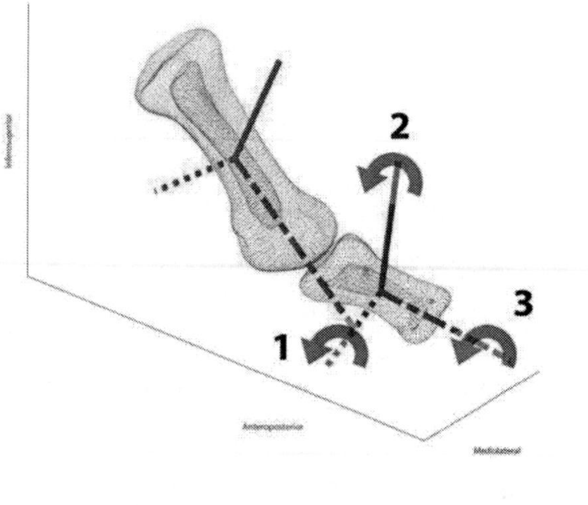

Figure 3: Coordinate system triads were calculated for each bone. The resulting rotation matrix calculated to align the triads was decomposed to find the set of angles to measure pronation. The axes were defined as X – mediolateral, Y - anteroposterior and Z - inferosuperior directions.

RESULTS

The average pronation of the first metatarsal relative to the second metatarsal was 19.8 ± 7.1 and 28.3 ± 10.8 in the normal and HV groups respectively ($p < 0.05$) (Figure 4). The differences in HV angles and IM angles measured by dot product between normal and HV patients were 22° and 7° respectively (Table 1). The measures from the 3-D reconstructions were well correlated with the standard X-ray measures as well as with measures directly from the CT scans. The influence of weightbearing demonstrated pronation angle differences that were greater in the normal group than in the HV group for both the IM pronation and the HV pronation ($p < 0.05$) (Figure 1). The regression analysis of the pronation and intermetatarsal angle was not found to be significant, with a weak correlation ($r^2 = 0.26$).

Table 1: The group averages are shown below for the correlational analysis of the conventional methods of angle measurement and dot product (DP) of our landmark selection data. For all angles, r > 0.75.

HV Angle	**X-ray**	**CT**	**DP**
HV Group	32.1 ± 10.5	31.9 ±10.3	36.0 ± 10.8
Normal Group	12.7 ± 3.6	15.3 ± 4.5	13.2 ± 3.0
IM Angle			
HV Group	16.0 ± 2.5	15.9 ± 4.1	17.7 ± 2.5
Normal Group	9.9 ± 1.4	9.2 ± 2.3	10.9 ± 1.7

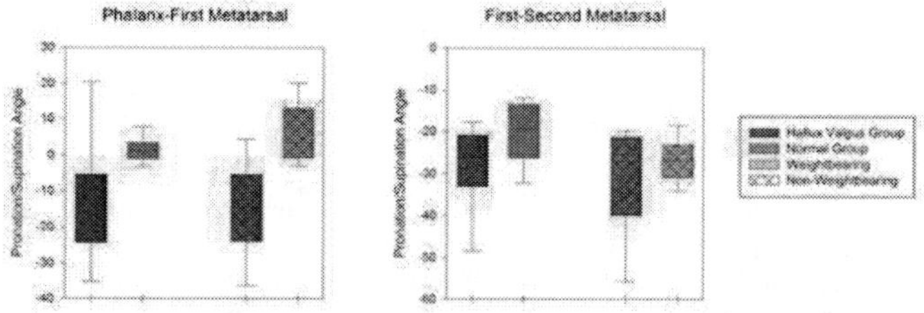

Figure 4: The group averages are shown above for the HVA Pronation and IMA Pronation. The groups (HV or normal) are denoted by color and the conditions (weightbearing or non-weightbearing) are denoted by fill.

DISCUSSION

This study is the first to report a value of pronation in HV feet from 3D based analysis. The pronation of the first metatarsal relative to the second metatarsal between normal and HV patients is significantly different. While the first metatarsal had measurable pronation in patients with hallux valgus, that value was not predicted by the IMA regression. This measure has some consequence as current clinical imaging techniques include X-ray only capabilities. Therefore, it is not possible for clinicians to estimate the degree of pronation by the IMA measured on plan film X-rays. The findings of this study indicate pronation should be considered in any surgical technique that seeks to restore native configurations.

ACKNOWLEDGEMENTS

This work was completed through funding from the Orthopaedic Practices group at the University of Pittsburgh Medical Center.

REFERENCES

[1] Perera, et al., J *Bone Joint Surg Am* 93.17, 2011.
[2] Grood, E. S., & Suntay, W. J., *J Biomech Eng*, 105.2, 1983.

POST-OPERATIVE EFFECTS OF ALTERING FLEXION AND EXTENSION GAPS DURING TOTAL KNEE ARTHROPLASTY: A FINITE ELEMENT STUDY

Ruth A. Solomon, Andrew L. Sori, Shihab Asfour, Francesco Travascio

Department of Industrial Engineering
University of Miami
Coral Gables, FL, USA

INTRODUCTION

Total knee arthroplasty (TKA) is a procedure that is frequently used to relieve pain and restore function in severely arthritic and injured knees. A major goal of modern TKA is to improve range of motion (ROM) while preserving joint stability. Maintaining equal flexion and extension gaps (see fig. 1)[1] has been a commonly recognized surgical goal, as it has been believed to reduce instability or incidence of stiffness. Nonetheless, a recent study on 404 post-operative TKA patients showed that controlled flexion gap increases of 2.5 mm or could increase postoperative flexion and patient satisfaction.[2] In addition, several comparative clinical studies have indicated that slightly lax knees may result in increased patient satisfaction, while a biomechanical cadaver study showed that a 2mm controlled increase in flexion gap lead to a reduction in tibiofemoral force beyond 90 degrees of passive flexion, with no significant affect to medial and lateral ligament strain. [2,3] Because little research has been done to confirm the cadaver results and most of the studies conducted did not intentionally increase flexion gap, further evaluation is necessary to determine whether or not deliberately increasing flexion gaps is worthwhile, and if so, by what margin the flexion gaps should be increased. The objective of this study is to perform computational analysis on a generic bone model in order to compare the distribution of stresses in the distal femur after a TKA with equal flexion and extension gaps to those of a TKA with 1 mm controlled increases in flexion gap.

METHODS

A CT scan of a generic femur bone model was converted into a 3D geometry, including both cortical and cancellous bone using Mimics Innovation Suite (Materialise, Belgium). With the help of a surgeon, the femur was fitted with a modeled TKA implant and aligned with flexion gaps 0, 1, 2, and 3mm larger than the extension gap. Both implant and bone mechanical properties were derived from literature. [4-6] With the distal tibia fixed, a static load of approximately 700N was applied to the top of the femur to simulate a body load. A tied contact was applied to the tibial-femoral implant interface and 2 scenarios were simulated for each flexion gap configuration: (1) with the leg at full extension; (2) with the leg at 90 degrees flexion (fig. 2 shows the contact between the tibial and femoral implants at full extension). The computational analysis was conducted using FEBio (University of Utah) to determine the resulting peak effective stresses in the distal femur.

RESULTS

When a body load was applied at 90 degrees flexion, the peak effective stresses in the distal femur consistently decreased with increased flexion gap (see figure 3). Increasing the flexion gap to 3mm larger than the extension gap resulted in a peak effective stress nearly three times lower (61.33MPa) than when the flexion gap was held equal to the extension gap (176.6MPa). Additionally, these peak effective stresses consistently occurred in the intercondyloid fossa region of the distal femur.

On the other hand, when the load was applied with the knee model at full extension, the peak effective stresses within the distal femur also initially decreased with increased flexion gap up until the flexion gap became 2 mm larger than the extension gap (from 280.6MPa to 42.13MPa), at which point the peak stress then began to increase (to 61.33MPa at 3mm increased flexion gap).

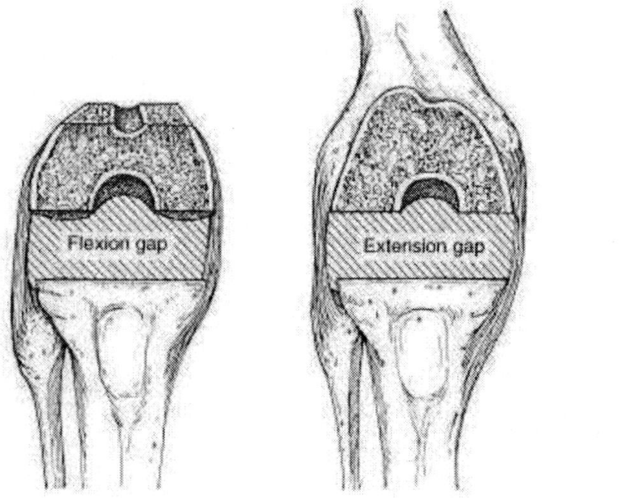

Figure 1: Sketch of the flexion gap and the extension gap cuts for a TKA [1].

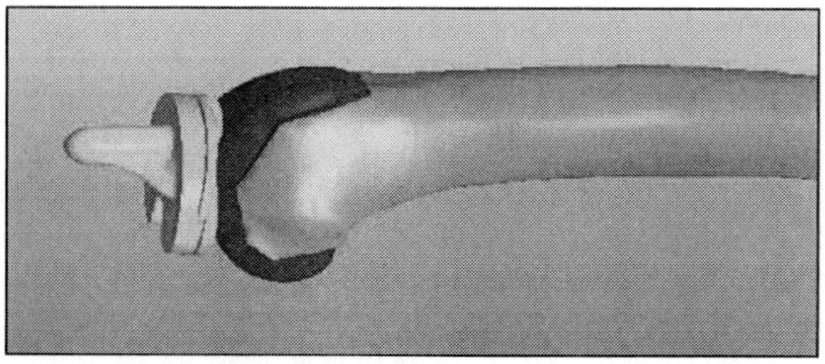

Figure 2: Femur (green) with femoral implant (blue) in alignment with the tibia implant (yellow) when the knee is in full extension.

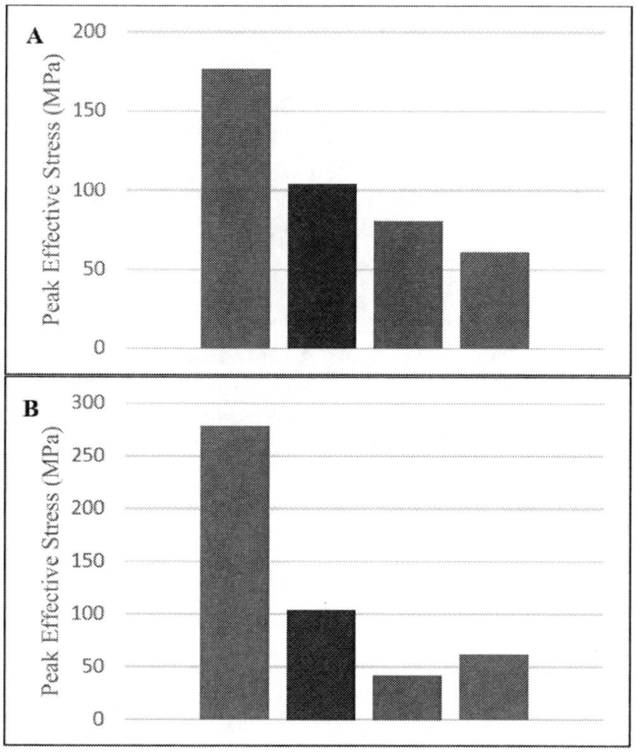

Figure 3A: Peak effective stress of different flexion gaps during flexion. Figure 3B: Peak effective stress of different flexion gaps during extension. Flexion gaps were evaluated at 0 mm (blue), 1 mm (red), 2 mm (green) and 3 mm (orange) greater than extension.

DISCUSSION

A common goal when performing a total knee arthroplasty is to produce equally balanced flexion and extension gaps; however, recent studies indicate that an increase in flexion gap may result in an increased post-operative flexion and improved patient satisfaction, without significantly effecting medial and lateral ligament strain.[2,3] The results of this study indicate that controlled increases in flexion gap may result in decreased peak effective stresses in the distal femur under a body load when the knee is in 90-degree flexion. Additionally, lower peak stresses were found in the distal femur when a body load was applied to a knee in full extension up to 2mm, after which the peak effective stress began to increase. This information may be necessary when considering future studies on increasing flexion gap and the margin of said increase, as larger peak stresses may result in polyethylene wear or an increased prominence of femoral fracture, overall contributing to a decrease in the success of long term TKAs.

REFERENCES

[1] Yasgur, D, et al. Knee Arthroplasty. 2016
[2] Ismailidis, P et al. *Knee Surg Sports Traumatol Arthrosc*, 2016. p. 1-7.
[3] Jeffcote, B et al., *J Bone Jt Surg Br*, 2007. 89(11): p. 1528-2533.
[4] Cuppone, M et al. *Calcif Tissue Int,* 2004. 74(3): p. 302-309.
[5] Wirtz, D et al. *J Biomech*, 2000. 33(10): p. 1325-1330.
[6] Zalzal P, et al. *J of Arthro*, 2006. 21(5): p. 737-743.

SB³C2017
Summer Biomechanics, Bioengineering and Biotransport Conference
June 21 – 24, Tucson, AZ, USA

COMPUTATIONAL ANALYSIS OF THE CHANGES IN INTRADISCAL PRESSURE AT ADJACENT SEGMENTS AFTER POSTERIOR FIXATION FOR BURST FRACTURE

S. Elmasry, S. Asfour, F. Travascio

Department of Industrial Engineering
University of Miami
Coral Gables, FL, USA

INTRODUCTION

Approximately 90% of spinal fractures occur at the thoracolumbar junction. Among these fractures, 10% to 20% are considered burst fracture (BF) [1]. In such cases, posterior fixation is commonly performed as a surgical treatment, especially when the fracture is moderate and associated with partial neurologic dysfunction [2]. Different posterior fixation constructs can be utilized such as short and long constructs with/without the inclusion of the level of fracture [3]. However, the knowledge of the biomechanical implications of such constructs on the adjacent segments is limited. It is believed that spinal fusion is associated with signs of disc degeneration at the adjacent segments [4]. Moreover, both *in-vitro* and numerical studies have shown that spinal fusion lead to increased intradiscal pressure (IDP) at the adjacent segments [5,6]. To our best knowledge, no study compared the IDP alteration at the adjacent segments due to long or short posterior fixation used in the treatment of BF. Also, the consequence of adding pedicle screws at the level of fracture, in both procedures, on the IDP at the adjacent segments is unknown. Accordingly, the objective of this study is to measure the changes of IDP at the adjacent segments after performing short/long posterior fixation with/without the inclusion of the level of fracture.

METHODS

A 3D nonlinear finite element (FE) model of the thoracolumbar spine (from T11 to L4) was reconstructed from a CT scan of a healthy male subject. Lumbar vertebrae were modeled as inner trabecular bone enclosed by an outer cortical shell. The intervertebral discs were assumed inhomogeneous biphasic media and composed of nucleus pulposus and fiber-reinforced annulus fibrosis [7]. Major spine ligaments were included in the model. Standard pedicle screws, posterior rods, and cross-links were defined as reported in the literature [8]. Four different fixation constructs were modeled: two short-segment constructs (T12-L2); one without pedicle screws at L1 (SSPF) and the other with pedicle screws at L1 (SSPFI), and two long-segment constructs (T11-L3); one without pedicle screws at L1 (LSPF) and the other with pedicle screws at L1 (LSPFI), see Figure 1. For all the simulated cases, the BF was simulated at the vertebral body of L1. All the constructs' models and the intact one were tested by applying flexion, extension, axial and lateral bending moments as loading conditions. The models were validated by comparing their predicted ROM of T12-L2 to data reported from an *in vitro* study [9]. The IDP at T12-L1 and L1-L2 discs was measured after applying each loading condition and compared.

Poster Presentation #P150

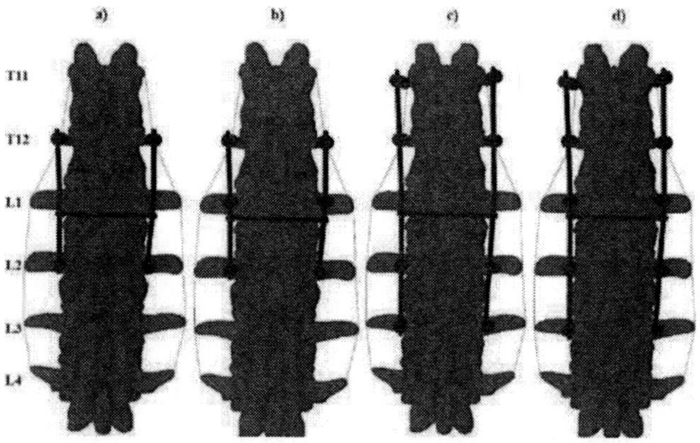

Figure 1: FE models of the fixation constructs: (a) SSPF; (b) SSPFI; (c) LSPF; (d) LSPFI

RESULTS

Models' predictions of angular ROM of T12-L2 junctions were in agreement with the experimental data with the only exception of LSPF and LSPFI in lateral bending (Figure 2).

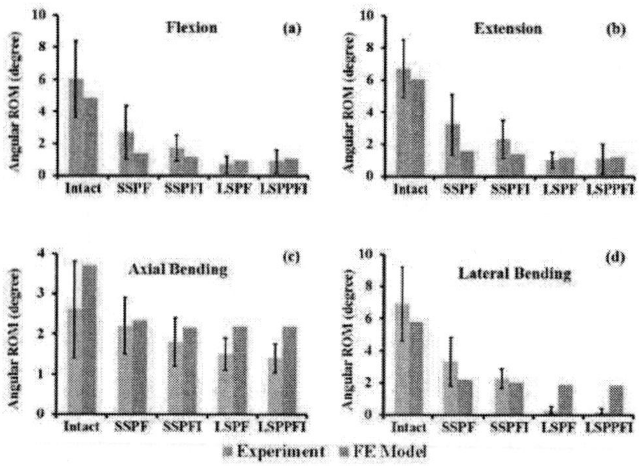

Figure 2: Models validation by comparing angular ROM of T12-L2 to in vitro results

The magnitudes and the distribution of the intradiscal pressure were analogous across the constructs, with only two exceptions during extension and axial bending: at T12-L1, SSPF generated 10% higher pressure than SSPFI (Figure 3), and at L1-L2, SSPF generated 15% lower pressure than SSPFI (Figure 4). When compared to the values attained in the intact spine, the IDP at T12-L1 increased up to 500% during flexion and 60% during axial bending, and decreased to 50% in lateral bending; no major changes were observed in extension (Figure 3). Also, at L1-L2 disc, the intradiscal pressure increased in flexion (~300%), axial bending (~200%) and extension (~150%), and decreased in lateral bending (75%) when compared to the intact case (Figure 4).

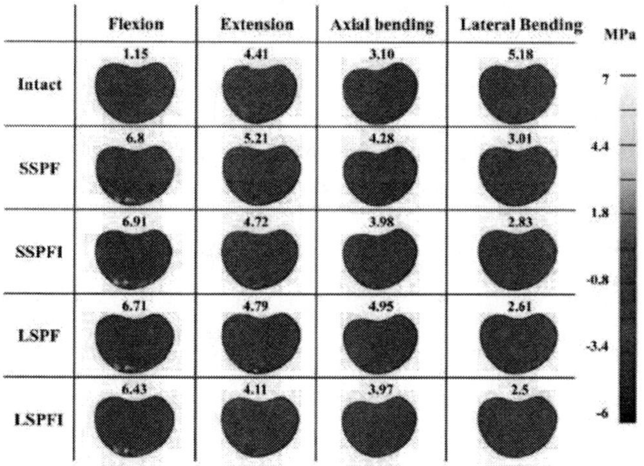

Figure 3: Intradiscal pressure distribution at T12-L1 IVD. The maximum pressure is reported in MPa.

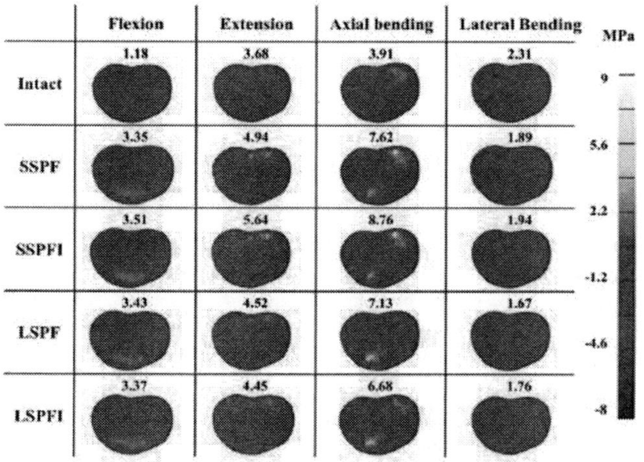

Figure 4: Intradiscal pressure distribution at L1-L2 IVD. The maximum pressure is reported in MPa.

DISCUSSION

Deleterious biomechanical changes on the adjacent segments after fusion may lead to adjacent segment disease [10]. The increase in magnitude of the IDP at adjacent segments provides an indication of stability of the anterior column at the level of fracture [11]; however, it may trigger a degenerative process due to overloading the adjacent discs [12]. In this analysis, the IDP magnitudes and distributions at the adjacent segments after performing four different fixation procedures for treatment of BF were investigated. The results show no major changes of the IDP magnitude across the investigated constructs. However, they were all larger than that attained in the intact spine for all tested loading conditions with the exception of lateral bending. These findings suggest that long and short segment constructs demonstrate similar risk on the health of the adjacent segments by increasing their IDP. Moreover, it indicates that adding pedicle screws at the fracture level induces minor changes on the IDP at the adjacent segments.

ACKNOWLEDGEMENTS

Study supported by funds donated to the Biomechanics Research Group of the University of Miami.

REFERENCES

[1] Dai, L et al., *Surg neurol,* 67: 221-231, 2007.
[2] Scholl, B et al., *Orthop*, 29, 2006.
[3] McLain, R *Spine*, 31:70-79, 2006.
[4] Mannion, A et al., *Spine*, 39:1373-83, 2014.
[5] Hilibrand, A & Robbins, M, *Spine J*, 4:190-194.
[6] Park, P et al., *Spine*, 29:1938-44, 2004.
[7] Travascio, F et al, *J Spine*, 4, 2015.
[8] Park, W et al., *J Orthop Sci*, 14: 443-49, 2009.
[9] Baaj, A et al., *J Neurosurg: Spine*, 14: 192-197, 2011.
[10] Rahm, MD & Hall, BB, *J Spinal Dis*, 9:392-400, 1996.
[11] Mahar, A et al., *Spine*, 32: 1503-1507, 2007.
[12] Stokes, I & Iatridis, J, *Spine*, 29:2724-32, 2004.

AN EFFICIENT NUMERICAL INTEGRATION METHOD FOR NON-LINEAR VISCOELASTIC MODELING

Nicole L. Ramo (1), Kevin L. Troyer (2), Christian M. Puttlitz (1,3)

(1) School of Biomedical Engineering
Colorado State University
Fort Collins, CO, USA

(2) Component Science and Mechanics
Sandia National Laboratories
Albuquerque, NM, USA

(3) Department of Mechanical Engineering
Colorado State University
Fort Collins, CO, USA

INTRODUCTION

Viscoelastic theory describes the time-dependent relationship between stress and strain. For viscoelastic materials, the current stress is dependent upon all previous loading events. This history-dependent behavior complicates numerical analyses of viscoelastic materials because the stress at each step throughout the load history must be computed and stored in order to obtain the current stress. For three-dimensional finite element models, computing and storing the stress tensor at each integration point and time step quickly becomes intractable. To simplify numerical analyses of viscoelastic materials, it is common to use a discrete series of exponentials (such as Prony series) to describe the time-dependent relationship between stress and strain.

Accordingly, we have developed a novel numerical integration technique that computes the current stress from a state variable stored from the preceding time step, avoiding the need to store the stress at each time point of the analysis. In this study, we derive the integration technique, implement it using computational methods, and verify its ability to fit non-linear viscoelastic data by recovering the parameters used to create input data.

METHODS

Uniaxial non-linear viscoelastic material behavior may be represented by the hereditary (or convolution) integral:

$$\sigma(\mathrm{t}, \varepsilon) = \int_0^t E(t-\tau, \varepsilon)\frac{d\varepsilon(\tau)}{d\tau}d\tau \qquad (1)$$

where σ is stress, ε is strain, t is time, $E(t,\varepsilon)$ is the material's relaxation modulus, and τ is a time variable of integration representing the history effect. The relaxation modulus for non-linear viscoelastic materials is time and strain-dependent, and must be continuous and monotonically decreasing. When modeling biological materials, we may approximate the non-linear relaxation modulus using a discrete Prony series:

$$\mathrm{E}(\mathrm{t}, \varepsilon) = E_\infty(\varepsilon) + \textstyle\sum_{i=1}^N E_i(\varepsilon)e^{\frac{-t}{\tau_i}} \qquad (2)$$

where $E_i(\varepsilon)$ is the strain-dependent Prony weight corresponding to time constant τ_i, $E_\infty(\varepsilon)$ represents the long-term strain-dependent modulus, and N defines the finite number of exponential Prony terms. Combining the above equations and simplifying the expression yields:

$$\sigma(\mathrm{t}, \varepsilon) = \int_0^t \left(\textstyle\sum_{i=1}^N E_i(\varepsilon)e^{\frac{-(t-\tau)}{\tau_i}}\right)\frac{d\varepsilon(\tau)}{d\tau}d\tau + E_\infty(\varepsilon)\cdot\varepsilon(t) \qquad (3)$$

A history state variable at the current time t can then be defined as:

$$h_i(\mathrm{t}, \varepsilon) = \int_0^t \left(E_i(\varepsilon)e^{\frac{-(t-\tau)}{\tau_i}}\right)\frac{d\varepsilon(\tau)}{d\tau}d\tau \qquad (4)$$

such that equation (3) becomes:

$$\sigma(\mathrm{t}, \varepsilon) = \textstyle\sum_{i=1}^N h_i(t,\varepsilon) + E_\infty(\varepsilon)\cdot\varepsilon(t) \qquad (5)$$

The history state variable can then be used to recursively update the stress over time. The updated history state variable at the next time step, $t+\Delta t$, is:

$$h_i(\mathrm{t}+\Delta\mathrm{t}, \varepsilon) = \int_0^{t+\Delta t} \left(E_i(\varepsilon)e^{\frac{-(t+\Delta t-\tau)}{\tau_i}}\right)\frac{d\varepsilon(\tau)}{d\tau}d\tau \qquad (6)$$

Using the summation rule for definite integrals and combining the result with equation (5) yields the following expression for the updated stress at the next time step:

$$\sigma(t+\Delta t,\varepsilon) = \int_0^t \left(\sum_{i=1}^N E_i(\varepsilon) e^{\frac{-(t+\Delta t-\tau)}{\tau_i}} \right) \frac{d\varepsilon(\tau)}{d\tau} d\tau + \int_t^{t+\Delta t} \left(\sum_{i=1}^N E_i(\varepsilon) e^{\frac{-(t+\Delta t-\tau)}{\tau_i}} \right) \frac{d\varepsilon(\tau)}{d\tau} d\tau + E_\infty(\varepsilon) \cdot \varepsilon(t+\Delta t) \qquad (7)$$

where the first integral term represents the effect of the past loading events, the second integral term represents the effect of the current loading event, and the final term represents the effect of the material's long-term (equilibrium) response. The preceding history state variable, $h_i(\mathrm{t},\varepsilon)$, can be factored out of the first term of (7) to yield:

$$\int_0^t \left(\sum_{i=1}^N E_i(\varepsilon) e^{\frac{-(t+\Delta t-\tau)}{\tau_i}} \right) \frac{d\varepsilon(\tau)}{d\tau} d\tau = \sum_{i=1}^N h_i(\mathrm{t},\varepsilon) e^{\frac{-\Delta t}{\tau_i}} \qquad (8)$$

Application of the second mean-value theorem for integrals and the central difference rule, results in the following expression for the second term of (7):

$$\int_t^{t+\Delta t} \left(\sum_{i=1}^N E_i(\varepsilon) e^{\frac{-(t+\Delta t-\tau)}{\tau_i}} \right) \frac{d\varepsilon(\tau)}{d\tau} d\tau = \sum_{i=1}^N \frac{E_i(\varepsilon)\tau_i \Delta\varepsilon}{\Delta t} \left(1 - e^{\frac{-\Delta t}{\tau_i}}\right) \qquad (9)$$

Combining (8) and (9) results in the final expression for the updated stress:

$$\sigma(\mathrm{t}+\Delta\mathrm{t},\varepsilon) = \sum_{i=1}^N \left[h_i(\mathrm{t},\varepsilon) e^{\frac{-\Delta t}{\tau_i}} + \frac{E_i(\varepsilon)\left(1 - e^{\frac{-\Delta t}{\tau_i}}\right)}{\Delta t / \tau_i} \Delta\varepsilon \right] + E_\infty(\varepsilon) \cdot \varepsilon(t+\Delta t) \qquad (10)$$

From equation (10), it is important to note that: (i) only the history state variable $h_i(\mathrm{t},\varepsilon)$ from the previous time step needs to be stored in order to compute the updated stress, and (ii) this formulation may be fit to an arbitrary strain history (e.g. stress relaxation or cyclical).

RESULTS

In order to verify the model described above, the ability of equation (10) to recover input parameters was examined. Artificial experimental data were created by prescribing the form and coefficients of the relaxation modulus $\mathrm{E}(\mathrm{t},\varepsilon)$. For this study, a 4-term Prony series was used to approximate the relaxation modulus where the strain-dependent weights were represented by a 2-term polynomial function:

$$E_\infty(\varepsilon) = C_1^\infty \varepsilon + C_2^\infty \varepsilon^2 \qquad (11)$$

$$E_i(\varepsilon) = C_1^{\tau_i} \varepsilon + C_2^{\tau_i} \varepsilon^2 \qquad (12)$$

Decadal time values were chosen for the time constants ($\tau_1 = 0.1\ \mathrm{s}, \tau_2 = 1\ \mathrm{s}, \tau_3 = 10\ \mathrm{s}, \tau_4 = 100\ \mathrm{s}$). Therefore, to create experimental data curves, a total of 10 fitted parameters along with the specific strain-history for each curve must be set. The input fitted parameters chosen for this study were those obtained from viscoelastic modeling of ovine Achilles tendon [1]. Once the artificial experimental stress-strain data were created, Matlab's fmincon function was used to find the 10 coefficients that minimized the sum of squared residuals.

Stress-Relaxation Data: Six stress-relaxation data curves were created based on experimental protocols previously utilized by our laboratory (10% ε/s ramp rate; a 100 s total test time; and strain magnitudes: 1-6%; [1,2]). In order to facilitate efficient fitting of stress-relaxation data, a Heaviside step assumption (HSA) was used to provide initial guesses for the fit of the entire curve (Table 1).

Table 1: Fitting results for stress-relaxation data (units MPa).

	Input Parameters from [1]	HSA Fitted Parameters	Final Fitted Parameters	% Error of Final Parameters
$C_1^{\tau_1}$	901.1	325.8	901.0	**0.01**
$C_2^{\tau_1}$	8437.0	- 2282.6	8439.6	**0.03**
$C_1^{\tau_2}$	343.5	182.3	343.6	**0.03**
$C_2^{\tau_2}$	- 684.1	- 811.9	- 686.48	**0.35**
$C_1^{\tau_3}$	331.2	179.2	331.1	**0.03**
$C_2^{\tau_3}$	- 1201.2	- 608.5	- 1198.8	**0.20**
$C_1^{\tau_4}$	476.5	261.0	476.6	**0.03**
$C_2^{\tau_4}$	- 363.4	- 388.2	- 366.6	**0.88**
C_1^∞	4403.1	- 948.2	4403.1	**0.0008**
C_2^∞	- 9959.3	4089.8	- 9958.6	**0.01**

Cyclic Data: Four single-cycle curves were created to examine the model's ability to fit dynamic data, another common experimental test for evaluating viscoelastic tissues. Each curve followed a cosine pattern to a maximum of 6% strain at a frequency given by $f_i = 1/2\pi\tau_i$. As shown in Figure 1, the viscoelastic modeling method fit the data very well and recovered the input coefficients with relatively small errors.

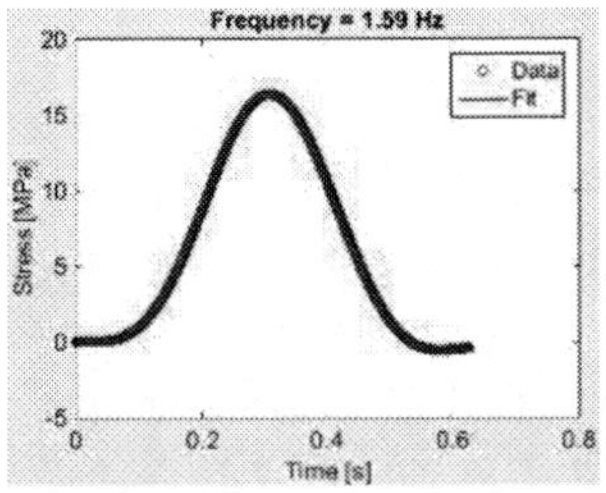

	% Error		**% Error**
$C_1^{\tau_1}$	**0.0003**	$C_1^{\tau_3}$	**0.0056**
$C_2^{\tau_1}$	**0.0007**	$C_2^{\tau_3}$	**0.0422**
$C_1^{\tau_2}$	**0.0025**	$C_1^{\tau_4}$	**0.0150**
$C_2^{\tau_2}$	**0.0345**	$C_2^{\tau_4}$	**0.5561**
C_1^∞	**0.0007**	C_2^∞	**0.0058**

Figure 1: Fitting results for cyclic data (representative plot shown).

DISCUSSION

This work presents a new method of modeling non-linear viscoelasticity which does not require storage of the stress at each time step of the loading history. Very strong recovery of all ten input parameters validates this new method for both stress-relaxation and cyclic experimental data. In addition, it is interesting to note that recovery of the input parameters is consistently better for cyclic data compared to stress relaxation data. This finding suggests that cyclic experiments may be better posed for fitting this viscoelastic formulation. In the future, we will: (i) investigate parameter sensitivity to different experimental loading conditions, and (ii) apply the technique developed herein to characterize the viscoelastic behavior of spinal cord and meningeal tissues to improve spinal cord injury models.

ACKNOWLEDGEMENTS

Sandia National Laboratories is a multi-mission laboratory managed and operated by Sandia Corporation, a wholly owned subsidiary of Lockheed Martin Corporation, for the U.S. Department of Energy's National Nuclear Security Administration under contract DE-AC04-94AL85000.

REFERENCES

[1] Troyer, KL et al. *J Biomech Eng,* 134:11, 2012.
[2] Shetye, SS et al. *Acta Biomaterialia.* 10:2 797-7, 2014.

INFLUENCE OF THE DISC HEIGHT AND ANNULUS FIBROSUS AREA OVER THE RANGE OF MOTION OF THE HUMAN SPINE, A PROBABILISTIC ANALYSIS

H. E. Jaramillo (1)

(1) Departmento de Energética y Mecánica
Universidad Autonoma de Occidente
Cali, Valle, Colombia

INTRODUCTION

Loads, geometry and material properties are uncertainties present in the biomechanics of the spine and many deterministic finite elements analysis has been carried out to understand their influence. However, few studies [1]–[9] have accomplished probabilistic analysis in order to obtain a more accurate estimate [10] of its influence on the final response [11]. In this direction, Spilker et al. [1]–[3] used a model of the spine under axial compressive load to find that increases in the disc height and the ratio of the nucleus area to disc area causes a reduction of the intradiscal pressure (IDP), the bulge, and the vertical displacement of the disc. Fagan et al. [5] found that the properties of the fibers of the annulus fibrosus have no significant effect on the disc stiffness under compression but it affects the flexural and torsional properties. Malandrino et al. [6] considered the pore-elastic effect between the annulus, nucleus pulposus, articular cartilage, and trabecular bone to conclude that the permeability is an important factor in the vertical displacements during compression, the IDP in compression and torsion, the pore pressure in torsion and the speed of fluid leaving through the vertebral bodies. Niemeyer et al. [8] used a simplified model of the L3-L4 segment to determine that the disc height, width, thickness and position of the facet joints have a great impact over IDP, range of motion (ROM) and facet joint contact forces. Finally, Thacker et al. [9] used a model of the C4-C6 segment, they found that the most important variables in the probabilistic response were the modulus of elasticity of the annulus fibrosus and the force-displacement curves of the flavum ligament, interspinous ligament and the capsular ligaments.

Considering that the intervertebral disc (IVD) have substantial variations of the geometrical properties among individuals and between levels of the spine, this paper presents probabilistic finite element (FE) analyses intended to assess the effects of the average disc height (h_p) and the area of the annulus fibrosus (AF) on the biomechanics of the lumbar spine. In particular, a finite element model of the annulus fibrosus was loaded under flexion, extension, lateral flexion and axial rotation was analyzed for different combinations of the average disc height and AF area (A) in order to obtain their effect over the ROM.

METHODS

We extracted the geometry of the L4-L5 annulus fibrosus from a realistic FE model [12]. To take into account the high stiffness of the vertebrae with respect to that of the AF, the endplates were considered to be rigid. The AF was represented as a homogeneous hyperelastic matrix reinforced with two families of fibers [13], [14]. The directions of the fibers varied with the radial coordinate from 25° on the outer boundary to 46° on the inner boundary, where this angle was measured with respect to the transverse plane [12], [13], [15]. Consistent with previous studies, a Yeoh energy function W_m was used for the matrix while a the hyperelastic function W_f was used for the fibers, as follows,

$$W_m = c_1(I_1 = 3) = c_2(I_1 = 3)^2 = c_3(I_1 = 3)^3 \tag{1}$$

$$W_f = \frac{a_1}{a_2}\left[e^{a_2(I_4=1)^2} = e^{a_2(I_6=1)^2}\right], \tag{2}$$

where c_1, c_2, c_3, a_1 and a_2 are material constants; I_1 is the first invariant of the deviatoric Green deformation tensor, and I_4 and I_6 are the deviatoric invariants associated with the two families of fibers.

A subroutine Uanisohyper of the program Abaqus 6.14.2 (Dassault Systemes SA, USA) was developed to implement the Equations (1) and (2). A set of 50 combinations of h_p (mean=18.1 mm, DS=3.5 mm) and A (mean=49.8%, DS=4.6%) was determined randomly using Excel (Microsoft, USA) according to a normal distribution. The h_p was calculated as the average of the anterior and posterior disc height and A as the percentage ratio between the area of the annulus fibrosus versus the total area of the disc. The constants a_1, a_2, c_1, c_2, and c_3 were equal to 2.1 MPa, 25.0 unitless, 0.035 MPa, 0.00065 MPa and 0.00045 MPa, respectively [12].

For each AF, the caudal endplate was fully restricted while a pure moment ramped from 0 to 8 N-m was applied to the cranial endplate in each of flexion, extension, lateral flexion, and axial rotation movements. A convergence analysis was performed to develop a final mesh of 9472 eight-node hybrid hexahedral elements (C3D8H).

The ROM was selected as the output variable. The ROM was plotted as a function of h_p and A. The ROM sensitivity to each of the input random variables (h_p, A) was calculated according to the equation proposed by Thacker et al. [9].

RESULTS

An increase of the average disc height (h_p) produces a increase of the ROM, but the ROM decreases when the annulus area (A) increase (Figures 1 and 2). In this direction, the correlation factors (CF) between the ROM and h_p were 0.999, 1.0, 1.0 and 0.975 for flexion, extension, lateral flexion and axial rotation respectively. The CF between the ROM and annulus area were -0.998, -0.993, -0.998, -0.976 for flexion, extension, lateral flexion and axial rotation respectively.

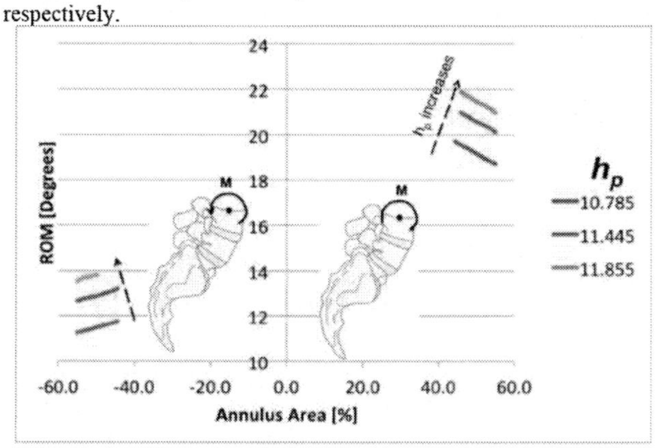

Figure 1: ROM versus annulus area (A) and the average disc height (h_p) in Flexion (left) and extension (right) at 8 N-m

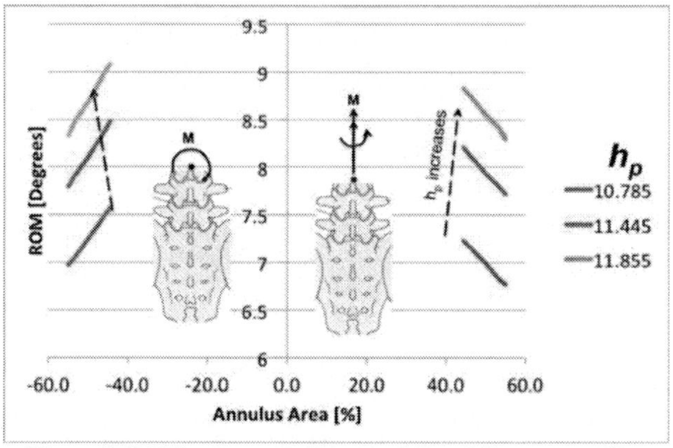

Figure 2: ROM versus annulus area (A) and the average disc height (h_p) in lateral flexion (left) and axial rotation (right) at 8 N-m

The ROM is most sensitive to h_p that A. The sensitivity factor calculated with h_p was between 1.20 and 1.90, while the ROM calculated with A was between -0.05 and -0.10 with A (Figure 3).

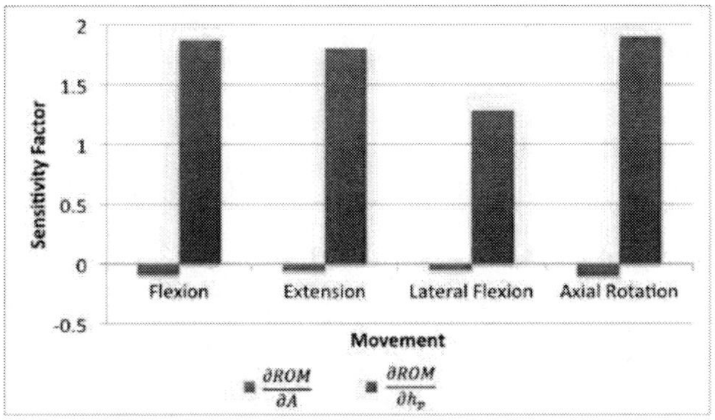

Figure 3: Sensitivity factor for each movement

DISCUSSION

The high level of uncertainty of the geometrical properties of the AF produced a significant dispersion in the biomechanical response, which cannot be fully described with a deterministic model [4], and also we can not generalize the results obtained. Particularly with the intervertebral disc, generally the geometrical properties (annulus area, anterior disc height, posterior disc height, et.) do not reported in the finite elements models of the human spine and these plays an important role in the biomechanics of the spine [8].

The strong influence the h_p over the ROM found in these analysis is agree with the study reported by Niemeyer et al. [8]. However, the influence of annulus area over the ROM could not compared with other study because these not found.

ACKNOWLEDGEMENTS

The author appreciates the support of the Universidad Autónoma de Occidente (Cali-Colombia).

REFERENCES

[1] Spilker, R. L. *J. Biomech.*, 13,10: 895–901, 1980.
[2] Spilker, R. L. et al. *J. Biomech.*, 17, 2:103–112, 1984.
[3] Spilker, R. L. et al. *J. Biomech. Eng.*, 108,1: 1–11, 1986.
[4] Rao, A. A. et al. *J. Biomed. Eng.*, 13, 2: 139–151, Mar. 1991.
[5] Fagan, M. J. et al. *Proc. Inst. Mech. Eng. [H]*, 216, 5: 299–314, 2002.
[6] Malandrino, A. et al. *J. Biomech.*, 42,16: 2780–2788, 2009.
[7] Rohlmann, A. et al. *Eur. Spine J.*, 18, 1: 89–97, 2009.
[8] Niemeyer, F. et al. *J. Biomech.*, 45, 8:1414–1423, 2012.
[9] Thacker, B. et al. *Abaqus*, 1, 1: 1–12, 2013.
[10] Madsen, H. O., et al. *Methods of Estructural Safety*. USA: Dover Publications Inc., 1986.
[11] Laz, P. J. et al. *Proc. Inst. Mech. Eng. [H]*, 224, 8: 927–943, 2010.
[12] Jaramillo, H. E. et al. *ACTA Bioeng. Biomech.*, 17, 2, 2015.
[13] Ayturk, U. M. et al. *J. Biomech. Eng.*, 132, 6: 061007–061007, 2010.
[14] O'Connell, G. D. *J. Biomech. Eng.*, 131, 11: 111007, 2009.
[15] Schmidt, H. *et al.*, *Clin. Biomech.*, 21, 4: 337–344, 2006.
[16] Cortes, D. H. et al. *J. Orthop. Res.*, p. n/a–n/a, 2013.

SB3C2017
Summer Biomechanics, Bioengineering and Biotransport Conference
June 21 – 24, Tucson, AZ, USA

CERVICAL SPINE FINITE ELEMENT MODEL WITH ANATOMICALLY ACCURATE ASYMMETRIC INTERVERTEBRAL DISCS

Jobin D. John (1,2), Mike W. J. Arun (1), Saravana Kumar G. (2), Narayan Yoganandan (1)

(1) Department of Neurosurgery
Medical College of Wisconsin
Mil□au□ee, Wisconsin, USA

(2) Department of Engineering Design
Indian Institute of Technology Madras
Chennai, Tamil Nadu, India

INTRODUCTION

Finite element (FE) models of the human cervical spine have been used to study the response under physiological and traumatic loading conditions. They include studies on the effects of degeneration, surgical intervention, artificial disc replacement, and rear impacts induced whiplash associated disorders. Some of them have also been incorporated into whole body FE models for simulating automotive crashes [1], [2]. It is important to accurately represent the geometry of spinal components and use appropriate material models to obtain realistic outputs from FE models. Considerable efforts have been advanced to better characterize the material definitions [3]; however, relatively little research has been conducted to incorporate accurate geometry of the components such as the intervertebral disc, the main transmitter of the load between vertebral bodies. Numerous existing FE models of the cervical spine have relied on definitions similar to the lumbar spine. These models have represented the annulus fibrosus as concentric layers of laminae.

Microdissection and morphological studies of the human cervical spine have shown that the annulus has a crescent shape in the anterior region which tapers laterally and forms a thin layer posteriorly [4]–[6]. The crescent shape of annulus imparts asymmetry to the disc. The effects of the asymmetry on flexion and extension responses have not been addressed. Therefore, the objective of the current study was to develop a cervical spine FE model with accurate geometrical definitions of the disc, validate its moment-rotation characteristics under flexion-extension using human cadaver experimental data, and determine the effects of asymmetry due to disc geometry.

METHODS

The FE model of the C2-C7 spinal column was developed using a mapping block-based hexahedral meshing technique [7]. The mesh was generated on geometry segmented from computed tomography images of a mid-size male spine. The soft tissues definitions included intervertebral discs (annulus fibrosus and nucleus pulposus), facet joints (articular cartilage, capsular ligaments, fluid), and ligaments (anterior longitudinal, posterior longitudinal, ligamentum flavum, and interspinous). Mapping block techniques were used to generate meshes for soft tissues also. Element types and material properties are given (Table 1). The mesh quality was evaluated using the aspect ratio and Jacobian. The thresholds for the aspect ratio of hexahedral and quadrilateral elements were 5 and 3. The threshold for the Jacobian was 0.6 for solid and shell elements.

The anterior heights of discs were greater than posterior heights. The spinal column had normal lordosis, typical of a healthy spine. The extent of the annulus and nucleus boundaries was defined based on measurements at two sagittal (SP1 and SP2), and two coronal planes (CP1 and CP2) (Fig.1) [5]. The resulting asymmetric mesh was compared with in-house cryomicrotome images of sagittal and coronal sections of healthy cervical spines. A concentric-symmetric disc was meshed to compare the responses between the two geometries.

Table 1: Element Types and material definitions

Component	Element type	Material
Cancellous	Hexahedral, under integrated	Linear elastic[8]
Cortical	Quadrilateral, shell	Linear elastic[9],[10]
AF ground	Hexahedral,	Hill foam[11],[12]
AF fibers	Quadrilateral, membrane	Nonlinear elastic [2], [13]
Nucleus	Hexahedral, constant stress	Viscoelastic fluid [14]
Facet fluid	Hexahedral, constant stress	Fluid [15]
Ligaments	Quadrilateral, membrane	Nonlinear elastic[2]

Flexion-extension loads (up to 2 Nm, each mode) were applied to the spinal column and moment-rotation responses were obtained for every segment of the asymmetric disc model. They were compared with experimental data [16]. The responses of the symmetric disc model under flexion-extension for C5-C6 disc segment were compared with the asymmetric disc model. FE preprocessor ANSA 16.1.0 (BETA CAE Systems S. A.) was used to generate the model and LS-DYNA R8.1.0 (LSTC) was used to solve the model.

RESULTS

The thickness of the anterior annulus in the disc FE mesh, on an average, was three times the posterior thickness at the mid-sagittal section (Fig. 2). The asymmetric and crescent shape of the annulus was modeled by implementing a C-type mesh transition in the lateral sides (Fig. 1). The mesh quality evaluations showed that 95% of hexahedral elements and 91.5% of quadrilateral elements had an aspect ratio less than 5 and 3, respectively, and 94.9% of hexahedral elements and 99.65% of quadrilateral elements had a Jacobian value of higher than 0.6.

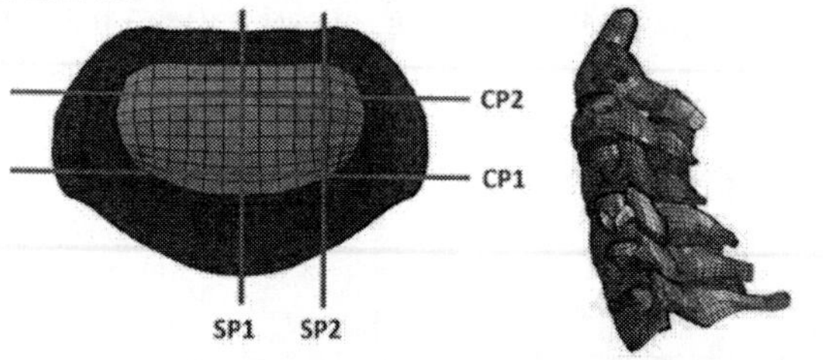

Figure 1: (Left) Top view of IVD, showing sagittal and coronal planes at which the annulus fibrosus boundaries were measured (Right) C2-C7 FE model

The moment-rotation responses of all the segments (Fig. 3) for the asymmetric model were within the experimental corridors (mean ±1 standard deviation), exhibiting non-linear response. The responses were stiffer in extension than flexion. The response of the asymmetric disc model was stiffer than the symmetric model by approximately 30% in extension, without considerable change in flexion. The patterns of the von Mises stresses in the endplates were also different between the two disc models (Fig. 4).

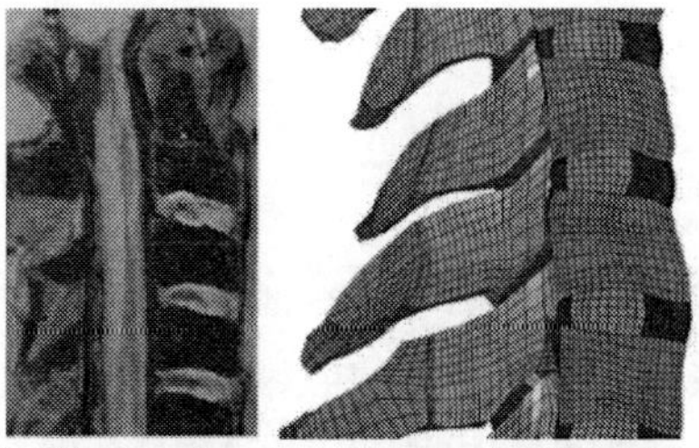

Figure 2: (Left) Crymicrotome image at mid-sagittal plane, and (Right) Mid-sagittal cross section of FE model

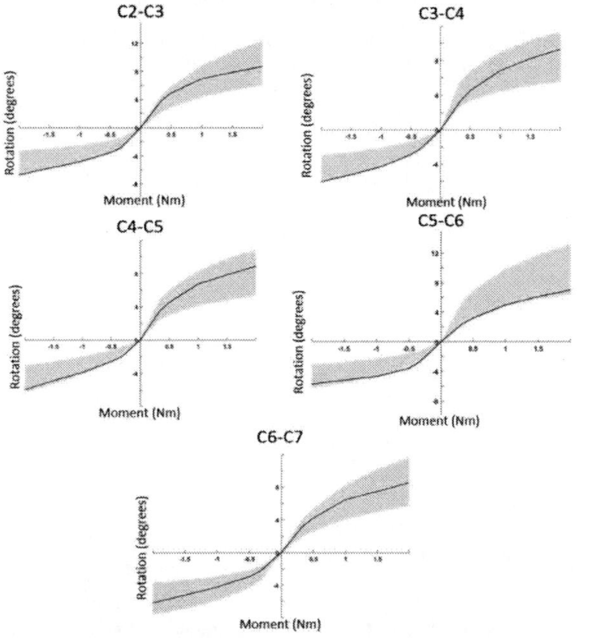

Figure 3: Moment rotation FE response curves (red), compared with experimental response corridors (shaded)

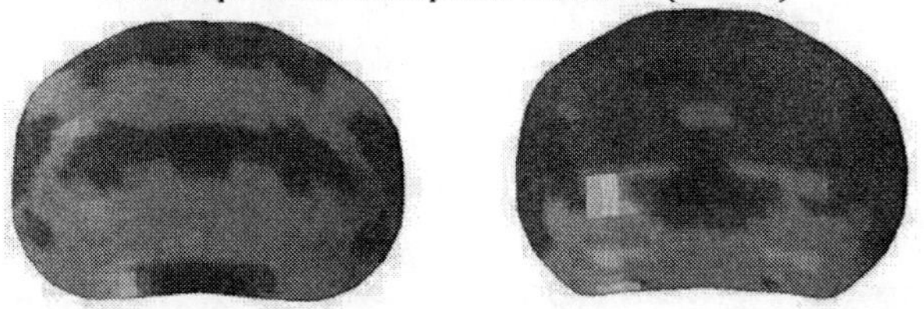

Figure 4: Comparison of von-Mises stress in the endplates during extension: (Left) symmetric disc (Right) asymmetric disc

DISCUSSION

It is not uncommon to extrapolate findings from studies of lumbar disc to cervical disc. However, loading conditions on the cervical spine and the pathologies that affect it are different from that from the lumbar spine. As relative motions between two vertebrae are dictated by the intervening soft tissues, their accurate geometrical representations are necessary to obtain realistic outputs. It is known from cadaver experiments that the response of a healthy cervical segment in extension is stiffer than in flexion [16]. This can be attributed to two factors: the asymmetry of the disc and engagement of facet joints. The moment-rotation response curves of previous models lack the asymmetric response [1], [3], especially at lower magnitudes of extension moments where disc is the major load-sharing component [12]. The comparison between asymmetric and symmetric annulus fibrosus demonstrated a considerable difference in the extension response. Therefore, it is important to simulate the asymmetry of the annulus fibrosus in future FE models of the cervical spine.

In vivo studies have shown that strains and other internal responses are region-specific: posterior-lateral region usually exhibits higher magnitudes [17]–[19]. This pattern was observed in the current study. The displacement of the nucleus pulposus posteriorly in the asymmetric disc also changed the endplate stress distribution pattern. Anatomically accurate material distributions in the different disc regions would ensure realistic outputs from models.

Using mapping block-based mesh generation technique for both hard and soft tissues ensured common nodal connectivity between the IVDs and adjacent vertebrae. This meshing technique eliminated the need for contact definitions, reducing non-linearity and resulting in better model stability. Hexahedral mesh generated using mapping blocks also had good mesh quality, as seen from the element quality measures. The use of the C-type mesh transition to implement thinner posterior annulus fibrosus helped to avoid small elements in the posterior, increasing computational efficiency.

In summary, a FE model of the human C2-C7 spinal column with anatomically accurate disc geometry was developed using a mapping block-based meshing technique. Moment-rotation responses of all segments were within the human cadaver experimental corridors, and non-linear under loading. The responses of all spine segments were asymmetric in flexion and extension modes. This study has shown the need to incorporate accurate disc geometry in cervical spine FE models to obtain realistic responses. Such models will assist better delineation of load-sharing under different modes, and for different components.

ACKNOWLEDGEMENTS

This material is the result of work supported with resources and use of facilities at the Zablocki VA Medical Center (ZVAMC), Milwaukee, Wisconsin; the Department of Neurosurgery at the Medical College of Wisconsin (MCW), W81XWH-16-1-0010. The authors are part time employees of the ZVAMC. Views expressed in this article are those of the authors and do not necessarily represent the funding organizations.

REFERENCES

[1] M. B. Panzer et al., *Med. Eng.Phy.* vol. 33, no. 9, 2011.
[2] J. Östh et al., *J. Biomech. Eng.*, vol. 138, no. 6, 2016.
[3] T. Mustafy et al., *J. Biomech.*, vol. 47, no. 12, 2014.
[4] S. Mercer et al., *Spine*, vol. 24, no. 7, Apr. 1999.
[5] J. Tonetti et al., *Surg. Radiol. Anat.*, vol. 27, no. 3, Aug. 2005.
[6] K. M. Browne et al., *J. Neurosurg. Spine*, vol. 12, no. 3, 2010.
[7] K. H. Shivanna et al., *Comput. Aided Des.*, vol. 42, no. 12,
[8] D. L. Kopperdahl et al., *J. Biomech.*, vol. 31, no. 7, 1998.
[9] D. T. Reilly et al., *J. Biomech.*, vol. 8, no. 6, 1975.
[10] N. Yoganandan et al., *Bone*, vol. 39, no. 2, 2006.
[11] S. Ebara et al., *Spine*, vol. 21, no. 4. 1996.
[12] M. B. Panzer et al., *J. Biomech.*, vol. 42, no. 4, Mar. 2009.
[13] J. J. Cassidy et al., *Connect. Tissue Res.*, vol. 23, no. 1,
[14] J. C. Iatridis et al., *J. Orthop. Res.*, vol. 15, no. 2, Mar. 1997.
[15] S. Kumaresan et al., *J. Biomech.*, vol. 31, no. 4, Apr. 1998.
[16] J. A. Wheeldon et al., *J. Biomech.*, vol. 39, no. 2, Jan. 2006.
[17] W. Anderst et al., *Ann. Biomed. Eng.*, vol. 44, no. 5, 2016.
[18] S. Matsunaga et al., *Spine*, vol. 24, no. 7, Apr. 1999.
[19] S. Ito et al., *Eur. Spine J.*, vol. 14, no. 4, 2005.

BIOMECHANICAL DIFFERENCES BETWEEN MALE AND FEMALE SACROILIAC JOINTS IMPLANTED WITH THREE DIFFERENT SACROILIAC IMPLANT SYSTEMS: RANGE OF MOTION STUDY

Amin Joukar (1), Anoli Shah (1), Ali Kiapour (1), Ardalan Seyed Vosoughi (1), Anand K. Agarwal (1), Hossein Elgafy (1), Nabil Ebraheim (1), Vijay K. Goel (1)

(1) Engineering Center for Orthopaedic Research Excellence (E-CORE), University of Toledo, Toledo, Ohio

INTRODUCTION

Low back pain (LBP) is a common reason for primary care with approximately 90% of adults being impacted by this condition at some time in their lives. Majority of LBP originates from the lumbar spine, while more recent studies have estimated that the sacroiliac joint (SIJ) is the actual source of pain in 15-30% of cases [1]. Much like the other regions of the spine, it is essential that we understand the SIJ mechanics. There are anatomical differences between male and female SIJs [1].
Increased awareness of the prevalence of the SIJ as a source of LBP has given rise to the importance of planning of the treatment related to SIJ. When non-surgical methods (NSM) failed to reduce the pain and discomfort of patients with suspected SIJ dysfunction, surgical measures become an option, beginning with open arthrodesis, or fusion of the SIJ. There are various implant devices to fuse SIJ using minimally invasive surgery (MIS).
Several *in vitro* and *in silico* studies have been done to determine the efficiency of different implants in SIJ motion reduction. Soriano-Baron et al. [2] conducted a cadaver study to investigate the effect of placement of sacroiliac joint fusion implants which were triangular implants. They showed that placement of three implants in trans-articular and inline approaches significantly reduced the ROM in all motions.
Lindsey et al. [3] also showed that SIJ fusion using three triangular implants provided significant reduction in SIJ motion in all six motions of spine. Bruna-Rosso et al. used finite element method to analyze SIJ biomechanics under RI-ALTO fusion implant under compression load. The ROM of SIJ reduced significantly and proximally insertion of implant which was farther from SIJ center of rotation was more efficient than distally insertion of implant.
Different implant designs exist in the market and there is a lack of biomechanical study to prove their ability in providing SIJ stability. The objective of this study is to assess the biomechanics of the three types of SIJ fusion implants, triangular implants, half threaded screw, and fully threaded screw, in terms of motion reduction and their effect on each gender using finite element simulations.

METHODS

The ligamentous L1- pelvis of male and female FE models were developed from the CT scans of a male and a female patients. Isotropic elastic, hyperelastic, and hypoelastic material properties were used for cortical and cancellous bone and nucleus, annulus, and ligaments, as detailed for the male model elsewhere [3]. All three implants are made of titanium. In each surgery, three implants were placed across SIJ. Triangular implant and half threaded screw were inserted laterally to the SIJ, but fully thread screw was located posteriorly. The interaction between implants and bone were defined as tie. The model was fixed at the hip joint and subjected to 10 Nm moment and 400 N follower load to simulate flexion-extension, lateral bending and axial rotation. In all six loading modes, the range of motion of sacroiliac joint was calculated for both models under all three fusion devices and compared to each other, Figure 1 [3]. Female and male intact spinopelvic models were validated against the previous literature data. All implants were placed into the left sacroiliac joint of both male and female models fusing the left SIJ and leaving right SIJ unfused.

RESULTS

The range of motion (ROM) of the intact female and male SIJs were calculated (Table 1).

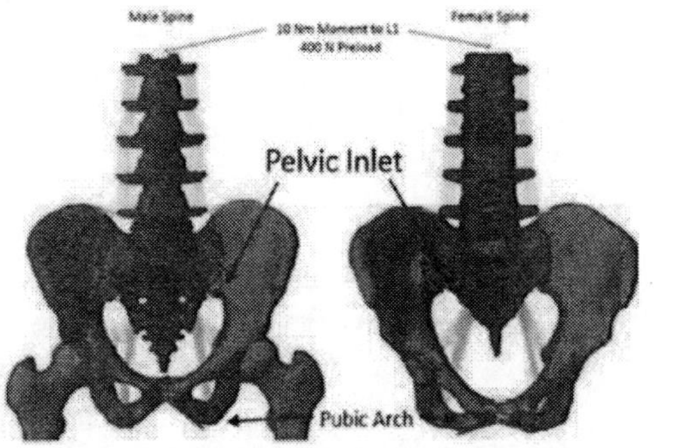

Figure 1: Finite element models of male and female sacroiliac joints.

The range of motion of the female SIJ was higher than the male SIJ range of motion in flexion, extension, and lateral bending up to 41% (from Table 1). In right rotation, the range of motion of male and female were similar, but the range of motion of the female model was about three times higher than the male model in left axial rotation. The motion reduction at left SIJ under all three implants fusion for male and female models are shown (Table 2 and 3). It was shown that the half threaded screw provided better stability at the fused SIJ compared to two other implants for the male model in all loading conditions. In the female model, the triangular implant provided maximum stability at the left SIJ compared to two other implants in all loading conditions.

Table 1: Range of motion of intact male and female SIJs

	Range of Motion (Degree)			
Motion	Male Left SIJ	Female Left SIJ	Male Right SIJ	Female Right SIJ
Flexion	0.92	1.2	0.89	1.22
Extension	1.76	2.48	2.02	2.63
Right Bending	0.7	0.8	0.6	0.8
Left Bending	0.32	0.97	0.49	0.97
Right Rotation	0.58	0.53	0.526	0.53
Left Rotation	0.46	1.37	0.526	1.45

Table 2: Motion reduction at left SIJ for implanted male models

Motion	Motion Reduction @ Left SIJ		
	Triangular Implant	Fully threaded screw	Half threaded screw
Extension	97.9%	98.1%	99%
Flexion	94.3%	93.4%	97.2%
Right Bending	92.5%	91.6%	98.3%
Left Bending	91.8%	91.8%	98%
Right Rotation	83%	79.2%	90.5%
Left Rotation	88.4%	82.7%	92.3%

Table 3: Motion reduction at left SIJ for implanted female models

Motion	Motion Reduction @ Left SIJ		
	Triangular Implant	Fully threaded screw	Half threaded screw
Extension	97%	96%	96%
Flexion	96%	94%	94%
Right Bending	96%	86%	92%
Left Bending	94%	82%	84%
Right Rotation	91%	75%	76%
Left Rotation	93%	81%	87%

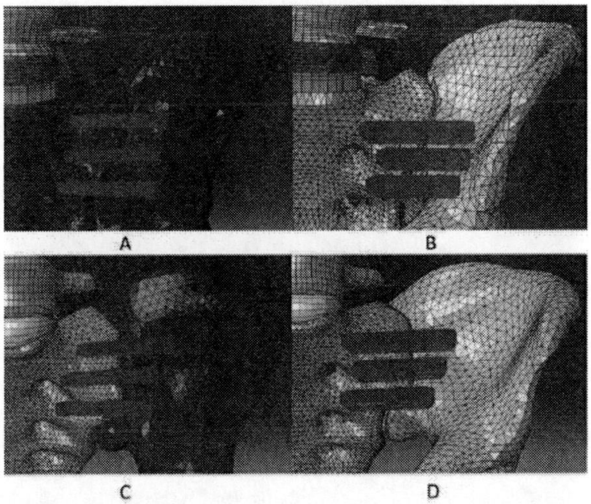

Figure 2: A) Half threaded screw placement in the male model, B) Triangular implant placement in the male model, C) Half threaded screw placement in the female model, D) Triangular implant placement in the female model

DISCUSSION

Our Finite element analysis showed that sacroiliac joint had relatively higher mobility in the females compared to males. This can be attributed to certain anatomical correlations such as curvature of the SIJ surfaces and a greater pubic angle compared to those of males.
Half threaded screw provided better stabilization for the left SIJ ROM compared to the triangular implant and fully threaded screw for the male model in all motions. This might be explained by the different placement of implants, since the triangular and half threaded screw were placed laterally while fully threaded screw was placed posteriorly. This means the implants with lateral placement showed better stability than the posterior placement. For the male model, one of the reasons that caused a better stabilization for half treaded screw compared to triangular implant was more penetration of this screw to the sacrum (Figure 2. A & B).
In the female model, triangular implants reduced fused SIJ motion more than two other implants. Triangular implants were inserted deeper compared to half threaded screw; therefore, less mobility at SIJ by using this implant was observed for the female model (Figure 2. C & D). As can be seen, the appropriate implant for each gender was different. Because the implants which were used for each gender had the same length, so we could see the effect of anatomy in how far an implant using similar placement can be inserted in each gender.
One limitation of the present study was use of the material properties for the female model as is male model due to lack of experimental data.

ACKNOWLEDGEMENTS

The work was supported by NSF Industry/University Cooperative Research Center at The University of California at San Francisco, CA and The University of Toledo, Toledo, OH (www.nsfcdmi.org).

REFERENCES

[1] Vleeming, A et al, *J. Anat. Journal of Anatomy*, 221(6):537-567, 2012.
[2] Bruna-Rosso, C et al, *Int J Spine Surg*, 10:16, 2016.
[3] Lindsey, D et al, *Int J Spine Surg*, 9:64, 2015.

CHARACTERIZATION OF THE AVERAGE LUMBAR SPINE INTERVERTEBRAL DISC ANNULUS PROPERTIES BASED ON RAW DATA SETS

Jessica S. Coogan (1), Brian D. Stemper (2), Daniel P. Nicolella (1)

(1) Mechanical Engineering
Southwest Research Institute
San Antonio, TX, USA

(2) Department of Neurosurgery
Medical College of Wisconsin
Zablocki Veterans Affairs Medical Center
Milwaukee, WI, USA

INTRODUCTION

Finite element models of the spine have been increasingly used to investigate the biomechanics associated with spinal injuries. One key component of finite element modeling is the assignment of material properties to the structures of the model, including the ligaments and the intervertebral disc. The annulus fibrosus (AF) of the intervertebral disc has a highly organized, lamellar architecture with collagen fibers oriented at alternating angles in each lamella. This structure leads to a non-linear force versus deformation response when the AF is loaded in tension, with a toe, linear, traumatic, and post-failure region. Several experimental studies have determined the non-linear properties of the AF [1-4]. However, these studies only provide representative force versus deformation curves for single experiments, whereas utilizing average curves with standard deviation bars would allow finite element simulations to represent the average population. Also, since individual lamella can have different inflection points for the various regions of the response, averaging the forces at common elongation points does not accurately capture the characteristic response. A normalization procedure was previously applied to spinal ligaments to calculate an average response curve that maintains the characteristic non-linear shape of the curve [5]. The goal of this work was to apply the same procedure to determine the average response and standard deviation bars for a set of experimental data characterizing the tensile response of the lumbar spine AF specimens [6]. The average curve was used to validate a finite element model of the full lumbar AF.

METHODS

Experimental data were obtained from young human lumbar annulus coupon specimens as described in a separate study [6]. A total of 145 test specimens were obtained from five post-mortem human subject (PMHS) lumbar spines (T12-L1 through L4-L5) with a mean age of 36±6 years. Test specimens consisted of a single layer of annular fibers (approx. 5 mm length, 2.7 mm width, 0.8 mm thickness), with the fibers oriented in the direction of loading. Tensile testing was performed using an electrohydraulic testing system, and the specimens were loaded at a quasi-static rate (0.005 mm/s) until failure. The current study investigated the curve fits for each experimental dataset.

Based on methods used to create average cervical ligament properties [5], curve fits were undertaken for each individual lumbar annulus specimen. The toe and linear regions were fit using functions that described the toe region using an exponential curve, and the linear region with a linear curve [5].

$$\text{Toe: } F = A_3(\exp(A_4 \cdot d) - 1)d^{(m^*)} \text{ for } d < d^* \quad (1)$$
$$\text{Linear: } F = F_{d^*} + (d - d^*)A_5 \text{ for } d > d^* \quad (2)$$

Force (F) is described as a function of displacement (d), where A_3, A_4, and m* are toe region constants, d* is the transition point between the toe and linear region, F_{d^*} is the force at the transition point, and A_5 is the slope or stiffness of the linear region. Constraints were applied to limit m*≥1 and slope and force at d* to be equal in both exponential and linear curves. The curves were fit to maximize the r^2 value through adjusting the variables A_3, A_4, m*, and d* using a commercial nonlinear optimization solver (Excel 2010, Microsoft). The traumatic and post-failure regions were not considered in this study.

Using the optimized fit parameters, each experimental curve was interpolated in order perform normalization. For each curve, 30 equally spaced points between 0 and d* represented the toe portion, and the corresponding force was determined using equation (1). Equation (2) was then used to determine the linear portion of the curve, where the endpoint of the linear portion was assumed to be at

2d* since the traumatic and post-failure regions were not considered. Each curve was then normalized to d*=1 and F_{d*}=1. The normalized curves were averaged to create a normalized average curve. The original d* and F_{d*} values for each experiment were averaged and used to scale-up the normalized average curve. In addition, a new method was devised to determine the standard deviation (SD) bars for the average curve. The SD of d* and F_{d*} was calculated. Both variables contribute to the variability of the total curve, so the combined variation was represented by an ellipse centered at (d*, F_{d*}) with minor axis equal to the SD of d* and major axis equal to the SD of F_{d*}. With the angle being defined from the positive x-axis, the +1 SD curve was defined by scaling up the normalized curve using the ellipse values at 135° and the -1 SD curve was defined by scaling up using the ellipse values at 315°. Equations (1) and (2) were determined for the average ± 1 SD curves.

A finite element model of the lumbar spine AF was created based on previous methods [7]. The L4-L5 motion segment was isolated, and the ligaments, nucleus, and vertebral arches were removed in order to replicate the experimental design of [8], where the full lumbar AF was tested in isolation. The AF was modeled with a fiber-reinforced transversely isotropic hyperelastic material model (LS-Dyna, LSTC, Livermore, CA). The parameters of the material model were optimized to fit the average ±1 SD force-displacement response. The material model is characterized by hyperelastic coefficients (C1-C5), a bulk modulus (XK), a stretch ratio at which the fibers are straightened (XLAM), and a fiber orientation vector. C2 was assumed to be 0, which simplifies the material model into the neo-Hookean constitutive model. C4 was chosen to be 25 in order to avoid overfitting the data, and XK was chosen to be 1e8 Pa in order to enforce near incompressibility. The alternating fiber orientation was generated by creating two different parts for the AF that alternated from the superficial to deep location, where the fiber orientation vector was assumed to be ±35° with respect to the transverse plane [9]. The model was loaded with a pure moment in flexion and extension to 10 Nm.

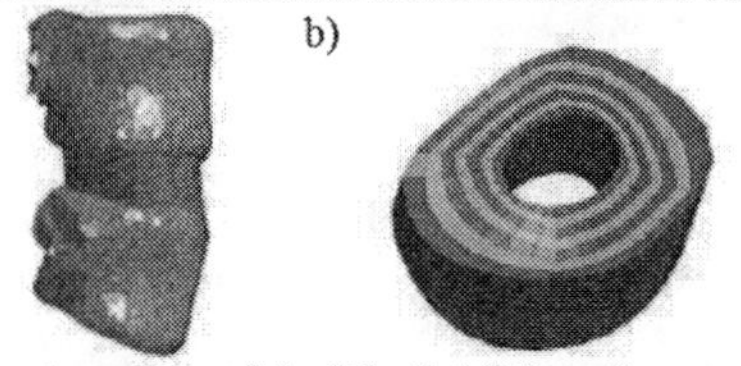

Figure 1: a) Finite element model of the L4-L5 motion segment b) Model of the AF with different parts shown in blue and green.

RESULTS

From the 145 experiments, those that showed only a linear force versus displacement response without the toe region were excluded, leaving a total of 111 experiments. The average ±1 SD force-displacement response of the lumbar AF specimens is shown in Figure 2, the fit parameters are given in Table 1, and the material model parameters are given in Table 2.

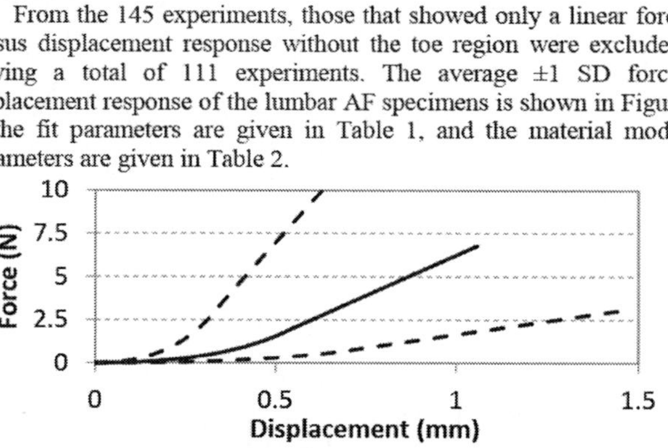

Figure 2: Average (solid line) ±1 SD (dotted lines) force vs. displacement response of the lumbar AF specimens

Table 1: Fit parameters for the average ± 1 SD curves.

Parameter	A_3	A_4	A5	m*	d*	F_{d*}
Average	5.01	1.21	9.30	1.39	0.53	1.86
-1 SD	3.09	0.50	3.09	1.54	0.74	0.86
+1 SD	4.24	3.59	23.41	1.06	0.32	2.85

Table 2: LS-Dyna material model parameters for the lumbar AF.

Parameter	Average	-1 SD	+1 SD
C1	2.65e5 Pa	1.01e5 Pa	6.29e5 Pa
C2	0	0	0
C3	5.33e4 Pa	8.75e3 Pa	2.66e5 Pa
C4	25	25	25
C5	2.48e7 Pa	8.22e6 Pa	6.67e7 Pa
XK	1e8 Pa	1e8 Pa	1e8 Pa
XLAM	1.095	1.135	1.06

The flexion and extension response of the finite element model was compared against the experimental results of [8]. In the experimental study, the range of motion (ROM) at 10 Nm applied moment was 15.8° ± 1.7° for flexion and 16.3° ± 0.5° for extension. The finite element model had very similar average response to the experimental study, but larger standard deviation. The simulated flexion response was 15.8°±6.6°, and the extension response was 15.8°±6.6°.

DISCUSSION

This study utilized a normalization technique in order to provide an average force-deformation response for the lumbar AF from individual experimental datasets. In addition, ±1 SD bars for the experimental data were determined. These data were incorporated into a finite element model of the full lumbar AF in order to validate the response. The average simulated response was very similar to the experimental response, but the variability was higher for the simulated response. This indicated that the variability seen in individual lumbar annulus coupons is larger than that of the full lumbar AF. Intuitively, as individual annulus specimens are summed together to make the entire lumbar AF, the highs and lows will cancel each other out to reduce the variability of the structure as a whole. In order to more accurately represent the variability of the full lumbar AF, the same procedure could be applied to experimental datasets utilizing the entire annulus fibrosus.

ACKNOWLEDGEMENTS

This work was supported by the Fire & Cabin Safety Technical Community Representative Group 2011/12 Project 2: Injury Criteria for Obliquely-oriented seats through FAA Aviation Research Grant 16-G-001.

REFERENCES

[1] Acaroglu ER, et al., *Spine*, 20: 2690-2701, 1995.
[2] Ebara S, et al., *Spine* 21: 452-261, 1996.
[3] Fujita Y, et al., *J Ortho Res*, 15: 814:819, 1997.
[4] Holzapfel GA, et al., *Biomechan Model Mechanobiol*, 3:125-140, 2005.
[5] Mattucci SFE, et al., *J Mech Beh Biomed Matl* 41: 251-260, 2015.
[6] Stemper BD, et al., *J Neurosurg: Spine* 21: 608-613, 2014.
[7] Coogan JS, et al., *Frontier Bioeng Biotech* 4, 2016.
[8] Schmidt H, et al., *Clin Biomech* 22: 377-384, 2007.
[9] Cassidy, et al., *Connect Tissue Res* 23, 75-88, 1989.

SB^{3}C2017
Summer Biomechanics, Bioengineering and Biotransport Conference
June 21 – 24, Tucson, AZ, USA

FINITE ELEMENT METHOD FOR PREDICTING FAILURE LOCATION OF ANNULUS FIBROSUS IN UNIAXIAL TENSION

Benjamin Werbner*, Minhao Zhou*, Grace O'Connell

Mechanical Engineering
University of California, Berkeley
Berkeley, CA, United States

INTRODUCTION

Fiber-reinforced tissues experience large, complex loads during daily activities. Failure of these tissues often results in debilitating pain and reduced mobility. Understanding the failure behavior of tissues with limited self-healing capabilities, such as annulus fibrosus (AF) in the intervertebral disc, is of particular importance, as their failure may transfer stresses to adjacent tissues, resulting in a cascade of damage and degeneration [1].

Unfortunately, the failure mechanics of fiber-reinforced tissues, particularly those with fibers oriented off-axis from the applied load, is not well understood. Previous studies have suggested that this is due to limited specimen sizes and high strains at the testing grips, which may cause unpredictable tissue failure [2, 3]. Recent work by Peloquin *et al.* identified five different failure modes in meniscus specimens, with less than 25% of specimens failing at the mid-substance [3]. Such unpredictable failure behavior may be responsible for the large variation of failure properties reported in the literature.

Finite element models (FEMs) are powerful tools that can provide insight into local tissue deformations that are difficult or impossible to measure experimentally [4]. Moreover, FEMs are valuable tools for directing and guiding time-intensive experimental studies. By utilizing both finite element models and mechanical testing, we aimed to develop and validate a robust testing protocol to investigate failure properties of annulus fibrosus. In particular, the geometries of experimental test specimens were designed based on simulation results. This study focused on AF tissue, a complex angle-ply laminate structure composed of alternating layers of fibers oriented between ±30° and ±45° to the transverse plane [5]. The findings presented here, however, are applicable to other fiber-reinforced materials as well.

METHODS

Modeling: FEMs were developed representing rectangular AF specimens oriented in the circumferential-axial direction (circ.-ax., Fig 1A) (PreView 1.19.0; FEBio 2.5.2; ~180k nodes, ~170k hexahedral elements). Bulk tissue dimensions were defined to be 4.8 mm x 2 mm x 0.8 mm, which is proportional to experimental samples from our previous work [5]. Four 200 μm-thick, welded-interface lamellae were included with fibers oriented at ±30° to the transverse plane to represent outer AF, and with fibers oriented at ±45° to represent inner AF [6]. Full-width sandpaper grips were simulated to cover 0.4 mm of the tissue at both ends. A homogeneous, hyperelastic material description was used [7], with material coefficients based on preliminary tests of circ.-ax. AF in tension (ρ= 1 g/cm^3, C_1= C_2 = 0.1 MPa, C_3= 0.02 MPa, C_4= 30, C_5= 25 MPa, k= 50 MPa, λ= 1.15).

In addition to the 'Intact' model, two notch geometries were evaluated. For the first notch geometry, a 2 mm x 0.04 mm x 0.04 mm block of material was removed at the mid-substance (MS), such that half of the original cross-sectional area remained (Fig. 1B- 'Half'). The second notch geometry was created by removing material from the remaining region, such that one quarter of the original cross-sectional area remained (Fig. 1B- 'Quarter').

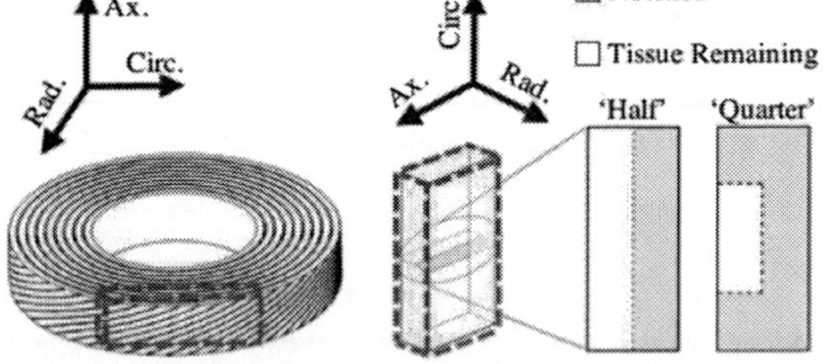

Fig. 1: A. Specimen orientation and B. notch geometries

Simulated loading was applied in two stages: a 10% compressive strain was applied at the gripped region, followed by uniaxial tension to 30% global engineering strain along the specimen length. Boundary conditions at the grips were fixed to represent no slipping between the grips and the sample surface.

Local strains for each node in the mid-substance and at the grip-line were sorted in descending order, and the average strain for each region was calculated as the average of the top 35 nodes. Then, the MS:grip strain ratio was calculated as the average mid-substance local strain divided by the average grip-line local strain. An MS:grip strain ratio greater than 1.0 indicates that the average strain is higher at the mid-substance than at the grip. Average local strain of 65% served as an estimate of bulk failure initiation, and failure was predicted at the location reaching this strain threshold earliest in the simulation.

Experimental: Healthy intervertebral discs were dissected from mature bovine caudal spines. Rectangular tissue sections oriented in the circ.-ax. direction were isolated from the outer-middle annulus, and parallel surfaces were created using a freezing stage microtome (thickness = 2 mm; Intact, n = 6). For Half notch specimens, a full-width (5.2 ± 0.6 mm) notch was created using a scalpel and depth-stop, resulting in a final thickness of 1 mm at the mid-substance (n = 8). For Quarter notch samples, additional cuts were made to further reduce the cross-sectional area to a final width of 2.9 ± 0.3 mm (n = 8).

Each sample was superglued into sandpaper, which was secured into a custom saline water bath attached to an Instron testing machine. A 0.05 N pre-load was used to remove slack before imaging the specimens with a scale bar to determine the exact width and length before testing. Uniaxial tension was applied at a rate of 50 mm/min. Lagrangian strain was calculated as the change in displacement divided by the initial grip-to-grip distance. Stress was calculated by dividing the measured force by the mid-substance cross-sectional area. Failure stress was defined as the maximum stress recorded during the test, and failure strain was defined as the strain corresponding with the failure stress. Toe- and linear-region moduli were calculated using a custom bi-linear fit to the stress-strain response.

RESULTS

Modeling: The model fit well to the nonlinear, experimental stress-strain data (R = 0.99). Average local strains reached 65% at either the mid-substance or at the grip-line for all simulations. For Intact and Half notch simulations, the average local strain was always higher at the grip-line than at the mid-substance by at least 15% (Fig. 2- black arrows; Fig. 3- red and green lines), so the model predicts grip-line failure in these samples. In Quarter notch simulations, maximum local strain at failure occurred at the notch site (Fig. 2- white arrows; Fig. 3- black and blue lines), so the model predicts mid-substance failure in these samples.

For inner-middle AF tissue, which has fibers oriented at ±45° from the loading direction, failure was strongly predicted to occur at the mid-substance (Fig. 3- black line, MS:grip ratio = 2.52). Outer AF, which has fibers oriented at ±30° from loading, was also predicted to fail at the mid-substance, but with less certainty, due to the low MS:grip ratio (1.04) at the time of predicted failure (Fig. 3- blue line).

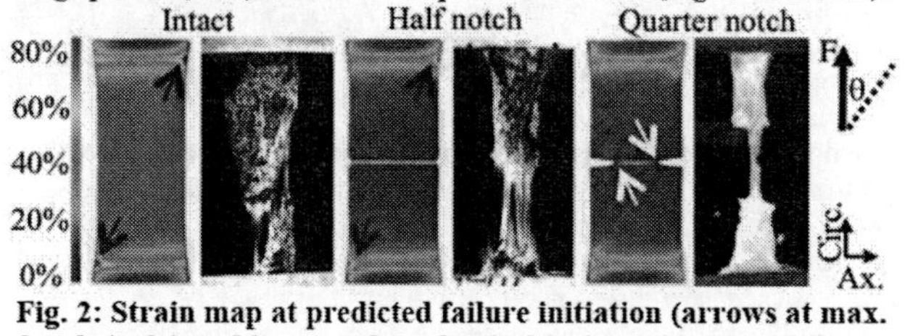

Fig. 2: Strain map at predicted failure initiation (arrows at max. local strain) and images of mechanical test specimens at failure.

Experimental: All samples tested exhibited a nonlinear stress-strain response. Bulk tissue failure originated at the grip-line in 100% of Intact samples and in 75% of Half notch samples. Quarter notch samples failed robustly within 2 mm of the mid-substance for all but one specimen (88%). Failure stress, failure strain, and toe- and linear-region moduli for the Quarter notch samples that failed at the mid-substance were 4.93 ± 1.07 MPa, 52% ± 7%, 1.55 ± 0.17 MPa, and 19.25 ± 3.27 MPa, respectively.

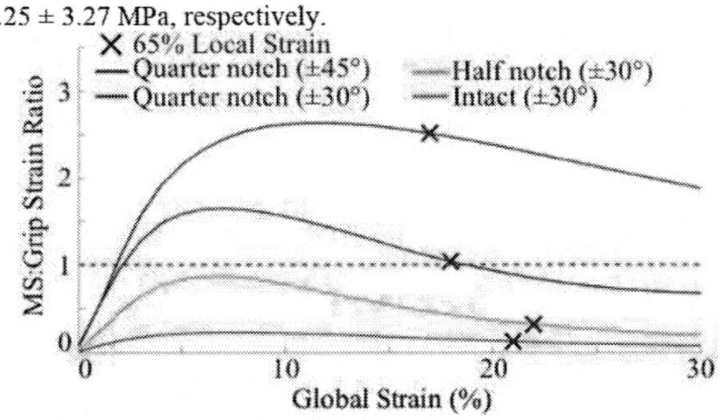

Fig. 3: MS:grip strain ratio vs. global strain for three notch geometries, including two fiber angles. X denotes predicted failure.

DISCUSSION

We investigated the effectiveness of using an FEM to predict the failure location of annulus fibrosus tissue samples with different mid-substance notch geometries. Repeatable mid-substance failure was achieved only in Quarter notch samples, which represent a modified dog-bone geometry. FEM results were validated with experimental tests. The standard deviations for material properties of Quarter notch specimens were less than 22% of the mean, which is low compared to data in the literature (>25%) [8-10]. The simulations indicate that grip-line failure was the result of higher compressive strains from gripping outweighing tensile strains at the mid-substance. Therefore, it is essential to model grip effects when investigating failure properties.

Our previous work used a Half notch geometry to induce repeatable mid-substance failure in circumferential-radial AF samples [5], suggesting that specimens with shorter fiber bundles tend to have higher MS:grip strain ratios. Additionally, the results of this study indicate that specimens with fibers more aligned with the loading direction are more likely to fail at the grip-line, thus mid-substance failure is more likely in inner- and middle-AF specimens.

We used a pre-defined strain threshold of 65% to predict tissue failure due to the lack of experimental data providing local failure strains. Furthermore, our model did not include a description for damage, resulting in a significant difference between the models' predicted bulk failure around 20% global strain and experimentally measured bulk failure at ~50% global strain. Future work will utilize a heterogeneous model that incorporates the tissue's micro-mechanical structure. This model will allow us to include corresponding damage descriptions to better investigate mechanisms of AF tissue failure [7].

This study used a combined approach of computational modeling and mechanical testing to develop and validate a method for sample preparation. Our approach focused on ensuring mid-substance failure of AF, which is a complex fiber-reinforced tissue with fibers aligned off-axis from the applied load. In conclusion, the combined method facilitates efficient and accurate evaluation of AF failure location.

ACKNOWLEDGEMENTS *Both authors contributed equally.

REFERENCES [1] Adams, M.+, *Spine*, 2006; [2] Jacobs, N.+, *J Biomech Eng*, 2013; [3] Peloquin, J.+, *J Biomech Eng*, 2016; [4] Schmidt, H.+, *J Biomech*, 2013; [5] Werbner, B.+, *ORS*, 2017; [6] Cassidy, J.+, *Conn Tiss*, 1989; [7] Yang, B.+, *SB3C*, 2016; [8] Green, T.+, *Eur Spine J*, 1993; [9] Acaroglu, E.+, *Spine*, 1995; [10] O'Connell, G.+, *J Biomech Eng*, 2009.

SB^{3}C2017
Summer Biomechanics, Bioengineering and Biotransport Conference
June 21 – 24, Tucson, AZ, USA

A SEMI-AUTOMATED APPROACH FOR CREATING A SUBJECT-SPECIFIC FINITE ELEMENT MODEL OF THE INTERVERTEBRAL DISC

Bo Yang (1), Yeabsra B. Habtegebriel (1), Yu Ma (2), Michael F Wendland (3), Grace D. O'Connell (1)

(1) Mechanical Engineering
University of California, Berkeley
Berkeley, CA, United States

(2) Mathematics
University of California, Berkeley
Berkeley, CA, United States

(3) I□BBB - □B3 Institute
University of California, Berkeley
Berkeley, CA, United States

INTRODUCTION

Over one-third of the U.S. population suffers from lower back pain and a high proportion of these cases are associated with intervertebral disc malfunction [1]. Finite element modeling (FEM) is a powerful tool for understanding disc joint mechanics, providing detailed information of stress distribution within the disc and surrounding tissues. The accuracy of FEMs is dependent on the material properties used and the ability to accurately describe the disc's complex geometry and tissue architecture.

The disc is made of fibrocartilage tissue and consists of three distinct regions: nucleus pulposus (NP), annulus fibrosus (AF), and cartilage endplate (CEP). Many of the current disc models have been generated using mesh based on computer-aided design (CAD) [2, 3]. This approach first obtains 3D image data (computer tomography or magnetic resonance images (MRI)), and then surfaces are reconstructed through segmentation. Finally, elements are generated within the disc boundary. Without further processing, image data can only provide whole disc geometry, with NP segmentation from the AF set based on anatomical data in the literature. Generally, generating a CAD-based mesh is a time-consuming manual process, because the preferred hexahedral elements cannot be generated automatically like tetrahedral elements. Furthermore, the manual process can lead to user variability in tissue geometry and position.

In contrast, voxel meshing, or an image-based meshing, is an alternative technique to CAD-based meshing. This approach directly converts imaging data to brick elements and has been widely used in bone modeling [4, 5], but rarely for disc modeling. Therefore, to overcome aforementioned disadvantages in CAD-based meshing, the objective of this study was to develop algorithms to generate a voxel mesh of a bovine disc from MRI scans. MRI scans were turned into a sequence of images to create a 3D solid. Then, individual parts of the disc were identified, including the AF, NP, and vertebral bodies using custom algorithms in Matlab (Mathworks R2016b). The outputted mesh file is widely accepted by finite element programs (e.g., ABAQUS and FEBio).

METHODS

Bovine caudal spine sections were obtained from a local abattoir. Bone-disc-bone motion segments were prepared by removing surrounding musculature and cutting through the superior and inferior vertebrae (n = 11). Each sample was imaged using a 3D Fast Low Angle Shot (FLASH) sequence (7T, Bruker MRI machine; FOV = 5 X 5 X 2.5 mm, 128 slices, isotropic resolution = 0.195 mm/pixel). Then, the sample underwent a 16-slice 2D scan using a T2 RARE sequence (TR = 6 s, 5 echoes with TEs of 11 ms, 33 ms, 55 ms, 77 ms, and 99 ms, thickness = 0.585 mm, in-plane resolution = 0.195 mm/pixel). Compared to the 3D scan, the 2D T2 weighted image had a thicker slice to increase signal strength, resulting in the NP and AF being represented separate values on the grey scale.

Raw datum were imported into ImageJ to generate a 3D image stack (Fig. 1) [2]. First, image slices from the 3D scan were analyzed to generate the 3D disc boundary. Non-disc tissue surrounding the disc (*e.g.*, surrounding ligaments) had the same signal intensity as the disc, requiring manual selection of the disc boundary (Fig. 2A). Therefore, every 6th image slice (~25% of acquired slices) was used to select the disc boundary. Once the disc boundary was selected, a signed distance function (negative inside and positive outside) was used to determine the disc boundary between image slices through interpolation [6]. The output of this process was a 3D matrix of signed distance function.

Fig. 1: 3D scan in ImageJ.

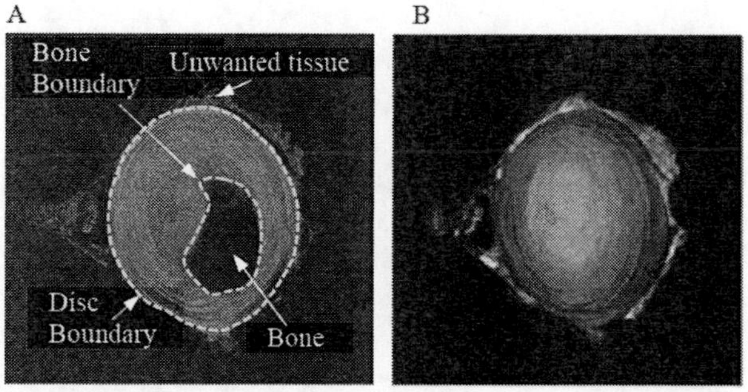

Fig. 2: (A) A 3D slice close to the bone. (B) A 2D T2 weighted image at the middle disc height.

At the superior and inferior ends of the disc, vertebral bodies were observed in MRI scans as a dark region (Fig. 2A). To obtain a disc-only mesh geometry for finite element modeling, this region was subtracted. To detect the bony region, 3D slices were binarized with a low threshold (0.25) making the bone area in the middle black and the surrounding disc tissue gray (Fig. 2A). The bone boundary was detected using the Sobel edge-detection method [7], which resulted in two regions being selected (Fig. 2A – dashed line). The outer edge selection was removed, leaving the bone boundary (Boolean operation). To enable interpolation, the boundary curve was turned into signed distance function.

In the third step, the boundary between NP and AF was extracted automatically from 2D T2 weighted scans, where the signal intensity in the NP is greater than AF due to a higher water content (Fig. 2B). The image was binarized and a threshold of 0.55 was selected for robust and repeatable results, based on preliminary experiments. Unwanted tissue surrounding the disc was as bright as NP, and, thus, remained in the binarized image. Therefore, unconnected regions of white pixels were removed using a built-in Matlab function (Matlab – bwareaopen). Finally, then NP boundary was converted into a signed distance function and interpolated for all heights throughout the disc's cross-sectional area.

CEP and bony endplate were automatically added to the superior and inferior sections of the segmented disc and was based on anatomy reported in the literature. CEP had a thickness of 0.6 mm, covering the entire NP and inner AF, while the bony endplate had a thickness of 1.5 mm [8, 9]. The last step was to write NP, AF, CEP, and bony endplate geometries into an Abaqus mesh file (.inp), which could then be imported into FEM software, such as FEBio. Finally, a Matlab GUI was developed to conduct all processes described above.

RESULTS

The disc boundary edge was obtained following boundary selection and interpolation (Fig. 3A). Fig. 4 A-D show binarization, cleaning, and edge detection of a vertebral body. Once separated, the bone was subtracted to obtain disc-only tissue at the superior and inferior ends of the disc. Fig. 4 E-D show the automated process for detecting the NP boundary from the surrounding AF tissue. The final voxel mesh contained four parts, including the NP, AF, CEP, and bony endplate (Fig. 3B).

DISCUSSION

We developed algorithms to create a semi-automated approach for converting MRI scans of the intervertebral disc into a voxel mesh for finite element analysis. While the process to determine the disc boundary required manual selection, detection of the vertebral body and NP boundaries were automated with pre-defined thresholds. Furthermore, disc segmentation and mesh generation processes were automated. In general, generation of a voxel mesh for a single disc was generated within 15 minutes (MRI scan time not included), which is much lower than the commonly used CAD-based approach (a couple of ***days***).

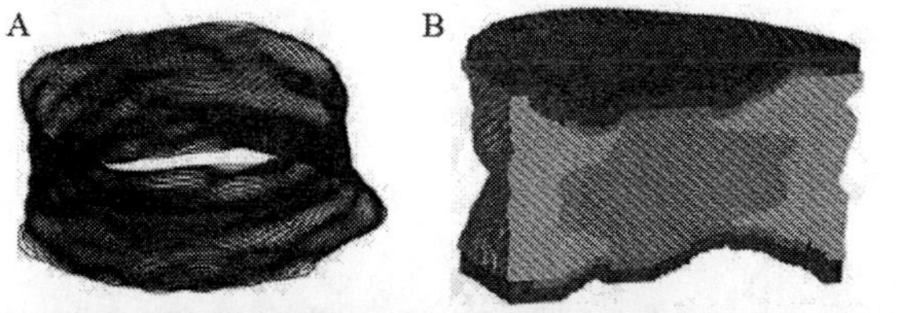

Fig. 3: (A) Disc boundary. (B) Disc meshes shown in Preview.

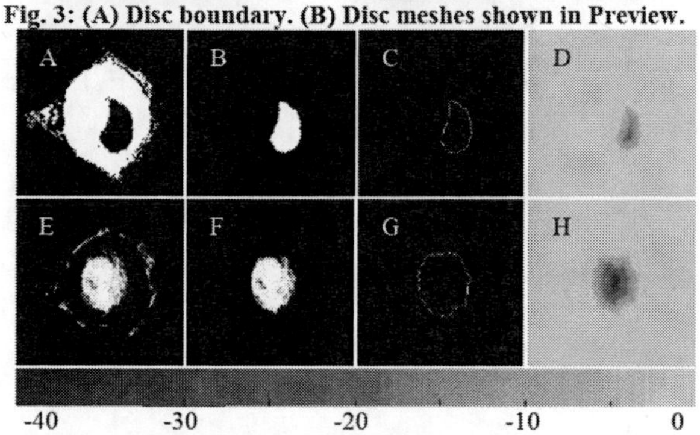

Fig. 4: (A-D) shows the process of binarization, cleaning, edge detection, and obtaining signed distance function for bone bound, original figure is Fig. 2 (A). (E-D) shows the process of detecting NP boundary, original figure is Fig. 3 (B).

Brick elements from the voxel mesh approach have higher quality than manually generated hexahedral meshing [10]. However, a disadvantage of the voxel mesh approach is the large number of elements that are generated, requiring more computational power for finite element analysis. For example, the bovine intervertebral disc model contained ~$1x10^6$ elements, while a CAD-based mesh approach would result in less than 10,000 elements. Therefore, voxel-mesh models require a super computer for analysis.

Auto-segmentation provided subject-specific disc geometry, and NP position and geometry (Fig. 3B). We were able to visualize AF lamellae in MRI scans (lamellae thickness ~0.3-0.7 mm; Fig. 2B) [11]; however, the scan resolution was not sufficient to determine boundaries between lamellae. In conclusion, the semi-automated voxel mesh approach presented here allowed us to create subject-specific models, with the NP, AF and vertebral bodies defined as separate tissues. The methods described here will be valuable for developing patient-specific models of human discs that can be used for diagnosis, surgical planning, or treatment evaluation.

ACKNOWLEDGEMENTS

This study was supported by the Hellman Fellows Fund.

REFERENCE

[1] Health, United States, 2015 retrieved from: https://www.cdc.gov; [2] Jacobs, NT, et al., *J. Biomech.*, 47.11: 2540-2546, 2014; [3] Yao, H, et al, *J. Biomech*, 40.9: 2071-2077, 2007; [4] Hollister, SJ, et al., *Biotechnol Bioeng*, 43.7: 586-596, 1994; [5] Morgan, EF, et al., *BoneKEy-Ostevision*, 2.12: 8-19, 2005; [6] Tsai, A, et al., *IEEE Trans Med Imaging*, 22.2: 137-154, 2003; [7] Sobel, I, et al., *Comput Vision Graph*, 8.1: 127-135, 1978; [8] Wu, Y, et al., *J Biomech*, 48.12: 3185-3191, 2015; [9] Davies, J, John Wiley & Sons, 2012; [10] Ghosh, S, et al., New York: Springer, 2011; [11] Holzapfel, G.A. et al., *Biomech Model Mechan*, 3.3 (2005): 125-140, 2005.

Collagen Fiber Orientation of Tendon-Bone Insertion Tissues

Sandhya Chandrasekaran, Mark Pankow, Kara Peters, Hsiao-Ying Shadow Huang (1)

(1) Mechanical and Aerospace Engineering
North Carolina State University
Raleigh, NC, USA

INTRODUCTION: Tendon-bone insertion is a highly graded tissue, transitioning from 200MPa tensile modulus at the tendon end to 20MPa tensile modulus at the bone, across just a few hundred micrometers. In this study, we examine the insertion tissue macrostructure across its width and depth and provide a quantitative description of its collagen orientation and mineral concentration by using image analysis and mass spectrometry, respectively. Histological results revealed uniformity in global collagen orientation at all depths, indicative of mechanical anisotropy, although at mid-depth, the highest fiber density, least amount of dispersion, and least cellular circularity were evident. Collagen orientation distribution obtained through two-dimensional fast Fourier transformation (FFT) of histological imaging data from fluorescent microscopy agreed with past results based on polarized light microscopy. Results revealed global fiber orientation across the tendon-bone insertion to be preserved along physiologic tension. Gradation in the fiber distribution orientation index across the insertion was reflective of a decrease in anisotropy from the tendon to the bone.

The merit of this study lies in the image-based simplified approach to fiber distribution quantification and in the high spatial resolution of the compositional analysis. In conjunction with the mechanical properties of the insertion tissue, fiber and mineral distribution results for the insertion from the current study may potentially be incorporated into the development of a structural constitutive approach toward computational modeling. Characterizing the properties of the native insertion tissue would provide the microstructural basis for developing biomimetic scaffolds to recreate the graded morphology of a fibrocartilaginous insertion.

METHODS: Digital flexor tendon-bone units were removed from porcine forelimbs procured from the local abattoir. Two samples were preserved for microscopy in 10% buffered formalin under zero tension over two to three days prior to dehydration and decalcification. Specimens were processed for standard paraffin embedded histology and sectioned in both the horizontal and sagittal planes. Brightfield images at 400× magnification (water immersion) were obtained using the Zeiss Axioimager (Zeiss Inc., Germany) microscope at the Cellular and Molecular Imaging Facility at North Carolina State University. The collagen structures of both planes were imaged using Hematoxylin and Eosin (H&E) staining and Verhoeff-Van Geison (VVG) staining, respectively. For image analysis using fast Fourier transformation and utilizing the fluorescence of the dye, H&E stained sections at mid-depth in the horizontal plane were imaged using a GFP fluorescence filter set (excitation wavelength = 450-490 nm and emission wavelength = 500-550 nm) at a 40× objective (water immersion) (**Figure 1a**). Image analysis was performed after cropping out the three regions of interest, namely pure bone, insertion, and pure tendon, as shown in **Figure 1b** where α represents the mean fiber orientation in each image.

Images from fluorescent microscopy were converted from RGB to grayscale before converting to the frequency domain by applying 2D discrete Fourier transformation as shown in **Equation (1)** by using the function in a custom MATLAB (2016b, MathWorks, Natick, MA) script [1]:

$$F(u,v)=\sum_{x=1}^{M}\sum_{y=1}^{N} f(x,y)e^{-i2\pi(\frac{ux}{M})}e^{-i2\pi(\frac{vy}{N})} \quad (1)$$

In **Equation (1)**, and denote the pixel coordinates on the image while and denote the frequencies corresponding to changes in the pixel intensity across the image spatial coordinates. Total pixel count is given by M×N for each of the cropped images in **Figure 1b**. For example, in the tendon region (T) the pixel count is 112×112. The simplistic case of the regularly spaced alternating black and white stripes on an image would correspond to a single harmonic frequency. The actual image being analyzed is thus the inverse transformation of the superposition of such frequencies in the x and y-directions. The function is applied to the Fourier transform to cluster the zero-frequency component in the middle of the spectrum. The transform is converted to a log spectrum of its magnitude to eliminate its imaginary component. By means of noise filtering, the spectrum is shifted by its minimum value for the initial value to be zero and scaled down by its highest value to produce the FFT as shown in **Figure 1c**. The angle of orientation of white pixels on the FFT image is represented by Θ.

Fiber orientation distribution is interpreted based on the angular distribution of FFT image pixels. By fitting the FFT image pixel data onto a circular distribution, the number of pixels oriented along each incremental angle from 0 to π is obtained. The count of FFT image pixels associated with each incremental angle is baseline corrected using the minimum value and the result is normalized as a fraction of the maximum value. By numerical integration, the total area under angular distribution of the normalized pixel count is obtained. Each individual pixel count is further normalized by the total area to obtain the unit area orientation distribution as shown in **Figure 1d**. It should be noted that the orientation distribution corresponds to the angular orientation of the FFT image pixels with respect to the indicated coordinate system (**Figure 1c**). Mean angles of orientation associated with each of the three regions of interest are calculated as the angles corresponding to the centroids of the respective distributions. Hence, the peak angle in each distribution corresponds to the direction of highest frequency content and is presumably 90° out of phase with the orientation of the fibers in the cropped images in **Figure 1b**. For example, in the tendon region (T), the angle on the FFT (Θ) corresponding to a peak value of orientation distribution is 177° as seen in **Figure 1d**. This corresponds to a fiber orientation angle (α) of 87° (177° - 90° = 87°) in **Figure 1b**. Similarly, in the insertion region (I), the angle on the FFT (Θ) corresponding to a peak value of orientation distribution is 173° in **Figure 1d**. This corresponds to a fiber orientation angle (α) of 83° in **Figure 1b**. Orientation index (OI), used as a measure of fiber dispersion, is calculated as the difference between the angles bounding the middle 50% area under the orientation distribution curve. Hence, an increase in randomization of fiber direction would correspond to shorter peaks and wider orientation indices.

RESULTS

Fiber orientation gradation across insertion regions is studied quantitatively through image analyses of histological data as summarized in **Figure 1**. Fluorescent imaging (**Figure 1a**: H&E staining, GFP fluorescence filter set with excitation wavelength = 450-490 nm and emission wavelength = 500-550 nm, 40× objective, water immersion) was used to selectively crop out images from regions of interest, namely pure bone, insertion, and pure tendon, as shown in **Figure 1b**, and FFT was performed using a custom MATLAB code [1]. Results of FFT shown in **Figure 2c** clearly indicate gradation in the directionality of fiber orientation from the bone to the tendon region. Although the angle corresponding to peak values of orientation distribution is maintained across insertion, the progressive increase in the circularity of the FFT image from the tendon to the bony end is a reflection of the gradual decrease in the degree of anisotropy. This is in clear agreement with previous results from polarized light microscopy which also concluded that there is increased dispersion in fiber orientation at the bony end [2]. Angular distribution of FFT image pixels (indicated as Θ in **Figure 1c**) is obtained as a measure of the collagen fiber orientation distribution in each of the three regions of bone, insertion, and tendon, and the results are presented in **Figure 1d**. Thus, the orientation distribution of FFT at a given angle (Θ) indicates the associated number of frequencies in the frequency domain (or number of white pixels in the FFT oriented at Θ). The direction perpendicular to fiber orientation on the image would correspond to a higher number of associated frequencies and hence higher "orientation distribution." For example, in the cropped image from the tendon region in **Figure 1b**, the mean fiber direction (indicated as α) is oriented at nearly 177°. This indicates an angular orientation for the FFT at Θ = 87°. The orientation index predictably increases from the tendon to the insertion to the bone, suggesting a decrease in the directionality, i.e., decreases in the anisotropy of the fiber direction.

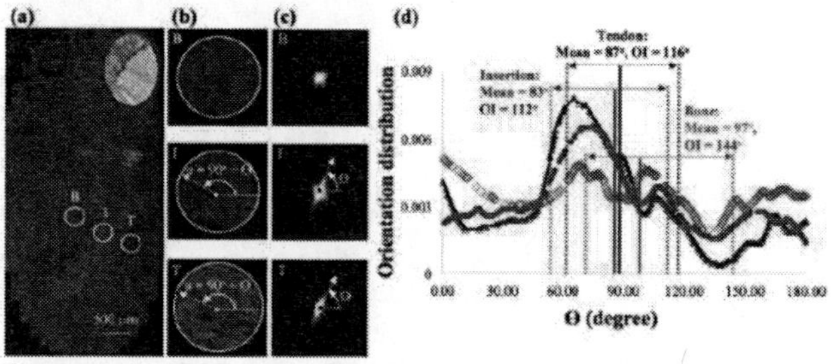

Figure 1 (a) H&E stained section was used for measurement of collagen fiber orientation in the bone (B), insertion (I), and tendon (T) regions. (b) Mean fiber orientation is represented by α in the cropped images, (c) FFT was applied to the cropped images. Angle of orientation of the white pixels on the FFT image is represented by Θ. (d) Orientation index decreases from bone to tendon, suggesting a decrease in fiber dispersion and an increase in anisotropy.

CONCLUSION

The insertion forms a functionally-graded interface between tendon and bone. This study addresses quantification of gradual changes in: (1) the orientation distribution of collagen fibers on two orthogonal planes; and (2) the relative concentration of the organic and inorganic tissue components. In this study, we have implemented an FFT-based method for determining collagen organization from histological sections which is shown to be an efficient replacement for the elaborate polarized light microscopy method. This quantification of collagen fiber orientation and dispersion across an insertion in the present study is validated against measurements from polarized light microscopy [3-4]. Features at the micrometer scale are important factors in the axial and shear moduli of tissue within the tendon-to-bone insertion [5]. Similar examination of the biocomposition across a healing tissue could help with understanding the contrast from results in a healthy tissue. The FFT provides complementary information about the complex structural properties of the inhomogeneous tendon-bone insertion tissue, and these results overall can pose important considerations in developing structure- and constituent-based constitutive models and injury mechanisms under super-physiological loading conditions.

ACKNOWLEDGEMENTS

This research was supported by NSF BMMB 1400018.

REFERENCES

[1] Masoumi N, Johnson KL, Howell MC, Engelmayr GC, Jr. Acta Biomater 2013;9(4):5974-88.

[2] Benjamin M, et al. Journal of Anatomy 1986;149:89-100.

[3] Thomopoulos S, et al. Journal of Orthopaedic Research 2003;21(3):413-419.

[4] Thomopoulos S et al. Journal of Biomechanics 2006;39(10):1842-1851.

[5] Genin GM, et al. Biophysical Journal 2009;97(4):976-985.

AGE-DEPENDENT FUNCTION OF THE ANTERIOR CRUCIATE LIGAMENT DURING POST-NATAL SKELETAL GROWTH IN THE PORCINE MODEL

Stephanie G. Cone (1), Emily P. Lambeth (1), Paul B. Warren (1), Stephanie D. Teeter (1), Jorge A. Piedrahita (2), Jeffrey T. Spang (3), Matthew B. Fisher (1,3)

(1) Department of Biomedical Engineering
North Carolina State University and the
University of North Carolina – Chapel Hill
Raleigh, NC, United States

(2) Department of Molecular Biomedical Sciences
North Carolina State University
Raleigh, NC, United States

(3) Department of Orthopaedics
University of North Carolina – Chapel Hill
Chapel Hill, NC, United States

INTRODUCTION

Anterior cruciate ligament (ACL) injuries of the knee occur in over 200,000 people annually in the United States, and incidence rates in children and adolescents have recently been rapidly increasing [1]. Treatment of ACL injuries in these young populations are complicated by previous findings showing that the anatomic orientation and the rate of volume change of the ACL vary with age prior to skeletal maturity [2,3]. However, little is known about the functional properties of the ACL during growth. The ACL features additional complexity due to the presence of its anteromedial (AM) and posterolateral (PL) bundles, that respond differently to various loading regimes and joint positions in adults [4]. Again, the function of these individual bundles during growth is unknown. Previous studies have established the pig as an acceptable pre-clinical model for studying the structure and function of the human ACL, and recent work in our laboratory has shown that the skeletally immature porcine ACL exhibits orientation changes that parallel those found in human growth [5-8]. Thus, the objective of the current work was to study changes in the contribution of the ACL and its bundles to joint function throughout skeletal growth. Our hypothesis was that the ACL functions to stabilize the joint against both excessive anterior tibial translation and varus-valgus rotation across all age groups, but that the individual contributions of the AM and PL bundles of the ACL in response to applied loads would vary with age.

METHODS

Stifle joints (knees) were collected from the hind limbs of 25 female Yorkshire pigs from a range of ages (1.5, 3, 4.5, 6, and 18 months; n=5/age group) from the Swine Educational Unit at NC State University following protocols approved by the Institutional Animal Care and Use Committee. These ages are equivalent to early juvenile through late adolescent ages in humans. All joints were stored in saline soaked gauze at -20° C between experimental procedures.

For functional testing, the joints were loaded into a 6 degree-of-freedom (DOF) universal force sensing robotic testing system (Kuka KR210 & SimVitro). The femur was rigidly fixed to the ground in custom clamps, and the tibia was attached to the end effector of the robotic arm in series with a 6-DOF force/moment sensor (ATI) [9,10]. Age-specific loading parameters were established by scaling selected loads for the 6 month age group (100 N anterior-posterior drawer, 5 N•m varus-valgus moment) relative to the cross-sectional area of the femoral cortical bone previously measured via magnetic resonance imaging.

A passive path with minimal forces and moments was established for each specimen from full extension (~40° in the pig) to 90° of flexion. An anterior-posterior (AP) load was applied to each joint at 40°, 60°, and 90°, as well as an isolated varus-valgus load at 60°, and the resulting kinematics were recorded. The AM and PL bundles of the ACL were transected sequentially, and the intact kinematics were repeated under each state while recording new forces. The scaled loading parameters were applied on the ACL-deficient joint, and the resulting ACL-deficient kinematics were recorded.

The individual force contributions of the AM and PL bundles of the ACL as well as the overall force carried by the ACL were calculated within a custom Matlab code using the principle of superposition [11]. Anterior-posterior tibial translation (APTT) and varus-valgus rotation data were calculated from the resulting kinematics for the ACL intact and ACL-deficient states. APTT values were normalized to the anterior-posterior length of the tibial plateau of each specimen previously measured via magnetic resonance imaging. Two-way ANOVA tests with Bonferroni post-hocs were performed to determine statistical significance. An overall alpha value of 0.05 was maintained.

RESULTS

Data for APTT was normalized to the length of the tibial plateau, presented in **Figure 1A** (data at 60° of flexion). Normalized APTT increased 2-3 fold after ACL transection (**Figure 1A**) across all flexion angles ($p<0.05$). At 40° and 60° of flexion, APTT decreased with increasing age ($p<0.05$); however, there was no statistically significant effect due to age at 90° of flexion ($p>0.05$). Under an applied varus-valgus moment (performed only at 60° of flexion), ACL transection resulted in increased varus-valgus rotations of 3-6° on average at all ages ($p<0.05$) (**Figure 1B**). Older age groups had decreased varus-valgus laxity in both the intact and ACL deficient states, with an average decrease of 15° and 16° between the 1.5 and 18 month age groups between the intact and ACL-deficient states, respectively ($p<0.05$). As shown in **Figure 1B**, these age-dependent changes were statistically significant between the 1.5 month age group and the 4.5, 6, and 18 month groups, as well as between 3 month age group and the 18 month old specimens ($p<0.05$).

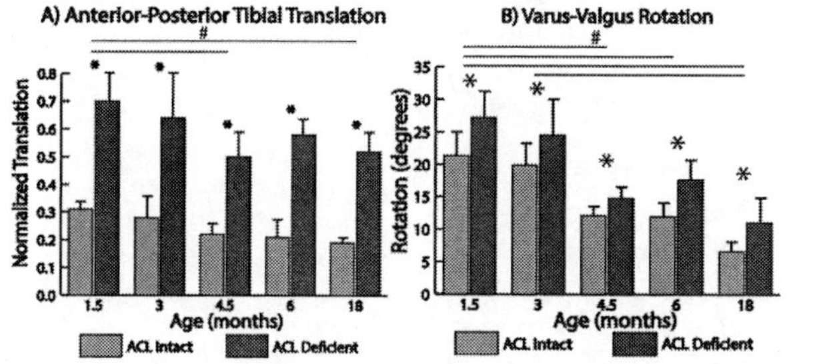

Figure 1: ACL transection and increasing age both impact joint laxity, here measured as anterior-posterior tibial translation (A) and varus-valgus rotation (B) at 60° of flexion. AP tibial translation was normalized to the A-P length of the tibial plateau. * represents statistical significance between states, # represents statistical significance between age groups (p=0.05).

In terms of forces in individual bundles under an anterior tibial load, significant interactions were found between age and bundle type (AM or PL) at all flexion angles ($p<0.05$). More specifically older age groups featured an increasing contribution from the AM bundle, paired with a decreasing contribution from the PL bundle. At 60°, the two bundles carried similar resultant forces under an anterior load across all flexion angles from the early juvenile (1.5 months) stage through the onset of adolescence (6 months), with the AM bundle carrying significantly more of the resultant force in the late adolescent (18 months) group ($p<0.05$) (**Figure 2B**).

A significant interaction ($p<0.05$) between age and bundle also existed under an applied varus moment ($p<0.05$). The contributions of the AM and PL bundles were significantly different under varus loading beginning in the early adolescent age groups (4.5 and 6 months) ($p<0.01$) (**Figure 2C**). The contribution of the AM bundle under varus loading increased with increasing age, from an average of 55% at 1.5 months to an average of 82% in the 18 month group, with changes reaching statistical significance between 1.5 and 6, and 1.5 and 18 month groups ($p<0.05$) (**Figure 2C**). However, for valgus moment, neither age nor bundle type had a significant impact on relative bundle function ($p>0.05$) (**Figure 2D**).

DISCUSSION

These findings suggest that the function of the ACL as a primary soft tissue stabilizer under anterior tibial drawer and varus-valgus moment is conserved throughout skeletal growth, but the specific contributions of its AM and PL bundles vary with age in the porcine model. The finding that the AM bundle carries the majority of the load in the ACL under anterior drawer in the mature age groups is consistent with previous literature [4]. Interestingly, the contributions of the AM and PL bundles vary with an applied varus moment across ages, while there are no significant differences in the bundle behavior under a valgus load. These findings suggest that while a primary function of the ACL is conserved during post-natal skeletal growth, the function of the ACL bundles evolve in an age dependent manner. While this study is limited by the use of only a limited number of female specimens, future work will look to replicate this project in male animals and to increase samples sizes. These age-specific changes may have implications in elucidating injury mechanisms, surgical reconstruction, and design of tissue engineered constructs in pediatric and adolescent patients after ACL injury.

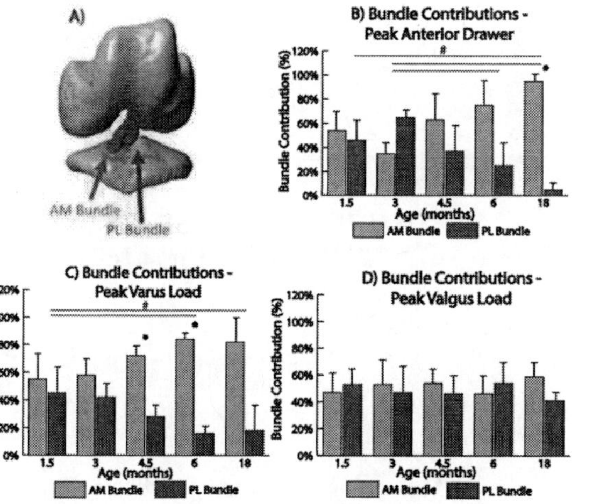

Figure 2: Function of the AM and PL bundles depicted in (A) varies with age. Increasing age results in an increased contribution of the AM bundle to the overall ACL behavior under an applied anterior drawer at 60° of flexion (B). Additionally, the AM bundle is increasingly dominant with age under a varus moment (C), while the AM and PL bundles contribute similarly across age groups under a valgus moment (D). * represents statistical significance between bundles, # represents statistical significance between ages (p=0.05).

ACKNOWLEDGEMENTS

We would like to acknowledge the Swine Educational Unit at NC State for their contributions to this work. Funding provided by NIH (R03 AR068112) and NSF GRFP (DGE-1252376).

REFERENCES

[1] Paterno, MV et al. *Am J Sports Med,* 42(7): 1567-1573, 2014; [2] Kim, HK et al. *Radiology,* 826-835, 2008; [3] Tuca, M et al. *Knee Surg Sports Traumatol Arthrosc,* 24(3): 780-787, 2016; [4] Livesay, GA et al. *J Orthop Res,* 278-284, 1997; [5] Xerogeanes, JW et al. *Annals Biomed Engr,* 26(3): 345-352, 1998; [6] Proffen, BL et al. *Knee,* 19(4): 493-499, 2012; [7] Cone, SG et al. SB3C 2016; [8] Cone, SG et al. ORS 2016; [9] Fisher, MB et al. *J Orthop Res,* 1373-1379, 2010; [10] Zamarra, G et al. *Knee Surg Sports Traumatol Arthrosc,* 11-19, 2010; [11] Rudy, TW et al. *J Biomech,* 29(10): 1357-1360, 1996.

THREE DIMENSIONAL STRAIN ANALYSIS OF THE HUMAN ANTERIOR CRUCIATE LIGAMENT DURING ANTERIOR TIBIAL TRANSLATION

Satoshi Yamakawa (1), Richard E. Debski (2), Hiromichi Fujie (3)

(1) Division of Human Mechatronics Systems,
Graduate School of System Design,
Tokyo Metropolitan University
Tokyo, Japan

(2) Bioengineering Department,
Faculty of Engineering,
University of Pittsburgh
Pittsburgh, PA, USA

(3) Department of Intelligent Mechanical Systems,
Faculty of System Design,
Tokyo Metropolitan University
Tokyo, Japan

INTRODUCTION

Quantitative biomechanical data on the strain behavior of the anterior cruciate ligament (ACL) during physiological knee motion are necessary for better understanding of the mechanical function of the ACL and the improvement of ACL reconstruction. Although strain distribution in the ACL was obtained in previous studies [1, 2], physiological knee motion and loading were not simulated due to experimental limitations. In addition, it was impossible to keep the intact state of ACL surface because of physical markers and chemical liquid for measurement. To solve these problems, both a robotic joint test system and a novel three dimensional image analysis method were applied to the determination of strain distribution in the ACL during anterior tibial translation in the present study.

METHODS

A 6-DOF robotic system consisting of a custom-made 6-axis manipulator with a 6-DOF universal force/moment sensor (FR-2010, Technology Service, Japan) was used. In the system, a LabView-based control program runs on a windows PC to control the displacement of, and force/moment applied to, the cadaveric knee joints with respect to the knee joint coordinate system [3]. Human knees (n=3) were dissected down to the joint capsule and fixed to the robotic system, and were subjected to the anterior drawer test up to 100 N at maximum knee extension and 30 degree of knee flexion. During the test, the anterior-posterior DOF was translated under displacement control while all the DOFs except the AP and flexion-extension DOFs were set under force control with prescribed force/moment at zero. After the lateral condyle was removed the ACL surface was recorded from two directions using a CCD camera. Then, the intact knee motion was reproduced to the knee using the robotic system while recording the ACL surface using the CCD camera. At 100 N of anterior loading, the ACL surface was recorded again from the two directions. In the recorded image, the ACL was transversely divided into 5 portions; 0-10%, 10-20%, 20-80%, 80-90% and 90-100% of the length from the femoral attachment, and was longitudinally divided into 4 portions from anterior to posterior fibers (Figure 1).

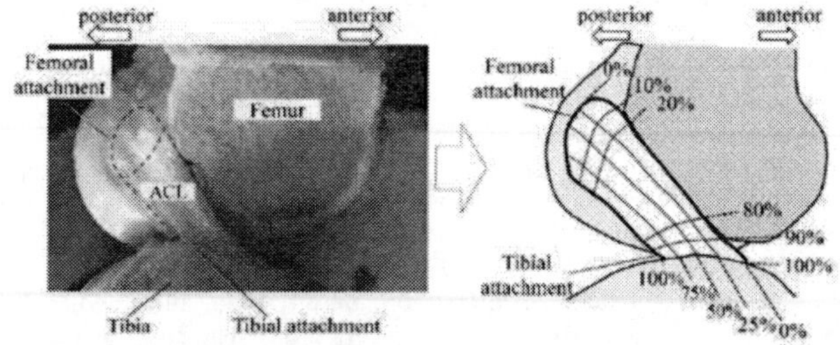

Figure 1: Recorded image of the medial surface of the ACL (left) and twenty portions on the ACL medial surface for strain calculation (right)

Multiple markers were fixed to the proximal and distal edges of each portion on the recorded image using a motion tracking system based on image correlation method (Motion Analyzer, Keyence, Japan). Two-dimensional positions of all the markers with respect to the two camera coordinate system at 0 N and 100 N of anterior drawer load were determined using the motion tracking system. The three-dimensional positions of all the markers were, then, calculated from the two sets of the two-dimensional data using geometric data of the CCD camera with respect to the knee. The above-described three-dimensional position calculation, called rotational stereo imaging method, was developed in the present study. Using the position data of the makers, length changes in all portions were used to analyze local strains in the ACL in response to 0 N and 100 N of anterior loads. Note that zero strains (no deformations) were defined when the knee was located at maximum extension with 0.5 Nm of extension moment. In addition, strain-anterior force relationship during anterior loading was also determined using a normal 2-D strain analysis.

RESULTS

Table 1 and 2 indicated that strain distribution in the ACL in response to 100 N of anterior load at maximum extension and 30 degree of knee flexion. Strain in the ACL was distributed throughout the entire ACL and large strains were observed in the ACL attachments area (0-20% and 80-100%) at maximum extension. At 30 degree of flexion, ACL strain shifted to anterior regions with no strained regions found in posterior and mid-posterior regions. Note that ACL strain in the anterior fibers was higher at 30 degree of flexion than at maximum extension. Regarding the relationship between ACL strain and anterior force during anterior loading, strain increased constantly in anterior and posterior fibers in the femoral attachment area at maximum extension. In contrast, at 30 degree of knee flexion, strain increased rapidly at the beginning of anterior load application, and gradually thereafter in the anterior fiber (Fig.2). Note that the results from one specimen at 30 degree of knee flexion were excluded because the direction of a line passing adjacent two markers was rotated more than 15 degree.

DISCUSSION AND SUMMARY

In the present study, the strain distribution in the human ACL during anterior translation was determined using the 6-DOF robotic system and the rotational stereo imaging method. In previous studies, video tracking systems were often involved to analyze deformation and strain in soft tissues. Although high resolution analysis was possible, there were limitations in position and number of markers with the systems. It was also problematic in video tracking systems that relative motion between markers and tissues affected experimental results. These problems are solved in the present method by the use of the rotational stereo imaging method. More importantly, it was possible to determine the three-dimensional position of markers with the method, which allows us to analyze the 3-D strain distribution of undulated structures such as the human ACL at flexion. Sakane et al, and Fujie et al reported that in situ force in the anterior fiber in the ACL in response to anterior drawer force is increased with the increase of knee flexion, while in situ force in the posterior fiber is decreased [4, 5]. In the present study, a similar trend was found in the strain distribution. More interestingly, we found a load-dependency in the relationship between strain and anterior force. It is well known that ligament and tendon tissues have crimp structure at the sub-ligament and sub-tendon level. Franchi et al reported that the crimp structure is more located in both ends of ligaments and tendons than in mid substance [6]. Weiss et al reported that a wavy crimp structure was found in the proximal one quarter of the ACL [7]. Considering the obtained results together with the previous reports, it is suggested that the strain on the ACL is site- and force-dependent probably due to unevenness in both distribution and property/function of crimp structures. The unique and complex characteristics of the ACL strain distribution are important for maintaining the mechanical function and stability of the knee joint in response to various physiological loadings.

Table 1 Strain distribution in the ACL in response to 100 N of anterior force to the knee at maximum extension

Length \ Width	Posterior 75-100%	50-75%	25-50%	Anterior 0-25%
0-10%	4.6	3.6	1.1	3.0
10-20%	2.2	1.7	1.9	3.9
20-80%	2.3	1.7	1.4	1.0
80-90%	5.2	3.7	1.2	3.2
90-100%	3.1	2.3	2.3	0.3

Table 2 Strain distribution in the ACL in response to 100 N of anterior force to the knee at 30 degree of flexion

Length \ Width	Posterior 75-100%	50-75%	25-50%	Anterior 0-25%
0-10%	1.9	2.6	4.8	5.0
10-20%	-	3.1	5.1	4.7
20-80%	-	-	0.9	2.6
80-90%	-	3.1	0.8	3.8
90-100%	-	-	-	1.9

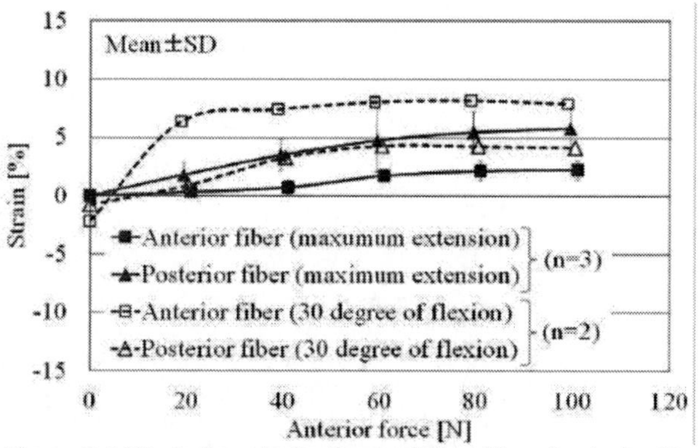

Figure 2 ACL strain-anterior force relationships of anterior (0-25%) and posterior (75-100%) fibers in the femoral attachment area (0-10%) at maximum knee extension and 30 degree of knee flexion

ACKNOWLEDGEMENTS

The present study was financially supported by the JSPS KAKENHI (#16J06537) and the Grant-In Aid Scientific Research B (#16H03172) from the MEXT.

REFERENCES

[1] Yamamoto K., et al., Medical Engineering & Physics, 1998
[2] Spalazzi J.P., et al., Wiley InterScience, 2006
[3] Grood, E.S. and Suntay, W.J., J. Biomech. Eng. (ASME), 1983.
[4] Sakane M., et al., JOR, 1997
[5] Fujie H., et al., Knee Surg. Sports Traumatol Arthrosc., 2011
[6] Franchi M., et al., Connect Tissue Res., 2012.
[7] Weiss M., et al., Knee Surg. Sports Traumatol Arthrosc., 2012.

SB³C2017
Summer Biomechanics, Bioengineering and Biotransport Conference
June 21 – 24, Tucson, AZ, USA

EVALUATING THE APPROPRIATENESS OF TRANSVERSELY ISOTROPIC CONSTITUTIVE THEORIES FOR STRUCTURAL LIGAMENTS

Benjamin C. Marchi (1), Callan M. Luetkemeyer (1), Ellen M. Arruda (1,2,3)

(1) Department of Mechanical Engineering
University of Michigan
Ann Arbor, MI, USA

(2) Department of Biomedical Engineering
University of Michigan
Ann Arbor, MI, USA

(3) Program in Macromolecular Science and Engineering
University of Michigan
Ann Arbor, MI, USA

INTRODUCTION

The mechanical responses of soft tissues are complex and specific, potentially incorporating components of nonlinearity, anisotropy, and viscoelasticity. The challenges associated with modeling these inherent constitutive nuances can be aggravated by nontrivial tissue geometries, as well as possible regional heterogeneity within the tissue bulk—manifesting from either microstructural or compositional spatial variation. Yet, despite these obstacles, computational models of soft tissues, particularly at the continuum level, have proven to be of tremendous clinical value. The knee has been especially amenable to continuum modeling approaches. Computational methods in structural biomechanics have been successfully employed to investigate various healthy[1-2] and diseased knee pathologies[3-4]. Additionally, computational models, in particular finite element (FE) models, are well suited to study the efficacy of reconstructive ligament procedures[5-6] and are easily adaptable to parametric investigations[7-8].

However, given the convenience and availability of continuum level formulations of biomechanical problems, differentiating between various constitutive theories can prove challenging. It is important to recall that these constitutive descriptions are at the center of continuum mechanics, and they can be postulated in myriad ways. The fundamental differences among theories play a crucial role in dictating the predicted response of a system, potentially with meaningful clinical consequences, and the conclusions generated from computational biomechanics models are entirely predicated on the veracity of the underlying constitutive theory—i.e., a model built with one description could predict normal motion, while another may predict injury or disease. Therefore, constitutive theories must be judged on how faithfully they describe observed deformation, and the fidelity with which they predict material responses not used in their formulation.

It would be a difficult problem to evaluate constitutive theories in a world with perfect and complete information, but physiological reality does not afford such luxuries. Experiments are costly and challenging, and the data generated from them is often inconsistent and at times conflicting. Therefore, approximations and simplifications are made to account for limited characterization data. For supporting ligaments in the knee, constitutive theories have been developed to address these issues, predominately assuming transverse isotropy with the preferred material direction aligned with structural collagen. This class of constitutive theories is well motivated by the microstructure of ligaments, but experimental evidence mechanically supporting the claim of transverse isotropy can be categorized as sparse at best[9-11]. Generally, wide variation in the expected response of a given ligament type is accepted as intrinsic patient-to-patient variability. This phenomenon is intensified by the limited availability of transverse quasi-static uniaxial behavior, which has shown orders of magnitude variation for the medial collateral ligament (MCL)[10-11].

This work attempts to judiciously describe the process by which candidate constitutive theories can be evaluated in the context of transverse hyperelasticity using quasi-static stress-strain experimental data of ligament structures in the knee. In particular, the anteromedial (AM) and posterolateral (PL) bundles of the anterior cruciate ligament (ACL) and the MCL are examined. We show how recent data for the MCL, both from the literature[11] and obtained herein, as well as from both bundles of the ACL, fail to be accurately represented assuming the form of previously validated and generally accepted constitutive theories. We widen the set of potential constitutive models to include chain-based statistical mechanics and novel hybrid theories to explore the range of representative models for ligaments, showing how the shape of the transverse stress-strain behavior displayed by ligaments is inconsistent with commonly employed constitutive descriptions.

METHODS

In addition to relevant quasi-static stress-strain behavior the MCL both along and perpendicular to the tissue average collagen orientation available in the literature[10-11], each ACL bundle and the MCL were tested in multiple configurations. Instead of assuming applied displacements at the boundary mapped exactly to the tissue level, full-field, noncontact strain fields were calculated throughout the loading using digital image correlation (DIC). DIC provides a complete surface description of deformation, which for ligaments is generally inhomogeneous. For the purposes of evaluating continuum models at the tissue length scale, surface strains were homogenized to obtain a representative, yet accurate, strain at each increment of displacement.

The suitability of numerous transversely isotropic, hyperelastic constitutive models containing a single material direction were studied in the context of describing the direction-dependent, quasi-static stress-strain behavior of each ligament structure. Transversely isotropic, exponential-based material models, commonly employed in continuum descriptions of ligaments[3,6-7], were evaluated with respect to the ease of their construction, effectiveness, and respective FE implementation. Additionally, chain-based statistical mechanics models, which have recently been successful in describing ligament deformation[1], were considered. Motivated by the microstructure of ligaments, a new hybrid model—consisting of a compressible form of an isotropic, eight-chain MacKintosh network model with Holzapfel-Gasser-Odgen (HGO)[12] type behavior along the preferred material direction—was proposed (referred to as MAC/HGO) as an alternative continuum description.

RESULTS

Stress-strain responses of the MCL are shown perpendicular to the averaged collagen orientation in Fig. 1. It is interesting to note the presence of nonlinearity—manifesting as initial concavity—in both previous studies[10-11], though there is significant variation in the extent of this behavior. There are also orders of magnitude differences between the axial (not shown) and transverse (Fig. 1) stress-strain responses, indicating a high degree of tissue level, bulk anisotropy.

The appropriateness of a particular constitutive form was predominately determined by the shape of the stress-strain behavior normal to the preferred material orientation—shown in Fig. 1. Constitutive theories performed adequately independent from model type assuming transverse data that was nearly linear in the preferred direction of loading, like that presented previously for the MCL[10] (with $R^2 > 0.97$ both along and normal to the primary collagen direction). Figure 1 illustrates the ability of neo-Hookean type behavior in the transverse direction, typical of the HGO type models, in generally capturing the stress-strain response in certain situations; however, note that even in the best case, there is a discrepancy between the shape of the stress-strain data (slightly concave) and the model fitting (slightly convex).

This inconsistency is exaggerated with data of the form presented recently[11], as well as obtained in our concurrent work (not shown). Data were not well represented by most constitutive model forms, save for the MAC/HGO theory (Fig. 1). Though there is almost no reduction in predictive power along the preferred material direction— $R^2 > 0.95$ for recent MCL data[11] (not shown)—it is accompanied by a loss in representativeness of the transverse response—from $R^2 > 0.97$ assuming nearly linear data[10] to $R^2 < 0.78$ with nonlinear behavior[11] (Fig. 1 shows a representative predictive inability of typically employed ligament constitutive theories). The hybrid MAC/HGO model outperforms its peers in describing the data that are nonlinear in the transverse direction[11] (Fig. 1), with $R^2 = 0.98$ and $R^2 = 1.0$ along and normal to the preferred material orientation, respectively.

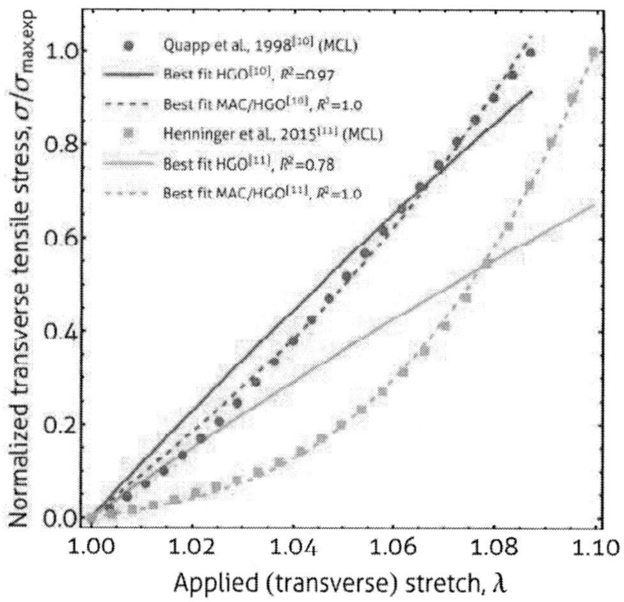

Figure 1: Normalized stress-stretch behavior of the MCL normal to the preferred material direction with a sampling of bit fit constitutive theory responses. Stresses are normalized with respect to their corresponding maximum observed experimental value.

DISCUSSION

The shape and magnitude of the axial and transverse data obtained in the current work are largely consistent with that of recent observations for the MCL[11]. Moreover, the homogenized surface strain descriptions provided by DIC also yield stretch information normal to the direction gauging the predictive power of constitutive theories. For ligaments, the performance of constitutive theories is largely tied to the form of the transverse stress-strain behavior. Initial concavity in the transverse stress-strain response is impossible to faithfully model when the isotropic response of the material is assumed to be linear or nearly linear, as is commonly the case currently. In particular, models which are exponential in I_4 and transversely isotropic chain-based models struggle to capture the initial nonlinearity of the transverse material response as they are traditionally posed.

ACKNOWLEDGEMENTS

This material is based upon work in part supported by the National Science Foundation Graduate Research Fellowship under Grant No. DGE 1256260.

REFERENCES

[1] Marchi, BC et al., *Biomech model mechanobiol*, 1-12, 2016.
[2] Mootanah, R et al., *Comp in biomech and bio eng*, 17:1502-1517, 2014.
[3] Peña, E et al., *Computers in biol and med*, 37:376-387, 2007.
[4] Shirazi, R et al., *J biomech*, 42:2458-2465, 2009.
[5] Shirazi, R et al., *Clin biomech*, 24:755-761, 2009.
[6] Westermann, RW et al., *J knee surg*, 2016.
[7] Kiapour, A et al., *J biomech eng*, 136:011002, 2014.
[8] Marouane, H et al., *J biomech*, 47:1353-1359, 2014.
[9] Butler, DL et al., *J of biomech engin*, 112:38-45, 1990
[10] Quapp, KM et al., *J of biomech engin*, 120:757-763, 1998
[11] Henninger, HB et al., *Acta biomat*, 25:304-312, 2015.
[12] Gasser, CT et al., *J royal society inter*, 3:15-35, 2006.

ESTABLISHING THE PROPER REFERENCE CONFIGURATION FOR FINITE ELEMENT MODELS OF THE SUPRASPINATUS TENDON

**R. Matthew Miller (1), James Thunes (1),
Volker Musahl (2,1), Spandan Maiti (1), Richard E. Debski (1,2)**

(1) Department of Bioengineering
University of Pittsburgh
Pittsburgh, PA, USA

(2) Department of Orthopaedic Surgery
University of Pittsburgh
Pittsburgh, PA, USA

INTRODUCTION

Accurate calculations of tissue strains are dependent upon the reference configuration used [1-4]. For investigation of musculoskeletal connective tissues, the reference configuration is typically determined for the joint position and preload at which all collagen fibers in the tissue start to be loaded (i.e. all "slack" within the tissue is removed). Failure to adequately define an appropriate reference configuration can lead to over- or under-estimation of tissue strains. The reference configuration must be rigorously defined to ensure that all experimental measures and computational predictions accurately represent the mechanical behavior of the tendon. The tendons of the rotator cuff in particular inherently experience uneven loading due to their inhomogeneous tensile properties, as well as from the natural curvature of the tendon at its insertion to the humerus. Preliminary experiments to measure supraspinatus tendon surface strains using a preload of 5 N as the reference configuration resulted in inaccurate calculations of strain for validation of a finite element model of the supraspinatus tendon. Therefore, in order to account for the inhomogeneous mechanical behavior of the supraspinatus tendon, rigorous analysis of experimentally measured tendon surface strains for multiple loading magnitudes was performed to establish an appropriate reference configuration.

METHODS

An intact, fresh-frozen cadaveric shoulder (male, 70 years old) was dissected to remove all soft tissue except for the rotator cuff tendons, which were left attached to their insertions on the humerus. The presence of any tendon thickenings (such as the rotator cable) was noted and their dimensions were recorded. A 6 x 4 array of black, plastic markers was affixed to the bursal and articular surfaces of the tendon using cyanoacrylate glue (Figure 1) for determination of tendon surface strains during experimental testing. This resulted in 30 surface elements (15 on each side) constructed from sets of four surface markers (Figure 1-C). The 2 mm-diameter markers were placed 5 mm apart, with the bottom row located just medial to the tendon insertion on the greater tuberosity of the bursal side. On the articular side, the bottom row of markers was placed approximately 1 cm medial to the tendon insertion to allow marker visibility.

The humerus was then secured to a materials testing machine (Model 5965, Instron Corp, Norwood, MA), and the supraspinatus tendon was then clamped to the crosshead for cyclic loading at 70° of glenohumeral abduction. This joint position allowed visualization of the markers on the bursal and articular surfaces of the tendon and minimized wrapping of the tendon around the humeral head. A 22N load was placed on the infraspinatus via a pulley system to simulate the load-sharing interaction with the supraspinatus tendon [5]. A four-camera optical tracking system (DMAS7 software, Spica Technology, Kihei, Maui HI) was used to track marker positions on the articular and bursal sides of the tendon throughout testing [6]. This system allows for calculation of three-dimensional surface strains on both tendon surfaces simultaneously.

A 5N preload was first applied to the supraspinatus tendon at 70° of glenohumeral abduction, followed by preconditioning from 5-50N for 10 cycles. Next, uniaxial cyclic tensile loading from 5-200N was performed for 50 cycles at 20 mm/min. During the 5-200N cyclic loading sets, bursal and articular surface strains were calculated at peak loads ranging from 10-200N in increments of 10N (i.e. 10N, 20N, 30N, …, 200N). For each loaded configuration, strain was calculated as the deformation relative to the recorded 5N preload state for each of the 15 elements on the bursal and articular surfaces (Figure 1-C). The strain for all elements on each surface was then averaged. Additionally, the difference in strain between loaded configurations was calculated for each load increment (i.e. strain at 20N minus strain at 10N, strain at 30N minus strain at 20N, etc.). The calculated difference in strain was plotted versus applied load, representing the first derivative of the strain versus load relationship. The updated preloaded state (i.e. reference configuration) was defined as the load at which the minimum inflection point was located on this plot, representing the point at which collagen fibers begin to be loaded (Point 3, Figure 2-B).

RESULTS

Average strains versus load at 70° of abduction were similar for bursal and articular surfaces. From 5-20N, strain increased non-linearly with load as collagen fibers in the tendon began to be stretched (Point 1 to 2, Figure 2-A). At ~20N, the curve reached a plateau for which load increased but the average strain was constant (Point 2 to 3, Figure 2-A). At the end of this plateau region (40N on the bursal side and 30N on the articular side), strain again began to increase. These loads corresponded with the minimum inflection point on the plot of difference in strain versus load (Point 3, Figure 2-B).

DISCUSSION

The difference in position of the minimum inflection point between articular and bursal surfaces indicates uneven loading between surfaces, where the collagen fibers on the articular side reach a "preloaded" state earlier. The presence of the rotator cable and the natural curvature of the tendon, especially near the enthesis, could cause this phenomenon. It is therefore important to consider the uneven loading between the bursal and articular surfaces when establishing a reference configuration and when making comparisons between surfaces. This observation is supported by previous studies that found the articular surface to have higher strains [7,8] and incidence of partial tears compared to the bursal surface [9,10].

Based on these considerations, a 40N preload at 70° of glenohumeral abduction was chosen as the reference configuration to ensure that slack is fully removed from the entire tendon. This reference configuration was used for experimental strain calculations as well as for computational model simulations. It is important to consider that this reference configuration is specific to this specimen, loading conditions, and experimental protocol. Since the reference configuration can drastically affect strain calculations, future experiments and computational models of soft tissues should rigorously establish the reference configuration on a specimen by specimen basis.

ACKNOWLEDGEMENTS

Support from The Albert B. Ferguson, Jr., M.D. Orthopaedic Fund (AD2015-1765-23), the Department of Orthopaedic Surgery, and Pittsburgh Chapter of the ARCS Foundation is gratefully acknowledged.

REFERENCES

[1] Debski et al. JSES. 2005;14:24S-31S.
[2] Ellis et al. JOR. 2006;24(4):800-810.
[3] Wagner et al. JOR. 2004;22(4):901-909.
[4] Lake et al. JOR. 2009;27(12):1596-1602.
[5] Miller et al. JOR. 2014;32(10):1283-1289.
[6] Moore et al. Ann Biomed Eng. 2010;38(1):66-76.
[7] Reilly et al. JBJS Br. 2003;85(4): 594-599.
[8] Huang et al. JOR. 2005;23(4):924-930.
[9] Payne et al. AJSM. 1997; 25(3):299-305.
[10] Sakurai et al. JSES. 1998;7(5):510-515.

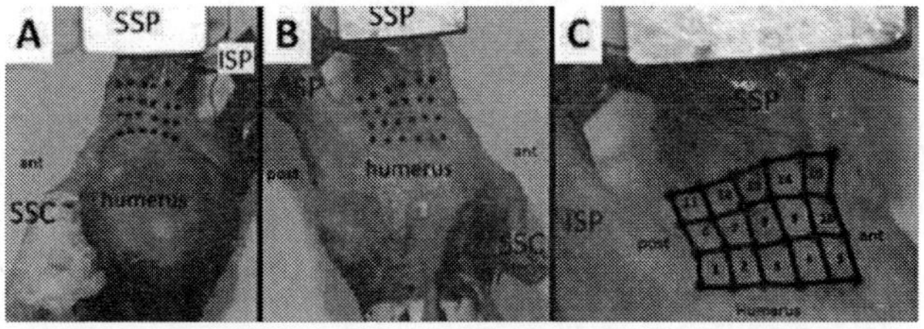

Figure 1: Mechanical testing and geometry of supraspinatus tendon. A) Articular and B) bursal surfaces of the supraspinatus tendon. C) 6 X 4 array of markers for strain calculations during the experiment. (SSP: Supraspinatus tendon; ISP: Infraspinatus tendon; SSC: Subscapularis tendon.)

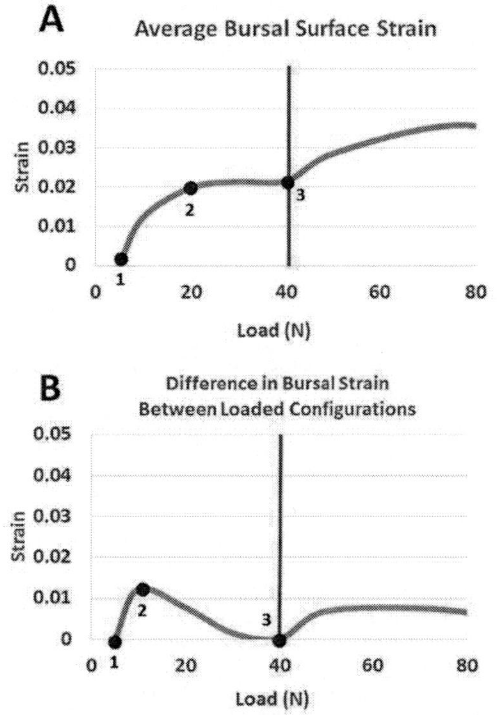

Figure 2: A) Average strain and B) Average difference in strain versus load for the bursal tendon surface at 70° of glenohumeral abduction. From 0-20N, measures of strain increased quickly due to "unfolding" of the tendon tissue as collagen fibers began to be loaded (points 1 to 2). The minimum inflection point on the bursal side was located at 40N (marked by red lines, point 3). This inflection point denotes the end of a plateau in the curve of strain versus load that is indicative of rigid body motion and the tissue taking up load without deforming uniformly (points 2 to 3).

SB³C2017
Summer Biomechanics, Bioengineering and Biotransport Conference
June 21 – 24, Tucson, AZ, USA

A CELL-BASED CROSS-CORRELATION IMAGING ANALYSIS METHOD FOR QUANTIFICATION OF 3-D TENDON STRAINS

Ashley K. Fung (1), J. J. Paredes (2), Rebecca Bell (1), Nelly Andarawis-Puri (1,2,3)

(1) Sibley School of Mechanical and Aerospace Engineering
Cornell University
Ithaca, NY, USA

(2) Nancy E. and Peter C. Meinig School of Biomedical Engineering
Cornell University
Ithaca, NY, USA

(3) Hospital for Special Surgery
New York, NY, USA

INTRODUCTION

Tendons are load-bearing musculoskeletal tissues that play an essential role in allowing joint motion with stability. Macroscopic mechanical loads modulate cellular responses that can lead to disparate outcomes such as maintenance of homeostasis, degeneration, repair, or remodeling. However, it remains unknown how macroscopic deformations to the tendon modulate the cellular response on a microscopic level, leaving a knowledge gap in our understanding of the resulting mechanotransduction pathways. We hypothesize that the unique relationship between matrix and cell deformation will dictate biological outcomes. A first step towards addressing this hypothesis is the development of a tool that uses the innate microstructure of the tendon – a 3-D speckle pattern of stained cells – which has the potential for longitudinal *in vivo* measurements during healing or degeneration. Therefore, the objective of this study was to develop and validate a cross-correlation based method for quantification of matrix strain through the use of tetrahedral finite elements that can be used to assess the *in vivo* temporal response of the tendon.

METHODS

Image Acquisition: Patellar tendons from freshly sacrificed C57BL/6 mice were stained with Syto 9 (Thermo Fisher) nucleic acid stain and subjected to a nominal load of 100g. Under a 20x objective lens, Z-stack images at 1 μm intervals through a depth of 80 μm were acquired using an upright multiphoton microscope tuned to a wavelength of 960 nm. Images had a resolution of 2048x2048 spanning an area of 425.1μm x 425.1μm.

Cross-Correlation Algorithm: A custom MATLAB program was developed to calculate strain between an unloaded and loaded image. This algorithm is unique from other existing methods for strain calculation in that it accounts for out-of-plane motion in a 3-D volume, making it ideally suited for analysis of 3-D *in vivo* strains, and these considerations are necessary in damaged and diseased tendons. Morphological top-hat filtering was used to correct the uneven background thus removing noise and highlighting the stained nuclei. Four additional slices were linearly interpolated between images such that the resolution in any plane was equal to 0.2μm /pixel.

A 7x7x3 grid of nodes was established on both images with spacing of 200, 200 and 100 pixels in the X, Y, and Z directions. On the unloaded image, each node served as the center of a 200x200x100 pixel template volume. A 400x400x150 search volume was established, and a template matching, cross-correlation algorithm, which makes use of the convolution theorem, was performed to find the location that best correlates with the template. A normalized correlation coefficient matrix was identified for each node utilizing fast Fourier transforms of the template and search volumes [1], and the location of highest correlation was noted as the initial displacement.

To ensure confidence in the identified displacement values, a second search iteration was implemented, which moved the center of the search volume to the location identified in the first iteration. This was followed by a third, which continually reduced the volume of the template until the solution converged to a single location. Final displacements in the X, Y, and Z-direction were recorded.

Strain Calculation: Delaunay triangulation was utilized to form a mesh on the 3-D volume consisting of 147 nodes and 432 tetrahedral elements. Tetrahedrals were treated as elements of constant strain, and the Lagrangian strain tensor was calculated for each to find average and standard deviations of ε_{xx}, ε_{yy}, and ε_{zz} (Fig. 1).

Validation: To test the accuracy of the analysis, images were scaled using bilinear interpolation in ImageJ to simulate local tissue strains ranging from 0 to +10%, -5 to 0% and -10 to 0% in the X, Y, and Z-directions, respectively. The direction parallel to uniaxial loading and collagen fiber orientation was denoted as the X-direction. The cross-correlation algorithm was then used to conduct the strain analysis on the original "unloaded" Z-stack and scaled "loaded" images.

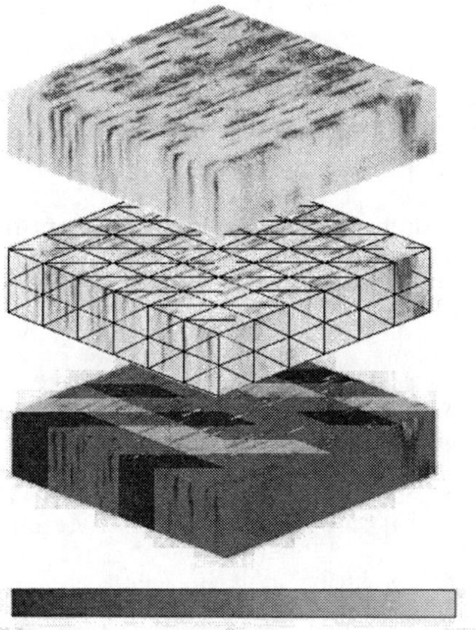

Figure 1: A finite element tetrahedral mesh was imposed on the 3-D volume [2], and an example strain map for an expected normal strain value of 1%.

RESULTS

Strains, ranging from 0.5% to 10%, were first applied to the 3-D image in the X-direction. Percent error between the expected values and the average normal strain, ε_{xx}, was calculated for each level of applied strain (Fig. 2). In the X-direction, error was found to be independent of applied strain ($p = 0.7663$) with an average error of 3.3%. Because the resolution of the image in the X direction equals that in the Y, we expect errors in this direction to be largely the same. For values of strain less than 7%, the algorithm detected no strain in the Y and Z directions, as expected, whereas large strain values resulted in non-zero determinations of strain in these two directions.

This process was repeated for the Z-direction (with no strain along the X and Y directions) to assess the resolution on output strain values. In contrast to the X-direction, errors were observed to be greatest at lower values of applied strain ($p = 0.0065$). Percent errors, however, reached values consistent with those observed in the X-direction at strain values greater than 2% (Fig. 2). Further analysis showed that while average errors in the X-direction were consistently lower than in the Z-direction for small strains, smaller standard deviations suggest that the minimum one pixel displacement allowed was a large contributor to the percent error observed.

DISCUSSION

We have developed an image analysis tool to determine local tissue strains in mouse patellar tendons using a speckle pattern of stained cells and by implementing a cross-correlation algorithm to calculate strain tensors across a 3-D tetrahedral finite element mesh. Characterizing the 3-D strain environment can provide insight as to how the relationship between the macro and microenvironment influences function in healthy, diseased, and injured tendons. Results support that multiphoton microscopy can be applied to transgenic reporter mice, in which cells are fluorescent, for *in vivo* temporal assessment of cell and matrix deformations.

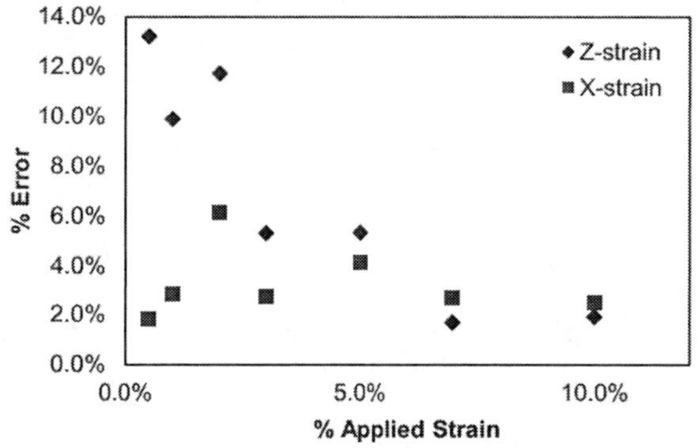

Figure 2: Percent errors in the X- and Z-directions differed at small strain values but appear to converge as applied strain levels increased.

The use of interpolation tools and a smaller overall dimension in the Z-direction are likely causes of greater error in this direction as compared to the X and Y directions. Small standard deviations suggest that these results can be improved through Z-stacks acquired at smaller intervals, which would minimize the use of interpolated data and increase the resolution in this dimension. This method, however, would require greater image acquisition and processing times.

Previous studies have used surface markers to determine strain or cell nuclei to track relative strain between cells [3-4]. These analyses are limited in information, as they capture strain in specific regions – those along the top surface of the tendon or where cells are abundant. The benefit of a 3-D analysis, as conducted here, not only provides a more comprehensive picture of the local strain environment but also includes the ability to consider out-of-plane motion of the cells which can occur as a result of misalignment with the direction of uniaxial loading or anisotropic material properties.

While the validation of this analysis method was conducted using naïve mouse patellar tendons, the results support that the cross-correlation algorithm can be extended to assess strain measurements in diseased and damaged tendons where the organization may be different than the microstructure observed here. The development of methods to quantify cell deformation using the innate tendon structure of stained cells is ongoing. Combined with the matrix strain method described, a successful method for determining cell deformation will draw us closer to understanding how the relationship between matrix and cell deformation influences the biological response.

ACKNOWLEDGEMENTS

Imaging data was acquired through the Cornell University Biotechnology Resource Center with NIH S10OD018516 funding.

REFERENCES

[1] Kroon, D. Fast/Robust Template Matching, MATLAB Central File Exchange, 2009. [2] Woodford, O. vol3d, MATLAB Central File Exchange, 2009. [3] Andarawis-Puri, N et al. *J Biomech*, 42(2):158-163, 2009. [4] Screen, H et al. *J Strain Analysis*, 44(5):327-335, 2009.

A NEW METHOD TO DETERMINE SUBJECT-SPECIFIC PROPERTIES OF KNEE LIGAMENTS USING BAYESIAN CALIBRATION

Mohammad Kia (1), Jonathan A. Race (3), Po-Hsu Chen (3), Thomas L. Wickiewicz (2), Andrew D. Pearle (2), Thomas Santner (3), Carl W. Imhauser (1)

(1) Department of Biomechanics
Hospital for Special Surgery
New York, NY, USA

(2) Department of Orthopedic Surgery
Hospital for Special Surgery
New York, NY, USA

(3) Department of Statistics
The Ohio State University
Columbus, OH, USA

INTRODUCTION

Present-day surgical guidelines for treating knee ligament injures are essentially one-size-fits-all. They do not account for immense variations in knee laxity between individuals despite the important role that such variations play in risk of ligament and graft rupture[1]. The heterogeneity in knee laxity is likely due to interpersonal variations in the articular and ligamentous stabilizers [2]. However, the numerous soft tissues of the knee, the interplay of ligament slack length (i.e., the length at which the ligament begins to carry force) and stiffness, and the wide variability in these properties makes identifying the mechanisms controlling knee laxity challenging. Without knowledge of the stabilizing mechanisms of each patient's knee, personalized guidelines for how to best restore laxity cannot be generated.

Computational models hold tremendous promise for identifying subject-specific mechanisms of knee laxity. Subject-specific articular anatomy and soft tissue attachment sites can be easily incorporated into such models via medical imaging. However, in many cases, knee models include only standardized slack lengths and average stiffnesses for the ligaments [3, 4], which severely limit the models' ability to be used to develop patient-specific treatments for ligament injury.

Therefore, we have integrated Bayesian statistical methods to estimate ligament properties of our computational knee models. By utilizing knowledge on the population distributions of the ligament properties, the Bayesian approach enables a more informed estimate of these properties relative to commonly-used optimization strategies[5]. Despite its advantages, this approach has not been used to obtain subject-specific ligament properties. Thus, the goal of this work is to develop a statistically-calibrated computational model of the knee to identify critical combinations of patient-specific ligaments and ligament properties that control knee laxity.

METHODS

Statistical calibration of ligaments of the computational model of the knee involved three steps. First, we developed a multibody dynamics model of a cadaveric knee from a 20-year-old male. We will refer to this as the knee simulator. The knee simulator consisted of subject-specific geometries of the bones, articular cartilage, menisci, and ligament insertions obtained using computed tomography (CT) (Fig. 1)[6]. The collaterals, cruciates, capsule, and meniscal attachments were represented with 42 force elements with tension-only, nonlinear force-elongation responses. Initially, mean structural properties were utilized for all the ligaments, while slack lengths were determined using a previously-defined algorithm[6]. These ligament stiffnesses and slack lengths were considered our 'default' values. We simulated clinical tests of rotatory laxity by rigidly fixing the femur and maintaining the tibia in 15° flexion, leaving it with five degrees of freedom. Then, multiplanar torques consisting of 8 Nm of valgus and 4 Nm of internal rotation were applied under 10 N of compression. In each loading phase, the anterior-posterior (AP) translations of the medial and lateral compartments of the tibia were measured to assess knee laxity.

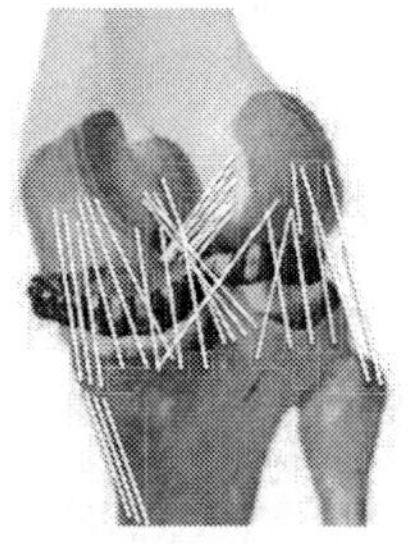

Figure 1: Multibody model (i.e., knee simulator) of the tibiofemoral joint.

Second, sensitivity analyses were conducted to identify the most active ligaments and ligament parameters to be included in Bayesian calibration. A total of 900 simulations were run in which the knee simulator inputs were varied over at least ±2 standard deviations of their mean values reported in the literature. These inputs consisted of slack length, toe region, and linear stiffness of the anterior cruciate ligament (ACL), posterior cruciate ligament (PCL), medial collateral ligament (MCL), posterior oblique ligament (POL), lateral collateral ligament (LCL), anterolateral ligament (ALL), and lateral meniscal fascicle (LMF). The sensitivity of AP translations of the medial and lateral compartments to variations in the knee simulator inputs in response to the multiplanar torques was determined by calculating main effect sensitivity indices (SI) for each input. The main effect SI of an input is the proportion of the total variation in the outputs (i.e., the compartmental translations) due to changes in said input.

In the third and final step, we conducted a Bayesian calibration using the soft tissue properties that had the largest influence on the AP translations of the medial and lateral compartments as calibration parameters. Bayesian calibration produces a distribution of likely values for the calibration parameters, and it requires data from both the knee simulator and from a physical experiment. Thus, we conducted biomechanical testing (i.e., a physical experiment) of rotatory laxity of the same cadaveric knee from which the knee simulator was developed using a robotic manipulator[7]. That is, we used the same test of rotatory laxity *in vitro* that we used *in silico*. Superposition was also used to determine the force carried by the ACL. All loads, boundary conditions, and kinematic definitions were maintained between the knee simulator and physical experiment. Our Bayesian statistical calibration combines outputs from the knee simulator and cadaver experiment using three strategies to assign values of the true calibration parameters: (1) knowledge of each calibration parameter from the literature and subject matter experts in the form of a prior distribution; (2) estimation of the bias between the knee simulator and the physical experiment where bias is the difference between a run of the knee simulator at the true values of the calibration parameters and that measured in the cadaver experiment; and (3) use of a pooled or average estimate of the knee simulator output over the updated prior distribution of "likely" calibration parameters[8].

The results of this study will be presented as follows. First, we report the ligaments that had the largest influence on the AP translations of the medial and lateral compartments via the main effect SI's. Second, we report the posterior distributions obtained from the Bayesian calibration using the most active ligament parameters. Third, we identify the median of the posterior distribution of each ligament parameter. Then, to assess whether this single set of calibrated ligament properties yields AP compartmental translations that track those measured in the physical experiment, we substitute them back into the knee simulator and calculate a root mean square (RMS) difference of predicted and experimentally measured compartmental translations. As a final, independent check, we compare the difference in the ACL force at the peak applied load predicted by the calibrated knee simulator with that measured in the physical experiment.

RESULTS

The ligaments with the largest influence on the AP translations of the lateral compartment were the ACL and the LCL, while medial compartment translations were controlled by the ACL, MCL, POL, and PCL. Thus, the properties of these ligaments were included in the Bayesian calibration. In terms of ligament parameters, slack length was dominant. For example, with 10 N of compression compared to combined moments, ACL slack was responsible for 77% and 32% of the variation in AP translation of the lateral compartment, respectively.

Stiffness played a larger role with combined valgus and internal rotation moments. For example, POL stiffness accounted for 26% of the variations in medial compartment AP translations with these loads.

The posterior distributions of each ligament parameter indicate a preference for decreasing slack length and increasing stiffness (Fig. 2). Narrower boxplots such as for MCL slack indicate that the Bayesian calibration yields a more certain set of ligament parameters (Fig. 2).

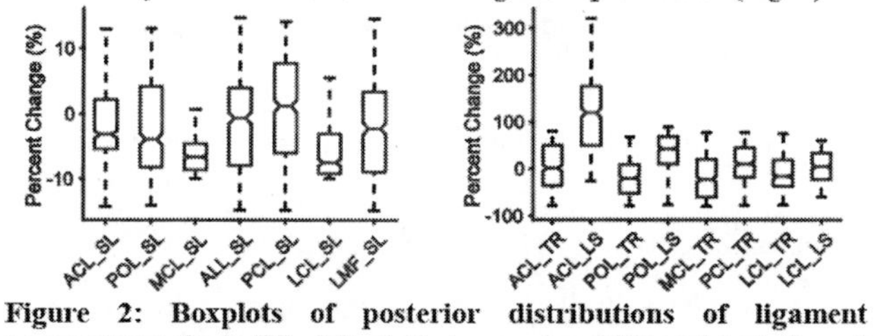

Figure 2: Boxplots of posterior distributions of ligament parameters: slack (SL, left plot), toe region (TR, right plot), and linear stiffness (LS, right plot). Data are presented as a percent change relative to the default value of each ligament parameter. The red line is the median, box ends denote 25% and 75% quartiles and whiskers include 150% of the interquartile range.

The median of the posterior distributions from the Bayesian calibration indicate that the ACL slack and stiffness should be decreased by 3% and increased by 120% of their default values, respectively (Fig. 2). Similarly, the POL slack and stiffness should be decreased by 4% and increased by 43%, respectively. Lastly, the LCL slack and stiffness should be decreased by 7% and increased by 5%, respectively.

Setting the ligament parameters in the knee simulator from the default values to the median of the posterior distributions decreased RMS error between AP translations predicted by the knee simulator and those measured in the physical experiment (Fig. 3). Specifically, RMS error decreased from 1.8 to 0.6 mm in the medial compartment and from 4.2 to 2.4 mm in the lateral compartment (Fig. 3). Similarly, differences in ACL force between simulator predictions and cadaveric measurements decreased from 41 to 12 N after including Bayesian-calibrated ligament parameters.

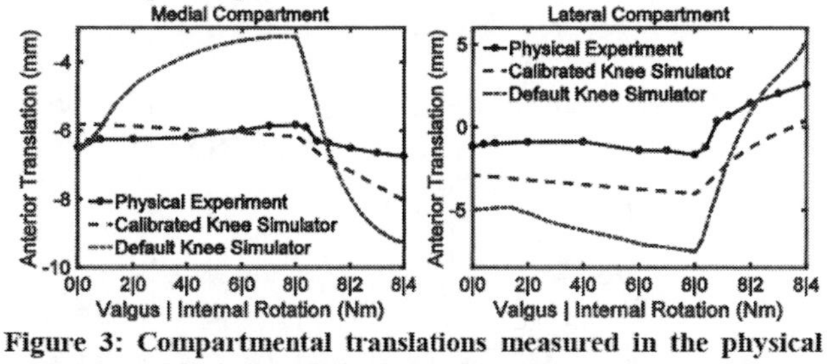

Figure 3: Compartmental translations measured in the physical experiment (black line with circles at each increment of applied load), and as predicted by the knee simulator with ligament properties set to their default values (red fine dash), and set to the median of the posterior distribution (blue dash).

DISCUSSION

We have established a new approach that combines a physics-based simulator of the tibiofemoral joint, the measurements from a physical experiment of the same knee that was modeled, and a Bayesian statistical method to identify likely subject-specific properties of knee ligaments. The approach yielded improved predictions of knee kinematics and ACL force.

The sensitivity analysis and the Bayesian calibration revealed that decreasing ligament slack and increasing ligament stiffness was important for controlling rotatory laxity in this specimen. These findings have high potential for clinical translation. Specifically, ligament slack length can be adjusted intraoperatively by changing the flexion angle, attachment points, and pretension at which ligament grafts are fixed. Stiffness can be adjusted intraoperatively by changing graft type or doubling over graft tissue.

The measurements of AP translations in this cadaveric knee were less than those measured in the knee simulator with default ligament parameters (Fig. 3). In keeping with biomechanical intuition, the Bayesian calibration predicted that a decrease in slack length and an increase in stiffness were needed to reduce this difference. Thus, these findings are a proof of concept. This new approach must also be corroborated with data from a broader population of knees, especially those with high laxity given their elevated injury risk[1]. Structural characterization is also needed to determine if the actual ligament stiffnesses agree with those predicted using Bayesian calibration.

We have developed a novel method to identify critical combinations of patient-specific ligaments and ligament properties (i.e., slack length and stiffness) that control rotatory laxity. This will enable personalized treatments of ligament injury to improve knee function and reduce risk of reinjury.

ACKNOWLEDGEMENTS

The Clark, Kirby, and Gosnell Family Foundations

REFERENCES

[1] Magnussen, et al. Am J Sports Med, 44:3077-82, 2016. [2] Hsu, et al. Am J Sports Med, 34:765-70, 2006. [3] Guess, et al. Med Eng Phys, 32:505-15, 2010. [4] Thelen, et al. J Biomech Eng, 136:021033, 2014. [5] Stone. Bayes' Rule. Sebtel Press, 2013. [6] Kia, et al. J Biomech Eng, 138:051010, 2016. [7] Imhauser, et al. Am J Sports Med, 41:815-25, 2013. [8] Higdon, et al. SIAM J Sci Comput, 26:448-66, 2004.

SB[3]C2017
Summer Biomechanics, Bioengineering and Biotransport Conference
June 21 – 24, Tucson, AZ, USA

DETERMINING THE POTENTIAL ROLE OF GLYCOSAMINOGLYCAN CLUSTERS IN TENDON MECHANICAL HOMEOSTASIS

Cody M. O'Cain (1), Wendell M.R. Heard (2), Felix H. Savoie (2), Sara Roccabianca (3), Ronald C. Anderson (1), Kristin S. Miller (1)

(1) Biomedical Engineering
Tulane University
New Orleans, LA, United States

(2) Orthopaedic Surgery
Tulane University
New Orleans, LA, United States

(3) Mechanical Engineering
Michigan State University
East Lansing, MI, United States

INTRODUCTION

Tendons permit locomotion by transferring muscle-generated forces to bone. Tendinopathy is a general term referring to a pathologic tendon, which may be induced by mechanical overuse and/or other pathological factors. Tendons are composed primarily of type I collagen fibers aligned along the longitudinal axis, however, smaller amounts of glycosaminoglycans (GAGs), proteoglycans, elastic fibers, minor collagens, and other glycoproteins are also present. While GAGs compose a small portion of tendon, GAG content has been observed to increase in tendons subjected to increased mechanical load and/or pathological factors, often in the form of aggregate clusters [1-4]. Further, GAG clusters have been suggested to contribute to tendinopathy onset and/or compromised structural integrity [2-4]. Previous work in vascular mechanics has shown that abnormal increases in GAG clusters may over-pressurize the inner layers of arteries leading to compromised structural integrity and potential mechanobiological consequences [5]. The role of GAGs in tendon mechanics, however, remains unclear. Therefore, the objective of this study was to develop a finite element model, which considered the size and shape of GAG clusters on the stress distributions within the surrounding extracellular matrix to examine a potential etiology of tendinopathy. We hypothesize that larger GAG clusters will increase local stresses, which may be sufficient to damage adjacent collagen fibers compromising structural integrity, and preventing proper homeostatic conditions.

METHODS

The finite element model consisted of two components: the tendon and GAG cluster. The highly aligned collagen fibers within tendon motivated the use of a one-fiber family model, defined by a strain energy function W of the form

$$W = \frac{c}{2}(\mathrm{I}_1 - 3) - \frac{k_1}{k_2}\{\exp[k_2(\mathrm{I}_4 - 1)^2] - 1\} \quad (1)$$

where material parameter c denotes the ground matrix shear modulus, k_1 and k_2 are the collagen-related stress-like and dimensionless parameters respectively, I_1 and I_4 are the first and forth invariant of the right Cauchy-Green tensor $\mathbf{C}$. The GAG cluster was assumed to resist loading in an isotropic manner; hence a neo-Hookean form of W was considered

$$W = \frac{\mu}{2}(\mathrm{I}_1 - 3) - \mu \ln \mathrm{J} \quad (2)$$

where μ represents the shear modulus, and J represents the determinant of the deformation gradient $\mathbf{F}$ [6]. To initialize the model, the best-fit material parameters were calculated from uniaxial tensile data obtained in the rat Achilles tendon via nonlinear regression [7]. After initialization, the model was granted a donnan osmotic pressure to allow for the swelling potential of the GAG cluster [6]. Both components were created using Solidworks, meshed within Abaqus, and constructed in the open source finite element software FEBio [8]. To consider the role of GAGs in tendon mechanical response, parametric studies were performed to evaluate the role of GAG geometry and fixed charge density while subjected to 1–4% axial strain. Motivated by histological observations [1], GAG clusters were modeled as ellipsoids with a minor axis "A" running radially, and a major axis "B" running axially with $A \in$ [1.5–6 μm] and $B \in$ [8–65 μm]. These bounds and cluster sizes were calculated from histological images from rat supraspinatus tendon with and without overuse pathology using ImageJ [1]. GAG fixed charge density was evaluated between [0–160 mEq/l] [9]. Data was exported from FEBio's Postview, and analyzed in a custom Matlab script.

RESULTS

The one-fiber family model described the uniaxial rat Achilles tendon data reasonably well ($R^2 = 0.975$). The average GAG cluster size from histological observation was $A = 3.2 = 0.92\mu m$, $B = 21.1 = 5.37\mu m$ for the control, and $A = 3.0 = 0.92\mu m$, $B = 40.2 = 11.4\mu m$ for the overuse group. The value of A for the overuse group was varied ($A = 1.5, 3, 6\ \mu m$) while the value of B was held constant ($B = 40.2\ \mu m$) to create three ellipsoids bounded by the overuse pathologies observed (Figure 1). As the ratio A/B increased, an increase in radial stress and decrease axial stress was observed (Figure 2).

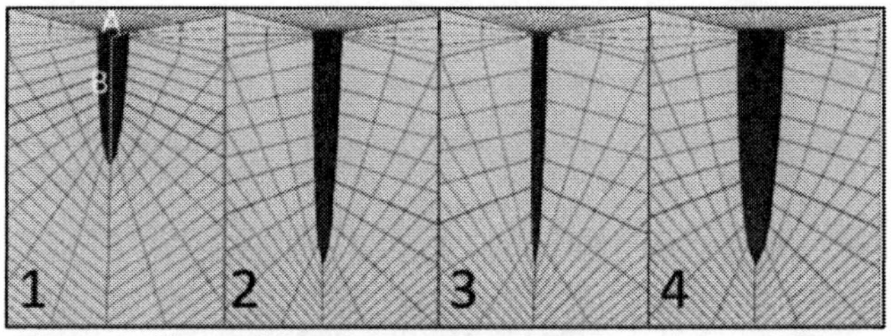

Figure 1: View of the undeformed local GAG cluster mesh from the four models: (1) Control, (2) Average Overuse, (3) Narrow Overuse, (4) Wide Overuse.

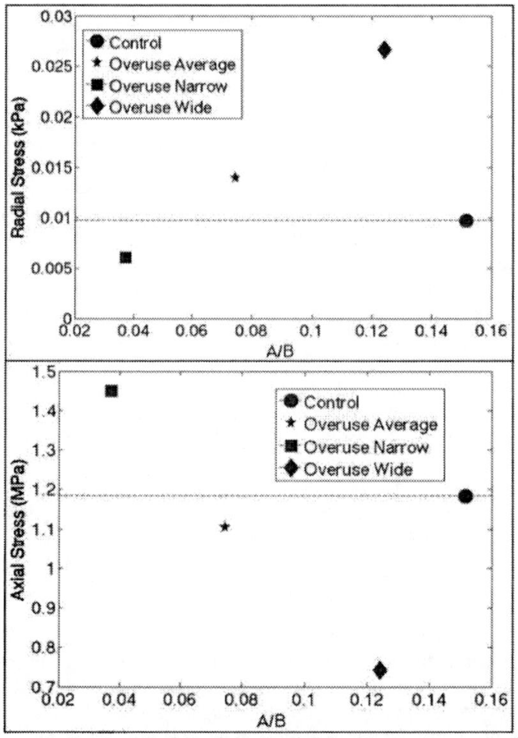

Figure 2: Peak radial (top) and axial (bottom) stresses in local region of GAG cluster as a function of increasing aspect ratio A/B (0.037 – 0.152). Tendon placed under 2.6% axial strain.

Increasing fixed charge density was observed to increase radial stress in all cases; however, the effect on axial stress was minimal (Figure 3). A greater influence was observed on the overuse models compared to the control for radial stress (Figure 3).

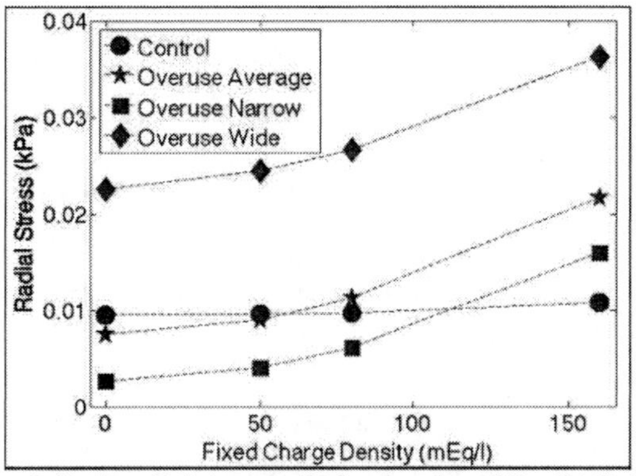

Figure 3: Peak radial stress in the local region of the GAG cluster as a function of increasing fixed charge density (0 – 160 mEq/l). Tendon placed under 2.6% axial strain.

DISCUSSION

This study supports the hypothesis that GAG clusters impact local tendon mechanics. Specifically, increasing GAG cluster size was shown to increase radial stress in tendons. While tendons are primarily loaded in the longitudinal direction, altered radial loading may alter collagen fiber microstructure, such as organization and undulation, which may impact cytoskeletal homeostasis and gene expression by disrupting cell-matrix interactions [10]. Further, increased local radial stresses may damage collagen fibers and cell-matrix interactions making them less sensitive to extracellular matrix strain [10-11]. Hence, we further hypothesize that the less anticipated local changes in radial stress may alter the surrounding collagen microstructure, which may induce altered gene expression by resident cells [2-4,10-11]. The question of whether GAGs are helpful or detrimental to tendinopathy recovery still remains [11], and it may be that it is a matter of proper size and number. Consequently, further study is needed to understand the implications of these differences on global and local tendon function and mechanobiology. Limitations of this study include simplified GAG geometry and lack of experimental data on fixed charge density values in tendons. These limitations can be addressed as additional experimental data becomes available. In conclusion, GAG clusters may play a significant role in tendon mechanics and merit further consideration in tendon mechanobiology and growth and remodeling in healthy and pathological states.

ACKNOWLEDGEMENTS

This work was supported by a grant from the Oscar Lee Putnam Cultural and Intellectual Enrichment Program. The author would like to acknowledge Akinjide Akintunde and Jason Schuster for their assistance.

REFERENCES

[1] Attia, M et al., JOR, 30:61-71, 2011.
[2] Attia, M et al., BJSM, 48.6:469-475, 2014.
[3] Riley, GP et al., SJMSS, 15.4:241-251, 2005.
[4] Fu, SC et al., CJSM, 17.2:129-134, 2007.
[5] Roccabianca, S et al., JR Soc Inter, 11, 20140397, 2014.
[6] Roccabianca, S et al., Biomech Model Mech, 13(1):13-25, 2014.
[7] Eliasson, P et al., JAP, 107:399–407, 2009.
[8] Maas, SA et al., JBME, 134(1):011005, 2012.
[9] Chahine, NO et al., Biophysical Journal, 89.3:1543-1550, 2005.
[10] Arnoczky, SP et al., CORR, 466.7:1583-1591, 2008.
[11] Woojin, MH et al., Nature materials, 15:477-484, 2016.

MULTIAXIAL MECHANICAL RESPONSES OF ANTERIOR CRUCIATE LIGAMENT BUNDLES REFLECT DIFFERENCES IN MICROSTRUCTURE

Callan M. Luetkemeyer (1), Benjamin C. Marchi (1), Ellen M. Arruda (1,2,3)

(1) Department of Mechanical Engineering
University of Michigan
Ann Arbor, MI, USA

(2) Department of Biomedical Engineering
University of Michigan
Ann Arbor, MI, USA

(3) Program in Macromolecular Science and Engineering
University of Michigan
Ann Arbor, MI, USA

INTRODUCTION

Researchers and clinicians have become increasingly interested in identifying anatomical risk factors for anterior cruciate ligament (ACL) injury due to the growing rate of ACL tears and the tendency for this injury to be associated with increased susceptibility to osteoarthritis[1]. Most studies with this focus have found statistical correlations between geometric measurements and injury rates[2]. However, the correlation between injury rate and intercondylar notch width, for example, may not imply a causal relationship. In order to determine if a given anatomical feature directly increases the risk of injury, researchers must be able to test one variable at a time – a practically impossible task for experimental studies.

Where experiments struggle, computational models have the ability to excel. A finite element model of the ACL and its surrounding structures would provide a platform from which parametric studies could evaluate the consequences of varying a single anatomical feature on the predicted strain field. However, the credibility of finite element models relies heavily on the accuracy of the constitutive models used. Because the ACL is highly anisotropic and experiences more than just longitudinal deformation *in vivo*, the constitutive model implemented must capture its mechanical response to all possible types of deformation.

The medial collateral ligament (MCL) has been characterized in multiple loading configurations using grip-to-grip strains[3-5]. These studies showed that the matrix (non-collagenous) stress-strain response is nonlinear with initial concavity. This suggests that a neo-Hookean form for the matrix contribution to the strain energy density function, such as in the popular Holzapfel-Gasser-Ogden constitutive model[6], may not accurately describe the matrix mechanical response.

Additionally, recent experimental studies have shown that the anteromedial (AM) and posterolateral (PL) bundles of the ACL have different microstructural and mechanical properties; the AM bundle displays a higher degree of collagen alignment and is consequently stiffer in the mean fiber direction than the PL bundle[7-9]. Thus, separate constitutive models (or at least different fitting parameters) should be used to describe each bundle individually. However, the mechanical characterization of the ACL bundles is incomplete, as their mechanical responses have been investigated in the mean fiber direction only.

The main objective of this study is to determine the average quasi-static tensile mechanical responses along, and orthogonal to, the mean fiber direction for the AM bundle, PL bundle, and MCL of ovine specimens for the purpose of constitutive modeling. Digital image correlation (DIC) is used to accurately measure displacements, from which strain fields are computed. It is hypothesized that there will be differences in the mechanical responses of the bundles in both directions, indicative of their differences in collagen orientation.

METHODS

Eight ovine knee specimens were obtained from a local abattoir. Surrounding tissues were removed and each ligament or bundle of interest was first cut in half in the collagen fiber direction on the thinnest side to provide two thin specimens from each tissue. This was done in order to acquire flatter, more uniform specimens. The two halves were then cut at the entheses to remove them from the bone. One half was used for uniaxial testing along the fiber direction and the other was used for plane strain testing normal to the mean fiber direction. All specimens were wrapped in phosphate buffered saline-soaked gauze and used within two hours.

Superglue was used to attach the ends of each tissue to custom-made grooved grips. These grips attached to a custom horizontal mechanical testing setup under a microscope with a CCD camera.

Images of each specimen's thinnest dimension were taken before mounting the specimen to the setup with the widest dimension facing up, so that the cross-sectional area could be estimated from the images following testing. Each specimen was then pre-loaded to 0.01 N, speckled with micronic graphite particles and stretched at a constant displacement rate of 0.1 mm/s until failure. Images and force readings were acquired at 10 Hz for axial specimens and 20 Hz for transverse specimens. DIC software (Correlated Solutions, Inc.) was used to track the speckle pattern and determine displacement fields from which strain fields were calculated.

RESULTS

Preliminary results show differences in the mechanical responses of the AM and PL bundles, with the AM being stiffer in the fiber direction but more compliant in the direction normal to the fibers (Figure 1a-b). All three ligaments appear to be at least one order of magnitude more compliant in the direction normal to the fibers, supporting previous work that shows ligaments are highly anisotropic. Additionally, the bundles appear to have distinct transverse strain responses when stretched along the mean fiber direction (Figure 2a). However, both bundles and the MCL appear to deform in approximately plane strain when stretched in the direction orthogonal to the fibers (Figure 2b).

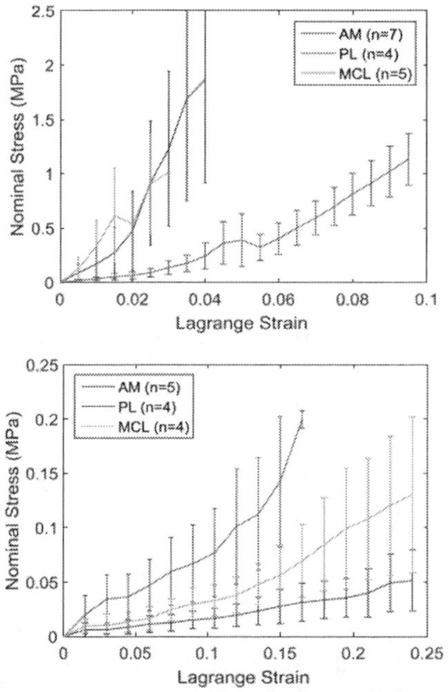

(a) (b)

Figure 1: Mean stress-strain data (a) along and (b) normal to the mean fiber direction for the MCL and AM and PL bundles.

Though not shown here, grip-to-grip strains were found to be significantly higher than tissue-level strains. Several specimens were not used in the calculation of mean responses due to a loss of image correlation at small strains. This was the result of poor speckle patterning, which has continually improved with time.

DISCUSSION

This work in progress presents, for the first time, the stress-strain relationships for both ACL bundles in multiple loading directions. This information is essential for the construction of accurate constitutive models. These results suggest that the AM and PL have distinct mechanical responses to tensile stretching both along and normal to the mean fiber direction. Moreover, the differences in these responses are consistent with the differences found in the degree of collagen alignment found by Skelley, Castile and colleagues[7-9]. Collagen fibers in the PL are less aligned than in the AM, and thus, contribute more to the mechanical response of the PL than the AM when the bundles are stretched normal to the mean fiber direction, and more to the response of the AM when pulled along the fiber direction.

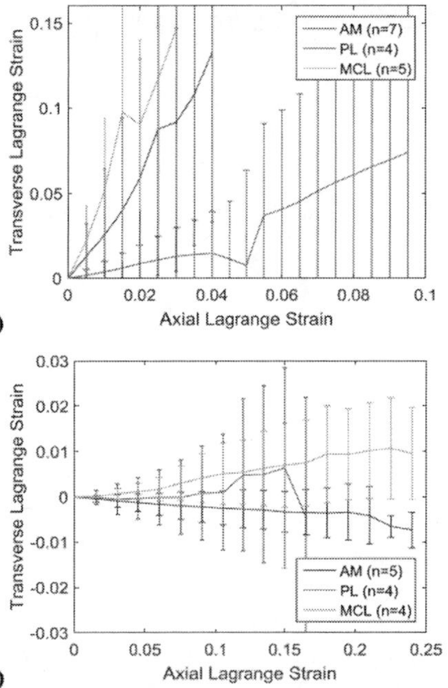

(a) (b)

Figure 2: Mean transverse strain response for testing (a) along and (b) normal to the mean fiber direction for the MCL and AM and PL bundles.

The implementation of DIC allowed for not only a more accurate measurement of the strain in the direction of the applied stretch, but also transverse strains. These transverse strains have the potential to be important in evaluating the suitability of various hyperelastic constitutive theories. Concurrent work includes an evaluation of the aptness of common constitutive models to describe these results, and future work will consider viscoelasticity.

ACKNOWLEDGEMENTS

This material is based upon work supported by the National Science Foundation Graduate Research Fellowship under Grant No. DGE 1256260.

REFERENCES

[1] Kessler, MA et al., *Knee Surg*, 16 (5): 442-448, 2008.
[2] Vacek, PM et al., *Am J Sports Med*, 44 (6): 1492-1501, 2016.
[3] Quapp, KM et al., *J Biomech Eng*, 120: 757-763, 1998.
[4] Bonifasi-Lista, C et al., *J Orthop Res*, 23: 67-76, 2005.
[5] Henninger, HB et al., *Acta Biomat*, 25: 304-312, 2015.
[6] Holzapfel, GA et al., *J Elasticity*, 61: 1-48, 2000.
[7] Skelley, NW et al., *Am J Sports Med*, 43 (4): 928-936, 2015.
[8] Skelley, NW et al., *Am J Sports Med*, 44 (11): 2892-2899, 2016.
[9] Castile, RM et al., *J Biomech*, 49 (1):87-93, 2016.

SB³C2017
Summer Biomechanics, Bioengineering and Biotransport Conference
June 21 – 24, Tucson, AZ, USA

PERIODONTITIS'S AFFECT ON THE SHEAR MECHANICAL BEHAVIOR OF THE FIBROUS PERIODONTAL LIGAMENT. A TRANSVERSELY ISOTROPIC HYPERELASTIC MODEL.

David S. Nedrelow (1), Victor H. Barocas (1)

(1) Biomedical Engineering
University of Minnesota
Minneapolis, MN, USA

INTRODUCTION

Periodontitis (PDT) is an inflammatory disease that causes damage to the tooth's supporting tissues and is a common cause of tooth loss. The periodontal ligament (PDL) that attaches the tooth to bone undergoes changes in both composition and architecture during the inflammatory process[1]. Although we know that the PDL structure changes, and we know from many other tissues that collagen architecture plays a major role in determining the mechanical properties[2,3,4], to our knowledge, no work has been done to analyze the effect of PDT on the PDL's mechanical anisotropy. We hypothesize that the observed changes in fiber orientation due to PDT could result in a PDL that is more compliant than healthy PDL even if one ignores compositional changes. Hyperelastic models, now used routinely in advanced finite element analysis of other tissues[5,6], as well as in some studies of PDL mechanics[7] make it possible to analyze the problem in terms of structure-stress effects.

METHODS

The collagen microstructure of the PDL was modeled as a transversely isotropic material, with planes of radially aligned fibers. The fiber directions varied in the vertical zr-plane of the cylinder, and did not change in the third angular θ-axis. Stiffness parameters were derived from a fit of the Holzapfel-Gasser-Ogden hyperelastic model (Eq. 1) to previously reported shear stress data[8] and assigned healthy collagen architecture parameters.

$$W = C_0(I_1 - 3) + \frac{k_1}{2k_2}\{exp[k_2\langle E_\alpha\rangle^2] - 1\} \quad (1)$$

Where C_o, the neo-Hookian matrix stiffness, k_1, a measure of fiber stiffness, and k_2 for fiber nonlinearity, and I_1 is the first Invariant, Fiber direction and distribution parameters were in the following equation.

$$E_\alpha = \kappa(I_1 - 3) + (1 - 3\kappa)(I_4 - 1) \quad (2)$$

Where K, the degree of anisotropy, is the fiber direction distribution term ranging from 0 (totally anisotropic) to 1/3 (totally isotropic), and I_4 is the fourth invariant expressing fiber direction. Theoretical stresses for PDL with PDT were generated by changing only the architecture parameters, while keeping the stiffness parameters (in Eq. 1) unchanged. In this way, an *in silico* comparison was made between the PDL's healthy and PDT states based only on reported pathological differences in collagen fiber orientation. All calculations were performed in Matlab.

RESULTS

The stresses in the coronal regions of the PDL (especially mesial-coronal) were lowered by PDT; however, in the apical regions, the stresses were increased (Fig). There was a very large increase in stress in the distal-apical region by PDT. As previously published, the angle of PDL fibers from the surface of the tooth (perpendicular is zero degrees) and distribution approached the perpendicular (zero angle) for all regions as a result of inlfammation.

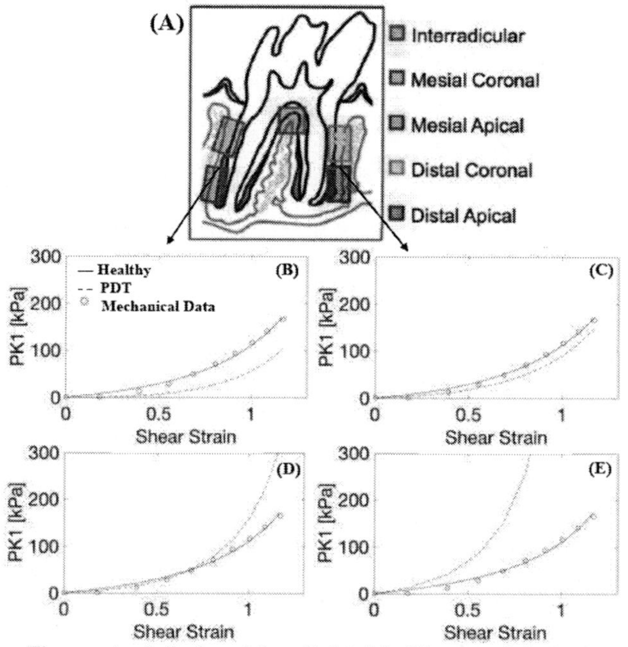

Figure: Analytical model applied to birefringence data from healthy and inflamed PDL in (A) four anatomical regions of Rat 2nd Molar PDL. (B) Mesial Coronal, (C) Distal Coronal, (D) Mesial Apical, and (E) Distal Apical plots of 1st Piola-Kirchhoff shear stress vs. shear strain.

DISCUSSION

PDT-related structural changes had a significant effect on PDL stiffness in the apical region but did not have a significant effect in the coronal region. Pure shear mechanics were applied in both the apical and coronal analysis, but apical PDL likely absorbs a slightly more compressive load than the coronal PDL during normal masticatory function. The lower coronal stresses are in agreement with anecdotal evidence that PDT-involved teeth are easier to extract. There are, however, other compositional changes (*e.g.* bone degradation) that contribute to tooth-loosening in patients with PDT.

It is well known that periodontitis causes an influx in collagen-cleaving enzymes, some of which have been suggested to be less effective at cleaving strained fibers[9]. Because the stress increased in the apical regions, this fiber direction may hve been more aligned with the direction of principal stress. Taken together, this suggests more of the shear load is borne by the apex of the root than the coronal root after inflammation of the PDL has occurred.

ACKNOWLEDGEMENTS

This study was not supported by any funding source.

REFERENCES

[1] Lin, J D et al., *J. Biomech*, 47(9):2102-2114, 2014.
[2] Brum, J et al., *Phys. Med. Biol*, 59:505–523, 2014.
[3] Kelleher, J E et al.,. *J. Acoust. Soc. Am*, 133(3):1625–1636, 2013.
[4] Gasser, T C et al., *J. R. Soc. Interface*, 3(6):15–35, 2006.
[5] Henak, C R et al., *J. Biomech. Eng.*, 136(2):21021, 2014.
[6] Wittek, A et al., *J. Mech. Behav. Biomed. Mater*, 1–17, 2015.
[7] Limbert, G. et al., *Comput. Methods Biomech. Biomed. Engin*, 6(5–6), 337–345, 2006.
[8] Komatsu, K et al., *Arch. Oral Biol*, 49(10), 817–824 2004.
[9] Ruberti, J W et al., *Biochem. Biophys. Res. Commun*, 336(2), 483–489, 2005.

SB3C2017
Summer Biomechanics, Bioengineering and Biotransport Conference
June 21 – 24, Tucson, AZ, USA

ROLE OF VASCULAR RECESSION ON THE NUTRITION OF THE HUMAN MENISCUS: A COMPUTATIONAL ANALYSIS

F. Travascio (1), A. Jackson (2), S. Asfour (1)

(1) Department of Industrial Engineering
University of Miami
Coral Gables, FL, USA

(2) Department of Biomedical Engineering
University of Miami
Coral Gables, FL, USA

INTRODUCTION

A healthy meniscus tissue is necessary to maintain functioning of the knee and to protect the underlying cartilage. It is estimated that degeneration of the meniscus affects approximately 35% of the population [1]. Such pathological condition is believed to be a precursor of knee osteoarthritis. Most likely, meniscal degeneration is multi-factorial, including effects of aging, joint mechanical loading and presence of inflammatory cytokines in synovial fluid [2,3]. Accordingly, many studies have investigated meniscus composition, morphology, mechanical properties, and response to pro-catabolic stimuli [2-4]. However, in order to maintain a healthy tissue, an adequate turnover of the extracellular matrix (ECM) is necessary. This is accomplished by tissue cells with the condition that adequate nutritional supply is provided. So far, the role of nutrition in meniscus homeostasis and degeneration has been overlooked.

The human meniscus modifies its structure as the musculoskeletal structure of the individual matures. More specifically, in adult menisci, the blood vessels progressively recede to the outer 10% of the tissue (red zone), leaving the majority of the tissue avascular (white zone) [5]. This may cause major changes in the mechanisms of transport and amount of nourishment the meniscus receives: solutes' delivery to the white zone must come from either the peripheral vascularized regions (red zone), or from the synovial fluid bathing the meniscus in the knee joint. In either case, nutrients and other important molecules for tissue homeostasis must diffuse from these surrounding areas through the tissue ECM to reach meniscal cells. Diffusion is a transport mechanism much less efficient than transvascular delivery from adjacent blood vessels in the red zones. Therefore, we hypothesize that vascular recession may cause a major reduction of the nutritional supply to the meniscal cells of the white zone. In order to test our hypothesis, we performed an assessment on the changes of nutritional levels in the meniscus upon the natural process of vascular recession. This was done by developing a computational model of meniscus homeostasis based on previously reported physical parameters and experimentally determined relations regulating transport and metabolism of nutrients.

METHODS

A three-dimensional diffusive-reactive finite element model was developed to describe the homeostasis of the meniscus. The tissue was modeled as composed of two regions: an external vascular portion (red zone), and an internal non-vascular one (white zone), see Figure 1a-c. Cells were also included into the tissue, and their density (ρ^{cell}) varied from region to region [6]. Nutrients and metabolism byproducts trafficking the meniscus (oxygen, glucose and lactate) followed a diffusive-reactive law:

$$\frac{\partial c^{\alpha}}{\partial t} = D^{\alpha}\nabla^2 c^{\alpha} + Q^{\alpha}, \quad (1)$$

where c^{α} is the concentration (per unit water volume) of solute α and D^{α} its diffusivity. The term Q^{α} accounted for solutes' consumption/production due to cell metabolism [7]:

$$Q^{oxygen} = \frac{V'_{max}(pH-4.95)c^{oxygen}}{K'_m(pH-4.59)+c^{oxygen}}\rho^{cell}, \quad (2)$$

$$Q^{lactate} = exp(-2.47 + 0.93pH + 0.1c^{oxygen} - 0.0058c^{oxygen})\frac{c^{glucose}}{K_m+c^{glucose}}\rho^{cell}, \quad (3)$$

$$Q^{glucose} = -0.5Q^{lactate}, \quad (4)$$

where V'_{max} is the maximum consumption rate of oxygen, and K'_m and K_m are the Michaelis–Menten constants for oxygen and lactate, respectively. Also, the term pH is computed as:

$$pH = -0.1c^{lactate} + 7.5. \quad (4)$$

Diffusivity was modeled as anisotropic, being faster in the direction parallel to collagen fibers [8]. Solutes were provided by the synovial fluid surrounding the tissue, and from the vascular network present into the meniscal red zone. Accordingly, to account for molecular inflow from the blood vessels, the solutes' mass balance equation (1) in the red zone included a dedicated source term as previously reported [9].

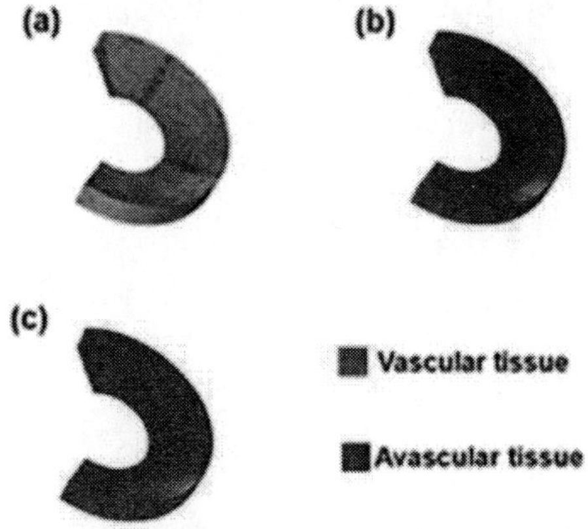

Figure 1: Finite element model of human meniscus: (a) computational domain; (b) young meniscus (33% vascular); (c) adult meniscus (10% vascular).

Two cases were simulated and compared: (1) a young meniscus, characterized by a red zone comprising the outer 33% of the tissue, and (2) an adult meniscus, whose red zone only accounted for the outer 10% of the tissue. Steady state levels of oxygen, glucose and lactate in the tissues were determined and compared. The results were expressed in terms of percent change of solute concentration from young to adult meniscus.

RESULTS

Changes of nutrients levels in human meniscus upon the natural process of vascular recession are reported in Figure 2. Major changes occurred in the avascular meniscal region immediately adjacent to the red zone. It was found that, due to the vascular recession, glucose and oxygen levels in the meniscus reduced to 50% and 60% of the values attained in a young tissue. In contrast, lactate levels in the adult meniscus were found to be up to 130% higher than those calculated for the young tissue.

DISCUSSION

A new computational finite element model for human meniscus homeostasis was developed to investigate the effects of the natural process of vascular recession on the availability of nutrients to meniscal cells.

The results of this study indicate that changes in vascularity of the meniscus, as observed during the maturation from young to adult tissue, cause a reduction of tissue levels of glucose and oxygen up to 50% and 60%, respectively. Both glucose and oxygen are necessary to meniscal cells in order to biosynthesize new extracellular matrix used to maintain and restore the meniscus tissue. Shortage of these nutrients may compromise the homeostasis of the meniscus. The results of this study suggest that this may be the case for the avascular meniscal region immediately adjacent to the red zone, see Figure 2.

The findings hereby reported are consistent with previous studies: It has been proposed that the inefficient mechanism of nutrients supply occurring in the white zone causes the central core of the meniscus to deteriorate with age, making the tissue more susceptible to cleavage tears [10]. Also, experimental studies reported that avascularity significantly hinders the ability for self-repair within the white zones, making the meniscus susceptible to permanent degenerative lesions [6].

The above considerations and the computational analysis performed in thisstudy, suggest that nutrition plays a crucial role in the homeostasis of the meniscus, and a reduced nutritional supply due to the progressive vascular recession may constitute a major cause of meniscal tissue degeneration.

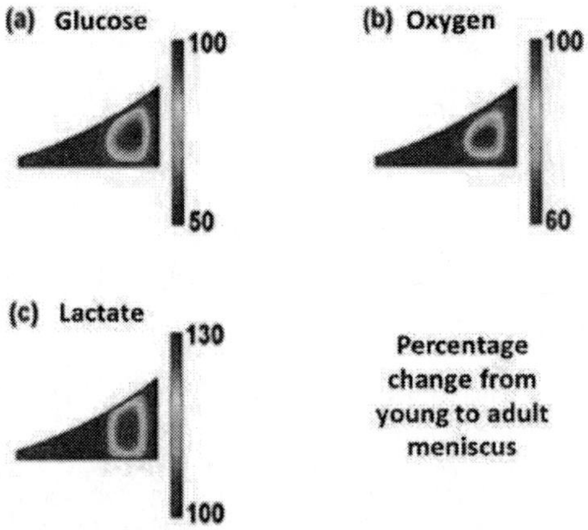

Figure 2: Changes in nutrients levels in meniscus upon vascular recession. Data are reported as percentage change from young to adult meniscus: (a) glucose; (b) oxygen; (c) lactate.

ACKNOWLEDGEMENTS

This study was supported by funds donated to the Biomechanics Research Group at the University of Miami.

REFERENCES

[1] Englund, M et al., *Osteoarthrit Cart,* 24:270-3, 2016.
[2] McNulty, AL Guilak, F, *J Biomech*, 48:1469-78, 2015.
[3] McAlinden, A et al., *Osteoarthrit Cart,* 9:33-41, 2001.
[4] McNulty, AL et al., *J Orthop Res*, 31:1039-45, 2013.
[5] Sweigart, MA Athanasiou, KA *Tissue Eng*, 7:111-29, 2001.
[6] Markis, EA et al., *Biomaterials*, 32:7411-31, 2011.
[7] Huang, CY et al., *J Biomech*, 47:1149-55, 2012.
[8] Travascio, F et al., *Ann Biomed Eng*, 37:813-23, 2009.
[9] Danzig, LA et al., *J Orthop Res*, 5:409-13, 1987.
[10] Smillie, IS. *Injuries of the Knee Joint*, Edinburgh: Churchill Livingston, 1978

SB^{3}C2017
Summer Biomechanics, Bioengineering and Biotransport Conference
June 21 – 24, Tucson, AZ, USA

CHARACTERIZATION OF THE IN-VIVO INFLAMMATORY RESPONSE TO POLYCARBONATE-URETHANE WEAR DEBRIS

Maoz Shemesh (1), Jonathan J. Elsner (1), Noa Cohen (1) Lotem Mahluf (1), Roni Noyvirt (1), Shmuel Israeli (2), Judit Krausz (2), Natalia Edison (2), Nimrod Rozen (2), Eran Linder-Ganz, (1)

(1) Active Imlpants LLC,
Memphis, TN, 38120

(2) Emek Medical Center,
Afula, Israel

INTRODUCTION

The surveillance of the clinical performance of a synthetic Polycarbonate-urethane (PCU) meniscus implant (NUsurface®, Active Implants LLC., Memphis TN, Linder-Ganz et al., 2010; Elsner et al., 2010) has thus far confirmed that the device can provide pain relief up to eight years post implantation. Ongoing clinical research has also established that symptoms associated with synovitis – pain and swelling, secondary to implant wear, are the main cause of revision surgery in the short term (1-4 years). The occurrence of synovitis has typically been noted in knee joints where the implant came in direct contact with exposed bone or other irregularities of the articulating surfaces (e.g. osteophytes) which were not addressed during implantation. In these cases, the presence of abrasion on the surface of the retrieved implant was confirmed by scanning microscopy.

The inflammatory response to wear debris has been studied and is understood for traditional bearing materials such as polyethylene, but limited information exists on the response to new elastomeric bearing materials such as PCU (Richard 2010, Wippermann 2008). Our study aimed to provide a systematic analysis of synovial tissues collected during revision surgeries, in order to understand the interaction between PCU and the joint environment, when it used as a bearing material, and to refine the indications for its use. It is important to evaluate the body's response to the material not only with respect to the safety and longevity of the device, but also with respect to the condition of the joint, post-removal, and any effect it may exert on other future treatment options.

METHODS

Between 2012 and 2015, synovium tissue samples were collected for histology in 10 retrieval cases (Ethical Committee Approval #0054-11-EMC). It was confirmed by microscopy that the reason for the damage and removal of the implant was abrasion as a result of direct contact with a rough surface (Grade IV lesions according to Outerbridge Grading System) or bony protrusions (osteophytes). Tissue samples were collected during the retrieval procedure from a location adjacent to the implant or from other sites where synovitis was detected. Synovitis was mostly present in the suprapatellar pouch, as well as in the lateral compartment. The average retrieval time was 30.4 months after implantation (Range of 7-50 months). The samples were stained routinely with Hematoxylin & Eosin (H&E). Tissue sections were examined and photographed by an experienced pathologist (JK) using a plain and polarized light microscope (BX53F, Olympus). General pathologic evaluation of the tissue was performed in order to characterize the type of cells exist in the synovial fragments, as well as the type of the biological response post-implantation.

PCU wear particles were traced from images using an image processing software (SigmaScan Pro) and their effective diameter and shape factor were calculated. The shape factor is a dimensionless quantity that reflects the deviation of the feature's outline from a circle shape. The articulating surfaces of the retrieved implants were photographed by Environmental scanning electron microscopy (ESEM quanta 200 FEG, FEI) to characterize the location and severity of the abrasion.

RESULTS

Macroscopic examination of the retrieved implants demonstrated wear mostly along the outer rim of the device. ESEM observations of the abraded areas showed a uniform morphology with increased roughness of the PCU surface, compared to the usual smooth appearance (Fig. 1). In most cases, the direction of wear was consistent with the principal direction of articulation (Anterior-Posterior).

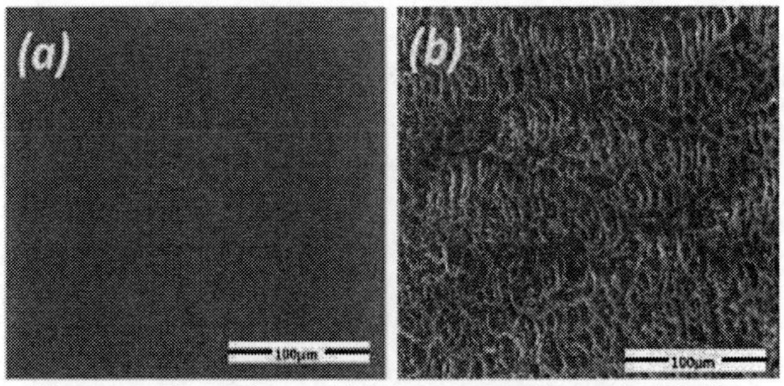

Figure 1. Abraded area (right) compared to control (left)

Microscopic examination of tissue samples revealed fragments of synovial tissue showing papillary hyperplasia with proliferation of macrophages, fibroblasts and lymphocytic infiltrate. Some of the macrophages and giant cells contained hemosiderin in the cytoplasm. This finding is compatible with old hemorrhage. In addition, a proliferation of small blood vessels was observed with mixoid matrix in the papillary formations, consistent with granulation tissue. Multi-nucleated giant cells containing transparent foreign body material with positive birefringence in polarized light were observed in the samples (Fig. 2). These findings are consistent with chronic inflammation due to the presence of foreign body.

The PCU particles, traces from the polarized light photographs, were of variable forms with a non-smooth surface structure. Quantitative analysis of the particles demonstrated an effective particle diameter of 8.47±8.81 µm. The PCU particles were generally of globular shape, as portrayed by average shape factor value of 0.71.
The distribution of particle sizes is demonstrated in Fig. 3 with a diameter range of 1.02 to 229.31 µm. It can be noticed that the majority of particles are lying between 4-16 µm. In addition, no significant variability was found between different patients in term of size distribution.

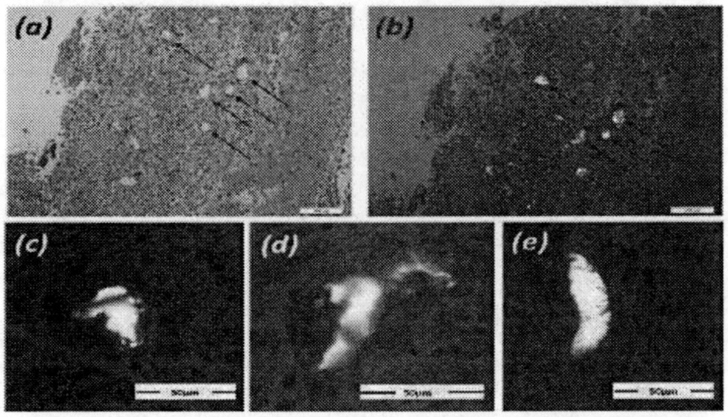

Figure 2. Plain and polarized light microscopy of the tissue

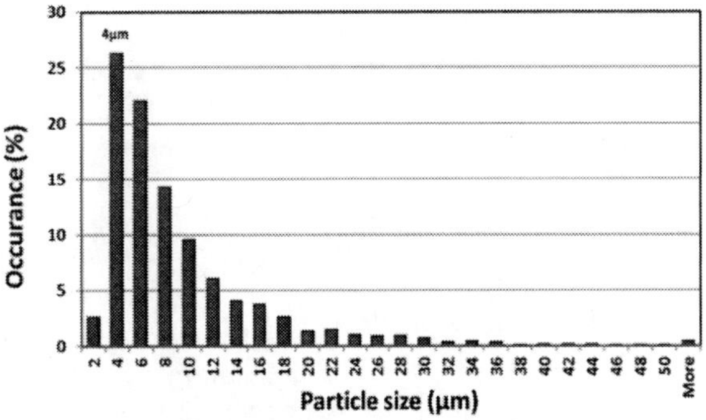

Figure 3. Particle size distribution

DISCUSSION

Implant wear was found to be the main cause of short-term revisions due to inflammatory response. We hypothesized a clinical effect is only exerted when particles accumulate beyond a threshold level, which is currently unknown. The particles which were embedded in the tissue varied in size and shape but the typical morphology and sizes were consistent with those previously reported from laboratory wear testing (Elsner, 2015). At time of revision, the macroscopic inflammatory response was limited to specific locations, which were not always in direct contact with the implant. Synovitis was most commonly found in the lateral and the supra paterllar pouch, thus confirming that the response is to wear particles and not to contact between the synovium and the implant itself. The overall joint condition upon revision was considered acceptable. It did not jeopardize or limit subsequent treatment selection, including re-implantation of the same device. Intra-operative shaving of the synovitis (synovectomy) typically eliminated the symptoms that were reported prior to surgery. Microscopic examination of synovial tissue indicated evidences of chronic inflammation in the presence of PCU particles.
Our finding contributes to the improvement and the understanding of indications-for-use for PCU as a bearing material in orthopedics. Direct articulation between the implant and exposed bone should be avoided at the time of implantation, as this can accelerate wear and lead to an inflammatory response. However, the latter effect is treatable and can be addressed easily by synovectomy in a revision surgery, followed by the next treatment of choice.

ACKNOWLEDGEMENTS

This Study was funded by Active Implants LLC.

REFERENCES

[1] Linder-Ganz, E., et al. J Biomech Eng. (2010) 132 (2).
[2] Elsner, J.J., et al. J Biomech Eng. (2010) 132 (9).
[3] Wippermann, B., et al. J long-term Eff. of Med Imp (2008) 18 (1).
[4] Smith, R.A., et al. J Biomed Mat Res Part A 93.1 (2010): 227-234.
[5] Elsner, J.J., et al. J Mech Behav Biomed Mat 49 (2015): 310-320.

SB³C2017
Summer Biomechanics, Bioengineering and Biotransport Conference
June 21 –24, Tucson,AZ, USA

NANOINDENTATION BASED APPROACH FOR THE MECHANICAL CHARACTERIZATION OF POLYMERIC MICROSPHERES FOR DRUG DELIVERY

G. Serino(1), V. Crognale(2), C. Del Gaudio(2), U. Morbiducci(1)A. Audenino (1)

(1) Department of Mechanical and Aerospace Engineering
Politecnico di Torino
Turin, Italy

(2) Department of Enterprise Engineering
University of Rome "Tor Vergata"
Rome. Italy

INTRODUCTION

Polymeric microspheres are largely studied for biomedical applications as, e.g., embolic agents to treat hyper-vascular tumors [1], or in tissue engineering [2]. The possibility to produce an ad hoc drug delivery system is the mostly studied characteristic of these type of microspheres.

The class of polymeric microspheres used for biomedical applications are mainly composed by hydrophilic polymers, because of their biocompatibility. The polymers chosen for this study are Gelatin and Chitosan. Due to the instability of these two polymers in aqueous environment a natural cross-liker (genipin) is added.

Here the mechanical properties of a new design of polymeric microspheres to be used as carriers for local release of drugs and grow factors, are investigated. The rationale of the study is understand how the used polymers and the presence of the cross-linker influence the mechanical properties of the microspheres and therefore the effectiveness in properly release drugs [3][4].

Technically, nanoindentation was applied on microspheres of size in the range 20-70 μm. The mechanical characterization highlighted a viscous-elastic behavior of microspheres, with an increasing area of the characteristics hysteresis loops when the genipin concentration increases. Moreover, on measured load-displacement data, the Hertz model was applied to estimate the Young's Modulus.

A protocol for the mechanical characterization of polymeric microspheres used for drug delivery will allow: (1) to support their design phase and (2) to improve their effectiveness in targeting the release of drugs.

METHODS

To produce microspheres the water-in-oil emulsion method was applied to gelatin type A (derived from the porcine skin), chitosan and to a blend of these two polymers. Moreover three different values of genepin (a natural cross-linker) concentration: 0.1%; 0.5%; and 1% w/v were used.

Nanoindentation combined with the Hertz model was applied to estimate the mechanical properties of the microspheres.

In detail, the Nanoindenter XP (Agilent/MTS company), characterized by a theoretical force resolution of 50 nN and a theoretical displacement resolution lower than 0.01 nm, was adopted and a specific protocol was developed for characterize the microspheres through a micro-compression test. Here a flat end punch with a diameter of 500 μm was selected.

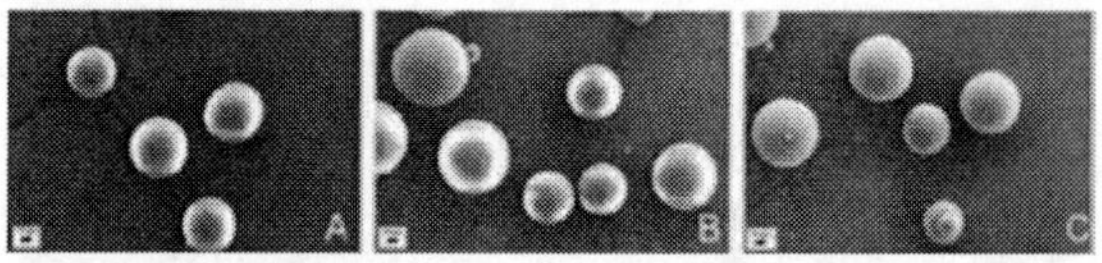

Figure 1: SEM micrographs of cross-linked gelatin (A), chitosan (B), and blend 5/1 (C) microspheres. Scale bar 10 μm.

The compression test is characterized by three steps: loading, hold and unloading. During the loading and unloading phases the velocity of the indenter punch was set at a constant value of 10 nm s^{-1}. In the hold phase, consisted of a stabilization period, the maximum value of the load was maintained for a period of 5 s. The maximum value of the load was reached in correspondence of an indentation depth equal to the 5% of the initial diameter of the microspheres. The initial diameter of microspheres was evaluated before the beginning of the test, using the optical system embedded in the Nanoindenter XP. Load and displacement signals were acquired at a sampling rate of 5 Hz.

On measured load-displacement curves as obtained from the compression tests, the Hertz model [5] was applied on the loading phase and the Young moduls was estimated according to [6]:

$$F=\frac{4\sqrt{R}}{3}\frac{E}{1-\upsilon^{2}}\sqrt{\left(\frac{H}{2}\right)^{3}}=\frac{4\sqrt{R}}{3}E^{*}\sqrt{\left(\frac{H}{2}\right)^{3}} \quad (1)$$

where F is the load, H is the displacement, E is the Young modulus, υ is the Poisson's ratio and R is the radius of the microsphere. The value of the Poisson's ratio was set equal to 0.5 [2][7]. Error analysis, in

terms of mean percentage error, was performed to evaluate the quality of the fitting [8].
For each group of microspheres,40 indentation tests were carried out. Outliers have been excluded by applying the modified Thompson's Tau method.

RESULTS

Explanatory compression load-displacement curves as obtained from nanoindentation, representative of all the tests performed on single microspheres, are presented in Figure 2. The curve is characterized by an hysteresis loop typically characterizing viscous-elastic materials.

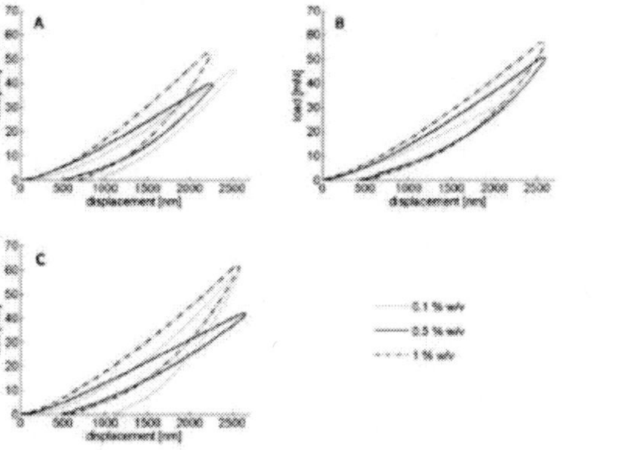

Figure 2: Explanatory load-displacement curves as obtained from nanoindentation tests performed on(A) gelatin, (B) chitosan, and (C) blend 5/1microspheres. For each polymer-based microsphere, data related to three different concentrations of cross-linking agent are presented.

The result of a more quantitative characterization of the mechanical properties of the microspheres is summarized in Fig. 3, where the average values of the Young's modulus for gelatin, chitosan and blend cross-linked microspheres are shown.
Statistical analysis highlights that polymers and cross-linker at different concentrations can result in significantly different (even if moderate) mechanical behavior of microspheres. In particular, it can be noticed from Fig. 3 that: (*i*) in gelatin microspheres, the progressive increase in the percentage of genipin over 0.5% results in a statistically significant stiffening, (*ii*) chitosan microspheres become significantly stiffer for genipin concentration higher than 0.1%, thus highlighting a more marked role of cross-linking, with respect to gelatin, in determining microspheres mechanical properties, and (*iii*) increasing the percentage of genipin from 0.1 to 1% in blend 5/1 composition leads to statistically significant differences, thus highlighting that blending could be used to appropriately modulate microspheres mechanical properties, even within a moderate range of variation.

DISCUSSION

Characteristics hysteresis loops of the curves indicate that part of the energy accumulated by the deformed microsphere during the loading phase is dissipated: as reported elsewhere [9], this higher energy loss is associated with a more marked cross-linking of the polymers chains. According to this, it can be observed that in general microspheres exhibit an hysteresis loop with an increasing area and the highest strength, as a consequence, when genipin concentration increases from 0.1 to 1%, proving, the influence of cross-linking on the mechanical behavior of the microspheres.

Error analysis highlights that there is an overall satisfactory agreement between Hertz model and experimental data (the obtained mean percentage errors are below the 6.9%), thus confirming that Eq. 1 can provide a robust estimation of microspheres Young modulus.
The composition of the polymeric microspheres, does influence not only their mechanical properties, but also the kinetic of release of drugs the microspheres are loaded with. Drug release experiments, performed by using methylene blue (data not shown)clearly indicate that the time course of the release of the therapeutic agent strongly depends on the used polymer(s): the stiffer microspheres, made of chitosan and 0.5 % w/v of genipin, release the loaded drug slower than gelatin microspheres (made of the same genipin concentration), these last being characterized by the lowest values of Young's modulus. Furthermore, blend 5/1 microspheres (made with the same genipin concentration) show an intermediate release profile.

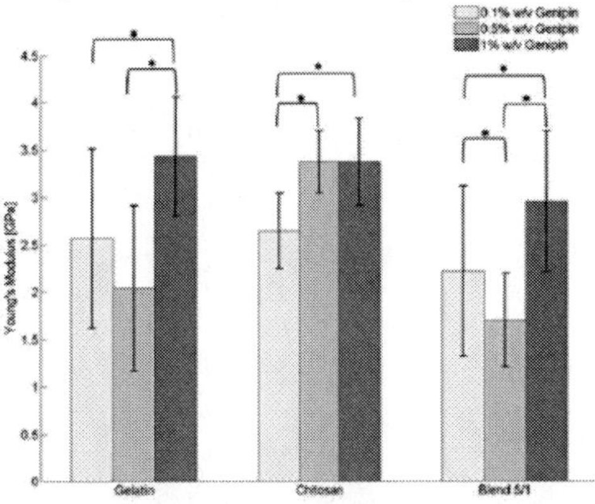

Figure 3:Average values of the Young's modulus as obtained for gelatin, chitosan and blend 5/1 microspheres at three different concentrations of genipin. Statistical significant differences ($p<0.05$) between single microsphere preparations with different genipin concentration are also reported.

The results here presented indicate a correlation between the mechanical behavior of the microspheres and their release kinetic.
Otherwise more experiments have to be perfomed, e.g., varying the concentration of genipin, in way to better understand how the degree of cross-linking influence the mechanical properties and consequently the release of drugs.
In conclusion, here we report that blending natural polymers and adding genipin as natural cross-linker could lead the production of natural microspheres with adjustable mechanical properties, suitable for drug transport and delivery.

REFERENCES

[1] Kuniyuki H et al., *J of the Mechanical Behavior of Biomedical Materials,* 3:497-503, 2010.
[2] Wang C et al., *Chimical Engineering Science*, 60:6649-6657, 2005.
[3] Habraken W et al., *J Biomed Mater Res,* 87A:643-655, 2008.
[4] Yan Y et al., *Powder Technology,* 192:122-130, 2009.
[5] Hertz H., *On the contact of elastic solids,* 318, 1896.
[6] Wheeler A. Ganji A., *Introduction to Engeneerin Esperimentation.*
[7] Hidaka K et al., *J of the Mechanical Behavior of Biomedical Materials,* 4:2161-2167, 2011.
[8]A.J. Wheeler et al.,*Introduction to engineering experimentatio.*
[9] O. Lieleg, et al.,*Phys. Rev. Lett.*, 101 (2008) 108101.

MULTI-SCALE MECHANICAL PROPERTIES OF COLLAGEN MATRIX

Haiyue Li[1], Bin Xu[1], Enhua Zhou[3], Raimon Sunyer[4,5], and Yanhang Zhang[1,2]

[1] Department of Mechanical Engineering
[2] Department of Biomedical Engineering
Boston University
Boston, MA, USA

[3] Ophthalmology, Novartis Institutes for
BioMedical Research,
Cambridge, MA, USA

[4] Institute for Bioengineering of Catalonia
Barcelona, Spain

[5] Centro de Investigación Biomédica en Red
en Bioingenienía,
Biomateriales y Nanomedicina,
Madrid, Spain

INTRODUCTION

The Extracellular Matrix (ECM) provides the principal avenue for mechanochemical communication between tissue and cells[1]. These signals are communicated in a reciprocal manner, playing critical roles in establishing tissue structure-function relationships and controlling cell fate. For example, the stiffness of ECM not only provides structural support but also plays an important role in regulating a series of cell activities such as differentiation, proliferation and migration. On the other hand, cells sense the local forces exerted by ECM and then modulate the structure and composition of the ECM. However, the mechanisms by which ECM mechanics influence cell and tissue function remain to be elucidated since the events associated with these processes span size scales from the tissue to molecular level[1]. Furthermore, the ECM has extremely complex hierarchical three-dimensional (3-D) structures and there exists a tremendous interdependence of ECM compositional, structural, and mechanical properties[1]. To date it is still not well understood how mechanical forces are translated within ECM, or from the tissue- to cellular level. Such information is imperative since the magnitude of the load at the tissue level is different from that experienced by the cell, and the differential load distribution is highly dependent on the architecture and mechanical properties of the ECM[1].

In the present study, the macro and microscale mechanical properties of type I collagen matrix were studied. Optical Magnetic Twisting Cytometry (OMTC) method was used to measure the local mechanical properties of the collagen matrix. Streptavidin coated ferromagnetic beads were used to achieve beads-matrix binding. The macroscopic mechanical properties of collagen gel were measured using a parallel plate rheometer.

MATERIALS AND METHODS

Sample preparation: Neutralized collagen solution was prepared by quickly mixing Nutragen collagen solution (Advanced BioMatrix), 10× PBS (Fisher Scientific) and 0.1M NaOH (Fisher Scientific) solution with a ratio of 8:1:1 at 4°C with a final collagen concentration of 2.0mg/ml and 4.8mg/ml.

Rheological testing: Rheological testing was performed by using a parallel-plate (R=2cm) AR-2000 Rheometer (TA Instruments). Collagen gel was polymerized in situ by quickly loading 1ml neutralized collagen solution on a precooled Peltier stage and then the temperature of the stage was quickly raised to 37°C for gelation. After collagen gel was formed, frequency sweep tests were performed. Oscillating shear strains of 1% were applied to the samples with frequencies ranging from 0.05 to 30Hz. Results from rheological testing were averaged from four samples.

Optical magnetic twisting cytometry: Neutralized collagen solution was transferred into a 3cm-diameter petri dish and incubated at 37°C for 2 hours for gelation. Streptavidin coated ferromagnetic beads (Spherotech Co., 4.7um in diameter) with a bead solution of 0.05mg/ml was poured on collagen gel surface followed by incubation at 37°C for 20 min to allow binding between beads and collagen fibrils. Ferromagnetic microbeads were first magnetized by a brief (0.1ms) strong external magnetic pulse (>0.1T) in the horizontal plane of the collagen gel surface. A homogeneous, oscillatory weak (<0.01T) magnetic twisting field was then applied to the beads in the vertical direction. This, in turn, induces a mechanical "twisting" torque that causes the beads to rotate[2]. The so-called specific twisting torque T with a unit of Pa characterizes the twisting torque per unit volume. The lateral translation d as a result of the rotation was detected optically (Figure 1). The apparent modulus G_d^* is a function of oscillating frequency ω, and can be calculated as:

$$G_d^*(\omega) = \frac{T(\omega)}{d(\omega)} = G_d'(\omega) + iG_d''(\omega) \quad (1)$$

where G_d' and G_d'' represent the apparent storage and loss modulus, respectively, and have a unit of Pa/nm. Since the collagen gel is a viscoelastic material, a phase angle, δ, exists between the twisting torque and bead displacement response, and can be calculated as:

$$\delta = arctan\frac{G_d''(\omega)}{G_d'(\omega)} \quad (2)$$

A geometric factor β is required to transfer the apparent shear modulus (G_d) to shear modulus (G) [3].

$$G_d = \frac{\kappa\beta}{R}G \quad (3)$$

where κ is shape factor (6 for spherical beads) and R is the radius of the bead.

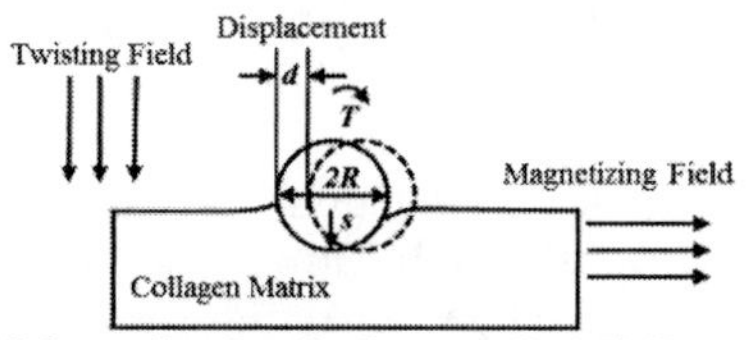

Figure 1: Schematic of optical magnetic twisting cytometry.

RESULTS AND DISCUSSION

By comparing the microscopic with the macroscopic frequency sweep measurements in Figure 2, there are several striking differences: first of all, the local mechanical measurements do not show obvious dependency on collagen concentration, i.e., the apparent storage modulus appears to be similar for all collagen concentrations. Further, there is no obvious frequency dependency of the apparent storage modulus. The apparent loss modulus, however, has a much stronger dependency on frequency, especially at higher frequencies (>10Hz). In local mechanical property measurements, micro-scale ferromagnetic beads (averaged diameter 4.7μm) were used as probes to locally deform the attached collagen fibrils. The diameter of the beads is comparable to the averaged mesh size of the collagen matrix, which was reported to range from 2.3-3.2 μm for the collagen concentrations used in our study[4]. Since the averaged mesh size characterizes the averaged distance between the neighboring collagen fibrils, the number of collagen fibrils attached by single bead is nearly the same among the three collagen concentrations at microscale, thus the dependency on collagen concentration is reduced when decreasing the measurement scales from macro to microscale.

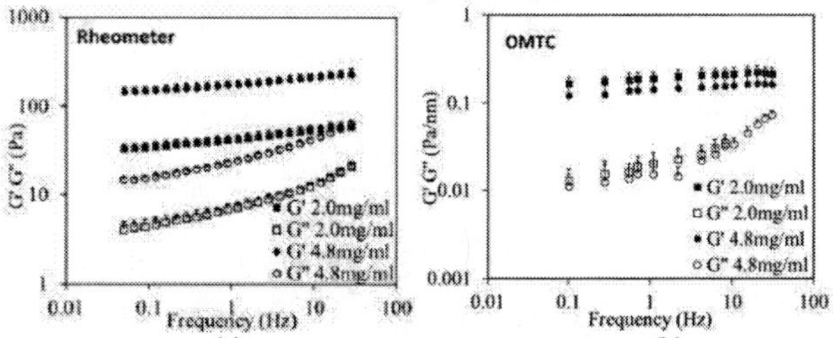

(a) (b)

Figure 2: Averaged storage and loss modulus of 2.0mg/ml, and 4.8mg/ml collagen matrix from (a) macroscopic frequency sweep tests, and (b) microscopic OMTC measurement.

Further comparisons between macroscopic and microscopic measurements were made by comparing the dimensionless quantity, phase lag angle, from both the rheometer and OMTC measurements (Figure 3). The non-dimensional quantity phase angle provides a more direct comparison between the measurements at macro and microscopic scales. The phase angle measured using a rheometer, which captures the macroscopic material behavior, shows a decreasing trend as the collagen concentration increases (Figure 3a), which implies that there is more energy stored relative to that dissipated, resulting from an increase in fibril density. However, this trend is absent from the OMTC measurements (Figure 3b). The phase angle from OMTC measurements shows independency on collagen concentration (Figure 3b). This further proves that the dependency on collagen concentration is reduced in the microscopic mechanical property measurements. Interestingly, when the twisting frequency increases to above 10Hz, the viscous contribution becomes more significant, resulting in a rapid increase in the phase angle from the OMTC measurements. This suggests that the contribution of the viscous effect is more significant in the OMTC measurements since the beads bound to collagen fibrils are surrounded by matrix interstitial fluid and are more exposed/sensitive to the high frequency dependency of the microscopic local fluid environment. The local measurements tend to have a smaller phase angle compared to that at macro scales at low frequencies. This is likely due to the more immediate/direct deformation of collagen fibrils in OMTC measurements through forces exerted by locally attached ferromagnetic beads, as opposed to the nonaffine fiber network deformation in macroscopic measurements that typically involves fiber reorientation, which precludes the development of fibril strain[5]. The discrepancy in phase angle between the macro and microscopic measurements at low frequencies diminishes as the collagen concentration increases, as the nonaffinity decreases with higher collagen concentration due to the formation of a denser network that limits the room available for fiber reorientation.

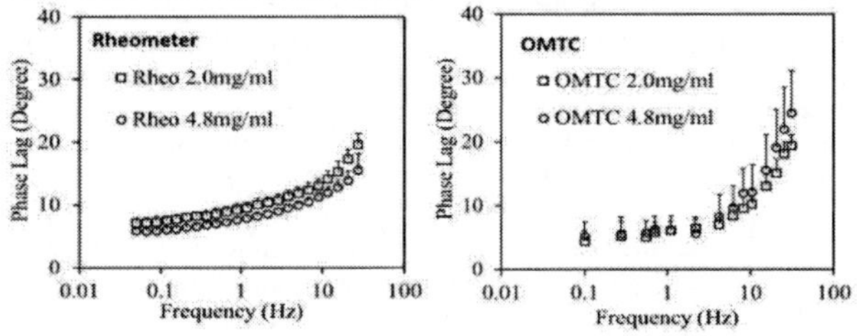

(a) (b)

Figure 3: Averaged phase angle based on measurements from (a) rheometer and (b) OMTC for collagen matrix with collagen concentrations of 2.0mg/ml and 4.8mg/ml.

Our study on the multiscale mechanics of collage matrix suggested several interesting differences between the macro and microscopic mechanical properties originated from the scales of measurements. In the macroscale measurements, the storage and loss modulus increases with collagen concentration. The nonaffine deformation of the fibril structural network plays an important role. At the microscopic scale, the local properties are less sensitive to changes in collagen concentration. However, the loss modulus is more affected by the local interstitial fluid environment, leading to an increase in viscosity at high frequencies.

ACKNOWLEDGEMENTS

Funding support from National Science Foundation to YZ (NSF CMMI 0954825) is acknowledged.

REFERENCES

[1]Pizzo A, et al. J. Appl. Physiol. 205; 98(5): p. 1909-1921.
[2] Fabry B, et al. J. Appl. Physiol. 2001; 91: p. 986-994.
[3] Mijailovich SM et al. J. Appl. Physiol. 2002; 93: p. 1429-1436
[4] Yang Y, Kaufman L. Biophys. J. 2009; 96(4): p. 1566-1585.
[5] Onck P, et al. Phys. Rev. Lett. 2005; 95(17): p. 178102.

SB³C2017
Summer Biomechanics, Bioengineering and Biotransport Conference
June 21 – 24, Tucson, AZ, USA

MODELING MECHANICAL PROPERTY CHANGES OF COLLAGEN FIBRILS FOLLOWING CYCLIC LOADING

Michelle L. Chen (1), Monica E. Susilo (2), Jeffrey A. Ruberti (2), Thao D. Nguyen (1)

(1) Mechanical Engineering
Johns Hopkins University
Baltimore, MD, USA

(2) Bioengineering
Northeastern University
Boston, MA, USA

INTRODUCTION

Fibrillar collagen is the main contributor to the stiffness and strength of soft connective tissues. The collagen fibrils provide the tissues with their stiffness and strength, both of which can be anisotropic depending on the fibril arrangement and interactions. The collagen architecture can adapt (remodel) to mechanical load by changing the arrangement, connectivity, and intrinsic properties of the collagen fibrils, all of which can significantly alter the mechanical properties at the tissue level.

In a previous study, we showed that cyclic loading at 13% and 26% strain increased the stiffness and strength of decellularized collagen constructs [1]. A greater stiffening effect was measured for the larger strain amplitude. The stiffening effect occurred both as an increase in the slope of the linear portion of the strain-stiffening stress response and as a left-ward shift of the stress-strain curve accompanied by an increase in the reference length of the specimen. The permanent elongation and softening of the stress-strain curve are commonly observed as preconditioning effects caused by the breaking of crosslinks and the reorientation of fibrils along the loading direction [2, 3]. However, we did not measure significant changes in the orientation distribution of the collagen fibrils with cyclic loading. Furthermore, the failure stress increased with cyclic loading, which is opposite to the softening effect observed in precondition [2].

Based on these findings, we hypothesized that cyclic loading increased the intrinsic stiffness and strength of the collagen fibrils in the decellularized collagen constructs. To test this hypothesis, we applied a distributed fiber model with fibril damage and fit the model parameters for the fibril strength and stiffness to the stress-strain curve of the constructs after cyclic loading. Preliminary results confirmed that the stiffening effect observed in the stress strain response was caused by an increase in the modulus of the fibril rather than by a change in the fibril microstructure.

METHODS

EXPERIMENTAL PROCEDURE Susilo et al. [1] prepared thin dense disorganized collagen substrate (DDCS) sheets from bovine collagen type I fibrils suspended in a poly(ethylene) glycol (PEG) matrix. Each sheet was divided into two and then pre-conditioned by uniaxial tension loading for either 1 cycle or 50 cycles and with a strain amplitude of either 13% or 26% strain, forming four experimental groups. After pre-conditioning, the specimens were either mechanically loaded to failure, or used for TEM imaging. The TEM images were used to determine the planar fibril orientation distribution via Fourier transform analysis [1].

CONSTITUTIVE MODEL A planar orientation distribution function, $\rho(\theta)$, was used to describe the collagen fibril arrangement, where θ is the fibril angle relative to the dominant fibril orientation. The elastica model of Comninou and Yannas [4] was applied to describe the strain-stiffening behavior of the collagen fibrils. In the elastica model, the fibrils are represented as planar sinusoidal Euler-Bernoulli beams. The midline of the undeformed fibril is given by $X_2 = a\sin(bX_1)$, where (X_1, X_2) are coordinates along the horizontal and vertical axes, $\Theta_0=ab$, is the crimp angle, which is defined as the maximum angle that fibril makes with the horizontal axis, and $L=2\pi/b$ is the length of the fibril projected onto the horizontal axis. The beam is assumed to be simply supported and subjected to a horizontal force F applied at $X_1=L$. The applied force is described by the non-dimensionalized parameter $\alpha=F/EA$, where A is the cross-section area, E is the Young's modulus of the fibril, and EA describes the axial stiffness of the fibril. The bending stiffness of the fibril is described by $E\beta$, where $\beta=b^2r^2$ is the slenderness ratio, and r is the fibril radius.

We assumed that mechanical loading beyond a stress threshold, σ_{min}, can induce damage by degrading the fibril modulus as $E=(1-\xi)E_0$ where E_0 is the Young's modulus of the undamaged fibril and $0 < \xi < 1$ is the damage factor. We further assume that damage evolves as,

$$\frac{d\xi}{dt} = \begin{cases} C\left(\frac{(1-\xi)(\lambda_f-1)}{H} - 1\right), & \text{if } \frac{(1-\xi)(\lambda_f-1)}{H} > 1 \\ 0, & \text{otherwise.} \end{cases} \quad (1)$$

where $H = \sigma_{min}/E_0$ is a non-dimensionalized stress threshold and C the rate of damage at initiation. The deformed configuration $\Theta(X_1)$ and fibril stretch, λ_f can be solved analytically for the case of small crimp angle and assuming small fibril strain [4].

$$\Theta(X_1) = \frac{\beta}{4\alpha(1+\alpha)+\beta}\Theta_0(X_1) \qquad \lambda_f(X_1) = 1+\alpha\cos(\Theta(X_1)) \quad (2)$$

Following the derivation shown in Tonge et al. [5], the fiber strain energy density can be written as the sum of a bending component that describes the contribution from the wavy fibril straightening and an axial component from the fibril stretch,

$$\Psi_{fib} = \underbrace{\frac{1}{L}\int_0^L \left[\frac{(1-\xi)E_0\beta}{8}\left(\Theta_0(X_1)-\Theta(X_1)\right)^2\right]dX_1}_{\Psi_{bending}} + \underbrace{\frac{1}{L}\int_0^L \left[\frac{(1-\xi)E_0}{2}\left(\lambda_f(X_1)-1\right)^2\right]dX_1}_{\Psi_{axial}} \quad (3)$$

We assume at the tissue level that the fibrils deform affinely with the macroscopic deformation gradient. This allows the fibril to be represented by a material line with orientation θ and the macroscopic fibril stretch to be computed from the deformation gradient as $\lambda(\theta) = \sqrt{\mathbf{C} : (\mathbf{a}_0(\theta) \otimes \mathbf{a}_0(\theta))}$, where $\mathbf{a}_0$ is the fibril orientation vector. At the tissue level, the strain energy density is assumed to be additively decomposed into an isotropic, incompressible component for the matrix and an anisotropic component from the fibrils,

$$\Psi = \frac{\mu}{2}(I_1 - 3) - \frac{p}{2}(I_3 - 1) + \int_{-\pi}^{\pi} \Psi_{fib}\left(\lambda(\theta)\right)\rho(\theta)\,d\theta \quad (4)$$

where μ is the shear modulus of the matrix, p the incompressibility constraint, and I_1 and I_3 are the first and third invariants of $\mathbf{C}$, the right Cauchy-Green deformation tensor.

PARAMETER FITTING For each experimental group, the parameters of the constitutive model were fit to the average uniaxial tension stress-stretch curve of the specimens after preconditioning. The orientation measurements obtained from the TEM images taken after preconditioning were used to describe $\rho(\theta)$. The fibril modulus, E, slenderness ratio β, and crimp angle Θ_0 were fit to the average stress strain response for each $\rho(\theta)$.

To simulate uniaxial tension we applied a stretch in the vertical direction $\lambda_{22} = \sqrt{(F_{12})^2 + (F_{22})^2}$ and solved for the vertical component of the first Piola-Kirchoff stress and components of the deformation gradient $\mathbf{F}$. We assumed that $F_{21}=0$ because the specimens were tightly gripped during testing. We also assumed $F_{13} = F_{31} = F_{23} = F_{32} = 0$ because there is no out-of-plane shear. The first Piola-Kirchhoff stress response is,

$$\mathbf{P} = 2\mathbf{F}\frac{\partial \Psi}{\partial \mathbf{C}} = 2c_1\mathbf{F} - p\mathbf{F}\mathbf{C}^{-1} + \int_{-\pi}^{\pi} 2\rho(\theta)\,\mathbf{F}\frac{\partial \Psi_{fib}}{\partial \mathbf{C}}\,d\theta \quad (5)$$

where $\mathbf{C} = \mathbf{F}^T\mathbf{F}$ is the right Cauchy-Green. The material parameters were determined by minimizing the absolute difference between the computed P_{22} and the uniaxial engineering stress using fmincon (Matlab 9.1). To simplify matters, we set the matrix shear modulus to μ = 0.01 MPa [5], because it primarily affects the unknown out-of-plane deformation. Preliminary parameter study showed that μ had a negligible impact on the in-plane stress response for the fibril parameters obtained in Table 1. We neglected damage parameters H and C, fitting to the part of the stress-stretch curve where $d^2P_{22}/d\lambda_{22}^2 >$ 0.

RESULTS

Table 1 lists the material parameters for the 13% and the 26% strain experimental groups. The material parameters for each fibril distribution are listed, with the average value bolded. Figure 1a shows the stress-strain curve comparing the measured and model fit for the 1 and 50 cycle cases at 13% and 26% strain.

Table 1 - Model Parameters

Material Para-meter	13% strain		26% strain	
	1 cycle	50 cycle	1 cycle	50 cycle
E (MPa)	20.65 21.71 20.45 Avg = **20.94**	22.89 24.87 23.70 Avg =**23.82**	17.60 16.77 18.76 Avg =**17.71**	29.70 24.77 28.34 Avg =**27.60**
Θ_0	31.32° 31.28° 31.32° Avg =**31.31°**	20.89° 20.92° 20.91° Avg =**20.91°**	38.45° 38.43° 38.47° Avg =**38.45°**	30.08° 30.09° 29.98° Avg =**30.05°**
β	0.0213 0.0209 0.0212 Avg =**0.0212**	0.0129 0.0130 0.0131 Avg =**0.0130**	0.0203 0.0207 0.0203 Avg =**0.0204**	0.0076 0.0076 0.0072 Avg =**0.0075**

In a paired t-test, the materials exposed to cyclic loading for 50 cycles had a larger E (p=0.0123 for 13%, n=3 and p=0.0142 for 26%, n=3), a smaller crimp angle Θ_0 (p=3.91E-06 for 13%, n=3 and p=2.93E-05 for 26%, n=3), and smaller fibril slenderness ratio β (p=2.67E-04 for 13%, n=3 and p=9.77E-05 for 26%, n=3). A large change in the parameters was obtained for the larger applied strain amplitude.

We next performed a parameter study to understand the implications of the alterations in material parameters. The material parameters of the 26% strain group were varied one-by-one from the averaged values to the corresponding 50 cycle groups. Figure 1b plots the stress-stretch curve of the baseline 26% strain group along with the results of the parameter study. Also plotted for comparison in Figure 1b are the average and standard deviation of the stress-stretch curves for the 1 cycle and 50 cycle 26% strain amplitude, measured during the load-to-failure uniaxial tension testing and the range of stress-stretch curves calculated for different $\rho(\theta)$. A decrease in the crimp angle caused the fibril to transition from a bending dominated to stretch dominated deformation at a smaller strain. This caused the stress-stretch curve to shift left. The decrease in the slenderness ratio caused a more compliant bending response, which shifted the stress-stretch curve slightly to the right. Axial stiffness increased from the 1 cycle to the 50 cycle which increased the slope in the linear region of the stress-stretch curve. E is the only parameter to affect the slope in the linear region of the stress-stretch curve. These alterations to the stress-stretch curve from the parameter study were larger than the variations in the stress-stretch curves caused by the orientation distributions of the collagen fibrils.

DISCUSSION

We showed that the axial stiffness of the collagen fibril increased with cyclic loading. Larger increase was obtained for a larger applied strain. Furthermore, the increase in the axial stiffness had a larger effect on the uniaxial stress-stretch response than the measured alterations in the fibril orientation distribution caused by cyclic loading. Our findings imply that cyclic loading can increase the intrinsic material stiffness of the collagen fibrils and may be an important mechanism of the remodeling process.

We are currently working to implement the distributed fibril model with fibril damage and will repeat the modeling study to determine the effect of cyclic loading on the intrinsic strength stiffness the collagen fibril.

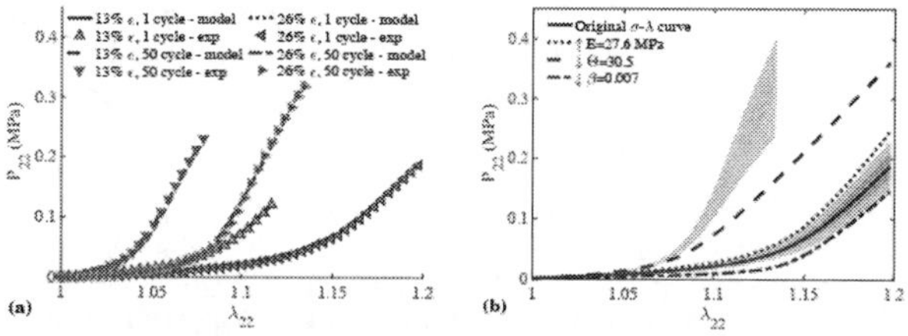

Figure 1: (a) Fitted load-to-failure P_{22}-λ_{22} curves (b) Variation in P_{22}-λ_{22} response across different fibril distributions, and different load-to-failure samples for substrates preconditioned to 26% strain. Blue bands represent 1 standard deviation from the average of P_{22}-λ_{22} curves calculated for the different $\rho(\theta)$. Red (pink) bands represent 1 standard deviation from the average of P_{22}-λ_{22} curves. Solid line represents the averaged 1 cycle P_{22}-λ_{22} response

REFERENCES

[1] Susilo, M et al., *Interface Focus*, 6:20150088, 2015.
[2] Eshel H et al., *Ann Biomed Eng*, 29:164–172, 2001.
[3] Tower T et al., *Ann Biomed Eng*, 30:1221–1233, 2002.
[4] Comninou, M et al., *J Biomech*, 9:427-433, 1976.
[5] Tonge, T et al., *Biophysical Journal*, 109:2689–2700, 2015.

SB^{3}C2017
Summer Biomechanics, Bioengineering and Biotransport Conference
June 21 – 24, Tucson, AZ, USA

RACE RELATED DIFFERENCES IN SCLERAL THICKNESS USING MICROCOMPUTED TOMOGRAPHY

Kenneth J. Furdella (1), Ehab A. Tamimi (1), Jonathan P. Vande Geest (1,2,3)

(1) Department of Bioengineering
University of Pittsburgh
Pittsburgh, PA, USA

(2) Louis J. Fox Center for Vision Restoration
University of Pittsburgh
Pittsburgh, PA, USA

(3) 3McGowan Institute for Regenerative Medicine
University of Pittsburgh
Pittsburgh, PA, USA

INTRODUCTION

The sclera is the primary load-bearing tissue of the eye [1]. Therefore, the eye's mechanical strength is greatly affected by the sclera. Differences in scleral properties can increase or decrease scleral deformation based on intraocular pressure, microstructure, or disease.

In order to evaluate these properties, scleral deformation and geometry must be measured. Previous work in our lab has evaluated sclera deformation using Sequential-Digital Image Correlation (S-DIC) [2].

To evaluate scleral thickness, previous work has shown that it can be measured using high frequency ultrasound, ultrasonic pachymeter, microcomputer tomography, and several other techniques [3, 4]. The ultrasonic pachymeter is a common technique to measure scleral thickness, but provides limited sampling that could be problematic in determining scleral regional differences. To increase sampling, other groups have used high frequency ultrasound. This technique can provide high resolution images. One possible down fall of high frequency ultrasound is the difficulty reconstructing the optic nerve region. In order to deal with these problems, microcomputer tomography can provide a solution for scleral thickness measurements. Specifically, it can provide high resolution cross sectional scleral images for thickness calculation.

It is also known that there is a higher prevalence of primary open angle glaucoma in those of European decent (ED), African Decent (AD), and Hispanic Ethnicity (HE). The primary goal of this work is to obtain regional thickness maps in racioethnic groups of ED, AD, and HE. Furthermore, the results from this study in combination with our previous work in scleral deformation mapping can be used to determine the material properties of the scleral shell.

In this study, eight posterior scleral shells human donor samples previously used for S-DIC were imaged using microcomputer tomography and reconstructed. The reconstructed surfaces were used to determine scleral thickness globally and regionally.

METHODS

Posterior scleral shells from 8 donors were obtained from Alabama Eye Bank in Birmingham, AL, USA, the San Diego Eye Bank in San Diego, CA, USA and Michigan Eye Bank in Ann Arbor, USA. Each posterior sclera sample was previously used in S-DIC testing and was subdivided into racioethnic groups: ED (n=3), AD (n=3), and HE (n=2). The donors ranged in age from 57 to 84 years old.

The scleral samples were soaked in 1% Phosphotungstic Acid (PTA) for 35 minutes. The samples were imaged using the Siemens Inveon scanner (microcomputer tomography) with the following parameters: 80 kVp, 450 μA, 1650 ms exposure, 220 degrees of rotation, a binning of 2, and a 53.79 transaxial by 53.79 axial mm field of view. The isotropic voxel size was 35.6 microns.

The posterior sclera and optic nerve head were isolated using image segmentation in MATLAB™. Briefly, the image sets were thresholded to remove background noise. Next, Chan-Vese segmentation was used to isolate the eye from the specimen holder. The top red line in Figure 1A represent image segmentation with Figure 1B showing reconstructed outer eye surface. The top and bottom scleral surfaces were isolated and used to calculate the scleral thickness. Finally, the optic nerve head was removed and sclera was divided into four regions (superior, inferior, nasal, and temporal).

Each region in the eye was compared using a statistics script in R. The linear mixed model was used to model thickness assuming race and regions as fixed effect terms. The eye donor was considered a random effects term. The thickness estimates were calculated for reach race and region and post hoc piecewise comparisons were performed using least square means.

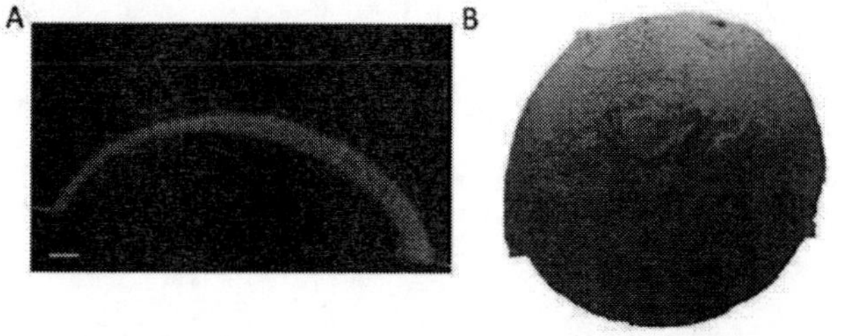

Figure 1: (A) Scleral cross sectional view. Red lines highlight the areas selected for thickness calculations. The scale bar represents 2mm. (B) Top surface selected from Chan-Vese segmentation.

RESULTS

Scleral thickness was calculated for each racioethnic group. The average sclera thickness for HE, ED, and AD was 821μm ±359μm, 1022μm ± 263μm, and 743μm ± 258μm, respectively. Representative thickness map distribution for each racioethnic group is shown in Figure 2. In general, thickness decreased with distance from the optic nerve head.

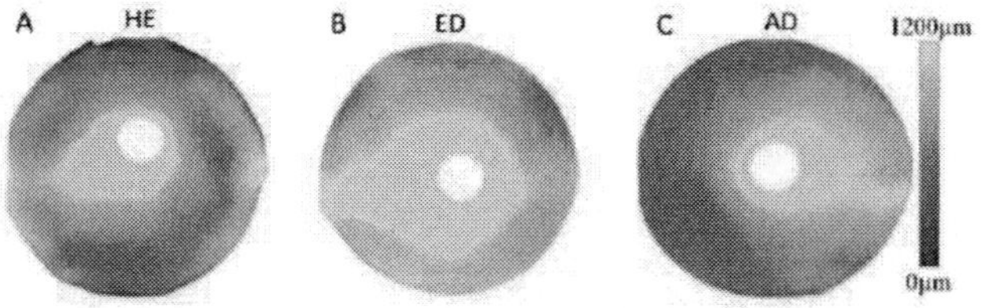

Figure 2: (A) HE, (B) ED, and (C) AD thickness distribution map.

Next, the sclera regions for each racioethnic group were compared and are shown in Figure 3. Each group was divided into 4 sub-regions: superior, nasal, inferior, and temporal. The temporal region in the AD group was significantly smaller than the ED with a p-value < 0.05. The superior regions in the AD and HE were found to be significantly smaller than ED with a p-value < 0.05.

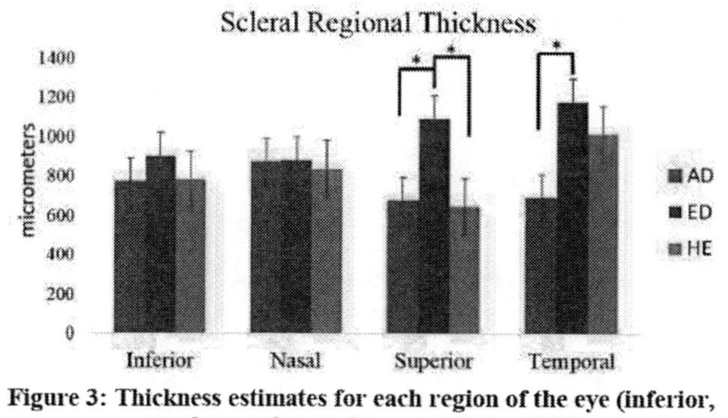

Figure 3: Thickness estimates for each region of the eye (inferior, nasal, superior, and temporal). * p<0.05

DISCUSSION

Our results demonstrated that microcomputed technology can determine scleral thickness between racioethnic groups. Specifically, our data demonstrates that scleral regional differences exist in the temporal region between AD and ED. There were also regionals differences seen in AD, HE, and ED in the inferior region.

Our sclera thickness measurements are similar to previously recorded work [1, 5-9]. First our results agree that thickness decreases with distance from the optic nerve head. Second, our average scleral thickness is similar with the previous recorded results, around 700μm. The only discrepancy between our results and the literature values is that the temporal and nasal regions tend to be a slightly thinner.

One potential limitation is the inability of our image segmentation process to differentiate sclera tissue from extraneous connective tissue, seen in Figure 1B. One possible solution would be to create a fitted surface to the segmented data to get rid false thickness measurements. Furthermore, this limitation could potentially account for our slightly larger thicknesses values.

Future work will divide the data into peripapillary sclera and nonperipapillary sclera. This could help reduce the mean thickness in nonperipapillary sclera regions from the optic nerve head. Our thickness maps and previously recorded scleral deformation data will be used in a subdomain inverse finite element algorithm in the future to estimate scleral material properties.

ACKNOWLEDGEMENTS

Research reported in this abstract was supported by the National Eye Institute of the National Institutes of Health under award number R01EY020890 to JPVG.

REFERENCES

1. Norman, R.E., et al., *Dimensions of the human sclera: Thickness measurement and regional changes with axial length.* Exp Eye Res, 2010. **90**(2): p. 277-84.
2. Pyne, J.D., et al., *Sequential-digital image correlation for mapping human posterior sclera and optic nerve head deformation.* J Biomech Eng, 2014. **136**(2): p. 021002.
3. Coudrillier, B., et al., *Effects of age and diabetes on scleral stiffness.* J Biomech Eng, 2015. **137**(7).
4. Tang, J. and J. Liu, *Ultrasonic measurement of scleral cross-sectional strains during elevations of intraocular pressure: method validation and initial results in posterior porcine sclera.* J Biomech Eng, 2012. **134**(9): p. 091007.
5. Olsen, T.W., et al., *Human sclera: thickness and surface area.* Am J Ophthalmol, 1998. **125**(2): p. 237-41.
6. Buckhurst, H.D., et al., *Measurement of Scleral Thickness in Humans Using Anterior Segment Optical Coherent Tomography.* PLoS One, 2015. **10**(7): p. e0132902.
7. Vurgese, S., S. Panda-Jonas, and J.B. Jonas, *Scleral thickness in human eyes.* PLoS One, 2012. **7**(1): p. e29692.
8. Shen, L., et al., *Scleral Thickness in Chinese Eyes.* Invest Ophthalmol Vis Sci, 2015. **56**(4): p. 2720-7.
9. Ebneter, A., N.U. Haner, and M.S. Zinkernagel, *Metrics of the normal anterior sclera: imaging with optical coherence tomography.* Graefes Arch Clin Exp Ophthalmol, 2015. **253**(9): p. 1575-80.

SB3C2017
Summer Biomechanics, Bioengineering and Biotransport Conference
June 21 –June 24, Tucson, AZ, USA

DETERMINATION OF PROPER STORAGE CONDITION AND CONSTITUTIVE MODEL FOR PORCINE URINARY BLADDER WALL MECHANICAL PROPERTIES

T. Tuttle, T. R. Bush and S. Roccabianca

Mechanical Engineering Department
Michigan State University
East Lansing, Michigan, U.S.A.

INTRODUCTION

The urinary bladder (UB) is a musculomembranous hollow organ whose main function is to store and periodically void urine in a controlled manner, which is achieved through a regulated cycle of mechanical relaxation and contraction (i.e., micturition cycle). As urine is collected in the bladder, the compliance of the tissue (i.e., the tissue elasticity) allows the bladder to enlarge so that the internal pressure stays low and relatively constant (**Figure 1, Bottom**, between points A and B). Then, during the voiding phase, the muscle within the bladder wall (i.e., the detrusor) contracts and both sphincters relax allowing urine to flow from the bladder (**Figure 1, Bottom**, between points C and D). Alterations to this cycle, or obstruction of the urethral outlet, can give rise to a number of storage or voiding disorders, e.g., underactive / overactive bladder syndromes.

Clinically, the parameter that urologists use to diagnose loss of functionality in the bladder is **compliance** [1]. The medical definition of bladder compliance is the ratio of changes in volume over changes in pressure during filling, and it is expressed in ml/cmH_2O (**Figure 1, Bottom**). These data are obtained by performing a urodynamic test, where the bladder volume, pressure, and urine flow are recorded during a whole micturition cycle. Bladder compliance values have, however, a wide range of "normality" reported in the literature: some authors report normal values to be between 30-100 ml/cmH_2O, others say anything above 40 ml/cmH_2O is normal, others consider values higher than 100 ml/cmH_2O still in the normal range [2]. Additionally, a cough, sneeze, or movement can affect these measures significantly and it is not standardized. Further, the key points one should consider when recording a urodynamic test is highly dependent on the operator (i.e., the positioning of points A, B, C, and D in **Figure 1, Bottom**). A more robust approach for identification of compliance is needed so that normality ranges can be determined. To this extent, we submit that a measure of the stress in urinary bladder wall could be a better predictor of the normal functionality of the bladder. For these reason, there is a need for more accurate mechanical models of the UB.

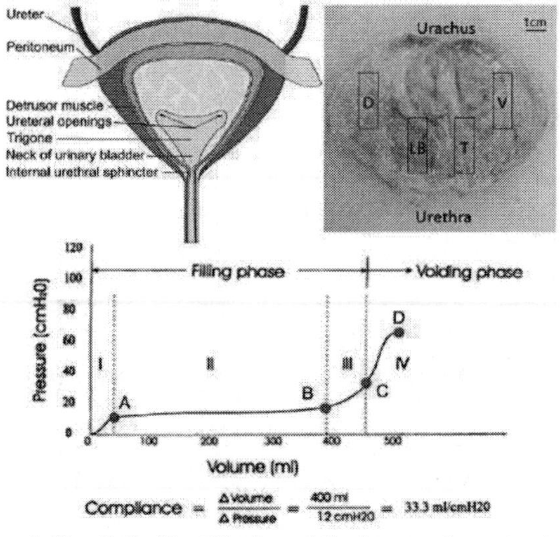

Figure 1. Top Left: Identification of the lower urinary tract. Top Right: Dissected pig UB, luminal side: dorsal (D), lower Body (LB), trigone (T), and ventral (V) locations shown. Bottom: Representative plot of urodynamics data with key measurement points. These points are identified in a subjective fashion by the clinician or technician who conducted the test.

Poster Presentation #P174

METHODS

Four intact pig UB (all male) were collected from the Michigan State University Meat Lab, then cleaned of any contents and adjacent tissues. For biomechanical characterization, four samples from each UB were harvested along the apex-to-base axis from four anatomical regions: dorsal (D), ventral (V), trigon (T) and lower body (LB), respectively (see **Figure 1**). We randomized anatomical location and storage protocols to avoid bias.

Storage Study. Following isolation, specimens were tested fresh (i.e., Control), or stored following three different storage protocols and tested afterward. The storage protocols were: (a) 24 hours in refrigeration at 4°C (i.e., Protocol #1), (b) flash-freeze (-80°C) followed by thawing for 6 hours at 4°C (i.e., Protocol #2), and (c) flash freeze (-80°C) followed by thawing for 24 hours at 4°C (i.e., Protocol #1). This study is important because the tissue is available in bulk (i.e., pigs are slaughtered once a month in groups of at least 10 pigs) or without previous notice (i.e., human tissue). For these reasons, there is a necessity to store the tissue to be able to maximize the number of samples collected and tested.

Mechanical Testing. Samples were then subjected to a uniaxial tensile test along the apex-to-base direction. The tissue was first preconditioned for 8 cycles using a displacement rate of 0.3 mm/s and a stretch ratio of $\lambda = 1.6$, where λ is the deformed length of the sample divided by the preloaded sample length. After preconditioning, a sample-specific stretch ratio λ_{test} was estimated to ensure a maximum stress of at least ~200 kPa. Finally, the tensile test consisted of 3 loading cycles performed at a displacement rate of 0.15 mm/s reaching the stretch of λ_{test}.

Constitutive description. We also modeled the data collected using the constrained mixture theory [3-4]. We employed a two-fiber family model, motivated by the assumption that the primary constituents that bear load under tension are: (1) an amorphous matrix, mostly residing in the submucosa and consisting of an isotropic matrix of collagen and elastin fibers; and (2) families of locally parallel smooth muscle cells and collagen fibers distributed along two main preferential directions, apex-to-base and circumferential, residing in the detrusor. The constitutive model is described in terms of a strain energy function W of the form:

$$W = \frac{c^{am}}{2}(I_C - 3) + \sum_{k=1}^{2} \frac{c_1^k}{4c_2^k}\left(exp[c_2^k((\lambda^k)^2 - 1)^2] - 1\right), \qquad (1)$$

where c^{am}, and c_1^k are material parameters with the dimension of a stress, c_2^k is a dimensionless material parameter., I_C is the first invariant of the right Cauchy–Green tensor $\boldsymbol{C}$ (i.e., $I_C = tr\boldsymbol{C}$), and λ^k is the stretch experienced by the kth fiber family oriented in direction $\boldsymbol{M}^k = [0, sin\,\alpha_0^k, \cos\alpha_0^k]$ in an appropriate reference configuration. We let $\alpha_0^1 = 0$ [deg] (apex-to-base family) and $\alpha_0^1 = 90$ [deg] (circumferential family). We then used nonlinear regression to determine best-fit values of the three model parameters (**Table 1**).

RESULTS

Figure 2 shows the effects of storage protocol and location on mechanical behavior of the UB. First, this study indicated that the four storage protocols yield results that are not significantly different, giving the option of storing the tissue at -80°C for longer periods. Second, the mechanical tests performed indicated that the inter-subject variability of the mechanical behavior were not significantly different, the animals were all the same age at time of slaughter and same sex (data not shown). Our data show that the mechanical responses are highly reproducible across pigs. **Figure 2** also shows differences between anatomical locations in pig bladders that are confirmed by previously published data [5]. These results confirm the existence of location-specific mechanical behavior in pigs' UB, and the ability of the conducted tests to detect these differences. Finally, the model show the good predictive capabilities in pigs UB (i.e., root mean square error is never higher than 0.07 as shown in **Table 1**).

DISCUSSION

This work represents the first step toward the development of a mechanical model that will have the capability of describing the changes in the mechanical property of the UBW resulting from changes in the urodynamics.

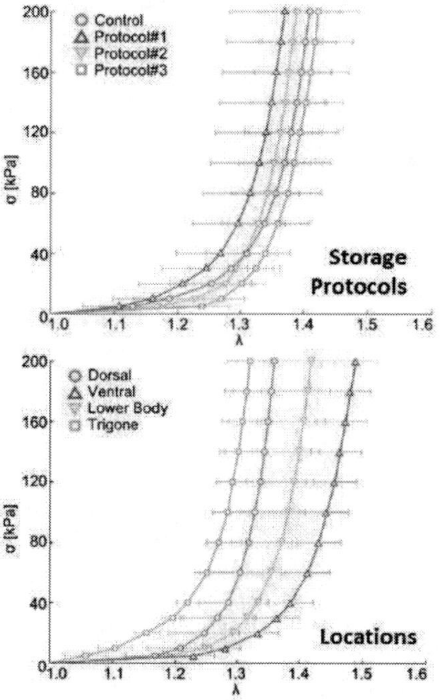

Figure 2. Top: Cauchy stress-stretch curves averaged by protocols, each curve represents the average of four tests. Bottom Cauchy stress-stretch curves averaged by locations, each curve represents the average of four tests.

Table 1. : Best-fit material parameters of the constitutive model.

	c^{am} [MPa]	c_1^k [MPa]	c_2^k	RMSE
Pig#1	3.8	6.2	2.7	0.04
Pig#2	11.4	4.4	4.4	0.07
Pig#3	6.4	5.2	2.4	0.04
Pig#4	9.6	5.1	4.3	0.06

REFERENCES

[1] Nitti, V. W. *Reviews in urology*, 7 (Suppl 6), S14, 2005.
[2] Wyndaele, J. J. *The Journal of urology*, 161(3), 899-902, 1999.
[3] Roccabianca S. et al. *J. Mech. Behavior Biomed.*, 19: 618-634, 2014.
[4] Humprey J. D. et al. *Biomech. Model. Mechan.*, 2:109-126, 2003.
[5] Korossis S. et al. *Biomaterials*, 30 (2). pp. 266-275, 2009.

COMPARATIVE STUDY OF THE TENSILE STRENGTH OF COMMONLY UTILIZED SUTURE MATERIALS

Sourav S. Patnaik (1, 3), James R. Butler (2), Bryn Brazile (1, 4), Margot Damaser (5), Jun Liao (1, 6)

(1) Department of Agricultural and Biological Engineering, and (2) Center for Veterinary Medicine, Mississippi State University, Mississippi State, MS, USA

(3) Department of Mechanical Engineering, University of Texas at San Antonio, San Antonio, TX, USA

(4) Eye and Ear Institute University of Pittsburgh, Pittsburgh, PA, USA

(5) Department of Biomedical Engineering, Cleveland Clinic, Cleveland, OH, USA

(6) Department of Bioengineering University of Texas at Arlington Arlington, TX, USA

INTRODUCTION

Pelvic floor disorders, including stress urinary incontinence and pelvic organ prolapse, results from the failure of DeLancey's three levels of structural support. To restore this support such that it holds the organs in place, clinicians employed various surgical meshes (artificial, biological, and hybrid) during pelvic reconstructive surgeries. There is no unique or widely accepted "gold standard" surgical technique for correcting pelvic organ prolapse, and hence, several different types of gynecological meshes are utilized. More than 30% of the prolapse or incontinence surgeries are repeat procedures [1], and the use of permanent sutures (e.g. Prolene™, Silk, Nylon,) in these gynecological surgeries often lead to complications such as granulation, fistula, bleeding and infections [2]. Complications arising from implanted gynecological meshes can be due to surgical issues, patient characteristics, or the material properties of the mesh [3]. In addition, the mechanical integrity of the tissue-mesh or tissue-tissue interface, and the type of suture used, can influence the outcome of the surgical procedure. Hence, it is important to understand the mechanical behavior of these surgical suture materials. For this study, we evaluated the tissue-suture integrity by comparing the suture retention strength of two commonly utilized suture materials (monofilament vs. polyfilament); polydioxanone (PDS™) material was selected for the monofilament group and Vicryl™ material was chosen for the polyfilament group. To evaluate the mechanical integrity of these suture materials, suture retention testing method was employed which involves the tensile testing of two tissue strips united by suturing them using surgical knots. Two of the most commonly utilized surgical knots in pelvic reconstructive surgeries – granny knot and square knot – were utilized for evaluation of each material type. We hypothesize that the PDS suture materials will exhibit more load bearing capacity as compared to the Vicryl suture materials. Additionally, we hypothesize that granny knots will be superior in load bearing strength as compared to square knots.

METHODS

Suture retention testing was performed as per previously established protocol [4]. Two types of knots (granny or square) and two types suture materials (PDS*II size 2-0 or Vicryl size 2-0) were used for comparison in this study for a total of four groups: PDS-Granny, PDS-Square, Vicryl-Granny & Vicryl-Square. Sheep vaginal wall tissues were obtained from a commercial abattoir (N=6 tracts) and the fat and other excess tissues were carefully removed. Each tract was dissected along the longitudinal axis into strips of about 20 mm long & 12 mm wide. Two vaginal tissue strips were then sutured together (length to length) using 3 simple interrupted stitches (either square or granny knots). Each square or granny stitch had four throws (2 knots each), and all stitches were performed by a Diplomate of the American College of Veterinary Surgeons (James R. Butler). The resulting samples (n=5 per group) were about 40 mm long & 12 mm wide in size. Each specimen was then loaded onto the uniaxial tension machine (Mach-1™,

Biomomentum Inc., Canada) with a grip to grip length of about 30 mm and subjected to uniaxial tension testing at a rate of 20 mm per minute. Figure 1(A) shows an example of uniaxial tensile testing setup with a PDS square knot sample. For each of the samples tested, the frequency of suture knot failure (i.e. the number of suture sites that contained tissue tears were recorded) and the mechanical parameters (maximum load, maximum displacement, and tensile modulus) were evaluated. Statistical analyses were performed using SPSS, and pairwise comparison was performed by Fisher LSD test. Data was considered significant at $p < 0.05$.

RESULTS

Data was significantly different across the groups for maximum displacement ($P<0.05$) and tensile modulus ($p<0.05$), but not for maximum load (Fig 1B-D). The average displacement at failure, in mm, was 16.33 ± 3.31 for PDS-Granny, 13.94 ± 1.49 for PDS-Square, 18.37 ± 2.78 for Vicryl-Granny, and 17.57 ± 3.24 for Vicryl-Square, shown in Figure 1C. In pairwise comparisons, displacement characteristics of PDS square group was significantly different from the Vicryl groups; the order was as follows – Vicryl-Granny >>Vicryl-Square>>PDS-Square (Fig 1C). However, in case of modulus, it was significantly different for PDS –granny only. The tensile modulus, in Pa, for PDS-Granny (135.56 ± 76.42) was lower by 36% as compared to PDS-Square (212.16 ± 109.74) ($p<0.05$); whereas the Vicryl groups exhibited similar modulus ranges ($p>0.5$) (Fig 1D). The PDS square group experienced the lowest max load in addition to having the highest number of suture knot failures (Fig 2). The Vicryl-Granny group experienced the largest max load but Vicryl-Square group experienced the fewest suture knot failures, shown in Figure 1B.

DISCUSSION

Usually, the selection of the suture material, the kind of surgical knots to use or the type of suture itself is at the discretion of the surgeon. This is typically defaulted to what the surgeon is most comfortable using given the surgical needs. With the given complex mechanical requirements of the female pelvic floor, no single suture knot or suture material is adequate to fit all meshes options or every individual. Vicryl sutures are known for better post-operative effects; whereas, PDS is preferred for wound closure tasks owing to their lower creep behavior [4]. Overall, our study demonstrated that the mechanical properties of granny knots were better suited for higher loads and longer displacements than square knots when using the same suture material. Thus, instances in which high loads or large displacements are to occur, surgeons may choose to use the granny knot instead of the square knot for more reliable tensile strength. The Vicryl suture material groups demonstrated a lower frequency of knot failure with increasing load when compared to the PDS groups. Therefore, surgeons might choose Vicryl over PDS suture materials for high load situations.

Along with the physical characteristics (braided or monofilament), biomechanical properties of the suture materials also play a vital role on the integrity of surgical interface and ultimately the union and performance of the implanted gynecological mesh. In this study, we report the tissue-tissue suture retention ability of commonly used suture materials. However, biomechanical evaluation of tissue-mesh interface will be beneficial for pelvic floor surgeries. Furthermore, additional investigations with animal models are necessary to look at different suture materials as well as other suture knot types to determine the optimum combination of surgical material and knot type.

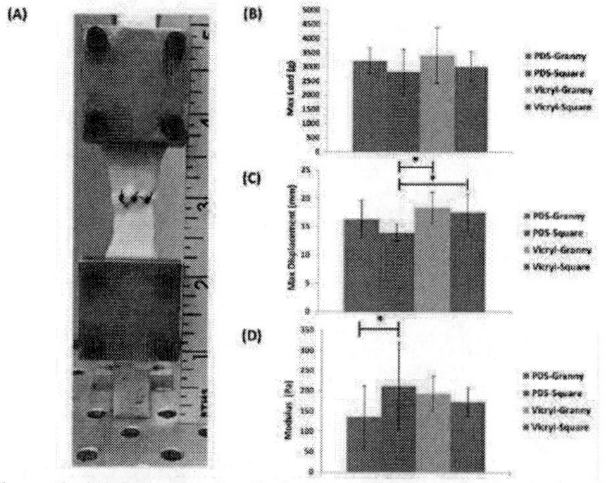

Figure 1: (A) Setup for tensile testing of PDS square knot sample. Biomechanical data from suture retention tests (B) max. load, (C) max. displacement and (D) tensile modulus data are shown here. * indicated p <0.05.

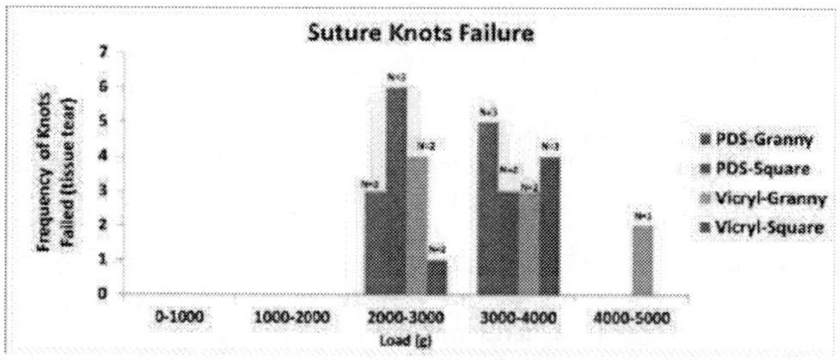

Figure 2: Frequency of tissue tear at a given load (knot failure)

REFERENCES

1. Olsen, AL., et al. *Obset Gynecol.* 89:501-6, 1997.
2. Given, FT., et al, Posterior culdeplasty revisited. *Am J Obstet Gynecol.* 153:135-8, 1985.
3. Alvarez, J., et al., *Minerva Ginecologica*, 65(1), 53-67, 2013.
4. Law, AY, et al. *Veterinary Surgery*, 46: 59-66, 2017.
5. Madhuvrata, P., et al. *J Obstet Gynaecol,* 31(5), 429-435, 2011.

SB^{3}C2017
Summer Biomechanics, Bioengineering and Biotransport Conference
June 21 – 24, Tucson, AZ, USA

A FINITE-ELEMENT MODEL OF PACINIAN CORPUSCLE CLUSTERING IN HUMAN SKIN

Julia C. Quindlen (1), Victor H. Barocas (1)

(1) Department of Biomedical Engineering
University of Minnesota
Minneapolis, MN, USA

INTRODUCTION

Cutaneous mechanoreceptors are mechano-neural complexes that transduce mechanical deformation of the skin into neural signals and ultimately produce the sensation of the applied tactile stimulus. The Pacinian corpuscle (PC) is the mechanoreceptor that encodes the high-frequency components of mechanical stimuli [1]. The PC structure consists of a single nerve fiber (neurite) containing mechanogated ion channels surrounded by a capsule of approximately 30 lamellae separated by fluid-filled spaces [1,2]. The capsule's structure is critical to the PC's role as a bandpass filter to mechanical vibrations [3,4].

PCs are located within the dermis and subcutaneous tissue of skin and are often found in clusters of receptors with overlapping receptive fields [5]. PC clustering could provide a more nuanced tactile encoding of a vibrational stimulus. Size variation within a cluster can affect frequency tuning, as the number of lamellae and the outer radius are important factors in determining a PC's frequency sensitivity [3]. PCs within a cluster would also be exposed to different components of a mechanical stimulus depending on the location of the PC with respect to the vibration. The goal of this study was to determine (1) whether the presence of one PC affects a second PC and (2) how the location of PCs within a cluster affects the mechanical stimulus that reaches each PC. Elucidating the mechanisms that govern the mechanoreceptors involved in somatosensation is a necessary step towards the development of haptic feedback-enabled prosthetics and understanding the vibrotactile sensitivity associated with aging and diseases, such as diabetic peripheral neuropathy [6,7].

METHODS

Previously, we developed a computer model to simulate the process by which a mechanical stimulus is transduced into neural spikes [3]. This model is a series of three sub-models that each address a specific stage of the mechano-neural transduction process: (1) A MATLAB finite-element mechanical model that uses shell and lubrication theories to model fluid-lamellar interaction in the outer core (henceforth "shell model"), (2) A mechanical inner core and neurite model, and (3) An electrochemical neurite model.

For the current study, we developed a finite-element model of the PC embedded within human skin in COMSOL. As a first step, we created a homogeneous viscoelastic model in COMSOL to match the behavior of our shell model described above. The shell model was run at indentation frequencies spanning 10-10,000 Hz with constant applied pressure. The shell model was run containing 30 lamellae with a Young's modulus of 3 kPa and an inter-lamellar viscosity of 1.4 mPa-s. The maximum displacement at each indentation frequency was recorded. A viscoelastic model was created in COMSOL to match the displacement profile obtained from the shell model (Fig. 1). The PC was modeled as a Standard Linear Solid with a shear modulus of 500 kPa, a relaxation time of 11 μs, and long-term static stiffness.

Skin was modeled using parameters published in the literature based on experimental surface deformation profiles for the human finger [8]. The five layers were modeled with the following Young's moduli (Fig. 2A): Epidermis (E): 0.18 MPa, Dermis (D): 18 kPa, Adipose Tissue (AT): 18 kPa, Fibrous Matrix (FM): 1.8 kPa, Bone (B): 1.8 GPa; all layers were modeled with Poisson's ratio 0.48. Rayleigh damping was introduced into the layers based off of published finger models [9]: mass damping of 1×10^{-7} s^{-1} and viscous damping of 0.4 ms were applied for the soft tissues; mass damping of 1×10^{-4} s^{-1} and 0.1 μs were used for bone. The epidermis was then vibrated with a simulated indenter (Fig. 2A) of radius 1 mm at 150 Hz with a maximum displacement of 30 μm. Two PCs with a 255.6 μm radius were modeled in the dermis and adipose tissue layers to observe how PCs in a cluster interact in different orientations [2]. The following two cases were analyzed (Fig. 2B): (1) One PC (Top) was

Poster Presentation #P176

embedded 2 mm below the surface of the skin. A second PC (Bottom) was embedded directly below the Top PC along the y-axis and the center-to-center distance (D_1) between the PCs was increased. (2) One PC (Center) was embedded 2 mm below the surface of the skin. A second PC (Lateral) was also embedded 2 mm below the skin surface and was moved along the +z axis away from the indenter, increasing the center-to-center distance (D_2) between the PCs. Simulations were also run for just one PC in each case. The displacement and phase shift from the applied pressure that reached the PC surfaces were recorded. The entire structure was meshed with 7846 tetrahedral elements.

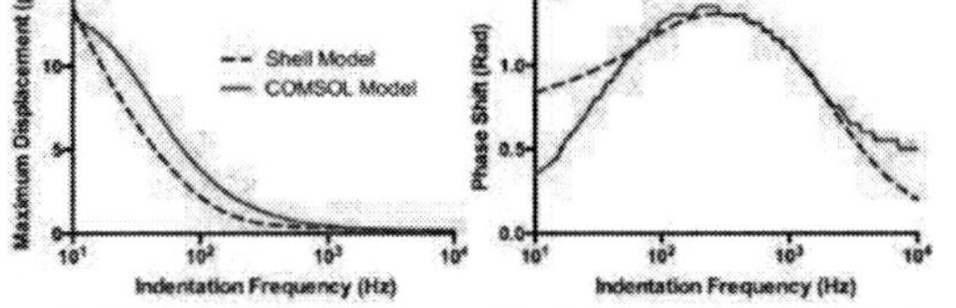

Figure 1: The displacement amplitude and phase shift between maximum pressure and maximum displacement on the PC outer surface for the shell and COMSOL models.

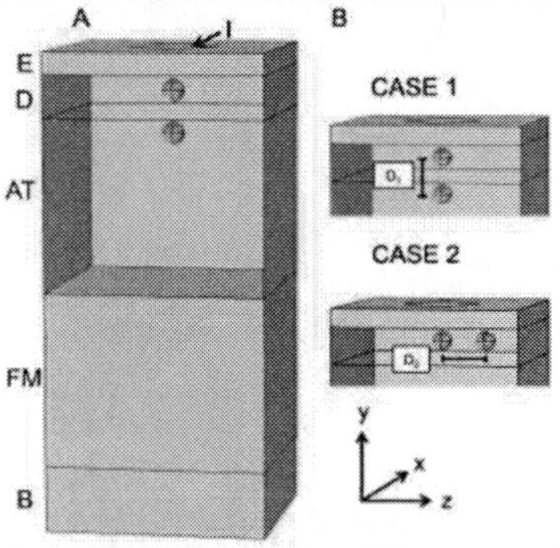

Figure 2. (A) Five skin layers and two PCs were modeled. 150 Hz vibrations were applied via a 1 mm radius indenter (I). (B) The distance between PCs was increased for two different orientations.

RESULTS

Case 1: When two PCs were aligned along the y-axis, the top PC experienced almost no effect due to the bottom PC, but the bottom PC experienced a slight increase in oscillation amplitude and a decrease in phase shift relative to the stimulus (Fig. 3). This effect was most pronounced at the smallest PC separations, as one might expect. The oscillation amplified and the phase shift increased with PC depth.

Case 2: When two PCs were aligned along the z-axis, the lateral PC had a minimal effect on the center PC, and the center PC led to a very slight increase in oscillation amplitude with negligible change in phase shift (Fig. 4). The phase shift increased strongly with distance from the indenter, an effect due to the displacement from the indenter site, not due to interaction between the two PCs.

DISCUSSION

PCs are often localized to clusters within skin, but the reason for this grouping remains unknown. Previous computer models have characterized the mechanical and/or neural responses of single PCs, but no published study has investigated the behavior of a cluster [3,4].

A single PC can detect low amplitude vibrations with low spatial resolution [1]. The results of this study suggest that clustering may enhance spatial discriminability, as PCs at various distances from the stimulus will receive different oscillation amplitudes at different phase shifts. The summation of unique spike trains produced by a PC cluster may allow for better localization of a stimulus on the skin. In addition, the time lag between the responses of PCs within a cluster may contribute to the directional tuning of PCs, drawing parallels to sound localization in the auditory system which relies on the time lag between the sound's arrival at each ear. Though the PCs modeled in this study were of equal diameter, changing the size of PCs within a cluster will affect the cluster's overall response to vibration due to the dependence of frequency sensitivity on PC size [3]. Size and positional variations of PCs within a cluster therefore enable clusters to provide more information about a stimulus than can be given by a single PC.

Previous studies have investigated vibration propagation across the surface of skin [10,11]. The current study instead analyzes wave transmission through skin and the resulting receptor deformation. Our model predicts wave propagation through skin at a speed of 14 m/s, which is comparable to experimental propagation speed measurements of 15 m/s across fingertip skin [11]. The presence of PCs did not cause large deformations of the pressure wave as it transmitted through the skin. Our results show some shielding of the pressure due to the presence of the PCs, especially in Case 1, but the primary cause of stimulus time lag or attenuation was PC distance from the indenter.

This study is the first to investigate the behavior of a cluster of PCs within skin and analyze the transmission of a vibration through the skin to the PCs. In future work, the mechanical model created in this study can be combined with our neural model of the PC neurite [3] to simulate action potentials that would originate from a cluster.

ACKNOWLEDGEMENTS

The research was supported the University of Minnesota Doctoral Dissertation Fellowship. Computational resources were provided by the Minnesota Supercomputing Institute.

REFERENCES

[1] Bell, J et al. *Prog Neurobiol*, 42:79-128, 1994. [2] Loewenstein, W et al. *J Physiol*, 182: 346-378, 1966. [3] Quindlen, J et al. *Integr Biol*, 8, 1111-11125, 2016. [4] Biswas et al., *IEEE Trans Haptics*, 8, 31-42, 2015. [5] Wu, G et al. *Exp Brain Res*, 126: 399-409, 1999. [6] Verillo, R. *J Gerontol*, 35:185-193. [7] Vinik, A et al. *Endocrinol Metab Clin North Am*, 42:747-787, 2013. [8] Dandekar, K et al. *J Biomech Eng*, 125:682-691, 2003. [9] Wu, J et al. *Med Eng Phys*, 28:816-826, 2006. [10] Shao, Y et al. *PNAS*, 113:4188-4193, 2016. [11] Manfredi, L et al. *PLoS ONE*, 7:e31203, 2012.

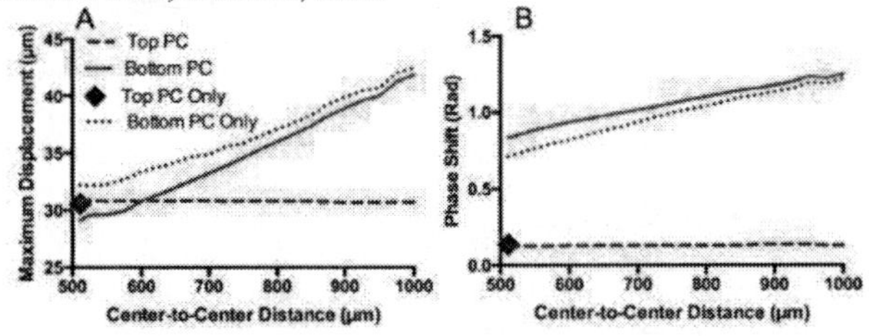

Figure 3: Case 1. (A) Maximum displacement at the PC surface when both (dashed and solid lines) or one PC was present. (B) The phase shifts between applied pressure and PC displacement.

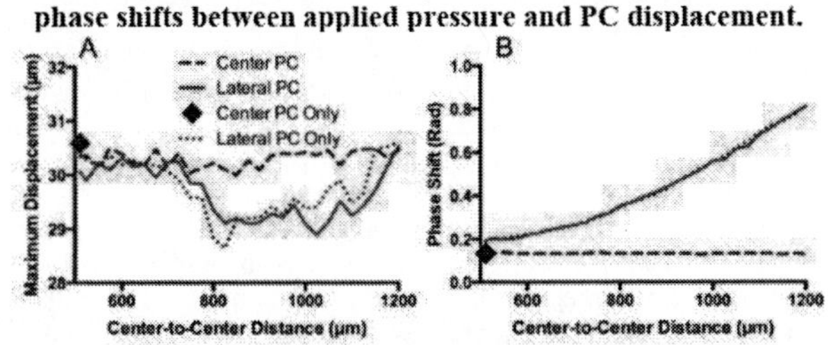

Figure 4: Case 2. (A) Maximum displacement at the PC surface when both (dashed and solid lines) or one PC was present. (B) The phase shifts between applied pressure and PC displacement.

SB³C2017
Summer Biomechanics, Bioengineering and Biotransport Conference
June 21 – 24, Tucson, AZ, USA

MODELING CREEP INDENTATION OF BRAIN SLICES AS A FIBER-REINFORCED BIPHASIC MATERIAL

Ruizhi Wang (1) and Malisa Sarntinoranont (1)

(1) Department of Mechanical and Aerospace Engineering
University of Florida
Gainesville, FL, USA

INTRODUCTION

Experiments have shown that brain can be described as a biphasic continuum, where a porous solid matrix is saturated with interstitial fluid [1,2]. Biphasic models can provide a mechanistic description of deformation and transport phenomena in soft tissues, and have been widely used in computational investigations of clinical applications such surgical planning [3,4] and convention-enhanced drug delivery [5-7]. Accuracy of corresponding mechanical properties, i.e., stiffness and compressibility of the solid matrix as well as hydraulic permeability, is essential for the reliability of simulations. To this end, direct mechanical testing [1,2] and imaging methods [8] have been employed to measure such properties in brain. However, because of different sample preparation and loading conditions used in these experiments, properties have been found to vary in a relatively large range. The appropriate constitutive relation for the solid matrix is also open to debate.

It is well established by biphasic theory that the intrinsic interaction between the porous solid matrix and the interstitial fluid explains explicit viscoelastic behavior of soft tissues [9]. Features in the time-dependent deformation of tissues can be related to the biphasic mechanical properties [10,13]. In a previous study of our group, the creep deformation of the cerebral cortex, hippocampus and caudate/putamen in acute rat brain tissue slices under micro-indentation was recorded using an optical coherence tomography (OCT) system [11]. In the current study, a biphasic finite element model of creep indentation was created to fit these data. Since these brain regions are all intricate and heterogeneous, it is necessary to take the contributions of different microstructural components into consideration. Cerebral gray matter is mainly composed of nerve cell bodies (soma), neuropil and glial cells, while white matter mainly consists of myelinated axons which make it a fibrous structure. Given this tissue structure, the solid matrix in our model was defined as a tension-compression nonlinear solid mixture where a compressible, isotropic, neo-Hookean ground matrix is reinforced by fibers. A sensitivity analysis was performed. The Young's modulus E, fiber modulus ξ and permeability k in each anatomical region were estimated for Poisson's ratio ν ranging from 0.35 to 0.49.

METHODS

The numerical simulation was performed under the biphasic continuum framework [9]. A constant, homogeneous and isotropic hydraulic permeability was assumed. The solid matrix of brain tissue was treated as a compressible, isotropic, neo-Hookean ground reinforced by continuously distributed fibers, which can only sustain tension and have zero response to compression. The fiber strain energy density function was defined as [12]

$$\psi = \frac{\xi}{\alpha\beta}\left(exp\left[\alpha(I_n - 1)^{\beta}\right] - 1\right) \tag{1}$$

where fiber modulus ξ represents the stiffness of fiber bundles. α and β were set to 0 and 2. I_n is the square of the fiber stretch. The Cauchy stress of the fiber bundles is given by

$$\boldsymbol{\sigma}(\boldsymbol{n}) = H(I_n - 1)\frac{2I_n}{J}\frac{\partial\psi}{\partial I_n}\boldsymbol{n}\otimes\boldsymbol{n} \tag{2}$$

where H is a unit step function. J is the Jacobian of the deformation. $\boldsymbol{n} = \mathbf{F}\cdot\mathbf{N}/I_n$, in which $\mathbf{F}$ is the deformation gradient, and $\mathbf{N}$ is the unit vector along the fiber direction. Under the assumption of spherical fiber angular distribution, the Cauchy stress tensor was obtained by integrating $\boldsymbol{\sigma}(\boldsymbol{n})$ over all possible fiber directions.

$$\boldsymbol{\sigma} = \int_0^{2\pi}\int_0^{\pi}\boldsymbol{\sigma}(\boldsymbol{n})\, sin\varphi d\varphi d\theta \tag{3}$$

Biphasic contact problems were solved using the FEBio software suite (version 2.5.2, University of Utah, Salt Lake City, UT) using the geometry shown in **Figure 1**. A 37 μN constant force was prescribed on the indenter. Each simulation was for 10 minutes, and the vertical displacement of the tissue surface was monitored to generate creep curves.

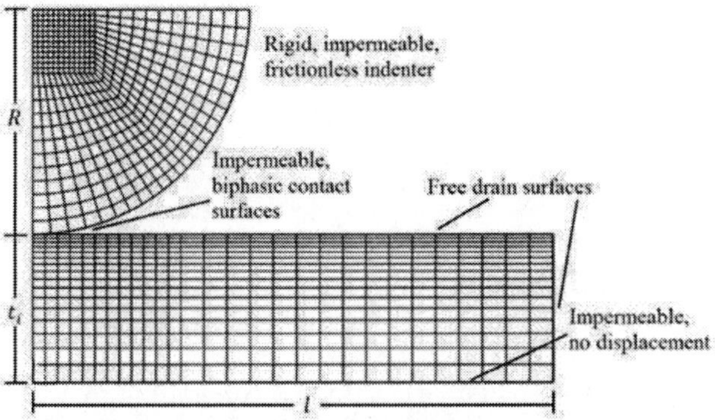

Figure 1: Finite element mesh and boundary conditions for indentation of rat brain tissue slices. Tissue radial boundary, $l = 1.2mm$, initial tissue thickness, $t_i = 330\mu m$, and spherical indenter radius, $R = 500\mu m$.

RESULTS

To be consistent with literature [1,2,11,14], the range of Poisson's ratio ranged from 0.35-0.49. The distinct role that ξ, E and k plays in shaping the creep curve reported by Chen et al. [13] is also observed in this study with an impermeable spherical indenter. ξ determines the instantaneous displacement while the infinite deformation is dominated by E. The time required to reach equilibrium is governed by k. Estimated biphasic properties data in each anatomical region are listed in **Table 1**.

Table 1: Estimates of biphasic brain tissue properties at different Poisson's ratios.

	ν	E (Pa)	ξ (Pa)	k (m^4/N s)
Cerebral Cortex	0.35	1080	16000	1.36×10^{-13}
	0.4	820	17100	1.36×10^{-13}
	0.45	500	18200	1.37×10^{-13}
	0.49	141	12500	1.38×10^{-13}
Hippocampus	0.35	545	5900	2.03×10^{-13}
	0.4	432	6120	2.05×10^{-13}
	0.45	265	6210	2.04×10^{-13}
	0.49	71	5297	2.04×10^{-13}
Putamen	0.35	323	3670	2.85×10^{-13}
	0.4	255	3820	2.87×10^{-13}
	0.45	157	3910	2.87×10^{-13}
	0.49	45	3430	2.88×10^{-13}

E was found to decrease as ν increased from 0.35 to 0.49, while fiber modulus showed a non-monotone tendency as a slow increase in ξ was followed by a dramatic drop as ν reached 0.49. The sensitivity of both E and ξ to ν increased with a nearly incompressible ground matrix, although Young's modulus was more sensitive than fiber modulus. Furthermore, taking ξ/E as a rough measure of the tension-compression nonlinearity of the solid matrix, it can be observed in **Figure 2** that the contrast between tensile and compressive modulus is enhanced by a larger Poisson's ratio. It should also be noted that within a certain region, k changed only slightly with different properties of the solid matrix.

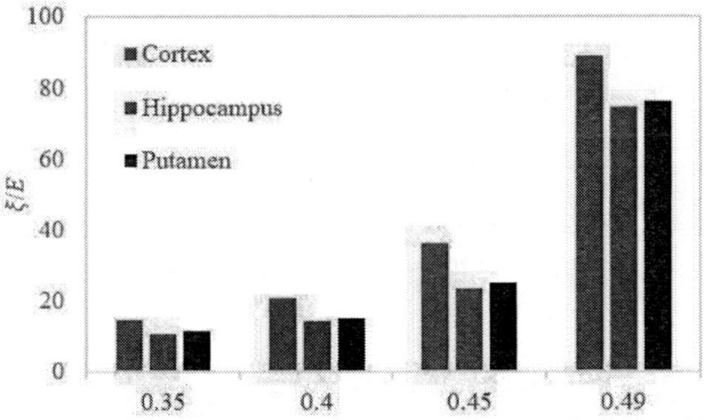

Figure 2: Comparison of tension-compression nonlinearity with increasing Poisson's ratio.

DISCUSSION

Simulated Young's modulus of the neo-Hookean ground and the hydraulic permeability fell within an acceptable range compared with those in literature [1,2,8,14]. Their relatively low values might be reflective of damaged vasculature due to tissue slicing. In this study, fibers greatly increased the effective tensile modulus. Since Poisson's ratio at the instant when time $t = 0_+$ is effectively 0.5, a larger ν results in a smaller deformation ratio (i.e., the ratio of the equilibrium displacement to the instantaneous displacement), as well as an enhanced tension-compression contrast required to fit the creep curve. Moreover, when ν is fixed, greater tension-compression nonlinearity could introduce a larger deformation ratio under indentation, as exhibited by comparison between different anatomical regions in **Figure 2**. Actual fiber distribution pattern is likely more organized than in our model, and fiber orientation could lead to anisotropy in micro-scale anatomical regions. The estimated biphasic properties indicate the potential tension-compression nonlinearity in the porous solid matrix of rat brain tissues. Realistic tissue compression might be better accounted for using this model. Given that permeability was nearly independent of the properties of the solid matrix, this might provide a simpler way to determine permeability from creep indentation.

ACKNOWLEDGEMENTS

We would like to acknowledge the FEBio software suite.

REFERENCES

[1] Franceschini, G et al., *J Mech Phys Solids*, 54:2592-2620, 2006.
[2] Cheng, S et al., *J Biomech*, 40:117-124, 2007.
[3] Platenik L.A et al., *IEEE Trans Biomed Eng*, 49:823-835, 2002.
[4] Lunn, K.E et al., *IEEE Trans Biomed Eng*, 53:1893-1900, 2006.
[5] Morrison, P.F et al., *J Neurosurg*, 107:74-83, 2007.
[6] Chen, X et al., *Ann Biomed Eng*, 35:2145-2158, 2007.
[7] Astary, G.W et al., *J Neurosci Methods*, 187:129-137, 2010.
[8] Weaver, J.B et al., *Phys Med Biol*, 57:7275-7287, 2012.
[9] Mow, V.C et al., *J Biomech Eng*, 102:73-84, 1980.
[10] Hu, Y et al., *Appl Phys Lett*, 96: 121904, 2010.
[11] Lee, S.J et al., *J Mech Behav Biomed Mater*, 29:213-224, 2014.
[12] Ateshian, G.A et al., *J Biomech Eng*, 131:061003, 2009.
[13] Chen, X et al., *J Biomech Ann Biomed Eng*, 44:1148-1158, 2016.
[14] Kyriacou, S.K et al., *Biomechan Model Mechanobiol*, 1:151-164, 2002.

POSTURAL INFLUENCE ON THORACOABDOMINAL ORGANS OF 5TH, 50TH, AND 95TH PERCENTILE MALE SUBJECTS

James P. Gaewsky (1,2), Katelyn A. Greene (2,3), F. Scott Gayzik (1,2), Ashley A. Weaver (1,2)

(1) Wake Forest University School of Medicine, Winston-Salem, NC, USA

(2) Virginia Tech - Wake Forest Center for Injury Biomechanics, Winston-Salem, NC, USA

(3) University of California, Berkeley, Berkeley, CA

INTRODUCTION

Changes in posture, such as sitting down or lying supine induce alterations of the skeleton and soft tissue that impact the morphology and position of organs within the abdominal region [1]. Understanding the extent of these postural changes on thoracoabdominal organs is crucial in the design of implantable medical devices and in device evaluation with computational human body modeling [2,3]. Preceding studies have demonstrated the influence of posture on relative organ location and rib coverage [2,4,5]. This study extends the investigation of postural changes to include thoracoabdominal organ volume and morphology through development and examination of 3D models of the 5th, 50th and 95th percentile male subjects in both the supine and seated postures.

METHODS

Medical images of the three male subjects representing the 5th, 50th, and 95th percentile by height and weight were obtained in the form of standard thoracic CT scans and both supine and seated thoracic MRI scans. The CT scan data was utilized to segment skeletal structures including the clavicles, vertebrae, and ribs (Figure 1a) using semi-automated tools in Mimics v.19 (Materialise, Leuven, Belgium). Segmented bones were then co-registered in the supine and seated MRI scans and manually positioned. Solid organs of interest (lungs, heart, kidneys, liver, and spleen) were segmented from the MRI scans using a combination of thresholding, dynamic region growing, and manual techniques (Figure 1b). All segmented structures were post-processed to make fine adjustments that resulted in smoothed stereolithography (STL) structures, free from holes (Figure 1c). This process created six 3D anatomical CAD models (three subjects in supine and seated positions).

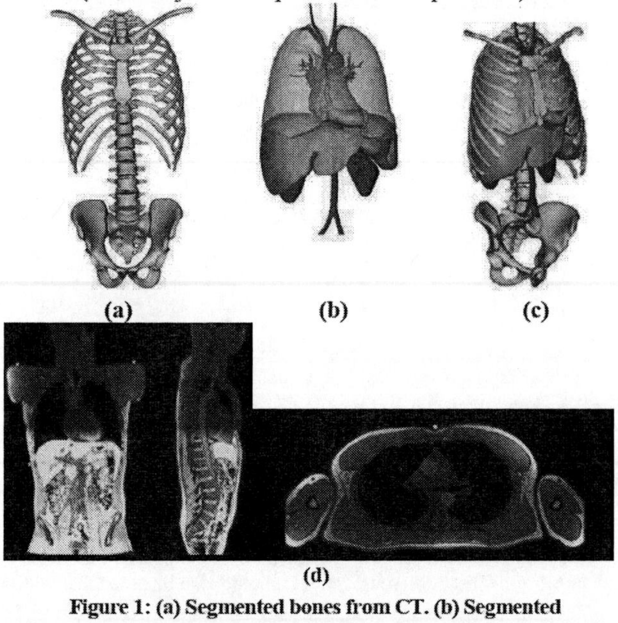

Figure 1: (a) Segmented bones from CT. (b) Segmented thoracoabdominal organs from MRI. (c) Co-registered model. (d) example of aorta segmented in MRI.

Volumes of the individual organs were then calculated and paired t-tests were used to analyze the significance of volume changes among the subjects. Morphological organ changes were quantified by conducting an automatic best fit alignment and 3D comparison of the reference object (supine organ) to the test object (seated organ). The maximum and minimum deviations experienced by each organ represent the extent of compression and expansion experienced by the organs in the supine versus seated position (Figure 2).

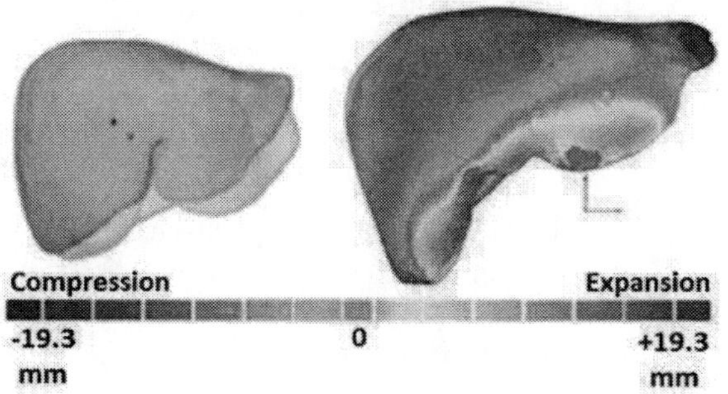

Figure 2: The supine and seated liver structures are overlaid (Left) and the spatial deviation between the surfaces is calculated. The dark blue represents areas of compression whereas dark red spots indicate areas of expansion (Right).

Organ morphology changes were quantified by normalizing the deviation results with organ volume. The normalized deviation value was calculated using the root mean square error (RMS) of the deviation between supine and seated organs according to Eq. (1).

$$\frac{\text{RMS}}{(\text{Supine Organ Volume})^{1/3}} \quad (1)$$

RESULTS

The postural changes can be visualized in the side views of the six models developed shown in Figure 3. The volumetric thoracoabdominal results indicated that there is no statistically significant change in volume ($p>0.05$) between the supine and seated positioning among the 5th, 50th and 95th percentile males. The liver experienced the most significant change in morphology given its volume while the kidneys experienced the smallest change in morphology (Figure 4).

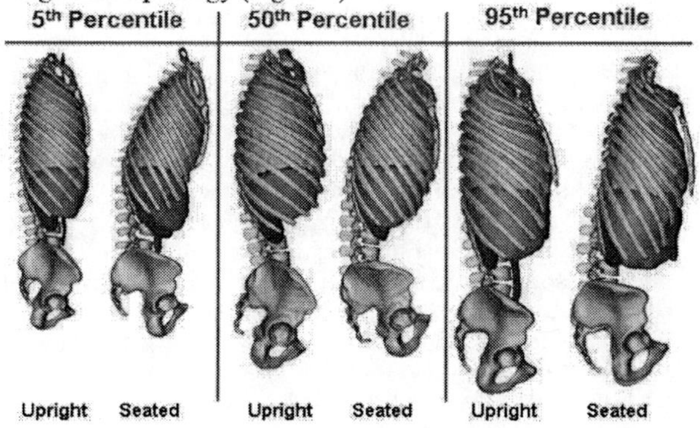

Figure 3: Six models were developed and analyzed.

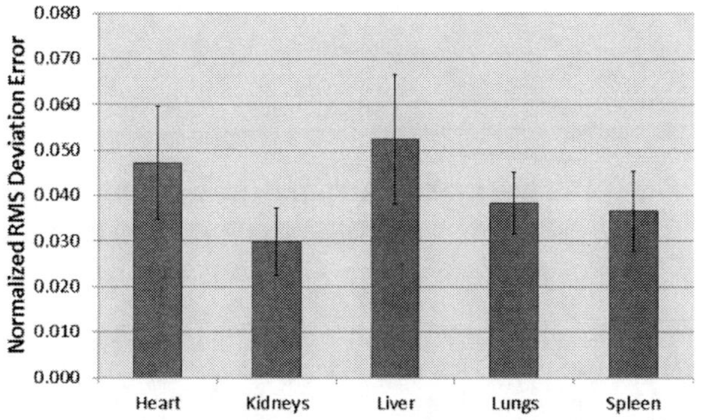

Figure 4: Graph exemplifying morphology change resulting from posture change. Analysis indicates the liver displayed the largest change while the kidneys experienced the smallest change.

DISCUSSION

The results from this study indicate that organ morphology is influenced by changes in posture, while organ volume stays relatively constant. Given that the subjects in the study reflect the characteristics of the general male population, quantification of the movement and morphology of the heart, lungs, and other soft tissue organs in relation to the skeletal structures in a seated versus supine posture has important implications in the design of medical devices. These devices include implantable cardioverter defibrillators or pacemakers which require placement of leads within the heart chambers. Anatomical variation within the population must also be considered when designing devices for individuals of different sizes and shapes (i.e. 5th, 50th, versus 95th percentile).

The data from this study may also benefit the injury biomechanics field. With knowledge of the anatomical variation in supine and upright postures, intermediate postures may be interpolated. These interpolated models could be useful for studying blunt trauma injury to subject-specific models in an array of postures. The results obtained suggest that anatomical variations along with postural changes and the corresponding organ morphology changes should be considered in future device designs and computational human body model development.

ACKNOWLEDGEMENTS

This project was partially supported by the National Science Foundation Research Experience for Undergraduates (REU) under Award No.1559700 and Medtronic.

REFERENCES

[1] Howes MK. Ann Adv Automot Med. (2013) 57: 209-224.
[2] Lafon Y. J Biomech. (2010) 43: 1269-1278.
[3] Hayes AR. Ann Adv Automot Med. (2013) 111-121.
[4] Hayes AR. Comput Math Methods Med. (2013) 1-12.
[5] Beillas P. Stapp Car Crash J. (2009) 53: 1-21.

DETERMINATION OF PROPER STORAGE CONDITION AND CONSTITUTIVE MODEL FOR RAT BACK SKIN MECHANICAL PROPERTIES

S. Chen (1), S. Roccabianca(1)

(1) Mechanical Engineering
Michigan State University
East Lansing, MI, US

INTRODUCTION

The determination of biomechanical properties of soft tissues like skin involves mechanical tensile test. In practice, skin samples cannot always be tested right after removal from the subjects. Hence, a storage protocol that can preserve most of the mechanical properties of skin tissue plays an important role in studies that involve mechanical tensile test.

The purpose of this work is to study the effect of different storage conditions on rat skin by evaluating mechanical characteristics using several parameters, and to determine the storage protocol that preserve skin mechanical properties the most. Also, the differences due to orientation, location and gender are studied in this work. Finally, a micro-structurally motivated constitutive model is used to fit the averaged data for each location on rat back.

METHODS

Sixteen Sprague Dawley rats, eight males and eight females, aged between 10 to 12 weeks were used in the study. The dorsal skin was shaved and the whole piece of back skin flap was excised. A custom made dog bone shape die was used to punch specimens from the flap. Four specimens were collected from each animal, two from the cranial location, on each side of the spine, and two from the caudal location, on each side of the spine. The animals were divided into two groups of eight animals, four males and four females: in Group I, the samples were cut along the spine direction (axial); in Group II, specimens were cut in the direction perpendicular to the spine (transverse).

Specimens from Group I were used for storage effect study. They were separated into four storage protocols randomly. Protocol I: specimens were tested right after they were cut from back skin flap. Protocol II: specimens were stored in 4°C for 24 hours before testing. In Protocol III: specimens were flash frozen in isopentane that submerged in dry ice (-78.5°C), then stored at -80°C, and thawed at 4°C for 6h before testing. Protocol IV: flash-freezing and storing process were the same as Protocol III, then specimens were thawed at 4°C for 24h before testing. Specimens from Group II were stored using the storage protocol that proved to be the best (Protocol III) determined by Group I specimens before testing.

Quasi-static uniaxial tensile test was performed on skin specimens to estimate mechanical properties. 4 markers were placed on the central part of the sample to track the length changing by Digital Image Correlation method using a CCD camera (i.e., Hitachi model KP-M2AN). The whole tensile test had three parts. Part I: 10 cycles of preconditioning at 20% strain; Part II: strain-incremental cyclic loading aiming for ideal stress level; Part III: rupture loading. Before each part of the test, a preload of 10g was applied to the specimen and the geometrical dimensions of the specimen were recorded to serve as reference configurations for corresponding stress-stretch curve. Testing speed was 0.83 mm/s.

An ideal curve from cyclic loading (the last one from Part II of the test, **Fig. 2**) of each sample was used to estimate the mechanical parameters of the micro-structurally motivated constitutive model. To describe the nonlinear, orthotropic behavior of skin, we employed a mass averaged strain energy function (SEF) of the form

$$W = \phi^e W^e(\mathbf{F}^e) + \sum_{k=1}^{3} \phi^k W^k(\lambda^k) \tag{1}$$

where ϕ^α and W^α represent the mass fraction and the SEF for each constituent, respectively, with $\alpha = e, k$ for elastic fibers and families of collagen fiber ($k = 1,2,3 \ldots$), respectively. We considered 2 collagen fibers families in this work, namely $k = 1, 2$ for symmetric diagonal fibers with respect to the spine. Elastin has been described by using a neo-Hookean constitutive model,

$$W^e(\mathbf{F}^e) = \frac{c^e}{2}(tr\mathbf{C}^e - 3) \tag{2}$$

where c^e represents a material parameter with the dimensions of stress and $\mathbf{C}^e$ represents the right Cauchy-Green tensor describing the deformation that affects the elastic network, i.e., $\mathbf{C}^e = (\mathbf{F}^e)^T\mathbf{F}^e$. Collagen fibers mechanical behavior has been described by using a Fung-type exponential SEF,

$$W^k(\lambda^k) = \frac{c_1^k}{4c_2^k}\{\exp[c_2^k((\lambda^k)^2 - 1)^2] - 1\}, for\ \lambda^k > 1 \quad (3)$$

where c_1^k and c_2^k are material parameters, with the dimension of stress and dimensionless respectively, λ^k describes the stretch in the direction of kth collagen fiber family defined as $\lambda^k = \sqrt{\mathbf{M}^k \cdot \mathbf{C}\mathbf{M}^k}$, where $\mathbf{M}^k = (0, \sin\alpha_0^k, \cos\alpha_0^k)$ with angles, $\alpha_0^1 = \alpha_0 + \gamma$ and $\alpha_0^2 = -\alpha_0 + \gamma$.

RESULTS

Mechanical properties of rat back skin are evaluated by using five parameters, defined as follows: (1) **initial slope,** represents the beginning 2% deformation of the rupture stress-stretch curve, we hypothesize that this response is dominated by the elastin content and approximately linear; (2) **maximum slope**, describes the quasi – linear part of the rupture stress-stretch curve at high stress level, which we hypothesize is dominated by the mechanical behavior of collagen fibers; (3) **rupture strain and** (4) **ultimate tensile strength (UTS)**, represent the strain and stress value at rupture, respectively; and (5) **toughness**, represents the work input done per unit volume of a specimen during rupture loading, defined as the area under the rupture stress-stretch loading curve .

ANOVA results show that there are no statistically significant (5%) differences between the four storage protocols on the five evaluated parameters. Figure 1 shows that Protocol III is the one that preserves the stress-stretch curve characteristics the most in both cyclic loading and rupture loading curves, compare to the fresh ones in Protocol I.

Since no significant effects from different storage protocols were found, data from Group I specimens, along with data from Group II, are used to analyze the effects of orientation, location and gender on rat back skin. By running Student's t-test, no statistically significant differences are found between male & female, left & right back. Caudal location has significant higher mean values than cranial location on all parameters, except for rupture strain, on which caudal location is also higher, just not statistically significant.

As to the differences between axial and transverse directions, axial specimens show significantly higher mean values on rupture strain and toughness at both cranial and caudal locations; moreover, axial specimens have significantly higher initial slope and lower maximum slope than transverse specimens at caudal location, which is the same trend at cranial location too, just not statistically significant.

Table 1 shows the best-fit material parameters of the constitutive model for different locations on rat back. **Figure 2** shows the averaged experimental data of upper left and lower left back, and corresponding model fitting curves. Fitting results for c^e show that the material parameter associated with elastin is higher in caudal locations when compared to cranial locations.

DISCUSSION

The study shows that there are no statistically significant differences between our 4 storage protocols. The study also reports no differences between male and female animals, but statistically significant differences on orientation and location.

The effect of storage conditions on rat skin has been studied by previous researchers, but not extensively. T. L. Foutz et al., 1992 reported that tissue freezing significantly increased fracture strength [1], but didn't affect yield strength, ultimate strength. However, their study didn't give clear definitions of evaluated parameters. R. D. Marangoni et al., 1966 reported that flash freezing plus -92.8°C storage condition preserved guinea pig skin mechanical properties the most, compared to 1.6°C and 37.8°C storage conditions [2], which supports the results in current study. When it comes to orientation differences, A. Karimi et al., 2015 [3] reported bigger maximum slope values of transverse specimens than axial specimens on mice back skin, as reported in current study.

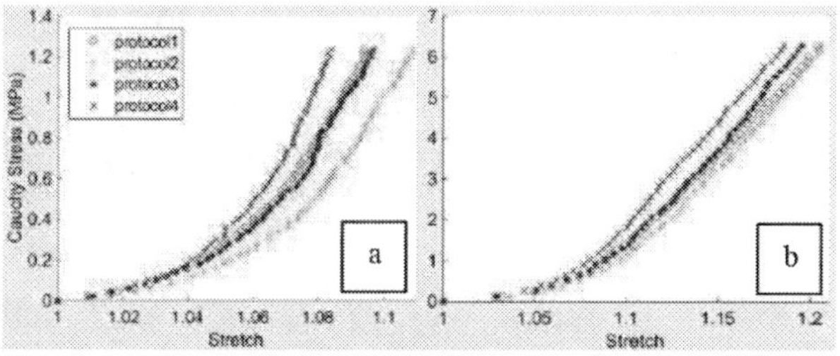

Figure 1: (a) Averaged cyclic loading curves based on different storage protocols using the last cycle from tensile test Part II. (b) Averaged rupture loading curves based on different storage protocols.

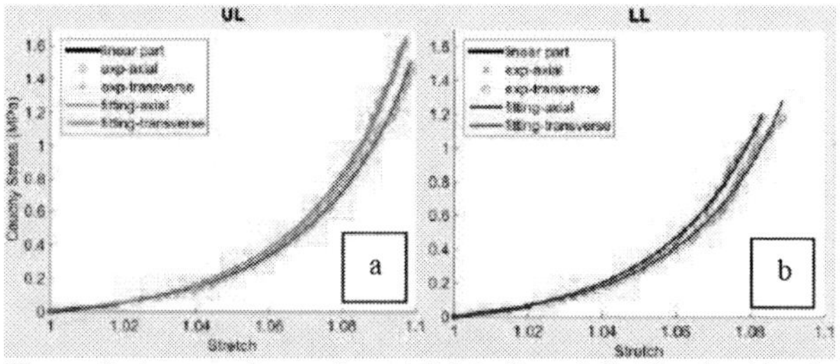

Figure 2: Averaged experimental data from left side of rat back, cranial (a) and caudal (b) locations, and corresponding model fitting curves.

Table 1: Best-fit material parameters of the constitutive model for different locations on rat back

	c^e(MPa)	c_1^{12}(MPa)	c_2^{12}	α_0 (°)	RMSE
UL	0.587	130.2	1463.9	45.10	0.036
UR	0.584	157.8	878.4	45.11	0.061
LL	0.734	153.3	1994.3	44.89	0.045
LR	0.623	134.9	1124.7	44.93	0.048

*UL: upper(cranial) left; UR: upper(cranial) right;
LL: lower(caudal) left; LR: lower(caudal) right.

ACKNOWLEDGEMENTS

We gratefully acknowledge Christina Chan, PhD and Ryan Thompson for the technical assistance.

REFERENCES

[1] Foutz, TI et al., *American Journal of Veterinary Research*, **53**: 788-792, 1992.

[2] Marangoni, RD et al., *Annal of the New York Academy of Sciences*, **136:** 441-453, 1966.

[3] Karimi, A et al., *Computer Methods in Biomechanics and Biomedical Engineering*, **18**: 1768-1774, 2015.

SB3C2017
Summer Biomechanics, Bioengineering and Biotransport Conference
June 21 – 24, Tucson, AZ, USA

CHARACTERIZATION OF PEDIATRIC BRAIN VISCOELASTICITY USING MULTI-FREQUENCY MAGNETIC RESONANCE ELASTOGRAPHY

Mehmet Kurt (1), Fabiola Macruz (2), Efe Ozkaya (1), Kim B. Pauly (2,3), Max Wintermark (2)

(1) Department of Mechanical Engineering
Stevens Institute of Technology
Hoboken, NJ, USA

(2) Department of Radiology
Stanford University
Stanford, CA, USA

(3) Department of Bioengineering
Stanford University
Stanford, CA, USA

INTRODUCTION

The developing brain is a unique physiological substrate with ever-changing physiology. The pediatric brain undergoes dramatic changes and significant growth of neural networks throughout development. Some of the key structural features of the pediatric brain change the brain tissue biomechanics when compared with the adult brain: **1)** Brain water content in the brain tissue is higher, resulting in reduced brain compliance [1] **2)** Axons in the pediatric brain goes through rapid myelination (with a slower rate after 18 months), which dramatically changes the mechanical properties [1,2].

Mechanical assessment of human pediatric brain has been largely limited. Pediatric brain tissue is difficult to obtain and *in vitro* studies have not been extensive. Several animal studies have been conducted for use in pediatric brain modeling, with a particular focus on infants [3-4]. Most of the literature on the pediatric brain tissue characterization is focused between birth and 18 months old since the myelination of axons during this period is rapid [5]. However, the volume growth of white matter and myelination continues through the adolescence, albeit with a slower pace. Nevertheless, this can potentially result in significant age differences in the local tissue properties of the pediatric brain, even at latter stages of childhood.

Magnetic resonance elastography (MRE) has been gaining attention as a non-invasive means to measure material properties of biological tissues, especially brain tissue. It can now be used as a diagnostic tool for detecting the onset of neurological diseases such as Alzheimer's [6]. The calculations of material properties of brain tissue via MRE are dependent on the actuation and image acquisition frequencies, yet to date; these parameters have not been rigorously investigated for studying brain material properties for large frequency ranges (e.g., 25-62.5 Hz frequency range in [7]). In this study, we used several actuation frequency values to investigate the in vivo viscoelasticity of the pediatric human brain in healthy volunteers. We used the experimental results to fit linear viscoelastic material models. To our best knowledge, this is the **first study** that investigated the mechanical properties of the pediatric brain *in vivo*.

METHODS

In this study, multi-frequency magnetic resonance elastography (Mayo Clinic; Rochester, MN) was used to investigate the *in vivo* viscoelasticity of healthy human brain in 12 volunteers (6 males and 6 females) ranging in age from 7 to 17 years. We categorized the volunteers into 3 groups (4 volunteer in each group): Children (7-9 years old), early adolescents (10-13 years old) and middle adolescents (14-17 years old). We varied the vibration frequencies of the MRE actuator in an acoustic range from 40 to 80 Hz to create a database of brain viscoelasticity with respect to varying frequencies of shear vibrations in the brain tissue. We fit a simple spring-dashpot model (standard linear solid model, *i.e.*, SLS) to the experimental data, wherein the viscosity and elasticity are combined in a parameter η that describes the solid-fluid behavior of the tissue, and E_1 and E_2 which describe the elastic solid behavior (Figure 1A). We examined the mid-axial slice of the brain for each subject to calculate the respective average storage and loss moduli (Figure 1B).

In order to capture the average brain moduli of each volunteer, the whole gray and white matter regions were selected as regions of interest (ROI) and the corresponding mean storage and loss moduli were recorded. Then, mean values and standard deviations for each age group were calculated.

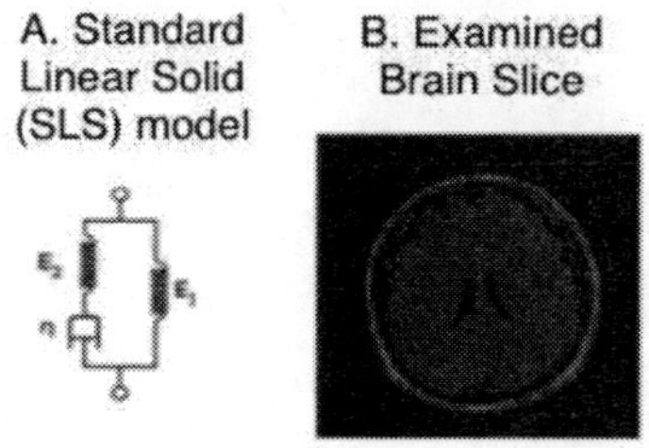

Figure 1 A. Simple spring-dashpot model (SLS) to model brain viscoelasticity B. A representative anatomic image of the examined mid-axial slice

RESULTS

We studied the change in storage and loss moduli in each age group as a function of MRE frequency (Figure 2). We found no statistical significance between children (7-9 years old) and early adolescents (10-13 years old). Storage moduli of middle adolescents (14-17 years old) were 12% lower than children and early adolescents. Storage moduli at 80 Hz (~2400 Pa) and loss moduli at 40 Hz (~550 Pa) were the closest values among 3 groups.

We fitted SLS models (Figure 1A) to the corresponding storage and loss moduli data for each age group (Figure 2). The SLS fits to the average brain storage and loss moduli in each age group yielded R^2=0.81, 0.74 and 0.72 respectively, which suggests that this mechanical model is a good representation of the average brain mechanical properties (Table 1).

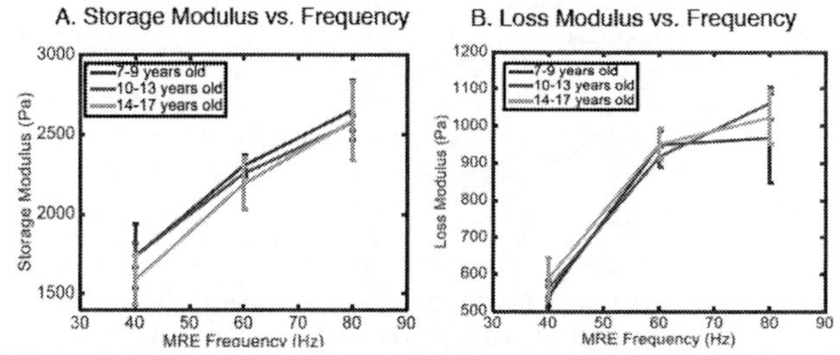

Figure 2 A. Storage modulus and B. Loss modulus as a function of MRE actuation frequency and age range.

Table 1: Frequency-independent SLS parameters for 3 age groups

Age Range	E_1 (kPa)	E_2 (kPa)	η (Pa)
7-9	1.61 (0.24)	1.94 (0.29)	3.55 (0.26)
10-13	1.63 (0.18)	2.08 (0.31)	3.35 (0.29)
14-17	1.46 (0.12)	2.01 (0.25)	3.76 (0.19)

DISCUSSION

Due to the sample size considered in this pilot study, we did not see significant differences in the mechanical properties of the brain among children and early adolescents. However, we were still able to obtain a few important insights regarding the age-dependent viscoelasticity changes in the pediatric brain. There was a statistically significant softening in the steady-state stiffness (E_1) of the brain for the later age group (14-17 years old), which is consistent with the findings of previous studies on the effect of aging in adults [7]. For these age groups the viscosity term, $\boldsymbol{\eta}$, is much lower than that of adults reported in previous studies (*e.g.*, 4.49 Pa [8],) which could be related with the high water content in the pediatric brain.

Another interesting observation was the change in the frequency-dependent behavior of the brain compared with adults. In a previous multi-frequency MRE study with adults, we found that the storage modulus converged to an asymptote around 80 Hz and loss modulus peaked before 80 Hz [8]. Interestingly, in our MRE data with the pediatric brain, we did not detect the peak of the loss modulus as a function of frequency and observed a continuous increase within the frequency range of interest. Similarly, storage moduli values increased almost linearly within this frequency range. This suggests that rate-dependent loading of the pediatric brain is substantially different than that of the adult brain, which is of interest to many communities, such as traumatic brain injury (TBI).

We are currently obtaining a larger dataset in order to study statistically significant age-dependent changes in brain viscoelasticity during childhood and adolescence, if any.

ACKNOWLEDGEMENTS

The authors would like to acknowledge Lucile Packard Children's Hospital for the support through a postdoctoral research grant (Recipient: Dr. Kurt) and GE Healthcare.

REFERENCES

[1] Lenroot, R et al., *Neuroscience & Biobehavioral Reviews* , 34:859-871, 2001.
[2] Weickenmeier, J et al., *Acta Biomateriala*, 42:265-272, 2016.
[3] Prange, M. et al., *J. Biomech. Eng.*, 124.2: 244-252. 2002.
[4] Thibault, K. et al., J. Biomech., 31:1119-1126, 1998.
[5] Duhaime et al., *J. Neurosurgery*, 93:455-462, 2000.
[6] Murphy et al., *JMRI*, 3,2011.
[7] Sack et al., *Neuroimage*, 46.3 :652-657,2009
[8] Kurt et al., *In: Proceedings of BMES*, 2016

LUNG MICROMECHANICS OF PULMONARY FIBROSIS: A FINITE ELEMENT ANALYSIS

Bo Yang (1), Abdulrahman Jbaily (1), Yintong Lu (2), Andrew J. Szeri (1), Grace D. O'Connell (1)

(1) Mechanical Engineering
University of California, Berkeley
Berkeley, CA, United States

(2) Mathematics
University of California, Berkeley
Berkeley, CA, United States

INTRODUCTION

Human lungs consist of a branching network of airways that become narrower, shorter and more abundant from the trachea to the bronchiole, eventually terminating at the alveolus. There are approximately 300 million alveoli, which are microscopic balloon-like sacs found in the form of small clusters. This offers a large surface area for gas exchange estimated at about 1 m^2/kg of body mass or more [2].

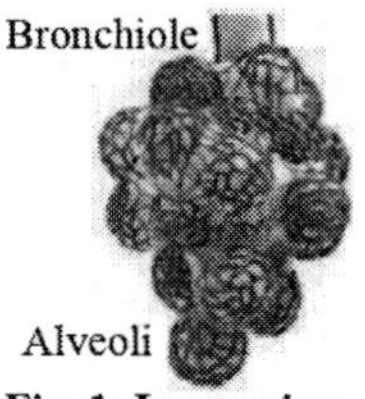

Fig. 1: Lung microstructure [1].

The alveolar wall has a complex composition, containing elastin and collagen fibers, and is coated on the inside with a layer of fluid. Surface tension at the air-fluid interface is regulated by surfactants secreted from epithelial cells that line the alveolar wall. Tension in the fibrous tissue in combination with surface tension at the air-fluid interface maintains the internal lung structure by exerting a deflating force, often referred to as elastic recoil [3]. However, it has been shown that the contribution of each type of tension force to the total elastic recoil is not necessarily equally distributed [4].

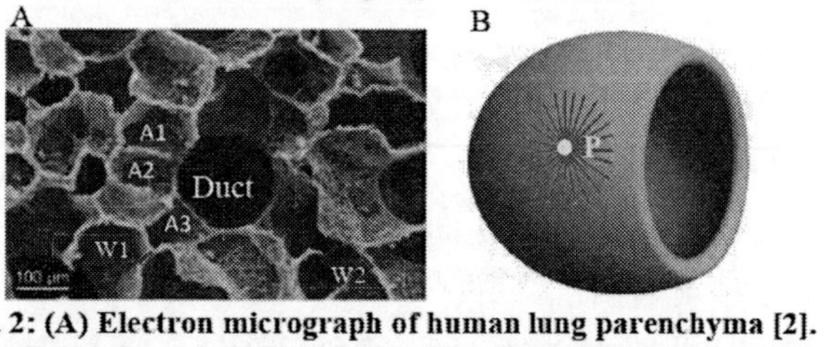

Fig. 2: (A) Electron micrograph of human lung parenchyma [2]. Alveoli (A1-3) and wall thickness (W1-2) used to define model geometry. (B) Alveolus model and an arbitrary point (P) with 12 groups of fibers.

Lung function is often characterized in the literature through Pressure-Volume (PV) curves. Specific aspects about alveoli function, including pulmonary tissue mechanics, cannot be elucidated from PV curves. A better understanding of pulmonary tissue mechanics is critical for improving treatment strategies for certain lung diseases. For example, a recent study utilized a multi-scale computational model to investigate ventilator-induced lung injuries [5]. In this paper, we develop a finite element model that specifically determines the effect of pulmonary wall fibrous tissue mechanics on alveolus function. After validation with PV curves in the literature, the model is used to study pulmonary fibrosis (PF), which is a disease where both the alveolar wall thickness and stiffness increase [6].

METHODS

The alveolus geometry was obtained from an electron micrograph of the human lung (Fig. 2A) [2]. A single alveolus formed an ellipsoidal shell, with a major axis of 122 ± 6.1 μm, a minor axis of 93 ± 3.5 μm, a circular opening of 70 ± 3.0 μm, and a wall thickness of 7 μm (Fig. 2B). The representative alveolus geometry was processed with a custom-written algorithm to create a 3D mesh (53760 hexahedral elements; MATLAB R2016b, Mathworks Inc.).

The mesh was imported into Preview (preprocessor of FEBio Package [7]), where boundary and loading conditions were specified. The alveolar wall was composed of collagen fibers embedded in a ground matrix that encompassed elastin, different types of cells, capillaries, etc. In order to evaluate the importance of collagen fibers in lung diseases, such as PF, we added tension-only fibers. Fibers were described with an exponential stress-strain response (three material coefficients: ξ, α, and β), while the ground matrix was described as a Neo-Hookean material (two material coefficients: E and ν). Due to the paucity of data on the alignment of collagen fibers in the alveolar wall, the following fiber arrangement was assumed: twelve groups of fibers (15° between two adjacent groups) of equal strength were laid in the tangential plane of the alveolar wall (Fig. 2B). The opening of the alveolus was constrained in all directions, and varying pressure was exerted on the inner surface, resulting in cycles of inflation and deflation.

The material coefficients were determined by parameter identification of the PV response from our simulation with data in the literature [8]. Because our specific interest is alveolar tissue mechanics, we used PV curves determined quasi-statically from saline-filled lungs, which eliminates the contribution of surface tension forces during lung recoil [3]. Furthermore, we assumed that the PV curve of the whole lung applies to a single alveolus, due to the fact that major changes in lung volume are attributed to changes in the alveolar configuration [3]. Results were validated with PV curves of saline-filled cat lungs, due to the absence of saline-filled human lung data. The model was considered valid if the coefficient of the determinant (R^2) between the model simulation and data was greater than 0.9.

Once the material coefficients were determined, stresses and strains on the inner and outer wall were evaluated with an applied internal pressure of 8 cmH_2O, which is pressure at total lung capacity for the saline-filled lung experiment [8]. Then, the alveolar wall thickness (t) and fiber modulus (ξ) were increased by 30%, to represent changes noted with pulmonary fibrosis (three simulated cases).

RESULTS

After parameter identification of the matrix and fiber material coefficients ($E = 1$ KPa, $\nu = 0.45$, $\xi = 0.03$ KPa, $\alpha = 0.1$, and $\beta = 2.5$), we obtained a PV curve that matched well with experimental data (Fig. 3A; control, R^2=0.97). The alveolar wall thickness decreased at different rates with pressure (Fig. 3B). At low pressures that represent normal inhalation (~1 cmH_2O), a steep descent in wall thickness was observed, while at higher pressures, the thickness plateaued near 2 μm. The inner wall of the alveolus had an effective strain of ~200%, corresponding to an effective stress of 40 kPa. The outer surface had a lower effective strain at ~160%, with an effective stress of 11 kPa (Figs. 3C&D).

All three simulated cases of PF resulted in lower PV curves than the healthy lung (Fig. 3A). Furthermore, total lung capacity decreased by 10% for a thicker alveolus with stiffer fibers (red line). The change in alveolus capacity was more pronounced at lower pressures, where the volume decreased by 25% at 2 cmH_2O.

DISCUSSION

A FEM model of a single alveolus was developed to study pulmonary tissue mechanics. The material properties of the alveolar wall were described by a combination of constitutive relations for the matrix and the fibers.

PV curves from the model allowed us to investigate the effect of pulmonary fibrosis on lung behavior. The two symptoms associated with PF, thicker wall and stiffer fibers, both caused the lung to become less compliant. Moreover, a combination of the two symptoms, which represents a severe case of PF, had a more drastic effect on tissue compliance. The decrease in alveolar volume during inhalation suggests a lower surface area available for gas exchange, which agrees well with observations in patients suffering from PF [6, 9,10].

The strain and stress distributions revealed that an alveolus undergoes asymmetric expansion and contraction during breathing. This can be mainly attributed to the boundary conditions at the airway opening. Future work will incorporate the surfactants that line the alveoli wall to investigate the effect of surface tension due to surfactant concentration on alveoli mechanics. Also, it is apparent that the inner and outer walls experienced different strain and stress values, which suggests that a thin wall assumption may not be an accurate approach in modeling.

It is worth mentioning that the results of our preliminary model are limited by the lack of necessary information and data in the literature. A notable improvement to our model would be the replacement of the homogeneous material description with a more specific one that contains information on fiber arrangements, and elastin and collagen densities. Also, the implications of assuming equivalence between the PV curve of a single alveolus and that of a whole lung are not well understood; nor are the limitations of the quasi-static PV curves determined from saline experiments clear.

This study is part of an ongoing project that aims to estimate the work of breathing and gas exchange rates in human lungs as a function of multiple biological parameters, such as breathing cadence, pulmonary surfactant properties, and pulmonary tissue mechanics. The model presented here demonstrates that finite element analyses of the alveolus can appropriately describe tissue compliance of both healthy and unhealthy lungs, which will be important for understanding disease mechanisms.

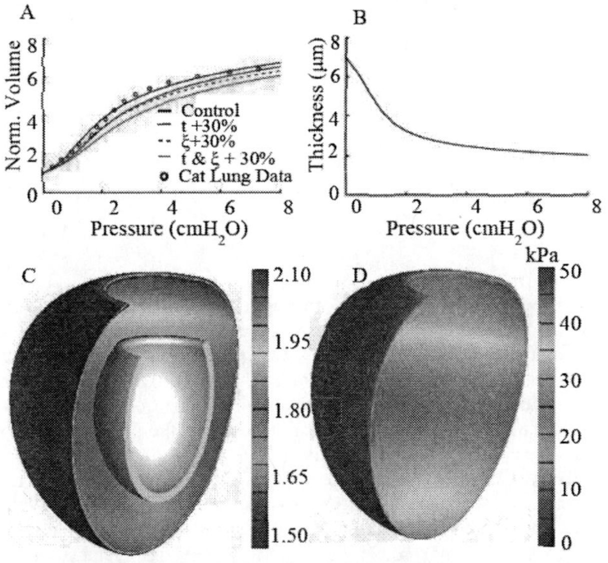

Fig. 3: (A) Volume (normalized by Residual Volume) vs. Pressure for the model, cat lung experiment, and different cases modeling PF. (B) Alveolar wall thickness vs Pressure. (C) Cross sectional view of alveolar reference configuration (grey), effective Lagrangian strain and (D) stress distribution at 8 cmH_2O.

ACKNOWLEDGEMENTS

This study was supported by the Hellman Foundation.

REFERENCES

[1] MedicalGraphics, http://www.medicalgraphics.de/en/; [2] Hsia, CCW et al., *Compr Physiol*, 6.2:827–895, 2016; [3] Hoppin, FG et al., *Handbook of Physiol*, ch.13:195-215, 1986; [4] Von Neergaard, K et al., *Z. Gesamte Exp. Med*, 66:373-394, 1929; [5] Wall, WA et al., Int. J. Numer. Meth. Biomed. Engng, 26:807-827, 2010; [6] Muhlfeld C et al., *Am J Physiol Lung Cell Mol Physiol*, 305:205-221, 2013; [7] Maas, SA, *J Biomech Eng.*, 134.1:011005, 2012; [8] Harris, RS, *Respir. Care*, 50.1:78-99, 2005; [9] Selman, M et al., *Ann. Intern. Med.*, 134.2, 136-151, 2001; [10] Gunther, A et al., Am. J. Respir. Crit. Care Med., 168.11:1358-1365, 2003

SB³C2017
Summer Biomechanics, Bioengineering and Biotransport Conference
June 21 – 24, Tucson, AZ, USA

NONLINEAR VISCOELASTIC RESPONSES OF PLGA FIBERS UNDER PHYSIOLOGIC CONDITIONS

Andrew B. Robbins (1), Hunter W. Storaci (1), Michael R. Moreno, (1,2) Anastasia Muliana (2)

(1) Biomedical Engineering
Texas A&M University
College Station, Texas, US

(2) Mechanical Engineering
Texas A&M University
College Station, Texas, US

INTRODUCTION

Biodegradable polymers are appealing in many biomedical applications. For example, scaffolds for engineered tissues are needed to provide temporary mechanical support for the tissue until cells can produce their own extra-cellular matrix. Another advantage of using biodegradable polymers is that it is possible to control the degradation rate by altering the macromolecular structure of the polymers through adding catalysts or other additives. Since one of the purposes of implants and scaffolds is to provide mechanical support, it is then imperative to understand the responses of biodegradable polymers under mechanical stimulus prior to the design of these implants and scaffolds. Of equal importance is to understand changes in the mechanical responses of biodegradable polymers as they degrade.

There have been many experimental studies conducted on examining the mechanical responses of biodegradable polymers [1,2,3,5,6,8,9,10,11]. In the above studies, uniaxial tensile loading was considered and the tested biodegradable polymers were shown to exhibit nonlinear and time-dependent behaviors when subjected to mechanical loading.

To describe the mechanical responses of biodegradable polymers several constitutive material models have been formulated. Some of the constitutive models of biodegradable polymers consider the nonlinear elastic response which allows for strain assisted degradation, i.e., the rate of degradation depends on the deformation gradient [8,9]. Muliana and Rajagopal [7] presented a quasi-linear viscoelastic model incorporating material degradation due to the diffusion of fluid. A nonlinear strain measure of a linearized strain is considered to incorporate the nonlinear elastic responses. A rate of degradation model that depends on fluid content originally proposed by Soares [7] is adopted and incorporated to the constitutive model. The model was shown capable of capturing relaxation responses in PLLA fibers.

In the present study, we investigate the response of PLGA biodegradable polymer fibers subject to a uniaxial tension while being immersed in PBS at 37C. The following experimental tests were conducted: stress relaxation at constant deformations and quasi-static ramps at different deformation rates. In addition we also conducted a limited test on PLGA fibers at room temperature under stress relaxation, and compared the responses to the ones immersed in PBS at 37C. To describe the mechanical behaviors of the tested fibers, a uniaxial nonlinear viscoelastic model based on a single integral form is considered.

METHODS

PLGA Polymers

PLGA polymers (Poly(lactic-co-glycolic acid) acid endcap (1:1 LA:GA) (Mn:25,000-35,000 Da)) were provided by Akina Inc. (West Lafayette, Indiana). Preliminary analysis of the polymer prior to extrusion was provided by Akina. The number average molecular weight (Mn) was calculated as 26,175 ± 551 Da and the weight average molecular weight (Mw) was calculated as 55,253 ± 11911 Da. The Tg of the polymer was reported as 40C. All dimensions were measured manually with calipers.

Mechanical Testing

Polymers were submerged in phosphate buffered saline (PBS) at 37C for 12-15 minutes prior to testing. A constant stretch rate was applied to the polymers using the Universal Test Machine (model 120R225, Test Resources, Shakopee, MN). Load, position, and time data was recorded until failure or slipping occurred, or the polymer had stretched so far as to come above the level of the saline solution. The stretch rates used ranged from 10 - 2000%/min, with a 1.0 in gage length for all uniaxial tests. For stress relaxation testing the same fixturing,

Poster Presentation #P182

gage length, and setup was used. The polymers were uniaxially stretched at the max rate (2000%/min) under load control until the load reached the specified load value (30% or 80% of the maximum load that was found during the uniaxial tensile tests) then held at a constant stretch until no more relaxation was observed.

Polymer Analysis

The polymer molecular weight and molecular weight distribution were determined by gel permeation chromatography (GPC) performed on an EcoSEC HLC-8320GPC (Tosoh America, Inc.). Polymer solutions were prepared at a known concentration (2.5-5 mg/mL) and an injection volume of 20 μL was used. The polymer molecular weight and molecular weight distribution GPC analysis of the polymer was conducted on samples in three different conditions: dry, soaked in PBS for 10-15 minutes at 37C, and samples soaked in 10-15 minutes in PBS at 37C and then uniaxially stretched at a rate of 1000%/min and stretched to 400% of the original gage length. Differential Scanning Calorimetry (DSC) testing was performed on four samples of the polymer using aluminum crucibles with about 5mg of samples, under dynamic air atmosphere (50mL/min), and a heating rate of 20C/min at a temperature range of 0-100C.

Non-Linear Viscoelastic Constitutive Model

In this study, we model the PLGA fiber as a nonlinear viscoelastic material. We adopt the quasi-linear viscoelastic (QLV) model, introduced by Fung [4] for modeling response of biological and polymeric solids. Fung expressed the second Piola-Kirchhoff stress as function of the Green-St Venant strain and the nonlinear stress relaxation function, which for one-dimensional relation is expressed as:

$$S(t) = \int_0^t D(t-s) \frac{dS^e}{dE}\frac{dE}{ds} ds \qquad (1)$$

where D(t) is the normalized relaxation function that should be positive definite, continuous, and monotonically decreasing with time, Se(E) is the nonlinear elastic stress function that depends on the Green-St Venant strain measure E, t is the current time, and s is the time history. The following function is used for the normalized relaxation function:

$$D(t) = D_\infty + \sum_{n=1}^{N} D_N e^{-t/\tau_N} \qquad (2)$$

RESULTS

Mechanical Testing

The polymer exhibited a non-linear viscoelastic response in constant stretch rate tests. Selected results are shown.

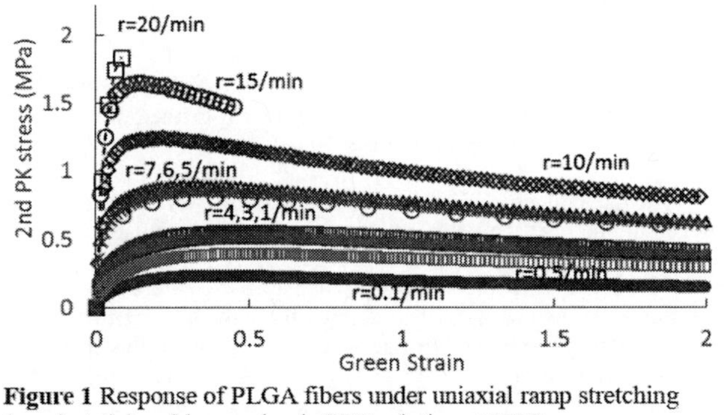

Figure 1 Response of PLGA fibers under uniaxial ramp stretching (quasi-static) and immersion in PBS solution at 37°C.

Polymer Analysis

Results from the GPC analysis of the PLGA polymers are as follows: Dry samples had a Mn 13,374 ± 759 Da, and Mw of 22,007 ± 734 Da. Samples soaked in PBS had a Mn of 12,341 ± 469 Da, and a Mw of 21,450 ± 370 Da. The Tg, calculated from the DSC results, was found to be 28.9 ± 2.2C.

DISCUSSION

After extrusion, the PLGA fibers had a Tg lower than originally reported by the manufacturer. Importantly, the Tg was lower than 37C at which the experiments were performed. This resulted in viscoelastic fluid-like behavior with rapid relaxation and large permanent deformations after testing. GPC analysis demonstrated that the experiments did not last long enough to significantly degrade the polymer, and even after mechanical loading, the molecular weight and Tg did not change; indicating little to no degradation during the experiment.

CONCLUSIONS

The fluid-like behavior of this extruded PLGA material is not exhibited at room temperature, with the polymer being much stiffer and more brittle. These temperature dependent properties could be exploited to produce devices that become flexible when exposed to physiological temperatures.

ACKNOWLEDGEMENTS

We would like to acknowledge NSF and ONR for supporting part of this work.

REFERENCES

[1] Breche Q, et. al., *Polymer Degradation and Stability*, 131:145-156, 2016.

[2] Dreher ML, et. al., *Journal of the Mechanical Behavior of Biomedical Materials*, 29:470-479, 2014.

[3] Dusunceli N, et. al., *Polymer Engineering and Science*, 2016.

[4] Fung YC, Biomechanics: Mechanical Properties of Living Tissues 2nd Ed., Springer, New York, 2010.

[5] Hayman D, et. al., *ASME J Biomechanical Engineering* 136:081006-1, 2014.

[6] Miller ND and Williams DF, *Biomaterials*, 5:365-368, 1984.

[7] Muliana A, et. al., *mechanics research communications*. 39:51-58, 2012.

[8] Soares, JS, *PhD Dissertation, Mechanical Engineering Texas A&M University*, 2008.

[9] Soares, JS, et. al., *Biomech. Model Mechanibiol*, 9:177-186, 2010.

[10] Vieira AC, et. al., *Materials Science Forum*, 730-732, pp. 56-61, 2011.

[11] Vieira AC, et. al., *International J. Solids and Structures*, 51:1164-1174, 2014.

SB^{3}C2017
Summer Biomechanics, Bioengineering and Biotransport Conference
June 21 – 24, Tucson, AZ, USA

EX-VIVO BIOMECHANICAL CHARACTERIZATION OF ARTERIOVENOUS FISTULAS

Aman Mahipat (1), Mirunalini Thirugnanasambandam (1), Sourav Patnaik (2), Roberto Vazquez (3), Ender A. Finol (2)

(1) Department of Biomedical Engineering
University of Texas at San Antonio
San Antonio, TX, USA

(2) Department of Mechanical Engineering
University of Texas at San Antonio
San Antonio, TX, USA

(3) Leonard M. Miller School of Medicine
University of Miami
Miami, FL, USA

INTRODUCTION

Hemodialysis vascular access dysfunction is a major cause of renal failure, and the major complications include stenosis and thrombosis, infections, digital ischemia, cardiac failure and aneurysm [1]. The three primary ways to gain vascular access are the native arteriovenous fistula (AVF), using synthetic PTFE graft and with cuffed catheters. AVFs, formed by joining an artery and a vein, are considered the gold standard method for vascular access. Despite this, AVFs fail in the early stages of maturation or require secondary interventions. These failures are believed to be caused by changes in the biomechanical properties of the fistulas over time. Using a pre-clinical model, we hypothesized that treating AVFs with the anti-fibrotic agent β-aminopropionitrile (BAPN) results in a decrease in AVF stiffness. This hypothesis was tested by conducting mechanical pressure-inflation testing on epigastric veins, AVFs formed by joining the right epigastric vein to the femoral artery, and AVFs treated with BAPN.

METHODS

The veins and AVFs were surgically extracted from Sprague-Dawley male rats (Harlan, Indianapolis, IN) and stored in vials filled with Dulbecco's Modified Eagle Medium for subsequent mechanical characterization. Three groups of specimens were tested: (1) left epigastric veins (EVs - sham controls); (2) AVFs created by joining the right epigastric vein to the femoral artery (controls); and (3) AVFs treated with a lysyl oxidase inhibitor (BAPN), to decrease collagen cross-linking. Pressure-inflation tests, as shown schematically in Fig. 1, were performed using the one-end free technique to analyze the deformation of the vessels when subjected to incremental intraluminal pressures [2]. The experimental setup consisted of a camera mounted on an inverted microscope (Olympus, Center Valley, PA) adjusted to capture images of the specimens at varying pressures, 500-µm-diameter stainless steel cannulas with machined grooves, 6-0 braided silk suture (Teleflex, Coventry, CT). The cannulas were placed in a vessel chamber (Living System Instrumentation, Burlington, VT) filled with PBS at room temperature.

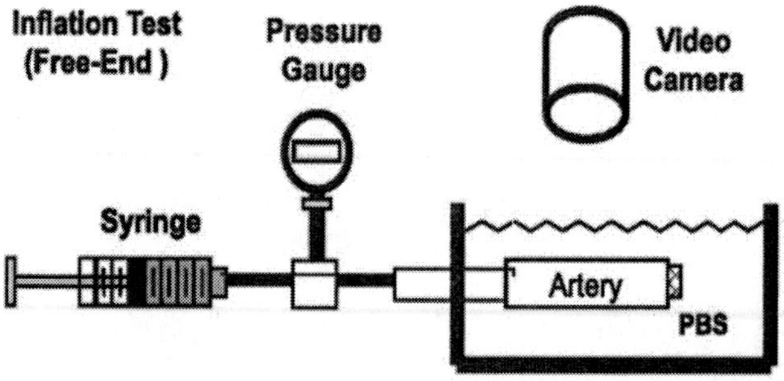

Figure 1: Schematic representation of the experimental setup of one-end free mechanical testing.

All specimens (n = 10-15 per group) were mechanically preconditioned before testing through pressurized inflation of 2-3 cycles at 40 mmHg and 60 mmHg, respectively. Once the specimens were mounted onto the cannula, the suture to suture length was measured. Perfusions were then applied through the proximal end of the cannula by a 60 mL syringe and simultaneously the change in pressure was measure by the pressure gauge (Ashcroft D1005PS). The distal ends of the specimens were sutured and left free to float on the PBS. Controls and the AVFs (untreated) were able to sustain a

Poster Presentation #P183

maximum pressure of 40 mmHg while the AVFs treated with BAPN could inflate circumferentially and axially to a maximum pressure of 120 mmHg – 150 mmHg. During mechanical testing, the specimens were pressurized and the change in outer diameter and length of the specimens were calculated by still images captured from the microscope camera (Lumenera Infinity 1) and saved for further analysis (Image-Pro Plus). Mean circumferential and axial Cauchy stresses along with Green's strains (circumferential and axial) for finite deformation were calculated using Eqs. (1) – (4),

$$\sigma_\theta = \frac{Pr_i}{r_e - r_i} \quad (1)$$

$$\sigma_z = \frac{Pr_i}{2(r_e - r_i)} \quad (2)$$

$$E_\theta = \frac{1}{2}(\lambda_\theta^2 - 1) \quad (3)$$

$$E_z = \frac{1}{2}(\lambda_z^2 - 1) \quad (4)$$

where P is the applied transmural pressure. Mean stresses and strains were used so that the unloaded state can be the reference state instead of the zero-stress state obtained from opening angles. These data was used to fit a two-dimensional Fung model, Eq. (5), resulting in an assessment of their mechanical behavior *ex vivo*,

$$w = \frac{1}{2}c_0(e^Q - 1), \qquad Q = c_1E_\theta^2 + c_2E_z^2 + 2c_4E_\theta E_z \quad (5)$$

where c_0, c_1, c_2, and c_4 are unknown material constants. These were obtained by multi-objective optimization using a genetic algorithm. The experimental data was also used to calculate the maximum tangential modulus (MTM), and the circumferential and axial stretch ratios. The latter were compared between the groups by two-way analysis of variance (ANOVA) with Tukey's honest significance test for multiple comparisons to determine the significance of the material model parameters. Data was considered significant at when $p < 0.05$.

RESULTS

Mechanical testing revealed that AVFs treated with BAPN exhibited lower circumferential stretch ratios (mm/mm) (1.02 ± 0.09 vs. 1.19 ± 0.05 vs. 1.07 ± 0.07) and lower axial stretch ratios (mm/mm) (1.08 ± 0.05 vs. 1.23 ± 0.07 vs. 1.20 ± 0.12) compared to the untreated AVFs and EVs, respectively. The MTM was evaluated to quantify the material stiffness, resulting in a lower MTM for the BAPN treated AVFs compared to the untreated AVFs, as seen in Fig. 2. Interestingly, the EVs had the smallest MTM in both directions compared to the other groups, showing the least amount of stiffness. Comparing the BAPN treated specimens with the untreated group, circumferentially and axially we observe a reduction in MTM of 26% and 20%, respectively.

The EVs (controls) possess low values of c_0, c_1, c_2 and c_4 with respect to AVF (untreated) and AVF (treated), which are nearly similar. Since material constants are obtained from fits of the data and the equations used here are primarily phenomenological, it could be seen that there is significantly low material stiffness in the control group at low strains compared to the AVF (untreated) and the AVF (treated with BAPN) groups in both the axial and circumferential directions.

DISCUSSION

This research emphasized the correlation of biomechanical properties and biological changes in the AVFs when treated with the lysyl oxidase inhibitor BAPN. Changes in the biomechanical parameters were analyzed and it was seen that the specimens treated with BAPN had a decrease in the mean outer circumferential and axial stretch ratios as a result of inhibition of collagen cross-linking. Stress-strain curves calculated using the 2D Fung model and to quantify the material stiffness and MTM was calculated which showed that BAPN prevents the formation of new cross-linking and reduced the AVF stiffening and weakened the relationship between collagen crosslinking and the AVF mechanics. The changes in structural and biological parameters suggest that BAPN plays a crucial role in reduction of AVF stiffening, which is the major cause of concern for failure of vascular access due to thrombosis, stenosis and intimal hyperplasia. Our results suggest that BAPN treatment affects the mechanical strength and structural architecture of AVFs, possibly resulting in decreased crosslinking of collagen, which may play an important role during thrombosis development and stenosis of the fistula during hemodialysis.

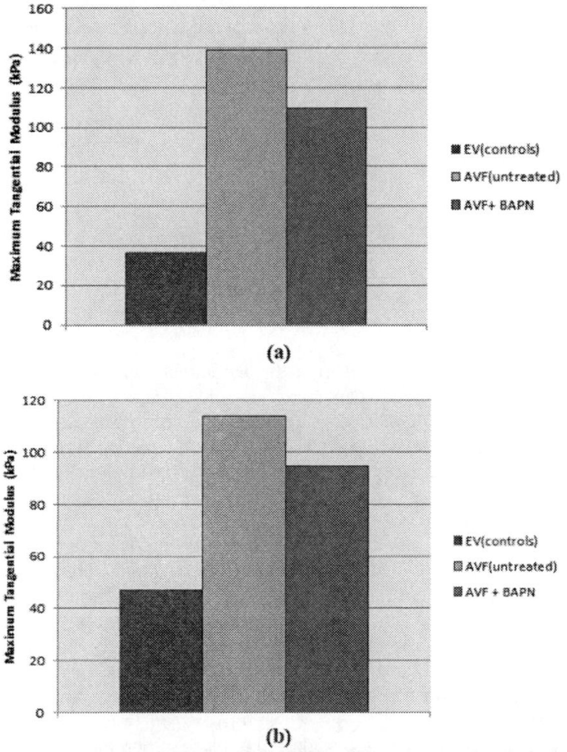

Figure 2: (a) Circumferential MTM for EV (controls), AVF and BAPN treated AVF, which is evaluated by calculating the peak slope of the stress-strain curves (p < 0.0.5; n = 10-15 per group). (b) Axial MTM for EV (controls), AVF and BAPN treated AVF (p < 0.0.5; n = 10-15 per group).

ACKNOWLEDGEMENTS

This work was supported in part by The University of Texas System Board of Regents - Science and Technology Acquisition and Retention (STARS) program.

REFERENCES

1. Gottmann, U et al., *J. Med. Case Rep.*, 6(1): 189-192, 2012.
2. Lee, AY et al., *Cardiovasc Eng Tech*, 1(4): 282-9, 2010.

ASSESSMENT OF MATERIAL PROPERTIES OF THIN FILM WOUND-TREATMENT POLYMERS

Krysta-Lynn H. Amezcua (1), Sourav Patnaik (2), Mirunalini Thirugnanasambandam (1), Matthew Reilly (3), Ender A. Finol (2)

(1) Department of Biomedical Engineering
University of Texas at San Antonio
San Antonio, TX, U.S.A.

(2) Department of Mechanical Engineering
University of Texas at San Antonio
San Antonio, TX, U.S.A.

(3) Department of Biomedical Engineering
The Ohio State University
Columbus, OH, U.S.A.

INTRODUCTION

Wounds and burns disrupt the natural barrier of the skin and expose the body to pathogens. Liquid bandage polymers are commonly used for soft tissue repair as they can be applied quickly and provide a protective barrier. One problem commonly encountered with polymeric dressings is the tendency for cured polymers to adhere to external surfaces [1], inflicting further damage to the wound upon removal. Curing of polymeric dressings involves complex mechanisms of polymer-substrate adhesion, which may be influenced by the material properties of the polymer [2]. Hence, biomechanical evaluation of these liquid bandage polymers is required to understand their structural and functional roles in wound healing applications. In the present work, the mechanical behavior of three wound-treatment polymeric films obtained from commercially available liquid bandages (New-Skin®, Skin Shield®, and Cavilon®) was characterized through uniaxial tensile testing and constitutive modeling. The ensuing mechanical properties were compared to those of porcine skin.

METHODS

Uniaxial tensile tests were used to assess the mechanical properties of each wound-treatment polymer. Ten specimens were tested for each polymer group (New-Skin®, Skin Shield®, or Cavilon®). Specimens were cut in dog-bone shape (length to width ratio of 2:1) with a thickness of approximately 0.25 mm. Mechanical tests were carried out in a uniaxial tension machine (Electroforce T3200, TA Instruments, Inc., New Castle, DE) and the specimens were subject to maximum extension at a rate of 0.03 mm/s. Tensile force and extension data were collected at 50 Hz with a sampling rate of 0.5 data points per second and post-processed utilizing a custom MATLAB script (Mathworks, Inc., Natick, MA). From these, the Cauchy stress and instantaneous stretch ratio were calculated. For each tensile test, a stress-stretch curve was generated for a total of ten stress-stretch curves per polymer. Young's moduli (E) were calculated for each specimen by estimating the slope of the curve in the linear elastic region (fitted to the straight portion of the curve at lower strains) using linear least-squares regression. From each stress-stretch plot, the area under the curve (strain energy density) was calculated using numerical integration with the trapezoidal method.

The stress-stretch behavior of each polymer was characterized using a hyperelastic, incompressible Mooney-Rivlin model mathematically described by Eq. (1) as follows:

$$\sigma = 2C_1\left(\lambda^2 - \frac{1}{\lambda}\right) + 2C_2\left(\lambda - \frac{1}{\lambda^2}\right) \quad (1)$$

where σ is the Cauchy stress, λ is the stretch ratio and C_n are the material model constants.

For each stress-stretch curve, Mooney-Rivlin model-predicted stresses were computed, and C_1 and C_2 were estimated accordingly. Fitting was achieved using the Levenberg-Marquadt algorithm, which allows specification of an arbitrary, nonlinear function for the stress in terms of the stretch ratio. The relative error was then calculated to determine how well the estimated C_1 and C_2 constants fit the experimental data. A one-way analysis of variance (ANOVA) was conducted to compare the Young's modulus (E), the stress at the maximum stretch ratio, the strain energy at the maximum stretch ratio, and C_1 and C_2 constants among the three polymers. Pairwise comparisons for each variable was performed using Tukey's test. Variances were considered significant for $p < 0.05$.

RESULTS

Poster Presentation #P184

All three wound-treatment polymers remained structurally intact for large deformations up to 78% engineering strain, which is the average failure strain of porcine skin [3]. Hence, for a comparative analysis all the data reported calculated for New-Skin®, Skin Shield® and Cavilon® in this study is upto a stretch ratio of 1.78. The New-Skin® specimens exhibited a mean stress of 53.2 MPa at maximum stretch ratio, which was statistically different ($p < 0.0001$) than the other polymers (see Table 1). Skin Shield® and Cavilon® exhibited lower mean tensile stresses (0.7 MPa vs. 14.3 MPa, respectively) and were not significantly different from each other ($p = 0.13$) at a stretch ratio of 1.78. New-Skin® displayed the highest mean strain energy density (67.2 MJ/m^3) at $\lambda = 1.78$ ($p < 0.0001$). Similarly, Skin Shield® and Cavilon® displayed lower mean strain energies (0.5 MJ/m^3 vs. 8.5 MJ/m^3, respectively) and were not significantly different ($p = 0.77$) from each other. The Young's moduli of the polymers are also described in Table 1, indicating that New-Skin® was the stiffest material tested.

Table 1: Mechanical properties of New-Skin®, Skin Shield®, Cavilon®, and porcine skin [3].

Material	Stress at λ = 1.78 (MPa)	Strain Energy Density at λ = 1.78 (MJ/m^3)	E (MPa)
New-Skin®	53.2 ± 25.9	67.2 ± 45.1	270.7 ± 84.4
Skin Shield®	14.3 ± 2.6	8.5 ± 1.6	71.4 ± 19.4
Cavilon®	0.7 ± 0.1	0.5 ± 0.1	4.0 ± 1.3
Skin	1.0	0.4	1.5

Compared to porcine skin, Cavilon® exhibited similar mechanical characteristics, such as the mean modulus of elasticity (4.0 MPa vs. 1.5 MPa, respectively), mean stress (0.7 MPa vs. 1.0 MPa, respectively) and mean strain energy density at $\lambda = 1.78$ (0.5 MJ/m^3 vs. 0.4 MJ/m^3, respectively). New-Skin® was found to be the stiffest of the polymers compared to porcine skin. Tensile stress and strain energy density of Cavilon® was found to be closest to porcine skin.

The behavior of each polymer was further characterized through constitutive modeling. Mooney-Rivlin material constants for each polymer were compared to the C_1 and C_2 constants for porcine skin reported by Shergold et al. [3]. Table 2 reveals that the constitutive parameters for Cavilon® ($C_1 = 1.2$ MPa and $C_2 = -0.6$ MPa) were most similar to porcine skin. These constants were analyzed with ANOVA, which revealed statistically significant differences in the mean C_1 ($p < 0.0001$) and C_2 ($p < 0.0001$) of each polymer's Mooney-Rivlin strain energy function. The differences in material behavior amongst the three polymers and the similarities of Cavilon® with porcine skin are represented graphically in Fig. 1.

DISCUSSION

In the present work we assessed the efficacy of polymeric films as wound treatment barriers by characterizing the mechanical behavior of three commercially available liquid polymer wound dressings. A Mooney-Rivlin strain energy function was chosen to model the stress-stretch behavior of porcine skin using experimentally derived function constants reported in [4]. By comparing the mechanical properties of three polymeric films to the properties of porcine skin, Cavilon® was revealed to be most similar to porcine skin in terms of modulus of elasticity, maximum stress, and strain energy. The Young's moduli revealed statistical differences among the three polymers, while the maximum stress and strain energy density at a stretch ratio of 1.78 were statistically the same for Skin Shield® and Cavilon®. Future work in this field should include rheological testing of polymeric films to assist with optimization of wound-treatment polymer designs. Future studies will focus on understanding the structure-function relationship of wound-treatment polymers that will help us improve the quality of biomaterials suitable for patients' clinical needs.

Table 2: Mooney-Rivlin C_1 and C_2 constants for New-Skin®, Skin Shield®, Cavilon® (with goodness of fit measures), and porcine skin [3].

Material	C_1 (MPa)	C_2 (MPa)	Goodness of fit
New-Skin®	35.2 ± 18.1	-13.5 ± 8.0	88% ± 2%
Skin Shield®	15.7 ± 4.5	-6.3 ± 2.3	92% ± 3%
Cavilon®	1.2 ± 0.3	-0.6 ± 0.1	91% ± 4%
Skin	0.3	0	-------

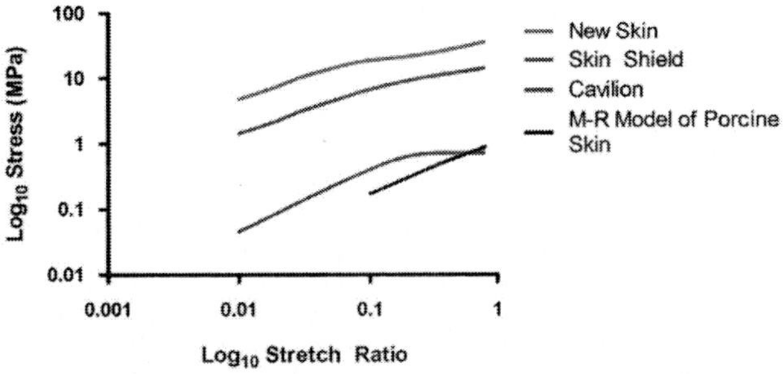

Figure 1: Mooney-Rivlin model for porcine skin and stress vs. stretch data for New-Skin®, Skin Shield, and Cavilon.

ACKNOWLEDGEMENTS

Funding for this work was provided in part by the UTSA MARC U*STAR GM 007717 Research Training Program.

REFERENCES

1. Salamone, J.C., et al. *Regen Biomater.* 3(2) 127–128, 2016
2. Kheyfets, V.O., et al. *J Biomech Eng*, 136(7): 0745011–0745015, 2014.
3. Shergold, O.A., et al. *Int J Impact Eng.* 32(9)1384-1402, 2006.
4. Yamada H., Williams and Wilkins, Baltimore, MD, pp. 226-231, 1970.

SB³C2017
Summer Biomechanics, Bioengineering and Biotransport Conference
June 21 – 24, Tucson, AZ, USA

PRESSURE INDUCED DAMAGE OF PULMONARY ARTERY

S. Baek (1), A. Rao (2), Y. Wang (1), L. Alison (1), S. Roccabianca (1)

(1) Department of Mechanical Engineering
Michigan State University
East Lansing, MI, US

(2) Department of Mechanical Engineering
National Institute of Technology Karnataka
Mangalore, Karnataka, India

INTRODUCTION

Pulmonary hypertension (PH) is associated with elevated pulmonary arterial pressure. PH prognosis remains poor with 15% mortality rate within 1 year even with modern clinical managements, while pharmacological treatments of high blood pressure have been effective in systemic hypertension patients for decades. Over the past decades, wall shear stress (WSS) has been considered an important hemodynamic factor, affecting cell mechanotransduction and growth and remodeling in several vascular diseases. Recent studies in PH proposed that a low WSS attributes to vascular adaptation and causing inflammatory and proliferative cell expression [1]. Valentín et al. [2] showed, however, that a 30% increase in flow rate produces only a 9% increase in arterial diameter. Moreover, it has been showed that the caliber of human pulmonary artery is much larger in PH patients than those of normal, e.g. more than 50% larger diameters in PH compared to normal subjects in young patients [3] and about 30% larger in adult PH [4]. For these reasons, we suggest that low WSS alone could not be responsible for the large increase of arterial diameter in PH. In this study, consequently, we investigated the hypothesis that increased transmural pressure can induce irreversible damage in a large animal's pulmonary artery. The objectives of this study are, first, to describe the mechanics and the damage behavior of pulmonary arteries, and second, to shed light on the relevance of pressure-induced damage mechanism in human PH. Specifically, we performed an in vitro cyclic inflation test to characterize the mechanical behavior, in homeostatic and high-pressure conditions, of porcine pulmonary arteries.

METHODS

The porcine pulmonary artery from two pigs, aged approximately six months, is obtained from a local vendor. The porcine specimen is dissected and the peripherally loose connective and adipose tissues are carefully removed from the pulmonary artery, trunk to one main branch. After suturing small branches both ends are mounted on cannulas and the specimen is preconditioned through 5 cyclic inflation – deflation loading cycles over the homeostatic pressure range 0 – 30 mmHg (labeled '**1**' in **Fig. 2** top and center). Then, the specimen is loaded with 5 loading protocols each one of them consisting in 5 inflation – deflation cycles, as follows: from 0 to 50 mmHg (two times; labeled '**2**' and '**3**'), from 0 to 100 mmHg (two times; the labeled '**4**' and '**5**'), and from 0 to 50 mmHg (labeled '**6**', all labels referred to **Fig. 2** top and center). One CCD camera records digital images of a selected region of one branch (a red box shown in **Fig. 1**), the outer diameter is measured as an average over the selected region. Simultaneously, pressure and outer diameter values are recorded as well. The other main branch is separated, excised and fixed as control for histological study.

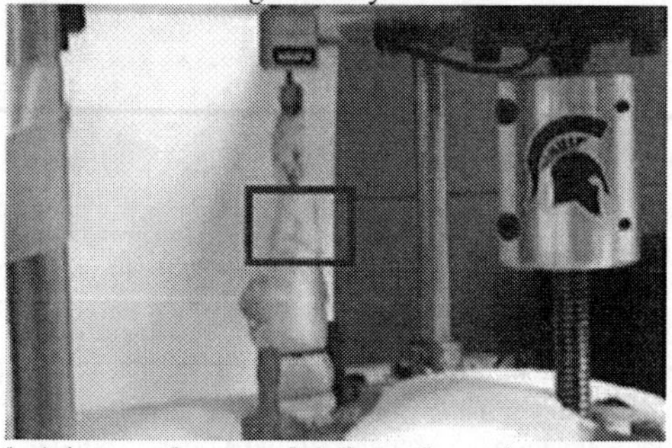

Figure 1: A image of a cannulated porcine pulmonary artery during cyclic inflation and a region shown in red box is used to calculate the averaged outer diameter

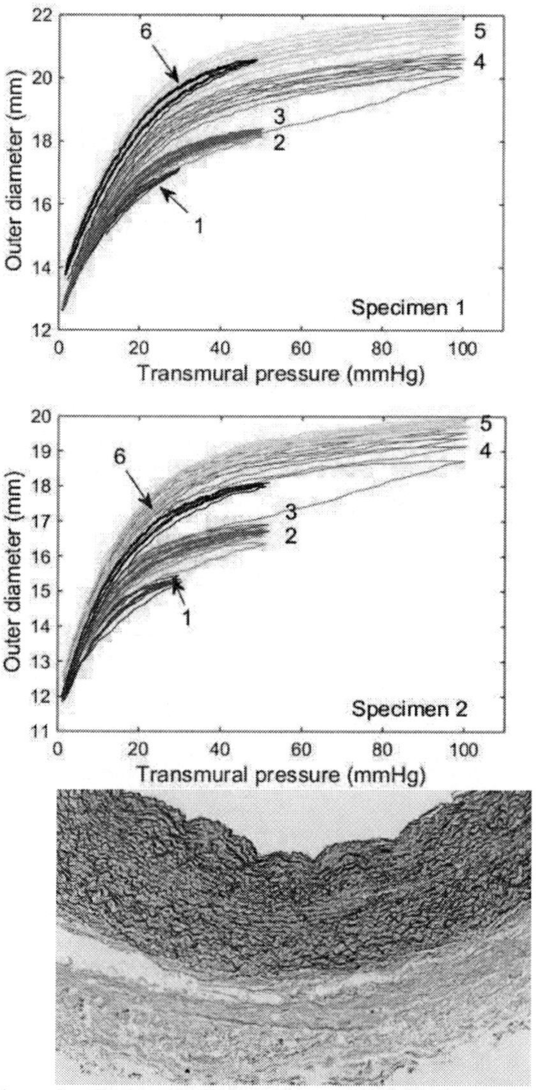

Figure 2: (Top) outer diameter-pressure plots of cyclic loading of Specimen 1, (Middle) outer diameter-pressure plots of cyclic loading of Specimen 2 and (Bottom) histology of Specimen 1.

RESULTS

The cyclic pressure-diameter graphs for the two specimens tested are shown in **Fig. 2**. During cyclic inflation tests, most arteries in systemic circulation exhibit typical pseudo-elastic behavior, where the second cyclic curve shifts with respect to the first, however with multiple cycles the loading and unloading curves becomes identical. So it is not clear if there is damage during cyclic loading of protocols **2** and **3** (i.e., 0 to 50 mmHg) and the shifting of pressure-diameter might be due to viscoelastic behavior, albeit Specimen 2 shows a visible shifting. The first loading cycle of protocol **4** (i.e., 0 – 100 mmHg), however, markedly shows a softening behavior, staring at about 50 mmHg, but the followed loading curves are shown only gradual shifting. Similarly, in the following loading cycles for pressure range of 0 – 100 mmHg (protocol **5**), the softening behavior is not visible, and rather the gradual shifting of the curves is shown. Finally, in protocol **6**, we repeat the cyclic loading 0 – 50 mmHg confirming that the artery underwent damage compromising the mechanical behavior, highlighted by the difference between the curves of protocols **2** and **3** compared to **6**. Overall, the pressure-diameter curves are similar between the two specimens, but the graphs of protocol **6** for Specimen 2 that shows some level of recovery of mechanical behavior, which is not shown in Specimen 1.

DISCUSSION

If the mean pressure of the pulmonary artery is greater than 25 mmHg in children, pediatric PH is diagnosed. Based on the in vitro tests of two porcine specimens, we showed that a high pressure could induce irreversible damage of mechanical behavior (i.e., during cyclic loading from 0 to 100 mmHg), while it is not certain however if mechanical damage were present for a pressure lower than 50 mmHg. Nonetheless, in PH, pathological conditions (e.g., inflammatory condition and low wall shear stress) may cause weakening of vessel wall and it is feasible that both of mechanical and biological cues induce the arterial damage. In pulmonary arteries, the physiological tissue strain is largely dependent on the stiffness of elastin at diastolic pressure [5], which determines the elastic recoiling of artery in the low stretch region. Interestingly, the outer diameter at 0 mmHg pressure is different between protocol **1** and protocol **6** for Specimen 1 (i.e., increase of outer diameter), in contrast Specimen 2 shows a similar diameter at 0 mmHg pressure between protocol **1** and **6**. We propose that the amount of elastin damage can be estimated using two loading protocols, namely **3** and **6**. Its validation, however, requires more biomechanical modeling and eventually from further analysis of histology (**Fig. 2** (Bottom)).

Based on understandings from animal models of PH, multiple studies have hypothesized that smooth muscle hypertrophy, endothelial dysfunction, deposition of collagen, loss of capillaries and elastin fragmentation contribute to the pathogenesis of PH through remodeling of the extracellular matrix. It has been reported in 2000 [6] that PH was completely reversible in a rodent models by a serine elastase inhibitor, whereas still clinical management of PH is mostly palliative, not curative, and it relies on vasodilators such as oxygen and nitric oxide yet. Surprisingly, few recognized the importance of the caliber of main pulmonary arteries in biomechanics and most of biomechanical studies were focused on arterial stiffness and WSS is considered as an inducer. So this study submit the possible idea that small animal models were able to successfully reverse the pathology because the increase of calibers in the pulmonary artery in small animal model were not large. This study brings attention to the role of main pulmonary artery caliber and to the pressing need of understanding of mechanical damage of artery. There are clear limitations of study, though (e.g., the number of specimen) and the ongoing study will increase the number of specimens and further analyze histology and the constitutive modeling.

ACKNOWLEDGEMENTS

This study was supported by, in part, inGEAR program in College of Engineering at Michigan State University and NSF CAREER (CMMI-1150376).

REFERENCES

[1] Kheyfets, V. et al., *J, Biomech. Eng.* 135:061011, 2013
[2] Valentín, A. et al., *J.R. Soc. Interface* 6:293-306, 2009
[3] Truong. U. et al., *J. Cardiovasc. Magn. Resonance*, 15:81, 2013
[4] Edwards, P.D. et al., *Br J. Radiol.* 71: 1018-1020, 1998
[5] Lammers, S. R. et al., *Am J. Physiol. Heart Circ. Physiol.*, 295: H1451-H1451, 2008
[6] Cowan, K.N. et al., *Nature Medicine*, 6:698-702, 2000

TRANSESOPHAGEAL ECHOCARDIOGRAPHY ENABLES REGIONAL QUANTIFICATION OF LEFT VENTRICULAR STRAIN IN A PORCINE MODEL OF MYOCARDIAL INFARCTION

William M. Torres (1,2), Alison T. Thames (2), Tarek Shazly (1), Francis G. Spinale (1,2,3,4)

(1) College of Engineering and Computing
University of South Carolina
Columbia, SC, USA

(2) Cardiovascular Translation Research Center
University of South Carolina School of Medicine
Columbia, SC, USA

(3) Department of Cell Biology and Anatomy
University of South Carolina School of Medicine
Columbia, SC, USA

(4) WJB Dorn Veteran Affairs Medical Center
Columbia, SC, USA

INTRODUCTION

The National Heart, Lung and Blood Institute reports that approximately 15.5 million Americans over the age of 20 have coronary artery disease (CAD).[1] Acute myocardial infarction (MI) occurs in a subset of CAD patients, and is characterized by localized necrosis of the myocardium following prolonged ischemia. Though the survival rate of patients who suffer MI is 35-65%, life expectancy is significantly reduced and the chances of a second MI are greater than 10%.[2,3] Elevation of risk can be attributed to mechanically-mediated changes in left ventricle (LV) geometry as well as myocardial composition and properties in a process generically termed adverse post-MI remodeling.

Adverse post-MI remodeling includes inflammatory cascades that impact both the micro- and macro-structures of the myocardium. A local proliferation of fibroblasts and fibrillar collagen deposition occurs in the infarct region. The altered extracellular matrix of the infarct is coupled with geometrical changes, including a localized wall thinning (MI region thinning) and gradual expansion of the MI area (MI region expansion) – these processes compromise LV function and promote the progression to heart failure.[4]

Adverse post-MI remodeling can be understood as a mechanobiological process that seeks to restore homeostatic wall stress in the LV following deleterious changes in myocardial properties and geometry. Therefore, techniques to quantify LV mechanics in-vivo are critical for understanding the progression of adverse post-MI remodeling and evaluating strategies to favorably curtail its trajectory. Recent advances in ultrasound echocardiography hardware and software provide the capacity to temporally track wall deformation throughout the LV. Local segmental strains (1) and strain rates (2) are computed as:

$$\varepsilon = \frac{L-L_0}{L_0} = \frac{\Delta L}{L_0} \qquad (1)$$

$$\gamma = \frac{(\Delta L/\Delta t)}{L_0} = \frac{\Delta V}{L_0}, \qquad (2)$$

where ε is segmental strain, L_0 is the baseline segmental length, L is the instantaneous segmental length, γ is strain rate, Δt is the elapsed time, and ΔV is the velocity gradient in the segment studied.[5] Two-dimensional speckle tracking echocardiography (STE) entails the definition of "speckle clusters" (natural acoustic markers) and continuous tracking of their relative position over the cardiac cycle. Clusters are used to define segmental lengths that enable quantification of myocardial strain and strain rate in accordance with Eqs. (1) and (2).

Our study utilizes 2-D STE to characterize the spatial variation in strain/strain rates throughout the LV wall in a porcine model of MI. Our findings demonstrate significant changes in wall strain profiles at 28 days post-MI, including both within the MI region and adjacent/remote regions of the viable myocardium.

METHODS

Porcine model: All animals were treated and cared for in accordance with the National Institutes of Health Guide for the Care and Use of Laboratory Animals, and all protocols were approved by the University of South Carolina's Institutional Animal Care and Use Committee. MI was induced by occluding the proximal left anterior descending artery in Yorkshire pigs ($n=5$) with a balloon catheter accessed through the femoral artery. The occlusion was held in place for 90 minutes and then removed to simulate injury reperfusion. Due to the lack of collateral circulation in pigs, the geometry and location of the resultant infarct region are highly repeatable.

Echocardiography: Standard transthoracic echocardiographic studies were performed (GE, VIVID E9) to obtain end diastolic and end systolic measurements of LV posterior wall thickness (PW), interventricular septum thickness (IVS), and volume (EDV/ESV) via Simpson's method of disk. Transesophageal echocardiographic studies were performed to quantify myocardial deformation through 2D STE. These studies were completed serially prior to the surgical induction of the MI and at 28 days after injury reperfusion. The same protocol was utilized on age/gender matched control animals ($n=10$) to provide a referent normal basis for comparison.

Strain quantification: The reference length L_0 for each segment was determined by using the peak of the R-wave on the superimposed ECG as a marker of end-diastole. Temporal maps of the longitudinal strain of the LV mid-wall were attained from the parasternal long axis. In each group, the the average region-specific curves were fit with a third-order polynomial regression after normalization of heart rate. Statistical analysis software, Matlab (Mathworks, 2016), was then used to quantify the following in-vivo response variables: segmental strain at end systole, peak segmental strain, systolic segmental strain rate, and mechanical dispersion. End systole was defined by the moment at which the aortic valve was fully closed after ejection. Peak segmental strain was quantified as the maximum relative strain value for each region of the myocardium. Systolic segmental strain rate was defined as the average rate at which the myocardium was deformed between the peak positive segmental strain value (tension) and the peak negative segmental strain value (compression). Mechanical dispersion was defined by calculating the standard deviation of the times at which peak negative segmental strain was achieved at each region of the myocardium.[6] All data was analyzed and recorded as the mean plus or minus the standard error of the mean.

RESULTS

Transesophageal echocardiograms were collected at heart rates of 112.88±5.79 bpm (referent normal group) and 98.80±11.50 bpm (post-MI group), which corresponds to no statistical difference between groups. Region-specific segmental strains were normalized/averaged to generated representative profiles for the two study groups (Figure 1).

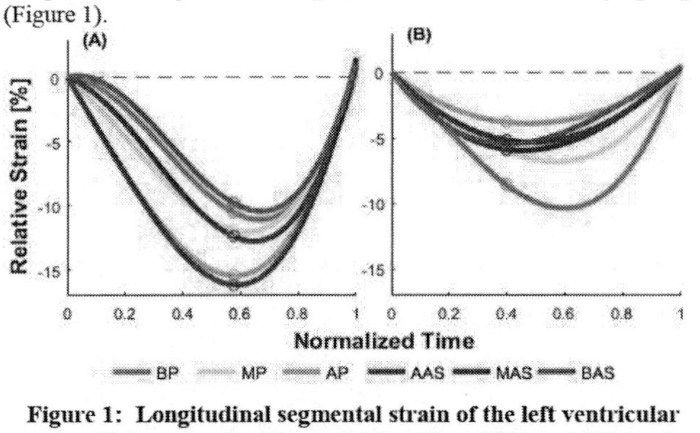

Figure 1: Longitudinal segmental strain of the left ventricular mid-wall for (A) referent normal (*n=10*) and (B) post-MI (*n=5*) groups. Region-specific strain was quantified in the basal posterior (BP), mid posterior (MP), apical posterior (AP), apical anteroseptal (AAS), mid anteroseptal (MAS), and basal anteroseptal (BAS) regions of the LV. End systole is demarcated with a circle imposed on the curve for each segment.

Quantitative comparison between the referent normal and MI groups are summarized in Table 1. Most notably, MI induced a 77% reduction of end systolic segmental strain (p=0.0047) and 72% reduction of systolic segmental strain rate (p=0.0116) in the area most proximal to the infarct, the apical posterior. In contrast, no significant differences in strain/strain rate were observed in the basal posterior region. Significant mechanical differences also emerged in other wall regions, with the general trend being a reduction in both strain magnitudes and rates post-MI. Although mechanical dispersion was not significantly altered, the observed 50% increase post-MI is reflective of compromised myocardial function.

Table 1: Quantification of echocardiographic response variables, including segmental strains and strain rates, for referent normal and 28 day post-MI experimental groups.

	Normal (n=10)	MI (n=5)	p-Value
End Systolic LS [%]			
Basal Posterior	-10.87±1.11	-8.98±0.99	n.s.
Mid Posterior	-12.15±0.69	-5.98±0.78	n.s.
Apical Posterior	-15.72±0.67	-3.59±0.86	0.0047
Apical Anteroseptal	-16.42±0.93	-5.27±0.98	0.0353
Mid Anteroseptal	-12.51±0.48	-6.04±1.01	n.s.
Basal Anteroseptal	-9.92±0.47	-5.49±0.92	n.s.
Systolic Strain Rate [s^{-1}]			
Basal Posterior	-25.35±2.73	-32.53±1.79	n.s.
Mid Posterior	-26.74±1.49	-17.18±2.44	n.s.
Apical Posterior	-36.88±1.53	-10.36±2.68	0.0116
Apical Anteroseptal	-38.27±2.06	-18.65±1.53	n.s.
Mid Anteroseptal	-22.39±2.07	-11.89±3.26	n.s.
Basal Anteroseptal	-29.16±0.63	-18.03±1.44	0.0206
Mechanical Dispersion [s]	0.10±0.01	0.15±0.01	n.s.
LVEDV [mL]	48.82±2.42	59.82±2.39	0.0421
EF [%]	56.63±1.63	37.68±6.99	0.0006

Values are mean ± SEM; n.s. = not significant
MI = myocardial infarction. LS = longitudinal strain; LVEDV = left ventricular end diastolic volume; LVESV = left ventricular end systolic volume; EF = ejection fraction

DISCUSSION

2D STE provides a noninvasive and effective method to assess LV mechanics in a preclinical model of MI. Our findings demonstrate that adverse post-MI remodeling differentially affects strains/strain rates throughout the myocardium, underscoring the complexity of mechanical changes that cumulatively result in a loss of LV function. When viewed together, post-MI trends in mechanical dispersion and the significant reduction in ejection fraction suggests that a normal sinus rhythm has been largely reestablished, albeit at a reduced level of LV function. Future studies will utilize 2D STE to evaluate emergent strategies for attenuating adverse post-MI remodeling based on mechanical augmentation of the infarct region, including the direct injection of biomaterials into the myocardium.

ACKNOWLEDGEMENTS

This work was supported by the National Institute of Health grant 5R01HL111090 (to F.G.S).

REFERENCES

1. Mozaffarian , D et al., *Circulation*. 2015;131:00-00.
2. Mann, D et al., *Circulation*. 2005;111(21):2837-2849.
3. Smolina, K et al., *Circ Cardiovasc Qual Outcomes*. 2012;5(4): 532-540.
4. Andreu, I et al., *Acta Biomater*. 2014;10(7):3235-3242.
5. Dandel, M et al., *Curr Cardiol Rev*. 2009;5(2):133-148.
6. Haugaa, K et al., *Eur Heart J*. 2009;30(3):330-337.

SB[3]C2017
Summer Biomechanics, Bioengineering and Biotransport Conference
June 21 – 24, Tucson, AZ, USA

EARLY REMODELING OF PULMONARY AUTOGRAFT AFTER ROSS PROCEDURE: WALL STRESS ANALYSIS

Yue Xuan[1], Ismail El-Hamamsy[2], Francois-Pierre Mongeon[3], Richard L. Leask[4], Alexander Emmott[4], Aly Ghoneim[5], Elaine E. Tseng[1], Liang Ge[1].

[1]Department of Surgery, University of California San Francisco Medical Center and San Francisco VA Medical Center, San Francisco, CA USA

[2]Division of Cardiac Surgery and [3]Department of Medicine, Montreal Heart Institute, Montreal, Canada

[4]Department of Chemical Engineering and [5]Division of Cardiac Surgery, McGill University, Montreal, Canada

INTRODUCTION

Ross procedure is one of the best options for children and young adults who need aortic valve replacement. The Ross procedure provides excellent hemodynamics, freedom from oral anticoagulation and a living tissue that can grow. However, the aneurysmal remodeling of the autograft would lead to complex reoperation because of the autograft failure. The remodeling of autograft can only be passively monitored and the biomechanics of dilatation is unclear. We previously demonstrated elevated autograft wall stress from pulmonic pressure to aortic pressure. Increased wall stress is an important risk factor for evaluating the biomechanical remodeling that can lead to autograft failure. In this study, we investigated the patient-specific stress distribution on the autograft at one year after Ross surgery.

METHODS

14 patients underwent Ross procedure were consented for the study. Intraoperative extra pulmonary root tissues were collected when available. Cine magnetic resonance imaging was performed at one-year postoperation and lumen geometry of the autografts and aortas were reconstructed based on the MRI images. Material properties and wall thickness of autografts and aortas measured from biaxial stretching were incorporated if available and averaged material property and thickness were used for others. The multiplicative approach was used to account for the prestress from in-vivo imaging. Fiber-embedded hyperelastic material model were developed and implemented to determine autograft wall stress on autografts at systematic pressure.

RESULTS

At systematic systole, the averaged first principal stress on autograft wall were 773±342KPa, 427±190KPa, and 552±247Kpa at sinotubular junction, sinuses, and annulus, respectively (figure 1). The mean autograft diameter was 35.8±1.7mm, 37.9±3.2mm, and 35.3±2.25mm at the sinotubular junction, sinuses, and annulus, respectively.

Poster Presentation #P188

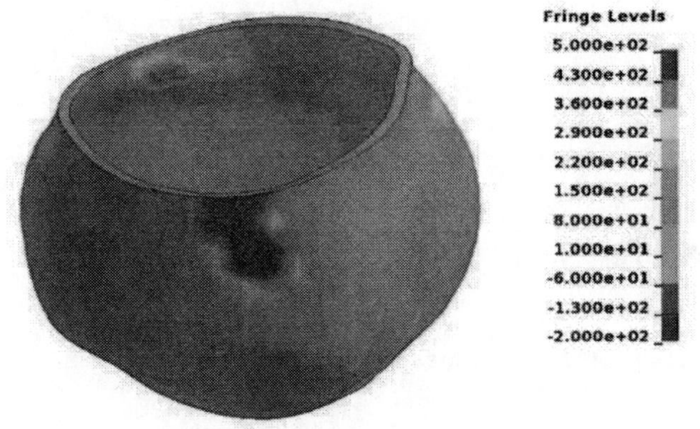

Figure 1: Typical example of wall stress distribution on autograft at one year postoperation.

DISCUSSION

The peak first principal stresses were mainly located at the sinotubular junction and/or annulus. Our finding explains the clinical observation that the autograft failure was usually accompanied by the dilation of sinotubular junction and/or annulus. The dilatation of sinus itself, which had the largest diameter, may not lead to aortic regurgitation. Future clinical data that follow the Ross patients will be incorporated to better understand the aneurysmal remodeling of autograft.

ACKNOWLEDGEMENTS

The current study is partly supported by the American Heart Association Postdoctoral Fellowship 16POST31420013.

SPATIAL SCALING IN MULTISCALE MODELS: A METHOD FOR COUPLING AGENT-BASED AND FINITE-ELEMENT MODELS OF TISSUE REMODELING

J.J. Lee (1), L. Talman (1), S. M. Peirce (1, 2), J. W. Holmes (1, 2, 3)

(1) Department of Biomedical Engineering
University of Virginia
Charlottesville, Virginia, U.S.A

(2) Berne Cardiovascular Research Center
University of Virginia
Charlottesville, Virginia, U.S.A

(3) Department of Medicine
University of Virginia
Charlottesville, Virginia, U.S.A

INTRODUCTION

Cellular activities, extracellular matrix (ECM) homeostasis and remodeling, and tissue mechanics form a mechanobiological feedback loop. For example, during wound healing, cell migration and ECM remodeling are both sensitive to the mechanical environment, even as cell-driven ECM remodeling dynamically modifies tissue mechanical properties. Multi-scale models that couple agent-based modeling (ABM) to represent cell behaviors and finite element modeling (FEM) to represent tissue mechanics have been used to explore cell-ECM-mechanics interactions in healing myocardial infarcts [1], muscular dystrophy [2], and vascular tissue engineering [3].

One of the challenges in coupling an ABM to an FEM is spatial scaling. ABMs typically require grid sizes on the micron scale to adequately represent cell migration, chemokine diffusion, and other relevant processes, while FEMs of tissue mechanics generally employ element sizes on the order of millimeters or centimeters, consistent with the assumptions of continuum mechanics. Furthermore, the two modeling frameworks typically employ different coordinate systems that may be related through nonlinear mappings. Thus, the present study aims to develop a general approach to ABM-FEM coupling that accounts for different spatial scaling in the two model components while allowing the user flexibility to adjust the mesh density of each component independently.

METHODS

Agent-based model: The agent-based model of cell migration and ECM remodeling was constructed in Repast Simphony (2.3.1), an open source, Java-based, agent Simulation Toolkit [4]. The overall framework was based on our previously published ABM [1,5] of infarct healing. Fibroblasts were modeled as agents with a radius of 5-μm that each occupy one discrete point on a regular grid, (referred to in Repast Simphony as a "GridPoint"). They migrate and replicate without overlapping deposit and degrade collagen, and apoptose after a specified lifetime. Users specify model dimensions, initial cell density, cell division and apoptosis rates, and collagen deposition and degradation rates through a dialog box. For this study, a 1-cm square was simulated as a two-dimensional 100-by-100 array of GridPoints. The amount of collagen at each GridPoint was simulated as a 6-bit grayscale image with intensity ranging from 0 to 63, and displayed using pseudo red colormaps. We employed the following parameters to approximately match the time course of collagen accumulation in our previously published ABM [4]: 1 hour per simulated time step (tick) with a total simulation time of 42 days, cell migration speed 10-μm/tick; apoptosis every 240 ticks, mitosis every 96 ticks, collagen deposition every 30 ticks, and collagen degradation every 3 ticks.

Finite element model: The 1-cm square and 2.5-μm thick slab of tissue was simulated as a neo-Hookean material with variable meshing in FEBio (version 2.4.2). The slab was loaded in the x direction with a prescribed traction force. The material properties of each element varied with local collagen fraction (F_{cf}) according to equations (1). After every 7 days of ABM simulation, new material parameters were computed and FE simulations repeated, with strains passed back to the ABM.

$$W = C_1(I_1 = 3)\,,\ C_1 = 2.6\big(1 = 4(F_{cf} = 0.03)/0.27\big). \qquad (1)$$

Registration of ABM grid to FE mesh: Spatial positions within an FE mesh are identified by element coordinates, typically a normalized coordinate system running from 0 to 1 in each direction; the mapping between element coordinates (ζ_1, ζ_2, ζ_3) and physical (i.e., 'real-world') coordinates (x_1,x_2,x_3) is specified by interpolation functions and nodal parameters n_{jk}:

$$x_1 = f_1(\xi_i, u_{jk}),\ x_2 = f_2(\xi_i, u_{jk}),\ x_3 = f_3(\xi_i, u_{jk}). \qquad (2)$$

In general, points that are evenly spaced in physical coordinates will not be evenly spaced in element coordinates (Figure 1A). Thus, when mapping between an ABM grid and a FE mesh, there are two possible approaches. One possibility (employed in this study, see details below) is to establish an ABM grid with GridPoints that are evenly spaced in physical coordinates. Within the ABM, this simplifies calculations of cell migration, chemokine diffusion, and other processes that involve distances; however, when mapping to the FEM it is necessary to invert equations (2) to identify the element coordinates corresponding to each ABM GridPoint. The other option would be to first select an array of points within the FEM, then treat them as GridPoints and account for their (known) nonuniform physical spacing in the ABM calculations.

Case study of FE mesh refinement in coupled model: To demonstrate the ability to establish spatial correspondence between an ABM and FEM with variable grid/mesh densities, we performed a mesh refinement study in our simulated tissue slab while holding ABM grid density constant. We varied the FE mesh (Figure 1B) from 2-by-2 to 50-by-50 elements. For each FE mesh, we established the element coordinates ξ_1, ξ_2 (properties were assumed to be uniform through the thickness of the tissue slab) of each ABM GridPoint and stored the FE element number and coordinates as variables associated with that GridPoint in Repast. We then ran the coupled simulations as described above, exporting collagen amount at each GridPoint to a file that was used to compute new material parameters for each element, then returning strain values at each grid point from the FEM to ABM. Because we used linear elements, in these simulations the material coefficient c1 for each element was based on the average collagen content across all GridPoints within that element, and the strains returned to the ABM were identical at all GridPoints within one element. As a measure of variability in collagen content and strain across the elements in the FE mesh, we calculated the coefficient of variation (CV) by dividing the standard deviation by the mean.

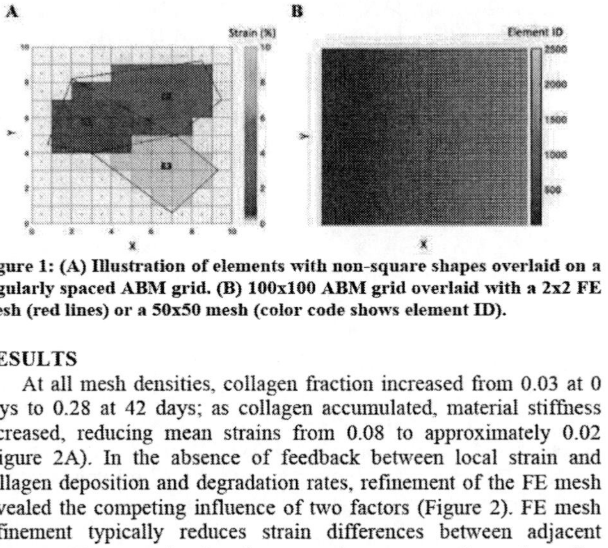

Figure 1: (A) Illustration of elements with non-square shapes overlaid on a regularly spaced ABM grid. (B) 100x100 ABM grid overlaid with a 2x2 FE mesh (red lines) or a 50x50 mesh (color code shows element ID).

RESULTS

At all mesh densities, collagen fraction increased from 0.03 at 0 days to 0.28 at 42 days; as collagen accumulated, material stiffness increased, reducing mean strains from 0.08 to approximately 0.02 (Figure 2A). In the absence of feedback between local strain and collagen deposition and degradation rates, refinement of the FE mesh revealed the competing influence of two factors (Figure 2). FE mesh refinement typically reduces strain differences between adjacent elements. However, the fact that averaging over smaller and smaller regions of the ABM produced spatial heterogeneity in collagen content and material properties dominated the coupled problem, increasing strain differences between adjacent elements as the number of elements increased (Figure 2C,D).

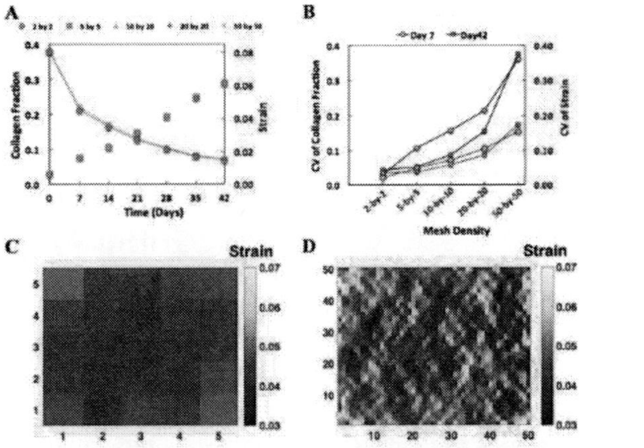

Figure 2: Coupling models with varying FE mesh densities yielded (A) similar overall collagen fraction (left axis) and strain (right axis; blue dash lines) over a time course of 0-42 days but produced higher variability in collagen content and strain (B) and more spatial heterogeneity in strain (C, 5x5 mesh, D 50x50 mesh at Day 7) as the number of elements increased.

DISCUSSION

Here, we introduced a general approach to establishing spatial correspondence between an agent-based model grid and a finite-element model mesh. We demonstrated this approach by performing a mesh refinement study on a simple FE model coupled to an ABM of scar formation. We found that contrary to expectations for a standard FE mesh refinement study, refining the FE mesh actually increased strain differences between adjacent elements in the FEM. We expect this heterogeneity to have significant nonlinear effects in our coupled model once strain is allowed to feed back on collagen deposition and degradation in the ABM. Thus, although the ABM and FEM literature provide methods for selecting grid or mesh densities when using either framework alone, new approaches for determining grid and mesh densities may be required for coupled models.

One of the challenges in mapping between an ABM grid and FE mesh is that points evenly spaced within an element are not necessarily evenly spaced in physical coordinates. We chose to establish an ABM grid with points that are regularly spaced in physical coordinates, then invert the FE interpolation functions to identify corresponding points in the FE mesh. This approach was straightforward and fast when using linear finite-elements, but will be more costly when using nonlinear elements; its cost will further increase in simulations of growth and remodeling where the grid-to-mesh correspondence must be recomputed. In some cases, the converse approach of employing non-uniformly spaced ABM grid points and accounting for their spacing within the ABM may prove more efficient.

ACKNOWLEDGMENTS

This study was supported by NIH grants R01 HL-116449 (JWH, JJL) and U01 HL-120862 (JWH, SPC, LT).

REFERENCES

[1] Rouillard, A et al., *Prog Biophys Mol Biol*, 115(2):235-243, 2014.
[2] Virgilio, K et al., *Interface Focus*, 5(2):20140080, 2015.
[3] Zahedmanesh, H et al., *Biomech Model Mechanobiol*, 11(3):363-377, 2012.
[4] North, M. et al., *Complex Adapt Syst Model*, 1(1):3, 2013.
[5] Rouillard, A et al., *J Physiol*, 590(18):4585-4602, 2012.

REGION-SPECIFIC ORTHOTROPIC GROWTH OF THE PEDIATRIC THORACIC SPINE THROUGH FINITE ELEMENT METHODS

John Dougherty, James Peters, Sriram Balasubramanian

School of Biomedical Engineering, Science and Health Systems
Drexel University
Philadelphia, PA, US

INTRODUCTION

Adolescent Idiopathic Scoliosis (AIS) is a three-dimensional deformity characterized by a progressive lateral curvature and axial rotation of the spine combined with structural changes in the rib cage, and affects five to nine million children ages 10-18 years in the US. Currently, the most common intervention to correct AIS deformity is spinal fusion. However, this method restricts the patient's future growth at the levels of instrumentation. Knowledge of the growth remaining in the spine is critical for predicting progression and surgical planning. Finite element (FE) methods have previously been used to create computational models of the growing spine. However, these models have been limited to using a singular growth strain on only the vertebral bodies for the entire spine. The objectives of this study are to apply region-specific growth strain data to a pediatric FE model of the osteoligamentous thoracic spine.

METHODS

Based on our previous work, region-specific growth strains showed differences not only in the direction of growth, but also with age, gender, and level of the spine (Peters et al. 2015). To apply these strains, an approach was taken similar to Fok et al. 2010 using thermal loads. The growth strains were applied for the various vertebral features such as facets, pedicles etc. A previously validated hexahedral meshed 10 year-old male thoracic spine model was used for this analysis. Elements were manually selected in Hypermesh to comprise each region at each level of the spine. A uniform thermal load was applied on the model using ANSYS 17.0. New thermal expansion coefficient values (α) in the X, Y and Z directions were calculated to account for the orthotropic growth.

Developing new orthotropic and anisotropic thermal growth equations began with Hooke's law for orthotropic materials.

$$\varepsilon_x = \frac{\sigma_x}{E_x} - \frac{v_{xy}\sigma_y}{E_y} - \frac{v_{xz}\sigma_z}{E_z} \tag{1}$$

$$\varepsilon_y = -\frac{v_{xy}\sigma_x}{E_x} + \frac{\sigma_y}{E_y} - \frac{v_{yz}\sigma_z}{E_z} \tag{2}$$

$$\varepsilon_z = -\frac{v_{xz}\sigma_x}{E_x} + \frac{v_{yz}\sigma_y}{E_y} + \frac{\sigma_z}{E_z} \tag{3}$$

where ε is strain, E is the Young's modulus, σ is the stress, and ν is the Poisson's ratio. Previously developed growth equations used thermal loads as a means to create strains.

$$\varepsilon = \dot{\varepsilon}\Delta t \tag{4}$$

$$\varepsilon = \alpha\Delta T \tag{5}$$

where $\dot{\varepsilon}$ is the strain rate, Δt is the time period, α is the thermal expansion coefficient of cortical bone, and ΔT is the thermal load. Combining Eqs. 1-3 with Eqs. 4-5, new orthotropic and anisotropic thermal growth equations were determined.

$$\alpha_x T = \frac{\sigma_x}{E_x} - \frac{\nu_{xy}\sigma_y}{E_y} - \frac{\nu_{xz}\sigma_z}{E_z} \quad (6)$$

$$\alpha_y T = -\frac{\nu_{xy}\sigma_x}{E_x} + \frac{\sigma_y}{E_y} - \frac{\nu_{yz}\sigma_z}{E_z} \quad (7)$$

$$\alpha_z T = -\frac{\nu_{xz}\sigma_x}{E_x} - \frac{\nu_{yz}\sigma_y}{E_y} + \frac{\sigma_z}{E_z} \quad (8)$$

A uniform thermal load was applied on the model using ANSYS 17.0 (ANSYS Inc, PA, USA). Thermal expansion coefficient values (α) in the X, Y and Z directions were calculated to account for

An iterative process was used to compute thermal growth. Each vertebrae was angled differently at each level, and orthotropic properties in ANSYS are uniquely based on each element's orientation, leading to discrepancies even after using the equations. Factor errors were used to scale the α-values, controlling the growth in each direction to improve accuracy. This process was repeated until growth strains (ε) were within 5% of the published literature.

RESULTS

The ε-values from one year of growth in the FE thoracic spine model were compared with the regional normative ε-values computed using linear regression equations for 10 and 11 year-old male subjects from T1-T12 (Table 2). All ε-values have errors below 5%.

Table 2. Growth data for 10 year-old male from the FE model. FE calculated ε-values from one year of growth are in the first row. The percent error with normative ε-values from retrospective CT scans (Peters et al. 2015) are shown in the second row.

	T1	T2	T3	T4	T5	T6	T7	T8	T9	T10	T11	T12
Facets Height	0.0387 0.48%	0.0365 0.11%	0.0352 0.73%	0.0398 1.46%	0.0375 0.21%	0.0399 2.06%	0.0412 2.80%	0.0399 2.45%	0.0412 1.02%	0.0438 0.19%	0.0452 2.44%	0.0458 1.01%
Left Pedicle Height	0.0472 2.88%	0.0479 1.26%	0.0439 0.10	0.0455 0.62%	0.0453 3.19%	0.0455 0.94%	0.0489 4.11%	0.0488 3.43%	0.0452 1.44%	0.0451 1.33%	0.0478 2.56%	0.0483 1.03%
Right Pedicle Height	0.0518 2.68%	0.0496 1.10%	0.0433 0.13%	0.0446 0.31%	0.0465 2.28%	0.0479 0.87%	0.0498 4.08%	0.0491 3.32%	0.0468 1.42%	0.0455 0.97%	0.0511 2.46%	0.0482 0.98%
Left Pedicle Width	0.0384 0.58%	0.0339 0.53%	0.0324 0.17%	0.0302 3.19%	0.0279 2.91%	0.0313 0.06%	0.0347 2.32%	0.0293 0.12%	0.0308 0.02%	0.0325 3.29%	0.0390 0.03%	0.0424 2.06%
Right Pedicle Width	0.0360 1.00%	0.0353 0.36%	0.0300 0.17%	0.0283 3.26%	0.0299 1.02%	0.0328 0.21%	0.0319 1.06%	0.0341 0.53%	0.0317 0.06%	0.0369 2.50%	0.0393 0.12%	0.0439 1.93%
Spinous Process Length	0.0404 3.46%	0.0424 0.55%	0.0407 1.13%	0.0410 1.26%	0.0402 0.15%	0.0386 2.47%	0.0426 1.50%	0.0421 1.94%	0.0424 3.05%	0.0410 3.11%	0.0355 0.71%	0.0380 2.10%
Left Transverse Process Width	0.0451 0.91%	0.0411 1.99%	0.0347 3.38%	0.0373 3.37%	0.0448 0.02%	0.0425 1.15%	0.0414 0.30%	0.0433 0.18%	0.0413 2.27%	0.0382 0.02%	0.0343 0.01%	0.0316 0.91%
Right Transverse Process Width	0.0451 0.91%	0.0411 1.99%	0.0347 3.38%	0.0373 3.37%	0.0448 0.02%	0.0425 1.15%	0.0414 0.30%	0.0433 0.18%	0.0413 2.27%	0.0382 0.02%	0.0343 0.01%	0.0316 0.91%
Vertebral Body Height	0.0643 2.08%	0.0669 0.42%	0.0608 0.48%	0.0603 2.87%	0.0607 2.05%	0.0610 1.92%	0.0615 1.25%	0.0585 1.42%	0.0636 0.11%	0.0629 0.41%	0.0637 0.79%	0.0629 0.53%

DISCUSSION

Using region-specific strain data from retrospective CT scans, new methods were developed to generate comprehensive growth in the thoracic spine through FE analysis. Orthotropic thermal expansion allowed for improved control of the output strains within regions, allowing for a more accurate representation of spine growth. Though this analysis was for a template pediatric FE model over a one-year period of time, this strain data exists for ages 1-19 years, for both males and females. Such a method can be utilized for all patient-specific geometries and can be translated for any FE growth simulation.

Our FE model demonstrates new methods to accurately apply region-specific normative growth, which will be expanded to account for the growth of scoliotic spines and allow clinicians to better understand the spinal curvature progression of AIS with patient-specific models. These advanced methods would aid in evaluating deformity-specific interventions and with preoperative planning for AIS patient-specific correction.

ACKNOWLEDGEMENTS

I would like to thank Prasannaah Hadagali and members of the Drexel Orthopedic Biomechanics Laboratory for their feedback throughout this study. I would also like to thank the Drexel iCARE GAANN program for supporting my work.

REFERENCES

[1] Peters, J.R., et al., *Age-and gender-related changes in pediatric thoracic vertebral morphology*. The Spine Journal, 2015.

[2] Fok, J., S. Adeeb, and J. Carey, *FEM simulation of non-progressive growth from asymmetric loading and vicious cycle theory: scoliosis study proof of concept*. The Open Biomedical Engineering Journal, 2010. **4**: p. 162.

SB[3]C2017
Summer Biomechanics, Bioengineering and Biotransport Conference
June 21 – 24, Tucson, AZ, USA

BIOMECHANICS OF EARLY EMBRYONIC BRAIN MORPHOGENESIS

Hannah Grover (1), Wei Zeng (1), Shicheng Huang (1), Lina Zhang (1, 2), Yan Li (1), Nan Hu (1), Zi Chen (1)

(1) Thayer School of Engineering
Dartmouth College
Hanover, NH, United States

(2) Department of Engineering
Mechanics, Shanghai Jiao Tong University,
Shanghai 200240, China

INTRODUCTION

Morphogenesis is the process by which tissue and organs form[1,2]. The proper development of the neural tube (NT) is an essential part of embryonic morphogenesis. Improper formation of the NT can lead to abnormalities in brain development known as neural tube defects (NTDs) which include: spina bifida, anencephaly, situs inverses, and heterotaxia[1,3,4]. Flexure and torsion of the developing NT are one of the first events that establish the left-right asymmetry in the developing embryo. The biomechanics that drive this left-right asymmetrical event remain unclear[1,4]. The complexity of this process has resulted in the adaptation of computational modeling to biological events to better understand the role of mechanical forces in NT morphogenesis. Modeling of embryonic tissue formation is not a trivial endeavor as soft tissue has been shown to undergo large elastic deformations under physiological loads and contain residual stress and strain[2,5,6]. We employ computational modeling to further elucidate the mechanical forces that are involved in brain morphogenesis.

We focus on the early brain development of chick embryo, which is a popular tool for human embryonic morphogenesis research due to their similarity of development. In chick embryos, flexure and torsion of the NT increase during 45 to 56 hours of cultivation beginning at Hamburger-Hamilton stage 12 (HH12) and concluding at HH16[1,2,7]. Comparatively, in a human embryo these processes do not begin until after day 17[7]. The NT is initially straight at HH12 and proceeds to develop cranial, cervical, and thoracic curvatures as the embryo develops to HH16. There are multiple potential driving factors in the flexure and torsion of the NT during embryonic growth, the alteration of which could illicit improper development[1]. The goal of this research is to identify sources of the mechanical forces that aid in the development of the NT to provide new insight into the roles of mechanical forces behind early brain morphogenesis. The first step in this process is to model the growth of the NT based on experimental data. The results could shed new light on the causes of NTDs in the hope that they can become more preventable or correctable.

METHODS

Optical coherence tomography (OCT) imaging is a non-invasive, non-destructive imaging method which can generate 2D cross-sectional and 3D volumetric images[8] of live embryos. This imaging method is used to probe and measure internal structures of the embryos *in vivo* without harming them. OCT scans were taken of control embryos at HH12-16 to build a control model for computational simulations. The torsional angle of the NT was found by reconstructing the 3D images of the embryos, in ImageJ, and determining where the cranial portion of the NT was located. The perimeter of the NT was marked in ImageJ and an ellipse fitting program in MATLAB was run with the obtained XY coordinates as shown in Fig. 1. The orientation angle of the ellipse was used to assess the torsional angle of the NT.

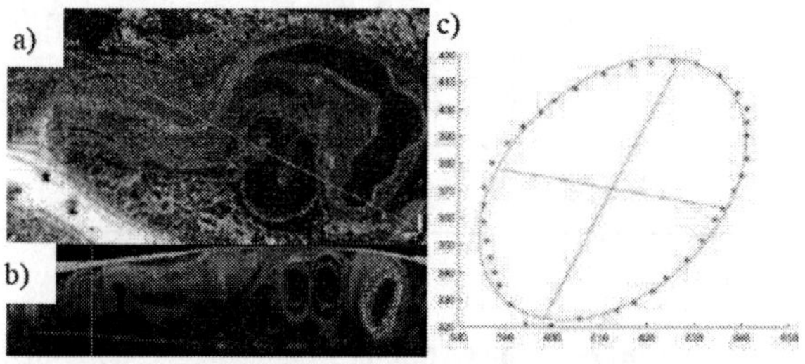

Figure 1. a) XY-Plane OCT image of HH15 embryo. b) YZ-plane OCT image of HH15 Embryo with perimeter of NT marked. c) Ellipse fit to NT markers in MATLAB.

Using the program ImageJ, markers were placed along the cranial, cervical, and thoracic curvatures as shown in Fig. 2. The XY coordinates of these markers were run through a MATLAB program and fit to a circle using the least mean squares minimization method. The radii of these curvatures are used to characterize the flexure of the neural tube.

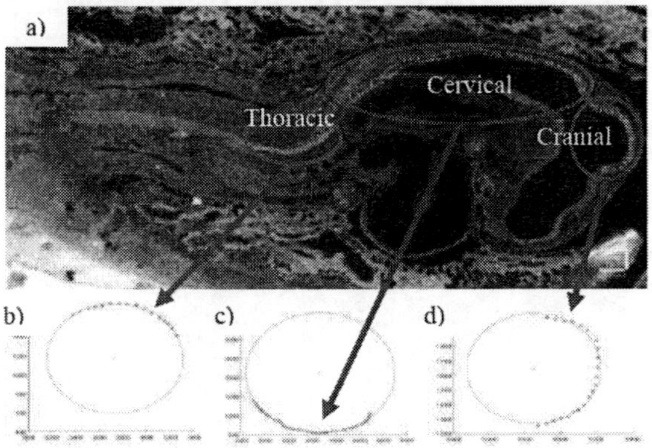

Figure 2. a) Markers placed on OCT image of HH15 embryo. Circle fitting of: b) Thoracic curvature, c) Cervical curvature, d) Thoracic curvature.

To help interpret our experimental results, a 3D finite-element model of an embryo during HH stages 14-17 was constructed using the commercial finite element software, Abaqus (version 6.10) and is shown in Fig. 3. The model for the brain consists of four segments: a straight cylinder representing the posterior part of the spinal cord and three curved segments representing the rest of the spinal cord and the brain ventricles (see Fig. 3a).

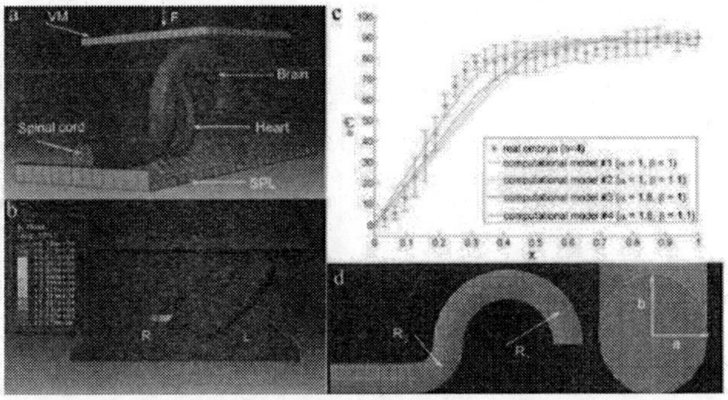

Figure 3. (a) Simplified geometry of a chick embryo at HH14-17. (b) Rightward torsion of brain (head-on view) in an embryo model with a rightward-looped heart. (c) Torsional angle as a function of position in a real embryo and four computational models, which were employed to explore how α and β (both defined in (d)) affect torsion. The position (the X-coordinate) is normalized by the arclength along the embryo. (d) The yellow line indicates the centerline of the embryonic neural tube in computational models (α ≡ R1/R2). β ≡ b/a defines the aspect ratio of the elliptic cross-section.

Moreover, we measured the torsional angle as a function of position to study the effects of the flexure radius ratio ($\alpha = R1/R2$) and the aspect ratio of the cross section ($\beta = b/a$) on torsion (Fig. 3d). In all cases, the angle increases asymptotically approaching nearly 90 degrees (Fig. 3c). The initial slope, however, depends on both the flexure radius and aspect ratios. The slope is smallest when $\alpha = \beta = 1$, and it increases when α increases to 1.8 or β increases to 1.1. When $\alpha = 1.8$ and the aspect ratio $\beta = 1.1$, as is the case in a real chick embryo at HH stage 16, the slope becomes largest and the torsional angles agree well with those in the real embryo. These results are also consistent with physical intuition. When β increases, the ratio between torsional rigidity and flexural rigidity decreseases as shown by equation (1).

$$\gamma = \frac{GJ}{EI} \sim \frac{1}{(1+\beta^2)} \tag{1}$$

As a result, it becomes easier for the brain to twist, and hence the slope is expected to increase. On the other hand, when the flexure radius ratio increases, twisting can occur over a relatively shorter range in the spinal cord and the hind brain, also resulting in an increase of the slope.

Using experimentally gathered curvatures as inputs, this previous model will be updated to more accurately reflect embryonic growth. Further, the model will be used to verify whether the degree of brain torsion during HH12-16 reflects that seen in embryos.

RESULTS

Torsional angle data was gathered for: HH12, n=4; HH13, n=5; HH14, n=7; HH15, n =7; and HH16, n=4. The results are graphed in Fig. 4a. From HH12 to HH14 there is a change of about 30 degrees but the rate of change gradually decreases during development.

Cranial, cervical, and thoracic curvature data has been gathered for: HH13, n=6; HH14, n=7; HH15, n =7; and HH16, n=4. Fig. 4b shows the average radius of curvature at each stage. HH12 was neglected as the NT is nearly straight.

DISCUSSION

The results of the research so far indicate that: 1) The cranial and thoracic curvatures both change during growth, i.e. increase and decrease respectively and 2) the cervical curvature remains mostly unchanged during growth. This is significant as the previous model, which used estimates and a combined cranial/cervical curvature, may not accurately reflect the changes in radius of curvature that occurs during embryonic growth. In the future, modeling of these earlier embryonic stages may prove beneficial to further understanding the process of growth.

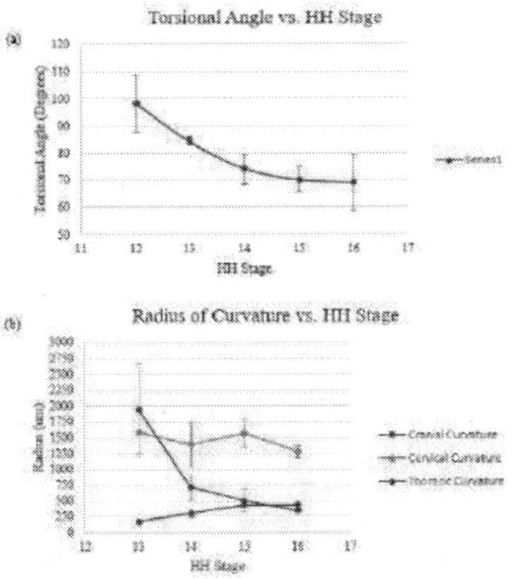

Figure 4. (a) Average torsional Angle for HH12-16. (b) Average cranial, cervical, and thoracic curvatures for HH13-16.

ACKNOWLEDGEMENTS

Z.C. acknowledges the support by the Society in Science-Branco Weiss Fellowship, administered by ETH Zürich.

REFERENCES

[1] Chen, Z. et al (2016). *J. R. Soc. Interface*, 13(124). **[2]** Ramasubramanian, A., & Taber, L. A. (2008). *Biomech Model Mechanobiol* 7(2), 77–91. **[3]** Greene, N. D. E., & Copp, A. J. (2009). *Prenatal Diagnosis*, 29(4), 303–311. **[4]** Wyczalkowski, M. A. et al. (2012). *Birth Defects Res C*, 96(2), 132–152. **[5]** Rodriguez, E. K. et al. (1994). *J Biomech*, 27(4), 455–467. **[6]** Taber et al., 1993. *Circ. Res* 72(2):455-462. **[7]** Taber, L. A. (2006). Int. J. Dev. Biol., 50, 323–332. **[8]** Shi, Y. et al. (2014). *Front Physiol*, 5(August), 1–14.

VASCULAR GROWTH AND REMODELING WITH STOCHASTIC OPTIMAL STRESS-DRIVEN FIBER DEPOSITION

Jiacheng Wu and Shawn C. Shadden

Department of Mechanical Engineering
University of California, Berkeley
Berkeley, CA, USA

INTRODUCTION

The progression of cardiovascular disease is closely related to the functional adaptions of arteries to mechanical stimuli. To better understand this adaptive behavior, a mathematical framework of vascular growth and remodeling (G&R) was proposed by Humphrey *et al.* based on a constrained mixture theory [1] in which a vessel is assumed to be able to adapt to mechanical stress to recover a homeostatic stress state. This framework was first applied to idealized geometries to simulate aneurysm expansion and later extended to 3D geometries, and has been coupled with computational fluid dynamics to incorporate the influence of the evolving hemodynamics.

In the vascular G&R, the vessel wall is modeled as fiber-reinforced material with collagen (and potentially smooth muscle) treated as fiber and elastin is modeled as ground matrix. The angle defining the collagen fiber direction is very important to the overall mechanical property of the vascular mixture. In prior works, newly produced collagen fiber is only allowed to be deposited in predefined directions, while in reality the fibers are not so rigidly constrained. Indeed, it is known that fibers within one collagen fiber family are not precisely aligned, but rather fiber dispersion is observed, and this dispersion can be important for accurately characterizing vessel wall mechanics [2]. In the work herein, we implement a vascular G&R computational framework in which newly produced collagen fibers can be deposited more freely based on the solution of an optimization problem inspired by [3]. The directions depend on the ratio of the two largest principal stresses, and thus allow fiber deposition to adapt to the stress state. In addition, fiber dispersion is also controlled by a stress-dependent stochasticity to account for sources of dispersion not explicitly modeled.

METHODS

Constrained mixture theory of growth and remodeling

The vessel wall is modeled as a constrained mixture made of two constituents: elastin and collagen, in a 3D continuum geometry. Smooth muscle fibers are ignored to focus on the passive mechanical stress generated by vessel wall. The kinetics of G&R is characterized by the mass density evolution equations for elastin and collagen fibers:

$$M_e(t) = M_e(0)Q_e(t) \quad \text{(elastin)}\,, \tag{1}$$

$$M_i(t) = M_i(0)Q_i(t) + \int_0^t m_i(\tau)q_i(t-\tau)d\tau, \quad i=1,2 \quad \text{(collagen)}\,, \tag{2}$$

where i is the index for two collagen fiber families oriented in two helical directions. $m_i(\tau)$ is the mass production rate for collagen fiber family i at time τ. Functional elastin is thought to be only produced in early development, therefore there is no second term to take account production new elastin. $Q_e(t)$, $Q_i(t)$ and $q_i(t)$ are functions characterizing the natural decay of elastin and collagen.

The constitutive relation for collagen is assumed to follow Fung-type exponential form, and the strain-energy density function is given by

$$W(\mathbf{C}^i_{n(\tau)}(t)) = \frac{c_2}{4c_3}\left\{\exp\left[c_3\left(I^i_{n(\tau)}(t)-1\right)^2\right]-1\right\} \tag{3}$$

where $I^i_{n(\tau)}(t)$ is the fiber invariant in the direction of collagen family i. On the other hand, elastin is modeled using Neo-Hookean material,

$$W_e(\mathbf{C}(t)) = \frac{c_1}{2}\left(I_1(t)-1\right)\,, \tag{4}$$

where I_1 is the first invariant of the right Cauchy-Green deformation tensor. The overall strain-energy function of the vessel wall mixture is modeled based on the Holzapfel model [6] for nearly incompressible material

$$\begin{aligned}\Psi(\mathbf{C}) =& U(J) + M_e(t)W_e(\mathbf{C}) \\ &+ \sum_{i=1,2} M_i(0)Q_i(t)W(\mathbf{C}^i_{n(0)}(t)) \\ &+ \int_0^t m_i(\tau)q_i(t-\tau)W(\mathbf{C}^i_{n(\tau)}(t))d\tau\,,\end{aligned} \tag{5}$$

where $U(J)$ is a penalty term ensures nearly incompressibility. At each G&R step, the solid mechanics problem is solved using finite element method to obtain local stress.

Optimal fiber deposition

In the current work, collagen fiber is deposited based on the local stress tensor, instead of being restricted to a pre-defined collagen fiber family direction. The optimal collagen fiber deposition angle is given by the solution from the following optimization problem

$$\begin{aligned}&\min_{M,\theta_f} M \\ &s.t. \\ &\sigma_f M\pi\cos^2\theta_f \geq T_{11} \\ &\sigma_f M\pi\sin^2\theta_f \geq T_{22}\end{aligned} \tag{6}$$

where T_{11} and T_{22} are the two largest principal values of the stress tensor T, and usually these two directions correspond to local circumferential

Poster Presentation #P192

and axial directions of blood vessel. σ_f is the stress a fiber can provide in normal condition. M is the total mass density of the collagen fibers. θ_f is the direction of collagen fiber with respect to circumferential direction. The idea behind this optimization problem is that our body want to minimize the mass needed for the fibers while still sustaining the force in both circumferential and axial directions. The solution for (6) is

$$M^* = \frac{T_{11} + T_{22}}{\sigma_f \pi}, \quad \theta^* = \arctan\sqrt{\frac{T_{22}}{T_{11}}}, \tag{7}$$

which defines the direction for newly produced collagen fiber.

The mass production rate of collagen fiber at time t is

$$m_i(t) = \frac{M(t)}{M(0)}\left(K_\sigma(\sigma_i(t) - \sigma^h) + f_h\right) \tag{8}$$

where K_σ is a feedback gain for stress deviation. f_h is the baseline value for collagen mass production rate. $\sigma_i(t)$ is the stress in the nominal direction $e_{fi}(t)$ of collagen fiber family i

$$\sigma_i(t) = e_{fi}(t) \cdot T(t) e_{fi}(t) \tag{9}$$

where the nominal direction $e_{fi}(t)$ is defined as

$$e_{fi}(t) = \frac{\int_0^t m_i(\tau) q(t-\tau) e_i(\tau) d\tau}{M_i(t)} \tag{10}$$

which is the average collagen fiber direction over the collagen family i weighted by mass.

While the above framework leads to some fiber dispersion, additional dispersion must be added to account for other influences that are not explicitly modeled. This is accomplished by imposing a stochastic deposition angle, which is assumed to follow a normal distribution

$$\theta \sim \mathcal{N}(\theta_f, \Sigma(\sigma_i))$$

where the mean deposition angle θ_f is set to be the optimal deposition angle defined in (7). The standard deviation $\Sigma(\sigma_i)$ is defined as a function of the fiber stress σ_i

$$\Sigma(\sigma_i) = \Sigma_0 \frac{\sigma_{scale}}{\sigma_i} \tag{11}$$

where σ_{scale} is a scale parameter for vessel wall stress and Σ_0 is the nominal value for the standard deviation. $\Sigma(\sigma_i)$ a decreasing function with respect to σ_i, which is consistent with the fact that fibers are aligned more coherently (less dispersion) when the stress σ_i is higher [4].

RESULTS

Numerical simulations are carried out in an idealized cylindrical geometry with radius of 10 mm and thickness of 1mm, which represents the healthly human abdominal aorta. The pressure imposed is set to 13332 Pa representing physiological blood pressure. For the parameters for stochastic fiber deposition, Σ_0 is set to be 0.1 and σ_{scale} is set to be 200 kPa to represent physiological level of stress in human aorta.

The results show that collagen fiber dispersion can be reproduced by combining constrained mixture theory of G&R with stochastic optimal fiber deposition. Figure 1 shows the fiber dispersion patterns in the two helical directions and the vessel wall stress distribution generated by the proposed framework. It can be noticed that the generated fiber dispersion is smaller in the inner layer and larger in the outer layer (see Figure 2). This is because higher stress in the inner layer yields smaller standard deviation Σ for stochastic fiber deposition, and thus the actual deposition angles are more concentrated around the mean deposition angle given by θ_f. This fact matches with the experimental observations in [5] that dispersion is larger in the adventitia (outer layer) than in the media (inner layer).

REFERENCES

[1] Humphrey, J et al., Math Models Methods Appl. Sci. 12(03):407-430, 2002.

[2] Gasser, T et al., J R Soc. Interface, 3(6), pp.15-35., 2006.

[3] Cyron, C et al., Math Mech Solids, 20(6), pp.680-696., 2015.

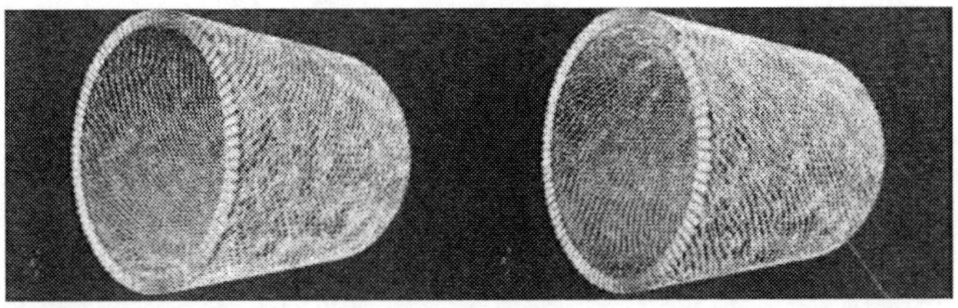

Figure 1: Collagen fiber distribution in two helical directions. Color indicates the magnitude of stress: high stress is red, low stress is blue.

Figure 2: Fiber dispersion in inner layer (left) and outer layer (right) of vessel wall

[4] Finlay, H et al., J Vasc Res, 32(5), pp.301-312., 1995.

[5] Niestrawska, J et al., J R Soc. Interface, 13(124), p.20160620. 2016

[6] Holzapfel, G et al., J Elasticity, 61(1-3), pp.1-48. 2000

SB3C2017
Summer Biomechanics, Bioengineering and Biotransport Conference
June 21 – 24, Tucson, AZ, USA

THE EFFECTS OF ALIGNMENT AND MISALIGNMENT OF AUTOGRAFTS IN THE REPAIR OF VOLUMETRIC MUSCLE LOSS INJURIES

J. Kim (1), B. Kasukonis (1), T. Washington (2), J. Wolchok (1)

(1) Department of Biomedical Engineering
University of Arkansas
Fayetteville, AR, United States

(2) Department of Exercise Science
University of Arkansas
Fayetteville, AR, United States

INTRODUCTION

When provided the appropriate cues, muscle has a healthy capacity for self-repair. Following mild muscle damage (ex. strains and contusions) myocytes and progenitor (satellite) cells are typically damaged but the underlying extracellular matrix (ECM) is largely intact. The regenerative response to mild injury is robust and surgical intervention is rarely warranted. However, when significant muscle volume is lost (trauma, infection, or surgical resection) both the cells *and* the underlying ECM are lost. Termed volumetric muscle loss (VML), the bulk loss (>20%) of muscle tissue overwhelms the capacity for regeneration.

The differential response to mild versus severe muscle injury suggests a vital role for ECM during muscle regeneration. We hypothesize that the preservation of ECM following mild muscle injury allows force and alignment cues to work in unison during regeneration, which promotes progenitor cell myogenesis. Alternatively, the loss of ECM organization following VML injury leads to cue (force and alignment) incongruence, which inhibits myogenesis. This study was designed to test the hypothesis that *congruent mechanical and architectural cues (strain applied in the direction of alignment) trigger a pro-myogenic response*. The following abstract describes both *in-vitro* and *in-vivo* experiments designed to explore this hypothesis.

METHODS

In-vivo Muscle Regeneration

Muscle regeneration experiments were performed on 12-week old male Fischer 344 rats (Envigo, Indianapolis, IN) in accordance with University of Arkansas IACUC protocols (#14044) and guided by published studies [1]. The tibialis anterior (TA) muscle was identified and a VML defect (size = 20% of TA muscle mass) was created using a biopsy punch (d=8mm) inserted to a depth of 3mm (Figure 1A). The autograft muscle plug was weighed and immediately re-implanted back into the defect in its native orientation (aligned; rotation=0°) or rotated either 45° or 90° with respect to the surrounding tissue (unaligned). Muscle plug autografts were sutured to the surrounding muscle using a series of interrupted stitches (6-0 Prolene, Ethicon, MA). The untreated contra-lateral limb served as a positive control. All animals were housed for either a 2 (n=6/group) or 12-week (n=10/group) recovery period. At the completion of the recovery period, peak contractile force, mass, repair site gene expression, and repair site tissue histology was assessed using techniques familiar to our group [2].

Figure 1: Muscle plugs were created with a biopsy punch (A) and re-implanted into the defect aligned to the surrounding muscle or rotated either 45° or 90°. A bioreactor (B) will be used to apply cyclic strain (mechanical cues) to microcontact printed lanes of myoblast cells (B: inset).

In-vitro Myoblast Mechanobiology

Microcontact printing was utilized to create elastic substrates patterned with fibronectin lanes (100μm wide). Fibronectin lanes were

Poster Presentation #P193

seeded with muscle myoblasts cells (C2C12, ATCC) to create aligned muscle mimetic cell patterns for *in-vitro* testing. C2C12 cells were maintained in static culture for a 72-hour attachment period. At the completion of the attachment period, mechanical cues, in the form of cyclic strain conditioning (1Hz, 10%, 4 hrs/day, 7 days), were applied to myoblast cell samples with the aid of a custom built bioreactor. Cells seeded lanes were oriented at 0°, 45°, or 90° with respect to the direction of stretch in order to mimic the conditions of the *in-vivo* study. Unaligned and unconditioned cells were maintained as comparative controls.

At the completion of the 7-day conditioning period, treated and control samples were fixed (10% formalin) and immuno-stained for the presence of myosin heavy chain (MHC), a muscle specific cytoskeletal element, and counterstained with the nuclear dye DAPI. For each sample, myogenesis was calculated as the percentage of MHC positive cells relative to total nuclei count (termed the fusion index) using a published protocol [3]. The average fusion index was calculated for all treated (0°, 45°, 90°) and control samples (n=4-5/sample group).

RESULTS

In-vivo Myogenesis Experiments: The repair of VML injuries using aligned implants significantly increased recovery of peak contractile force (77±11% of normal contralateral muscle) when compared to defect repair with either 45° (p=0.03) or 90° (p=0.005) rotated implants (65±10% and 61±11% respectively). Similar improvements were measured for restoration of muscle mass in response to repair with aligned implants (91±8% of normal contralateral) when compared to both unaligned groups (88±5% and 84±7% respectively). Histological findings revealed a notable decrease in collagen dense repair tissue formation at the implant site for aligned samples. Alignment significantly increased the expression of both Pax7 (satellite cells) and MyoD (myogenesis) within tissue harvested at the implant site, when compared to 45° and 90° rotated samples.

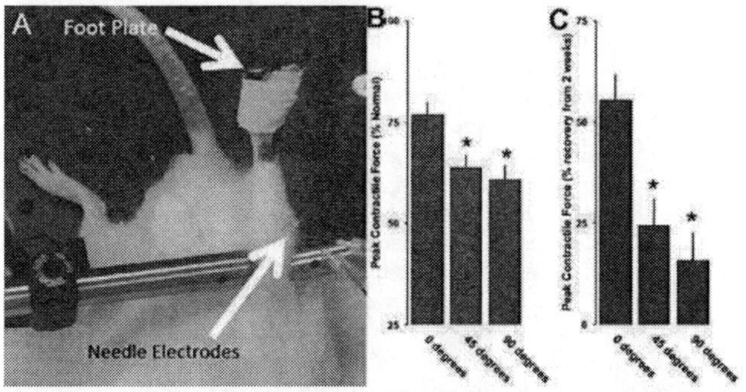

Figure 1: Peak tetanic TA contractions were generated via stimulation of the peroneal nerve and measured with a force plate (A). TA muscles repaired with aligned implants were significantly stronger than either 45° or 90° rotated implants. Muscle strength recovery, measured as the percent increase from 2 weeks, was significantly improved in the aligned group when compared to rotated implant repair (C). * = p<0.05 compared to 0 degrees.

In-vitro Myogenesis Experiments: The application of cyclic strain (10%) in the direction of alignment (0°) significantly increased myoblast myogenesis when compared to the unaligned groups. The average fusion index increased from a low of 3.7±3.1% for the unaligned/unstrained control group, to 18.4±8.8%, 13.4±7.4%, and 8.5±5.1% for the 0°, 45°, and 90° groups respectively. The increase in the fusion index was significant for both the 0° and 45° treatment groups. Although elevated, the difference between controls and 90° were not significant. Additionally, the 0° samples exhibited a higher fusion index compared to 90° samples.

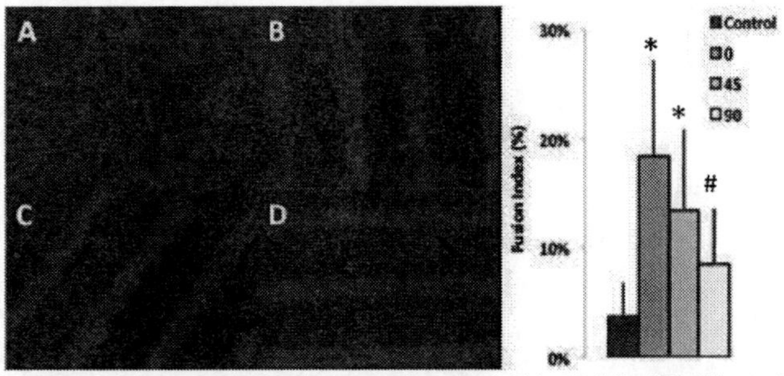

Figure 2: Representative images of control (A), 0° (B), 45° (C), and 90° (D) samples immune-stained with MHC (red) and DAPI (blue) Myogenesis (as measured via the fusion index) was elevated when strain was applied in the direction of cell alignment and decreased when mechanical and alignment cues were incongruent. * = p<0.05 compared to controls. # = p<0.05 compared to 0 degrees.

DISCUSSION

The results from the combined *in-vitro* and *in-vivo* studies provided strong evidence in support of the central hypothesis, and suggest that *congruent mechanical and architectural cues do trigger a pro-myogenic response*. Although there are previous studies that have investigated the response to stretch by myoblasts and how downstream events such as gene expression and mechanotransduction pathways are affected, there still remains a gap in knowledge when it comes to the effects of misalignment on myogenesis. We believe this study is the first to explore the role of cue misalignment on muscle myogenesis. These findings are significant clinically when one considers that the importance of alignment (scaffold architecture) and mechanical stimulation (rehabilitation protocols) as part of a VML repair strategy is still unsettled. These findings suggest that scaffolds that recapitulate the native ECM architecture when combined with a postoperative rehabilitation program could be utilized to enhance muscle regeneration in a clinical setting.

How cross-talk between force and alignment cues might promote or inhibit myogenesis is unknown, however, we know that the oriented arrangement of cytoskeletal elements in the direction of cell alignment has been shown to increase longitudinal stiffness and reduce transverse stiffness. Consequently, strains applied to cells incongruent to alignment would lead to the overloading of transverse cytoskeletal elements triggering the activation of strain sensitive focal adhesion kinase (FAK) mechanotransduction pathways, including p38 signaling. Activation of the p38 mechanotransduction pathway has been shown to increase the synthesis of the cytokine transforming growth factor beta 1 (TGFβ1), which is a known inhibitor of myogenesis.

ACKNOWLEDGEMENTS

This research was supported by the National Institute of Arthritis and Musculoskeletal and Skin Diseases of the NIH under Award Number R15AR064481 as well as the Arkansas Biosciences Institute.

REFERENCES

[1] Walters, T. J., *Biores Open Access*, 1(6):280-290, 2012.
[2] Wolchok, J., *Tissue Eng Part A*, 22(19-20):1151-1163, 2016.
[3] Elvassore, N., *Langmuir*, 28(5):2718-2726, 2012.

SB³C2017
Summer Biomechanics, Bioengineering and Biotransport Conference
June 21 – 24, Tucson, AZ, USA

A BIOMECHANICAL COMPARISON OF TWO METHODS OF SCAPULAR NECK FRACTURE FIXATION

Hema J. Sulkar (1,2), Robert Z. Tashjian (2), Heath B. Henninger (1,2)

(1) Department of Bioengineering
University of Utah
Salt Lake City, UT, USA

(2) Department of Orthopaedics
University of Utah
Salt Lake City, UT, USA

INTRODUCTION

Traumatic injuries to the shoulder can result in fractures of the glenoid and scapula. While scapular fractures constitute only 1% of all fractures and 5% of all shoulder fractures,[1] and they may be treated non-operatively, unstable scapular neck fractures are often indicated for operative treatment. Scapular neck fractures that involve greater than 1 cm of bony displacement or 45° of fragment misalignment are recommended for surgical fixation. Plating restores the anatomical relationship of the functionally imbalanced glenohumeral joint.[2] Favorable clinical outcomes following open reduction internal fixation of scapular neck fractures have been reported.[2-4]

The standard for scapular neck fracture fixation is a single plate extending along the posterior surface of the lateral scapula border up to engage the glenoid fragment. The lateral scapular border is the thickest segment of the scapular body, and has traditionally been used for fixation.[5] Typically, a reconstruction plate is affixed with at least 3 screws in the lateral border and 3 screws engaging the glenoid vault. From a surgical standpoint there are several problems with the single plate construct. First, the screws in the lateral border are short (<10 mm). Consequently, the inferior fixation can be limited, especially in patients with poor bone quality. Second, it can be difficult to position the 3 screws through the plate into the glenoid. This is especially pronounced in smaller patients. Placement of a plate may compromise the suprascapular nerve and artery via over-retraction of soft tissues required to place the plate. Finally, the limited Judet approach commonly used for fixation splits the infraspinatus/teres minor interval and it can be difficult to extend the plate inferiorly along the lateral scapula without excessive tissue retraction or muscle stripping.

An alternative fixation method was developed to improve biomechanical stability with smaller amounts of soft dissection and less risk for neurovascular injury. Two short plates are placed in parallel, extending from the glenoid directly medial onto the scapula. Because the plates are small, 4 screws are easily placed into the glenoid fragment. Because the plates extend directly medial from the glenoid there is still significant bone for screw purchase (~14-16 mm) before the scapula becomes too thin. Now using the modified Judet approach, no significant stripping or retraction is required which minimizes soft tissue and neurovascular compromise. Consequently, the limitations of the single plate/lateral border technique are avoided.

Very limited data exists regarding scapula neck fracture fixation. Most studies are the result of small clinical series evaluating surgically treated patients.[6, 7] Biomechanical testing has only been performed on the canine scapula.[8] No biomechanical studies have evaluated the initial fixation strength of glenoid neck or scapula fracture constructs in the human shoulder. The purpose of this study was to compare the initial biomechanical strength of a glenoid neck fracture repaired with a single lateral border plate (standard technique) and a new construct with two smaller medially oriented parallel plates.

METHODS

Composite human scapulae (Sawbones, Pacific Research Laboratories, Vashon, WA) were used to eliminate bone quality and bone shape as confounding variables. A 2.5 mm wide scapular neck fracture was created in each specimen with a cutting jig to ensure a repeatable size and location of the defect (Fig 1). The osteotomy along the anatomical neck extended from the lateral scapula border proximally through the scapular body at the base of the spine (spinoglenoid notch) up to the base of the coracoid process. The osteotomy was not extended completely to the superior border in order to maintain the gap and apply consistent loading on the fixations. This also preserved the orientation of the glenoid in relation to the scapula.

Poster Presentation #P194

The 1-plate repairs were fixed with a 3.5 mm 6-hole recon plate, from the lateral border extending superiorly across the posterior glenoid (Fig. 1, left). Three bicortical screws were placed in the scapula, and three in the glenoid vault. The 2-plate repairs were fixed with two 4-hole plates (superior 2.4 mm, inferior 2.7 mm) each with two bicortical screws in the scapula and two in the vault (Fig. 1, right). Plates and screws were obtained from Depuy Synthes (Warsaw, IN).

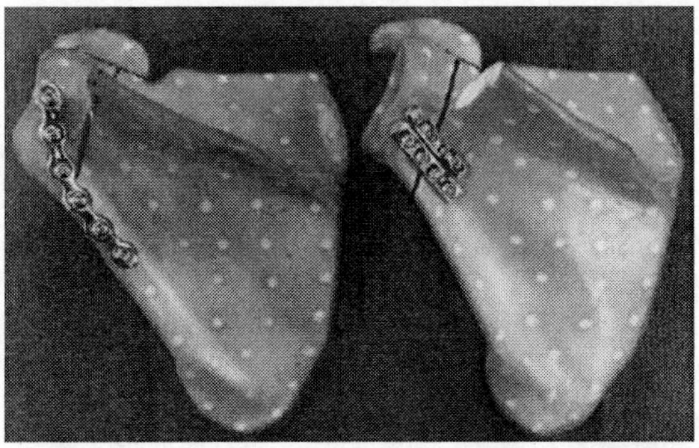

Figure 1. Sample specimens with 1-Plate (left) and 2-Plate (right) constructs securing a fracture gap. Note that the acromion has been removed in order to make the fracture cut.

Eight pairs of scapulae were tested. Seven pairs underwent fatigue followed by failure and 2 scapulae were tested without any fixation to baseline the strength of the bone bridge at the base of the coracoid. A uniaxial servo-hydraulic materials testing machine (Instron 1331 Load Frame, Model 8800 controller, Instron Corp., Norwood, MA, USA) equipped with a 5 kN tension/compression load cell (Model 2518-103, Instron Corp., Norwood, MA, USA) applied loads directly on the glenoid surface. A 40 mm diameter steel ball replicated the contact of the humeral head centered on the glenoid. Optotrak Certus rigid bodies (NDI, Waterloo, ON, Canada) were affixed to the specimens to track displacement of the glenoid relative to the scapula. For the fatigue testing, specimens were loaded from 20–150 N for 1000 cycles at 1 Hz. The 150 N load was determined from pilot failure tests to ensure fatigue testing occurred on the elastic portion of the load-displacement response of the construct. For the failure, a 10 mm displacement was applied at 1 mm/s. Forces were compared at a 1 mm gap closure point in the failure test. This allowed evaluation at a point prior to complete fracture gap closing and bone-on-bone loading, which might mask the contributions of the plates. Statistical analysis was performed using unpaired t-tests with significance at $P \leq 0.050$. Data are presented as mean ± SD.

RESULTS

Fatigue loading over 1000 cycles showed higher displacement for the 2-Plate construct compared to the 1-Plate construct at 100 and 200 cycles ($P \leq 0.038$), but both constructs stabilized by 100 cycles and further micromotion of the bone segments was not detected (Fig. 2). No differences were detected between cycles within a group (1-Plate: $P \geq 0.757$, 2-Plate: $P \geq 0.318$).

The compressive loads upon reaching 1 mm of gap closure were comparable for the two constructs (Fig. 3) ($P = 0.930$), but each construct was significantly stronger than the specimens without any fixation ($P \leq 0.001$). The stiffness values measured at the 1 mm gap reduction point in the failure tests were 436 ± 116, 355 ± 67 and 174 ± 13 N/mm for the 1-Plate, 2-Plate and no-construct tests, respectively. Significant differences were detected between each of the constructs and the no-construct scapulae (both $P \leq 0.008$) but not between the two constructs ($P = 0.164$).

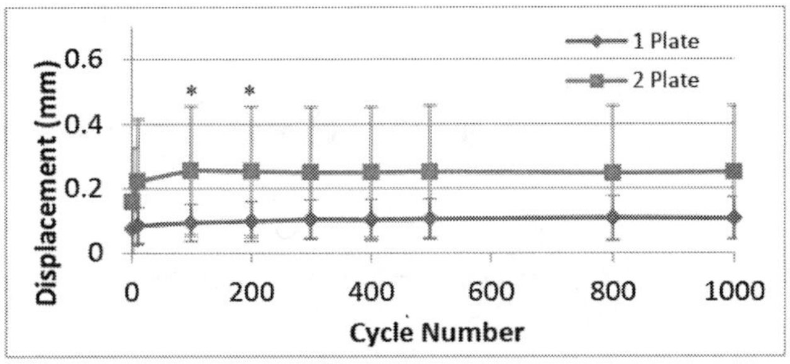

Figure 2. Displacement of the glenoid relative to the scapula for 1000 cycles of fatigue testing loaded from 20-150 N. (*P ≤ 0.038). Mean ± SD

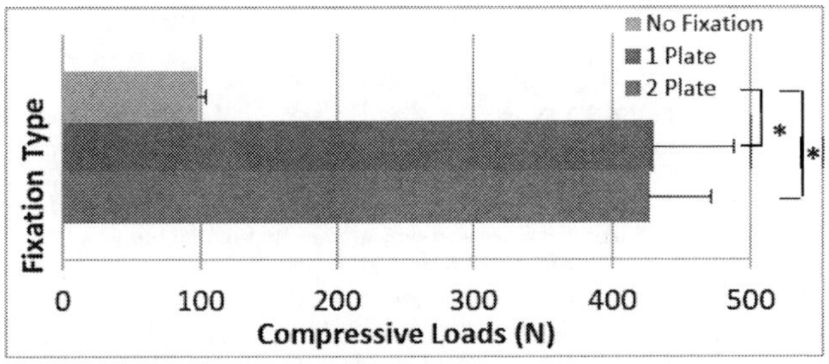

Figure 3. Compressive loads sustained by the fixed specimens at 1 mm fracture gap reduction (*P ≤ 0.001). Mean ± SD

DISCUSSION

The fatigue and failure testing lacked statistical difference between the two fixation types. Displacement ($\Delta < 0.2$ mm), stiffness ($\Delta < 80$ N/mm) and compressive loads ($\Delta < 10$ N) confirm that both constructs performed comparably. The compressive loads sustained by specimens fixed constructs were, as expected, significantly greater than the no-construct specimens. This showed that the fixations were loaded significantly and the bone bridge only served to stabilize alignment of the repairs.

The experimental model used a 2.5 mm fracture gap which is a deviation from clinical practice where the surgeon would reduce the fracture and approximate the bone fragments to allow for healing. However, the gap was required so that test loads were distributed over the construct fixation, rather than a bone-on-bone interface. Although the Sawbones replicate human bone with a hard cortical shell and foam core approximating cancellous bone, in the future, these results should be validated using human cadaver specimens to capture the natural variation in shape and mechanical properties of human bone.

In conclusion, the 1- and 2-plate fixation techniques performed comparably, but the 2-plate construct has favorable advantages in placement and surgical technique.

ACKNOWLEDGEMENTS

Funding from the University of Utah Faculty Research and Creative Grant mechanism is gratefully acknowledged.

REFERENCES

[1] Goss, *JAAOS*, 3, 22–33 (1995). [2] Hardegger, *JBJS Br*, 66, 725–31 (1984). [3] Khallaf, *Med Princ Pract*,15, 443–8 (2006). [4] Schroder, *JBJS*, 1623–30 (2016). [5] Burke, *JSES*, 15, 645–8 (2006). [6] Bahk, *JBJS, Am* Oct 2009; 91(10):2492-2510. [7] Jones, *JOT*, 23, 558–64 (2009). [8] Acquaviva, *JAAHA*, 48, 372–8 (2012).

SB³C2017
Summer Biomechanics, Bioengineering and Biotransport Conference
June 21 – 24, Tucson, AZ, USA

DEVELOPMENT AND APPLICATION OF A SIX-YEAR-OLD CHILD PEDESTRIAN FINITE ELEMENT MODEL

Haiyan. Li(1), Wenle. Lv(1), Shihai. Cui(1), Lijuan. He(1), Shijie. Ruan(1), Chunxiang. Wang(2)

(1) College of Mechanical Engineering
Tianjin University of Science and Technology
Tianjin, China

(2) Department of Radiology
Tianjin Children's Hospital
Tianjin, China

INTRODUCTION

Finite element (FE) models of the human body are becoming increasing popular and are used to study the impact biomechanics and injury analysis during vehicle crashes and sports for recently two decades. Many developed models are devoted to body regions such as the head, neck, thorax, abdomen, and upper and lower extremities, but the whole body FE models are rare and have many limitations. In recent year the Global Human Body Models Consortium and Toyota Central R&D Labs developed the GHBMC 50th percentile male seated occupant model [1] and the Total Human Model (THUMS) [2] to understand the occupant injury mechanisms in pre-crash and during crash.

The children are not small adults has been known for decades and has been considered during the development of current restraints that protect the child in the crash phase [3]. They are important for future research and much research is currently ongoing because the age-specific anatomical changes should be captured in research. Mizuno et al [4] developed the first 3-year old child human body FE model from a 50th percentile human body model using scaling techniques. The Wayne State University (WSU) 10-year old FE model was developed and was used to analysis the related traffic injury[5]. A whole body FE model of 6-year old child is urgent to develop based on the age anatomic structure.

In this study, a whole body finite element model of 6-year-old pediatric pedestrian was developed with detailed anatomical characterization based on the CT images. Also, the model was simulated the traffic accident to study its injury mechanism in vehicle-pedestrian collisions.

METHODS

6-year-old child pedestrian finite element model development

The model development procedures are briefly outlined in Fig. 1. The geometric data of the proposed six-year-old child pedestrian finite element model was based on CT scan images. The FE model of the full body of 6-year-old pediatric pedestrian was completed by integrating all the component models through constructing the skin, soft tissues and joint ligaments, according to CT images of a 6-year-old child who has a height of 113.5cm and weighs 20kg.

6-year-old child pedestrian finite element model validation

Each component-level (body part) of the 6-year-old pedestrian finite element model was validated against the test data reported in the literature. Table 1 summarizes these validation studies, including the body part, loading condition, basic information of sample and references documenting, model development and validation.

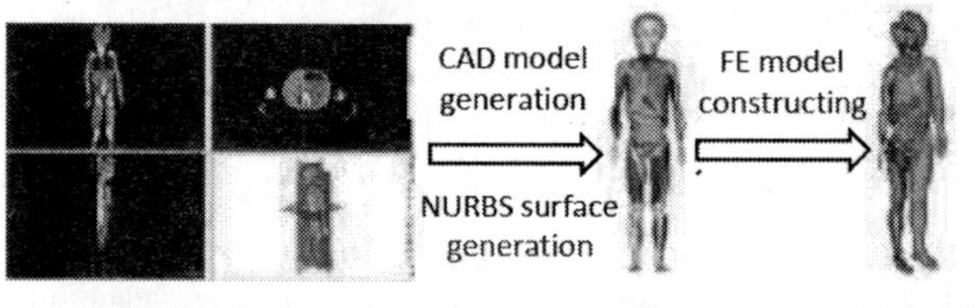

Figure 1: 6-year-old child pedestrian finite element model development procedures.

Table 1: Summary of the component-level model validation

Part	Loading condition	Sample experiment	Reference	Model development and validation
Head	static compression	child cadaver	Prange et al.(2004)	Cui et al(2015)
	impact experiment	adult cadaver	Nahum et al.(1977)	
Neck	dynamic tension of vertebral segments	child cadaver	Luck et al. (2013)	Lv et al.(2016)
	tension of full cervical vertebra	child cadaver	Ouyang et al.(2005)	
	low speed impact	child volunteer	Arbogast et al. (2009)	
Thorax	front impact	child cadaver	Ouyang et al. (2006)	Lv et al.(2016)
	lateral impact	adult cadaver	Shaw et al. (2006)	Lv et al.(2016)
Abdomen	front impact	child cadaver	Ouyang et al. (2015)	Lv et al.(2016)
	lateral impact	adult cadaver	Shaw et al. (1989)	Lv et al.(2016)
Lower Extremities	Shear/bending test	adult cadaver	Kajzer et al.(1999)	Li Bei(2013)

6-year-old child pedestrian finite element model application

Collision Simulation of vehicle and child pedestrian were set up (as Figure 2.) using the validated 6-year-old finite element model and the car finite model obtained from the vehicle model database by the National Crash Analysis Center (NCAC). The collision speeds of simulations were defined as 30 and 50km/h, respectively.

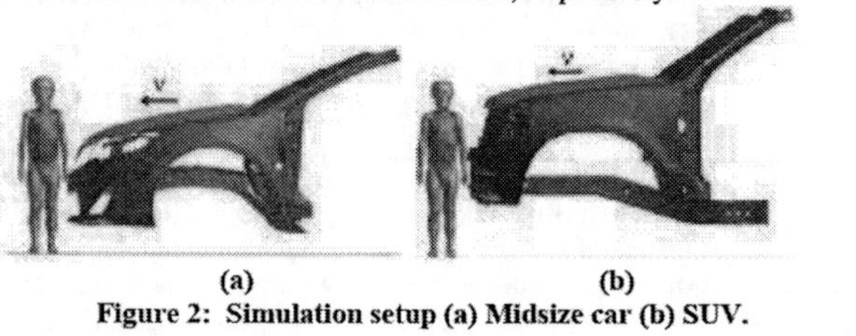

(a) (b)

Figure 2: Simulation setup (a) Midsize car (b) SUV.

RESULTS

The total 6-year-old FE pedestrian model contains 1,050,835 nodes and 1,339,386 elements. Each body part has the detail anatomical structure (Figure 3) and was validated effectively. This detailed validation studies were presented in the papers as shown in Table 1.

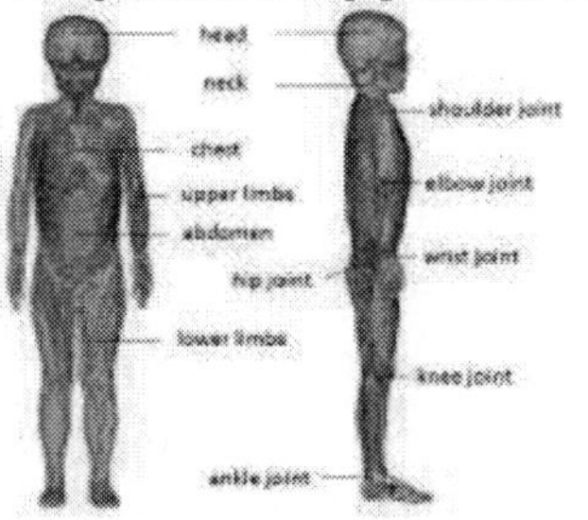

Figure 3: 6-year-old child pedestrian finite element model with detail antimony.

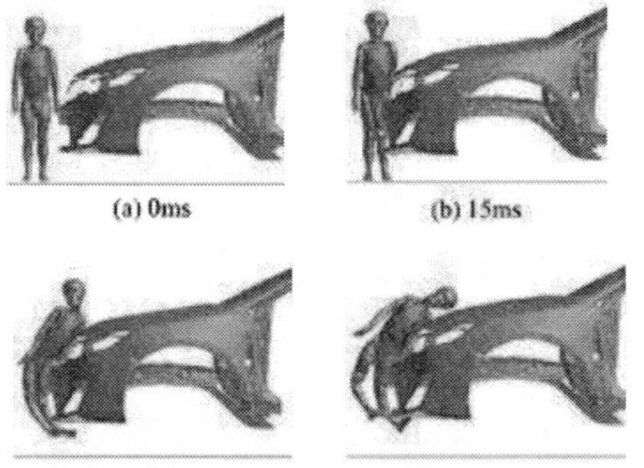

(a) 0ms (b) 15ms

(c) 30ms (d) 45ms

Figure 4: Dynamic responses of the pediatric pedestrian at impact speed 50 km/h with midsize car.

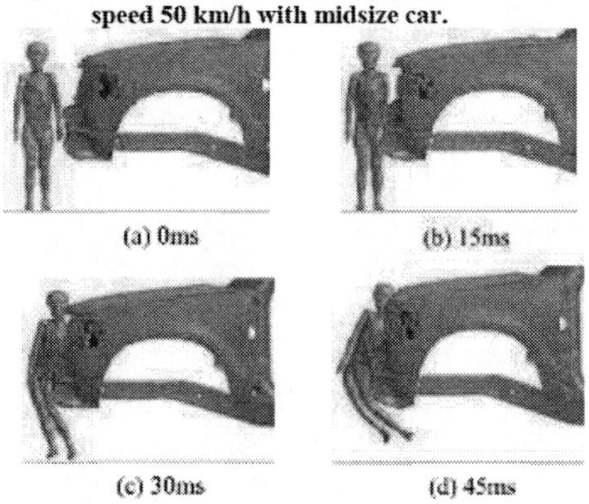

(a) 0ms (b) 15ms

(c) 30ms (d) 45ms

Figure 5: Dynamic responses of the pediatric pedestrian at impact speed30 km/h with SUV.

Figure 4 and 5 shows the dynamic responses of the pediatric pedestrian at different time stages in midsize car/SUV-pedestrian collisions at impact speeds of 50km/h and 30km/h, respectively.

As shown in Figure 6, the deformations and VCmax of pediatric chest and abdomen were proportional to collision speed. Maximum deformations and VCmax of chest in SUV-pedestrian simulations were greater than that in midsize car-pedestrian simulations at the same collision speed, and it was contrary to abdominal maximum deformation and VCmax.

As shown in Figure 7(a), it was found that maximum first principal strains of lung and heart were proportional to collision speed. Maximum first principal strains of lung and heart in SUV-pedestrian simulations were larger than that in midsize car-pedestrian simulations at the same collision speed. Also, the maximum first principal strain of lung was larger than that of heart at the same collision speed. In Figure 7(b), the maximum first principal strains of liver, kidney and spleen increased with the increase of collision speed, and the strains in midsize car- pedestrian collisions were greater than that in SUV-pedestrian collisions at the same speed. The influence of collision speed on the maximum first principal strain of kidney was larger than that on liver and spleen maximum first principal strain.

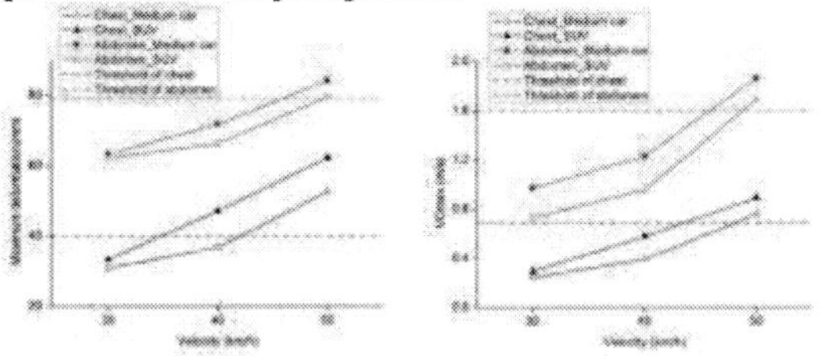

(a)Maximum deformation (b) Maximum VC values

Figure 6: Maximum deformations and VCmax of chest/abdomen.

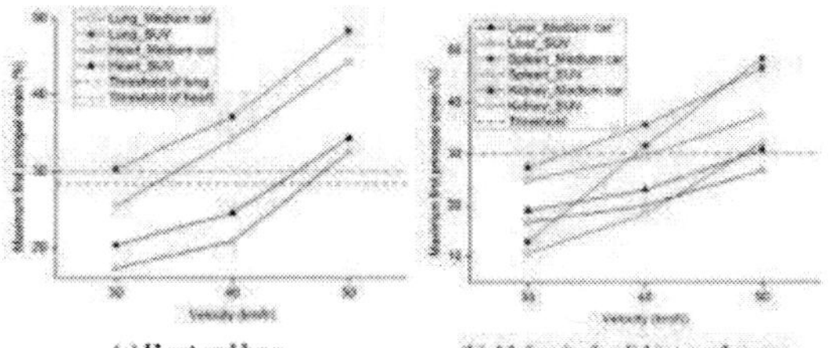

(a) Heart and lung (b) Abdominal solid internal organs

Figure 7: Maximum first principal strains of lung, heart and abdominal solid internal organs.

DISCUSSION

The injury of pediatric pedestrian chest/abdomen related to the contact positions on midsize car/SUV because of the change of height and the structure of midsize car and SUV.

Injuries of chest and abdomen in midsize car-pedestrian collision simulations predicted by first principal strain were consistent with those predicted by compression/viscous criteria.

The TUST 6-year-old child pedestrian finite element model can be used to simulate collisions in order to study the injury mechanisms and protection.

ACKNOWLEDGEMENTS

This study is supported by the National Natural Science Foundation of China (81371360, 81471274, and 81201015) and American Ford Automobile Research Institute.

REFERENCES

[1] Schwartz D et al., Traffic Injury Prevention, 16: 49-56, 2015.
[2] Iwamoto M et al., Traffic Injury Prevention, 16:36-48, 2015.
[3]Brolin K et al., IATSS Research, 38:92-102, 2015.
[4] Mizuno K et al., Traffic Injury Prevention, 6:361–371, 2005.
[5] Zhu F et al., Annals of Biomedical Engineering, 44:258-271, 2016.

All Authors:
Haiyan,Li lihaiyan@tust.edu.cn
Wenle, Lv lvwenle@tust.edu.cn
Shihai, Cui shihaicui@tust.edu.cn
Lijuan, He helijuan@tust.edu.cn
Shijie, Ruan jruan36@yahoo.com
Chunxiang, Wang tjchct@163.com

SB³C2017
Summer Biomechanics, Bioengineering and Biotransport Conference
June 21 – 24, Tucson, AZ, USA

MEASURING OLIGODENDROCYTE MECHANICS FOLLOWING SIMULATED TRAUMATIC BRAIN INJURY

Nicholas J. Braun (1), Zaw Win (1), Dezhi Liao (2), Patrick W. Alford (1)

(1) Biomedical Engineering
University of Minnesota – Twin Cities
Minneapolis, MN, USA

(2) Neuroscience
University of Minnesota – Twin Cities
Minneapolis, MN, USA

INTRODUCTION

The brain is a complex assembly of different cell types with highly specialized roles. Neurons are the basic functional unit of the brain, transmitting signals and information to other neurons or regions of the body. Glial cells, however, constitute the majority of cells in the brain, and are tasked with supporting and protecting neurons, chemically and mechanically [1]. Oligodendrocytes extend processes that sheath and insulate neuron axons with myelin, forming physical interactions between the two cell types for the purpose of electrical signal propagation [2].

During a traumatic brain injury (TBI), mechanical forces damage tissue on the cellular scale [3]. Diffuse axonal injury (DAI) is a hallmark of TBI, and encompasses a spectrum of effects in the myelin-rich white matter of the brain, including axonal cytoskeleton damage, disruption of axonal transport, and in extreme cases, Wallerian degeneration [4]. Since TBI readily affects the white matter tracts where oligodendrocytes are most prominent, and oligodendrocytes form physical myelin linkages with as many as 50 neurons at a time, the mechanics of oligodendrocytes likely play a role in determining the severity of TBI repercussions [2].

If oligodendrocyte mechanics are altered by a TBI, a subsequent injury could have amplified consequences or more severe DAI. **Here, we present a method for simulating TBI on oligodendrocytes with a high-velocity cell stretcher (HVCS) and subsequently measuring their material properties using cellular micro-biaxial stretching (CμBS). The results will provide insight to TBI pathophysiology and improve the understanding of the role of glial cells in TBI.**

METHODS

Cell Harvest, Culture, and Oligodendrocyte Isolation. Dissociated cortical mixed glial cultures were harvested from P0 newborn rat pups and seeded in flasks that had been coated with 100 μg/mL poly-D-lysine in a 37°C incubator overnight. Media changes (DMEM, 20% serum, 1% penicillin-streptomycin-glutamine, 1% sodium pyruvate) occurred every 3 days. At DIV 10, flasks were sealed and shaken at 200 rpm, 37°C, for 1 hour to remove microglia. Media was changed and cells were shaken again, overnight at 200 rpm, 37°C, to dislodge oligodendrocytes. The cell suspension was then incubated in an untreated petri dish for 1 hour to selectively remove contaminate cells (residual microglia and astrocytes can adhere to untreated plastic). The remaining oligodendrocytes in suspension were then seeded in flasks coated with poly-D-lysine as described previously. Stretching constructs were seeded at low density 7 days after oligodendrocyte isolation [5].

Construct Fabrication. Fibronectin (FN) was micropatterned onto polyacrylamide (PA) gels doped with fluorescent microbeads. The gels were bonded to elastomer membranes suspended in stainless-steel clamps as shown in Figure 1. FN was incubated on PDMS stamps featuring raised 60 μm radius circles for 1 hour. Stamps were then placed feature-side down onto glass coverslips for 30 minutes. Suspended elastomer membranes were treated with 10% benzophenone [6] and a PA solution of 10% acrylamide, 0.13% bis-acrylamide, and 1% 0.2 μm fluorescent microbeads was degassed. To polymerize gels, 0.2% tetramethylethylenediamine, 1M hydrochloric acid to a pH of 7.0-7.4, 10 μg/ml N-hydroxysuccinimideacrylic ester, and 0.05% aminopropylsilane were quickly added [7]. 10 μL droplets of this solution were rapidly placed on suspended membranes and the stamped coverslips were placed face-down on the PA droplet. Gels were then cured via UV exposure and rehydrated, allowing for removal of the glass coverslip. PA gels were passivated in 4% bovine serum albumin and rinsed with PBS before use.

Cell Stretching. A linear motor was used to apply uniaxial high-speed stretches at a strain rate of approximately 400% per second to simulate TBI. Varying stretches can be performed at this rate to simulate a spectrum of TBI severities. CμBS was then used to complete traction force microscopy experiments at 0-20% strain in 5% strain increments (Figure 2A).

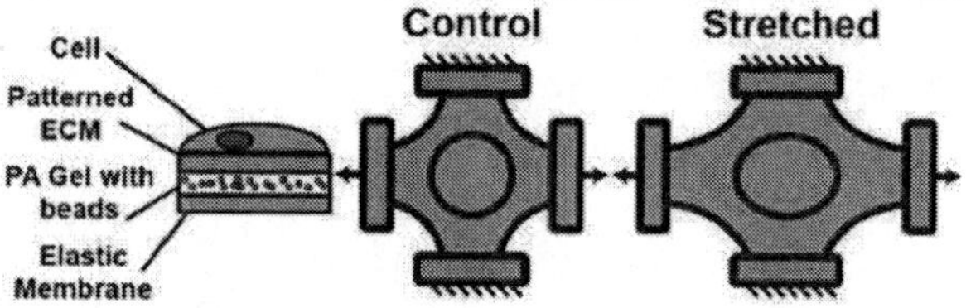

Figure 1: Illustration of micropatterned PA gel bonded to elastic membrane and uniaxial stretching of a cell seeded on the assembled construct.

Traction Force Microscopy. Stretching constructs were seeded with cells and then mounted on CμBS in an Olympus IX81ZDC confocal microscope. Substrates were stretched from 0-20% in increments of 5% at a slow rate so that viscous effects could be ignored. Stretching occurred in the same direction as TBI simulation. Bright field images of the cells and images of the fluorescent bead layer just below the cells were collected at each strain. Cells were then lysed with sodium dodecyl sulfate and reference images of the bead layer were collected.

Cell Force Analysis. Bead displacements were found by comparing bead images with cells present to cell-free bead images by using a published particle image velocimetry ImageJ plugin [8]. Known PA material properties were combined with bead displacements to yield traction forces via an ImageJ unconstrained FTTC algorithm [8].

RESULTS

We applied stains of 0%-20% to oligodendrocytes seeded on FN circles. At each strain step, we recorded traction forces as described previously [9, 10]. As expected, cells exhibited strain-dependent traction that increased by nearly a factor of 2 when comparing the 0% control to higher strains. Similar results were exhibited between the example single cell (Figure 2B) and the total experimental population (Figure 2C).

DISCUSSION

Here, we have demonstrated the utility of a system used to measure the material properties of single oligodendrocytes seeded on FN circles. Our data demonstrate that traction forces produced by oligodendrocytes increase with applied substrate strain, yielding baseline values that oligodendrocytes exposed to TBI simulations will be compared to. Previous studies have shown that TBI simulations can elicit phenotypical changes in cells that alter their mechanics [11]. Preliminary experiments in our lab using mixed glial cell types have corroborated similar studies by demonstrating increased contractility of glial cells following TBI simulation.

Moving forward, experiments will be repeated with the addition of TBI simulations of increasing severities. We hypothesize that oligodendrocytes will exhibit some degree of increased contractility or resistance to strain if treated with a TBI simulation below the threshold of cellular death. An alteration in oligodendrocyte mechanical properties in this *in vitro* experiment could be the first step in demonstrating a change in brain cell mechanics in TBI patients and provide insight as to why recurrent TBIs can lead to catastrophic outcomes or long-term neurodegeneration.

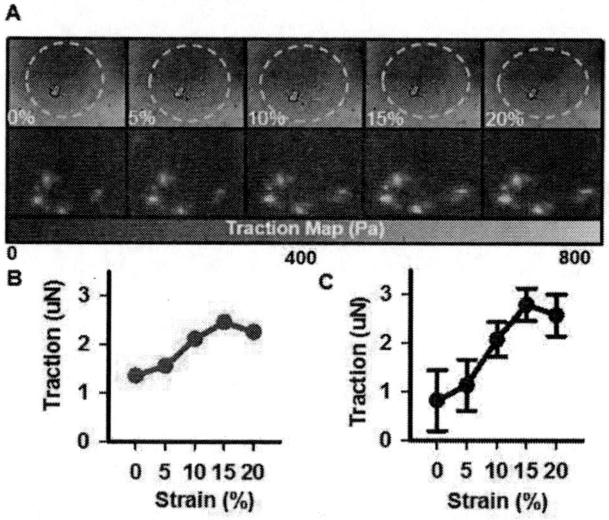

Figure 2: Oligodendrocyte traction force output is strain-dependent. A. *Top* – Brightfield images of a single oligodendrocyte on a 60 μm FN circle (yellow) being stretched from 0%-20% strain. *Bottom* – Traction maps for the oligodendrocyte at each stage of the stretching protocol. **B.** Resultant total traction force produced by the cell pictured in A. **C.** Total traction forces for whole sample of oligodendrocytes. *Error bars;* SEM, n=9.

ACKNOWLEDGEMENTS

We wish to thank our funding from the University of Minnesota Institute for Engineering in Medicine Seed Grant (DL, PWA), NIH-NINDS R21 (DL), NSF CMMI-1553255 (PWA), and the 3M Science and Technology Graduate Fellowship (NB).

REFERENCES

[1] Temburni, MK and Jacob, MH, *Proc Natl Acad Sci,* 98(7):3631-3632, 2001.
[2] Baumann, N and Pham-Dihn, D, *Physiol Rev,* 81(2):871-927, 2001.
[3] Magnuson J, *Curr Neurol Neurosci Rep,* 12(5):570-9, 2012.
[4] Johnson, VE et al., *Exp Neurol,* 246:35-43, 2013.
[5] Chen, Y et al., *Nat Protoc,* 2(5):1044-51, 2007.
[6] Simmons, CS et al., *Lab Chip,* 13:646-649, 2013.
[7] Polio, SR et al., *Methods in Cell Bio,* 17-31, 2014
[8] Tseng, Q et al., *Proc Nat Acad Sci,* 109:1506-1511, 2012.
[9] Win, Z et al., *SB3C Conference Proceedings,* 464-465, 2016.
[10] Win, Z et al, *J Biomech Eng,* under review.
[11] Alford, PW et al., *Proc Nat Acad Sci,* 108(31):12705-12710, 2011.

SB³C2017
Summer Biomechanics, Bioengineering and Biotransport Conference
June 21-24, Tucson, Ari□ona, USA

MOPHOMETRIC ANALYSIS OF CEREBELLUM IN TYPE I CHIARI MALFORMATION

Dipankar Biswas (1), Maggie S. Eppelheimer (2), James R. Houston (3), Audrey Braun (1), Richard Labuda (4), Francis Loth (1,2)

(1) Department of Mechanical Engineering
The University of A□ron
A□ron, Ohio, USA

(2) Department of Biomedical Engineering
The University of A□ron
A□ron, Ohio, USA

(3) Department of Psychology
The University of A□ron
A□ron, Ohio, USA

(4) Con□uer Chiari
We□ford, Pennsylvania, USA

INTRODUCTION

Type I Chiari malformation (CMI) is a neurological disorder of the hindbrain in which the position of cerebellum is lower in the brain which could serve as an obstruction to cerebrospinal fluid motion. CMI patients present a wide range of symptoms that can vary from pain in the head and neck, dizziness or impaired cognitive functions. It is radiologically identified by the position of the cerebellar tonsils to be three to five millimeters below the foramen magnum. The goal of the present study is to determine what cerebellum morphometric parameters beyond the tonsillar herniation are different for CMI patients as compared to healthy controls. A few studies looked at the posterior cranial fossa osseous (PCF) volume in CMI patients as compared to healthy controls [1, 2]. Karagöz et. al. [1] concluded that PCF is underdeveloped in CMI patients compared to healthy controls while Roller et. al. [2] concluded that the difference is insignificant when controlled for demographics. In another study Taştemur et. al. [3] showed that the diameter of foramen magnum, and the height and diameter of the cerebellum increases significantly in CMI patients as compared to healthy controls. They also showed that maximum cranial height, supraocciput length, clivus length, and PCF height decreases significantly in CMI patients as compared to healthy volunteers. None of these studies, however looked into the shape or orientation of the cerebellum of CMI patients and compared them to healthy controls. Additional markers regarding the shape of the cerebellum and its position and orientation with respect to other structures in the brain may lead to better understand the symptomatology and prognosis of the disease.

METHODS

The goal of this research is to semi-automatically segment the cerebellum from the mid-sagittal MRI images of 20 CMI patients and 20 healthy controls and compare them with respect to 11 different parameters. These include: total cerebellum area, tonsillar area, cerebellum area inside the PCF, perimeter, aspect ratio, cerebellum to PCF area ratio, and the cerebellum angles with respect to the McRae's line, spinal cord, clivus bone, anterior base of the brain and the anterior dural line. The images were selected from a voluntarily submitted patient image database, known as Chiari1000, and stored on an OsiriX DICOM PACS server at the Conquer Chiari research center at The University of Akron. The MR images of healthy controls were obtained through the Washington University, University of Minnesota, Oxford University Human Connectome Project consortium and Akron General Medical Center. The subjects were restricted to be all white females, age between 18 and 38 years and body mass index between 18.5 and 40.

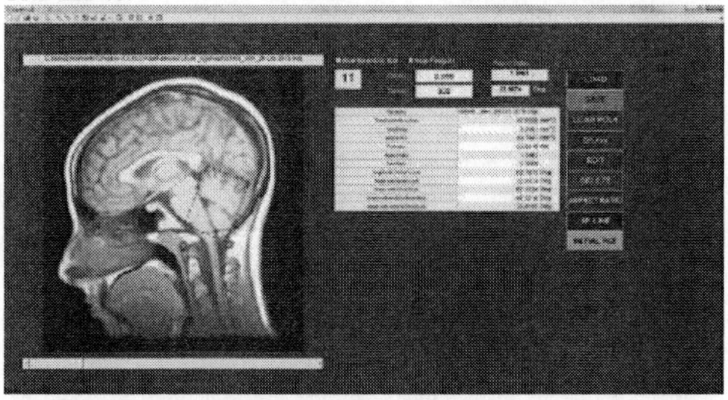

Figure 1: ***CerePro2D*** **cerebellum analysis software. Cerebellum inside PCF identified in cyan, tonsillar area in red, best fit minimum area rectangle in dotted red, and PCF in dotted green.**

Poster Presentation #P197

The mid-sagittal MRI image was obtained by meeting three out of four criteria namely, (1) the genu of the corpus callosum, (2) the splenium of the corpus callosum, (3) the pituitary infundibulum, and (4) the cerebral aqueduct. An algorithm was developed in MATLAB (Mathworks, Natick, MA) to segment the cerebellum geometry using the MRI (*CerePro2D*), shown in Figure 1. A manually selected point inside the cerebellum was used as a seed point to identify the cerebellum using the Chan-Vese region growing algorithm. A rectangle with minimum area was fit around the cerebellum and the major and minor axis of the rectangle identified. The other features of the brain with which the cerebellum shape, position and orientation were compared were identified manually using another in-house software (*MorphPro*) also written in MATLAB. The direction of spinal cord was identified as a line along the posterior edge of the spinal cord from the bottom of C2 to the bottom of C4 vertebrae. PCF area was marked by the tentorium, the supraoccipital portion of the occipital bone, McRae's line, and clivus. The McRae's line was identified as the line joining the basion and the opisthion. The line representing the clivus was obtained by joining the top of clivus and the basion. The line representing the anterior base of the brain was a line extending across the anterior cranial fossa to the tip of the dorsum sallae. The aspect ratio was defined as the ratio of the major and minor axis of the best fit minimum area rectangle. Two tailed equal variance sample t-test was used to compare between the CMI patient and the control images. In addition, the cerebellum shapes (Figures 2 and 3) of the two sets were compared using a cross-correlation algorithm (normxcorr2). Each CMI cerebellum shape was compared to each of the 20 control cerebellum shapes and the mean of the 20 correlation coefficients were calculated. This was repeated for the control cerebellum shapes where the shape of each control cerebellum was correlated to the other 19 and the mean of the 19 coefficients were calculated. A two tailed unequal variance sample t-test was then conducted on these set of means. A modified Levene's test was conducted in Microsoft Excel to determine whether homoscedastic or heteroscedastic t-test should be used.

RESULTS

A total of 11 parameters were compared with respect to the cerebellum between the Chiari and the healthy controls and the results shown in Table 1. The 11 parameters examined were not statistically significantly different between the Chiari and the healthy control cases except for the tonsillar area which is expected. Interesting to note was that the cerebellum orientation with respect to the McRae's line was not different from that of healthy controls (means were 62.7° and 62.6° for CMI and controls, respectively, $p=0.99$).

The mean of the correlation coefficients when (A) Chiari cerebellum shapes were compared to the healthy controls, (B) the healthy controls with themselves, and (C) Chiari with other Chiari, were 0.80 ± 0.05, 0.84 ± 0.02 and 0.78 ± 0.04, respectively. Thus, on average the shape of the Chiari cerebellums were found to be significantly different from those of the healthy cerebellums (A vs. B, $p<0.01$). When groups were correlated to themselves, the healthy controls were significantly more similar in shape than the Chiari cases (B vs. C, $p<0.0001$).

DISCUSSION

A concern for CMI patients by some neurosurgeons is that the cerebellum is undergoing stresses that may move or deform the cerebellum due to insufficient space in the PCF and an abnormal pressure environment. The present results, however, implied that the crowding inside the PCF of CMI patients were similar compared to the healthy controls. Cerebellum shapes were shown to vary more for CMI patients than healthy controls implying CMI cerebellums are more dysmorphic than those of healthy controls. This research may help understand the stresses experienced on the cerebellum by CMI patients. One limitation of the present study was the variety of MR scanners from which the images were obtained. This difference in quality might have some effect on the results and need further study.

Table 2: Mean results for the Chiari and control groups. Only parameter significantly different between the two groups is tonsillar area.

Parameters	Chiari	Control	p
Total cerebellum area	48.7±5.6mm^2	45.9±4.9mm^2	0.10
Tonsil area	1.6±1.7mm^2	0.04±0.1mm^2	0.0003
Cerebellum area in PCF	47.1±6.0mm^2	45.9±4.8mm^2	0.47
Perimeter	52.1±13.8mm	57.4±11.8mm	0.20
Aspect ratio	1.5±0.4	1.3±0.2	0.21
Angle with spinal cord	20.1±31.1°	18.6±33.0°	0.88
Area ratio	0.4±0.1	0.3±0.1	0.07
Angle with McRae's Line	62.7±15.6°	62.6±20.0°	0.99
Angle with clivus Bone	47.1±29.9°	41.6±31.4°	0.58
Angle with anterior brain base	-73.4±27.6°	-69.7±44.0°	0.75
Angle with anterior dural line	18.8±31.3°	18.0±33.1°	0.94

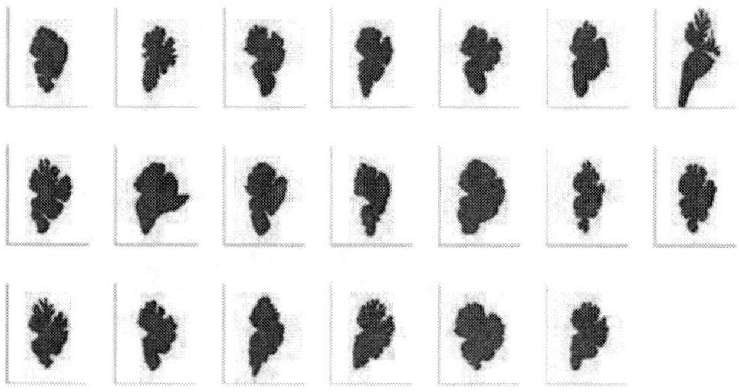

Figure 3: Cerebellum shapes of 20 Chiari patients segmented from their MR images using *CerePro2D*.

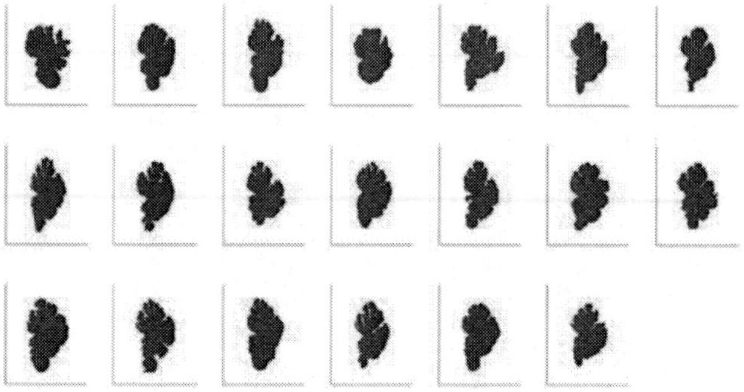

Figure 4: Cerebellum shapes of 20 healthy controls segmented from their MR images using *CerePro2D*.

REFERENCES

[1] Karagöz, F. et al., *Acta Neurochir*, 144:165-171, 2002.
[2] Roller, L.A. et al., *Neuroradiology*, 204:835-841, 2015.
[3] Taştemur, Y. et al., *Iran J Radiol*, e24436 ,2016.

COMPOSITE HYDROGEL: A NEW TOOL FOR REPRODUCING THE MECHANICAL BEHAVIOUR OF SOFT HUMAN TISSUES

Z. Tan (1), A. Forte (1), C. Parisi (2), F. Rodriguez Y Baena (3) and D. Dini (1)

(1) Tribology group
Department of Mechanical Engineering
Imperial College London
London, United Kingdom

(2) Tissue Engineering and Biophotonics Division
King's College London
London, UK

(3) Mechatronics in Medicine
Department of Mechanical Engineering
Imperial College London
London, United □ingdom

INTRODUCTION

The characterisation of the mechanical response of real soft tissues, such as brain, liver and cartilage, is immensely important as it allows us to understand the way they respond under a variety of different loads. Hence, ways to reduce damage to living tissues during real life scenarios can be identified and developed. However, real tissue is difficult to obtain and test due to accessibility. Therefore, there is a huge advantage in developing an accurate synthetic tissue phantom that is easier to procure and produce. This has led to the popularity of hydrogels, which have been developed into tissue mimicking materials due to their biocompatibility and stiffness tunability.

A composite hydrogel (CH) constituting of poly(vinyl) alcohol (PVA) and phytagel has been developed and is able to match the complex viscoelastic behaviour of brain [1]. The CH can be tuned to achieve different stiffness and relaxation responses by varying the concentrations of each hydrogel component. Therefore, a map showing the concentrations of PVA and phytagel required to match the mechanical responses of different soft tissues can be built. The stress mapping at 30% strain is compared to various soft tissue properties found in literature and reported here.

In addition, a mechanically accurate tissue phantom material opens the doors to many applications in the study of mechanobiology and regenerative medicine. This is because the stiffness of a cell seeded substrate has been shown to affect cell differentiation [2]. Furthermore, other mechanical loading conditions, such as tension and cyclic loading have been shown to affect cell growth and ECM synthesis [3], [4]. This study also demonstrates the cell viability of the CH substrate, therefore extending its range of applicability to investigate how cells are affected by impact and fracture loads using cell-seeded CH phantoms.

METHODS

Various combinations of different concentrations of PVA and phytagel were tested to observe the effect of each constituent hydrogel on the overall properties of the CH. PVA (molecular weight 146,000-186,000 Da), phytagel and deionised water were supplied by Sigma Aldrich, USA. The CH was prepared by dissolving various amounts of the PVA and phytagel powders, according to the concentrations marked in Fig. 1, separately in deionised water for 1h at 90 °C. The separate solutions were combined in a 1:1 weight ratio and stirred at 70 °C for an additional 30 min. The mixed solution was then poured into a petri dish and allowed to cool to room temperature.

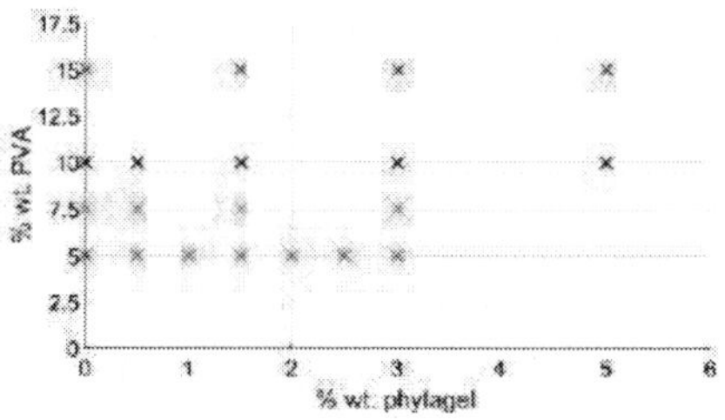

Figure 1: Mapping matrix of CH compositions tested.

The CH is physically cross-linked by hydrogen bonds that form during the thawing process when the hydrogel undergoes a freeze-thaw cycle. Therefore, the solution is frozen and stored at -25 °C for 15 hrs. Before testing, the samples are thawed to room temperature. Cylindrical samples of 10 mm diameter and 7 mm height were cut from the petri dish bulk material using a biopsy punch. The mechanical behaviour of the CH samples was then analysed.

Poster Presentation #P198

In this study, unconfined uniaxial compression tests at 30% strain and 1 s-1 strain rate on CHs of varying compositions were carried out to obtain the changes in mechanical behaviour with respect to hydrogel composition. The tests were performed using a Mach-1™ mechanical testing system (Biomomentum, Canada) and a 1.5 N load cell (Honeywell, USA). Samples were kept hydrated until tested. The true stress was calculated at 30% engineering strain under an incompressible assumption, which is commonly used to model soft tissues [5],[6]. Once the stress map was completed, the mechanical characteristics of different soft tissues found in literature were compared and the best matched compositions were determined.

Cell viability on PVA/phytagel 5/0.59 wt% CH samples was also explored. Primary Normal Human Dermal Fibroblasts (NHDF, Promocell, Germany) were seeded onto the CH hydrogels, which were coated with collagen, poly-L-lysine or gelatin to enhance cell attachment, a technique recently proposed and widely adopted in the literature [2]. Cell viability was assessed using Live/Dead staining where the samples were stained with calcein for live cells and ethidium homodimer-1 (EthD-1) for dead cells after 72 hrs in culture. Fluorescence microscopy was used to observe the Live (green, FITC) and dead (red, TRITC) cells.

RESULTS

The effect of the constituent PVA and phytagel on the stiffness of the overall composite hydrogel is shown on Fig 2. Increasing the concentration for both PVA and phytagel increases the maximum stress at 30% compressive strain. The compositions found to match the stiffness and relaxation responses of brain, liver and posterior cartilage were, respectively: PVA/phytagel 5/1.81 wt%, PVA/phytagel 10/3.33 wt% and PVA/phytagel 15/4.45 wt%.

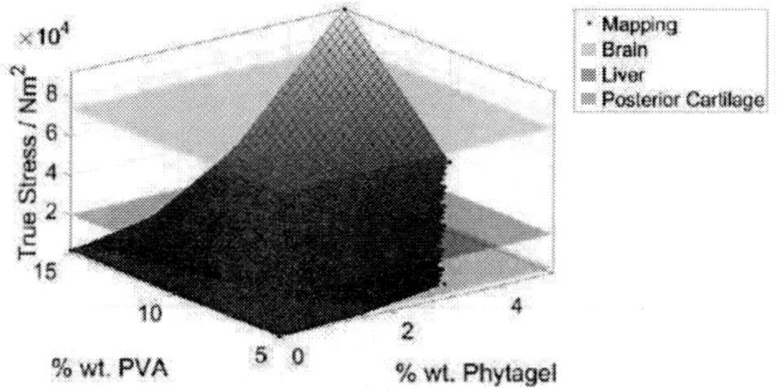

Figure 2: Maximum stress at 30% strain dependency on the constituent PVA and phytagel concentrations of the Composite Hydrogel (CH).

The results from the cell seeding reported in Fig. 3 show that the collagen coating achieves the greatest cell viability with over 97% live cells. In comparison to the number of cells attached onto the collagen coating, the poly-L-lysine and gelatin coating achieved an attachment rate of respectively, 76% and 8%. Cell morphology was different for each coating. Cells on the gelatin-coated substrate were all rounded. On the collagen-coated materials, all cells were well-spread and on poly-L-lysine-coated samples, 80% of the live cells exhibited good spreading. The remaining 20% live cells found on the poly-L-lysine coated hydrogel formed regions of tightly packed cell clusters, hence the density of cell attachment is less homogeneous than the collagen coated hydrogel.

Figure 3: Cell viability on (A) collagen, (B) poly-l-lysine and (C) gelatin coated 3D printed CH. All scale bars are 100 μm.

DISCUSSION

The study found that specific compositions of the CHs were able to reproduce the viscoelastic responses of various soft tissues with stiffness ranging from orders of magnitude 1 to $1x10^5$; this covers very soft tissues, such as brain and lung, to stiffer tissues, such as cartilage.

The apparent scarcity of attached live cells on the poly-L-lysine is due to the fact that the cells form densely packed clusters in localised regions, which may be due to the inhomogeneity of the coating. A higher cell attachment could be achieved using this coating with an optimized deposition process, which is currently under investigation.

With the encouraging results for cell viability on the CH substrate, the CH has showed great biological potentials as alternative and realistic tissue mimicking materials for accurate in vitro investigations of a wide range of bioengineering and mechanobiology applications. Further mechanical tests to investigate cellular response to a variety of mechanical loads are currently under investigation. These experiments include impact tests, needle insertion tests and tribological tests to match, respectively, the high strain rate behaviour of brain in traumatic injury, the fracture behaviour of liver during surgery and the tribological behaviour of cartilage.

ACKNOWLEDGEMENTS

The authors would like to acknowledge the support of the EPSRC via the Established Career Fellowship EP/N025954/1 awarded to Prof. Dini and funding from the European Union's EU Research and Innovation programme Horizon 2020 under grant agreement no 688279.

REFERENCES

[1] A. E. Forte, S. Galvan, F. Manieri, F. Rodriguez y Baena, and D. Dini, "A composite hydrogel for brain tissue phantoms," *Mater. Des.*, vol. 112, pp. 227–238, 2016.

[2] A. J. Engler, S. Sen, H. L. Sweeney, and D. E. Discher, "Matrix Elasticity Directs Stem Cell Lineage Specification," *Cell*, vol. 126, no. 4, pp. 677–689, 2006.

[3] H.-S. Yu, J.-J. Kim, H.-W. Kim, M. P. Lewis, and I. Wall, "Impact of mechanical stretch on the cell behaviors of bone and surrounding tissues.," *J. Tissue Eng.*, vol. 6, pp. 1–24, 2015.

[4] V. Gupta and K. J. Grande-Allen, "Effects of static and cyclic loading in regulating extracellular matrix synthesis by cardiovascular cells," *Cardiovasc. Res.*, vol. 72, no. 3, pp. 375–383, 2006.

[5] M. S. Estes and J. H. MacElhaney, *Response of brain tissue of compressive loading*. New York: ASME, 1970.

[6] A. E. Forte, S. M. Gentleman, and D. Dini, "On the characterization of the heterogeneous mechanical response of human brain tissue," *Biomech. Model. Mechanobiol.*, no. 2003, pp. 1–14, 2016.

STRAIN RESPONSES OF THE HUMAN BRAIN WITH MORPHOLOGICALLY AGE-APPROPRIATE HEAD MODELS

Bei Li (1,2), Wei Zhao (2), Haiyan Li (1), Songbai Ji (2), Shijie Ruan (1)

(1) Center for Injury Biomechanics and Vehicle Safety Engineering
Tianjin University of Science and Technology
Tianjin, 300222, China

(2) Department of Biomedical Engineering
Worcester Polytechnic Institute
Worcester, MA 01605, USA

INTRODUCTION

Traumatic brain injury (TBI) poses a special challenge in pediatric population [1]. In particular, the number of TBI in children aged from 0–19 years is reported to be underestimated [1]. Thus, a reliable injury assessment is required to prevent and mitigate this prevalent disease.

Finite element (FE) models have been increasingly important in brain injury studies. They have been extensively employed in adult brain injuries studies over decades [2, 3]. Unfortunately, only a few studies are focused on pediatric brain injury, among which most pediatric head FE models aged from 0–3 years [4-7] involving those scaled from adult models [8,9].

To understand the biomechanics of pediatric brain injury, it is important to recognize, at the first step, that differently-aged pediatric heads differ structurally and dimensionally. In particular, the morphology of the brain does not linearly grow with age [8]. Hence, it may not be appropriate to assume pediatric head to be a miniature adult head [9].

Given these considerations, it is necessary to develop age-appropriate head FE models for the pediatric population, including brain material properties, size and shape. In this study, we specifically investigated the importance of head/brain shape and size on model-estimated strains. As data for the material properties of the brain in children are lacking, the same material properties of the adult brain were used. This study may serve as a first step to better understand the importance of developing an age-appropriate head FE model for brain injury studies in the pediatric population.

METHODS

In this study, FE models of a 50th percentile adult, 6-year-old (6yo) child and 3-year-old (3yo) child head models were used. The models were created with high mesh quality and geometrical accuracy based on CT scans. The brain material properties and brain-skull boundary conditions were identical to that of Simulated Injury Monitor (SIMon) [2]. The width of the 50th percentile adult model was slightly smaller than that of the 6yo and 3yo models (**Table 1**). This was because the geometry of the 50th percentile adult head model was based on statistical measurements of CT scans averaged from a population group, while CT scans for the other two models were from a specific subject (due to the paucity in image data availability for the pediatric population).

Table 1. Characteristics of the three head FE models

	Circumference (cm)	Width (cm)	Length (cm)	Head mass (kg)	Brain mass (kg)
adult	54.55	14.32	17.87	4.31	1.77
6yo	51.64	15.44	16.40	3.31	1.65
3yo	47.01	15.38	16.06	2.66	1.36

The three FE models were first validated against relative brain-skull displacement data measured in three cadaveric head impacts (frontal (C383-T1), occipital (C755-T2) and parietal (C394-T4) [10, 11]. In the validations, all the FE models were linearly scaled to match the sizes of cadaver heads based on the following scaling law proposed by Mertz et al. [12]:

$$R = (C + W + L)_s/(C + W + L)_{50th} \qquad (1)$$

where C, W, and L are circumference, width and length of subject (s) and the 50th percentile head (**Table 1**). Six degrees-of-freedom accelerations were prescribed onto the rigid skull. Model-estimated relative brain-skull displacements at selected neutral density targets (NDTs) were compared with the experimental measurements.

For each model, a triangulated angular acceleration with a peak magnitude of 4500 rad/s^2 and duration of 10 ms (with another 25 ms holding period to maintain the same angular velocity) was then prescribed to the rigid skull to simulate uniaxial coronal head rotations. Their resulting cumulative brain strains were compared.

Lastly, the 50th percentile adult model was linearly scaled to match the sizes of 6yo and 3yo models using the scaling law (Eqn.1), while maintaining all other model configurations. This was to assume that the child and adult heads have similar geometry [12]. The same angular acceleration was prescribed to the rigid skull of the scaled models.

Element-wise peak maximum principal strains during the entire simulation, regardless of the time of occurrence, were extracted for each simulation. The 95th percentile peak strain magnitude of all elements (ε_p) was calculated to avoid potential numerical issues [13]. The cumulative strain damage measure (CSDM) at a strain threshold of 0.1 was computed.

RESULTS

Fig. 1 compares simulated relative brain-skull motions and their experimental counterparts for selected NDTs.

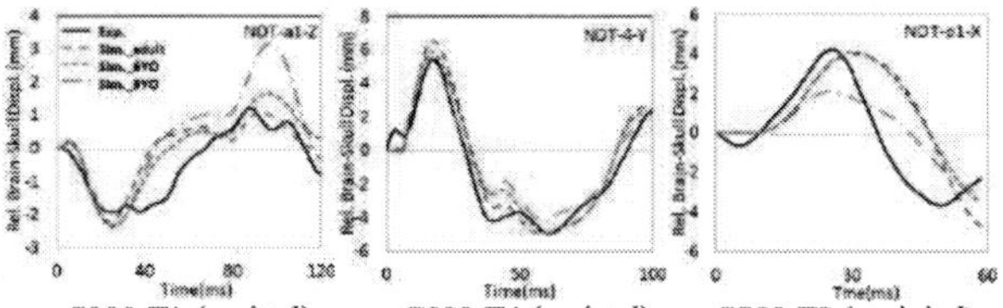

C383-T1 (sagittal) C393-T4 (parietal) C755-T2 (occipital)

Figure 1 Comparison between simulated and experimental time histories of relative brain-skull displacement for selected NDTs.

Table 2 Pearson correlation coefficients between simulation results and experimental data for each test and selected NDTs (p<0.05 for all tests, except for NDT-P5-Z in C755-T2 and NDT-4-Y in C393-T4)

	adult	6yo	3yo
C383-T1	0.720	0.716	0.724
C393-T4	0.630	0.640	0.657
C755-T2	0.463	0.582	0.435

For strain magnitudes of the three actual models subjected to the identical angular acceleration, the 50th percentile adult and 6yo child head models yielded nearly identical ε_p (0.25 and 0.24, respectively) and CSDM (54.38% and 54.19%, respectively) (**Table 3**). However, for the 3yo child head model slightly lower ε_p and CSDM were found (0.21 and 44.28%, respectively). The whole brain distributions of the three actual models were largely similar in the cerebrum with large strains occurred in the peripheral and central brain, but different in the brainstem and cerebellum with substantial strains presented in 6yo model (**Fig. 2**).

Table 3 Comparison of ε_p and CSDM at a strain threshold of 0.1 among the actual and scaled models.

	adult	6yo	3yo	6yo-scaled	3yo-scaled
ε_p	0.25	0.24	0.21	0.22	0.20
CSDM (%)	54.38	54.19	44.28	45.90	38.50

With identical angular acceleration, the scaled models had ε_p similar to that from actual models. However, CSDM from scaled models were underestimated relative to their actual counterparts (15.3% and 13.1% for 6yo and 3yo models, respectively; **Table 3**). The strain distributions of the scaled models were largely similar to their actual baseline counterparts with large strains occurred in the peripheral and central brain (**Fig. 3**). However, the strain distributions dramatically differed in the cerebellum and brainstem regions with scaled models yielding an underestimation (**Fig. 3b** and **d**).

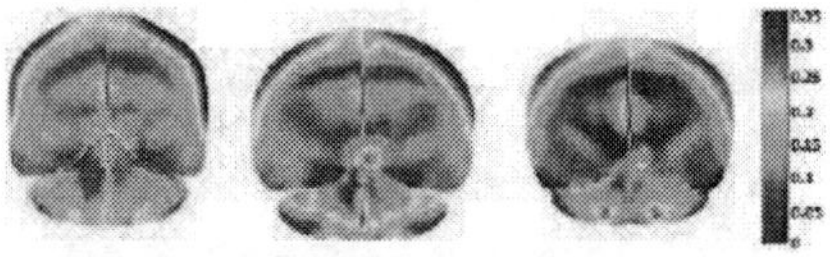

(a) (b) (c)

Figure 2 Comparison of element-wise strain distributions for the 50th percentile adult (a), 6yo child (b), and 3yo child (c) models.

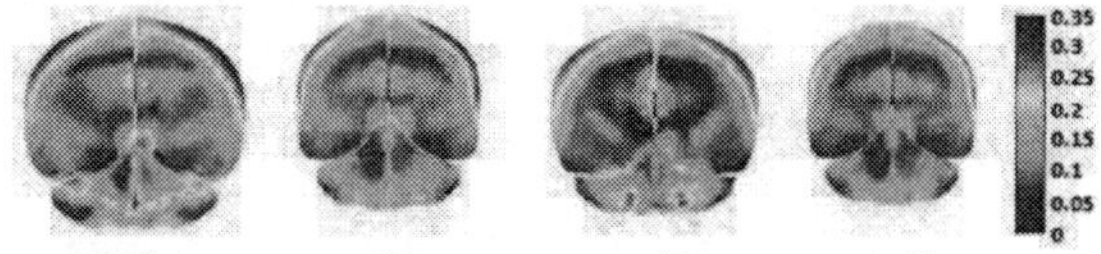

(a) (b) (c) (d)

Figure 3 Comparison of strains between scaled models and their actual counterparts. (a) and (c): actual 6yo and 3yo child models, respectively; (b) and (d): 6yo and 3yo models scaled from a 50th percentile adult model.

DISCUSSION

Comparing strains from the three different head models experiencing the same angular acceleration, we found similar strain pattern for the majority of the brain. This suggested that similar brain regions experience high injury risk across different ages. For strain magnitude, the 6yo and adult heads (in similar size) had comparable ε_p and CSDM. The 3yo child head sustained the lowest strain responses, due to a smaller size. However, it must be noted that the growing brain may be more vulnerable to mechanical loads due to its softer material properties [14], which was not investigated here and was a limitation.

The scaled models (from the 50th percentile adult head) predicted strain patterns similar to the actual models, but with substantial underestimation in magnitudes. This was because the brain morphology does not grow linearly with age [8]. The scaling law assumed similar geometry between the child and the adult heads, but did not account for their shape differences. Therefore, this suggested a need to establish child FE models from their actual brain scans, rather than by directly scaling from an adult head model.

Although this study was subject to the limitation of not considering the difference in material properties in the growing brain, still, it provided in-depth biomechanical understanding of the pediatric brain.

ACKNOWLEDGEMENTS

Funding is provided by the National Natural Science Foundation of China grant 81471274 (SR), the China Scholarship Council (BL).

REFERENCES

[1] CDC, 2015. [2] Takhounts et al., *Stapp Car Crash*, 52:1–31, 2008. [3] Ji et al., *J Neurotrauma*, 32(7): 441-454, 2015. [4] Coats et al., *Stapp Car Crash J.* 51, 1–15,2007. [5] Klinich et al., *Stapp Car Crash*, 165–194, 2002. [6] Roth et al., *Comput Meth Prog Bio*, 25–33,2010. [7] Roth et al., *Int. J. of Legal Med.*, 223–228, 2007. [8] Roth., *Comput Meth Prog Bio*, 90(3)262–274, 2008. [9] Roth,. *Int. J of vehicle safety*, 2(4): 404-421. 2007. [10] Hardy et al., *Stapp Car Crash*, 45:337–368, 2001. [11] Hardy et al., *Stapp Car Crash J.*, 17–80, 2007. [12] Merz et al., *SAE Paper 890756*, 1989. [13] Gabler et al., *Ann Biomed Engineering*, 44(12), 3705-3718, 2016. [14] Kennard, *Arch NeurPsych*, 44(2): 377-397, 1940.

INTERRUPTED HIGH-RATE COMPRESSION OF PORCINE BRAIN TISSUE UTILIZING THE SPLIT-HOPKINSON PRESSURE BAR METHOD

Haden A. Johnson (1,2), Jonathon Miller (2), Wilburn R. Whittington (2), Alicia K. Olivier (3), Michael D. Jones (4), Rajkumar Prabhu (1,2), Lakiesha N. Williams (1,2)

(1) Department of Agricultural and Biological Engineering
Mississippi State University
Starkville, MS, USA

(2) Center for Advanced Vehicular Systems
Mississippi State University
Starkville, MS, USA

(3) Department of Pathobiology and Population Medicine
Mississippi State University
Starkville, MS, USA

(4) School of Engineering
Cardiff University
Cardiff, Wales, UK

INTRODUCTION

The Center for Disease Control reports approximately 1.7 million traumatic brain injuries (TBIs) in the United states each year.[1] Of these cases, approximately 52,000 result in death, and more than 5 million people are currently living with a TBI related disability.[1]

There have been many studies of the mechanical response of brain tissue under dynamic loading, however, testing methods have necessitated the complete destruction of the test sample.[2] Thus, there is limited understanding of the strain rate related microstructural tissue deformation at the time of loading. Other studies have been able to evaluate the microstructural response of brain tissue, but these tests only analyzed damage at quasi-static loading rates.[3] These previous studies do demonstrate that brain tissue is very strain rate sensitive and exhibits a nonlinear viscoelastic behavior.[3,4]

By studying the response of brain tissue at various strain rates, the relative severity of TBI can be evaluated by analyzing the microstructure of the tissue. The aim of this research is to develop a testing method capable of quantifying the microstructural damage that brain tissue acquires at high strain rates.

METHODS

Adult pig heads were obtained from a local abattoir (Sansing Meat Service, Maben, MS), and they were transports to the Agricultural and Biological Engineering building at Mississippi State University. Intact porcine brains were extracted from the heads, and cylindrical (20 mm diameter) samples of mixed gray and white matter, example shown in Figure 1, were resected from the superior cortical region of each hemisphere using a custom stainless steel punch. The cylindrical samples were cut into 10 mm lengths to provide a 2:1 aspect ratio. The samples were then wrapped in gauze soaked in Phosphate Buffered Saline (PBS) and transported to the Center for Advanced Vehicular Systems (CAVS) for testing. All tests were completed within three hours post mortem to limit any effects of deterioration.

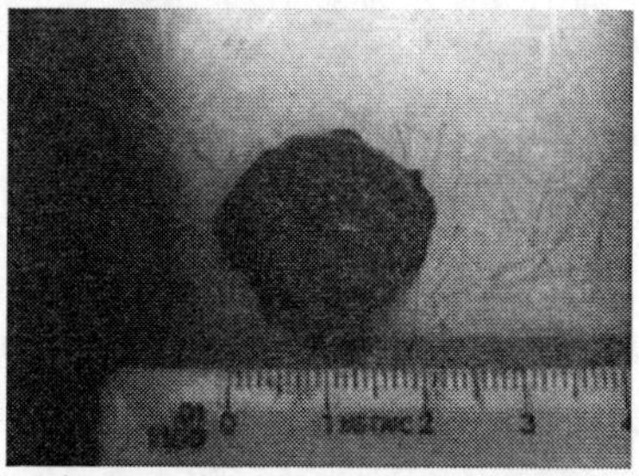

Figure 1: Cylindrical porcine brain tissue sample of mixed gray and white matter with a diameter of 20 mm.

A polymeric Split-Hopkinson Pressure Bar (SHPB) was chosen for compression testing due to its capability in achieving high strain rates. For a detailed description of the SHPB testing methodology for soft biological materials, the reader is referred to previous discussions by Prabhu *et al.* (2011).[2] Strain waves captured during testing were analyzed with the Split Hopkinson Pressure Bar Graphical Analysis Tool to produce stress-strain data.[4] Full compression tests, to failure, were initially performed to evaluate the mechanical response of the brain tissue. Tests were conducted at strain rates of 650, 800, and 900 s^{-1}.

To perform the interruption tests necessary for histological analysis of the brain tissue, custom attachments were designed and fabricated. Cylindrical disks, 60 mm in diameter and 13 mm in thickness, were cut from a sheet of polycarbonate, similar to the

Hopkinson bars. Wells were machined on either side of the disks to fit onto the ends of the bars and allow for the brain samples to fit inside. One pair of disks was necessary for each interruption test. The sides that hold the tissue samples were machined to an appropriate thickness, necessary to allow for compression to strain levels of 15, 30 and 40%. Slots were cut out of one of the disks from each pair to allow formalin to enter and penetrate the tissue after completion of each test. A pair of attachments can be seen below in Figure 2.

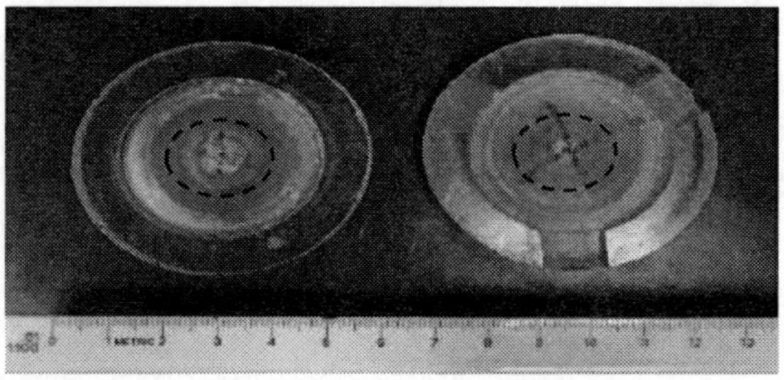

Figure 2: A pair of interruption attachments with the one on the right having slots for formalin to enter and penetrate the sample after testing is complete. Black, dashed circles represent approximate sample size in comparison to attachments.

To perform the interruption tests, double-sided adhesive tape was placed on the rings of the attachments surrounding the specimen wells. The attachments were then fitted to the incident and transmitted bars and the sample placed in the wells between them. The SHPB was then pressurized to predetermined (psi) levels to achieve the desired strain rates of 650, 800, and 900 s^{-1}. Upon releasing the gas, the samples were compressed until the attachments made contact with each other and interrupted the test. The double-sided adhesive tape ensured that the attachments would remain in contact with one another and hold the samples in the partially compressed state. After each test, the attachments, with the sample inside, were removed from the SHPB and placed into a formalin bath to fix the tissue. After fixation, the samples were sectioned and stained with Hematoxylin and Eosin (H&E) and Luxol Fast Blue (LFB).

RESULTS

Experimental test results from the full compression tests at strain rates of 650, 800, and 900 s^{-1} are shown in Figure 3. SHPB results show a nonlinear response with an increasing maximum stress as the strain rate increases.

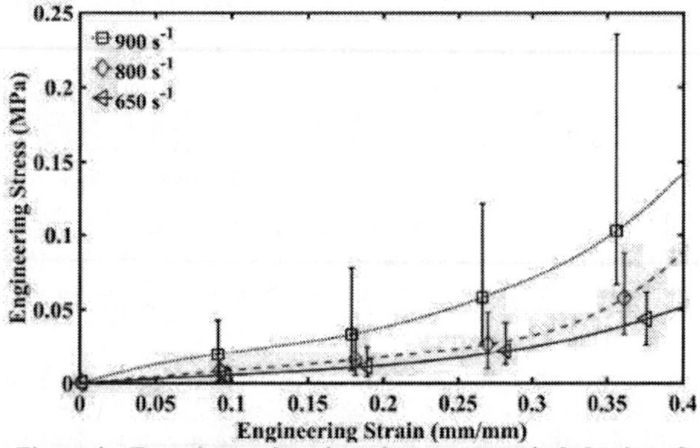

Figure 3: Experimental engineering stress-strain behavior of porcine brain tissue at strain rates of 650, 800, and 900 s^{-1}.

Images from three different tissue samples can be seen below in Figure 4. The samples were compressed at a rate of 650 s^{-1} and the tests were interrupted at 15, 30, and 40% strain. Initial observations reveal that as the strain levels increase, so does the density of neural and glial cells in each image.

Figure 4: Gray matter stained with LFB. Tissue was compressed at a strain rate of 650 s^{-1} to strain levels of (A) 15%, (B) 30%, and (C) 40%.

DISCUSSION

The stress-strain curves in Figure 3 show the brain's strain rate dependency, remaining consistent with results from previous studies.[2,3] However, more deviation can be seen in the results as the strain rate increases. This is likely due to the range of high rates associated with SHPB testing and the tissue's response threshold. Thus rates at 650 and 800s^{-1} appear to produce similar results.

The images in Figure 4 show the brain tissue cells becoming noticeably more compacted with increasing strain levels. Further analysis will be performed using ImageJ image analyzing software to determine area fraction and nearest neighbor distance of the neural cells.[5] This data will be compared across the different strain rates and levels of strain that were tested. This information will lead to another means of quantifying the damage that occurs to brain tissue under high-rate compressive loading.

ACKNOWLEDGEMENTS

Material presented in this paper is a product of the CREATE-GV Element of the Computational Research and Engineering Acquisition Tools and Environments (CREATE) Program sponsored by the U.S. Department of Defense HPC Modernization Program Office. This effort was sponsored under contract number W912HZ-13-C-0037.

I would also like to thank Mississippi State University's College of Veterinary Medicine Diagnostic Laboratory Services (CVM-DLS) for histological processing and staining and Sansing Meat Processing for their assistance with acquiring adequate test specimens.

REFERENCES

[1] Faul, M. et al., "Traumatic brain injury in the United States: emergency department visits, hospitalizations, and deaths, (2002-2006)," (2010).

[2] Prabhu, R. et al., "Coupled experiment/finite element analysis on the mechanical response of porcine brain under high strain rates," Journal of the Mechanical Behavior of Biomedical Materials, 4 (7), 1067-1080, (2011).

[3] Begonia, M.T., "The influence of strain rate dependency on the structure—property relations of porcine brain," Annals of biomedical engineering, 38 (10), 3043-3057, (2010).

[4] Francis, D.K., Whittington, W.R. et al., "Split Hopkinson Pressure Bar Graphical Analysis Tool," Experimental Mechanics, (2016).

[5] Rasband, W.S., ImageJ, U. S. National Institutes of Health, Bethesda, Maryland, USA, http://imagej.nih.gov/ij/, (1997-2016).

A METHOD TO LEVERAGE DETAILED AND SIMPLIFIED OCCUPANTS FOR COMPUTATIONAL EFFICIENCY IN PRE-CRASH SIMULATIONS

B. Guleyupoglu (1,2) and F.S. Gayzik (1,2)

(1) Department of Biomedical Engineering
Wa□e □orest School of Medicine
Winston-Salem, NC, USA

(2) □irginia Tech – Wa□e □orest
Center for In□ury Biomechanics
Winston-Salem, NC, USA

INTRODUCTION

Computational human body models are used to evaluate vehicle performance during crash events. These models provide important information on how the human body may interact with safety systems that are present in the vehicle and what risks are present to the occupant [1]. With the addition of active safety systems, which activate prior to a crash event or the pre-crash phase, the required simulation time to capture the full event may be on the order of 1 to 2 seconds, whereas crash-event only simulations are typically a fraction of that, typically 150 msec. While output from a detailed human body model is useful during the pre-crash phase, it is largely kinematics based and drastically increases computational cost. The use of a simplified model for the pre-crash phase and substituting the detailed model for the crash phase of a crash event is a sensible approach to reduce this cost. The objective of this study is to develop a method to switch between models of different levels of detail. To assess this method, we leverage two previously published computational models, the Global Human Body Models Consortium (GHBMC) 50th percentile male (M50) simplified occupant (-OS) and the detailed occupant (-O).

METHODS

The GHBMC M50-O (v. 4.3, 1.3M nodes, 2.2M elements, 77.0 kg) and M50-OS (v. 1.8, 295k nodes, 357k elements, 79.6 kg), were used in this study. The models were gravity settled into the driver position of a generic interior designed for studies in a frontal crash. The models are belted into the generic interior with an advanced three point belt with properties based on literature [2]. A severe braking case will be used to bring the M50-OS out of position. The specific case that will be used will be the pre-crash portion of the NCAP 0.8 G braking case (FCW+PBA+PB) which occurs over a period of about 670 msec. During the pre-crash simulation gravity will not be turned on to reduce any slumping that may occur over a long period of time. The occupant will remain belted for the out of position simulation; however none of the passive safety systems will be active during this time.

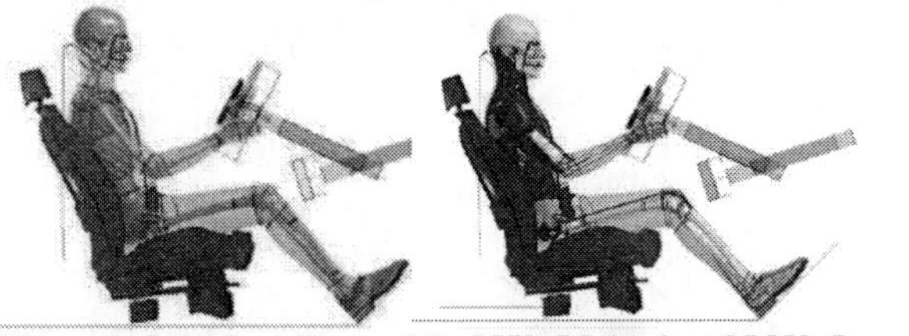

Figure 1: Seated positions of the M50-OS (top) and M50-O (bottom) models in the occupant compartment.

Once the M50-OS model has been brought into an interim position through a deceleration simulation, the model will be iteratively morphed using a Kriging method [3]. The method relies an assumed random process z with mean zero and a covariance of (1) between $z(w)$ and $z(x)$ where σ^2_l is a process variance and $R(\theta,w,x)$ is the correlation model with parameter θ.

(1) $$E[z_l(w), z_l(x)] = \sigma_l^2 R(\theta,w,x) \qquad l = 1, \dots, q$$

It also requires a kriging predictor y where the true value can be written as (2) where F is a regression function, β is a regression parameter and α is an approximation error.

(2) $$y_l(x) = F(\beta_{:,l}, x) + \alpha(\beta_{:,l}, x)$$

In the method used here, a Gaussian correlation function as well as a second order regression function was used to morph the model. The homologous landmarks that were used for the morph were based on anatomical landmarks previously used in literature [4]. The automated process for iterative morphing follows the flow chart in Figure 2. The states that are used are not a single time step but multiple time steps in between states.

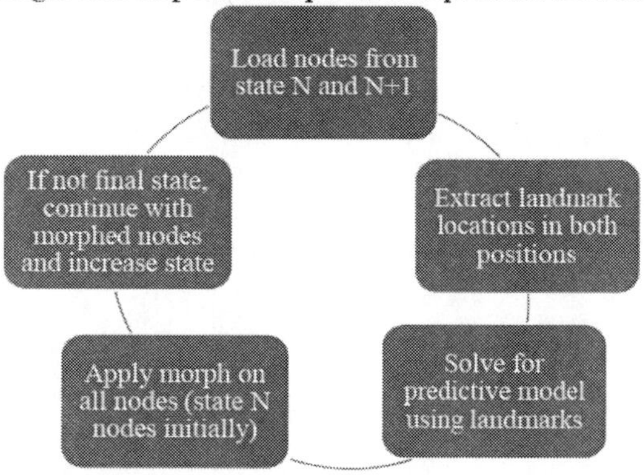

Figure 2: Flowchart of iterative kriging morphing method. Method starts with loading nodes from state N.

RESULTS

Simulating the M50-OS model in the pre-crash deceleration pulse had an average solution time of 0.5 min/msec while the M50-O model had an average solution time of 14.4 min/msec. Use of the M50-OS model saved an average of 13.9 min/msec for the interim position solution.

The morphed model demonstrated good agreement with the positioning of the target state (Figure 3). The automated morphing method averaged 28 minutes for nodal data extraction and 25 consecutive morphs. The final mass of the morphed model is 76.3 kg. Even though there were 25 morphs, the geometry within the model has also been largely preserved without any large distortions occurring (Figure 4).

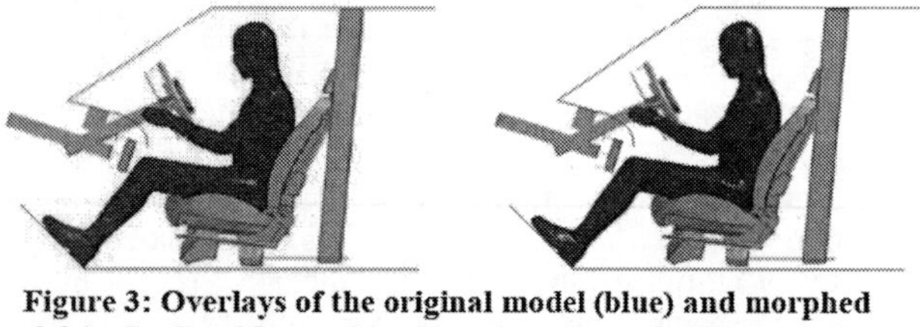

Figure 3: Overlays of the original model (blue) and morphed model (red) after 25 morphing iterations through 671 msec of pre-crash time. Morphed on top (left) and original on top (right).

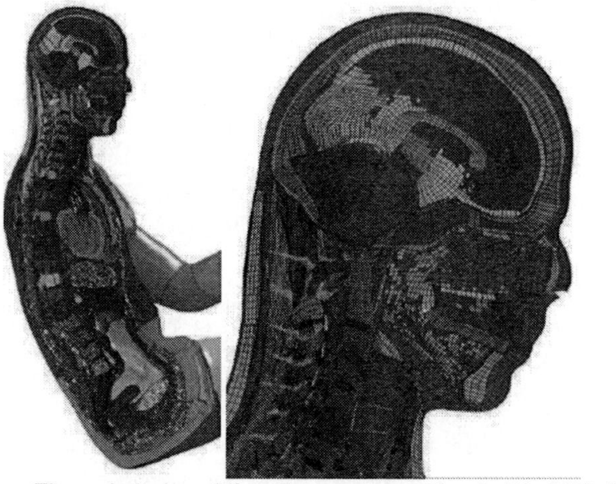

Figure 4: A view of the spine, organs and head of the M50-O model after morphing. Shapes of the structures have been largely preserved and element quality is high.

DISCUSSION

The method developed here to iteratively morph the M50-O model based on an interim position achieved by the M50-OS model achieved a time savings of around 154.5 hours. The element quality of the model also remained high at the end of the entire morphing algorithm. The mass of the model did decrease by 0.7 kg however, this only a 0.9% change in mass and may be further optimized through additional landmark selection.

There are limitations to this method however. Currently, the model does have some distortions from the limited set of landmarks used such as a minor dent in the back of the model. There are also areas around where the seatbelt interacts with the model that did not retain all of the deformation that occurred during the pre-crash simulation.

ACKNOWLEDGEMENTS

The authors would like to acknowledge the Global Human Body Models Consortium, LLC for funding and support. Wake Forest University is the Integration Center of Expertise of the GHBMC. The authors gratefully acknowledge the meshing and validation efforts by our academic partners in the GHBMC. All simulations were run on the DEAC cluster at Wake Forest University, with support provided by Drs. Timothy Miller and Damian Valles.

Dr. F. Scott Gayzik is a member of Elemance, LLC., which is the sole licensor and distributor of academic and commercial licenses for use of the GHBMC, LLC. – owned M50 occupant models.

REFERENCES

[1] Guleyupoglu, B et al. *Traffic Injury Prevention*, 2017
[2] Forman, JL et al., *Stapp Car Crash J*, 53:49-74, 2009.
[3] Lophaven, SN et al., *DACE Manual v2.0*, 2002.
[4] Jolivet, E et al., *Stapp Car Crash J*, 59:337-57, 2015.

SB^{3}C2017
Summer Biomechanics, Bioengineering and Biotransport Conference
June 21 – 24, Tucson, AZ, USA

PREDICTING STENT GRAFT ROTATION IN PATIENT SPECIFIC ABDOMINAL AORTIC ANEURYSM REPAIR USING FINITE ELEMENT ANALYSIS

Ryan M. Sanford (1), Sean A. Crawford (2, 3), Matthew G. Doyle (1, 2), Thomas L. Forbes (2), Cristina H. Amon (1, 3)

(1) Department of Mechanical and Industrial Engineering
University of Toronto
Toronto, Ontario, Canada

(2) Division of Vascular Surgery, Department of Surgery
University of Toronto
Toronto, Ontario, Canada

(3) Institute of Biomaterials and Biomedical Engineering
University of Toronto
Toronto, Ontario, Canada

INTRODUCTION

Abdominal aortic aneurysms (AAA) are pathological dilations of the arterial wall, which, if left untreated, can lead to aortic rupture and potentially patient death. AAAs can be repaired either through an open procedure, or through a minimally invasive surgical procedure called endovascular aneurysm repair (EVAR), which utilizes a medical device called a stent graft. During EVAR, a stiff guidewire is first inserted through the iliac artery and into the aneurysm sac. Next, the stent graft, which is enclosed within a sheath, is inserted and advanced through the iliac artery, following the path of the guidewire. When it is in the desired located in the aorta, the sheath is withdrawn, and the stent graft expands to its uncompressed size.

Depending on the location of the aneurysm along the aorta, it may be necessary to use a stent graft with fenestrations at the locations of the branch arteries in order to preserve blood flow to these vessels. These fenestrated stent grafts are custom-made for a specific patient to ensure that the patency of the visceral vessels is maintained following device deployment. However, occasionally during stent graft deployment, the device may rotate, in which case the fenestrations no longer line up with their intended vessels [1]. This can lead to longer surgical times, increased contrast use, kidney failure, and death. Rotation is thought to be caused by tortuosity, stenosis, and/or local areas of calcification in the iliac artery through which the device is delivered prior to deployment.

Several groups [2-4] have used finite element analysis to simulate stent graft deployment for AAA repair; however, none have included delivery within a sheath, nor have they investigated rotational effects during stent graft deployment.

The objective of this study was to use finite element analysis to simulate stent graft deployment within a patient-specific AAA to predict whether graft rotation during deployment was likely to occur.

METHODS

This computational study was performed using the explicit finite element solver LS-DYNA (LSTC, Livermore, CA, USA). As a proof of concept, pre-operative CT scans of a selected patient who had undergone fenestrated stent graft implantation were obtained. Geometries of the AAA and the iliac arteries were extracted from the CT scans using the open-source software The Vascular Modelling Tool Kit. A stent graft model and sheath were developed using SolidWorks (Dassault Systèmes, Waltham, MA, USA). Using HyperMesh (Altair, Troy, MI, USA), the graft fabric, sheath, and vessel walls were meshed using 50,626, 5174, 11,070 shell elements, respectively, while the stents and the guidewire were meshed using 3330 and 749 beam elements, respectively. Material properties for the sheath, stent graft, and stents were taken from the literature [3]. Representative material properties were also used for the guidewire. For the iliac artery and the aorta, a single two-parameter Mooney-Rivlin hyperelastic material model was developed using biaxial tensile test data of aged human iliac arteries taken from the literature [5]. Friction was included in the model between the graft and the sheath, and between the sheath and the vessel wall, using coefficients of friction from our previous work [6].

The first step in the stent graft delivery and deployment simulations was to simulate the insertion of the guidewire. To do so, a straight guidewire was displaced to follow the path of the iliac artery centerline. This imposed displacement condition was then removed, which initiated contact between the wire and the vessel wall and caused the wire to attempt to deform back to straight. After equilibrium was reached, stent graft delivery and deployment were simulated as follows. First, the stent graft model was crimped and inserted within the sheath. Then, it was delivered through the distal end of an iliac artery. The sheath and the stent graft were then prescribed a velocity, which caused them to traverse the iliac artery, following the path of the guidewire,

Poster Presentation #P202

until they were in the correct position within the aorta. The sheath was then withdrawn, allowing the graft to expand to its normal size. Rotation of the stent graft was then determined by calculating the rotation of two opposing nodes on the graft via a comparison between their initial and final position.

RESULTS

Figure 1 shows three of the stages of the stent graft delivery and deployment simulations, namely, the stent graft prior to crimping (a), the stent graft compressed within the sheath during delivery (b), and the stent graft after deployment (c).

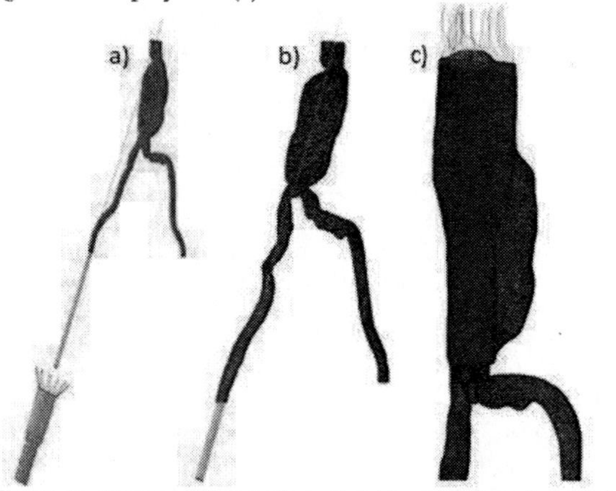

Figure 1: Stages of the stent graft delivery and deployment simulations: the stent graft prior to crimping (a), the stent graft during delivery (b), and post-deployment (c).

Figure 2 is a map of the nodal tracking of two points, one on either side of the stent graft during the delivery and deployment simulations. The initial and final positions of the nodes are highlighted in red, and the amount of rotation of the graft is the angle between the initial and final nodal positions. For clarity, the only the graft has been shown. The stent graft has rotated approximately 35° clockwise upon deployment.

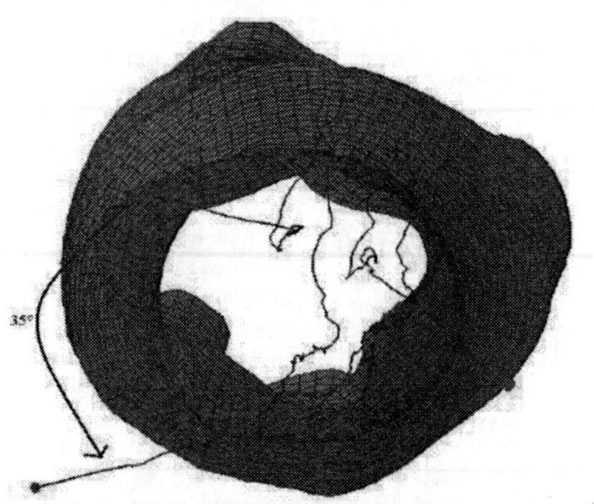

Figure 2: Results of deployment showing rotation of graft via the tracking of two opposing nodes.

DISCUSSION

Simulations of the delivery and deployment of a fenestrated stent graft have been successfully performed. As has been observed clinically, the device rotated upon deployment, and, in this case, that rotation was approximately 35°, which would likely lead to significant complications in cannulating the branch vessels to complete the surgical procedure.

These simulations improve upon previous work by including, for the first time, rotational effects during delivery and deployment. Additionally, previous stent graft simulations have simulated deployment in the absence of the guidewire [2], which, as Figure 1 shows, can severely deform the iliac artery. Deformed iliac arteries are present clinically, and could potentially influence rotation via further frictional interactions with the sheath, and by extension, the stent graft.

There are several limitations to this study. Localized calcium, which is thought to be important in the influence of rotation, is not present in this model. Furthermore, frictional coefficients were measured between a stent graft and a synthetic material, instead of between the stent graft and the aorta. Additionally, the same material model is used for both the iliac arteries and the aorta, and this material is not specific to this particular patient, though as the material model was derived from an elderly patient with significant comorbidities, it is representative of an EVAR patient's arteries. Finally, this case has not yet been validated against clinical data. Ongoing work is being conducted to address these limitations in future simulations, and further patient specific models are being developed to validate the model against clinical data.

In conclusion, the model developed in this work is the first step towards predicting stent graft rotation prior to surgery. Once validated against clinical data, it will allow the attending surgeon to identify patients at risk for this adverse outcome prior to device implantation and modify their surgical approach accordingly.

ACKNOWLEDGEMENTS

We are grateful to Hossein Mohammadi for helping with LS-DYNA and to Olena (Helen) Genis for creating the stent graft models in SolidWorks. Funding for this work was provided by a Physicians' Services Incorporated Foundation Health Research Grant.

REFERENCES

[1] Crawford, S et al., *J Vasc Surg*, 64:1547, 2016.
[2] De Bock, S et al., *J Mech Behav Biomed Mater*, 13:129-139, 2012.
[3] Roy, D et al., *Int J Numer Method Biomed Eng*, 32:e02769, 2016.
[4] Auricchio, F et al., *Comp Biol Med*, 43:386-394, 2013.
[5] Kamenskiy, AV et al., *Biomech Model Mechanobiol*, 13:1341-1359, 2014.
[6] Doyle, MG et al., *Proceedings of SB^{3}C2016, National Harbor, Maryland, USA*, SB3C2016-60, 2016.

MECHANICAL CHARACTERISATION AND MODELLING OF THROMBUS MATERIAL

S. Johnson (1), S. Duffy (2,3), M. Gilvarry (3), J.P. McGarry (1), P.E. McHugh (1)

(1) Department of Biomedical Engineering,
College of Engineering and Informatics,
National University of Ireland, Galway, Ireland

(2) GmedTech,
Galway-Mayo Institute of Technology,
Galway, Ireland

(3) Neuravi Ltd, □al□ay, Ireland

INTRODUCTION

A thrombus or blood clot is a solid mass, made up of a network of fibrin, platelets and other blood components. Blood clots can form in the heart or blood vessels through various pathways, for example as a result of exposed tissue factor from vascular injury, as a result of low flow/stasis, or in very high shear flow conditions. Embolization of cardiac or vascular originating blood clots causing an occlusion of the neurovasculature is the major cause of stroke, accounting for 85% of all stroke. The composition of blood clots (e.g. the fibrin:cellular ratio) will vary depending on the conditions in which the clot is formed, or the age of the clot before it embolizes.

The enormous prevalence of vascular disease has led to the development of minimally invasive treatments such as stent angioplasty, where the implanted stent scaffolds the occluded artery and restores patency [2], and embolic filtration, where the filter captures embolic (thrombus) material in the blood stream and thereby minimises stroke risk. Mechanical thrombectomy, where the thrombus is removed from a vessel, is becoming the new standard of care in stroke therapy [2]. Therefore, defining the material properties of thrombus has obvious implications for the development of treatment devices. To-date, this material definition has not been adequately investigated.

Thrombus material has been found to be incompressible, nonlinearly elastic and inhomogeneous[3]. It has been shown to be highly heterogeneous, with both fibrous regions and cellular regions [4]. Although some experimentation has been performed, the viscoelasticity and rupture behaviour of the material is still not fully understood and model development has been extremely limited. Therefore, the aim of this research is to generate a better understanding of the biomechanical properties and viscoelastic behaviour of thrombus material. This data will be used to formulate an accurate material (constitutive) model for thrombus.

METHODS

The research involves linked experimental and computational modelling aspects.

Experimental Methods

Thrombus analogue material was generated (performed in collaboration with Neuravi Ltd, using their recently developed thrombus analogue generation protocols) [5]. The analogue material samples were generated to represent a range of overall thrombus tissue types, with varying red blood cell content.

Uniaxial unconfined compression testing was used to carry out a range of experiments to determine the viscoelasticity and rate-dependence of the material (Figure 1). The testing was performed using a 10N load cell on a Zwick biaxial testing machine.

Figure 1: Image illustrating the compression test set-up a) before and b) during loading.

Firstly, stress-relaxation testing of the material was carried out to determine if the material was viscoelastic. A strain of 30% was applied to the material and it was held constant for 1000s. Secondly, the rate-dependence of the material was investigated by applying a strain of 30% at strain rates of 1%, 5%, 10% and 15% per second (Figure 2).

Uniaxial tensile testing was also carried out. Dogbone-shaped specimens of the analogue clot material were produced. A 100N load cell was used and samples were loaded at a strain rate of 10mm/min. The test was run until failure occurred (Figure 4).

Computational Methods

The experimental results were used to develop a constitutive model for the material for finite element analysis (FEA). A two-arm generalized Maxwell model, also known as the Wiechert model, was selected to represent the material's viscoelasticity. The stiffness of the material, E(t), is defined as the sum of the stiffness of the isolated spring plus the stiffness of each of the Maxwell spring-dashpot arms:

$$E(t) = E_{\infty} + \sum_j E_j e^{\frac{-t}{\tau_j}} \qquad (1)$$

where E_{∞} is the long term modulus of the material once it is completely relaxed, and τ_j is the relaxation time.

The model was calibrated to the experimental stress-relaxation results by determining appropriate values for the material constants E_0, E_1, E_2, τ_1 and τ_2. The model was then related to a Prony series, so that stress relaxation experiment could be simulated in Abaqus using Prony series viscoelasticity.

A two-parameter Mooney-Rivlin model was also used to capture the hyperelasticity of the material. The Cauchy stress for a Mooney-Rivlin hyperelastic material under uniaxial compression is given by:

$$\sigma = C_{10}(\lambda^2 - \lambda^{-1}) + C_{01}(\lambda - \lambda^{-2}) \qquad (2)$$

where C_{10} and C_{01} are the material constants and λ is the stretch.

An inverse FE method, as proposed in [6], was used to optimise the constitutive parameters. An FE simulation replicating the experiment, using a single-element unit cube model, was performed and the net force–displacement curve was computed after each simulation. A least-squares regression method was implemented in MATLAB to minimize the error between the FE and experimental force-displacement curves until a good fit was achieved.

RESULTS

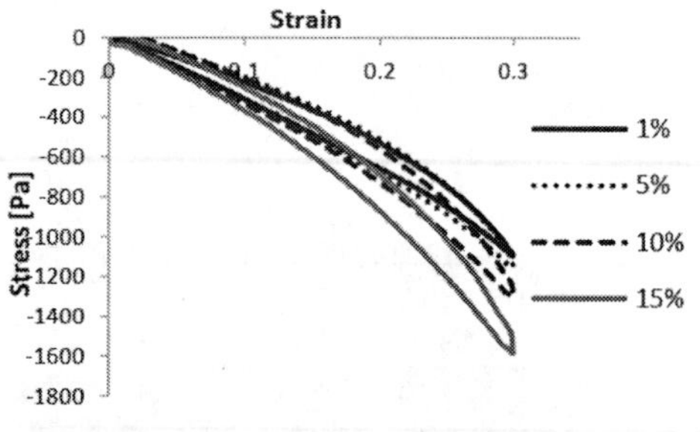

Figure 2: Plot of Stress versus Strain for the compressive loading and unloading experiments at strain rates of 1%, 5%, 10% and 15%.

Figure 2 shows that the clot material exhibited different behavior when loaded at various strain rates, as the material becomes stiffer at higher strain rates. The material also had a different loading and unloading curve, further highlighting its viscoelasticity.

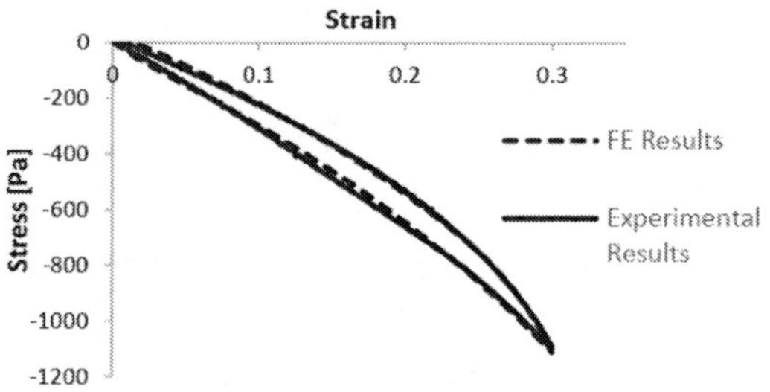

Figure 3: Plot of Stress versus Strain comparing the experimental results with the results from the FE simulation.

Figure 3 above shows that the optimised constitutive parameters accurately captured the experimental results.

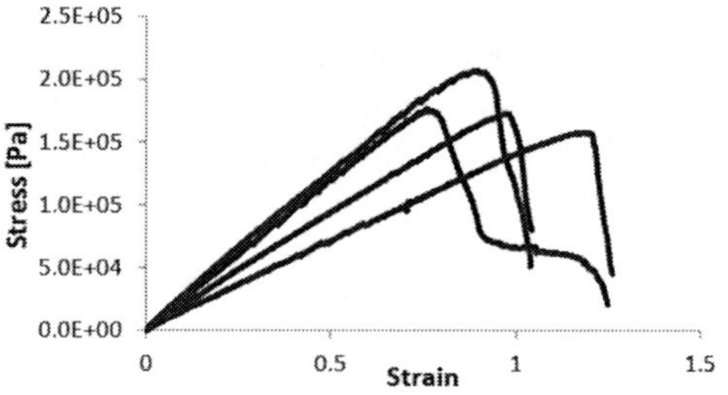

Figure 4: Stress versus Strain plot for the uniaxial tensile testing of samples tested at a strain rate of 10mm/min.

The results from the uniaxial tensile testing in Figure 4 indicate that the material is very elastic as it can reach very high strains of up to 1.2 before failure.

DISCUSSION

The thrombus material was found to exhibit a viscoelastic response. Therefore, a combined hyperelastic and viscoelastic constitutive model was implemented to predict the material behaviour. The optimised material parameters accurately captured the material response under compressive loading and unloading conditions. This indicates that the constitutive model is suitable for implementation in other geometries.

The results from the uniaxial tensile testing indicate that the material is highly elastic as it can reach very high strains before failure. There is also little plasticity indicated.

Clot types also vary depending on their constituents. The results above illustrate the behavior of one type of clot analogue. In the future the properties of a range of clot types with varying constituents will be investigated and compared.

ACKNOWLEDGEMENTS

The authors would like to acknowledge the support of Neuravi Ltd, the Irish Research Council Enterprise Partnership scheme (Grant No. EPSPG/2015/92) and the NUI Hardiman Research Scholarship for this research.

REFERENCES

[1] Furie, B et al., N. Engl. J. Med., vol. 359, no. 9, pp. 938–949, 2008.
[2] Berkhemer O.A, et al., N. Engl. J. Med., no. 372, pp. 11–20, 2014.
[3] Wang, D.H et al., J. Biomech. Eng., vol. 123, no. 6, pp. 536–9, 2001.
[4] Marder, V.J et al., Hemostasis and Thrombosis: Basic Principles and Clinical Practice. 2012.
[5] Duffy, S et al., J. Neurointerv. Surg., vol. 2010, no 6, pp: 1-8, 2016.
[6] Nolan, D.R et al. Ann. Biomed. Eng., vol. 44, no. 4, pp. 993–1007, 2016.

MOUSE AORTIC MECHANICS ASSESSED FROM FINITE ELEMENT SIMULATION USING RING PULL TEST PROPERTIES

Ryan R. Mahutga (1), Neeta Adhikari (2), Jennifer L. Hall (2), and Victor H. Barocas (1)

(1) Department of Biomedical Engineering
University of Minnesota
Minneapolis, MN, USA

(2) Department of Medicine
University of Minnesota
Minneapolis, MN, USA

INTRODUCTION

Cardiovascular disease continues to be the leading cause of death in the U.S., accounting for 23.5% of all deaths in 2013 according to the Centers for Disease Control and Prevention (CDC)[1]. In addition, 17,000 deaths were a result of aortic aneurysm[2]. Therefore, it remains important to study and understand vascular mechanics, how properties change in disease, and how intervention affects function. Many aortic complications arise from disease or genetic predisposition. Mice make an excellent analog to study disease and genetic involvement in passive, static vascular mechanics because mice have similar vascular structure and blood pressure to humans[3]. To date, much research on the murine aorta have focused on mechanics of straight, unbranching vessels using complex systems of cameras, mirrors, pumps, and pressure regulators[4-7]. While there is much to be learned from such studies, we posit that there could be a simpler more expeditious way to understand arteries in mice, especially if one is concerned with non-uniform geometries.

The focus of this study was to examine the properties of the wild-type mouse aorta from the aortic root to the renal branches using ring pull tests. The properties ascertained from the ring pull tests were then used in finite element (FE) simulations to approximate the behavior of the aorta under pressurization. The FE simulations were then compared to inflation tests to determine the efficacy of using properties approximated from ring pull tests in evaluating aortic function.

METHODS

Experimental Specimens

Aortas were excised from healthy, adult male wild-type mice aging from 8 months to over 1 year. The aorta was excised from the iliac bifurcation to the top of the arch, keeping the heart attached at the proximal end. Samples were then prepared for either inflation or ring pull tests as outlined below.

Ring Pull Test Protocol

Using the intact aorta, the heart was first excised from the aortic root. Ring samples of ~0.5mm and ~2mm width were removed in an alternating manner starting at the aortic root and continuing to the renal bifurcation (Fig. 1). The 0.5mm rings were used to measure the average diameter and wall thickness of the aorta, and were cut open to give an opening angle as an indicator of residual stress. The 2mm rings were used in the ring pull tests.

The ring pull tests were completed by placing rings on a custom fixture consisting of two thin wires passing through the center of the rings[8]. The samples were oriented so that the pull would be uniaxial. The rings were then prestretched by a 0.01N load, and the testing performed at a pull rate of 10mm/min until failure. The ring pull tests were treated as uniaxial tension tests assuming the wire was sufficiently thin and relatively frictionless.

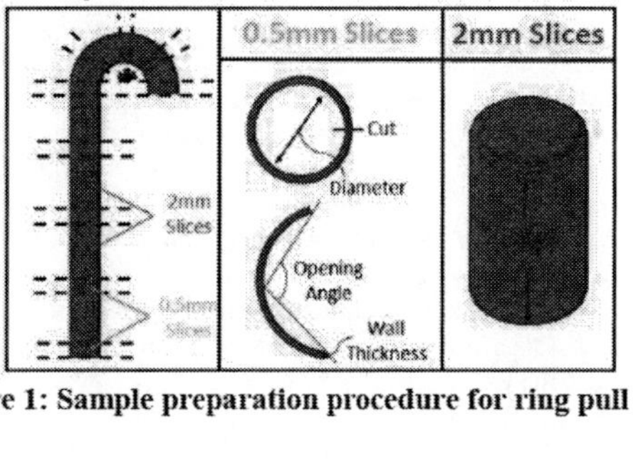

Figure 1: Sample preparation procedure for ring pull tests

Inflation Protocol

Inflation testing was performed by taking the intact aorta and cannulating it on a glass micro-pipette. The aorta was then mounted and submerged in 1x PBS. A stitch was placed in the apex of the heart to tie the heart in place. A suture was placed around the top vessels (brachiocephalic, carotid, subclavian, renal) to seal them. Pressure was measured upstream of the aorta as saline was pumped at a constant rate into the aorta. Imaging was performed along the length of the aorta using a VEVO2100 small animal ultrasound system with 55MHz probe. A z-stack was created with slices every 70um along ~12mm of length.

FE Simulation

An FE simulation geometry was created using images from an aorta prior to testing (Fig. 2). The vessel wall thickness was specified from the 0.5mm ring data. The solid model was then meshed and imported into FEBio for simulation[9] (Fig. 2).

The aorta was partitioned into six regions in this study: root, ascending, descending, upper thoracic, lower thoracic, and abdominal. The material properties were changed based on regional averages from ring pull testing. The strain energy density model was consistent across all regions, and took the form of an uncoupled solid mixture enforcing near incompressibility. The mixture consisted of a neo-Hookean ground matrix and two fiber families, one oriented circumferentially and the other oriented axially. These fibers were represented by the exponential-power strain energy density function (1)[9,10].

$$W_{Fibers} = \frac{\xi}{2\alpha}(\exp[\alpha(I_n^2 - 1)^2] - 1) \quad (1)$$

where: W_{Fibers} is the strain energy density of the fibers
I_n is the square of the fiber stretch
ξ is the bulk fiber bundle modulus
α is the exponent representing nonlinearity

RESULTS

Ring pull tests were performed on three aortas. The neo-Hookean parameter was fit to these tests assuming the neo-Hookean component dominated the stress-strain curve in the small strain regime (strain < 10%), with the averaged parameter being 25.24±7.131 [kPa]. The circumferential fiber parameters, ξ and α, were then fit to the ring test data. The parameters for each region are shown in Table 1. To get a stable solution for this mixture model, fiber properties were assumed for the axial direction (Table 1). These properties represent a significantly stiffer fiber with similar nonlinearity to those oriented circumferentially. For boundary conditions, the proximal end at the aortic root and the distal end at the abdominal aorta were fixed in x,y,z space.

Table 1: Model Constitutive Parameters

Region	α	ξ [kPa]
Root	0.338±0.057	20.56±2.65
Ascending	0.198±0.018	24.83±9.33
Descending	0.201±0.001	29.61±0.707
Upper Thoracic	0.175±0.008	27.12±8.89
Lower Thoracic	0.216±0.085	35.37±6.07
Abdominal	0.138±0.043	78.51±4.19
Axial	0.200	100

Simulation results performed as expected from experimentation. The aorta inflated circumferentially eventually reaching a point where it bowed outward due to the axial strain. Figure 2 shows the simulation results for surface strain and average circumferential stretch at a representative pressure. Figure 3 shows the comparison of data from inflation of an aorta to this model. Stress-stretch plots are shown where dashed lines in brighter colors show experimental data and dotted lines in duller colors represent the model prediction. Figure 4 shows the difference in exterior and luminal stretch for the aorta under pressurization. Dash-dot lines represent the luminal surface and dashed lines represent the exterior surface.

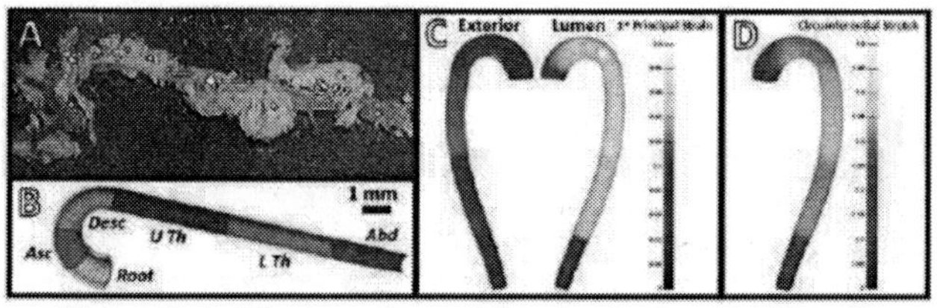

Figure 2: A. Mouse aorta used for ring testing. B. FE mesh recreated from the aorta in A. C. Lagrange strain on the aorta exterior surface (left) and lumen wall (right) at 120 mmHg. B. The average circumferential stretch of the aorta at 120 mmHg.

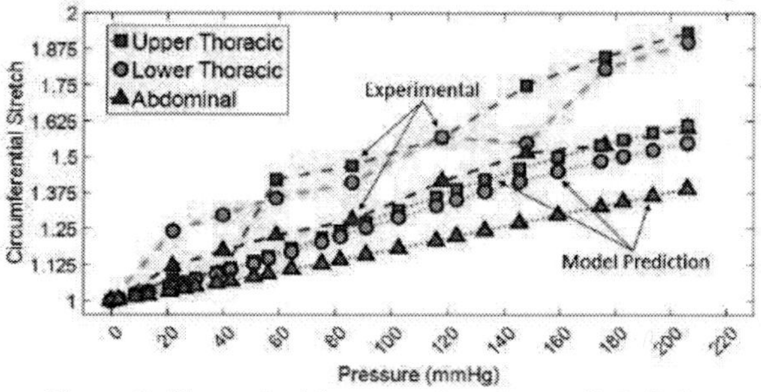

Figure 3: Stress-stretch comparison for aortic inflation

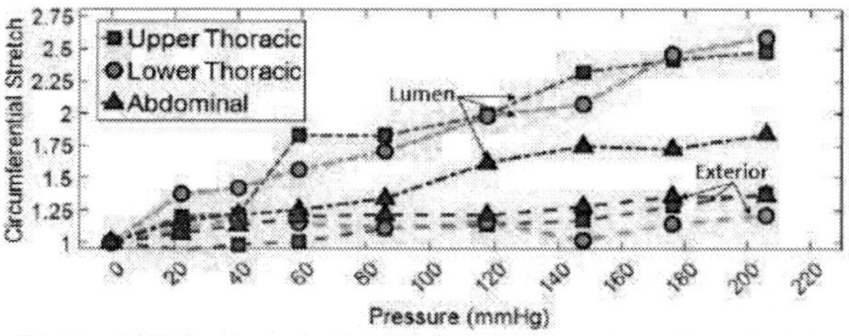

Figure 4: Experimental luminal and exterior surface stretch

DISCUSSION

The simulations show reasonable agreement with the inflation test results both qualitatively and quantitatively. The simulation shows that the artery begins bowing outward around 100mmHg, which had been observed experimentally. The exterior surface and luminal surface show differing strains in the model, a phenomenon that is backed up by the results obtained from experimentation. The pressure-stretch relationship from the model was approximately a 20% underestimation from experimental for the regions compared. This error could be due partly to assuming axial fiber properties, and partly due to not accounting for the residual stress within the material.

Future works will remove the two major sources of error by performing axial testing to get fiber properties, and by adapting this model to account for residual stresses based on experimentally obtained opening angles. The addition of experimental measurements of stress-stretch relations for the root, ascending, and descending parts of the arch should give a full picture of the aortic behavior to compare to this model.

REFERENCES

1. Xu J, et al. *National Vital Statistics Reports*. 2016; 64(2). 2. CDC. *Aortic aneurysm fact sheet*. 2016. 3. Trueting P, Dintzis S. *Comparative anatomy and histology;* 2012. 4. Gleason R, et al. *J. BME,* 2004; 126. 5. Ferruzi J, et al. *Annals of BME*. 2013; 41(7). 6. Bersi M, et al. *Annals of BME*. 2014; 42(6). 7. Wagenseil J, et al. *Am J Physiol Heart Circ Physiol*. 2010; 299. 8. Humphrey J. *Cadiovascular solid mechanics;* 2002. 9. Maas, S. et al. *FEBio Theory Manual*, 2016; (v2.5). 10. Holzapfel, G. et al. *J. BME,* 2004; 126.

SB³C2017
Summer Biomechanics, Bioengineering and Biotransport Conference
June 21 – 24, Tucson, AZ, USA

A COMPARISON OF MORPHOLOGICAL PARAMETERS IN ASIAN AND CAUCASIAN ABDOMINAL AORTIC ANEURYSM PATIENTS USING BIOMECHANICAL AND MACHINE LEARNING METHODS

Tejas Canchi (1, 2, 3), Hong N. Nguyen (1), Sourav S. Patnaik (4), Eddie Y.K. Ng (2), Dinesh K. Srinivasan (3), Sriram Narayanan (5), Ender A. Finol (4)

(1) Department of Biomedical Engineering
University of Texas at San Antonio
San Antonio, Texas, U.S.A.

(2) School of Mechanical and Aerospace Engineering
Nanyang Technological University
Singapore

(3) Lee Kong Chian School of Medicine
Nanyang Technological University
Singapore

(4) Department of Mechanical Engineering
University of Texas at San Antonio
San Antonio, Texas, U.S.A.

(5) Department of General Surgery
Tan Tock Seng Hospital
Singapore

INTRODUCTION

Abdominal Aortic Aneurysm (AAA) is a vascular condition that occurs in the infrarenal region of the aorta. It involves the dilation of the healthy aorta to about 2-2.5 times its original diameter. Without intervention, this dilation leads to rupture over time. The treatment of this condition when diagnosed is through a surgical procedure known as Endovascular Aneurysm Repair (EVAR) where a stent graft is inserted in the diseased abdominal aorta to restore its structural integrity. The geometry of the Asian aorta is suspected to be different from the Caucasian aorta in terms of neck length, common iliac artery diameter and the distance between the lowest renal artery and the iliac bifurcation. To this end, in a clinical investigation to compare 3 commercially available stents, Cheng et al. [1] carried out an assessment of 65 Chinese patients who underwent EVAR in comparison with European and American patients. It was found that the common iliac arteries (CIA) were significantly shorter in Asians, particularly the right common iliac artery. This outcome was corroborated by Mladenovic et al. [2] who found the CIAs to be shorter, along with the mean neck length of Asian AAA being significantly lower than Europeans (22.73 mm vs. 38.72 mm). These differences have caused difficulties with the graft delivery during EVAR and subsequent follow up of AAA Asian patients. The clinical significance of these differences can be brought out through a comparison of the geometric parameters in each cohort along with the mechanical stresses developed in these aneurysms. Therefore, the objective of this work is to analyze, in a cohort of 10 Asian and 10 Caucasian AAA patients, the wall stress and geometric indices that characterize each group and quantify any differences that may exist.

METHODS

Anonymized patient-specific abdominal computed tomography angiographies were obtained retrospectively following appropriate IRB approvals obtained from Allegheny General Hospital, in Pittsburgh, Pennsylvania and the Tan Tock Seng Hospital, Singapore vascular database. The 10 AAA for each group were chosen such that the range of maximum diameters (Dmax) was similar for both groups in the cohort. The images were subjected to segmentation using the in-house script AAAVasc following the method and validation described in [3].

Meshing of the 3D geometries was done using AAAMesh, an in-house script developed exclusively for AAA quadratic hexahedral meshing. The meshes were then exported to the finite element analysis solver ADINA (Adina R&D Inc., Watertown, MA) where the biomechanical stresses were computed using a 2-parameter Mooney-Rivlin constitutive model for the properties of the arterial wall [4]. The finite element equations were solved in 24 time steps incrementally to achieve an intraluminal pressure loading of 120 mmHg. The patient-specific non-uniform wall thickness resulting from the image segmentation was incorporated in the FEA model. The first principal stress distribution on the AAA wall was calculated from which the following biomechanical parameters were computed: maximum of the first principal stress (PWS), spatially averaged wall stress (AWS), and the 99th percentile wall stress (99th–WS).

Sixty-four geometric indices were estimated for both patient groups using several Matlab (The Mathworks Inc., Natick, MA) scripts. These were divided into eleven 1-D size indices, nine 2-D shape indices, five thrombus-related indices, four 3-D size indices, two 3-D shape indices and eleven wall thickness-related indices, in addition to ten surface curvature-based indices. The complete mathematical formulation of thee geometry metrics can be found in [5].

Poster Presentation #P206

The geometric indices and biomechanical parameters were subject to a machine learning algorithm (MLA) using the open source application WEKA [6]. The class variable was chosen to be race with Caucasian and Asian being the two states of the variable. The decision tree machine learning analysis yielded values of classification accuracy and a Receiver Operating Characteristic (ROC) curve. The geometric and biomechanical data were also analyzed using an independent t-test (2-tailed) using SPSS (IBM Corporation., Armonk, NY) to determine if there are statistically significant parameters that differentiate each group.

RESULTS

The range of Dmax for the patient cohort was 29.1 – 67.4 mm. The mean PWS in the Asian group was 0.90 MPa compared to 0.92 MPa for the Caucasian group ($p = 0.94$). The mean AWS were 0.18 MPa and 0.21 MPa for the Asian and Caucasian groups, respectively ($p = 0.50$). The mean 99th–WS were 0.36 MPa for the Asian AAA and 0.39 MPa for the Caucasian AAA ($p = 0.69$). However, differences were observed in size-matched comparisons. An Asian AAA with a Dmax of 29.1 mm exhibited a PWS of 0.9 MPa while a Caucasian AAA with a Dmax of 27.3 mm resulted in a PWS of 0.2 MPa. Likewise, a 55.4 mm Asian AAA had a PWS of 1.13 MPa while a 56.4 mm Caucasian AAA had a PWS of 0.77 MPa.

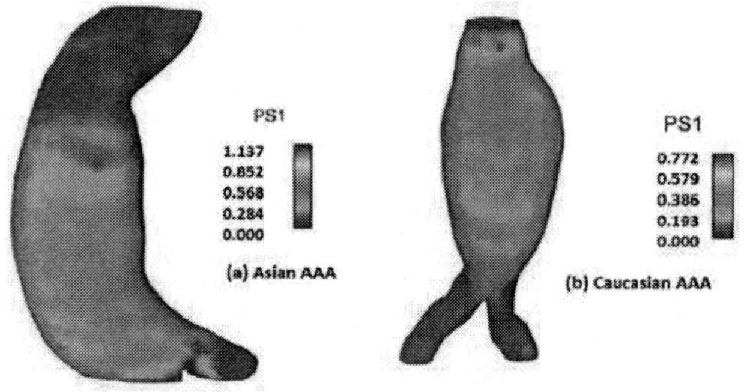

Figure 1: Spatial distribution of first principal stress in an (a) Asian and (b) Caucasian AAA (MPa).

Figure 1 shows the wall stress distributions for an Asian AAA with a Dmax of 55.4 mm and a Caucasian AAA with a Dmax of 56.4 mm. The distributions are similar while the Asian AAA tends to be more tortuous than the Caucasian AAA.

The MLA generated a decision tree that contained the geometry index GLN (the L2-norm of the Gaussian curvature). This was the determining variable when deciding the classification of an aneurysm belonging to either the Asian or Caucasian groups, as seen in Fig. 2. The classification accuracy was 95% with a ROC area under the curve of 0.915, as seen in Fig. 3.

The independent t-test yielded 15 parameters that were significant between the two groups. The test was conducted assuming non equal variances and a significance level of $\alpha = 0.05$. These included neck height ($p < 0.002$), sac height ($p < 0.006$), maximum compactness ($p < 0.006$), minimum thrombus thickness ($p < 0.004$), and GLN ($p < 0.002$). Noteworthy is that the latter was also seen in the machine learning analysis as the determining variable of classification between the two groups. Other curvature-based indices that were also significant include: the Gaussian (KG, $p < 0.003$) and Mean curvatures (KM, $p < 0.010$), the first (K1AA, $p < 0.001$) and second principal curvatures (K2AA, $p < 0.001$) and the L2-norm of the second principal curvature (K2LN, $p < 0.008$).

DISCUSSION

This is the first known work on comparing the geometric and biomechanical characteristics of Asian and Caucasian AAA. The biomechanical comparison resulted in Asian AAA having higher peak wall stresses for small maximum diameters. However, there were no statistical differences in PWS, AWS, or 99th–WS between the two groups when comparing them without size-matching. The classification analysis by means of a MLA resulted in AAA wall surface curvature being the most important parameter in the classification, in agreement with the work of Lee et al [6]. Curvature is a spatially distributed metric highly correlated with wall stress [7]. This finding was corroborated through a student t-test where the geometry metrics that were significant between the two groups included the surface curvatures and neck height, which has been reported to be one of the important morphological parameters that differentiate Asian from Caucasian AAA [2]. A larger patient cohort is required to corroborate these findings.

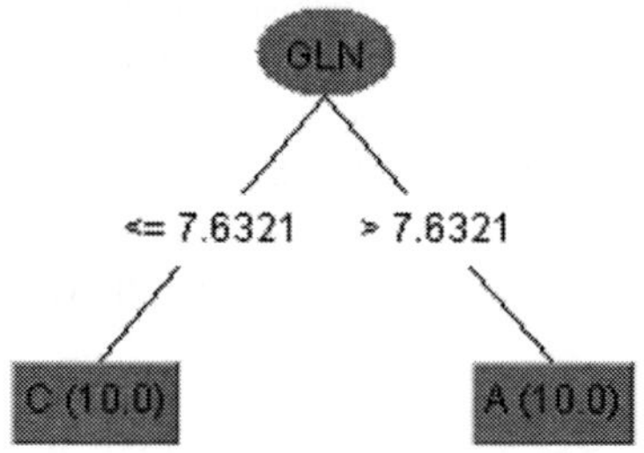

Figure 2: Decision Tree visual ensuing from the MLA application.

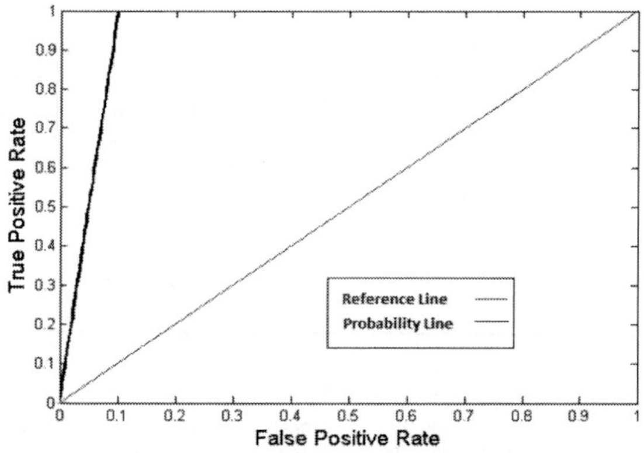

Figure 3: Receiver Operator Characteristic (ROC) curve for the decision tree machine learning analysis.

ACKNOWLEDGEMENTS

Tejas Canchi acknowledges receipt of the Research Student Scholarship from the Lee Kong Chian School of Medicine, Nanyang Technological University, Singapore.

REFERENCES

1. Cheng, SWK et al., *J. Endovascular Therapy*, 11:605-612, 2004.
2. Mladenovic, AS, et al., *Eur J Radiol*, 81:1990-1997, 2012.
3. Shum, J, et al., *Medical Physics*, 37(2):638-648, 2010.
4. Raghavan, ML and Vorp, DA, *J Biomech*, 33(4):475-482, 2000.
5. Shum J, et al., *Ann Biomed Eng.*, 39(1):277-286, 2011.
6. Lee, K, et al., *Ann Biomed Eng*, 41(3):562-576, 2013.
7. Giannoglou, G, et al., *Angiology*, 57(4):487-494, 2006.

MULTIDIMENSIONAL ANEURYSM GROWTH: A BIOENGINEERING APPROACH TO ASSESS THORACIC AORTIC ANEURYSMS

Alina Ismaguilova (1), Giampaolo Martufi (1), Jehangir J. Appoo (2), Eric J. Herget (3), Amy Bromley (4), Lorraine T. Royall (4), Naeem Merchant (3), Elena S. Di Martino (5).

(1) Biomedical Engineering
University of Calgary
Calgary AB Canada

(2) Dept. of Cardiac Surgery
Libin Cardiovascular Institute of Alberta
Calgary AB Canada

(3) Dept. of Radiology
Foothills Medical Centre
Calgary AB Canada

(4) Dept. of Pathology and Laboratory Medicine
Foothills Medical Centre
Calgary AB Canada

(5) Dept. of Civil Engineering
University of Calgary
Calgary AB Canada

INTRODUCTION

An aortic aneurysm is the pathological dilation of the largest artery in the body: the aorta. A healthy aorta exhibits low stiffness at low strain and high stiffness at supra-physiologic strain, which enables the aorta to cyclically stretch without dilating past its physiological threshold. An aneurysm is a dilatation of the aorta accompanied by stiffening and weakening of the tissue [1]. Aortic aneurysms are generally indolent until they become symptomatic, at which point they present catastrophically with intimal dissection or rupture, resulting in massive internal haemorrhage and/or death [2,3]. At time of rupture, patient mortality approaches 100% with over half of patients dying before reaching the hospital [4]. There is thus a clear need to develop a method that helps better prioritize treatment options for such a life-threatening prognosis.

Variable incidences of aneurysmal growth in the descending aorta following the surgical repair of acute Stanford Type A aortic dissection have been observed. Such a pathology is unique as it involves the repair of one problem and the consequent presentation of another problem distally along the aorta. The development of an aortic pathology distally from the initial Type A dissection is an occurrence that has not yet been well explored. Although it has been observed that dissection is the most common cause of aneurysmal growth, such studies are restricted to the same aortic segment [5]. In addition, the monitoring of diameter in an aneurysm is normally performed by monitoring the changes in diameter at the location of the maximum diameter of the aneurysm. This method, however, only monitors aneurysmal growth at one cross-section, thus areas of fast growth in other areas of the aorta will be missed and axial growth cannot be quantified [6]. With the use of three-dimensional remodeling techniques and histochemical analysis of surgical samples, it is possible develop a more quantitative, non-invasive method of risk assessment of TAA growth in patients that have undergone surgical resection of the ascending aorta.

The current standard of care considers maximal aortic diameter to guide decisions for elective repair of thoracic aortic aneurysms, herein referred to as the "surgical standard" [7]. Such a method of risk assessment has faced strong challenge because of its inability to accurately predict rupture for all cases. It has been commonly accepted by vascular surgeons in literature that aneurysms can rupture at virtually any size, and the consequences of aortic rupture are often fatal [8]. Both the clinically established size criterion as well as novel approaches of computational engineering stress analysis for rupture risk of aneurysms have limitations. In particular, the existence of inter- and intra-patient variations in material properties is known, but has so far not been addressed.

With these limitations in mind, the primary goal of this study is to explore the growth patterns of the entire descending aorta in patients following acute DeBakey Type I surgery. A multidimensional investigation of descending aorta growth was applied to CTA follow-up data from 37 patients. the applied analysis method uses multiple centerline-based diameter measurements between the aortic arch and the iliac arteries. From these geometric measurements, we aim to compare the the progression of multidimensional growth across the entire descending aorta with the surgical standard.

METHODS

Data from 130 patients who have undergone elective repair of an acute DeBakey Type I dissection was retrospectively collected from the

Foothills Medical Center in Calgary, Canada. Criteria for inclusion involved the availability of at least two sufficiently high-resolution CTA scans that allowed for accurate reconstructions of the vessel at two different time points. CTA scans with strongly inhomogeneous and inconsistent lumen intensity were rejected. Other exclusion criteria were patients with: intramural hematoma (IMH), penetrating aortic ulcer (PAU), contrast agent intolerance, and descending aorta surgery.

Ascending aorta pathology samples were collected for all the patients during the Type I dissection surgery. The samples were stained with Movat's pentachrome stain. The image of the stained sections obtained by microscopy was captured by a DP73 Olympus video camera (Olympus, Massachusetts, USA) for automatic colorimetric analysis to quantify elastin and proteoglycan content.

The descending aorta of 40 patients (31 male and 9 female) was systematically reconstructed from CTA data (A4clinics, VASCOPS) to monitor three-dimensional growth over time. Inclusion criteria for this study involves Type A dissection patients with at least 1 year of follow-up and the availability of at least two sufficiently high-resolution CTA scans that allowed an accurate three-dimensional (3D) reconstruction. Aneurysm growth was monitored at 100 cross-sections perpendicular to the centerline per single reconstruction. Histological colorimetric analysis was done on surgically removed specimens from the dissection site in the ascending aorta to measure the amount of proteoglycans, smooth muscle, and elastin, and to quantify medial thickness, degree of cystic medial degeneration, and inflammation.

The descending aorta from 106 CTA scans was systematically reconstructed from the origin of the left subclavian artery to the upper renal artery using deformable contour model implemented in the A4Clinics 4.0 diagnostic software (VASCOPS GmbH, Graz, Austria) by a single observer. The developed reconstructions were then verified by another co-author.

The 3D geometry of the descending aorta was automatically segregated into cross-sections orthogonal to the lumen centerline such that the aorta's luminal and outer wall surfaces could be represented through 100 cross-sections (or "splines"). The combination of these splines and the vessel centerline provided a simplified 3D representation of the patient's descending aorta. Fully automatic analysis of the final geometries was then performed with a locally developed algorithm using MatLab software (The MathWorks, Natick, Mass).

RESULTS

Analysis of the 3D growth data provided results that greatly contrasted the current clinical standard of care. On average, we found that the cross-section of the aorta that exhibited the fastest rate of growth (or, radial expansion in millimeters per year) did not coincide with surgical standard, or maximum diameter (Spearman's rho = 0.13, $p = 0.36$). As observed in Figure 1, the most rapidly growing cross-section of the descending aorta had significantly greater growth when compared to that of the surgical standard (3.4 mm/yr vs. 0.6 mm/yr, $p<0.001$). It is also important to note that axial growth was also observed along the aorta, with an average lengthening in aneurysm sac of 8.9 mm/yr. No statistical correlation was found between maximal growth and elastin or proteoglycan content from histological analysis, although cases with focal cystic medial degeneration in the ascending segment exhibited the highest maximum growth in the descending aorta, when compared with aortas showing diffuse inflammation.

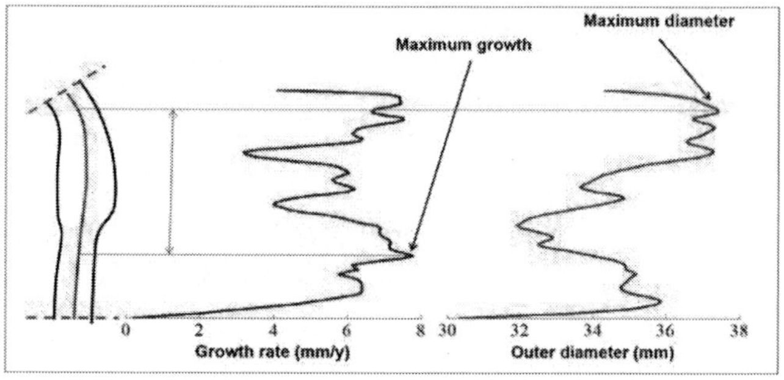

Fig. 1. Comparison between growth rate (mm/year) and outer diameter (mm) of a descending TAA.

DISCUSSION

The results indicate that the descending aorta grows haphazardly at various locations along its length with areas of fast diameter expansion as well as, remarkably, areas of shrinkage (**Fig. 1**). More specifically, maximum aneurysmal growth does not necessarily correspond to maximum diameter. When compared with aortas showing diffuse inflammation, cases with focal cystic medial degeneration in the ascending aortic segment exhibited the highest maximum growth in the descending aorta.

The current clinical standard of monitoring aneurysm growth through changes in the surgical standard maximum diameter ultimately does not capture the state of the aneurysm. Due to the heterogeneous nature of vessel dilatation, localized spots of fast diameter growth can be detected through multiple centerline-based diameter measurements over the whole aneurysm. Such a novel discovery further reinforces the quality of using multidimensional bioengineering techniques for quantitative aneurysm surveillance.

We believe that this project is a step forward in aortic aneurysm assessment. These findings can further be utilized in clinical practice in order to assess the risk of rupture of each aneurysm on a patient-specific basis. Globally, an improvement in TAA risk assessment using such bioengineering techniques will avoid aneurysm-related mortality without increasing the number of unnecessary surgeries and their consequent complications.

ACKNOWLEDGEMENTS

Support from the NSERC Discovery Grant to EDM is acknowledged.

REFERENCES

[1] García-Herrera, C, *Med Biol Eng Comput,* 50(6):559-66, 2012.

[2] Isselbacher, E, *Circulation,* 111(6):816-28, 2005.

[3] Vorp, D et al., *Ann Thorac Surg,* 75:1210-4, 2003.

[4] Johansson, G, *J Vasc Surg,* 21(6):985-8.

[5] LeMaire, S et al.,. *Rev Cardiol,* 8(2):103-13, 2011.

[6] Martufi, G et al., *Biomech Model Mechanobiol,* 13(5):917-28, 2014.

[7] Elefteriades, J, *Yale J Biol Med,* 81(4): 175-86, 2008.

[8] Elefteriades, J, *Ann Thorac Surg,* 74:S1877-80, 2002.

CHARACTERISATION OF MECHANICAL PROPERTIES OF VASCULAR TISSUE IN A QUASI-2D SETTING

Stefan N. Sanders, Frans N. van de Vosse, Marcel C.M. Rutten

Biomedical Engineering
Eindhoven University of Technology
Eindhoven, The Netherlands

INTRODUCTION

Rupture of atherosclerotic plaques in the carotid artery is a major cause for stroke. Rupture is a dangerous event, since it causes total vessel obstruction through thrombosis, and as such may be a lethal condition. Currently, the level of lumen occlusion caused by the stenosis is used to estimate the risk of plaque rupture. However, plaque rupture occurs when the mechanical stresses in the cap of the plaque exceed the local tissue strength, not necessarily when the luminal occlusion exceeds a certain level. Therefore, a biomechanical model of the plaque and its mechanical properties may help to better assess rupture risk.

We developed a method to assess material properties of (diseased) vascular tissue in a quasi-2D setting. Comparing traditional whole-vessel inflation experiments [1] to our inflation method on thin slices of the same carotid artery, revealed different material behaviour. However, a difference in pre-stretch is not expected to alter material properties [2]. Possibly, the thickness of the slice effects the integrity of the sample, changing the material behaviour. In this study, we looked into the effect of slice thickness on material properties and compared our method to whole-vessel inflation experiments.

METHODS

To look into the effect of slice thickness on the material behaviour, healthy porcine carotid arteries were cut into slices of 0.3mm, 0.5mm and 0.7mm. The samples were slightly compressed (6-7%) between two glass plates and inflated up to 100mmHg by injecting fluid into the lumen, through a hole in the bottom plate. A pressure sensor monitored the intraluminal pressure, while a high speed camera recorded the displacement of the sample. The lumen diameter was calculated from these images, the resulting pressure-diameter curves were compared.

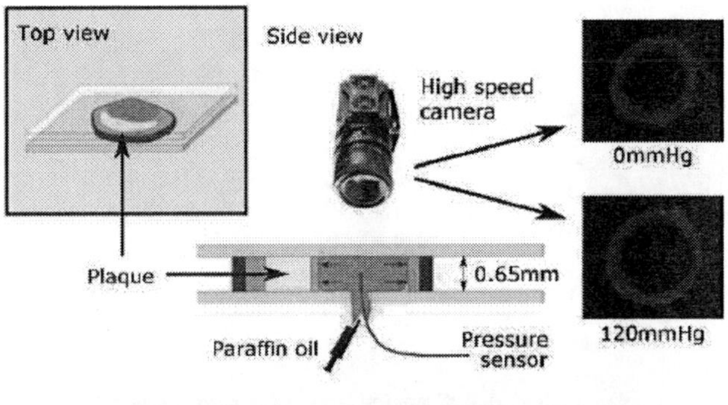

Figure 1: Schematic overview of the experiment.

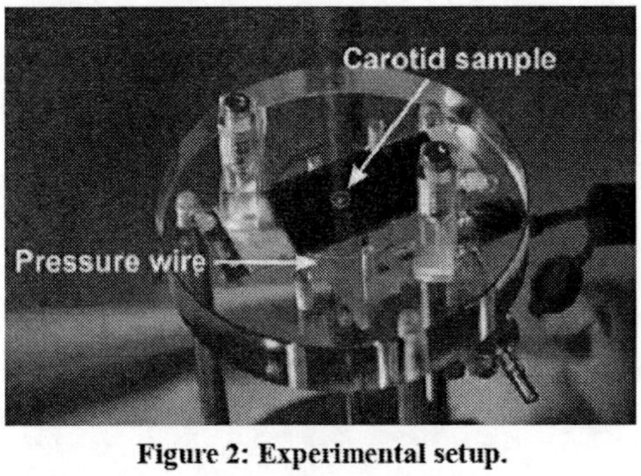

Figure 2: Experimental setup.

For validation of our ring-inflation method, whole-vessel inflation experiments were performed. A healthy porcine carotid artery was mounted in a water bath, with 60% pre-stretch to mimic the in-vivo situation [3]. An ultrasound probe, placed over the vessel, recorded the displacements while the artery was pressurized (figure 3). Next, slices of 0.7mm thickness were cut from the artery, mounted in the ring-inflation setup and slowly pressurized to at least 120mmHg. For reproducibility reasons, the experiments were repeated three times.

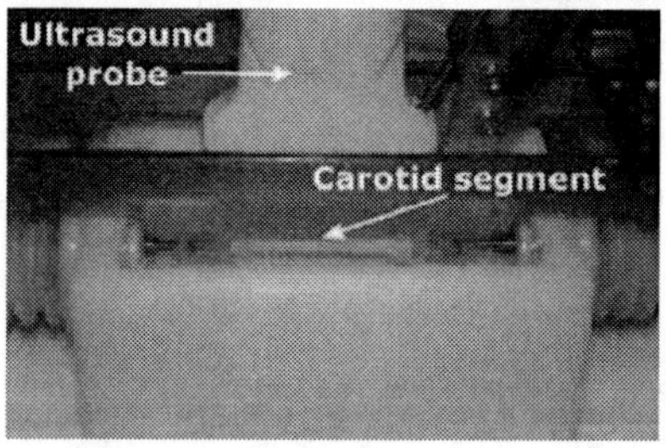

Figure 3: Ultrasound experiment setup.

For comparison, a 1D axisymmetric finite element model was used to estimate the material properties from the pressure-diameter relations obtained through both methods. A Holzapfel-Gasser-Ogden [4] material model was used to obtain matrix stiffness, collagen fibre stiffness and the fibre stiffening parameters, c, k_1 and k_2, respectively.

RESULTS

Sample thickness appears to affect material behaviour (figure 4). Increasing the slice thickness results in stiffer samples.

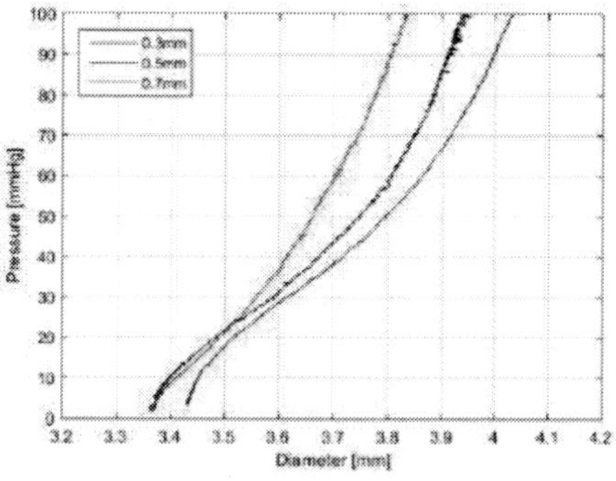

Figure 4: Pressure-diameter curves for samples of different thickness.

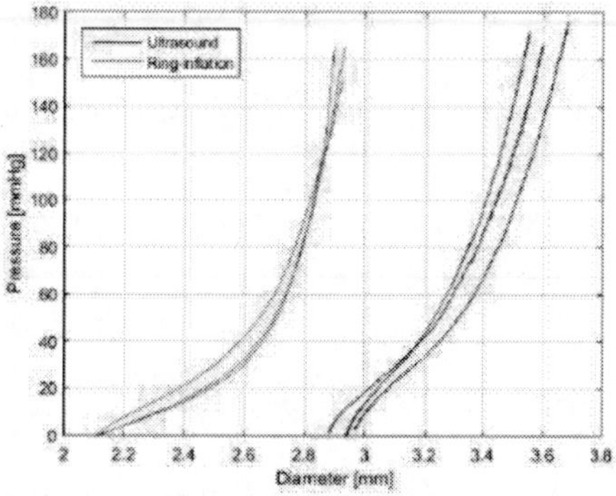

Figure 5: Pressure diameter curves of the ultrasound (red) and ring-inflation (blue) experiments .

Table 1: Material parameter estimates for the ultrasound and ring-inflation experiment, depicted as mean±standard deviation.

Ultrasound experiment			Ring-inflation experiment		
c [kPa]	k_1 [kPa]	k_2 [-]	c [kPa]	k_1 [kPa]	k_2 [-]
18.4±5.7	12.1±0.9	19.1±0.6	31.3±5.9	5.0±1.7	19.1±0.6

Typical strain-stiffening pressure-diameter curves are obtained with both the ultrasound and ring-inflation measurements (figure 5). For both experiments the pressure-diameter curves and resulting material parameters show little variation, indicating reproducibility (table 1). The matrix stiffness appears higher in the ring-inflation experiment, whereas the fibre stiffness parameters are higher in the ultrasound experiment. As the matrix and fibre stiffness are dependent on each other, the model was rerun, interchanging the parameters. The mean ultrasound material parameters were used as input with the boundary conditions of the ring-inflation experiment (figure 6). The resulting curve is in the same diameter range as the ring-inflation data.

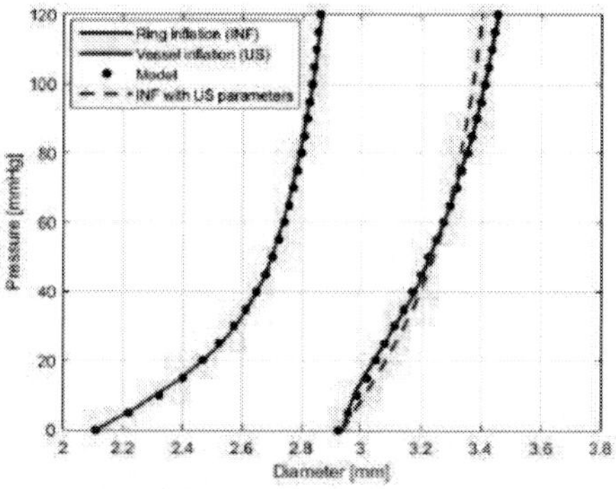

Figure 6: Comparison ultrasound and ring-inflation experiment. The raw ultrasound (red) and ring-inflation (blue) data are shown with the model results (black dots) as well as the model result with interchanged parameters (red dashed line).

DISCUSSION

Our results indicate the need for a minimum size of sample to be taken before it can be considered representative, enabling the collection of accurate data on material properties during testing. More experiments are needed to find this minimal thickness, in which the properties of the vessel are preserved. A good agreement between the ring-inflation and whole-vessel ultrasound experiment was found, using 0.7mm thick samples.

The results so far hint towards a method that may successfully be applied for assessment of vascular properties, in a quasi 2D experiment. In future applications, heterogeneous properties, like in atherosclerotic plaque material, may be assessed as well, using vital staining techniques to distinguish different tissue components without affecting their mechanical behaviour.

ACKNOWLEDGEMENTS

This project is funded by the STW open technology program (STW #12548).

REFERENCES

[1] van den Broek, CN et al., *Biomech. Mod. Mech.*, 10:249-258, 2011.
[2] Schulze-Bauer, CAJ et al., *J. Biomech. Eng.*, 125: 395-406, 2003.
[3] Boekhoven, RW et al., *Ultrasound. Med. Biol.*, 40:1329-1342, 2014.
[4] Holzapfel, GA et al., *J. Elasticity*, 61:1-48, 2000.

SB^{3}C2017
Summer Biomechanics, Bioengineering and Biotransport Conference
June 21 – 24, Tucson, AZ, USA

TOWARDS ACCURATE ATHEROSCLEROTIC PLAQUE FAILURE MODELS: INVESTIGATING MATRIX-CALCIFICATION DELAMINATION

Brían L. O'Reilly (1), Peter E. McHugh (1), Patrick McGarry (1)

(1) Biomedical Engineering
National University of Ireland, Galway
Galway, Ireland

INTRODUCTION

Atherosclerosis is a major cause of heart failure and strokes worldwide. This disease entails the formation of an occlusion within an artery by a lesion of atherosclerotic plaque. Clinical intervention typically involves the widening of the blocked section of the vessel by the deployment of a stent.

Histological studies of atherosclerotic plaque have provided extensive characterization of the composition of the diseased tissue [1]. However, the mechanical behavior of atherosclerotic plaque has not been rigorously investigated to date. The mechanical contributions of calcifications and soft tissue anisotropy have not been parsed. Additionally, failure mechanisms of composite calcified plaques have not yet been uncovered. As the mechanical deformation and rupture of plaque is the key factor that governs the success of angioplasty procedures, an improved understanding of plaque mechanics must be developed in order to improve patient outcomes and avoid stent fracture [2].

In this paper we develop experimental and computation techniques to provide new insight into the mechanical behavior and failure of calcified plaques.

METHODS

Experimental Testing

Excised human atherosclerotic plaque samples were acquired from University College Hospital Galway as approved by the hospital's ethical research committee. The plaque samples were scanned using a 14.8 μm resolution micro-computed topography (μCT) scanner. Mineral oil was used as a medium during scanning. This technique allows for imaging of the surface of the low-density fibrous tissue in addition to the high density calcifications. Scans provide high resolution 3D digitized images of bi-phasic plaques in which irregular calcification morphologies are embedded in irregular soft fibrous tissue morphologies, as shown in Figure 1(a).

Composite sections of the plaque, consisting of calcifications embedded in soft fibrous tissue, were subjected to uniaxial stretching to failure. Following mechanical testing, the fractured composite plaques are subjected to scanning electronic microscopy (SEM) to identify potential failure mechanisms. In addition to the testing of composite sections, homogeneous sections of plaque without any calcifications were biaxially tested to parse anisotropic mechanical behavior of the fibrous tissue.

Model Development

The complexity of plaque composite morpholgies combined with non-linear material behavior and damage evolution requires the development of finite element models in order to interpret mechanical test data. In addition to accurately representing the composite morphology, models must also incorporate appropriate anisotropic constitutive laws and damage mechanisms.

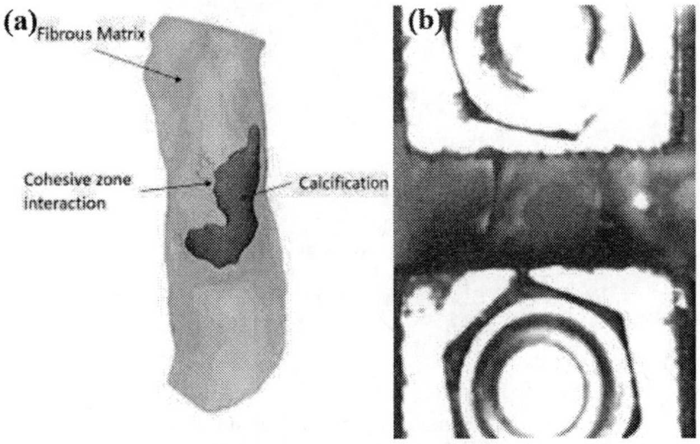

Figure 1: (a) CT generated atherosclerotic plaque geometries. The plaque consists of macrocalcifcation embedded in matrix material. (b) Experimental testing of an atherosclerotic plaque sample to failure.

A compressible form of the HGO anisotropic hyperelastic model [3, 4] is calibrated using the biaxial test results, whereby the stress tensor is given as

$$\sigma = \kappa_0(J-1)\mathbf{I} + \mu_0 J\left(\overline{\mathbf{B}} - \frac{1}{3}\bar{I}_1\mathbf{I}\right) + 2J^{-1}k_1 \sum_{i=4,6}(I_i - 1)\exp[k_2(I_i - 1)^2](\boldsymbol{a}_i \otimes \boldsymbol{a}_i) \quad (1)$$

where κ_0 is the isotropic bulk modulus, μ_0 is the isotropic shear modulus, $k_1$and k_2 are the anisotropic material constants, $\boldsymbol{a}_i$ is the fiber directional vector, J is the determinant of the deformation gradient, I_1 is the first invariant of the right Cauchy-Green tensor and I_4 and I_6 are the anisotropic invariants describing the deformation of reinforcing fibres.

CT scans are used to develop finite element meshes of soft tissue and calcifications. Based on SEM imaging results (discussed below) fracture at the soft tissue-calcification interface is simulated using a mixed mode cohesive zone model proposed by McGarry et al. [5] whereby the work of separation given as

$$\phi(\Delta_n, \Delta_t) = \phi_0 + \sigma_{max}\exp(1)\left(1 + \sqrt{\frac{\Delta_n^2}{\delta_n^2} + \frac{\Delta_t^2}{\delta_t^2}}\right)\exp\left(-\sqrt{\frac{\Delta_n^2}{\delta_n^2} + \frac{\Delta_t^2}{\delta_t^2}}\right) \quad (2)$$

Expressions for normal and tangential traction are derived from equation (2), resulting in fully coupled mixed mode behavior.

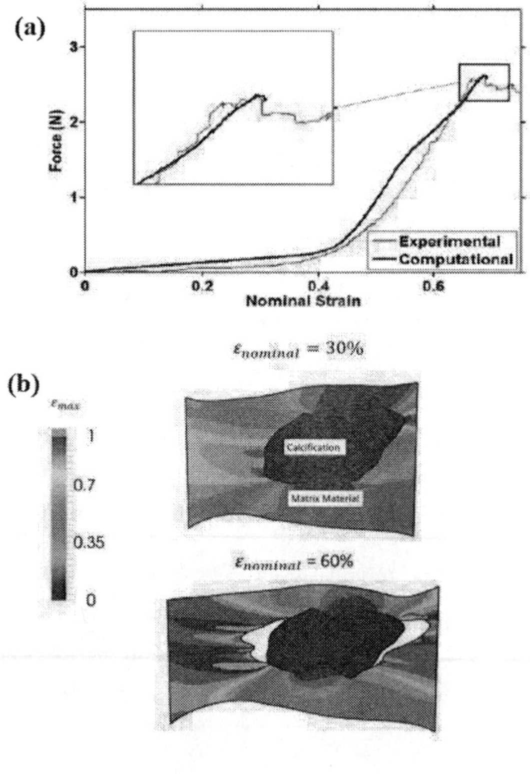

Figure 2: (a) Experimentally measured and computationally predicted force evolution and ultimate failure of composite calcified plaque. (b) Computed distribution of strain in composite plaque prior to failure (top) and at the point of failure (bottom). Significant delamination of soft tissue from calcification surface is predicted at failure.

Figure 3: SEM image of the fracture surface along the delaminated calcification. The white box highlights the transition from a damaged region (fibers ruptured/removed) to an undamaged region (with interdigitated fibers).

RESULTS

Mechanical testing of the composite plaque sample results in a non-linear force-strain relationship, as shown in Figure 2(a). The structure is initially compliant, but at a nominal applied strain of ~0.4 a significant increase in stiffness occurs as collagen fibres in the soft tissue become highly stretched. The sample fractures at a nominal strain of 62% force.

SEM imaging reveals that fracture occurs at the interface between the soft tissue and calcified particles. As highlighted in Figure 3, fibers that are initially connected to the stiff calcification through inter-digitation are observed to pull-out and rupture when the sample fails.

As shown in Figure 2(a), the computational model accurately predicts the non-linear behavior of the composite plaque, in addition to correctly predicting sample failure. The strain distribution in the sample prior to failure is shown in Figure 2(b) revealing three areas of localized deformation in the soft tissue. At an applied nominal strain of 60% significant delamination propagates along the calcification surface, providing an accurate prediction of sample failure.

DISCUSSION

This study presents the first investigation into plaque rupture models using realistic geometries derived from CT scans. Experimental testing uncovers non-linear material behavior and damage propagation at the soft tissue-calcification interface, leading to mechanical failure of the composite structure. Cohesive zone modelling is shown to accurately predict the mode of failure, in addition to predicting the applied load at which ultimate structural failure occurs. The experimental-computational analyses performed in this study provide new insight into the mechanical behavior and failure modes of calcified plaques. The modelling approach can potentially be extended to guide clinical treatment by simulating patient specific valuable design tool for evaluation and development of implant devices.

ACKNOWLEDGEMENTS

The authors would like to thank the clinicians at the Western Vascular Institute in University College Hospital Galway for provision of plaque samples. BOR acknowledges a NUIG College of Engineering and Informatics PhD Scholarship.

REFERENCES

[1] Saam, T et al., *Arterioscler. Thromb. Vasc. Biol.*, 25:234-239, 2005
[2] Morlacchi, S et al., *J. Biomech,*, 47:899-907, 2014
[3] Holzapfel, GA et al., *J. Elasticity*, 61:1-48, 2000
[4] Nolan, DR et al., *J. Mech. Behav. Biomed.*, 39:48-60, 2014
[5] McGarry, JP et al., *J. Mech. Phys. Solids.*, 63:336-362, 2014

SB^{3}C2017
Summer Biomechanics, Bioengineering and Biotransport Conference
June 21 – 24, Tucson, AZ, USA

NON-CONTACT MEASUREMENT OF CAROTID ARTERY PULSEWAVE VELOCITY: NECK PHANTOM AND PRELIMINARY IN-VIVO RESULTS

Jon Reeves (1&2), Shyam Thacker (1), Awais Yousaf (1), Malcolm Birch (1&2), Viviana Mancini (3), Daniela Tommasin (3), Louise Marais (4), Pierre Boutouyrie (4), Patrick Segers (3), Steve Greenwald (1),

(1) Blizard Institute
Queen Mary University of London
London, UK

(2) Barts Health Trust
London, UK

(3) Institute Biomedical Technology
Ghent University
Ghent, Belgium

(4) INSERM U970
56 rue Leblanc
75015 PARIS, France

INTRODUCTION

Aortic pulse wave velocity (PWV), measured from the time delay of the pulse between carotid and femoral measuring sites, is a powerful prognostic indicator for cardiovascular disease (CVD). Currently, no tools are available to measure PWV to screen large populations for CVD risk factors, because all commercially available devices require skilled operators and most are prone to movement artefacts. As part of a larger consortium, we aim to build a non-contact device to measure (PWV) between the carotid and femoral arteries as well as measuring PWV in the carotid artery itself by detecting movement of the skin due to the underlying pressure pulse. Our immediate objective is to prove the principle that PWV can be measured in a simple model (phantom) of the neck containing a tissue mimicking gel, using a commercially available dual-beam laser Doppler vibrometer (LDV). The LDV measures displacement at the surface of the model at two sites a known distance apart. Displacement of the surface is due to the passage of a pressure pulse wave in a compliant tube embedded within the model.

A secondary aim of this work is to test the hypothesis that the same equipment can detect low-amplitude displacement of the skin (amplitude $\approx 1\mu m$, frequency range 100 - 1000 Hz) due to disturbed flow distal to a stenosis [1]. The data from the model measurements provides input for computational models, currently under development, of the processes by which energy is transferred from the flowing blood to the arterial wall and thence, through soft tissue, to the skin.

Finally, we report the results of preliminary measurements of carotid artery pulse transit time in healthy volunteers, confirming an earlier report [2] that a LDV can be used to detect the pulse wave at two sites in the carotid artery and thus measure carotid artery PWV.

METHODS

The model consists of a cuboidal acrylic container, open at the top, filled with a soft-tissue-mimicking viscous gel (Aquasonic, Parker Labs, Fairfield, NJ. Length 400, width150, height 100 mm), with an embedded latex tube (i.d. 6.5mm, wall thickness 0.25 mm) representing the common carotid artery (depth below gel surface 10 mm). The upper surface of the gel is covered with a thermoplastic polyurethane sheet 0.1mm thick, (Platilon, Epurex Films, Bomlitz, Germany), to mimic the skin. The system is water-filled, pressurized from a header tank and pulsatile flow is imposed with an in-house programmable piston pump. Pressure in the tube is measured with 2 catheter-tip manometers (6f gauge, Gaeltec, Dunvegan, Scotland) and surface movement, with a dual beam laser-Doppler vibrometer (SIOS Meβtechnik, Ilmenau, Germany) and/or up to 6 miniature 3-axis MEMS accelerometers (ADXL 337, Analog Devices, Norwood MA.) to provide an independent measure of skin movement and to map the displacement field.

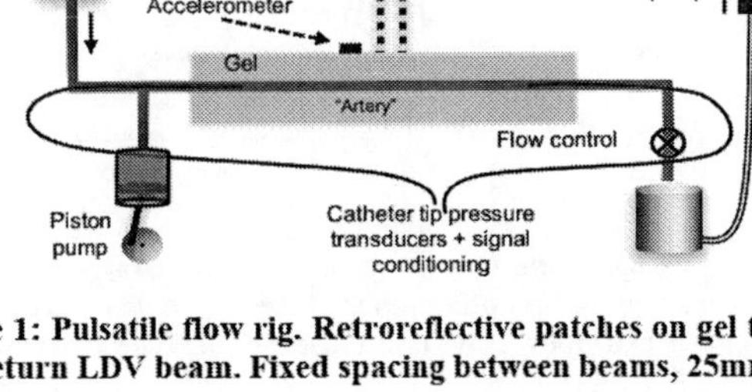

Figure 1: Pulsatile flow rig. Retroreflective patches on gel top surface return LDV beam. Fixed spacing between beams, 25mm

Poster Presentation #P210

The in-vivo measurements were conducted on subjects lying supine with the LDV beams directed onto the skin overlying the right common and external carotids (figure 4).

Data were sampled at 10kHz (Powerlab 16/35, AD Instruments, Oxford, UK), typically for 10 seconds and displayed in real time with associated *Labchart* software. Pulse wave transit time between measurement sites was calculated by cross correlation of the surface velocity from the first derivatives of the paired LDV or pressure signals and from the first integral of paired accelerometer signals.

RESULTS

Under pulsatile flow, LDV displacement measurements normal to the surface (≈100μm) at two sites 25 mm apart, gave PWV values in the range 4-10 ms^{-1}, agreeing well with those obtained from pressure measurements within the embedded tube. Typical pressure and displacement signals are shown in figure 2. The mean PWV (±SD) for a typical run of 10 pulses was 6.00 ± 2.56 for pressure and 5.68 ± 0.91 for displacement.

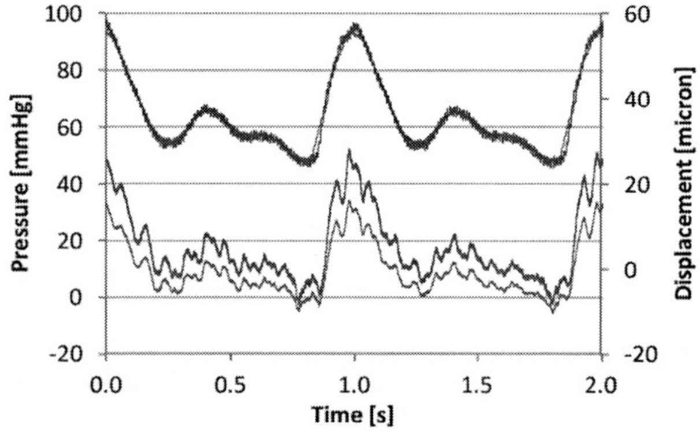

Figure 2. Pressure within embedded tube (upper 2 traces) and displacement of phantom's surface (lower 2 traces) when subjected to pulsatile flow.

In a stenosed tube (70% by area), the frequency spectra of the surface movement measured with accelerometers showed peaks in the range 100 – 600 Hz. At frequencies above 150 Hz the magnitude of the peaks increased with mean flow rate, although in some cases this increase was not monotonic (figure 3).

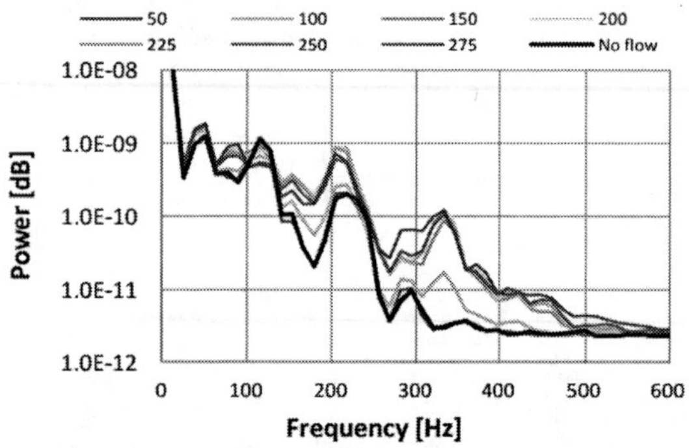

Figure 3. FFT spectrum for a range of mean flow rates of accelerometer placed on phantom surface 20mm downstream from stenosis exit. Flow rates in ml/minute. LDV measurements on this system in progress.

We have carried out in-vivo measurements of skin displacement over the carotid artery in 10 subjects ranging in age from 23 to 70 with the set up in figure 4 and obtained pulsatile waveforms with similar characteristics (figure 5) to those derived by, for instance, tonometry [3], although this depended critically on the beam position, with the most repeatable signals being obtained from the more proximal positions. PWV values in the range 4.6 to 10.1 ms^{-1} were recorded with the older subjects having the higher values, as expected.

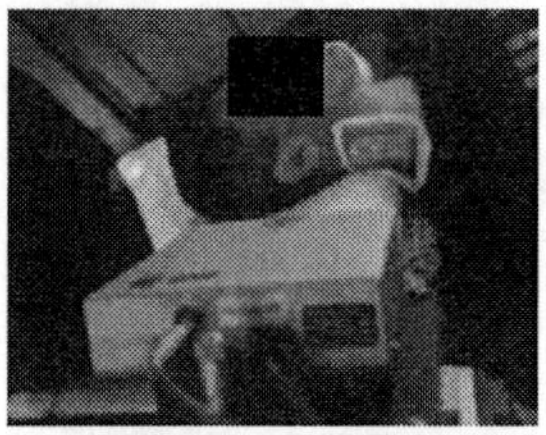

Figure 4. Set up for carotid artery PWV measurement. LDV in foreground, laser-subject distance ≈ 400 mm, beam separation, 25 mm. Array of retro-reflective patches to map displacement field.

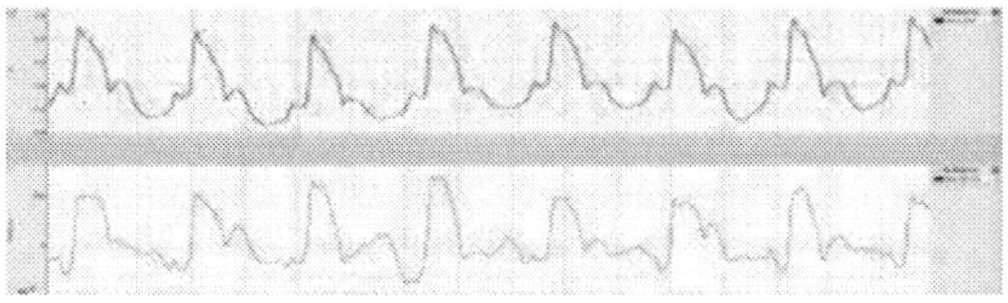

Figure 5. LDV signal from mid neck region (upper trace) and displacement derived from accelerometer at same position (arbitrary units).

DISCUSSION

We have shown that in a simple neck phantom containing an embedded compliant tube representing the common and external carotid arteries, measurements of surface displacement allow reliable estimation of PWV in the tube. Currently, the phantom does not include other structures, so their effect on signal strength shape and timing remains to be assessed. Furthermore, the 'artery' is simply a uniform tube of constant diameter and does not yet include the internal/external carotid bifurcation. A 40% by volume glycerol/water solution will shortly replace the water to more closely mimic blood viscosity.

We have also observed that low amplitude acoustic signals in the frequency range 100 – 500 Hz are associated with the presence of an axisymmetric stenosis at mean flow rates similar to those seen in the carotid artery [4]. We have also shown that, in-vivo, plausible values of carotid artery PWV may be obtained, requiring no direct contact with the subject, and that the shape and timing of displacement signals derived from accelerometers are similar in shape and timing to those from the LDV. Repeatability measurements are in progress. Other consortium members are developing miniaturized optics and electronics to build a multi-beam hand-held device which will be used in a forthcoming clinical validation trial.

ACKNOWLEDGEMENTS

Funding: EU Horizon 2020 programme (CARDIS; ICT-26-2014)

REFERENCES

[1] Semmlow, J. & Rahalkar, K. *Ann Rev Biomed Eng, 9*:449-469, 2007.

[2] De Melis M. et al., *Am J Hypertens*, 21:1280-1283, 2008.

[3] Chen C-H et al. *Hypertension,* 27:168-175, 1996

[4] Likittanasombut P. et al. *J. Neuroimag,* 16:34-38, 2006

NEW IMPLANTABLE FORCE TRANSDUCER FOR THE AORTIC ANNULUS

Tommy Bechsgaard (1,2), Hans Nygaard (1), Sten L. Nielsen (1), Peter Johansen (1,2)

(1) Department of Cardiothoracic & Vascular Surgery, Aarhus University Hospital
Aarhus N, Denmark

(2) Department of Engineering, Faculty of Science and Technology, Aarhus University
Aarhus N, Denmark

INTRODUCTION

In recent years the use of valve sparing repair techniques for patients with insufficient aortic valves has become more common. Overall, the repair procedures can be divided into two groups; a reimplantation as proposed by David et al [1] and the remodeling as proposed by Yacoub et al [2]. Several modifications have been proposed to the original procedures introduced more than 20 years ago. However, there are still limited experimental data, documenting the biomechanical impact of the different interventions. More specific, how the different techniques alter the force distribution in the aortic root. Such data is important as the forces could impact the longevity of the repair as well as the functionality of the aortic valve.

Hence, a new force transducer that allows regional force measurements at the aortic root has been designed, manufactured and tested.

METHODS

The design of the force transducer for the aortic annulus was based on transducers previously developed at our institution [3]. The geometry of the transducer was derived from the anatomy of a pig with a body weight of 80 kg. This resulted in a circular design with three equidistant placed spring elements (arms) (Figure 1). The base ring of the transducer had a diameter of 20 mm and the diameter of the spring elements was 19 mm and all three arms were 5 mm high. The transducer was designed to be implanted within the left ventricular outflow tract, with the ring base proximal to the aortic valve, and the three arms sutured to the interleaflet triangles of the aortic root, at the annular level.

The lead wires were routed within the base of the transducer and only externalized when connected to the strain gauges. For mechanical protection, lead wires and strain gauges were coated with two-component epoxy. The lead cable was attached to the transducer with a strain relief and heat shrinkable tubing.

The transducer frame was designed and simulated using CAD software (Solidworks 2014, Dassault Systèmes, Waltham, USA), where the spring element was optimized to match the maximal recommended strain of the chosen miniature strain gauge (Model S5024, Vishay Precision Group, Basingstoke, UK). The frame was manufactured using rapid prototyping techniques in a nylon material. Two strain gauges where mounted on each spring element and setup in a half bridge configuration to minimize the temperature drift and optimize the stability and signal-to-noise ratio.

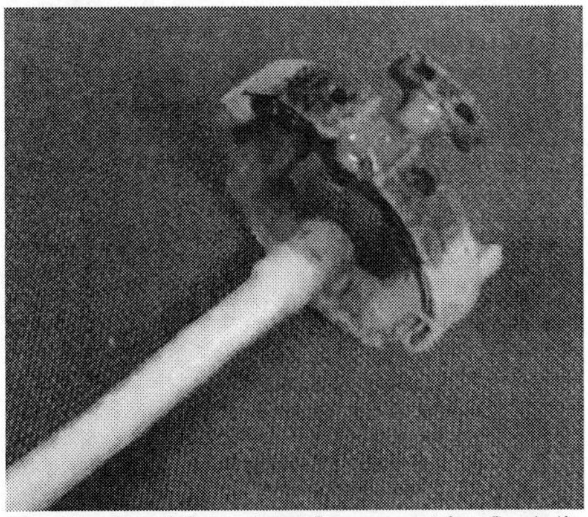

Figure 1: The developed force transducer, post implantation in a pilot experiment. The arms are color coded for easy identification of their anatomical positions during implantation.

Prior to implantation, the transducer was calibrated using a digital force meter setup. The calibration values were obtained using a linear regression between the digital force meter output and the output of the strain gauges on the arm under calibration.

The pig used in the pilot experiment was provided by the University of Aarhus Animal Experimental Farm and was a mix of Danish Landrace and Duroc with a bodyweight of 80 kg. The anesthetic protocol has previously been described [4]. The handling of the animal in this experiment was conducted in compliance with the Danish law on animal experimentation and the study was approved by the Danish Inspectorate of Animal Experimentation.

The hemodynamic and force data were acquired with dedicated hardware and software (NI 9237, NI 9215 and LabVIEW 2015, National Instruments, Austin Texas, USA).

RESULTS

The calibration of the three arms is shown in Figure 2. All fitted values had an R^2 value better than 0.99.

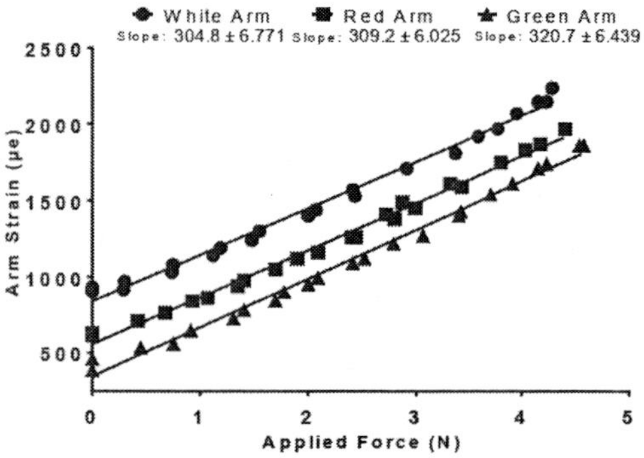

Figure 2: Static calibration of all three arms of the transducer. A small offset have been added to the graphs to make the regression lines more visible.

The pilot data were acquired 45 min after the last dose of cardioplegia during hemodynamically stable conditions (Figure 3). The mean heart rate was 68 BPM and the peak ventricular pressure was 84 mmHg.

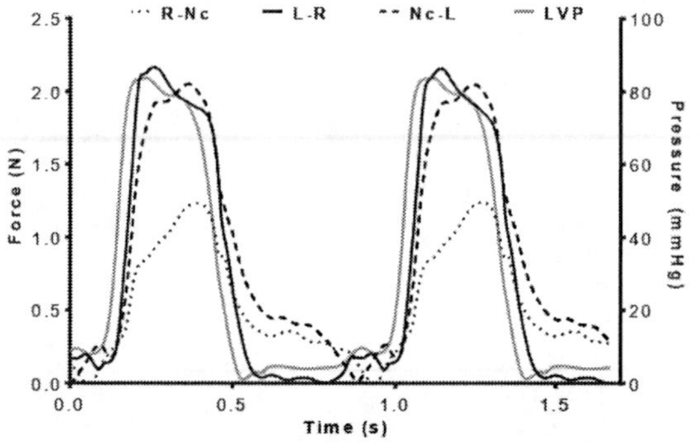

Figure 3: Force and pressure data from the pilot experiment. R-Nc is the force at the interleaflet triangle at annular level between Right and NonCoronary, L-R is Left to Right and Nc-L is NonCoronary to Left. LVP is the left ventricular pressure.

DISCUSSION

The study was successful with regard to developing a new force transducer for quantifying forces of the aortic annulus. Furthermore, the measured forces are within the expected range and similar to other studies focusing on the aortic valve [5].

It seems that there is a small delay between the increase in ventricular pressure and the increase in all three forces (Figure 3). This could be interpreted as the forces were governed by the displacement of fluid as this is delayed due to fluid inertia. It is important to stress that the current experiment was conducted as an acute experiment on a healthy pig and therefore may not reflect the actual forces during a pathological state.

In conclusion, we believe that the proposed force transducer will aid in the refinement of the existing surgical techniques and will guide the development of new techniques. Furthermore, a more detailed description of surgical procedures that include force measurements will also help the surgeon in selecting the optimum technique for the pathology at hand, thus optimizing the treatment of each individual patient.

ACKNOWLEDGEMENTS

The subproject was partly funded by The Danish Heart Foundation Grant #15-R99-A5942-22906, The Danish Council for Independent Research Grant #DFF -4004-00317, Direktør Jacob Madsens & Hustru Olga Madsens Fond, Direktør Kurt Bønnelycke og hustru fru Grethe Bønnelyckes Fond, Edith og Olfert Dines Hansens Legat, Helga og Peter Kornings fond, Holger Rabitz og hustru Doris Mary født Philipp's Mindefond, Snedkermester Sophus Jacobsen og hustru Astrid Jacobsens Fond, Karen Elise Jensens Fond.

REFERENCES

1. David TE, Feindel CM (1992) An Aortic-Valve Sparing Operation for Patients with Aortic Incompetence and Aneurysm of the Ascending Aorta. J Thorac Cardiovasc Surg 103:617–622.
2. Sarsam MAI, Yacoub M (1993) Remodeling of the Aortic-Valve Anulus. J Thorac Cardiovasc Surg 105:435–438.
3. Kragsnaes ES, Honge JL, Askov JB, et al (2013) In-plane Tricuspid Valve Force Measurements: Development of a Strain Gauge Instrumented Annuloplasty Ring. Cardiovasc Eng Technol 4:131–138. doi: 10.1007/s13239-013-0135-9
4. Jensen MØ, Jensen H, Nielsen SL, et al (2008) What forces act on a flat rigid mitral annuloplasty ring? J Heart Valve Dis 17:267–75; discussion 275.
5. Goetz W a, Lim KH, Ibled R, et al (2006) Forces at single point attached commissures (SPAC) in pericardial aortic valve prosthesis. Eur J Cardiothorac Surg 29:150–5. doi: 10.1016/j.ejcts.2005.12.002

SB³C2017
Summer Biomechanics, Bioengineering and Biotransport Conference
June 21 – 24, Tucson, AZ, USA

CHARACTERISTICS OF CLOTS FROM ACUTE ISCHEMIC STROKE AND LABORATORY ANALOGS

J.Y. Chueh (1), J. Litchman (1), R.A. Arslanian (1), S.L. Carniato (2), D.E. Rex (1), A.S. Puri (1), M.C. Howk (1), M.J. Gounis (1)

(1) University of Massachusetts Medical School
Worcester, MA, USA

(2) Department of Preclinical Science
Stryker Neurovascular
Fremont, CA, USA

INTRODUCTION

The findings of recent randomized control trials have shown a convincing benefit from stent retriever-mediated mechanical thrombectomy for selected acute ischemic stroke (AIS) patients with large vessel occlusion in the anterior circulation.[1-2] The ability of the stent retriever to retain the clot in the stent struts during clot retrieval is an important predictor of successful recanalization and is affected by the clot type. Loss of interaction between clot and stent retriever when passing the tortuous cerebral arteries can result in partial or failed thrombectomy, which increases the need for multiple passes and additional rescue therapy. Multiple passes may be associated with reduced good clinical outcomes due to delay of recanalization or injury to the vessel.[3]

It is our hypothesis that calcified and fibrin-rich clots are stiffer than erythrocyte–rich clots; therefore, they may serve as the challenging clot models to evaluate clot-device interactions, and further aid the design and development of future generations of stent retrievers. We previously used the "clot integration factor (CIF)" to quantitatively evaluate the integration of the clot into a stent retriever in a controlled in vitro experiment. The CIF is defined as the amount of clot that migrates into the inner volume of the expanded stent.[4]

In phase 1 of this proposed project, clinical specimens were collected and analyzed with histological staining and dynamic mechanical analysis (DMA). Preparation of experimental clot models was conducted in phase 2 of the study, and the laboratory analogs underwent the same characterization process as the clinical specimens. A clot preservation study was performed to investigate how freeze treatment and storage temperatures affect the clot mechanical properties with a goal to extend clot shelf life.

METHODS

Characterization of clot specimens retrieved from patients with acute ischemic stroke (AIS)

Fifty clot specimens (length >2mm, diameter >1mm) from 28 patients with AIS were collected for characterization. All human specimens were collected with approval from our Institutional Review Board. Stiffness and elasticity of each clot specimen were measured.

- **Mechanical characterization:** DMA (Q800; TA Instruments, New Castle, DE) was used to explore the stiffness and elasticity of the clots.[5] In the controlled force mode, stress variation over 0-75% strain (E1) and 75-100% strain (E2) was calculated, respectively, as an indication of clot stiffness. In the stress-relaxation mode, strain recovery, a measure of material elasticity, was acquired.
- **Histologic assessment:** Retrieved clot materials were fixed in 10% buffered formalin immediately after retrieval. Specimens were then embedded in paraffin wax and cut into 5μm sections. Sections were dewaxed and hydrated with distilled water in preparation for a modified version of the Martius scarlet blue (MSB) method for staining fibrin, collagen, and erythrocytes. Each slide was subjected to quantitative analysis and segmentation using Image Pro Plus (Media Cybernetics, Atlanta, GA) software following the MSB staining. Tiled images were taken at 2× magnification. The clot components analyzed were fibrin, erythrocytes, and a mixed region of both fibrin and erythrocytes. Each component was assigned a color and manually highlighted to determine physical composition and differentiate each constituent.

Preparation and characterization of laboratory analogs

Three different clot models were prepared, and subjected to the mechanical and histological characterization described previously.

- **Calcium apatite-rich clots:** Bovine blood was first mixed with fibrinogen and hydroxyapatite. The mixture was then added into thrombin (2.5NIHU/mL) solution and injected into a silicone tube to facilitate clotting.
- **Fibrin-rich clots:** Thrombin-induced clotting was conducted by simultaneously injecting whole bovine blood and calcium chloride /thrombin/fibrinogen solution into a silicone tube to give blood mixtures with final thrombin concentration of 5 NIHU/mL blood.
- **Barium sulfate-doped clots:** Similar thrombin-induced clotting method described above was repeated to prepare the barium sulfate-doped clots except that 1 g of barium sulfate was added into 10mL whole blood, and no fibrinogen was introduced during the preparation process.

Development of a clot preservation method

A clot preservation study was performed with the use of the experimental fibrin-rich clot. The fibrin-rich clot was stored at 4°C and -80°C, respectively, prior to the clot stability tests. Clot stability/shelf life was determined by 1) clot appearance including weight and shape, and 2) clot stiffness and elasticity. Mechanical examination was conducted every day for 4 days.

RESULTS

Figure 1 shows a 60 year-old male presented with an AIS due to right ICA and M1 occlusions. A clot (Figure 1, left) was removed by using the thrombectomy device for characterization, and was trimmed to the desired size and shape prior to the DMA test (Figure 1, right).

Figure1. A clot is successfully retrieved via mechanical thrombectomy.

Comparisons of human specimens and experimental clot models in terms of S-S curves are presented in Figure 2. Overall, the experimental clot models were slightly stiffer than the human specimens. It should be noted that the calcium apatite-rich clot had the highest E1, as compared to the fibrin-rich clot and the barium sulfate-doped clot (Figure 2, inset).

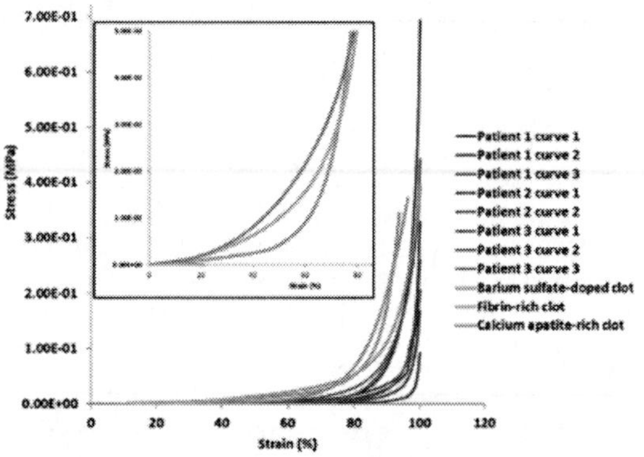

Figure 2. S-S curves of clinical samples and experimental clot models. Inset: Stress variation over 0-75% strain.

MSB results (Figure 3) revealed the heterogeneous structure of a human clot, which was mainly composed of fibrin (41%) with erythrocyte aggregates (30%) dispersed in it.

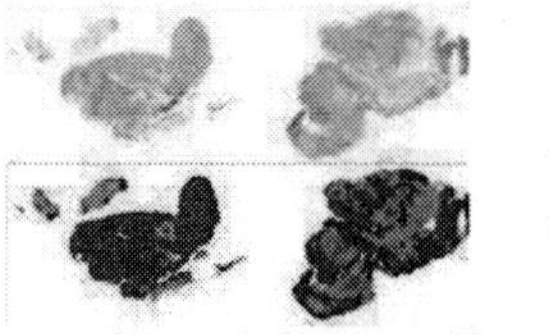

COMPONENT	TOTAL AREA
RBC	27899
FIBRIN	38595
RBC and FIBRIN	28004
COMPONENT	PERCENTAGE (CALC)
RBC	29.6
FIBRIN	40.8
RBC and FIBRIN	29.6

Figure 3. (Upper left) A paraffin-embedded section of human clot is stained with MSB to the detailed structure. (Lower left) The corresponding segmented mask shows erythrocytes in green, fibrin in blue, and fibrin/erythrocyte mixed area in purple. (Table) Quantitative analysis of the percentage of erythrocytes, fibrin, and mixed erythrocytes and fibrin by area.

Our preliminary results suggested that clot stability, the resistance of the clot to mechanical stress and fibrinolytic dissolution, was affected by storage temperature (Figure 4). An increase in stiffness was observed for clots stored at 4°C on day 2 and 3. E1 was increased from 0.0001MPa on day 1 to 0.00023MPa on day 3, while E2 was increased from 0.006MPa on day 1 to 0.016MPa on day 3. On day 4, both E1 and E2 were reduced to the original values. For clots stored at -80°C, a continuous reduction in clot stiffness over time was recorded (reduction in E2 was greater than 50%; data not shown).

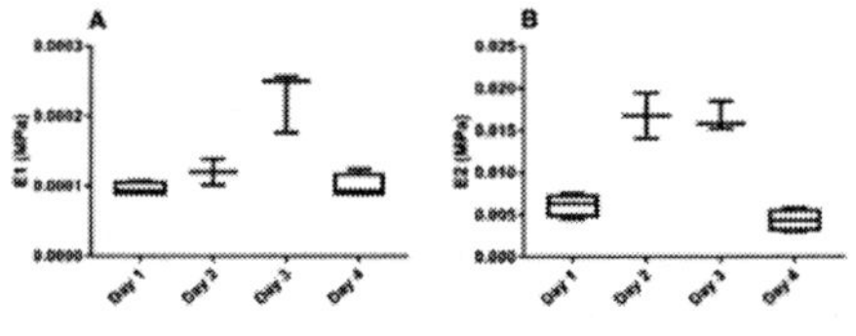

Figure 4. Clot stability over time (4 degree storage). (A) Changes in E1: stress variation over 0-75% strain (B) Changes in E2: stress variation over 75-100% strain.

DISCUSSION

Clinically representative clot models were successfully developed in this study. Our results support the hypothesis that calcified and fibrin-rich clots are stiffer than erythrocytes–rich clots.

The stability results of this study agree with the previous findings that cooling rate disparities may affect ice formation during the freezing of biological tissue that may result in important mechanical changes.[6]

The isolation of each respective target area acquired by the segmentation method during histologic assessment allowed for precise measurements that enabled the depiction of area and percent composition per component. Completion of the histologic assessment was instrumental in determining the quantity of each key component of the clots, and will be correlated with technical metrics of the procedure such as number of passes, final TICI, and time to recanalization from groin puncture.

ACKNOWLEDGEMENTS

This work was supported by Stryker Neurovascular. The contents are solely the responsibility of the authors and do not necessarily represent the official views of Stryker Neurovascular.

REFERENCES

[1] Saver, JL et al., *N Engl J Med*, 372: 2285-2295, 2015.
[2] Campbell, BC et al., *N Engl J Med*, 372: 1009-1018, 2015.
[3] Saver, JL et al., JAMA, 316:1279-1289, 2016.
[4] van der Marel, K et al., *J Neurointerv Surg*, 8: 1278-1282, 2015.
[5] Chueh, JY et al., AJNR Am J Neuroradiol, 32:1237-1244, 2011.
[6] Muldrewh, K et al., Cryobiology, 40:102-109, 2000.

THREE PART HYPERELASTIC LAW FOR ANISOTROPIC AORTIC TISSUE: MODEL DEVELOPMENT AND EXPERIMENTAL VALIDATION

Catherine A. O'Connor (1), David R. Nolan (2), Eóin McEvoy (1), Patrick McGarry (1)

(1) Department of Biomedical Engineering
National University of Ireland, Galway
Galway, Ireland

(2) Trinity Centre for Bioengineering,
Trinity Biomedical Sciences Institute
Trinity College, Dublin
Dublin, Ireland

INTRODUCTION

Several hyperelastic formulations for soft tissue use an exponential form of the strain energy density to simulate strain stiffening due to extension of embedded collagen fibers [1, 2, 3]. Such fiber based formulations also incorporate in-plane fiber dispersion [4], in addition to out-of-plane fiber dispersion [5]. Such anisotropic exponential strain stiffening hyperelastic laws with two fiber families have been shown to accurately simulate muscular arteries, e.g. iliac [4, 6] and coronary [7] arteries.

In the current paper we perform mechanical tests to characterize the anisotropic behavior of aortic tissue. Results suggests that exponential strain hardening formulations previously developed for "muscular" arteries are not suitable for "elastic" arteries. We propose a quasi-bilinear anisotropic hyperelastic formulation and provide and demonstrate that this new material model accurately simulates our experimental test data.

MODEL DEVELOPMENT

Building upon previous constitutive laws for soft tissue we use an additive decomposition of the strain energy density function into an isotropic component, which represents the ground matrix, and an anisotropic component, representing two families of reinforcing fibers,

$$\boldsymbol{\Psi} = \boldsymbol{\Psi_{iso}} + \textstyle\sum_{i=4,6} \boldsymbol{\Psi_a} \quad (1)$$

For simplicity we assume that the ground matrix behaves as a slightly compressible neo-Hookean material [8, 9]. We propose an anisotropic component that consists of three distinct regimes (Figure 1A): (i) an initial compliant quasi-linear region to model the stretching of coiled collagen fibers; (ii) a non-linear transition region to model the uncoiling of fibers; (iii) a stiff quasi-linear region to model fully uncoiled fibers.

The piece-wise strain-energy density function for the anisotropic component is given as

$$\boldsymbol{\Psi_a} = \begin{cases} \sum_{i=4,6} K_1 I_i + 2K_1 I_i^{-\frac{1}{2}} + \Psi_0 & (\varepsilon_f) \leq D_1 \\ \sum_{i=4,6} T_1 I_i + T_2 I_i^{-\frac{1}{2}} - T_3 I_i^{-2} + \Psi_0 & D_1 \leq (\varepsilon_f) \leq D_2 \\ \sum_{i=4,6} K_2 I_i - 2 I_i^{-\frac{1}{2}} + \Psi_0 & (\varepsilon_f) \geq D_2 \end{cases} \quad (2)$$

where ε_f is the axial strain in a fiber and parameters D_1 and D_2 define the boundaries of the three fiber deformation regimes, as shown in Figure 1A. K_1 and K_2 govern the effective stiffness of the two quasi-linear regions. Two fiber families are considered; *i=4* and *i=6*. The anisotropic invariant I_i is the square of the stretch of fibre *i*, given as

$$I_i = \boldsymbol{C}: \boldsymbol{M_i} \otimes \boldsymbol{M_i} \quad (3)$$

where $\boldsymbol{C}$ is the right Cauchy-Green deformation tensor and $\boldsymbol{M_i} = \boldsymbol{F}\boldsymbol{M_{i0}}$, where $\boldsymbol{F}$ is the deformation gradient and $\boldsymbol{M_{i0}}$ is a vector describing the undeformed orientation of fibre *i*.

The Cauchy stress is obtained from the strain energy density function using

$$\boldsymbol{\sigma} = J^{-1}\boldsymbol{F}\frac{\partial \boldsymbol{\Psi}}{\partial \boldsymbol{F}} \quad (4)$$

where *J=det(F)*.

Fiber dispersion, as illustrated in Figure 1B is incorporated into the constitutive model through the generalized structure tensor $\boldsymbol{H}$ proposed by Holzapfel et al. [5], whereby

$$\boldsymbol{H} = A\boldsymbol{I_1} + B\boldsymbol{M_i} \otimes \boldsymbol{M_i} + C_{op}\boldsymbol{M_n} \otimes \boldsymbol{M_n} \quad (5)$$

where $\boldsymbol{M_n}$ is a vector orthogonal to the plane formed by the two fiber families. Parameters A and B are functions of the in-plane and out-of-plane fiber dispersions and $C_{op} = (1\text{-}3A\text{-}B)$.

The dispersed fibre strain for family *i* is given as

$$E_i = \boldsymbol{H_i}: (\boldsymbol{C} - \boldsymbol{I}) \quad (6)$$

leading to

$$E_i = AI_1 + BI_i + C_{op}I_n \quad (7)$$

Fiber dispersion is then incorporated into the anisotropic strain energy density function by substitution of E_i for I_i in equation (2).

The Cauchy stress, obtained from equation (4) is then given as

$$J\sigma_A = \begin{cases} \sum_{i=4,6}\left[E_1\left(1-E_i^{-\frac{3}{2}}\right)\left(AB + BM_i\otimes M_i + C_{op}M_n\otimes M_n\right)\right] \\ \sum_{i=4,6}\left[\left(S_1 + PE_i^{-3} + \frac{1}{2}S_2E_i^{-\frac{3}{2}}\right)\left(AB + BM_i\otimes M_i + C_{op}M_n\otimes M_n\right)\right] \\ \sum_{i=4,6}\left[\left(S_3 + E_i^{-\frac{3}{2}}\right)\left(AB + BM_i\otimes M_i + C_{op}M_n\otimes M_n\right)\right] \end{cases} \quad (8)$$

This anisotropic hyperelastic formulation is implemented in a user defined material subroutine (UMAT) in Abaqus 6.14 (Simula, Providence, RI, USA).

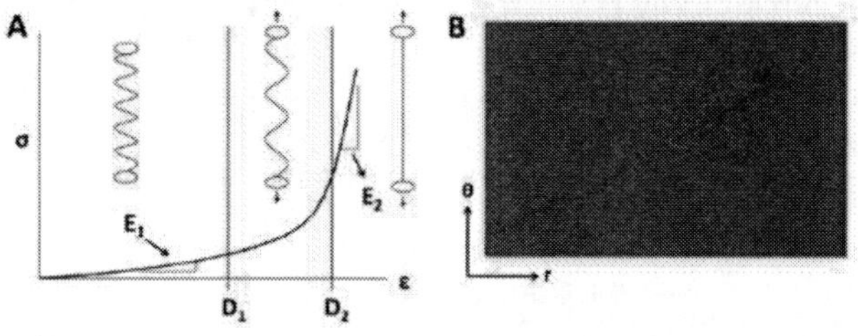

Figure 1: (A) Schematic showing the in-plane fiber dispersion cone for one fiber family. (B) Illustration of stress-strain curve showing the three regimes of fiber stretching.

In order to illustrate the importance of fiber dispersion in our new formulation we simulate stretching of aortic tissue in the 1-direction, with the tissue free to expand or contract in the lateral directions ($\sigma_{22} = \sigma_{33} = 0$). In Figure 2 we show that an auxetic expansion in the out-of-plane direction, $\varepsilon_3>0$, occurs for high ratios of fiber stiffness to matrix stiffness (R) when dispersion is not included in the model. Such auxetic behavior is not observed in our experiments on aortic tissue and has not been reported in any previous mechanical tests on arterial tissue. When dispersion is included in the model the out-of-plane dispersion parameter C_{op} must be positive in order to ensure that $\varepsilon_3<0$.

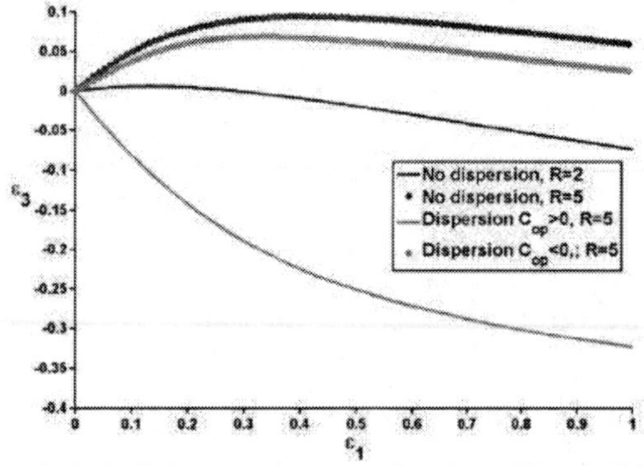

Figure 2: Out-of-plane strain ε_3 as a function of the applied strain ε_1. Note that $\varepsilon_3>0$ signifies auxetic behavior, which is not observed experimentally for aortic tissue. R is the ratio of fiber stiffness to matrix stiffness.

EXPERIMENTAL TESTING OF AORTIC TISSUE

Dog-bone specimens of descending ovine aorta in the axial (n=8) and circumferential (n=9) directions were subjected to uniaxial tensile testing to failure at a nominal strain rate of 0.0125^{-1}. Preconditioning was performed by loading and unloading samples to a nominal strain of 0.02 for 10 cycles prior to the commencement of testing. Sample deformation during testing was measured using a video extensometer.

Figure 3 shows the nominal stress-strain behavior (mean±SD) of the aortic tissue when stretched in the circumferential and axial directions. Clearly the tissue is highly anisotropic for the entire applied strain range, with the circumferential direction exhibiting a significantly higher stiffness. The three regions of deformation outlined in Figure 1A are clearly evident. In the case of the circumferential direction the transition region between the two linear regions is observed in the strain range from 0.65 to 0.75. A quasi-linear stress-strain relationship is observed before and after the transition region. The proposed piecewise hyperelastic model provides a good prediction of the material anisotropy, particularly in the transition and stiff quasi-linear regimes.

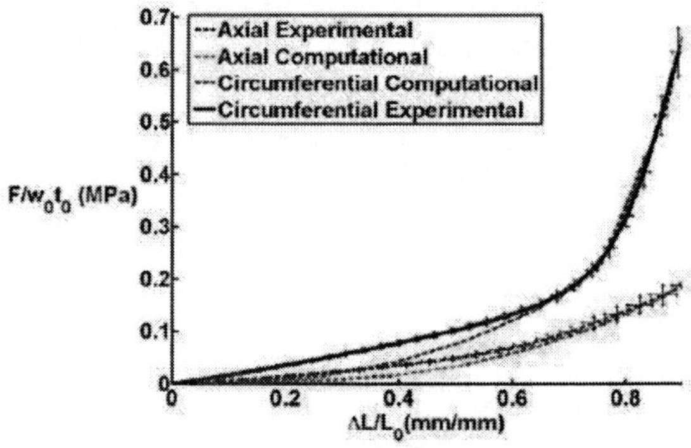

Figure 3: Experimental test results (mean±SD) for stretching of the aortic tissue in the circumferential and axial directions. Model predictions are also shown.

DISCUSSION

Experimental testing of anisotropic aortic tissue to high strains reveals that the tissue exhibits three deformation regimes: a low stiffness linear regime; a non-linear transition regime; a high stiffness linear regime. A novel hyperelastic formulation is proposed to model this behavior. Incorporation of fiber dispersion is required to eliminate non-physical auxetic behavior. An accurate model of aortic hyperelasticity is critical for the design of implant grafts for the treatment of aortic aneurysms where a key challenge is the maintenance of physiological levels of effective vessel compliance.

ACKNOWLEDGEMENTS

The authors wish to acknowledge funding from Science Foundation Ireland (SFI-12/IP/1723).

REFERENCES

1. Holzapfel, GA et al., *J. Elasticity*, 61:1-48, 2000.
2. Fung, Y, *Am. J. Physiol.-Legacy Content*, 213:1532-1544, 1967.8
3. Weiss, JA et al. *Comput. Methods in Appl. Mech. Eng.*, 135:107-128, 1996.
4. Gasser, TC et al., *J. R. Soc. Interface*, 3:15-35, 2006.
5. Holzapfel, GA et al., *J. R. Soc. Interface*, 12:20150188, 2015.
6. Schulze-Bauer, CAJ, *J. Biomech. Eng.*, 125:395-406, 2003.
7. Holzapfel, GA et al., *Am. J. Physiol-Heart C*, 289:H2048-H2058, 2005.
8. Nolan, DR at al., *J. Mech. Behav. Biomed. Mater.*, 39:48-60, 2014.
9. Nolan and McGarry, *Ann. Biomed. Eng.*, 44:993-1007, 2016.

Poster Presentation #P213

A SEMIAUTOMATIC METHOD FOR THE DETECTION OF PATIENT SPECIFIC AORTIC GEOMETRIES AND MECHANICAL PROPERTIES USING 4D FLOW MRA

Jamie Concannon (1,2), Niamh Hynes (2), Sherif Sultan (2), Patrick McGarry (1), Peter E. McHugh (1)

(1) Discipline of Biomedical Engineering,
College of Engineering and Informatics,
National University of Ireland, Galway, Ireland.

(2) Western Vascular Institute,
Department of Vascular and Endovascular Surgery,
Galway University Hospitals, Galway, Ireland

INTRODUCTION

Quantification of regional mechanical heterogeneity in the human aorta is currently lacking. Notwithstanding a handful of studies on the variance in aortic cyclic strain or associated mechanical properties in animal models [1,2], such techniques have yet to be transferred into clinical practice for the benefit of high risk patients.

The importance of quantifying regional aortic mechanical properties has surfaced in recent years due to complications arising from a paradigm shift from open surgery to endovascular repair techniques for the treatment of aortic pathologies. The Endovascular Aortic Repair (EVAR) technique has provided high risk patients with an alternative intervention, however the technique is not without limitations. Many complications remain particularly apparent following the treatment of Thoracoabdominal Aortic Aneurysms (TAAAs) which, by definition involve one or more origins of the coeliac, superior mesenteric or renal arteries. Emergency repair for such aneurysms is associated with 30-day mortality rates of up to 48%, alongside high rates of paraplegia (8.0%) and renal failure (8.3%) [3,4].

It is critical, therefore, to understand the regional mechanical properties of the aorta in order to inform patient intervention strategies through computational analyses instead of the currently used diameter criterion, when the repair of TAAAs is associated with dismal post-operative success, and rupture results in mortality in up to 100% of cases [5]. Following the decision to eventually intervene, stent graft placement in the abdominal aorta has shown to reduce aortic compliance by 10-21% [6], which in itself has been shown to induce devastating outcomes such as neointimal hyperplasia, graft thrombosis and ultimately, failure [7].

This poses the question: What are the levels of compliance mismatch in the thoracic section of the aorta of aneurysmal patients if cyclic strain and mechanical properties are heterogeneous? And additionally, would quantifying regional aortic properties result in improved rupture prediction simulations, stent design and in turn, patient outcomes?

METHODS

The axial phase images of a 4D Flow MRA scan were isolated for a healthy subject (HR = 63 bpm) and 8 planes along the descending aorta originating distal to the left subclavian artery to the infra-renal aorta were analyzed in order to quantify regional Green-Lagrange circumferential cyclic strain (E) according to (1). An automatic vessel detection algorithm (VDA) was used to capture the boundary of the inner lumen in each plane throughout the cardiac cycle (949 ms) based on thresholding of the gadolinium dye arriving in the plane. Each plane was then further analyzed in both diastole and systole to investigate the change in area and circumference due to the cyclic normotensive loading conditions of the subject (BP = 120/80 mmHg).

$$E = \frac{1}{2}\left[\left(\frac{C_{Systole}}{C_{Diastole}}\right)^2 - 1\right] \text{ x } 100\% \qquad (1)$$

The points along the lumen boundary were subsequently translated into Cartesian coordinates (mm) and imported into Abaqus 6.14-1 (DS, Simulia, USA) using a point cloud reconstruction technique. Each planar lumen boundary was given a uniform wall thickness of 1.39 mm and meshed using C3D8RH elements. An anisotropic hyperelastic constitutive law was implemented for each plane in order to mimic healthy aortic properties as seen in Table 1 [8].

$$\psi = \frac{c}{2}(I_1 - 3) + \sum_{n=4,6}\frac{k_1}{k_2}\{exp[k_2(I_n^* - 1)^2] - 1\} \qquad (2)$$

$$I_n^* = \kappa I_1 + (1 - 3\kappa)I_n \qquad (3)$$

Table 1: Anisotropic hyperelastic constitutive law parameters.

μ (MPa)	D (MPa)	k1 (MPa)	k2	κ	γ (°)
0.017	0	0.56	16.21	0.18	51.0

Poster Presentation #P214

Finally, a custom Zero-Pressure-Geometry (ZPG) algorithm was implemented in order to converge upon the unloaded lumen geometry of each plane to an accuracy of 1E-6 mm (Figure 1).

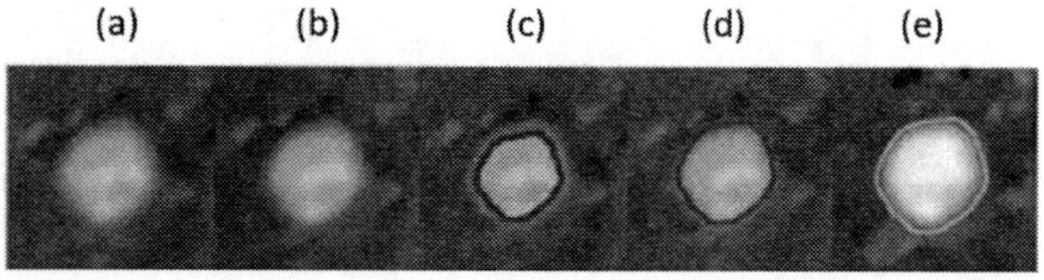

Figure 1: (a) Raw axial plane image, (b) lumen boundary isolated post VDA, (c) Converged ZPG, (d) Loaded Diastolic Geometry (80mmHg), (e) Loaded Systolic Geometry (120mmHg).

RESULTS

High levels of Green-Lagrange strain are evident in the proximal descending aorta whereas levels are significantly lower in the para-renal Planes 6-8 (Figure 2). Note that Plane 2 has a low cyclic strain that can be attributed to aortic compression from the heart and thoracic vertebrae in this specific subject. The average residual is defined as the distance between each node on the loaded systolic geometry (odb geometry) and its corresponding node on the scan derived systolic geometry for each plane (Figure 2). Plane 7 resulted in the lowest average residual of 0.29mm and maximum distance of 1.14mm while Plane 3 resulted in an average residual of 1.23mm and a maximum of 2.59mm.

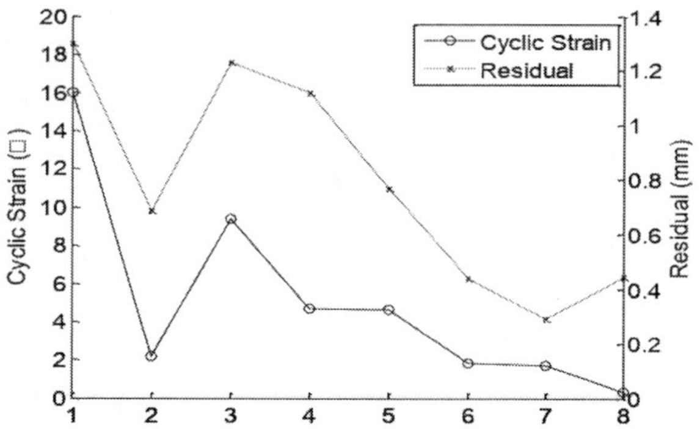

Figure 2: Cyclic Strain levels based upon VDA and Equation 1 with average residual superimposed to show accuracy of the given constitutive law in predicting the systolic geometry.

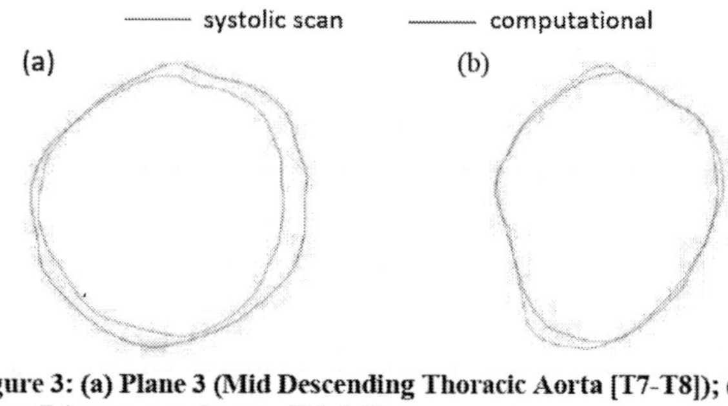

Figure 3: (a) Plane 3 (Mid Descending Thoracic Aorta [T7-T8]); (b) Plane 7 (supra-renal aorta [L1-L2]).

The anisotropic hyperelastic constitutive parameters in Table 1 result in the under prediction of strain in the thoracic Plane 3 whereas the chosen parameters in Table 1 are far more accurate in predicting the actual systolic geometry in the abdominal Plane 7 (Figure 3).

DISCUSSION

The healthy subject analyzed in this study illustrates the heterogeneity in cyclic strain experienced by the aorta during the cardiac cycle under normotensive conditions. Green-Lagrange strain varied from 16% in the mid descending thoracic aorta to 0.3% in the infra-renal aorta, levels of which are comparable to those reported by Morrison & Taylor [9].

The nature of the step changes in strain between planes presented here, although on average 20mm apart, suggest that there may be significant merit in investigating further the effects of stenting in particular zones during EVAR procedures. Since the level of cyclic strain is low in the para-renal segment of the aorta (<2%), stenting may not have such a drastic effect on the compliance of the abdominal wall as if the stent was deployed in Plane 3, where the cyclic strain levels exceed 9%. The effects of reduced aortic compliance, particularly proximal, are widely published and result in increased left ventricular afterload and subsequent cardiac failure through raised aortic Pulse Wave Velocities [6]. This paired with existing aneurysm related comorbidities, further increases the risk of mortality of the patient and decreases their chances of being offered a reintervention due to comorbidity risks.

The constitutive law implemented here resulted in an average accuracy in reproducing the loaded systolic geometry of 0.29mm for Plane 7, indicating that it may be feasible to model the infra-renal aorta provided the ZPG algorithm is used concomitantly. The use of the same constitutive parameters in more proximal planes of the aorta, however, grossly underestimates the level of strain experienced in vivo and may potentially result in an underestimation of wall stresses and subsequently, potential rupture points in the thoracoabdominal aorta.

ACKNOWLEDGEMENTS

Funding for this research was provided by The Irish Research Council Enterprise Partnership Scheme (RCS1564).

REFERENCES

[1] Draney, M. T., et al., *Magnetic Resonance in Medicine*, 52(2), 286–95, 2004.

[2] Moriwaki, T., et al., *Journal of Artificial Organs*, 14(4), 276–83, 2011.

[3] Johns, N., et al., *Journal of Cardiothoracic Surgery*, 9(1), 195, 2014.

[4] Coselli, J. S., & LeMaire, S. A. *The Annals of Thoracic Surgery*, 83(2), S862-4–2, 2007.

[5] Kheirelseid, E. A. H., & Madhavan, P. *Irish Journal of Medical Science*, 183(2), 153–60, 2014.

[6] Morris, L., & Sultan, S. *European Journal of Vascular and Endovascular Surgery*, 51(1), 44–55, 2016.

[7] Singh, C., & Wang, X, *Journal of Functional Biomaterials*, 6(3), 500–25, 2015.

[8] Weisbecker, H., & Holzapfel, G. A. *Journal of Mechanical Behaviour of Biomedical Materials*, 2012.

[9] Morrison, T. M., & Taylor, C. A. *Journal of Vascular Surgery*, 49(4), 1029–36, 2009.

SB³C2017
Summer Biomechanics, Bioengineering and Biotransport Conference
June 21 – 24, Tucson, AZ, USA

MODELING OF MYOCARDIUM COMPRESSIBILITY AND ITS IMPACT IN COMPUTATIONAL SIMULATIONS OF THE FUNCTIONING HEART

João S. Soares (1), David S. Li (1), Eric Lai (2),
Joseph H. Gorman III (2), Robert C. Gorman (2), and Michael S. Sacks (1)

(1) Center for Cardiovascular Simulation
Institute for Computational Engineering and Sciences
Department of Biomedical Engineering
University of Texas at Austin
Austin, Texas, USA

(2) Gorman Cardiovascular Research Group
Perelman School of Medicine
University of Pennsylvania
Philadelphia, Pennsylvania, USA

INTRODUCTION

Myocardial infarction (MI) induces maladaptive remodeling of the left ventricle (LV), causing dilation, wall thinning, change in mechanical properties, and loss of contractile function [1]. Simulation technologies can potentially lead the way for in-silico based models of the heart for many therapeutic applications. Specifically, a computational platform to accurately evaluate the effect of MI on cardiac function impairment, acutely and chronically, would be extremely valuable to understand etiology and pathophysiology of myocardium remodeling, its impact on tissue-level properties and organ-level cardiac function, and ultimately, to improve virtual surgery technologies, medical device development, as well as to provide quantitative risk stratification tools for these interventions.

However, material modeling of myocardium, and its critical numerical implementation, remains an area where much progress is required. One particular challenging feature of functioning myocardium is the interaction of coronary flow with myocardial contractility and compressibility. While volumetric changes in the myocardium during the cardiac cycle have been known to occur for at least two decades [2], their incorporation into cardiac simulations and realization of this important effect in cardiac function has yet to be fully accomplished. In the following study we present novel experimental studies, material modeling, and numerical simulations as first step in developing more realistic cardiac models.

METHODS

We have developed an in-silico model of MI based on extensive datasets from a single ovine heart, using a comprehensive model pipeline (Fig. 1). All data collection of this in silico heart model was performed at the Visible Heart Laboratory (VHL, University of Minnesota). Magnetic resonance images (MRI) at end-diastole were

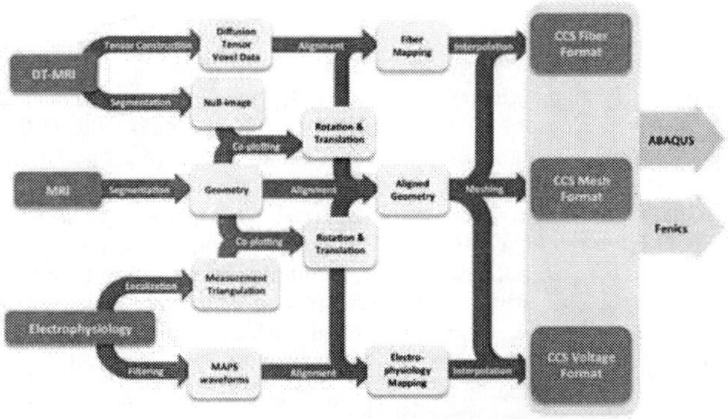

Figure 1. Complete modeling pipeline from the single heart data source.

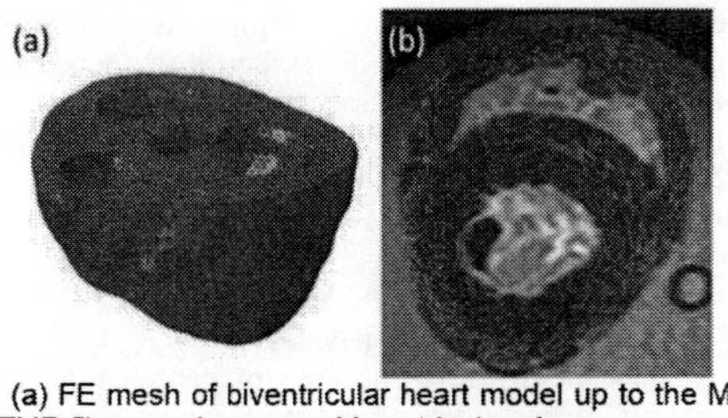

Figure 2. (a) FE mesh of biventricular heart model up to the MV plane; (b) MR-DTMR fiber overlay on a mid-ventricular plane.

segmented to create a finite element (FE) mesh (Fig. 2a). Diffusion tensor MRI (DTMRI) data was employed to prescribe principal fiber direction (Fig. 2b). As a first step, we utilized a conventional incompressible transversely isotropic Fung-based hyperelastic model for the passive mechanical properties of myocardium of the form

$$\mathbf{T} = =\frac{\partial W^{\mathrm{vol}}}{\partial J} = \frac{1}{J}\tilde{\mathbf{F}}\frac{\partial W^{\mathrm{dev}}}{\partial \tilde{\mathbf{E}}}\tilde{\mathbf{F}}^{\mathrm{T}} = \frac{1}{J}\mathbf{F}\mathbf{S}^{\mathrm{act}}\mathbf{F}^{\mathrm{T}}$$

where $J = \det \mathbf{F}$, the deviatoric deformation gradient is defined as $\tilde{\mathbf{F}} = J^{=1/3}\mathbf{F}$, and $\tilde{\mathbf{E}} = \frac{1}{2}(\tilde{\mathbf{F}}^{\mathrm{T}}\tilde{\mathbf{F}} = \mathbf{I})$. The passive mechanics of myorcardium is described by the volumetric and deviatoric responses

$$W^{\mathrm{vol}} = \frac{K}{2}\left(\frac{J^2 = 1}{2} = \ln J\right)$$

$$W^{\mathrm{dev}} = \frac{c}{2}\left[\exp(\alpha \tilde{\mathbf{E}} \cdot \mathbb{C}\tilde{\mathbf{E}}) = 1\right]$$

where K is the bulk modulus (with $K \gg c$ enforcing material incompressibility), c and α are passive material parameters, and 4th order tensor $\mathbb{C}$ characterizes the transverse isotropy of the material [3]. The passive model is calibrated to match the general end-diastolic pressure volume relationship obtained by Klotz et al. [4]. The active stress responsible for myocardium contraction follows Hunter-McCulloch-Ter Keurs model [5] and is given by

$$\mathbf{S}^{\mathrm{act}} = T_{Ca^{2=}}(\mathbf{x}, t)[\mathbf{1} = \boldsymbol{\beta}(\boldsymbol{\lambda} = \mathbf{1})]\mathbf{m} \otimes \mathbf{m}$$

where $\mathbf{m}$ is the fiber direction and local fiber stretch is $\lambda = \mathbf{m} \cdot \mathbf{F}^{\mathrm{T}}\mathbf{F}\mathbf{m}$. Time- and space-dependent active contraction $T_{Ca^{2=}}(\mathbf{x}, t)$ is driven by epicardial electrical activity measured with monophasic action potentials (MAP). The pressure-volume (PV) loop, measured through catheterization and sonomicrometry, was used to best-fit the modulation of active contraction.

In the Gorman Cardiovascular Research Group, we conducted novel regional experiments using an ovine model with a 27 sonocrystals implanted in a 3D array in the LV free wall (Fig. 3) [6]. A prolate spheroidal coordinate wedge is defined in the region delimited by the sonocrystals, and the regional strain field is determined from their motion using FE techniques with Sobolev norm regulation. From the strain field we determine the regional dilatation (normalized volume change). Experimental data demonstrated a progressive decrease in dilatation from the epi- to the endocardial surfaces, and from the direction of the base to the apex. These results indicate that myocardium is compressible during systole only. Myocardial tissue volume can only decrease (with dilatation ranging from 1.0 to ~ 0.75), so that conventional compressible material models cannot be used, as $J \leq 1$ still needs to be enforced. To account for this distinctive type of compressible material behavior (i.e. incompressible to volume increase but compressible while actively contracting), we have developed a specialized material model that utilized a penalty term allowing for a reduction in volume only and was modulated by the active contraction.

RESULTS

MRI, DTMRI, MAP, and in vivo PV loops allow the construction of a physiologically significant in silico heart model. Passive mechanical properties compared favorably with previous literature, time-modulation of active contraction shows good agreement with typical calcium-activated contraction, and qualitative comparison with in vivo 2D echo validated the baseline healthy model [7]. Testing of the material model was done in several steps, starting with a basic 3D cube to investigate how axial contraction produces transverse expansion (Fig. 4). Note that as the level of compressibility increases, the level of transversal expansion reduces.

We simulated stiffening and loss of contractility in infarcted myocardium by inversely determined matching an infarcted PV loop. In vivo function was validated with the transmural strain smooth fields determined with sonocrystal arrays (Fig. 3) validated quantitatively the baseline and infarct models. Changes in properties (passive mechanics and active contraction) in infarcted and border-zone regions around the

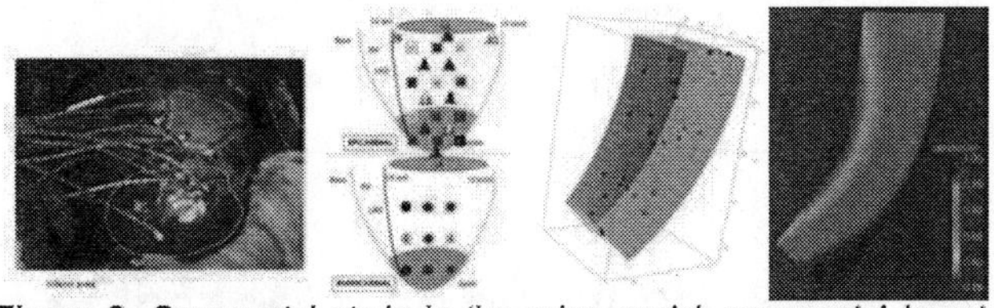

Figure 3. Sonocrystal study in the ovine model, sonocrystal layout, prolate spheroid "wedge" used for strain analysis, and regional dilatation for normal myocardium, showing regional variations of volume change.

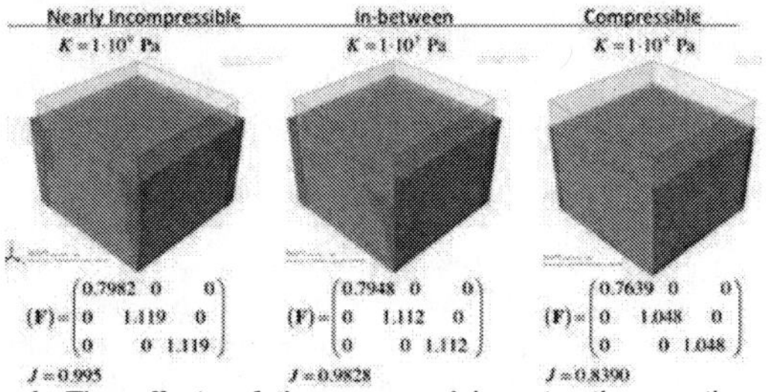

Figure 4. The effects of the same axial contraction on the corresponding transverse strain and dilatation. Material compressibility impacts the relationship between active contraction and wall thickening.

LV apex to best-fit the in vivo MI PV loop resulted in good agreement with transmural strain field observed in vivo and with previous findings.

DISCUSSION

The employment of a complete dataset from a single heart to develop an in-silico model avoids unnecessary image registration between MRI and DTMRI images. The coupling between electrical activity and active contraction is significant to heart biomechanics and is needed to obtain physiological realistic heart function. Myocardium is extensively perfused with distensible vessels (approximately 10-20% of the total volume), thus increasing the stiffness of the surrounding tissue, as during muscle contraction, will cause the fluid to be extruded from these vessels [2] – the finding that perfused myocardium is compressible implies that results from analyses that assume incompressibility are not realistic. Not only the ventricular walls modeled with incompressible myocardium will be thicker at peak systole (as the relationship between fiber shortening and wall thickening is highly sensitive to volumetric changes) but also ventricular stresses will be substantially overestimated – the former is crucial for accurate representation of ventricular kinematics and the latter is a key factor in myofiber stress and ventricular remodeling. To simulate MI, we have found that both impairment of contraction and stiffening of myocardium were necessary to replicate appropriately MI biomechanics. Subsequent remodeling mechanisms of collagen fiber re-orientation may be important and are subject of future studies.

ACKNOWLEDGEMENTS

NIH (R01 HL068816, HL089750, HL070969, HL108330, HL063954)

REFERENCES

[1] Gorman, R.C., et al., J Cardiovasc Transl Res, 2011. 4: 73-81;
[2] Yin, F.C., et al., Am J Physiol, 1996. 271: H1864-70;
[3] Wang, V.Y., et al., Med Image Anal, 2009. 13: 773-84;
[4] Klotz, S., et al., Nat Protoc, 2007. 2: 2152-8;
[5] Hunter, P.J., et al., Prog Biophys Mol Biol, 1998. 69: 289-331;
[6] Jackson, B.M., et al., J Am Coll Cardiol, 2002. 40: 1160-71;
[7] McGarvey, J.R., et al., Ann Thorac Surg, 2015. 100: 582-9;

TOPOLOGICAL AND GEOMETRICAL ANALYSES OF 3D EPICARDIAL ELASTIN FIBER NETWORK

Xiaodan Shi (1), Song Zhang (2), Katherine M. Copeland (1), Yue Liu (3), Huajian Gao (3), Jun Liao (1)

(1) Department of Bioengineering, University of Texas at Arlington, Arlington, TX 76019, USA

(2) Department of Computer Science and Engineering, Mississippi State University, Mississippi State, MS 39762, USA

(3) School of Engineering, Brown University, Providence, RI 02912, USA

INTRODUCTION

Heart disease is the leading cause of death in the United States each year. Among the components of the cardiac extracellular matrix (ECM) in the left ventricle (LV), cardiac elastin is believed to play a vital role in the overall cardiac biomechanics. However, to the best of our knowledge, there is no thorough study that details the 3D geometrical configuration of cardiac elastin fiber network. In this study, we have applied laser scanning confocal microscopy (LSCM) to obtain the elastin fiber images from the epicardium layer of the porcine LV. We found that the epicardial elastin network has location- and depth-dependency. We further reconstructed the rich 3D information from the 2D confocal images and performed the topological and geometrical analyses of 3D elastin network. The intricate interconnectivity, geometry, and topology of the 3D epicardial elastin fibers are largely unknown on a whole heart scale and are crucial to understand the structural-functional design the cardiac elastin.

METHODS

Confocal microscopy images are noisy. A simple threshold based method that identified pixels with high intensities as fibers generates spurious features and miss true fibers. We first used the multi-scale vessel enhancement filtering [1] to segment the elastin fibers and create a "fiberness" value at each pixel. To acquire an abstract representation of the structure and connectivity of the network, each fiberness image was then binarized and removed outside surface fibers if no homotopy changed. We used the homotopic thinning algorithm [2] to produce a one-voxel-thick centerline skeleton for the elastin fibers. Spurious fiber branches were removed based on the length of the branches. From the elastin fiber skeleton, network metrics were built by (1) removing the spurious fibers using length threshold; (2) identifying nodes which connect two or more fibers; and (3) identifying links between a pair of nodes and terminal links where at least one of its end nodes has only one fiber connecting to it.

RESULTS

By means of fiber segmentation, skeletonization, and network-based metrics reconstruction, we were able to create a fiberness map of elastin network in the epicardium of the porcine LV. Elastin fibers were found to show different shape, connectivity, orientation, and scale at multiple regions in the heart. As an example, Figure 1 (A) reveals the intricate structure of 3D rendering of elastin fibers in the apex region of the heart. Figure 1 (B) shows the skeletonized centerlines of the fibers

Poster Presentation #P216

shown in figure 1 (A). Fiberness biometrics was then computed in Figure 1 (C) to further reveal the topological structure of the elastin fibers and cordless cycles in the network. Topological analysis shows the distribution of cordless cycles in the network. Red, green, blue, magenta, and yellow fibers are part of the cordless cycles with 3 to 7 edges, respectively.

DISCUSSION

In this study, fiber-based metrics reconstruction from epicardial elastin network was achieved for the first time. Fiber-based biometrics is more specific and more robust to the disturbance to fiber locations across subjects. This fiberness map can be rendered for display or used as mask for further processing. The density and complexity of the fibers suggest that a rich amount of information can be mined from a 3D topological and geometrical study, which are highly correlated to their anatomically locations. The cycle structure can help reveal the biological design of the elastin fibers. Potential metrics include fiber length, average fiber curvature, average scale along fiber, 3D orientation distribution in the network, numbers of branching points in the network, number of cliques in a network, etc.

ACKNOWLEDGEMENTS

This works is supported by NSF EPS-0903787 and NIFA CRIS (#MIS-351070).

REFERENCES

[1] Frangi, A.F., et al. Medical Image Computing and Computer-Assisted Interventation, 1998: p. 130-137.
[2] Lee, T.C., et al. CVGIP: Graphical Models and Image Processing, 1994. **56**(6): p. 462-478.

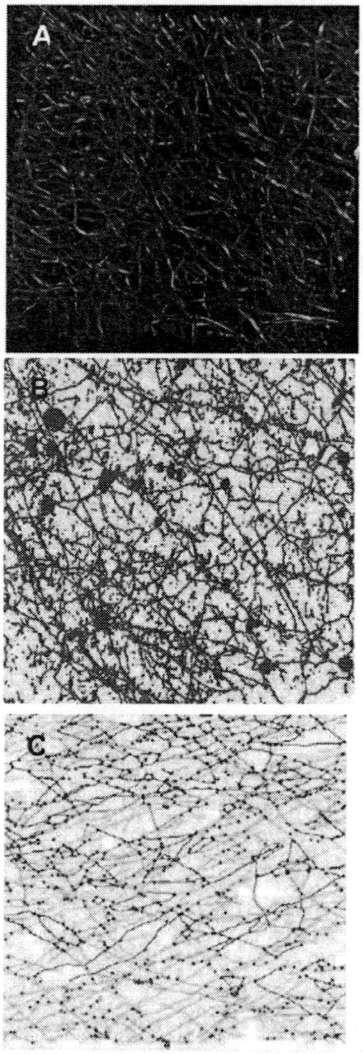

Figure 1: 3D elastin network structure of the apex region in porcine heart. (A) The 3D rendering of elastin fibers; (B) The skeletonization of the elastin fibers; (C) Reconstruction of 3D biometrics of elastin fibers

SB³C2017
Summer Biomechanics, Bioengineering and Biotransport Conference
June 21 – 24, Tucson, AZ, USA

BIAXIAL MECHANICAL PROPERTIES OF PORCINE TRICUSPID VALVE LEAFLETS

Keyvan Amini Khoiy (1), Rouzbeh Amini (1)

(1) Department of Biomedical Engineering
The University of Akron
Akron, OH, USA

INTRODUCTION

The tricuspid valve (TV) prevents blood backflow from the right ventricle to the right atrium during ventricular systole [1]. TV malfunctions such as valvular regurgitation or stenosis may require surgical interventions such as total valve replacement or valvular repair. TV repair has better outcomes when compared to the total TV replacement [2-4]. However, the outcomes of TV surgeries are still not completely satisfying and TV surgery is one of the most high-risk valve procedures [6]. To improve the existing surgical techniques and develop more successful valve repair procedures, accurate quantification of the biomechanical properties of native TV tissues is essential. Such data are necessary both in development of tissue equivalent biomaterials and as input parameters for generating computational models of healthy and diseased TVs.

Considering the native loading environment of the cardiac valves, we have previously used biaxial tensile testing technique to evaluate the stress-strain response of the porcine TV leaflets [5]. In the current study, we utilized a phenomenological constitutive model to quantify the anisotropic and nonlinear material properties of the porcine TV leaflets. The material constants will be used in our future computer simulations of TV to mimic its deformation in vivo.

METHODS

The sample preparation and biaxial tensile testing procedures have been described in detail in our previous publication [5]. Briefly, a total of ten square-shaped ($11\,mm$ by $11\,mm$) freshly excised samples per TV leaflet were used. Specimens were mounted to the biaxial tensile testing equipment with the circumferential direction aligned with one stretching axis and the radial direction aligned with the other stretching axis. Each leaflet then experienced five different tension-controlled loading protocols. For each protocol, the ratio of axial tensions $T_c{:}T_r$ was kept constant during the test ($T_c{:}T_r = 1{:}1, 1{:}0.75, 0.75{:}1, 1{:}0.5, 0.5{:}1$ for the five protocols). The strains and stresses were subsequently calculated from tension-deformation data as described previously [5].

Similar to the other cardiac valves, within the small specimen region, TV leaflet tissue was assumed to be incompressible, homogenous, and hyperelastic undergoing finite deformations [6]. As such, we could describe the mechanical response of the TV leaflets by a strain energy function W. Therefore, the elements of the second Piola-Kirchhoff stress tensor were determined by the following equation:

$$S_{ij} = \frac{\partial W}{\partial E_{ij}} \qquad (1)$$

where $\mathbf{E}$ was the Green-Lagrangian strain tensor. To investigate the pseudo-elastic response of the TV leaflets and choose the suitable form of the strain energy function, all the stress-strain data were independently fit to the following response functions for each directional component as described by Vande Geest et al. [7]:

$$\begin{aligned} S_{cc} &= c_0(c_1E_{cc} + c_3E_{rr} + c_4E_{cc}E_{rr} \\ &\quad +\tfrac{1}{2}c_5E_{rr}^2 + c_6E_{cc}E_{rr}^2 + 2c_7E_{cc}^3)e^Q \\ S_{rr} &= c_0(c_2E_{rr} + c_3E_{cc} + c_5E_{cc}E_{rr} \\ &\quad +\tfrac{1}{2}c_4E_{cc}^2 + c_6E_{cc}^2E_{rr} + 2c_8E_{rr}^3)e^Q \end{aligned} \qquad (2)$$

where c and r denoted circumferential and radial directions, respectively and

$$\begin{aligned} Q = &(c_1E_{cc}^2 + c_2E_{rr}^2 + 2c_3E_{cc}E_{rr} + c_4E_{cc}^2E_{rr} + c_5E_{rr}^2E_{cc} \\ &+ c_6E_{cc}^2E_{rr}^2 + c_7E_{cc}^4 + c_8E_{rr}^4) \end{aligned}$$

The specific form of the resulting stress surfaces led to selection of an anisotropic Fung-type strain energy function [8]. As shear stresses were negligible [5], the strain energy function W was expressed as:

$$W = \frac{c}{2}\left(e^{a_1 E_{cc}^2 + a_2 E_{rr}^2 + 2a_3 E_{cc} E_{rr}} - 1\right) \quad (3)$$

where c and a_i were material constants. Anisotropy index (AI, a measure of anisotropy) was also calculated as:

$$AI = min\left(\frac{a_1+a_3}{a_2+a_3}, \frac{a_2+a_3}{a_1+a_3}\right) \quad (4)$$

RESULTS

The biaxial stress-strain data fitted the response functions of equation (2) well with an average R-squared of 0.94. The graphs of constant stresses were plotted in the strain field (versus E_{cc} and E_{rr}) to help investigate the selection of strain energy function. For an ideal isotropic material the constant stress contours are symmetric with respect to $E_{cc} = E_{rr}$ line. As shown in Fig. 1 for typical leaflets, the stress contours are asymmetric for all three leaflets indicating an anisotropic mechanical response. Therefore, the anisotropic Fung-type strain energy function expressed in equation (3) was chosen [8].

The five-protocol stress-strain data fitted the Fung-type model well with average R-squared of 0.94, 0.92, and 0.92 for anterior, posterior, and septal leaflets, respectively (Fig. 2). Table 1 shows the model parameters for the three representative leaflets. The mean and standard deviation of anisotropy indices were 0.52 ± 0.24, 0.39 ± 0.20, and 0.51 ± 0.25 for the anterior, posterior, and septal leaflets, respectively. Statistical analysis showed that there was no significant difference between the anisotropy indices of different leaflets ($p > 0.05$ for all comparisons). In all three leaflets, the anisotropy index was significantly smaller than 1 indicating that all leaflets were stiffer in the circumferential direction in comparison to the radial direction.

Table 1: Material constants of Fung-type model calculated for representative anterior, posterior, and septal leaflets.

Leaflet	$c[kpa]$	a_1	a_2	a_3	R^2	AI
Anterior	1.02	47.81	14.06	5.99	0.964	0.37
Posterior	0.47	58.15	33.54	1.09	0.988	0.58
Septal	0.85	123.25	40.62	8.04	0.985	0.37

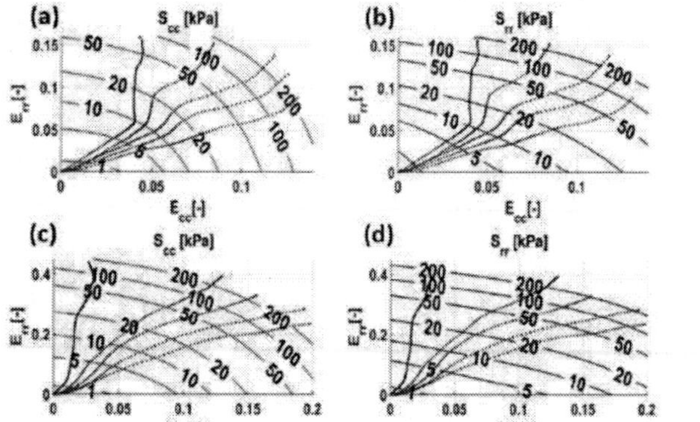

Figure 1: Contours of constant stresses superimposed over the strain field for typical (a and b) anterior, (c and d) posterior, and (e and f) septal leaflets.

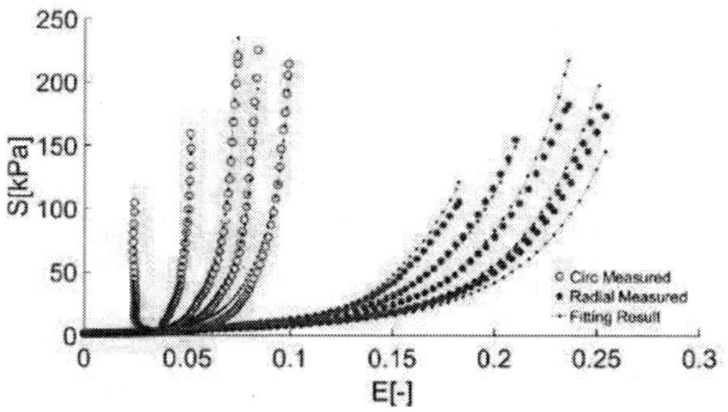

Figure 2: The constitutive model fit (dot-line) plotted over the experimental data from all five protocols for circumferential (circles) and Radial (stars) directions for a typical septal leaflet.

DISCUSSION

After independently fitting the planar biaxial stress-strain data from porcine TV leaflets to the response functions of equation (2), we plotted the constant stress contours versus the strain field. Asymmetric contours showed that all three leaflets are anisotropic. The dramatic changes in the stress contours from small strain values to larger strain values indicated that an exponential form for the strain energy function is appropriate for TV leaflets. As such, an anisotropic Fung-type strain energy function was selected.

Although the constitutive model used in this study predicted the response of the tissue with high confidence levels, its implementation in finite element models of the TV should be accompanied with other considerations. For example, considering the high degree of complexity at the tissue microstructural level, the assumption of homogeneity of the tissue will not be accurate for all different regions of the valve leaflets. As such, further research is necessary to quantify microstructural architecture of the TV leaflets and the variation of tissue material parameters over the entire leaflet.

Unlike multi-scale models [9], phenomenological models similar to the Fung-type model used in this study, do not provide comprehensive information about potential changes in tissue microstructure. Nevertheless, implementation of phenomenological strain energy functions in commercial finite element software is often less complicated. In addition, with their large number of degrees of freedom, the multi-scale models are much more computationally costly. As such, if one's intention is only to investigate the tissue level mechanical responses, using the proposed model with additional correction for shear-related numerical stability [10] is advantageous.

ACKNOWLEDGEMENTS

Funding for this project was provided in part by American Heart Association Grant No. 16SDG27770088.

REFERENCES

[1] Hall, J. E., *Elsevier Health Sciences*, 2010.
[2] Singh, J. P., *Am J Cardiol*, 83:897-902, 1999.
[3] Stuge, O., *J Thora Cardiovasc Surg*, 132:1258-1261, 2006.
[4] Taramasso, M., *J Am Coll Cardiol*, 59:703-710, 2012.
[5] Amini Khoiy, K., *J Biomech Eng*, 138:104504, 2016.
[6] May-Newman, K., *J Biomech Eng*, 120:38-47, 1998.
[7] Vande Geest, J. P., *J Biomech Eng*, 126:815-822, 2004.
[8] Tong, P., *J Biomech*, 9:649-657, 1976.
[9] Stylianopoulos, T., *Comput Methods Appl Mech Eng*, 196:2981-2990, 2007.
[10] Vande Geest, J. P., *Ann Biomed Eng*, 36:921-932, 2008.

SB³C2017
Summer Biomechanics, Bioengineering and Biotransport Conference
June 21 – 24, Tucson, AZ, USA

EFFECT OF LAYER- AND ORGAN-SCALE RESIDUAL STRAIN ON THE OPENING-CLOSING BEHAVIOR OF A HEART VALVE

Rogelio Ortigosa[1], Antonio J. Gil[1], Ankush Aggarwal[1]

(1) Zienkiewicz Centre for Computational Engineering,
Swansea University, Swansea, UK

INTRODUCTION

Heart valves play an important role in the normal functioning of heart by ensuring a unidirectional blood flow during both systole and diastole. Hence, the geometric and mechanical properties of heart valves have evolved to provide a highly efficient performance. Any deviation from the optimal properties, either because of disease or surgery, results in a lower cardiac output or remodeling of tissue with long term complications.

Residual strains have been reported in the aortic valve leaflets both at the organ and layer level [1,2] (Fig. 1). The functional importance of this residual strain characteristic is speculated, but has not been established or quantified. In this study, we investigate the role residual strain plays in the heart valve function by performing a detailed computational analysis.

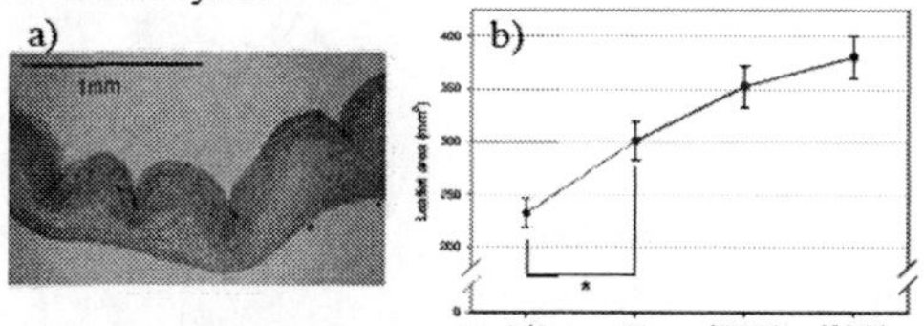

Figure 1: (a) Layer-scale (adapted from [7]) and (b) organ-scale residual strain (adapted from [2]) for aortic valve leaflet

METHODS

We use aortic valve as an example of the heart valve, which has three semi-lunar shaped leaflets. A new shell model of a single aortic valve leaflet is created using triangular elements, which allows us to incorporate layer-level details [3] (Fig. 2). An average thickness of 0.4mm is used. Three layers are modeled through the thickness of the valve – fibrosa, spongiosa, and ventricularis – each with different mechanical properties (Fig. 2). The residual strain is introduced in the constitutive law via multiplicative decomposition [4]

$$F = F_e Q \hat{F} Q^T$$

where F_e is the elastic deformation, Q is the rotation matrix that defines local material coordinate, and $\hat{F}$ is the residual strain tensor in the material coordinate system

$$\hat{F} = \begin{bmatrix} \lambda_c & 0 & 0 \\ 0 & \lambda_r & 0 \\ 0 & 0 & 1/\lambda_c\lambda_r \end{bmatrix}.$$

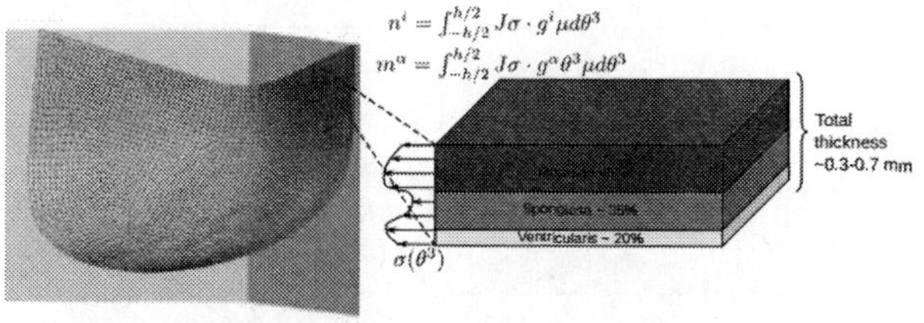

Figure 2: A new shell model formulation allows us to incorporate layer-specific properties

The total deformation gradient F is used in the constitutive law. The constitutive law for fibrous tissue (fibrosa and ventricularis) is taken as the simplified structural model

$$\psi_f = \int_\theta \Gamma(\theta)\, \psi_{ens}(\mathbf{E}: \boldsymbol{N} \otimes \boldsymbol{N}) d\theta$$

where ψ_{ens} is the ensemble response and is approximated using an exponential function. This model has been shown to accurately capture the mechanical response of valve tissue [5]. For spongiosa, which is a gel-like structure, neo-Hookean model is used. Contact with other leaflets is approximated as with a rigid wall [6]. Each surface node has six degrees of freedom (three translational and three rotational). In addition, the thickness stretch varies along the thickness. Thus, two degrees of freedom for the stretch are assigned per layer to each node. A normal pressure load approximating the cardiac pressure is applied and the dynamic equation of the system is solved using an implicit method to guarantee stability. The resulting deformation is analyzed to obtain the opening area as a function of time (Fig. 3).

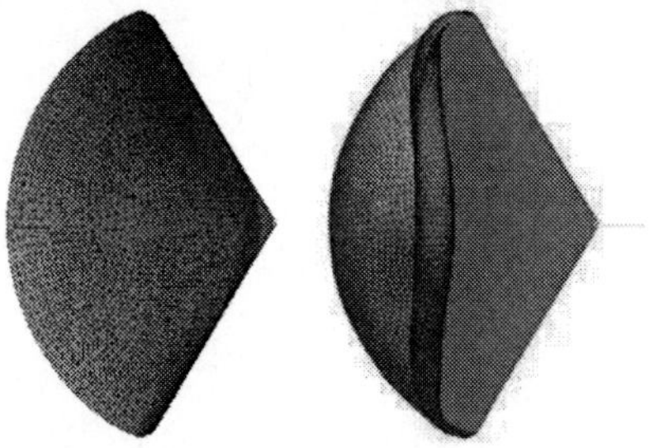

Figure 3: From the simulated valve shape, open area is calculated by integrating its projection

RESULTS

The residual strains observed experimentally are used (Table 1). The resulting opening-closing behavior is simulated and analyzed for a range of residual strains (from no residual strain to experimental values). The results show that the in-vivo residual strains provide an improved opening-closing behavior, as the leaflet opens much faster (Fig. 4).

Table 1: The residual strain's principal stretches (circumferential and radial) in the top and bottom layer of aortic valve.

Residual strain	Layer	In-vivo value
λ_c	Fibrosa	0.95
λ_r	Fibrosa	0.78
λ_c	Ventricularis	0.8
λ_r	Ventricularis	0.55

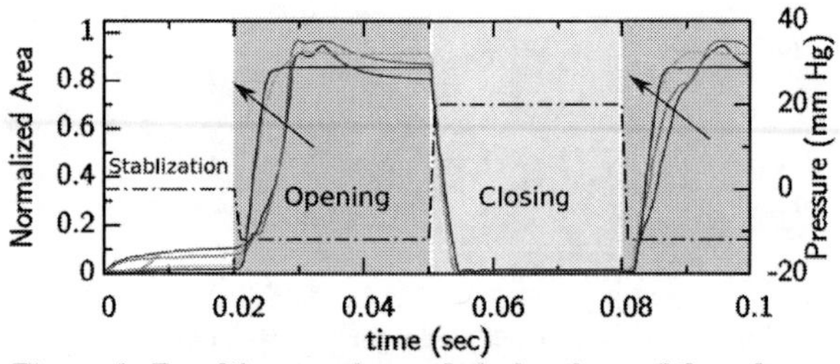

Figure 4: Resulting opening and closing times of the valve as the layer-scale residual strain is varied

DISCUSSION

In this study, we present a computational model that describes the mechanical properties of heart valve while taking into account the layer-specific mechanics. This is the first time such a multi-scale model has been formulated, where the layer and organ scales are coupled two-ways. In the previous study by Mofrad et al. [7], only one way coupling was established. We focus on the role on residual strains, and use aortic valve as an example. We simulate the opening-closing behavior of the valve with varying residual strains in the layers, and observe that the leaflet opens quicker with residual strain. The opposite residual strain did not affect the leaflet dynamics. Moreover, the closing times were relatively insensitive to the residual strain.

Overall, these findings provide an important insight into the mechanics of heart valves. The results indicate that the valve leaflets' residual strains may have a functional role in the cardiac system, and might be a mechanism of providing an improved mechanics. Thus, there is a need for consideration of residual strains while designing artificial heart valves and valve repair procedures.

In the future, this model will be further extended to elucidate the role of each layer in the valve function. This would require more detailed description of the leaflet mechanics. The results would also accurate define the stress state of valve interstitial cells during cardiac cycle, which are responsible for the remodeling and disease development [8].

ACKNOWLEDGEMENTS

The authors gratefully acknowledge the financial support provided by the Welsh Government and Higher Education Funding Council for Wales through the Sêr Cymru National Research Network in Advanced Engineering and Materials (Grant No. F28) and the Engineering and Physical Sciences Research Council of the UK (Grant No. EP/P018912/1).

REFERENCES

[1] Stella and Sacks, J Biomech Eng 129, 757-766, 2007
[2] Aggarwal et al., J Biomech, 49(12):2481-2490, 2016
[3] Ortigosa and Gil, CMAME (under review)
[4] Rodriguez, Hoger, and McCulloch, J Biomech 27, 455, 1994
[5] Fan and Sacks, J Biomech 47: 2043-2054, 2014
[6] Aggarwal and Sacks, BMMB, 15(4):909-932, 2016
[7] Weinberg and Mofrad, Cardivasc Eng, 7:140-155, 2007
[8] Rego et al., J. Royal Society Interface, 13(125), 2016

SB[3]C2017
Summer Biomechanics, Bioengineering and Biotransport Conference
June 21 – 24, Tucson, AZ, USA

A BIOLOGICALLY MOTIVATED COMPUTATIONAL MODEL TO EXPLAIN ANISOTROPIC PRESTRETCH IN THE MITRAL VALVE

Mathieu A. J. van Kelle (1,2,3), Manuel K. Rausch (4), Carlijn V.C. Bouten (1,2), Ellen Kuhl (3), Sandra Loerakker (1,2,3)

(1) Department of Biomedical Engineering
Eindhoven University of Technology
Eindhoven, The Netherlands

(2) Institute for Complex Molecular Systems
Eindhoven University of Technology
Eindhoven, The Netherlands

(3) Mechanical Engineering
Stanford University
Stanford, CA, United States

(4) Department of Aerospace Engineering & Engineering Mechanics
University of Texas,
Austin, TX, United States

INTRODUCTION

In soft biological tissues, the processes of growth, development, remodeling and repair lead to the development of tissue prestretch. Neglecting these tissue prestretches when studying mechanical properties of biological tissues, can lead to great discrepancies in obtained results[1]. However, accurately determining de degree of prestretch in tissues in their *in vivo* situation is often hard, if possible at all, due to the complex geometry and loading conditions. Therefore, computational models are necessary to accurately predict tissue prestretch, but also to gain mechanistic understanding of the development of these prestretches.

In a recent study, Rausch *et al*[1,2] developed a computational model to describe prestretch in the mitral valve leaflet. While *a priori* assuming a certain degree of prestretch, they investigated the actual *in vivo* leaflet stresses and strains and intrinsic material properties. By doing so, this model could nicely predict the degree of prestretch in the mitral valve. However, in this model it was assumed that prestretch is mapped homogeneously and isotropically over the valve, while actually there is experimental evidence suggesting that prestretch in the mitral valve is rather anisotropic[3]. Moreover, the mechanism through which prestretch is induced was not addressed, leaving room for improvement of the current framework.

In the present study, the framework by Rausch *et al*[1,2] was adapted to allow for inhomogeneous and anisotropic prestretch development, where the local magnitude and direction of this prestretch depends on a biologically motived algorithm for cell-mediated prestretch[4].

METHODS

Constitutive Framework

To model the constitutive behavior of the ground matrix and collagen fibers in the mitral valve, a Fung-type hyperelastic model[2,5] was used that uses two decoupled terms to describe the isotropic and anisotropic tissue response respectively. To account for anisotropic prestretch, a second fiber family was introduced. Consequently, the isochoric strain energy function is as follows:

$$\overline{\psi} = c_0(\overline{I_1} - 3) + \frac{c_1}{2c_2}(\exp(c_2(\varkappa_1\overline{I_1} + (1 - 3\varkappa_1)\overline{I_4'}_1 - 1)^2 + \frac{c_3}{2c_4}(\exp(c_4(\varkappa_2\overline{I_1} + (1 - 3\varkappa_2)\overline{I_4'}_2 - 1)^2 \quad (1)$$

With stiffness parameters c_0, c_1, c_2, c_3, and c_4, fiber dispersities $\varkappa_1$ and $\varkappa_2$ and isochoric invariants $\overline{I_1}$, $\overline{I_4}_1$ and $\overline{I_4}_2$ of the isochoric part of the right Cauchy-Green tensor:

$$\overline{I_1} = \overline{\boldsymbol{C}}:\boldsymbol{I} \quad \text{and} \quad \overline{I_{4_i}} = \overline{\boldsymbol{C}}:\boldsymbol{N_i} \quad (2)$$

With unity tensor $\boldsymbol{I}$ and structural tensor $\boldsymbol{N_i}$ corresponding with each fiber direction.

Prestretch

Prestretch was induced through manipulation of the fourth invariants by multiplication with the square of a preferred collagen fiber prestretch λ_p, in a similar manner as done by Nagel & Kelly (2012)[6]:

$$\overline{I_4'}_i = \lambda_p^{\,2}\,\overline{I_{4_i}} \quad (3)$$

The previously introduced cell-mediated tissue compaction model by Loerakker *et al (2014)*[4] will be used to hypothesize the preferred collagen prestretch, based on the strain and strain rate of the leaflet, and the contractility of the cells.

Implementation

The entire framework was implemented in ABAQUS (version 6.14) using the user defined subroutine UANISOHYPER_INV, in which the first and second derivatives of the strain energy function (equation 1) with respect to its invariants were specified.

RESULTS

First, the proposed framework was tested on a circular membrane (Figure 1a), by explicitly prescribing the preferred prestretch. In the case of a single fiber direction, prestretch can only develop in one direction (Figure 1b). By the introduction of a second, perpendicularly oriented fiber direction, prestretch can develop in two directions, either isotropically (Figure 1c), or anisotropically (Figure 1d).

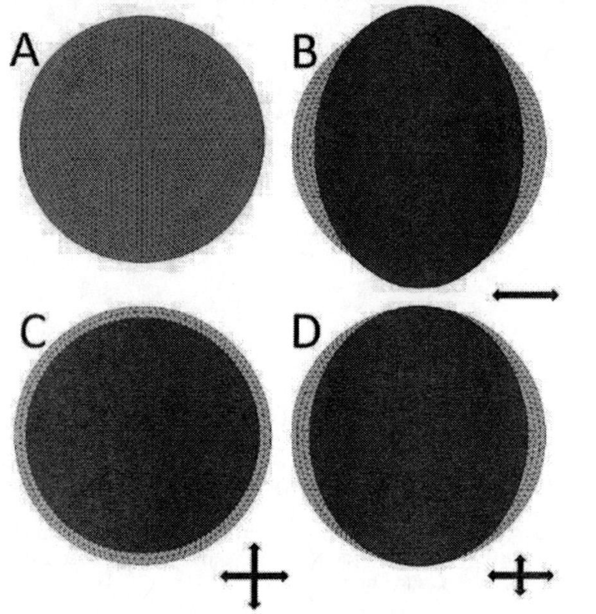

Figure 1: Circular membrane at the start of the simulation (A). Deformed membranes for prestretch (arrows) in one direction (B), isotropic prestretch in two directions (C), anisotropic prestretch in two directions (D).

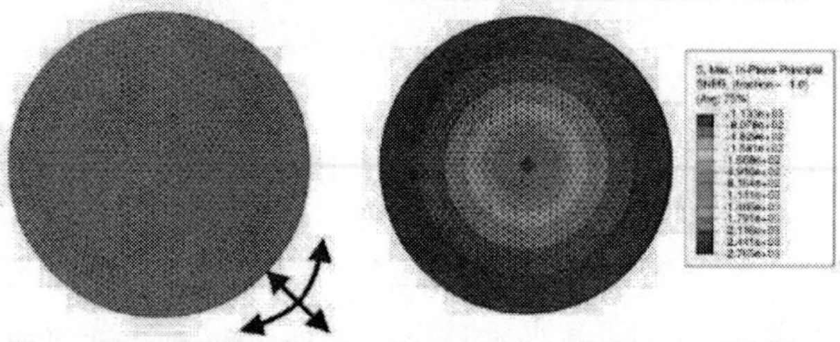

Figure 2: Left) Circular membrane with radial-tangential fiber directions (arrows), Right) Inhomogeneous prestress development.

Next, a radial-tangential fiber direction was introduced in the circular membrane (Figure 2, left). This shows that although the prestretch can be equal in the two (radial-tangential) fiber directions, this can still lead to inhomogeneous prestress development (Figure 2, right).

Finally, the framework was applied to a mesh representing the anterior mitral valve leaflet, as constructed by Rausch *et al*[2]. The prestretch was again explicitly prescribed and the fiber orientation varied locally (Figure 3 top). Again, a clearly anisotropic, inhomogeneous prestress development can be observed, even though the local prestretch with respect to the (local) fiber directions is the same (Figure 3 bottom).

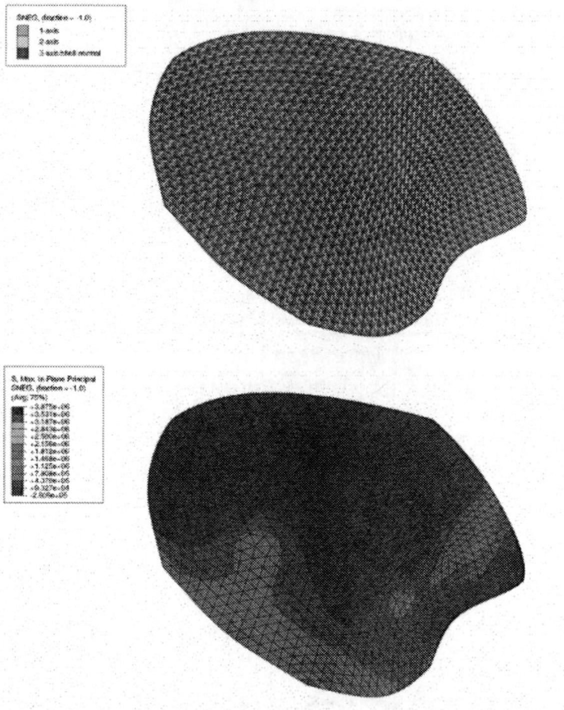

Figure 3: Top) Model of anterior mitral leaflet with fiber orientations. Bottom) Inhomogeneous prestress development.

DISCUSSION

In the present study, a computational model for anisotropic, inhomogeneous prestretch development was developed. The current results illustrate the feasibility of this model. As a next step, the cell-mediated tissue compaction model[4] will be incorporated to obtain the preferred collagen prestretch. Future research will focus on validating the predicted prestretches in the mitral valve leaflet. First, results will be compared to the predictions by Rausch *et al*[2]. Second, upon release of the constraints in the FE simulations, leaflet shrinkages will be quantitatively compared to previously reported mitral valve contraction after excision from the heart[3].

ACKNOWLEDGEMENTS

We acknowledge the support from the Netherlands Cardiovascular Research Initiative (CVON 2012-01): The Dutch Heart Foundation, Dutch Federation of University Medical Centers, the Netherlands Organization for Health Research and Development and the Royal Netherlands Academy of Sciences.

REFERENCES

[1] Rausch, MK *et al.*, *BMMB*, 12:1053-1071, 2013.
[2] Rausch, MK *et al.*, *J Mech Phys Solids*, 61(9):1955-1969, 2013.
[3] Amini, R *et al.*, *Ann Biomed Eng*, 40(7), 1455-1467, 2012.
[4] Loerakker, S *et al.*, *BMMB*, 13:985-1001, 2014.
[5] Gasser, TC *et al.*, *J R Soc interface*, 3:15-35, 2006.
[6] Nagel, T. & Kelly, DJ., *BMMB*, 11: 325-339, 2012

FINITE ELEMENT MODELING OF MITRAL VALVE PATCH AUGMENTATION AND EFFECTS ON CHORDAL FORCE DISTRIBUTION

Jonathan F. Wenk (1,2), Morten O. Jensen (3)

(1) Department of Mechanical Engineering
University of Kentucky
Lexington, K□, USA

(2) Department of Surgery
University of Kentucky
Lexington, K□, USA

(3) Department of Biomedical Engineering
University of Arkansas
Fayetteville, AR, USA

INTRODUCTION

The mitral valve (MV) is the primary structure that controls the one-way flow of blood from the left atrium to the left ventricle. During various disease states the valve can become compromised, leading to mitral regurgitation (MR). If left unchecked, MR can significantly accelerate the adverse effects of heart failure. Several surgical interventions have been proposed in order to repair and prevent MR. One such approach involves the implantation of a patch in the anterior leaflet of the MV (**Fig 1**) [1]. The goal is to augment the leaflet area in order to increase coaptation between the two leaflets of the MV, thus reducing MR. However, there remain open questions about the effects of this procedure on chordal force distributions and leaflet stress.

As a compliment to in vivo and in vitro experimental studies, computational models of the MV can provide additional insights on treatment options for MR, as well as guide device design. The goal of the current study was to employ a finite element (FE) model of the mitral apparatus in order to investigate the effects of patch augmentation on chordal forces and leaflet coaptation. The results of the numerical study were then compared to previous studies in the literature.

METHODS

Inspired by previous in vivo [2] and in vitro [3] studies of MV patch augmentation, a previously develop FE model [4] of the mitral apparatus was modified to mimic this procedure. Specifically, the FE model of a MV (both anterior and posterior leaflets), which included the chordae tendineae and papillary muscle attachments, was reconstructed using the geometry from previously collect MRI data of a sheep heart. The papillary muscles were meshed with 8-node trilinear brick elements, the mitral valve leaflets with 4-node bilinear membrane elements, and the chordae with 2-node beam elements (**Fig 2A**). A nearly incompressible, transversely isotropic, hyperelastic constitutive law was used for papillary muscles [4], the MV leaflets were modeled with an anisotropic layered shell material model [5], and the chordae tendineae were modeled with a cable element formulation [4]. The edge and strut chordae were attached to the leaflets and papillary muscles in a configuration that mimicked in vivo data. In order to represent the presence of a pericardial patch (**Fig 1**), the same region of the anterior leaflet was modified in the FE model to have different properties and thickness (**Fig 2B**). The material properties were adjusted until the bulge in the patch was within one Std Dev of in vivo measurements [2].

The annulus of the MV was assumed to be completely fixed, along with the mid-myocardial surface of the papillary muscles. The valve leaflets were loaded to end-systole using a boundary condition that was based on in vivo measurements of ventricular and atrial pressures [4]. Both the baseline and patch augmented models were simulated using the nonlinear FE solve LS-DYNA (Livermore Software Technology Corporation, Livermore, CA).

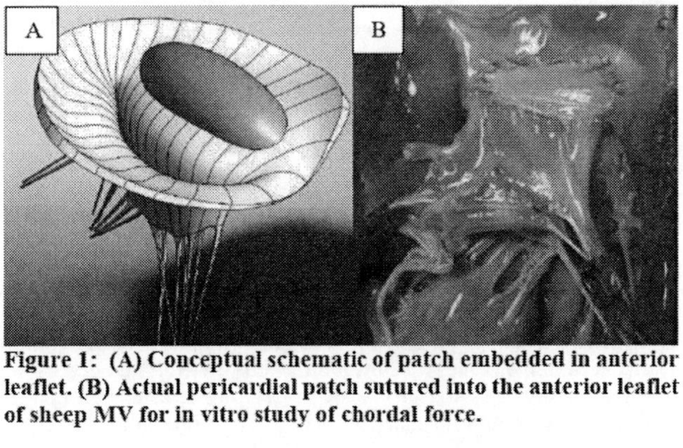

Figure 1: (A) Conceptual schematic of patch embedded in anterior leaflet. (B) Actual pericardial patch sutured into the anterior leaflet of sheep MV for in vitro study of chordal force.

Poster Presentation #P220

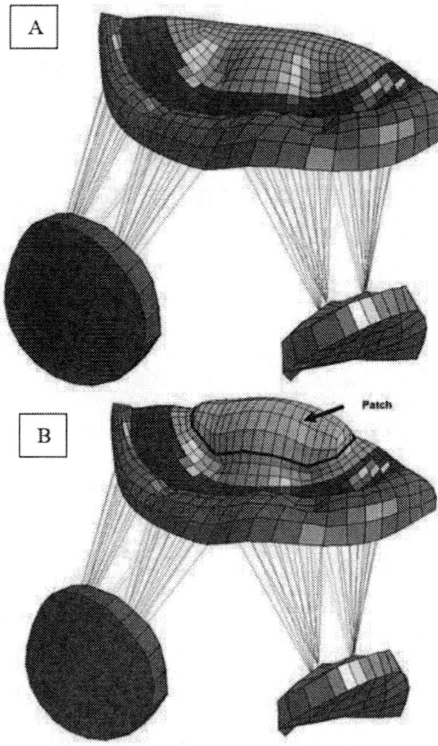

Figure 2: (A) Baseline FE model of mitral apparatus loaded to end-systole. (B) FE model with patch embedded in the anterior leaflet. Note: Posterior leaflet (blue), anterior leaflet (green/yellow).

RESULTS

The deformed shape of the mitral apparatus (without and with patch) can be seen in **Fig 2**. In order to further visualize the effects of the patch on the anterior leaflet, the deformed shape of the leaflet meshes are overlapped (without and with patch) in **Fig 3**. It can be seen that the strut chord insertion points move towards the edge of the leaflet (towards the coaptation zone), as well as downward towards the papillary muscles due to the presence of the patch (**Fig 3A**). In addition, the edge of the anterior leaflet moves inward towards the coaptation zone and downward towards the papillary muscles due to the presence of the patch (**Fig 3B**). This led to an overall increase in the coaptation area between the anterior and posterior leaflets. The average force in the chordae along the edge of the anterior leaflet decreased by approximately 25% due to the presence of the patch. This is likely due to the downward displacement of the leaflet edge, which reduces the amount of stretch in the edge chords.

As an initial sensitivity study, the bulge in the pericardial patch region was altered over a range of values. It was found that when the bulge in the patch increases, the force in the edge chordae decreases. This was caused by an increase in downward displacement of the leaflet edge as the bulge in the patch increases.

DISCUSSION

The results of this study indicate that the implantation of a patch in the anterior leaflet could increase coaptation between the leaflets, potentially reducing or eliminating MR. The deformation of the anterior leaflet seen in the model (i.e., inward towards the edge and downward towards the papillary muscles) shows strong agreement with both in vivo and in vitro experimental studies of patch augmentation [1, 2].

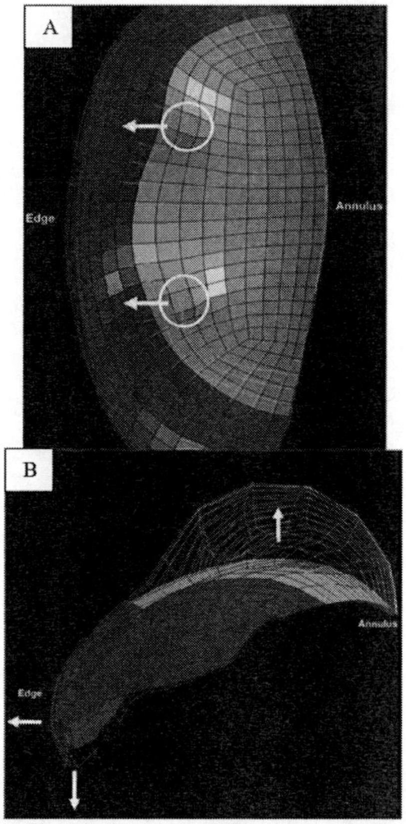

Figure 3: (A) Ventricular view of the anterior leaflet. The white circles indicate the strut chord insertion points. (B) Side view of the anterior leaflet. Note: Filled in elements represent the baseline model, while the wireframe elements represent the model with a pericardial patch. Arrows indicate the direction of altered deformation when the patch is implanted.

Additionally, the model predicted a reduction in average force transmitted through the edge chords, which is also in strong agreement with previous experimental studies [2]. The results of the current study indicate that the initial modeling approach has the potential to replicate the actual in vivo environment and identify how chordal forces are redistributed. Thus, the model could be used as a means of testing various treatment approaches for eliminating MR in a patient-specific way using imaging data for model construction.

ACKNOWLEDGEMENTS

Funding to support this work was provided by the Arkansas Research Alliance (Jensen) and NSF grant CMMI-1538754 (Wenk).

REFERENCES

[1] Aubert, SE et al., *Ann Thorac Surg*, 83:1560-1561, 2007.
[2] Granier, M et al., *J Thorac Cardiovasc Surg*, 142:1585-1587, 2011.
[3] Ostli, B et al., *Cardiovasc Eng Tech*, 3:263-268, 2012.
[4] Wenk, JF et al., *Ann Thorac Surg*, 89:1546-1554, 2010.
[5] Wenk JF et al, Med Biol Eng Comput, 50:1071-1076, 2012.

A PARAMETRIC STUDY OF THE OPTIMAL SHAPE AND LEAFLET PROPERTIES IN BIOPROSTHETIC HEART VALVES

Rana Zakerzadeh (1), Fei Xu (2), Michael C.H Wu (2), Ming-Chen Hsu (2), Michael S. Sacks (1)

(1) Institute for Computational Engineering and Science (ICES)
University of Texas at Austin
Austin, Texas, USA

(2) Department of Mechanical Engineering
Iowa State University
Ames, Iowa, USA

INTRODUCTION

Replacement heart valves fabricated from biologically derived materials are referred to as bioprosthetic heart valves (BHVs). BHV have important advantages in that they do not require permanent anticoagulation therapy, and have blood flow characteristics like the native valve. However, BHV durability continues to remain limited to the range of 10-15 years; and therefore, they are typically only used in patient ages 57 years or older. Device failure continues to be the result of leaflet structural deterioration mediated by fatigue and/or tissue mineralization. Also; valve performance is highly dependent on the geometry of the leaflets. There is thus a profound need for the development of computational models for simulating BHV in the design context. These numerical models can provide information that would be valuable for clinical diagnosis and treatment. Built upon the computational approach presented in [1, 2], In this work, we developed a framework for parametric study of bioprosthetic heart valve, to identify the geometry for the cusps that would yield in normal valve dynamics. We also simulate the valve deformations with various perturbations of the tissue properties, to investigate the effect of alteration of the material model of the leaflets, including the collagen fiber orientation, as well as the fiber stiffness. These results can provide guidelines for designing leaflet tissues to improve valve durability.

METHODS

We consider the pressure-driven structure dynamics of a deformable valve representing a BHV. We use a nonlinear anisotropic material model for the leaflets [3] to enhance the physical realism of the BHV dynamics simulation and to evaluate the impact of leaflet properties on the deformation and dynamics of the BHV. The added nonlinear Neo-Hookean term represents the low-strain responses and is needed also for numerical stability [4]. Moreover, the exponential part captures the important exponential nature of the soft tissue behavior. Therefore, strain energy is a function of two invariants as follows:

$$W(I_1, I_4) = \frac{c_0}{2}(I_1 - 3) + c_1\left(e^{c_2(I_1-3)^2+c_3(\sqrt{I_4}-1)^4} - 1\right) \quad (1)$$

Where I_1 is the first invariant of the left Cauchy–Green deformation tensor C, and I_4 is the fourth invariant for a transversely isotropic material. and M is the unit vector defining the collagen fiber direction in the undeformed configuration.

$$I_1 = tr\,C \quad I_4 = M \cdot C \cdot M \quad (2)$$

The values of the coefficients c_1 c_2 and c_3 are obtained from the data presented in [3]. The coefficient of the Neo-Hookean term is set to 100 kPa. In order to approximate this problem, we use a thin shell isogeometric structure dynamic finite element solver developed in [1]. We apply a physiological transvalvular pressure load to the leaflets to enforce the pressure difference between left ventricle and aorta.

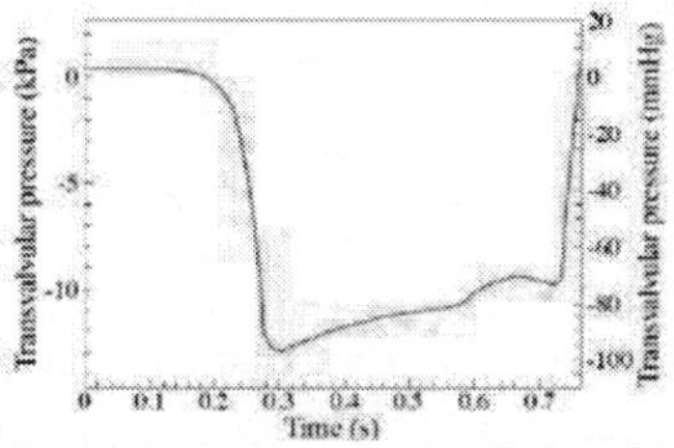

Figure 1: Transvalvular pressure applied to the leaflets as a function of time.

RESULTS

In Figure 2, the sensitivity results for different values of fiber stiffness c_3 (top panel) and fiber orientation (bottom panel) are presented. It is assumed that initially the fibers are aligned in the circumferential direction through the leaflet and we compute the 30-degree and 45-degree deviation of unit vectors in the tangent space of a reference configuration. Our observation suggests that perturbing fiber stiffness coefficient and fiber orientation, has a significant influence on the valve deformation only at the closing stage. Namely, the strain distribution of the leaflets is the same between all the cases for the opening phase of the cardiac cycle. The reason is that in dynamic simulation of heart valve in a full cardiac cycle, the opening phase is in the small strain regime, when tissue does not behave exponentially and the Neo-Hookean term dominates. However, when the valve is closed, the exponential terms dominate that representing the fiber action.

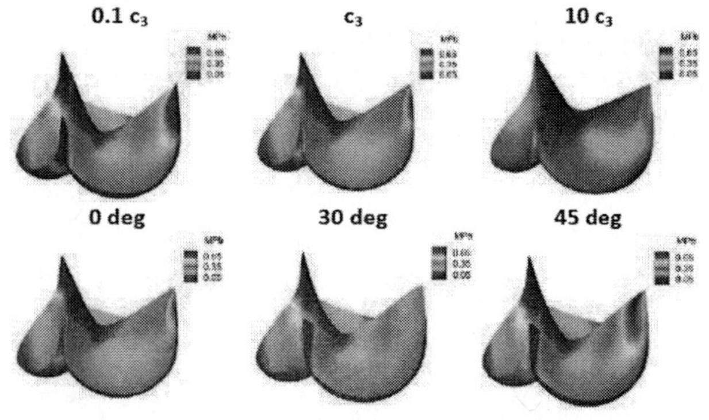

Figure 2: Deformation of the valve colored by Maximum in-plane principal Green-Lagrange strain. The chosen time point is at the end of cardiac cycle when the valve is fully closed

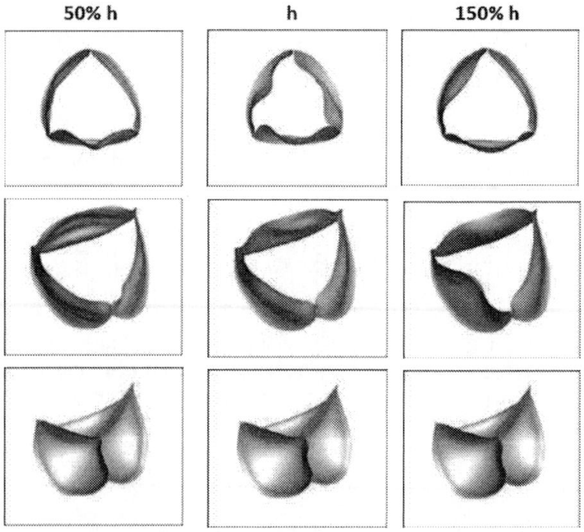

Figure 3: Comparison between deformation of the valve at several points during the cardiac cycle, for different values of leaflet's thickness (h)

Parametric study for geometry has been performed using our FSI solver [2]. The influence of leaflet's thickness on valve dynamics has been presented in Figure 3. We use the valve geometry discussed in [2] where the thickness of the leaflets (h) is 0.0386 cm; and we qualitatively compare the leaflets deformation for different values of thickness. The results show that original thickness case provided the best result. 150% thickness case has a smaller opening area and one leaflet pops open abruptly during peak of the flow. 50% thickness case introduces more fluttering in the leaflets but has a larger opening area than the original thickness case. We also examine the effect of changing leaflet shape on valve dynamics by adjusting some geometrical parameters. The geometric effect of varying surface curvature of the leaflets is shown in upper panel of Figure 4. In the lower panel, deformation of the leaflets for each surface curvature is provided. We observe a noticeable influence of the surface curvature on the overall tissue deformation.

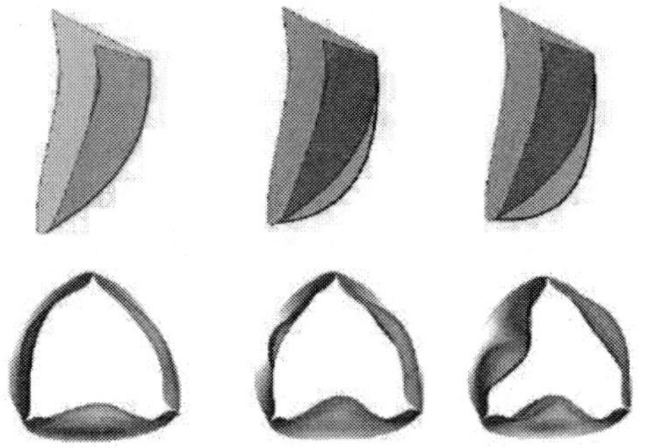

Figure 4: Comparing valve deformation for different values of surface curvature of the leaflets, at the same time point in the cardiac cycle.

DISCUSSION

The change of parameters (geometric and material) can greatly affect the valve performance. Our objective is performing a comprehensive parametric study to help understand the physics of valves. The goal is to provide a framework which enables us to use more physically realistic soft tissue material model in heart valve simulations and to investigate the influence of the geometry on heart valve performance indicated by the effective orifice area during the opening and the coaptation area during the closing. Starting from the microstructural material model presented in [5] we are currently developing a phenomenological model that covers the interaction between the tissue's collagen fibers and its ground matrix as well as the interaction between different fibers. We use this computationally tractable material model to facilitate the integration of the model into our FSI solver [2].

ACKNOWLEDGEMENTS

This work has been supported by NHBLI grant R01 HL129077. HPC resources that have contributed to the results reported in this paper are provided by the Texas Advanced Computing Center (TACC).

REFERENCES

1. Hsu, M.-C., et al. Computational mechanics, 2015. **55**(6): p. 1211-1225.
2. Kamensky, D., et al. Computer methods in applied mechanics and engineering, 2015. **284**: p. 1005-1053.
3. May-Newman, K., et al. Journal of biomechanical engineering, 2009. **131**(8): p. 081009.
4. Lee, C.H., et al. J Biomech, 2014. **47**(9): p. 2055-63.
5. Sacks, M.S., et al. Interface focus, 2016. **6**(1): p. 20150090.

SB³C2017
Summer Biomechanics, Bioengineering and Biotransport Conference
June 21 – 24, Tucson, AZ, USA

IN-VIVO STRESS ESTIMATION OF THE FUNCTIONING HEART VALVES AND ITS IMPLICATIONS FOR ANNULOPLASTY RING-BASED VALVE SURGICAL REPAIR

Chung-Hao Lee (1), and Michael S. Sacks (2)

(1) School of Aerospace and Mechanical Engineering
The University of Oklahoma
Norman, OK, USA

(2) Institute for Computational Engineering and Sciences
Department of Biomedical Engineering
The University of Texas at Austin
Austin, TX, USA

INTRODUCTION

Several recent studies [1-3] have shown that excessive tissue stress and the resulting strain-induced tissue remodeling and damage may be primary etiological factors responsible for the significant number of mitral valve repair failures. Moreover, these undesirable stress-induced changes in the valve interstitial cells (VICs) and the consequent collagen biosynthesis are likely associated with the limited durability of valve repair. Thus, there is a critical need for developing reliable and predictive computer simulation tools, which allow for systematic evaluations of the valve function at the normal, pathophysiological, and surgically repaired states by enabling the connection of the organ-level biomechanical performance to *in vivo* tissue stresses and the phenotypic and biosynthetic responses of the VICs. The overall goal of this study was to make such an important quantitative connection for our future development of simulation-assisted therapeutics for heart valve disease.

METHODS

Sonocrystal data: 3D echocardiography methodology was employed for acquisition of the positional data of the MV anterior leaflet, and the *in vivo* sonocrystal data acquired from the previous study [4] were utilized for valve leaflet deformation analysis as well as the subsequent *in vivo* valve tissue stress estimations. In brief, 9 male Dorsett sheep (35-45 kg) were used in compliance with NIH guidelines for the care and use of laboratory animals. With a sterile left lateral thoracotomy, 2-mm hemispherical piezoelectric transducers were sutured to the central region of the anterior leaflet in a 10-mm square array (Fig. 1a), and the 3D coordinate of each transducer was determined at an interval of 5 ms with transvalvular pressure continuously monitored (Fig. 1b).

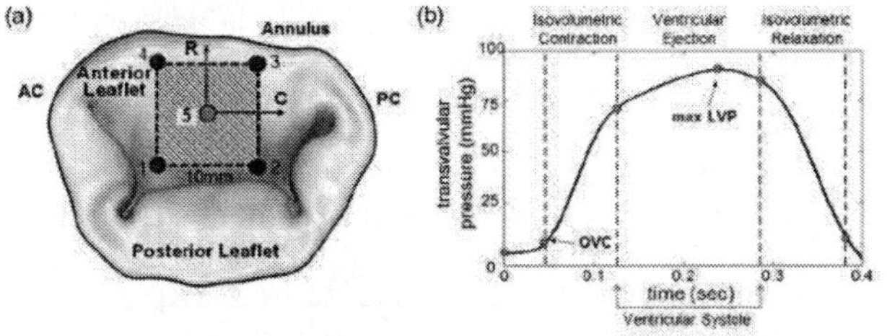

Figure 1: (a) Schematic of mitral valve leaflets and the arrangement of 5 sonocrystal transducers, (b) an illustration of the measured transvalvular pressure over one representative cardiac cycle, applied as pressure loading for the inverse FE simulations.

Valve leaflet deformation & annuloplasty ring effect: In order to calculate the in-surface deformations and strains of the MVAL tissue, a 4-node bilinear finite element was used based on the average sonocrystal marker positions, following previously developed approaches [5]. Due to the lack of available sonocrystal data for inverse finite element (FE) modeling, we simply adjusted the tissue-level deformations based on the information from [4] in order to investigate the effect due to the implanted annuloplasty ring (Fig. 2a-c).

Poster Presentation #P222

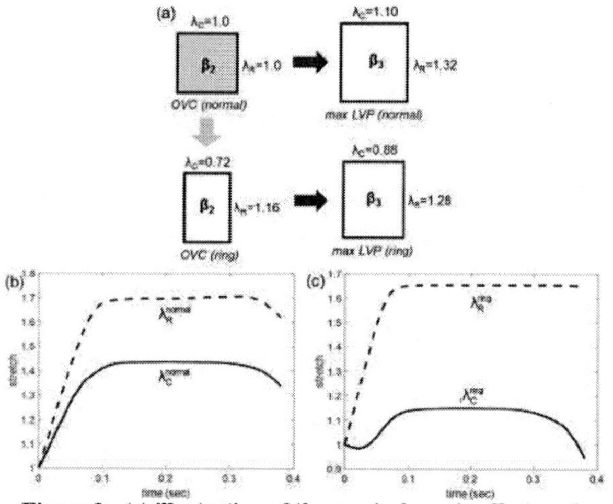

Figure 2: (a) illustration of the surgical repair effect on the functional strains in the central region of the MVAL previously quantified in [4], as utilized in the simulations of MVIC behaviors for both normal and surgically repaired MVs. Quantified valve leaflet stretches over a cardiac cycle: (b) normal, (c) ring repaired.

Inverse FE modeling for in vivo stress estimation: We employed the previously developed inverse FE modeling framework [6] to quantify the *in vivo* tissue stresses with respect to both *in vivo* and *in vitro* reference configuration by incorporating the quantified pre-strains in [4], using a two-step simulation procedure: (i) dimensional change simulation, and (ii) *in vivo* functioning simulation. A microstructural constitutive model was used in these inverse modeling-based calculations:

$$\mathbf{S}_{SSM}(\mathbf{C}) = \mu_m\left(\mathbf{I} - C_{33}\mathbf{C}^{-1}\right) + \int_{-\frac{\pi}{2}}^{\frac{\pi}{2}} \Gamma(\theta) c_0 \left(e^{c_1 E_{ens}} - 1\right) \mathbf{n} \otimes \mathbf{n} d\theta \quad (1)$$

RESULTS

The estimated in vivo stresses at peak pressure loading of the representative tissue with respect to *in vivo* reference configuration were σ_{CC}=~360kPa and σ_{RR}=~450kPa in the circumferential and radial directions, respectively, whereas the estimated stresses with respect to *in vitro* reference configuration were σ_{CC}=~510kPa and σ_{RR}=~740kPa (Fig. 3a-b). The amount of the inherent pre-stresses was determined from FE calculations with the optimal parameter set (σ_{CC}=30.4kPa, and σ_{RR}=20.5kPa), which is relatively small compared to the functional stresses.

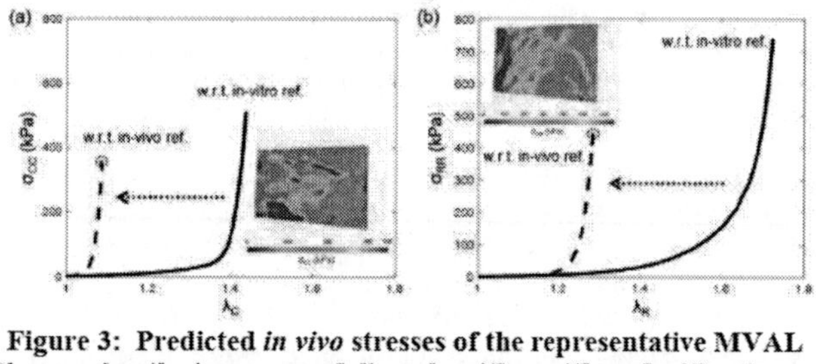

Figure 3: Predicted *in vivo* stresses of the representative MVAL tissue using the inverse modeling algorithm with and without pre-strain effects: (a) in the circumferential direction, and (b) in the radial direction. The inlets show the corresponding principal stresses and their directions at the maximum pressure loading.

By further incorporating previously developed down-scale modeling framework [7], we were able to quantify the underlying *in vivo* VIC deformation, in terms of a nucleus aspect ratio (NAR) defined as the ratio between the major and minor dimensions of the VIC ellipsoid, for both the normal and repaired mitral valves (Fig. 4).

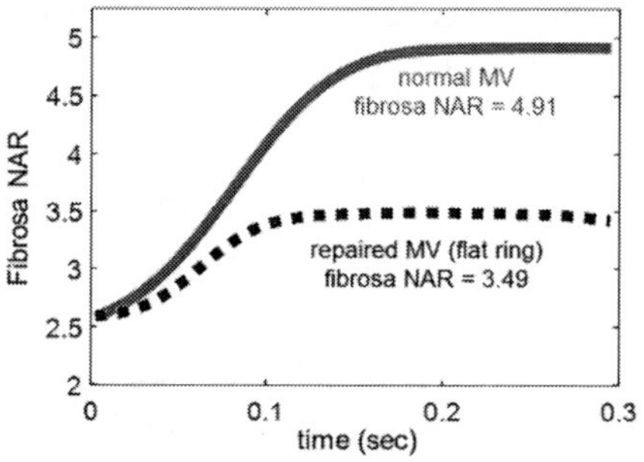

Figure 4: Comparison of the predicted fibrosa VIC deformation between the normal and the repaired valves, showing the remarked reduction of the NAR value (from 4.91 to 3.49) due to ring-induced changes in the annular geometry/dimension.

DISCUSSION

Our pre-stress results, while small in magnitude, indicated that the long toe-region of the valve tissue stress-strain behavior, typically observed *in vitro*, is completely absorbed by the pre-strains (Fig. 3a-b). This implies that the elastin fiber network is primarily responsible for the pre-strain, whereas collagen fibers subsequently engage in bearing the majority of *in vivo* physiological loading. Moreover, it is interesting to note that such surgical ring-induced geometry changes (28% circumferential contraction and 16% radial expansion at onset volumetric contraction and 12% circumferential contraction and 28% radial expansion at the maximum pressure loading, Fig. 2b-c) resulted in notable reduction of the fibrosa NARs from 4.91 to 3.49 (Fig. 4). The findings in this study could be beneficial for the design of individual-optimized annuloplasty surgical repair that could not only best restore the functional geometry of the valve annulus but could also achieve the valve homeostasis at the organ, tissue and cellular levels for improving the long-term surgical outcomes associated with the treatments of patients with functional or ischemic mitral regurgitation.

ACKNOWLEDGEMENTS

This work is supported by the American Heart Association Scientist Development Grant 16SDG27760143 (CHL) and the NIH R01 grant HL119297 (MSS).

REFERENCES

[1] Braunberger, E. et. al., *Circulation*, 104:I8-11, 2001.
[2] Flameng, W. et al., *Circulation*, 107:1609-13, 2003.
[3] Flameng, W. et al., *J Thorac Cardiovas Surg*, 135:274-82, 2008.
[4] Amini, R. et al., *Ann Biomed Eng*, 40:1455-67, 2012.
[5] Sacks, M.S. et al., *Ann Thorac Surg*, 82: 1369-77, 2006.
[6] Lee, C.H. et al., *J Biomech*, 47:2055-63, 2014.
[7] Lee, C.H. et al., *J Theor Biol*, 373:26-39, 2015.

SB³C2017
Summer Biomechanics, Bioengineering and Biotransport Conference
June 21 –June 24, Tucson, AZ, USA

PHENOMENOLOGICAL BASED CONSTITUTIVE MODELING OF JUGULAR VENOUS TISSUE

Nayyan Kaul (1) and Hsiao-Ying Shadow Huang (2)

(1) PhD Candidate
Mechanical and Aerospace Engineering
North Carolina State University
Raleigh, NC, USA

(2) Associate Professor
Mechanical and Aerospace Engineering
North Carolina State University
Raleigh, NC, USA

INTRODUCTION

Venous valve tissues, while used in vein reconstruction surgeries and bio prosthetic valves with moderate success, does not have extensive studies on their structure and modelling. Their inherent anisotropic, non-linear behavior combined with severe diseases inflicting the veins like chronic venous insufficiency warrants understanding the structure and material behavior of venous valve tissues. Hence, before any bio-prosthetic grafts can be used in place of tissues, it is of utmost importance to know about the mechanical and structural properties of the tissue which can lead to an increase in success rates of valve replacement surgeries. The longevity of the bio-prosthetics would also increase if the manufactured grafts behave exactly similar to the native valves.

With the scant information on uniaxial and biaxial mechanical properties of jugular venous valve and wall tissue from any previous studies, the current focus of our study was to understand the material behavior by determining an established phenomenological strain energy based constitutive relation for the tissues. We used bovine veins to study the behavior of valve leaflet tissue and adjoining wall tissue (from proximal and distal end of the vein) under different biaxial testing protocols. We looked at the behavior of numerical partial derivatives of strain energy to select a suitable functional form of strain energy for wall and valve tissue from the literature. Using this strain energy descriptor, we determined Cauchy stress and compared it with experimental results under displacement controlled biaxial testing protocols to find material specific model parameters by Powell's method algorithm. Results indicated that whereas wall tissue strain energy can be explained using a polynomial nonlinear function, the valve tissue, due to higher nonlinearities, requires an exponential function. This study can prove helpful in primary stages of bio-prosthetic designs; replacement surgeries; can be of support to any future studies investigating structural models and to study valvular diseases by giving a way to understand material properties, behavior and to form a continuum model when required for numerical analyses and computational simulations.

METHODS

Structural exploration studies on jugular venous tissues have shown that tissue has incompressible homogeneous structure. Valve tissues are known to have bundles of aligned collagen fibers. The belly region of the valve has more homogeneous structure compared to the commissures or attachment area. Few studies done towards the mechanical behavior aspect of the Jugular venous tissue indicates non-linear and anisotropic behavior for the valve and wall tissue, valve having more nonlinearities. This information enables us to safely assume that the tissue under study, valve and wall alike, are pseudo-elastic, incompressible, homogeneous and locally transversely isotropic with respect to fiber axis.

Behavior of such hyper-elastic soft tissues can be explained using strain energy expression - 'W' which is a function of two invariants of the deformation [1] i.e. I_1 and I_4 ($= \alpha^2$), along with the constitutive equation given as follows, where all symbols have their standard meanings.

$$\boldsymbol{t} = -p\boldsymbol{I} + 2W_1\boldsymbol{B} + \left(\frac{W_\alpha}{\alpha}\right)\boldsymbol{F}\cdot\boldsymbol{N}\otimes\boldsymbol{N}\cdot\boldsymbol{F}^T \quad (1)$$

The strain energy descriptor was decided on the basis of experimentation which involved biaxial testing where alternatively each invariant was held at a constant value, called "constant invariant testing." The plots of $W_1 = \dfrac{\partial W}{\partial I}$ and $W_\alpha = \dfrac{\partial W}{\partial \alpha}$ with respect to the invariants, termed as "response curves," provides an empirical basis to decide a suitable function for the strain energy descriptor.

Tissue Procurement & Preparation: Due to issues associated with procurement and testing human tissues and the fact that in practice, bovine tissues are used on a large scale as a suitable substitute from medical point of view in bio prosthetic grafts; in the current study, we used tissue samples from a mature bovine of Holstein breed, 10+ years old with an approx. weight of 1250 lbs. Valve and wall tissue samples were extracted from jugular veins of the animal. Vein samples were shipped overnight on ice, ~24 hours post-slaughter. Test specimens were cut from the belly region of the valve and avalvular region of the wall in square section of side 7-10 mm and stored in Hanks balanced salt solution (HBSS).

Testing Protocols: Discussed in detail in a previous study [2], we mount the tissue test specimens on the biaxial testing apparatus using rakes in a bath of HBSS pre-heated to 37 degrees Celsius to imitate the physiological conditions. The effective central region subject to stretching is 4.5 mm x 4.5 mm. The test rig has two load cells (10N±0.02N) mounted perpendicular to each other and two actuators opposite to load cells to simulate the loading or stretching. The displacement of the actuators is recorded along with the load cells readings. Test Specimen is preconditioned with a preload of 10mN for 8 loading-unloading cycle up to 30% strain at a strain rate of 1% per sec followed by a rest period of 300 secs. Thickness of the tissue samples is recorded before testing 4-5 times by using a dial gauge (±0.0001 inch; The L.S. Starrett Co., Athol, MA) and then an average value was used for stress calculations. Based on an approximate 60% true strain breaking strength, we tested a total of 16 wall samples along with 8 samples of valve tissue for $3.2 < I_1 < 4.4$ and $1.1 < \alpha < 1.7$.

Model and Parameters: Based on the behavior of the response curves (linear for wall tissue and exponential for the valve tissue), we selected suitable descriptors from the literature: 5-parameter polynomial type strain energy descriptor [1, 3] for the wall tissue and a 3-parameter exponential type function [4] for the valve tissue as shown.

$$W^{wall} = c_1(\alpha-1)^2 + c_2(\alpha-1)^3 + c_3(I_1-3) + c_4(I_1-3)(\alpha-1) + c_5(I_1-3)^2$$

$$W^{valve} = c_0(\exp(c_1(I_1-3)^2 + c_2(\alpha-1)^4)-1)$$

Note that c's in the above two strain energy functions are different parameters. After selecting the descriptor function, the next step of the process required us to find material parameter values (5 for W^{wall} and 3 for W^{valve}). The expressions for model predicted stress for the wall and valve tissues were found by substituting W_I and W_α in the constitutive relation in **Equation 1**, and as below:

$$t_{11}^{wall} = 2(\lambda_1^2-\lambda_3^2)[c_3+c_4(\alpha-1)+2c_5(I_1-3)] + \lambda_1[2c_1(\alpha-1)+3c_2(\alpha-1)^2+c_4(I_1-3)] \quad (2)$$
$$t_{22}^{wall} = 2(\lambda_2^2-\lambda_3^2)[c_3+c_4(\alpha-1)+2c_5(I_1-3)]$$

$$t_{11}^{valve} = (4c_0c_1(\lambda_1^2-\lambda_3^2)(I_1-3) + 4c_0c_2\lambda_1(\alpha-1)^3)\exp(c_1(I_1-3)^2+c_2(\alpha-1)^4), \quad (3)$$
$$t_{22}^{valve} = 4c_0c_1(\lambda_2^2-\lambda_3^2)(I_1-3)\exp(c_1(I_1-3)^2+c_2(\alpha-1)^4)$$

where λ's are stretches. For parameter estimation, a separate set of biaxial testing was conducted in which separate preconditioned tissues were subjected to 4 off-axial and 1 equi-biaxial stretching protocol consecutively (in the order of A: C for wall tissue samples or C: R for valve tissue sample = 2:1, 1:2, 1.5:1, 1:1.5, 1:1) where the ratio indicates the ratio of strains in the orthogonal directions. Best fit values were found for each sample separately by minimizing the sum of the squares of the residuals using the Powell's method algorithm. The residual was defined as the error between experimental measured and theoretically predicted Cauchy stress. A maximum of 70% true strain was selected for this phase of biaxial testing.

RESULTS AND DISCUSSION

Parameter fitting was done for three testing ratios (i.e., 1:1, 1.5:1, 1:1.5) for every wall and valve samples. The material parameters found were used to predict the experimental data for the two remaining test ratios to verify the usefulness of the model. Stress vs. stretch data along with the model fitting for the wall and valve tissue for equi-biaxial test ratio (1:1) is shown in **Figure 1**. The correlation coefficients for all the wall tissues were found above 0.9; whereas for the valve tissue, they were as low as 0.7, indicating some prediction error. The response curves for the wall matched very well to the ones in the myocardium study [1, 3], but the response curves for the valve, while exponential in nature, had some notable differences from the ones presented in mitral valve study [4], and hence the prediction error was expected. It was also observed that the distal wall tissue samples have more strain energy density per unit volume indicating higher forces at distal end compared to proximal end at any instant; which could possibly be attributed to the higher hydrostatic pressure at the distal end of the vein.

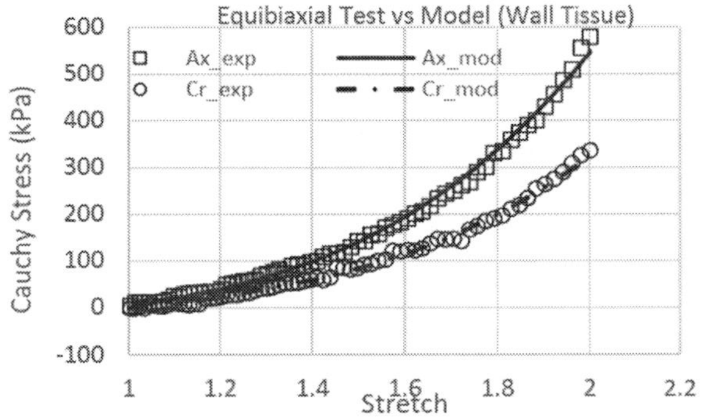

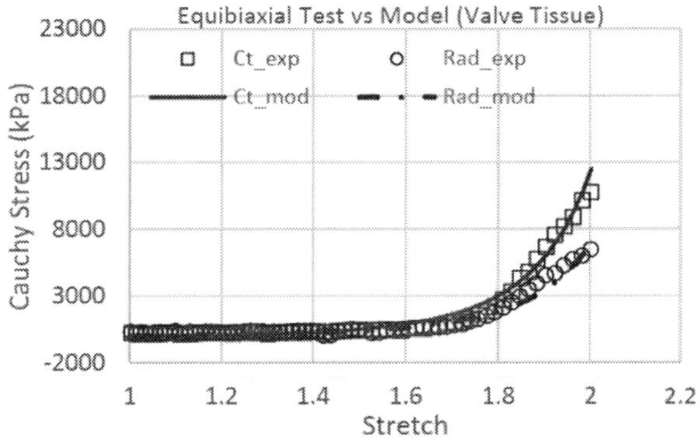

Figure 1: Top: The model fits using the individual material parameters for the wall tissue (Equation 2, c1 = 29.7, c2 = 33.7, c3= 12.3, c4 = -9.90, c5 = 3.82). Bottom: The model fits using the individual material parameters for the valve tissue (Equation 3, c0 = 472.7, c1 = 0.043, c2= 0.34).

ACKNOWLEDGEMENT

This research was supported by NSF CAREER GARDE 1553430.

REFERENCES

[1] Humphrey JD, Strumpf RK, Yin FCP. Journal of Biomechanical Engineering. 1990; 112(3):333-339. doi: 10.1115/1.2891193.

[2] Huang HYS and Lu J. *Biomechanics and Modeling in Mechanobiology*. in press.

[3] Humphrey JD, Strumpf RK, Yin FCP. Journal of Biomechanical Engineering. 1990; 112(3):340-346. doi: 10.1115/1.2891194.

[4] May-Newman K, Yin FCP. Journal of Biomechanical Engineering. 1998; 120(1):38-47. doi: 10.1115/1.2834305.

SB^{3}C2017
Summer Biomechanics, Bioengineering and Biotransport Conference
June 21 – 24, Tucson, AZ, USA

MECHANICAL AND STRUCTURAL CHARACTERIZATION OF PULMONARY ARTERIES IN TWO PAH ANIMAL MODELS

Daniela Vélez-Rendón (1), Erica R. Pursell (1), Daniela Valdez-Jasso (1)

(1) Department of Bioengineering
University of Illinois at Chicago
Chicago, IL, USA

INTRODUCTION

Pulmonary arterial hypertension (PAH) is an aggressive disease of the pulmonary vasculature defined by a mean pulmonary arterial pressure (mPAP) greater than 25 mmHg. During PAH, a pressure overload is imposed on the right ventricle (RV) affecting its performance. Likewise, the pulmonary vessels undergo remodeling, altering their ability to store and deliver blood [1- 2]. This remodeling process and the underlying mechanisms governing the disease remain poorly understood. Here in this study, we aim to elucidate the structural and mechanical changes undergone by the pulmonary vasculature. Using two well-established murine models of PAH, left and right pulmonary segments are tested biaxially and scanned under a multiphoton microscope.

METHODS

The experimental protocol used for this study was approved by the University of Illinois at Chicago Animal Care and Use Committee. 20 male Sprague Dawley rats were used in the study, out of which 7 were injected with a placebo (PL) solution to account for aging effects. PAH was induced in 14 animals. 5 were injected with a single dose of 20mg/kg of Sugen (SU5416, Sigma-Aldrich) and were placed on a hypoxia chamber for 3 weeks followed by 3 weeks of normoxia (SuHx group). The 8 MCT-treated animals were subcutaneously injected with a dose of 60mg/kg of monocrotaline (crotaline, Sigma-Aldrich) and kept in normoxia for 4 weeks. The animals underwent open chest surgery to obtain RV and mPAP pressures. At the end of the surgery, intralobular left and right pulmonary arteries (LPA and RPA) were harvested and cannulated for tubular biaxial testing and imaging.

A proximal segment of the LPA and a distal portion of the RPA were placed on a BioDynamic chamber (ElectroForce Systems Group, Eden Prairie, MN) and phosphate buffered saline circulated through the vessels at 37°C. Two tests were performed to investigate the circumferential and axial mechanical properties. Intramural pressure, outer diameter and axial forces were recorded. In the circumferential test, the vessels were subjected to dynamic pressure matching the *in-vivo* measured values and stretched to their *ex-vivo* length. During the axial test, the vessels were cyclically stretched from 40 to 45% of their *ex-vivo* length and their corresponding mPAP was kept constant. Using the quasilinear viscoelasticity theory approach, strain ϵ at any given time t is the convolution of a static elastic response s_j with time-invariant dynamics component or creep function K [3]. s_j is a function of the intramural pressure in the circumferential direction (θ) and in the axial direction (z), F is the axial load experienced by the vessel. The parameter C_1 is the amplitude associated with the relaxation time (b_1) or frequency $(1/b_1)$.

$$\epsilon_j(t) = \mathrm{K}(0)s_j[F_j(t)] + \int_{t_0}^{t} s_j[F_j(\gamma)]\frac{d\mathrm{K}(t-\gamma)}{d\gamma}d\gamma\,, j = \theta, z \qquad (1)$$

where $\mathrm{K}(t) = 1 - \sum_{i=1}^{n} C_i e^{-t/b_i}$

The elastic response relates to pressure and axial force as

$$s_\theta[p(t)] = \frac{r_0(1-\nu_{z\theta}\nu_{\theta z})}{E_\theta h}p \qquad s_z[f(t)] = \frac{(1-\nu_{\theta z}\nu_{z\theta})}{E_z}\sigma_z \qquad (2)$$

And strain (ϵ) relates to area (A) and length (L) as

$$A(t) = \frac{A_0}{(1-\epsilon(t))^2} \qquad L(t) = L_0(\epsilon_z(t)+1) \qquad (3)$$

Here, ν is Poisson's ratio, E is Young's elastic modulus, σ is the normal stress tensor, p intramural pressure, r_0 is the radius at zero pressure at *ex-vivo* length, h is the measured vessel wall thickness and L_0 is the length of the vessel L before mechanical testing. Data are fitted by optimizing the parameters $E_j, C_{1,j}, b_{1,j}, L_0$ and A_o in MATLAB.

Poster Presentation #P224

For the structural analysis, the proximal segment of the RPA and distal of the LPA were cannulated and imaged under a multiphoton microscope (MPM). Collagen fiber distributions were obtained by using the directionality function in Fiji [4]. Histograms of the angle distributions per vessel type and treatment were generated. Here, the fiber angles ranged from -90° to 90°, with 0° corresponding to the axial direction and ± 90° to the circumferential direction.

RESULTS

Severe PAH was confirmed with the increase in the average RV systolic pressure (RVSP) from the PL (31.6 ± 2.8 mmHg) to the MCT (68.7 ± 11.2 mmHg) and SuHx (76.6 ± 9.8 mmHg) groups. There was also an increase in the mPAP from the PL (24.6 ± 3 mmHg) to the MCT (46.9 ± 5.3 mmHg) and SuHx (55.1 ± 0.1 mmHg).

Pressure-area relation and load-vessel length data measurements for each treatment group are shown in Figure 1 and are plotted against the model predictions. The shape of the pressure-area relation changed from a nonlinear to a linear form from the PL to both MCT and SuHx groups, while the amplitude of the force-vessel length relation changed decreased with PAH.

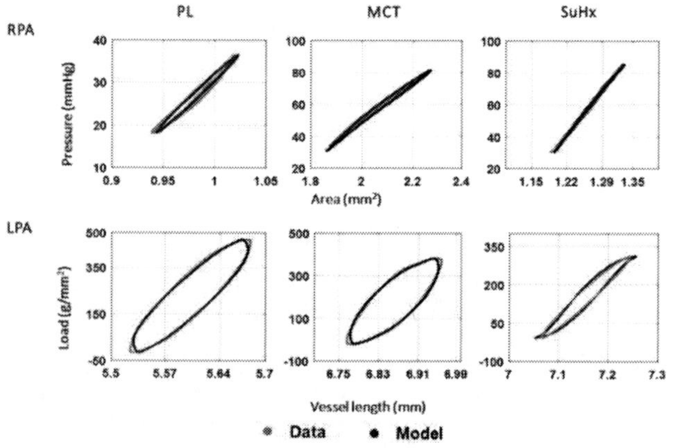

Figure 1: Representative pressure-area (top) and load-vessel length (bottom) plots of model fitted to data from the RPA (top) and LPA (bottom) of a PL animal, MCT animal and SuHx animal.

In the circumferential direction, E_θ statistically significantly increased in the RPA from the PL to MCT ($3.4\pm3.1\times10^3$ to $26.2\pm12.1\times10^3$ mmHg), while the SuHx group it remained unaffected ($2.9\pm2.7\times10^3$ mmHg). A_0 significantly increased in the LPA of both PAH animal models when compared with the PL vessels. The viscoelastic parameters $C_{1,\theta}$, remained overall constant with the progression of PAH. However, the time relaxation when transformed $\ln(b_{1,\theta})$ statistically significantly decrease in the MCT group.

Vessel stiffness in the axial direction decreased from PL to MCT and SuHx, without reaching any statistical significance. The amplitude C_1 in the axial stress-strain relation decreased in the LPA and RPA from PL to SuHx. In contrast, the relaxation time $b_{1,z}$ did not change in any of the diseased animals.

Results from MPM imaging also indicated a variation on the structural remodeling depending on the PAH animal model. Qualitatively from the MPM images one finds the collagen fibers to be crimped (Figure 2). However, fiber orientation of the vessels from the PL and SuHx group are more randomly distributed. When these fibers angle orientation were quantified, the PL group was aligned to 0° and 13°, the SuHx group to 13° and -6°, for the LPA and RPA respectively. In the MCT-treated animals, two preferred fiber directions were found. For the LPA, the preferred fiber orientation was towards the -9° and 14°, while the RPA -28° and 14°.

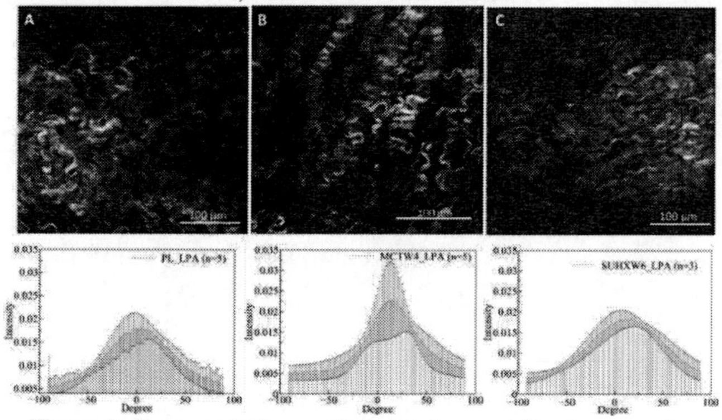

Figure 2: Representative multiphoton images of a given vessel (top) and the corresponding histogram of collagen fiber orientation (bottom) from the LPA of PL, MCT and SuHx groups. Note for the MCT group just one preferred fiber orientation histogram is shown.

DISCUSSION

Even though the severity of PAH is more pronounced in the SuHx model (higher mPAP and RVSP, and formation of plexiform lesions in patients with PAH) [5], the remodeling undergone by the vessels was not as marked as in the MCT group. In particular, the LPA and RPA did not remodel in the same matter. In the MCT treated animals, the circumferential stiffness of the RPA was significantly larger. While in the SuHx treated animals there were also an increase in circumferential stiffness, this occurred in the LPA. In the axial direction vessels became more compliant in both PAH models, but it was in the SuHx group that the decrease was more pronounced.

Mechanical changes of the pulmonary vessels were corroborated in the structural analysis. The collagen fiber distribution of the SuHx treated animals was tighter and it was similar to the distribution of the placebo group. With the MCT group, the arrangement of the collagen fibers was more drastic and presented with two different directions for each vessel type.

Future work will require histological investigation to confirm the formation of the plexiform lesions in animals and to be related to human explant tissues. These findings will be analyzed in conjunction with 1-dimensional fluid model and incorporate impedance analysis. Finally, the SuHx animals will be kept in normoxia for more weeks to reach acute PAH levels.

ACKNOWLEDGEMENTS

DVJ would like to acknowledge funding from the American Heart Association Scientist Development Grant 16SDG29670010 and UIC College of Engineering Seed Fund, and ERP the Lincoln Graduate Student Fellowship.

REFERENCES

[1] Wang, Z et al., *PLoS One*, vol. 8, no. 11, p. e78569, 2013.
[2] Bellofiore, A. et al., *Ann. Biomed. Eng.*, vol. 41: 1384–98, 2013.
[3] Pursell, E. R. et al., *J. Biomech. Eng.*, 138, no: 111004, 2016.
[4] Schindelin, J. et al., *Nat. Methods*, vol. 9: 676–682, 2012.
[5] Gomez-Arroyo, J. G. et al., *Am. J. Physiol. Lung Cell. Mol. Physiol.*, vol. 302, no. 4, pp. L363-9, 2012.

Poster Presentation #P224

SB^{3}C2017
Summer Biomechanics, Bioengineering and Biotransport Conference
June 21 – 24, Tucson, AZ, USA

STATICS AND DYNAMICS OF AORTIC SEGMENT WITH RESIDUAL STRESSES

Ivan Breslavskyi, Marco Amabili

Department of Mechanical Engineering
McGill University
Montreal, □uebec, Canada

INTRODUCTION

The present study is devoted to the analysis of static and dynamic behavior of the three-layered circular cylindrical shell representing the intima, media and adventitia of the aorta. The shell is loaded by pulsatile internal pressure.

The study presents a method for nonlinear dynamic analysis of the human aorta model. Two types of residual stress in the aorta are taken into account.

METHODS

Static and dynamic analysis of circular cylindrical shell that models a segment of human aorta is carried out in this study. The shell is assumed to be three-layered, with each layer being hyperelastic, fiber-reinforced, viscoelastic material. Following the paper [1], we chose a material model in form of combination of Neo-Hookean and Fung models. The strain energy density for this case has the following form:

$$W = \frac{E}{6}(I_1 - 3) + \frac{\mu_1}{\mu_2}\left(Exp\left(4\mu_2(\varepsilon_x^2 cos^4\beta + 2\varepsilon_x\varepsilon_\theta cos^2\beta sin^2\beta + \varepsilon_\theta^2 sin^4\beta)\right) - 1\right), \quad (1)$$

where E is the Young's modulus of the shell material; I_1 is the first invariant of the right Cauchy-Green deformation tensor $\mathbf{C} = \mathbf{2E} + \mathbf{I}$, and $\mathbf{E}$ is Green-Lagrange strain tensor, μ_1, μ_2, β are material parameterts. The material (1-3) is anisotropic material due to Fung term, which is included to describe anisotropic response of the fibers and their sharp increase in stiffness. Neo-Hookean term describes the isotropic part of the tissue response. The geometrical nonlinearity is described with the higher-order shear deformation theory [2].

The viscoelasticity is taken into account through employment of Kelvin-Voigt model [3].

The shell is filled with quiescent fluid [4]. Since we deal with large deflections and large strains, the displacement-dependent expression for pressure is used [5]. Such load is proportional to the area of the deformed surface and is orthogonal to it.

Strain energy density in the form (1-3) is not polynomial in strain components, which significantly complicates the investigation of the dynamics. The analysis is simplified by expansion of strain energy density into truncated after order 4 series in strain components. Such model is able to describe the behavior of the vessel only locally, in the vicinity of a given configuration of interest around which the expansion is made. To reach highly deformed configurations, a sequence of successive local models has to be built.

Residual stresses that are stored in the aorta without pressure load are of two different sources. The first one is a simple axial stretch. The second type of stress is associated with the fact that the stress-free configuration for the aorta is not a closed circular cylindrical shell, but rather an open cylindrical shell. Moreover, the opening angle and the radius are all different for three aortic layers. This type of residual stress is taken into account as a stress resulting from the "closing" of the open cylindrical shell with additional condition of the same radii at the intima-media and media-adventitia connecting surfaces. Such deformation is not unique, so as additional conditions we used the experimentally observed feature that internal layer (intima) is fully in compression and the outer layer, adventitia, is fully in tension [6] and circumferential stress resultant is zero. The distribution of this type of residual stress is shown in (Fig. 1). The geometry of the shell in open and closed configuration is taken from [6, 7].

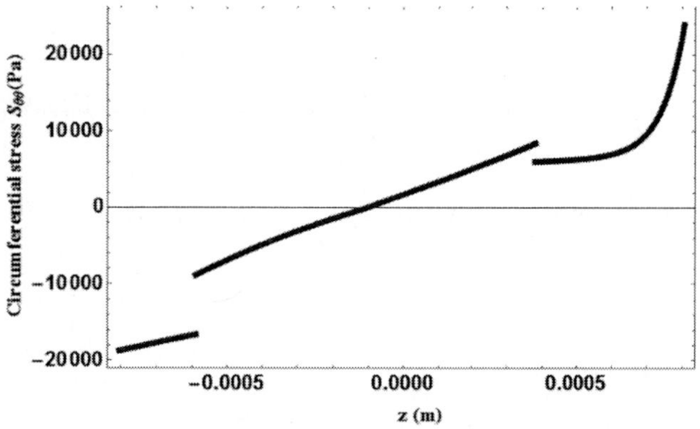

Figure 1: Circumferential stress distribution through the thickness.

RESULTS

We consider the shell with geometrical parameters that are similar to the parameters of the straight thoracic aortic segment of human aorta. The geometrical parameters are taken from [7, 8]. The material parameters are partially determined based on experimental data from [9, 10]. We solved the problem both with and without residual stresses.

Initially, the static problem of inflation of the shell under the pressure load was studied. The pressure-defection curve is shown in (Fig. 2). As can be seen, the stiffness drastically increases after certain strain threshold is reached.

Afterwards we analyzed free and forced axisymmetric vibrations around pre-loaded state of the entire artery as a composite shell and also single-layered shell to understand the contribution of different layers. For the three-layered artery model the configuration in the middle of physiological range (12.5 kPa pressure) was chosen for the expansion of the strain energy density around it. Vibrations with dominant first axisymmetric bending mode and with dominant mode with two half-waves in circumferential direction of the empty and fluid-filled shells are studied. The frequency response for the empty shell made of adventitial material is shown in (Fig. 3). We can see that the nonlinearity is weak enough for this case, however, in the case of fluid-filled shell it is stronger.

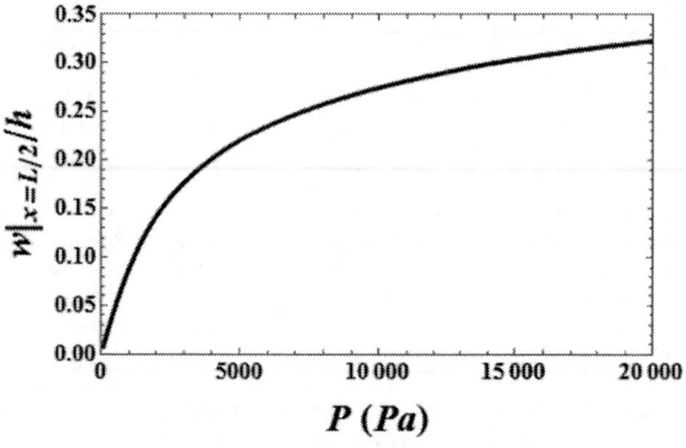

Figure 2: Static pressure-deflection curve. The abscissa shows central deflection.

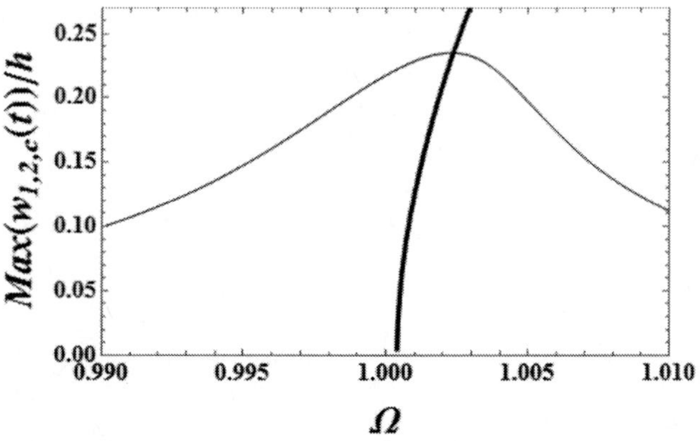

Figure 3: Frequency response of empty shell vibrations around pre-loaded configuration (adventitial material). Principal bending mode amplitude vs. frequency. Frequency response (thin line) and backbone curve (thick line).

DISCUSSION

The method for the analysis of nonlinear vibrations of arteries is presented. The method is able to deal with hyperelastic, viscoelastic, multilayered, fiber-reinforced shells that model blood vessels. The method is tested on the human aortic segment model. Under the static pressure initially soft shell becomes much stiffer with the pressure growth, which is the common feature for soft biological tissues. The dynamic nonlinearity is not very strong, but still nonlinear effects, like change in stiffening type, are observed.

ACKNOWLEDGEMENTS

The authors acknowledge the financial support of the NSERC Discovery Grant, Canada Research Chair and the Qatar National Research Fund NPRP 7-032-2-016.

REFERENCES

[1] Holzapfel, G.A., et al., *J Elast,* 61:1–48, 2000.

[2] Amabili, M., Reddy, J. N. *Int J Non-linear Mech* 45(4):409-418, 2010.

[3] Lakes, R., Viscoelastic Materials. New York, USA 2009.

[4] Amabili, M.: Nonlinear vibrations and stability of shells and plates. New York, USA 2008.

[5] Amabili, M., Breslavsky, I.D., *Int J Non-linear Mech* 77:265-273, 2015.

[6] Holzapfel, G.A., Ogden, R.W., *J R Soc Interface,* 7:787–799, 2010.

[7] Labrosse, M.R., et al., *J Mech Behavior Biomed Mat*, 17:44–55, 2013.

[8] Amabili, M., et al., *Int J Numer Methods Biomed Eng* 28(5):495-512, 2012.

[9] Holzapfel, G.A., et al. *J R Soc Interface* 12:20150188, 2015.

[10] Sassani, S.G., et al., *J Mech Behavior Biomed Mat*, 49:141-161, 2015.

Poster Presentation #P225

PHYSIOLOGY-MODELING COUPLED EXPERIMENT: A HIGH FIDELITY HARDWARE-IN-THE-LOOP HYBRID MODEL FOR THE CIRCULATION

Ethan O. Kung (1, 2), Masoud Farahmand (1), Akash Gupta (1)

(1) Dept. of Mechanical Engineering
Clemson University
Clemson, SC, USA

(2) Dept. of Bioengineering
Clemson University
Clemson, SC, USA

INTRODUCTION

Computational and in-vitro approaches, each with its own strengths and weaknesses, have provided useful tools for investigating cardiovascular biomechanics. While the hardware-in-the-loop (HIL) technique has been widely used in the automotive and aviation industries to extract the best of both of these approaches, this powerful technique has not been widely employed for cardiovascular modeling due to the physical limitations of real-time hydraulic control and measurements. Previous attempts on cardiovascular hybrid models have been limited to scenarios with high amplitudes of hydraulic signals in order to tolerate input and output errors and delays in the physical experiment. The quality and application of these hybrid models were severely limited by the hardware bandwidth and signal qualities in the physical experiment. Furthermore, there has not been robust verification of previous hybrid models against a well-defined standard, thus the detriments of experimental errors propagating in an HIL setup could not be quantitatively evaluated.

In this study, we developed a novel HIL framework called the Physiology-Modeling Coupled Experiment (PMCE), which leverages cardiovascular periodicity to implement a robust coupling between the numerical and experimental domains in a cardiovascular hybrid model. This mitigates physical limitations in the hydraulic experiment including sensor response time, measurement noise, actuator non-linearity, and signal delays due to hardware bandwidth or physical wave propagation. Furthermore, the PMCE method guarantees the convergence of the solution below a predefined residual, ensuring a high-quality hybrid model. The iterative nature of the PMCE also both relieves hardware bandwidth requirements and provides compatibility with sophisticated numerical methods that cannot be performed in real-time.

We present verification of the PMCE against state-of-the-art multiscale computational fluid dynamic (CFD) simulations that have been previously validated. The verification was performed in a challenging context where the PMCE incorporated a sophisticated numerical physiology model of the single-ventricle circulation, and a hydraulic experiment exhibiting small pressure and flow signals requiring high measurement and actuation precision.

METHODS

Multi-domain Coupling in the PMCE Making use of the periodic nature of cardiovascular flow, we implemented an iterative framework to interface signals of pressure and flow between the experimental and numerical domains with the ability to perform signal cleanup and correction, thus achieving a high-fidelity coupled system (Figure 1). Using derivative-based optimization and a proportional controller, we developed a coupling algorithm to identify the correct flow waveform through the numerical-experimental interface which would result in identical pressure drop across the experimental section and the corresponding locations in the physiology model. The iterative algorithm continues until the error between the two pressure drops is smaller than a user specified tolerance.

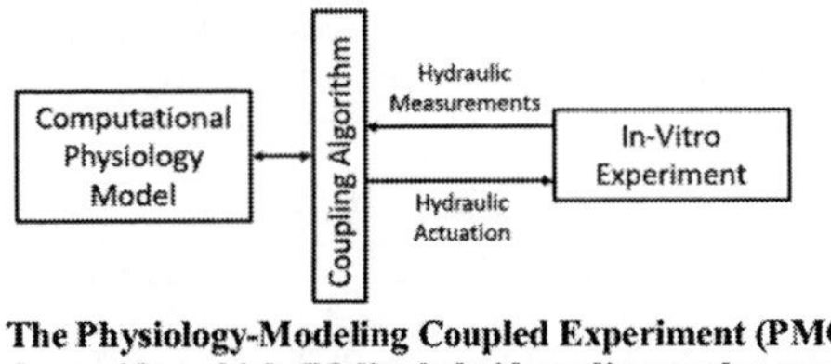

Figure 1: The Physiology-Modeling Coupled Experiment (PMCE) framework provides a high-fidelity hybrid cardiovascular model

Verification To verify the solution accuracy of the PMCE against multiscale CFD simulations, we constructed parallel PMCE and multiscale CFD models each describing a mid-sized Fontan patient (weight=50kg and height=150 cm) with an 85% area left pulmonary artery (LPA) stenosis under two different metabolic states (1 MET and 5 MET). A previously published computational physiology model of the Fontan circulation [1] was used in both the PMCE and the multiscale simulations. The in-vitro segment of the PMCE contained a 3D printed physical phantom of the LPA stenosis; the same stenosis geometry was used as the 3D model for the CFD portion of the multiscale simulations.

In-Vitro Setup We placed the physical phantom in a flow loop where the flow waveform is produced by a computer programmable pump [2]. We used a camp-on flow sensor (Transonic Systems, NY) to measure flow through the phantom and two catheter pressure transducers (Millar Inc., TX) placed at the phantom's inlet and outlet to measure pressure drop. All tubing in the flow loop was rigid and the working fluid was a 40% glycerol solution. The pressure and flow data were processed through a 60Hz band-stop filter and averaged over 13 cycles to strengthen signal to noise ratio.

Multiscale Simulations The 3D stenosis model was discretized into an anisotropic finite element mesh with maximum edge size = 0.5mm near the stenosis region and 0.9mm elsewhere. Simulations were performed using a previously validated in-house solver [3,4]. A time step size of 0.001s was used and the simulations were run for a minimum of 30s to establish stable periodicity.

RESULTS

The results produced by the PMCE agree extremely well with those produced by the multiscale CFD simulations. Comparisons of flow waveforms through the LPA and pulmonary pressure waveforms revealed R^2 values > 0.98 in all cases (Figure 2). The RMS errors of the flow waveforms were less than 2.8% of the mean flow, and of the pressure waveforms were less than 1.2% of the mean pressure at both 1 and 5 MET. Comparisons of parameter values averaged across one period also show extremely good agreements between PMCE and multiscale simulation (Table 1). All comparisons except one show agreement to within 0.4%, with the largest discrepancy being 1.64%.

DISCUSSION

Unique to previous hybrid model approaches, the PMCE method leverages cardiovascular periodicity to mitigate physical limitations in the hydraulic experiment, allowing for averaging and filtering of signals and iterative correction of actuator outputs. This framework not only enables high-fidelity coupling between the experimental and numerical domains, it also allows for the direct evaluation and continual improvement of the solution quality until the desired residual is satisfied.

Table 1: Comparisons of cardiac output (CO), aortic pressure (P_{ao}), pulmonary pressure (P_{pul}), and LPA flow fraction (LPA Q_{frac}) between PMCE and multiscale simulation. The table presents mean values of these parameters across one period.

		CO (L/min)	P_{ao} (mmHg)	P_{pul} (mmHg)	LPA Q_{frac}
1 MET	Multiscale Sim	3.405	86.93	13.10	0.3024
	PMCE	3.413	86.99	13.08	0.3074
	Difference	0.24%	0.07%	0.15%	1.64%
5 MET	Multiscale Sim	6.593	96.42	16.994	0.2291
	PMCE	6.618	96.45	16.987	0.2288
	Difference	0.38%	0.035%	0.040%	0.18%

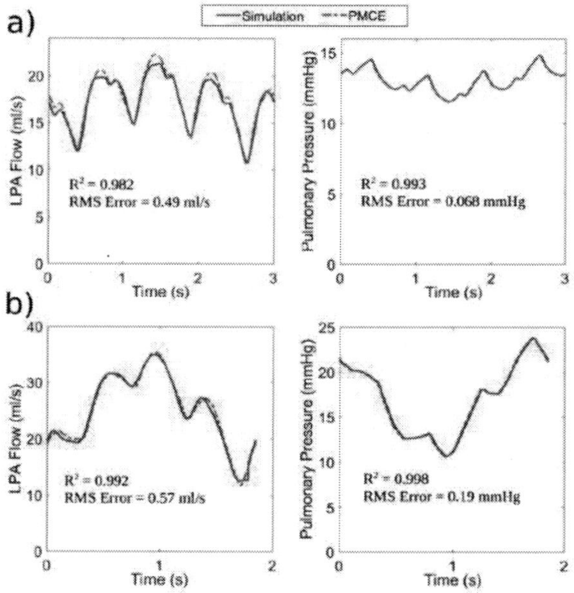

Figure 2: Flow and pressure waveform comparisons between PMCE and multiscale simulation. a) 1 MET, and b) 5MET

A limitation of the PMCE framework is that its iterative nature implies potentially significant time for model execution. This factor would be especially important when a multiple-outlet or compliant experimental section is used, since a large number of iterations may be required to identify multiple flow waveforms. Similar to current CFD simulations, the PMCE execution could require tens of hours for a highly complex scenario.

In this study, we used a scenario involving a rigid 3D section where the multiscale CFD simulations can provide the actual model solution to verify the accuracy of the PMCE. The verified PMCE framework can now provide a tool that is much more powerful than current multiscale CFD simulations in the area of modeling complex fluid-structure interactions in closed-loop with a systemic physiology model. The capability to investigate a real physical system in the context of patient physiology has the potential to transform the design and testing of many cardiovascular devices and procedures.

ACKNOWLEDGEMENTS

This work was supported by an Award from the American Heart Association and The Children's Heart Foundation.

REFERENCES

[1] Kung, E et al., *J Biomech Eng*, 136(8):081007, 2014.
[2] Mechoor, R et al., *J Biomech Eng*, 138(11):111002, 2016
[3] Esmaily-Moghadam, M et al., *Comp Meth App Mech*, 286(1):40-62, 2015
[4] Steinman, D. A., et al, *J Biomech Eng*, 135(2):21016, 2013

IN-VITRO VALIDATION OF A LUMPED-PARAMETER MODEL FOR A FONTAN RIGHT-SIDE ASSIST DEVICE

Mitra Shabanisamghabady (1), Ehsan Mirzaei (1), Minoo N. Kavarana (2), Dimitrios Georgakopoulos (3), Ethan O. Kung (1,4)

(1) Department of Mechanical Engineering
Clemson University
Clemson, South Carolina, USA

(2) Division of Cardiothoracic Surgery
Medical University of South Carolina
Charleston, South Carolina, USA

(3) Sunshine Heart Inc.
Eden Prairie, Minnesota, USA

(4) Department of Bioengineering
Clemson University
Clemson, South Carolina, USA

INTRODUCTION

Children born with uni-ventricular hearts typically undergo a three-stage surgery with the third stage being the Fontan procedure. It involves diverting the venous blood from the IVC and SVC to the pulmonary arteries without passing through heart (total cavo-pulmonary connection) commonly by the way of an extra cardiac Gore-Tex graft [1,2]. Fontan procedure provides excellent short-term palliation but a large proportion of these children face severe life-threatening problems from elevated venous pressures to ventricular dysfunction [3,4]. An ideal solution is restoring a cavo-pulmonary power source which can convert the uni-ventricular system back into a normal bi-ventricular circulation. Due to the size constraints in these children mechanical support is very high risk [5]. We propose a novel design for Fontan right-side support that applies the external aortic balloon concept (C-Pulse Heart Assist System, Sunshine Heart Inc., MN) to the Fontan Extra-cardiac graft. This attractive approach provides benefits with minimum risk to the patients. In this research, our goal is to experimentally validate a computational lumped-parameter network (LPN) for investigating the feasibility of this concept.

METHODS

Lumped Parameter Model Lumped-parameter network (LPN) comprising electrical circuit elements such as resistors, capacitors and diodes can be used to model Fontan physiology [6]. In this research, we implemented an LPN model of the C-Pulse heart assist device (Figure 1) and validated the model via equivalent in-vitro experiments. The C-Pulse system includes a non-blood contacting cuff, ECG sensing leads, Percutaneous Interface Lead (drive line), and an external driver (pump). In the Fontan right-side assist application, the cuff is wrapped around the extra cardiac Gore-Tex graft and is inflated and deflated by the driver. To model the function of the C-Pulse device with an LPN, the pressure waveform generated by the C-Pulse system (Figure 2) is prescribed to the Fontan graft. We used a basic three element Windkessel model consisting of a proximal resistor (R), capacitor (C), and distal resistor (R_d) to mimic the pulmonary vascular impedance in the validation setup. The parameter values for the electrical components of the LPN were derived from a previously published Fontan LPN [6].

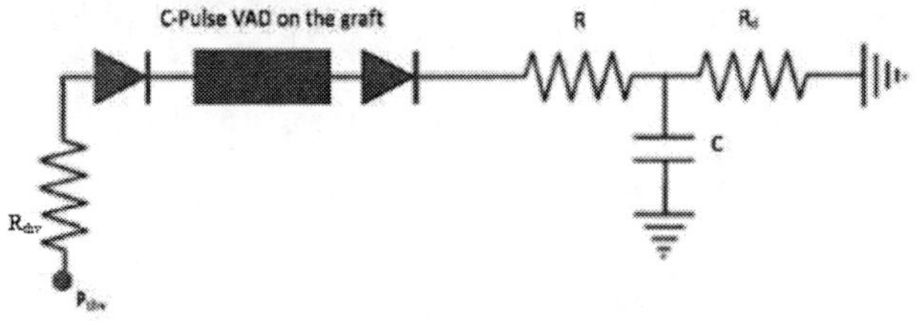

Figure 1: Open-loop lumped-parameter Network containing the C-Pulse device model

In-Vitro Setup To mimic the same scenario with an in-vitro experimental setup (Figure 3), we placed the extra cardiac Gore-Tex graft and two one-way valves before and after it in a flow loop. The C-Pulse balloon pump was mounted on the graft. Using an elevated reservoir before the graft and a real-time programmable pump [8] the upstream pressure was kept constant. To recreate the downstream impedance, an air-pocket capacitor was used [7]. The C-Pulse balloon pump driver settings were fixed but the pump heart rate was changed during the experiment using the controller. All tubing was rigid and the working fluid was a 40% glycerol solution. Data acquired from the pressure transducers and flow sensor was post processed using MATLAB.

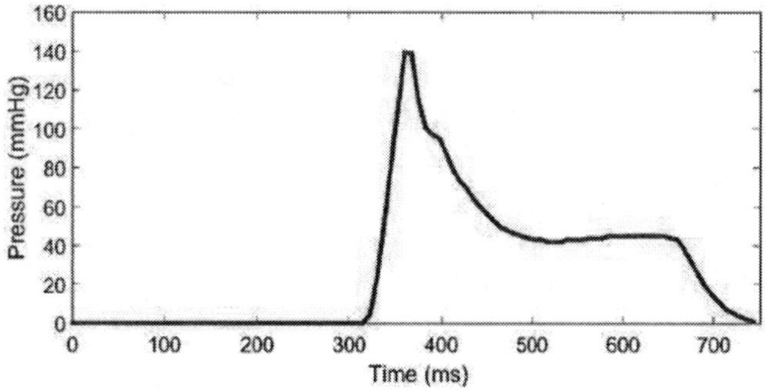

Figure 2: C-Pulse pressure waveform

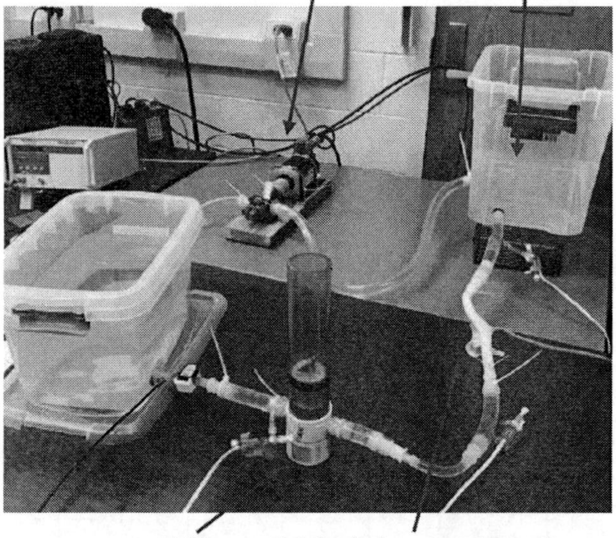

Figure 3: Experimental setup for the LPN validation

RESULTS AND DISCUSSION

The purpose of this study was to use in-vitro experimental data to validate an LPN model of the balloon pump concept for Fontan right-side assist. Comparisons of the averaged parameter values show good agreement between the LPN and in-vitro results (Table 1). The difference observed in the averaged flow results was partially due to the small backflow in the one-way valves used in the experimental setup. The downstream resistances and capacitance in the experiment were non-ideal components which could also contribute to the experimental error.

The operation of the C-Pulse device only produced improved flow rate at the highest heart rate setting (120 beats/min), where the improvement was 7.7%. Results from both the LPN and the experimental set-up for different pump heart rates show that the balloon pump right-side assist concept is potentially feasible at higher pump rates. Due to the limitation of the C-Pulse pump the pump rate was only increased to 120 beats/min in this study.

Further modeling and experimentation can lead to the optimal design, implementation, and operation settings for the C-Pulse in the Fontan right-side assist application. Using the validated LPN model, assessment of physiologic impacts of device implementation can be performed with considerations to individual patient characteristics.

ACKNOWLEDGEMENTS

This work was supported by an Award from the American Heart Association and The Children's Heart Foundation and a grant by Saving tiny Hearts Society. We also acknowledge Sunshine Heart for providing the C-Pulse balloon pump.

REFERENCES

[1] Starnes, S. L. et al., *J Thorac Cardiovasc Surg*, 122:518-523, 2001.
[2] Ericken, E. et al., *Int J Cardiol*, 116:212-218, 2007.
[3] Khairy, P. et al., *J Circulation*, 117:85-92, 2008.
[4] de Leval, M. R. et al., *Nat Clin Pract Cardiovasc Med*, 2:202-208, 2005.
[5] Horne, D. et al., Semin Thorac Cardiovasc Surg Peditr Card Surg Annu, 18:17-24, 2015.
[6] Kung, E. et al., *J Biomech Eng*, 136, 2014.
[7] Kung, E. et al., *Cardiovascular Engineering and Technology*, 2(1):2-14, 2011.
[8] Mechoor, R. et al. Journal of Biomechanical Engineering, 138(11), 2016.

Table 1: Comparison of flow rate results (L/min) between LPN and in-vitro experiment

	LPN	In-vitro	Difference
Reference	2.230	2.200	1.36%
C-Pulse (66 beats/min)	2.042	2.037	0.25%
C-Pulse (80 beats/min)	2.045	2.040	0.24%
C-Pulse (120 beats/min)	2.400	2.370	1.26%

SB³C2017
Summer Biomechanics, Bioengineering and Biotransport Conference
June 21 – 24, Tucson, AZ, USA

TOWARDS A MULTI-FIDELITY HEMODYNAMIC MODEL PIPELINE FOR THE ANALYSIS OF CARDIOVASCULAR FLOW UNDER UNCERTAINTY

Casey M. Fleeter (1), Daniele E. Schiavazzi (2), Alison L. Marsden (3)

(1) Institute for Computational and Mathematical Engineering
Stanford University
Stanford, CA, USA

(2) Department of Applied and Computational Mathematics and Statistics
University of Notre Dame
Notre Dame, IN, USA

(3) Department of Pediatrics
Stanford University
Stanford, CA, USA

INTRODUCTION

Hemodynamic modeling is increasingly being adopted in the diagnosis and treatment of cardiovascular disease and surgical planning. Hemodynamic modeling generally consists of anatomic model construction from medical image data, followed by solution of the incompressible Navier-Stokes equations governing blood flow in elastically deformable vessels. However, various simplifying assumptions can be made to generate models of "intermediate complexity."

Further simplifications may be related to the flow dimensionality and patient specific geometry of the vasculature. The so-called *one-dimensional formulation of hemodynamics*, for example, integrates the equations across the vessel cross section, thereby reducing the dimensionality to solve for flow and cross sectional area as a function of time and axial distance. This is typically done for Newtonian flow with deformable, impermeable and elastic walls, allowing for a range of material models. The one-dimensional formulation of hemodynamics adopted in this project is adapted from Hughes and Lubliner [2], with the constitutive relationship proposed by Olufsen et al. [4]. Details of the numerical solution method are explained in [5]. A wide range of boundary conditions are available for the one-dimensional formulation, with implementation specifics presented in [3].

In contrast to many other branches of computational mechanics, computations of biological and biomedical systems are intrinsically affected by multiple sources of uncertainty. Under these conditions, a deterministic description of the hemodynamic response provides limited information of the process under study. This motivates the transition to a stochastic framework where all parameters affecting boundary conditions, material constitutive behavior and model geometry are defined in probability with distributions either assumed or assimilated from available patient-specific data.

We discuss our efforts to create an automatic pipeline generating a cascade of varying-fidelity cardiovascular models ranging from computationally inexpensive, patient agnostic lumped parameter models to computationally intensive, patient-specific models of blood flow. Due to the computational cost of performing stochastic analysis (parameter estimation and uncertainty propagation) directly on the multi-scale models, this pipeline facilitates the statistical characterization of quantities not significantly affected by the three-dimensionality of the local flow field and whose statistical characterization can be performed at a reduced computational cost.

We also show the integration of the proposed pipeline in the SimVascular open source framework [1] (www.simvascular.org), in which the one-dimensional model is automatically generated based on the vessel centerline paths and segmentations created in SimVascular from the medical images acquired on the patient. We also present a validation study, comparing the pressures and flows obtained from the one-dimensional and three-dimensional models and critically discuss their difference considering the underlying assumptions.

Finally, we discuss the initial steps towards the development of a framework for uncertainty propagation that takes advantage of a cascade of models with increasing fidelity.

METHODS

Our one-dimensional solver accepts input files divided into several *cards*. *Joints* are defined at the intersection of straight vascular segments, specifying the segments entering and exiting the joint. *Segments* are straight cylindrical tubes connected through joints, representing a blood vessel with axis-symmetric deformation. *Input waveforms* (flow or pressure) are prescribed at the inlet joint, while *boundary conditions* can be applied to the model outlets. Both initial and boundary conditions use *data tables* to specify temporal or frequency dependence. Linear elastic and material models can be described through *material* cards, for example using the constitutive models introduced by Olufsen and others [4]. Work is underway to extend the code with the ability to couple the inlets/outlets with a zero-dimensional lumped parameter model.

We created a python script pipeline to automatically convert SimVascular segmentation group files into an input file compatible with the one-dimensional solver. The geometry of the model, consisting of path centerlines and cross-sectional segmentations, is extracted and converted into joints and segments by automatically detecting the intersections between different Simvascular segmentation groups. The script handles simple joints (two segments connected linearly), as well as more complex configurations with multiple inlets and outlets, such as a parent vessel bifurcating into two child branches. All intersections are recorded and their cards are written to the input file, along with the specified boundary condition, initial condition and material property cards.

The results from the one-dimensional solver were validated by comparing the model outputs generated from a three-dimensional model with those converted to a one-dimensional model through the proposed pipeline for two patient models.

The first is a patient specific model of the aorto-iliac bifurcation. This model was chosen for the length of the vessel, to demonstrate the change in pressure and flow waveforms along the vessel length to confirm these results are replicated in both solvers.

This model is illustrated in Figure 1. The second is a patient specific model including the thoracic aorta and the coronary arteries. This model was selected to perform model validation under a wide range of vessel diameters, as well as to ensure proper functioning for future uncertainty quantification work using coronary model databases. This model is illustrated in Figure 2.

For both models, resistance boundary conditions at the outlets and both steady and pulsatile inlet waveforms were utilized. Outlet resistance values were tuned to produce physiologic pressure and flow output waveforms, with an average pressure of 80 mmHg and the Womersley flow profile resulting in 5 L/min of blood flow. The three-dimensional simulations were run with a rigid wall assumption, and the one-dimensional model was run with an Olufsen material model, where parameters defining the elastic modulus (k1, k2, and k3) were given by 2×107, −22.5267, and 8.65×105 respectively.

Perturbations in boundary conditions, constitutive model and geometrical parameters can be easily implemented using a one-dimensional formulation, by direct interaction with the input file format. Ease of parameterization together with limited solution cost make a one-dimensional hemodynamics solver ideal to perform comparison between the performance of various non-intrusive methods for uncertainty quantification. Moreover, by systematically exploring the solution associated with various parameter realizations is is possible to further study the robustness of the implemented solver.

In this talk, a preliminary investigation is presented that considers uncertainty in the outlet resistances in a model of the aorto-iliac bifurcation. Several methods were compared, including Monte Carlo and Quasi-Monte Carlo simulation, generalized polynomial chaos (gPC) and a recently proposed generalized multi-resolution expansion for uncertainty propagation.

RESULTS

The aorto-iliac bifurcation (Figure 1) and coronary models (Figure 2) are analyzed under steady state and pulsatile flow conditions. Flow profiles generated with the one-dimensional and three- dimensional solvers, illustrated for the aorto-iliac model (Figure 1), respectively, agree well both in shape and amplitude. Conversely, the two solvers generate slightly different pressure profiles as shown for the coronary model (Figure 2). These differences are, however, justified due to the respective underlying assumptions for the deformability of the vessel walls. We expect these differences to be minimized by including fluid-structure interaction capabilities to the three-dimensional solver.

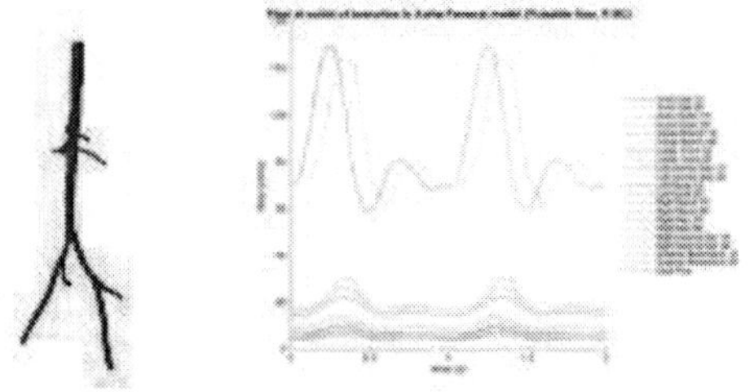

Figure 1: Aorta-iliac model flows, resistance BCs.

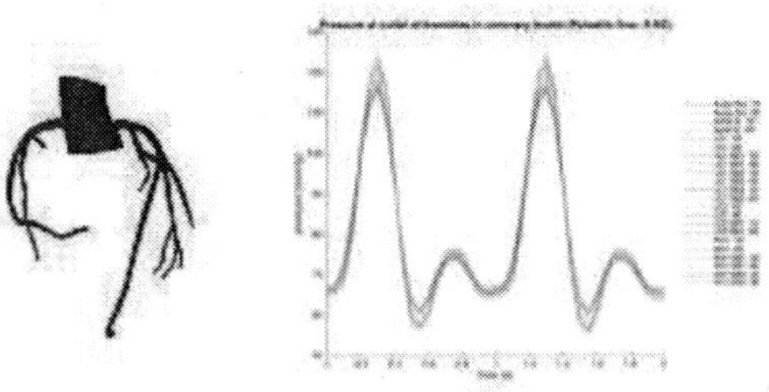

Figure 2: Coronary model pressures, resistance BCs.

DISCUSSION

The results of our validation study confirm the consistency of three-dimensional and one-dimensional model outputs. The results are also consistent with the formulation of the two underlying solvers where the difference in vessels model deformability justified the observed differences in output pressures and flows. This relates to the wall material model implemented in the one-dimensional solver and the rigid wall formulation considered instead in the three-dimensional case.

Our parametric studies confirmed the robustness of the one-dimensional hemodynamic solver to a variation in the boundary conditions, material and geometrical parameters. In our uncertainty study, convergence to the output statistics confirmed the expectations, with Monte Carlo approaches requiring the larger number of model realizations followed by Quasi-Monte Carlo and gPC. Multiresolution uncertainty propagation required the least number of model evaluations taking advantage of built-in adaptivity.

Future work will be devoted to perform additional validation tests using the SimVascular's fluid structure interaction capability, and on investigating how to take advantage of a cascade of automatically generated multi-fidelity cardiovascular models to speed up the computation of the response statistics.

REFERENCES

[1] Updegrove, A et al., *Ann Biomedical Engg*, 44:1–17, 2016.

[2] Hughes, T et al., *Math Biosci*, 18:161–170, 1973.

[3] Vignon, I, Ph.D. thesis, Stanford University, 2006.

[4] Olufsen, M, *Am J Physiol*, 276:H257-268, 1999.

[5] Wan, J et al., *Comput Method Biomec*, 5:195–206, 2002.

Poster Presentation #P228

SB³C2017
Summer Biomechanics, Bioengineering and Biotransport Conference
June 21 – 24, Tucson, AZ, USA

A CFD-BASED GENETIC ALGORITHM APPLIED TO THE DESIGN OF FLOW-DIVERTING STENT FOR IDENTIFYING THE WIRE CONFIGURATION THAT MAXIMALLY DISRUPTS THE BUNDLE OF ANEURYSM INFLOW

Mingzi Zhang (1,2), Hitomi Anzai (3), Bastien Chopard (4), Yi Qian (2), Makoto Ohta (5)

(1) Graduate School of Engineering
Tohoku University
Sendai, Miyagi, Japan

(2) Faculty of Medicine and Health Sciences
Macquarie University
Sydney, N.S.W., Australia

(3) Frontier Research Institute for Interdisciplinary Sciences
Tohoku University
Sendai, Miyagi, Japan

(4) Department of Computer Science
University of Geneva
Carouge, Geneva, Switzerland

(5) Institute of Fluid Science
Tohoku University
Sendai, Miyagi, Japan

INTRODUCTION

Flow-diverting (FD) stents, as an emerging endovascular device, can be used to treat intracranial aneurysms, a life-threatening disease which can lead to subarachnoid haemorrhage due to aneurysm rupture. While FD treatments were demonstrated to be effective in inducing the blood clotting and thrombotic occlusion of the aneurysm, a number of patients may still suffer from delayed aneurysm rupture, incomplete aneurysm occlusion, or post-treatment in-stent stenosis [1].

In FD treatments, an FD diverts much of the aneurysmal inflow away, such that the amount of blood entering the aneurysm decreases. To achieve an improved flow diversion, a denser configuration of FD wires is desired to maximally disrupt the bundle of aneurysmal inflow (BOI); however, an in-stent stenosis complication may occur when the high flow diversion is pursued by using FDs with low device porosities.

To improve the flow diversion of a stent without reducing its nominal porosity, we report a computational approach to identify the stent wire configuration with the maximal BOI disruption in a given aneurysm model geometry, by using an automated genetic algorithm based on a series of computational fluid dynamics (CFD) simulations of the aneurysmal local haemodynamics.

METHODS

The patient aneurysm model we studied is an internal carotid artery (ICA) aneurysm, with an inlet diameter of 3.8 mm. A model FD geometry was constructed as a helix-like structure that mimics those FD stents available in the markets. The model FD was assumed to be composed of a total of 8 wires: 4 clockwise and 4 anti-clockwise wires with a thickness of 50 μm. We defined the trajectory of each helix wire based on the centreline of the aneurysm parent artery, and the maximum inscribed sphere radius (MISR) corresponded to the discrete points along the centreline.

We employed a random modification procedure to assign a slight structural alteration to the helical wires at each step, and evaluated this alteration's effect on the aneurysmal haemodynamic modifications. To perform the random modification, one of the eight helices was arbitrarily chosen, and a stochastic variable ε was then added to its starting phase, which finally resulted in a tiny longitudinal displacement of the randomly selected helix. A genetic algorithm (see Fig. 1) was developed to find the optimal solution out of the randomly created candidates, and to control the random modification progressing towards a global optimum. We used the aneurysmal average velocity as the objective function in the genetic algorithm, a sensitive parameter thought to be associated with the aneurysmal inflow reduction [2].

An open-source, public-available CFD solver *Palabos* (http://www.palabos.org/) was applied in haemodynamic simulations. *Palabos* is based on lattice Boltzmann method (LBM), a mesoscopic approach in which a fluid is described in terms of the density distribution $f_i(r,t)$ of idealised fluid particles moving and colliding on a lattice structure:

$$f_i(r + \Delta t v_i, t + \Delta t) = f_i(r,t) + \frac{1}{\tau}\left(f_i^{eq} - f_i\right) \qquad (1)$$

where f^{eq} and τ are respectively the local equilibrium distribution and relaxation time [3]. Since the Cartesian mesh applies to LBM, the stent wire trajectories were defined as bounce-back lattice at the beginning of each simulation, which enabled the automation of the proposed genetic algorithm. The blood was assumed to be a Newtonian and incompressible fluid, with density and kinematic viscosity fixed to 1040 kg/m^3 and $4.0e^{-6}$ m^2/s. We prescribed a parabolic velocity profile (average velocity of 0.21 m/s, corresponding to a Reynolds Number of 200) at the inlet, and a constant pressure 0 Pa at the outlet of the

Poster Presentation #P229

aneurysm model geometry. A convergence was assumed when the change in average energy of fluid cells was no more than 10^{-6} kg m^2/s^2.

RESULTS

To quantify the differences of aneurysmal inflow of each randomly generated FD wire configuration, we defined a flow reduction index:

$$R_f = \frac{V_{w/o} - V_{current}}{V_{w/o}} \quad (2)$$

to measure the flow-diversion efficacy. The $V_{w/o}$ is the aneurysmal average velocity in the untreated condition, and the $V_{current}$ denotes the average velocity at a 'current' stage. Figure 2 demonstrates the change of R_f with respect to the running step in the genetic algorithm. As can be observed, starting from the initial structure (R_f at around 70%), the flow diversion improved noticeably during the initial hundreds of iterations, and gradually stabilised at around 95%.

Figure 3 shows the streamlines (colour-coded with velocity magnitude) and iso-velocity surfaces (corresponding to 0.12 m/s) generated inside the aneurysm geometry with FD stent before and after this genetic procedure. In the FD treatment with the initial uniformed wire configuration, a strong jet flow can be observed entering the aneurysm through its ostium. Whereas the aneurysm inflow cannot be seen clearly with the wire configuration identified by the genetic algorithm. It is worth noting that, in the latter scenario, a concentration of FD wires inside the BOI area can be observed.

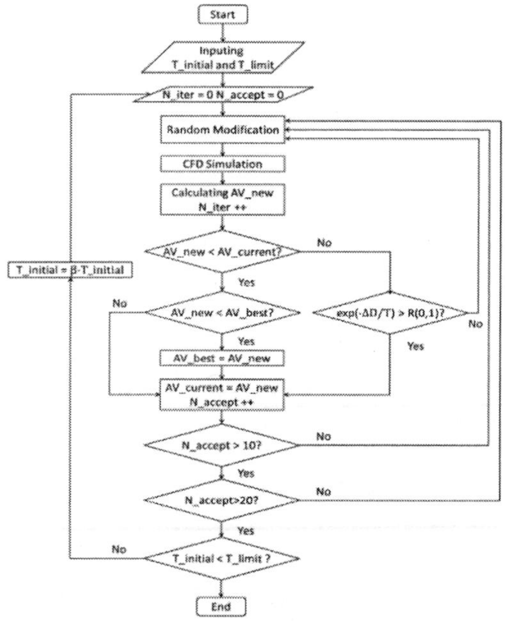

Figure 1: The flow chart of the proposed the genetic algorithm.

DISCUSSION

The CFD-based genetic algorithm developed in this study modified the FD wire configuration to enable it better disrupt the BOI area, thereby improving the flow-diversion efficacy without altering the original device porosity. To insure that the genetic algorithm could find the optimal solution of all the stent configuration candidates, a configuration with an inferior flow diversion might also be accepted according to our algorithm. In addition, the developed algorithm, such as a study we previously reported [4], can accept other haemodynamic parameter as the objective function(s), as long as the Input parameters (see Fig. 1) are well established.

A possible limitation of this method from the viewpoint of clinical practice is that sending such an FD stent to the desired location could be a challenge for interventionists, since a discrepancy of FD position in longitudinal direction during device deployment might comprise the flow-diversion efficacy. In the future work, this problem might be solved by embedding a robustness test into the genetic algorithm to improve the reliability of the identified wire configuration.

In conclusion, a practical genetic algorithm was developed based on the aneurysmal haemodynamic simulations to identify the FD wire structure that maximally disrupt the BOI area in this study. This algorithm potentially contributes to the study of the patient-specific design of FD stents.

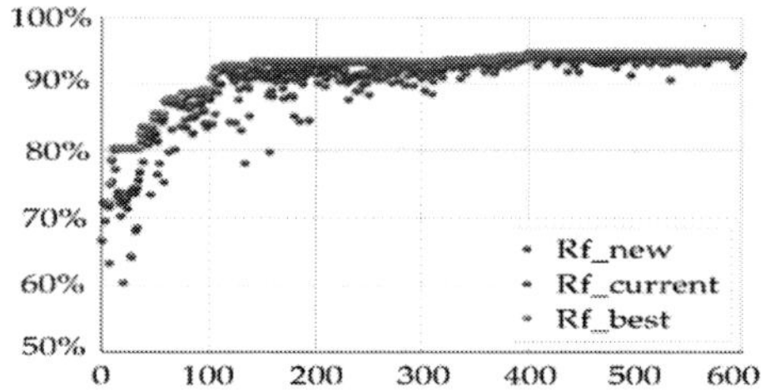

Figure 2: The change of flow-diversion efficacy with respect to the iterations of the applied genetic algorithm. (Rf_new, Rf_current, and Rf_best *refer to the flow reductions calculated for the* AV_new, AV_current, *and* AV_best *in Fig. 1)*

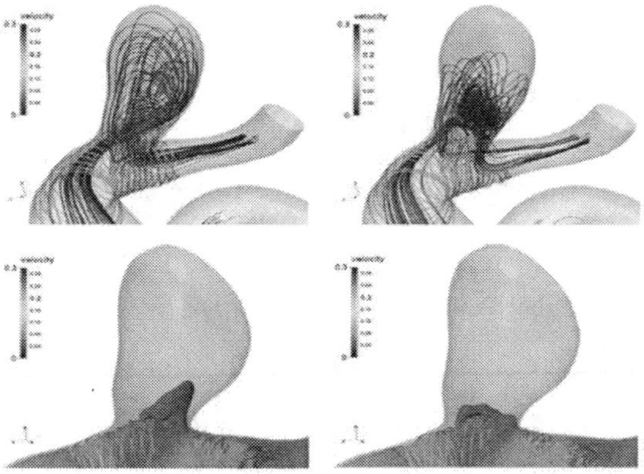

Figure 3: Visualisations of aneurysmal flow patterns and iso-velocity surfaces with FD treatments of the initial structure and the algorithm-chosen structure. *(Top: streamlines colour-coded by velocity magnitude; bottom: iso-velocity surfaces)*

ACKNOWLEDGEMENTS

This study was supported by the ImPACT Program of Council for Science, Technology and Innovation (Cabinet Office, Government of Japan), the Macquarie University Research Excellence Scholarship (iMQRES, No. 2015257), and Australian Research Council Linkage Projects (Grant ID: LP130100423).

REFERENCES

[1] Cohen, J *et al.*, *J. Clin. Neurosci.*, 21(7): 1116–1122, 2014.
[2] Chung, B *et al.*, *J. Neurointerventional Surg.*, 0:1-5 2014.
[3] Succi, S, *The Lattice-Boltzmann Equation*. Oxford, 2001.
[4] Anzai, H *et al.*, *J. Biomech. Eng.*, 136(6):61006, 2014.

OPTIMIZATION OF SYSTEMIC-TO-PULMONARY SHUNT DESIGN IN THE ASSISTED BI-DIRECTIONAL GLENN

A. Verma (1), M. Esmaily-Moghadam (1), J.K. Shang (2), R. Figliola (3), T.Y. Hsia (4), A. Marsden (5)

(1) Department of Mechanical Engineering
Stanford University
Stanford, CA, USA

(2) Department of Mechanical Engineering
University of Rochester
Rochester, NY, USA

(3) Department of Mechanical Engineering
Clemson University
Clemson, SC, USA

(4) Great Ormond Street Hospital for Children, and University College London Institute of Cardiovascular Science, London, UK

(□) Department of Pediatrics
Stanford University
Stanford, CA, USA

INTRODUCTION

Palliative treatment for patients with single ventricle congenital heart diseases remains clinically challenging, with mortality rates after first stage surgery approaching 30% [1] in some centers. The first stage surgery, typically the Norwood procedure [2] for patients with hypoplastic left heart syndrome, is performed within the first few days of life. This involves establishing the functioning ventricle as the main driver of systemic circulation and providing flow to the pulmonary arteries (PAs) through a systemic-to-pulmonary shunt. Recently, Esmaily-Moghadam et.al proposed an alternative to the current stage 1 surgery called the Assisted Bi-direction Glenn (ABG) [2], which modifies the conventional stage-2 surgery, the Bi-Directional Glenn (BDG) by adding a systemic-to-pulmonary shunt between the brachiocephalic artery and the Superior Vena Cava (SVC) (Figure 1). Since the eventual ligation of the shunt results in stage-2 BDG physiology, this approach has the potential to reduce the number of surgical stages from 3 to 2. Other potential benefits of this procedure include reduced ventricular load and reduced risk of shunt thrombosis.

The objective of this study is to determine the effect of varying systemic-to-pulmonary shunt geometry on pulmonary hemodynamics, particularly the pressure in Superior Vena Cava and flow in the PAs. Using optimization, we take a systematic approach to shunt design, aimed at increased pulmonary flow and maintaining moderate SVC Pressure.

METHODS

Cylindrical idealizations of the aorto-pulmonary geometry are constructed, based on our former ABG study [2]. Each model includes a circular aortic arch, 4 upper branches, 2 coronary arteries, the SVC, the two pulmonary arteries and a converging systemic-to-pulmonary shunt (Figure 1). The shunt is divided into two parts: the graft, which connects to the brachiocephalic artery, and the nozzle, which injects flow from the systemic circulation into the SVC. The shunt geometry was parametrized by selecting a total of 6 design parameters for the shunt that could be varied using automated scripts (Figure 2).

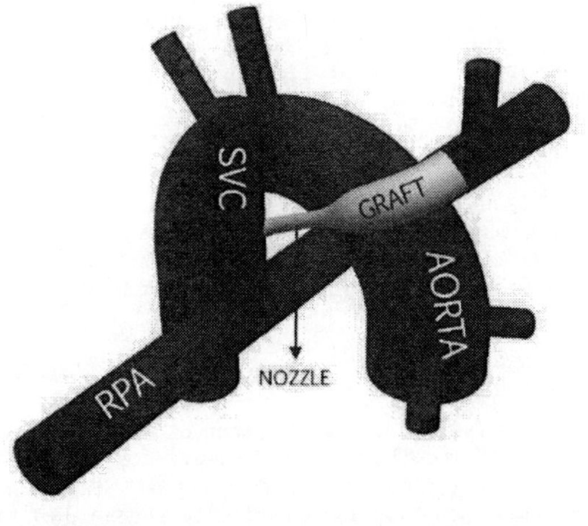

Figure 1: Model geometry used for the 3D hemodynamics simulations. Note that shunt shapes were used in the study.

SimVascular, an open source software package for cardiovascular modeling [4], was used to construct the 3D geometries. An in-house three-dimensional finite element flow simulation solver is used for solving the unsteady, incompressible Navier-Stokes equations. A closed loop 0D lumped parameter network, which represents the full body, single-ventricle circulation, was used to provide boundary conditions for the 3D domain. The parameters of this lumped parameter network are based on our previous studies [3,5]. The pipeline of modeling and meshing to simulation was automated with the use of scripts.

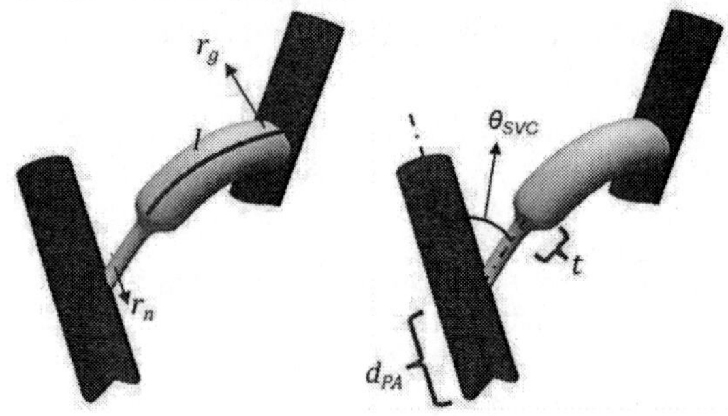

Figure 2: Design parameters of the systemic-to-pulmonary shunt.

Parameter	Lower Bound	Upper Bound
r_g	1mm	2mm
r_n	0.3mm	1mm
θ_{SVC}	20°	80°
d_{PA}	6mm	12mm
l	60% graft length	90% graft length
t	10% graft length	50% graft length

Table 1: Bounds on design parameters for the graft. R_v is the radius of the parent vessel.

We pose the design space exploration process as an optimization problem for maximizing pulmonary flow with a physiological constraint on the SVC pressure. We utilize the Surrogate Management Framework (SMF), a derivative-free optimization method, for the optimization process [6] and we use a large initial sampling to ensure sufficient coverage. Bounds on the design parameters are specified in Table 1. A parallel framework was employed to increase efficiency of the optimization runs.

RESULTS

The SVC Pressure and the pulmonary flow both increase nearly linearly with an increase in nozzle radius r_f (Figures 3 & 4). There is also an observed increase in the scatter of the responses at larger values of nozzle radius. We do not observe any significant trends with the other design parameters.

DISCUSSION

The effect of the systemic-to-pulmonary shunt on SVC Pressure and the total pulmonary flow can be represented almost entirely by the effect of the nozzle radius. This allows for predictability of SVC Pressure and total pulmonary flow, at low nozzle radii r_f. From Figures 3&4, we can see an increase in variance at larger values of pulmonary flow and SVC pressure. This indicates a sensitivity to design parameters at larger values of nozzle radius, which implies an interaction between the nozzle radius and the other design parameters. Based on this data, a linearized model for effect of these parameters can be developed, which can be used towards designing shunts that achieve a target pulmonary flow rate for the ABG physiology. Alternatively, it can be used to find an upper bound on the nozzle radius based on a target SVC pressure. These trends also indicate that a tight coupling between pulmonary flow and SVC pressure. Hence, in general, there is a trade-off between increasing pulmonary flow and decreasing SVC pressure.

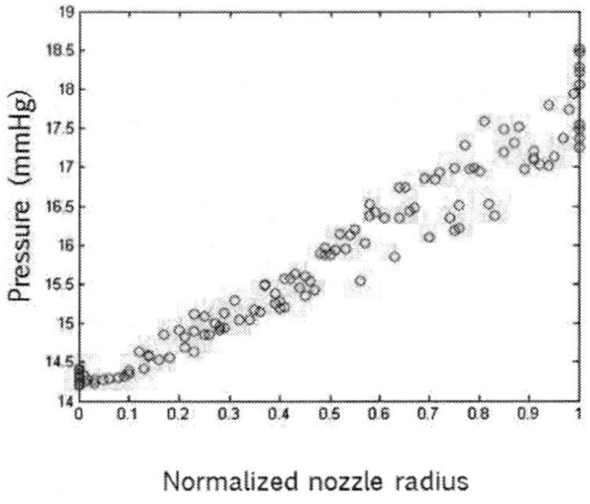

Figure 3: Variation of SVC pressure with the nozzle radius r_f

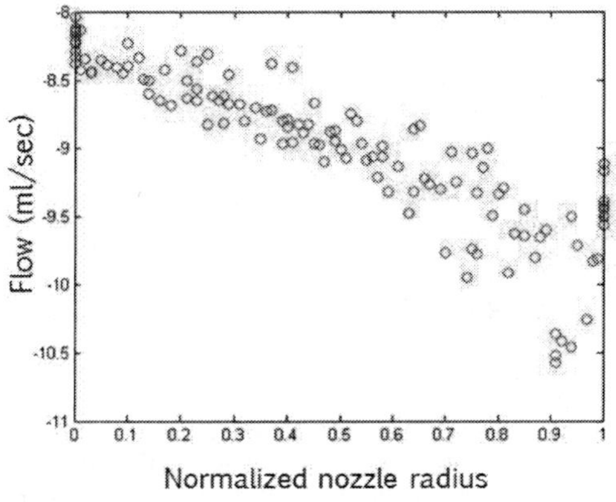

Figure 4: Pulmonary flow with the nozzle radius r_f

ACKNOWLEDGEMENTS

This work was supported by a Burroughs Wellcome Fund CASI Award, an NSF Career Award (OCI1150184) and a Leducq Foundation Network of Excellence.

REFERENCES

[1] Tamisier, D et al, *Ann Thorac Surg*, 49: 797-801, 1990.

[2] Norwood,WI et al, *J Thorac Cardiovasc Surg*, 82(4):511-9, 1981.

[3] Esmaily-Moghadam, M et al., *J Thorac Cardiovasc Surg*, 149(3):699-705, 2015.

[4] Updegrove, A. et al., *Ann Biomed Eng*, 1-17, 2016.

[5] Migliavacca, F, *Am J Physiol Heart Circ Physiol*, 280: 2076-86, 2001.

[6] Booker, AJ, *Boeing Tech Rep*, M&CT-TECH-00-002, 2000.

SB^{3}C2017
Summer Biomechanics, Bioengineering and Biotransport Conference
June 21 – 24, Tucson, AZ, USA

HEMODYNAMICS CONSEQUENCES OF DIFFERENT DESIGNS OF THE MONALSA STENT GRAFT

R. Tricarico (1), Y. He (2,3), R. Tran-Son-Tay (1,4), S. Scali (2.3), T. Lee (3), S. A. Berceli (1,2,3)

(1) Department of Biomedical Engineering
University of Florida,
Gainesville, FL

(2) North Florida/South Georgia Veterans
Health System
University of Florida,
Gainesville, FL

(3) Department of Surgery
University of Florida,
Gainesville, FL

(4) Department of Mechanical and Aerospace
Engineering
University of Florida,
Gainesville, FL

INTRODUCTION

Aortic aneurysms are degenerative diseases with multifactorial pathogenesis that claim more than 15,000 lives every year in the United States alone. Since the mid-nineties[1], endovascular procedures have been increasingly utilized in clinical practice for treating similar pathologies, offering a minimally invasive solution, and a shorter patient recovery[2,3]. In cases of aneurysm involving the aortic arch, advanced thoracic endovascular repairs (adTEVAR) procedures can be performed. AdTEVAR techniques aim to preserve blood perfusion into one or more aortic arch branches, while excluding diseased area from the aortic lumen.

Nowadays, an increasing number of stent devices are becoming available for adTEVAR in high-risk cases, and stent graft designs along with their endovascular delivery systems are becoming technologically more sophisticated. However, adTEVAR procedures are still fairly complicated, and highly specialized surgical skills are required. Furthermore, the impact of different stent graft designs on the aortic hemodynamics is still under investigation.

The objective of this study is to computationally investigate the hemodynamics of different configurations of one of the newest devices that is approaching the market. The Valiant MonaLSA device (Medtronic Inc., Dublin, Ireland) consists of an aortic stent graft with a manufactured funnel-shape cuff at level of the left subclavian artery (LSA) orifice which allows the perfusion of the left extremity, through a branch stent graft deployed inside the LSA. We perform *in-silico* modifications on a post-operatively patient-specific 3D aortic geometry treated with MonaLSA device, to analyze how geometrical characteristics affect the arterial hemodynamics.

METHODS

A post-operative aortic geometry of a patient treated with MonaLSA device, including the aortic arch, main aortic branches, and vertebral arteries, was extracted from computed tomography (CT) scan through segmentation in Mimics (Materialize, Leuven, Belgium).
In-silico modifications, on the portion representing the stent graft placed inside the LSA, were performed to recreate various stent graft designs and placements (**Figure 1**) in Ansys Spaceclaim (Ansys Inc., Canonsburg, PA). Stent graft length, diameter, curvature, and orifice misalignment were among the modified parameters, a total of 9 configurations was analyzed.

The surfaces of the nine models were smoothed, and their outlets extended. Tetrahedral meshes were used for the computational fluid dynamic (CFD) simulations in Ansys Fluent. Stationary and no-slip conditions at the walls were imposed; a pulsatile profile of blood flow, extracted from literature[4] and scaled based on the patient's estimated cardiac output[5] (4.2 L/min), was imposed at the inlet and implemented to obtain the solution of the transient simulations under laminar flow conditions. Patient-specific flow rate waveforms ($T = 0.88$ s) at the branches, and pressure waveform at the descending aorta were imposed as outlet boundary conditions.

Cross-sectional areas (CSA), time-average and maximum wall shear stress (aveWSS and maxWSS), time-average of WSS gradient (aveWSSG), oscillatory shear index (OSI), and relative resident time (RRT) have been analyzed. Hemodynamic differences between the modified designs and optimal stent graft deployment along the stented portion of the LSA have been investigated.

RESULTS

Four configurations were obtained by modifying the stent graft length, and four by variating the stent graft curvature; one configuration with LSA orifice misalignment was also tested (**Figure 1**). The analyzed tract of the LSA was 160 mm long with an average diameter of 7 mm; a stent graft of 12 mm of diameter was deployed inside the first 30 mm of the artery during adTEVAR (**Figure 2**). The configuration with the

Poster Presentation #P231

shortest stent graft design (*0L*) represents the optimal deployment configuration, and has been used as baseline for the calculations of the variables' differences. **Figure 3** shows the maximum differences from baseline of the WSS-related variables for each stent graft design on the stented portion of the LSA (30 mm). In particular, most of the configurations presents both maxWSS and/or aveWSSG values lower than the optimal deployment (**Figure 3**, **a** and **b**). The curved designs of 60 degrees, in both forward and backward configurations, show the lowest maxWSS compared to the optimal deployment. On the other hand, all straight designs show peaks of OSI higher than the optimal deployment (**Figure 3, c**). Lastly, the configurations with the highest RRT differences from the optimal deployment of the stent graft, have curved designs (*30B, 60F/B*), while the all straight stent graft designs maintained lower difference ranges (**Figure 3, d**).

Worthy of attention are the first 5-15 mm of stent graft length deployed inside the aortic arch, which have been omitted in the previous analysis due to shorter length of the baseline. However, the highest values of WSS have been observed on this area of the stent, as shown in **Figure 4** (LSA length < 0).

On the native vessel (LSA length > 30 mm), all variables show negligible variation, similarly to the maxWSS (**Figure 4, a**), except for the RRT, which ranges change consistently depending on stent graft design (**Figure 4, b**).

The mean (M$\pm$*SD*) of aveWSS on the entire LSA length is 1.26$\pm$0.8 Pa, with variation of less than $\pm$0.1 Pa among stent graft designs. On the other hand, the hemodynamic changes among different configurations on the entire length of bilateral carotid and vertebral artery as well as right subclavian result negligible. The longitudinal means of aveWSS on the carotid and the vertebral arteries are in the range of 2.2-2.8 Pa, about 3-fold lower than the ones on the right subclavian artery (6.5$\pm$4.2 Pa).

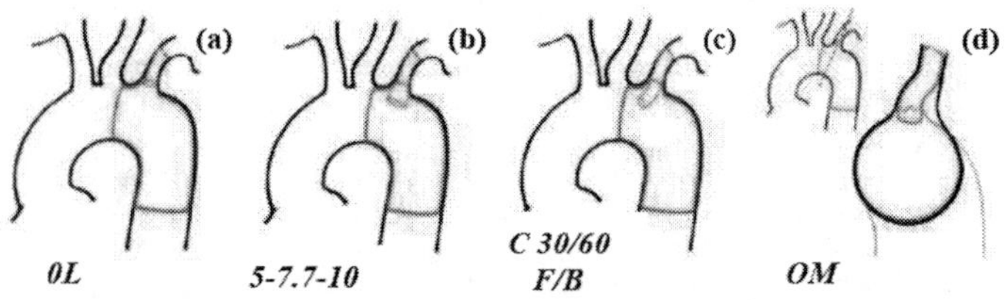

Figure 1. Aortic configurations obtained by in-silico modifications of a MonaLSA device (a, 0L). Variation of the length of the branch stent graft have been applied (5 mm (*5L*), 7.7 mm (*7.7L*), 10 mm (*10L*)) to obtain different downward facing stent graft designs (b). The stent graft curvature has been modified (c) with 30 and 60 degrees in forward and backward directions (*C30/60 F/B*). An orifice misalignment (*OM*) of the aortic stent graft has been recreated (d) with 30 degrees of rotation to the left.

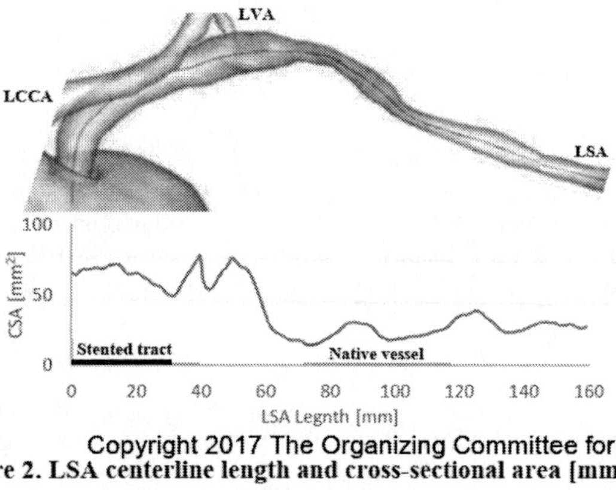

Figure 2. LSA centerline length and cross-sectional area [mm²].

DISCUSSION

To the best of our knowledge, this is the first computational modeling of realistic stent graft configurations in a patient-specific aortic arch. We observed that different stent graft designs as well as stent graft suboptimal placement influence the *in-situ* hemodynamics of adTEVAR. Among the analyzed designs, the C60F and C60B presented the highest RRT compared to all other configurations, suggesting potential abnormal flow as result of low and oscillatory shear. Larger residence time of the blood particles along the arterial wall may cause leucocyte activation and lead to higher risk of atherosclerosis. Analyses are ongoing; pressure drops along the LSA, and WSS-related variables on the aortic wall area around the LSA stent orifice are being investigated.

Among the limitations, mechanical characteristics of the stent graft and its interaction with the blood flow were neglected. Moreover, we hypothesized standard boundary conditions for the various stent graft configurations. Further investigation is needed to confirm our findings.

ACKNOWLEDGEMENTS

Kind acknowledgments to Medtronic Inc. for granting our research.

REFERENCES

[1] DUKE, *N Engl J Med* 331:1729–34, 1994.
[2] Bavaria et al., *J Thorac Cardiovasc Surg* 133 (2): 369–377 2007.
[3] Scali et al., *J Vasc Surg*, 53 (6): 1499–1505, 2011.
[4] Gallo, D. et al., *Ann Biomed Eng* 40 (3): 729–41, 2011.
[5] DeSimone et al., *Circulation* 95 (7): 1837–43, 1997.

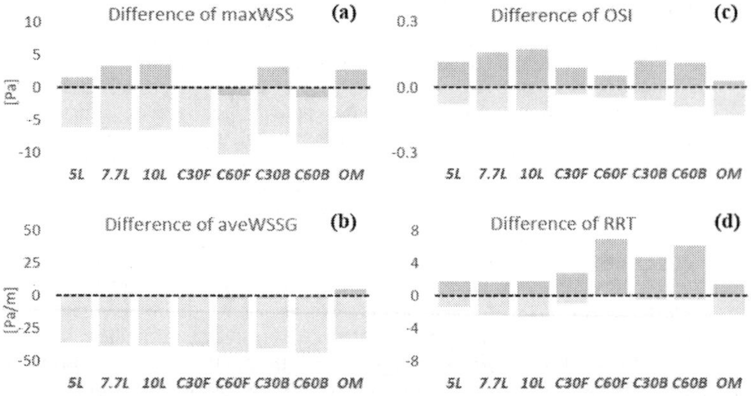

Figure 3. Hemodynamic analysis. Maximum differences between each stent graft design and the *0L* configuration in terms of maxWSS (a), average WSSG (b), OSI (c), and RRT (d) along the stented LSA. Positive differences indicate larger values on the configuration of interest compared to the *0L*, vice versa, negative differences refer to larger values on the *0L* design compared to the configuration of interest.

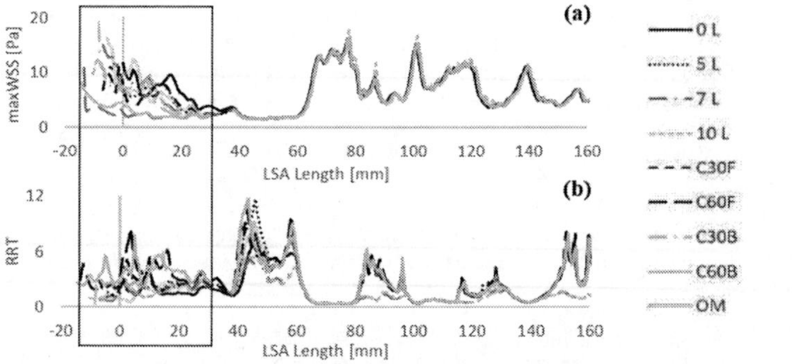

Figure 4. Absolute values of maximum WSS (a) and RRT (b) of the nine stent graft designs along the LSA length. Negative values are inside the aortic lumen. Values larger than 30 mm represent the native artery.

NONLINEAR DYNAMICS OF DACRON AORTIC PROSTHESES CONVEYING PULSATILE FLOW

E. Tubaldi, M. Amabili, M.P. Paidoussis

Department of Mechanical Engineering,
McGill University
Montreal, Quebec, Canada

INTRODUCTION

Dacron and ePTFE represent the standard materials for large-diameter vascular grafts used in various circumstances of vascular maladies requiring replacements of components of the cardiovascular system, such as vessel patches for aneurysms. Synthetic grafts function well in this high-flow, low resistance circuits with high long-term patencies. The main disadvantages of such grafts are reduced compliance and reduced tissue incorporation. The compliance mismatch between arteries and grafts can cause flow disruption, increase the risk of thrombosis and it is known to reduce graft patency. Wide knowledge about the distinctly different mechanical properties of the Dacron implants with respect to the native aorta is available in literature while very little is known about the dynamic behavior of these prostheses.

Artificial blood vessels can be modeled as thin-walled shells conveying pulsating flow. In this study, a Dacron thoracic aortic replacement is modeled as an orthotropic cylindrical shell and geometrically nonlinear vibration response to pulsatile flow is here presented via frequency-response curves in the physiological range. It is shown that during exercise (i.e. high pulsatile flow velocity and high heart rate) the frequency-response presents several superharmonic resonance peaks. In the limit case of low modal damping values, the prosthesis can lose stability with deformation in the cross-section by ovalization modes of the shell (shell-like buckling). In this deformed condition, high stress localized regions appear in the inner wall and flow separation or turbulence can develop eventually causing the failure of the prosthesis.

METHODS

In this study, Dacron vascular prostheses currently used in thoracic aortic replacement surgery are modeled as orthotropic linearly elastic shells via the nonlinear Novozhilov shell theory. As shown in Fig. 1, the circular cylindrical shell under consideration is characterized by a mean radius R, a shell thickness h and a shell length L. Flexible constraints to simulate connection with the remaining tissue, are applied at the shell ends [1].

In deformable shells conveying pulsatile flow, oscillatory pressure changes cause local movements of the fluid and deformation of the shell wall, which propagate downstream in the form of a wave.

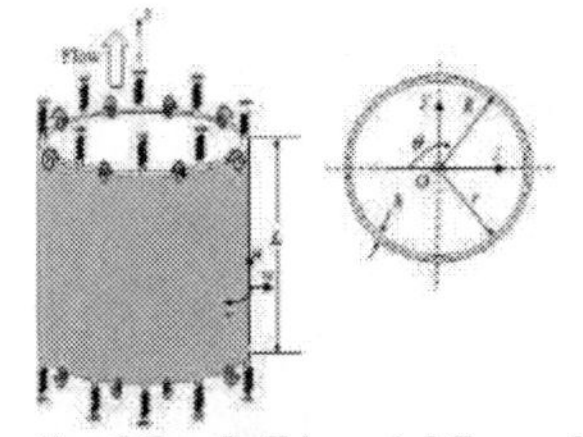

Figure 1: Schematic of the shell in axial flow with boundary conditions at the shell ends.

Pressure and flow velocity distributions within the shell are oscillatory both in space and time and they can be represented by a Fourier series as follows:

$$U(x,t) = \bar{U} + \sum_{n=1}^{N}\left(a_{v,n}\cos\left(n\,\Omega\,(t - x/c_0)\right) + b_{v,n}\sin\left(n\,\Omega\,(t - x/c_0)\right)\right), \qquad (1)$$

$$p_m(x,t) = \bar{p}_m + \sum_{n=1}^{N}\left(a_{p,n}\cos\left(n\,\Omega\,(t - x/c_0)\right) + b_{p,n}\sin\left(n\,\Omega\,(t - x/c_0)\right)\right), \qquad (2)$$

where Ω is the heart rate, $\bar{p}_m$ and $\bar{U}$ represent the steady component of the pressure field and the pulsatile mean flow velocity, respectively, and N represents the number of terms in the series expansion. The coefficients a_n and b_n are the Fourier cosine and sine coefficients, respectively. The physiological waveforms of velocity and pressure [2] in the thoracic aorta are approximated with the first eight harmonics ($N=8$) of the Fourier expansion, as shown in Fig. 2.

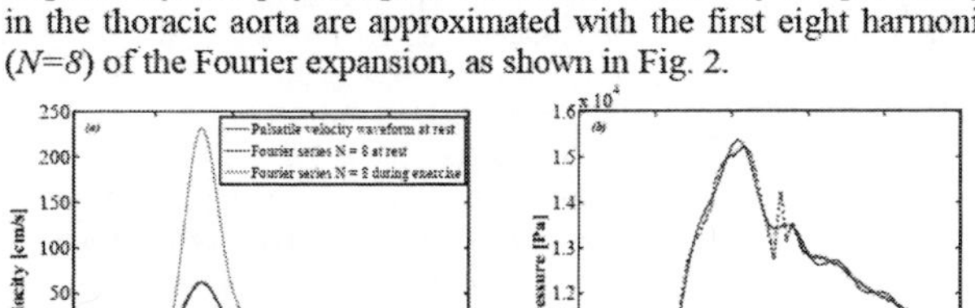

Figure 2: (a) Flow velocity at rest and during exercise and (b) transmural pressure values in the Dacron prosthesis.

If the wall thickness is small compared with the shell radius and if the effects of viscosity can be neglected, the wave speed c_0 is given approximately by the so called Moens-Korteweg formula.

The fluid is modeled as a Newtonian pulsatile flow, and it is formulated using a hybrid model that contains the unsteady effects obtained from the linear potential flow theory and the pulsatile viscous effects obtained from the unsteady time-averaged Navier-Stokes equations [1]. Coupled fluid-structure Lagrange equations for a non-material volume with wave propagation in case of pulsatile flow have been applied [3].

RESULTS

Direct integration of the equations of motion by means of IMSL DIVPAG Fortran routine, which uses the Adams-Gear integration scheme, has been used to solve the nonlinear equations of motion. The mechanical properties and the characteristics of the vessel and of the fluid here studied are: L = 0.18 m, h = 0.361 mm, R = 0.015 m, Poisson ratio $\nu_{\theta x}$ = 0.3, E_x = 0.87 MPa, E_θ = 12 MPa, Young moduli in x and θ direction, respectively, mass density of the shell ρ_S = 1247 kg/m^3, fluid density ρ_F = 1050 kg/m^3. The time response of the Dacron implant subjected to the physiological waveforms of velocity and pressure during the heart beating period at rest and in exercise conditions is presented in Fig. 3; the shell radial displacement w, non-dimensionalized with respect to the thickness h, is shown at the angular coordinate $\theta = 0$ versus time and at three different axial positions x=L/4, L/2, 3L/4. The shell transverse sections are circular at any point since only axisymmetric modes are activated in this case.

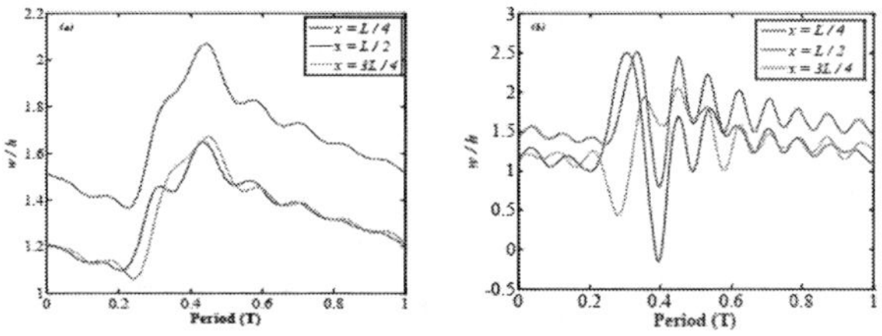

Figure 3: Time response (a) at rest (60 bpm) and (b) in exercise conditions (180 bpm).

The time response at 180 bpm shows the presence of high frequencies while the response at rest reproduces the pressure behavior. Fig. 4 presents the maximum amplitude of the response versus the heart rate for the most significant axisymmetric modes $w_{1,0}/h$ and $w_{2,0}/h$ and asymmetric modes $w_{1,2}/h$ and $w_{2,2}/h$ (non-dimensionalized with respect to the thickness h of the shell) obtained by increasing the pulsating frequency for different damping values ζ. When the vessel is pressurized with the steady transmural pressure $\bar{p}_m$ = 12253 Pa, which is the mean value obtained in Fig. 2(b), the fundamental mode is the first axisymmetric mode whose frequency is $\omega_{1,0}$ = 35.75 Hz (notation: $\omega_{m,n}$ where m is the number of longitudinal half-waves, n is the number of circumferential waves). Resonance peaks shown in Fig. 4 correspond to the superharmonic response of the system under harmonic excitation with frequency $\Omega = \omega_{1,0}/k$ where k = 2, 3,.., 2N. Resonant peaks for different values of damping ζ correspond to a circumferential dilation of the prosthesis that can compromise its long term function. It is well known that graft dilation is one of the main reasons of graft failure since it can induce yarn breakages and eventually tears. The sudden change in dimeter of the circular cross-section for different heart rates represented by the resonant peaks may also provoke fluctuations in the shear stress and disturbed blood flow. As shown in Fig. (4c-d), for the limit case of damping ζ = 0.01 and high heart rates, the prosthesis loses stability by buckling with deformation of the cross-section compromising its proper functioning. In this case, chaos regime is detected (several dots correspond to every frequency in the bifurcation diagrams).

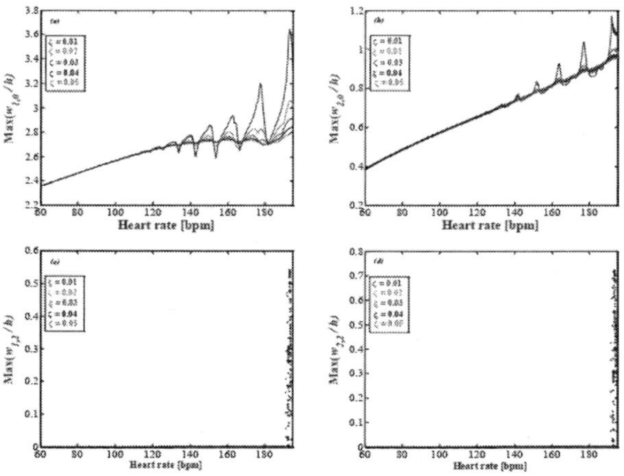

Figure 4: Frequency response of the modes (a) $w_{1,0}/h$ (b) $w_{2,0}/h$ (c) $w_{1,2}/h$ (d) $w_{2,2}/h$ for different damping values ζ.

DISCUSSION

This study represents one of the first researches providing a deep insight in the dynamic behavior of textile structures used in today vascular prosthetic grafts. Higher harmonics contributions are observed in the time responses at 180 bpm. Frequency-responses show resonance peaks due to the superharmonics of the pulsatile flow excitation. This can facilitate the graft dilative characteristic and disturb the flow. Different damping values ζ are considered since experimental characterization of this parameter has never been addressed in literature. The authors are currently involved in the experimental identification of the damping value of Dacron prosthesis used for thoracic aortic replacement. For the limit case of ζ = 0.01, flow-induced buckling of the aortic prosthesis is possible. This would induce high stress regions which, combined with the fatigue cycles of the heart beats, could contribute to the initiation of a rupture. With respect to the limitations of the present work, a more accurate structural model should include the viscoelasticity of the graft material. Despite the limitations, this research represents the first study of nonlinear vibrations of aortic replacements and it provides a substantial progress in the understanding of the dynamics of installed implants. Making surgeons aware of the effects in the differences between dynamical behaviors of prostheses with respect to the human arteries can aid in surgical decision making and eventually inspire the design of new materials or techniques for the fabrication of a next generation prostheses.

ACKNOWLEDGEMENTS

The authors would like to thank NSERC Discovery Grant program, Qatar Foundation grant NPRP 7-032-2-016 and the Canada Research Chair program for their financial support.

REFERENCES

[1] Amabili, M et al, A three-layer model for buckling of a human aortic segment under specific flow-pressure conditions, International Journal for Numerical Methods in Biomedical Engineering, 28 (2012) 495-512.

[2] Mills, C et al, Pressure-flow relationships and vascular impedance in man, Cardiovascular Research, (1970) 405-417.

[3] Tubaldi, E et al, Nonlinear response of shells conveying pulsatile flow, in: XXIV ICTAM, Montreal, 2016.

SB[3]C2017
Summer Biomechanics, Bioengineering and Biotransport Conference
June 21 – 24, Tucson, AZ, USA

A NOVEL MODELING APPROACH TO QUANTIFY CORONARY PERFUSION AFTER TRANSCATHETER AORTIC VALVE REPLACEMENT

Harkamaljot S. Kandail (1), Setu Trivedi (2), John F. LaDisa (1, 3)

(1) Department of Biomedical Engineering
Marquette University and Medical College of Wisconsin, Milwaukee, WI, USA

(2) Aurora St. Luke's Medical Center
Aurora Healthcare
Milwaukee, WI, USA

(3) Departments of Medicine and Physiology
Medical College of Wisconsin
Milwaukee, WI, USA

INTRODUCTION

The advent of minimally-invasive transcatheter aortic valve replacement (TAVR) procedures now enables high risk patients with aortic stenosis to be treated endovascularly. During TAVR, a bioprosthetic valve is guided fluoroscopically to the diseased native aortic valve and once deployed, the tissue of the bioprosthetic valve assumes the workload of the native valve by pushing it against the arterial wall. One of the most commonly used bioprosthetic devices for TAVR procedures is CoreValve and this device shows promising post-operative results. However, there is a paucity of detailed knowledge regarding post-TAVR coronary hemodynamics in patients treated with this device. The objective of the current research is to address this gap in the literature by simulating patient-specific aortic and coronary hemodynamics in subjects who underwent TAVR with the CoreValve. While important related work has been conducted [1, 2, 3], novelty is derived in the current work through the use of an advanced fluid structure interaction (FSI) approach that has not previously been applied with TAVR.

METHODS

Patient-specific aortic geometries were obtained using the CRIMSON open-source software package (Cardiovascular Integrated Modeling and Simulation, available at http://www.crimson.software/). An accurate model of the 29 mm CoreValve device, which is the most commonly used device for the TAVR procedure was created using a commercially available computer aided design (CAD) software (Solidworks; Dassault Systèmes, Vélizy-Villacoublay, France). The CAD model was then meshed, crimped and released into patient-specific aortic geometries using Abaqus/Explicit (Dassault Systèmes, Vélizy-Villacoublay, France). The porcine tissue of the CoreValve was modelled as a linear elastic material with Young's modulus of 8 MPa and Poisson's ratio of 0.45 [2, 3]. CoreValve's Nitinol frame was modelled as a super elastic shape memory alloy using VUMAT subroutine with material properties obtained from Prasad et al. [4]. Contact between model parts was defined according to the robust general contact algorithm of Abaqus/Explicit which enforces the contact constraints using a penalty based contact method. In order to reduce the computational time associated with the deployment simulations, a variable mass scaling technique was employed. The kinetic and internal energies were also monitored throughout the deployment simulation to keep their ratio below 5% in order to ensure a quasi-static simulation.

Once deployed, the deformed geometry of the CoreValve device was extracted and FSI flow simulations were performed. Valve leaflets undergo extreme deformation during a cardiac cycle. This limits the applicability of traditional FSI approaches such as arbitrary Lagrangian Eulerian and immersed boundary methods due to issues associated with re-meshing and loss of void, respectively [3]. As a result, a novel FSI procedure called Smoothed Particle Hydrodynamics (SPH) was employed. SPH is an advanced meshless Lagrangian method where the fluid domain is modelled as a set of particles instead of a conventional mesh. Briefly, the individual physical properties of each SPH particle, A, are interpolated using the properties of the neighboring particles through a kernel function, W, as shown below:

$$A(\mathbf{r}_a) = \sum_b m_b \frac{A_b}{\rho_b} W(\mathbf{r}_a - \mathbf{r}_b, h) \quad (1)$$

where subscripts a and b refer to neighboring particles 'a' and 'b' at position $\mathbf{r}_a$ and $\mathbf{r}_b$, respectively while h refers to the smoothing radius. Particle 'b' has mass m_b and density ρ_b and, using equation (1), mass and momentum conservation equations can be written as equations 2 and 3, respectively.

$$\frac{d\rho_a}{dt} = \sum_b m_b \mathbf{V}_{ab} \cdot \nabla_a W_{ab} \tag{2}$$

where $\mathbf{V}_{ab} = \mathbf{V}_a - \mathbf{V}_b$ denotes the relative velocity vector between particles 'a' and 'b' and $\nabla_a W_{ab}$ represents the gradient of the kernel function.

$$\frac{d\mathbf{V}_a}{dt} = -\sum_b m_b \left(\frac{p_a + p_b}{\rho_a \rho_b} \right) \nabla_a W_{ab} + \sum_b m_b \frac{(\mu_a + \mu_b)\mathbf{V}_{ab}}{\rho_a \rho_b \mathbf{r}_{ab}^2} . \nabla_a W_{ab} \tag{3}$$

where p and μ are pressure and dynamic viscosity of the blood (0.004 Pa s), respectively, and the particle movement is governed by:

$$\frac{d\mathbf{r}_a}{dt} = \mathbf{V}_a \tag{4}$$

Finally, the fluid pressure and density are coupled using the linear $U_s - U_p$ Hugoniot form of the Mie-Grüneisen equation of state:

$$p = \rho_0 c_0^2 \mu \tag{5}$$

The artificial speed of sound c was assumed to be 145 m/s to maintain the incompressible behavior of blood flow [5] while the reference blood density was assumed to be 1060 kg/m^3. Left ventricle, coronary and aorta pressure waveforms were applied as boundary conditions for the SPH simulations. These pressure waveforms were obtained by running a conventional closed-loop flow simulation using CRIMSON [6]. All SPH simulations were performed again using Abaqus/Explicit while post-processing of simulation results was performed using Abaqus/CAE and Ensight (CEI; Apex, NC, USA).

RESULTS

Preliminary results from work conducted with SPH simulations to date suggest that if the CoreValve device is deployed at the optimum landing zone (as specified by the manufacturer), it can restore normal physiologic coronary perfusion by channeling approximately 5% of the cardiac output into the left and right coronary arteries (under resting conditions). Clinical studies suggest there is a considerable decrease coronary perfusion (only 2.5-3% of the cardiac output) when the device was implanted too high [7]. The implications of these changes in flow and resulting patterns of wall shear stress (WSS) are now being studied using the current SPH approach. Clinical studies have also shown that physiologic time variance of coronary flow rate waveforms were observed, with the majority of coronary perfusion occurring during diastole when the left ventricle was relaxing. These details are also now being incorporated using the current methods. Due to the supra-annular nature of the valve leaflets in CoreValve, indices of WSS such as time-averaged WSS are expected to differ as compared to forthcoming normal subjects.

DISCUSSION

The current work showcases the utility of the SPH method to perform FSI simulations in scenarios where the fluid domain undergoes extreme deformation. Preliminary results from this research using the SPH approach indicate that if deployed correctly, CoreValve may alleviate the adverse effects of aortic stenosis. Although potentially rare, recent clinical studies have demonstrated decreased coronary blood flow in subjects with improperly deployed CoreValve, i.e. the device deployed too high. The local hemodynamics implications of such scenarios will be revealed by completion of the current work. The curvature of the aortic root helps to guide the blood to the coronary ostia during diastole when there is flow reversal in the aorta. As a result, if less blood is directed to the aortic root during diastole, the coronary perfusion will also reduce, and this is manifested in CoreValve devices that are implanted too high. This will be also be verified computationally in the forthcoming simulations. While only resting scenarios were simulated to date, the efficacy of CoreValve will also be analyzed during exercise conditions at three different heart rates [8] in this ongoing research.

ACKNOWLEDGEMENTS

This research project is funded by a grant from the Aurora Research Institute and the Marquette University Strategic Innovation Fund.

REFERENCES

[1] Tan FPP, Xu XY, Torii R, et al. Comparison of Aortic Flow Patterns Before and After Transcatheter Aortic Valve Implantation. *Cardiovasc Eng Technol.* 3:123-135, 2012.

[2] Wu W, Pott D, Mazza B, et al. Fluid-structure interaction model of a percutaneous aortic valve: Comparison with in vitro and feasibility study in a patient-specific case. *Ann Biomed Eng*, 44:590-603, 2016.

[3] Bavo AM, Rocatello G, Iannaccone F, et al. Fluid-structure interaction simulation of prosthetic aortic valves: Comparison between immersed boundary and arbitrary Lagrangian-Eulerian techniques for the mesh representation. *PLoS ONE*, 11:e0154517, 2016.

[4] Prasad A, Xiao N, Gong XY, et al. A computational framework for investigating the positional stability of aortic endografts. *Biomech Model Mechanobiol*, 12:869-887, 2013.

[5] Mao W, Li K and Sun W. Fluid-structure interaction study of transcatheter aortic valve dynamics using smoothed particle hydrodynamics. *Cardiovasc Eng Technol*, 7:374-388, 2016.

[6] Kim HJ, Vignon-Clementel IE, Coogan JS, et al. Patient-specific modeling of blood flow and pressure in human coronary arteries. *Ann Biomed Eng*, 38:3195-3209, 2010.

[7] Laborde JC, Brecker SJD, Roy D, et al. Complications at the Time of Transcatheter Aortic Valve Implantation. *Methodist Debakey Cardiovasc* J, 8:38-41, 2012.

[8] Ellwein L, Samyn MM, Danduran M, et al. Toward translating near-infrared spectroscopy oxygen saturation data for the non-invasive prediction of spatial and temporal hemodynamics during exercise. *Biomech Model Mechanobiol*, doi:10.1007/s10237-016-0803-4, 2016.

PREDICTING CALCIFIC AORTIC VALVE DISEASE PROGRESSION AND ITS EFFECT ON TRANSCATHETER AORTIC VALVE DEPLOYMENT IN BICUSPID VALVES

Gil Marom (1), Karin Lavon (2), Matteo Bianchi (1), Rotem Halevi (2), Ashraf Hamdan (3), Ehud Raanani (3), Rami Haj-Ali (1,2), Danny Bluestein (1)

(1) Department of Biomedical Engineering
Stony Brook University
Stony Brook, NY, USA

(2) School of Mechanical Engineering
Tel Aviv University
Tel Aviv, Israel

(3) Department of Cardio- thoracic Surgery
Chaim Sheba Medical Center
Tel Hashomer, Israel

INTRODUCTION

Calcific aortic valve disease (CAVD) is a formation of tissue similar to bone on the leaflets of the aortic valve (AV). The presence of such calcifications rapidly leads to aortic stenosis (AS), a condition which requires an early intervention; however, patients usually do not present symptoms until the disease has progressed to an advanced stage [2]. It is therefore rare to have a CT scan acquired at the early stages of the disease. Bicuspid Aortic Valve (BAV) is the most common type of congenital heart disease where the valve has only two leaflets rather than the normal three. The most common configuration of BAV, type 1, includes one fused leaflet with two fused segments. BAV is associated with secondary pathologies of the valve and the aorta, such as aortic regurgitation, AS, and ascending aorta dilatation [3]. The progression of CAVD leading to AS in BAV patients is accelerated compared to TAV patients [4]. Transcatheter aortic valve replacement (TAVR) is a relatively new procedure that provides an alternative to open-heart surgical replacement for inoperable patients with severe AS. In this minimally invasive intervention, a stent with a mounted bioprosthetic valve is delivered through the arteries and deployed on the stenotic native valve. Currently, the latest generation FDA-approved TAVR devices are the Edwards Lifesciences Sapien 3 and Medtronic Corevalve Evolut R. In recent years, more centers around the world started to perform TAVR also in BAV patients which raise concerns stemming from the BAV asymmetrical structure and its elliptical annulus and valve orifice. The outcomes of TAVR in BAVs are also highly dependent on the type of device, namely Edwards Sapien valve might require post-implantation balloon dilation to minimize paravalvular leakage (PVL) [5, 6] while Medtronic CoreValve has higher occurrence of post-implantation central regurgitation [6].

CAVD has been studied using numerical tools to investigate the mechanical stress, hemodynamics and kinematics of the valve. One of the main use for patient-specific modeling of CAVD is TAVR procedural planning [7] with studies examining prosthesis deployment and the device design. However, previous TAVR studies did not account for the progression of the disease. Our group studied BAV with Fluid Structure Interaction (FSI) methods [8] to compare tricuspid valve with three morphologies of BAV while other groups also modeled BAVs for similar aims.

Detailed information on CAVD progression is currently limited because the disease is diagnosed during its advanced stages. We aim to expand our methods of retrospectively calcification growth evaluation techniques [1] to BAVs and to assess the deployment of the latest generation TAVR devices, both balloon and self-expandable, with numerical models. The PVLs will be compared using hemodynamics models of the diastole in the resulted deployed configurations.

METHODS

CT scans of pre-TAVR patient with BAV were collected from our existing database of CAVD scans. A severe AS patient was chosen because of the heavily calcified raphe region, contributing to the assumed TAVR elliptical deployment. Our Reverse Calcification Technique (RCT) was employed for this BAV patient in a similar manner to our previously suggested RCT for TAVs [1]. This technique is based on using pre-TAVR CT scans of AVs to study the calcification progression that leads to the current state. Spatial gradients of the densities in the calcified regions are analyzed at the current late stage (identified from the HU range), and are utilized to predict the progression [1]. For TAV, this technique was validated by obtaining two scans from the same patient during both severe and mild stages (Fig. 1, left). Close agreement was found between the old scan and the RCT prediction from a recent scan.

Idealized geometry of the BAV type 1 anatomy was created for non-fused leaflet angle of 140° and symmetrical fused leaflet. For this

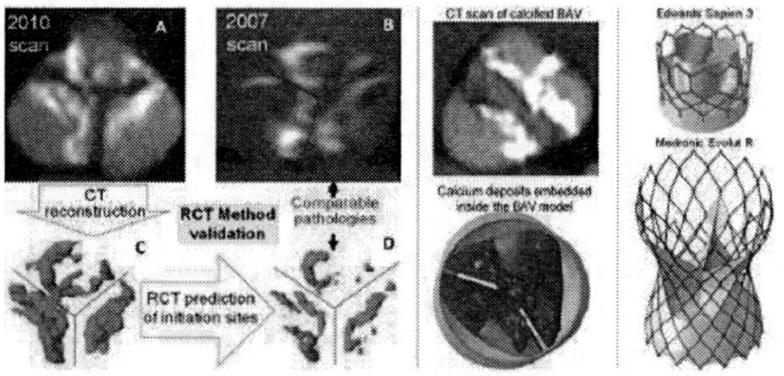

Figure 1: Left - A. 2010 and B. 2007 CT scans of the same patient; C. 3D image from the 2010 scan; D. RCT prediction similar to B [1]. **Center** - BAV CAVD patterns and their reconstructed deposits embedding in a parametric BAV geometry. **Right** - FE models of new generation TAVR valves.

purpose, our existing parametric geometry and mesh generation method [9] was modified for asymmetric BAVs. The fusion of the leaflets is defined as circular arcs that follow the leaflet's curvature and Bezier curve in the coaptation-belly connection. The native leaflets and root are meshed with shell elements in this model and have native tissue properties. The calcium deposits are tied to the leaflets and have calcium material properties (Fig. 1, center).

Deployment of the two TAVR devices, Sapien 3 and Evolut R (Fig. 1, right), are modeled in the CAVD-BAV anatomy. For Sapien 3, the leaflets and stent were drawn based on publicly available figures from Edwards. The 3D shape of the leaflets was calculated in finite elements (FE) by suturing them to the cuff. The cuff, that was also drawn based on public images, is necessary to estimate the leakage and connect the leaflets and the stent. The stent of the CoreValve was generated based on Bezier curves, resulted in a structured hexahedral mesh. Both stents have ~70,000 elements. The leaflets of the CoreValve were generated based on the stent mesh and meshed with shell elements. The cuffs in both models were modeled with membrane elements. The materials used to model the stents are MP35N alloy and purely superelastic NiTi for the Sapien 3 and Evolut R, respectively. The stents were initially crimped with a cylindrical crimper. The Evolut R deployment is a result of the residual stresses present in the stent after the crimping while gradually pulling the sleeve toward the aorta. The Sapien 3 is deployed by balloon (NovaFlex+) inflation as we previously used for first generation Sapien [10]. The FE solver is SIMULIA Abaqus (Dassault Systèmes, Providence, RI).

The FE analyses of the deployments will be used as predicative models of possible complications. The migration of the stents as a result of insufficient anchoring forces will be modeled by recoil phase after the deployment [10]. Post-TAVR diastolic regurgitation in BAVs is compared between Sapien 3 and Evolut R. The leaflets and cuffs are deformed to the final configuration of the deployed stents and the leaflets are then closed by applying transvalvular pressure. Capvidia FlowVison will be used for regurgitation estimation.

RESULTS

The RCT subtraction algorithm was commenced and used to generate various stages of the CAVD disease in BAV. The pre-TAVR stage (Fig. 1, center) was used as the input for this technique and the reverse configurations were generated until reaching the initiation nodule sites. The left column of Fig. 2 shows three of the calculated configurations. An initiation nodule of the calcification growth appears on the raphe region, a location that is subjected to higher stresses in healthy BAV type 1. The non-fused leaflet has similar arc shaped pattern as in TAVs [1] while the arcs are connected in the fused leaflet.

Preliminary deployment of the Evolut R stent in healthy BAV is shown in the right column of Fig. 2. The resulting deployed

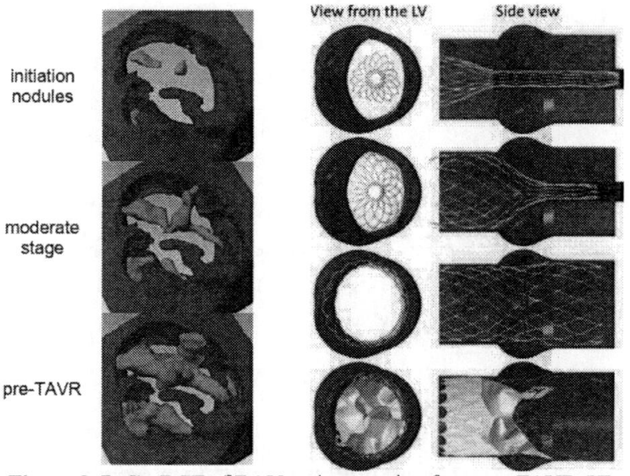

Figure 2: **Left** - RCT of BAV patient starting from pre-TAVR CT scan. **Right** - Evolut R stent deployment in BAV and the deformed prosthetic leaflets and cuff in the final configuration (bottom).

configuration was used to deform the leaflets and cuff to the final configuration in the same method that will be used to create the CFD domain. The lack of the calcifications is probably the reason for the more circular than expected deployment. This circular configuration enabled the bioprosthetic leaflets to be centrally closed. Detailed prediction of the leakage locations, including in the small gaps between the native leaflets and the cuff, will be found from the following CFD model.

DISCUSSION

In this study we expanded our previously published models of retrospective estimation of CAVD progression and parametric geometry of the aortic root from TAV to BAV morphologies. The RCT was able to generate calcific deposits geometries in various stages but further validation of this technique in BAVs is still required. Similarly, the parametric geometry algorithm needs to be compared with 3D Echo scans. The balloon expandable Sapien 3 assumed the circular configuration of the balloon while the Evolut R adapted to the elliptical shape of the BAV opening, leading to very different hemodynamic during the diastole. The circular Sapien 3 deployment in an elliptical opening may leave gaps between the stent and the root but the outer skirt can help in reducing the severity of the PVL. The highly elliptical deployment of the Evolut R in the calcified BAV might prevent proper coaptation between the bioprosthetic leaflets and a full closure but the 'fish mouth' opening of the BAV could be narrow enough to prevent it. Similar deployment models will be solved for previous stages of the disease to estimate the desired occasion for intervention.

ACKNOWLEDGEMENTS

This work was partially supported by grants from the Nicholas and Elizabeth Slezak Super Center for Cardiac Research and Biomedical Engineering at Tel Aviv University (RHA) and Simor Feffer (AH).

REFERENCES

[1] Halevi, R, et al., *J Biomech*, 48:489-97, 2015.
[2] Rosenhek, R, et al., *N Engl J Med*, 343:611-7, 2000.
[3] Cedars, A, and Braverman, AC, *Prog Pediatr Cardiol*, 34:91-96, 2012.
[4] Beppu, S, et al., *Am J Cardiol*, 71:322-7, 1993.
[5] Hayashida, K, et al., *Circ Cardiovasc Interv*, 6:284-91, 2013.
[6] Mylotte, D, et al., *J Am Coll Cardiol*, 64:2330-9, 2014.
[7] Capelli, C, et al., *Ann Biomed Eng*, 40:2663-73, 2012.
[8] Marom, G, et al., *Med Biol Eng Comput*, 51:839-48, 2013.
[9] Haj-Ali, R, et al., *J Biomech*, 45:2392-7, 2012.
[10] Bianchi, M, et al., *Artif Organs*, 40:E292-E304, 2016.

IMPLICATION OF FLOW DEPENDENCE OF VALVE AREA ON THE FORMULATION OF A SEVERITY INDEX FOR CALCIFIC AORTIC VALVE STENOSIS

Megan K. Heitkemper (1), Hoda Hatoum (1), Jennifer Dollery (2), Juan A. Crestenello (2), Lakshmi P. Dasi (1)

(1) Biomedical Engineering Department
The Ohio State University
Columbus, Ohio, USA

(2) Division of Cardiac Surgery
The Ohio State University
Columbus, Ohio, USA

INTRODUCTION

Cardiovascular disease is the leading cause of death in the United States and heart valve replacement, usually for aortic valve disease, is the second most common cardiac operation performed in the US [1]. The parameters most frequently used to assess the severity of aortic valve stenosis are the transvalvular pressure gradient (TPG) and the effective orifice area (EOA), which represents the minimum cross-sectional area of the jet downstream of the valve [2]. More recent indices [3] focus on net energy losses, however with the assumption that the effective orifice area is independent of flow.

Currently the drawback of using indices that inherently depend on flow to characterize the severity of aortic stenosis continues to plague the timing of interventions in patients, particularly those with low flow and low TPG. From a fundamental biomechanics standpoint, the dependence of valve area and pressure drop with flow, arises due to geometric as well as fluid dynamic factors corresponding to: (1) the variable opening of the valve due to the inherently flexible nature of the leaflets, which is different between severely calcified and non-calcified valves, and (2) the changing flow regimes with corresponding changes in Reynolds number through systole. Better characterizing and modeling these dependences in real calcified aortic valves can help formulate a superior index to assess the severity of aortic stenosis compared to current clinical standards.

Towards this goal, in this study we aim to investigate the influence of flow and cardiac output on the measures of EOA, pressure drop, and the actual aortic valve area for the cases of non-calcified and severely calcified aortic leaflets.

METHODS

The hemodynamic performances of a non-calcified 23 mm St. Jude Medical Trifecta aortic SAV and a 23 mm severely stenosed Carpentier-Edwards Perimount Magna Ease SAV extracted from a patient who underwent a redo aortic valve replacement were assessed in a left heart simulator flow loop under physiological pressure (120/80 mmHg) and heart rate (60bpm). Cardiac output of (1, 3, 5, and 7 L/min) was varied by changing the stroke volume of the pump. The flow loop consists of a reservoir tank, mitral valve, pump, valve holder with flow development region, systemic compliance chamber, and resistance valve. A 60/40 water/glycerin mixture by volume with density of 1090 Kg/m^3 and a kinematic viscosity of 3.88 cSt was used as the working fluid. The pressure and flow rate across the aortic valve were regulated by an in-house LabVIEW program. The flow rate was measured with an ultrasonic Transonic flowprobe located immediately upstream of the aortic valve, and the aortic and ventricular pressures were measured using Validyne pressure transducers. A high speed camera was used to record valve leaflet motion at 1000 fps throughout the cardiac cycle for each of the four chosen cardiac outputs. Time dependent EOA was computed for each case using Equation (1),

$$EOA(t) = \frac{Q(t)}{51.6\sqrt{|\Delta P(t)|}} \qquad (1)$$

and then normalized by the inflow annular area, $A_{Annular}$, of each valve. In addition, averaged EOA using the Gorlin equation and energy loss index (ELI) was also calculated. Time resolved particle image velocimetry was performed for all cases to capture the spatio-temporal evolution of the central jet throughout systole both qualitatively and quantitatively.

Poster Presentation #P235

RESULTS

The normalized average EOA with respect to $A_{Annular}$ as a function of cardiac output in L/min for the 23 mm stenosed SAV and the 23 mm non-calcified SAV is shown in Figure 1(a). The calcified valve had a normalized EOA that ranged from 0.112 at 1 L/min to 0.214 at 7 L/min, and the non-calcified valve ranged from 0.165 at 1 L/min to 0.582 at 7 L/min. High speed images of maximum valve openings during peak systole are shown in Figure 1(b), with cardiac outputs increasing from left to right.

(a)

(b)

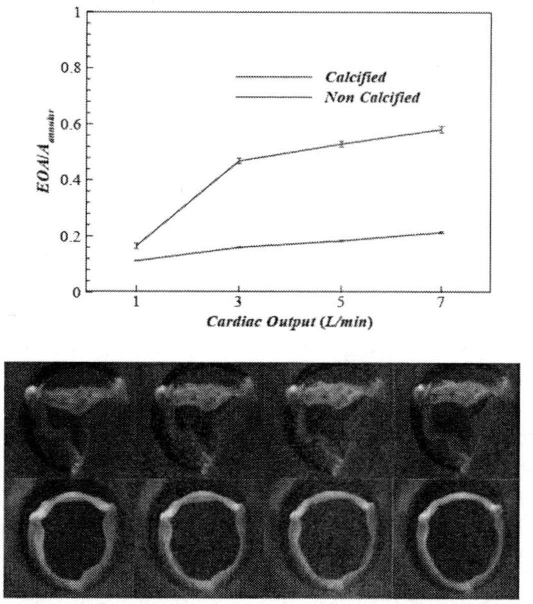

Figure 1: (a) EOA/$A_{Annular}$ vs. cardiac output in L/min for the calcified and the non-calcified valves (b) Top: High speed images of the calcified valve at peak systole for cardiac outputs of 1, 3, 5, and 7 L/min from left to right. Bottom: High speed images of the non-calcified valve at peak systole for cardiac outputs of 1, 3, 5, and 7 L/min from left to right.

The normalized instantaneous EOA (using equation 1) versus time is presented for each of the tested cardiac outputs using the calcified and non-calcified valves in Figures 2(a) and 2(b) respectively. The calcified valve shows less variation in EOA throughout systole than the non-calcified valve. The calcified valve, Figure 2(a), also shows that the EOA during peak systole at flowrate 7 L/min is almost double the EOA at the same point in time for the 1 L/min flow rate. Similarly, the peak EOAs for the non-calcified valve change by roughly 200% from the 1 L/min case to the 7 L/min case. We have also computed the ELI using methods described in [4] and have found that, as expected, ELI also strongly depends on flow.

DISCUSSION

This study illustrates that flow dependency characteristics of a non-calcified aortic valve are significantly different than that of a calcified aortic valve. While the two curves shown in Figure 1(a) may seem to be self-similar, this behavior is not supported by Figure 2.

(a)

(b)

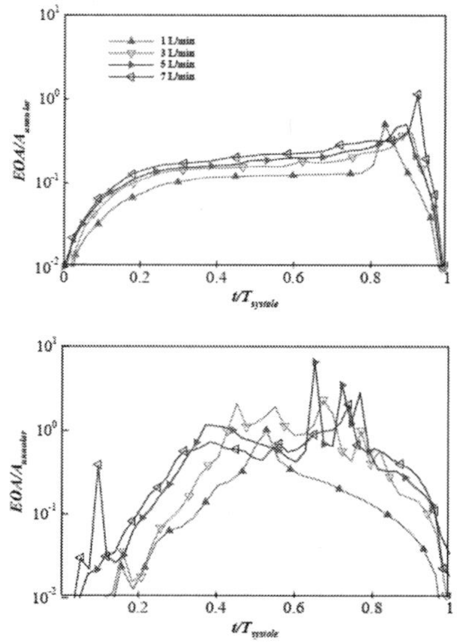

Figure 2: Normalized EOA as a function of time spent in systole for the (a) calcified valve and (b) non-calcified valve.

Instantaneous EOA behavior of the calcified valve follows a waveform shape distinct from that followed by the non-calcified aortic valve. In particular, the non-calcified aortic valve's instantaneous EOA fluctuates for a much larger portion of systole owing to self-excited oscillations between the flexible leaflets and the turbulent central jet. In contrast, there are no fluctuations in leaflets or the instantaneous EOA for the calcified valve. Instantaneous EOAs appear to produce an artificial boost just before valve closure due to an apparent reduction in instantaneous TPG due to flow deceleration. Nevertheless, the shape of the instantaneous EOA curves across the four cardiac outputs appears to follow some self-similar characteristics leading to the hypothesis that it is indeed possible to formulate a single flow independent index of severity for calcific aortic stenosis.

SUMMARY

This study introduces an attempt to formulate an index that better evaluates the severity of calcific aortic stenosis. Calcification of the SAV leads to a distinct pattern in instantaneous EOA over the systolic portion of the cardiac cycle when compared with the observations made with a non-calcified valve. The average EOAs for a calcified valve change significantly less than those of the non-calcified valve.

ACKNOWLEDGEMENTS

The research done was supported by National Institutes of Health (NIH) under Award Number R01HL119824.

REFERENCES

[1] Mozaffarian, D., *et al., Circulation,* 339-410, 2015.
[2] Shavelle, D.M., et al., *Cardiology,* 9.1–9.10, 2000.
[3] Garcia D., et al. *Journal of biomechanics,* 1303-1311, 2005.
[4] Seaman C., et al., *Journal of Biomechanical Engineering*, 1-11, 2014.

SUB-ANNULAR DEPLOYMENT IN VALVE-IN-VALVE IS MOST OPTIMAL WHEN CONSIDERING BOTH PRESSURE GRADIENTS AND LEAFLET THROMBOSIS RISK

Hoda Hatoum (1), Atieh Yousefi (1), Jennifer Dollery (2), Pablo Maureira (3), Juan A. Crestanello (2), Lakshmi P. Dasi (1)

(1) Department of Biomedical Engineering
The Ohio State University
Columbus, Ohio, USA

(2) Division of Cardiac Surgery
The Ohio State University
Columbus, Ohio, USA

(3) Department of Cardiovascular Surgery
Lorraine University Hospital of Nancy
Nancy, France

INTRODUCTION

Transcatheter aortic valve (TAV) implantation inside a failed bioprosthetic surgical aortic valve (SAV) – also known as valve-in-valve (ViV) – offers an alternative to redo surgery for high-risk patients. Unfortunately, adverse consequences such as elevated pressure gradients, residual stenosis and paravalvular leakage persist after ViV implantation. Several studies have highlighted the importance of depth of implantation and angular orientation of the TAV with respect to the bioprosthesis on the clinical outcomes and the hemodynamic parameters [1-3] demonstrating that supra-annular deployment and the alignment of the TAV commissures with the bioprosthetic SAV commissures yielded the lowest pressure gradients. This study represents a comprehensive analysis that aims to weigh the benefits in terms of low pressure gradients against sinus flow stasis leading to prosthetic leaflet thrombus formation due to different axial positions and angular orientations in ViV with and without coronary flow. This study shows that any gain achieved with lower pressure gradients due to supra-annular deployment of the TAV in ViV is negated with lower sinus washout and therefore higher risk of leaflet thrombosis.

METHODS

A 23 mm Medtronic Evolut TAV was implanted in a severely degenerated 23 mm Carpentier-Edwards Perimount Magna Ease SAV extracted from a patient who underwent a redo surgery. The TAV was deployed at 16 different combinations composed of 4 axial positions namely +6.0, 0, -6.2 and -9.8 mm relative to the bioprosthesis and 4 angular orientations 0°, 30°, 60°, and 90° (Figure 1a). The negative axial positions denote a sub-annular deployment and the positive a supra-annular deployment relative to the bioprosthesis annulus. The hemodynamic performance of these ViV configurations was assessed in a flow loop that simulates the left heart under physiological pressure and flow with and without coronary flow. The flow loop includes a reservoir tank, mitral valve, pump, flow development region, systemic compliance chamber and resistance valve. The coronary circuit was designed to provide physiological coronary flows throughout the cardiac cycle through the use of a pneumatically controlled Starling resistor. Namely, this resistor was collapsed or expanded during specific intervals to match the changes in coronary flow during myocardial isovolumetric contraction and relaxation, respectively. A small compliance chamber was used to modulate the compressive forces. The working fluid used was approximately 60/40 water/glycerin mixture by volume, a density of 1090 Kg/m^3 and a kinematic viscosity of 3.88 cSt. The pressure and the flow rate across the aortic valve were regulated by a compliance chamber, a resistance valve, and a pump that is controlled by an in-house LabVIEW program. The flow rate was measured with an ultrasonic Transonic flow probe located upstream of the aortic valve, and aortic and ventricular pressures were measured using Validyne pressure transducers. Flow and pressure waveforms are recorded in LabVIEW. The flow was seeded with fluorescent PMMA-Rhodamine B particles of sizes ranging from 1 to 20 μm. The sinus hemodynamic assessment of the ViV was performed using high resolution particle image velocimetry (PIV) which includes illuminating a region of interest in the flow domain using a laser sheet created by a Nd:YLF Single Cavity Diode Pumped Solid State, High Repetition Rate Laser coupled with external spherical and cylindrical lenses. The spatial and temporal resolutions were 0.159 mm/pixel and 4000 Hz respectively.

RESULTS

The axial and angular deployments of the TAV in the ViV influenced the average pressure gradients (Figure 1b). For all the angular orientations, supra-annular deployment at +6 mm yielded the lowest values for the average pressure gradients in comparison with the sub-annular ones at -6.2 and -9.8 mm and the recommended deployment at 0 mm in accordance with published literature [2,4]. Every angle had a different effect on the pressure gradient at a specific axial position. The most optimum hemodynamic results were obtained at 60° angular orientation. The pressure gradient for a given axial position varied between 1.5 mmHg and 5 mmHg due to angular orientation variations. The pressure difference for a given angular position varied between 2.5 mmHg and 6.2 mmHg due to axial position changes. From a leaflet washout standpoint with and without coronary flow (Figure 1c) at peak systole where sinus washout is the least compared to other phases of the cardiac cycle, washout was maximum in the -9.8 mm axial deployment case (sub-annular) and minimum in the supra-annular deployment case at +6 mm. For a particular axial position at 0 mm, having an angle of 60° lowers the vortex rotation with and without coronary flow compared to 0°. The sinus velocity vectors and vorticity contours showed that the presence of coronary flow improves significantly the rotation inside the sinus which making the non-coronary sinus the conservative one.

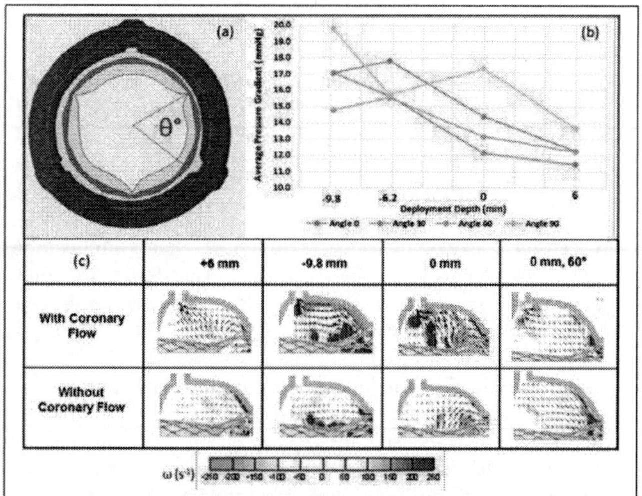

Figure 1: (a) Angular orientations of the TAV in the SAV (b) Average pressure gradients versus axial deployment depths and angular orientations (c) Sinus velocity and vorticity fields at peak systole with and without coronary flow

The axial deployments and angular orientations of the TAV with respect to the SAV in ViV were tested and their hemodynamic parameters were assessed. Although supra-annular deployment regardless of any angular orientation provided the lowest average pressure gradients, it yielded the lowest sinus washout in the presence and absence of coronary flow. Lowest sinus washout is associated with leaflet thrombosis development [5]. In the absence of coronary flow, stagnation was more likely to occur (Figure 1c) [6,7]. Clinical data have shown that developing leaflet thrombosis is more prevalent in the non-coronary cusp compared to the coronary cusp [8]. This observation comes in accordance with the findings of this study. In addition, while it seems that supra-annular deployment maximizes the chance of achieving lower pressure gradients, this improvement may very well be negated by an incorrect angular orientation of the TAV in the ViV. The gains achieved in lower pressure gradients at a particular axial position are offset by a variation of angular orientation of the TAV with respect to the SAV. Further, looking at sinus washout decreased as a direct consequence of supra-annular deployment places supra-annular positions at a disadvantage in terms of long-term benefits of ViV.

SUMMARY

In-vitro ViV hemodynamic studies using a TAV deployed in different axial positions with respect to a calcified SAV annulus showed that supra-annular axial deployment advantages of achieving lower pressure gradients are negated by a low sinus washout that leads to a high risk of leaflet thrombosis. In addition, this study showed that any variation in angular orientation at a particular axial position can compromise the pressure gradients values.

ACKNOWLEDGEMENTS

The research done was supported by National Institutes of Health (NIH) under Award Number R01HL119824.

REFERENCES

[1] Azadani, A.N., et al., *The Annals of Thoracic Surgery*, 92(2):548-555, 2011
[2] Simonato, M., et al., *Euro Intervention*, 12(7):909-917, 2016
[3] Midha, P.A., et al., *JACC: Cardiovascular Interventions*, 9(15):1618-1628, 2016
[4] Simonato, M., et al., *Circulation: Cardiovascular Interventions*, 9(6):e003651
[5] Hatoum, H., et al., *New England Journal of Medicine*, 374:1591-91, 2016
[6] Moore, B.L., et al., *Annals of Biomedical Engineering*, 43:2231-41, 2015
[7] Wootton, D.M., et al., *Annual Review of Biomedical Engineering*, 1:299-329, 1999
[8] Makkar, R.R., et al., *New England Journal of Medicine*, 373:2015-24, 2015

IN VITRO ASSESSMENT OF PROSTHETIC VALVE FLUID MECHANICS IN THE PEDIATRIC PULMONARY OUTFLOW TRACT

N. Schiavone (1), C. Elkins (1), J. Feinstein (2), D. McElhinney (3), J. K. Eaton (1), A. Marsden (4)

(1) Department of Mechanical Engineering
Stanford University
Stanford, CA, USA

(2) Department of Pediatric Cardiology
Stanford University
Stanford, CA, USA

(3) Department of Cardiothoracic Surgery
Stanford University
Stanford, CA, USA

(4) Departments of Pediatrics and Bioengineering
Stanford University
Stanford, CA, USA

INTRODUCTION

Tetralogy of Fallot (ToF), the most common type of cyanotic congenital heart defect, affects 1 in every 2500 newborns annually. Typical surgical repair for ToF requires an initial surgery at a few months of age to widen the pulmonary outflow tract and close the ventricular septal defect; often it leaves the patient with pulmonary regurgiation. The patient's right ventricular size is monitored and a subsequent procedure may be necessary to place a pulmonary valve. The prosthetic valve can be implanted surgically or delivered by catheter percutaneously. These valves become subject to degradation and dysfunction, resulting in variations in valve longevity, often leading to additional surgical interventions [1]. However, there is little understanding of structure of the blood flow local to the valve and what factors of the hemodynamic environment may lead to valve dysfunction. More comprehensive knowledge of prosthetic valve performance and leaflet dynamics may aid clinicians in selecting the best location and orientation for the valve to prevent dysfunction and eventual failure.

This work aims to experimentally explore the flow structure local to a St. Jude Medical Epic valve (a porcine bioprosthetic) in a 3D printed model of the pulmonary anatomy. Variations of the right ventricular outflow tract (RVOT) and the pulmonary anatomy can be printed to represent realistic anatomies found in both healthy and diseased patients. Using magnetic resonance velocimetry (MRV), the 3D, three-component phase-averaged velocity fields of the flow can be measured over the cardiac cycle [2]. This work will focus on key features of the flow field identified local to a replacement pulmonary valve in a 3D model of a healthy RVOT that will serve as a baseline for future experiments.

METHODS

The experimental method consists of two main steps: 1) using patient data to construct a physical model of the RVOT using 3D printing, and 2) using MRV to measure the velocity fields in the model using a blood-analog fluid. An idealized RVOT was designed based on magnetic resonance imaging (MRI) measurements of pulmonary anatomies from six healthy subjects ranging in age from 11-13 years old. For this experiment, we chose a 25mm diameter valve and scaled our model to match, resulting in a main pulmonary artery (MPA) diameter of 25mm, a right pulmonary artery (RPA) diameter of 14mm, and a left pulmonary artery (LPA) diameter of 12mm. Additional key design features included an MPA radius of curvature of 41mm in the sagittal plane and 38mm in the axial plane and a 108 degree bifurcation angle between the LPA and the RPA. The resulting computer aided design (CAD) model of the anatomy is shown below, along with a St. Jude Medical Epic valve. The valve is tri-leaflet and was oriented with one of valve commissures aligned with the anterior curvature of the vessel (Figure 1).

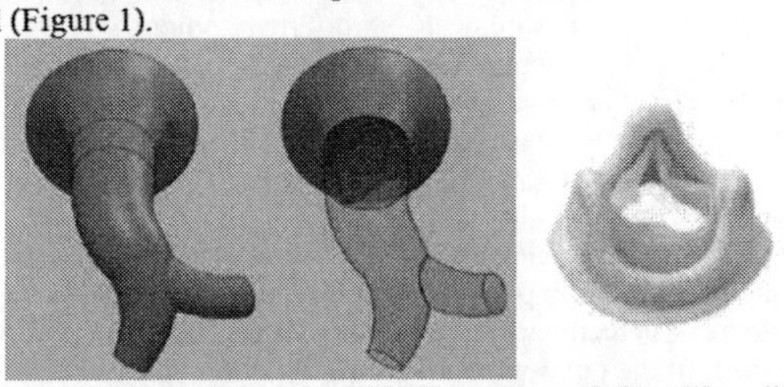

Figure 1: CAD model of the RVOT anatomy (left) with artificial valve location shown in red and a St. Jude Medical Epic valve, a porcine bioprosthetic replacement valve (right)

The 3D model of the anatomy (Figure 2), with the valve in place, was attached to a flow loop powered by a centrifugal pump with a steady flow rate of 3.5 liters per minute. An actuated cylindrical-piston system was connected downstream of the centrifugal pump to create a pulsatile flow rate waveform representative of a healthy heart pumping at 70 beats per minute. This flow rate waveform (Figure 3) was measured using a Transonic Systems flow probe connected to flexible tubing before the entrance to the anatomical model. The outlets from the left and right pulmonary arteries of the 3D model were returned to a reservoir in separate tubing, with flow valves used to vary resistances in order to achieve a physiological 55/45 RPA/LPA flow split. The blood analog fluid used in the system was a water-glycerin solution with 40% glycerin resulting in a viscosity of 3.9 centipoise at 22 degrees C.

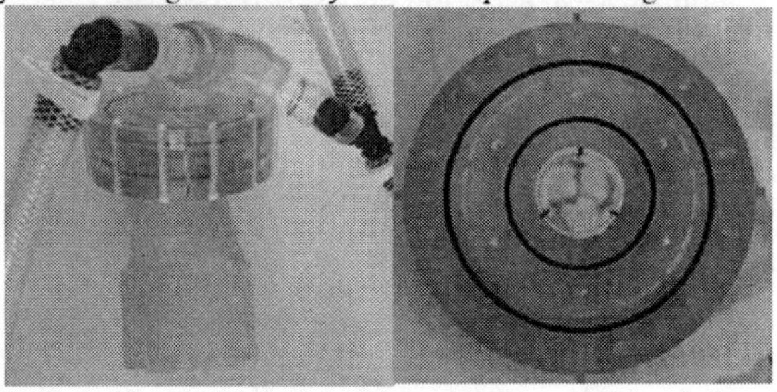

Figure 2: 3D printed model of the pulmonary anatomy (left) and section showing insertion of surgical replacement valve (right)

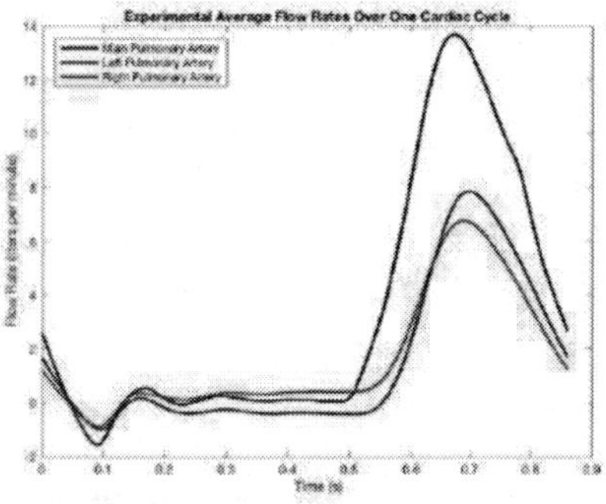

Figure 3: Experimental flow rates over the cardiac cycle

The 3D, three-component, phase-averaged velocity data was obtained using magnetic resonance velocimetry (MRV). The model was placed in a whole body, 3 Tesla General Electric MRI scanner at the Richard M. Lucas Center for Imaging. The scan was performed in the sagittal direction, with a spatial resolution of 0.9mm in the sagittal, axial, and coronal directions. The MRV experiment was phase-locked, which allows for measurements at different points throughout the cardiac cycle. The temporal resolution was 0.092 seconds, which allowed for direction measurement of 9 phases; this data was used to interpolate to 20 total phases in the cardiac cycle.

RESULTS

The MRV experiment provided velocity fields throughout the entire model at 20 time points during the cardiac cycle. Shown below are contours of velocity normal to a plane slicing through the vessel just downstream of the pulmonary valve. The location of the slice is shown in Figure 4, while the velocity contours with in-plane velocity vectors are shown at two different time points (Figure 5). The time of each image corresponds to the time in the input flow rate waveform (Figure 3). The images in Figure 5 occur right before and after the systolic peak in flow rate.

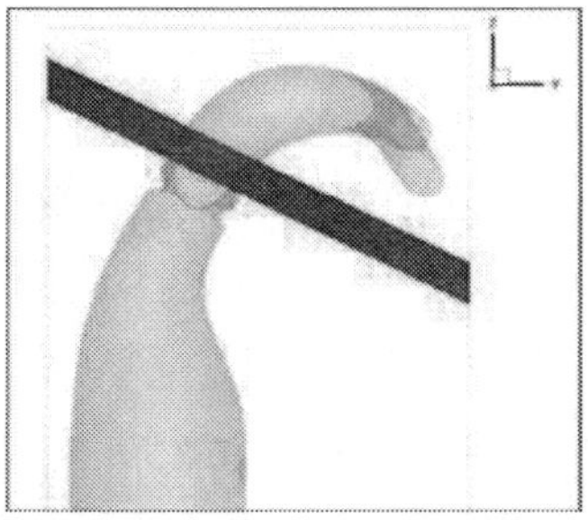

Figure 4: Location of velocity contour slice in Figure 5

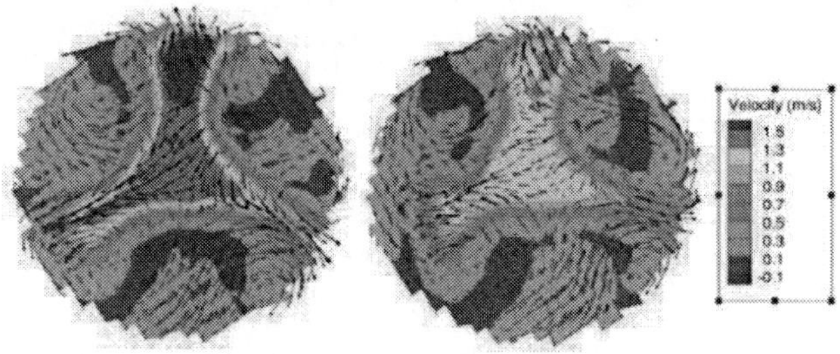

Figure 5: Velocity magnitude contour with in-plane velocity vectors. Shown phases are at 279° (left) and 315° (right)

DISCUSSION

The asymmetry of the flow is demonstrated in Figure 5. Although the valve itself has three-way symmetry due to its three leaflets, the velocity magnitude is higher on the outside of the vessel, likely due to the curvature and twist of the vessel distal to the valve. Further examination of this asymmetry, particularly in the setting of varying outflow tract geometries, may reveal negative effects on valve performance.

In addition, the in-plane vectors reveal secondary flow structures. The structure of the valve that helps the leaflets close creates blockages in the flow. As the flow continues through the valve opening, areas of recirculating flow begin to form downstream of these blockages. This creates vortices that persist in the flow field after systole. These vortical structures and the asymmetry of the flow are key features for further exploration that may contribute to eventual valve dysfunction.

Future work will focus on adjusting the valve's position and orientation within the anatomy to discover flow conditions with the least negative impact on valve performance. In addition, we will measure the flow in a ToF patient specific anatomy to better understand how anomalies in the diseased vessel geometries may lead to fluid structures that affect valve performance.

ACKNOWLEDGEMENTS

Nicole Schiavone is funded by the Benchmark Capital Fellowship in Congenital Cardiovascular Bioengineering. Additional support is provided by the Stanford Child Health Research Institute.

REFERENCES

[1] McElhinney, D. et al., *Circulation – Cardiovascular Interventions*, 4:602-U130, 2011.

[2] Elkins, C. et al., Magnetic resonance velocimetry, *Experiments in Fluids*, 43:6:823-858, 2007.

SB³C2017
Summer Biomechanics, Bioengineering and Biotransport Conference
June 21 – 24, Tucson, AZ, USA

NOVEL TECHNIQUE FOR OPTICAL STRAIN MEASUREMENTS ON THE MITRAL VALVE ANTERIOR LEAFLET

Søren N. Skov (1,2), Oliver Blome (2), Mariam A. Noor (2), Peter Johansen (1,2)

(1) Department of Cardiothoracic Surgery
Aarhus University
Aarhus, Denmark

(2) Department of Engineering
Aarhus University
Aarhus, Denmark

INTRODUCTION

Several biomechanical approaches have been used in-vitro and in-vivo for strain measurements both directly on the valve leaflets [1-2] and in the mitral annulus [3]. In-depth knowledge about the mitral valve dynamics is relevant for the development of surgical devices used for valve reconstruction and replacement.

Reconstruction of the mitral valve apparatus is preferred instead of performing a total valve replacement, especially for patients with degenerative valve disease [4]. The mitral valve consists of two leaflets, the anterior and posterior leaflet. In reconstruction surgery these leaflets are repaired if necessary and typically followed by an annuloplasty ring implantation and sometimes artificial chords to assure proper function of the mitral leaflets. Knowledge about the leaflet strain distribution are important to understand both diseased and non-diseased valves and the impact of different surgical devices such as annuloplasty rings.

The aim of this study was to establish a new experimental model for in-vitro 3D strain measurements directly on the mitral valve anterior leaflet using digital image correlation technology. We hypothesized that the strain distribution of the anterior leaflet increases with elevated left ventricular pressure.

METHODS

Measurements were performed in-vitro in an experimental setup as illustrated in Figure 1. A fresh heart from a 100 kg pig was brought to the lab from a slaughterhouse nearby. The left atrium was removed to provide a visual access to the mitral valve and the heart was fixated in a dedicated chamber with 10-12 sutures placed around the mitral annulus and fastened to the chamber with peans. The chamber was designed with an angle of 45° for optimal position of the cameras. Left ventricular pressure was provided through a nozzle placed in the aorta blowing air into the left ventricle. Air was distributed from a Jun-Air compressor and fine-tuned with a homemade pressure control unit to provide different pressures in the left ventricle. Left ventricular pressure was monitored with a microtip pressure catheter (SPR-350S, Millar Instruments, Houston, TX, USA) and amplified with a two-channel pressure control unit (PCU-2000, Millar Instruments).

For optical strain measurements we used a complete ARAMIS 4M system with dedicated ARAMIS software (ARAMIS v. 6.3.0, GOM, Braunschweig, Germany). The optical sensors consisted of two 4 megapixel cameras, each mounted together with a lamp to an adjustable holding stand (Figure 1).

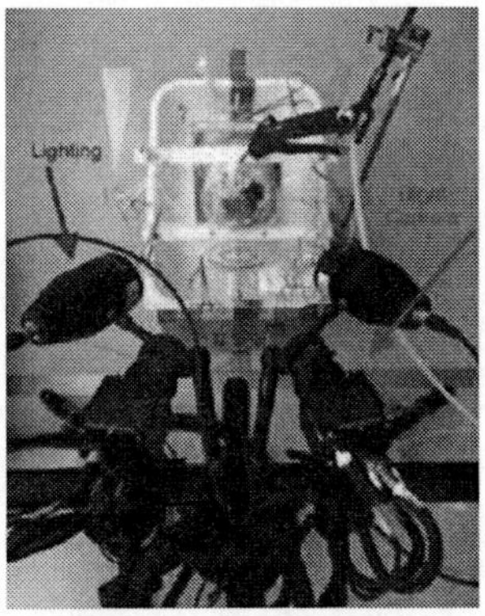

Figure 1: Experimental setup with two cameras and the heart mounted in a dedicated chamber for optical access of the valve.

The cameras were placed in a distance of 345 mm from the valve and oriented with an angle of 25° (seen from the valve). The ARAMIS software is able to measure 3D dimensional strain maps using digital image correlation between images recorded simultaneously from each camera. The ARAMIS software combines neighboring pixels to create small facets (19x19 pixels) in the images. The distortion in these facets are analyzed relative to the previous image to determine displacement and strain of the specific facet area. To improve the facet and distortion detection a stochastic ink pattern was applied to the mitral valve leaflets using black spray paint. Calibration was performed with dedicated calibration objects in order to adjust the alignment and focus of both cameras. Measurements were performed with a frame rate of 60 fps and a duration of 2 seconds creating 120 images from each camera.

Before data acquisition the left ventricular pressure was set to around 20 mmHg in order to close the mitral leaflets. After closure and initiation of data acquisition the pressure was increased in the left ventricular to around 150 mmHg and simultaneous data from both cameras and the pressure catheter was recorded for analysis.

RESULTS

Representative images from the left camera with a color strain map of the anterior leaflet are illustrated in Figure 2. It is clear that the major principle strain distribution is heterogenic and that the strain increases with an elevated left ventricular pressure.

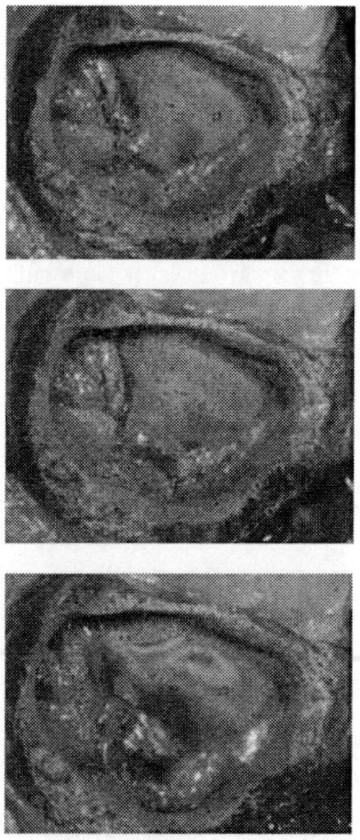

Figure 2: Major principal strain map of the anterior leaflet at different left ventricular pressure. Blue areas represent low strain and red areas with a high strain.

The development in maximum major principle strain with increased left ventricular pressure (imitating the early and end-systolic phase) indicated a linear tendency as illustrated in Figure 3. The measurements performed on a single heart suggested an average 0.34 % strain increase per mmHg.

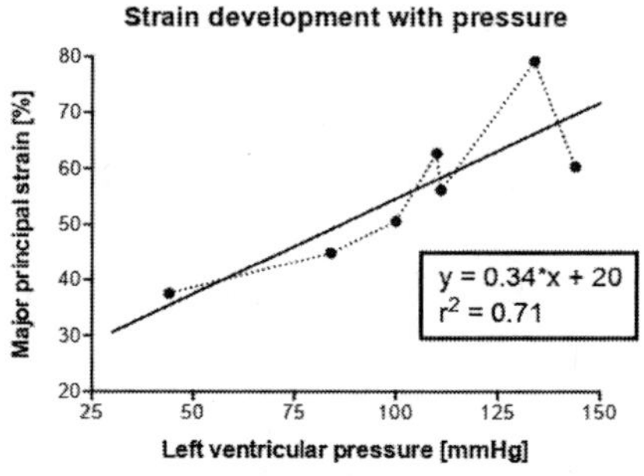

Figure 3: Maximum major strain as a function of left ventricular pressure.

DISCUSSION

The heterogenic strain distribution on the anterior mitral leaflet with high strain accumulation close to the annulus and at the edge near to the coaptation zone may reflect differences in thickness and underlying chordae tendineae distribution. The "strain-valley" configuration with lower strain in the middle of the anterior leaflet was identified in other pilot experiments, but further measurements and refinement is necessary to explain this phenomenon.

We chose only to include the anterior leaflet since the posterior leaflet is less well defined with several clefts (scallops between P1, P2 and P3). We hope that we will be able to include the posterior leaflet in further studies to achieve a complete understanding of the mitral valve.

In this study we have established a new platform for measuring high resolution field strain of the anterior mitral valve. As hypothesized, the preliminary data indicated a linear increase in major principle strain with elevated left ventricular pressure. We believe that this tool can be used for instance to differentiate the impact on leaflet strain using different mitral annuloplasty rings. The technique could also be applied to other valves and several other surgical devices such as Transcatheter aortic valve implantation (TAVI).

ACKNOWLEDGEMENTS

The authors wish to thank the following foundations for their grateful financial support: Snedkermester Sophus Jacobsen og Hustru Astrid Jacobsens Fond, Arvid Nilssons Fond, Karen Elise Jensens fond. We also wish to thank Zebicon A/S, Billund, Denmark for supplying the Aramis 4M system in a two-week trial period.

REFERENCES

[1] Jimenez JH, Liou SW, Padala M, et al. A saddle-shaped annulus reduces systolic strain on the central region of the mitral valve anterior leaflet. J Thorac Cardiovasc Surg 2007;134:1562–8.

[2] Heide-Jørgensen S, Kumaran Krishna S, Taborsky J, et al. A Novel Method for Optical High Spatiotemporal Strain Analysis for Transcatheter Aortic Valves in Vitro. J Biomech Eng 2016;138:034504.

[3] Skov SN, Røpcke DM, Ilkjær C, et al. New mitral annular force transducer optimized to distinguish annular segments and multi-plane forces. J Biomech 2016;49:742–8.

[4] David TE, Armstrong S, McCrindle BW, et al. Late outcomes of mitral valve repair for mitral regurgitation due to degenerative disease. Circulation 2013;127:1485–92.

SB[3]C2017
Summer Biomechanics, Bioengineering and Biotransport Conference
June 21 – 24, Tucson, AZ, USA

SIMULATION OF BLOOD AS A PARTICULATE FLOW IN THE HINGE GAP REGION OF A MECHANICAL HEART VALVE

Fazlolah Mohaghegh (1), H.S. Udaykumar (1)

(1) Department of Mechanical and Industrial Engineering
University of Iowa
Iowa City, IA, USA

INTRODUCTION

Approximately 90,000 heart valve replacement surgeries are performed annually in the US and 280,000 worldwide [1]. Depending on the diagnosis if a corrective surgery does not help, a substitute valve should be implanted. Almost half of the prosthetic valves are Mechanical heart valves (MHVs). Unlike bio-prosthetic-heart valves, mechanical heart valves cause thrombus formation because of non-physiological nature of the prosthesis in the pass of blood flow [2]. The contact of blood elements with the hard parts of the valve's surface and the exposure of the cells to the very high fluid stress in the systole leads to damage to the blood elements [3] or more importantly platelet activation [4] in the hinge. Anti-coagulant therapy is the popular way to prevent thrombus formation due to platelet activation; however, it comes with multiple complications such as hemorrhage and infection. Modification to the design in the gap region will prevent the activation of platelets.

This work represents a computational framework for the simulation of blood as a particle-laden flow. We focus specially on the meso-scale simulation of blood flow in the hinge gap region of a mechanical heart valve (MHV). The aim of the study is to find out the effect of flow near the gap region on the particles' motion, especially as the high shear stress happens at the region.

METHODS

To perform the simulation, we implemented the smoothed profile method (SPM) [5-8] to apply the fluid-structure interaction of blood cells as rigid particles with the carrier fluid. SPM has two stages: at the first stage, the particle-free field is solved. At the second stage, the effect of the particles is added to the domain by using an extra projection. The second projection uses an indicator function field to represent each particle. Considering the ith particle, the indicator function is defined as:

$$\varphi_i(\boldsymbol{x},t) = \frac{1}{2}\left[\tanh\left(\frac{d_i(\boldsymbol{x},t)}{\xi_i}\right) + 1\right] \quad (1)$$

Here, $d_i(\boldsymbol{x},t)$ is the signed normal distance to the particle surface and ξ_i represents the specified interface thickness. The total indicator function field $\varphi(\boldsymbol{x},t)$ is the linear summation of reach particle's field. The velocity of the rigid particles is convolved with the indicator function field to form the particle velocity field which is defined everywhere in the computational domain as:

$$\boldsymbol{u}_p^{n+1}(\boldsymbol{x},t) = \left[\sum_{i=1}^{N_p}\left(\boldsymbol{V}_i^{n+1} + \boldsymbol{\Omega}_i^{n+1}\times\left[x - \boldsymbol{X}_i^{n+1}\right]\right)\varphi_i\right]/\varphi \quad (2)$$

$\boldsymbol{X}_i^{n+1}$, $\boldsymbol{V}_i^{n+1}$ and $\boldsymbol{\Omega}_i^{n+1}$ are the updated i^{th} particle position, velocity and angular velocity, respectively. Then the second projection updates the velocity field to add the effect of particle presence to the domain.

$$\boldsymbol{u}^{n+1} = \boldsymbol{u}^* + \varphi\left(\boldsymbol{u}_p^{n+1} - \boldsymbol{u}^*\right) - \frac{\nabla p_p}{\rho}\Delta t \quad (3)$$

where $\boldsymbol{u}^*$ is the provisional velocity from the first stage. p_p is an extra pressure field which is added to the fluid pressure to enforce the divergence free field. It is obtained from a second pressure Poisson equation as

$$\frac{\Delta t}{\rho}\nabla^2 p_p = \nabla.\varphi\left(\boldsymbol{u}_p^{n+1} - \boldsymbol{u}^*\right) \quad (4)$$

The forces and moments on the particles are calculated using a volumetric integration:

$$\boldsymbol{F}_{h_i}^{n+1} = \frac{1}{\Delta t}\int \rho\,\varphi_i\left(\boldsymbol{u}^* - \boldsymbol{u}_p^n\right)d\boldsymbol{x} \quad (5)$$

$$\boldsymbol{T}_{h_i}^{n+1} = \frac{1}{\Delta t}\int \boldsymbol{r}_i^{n+1}\times\left[\rho\varphi_i\left(\boldsymbol{u}^* - \boldsymbol{u}_p^n\right)\right]d\boldsymbol{x} \quad (6)$$

Since the density of the blood cells is similar to the fluid, the high added mass effect can lead to numerical instabilities. To stabilize the solution, we implemented strongly coupled FSI coupling by sequentially updating fluid and structure to reach the desired convergence. By using Aitken relaxation as described in [9,11] reaching

Poster Presentation #P239

the convergence is accelerated. An additional external repulsion force prevents the overlapping of the particles when they are very close.

RESULTS

At first the robustness of the code is checked for a single RBC located inside a horizontal channel at $\theta = \pi/4$. Placing the particle with the same density as fluid, with non-dimensional diameter $D_p = (2.0,0.5)$ and $Re = 1500$ (with respect to the channel width), a uniform unit velocity moves the particle as shown in figures (1) and (2). The particle has a sudden motion at the start, but the acceleration rate declines with the increase in velocity. The particle starts to rotate just after the start of motion and gradually aligns itself with the flow. As the particle and fluid have similar densities, the gravitational force is not considered.

Figure (3) shows snapshots of RBCs position, velocity and vortex at $t = 40$s for a flow over a constriction at $Re = 800$. While most of the particles escape the constriction at significantly higher velocities than the entrance velocity, a few of them are trapped in the recirculation zone. Moreover, the elevated velocity at the constriction due to conservation of mass continues to form a shear flow at the downstream. We have shown the vortex contours as a criterion for the shear stress. The maximum shear flow happens near the walls at the constriction as well as in the recirculation zone.

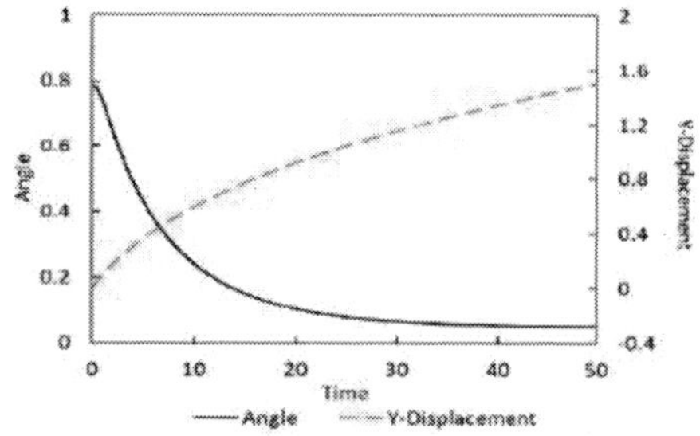

Figure 1: Variation of angle and counter stream-wise displacement with time for an RBC in a horizontal channel.

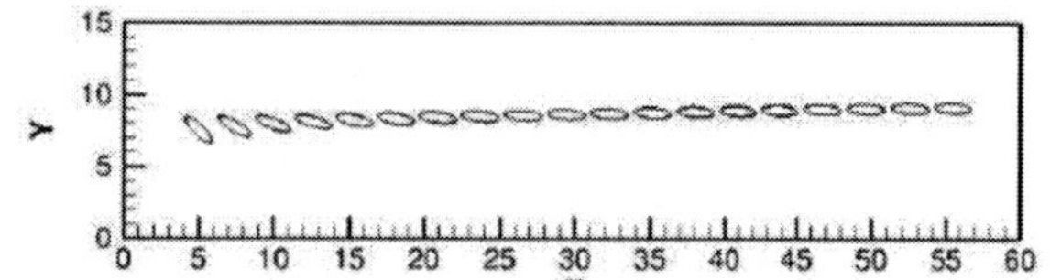

Figure 2: Sequence of a single RBC position and angle in a horizontal channel at Re=1500.

DISCUSSION

We were able to successfully model the motion of blood cells in a horizontal channel using strongly coupled fluid structure interaction. Especially the drift of the particle from the center and the final alignment of the particle with the flow is observed. The velocity of the particle approaches the bulk velocity asymptotically.

When the blood passes a constriction e.g. the hinge gap region of a mechanical heart valve, the particles very close to the walls are subject to higher shear stresses. The high shear stresses also happen locally near the particle interfaces especially when two particles collide. Since the collision is of short time, we conclude that platelet activation can happen in places near the walls and the recirculation zone behind the constriction.

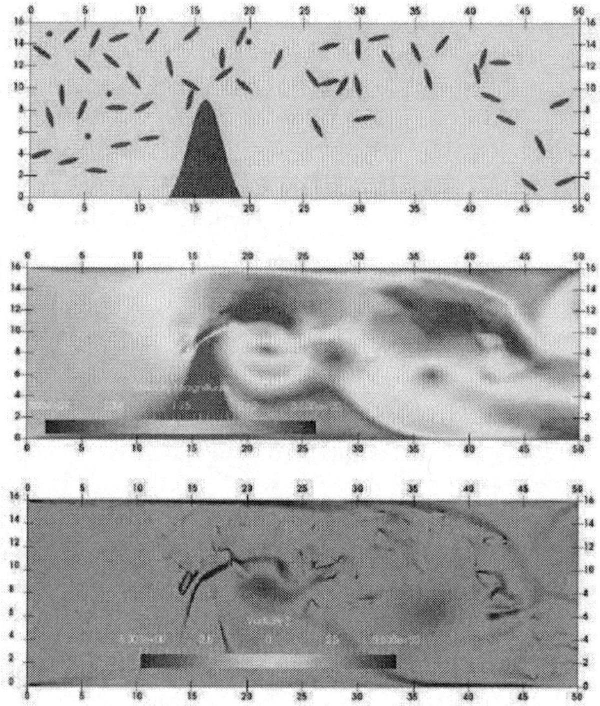

Figure 3: A snapshot of flow in a shear flow at T=40. Top) The position of the particles in the domain. Middle) Contour of velocity magnitude. Bottom) The strain contours.

REFERENCES

[1] Pibarot, P. and J.G. Dumesnil, *Circulation*, 2009 119(7): p. 1034-1048.

[2] Körfer R., et al., *The Journal of heart valve disease*, 2006 15(3): p. 404-12; discussion 413.

[3] Bluestein D., Li Y., and Krukenkamp I., *Journal of biomechanics*, 2002 35(12): p. 1533-1540.

[4] Simon H.A., et al., *Annals of biomedical engineering*, 2010 38(11): p. 3295-3310.

[5] Mohaghegh F., Udaykumar HS., *Computers & Fluids*, 2016 Nov 25;140:39-58.

[6] Mohaghegh F., Mousel J., Udaykumar HS., *ASME 2014 4th Joint US-European Fluids Engineering Division Summer Meeting collocated with the ASME* 2014.

[7] Luo X., Maxey MR., Karniadakis GE., *Journal of Computational Physics*, 2009 Mar 20;228(5):1750-69.

[8] Nakayama Y., Yamamoto R., *Physical Review E.* 2005 Mar 25;71(3):036707.

[9] Mohaghegh F., Udaykumar HS., *ASME 2016 Fluids Engineering Division Summer Meeting collocated with the ASME* 2016.

[10] Mohaghegh F., Udaykumar HS., *Computers & Fluids.* 2017 Jan 17;143:103-19.

[11] Baek H., Karniadakis GE., *Journal of Computational Physics* 2011 Jun 1;230(12):4384-402.

SB^{3}C2017
Summer Biomechanics, Bioengineering and Biotransport Conference
June 21 – 24, Tucson, AZ, USA

EFFECT OF LEFT VERSUS RIGHT CORONARY FLOW WAVEFORMS ON AORTIC SINUS HEMODYNAMICS

Dorma C. Flemister (1), Ryan W. Oba (1), Atieh Yousefi (1), Hoda Hatoum (2), Juan A. Crestanello MD (3), Lakshmi P. Dasi PhD (1)

(1) Department of Biomedical Engineering
The Ohio State University
Columbus, Ohio, USA

(2) Department of Mechanical Engineering
The Ohio State University
Columbus, Ohio, USA

(3) Division of Cardiovascular Surgery -
Wexner Medical Center,
The Ohio State University
Columbus, Ohio, USA

INTRODUCTION

Cardiovascular disease is the major cause of death worldwide. A large subset of individuals with cardiovascular disease develop calcific aortic valve disease (CAVD) whose long-term progression leads to stenotic valves that impair the cardiovascular system's pumping function. For highly stenotic aortic valves, surgical intervention is necessary but complications remain including restenosis of the prosthesis.

While evidence suggests that hemodynamics regulate stenosis, the etiology of valve stenosis remains incomplete. Previous studies of sinus vortex dynamics have identified novel fine-scale hemodynamics within the aortic sinus which may play a role in the onset of CAVD [1]. Additionally, investigations of the coronaries throughout the cardiac cycle have been modeled but left unspecified which coronary was modeled [2]. Moreover, recent studies have developed systems for the assessment of coronary flow but these studies were not conducted in the context of assessing aortic sinus hemodynamics [3].

The objective of this study is to elucidate a substantial difference in the hemodynamics between the left and right sinuses during normal physiological conditions. Investigating in detail the difference in hemodynamic parameters such as the magnitude and shape of the coronary flow waveforms between the left and right coronary sinuses simultaneously has not been performed. Knowledge of the aortic sinus hemodynamics in the coronary can provide insight into some causes of these adverse effects.

METHODS

A left heart flow loop system described previously [2], which simulates flow from the left ventricle to the aorta, was modified through the addition of a coronary circuit. This circuit was designed to provide physiological coronary flows throughout the cardiac cycle through the use of a pneumatically controlled Starling resistor. Namely, this resistor was collapsed or expanded during specific intervals to match the changes in coronary flow during myocardial isovolumetric contraction and relaxation, respectively. A small compliance chamber was used to modulate the compressive forces. The working fluid used was a 60:40 water-glycerin blood viscosity matching analog solution, and a porcine bioprosthetic aortic valve (23mm Medtronic Hancock II) was mounted inside the acrylic aortic sinus chamber. The port for the coronary ostia was a 4mm diameter hole machined 5mm from the base of the sinus. A flow loop specific Lab-VIEW program was utilized at average resting conditions (HR ~ 60 bpm; BP ~ 120/80 mmHg). Average cardiac output of 3.9 L/min was calculated for this valve size based upon existing relationships between body size area and cardiac output [4]. Coronary flow measurements were based on a ~250 mL/min average with 70% of the flow going to the left coronary (~175 mL/min) and 30% to the right (~75 mL/min) [5]. Aortic and coronary flow rates were measured with Transonic ultrasonic flow probes. Pressure measurements were measured downstream and upstream the valve using Validyne pressure transducers. Flow visualization was performed using two-dimensional time-resolved particle image velocimetry (PIV), according to procedures by Moore *et al* [2].

RESULTS

During the cardiac cycle there are differences in the flow waveforms for the left and right coronaries (Figure 1). The left coronary case averaged 155 mL/min with a maximum 239ml/min and minimum of -11mL/min. The right coronary case averaged 76 mL/min with maximum of 95 mL/min and minimum of 65 mL/min. These values being within the physiological range [5]. When systole begins and the

Poster Presentation #P240

aortic valve pressure increases, the left coronary flow begins to plummet. The left coronary's fall ends once the flow briefly reverses. The left coronary then rises and peaks just before aortic deceleration where it briefly falls. At the closure of the aortic valve, the dicrotic notch briefly spikes aortic pressure which in turn spikes the left coronary flow to a second peak. Once diastole begins, the left coronary flow falls with the aortic pressure. The right coronary case follows the aortic pressure more closely than the left due to the lack of contractile forces squeezing the vessel. For diastole, both follow the fall in aortic pressure after reaching peak flow rates.

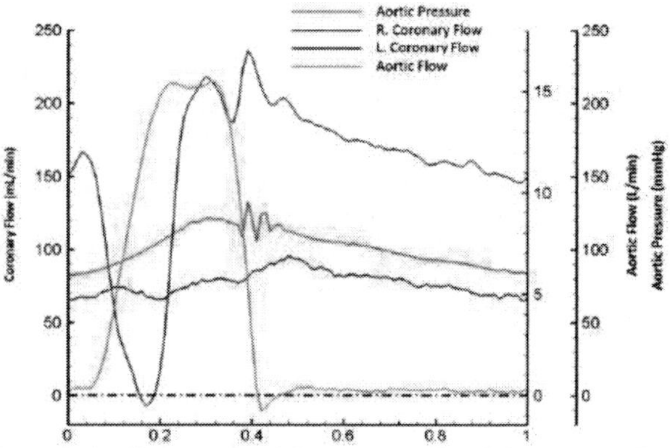

Figure 1. Aortic pressure, right coronary flow, left coronary flow, and aortic flow during one cardiac cycle.

Velocity vectors and vorticity maps within the aortic sinus region were calculated from high speed-camera PIV images (Figure 2). In both coronary cases, there is a significant presence of sinus vorticity during early systole. In the left coronary flow case, during systole the flow begins to slow until a brief period of reversal which diminishes the strength of the sinus vortex. In the right coronary case, since the flow remains positive throughout the cardiac cycle the sinus is subject to stronger sinus vorticities in systole. The velocity vectors reiterate that the flow rate is higher in diastole. Vortices are at their peak at mid systole before tapering off at the end of systole and during diastole.

DISCUSSION

Overall this study highlights the very sensitive nature of coronary flow on sinus hemodynamics and also indirectly leaflet opening. The coronary circuit appears to successfully mimic physiological left and right coronary flows. While both coronary cases are subject to the influences of the aortic flow and pressure, there are major differences in the sinus flow waveforms during systole. Specifically, the left coronary case has a multi-phasic waveform with the major peak occurring during early diastole [6]. Conversely, the right coronary case has reduced flow throughout and follows the aortic pressure. The right coronary's relatively constant flow is able to better open the leaflet during systole in contrast to left coronary case where the coronary flow is slowing down (meaning there is higher pressure at the ostia relative to the leaflet). A slight shift in leaflet vertical position changes the re-attachment point of the jet which explains why there is more vorticity in the right coronary sinus case. The left coronary flow's more chaotic nature diminishes the effects of vorticity and also exerts pressure onto the leaflet pushing the jet downward. The coronaries' effect within the sinus showcases how the left and right cases distinctly impact sinus vorticities. These effects may provide further insights into the onset and progression of CAVD and merits further investigation.

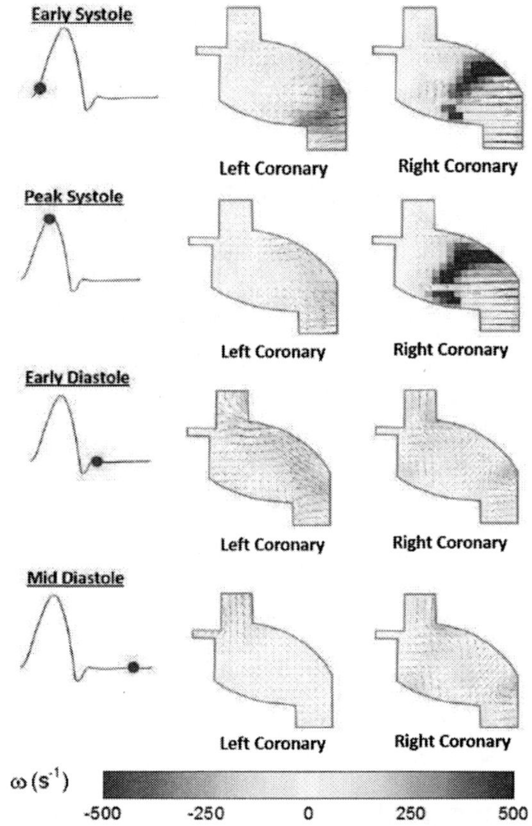

Figure 2: Aortic sinus velocity vectors and vorticity maps within the coronary sinus across cardiac cycle.

SUMMARY

This study showed a pneumatically controlled coronary circuit could successfully mimic physiological left and right coronary flows. There exists a tangible difference in the sinus hemodynamics between the left and right sinuses during normal physiological conditions.

ACKNOWLEDGEMENTS

The Ohio State University Department of Biomedical Engineering for GTA support

The Ohio State University Wexner Medical Center

Davis Heart & Lung Research Institute

This research was supported by the National Institutes of Health under Award Number 3R01HL118924 and 5R01HL119824

REFERENCES

[1] Moore B, Dasi LP. *Exp Fluids*. 55:1770, 2014.

[2] Moore, B., Dasi LP, *Ann. Biomed Eng*, 43:2231–2241, 2015

[3] Piola, M et al., *Physiol. Meas*., 37:2274–2285, 2016

[4] Sluysmans, T., Colan S., *J Appl Physiol* 99:445– 457, 2005

[5] Sankaran S et al., *Ann. Biomed. Eng*. 40:2228–42, 2012

[6] Olgac U et al., *Am J Physiol Heart Circ Physiol*. 296:H1969–H1982. 2009

SB³C2017
Summer Biomechanics, Bioengineering and Biotransport Conference
June 21 – 24, Tucson, AZ, USA

SUPPRESSING UNSTEADY FLOW IN ARTERIO-VENOUS FISTULAE

L. Grechy (1), F. Iori (1), R. W. Corbett (2), W. M. W. Gedroyc (3), N. Duncan (2), C. G. Caro (4), P. E. Vincent (1)

(1) Department of Aeronautics, Imperial College London, London, UK

(2) Imperial College Renal and Transplant Centre, Hammersmith Hospital, London, UK

(3) St. Marys Hospital, Praed Street, London, UK

(4) Department of Bioengineering, Imperial College London, London, UK

INTRODUCTION

Haemodialysis is a common form of renal replacement therapy for patients who suffer from End-Stage Renal Disease (ESRD) [1]. Efficient haemodialysis requires high-quality vascular access, allowing the rapid removal of a patient's blood, which passes through a dialyser before being returned to the body. The `gold-standard' method of vascular access is via creation of an Arterio-Venous Fistula (AVF) [2], formed by surgically connecting an artery and vein, usually in the patient's arm. The connection or `anastomosis' can take a number of configurations. However, an end-to-side layout, where the end of the vein is connected to the side of the artery, has emerged as the clinically preferred configuration [3].

After an AVF is formed, high venous flow rates should stimulate vascular dilation and re-modeling, such that the AVF `matures', and can be used for haemodialysis [4]. Unfortunately, however, up to 60% of AVF do not mature as anticipated [5] resulting in unacceptable patient morbidity and a significant burden on healthcare expenditure. A major cause of AVF failure is the formation of stenosis due to development of Intimal Hyperplasia (IH) in the juxta-anastomotic area and the vein [3,4,6]. Whilst the exact mechanisms underlying development of IH in AVF are unknown, there is considerable evidence to suggest that the inherently unphysiological flow patterns within AVF play an important role [4,7]. In particular regions of low Wall Shear Stress (WSS) [8,9], low lumen-to-wall oxygen flux (leading to wall hypoxia)[10] and high-frequency flow unsteadiness [11,12,13,14] have all been implicated in the initiation and development of IH.

The present study is based on the latter hypothesis, that high-frequency flow unsteadiness plays an important role in development of IH. Specifically, a Kriging based surrogate model [15] and a Mesh Adaptive Direct Search (MADS) optimisation framework is employed [16,17], with CFD simulations, and a new cost function, to design a novel non-planar AVF configuration that suppresses high-frequency flow unsteadiness.

METHODS

Geometric Parametrisation

The artery was constructed by sweeping a circular section along a plane-arc centerline, as described in [12]. The vein was constructed by sweeping a circular section along a two-piece cubic hermitian spline, connected to the mid-point of the plane-arc artery. The resulting configuration was parameterised by a set of six quantities:

- the bifurcation angle between the artery and the vein (two parameters);
- the position of the middle of the vein (three parameters);
- the position of the distal end of the vein (one parameter).

Cost Function

We introduce a new cost function that indicates whether a particular AVF configuration is likely to induce unsteady flow. Evaluating the cost function for a given configuration comprised of two stages. Firstly, the steady flow solver of Star-CCM+ was used to find a steady-state flow solution in the configuration. This solution was then used as the initial condition for the segregated implicit unsteady solver

Poster Presentation #P241

of Star-CCM+, which was run for a fixed time-period. The cost function itself was then calculated based on the deviation of the unsteady solution from the steady-state initial condition in finite time, with large deviations resulting in increased values. All simulations were run with steady-state inflow conditions, similar to those used in [8].

Optimisation Framework

The MATLAB DACE package [18] was used for building the Kriging surrogate model, which was seeded with a cost function for a Latin Hypercube Sampling of 91 valid configurations. After the first Kriging model was built, the MADS algorithm [16,17] was employed to find an optimal configuration that minimized high frequency unsteadiness.

Validation

High-resolution pulsatile simulations of flow were undertaken in the optimal configuration in order to check that high-frequency flow unsteadiness was indeed suppressed. Such a check was important; since the optimization framework itself was based on a cost function derived with non-pulsatile inflow conditions.

RESULTS

The optimal configuration is shown in Fig. 1. Salient features of the optimal configuration include:

- an anastomosis on the outer curvature of the arterial bend, in line with the results of Iori et al. [12];
- an inherently non-planar looped venous section;
- an obtuse anastomotic angle, more similar to the ones found in physiological vascular branching than typically formed in AVF.

High-resolution pulsatile simulations of flow in the optimal configuration confirmed no significant unsteadiness above 20Hz, whereas the frequencies registered in [11,12,13,14] were all above 70Hz, and up 1000 Hz. The configuration therefore appears to suppress high-frequency unsteadiness as designed.

DISCUSSION

High-frequency flow unsteadiness has been associated with development of vascular pathology in AVF. In this study we employed a MADS optimisation framework with CFD simulations and a new cost function, to design a novel non-planar AVF configuration that suppresses high-frequency flow unsteadiness. Salient aspects of the proposed configuration include an anastomosis on the outer curvature of an arterial bend, and a looped non-planar venous section. Current work is underway to deploy a prototype device in a porcine model that can hold AVF in the optimal configuration.

ACKNOWLEDGEMENTS

The authors are grateful for support from the National Institute for Health Research Imperial Biomedical Research Centre, CD-Adapco, the British Heart Foundation, and the Engineering and Physical Sciences Research Council. Outcomes of this work have been submitted for protection under patent number 1603501.6.

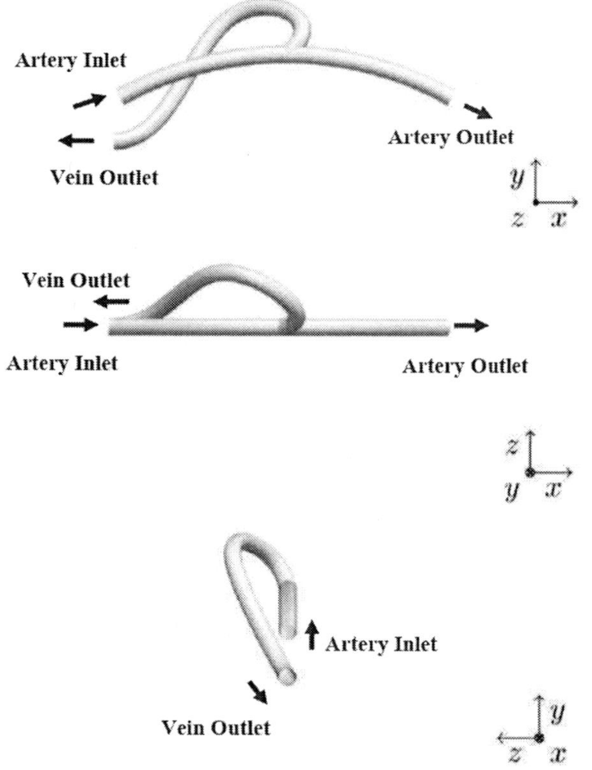

Figure 1: Optimal configuration.

REFERENCES

[1] US Renal Data System, *AJKD,* 67(3):S197-S2010,2015.
[2] Tordoir, J et al., *Nephrol Dial Transplant*, 22(2):88–117, 2007.
[3] Konner, K, *JASN,* suppl 1(48):S176–S47, 2003.
[4] Dixon, B S, *Kidney Int*, 70(8):141322, 2006.
[5] Al-Jaishi, A et al., *AJKD,* 63(3):464–478,2014.
[6] Riella, M C et al., *Nature Rev, Nephrol*, 9(6):348–357, 2013.
[7] Qiu, J et al., *JRSI*, 11(90):20130852, 2014.
[8] Caro, C G et al., *Nature*, 23(5211):1159–1161, 1969.
[9] Peiffer, V et al, *Cardiovascular Res.*, 99(2):242–50, 2013.
[10] Tarbell, J M et al., *Ann Review Biomed Eng*, 5:79–118, 2003.Loth, F et al., *J Biomech Eng*, 124(1):44–51, 2002.
[11] Nathan, I et al., *Bulletin of the New York Academy of Medicine*, 53(10):849–868, 1977.
[12] Iori, F et al., *Phys Fluids*, 27(3):031903, 2015.
[13] Bozzetto, M et al., *Ann Biomed Eng*, 44(8):2388-401, 2015.
[14] Browne, L D et al., *Cardiovasc Eng Technol*, 6(4):450–462, 2015.
[15] Sacks, J et al. *Stat Science*, 4(4):409–423, 1989.
[16] Audet, C et al., *SIAM J Optim*, 18(4):1501–1503, 2008.
[17] Marsden, A L et al., *Comput Method Appl M*, 197(24):1890–1905, 2008.
[18] Lophaven, S et al., *DACE: A MATLAB Kriging toolbox version* 2.0, Technical Report IMM-REP-2002.

SB³C2017
Summer Biomechanics, Bioengineering and Biotransport Conference
June 21 – 24, Tucson, AZ, USA

QUANTIFICATION OF CHANGES IN BLOOD FLOW DYNAMICS IN LEFT VENTRICLES OF PORCINE HEARTS BEFORE AND AFTER MYOCARDIAL INFARCTION

Vivek Vasudevan (1), Adriel J. J. Low (1), Sarayu P. Annamalai (2), Smita Sampath (2), Kian K. Poh (3), Teresa Totman (4), Muhammad Mazlan (4), Mark A. Richards (5, 6), Dominique P. V. de Kleijn (4), Chih-Liang Chin (2), Choon-Hwai Yap (1)

(1) Department of Biomedical Engineering
National University of Singapore
Singapore

(2) Translational Biomarkers
Merck Research Laboratories, MSD
Singapore

(3) Department of Cardiology
National University Health Systems
Singapore

(4) Department of Surgery
National University Health Systems
Singapore

(5) Cardiovascular Research Institute
National University Heart Centre
Singapore

(6) The Christchurch Heart Institute
University of Otago
New Zealand

INTRODUCTION

Cardiovascular disease is the leading cause for death in the world, of which, myocardial infaction is a major category. Consequently, advancing our knowledge of heart function during & after an infarction is an on-going research endeavour, as they may lead to improved prognosis, diagnosis & treatment strategies.

The spatiotemporal characteristics of LV fluid dynamics are expected to alter in heart diseases. For example, changes to the diastolic ventricular vortex ring dynamics during cardiomyopathy have been proposed as a marker for the disease [1]. In Myocardial Infarction (MI), parts of the myocardium perish, exposing rest of the heart to high stresses, leading to decreased ejection fraction, elevated heart rates, and reduced cardiac output [2]. These changes are likely to produce flow changes which can be used as disease markers. Thus far, however, effects of these pathological changes on LV flow & energy dynamics have not been comprehensively characterized.

In this work, MI was induced in a porcine animal model, CINE MRI images of the porcine heart were collected, & 3D dynamic mesh CFD was performed to charatcerise the changes in flow & energy dynamics of the LV caused by MI. Effects of key physiological parameters were investigated to understand their relative contributions to the observed changes during infarction. Further, a non-dimensional number that could effectively represent the energy efficiency of flow in the LV was proposed.

METHODS

MI was induced in two porcine subjects (#68 & #70) through permanent ligation of left circumflex coronary artery (LCx). MRI scans were performed at 0 week (baseline, before infarction), 1 week (acute infarction) & 4 weeks (chronic infarction) time-point. Cine images were acquired for several short-axis slices to cover the porcine hearts. LV twisting motion was imaged via SPAMM tagging, which encoded the 2-D displacement in respective short axis slices.

Mathematical modelling of LV wall motion & dynamic mesh CFD were performed using previously established methods [3]. The LV was reconstructed from MRI images at every time point, & used to construct a mathematical model of wall motion, which composed of a radial wall motion about the centroid & an azimuthal LV twisting motion. CFD was performed with ANSYS FLUENT 15.0, with a User-defined function to specify LV wall motions. Simulations were performed for four heart beats, & only results from the last cycle were analysed. Boundary conditions chosen were such that during diastole, the mitral opening was a zero pressure inlet & the outflow tract (LVOT) was a wall. Conversely, during systole, the mitral opening was a wall, while the LVOT was a zero pressure outlet.

To understand the influence of key physiological parameters on the LV fluid & energy dynamics, simulations were conducted where the stroke volume (SV), ventricular size & heart rate (HR) were varied in porcine subject #68. The LV energetics was quantified via two energy indices: specific energy loss (i.e. viscous flow energy loss per unit volume of blood pumped by the heart); & kinetic energy (KE) density (i.e. average KE per unit LV volume) during the diastolic phase.

RESULTS

The physiological parameters of the simulated porcine hearts based on MRI scans (Table 1) show that acute response of the heart for both subjects was ventricular dilation & elevated HR. There was a gradual decrease in ejection fraction (EF) from Week 0 to Week 4, suggesting development of systolic dysfunction with disease progression.

Additional simulations were performed, based on scans from subject #68, but with a single physiological parameter altered, such as heart rate (HR), end-diastolic volume (EDV) or stroke volume (SV), in a controlled manner, to understand the effects of these individual parameters. The details are provided in Table 2.

Tables 1 & 2 also show the simulation results, demonstrating the diastolic vortex rings, & wall shear stress color contour. These images show the extent of vortex ring penetration into the LV chamber as well the relative dynamics in the LV chambers.

Quantification of specific energy loss & diastolic KE density are shown in figure 1. The lines joining data points indicate that these two data points have identical physiological parameters, except for the parameter being varied (in the x-axis).

Poster Presentation #P242

Table 1: Physiological parameters of the MRI-scanned hearts

	Subject #68			Subject #70		
	W 0	W 1	W 4	W 0	W 1	W 4
HR (bpm)	63	93	95	66	94	90
EDV (ml)	96	139	95	90	109	106
SV (ml)	54	63	34	46	53	49
EF (%)	56	45	36	51	49	47
Vortex rings @ EDV						

Table 2: Parameters of additional simulations (on subject #68) to study effects of HR, chamber dilation & SV

	W_0 fast	W_1 slow	W_0 bigEDV	W_1 smallSV	W_1 smallerSV
HR (bpm)	93	63	63	93	93
EDV (ml)	96	139	139	140	140
SV (ml)	54	63	54	54	34
EF (%)	56	45	38	39	24
Vortex rings @ EDV					

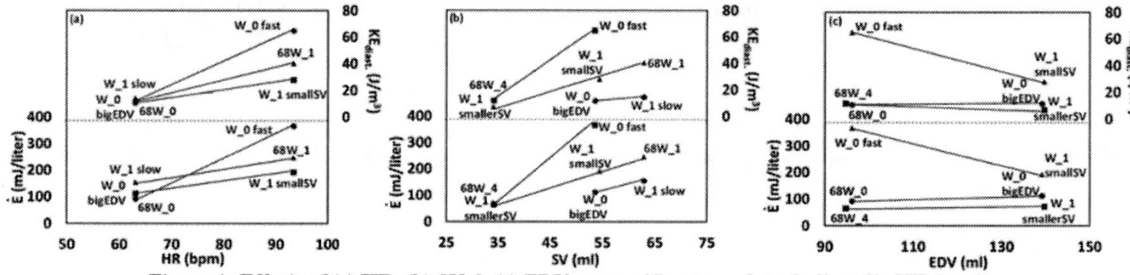

Figure 1: Effects of (a) HR, (b) SV & (c) EDV on specific energy loss & diastolic KE density

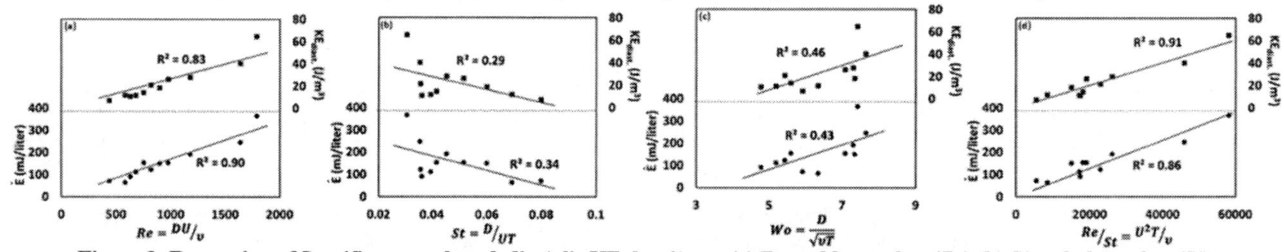

Figure 2: Regression of Specific energy loss & diastolic KE density vs. (a) Reynolds number (*Re*) (b) Strouhal number (St) (c) Wormersley number (*Wo*) & (d) *Re/St*

Generally, increasing HR increased energy loss & KE density, but the effect was more prominent at lower EDV & higher SV. Increasing SV generally increased energy loss & KE density as well, & this effect was also more prominent at lower EDV. The effect of increasing EDV was not consistent. In some cases energy loss & KE density decreased, while in other cases they increased slightly. This complex interplay between the three parameters, made it difficult to understand the contributions of these physiological parameters on energy dynamics.

This was the motivation for devising a non-dimensional number that incorporated all 3 physiological parameters to describe LV energetics. Regression analysis of the energy indices with several non-dimensional numbers was performed to determine which number effectively represented the energy dynamics. Results showed that energy loss & KE density had positive correlations with *Re*, *Re/St* & *Wo*, but negative correlation with *St* (Figure 2). *Re* & *Re/St* showed very good linear relationships with the high R^2 values, & hence, could most effectively represent the energy efficiency of flow in the LV.

DISCUSSION

Inter-subject variabilities in comparing pre- & post-MI conditions were obviated via use of the same porcine subject before & after MI. Drastic changes were observed in energy dynamics of the intraventricular flows due to induced MI. At week 1, ventricular flows always had significantly higher vortex dynamics & energy losses, most likely due to the trauma of infarction. At week 4, the ventricle generally had a reduced KE and lower energy losses, but magnitude of this reduction depended on how well the LV recovered to its pre-infarct size & cardiac output.

While attempting to understand what physiological parameter contributed to the drastic changes in LV energy dynamics over the infarction, it was discovered that the effects of various physiological parameters have an intricate interplay, & could not be studied in isolation. To this end, the non-dimensional number *Re/St* was proposed, which included all 3 physiological parameters, and could represent the state of energy efficiency in the LV very well.

We believe that this non-dimensional number, *Re/St*, can easily be measured in a clinical setting with currently available equipment (e.g., pulsed-wave Doppler for peak E-wave velocity, *U*; ECG for cardiac duration, *T*; standard value for kinematic viscosity of human blood, *v*), & hence, can be used as a potential prognostic or diagnostic biomarker.

ACKNOWLEDGEMENTS

The study was funded by Singapore Millennium Foundation Grant: "*Minimally-invasive, Thoracoscopically Implantable Cardiac Assist Device for Palliative Care of Elderly with Failing Hearts*" (PI: Yap).

REFERENCES

[1]. Gharib, M. et al., *Proc Natl Acad Sci USA*, 103:6305, 2006
[2]. Kupper, W. et al., *Am J Cardiol*, 40:900, 1977
[3]. Wiputra, H. et al., *Am J Physiol Heart and Circ Physiol*, 2016

SB^{3}C2017
Summer Biomechanics, Bioengineering and Biotransport Conference
June 21 – 24, Tucson, AZ, USA

A SIMULATION STUDY OF THE AGE-RELATED CHANGES IN THE CARDIOVASCULAR SYSTEM

Stamatia Z. Pagoulatou (1), Nikolaos Stergiopulos (1)

(1) Laboratory of Hemodynamics and Cardiovascular Technology
Swiss Federal Institute of Technology - École Polytechnique Fédérale de Lausanne (EPFL)
Lausanne, Switzerland

INTRODUCTION

Age-related changes are noted mainly in the big elastic vessels, i.e. the aorta and its major branches, which undergo gradual stiffening and dilation. These alterations lead to a continuous increase of systolic pressure (SBP) and pulse pressure (PP) over time, which affects in turn the cardiac structure and function.

The increase in PP is attributed to the change of both the forward and backward travelling waves. The contribution of both components, however, still remains a controversial topic. One possible theory is that global arterial stiffening causes the augmentation and earlier arrival of the reflected waves. A second mechanism proposed by Mitchell et al.; [1]; suggests that the non-uniform aortic stiffening, whereby proximal aorta stiffens more than the distal aorta and peripheral vessels, leads to an increase in the characteristic impedance predominantly in the proximal aorta, causing an increase in amplitude of the forward pressure wave.

The use of mathematical models of the cardiovascular system to simulate aging could shed light to these debates and broaden our understanding on the aging process. Previous work from Maksuti et al.; [2]; has tried to do so by employing a lumped parameter model of the arterial tree. Although this aging model did manage to predict systolic (SBP) and diastolic (DBP) pressure evolution with certain accuracy, it was unable to capture successfully the wave transmission effects.

The objective of this study is to simulate the major aging mechanisms using a one-dimensional model of the arterial tree. We expect that our model will be able not only to predict pressure evolution but also to capture the characteristics of the pressure waveform with accuracy. Furthermore, we can analyze the contribution of the earlier-arriving and amplified reflection wave and the augmented forward wave to the development of isolated systolic hypertension and thereby assess the relative importance of the two wave components.

METHODS

The aging cardiovascular system was simulated using a detailed, already validated, one-dimensional model of the arterial tree, a thorough description of which can be found in [3]. The model was modified in order to include the opening of the aortic valve and the phenomena of inertia and turbulence in the flow exiting the ventricle, according to Mynard et al.; [4] :

$$\Delta p = P_{lv} - P_a = BQ^2 + L\frac{dQ}{dt} \qquad (1)$$

where B and L are blood resistance and inertance, respectively.

The major age-related change in the arterial system is the stiffening of the aorta and the proximal arteries. Although most studies assess total aortic stiffening through measurement of the carotid-to-femoral PWV, it has been shown that the elastic properties of different aortic regions do not undergo uniform alterations. Therefore we chose to use values of regional PWV as a function of the mean arterial lumen diameter. Aging was introduced by changing the local PWV values according to literature data, [5-6]. Despite some dispersion, the empirical inverse relations between artery size and PWV, similar to the one proposed by Reymond et al.; [3];,were well fitted to literature data for all decades (R^2>0.75).

Total peripheral resistance and terminal resistances were estimated from the expected mean arterial pressure for each decade (according to the Framingham Heart Study; [7];) and assuming that cardiac output remains almost unchanged.

Regarding cardiac aging, we employed the physiological rules developed by Maksuti et al.; [2]; in order to determine cardiac function. More specifically, elastance parameters were modified in proportion to the ventricular wall thickness in order to normalize ventricular wall stress. Heart rate was changed for each decade according to literature values.

RESULTS

The simulation-derived evolution of central and brachial SBP and DBP with age is depicted in Figure 1 in comparison to the clinical pressure data from the Framingham Heart Study; [7]. Goodness of fit parameters were also calculated; nRMSE was 2.2%.

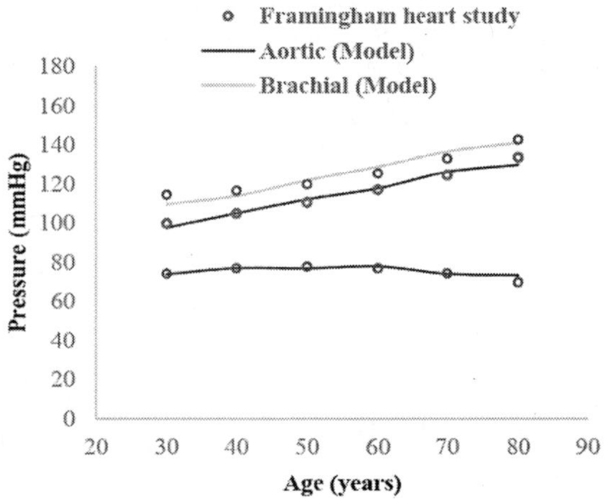

Figure 1: Pressure evolution during aging: Model vs Prediction.

Aortic pressure waves were separated into their forward and backward components. Augmentation Index (AI), which is a wave shape index that reflects the contribution of the backward wave to central pulse pressure, was calculated as described in [8] and compared to the expected values from Torjesen et al.; [9]; as shown in Figure 2.

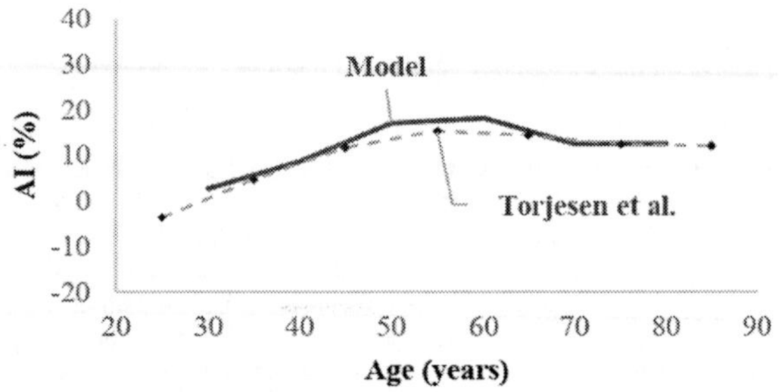

Figure 2: Augmentation index during aging: Model vs Prediction.

The contribution of the forward pressure wave to systolic pressure augmentation during aging was quantified as the ratio of the forward wave pulse pressure to the central pulse pressure, fPP/cPP. The model results were compared to those of the 0-D model, as depicted in Figure 3. Between 30 to 80 years old, the ratio increased from 0.78 to 0.83, while the 0D model of Maksuti et al.; [2]; predicted a decline.

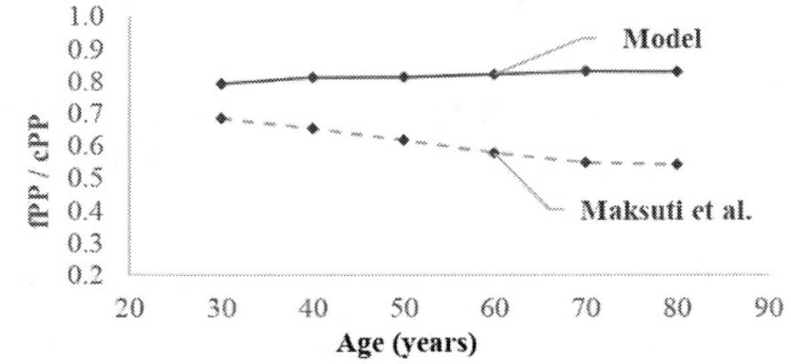

Figure 3: Contribution of forward pressure wave to pressure augmentation during aging: Model results vs 0D model.

DISCUSSION

The present study aimed to achieve three goals. First, to develop rules and accurately simulate the aging process using a realistic 1-D model of the arterial tree. In doing so, it became imperative to improve the initial 1D model of Reymond et al.; [3]; by including flow phenomena at the aortic valve that strongly influence the shape of the aortic pressure waveforms.

The second goal was to validate the predictions of the model by comparison to published data from large-scale clinical studies. Despite the simplicity of the rules we applied and the fact that we did not take into consideration how the arterial geometry evolves during aging, our simulations were very accurate; SBP and DBP prediction was excellent when compared to the findings of the Framingham heart study. Furthermore, the pressure waveforms produced were of the anticipated shape and predicted AI accurately for each decade.

The third goal of our study was to examine whether the augmented forward pressure waves contribute significantly to the development of essential hypertension. Our results showed that the forward wave amplitude increases with age and, thus, becomes a major determinant of the increase in PP with age, as suggested by Mitchell et al., [1]. This does not hold true for the 0-D model that failed to capture the wave transmission effects.

This study lays the ground for further investigation of the aging mechanisms and their contribution to the development of essential hypertension. New detailed information about age-related changes in aortic geometry and properties would help further improve the accuracy of the mathematical models of the aging cardiovascular tree.

REFERENCES

[1] Mitchell, G et al., *Hypertension*, 51(1):105-111, 2008.
[2] Maksuti, E et al., *PLOS ONE*, 11(6): e0157493, 2016.
[3] Reymond, P et al., *Am J Physiol Heart Circ Physiol*, 297: H208–H222, 2009.
[4] Mynard, J et al., *Int. J. Numer. Meth. Biomed. Engng.*; 28:626–641,2012.
[5] Mitchell, G et al., *Hypertension*, 43:1239-1245, 2004.
[6] Rogers, W et al., *J Am Coll Cardio*, 38(4):1123-9, 2001.
[7] Franklin, S et al., *Circulation*, 96:308-315, 1997.
[8] Murgo, J et al., *Circulation*, 62:105-116, 1980.
[9] Torjesen, A et al., *Hypertension*, 64(2):259-265, 2014.

SB³C2017
Summer Biomechanics, Bioengineering and Biotransport Conference
June 21 – 24, Tucson, AZ, USA

CFD-SHAPE OPTIMIZATION COUPLING EXPLAINS PARTIAL RESTORATION OF HOMEOSTATIC WSS IN VENOUS NEOINTIMAL HYPERPLASIA

S. M. Javid Mahmoudzadeh Akherat (1), Kevin W. Cassel (1), Marta P. Wlodarczyk (1), Mary S. Hammes (2)

(1) MMAE Department
Illinois Institute of Technology
Chicago, IL, USA

(2) Department of Medicine
University of Chicago
Chicago, IL, USA

INTRODUCTION

A vascular access is necessary to provide adequate blood flow rate to provide hemodialysis for end-stage renal disease (ESRD) patients. The access that provides the best outcomes is the arteriovenous fistula (AVF), which is a direct surgical connection between the arterial and venous systems at an anastomosis. As a result of diversion of the arterial flows into the venous system and a dramatic increase in the flow, multiple mechano-transduction stimuli are triggered that result in hyperplastic growth in the medial layer of the veins. Excessive neointimal hyperplasia (NH) is an endemic condition in the renal failure patient on chronic dialysis that eventually leads to venous stenosis, excessive venous pressure, aneurysmal growth, and heart failure (Figure 1). It is believed that non-homeostatic wall shear stress (WSS) distribution, brought about by very high flow rates through curved geometries (the cephalic arch or the anastomosis) is the main triggering factor of NH and stenotic growth. Growth and remolding (G&R) is the first-principles, continuum mechanics approach to describe the soft tissue growth, and it can theoretically be applied to the case of NH growth. However, it has limited application in the renal failure patient's NH as the human data required to define constitutive relation of wall constituents' production, removal, and turnover are very scarce [1].

In the absence of an in-depth understanding of the origins of the stress-mediated proliferative growth, we have developed a completely novel method for elucidating NH growth in dialysis patients. We have hypothesized that a strong coupling of CFD and shape optimization can be utilized that seeks to restore homeostatic WSS levels through imposition of physiological definitions of WSS target values and evolution of the shape in a well-posed detailed mathematical optimization. It is shown that the CFD-shape optimization framework accurately predicts the patient-specific mode of failure and exhibits a partial restoration of homeostatic WSS distribution and hemodynamic environment. It is also shown that pressure-mediated growth is a secondary effect present in the fistula environment that causes dilative growth downstream of the anastomosis.

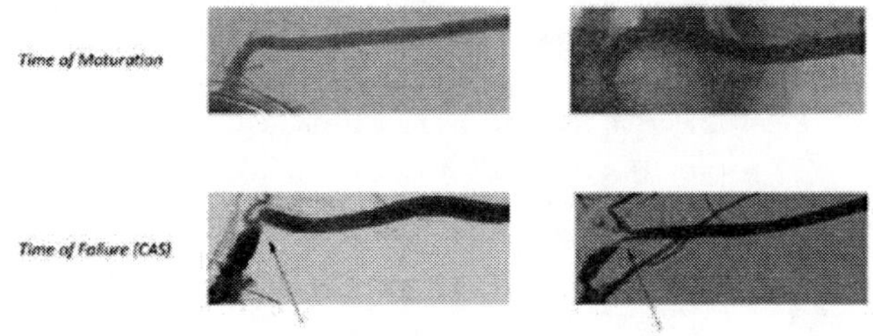

Figure 1 Cephalic arch stenosis (CAS) developed due to excessive neointimal hyperplasia in dialysis patients.

METHODS

A large cohort of end-stage renal disease patients were registered at the University of Chicago hospital and were followed for up to three years while receiving dialysis via BCF access. Patient-specific Doppler velocimetry, whole blood viscosity, X-ray venography, and other clinical measurements were performed on each patient at the time of maturation and in every follow up session. This data forms the basis for patient-specific hemodynamics simulations, where rigid impermeable walls with no-slip boundary condition are considered for a domain with patient-specific fixed pressure Dirichlet imposed at the

Poster Presentation #P244

inlet and a fixed zero-pressure zero-divergence at the outlet. Thereafter, a shape optimization module (COMSOL Multiphysics 5.2a) was coupled with the CFD simulation. The objective functional for the shape optimization problem reads,

$$J = \int |\tau_{w|actual}(q) - \tau_{w,homeostatic}| dx, \quad (1)$$

where $\tau_{w,homeostatic}$ is the homeostatic WSS target value, taken to be the mean value of the physiological range of WSS in the cephalic vein ($0.076 \leq \tau_w \leq 0.76$ Pa [2]). $\tau_{w|actual}(\mathbf{q})$ is obtained from the patient-specific CFD solution at the time of maturation and $\mathbf{q}$ is a vector quantity of parameters that controls the geometrical deformations of interest [1], provided by Bernstein basis function shape parameterization scheme. In other words, a shape is sought that best reregulates the WSS distribution subject to the constraint that Navier-Stokes equations and the fixed venous pressure difference are satisfied.

RESULTS

2D Simulations:

As can be seen from Figure 2, the shape that has an improved physiological WSS distribution (solid-blue lines) is the one that is constricted in the arch section of the vein. It also can be seen that

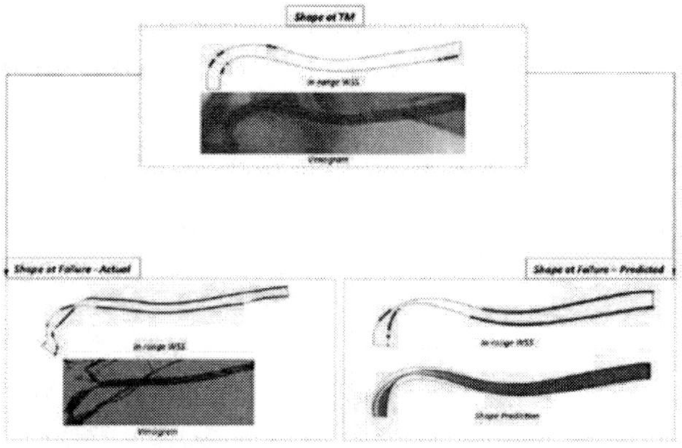

Figure 2 Partial restoration of homeostatic WSS (solid blue lines)

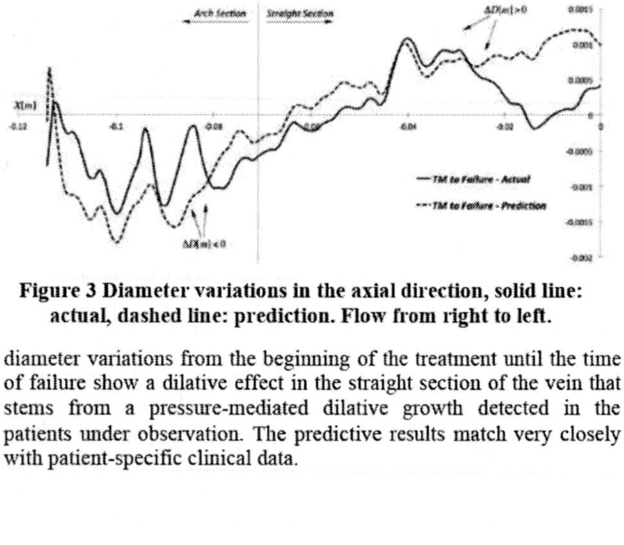

Figure 3 Diameter variations in the axial direction, solid line: actual, dashed line: prediction. Flow from right to left.

diameter variations from the beginning of the treatment until the time of failure show a dilative effect in the straight section of the vein that stems from a pressure-mediated dilative growth detected in the patients under observation. The predictive results match very closely with patient-specific clinical data.

3D Simulations:

These predictive simulations were also performed in three-dimensional patient-specific domains. The results still indicate restoration of in-range WSS and a drop in flow rate to maintain the fixed inlet pressure constraint imposition, elucidating the hypertensive pressure build up that explains the aneurysmal growth in the straight section of the vessel (see Figure 3). Figure 4 illustrates a partial restoration of homeostatic WSS (blue), and a strong regulation of high (green, $\tau_w \geq 0.76$Pa) and low (red, $\tau_w \leq 0.076$Pa) wall stresses in the 3D domain.

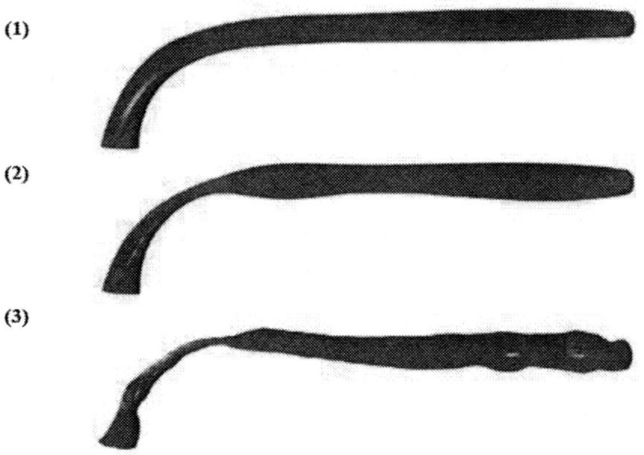

Figure 4 Patient-specific 3D simulations, (1) at time of maturation, (2) predicted failure, (3) actual failure.

DISCUSSION

It is shown that a strong coupling of CFD-shape optimization in patient-specific hemodynamics simulations of renal failure patients can accurately predict the onset and development of stenosis in the cephalic arch. It is also concluded that stress-mediated growth causes inward stenosis in the arch section and pressure-mediated growth triggers outward dilation in the straight section of the vein. It is also concluded that NH and the subsequent stenosis serve to restore homeostatic conditions in the fistula environment. This method can be employed for individualized patient-specific treatment planning and for monitoring and maintaining chronic dialysis treatment. Controlling the flow on a subject-specific scale holds promise to prevent cephalic arch stenosis [3], which is one of the application of the computational method presented that is currently under investigation.

ACKNOWLEDGEMENTS

Research reported in this publication is supported by the National Institute of Diabetes and Digestive And Kidney Diseases of the National Institutes of Health under Award Number R01DK090769.

REFERENCES

[1] Akherat, S.J.M., 2016. *Development of a Predictive Framework to Forecast Venous Stenosis* (Doctoral dissertation, Illinois Institute of Technology).

[2] Hammes M., et al. (2016) Increased Inlet Blood Flow Velocity Predicts Low Wall Shear Stress in the Cephalic Arch of Patients with Brachiocephalic Fistula Access. PLoS ONE 11(4): e0152873. doi:10.1371/journal.pone.0152873

[3] Miller, G.A, et al., (2010). The MILLER banding procedure is an effective method for treating dialysis-associated steal syndrome. *Kidney international*, *77*(4), pp.359-366.

SB³C2017
Summer Biomechanics, Bioengineering and Biotransport Conference
June 21 – 24, Tucson, AZ, USA

COMPUTATIONAL ACCESS FLOW REDUCTION EFFECT ON WALL SHEAR STRESS IN BRACHIOCEPHALIC FISTULAE

Marta P. Wlodarczyk (1), S. M. Javid Mahmoudzadeh Akherat (1), Kevin W. Cassel (1), Marry Hammes (2)

(1) Department of Mechanical, Materials, and Aerospace Engineering
Illinois Institute of Technology
Chicago, IL, USA

(2) Department of Medicine
University of Chicago
Chicago, IL, USA

INTRODUCTION

Brachiocephalic fistulae (BCF), which are one of the recommended access sites for hemodialysis patients result in 38% of failures with the primary leading cause of failure due to cephalic arch stenosis (19.5-77%) (CAS) [1]. The creation of the BCF for dialysis produces an anatomically and physiologically abnormal situation. High volume flow is diverted from a thick-walled muscular pulsatile artery into a thin-walled vein. Mean brachial artery flow increases from 56.2±20.0 mL/min before creation of an arteriovenous fistula (AVF) to 365.0±129.3 mL/min one day after and to 720.4±132.8 mL/min 28 days after creation. BCF anastomosis creates excessive flow through the curved cephalic arch that leads to non-physiological hemodynamics and creates regions of non-homeostatic wall shear stress (WSS) [2]. The extreme hemodynamic environment that BCF creates triggers the development of neointimal hyperplasia (NH) leading to CAS in most of the patients, which leads to access site failure. Because a high flow rate through a naturally curved vein creates low-WSS regions, reduction of the flow rate is a potential mitigation. However, this must be balanced by the minimum necessary flow rate required to perform effective dialysis. The effect of fistula flow reduction via banding of the juxta-anastomotic segment on the recurrence of CAS has been clinically investigated by Miller et al [1]. The Miller's procedure requires the diameter of the access lumen to be reduced by 60-80% in order to significantly affect flow [3]. As a result, the cephalic arch intervention rate was reduced from 3.34 to 0.9 per access-year. Flow reduction of a BCF arteriovenous fistula could effectively diminish the incidence of symptomatic CAS. Even though the method of banding holds promise, its effectiveness on overall homeostatic hemodynamics relaxation has not been thoroughly evaluated. Therefore, the objective of this study is to computationally assess the efficacy and level of patient-specific access flow reduction to restore physiological WSS.

METHODS

Patient specific 3-D cephalic vein and Venturi geometries: Patient specific venograms, Doppler velocimetry measurements, and whole blood viscosity (WBV) were obtained at the fistula time of maturation. Intravascular ultrasound (IVUS) images are used to form the patient-specific three-dimensional geometries used herein. Two patient geometries (patient 5 and patient 8) were used in this study. To replicate the Miller's banding zone, a five centimeter Venturi was modeled and attached to each geometry in COMSOL Multiphysics, which was the starting geometry, reference Figure 1. This allowed for a fully-developed inlet velocity profile imposition at the domain's inlet. A deformable constriction site has been created in the Venturi, three centimeters prior to the inlet of the cephalic vein. The deformable constriction was allowed to constrain a maximum of 60% of its original shape.

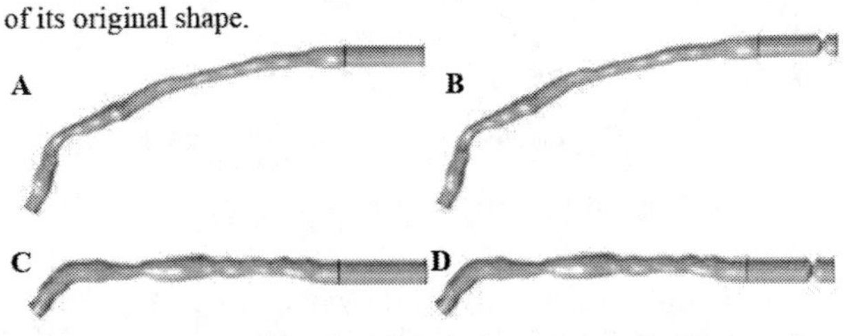

Figure 1: (A) Cephalic vein with unconstricted Venturi for patient 5; (B) cephalic vein with constricted Venturi for patient 5; (C) Cephalic vein with unconstructed Venturi for patient 8; (D) cephalic vein with constricted Venturi for patient 8.

Poster Presentation #P245

CFD Simulations: Walls were taken to be impermeable and rigid with no-slip boundary conditions. Patient-specific pressure inlet condition obtained from the Doppler velocimetry along with zero pressure Dirichlet boundary condition with a zero traction was applied at the outlet to obtain the patient-specific flow rate. Blood flow in the cephalic vein is laminar and can be taken to be almost steady under physiological conditions. WBV for the CFD simulation is taken to be the patient-specific high-shear rate asymptotic viscosity values from the viscometry tests. The steady, incompressible, Newtonian Navier-Stokes equations are solved for all the CFD-shape optimization simulations with a density taken to be $\rho = 1050$ kg/m^3.

Wall Shear Stress: A physiologic range of WSS distribution in the cephalic vein pre-surgery was statistically and computationally identified to be $0.076 \leq \tau_w \leq 0.76$ Pa [4].

Objective functional and shape optimization: The CFD simulations were completed at time of maturation with the patient-specific inlet venous pressure. The objective functional J that is being minimized is defined as

$$J = \int | \tau_{wall|Actual}(q) - \tau_{wall|Physiological,Mean} | dx \qquad (1)$$

where $\tau_{wall|Physiological,Mean}$ is the desirable physiological WSS distribution sought by the virtue of the physiological definition of the target WSS value in the cephalic vein, which we take to be equal to 0.418 Pa, the mean value of the mentioned physiological range, **q** is a vector quantity of parameters that controls the geometrical deformations of interest [4]. The objective functional and WSS distribution were monitored as the geometry was deformed.

RESULTS

Figures 2 and 3 show the WSS distribution prior to and post Miller's Venturi constriction for patients 5 and 8. The objective function value prior to constriction was 0.014936 and 0.0050366 for patients 5 and 8, respectively. After the shape optimization and the resulting banding, the objective functional was reduced to 0.012368 and 0.0028162 with the patient-specific constriction levels of 44% and 56% of the original inlet area, respectively. The velocity was reduced from 0.551 to 0.466 m/s for patient 5 and from 0.209 to 0.135 m/s for patient 8. Figure 4 shows the velocity contours at three different locations pre-and post-constriction. Note that the post constriction flows for patient 8 involve secondary flows with WSS below the 0.076 Pa threshold in the inner portion of the arch.

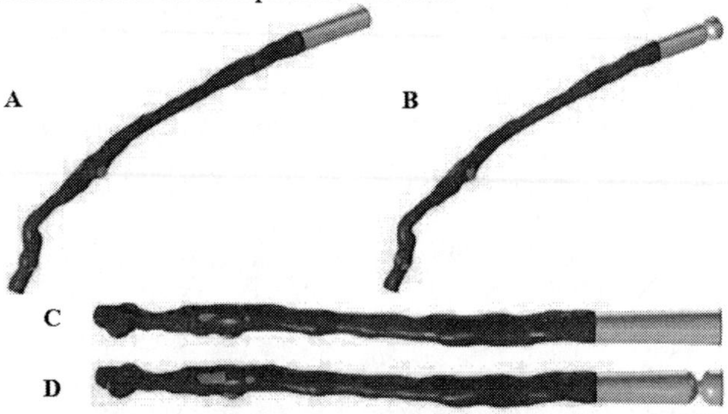

Figure 2: The three-dimensional WSS distribution at TM, Patient 5 prior to constriction (A) (C) and post constriction (B) (D), Green: $0.076 \leq \tau_w \leq 0.76$ Pa, Red: $\tau_w \leq 0.076$ Pa, Blue: $\tau_w \leq 0.76$ Pa

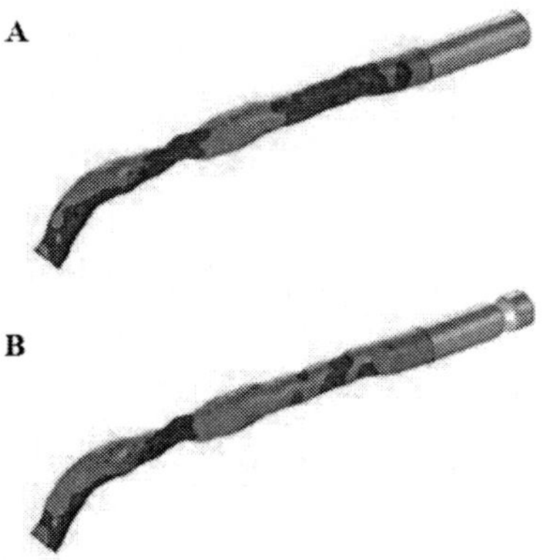

Figure 2: The three-dimensional WSS distribution at TM, Patient 5 prior to constriction (A) and post constriction (B), Green: $0.076 \leq \tau_w \leq 0.76$ Pa, Red: $\tau_w \leq 0.076$ Pa, Blue: $\tau_w \leq 0.76$ Pa

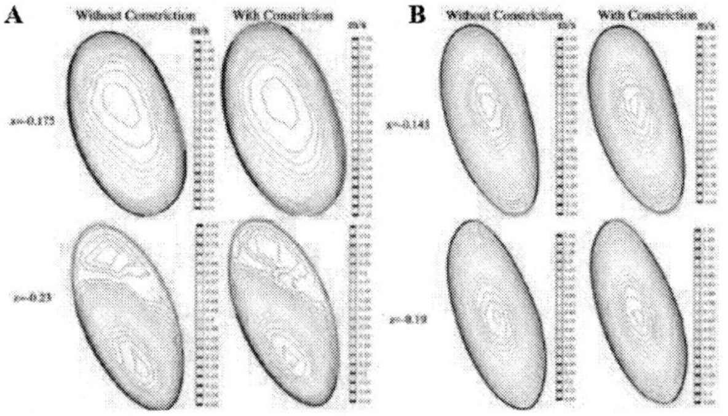

Figure 4: Cross-sectional velocity contours superimposed on WSS line plots at three different locations for patient 5 (A) and patient 5 (B) Green: $0.076 \leq \tau_w \leq 0.76$ Pa, Red: $\tau_w \leq 0.076$ Pa, Blue: $\tau_w \leq 0.76$ Pa

DISCUSSION

This study reveals that BCF access results in non-homeostatic WSS distribution that triggers the growth of neointimal hyperplastic lesions followed by venous stenosis and thrombotic incidents. The presented computational method can be clinically utilized to determine the level of banding necessary to restore the WSS closer to homeostatic values on a patient-specific scale via Miller's banding method. Results demonstrate that the level of banding required for treatment planning is, in fact, a patient-specific value that needs to be non-invasively investigated prior to the banding surgery.

ACKNOWLEDGEMENTS

Research reported in this publication is supported by the National Institute of Diabetes and Digestive and Kidney Diseases of the National Institutes of Health under Award Number R01DK090769.

REFERENCES

[1] Miller, G. A., et al., J Vasc Access, 11(4):281-7, 2010 Oct-Dec;
[2] Albayrak, R., et al., J Clin Ultrasound 35(3): 133–137, 2007.
[3] Miller G.A., et al., Kidney Int.,77(4):359-66, 2010.
[4] Mahmoudzadeh Akherat, S.J.M., 2016. Development of a Predictive Framework to Forecast Venous Stenosis (Doctoral dissertation, Illinois Institute of Technology).

SB³C2017
Summer Biomechanics, Bioengineering and Biotransport Conference
June 21 – 24, Tucson, AZ, USA

DYNAMIC MESH COMPUTATIONAL FLUID DYNAMICS OF THE CHICK EMBRYONIC HEART BASED ON 4D HIGH-FREQUENCY ULTRASOUND IMAGING

Sheldon Ho (1,2), Yue Yin Loh (1), Hadi Wiputra (1), Choon Hwai Yap (1)

(1) Department of Biomedical Engineering. National University of Singapore, Singapore

(2) NUS Graduate School for Integrative Science and Engineering, National University of Singapore

INTRODUCTION

Congenital cardiovascular malformations are the most common type of birth defect, affecting approximately 0.6–1.9% of newborns each year, and are the leading causes of birth defect related deaths[1, 2]. The only treatment is post-natal surgeries which are drastic and expensive, yet yield imperfect outcome despite advances over the years. Recent investigation has provided evidence that the mechanical forces of blood flow can influence the development of the embryonic heart to lead to congenital malformations[3-5]. Thus, further studies in embryonic cardiovascular fluid mechanics are warranted to understanding the role of mechanics in congenital heart diseases.

A good way to perform detailed investigation of embryonic cardiovascular mechanics is to perform high resolution imaging on followed by computational fluid dynamics (CFD). Current embryonic imaging techniques either have insufficient penetration depths[6], resolution[7] or are destructive[8], which prevent repeated scans of the same subjects. It is thus difficult to get 4D images of embryonic hearts for CFD.

In the current study, we first present a novel technique for 4D high-frequency ultrasound imaging of the chick embryo heart, which enables non-invasive quantification of the cardiac dynamics, and repeated imaging of the same subject. We then demonstrate that these images could support dynamic mesh CFD simulations of the HH25 (4.5 days old) chick embryonic heart. The fluid mechanics of the HH25 chick embryonic ventricle is also discussed.

METHODS

Scans and simulations for 2 chick embryos are presented. Fertilized Bovans Goldline chicken eggs at Hamburger-Hamilton (HH) stage 25 had their shell at the top removed, to expose the embryos for imaging.

High-frequency (40 MHz) 2D B-mode cine-ultrasound imaging was performed at multiple planes for a chick embryo at Hamburger-Hamilton (HH) stage 25. Spatial and temporal correlation was used to determine matching phases in the cardiac cycle between neighboring imaging planes, and across different heart beats within the same imaging plane. Images from 20-25 cardiac cycles within each plane were averaged into one ensemble-averaged cardiac cycle, using quadratic mean instead of arithmetic mean thus resulting in good contrast between blood and tissue spaces, which could not be achieved with arithmetic mean. Images from different imaging planes were adjusted to match each other in cardiac cycle phase. This creates a 4D image of the blood space within the embryonic cardiovascular system. Vascular Modeling Toolkit (VMTK, www.vmtk.org), was used for 3D segmentation of the ventricle over 20-30 time points.

The chick embryonic heart ventricular wall motion was described mathematically with equation 1. The wall location, $r_{model}(\theta,\varphi,t)$, from the centroid was expressed in spherical coordinates as a function of the initial wall location, $r_0(\theta,\varphi)$, the characteristic wall motion waveform, $\Omega(t)$, and the amplitude of wall motion, $\alpha(\theta,\varphi)$:

$$r_{model}(\theta,\varphi,t) = \alpha(\theta,\varphi)\,\Omega(t) + r_0(\theta,\varphi) \qquad (1)$$

where θ and φ were the angle coordinates, $\alpha(\theta,\varphi)$ was expressed as a the 3rd order spherical harmonics function with the least square best fit to the segmented images, and $\Omega(t)$ was expressed as a 5th order Fourier series with a least square best fit to the cube root of the volume over time curve.

Pulsewave Doppler was used to measure the flow velocity waveform at the proximal outflow tract of the ventricle for four sets of embryos, and ultrasound was used to measure the outflow tract sizes. Data from the 4 embryos were averaged and used as CFD boundary conditions.

Dynamic mesh CFD simulations were performed with ANSYS FLUENT, using a UDF to enforce the ventricular wall motion. Both inlet and outlet was also extruded by at least 2x of their diameters length to remove boundary conditions artefacts. The simulation was run with more than 10^6 elements over 3 cycles, and data from the last cycle was used for analysis. Blood was modeled as an incompressible Newtonian fluid with density of 1060 kg/m^3 and viscosity of 0.0015 $Pa{\cdot}s^2$.

RESULTS

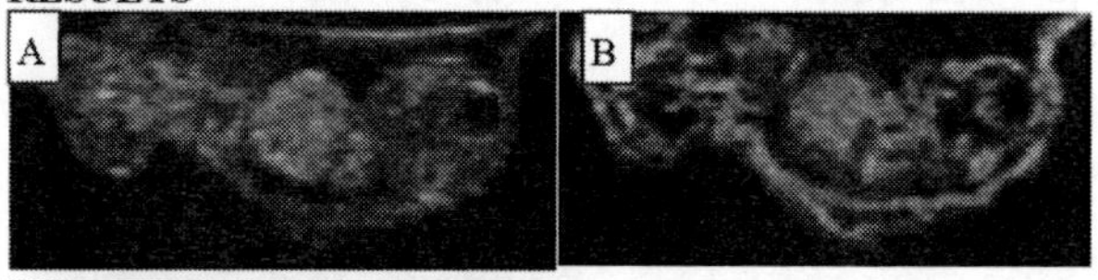

Figure 1: a) Raw ultrasound image. b) Image after ensemble averaging via spatial and temporal correlation. d) Image after speckle filter.

Figure 1a shows the raw ultrasound image, demonstrating that tissue and blood spaces could not be distinguished. With Figure 1b, we demonstrate that using quadratic averaging to ensemble average more than 20 cardiac cycles into one average cycle (via spatial-temporal correlation) enabled excellent delineation of the heart chambers and blood vessels. Applying a speckle to figure 1b gave a clean image (Figure 1c). The hearts were reconstructed at all time points from these ensemble-averaged, filtered images.

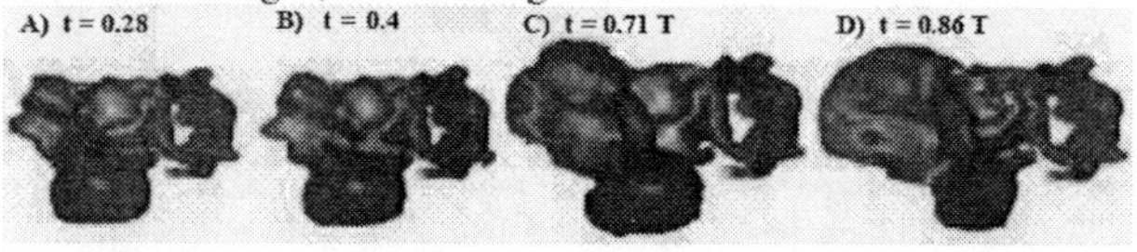

Figure 2: 3D reconstructions of chick embryonic heart at different time point based on high frequency ultrasound imaging. Red, yellow – left, right atria; green – ventricle; purple – outflow tract.

The reconstructed heart (figure 2) revealed quantitative information on the dynamics of the embryonic heart, such as the stroke volumes of the atria and ventricle, and the peristalsis-like motion of the outflow tract, which at times was completely collapsed. With our imaging technique, for the first time, fine anatomic structures such as the pharyngeal aortic arches could be non-invasively imaged at high resolution. The Doppler velocity measurements at the outflow tract are shown in figure 3.

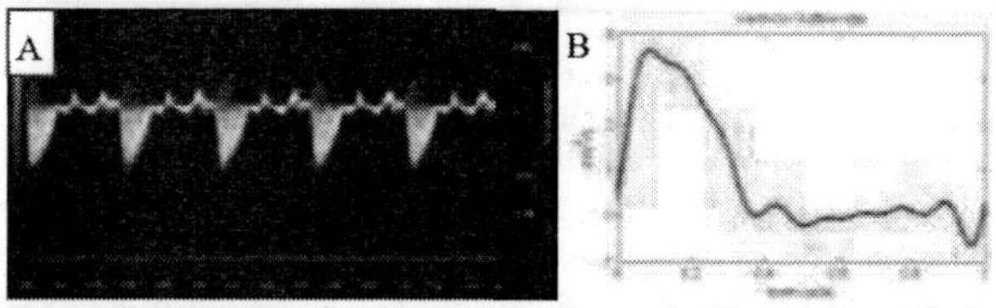

Figure 3: a) Pulsewave Doppler at the outlet tract of ventricular of 4.5 days old chick. b) Average waveform from 4 subjects.

Figure 4 shows the wall shear stress magnitudes (WSS) and streamlines of the ventricle at different phase of the cardiac cycles. Elevated WSS were concentrated around the entry and exit region of the ventricle, which has smaller cross-sectional area, and decrease towards the apex of the heart. WSS in the right side of the ventricle were higher than the left side of the ventricle, and experienced different shear stress waveform over time. Due to the low Reynolds number, flow streamlines were generally well behaved and stable.

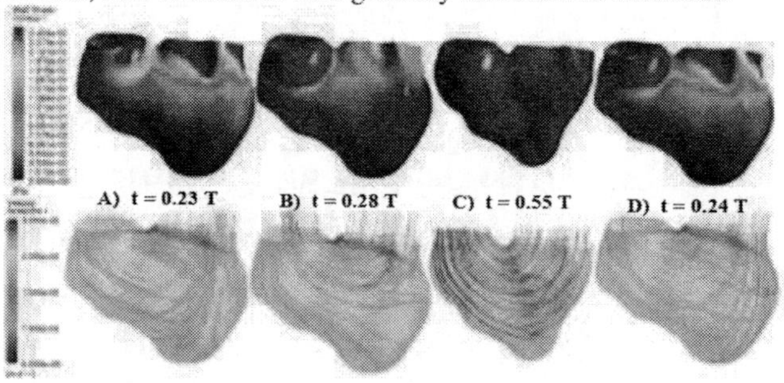

Figure 4: Wall shear stress and streamline flow velocity of the ventricle during a)maximum inflow, b)maximum outflow, c)end systole, d)end diastole. Outlet is on the left and inlet is on the right.

DISCUSSION

We have developed an imaging technique on small animal embryonic hearts with sufficient resolution to support CFD simulations of flow within the heart. Our technique uses high-frequency ultrasound, and as such, has higher resolution than MRI, larger depth than OCT, and is completely non-invasive, requiring no contrast agent. The technique is suitable for later stages chick embryos, which are larger and optical access to the heart is low. A further advantage of using ultrasound is the ready availability of Doppler measurements to obtain boundary conditions.

Due to the small diameter and low flow velocity of the HH215 chick embryos, flow manifested as highly orderly laminar flow with low Reynolds number of ~50. WSS over time waveforms generally adopted the shape of the flow velocities nearby. In the embryonic heart, however, flow moves linearly from the left atria into the left ventricle before moving into the right ventricle. Due to the contraction of the ventricular wall, velocity over time waveforms in the left ventricle and right ventricle were different, resulting in differences in the WSS waveforms. The left ventricle also experiences lower WSS than the right ventricle. Highest WSS were observed around the inlet and outlet regions, due to the small cross-sectional area of these regions, while the lowest were at the apex and in the left ventricle due to low velocities.

Our imaging and CFD techniques can be further extended to other parts of the heart, such as the atria and the outflow tract. Future work could also include using our technique to compare the fluid mechanics of normal chick hearts versus that in left-atria-ligated chick hearts, which are models for the Hypoplastic Left Heart Syndrome disease.

ACKNOWLEDGEMENTS

We thank the National University of Singapore Young Investigator award, entitled "Fluid Mechanics and Mechanobiology of the Congenital Outflow Tract Malformation" (PI: Yap) for funding.

REFERENCES

[1] M. D. Reller et al, J Pediatr 153 (6), 807-813 (2008).
[2] J. I. E. Hoffman and S. Kaplan, JACC 39 (12), 1890-1900 (2002).
[3] J. L. Lucitti et al, J Exp Biol 208 (Pt 10), 1877-1885 (2005).
[4] K. Tobita et al, Am J Physiol - Heart Circ 279 (3), H959 (2000).
[5] N. Hu et al, Anat Rec (Hoboken) 292 (5), 652-660 (2009).
[6] J. Männer, The Anatomical Record 259 (3), 248-262 (2000).
[7] A. R. Kherlopian et al, BMC Syst Biol 2, 74 (2008).
[8] J. T. Butcher et al, Dev Dyn 236 (3), 802-809 (2007).

A FLUID-STRUCTURE INTERACTION MODEL OF A CUFFED CAROTID ARTERY OF AN APOE$^{-/-}$ MOUSE

Ryan M. Pedrigi (1,2), Miten B. Patel (1), Vikram V. Mehta (1), Fotios Savvopoulos (1), Avinash Kondiboyina (1), Lucas H. Timmins (3), Rob Krams (1)

(1) Department Bioengineering
Imperial College London
London, UK

(2) Department of Mechanical and Materials Engineering
University of Nebraska
Lincoln, NE, USA

(3) Department of Bioengineering
University of Utah
Salt Lake City, UT, USA

INTRODUCTION

Atherosclerotic plaques tend to form in vessels with high curvature or near bifurcations that cause changes in blood flow magnitude and direction compared to straight, non-bifurcating vessels. These blood flow profiles are often denoted disturbed or atherogenic. To study the causal relationship between disturbed blood flow and plaque development, we previously developed a blood flow-modifying cuff for placement around one of the carotid arteries of ApoE$^{-/-}$ mice (1). We recently demonstrated an increase in several wall shear stress (WSS) metrics that describe different types of atherogenic flow in the cuffed vessel using computational fluid dynamics (CFD) (2).

Although CFD is the commonest method to quantify flow in diseased arteries, it is limited by the assumption of a rigid wall. Consideration of the dilatation and contraction of the elastic artery over the cardiac cycle will influence the calculation of WSS and, in addition, the wall mechanics may play a mechanobiological role in disease development. Therefore, in this study, we developed a fluid-structure interaction (FSI) model of the instrumented and control carotid artery of an ApoE$^{-/-}$ mouse from in vivo micro-CT images and demonstrate physiologically realistic values for wall stress/strain and shear stress that are perturbed with cuff placement.

METHODS

Animals and Imaging. All experimental procedures complied with the Animals Act of 1986. One female ApoE$^{-/-}$ mouse was placed on a high-fat diet at 11 weeks and, 2 weeks later, instrumented with a blood flow-modifying cuff around the left carotid artery. Nine weeks after cuff placement, the mouse was scanned with micro-CT to obtain an accurate 3-D reconstruction of the in vivo geometry of each carotid artery. Micro-CT images were loaded into the Vascular Modeling Toolkit to segment each carotid artery and smooth the 3-D reconstruction. In addition, ultrasound measurements of blood velocity were acquired at the inlet of each carotid artery 3 days after μCT imaging.

Fluid-structure interaction modeling. The wall thickness and composition of the 3D reconstructed carotid artery geometry was obtained from co-registered histology and meshed with linear hybrid hexahedral elements using custom MATLAB software. This wall mesh was imported into Abaqus v6.14 where the fluid domain was meshed with linear hexahedral elements and a FSI simulation was performed. Optimal mesh density was determined via a convergence test. The final FSI model consisted of the blood, vessel wall, perivascular tissue to constrain the motion of the vessel, an inlet extension to allow the idealised parabolic profile to adapt to the vessel geometry and an outlet extension to simulate the downstream vasculature (3).

Blood was assumed to be a non-Newtonian fluid described by the Carreau-Yasuda model, the vessel wall was modeled as a hyperelastic incompressible isotropic material using the Ogden form of the strain energy function, and the perivascular tissue was modeled as a compliant and linear elastic material.

The outlet extension consisted of two parts; a linearly elastic part to simulate the compliance of the downstream vasculature and a rigid part with a reduced diameter to model its resistance. Reported values for compliance and resistance (4) determined the length and Young's modulus of the elastic part and the contraction ratio of the rigid part.

FSI simulations imposed a pulsatile flow of the blood at the inlet of the vessel by assigning a parabolic flow profile measured from ultrasound. To account for residual stress and pre-stress of the vessel due to diastolic pressure, the vessel wall was initialized with circumferential residual stress computed from computationally opening the vessel to 92° (5) and then closing followed by incrementally pressurising the vessel to 66.5 mmHg using the backward integration method (6); these simulations employed a solids-only approach. Additional boundary conditions included fully constraining movement of the outer-most layer of the perivascular tissue and, in the case of the instrumented vessel, of the cuff region. Every simulation was run over three cardiac cycles. Shear stress obtained from the second cycle was found to be within 1% of that in the third cycle, demonstrating periodic convergence. During the third cycle, blood velocity, shear stress, wall (Cauchy) stress, and finite strain were exported at 50 points over the cardiac cycle.

RESULTS

The control carotid FSI simulations exhibited a physiologically realistic pressure waveform that ranged from 51 to 82 mmHg to drive the prescribed velocity profile over the cardiac cycle (Figure 1). Resultant mean circumferential vessel wall Cauchy stress and finite strain in the control vessel were 40.6 kPa and 26.3%, respectively, which aligns with previous mechanical behaviour studies of the mouse carotid artery (7). The nominal vessel dilatation was 52 μm (across the diameter) over the cardiac cycle. WSS in the control vessel had a mean magnitude of 8.6 Pa with near homogeneity over the length and cross-section of the vessel.

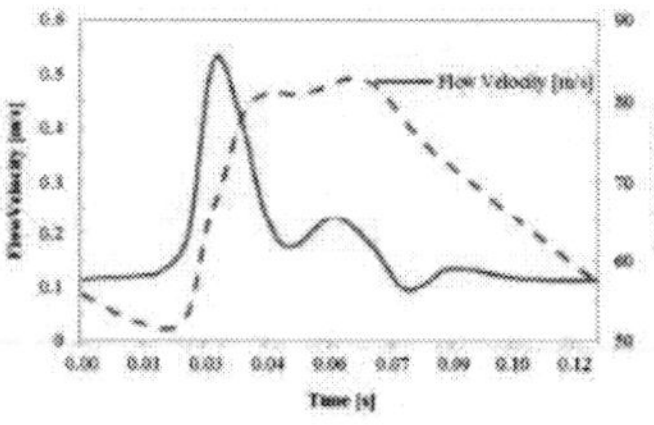

Figure 1: Control carotid flow velocity and pressure waveforms

In contrast to the control, the instrumented vessel exhibited reduced WSS upstream of the cuff, high WSS within the cuff, and a combination of low and high WSS downstream of the cuff depending on position (Figure 2). In addition, stress concentrations of 88.7 kPa were observed within the wall of the vessel immediately outside of the inlet and outlet of the cuff (note: not shown in Figure 2 which shows the inner wall only).

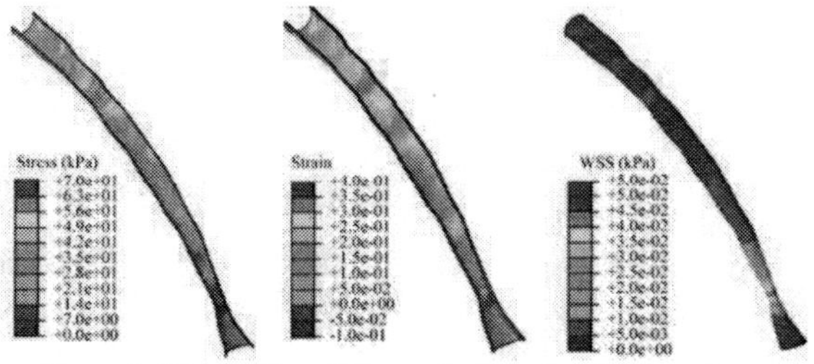

Figure 2: Distribution of stress, strain and wall shear stress over the endothelium of the instrumented vessel

Simulations were also performed without pre-stress, which caused a much larger vessel dilatation of 100 μm (across the diameter) to accommodate the diastolic pressure, followed by a further dilatation of 60 μm from diastole to systole, and reduced mean circumferential wall Cauchy stress and finite strain of 22.6 kPa and 17.4%, respectively at systole

DISCUSSION

The mouse has long been used to study the biomechanics of atherosclerosis, with virtually all studies employing CFD to examine flow within diseased and non-diseased vessels. To our knowledge, only one group has utilised an FSI model to report blood flow profiles in the mouse carotid artery and aorta, but with no consideration of the wall mechanics (8). Our study is the first to quantify the vessel wall mechanics and the hemodynamics in both a non-diseased and atherosclerotic artery of the mouse. This is important as we recently hypothesized that disturbed vessel wall mechanics may act synergistically with disturbed blood flow to promote different advanced plaque types (9). Such disturbances to the wall mechanics occur as the plaque develops and as the result of instrumentation of the carotid artery with a cuff, which may exacerbate the atherogenic effect of the cuff particularly at the interface with the surrounding vessel where plaques are known to develop. Our results herein provide initial support for this hypothesis. We also demonstrate the importance of incorporating the residual stress and pre-stress of the vessel for accurate quantification of the wall mechanics.

ACKNOWLEDGEMENTS

This work was supported by the British Heart Foundation (grants RG/11/13/29055 and PG/15/49/31595).

REFERENCES

[1] Cheng, C et al., *Circulation*, 113:2744-2753, 2006.
[2] Pedrigi, R et al., *R Soc Open Sci.*, 3:160588, 2016.
[3] Pahlevan, NM et al., *Ann Biomed Eng.*, 39:1470-81, 2011.
[4] Cuomo, F et al., *Ann Biomed Eng.*, 43:1555-70, 2015.
[5] Hansen, L et al., *Ann Biomed Eng.* 41:682-693, 2013.
[6] Speelman, L et al., *J.Biomech.*42:1713-1719, 2009.
[7] Eberth, JF et al., *Ann Biomed Eng.* 37:1093-104, 2009.
[8] De Wilde, D et al., *Ann Biomed Eng.* 44:2327–2338, 2016.
[9] Pedrigi, RM et al., *ATVB*, 34:2224-31, 2014.

A MULTISCALE MODEL OF THE ENDOTHELIAL GLYCOCALYX AS MECHANOSENSOR OF HEMODYNAMIC SHEAR FORCES

Diego Gallo (1), Pablo Sáez (2), Umberto Morbiducci (1)

(1) Department of Mechanical and Aerospace Engineering
Politecnico di Torino
Turin, Italy

(2) Laboratori de Calcul Numeric (LaCaN)
Universitat Politecnica de Catalunya
Barcelona, Spain

INTRODUCTION

a multiscale model of the endothelial glycocalyx as mechanosensor of hemodynamic shear forces

Atheroma plaque formation is one of the leading causes of cardiovascular disorders in the developed countries. The glycocalyx (GC), a bush-like layer coating the endothelial cells (ECs), plays a key role in the mechano-sensing and -transduction processes promoting atherosclerosis [1]. In fact, GC has been identified as the mechanosensor of the forces exerted by the streaming blood onto the vascular endothelial lining [2,3,4].

The focal development of atherosclerosis has been attributed to disturbed shear forces (see, e.g., ref. [5]). However, these forces, exerted by the flowing blood, are transmitted to the endothelial cells (ECs) cytoskeleton through the transmembrane proteins, i.e. the terminal structures of the GC layer on the ECs side. Although the ECs gene expression and protein production as a consequence of disturbed shear has been extensively studied [6], the mechanisms by which fluid shear forces are transmitted to the subcellular elements through the endothelial transmembrane anchoring structures is still not fully explored.

In this study a multiscale approach is applied to analyze how fluid shear forces are transmitted to the ECs transmembrane anchors through the mechanical response of the GC layer. To do it, wall shear stress (WSS) vector time histories are calculated at the luminal surface of a carotid bifurcation, a vascular district prone to atherosclerotic lesions development, through image-based computational hemodynamic modelling. Then, a mechanical model of the GC structure, based on the Timoshenko beam theory [7], is applied to simulate the mechanical behavior of the GC layer. Realistic atheroprone and atheroprotective WSS vector time histories, as obtained from computational hemodynamics, are prescribed to the idealized GC structure, and the dynamic forces transmitted to the anchoring elements on the EC membrane are evaluated. Finally, the comparison of the distribution of WSS phenotypes and of the corresponding mechanical forces transmitted to the anchors at the membrane level is performed, and the possible scenario opened by the presented findings is discussed.

METHODS

An ostensibly healthy carotid bifurcation geometry was reconstructed from angiographic computed tomography images [8]. The finite volume method was applied to solve the governing equations of the fluid motion under unsteady flow conditions. The general purpose computational fluid dynamics (CFD) code Fluent (ANSYS Inc., USA) was used on a tetrahedral computational mesh-grid of cardinality 1400000. Exhaustive details on the computational setting can be found elsewhere [8]. From the CFD simulation, the near-wall hemodynamics was characterized in terms of time-averaged WSS (TAWSS), and oscillatory shear index (OSI).

The endothelial GC presents a 3D bush-like structure with a constant spacing of 20 nm in all directions and a diameter of 10-12 nm [9; 10]. The anchoring locations present an hexagonal-like distribution and are spaced each other 100 nm [11; 12]. Based on these observations, Weinbaum and co-workers [13] proposed a continuum mechanics model of the endothelial GC, that has been adapted here. In detail, the hexagonal symmetric structure was simplified to a single branch centered in the anchor. This idealization relies on the assumption that the same fluid shear stress acts on all the elements of the hexagonal structure. The Timoshenko beam theory [7] was applied in 3D to model GC deflection and to gather the reaction forces in the anchoring structures. In order to solve the Timoshenko beam theory, the finite element method-based commercial software ABAQUS was used. The shear stress was applied to 1/6 of the upper segment of the beams [13]: in fact, the GC layer acts as a dense porous medium to the blood. As a consequence, blood flow is not able to penetrate the GC layer in depth, so that the velocity profile within the layer rapidly decreases from the GC tip to the lumen surface, vanishing close to the EC membrane surface [13]. The attachment of the GC to the anchoring structures was modeled in terms of Dirichlet boundary condition, with zero imposed displacements. The fluid stimuli $\mathbf{F}_{shear}$ (as the force derived from the WSS vector applied to the surface of the upper segment of the GC), and the mechanical forces transduced by the glycocalyx $\mathbf{F}_{mem}$ (obtained at the GC anchor point on the EC membrane) were calculated at each node on the luminal surface of the carotid bifurcation. For a quantitative description of the role played by the GC layer in transducing the fluid stimuli to the EC membrane, we introduce here a new indicator, the Oscillatory Force Index (OFI):

$$\mathrm{OFI} = 0.5\left[1-\left(\frac{\left|\int_0^T \mathbf{F}_{mem}\,\mathrm{d}t\right|}{\int_0^T \left|\mathbf{F}_{mem}\right|\cdot \mathrm{d}t}\right)\right] \qquad 0 \le \mathrm{OFI} \le 0.5$$

Inspired by the OSI, OFI is a measure of the oscillations of the mechanical force $\mathbf{F}_{mem}$ transmitted by the GC to the EC membrane.

RESULTS

A snapshot of the deformed shape of the Timoshenko beam-based model of one GC is presented in Fig. 1, where the deformation of the structure is obtained by applying an instantaneous fluid shear force $\mathbf{F}_{shear}$ as obtained from CFD simulations.

Among the main findings of the study, we report that: (1) discrepancies between the OSI distribution and OFI distribution at the luminal surface can be observed, in particular in the bifurcation region (Fig. 2, top row); (2) the cycle-averaged value of the magnitude of the force transmitted to the transmembrane anchor $\mathbf{F}_{mem}$ is, in general, similar to the distribution of the fluid shear force $\mathbf{F}_{shear}$, in particular in the bifurcation region, where the lowest forces values can be observed (Fig. 2, bottom row). These findings suggest that the GC layer does not merely transmits near-wall forces to the EC membrane as they are: the GC mechanical transduction could lead to a modification of the sensed pattern of blood shear forces, especially in terms of direction.

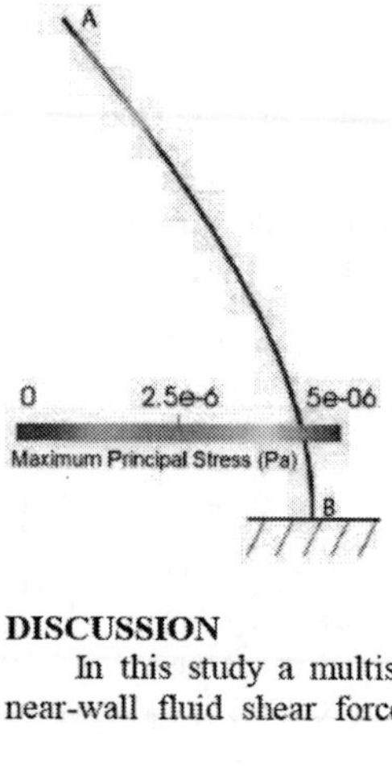

Figure 1: explanatory example of GC deformed by applying an instantaneous fluid shear force $\mathbf{F}_{shear}$ as obtained from CFD simulations. $\mathbf{F}_{shear}$ is applied in A, $\mathbf{F}_{mem}$ in B. The deformed GC structure is color-coded with the instantaneous local value of the maximum principal stress.

DISCUSSION

In this study a multiscale approach is applied to analyze how near-wall fluid shear forces are transmitted to the transmembrane anchors of ECs, through the mechanical response of the GC layer. The presented analysis, although preliminary, suggests that the presence of the GC layer alters focally both the magnitude and directionality patterns of the real forces acting on the membrane of ECs, with respect to WSS stimuli, widely considered as a localizing factor of vascular disease. This study is intended to investigate the mechanical role of the GC layer in transmitting hemodynamic shear forces to the EC membrane assuming that the acting near-wall fluid forces are not disruptive of the GC layer itself. This is like to say that the study is intended to investigate the role of GC mechanosensors at a pre-disease stage. In conclusion, the approach proposed here could contribute to clarify the mechanisms of transmission of local near-wall fluid forces sensed by the GC to the EC membrane, thus bridging the gap of knowledge still existing. In particular, mapping near-wall forces distribution at the luminal surface *versus* the forces transmitted to the EC membrane could represent a powerful tool to link hemodynamics to the mechanobiology of the endothelium.

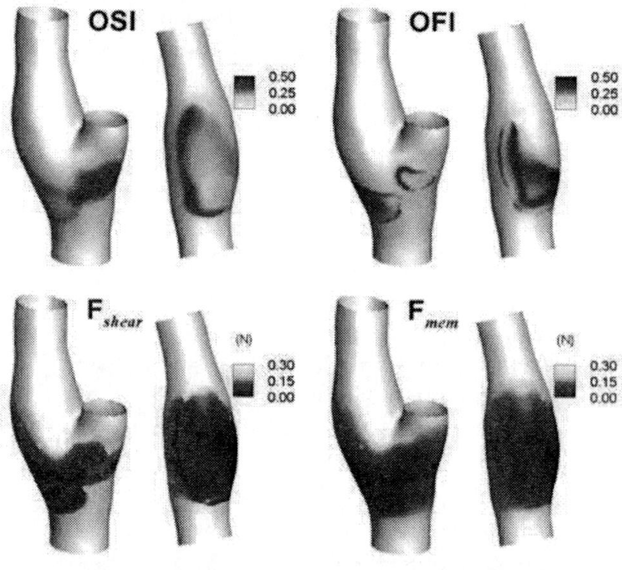

Figure 2: OFI *vs* OSI distributions (upper panel), and cycle-average $\mathbf{F}_{shear}$ *vs* $\mathbf{F}_{mem}$ distributions (lower panel) at the luminal surface of the carotid bifurcation. The role played by GC in modifying the near-wall hemodynamic forces direction, when they are transmitted to the transmembrane anchors, can be appreciated by visual inspection.

REFERENCES

[1] Tarbell, JM, *Annu Rev Biomed Eng*, 5:79-118, 2003.
[2] Davies, PF et al., *Ann Biomed Eng*, 33(12):1714-8, 2005.
[3] Tarbell, JM et al., *Ann Biomed Eng*, 33(12):1719-23, 2005.
[4] Chien, S, *Am J Physiol Hear Circ Physiol*, 292(3):H1209-24, 2007.
[5] Morbiducci, U et al., *Thromb Haemost*, 115(3):484-92, 2016.
[6] Chiu, JJ, Chien S, *Physiol Rev*, 91(1):327-87, 2011.
[7] Timoshenko, SP, *Philos Mag Ser*, 43:125-131, 1922.
[8] Morbiducci, U et al., *J Biomech*, 44:2427-2438, 2011.
[9] Squire JM, et al., *J Struct Biol*, 136(3):239-55, 2001.
[10] Pries AR, et al., *Pugers Arch*, 440(5):653-66, 2000.
[11] Rostgaard J., et al., *Microvasc Res*, 53(1):1-13 1997.
[12] Arkill KP, et al., *Biophys J*, 101(5):1046_56, 2011.
[13] Weinbaum, S et al., *Proc Natl Acad Sci*, 100(13):7988-95, 2003.

QUANTIFICATION OF VENTRICULAR HEMODYNAMIC AND WALL SHEAR STRESS ABNORMALITIES IN DISCRETE SUBAORTIC STENOSIS

Jason A. Shar (1), Philippe Sucosky (1)

(1) Mechanical and Materials Engineering
Wright State University
Dayton, Ohio, United States of America

INTRODUCTION

Discrete subaortic stenosis (DSS) is an acquired lesion that obstructs blood flow through the left ventricular outflow tract (LVOT) [1]. It results from either the formation of a circular fibromuscular rim of tissue, or muscular narrowing [1]. Risks associated with the lesion include left ventricle hypertrophy, aortic regurgitation, ventricular septal defects, and death [1]. DSS can form in isolation or concurrently with other congenital defects such as a bicuspid aortic valve [1]. The disease progresses rapidly, especially in children during the early stages of life [1]. DSS affects approximately 6.5% of the adult population and constitutes over 20% of reported LVOT obstruction in children [1], [2]. The only treatment consists of the surgical resection of the membrane [1]. It has been reported, however, that recurrence in patients is high (26-37%) over a period of thirteen years [1]. The precise etiology of the lesion is not known, but thought to have multifactorial causes ranging from inflammatory responses, secondary genetic factors, and cell proliferation due to shear forces imposed by morphological abnormalities [1].

Wall shear stress (WSS) can be measured *in vivo* by echocardiography via calculation of the spatial blood flow velocity gradients. However, this modality is limited by poor spatial resolution. Computational fluid dynamics has the potential to provide a more accurate characterization of the hemodynamic environment within the left ventricle. To date, there have been few computational studies that have characterized blood flow abnormalities due to DSS at the LVOT. One such study utilized a simple ventricle geometry and altered the aortoseptal angle to determine the resulting WSS at the LVOT and better understand the conditions in which DSS forms [2]. Other studies have accurately modeled blood flow in the left ventricle by utilizing more anatomically correct geometries and blood flow conditions, and have characterized velocity streamlines and vorticity [3]. However, no study has characterized hemodynamics of the LVOT in the presence of DSS. The objective of this study was to computationally quantify WSS near the LVOT in a contracting left ventricle, in both a normal and DSS geometry. The hypothesis is that DSS creates higher WSS conditions at the LVOT.

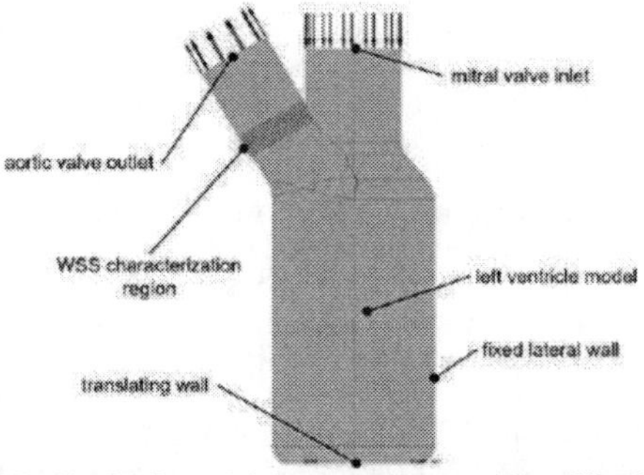

Figure 1 Idealized left ventricle geometry considered in the model.

METHODS

As a first approximation, the left ventricle geometry was modeled in Solidworks as an axially-contracting cylinder (Figure 1). The height and diameter of the cylinder (7.2 and 4.2 cm, respectively) were based on patient measurements taken via echocardiography at end-diastole [4]. This methodology resulted in an end-diastolic left ventricular volume of 99.6 mL, which falls within the acceptable range of published values [4]. A fluid inlet (2.5 cm in diameter) and an outlet (1.9 cm in diameter) were added at the locations of the mitral and aortic valves, respectively. A second geometry was created with DSS imposed at the

LVOT approximately 20 mm away from the outlet. Due to the lack of DSS dimensional data in the literature, the lesion was approximated as a 0.6 cm-wide rim of tissue narrowing the LVOT by 32%.

Ansys CFX was used to compute the fluid flow. Blood was approximated as Newtonian, with a density ρ of 1050 kg/m^3 and a dynamic viscosity μ of 0.0035 kg/s [3]. Ventricular contraction was simulated using a displacement profile, and the maximum stroke length (5.2 cm) was determined to enforce a physiologic end-systolic volume of 72.0 mL [4]. The left-ventricle geometry was meshed using tetrahedral elements and consisted of 325,079 and 353,943 elements, in the normal and stenosed anatomy, respectively. The maximum aspect ratio for the normal and stenosed anatomy was 1.98892e+03 and 1.98892e+03, respectively. Mass flow rate profiles were prescribed at the mitral valve inlet and aortic valve outlet to simulate fluid ejection during systole and ventricular refill during diastole under a physiologic cardiac ouput of 5.0 L/min. The cardiac cycle (0.86 s) simulated a physiologic heart rate of 70 beats per minute and was discretized in 100 time steps. The simulations were run for two cardiac cycles. The unsteady flow solution was computed assuming incompressible laminar flow using the Navier-Stokes equations,

$$\rho \frac{D\underline{\mathbf{V}}}{Dt} = \rho\underline{\mathbf{g}} - \nabla P + \mu\nabla^2\underline{\mathbf{V}}, \quad (1)$$

and the continuity equation,

$$\nabla \cdot \underline{\mathrm{V}} = 0, \quad (2)$$

where P is the pressure, $\underline{\mathbf{V}}$ is the fluid velocity, and $\underline{\mathbf{g}}$ is the gravitational acceleration. The analysis consisted of the characterization of the flow streamlines, vorticity, WSS, pressure, and velocity fields. WSS was computed in the same data collection region for both geometries (see Figure 1).

RESULTS

Average and maximum WSS values predicted at the LVOT were 5.7 and 19.7 Pa, respectively, in the normal anatomy, and 7.6 and 22.7, respectively, in the stenosed anatomy (Figure 2). Maximum and average velocity values at the aortic outlet in the normal ventricle were 0.82 and 0.62 m/s, respectively, which agreed with previous reports [3]. The presence of DSS resulted in a 16% and 32% increase in average and maximum velocity at the model outlet, respectively. Pressure drop was captured in the stenosed geometry between two points located immediately before and after the DSS lesion (Figure 3). The maximum pressure drop for the normal and stenosed geometries were 17.9 and 28.4 mmHg, respectively, which is consistent with published values [1]. Flow streamlines taken during systole at the LVOT (Figure 3) show increased helicity in the stenosed geometry, as well as evidence of jet

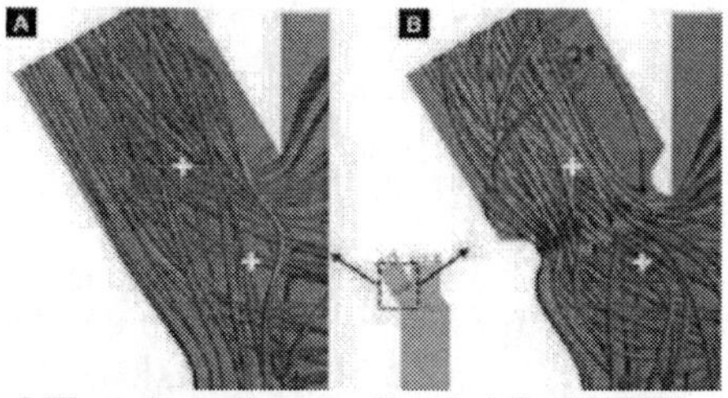

Figure 3 Flow streamlines predicted at the LVOT in: A) the normal ventricle; and B) the DSS ventricle (white crosses: pressure drop measurement points).

formation. The preliminary assessment of this idealized model was performed by qualitatively comparing the streamline fields captured in both geometries to those described in previous computational and in vivo studies (Figure 4). The model predicted the same vortex formation

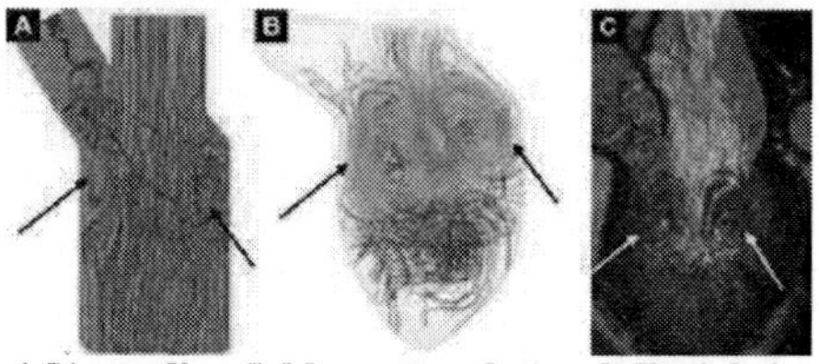

Figure 4 Streamline fields captured at end-diastole in: A) the present model; B) a previous model [6]; and C) in vivo using PC-MRI [5] (arrows: vortex formation).

in the medial and lateral section of the ventricle, as that reported in other studies [5], [6].

DISCUSSION

In this study, a simple computational approach implementing idealized geometries of a normal left ventricle with and without DSS as developed to provide a first approximation of the impact of a DSS lesion on LVOT hemodynamics. As hypothesized, average and maximum WSS were significantly higher (by 33% and 15.2 %, respectively) in the stenosed geometry than in the normal geometry.

While the WSS overload imposed by the DSS lesion near the LVOT combined with the increased flow recirculation may trigger some biological cascades involved in DSS pathogenesis, a better hemodynamic characterization using a more physiologic left ventricle model and accompanying mechanobiological experiments are needed to assess this hypothesis.

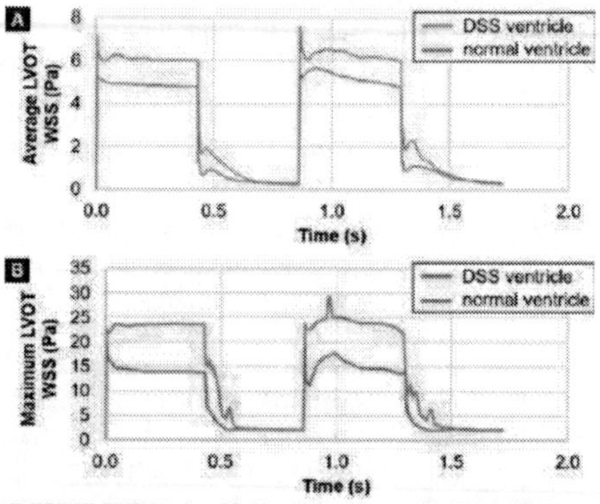

Figure 2 LVOT WSS predictions: A) average; and B) maximum WSS captured in the characterization region.

ACKNOWLEDGEMENTS

This project was funded by the College of Engineering and Computer Science at Wright State University.

REFERENCES

[1] Foker, J. E., *Circulation*, 127:1447–1451, 2013.
[2] Cape, E. G. et al, *J. Am. Coll. Cardiol.*, 30:247–254, 1997.
[3] TRIULZI, M. et al., *Echocardiography*, 1:403–426, 1984.
[4] Boyang Su et al., *2014 IEEE Conf. Biomed. Eng. Sci.*,276–280,
[5] Pedrizzetti, G. et al., *Ann. Biomed. Eng.*, 43:26–40, 2014.
[6] S. N. Doost, S. N. et al., *Biomed. Eng. Online*, 15:101, 2016.

Inconsistent Application of the Scalar Stress Concept in the Power-Law Hemolysis Model

Mohammad M. Faghih, M. Keith Sharp

Department of Mechanical Engineering
University of Louisville
Louisville, KY, USA

INTRODUCTION

A surprising outcome from decades of development of cardiovascular devices is how little hemolysis (damage to red blood cells) occurs, for instance, in high-speed centrifugal blood pumps and hard, tilting-disc valves. Limited plasma-free hemoglobin is not by itself life threatening, but rather represents a homeostatic stress that may, in combination with other conditions, precipitate a decline in health status. Therefore, analysis of the hemolysis caused by any new device remains a part of the device development cycle. Such analysis, which includes simulations and testing, represents a considerable expense in terms of time and money that could be reduced with an accurate numerical technique for predicting hemolysis. An accurate model could be used during the design stage to refine device geometry and operational protocols to minimize hemolysis and reduce the need for physical testing of multiple prototypes.

The power-law model is the most popular model for predicting hemolysis

$$\Delta PfHb / Hb = C\tau^a t^b \quad (1)$$

where *PfHb* is the plasma-free hemoglobin, *Hb* is the total hemoglobin in the blood (intracellular and extracellular), τ is fluid shear stress, t is exposure time to shear stress and a, b and C are empirical constants [Giersiepen, *et al.* 1990]. The power-law curve fit was matched to data for pure, laminar shear stress, however it has been hypothesized that a scalar, resultant stress, analogous to the von Mises stress in solid mechanics, may allow the extension of the power-law model to more complex flows that include multiple components of shear and normal stresses [Bludszuweit 1994, 1995]. The von Mises stress, which is related to deviatoric strain energy, is

$$\sigma_{vM} = \left[\frac{3}{2}\sigma'_{ij} : \sigma'_{ij}\right]^{1/2} \quad (2)$$

where $\sigma'_{ij} = \sigma_{ij} - \frac{1}{3}\delta_{ij}\sigma_{kk}$ is the deviatoric stress and σ_{ij} is the full stress. The second term on the right hand side represents the pressure, in which δ_{ij} is the Kronecker delta. *i, j, k* are tensor indices, each representing all three orthogonal coordinate axes. The von Mises stress can be expanded to

$$\sigma_{vM} = \left\{\frac{1}{2}\left[\left(\sigma_{xx}-\sigma_{yy}\right)^2 + \left(\sigma_{yy}-\sigma_{zz}\right)^2 + \left(\sigma_{zz}-\sigma_{xx}\right)^2\right] + 3\left[\sigma_{xy}^2 + \sigma_{yz}^2 + \sigma_{zx}^2\right]\right\}^{1/2} \quad (3)$$

(1)

The constant 3/2 in equation 2 is somewhat arbitrarily chosen to cause the von Mises stress to equal the normal stress for a purely extensional stress condition $\sigma_{vM} = \sigma_{xx}$.

To apply the von Mises concept to fluid stresses, it is important to recall that the stress tensor is symmetric, i.e., $\sigma_{xy} = \sigma_{yx}$. Thus for the pure shear flow for which the power-law model was derived, the shear stress has two components $\tau = \sigma_{xy} = \sigma_{yx}$ and the von Mises stress becomes $\sigma_{vM} = \sqrt{3}\sigma_{xy}$. Inserting the von Mises stress into equation 1, therefore, results in a prediction of hemolysis that is $3^{a/2}$ greater than using τ. To create a scalar stress that is compatible with the power-law model, we simply choose a different constant that causes the scalar stress to equal the shear stress when the flow is a pure shear flow. This scalar stress is

$$\sigma = \left[\frac{1}{2}\sigma'_{ij} : \sigma'_{ij}\right]^{1/2} \quad (4)$$

which can be expanded to

$$\sigma = \left\{\frac{1}{6}\left[\left(\sigma_{xx} - \sigma_{yy}\right)^2 + \left(\sigma_{yy} - \sigma_{zz}\right)^2 + \left(\sigma_{zz} - \sigma_{xx}\right)^2\right] + \left[\sigma_{xy}^2 + \sigma_{yz}^2 + \sigma_{zx}^2\right]\right\}^{1/2} \quad (5)$$

THE BLUDSZUWEIT SCALAR STRESS

The scalar stress as originally presented by Bludszuweit [1994, 1995] is given by

$$\sigma = \left[\frac{1}{6}\sum\left(\sigma_{ii} - \sigma_{jj}\right)^2 + \sum\sigma_{ij}^2\right]^{1/2} \quad (6)$$

where the summations do not include terms for which $i = j$. Since $\sigma_{ii} - \sigma_{jj} = \sigma_{jj} - \sigma_{ii}$ and $\sigma_{ij} = \sigma_{ji}$ for fluid flows, the expanded formula becomes

$$\sigma = \left\{\frac{1}{3}\left[\left(\sigma_{xx} - \sigma_{yy}\right)^2 + \left(\sigma_{yy} - \sigma_{zz}\right)^2 + \left(\sigma_{zz} - \sigma_{xx}\right)^2\right] + 2\left[\sigma_{xy}^2 + \sigma_{yz}^2 + \sigma_{zx}^2\right]\right\}^{1/2} \quad (7)$$

Comparing equations 5 and 7, it is evident that the Bludszuweit derivation contains an error that results in a hemolysis prediction that is $2^{a/2}$ too large when the scalar stress of equation 7 is used in the power-law model for a simple shear flow [Faghih & Sharp 2016].

Inconsistent application of scalar stress

The power-law model has been broadly applied to many simulations of cardiovascular flows, as summarized by Faghih & Sharp [2016]. Many investigators have apparently checked the Bludszuweit derivation and have used the corrected scalar stress of equations 4 and 5. However, unfortunately, many have used the original formula as well. Some have also not specified how scalar stress was calculated.

The impact of using the incorrect scalar stress formula depends on the stress and exposure time histories in the flow within the device. If the flow is characterized by passage through a channel in which shear is the dominant term, for instance, then the hemolysis prediction would be $2^{a/2} = 2.31$ (using the Giersiepen, *et al.* [1990] value of $a = 2.416$) times higher than that resulting from using the shear stress. While the trends of hemolysis among different designs or conditions may be similar, the 2.31X difference may impact conclusions regarding the acceptability of the predicted hemolysis, particularly if the criterion includes a threshold of hemolysis. Another type of investigation for which conclusions might change considerably is that of comparisons of the quality of fit of different power-law models (of which there are at least three [Giersiepen *et al.* 1990, Heuser and Opitz 1980 and Zhang, *et al.* 2011]) to experimental data. Each model has a different value of the exponent a, thus the change in hemolysis prediction varies among the models.

DISCUSSION

This report is offered in the hope that future investigations will apply the scalar stress that is consistent with its development from measurements in pure shear flow. However, from a broader and more fundamental perspective, validity of the von Mises-like scalar stress for hemolysis prediction remains to be verified. Alternative forms of a scalar stress, such as the maximum principle stress difference [Farinas, *et al.* 2006], have been suggested. Further, it has been demonstrated that extensional stress deforms red cells more readily than does shear stress [Lee, *et al.* 2009], which suggests that hemolysis may be larger in extensional than shear flow of the same scalar stress [Down, *et al.* 2011]. This phenomenon is plausible, since pure extensional flow causes stretching of the cell without tank-treading, thus the distribution of tension around the membrane is nearly constant (discounting viscoelastic effects and time-dependent deformation of the cell), but shear causes tank-treading or tumbling, for which the local membrane tensions are periodic. A steady, concentrated tension of longer duration (extensional flow) might more likely lead to membrane rupture than a tension briefly applied locally as the membrane circulates through the region of high tension (shear flow). The scalar stress has also been applied to turbulent flows. Several theories have been forwarded for how turbulent flows exert stresses on cells [Jones 1995, Quinlan and Dooley 2007, Antiga & Steinman 2009]. However, no experimental validation of any relationship of cell membrane tension to turbulence characteristics is yet available. In spite of several decades of work on hemolysis modeling, fundamental mechanisms remain to be discovered and quantified to produce accurate and universal hemolysis predictions for complex flows. Particular focus should be placed on cell and molecular-level modeling to relate fluid stresses to membrane tensions and membrane tensions to hemoglobin release.

REFERENCES

Antiga, L & Steinman DA. *Biorheology* 46(2):77-81, 2009.

Bludszuweit C. Doctoral Thesis, Bioengineering Unit, University of Strathclyde, 1994.

Bludszuweit C. *Artif Organs* 19(7):590-96, 1995.

Down LA, Papavassiliou DV & O'Rear EA. *Ann Biomed Eng* 39:6:1632–42, 2011.

Faghih MM & Sharp MK. *J Biomech Eng* 138:12, 2016.

Farinas MI, Garon A, Lacasse D, & D N'Dri D. *J Biomech Eng* 128(5), pp. 688-96, 2006.

Giersiepen M, Wurzinger LJ, Opitz R, & Reul H. *Int J Artif Organs* 13(5):300-306, 1990.

Heuser G, & Opitz R. *Biorheology* 17(1-2):17-24, 1980.

Jones SA. *Ann Biomed Eng* **23**(1): 21-28, 1995.

Lee SS, Yim Y, Ahn KH & Lee SJ. *Biomed Microdevices* 11:5:1021–27, 2009.

Quinlan NJ & Dooley PN. *Ann Biomed Eng* **35**(8):1347-56, 2009.

Zhang T, Taskin ME, Fang H-B, Pampori A, Jarvik R, Griffith BP, & Wu ZJ. *Artif Organs* 35(12):1180-86, 2011.

SB³C2017
Summer Biomechanics, Bioengineering and Biotransport Conference
June 21–24, 2016, Tucson, AZ, USA

FICTITIOUS DOMAIN PARTICLE-BASED MODELING FOR THROMBOSIS

Debanjan Mukherjee and Shawn C. Shadden

Department of Mechanical Engineering
University of California, Berkeley
Berkeley, CA, USA

INTRODUCTION

Thrombosis is a key cause of severe cardiovascular diseases like heart attack and stroke [1]. In addition, it leads to major complications during surgical procedures, and medical device deployment. Blood flow and hemodynamic loading plays a crucial role in thrombosis. The evaluation of blood flow patterns and flow-induced loading on a realistic blood clot within an anatomically realistic vasculature is a complicated task. The complication arises primarily due to the complexity of interactions between unsteady, pulsatile flow of blood with an arbitrarily shaped aggregated blood clot (thrombus). Realistic thrombi in large arteries possess arbitrary aggregate morphology and microstructure varying with time as the thrombi grow (or embolise). Efficient mesh or lattice-based description of the blood-thrombus interface, that can grow or shrink with time, is difficult. Furthermore, understanding intra-thrombus flow and transport requires resolution of thrombus internal microstructure, which is heterogeneous and non-uniform. Resolving these characteristic microstructures using conventional mesh based methods is difficult, and often a homogenized porous medium approach is employed which does not use an explicit description of the internal microstructure.

Here, we address this challenge by proposing a modeling framework using a combination of fictitious domain finite element method and discrete element method. The central premise here is to replace the complex blood-thrombus domain by a simpler, regular, "fictitious" domain that constitutes a background mesh, and numerically embed the blood (fluid), and thrombus (solid) domains. The thrombus domain is further described using a mesh-free off-lattice discrete element method which renders substantial flexibility in representation of complex thrombus geometry and microstructure. Two illustrative numerical examples are provided to demonstrate the efficacy of the proposed approach.

METHODS

We consider a single computational domain, referred to as fictitious domain as above, and assumed blood to be an incompressible, Newtonian fluid whose motion across this domain was modeled using a Petrov-Galerkin stabilized finite element formulation. The velocity and pressure boundary conditions on the original blood domain were imposed on the corresponding boundaries in the fictitious domain. The influence of the clot sub-domain on blood flow was then accounted for by imposing a source function within the Petrov-Galerkin variational formulation for the flow. For the present study, this function was chosen in form of a penalty term that enforces the flow velocity to match the local velocity within the thrombus sub-domain. The penalty constant for this added penalty term was scaled with element size to ensure the additional penalization does not induce ill-conditioning of the resultant matrix system of equations generated from the finite element formulation. The overall discretized system was integrated over time using a one-step midpoint integration scheme. The implementation of this penalty coupling implicitly assumed a definition of the thrombus sub-domain. The arbitrary shape and microstructure of the thrombus sub-domain was handled here by representing the thrombus as a collection of discrete particles. This involved identifying the thrombus manifold geometry information from medical images (for macro-scale thrombi) or platelet centroidal locations from fluorescent microscopy images (for micro-scale thrombus geometry), and generating an ensemble of particles based on this information. Illustrative examples of this particle based representation procedure has been presented in Fig: 1. Within the ensemble, each discrete particle was modeled as a superquadric

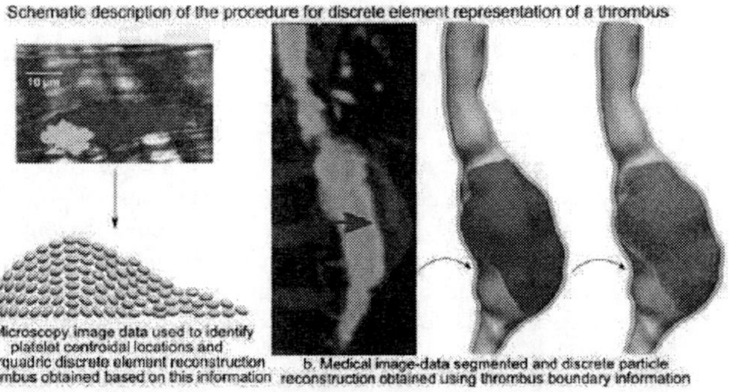

FIGURE 1: Illustration of the reconstruction of a thrombus using discrete particles for micro- (a) and macro-scale (b) clots.

object, which can be uniquely identified using three size parameters and two shape parameters (in 3D) [2]. Each superquadric discrete particle has a uniquely defined in-out function that helps identify computational nodes that lie inside or outside of a particle. This was used to identify the interior locations of the overall thrombus sub-domain for imposing the penalty term. Additionally, the flexibility of superquadric discrete particle representation enabled easy manipulation of individual particle geometries by modifying the shape and size parameters. This enabled rapid variations in microstructure and morphology for a given thrombus sub-domain configuration. The complete formulation was implemented as a solver using the open-source finite element library FEniCS [3]. Here we present results from two representative numerical case-studies that demonstrate the efficacy of the proposed fictitious domain discrete particle method in resolving macro-scale flow around a clot in a large artery, as well as micro-scale intra-thrombus flow.

Poster Presentation #P252

RESULTS

Microscale interactions

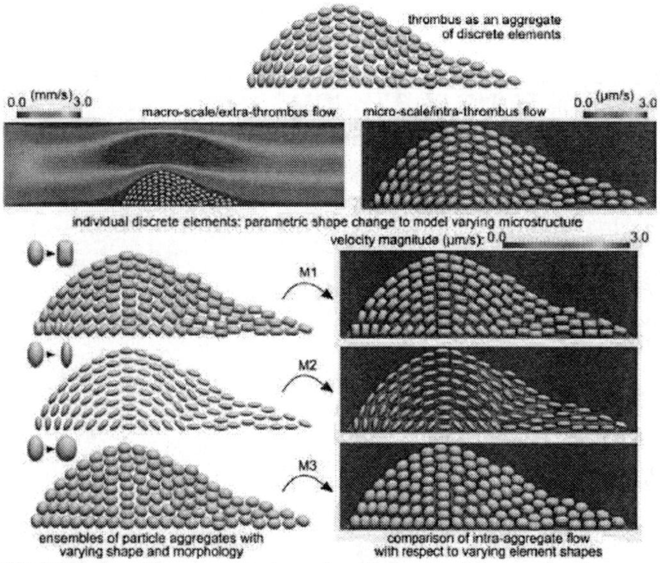

FIGURE 2: Flow around and within a micro-scale thrombotic platelet plug, alongwith illustration of variations in intra-thrombus flow for varying internal thrombus microstructure.

For the micro-scale flow-thrombus interaction simulation, we chose the model system described in [4]. Specifically, a thrombus formed within an injured vessel in a mouse cremaster muscle was imaged, and reconstructed using discrete particles. Each particle was a superquadric with planar aspect ratio 1.0:0.6. The reconstructed thrombus was embedded within a rectangular channel with width representative of a small arteriole, and a parabolic flow profile with peak centerline velocity 2 mm/s was imposed at the vessel inlet. The background fluid was plasma (density: 1.025 g/cc, viscosity: 1.7 cP), and the corresponding flow velocity fields around and within the reconstructed thrombus is presented in Fig: 2. The observed maximum flow velocity within the channel was 3.2 mm/s, while peak intra-thrombus velocity was 3.29 μm/s. These values, alongwith the spatial variation of flow, is in excellent agreement with the results obtained from simulations with explicitly meshed clots (i.e. no embedded domains) as reported in [4]. The discrete particle approach enables modeling varying clot microstructures by further parametric modifications of individual discrete element shape/size. This has been illustrated in Fig: 2 for three different variations (M1-M3) in thrombus morphology with respect to the original thrombus. Corresponding intra-thrombus flow velocities are observed to be influenced by thrombus interstitial space and connectivity. Microstructural variations are also observed to have small but noticeable influence on extra-thrombus flow. For example, when the peak extra-thrombus velocities are compared, the highest is observed for model M3 (3.33mm/s) and the lowest for model M2 (3.26 mm/s). The difference between the two is 70 μm/s, likely to increase with higher incoming flow at the inlet, and caused primarily by variations in interstitial space morphology.

Macroscale interactions

For demonstrating the efficacy of the proposed fictitious domain method in resolving macro-scale thrombus-hemodynamics interactions, a model system comprising an occlusion within an idealized channel was devised. As described in Fig: 3.a., the channel width (d_v) was assumed to be that of the common carotid artery ($\approx$ 6.0 mm), with a measured inflow profile imposed at the channel inlet to drive the flow (Fig: 3.b.). An idealized hemispherical occlusion was embedded within the channel, with a radius d_c. The flow was started from rest, and computed over three cardiac cycles, and the flow field obtained from the final cardiac cycle was compared across varying occlusion sizes (in comparison with channel width). The corresponding flow fields obtained for three different occlusions (d_c/d_v = 0.25, 0.33, 0.42) have been presented across five successive instants during a cardiac cycle in panels c1-c3 in Fig: 3. All velocity magnitudes have been scaled to 700.0 cm/s. The results capture the temporally varying recirculation region and flow reattachment distal to the occlusion, and the resultant vortical structures in the vicinity of the clot. This is relevant for progression of thrombotic activity in large artery thrombosis, since the resultant temporally varying, oscillating, shear loading on platelets, and the extent of recirculatory flow have an intimate role to play in mechanical platelet activation and subsequent aggregation of platelets. The results from the simulations presented here clearly indicate that the proposed numerical method is capable of capturing such spatiotemporally varying flow in the proximity of a large-artery thrombus. We remark here, in addition, that while the model system here involved an idealized circular/hemispherical occlusion, following the flexibility of shape representation as described in the previous sections, this can now be extended to model macroscopic occlusions of any shape.

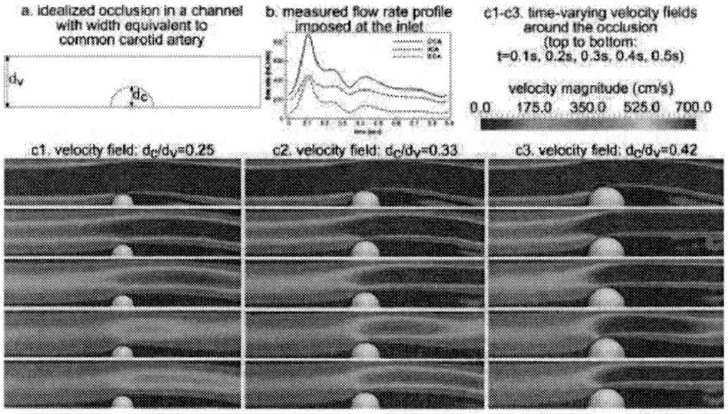

FIGURE 3: Macroscale interactions between an idealized thrombotic occlusion and pulsatile hemodynamics, in a channel of width equivalent to common carotid artery.

DISCUSSION

The results presented here establish the ability of the proposed fictitious domain discrete particle framework to resolve thrombus-hemodynamics interactions and potentially enable modeling of both intra-thrombus and extra-thrombus flow. While the former has broad implications in understanding transport of coagulation agonists and thrombolytic drugs within an existing thrombus, the latter is of key interest in the biomechanics of thrombotic disease progression and embolization. The premise of parametrically defined discrete particle aggregates embedded within a simplified mesh addresses the challenge of modeling realistic thrombus morphologies. However, when compared with an explicitly meshed blood-thrombus interface, in the proposed fictitious domain approach, we recover only a discretized approximation of that interface. This may be associated with numerical errors, and a closer inspection of the bounds of this approximation error is underway. The imposition of a single simplified background mesh, also enables this framework to directly use the native parallelization of existing parallel flow solvers, thus allowing for potentially large-scale patient-specific simulations. This is also currently an area of active research interest.

ACKNOWLEDGMENTS

This research was supported by the American Heart Association, Award No. 16POST27500023.

REFERENCES

[1] Raskob, G.E., et.al. *Arterioscler. Throm. Vas.*, 34(11):2363-71, 2014.
[2] Barr, A.H. *IEEE Comput. Graph.*, 1.1:11-23, 1981.
[3] Alnaes, M.S., et.al. *Arch. Numer. Software*, 3(100), 2015.
[4] Tomaiuolo, M., et.al. *Blood*, 124(11):1816-1823, 2014.

A VIRTUAL INTER-LABORATORY COMPARISON OF PREDICTED HEMODYNAMIC INDICES IN INTRACRANIAL ANEURYSMS: CONSISTENT OR NOT?

Aslak W. Bergersen (1), Kristian Valen-Sendstad (1)

(1) Scientific Computing Department
Simula Research Laboratory
Oslo, Norway

INTRODUCTION

Computational fluid dynamic (CFD) simulations can provide plausible hemodynamic stimuli, which besides biological effects, are believed to play a role in vascular remodeling [1]. More specifically, an abnormal hemodynamic stimulus is hypothesized to be a surrogate of aneurysm wall weakening and therefore potential risk of rupture. However, results are equivocal and an example of inter-laboratory discrepancy in the literature is whether physiologically too high or too low wall shear stress (WSS) is associated with aneurysm rupture status in large retrospective studies, frequently exemplified by the conflicting results from Cebral et al. [2] and Xiang et al. [3].

Patient-specific flow rates are rarely available and one has to make certain assumptions of how the flow relates to the inlet diameter. There is a natural variation of vessel diameter, but how the flow scales with the diameter is not necessarily obvious. We hypothesized that flow rate assumptions might be a contributing factor for why CFD-based results differ. The aim of the study was to clarify the impact of the two most common flow rate assumptions; constant parent artery WSS [2] versus constant flow rates [3], on commonly computed hemodynamic indices, and more specifically, how well they correlate. Mathematically, this can be expressed as:

$$Q_i \sim Q_{avg} \times \frac{D_i^n}{\mathrm{mean}_i(D_i^n)} \qquad (1)$$

where the subscripts ***avg*** and ***i*** refers to average and individual, respectively, and D is the measured inlet diameter and Q the imposed flow rate. Prescribing *constant flow rates and constant parent artery WSS* corresponds to setting $u = 0$ and $u = 3$ in equation 1, respectively.

METHODS

We included the first five consecutive internal carotid artery (ICA) and middle cerebral artery terminal aneurysms that extended down to the cervical segment of the ICA from the open-source Aneurisk database. Two of the models harbored two aneurysms, giving a total of 12 aneurysms. The mean ICA inlet diameter was 4.38(+/-.67) mm. Meshes were generated using the Vascular Modeling Toolkit and final models averaged 3 million tetrahedron cells, ranging 1.8-3.7 million depending on the extent of the computational domain. Two cardiac cycles were simulated taking 3,000 time steps per cycle with a first order accurate CFD solver designed [4] and demonstrated [5] to mimic the default and recommended solver in ANSYS Fluent.

We specified a Womersley profile at the inlet using an older adult waveform and applied a resistance boundary condition at the outlets. We computed the following proposed metrics of rupture risk; WSS, maximum WSS (MWSS), WSS gradients (WSSG), and oscillatory shear index (OSI). The above-mentioned solution strategy was performed on the ten chosen cases using the *exact same mean* flow rates (Q_{avg} = 4.1 ml/s), but scaled according to the two most common assumptions.

RESULTS

Linear regressions of the space/time reduced hemodynamic indices for *constant WSS* versus *constant flow rates* are shown in Figure 1. We can conclude that there was weak or no correlation for WSS, MWSS, and WSSG, with R^2 values of 0.0004, 0.1845, and 0.0789, respectively. There was a stronger correlation for OSI with an R^2 of 0.8651. Looking now on the slope of the regression lines, it is close to zero for WSS, whereas 0.52, 0.13, and 0.93 for MWSS, WSSG, and OSI, respectively.

Poster Presentation #P255

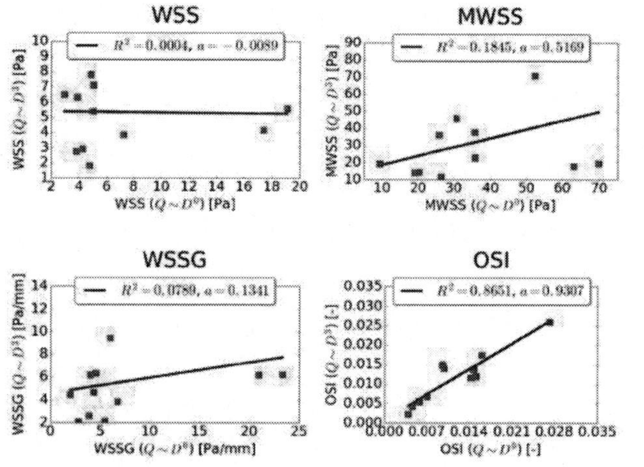

Figure 1: Comparing reduced hemodynamic indices for the two most common flow rate assumptions; constant parent artery wall shear stress [2] versus constant flow rate [3].

DISCUSSION

We have demonstrated that flow rate assumption has a profound impact on derived hemodynamic indices, for *identical mean flow rates*, but scaled according to the two most common flow rate assumptions in the literature. More importantly, these presumably equally valid assumptions led to weak or no correlations for the most commonly computed hemodynamic indices. The reported differences are not spurious differences in single cases, but consistent and persistent differences ("errors") in the 100 - 200% range. The impact of flow rates assumptions is larger than previously investigated modeling uncertainties, such as imaging modality, compliant walls, and numerical solution strategy, not to mention non-Newtonian effects.

The reported weak or no correlation between hemodynamic indices is indeed true, but admittedly only half of the story. Figure 2 shows iso-surface of instantaneous velocity fields during mid systole for the constant flow rate assumption ($Q \sim D^0$) in the left column, and constant WSS assumption ($Q \sim D^3$) in the right column. The figure is sorted on inlet diameter, with -1 standard deviation on the left hand side, mean in the middle, and +1 standard deviation to the right. In the model with smaller inlet diameter, the constant flow rate assumption results in higher velocities compared to the constant WSS assumption, and opposite for the model with larger inlet diameter.

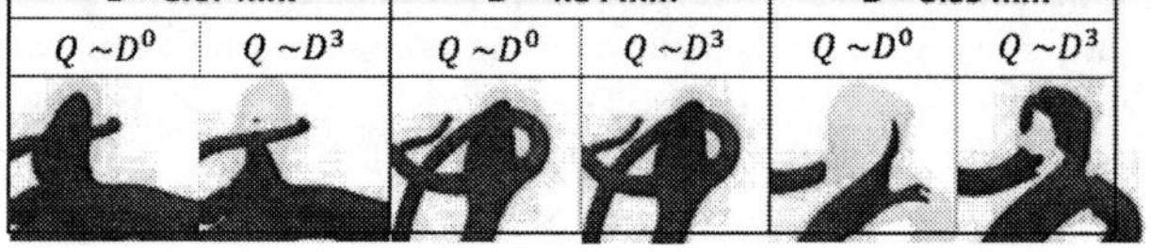

Figure 1: Iso-velocity surface in three aneurysms: Left frame shows a model with an inlet diameter of one standard deviation below the mean, as mean in the middle, and plus one standard deviation to the right.

Now plotting the *WSS ratio* (WSS($Q \sim D^0$)/WSS($Q \sim D^3$)) versus diameter Figure 3 (left), we observe a strong inverse correlation in the ratio of the computed WSS. The WSS ratios reflect the observations of Figure 1, but reveals ratios of 200-400% in predicted WSS levels in models with small vs. large inlet diameters, respectively, but also zero around for inlet diameters close to the mean. Furthermore, if we rank WSS levels from low to high for the two most common flow rate assumptions, Figure 3 (right) reveals that rank-ordering is uncorrelated, again highlighting current difficulties in deciphering the aneurysm literature.

A natural step for inter-laboratory comparison of results would be to normalize the sac/time averaged WSS levels to the time averaged parent artery WSS. This provides statistically correlated WSS values and rank ordering to be consistent (results not displayed). However, this only measures the WSS ratio between the aneurysm and parent artery, and comes at the cost of losing information on the absolute values. Furthermore, normalized values are actually not commonly reported in the literature.

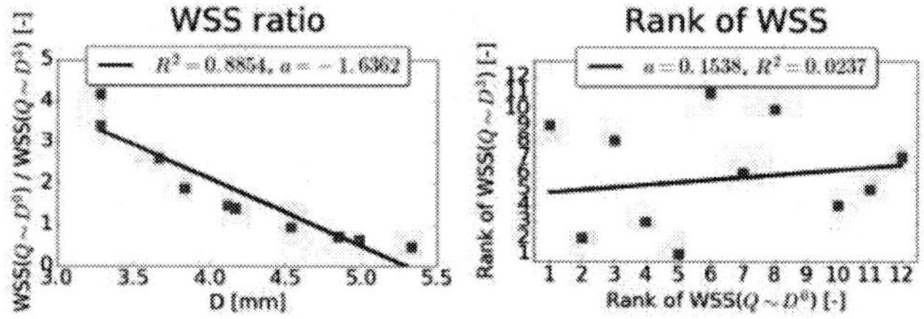

Figure 3: Comparing WSS ratio and rank ordering from low to high for the two most common flow rate assumptions; constant parent artery wall shear stress [2] versus constant flow rate [3].

It should be mentioned that the study design and reported differences in hemodynamic indices are indeed conservative. The prescribed mean flow rates were 4.1 ml/s, based on gold-standard flow rate measurements in 94 older adults from Hoi et al. [6]. These flow rates do not perfectly match either of the exemplified groups' flow rates, but reflect a conservative compromise between the exemplified groups' approaches. First comparing flow rates, 4.1 ml/s is less than what Xiang et al. prescribed (4.6 ml/s), but more than the 3.54 ml/s that Cebral et al.'s approach would have led to for the current dataset. Now comparing WSS at the inlet, the resulting mean WSS became 1.64 Pa, which is slightly above the constant WSS of 1.5 Pa which Cebral et al. applied, but considerably lower than 1.95 Pa which the constant flow rate assumption by Xiang et al. would have given for the current dataset. Either way, our virtual inter-laboratory experiment is nevertheless conservative, measured either in flow rates or WSS, with current WSS and flow rate values falling in between the "extremes".

Finally, while not claiming to be experts in cerebral blood flow and metabolism, a simple back-of-the-envelope calculation of the WSS imposed on the parent artery for this data (sub-) set reveals that the constant flow rate assumption leads un relatively un-physiological variability of WSS in the ICA; both below 1.0 Pa and above 4.0 Pa. At the same time, the flow rates imposed assuming constant WSS are also relatively "extreme" falling below 2 ml/s and approaching 7 ml/s; far beyond the standard deviation of 1ml/s reported by Hoi et al. [6].

In conclusion, flow rate assumptions may currently not be excluded as a contributing factor to the reported inter-laboratory discrepancies of hemodynamic indices.

REFERENCES

[1] Malek AM et al. *JAMA* 282.21 (1999): 2035-2042.
[2] Cebral JR et al. *AJNR*. 2005 Nov 1;26(10):2550-9.
[3] Xiang J et al. *Stroke*. 2011 Jan 1;42(1):144-52.
[4] Valen-Sendstad K et al. *AJNR*. 2014 Mar 1;35(3):536-43.
[5] Khan MO et al. *AJNR*. 2015 Jul 1;36(7):1310-6.
[6] Hoi Y et al. Physiol Meas. 2010 Mar;31(3):291-302

SBC^{3}C2017
Summer Biomechanics, Bioengineering ans Biotransport Conference
June 21 - 24, 2017, Tucson, AZ, USA

INTRACRANIAL VASCULAR DISEASE EVALUATION WITH COMBINED VESSEL WALL IMAGING AND PATIENT SPECIFIC HEMODYNAMICS

Kurt R. Sansom (1), Mahmud Mossa-Basha (2), Chun Yuan (2,3), Alberto Aliseda (1), Gador Canton(1,3)

(1) Dept. of Mechanical Engineering
University of Washington
Seattle, Washington, USA

(2) Department of Radiology
University of Washington
Seattle, Washington, USA

(3) Vascular Imaging Laboratory
University of Washington
Seattle, Washington, USA

INTRODUCTION

Intracranial vascular pathologies are currently evaluated with angiography, whether conventional digital subtraction angiography (DSA) or non-invasive (MR angiography, CT angiography) techniques. Imaging with these techniques provides high quality anatomic information about the lumen in the vasculature, but does not resolve enough information about the vessel wall disease characteristics or evaluate any functional hemodynamics. Magnetic resonance angiography (MRA) provides accurate anatomic information, but current techniques present limitations on the resolution with which the vessel wall characteristics can be measured, making differential diagnostics of cerebral vasculopathies difficult or impossible [1]. High-resolution MR black-blood vessel wall imaging (VWI) was developed to subtract the signal from flowing blood which allows for direct visualization of the vessel wall at higher resolution and contrast, enabling differentiation between various intracranial vasculopathies [1]. However, these imaging modalities cannot, fundamentally, provide quantitative functional assessments of vasculature disease state or prognosis. Additional insights must to be extracted from the imaging, beyond anatomical information.

This study incorporates computational fluid mechanics (CFD) models based on MR angiography (MRA) and PC-MRI velocimetry with intracranial vessel-wall MRI (VWMRI) to correlate histologically unstable wall changes in unruptured intracranial aneurysms (UIA). The addition of modeling provides physiological information about the hemodynamic forces that the vessel wall is subject to, such as wall shear stress, and metrics, such as oscillatory shear index. Each VWMRI has been evaluated for pattern and presence of wall enhancement and thin-walled regions and is compared with CFD models of the 3D-vessel reconstruction. Histology of each resected aneurysm evaluated for inflammatory infiltration and wall thickness features are incorporated to validate the analysis determined from VWMRI and CFD.

METHODS

Five patients were consecutively selected from a subset of a larger study where the patient underwent multi-contrast 3D VWMRI within one week prior to aneurysm clipping. MR sequences included contrast-enhanced 3D MRA, pre- and post-contrast, and 2D phase contrast MRA to measure inflow at the parent vessel for CFD analysis. An example aneurysm is depicted in Figure 1. The anatomical information from the 3D MRA sequence was segmented with from a combination of inhouse and open source tools (http://www.slicer.org) [2]. Vesselness and Ranking Orientation Responses of Path Operators (RORPO) filters were used to delineate tubular structures in the image volumes [3] [4]. Segmentation was performed using semi-automated colliding fronts level set segmentation. A surface mesh representation was extracted, smoothed, and post processed to add vessel extensions for Womersley inflow boundary conditions with the Vascular Modeling Toolkit(VTMK v1.3).

Coarse CFD volume meshes were generated in StarCCM+ (Release 11.04, CD-adapco) and spherigon patches applied to the volume surface to efficiently smooth polygon surfaces, generating high order volume mesh information. Blood flow was assumed to be incompressible and Newtonian with a density at $1050 kg/m^3$ and dynamic viscosity of 3.5 cP. A time dependent Womersley velocity profile was applied at the inlet, based on velocity information extracted from 2D phase contrast MRA scaled to the model inlet area. The outlets were modeled with a reduced order RCR pressure boundary condition to emulate downstream vascular resistance and compliance, and rigid wall assumption is used for intracranial vasculature. A high order spectral/hp finite element Navier-Stokes solver, Nektar++ [5] was used for the simulations. Each simulation was run for three cardiac cycles, where the first two cycles were discarded to ensure transient initial condition artifacts are eliminated.

Values of wall shear stress (WSS), WSS gradient, (WSSG), and oscillatory shear index (OSI) were averaged over the entire area of the aneurysm sac. Blood flow into the aneurysm ($Q_{Aneurysm}$, net positive and net negative) and viscous dissipation (ε), a measure of energy loss, were integrated over the aneurysm neck area and sac volume, respectively. All hemodynamic parameters were calculated at both peak systole, and averaged over one cardiac cycle, with the exception of OSI which was calculated from the average cardiac cycle, as defined.

RESULTS

The calculated time averaged wall shear stress for a 4 mm anterior communicating aneurysm is depicted in Figure 2. CFD provides easy access to accurate wall shear stress and the metrics listed above that cannot be provided solely by VWMRI or Time of Flight MRA (TOFMRA). While VWMRI provides information about vessel enhancement and TOFMRA provides accurate anatomical information, all of these combined data-sets provide a foundation to enable differentiation between various intracranial vasculopathies and provide quantitative functional assessments of vasculature disease. VWMRI combined with CFD could provide a more accurate assessment and have the potential to become a clinical tool for predicting disease progression.

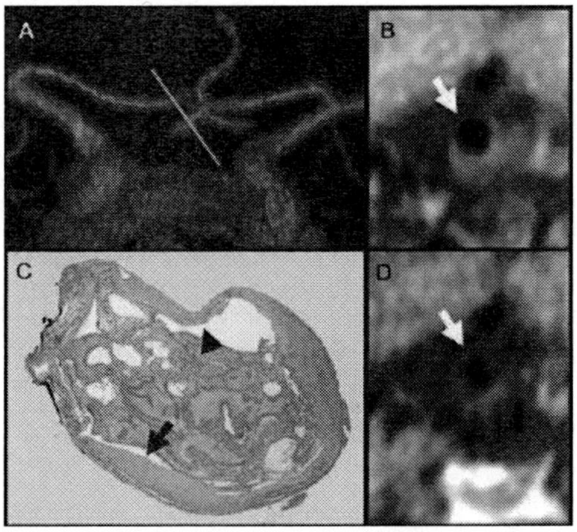

FIGURE 1. Intracranial vessel-wall MRI (VWMRI) with histological correlation for a incidental 4 mm anterior communicating artery aneurysm: 3D-Surface rendered CE-MRA with histology plane shown by the green line (A). Pre- and post-contrast VWMRI through mid-wall (B and D respectively) demonstrates wall thinning (arrows) without areas of enhancement. Histology shows the same areas (arrowheads) as noted on VWMRI without evidence of inflammation. (C)

DISCUSSION

This preliminary data is a solid foundation on which to build a new framework for combined VMWI-CFD development. The next steps include completing the analysis on additional aneurysm cases and provide analysis for the quantitative metrics described above. Additionally, providing the mapping between the CFD, VWMRI and histology for each case can provide quantitative cross-validation for the various modalities against the gold-standard of histology.

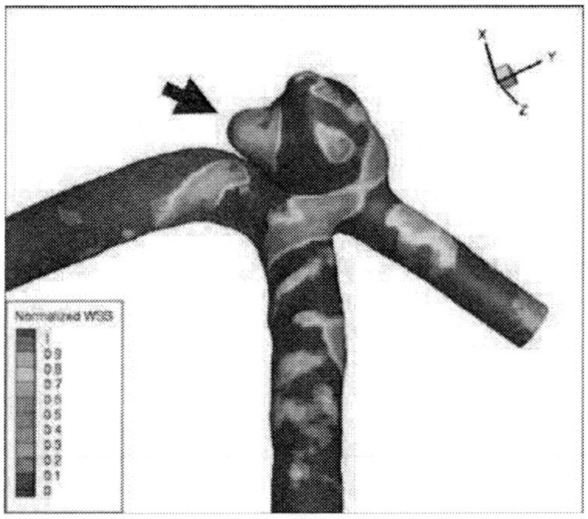

FIGURE 2. A patient with 4 mm anterior communicating aneurysm. CFD Time averaged wall shear stress map shows a prominent region of reduced wall shear stress involving one of the aneurysm lobes (arrow).

ACKNOWLEDGMENT

This work has been financially supported by a NINDS R01 grant (1R01NS088072-01A1) and an NIBIB R43 grant (2 R42 EB018124 02).

REFERENCES

[1] Mossa-Basha, M., Alexander, M., Gaddikeri, S., Yuan, C., and Gandhi, D., 2016. "Vessel wall imaging for intracranial vascular disease evaluation". *Journal of NeuroInterventional Surgery,* **8**(11), Nov., pp. 1154–1159.

[2] Fedorov, A., Beichel, R., Kalpathy-Cramer, J., Finet, J., Fillion-Robin, J.-C., Pujol, S., Bauer, C., Jennings, D., Fennessy, F., Sonka, M., Buatti, J., Aylward, S., Miller, J. V., Pieper, S., and Kikinis, R., 2012. "3d Slicer as an image computing platform for the Quantitative Imaging Network". *Magnetic Resonance Imaging,* **30**(9), Nov., pp. 1323–1341.

[3] Frangi, A. F., Niessen, W. J., Vincken, K. L., and Viergever, M. A., 1998. "Multiscale vessel enhancement filtering". In Medical Image Computing and Computer-Assisted Intervention MICCAI98, W. M. Wells, A. Colchester, and S. Delp, eds., Lecture Notes in Computer Science, Springer Berlin Heidelberg, pp. 130–137. DOI: 10.1007/BFb0056195.

[4] Merveille, O., Talbot, H., Najman, L., and Passat, N., 2014. "Tubular Structure Filtering by Ranking Orientation Responses of Path Operators". In Computer Vision ECCV 2014, D. Fleet, T. Pajdla, B. Schiele, and T. Tuytelaars, eds., Lecture Notes in Computer Science, Springer International Publishing, pp. 203–218. DOI: 10.1007/978-3-319-10605-2_14.

[5] Cantwell, C. D., Moxey, D., Comerford, A., Bolis, A., Rocco, G., Mengaldo, G., De Grazia, D., Yakovlev, S., Lombard, J. E., Ekelschot, D., Jordi, B., Xu, H., Mohamied, Y., Eskilsson, C., Nelson, B., Vos, P., Biotto, C., Kirby, R. M., and Sherwin, S. J., 2015. "Nektar++: An open-source spectral/ element framework". *Computer Physics Communications,* **192**, July, pp. 205–219.

ONTOLOGY FOR CEREBRAL ANEURYSM MORPHOMETRICS

Benjamin M. Berkowitz (1), Elizabeth Niedert (1), Suresh M.L. Raghavan (1)

(1) Department of Biomedical Engineering
University of Iowa
Iowa City, IA, USA

INTRODUCTION

Is the morphology of a cerebral aneurysm a factor in its clinical outcome? Clinicians have surmised that aneurysm morphology could be used to assess risk of rupture under observation, during flow diverter deployment, or recurrence of an aneurysm that is coil embolized. But how should one go about testing whether this is possible? First, we need to find ways to quantify morphology and then test the prognostic capability of the metrics in an adequately powered patient population. However, morphological quantification is easier said than done. What exactly constitutes morphology? How many truly independent morphological features are there? How do we mathematically capture them with a set of metrics? Can we ensure each of these metrics is physically meaningful, intuitive, unique, and exhaustively describes morphology? Based on a study of 276 unruptured cerebral aneurysms from three independent populations, we propose a minimal set of independent, meaningful metrics that quantitatively captures cerebral aneurysm morphology.

METHODS

An automated protocol was developed using libraries from the Vascular Modeling Toolkit (VMTK) and Visualization Toolkit (VTK) to process segmented medical images and extract an exhaustive set of reported [1-6] and novel (newly developed by us) [7, 8] morphological indices that quantified all relevant morphological features. This set of morphological indices included the following (many of the features have numerous definitions which were included): height and maximum diameter, maximal dimensions, volume of the aneurysm using planar and non-planar isolations, volume of the Voronoi Diagram Core (VDC), size ratio, bottleneck, tissue stretch ratio, volume-to-ostium ratio using aneurysm volume and VDC volume, aspect ratio, prolateness, oblateness, ellipticity, irregularity, non-sphericity and fusiformity.

Several of the novel set of morphological indices utilized a concept similar to the VDC, as defined by Piccinelli et al. as the envelope of the 75th percentile and larger Voronoi spheres (VS) of an aneurysm sac geometry [6], creating a surface in which only larger morphological features of an aneurysm are present. We sought to create a new set of metrics using this concept. Instead of estimating an optimal VS percentile cutoff like Piccinelli et al., we varied VS percentile cutoff from 0 to 99. We concurrently measured the volume of the VS envelope as increasingly larger shape features were smoothed out. The VS envelope volume in relation to the shape feature size was used in the definition of a metric to quantify non-spherical shape (as NSIvdc). The same process can be applied to the aneurysm's best-fit ellipsoid to quantify ellipticity (as EIvdc) and irregularity (as IRRvdc).

Each index was analyzed for robustness and inter-user variability by assessing the results of analyses performed by two users on eight patient-specific aneurysm data sets. Each index was then assessed for sensitivity to the morphological characteristic that it was intended to quantify by building hypothetical aneurysm models where that shape feature was perturbed in a controlled manner. The metrics that were most robust, least prone to user-variability and adequately feature sensitive were used in a factor analysis with a population of unruptured aneurysms from five data centers and 276 aneurysms. Factor analysis is a statistical process that analyzes variables, in this case morphological indices, for co-variance in a population. Variables that vary together strongly are grouped into factors, which uncover underlying data constructs and can be interpreted to uncover meaning in the data.

Data for this study was sourced from Ramachandran et al. 2016 [3] and the Aneurisk project [9]. The results from the factor analysis were utilized to ascertain what morphological features those metrics truly described, if there were any redundancies in definition, and the variance each morphological trait described in the population. An ideal set of metrics from this analysis was identified and recommended for use in future morphological analyses of cerebral aneurysms.

RESULTS

Of the 44 original morphological indices, 18 were eliminated because of poor inter-user variability performance, poor sensitivity to the morphological trait that each was intended to measure, or logical deficiency in definition. Five underlying morphological constructs were uncovered through the factor analysis. The results of the factor analysis are summarized in Figure 1.

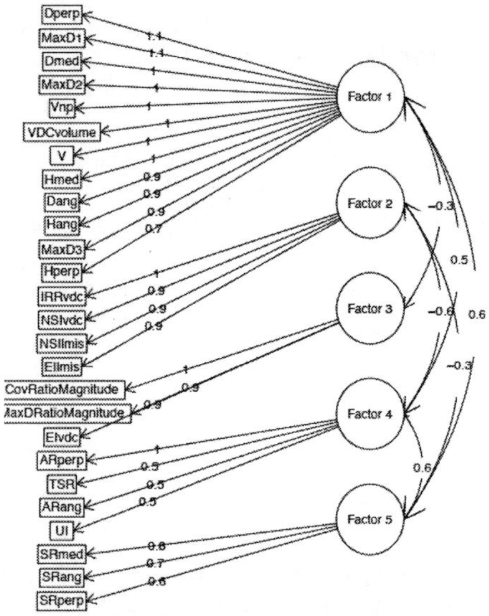

Figure 1: Factor diagram of the morphological indices (boxed), factor scores (overlaid on arrows pointing to morphological indices), and factor inter-correlations (overlaid on arrows between factors).

The first factor, sac size, was highly aligned with morphological indices that measured volume and one-dimensional size measurements. Sac size described 48% of the variance in the data set. The second factor, sac irregularity, was highly aligned with morphological indices that described various aspects of irregular shape, and described 16% of the variance in the data set. The third factor, sac ellipticity, aligned highly with morphological indices that described the overarching ellipticity of the aneurysm sac independent of other non-spherical characteristics. Sac ellipticity described 13% of the variance in the data set. The fourth factor, neck-sac size relation, aligned mainly with morphological indices that described the size relationship between an aneurysm's sac and its neck. Neck-sac size relation described 9% of the variance in the data set. The fifth factor, sac-vessel size relation, aligned highly with morphological indices that described the size of the aneurysm sac in relation to its parent vessel. Sac-vessel size relation described 8% of the variance in the data set. All five of these factors in combination described 93% of the variance in the data set. Five morphological indices – non-planar isolation sac volume (Vnp), VDC evolution irregularity index (IRRvdc), tissue stretch ratio (TSR), VDC evolution ellipticity index (EIvdc) and size ratio (SRang) were determined to be the key indices to describe aneurysm morphology.

Vnp is the volume of a non-planar Boolean isolation [7] of the aneurysm from its parent vessel reconstruction [10]. IRRvdc and EIvdc, as described above, quantify sac irregularity and sac elongation [8]. TSR quantifies the size of the ostium with relation to the dome as the ratio of the surface areas of each respective surface [7]. Finally, SRang is defined as the ratio of the aneurysm height (the distance from the centroid of the ostium to the furthest point on the dome) to the parent vessel diameter [8].

It is also important to address these morphological factors in a manner such that they can be quantified by clinicians or scientists who lack the resources to perform complex 3-dimensional morphological calculations. In many instances it may be necessary to assess aneurysm morphology from manual measurements. In this case, the key indices would be, respectively, primary maximum dimension (MaxD1) [8], a subjective categorical assessment of regular/irregular shape, aspect ratio using a height from the ostium centroid to the furthest point on the dome (ARang) [8], a characterization of ellipticity using ratios of the maximum dimension and its two perpendicular maximum dimensions (MaxDRatioMagnitude), and SRang [8]. These morphological indices, as well as the higher dimensional metrics, are illustrated in Figure 3.

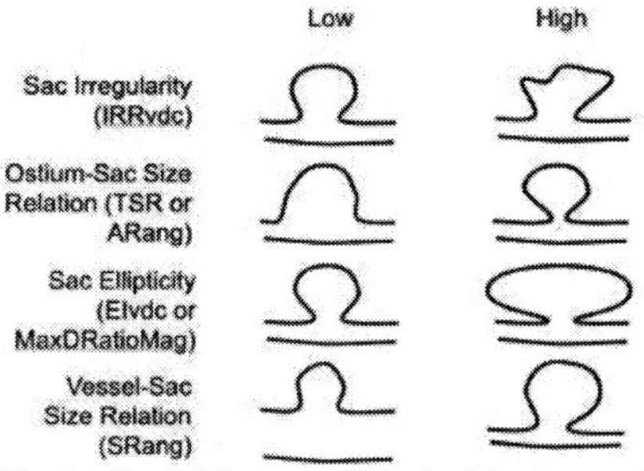

Figure 3: Illustration of the morphological factors and metrics.

DISCUSSION

Many studies have introduced novel morphological indices. However, there has been no consolidation of such metrics into a minimal set of physically meaningful, independent metrics that fully describe aneurysm morphology. This study defined such an optimal set of metrics, and laid out methods to fully assess the relevant morphological features of aneurysms using an optimal set of morphological indices – a necessary step prior to performing the sweeping longitudinal cohort study that the field requires.

ACKNOWLEDGEMENTS

Dr. Luca Antiga assisted with the implementation of VMTK. Dr. David Hasan and Dr. Bob Harbaugh assisted in study subject recruitment and data collection. Many of the aneurysm models used in this study were originally segmented and organized from image data by Dr. Manasi Ramachandran, Dr. Rohini Retarekar, Steve Lin, Ben Dickerhoff, Tatiana Correa and Kevin Johnson. Data available from the Aneurisk study was also utilized.

REFERENCES

[1] Raghavan, ML et al., *Neurosurgery*, 2005.
[2] Dhar, S et al., *AJNR*, 2008.
[3] Ramachandran, M et al., *Neurosurgery*, 2016.
[4] Yasuda et al., *Neurosurgery*, 2011.
[5] Lauric et al., *IEEE Trans. Biomed. Eng.*, 2011.
[6] Piccinelli et al., *Neurosurgery*, 2012.
[7] Berkowitz, B, *U. of Iowa*, 2012.
[8] Berkowitz, B, *U. of Iowa*, 2016.
[9] Emory U. Dept. Math & CS, *AneuriskWeb project website*, 2012.
[10] Ford et al., *Br J Radiol*, 2009.

SB^{3}C2017
Summer Biomechanics, Bioengineering and Biotransport Conference
June 21 – 24, Tucson, AZ, USA

FLUID MECHANICS AND EVOLUTION OF COOPERATION

D.C. Vural, G. Uppal

Department of Physics
University of Notre Dame
Notre Dame, IN, USA

INTRODUCTION

Bacteria are social species. They produce and share goods vital for nourishment, accommodation, communication and defense. However, strains that benefit from these without contributing reproduce quicker and compromise the cooperative nature of the group. How cooperation emerges and persists despite its evolutionary instability is an important open problem in biology. Here we approach this problem from continuum mechanics point of view. Using computer simulations and analytical theory, we study the role of fluid dynamics on the social evolution of bacteria.

The theory of inter-species interactions is traditionally rooted in first order chemical reaction kinetics, also known as Lotka-Volterra equations (LV). In this framework, the concentration of, say, a predator increases as it "collides" with a prey, and the reaction rate is proportional to the concentration of both species. An interaction matrix fully determines the dynamics of the ecosystem, and the formalism has close connections with game theory [1]. As mathematically elegant this approach may be, life does not take place in a reaction chamber. Most organisms evolve in fluid media. In some cases cells couple to the flow in their habitat, and in other cases they themself constitute a fluid. Phytoplankton patches for example, are shaped by turbulent eddies, and form fractal filaments [2,3]. Vortices cause localized population blooms that would otherwise not exist [4]. Bacteria in biofilms, and cells in developing tissues and tumors constitute fluids with interesting mechanical properties [5,6,7].

The fundamental question addressed here is the effect of fluidic forces on the emergence of sociality within and across species. Specifically, we determine how a flow influences the formation of cooperation across and within species.

METHODS

Our work is computational and analytical. We consider the population dynamics of bacteria embedded in a flowing viscous fluid. The bacteria are assumed to secrete one waste and one "public good". The public good is beneficial to those near by, and has an evolutionary cost, i.e. those secreting more of it reproduce slower.

The flow field $\vec{u}(x,t)$ is determined by the Navier-Stokes equations for incompressible viscid fluids for a number of simple geometries. $\vec{u}(x,t)$ influences the motion of bacteria population and their secretions according to the diffusion advection reaction equations

$$\frac{\partial u_i}{\partial t} = \delta_i \nabla^2 u_i - \vec{u}\cdot\vec{\nabla} u_i = u_i \left[\sum_\alpha B_{i\alpha} \frac{C_\alpha}{C_\alpha = K_\alpha} = D_{i\alpha} S_{i\alpha} \right] = \varepsilon \sum_\alpha \frac{\partial^2 u_i}{\partial S_{i\alpha}^2}$$

$$\frac{\partial c_\alpha}{\partial t} = d_\alpha \nabla^2 c_\alpha = \vec{u}\cdot\vec{\nabla} c_\alpha = \textstyle\sum_i \int S_{i\alpha} u_i \, dS_\alpha = \lambda_\alpha c_\alpha. \qquad (1)$$

Here $u_i(x,t,S_{i\alpha})$ is the number density of the ith species at position x, time t, that secretes a chemical α at a rate $S_{i\alpha}$, indexed by α. $C_\alpha(x,t)$ is the concentration of the αth chemical. The first and second terms in both equations describe diffusion and advection of chemicals and species. $B_{i\alpha}$ quantifies the benefit of a chemical α to a species i. The two terms in the square bracket quantifies fitness, which depends on the benefit due to the chemicals $C_{i\alpha}$ minus the cost $D_{i\alpha}S_{i\alpha}$ of secretion. Note that the benefit of chemical $C_{i\alpha}$ saturates with concentration: this "Monod term" is common in the literature, e.g. bacteria cannot grow infinitely fast even if resources are infinite. The cost on the other hand, is proportional to the secretion amount. The final term in the first equation, represents mutation, which can be thought as a "random walk in phenotype space", i.e. diffusion in $S_{i\alpha}$. The last two terms in the second equation describe the chemical production and decay.

RESULTS

First, we observed that randomly placed bacteria cluster into communities, since bacteria far away from others have lesser access to the secreted beneficial goods and will die with higher likelihood (**Fig.1**). We furthermore observed, as expected, that when mutations were turned on, cheating mutants take over clusters.

Our central result concerns the effect of flow on social evolution. With no flow, all clusters are eventually taken over by mutants. Furthermore, lacking public goods the clusters die.

When a large flow was introduced however, clusters became strained, and at started fragmenting at a certain rate. Fragmentation limits the spread of cheaters: as long as a group fragments before a cheating mutant appears, it will be able to pass on its cooperative nature. Thus, the evolutionary fate of bacteria depends on the local shear. We have also derived a condition for cooperation to persist at steady state:

$$\mu N = m. du/dr = \omega_0. \quad (2)$$

Here μ is the mutation rate, N is the population of a cluster, m is a dimensionless constant, du/dr is the shear, and ω_0 is a constant that depends on the diffusion constant of the bacteria and their secretions.

Naturally, shear is seldom homogeneous in a natural habitat. Even a simple steady laminar flow may give rise to a diversity of social behaviors. For a laminar flow through a cylindrical pipe for example, we find that bacteria are maximally cooperative near the flow boundary, and decreasingly so as one moves towards the bulk. Beyond a critical distance from the boundary, the fluid permits no life, social or otherwise, because all groups were taken over and annihilated by cheaters (**Fig.1**). In a Rankine vortex, we were able to localize cooperation to a ring, or depending on the size of the vortex, a circular domain (**Fig.2**).

DISCUSSION

Social evolution is a fundamental question, with implications ranging from the origins of life to the treatment of pathogenic biofilms. Present theoretical understanding of this subject is rooted in game theory, a mathematically elegant and insightful framework, but typically, with very little physical foundation. Our work is distinguished by the view that social evolution is a mechanical phenomenon. The laws of fluid dynamics and reaction kinetics that govern organisms and their environment are the ultimate determinants of evolutionary outcomes. Our approach can also be considered as the beginnings of evolutionary engineering, since our original question can also be framed as, "what flow pattern gives rise to a particular (desirable) ecological structure". Understanding the formation of social interactions in microbes is essential to control these interactions.

We conclude that flow patterns crucially influence the fate of bacteria, and determine whether social cooperation can evolve and persist.

ACKNOWLEDGEMENTS

This work was supported by the Defense Advanced Research Projects Agency, Contract No. HR0011-16-C-0062

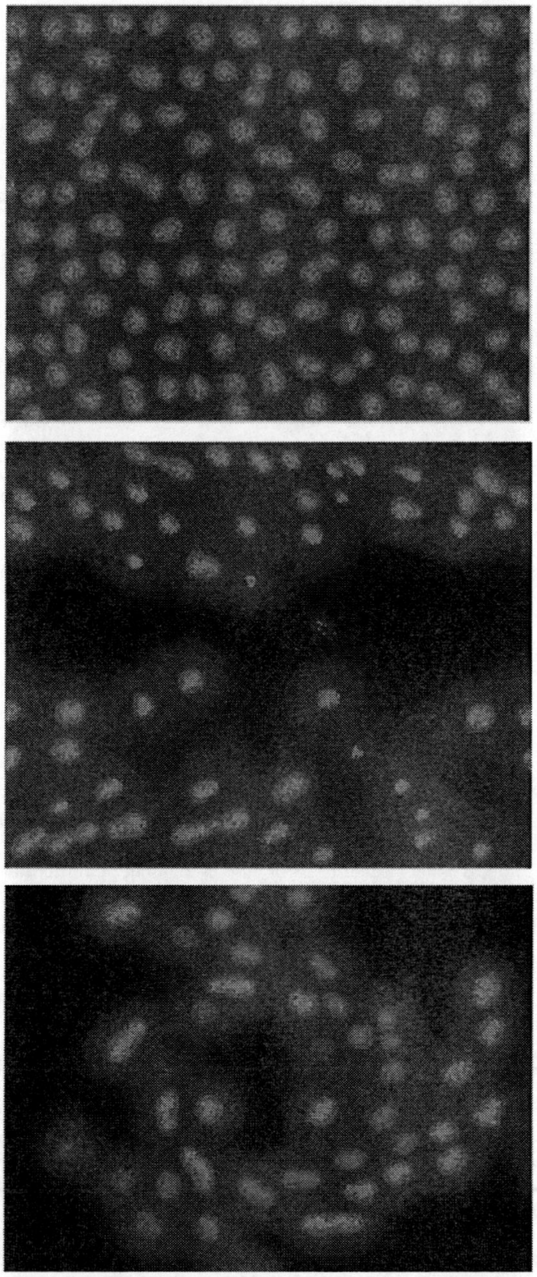

Fig.1. Evolution of Cooperation. No Flow (Top), Laminar Flow through a Pipe (Center). Rankine Vortex (Bottom). Green dots: Bacteria. Red: Public good. Blue: Waste. Shear fragments groups and limits the spread of cheaters. If the shear-induced fragmentation is larger than the mutation rate, cooperation persists in steady state. In regions below a critical shear cheaters will take over and kill all.

REFERENCES

[1] Hofbauer, J, Karl S. Bull Am Math Soc 40.4, 479-519, 2003.
[2] Koshel, K.V, and Sergei V.P., Physics-uspekhi 49.11, 2006.
[3] Tél, T, et al., Phys. Rep., 413(2), 91-196, 2005.
[4] Sandulescu, M, et al. arXiv preprint arXiv:0802.3973, 2008.
[5] Angelini, TE, et al. Proc Nat Acad Sci 108(12), 4714-4719, 2011.
[6] Marchetti, MC, et al. Rev Mod Phy, 85(3), 1143, 2013.
[7] Basan, M et al. HFSP journal, 3(4), 265-272, 2009.

SB³C2017
Summer Biomechanics, Bioengineering and Biotransport Conference
June 21 – 24, Tucson, AZ, USA

FLUID-STRUCTURE INTERACTION OF THE NON-CONTACT TONOMOTERY TEST

Wei Wu (1), Miguel A. Ariza (2), Mauro Malve (3), Ender A. Finol (1), Begoña Calvo (2), Jose F. Rodriguez (4)

(1) Department of Mechanical Engineering
University of Texas at San Antonio
San Antonio, Texas, USA

(2) Department of Mechanical Engineering
University of Zaragoza
Zaragoza, Spain

(3) Department of Mechanical Engineering, Energetics and Materials
Public University of Navarra
Pamplona, Navarra, Spain

(4) Department of Chemistry, Materials, and Chemical Engineering "Giulio Natta"
Politecnico di Milano
Milan, Italy

INTRODUCTION

The corneal shape is the result of the equilibrium between its mechanical stiffness (related to the corneal geometry and the intrinsic stiffness of the corneal tissue), intraocular pressure (IOP) and the external forces acting upon it such as an external pressure. An imbalance between these parameters, e.g. an increment of IOP, a decrease of the corneal thickness induced by refractive surgery or by a corneal material weakening due to a disruption of collagen fibers, can produce ocular pathologies which seriously affect a patient's sight. Consequently, it is important to understand how ocular factors such as IOP, geometry and corneal material are related to pathologies in order to improve treatments.

Non-contact tonometry has recently gained interest as a diagnostic tool in ophthalmology and as an alternative method for characterizing the mechanical behavior of the cornea. In a non-contact tonometry test, a high-velocity air jet is applied to the cornea for a very short time (< 30 ms), causing the cornea to deform, while the corneal motion is recorded by a high-speed camera. A number of biomarkers associated with the motion of the cornea, i.e. maximum corneal displacement and time between first and second applanations, among others, have been proposed to characterize pre- and post-operative biomechanical changes [1–4]. Identification of the material parameters associated with the corneal mechanical behavior by means of a non-contact tonometry test is performed using the inverse finite element method [5-6]. Hence, an accurate simulation of the non-contact tonometry test is required. In this regard, most approaches model the non-contact tonometry test as a quasi-static problem, and neglect the inner structures of the eye, i.e. the lens, ciliary muscles, and, most importantly, the aqueous and vitreous humors. Considering that during a non-contact tonometry test the dynamic pressure on the anterior surface of the cornea varies from 0 to 9-15 kPa in approximately 15 ms, implying a loading rate of 1 MPa/s, neglecting inertial effects may result in inaccurate results. In this work, we perform fluid-structure interaction (FSI) simulations in a two-dimensional model of the eye accounting for inertial effects and evaluating the effect of considering or not the presence of the humors in the model of the eyeball. Our results indicate that when inertial effects are accounted for in the simulations, neglecting the humors in the model will lead to non-physiological results.

METHODS

A two-dimensional model of the eyeball including the crystalline and ciliary muscles as inner structures as well as both humors –vitreous and aqueous– were considered as shown in Figure 1.

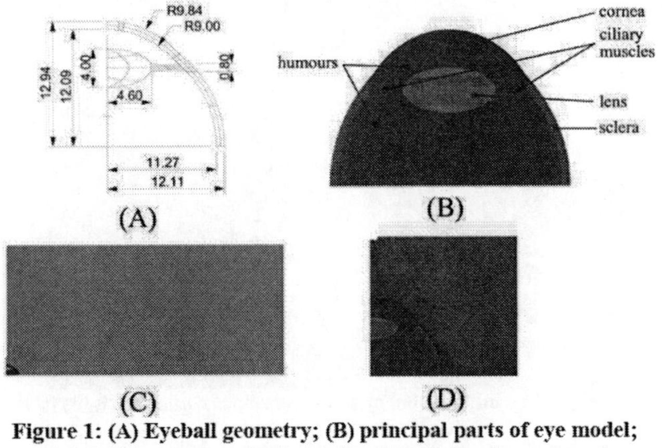

Figure 1: (A) Eyeball geometry; (B) principal parts of eye model; (C) full FSI model; and (D) details of the finite element mesh.

The cornea was modeled as an isotropic material described by a strain energy function of the form [5]

$$W_{cornea} = D_1\left(e^{D_2(\bar{I}_1-3)} + 1\right) + \kappa(J-1)^2 \quad (1)$$

where $\bar{I}_1$ is the first invariant of the modified right Cauchy-Green tensor, $\bar{C}$, $J = \sqrt{\det C}$ is the elastic volume ratio, $D_1 = 120.9$ kPa and $D_2 = 132.22$ are material constants chosen such that they fit inflation test experiments [7], and κ is the bulk modulus. The sclera and ciliary muscles were also modeled as an isotropic material with the following strain energy function

$$W_{sclera} = C_{10}(\bar{I}_1 - 3) + C_{20}(\bar{I}_1 - 3)^2 + C_{30}(\bar{I}_1 - 3)^3 + \kappa(J-1)^2 \quad (2)$$

where $C_{10} = 810$ kPa, $C_{20} = 56{,}050$ kPa, and $C_{30} = 2{,}332{,}260$ kPa. The lens was modeled as a linear elastic material with $E = 7$ MPa and $\nu = 0.33$. The humors were considered to be water.

The air-jet was applied as a ramp from 0 to 120 m/s in 15 ms and then back to zero velocity in 15 ms. Zero pressure was imposed as the outflow condition and a variational multiscale based turbulent flow model was used for the simulation. An intraocular pressure of 2 kPa was applied before initiating the air-jet (Figure 1D). A mesh sensitivity analysis was performed in the structural part to determine the minimum number of elements required for modeling the bending mode of the cornea. All simulations were performed using LS-DYNA Release 9.0 (LSTC, Livermore CA, USA and ANSYS, Inc., Canonsburg PA, USA) with ICFD as fluid solver.

RESULTS

The air-jet caused a deflection in the cornea that reached a maximum of 0.6 mm in correspondence with the peak velocity of the air-jet (see Figure 2), as observed in the actual application of the CorVis-ST non-contact tonometer [4]. During loading, the pressure in the anterior and posterior cameras of the eye (humor aqueous and humors vitreous) increased three-fold (from 2 kPa to 6 kPa), indicating that IOP changed because of the air-jet loading. In addition, as shown in Figure 2, the dynamic pressure exerted by the air-jet is concentrated on an area of 6 mm in diameter with a Gaussian type spatial distribution, and reaches a maximum of 8.99 kPa, which is comparable to the theoretical pressure of 9.0 kPa.

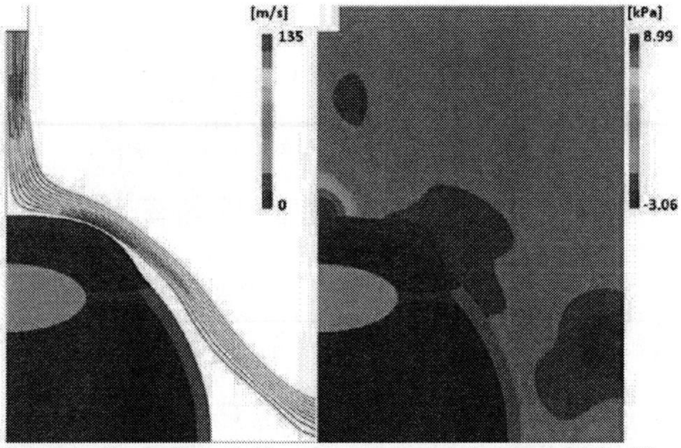

Figure 2: Streamlines (left) and pressure distribution (right) of the air-jet when the maximum velocity (120 m/s) is reached. At this instant, the cornea has the maximum concavity.

The displacement experienced by the cornea is sensitive to the presence of the humor in the model. If the simulation is performed without humor and with a constant IOP of 2 kPa, corneal displacement is non-physiological (about 18 times larger than the model with humors, as shown in Figure 3), indicating humors are a necessary feature in the model.

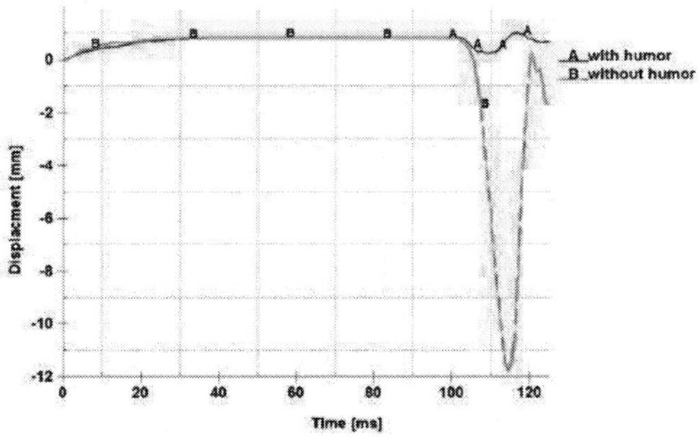

Figure 3: Displacement of the apex of the cornea with and without both humors. From 0 to 95 ms, both models are pressurized and stabilized at the IOP, after that the air-jet lasted 30 ms.

DISCUSSION

Non-contact tonometry is gaining popularity as a diagnosis tool in ophthalmology and a number of studies are considering this test as an alternative to characterize the mechanical behavior of the corneal tissue [3-5]. However, to understand the contribution of the cornea to the mechanical response measured due to the air-jet, accurate simulations of the tests are required. Results from this study indicate that FSI simulations accounting for the presence of the internal humors in the eye are required to simulate accurately the non-contact tonometry test. Neglecting the humors and imposing a constant IOP during simulation lead to non-physiological displacements of the cornea. In addition, these results indicate that most of the dynamic loading imposed by the air-jet is absorbed by the humors with minimal contribution of the cornea. Additional tests are required to study the sensitivity of the results (e.g., maximum corneal displacement) on the mechanical properties of the cornea and corneal thickness.

ACKNOWLEDGEMENTS

This research was funded in part by the European Union's Seven Framework Program managed by REA Research Executive agency http://ec.europa.eu/research/rea (FP7/2007–2013) under Grant Agreement FP7-SME-2013 606634 and the Spanish Ministry of Economy and Competitiveness (DPI2011-27939-C02201 and DPI2014-54981R).

REFERENCES

[1] Lanza, M et al., *J Biophotonics*, 9(5):467-477, 2016.
[2] Ariza, MA et al., *Plos One* 10(3): e0121486, 2016.
[3] Roy, AS et al., *J Mech Behav Biomed Mater*, 48: 173-182, 2015.
[4] Hong, Y et al., *Optom Vis Sci*, 90(1): e1-e8, 2013.
[5] Ariza MA et al, *Comput Methods Appl Mech Eng*, 2016 (in press).
[6] Lago M et al, *J Biomech*, 48(1): 38-43, 2015.
[7] Elsheikh A, et al. *Curr Eye Res*. 40(11): 1111-1119, 2015.

SB^3C2017
Summer Biomechanics, Bioengineering and Biotransport Conference
June 21 – 24, Tucson, AZ, USA

THE EFFECTS OF PRECONDITIONING ON UNIAXIAL TENSILE TESTS OF PORCINE CORNEA

Hamed Hatami-Marbini, Sandeep J. Mysore

Department of Mechanical & Industrial Engineering,
University of Illinois at Chicago
Chicago, Illinois, USA

INTRODUCTION

The cornea is a transparent dome-shaped tissue at the front of the eye. It is mainly composed of layers of collagen fibers and protects the inner parts of the eye while transmitting the incident light [1-2]. The mechanical properties of the cornea have been widely studied using uniaxial tensile experiments. In this testing procedure, a stress or strain controlled strain experiment is performed on a rectangular corneal strip. It is well developed that the tensile response of the cornea is rate dependent [3-4]; thus, it is expected that the prior history of the samples affects the experimental measurements. Factors such as tissue extraction from the organism and tissue handling during sample preparation will influence the stress history of the specimens. Pre-conditioning is a well-accepted method to remove any stress history, abet in obtaining repeatable results, and create a recoverable reference state [3-5].

Although there exist few studies that have been done without a preconditioning step, preconditioning is often an integral part of the uniaxial testing technique. Pre-conditioning is analogous to the warm up session of an athlete before the actual training which enables the athlete to repetitively perform a certain action efficiently. In preconditioning, specimens are subjected to a given cyclic load prior to the actual experiment in order to create a consistent loading history and to investigate more accurately the possible effects of various loading conditions such as strain rate. Despite the importance of the preconditioning, there does not exist a consistent protocol for the corneal tissue. The primary objective of the present study is to characterize the influence of various pre-conditioning protocols on the repeatability of stress values of in strain-controlled uniaxial tensile test of porcine corneas.

METHODS

Specimen preparation
Fresh porcine eyes were obtained from a local abattoir within 6 hours of slaughtering. The corneal scleral ring with at least 2 mm of sclera, was carefully excised from the ocular globe using surgical scissors. A custom-made double bladed device was used to slice a 5 mm wide corneal strips with about 2 mm scleral tissue. The width of the sliced strip was verified using a digital calliper (Mitutoyo, Japan) and their thickness was measured using a digital Pachymeter (DGH Technology Inc., Pennsylvania). The samples were allowed to reach a constant thickness prior mounting them into the testing device.

Experiments
The tensile experiments were conducted using a Dynamic Mechanical Analyser (RSAG2, TA Instruments USA). An immersion chamber attachment was used to run the experiments in mineral oil (Fisher Science Education, Illinois) in order to prevent any change in the thickness of the specimen during the test [2]. The corneal strip was loaded on to the tension geometry at a given loading gap using serrated grips. The samples were

then preconditioned by stretching them up to 5% strain for ten cycles followed by a five-minute relaxation period. A tare load of 100 mN was applied to the pre-conditioned strips to remove any slack and straighten them. The length at this force was noted down as the initial length L_0. A repeatability test, consisting of three loading cycles with a five-minute relaxation time in between the cycles was performed to check the repeatability of stress values after preconditioning. The strips were then stretched to 5% strain and immediately brought back to its initial length L_0, where it rested for five minutes. This cycle was performed three times in total in order to measure the consistency of the experimental measurements. Figure 1 shows the preconditioning protocol that was used.

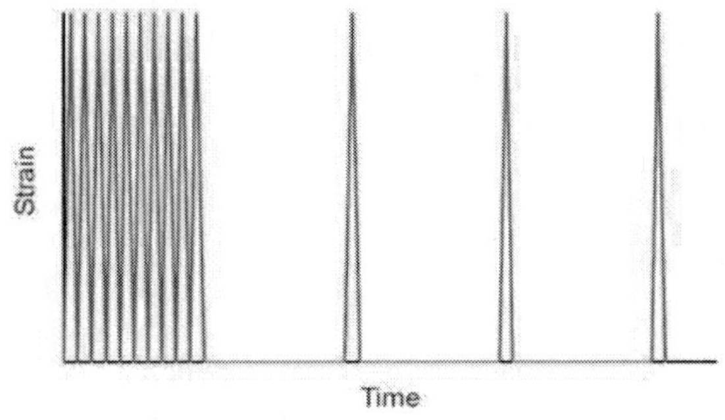

Fig. 1 The preconditioning procedure which was used at the beginning of the experiments. The samples were preconditioned by stretching them up to 5% strain for ten cycles followed by a five-minute relaxation period. Then, they were tested three times by applying 5% strain with 5 minutes relaxation in between.

RESULTS

Figures 2-3 show how the preconditioning step affects the repeatability of the experimental measurements for typical corneal samples. In these plots, the normalized uniaxial stress is plotted as function of time. In Figure 2, it is seen that despite the high value of maximum tensile stress, it reaches a constant value following the completion of the preconditioning procedure. Nevertheless, the results shown in Figure 3 shows that the maximum tensile stress during the actual experiment is different than the constant value at the end of the preconditioning step.

DISCUSSION

The main goal of this study was to characterize the effects of preconditioning protocol on the repeatability of the stress values obtained from strain controlled uniaxial tensile tests of porcine cornea. During the course of this study, the relation between the loading rate during the preconditioning step and the rate of loading during the actual tensile was also characterized.

The findings of this study is important because we have observed that obtaining a repeatable experimental measurement is highly dependent on the accurate selection of the preconditioning protocol. In other words, although an arbitrary selected preconditioning step will remove some inconsistency from the experimental data, it does not warrant obtaining a uniform and repeatable mechanical properties. For example, the stress response obtained after the preconditioning may not have yet reached its constant value (Figure 3). In a future publication, the effects of preconditioning, i.e. the number of cycles, strain rate, and strain magnitude, on measured tensile properties of the corneal tissue will be presented.

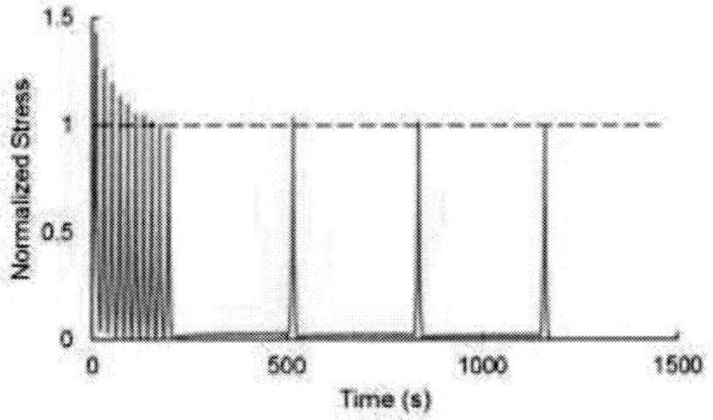

Fig. 2 The variation of normalized tensile stress as a function of time for a typical sample. It is seen that the stress value reaches a plateau after ten cycles of preconditioning and remain like that during the actual experiment.

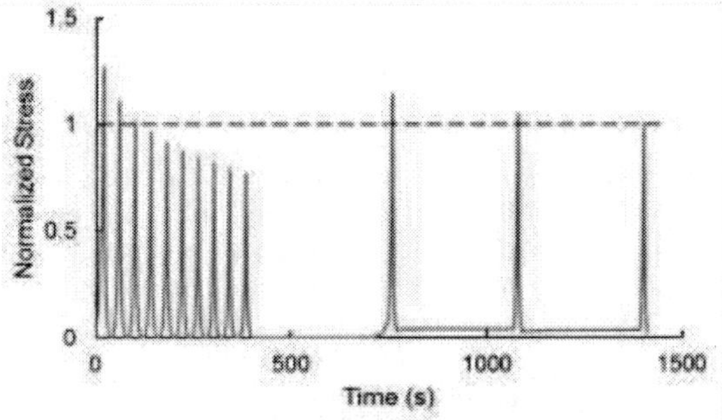

Fig. 3 The variation of normalized tensile stress as a function of time for a typical sample. It is seen that the stress value reaches a plateau after ten cycles of preconditioning. Nevertheless, there still exist significant variation in the stress value when preforming the actual experiment.

ACKNOWLEDGEMENTS

This study was supported in part by National Science Foundation (1635290) and UIC College of Engineering.

REFERENCES

[1] Maurice, DM, in The Eye, Academic Press, 1-184, 1984
[2] Hatami-Marbini, H, et al. Exp Eye Res, 116:47-54, 2013
[3] Boyce, BL et al. J Biomech., 40(11):2367-2376, 2007
[4] Hatami-Marbini, H. Ann Biomed Eng., 42(8):1740-1748, 2014
[5] Hatami-Marbini, H et al. Exp Eye Res., 120:103-108, 2014

INTRAOCULAR PRESSURE MEASUREMENT THROUGH THE LASER INDUCED CAVITATION BUBBLES DYNAMICS.

Luis F. Devia-Cruz[1], Carlos A. Zuniga-Romero[1]
Guillermo Aguilar[2], Santiago Camacho-López[1]

(1) Departamento de Óptica, Centro de Investigación Científica y de Educación Superior de Ensenada, Carretera Ensenada-Tijuana No. 3918, Zona Playitas, C.P. 22860, Ensenada, B. C., Mexico.
(2) Department of Mechanical Engineering, University of California, Riverside, 92521, Riverside, CA, USA.

INTRODUCTION

The Intraocular pressure (IOP) is a mechanical property of the intraocular fluids in the eyeball. One of the main eye diseases that affect the IOP is the so called Glaucoma, it alters the intraocular aqueous humor balance, causing an elevation of the intraocular pressure, and therefore irreversible damage to the optic nerve terminal fibers. Currently, the IOP measurement is performed by using techniques based on the Imbert-Fick principle (Chihara 2008); which states that the pressure of a liquid inside a thin-walled sphere is proportional to the force required to flatten part of its surface, divided by the area of flattening. This type of measurement requires physical contact to directly induce a mechanical deformation of the human eyeball. Several ophthalmologist reports point out the fact that patient movements, blood pressure changes, and the cornea physical characteristics which are distinct person to person, directly affect the IOP measure accuracy and precision.

The study of the laser-induced cavitation phenomenon in biological media began with the use of Q- switched lasers in therapeutic applications in the mid 60's Lee &Gojani [1]. With the advent of lasers in medicine, it is possible to cut and remove material at a few microns scale; these new techniques were based on the mechanism of breaking of molecular bonds due to high energy lasers that provided UV ranges and high absorption of the tissue to such wavelengths Niemz [2].

Cavitation bubbles are well characterized through their maximum radius and collapse time; however them both are elaborated and expensive to determine. The traditional laser-photographic techniques require several laser shots and bubbles generated in order to recreate a whole bubble event and its dynamics Gregorčič[3], Evans [4]. An immediate disadvantage is the fact that even when the shot to shot energy per pulse does not fluctuate much, the stochastic nature of the laser-induced plasma produces a slightly different cavitation bubble from shot to shot making it somehow hard to fully recreate a single cavitation bubble event from a series of many laser shots. The use of high-speed cameras, say over 300.000 frames per second,is expensive and no matter this speed it still happens many times to miss either the maximum radius or the exact time collapse frames. This is a significant inconvenience since those data are quite important for the study and applications of the cavitation bubble phenomenon.

METHODS

The experimental part of this work is based on a recently reported cavitation characterization technique called Transmittance Spatial Modulation (TSM), Devia et Al. 2012 [5] , where the modulation of the spatial intensity profile of a probe CW beam is generated by the cavitation bubble itself. A HeNe laser (@632.8 nm) was used as probe beam, it passed co-linearly (see Figure 1) with the pump (@532nm) beam and then it is collimated throughout the sample. When the pump pulse arrives and the cavitation process starts, the bubble grows to its maximum size, and then it collapses to start it over again.

The bubble acts as a negative lens, therefore the (collimated probe) light incident onto the bubble is deflected or/and scattered sideways and the total amount of light detected is inversely proportional to the bubble size. All the transmitted light was collected through a lens onto a fast (1 ns rise time) photo diode, whose signal was visualized and recorded using an oscilloscope (1GHz, Tektronix DPO). Figure 2 shows a typical trace of the time evolution of the whole cavitation event. The collapse time for the first bubble depends on both the pump pulse energy and on the actual liquid pressure. It is then possible, to infer the actual liquid pressure by just measuring the first collapse time for a given pump pulse energy.

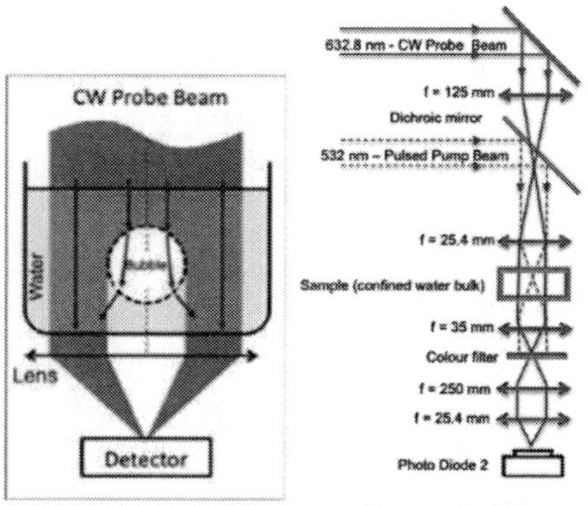

Figure 1.STM technique operation principle and experimental setup.

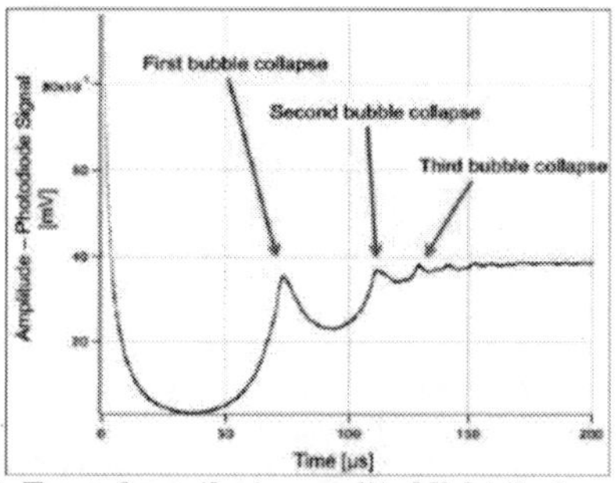

Figure 2. Trace from the transmitted light that propagates through the cavitation Bubble plane. The collapse times can be measured off from this trace with a high time resolution, since the trace is obtained by using a very fast photo diode.

RESULTS

The collapse time was measured under several liquid pressure values and pulse energy conditions. Parametric curves of the collapse time dependence on the liquid pressure were obtained and are shown in Figure 3.

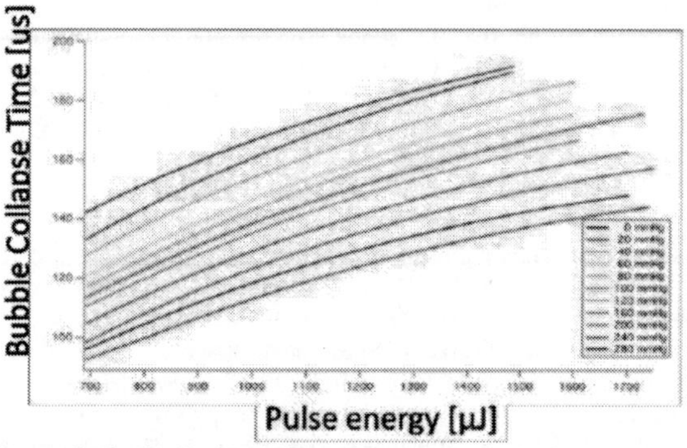

Figure 3. Cavitation bubble collapse time dependence on the liquid pressure and pump pulse energy.

DISCUSSION

The first collapse time in a laser-induced cavitation bubble event depends clearly on both the external liquid pressure and on the laser pulse energy. For well-known pulse energy, it is possible to infer the actual liquid pressure by precisely measuring the bubble collapse time. The presented STM technique seems to be a feasible way of implementing an IOP sensor. By using this technique the IOP measurement is expected to be more accurate than traditional tonometry measurements. Some advantages of the laser- induced cavitation and STM based sensor, among other, are that the in this case the IOP measurements would not depend on person to person factors as: patient age, gender, patient eye ball movements during measuring and/or corneal thickness. This is so due to the fact that the through the scheme proposed here the IOP measurement takes only a few tens of microseconds and the sensor measures the IOP directly inside the anterior chamber, avoiding all the above mentioned factor as sources of uncertainty on the measurements.

ACKNOWLEDGEMENTS

LFDC acknowledges CONACyT support through PhD scholarship. Authors acknowledge partial support to this research work by Oftálmica Internacional.

REFERENCES

[1] Lee, H., et al., Dynamics of laser-induced bubble collapse visualized by time-resolved optical shadowgraph. Journal of Visualization, 2011. 14 (4): p. 331-337.

[2] Laser Tissue Interactions Fundamentals and Applications. Markolf H.Niemz. Third ed. Biological and medical physics, biomedical engineering. 1996, Springer: Germany.

[3] Gregorčič, P., et al., Measurements of cavitation bubble dynamics based on a beam-deflection probe. AppliedPhysics A, 2008. 93 (4): p. 901-905.

[4] Evans, R. and S. Camacho-López, Pump-probe imaging of nanosecond laser-induced bubbles in distilled watersolutions: Observations of laser-produced-plasma. Journal of Applied Physics, 2010.108(10): p. 103106.

[5] Devia-Cruz, Luis F., Santiago Camacho-López, Rodger Evans, et al. 2012. Laser-induced cavitation phenomenon studied using three different optically-based approaches – An initial overview of results. Photonics and Lasers in Medicine. 1(3): 143-229.

INFANT ORAL CAVITY PRESSURE DATA PROCESSING

Lin Jiang, Fatemeh Hassanipour

Department of Mechanical Engineering
University of Texas at Dallas
Richardson, TX, USA

INTRODUCTION

It is well accepted that infants obtain both nutrients and immunities required for growth from breast milk. Previous work has explored and modeled the vacuum used by infants while breastfeeding. However, the total amount of pressure from infant's oral cavity applied on the nipple-areola complex during time-varying suckling patterns is unclear.

In this paper, the raw data extracted from the vacuum pressure, ultrasound images of the oral cavity and time dependent motion of the wave-like tongue movement during breastfeeding are de-noised, compared and analyzed. Flexible resistance pressure sensor strips were used for oral pressure. To evaluate the relationship between the oral pressure and vacuum applied by the infant, we first measure the nipple dimension changes seen on the ultrasound image. Then the noise and unexpected outliers from the pressure data were removed and a processed oral pressure for both infant's upper gum and lower jaw is analyzed. A natural-looking oral movement of infants during breast-feeding was derived from captured sensor data.

METHODS

For eight groups of participants, the vacuum, upper gum and lower jaw pressures from infants during breastfeeding were measured by collaborators at University of Western Australia, Perth, WA, Australia. Flexible resistance pressure sensor strips I-Scan 9801/9830 (Tekscan Inc., Boston, MA, USA) were used for oral pressure. The position of sensors during experiments can be seen in Fig.1, where U1-U7 represents the displacements of the sensor cells that were under the upper gum of infants while L1-L7 indicated that the sensors cells were placed under infants' lower jaw. The ultrasound images and vacuum pressure were obtained by an endocavity convex transducer Acuson, XP10 (Siemens, Mountain View, California, USA) placed under infant chin and a silicon tube (650×4mm) connected to a disposable pressure transducer (Cobe Laboratories, Frenchs Forest, NSW2086, Australia), respectively.

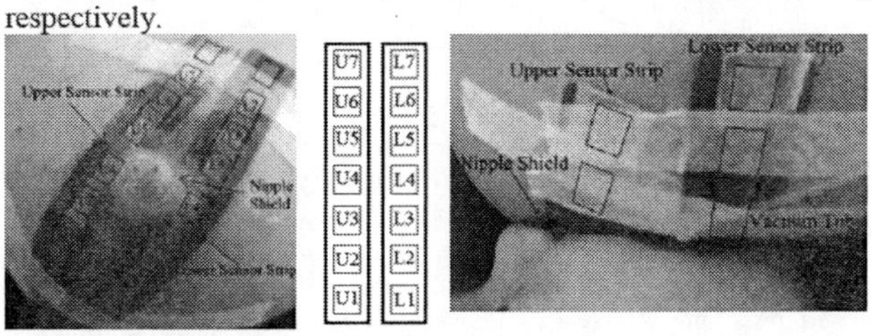

Figure 1 (left) Sensor placement on mum's breast (middle) Schematic description of the sensor cells (right) Vacuum tube and sensor position during clinic experiment

All image processing analyses were performed using MATLAB. The suck cycle showing milk ejection was captured in the ultrasound images which are known as Nutritive Suckling (NS) [1]. Fig. 2a shows an ultrasound image when vacuum was at its peak value. The tongue and upper palate are wrapped around the nipple at its lowest position. Fig. 2b represents a minimum vacuum value when the infant compresses the nipple with his upper palate, tongue, and jaw, causing the nipple to stretch and elongate significantly inside infant's oral cavity (see Fig. 2c). The pressure from both infant's inner vacuum and oral pressure created this phenomenon in a NS cycle.

Calibration is made with a known dimension in ultrasound images. A manual design-nation of the approximate edge on each frame was used to outline the visualized boundaries for nipple, upper palate and tongue. A self-programmed measurement system was achieved in MATLAB to get average dimensions of the nipple width and length with tongue moving up (TU) and tongue moving down (TD).

Figure 2 (a) Ultrasound image when tongue moves down (b) Ultrasound image when tongue moves up (c) Schematic description of nipple dimension measurement method

The average nipple width and elongation with observable tongue up (TU) and tongue down (TD) inner oral pattern during 10 seconds of NS cycles were measured using image processing toolbox in MATLAB for one of the Participates.

Data De-noising and Statistical Analysis Files including raw sensor readings were derived using I-Scan 7.60 compiling software from Tekscan and was imported into MATLAB with user defined programs. After loading the linear calibration and applying unit conversion, all sensor data were organized and stored in MATLAB files. The original sensor signal contained an arbitrary range of white noise and some unpredictable outliers as can be seen in 10 seconds of NS from this infant in Figure 3. A post-processing filter based polynomial de-noising algorithm known as BUTTERWORTH with a MATLAB command BUTTER [2] are applied for the overlapped experimental and the noise signals.

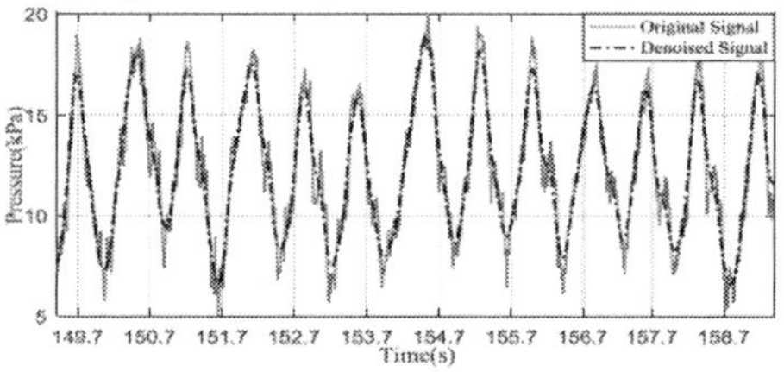

Figure 3 Signal de-noising sample from Pressure Sensor U4

RESULTS

The average nipple width was smaller during tongue moving up then when the tongue was at its lowest position in Table 1. Meanwhile, nipple length was longer when infant's tongue moved up then when the tongue moved down. We observed a vacuum peak and milk spray from the nipple when the tongue moves down. This result shows that when the infant was creating the vacuum, its oral cavity opens to allow the nipple to expand widthwise and shorten lengthwise. Conversely, when the vacuum is at its lowest is when the tongue and upper palate are compressing the nipple causing the nipple dimensions to shorten the width and increase the length. Also, the change of nipple length would be as long as 3.49 mm and the compression of the nipple in the width was almost 1.18 mm.

Table 1 Nipple Dimension Measurement

Nipple Dimension (mm)	Tongue Up	Tongue Down	Deviation
Average Nipple Width	9.13±0.11	9.85±0.57	1.18
Nipple Length	23.42±1.14	22.11±1.37	3.49

As we observed from Fig. 1C, sensor 7 was not placed wholly under infant's mouth. The pressure values from the sensor cells not under the infant's mouth were most likely from infant's hand compressing mum's breast. Sensor cells not located under infant's mouth were ignored. Fig. 4 shows 14 suck cycles during 10 seconds of NS while effective oral pressure sensor data was averaged for both the upper and lower gum/jaw after applying a BUTTERWORTH filter to the recorded raw data. The filtered data gave us a smooth sinusoidal signal as well as kept the original peaks of the waves. For Infant 6, a total pressure of approximate 9 kPa was applied on mum's breast from the upper part of infant's mouth and 5 kPa was from the lower jaw motion of the infant during breastfeeding. Furthermore, the upper palate moved steadier than the lower jaw, the deviation for upper palate pressure movement was within 0.5 kPa while the change for lower motion was around 2 kpa during 10 seconds NS. Meanwhile, the pattern of upper and lower oral pressures matched the vacuum peaks during these suck cycles. When the vacuum experienced a local minimum (around -10 kPa) the upper and lower pressures would simultaneously reach a local minimum value. This phenomenon matches what we observed in the ultrasound image; that is the infant's jaw dropped to create a vacuum.

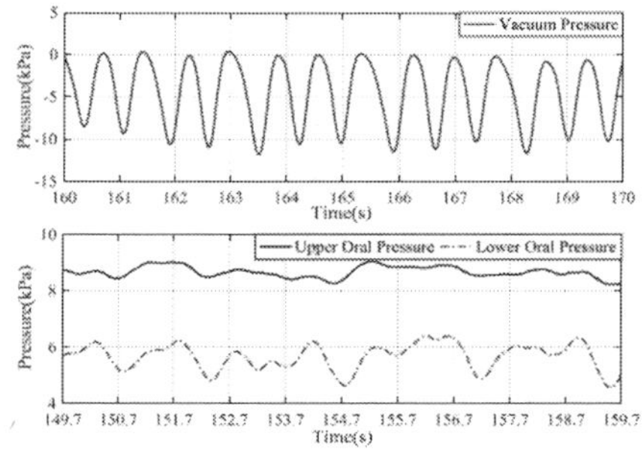

Figure 4 Vacuum Pressure vs. Total Oral Pressure during 10 seconds Nutritive Suckling Breastfeeding

DISCUSSION

The results from this study show that pressure change from infant's oral cavity was moving correspondingly with the suckling pattern during breastfeeding. Furthermore, we provided the exact value about the pressure applied on mum's breast from infants' mouth instead of just demonstrating the importance of oral pressure as previous studies. Among all the eight participants (see Fig. 5), three participants exerted average pressures that exceeded 10 kPa. These three participant mums complained about nipple pain. This finding may indicate that the nipple pain of these mums may be attributed to, at least in part, infants' upper or lower oral pressure during breastfeeding.

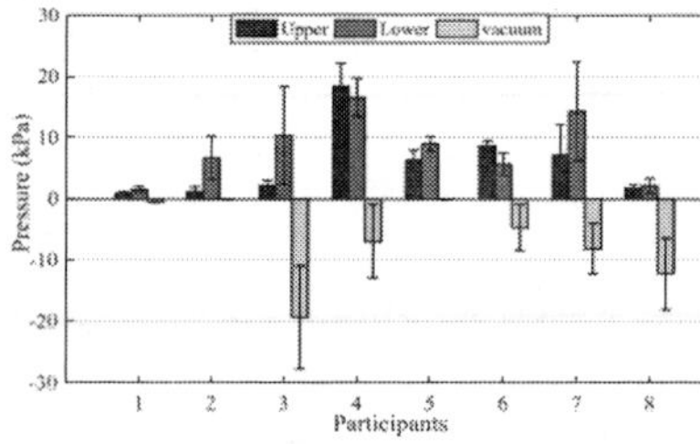

Figure 5 Oral Cavity Pressure on the Breast during 10 Seconds Nutritive Suckling of Breastfeeding for eight infants

ACKNOWLEDGEMENTS

The authors would like to thank Ms. Diana Alatalo, Dr. Donna Geddes, and the Hartmann Human Lactation Research Group in Perth, WA, Australia for their assistance with capturing these data. This research was funded by NSF Grant #1454334.

REFERENCES

[1] Geddes, DT et al., Early Human Dev, 84: 471-477, 2008.
[2] Butterworth, S, *Wireless Eng*, 7: 536-541, 1930.

SB³C2017
Summer Biomechanics, Bioengineering and Biotransport Conference
June 21 – 24, Tucson, AZ, USA

AN IN VITRO MODEL OF INTRATHECAL CEREBROSPINAL FLUID DYNAMICS WITH DORSAL AND VENTRAL SPINAL CORD NERVE ROOTLETS

Lucas R. Sass[1], Mohammadreza Khani[1], Olivier Baledent[2], Bryn A. Martin[1]

(1) Department of Biological Engineering
University of Idaho
Moscow, Idaho, U.S.A.

(2) BioFlow Image
University of Picardy Jules Verne
Amiens, France

INTRODUCTION

The dynamic motion of cerebrospinal fluid (CSF) around the brain, spinal cord, and within the ventricles is thought to be an important factor contributing to a number of central nervous system (CNS) diseases and can be leveraged for use in CSF-system based therapies such as intrathecal drug and gene vector delivery to the CNS. Dynamic MRI measurements have shown that CSF moves in an oscillatory manner within the intrathecal space surrounding the spinal cord [1]. It is a clear fluid with the viscosity of water at body temperature. Subject specific computational fluid dynamics modeling has been used to investigate the role of CSF dynamics in CNS diseases [2] and the impact of microanatomical features, such as spinal cord nerve rootlets and denticulate ligaments, on CSF dynamics [3]. Idealized in vitro models have been used to help understand how spinal cord fluid structure interaction can contribute to syringomyelia [4]. However, at present an in vitro model of the complete spine has not been designed nor constructed. Such an in vitro model would allow characterization and optimization of intrathecal-based therapeutic devices and validation of numerical modeling approaches. Herein, we describe the design and hydrodynamic characterization of an in vitro intrathecal CSF system model (Figure 1).

METHODS

Unsteady CSF flow quantification: CSF flow was measured at the C2-C3, C7-T1, and T10-T11 vertebral levels by phase-contrast MRI with retrospective ECG gating with 32 cardiac phases. A slice thickness of 5.0 mm and in-plane resolution of 0.54 x 0.54 mm were used. Slice orientation was perpendicular to the CSF flow direction and intersected vertebral discs. Unsteady CSF flow was quantified at each axial location using methods previously published by our group [2]. To obtain a smooth distribution of CSF flow along the spine, CSF flow was interpolated between each measurement location.

Intrathecal CSF space geometry: High-resolution T2-weighted MRI anatomic measurements were obtained for a healthy 23-year-old female subject in three volumes (30 x 30 x 7 cm) with a General

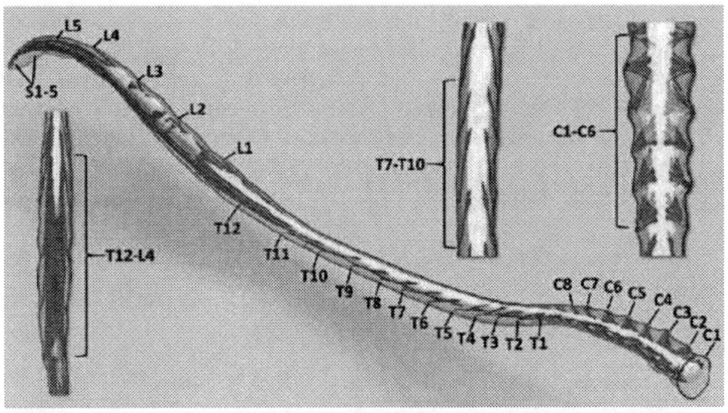

Figure 1: Intrathecal CSF system model with nerve roots labeled.

Electric 3T scanner (Signa HDxt, software 15.0_M4_0910.a). Each MRI data volume was segmented using ITK-SNAP (Version 3.4.0, University of Pennsylvania, U.S.A.), exported as a .STL file and then imported into Blender (Version 2.77a, Amsterdam, Netherlands). In brief, the model was sliced by a series of axial planes to define the spinal cord and dura surface. Resulting edge loops were extruded (~1.0 mm) to create a series of profiles. Simple circular mesh objects were fitted to every profile using the shrinkwrap modifier in Blender and new spinal cord and dural surfaces were created by bridging the edge loops. Manual sculpting was performed in Blender to improve the overlay fit of the original and remeshed surfaces to a ~50% interference by visual inspection with equal overall model volumes. Finally, shrinkwrap (with offset to maintain correct volume) and smoothing modifiers were applied to each surface to achieve the final fit of the remeshed surface to that of the original segmentation mesh.

Spinal cord nerve rootlet addition: The 30 pairs of dorsal and ventral spinal cord nerve rootlets were manually added to the geometric model as these small features were only partially visible in the high-resolution anatomic MRI measurements. In brief, the nerve root

Poster Presentation #P263

insertion location along the spinal cord was visually determined using the MRI measurements. Each individual nerve rootlet was then designed based on an organic modeling process that consisted of cross-sectional extrusion, tapering and bending of individual nerve rootlets. Similar to our previous study [3], we took into account the radicular line, descending angle and nerve root thickness as design parameters. In addition, we created individual nerve rootlets and took into account dimensions based on subject-specific nerve root location, anatomic reference imagery and cadaveric measurements based on studies by Bozkurt et al., Zhou et al., and Hauck et al. [5-7]

Hydrodynamic and geometric quantification: CSF hydrodynamics and geometry were quantified along the entire spine based on our previously published methods [2]. In summary, the axial distribution of cross-sectional area and wetted perimeter of the spinal cord, dura and spinal cord nerve rootlets were determined at 1 mm intervals. These measurements were then used to compute axial distribution of hydraulic diameter, Reynolds number (Re) based on hydraulic diameter and Womersley number (Wo). Axial distribution of mean CSF velocity was computed based on an axial smoothing of the unsteady CSF flows (not shown) and local cross-sectional area.

RESULTS

The final model included the 30 pairs of dorsal and ventral nerve rootlets, spinal cord with coccygeal nerve and dural wall (Figure 1). Total intrathecal CSF volume below the foramen magnum was 96.8 ml. Volumes of the dura, spinal cord and nerve rootlets were 122.4, 19.8 and 5.8 cm^3, respectively. Volumetric error between the remeshed surfaces compared to the original segmentations was calculated at 1.02%, with an original segmentation volume of 100.5 cm^3. The surface areas for the dura, spinal cord and nerve rootlets were 317.4 cm^2, 113.0 cm^2 and 232.3 cm^2 respectively. The average cross sectional areas of the dura, spinal cord and nerve roots were 2.02 cm^2, 0.33 cm^2 and 0.10 cm^2 respectively (Figure 2). Average Reynolds and Womersley number was 69.4 and 9.6, respectively.

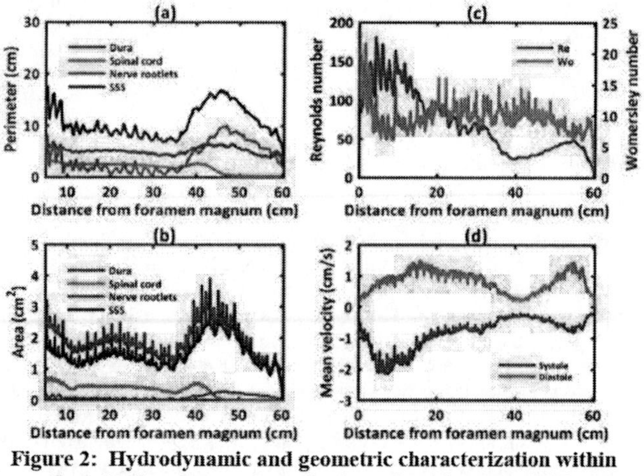

Figure 2: Hydrodynamic and geometric characterization within the intrathecal space (spinal subarachnoid space = SSS).

DISCUSSION

The presented intrathecal CSF system model provides a foundation for testing of intrathecal therapeutic systems and validation of numerical modeling approaches. In addition, it provides detailed information on subject-specific axial distribution of geometric and hydrodynamic parameters within the intrathecal space.

Geometric analysis of the intrathecal CSF space showed that the spinal cord nerve rootlets contribute 35% of the overall intrathecal CNS tissue surface area (based on SSS perimeter in Figure 2a). This information could be important for calculation of pharmacokinetic absorption rate into the CNS tissue.

Intrathecal CSF volume below the foramen magnum was ~100 ml. Studies in the literature have considered the entire CSF volume within the intracranial space and spine to be on the order of ~125 to 150 ml with a relatively small contribution of intrathecal CSF. To our knowledge, no previous studies have collected high-resolution MRI data over the entire intrathecal space and used this information to quantify CSF volume in the human. Previous studies have primarily relied on cadaveric drainage of CSF to estimate the CSF volume, a technique possibly prone to error since the delicate meninges of the CNS collapse post-mortem and many fine membranes are present in the intrathecal space that may restrict CSF drainage. Future studies should be conducted to characterize intrathecal CSF geometry.

The hydrodynamic results show that CSF flow is laminar throughout the spine with a peak Reynolds number of ~181. The region of the spine with most hydrodynamic activity in terms of Reynolds number and mean CSF velocities was at C3-C4. Reynolds number had local minimum value near T12. This information could help optimize intrathecal drug infusion location with interest towards increasing drug spread or possibly decreasing it to elevate drug resonance time and uptake into the CNS tissue.

In this study the semi-automatic tool in ITK-SNAP was used for CSF space segmentation. After the first versions of the remeshed dura and spinal cord were completed, the enclosed CSF volume was checked against that of the original non-smoothed segmentation. Despite a close surface interference, the error was found to be >25%. Our process mitigated this error by applying a shrinkwrap modifier in Blender. We found that a consistent surface bias of ~25μm added ~1% to the volumetric error. With anatomic MRI data typically collected at a resolution well above this level, it is easy to accumulate a significant error in the segmentation process highlighting the need for improved CSF space segmentation methods.

ACKNOWLEDGEMENTS

This work was supported by an Intuitional Development Award (IDeA) from the National Institute of General Medical Sciences (NIGMS) of the National Institutes of health (NIH) under Grant #P20GM1033408, NIH NIGMS Grant #4U54GM104944-04TBD and University of Idaho Vandal Ideas Project.

REFERENCES

1. Loth F. et al. Hydrodynamic modeling of cerebrospinal fluid motion within the spinal cavity. J Biomech Eng. 2001;123(1):71-9.
2. Martin BA, et al. Hydrodynamic and longitudinal impedance analysis of cerebrospinal fluid dynamics. PMCID: PMCPMC3794956.
3. Heidari Pahlavian S, et al. The impact of spinal cord nerve roots and denticulate ligaments. PMCID: PMC3977950.
4. Martin BA, et al. Spinal Subarachnoid Space Pressure Measurements in an In Vitro. J Biomech Eng-T Asme. 2010;132(11).
5. Bozkurt M, et al. Microsurgical anatomy of the dorsal thoracic rootlets and dorsal root entry zones. Acta Neurochir (Wien). 2012;154(7):1235-9.
6. Zhou MW, et al. Microsurgical anatomy of lumbosacral nerve rootlets for highly selective rhizotomy in chronic spinal cord injury. PubMed PMID: 21089050.
7. Hauck EF, et al. Intradural microanatomy of the nerve roots S1-S5 at their origin from the conus medullaris. J Neurosurg Spine. 2008;9(2):207-12.

Poster Presentation #P263

SB³C2017
Summer Biomechanics, Bioengineering and Biotransport Conference
June 21 – 24, Tucson, AZ, USA

ACCELERATING CARDIOVASCULAR SEGMENTATION WITH CONVOLUTIONAL NEURAL NETWORKS

Gabriel D. Maher (1), Jameson M. Merkow (2), Alison L. Marsden (3)

(1) Institute for Computational and Mathematical Engineering,
Stanford University
Palo Alto, CA, USA

(2) Electrical and Computer Engineering,
University of California San Diego,
San Diego, CA, USA

(3) Bioengineering, Pediatrics,
Stanford University
Palo Alto, CA, USA

INTRODUCTION

In recent years, numerical simulations of cardiovascular hemodynamics have improved our understanding of cardiovascular biomechanics relevant to a range of clinical applications such as atherosclerosis [1] and coronary artery disease [2]. However, cardiovascular simulations require construction of accurate three dimensional patient-specific models of the cardiovascular anatomy based on medical image data [3]. Models are typically constructed from magnetic resonance (MR) or computed tomography (CT) scans.

SimVascular is an open-source software package with which users can perform patient-specific blood flow simulations [4] [5]. SimVascular contains functionality to visualize and segment medical images, construct meshes for numerical simulations and perform cardiovascular fluid dynamics and fluid-structure interaction simulations. At present, to segment medical images with SimVascular, users often perform manual segmentation or use classical image processing methods such as thresholding [5] or active contours [6][7].

The image segmentation process often requires laborious user intervention, making the modeling process cumbersome and time-consuming, and preventing high-throughput model generation for large patient cohorts. It also introduces user variability. Medical image segmentation is thus a significant bottleneck in performing cardiovascular simulations. It is therefore desirable to find, or develop, cardiovascular medical image segmentation methods that can be used without significant user intervention across a range of medical images.

Convolutional Neural Networks (CNNs) are machine learning models tailored for processing visual data. Example medical image processing methods developed with CNNs include, among others, pancreas segmentation [8], brain lesion detection [9], brain tumor segmentation [10], kidney segmentation [11] and cardiovascular edge detection [12]. Segmentation methods developed with CNNs are typically parameter-free making them a promising approach for improving the patient-specific modeling process. Furthermore the Vascular Model Repository [14] is a repository that contains over 100 medical images and cardiovascular models segmented by users of SimVascular, and is a suitable source of data with which to train machine learning models for cardiovascular segmentation.

In this work we propose a CNN-based medical image segmentation method that can be used to construct accurate 3D cardiovascular models for numerical blood flow simulations. The method consists of (1) using CNNs to segment image patches extracted from medical images with user-annotated vessel centerlines, (2) extracting vessel boundaries from the segmentations, and (3) combining the extracted boundaries to form a solid model. We demonstrate that the CNN approach outperforms an active contour method when compared to manually segmented vessel boundaries.

METHODS

The proposed method for cardiovascular model construction in this work is based on the current model construction process available through SimVascular [5][7]. An outline of the model construction process is shown in Figure 1. In the first step users load a 3D medical image and annotate vessel centerlines by selecting (x,y,z) coordinates in the image. For the second step, segmentations can be produced along individual vessels, using an intensity probe that displays the image intensity along the vessel. During the third step solid models for each vessel are constructed by lofting the segmentations from step 2.

Centerline annotation and vessel segmentation are the most time consuming parts of the model construction process. The proposed method improves the vessel segmentation step. Our method takes a 2D cross sectional image along a centerline path as input to a CNN that has been trained for vessel segmentation. Vessel boundaries are then extracted from the segmentation with the marching squares algorithm. For CNN architectures we build a simple fully-convolutional CNN with both convolutional and fully-connected layers. Our CNN architecture localizes vessels close to the centerline, hence we denote it as Region-Selection Network (RSN). Further, we apply the object-boundary guided semantic segmentation method [14] to RSN and call the resulting CNN Object-Boundary Guided RSN (OBG_RSN). We compare our result to the Holistically-Nested edge-detection (HED) [15] and I2INet [12] architectures as these are state of the art for edge-detection and cardiovascular edge-detection respectively. In this case we have adapted the I2INet architecture to work on 2D image patches. The training, validation and test datasets consist of 17331, 3036 and 1831 64x64 image patches respectively. The data is split so that image patches from the same medical image are always in the same set. As a baseline classical image processing method we use the level set method available in SimVascular [7], with a fixed set of parameters tuned to produce accurate segmentations on a reference aorta model.

To assess the quality of vessel boundaries produced by each method we computed the Jaccard distance error metric [16] (equation 1) with respect to user-produced vessel boundaries. *A* and *B* denote sets and $|A|$ denotes cardinality of set *A*. *J(A,B)* is *0* if *A* and *B* are non-overlapping and 1 if they are the same. As such the Jaccard distance allows the similarity between vessel boundaries to be quantified.

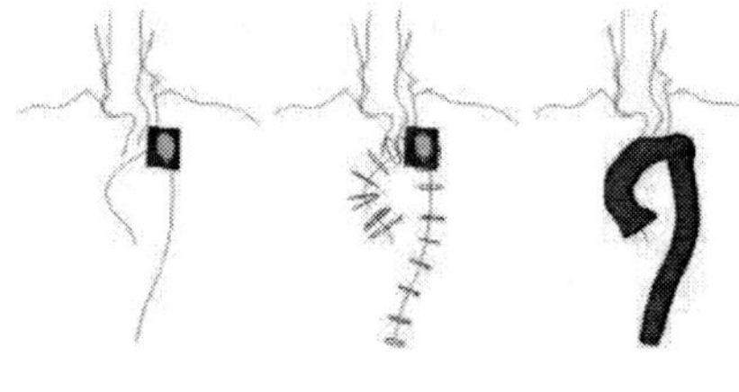

Figure 1: Cardiovascular model construction process in Simvascular (from [5]). Left: Vessel centerlines and medical image intensity probe, Middle: Vessel centerlines with vessel boundaries, Right: Solid model constructed from vessel boundaries.

$$J(A,B)=\frac{|A\cap B|}{|A\cup B|} \qquad (1)$$

RESULTS

Example vessel boundaries and error metric results for each method are shown in Figure 2 for example image patches. For large and clearly visible vessels, all methods produce a vessel boundary that resembles that segmented by the user. When the vessels are small, or ambiguous, the level set method is unable to produce the desired boundary. All CNN architectures are able to segment small vessels and infer reasonable boundaries in the ambiguous fourth image patch.

To quantify the performance of each method, we define the cumulative error distribution *F(x)* (equation 2), where *N* is the total number of image patches. For a given method, *F(x)* denotes the fraction of vessel boundaries with a Jaccard distance below *x*. Figure 3 shows the cumulative error distribution for each method. A larger area under a curve approximately indicates better performance. It is evident that all CNN-based methods have similar performance and improve on the level set. The level set underperforms due to failing to produce a vessel boundary on approximately 60% of image patches.

DISCUSSION

Our results provide evidence that CNN-based segmentation methods can be used to improve the patient-specific modeling process for cardiovascular models. However, the CNN-based method is conditional on the availability of medical images with accurate annotated vessel centerlines. The fact that all CNN-based methods performed similarly indicates that the precise CNN architecture is not crucial, as long as the CNN has enough capacity to model the data.

The underperformance of the level set method in this case is due to the fact that only one set of level set parameters was used for all images. The performance of the level set could be improved by using multiple parameter sets. This illustrates why the level set requires significant user intervention to be used across many images. Once the CNNs have been trained, the CNN-based method is parameter free and therefore requires significantly less user intervention.

$$F(x)=\frac{|\{J_i : J_i \leq x\}|}{N} \qquad (2)$$

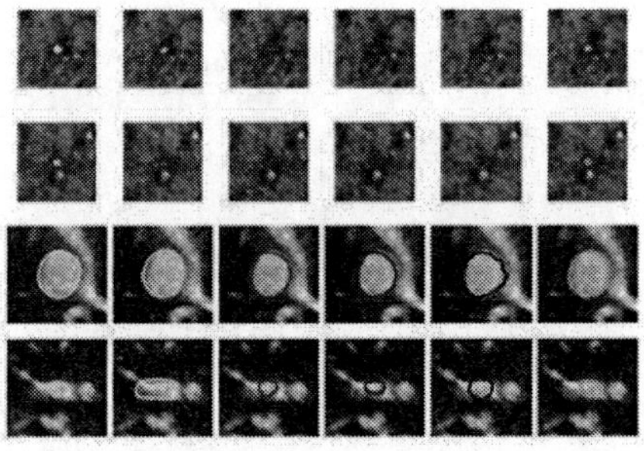

Figure 2: Example image patches and corresponding vessel boundaries produced. Yellow: Manual segmentation, Pink: Level set, Red: RSN, Blue: OBG_RSN, Black: HED, Orange: I2INet

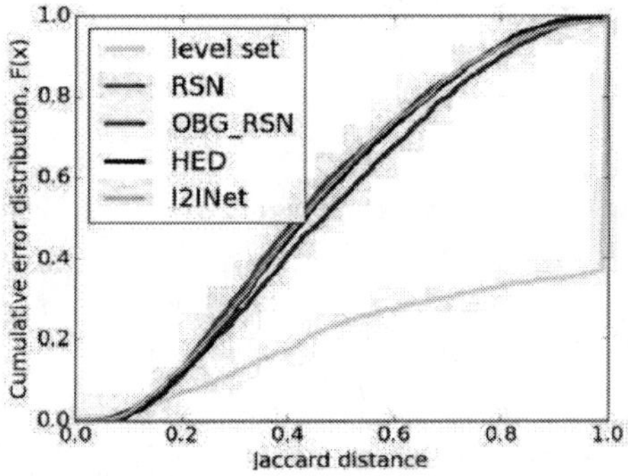

Figure 3: Cumulative error distribution for CNNs and level set.

REFERENCES

[1] Samady, H et al., Circulation, 124:779-788, 2011.
[2] Taylor, C A et al., JACC, 61:2233-2241, 2013.
[3] Taylor, C A et al., Annu Rev Biomed Eng., 11:109-134, 2009.
[4] http://simvascular.github.io/, accessed on 01/10/2017.
[5] Updegrove, A et al., Annals of Biomed Eng., 1-17, 2016.
[6] Osher, S and Sethian J A, JCP, 79:12-49, 1988.
[7] Wang, K C, PhD Thesis, Stanford University, 2001.
[8] Cai, J et al., MICCAI, 442-450, 2016.
[9] Kamnitsas, K et al., Medical Image Analysis, 36:61-78, 2017.
[10] Menze, B H et al., MICCAI-BRATS, 2015.
[11] Çiçek, Ö et al., MICCAI, 424-432, 2016.
[12] Merkow, J et al, MICCAI, 371-379, 2016
[13] https://simtk.org/projects/cv-gmodels, accessed on 01/10/2017.
[14] Huang, Q et al., arXiv:1603.09742, 2016
[15] Xie, S and Tu, Z, ICCV, 1395-1403, 2015
[16] Everingham, M, IJCV, 2:303-338, 2010

SB^{3}C2017
Summer Biomechanics, Bioengineering and Biotransport Conference
June 21 – 24, Tucson, AZ, USA

HOW TEMPERATURE INFLUENCES THE VISCOSITY OF HORNWORM HEMOLYMPH

Melissa C. Kenny (1), Matthew N. Giarra (2), and John J. Socha (1)

(1) Department of Biomedical Engineering and Mechanics
Virginia Tech
Blacksburg, VA, USA

(2) Department of Mechanical Engineering
Virginia Tech
Blacksburg, VA, USA

INTRODUCTION

The insect circulatory system is responsible for the circulation of nutrients, hormones, and metabolic waste via the transport of hemolymph. The hemolymph also functions to close off wound sites, destroy internal parasites, and, in some cases, aid in thermoregulation [1,2]. The hemolymph is crucial in the growth and development of holometabolous insects and can change throughout the life cycle [2]. Insect hemolymph is composed primarily of water with added inorganic salts, amino acids, proteins, lipids, carbohydrates/sugars, metabolic waste, and hormones [1]. While much of the composition and function of hemolymph is known, many of its physical properties, such as the viscosity, are still unknown. These properties are integral in understanding the behavior of the hemolymph for circulation and flow.

As ectotherms, insects are significantly affected by environmental temperatures, and have evolved multiple mechanisms to respond to temperature extremes. For example, some insects upregulate heat shock proteins in response to heat to minimize protein denaturation, and some increase glycerol production and purge water in response to cold to minimize ice crystal formation [3]. However, the effect of temperature on the physical properties of hemolymph has not been studied, and changes in viscosity in particular may greatly influence an insect's ability to circulate hemolymph. Here, we address the questions: What is the viscosity of the insect hemolymph? What is the effect of temperature on the viscosity of insect hemolymph?

METHODS

We used *Manduca sexta* larvae to determine the viscosity of hemolymph across a range of temperatures. These larvae were chosen due to their extensive use in research and knowledge of their life cycle and growth patterns [4,5]. Measurements were taken using a cone and plate viscometer (Brookfield Engineering DV-II+ Pro) attached to a water circulator (Lauda RE206) to regulate temperature. Trials were performed at temperatures ranging from 0-45°C at 5°C internals. This range was chosen due to experimental limitations (0°C) and the upper lethal temperature for this species (45°C) [4]. To minimize oxygen-induced clotting (Fig. 1), experiments were performed in a sealed glove box flooded with dry nitrogen gas. Oxygen concentration within the box was measured to be < 2% for all trials. A small humidifier (Pure Enrichment Ultrasonic Cool Mist) was placed within the glove box and maintained 60-80% humidity for all trials. All experiments were performed at nearly constant shear rates of 450-460.8 s^{-1}.

To extract hemolymph (Fig. 1), larvae were first anesthetized via exposure to nitrogen gas for 5-10 minutes. A small incision was made on the ventral surface of the animal between the 2nd and 3rd set of prolegs. Hemolymph was collected on Parafilm and 0.7 mL was transferred to the cup of the viscometer. To minimize oxygen exposure, the time from incision to the start of the viscometer was <2 minutes.

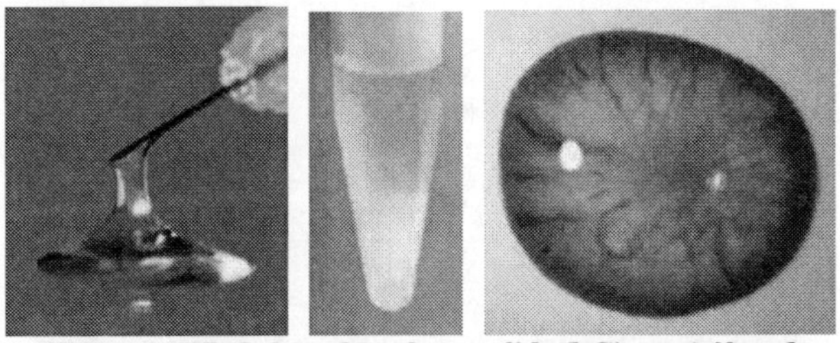

Figure 1: Whole hemolymph on a slide (left); centrifuged hemolymph showing separation of plasma (center); hemolymph after approximately 11 minutes of exposure to ambient air (right).

Additional trials were conducted with the plasma separated from the hemolymph (Fig. 1). To separate the plasma from the hemocytes, hemolymph was spun via centrifuge (Miulab Mini Centrifuge, Mini-10k+) for 5 minutes at 10,000 RPM. The supernatant was isolated and centrifugation was repeated once more.

Viscosity measurements were recorded at a rate of 2 Hz. Data was processed using a running window standard deviation in a custom MATLAB (MathWorks) code. Viscosity values were considered steady when the standard deviation was below a 0.05 SD threshold. The final viscosity value for each trial represents the average of the steady values.

RESULTS

Preliminary results revealed the viscosity at 20°C of whole hemolymph was 2.82 ± 0.23 cP and plasma was 1.84 ± 0.12 cP. Whole hemolymph viscosity increased with decreasing temperature, and showed an increase in average viscosity from 1.83 to 14.11 cP from 45°C to 0°C (Fig. 2). Plasma showed the same trend, with an increase in average viscosity from 1.07 to 3.62 cP from 45°C to 0°C.

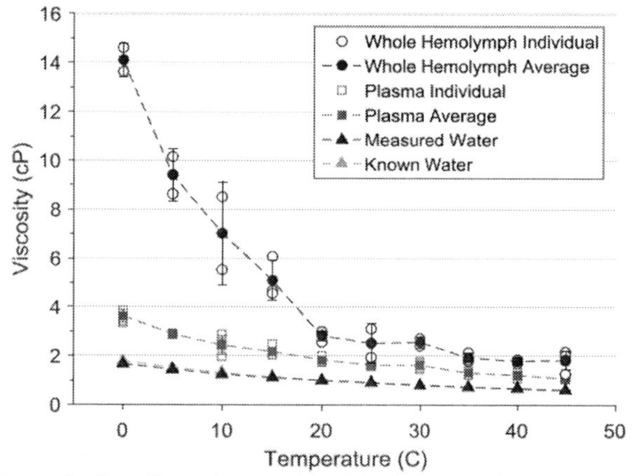

Figure 2: Viscosity vs temperature for whole hemolymph, plasma, and water. Individual trials and averages are shown for each temperature. Bars indicate standard deviation of the mean.

Relative to water, the viscosity of plasma was 1.91 ± 0.13 times the viscosity of water at all temperatures, with a range of 1.77 to 2.18. The variance in viscosity of whole hemolymph changed with temperature (Fig. 2). At 20-45°C, the viscosity of whole hemolymph was 2.85 ± 0.20 times that of water with a range of 2.66-3.16. From 0-15°C, whole hemolymph viscosity had greater variation relative to water, with a range of 4.61 to 8.49 times the viscosity of water.

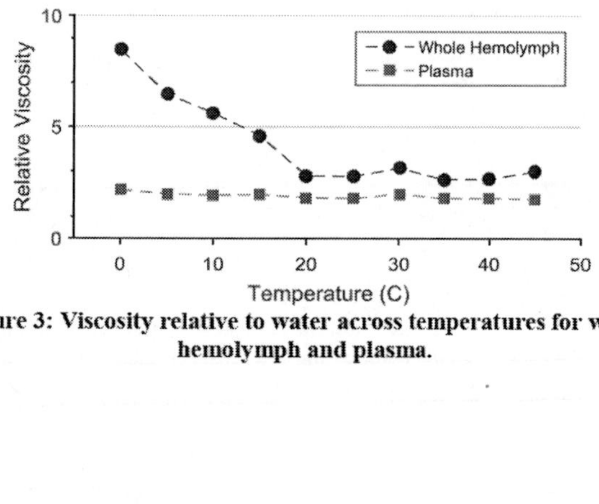

Figure 3: Viscosity relative to water across temperatures for whole hemolymph and plasma.

Viscosity of whole hemolymph, plasma, and water were measured at 25°C with varying shear rates by changing the speed of the viscometer spindle. The viscosity of water did not vary significantly with changes in shear rate (as expected), showing only a 0.02 cP increase when shear rate varied from 90 to 900 s^{-1}. However, the viscosity of both whole hemolymph and plasma decreased with increasing shear rate, showing a decrease of approximately 1.09 cP when shear rate varied from 225 to 1125 s^{-1} for whole hemolymph and 90 to 900 s^{-1} for plasma.

DISCUSSION

Although it is composed primarily of water, insect hemolymph has a significantly higher viscosity. At 20°C, the viscosity of hemolymph is 2.82 ± 0.23 cP as compared to the 1.00 cP viscosity of water at the same temperature. This difference may result from the added inorganic salts, nutrients, hemocytes, and other components of the hemolymph [1]. Additionally, whole hemolymph and plasma are shear-thinning non-Newtonian fluids, showing an increase in viscosity with decreasing shear rate. This is similar to both vertebrate whole blood and blood plasma [6]. Conversely, water is a Newtonian fluid, maintaining a constant viscosity regardless of shear rate. Understanding the rheological properties of insect hemolymph is necessary to study the circulatory system function in these animals.

Hemolymph viscosity also changes with temperature, showing an increase of 7.7 times for whole hemolymph from 45°C to 0°C, with a 3.4 times increase for plasma alone (Fig. 2). Water, by contrast, only experiences a three-fold increase over the same range. Cold and heat tolerance in insects are well studied. Changes in temperature are known to affect activity, locomotion, feeding, and other behaviors in insects [3-5]. Development of *Manduca sexta* larvae specifically is known to be optimum at 20-25°C, with only 12% survival at 15°C and 50% at 35°C [5]. Feeding behavior has been measured as well, showing minimal feeding in larvae below 12°C and a significant decrease in bites/second during feeding as temperature decreases [4]. Changes in hemolymph viscosity may represent an underappreciated factor in the reduction of activity with decreasing temperature. It could affect hydrostatic pressure for proleg usage or the ability of the dorsal vessel to move hemolymph, thus affecting the circulation of nutrients, waste, and other chemicals necessary for growth and survival.

Significance: Measuring the rheological properties of insect hemolymph is an important step towards a greater understanding of the mechanics of the insect circulatory system. Hemolymph must travel not only through an insect's open body cavity, but also through the length of thin legs and antennae, and through the delicate vessels in the wings. From an engineering perspective, the insect circulatory system functions using micro-sized spaces, utilizing several micro-pumps, diaphragms, and other organs. Understanding the mechanics of this system at the microscale could be used as inspiration for miniaturization and scaling of design, most notably in microfluidics.

ACKNOWLEDGEMENTS

We would like to acknowledge the help of Ellen Granata, Patrick Roagers, Amanda Barnes, Brittany Horton, and Katie Johnston. Supported by NSF 1558052 and 1301037

REFERENCES

[1] Chapman, R.F. "Insects: structure and function" *In Print*, 2013.
[2] Snodgrass, R.E. "Principles of Insect Morphology" *In Print*, 1935.
[3] Clark, M.S., Worland, M.R. *J Comp Physiol B*, 178:917-933, 2008.
[4] Casey, T.M. *Ecology*, 57(3):485-497, 1976.
[5] Reynolds, S.E., *J. Insect Phys.*, 31:129-134, 1985.
[6] Brust, M. et al. *Phys. Rev. Lett.*, 110:078305, 2013.

SB^{3}C2017
Summer Biomechanics, Bioengineering and Biotransport Conference
June 21 – 24, Tucson, AZ, USA

COMPUTATIONAL MODEL OF AN INITIAL LYMPHATIC NETWORK

Bernard O. Ikhimwin (1), Samira Jamalian (2), Charlie Macaskill (1), Christopher D. Bertram (1)

(1) School of Mathematics and Statistics
University of Sydney, New South Wales,
Australia

(2) Department of Bioengineering,
Imperial College, London, UK

INTRODUCTION

The lymphatic vascular system consists of networks of numerous vessels which play a key role in immune surveillance by transporting lymph (a colourless liquid that contains white blood cells that helps to purge undesirable materials and toxins from the body) and protein from the tissue space back to the circulatory system. In contrast to the cardiovascular system which has a central pump, the lymphatic vascular system has no central pump hence the transport of fluid against gravity relies on local extrinsic and intrinsic pumping. However, since initial lymphatics lack smooth muscle cells, they cannot perform intrinsic pumping. In this work we used a lumped-parameter model to describe the transport of lymph through an initial lymphatic network of the mesentery by extrinsic pumping. We used an image of the initial lymphatics from [1], from which we obtained the length and diameter of the branches which we fed into our model to calculate the flow-rate across the network against adverse pressure gradients.

METHODS

We adapted an existing lumped-parameter model to describe the transport of lymph through the network under the impulsion of time-varying external pressure. We used a sigmoidal function [2,3] to model the resistance of the secondary valves which we found in the network vessels; see Fig. 1. Initial lymphatics have primary valves to regulate the intake of lymph from the interstitial space to the lumen. We assumed all inlets of the network to be primary valves. Following [4], we assumed perfect primary-valve function (i.e. no retrograde flow).

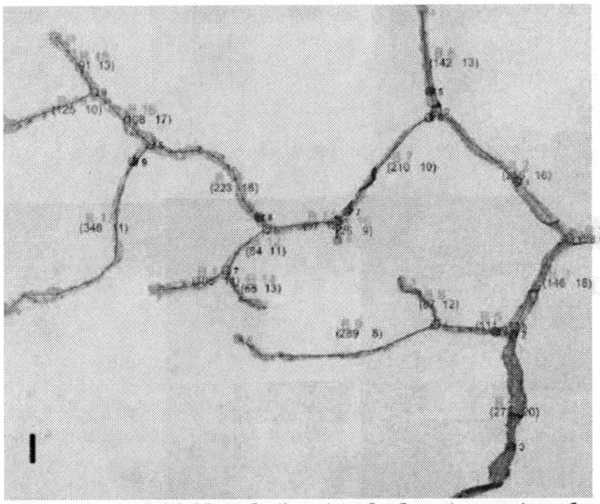

Figure 1: An initial lymphatic network of a rat mesentery, from fig. 1 of [1]. The network contains secondary valves labeled V1 to V9 (blue) and primary valves at the inlets of the network (labeled E2 to E11 in green, E1 is the outlet). B1-B19 (cyan) are the branch numbers of the network and the length and average diameter are printed (in red) on each of the branches. Lines of dark pixels that run across the network are blood vessels.

MATHEMATICAL MODEL

The equations of the model are as follows.

$$\frac{dV}{dt} = Q_n - Q_{n+1} \quad (1)$$

$$Q_n = \frac{P_{n-1,2} - P_{n,1}}{R_n} = \frac{P_{n,1} - P_{n,m}}{R_{ves}/2} \quad (2)$$

$$P_{n,m} = P_{ext}(t) + sD \quad (3)$$

$$R_n = f(P_{n-1,2} - P_{n,1}) \quad (4)$$

Eq. (1) describes conservation of mass, where Q_n and Q_{n+1} are flow-rate into and flow-rate out of a lymphangion, both through the appropriate valve. Eq. (2) describes conservation of momentum where $P_{n-1,2}$ and $P_{n,1}$ are the pressures upstream and downstream of a given valve respectively; $P_{n,m}$ is a mid-lymphangion pressure. Eq. (3) represents a balance of forces at the vessel wall. We model the time-varying external pressure with a sinusoidal function; s represents the slope of the linear relation between transmural pressure and diameter. Eq. (4) represents the valve resistance as a function of transvalvular pressure gradient. The resistance differs for primary and secondary valves; we have used an exponential function fitted to the data of [4] for the resistance of the primary valves, and a sigmoidal function adopted from [2] for the resistance of the secondary valves.

We have made the vessel inlet pressures equal to the time-varying external pressure, and we keep the network outlet pressure constant.

RESULTS

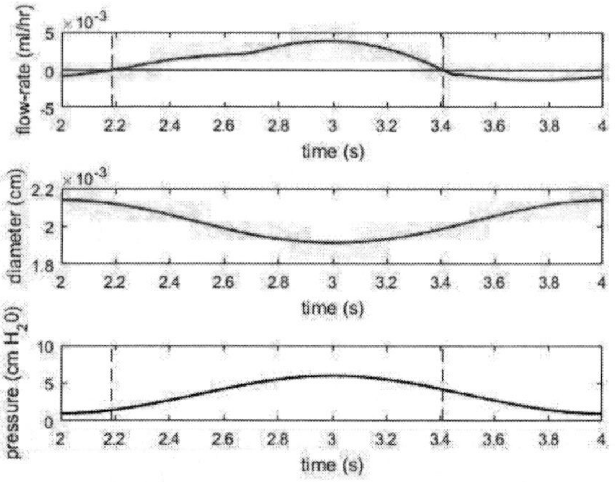

Figure 2: A plot of the outlet flow-rate, outlet diameter and external pressure as a function of time. The period of oscillation is 2s, P_{out} = 4 cmH$_2$O, P_{in} = $P_{ext}(t)$, which varies between 1 and 6 cm H$_2$O. Note that P_{ext} is higher when Q descends through zero than when it rises through zero.

Fig. 2 shows the instantaneous flow-rate, diameter and pressure for a given cycle. As the external pressure increases, the flow-rate increases while the diameter decreases; this is because of the lymphatic vessel compression due to motion of adjacent tissues, which leads to expulsion of lymph. During the early contraction and late relaxation phases of the external pressure, it is observed that there is backward flow since the network is up against an adverse pressure gradient (i.e. the inlet pressures oscillates between 1 cmH$_2$O and 6 cmH$_2$O while the outlet pressure is set to be 4 cmH$_2$O) during these phases. Early in the compression phase there is backward flow; this flow only fills the lymphatic network, because the ends of the network contain assumed-perfect primary valves which prevent backward flow. This increases the pressure in the vessels of the network. As the external pressure increases, the backward flow becomes forward flow while the pressure gradient is still adverse. As the external pressure increases further, the flow-rate peaks due to the favourable pressure gradient. The cycle-mean flow-rate was 0.0010 ml/hr.

Setting the external pressure and inlet pressures constant at 3.5 cmH$_2$O while keeping the outlet pressure at 4 cmH$_2$O resulted in no flow-rate in the steady state. The reason for this is the primary valves, which do not allow retrograde flow (i.e. the backward flow of lymph through the network outlet only fills the network, thereby increasing the pressure until it reaches that at the outlet). On the other hand, the secondary valves are biased to the open state, and thus permit some backflow. The inlet pressures and the external pressure were also set to 5 cmH$_2$O while the outlet pressure remained at 4 cmH$_2$O, this resulted in a forward flow-rate of 0.0019 ml/hr. We also varied $P_{ext}(t)$ between 1 cmH$_2$O and 9 cmH$_2$O, so that the average external pressure was 5 cmH$_2$O; the cycle-mean flow-rate through the inlet and outlet branches was greater than the steady-state flow-rate with the inlet and external pressures set to 5 cmH$_2$O.

DISCUSSION

We have formulated a model to describe lymph transport in a non-symmetric initial-lymphatic network which has both primary and secondary valves. A time-varying external pressure was used to model the extrinsic pumping mechanism, which depends on relative motion of external tissues.

The primary valves were assumed to be at the inlets of the network; later we plan to include disseminated inflow through initial lymphatic vessel walls everywhere. The secondary valves were placed where they appear in the imaged network; they are relatively sparse, and do not seem to follow a systematic placement scheme.

A linear relationship between transmural pressure and diameter was assumed in the absence of better information. The inlet pressures were set to equal the time-varying external pressure. The instantaneous flow-rate showed both forward and backward flow, with more backward flow during the reduction of external compression.

The backward flow is abated due to the primary valves in the network, this is seen when we set the external and inlet pressures to a constant ($P_{ext}(t)$ = 3.5 cmH$_2$O) while the outlet pressure remained 4 cmH$_2$O, this resulted in no flow-rate at steady state.

The result showed that periodic compression of initial lymphatic vessels by adjacent external tissues increases flow-rate compared to a constant external pressure.

ACKNOWLEDGEMENTS

CDB gratefully acknowledges generous assistance from Dr. Walter L. Murfee of Tulane University. All authors acknowledge funding from NIH grant U01-HL-123420.

REFERENCES

[1] Murfee, W L, et al., *Lymphatic Res Biol* 5(2): 81-90, 2007.
[2] Bertram, C D, et al., *J Biomech Eng* 133(1): 011008, 2011.
[3] Jamalian, S, et al., *PloS One* 11(2): e0148384, 2016.
[4] Galie, P, and R L Spilker, *J Biomech Eng* 131(11): 111004, 2009.

SB^3C2017
Summer Biomechanics, Bioengineering and Biotransport Conference
June 21 – 24, Tucson, AZ, USA

CAPTURING THE ORAL PERIPHERAL PRESSURE OF INFANTS WHILE BREASTFEEDING

Diana L. Alatalo, Fatemeh Hassanipour

Department of Mechanical Engineering
The University of Texas at Dallas
Richardson, TX, USA

INTRODUCTION

The importance of breastfeeding for the health and well-being of both mother and infant are well documented. The breast contains a unique ductal system for transporting milk to infants after birth. This bio-fluid transport phenomenon involves multiple mechanisms, both physical and hormonal [1]. Modeling the mechanisms requires obtaining the physical forces exerted by the infant while breastfeeding.

The dynamics of milk removal from the breast by the infant have been studied for decades. Ultrasound studies by Geddes et al. identified four stages of sucking that alter the shape of the breast and recorded normal intra-oral vacuum patterns [1]. The authors related the positioning of the tongue and changes in the nipple dimensions to the intra-oral vacuum pressure. Their study suggested that vacuum pressure played a significant role in milk removal from the breast by the infant. In recent years the use of simulation software has been used to further explore the mechanics of milk removal with heavy emphasis on negative vacuum pressure [2] [3].

The positive oral pressure exerted by the infant's mouth on the breast has received limited attention in recent decades. However, both breast massage and infant suckling are known to positively affect milk flow [4] [5]. Thus, the questions regarding the biomechanics of milk removal by infants during breastfeeding remain unresolved. A primary piece of information is missing from all previous studies, namely the amount of pressure exerted on the nipple-areola complex by the infant. The objective of this study was to capture the oral peripheral pressure of infants while breastfeeding in combination with the negative vacuum pressure.

METHODS

Participants. After obtaining IRB and ethics approval, positive oral pressure data were successfully obtained from eight dyads, with one dyad providing two sets of data on separate days. Ages of infants ranged from 6 days to 21 months. All infants were successfully breastfed with the use of a nipple shield.

Peripheral oral pressure: The I-Scan System by Tekscan™, Inc. was used to capture the peripheral oral pressure. Two pressure mapping sensors were used, model numbers 9801 and 9830. Both sensors are designed to be cut into strips for use over curved surfaces. The maximum pressures are 34 kPa and 69 kPa for these two sensors respectively. Each sensor was cut and edges smoothed to prevent the sensor edge from cutting into the mother's skin. Two strips from the sensor were attached to the breast with tape and covered with a breast shield to minimize moisture exposure to the strips and prevent the strips from entering the mouth of the infant (see Figure 1). For infants who regularly breastfed without a nipple shield, the tip of the shield was cut off to allow the nipple to move normally within the oral cavity. Since the width of the strips was too wide to comfortably fit around the nipple under a shield, the strips were placed along the top and bottom of the areola as close to the base of the nipple as possible. The strips were oriented to approximate the location of the upper and lower gums and lips of the infant while suckling. The sensors were zeroed and calibrated according to manufacturer guidelines using the I-Scan built in multi-point calibration algorithm. The sensitivity level was varied on some participants to avoid saturation of individual sensels and maximize data acquisition.

Poster Presentation #P267

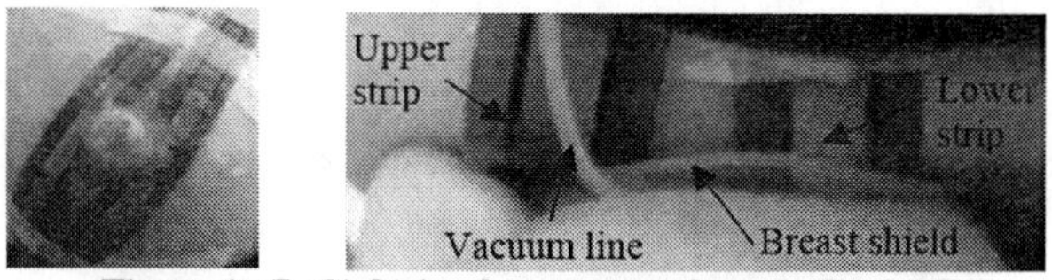

Figure 1: (Left) Strip placement on breast, (Right) Data acquisition while infant nursing

Intra-oral pressure, ultrasound imaging, and milk intake: Intra-oral pressure, ultrasound imaging, and milk intake were captured as outlined by Geddes et al. [1], except the supply line for the pressure transducer was filled with water instead of mother's expressed breastmilk and placed alongside the nipple on top of the nipple shield, when in use. The ultrasound imaging was used to determine periods of nutritive and non-nutritive suckling, as well as to visualize the changes in the nipple dimensions from infant pressures.

RESULTS

The peripheral oral pressure varied considerably among participants. The raw data, before denoising, demonstrated a rhythmic pattern with pressure changes for both the upper and lower lips and gums. Prior to the infant fully attaching to the breast and beginning to suck, the nipple experienced variations in pressure (see Figure 2). Four infants displayed step-like changes in their oral pressures for one or both gums as nursing progressed (see Figure 3). Both sensors recorded large variations in oral pressure values (see Figure 4).

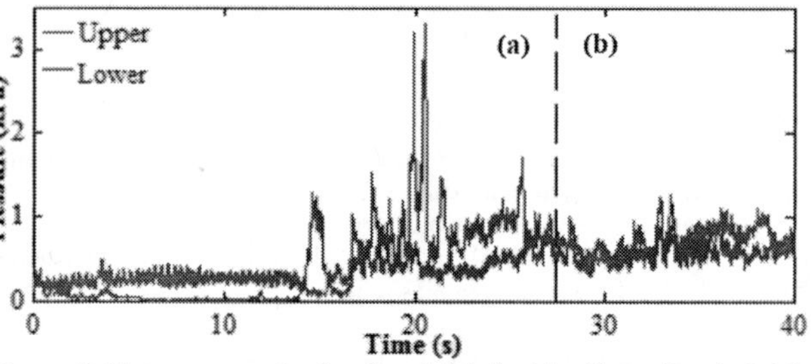

Figure 2: Pressure on nipple when (a) infant is attempting to latch on and (b) when infant is latched on and sucking.

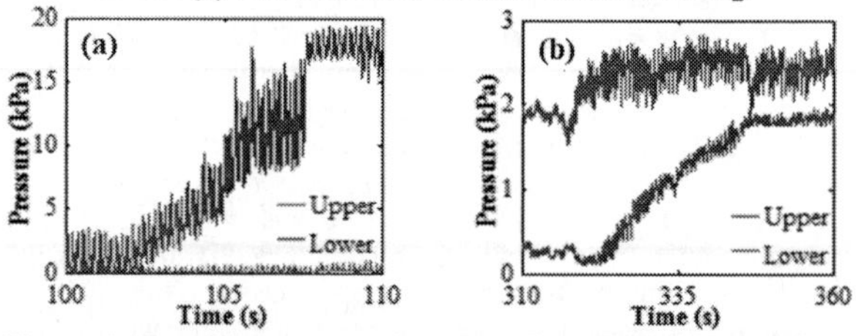

Figure 3: Step-like pressure increases for infant 5 (a) & 7 (b).

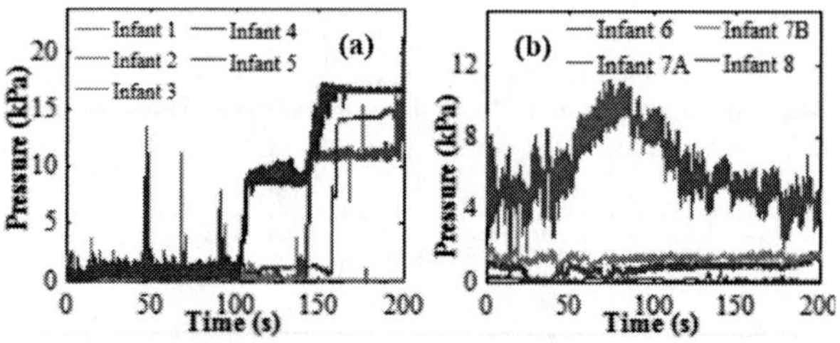

Figure 4: Total pressure on sensor #9801 (a) and #9830 (b).

DISCUSSION

The Tekscan™ I-Scan System successfully captured pressure exerted on the breast from an infant's upper and lower lips/gums while breastfeeding. The level of accuracy is unknown as, after a thorough literature review, this was the first attempt by any researcher to actually measure these pressure forces. While the chosen sensors are designed to increase accuracy when measuring on curved surfaces, the radius of curvature of the breast differs for each woman. Additionally, the density of the breast changes during breastfeeding due to the transfer of milk to the infant and increase in blood flow to the breast. The geometry of the sensor strips limited the area of sensing to the areola. To measure the pressures applied on the nipple will require narrower sensors with greater flexibility.

The pressure data consistently demonstrated a sinusoidal pattern that mimicked the sinusoidal vacuum pattern. Previous clinical research found women begin to experience discomfort and pain when compression on the nipple-areola complex exceeds 35-40 kPa [5]. Although some of the mothers who participated in this study frequently utilized nipple shields for pain, only Infant 1 attained pressure values above 40 kPa towards the end of the feed. Thus, compression of the nipple-areola complex is not likely the source of pain.

The source of the step-like increases in pressure is not clear. All of the steps occurred after the infants had been suckling for some time. The flow of milk through the breast ductal system varies in response to hormones affecting milk ejection. Infants alter their own suckling patterns in response to milk flow [1]. Thus, the increases in oral pressures are likely in response to the infant's attempt to regulate flow.

Periodic spikes in the pressure readings occurred with values that far exceeded expectations (see Figure 4). As can be seen in Figure 1, not all the recording sensels were located directly under the infant's mouth. In the raw data discussed here, those sensels were not excluded since evidence shows that pressure applied to the breast by mothers affects milk flow [4]. During breastfeeding, it is common for both the mother and infant to apply pressure in response to infant positioning, among other things. These spikes may be attributed to mother or infant hand movements.

These preliminary findings provide insight into the amount of positive pressure the areola experiences during breastfeeding. Additional work is underway to capture the pressure exerted on the nipple by the tongue and upper palate. Also, the processed data will be further examined to find correlations between the vacuum and oral pressure amplitudes and milk flow patterns. This data will be used to modify simulations of breastfeeding and provide improved understanding of the mechanics involved in milk extraction from the breast.

ACKNOWLEDGEMENTS

The authors would like to thank Dr. Donna Geddes and the Hartmann Human Lactation Research Group in Perth, WA, Australia for their assistance with capturing this data. This research was funded by NSF Grant #1454334 and NSF EAPSI Grant #1614350.

REFERENCES

[1] Geddes, DT et al., *Early Human Dev.*, 84:471-7, 2008.
[2] Mortazavi, N et al., *ASME J Biomed Eng. (Accepted)*, 2014.
[3] Elad, D et al., *PNAS*, 111:5230-5, 2014.
[4] Morton, J et al., *J Perinatology*, 29:757-64, 2009.
[5] Alekseev, N et al., *Russian J Physiology*, 86:711-9, 2000.

INTRODUCTION OF A RE-ENGINEERED USER INTERFACE AND MODULAR ARCHITECTURE FOR THE SIMVASCULAR OPEN SOURCE PIPELINE FOR CARDIOVASCULAR MODELING

Hongzhi Lan (1), Adam Updegrove (2), Nathan M. Wilson (3), Shawn C. Shadden (2), Alison L. Marsden (1)

(1) Pediatrics and Bioengineering
Stanford University
Stanford, California, USA

(2) Mechanical Engineering
University of California - Berkeley
Berkeley, California, USA

(3) Open Source Medical Software Corporation
Santa Monica, California, USA

INTRODUCTION

Modeling and simulation of cardiovascular hemodynamics can provide invaluable data to predict outcomes of surgical interventions and optimize medical devices for individual patients [1]. It is a complex process involving multiple procedures such as image segmentation, meshing, flow simulation, and scientific visualization. SimVascular (www.simvascular.org) is currently the only fully open source software package enabling the entire process in a unified platform [2]. However, the core of the original graphical user interface (GUI) of SimVascular was designed in the late 1990's using very popular design paradigms and toolkits (e.g. Tcl/Tk) at the time. Due to limitations of computer memory and hardware, many steps in the process took significant amounts of computational time and stretched the limits of accessible memory so the interface was file based and relied more on scripting than is commonplace today. The factors contributed to the GUI being less intuitive and interactive which decreased efficiency and increased the learning curve.

Here, we introduce a major SimVascular release with a completely overhauled modern user interface, new modular software architecture, with major improvements including enhanced data/project management, intuitive GUI, efficient user interaction, a cleaner pipeline, plugin mechanism. In addition, we review the major new features in each of the steps in the modeling pipeline.

METHODS

Software Architecture The new architecture of SimVascular is summarized in Fig. 1. SimVascular is now fully open source and leverages several existing open source packages including: OpenCASCADE, MMG, Vascular Modeling Toolkit (VMTK), Tetrahedral Mesh Generator (TetGen), and the SimVascular Linear Solver (svLS).

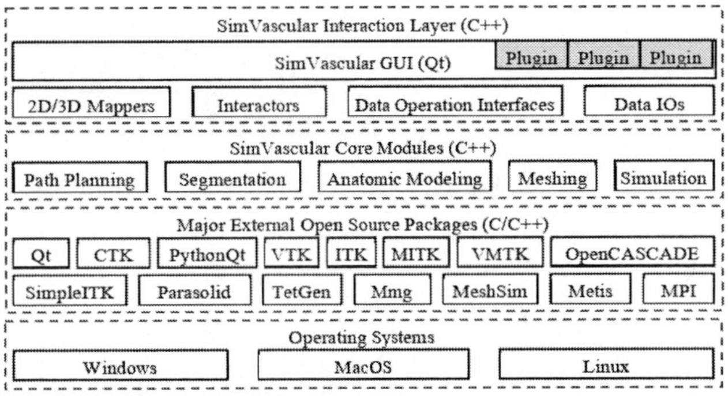

Figure 1: SimVascular Software Architecture

The new architecture has two clear layers for user interaction and data processing, respectively. Its structure is more modular and avoids unnecessary couplings between two layers. The data processing layer is composed of SimVascular core modules, only for definition and computation of various types of data.

Interaction Layer The SimVascular interaction layer is developed using Qt and the Medical Imaging Interaction Toolkit (MITK). It employs Common Toolkit (CTK) and MITK plugin frameworks to make functions in GUI more separate. With consistent plugin interfaces, developer and users are able to create their own custom plugins independently. The GUI is intuitive and clean (Fig. 2). It loads SimVascular and MITK plugins dynamically during launching. The plugins contain tools for the entire modeling pipeline, including image visualization/processing, 2D/3D segmentation, modeling, meshing, and simulation. The newly implemented Data Manager uses a project format to organize all the data for one simulation case and provides the user with easy access to the associated data and tools. SimVascular can automatically switch to the corresponding tool and update the tool GUI as users move from one data type to another.

To interface between the GUI and data processing layer, custom classes have been developed specifically for SimVascular's communication with users (Fig. 1). Specialized 2D/3D mappers are created for various types of data to enhance the visualization in 2D/3D views. Interactors use XML-based scripts to define and control the user's interaction with data through the GUI. By implementing data operation interfaces, SimVascular can track data changes and enable undo/redo features to restore data to the previous states. The data IOs provide flexible ways to read/write data of various types in a number of file formats, such as XML, VTK, images and CAD models. SimVascular also supports importing/exporting data in legacy SimVascular file formats to facilitate users' migration to the latest version. In general, the interaction layer greatly enhances the interactive environment to let users handle data more efficiently, and simplifies the procedures required for the modeling/simulation pipeline.

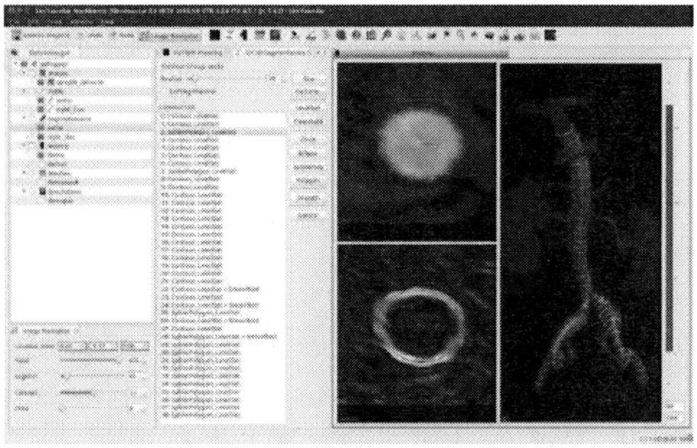

Figure 2: SimVascular GUI

SIMVASCULAR MODELING/SIMULATION PIPELINE

SimVascular provides a complete pipeline for the process of cardiovascular modeling and simulation.

Path Planning A vessel path (centerline) is defined by a set of control points. Users are able to easily locate center points based on vessel cross section using the multi-view Display. The image can also be resliced along a vessel path to enable users to check the quality of the path and modify if needed.

Image Segmentation SimVascular enables creating contours of the vessel lumen along the paths via algorithm-based methods (level set, threshold), and interaction-based or manual ways (circle, ellipse, polygon, spline polygon). Fourier smoothing, interactive shifting/scaling are incorporated. Batch segmentation by level set or threshold is implemented for automated segmentation of an entire vessel with minimal user intervention. Integrated direct 3D image segmentation based on level set methods is also available.

Anatomic Modeling SimVacular can loft 3D solid models for vessels based on 2D segmentation and unions the vessels to create a complete 3D model of the vasculature of interest. There are three types of models: analytic (OpenCASCADE, Parasolid) and discrete (vtkPolyData). For discrete models, SimVascular provides robust Boolean operations for discrete triangulated surfaces, algorithms to combine 2D and 3D image segmentations in a single model, and global/local surface operations like smoothing, remeshing, decimation, subdivision, trimming, etc. [3]. For analytic models, typical CAD procedures such as blending, cutting, and Boolean operations are available. The evolving solid model can be visualized as new segmentations are added, providing the user an opportunity for model assessment and refinement on the fly.

Meshing SimVascular uses TetGen to generate tetrahedral unstructured meshes for the 3D anatomic model that are suitable for computational hemodynamics modeling. To enhance mesh quality, advanced options are provided, including local refinement near selected surfaces, boundary layer meshing, spherical refinement, and radius-based meshing. Users can check the quality of the surface mesh and inner volume mesh. Mesh adaption tools utilizing a-posteriori error estimates are also integrated. SimVascular continues to support meshing with the commercial package MeshSim.

Simulation The SimVascular pipeline enables users to perform hemodynamic simulations at physiologic flow rates and pressure levels. Time dependent flow rates in a specified profile (plug, parabolic or Womersley) or lumped parameter heart models may be defined for the inlet(s). Impedance boundary conditions (BCs), Windkessel-type BCs, and more complicated lumped parameter network (LPN) models like coronary BCs and closed loop BCs can be specified at the outlets by tuning model parameters to match available clinical data [4]. For a deformable wall, fluid structure interaction (FSI) is incorporated using the coupled momentum method (CMM) [5], with the wall modeled as a linear elastic material with uniform or variable elastic modulus and thickness along the vessel. SimVascular can post-process the simulation results and export relevant hemodynamic quantities in VTK file formats that are compatible with the open source visualization package ParaView.

RESULTS AND DISCUSSION

SimVascular has been significantly improved throughout the pipeline with expanded functionality and increased ease of use. In particular, the most recent version provides a newly-engineered modular software architecture and improved graphical user interface with increased image handling capabilities and ease of use. Since the inception of the revitalization project in 2013, SimVascular has attracted up to 1,600 unique users worldwide and has been utilized in over 80 abstracts and journal publications. Active research and development continues on new image segmentation techniques based on machine learning and neural networks as well as on algorithms to convert discrete surface triangulations directly to analytic surface models. Work on an optimization module to enable users to automate device design and perform virtual surgical planning is also underway.

ACKNOWLEDGEMENTS

Support was provided by the NSF SSI program (Award No. 1562450 and 1407834) and an NSF CAREER Award. Software hosting is provided by simtk.org at Stanford University.

REFERENCES

[1] Taylor, CA and Figueroa, CA, *Annual Review of Biomedical Engineering*, 11:109-134, 2009.

[2] Updegrove A et al., *Annals of Biomedical Engineering*, online, 2016.

[3] Updegrove A et al., *Advances in Engineering Software*, 95:16-27, 2016.

[4] Moghadam, ME et al., *Journal of Computational Physics*, 244:63-79, 2013.

[5] Figueroa, CA et al., *Computer Methods in Applied Mechanics and Engineering*, 195:5685-5706, 2006.

HEMODYNAMIC EFFECTS OF STENOSIS IN THE INFERIOR VENA CAVA CONDUIT AND LEFT PULMONARY ARTERY OF THE FONTAN CIRCULATION

Masoud Farahmand (1), Ethan O. Kung (1,2)

(1) Department of Mechanical Engineering
Clemson University
Clemson, SC, USA

(2) Department of Bioengineering
Clemson University
Clemson, SC, USA

INTRODUCTION

Aortic compression, space reduction in the area of anastomosis, torsion of the graft during growth, and calcification are the main reasons for the obstruction of left pulmonary artery (LPA) and inferior vena cava (IVC) conduit in the Fontan circulation [1]. Long term stability and hemodynamic performance of Fontan patients depend on normal blood flow in the LPA and systemic veins. Currently the impacts of LPA and IVC conduit obstructions on adult Fontan hemodynamic performance during exercise conditions is still unclear.

Many previous studies have developed numerical models to evaluate the impact of LPA and IVC conduit stenoses on Fontan hemodynamics based on examining local energy-loss. Few studies have also considered the interactions between the local hemodynamics and global physiology using multi-scale computational simulations. Recently, we have developed a novel computational-experimental technique which we refer to as the Physiology-Modeling Coupled Experiment (PMCE) for testing a variety of medical devices and anatomical models. This novel technique has the advantage of coupling a physical experiment to a computational global physiology model. In this study, we utilized the PMCE to investigate the effects of an 85% (area-based) LPA and IVC conduit stenoses on an adult Fontan patient under 1, 3, 5 metabolic equivalents (MET).

METHODS

Numerical-Experimental Interface coupling The PMCE is a hybrid model which couples a computational physiology model to an in-vitro experiment. In summary, the PMCE passes flow and pressure information between the computational and physical domains at a periodic basis. The coupling approach utilizes a proportional controller to modify the flow waveform at the inlet of the anatomical model such that the measured pressure drop across the experiment matches the pressure difference across the corresponding points in the computational physiology model. In this study, the stenosis region is represented by the in-vitro experiment, and the patient global physiology is represented by a computational physiology model.

Computational Physiology Model Based on prior works [2], a lumped parameter network (LPN) was employed to model the physiology of an adult Fontan patient. The LPN has the capability to adjust model parameters such as cardiac function, vascular resistances, and respiratory parameters based on the metabolic state and patient body-size. The physiology we modeled in this study represents a Fontan patient with weight=50kg and height=150 cm.

In-vitro Experiment For the experimental section, we constructed two physical phantoms representing IVC and LPA stenosis. We placed the physical phantoms in a flow loop where the flow waveform is produced by a computer programmable pump [3]. We used an external flow meter (Transonic Systems, NY) to capture the output flow waveform from the stenotic model. All tubing in the flow system were rigid and two catheter pressure transducers (Miller Instruments, TX) were placed at the inlet and outlet of the phantom to monitor the pressure waveforms. The working fluid was a 40% glycerol solution with density of $1.1 \frac{gr}{cm^3}$ and dynamic viscosity of 0.0041 Pa.s. The pressure and flow data was averaged over 13 cycles to strengthen signal quality and signal to noise ratio. We used a band stop filter to remove 60Hz power-line disturbance from the acquired data before the averaging.

Reference Normal Cases In addition to the PMCE models, we performed LPN-only simulations for the example patient to obtain the hemodynamics of the reference normal case with no vascular obstructions.

RESULTS

Comparing the results with the reference normal case, the cardiac output and pulmonary pressure are significantly affected by the presence of an 85% LPA or IVC conduit stenosis. At 5 MET, the IVC conduit stenosis led to a 41% (6mmHg) and 22% (1.55 L/min) decrease in the pulmonary pressure and cardiac output, respectively (Table 1). The IVC conduit stenosis also resulted in a 43% (6.68 mmHg) increase in the central venous pressure at 5 MET.

In the presence of an LPA stenosis, the LPA flow fraction is affected by the intensity of exercise. Comparing the results between 1 MET and 5 MET, the LPA flow fraction decreased from ~30% to ~23%.

We found that an 85% LPA or IVC conduit stenosis has noticeable effects on the left-ventricular PV loop especially during exercise (Figure 1). At 5 MET, a severe IVC conduit stenosis resulted in approximately 37% decrease in the ventricular power compared to the reference case.

DISCUSSION

In this study we used a coupled numerical-experimental technique to quantify the consequences of the IVC conduit stenosis and LPA stenosis on hemodynamics of a Fontan patient. In agreement with previous literature, our results show that the local power-loss across the stenosis is a small fraction of total ventricular power [4]. Nonetheless, the left-ventricular PV loop is considerably altered by the presence of a severe IVC or LPA stenosis compared to the reference normal case.

The LPA stenosis is considered to be clinically significant if the LPA flow fraction is less than 30% and (or) pressure drop across the stenosis is higher than 3mmHg [5]. An examination of results between different MET levels shows that an 85% obstruction in the LPA led to an LPA flow fraction < 30% and pressure drop > 3mmHg.

The PMCE method combines an in-vitro experiment with a systemic physiology model to assess the impact of a local anatomy anomaly on the global circulation. This study serves as an example of how the PMCE can be used to obtain quantitative measures of fluid dynamics and physiology in a fully-coupled cardiovascular hybrid model.

Table 1. Mean values of physiological parameters for an example patient (weight=50kg and height=150 cm) at 3 different MET levels. Parameters are shown for the normal case without stenosis, case with IVC conduit stenosis, and LPA stenosis. Parameter values are cardiac output (CO) in L/min, Inferior vena cava flow (Q_{IVC}) in L/min, local power-loss across the stenosis inmW, pulmonary pressure (P_{pul}) and pressure drop across stenosis (ΔP_{st}) in mmHg.

MET		CO	Q_{IVC}	P_{pul}	ΔP_{st}	Power-loss
1	Normal Case	3.54	2.07	12.30	0	0
	IVC Stenosis	3.32	1.91	10.91	-3.18	13.49
	LPA Stenosis	3.41	2.01	13.08	-3.15	7.34
3	Normal Case	5.40	3.84	14.02	0	0
	IVC Stenosis	4.57	3.13	10.10	-8.40	58.42
	LPA Stenosis	5.14	3.65	15.13	-5.01	14.83
5	Normal Case	6.98	5.38	15.60	0	0
	IVC Stenosis	5.43	3.96	9.18	-13.40	117.91
	LPA Stenosis	6.61	5.10	16.98	-6.45	21.70

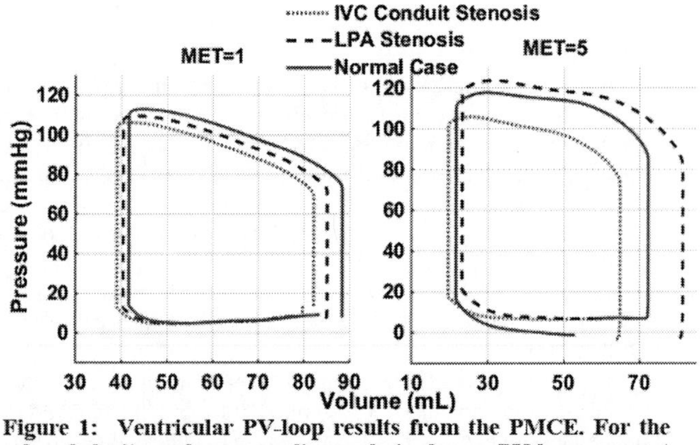

Figure 1: Ventricular PV-loop results from the PMCE. For the sake of clarity, only one cardiac cycle is shown. PV loops are not enclosed due to respiration.

ACKNOWLEDGEMENTS

This work was supported by an Award from the American Heart Association and The Children's Heart Foundation.

REFERENCES

[1] Giannico, S et al., *J Am Coll Cardiol*, 47: 2065–2073, 2006.
[2] Kung, E et al., *J Biomech*, 136:081007, 2014.
[3] Mechoor, R. R et al., *J Biomech*, 138:111002, 2016.
[4] Schiavazzi, D.E. et al., *J Thorac Cardiovasc Surg*, 149:689-696, 2015.
[5] Azakie, A et al., *J Biomech*, 71:1541-6, 2001.

SB^{3}C2017
Summer Biomechanics, Bioengineering and Biotransport Conference
June 21 – June 24, Tucson, AZ, USA

SICKLE RED BLOOD CELL ADHESION TO HEME ACTIVATED ENDOTHELIAL CELLS IN MICROSCALE FLOW

E. Kucukal (1), A. Ilich (2), J. A. Little (3), N. S. Key (2), U. A. Gurkan (1, 4)

(1) Department of Mechanical and Aerospace Engineering,
Case Western □eserve University
Cleveland, O□, USA

(2) Division of □ematology□Oncology,
University of North Carolina Medical Center
Chapel □ill, NC, USA

(3) Division of □ematology□Oncology, Case Western □eserve University, University □ospitals Seidman Cancer Center,
Cleveland, O□

(4) Department of Biomedical Engineering,
Case Western □eserve University
Cleveland, O□, USA

INTRODUCTION

Episodic vascular occlusion due to abnormal adhesion of red blood cells (RBCs) is a hallmark of sickle cell disease (SCD). These episodes are due to the adhesive interplay between different cell types in the human microvasculature, further mediated by plasma proteins. A large number of studies have shown elevated sickle RBC adhesion to various endothelial proteins, including fibronectin, thrombospondin, and VCAM-1 using in vitro platforms [1]. Moreover, adhesion is enhanced by overexpressed vascular adhesion molecules, such as ICAM-1, P-selectin, and E-selectin, on the highly activated endothelium in SCD patients. Activation and resulting dysfunction of the endothelium have also been implicated in white blood cell and platelet adhesion, which facilitate RBC adhesion through a secondary pathway.

Heme is a metal complex located inside the hemoglobin (Hb) protein that is responsible for carrying O_2 by RBCs. In healthy individuals, heme is confined within RBCs where it is isolated from the surrounding cells and tissues. With hemolysis, Hb and thus heme are released into the plasma where it may circulate in its free form. High rates of hemolysis are typical in patients with SCD. Heme has recently been shown to induce endothelial activation and vaso-occlusion in a murine model of SCD [2]. Furthermore, when activated with heme, human umbilical vein endothelial cells (HUVECs) increased nitrite, endothelial nitric oxide synthase (eNOS), and hypoxia-inducible factor (HIF)-α in a dose dependent manner [3]. In this study, we developed and utilized an endothelialized microfluidic platform to assess the contribution of heme to sickle RBC adhesion *in vitro*. We further assessed the adhesion of RBCs to ICAM-1 and VCAM-1, adhesion molecules that are upregulated on activated endothelium, using whole blood from a clinically diverse population of SCD patients.

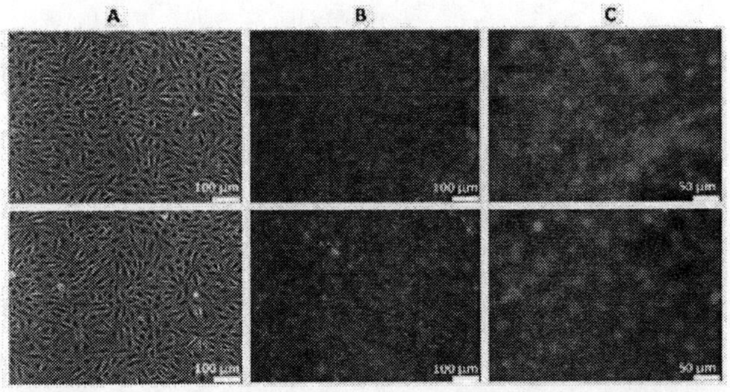

Figure 1. Representative images of HUVECs immobilized in microchannels. (A) 10X phase-contrast images of cultured cells in static conditions after a 4h incubation period. Top and bottom images are from two different fields of view within the same channel. (B) Fluorescent images are shown at 10X magnification. The nuclei and actin filaments were stained with DAPI (blue) and rhodamine phalloidin (pink) respectively. (C) Formation of cell-cell junctions can be seen more clearly at 20X magnification.

METHODS

Microfluidic channels were fabricated using a lamination technique. The channel geometry was determined by laser-cut double-sided adhesive (DSA) films that were placed between a top PMMA cap and bottom microscope glass slide. The adhesion proteins were then perfused into the surface treated devices and allowed to incubate for 1.5 hours at room temperature. To eliminate non-specific binding, the channels were filled with 2% bovine serum albumin (BSA) and incubated at 4°C overnight. To prepare endothelialized devices,

Poster Presentation #P271

HUVECs were cultured and seeded into fibronectin-coated microchannels at a density of 8x10^6 cells/mL and incubated for 4 hours at 37°C, 5% CO_2 for complete cell attachment to the channel surface. Representative phase-contrast and fluorescent images of immobilized cells from two different fields of view (FOV) are shown in Figure 1. Endothelialized channels for adhesion experiments were connected to syringes with either culture medium, or heme solutions (20 and 40 μM) that were perfused into microchannels at a constant shear stress of 5.4 dyne/cm2 for 1 hour using a constant displacement syringe pump. The device temperature was maintained at 37 °C using a heating plate. Thereafter, the channels were rinsed with fresh culture medium and connected to syringes containing whole blood from patients with SCD. A total sample volume of 15 μl was perfused in the channels at a shear stress of 1 dyne/cm^2. The channels were washed with culture medium again to remove non-adherent cells.

RESULTS

We first examined sickle RBC adhesion to VCAM-1 and ICAM-1 in protein-immobilized channels given that their expression on endothelium has been observed with heme activation. Figure 2 shows the adherent cell numbers to ICAM-1 and VCAM-1 for a different population of SCD patients. We separated the RBCs into two categories: 1- deformable cells, 2- non-deformable cells as previously defined by Alapan et al [4, 5]. Both RBC types displayed increased adhesion to both proteins in a patient-dependent fashion. Notably, we observed an elevation in adherent RBC numbers to ICAM-1 for patients with SCD who had self-referred for management of acute pain to the acute care clinic. The adhesion experiments in VCAM-1 channels were only performed on patients at clinical baseline, and these show a heterogeneous adhesion profile.

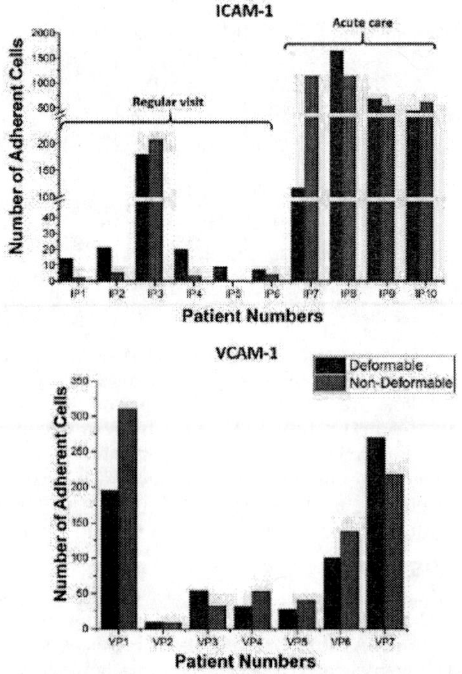

Figure 2. Sickle RBCs adhere to ICAM-1 and VCAM-1 in a heterogeneous manner. (A) Patients with SCD being managed for acute pain showed significantly greater RBC adhesion to ICAM-1 than did patients at clinical baseline. (B) Both deformable and non-deformable cells displayed variable adhesion to VCAM-1 in samples from.

Our results also indicated a significant change in adherent cell numbers when HUVECs within channels were activated with 20 and 40 uM heme solutions (Figure 3). At 20 and 40 uM Heme, compared with incubation with media alone, adhesion rose in all patients relative to baseline, although the degree of the effect was variable. These data support the hypothesis that heme activation of HUVECs enhances RBC adhesion.

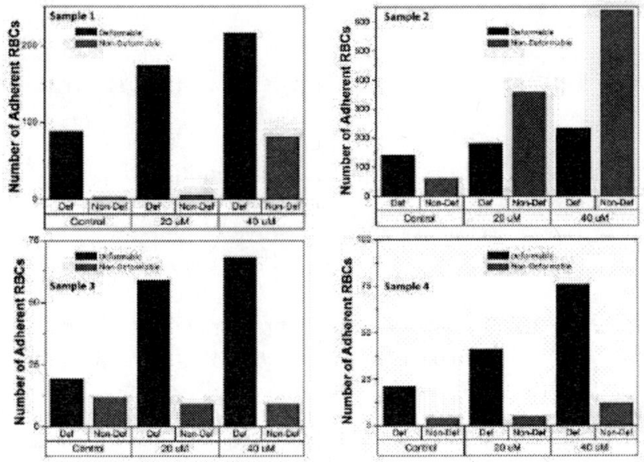

Figure 3. Heme activation of HUVECs promotes sickle RBC adhesion in 4 different patients in a concentration dependent fashion. The response of deformable and non-deformable RBCs to increasing heme concentrations tends to be heterogeneous and patient specific.

DISCUSSION

We present microfluidic devices through which to study sickle red cell adhesion to isolated proteins (VCAM-1 and ICAM-1) or to endothelialized surfaces following activation by heme. We have shown that both VCAM-1 and ICAM-1, present on the activated endothelial lining, led to increased RBC adhesion that varies amongst patients with SCD, but indicated higher adhesion rates to ICAM-1 in patients being managed for acute pain compared to those who are not; this finding supports a link between *in vitro* adherent cell numbers and VOC episodes. We also observed rising RBCs adherence to HUVECs in a heme concentration-dependent fashion.

REFERENCES

[1] Kaul, Dhananjay K., Eileen Finnegan, and Gilda A. Barabino. "Sickle red cell–endothelium interactions." *Microcirculation* 16.1 (2009): 97-111.

[2] Belcher, John D., et al. "Heme triggers TLR4 signaling leading to endothelial cell activation and vaso-occlusion in murine sickle cell disease." *Blood* 123.3 (2014): 377-390.

[3] Da Guarda, Caroline C., et al. "Heme changes HIF-α, eNOS and nitrite production in HUVECs after simvastatin, HU, and ascorbic acid therapies." *Microvascular research* 106 (2016): 128-136.

[4] Alapan, Yunus, Jane A. Little, and Umut A. Gurkan. "Heterogeneous red blood cell adhesion and deformability in sickle cell disease." *Scientific reports* 4 (2014): 7173.

[5] Alapan, Yunus, et al. "Sickle cell disease biochip: a functional red blood cell adhesion assay for monitoring sickle cell disease." *Translational Research* 173 (2016): 74-91.

SB^{3}C2017
Summer Biomechanics, Bioengineering and Biotransport Conference
June 21 – 24, Tucson, AZ, USA

PARAMETRIC ANALYSIS OF NEW CORONARY ARTERY BYPASS CONFIGURATIONS

Gokce N. Oguz (1), Senol Piskin (1), Tijen Alkan Bozkaya (2), Mehmet Sanser Ates (2), Haldun Karagoz (3) Kerem Pekkan(1)

(1) Department of Mechanical Engineering
Koc University
Sariyer, Istanbul, Turkey

(2) Department of Cardiovascular Surgery, Koç University Hospital,
Topkapi, Istanbul, Turkey

(2) Department of Cardiovascular Surgery, American Hospital, Nisantasi,
Istanbul, Turkey

INTRODUCTION

Coronary artery diseases (CAD) are statistically identified as the principal cause of morbidity in both men and women by American Heart Association [1]. CAD develops as a result of plaque buildup in arteries that supply blood to heart muscle tissue (myocardium). The elastic and smooth characteristics of arteries becomes rigid and narrowed (or even blocked) which can lead to myocardium dysfunction and heart attack eventually due to insufficient blood perfusion in myocardium. CAD can occur in a single vessel or in multiple vessels concurrently. Coronary artery bypass grafting (CABG) is a surgical intervention to create an alternative route around the blocked arteries by using a section of blood vessel from leg, chest or another part of the body and it is the most common type of open heart surgery in United States. There has been progress in methodology of CABG operations in terms of distal target selection, choice of conduit and sequential anastomotic model [2] since the early CABG surgeries.

In sequential grafting technique, more than one distal anastomosis is constructed per segment of conduit and results two or more distal anastomosis per single proximal anastomosis [2]. According to intraoperative studies, sequential grafting provides higher blood flow and higher velocity in the proximal segment compared to a single graft with an end-to-side anastomosis [3]. Furthermore, sequential grafting has better patency rate than individual grafting in general [4]. The hemodynamics of sequential grafts has been studied widely using computational fluid dynamic (CFD) simulations [3-5]. Hemodynamic parameters such as wall shear stress, residence time of the blood, pressure drop and energy loss have been compared between different sequential grafting geometries as well as individual grafting. The parameters that define the geometry of sequential grafting and studied in literature include the anastomosis angle, shape, and diameter. However, the effect of distance between anastomosis sites of sequential grafts has not been studied before, to the best of our knowledge.

In a previous study, we have investigated the hemodynamic efficiency of patient-specific complete revascularization after stenosis for 3 different scenarios: right coronary artery (RCA), left interior descending artery (LAD) and left circumflex artery (LCX) bypasses [5]. We have also compared sequential grafting with individual grafting in terms of blood perfusion. The simulations have been conducted using an experimentally validated commercial CFD software FLUENT (ANSYS Inc., PA) and the results have been validated.

In this study, we investigate the effect of anastomosis site distance in sequential grafting by using our previously constructed patient-specific model. Furthermore, through a new validation campaign, all of the simulations are conducted in an open-source CFD software OpenFOAM (Inc). Effects of the distance between anastomosis sites of sequential grafts are analyzed in detail and the results of our previous study [5] are used for validation of the current study.

METHODS

The model used in our previous study was based on patient-specific anatomy of a 54-year-old patient with CAD. It was acquired from post-op computed tomography (CT) images and reconstructed using Simpleware-Scan IP (Simpleware Ltd., Innovation Centre, Exeter, UK) and refined for CFD meshing in in Geomagic (Geomagic Inc., NC, USA). The healthy geometry was reconstructed by modifying the patient-specific data and was used for validation of the study. High quality tetrahedral grids were created using Gambit 2.3.6 (ANSYS Inc., PA) with ~2 million elements.

Sequential grafts, in the current study, are generated virtually based on the post-op CT images of the patient (see Figure 1) for four different anastomosis site distances: 0.5, 1.0, 1.5 and 2.0 mm with radii of 0.5, 1.0 and 1.5 mm.

The simulations are performed using 3D steady, laminar, incompressible, and non-Newtonian Navier-Stokes equations using a second-order solver (nonNewtonianIcoFoam). A constant, plug velocity profile with 0.091 m/s is defined for the inlet boundary condition and constant pressure boundary conditions are specified at all artery outlets.

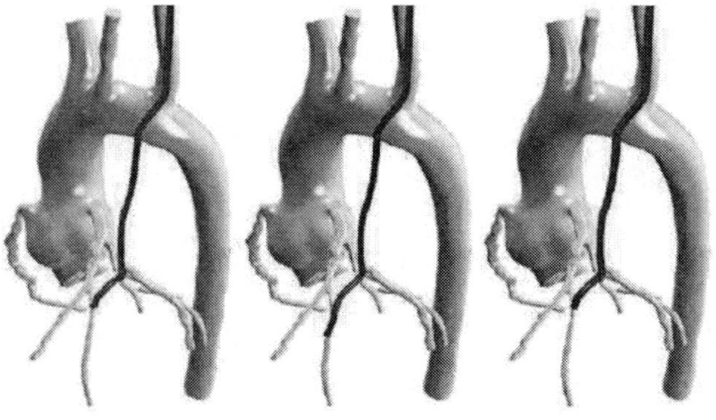

Figure 1: Two distal anastomosis sequential grafting with (Left) distance 1 mm and radius 1 mm (Middle) distance 2 mm and radius 1 mm (Right) distance 1 mm and radius 2 mm

Generalized power law model is implemented in to capture the non-Newtonian behavior in low shear regions (Eqn. 1).

$$\mu = \lambda|\dot{\gamma}|^{n-1}\ (units\ in\ poise)$$

where

$$\lambda(\dot{\gamma}) = \mu_\infty + \Delta\mu\, e^{\left[-\left(1+\frac{|\dot{\gamma}|}{a}\right)e^{\left(\frac{-b}{|\dot{\gamma}|}\right)}\right]} \quad (1)$$

$$n(\dot{\gamma}) = n_\infty - \Delta n\, e^{\left[-\left(1+\frac{|\dot{\gamma}|}{c}\right)e^{\left(\frac{-d}{|\dot{\gamma}|}\right)}\right]}$$

where $\mu_\infty = 0.035$, $n_\infty = 1.0$, $\Delta\mu = 0.25$, $\Delta n = 0.45$, $a = 50$, $b = 3$, $c = 50$ and $d = 4$. Blood density is specified as 1060 kg/m^3. Time-step is set to 0.1 milliseconds and last 100 time-step data is averaged to reduce the effect of oscillation in residuals.

RESULTS

OpenFoam results with healthy geometry are verified using previous Fluent results for both Newtonian and non-Newtonian simulations. A representative velocity field for a slice taken from the descending aorta is compared in Figure 2 for non-Newtonian cases in OpenFOAM and Fluent. Results show that the non-Newtonian model implementation in OpenFOAM is successful.

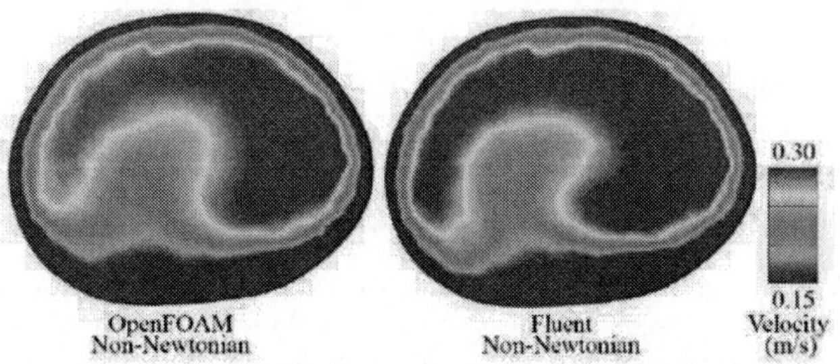

Figure 2: Velocity distribution on a cross-section at the proximal ascending aorta non-Newtonian OpenFOAM and non-Newtonian Fluent results, respectively.

Pressure fields are presented in Figure 3. As expected, there is a visual difference between Newtonian case and non-Newtonian case while OpenFOAM results and Fluent results agree well for pressure field. The same characteristics are observed in the velocity field comparison.

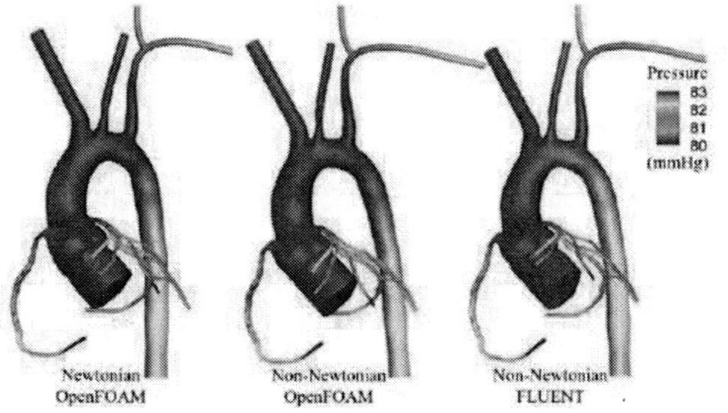

Figure 3: Pressure field on the surface is visualized for Newtonian OpenFOAM, non-Newtonian OpenFOAM and non-Newtonian Fluent results respectively

DISCUSSION and SUMMARY

A framework for studying complex by-pass operation scenarios is established using open CFD software, which will be used in evaluating new by-pass configurations. Overall, the nonNewtonianIcoFoam solver of OpenFOAM provides acceptable results for this problem. In the future realistic (physiological) blood flow behavior through resistance boundary conditions will be implemented to inlets and outlets as it in our previous study. Sequential graft configurations covering different anastomosis site distances (0.5, 1.0, 1.5 and 2.0 mm) with radii of 0.5, 1.0 and 1.5 mm will be simulated with these advanced boundary conditions to get accurate results and to find the effect of graft anastomosis distance.

The simulations for the cases with different distances between anastomosis sites for different radius will give further information about the radius effect. Different anastomosis angles will also be analyzed in addition to current geometries whereby it would give us a brief understanding of the sequential grafting methodology. Thus, this study will give us insight about the relation between three sequential grafting parameters: distance between anastomosis site, graft diameter and anastomosis angle, which have not been studied (and related) in the literature. The clinical outcome of the current study may be crucial and supportive for the CABG surgeries.

ACKNOWLEDGEMENTS

Funding was provided by grants EU FP7 293931-CardioFluidMechanics and TUBITAK 1003

REFERENCES

[1] Rosamond W et al., *Circulation*, vol. 117, pp. e25-146, 2008.
[2] Ouzounian M et al., *J Cardiothorac Surg*, ol. 38, pp. 579-84, 2010.
[3] Kabinejadian F et al., *Ann Biomed Eng*, vol. 38, pp. 3135-50, 2010.
[4] Hajati O et al., *Conf Proc IEEE Eng Med Biol Soc*, vol. 2012, pp. 641-4, 2012.
[5] Dur O et al., *Cardiovasc Eng Technol*, vol. 2, pp. 35-47, Mar 2011.

DART BRACE: DAILY ADVANCED RANGE OF MOTION THERAPY FOR MAXIMIZING FUNCTION

Anthony D. Anderson, II
Clarkson University
Potsdam, NY, USA

Bridgette Bousquet
Clarkson University
Potsdam, NY, USA

Megan M. Curry
Clarkson University
Potsdam, NY, USA

Brian G. Davis
Clarkson University
Potsdam, NY, USA

Catherine R. Pelton
Clarkson University
Potsdam, NY, USA

Christopher Robinson
Clarkson University
Potsdam, NY, USA

Dr. Laurel Kuxhaus
Clarkson University
Potsdam, NY, USA

Faculty Advisor(s)
Margret Shea, PA-S2
Clarkson University
Potsdam, NY, USA

Dr. Victoria Priganc
Clarkson University
Potsdam, NY, USA

INTRODUCTION

Injury to the scapho-lunate (SL) ligament, the primary stabilizer of the wrist, is one of the most common wrist traumas [1,2]. Tears or defects to this ligament occur in 29% of the population (a cadaver study) and, according to the U.S. Census Bureau, wrist injuries are accountable for 10 days of work loss each year per worker sustaining an injury [3]. This injury affects people of all ages and often requires surgery [4,5]. Minor injury, such as a ligament tear, often results from overuse of the wrist, while a traumatic dislocation of the ligament often occurs from falling on the outstretched hand [1,2,5]. Injuries to the SL ligament can be treated by restricting motion or by surgical repair [2]. Typical recovery from the SL ligament surgery includes 6-8 weeks of immobilization. It is possible to shorten the immobilization time with a limited motion brace that will restrict motion to the dart thrower's plane (DTP) [2]. Limited motion braces promote bone growth, healing of associated ligaments and joint mobility [6]. Movements in the DTP will minimize the risk of reinjuring the SL ligament via strain while allowing the patient to regain limited mobilization earlier than with traditional motionless braces, maximizing function during recovery. This will improve quality of life and shorten total recovery time [1,6]. Furthermore, post-injury joint activity promotes blood flow and contributes to a faster recovery time [7]. The dart thrower's motion (DTM) is defined by wrist motion starting at 30° extension with 10° radial deviation and moves through 30° flexion with 10° ulnar deviation [6]. The objective of this project is to design, manufacture, test and commercialize a wrist brace that allows movement only in the DTP.

Current DTP wrist braces are not commercially available [8,9]. The nature of the slightly oblique DTP and the establishment of the axis of rotation poses the greatest design challenge [8]. The specifications for a new, inventive design for this Class I Medical Device are similar to existing braces in the sense of function and safety. However, the new design is more user friendly, comfortable, and will be readily available. In function, the brace restricts motion to the DTP while delivering accurate, smooth movement. The brace does not hinder the healing process of the SL ligament or the incision area. Secondary design objectives include: optimizing cost and manufacturing efficiency; scalable sizing options; improving the ability to don and doff; and making it both waterproof and unburdensome.

PRODUCT DESIGN

The SL brace ensures a limited range of motion in the DTP. The brace (Autodesk Inventor 2017, Mill Valley, CA) is composed of hand and forearm pieces connected with two pins (double sided rivets). The brace is 3D printed and thermo-molded to the patient.

The dorsal pin is located over the dorsal surface of the base of the 4th metacarpal. This pin, in conjunction with a slider-slot combination, allows pivoting action with translation (Figure 1). The palmar pin is located over the tubercle of the scaphoid. The pinhole on the palmar side is 1.5mm larger in diameter than the pin. This allows the palmar pin and axis of rotation to shift slightly in conjunction with the dorsal slider. These considerations allow the wrist to rotate through the full oblique DTP. Initial designs without a slider revealed the pivoting system allowed radial-ulnar deviation (RUD) but prohibited flexion and extension. The inclusion of the slider on the dorsal side mitigates this problem. To ensure that RUD is limited within the DTP, a stopping mechanism (not shown) will be implemented.

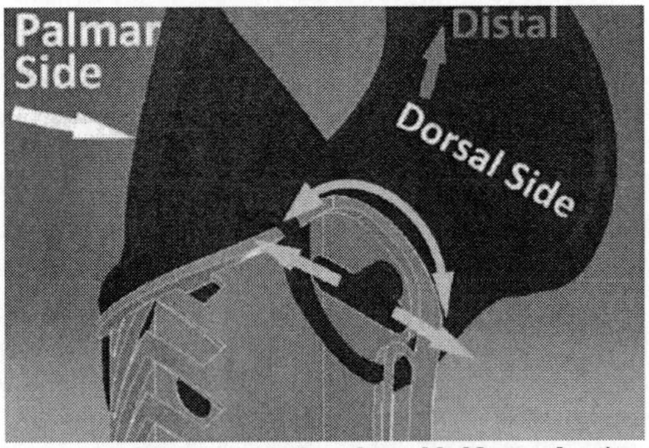

Figure 1: A 3D representation of a molded brace showing a detailed view of the pin-slider combination.

The brace design follows good splinting practices by supporting two-thirds of the forearm and avoids interference with anatomical landmarks such as the ulnar styloid [6,11]. The hand piece is entirely proximal to the transverse palmar crease, permitting unrestricted phalangeal movement. A mesh design simplifies molding and gives superior breathability. Thickened segments along the palmar and dorsal sides of the forearm and hand ensure the rigidity and conformity of the brace. The inside of the brace will be lined with a silicone padding to provide all-day comfort while maintaining water resistance. To protect postoperative incisions from irritation, a simple bandage or wrap will act as a boundary. Velcro straps ensure a snug fit while making it easy to don and doff.

The pieces of the brace are printed flat using a Raise3D N2 Plus printer (Santa Clara, CA). Then the printed pieces are thermoformed to ensure a custom fit (Figure 2). Initially polylactic acid (PLA) printer filament was used; polyethylene terephthalate (PETG) will produce a stronger brace. The thermoforming characteristics of PLA/PETG allow the brace to be custom to each patient. Both materials are 100% recyclable and biodegradable.

The thermoforming process (in a practitioner environment) includes aligning the pins by placing them over anatomical landmarks while thermo-molding; this establishes the axis of rotation for the DTP. The brace is heated in a hot water bath (between 195°F and 200°F) and then molded to the patient. The heating and molding process is repeated as needed to custom fit the brace to each individual (Figure 2).

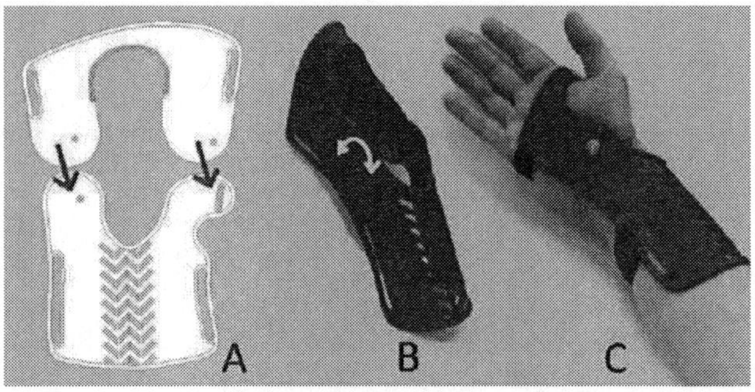

Figure 2: The two flattened pieces (A). An early brace prototype thermoformed (B) around user's arm (C).

The SL Brace motion will be analyzed using motion capture (Qualysis) and MATLAB to show the movement is limited to the defined DTP. Future prototype developments will modify the shape to increase comfort, simplicity, conformity, and strength. Further materials (other 3D printing plastics and more traditional thermoplastics) and fabrication techniques will also be tested to make the best selection for strength and moldability. Upon Institutional Review Board approval, surveys will be sent to medical professionals to evaluate clinical viability and value, while ensuring all design requirements have been met. This will aid in future developments.

BUDGET & MARKET ANALYSIS

To bring the SL Brace to market, commercial production will be performed by a startup LLC created with the guidance of the Clarkson University Shipley Center for Innovation. The two flat pieces will be shipped to medical offices, which will then be individually molded to each patient at their physician's practice. Six standard sizes, available for the left or right wrist, will be available for order in bulk or individually by clinicians. Additionally, clinicians with access to a 3D printer may pay a subscription fee to print the design onsite.

Thus far, seven full size SL Brace prototypes have been produced at a total cost of $84 in materials. Each print requires about 30 minutes of set-up and clean-up time in addition to a 3-hour print cycle. Production operators do not need to be present as the brace is 3D printed, thus labor over the print time has not been included in cost analysis. Current cost projections show the direct cost of commercial production would be less than $35 per brace. Further anticipated costs include a 3D printer and a computer for a cost of $3,900 [Raise3D N2 FFF and HP Envy 15t]. These purchases would be depreciated evenly over a 5-year period at $780 annually. Given a single printer should produce a minimum of 1,000 braces annually, more printers with increased printing capacity will be purchased to accommodate increases in sales.

To market this product direct connections will be made with regional OT/PT practices and surgeons. Conference and exposition attendance could further increase sales. Typical wrist braces prescribed to patients cost between $150 and $200. Given this range, it is reasonable to assume a sale price of $150 plus shipping [11]. According to the 2014 census, the population of New York and the New England region is over 34 million, meaning that at an injury rate of 29% about 9.9 million people could sustain an SL injury in their lifetime [3]. If these injuries are sustained at different times over the typical lifespan of a person, currently 78, then there is an average of 127,000 SL injuries per year [12]. Assuming the SL Brace is prescribed in 1% of these cases in our first year, a total of 1,270 sales is expected, generating over $190,000 in sales annually. If market share increases by 1% each year, sales will grow to $952,000 in the 5th year. With a projected minimum profit of $100 per unit sold, only 8 braces would need to be sold annually to break even on the initial investment. This conservative estimate shows that the SL Brace is both a widely useful and a potentially profitable venture with high margins that will support labor cost and fund company growth.

The existing SL Brace has a new, innovative design that augments wrist injury rehabilitation by virtue of precise, limited motion. Looking forward, additional models could be created for activities that subject the brace to intense stresses and strains such as heavy lifting, contact sports, or high-impact activities. This design could also be adapted to accommodate other common carpal injuries; simple changes to to the design would allow movements about different axes of motion. Post-surgery patients are allowed to regain limited mobility of their wrist faster than traditional rigid braces, offering improved patient quality of life by enabling some normality in basic functions such as drinking from a mug, cutting with a knife, or tying shoes. Returning to daily activities earlier during the recovery process can prevent compensatory injuries due to coping strategies while improving patient quality of life. The SL Brace allows wrist motion in the DTP, ultimately encouraging healing while improving quality of life during recovery.

ACKNOWLEDGEMENTS

This material is based upon work supported by the National Science Foundation under Grant No. GARDE-1510367.

REFERENCES

[1] Weiss, A P C J Hand Surg Am, 23(2): 205-215, 1998.
[2] Kitay, A et al J Hand Surg Am, 37(10): 2175-2196, 2012.
[3] https://www.bls.gov/news.release/pdf/osh2.pdf
[4] Wright, T W et al. J Hand Surg Am, 19(2): 313-318, 1994.
[5] Cooper, C. "Fund. of hand therapy" (2nd ed.) St. Louis: Mosby Elsevier. 2013.
[6] Jacobs, M. et al. "Orthotic Intervention" 2nd Ed. Philadelphia: Lippincott Williams & Wilkins. 2014.
[7] Hauser, R. A., et al. TOREHJ 6(1) 2013.
[8] Werner, F W et al., J Hand Ther 29(2):175-182, 2016.
[9] Braidotti, F et al., J Hand Ther, 28(3): 329-335, 2015.
[10]https://blog.orfit.com/blog/splint-in-the-spotlight-dart-throwers-motion-orthosis-also-known-as-the-dart-splint/
[11]Personal contact: Victoria Priganc, PhD,OTR,CHT,CLT, 2/6/17
[12] https://www.cdc.gov/nchs/fastats/life-expectancy.htm

SB³C2017
Summer Biomechanics, Bioengineering and Biotransport Conference
June 21- 24, Tucson, AZ, USA

WEARABLE REHABILITATION: A CUSTOMIZABLE CONTINUOUS PASSIVE MOTION DEVICE FOR EARLY PHALANGEAL MOBILIZATION

Sydney D. Crady
Clarkson University
Potsdam, NY, USA

Kiersten E. Drapeau
Clarkson University
Potsdam, NY, USA

Thomas C. Piersall
Clarkson University
Potsdam, NY, USA

Elizabeth A. Cassady
Clarkson University
Potsdam, NY, USA

Alyson A. Weisner
Clarkson University
Potsdam, NY, USA

Deja A. Robinson
Clarkson University
Potsdam, NY, USA

Faculty Advisor(s)

Dr. Laurel Kuxhaus
Clarkson University
Potsdam, NY, USA

Dr. Kevin B. Fite
Clarkson University
Potsdam, NY, USA

Dr. Victoria Piganc
Clarkson University
Potsdam, NY, USA

John LaRue
Clarkson University
Potsdam, NY, USA

Molly Kelso
Clarkson University
Potsdam, NY, USA

INTRODUCTION

While the index finger appears to be small and simple, it is very important in the execution of everyday tasks. Three joints in the index finger allow these tasks to be completed with ease: the distal interphalangeal (DIP), proximal interphalangeal (PIP), and metacarpophalangeal (MP) joints. [1] Unlike other fingers, the index finger has an independent extensor indicis tendon and muscle. This allows the index finger to aid in tasks such as grasping, pinching, and many fine motor skills. However, if the index finger is injured, this can cause a large hindrance when carrying out daily tasks.

About 40% of cases in emergency treatment of surgery and orthopedics are hand trauma. [2] Finger injuries appear to be very straightforward from a distance, but the anatomy and biomechanics behind them are incredibly complex. Improper treatment of finger injuries can cause physical deformity and permanent functional impairment over time.

Currently, the main choice in healing severe traumatic finger injuries is surgery. However, a common sequela of finger injury is joint stiffness. There are four stages of joint stiffness: bleeding, edema, granulation tissue, and fibrosis. Rehabilitation should be started in the first two stages to force accumulated periarticular interstitial fluid away from joints. If this is not accomplished then, during stages three and four the finger will become especially stiff leading to complete loss of compliance, severe pain, and permanent damage. [3]

Continuous passive motion (CPM) is crucial in the maintenance of periarticular tissue and joint range of motion leading to an efficient and effective recovery. [3] Current devices to prevent joint stiffness for the hand are inadequate and often expensive, time-consuming, and cumbersome. They do not target the specific injury efficiently because they focus on the whole hand rather than one specific finger. The alternative rehabilitation choice for the finger is to recommend that the patient regularly squeeze objects such as tennis or stress balls. This technique is not easily regulated and is very time consuming as well. Both of these methods are very inconvenient for the patient, and as a result, patients opt not to use them and instead can cause permanent damage to the finger. To resolve these problems, a specialized, affordable, and portable device is necessary.

Such a device will use CPM to combat joint stiffness, reducing and ridding of blood and fluid by a flexion and extension motion known as pumping. The fluid will be forced out of the region, restoring mobility and accelerating clearance of hemarthrosis. [3] Clinicians would recommend its use from four to eight hours per day for two to three weeks, based on the severity of the injury. [4]

The end goal of the CPM machine is to create a portable, customizable rehabilitation device for adults recovering from finger injuries to their joints, tendons, or ligaments. While existing devices are expensive, bulky, and must be plugged into the wall, this device will be the first affordable and portable machine of its kind.

PRODUCT DESIGN

The single digit CPM machine will rehabilitate the index finger effectively while also allowing the patient to carry out everyday activities. For simplicity and efficiency, a single attachment method was selected. The movement produced will be located at the distal phalanx.

To produce an elliptical fingertip motion, the following components will be used (**Figure 1** and **Table 1**): 3D printed guide piece (Part No. 5), 3D printed articulator (No. 6), 3D printed finger cuff (No. 7), 3D printed base (actuator unit) (No. 1), motor (No. 2), microchip (No. 3), battery (No. 4), and necessary electronics. An on-off switch will be located on the top of the actuator unit for easy start and stop.

The guide piece attaches to the shaft of the motor on one end and to the articulator at the other end. The motor moves in a circular motion. Because the attachment point of the guide piece to the articulator is offset, the circular motion is transformed, causing the distal end of the finger to move in an elliptical motion.

The articulator attaches to the guide piece at one end and to the finger cuff at the other end. The attachment point of the articulator and the guide piece will be adjusted based on the length of the finger to ensure proper motion. The finger cuff includes a clamp that can be adjusted by the patient to ensure proper fit, and also allows them to remove the product with ease.

Using a hook and loop strap, the actuator unit attaches to the dorsal side of the hand and will provide the power source for the machine. A small, lightweight 2"x2"x0.75" box (**Figure 1**) houses the motor/gearbox, a 3.7V lithium-ion-polymer battery with a lifespan of 8 hours, and a dual motor processor.

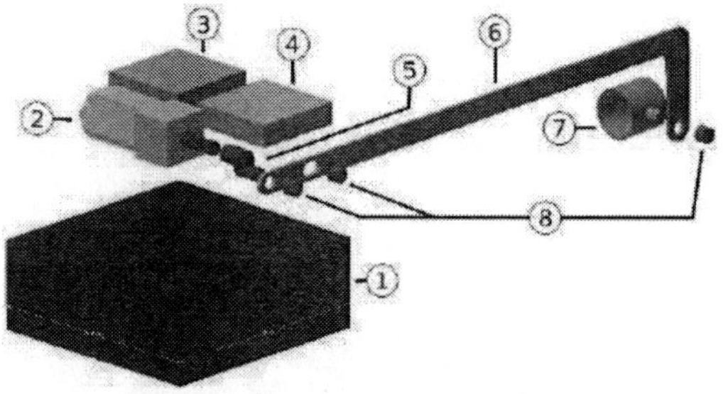

Figure 1 - Exploded Diagram (Autodesk Inventor) - Part Numbers Correspond to Table 1

Figure 2 - Rendered Main Device Housing Attached to a Hand

Since the rotational rate of the CPM machine is not as much of a concern for the rehabilitation of the finger as the range of motion, the motor provides continuous rotation of the user's finger at a low rate of 0.6 RPM. This is fast enough to ensure finger movement, yet slow enough that it will not interfere with the user's daily activities. The amplitude will be patient specific and equal to the maximum range of motion of the index finger as measured by the clinician. The articulator is normal to the plane of the index finger and will allow for smooth, consistent motion, while allowing wrist mobility so they can easily perform daily activities. The electronics will also include a fail-safe feature that will turn off the motor if the user resists the passive motion. This not only will keep the motor from burning out, but will also allow the user to temporarily use the affected finger (i.e. to lift a cup of coffee).

To 3D print the customizable parts, a user-friendly software program will allow clinicians to input finger dimensions (lengths between joints) to ensure a custom fit. Custom parts (articulator, finger cuff, and guide piece) will be 3D printed based on these measurements.

BUDGET & MARKET ANALYSIS

This new solution for clinicians and their patients will decrease the cost and time for treating digit injuries. Our product has the potential to dramatically increase quality of life during rehabilitation time. As our team enters the market and initiates contact with clinical offices, we will submit a patent for our customizable single digit CPM machine and its associating software.

Table 1 lists the total per unit cost of prototyping and producing a single digit CPM machine at 33,500 units. Though the current prototype can only rehabilitate the index finger, future iterations of the device will be able to work with each finger. Initial startup costs are estimated to be approximately $10,000.00. This includes the cost of utilities, initial equipment purchases, sales and product material, packaging material, design time for the customizable software and our CPM machine, as well as an advertisement cost. The actuator unit will be manufactured and sold to clinicians.

Current hand CPM machines cost between $2,000.00 and $6,000.00 for clinical offices to purchase and then patients pay up to $800.00 for a two-week rental. Selling our standard device outright for only $300.00 each will reduce the overall burden of therapy expenses on a patient. Clinicians will benefit by having a shorter turnover time for clients and will no longer have to deal with the extra steps of filling out rental contracts. In addition, the clinicians will buy a subscription for a software package, which allows them to design the patient-specific 3D elements. The software required to personalize the patient's fittings will be sold to clinical offices as an initial cost of $300.00 with an annual subscription of $100.00 for the software. These elements can then be printed in-house or outsourced to a 3D printing facility. Overall, our CPM machine and software package will reduce patient and clinician's costs by 75% in comparison to the least expensive CPM machine currently on the market.

Table 1 - Part List with Production & Prototype Cost

Part Name	Prototype Cost	Production Cost
(1) Housing Material	$7.00	$4.50
(2) Motor	$10.00	$6.40
(3) Microchip	$5.00	$3.20
(4) Battery - MicroUSB	$10.00	$6.40
(5) Guide Piece	$1.00	$0.65
(6) Articulator	$5.00	$3.20
(7) Finger Cuff	$1.00	$0.65
(8) Hardware	$1.00	$0.65
Hook and Loop Strap	$2.00	$1.30
Total	$42.00	$26.30

The single digit CPM machine has a large potential market, with 3.37 million wrist, hand, and finger injuries reported in emergency rooms in the United States annually. [5] Within the first year of the introducing the product to the market, we can capture 33,500 of these injuries and treat them with the single digit CPM machine. These machines can be produced, packaged, sold with the necessary software, and delivered with limited production costs. Sales exiting the first year leads to a profit of $9.2 million. This would enable the hiring of new employees such as sales representatives, software engineers, design engineers, quality analysts, and financial managers. With these new employees, along with additional capital, research and development for a newer, more versatile CPM machine will be possible.

ACKNOWLEDGEMENTS

This work is supported by the National Science Foundation under Grant No. GARDE-1510367.

REFERENCES

[1] Brook, N. et al., 1995, Med Eng Phys, 17:54-63

[2] Yili, F. et al., 2007, Robotics and Biomimetics

[3] O'Driscoll, S. et al., 2000, Rehab, Res., 37:179-188

[4] Chiarello, Cynthia. et al., 1997, Orthop & Sports Phys Ther, 25: 119-127

[5] CDC., 2013, Emergency Dept. Summary

DESIGN AND FABRICATION OF A SMALL *IN VIVO* BIOMECHANICAL TESTING DEVICE: THE PORTABLE *IN VIVO* TISSUE TESTER (PIVTT)

Thomas Zamorski
Widener University
Chester, PA, USA

Bridgette M. Saverine
Widener University
Chester, PA, USA

Tobi Odesanya
Widener University
Chester, PA, USA

Veronica E. Schimpf
Widener University
Chester, PA, USA

Michael Tarquini
Widener University
Chester, PA, USA

Faculty Advisor
Dr. Anita Singh
Widener University
Chester, PA, USA

INTRODUCTION

The brachial plexus (BP) is a large bundle of nerves that originates in the neck and runs through the shoulder into the arm [1]. Neonatal brachial plexus palsy (NBPP) is an injury that can result in a loss of movement or weakness of the arm. During a difficult childbirth, the infant's head and neck may be pulled to the side by maternal (endogenous) and clinician-applied (exogenous) forces and stretch the brachial plexus nerve, causing damage; the extent of the damage depends on how far the nerve is stretched [1]. Despite considerable research and improvements in obstetrical care, birth-related brachial plexus injuries occur in 0.13 to 3.6 per 1000 live births [2] and permanent NBPP continues to occur in 1.1 to 2.2 out of 10,000 births remaining a regular challenge for the affected families and treating physicians [3]. Currently, there is no real form of prevention for the incidence of brachial plexus palsy. According to the few studies conducted on the brachial plexus, the effects of the forces on the brachial plexus are directly related to: magnitude, loading rate, surrounding tissue properties, and how the applied force is transmitted to the brachial plexus. Since *in vivo* measurements of the forces, fetal shoulder deformation, and the response of the brachial plexus are difficult to assess during delivery, computational and physical models are used to simulate these events using *in vitro* biomechanical data for the brachial plexus. In order to provide reliable data for this issue, *in vivo* tests need to be performed in brachial plexus nerve tissue that closely resembles that of an infant.

PRODUCT DESIGN

Based on the lack of *in vivo* biomechanical data on the brachial plexus, a goal was established to design, build, and validate a small *in vivo* biomechanical testing device that would measure the biomechanical properties of brachial plexus nerve tissue in a piglet model. This device: Portable In Vivo Tissue Tester (PIVTT) must be small enough to fit within the operating space of a piglet, it must be biocompatible, and it must measure several biomechanical properties such as the load applied and the change in length of the nerve segment.

In order to identify the design constraints and components needed for the rig, literature review and preliminary testing on a piglet were completed. The following design criteria were identified from literature review: the device materials must be biocompatible and the non-slip clamps must not harm the nerve tissue [4], the device must apply tensile loading rates between 0.01 mm/s and 10 mm/s to the nerve tissue to simulate the quasistatic and dynamic rates experienced during brachial plexus injury [4,5], the data acquisition system must collect data at a sampling rate of at least 25 Hz [4], a camera must be used to capture the *in vivo* tissue strain and follow a timestamp that coincides with the data collection [5], and tissue biomarkers must be used to follow the stretch of the brachial plexus on the camera [5]. After performing the literature review, it was clear that the rig would require the use of a load cell, linear actuator, camera, and data acquisition system. Preliminary measurements were then performed on a piglet to find the size of the operating space, loads that would be applied to the nerve, and length of stretch. After performing preliminary measurements on 2 separate cadaveric neonatal piglets, the following design criteria were identified for the device: an operating space of 1" by 1.5", the ability to apply at least 100 N of tensile force, the ability to stretch the brachial plexus a distance of 10", and a camera capable of recording at 120 frames per second.

After taking all of these design criteria into consideration, the device was designed and fabricated as shown in Figure 1. Main components of the device are numbered and labeled in the legend. The linear actuator and load cell (including a load cell amplifier) were controlled using an Arduino Uno Rev3 with a MegaMoto Plus H-Bridge Control Shield attachment with a user interface created in LabView. This user interface allowed the user to control the speed of the actuator and measure the force applied to the brachial plexus in real time while recording the test for analysis and playback using the camera.

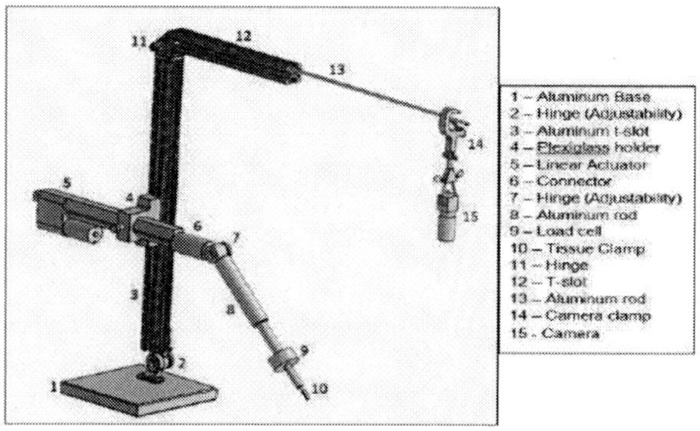

Figure 1: PIVTT Schematic

PRODUCT TESTING AND VALIDATION

For validation of the completed rig and components, four test runs were performed on a rubber band model and one test was conducted on a 6-8 weeks old piglet. For the rubber band tests, the position and load applied were recorded in Excel through the user interface. This data was then used to create a position vs. time and load applied vs. time graph. The length and cross sectional area of the rubber band were measured prior to testing and used to calculate the stress applied to the rubber band over time. These values were then used to create a stress-strain curve. The slope of the linear portion of this stress-strain curve could then be calculated to find the Young's Modulus of the rubber band. Important values calculated from this test are also shown in Table 1. The purpose of the rubber band model was to validate that the components were working properly.

Table 1: Significant Values from Rubber Band Testing

Simulation	Stretch Rate (mm/s)	Max Stress (MPa)	Max Strain (mm/mm)
1	10.89	2.54	1.97
2	10.81	2.48	1.97
3	2.04	1.72	1.04
4	2.06	2.06	1.15

For the piglet test, a brachial plexus nerve segment was isolated within a freshly euthanized piglet and attached to the specialized clamp. The same properties measured during the rubber band tests were measured during the piglet test. The values of force and position of the nerve over time were used to generate the stress-strain curve as shown in Figure 2. This curve was used to calculate the Young's Modulus (E), ultimate stress, and the ultimate strain. The Young's Modulus was calculated from the slope of the linear portion of the graph. The values measured from the piglet test are shown in Table 2.

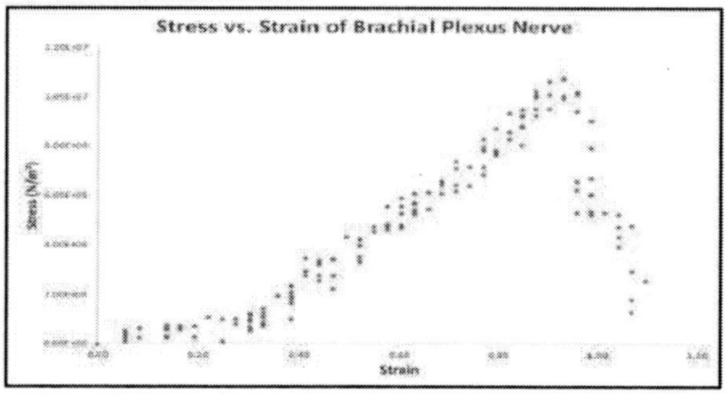

Figure 2: Stress vs. Strain Curve Produced From Piglet BP Testing

By successfully fabricating and validating this testing device, the *in vivo* properties of brachial plexus will be measured and the properties obtained from this testing can be used to create an accurate, biofidelic computational model.

Table 2: Biomechanical Properties of Piglet Brachial Plexus Nerve

BP Test	Stretch Rate (mm/s)	Max Stress (MPa)	Max Strain (mm/mm)	Young's Modulus (MPa)
1	0.63	10.8	1.1	15.19

BUDGET AND MARKET ANALYSIS

At the start of the project, the device was allocated a budget of $1,000. After the design, fabrication, and validation of the device, we spent a total of $905.39, keeping the total cost of the device under budget. The PIVTT was designed and built on a tight budget for academic research projects. Thus, although the market for this device might appear small, the impact of the research using this device is high. Also the device can be utilized to test various *in vivo* tissue properties in animals and humans. The low cost, ease of fabrication and portability while being efficient can benefit several labs and universities across the country. We anticipate fifty of these units to be fabricated and sold per year. The manufacturing cost at this volume would be kept under $800 per unit. Each unit will be sold for $1,500 due to the labor needed to assemble and troubleshoot the device during fabrication. Revenue from the sale of each unit will be placed towards improving the post-test processing of data and generating a biofidelic computational model which will be used in Ob/Gyn training and education for preventing the occurrence of NBPP during childbirth.

ACKNOWLEDGEMENTS

The team would like to thank Ishan Ali and Progressive Automations for their assistance in the purchase and use of the linear actuator, Arduino Uno, and MegaMoto, Chris Lange and Tacuna Systems, Inc. for their assistance in the purchase and use of the load cell and amplifier, Dr. Arun Ramakrishnan from Drexel University for his input in the circuit configuration, and Jacob Fenstermaker from Widener University for his help in building the PIVTT.

REFERENCES

[1] - N. Zhang, B. Gonik, and M. Grimm, Summer Bioengineering Conference, 2003.

[2] - "Brachial Plexus and Peripheral Nerve Injuries." The Children's Hospital of Philadelphia. Information Site.

[3] - Chauhan, SP, Blackwell SB, Ananth CV. "Seminars in Perinatology." 2014. Vol. 8(4): 210-218. PMID:24863027

[4] - Zamorski, T., et al. "Investigating the Mechanism of Neonatal Brachial Plexus Palsy Using Biomechanical Testing." Dept. of Biomedical Eng. Widener University. 2015.

[5] - Kawai, H. et al. "Stretching of the brachial plexus in rabbits." Acfa Orthop Scand 1989; 60(6): 635-638.

AUTHOR INDEX

AUTHOR INDEX

AUTHOR INDEX

AUTHOR INDEX

AUTHOR INDEX

AUTHOR INDEX

AUTHOR INDEX

AUTHOR INDEX

AUTHOR INDEX

AUTHOR INDEX

AUTHOR INDEX

AUTHOR INDEX

AUTHOR INDEX

AUTHOR INDEX

AUTHOR INDEX

AUTHOR INDEX

AUTHOR INDEX

AUTHOR INDEX

AUTHOR INDEX

Summer Biomechanics, Bioengineering and Biotransport Organizing Committee
201 Waterfront St
National Harbor, MD 20745
USA

ISBN 978-1-5108-5993-7
9 781510 859937

CURRAN ASSOCIATES INC.
proceedings.com